Organic Chemicals in the Environment
Mechanisms of Degradation and Transformation

Second Edition

Organic Chemicals in the Environment

Mechanisms of Degradation and Transformation

Second Edition

Alasdair H. Neilson and Ann-Sofie Allard

CRC Press
Taylor & Francis Group
Boca Raton London New York

CRC Press is an imprint of the
Taylor & Francis Group, an **informa** business

CRC Press
Taylor & Francis Group
6000 Broken Sound Parkway NW, Suite 300
Boca Raton, FL 33487-2742

First issued in paperback 2021
First issued in hardback 2019

ISBN 13: 978-1-03-209917-0 (pbk)
ISBN 13: 978-1-4398-2637-9 (hbk)

Library of Congress Cataloging-in-Publication Data

Neilson, Alasdair H.
    Environmental degradation and transformation of organic chemicals / Alasdair H. Neilson, Ann-Sofie Allard. -- 2nd ed.
        p. cm.
    Includes bibliographical references and index.
    ISBN 978-1-4398-2637-9 (hardback)
    1. Aquatic organisms--Effect of water pollution on. 2. Organic water pollutants--Environmental aspects. I. Allard, Ann-Sofie. II. Title.

QH545.W3N45 2012
578.76--dc23

2012006803

Visit the Taylor & Francis Web site at
http://www.taylorandfrancis.com

and the CRC Press Web site at
http://www.crcpress.com

Contents

Section II   Experimental Procedures

Section III   Pathways and Mechanisms of Degradation and Transformation

Preface

The layout of this volume follows essentially that of the previous edition *Environmental Degradation and Transformation of Organic Chemicals* by the same authors, with the omission of Chapters 12 through 14 that dealt with bioremediation and which are fully covered elsewhere. As before, this volume deals primarily with organic chemicals in the aquatic and terrestrial environments, including products from atmospheric transformation.

Biochemical reactions parallel those in organic chemistry and a mechanistic approach has proved valuable for both. Most of the principles that have emerged apply equally to the aquatic, atmospheric, and terrestrial environments, and the approach throughout this volume is, therefore, essentially mechanistic and biochemical. The mechanisms of degradation are valuable since they may be generalized to provide guidance for structurally similar contaminants. This is particularly important in view of the widening spectrum of contaminants including agrochemical and pharmaceutical products. In addition, metabolites produced by biochemical transformation of the substrate—rather than by degradation—or from partial abiotic reactions may be terminal and persistent or toxic to other components of the ecosystem including the microorganisms that produced them.

Although phase partition is not discussed, this is an important factor in determining biodegradability since a contaminant will seldom remain in the phase to which it is initially discharged. This is illustrated by the following examples:

1. Compounds deposited as solids (including agrochemicals and contaminants) may reach groundwater and watercourses as a result of partition and leaching.
2. Substances with even marginal volatility will enter the atmosphere, and after transformation may then reenter the aquatic and terrestrial environments through precipitation.
3. Aquatic biota may bring about transformation to metabolites that are then disseminated to a considerable distance from their source.
4. Particularly, polar contaminants may associate with polymeric humic components of soil, water, and sediment. Their biodegradation then depends on the degree to which these processes are reversible and the contaminants become accessible to microorganisms (bioavailable). This is especially significant after weathering (aging), even in the case of nonpolar compounds.

There are omissions in the material that is covered, and the following are not discussed systematically:

1. Naturally occurring polymers: polysaccharides including cellulose and chitin; polyisoprenoids and lignin; synthetic polyamides, polyurethanes, and polysiloxanes.
2. Natural polypyrroles such as hemin and chlorophylls.
3. Metabolites produced by plants and microorganisms although a plethora of halogenated metabolites is produced by marine biota such as halomethanes, polybrominated phenols, and polybrominated diphenyl ethers may be undesirable contaminants.
4. Agrochemicals and veterinary chemotherapeutic agents that have been widely used in large-scale agriculture and animal husbandry.
5. Pharmaceuticals that encompass a wide range of structures including hormone disrupters which have raised serious concern.

The genetics of degradation pathways and metabolic regulation are not rigorously discussed, and there is no discussion of models either for the analysis of degradation kinetics, or for the prediction of biodegradability. Only limited attention is drawn to metabolism and synthesis by genetically or meta-bolically manipulated bacteria that have achieved prominence in biotechnology.

The number of references that are cited is necessarily restrictive and, although numerous, they repre-sent merely an eclectic selection from the vast literature. As far as possible, the references after each section are given to the primary literature that has been subjected to the scrutiny of peer review, and, therefore, provides a solid and reproducible basis for the critical reader. Some earlier work has been cited when this has led to lasting concepts, though it may be difficult to evaluate the nomenclature by current standards.

Nomenclature has presented serious problems since the numbers of new taxa that are being described appear to be increasing exponentially. Although no attempt has been made to provide the current assign-ment for all taxa, we have used those employed by the authors and the designations that occur most fre-quently in the current literature. A word of caution is appropriate: the reactions that are used for illustration are seldom specific for a single taxon, and it is not generally possible to establish the range of organisms which will be able to carry out that reaction.

The basic work that provides a detailed coverage of the nomenclature of bacteria and archaea is the seven-volume *The Prokaryotes: A Handbook on the Biology of Bacteria, Archaea* (Ed. Stanley Falkow), Springer, New York. In addition, references in the text frequently cite taxonomic articles from the *International Journal of Systematic and Evolutionary Microbiology.*

Whereas the distinction has been made in this book between the activities of bacteria and archaea, the terms prokaryote and eukaryote have been retained for convenience, and further division of the archaea has not been attempted.

The contents of the chapters are interdependent. In summary, they deal with the following for which pathways have been illustrated with figures that provide details of the intermediates involved.

Chapters 1 through 3 provide a broad perspective on abiotic and biotic reactions, including the basic chemical and biochemical reactions.

Chapter 4 describes the reaction of bacteria to stress induced by the physical or chemical environment, and discusses the various types of microbial interactions. Contaminants seldom consist of single sub-strates and the pathways used for biodegradation of some components may be incompatible with that for others that are present.

Chapter 5 provides a brief introduction to the range of experimental procedures that have been used, and includes a section dealing with the study of microbial populations that is discussed in detail in Chapter 6.

Chapter 6 describes procedures for establishing the details of metabolic pathways using isotopes, and includes the application of stable isotope probes and stable isotope fractionation to microbial popula-tions. The application of physical methods to structure determination using NMR, EPR, and x-ray crys-tallography is widely used and is briefly outlined.

Chapters 7 through 11 provide details of the biochemical reactions involved in the biodegradation of the major groups of aliphatic, alicyclic, carbocyclic aromatic, and heterocyclic compounds, including halogenated substrates. Although emphasis is placed on the pathways, general accounts of the enzymes involved and the genetics are provided.

Acknowledgment

We thank Östen Ekengren, director of Environmental Technology, for the generous provision of library facilities.

Authors

Alasdair H. Neilson was a principal scientist until retirement from IVL Swedish Environmental Research Institute in Stockholm. He studied chemistry at the University of Glasgow and obtained his PhD in organic chemistry in Alexander Todd's laboratory at Cambridge. He carried out further research at Cambridge in organic chemistry and in theoretical chemistry with Charles Coulson at Oxford. He held academic positions in the universities of Glasgow and Sussex and obtained industrial experience in the pharmaceutical industry. He consolidated his experience by turning to research in microbiology during a prolonged stay with Roger Stanier and Mike Doudoroff in Berkeley. His interests have ranged widely and included studies on nitrogen fixation, carbon and nitrogen metabolism in algae, and various aspects of environmental science, including biodegradation and biotransformation, chemical and microbiological reactions in contaminated sediments, and ecotoxicology. With his group of collaborators, these studies have resulted in publications in *Applied and Environmental Microbiology, Journal of Chromatography, Environmental Science and Technology*, and *Ecotoxicology and Environmental Safety*, and in chapters contributed to several volumes of *The Handbook of Environmental Chemistry* (editor-in-chief O. Hutzinger). He is a member of the American Chemical Society, the American Society for Microbiology, the American Society of Crystallography, and the American Association for the Advancement of Science.

Ann-Sofie Allard was a chemical microbiologist and is currently a senior microbiologist at IVL Swedish Environmental Research Institute in Stockholm. She has carried out research in a wide range of water quality issues including the distribution of *Yersinia enterocolitica* in Swedish freshwater systems, and processes for the removal of hormone disrupters. She has also carried out extensive studies on the biodegradation and biotransformation of organic contaminants and in ecotoxicology, and she has implemented independent studies on the uptake and metabolism of organic contaminants and metals in higher plants in the context of bioremediation. Her studies have been published in *Applied and Environmental Microbiology, International Biodeterioration and Biodegradation, Environmental Chemistry and Ecotoxicology, Journal of Environmental Science and Health*, and in chapters contributed to several volumes of *The Handbook of Environmental Chemistry* (editor-in-chief O. Hutzinger). She is a member of the American Chemical Society.

Section I

Degradation and Transformation Processes

1

Abiotic Reactions: An Outline

1.1 Introduction

Virtually any of the plethora of reactions in organic chemistry may be exploited for the abiotic degradation of xenobiotics. These include nucleophilic displacement, oxidation, reduction, thermal reactions, and halogenation. Hydrolytic reactions may convert compounds such as esters, amides, or nitriles into the corresponding carboxylic acids, or ureas and carbamides into the amines. These abiotic reactions may, therefore, be the first step in the degradation of such compounds. The transformation products may, however, be resistant to further chemical transformation so that their ultimate fate is dependent upon subsequent microbial reactions. For example, for some urea herbicides, the limiting factor is the rate of microbial degradation of the chlorinated anilines that are the initial products of hydrolysis. The role of abiotic reactions should, therefore, always be taken into consideration, and should be carefully evaluated in laboratory experiments on biodegradation and biotransformation. The results of experiments directed to microbial degradation are probably discarded if they show substantial interference from abiotic reactions. A good illustration of the complementary roles of abiotic and biotic processes is offered by the degradation of tributyltin compounds. Earlier experiments (Seligman et al. 1986) had demonstrated the degradation of tributyltin to dibutyltin primarily by microbial processes. It was subsequently shown, however, that an important abiotic reaction mediated by fine-grained sediments resulted in the formation of monobutyltin and inorganic tin also (Stang et al. 1992). It was, therefore, concluded that both processes were important in determining the fate of tributyltin in the marine environment.

A study of the carbamate biocides, carbaryl, and propham illustrates the care that should be exercised in determining the relative importance of chemical hydrolysis, photolysis, and bacterial degradation (Figure 1.1) (Wolfe et al. 1978). For carbaryl, the half-life for hydrolysis increased from 0.15 d at pH 9 to 1500 d at pH 5, while that for photolysis was 6.6 d: biodegradation was too slow to be significant. In contrast, the half-lives of propham for hydrolysis and photolysis were $>10^4$ and 121—so greatly exceeding the half-life of 2.9 d for biodegradation that abiotic processes would be considered to be of subordinate significance. Close attention to structural features of xenobiotics is, therefore, clearly imperative before making generalizations on the relative significance of alternative degradative pathways.

1.2 Photochemical Reactions in Aqueous and Terrestrial Environments

Photochemical reactions are important in atmospheric reactions, in terrestrial areas of high solar irradiation such as the surface of soils, and in aquatic systems containing ultraviolet (UV)-absorbing humic and fulvic acids (Zepp et al. 1981a,b). They may be relevant especially for otherwise recalcitrant

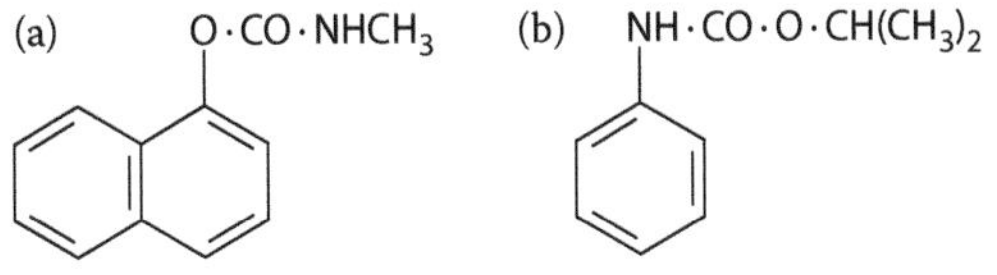

(a) $O \cdot CO \cdot NHCH_3$ (b) $NH \cdot CO \cdot O \cdot CH(CH_3)_2$

FIGURE 1.1   Carbaryl (a) and propham (b).

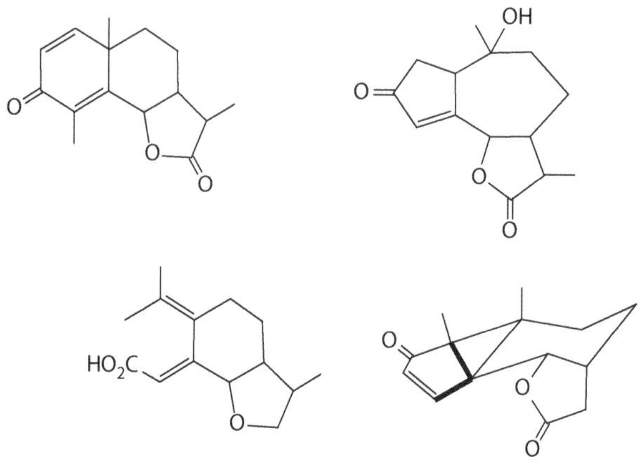

FIGURE 1.2 Photochemical transformation of santonin.

compounds. It has also been shown (Zepp and Schlotzhauer 1983) that although the presence of algae may enhance photometabolism, this is subservient to direct photolysis at the cell densities likely to be encountered in rivers and lakes. It should be noted that different products may be produced in natural river water and in buffered medium. For example, photolysis of triclopyr (3,5,6-trichloro-2-pyridyloxy-acetic acid) in sterile medium at pH 7 resulted in hydrolytic replacement of one chlorine atom, whereas in river water the ring was degraded to form oxamic acid as the principal product (Woodburn et al. 1993). Particular attention has, therefore, been understandably directed to the photolytic degradation of biocides—including agrochemicals—that are applied to terrestrial systems and enter the aquatic system through leaching. There has been increased interest in their phototoxicity toward a range of biota (references in Monson et al. 1999), and this may be attributed to some of the reactions and transformations that are discussed later in this chapter. It should be emphasized that photochemical reactions may produce molecules structurally more complex and less susceptible to degradation than their precursors, even though the deep-seated rearrangements induced in complex compounds such as the terpene santonin during UV irradiation (Figure 1.2) are not likely to be encountered in environmental situations.

1.3 The Diversity of Photochemical Transformations

In broad terms, the following types of reactions are mediated by the homolytic fission products of water (formally, hydrogen, and hydroxyl radicals), and by molecular oxygen including its excited states—hydrolysis, elimination, oxidation, reduction, and cyclization.

1.3.1 The Role of Hydroxyl Radicals

The hydroxyl radical plays two essentially different roles: (a) as a reactant mediating the transformations of xenobiotics and (b) as a toxicant that damages DNA. They are important in a number of environments: (1) in aquatic systems under irradiation, (2) in the troposphere, which is discussed later, and (3) in biological systems in the context of superoxide dismutase and the role of iron. Hydroxyl radicals in aqueous media can be generated by several mechanisms:

1. Photolysis of nitrite and nitrate (Brezonik and Fulkerson-Brekken 1998)
2. Fenton reaction with H_2O_2 and Fe^{2+} in the absence or presence of light (Fukushima and Tatsumi 2001)
3. Photolysis of fulvic acids under anaerobic conditions (Vaughan and Blough 1998)
4. Reaction of Fe(III) or Cu(II) complexes of humic acids with hydrogen peroxide (Paciolla et al. 1999).

1.3.2 Illustrations of Photochemical Transformations in Aqueous Solutions

1.3.2.1 Chloroalkanes and Chloroalkenes

1. The photolysis of chloroalkanes and chloroalkenes has received considerable attention and results in the formation of phosgene as one of the final products. The photodegradation of 1,1,1-trichloroethane involves hydrogen abstraction, and oxidation to trichloroacetaldehyde that is degraded by a complex series of reactions to phosgene (Nelson et al. 1990; Platz et al. 1995). Tetrachloroethene is degraded by reaction with chlorine radicals and oxidation to pentachloropropanol radical that also forms phosgene (Franklin 1994). Attention is drawn to these reactions in the context of the atmospheric dissemination of xenobiotics.

2. Heptachlor and *cis*-chlordane, both of which are chiral, produce caged or half-caged structures (Figure 1.3) on irradiation, and these products have been identified in biota from the Baltic, the Arctic, and the Antarctic (Buser and Müller 1993).

1.3.2.2 Arenes

1. Photochemical transformation of pyrene in aqueous media produced the 1,6- and 1,8-quinones as stable end products after initial formation of 1-hydroxypyrene (Sigman et al. 1998).

2. The transformation of arenes and alkanes by hydroxyl radicals produced photochemically from H_2O_2 in the range $-20°C$ to $-196°C$ yielded the products given in Table 1.1 (Dolinová et al. 2006).

1.3.2.3 Nitroarenes

In the presence of both light and hydrogen peroxide, 2,4-dinitrotoluene is oxidized to the corresponding carboxylic acid; this is then decarboxylated to 1,3-dinitrobenzene, which is degraded further by hydroxylation and ring fission (Figure 1.4) (Ho 1986). Analogous reaction products were formed from 2,4,6-trinitrotoluene and hydroxylated to various nitrophenols and nitrocatechols before fission of the aromatic rings, and included the dimeric 2,2′-carboxy-3,3′,5,5′-tetranitroazoxybenzene (Godejohann

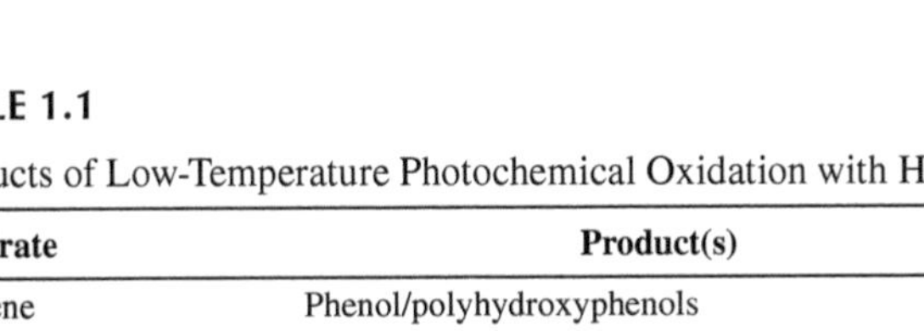

FIGURE 1.3 Photochemical transformation of atrazine.

TABLE 1.1

Products of Low-Temperature Photochemical Oxidation with H_2O_2

Substrate	Product(s)
Benzene	Phenol/polyhydroxyphenols
Naphthalene	Naphth-1-ol, Naphth-2-ol, Naphtho-1,4-quinone
Anthracene	Anthracene-9,10-dione
Cyclohexane	Cyclohexanol, Cyclohexanone
Butane	Butan-1-ol, Butan-2-ol, Butyrate

Source: Adapted from Dolinová N et al. 2006. *Environ Sci Technol* 40: 7668–7674.

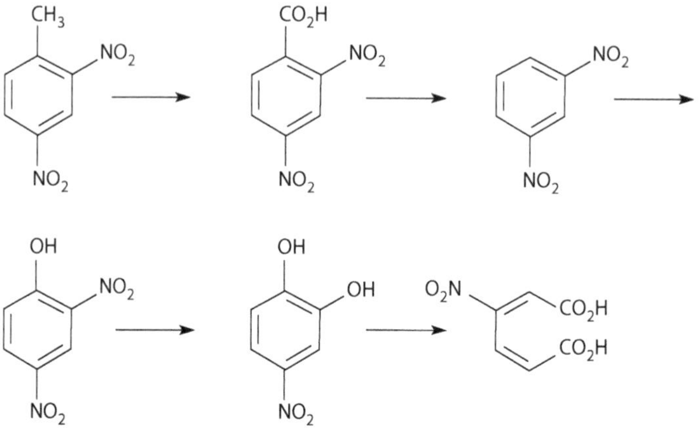

FIGURE 1.4 Photochemical transformation of pentachlorophenol.

et al. 1998). Nitrobenzene, 1-chloro-2,4-dinitrobenzene, 2,4-dinitrophenol, and 4-nitrophenol were degraded with the formation of formate, oxalate, and nitrate (Einschlag et al. 2002).

1.3.2.4 Heteroarenes

1. Atrazine is successively transformed to 2,4,6-trihydroxy-1,3,5-triazine (Pelizzetti et al. 1990) by dealkylation of the alkylamine side chains and hydrolytic displacement of the ring chlorine and amino groups (Figure 1.5). A comparison has been made between direct photolysis and nitrate-mediated hydroxyl radical reactions (Torrents et al. 1997): the rates of the latter were much greater under the conditions of this experiment, and the major difference in the products was the absence of ring hydroxylation with loss of chloride.

2. A potential insecticide that is a derivative of tetrahydro-1,3-thiazine undergoes a number of reactions resulting in some 43 products of which the dimeric azo compound is the principal one in aqueous solutions (Figure 1.6) (Kleier et al. 1985).

3. The photolytic degradation of the fluoroquinolone enrofloxacin involves a number of reactions that produce 6-fluoro-7-amino-1-cyclopropylquinolone 2-carboxylic acid that is then degraded to CO_2 via reactions involving fission of the benzenoid ring with loss of fluoride, dealkylation, and decarboxylation (Burhenne et al. 1997a,b) (Figure 1.7).

4. The transformation of isoquinoline has been studied both under photochemical conditions with hydrogen peroxide, and in the dark with hydroxyl radicals (Beitz et al. 1998). The former resulted in fission of the pyridine ring with the formation of phthalic dialdehyde and phthalimide, whereas the major product from the latter reaction involved oxidation of the benzene ring with the formation of isoquinoline-5,8-quinone and a hydroxylated quinone.

5. The photodegradation of the contact herbicide paraquat yielded many degradation products, but the major pathway produced 1,2,3,4-tetrahydro-1-ketopyrido[1,2-*a*]-5-pyrazinium that was further degraded to pyridine-2-carboxamide and pyridine-2-carboxylate (Figure 1.8) (Smith and Grove 1969).

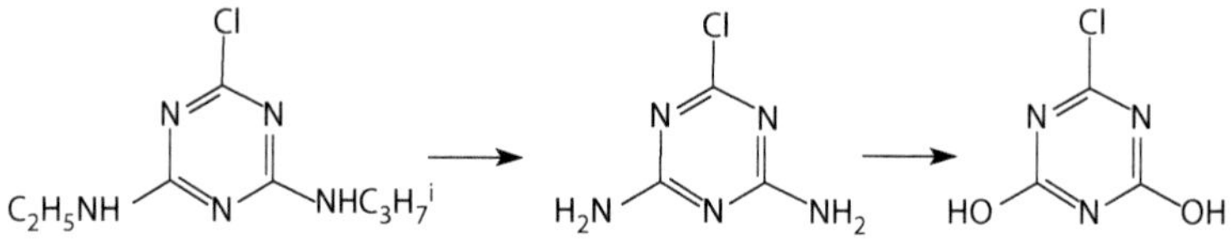

FIGURE 1.5 Photochemical transformation of 3-trifluoromethyl-4-nitrophenol.

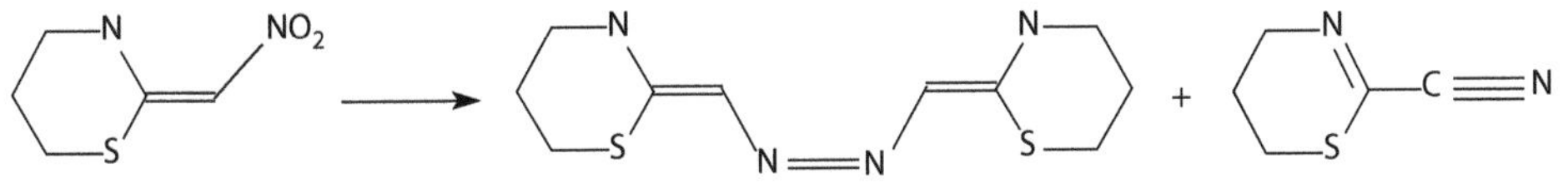

FIGURE 1.6 Photochemical transformation of a tetrahydro-1,3-thiazine.

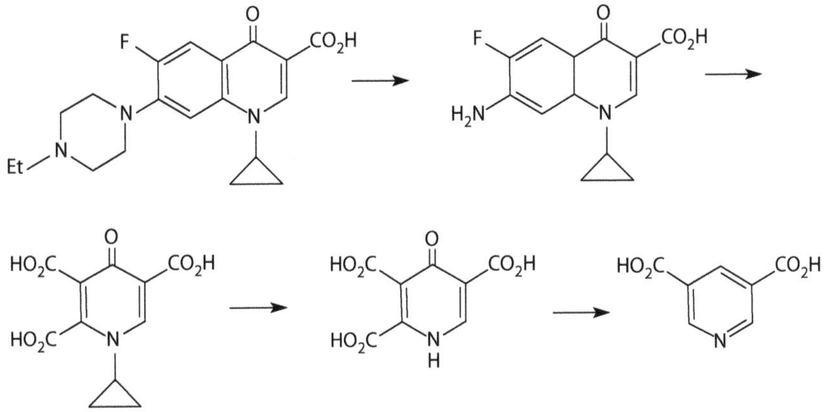

FIGURE 1.7 Photochemical transformation of trifluralin.

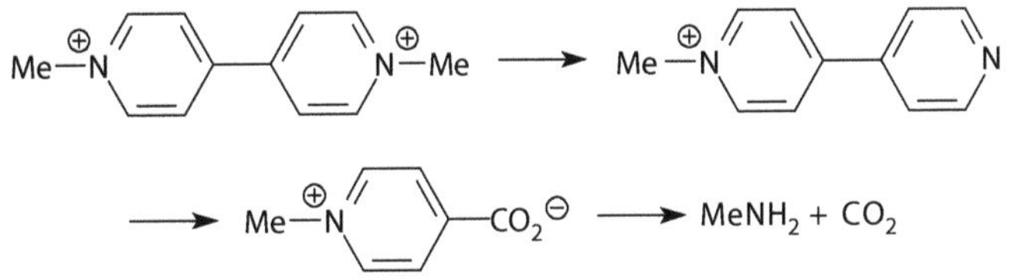

FIGURE 1.8 Photochemical transformation of chlordane.

1.3.2.5 Organofluorines

1. The main products of photolysis of 3-trifluoromethyl-4-nitrophenol are 2,5-dihydroxy-benzo-ate produced by hydrolytic loss of the nitro group and oxidation of the trifluoromethyl group, together with a compound identified as a condensation product of the original compound and the dihydroxybenzoate (Figure 1.9) (Carey and Cox 1981).

2. The herbicide trifluralin undergoes a photochemical reaction in which the *n*-propyl side chain of the amine reacts with the vicinal nitro group to form the benzopyrazine (Figure 1.10) (Soderquist et al. 1975).

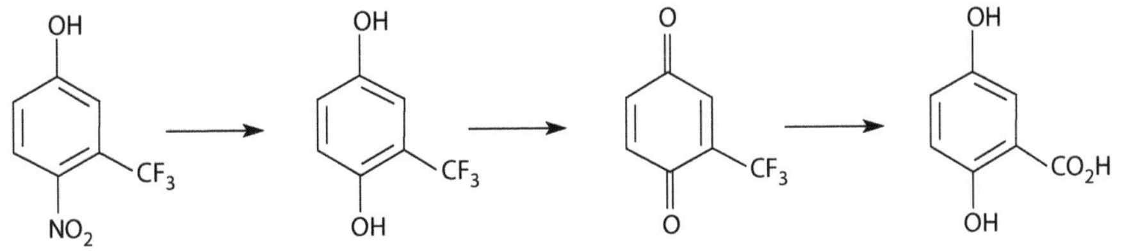

FIGURE 1.9 Photochemical degradation of enrofloxin.

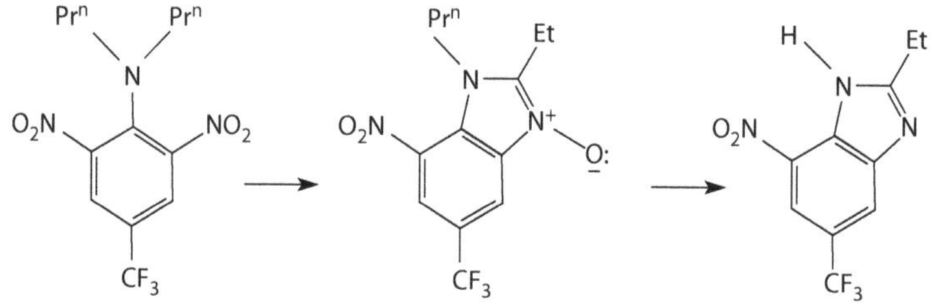

FIGURE 1.10   Photochemical transformation of 2,4-dinitrotoluene.

3. The psychopharmaceutical drug fluoxetine (Prozac) is degraded both directly and by the faster reaction with OH radicals (Lam et al. 2005). In both reactions, the ring bearing the CF_3 group was degraded in high yield to 4-(difluoromethylene)-cyclohexa-2,5-diene-1-one.

1.3.2.6 Organochlorines

1. Pentachlorophenol produces a wide variety of transformation products, including chloranilic acid (2,5-dichloro-3,6-dihydroxybenzo-1,4-quinone) by hydrolysis and oxidation, a dichlorocyclopentanedione by ring contraction, and dichloromaleic acid by cleavage of the aromatic ring (Figure 1.11) (Wong and Crosby 1981).
2. It has been suggested that the photochemical reaction of pentachlorophenol in aqueous solution to produce octachlorodibenzo[1,4] dioxin and some of the heptachloro congener could account for the discrepancy between values for the emission of chlorinated dioxins and their deposition, which is significant for the octachloro congener (Baker and Hites 2000).
3. Whereas photolysis of 2- and 4-chlorophenols in aqueous solution produced catechol and hydroquinone, in ice the more toxic dimeric chlorinated dihydroxybiphenyls were formed (Bláha et al. 2004).
4. Photodegradation of the nonsteroidal anti-inflammatory drug diclofenac produced carbazole-1-acetate as the major product (Figure 1.12) (Moore et al. 1990). In a lake under natural conditions, it was rapidly decomposed photochemically though none of the products produced in laboratory experiments could be detected (Buser et al. 1998).

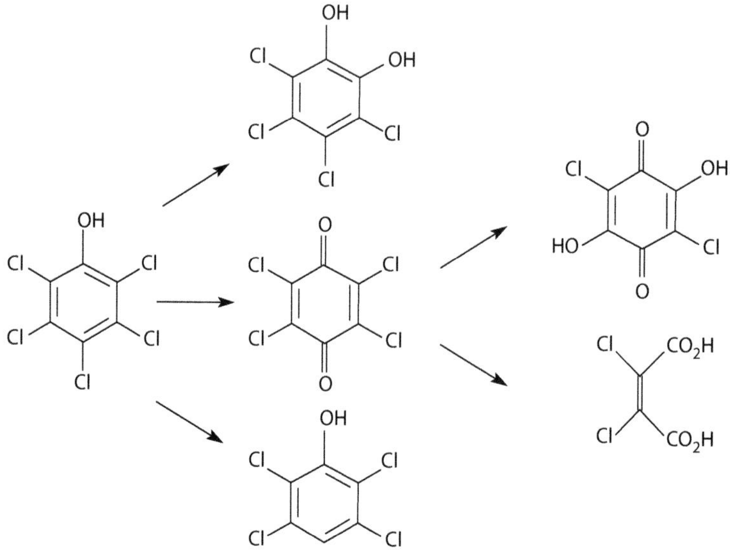

FIGURE 1.11   Photochemical transformation of paraquat.

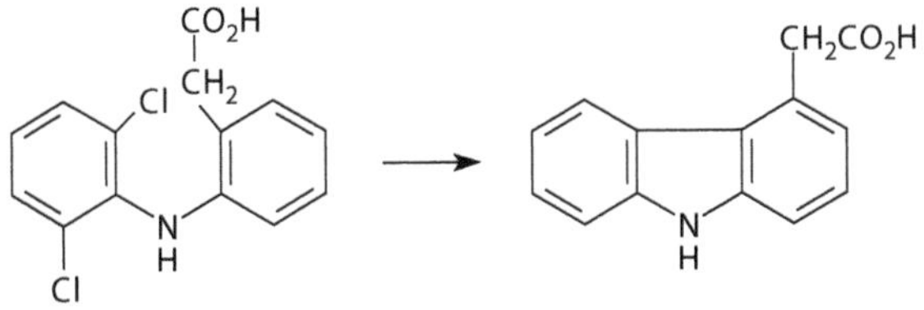

FIGURE 1.12 Photochemical transformation of diclofenac.

5. The photochemical dechlorination of PCB congeners was examined in the solid phase at −25°C. Reductive dehalogenation was the principal reaction, presumptively involving trace organic molecules as the source of the hydrogen atoms, since the possibility of using water was excluded from experiments using D_2O (Matykiewiczová et al. 2007). 2,4-Dichlorobiphenyl (PCB-7) formed biphenyl, and 2,4,5,2′,4′,5′-hexachlorobiphenyl (PCB-153) produced 2,2′,5,5′-tetrachlorobiphenyl (PCB-52), in greater yield than 2,4,5,2′,5′-pentachlorobiphenyl (PCB-101).

1.3.2.7 Organobromines

Photochemical debromination of polybrominated diphenyl ethers was examined in radiation from an UV lamp (Eriksson et al. 2004). Decabromodiphenyl ether was rapidly debrominated with the production of nano-, octa-, hepta- and di- and tri-brominated products. There were two issues that are worth noting: (a) Rates of debromination were greater for the highly brominated congeners and lowest for less highly substituted congeners, so that di- and tri-brominated congeners were most resistant. (b) A metabolite tentatively identified as a methoxy tetrabromodibenzofuran was produced by nucleophilic reaction with the methanol solvent.

1.3.2.8 Miscellaneous

1. Although ethylenediaminetetraacetic acid (EDTA) is biodegradable under specific laboratory conditions (Belly et al. 1975; Lauff et al. 1990; Nörtemann 1992; Witschel et al. 1997), the primary mode of degradation in the natural aquatic environment involves photolysis of the Fe complex (Lockhart and Blakeley 1975; Kari and Giger 1995). Its persistence is critically determined not only by the degree of insolation but also on the concentration of Fe in the environment, since complexes with other metals including Ca and Zn are relatively resistant to photolysis (Kari et al. 1995). The available evidence suggests that in contrast to nitrotriacetic acid (NTA) that is more readily biodegradable, EDTA is likely to be persistent except in environments in which concentrations of Fe greatly exceed those of other cations.

2. Stilbenes that are used as fluorescent whitening agents are photolytically degraded by reactions involving *cis–trans* isomerization followed by hydration of the double bond, or oxidative fission of the double bond to yield aldehydes (Kramer et al. 1996).

3. A reaction of clinical significance is the photolysis of metallo organic complexes with carbonyl or nitrosyl ligands to establish the benign release carbon monoxide or nitric oxide as cytoprotective agents. Carbon monoxide is released under low-energy visible illumination from a range of tricarbonyls in which manganese is ligated to the tripodal ligands tris(2-pyridyl)amine and (2-pyridylmethyl)(2-quinolylmethyl)amine (Gonzalez et al. 2011), or tris(pyrazolyl)methane (Schatzschneider 2010). Analogously, nitrosyl complexes have been synthesized in which the potential NO donor may be released by illumination. A manganese complex containing a quadridentate polypyridyl ligand with a singe NO *trans* to the pyridone nitrogen was activated by near-infrared light with the release of NO (Hoffman-Luca et al. 2009), and ruthenium complexes with quadridentate ligands containing carboxamido and phenolato donors displayed increased photolability (Fry et al. 2010).

1.4 Hydroxyl Radicals in the Destruction of Contaminants

The destruction of contaminants has directed attention to the use of hydroxyl radical-mediated reactions. These reactions should be viewed against those with hydroxyl radicals that occur in the atmosphere.

1.4.1 Fenton's Reagent

Hydrogen peroxide in the presence of Fe^{2+} or Fe^{3+} (Fenton's reagent) has been used in a range of configurations including irradiation, electrochemical, and both cathodic and anodic conditions (references in Wang et al. 2004). In all of these, the reaction involves hydroxyl radicals and systematic investigations have been carried out on the effect of pH, the molar ratio of H_2O_2/substrate, and the possible complications resulting from the formation of iron complexes. As an alternative to the Fe(III)/Fe(II)–H_2O_2 reaction, the Ce (IV)/Ce(III)–H_2O_2 reaction produced the hydroxyl radical and the superoxide radical, both of which were identified by EPR (Heckert et al. 2008).

The Fenton reaction has been studied particularly intensively for the destruction agrochemicals including chlorinated phenoxyacetic acid (Sun and Pignatello 1993) and chloroacetanilide herbicides (Friedman et al. 2006). It is worth noting an unusual reaction in which carbon suboxide (C_3O_2) was produced from the reaction of Fe(III) and H_2O_2 on catechol, 3,5-dichlorocatechol, 4,5-tetrachlorocatechol, or tetrachlorocatechol. The fact that C_3O_2 was produced from the chlorinated compounds indicates the complexity of the reaction in addition to the basic requirement for the O–C–C–C–O unit (Huber et al. 2007).

Although this reaction may have limited environmental relevance except under rather special circumstances, it has been applied in combination with biological treatment of polycyclic aromatic hydrocarbons (PAHs) (Pradhan et al. 1997). Attention is drawn to it here since, under conditions where the concentration of oxidant is limiting, intermediates may be formed that are stable and that may possibly exert adverse environmental effects. Some examples that illustrate the formation of intermediates are given, although it should be emphasized that total destruction of the relevant xenobiotics under optimal conditions can be successfully accomplished. The structures of the products that are produced by the action of Fenton's reagent on chlorobenzene are shown in Figure 1.13a

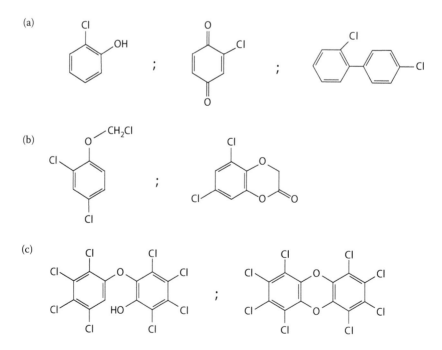

FIGURE 1.13   Transformation products from (a) chlorobenzene, (b) 2,4-dichlorophenoxyacetate, (c) pentachlorophenol.

(Sedlak and Andren 1991), on 2,4-dichlorophenoxyacetate in Figure 1.13b (Sun and Pignatello 1993), and on pentachlorophenol in Figure 1.13c (Fukushima and Tatsumi 2001). The UV-enhanced Fenton transformation of atrazine produced 2,4-diamino-6-hydroxy-1,3,5-triazine by a series of interacting reactions (Chan and Chu 2006). Whereas the degradation of azo dyes by Fenton's reagent produced water- and CH_2Cl_2-soluble transformation products including nitrobenzene from Disperse Orange 3 that contains a nitro group, benzene was tentatively identified among volatile products from Solvent Yellow 14 (Spadaro et al. 1994).

1.4.2 Polyoxometalates (Heteropolyacids)

These are complexes formed between tungstates and molybdates, and silicate or phosphate, and have been used to generate hydroxyl radicals photochemically. The tungstates $PW_{12}O_{40}^{3-}$ and $SiW_{12}O_{40}^{4-}$ have been most frequently used.

1. Degradation of 2,4,6-trichlorophenol by $PW_{12}O_{40}^{3-}$ formed a number of products as intermediates, including 2,6-dichlorohydroquinone followed by fission of the ring to maleate, oxalate, acetate, and formate (Androulaki et al. 2000).
2. Nonafluoropentanoic acid was decomposed in aqueous solution to fluoride and CO_2 catalyzed by $H_3PW_{12}O_{40}$ under UV–visible light radiation. The reaction was initiated by decarboxylation followed by a series of reactions involving oxidations (Hori et al. 2004b).
3. The polyoxometalate $PW_{12}O_{40}^{3-}$ was immobilized on an anion-exchange resin, and used to demonstrate the degradation in the presence of H_2O_2 of the phthalein dye rhodamine B to phthalate and a number of short-chain aliphatic mono- and dicarboxylates (Lei et al. 2005).

They have also been used to bring about photochemical reduction of Hg^{2+} via Hg_2^{2+} Hg_2^{2+} to Hg^0 (Troupis et al. 2005).

1.4.3 Photolytic Degradation on TiO_2

The mechanism involves photochemical production of a free electron in the conduction band (e_{cb}^-) and a corresponding hole (h_{vb}^+) in the valence band. Both of these produce H_2O_2 and thence hydroxyl radicals.

1. In the presence of slurries of TiO_2 that served as a photochemical sensitizer, methyl *tert*-butyl ether was photochemically decomposed at wavelengths <290 nm. The products were essentially the same as those produced by hydroxyl radicals under atmospheric conditions (Baretto et al. 1995): *tert*-butyl formate and *tert*-butanol were rapidly formed and further degraded to formate, acetone, acetate, and but-2-ene.
2. The photocatalytic oxidation of various EDTA complexes has been examined (Madden et al. 1997). The rates and efficiencies were strongly dependent on the metal and the reactions are generally similar to those involved in electrochemical oxidation (Pakalapati et al. 1996).
3. A number of products are formed from trichloroethene including tetrachloromethane, hexachloroethane, pentachloroethane, and tetrachlororethene, although the last two were shown to be degradable in separate experiments (Hung and Marinas 1997). In TiO_2 slurries, the photochemical degradation of chloroform, bromoform, and tetrachloromethane involves initial formation of the trihalomethyl radicals. In the absence of oxygen, these are further decomposed via dihalocarbenes to CO. Dichlorocarbene was found as an intermediate in the degradation of trichloroacetate (Choi and Hoffmann 1997).
4. The degradation of the herbicide 3-amino-1*H*-1,2,4-triazole (amitrole) produced a number of transformation products after fission of the triazole ring. These reacted together to form 2,4,6-trihydroxy-1,3,5-triazine that was a stable end product (Watanabe et al. 2005).
5. The oxidation of tetrachlorobiphenyl congeners (23-34, 25-34, 345-4) with only one *ortho*-chlorine substituent was examined using nanostructured TiO_2 immobilized on quartz beads.

Initial reactions were hydroxylations by OH radicals that were followed by fission of a single ring to chlorobenzoates and low yields of succinate and glycolate (Nomiyama et al. 2005).

6. The TiO_2-mediated photocatalytic debromination of decabromodiphenyl ether has been examined (Sun et al. 2009). Successive debrominations resulted principally in the formation of nanobromo (BDE 206), octobromo (BDE 203) + (BDE 196), heptabromo (BDE 183), hexabromo (BDE 153), pentabromo (BDE 99) congeners and, finally 1,1′,3,3′-tetrabromodiphenylether (BDE 47) (Sun et al. 2009).

1.5 Other Photochemically Induced Reactions

1. Two types of reactions are important in the photochemical transformation of PAHs, those with molecular oxygen and those involving cyclization. Illustrative examples are provided by the photooxidation of 7,12-dimethylbenz[*a*]anthracene (Lee and Harvey 1986) (Figure 1.14a) and benzo[*a*]pyrene (Lee-Ruff et al. 1986) (Figure 1.14b), and the cyclization of *cis*-stilbene (Figure 1.14c).

2. In nonaqueous solutions, two other types of reactions have been observed with polycyclic arenes: condensation via free-radical reactions and oxidative ring fission.

   a. Irradiation of benz[*a*]anthracene in benzene solutions in the presence of xanth-9-one or vanillin produced a number of transformation products tentatively identified as the result of oxidation and cleavage of ring A, ring C, ring D, and rings C and D, and rings B, C, and D, respectively (Jang and McDow 1997).

   b. 1-Nitropyrene is a widely distributed contaminant produced in the troposphere by reaction of nitrate radicals with pyrene, which is discussed later. A solution in benzene was photochemically transformed into 9-hydroxy-1-nitropyrene that is less mutagenic than its precursor (Koizumi et al. 1994).

3. The photochemical transformation of phenanthrene sorbed on silica gel (Barbas et al. 1996) resulted in a variety of products including *cis*-9,10-dihydrodihydroxyphenanthrene, phenanthrene-9,10-quinone, and a number of ring fission products including biphenyl-2,2′-dicarboxaldehyde, naphthalene-1,2-dicarboxylic acid, and benzo[*c*]coumarin.

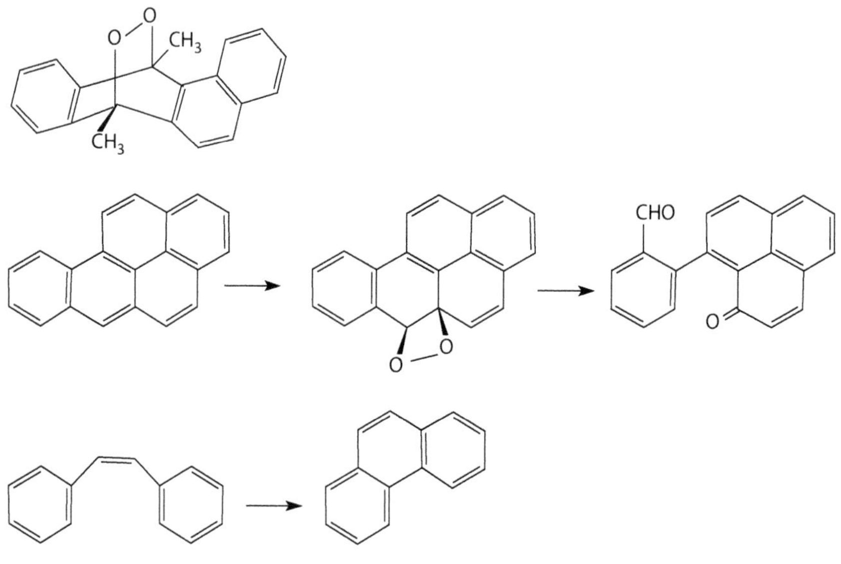

FIGURE 1.14 Photooxygenation of (a) 7,12-dimethylbenz[*a*]anthracene, (b) benzo[*a*]pyrene, (c) photocylization of *cis*-stilbene.

4. The products of the photooxidation of naphthylamines adsorbed on particles of silica and alumina were putatively less toxic than their precursors (Hasegawa et al. 1993).

5. Plant cuticles can play diverse roles.

   a. It has been suggested that photochemically induced reactions may take place between biocides and the biomolecules of plant cuticles. Laboratory experiments have examined addition reactions between DDT and methyl oleate, and were used to illustrate reactions, which resulted in the production of "bound" DDT residues (Figure 1.15) (Schwack 1988).

   b. The herbicide sulcotrione that is an alternative to atrazine is transformed photochemically on maize cuticular wax to a tetrahydroxanthone (ter Halle et al. 2006) (Figure 1.16).

6. There has been considerable interest in the photochemical decomposition of perfluorinated compounds that have become ubiquitous.

   a. Perfluorooctanoate can be successively degraded to C_6, C_5, C_4, and C_3 perfluoro acids, and to fluoride and CO_2 (Hori et al. 2004a). The reaction is initiated by loss of CO_2 with the formation of C_7F_{15} radical that reacts with water to form $C_7F_{15}OH$. The following sequence of reactions is based on results from reactions in $H_2^{18}O$:

   $$C_7F_{15}OH \rightarrow C_6F_{13}COF \rightarrow H^+ + F^-$$

   $$C_6F_{13}COF + H_2O \rightarrow C_6F_{13}CO_2H + H^+ + F^-$$

   b. The decomposition of perfluorocarboxylic acids $(C_4–C_8)$ in the presence of the sulfate radical anion produced photochemically from persulfate $(S_2O_8^{2-})$ in the presence of oxygen resulted in the production of F^-, CO_2, and shorter-chain perfluoro fatty acids (Hori et al. 2005a). To facilitate mixing of the poorly water-soluble perfluorocarboxylic acids, the reaction has been carried out in a two-phase aqueous/liquid CO_2 system (Hori et al. 2005b). High rates of degradation were observed for $C_9–C_{11}$ substrates, with the formation of fluoride, and only minor production of shorter-chain perfluorocarboxylic acids.

7. Photolysis of the potential fuel additive methylcyclopentadienyl manganese tricarbonyl resulted in the loss of CO and formation of a compound tentatively identified as the dicarbonyl (Wallington et al. 1999). In addition, there is concern over the particulates that are emitted during normal vehicle running. Speciation of the inorganic products has been carried out using manganese K-edge x-ray absorption fine structure (XANES) and showed that a number of products were formed including Mn_3O_4, $MnSO_4$, and $MnPO_4$ (Ressler et al. 2000).

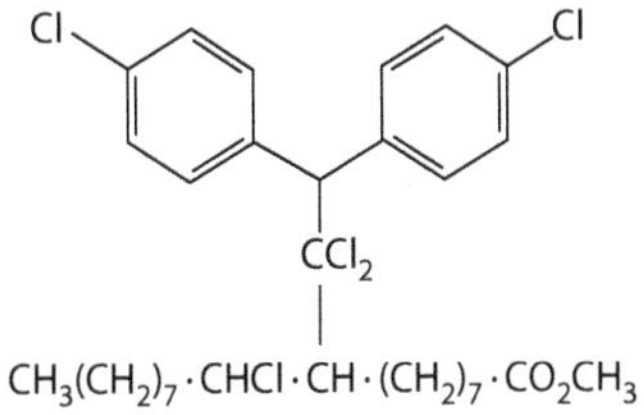

FIGURE 1.15 Product of reaction between DDT and methyl oleate.

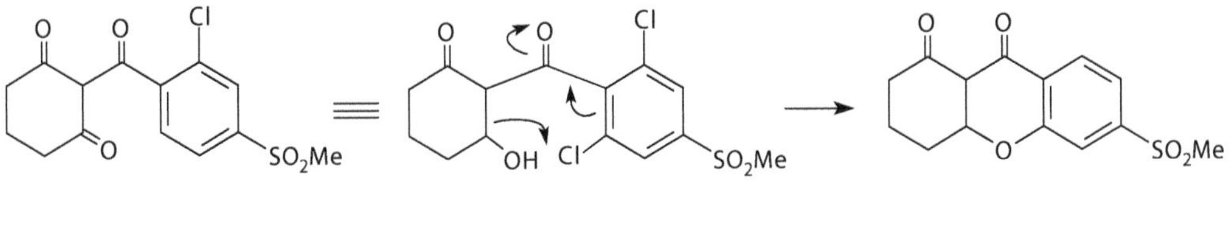

FIGURE 1.16 Photochemical transformation of sulcotrione.

1.6 The Role of Humic Matter: Singlet Dioxygen

It is well established that important photochemical reactions are mediated by humic material in the aquatic environment (Zepp et al. 1981a,b), and that these are particularly significant for hydrophobic contaminants. Partial reductive dechlorination of the persistent insecticide mirex associated with humic matter has been reported (Burns et al. 1996, 1997; Lambrych and Hassett 2006), and it was shown that a kinetic model required differentiation of the bound and dissolved phases (Burns et al. 1996). The fundamental mechanism for the production of singlet dioxygen in aqueous solution was elucidated by Zepp et al. (1985). In the presence of the excited triplet states of chromophores such as humic acids, reaction with the triplet ground state of dioxygen ($^3O_2$) can produce singlet dioxygen ($^1O_2$) whose reaction with dienes has been traditionally used as a specific marker. Singlet $^1O_2$ is highly reactive with a half-life of only 4 μs, but is significant in the aquatic environment in the neighborhood of the exciting chromophore. Using a trap-and-trigger chemiluminescent probe, it has been shown that within a volume containing the chromophore, high densities of $^1O_2$ may be produced, and that this decreases during the transition to the true aquatic phase (Latch and McNeill 2006). Intrahumic reactions are, therefore, significant primarily for contaminants that are hydrophobic and that associate with the chromophoric groups of humic acids.

1.7 Interactions between Photochemical and Other Reactions

It has been shown that a combination of photolytic and biotic reactions can result in enhanced degradation of xenobiotics in municipal treatment systems, for example, of chlorophenols (Miller et al. 1988a) and benzo[*a*]pyrene (Miller et al. 1988b). Two examples illustrate the success of a combination of microbial and photochemical reactions in accomplishing the degradation of widely different xenobiotics in natural ecosystems. Both of them involved marine bacteria, and it, therefore, seems plausible to assume that such processes might be especially important in warm-water marine environments.

1. The coupled degradation of pyridine dicarboxylates (Amador and Taylor 1990).
2. The degradation of 3- and 4-trifluoromethylbenzoate: microbial transformation resulted in the formation of catechol intermediates that were converted into 7,7,7-trifluoro-hepta-2,4-diene-6-one carboxylate, but this was subsequently degraded photochemically with the loss of fluoride (Taylor et al. 1993) (Figure 1.17). This degradation may be compared with the purely photochemical degradation of TFM that has already been noted, and contrasted with the resistance of trifluoromethylbenzoates to microbial degradation (Chapter 9, Part 3).

Collectively, these examples illustrate the diversity of transformations of xenobiotics that are photochemically induced in aquatic and terrestrial systems. Photochemical reactions in the troposphere are extremely important in determining the fate and persistence of not only xenobiotics but also of naturally occurring compounds. A few illustrations are given as introduction:

1. The occurrence of C_8 and C_9 dicarboxylic acids in samples of atmospheric particles and in recent sediments (Stephanou 1992; Stephanou and Stratigakis 1993) has been attributed to photochemical degradation of unsaturated carboxylic acids that are widespread in almost all biota.

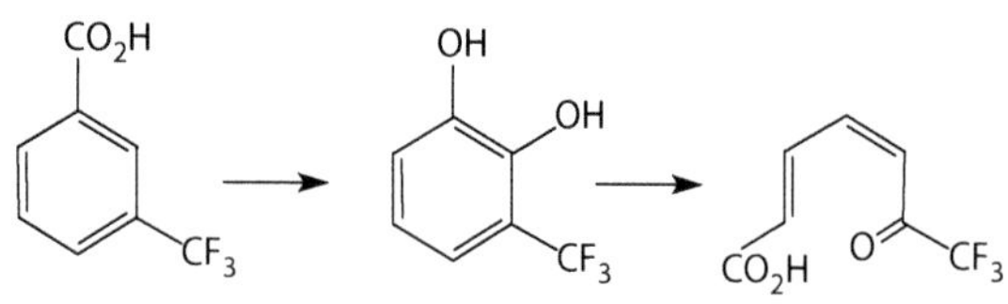

FIGURE 1.17 Microbial followed by photochemical degradation of 3-trifluorobenzoate hydroxyl radical transformation of isoprene.

2. The formation of peroxyacetyl nitrate from isoprene (Grosjean et al. 1993a) and of peroxypropionyl nitrate (Grosjean et al. 1993b) from *cis*-3-hexen-1-ol that is derived from higher plants, illustrate important contributions to atmospheric degradation (Seefeld and Kerr 1997).

3. Attention has been given to possible adverse effects of incorporating *tert*-butyl methyl ether into automobile fuels. *tert*-Butyl formate is an established product of photolysis, and it has been shown that photolysis in the presence of NO can produce the relatively stable *tert*-butoxyformyl peroxynitrate. Its stability is comparable to that of peroxyacetyl nitrate, and may, therefore, increase the potential for disseminating NO_x (Kirchner et al. 1997).

1.8 Reactions in the Troposphere

Although chemical transformations in the atmosphere may seem peripheral to this discussion, these reactions should be considered since their products may subsequently enter the aquatic and terrestrial environments: the persistence and the toxicity of these secondary products are, therefore, relevant to this discussion.

Details of the relevant principles and details of the methodology are covered in a comprehensive treatise (Finlayson-Pitts and Pitts 1986), and reference should be made to a review on tropospheric air pollution (Finlayson-Pitts and Pitts 1997), and atmospheric aerosols (Andreae and Crutzen 1997). Attention has been directed to the potentially adverse reactions including ozone depletion induced by a range of anthropogenic compounds. The atmospheric chemistry of hydrofluorocarbons and hydrofluoroethenes is discussed by Wallington and Nielsen (2002), and organic bromine and iodine compounds by Orlando (2003). Although reactions with free radicals generally dominate, direct photolysis is important for alkyl iodides (Orlando 2003).

There are a number of important reasons for discussing the reactions of organic compounds in the troposphere:

1. After emission, contaminants may be partitioned among the terrestrial, aquatic, and various atmospheric phases, and those of sufficient volatility or associated with particles may be transported over long distances. This is not a passive process, however, since important transformations may take place in the troposphere during transit so that attention should also be directed to their transformation products.

2. Considerable attention has been given to the persistence and fate of organic compounds in the troposphere, and this has been increasingly motivated by their possible role in the production of ozone by reactions involving NO_x.

3. Concern has been expressed over the destruction of ozone in the stratosphere brought about by its reactions with chlorine atoms produced from chlorofluoroalkanes that are persistent in the troposphere, and that may contribute to radiatively active gases other than CO_2.

Reactions in the troposphere are mediated by reactions involving hydroxyl radicals produced photochemically during daylight, by nitrate radicals that are significant during the night (Platt et al. 1984), by ozone and, in some circumstances by $O(^3P)$.

The overall reactions involved in the production of hydroxyl radicals are

$$O_3 + h\nu \rightarrow O_2 \rightarrow O\ (^1D) : O(^1D) + H_2O \rightarrow 2\ OH$$

$$O(^1D) \rightarrow O(^3P) : O(^3P) + O_2 \rightarrow O_3$$

Note: Roman capitals (S, P, D, etc.) are used for the states of atoms, and Greek capitals (Σ, Π, Δ, etc.) for those of molecules, and that the ground state of O_2 is a triplet $O_2(^3\Sigma)$.

The reaction $O_3 + h\nu \rightarrow O(^1D) + O_2(^1\Delta)$ has an energy threshold at 310 nm, and the other possible reaction $O_3 + h\nu \rightarrow O(^1D) + O_2(^3\Sigma)$ is formally forbidden by conservation of spin. Increasing evidence has, however, accumulated to show that the rate of production of $O(^1D)$, and, therefore, of hydroxyl radicals at wavelengths >310 nm is significant and that, therefore, in contrast to previous assumptions, the latter reaction makes an important contribution (Ravishankara et al. 1998).

Nitrate radicals are formed from NO that is produced during combustion processes and these are significant only during the night in the absence of photochemically produced OH radicals. They are formed by the reactions

$$NO + O_3 \rightarrow NO_2 \rightarrow O_2 : NO_2 + O_3 \rightarrow NO_3 + O_2$$

The concentrations of all these depend on local conditions, the time of day, and both altitude and latitude. Values of ca. 10^6 molecules/cm^3 for OH, 10^8–10^{10} molecules/cm^3 for NO$_3$, and ca. 10^{11} molecules/cm^3 for ozone have been reported. Not all of these reactants are equally important, and the rates of reaction with a substrate vary considerably. Reactions with hydroxyl radicals are generally the most important, and some illustrative values are given for the rates of reaction (cm^3/s/molecule) with hydroxyl radicals, nitrate radicals, and ozone (Atkinson 1990; summary of PAHs by Arey 1998).

1.8.1 Survey of Reactions

The major reactions carried out by hydroxyl and nitrate radicals may conveniently be represented for a primary alkane RH or a secondary alkane R$_2$CH. In both, hydrogen abstraction is the initiating reaction.

1. Hydrogen abstraction

$$RH + HO \rightarrow R + H_2O$$

$$RH + NO_3 \rightarrow R + HNO_3$$

2. Formation of alkylperoxy radicals

$$R + O_2 \rightarrow R\text{–}O\text{–}O$$

3. Reactions of alkylperoxy radicals with NO$_x$

$$R\text{–}O\text{–}O + NO \rightarrow R\text{–}O + NO_2$$

$$R\text{–}O\text{–}O + NO \rightarrow R\text{–}O\text{–}NO_2$$

4. Reactions of alkyloxy radicals

$$R_2CH\text{–}O + O_2 \rightarrow R_2CO + HO_2$$

$$R_2CH\text{–}O + NO \rightarrow R_2CH\text{–}O\text{–}NO_2 \rightarrow R_2CO + HNO$$

The concentration of NO determines the relative importance of reaction 3, and the formation of NO$_2$ is particularly significant since this is readily photolyzed to produce O(^{3}P) that reacts with oxygen to produce ozone. This alkane–NO$_x$ reaction may produce O$_3$ at the troposphere–stratosphere interface:

$$NO_2 \rightarrow NO + O(^3P) : O(^3P) + O_2 \rightarrow O_3$$

This is the main reaction for the formation of ozone although, under equilibrium conditions, the concentrations of NO$_2$, NO, and O$_3$ are interdependent and no net synthesis of O$_3$ occurs. When, however, the equilibrium is disturbed and NO is removed by reactions with alkylperoxy radicals (reactions 1 + 2 + 3), synthesis of O$_3$ may take place.

$$RH + OH \rightarrow R + H_2O : R + O_2 \rightarrow RO_2 : RO_2 + NO \rightarrow RO + NO_2$$

The extent to which this occurs depends on a number of issues (Finlayson-Pitts and Pitts 1997), including the reactivity of the hydrocarbon that is itself a function of many factors. It has been proposed that the possibility of ozone formation is best described by a reactivity index of incremental hydrocarbon reactivity (Carter and Atkinson 1987, 1989) that combines the rate of formation of O$_3$ with that of the reduction in the concentration of NO. The method has been applied, for example, to oxygenate additives to automobile fuel (Japar et al. 1991), while both anthropogenic compounds and naturally occurring hydrocarbons may be reactive.

TABLE 1.2

Degradation of Xenobiotic under Simulated Atmospheric Conditions

Xenobiotic	Reference
Aliphatic and aromatic hydrocarbons	Tuazon et al. (1986)
Substituted monocyclic aromatic compounds	Tuazon et al. (1986)
Terpenes	Atkinson et al. (1985b)
Amines	Atkinson et al. (1987c)
Heteroarenes	Atkinson et al. (1985a)
Chlorinated aromatic hydrocarbons	Kwok et al. (1995)
Volatile methyl-silicon compounds	Tuazon et al. (2000)

Clearly, whether or not ozone is formed depends also on the rate at which, for example, unsaturated hydrocarbons react with it. Rates of reactions of ozone with alkanes are, as noted above, much slower than for reaction with OH radicals, and reactions with ozone are of the greatest significance with unsaturated aliphatic compounds. The pathways plausibly follow those involved in chemical ozonization (Hudlický 1990).

The kinetics of the various reactions have been explored in detail using large-volume chambers that can be used to simulate reactions in the troposphere. They have frequently used hydroxyl radicals formed by photolysis of methyl (or ethyl) nitrite, with the addition of NO to inhibit photolysis of NO_2. This would result in the formation of $O(^3P)$ atoms, and subsequent reaction with O_2 would produce ozone, and hence NO_3 radicals from NO_2. Nitrate radicals are produced by the thermal decomposition of N_2O_5, and in experiments with O_3, a scavenger for hydroxyl radicals is added. Details of the different experimental procedures for the measurement of absolute and relative rates have been summarized, and attention drawn to the often considerable spread of values for experiments carried out at room temperature (~298 K) (Atkinson 1986). It should be emphasized that in the real troposphere, both the rates—and possibly the products—of transformation will be determined by seasonal differences both in temperature and the intensity of solar radiation. These are determined both by latitude and altitude.

The kinetics of the reactions of several xenobiotics with hydroxyl or nitrate radicals have been examined under simulated atmospheric conditions and include those given in Table 1.2.

For polychlorinated biphenyls (PCBs), rate constants were highly dependent on the number of chlorine atoms, and calculated atmospheric lifetimes varied from 2 d for 3-chlorobiphenyl to 34 d for 236-25 pentachlorobiphenyl (Anderson and Hites 1996). It was estimated that loss by hydroxylation in the atmosphere was a primary process for the removal of PCBs from the environment. It was later shown that the products were chlorinated benzoic acids produced by initial reaction with a hydroxyl radical at the 1-position followed by transannular dioxygenation at the 2- and 5-positions followed by ring fission (Brubaker and Hites 1998). Reactions of hydroxyl radicals with polychlorinated dibenzo[1,4]dioxins and dibenzofurans also play an important role for their removal from the atmosphere (Brubaker and Hites 1997). The gas phase and the particulate phase are in equilibrium, and the results show that gas-phase reactions with hydroxyl radicals are important for the compounds with fewer numbers of chlorine substituents, whereas for those with higher numbers of substituents particle-phase removal is significant.

Considerable attention has been directed in determining the products from reactions of aliphatic hydrocarbons, aromatic compounds, and unsaturated compounds including biogenic terpenes that exhibit appreciable volatility. These studies have been conducted both in simulation chambers and using natural sunlight in the presence of NO.

1.8.2 Survey of Reactants

1.8.2.1 Aliphatic Hydrocarbons

A range of transformation products has been identified from simulated reactions of alkanes. These include alkyl nitrates by reactions that have already been given, but also include a range of hydroxycarbonyls that are summarized in Table 1.3 (Reisen et al. 2005).

TABLE 1.3

Products from Hydroxyl-Radical Initiated Simulation Reaction with
n-Alkanes

Alkane	Product	Alkane	Product
n-Pentane	5-Hydroxypentan-2-one	*n*-Octane	5-Hydroxyoctan-2-one
	5-Hydroxypentanal		6-Hydroxyoctan-3-one
n-Hexane	5-Hydroxyhexan-2-one		7-Hydroxyoctan-4-one
	6-Hydroxyhexan-3-one		4-Hydroxyoctanal
n-Heptane	5-Hydroxyheptan-2-one		
	6-Hydroxyheptan-3-one		
	1-Hydroxyheptan-4-one		
	4-Hydroxyheptanal		

Source: Adapted from Reisen F et al. 2005. *Environ Sci Technol* 39: 4447–4453.

1.8.2.2 Aromatic Hydrocarbons

On account of their volatility, considerable effort has been directed to the atmospheric chemistry of monocyclic arenes. It has been shown that phenols represent a major group of transformation products, and that further reactions result in fission of the aromatic ring. For example, *o*-xylene forms diacetyl, methylgly-oxal, and glyoxal (Tuazon et al. 1986), which are also produced by ozonolysis (Levine and Cole 1932). The photooxidation of alkyl benzenes that are atmospheric contaminants with high volatility has been studied in detail: the important role of epoxides was demonstrated (Yu and Jeffries 1997) and reaction pathways were delineated (Yu et al. 1997). Products included both those with the ring intact, such as aromatic alde-hydes and quinones, together with a wide range of aliphatic compounds containing alcohol, ketone, and epoxy functional groups resulting from ring fission. A study using a large volume photoreactor at 298 K with an optical path of 579.2 m examined the role of benzene oxide and its tautomer oxepin since the equilibrium between these was probably shifted towards the oxepin (Klotz et al. 1997). Further hydrox-ylation of this followed by ring fission produced hexa-2,4-diene-1,6-dial (muconaldehyde). On account of interest in the role of nitrate radicals, the reaction with benzene oxide/oxepin was examined and yielded a 2,3-epoxide, from which hexa-2,4-diene-1,6-dial could be produced. On confirmation, it has been shown that chemical oxidation of benzene/oxepin with dimethyldioxirane produced muconaldehyde and, although the 2,3-epoxide could not be identified, the isomeric 4,5-epxyoxepin was a minor product (Bleasdale et al. 1997).

The atmospheric chemistry of naphthalene has been examined intensively since it is the most volatile PAH. The reaction with hydroxy radicals is dominant and produces naphthalene 2-formylcinnamal-dehyde (Arey 1998). The atmospheric transformation of naphthalene and alkylnaphthalenes has been examined in a Teflon chamber using hydroxyl radicals generated by the photolysis of methyl nitrite. These reactions were initiated by ring-scission to yield dicarbonyls that were subsequently converted into phthalic anhydrides by fission of one of the rings (Wang et al. 2007).

1.8.2.3 Biogenic Terpenes

Monoterpenes are appreciably volatile and are produced in substantial quantities by a range of higher plants and trees. Only a brief summary of examples is given here.

1. The carbonyl products from the transformation of isoprene in the presence of NO have been identified as methyl vinyl ketone, methacrolein, hydroxyacetaldehyde and hydroxyacetone (Grosjean et al. 1993a). Reactions with OH radicals and oxidation have been shown to produce methyl vinyl ketone, methacrolein and 3-methylfuran, and the kinetics of their production have been examined (Lee et al. 2005). A further examination of the reaction with hydroxyl radicals has added important details (Paulot et al. 2009). The initial reaction involves attack of hydroxyl

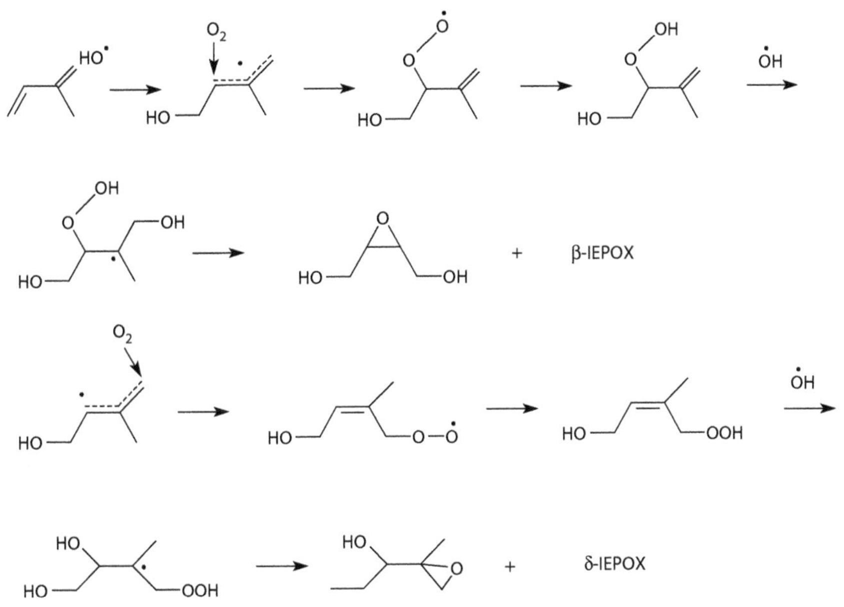

FIGURE 1.18 Reaction of isoprene with hydroxyl radicals. (Modified from Paulot F et al. 2009. *Science* 325: 730–733.)

radicals at C_1 followed by reaction with O_2 to produce the 1-hydroxy-2-hydroperoxide that may give rise to methacrylate and methyl vinyl ketone. It has been shown that the initial C_2 hydroperoxide may also react with an OH radical followed by an intramolecular reaction to produce a 2,3-epoxide with hydroxyl groups at C_1 and C_4. As an alternative, the C_4 peroxide may produce the analogous 3,4-epoxide (Figure 1.18). Both these reactions produce hydroxyl radicals whose reformation contributes to the stability of HO_x levels in remote regions, and both the hydroperoxides and the epoxides have been detected in and contribute to aerosols.

2. Products from reaction of α-pinene with ozone that produced a range of cyclobutane carboxylic acids (Kamens et al. 1999).

3. Rapid reactions of linalool with OH radicals, NO_3 radicals, and ozone in which the major products were acetone and 5-ethenyldihydro-5-methyl-2(3*H*)-furanone (Shu et al. 1997).

4. Plant metabolite *cis*-hex-3-ene-1-ol that is the precursor of peroxypropionyl nitrate (Grosjean et al. 1993b) analogous to peroxyacetyl nitrate.

5. Degradation of many terpenes has been examined including β-pinene, d-limonene, and *trans*-caryophyllene (Grosjean et al. 1993c).

6. Products formed by reaction of NO_3 radicals with α-pinene have been identified and include pinane epoxide, 2-hydroxypinane-3-nitrate, and 3-ketopinan-2-nitrate formed by reactions at the double bond, and pinonaldehyde that is produced by ring fission between C_2 and C_3 (Wängberg et al. 1997). These reactions should be viewed in the general context of "odd nitrogen" to which alkyl nitrates belong (Schneider et al. 1998).

7. Gas-phase products from the reactions of ozone with the monoterpenes (–)-β-pinene and (+)-sabinene included the ketones formed by oxidative fission of the exocyclic C=C bonds as well as ozonides from the addition of ozone to this bond (Griesbaum et al. 1998).

1.9 Reentry of Tropospheric Transformation Products

An important aspect of atmospheric reactions is the possibility that tropospheric transformation products subsequently enter aquatic and terrestrial ecosystems through precipitation or by particle deposition.

1.9.1 Halogenated Alkanes and Alkenes

The stability of perchlorofluoroalkanes is due to the absence of hydrogen atoms that may be abstracted by reaction with hydroxyl radicals. Attention has, therefore, been directed to chlorofluoroalkanes containing at least one hydrogen atom (Hayman and Derwent 1997). Considerable effort has also been directed to the reactions of chloroalkanes and chloroalkenes, and this deserves a rather more detailed examination in the light of interest in the products that are formed.

1. There has been concern over the fate of halogenated aliphatic compounds in the atmosphere, and a single illustration of the diverse consequences is noted here. The initial reaction of 1,1,1-trichloroethane with hydroxyl radicals produces the $Cl_3C–CH_2$ radical by abstraction of H and then undergoes a complex series of reactions including the following:

$$Cl_3C–CH_2 + O_2 \rightarrow Cl_3C–CH_2O_2$$

$$2Cl_3C–CH_2O_2 \rightarrow 2Cl_3C–CH_2O$$

$$Cl_3C–CH_2O + O_2 \rightarrow Cl_3C–CHO$$

   In addition, the alkoxy radical $Cl_3C–CH_2O$ produces highly reactive $COCl_2$ (phosgene) (Nelson et al. 1990; Platz et al. 1995) that has been identified in atmospheric samples and was attributed to the transformation of *gem*-dichloro aliphatic compounds (Grosjean 1991).

2. The atmospheric degradation of tetrachloroethene is initiated by reaction with Cl radicals and produces trichloroacetyl chloride as the primary intermediate which is formed followed by the following reactions (Franklin 1994):

$$Cl_2C=CCl_2 + Cl \rightarrow Cl_3C–CCl_2$$

$$Cl_3C–CCl_2 + O_2 \rightarrow Cl_3C–CCl_2O_2$$

$$Cl_3C–CCl_2O_2 + NO \rightarrow Cl_3C–CCl_2O + NO_2$$

$$Cl_3C–CCl_2O \rightarrow Cl_3C–COCl + Cl \rightarrow COCl_2 + CCl_3$$

$$CCl_3 + O_2 + NO \rightarrow COCl_2 + NO_2 + Cl$$

An overview of the reactions involving trihalomethanes (haloforms) CHXYZ, where X, Y, and Z are halogen atoms, has been given in the context of ozone depletion (Hayman and Derwent 1997). Interest in the formation of trichloroacetaldehyde formed from trichloroethane and tetrachloroethene is heightened by the phytotoxicity of trichloroacetic acid (Frank et al. 1994), and by its occurrence in rainwater that seems to be a major source of this contaminant (Müller et al. 1996). The situation in Japan seems, however, to underscore the possible significance of other sources including chlorinated wastewater (Hashimoto et al. 1998). Whereas there is no doubt about the occurrence of trichloroacetic acid in rainwater (Stidson et al. 2004), its major source is unresolved since questions remain on the rate of hydrolysis of trichloroacetaldehyde (Jordan et al. 1999).

Low concentrations of trifluoroacetate have been found in lakes in California and Nevada (Wujcik et al. 1998). It is formed by atmospheric reactions from 1,1,1,2-tetrafluoroethane and from the chlorofluorocarbon replacement compound $CF_3–CH_2F$ (HFC-134a) in an estimated yield of 7–20% (Wallington et al. 1996). CF_3OH that is formed from CF_3 in the stratosphere is apparently a sink for its oxidation products (Wallington and Schneider 1994).

$$CF_3–CH_2F + OH \rightarrow CF_3–CHF$$

$$CF_3–CHF + O_2 \rightarrow CF_3–CHFO_2$$

$$CF_3–CHFO_2 + NO \rightarrow CF_3–CHFO + NO_2$$

$$CF_3–CHFO + O_2 \rightarrow CF_3–COF$$

$$CF_3–CHFO \rightarrow CF_3 + H–COF$$

It has been suggested that an alternative source of trifluoroacetic acid in the environment may be provided by the aqueous photolysis of TFM and related trifluoromethylphenols (Ellis and Mabury 2000). Although trifluoroacetate is accumulated by a range of biota through incorporation into biomolecules (Standley and Bott 1998), unlike trichloroacetate, it is only weakly phytotoxic, and there is no evidence for its inhibitory effect on methanogenesis (Emptage et al. 1997). Attention has been directed to the transformation of perfluorotelomer alcohols $CF_3(CF_2)_7–CH_2CH_2OH$ that could be the source of the ubiquitous perfluorinated carboxylic acids (PFCAs), and that have been produced during simulated atmospheric reactions with chlorine radicals as surrogate for hydroxyl radicals (Ellis et al. 2004).

1.9.2 Arenes and Nitroarenes

The transformation of arenes in the troposphere has been discussed in detail (Arey 1998). Their destruction can be mediated by reaction with hydroxyl radicals, and from naphthalene a wide range of compounds is produced, including 1- and 2-naphthols, 2-formylcinnamaldehyde, phthalic anhydride, and with less certainty 1,4-naphthoquinone and 2,3-epoxynaphthoquinone. Both 1- and 2-nitronaphthalene were formed through the intervention of NO_2 (Bunce et al. 1997). Attention has also been directed to the composition of secondary organic aerosols from the photooxidation of monocyclic aromatic hydrocarbons in the presence of NO_x (Forstner et al. 1997); the main products from a range of alkylated aromatics were 2,5-furandione and the 3-methyl and 3-ethyl congeners.

Considerable attention has been directed to the formation of nitroarenes that may be formed by several mechanisms: (a) initial reaction with hydroxyl radicals followed by reactions with nitrate radicals or NO_2 and (b) direct reaction with nitrate radicals. The first is important for arenes in the troposphere, whereas the second is a thermal reaction that occurs during combustion of arenes. The kinetics of formation of nitroarenes by gas-phase reaction with N_2O_5 has been examined for naphthalene (Pitts et al. 1985a) and methylnaphthalenes (Zielinska et al. 1989); biphenyl (Atkinson et al. 1987a,b); acephenanthrylene (Zielinska et al. 1988); and for adsorbed pyrene (Pitts et al. 1985b). Both 1- and 2-nitronaphthalene were formed through OH radical-initiated reactions with naphthalene by the intervention of NO_2 (Bunce et al. 1997). The major product from the first group of reactions is 2-nitronaphthalene, and a number of other nitroarenes have been identified including nitropyrene and nitrofluoranthenes (Arey 1998). The tentative identification of hydroxylated nitroarenes in air particulate samples (Nishioka et al. 1988) is consistent with the operation of this dual mechanism. Reaction of methyl arenes with nitrate radicals in the gas phase gives rise to a number of products. From toluene, the major product was benzaldehyde with lesser amounts of 2-nitrotoluene > benzyl alcohol nitrate > 4-nitrotoluene > 3-nitrotoluene (Chlodini et al. 1993). An important example is the formation of the mutagenic 2-nitro- and 6-nitro-6*H*-dibenzo[*b,d*]pyran-6-ones (Figure 1.19) from the oxidation of phenanthrene in the presence of NO_x and methyl nitrite as a source of hydroxyl radicals (Helmig et al. 1992a). These compounds have been identified in samples of ambient air (Helmig et al. 1992b), and analogous compounds from pyrene have been tentatively identified (Sasaki et al. 1995). These compounds add further examples to the list of mononitroarenes that already include 2-nitropyrene and 2-nitrofluoranthene, and it appears plausible to suggest that comparable reactions are involved in the formation of the 1,6- and 1,8-dinitroarenes that have been identified in diesel exhaust. 3-Nitrobenzanthrone that is formally analogous to the dibenzopyrones noted earlier has also been identified in diesel exhaust and is also highly mutagenic to *Salmonella typhimurium* strain TA 98 (Enya et al. 1997).

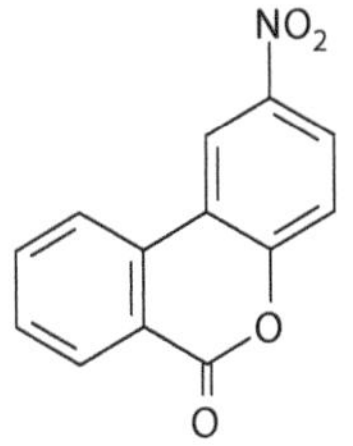

FIGURE 1.19 Product from the photochemical reaction of phenanthrene and No$_x$.

Many nitroarenes including 2-nitronaphthalene are direct-acting, frameshift mutagens in the Ames test (Rosenkranz and Mermelstein 1983). Although the mechanism has not been finally resolved, it appears to involve metabolic participation of the test organisms, for example, by reduction. In addition, nitroarenes may be reduced microbiologically to the corresponding amino compounds, and in terrestrial and aquatic systems these may have undesirable properties: (a) some including 2-aminonaphthalene are carcinogenic to mammals and (b) they react with components of humic and fulvic acids, which makes them more recalcitrant to degradation and, therefore, more persistent in ecosystems.

Although nitrofluoranthenes and nitropyrenes are established atmospheric products from the nitration of the respective PAHs, their direct mutagenic activity is unable to account for the total mutagenic activity of airborne samples. It has been shown that the concentrations of 1- and 2-nitrotriphenylenes were higher than those of 1-nitropyrene, and that 1-nitrotriphenylene was a highly active direct-acting mutagen in *S. typhimurium* strains TA98 and YG 1024 (Ishii et al. 2000). Nitrated phenols have been found in rainwater, and are secondary products from atmospheric oxidation of benzene and alkylated toluenes followed by nitration (Kohler and Heeb 2003). Plausible mechanisms for the formation of 2,4-dinitrophenol from 2- and 4-nitrophenol in the atmospheric liquid phase have been proposed (Vione et al. 2005), and include nitration by NO_2 or NO_3 radicals. A wide range of azaarenes is produced during combustion (Herod 1998), and these may enter the troposphere, with formation of the corresponding nitro derivatives.

1.9.3 Alkylated Arenes

The products from the oxidation of alkylbenzenes under simulated atmospheric conditions have been examined. Both ring epoxides that were highly functionalized, and aliphatic epoxides from ring fission were tentatively identified (Yu and Jeffries 1997). Formation of the epoxides, many of which are mutagenic, may cause concern over the transformation products of monocyclic aromatic hydrocarbons in the atmosphere. Toluene is an important atmospheric contaminant and its photochemistry has been explored using EUROPHORE, which is described in Chapter 5 (Klotz et al. 1998). This used natural illumination in a large-volume chamber equipped with differential optical spectroscopy (path length 326.8 m) and Fourier transform spectroscopy (path length 128 m) to minimize artifacts from sampling or treatment for analysis. The photochemistry of toluene has been explored using concentrations of NO_x resembled those found in the atmosphere, while concentrations of ozone were low at the beginning of the experiment. The major reaction involved OH radicals that initiated two reactions: (a) removal of a hydrogen atom from the methyl group to form benzaldehyde and (b) formation of a toluene-hydroxyl adduct (methyl-hydroxy-cyclohexadienyl radical) that reacted with oxygen to produce *o*-, *m*-, and *p*-cresol that reacted further with nitrate radicals (Alvarez et al. 2007).

1.9.4 Sulfides and Disulfides

Sulfides and disulfides can be produced by bacterial reactions in the marine environment. 2-Dimethylthiopropionic acid is produced by algae and by the marsh grass *Spartina alternifolia*, and may then be metabolized in sediment slurries under anoxic conditions to dimethyl sulfide (Kiene and Taylor 1988), and by aerobic bacteria to methyl sulfide (Taylor and Gilchrist 1991). Further details are given in Chapter 11, Part 2. Methyl sulfide can also be produced by biological methylation of sulfide itself (HS^-). Carbon radicals are not the initial atmospheric products from organic sulfides and disulfides, and the reactions also provide an example in which the rates of reaction with nitrate radicals exceed those with hydroxyl radicals. Dimethyl sulfides—and possibly methyl sulfide as well—are oxidized in the troposphere to sulfur dioxide and methanesulfonic acids:

$$CH_3\text{–}SH \rightarrow CH_3\text{–}SO_3H$$

$$CH_3\text{–}S\text{–}S\text{–}CH_3 \rightarrow CH_3\text{–}SO_3H + CH_3\text{–}SO$$

$$CH_3\text{–}SO \rightarrow CH_3 + SO_2$$

It has been suggested that these compounds may play a critical role in promoting cloud formation (Charleson et al. 1987), so that the long-term effect of the biosynthesis of methyl sulfides on climate alteration may be important. And yet, at first glance this seems far removed from the production of an osmolyte by marine algae and its metabolism in aquatic systems. The occurrence of methyl sulfates in atmospheric samples (Eatough et al. 1986) should be noted, although the mechanism of its formation appears not to have been established. These reactions provide a good example of the long chain of events that may bring about environmental effects through the subtle interaction of biotic and abiotic reactions in both the aquatic and atmospheric environments.

Appreciation of interactive processes that have been outlined has been able to illuminate discussion of mechanisms for reactions as diverse as the acidification of water masses, climate alteration, ozone formation and destruction, and the possible environmental roles of trichloroacetic acid and nitroarenes.

1.10 Chemically Mediated Transformations

Only a limited number of the plethora of known chemical reactions has been observed in transforming xenobiotics in the environment. An attempt is made merely to present a classification of the reactions that take place with illustrations of their occurrence.

1.10.1 Hydrolysis

Organic compounds containing carbonyl groups flanked by alkoxy groups (esters) or by amino- or substituted amino groups (amides, carbamates, and ureas) may be hydrolyzed by purely abiotic reactions under appropriate conditions of pH. The generally high pH of seawater (ca. 8.2) suggests that chemical hydrolysis may be important in this environment. In contrast, although very few natural aquatic ecosystems have pH values sufficiently low for acidic hydrolysis to be of major importance, this could be important in terrestrial systems when the pH is lowered by fermentation of organic substrates. It is, therefore, important to distinguish between alkaline or neutral, and acidic hydrolytic mechanisms. It should also be appreciated that both hydrolytic and photolytic mechanisms may operate simultaneously and that the products of these reactions may not necessarily be identical.

It has already been noted that substantial numbers of important agrochemicals contain carbonyl groups, so that abiotic hydrolysis may be the primary reaction in their transformation: the example of carbaryl has already been cited (Wolfe et al. 1978). The same general principles may be extended to phosphate and thiophosphate esters, although in these cases it is important to bear in mind the stability to hydrolysis of primary and secondary phosphate esters under the neutral or alkaline conditions that prevail in most natural ecosystems. In contrast, sulfate esters and sulfamides are generally quite resistant to chemical hydrolysis except under rather drastic conditions, and microbial sulfatases and sulfamidases generally mediate their hydrolysis. Some examples illustrate the range of hydrolytic reactions that involve diverse agrochemicals that may enter aquatic systems by leaching from the soil.

1. The cyclic sulfite of α- and β-endosulfan (Singh et al. 1991).
2. The carbamate phenmedipham that results in the intermediate formation of *m*-tolyl isocyanate (Figure 1.20) (Bergon et al. 1985).
3. 2-(Thiocyanomethylthio)benzthiazole initially forms 2-thiobenzthiazole, which is transient and then rapidly degrades photochemically to benzothiazole and 2-hydroxybenzothiazole (Brownlee et al. 1992).

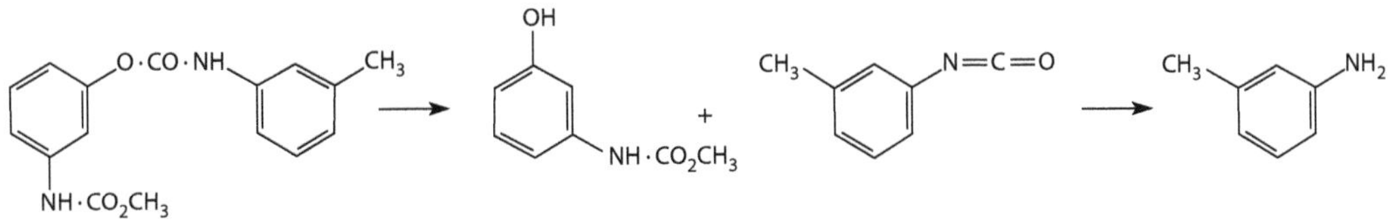

FIGURE 1.20 Hydrolysis of phenmedipham.

4. Aldicarb undergoes simple hydrolysis at pH values above 7, whereas at pH values <5 an elimination reaction intervenes (Figure 1.21) (Bank and Tyrrell 1984).

5. The sulfonyl urea sulfometuron methyl is stable at neutral or alkaline pH values, but is hydrolyzed at pH 5 to methyl 2-aminosulfonylbenzoate that is cyclized to saccharin (Figure 1.22) (Harvey et al. 1985). The original compound is completely degraded to CO_2 by photolysis.

6. The pyrethroid insecticides fenvalerate and cypermethrin are hydrolyzed under alkaline conditions at low substrate concentrations, but at higher concentrations the initially formed 3-phenoxybenzaldehyde reacts further with the substrate to form dimeric compounds (Figure 1.23) (Camilleri 1984).

7. The sulfonyl urea herbicide rimsulfuron is degraded increasingly rapidly in the pH range 5–9. The main degradation pathway is by rearrangement of the sulfonyl urea group followed by hydrolysis (Schneiders et al. 1993) (Figure 1.24).

8. The thiophosphate phorate is degraded in aqueous solutions at pH 8.5 to yield diethyl sulfide and formaldehyde, which are formed by nucleophilic attack either at the P=S atom or the methylene dithioketal carbon atom (Hong and Pehkonen 1998).

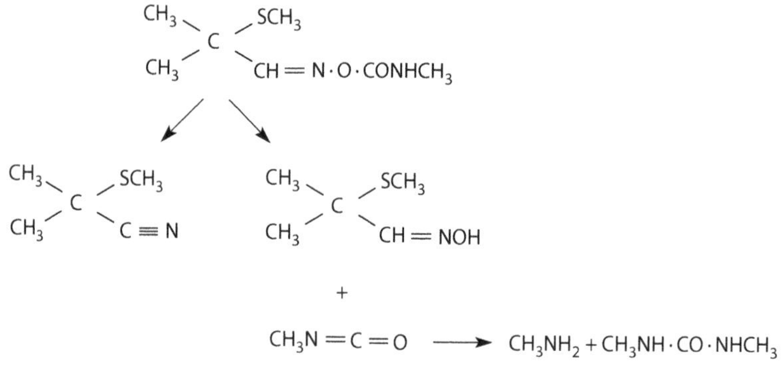

FIGURE 1.21 Hydrolysis of aldicarb.

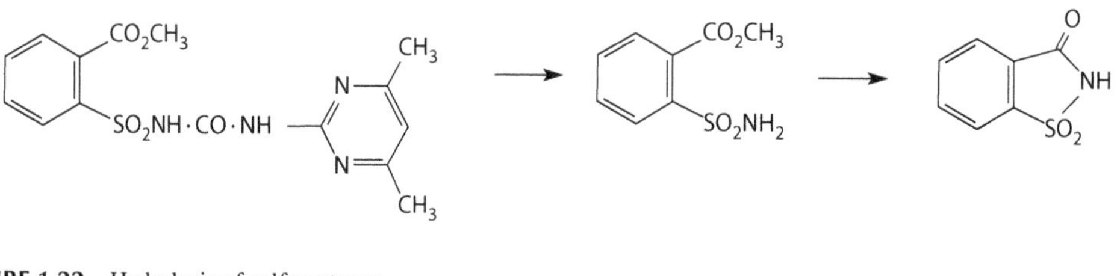

FIGURE 1.22 Hydrolysis of sulfometuron.

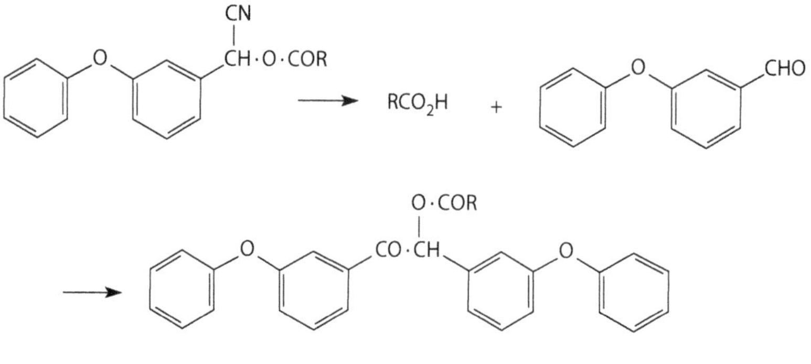

FIGURE 1.23 Hydrolysis of the pyrethroid insecticides fenvalerate and cypermethrin.

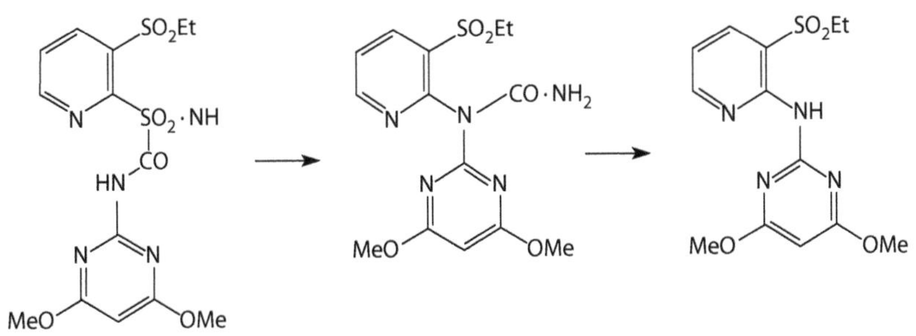

FIGURE 1.24 Hydrolysis of rimsulfuron.

9. Hydrolysis of the isoxazole herbicide isoxaflutole rapidly produces a biologically active diketonitrile that is further degraded to products, which are partly bound to soil components and partly degraded to CO_2 (Beltran et al. 2000; Lin et al. 2002; Taylor-Lovell et al. 2002) (Figure 1.25).
10. In alkaline solution, 3-bromo-2,2-bis(bromomethyl)propanol undergoes successive loss of bromide to produce 2,6-dioxaspiro(3,3)-heptane (Ezra et al. 2005) (Figure 1.26).
11. The reaction of brominated diphenyl ethers with methoxide in dimethyl formamide has been examined, and suggested as a ranking of their susceptibility to hydrolytic reactions under natural conditions (Rahm et al. 2005). The nature of the products was not apparently systematically examined.

Three important comments should be added:

1. Abiotic hydrolysis generally accomplishes only a single step in the ultimate degradation of the compounds that have been used for illustration. The intervention of subsequent biotic reactions is, therefore, almost invariably necessary for their complete mineralization.
2. The operation of these hydrolytic reactions is independent of the oxygen concentration of the system so that—in contrast to biotic degradation and transformation—these reactions may occur effectively under both aerobic and anaerobic conditions.
3. Rates of hydrolysis may be influenced by the presence of dissolved organic carbon, or organic components of soil and sediment. The magnitude of the effect is determined by the structure of the compound and by the kinetics of its association with these components. For example, whereas the neutral hydrolysis of chlorpyrifos was unaffected by sorption to sediments, the

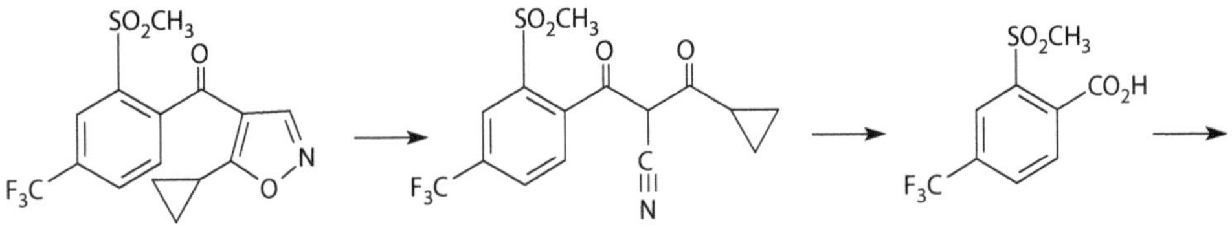

FIGURE 1.25 Degradation of isoxaflutole.

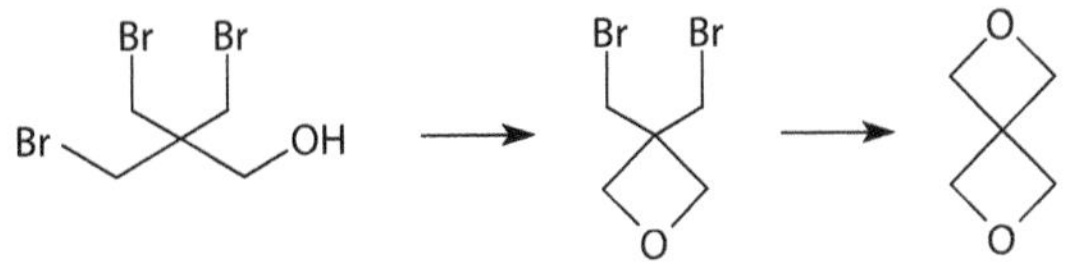

FIGURE 1.26 Degradation of 3-bromo-2,2-bis(bromomethyl)propanol.

rate of alkaline hydrolysis was considerably slower (Macalady and Wolfe 1985); humic acid also reduced the rate of alkaline hydrolysis of 1-octyl 2,4-dichloro-phenoxyacetate (Perdue and Wolfe 1982). Conversely, sediment sorption had no effect on the neutral hydrolysis of 4-chlorostilbene oxide, although the rate below pH 5 where acid hydrolysis dominates was reduced (Metwally and Wolfe 1990).

1.10.2 Reductive Displacement: Dehalogenation and Desulfurization

There has been considerable interest in the abiotic dechlorination of chlorinated ethenes at contaminated sites. Reductive dehalogenation has, therefore, been examined using a range of reductants, many of them involving reduced complexes of porphyrins or corrins.

1.11 Reductive Dehalogenation

1.11.1 Zero-Valent Metals

Zero-valent iron has been used for diverse dehalogenations:

1. The reductive dechlorination of DDT with zero-valent iron in an anaerobic aqueous medium produced DDD by reductive dechlorination, and this was further transformed at higher concentrations of Fe (Sayles et al. 1997).
2. Zero-valent iron was able to degrade octahydro-1,3,5,7-tetranitro-1,3,5,7-tetrazocine (HMX) to hydrazine and formaldehyde, putatively via the intermediate methylendinitramine (Monteil-Rivera et al. 2005).
3. Zero-valent metals in subcritical water have been applied to the dechlorination of PCBs (Kubátova et al. 2003; Yak et al. 1999), and is noted again later.
4. The dechlorination of polychlorinated dibenzo[1,4]dioxins was examined using zero-valent iron that was palladized electrochemically. 1,2,3,4-Tetrachlorodibenzo[1,4]dioxin was dechlorinated successively to 1,2,3-trichloro, 1,2-dichloro, and 1-monochloro congeners (Kim et al. 2008).

Attention has been particularly devoted to the application of nanoscale particles of zero-valent iron, and a range of halogenated aliphatic compounds has been examined.

1. For trichloroethene (TCE), the stoichiometric amount of iron and the effect of different preparations determine the outcome of the several competing reactions. Coupling products such as butenes, acetylene and its reduction products ethene and ethane, and products with five or six carbon atoms were formed (Liu et al. 2005). Although a field-scale application successfully lowered the concentration of TCE, there was evidence for the formation of the undesirable *cis*- 1,2-dichloroethene and 1-chloroethene (vinyl chloride) in the groundwater (Quinn et al. 2005).
2. A range of chlorinated ethanes has been examined and several transformations were observed (Song and Carraway 2005):
   a. Combined β-elimination and reductive dechlorination:
      Hexachloroethane → tetrachloroethene
      Pentachloroethane → tetrachloroethene and trichloroethene
      1,1,2,2-Tetrachloroethane → *cis*- and *trans*-dichloroethene
   b. Reductive dechlorination: 1,1,1-trichloroethane → 1,1-dichloroethane
   c. α-Elimination: 1,1-dichloroethane → ethane
3. Experiments have been carried out to compare nanoscale catalysts composed of Fe-, Ni-, and Co-complexes of several porphyrins or cyanocobalamin (Dror et al. 2005). A cobalt–porphyrin

complex and cyanocobalamin in the presence of Ti(III)citrate reduced the initial concentrations of tetrachloromethane and tetrachloroethene by ⊕99.5%, and the porphyrin was equally effective with trichloroethene. The advantage of using heterogeneous catalysts was shown by experiments in repetitive cycling of tetrachloromethane. Zero-valent metals degrade vicinal dichlorides such as tetrachloroethene by α-elimination to produce dichloroacetylene and finally acetylene (Roberts et al. 1996).

1.11.2 Chromous Chloride

1. Cr(II) has been used to bring about dehalogenation of alkyl halides involving the production of alkyl radicals, and details have been provided in a substantive review (Castro 1998). The ease of reduction is generally: iodides > bromides > chlorides, while tertiary halides are the most reactive and primary halides the least (Castro and Kray 1963, 1966).

2. The stereospecificity of dehalogenation of vicinal dibromides to olefins was examined for reducing agents including Cr(II), iodide, and Fe^0 (Totten et al. 2001). For dibromostilbene, the (*E*)-stilbene represented >70% of the total olefin that was produced, and for *threo*-dibromopentane reduction by Cr(II) produced ca. 70% of (*E*)-pent-2-ene, whereas values for iodide and Fe^0 were <5% of this.

1.11.3 Porphyrins and Corrins

1. Attention has been directed to the role of corrins and porphyrins in the absence of biological systems. A range of structurally diverse compounds can be dechlorinated: these include DDT (Zoro et al. 1974); lindane (Marks et al. 1989); mirex (Holmstead 1976); C$_1$ chloroalkanes (Krone et al. 1989); C$_2$ chloroalkenes (Gantzer and Wackett 1991), and C$_2$ chloroalkanes (Schanke and Wackett 1992). Detailed mechanistic examination of the dehydrochlorination of pentachloroethane to tetrachloroethene reveals, however, the potential complexity of this reaction, and the possibly significant role of pentachloroethane in the abiotic transformation of hexachloroethane (Roberts and Gschwend 1991).

2. Considerable attention has been directed to dehalogenation mediated by corrinoids and porphyrins in the presence of a chemical reductant (references in Gantzer and Wackett 1991; Glod et al. 1997; Workman et al. 1997). Illustrations are provided by the dechlorination and elimination reactions carried out by titanium(III) citrate and hydroxocobalamin (Bosma et al. 1988; Glod et al. 1997). The involvement of corrinoids and porphyrins is consistent with the occurrence of analogous mechanisms for biological reactions that are dechlorination generally transformations involving dehydrochlorination (elimination) and reductive.

   a. Chlorofluoroalkanes—interest in their adverse environmental effects has stimulated interest in their anaerobic transformation. Although this is probably mediated by abiotic reactions involving porphyrins (Lovley and Woodward 1992; Lesage et al. 1992), it should be noted that the C–F bond was apparently retained in the products (Lesage et al. 1992).

   b. Dechlorination and elimination reactions of hexachlorobuta-1,3-diene took place in the presence of titanium(III) citrate and hydroxocobalamin (Bosma et al. 1994). Hexachlorobuta-1,3-diene was dechlorinated to the pentachloro compound and, by dechlorination and elimination, successively to trichlorobut-1-ene-3-yne (probably the 1,2,2-trichloro isomer), and but-1-ene-2-yne. The specificity of corrins and porphyrins is of particular interest since it seems to be significantly less than that of the enzymes generally implicated in microbial dechlorination. In aqueous solution containing cobalamin and titanium(III) as reductant, tetrachloroethene was dehalogenated to the trichloroethene radical, and thence to the dichlorovinyl radical, chloroacetylene, and acetylene (Glod et al. 1997).

   c. Reductive dechlorination of the α-, β-, γ-, and δ-isomers of hexachlorocyclohexane was examined under several conditions. The most effective was the combination of titanium(III) citrate and hydroxocobalamin, when all the isomers were degraded in the order of rates $\gamma > \alpha > \delta > \beta$. Tetrachlorocyclohexene was formed as a transient intermediate with chlorobenzene as the ultimate product (Rodríguez-Garrido et al. 2004).

d. The reductive defluorination of PFOS isomers was examined under anaerobic conditions using vitamin B_{12} and Ti(III)-citrate (Ochoa-Herrera et al. 2008). Fluoride was produced, and the branched-chain isomers with CF_3 groups at several positions were preferentially defluorinated.

3. An additional aspect of these dehalogenations that elucidates the role of vitamin B_{12} is provided by experiments with *Shewanella alga* strain BrY (Workman et al. 1997). This organism carries out reduction of Fe(III) and Co(III) during growth with lactate and H_2, and was used to reduce vitamin B_{12a} anaerobically in the presence of an electron donor. The biologically reduced vitamin B_{12} was then able to transform tetrachloromethane to CO.

4. The reactions of alkyl halides with Fe(II) deuteroporphyrin IX have been examined (Wade and Castro 1973). Three classes of reaction were observed (i) hydrogenolysis, (ii) elimination to alkenes, and (iii) coupling of alkyl-free radicals. Further discussion has been given in Castro (1998).

5. Experiments have been carried out to mimic the reactions of model systems for coenzyme F_{430} that is involved in the terminal step in the biosynthesis of methane, and that is able to dechlorinate CCl_4 successively to $CHCl_3$ and CH_2Cl_2 (Krone et al. 1989). Nickel(I) isobacteriochlorin anion was generated electrolytically and used to examine the reactions with alkyl halides in dimethylformamide (Helvenston and Castro 1992). The three classes of reaction were the same as those observed with Fe(II) deuteroporphyrin IX that have already been noted.

6. Experiments that were carried out to compare nanoscale catalysts composed of Fe, Ni, and Co complexes of several porphyrins, or cyanocobalamin have already been noted (Dror et al. 2005).

1.12 Thiol and Sulfide Reductants

Dehalogenation mediated by sulfides has been observed both in homogeneous solutions of sulfides in the presence of quinone electron acceptors, and in heterogeneous systems using various forms of iron sulfide without additional electron acceptors.

1.12.1 Homogeneous Dechlorination

The dehalogenation of chlorinated alkanes and alkenes has been examined using sulfide and its soluble derivatives in the presence of quinones: illustrative examples include the following.

1. The formation of tetrachloroethene from hexachloroethane with H_2S in the presence of 5-hydroxy-naphtho-1,4-quinone (juglone) (Perlinger et al. 1996).

2. The reduction of hexachloroethane by both HS^- and polysulfides (S_4^{2-}) to tetrachloroethene and pentachloroethane (Miller et al. 1998).

3. The transformation of tetrachloromethane by HS^- and cysteine in the presence of quinones as electron-transfer mediators (Doong and Chiang 2005). 5-Hydroxy-naphtho-1,4-quinone (juglone) was the most effective, and much greater than anthraquinone 2,6-disulfate. The enhancement was the result of the formation of mercaptoquinones or, from anthraquinone 2,6-disulfate semiquinone radicals.

1.12.2 Heterogeneous Dechlorination

Considerable attention has been given to the reductive dechlorination in the *absence* of quinones by various forms of insoluble iron sulfide. These can be formed by the reaction between sulfide produced from sulfate by anaerobic bacteria and sedimentary metal oxide/hydroxide minerals. Representative examples include the following.

1. *Hexachloroethane:* Tetrachloroethene was produced from hexachloroethane putatively via free radical reactions (Butler and Hayes 1998), and the rates of dechlorination of hexachloroethane by the iron sulfide Mackinawite decreased in the presence of Cr(III) and Mn(II), and increased in the presence of Co(II), Ni(II), Cu(II), Zn (II) (Jeong and Hayes 2003).

2. *Trichloroethene and tetrachloroethene:* The dehydrochlorination of trichloroethene and tetrachloroethene produced acetylene as the final product—putatively by free radical reaction (Butler and Hayes 1999), and it was shown that there were two competing reactions: (a) trichloroethene → *cis*-dichloroethene and (b) HC≡C–Cl→HC≡CH (Butler and Hayes 2001). The rates of reductive dechlorination of tetrachloroethene and trichloroethene were increased by the presence of Co(II) and Hg(II) and decreased by Ni(II) (Jeong and Hayes 2007).

1.13 Reductive Desulfurization

In the context of diagenesis in recent anoxic sediments, reduced carotenoids, steroids, and hopanoids have been identified, and it has been suggested that reduction by sulfide, produced for example, by the reduction of sulfate could play an important part (Hebting et al. 2006). The partial reduction of carotenoids by sulfide has been observed as a result of the addition of sulfide to selected allylic double bonds, followed by reductive desulfurization. This is supported by the finding that the thiol in allylic thiols could be reductively removed by sulfide to produce unsaturated products from free-radical reactions (Hebting et al. 2003).

1.13.1 Reductions Other than Dehalogenation

Aromatic nitro compounds include both important explosives and a number of agrochemicals. Concern with their fate has motivated extensive examination of their reduction to amines under a range of conditions.

1. Reduction of monocyclic aromatic nitro compounds has been demonstrated: (a) with reduced sulfur compounds mediated by a naphthoquinone or an iron porphyrin (Schwarzenbach et al. 1990), and (b) by Fe(II) and magnetite produced by the action of the anaerobic bacterium *Geobacter metallireducens* (Heijman et al. 1993). Quinone-mediated reduction of monocyclic aromatic nitro compounds by the supernatant monocyclic aromatic nitro compounds has been noted (Glaus et al. 1992), and these reactions may be significant in determining the fate of aromatic nitro compounds in reducing environments (Dunnivant et al. 1992).

2. The reduction of 2,4,6-trinitrotoluene with Fe⁰ has been extensively studied (references in Bandstra et al. 2005), and it has finally produced 2,4,6-triaminotoluene that could undergo polymerization.

3. The degradation of trifluralin [2,6-dinitro-*N,N*-dipropyl-4-(trifluoromethyl)benzenamine] by Fe(II)/goethite has been examined under anaerobic conditions as a model for reactions in flooded soils (Klupinski and Chin 2003). A range of transformation products was found including those produced by reduction of the nitro groups, dealkylation, and formation of benzimidazoles presumably via intermediate nitroso compounds (Figure 1.27).

A number of other abiotic reductions have been described:

1. The aerobic biodegradation of monocyclic azaarenes frequently involves reduction (Chapter 10, Part 1), but purely chemical reduction may take place under highly anaerobic conditions. This has been encountered with the substituted 1,2,4-triazolo[1,5*a*]pyrimidine (Flumetsulam) (Wolt et al. 1992) (Figure 1.28).

2. Cell-free supernatants may mediate reductions. The reduction of aromatic nitro compounds by SH⁻ was mediated by the filtrate from a strain of *Streptomyces* sp. that is known to synthesize

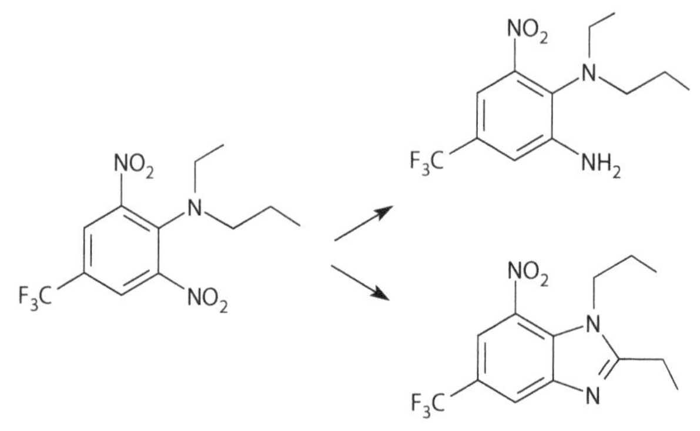

FIGURE 1.27   Degradation of 2,6-dinitro-*N,N*-dipropyl(trifluoromethyl)benzenamine.

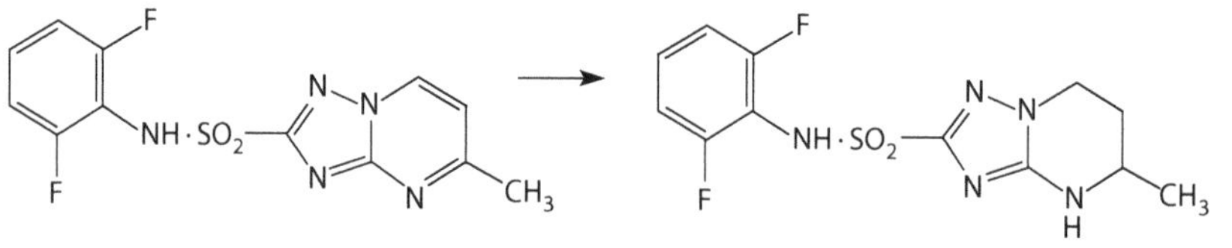

FIGURE 1.28   Reductive degradation of 1,2,4-triazolo[1,5*a*]pyrimidine.

2-amino-3-carboxy-5-hydroxybenzo-1,4-quinone (cinnaquinone) and the 6,6′-diquinone (dicin-naquinone) as secondary metabolites (Glaus et al. 1992). The quinones presumably function as electron-transfer mediators.

3. Dimerization of a number of arylalkenes catalyzed by vitamin B_{12} and Ti(III) as reductant has been examined (Shey et al. 2002). Mechanisms were examined including the requirement of a reductant, and a reaction was proposed that involved the formation of radicals at the benzylic carbon atoms.

1.13.2 Nucleophilic Reactions

Although the foregoing reactions involve dehalogenation by reduction or elimination, nucleophilic displacement of chloride may also be important. This has been examined with dihalomethanes using HS^- at concentrations that might be encountered in environments where active anaerobic sulfate reduction is taking place. The rates of reaction with HS^- exceeded those for hydrolysis and at pH values above 7 in systems that are in equilibrium with elementary sulfur, the rates with polysulfide exceeded those with HS^-. The principal product from dihalomethanes was the polythiomethylene $HS-(CH_2-S)_nH$ (Roberts et al. 1992).

Attention is briefly drawn to hydrolytic procedures that have been considered for the destruction of xenobiotics. Although these are carried out under conditions that are not relevant to the aquatic environment, they may be useful as a background to alternative remediation programs. Three examples involving CaO and related compounds may be used as examples of important and unprecedented reactions:

1. The destruction of DDT by ball milling with CaO resulted in substantial loss of chloride and produced a graphitic product containing some residual chlorine. In addition, an exceptional rearrangement occurred with the formation of bis(4-chlorophenyl)ethyne that was identified by ^{1}H NMR (Hall et al. 1996) (Figure 1.29).

2. Treatment of 1,2,3,4-tetrachlorodibenzo[1,4]dioxin on Ca-based sorbents at temperatures of 160–300°C resulted in its conversion into products with molecular masses of 302 and 394 that

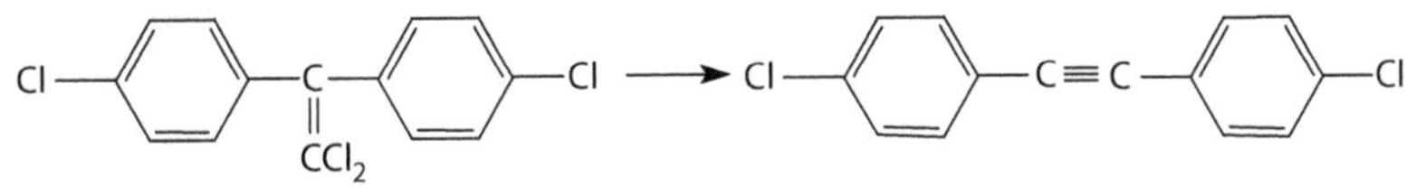

FIGURE 1.29 Alkaline destruction of DDT.

were tentatively identified as chlorinated benzofurans and 1-phenylnaphthalene or anthracenes (Gullett et al. 1997).

3. CaO was activated by the addition of Ni (acetylacetonate)$_2$ followed by calcining at 973 K. This resulted in the production of nickel on the carbonate matrix, and was effective in decomposing dichlorodifluoromethane (CFC-12) to CO_2 at 723 K—with low amounts of CO—while the halogen was combined as CaFCl (Tamai et al. 2006).

1.13.3 Oxidations

These have already been noted in the context of hydroxyl radical-initiated oxidations, and reference should be made to an extensive review by Worobey (1989) that covers a wider range of abiotic oxidations. Some have attracted interest in the context of the destruction of xenobiotics, and reference has already been made to photochemically induced oxidations.

1.13.3.1 Anodic Oxidation

An interesting study examined the anodic oxidation of EDTA at alkaline pH on a smooth platinum electrode (Pakalapati et al. 1996). Degradation was initiated by removal of the acetate side chains as formaldehyde, followed by deamination of the ethylenediamine that formed glyoxal and oxalate. Oxalate and formaldehyde are oxidized to CO_2, and adsorption was an integral part of the oxidation.

1.13.3.2 Ozone

Although reactions carried out by ozone have attracted enormous attention in the atmospheric environment, ozone has also been used extensively in the treatment of drinking water without the production of undesirable trihalomethanes from the use of molecular chlorine (Richardson et al. 1999). It has been examined for the removal of a number of contaminants, and ozone is considered to be a selective oxidant, even though quite complex reactions may occur.

1. Laboratory experiments on the ozonization of the dibenzoazepine drug carbamazepine (McDowell et al. 2005) showed the occurrence of a number of transformation products that were initiated by reaction at the olefinic double bond between the rings (Figure 1.30).
2. The degradation of aminodinitrotoluenes has been examined, and destruction of the rings took place with the release of NO_3^- and NO_2^-. Use of ^{15}N amino compounds revealed that pyruvamide was formed from the 2-amino compound and oxamic acid from the 4-amino compound (Spanggord et al. 2000).
3. Ozone attacks the rings of PAHs rather indiscriminately with fission of the rings to produce aldehyde groups. There has been concern, however, since the products may be more harmful than

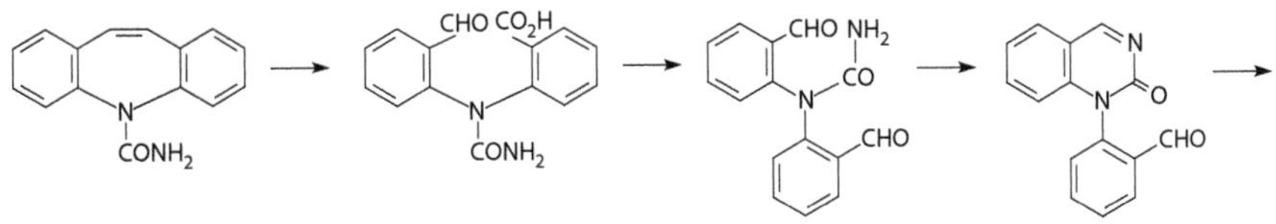

FIGURE 1.30 Ozonization of carbamazepine.

their precursors. In the studies that are used as illustration, *in vitro* gap junctional intracellular communication (GJIC) was used to assess adverse alteration on the expression of genes at the transcription, translational, or posttranslational level:

a. Pyrene was degraded to two groups of compounds: phenanthrene rings with one or more carbonyl groups by oxidation and fission of Δ^{9-10}, and biphenyls with four carbonyl groups *ortho* to the ring junction by fission of both Δ^{9-10} and Δ^{4-5} (Herner et al. 2001).

b. The product from chrysene was produced by oxidation and fission of Δ^{11-12} to the dialdehyde (Luster-Teasley et al. 2002).

4. Effective removal of the estrogens 17β-estradiol, estrone, and 17α-ethynylestradiol has been achieved using ozone under conditions that simulated those used for water treatment (Deborde et al. 2005). Analysis of the products showed the occurrence of two reactions:

a. Fission of rings A and B in estrone and 17β-estradiol to yield products with a carboxyl group at C_9 and a carboxyethyl group at C_8 (Figure 1.31a).

b. Oxidation of ring A to a 1,2-dione that underwent a benzylic acid rearrangement with ring contraction to a cyclopentane carboxylate (Figure 1.31b).

Analogous reactions were postulated to occur for 17α-ethynylestradiol and, on the basis of experiments with model compounds, oxidation of the ethynyl group to –COCHO (Huber et al. 2004).

5. Combined treatment of atrazine with ozone and H_2O_2 resulted in retention of the triazine ring, and oxidative dealkylation with or without replacement of the 2-chloro group by hydroxyl (Nélieu et al. 2000). Reaction with ozone and hydroxyl radicals formed the analogous products with the additional formation of the acetamido group from one of the *N*-alkylated groups (Acero et al. 2000).

6. A range of products was obtained from aniline including those from oxidative coupling (azozybenzene, azobenzene, and benzidine), and phenazine by dimerization (Chan and Larson 1991). Oxidation of *m*-phenylenediamine was initiated by the oxidation of two molecules to produce an *N*-phenyl-2-aminoquinone-imine that reacted with *m*-phenylenediamine to produce 2-amino-5-phenylaminoquinone-imine after further oxidation (Kami et al. 2000).

7. Two principal reactions were observed from the ozone treatment of ciprofloxacin: (a) fission of the piperazinyl ring by oxidation and ring fission, and (b) fission of the quinolone ring to produce a 4-fluoro-*N*-cyclopropyl-isatin with retention of the residual piperazinyl at C_5 (De Witte et al. 2008). It is worth noting two issues: these reactions differ from those that have been observed for the photochemical degradation which has already been discussed, and they may be compared with the fungal degradations of enrofloxacin (Wetzstein et al. 1999) and ciprofloxacin (Wetzstein et al. 1997) that are noted in Chapter 10.

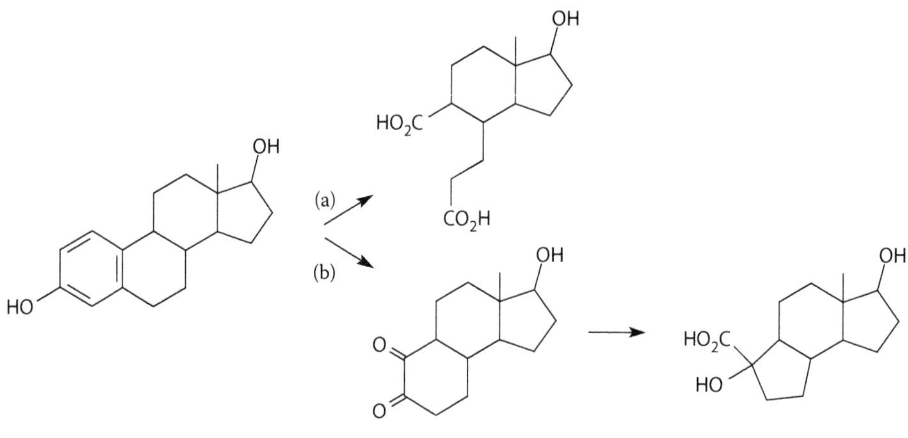

FIGURE 1.31   Ozonization of estrone/estradiol.

1.13.3.3 Hydrogen Peroxide

The use of hydrogen peroxide in conjunction with Fe(II) (Fenton's reagent) or ozone has already been noted. It has been used alone to examine the products from *o*- and *m*-phenylenediamines in the context of their mutagenicity (Watanabe et al. 1989). Successive reactions produced 3,4-diaminophenazine from *o*-phenylenediamine, and 3,7-diaminophenazine from *m*-phenylenediamine.

1.13.3.4 Manganese Dioxide

Manganese dioxide has been used to carry out a range of chemical oxidations, and is the reagent of choice for the oxidation of allylic alcohols (Hudlický 1990). There are several methods for its preparation that may account for differences in its activity. Although it has seldom been exploited for the oxidation of contaminants, two widely different applications have been described:

1. The transformation of several fluoroquinolone antibacterials was examined and a number of products from ciprafloxacin were tentatively identified (Zhang and Huang 2005). The quinolone ring was unchanged, and the major product was produced by fission of the piperazine ring to a phenamine.
2. The oxidation of triclosan was initiated by oxidation to the PhO radical. In analogy with reactions established in biomimetic synthesis, this underwent coupling to produce biphenyl ethers and biphenyls that were oxidized further to diphenoquinones (Zhang and Huang 2003).

1.13.3.5 Sulfate Radicals

The photochemical generation of sulfate radicals for the degradation of perfluorocarboxylic acids has already been noted (Hori et al. 2005a,b). Sulfate radicals are powerful oxidants, and can be produced by the co-mediated degradation of peroxymonosulfate (HSO_5^-). They have been shown to be effective for the degradation of 2,4-dichlorophenol and atrazine with the formation of sulfate (Anipsitakis and Dionysiou 2003), and it was shown that they were more effective than conventional Fenton oxidation with hydroxyl radicals, and were effective over a wider range of pH. Sulfate radicals are powerful oxidizing agents and can oxidize chloride to chlorine radicals and molecular chlorine that may be incorporated into the reactant. For example, reaction of peroxymonosulfate with 2,4-dichlorophenol in the presence of chloride produced 2,4,6-trichlorophenol; 2,3,5,6-tetrachlorohydroquinone; and the ring fission products penta- and tetrachloroacetone, and tetrachloromethane (Anipsitakis et al. 2006). Dioxirane that is produced *in situ* by the reaction of methylpyruvate and peroxymonosulfate has been evaluated as (a) an effective disinfecting agent (Wong et al. 2006) and (b) oxidant of alkenes, alkanes, and phenols (references in Wong et al. 2006).

1.13.4 Halogenation

Halogenation is important for disinfecting drinking water supplies, generally using molecular chlorine. Most attention has been directed to the adverse production of haloforms and haloacetates from reactions of chlorine with natural substrates, although in water containing bromide/iodide, a number of other reactions may occur.

1. While attempts to decrease the formation of haloforms have used chloramination, the undesirable formation of iodoform may take place in the presence of iodide (Leitner 1998).
2. The formation of iodoacetate and iodinated propenoates has been demonstrated (Plewa et al. 2004), and the cytotoxic and genotoxic properties of the former (Cemeli et al. 2006) has aroused concern.
3. In water containing particularly high concentrations of bromide, there is a rather special situation; chlorination with a mixture of chlorine and ClO_2 produced 2,3,5- tribromopyrrole—putatively from humic acids—that is strongly cytotoxic and genotoxic (Richardson et al. 2003).

In addition, there is interest in the halogenation of a wide range of anthropogenic contaminants, some of which may occur in raw water before treatment.

1. Pyrene is a common PAH contaminant and may occur in drinking water. Chlorination of water with or without bromide that may be present in coastal environments has been examined. Both chlorinated and brominated pyrenes with halogen substituents at the 1,3-, 1,6-, and 1,8-positions were found, and could putatively be produced by reaction of pyrene with hypochlorous acid and hypochlorite (Hu et al. 2006).

2. Degradation of the insecticide chlorpyrifos has been examined in solutions of aqueous chlorine in which the primary oxidant is hypochlorous acid/hypochlorite. The final product was 2,3,5-trichloropyrid-2-one produced either directly, or via initial oxidative conversion of the thioate ester by replacement of the sulfur with oxygen (Duirk and Collette 2006). An analogous oxidation was found with diazinon (Zhang and Pehkonen 1999).

3. Transformation of the widely used over-the-counter analgesic acetaminophen (paracetamol) during chlorination produced the toxic 1,4-benzoquinone via the *N*-acetylquinone-imine and minor amounts of products from chlorination of the phenolic ring (Bedner and Maccrehan 2006).

4. The chlorination of the antibacterial sulfonamide sulfamethoxazole was initiated by N-chlorination of the primary amine. Further reaction of the *N,N*-dichlorinated compound resulted in the final production of 3-amino-5-methyloxazole and 1,4-benzoquinone-imine (Dodd and Huang 2004).

5. The effect of chlorinating estrone in aqueous solution has been examined. Three reactions involving chlorination of different rings emerged:
   i. Ring A to produce 2- and 4-chlorinated products (Figure 1.32a)
   ii. Ring D at C_{16} followed by hydrolytic ring fission of ring D (Figure 1.32b)
   iii. Ring A at C_{10} followed by fission of ring B with the formation of a ketone in ring C at C_9 (Figure 1.32c) (Hu et al. 2003)

6. Amino acids have been found in samples of river water (Lee et al. 2006), and chlorination of glycine may produce cyanogen chloride, via *N,N*-dichloroglycine. At pH values >6, this is converted into cyanogen chloride, whereas at lower pH values *N*-chloromethylimine is formed (Na and Olson 2006).

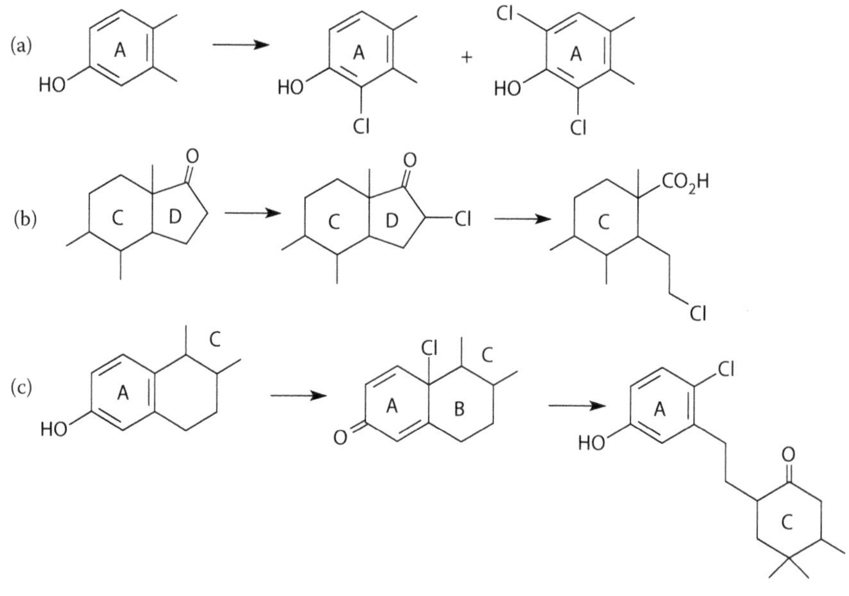

FIGURE 1.32 Chlorination of estrone.

1.13.5 Thermal Reactions

1.13.5.1 Gas Phase

The products of incomplete combustion may be associated with particulate matter before their discharge into the atmosphere, and these may ultimately enter the aquatic and terrestrial environments in the form of precipitation and dry deposition. It is, therefore, essential to ensure total destruction of the contaminants, generally by raising the temperature. The spectrum of compounds that have been examined is quite extensive, and several of them are produced by reactions between hydrocarbons and inorganic sulfur or nitrogen constituents of air. Some illustrative examples involving other types of reaction include the following:

1. The pyrolysis of vinylidene chloride produced a range of chlorinated aromatic compounds including polychlorinated benzenes, styrenes, and naphthalenes (Yasahura and Morita 1988), and a series of chlorinated acids including chlorobenzoic acids has been identified in emissions from a municipal incinerator (Mowrer and Nordin 1987).

2. Nitroaromatic compounds have been identified in diesel engine emissions (Salmeen et al. 1984), and attention has been directed particularly to 1,8- and 1,6-dinitropyrene that are mutagenic, and possibly carcinogenic (Nakagawa et al. 1983).

3. A wide range of azaarenes including acridines and benzacridines, 4-azafluorene, and 10-azabenzo[*a*]pyrene (Figure 1.33) has been identified in particulate samples of urban air, and some of them have been recovered from contaminated sediments (Yamauchi and Handa 1987).

4. Ketonic and quinonoid derivatives of aromatic hydrocarbons have been identified in automobile (Alsberg et al. 1985) and diesel exhaust particulates (Levsen 1988), and have been recovered from samples of marine sediments (Fernández et al. 1992).

5. Halogenated phenols, particularly 2-bromo-, 2,4-dibromo-, and 2,4,6-tribromophenol, have been identified in automotive emissions and are the products of thermal reactions involving the dibromoethane fuel additive (Müller and Buser 1986). It could, therefore, no longer be assumed that such compounds are exclusively the products of biosynthesis by marine algae.

6. Complex reactions occur during high-temperature treatment of aromatic hydrocarbons. An important class of reactions involves the cyclization and condensation of simpler PAHs to form highly condensed polycyclic compounds. This is discussed more fully by Zander (1995).

   a. A number of pentacyclic aromatic hydrocarbons have been identified as products of the gas-phase pyrolysis of methylnaphthalenes. These were formed from 1-methyl- and 2-methylnaphthalene by dimerization (Lang and Buffleb 1958) at various positions, whereas direct coupling with loss of the methyl group was found to be dominant with 2-methylnaphthalene (Lijinsky and Taha 1961) (Figure 1.34).

   b. A hypothetical scheme involving 2-carbon and 4-carbon additions has been used to illustrate the formation of coronene (circumbenzene) and ovalene (circumnaphthalene) from phenanthrene (Figure 1.35).

7. Concern has been expressed over the formation of chlorinated dibenzo[1,4]dioxins and dibenzofurans during the thermal transformation of organic material in the presence of chloride or organochlorine compounds.

   a. The high-temperature oxidation of 2-chlorophenol yielded a number of products. At 600°C these included 4,6-dichlorodibenzofuran, 1-chlorodibenzo[1,4]dioxin, and low concentrations of 2,4- and 2,6-dichlorophenol that were produced by reactions with chloride radicals

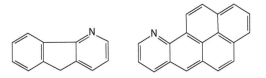

FIGURE 1.33 Azaarenes identified in particulate samples of urban air.

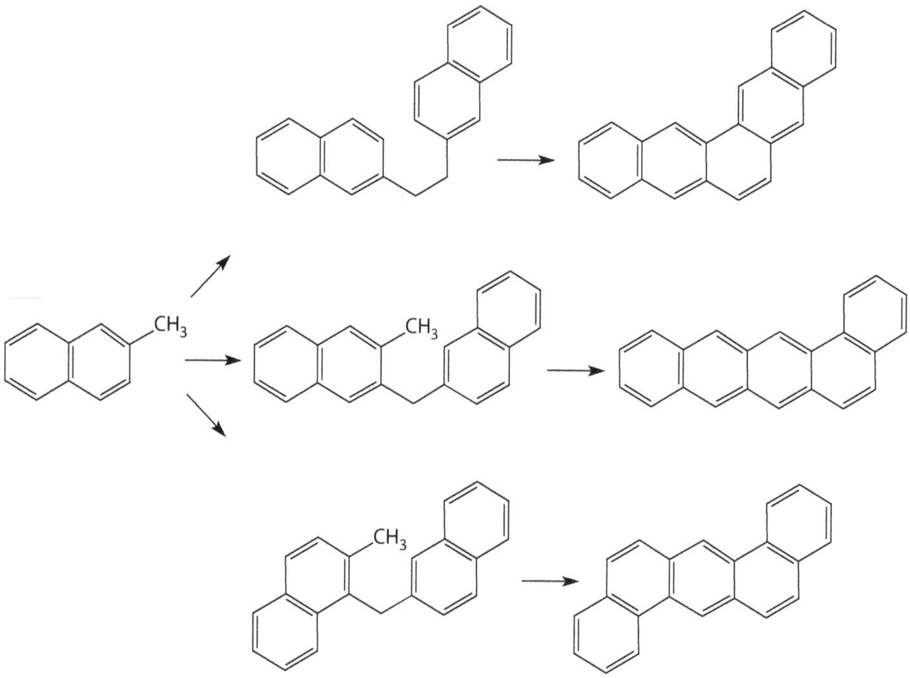

FIGURE 1.34   Products from the pyrolysis of 2-methylnaphthalene.

and hydroxyl radicals. They were, however, essentially eliminated at temperatures above 800°C (Evans and Dellinger 2005a). In contrast, under comparable conditions, although a mixture of 2-chloro- and 2-bromophenol also yielded a range of halogenated dibenzofurans, 4,6-dibromodibenzofuran was not produced (Evans and Dellinger 2005b).

b. In experiments to clarify the ameliorating effect on the formation of these undesirable products, the effect of adding urea was studied. Reaction of graphite, $CuCl_2$, and urea at 300°C resulted in substantial reduction in the amounts of chlorinated dioxins and furans that were produced, and the formation of 2,6-dichlorobenzonitrile, 4-chloro-2-methylaniline, and pentachloropyridine (Kuzuhara et al. 2005).

1.13.5.2 In Water

There are two distinct conditions that have been used: above the critical temperature and pressure (374°C and 218 atm) water becomes a supercritical fluid in which the distinction between the liquid and gaseous states disappears. Since supercritical water can dissolve nonpolar compounds, it has been examined for the degradation of such contaminants. Subcritical water in which the liquid state is maintained by the pressure of the containing vessel has also achieved attention.

1. Destruction of the explosives RDX, HMX, and TNT has been examined using subcritical water in both laboratory- and pilot-scale experiments. In contaminated soils at 150°C, considerable amounts of TNT remained in the soil after reaction for 5 h, and of HMX for 2.5 h. In the pilot-scale experiments, heating at 275°C for 1 h accomplished complete destruction of RDX and TNT, and ca. 98% destruction of HMX (Hawthorne et al. 2000).

2. PCBs

   a. Destruction of PCBs containing congeners with two to eight chlorine substituents was examined in supercritical water under oxidizing or alkaline reducing conditions. The latter were more effective and, although chlorinated dibenzofurans with one to six chlorine substituents

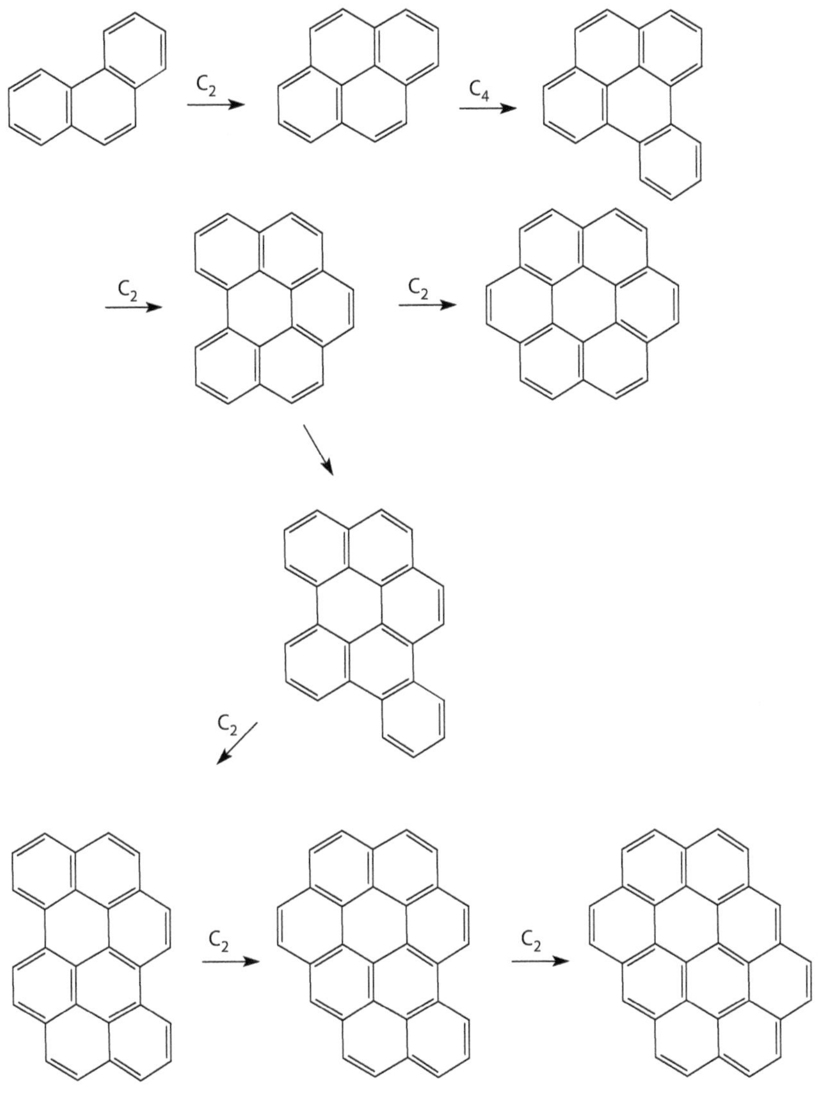

FIGURE 1.35 Successive C_2 and C_4 addition reactions.

were formed at temperatures of 250°C they were removed at 350°C. Experiments with the 245–245 hexachloro congener were used to illustrate the formation of chlorinated dibenzo-furans (Weber et al. 2002).

b. Examination of the destruction of Arochlor 1254 in paint scrapings at 350°C revealed the significance of the metals. Experiments with zero-valent metal additions in the absence of paint showed their effectiveness in the order Pb≡Cu > Al > Zn > Fe. With Pb and Cu destruction of tetra to heptachloro congeners to predominantly monochlorobiphenyl occurred (Kubátova et al. 2003).

c. Zero-valent iron has been used for the reductive dechlorination of PCBs in subcritical water at 250°C (Yak et al. 1999). Extensive loss of the more highly chlorinated congeners in Arochlor 12609 took place with the formation of congeners having one to four chlorine substituents, while with longer exposure times these were virtually eliminated.

3. At 200°C destruction of γ-hexachlorocyclohexane in subcritical water occurred in 1 h, and for dieldrin in 1 h at 300°C. The pathway for γ-HCH involved successive formation of

1,2,4-trichloro-, 1,4-dichloro-, and monochlorobenzene that was hydrolyzed to phenol (Kubátova et al. 2002).

4. The use of subcritical water at 275°C was successful in removing PAHs with two to six rings (including the carcinogenic benzo[*a*]pyrene) from soil at a contaminated site, and the wastewater that was produced appeared to be suitable for further exploitation (Lagadec et al. 2000).

1.13.6 Electrocatalytic Reactions

The energy for the fission of the covalent bond in organic contaminants is normally supplied thermally using thermodynamically accessible chemical or biochemical reactions, or by the introduction of catalysts to lower the activation energy of the reactions. There has been interest, however, in using electrical energy in a number of forms to carry out these reactions. A selection of processes for the destruction of contaminant is noted with some illustrative examples.

1. The simplest applications involve direct anodic oxidation in aqueous media.
    a. During the anodic oxidation of EDTA at alkaline pH on a smooth platinum electrode (Pakalapati et al. 1996), degradation was initiated by removal of the acetate side chains as formaldehyde, followed by deamination of the ethylenediamine that is formed to glyoxal and oxalate. Oxalate and formaldehyde were oxidized to CO_2.
    b. Using a platinum electrode coated with a metal oxide film (Ti, Ru, Sn, and Sb) in conjunction with a solid Nafion membrane, complete destruction of 4-chlorophenol could be achieved in reactions involving the formation of benzoquinone followed by ring fission to succinate acetate and CO_2 (Johnson et al. 1999). On account of the products this was termed "electrochemical incineration."
2. Electrohydraulic discharge plasmas have been used in different configurations.
    a. The degradation of 2,4,6-trinitrotoluene has been carried out in the presence of ozone. The hydroxyl radicals that were produced brought about oxidation to an intermediate that was identified as 2,4,6-trinitrobenzaldehyde (Lang et al. 1998).
    b. The degradation of 4-chlorophenol and 3,4-dichloroaniline was carried out by the shock wave that was developed in the cell (Willberg et al. 1996).
3. Variants of glow discharge plasmas have been used:
    a. The dechlorination of pentachlorophenol took place with the formation of C_1 and C_2 carboxylic acids (Sharma et al. 2000).
    b. In the presence of Fe(III), degradation of phenol which took place by hydroxylation followed by ring fission (Liu and Jiang 2005) in reactions are analogous to those in the electro-Fenton reaction that has been examined for the degradation of 4-nitrophenol (Oturan et al. 2000).
4. Electrolytic reduction has been carried out under several conditions.
    a. Using a porous nickel electrode and at voltages that were maintained by a potentiostat, a number of reductive reactions were observed: tetrachloroethene and trichloroethene → ethane and ethene, and penta-, tetra-, tri-, and dichloroethanes → ethane (Liu et al. 2000).
    b. Fuel cells have been used in various modifications to carry out gas-phase reductive dechlorination of trichloroethene:
        i. Using a polymer electrolyte membrane cell in which H_2 flowed through the anode chamber. The major intermediate chlorinated products from tetrachloroethene or tetrachloromethane were trichloroethene or trichloromethane, and these were finally reduced to a mixture of ethane and ethene, or methane (Liu et al. 2001).
        ii. Reduction of trichloroethene to ethane took place in a modified fuel cell to which H_2 was introduced, although the loss of catalytic activity with time could present a serious limitation (Ju et al. 2006).

1.14 Sonication

The transient collapse of vapor bubbles produced by ultrasonic pressure waves leads to average internal vapor temperatures of ca. 4000 K that result in pyrolytic reactions and the formation of hydroxyl radicals. This method is effective for contaminants that have a high air-to-water interface, and combines two reactions—pyrolysis and solvolysis. The following are given as illustration using contaminants of widely differing chemical structure.

1. In the degradation of CCl_4, two successive reactions were identified: $CCl_4 \rightarrow CCl_3$ and $CCl_3 \rightarrow CCl_2$, while CCl_3 produced low concentrations of hexachloroethane, and tetrachloroethene and Cl^- (Hua and Hoffmann 1996).

2. In the degradation of Triton X-100 three reactions were established (Destaillats et al. 2000):

   a. Pyrolysis of the 4-alkyl side chain with the successive production of

   $$HO-C(CH_3)_2-Ar- \text{ and } HO-Ar-$$

   b. Hydroxyl radical addition to the aromatic ring

   c. Hydroxyl radical attack of the polyethoxylated side chain

3. The complete degradation of PFOA and PFOS by sonication yielded CO_2, CO, and HF, and was proposed to involve the following stages (Vecitis et al. 2008):

   Stage 1: Pyrolytic loss of ionic functional groups to produce 1H-fluoroalkane or the perfluoro-1-alkene.

   Stage 2: Pyrolysis to the C_3 fluoro radical from which CF_2 carbene is produced, and pyrolysis of the C_5 fluoro radical to form the CF_2 carbene.

   Stage 3: Vapor-phase hydrolysis of CF_2 carbene with the production of CO, CO_2, and HF.

   $$\text{Sum: PFOS, PFOA} \rightarrow CO, CO_2, F$$

1.15 References

Acero JL, K Stemmler, U van Gunten. 2000. Degradation kinetics of atrazine and its degradation products with ozone and OH radicals: A predictive tool for drinking water treatment. *Environ Sci Technol* 34: 591–597.

Alsberg T et al.. 1985. Chemical and biological characterization of organic material from gasoline exhaust particles. *Environ Sci Technol* 19: 43–50.

Alvarez EG, J Viidanoja, A Munoz, K Wirtz, J Hjorth. 2007. Experimental confirmation of the dicarbonyl route in the photo-oxidation of toluene and benzene. *Environ Sci Technol* 41: 8362–8369.

Amador JA, BF Taylor. 1990. Coupled metabolic and photolytic pathway for degradation of pyridinedicarboxylic acids, especially dipicolinic acid. *Appl Environ Microbiol* 56: 1352–1356.

Anderson PN, RA Hites. 1996. OH radical reactions: The major removal pathway for polychlorinated biphenyls from the atmosphere. *Environ Sci Technol* 30: 1756–1763.

Andreae MO, Crutzen. 1997. Atmospheric aerosols: Biogeochemical sources and role in atmospheric chemistry. *Science* 276: 1052–1058.

Androulaki E, A Hiskia, D Dimotikali, C Minero, P Calza, A Papaconstantinou. 2000. Light induced elimination of mono- and polychlorinated phenols from aqueous solutions by $PW_{12}O_{40}{}^{3-}$. The case of 2,4,6-trichlorophenol. *Environ Sci Technol* 34: 2024–2028.

Anipsitakis GP, DD Dionysiou. 2003. Degradation of organic contaminants in water with sulfate radicals generated by the conjunction of peroxymonosulfate with cobalt. *Environ Sci Technol* 37: 4790–4897.

Anipsitakis GP, DD Dionysiou, MA Gonzalez. 2006. Cobalt-mediated activation of peroxymonosulfate and sulfate radical attack on phenolic compounds. Implications of chloride ions. *Environ Sci Technol* 40: 1000–1007.

Arey J. 1998. Atmospheric reactions of PAHs including formation of nitroarenes. *Handbook Environ Chem* 3I: 347–385.

Atkinson R. 1986. Kinetics and mechanisms of the gas-phase reactions of the hydroxyl radical with organic compounds under atmospheric conditions. *Chem Rev* 86: 29–201.

Atkinson R. 1990. Gas-phase troposphere chemistry of organic compounds: A review. *Atmos Environ* 24A: 1–41.

Atkinson R, J Arey, B Zielinska, SM Aschmann. 1987a. Kinetics and products of the gas-phase reactions of OH radicals and N_2O_5 with naphthalene and biphenyl. *Environ Sci Technol* 21: 1014–1022.

Atkinson R, SM Aschmann, AW Winer. 1987b. Kinetics of the reactions of NO_3 radicals with a series of aromatic compounds. *Environ Sci Technol* 21: 1123–1126.

Atkinson R, SM Aschmann, AM Winer, WPL Carter. 1985a. Rate constants for the gas-phase reactions of NO_3 radicals with furan, thiophene, and pyrrole at 295 1 K and atmospheric pressure. *Environ Sci Technol* 19: 87–90.

Atkinson R, SM Aschmann, AM Winer, JN Pitts. 1985b. Kinetics and atmospheric implications of the gas-phase reactions of NO_3 radicals with a series of monoterpenes and related organics at 294 2K. *Environ Sci Technol* 19: 159–163.

Atkinson R, WPL Carter. 1984. Kinetics and mechanisms of the gas-phase reactions of ozone with organic compounds under atmospheric conditions. *Chem Rev* 84: 437–470.

Atkinson R, EC Tuazon, TJ Wallington, SM Aschmann, J Arey, AM Winer, JN Pitts. 1987c. Atmospheric chemistry of aniline, *N,N*-dimethylaniline, pyridine, 1,3,5-triazine and nitrobenzene. *Environ Sci Technol* 21: 64–72

Baker JI, RA Hites. 2000. Is combustion the major source of polychlorinated dibenzo-*p*-dioxins and dibenzofurans to the environment? A mass balance study. *Environ Sci Technol* 34: 2879–2886.

Bandstra JZ, R Miehr, RL Johnson, PG Tratnyek. 2005. Reduction of 2,4,6-trinitrotoluene by iron metal: Kinetic controls on product distributions in batch experiments. *Environ Sci Technol* 39: 230–238.

Bank S, RJ Tyrrell. 1984. Kinetics and mechanism of alkaline and acidic hydrolysis of aldicarb. *J Agric Food Chem* 32: 1223–1232.

Barbas JT, ME Sigman, R Dabestani. 1996. Photochemical oxidation of phenanthrene sorbed on silica gel. *Environ Sci Technol* 30: 1776–1780.

Baretto RD, KA Gray, K Anders. 1995. Photocatalytic degradation of methyl-*tert*-butyl ether in TiO_2 slurries: A proposed reaction scheme *Water Res* 29: 1243–1248.

Bedner M, WA Maccrehan. 2006. Transformation of acetaminophen by chlorination produces the toxicants 1,4-benzoquinone and *N*-acetyl-*p*-benzoquinone imine. *Environ Sci Technol* 40:516–522.

Beitz T, W Bechmann, R Mitzner. 1998. Investigations of reactions of selected azaarenes with radicals in water 1. Hydroxyl and sulfate radicals. *J Phys Chem* A 102: 6760–6765.

Belly RT, JJ Lauff, CT Goodhue. 1975. Degradation of ethylenediaminetetraacetic acid by microbial populations from an aerated lagoon. *Appl Microbiol* 29: 787–794.

Beltran E, H Fenet, JF Cooper, CM Coste. 2000. Kinetics of abiotic hydrolysis of isoxaflutole: Influence of pH and temperature in aqueous mineral buffered medium. *J Agric Food Chem* 48: 4399–4403.

Bergon M, NB Hamida, J-P Calmon. 1985. Isocyanate formation in the decomposition of phenmedipham in aqueous media. *J Agric Food Chem* 33: 577–583.

Berndt T, O Boge. 2006. Formation of phenol and carbonyls from the atmospheric reaction of OH carbonyls with benzene. *Phys Chem Chem Phys* 8: 1205–1214.

Bláha L, J Klánová, P Klán, J Janosek, M Skarek, R Rúzicka. 2004. Toxicity increases in ice containing monochlorophenols upon photolysis: Environmental consequences. *Environ Sci Technol* 38: 2873–2878.

Bleasdale C, R Cameron, C Edwards, BT Golding. 1997. Dimethyldioxirane converts benzene oxide/oxepin into (*Z,Z*)-muconaldehyde and *sym*-oxepin oxide: Modeling the metabolism of benzene and its photooxidative degradation. *Chem Res Toxicol* 10: 1314–1318.

Bosma TNP, FHM Cottaar, MS Posthumus, CJ Teunis, A van Veidhuizen, G Schraa, AJB Zehnder. 1994. Comparison of reductive dechlorination of hexachloro-1,3-butadiene in Rhine sediments and model systems with hydroxocobalamin. *Environ Sci Technol* 28: 1124–1128.

Bosma TNP, JR van der Meer, G Schraa, ME Tros, AJB Zehnder. 1988. Reductive dechlorination of all trichloro- and dichlorobenzene isomers. *FEMS Microbiol Ecol* 53: 223–239.

Brezonik PL, J Fulkerson-Brekken. 1998. Nitrate-induced photoysis in natural waters: Controls on concentrations of hydroxyl radical photo-intermediates by natural scavenging agents. *Environ Sci Technol* 32: 3004–3010.

Brownlee BG, JH Carey, GA MacInnes, IT Pellizzari. 1992. Aquatic environmental chemistry of 2-(thiocyano-methylthio)benzothiazole and related benzothiazoles. *Environ Toxicol Chem* 11: 1153–1168.

Brubaker WW, RA Hites. 1997. Polychlorinated dibenzo-*p*-dioxins and dibenzofurans: Gas-phase hydroxyl radical reactions and related atmospheric removal. *Environ Sci Technol* 31: 1805–1810.

Brubaker WW, RA Hites. 1998. Gas-phase oxidation products of biphenyl and polychlorinated biphenyls. *Environ Sci Technol* 32: 3913–3918.

Bunce NJ, L Liu, J Zhu, DA Lane. 1997. Reaction of naphthalene and its derivatives with hydroxyl radicals in the gas phase. *Environ Sci Technol* 31: 2252–2259.

Burhenne J, M Ludwig, M Spiteller. 1997a. Photolytic degradation of fluoroquinolone carboxylic acids in aqueous solution. Primary photoproducts and half-lives. *Environ Sci Pollut Res* 4: 10–15.

Burhenne J, M Ludwig, M Spiteller. 1997b. Photolytic degradation of fluoroquinolone carboxylic acids in aqueous solution. Isolation and structural elucidation of polar photometabolites. *Environ Sci Pollut Res* 4: 61–67.

Burns SE, JP Hassett, MV Rossi. 1996. Binding effects on humic-mediated photoreaction: Intrahumic dechlorination of mirex in water. *Environ Sci Technol* 30: 2934–2941.

Burns SE, JP Hassett, MV Rossi. 1997. Mechanistic implications of the intrahumic dechlorination of mirex. *Environ Sci Technol* 31: 1365–1371.

Buser H-R, MD Müller. 1993. Enantioselective determination of chlordane components, metabolites, and photoconversion products in environmental samples using chiral high-resolution gas chromatography and mass spectrometry. *Environ Sci Technol* 27: 1211–1220.

Buser H-R, T Poiger, MD Müller. 1998. Occurrence and fate of the pharmaceutical drug diclofenac in surface waters: Rapid photodegradation in a lake. *Environ Sci Technol* 32: 3449–3456.

Butler EC, KF Hayes. 1998. Effects of solution composition and pH on the reductive dechlorination of hexa-chloroethane by iron sulfide. *Environ Sci Technol* 32: 1276–1284.

Butler EC, KF Hayes. 1999. Kinetics of the transformation of trichloroethylene and tetrachloroethylene by iron sulfide. *Environ Sci Technol* 33: 2021–2027.

Butler EC, KF Hayes. 2001. Factors influencing rates and products in the transformation of trichloroethylene by iron sulfide and iron metal. *Environ Sci Technol* 35: 3884–3891.

Camilleri P. 1984. Alkaline hydrolysis of some pyrethroid insecticides. *J Agric Food Chem* 32: 1122–1124.

Carey JH, ME Cox. 1981. Photodegradation of the lampricide 3-trifluoromethyl-4-nitrophenol (TFM) 1. Pathway of the direct photolysis in solution. *J Great Lakes Res* 7: 234–241.

Carter WPL, R Atkinson. 1987. An experimental study of incremental hydrocarbon reactivity. *Environ Sci Technol* 21: 670–679.

Carter WPL, R Atkinson. 1989. Computer modeling of incremental hydrocarbon reactivity. *Environ Sci Technol* 23: 864–880.

Castro CE. 1998. Environmental dehalogenation: Chemistry and mechanism. *Rev Environ Contam Toxicol* 155: 1–67.

Castro CE, WC Kray. 1963. The cleavage of bonds by low-valent transition metal ions. The homogeneous reduction of alkyl halides by chromous sulfate. *J Am Chem Soc* 85: 2768–2773.

Castro CE, WC Kray. 1966. Carbenoid intermediates from polyhalomethanes and chromium (II). The homoge-neous reduction of geminal halides by chromous sulfate. *J Am Chem Soc* 88: 4447–4458.

Cemeli E, ED Wagner, D Anderson, SD Richardson, MJ Plewa. 2006. Modulation of the cytoxicity and genotoxicity of the drinking water disinfection byproduct iodoacetic acid by suppressors of oxidative stress. *Environ Sci Technol* 40: 1878–1883.

Chan K-W, W Chu. 2006. Model applications and intermediates quantification of atrazine degradation by UV-enhanced Fenton process. *J Agric Food Chem* 54: 1804–1813.

Chan WF, RA Larson. 1991. Formation of mutagens from the aqueous reactions of ozone and anilines. *Water Res* 25: 15239–1538.

Charleson RJ, JE Lovelock, MO Andreae, SG Warren. 1987. Oceanic phytoplankton, atmospheric sulfur, cloud albedo and climate. *Nature* (London) 326: 655–661.

Chlodini G, B Rindone, F Cariati, S Polesello, G Restelli, J Hjorth. 1993. Comparison between the gas-phase and the solution reaction of the nitrate radical and methylarenes. *Environ Sci Technol* 27: 1659–1664.

Choi W, MR Hoffman. 1997. Novel photocatalytic mechanisms for $CHCl_3$, $CHBr_3$, and $CCl_3CO_2^-$ degradation and fate of photogenerated trihalomethyl radicals on TiO_2. *Environ Sci Technol* 31: 89–95.

De Witte B, J Dewulf, K Demeestere, V van de Vyvere, P de Wispelaere, H van Langenhove. 2008. Ozonation of ciprofloxacin in water: HRMS identification of reaction products and pathways. *Environ Sci Technol* 42: 4889–4895.

Deborde M, S Rabouan, J-P Duguet, B Legube. 2005. Kinetics of aqueous ozone induced oxidation of some endocrine disruptors. *Environ Sci Technol* 39: 6086–6092.

Destaillats H, H-M Hung, MR Hoffmann. 2000. Degradation of alkylphenol ethoxylate surfactants in water with ultrasonic irradiation. *Environ Sci Technol* 34: 311–317.

Dodd MC, C-H Huang. 2004. Transformation of the antibacterial agent sulfamethoxazole in reactions with chlorine: Kinetics, mechanisms, and pathways. *Environ Sci Technol* 38: 5607–5615.

Dolinová N, R Ruzicka, R Kurková, J Klánová, P Klán. 2006. Oxidation of aromatic and aliphatic hydrocarbons by OH radicals photochemically generated from H_2O_2 in ice. *Environ Sci Technol* 40: 7668–7674.

Doong R-A, H-C Chiang. 2005. Transformation of carbon tetrachloride by thiol reductants in the presence of quinone compounds. *Environ Sci Technol* 39: 7460–7468.

Dror I, D Baram, B Berkowitz. 2005. Use of nanosized catalysts for transformation of chloro-organic pollutants. *Environ Sci Technol* 39: 1283–1290.

Duirk SE, TW Collette. 2006. Degradation of chlopyrifos in aqueous chlorine solutions: Pathways, kinetics, and modeling. *Environ Sci Technol* 40: 546–551.

Dunnivant FM, RP Schwarzenbach, DL Macalady. 1992. Reduction of substituted nitrobenzenes in aqueous solutions containing natural organic matter. *Environ Sci Technol* 26: 2133–2141.

Eatough DJ, VF White, LD Hansen, NL Eatough, JL Cheney. 1986. Identification of gas-phase dimethyl sulfate and monomethyl hydrogen sulfate in the Los Angeles atmosphere. *Environ Sci Technol* 20: 867–872.

Einschlag FSG, J LOpez, L Carlos, AL Capparelli. 2002. Evaluation of the efficiency of photodegradation of nitroaromatics applying the UV/H_2O_2 technique. *Environ Sci Technol* 36: 3936–3944.

Ellis DA, SA Mabury. 2000. The aqueous photolysis of TFM and related trifluoromethylphenols. An alternate source of trifluoroacetic acid in the environment. *Environ Sci Technol* 34: 632–637.

Ellis DA, JW Martin, AO De Silva, SA Mabury, MD Hurley, MPS Andersen, TJ Wallington. 2004. Degradation of fluorotelomer alcohols: A likely atmospheric source of perfluorinated carboxylic acids. *Environ Sci Technol* 38: 3316–3321.

Emptage M, J Tabinowski, JM Odom. 1997. Effect of fluoroacetates on methanogenesis in samples from selected methanogenic environments. *Environ Sci Technol* 31: 732–734.

Enya T, H Suzuki, T Watanabe, T Hiurayama, Y Hisamatsu. 1997. 3-Nitrobenzanthrone, a powerful bacterial mutagen and suspected human carcinogen found in diesel exhaust and airborne particulates. *Environ Sci Technol* 31: 2772–2776.

Eriksson J, N Green, G Marsh, Å Bergman. 2004. Photochemical decomposition of 15 polybrominated diphenyl ether congeners in methanol/water. *Environ Sci Technol* 38: 3119–3125.

Evans CS, B Dellinger. 2005a. Mechanisms of dioxin formation from the high temperature oxidation of 2-chlorophenol. *Environ Sci Technol* 39: 122–127.

Evans CS, B Dellinger. 2005b. Formation of bromochlorodibenzo-*p*-dioxins and furans from the high tempareture pyrolysis of a 2-chlorophenol/2-bromophenol mixture. *Environ Sci Technol* 39: 7940–7948.

Ezra S, S Feinstein, I Bilkis, E Adar, J Ganor. 2005. Chemical transformation of 3-bromo-2,2-bis(bromomethyl) propanol under basic conditions. *Environ Sci Technol* 39: 505–512.

Fernández P, M Grifoll, AM Solanas, J M Bayiona, J Albalgés. 1992. Bioassay-directed chemical analysis of genotoxic compounds in coastal sediments. *Environ Sci Technol* 26: 817–829.

Finlayson-Pitts B, J Pitts. 1986. *Atmospheric Chemistry*. John Wiley and Sons, New York.

Finlayson-Pitts BJ, JN Pitts. 1997. Tropospheric air pollution: Ozone, airborne toxics, polycyclic aromatic hydrocarbons, and particles. *Science* 276: 1045–1052.

Forstner HJL, RC Flagan, JH Seinfeld. 1997. Secondary organic aerosol from the photoxodation of aromatic hydrocarbons: Molecular composition. *Environ Sci Technol* 31: 1345–1358.

Frank H, H Scholl, D Renschen, B Rether, A Laouedj, Y Norokorpi. 1994. Haloacetic acids, phytotoxic secondary air pollutants. *Environ Sci Pollut Res* 1: 4–14.

Franklin J. 1994. The atmospheric degradation and impact of perchloroethylene. *Toxicol Environ Chem* 46: 169–182.

Friedman CL, AT Lemley, A Hay. 2006. Degradation of chloroacetanilide herbicides by anodic Fenton treatment. *J Agric Food Chem* 54: 2640–2651.

Fry NL, MJ Rose, DL Rogow, C Nyitray, M Kaur, PK Marscharak. 2010. Ruthenium nitrosyls derived from tetradentate ligands containing carboxamido-N and phenolato-O donors: Syntheses, structures, photolability and time dependent density functional theory studies. *Inorg Chem* 49: 1487–1495.

Fukushima M, K Tatsumi. 2001. Degradation pathways of pentachlorophenol by photo-Fenton systems in the presence of iron (III) humic acid and hydrogen peroxide. *Environ Sci Technol* 35: 1771–1778.

Gantzer CJ, LP Wackett. 1991. Reductive dechlorination catalyzed by bacterial transition-metal coenzymes. *Environ Sci Technol* 25: 715–722.

Glaus MA, CG Heijman, RP Schwarzenbach, J Zeyer. 1992. Reduction of nitroaromatic compounds mediated by *Streptomyces* sp. exudates. *Appl Environ Microbiol* 58: 1945–1951.

Glod G, W Angst, C Holliger and RP Schwartzenbach. 1997. Corrinoid-mediated reduction of tetrachloroethene, trichloroethene,and trichlorofluoroethene in homogeneous solution: Reaction kinetics and reaction mechanisms. *Environ Sci Technol* 31: 253–260.

Godejohann M, M Astratov, A Preiss, K Levsen, C Mügge. 1998. Application of continuous-flow HPLC-proton-nuclear magnetic resonance spectroscopy and HPLC-thermospray spectroscopy for the structural elucidation of phototransformation products of 2,4,6-trinitrotoluene. *Anal Chem* 70: 4104–4110.

Gonzalez MA, MA Yim, S Cheng, A Moyes, AJ Hobbs, PK Mascharak. 2011. Manganese carbonyls bearing tripodal polypyridine ligands as photoactive carbon monoxide-releasing molecules. *Inorg Chem* 51: 601–608.

Griesbaum K, V Miclaus, IC Jung. 1998. Isolation of ozonides from gas-phase ozonolysis of terpenes. *Environ Sci Technol* 32: 647–649.

Grosjean D. 1991. Atmospheric chemistry of toxic contaminants. 4. Saturated halogenated aliphatics: Methyl bromide, epichlorhydrin, phosgene. *J Air Waste Manage Assoc* 41: 56–61.

Grosjean D, EL Williams, E Grosjean. 1993a. Atmospheric chemistry of isoprene and of its carbonyl products. *Environ Sci Technol* 27: 830–840.

Grosjean D, EL Williams II, E Grosjean. 1993b. A biogenic precursor of peroxypropionyl nitrate: Atmospheric oxidation of *cis*-3-hexen-1-ol. *Environ Sci Technol* 27: 979–981.

Grosjean D, EL Williams, E Grosjean, JM Andino, JH Seinfeld. 1993c. Atmospheric oxidation of biogenic hydrocarbons: Reaction of ozone with -pinene, D-limonene, and *trans*-caryophyllene. *Environ Sci Technol* 27: 2754–2758.

Gullett BK, DF Natschke, KRE Bruce. 1997. Thermal treatment of 1,2,3,4-tetrachlorodibenzo-*p*-dioxin by reaction with Ca-based sorbents at 23–300°C. *Environ Sci Technol* 31: 1855–1862.

Hall AK, JM Harrowfied, RJ Hart, PG McCormick. 1996. Mechanochemical reaction of DDT with calcium oxide. *Environ Sci Technol* 30: 3401–3407.

Harvey J, JJ Dulka, JJ Anderson. 1985. Properties of sulfometuroin methyl affecting its environmental fate: Aqueous hydrolysis and photolysis, mobility and adsorption on soils, and bioaccumulation potential. *J Agric Food Chem* 33: 590–596.

Hasegawa K et al. 1993. Photooxidation of naphthalenamines adsorbed on particles under simulated atmospheric conditions. *Environ Sci Technol* 27: 1819–1825.

Hashimoto S, T Azuma, A Otsuki. 1998. Distribution, sources, and stability of haloacetic acids in Tokyo Bay, Japan. *Environ Toxicol Chem* 17: 798–805.

Hawthorne SB, AJM Lagadec, D Kalderis, AV Lilke, DJ Miller. 2000. Pilot-scale destruction of TNT, RDX, and HMX on contaminated soils using subcritical water. *Environ Sci Technol* 34: 3224–3228.

Hayman GD, and RG Derwent. 1997. Atmospheric chemical reactivity and ozone-forming potentials of potential CFC replacements. *Environ Sci Technol* 31: 317–336.

Hebting Y, P Adam, P Albrecht. 2003. Reductive desulfurization of allylic thiols by HS^-/H_2S in water gives clue to chemical reactions widespread in natural environments. *Org Lett* 5: 1571–1574.

Hebting Y, P Schaeffer, A Behrens, P Adam, G Schmitt, P Schneckenburger, SM Bernasconi, P Ahlbrecht. 2006. Biomarker evidence for a major preservation pathway of sedimentary organic carbon. *Science* 312: 1627–1631.

Heckert EG, S Seal, WT Self. 2008. Fenton-like reaction catalyzed by the rare earth transition metal cerium. *Environ Sci Technol* 42: 5014–5019.

Heijman CG, C Holliger, MA Glaus, RP Schwarzenbach, J Zeyer. 1993. Abiotic reduction of 4-chloronitrobenzene to 4-chloroaniline in a dissimilatory iron-reducing enrichment culture. *Appl Environ Microbiol* 59: 4350–4353.

Helmig D, J Arey, WP Harger, R Atkinson, J López-Cancio. 1992a. Formation of mutagenic nitrodibenzopyranones and their occurrence in ambient air. *Environ Sci Technol* 26: 622–624.

Helmig D, J López-Cancio, J Arey, WP Harger, R Atkinson. 1992b. Quantification of ambient nitrodibenzopyranones: Further evidence for atmospheric mutagen formation. *Environ Sci Technol* 26: 2207–2213.

Helvenston MC, CE Castro. 1992. Nickel(I) octaethylisobacteriochlorin anion. An exceptional nucleophile. Reduction and coupling of alkyl halides by anionic and radical processes. A model for factor F-430. *J Am Chem Soc* 114: 8490–8496.

Herner HA, JE Trosko, SJ Masten. 2001. The epigenic toxicity of pyrene and related ozonation byproducts containing an aldehyde functional group. *Environ Sci Technol* 35: 3576–3583.

Herod A A. 1998. Azaarenes and thiaarenes. *Handbook Environ Chem* 3I: 271–323.

Ho PC. 1986. Photooxidation of 2,4-dinitrotoluene in aqueous solution in the presence of hydrogen peroxide. *Environ Sci Technol* 20: 260–267.

Hoffman-Luca CG, AA Eroy-Reveles, J Alvarenga, PK Mascharak. 2009. Synthesis, structures, and photochemistry of manganese nitrosyls derived from designed Schiff base ligands: Potential NO donors that can be activated by near-infrared light. *Inorg Chem* 48: 9104–9111.

Holmstead RL. 1976. Studies of the degradation of mirex with an iron (II) porphyrin model system. *J Agric Food Chem* 24: 620–624.

Hong F, S Pehkonen. 1998. Hydrolysis of phorate using simulated and environmental conditions: Rates, mechanisms, and product analysis. *J Agric Food Chem* 46: 1192–1199.

Hori H, E Hayakawa, H Einaga, S Kutsuna, K Koike, T Ibusuki, H Koatagawa, R Arakawa. 2004a. Decomposition of environmentally persistent perfluorooctanoic acid in water by photochemical approaches. *Environ Sci Technol* 38: 6118–6124.

Hori H, E Hayakawa, K Koike, H Einaga, T Ibusuki. 2004b. Decomposition of nonafluoropentanoic acid by heteropolyacid photocatalyst $H_3PW_{12}O_{40}$ in aqueous solution. *J Mol Cat* A 211: 35–41.

Hori H, A Yamamoto, E Hayakawa, S Taniyasu, S Kutsuna, H Kiatagawa, R Arakawa. 2005a. Efficient decomposition of environmentally persistent perfluorocarboxylic acids by use of persulfate as a photochemical oxidant. *Environ Sci Technol* 39: 2383–2388.

Hori H, A Yamamoto, S Kutsuna. 2005b. Efficient photochemical decomposition of long-chain perfluorocarboxylic acids by means of an aqueous/liquid CO_2 biphasic system. *Environ Sci Technol* 39: 7692–7697.

Hu J, S Cheng, T Aizawa, Y Terao, S Kunikane. 2003. Products of aqueous chlorination of 17β-estradiol and their estrogenic activities. *Environ Sci Technol* 37: 5665–5670.

Hu J, X Jin, S Kunikane, Y Terao, T Aizawa. 2006. Transformation of pyrene in aqueous chlorination in the presence and absence of bromide: Kinetics, products, and their aryl hydrocarbon receptor-mediated activities. *Environ Sci Technol* 40: 487–493.

Hua I, MR Hoffmann. 1996. Kinetics and mechanism of the sonolytic degradation of CCl_4: Intermediates and by-products. *Environ Sci Technol* 30: 864–871.

Huber MM, TA Ternesm U von Gunten. 2004. Removal of estrogenic activity and formation of oxidation products during ozonation of 17α-ethinylestradiol. *Environ Sci Technol* 38: 5177–5186.

Huber SG, G Kilian, HF Scholer. 2007. Carbon suboxide, a highly reactive intermediate from the abiotic degradation of aromatic compounds. *Environ Sci Technol* 41: 7802–7806.

Hudlický M. 1990. *Oxidations in Organic Chemistry*. American Chemical Society, Washington, DC, USA.

Hung C-H, BJ Marinas. 1997. Role of chlorine and oxygen in the photocatalytic degradation of trichloroethylene vapor on TiO_2 films. *Environ Sci Technol* 31: 562–568.

Ishii S, Y Hisamatsu, K Inazu, M Kadoi, K-J Aika. 2000. Ambient measurement of nitrotriphenylenes and possibility of nitrotriphenylene formation by atmospheric reaction. *Environ Sci Technol* 34: 1893–1899.

Jang M, SR McDow. 1997. Products of benz/*a*/anthracene photodegradation in the presence of known organic constituents of atmospheric aerosols. *Environ Sci Technol* 31: 1046–1053.

Japar SM, TJ Wallington, SJ Rudy, TY Chang. 1991. Ozone-forming potential of a series of oxygenated organic compounds. *Environ Sci Technol* 25: 415–420.

Jeong HY, KF Hayes. 2003. Impact of transition metals on reductive dechlorination rate of hexachloroethane by Mackinawite. *Environ Sci Technol* 37: 4650–4655.

Jeong HY, KF Hayes. 2007. Reductive dechlorination of tetrachloroethylene and trichloroethylene by Mackinawite (FeS) in the presence of metals: Reaction rates. *Environ Sci Technol* 41: 6390–6396.

Johnson SK, LL Houk, J Feng, RS Houk, DS Johnson. 1999. Electrochemical incineration of 4-chlorophenol and the identification of products and intermediates by mass spectroscopy. *Environ Sci Technol* 33: 2638–2644.

Jordan A, H Frank, ED Hoekstra, S Juuti. 1999. New directions: Exchange of comments on "The origins and occurrence of trichloroacetic acid". *Atmos Environ* 33: 4525–4527.

Ju X, M Hecht, RA Galhotra, WP Ela, EA Betterton, RG Arnold, AE Sáez. 2006. Destruction of gas-phase trichloroethylene in a modified fuel cell. *Environ Sci Technol* 40: 612–617.

Kamens R, M Jang, K Leach. 1999. Aerosol formation from the reaction of α-pinene and ozone using a gas-phase-kinetics-aeosol partitioning model. *Environ Sci Technol* 33: 1430–1438.

Kami H, T Watanabe, S Takemura, Y Kameda, T Hirayama. 2000. isolation and chemical–structural identification of a novel aromatic amine mutagen in an ozonized solution of *m*-phenylenediamine. *Chem Res Toxicol* 13: 165–169.

Kari FG, S Hilger, S Canonica. 1995. Determination of the reaction quantum yield for the photochemical degradation of Fe(III)—EDTA: Implications for the environmental fate of EDTA in surface waters. *Environ Sci Technol* 29: 1008–1017.

Kari FG, W Giger. 1995. Modelling the photochemical degradation of ethylenediaminetetraacetate in the River Glatt. *Environ Sci Technol* 29: 2814–2827.

Kiene RP, BF Taylor. 1988. Demethylation of dimethylsulfoniopropionate and production of thiols in anoxic marine sediments. *Appl Environ Microbiol* 54: 2208–2212.

Kim J-H, PG Tratnyek, T-S Chang. 2008. Rapid dechlorination of polylchlorinated dibenzo-*p*-dioxins by bimetallic and nanosized zerovalent iron. *Environ Sci Technol* 42: 4106–4112.

Kirchner F, LP Thüner, I Barnes, KH Becker, B Donner, F Zabel. 1997. Thermal lifetimes of peroxynitrates occurring in the atmospheric degradation of oxygenated fuel additives. *Environ Sci Technol* 31: 1801–1804.

Kleier D, I Holden, JE Casida, LO Ruzo. 1985. Novel photoreactions of an insecticidal nitromethylene heterocycle. *J Agric Food Chem* 33: 998–1000.

Klotz B et al. 1998. Atmospheric oxidation of toluene in a large-volume outdoor photoreactor: *In situ* determination of ring-retaining product yields. *J Phys Chem A* 102: 10289–10299.

Klotz B, I Barnes, KH Becker, BT Golding. 1997. Atmospheric chemistry of benzene oxide/oxepin. *J Chem Soc Faraday Trans* 93: 1507–1516.

Klupinski TP, Y-P Chin. 2003. Abiotic degradation of trifluralin by Fe(II): Kinetics and transformation pathways. *Environ Sci Technol* 37: 1311–1318.

Knitt LE, JR Shapley, TJ Strathmann. 2008. Rapid metal-catalyzed hydrodehalogenation of iodinated X-ray contrast media. *Environ Sci Technol* 42: 577–583.

Kohler M, NV Heeb. 2003. Determination of nitrated phenolic compounds in rain by liquid chromatography/ atmospheric pressure chemical ionization mass spectrometry. *Anal Chem* 75: 3115–3121.

Koizumi A, N Saitoh, T Suzuki, S Kamiyama. 1994. A novel compound, 9-hydroxy-1-nitropyrene, is a major photodegraded compound of 1-nitropyrene in the environment. *Arch Environ Health* 49: 87–92.

Kramer JB, Canonica, S, Hoigné, J Kaschig. 1996. Degradation of fluorescent whitening agents in sunlit natural waters. *Environ Sci Technol* 30: 2227–2234.

Krone UE, K Laufer, RH Thauer, HPC Hogenkamp. 1989. Coenzyme F_{430} as a possible catalyst for the reductive dehalogenation of chlorinated C-1 hydrocarbons in methanogenic bacteria. *Biochemistry* 28: 10061–10065.

Kubátova A, J Herman, TS Steckler, M de Veij, DJ Miller, EB Klunder, CM Wai, SB Hawthorne. 2003. Subcritical (hot/liquid) water dechlorination of PCBs (Arochlor 1254) with metal additives and in waste paint. *Environ Sci Technol* 37: 5757–5762.

Kubátova A, AJM Lagadec, SB Hawthorne. 2002. Dechlorination of lindane, dieldrin, tetrachloroethane, trichloroethene and PVC in subcritical water. *Environ Sci Technol* 36: 1337–1343.

Kuzuhara S, H Sato, N Tsubouchi, Y Ohtsuka, E Kasai. 2005. Effect of nitrogen-containing compounds on polychlorinated dibenzo-*p*-dioxin/dibenzofuran formation through *de novo* synthesis. *Environ Sci Technol* 39: 795–799.

Kwok ESC, R Atkinson, J Arey. 1995. Rate constants for the gas-phase reactions of the OH radical with dichlorobiphenyls, 1-chlorodibenzo-*p*-dioxin, 1,2-dimethoxybenzene, and diphenyl ether: Estimation of OH radical reaction rate constants for PCBs, PCDDs, and PCDFs. *Environ Sci Technol* 29: 1591–1598.

Lagadec AJM, DJ Miller, AV Lilke, SB Hawthorne. 2000. Pilot-scale subcritical water remediation of polycyclic aromatic hydrocarbon- and pesticide-contaminated soil. *Environ Sci Technol* 34: 1542–1548.

Lam MW, CJ Young, SA Mabury. 2005. Aqueous photochemical reaction kinetics and transformation of fluoxetine. *Environ Sci Technol* 39: 513–522.

Lambrych KL, JP Hassett. 2006. Wavelength-dependent photoreactivity of mirex in Lake Ontario. *Environ Sci Technol* 40: 858–863.

Lan Q, F Li, C Liu, X-Z Li. 2008. Heterogeneous photodegradation of pentachlorophenol with maghemite and oxalate under UV illumination. *Environ Sci Technol* 42: 7918–7923.

Lang KF, H Buffleb. 1958. Die Pyrolyse des α- and β-methyl-naphthalins. *Chem Ber* 91: 2866–870.

Lang PS, W-K Ching, DM Willberg, MR Hoffmann. 1998. Oxidative degradation of 2,4,6-trinitrotoluene by ozone in an electrohydraulic discharge reactor. *Environ Sci Technol* 32: 3142–3148.

Latch DE, K McNeill. 2006. Microheterogeneity of singlet oxygen distributions in irradiated humic acid solutions. *Science* 311: 1743–1747.

Lauff JL, DB Steele, LA Coogan, M Breitfeller. 1990. Degradation of the ferric chelate of EDTA by a pure culture of an *Agrobacterium* sp. *Appl Environ Microbiol* 56: 3346–3353.

Lee W, M Baasandori, PS Stevens, RA Hites. 2005. Monitoring OH-initiated oxidation kinetics of isoprene and its products using online mass spectrometry. *Environ Sci Technol* 39: 1030–1036.

Lee H, RG Harvey. 1986. Synthesis of the active diol epoxide metabolites of the potent carcinogenic hydrocarbon 7,12-dimethybenz[*a*]anthacene. *J Am Chem Soc* 51: 3502–3507.

Lee JH, C Na, RL Ramirez, TM Olson. 2006. Cyanogen chloride precursor analysis in chlorinated river water. *Environ Sci Technol* 40: 1478–1484.

Lee-Ruff E, H Kazarians-Moghaddam, M Katz. 1986. Controlled oxidations of benzo[*a*]pyrene. *Can J Chem* 64: 11297–1303.

Lei P, C Chen, Y Yang, W Ma, J Zhao, L Zang. 2005. Degradation of dye pollutants by immobilized polyoxometalate with H_2O_2 under visible-light irradiation. *Environ Sci Technol* 39: 8466–8474.

Leitner NKV. 1998. Chlorination and formation of organoiodinated compounds: The important role of ammonia. *Environ Sci Technol* 32: 1680–1685.

Lesage S, S Brown, KR Hosler. 1992. Degradation of chlorofluorocarbon-113 under anaerobic conditions. *Chemosphere* 24: 1225–1243.

Levine AA, AG Cole. 1932. The ozonides of ortho-xylene and the structure of the benzene ring. *J Am Chem Soc* 54: 338–341.

Levsen K. 1988. The analysis of diesel particulate. *Fresenius Z Anal Chem* 331: 467–478.

Li B, PL Gutierraz, NV Bloiugh. 1997. Trace determination of hydroxyl radical in biological systems. *Anal Chem* 69: 4295–4302.

Lijinsky W, CR Taha. 1961. The pyrolysis of 2-methylnaphthalene. *J Org Chem* 26: 3566–3568.

Lin C-H, RH Lerch, EM Thurmkan, HE Garrett, MF George. 2002. Determination of isoxaflutole (Balance) and its metabolites in water using solid-phase extraction followed by high-performance liquid chromatography with ultraviolet or mass spectrometry. *J Agric Food Chem* 50: 5816–5824.

Liu Z, RG Arnold EA Betterton, E Smotkin. 2001. Reductive dehalogenation of gas-phase chlorinated solvents using a modified fuel cell. *Environ Sci Technol* 35: 4320–4326.

Liu Z, EA Betterton, RG Arnold. 2000. Electrolytic reduction of low molecular weight chlorinated aliphatic compounds: Structural and thermodynamic effects on process kinetics. *Environ Sci Technol* 34: 804–811.

Liu YJ, XZ Jiang. 2005. Phenol degradation by a nonpulsed diaphragm glow discharge in an aqueous solution. *Environ Sci Technol* 39: 8512–8517.

Liu Y, SA Majetich, RD Tilton, DS Sholl, GV Lowry. 2005. TCE dechlorination rates, pathways, and efficiency of nanoscale particles with different properties. *Environ Sci Technol* 39: 1338–1345.

Lockhart HB, RV Blakeley. 1975. Aerobic photodegradation of Fe(III)-(ethylenedinitrilo)tetraacetate (Ferric EDTA). Implications for natural waters. *Environ Sci Technol* 9: 1035–1038.

Lovley DR, JC Woodward. 1992. Consumption of freons CFC-11 and CFC-12 by anaerobic sediments and soils. *Environ Sci Technol* 26: 925–929.

Luster-Teasley SL, JJ Yao, HH Herner, JE Trosko, SJ Masten. 2002. Ozonation of chrysene: Evaluation of byproduct mixtures and identification of toxic constituent. *Environ Sci Technol* 36: 869–776.

Macalady DL, NL Wolfe. 1985. Effects of sediment sorption and abiotic hydrolyses.1. Organophosphorothioate esters. *J Agric Food Chem* 33: 167–173.

Madden TH, AK Datye, M Fulton, MR Prairie, SA Majumdar, BM Stange. 1997. Oxidation of metal-EDTA complexes by TiO_2 photocatalysis. *Environ Sci Technol* 31: 3475–3481.

Marks TS, JD Allpress, A Maule. 1989. Dehalogenation of lindane by a variety of porphyrins and corrins. *Appl Environ Microbiol* 55: 1258–1261.

Matykiewiczová N, J Klánová, P Klán. 2007. Photochemical degradation of PCBs in snow. *Environ Sci Technol* 41: 8308–8314.

McDowell DC, MM Huber, M Wagner, U von Gunten, TA Ternes. 2005. Ozonation of carbamazepine in drinking water: Identification and kinetic study of major oxidation products. *Environ Sci Technol* 39: 8014–8022.

Metwally M E-S, NL Wolfe. 1990. Hydrolysis of chlorostilbene oxide. II. Modelling of hydrolysis in aquifer samples and in sediment–water systems. *Environ Toxicol Chem* 9: 963–973.

Miller RM, GM Singer, JD Rosen, R Bartha. 1988a. Sequential degradation of chlorophenols by photolytic and microbial treatment. *Environ Sci Technol* 22: 1215–1219.

Miller RM, GM Singer, JD Rosen, R Bartha. 1988b. Photolysis primes biodegradation of benzo[*a*]pyrene. *Appl Environ Microbiol* 54: 1724–1730.

Miller PL, D Vasudevan, PM Gschwend, AL Roberts. 1998. Transformatiom of hexachloroethane in a sulfidic natural water. *Environ Sci Technol* 32: 1269–1275.

Monson PD, DJ Call, DA Cox, K Liber, GT Ankley. 1999. Photoinduced toxicity of fluoranthene to northern leopard frogs (*Rana pipens*). *Environ Toxicol Chem* 18: 308–312.

Monteil-Rivera F, L Paquet, A Halasz, MT Montgomery, J Hawari. 2005. Reduction of octahydro-1,3,5,7-tetranitro-1,3,5,7-tetrazocine by zero-valent iron: Product distribution. *Environ Sci Technol* 39: 9725–9731.

Moore DE, S Roberts-Thomson, D Zhen, CC Duke. 1990. Photochemical studies on the anti-inflammatory drug diclofenac. *Photochem Photobiol* 52: 685–690.

Mowrer J, J Nordin. 1987. Characterization of halogenated organic acids in flue gases from municipal waste incinerators. *Chemosphere* 16: 1181–1192.

Müller MD, H-R Buser. 1986. Halogenated aromatic compounds in automotive emissions from leaded gasoline additives. *Environ Sci Technol* 20: 1151–1157.

Müller SR, H-R Zweifel, DJ Kinnison, JA Jacobsen, MA Meier, MM Ulrich, RP Schwarzenbach. 1996. Occurrence, sources, and fate of trichloroacetic acid in Swiss waters. *Environ Toxicol Chem* 15: 1470–1478.

Na C, TM Olson. 2006. Mechanism and kinetics of cyanogen chloride formation from the chlorination of glycine. *Environ Sci Technol* 40: 1469–1477.

Nakagawa R, S Kitamori, K Horikawa, K Nakashima, H Tokiwa. 1983. Identification of dinitropyrenes in diesel-exhaust particles. Their probable presence as the major mutagens. *Mutation Res* 124: 201–211.

Nélieu S, L Kerhoas, J Einhorn. 2000. Degradation of atrazine into ammeline by combined ozone/hydrogen peroxide treatment in water. *Environ Sci Technol* 34: 430–437.

Nelson L, I Shanahan, HW Sidebottom, J Treacy, OJ Nielsen. 1990. Kinetics and mechanism for the oxidation of 1,1,1-trichloroethane. *Int J Chem Kinet* 22: 577–590.

Nishioka MG, CC Howard, DA Conros, LM Ball. 1988. Detection of hydroxylated nitro aromatic and hydroxylated nitro polycyclic aromatic compounds in ambient air particulate extract using bioassay-directed fractionation. *Environ Sci Technol* 22: 908–915.

Nomiyama K, T Tanizaki, H Ishibashi, K Arizono, R Shinohara. 2005. Production mechanism of hydroxylated PCBs by oxidative degradation of selected PCBs using TiO_2 in water and estrogenic activity of their intermediates. *Environ Sci Technol* 39: 8762–8769.

Nörtemann B. 1992. Total degradation of EDTA by mixed cultures and a bacterial isolate. *Appl Environ Microbiol* 58: 671–676.

Novak PJ, L Daniels, GF Parkin. 1998. Rapid dechlorination of carbon tetrachloride and chloroform by extracellular agents in cultures of *Methanosarcina thermophila*. *Environ Sci Technol* 32: 3132–3136.

Ochoa-Herrera V, R Sierra-Alvarez, A Somogyi, NE Jacobsen, VH Wysocki, JA Field. 2008. Reductive defluorination of perfluorooctane sulfonate. *Environ Sci Technol* 42: 3260–3264.

Orlando JJ. 2003. Atmospheric chemistry of organic bromine and iodine compounds. *Handbook Environ Chem* 3R: 253–299.

Oturan MA, J peiroten, P Chartin, AJ Acher. 2000. Complete destruction of *p*-nitrophenol in aqueous medium by electro-Fenton method. *Environ Sci Technol* 34: 3474–3479.

Paciolla MD, G Davies, SA Jansen. 1999. Generation of hydroxyl radicals from metal-loaded humic acids. *Environ Sci Technol* 33: 1814–1818.

Pakalapati SNR, BN Popov, RE White. 1996. Anodic oxidation of ethylenediaminetetraacetic acid on platinum electrode in alkaline medium. *J Electrochem Soc* 143: 1636–1643.

Paulot F, JD Crounse, HG Kjaergaard, A Kürten, JM St. Clair, JH Seinfeld, PO Wennberg. 2009. Unexpected epoxide formation in the gas-phase photooxidation of isoprene. *Science* 325: 730–733.

Pelizzetti E, V Maurino, C Minero, V Carlin, E Pramauro, O Zerbinati, ML Tosata. 1990. Photocatalytics degradation of atrazine and other *s*-triazine herbicides. *Environ Sci Technol* 24: 1559–1565.

Perdue EM, NL Wolfe. 1982. Modification of pollutant hydrolysis kinetics in the presence of humic substances. *Environ Sci Technol* 16: 847–852.

Perlinger JA, W Angst, RP Schwarzenbach. 1996. Kinetics of the reduction of hexachloroethane by juglone in solutions containing hydrogen sulfide. *Environ Sci Technol* 30: 3408–3417.

Pitts, N Jr, R Atkinson, JA Sweetman, B Zielinska. 1985a. The gas-phase reaction of naphthalenes with N_2O_5 to form nitronaphthalenes. *Atmos Environ* 19: 701–705.

Pitts JN Jr, B Zielinska, JA Sweetman, R Atkinson, AM Winer. 1985b. Reactions of adsorbed pyrene and perylene with gaseous N_2O_5 under simulated atmospheric conditions. *Atmos Environ* 19: 911–915.

Platt UF, AM Winer, HW Biermann, R Atkinson, JN Pitts, Jr. 1984. Measurement of nitrate radical concentrations in continental air. *Environ Sci Technol* 18: 365–369.

Platz J, OJ Nielsen, J Sehested, TJ Wallington. 1995. Atmospheric chemistry of 1,1,1-trichloroethane: UV absorption spectra and self-reaction kinetics of CCl_3CH_2 and $CCl_3CH_2O_2$ radicals, kinetics of the reactions of the $CCl_3CH_2O_2$ radical with NO and NO_2, and the fate of alkoxy radical CCl_3CH_2O. *J Phys Chem* 99: 6570–6579.

Plewa MJ, ED Wagner, SD Richardson, AD Thurston, Y-T Woo, AB McKague. 2004. Chemical and biological characterization of newly discovered iodoacid drinking water disinfection byproducts. *Environ Sci Technol* 38: 4713–4722.

Pradhan SP, JR Paterek, BY Liu, JR Conrad, VJ Srivastava. 1997. Pilot-scale bioremediation of PAH-contaminated soils. *Appl Biochem Biotechnol* 63/65: 759–773.

Quinn J, C Geiger, C Clausen, K Brooks, C Coon, S ÓHara, T Krug, D Major, W-S Yoon, A Gavskar, T Holdsworth. 2005. Field demonstration of DNAPL dehalogenation using emulsified zero-valent iron. *Environ Sci Technol* 39: 1309–1318.

Rahm S, N Green, J Norrgran, Å Bergman. 2005. Hydrolysis of environmental contaminants as an experimental tool for indication of their persistency. *Environ Sci Technol* 39: 3128–3133.

Ravishankara AR, G Hancock, M Kawasaki, Y Matsumi. 1998. Photochemistry of ozone: Surprises and recent lessons. *Science* 280: 60–61.

Reisen F, SM Aschman, R Atkinson, J Arey. 2005. 1,4-hydroxycarbonyl products of the OH radical initiated reactions of C_5–C_8 *n*-alkanes in the presence of NO. *Environ Sci Technol* 39: 4447–4453.

Ressler T, J Wong, J Roos, IL Smith. 2000. Quantitative speciation of Mn-bearing particulates emitted from autos burning (methylcyclopentadienyl)manganese tricarbonyl-added gasolines using XANES spectroscopy. *Environ Sci Technol* 34: 950–958.

Richardson SD et al. 1999. Identification of new ozone disinfection byproducts in drinking water *Environ Sci Technol* 33: 3368–3377.

Richardson SD, AD Thurston, C Rav-Acha, L Groisman, I Popilevsky, O Juraev, V Glezer, AB McKague, MJ Plewa, ED Wagner. 2003. Tribromopyrrole, brominated acids, and other disinfection byproducts produced by disinfection of drinking water rich in bromide. *Environ Sci Technol* 37: 3782–3793.

Roberts AL, PM Gschwend. 1991. Mechanism of pentachloroethane dehydrochlorination to tetrachloroethylene. *Environ Sci Technol* 25: 76–86.

Roberts AL, PN Sanborn, PM Gschwend. 1992. Nucleophilic substitution of dihalomethanes with hydrogen sulfide species. *Environ Sci Technol* 26: 2263–2274.

Roberts AL, LA Totten, WA Arnold, DR Burris, TJ Campbell. 1996. Reductive elimination of chlorinated ethylene by zero-valent metals. *Environ Sci Technol* 30: 2654–2659.

Rodríguez-Garrido B, MC Arbestain, MC Monterroso, F Macías. 2004. Reductive dechlorination of α, β, δ, and γ-hexachlorocyclohexane isomers by hydroxocobalamin in the presence of either dithiothreitol or titanium(III) citrate as reducing agents. *Environ Sci Technol* 38: 5046–5052.

Rosenkranz HS, R Mermelstein. 1983. Mutagenicity and genotoxicity of nitroarenes. All nitro-containing chemicals were not created equal. *Mutat Res* 114: 216–267.

Salmeen IT, AM Pero, R Zator, D Schuetzle, TL Riley(1984) Ames assay chromatograms and the identification of mutagens in diesel particle extracts. *Environ Sci Technol* 18: 375–382.

Sasaki J, J Arey, WP Harger. 1995. Formation of mutagens from the photooxidation of 2–4-ring PAH. *Environ Sci Technol* 29: 1324–1335.

Sayles GD, G You, M Wang, MJ Kupferle. 1997. DDT, DDD, and DDE Dechlorination by zero-valent iron. *Environ Sci Technol* 31: 3448–3454.

Schatzschneider U. 2010. Photoactivated biological activity of transition-metal complexes. *Eur J Inorg Chem* 1451–1467, doi: 10.1002/ejic.201000003.

Schanke CA, LP Wackett. 1992. Environmental reductive elimination reactions of polychlorinated ethanes mimicked by transition-metal coenzymes. *Environ Sci Technol* 26: 830–833.

Schneider M, O Luxenhofer, A Deissler, K Ballschmiter. 1998. C_1–C_{15} alkyl nitrates, benzyl nitrate and bifunctional nitrates: Measurements in California and South Atlantic air and global comparison using C_2C_{14} and $CHBr_3$ as marker molecules. *Environ Sci Technol* 32: 3055–3062.

Schneiders GE, MK Koeppe, MV Naidu, P Horne, AM Brown, CF Mucha. 1993. Rate of rimsulfuron hydrolysis in the environment. *J Agric Food Chem* 41: 2404–2410.

Schwack W. 1988. Photoinduced additions of pesticides to biomolecules. 2. Model reactions of DDT and methoxychlor with methyl oleate. *J Agric Food Chem* 36: 645–648.

Schwarzenbach RP, R Stierliu, K Lanz, J Zeyer. 1990. Quinone and iron porphyrin mediated reduction of nitroaromatic compounds in homogeneous aqueous solution. *Environ. Sci. Technol.* 24: 1566–1574.

Sedlak DL, AW Andren. 1991. Oxidation of chlorobenzene with Fenton's reagent. *Environ Sci Technol* 25: 777–782.

Seefeld S, JA Kerr. 1997. Kinetics of reactions of propionylperoxy radicals with NO and NO_2: Peroxypropionyl nitrate formation under laboratory conditions related to the troposphere. *Environ Sci Technol* 31: 2949–2953.

Seligman PF, AO Valkirs, RF Lee. 1986. Degradation of tributytin in San Diego Bay, California, waters. *Environ Sci Technol* 20: 1229–1235.

Sharma AK, GB Josephson, DM Cameron, SC Goheen. 2000. Destruction of pentachlorophenol using glow discharge plasma process. *Environ Sci Technol* 34: 2267–2272.

Shey J, CM McGinley, KM McCauley, AS Dearth, BT Young, WA van der Donk. 2002. Mechanistic investigation of a novel vitamin B_{12}-catalyzed carbon–carbon bond forming reaction, the reductive dimerization of arylalkenes. *J Org Chem* 67: 837–846.

Shu Y, ESC Kwok, EC Tuazon, R Atkinson, J Arey. 1997. Products of the gas-phase reactions of linalool with OH radicals, NO_3 radicals, and O_3. *Environ Sci Technol* 31: 896–904.

Sigman ME, PF Schuler, MM Ghosh, RT Dabestani. 1998. Mechanism of pyrene photochemical oxidation in aqueous and surfactant solutions. *Environ Sci Technol* 32: 3980–3985.

Singh NC, TP Dasgupta, EV Roberts, A Mansingh. 1991. Dynamics of pesticides in tropical conditions. 1. Kinetic studies of volatilization, hydrolysis, and photolysis of dieldrin and a- and b-endosulfan. *J Agric Food Chem* 39: 575–579.

Smith AE, J Grove. 1969. Photodegradation of diquat in dilute aqueous solution and on silica gel. *J Agric Food Chem* 17: 609–613.

Soderquist CJ, DG Crosby, KW Moilanen, JN Seiber, JE Woodrow. 1975. Occurrence of trifluralin and its photoproducts in air. *J Agric Food Chem* 23: 304–309.

Song H, ER Carraway. 2005. Reduction of chlorinated ethanes by nanosized zero-valent iron; kinetics, pathways, and effect of reaction conditions. *Environ Sci Technol* 39: 6237–6245.

Spadaro JT, L Isabelle, V Renganathan. 1994. Hydroxyl radical mediated degradation of azo dyes: Evidence for benzene generation. *Environ Sci Technol* 28: 1389–1393.

Spanggord RJ, CD Yao. T Mill. 2000. Oxidation of aminodinitrotoluenes with ozone: Products and pathways. *Environ Sci Technol* 34: 497–504.

Standley LJ, TL Bott. 1998. Trifluoroacetate, an atmospheric breakdown product of hydrofluorocarbon refrigerants: Biomolecular fate in aquatic organisms. *Environ Sci Technol* 32: 469–475.

Stang PM, RF Lee, PF Seligman. 1992. Evidence for rapid, nonbiological degradation of tributyltin compounds in autoclaved and heat-treated fine-grained sediments. *Environ Sci Technol* 26: 1382–1387.

Stephanou EG. 1992. α,ω-dicarboxylic acid salts and α,ω-dicarboxylic acids. Photooxidation products of unsaturated fatty acids, present in marine aerosols and marine sediments. *Naturwiss* 79: 28–131.

Stephanou EG, N Stratigakis. 1993. Oxocarboxylic and α,ω-dicarboxylic acids: Photooxidation products of biogenic unsaturated fatty acids present in urban aerosols. *Environ Sci Technol* 27: 1403–1407.

Stidson RT, CA Dickey, JN Cape, KV Heal, MR Heal. 2004. Fluxes and reservoirs of trichloroacetic acid at a forest and moorland catchment. *Environ Sci Technol* 38: 1639–1647.

Sun Y, JJ Pignatello. 1993. Organic intermediates in the degradation of 2,4-dichlorophenoxyacetic acid by Fe^{3+}/H_2O_2 and Fe^{3+}/H_2O_2/UV. *J Agric Food Chem* 41: 1139–1142.

Sun C, D Zhao, C Chen, W Ma, J Zhao. 2009. TiO_2-mediated photocatalytic debromination of decabromodiphenyl ether: Kinetics and intermediates. *Environ Sci Technol* 43: 157–162.

Swanson MB, WA Ivancic, AM Saxena, JD Allton, GKÓBrian, T Suzuki, H Nishizawa, M Nokata. 1995. Direct photolysis of fenpyroximate in a buffered aqueous solution. *J Agric Food Chem* 43: 513–518.

Tamai T, K Inazu, K-I Aika. 2006. Dichlorodifluoromethane decomposition to CO_2 with simultaneous halogen fixation by calcium oxide based materials. *Environ Sci Technol* 40: 823–829.

Taylor BF, JA Amador, HS Levinson. 1993. Degradation of *meta*-trifluoromethylbenzoate by sequential microbial and photochemical treatments. *FEMS Microbiol Lett* 110: 213–216.

Taylor BF, DC Gilchrist. 1991. New routes for aerobic biodegradation of dimethylsulfoniopropionate. *Appl Environ Microbiol* 57: 3581–3584.

Taylor-Lovell S, GK Sims, LM Wax. 2002. Effects of moisture, temperature, and biological activity on the degradation of isoxaflutole in soil. *J Agric Food Chem* 50: 5626–5633.

ter Halle A, D Drncova, C Richard. 2006. Phototransformation of the herbicide sulcotrione on maize cuticular wax. *Environ Sci Technol* 40: 2989–2995.

Torrents A, BG Anderson, S Bilboulian, WE Johnson, CJ Hapeman. 1997. Atrazine photolysis: Mechanistic investigations of direct and nitrate-mediated hydroxyl radical processes and the influence of dissolved organic carbon from the Chesapeake Bay. 1997. *Environ Sci Technol* 31: 1476–1482.

Totten LA, U Jans, AL Roberts. 2001. Alkyl bromides as mechanistic probes of reductive dehalogenation: Reactions of vicinal stereoisomers with zerovalent metals. *Environ Sci Technol* 35: 2268–2274.

Troupis EGA, A Hiskia, E Papaconstantinou. 2005. Photochemical reduction and recovery of mercury by polyoxometalates. *Environ Sci Technol* 39: 4242–4248.

Tuazon EC, SM Aschmann, R Atkinson. 2000. Atmospheric degradation of volatile methyl-silicon compounds. *Environ Sci Technol* 34: 1970–1976.

Tuazon EC, H MacLeod, R Atkinson, WPL Carter. 1986. α-Dicarbonyl yields from the NO_x-air photooxidations of a series of aromatic hydrocarbons in air. *Environ Sci Technol* 20: 383–387.

Vaughan PP, NV Blough. 1998. Photochemical formation of hydroxyl radicals by constituents of natural waters. *Environ Sci Technol* 32: 2947–2953.

Vecitis CD, H Park, J Chamg, BT Mader, MR Hoffmann. 2008. Kinetics and mechanism of the sonolytic conversion of aqueous perfluorinated surfactants, perfluorooctanoate (PFOA) and perfluorooctane sulfonate (PFOS) into inorganic products. *J Phys Chem A* 112: 4261–4270.

Vione D, V Maurino, C Minero, E Pelizzetti. 2005. Aqueous atmospheric chemistry: Formation of 2,4-nitrophenol upon nitration of 2-nitrophenol and 4-nitrophenol in solution. *Environ Sci Technol* 39: 7921–7931.

Wade RS, CE Castro. 1973. Oxidation of iron(II) porphyrins by alkyl halides. *J Am Chem Soc* 95: 226–230.

Wallington TJ, MD Hurley, JM Fracheboud, JJ Orlando, GS Tyndall, J Sehested, TE Mögelberg, OJ Nielsen. 1996. Role of excited CF_3CFHO radicals in the atmospheric chemistry of HFC-134a. *J Phys Chem* 100: 18116–18122.

Wallington TJ, OJ Nielsen. 2002. Atmospheric chemistry and environmental impact of hydrofluorocarbons (HFCs) and hydrofluoroethers (HFEs). *Handbook Environ Chem* 3N: 85–102.

Wallington TJ, WF Schneider. 1994. The stratospheric fate of CF_3OH. *Environ Sci Technol* 28: 1198–1200.

Wallington TJ, O Sokolov, MD Hurley, GS Tyndall, JJ Orlando, I Barnes, KH Becker, R Vogt. 1999. Atmospheric chemistry of methylcyclopentadienyl manganese tricarbonyl: Photolysis, reaction with hydroxyl radicals and ozone. *Environ Sci Technol* 33: 4232–4238.

Wang L, R Atkinson, J Arey. 2007. Dicarbonyl products of the OH radical-initiated reactions of naphthalene and the C_1- and C_2-alkylnaphthalenes. *Environ Sci Technol* 41: 2803–2810.

Wang Q, EM Scherer, AT Lemley. 2004. Metribuzin degradation by membrane anodic Fenton treatment and its interaction with ferric iron. *Environ Sci Technol* 38: 1221–1227.

Wängberg I, I Barnes, KH Becker. 1997. Product and mechanistic study of the reaction of NO_3 radicals with α-pinene. *Environ Sci Technol* 31: 2130–2135.

Watanabe T, T Hirayama, S Fukui. 1989. Phenazine derivatives as the mutagenic reaction product from *o*- or *m*-phenylenediamine derivatives with hydrogen peroxide. *Mutation Res* 227: 135–145.

Watanabe N, S Horikoshi, A Kawasaki, H Hidaka, N Serpone. 2005. Formation of refractory ring-expanded triazine intermediates during the photocatalyzed mineralization of the endocrine disruptor amitrole and related triazole derivatives at UV-irradiated TiO_2/H_2O interfaces. *Environ Sci Technol* 39: 2320–2326.

Weber R, S Yoshida, K Miwa. 2002. PCB destruction in subcritical and supercritical water—Evaluation of PCDF formation and initial steps of degradation mechanisms. *Environ Sci Technol* 36: 1834–1844.

Wetzstein H-G, N Schmeer, W Karl. 1997. Degradation of the fluoroquinolone enrofloxacin by the brown-rot fungus *Gloeophyllum striatum*: Identification of metabolites. *Appl Environ Microbiol* 63: 4272–4281.

Wetzstein H-G, M Stadler, H-V Tichy, A Dalhoff, W Karl. 1999. Degradation of ciprofloxacin by basidiomycetes and identification of metabolitres generated by the brown-rot fungus *Gloeophyllum striatum. Appl Environ Microbiol* 65: 1556–1563.

Willberg DM, PS Lang, RH Höchemer, A Kratel, MR Hoffmann. 1996. Degradation of 4-chlorophenol, 3,4-dichloroaniline, and 2,4,6-trinitrotoluene in an electrohydraulic discharge reactor. *Environ Sci Technol* 30: 2526–2534.

Witschel M, S Nagel, T Egli. 1997. Identification and characterization of the two-enzyme system catalyzing the oxidation of EDTA in the EDTA-degrading bacterial strain DSM 9103. *J Bacteriol* 179: 6937–6943.

Wolfe NL, RG Zepp, DF Paris. 1978. Carbaryl, propham and chloropropham: A comparison of the rates of hydrolysis and photolysis with the rates of biolysis. *Water Res* 12: 565–571.

Wolt JD, JD Schwake, FR Batzer, SM Brown, LH McKendry, JR Miller, GA Roth, MA Stanga, D Portwood, DL Holbrook. 1992. Anaerobic aquatic degradation of flumetsulam [N-(2,6-difluorophenyl)-5-methyl[1,2,4] triazolo[1,5a]pyrimidine-2-sulfonamide]. *J Agric Food Chem* 40: 2302–2308.

Wong AS, DG Crosby. 1981. Photodecomposition of pentachlorophenol in water. *J Agric Food Chem* 29: 125–130.

Wong M-K, T-C Chan, W-Y Chan, W-K Chan, LLP Vrijmoed, DK ÓToole, C-M Che. 2006. Dioxiranes generated *in situ* from pyruvates and oxone as environmentally friendly oxidizing agents for disinfection. *Environ Sci Technol* 40: 625–630.

Woodburn KB, FR Batzer, FM White, MR Schulz. 1993. The aqueous photolysis of triclopyr. *J Agric Food Chem* 29: 125–130.

Workman SL Woods, YA Gorby, JK Fredrickson, and MJ Truex. 1997. Microbial reduction of vitamin B_{12} by *Shewanella alga* strain BrY with subsequent transformation of carbon tetrachloride. *Environ Sci Technol* 31: 2292–2297.

Worobey BL. 1989. Nonenzymatic biomimetic oxidation systems: Theory and application to transformation studies of environmental chemicals. *Handbook Environ Chem* 2E: 58–110.

Wujcik CE, TM Cahill, JN Seiber. 1998. Extraction and analysis of trifluoroacetic acid in environmental waters. *Anal Chem* 70: 4074–4080.

Yak HK, BW Wenclawiak, IF Cheng, JG Doyle, CM Wai. 1999. Reductive dechlorination of polychlorinated biphenyls by zerovalent iron in subcritical water. *Environ Sci Technol* 33: 1307–1310.

Yamauchi T, T Handa. 1987. Characterization of aza heterocyclic hydrocarbons in urban atmospheric particulate matter. *Environ Sci Technol* 21: 1177–1181.

Yasahura A, M Morita. 1988. Formation of chlorinated aromatic hydrocarbons by thermal decomposition of vinylidene chloride polymer. *Environ Sci Technol* 22: 646–650.

Yu J, HE Jeffries. 1997. Atmospheric photooxidation of alkylbenzenes II. Evidence of formation of epoxide intermediates. *Atmos Environ* 31: 2281–2287.

Yu J, HE Jeffries, KG Sexton. 1997. Atmospheric photooxidation of alkylbenzenes—I. Carbonyl product analysis. *Atmos Environ* 31: 2261–2280.

Zander M. 1995. *Polycyclische Aromaten*, pp. 213–217. BG Teuber, Stuttgart, Germany.

Zepp RG, GL Baugham, PA Scholtzhauer. 1981a. Comparison of photochemical behaviour of various humic substances in water I. Sunlight induced reactions of aquatic pollutants photosensitized by humic substances. *Chemosphere* 10: 109–117.

Zepp RG, GL Baugham, PA Scholtzhauer. 1981b. Comparison of photochemical behaviour of various humic substances in water: II Photosensitized oxygenations. *Chemosphere* 10: 119–126.

Zepp RG, PF Schlotzhauer. 1983. Influence of algae on photolysis rates of chemicals in water. *Environ Sci Technol* 17: 462–468.

Zepp RG, PF Schlotzhauer, RM Sink. 1985. Photosensitized transformations involving energy transfer in natural waters: Role of humic substances. *Environ Sci Technol* 19: 74–81.

Zhang H, C-H Huang. 2003. Oxidative transformation of triclosan and chlorophene by manganese dioxides. *Environ Sci Technol* 37: 2421–243.

Zhang H, C-H Huang. 2005. Oxidative transformation of fluoroquinolone antibacterial agents and structurally related amines by manganese dioxide. *Environ Sci Technol* 39: 4474–4483.

Zhang Q, SO Pehkonen. 1999. Oxidation of diazinon by aqueous chlorine: Kinetics, mechanisms, and product studies. *J Agric Food Chem* 47: 1760–1766.

Zielinska B, J Arey, R Atkinson, PA McElroy. 1988. Nitration of acephenanthrylene under simulated atmospheric conditions and in solution and the presence of nitroacephenanthrylenes in ambient particles. *Environ Sci Technol* 22: 1044–1048.

Zielinska B, J Arey, R Atkinson, PA McElroy. 1989. Formation of methylnitronaphthalenes from the gas-phase reactions of 1- and 2-methylnaphthalene with hydroxyl radicals and N_2O_5 and their occurrence in ambient air. *Environ Sci Technol* 23: 723–729.

Zoro JA, JM Hunter, G Eglinton, CC Ware. 1974. Degradation of *p,p'*-DDT in reducing environments. *Nature* 247: 235–237.

2

Biotic Reactions: An Outline of Reactions and Organisms

2.1 Introduction

The transformation of organic compounds by photochemical processes is discussed in Chapter 1. It is, however, generally conceded that biotic reactions involving microorganisms are of major significance in determining the fate and persistence of organic compounds in aquatic and terrestrial ecosystems. The role of higher biota in carrying out important transformations is addressed in Part 2 of this chapter, and that involving plants in Part 3.

Archaea, bacteria, cyanobacteria, fungi, yeasts, and algae comprise a large and diverse number of taxa. However, only a relatively small number of even the genera have been examined in the context of biodegradation, and there is no way of determining how representative these are. Care should, therefore, be exercised in drawing conclusions about the metabolic capability of the plethora of taxa included within these major groups of microorganisms. Attention is directed to the role of populations that are discussed in Chapter 5, Part 2.

2.2 Definitions: Degradation and Transformation

It is essential to make a clear distinction between biodegradation and biotransformation in the beginning.

Biodegradation: Under aerobic conditions, biodegradation results in the mineralization of an organic compound to carbon dioxide and water and—if the compound contains nitrogen, sulfur, phosphorus, or halogen—with the release of ammonium (or nitrite), sulfate, phosphate, or halide. These inorganic products may then enter well-established geochemical cycles. Under anaerobic conditions, methane may be formed in addition to carbon dioxide, and sulfate may be reduced to sulfide.

Biotransformation: In contrast, biotransformation involves only a restricted number of metabolic reactions, and the basic framework of the molecule remains essentially intact.

Illustrative examples of biotransformation reactions include the following, although it should be emphasized that other microorganisms may be able to degrade the substrates:

1. The hydroxylation of 14-chlorodehydroabietic acid by fungi (Figure 2.1) (Kutney et al. 1982)
2. The epoxidation of alkenes by bacteria (Patel et al. 1982; van Ginkel et al. 1987)
3. The formation of 16-chlorohexadecyl-16-chlorohexadecanoate from hexadecyl chloride by *Micrococcus cerificans* (Kolattukudy and Hankin 1968):

$$CH_3(CH_2)_{14}-CH_2Cl \rightarrow ClCH_2(CH_2)_{14}-CH_2-O-CO-(CH_2)_{14}-CH_2Cl$$

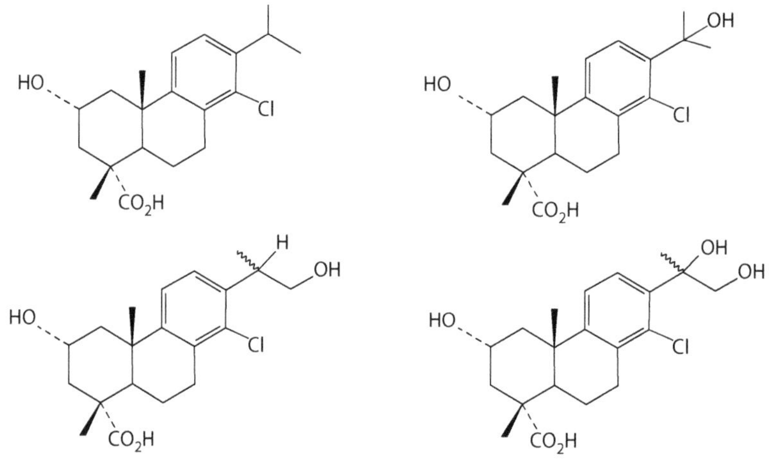

FIGURE 2.1 Biotransformation of dehydroabietic acid by *Mortierella isabellina.*

4. The *O*-methylation of chlorophenols to anisoles by fungi (Cserjesi and Johnson 1972; Gee and Peel 1974) and by bacteria (Suzuki 1978; Rott et al. 1979; Neilson et al. 1983; Häggblom et al. 1988)

5. The formation of glyceryl-2-nitrate from glyceryl trinitrate by *Phanerochaete chrysosporium* (Servent et al. 1991)

The initial biotransformation products may, in some cases, be incorporated into cellular material. For example, the carboxylic acids formed by the oxidation of long-chain *n*-alkyl chlorides were incorporated into cellular fatty acids by strains of *Mycobacterium* sp. (Murphy and Perry 1983). The fungus *Mortierella alpina*, in which hexadecene was oxidized by the fungus by ω-oxidation (Shimizu et al. 1991), formed lipids that contained carboxylic acids containing both 18 and 20 carbon atoms, including the unusual polyunsaturated acid 5*cis*, 8*cis*, 11*cis*, 14*cis*, 19-eicosapentaenoic acid. Metabolites of metolachlor that could only be extracted from the cells with acetone were apparently chemically bound to unidentified sulfur-containing cellular components (Liu et al. 1989).

Biodegradation and biotransformations are of course alternatives, but they are not mutually exclusive. For example, it has been suggested that for chlorophenolic compounds, the *O*-methylation reaction may be an important alternative to reactions that bring about their degradation (Allard et al. 1987). Apart from the environmental significance of biotransformation reactions, many of them have enormous importance in biotechnology for the production of valuable metabolites, for example, the synthesis of hydroxylated steroids and in reactions that take advantage of the oxidative potential of methanotrophic bacteria (Lidstrom and Stirling 1990) and rhodococci (Finnerty 1992).

It is also important to consider the degradation of xenobiotics in the wider context of metabolic reactions carried out by the cell. The cell must obtain energy to carry out essential biosynthetic (anabolic) reactions for its continued existence, and to enable growth and cell division to take place. The substrate cannot, therefore, be degraded completely to carbon dioxide or methane, for example, and a portion must be channeled into the biosynthesis of essential molecules. Indeed, many organisms will degrade xenobiotics only in the presence of a suitable more readily degraded growth substrate that supplies both cell carbon and the energy for growth: this is relevant to discussions of "cometabolism" and "concurrent metabolism," that is discussed in Chapter 4, Part 2.

Growth under anaerobic conditions is demanding both physiologically and biochemically, since the cells will generally obtain only low yields of energy from the growth substrate, and must, in addition, maintain a delicate balance between oxidative and reductive processes. Only a few examples of mechanisms for ATP generation in anaerobes are given below and are discussed in greater detail in later chapters:

1. Clastic reactions from 2-oxo acid-CoA esters produced in a number of degradations
2. Reactions involving carbamyl phosphate in the degradation of arginine in clostridia, and the fermentation of allantoin by *Streptococcus allantoicus*
3. The activity of formyl THF synthase during the fermentation of purines by clostridia
4. The reductive dechlorination of 3-chlorobenzoate by *Desulfomonile tiedjei* DCB-1
5. The proton pump in *Oxalobacter formigenes*
6. The biotin-dependent carboxylases that couple the decarboxylation of malonate to acetate in *Malonomonas rubra* to the transport of Na^+ across the cytoplasmic membrane

True fermentation implies that a single substrate is able to provide carbon for cell growth and at the same time satisfy the energy requirements of the cell. A simple example of fermentation is the catabolism of glucose by facultatively anaerobic bacteria to pyruvate that is further transformed into a variety of products, including acetate, butyrate, propionate, or ethanol, by different organisms. In contrast, a range of electron acceptors may be used under anaerobic conditions to mediate oxidative degradation of the carbon substrate at the expense of the reduction of the electron acceptors. For example, the following reductions may be coupled to oxidative degradation: nitrate to nitrogen (or nitrous oxide), sulfate to sulfide, carbonate to methane, fumarate to succinate, trimethylamine-*N*-oxide to trimethylamine, or dimethyl sulfoxide to dimethyl sulfide, which are discussed in Chapter 3, Part 2. The environments required by the relevant organisms are determined by the redox potential of the relevant reactions, so that increasingly reducing conditions are required for reduction of nitrate, sulfate, and carbonate.

Attention is drawn to the dechlorination by anaerobic bacteria of both chlorinated ethenes and chlorophenolic compounds that serve as electron acceptors with electron donors, including formate, pyruvate, and acetate. This is termed dehalorespiration and is important in the degradation of a range of halogenated compounds under anaerobic conditions, and is discussed further in Chapter 3, Chapter 7, and Chapter 9, Parts 1 and 2.

Probably, most of the microbial degradations and transformations that are discussed in this book are carried out by heterotrophic microorganisms that use the xenobiotic as a source of both carbon and energy, or by cometabolism (Chapter 4, Part 2). In addition, xenobiotics may serve only as sources of nitrogen, sulfur, or phosphorus. The attention is briefly drawn here to those groups of organisms many of whose members are autotrophic or lithotrophic. Discussion of the complex issue of organic nutrition of chemolithotrophic bacteria, and the use of the term "autotrophy" is given in classic reviews (Rittenberg 1972; Matin 1978). The groups of organisms that are discussed here in the context of biotransformation include the following:

1. Ammonia-oxidizing bacteria, including *Nitrosomonas europaea*
2. Facultatively heterotrophic thiobacilli that use a number of organic sulfur compounds as energy source
3. Oxygenic photolithotrophic algae and cyanobacteria

It is important to underscore the fact that carbon dioxide is required not only for the growth of strictly phototrophic and lithotrophic organisms. Many heterotrophic organisms that are heterotrophic have an obligate requirement for carbon dioxide for their growth. Illustrative examples include the following:

1. Anaerobic bacteria such as acetogens, methanogens, and propionic bacteria.
2. Aerobic bacteria that degrade propane (MacMichael and Brown 1987), the branched hydrocarbon 2,6-dimethyloct-2-ene (Fall et al. 1979), or oxidize carbon monoxide (Meyer and Schlegel 1983).
3. Cyanase that converts cyanate to NH_4^+ and CO_2 is dependent on bicarbonate, and is a recycling substrate rather than a hydrolase (Johnson and Anderson 1987). It can be used as a source of nitrogen for the growth for a few bacteria, including *Escherichia coli* and *Pseudomonas fluorescens* (Kunz and Nagappan 1989).

4. The anaerobic biotransformation of aromatic compounds may be dependent on CO_2, and a review by Ensign et al. (1998) provides a brief summary of the role of CO_2 in the metabolism of epoxides by *Xanthobacter* sp. strain Py2, and of acetone by both aerobic and anaerobic bacteria.

5. The lag after diluting glucose-grown cultures of *Escherichia coli* into fresh medium may be eliminated by the addition of $NaHCO_3$ and is consistent with the requirement of this organism for low concentrations of CO_2 for growth (Neidhardt et al. 1974).

2.3 Biodegradation of Enantiomers: Racemization

2.3.1 Enantiomers and Racemases

2.3.1.1 Biodegradation of Enantiomers

The molecules of a number of compounds are asymmetric. These include both naturally occurring compounds such as carbohydrates and amino acids, and agrochemicals such as chlorinated phenoxypropionates, certain *N*-substituted 2,6-dimethylanilines (e.g., metalaxyl) (Buser et al. 2002), nonplanar PCB congeners (i.e., IUPAC 91, 95, 132, 136, 149, 174, 176, 183) (Kaiser 1974), *o,p′*-DDT (Garrison et al. 2000), and α-hexachlorocyclohexane (Wiberg et al. 2001). It is worth adding that molecular dissymmetry may be imposed by the phosphonate atom in fonofos, and the sulfur atom in the sulfoxide of fipronil (Garrison 2006). These may, therefore, exist as pairs of mirror-image enantiomers that display differential biological activity and biodegradability. Only one of them may be encountered in environmental samples, and this may plausibly be attributed to the preferential destruction or transformation of one enantiomer that is consistent with observed significant differences in the biodegradability of enantiomers.

Different strategies for the biodegradation of enantiomers may be used, and are illustrated in the following examples, for which the pathways are discussed in the appropriate chapters.

1. Only the *R*(+) enantiomer of the herbicide 2-(2-methyl-4-chlorophenoxy)propionic acid was degraded (Tett et al. 1994), although cell extracts of *Sphingomonas herbicidovorans* grown with the *R*(+) or *S*(−) enantiomer, respectively, transformed selectively the *R*(+) or *S*(−) substrates to 2-methyl-4-chlorophenol (Nickel et al. 1997). The enantioselective 3-oxoglutarate-dependent dioxygenases have been purified and characterized from *Sphingomonas herbicidovorans* MH (Müller et al. 2006) that is able to degrade both enantiomers of the racemate (*R,S*)-4-chloro-2-methylphenoxypropionate (mecoprop) (Zipper et al. 1996), and from *Delftia acidovorans* (Schleinitz et al. 2004).

2. Degradation of linear alkylbenzenes may produce chiral 3-phenylbutyrate. The degradation of the *R*- and *S*-enantiomers has been examined in *Rhodococcus rhodochrous* strain PB1 (Simoni et al. 1996). Whereas the *S*-enantiomer undergoes dioxygenation to the catechol that is a terminal metabolite, the *R*-enantiomer is degraded to 3-phenylpropionate by the loss of the methyl group by an unresolved mechanism, followed by dioxygenation and ring fission.

3. The α-isomer of hexachlorocyclohexane exists in two enantiomeric forms, and both are degraded by *Sphingomonas paucimobilis* strain B90A by dehydrochlorination to 1,3,4,6-tetrachlorocyclohexa-1,4-diene that is spontaneously degraded to 1,2,4-trichlorophenol. In this strain, there are two enzymes LinA1 and LinA2 that specifically accept (+)-α-HCH or (−)-α-HCH to produce β-pentachlorocyclohexene 1 and 2. Both enzymes were 88% identical at the amino acid level, and LinA2 was identical to LinA of *S. paucimobilis* strain UT26. This implies that in strains such as UT26 that contain only a single dehydrochlorinase, enantioselective degradation may take place with enrichment of the nondegradable enantiomer (Suar et al. 2005).

4. Bis-(1-chloro-2-propyl)ether has two chiral centers, and exists in (*R,R*)-, (*S,S*)-, and a *meso* form. It is degraded by *Rhodococcus* sp. with a preference for the *S,S*-enantiomer with the intermediate formation of 1-chloro-propan-2-ol and chloroacetone (Garbe et al. 2006).

5. Iminodisuccinate that is a potential replacement for ETDA has two asymmetric centers and, since the C–N lyase that cleaves it to D-aspartate and fumarate is stereospecific, degradation is initiated by the activity of epimerases to the *R,S*-enantiomer. D-Aspartate is then isomerized to L-aspartate that undergoes elimination to fumarate (Cokesa et al. 2004).

Attention is drawn to additional aspects:

1. Redox reactions between ketones and the corresponding alkanols have been used to carry out the production of a single enantiomer. This is a cardinal reaction carried out by 3α-hydroxysteroid dehydrogenase/carbonyl reductases (Möbus and Maser 1998) which are members of the short-chain alkanol dehydrogenases/reductases (Jörnvall et al. 1995). For example, cells of *Acinetobacter* sp. NCIB 9871 grown with cyclohexanol carried out enantiomerically specific degradation of a racemic substituted norbornanone to a single ketone having >95% enantiomeric excess (Levitt et al. 1990).

2. Enantiomeric specificity is important in the microbial synthesis of enantiomers. This is exemplified by the stereospecific microbial oxidation of sulfides to sulfoxides (Holland 1988), and the application of Baeyer–Villiger-type microbial monooxygenases to a range of substrates, including sulfides to sulfoxides, amines to amine oxides, and olefins to epoxides (references in de Gonzalo et al. 2006; Rehndorf et al. 2009). Details of microbial hydroxylation—mainly by fungi—at specific positions of steroids and terpenoids are discussed in Chapter 7 dealing with cycloalkanes and steroids.

3. Fungal metabolism has been proposed as a model for mammalian metabolism and particular attention has been paid to the configuration and absolute stereochemistry of the products. A valuable example is the difference in the absolute configuration of the dihydrodiols produced by *Cunninghamella elegans* from anthracene and phenanthrene: these have the *S,S* configurations which is the opposite of the *R,R* configurations of the metabolites from rat liver microsomes (Cerniglia and Yang 1984). This may not, however, be true for other fungi.

4. Natural systems may be quite complex. For example, the enantiomerization of phenoxyalkanoic acids containing a chiral side chain has been studied in soil using ^2H_2O (Buser and Müller 1997). It was shown that there was an equilibrium between the *R*- and *S*-enantiomers of both 2-(4-chloro-2-methylphenoxy)propionic acid (MCPP) and 2-(2,4-dichlorophenoxy)propionic acid (DCPP) with an equilibrium constant favoring the herbicidally active *R*-enantiomers. The exchange reactions proceeded with both retention and inversion of configuration at the chiral sites. This important issue will certainly attract increasing attention in the context of the preferential microbial synthesis of intermediates of specific configuration.

2.3.2 Racemization

Racemization is the mechanism whereby a substrate of one configuration is converted into its optical enantiomer. This may be required for the degradation of a substrate, or for the provision of one enantiomer that is specifically required. There are two mechanisms for racemization when the asymmetric carbon atom carries a hydrogen substituent: one involves the cofactor pyridoxal-5′-phosphate and is generally used for amino acids, whereas the other that is cofactor independent involves a direct proton transfer involving pairs of amino acids.

2.3.2.1 D-Amino Acids

Although amino acids normally have the L-configuration, amino acids with the D-configuration are required to fulfill several functions: (a) cell-wall peptidoglycan biosynthesis, (b) synthesis of the poly-γ-D-glutamate capsule in *Bacillus anthracis*, and (c) the synthesis of a range of peptide antibiotics in *Bacillus brevis*, including the gramicidins and the tyrocidins. The synthesis of these D-amino acids is generally carried out by either of the two reactions both of which are dependent on pyridoxal-5′-phosphate: racemization or by the activity of a D-transaminase from the appropriate 2-oxoacid precursor and D-alanine. Analysis of the genome of several bacteria has revealed the presence of the appropriate racemases in a wide range of bacteria to provide these D-amino acids. In addition, it has emerged that D-amino acids play a specific and important role in cell morphology in stationary cultures and in the disruption of the structure of biofilms.

Alanine racemase is widely distributed in bacteria, and racemase activity has also been demonstrated for other amino acids, including threonine from *Pseudomonas putida* (Lim et al. 1993), arginine from *Pseudomonas graveolens* (Yorifuji and Ogata 1971), and ornithine from *Clostridium sticklandii* (Chen et al. 2000)—in all of which pyridoxal phosphate is bound to the enzyme. A number of bacteria possess two race-mases, for glutamate in *Bacillus sphaericus* (Fotheringham et al. 1998) and *Bacillus anthracis* (Dodd et al. 2007), and for alanine in *Salmonella typhimurium* (Wasserman et al. 1983), while *Staphylococcus haemo-lyticus* is able to synthesize D-glutamate both by the activity of a racemase and by a D-amino acid transami-nase (Pucci et al. 1995). The degradation of amino acids as sources of carbon and energy has generally been studied with the L-isomers: use of the D-isomer may, therefore, be dependent on the existence of a racemase, although different pathways may be used for the D- and L-isomers, for example, lysine in pseudomonads (Muramatsu et al. 2005). Although a racemase for lysine has been established (Ichihara et al. 1960), it has not been shown that the two pathways for L- and D-lysine are physiologically interconnected.

Some racemases do not, however, rely on the activity of pyridoxal-5′-phosphate. These include those, for example, L-proline racemase that is involved in the reductive segment of the fermentation of L-ornithine by *Clostridium sticklandii*. (Stadtman and Elliot 1957; Rudnick and Abeles 1975), 2-hydroxyproline in *Pseudomonas striata* (Adams and Noton 1964), glutamate in *Lactobacillus fermenti* (Gallo and Knowles 1993), and lysine racemase isolated from a soil metagenome (Chen et al. 2009). An alternative mechanism is discussed later.

2.3.2.2 Additional Roles for D-Amino Acids

1. It has been shown that D-amino acids play additional roles in the stationary phase of bacteria, and that they influence the composition of the peptidoglycan (Lam et al. 2009). In *Vibrio chol-erae*, D-Met and D-Leu accumulated in stationary cultures, whereas in *Bacillus subtilis* it was D-Tyr and D-Phe that accumulated. These D-amino acids brought about alteration in cell mor-phology in stationary-phase cultures, and this could be elicited by addition of the appropriate D-amino acids. Release of D-Leu and D-Met was greatest in *V. cholera*, *V. parahemolyticus*, and *Aeromonas hydrophila*, but essentially absent in the other Gram-negative and Gram-positive bacteria that were examined.

2. Biofilms have a finite lifetime and eventually undergo disassembly. This has been examined in *Bacillus subtilis* in which cells in biofilms are bound by an extracellular matrix composed of exopolysaccharide and amyloid fibers composed of the protein TasA. Medium from an 8-day culture of *B. subtilis* was concentrated on a C_{18} column, and it was shown that the eluate was able to inhibit pellicle formation, and that D-amino acids were specifically active in disassembly—in contrast to the L-amino acids that were inactive. The analogous effect was established for biofilms from *Staphylococcus aureus* and *Pseudomonas aeruginosa*, and it was suggested that D-amino acids might provide a general mechanism for disassembly of biofilms (Kolodkin-Gal et al. 2010).

2.3.2.3 Cofactor-Independent Racemization

Although the greatest attention has been devoted to the racemization of amino acids, there are additional examples of racemization to which brief attention is drawn, and in which cofactors are not involved. In these, a direct proton transfer mechanism operates involving pairs of amino acids, for example, lysine/ histidine, two lysines, two cysteines, or two aspartates. Cofactor-independent racemases include the following:

1. Interconversion of *R*- and *S*-isomers is necessary for the aerobic degradation of branched-chain alkane carboxylates and the C_{17} side chains of sterols and bile acids. This involves alteration in

the configuration of the α-hydrogen atom of the coenzyme A esters, and a two-base mechanism has been proposed that involves pairs of amino acids, for example, a lysine/histidine pair in mandelate racemase. This has been explored in mutants of *Mycobacterium tuberculosis* in which desired conserved amino acids were altered to alanine, and it was shown that Arg[91], His[136], Asp[156], and Glu[241] were determinants of catalytic activity for the α-methacyl coenzyme A racemase (Savolainen et al. 2005). Racemases are required for the degradation of pristane, phytane, and squalane in *Mycobacterium* sp. strain P101 (Sakai et al. 2004); in addition, the β-oxidation pathway has been used to illustrate the several stereochemical configurations that may be encountered in the degradation of pristane. The α-methacyl-CoA racemase has been characterized from *Gordonia polyprenivorans* VH2 (Arenskötter et al. 2008).

2. Both enantiomers of mandelate were degraded by *Pseudomonas putida* through the activity of a mandelate racemase (Hegeman 1966). The racemase (mdlA) is encoded in an operon that includes the following two enzymes in the pathway of degradation, *S*-mandelate dehydrogenase (mdlB) and benzoylformate decarboxylase (mdlC) (Tsou et al. 1990). The x-ray structures of mandelate racemase (MR) and muconate lactonizing enzyme (MLE) have been examined (Hasson et al. 1998). The structure of mandelate racemase showed the involvement of Lys[166] and His[297] assisted by Asp[270] and the carboxylate anion stabilized by both association with Lys[164] and a cation. It was suggested that in both of them the reactions involved a 1:1 proton transfer to produce R–C(R′)=C(O⁻)₂ where R′=OH for MR or H for MLE, followed by reversible reprotonation to either enantiomer.

3. Aspartate racemase from the archaeon *Pyrococcus horikoshii* OT3 is an example of a mechanism less commonly used among amino acids that involves a direct 1:1 proton transfer. The enzyme consists of a dimeric structure whose inter-subunits are held by a Cys[73]–Cys[73′] disulfide and, between the two domains, the Cys[82] and Cys[194] that carry out the two-base racemization and are strictly conserved (Liu et al. 2002).

2.4 Sequential Microbial and Chemical Reactions

Microbial activity may produce reactive intermediates that undergo spontaneous chemical transformation to terminal metabolites. This is quite a frequent occurrence, and its diversity is illustrated by the following examples:

1. A bacterial strain BN6 oxidizes 5-aminonaphthalene-2-sulfonate by established pathways to 6-amino-2-hydroxybenzalpyruvate that undergoes spontaneous cyclization to 5-hydroxyquinoline-2-carboxylate (Figure 2.2a) (Nörtemann et al. 1993).

2. The oxidation of benzo[*b*]thiophene by strains of pseudomonads produces the sulfoxide that undergoes an intramolecular Diels–Alder reaction followed by further transformation to benzo[*b*]naphtho[1,2-*d*]thiophene (Figure 2.2b) (Kropp et al. 1994).

3. The degradation of 4-chlorobiphenyl by *Sphingomonas paucimobilis* strain BPSI-3 formed the intermediates 4-chlorobenzoate and 4-chlorocatechol. Fission products from the catechol reacted with NH_4^+ to produce chloropyridine carboxylates (Davison et al. 1996) (Figure 2.2c).

4. 4-Nitrotoluene is degraded by a strain of *Mycobacterium* sp. via the corresponding 4-amino-3-hydroxytoluene (Spiess et al. 1998): this is dimerized abiotically to form a dihydrophenoxazinone, and after extradiol cleavage to 5-methylpyridine-2-carboxylate (Figure 2.2d).

5. The incubation of 3,5-dichloro-4-methoxybenzyl alcohol with methanogenic sludge produced the de-*O*-methylated compound that was transformed to 2,6-dichlorophenol, and abiotically dimerized to *bis*(3,5-dichloro-4-hydroxyphenyl)methane (Verhagen et al. 1998) (Figure 2.2e).

6. The transformation of aromatic amino acids to the 2-oxoacids was mediated by *Morganella morganii*, and these subsequently underwent a hemin-dependent chemical transformation with the production of CO (Hino and Tauchi 1987).

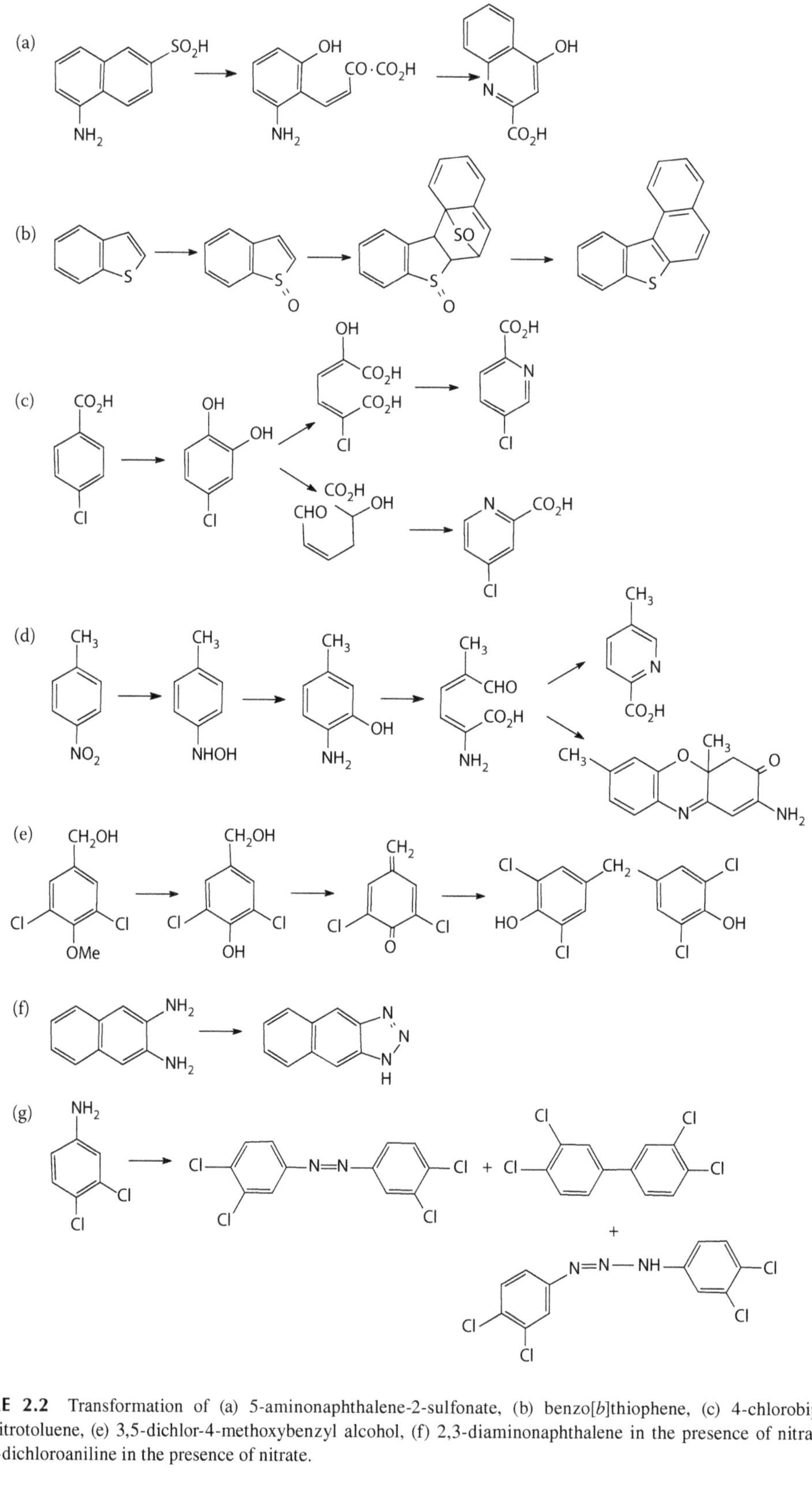

FIGURE 2.2 Transformation of (a) 5-aminonaphthalene-2-sulfonate, (b) benzo[*b*]thiophene, (c) 4-chlorobiphenyl, (d) 4-nitrotoluene, (e) 3,5-dichlor-4-methoxybenzyl alcohol, (f) 2,3-diaminonaphthalene in the presence of nitrate, and (g) 3,4-dichloroaniline in the presence of nitrate.

7. Tetrachloroethene may be degraded by bacteria via the epoxide, and chemical hydrolysis of this produces CO and CO_2 from oxalyl chloride as major products, whereas only low amounts of trichloroacetate were produced (Yoshioka et al. 2002).

Nitrite (or compounds at the same or lower oxidation level) is produced microbiologically from nitrate, and may then react with the substrate to produce stable end products. The production of nitrite is the sole metabolic function of the bacteria and, in view of concern over the presence of nitrate in groundwater, the following possible environmental significance of these or analogous reactions should not be overlooked.

1. A strain of *Escherichia coli* produces a naphthotriazole from 2,3-diaminonaphthalene and nitrite that is formed from nitrate by the action of nitrate reductase. The initial product is NO, which is converted by reactions with oxygen into the active nitrosylating agent that reacts chemically with the amine (Ji and Hollocher 1988). A comparable reaction may plausibly account for the formation of dimethylnitrosamine by *Pseudomonas stutzeri* during growth with dimethylamine in the presence of nitrite (Mills and Alexander 1976) (Figure 2.2f).

2. The formation of 3,3′,4,4′-tetrachloroazobenzene, 1,3-bis(3,4-dichlorophenyl)triazine and 3,3′,4,4′-tetrachlorobiphenyl from 3,4-dichloroaniline and nitrate by *E. coli* plausibly involved intermediate chemical formation of the diazonium compound by reaction of the amine with nitrite (Corke et al. 1979) (Figure 2.2g).

3. Nitro-containing metabolites have been isolated from a number of substrates when the medium contained nitrate. Reduction to nitrite is the putative source of the nitro group, and two mechanisms have been suggested: (a) nitrosation by NO^+ produced from HO–NO in a slightly acidic medium followed by oxidation or (b) nitration of an intermediate arene oxide by nitrite. Since all the primary metabolites have phenolic groups and the introduced nitro groups are *ortho* or *para* to these, the first is probably the more attractive general reaction. In support of the alternative, however, an NIH shift has been demonstrated in the transformation of 2-chlorobiphenyl to 2-hydroxy-3-chlorobiphenyl by a methylotrophic organism, so that the formation of an arene oxide cannot be excluded (Adriaens 1994). The following illustrate these reactions:

   a. The formation of nitro-containing metabolites during the degradation of 4-chlorobiphenyl by strain B-206 (Sylvestre et al. 1982).

   b. *Corynebacterium* sp. that utilizes dibenzothiophene as a sulfur source produced 2-hydroxybiphenyl, and subsequently nitrated this to form two hydroxynitrobiphenyls (Omori et al. 1992).

   c. The transformation of α-tocopherol by *Streptomyces catenulae* produced 5-nitro-tocopherol in addition to quinones (Rousseau et al. 1997).

   d. The transformation of 2-hydroxybenz[1,4]oxazin-3-one with *Gliocladium cibotii* produced the intermediate 2-hydroxyacetanilide that was transformed to nitro derivatives (Zikmundová et al. 2002).

2.5 The Spectrum of Organisms

2.5.1 Bacteria in Their Natural Habitats

Illustrations of the plethora of pathways used by bacteria for the degradation and biotransformation of xenobiotics are provided in Chapters 7 through 11. It is appropriate to say something here of the metabolic spectrum of specific groups of organisms, in particular, those that have hitherto achieved less prominence in discussions of biodegradation and biotransformation. The application of the methodology of molecular biology has revealed new dimensions in the range of naturally occurring organisms.

It should be appreciated that in natural situations, bacteria may be subjected to severe nutrient limitation so that they are compelled to reproduce at extremely low rates in order to conserve their metabolic energy (Kjellberg et al. 1987; Siegele and Kolter 1992). This does not necessarily mean, however, that these organisms have negligible metabolic potential toward xenobiotics. Other slow-growing organisms may be well adapted to the natural environment (Poindexter 1981), although they may not be numerically dominant among organisms isolated by normal procedures (Schut et al. 1993). An unusual situation has been observed in *Pseudomonas putida* strain mt-2 that contains the TOL plasmid pWWO. After growth with 3-methylbenzoate, cells were exposed to concentrations of toluene from 4 mg/L (growth supporting) to 130 mg/L (inhibitory) to 267 mg/L (lethal). Protein synthesis was rapidly inhibited with the concomitant production of new proteins, which were characteristic of cell starvation, and which could be suppressed by the addition of 3-methylbenzoate as carbon source. Cells exposed to 4 mg/L ceased to produce "starvation proteins" within 3 h and growth was initiated. At higher concentrations, these proteins persisted for increasing lengths of time and at 267 mg/L there was a rapid loss of viability (Vercellone-Smith and Herson 1997). An important example is provided by *Lactococcus lactis* that enters a resting phase after consumption of the available carbohydrates. The ensuing noncultivable state is characterized by the synthesis of 2-methylbutyrate that is derived from L-leucine, and provides the intermediate 3-hydroxy-3-methyl-glutarate (Ganesan et al. 2006).

Attention is directed to organisms that have hitherto evaded isolation or are represented by only a few cultivated examples. Such organisms may well outnumber those that have been isolated as pure cultures, frequently using elective enrichment. A few illustrations are given below:

1. Archaea belonging to the kingdom *Crenarchaeota* deserve attention. Although it has been assumed that these are extreme thermophiles, members of this group have been identified by molecular techniques in other habitats, for example, soils (Buckley et al. 1998), boreal forest soil (Jurgens et al. 1997), and in plant extracts (Simon et al. 2005).

2. Organisms belonging to the phylum Verrucomicrobia have been detected in soil throughout the world, although hitherto only a few organisms have been obtained in pure culture.

    a. *Chthoniobacter flavus* has been obtained in pure culture (Sangwan et al. 2004), and this was made possible by the use of a dilute complex medium solidified with gellan gum in place of agar (Janssen et al. 2002), that has since been extended (Sangwan et al. 2005). On the other hand, it has been shown that isolation on solid medium is more effective for members of *Acidobacteria*, *Verrucomicrobia*, and *Gemmatimonadetes* than the use of liquid serial dilution (Schoenborn et al. 2004).

    b. An organism Kam1 was isolated from an acidic hot spring in Kamchatka and was capable of methane oxidation at 55°C and pH 2 (Islam et al. 2008). On the basis of 16S rRNA gene sequences, it was most closely similar to organisms isolated from Yellowstone hot springs. Remarkably, Kam1 appeared to lack genes for the oxidation of methane, and cells contained polyhedral organelles resembling carboxysomes.

    c. Methanotrophs growing under highly acid conditions have been described, and one of them (SolV) was assigned to *Methylacidiphilum* (*Acidimethylosilex*) *fumariolicum* and contained the genes for the particulate methane monooxygenase and C_1 incorporation using a combination of the serine, tetrahydrofolate and ribulose-1,5-bisphosphate pathways (Pol et al. 2007). These organisms oxidize methane to CO_2, but the acidophilic thermophilic nitrogen-fixing strain of *Methylacidiphilum fumariolum* SolV is an autotroph that utilizes CO_2 as the sole source of carbon and contains a new type of ribulose-1,5-bisphosphate carboxylase/ oxygenase. This has been confirmed by the use of $^{13}CH_4$ or $^{13}CO_2$ (Khadem et al. 2011), and is relevant to the estimation of methanotrophs using stable isotope probing using labeled CH_4.

    d. Another extreme acidophile has been characterized that had an optimal pH for growth between 2.0 and 2.5, and was distinguished by the occurrence of genes for three particulate methane monooxygenases and the enzymes of the Benson–Calvin cycle for CO_2 assimilation (Dunfield et al. 2007).

Before discussing some of the larger groups of microorganisms that have been implicated in biodegradation and biotransformation, brief comments are made on other groups of organisms that have hitherto attracted somewhat limited attention, including

1. Typically clinical organisms that have been found in environmental samples can display degradative capability: these include, for example, *Klebsiella pneumoniae*, *Mycobacterium tuberculosis*, and *Ochrobactrum* spp.
2. Oligotrophic marine bacteria, including species of *Cycloclasticus*, *Neptunomonas*, *Marinobacter*, and *Sulfitobacter*
3. Gram-positive organisms belonging to the genera *Mycobacterium*, *Rhodococcus*, and *Gordonia*

2.5.2 Marine and Oligotrophic Bacteria

Although most illustrations have been taken from investigations of freshwater environments—lakes and rivers—fewer relate to the marine environment. In view of the area of the globe that is covered by open sea, this may seem remarkable, and the degradation of xenobiotics by marine bacteria has received somewhat limited attention. Roseobacters represent one of the major groups of marine bacteria and have been found not only in a range of geographical habitats but in various degrees of association. A review provides a valuable summary, and draws attention to some of their metabolic potential, including the oxidation of CO and the degradation of aromatic compounds (Buchan et al. 2005). Illustrative examples of degradation by marine bacteria are given below:

1. Strains of *Marinobacter* sp. that are able to degrade aliphatic hydrocarbons and related compounds have been isolated (Gauthier et al. 1992), and *Marinobacter* sp. strain CAB is capable of degrading 6,10,14-trimethylpentadecan-2-one under both aerobic and denitrifying conditions with the production of a range of metabolites (Rontani et al. 1997). It has been shown that *Alcanivorax borkumensis* is also able to degrade a number of hydrocarbons (Yakimov et al. 1998).
2. Organisms assigned to the genus *Cycloclasticus* have been isolated from a number of geographical locations and display a considerable metabolic versatility as follows:
    a. Strains were isolated from the Gulf of Mexico and from Puget Sound. In an artificial seawater medium, they were able to degrade a range of PAHs, including alkylated naphthalenes, phenanthrene, anthracene, and fluorene at concentrations ranging from 1 to 5 ppm (Geiselbrecht et al. 1998). Strains from both localities were numerically important, and were similar based on both 16S ribosomal DNA (rDNA) sequences and phylogenetic relationships determined from the sequences of arene dioxygenases.
    b. *Cycloclasticus oligotrophus* strain RB1 that harbors genes with a high degree of homology to that encoding xylene degradation in terrestrial pseudomonads is able to grow at the expense of toluene, xylenes, biphenyl, naphthalene, and phenanthrene (Wang et al. 1996).
    c. *Cycloclasticus* sp. strain A5 was able to grow at the expense of naphthalenes, phenanthrenes, and fluorenes—though not anthracene (Kasai et al. 2003).
3. Organisms isolated after a hydrocarbon spill in the marine environments included *Alcanivorax* that was able to degrade *n*-alkanes (Hara et al. 2003), and *Thalassolituus oleivorans* that was isolated from the Thames Estuary could degrade long-chain *n*-alkanes (McKew et al. 2007).
4. Bacteria isolated from marine macrofaunal burrow sediments and assigned to *Lutibacterium anuloederans* were able to degrade phenanthrene in heavily contaminated sediment (Chung and King 2001).
5. Marine bacteria were isolated from a creosote-contaminated sediment in Puget Sound by enrichment with naphthalene (Hedlund et al. 1999). It was shown that the gene encoding the naphthalene dioxygenase ISP from this strain was not closely related to those from naphthalene-degrading strains of *Pseudomonas* or *Burkholderia*. Although analysis of 16S rDNA suggested a close relation to the genus *Oceanospirillum*, the differences were considered sufficient to assign these strains to a new taxon—*Neptunomonas naphthovorans*.
6. Strains of *Sagittula stellata* and *Sulfitobacter* sp. that are members of the *Roseobacter* group of organisms were screened for the presence of 3,4-dihydroxybenzoate dioxygenase that is involved in the degradation of several aromatic carboxylates (Buchan et al. 2000). The enzyme

was widely distributed in this group, and the organisms were able to degrade a number of sub-strates, including benzoate, anthranilate, salicylate, 4-hydroxybenzoate, and vanillate.

7. Marine roseobacters that contain bacteriochlorophyll *a* have been described (Oz et al. 2005), and the bacteriochlorophyll *a*-containing marine bacterium *Porphyrobacter sanguineus* was able to degrade biphenyl and dibenzofuran and use them for growth (Hiraichi et al. 2002).

8. The roseobacters *Silicibacter pomeroyi* and *Roseovarius nubinhibens* were able to carry out the degradation of dimethylsulfoniopropionate to dimethylsulfide, and to methanethiol (González et al. 2003), and are discussed further in Chapter 11, Part 2.

It is necessary to take into account critical aspects of the physiology and biochemistry of these marine organisms and, although parenthetical comments on marine bacteria are made in various sections of this book, it is convenient to bring together some of their salient features. It should also be appreciated that terrestrial organisms that have a high tolerance to salinity may be isolated from inshore seawater samples. Among these are the yeasts that have been isolated from coastal marine sediments (MacGillivray and Shiaris 1993). Such organisms are excluded from the present discussion, which is restricted to oceanic water:

1. It is experimentally difficult to obtain numerical estimates of the total number of bacteria present in seawater, and the contribution of ultramicroorganisms that have a small cell volume and low concentrations of DNA may be seriously underestimated. Although it is possible to evaluate their contribution to the uptake and mineralization of readily degraded compounds such as amino acids and carbohydrates, it is more difficult to estimate their potential for degrading xeno-biotics at realistic concentrations. This is more extensively discussed in Chapter 5, Part 2.

2. The use of conventional plating procedures may result in the isolation of only fast-growing organisms that outgrow others—which may be numerically dominant and which are unable to produce colonies on such media. The substrate concentrations used for isolation may have been unrealistically high, so that obligately oligotrophic organisms were overgrown during attempted isolation.

3. There is indeterminacy in the term "oligotroph," and the dilemma is exacerbated by the fact that it may be impossible to isolate obligate oligotrophs by established procedures. The application of DNA probes should, however, contribute to an understanding of the role of these "nonculti-vable" organisms. Oligotrophic bacteria in the marine environment are able to utilize low substrate concentrations, and they may be important in pristine environments.

4. During prolonged storage in the laboratory under conditions of nutrient starvation, facultatively oliogotrophic bacteria may be isolated and these display transport systems for the uptake of amino acids and glucose that are coregulated.

5. Organisms in natural ecosystems may not be actively dividing but may, nonetheless, be metabolically active. This may be particularly important for ultramicro marine bacteria in their natural habitat.

2.5.3 Lithotrophic Bacteria

These are major groups of microorganisms that have achieved restricted prominence in discussions on biodegradation and biotransformation, and include both photolithotrophs and chemolithotrophs. Chemolithotrophs that metabolize elementary sulfur using CO_2 as a source of carbon are briefly noted in Chapter 11. Additional general comments include the following.

These organisms use CO_2 as their principal, or exclusive, source of carbon, and this is incorporated into cellular material generally by the Benson–Calvin cycle. This should be distinguished from heterotrophic organisms that may incorporate CO_2 during degradation, for example, CO_2 in the metabolism of epoxides by *Xanthobacter* sp. strain Py2, and of acetone by both aerobic and anaerobic bacteria. Energy for growth is obtained either from photochemical reactions (photolithotrophs) or by chemical oxidation of inorganic substrates such as reduced forms of nitrogen or sulfur (chemolithotrophs). In some organisms, organic carbon can be taken up and incorporated during growth even in organisms that are obligately chemoli-thotrophic or photolithotrophic. Organic carbon may, however, have an inhibitory effect on growth. Some species and strains of these organisms may also grow heterotrophically using organic carbon as sources of both energy and cell carbon. Attention is directed to reviews that cover the sometimes controversial aspects of lithotrophy and autotrophy (Kelly 1971; Rittenberg 1972; Whittenbury and Kelly 1977; Matin 1978). It is worth noting that many aerobic bacteria that belong to groups with well-established heterotrophic activity are also chemolithoautotrophic and use the oxidation of hydrogen as their source of energy (Bowien and Schlegel 1981). Attention is drawn to the reassignment of the *Pseudomonas* strains *P. flava*, *P. pseudoflava*, and *P. palleroni* to the genus *Hydrogenophaga*, and the degradative activity of species of *Xanthobacter*.

Attention has been directed to obligate lithotrophs such as those that oxidize carbon monoxide or hydrogen. Analysis of DNA extracts and PCR amplification for the *cbbL* gene that encodes the large subunit of ribulose-1,5-bisphosphate carboxylase/oxygenase that is typically found in obligate lithotrophs was carried out. The PCR products were cloned, sequenced, and analyzed, and the results showed that populations depended on the plant cover and land use, although the relevant populations of ammonia-oxidizing organisms were small in comparison with those of facultative lithotrophs (Tolli and King 2005).

Anaerobic lithotrophs that oxidize ammonium using nitrite as electron acceptor (anammox) are noted later in the section on anaerobic bacteria.

It has become increasingly clear that representatives of chemolithotrophic microorganisms may be effective in carrying out the transformation of xenobiotics. As an illustration, attention is directed to three groups of organisms: (1) the ammonia oxidizers, (2) the thiobacilli, and (3) algae and cyanobacteria. Habitats to which these organisms are physiologically adapted should, therefore, be considered in discussions on biodegradation and biotransformation.

2.5.4 Phototrophic Organisms

There are several large groups of phototrophic organisms: oxygenic algae and cyanobacteria, anaerobic purple nonsulfur bacteria, anaerobic sulfur bacteria, and aerobic nonoxygenic bacteria.

2.5.4.1 Oxygenic Algae and Cyanobacteria

The metabolic significance of oxygenic algae and cyanobacteria has received relatively limited attention in spite of the fact that they are important components of many ecosystems and may, for example, in the marine environment be of primary significance. Whereas the heterotrophic growth of algae at the expense of simple carbohydrates, amino acids, lower aliphatic carboxylic acids, and simple polyols is well documented (Neilson and Lewin 1974), the potential of algae for the metabolism of xenobiotics has been much less extensively explored. Among these metabolic possibilities, which have received less attention than they deserve, the following are used as illustration:

1. The transformation—though not apparently the degradation—of naphthalene has been examined in cyanobacteria and microalgae, including representatives of green, red, and brown algae, and diatoms (Cerniglia et al. 1980a, 1982). The transformation of biphenyl (Cerniglia et al. 1980b), aniline (Cerniglia et al. 1981), and methylnaphthalenes (Cerniglia et al. 1983) has been examined in cyanobacteria (Figure 2.3). Phenanthrene is metabolized by *Agmenellum quadruplicatum* to the *trans*-9,10-dihydrodiol by a monooxygenase system, and the 1-hydroxy-phenanthrene that was formed transiently was *O*-methylated (Narro et al. 1992). The biotransformation of benzo[*a*]pyrene has been demonstrated in a number of green algae, though this was not metabolized by a chlamydomonad, a chrysophyte, a euglenid, or a cyanobacterium (Warshawsky et al. 1995). The relative amounts of the products depended on the light sources and their intensity, and included 9,10-, 4,5-, 11,12-, and 7,8-dihydrodiols, the toxic 3,6-quinone, and phenols. The 11,12-, 7,8-, and 4,5-dihydrodiols produced by *Selenastrum capricornutum* had the *cis* configuration, which suggests that their formation was mediated by a dioxygenase (Warshawsky et al. 1988). It is worth noting that *Ochromonas danica* is able to degrade phenol by extradiol fission of the initially formed catechol (Semple and Cain 1996).

2. Some green algae are able to use aromatic sulfonic acids (Figure 2.4a) (Soeder et al. 1987) and aliphatic sulfonic acids (Figure 2.4b) (Biedlingmeier and Schmidt 1983) as sources of sulfur.

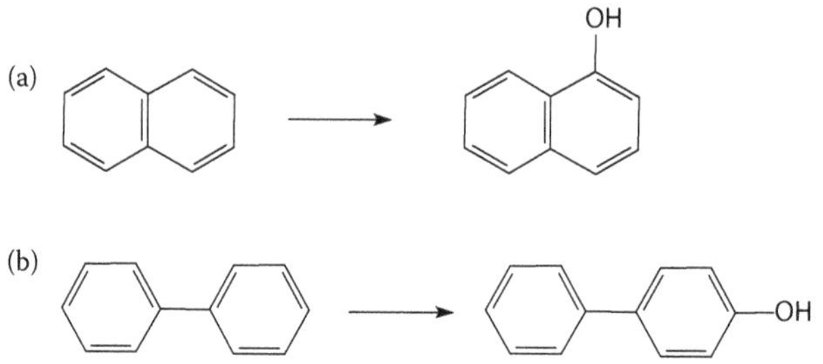

FIGURE 2.3 Hydroxylation by algae. (a) Naphthalene, (b) biphenyl.

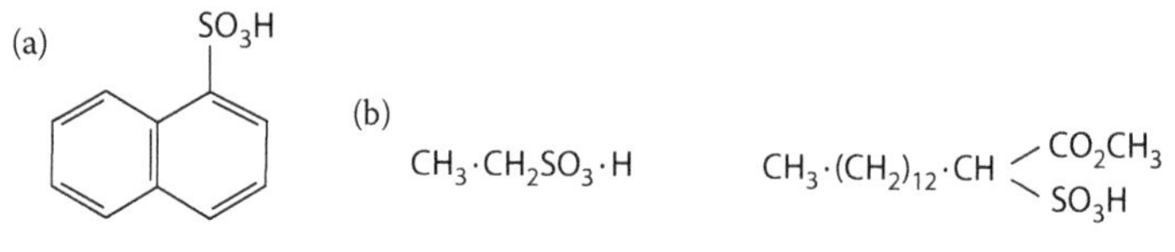

FIGURE 2.4 Organosulfur compounds used as sources of sulfur. (a) Naphthalene-1-sulsulfonic acid, (b) aliphatic sulfonic acids.

Cultures of *Scenedesmus obliquus* under conditions of sulfate limitation metabolized naphthalene-1-sulfonate to 1-hydroxy-naphthalene-2-sulfonate and the glucoside of naphth-1-ol (Kneifel et al. 1997). These results are consistent with the formation of a 1,2-epoxide followed by an NIH shift.

3. The cyanobacteria *Anabaena* sp. strain PCC 7120 and *Nostoc ellipsosporum* dechlorinated γ-hexachloro[*aaaeee*]cyclohexane in the light in the presence of nitrate to γ-pentachlorocyclohexene (Figure 2.5), and to a mixture of chlorobenzenes (Kuritz and Wolk 1995). The reaction is dependent on the functioning of the *nir* operon involved in nitrite reduction (Kuritz et al. 1997).

4. Alkanes are produced by fungi, algae, and bacteria, including cyanobacteria, and one of the pathways involving decarbonylation of fatty acid aldehydes has been examined in some.

   a. A survey of hydrocarbons in marine benthic algae included representatives of green, brown, and red algae in the Cape Cod area, USA. and revealed a complex overview that is briefly summarized: (a) *n*-pentadecane dominated in brown algae and *n*-heptadecane in red algae, (b) alkenes were dominant in some groups, and the degree of unsaturation ranged from 17:1 to 21:5 and 21:6 (Youngblood et al. 1971).

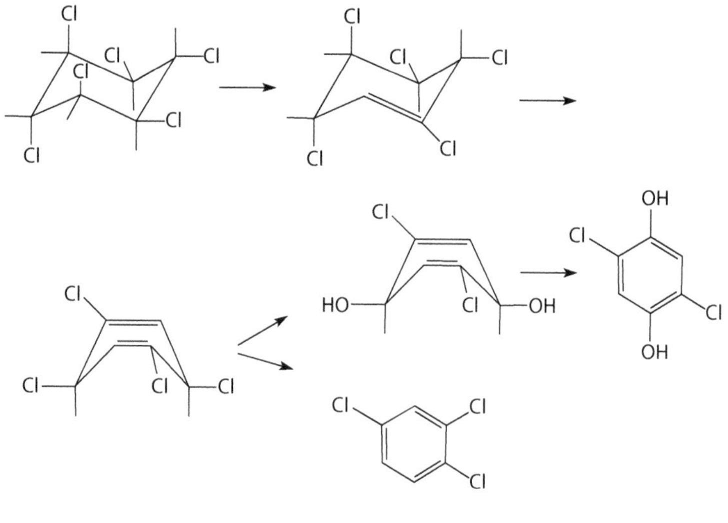

FIGURE 2.5 Transformation of γ-hexachlorocyclohexane by cyanobacteria.

b. There has been substantial interest in the synthesis of hydrocarbons by *Botryococcus braunii*. These comprise a plethora of structures that are linear, branched, and unsaturated, and with chain lengths from C_{25} to C_{40}. Since attention is directed to several extensive reviews (Banerjee et al. 2002; Metzger and Largeau 1999, 2005), it is sufficient to provide only a summary of highlight features.

Races of *Botryococcus braunii* and Summary of the Hydrocarbons Synthesized

Race	Hydrocarbons
A	Odd n-C_{25} to C_{31}; C_{25} Δ^1,Δ^{16}: C_{27} Δ^1,Δ^{18}: C_{29} Δ^1,Δ^{20}: C_{31} Δ^1,Δ^{22}
B	Triterpenoid botryococcenes C_{30} to C_{34} with methylation; squalene
L	Tetraterpenoid $C_{40}H_{78}$ lycopadiene

The mechanisms used by races of *Botryococcus braunii* include the following:

i. Microsomal preparations were able to produce heptadecane from octadecanal under anoxic conditions with the formation of CO, and were inhibited by metal-chelating agents (Dennis and Kolattukudy 1991).

ii. Long-chain hydrocarbons were produced by successive elongation of oleate with malonyl-CoA and formation of terminal Δ^1 by the reaction:

$-CH(OH)-CH_2-C(=O)X \rightarrow -CH=CH_2$ (Metzger and Largeau 1999). Note that these alkenes are not produced by head-to-head condensation.

iii. Dimerization of farnesyl diphosphate to presqualene diphosphate with a cyclopropane ring bearing a $-CH_2$-diphosphate followed by rearrangement to produce a quaternary C_{30} hydrocarbon with a methyl and an ethene group at C_{10} (Epstein and Rilling 1970; Poulter 1990). Triterpenoid hydrocarbons C_{30} to C_{37} are produced by this organism, and methylation by methionine of the precursor C_{30} hydrocarbon produces a series of botrycoccenes with methyl groups at the C_{30} at C_{37}, C_{16}, and C_{20} positions (Metzger et al. 1986). Although this pathway may, formally at least, produce squalene, the enzymes involved are probably different.

c. i. Alkanes dominated by $C_{17:0}$ and alkenes by $C_{19:1}$ were produced by a number of filamentous cyanobacteria (Winters et al. 1969), and the enzymology has been established for a group of them. The key reactions were reduction of the fatty acid acyl carrier proteins to the aldehydes followed by decarbonylation. A range of odd-numbered n-alkyl hydrocarbons—most frequently heptadecane or pentadecane—was observed both from unicellular species (*Synechococcus*, *Synechocystis*, *Cyanothece*) and filamentous species (*Anabaena*, *Nostoc*). It was shown by x-ray crystallography that the structure of the decarbonylase from *Prochlococcus marinus* that produced pentadecane was comparable to that of ribonucleotide reductase from *Escherichia coli* with two Fe atoms (or possibly one Mn and one Fe) at the active site. In addition, the purified aldehyde decarbonylase from *Nostoc punctiforme* was able to produce heptadecane from octadecanal though this required the addition of the reductants spinach ferredoxin, ferredoxin reductase, and NADPH (Schirmer et al. 2010).

ii. The synthesis of terminal alkenes has been described in the unicellular cyanobacterium *Synechococcus* sp. PCC 7220 and is initiated by Claisen-type condensation of the fatty acid acyl carrier protein with malonyl-CoA followed by decarboxylation with concomitant loss of the β-hydroxyl group to produce the alkene (Mendez-Perez et al. 2011):

$R-CO-S-ACP-1 + \text{malonyl-CoA} \rightarrow R-C(OH)-CH_2-CO-ACP-2 \rightarrow R-CH=CH_2 + CO_2$

5. The transformation of DDT to DDE—albeit in rather low yield—by elimination of one molecule of HCl has been observed in several marine algae (Rice and Sikka 1973).

2.5.4.2 Phototrophic Anaerobic Bacteria

Purple nonsulfur phototrophs are metabolically versatile and are capable of growth anaerobically in the light, heterotrophically in the dark under aerobic conditions, and in the dark by fermentation of, for example, pyruvate (Uffen 1973) and fructose (Schultz and Weaver 1982).

There has been interest in the metabolic potential of anaerobic phototrophic bacteria, particularly the purple nonsulfur organisms that can degrade aromatic compounds (Khanna et al. 1992). Such organisms are widely distributed in suitable ecosystems, and may, therefore, play a significant role in the degradation of xenobiotics. Less appears to be known, however, of the potential of other anaerobic phototrophs such as the purple and green sulfur bacteria to degrade xenobiotics. Some examples of metabolism by purple nonsulfur phototrophs include the following:

1. Dehalogenation of a number of halogenated alkanoic acids has been observed with *Rhodospirillum rubrum*, *R. photometricum*, and *Rhodopseudomonas palustris*. Substrates included 2- and 3-chloropropionic acid for all of these organisms, chloroacetate for *R. photometricum*, and 2-bromopropionate for *R. rubrum* and *Rhps. palustris*: acetate and propionate were produced and further metabolized (McGrath and Harfoot 1997).

2. Several nonsulfur phototrophs are able to synthesize C_4 and C_5 poly-3-hydroxyalkanoates, and *Rhodobacter capsulatus* is able to grow photoheterotrophically with C_2 to C_{12} fatty acids (Kranz et al. 1997). A study of mutants revealed that the synthesis of 3-hydroxyacyl-CoA was not dependent on the presence of PhaA (β-ketothiolase) and PhaB (acetoacetyl-CoA reductase).

3. *Rhodobacter capsulatus* was able to carry out the reduction of aromatic amines that included 2,4-dinitrotoluene, 3,4-dinitrobenzoate, and 2,4-dinitrophenol (Pérez-Reinado et al. 2005). Two reductases NprA and NprB were characterized that responded differently to the presence of ammonia.

4. The metabolism of 3-chlorobenzoate has been examined in a strain of *Rhodopseudomonas* (Egland et al. 2001). This was initiated by the formation of the coenzyme A ester that was dechlorinated to benzoate that was metabolized to acetyl-CoA by mechanisms that are discussed in Chapter 8.

5. The photometabolism of benzoate and ring-oxygenated benzoates has been examined in several organisms, and two processes were observed: (a) oxidation of side chains and (b) metabolism of the resulting benzoate by ring fission. Some representative results are given in Table 2.1.

2.5.4.3 Aerobic Nonoxygenic Phototrophs

Aerobic nonoxygenic phototrophic bacteria (AAB) belong to a highly heterogeneous group that contains both freshwater and marine representatives. They are widely distributed in eutrophic littoral zones and in the open oceans, as well as occurring as marine symbionts in a range of algae. Examples of their distribution in the oceans are given in Table 2.2.

These organisms generally—though not always—contain bacteriochlorophyll *a*, although they are unable to carry out photosynthesis, and many contain high concentrations of carotenoids that contribute to their intense color. For example, the genome of *Roseobacter denitrificans* does not possess

TABLE 2.1

Photometabolism of Aromatic Carboxylates by Purple Nonsulfur Phototrophic Bacteria

Organism	Substrates	Reference
Rhodopseudomonas palustris CGD 052	Benzoate, caffeate, 3-hydroxybenzoate, mandelate, vanillate	Harwood and Gibson (1988)
Rhodomicrobium vannielii	Benzoate, syringate, vanillate	Wright and Madigan (1991)
Rubrivivax benzoatilyticus	Benzoate, phenylalanine, sulfanilate	Ramana et al. (2006)

TABLE 2.2

Examples of the Marine Distribution of Aerobic Anoxygenic Phototrophs, and *Prochlococcus*

Area	Comment	Reference
South Pacific	An oligotrophic ocean with both AAB and *Prochlorococcus* in high concentrations	Lami et al. (2007)
Mid-Atlantic Bight	Concentrations of AAB exceeded those of *Prochlorococcus*, that were maximal in the photic zone	Cottrell et al. (2006)
Pacific, Indian, Atlantic	Concentrations of AAB greater in the coastal regions than oceanic	Jiao et al. (2007)
Pacific, Indian, China Sea	BChl a and Chl a greater in the open ocean	Jiao et al. (2010)
Central Baltic	The Baltic Sea has a low salinity that diminishes northward	Salka et al. (2008)

genes for the photosynthetic assimilation of CO_2, and specifically those for ribulose bisphosphate carboxylase and phosphoribulokinase are missing. This suggests the operation of a mixotrophic rather than autotrophic CO_2 assimilation (Swingley et al. 2007). Several marine strains, including *Roseobacter*, *Erythrobacter*, *Stappia*, and *Sulfitobacter*, have been examined as well as freshwater strains, including *Erythromonas* and *Erythromicrobium*. The *Roseobacter* group has been the subject of reviews (Allgaier et al. 2003; Buchan et al. 2005) and an overall review of aerobic nonoxygenic phototrophs (Yurkov and Beatty 1998). Procedures for the successful isolation of important representatives are discussed in Chapter 4; briefly, they incorporate extremely low substrate concentration, incubation at low temperature for several weeks, and dilution-to-extinction. It is not possible to include a discussion of the large number of genera in this group of bacteria and only a few brief comments on their metabolic capabilities are given.

2.5.4.3.1 Dimethylsulfoniopropionate

This is synthesized by marine phytoplankton as an osmolyte and its demethylation to dimethylsulfide and methanethiol has been observed in the roseobacters *Silicibacter pomeroyi* and *Roseovarius nubinhibens*. Whereas *S. pomeroyi* could grow at the expense of dimethylsulfoniopropionate and a number of putative degradation products, including acrylate, *R. nubinhibens* was unable to grow with acrylate, and neither strain could utilize 3-mercaptopropionate (González et al. 2003).

2.5.4.3.2 Carbon Monoxide

Carbon monoxide can be produced by photolysis of dissolved organic matter in the marine environment, and its oxidation has been demonstrated in species of the marine genus *Stappia*. Type I and type II of the gene *coxL* were found as well as *cbbL* that encodes the large subunit of ribulose bisphosphate carboxylase/oxygenase (Weber and King 2007). This strain was able to grow at the expense of a range of substrates that included propionate, malonate, ribose, sucrose, alanine, valine, and glutamate.

2.5.4.3.3 Aromatic Substrates

1. Arene carboxylates are products of the degradation of lignin and humic material. *Sagittula stellata* strains E-37 and Y3F were able to grow with 4-hydroxybenzoate and 3,4-dihydroxybenzoate and possessed the genes for the 3-oxoadipate pathway, while *Sulfitobacter* strain EE 36 was able to degrade a wider range of aromatic substrates that included anthranilate, benzoate, salicylate, and vanillate (Buchan et al. 2000).

2. A marine strain of *Porphyrobacter sanguineum* was able to grow at the expense of biphenyl and dibenzofuran (Hiraichi et al. 2002), and species of this genus of have been described from both freshwater thermal habitats and the ocean, and, on the basis of molecular evidence, were closely similar to *Erythromonas* and *Erythromicrobium*.

2.5.4.3.4 Alkanoates and Carbohydrates

After aerobic growth in the dark, activities of TCA cycle enzymes and the glyoxylate cycle were found in the freshwater *Erythromicrobium sibiricus*, although levels of succinate and 2-oxoglutarate were missing in *E. longus*. The high activities for enzymes of the Entner–Doudoroff pathway in *E. sibiricus* and *E. longus* were consistent with this as the primary mechanism for the catabolism of glucose (Yurkov and Beatty 1998).

2.5.5 Aerobic and Facultatively Anaerobic Bacteria

The well-established metabolic versatility of groups such as the pseudomonads and their numerous relatives, and the methanotrophs has possibly deflected attention from other groups that may be present in aquatic systems and which may play an important role in determining the fate of xenobiotics. This is increasingly being rectified with the isolation and description of new taxa. Although the potential of other Gram-negative groups, including the pseudomonads, acinetobacters, moraxellas, and species of *Alcaligenes* is well established, Gram-positive groups seem to have achieved generally somewhat less prominence in aquatic systems. In the succeeding paragraphs, some examples of the metabolic importance of a few of these groups of organisms are presented.

2.5.5.1 Gram-Positive Aerobic Bacteria

It is appropriate to summarize the genera of Gram-positive bacteria that are characterized by the existence of mycolic acids and menaquinones MK-8 or MK-9: *Corynebacterium*, *Dietzia*, *Gordonia*, *Mycobacterium*, *Nocardia*, and *Rhodococcus* (Rainey et al. 1995). These are explored in greater detail in the following paragraphs.

2.5.5.1.1 Mycobacteria

Mycobacteria have traditionally been divided into fast-growing and slow-growing strains. Strains of pathogenic significance typically belong to the latter group, whereas even those of clinical origin that are fast-growing are considered to be commensals, and environmental strains as rapid-growing saprophytes (e.g., Cheung and Kinkle 2001; Hennessee et al. 2009). A number of organisms considered as mycobacteria would now be classified as rhodococci or gordonia. Among these are the non-acid-fast organisms described by Gray and Thornton (1928) that degraded naphthalene. Care should, therefore, be exercised in the taxonomic description of the organisms.

The degradation of a broad range of substrates by mycobacteria has been described, including alkanes, haloalkanes, PAHs, nitroarenes, and chlorophenols. The growth of *Mycobacterium tuberculosis* in a defined medium is highly restricted by pH (Piddington et al. 2000), and *Mycobacterium bovis* is restricted to mildly acid conditions, whereas *Mycobacterium smegmatis* is tolerant of pH 5, although displaying only limited growth compared, for example, to that of *Salmonella enterica* at the same pH (Rao et al. 2001). Mycobacteria that were able to degrade PAHs have been isolated from soil with pH 2–5 (Uyttebroek et al. 2007), and the DGGE fingerprinting and sequence analysis showed that the dominant organism was closely similar to *Mycobacterium montefiorense* that belongs to the slow-growing group of mycobacteria (Levi et al. 2003).

2.5.5.1.1.1 Alkanes, Haloalkenes, Alkyl Ethers, Alkyl Nitrate Ester

1. *Alkanes   Mycobacterium paraffinicum* (Davis et al. 1956; Wayne et al. 1991; Toney et al. 2010) was able to degrade ethane, propane, and butane and its inability to grow on other substrates with the exception of their putative oxidation products. The name was validated by Toney et al. (2010) who provided a full description of this slow-growing scotochromogenic species that lacked amidase activity. The metabolism of propane and butane has been described for *Mycobacterium vaccae* strain JOB-5, although there were significant differences: propane is hydroxylated at C_2 (Vestal and Perry 1969) whereas butane undergoes terminal hydroxylation

(Phillips and Perry 1974). Organisms described as mycobacteria could use ethene and, less frequently, propene for growth, although no details were given (de Bont et al. 1980). The growth of some acyclic terpenoids has been observed, although without providing metabolic details. A strain SD4 assigned to *Mycobacterium ratisbonense* was able to grow at the expense of squalane, squalene, and pristane, and *Mycobacterium fortuitum* strain NF4 with farnesol, geranylacetone, and geranic acid while both strains were able to grow with C_{16} and C_{18} carboxylates (Berekaa and Steinbüchel 2000).

2. *Haloalkenes*   Some strains of mycobacteria grown with propane are able to oxidize the apparently unrelated substrate trichloroethene (Wackett et al. 1989). Strains of mycobacteria have been shown to degrade a range of halogenated alkenes. These include (a) vinyl chloride (Hartmans and de Bont 1992; Coleman and Spain 2003) and (b) *Mycobacterium* sp. strain GP1 that degrades 1,2-dibromoethane by a pathway that avoids the production of the toxic 2-bromoethanol and 2-bromoacetaldehyde by the formation of the epoxide (Poelarends et al. 1999). Possibly, more remarkable is the metabolic capacity of species of mycobacteria, including the human pathogen *M. tuberculosis* strain H37Rv (Jesenská et al. 2000). On the basis of amino acid and DNA sequences, the strain that was used contained three halohydrolases and the activity of the dehalogenase for debromination of 1,2-dibromoethane by other species of mycobacteria is given in Table 2.3. The haloalkane dehalogenase gene from *Mycobacterium avium* has been cloned and partly characterized (Jesenská et al. 2002).

3. *Tert-methyl butyl ether and morpholine*   *Mycobacterium austroafricanum* strain IFP 2012 transformed MTBE to *tert*-butanol, 2-methyl-1,2-propandiol, and 2-hydroxyisobutyrate that was degraded (Ferreira et al. 2006). A pathway for its degradation has been proposed that involved a cobalamin-dependent mutase, which converted 2-hydroxyisobutyrate-CoA into 3-hydroxybutyryl-CoA (Rohwerder et al. 2006). Morpholine can be degraded by several mycobacteria, including *Mycobacterium aurum* strain MO1 (Combourieu et al. 1998), *Mycobacterium* strain RP1 (Poupin et al. 1998), and *Mycobacterium chelonae* (Swain et al. 1991). The reaction is initiated by a cytochrome P450 monooxygenase that is also active against pyrrolidine and piperidine (Poupin et al. 1998).

4. *Alkyl nitrate ester*   2-Ethylhexyl nitrate is a major additive to diesel fuel and was degraded by *Mycobacterium austroafricanum*. This involved conversion to 2-ethylpentan-1,6-diol, followed by dehydrogenation to the 1-carboxylate and β-oxidation to 3(hydroxymethyl)pentanoate. This was then cyclized to the terminal metabolite 4-ethyldihydrofuran-2(3*H*)-one (Nicolau et al. 2008).

5. *Carbon monoxide*   The degradation of carbon monoxide has attracted recent attention, and a strain of *Mycobacterium* is able to degrade this unusual substrate as well as methanol (Park et al. 2003). The subunits of the carbon monoxide dehydrogenase *cutA*, *cutB*, and *cutC* correspond to the large, medium, and small subunits of the enzyme, and it was shown in *Mycobacterium* sp. strain JC1 DSM 3803 that CutR which is a member of the LysR-type transcriptional regulators was required for the expression of these subunits and effective growth with carbon monoxide (Oh et al. 2010).

TABLE 2.3

Specific Activity (μmol Bromide per mg Protein per min) of Dehalogenase for 1,2-Dibromoethane from Selected Species of Mycobacteria

Taxon	Activity	Taxon	Activity
M. bovis BCG MU10	99	*M. avium* MU1	36
M. fortuitum MU 8	76	*M. phlei* CCM 5639	22
M triviale MU3	61	*M. parafortuitum* MU 2	22
M. smegmatis CCM 4622	49	*M. chelonae*	20

2.5.5.1.1.2 Arenes

1. *PAHs* There has been increasing interest in the degradation of PAHs by mycobacteria (Brezna et al. 2003). These include naphthalene (Kelley et al. 1990), phenanthrene (Guerin and Jones 1988), anthracene (van Herwijnen et al. 2003), pyrene, and fluoranthene (Heitkamp et al. 1988; Grosser et al. 1991; Krivobok et al. 2003; Derz et al. 2004) (Table 2.4).

   The strain PYR-1 that degraded pyrene and fluoranthene (Govindaswami et al. 1995), the fluoranthene-degrading *Mycobacterium hodleri* (Kleespies et al. 1996), the fluoranthene- and pyrene-degrading *Mycobacterium frederiksbergense* (Willumsen et al. 2001) and the pyrene-degrading strains of *Mycobacterium pyrenivoransi* (Derz et al. 2004) belonged to the group of fast-growing scotochromogenic mycobacteria. The PAH-degrading *Mycobacterium gilvum* is able to degrade all the benzoquinoline isomers with the putative formation of 2-oxo-benzo-quinoline as an intermediate from 5,6-benzoquinoline (Willumsen et al. 2001).

2. *Phenols*

   a. Although *Mycobacterium goodii* is unable to grow at the expense of phenol, it is capable of hydroxylating phenol to hydroquinone after induction of the monooxygenase by acetone (Furuya et al. 2011). Sequence analysis of binuclear iron monooxygenase gene cluster revealed the presence of the *mimA*, *mimB*, *mimC*, and *mimD* genes that encode the oxygenase large subunit, a reductase, the oxygenase small subunit, and a coupling protein.

   b. Considerable interest has been expressed in the chlorophenol-degrading organism *Mycobacterium chlorophenolicum* (*Rhodococcus chlorophenolicus*) (Apajalahti et al. 1986; Briglia et al. 1994; Häggblom et al. 1994), partly motivated by its potential for application to bioremediation of chlorophenol-contaminated industrial sites (Häggblom and Valo 1995).

3. *Nitroarenes* The pathway used by a strain of *Mycobacterium* sp. for the degradation of 4-nitrotoluene is initiated by reduction to the hydroxylamine followed by rearrangement to 3-hydroxy-4-aminotoluene before further degradation (Spiess et al. 1998).

2.5.5.1.2 Rhodococci

Only a selection of examples is given below to illustrate the wide metabolic versatility of rhodococci: further details are given in the appropriate sections of Chapters 7 through 11:

1. Cyclopropanecarboxylate can be degraded by *Rhodococcus rhodochrous* (Toraya et al. 2004).
2. The degradation of isoprene by *Rhodococcus* sp. strain AD45 involves a glutathione-mediated reaction (van Hylckama Vlieg et al. 2000) and is noted again in Chapter 7, Part 1.

TABLE 2.4

Degradation of PAHs by Selected Strains of Mycobacteria

Substrate	Organism	Reference
Nap, Phe, 3-Me-Chol	*Mycobacterium* sp.	Heitkamp et al. (1988)
Phe, Pyr, Flu	*Mycobacterium gilvum* sp. BR1	Boldrin et al. (1993)
Pyr, BaPyr, BaAnth	*Mycobacterium* sp. RGII-135	Schneider et al. (1996)
Phe, Anth, Flu, Pyr	*Myco. frederiksbergense*	Willumsen et al. (2001)
Nap, Anth, Phe, Flu, Pyr	*Myco. vanbaalenii* PYR-1	Kim et al. (2006, 2007)
BaPyr	*Myco. vanbaalenii* PYR-1	Moody et al. (2004)
BaAnth	*Myco. vanbaalenii* PYR-1	Moody et al. (2005)
Phe, Flu, Pyr	*Myco. pyrenivorans*	Derz et al. (2004)

Note: 3-MeChol, 3-methylcholanthrene; Anth, anthracene; BaAnth, benz[*a*]anthracene; BaPyr, benzo[*a*]pyrene; Flu, fluoranthene; Nap, naphthalene; Phe, phenanthrene; Pyr, pyrene.

3. A number of aliphatic ethers can be degraded by *Rhodococcus* sp. strain DEE 5151 (Kim and Engesser 2004).

4. The aerobic degradation of acetylene has been accomplished by several strains of *Rhodococcus* and takes place by molybdenum-dependent hydration (Rosner et al. 1997).

5. A strain of *Rhodococcus* sp. was capable of degrading a number of chlorinated aliphatic hydrocarbons, including vinyl chloride and trichloroethene, as well as the aromatic hydrocarbons benzene, naphthalene, and biphenyl (Malachowsky et al. 1994).

6. A strain of *Rhodococcus opacus* isolated by enrichment with chlorobenzene was able to grow at the expense of a wide range of halogenated compounds. These included 1,3- and 1,4-dichlorobenzene, 1,3- and 1,4-dibromobenzene, 2-, 3-, and 4-fluorophenol, 2-, 3-, and 4-chlorophenol, 4-nitrophenol, 3- and 4-fluorobenzoate, and 3-chlorobenzoate (Zaitsev et al. 1995).

7. Several rhodococci have attracted interest for their ability to degrade PCBs, including an organism (*Acinetobacter* sp. strain P6) now assigned to *Rhodococcus globerulus* (Asturias and Timmis 1993), *R. erythropolis* (Maeda et al. 1995), and *Rhodococcus* sp. strain RHA1 (Seto et al. 1995b).

8. The degradation of 2,4,6-trichlorophenol by *Rhodococcus percolatus* has been described (Briglia et al. 1996).

9. 2,4,6-Trinitrophenol is degraded in a reaction involving ring reduction by hydride transfer from an NADPH-dependent F_{420} reductase (Hofmann et al. 2004).

10. *Rhodococcus jostii* has a rather unusual isolation history since the type strain was isolated after exhumation of a grave, and the specific name refers to the femur of a mediaeval Czech ruler in Moravia (Takeuchi et al. 2002). The RHA1 strain has been extensively examined and exhibited a remarkably broad spectrum of degradation; examples to illustrate the diversity are given in Table 2.5.

In addition, a number of biotransformations have been accomplished by rhodococci, including, for example, the hydrolysis of nitriles, including polyacronitriles (Tauber et al. 2000), and the reduction of the conjugated C·C double bond in 2-nitro-1-phenylprop-1-ene (Sakai et al. 1985).

2.5.5.1.3 Corynebacterium

This is a heterogeneous genus that covers a range of species encountered in clinical microbiology (Funke et al. 1997), including the human pathogen *Corynebacterium diphtheriae*, as well as important pathogens for wheat and corn, alfalfa, potato, and tomato (Carlson and Vidaver 1982). *Corynebacterium glutamicum* has achieved prominence in biotechnology for the synthesis of L-lysine and the food flavor sodium L-glutamate, and the biodegradable polyhydroxybutyrate, while various strains are also able to carry out a range of degradations (Table 2.6). *Corynebacterium glutamicum* is able to use a range of aliphatic sulfonates as sources of sulfur that are discussed in Chapter 11 (Koch et al. 2005), and is able to tolerate unusually high concentrations of both arsenite and arsenate (Ordóñez et al. 2005).

TABLE 2.5

Aerobic Degradations Carried Out by *Rhodococcus jostii* Strain RHA1

Substrate	Reference
Cholesterol	Rosłoniec et al. (2009)
Phenylacetate	Navarro-Llorens (2005)
Polychlorinated biphenyls	Seto et al. (1995a,b)
Brominated diphenyl ethers	Robrock et al. (2009)
Indole-3-acetic acid	Leveau and Gerards (2008)

TABLE 2.6

Degradations Carried Out by Strains of *Corynebacterium*

Species or Strain	Substrate	Reference
Corynebacterium sp. strain SY1	Dibenzothiophene	Omori et al. (1992)
Corynebacterium sp. strain C125	1,2,3,4-Tetrahydro-naphthalene	Sikkema and de Bont (1993)
Corynebacterium cyclohexanicum	4-Oxocyclohexane-carboxylate	Kaneda et al. (1993)
Corynebacterium sepedonicum	2,4-Dichlorobenzoate	Romanov and Hausinger (1996)
Corynebacterium glutamicum	Resorcinol	Huang et al. (2006)

2.5.5.1.4 Gordonia

A review has summarized the taxonomic status of the genus *Gordonia* (Arenskötter et al. 2004), and drawn attention to important degradations carried out by species of the genus some of which are summarized in Table 2.7. Other degradations include growth with alkanes from C_{13} to C_{22} by *Gordonia* sp. strain TY-5 (Kotani et al. 2003), and degradation of RDX (hexahydro-1,3,5-trinitro-1,3,5-triazine *Gordonia* sp. strain KTR9 (Thompson et al. 2005). The degradation of *cis*-1,4-polyisoprene is discussed in Chapter 3 on dioxygenases and in Chapter 7 on alkenes (Braaz et al. 2004).

2.5.5.1.5 Dietzia

This genus has been separated from *Rhodococcus* with the type strain *Dietzia maris* (Rainey et al. 1995). Several strains are able to grow at the expense of long-chain *n*-alkanes: *Dietzia maris* with C_6 to C_{17}, C_{19}, and C_{23} *n*-alkanes; the facultatively psychrophilic alkaliphile *Dietzia psychralcaliphila* with decanes, eicosane, and tetracosane—though not pristane (Yumoto et al. 2002); *Dietzia* sp. strain DQ 12-45-1b with C_6 to C_{40} *n*-alkanes. This strain contained the alkane integral-membrane monooxygenase (AlkB) fused to rubredoxin (Rd) encoded by *alkW1* and *alkW2* of which the first was active for the degradation of C_8 to C_{32} *n*-alkanes (Nie et al. 2011).

2.5.5.1.6 Actinomyces

The role of cytochrome P450 oxygenations in actinomycetes has been reviewed (O'Keefe and Harder 1991) and, although these organisms have been exhaustively explored as the source of clinically valuable antibiotics, their degradative capability has been less extensively examined. It has been shown that they have a virtually ubiquitous occurrence in tropical and subtropical marine sediments (Mincer et al. 2002), and all of these have been assigned to two taxa *Salinispora arenicola* and *S. tropica* (Mincer et al. 2005). Their occurrence has been explored in the marine environment as possible sources of biologically active metabolites (Magarvey et al. 2004).

TABLE 2.7

Degradations Carried Out by Species of *Gordonia*

Substrate	Organism	Reference
Benzothiophene	*G. desulfuricans*	Kim et al. (1999)
Dibenzothiophene	*G. amicalis*	Kim et al. (2000)
3-Alkylpyridines	*G. nitida*	Yoon et al. (2000)
Acetylene	*G. rubropertincta*	Rosner et al. (1997)
cis-1,4-Polyisoprene	*G. polyisoprenivorans*	Linos et al. (1999)
	G. westfalica	Linos et al. (2002)
Hexadecane/hexadecene	*G. alkanivorans*	Kummer et al. (1999)
Hexadecane/hexatriacontane	*G.* sp. strain SoCg	Piccolo et al. (2011)
Cholesterol > ergosterol > stigmasterol	*G. cholesterolivorans*	Drzyzga et al. (2011)

2.5.5.1.7 Streptomyces

These have become established biocontrol agents against fungal infection that are discussed in Part 3 of this chapter. They have been examined for the degradation of a number of contaminants, and several pathways for the introduction of oxygen have emerged.

1. Cytochrome P450 monooxygenation
   a. The transformation by *Streptomyces griseus* of benzene, toluene, naphthalene, biphenyl, and benzo[*a*]pyrene to the corresponding phenols has been observed (Trower et al. 1989). The oxidation of phenanthrene to (–)*trans*-[9*S*,10*S*]-9,10-dihydrodihydroxyphenanthrene by *Streptomyces flavovirens* with minor amounts of 9-hydroxyphenanthrene (Sutherland et al. 1990) is plausibly carried out by epoxidation and epoxide hydrolysis rather than by typical dioxygenation.
   b. The transformation of sulfonylureas by *Streptomyces griseolus* grown in a complex medium containing glucose when the methyl group of the heterocyclic moieties is hydroxylated, and for some substrates subsequently oxidized to the carboxylic acid (Romesser and O'Keefe 1986; O'Keefe et al. 1988).
2. Dioxygenation
   a. The metabolism of aromatic substrates has been examined in *Streptomyces setonii*, and the degradation of vanillate involved in decarboxylation to guaiacol followed by intradiol fission of the catechol that was produced (Pometto et al. 1981). This strain could also metabolize cinnamate and related carboxylic acids (Sutherland et al. 1983), and strains of *Amycolatopsis* sp. and *Streptomyces* sp. were able to metabolize benzoate and salicylate that were degraded by dioxygenation (Grund et al. 1990).
   b. The degradation of quercetin by 2,3-dioxygenation to 2-protocatechuoyl phloroglucinol carboxylate and carbon monoxide is carried out by *Streptomyces* sp. FLA (Merkens and Fetzner 2008). The enzyme activity was high when the medium was supplemented with Ni^{2+} or Co^{2+}—though not Fe^{2+}, Cu^{2+} or Zn^{2+}—and the ESR spectrum was consistent with the presence of high-spin $S^{3/2}$ Co^{2+} (Merkens et al. 2008).
   c. The degradation of 4-hydroxybenzoate and vanillate has been examined in *Streptomyces* sp. strain 2065, and the pathway involving intradiol dioxygenation of the initially formed 3,4-dihydroxybenzoate was characterized (Iwagami et al. 2000).
   d. The degradation of 2,4,6-trichlorophenol has been described for *Streptomyces rochei* strain 303, and is initiated by the formation of 2,6-dichlorohydroquinone that underwent 1,2-fission by a specific dioxygenase and has been characterized (Zaborina et al. 1995).
   e. The degradation of *cis*-poly-1,4-isoprene has been examined in *Streptomyces coelicolor* 1A, and a number of oxidation products identified (Bode et al. 2000). These were produced by oxidations of the subterminal double bonds followed by β-oxidation, hydroxylation, oxidation, and fission of the 3-oxoacids.
3. Several strains of *Streptomyces* sp. that were resistant to the herbicide alachlor [2-chloro-2′,6′-diethyl-*N*-(methoxymethyl)acetanilide] were able to carry out its degradation. Transformation was initiated by dechlorination of the acyl side chain and *N*-dealkylation, followed by cyclization with one of the aryl ethyl groups to form 8-ethylquinoline and, putatively, 7-ethylindole (Durães Sette et al. 2004).

2.5.5.2 Gram-Positive Facultatively Anaerobic Organisms

The genus *Staphylococcus* is traditionally associated with disease in humans. Dibenzofuran or fluorene served, however, as sole sources of carbon and energy for *Staphylococcus auriculans* strain BF63 that had no obvious clinical association. Salicylate and gentisate accumulated in the medium using dibenzofuran which is consistent with the angular dioxygenation at the 1,1a position followed by the formation of 2,3,2′-trihydroxybiphenyl before ring fission (Monna et al. 1993). This strain also carried out limited biotransformation of fluorene and dibenzo[1,4]dioxin, and these serve to illustrate the unsuspected metabolic potential of facultatively anaerobic Gram-positive organisms.

A satisfying evaluation of the metabolic potential of microorganisms in natural ecosystems should not, therefore, fail to consider all these organisms, which are certainly widespread, and to distinguish between rates of degradation and metabolic potential: slow-growing organisms may be extremely important in degrading xenobiotics in natural ecosystems.

2.5.6 Gram-Negative Aerobic Bacteria

There have been substantial developments in the taxonomy of pseudomonads based on the substantive defining studies of Palleroni (1984), and many new genera have been proposed, including, for example, *Sphingomonas*, *Comamonas*, *Burkholderia*, *Xanthomonas*, *Stenotrophomonas*, and *Variovorax* (Anzai et al. 2000). This and the issues surrounding assignment to the genera *Ralstonia*, *Wautersia*, and *Cupriavidus* (Vandamme and Coenye 2004) are not discussed here, although examples of these genera are given in Chapters 7 through 11 in the context of degradation.

Attention is briefly drawn to some groups that are not only widely distributed but have attracted attention for their ability to fix nitrogen.

2.5.6.1 Azotobacter

Burk and Winogradsky in the 1930s showed that azotobacters could readily be obtained from soil samples by elective enrichment with benzoate, and these provide the paradigm for aerobic nitrogen fixation. The degradative pathway for benzoate has been elucidated (Hardisson et al. 1969), and the range of substrates extended to 2,4,6-trichlorophenol (Li et al. 1992; Latus et al. 1995). The enzyme from *Azotobacter* sp. strain GP1 that catalyzes the formation of 2,6-dichlorohydroquinone from 2,4,6-trichlorophenol is a monooxygenase that requires NADH, FAD, and O_2 (Wieser et al. 1997) and, in the absence of a substrate, results in unproductive formation of H_2O_2. It is also able to accept other chlorophenols with the consumption of NADH, including 2,4-, 2,6-, 3,4-dichloro-, 2,4,5-trichloro-, and 2,3,4,5-, and 2,3,4,6-tetrachlorophenol.

2.5.6.2 Burkholderia

The *Burkholderia cepacia* complex includes genomovars that are significant and widely distributed pathogens in the clinical environment (Mahenthiralingam et al. 2005). *B. cepacia* genomovar III is the primary cause of cystic fibrosis, and there is increasing concern over its transmission. This genomovar has also been found in the rhizosphere of wheat and maize, and as endophytes on wheat and lupine (Balandreau et al. 2001). *B. cepacia* is the classical causative agent of soft-rot in onion, and strains have been isolated both as plant pathogens and from the rhizosphere, rhizoplane, and roots of plants. Among isolates of the complex from the maize rhizosphere, a variety of groups was isolated and showed a complex relation to the plants and to their environment (Ramette et al. 2005). By use of semiselective media, *B. cepacia* and *B. vietnamiensis* were consistently isolated from the rhizosphere of both maize and rice (Zhang and Xie 2006). In addition, some strains of *Burkholderia* are able to form effective symbiotic associations with, and nodulate species of the legume *Mimosa* (Chen et al. 2005), and *B. phytoformans* that was isolated from *Glomus vesiculiferum*-infected onion (Sessitsch et al. 2005) is able to produce ACC deaminase and the quorum-sensing 3-hydroxy-C8-homoserine lactone, and to establish epiphytic colonization on grapevine (*Vitis vinifera*) (Compact et al. 2005).

Other species have established roles in biodegradation, for example, *Burkholderia kururiensis* in the cometabolism of trichloroethene (Zhang et al. 2000); *Burkholderia phenoliruptrix* in the degradation of 2,4,5-trichlorophenoxyacetate (Coenye et al. 2004); and *Burkholderia xenovorans* in the degradation of PCBs (Goris et al. 2004) and diterpenoids (Smith et al. 2008) that are discussed in Chapters 7 and 9. Species of *Burkholderia* have been found in the rhizosphere and the rhizoplane of plants, and as endophytes in field crops; for example, in maize and sugar in Mexico and Brazil where they play a putative role in the fixation of nitrogen (Perin et al. 2006). These aspects and their function as biocontrol agents are discussed in Part 3 of this chapter.

2.5.6.3 Rhizobia

Rhizobia can form nodules on a range of crop and forage legumes, including *Lotus*, *Lupinus*, *Mimosa*, and *Leucaena*. They are able to fix nitrogen within the nodules, and have been divided into the fast-growing species of *Rhizobium* and the slow-growing *Bradyrhizobium*. The genera *Sinorhizobium* and *Azorhizobium* have been added. *Azorhizobium caulinodans* that can form stem and root nodules on the tropical legume *Sesbania rostrata* is distinguished by the ability of free-living strains to fix nitrogen (Dreyfus et al. 1988; Rinaudo et al. 1991), and it has been shown that root colonization in wheat (*Triticum aestivum*) was stimulated by the flavone naringenin (Webster et al. 1998).

Taxa belonging to both the genera *Rhizobium* and *Bradyrhizobium* are also capable of degrading simple aromatic compounds, including benzoate (Chen et al. 1984) and 4-hydroxybenzoate (Parke and Ornston 1986; Parke et al. 1991). It has been shown that 4-hydroxybenzoate hydroxylase is required for the transport of 4-hydroxybenzoate into the cell (Wong et al. 1994). In strains of *Rhizobium trifolium*, the metabolism of benzoate involves 3,4-dihydroxybenzoate (protocatechuate) 3,4-dioxygenase (Chen et al. 1984), or catechol 1,2-dioxygenase (Chen et al. 1985). The degradation of the toxic *N*-alkylpyrid-4-one mimosine by *Rhizobium* sp. strain TAL1145 has been described (Awaya et al. 2005). Rhizobia have quite a broad degradative capability that is illustrated by the following examples:

1. PCB congeners (Damaj and Ahmad 1996)
2. Flavones (Rao et al. 1991; Rao and Cooper 1994)
3. Dechlorination—though not the degradation—of atrazine (Bouquard et al. 1997)
4. Phosphonomycin (1,2-epoxypropylphosphonate) as a source of carbon, energy, and phosphorus (McGrath et al. 1998)

2.5.6.4 Sphingomonads

The description of the genus *Sphingomonas* has been revised and separated into four genera: *Sphingomonas*, *Sphingobium*, *Novosphingobium*, and *Sphingopyxis* (Takeuchi et al. 2001). Several species of *Sphingomonas* have been isolated from plants and plant roots, and include *Sph. azotofigens* from rice roots that is able to fix nitrogen (Xie and Yokota 2006), and others from the roots of apple, peach, and rose (Takeuchi et al. 1995). Many organisms with degradative capability have been assigned to species in the genus *Sphingomonas sensu lato*, and it has emerged that the organization of the genes for the degradation of aromatic substrates bears only a distant relation to those in pseudomonads, both in sequence homology and in gene organization (Pinyakong et al. 2003). Their metabolic activities are summarized in Table 2.8 and in Table 2.9 for species of *Novosphingobium*. It should be emphasized that the division between *Sphigomonas*, *Sphingobium*, and *Novosphingobium* that is used here is not rigorous, and taxonomic revisions are currently underway. A number of additional features are worth noting.

1. *Sphingobium* (*Sphingomomas*) *yanoikuyae* (Khan et al. 1996) that was originally assigned to *Beijerinckia* sp. strain B1 (Gibson 1999) was able to carry out dioxygenation and degradation of biphenyl, naphthalene, and phenanthrene (Zylstra and Kim 1997), and dioxygenation of benz[*a*]anthracene and benzo[*a*] pyrene (Jerina et al. 1984).
2. Cells of *Sphingomonas paucimobilis* strain EPA 505 grown at the expense of fluoranthene were able to degrade pyrene > benz[*a*]anthracene > benzo[*a*]pyrene ≈ chrysene > dibenz[*a,h*]anthracene (Ye et al. 1996).
3. The unusual degradation of chrysene by an undetermined species of *Sphingomonas* (Willison 2004; Demanèche et al. 2004), while *Sphingomonas* sp. strain CHY-1 that was isolated by enrichment with chrysene was capable of producing two ring-hydroxylating dioxygenases (Demanèche et al. 2004).
4. It has been shown in *Sphingobium yanoikuyae* strain B8/36 that the initially formed dihydrodiols may undergo further dioxygenation to *bis-cis*-dihydrodiols (*cis*-tetraols). Chrysene formed

TABLE 2.8

Degradation/Transformation of Substrates by Selected Species of Sphingomonads

Species	Substrate(s)	Reference
chlorophenolicus	Pentachlorophenol	Cai and Xun (2002)
herbicidovorans	Dichlorophenoxypropionate	Nickel et al. (1997); Müller et al. (2006)
indicum/japonicum/francense	Hexachlorocyclohexane isomers	Pal et al. (2005); Sharma et al. (2006)
paucimobilis	PAHs	Ye et al. (1996)
subarctica	Tetrachlorophenol, trichlorophenol	Nohynek et al. (1996)
terrae	Polyethylene glycol	Sugimoto et al. (2001)
wittichii	Dibenzo[1,4]dioxin	Yabuuchi et al. (2001)
	1,2,3-Trichloro- and 1,2,3,4,7,8-tetrachlorodibenzo[1,4]dioxin	Nam et al. (2006)
xenophaga	Naphthalene-2-sulfonate	Keck et al. (2002)
	Nonylphenol isomers	Gabriel et al. (2005)
yanoikuyae	Naphthalene, biphenyl, anthracene, phenanthrene	Jerina et al. (1976)
	Benz[*a*]anthracene	Jerina et al. (1984)

TABLE 2.9

Degradation/Transformation of Substrates by Species of *Novosphingobium*

Species	Substrate(s)	Reference
aromaticivorans	PCBs	Romine et al. (1999)
	Naphthalene, biphenyl	Balkwill et al. (1997)
naphthalenivorans	Naphthalene	Suzuki and Hiraishi (2007)
pentaromaticivorans	PAHs	Sohn et al. (2004)
ummariense	α-, β-, γ-, δ-hexachlorocyclohexane	Singh and Lal (2009)
yaihuense	Phenol, aniline, nitrobenzene	Liu et al. (2005)
cloacae	Nonylphenol	Fiji et al. (2001)
fuliginis	Phenanthrene	Prakash and Lal (2006)
	4-*tert*-Butylphenol	Toyama et al. (2010)

    3,4,8,9-*bis-cis*-dihydrodiol, and *bis*-dihydroxylation was demonstrated at the analogous positions for benzo[*c*]phenanthridine and benzo[*b*]naphtho[2,1-*d*]thiophene (Boyd et al. 1999).

  5. *Novosphingobium pentaromaticivorans* displays an unusually wide spectrum of PAHs that are degraded, including tetracyclic and pentacyclic representatives (Sohn et al. 2004).

2.5.6.5 *Gram-Negative Facultatively Anaerobic Bacteria*

2.5.6.5.1 *Enterobacteriaceae*

Although facultatively anaerobic bacteria and especially those belonging to the family Enterobacteriaceae have a long history as the agents of disease in man, there is increasing evidence for their importance in a wide range of environmental samples. Methods for their identification and classification have been extensively developed and have traditionally used the ability to ferment a wide range of carbohydrates as taxonomic characters. This has had the possibly unfortunate effect of deflecting attention from the capability of these organisms to degrade other classes of substrates. Their ability to utilize, for example, 3- and 4-hydroxybenzoates (Véron and Le Minor 1975), and nicotinate (Grimont et al. 1977) under aerobic conditions has, however, been quite extensively used for taxonomic classification. Some examples used to illustrate the metabolic capabilities of these somewhat neglected organisms that include species of the genera *Citrobacter*, *Enterobacter*, *Hafnia*, *Klebsiella*, *Salmonella*, and *Serratia*, as well as more recently described genera, are given:

1. The degradation of DDT by organisms designated *Aerobacter aerogenes* (possibly *Klebsiella aerogenes*) (Wedemeyer 1967) (Figure 2.6), and the partial reductive dechlorination of methoxychlor by *K. pneumoniae* (Baarschers et al. 1982).

2. The biotransformation of methyl phenyl phosphonate to benzene by *K. pneumoniae* (Cook et al. 1979) (Figure 2.7a). Further examples of the cleavage of the carbon–phosphorus bond by other members of the Enterobacteriaceae, and a discussion of the metabolism of phosphonates are given in Chapter 11.

3. The biotransformation of 2,4,6-trihydroxy-1,3,5-triazine and atrazine under anaerobic conditions by an unidentified facultative anaerobe (Jessee et al. 1983).

4. The biotransformation of γ-hexachloro[*aaaeee*]cyclohexane to tetrachlorocyclohexene by *Citrobacter freundii* (Figure 2.7b) (Jagnow et al. 1977).

5. The biotransformation of 3,5-dibromo-4-hydroxybenzonitrile to the corresponding acid by a strain of *K. pneumoniae* ssp. *ozaenae* which uses the substrate as sole source of nitrogen (Figure 2.7c) (McBride et al. 1986).

6. The decarboxylation of 4-hydroxycinnamic acid to 4-hydroxystyrene, and of ferulic acid (3-methoxy-4-hydroxycinnamic acid) to 4-vinylguaiacol by several strains of *Hafnia alvei* and *H. protea*, and by single strains of *Enterobacter cloacae* and *K. aerogenes* (Figure 2.7d) (Lindsay and Priest 1975). The decarboxylase has been purified from *Bacillus pumilis* (Degrassi et al. 1995).

7. Several taxa of Enterobacteriaceae, including *Morganella morganii*, *Proteus vulgaris*, and *Raoultella* (*Klebsiella*) *planticola*, are able to decarboxylate the amino acid histidine, which is abundant in the muscle tissue of scombroid fish (Yoshinaga and Frank 1982; Takahashi et al. 2003). The histamine produced has been associated with an incident of scombroid fish poisoning (Taylor et al. 1989).

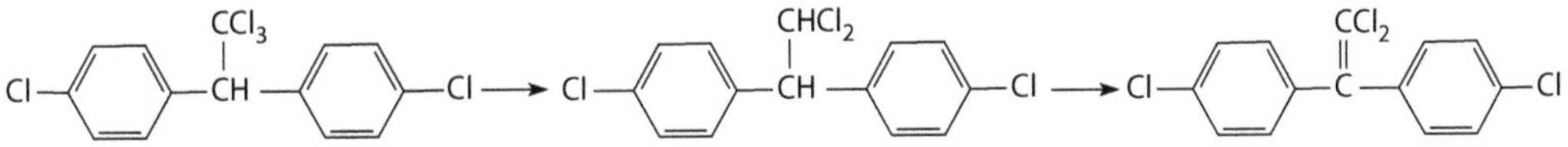

FIGURE 2.6 Degradation of DDT by *Aerobacter aerogenes*.

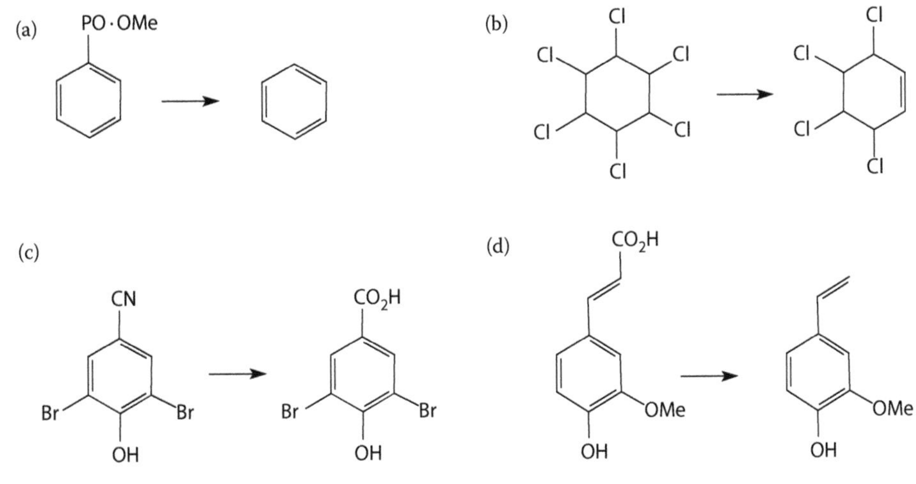

FIGURE 2.7 Transformation of (a) methylphenyl phosphonate, (b) γ-hexachlorocyclohexane, (c) 3,5-dibromo-4-hydroxybenzonitrile, and (d) decarboxylation.

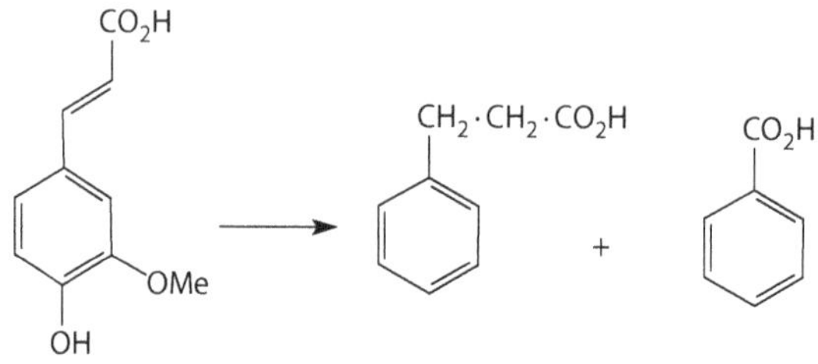

FIGURE 2.8 Metabolism of ferulic acid by *Enterobacter cloacae.*

8. The metabolism of ferulic acid (3-methoxy-4-hydroxycinnamic acid) by *Ent. cloacae* to a number of products, including phenylpropionate and benzoate (Figure 2.8) (Grbic-Galic 1986).

9. Utilization of uric acid as a nitrogen source by strains of *Aer. aerogenes, K. pneumoniae,* and *Serratia kiliensis* (Rouf and Lomprey 1968).

10. The sequential reduction of one of the nitro groups of 2,6-dinitrotoluene by *Salmonella typhimurium* (Sayama et al. 1992)—a taxon that is not generally noted for its metabolic activity.

11. Hexahydro-1,3,5-trinitro-1,3,5-triazine (RDX) is degraded by *Klebsiella pneumoniae* (Zhao et al. 2002) and, under conditions of oxygen limitation, by strains of *Morganella morganii* and *Providencia rettgeri* after initial reduction to the nitroso compounds (Kitts et al. 1994).

12. After growth in a medium containing suitable reductants such as glucose, a strain of *Enterobacter agglomerans* was able to reduce tetrachloroethene successively to trichloroethene and *cis*-1,2-dichloroethene (Sharma and McCarty 1996).

13. Degradation of pentaerythritol tetranitrate by *Ent. cloacae* (French et al. 1996) and of glyceryltrinitrate by *K. oxytoca* (Marshall & White 2001).

14. Degradation by *Enterobacter* sp. of chlorpyrifos with the formation of 3,5,6-pyridin-2-one and diethylthiophosphate that was used for growth and energy (Singh et al. 2004).

15. Degradation of naphthalene, fluoranthene, and pyrene by *Leclercia adecarboxylata* that hitherto had a clinical origin (Sarma et al. 2004), and of pyrene by an endophytic strain of *Enterobacter* sp. 12J1 (Sheng et al. 2008).

16. The taxonomic application of the ability of enteric organisms to grow with 4-hydroxyphenylacetate (Cooper and Skinner 1980) and 3-hydroxyphenylpropionic acid (Burlingame and Chapman 1983) has been established. In addition, it has been demonstrated that the enzyme that carries out the hydroxylation has a wide substrate range extending to 4-methylphenol, and even to 4-chlorophenol (Prieto et al. 1993).

2.5.6.5.2 Azoarcus and Thauera

A taxonomic examination of bacteria that have been classified as species of *Azoarcus* has been carried out. Many of them were isolated from roots of *Leptochloea fusca* that is an alkali- and salt-tolerant grass in the Punjab, or from the fungal sclerotium of *Oryza* sp. These plant-derived strains have been assigned to the genera *Azoarcus, Azovibrio, Azospira,* and *Azonexus,* belonging to the β-subclass of *Proteobacteria,* and were generally able to fix atmospheric nitrogen (Reinhold-Hurek and Hurek 2000).

1. Attention has been given to denitrifying organisms that could degrade toluene and that have been referred to the genera *Thauera* and *Azoarcus* (Anders et al. 1995). Particular interest has centered on strains of *Azoarcus evansii* (Anders et al. 1995) and *Azoarcus tolulyticus* both of which are able to degrade benzoate, 3-hydroxybenzoate and phenylalanine, whereas only the latter is able to degrade toluene under denitrifying or aerobic conditions (Zhou et al. 1995). This stimulated a comprehensive investigation of the toluene-degrading strains using a wide range of taxonomic characters (Song et al. 1999). On the basis of DNA–DNA hybridization, the strains could be divided into five groups: members of Groups 1 and 2 were assigned to *Azoarcus tolulyticus,* while those of Group 4 were assigned to the new species *Azoarcus toluclasticus* and

Group 5 to the new species *Azoarcus toluvorans*. All strains of Groups 1 and 2, and Group 5 were able to degrade toluene under both aerobic and denitrifying conditions while growth with phenylalanine was widely, though not uniformly, distributed.

2. Examination of denitrifying strains that could degrade fluoro-, chloro-, or bromobenzoates was carried out using a polyphasic approach. On this basis the assignment of strains to the new species *Thauera chlorobenzoica* was justified (Song et al. 2001).

3. *Azoarcus anaerobius* is a strictly anaerobic denitrifying organism that could carry out the degradation of resorcinol (1,3-dihydroxybenzene)—though not catechol or hydroquinone—and 16S rRNA sequence analysis showed it to be highly similar to *Azoarcus evansii* and *Azoarcus tolulyticus* (Springer et al. 1998). Simple alkanoates, cyclohexane carboxylate, and a range of aromatic substrates, including phenol, resorcinol, 3- and 4-hydroxybenzoate, phenylacetate, phenylalanine, and tyrosine, could be used as electron acceptors. The pathway used for anaerobic degradation of 1,3-dihydroxybenzene is completely different from that used for 1,2-dihydroxybenzene (catechol) and 1,4-dihydroxybenzene (hydroquinone) by other organisms since it is not initiated by carboxylation but by membrane-associated enzymes with the production of 2-hydroxybenzoquinone that undergoes ring fission to malate and acetate (Darley et al. 2007).

2.5.6.5.3 Vibrionaceae

Only a few examples have emerged to demonstrate the degradative capability of these organisms, and include the following:

1. An organism, tentatively identified as a strain of *Vibrio* sp., was able to degrade the 2-carboxylates of furan, pyrrole, and thiophene (Evans and Venables 1990).

2. A strain of *Aeromonas* sp. was able to degrade phenanthrene through *o*-phthalate (Kiyohara et al. 1976).

3. Bacteria isolated by growth with phenanthrene from a marine environment were subjected to numerical taxonomy, and the vibrios were assigned mostly to *Vibrio fluvialis* and only a few to *Vibrio parahaemolyticus* (West et al. 1984).

2.6 Microbial Metabolism of C_1 Compounds

2.6.1 Metabolism of C_1 Substrates

2.6.1.1 Organisms

Several groups of aerobic microorganisms are able to grow with C_1 substrates such as methanol and methylamine by mechanisms that produce formaldehyde from both of them, followed by the oxidation (glutathione-dependent or glutathione-independent) to formate and CO_2. For anaerobic methanogens, several C_1 substrates can serve as substrates in addition to CO_2. Methanotrophs are discussed in Chapter 3 in the context of monooxygenation and in Chapter 7, while further comments on the role of methylotrophs in the context of plant–microbe interactions are given in Part 3 of this chapter.

2.6.1.1.1 Bacteria

1. The range of aerobic methylotrophs is quite extensive, and includes the following genera of bacteria: *Methylococcus, Methylomonas, Methylobacter, Methylosinus, Methylocystis* and *Methylobacterium, Methylibium*, and *Methylophaga*, some of which can also utilize methane (references in Anthony 1982). In addition, methanotrophs belonging to the genus *Methylocella* are able to utilize a range of higher alkanoates, in addition to methane and methanol (Dedysh et al. 2005).

   a. Considerable effort has been given to the dehydrogenation of methanol and methylamine in *Methylobacterium extorquens*. Methanol dehydrogenase is probably the most thoroughly investigated quinoprotein, and pyrroloquinoline quinone (PQQ) and Ca^{2+} are used in *Methylobacterium extorquens* (Ramamoorthi and Lidstrom 1995). Dehydrogenation of methylamine involves, however, a different quinoprotein: tryptophan tryptophylquinone (TTQ) as the cofactor that has been demonstrated for methylamine in *Methylobacterium extorquens* (McIntire et al. 1991).

b. Organisms belonging to the genus *Methylocella* have been isolated from acidic environments such as acidic peat (*Methylocella palustris*), forest soil (*Methylocella silvestris*) and tundra soils (*Methylocella tundrae*) (references in Dedysh et al. 2005). They are unusual for methanotrophs since they can utilize not only the C_1 compounds (methane, methanol, and methylamine), but in addition C_2 compounds, including acetate, pyruvate, and succinate (Dedysh et al. 2005). They appear to possess only the soluble form of methane monooxygenase, and lack the intracytoplasmic membrane system to which particulate methane monooxygenase is generally bound.

c. The range of metabolic activities of *Methylibium petroleiphilum* is extensive and includes degradation of methyl *tert*-butyl ether, C_5 to C_{12} *n*-alkanes, and the monocyclic arenes benzene and toluene (Nakatsu et al. 2006; Kane et al. 2007).

d. Strains of *Methyloversatilis universalis* were obtained after enrichment either with methanol using contaminated soil (EHg5) or from lake water supplemented with formaldehyde (FAM5). They were metabolically versatile, and were able to grow at the expense of methanol, all methylamines, formate, glycerol, succinate, glucose, alanine, and toluene (Kaluzhnaya et al. 2006b).

e. Low GC-ratios species of *Methylophaga* belong to the γ-proteobacteria, are moderately halophilic and characterized by the use of the ribulose monophosphate pathway for the assimilation of C_1 substrates. *M. aminisulfidivorans* (Kim et al. 2007) is able to utilize dimethylsulfide, methanol and all the methylamines, and dimethylsulfide by *M. sulfidovorans* (de Zwart and Kuenen 1997). It has been shown that *M. thiooxidans* is able to degrade dimethylsulfide by demethylation to sulfide that was oxidized through thiosulfate to tetrathionate as the end product. Exogenous thiosulfate could be used during growth with dimethylsulfide or methanol when its oxidation was coupled to the generation of ATP (Boden et al. 2010).

f. *Methylobacterium populi* was isolated from tissue cultures of poplar (*Populus deltoides × nigra* DN34) (van Aken et al. 2004) and was, unusually in this genus, able to use methane as growth substrate. It was also able to utilize methylamine, while *M. aminovorans* can use all the methylamines as growth substrates, and *Methylotenera mobilis* is unusual among methylotrophs in its obligate growth at the expense of methylamine (Kalyuzhnaya et al. 2006a).

g. *Methylomirabilis oxyfera* is able to oxidize methane under anaerobic conditions (Ettwig et al. 2010). A particulate monooxygenase produced methanol that was degraded by dehydrogenation and tetrahydrofolate transfer reactions, while nitrate was successively reduced to nitrite and nitric oxide. In a novel reaction this underwent dismutation to N_2 and O_2 that was primarily used for methane oxidation by reactions used for aerobic bacteria, and involved a membrane-bound *bo*-type terminal oxidase (Wu et al. 2011). The dismutation may be considered formally analogous to the formation of oxygen and chloride from chlorite by chlorite dismutase (van Ginkel et al. 1996; Stenklo et al. 2001).

h. Exceptionally, a few bacteria are able to grow at the expense of halomethanes: (i) *Hyphomicrobium chloromethanicum* and *Methylobacterium chloromethanicum* with chloromethane, methanol, and methylamine (McDonald et al. 2001), (ii) the marine *Leisingera methylohalidivorans* with an extremely limited range of substrates was able to grow with chloromethane, bromomethane, methionine, and glycine betaine, though not with methylated amines, methanol, or formate (Schaefer et al. 2002). Other strains of roseobacters were, however, unable to utilize halomethanes (Martens et al. 2006).

2. The metabolic capability of the facultative methylotrophs *Paracoccus denitrificans* and *Pa. versutus* (formerly *Thiobacillus venustus*) (Katayama et al. 1995) enables them to grow at the expense of both methanol and methylamine. There are also a number of bacteria with more extended metabolic capacity that have been assigned to the genera *Pseudomonas*, *Bacillus*, and *Achromobacter*.

3. Verrucomicrobia

   Methane oxidation has been described for several organisms of the *Verrucomicrobia* phylum all of which were moderate extremophiles that are discussed in the following paragraphs:

   a. An organism (V4) that grew optimally at pH 2.0 to 2.5 was isolated from a thermal area in New Zealand (Dunfield et al. 2007). It grew robustly only on C_1 substrates (methane and methanol), contained the genes required for the operation of the Benson–Calvin cycle and for those encoding particulate methane monooxygenase that belonged to a new group of these enzymes.

   b. An organism (SolV) was isolated at pH 2.0 and 50°C from the Solfatara volcano mudpot (Pol et al. 2007), and the organism was assigned to *Methylacidiphilum* (*Acidimethylosilex fumariolicum*). Particulate methane monooxygenase activity was measured and sequencing of the DNA suggested that the organism used a combination of the serine, tetrahydrofolate, and ribulose-1,5-bisphosphate pathways for assimilation of carbon.

   c. An extremophile that carried out oxidation at 55°C and pH 2 lacked intracytoplasmic membrane, and most remarkably, lacked genes for methane monooxygenase (Islam et al. 2008).

4. Unique among methanotrophs is a filamentous organism that was described many years ago as *Crenothrix* and is additionally unusual in the methane-oxidizing system that is used (Stoecker et al. 2006).

2.6.1.1.2 Archaea

The structure and diverse functions of archaea have been described in a review (Jarrell et al. 2011). Archaea display a number of important metabolic features that include the following, and additional examples are given in the paragraphs dealing with thermophiles.

1. Archaea are the classical methanogens and, in addition, a number of them—*Methanosarcina barkeri* in particular—are able to utilize C_1 substrates, including methanol (Sauer et al. 1997), di- and trimethylamines (Paul et al. 2000), and dimethyl sulfide (Tallant et al. 2001), while *Methanomethylovorans hollandica* is able to utilize an unusually wide spectrum of substrates: methanol, methylamines, methanethiol, and dimethyl sulfide (Lomans et al. 1999).

2. Extremophilic archaea are involved in the metabolism of elementary sulfur: anaerobically by *Pyrococcus furiosus* (Blumentals et al. 1990), and *Sulfolobus brierleyi* (Emmel et al. 1986), and by *Acidianus infernus* that can carry this out lithotrophically under both aerobic conditions and anaerobically using H_2 (Segerer et al. 1986). In the dicarboxylate/4-hydroxybutyrate cycle that functions in *Thermoproteus neutrophilus* and *Ignicoccus hospitalis*, growth as strict anaerobes can be accomplished by reducing elementary sulfur with H_2 (Jahn et al. 2007).

3. Groups of anaerobes that have not been isolated hitherto are able to carry out the important anaerobic oxidation of methane, and belong to three clades of hitherto uncultured organisms, ANME1, ANME-2, and ANME-3, possibly by reversed methanogenesis (Hallam et al. 2004) and are discussed in Chapter 7.

4. Ammonia-oxidizing archaea are ubiquitous in both the marine (Wuchter et al. 2006) and the terrestrial (Leininger et al. 2006) environments. High-molecular-mass DNA was extracted from a marine archaeon *Nitrosopumilis maritimus* and the genome was sequenced. This revealed that the organism was able to carry out chemolithotrophic oxidation of ammonia to nitrite, but possessed only limited potential for the assimilation of organic carbon. Carbon dioxide assimilation was carried out by a modification of the 3-hydroxypropionate/4-hydroxybutyrate cycle, and the system for ammonia oxidation and the electron transport chain were dependent on copper, but were distinct from those used for bacterial ammonia oxidation. Genome sequences for the synthesis of the osmolyte ectoine hydroxylase were present (Walker et al. 2010).

5. Nitrogen fixation has been observed in a methanogenic archaeon from hydrothermal vent fluid. It was similar to *Methanocaldococcus jannaschii*, incorporated $^{15}N_2$, and expressed *nifH* mRNA (Mehta and Baross 2006).

6. Autotrophic carbon assimilation has been described in several *Crenarchaeota*. The 3-hydroxypropionate/4-hydroxybutyrate cycle is used by the anaerobic *Stygiolobus azoricus* (Berg et al. 2010) or microaerophilic *Sulfolobales* such as *Metallosphaera sedula* (Estelmann et al. 2011), whereas the alternative dicarboxylate/4-hydroxybutyrate cycle is used by the aerobic *Pyrolobus fumarii* (Berg et al. 2010). The distinction seems, however, to be correlated with the 16S rRNA-based phylogeny rather than the aerobic or anaerobic conditions of growth.

7. A range of archaea have been isolated from a bioleaching reactor and included phylotypes belonging to the *Sulfolabales* related to *Stygiolobus azoricus*, *Metallosphaera* sp., and *Acidianus infernus* (Mikkelsen et al. 2009).

8. Fe(III) reductases have been described from *Pyrobaculum islandicum* (Childers and Lovley 2001), and from the hyperthermophilic sulfate-reducing *Archaeoglobus fulgidus* (Vadas et al. 1999).

2.6.1.1.3 Yeasts

Several yeasts are able to degrade long-chain *n*-alkanes. This is carried out in two subcellular organelles: in microsomes, cytochrome P450 and the associated NADH reductase (Käppeli 1986) carry out hydroxylation, while dehydrogenation of the alkanols and β-oxidation take place in peroxisomes that are induced during growth with alkanes (Tanaka and Ueda 1993).

Further details of the metabolism of *n*-alkanes and alkylamines are given in Chapter 7.

Clearly then, yeasts possess metabolic potential for the degradation of xenobiotics that is little inferior to that of many bacterial groups, so that their role in natural ecosystems justifies the greater attention that has been directed to them (MacGillivray and Shiaris 1993). The metabolic potential of yeasts has attracted attention in different contexts, and it has emerged that, in contrast to many fungi, they are able to bring about fission of aromatic rings. Some illustrative examples include the following:

1. Ring fission clearly occurs during the metabolism of phenol (Walker 1973) by the yeast *Rhodotorula glutinis*, and of aromatic acids by various fungi (Cain et al. 1968; Durham et al. 1984; Gupta et al. 1986).

2. Analogous ring fission reactions have also been found in studies on the metabolism of other aromatic compounds by the yeast *Trichosporon cutaneum* whose metabolic versatility is indeed comparable with that of bacteria. Examples include the degradation of phenol and resorcinol (Gaal and Neujahr 1979), tryptophan and anthranilate (Anderson and Dagley 1981a), and aromatic acids (Anderson and Dagley 1981b).

3. *Trichosporon cutaneum* is able to grow at the expense of 4-methylphenol and degradation is initiated by hydroxylation to 4-methylcatechol. This undergoes intradiol fission to *cis,cis*-muconic acid followed by lactonization to 3-methylmuconolactone and rearrangement to 3-methyl-but-3-en-1,4-olide before hydrolysis to 4-methyl-3-oxoadipate and further metabolism to acetate and pyruvate (Powlowski and Dagley 1985; Powlowski et al. 1985).

4. The phenol-assimilating yeast *Candida maltosa* degraded a number of phenols, even though these were unable to support growth. Hydroxylation of 3-chloro- and 4-chlorophenol initially produced 4-chlorocatechol and then 5-chloropyrogallol (Polnisch et al. 1992). The yeast *Rhodotorula cresolica* was able to assimilate phenol, 3- and 4-methylphenol, catechol and 3- and 4-methylcatechol, resorcinol and hydroquinone, and a wide range of phenolic carboxylic acids (Middelhoven and Spaaij 1997).

5. The ability to grow at the expense of 4-hydroxy- and 3,4-dihydroxybenzoate has been used for the classification of medically important yeasts, including *Candida parapsilosis* (Cooper and Land 1979). This organism degrades these substrates by oxidative decarboxylation, catalyzed by a flavoprotein monooxygenase (Eppink et al. 1997).

6. Examples of hydroxylation followed by ring fission include the following:
   a. Transformation of diphenyl ether by *Trichosporum beigelii* is initiated by successive hydroxylation and extradiol ring fission of the resulting catechol (Schauer et al. 1995). In a formally analogous way, the metabolism of dibenzofuran by the yeast *Trichosporon*

mucoides involves initial hydroxylation of one of the rings followed by ring fission at the 2,3-position (Hammer et al. 1998).

b. Biphenyl is hydroxylated by *Trichosporon mucoides* to 2- and 4-hydroxy-, 2,3- and 3,4-dihydroxy-, and 2,3,4- and 3,4,5-trihydroxybiphenyl, and ring fission of 3,4-dihydroxybiphenyl and 3,4,5-trihydroxybiphenyl then occurred (Sietmann et al. 2001).

7. *Trichosporon adeninovorans* is able to grow at the expense of adenine, xanthine, and uric acid as well as primary amines as source of both carbon and nitrogen (Middelhoven et al. 1984). The presence of urate oxidase and allantoinase activities after growth on adenine and urate suggests the operation of conventional metabolism that is outlined in Chapter 10.

2.6.1.2 CO_2 Assimilation

It is appropriate to address briefly the mechanisms whereby CO_2 is assimilated since this is the simplest C_1 substrate and since this is accomplished by a range of both archaea and bacteria that include lithotrophs, phototrophs, and heterotrophs. Carbonic anhydrase plays a role in the physiology of bacteria of clinical importance, including *Streptococcus pneumoniae*, to minimize loss of internal CO_2 since HCO_3^- is unable to diffuse across membranes (Burghout et al. 2010), and *Helicobacter pylori* where the periplasmic α-carbonic anhydrase is essential for acid acclimation (Marcus et al. 2005). The metabolism of carbon monoxide, and methylamine and related compounds, is discussed in Chapter 7, Part 1, and of methanethiols and related compounds in Chapter 11, Part 2.

Although greatest prominence has been given to the Benson–Calvin (ribulose-1,5-bisphosphate) pathway that is used by plants, algae, and cyanobacteria for the assimilation of CO_2, there is a plethora of alternative pathways in archaea (Hügler et al. 2003a). These are briefly summarized, followed by a discussion in greater detail:

1. Ribulose-1,5-bisphosphate cycle used by phototrophs and lithotrophs
2. Reductive citric acid cycle used primarily by anaerobes
3. 3-Hydroxypropionate cycle by phototrophic anaerobes
4. Assimilation during heterotrophic growth
5. Wood pathway for acetogenesis
6. Reductive pathway used for methanogenesis

Ribulose-1,5-bisphosphate cycle: CO_2 is required for growth by phototrophs and lithotrophs, and is assimilated by the classic Benson–Calvin pathway using ribulose-1,5-bisphosphate carboxylase/oxygenase in algae and cyanobacteria. It is also used by lithotrophs, for example, *Pseudomonas oxaliticus* and *Ralstonia eutropha* during growth with formate.

CO_2 may also be produced from the substrate by the dehydrogenation of formate, carbon monoxide, or oxalate. In *Ralstonia eutropha*, these reactions are

$$\text{Formate} \rightarrow CO_2 + H_2$$
$$\text{Carbon monoxide: } CO + H_2O \rightarrow CO_2 + 2H^+ + 2\varepsilon$$
$$\text{Oxalate} \rightarrow H \cdot CO_2H + CO_2 \rightarrow 2CO_2 + H_2$$

followed by CO_2 assimilation by the ribulose-1,5-bisphosphate cycle. In *Pseudomonas oxaliticus*, oxalate is metabolized by the combined activity of oxalate decarboxylase and formate dehydrogenase: $H \cdot CO_2 \cdot CO_2H \rightarrow 2CO_2 + 2H^+ + 2e$. In this organism, however, CO_2 is assimilated not by the ribulose-1,5-bisphosphate cycle used for formate metabolism in this strain, but by dehydrogenation to glyoxylate, decarboxylative dimerization mediated by glyoxylate carboligase, and then entry into the phosphoglycerate cycle (Friedrich et al. 1979).

Reductive citric acid cycle: This involves the operation of the TCA cycle functioning in a reductive mode. There are two successive carboxylations—ferredoxin-mediated carboxylation of succinyl-CoA to 2-oxoglutarate by 2-oxoglutarate: ferredoxin oxidoreductase, and further carboxylation to isocitrate that is carried out by isocitrate dehydrogenase. Isocitrate is then converted to citrate by aconitase, and the activity of ATP citrate lyase produces oxalacetate that enters the cycle and acetyl-CoA that is carboxylated to pyruvate. This cycle is used by the anaerobic phototroph *Chlorobium limicola* (*thiosulfatophilum*) (Evans et al. 1966; Fuchs et al. 1980); by *Desulfobacter hydrogenophilus* (Schauder et al. 1987) and *Desulfobacter postgatei* (Möller et al. 1987); by the thiosulfate-oxidizing denitrifying autotroph *Thiomicrospira denitrificans* (Hügler et al. 2005); and by the extreme thermophile *Pyrobaculum islandicum* (Hügler et al. 2003a). This pathway is not restricted, however, to anaerobes since it is used for the assimilation of CO_2 by the obligately autotrophic, hydrogen-oxidizing *Hydrogenobacter thermophilus* (Shiba et al. 1985), and the thermophilic *Aquifex pyrophilus* (Beh et al. 1993).

Hydroxypropionate and related pathways: Several other pathways are used for CO_2 assimilation in anaerobes, all of them involving hydroxyacids. The steps in the hydroxypropionate pathway that involve carboxylation of propionyl-CoA to methylmalonyl-CoA and acetyl-CoA to malonyl-CoA are carried out by HCO_3^-.

1. The anaerobic 3-hydroxypropionate pathway has been elucidated for the green nonsulfur anaerobic bacterium *Chloroflexus aurantiacus* in which the ribulose-1,5-bisphosphate-carboxylase cycle does not operate. In this organism, CO_2 assimilation is accomplished by functioning of the 3-hydroxypropionate cycle (Herter et al. 2002). In brief, this operates by carboxylation of acetyl-CoA to malonyl-CoA, followed by sequential reduction by a multifunctional enzyme to 3-hydroxypropionate and propionate whose CoA ester undergoes further carboxylation to methylmalonyl-CoA and fission to glyoxylate. The carboxylation of both acetyl-CoA and propionyl-CoA is carried out by bicarbonate. The metabolism of glyoxylate involves the ethylmalonate cycle (Erb et al. 2007) in place of the classical glyoxylate cycle. Both pathways utilize the novel bifunctional enzyme mesaconyl-CoA hydratase (Zarzycki et al. 2008).

2. In *Metallosphaera sedula*, a variant of the 3-hydroxypropionate cycle is used. In this, the enzymes for the carboxylation of acetyl-CoA and propionyl-CoA have been characterized (Hügler et al. 2003b), and the reduction of 3-hydroxypropionate to propionyl-CoA requires the activity of at least two enzymes in place of the single one in *Chloroflexus aurantiacus* (Alber et al. 2008). Carboxylation of acetyl-CoA and propionyl-CoA is carried out with bicarbonate. This pathway has been identified additionally in the extreme thermophiles *Acidianus ambivalens* and *Sulfolobus* sp. strain VE 6 (Hügler et al. 2003a).

3. *Thermoproteus neutrophilus* is a facultative autotroph, and uses a dicarboxylic acid/4-hydroxybutyrate cycle for CO_2 assimilation in which oxalacetate is produced by carboxylation of phosphoenol pyruvate with bicarbonate (Ramos-Vera et al. 2009). Addition of acetate, pyruvate, succinate, and 4-hydroxybutyrate stimulated growth and were fed into the cycle at the appropriate places (Ramos-Vera et al. 2010).

Assimilation during heterotrophic growth: The initial proof for carboxylation during heterotrophic growth was found from the identification of the products from the fermentation of glycerol by *Propionibacterium pentosaceum*, when the label from CO_2 was found in the carboxyl group of succinate. This is discussed in detail in a classic review (Wood and Stjernholm 1962). This has since been amply verified for a range of degradative reactions in which CO_2 is required for heterotrophic growth. Carboxylation is involved—albeit by quite different mechanisms from the carboxylation by phototrophs and lithotrophs. Illustrative examples include degradation of both aliphatic and aromatic substrates.

1. CO_2 is incorporated during the degradation of a wide range of substrates, including the following:

   Geraniol by *Pseudomonas citronellosis* (Seubert and Remberger 1963; Fall and Hector 1977)

   Phytosterol by *Mycobacterium* sp. (Fujimoto et al. 1982)

   Acetone and epoxypropene by *Xanthobacter* strain Py2 (Sluis et al. 2002)

   Crotonate in the aerobic degradation of leucine and isoleucine by *Pseudomonas aeruginosa* (Massey et al. 1974; Höschle et al. 2005)

   Crotonate is metabolized by the anaerobe *Syntrophomonas acidotrophicus* to produce cyclohexane carboxylate (Mouttaki et al. 2007)

2. The anaerobic degradation of phenol by organisms of different taxonomic affiliation involves phosphorylation followed by carboxylation and dehydroxylation to benzoyl-CoA that undergoes

reduction and ring fission: *Thauera aromatica* (Schühle and Fuchs 2004), *Geobacter metallireducens* GS-15 (Schleinitz et al. 2009), and *Desulfobacterium anilini* (Ahn et al. 2009). An analogous pathway is used for aniline by *Desulfobacterium anilini* (Schnell and Schink 1991), and for catechol by *Thauera aromatica* with the formation of 3-hydroxybenzoate (Ding et al. 2008).

3. The metabolism of ketones may involve ATP-dependent carboxylation. This is used for (a) the aerobic degradation of acetone by *Xanthobacter autotrophicus* (Sluis et al. 1996), (b) the nitrate-dependent degradation of acetone by *Alicycliphilus* (Dullius et al. 2011) and (c) of acetophenone produced by *Aromatoleum aromaticum* from ethylbenzene by hydroxylation and dehydrogenation (Jobst et al. 2010). The carboxylase enzymes are generally similar although they differ in their metal requirement: Mn in acetone carboxylase (Boyd et al. 2004) and Zn in acetophenone carboxylase (Jobst et al. 2010).

Acetogenesis: It is convenient to summarize the range of C_1 substrates that can be utilized by anaerobic bacteria for the synthesis of acetate from glucose or fructose whose metabolism provides CO, CO_2, and formate by pathways given in a review by one of the pioneers in this area (Ljungdahl 1986).

The synthesis of acetate is carried out by a number of anaerobes (Table 2.10), and involves the following reactions in the Wood pathway for autotrophic fixation of CO_2 (Ljungdahl 1986). An outline of the enzymology has been given (Ragsdale 1991):

$$CO_2 \rightarrow H \cdot CO_2H \rightarrow THF\text{--}CHO \rightarrow THF\text{--}CH_3$$

$$THF\text{--}CH_3 + E\text{-}Co \rightarrow E\text{-}Co\text{--}CH_3$$

$$CO_2 + 2H \rightarrow CO + H_2O$$

$$E\text{-}Co\text{--}CH_3 + CO + HS\text{-}CoA \rightarrow CH_3\text{--}CO\text{-}SCoA$$

These involve the cooperation of a group of enzymes: formate dehydrogenase (FDH), tetrahydrofolate enzymes (THF), carbon monoxide dehydrogenase (CODH), carbon monoxide dehydrogenase/carbon monoxide dehydrogenase disulfide reductase (CODH-S_2 reductase), corrinoid proteins (Co^+, Co^{2+}, and Co^{3+}), and corrinoid methyltransferase.

Methanogenesis: This is accomplished by a range of archaea, that may utilize either CO_2/H_2 or a number of substrates, including formate, methanol, and acetate: the pathways are discussed in a review (Deppenmeier et al. 1996). The pathway for CO_2 utilization is, to some extent, analogous to that involved in acetogenesis from CO_2—albeit with some cardinal differences, including the use of formyl-methanofuran dehydrogenase to carry out the initial reaction, and tetrahydromethanopterin (H_4MPt) rather than tetrahydrofolate (H_4THF). In brief, the reactions are

$$CO_2 \rightarrow \text{formyl-methanofuran (formyl-MF)}$$

$$CHO\text{-}MF \rightarrow CHO\text{-}H_4MPt \text{ (formyl-tetrahydromethanopterin)}$$

$$CHO\text{-}H_4MPt \rightarrow CH_3\text{-}H_4MPt$$

$$CH_3\text{-}H_4MPt + HS\text{-}CoM \rightarrow CH_3\text{-}SCoM$$

$$CH_3\text{-}SCoM \rightarrow CH_4$$

TABLE 2.10

Utilization of C_1 Substrates by Anaerobes

Substrate	CO	H · CO$_2$H	CO$_2$/H$_2$	CH$_3$OH/CO$_2$
Acetobacterium woodii	+	+	+	+
Moorella (Clostridium) thermoacetica	+	+	+	+
Clostridium thermoautotrophicum	+	+	+	+
Sporomusa acidovorans	−	+	+	+
Desulfomaculum orientis	+	+	+	+

The metabolism of H_2 is carried out using F_{420}, and the resulting reductant is fed into the tetrahydro-methanopterin pathway. The metabolic activity of methane monooxygenase and related enzymes is explained as follows.

The enzymology of methane monooxygenase is discussed in Chapter 7, Part 1. Methane monooxygenase, cyclohexane monooxygenase, and the monooxygenase of *Nitrosomonas* have broad metabolic versatility. This is illustrated in Figures 2.9 through 2.11, and enzymological details of methane monooxygenase are given in Chapter 3. It is on account of this that methylotrophs have received attention for their technological potential (Lidstrom and Stirling 1990). An equally wide metabolic potential has also been demonstrated for cyclohexane monooxygenase, which has been shown to accomplish two broad types of reaction: one in which formally nucleophilic oxygen reacts with the substrate, and the other in which formally electrophilic oxygen is involved (Figure 2.10) (Branchaud and Walsh 1985).

In addition, it has emerged that the monooxygenase system in methanotrophs is similar to that in the nitrite-oxidizing bacteria, and that the spectrum of biotransformations is equally wide. There are also important groups of facultatively C_1-utilizing bacteria: (a) members of the genus *Xanthobacter* (Padden et al. 1997), and (b) *Methylocella silvestris* that was isolated from tundra soils and had the unusual capacity to use not only methane and methanol as energy sources, but also C_2 and C_3 substrates such as acetate and malate (Dedysh et al. 2005).

The following illustrate some of the biotransformations that have been observed with *Nitrosomonas europaea*. These are particularly interesting since this organism has an obligate dependency on CO_2 as carbon source, and has traditionally been considered to be extremely limited in its ability to use organic carbon for growth or transformation:

1. The oxidation of benzene to phenol and 1,4-dihydroxybenzene (Figure 2.11a) (Hyman et al. 1985), both side chain and ring oxidation of ethyl benzene, and ring hydroxylation of haloge-nated benzenes and nitrobenzene (Keener and Arp 1994).

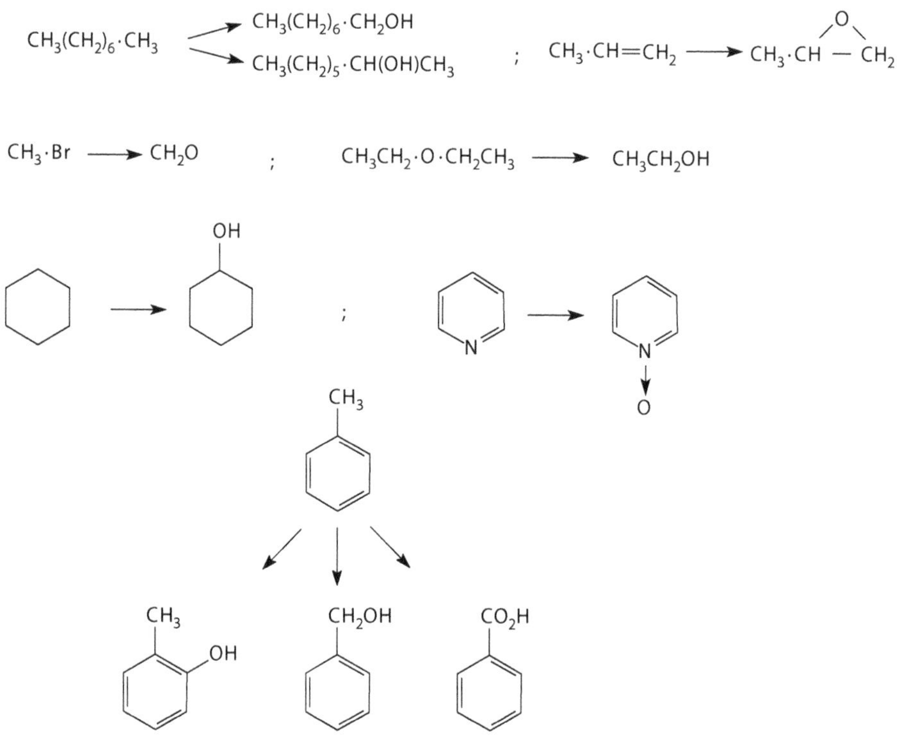

FIGURE 2.9   Reactions mediated by the monooxygenase system of methanotrophic bacteria.

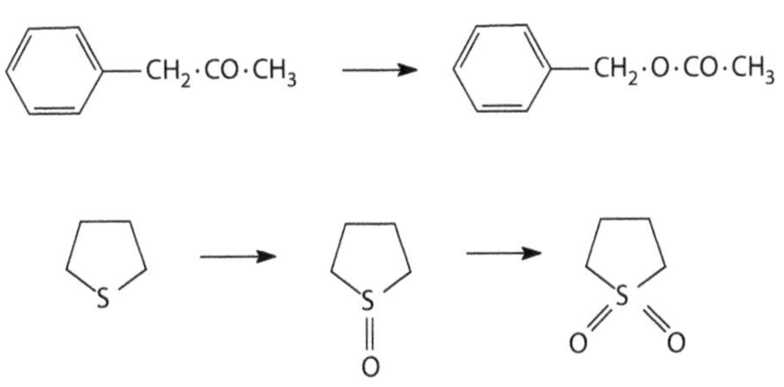

FIGURE 2.10 Reactions mediated by cyclohexane monooxygenase.

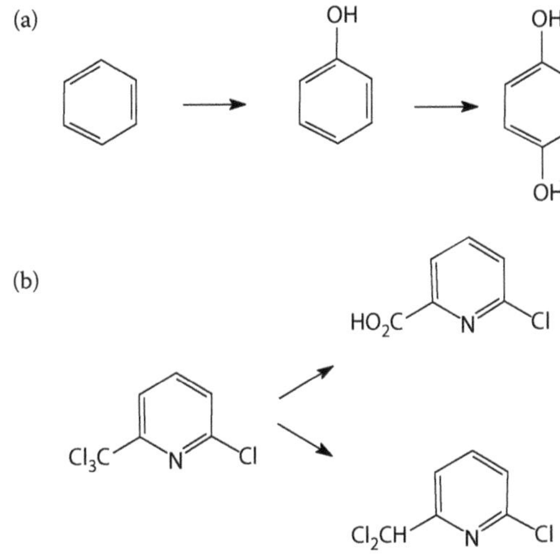

FIGURE 2.11 Metabolism by *Nitrosomonas europaea*. (a) Benzene and (b) 2-chloro-6-trichloromethylpyridine.

2. The oxidation of alkanes (C_1–C_8) to alkanols, and alkenes (C_2–C_5) to epoxides (Hyman et al. 1988).

3. The oxidation of methyl fluoride to formaldehyde (Hyman et al. 1994), and of chloroalkanes at carbon atoms substituted with a single chlorine atom to the corresponding aldehyde (Rasche et al. 1991).

4. The oxidation of a number of chloroalkanes and chloroalkenes, including dichloromethane, chloroform, 1,1,2-trichloroethane, and 1,2,2-trichloroethene (Vannelli et al. 1990). Although the rate of cometabolism of trihalomethanes increased with levels of bromine substitution, so did toxicity. Both factors must, therefore, be evaluated in the possible application of this strain (Wahman et al. 2005).

5. The oxidation of the trichloromethyl group in 2-chloro-6-trichloromethylpyridine to the corresponding carboxylic acid occurs at high oxygen concentrations during cooxidation of ammonia or hydrazine. At low oxygen concentrations, however, in the presence of hydrazine, reductive dechlorination to 2-chloro-6-dichloromethylpyridine occurs (Vannelli and Hooper 1992) (Figure 2.11b).

6. A range of sulfides, including methylsulfide, tetrahydrothiophene, and phenylmethylsulfide, are oxidized to the corresponding sulfoxides (Juliette et al. 1993).

As for methanotrophic bacteria, such transformations are probably confined neither to a single organism nor to strains of specific taxa within the group. For example, both *Nitrosococcus oceanus* and *Nitrosomonas europaea* are able to oxidize methane to CO_2 (Jones and Morita 1983; Ward 1987). The versatility of this group of organisms clearly motivates a reassessment of their ecological significance, particularly in the marine environment where they are widely distributed. The overall similarity of these transformations to those carried out by eukaryotic cytochrome P450 systems (Guengerich 1990) is striking.

2.7 Anaerobic Bacteria

There are several reasons for the increased interest in transformations carried out by anaerobic bacteria:

1. After discharge into the aquatic environment, many xenobiotics are partitioned from the aquatic phase into the sediment phase.
2. Sediments in the vicinity of industrial discharge often contain readily degraded organic matter, and the activity of aerobic and facultatively anaerobic bacteria then renders these sediments effectively anaerobic.

The fate of xenobiotics in many environments is, therefore, significantly determined by the degradative activity of anaerobic bacteria. No attempt is, however, made to cover the substantial literature on several aspects of anaerobic bacteria.

The terms "anaerobic" and "anoxic" are purely operational, and imply merely the absence of air or oxygen, and the absolute distinction between aerobic and anaerobic organisms is becoming increasingly blurred. The problem of defining anaerobic bacteria may, therefore, be best left to philosophers. Possibly, the critical issue is the degree to which low concentrations of oxygen are either necessary for growth or toxic. During the growth of bacteria in the absence of externally added electron acceptors, the term "fermentation" implies that a redox balance is achieved between the substrate (which may include CO_2) and its metabolites. An important group of syntrophic associations between anaerobic bacteria and H_2-utilizing organisms is discussed in Chapter 4, and in the context of specific degradations in Chapters 7 and 8.

A few examples are given to illustrate the apparently conflicting situations that may be encountered, and the gradients of response that may be elicited:

1. Although strictly anaerobic bacteria do not generally grow in the presence of high-potential electron acceptors such as oxygen or nitrate, an intriguing exception is provided by an obligately anaerobic organism that uses nitrate as electron acceptor during the degradation of resorcinol (Gorny et al. 1992). This isolation draws attention to the unknown extent to which such organisms exist in natural systems, since strictly anaerobic conditions are not always used for the isolation of organisms using nitrate as electron acceptor. Other formally similar organisms are either facultatively anaerobic or have a fermentative metabolism. In contrast, the normally strictly anaerobic sulfate-reducing organisms *Desulfobulbus propionicus* and *Desulfovibrio desulfuricans* may grow by reducing nitrate or nitrite to ammonia using hydrogen as electron donor (Seitz and Cypionka 1986). *Desulfovibrio vulgaris* Hildeborough is capable of growing at oxygen concentrations of 0.24–0.48 µM, and indeed it was suggested that this organism might protect anoxic environments from adverse effects resulting from intrusion of oxygen (Johnson et al. 1997). In the presence of nitrate, the acetogen *Clostridium thermoaceticum* oxidizes the *O*-methyl groups of vanillin or vanillate to CO_2 without production of acetate, which is usually formed in the absence of nitrate (Seifritz et al. 1993).
2. Some anaerobic organisms such as clostridia are appreciably tolerant of exposure to oxygen, whereas others such as *Wolinella succinogenes* that have hitherto been classified as anaerobes are in fact microaerophilic (Han et al. 1991). However, organisms such as *Nitrosomonas europaea* which normally obtain energy for growth by the oxidation of ammonia to nitrite, may apparently bring about denitrification of nitrite under conditions of oxygen stress (Poth and Focht 1985), or under anaerobic conditions in the presence of pyruvate (Abeliovich and Vonshak 1992).
3. It is important to note the existence of microaerotolerant or microaerophilic organisms. The example of *W. succinogenes* has already been noted, and another is provided by *Malonomonas*

rubra that uses malonate as the sole source of carbon and energy (Dehning and Schink 1989a). On the other hand, *Propionigenium modestum* that obtains its energy by the decarboxylation of succinate to propionate is a strictly anaerobic organism (Schink and Pfennig 1982).

Attention is directed to two groups of anaerobic bacteria that display metabolic versatility toward structurally diverse substrates—clostridia and anaerobic sulfate reducers.

2.7.1 Clostridia

The classical studies on the anaerobic metabolism of amino acids, purines, and pyrimidines by clostridia not only set out the relevant experimental procedures and thereby laid the foundations for virtually all future investigations, but they also brought to light the importance and range of what were later established as coenzyme B_{12}-mediated rearrangements. These reactions were mainly carried out by species of clostridia and indeed some of these degradations belong to the classical age of microbiology in Delft (Liebert 1909). The number of different clostridia investigated may be gained from the following selected examples:

1. a. The classic purine-fermenting organisms *Clostridium acidurici* and *Cl. cylindrosporum* ferment several purines, including uric acid, xanthine, and guanine.
   b. The degradation of pyrimidines such as orotic acid is accomplished by *Cl. oroticum*.
2. A large number of clostridia, including *Cl. perfringens*, *Cl. saccharobutyricum*, *Cl. propionicum*, *Cl. tetani*, *Cl. sporogenes*, and *Cl. tetanomorphum*, ferment a range of single amino acids, and some are able to participate in the Stickland reaction involving two amino acids that is discussed in Chapter 7.
3. Several clostridia, including *Cl. tyrobutyricum*, *Cl. thermoaceticum*, and *Cl. kluyveri*, are able to reduce the double bond of α,β-unsaturated aldehydes, ketones, and coenzyme A thioesters (Rohdish et al. 2001), and are noted in Chapter 3.

2.7.2 Anaerobic Sulfate-Reducing Bacteria

The spectrum of compounds degraded by anaerobic sulfate-reducing bacteria is continuously widening. It now includes, for example, alkanes (Aeckersberg et al. 1991, 1998); alkanols, alkylamines, alkanoic acids, and nicotinic acid (Imhoff-Stuckle and Pfennig 1983); indoles (Bak and Widdel 1986); methoxybenzoates (Tasaki et al. 1991); benzoate, hydroxybenzoate, and phenol (Drzyzga et al. 1993; Peters et al. 2004); catechol (Szewzyk and Pfennig 1987); and naphthalene (Musat et al. 2009). Further details of the pathways are provided in Chapter 7, Part 1 and Chapter 8, Part 3. It should also be noted that elementary sulfur (S^0) may serve as oxidant for organisms belonging to a number of genera, including *Desulfomicrobium*, *Desulfurella*, *Desulfuromonas*, and *Desulfuromusa* (references in Liesack and Finster 1994), that are noted in Chapter 11. The anaerobic oxidation of propionate may be accomplished in pure cultures of *Syntrophobacter wolinii* and *S. pfennigi* at the expense of sulfate reduction (Wallrabenstein et al. 1995).

2.7.3 Other Anaerobic Bacteria

Additional reactions carried out by some anaerobic bacteria are briefly summarized.

1. Compounds such as oxalate (Dehning and Schink 1989b) and malonate (Dehning et al. 1989; Janssen and Harfoot 1992) are degraded by decarboxylation with only a modest energy contribution. They are nevertheless able to support the growth of the appropriate organisms. The bioenergetics of the anaerobic *Oxalobacter formigenes* (Anantharam et al. 1989) and the microaerophilic *Malonomonas rubra* have been elucidated (Hilbi et al. 1993; Dimroth and Hilbi 1997).
2. Particularly extensive effort has been directed to the anaerobic dechlorination of alkenes and halogenated arenes, benzoates, and phenols that are discussed in Chapters 7 and 9. As a result

of this activity, the range of organisms that are able to carry this out is considerable and includes the following:

   a. A sulfate-reducing organism *Desulfomonile tiedjei* (DeWeerd et al. 1990).

   b. A sulfite-reducing organism *Desulfitobacterium dehalogenans* (Utkin et al. 1994).

   c. *Desulfitobacterium metallireducens* that can use tetrachloroethene and trichlororethene as electron acceptors during growth with lactate (Finneran et al. 2002).

   d. The chlororespiring *Anaeromyxobacter dehalogens* that can use a few 2-chlorinated phenols as electron acceptors (Sanford et al. 2002), and in addition Fe(III) (He and Sanford 2003).

   e. *Sulfurospirillum (Dehalospirillum) multivorans* that accomplishes the sequential reduction of tetrachloroethene to trichloroethene and *cis*-1,2-dichloroethene (Neumann et al. 1994).

3. Although they have not been obtained in pure culture, chemolithotrophic anaerobic bacteria (anammox) that oxidize ammonia using nitrite as electron acceptor and CO_2 as a source of carbon have been described, and are noted in Chapter 3 in the context of alternative electron acceptors. In addition, they can oxidize propionate to CO_2 (Güven et al. 2005) by a pathway that has not yet been resolved.

4. Important groups of organisms belonging to the genus *Geobacter* are widely distributed in sediments. They are able to couple the reduction of Fe(III) to the oxidation of low-molecular-weight organic carboxylic acids (Coates et al. 1996), and some are able to use the humic acid model anthraquinone 2,6-disulfonate as electron acceptor. In addition to acetate and benzoate, *G. metallireducens* and *G. grbiciae* are able to use toluene as an electron donor with anthraquinone 2,6-disulfonate as an electron acceptor (Coates et al. 2001). These organisms are of considerable ecological importance since some of them are able to use humic acid as an electron acceptor. Further comments are given in Chapter 3.

2.8 Organisms from Extreme Environments: Extremophiles

Increasing numbers of bacteria have been isolated that are able to grow under extremes of temperature, pH, salinity, and pressure—many of them obligately. There has been increasing interest in these bacteria and archaea on account of their potential for the synthesis of valuable enzymes that include proteolytic enzymes, and those that degrade starch, xylan, and cellulose (references in Antranikian and Egorova 2007). With increasing accessibility of thermophiles, psychrophiles, halophiles, acidophiles, alkaliphiles, and piezophiles, studies of these organisms have revealed significant features in the physiology of adaptation to extreme environments. This section provides only a small selection of examples from this burgeoning field, since metabolic aspects and those of clinical relevance are discussed in Chapter 4, Part 1, and since a book (Gerday and Glansdorf 2007) with an extensive bibliography has been devoted to these organisms, as well as the journal *Extremophiles*. Some additional details are given in Chapter 4, Part 1.

2.8.1 Psychrophiles

Interest in psychrophiles has been stimulated by the current interest in extraterrestrial life. There is an empirical difference between psychrotolerant and psychrophilic organisms, and the ambiguity in the definition has been discussed in reviews that cover the basic issues (Ingraham and Stokes 1959; Morita 1975). Broadly, the following definitions have been used:

   Psychrophiles—optimum growth temperature of <15°C, and do not grow above 20°C
   Psychrotrophs—optimum and maximum growth temperatures of >15°C and 20°C

A review that discusses the mechanisms of enzyme stability and activity at low temperatures provides valuable comment on the definition of psychrophiles (Feller and Gerday 2003). Only a few examples of low-temperature tolerance and growth are given here as illustration, while the response to the requirement for membrane fluidity is briefly discussed in Chapter 3, and metabolic aspects are given in Chapter 4.

1. Strains of *Thiobacillus* have been isolated from Lake Fryxell in the dry valleys of Victoria Land, Antarctica. They were able to grow at $-2°C$, optimally at $18°C$ with an upper limit at $31°C$, and were considered to be *psychrotolerant* (Sattley and Madigan 2006).

2. Strains of *Clostridium frigoris, Cl. lacusfryxellense, Cl. Bowmanii,* and *Cl. psychrophilum* from the same source had temperature optima (maxima) ($°C$) 5–7 [11], 8–12 [15], 12–16 [20], and 4 [10], and were, therefore, considered as *psychrophilic*.

3. Sulfur-oxidizing strains of *Thiomicrospira arctica* and *Th. psychrolphila* isolated from marine sediments in Svalbard had temperature ranges of $-2.0°C$ to $20.8°C$, and would be considered as *psychrophiles* (Knittel et al. 2005).

4. Strains assigned to *Psychrobacter pacificensis* were isolated from seawater at depths of 500–6000 m in the Japan Trench and grew on a complex medium at $4°C$ with an optimal at $25°C$ and a maximum of $38°C$ (Maruyama et al. 2000). They would be considered as *psychrotolerant*.

5. Strains of *Moritella profunda* and *M. abyssi* were isolated from sediments off the West African coast at a depth of 2815 m and grew on a complex medium between $2°C$ and $12°C$ and $2°C$ and $14°C$, respectively, and were moderately piezophilic (Xu et al. 2003). They were clearly *psychrophiles*.

6. A wide range of bacteria have been isolated from permafrost in the Russian arctic (Vishnivetskaya et al. 2006). More than half of them were Gram-positive rods comparable with *Arthrobacter*, while other dominant groups included *Firmicutes*, including *Bacillus* and *Exiguobacterium, Bacteroidetes, Gamma-Proteobacteria*, including *Psychrobacter* and *Pseudomonas* and *Alpha-Proteobacteria* affiliated with *Sphingomonas*. In a previous study in which broadly similar results were obtained, the metabolic competence of the organisms was measured by incorporation of ^{14}C-labeled acetate into lipids (Rivkina et al. 2000). The kinetics closely resembled sigmoidal growth curves and were monitored during 550 days. Acetate was incorporated at temperatures ranging from $+5°C$ to $-20°C$ and doubling times ranged from 20 days at $-10°C$ to ca. 160 days at $-20°C$. They would be classified as *psychrophiles*.

7. It has been established that bacteria isolated from ice-core samples were able to grow at temperatures below $0°C$.

   a. *Psychromonas ingrahamii* was isolated from Alaskan sea-ice at Point Barrow Alaska, and it was able to grow in the range $-12°C$ to $+10°C$ (Auman et al. 2006).

   b. *Psychrobacter arcticus* strain 273-4 was isolated from continuously frozen permafrost in Kolyma, and it was capable of growth in the range $-10°C$ to $+28°C$ (Bakermans et al. 2006).

   c. Halophilic bacteria similar to *Psychrobacter* on the basis of ribosomal RNA sequences were isolated from the deep ice of Lake Vida Antarctica, and strain LV40 was able to grow in the range $-8°C$ to $+18°C$ (Mondino et al. 2009).

8. Permafrost samples from 9 m were collected from Eureka, Ellesmere Island in the high Canadian arctic; they were examined by various procedures directed primarily to determine the diversity of culture-dependent and culture-independent microorganisms (Steven et al. 2007). They were described as *psychrotolerant*, although three isolates that belonged to the *Actinobacteria* were able to grow at $-5°C$ and one of them was unable to grow at $20°C$.

9. Bacteria in thawed ice samples from Ellesmere Island ($79°38'N$, $74°23'W$) produced CO_2 and CH_4 during extended incubation at $4°C$ in a diluted complex medium (yeast extract, casamino acids, starch, and glucose) (Skidmore et al. 2000). It was suggested that the results demonstrated that the subglacial environment beneath a polythermal glacier provided an acceptable habitat for microbial life.

10. Several investigations have examined a glacial ice core (GISP2) from Greenland:

    a. Anomalous concentrations of methane at 2954 and 3036 m were associated with the activity of methanogens that were characterized by the fluorescence of the F_{420} coenzyme (Tung et al. 2005).

    b. Thawed samples were treated by selective filtration for ultrasmall microorganisms, followed by plating and anaerobic incubation, or by cycles of filtration and enrichment for 2–8 months at −2°C or +5°C in media containing low concentrations of formate or acetate. Isolates were characterized by 16S rRNA sequencing (Miteva and Brenchley 2005), and several important conclusions could be drawn: (i) filtration increased the number of colonies that were obtained; (ii) the combination of filtration and cultivation increased both the number of colonies and their diversity; and (iii) spore-forming organisms were among the isolates.

    c. Thawed samples from a deep Greenland ice core were enriched anaerobically at 2°C to obtain *Herminiimonas glaciei* UMB49[T] that was able to grow between 1°C and 35°C, and was able to assimilate citrate, malate, succinate, and alanine (Loveland-Curtze et al. 2009).

Psychrophiles are not restricted to bacteria, and in cold oceanic water, algae play key roles in harvesting solar energy since they lie at the top of the food chain. Illustrative examples include the following:

1. Pure cultures of diatoms were obtained by enrichment from the ice-edge of the Bering Sea and were unable to grow in a defined mineral medium with seawater above 18°C. Strain KD-50 of *Chaetoceros* sp. had generation times of 20 h at 10°C and 165 h at 0°C and of strain K 3-3 of *Nitschzia* sp. 38 h at 10°C and 145 h at 0°C (Van Baalen and O'Donnell 1983).

2. Seawater samples from the Antarctic Ocean were obtained by growing the pellet from centrifugation in a mineral medium supplemented with penicillin G and streptomycin sulfate. Two strains were obtained that were classified as a bacillariophyte (strain A) and a prymnesiophyte (strain B). Upper temperature limits were 15°C for strain A and 10°C for strain B; both grew at 2.5°C, although strain B grew poorly at 7.5°C (Okuyama et al. 1992).

2.8.2 Thermophiles

It should be emphasized that whereas hyperthermophiles frequently belong to the archaea, valuable thermophilic bacteria carry out important hydrolytic reactions, for example, the endoxylanase from the anaerobe *Thermoanaerobacterium* strain JW/SL-YS 485 (Liu et al. 1996) and several xylanases from *Thermotoga maritima*, although only XynB showed functional activity and no loss of activity after 16 h at 85°C (Reeves et al. 2000).

There has been increased interest in hyperthermophiles for application to biotechnology and bioremediation; some are, in addition, tolerant of extremely low pH that are noted later as acidophiles.

1. The hyperthermophilic archaeon *Pyrococcus furiosus* is an anaerobe that is able to grow at temperatures up to 105°C and has a requirement for tungsten that is discussed in Chapter 3, Part 3. This organism contains several tungsten-containing ferredoxin oxidoreductases—for aldehydes, glyceraldehyde-3-phosphate, and formaldehyde (Mukund and Adams 1991, 1993, 1995; Roy et al. 1999). In addition, glutamate dehydrogenase from *Pyr. furiosus* is inactive below 40°C, and is able to use both NAD$^+$ and NADP$^+$ as cofactors (Klump et al. 1992).

2. The metabolism of elemental sulfur has been explored in a number of hyperthermophiles, many of which—with the exception of *Pyrococcus furiosus*—have an obligate requirement for S^0.

    a. The metabolism of elemental sulfur in the hyperthermophilic archaeon *Pyrococcus furiosus* takes place via polysulfides to produce sulfide (Blumentals et al. 1990). It is able to grow at the expense of peptides or carbohydrates, and it has been suggested that regeneration of oxidized ferredoxin is accomplished by three possible mechanisms: reduction of S^0 to sulfide, reduction of H$^+$ to H$_2$, or formation of alanine that is excreted into the medium in the absence of S^0. The hydrogenases exist in both soluble (Ma and Adams 2001) and particulate form (Sapra et al. 2000), and the [Ni–Fe] hydrogenase that also exhibits sulfur reductase activity has been termed a sulfurhydrogenase (Ma et al. 2000). The formation of alanine by transamination of pyruvate catalyzed by alanine aminotransferase is consistent with the high activity of the enzyme in cells grown with pyruvate (Ward et al. 2000).

b. The sulfur-oxidizing *Acidianus infernus* is capable of growth between 65°C and 96°C, is tolerant of pH from 5.5 to 2, and displays the unusual activity to carry this out under both aerobic and anaerobic conditions (Segerer et al. 1986).

c. Although strains of the strictly anaerobic genus *Ignicoccus* are able to reduce sulfur to sulfide using H_2, they were unable to use sulfate, sulfite, thiosulfate, dithionate, tetrathionate, or nitrate as electron acceptors (Huber et al. 2000).

Illustrations of the metabolic capacity for organic substrates include the following:

1. Methane-oxidizing thermophiles belonging to the phylum *Verrucomicrobia* have already been noted, and a single example is sufficient here: an organism Kam1 isolated from an acidic hot spring in Kamchatka and capable of methane oxidation at 55°C and pH 2 was assigned to the phylum *Verrucomicrobia* (Islam et al. 2008).

2. Several groups of moderately thermophilic alkane-oxidizing bacteria have been recovered from deep oilfields and display growth and degradation of a range of *n*-alkanes. These include (a) *Geobacillus uzensis* and *Geo. subterraneus* for C_{10} to C_{16} *n*-alkanes (Nazina et al. 2001); (b) *Geobacillus thermodenitrificans* for C_{15} to C_{36} (Feng et al. 2007); and (c) *Thermoleophilum album* for C_{12} to C_{19} *n*-alkanes (Zarilla and Perry 1984).

3. Oxidation of long-chain alkanoates has been observed for several hyperthermophilic anaerobic archaea: (a) *Archaeoglobus fulgidus* was able to use thiosulfate as electron acceptor for the oxidation of C_4 to C_{18} alkanoates (Khelifi et al. 2010) and (b) *Geoglobus ahangari* could use Fe(III)oxide as electron acceptor during oxidation of C_3–C_5 alkanoates, as well as with palmitate and stearate (Kashevi et al. 2002).

4. The hyperthermophilic *Ferroglobus placidus* was able to oxidize benzene stoichiometrically to CO_2 under anaerobic conditions using Fe(III) as the sole electron acceptor. Degradation took place putatively by carboxylation to benzoate followed by ring fission to 3-hydroxypimeroyl-CoA—and not via phenol by hydroxylation (Holmes et al. 2011).

5. Phenol hydroxylase has been characterized in the thermophilic *Bacillus stearothermophilus* strain BR219 (Kim and Oriel 1995), and the degradation of phenol involving phenol hydroxylase and catechol 2,3-dioxygenase has been analyzed in *Bacillus thermoleovorans* strain A2 (Duffner and Müller 1998) and strain FDTP-3 (Dong et al. 1992).

6. The thermophile *Chelatococcus* sp. strain MW10 was grown with glucose at 50°C in a cyclic fed-batch fermentor and was able to achieve a high level of poly-3-hydroxybutyrate (Ibrahim and Steinbüchel 2010).

7. Several hyperthermophilic archaea have been examined for the synthesis of short-chain alcohol dehydrogenases.

a. The short-chain alcohol dehydrogenase from the marine *Pyrococcus furiosus* that grows optimally at 100°C had a half-life of 22.5 h at 90°C and 25 min at 100°C. It showed high activity to propan-2-ol and butan-2-ol, and pyruvate in the reduction reaction (van der Oost et al. 2001). EPR revealed the presence of Fe in addition to Zn, it was oxygen sensitive and accepted C_2 to C_8 alkanols (Ma and Adams 1999).

b. The short-chain alcohol dehydrogenase from *Thermococcus sibircus* that was isolated from a high-temperature oil reservoir produced an enzyme with a half-life of 2 h at 90°C and 1 h at 100°C. It was able to accept a range of substrates for oxidation that included propan-2-ol, (*S*)-(+)-butan-2-ol, and D-arabinose, and retained activity in several solvents, including dimethylformamide, acetonitrile, and chloroform (Stekhanova et al. 2010).

c. The alcohol dehydrogenase from *Thermococcus guaymensis* is able to catalyze oxidation of C_2 to C_5 primary alkanols and, more effectively, secondary propan-2-ol, butan-2-ol, and pentan-2-ol, as well as the reverse reduction of butan-2-one and pentan-2-one. The enzyme also stereoselectively oxidized (2R,3R)(–)-butan,2,3-diol to (R)-2-acetoin and the reduction of racemic (R/S)-acetoin to (2R,3R)-butan-2,3-diol (Ying and Ma 2011).

8. Hyperthermophiles that synthesize hydrolytic enzymes for cellulose, xylan, chitin, and proteases have attracted substantial attention (references in Antranikian and Egorova 2007). Some examples illustrate the possibilities:

a. An endoglucanase from the anaerobe *Pyrococcus horikoshii* hydrolyzed the cross-linking in cellulose and displayed optimal activity at above 97°C (Ando et al. 2002).

b. *Pyrococcus furiosus* is able to grow at the expense of chitin, and displayed activities of two hydrolytic enzymes at 95°C: ChiA was an endochitinase that showed no activity toward oligomers smaller than chitotetraose, while ChiB was a chitibiosidase that successively hydrolyzed chitobiose from the nonreducing terminal of chitin (Gao et al. 2003). Both activities were, therefore, required for growth on chitin.

c. An amylolytic enzyme from *Pyrococcus furiosus* displayed an unusual range of hydrolytic activity with an optimal at 90°C (Yang et al. 2004). It could function as an amylase and cyclodextrin hydrolase to produce maltoheptaose and further maltotriose and maltotetraose, and it was able to hydrolyze pullulan to the trisaccharide panose and acarbose to acarviosin-glucose and glucose.

d. The growth of *Pyrococcus furiosus* and strains of *Thermococcus* was examined at 95°C and 85°C, respectively. Both groups could grow with peptides, maltose, and cellobiose in the presence of S^0, and *Pyro. furiosus* also in the absence of S^0 with maltose and cellobiose, although only some strains of *Thermococcus* had this ability. Growth with cellulose and production of H_2 in the absence of S^0 was observed for *Pyrococcus furiosus* that also produced H_2 from starch and maltose (Oslowski et al. 2011).

In addition, illustrative examples of their biotechnological applications based on analysis of the metagenomes include the following:

1. A gene (*erstE1*) encoding a thermostable esterase was isolated from *Escherichia coli* cells that had been transformed by DNA libraries with metagenomes from environmental samples isolated from thermal habitats. The enzyme belonged to the hormone-sensitive lipase family, could be overexpressed in *E. coli*, was active between 30°C and 95°C, and used 4-nitrophenyl esters with chain lengths of C_4–C_{16} (Rhee et al. 2005).

2. A metagenome was isolated from soil samples, a library was constructed, and was screened for amylase activity. The amlyase gene (*amyM*) was overexpressed and purified. Its properties suggested that it could be regarded as a type of maltogenic amylase, α-amylase, and 4-α-glucanotransferase (Yun et al. 2004).

2.8.3 Acidophiles

This section deals with organisms that are tolerant of pH < 2 and are sometimes, additionally, thermotolerant. These occur naturally in areas subject to thermal activity such as volcanoes where the microbial oxidation of discharged sulfur produces acid, and in drainage from mine tailings that contain sulfide minerals where bacterial oxidation produces acidity as a result of sulfide oxidation to sulfuric acid. Acidophilic organisms have been found among both archaea and several groups of bacteria, including autotrophs and heterotrophs (Johnson 2007). Acid tolerance in the clinical context is discussed in Chapter 4, and methods for isolating moderately acidophilic bacteria belonging to the phylum *Acidobacteria* that are widespread in the terrestrial environment are addressed in Chapter 5. Although the genomes of three species in the phylum have been explored, the difficulty in isolating these slow-growing heterotrophic organisms that are able to degrade plant polymers is one of the bottlenecks in their nutritional analysis (Ward et al. 2009).

Archaea of the Ferroplasmaceae family are characterized by tolerance of extreme pH, ability to uniformly oxidize Fe^{2+} in the presence of organic carbon, and resistance of some to Cu^{2+} and arsenite AsO_3H_3 (Golyshina 2011). Strains of the archaea *Ferroplasma acidarmus* and *Ferr. acidiphilum* are extreme acidophiles which have pH optima between 1.0 and 1.7. They have been isolated from acid mine drainage and are capable of anaerobic growth by reduction of Fe^{3+} and mixotrophically with a range of carbohydrates in the presence of Fe^{2+}. They are able to grow in the presence of Cd, Cu, As(III), and As(V) (Golyshina et al. 2000; Dopson et al. 2004). It is worth noting the nutritional ambiguities of these strains and the varying definition of autotrophy (Dopson et al. 2004). *Ferroplasma acidarmanus* has an obligate requirement for a minimum concentration of 100 mM sulfate and produced methanethiol from ^{35}S-labeled methionine, cysteine, and sulfate, while ^{3}H-labeled methionine suggested that this was used as the source of the methyl group (Baumler et al. 2007). The archaeon *Acidiplasma (Ferroplasma) cuprocumulans* was isolated from an industrial mineral sulfide bioleaching pile and had optimum growth between pH 1.0 and 1.2 (Hawkes et al. 2006), while *Acidiplasma aeolicum* (Golyshina et al. 2009) was isolated from a hydrothermal pool; both grew with yeast extract between pH 1.4 and 1.6.

Species of the genus *Picrophilus* that were most closely similar to species of *Thermoplasma* have been isolated from a dry highly acidic soil in Japan, and grew on yeast extract down to ca. pH 0 (Schleper et al. 1995).

Chemolithotrophs are able to oxidize reduced substrates such as Fe^{2+} and S^{2-} to produce ATP and produce protons from water during mineral oxidation by Fe^{3+}, while the initially produced thiosulfate is subsequently oxidized to sulfate by Fe^{3+}, and As^0 to arsenite. These organisms have achieved prominence in their application to solubilize mineral concentrates for the release of gold (and silver) from gold (silver)-bearing minerals such pyrite, arsenopyrite, and chalcopyrite before refining by cyanidation. They are employed in the commercial systems BIOX (Dew et al. 1997) and HIOX (d'Hugues et al. 2001) (references in Okibe et al. 2003), and the principle has been extended to other metals, including cobalt and copper (references in Norris et al. 2000). This has been termed bioleaching and has been applied both to mine tailings and in controlled reactor tanks (Rawlings and Johnson 2007). Only a brief summary of some salient issues is given:

1. a. Five strains of ore-leaching bacteria assigned to *Thiomonas (Thiobacillus) cuprinus* were examined for their ability to use sulfide ores, including arsenopyrite and chalcopyrite for growth, and to extract metals from them (Huber and Stetter 1990). They differed considerably in their extraction of Cu, Fe, U, and Zn, and this depended on the organism, the ore, and the pH.

   b. Organisms that were actively leaching pyrite, arsenopyrite, and chalcopyrite organisms at 45°C were analyzed by 16S rRNA sequence analysis. They were related to *Acidithiobacillus* and *Sulfobacillus* and were implicated in the release of Fe, As, and Cu (Dopson and Lindström 2004). A combination of the sulfide-oxidizing mixotrophic *Acidithiobacillus caldus* and the iron-oxidizing chemolithotroph *Leptospirillum ferriphilum*—both of which are acid-tolerant mesophiles—has been used in commercial mineral bioleaching. In addition to these, the iron-oxidizing *Ferroplasma acidiphilum* became dominant in a pilot-plant treating a combination of chalcopyrite, pyrite, and sphalerite (Okibe et al. 2003).

2. The archaeal community from a laboratory-scale bioleaching reactor at 78°C was examined by temperature gradient electrophoresis (TGGE) and fluorescence *in situ* hybridization (FISH) (Mikkelsen et al. 2009). The latter found phylotypes belonging to the *Sulfolabales* related to *Stygiolobus azoricus*, *Metallosphaera* sp., *Acidianus infernus*, and *Sulfurisphaera ohwakuensis*.

The efficiency of copper extraction from chalcopyrite depends, however, on a number of factors in addition to the specific organism: an examination of several factors underscored the role of Fe^{3+}, and both the pH and the redox potential of the system were paramount in determining the extraction (Vilcáez et al. 2008).

3. Arsenic as AsO_3^{3-} is released from arsenopyrite FeAsS, and Cu^{2+} from chalcopyrite; the organisms must, therefore, tolerate these. Resistance to arsenite that is produced during continuous mineral oxidation has been examined. In *Acidithiobacillus caldus*, resistance to As is carried on a transposon Tn*AtcArs* (Tuffin et al. 2005), while in *Leptospirillum ferriphilum*, the transposon Tn*LfArs* is involved (Tuffin et al. 2006). Although these were characterized in organisms from the same oxidation tank, these transposons were not apparently transferred between organisms. A facultative chemolithotroph *Thiomonas arsenivorans* isolated from gold mine samples in France, was able to use arsenite, sulfur, thiosulfate, and Fe(III) for autotrophic growth (Battaglia-Brunet et al. 2006). Anaerobic oxidation of arsenite has also been linked to chlorate reduction by strains identified by their 16S rRNAs genes as *Dechloromonas* and *Azospira*, of which pure cultures ECC1-pb1 and ECC1-pb2, respectively, were isolated, and under autotrophic conditions *Stenotrophomonas* (Sun et al. 2010).

2.8.4 Alkaliphiles

Bacteria that grow at pH values >9 include both obligate and facultative organisms. They have been discussed with extensive reference citations (Yumoto 2007), and occur naturally in soda lakes in East Africa, Central Asia, Europe, and North America (Sorokin and Kuenen 2005). A range of bacteria has been isolated from soda lakes in the Rift Valley, East Africa. Among aerobes, strains of *Bacillus* sp. have been most extensively examined (Horikoshi 1999), and among anaerobes, species of *Clostridium*, *Alkaliphilus*, and *Alkalibacter*. Especially noteworthy are isolations from the soda-depositing Lake Magadi of extremely halophilic saccharolytic, hydrogen-producing spirochaetes (Zhilina et al. 1996), and *Desulfonatronvibrio hydrogenovorans* that used only H_2 and formate as electron donors and forms of oxidized sulfur as acceptor (Zhilina et al. 1997).

Various species of the chemolithotrophic genera *Thioalkalimicrobium* and *Thioalkalivibrio* that are able to carry out oxidation of various sulfur oxyanions have been isolated from alkaline soda lakes (Sorokin et al. 2001) and are noted in Chapter 11.

Organisms, including *Alkaliphilus transvaalensis*, was isolated from the dam of a deep South-African gold mine (Takai et al. 2001), while strains of the obligate *Alkalibacterium iburiense* (Nakajima et al. 2005) and *Alkalibacterium psychrotolerans* (Yumoto et al. 2004) have been isolated from indigo-fermentation vats in Japan that were used to process the plant *Polygonum tinctorium* that contains indican as a glycoside of indoxyl: in traditional vat-dyeing, this was oxidized to insoluble indigo, reduced to soluble "indigo white," and reoxidized to indigo on the fabric.

Cyanobacteria are the dominant primary producer in less alkaline lakes, and *Arthrospira fisiformis* serves both as a carbon source for bacteria and as a food source for the Lesser Flamingo. The toxins microcystin and anatoxin-a have been detected in several lakes and in cultured strains of *Arthrospira fisiformis* from Lake Bogoria (Ballot et al. 2004). This provides support for the hypothesis that these toxins are responsible for the mass mortalities of Lesser Flamingos that have been observed.

2.8.5 Halophiles

Extreme halophiles have been discussed in several chapters of a book (Gerday and Glansdorf 2007) that discuss their response to and adaptation of extreme salinity. These organisms have been isolated from highly saline environments such as natural salt lakes and in evaporation ponds for the production of salt from seawater. Hypersaline microorganisms include the archaeal genera *Halobacterium*, *Haloferax*, *Natromonas*, and, among the bacteria, the red *Salinibacter ruber* that was isolated from saltern ponds (Antón et al. 2002). Its osmotic tolerance is noteworthy since this was dependent not on the synthesis of intracellular organic osmolytes (compatible solutes) that are discussed in Chapter 4, but on the concentration of KCl that was analogous to that of halophilic archaea (Oren et al. 2002). The organisms in a high-salinity evaporation pond were examined by plating on solid media and prolonged incubation, and by analysis of a library of PCR-amplified 16S rRNA genes (Burns et al. 2004a). The isolates were related to species of several genera of archaea, including *Haloferax*, *Halorubrum*, and *Natromonas*. The first of these was not, however, represented in the gene library, and it was suggested that this could be due to its ability to form colonies even though it was not a dominant group. A major group identified in the library was the square organisms of Walsby that have been isolated and cultivated (Bolhuis et al. 2004; Burns et al. 2004b).

1. a. A strain of the moderate halophilic archaeon assigned to the genus *Haloferax* was isolated from an oil-contaminated saline terrestrial site (Emerson et al. 1994). It was able to use a restricted range of aromatic substrates such as benzoate, cinnamate, and 3-phenylpropanoate as sole sources of carbon and energy. This strain was unable to use lactose or succinate, and this distinguished it from other members of the genus such as *Hf. volcanii* that were able to do so—though unable to use benzoate. This strain could be relevant for the degradation of metabolites produced from the primary degradation of PAHs.

   b. *Haloarcula* sp. strain D1 is an extreme halophile that was unusually able to grow aerobically at the expense of 4-hydroxybenzoate, and gentisate dioxygenase was expressed during growth at the expense of 4-hydroxybenzoate—though not on benzoate (Fairley et al. 2002).

4-Hydroxybenzoate was degraded using the gentisate pathway instead of via protocatechuate, hydroquinone, or catechol. In addition, it was shown using ^{2}H-labeled 2,6-dideutero-4-hydroxybenzoate that the reaction involved a hydroxylation-induced intramolecular migration (Fu and Oriel 1999). Further examples of this NIH migration are given in Chapter 3, Part 3. The degradation of 3-phenylpropionate by *Haloferax* sp. strain D1227 took place by dehydrogenation to 3-phenylcinnamate, chain shortening by β-oxidation to benzoate, and conversion to benzoyl-CoA followed by degradation via the gentisate pathway (Fairley et al. 2006).

2. Halophiles have achieved technical interest, for example, the synthesis of bacteriorhodopsin for photosensitizing devices (Margesin and Schinner 2001), poly-3-hydroxybutyrate by *Natrialba aegyptica* strain 56 (Hezayen et al. 2000, 2001) and *Haloarcula marismortui* (Han et al. 2007), and poly-γ-glutamic acid by *Natrialba aegyptica* strain 40 (Hezayen et al. 2001). Accumulation of poly-hydroxybutyrate and poly(3-hydroxybutyrate-*co*-3-hydroxyvalerate) was examined in a wide range of halophiles grown on a complex medium with glucose. The organisms included strains of *Haloferax, Haloarcula, Halorubrum, Natrialba, Natrinema,* and *Natronococcus,* and it was concluded that the PHA synthase subtype III A had features that distinguished it from the bacterial III B subtype (Han et al. 2010). Mechanisms for the synthesis of poly-3-hydroxyalkanoates are discussed in detail in Chapter 7.

2.9 Eukaryotic Microorganisms

2.9.1 Metabolism by Fungi

Although the cardinal importance of fungi in the terrestrial environment is unquestioned, and some examples of the synthesis of organohalogen metabolites are given in Chapter 3, Part 1, most attention in the aquatic environment has been directed to bacteria. An important exception is provided by studies in Lake Bonney, South Australia. The microflora of the lake contained a population of fungi, including *Trichoderma herzianum,* that was able to degrade the "free" but not the associated chloroguaiacols (van Leeuwen et al. 1997). In addition, *Epicoccum* sp., *Mucor circinelloides,* and *Penicillium expansum*—which are widely distributed soil fungi and may have entered the lake from run-off—were capable of bringing about association of tetrachloroguaiacol with organic components in the aqueous phase, so that this material could subsequently enter the sediment phase (van Leeuwen et al. 1997).

On account of the similarity of the metabolic systems of fungi to those of mammalian systems, it has been suggested that fungi could be used as models for screening purposes. Several relevant examples have been provided (Ferris et al. 1976; Smith and Rosazza 1983; Griffiths et al. 1992), and the fungus

Cunninghamella elegans has attracted particular attention. The reactions involved in the transformation of alachlor by this organism (Pothuluri et al. 1993) are similar to those encountered in other organisms: (a) mammalian systems, (b) the biotransformation of the analogous metalochlor by bluegill sunfish (*Lepomis macrochirus*), and (c) that carried out by a soil actinomycete, which is noted later in this chapter. A wide range of taxonomically diverse fungi has also been used for the synthesis of less readily available compounds, including hydroxylated steroids (Chapter 7, Part 2). Fungi—especially white-rot and brown-rot—have attracted considerable attention in the context of bioremediation of contaminated terrestrial sites, and examples are given in Chapter 14.

In the terrestrial environment—and possibly also in a few specialized aquatic ecosystems—fungi and yeasts play a cardinal role in biodegradation and biotransformation. The role of yeasts in the coastal marine environment is illustrated by results of their frequency and their potential for transformation of phenanthrene and benz[*a*]anthracene (MacGillivray and Shiaris 1993). The transformation of PAHs by certain fungi is analogous to that in mammalian systems so that fungal metabolism has been explored as a model for higher organisms, and extensive studies have been carried out, particularly with *Cunninghamella elegans*.

It has been suggested that the transformations accomplished by the brown-rot fungus *Gleophyllum striatum* may involve hydroxyl radicals, and this is supported by the overall similarity in the structures of the fungal metabolites with those produced with Fenton's reagent (Wetzstein et al. 1997).

Biotransformation (hydroxylation) of a wide range of PAHs and related compounds, including biphenyl, naphthalene, anthracene, phenanthrene, 4-, and 7-methylbenz[*a*]anthracene, and 7,12-dimethylbenz[*a*]anthracene, has been examined in a number of fungi, most extensively in species of *Cunninghamella*, especially *C. elegans* (McMillan et al. 1987). Hydroxylation of benzimidazole (Seigle-Murandi et al. 1986) by *Absidia spinosa*, and of biphenyl ether by *C. echinulata* (Seigle-Murandi et al. 1991) has also been studied on account of the industrial interest in the metabolites. The biotransformation of alachlor (2-chloro-*N*-methoxymethyl-*N*-[2,6-diethylphenyl]acetamide) by *C. elegans* has already been mentioned, and involves primarily hydroxylation at the benzylic carbon atom and loss of the methoxymethyl group (Pothuluri et al. 1993). In all these examples, the reactions are correctly described as biotransformations since the aromatic rings of these compounds remain unaltered.

Particular attention has been focused on the white-rot fungus *Phanerochaete chrysosporium* on account of its ability to degrade not only lignin but also to metabolize a wide range of unrelated compounds. These include PAHs (Bumpus 1989) and organochlorine compounds, including DDT (Bumpus and Aust 1987), PCBs (Eaton 1985), lindane, and chlordane (Kennedy et al. 1990), pentachlorophenol (Mileski et al. 1988), and 2,7-dichlorobenzo[1,4]dioxin (Valli et al. 1992b). The novel pathways for the degradation of 2,4-dichlorophenol (Valli and Gold 1991) (Chapter 9, Part 2) and 2,4-dinitrotoluene (Valli et al. 1992a) (Chapter 9, Part 5) are worth noting. The degradation of all these compounds is apparently mediated by two peroxidase systems—lignin peroxidases and manganese-dependent peroxidases—and a laccase that is produced by several white-rot fungi, though not by *P. chrysosporium*. A potentially serious interpretative ambiguity has, however, emerged from the observation that lignin peroxidase is able to polymerize a range of putative aromatic precursors to lignin, but that this is not the functional enzyme in the depolymerization of lignin (Sarkanen et al. 1991). The regulation of the synthesis of these oxidative enzymes is complex, and is influenced by nitrogen limitation, the growth status of the cells, and the concentration of manganese in the medium (Perez and Jeffries 1990). In addition, it seems clear that monooxygenase and epoxide hydrolase activities are also involved since the biotransformation of phenanthrene takes place even in the absence of peroxidase systems (Sutherland et al. 1991). *P. chrysosporium* is able to degrade anthracene by oxidation to anthra-9,10-quinone, followed by ring fission to produce *o*-phthalate. *Phanerochaete chrysosporium*, which is able to degrade simultaneously chlorobenzene and toluene (Yadav et al. 1995) has achieved some importance in the context of bioremediation since the bacterial degradation of these substrates is generally restricted by the incompatibility of the degradative pathways. Clearly, therefore, a number of important unresolved issues remain. In addition, attention is drawn to the role of fungal redox systems in the covalent linking of xenobiotics to aromatic components of humus in soils.

The metabolic activity of other white-rot fungi, including *Phanerochaete chrysosporium* and *Pleurotus ostreacus*, has been discussed in the context of polycyclic aromatic hydrocarbons. For example, the mineralization potential of the manganese peroxide system from *Nematoloma frowardii* for a number of substrates has been demonstrated (Hofrichter et al. 1998): the formation of CO_2 from labeled substrates ranged from 7% (pyrene) to 36% (pentachlorophenol), 42% (2-amino-4,6-dinitrotoluene), and 49% (catechol).

Several strains of white-rot fungi have been examined for their ability to degrade and mineralize selected PCB congeners (Beaudette et al. 1998). Mineralization of 2,4′,5 [U-^{14}C]trichloro-biphenyl as a fraction of the substrate added ranged from ca. 4.2% for a strain of *Pleurotus ostreacus*, 4.9% and 6.9% for two strains of *Bjerkandera adusta*, to 11% for a strain of *Trametes versicolor*, whereas *Phanerochaete chrysosporium* produced only ca. 2% of $^{14}CO_2$. There was no apparent correlation among levels of lignin peroxidase, manganese peroxidase, and degradative ability.

The degradation of phenolic compounds by fungi may involve rather unusual features of which the following three are given as illustration.

1. *Aspergillus fumigatus* degrades phenol using simultaneously two pathways: first, *ortho*-hydroxylation to catechol followed by ring cleavage to 3-oxoadipate, and second, successive hydroxylation to hydroquinone and 1,2,4-trihydroxybenzene before ring cleavage (Jones et al. 1995). The metabolism of 2-aminobenzoate (anthranilate) to 2,3-dihydroxybenzoate by *A. niger* is accomplished by apparent incorporation of one atom of oxygen from each of O_2 and H_2O, and is not a flavoprotein (Subramanian and Vaidyanathan 1984).

2. The degradation of 1,3,5-trihydroxybenzene (phloroglucinol) by *Fusarium solani* involves rearrangement to 1,2,3-trihydroxybenzene (pyrogallol) followed by ring cleavage to 2-oxo-hex-3-ene-1,6-dicarboxylate (Walker and Taylor 1983). This rearrangement is the opposite of that involved in the anaerobic degradation of 3,4,5-trihydroxybenzoate (Chapter 8, Part 4).

3. The imperfect fungus, *Paecilomyces liliacinus*, is able to produce successively mono-, di-, and trihydroxylated metabolites from biphenyl, and carry out intradiol ring fission of rings carrying adjacent hydroxyl groups (Gesell et al. 2001). Similarly, mono- and dihydroxylated metabolites are produced from dibenzofuran followed by intradiol fission of the ring with adjacent hydroxyl groups (Gesell et al. 2004).

Since the biosynthesis of hydrocarbons by algae and cyanobacteria has been discussed, it is relevant to note briefly their synthesis by a fungus. A structurally diverse array of volatile metabolites is produced by species of the epiphytic fungus *Ascocoryne* (Griffin et al. 2010). These included pent-2-ene, heptane, and octane—though not the long-chain branched alkanes that were previously reported—as well as alkanols such as 3-methylbutan-1-ol and esters, including ethyl acetate, heptyl acetate, and butan-2-ol propionate (Griffin et al. 2010).

2.10 References

Abeliovich A, A Vonshak. 1992. Anaerobic metabolism of *Nitrosomonas europaea*. *Arch Microbiol* 158: 267–270.

Adams E, IL Noton. 1964. Purification and properties of inducible hydroxyproline 2-epimerase from *Pseudomonas*. *J Biol Chem* 239: 1525–1535.

Adriaens P. 1994. Evidence for chlorine migration during oxidation of 2-chlorobiphenyl by a type II methanotroph. *Appl Environ Microbiol* 60: 1658–1662.

Aeckersberg F, F Bak, F Widdel. 1991. Anaerobic oxidation of saturated hydrocarbons to CO_2 by a new type of sulfate-reducing bacterium. *Arch Microbiol* 156: 5–14.

Aeckersberg F, FA Rainey, F Widdel. 1998. Growth, natural relationships, cellular fatty acids and metabolic adaptation of sulfate-reducing bacteria that utilize long-chain alkanes under anoxic conditions. *Arch Microbiol* 170: 361–369.

Ahn Y-B, J-C Chae, GJ Zylstra, MM Häggblom. 2009. Degradation of phenol via phenolphosphate and carboxylation to 4-hydroxybenzoate by a newly isolated strain of the sulfate-reducing bacterium *Desulfobacterium anilini*. *Appl Environ Microbiol* 75: 4248–4253.

Alber BE, JW Kung, G Fuchs. 2008. 3-Hydroxypropionyl-coenzyme A from *Metallosphaera sedula*, an enzyme involved in autotrophic CO_2 fixation. *J Bacteriol* 190: 1383–1389.

Allard A-S, M Remberger, AH Neilson. 1987. Bacterial *O*-methylation of halogen-substituted phenols. *Appl Environ Microbiol* 53: 839–845.

Allgaier M, H Uphoff, A Felske, I Wagner-Döbler. 2003. Aerobic anoxygenic photosynthesis in Roseobacter clade bacteria from diverse marine habitats. *Appl Environ Microbiol* 69: 5051–5059.

Anantharam V, MJ Allison, PC Maloney. 1989. Oxalate: Formate exchange. *J Biol Chem* 264: 7244–7250.

Anders H-J, A Kaetzke, P Kämpfer, W Ludwig, G Fuchs. 1995. Taxonomic position of aromatic-degrading denitrifying pseudomonad strains K 172 and KB 740, and their description as new members of the genera *Thauera*, as *Thauera aromatica* sp. nov., and *Azoarcus*, as *Azoarcus evansii* sp. nov., respectively, members of the beta subclass of the *Proteobacteria*. *Int J Syst Bacteriol* 45: 327–333.

Anderson JJ, S Dagley. 1981a. Catabolism of aromatic acids in *Trichosporon cutaneum*. *J Bacteriol* 141: 534–543.

Anderson JJ, S Dagley. 1981b. Catabolism of tryptophan, anthranilate, and 2,3-dihydroxybenzoate in *Trichosporon cutaneum*. *J Bacteriol* 146: 291–297.

Ando S, H Ishada, Y Kosugi, K Ishikawa. 2002. Hyperthermostable endoglucanase from *Pyrococcus horokoshii*. *Appl Environ Microbiol* 68: 430–433.

Anthony C. 1982. *The Biochemistry of Methylotrophs*. Academic Press, London.

Antón J, A Oren, S Benlloch, F Rodriguez-Valera, R Amann, R Rosello-Mora. 2002. *Salinibacter ruber* gen. nov., sp. nov., a novel extremely halophilic member of the *Bacteria* from saltern crystallizer ponds. *Int J Syst Evol Microbiol* 52: 485–491.

Antranikian G, K Egorova. 2007. Extremophiles, a unique resource of biocatalysts for industrial biotechnology, pp. 361–406. In *Physiology and Biochemistry of Extremophiles* (Eds. C Gerday, N Glansdorf), ASM Press, Washington, D.C.

Anzai Y, H Kim, J-Y Park, H Wakabayashi, H Oyaizu. 2000. Phylogenetic affiliation of the pseudomonads based on 16S rRNA sequence. *Int J Syst Evol Microbiol* 50: 1563–1589.

Apajalahti JH, P Kärpänoja, MS Salkinoja-Salonen. 1986. *Rhodococcus chlorophenolicus* sp. nov. a chlorophenol-mineralizing actinomycete. *Int J Syst Bacteriol* 36: 246–251.

Arenskötter M, D Bröker, A Steinbüchel. 2004. Biology of the metabolically diverse genus *Gordonia*. *Appl Environ Microbiol* 70: 3195–3204.

Arenskötter Q, J Heller, D Dietz, M Arenskötter, A Steinbüchel. 2008. Cloning and characterization of α-methylacyl coenzyme A racemase from *Gordonia polyisoprenivorans*. *Appl Environ Microbiol* 74: 7085–7089.

Asturias JA, KN Timmis. 1993. Three different 2,3-dihydroxybiphenyl-1,2-dioxygenase genes in the Gram-positive polychlorobiphenyl-degrading bacterium *Rhodococcus globerulus* P6. *J Bacteriol* 175: 4631–4640.

Auman AJ, JL Breeze, JJ Gosink, P Kämpfer, JT Staley. 2006. *Psychromonas ingrahamii* sp. nov., a novel gas vacuolated, psychrophilic bacterium isolated from Arctic polar sea ice. *Int J Syst Evol Microbiol* 56: 1001–1007.

Awaya JD, PM Fox, D Borthakur. 2005. *pyd* genes of *Rhizobium* sp. strain TAL1145 are required for degradation of 3-hydroxy-4-pyridone, an aromatic intermediate in mimosine metabolism. *J Bacteriol* 187: 4480–4487.

Baarschers WH, AI Bharaty, J Elvish. 1982. The biodegradation of methoxychlor by *Klebsiella pneumoniae*. *Can J Microbiol* 28: 176–179.

Bak F, F Widdel. 1986. Anaerobic degradatiion of indolic compounds by sulfate-reducing enrichment cultures, and description of *Desulfobacterium indolicum* gen. nov., sp. nov. *Arch Microbiol* 146: 170–176.

Bakermans C, HL Ayala-del-Rio, MA Pander, T Vishnovbetskaya, D Gilichinsky, MF Thomashow, JM Tiedje. 2006. *Psychrobacter cryohalolentis* sp. nov., and *Psychrobacter arcticus* sp. nov., isolated from Siberian permafrost. *Int J Syst Evol Microbiol* 56: 1285–1291.

Balandreau J, V Viallard, B Cournoyer, T Coenye, S Laevens, P Vandamme. 2001. *Burkholderia cepacia* genomovar III is a common plant-associated bacterium. *Appl Environ Microbiol* 67: 982–985.

Balkwill DL, et al. 1997. Taxonomic study of aromatic-degrading bacteria from deep-terrestrial-subsurface sediments and description of *Sphingomonas aromaticivorans* sp. nov., *Sphingomonas subterranea* sp. nov., and *Sphingomonas stygia* sp. nov. *Int J Syst Bacteriol* 47: 191–201.

Ballot A, L Krienitz, K Kotut, C Wiegand, JS Metcalf, GA Codd, S Pflugmacher. 2004. Cyanobacteria and cyanobacterial toxins in three alkaline Rift Valley lakes of Kenya—Lakes Bogori, Nakuru, and Elmenteita. *J Plankton Res* 26: 925–935.

Banerjee A, R Sharma, Y Chisti, UC Banerjee. 2002. *Botryococcus braunii*: A renewable source of hydrocarbons and other chemicals. *Crit Revs Biotechnol* 22: 245–279.

Battaglia-Brunet F, C Joulian, F Garrido, M-C Dictor, D Morin, K Coupland, DB Johnson, KB Hallberg, P Baranger. 2006. Oxidation of arsenite by *Thiomonas* strains and characterization of *Thiomonas arsenivorans* sp. nov. *Antonie van Leeuwenhoek* 89: 99–108.

Baumler DJ, K-F Hung, KC Jeong, CW Kaster. 2007. Production of methanethiol and volatile sulfur compounds by the archaeon "*Ferroplasma acidarmanus.*" *Extremophiles* 11: 841–851.

Beaudette LA, S Davies, PM Fedorak, OP Ward, MA Pickard. 1998. Comparsion of gas chromatography and mineralization experiments for measuring loss of selected polychlorobiphenyl congeners in cultures of white rot fungi. *Appl Environ Microbiol* 64: 2020–2025.

Beh M, G Strauss, R Huber, K-O Stetter, G Fuchs. 1993. Enzymes of the reductive citric acid cycle in the autotrophic eubacterium *Aquifex pyrophilus* and in the archaebacterium *Thermoproteus neutrophilus. Arch Microbiol* 160: 306–311.

Berekaa MM, A Steinbüchel. 2000. Microbial degradation of multiply branched alkane 2,6,10,15,19,23-hexamethyltetracosane (squalane) by *Mycobacterium fortuitum* and *Mycobacterium ratisbonense. Appl Environ Microbiol* 66: 4462–4467.

Berg IA, WH Ramos-Vera, A Petri, H Huber, G Fuchs. 2010. Study of the distribution of autotrophic CO_2 fixation cycles in *Crenarchaeota. Microbiology UK* 156: 256–269.

Biedlingmeier JJ, A Schmidt. 1983. Arylsulfonic acids and some *S*-containing detergents as sulfur sources for growth of *Chlorella fusca. Arch Microbiol* 136: 124–130.

Blumentals II, M Itoh, GJ Olson, RM Kelly. 1990. Role of polysulfides in reduction of elemental sulfur by the hyperthermophilic archaebacterium *Pyrococcus furiosus. Appl Environ Microbiol* 56: 1255–1262.

Bode HB, A Zeeck, K Plückhahn, D Jendrossek. 2000. Physiological and chemical investigations into microbial degradation of synthetic poly(*cis*-1,4-isoprene). *Appl Environ Microbiol* 66: 3680–3685.

Boden R, DP Kelly, JC Murrell, H Schäfer. 2010. Oxidation of dimethylsulfide to tetrathionate by *Methylophaga thiooxidans* sp. nov.: A new link in the sulfur cycle. *Environ Microbiol* 12: 2688–2699.

Boldrin B, A Tiehm, C Fritzsche. 1993. Degradation of phenanthrene, fluorene, fluoranthene, and pyrene by a *Mycobacterium* sp. *Appl Environ Microbiol* 59: 1927–1930.

Bolhuis H, EM te Poele, F Rodríguez-Valera. 2004. Isolation and cultivation of Walsby's square archaeon. *Environ Microbiol* 6: 1287–1291.

Bouquard C, J Ouazzani, J-C Promé, Y Michel-Briand, P Plésiat. 1997. Dechlorination of atrazine by a *Rhizobium* sp. isolate. *Appl Environ Microbiol* 63: 862–866.

Bowien B, HG Schlegel. 1981. Physiology and biochemistry of hydrogen-oxidizing bacteria. *Annu Rev Microbiol* 35: 405–452.

Boyd DR, ND Sharma, F Hempenstall, MA Kennedy, JF Malone, CCR Allen. 1999. *bis-cis*-Dihydrodiols: A new class of metabolites resulting from biphenyl dioxygenase-catalyzed sequential asymmetric *cis*-dihydroxylation of polycyclic arenes and heteroarenes. *J Org Chem* 64: 4005–4011.

Boyd JM, SA Elsworth, SA Ensign. 2004. Bacterial acetone carboxylase is a manganese-dependent metalloenzyme. *J Biol Chem* 46644–46651.

Braaz R, P Fischer, D Jendrossek. 2004. Novel type of heme-dependent oxygenase catalyzes oxidative cleavage of rubber (poly-*cis*-1,4-isoprene). *Appl Environ Microbiol* 70: 7388–7395.

Branchaud BP, CT Walsh. 1985. Functional group diversity in enzymatic oxygenation reactions catalyzed by bacterial flavin-containing cyclohexanone oxygenase. *J Am Chem Soc* 107: 2153–2161.

Brezna B, AA Khan, CE Cerniglia. 2003. Molecular characterization of dioxygenases froim polycyclic aromatic hydrocarbon-degrading *Mycobacterium* spp. *FEMS Microbiol Lett* 223: 177–183.

Briglia M, FA Rainey, E Stackebrandt, G Schraa, MS Salkinoja-Salonen. 1996. *Rhodococcus percolatus* sp. nov., a bacterium degrading 2,4,6-trichlorophenol. *Int J Syst Bacteriol* 46: 23–30.

Briglia M, RIL Eggen, DJ van Elsas, WM de Vos. 1994. Phylogenetic evidence for transfer of pentachlorophenol-mineralizing *Rhodococcus chlorophenolicus* PCP-I^T to the genus *Mycobacterium. Int J Syst Bacteriol* 44: 494–498.

Buchan A, LS Collier, EL Neidle, MA Moran. 2000. Key aromatic ring-cleaving enzyme, protocatechuate 3,4-dioxygenase, in the ecologically important marine *Roseobacter* lineage. *Appl Environ Microbiol* 66: 4662–4672.

Buchan A, JM González, MA Moran. 2005. Overview of the marine *Roseobacter* lineage. *Appl Environ Microbiol* 71: 5665–5677.

Buckley DH, JR Graber, TM Schmidt. 1998. Phylogenetic analysis of nonthermophilic members of the kingdom *Crenarchaeota* and their diversity and abundance in soils. *Appl Environ Microbiol* 64: 4333–4339.

Bumpus JA. 1989. Biodegradation of polycyclic aromatic hydrocarbons by *Phanerochaete chrysosporium. Appl Environ Microbiol* 55: 154–158.

Bumpus JA, SD Aust. 1987. Biodegradation of DDT [1,1,1-trichloro-2,2-bis(4-chlorophenyl)ethane] by the white rot fungus *Phanerochaete chrysosporium. Appl Environ Microbiol* 53: 2001–2008.

Burghout P, LE Cron, H Gradstedt, B Quintero, E Simonetti, JJE Bijlsma, HJ Bootsma, PWM Hermans. 2010. Carbonic anhydrase is essential for *Streptococcus pneumoniae* growth in environmental ambient air. *J Bacteriol* 192: 4054–4062.

Burlingame R, PJ Chapman. 1983. Catabolism of phenylpropionic acid and its 3-hydroxy derivative by *Escherichia coli. J Bacteriol* 155: 113–121.

Burns DG, HM Camakaris, PH Janssen, ML Dyall-Smith. 2004a. Combined use of cultivation-dependent and cultivation-independent methods indicates that members of most haloarchaeal groups in an Australian crystallizer pond are cultivable. *Appl Environ Microbiol* 70: 5258–5265.

Burns DG, HM Camakaris, PH Janssen, ML Dyall-Smith. 2004b. Cultivation of Walsby's square archaeon. *FEMS Microbiol Lett* 238: 469–473.

Buser H-R, MD Müller. 1997. Conversion reactions of various phenoxyalkanoic acid herbicides in soil 2 Elucidation of the enantiomerization process of chiral phenoxy acids from incubation in a D_2O/soil system. *Environ Sci Technol* 31: 1960–1967.

Buser H-R, MD Müller. 1997. Conversion reactions of various phenoxyalkanoic acid herbicides in soil 2 elucidation of the enantiomerization process of chiral phenoxy acids from incubation in a D_2O/soil system. *Environ Sci Technol* 31: 1960–1967.

Buser H-R, MD Müller, ME Balmer. 2002. Environmental behaviour of the chiral acetamide pesticide metalaxyl: Enantioselective degradation and chiral stability in soil. *Environ Sci Technol* 36: 221–226.

Cai M, L Xun. 2002. Organization and regulation of pentachlorophenol-degrading genes in *Sphingobium chlorophenolicum* ATCC 39723. *J Bacteriol* 184: 4672–4680.

Cain RB, RF Bilton, JA Darrah. 1968. The metabolism of aromatic acids by microorganisms. *Biochem J* 108: 797–832.

Carlson RR, AK Vidaver. 1982. Taxonomy of *Corynebacterium* plant pathogens, including a new pathogen of wheat, based on polyacrylamide gel electrophoresis of cellular proteins. *Int J Syst Evol Microbiol* 21: 315–326.

Cerniglia CE, JP Freeman, C van Baalen. 1981. Biotransformation and toxicity of aniline and aniline derivatives in cyanobacteria. *Arch Microbiol* 130: 272–275.

Cerniglia CE, JP Freeman, JR Althaus, C van Baalen. 1983. Metabolism and toxicity of 1- and 2-methylnaphthalene and their derivatives in cyanobacteria. *Arch Microbiol* 136: 177–183.

Cerniglia CE, DT Gibson, C van Baalen. 1980b. Oxidation of naphthalene by cyanobacteria and microalgae. *J Gen Microbiol* 116: 495–500.

Cerniglia CE, DT Gibson, C van Baalen. 1982. Naphthalene metabolism by diatoms isolated from the Kachemak Bay region of Alaska. *J Gen Microbiol* 128: 987–990.

Cerniglia CE, C van Baalen, DT Gibson. 1980a. Oxidation of biphenyl by the cyanobacterium, I sp. strain JCM. *Arch Microbiol* 125: 203–207.

Cerniglia CE, SK Yang. 1984. Stereospecific metabolism of anthracene and phenanthrene by the fungus *Cunninghamella elegans. Appl Environ Microbiol* 47: 119–124.

Chen YP, MJ Dilworth, AR Glenn. 1984. Aromatic metabolism in *Rhizobium trifolii*—protocatechuate 3,4-dioxygenase. *Arch Microbiol* 138: 187–190.

Chen, YP, AR Glenn, MJ Dilworth. 1985. Aromatic metabolism in *Rhizobium trifolii*—catechol 1,2-dioxygenase. *Arch Microbiol* 141: 225–228.

Chen I-C, W-D Lin, S-K Hsu, V Thiruvengadam, W-H Hsu. 2009. Isolation and characterization of a novel lysine racemase from a soil metagenomic library. *Appl Environ Microbiol* 75: 5161–5166.

Chen H-P, C-F Lin, Y-J Lee, S-S Tsay, S-H Wu. 2000. Purification and properties of ornithine racemase from *Clostridium sticklandii. J Bacteriol* 182: 2052–2054.

Chen W-M et al. 2005. Proof that *Burkholderia* strains form effective symbioses with legumes: A study of novel *Mimosa*-nodulating strains from South America. *Appl Environ Microbiol* 71: 7461–7471.

Cheung P-Y, BK Kinkle. 2001. *Mycobacterium* diversity and pyrene mineralization in petroleum contaminated soils. *Appl Environ Microbiol* 67: 2222–2229.

Childers SE, DR Lovley. 2001. Differences in Fe(III) reduction in the hyperthermophilic archaeon *Pyrobaculum islandicum*, versus mesophilic Fe(III)-reducing bacteria. *FEMS Microbiol Lett* 195: 253–258.

Chung WK, GM King. 2001. Isolation, characterization, and polyaromatic hydrocarbon potential of aerobic bacteria from marine macrofainal burrow sediments and description of *Lutibacterium anuloederans* gen. nov., sp. nov., and *Cycloclasticus spirillensis* sp. nov. *Appl Environ Microbiol* 67: 5585–5592.

Coates JD, VK Bhupathiraju, LA Achenbach, MJ McInerney, DR Lovely. 2001. *Geobacter hydrogenophilus, Geobacter chapellei* and *Geobacter grbiciae*, three new, strictly anaerobic, dissimilatory Fe(III) reducers. *Int J Syst Evolut Microbiol* 51: 581–588.

Coates JD, EJP Phillips, DJ Lonergan, H Jenter, DR Lovley. 1996. Isolation of *Geobacter* species from diverse sedimentary environments. *Appl Environ Microbiol* 62: 1531–1536.

Coenye T, D Henry, DP Speert, P Vandamme. 2004. *Burkholderia phenoliruprix* sp. nov., to accommodate the 2,4,5-trichlorophenoxyacetate and halophenol-degrading strain AC1100. *Syst Appl Microbiol* 27: 623–627.

Cokesa Z, H-J Knackmuss, P-G Rieger. 2004. Biodegradation of all stereoisomers of the EDTA substitute iminodisuccinate by *Agrobacterium tumefaciens* BY6 requires an epimerase and a stereoselective C–N lyase. *Appl Environ Microbiol* 70: 3941–3947.

Coleman NV, JC Spain. 2003. Epoxyalkane: Coenzyme M transferase in the ethene and vinyl chloride biodegradation pathways of *Mycobacterium* strain JS60. *J Bacteriol* 185: 5536–5546.

Combourieu B, P Besse, M Sancelme, H Veschambre, AM Delort, P Poupin, N Truffaut. 1998. Morpholine degradation pathway of *Mycobacterium aurum* MO1: Direct evidence of intermadiates by *in situ* ^{1}H nuclear magnetic resonance. *Appl Environ Microbiol* 64: 153–158.

Compact S, B Reiter, A Sessitsch, J Nowak, C Clément, AI Barka. 2005. Endophtic colonization of *Vitis vinifera* L. by plant growth-promoting bacterium *Burklholderia* sp. strain PsJN. *Appl Environ Microbiol* 71: 1685–693.

Cook AM, CG Daughton, M Alexander. 1979. Benzene from bacterial cleavage of the carbon–phosphorus bond of phenylphosphonates. *Biochem J* 184: 453–455.

Cooper BH, GA Land. 1979. Assimilation of protocatechuate acid and *p*-hydroxybenzoic acid as an aid to laboratory identification of *Candida parapsilosis* and other medically important yeasts. *J Clin Microbiol* 10: 343–345.

Cooper RA, MA Skinner. 1980. Catabolism of 3- and 4-hydroxyphenylacetate by the 3,4-dihydroxyphenylacetate pathway in *Escherichia coli*. *J Bacteriol* 143: 302–306.

Corke CT, NJ Bunce, A-L Beaumont, RL Merrick. 1979. Diazonium cations as intermediates in the microbial transformations of chloroanilines to chlorinated biphenyls, azo compounds and triazenes. *J Agric Food Chem* 27: 644–646.

Cottrell MT, A Mannino, DL Kirchman. 2006. Aerobic anoxic phototrophic bacteria in the Mid-Atlantic Bight and the North Pacific Gyre. *Appl Environ Microbiol* 72: 557–564.

Cserjesi AJ, EL Johnson. 1972. Methylation of pentachlorophenol by *Trichoderma virginatum*. *Can J Microbiol* 18: 45–49.

Damaj M, D Ahmad. 1996. Biodegradation of polychlorinated biphenyls by rhizobia: A novel finding. *Biochem Biophys Res Commun* 218: 908–915.

Darley PI, JA Hellstern, JI Medina-Bellver, S Marqués, B Schink, B Philipp. 2007. Heterologous expression and identification of the genes involved in anaerobic degradation of 1,3-dihydroxybenzene (resorcinol) in *Azoarcus anaerobius*. *J Bacteriol* 189: 3824–3833.

Davis JB, HH Chase, RL RFaymond. 1956. *Mycobacterium paraffiicum* n. sp., a bacterium isolated from soil. *Appl Microbiol* 4: 310–315.

Davison AD, P Karuso, DR Jardine, DA Veal. 1996. Halopicolinic acids, novel products arising through the degradation of chloro- and bromobiphenyl by *Sphingomonas paucimobilis* BPS1-3. *Can J Microbiol* 42: 66–71.

de Gonzalo G, G Ottolina, F Zambianchi, MW Fraaije, G Carrea. 2006. Biocatalytic properties of Baeyer–Villiger monooxygenases in aqueous solution. *J Mol Cat B Enzymatic* 39: 91–97.

de Bont JAM, SB Primrose, MD Collins, D Jones. 1980. Chemical studies on some bacteria which utilize gaseous unsaturated hydrocarbons. *J Gen Microbiol* 117: 97–102.

de Zwart JMM, JG Kuenen. 1997. Aerobic conversion of dimethylsulfide and hydrogen sulfide by *Methylophaga sulfidovorans*: Implication for modelling DMS conversion in a microbial mat. *FEMS Microbiol Ecol* 22: 155–165.

Dedysh SN, C Knief, PFDunfield. 2005. *Methylocella* species are facultatively methanotrophic. *J Bacteriol* 187: 4665–47670.

Degrassi G, PP de Laureto, CV Bruschi. 1995. Purification and chracterization of ferulate and *p*-coumarate decarboxylase from *Bacillus pumilis*. *Appl Environ Microbiol* 61: 326–332.

Dehning I, B Schink. 1989a. *Malonomonas rubra* gen nov. sp. nov., a microaerotolerant anaerobic bacterium growing by decarboxylation of malonate. *Arch Microbiol* 151: 427–433.

Dehning I, B Schink. 1989b. Two new species of anaerobic oxalate-fermenting bacteria, *Oxalobacter vibrioformis* sp. nov., and *Clostridium oxalicum* sp. nov., from sediment samples. *Arch Microbiol* 153: 79–84.

Dehning I, M Stieb, B Schink. 1989. *Sporomusa malonica* sp. nov., a homoacetogenic bacterium growing by decarboxylation of malonate or succinate. *Arch Microbiol* 151: 421–426.

Demanèche S, C Meyer, J Micoud, M Louwagie, JC Willison, Y Jouanneau. 2004. Identification and functional analysis of two aromatic-ring-hydroxylating dioxygenases from a *Sphingomonas* strain that degrades various polycyclic aromatic hydrocarbons. *Appl Environ Microbiol* 70: 6714–6725.

Dennis MW, PE Kolattukudy. 1991. Alkane biosynthesis by decarbonylation of aldehyde catalyzed by a microsomal preparation from *Botryococcus braunii*. *Arch Biochem Biophys* 287: 268–275.

Deppenmeier U, V Müller, G Gottschalk. 1996. Pathways of energy conservation in methanogenic archaea. *Arch Microbiol* 165: 149–163.

Derz K, U Klinner, I Schiphan, E Stackebrandt, RM Kroppenstadt. 2004. *Mycobacterium pyrenivorans* sp. nov., a novel polycyclic-aromatic-hydrocarbon-degrading species. *Int J Syst Evol Microbiol* 54: 2313–2317.

Dew DW, EN Lawson, JL Broadhurst. 1997. The BIOX® process for biooxidation of gold-bearing ores or concentrates, pp. 45–79. In *Biomining: Theory, Microbes and Industrial Processes* (Ed. Rawlings DE), Springer-Verlag, Berlin.

DeWeerd KA, L Mandelco, RS Tanner, CR Woese, JM Suflita. 1990. *Desulfomonile tiedjei* gen nov. and sp. nov., a novel anaerobic, dehalogenating, sulfate-reducing bacterium. *Arch Microbiol* 154: 23–30.

Dimroth P, H Hilbi. 1997. Enzymatic and genetic basis for bacterial growth on malonate. *Mol Microbiol* 25: 3–10.

Ding B, S Schmeling, G Fuchs. 2008. Anaerobic metabolism of catechol by the denitrifying bacterium *Thauera aromatica*—a result of promiscuous enzymes and regulators? *J Bacteriol* 190: 1620–1630.

Dodd D, JG Reese, CR LOuer, JD Ballard, MA Spies, SR Blanke. 2007. Functional comparison of the two *Bacillus anthracis* glutamate racemases. *J Bacteriol* 189: 5265–5275.

Dong F-M, L-L Wang, C-M Wang, J-P Cheng, Z-Q He, Z-J Sheng, R-Q Shen. 1992. Molecular cloning and mapping of phenol degradation genes from *Bacillus stearothermophilus* FDPT-3 and their expression in *Escherichia coli*. *Appl Environ Microbiol* 58: 2531–2535.

Dopson M, C Baker-Austin, A Hind, JF Bowman, PL Bond. 2004. Characterization of *Ferroplasma* isolates and *Ferroplasma acidarmanus* sp. nov., extreme acidophiles from acid mine drainage and industrial bioleaching environments. *Appl Environ Microbiol* 70: 2079–2088.

Dopson M, EB Lindström. 2004. Analysis of community composition during moderately thermophilic bioleaching of pyrite, arsenical pyrite, and chalcopyrite. *Microb Ecol* 48: 19–28.

Dreyfus B, JL Garcia, M Gillis. 1988. Characterization of *Azorhizobium caulinodans* gen. nov., sp. nov., a stem-nodulating n itrogen-fixing bacterium isolated from *Sesbania rostrata*. *Int J Syst Bacteriol* 38: 89–98.

Drzyzga O, LF de las Heras, V Morales, JM Navarro Llorens, J Perera. 2011. Cholesterol degradation by *Gordonia cholesterolivorans*. *Appl Environ Microbiol* 77: 4802–4810.

Drzyzga O, J Küver, K-H Blottevogel. 1993. Complete oxidation of benzoate and 4-hydroxybenzoate by a new sulfate-reducing bacterium resembling *Desulfoarculus*. *Arch Microbiol* 159: 109–113.

Duffner FM, R Müller. 1998. A novel phenol hydroxylase and catechol 2,3-dioxygenase from the thermophilic *Bacillus thermoleovorans* strain A2: nucleotide sequence and analysis of the genes. *FEMS Microbiol Lett* 161: 37–45.

Dullius CH, C.Y Chen, B Schink. 2011. Nitrate-dependent degradation of acetone by *Alicycliphilus* and *Paracoccus* strains and comparison of acetone carboxylase enzymes. *Appl Environ Microbiol* 77: 6821–6825.

Dunfield PF et al. 2007. Methane oxidation by an extremely acidophilic bacterium of the phylum Verrumicrobia. *Nature* 450: 879–882.

Durães Sette L, LA Mendonça Alves da Costa, AJ Marsaioli, GP Manfio. 2004. Degradation of alachlor by soil streptomyces. *Appl Microbiol Biotechnol* 64: 712–717.

Durham DR, CG McNamee, DP Stewart. 1984. Dissimilation of aromatic compounds in *Rhodotorula graminis*: Biochemical characterization of pleiotrophically negative mutants. *J Bacteriol* 160: 771–777.

Eaton DC. 1985. Mineralization of polychlorinated biphenyls by *Phanerochaete chrysosporium*, a lignolytic fungus. *Enzyme Microbiol Technol* 7: 194–196.

Egland PG, J Gibson, CS Harwood. 2001. Reductive, coenzyme A-mediated pathway for 3-chlorobenzoate degradation in the phototrophic bacterium *Rhodopseudomonas palustris*. *Appl Environ Microbiol* 67: 1396–1399.

Emerson D, S Chauhan, P Oriel, JA Breznak. 1994. *Haloferax* sp. D1227, a halophilic Archaeon capable of growth on aromatic compounds. *Arch Microbiol* 161: 445–452.

Emmel T, W Sand, WA König, E Bock. 1986. Evidence for the existence of sulfur oxygenase in *Sulfolobus brierleyi*. *J Gen Microbiol* 132: 3415–3420.

Ensign SA, FJ Smakk, JR Allen, MK Sluis. 1998. New roles for CO_2 in the microbial metabolism of aliphatic epoxides and ketones. *Arch Microbiol* 169: 179–187.

Eppink MHM, SA Boeren, J Vervoort, WJH van Berkel. 1997. Purification and properties of 4-hydroxybenzoate 1-hydroxylase (decarboxylating), a novel flavin adenein dinucleotide-dependent monooxygenase from *Candida parapilosis* CBS604. *J Bacteriol* 179: 6680–6687.

Epstein WW, HC Rilling. 1970. Studies on the mechanism of squalene biosynthesis. The structure of presqualene pyrophosphate. *J Biol Chem* 245: 4597–4605.

Erb TJ, IA Berg, V Brecht, M Müller, G Fuchs, BE Albert. 2007. Synthesis of C_5-dicarboxylic acids from C_2-units involving crotonyl-CoA carboxylase/reductase: The ethylmalonyl-CoA pathway. *Proc Natl Acad USA* 104: 10631–10636.

Estelmann S, M Hügler, W Eisenreich, K Werner, IA Berg, WH Ramos-Vera, RF Say, D Kockelkorn, N Gad'on, G Fuchs. 2011. Labeling and enzymatic studies of the central carbon metabolism in *Metallo-sphaera sedula*. *J Bacteriol* 193: 1191–1200.

Ettwig KF et al. 2010. Nitrite-driven anaerobic methane oxidation by oxygenic bacteria. *Nature* 464: 543–548.

Evans MCW, BB Buchanan, DI Arnon. 1966. A new ferredoxin-dependent carbon reduction cycle in a photosynthetic bacterium. *Proc Natl Acad Sci USA* 55: 928–933.

Evans JS, WA Venables. 1990. Degradation of thiophene-2-carboxylate, furan-2-carboxylate, pyrrole-2-carboxylate and other thiophene derivatives by the bacterium *Vibrio* YC1. *Appl Microbiol Biotechnol* 32: 715–720.

Fairley DJ, DR Boyd, ND Sharma, CCR Allen, P Morgan, MJ Larkin. 2002. Aerobic metabolism of 4-hydroxybenzoic acid in *Archaea* via an unusual pathway involving an intramolecular migration (NIH shift). *Appl Environ Microbiol* 68: 6246–6255.

Fairley DJ, G Wang, C Rensing, IL Pepper, MJ Larkin. 2006. Expression of gentisate 1,2-dioxygenase (*gdoA*) genes involved in aromatic degradation in two haloarchaeal genera. *Appl Microbiol Biotechnol* 73: 691–695.

Fall RR, JI Brown, TL Schaeffer. 1979. Enzyme recruitment allows the biodegradation of recalcitrant branched hydrocarbons by *Pseudomonas citronellolis*. *Appl Environ Microbiol* 38: 715–722.

Fall RR, ML Hector. 1977. Acyl-coenzyme A carboxylases. Homologous 3-methylcrotonyl-CoA and geranyl-CoA carboxylases from *Pseudomonas citronellolis*. *Biochemistry* 16: 4000–4005.

Feller G, C Gerday. 2003. Psychrophilic enzymes: Hot topics in cold adaptation. *Nature Revs Microbiol* 1: 200–208.

Feng L et al. 2007. Genome and proteome of long-chain alkane degrading *Geobacillus thermo-denitrificans* BG80-2 isolated fom a deep-subsurface oil reservoir. *Proc Natl Acad Sci USA* 104: 5602–5607.

Ferreira NL, D Labbé, F Monot, F Fayolle-Guichard, CW Greer. 2006. Genes involved in the methyl *tert*-butyl ether (MTBE) metabolic pathway of *Mycobacterium austroafricanum* IFP 2012. *Microbiology (UK)* 152: 1361–1374.

Ferris JP, LH MacDonald, MA Patrie, MA Martin. 1976. Aryl hydrocarbon hydroxylase activity in the fungus *Cunninghamella bainieri*; evidence for the presence of cytochrome P-450. *Arch Biochem Biophys* 175: 443–452.

Fijii K, N Urano, H Ushio, M Satomi, S KImura. 2001. *Sphingomonas cloaae* sp. nov., a nonylphenol-degrading bacterium from wastewater of a sewage-treatment plant in Tokyo. *Int J Syst Evolut Microbiol* 51: 603–610.

Finneran KT, HM Foirbush, CVG VanPraagh, DR Lovley. 2002. *Desulfitobacterium metallireducens* sp. nov., an anaerobic bacterium that couples growth to the reduction of metals and humic acids as well as chlorinated compounds. *Int J Syst Evol Microbiol* 52: 1929–1935.

Finnerty WR. 1992. The biology and genetics of the genus *Rhodococcus*. *Annu Rev Microbiol* 46: 193–218.

Fotheringham IG, SA Bledig, PP Taylor. 1998. Characterization of the genes encoding D-amino acid transaminase and glutamate racemase, two D-glutamate biosynthetic enzymes of *Bacillus sphaericus* ATCC 10208. *J Bacteriol* 180: 4319–4323.

French CE, S Nicklin, NC Bruce. 1996. Sequence and properties of pentaerythritol tetranitrate reductase from *Enterobacter cloacae* PB 2. *J Bacteriol* 178: 6623–6627.

Friedrich CG, B Bowien, B Fried rich. 1979. Formate and oxlate metabolism in *Alcaligenes eutrophus*. *J Gen Microbiol* 115: 185–192.

Fu W, P Oriel. 1999. Degradation of 3-phenylpropionic acid by *Haloferax* D1227. 1999. *Extremophiles* 3: 45–53.

Fuchs G, E Stupperich, G Eden. 1980. Autotrophic CO_2 fixation in *Chlorobium limicola*. Evidence for the operation of a reductive tricarboxylic acid cycle in growing cells. *Arch Microbiol* 128: 64–71.

Fujimoto Y, C-S Chen, AS Gopalan, CJ Sih. 1982. Microbial degradation of the phytosterol side chain. 2. Incorporation of $NaH^{14}CO_3$ into the C-28 position. *J Am Chem Soc* 104: 4720–4722.

Funke G, A von Graevenitz, JE Clarridge, KA Bernard. 1997. Clinical microbiology of coryneform bacteria. *Clin Microbiol Revs* 10: 125–159.

Furuya T, S Hirose, H Osanai, H Semba, K Kino. 2011. Identification of the monooxygenase gene clusters responsible for the regioselective oxidation of phenol to hydroquinone in mycobacteria. *Appl Environ Microbiol* 77: 1214–1220.

Gaal A, HY Neujahr. 1979. Metabolism of phenol and resorcinol in *Trichosporon cutaneum*. *J Bacteriol* 137: 13–21.

Gabriel FLP, W Giger, K Guenther, H-PE Kohler. 2005. Differential degradation of nonylphenol isomers by *Sphingomonas xenophaga* Bayram. *Appl Environ Microbiol* 71 1123–1129.

Gallo KA, JR Knowles. 1993. Purification, cloning, and cofactor independence of glutamate racemase from *Lactobacillus*. *Biochemistry* 32: 3981–3990.

Ganesan B, P Dobrowski, BC Weimer. 2006. Identification of the leucine-to-2-methylbutyric acid catabolic pathway of *Lactococcus lactis*. *Appl Environ Microbiol* 72: 4264–4273.

Gao J, MW Bauer, KR Shockley, MA Pysz, RM Kelly. 2003. Growth of hyperthermophilic archaeon *Pyrococcus furiosus* on chitin involves two family 18 chitinases. *Appl Environ Microbiol* 69: 3119–3128.

Garbe L-A, M Moreno-Horn, R Tressi, H Görisch. 2006. Preferential attack of the (*S*)-ether-linked carbons in bis-(1-chloro-2-propyl)ether by *Rhodococcus* sp. strain DTB. *FEMS Microbiol Ecol* 55: 113–121.

Garrison AW. 2006. Probing the enantioselectivity of chiral pesticides. *Environ Sci Technol* 40: 16–23.

Garrison AW, VA Nzengung, JK Avants, JJ Ellington, WJ Jones, D Rennels, NL Wolfe. 2000. Phytodegradation of *p,p'*-DDT and the enantiomers of *o,p'*-DDT. *Environ Sci Technol* 34: 1663–1670.

Gauthier MJ, B Lafay, R Christen, L Fernandez, M Acquaviva, P Bonin, J-C Bertrand. 1992. *Marinobacter hydrocarbonoclasticus* gen. nov., sp. nov., a new extremely halotolerant, hydrocarbon-degrading marine bacterium. *Int J Syst Bacteriol* 42: 568–576.

Gee JM, JL Peel. 1974. Metabolism of 2,3,4,6-tetrachlorophenol by micro-organisms from broiler house litter. *J Gen Microbiol* 85: 237–243.

Geiselbrecht AD, BP Hedland, MA Tichi, JT Staley. 1998. Isolation of marine polycyclic aromatic hydrocarbon (PAH)-degrading *Cycloclasticus* strains from the Gulf of Mexico and comparison of their PAH degradation ability with that of Puget Sound *Cycloclasticus* strains. *Appl Environ Microbiol* 64: 4703–4710.

Gerday C, N Glansdorf (Eds.) 2007. *Physiology and Biochemistry of Extremophiles*, pp. 223–253. ASM Press, Washington, D.C.

Gesell M, E Hammer, A Mikolasch. 2004. Oxidation and ring cleavage of dibenzofuran by the filamentous fungus *Paecilomyces lilacinus*. *Arch Microbiol* 182: 51–59.

Gesell M, E Hammer, M Specht, W Francke, F Schauer. 2001. Biotransformation of biphenyl by *Paecilomyces lilacinus* and characterization of ring cleavage products. *Appl Environ Microbiol* 67: 1551–1557.

Gibson DT. 1999. *Beijerinckia* sp. strain B1: A strain by any other name. *J Ind Microbiol Biotechnol* 23: 284–293.

Golyshina OV. 2011. Environmental, biogeographic, and biochemical patterns of archaea of the family Ferroplasmaceae. *Appl Environ Microbiol* 77: 5071–5078.

Golyshina OV, TA Pivovorova, GI Karavaiko, TF Kondrat'eva, ERB Moore, WR Abraham, H Lunsdorf, KN Timmis, MM Yakimov, PN Golyshin. 2000. *Ferroplasma acidiphilum* gen. nov. sp. nov., an acidophilic autotrophic, ferrous-oxidizing, cell-wall-lacking, mesophilic member of the *Ferroplasmaceae* fam. nov., comprising a distinct lineage of the Archaea. *Int J Syst Evol Microbiol* 50: 997–1006.

Golyshina OV, MM Yakimov, H Lünsdorf, M Ferrer, M Nimtz, KN Timmis, V Wray, BJ Tindall, PN Golyhshin. 2009. *Acidiplasma aeolicum* gen. nov., sp. nov., a euryarchaeon of the family *Ferroplasmaceae* isolated from a hydrothermal pool, and transfer of *Ferroplasma cupricumulans* to *Acidiplasma cupricumulans* comb. nov. *Int J Syst Evol Microbiol* 59: 2815–2823.

González JM et al. 2003. *Silicibacter pomeroyi* sp. nov. and *Roseovarius nubinhibens* sp. nov., dimethylsulfoniopropionate-demethylating bacteria from marine environments. *Int J Syst Evol Microbiol* 53: 1261–1269.

Goris J, P de Vos, J Caballero-Mellado, J-H Park, E Falsen, JM Tiedje, P Vandamme. 2004. Classification of the PCB- and biphenyl-degrading strain LB400 and relatives as *Burkholderia xenovorans* sp. nov. *Int J Syst Evol Microbiol* 54: 1677–1682.

Goris M, P de Vos, J Caballero-Mellado, J-H Park, E Falsen, JF Quensen III, JM Tiedje, P Vandamme. 1995. Classification of the PCB- and biphenyl-degrading strain LB400 and relatives as *Burkholderia xenovorans* sp. nov. *Int J Syst Evol Microbiol* 54: 1677–1681.

Gorny N, G Wahl, A Brune, B Schink. 1992. A strictly anaerobic nitrate-reducing bacterium growing with resorcinol and other aromatic compunds. *Arch Microbiol* 158: 48–53.

Govindaswami M, DJ Feldjake, BK Kinkle, DP Mindell, JC Loper. 1995. Phylogenetic comparison of two polycyclic aromatic hydrocarbon-degrading mycobacteria. *Appl Environ Microbiol* 61: 3221–3226.

Gray PHH, HG Thornton. 1928. Soil bacteria that decompose certain aromatic compounds. *Centbl Bakt (Abt 2)* 73: 74–96.

Grbic-Galic D. 1986. *O*-Demethylation, dehydroxylation, ring-reduction and cleavage of aromatic substrates by Enterobacteriaceae under anaerobic conditions. *J Appl Bacteriol* 61: 491–497.

Griffin MA, DJ Spakowicz, TA Gianoulis, SA Strobel. 2010. Volatile organic compounds produced by organisms in the genus *Ascocoryne* and a re-evaluation of myco-diesel production by NRRL 50072. *Microbiology (UK)* 156: 3814–3829.

Griffiths DA, DE Brown, SG Jezequel. 1992. Biotransformation of warfarin by the fungus *Beauveria bassiana*. *Appl Microbiol Biotechnol* 37: 169–175.

Grimont PAD, F Grimond, HLC Dulong de Rosnay, PHA Sneath. 1977. Taxonomy of the genus *Serratia*. *J Gen Microbiol* 98: 39–66.

Grosser RJ, D Warshawsky, JR Vestal. 1991. Indigenous and enhanced mineralization of pyrene, benzo[*a*] pyene, and carbazole in soils. *Appl Environ Microbiol* 57: 3462–3469.

Grund E, C Knorr, R Eichenlaub. 1990. Catabolism of benzoate and monohydroxylated benzoates by *Amycolatopsis* and *Streptomyces* spp. *Appl Environ Microbiol* 56: 1459–1464.

Guengerich P. 1990. Enzymatic oxidation of xenobiotic chemicals. *Crit Rev Biochem Mol Biol* 25: 97–153.

Guerin WF, GE Jones. 1988. Mineralization of phenanthrene by a *Mycobacterium* sp. *Appl Environ Microbiol* 54: 937–944.

Gupta JK, C Jebsen, H Kneifel. 1986. Sinapic acid degradation by the yeast *Rhodotorula graminis*. *J Gen Microbiol* 132: 2793–2799.

Güven D et al. 2005. Propionate oxidation by and methanol inhibition of anaerobic ammonium-oxidizing bacteria. *Appl Environ Microbiol* 71: 1066–1071.

Häggblom M, D Janke, PJM Middeldorp, M Salkinoja-Salonen. 1988. *O*-Methylation of chlorinated phenols in the genus *Rhodococcus*. *Arch Microbiol* 152: 6–9.

Häggblom MM, LJ Nohynek, NJ Palleroni, K Kronqvist, E-L Nurmiaho-Lassila, MS Salkinoja-Salonen, S Klatte, RM Kroppenstedt. 1994. Transfer of polychlorophenol-degrading *Rhodococcus chlorophenolicus* (Apajalahti et al. 1986) to the genus *Mycobacterium* as *Mycobacterium chlorophenolicum* comb. nov. *Int J Syst Bacteriol* 44: 485–493.

Häggblom MM, RJ Valo. 1995. Bioremedation of chlorophenol wastes, pp. 389–434. In *Microbial Transformation and Degradation of Toxic Organic Chemicals* (Eds. LY Young and CE Cerniglia), Wiley-Liss, New York, USA.

Hallam SJ, N Putnam, CM Preston, JC Detter, D Rokhsar, PM Richardson, EF DeLong. 2004. Reverse methanogenesis: Testing the hypothesis with environmental genomics. *Science* 305: 1457–1462.

Hammer E, D Krpowas, A Schäfer, M Specht, W Francke, F Schauer. 1998. Isolation and characterization of a dibenzofuran-degrading yeast: Identification of oxidation and ring clavage products. *Appl Environ Microbiol* 64: 2215–2219.

Han J, J Hou, H Liu, S Cai, B Feng, J Zhou, H Xiang. 2010. Wide distribution among halophilic archaea of a novel polyhydroxyalkanoate synthase subtype with homology to bacterial type III synthases. *Appl Environ Microbiol* 76: 7811–7819.

Han Y-H, RM Smibert, NR Krieg. 1991. *Wolinella recta, Wolinella curva, Bacteroides ureolyticus,* and *Bacteroides gracilis* are microaerophiles, not anaerobes. *Int J Syst Bacteriol* 41: 218–222.

Han J, Q Lu, L Zhou, J Zhou, H Xiang. 2007. Molecular characterization of the *phaEC*$_{Hm}$ genes, required for biosynthesis of poly(3-hydroxybutyrate) in the extremely halophilic archaeon *Halolarcula marismortui*. *Appl Environ Microbiol* 73: 6058–6065.

Hara A, K Syutsubo, S Harayama. 2003. *Alcanivorax* which prevails in oil-contaminated seawater exhibits broad substrate specificity for alkane degradation. *Environ Microbiol* 5: 746–753.

Hardisson C, JM Sala-Trapat, RY Stanier. 1969. Pathways for the oxidation of aromatic compounds by *Azotobacter*. *J Gen Microbiol* 59: 1–11.

Hartmans S, FJ Weber, BPM Somhorst, JAM de Bont. 1991. Alkene monooxygenase from *Mycobacterium* E3: A multicomponent enzyme. *J Gen Microbiol* 137: 2555–2560.

Hartmans S, JAM de Bont. 1992. Aerobic vinyl chloride metabolism in *Mycobacterium aurum* L1. *Appl Environ Microbiol* 58: 1220–1226.

Harwood CS, J Gibson. 1988. Anaerobic and aerobic metabolism of diverse aromatic compounds by the photosynthetic bacterium *Rhodopseudomonas palustris*. *Appl Environ Microbiol* 54: 712–717.

Hasson MS et al. 1998. Evolution of an enzyme active site: The structure of a new crystal form of muconate lactonizing enyme compared with mandelate racemase and enolase. *Proc Natl Acad Sci USA* 95: 10396—10401.

Hawkes R, P Franzmann, G O'Hara, J Plumb. 2006. *Ferroplasma cupricumulans* sp. nov., a novel moderately thermophilic, acidophilic archaeon isolated from an industrial-scale chalcocite bioleach heap. *Extremophiles* 10: 525–539.

He Q, RA Sanford. 2003. Characterization of Fe(III) reduction by chlororespiring *Anaeromyxobacter dehalogens*. *Appl Environ Microbiol* 69: 2712–2718.

Hedlund BP, AD Geiselbrecht, TJ Bair, JT Staley. 1999. Polycyclic aromatic hydrocarbon degradation by a new marine becterium, *Neptunomonas naphthovorans* gen. nov., sp. nov. *Appl Environ Microbiol* 65: 251–259.

Hegeman GD. 1966. Synthesis of the enzymes of the mandelate pathway by *Pseudomonas putida*. I. Synthesis of enzymes of the wild type. *J Bacteriol* 91: 1140–1154.

Heitkamp MA, JP Freeman, DW Miller, CE Cerniglia. 1988. Pyrene degradation by a *Mycobacterium* sp.: Identification of ring oxidation and ring fission products. *Appl Environ Microbiol* 54: 2556–2565.

Hennessee CT, J-S Seo, AM Alvarez, QX Li. 2009. Polycyclic aromatic hydrocarbon-degrading species isolated from Hawaiian soils: *Mycobacterium crocinum*, sp. nov., *Mycobacterium pallens* sp. nov., *Mycobacterium rutilum* sp. nov., *Mycobacterium rufum* sp. nov. and *Mycobacterium aromaticivorans* sp. nov. *Int J Syst Evol Microbiol* 59: 378–387.

Herter S, G Fuchs, A Bacher, W Eisenreich. 2002. A bicyclic autotrophic CO_2 fixation patway in *Chloroflexus aurantiacus*. *J Biol Chem* 277: 20277–20283.

Hezayen FE, BH Rehm, R Eberhardt, A Steinbuchel. 2000. Polymer production by two newly isolated and extremely halophilic archaea: Application of a novel corrosive-resistant bioreactor. *Appl Microbiol Biotechnol* 54: 319–325.

Hezayen FE, BH Rehm, R Eberhardt, A Steinbuchel. 2001. Transfer of *Natrialba asiatic* B1T to *Natrialba taiwanensis* sp. nov., and description of *Natrialba aegyptica* sp. nov., a novel, extremely halophilic, aerobic non-pigmented member of the *Archaea* from Egypt that produces extracellular poly(glutamic acid). *Int J Syst Evol Microbiol* 51: 1133–1142.

Hilbi H, R Hermann, P Dimroth. 1993. The malonate decarboxylase enzyme system of *Malonomonas rubra*: Evidence for the cytoplasmic location of the biotin-containing component. *Arch Microbiol* 160: 126–131.

Hino S, H Tauchi. 1987. Production of carbon monoxide from aromatic acids by *Morganella morganii*. *Arch Microbiol* 148: 167–171.

Hiraichi A, Y Yonemitsu, M Matsushita, YK Shin, H Kuraishi, K Kawahara. 2002. Characterization of *Porphybacter sanguineus* sp. nov., an aerobic bacteriochlorophyll-containing bacterium capable of degrading biphenyl and dibenzofuran. *Arch Microbiol* 178: 45–52.

Hofmann KW, H-J Knackmuss, G Heiss. 2004. Nitrite elimination and hydrolytic ring cleavage in 2,4,6-trinitrophenol (picric acid) degradation. *Appl Environ Microbiol* 70: 2854–2860.

Hofrichter M, K Scheibner, I Schneega, W Fritsche. 1998. Enzymatic combusion of aromatic and aliphatic compounds by manganese peroxidase from *Nematoloma frowardii*. *Appl Environ Microbiol* 64: 399–404.

Holland HL. 1988. Chiral sulfoxidation by biotransformation of organic sulfides. *Chem Rev* 88: 473–485.

Holmes DE, C Risso, JA Smith, DR Lovley. 2011. Anaerobic oxidation of benzene by the hyperthermophilic archaeon *Ferroglobus placidus*. *Appl Environ Microbiol* 77: 5926–5933.

Horikoshi K. 1999. Alkaliphiles: Some applications of their products for biotechnology. *Microbiol Mol Biol Rev* 63: 735–750.

Höschle B, D Jendrossek. 2005. Utilization of geraniol is dependent on molybdenum in *Pseudomonas aeruginosa*: Evidence for different metabolic routes for oxidation of geraniol and citronellol. *Microbiology (UK)* 151: 2277–2283.

Höschle B, V Gnau, D Jendrossek. 2005. Methylcrotonyl-CoA and geranyl-CoA carboxylases are involved in leucine/isovalerate utilization (Liu) and acyclic terpene utilization (Atu), and are encoded by *liuB /liuD* and *atuC/atuF*, in *Pseudomonas aeruginosa*. *Microbiology (UK)* 151: 3649–3656.

Huang Y, K-X Zhao, X-H Shen, MT Chaudhry, C-Y Jiang, S-J Liu. 2006. Genetic characterization of the res-orcinol catabolic pathway in *Corynebacterium glutamicum*. *Appl Environ Microbiol* 72: 7238–7245.

Huber H, S Burggraf, T Mayer, I Wyschkony, R Rachel, KO Stetter. 2000. *Ignicoccus* gen. nov., a novel genus of hyperthermophilic, chemolithotrophic archaea, represented by two new species, *Ignicoccus islandicus* sp. nov., and *Ignicoccus pacificus* sp. nov. *Int J Syst Evol Microbiol* 50: 2093–21200.

Huber H, KO Stetter. 1990. *Thiobacillus cuprinus* sp. nov., a novel facultatively organotrophic metal-mobilizing bacterium. *Arch Microbiol* 56: 315–322.

Hügler M, H Huber, KO Stetter, G Fuchs. 2003a. Autotrophic CO_2 fixation pathways in archaea (*Crenarchaeota*). *Arch Microbiol* 179: 160–173.

Hügler M, RS Krieger, M Jahn, G Fuchs. 2003b. Characterization of acetyl-CoA/propionyl-CoA carboxylase in *Metallosphaera sedula*. *Eur J Biochem* 270: 736–744.

Hügler M, CO Wirsen, G Fuchs, CD Taylor, SM Sievert. 2005. Evidence for autotrophic CO_2 fixation via the reductive tricarboxylic acid cycle by members of the ε subdivision of proteobacteria. *J Bacteriol* 187: 3020–3027.

Hyman MR, CL Page, DJ Arp. 1994. Oxidation of methyl fluoride and dimethyl ether by ammonia monooxy-genase in *Nitrosomonas europaea*. *Appl Environ Microbiol* 60: 3033–3035.

Hyman MR, IB Murton, DJ Arp. 1988. Interaction of ammonia monooxygenase from *Nitrosomonas europaea* with alkanes, alkenes, and alkynes. *Appl Environ Microbiol* 54: 3187–3190.

Hyman MR, AW Sansome-Smith, JH Shears, PM Wood. 1985. A kinetic study of benzene oxidation to phenol by whole cells of *Nitrosomonas europaea* and evidence for the further oxidation of phenol to hydroqui-none. *Arch Microbiol* 143: 302–306.

Ibrahim MHA, A Steinbüchel. 2010. High-cell-density cyclic fed-batch fermentation of a poly(3-hydroxybutyrate)-accumulating thermophile, *Chelatococcus* sp. strain MW10. *Appl Environ Microbiol* 76: 7890–7895.

Ichihara A, S Furiya, M Suda. 1960. Metabolism of ʟ-lysine by bacterial enzymes III. Lysine racemase. *J Biochem (Tokyo)* 48: 277–282.

Imhoff-Stuckle D, N Pfennig. 1983. Isolation and characterization of a nicotinic acid-degrading sulfate-reducing bacterium, *Desulfococcus niacini* sp. nov. *Arch Microbiol* 136: 194–198.

Ingraham JL, JL Stokes. 1959. Psychrophilic bacteria. *Bacteriol Revs* 23: 97–108.

Islam T, S Jensen, LJ Reigstad, Ø Larsen, N-K Birkeland. 2008. Methane oxidation at 55°C and pH 2 by a thermoacidophilic bacterium belonging to the *Verrucomicrobia* phylum. *Proc Natl Acad Sci USA* 105: 300–304.

Iwagami S, K Yang, J Davies. 2000. Characterization of the protocatechuate acid catabolic gene cluster from *Streptomyces* sp. strain 2065. *Appl Environ Microbiol* 66: 1499–1508.

Jagnow G, K Haider, P-C Ellwardt. 1977. Anaerobic dechlorination and degradation of hexachlorocyclohexane by anaerobic and facultatively anaerobic bacteria. *Arch Microbiol* 115: 285–292.

Jahn U, H Huber, W Eisenreich, M Hügler, G Fuchs. 2007. Insights into the autotrophic CO_2 fixation pathway of the archaeon *Ignicoccus hospitalis*: Comprehensive analysis of the central carbon metabolism. *J Bacteriol* 189: 4108–4119.

Janssen PH, CG Harfoot. 1992. Anaerobic malonate decarboxylation by *Citrobacter diversus*. *Arch Microbiol* 157: 471–474.

Janssen PH, PS Yates, BE Grinton, PM Taylor, M Sait. 2002. Improved cultivability of soil bacteria and isola-tion in pure culture of novel members of the divisions *Acidobacteria, Actinobacteria, Proteobacteria,* and *Verrucomicrobia*. *Appl Environ Microbiol* 68: 2391–2396.

Jarrell KF, AD Walters, C Bochiwal, JM Borgia, T Dickinson, JPJ Chong. 2011. Major players on the microbial stage: Why archaea are important. *Microbiology UK* 157: 919–936.

Jerina, DM, H Selander, H Yagi, MC Wells, JF Davey, V Mahadevan, DT Gibson. 1976. Dihydrodiols from anthracene and phenanthrene. *J Am Chem Soc* 98: 5988–5996.

Jerina DM, PJ van Bladeren, H Yagi, DT Gibson, V Mahadevan, AS Neese, M Koreeda, ND Sharma, DR Boyd. 1984. Synthesis and absolute configuration of the bacterial *cis*-1,2-, *cis*-8,9-, and *cis*-10, 11-dihydrodiol metabolites of benz[*a*]anthacene by a strain of *Beijerinckia*. *J Org Chem* 49: 3621–3628.

Jesenská A, M Bartos, V Czerneková, I Rychlík, I Pavlík, J Damborský. 2002. Cloning and expression of the haloalkane dehalogenase gene *dhmA* from *Mycobacterium avium* N85 and preliminary characterization of DhmA. *Appl Environ Microbiol* 68: 3224–3230.

Jesenská A, I Sedlácek, J Damborský. 2000. Dehalogenation of haloalkanes by *Mycobacterium tuberculosis* H37Rv and other mycobacteria. *Appl Environ Microbiol* 66: 219–222.

Jessee JA, RE Benoit, AC Hendricks, GC Allen, JL Neal. 1983. Anaerobic degradation of cyanuric acid, cysteine, and atrazine by a facultative anaerobic bacterium. *Appl Environ Microbiol* 45: 97–102.

Ji X-B, TC Hollocher. 1988. Mechanism for nitrosation of 2,3-diaminonaphthalene by *Escherichia coli*: Enzymatic production of NO followed by O_2-dependent chemical nitrosation. *Appl Environ Microbiol* 54: 1791–1794.

Jiao N, F Zhang, N Hong. 2010. Significant roles of bacteriochlorophyll *a* supplemental to chlorophyll *a* in the ocean. *ISME J* 4: 595–597.

Jiao N, Y Zhang, Y Zeng, N Hong, R Liu, F Chen, P Wang. 2007. Distinct distribution pattern of abundance and diversity of aerobic anoxygenic phototrophic bacteria in the global ocean. *Environ Microbiol* 9: 3091–3099.

Jobst B, K Schühle, U Linne, J Heider. 2010. ATP-dependent carboxylation of acetophenone by a novel type of carboxylase. *J Bacteriol* 192: 1387–1394.

Johnson DB. 2007. Physiology and ecology of acidophilic microorganisms, pp. 257–270. In *Physiology and Biochemistry of Extremophiles* (Eds. C Gerday, N Glansdorf), ASM Press, Washington, D.C.

Johnson WV, PM Anderson. 1987. Bicarbonate is a recycling substrate for cyanase. *J Biol Chem* 262: 9021–9025.

Johnson MS, IB Zhulin, M-E R Gapuzan, BL Taylor. 1997. Oxygen-dependent growth of the obligate anaerobe *Desulfovibrio vulgaris* Hildenborough. *J Bacteriol* 179: 5598–5601.

Jones KH, PW Trudgill, DJ Hopper. 1995. Evidence for two pathways for the metabolism of phenol by *Aspergillus fumigatus*. *Arch Microbiol* 163: 176–181.

Jones RD, RY Morita. 1983. Methane oxidation by *Nitrosococcus oceanus* and *Nitrosomonas europaea*. *Appl Environ Microbiol* 45: 401–410.

Jörnvall H, B Persson, M Krook, S Atrian, R Gonzàlez-Duarte, J Jeffery, D Ghosh. 1995. Short-chain dehydrogenases/reductases (SDR). *Biochemistry* 34: 6003–6013.

Juliette LY, MR Hayman, DJ Arp. 1993. Inhibition of ammonia oxidation in *Nitrosomonas europeae* by sulfur compounds: Thioethers are oxidized to sulfoxides by ammonia monooxygenase. *Appl Environ Microbiol* 59: 3718–3727.

Jurgens G, K Lindström, A Saano. 1997. Novel group within the kingdom *Crenararchaeota* from boreal forest soil. *Appl Environ Microbiol* 63: 803–805.

Khadem AF, A Pol, A Wieczorek, SS Mohammadi, K-J Francoijs, HG Stunnenberg, MSM Jetten, HJM Op den Camp. 2011. Autotrophic methanotrophy in Verrucomicrobia: *Methylacidiphilum fumariolicum* SolV uses a Calvin-Benson-Bassham cycle for carbon dioxide fixation. *J Bacteriol* 193: 4438–4446.

Kaiser KLE. 1974. On the optical activity of polychlorinated biphenyls. *Environ Pollut* 7: 93–101.

Kalyuzhnaya M, S Bowerman, J Laram, M Lidstrom, L Chistoserdova. 2006a. *Methylotenera mobilis* is an obligately methylamine-utilizing organism within the family Methylophilaceae. *Int J Syst Evol Microbiol* 56: 2819–2823.

Kaluzhnaya MG, P De Marco, S Bowerman, CC Pacheco, JC Lara, ME Lidstrom, L Chistoserdova. 2006b. *Methyloversatilis universalis* gen nov., sp nov., a novel taxon within the *Betaproteobacteria* represented by three methylotrophic isolates. *Int J Syst Evol Microbiol* 56: 2517–2522.

Kane SR et al. 2007. Whole-genome analysis of the methyl *tert*-butyl ether-degrading beta proteobacterium *Methylibium petroleiphilum* PM1. *J Bacteriol* 178: 1931–1945.

Kaneda T, H Obata, M Tokumoto. 1993. Aromatization of 4-oxocyclohexanecarboxylic acid to 4-hydroxybenzoic acid by two distinct desaturases from *Corynebacterium cyclohexanicum*. *Eur J Biochem* 218: 997–1003.

Käppeli, O. 1986. Cytochromes P-450 of yeasts. *Microbiol Rev* 50: 244–258.

Kasai Y, K Shindo, S Harayama, N Misawa. 2003. Molecular characterization and substrate preference of a polycyclic aromatic hydrocarbon dioxygenase from *Cycloclasticus* sp. strain A5. *Appl Environ Microbiol* 69: 6688–6697.

Kashevi K, JM Tor, DE HOlmes, CVG Van Praagh, A-L Reysenbach, DR Lovley. 2002. *Geoglobus ahangari* gen. nov., sp. nov., a novel hyperthermophilic archaeon capable of oxidizing organic acids and growing autotrophically on hydrogen with Fe(III) serving as sole electron acceptor. *Int J Syst Evol Microbiol* 52: 719–728.

Katayama Y, A Hiraishi, H Kuraishi. 1995. *Paracoccus thiocyanatus* sp. nov., a new species of thiocyanate-utilizing facultative chemolithotrophs, and transfer of *Thiobacillus versutus* to the genus *Paracoccus* as *Paracoccus versutus* comb. nov. with emendation of the genus. *Microbiology (UK)* 141: 1469–1477.

Keck A, J Rau, T Reemtsma, R Mattes, A Stolz, J Klein. 2002. Identification of quinonoid redox mediators that are formed during the degradation of naphthalene-2-sulfonate by *Sphingomonas xenophaga* BN6. *Appl Environ Microbiol* 68: 4341–4349.

Keener WK, DJ Arp. 1994. Transformations of aromatic compounds by *Nitrosomonas europaea*. *Appl Environ Microbiol* 60: 1914–1921.

Kelley I, JP Freeman, CE Cerniglia. 1990. Identification of metabolites from degradation of naphthalene by a *Mycobacterium* sp. *Biodegradation* 1: 283–290.

Kelly DP. 1971. Autotrophy: Concepts of lithotrophic bacteria and their organic metabolism. *Annu Rev Microbiol* 25: 177–210.

Kennedy DW, SD Aust, JA Bumpus. 1990. Comparative biodegradation of alkyl halide insecticides by the white rot fungus, *Phanerochaete chrysosporium* (BKM-F-1767). *Appl Environ Microbiol* 56: 2347–2353.

Khan AA, R-F Wang, W-W Cao, W Franklin, CE Cerniglia. 1996. Reclassification of a polycyclic aromatic hydrocarbon-metabolizing bacterium, *Beijerinckia* sp. strain B1, as *Sphingomonas yanoikuyae* by fatty acids analysis, protein pattern analysis, DNA-DNA hybridization, and 16S ribosomal pattern analysis. *Int J Syst Bacteriol* 46: 466–490.

Khanna P, B Rajkumar, N Jothikumar. 1992. Anoxygenic degradation of aromatic substances by *Rhodopseudomonas palustris*. *Curr Microbiol* 25: 63–67.

Khelifi N, V Grossi, M Hamdi, A Dolla, J-L Tholozan, B Ollivier, A Hirshler-Réa. 2010. Anaerobic oxidation of fatty acids and alkenes by the hyperthermophilic sulfate-reducing archaeon *Archaeoglobus fulgidus*. *Appl Environ Microbiol* 76: 3057–3060.

Kim SB, R Brown, C Oldfield, SC Gilbert, M Goodfellow. 1999. *Gordonia desulfuricans* sp. nov., a benzothiophere-desulfurizing actinomycete. *Int J Syst Bacteriol* 49: 1845–1851.

Kim SB, R Brown, C Oldfield, SC Gilbert, S Iliarionov, M Goodfellow. 2000. *Gordonia amicalis* sp. nov., a novel debenzothiophene-desulfurizing actinomycete. *Int J Syst Evol Microbiol* 50: 2031–2036.

Kim, HG, NV Doronina, YA Trotsenko, SW Kim. 2007. *Methylophaga aminisulsulfidivorans* sp. nov., a restricted facultatively methylotrophic marine bacterium. *Int J Syst Evol Microbiol* 57: 2096–2101.

Kim Y-H, K-H Engesser. 2004. Degradation of alkyl ethers, aralkyl ethers, and dibenzyl ether by *Rhodococcus* sp. strain DEE 5151, isolated from diethyl ether-containing enrichment cultures. *Appl Environ Microbiol* 70: 4398–4401.

Kim S-J, O Kweon, JP Ftreeman, RC Jones, MD Adjei, J-W Jhoo, Rd Edmonson, CE Cerniglia. 2006. Molecular cloning and expression of genes encoding a novel dioxygenase involved in low- and high-molecular-weight polycyclic aromatic hydrocarbon degradation in *Mycobacterium vanbaalenii* PYR-1. *Appl Environ Microbiol* 72: 1045–1054.

Kim S-J, O Kweon, RC Jones, JP Freeman, RD Edmonson, CE Cerniglia. 2007. Complete and integrated pathway in *Mycobacterium vanbaalenii* PYR-1 based on systems biology. *J Bacteriol* 189: 464–472.

Kim JC, PJ Oriel. 1995. Characterization of the *Bacillus stearothermophilus* BR219 phenol hydroxylase gene. *Appl Environ Microbiol* 61: 1252–1256.

Kitts CL, DP Cunningham, PJ Unkefer. 1994. Isolation of three hexahydro-1,3,5-trinitro-1,3,5-triazine-degrading species of the family Enterobacteriaceae from nitramine explosive-contaminated soil. *Appl Environ Microbiol* 60: 4608–4711.

Kiyohara H, K Nagao, R Nomi. 1976. Degradation of phenanthrene through *o*-phthalate by an *Aeromonas* sp. *Agric Biol Chem* 40: 1075–1082.

Kjellberg S, M Hermansson, P Mårdén, GW Jones. 1987. The transient phase between growth and nongrowth of heterotrophic bacteria with emphasis on the marine environment. *Annu Rev Microbiol* 41: 25–49.

Kleespies M, RM Kroppenstedt, FA Rainey, LE Webb, E Stackebrandt. 1996. *Mycobacterium hodleri* sp. nov., a new member of the fast-growing mycobacteria capable of degrading polycyclic aromatic hydrocarbons. *Int J Syst Bacteriol* 46: 683–687.

Klump H, J Di Ruggiero, M Kessel, J-B Park, MWW Adams, FT Robb. 1992. Glutamate dehydrogenase from the hyperthermophile *Pyrococcus furiosus*. Thermal denaturation and activation. *J Biol Chem* 267: 22681–22685.

Kneifel H, K Elmendorff, E Hegewald, CJ Soeder. 1997. Biotransformation of 1-naphthalenesulfonic acid by the green alga *Scenedesmus obliquus*. *Arch Microbiol* 167: 32–37.

Knittel K, J Kuever, A Meyerdierks, R Meinke, R Amann, T Brinkhoff. 2005. *Thiomicrospira arctica* sp. nov. and *Thiomicrospira psychrophila* sp. nov., psychrophilic, obligately chemolithoauthotrophic, sulfur-oxidizing bacteria isolated from marine Arctic sediments. *Int J Syst Evol Microbiol* 55: 781–786.

Koch DJ, C Rückert, DA Rey, A Mix, A Pühler, J Kalinowski. 2005. Role of the *ssu* and *seu* genes of *Corynebacterium glutamicum* ATCC 13032 in utilization of sulfonates and sulfonate esters as sulfur sources. *Appl Environ Microbiol* 71: 6104–6114.

Kolattukudy PE, L Hankin. 1968. Production of omega-haloesters from alkyl halides by *Micrococcus cerificans*. *J Gen Microbiol* 54: 145–153.

Kolodkin-Gal I, D Romero, S Cao, J Clardy, R Kolter, R Losick. 2010. D-Amino acids trigger biofilm disassembly. *Science* 328: 727–629.

Kotani T, T Yamamoto, H Yurimoto, Y Sakai, N Kato. 2003. Propane monooxygenase and NAD$^+$-dependent secondary alcohol dehydrogenase in propane metabolism by *Gordonia* sp. strain TY-5. *J Bacteriol* 185: 7120–7128.

Kranz RG, KK Gabbert, TA Locke, MY Madigan. 1997. Polyhydroxyalkanoate production in *Rhodobacter capsulatus*: Genes, mutants, expression and physiology. *Appl Environ Microbiol* 63: 3003–3009.

Krivobok S, S Kuony, C Meyer, M Louwagie, JC Wilson, Y Jouanneau. 2003. Identification of pyrene-induced proteins in *Mycobacterium* sp. strain 6PY1: Evidence for two ring-hydroxylating dioxygenases. *J Bacteriol* 185: 3828–3841.

Kropp KG, JA Goncalves, JT Anderson, PM Fedorak. 1994. Microbially mediated formation of benzonaphtho-thiophenes from benzo[*b*]thiophenes. *Appl Environ Microbiol* 60: 3624–3631.

Kummer C, P Schumann, E Stackebrandt. 1999. *Gordonia alkanivorans* sp. nov., isolated from tar-contaminated soil. *Int J Syst Bacteriol* 49: 1513–1522.

Kunz DA, O Nagappan. 1989. Cyanase-mediated utilization of cyanate in *Pseudomonas fluorescens* NCIB 11764. *Appl Environ Microbiol* 55: 256–258.

Kuritz T, LV Bocanera, NS Rivera. 1997. Dechlorination of lindane by the cyanobacterium *Anabaena* sp. strain PCC 7120 depends on the function of the *nir* operon. *J Bacteriol* 179: 3368–3370.

Kuritz T, CP Wolk. 1995. Use of filamentous cyanobacteria for biodegradation of organic pollutants. *Appl Environ Microbiol* 61: 234–238.

Kutney JP, E Dimitriadis, GM Hewitt, PJ Salisbury, M Singh. 1982. Studies related to biological detoxification of kraft mill effluent IV—The biodegradation of 14-chlorodehydroabietic acid with *Mortierella isabellina*. *Helv Chem Acta* 65: 1343–1350.

Kweon O, S-J Kim, RC Jones, JP Freeman, MA Adjei, RD Edmonson, CE Cerniglia. 2007. A polyomic approach to elucidate the fluoranthene-degradative pathways in *Mycobacterium vanbaalenii* PYR-1. *J Bacteriol* 189: 4635–4647.

Lam H, D-C Oh, P Cava, CN Takacs, J Clardy, MA de Pedro, MW Waldor. 2009. D-Amino acids govern stationary phase cell wall remodeling in bacteria. *Science* 325: 1552–1555.

Lami R, MT Cottrell. J Ras, O Ulloa, I Oberbosterer, H Claustre, DL Kirchman, P Lebaron. 2007. High abundance of aerobic anoxygenic photosynthetic bacteria in the South Pacific Ocean. *Appl Environ Microbiol* 73: 4198–4205.

Latus M, H-J Seitz, J Eberspächer, F Lingens. 1995. Purification and characterization of hydroxyquinol 1,2-dioxygenase from *Azotobacter* sp. strain GP1. *Appl Environ Microbiol* 61: 2453–2460.

Leininger S, T Ulrich, M Schloter, L Schwa rk, J Qi, GB Nicol, JI Prosser, SC Schuster, C Schleper. 2006. Archaea predominate among ammonia-oxidizing prokarytes in soils. *Nature* 442: 806–809.

Leveau JHL, S Gerards. 2008. Discovery of a bacterial gene cluster for catabolism of the plant hormone indole-3-acetic acid. *FEMS Microbiol Ecol* 65: 238–250.

Levi MH, J Bartell, L Gandolfo, SC Mole, SF Costa, LM Weiss, JK Johnson, G Osterhout, LH Herbst. 2003. Characterization of *Mycobacterium montefiorense* sp. nov., a novel pathogenic *Mycobacterium* from moray eels that is related to *Mycobacterium triplex*. *J Clin Microbiol* 41: 2147–2152.

Levitt MS, RF Newton, SM Roberts, AJ Willetts. 1990. Preparation of optically active 6'-fluorocarbocyclic nucleosides utilising an enantiospecific enzyme-catalysed Baeyer–Villiger type oxidation. *J Chem Soc Chem Comm* 619–620.

Li D-Y, J Eberspächer, B Wagner, J Kuntzer, F Lingens. 1992. Degradation of 2,4,6-trichlorophenol by *Azotobacter* sp. strain GP1. *Appl Environ Microbiol* 57: 1920–1928.

Liang Y, DR Gardner, CD Miller, D Chen, AJ Anderson, BC Weimer, RC Sims. 2006. Study of biochemical pathways and enzymes involved in pyrene degradation by *Mycobacterium* sp. strain KMS. *Appl Environ Microbiol* 72: 7821–7818.

Lidstrom ME, DI Stirling. 1990. Methylotrophs: Genetics and commercial applications. *Annu Rev Microbiol* 44: 27–58.

Liebert F. 1909. The decomposition of uric acid by bacteria. *Proc K Acad Ned Wetensch* 12: 54–64.

Liesack W, K Finster. 1994. Phylogenetic analysis of five strains of Gram-negative, obligately anaerobic, sulfur-reducing bacteria and description of *Desulfuromusa* gen. nov., including *Desulfuromusa kysingii* sp. nov., *Desulfuromusa bakii* sp. nov., and *Desulfuromusa succinoxidans* sp. nov. *Int J Syst Bacteriol* 44: 753–758.

Lim Y-H, K Yokoigawa, N Esaki, K Soda. 1993. A new amino acid racemase with threonine α-epimerase activity from *Pseudomonas putida*: Purification and characterization. *J Bacteriol* 175: 4213–4217.

Lindsay RF, FG Priest. 1975. Decarboxylation of substituted cinnamic acids by enterobacteria: The influence on beer flavour. *J Appl Bacteriol* 39: 181–187.

Linos A, MM Berekaa, A Steinbüchel, KK Kim, C Spröer, RM Kroppenstedt. 2002. *Gordonica westfalica* sp. nov., a novel rubber-degrading actinomycete. *Int J Syst Evol Microbiol* 66: 1133–1139.

Linos A, A Steinbüchel, C Spröer, RM Kroppenstedt. 1999. *Gordonia polyisoprenivorans* a rubber-degrading actinomycete isolated from an automobile tire. *Int J Syst Bacteriol* 49: 1785–1791.

Liu S-Y, FC Gherardini, M Matuschek, H Bahl, J Wiegel. 1996. Cloning, sequencing, and expression of the gene encoding a large S-layer-associated endoxylanase from *Thermoanaerobacterium* strain JW/SL-YS 485. *J Bacteriol* 178: 1539–1547.

Liu L, K Iwata, A Kita, Y Kawarabayasi, M Yohda, K Miki. 2002. Crystal structure of aspartate racemase from *Pyrococcus horikoshii* OT3 and its implications for molecular mechanism of PLP-independent racemization. *J Mol Biol* 319: 479–489.

Liu Z-P, B-J Wang, Y-H Liu, S-J Liu. 2005. *Novosphingobium taihuense* sp. nov., a novel aromatic-compound-degrading bacterium isolated from Taihu Lake, China. *Int Syst Evolut Microbiol* 55: 1229–12132.

Liu S-Y, Z Zheng, R Zhang, J-M Bollag. 1989. Sorption and metabolism of metolachlor by a bacterial community. *Appl Environ Microbiol* 55: 733–740.

Ljungdahl LG. 1986. Autotrophic acetate synthesis. *Annu Rev Microbiol* 40: 415–450.

Lomans BP, R Maas, R Luderer, HJP op den Camp, A Pol, C van der Drift, GD Vogels. 1999. Isolation and characterization of *Methanomethylovorans hollandica* gen. nov., sp. nov., isolated from freshwater sediment, a methylotrophic methanogen able to grow with dimethyl sulfide and methanethiol. *Appl Environ Microbiol* 65: 3641–3650.

Loveland-Curtze J, VI Miteva, JE Brenchley. 2009. *Herminiimonas glaciei* sp. nov., a novel ultramicrobacterium from 3041 m deep Greenland glacial ice. *Int J S yst Evol Microbiol* 59: 1272–1277.

Ma K, MWW Adams. 1999. An unusual oxygen-sensitive, iron- and zinc-containing alcohol dehydrogenase from the hyperthermophilic archaeon *Pyrococcus furiosus*. *J Bacteriol* 181: 1163–1170.

Ma K, MWW Adams. 2001. Hydrogenases I and II from *Pyrococcus furiosus*. *Methods Enzymol* 331: 208–216.

Ma K, R Weiss, MWW Adams. 2000. Characterization of hydrogenase II from the hyperthermophilic archaeon *Pyrococcus furiosus* and assessment of its role in sulfur reduction. *J Bacteriol* 182: 1864–1871.

MacGillivray AR, MP Shiaris. 1993. Biotransformation of polycyclic aromatic hydrocarbons by yeasts isolated from coastal sediments. *Appl Environ Microbiol* 59: 1613–1618.

MacMichael GJ, LR Brown. 1987. Role of carbon dioxide in catabolism of propane by "*Nocardia paraffinicum*" (*Rhodococcus rhodochrous*). *Appl Environ Microbiol* 53: 65–69.

Maeda M, S-Y Chung, E Song, T Kudo. 1995. Multiple genes encoding 2,3-dihydroxybiphenyl 1,2-dioxygenase in the Gram-positive polychlorinated biphenyl-degrading bacterium *Rhodococcus erythropolis* TA421, isolated from a termite ecosystem. *Appl Environ Microbiol* 61: 549–555.

Magarvey NA, JM Keller, V Bernan, M Dworkin, DH Sherman. 2004. Isolation and characterization of novel marine-derived actinomycete taxa rich in bioactive metabolites. *Appl Environ Microbiol* 70: 7529–7530.

Mahenthiralingam E, TA Urban, JB Goldberg. 2005. The multifarious, multireplicon *Burkholderia cepacia* complex. *Nat Revs Microbiol* 3: 144–156.

Malachowsky KJ, TJ Phelps, AB Teboli, DE Minnikin, DC White. 1994. Aerobic mineralization of trichloroethylene, vinyl chloride, and aromatic compounds by *Rhodococcus* species. *Appl Environ Microbiol* 60: 542–548.

Marcus EA, AP Moshfegh, G Sachs, DR Scott. 2005. The periplasmic α-carbonic anhydrase activity of *Helicobacter pylori* is essential for acid acclimatiom. *J Bacteriol* 187: 729–738.

Margesin R, E Schinner. 2001. Potential of halotolerant and halophilic microorganisms for biotechnology. *Extremophiles* 5: 73–83.

Marshall SJ, GF White. 2001. Complete denitration of nitroglycerin by bacteria isolated from a washwater soakaway. *Appl Environ Microbiol* 67: 2622–2626.

Martens T, T Heidorn, R Pukall, M Simon, BJ Tindall, T Brinkhoff. 2006. Reclassification of *Roseobacter gallaeciensis* Ruiz-Ponte et al. 1998 as *Phaeobacter gallaeciensis* gen. nov., comb. nov., description of *Phaeobacter inhibens* sp. nov., reclassification of *Ruegeria algicola* (Lafay et al. 1995) Uchino et al. 1999 as *Marinovum algicola* gen. nov., comb. nov., and emended descriptions of the genera *Roseobacter, Ruegeria* and *Leisingera*. *Int J Syst Evol Microbiol* 56: 1293–1304.

Massey LK, RS Conrad, JR Sokatch. 1974. Regulation of leucine catabolism in *Pseudomonas putida*. *J Bacteriol* 118: 112–120.

Maruyama A, D Honda, H Yamamoto, K Kitamura, T Higashihara. 2000. Phylogenetic analysis of psychrophilic bacteria isolated from the Japan Trench, including a description of the deep-sea species *Psychrobacter pacificensis* sp. nov. *Int J Syst Evol Microbiol* 50: 835–846.

Matin A. 1978. Organic nutrition of chemolithotrophic bacteria. *Annu Rev Microbiol* 32: 433–468.

McBride KE, JW Kenny, DM Stalker. 1986. Metabolism of the herbicide bromoxynil by *Klebsiella pneumoniae* subspecies *ozaenae*. *Appl Environ Microbiol* 52: 325–330.

McDonald IR, NV Doronina, YA Trotsenko, C McAnulla, JC Murrell. 2001. *Hyphomicrobium chloromethanicum* sp. nov. and *Methylobacterium chloromethanicum* sp.nov., chloromethane-utilizing bacteria isolated from a polluted environment. *Int J Syst Evol Microbiol* 51: 119–122.

McGrath JE, CG Harfoot. 1997. Reductive dehalogenation of halocarboxylic acids by the phototrophic genera *Rhodospirillum* and *Rhodopseudomonas*. *Appl Environ Microbiol* 63: 333–335.

McGrath JW, F Hammerschmidt, P Quinn. 1998. Biodegradation of phosphonomycin by *Rhizobium huakuii* PMY1. *Appl Environ Microbiol* 64: 356–358.

McIntire WS, DE Wemmer, A Chistoserdov, ME Lidstrom. 1991. A new cofactor in a prokaryotic enzyme: Tryptophan tryptophylquinone as the redox prosthetic group in methylamine dehydrogenase. *Science* 252: 817–824.

McKew BA, F Coulon, AM Osborn, KN Timmis, TJ McGenity. 2007. Determining the identity and roles of oil-metabolizing marine bacteria from the Thames estuary, UK. *Environ Microbiol* 9: 165–176.

McMillan DC, PP Fu, CE Cerniglia. 1987. Stereoselective fungal metabolism of 7,12-dimethyl-benz[a]anthracene: Identification and enantiomeric resolution of a K-region dihydrodiol. *Appl Environ Microbiol* 53: 2560–2566.

Mehta MO, JA Baross. 2006. Nitrogen fixation at 92°C by a hydrothermal vent archaeon. *Science* 314: 1783–1786.

Mendez-Perez D, MB Begemann, BF Pfleger. 2011. Modular synththase-encoding gene involved in α-olefin biosynthesis in *Synechococcus* sp. strain PCC 7002. *Appl Environ Microbiol* 77: 4264–4267.

Merkens H, S Fetzner. 2008. Transcriptional analysis of the *queD* gene coding for quercetinase of *Streptomyces* sp. FLA. *FEMS Microbiol Lett* 287: 100–107.

Merkens H, R Kappl, RP Jakob, FX Schmid, S Fetzner. 2008. Quercetinase QueD of Streptomyces sp. FLA, a monocuprin dioxygenase with a preference for nickel and cobalt. *Biochemistry* 47: 12185–12196.

Metzger P, M David, E Casadevall. 1986. Biosynthesis of triterpenoid hydrocarbons in the B-race of the green alga *Botryococcus braunii*. Sites of production and nature of the methylating agent. *Phytochemistry* 26: 129–134.

Metzger P, C Largeau. 1999. Chemicals of *Botryococcus braunii*, pp. 205–260. In *Chemicals from Microalgae* (Ed. Z Cohen), Taylor & Francis, London.

Metzger P, C Largeau. 2005. *Botryococcus braunii*: A rich source for hydrocarbons and related ether lipids. *Appl Microbiol Biotechnol* 66: 486–496.

Meyer O, HG Schlegel. 1983. Biology of aerobic carbon monoxide-oxidizing bacteria. *Annu Rev Microbiol* 37: 277–310.

Middelhoven WJ, F Spaaij. 1997. *Rhodotorula cresolica* sp. nov., a cresol-assimilating yeast species isolated from soil. *Int J Syst Bacteriol* 47: 324–327.

Middelhoven WJ, MC Hooglamer-Te Niet, NJW Kreger-Van Rij. 1984. *Trichosporon adeninovorans* sp. nov., a yeast species utilizing adenine, xanthine, uric acid, putrescine and primary *n*-alkylamines as the sole source of carbon, nitrogen and energy. *Antonie van Leeuwenhoek* 50: 369–378.

Mikkelsen D, U Kappler, AG McEwan, LI Sly. 2009. Probing the archaeal diversity of a mixed thermophilic bioleaching culture by TGGE and FISH. *Syst Appl Microbiol* 32: 510–513.

Mileski G, JA Bumpus, MA Jurek, SD Aust. 1988. Biodegradation of pentachlorophenol by the white rot fungus, *Phanerochaete chrysosporium*. *Appl Environ Microbiol* 54: 2885–2889.

Mills AL, M Alexander. 1976. *N*-nitrosamine formation by cultures of several microorganisms. *Appl Environ Microbiol* 31: 892–895.

Mincer TJ, W Fenical, PR Jensen. 2005. Culture-dependent and culture-independent diversity within the obligate marine actinomycete genus *Salinispora*. *Appl Environ Microbiol* 71: 7019–7028.

Mincer TJ, PR Jensen, CA Kauffman, W Fenical. 2002. Widespread and persistent populations of a new marine actinomycete taxon in ocean sediments. *Appl Environ Microbiol* 68: 5005–5011.

Miteva VI, JE Brenchley. 2005. Detection and isolation of ultrasmall microorganisms from a 120,000-year-old Greenland glacier ice core. *Appl Environ Microbiol* 71: 7806–7818.

Möbus E, E Maser. 1998. Molecular cloning, overexpression, and characterization of steroid-inducible 3α-hydroxysteroid dehydrogenase/carbonyl reductase from *Comamonas testosteroni*. *J Biol Chem* 273: 30888–30896.

Möller D, R Schauder, G Fuchs, RK Thauer. 1987. Acetate oxidation to CO_2 via a citric acid cycle involving an ATP-citrate lyase: A mechanism for the synthesis of ATP via substrate level phosphorylation in *Desulfobacter postgatei* growing on acetate and sulfate. *Arch Microbiol* 148: 202–207.

Mondino LJ, M Asao, MT Madigan. 2009. Cold-active halophilic bacteria from the ice-sealed Lake Vida, Antarctica. *Arch Microbiol* 191: 785–790.

Monna L, T Omori, T Kodama. 1993. Microbial degradation of dibenzofuran, fluorene, and dibenzo-*p*-dioxin by *Staphylococcus auriculans* DBF63. *Appl Environ Microbiol* 59: 285–289.

Moody JD, JP Freeman, CE Cerniglia. 2005. Degradation of benz[*a*]anthracene by *Mycobacterium vanbaalenii* PYR-1. *Biodegradation* 16: 513–526.

Moody JD, JP Freeman, PP Fu, CE Cerniglia. 2004. Degradation of benzo[*a*]pyrene by *Mycobacterium vanbaalenii* PYR-1. *Appl Environ Microbiol* 70: 340–345.

Morita RY. 1975. Psychrophilic bacteria. *Bacteriol Revs* 39: 144–167.

Mouttaki H, MA Nanny, MJ McInerney. 2007. Cyclohexane carboxylate and benzoate formation from crotonate in *Syntrophus aciditrophicus*. *Appl Environ Microbiol* 73: 930–938.

Mukund S, MWW Adams. 1991. The novel tungsten–iron–sulfur protein of the hyperthermophilic archaebacterium, *Pyrococcus furiosus*, is an aldehyde ferredoxin oxidoreductase. *J Biol Chem* 266: 14208–14216.

Mukund S, MWW Adams. 1993. Characterization of a novel tungsten-containing formaldehyde oxidoreductase from the extremely thermophilic archaeon, *Thermococcus litoralis*. *J Biol Chem* 268: 13592–13600.

Mukund S, MWW Adams. 1995. Glyceraldehyde-3-phosphate ferredoxin oxidoreductase, a novel tungsten-containing enzyme with a potential glycolytic role in the hyperthermophilic archaeon *Pyrococcus furiosus*. *J Biol Chem* 270: 8389–8392.

Müller TA, T Fleischmann, JR van der Meer, H-P E Kohler. 2006. Purification and characterization of two enantioselective α-ketoglutarate-dependent dioxygenases, RdpA and SdpA, from *Sphingomonas herbicidovorans*. *Appl Environ Microbiol* 72: 4853–4861.

Muramatsu H, H Mihara, R Kakutani, M Yasuda, M Ueda, T Kurihara, N Esaki. 2005. The putative malate/lactate dehydrogenase from *Pseudomonas putida* is an NADPH-dependent Δ^1-piperideine-2-carboxylate/Δ^1-pyrroline-2-carboxylate reductase involved in the catabolism of L-lysine and D-proline. *J Biol Chem* 280: 5329–5335.

Murphy GL, JJ Perry. 1983. Incorporation of chlorinated alkanes into fatty acids of hydrocarbon-utilizing mycobacteria. *J Bacteriol* 156: 1158–1164.

Musat F, A Galushko, J Jacob, F Widdel, M Kube, R Reinhardt, H Wilkes, B Schink, R Rabus. 2009. Anaerobic degradation of naphthalene and 2-methylnaphthalene by strains of marine sulphate-reducing bacteria. *Environ Microbiol* 11: 209–219.

Nakajima K, K Hirota, Y Nodasaka, I Yumoto. 2005. *Alkalibacterium iburiense* sp. nov., an obligate alkaliphile that reduces an indigo dye. *Int Syst Evol Microbiol* 55: 1525–1530.

Nakatsu CH, K Hristova, S Hanada, X-Y Meng, JR Hanson, KM Scow, Y Kamagata. 2006. *Methylibium petroleiphilum* gen. nov., sp. nov., a novel methyl *tert*-butyl ether-degrading methylotroph of the *Betaproteobacteria*. *Int J Syst Evol Microbiol* 56: 983–989.

Nam I-H, Y-M Kim, S Schmidt, Y-S Chang. 2006. Biotransformation of 1,2,3-tri- and 1,2,3,4,7,8-hexachlorodibenzo-*p*-dioxin by *Sphingomonas wittichii* strain RW1. *Appl Environ Microbiol* 72: 112–116.

Narro ML, CE Cerniglia, C Van Baalen, DT Gibson. 1992. Metabolism of phenanthrene by the marine cyanobacterium *Agmenellum quadruplicatum* PR-6. *Appl Environ Microbiol* 58: 1351–1359.

Navarro-Llorens JM, MA Patrauchan, GR Stewart, JE Davies, LD Eltis, WW Mohn. 2005. Phenylacetate catabolism in *Rhodococcus* sp. strain RHA1: A central pathway for degradation of aromatic compounds. *J Bacteriol* 187: 4497–4504.

Nazina TN et al. 2001. Taxonomic study of aerobic thermophilic bacilli: Descriptions of *Geobacillus subterraneus* gen. nov., sp. nov. and *Geobacillus uzenensis* sp. nov. from petroleum reservoirs and transfer of *Bacillus stearothermophilus, Bacillus thermocatenulatus, Bacillus thermoleovorans, Bacillus kaustophilus, Bacillus thermoglucosidasius*, and *Bacillus thermodenitrificans* to *Geobacillus* as the new combinations *G. stearothermophilus, G. thermocatenulatus, G. thermoleovorans, G. kaustophilus, G. thermoglucosidasius*, and *G. thermodenitrificans. Int J Syst Evol Microbiol* 51: 433–446.

Neidhardt FC, PL Bloch, DF Smith. 1974. Culture medium for enterobacteria. *J Bacteriol* 119: 736–747.

Neilson AH, RA Lewin. 1974. The uptake and utilization of organic carbon by algae: An essay in comparative biochemistry. *Phycologia* 13: 227–264.

Neilson AH, A-S Allard, P-Å Hynning, M Remberger, L Landner. 1983. Bacterial methylation of chlorinated phenols and guiaiacols: Formation of veratroles from guaiacols and high molecular weight chlorinated lignin. *Appl Environ Microbiol* 45: 774–783.

Neilson AH, T Larsson. 1980. The utilization of organic nitrogen for growth of algae: Physiological aspects. *Physiol Plant* 48: 542–553.

Neumann A, H Scholz-Muramatsu, G Diekert. 1994. Tetrachloroethene metabolism of *Dehalospirillum multivorans. Arch Microbiol* 162: 295–301.

Nickel K, MJ-F Suter, H-PE Kohler. 1997. Involvement of two α-ketoglutarate-dependent dioxygenases in enantioselective degradation of (*R*)- and (*S*)-mecoprop by *Sphingomonas herbicidovorans* MH. *Appl Environ Microbiol* 63: 6674–6679.

Nicolau E, L Kerhoas, M Lettere, Y Jouanneau, R Marchal. 2008. Biodegradation of 2-ethylhexyl nitrate by *Mycobacterium austroafricanum* IFP 2173. *Appl Environ Microbiol* 74: 6187–6193.

Nie Y, J Liang, Y-Q Tang, X-L Wu. 2011. Two novel alkane hydroxylase-rubredoxin fusion genes isolated from a *Dietzia* bacterium and the functions of fused rubredoxin domains in long-chain *n*-alkane degradation. *Appl Environ Microbiol* 77: 7279–7288.

Nohynek JL, E-L Nurmiaho-Lassila, EL Suhonen, H-J Busse, M Mohammadi, M Hantula, F Raney, Salkinoja-Salonen. 1996. Description of chlorophenol-degrading *Pseudomonas* sp. strains KF1, KF3, and NKF1 as a new species of the genus *Sphingomonas, Sphingomonas subarctica* sp. nov. *Int J Syst Bacteriol* 46: 1042–1055.

Norris PR, NP Burton, NAM Foulis. 2000. Acidophiles in bioreactor mineral processing. *Extremophiles* 4: 71–76.

Norris PR, DA Clark, JP Owen, S Waterhouse. 1996. Characteristics of *Sulfobacillus acidophilus* sp. nov., and other moderately thermophilic mineral sulfide-oxidizing bacteria. *Microbiology (UK)* 142: 775–783.

Nörtemann B, A Glässer, R Machinek, G Remberg, H-J Knackmuss. 1993. 5-Hydroxyquinoline-2-carboxylic acid, a dead-end metabolite from the bacterial oxidation of 5-aminonaphthalene-2-sulfonic acid. *Appl Environ Microbiol* 59: 1898–1903.

O'Keefe DP, JA Romesser, KJ Leto. 1988. Identification of constitutive and herbicide inducible cytochromes P-450 in *Streptomyces griseolus. Arch Microbiol* 149: 406–412.

O'Keefe DP, PA Harder. 1991. Occurrence and biological function of cytochrome P-450 monooxygenase in the actinomycetes. *Mol Microbiol* 5: 2099–2105.

Oh J-I, S-J Park, S-I Shin, I-J Ko, SJ Han, SW Park, T Song, YM Kim. 2010. Identification of *trans*- and *cis*-control elements involved in regulation of the carbon monoxide dehydrogenase genes in *Mycobacterium* sp. strain JC1 DSM 3803. *J Bacteriol* 192: 3925–3933.

Okibe N, M Gericke, KB Hallberg, DB Johnson. 2003. Enumeration and characterization of acidophilic microorganisms isolated from a pilot plant stirred-tank bioleaching operation. *Appl Environ Microbiol* 69: 1936–1943.

Okuyama H, K Kogame, M Mizuno, W Kobayashi, H Kanazawa, S Ohtani, K Wanatabe, H Kanda. 1992. Extremely psychrophilic microalgae isolated from the Antarctic Ocean. *Proc NIPR Symp Polar Biol* 5: 1–8.

Omori T, L Monna, Y Saiki, T Kodama. 1992. Desulfurization of dibenzothiophene by *Corynebacterium* sp. strain SY1. *Appl Environ Microbiol* 58: 911–915.

Ordóñez E, M Letek, N Valbuena, JA Gil, LK Mateos. 2005. Analysis of genes involved in arsenic resistance in *Corynebacterium glutamicum* ATCC 13032. *Appl Environ Microbiol* 71: 6206–6215.

Oren A, M Heldal, S Norland, EA Galinski. 2002. Intracellular ion and organic solute concentrations of the extremely halophilic bacterium *Salinibacter ruber. Extremophiles* 6: 491–498.

Oslowski DM, J-H Jung, D-H Seo, C-S Park, JF Holden. 2011. Production of hydrogen from α-1,4- and β-1,4-linked saccharides by marine hyperthermophilc archaea. *Appl Environ Microbiol* 77: 3169–3173.

Oz A, G Sabehi, M Koblízek, R Massana, A Béjà. 2005. *Roseobacter*-like bacteria in Red and Mediterranean Sea aerobic anoxygenic photosynthetic populations. *Appl Environ Microbiol* 71: 344–353.

Padden AN, FA Rainey, DP Kelly, AP Wood. 1997. *Xanthobacter tagetidis* sp. nov., an organism associated with *Tagetes* species and able to grow on substituted thiophenes. *Int J System Bacteriol* 47: 394–401.

Pal R et al. 2005. Hexachlorocyclohexane-degrading bacterial strains *Sphingomonas paucimobilis* B90A, UT26 and Sp+, having similar *lin* genes, represent three distinct species, *Sphingobium indicum* sp. nov., *Sphingobium japonicum* sp. nov., and *Sphingobium francense* sp. nov., and reclassification of [*Sphingomonas*] *chungbukensis* as *Sphingobium chungbukense* comb. nov. *Int J Syst Evol Microbiol* 55: 1965–1972.

Palleroni NJ. 1984. *Pseudomonas*, pp 141–199. In *Bergey's Manual of Systematic Bacteriology* Vol. 1 (Ed. NR Krieg), Williams and Wilkins, Baltimore, MD.

Park SW, EH Hwang, H Park, JA Kim, J Heo, KH Lee, T Song et al. 2003. Growth of mycobacteria on carbon monoxide and methanol. *J Bacteriol* 185: 142–147.

Parke D, LN Ornston. 1986. Enzymes of the β-ketoadipate pathway are inducible in *Rhizobium* and *Agrobacterium* spp. and constitutive in *Bradyrhizobium* spp. *J Bacteriol* 165: 288–292.

Parke D, F Rynne, A Glenn. 1991. Regulation of phenolic metabolism in *Rhizobium leguminosarum* biovar *trifolii. J Bacteriol* 173: 5546–5550.

Patel RN, CT Hou, AI Laskin, A Felix. 1982. Microbial oxidation of hydrocarbons: Properties of a soluble monooxygenase from a facultative methane-utilizing organisms *Methylobacterium* sp. strain CRL-26. *Appl Environ Microbiol* 44: 1130–1137.

Paul L, DJ Ferguson, JA Krzycki. 2000. The trimethylamine methyltransferase gene and multiple dimethylamine methyltransferase genes of *Methanosarcina barkeri* contain in-frame and read-through amber codons. *J Bacteriol* 182: 2520–2529.

Pellizari VH, S Bezborodnikov, JF Quensen, JM Tiedje. 1996. Evaluation of strains isolated by growth on naphthalene and biphenyl for hybridization of genes to dioxygenase probes and polychlorinated biphenyl-degrading ability. *Appl Environ Microbiol* 62: 2053–2058.

Perin L, L Martínez-Aguilar, R Castro-González, P Estrada-de los Santos, T Cabellos-Avelar, HV Guedes, VM Reis, J Caballero-Mallado. 2006. Diazotrophic *Burkholderia* species associated with field-grown maize and sugarcane. *Appl Environ Microbiol* 72: 3103–3110.

Perez J, TW Jeffries. 1990. Mineralization of ¹⁴C-ring-labelled synthetic lignin correlates with the production of lignin peroxidase, not of manganese peroxidase or laccase. *Appl Environ Microbiol* 56: 1806–1812.

Pérez-Reinado E, R Blasco, F Castillo, C Moreno-Vivían, MD Roldán. 2005. Regulation and characterization of two nitroreductase genes, *nprA* and *nprB* of *Rhodobacter capsulatus*. *Appl Environ Microbiol* 71: 7643–6549.

Peters F, M Rother, M Boll. 2004. Selenocysteine-containing proteins in anaerobic benzoate metabolism of *Desulfococcus multivorans. J Bacteriol* 186: 2156–2163.

Phillips WE, JJ Perry. 1974. Metabolism of *n*-butane and 2-butanone by *Mycobacterium vaccae. J Bacteriol* 120: 997–989.

Piccolo LL, C De Pasquale, R Fodale, AM Puglia, P Quatrini. 2011. Involvement of an alkane hydroxylase system of *Gordonia* sp. strain SoCg in degradation of solid *n*-alkanes. *Appl Environ Microbiol* 77: 1204–1213.

Piddington DL, A Kashkoulim, NA Buchmeier. 2000. Growth of *Mycobacterium tuberculosis* in a defined medium is very restricted by acid pH and Mg^{2+} levels. *Infect Immun* 68: 4518–4522.

Pinyakong O, H Habe, T Omori. 2003. The unique aromatic catabolic genes in sphingomonads degrading polycyclic aromatic hydrocarbons (PAHs). *J Gen Appl Microbiol* 49: 1–19.

Poelarends GJ, JET van Hylckama Vlieg, JR Marchesi, LM Freitas dos Santos, DB Janssen. 1999. Degradation of 1,2-dibromoethane by *Mycobacterium* sp. strain GP1. *J Bacteriol* 181: 2050–2058.

Poindexter JS. 1981. Oligotrophy. Fast and famine existence. *Adv Microb Ecol* 5: 63–89.

Pol A, K Heijmans, HR Harhangi, D Tedesco, MSM Jetten, HJM Op den Camp. 2007. Methanotrophy below pH 1 by a new Verrucomicrobia species. *Nature* 450: 874–878.

Polnisch E, H Kneifel, H Franzke, KL Hofmann. 1992. Degradation and dehalogenation of monochlorophenols by the phenol-assimilating yeast *Candida maltosa*. *Biodegradation* 2: 193–199.

Pometto AL, JB Sutherland, DL Crawford. 1981. *Streptomyces setonii*: Catabolism of vanillic acid via guaiacol and catechol. *Can J Microbiol* 27: 636–638.

Poth M, DD Focht. 1985. ^{15}N kinetic analysis of N_2O production by *Nitrosomonas europaea*: An examination of nitrifier denitrification. *Appl Environ Microbiol* 49: 1134–1141.

Pothuluri JV, JP Freeman, FE Evans, TB Moorman, CE Cerniglia. 1993. Metabolism of alachlor by the fungus *Cunninghamella elegans. J Agric Food Chem* 41: 483–488.

Poulter CD. 1990. Biosynthesis of non-head-to-tail terpenes. Formation of 1′-1 and 1′-3 linkages. *Acc Chem Res* 23: 70–77.

Poupin P, N Truffaut, B Combourieu, M Sancelelme, H Veschambre, AM Delort. 1998. Degradation of morpholine by an environmental *Mycobacterium* strain involves a cytochrome P-450. *Appl Environ Microbiol* 64: 159–165.

Powlowski JB, J Ingebrand, S Dagley. 1985. Enzymology of the β-ketoadipate pathway in *Trichosporon cutaneum. J Bacteriol* 163: 1136–1141.

Powlowski JB, S Dagley. 1985. β-ketoadipate pathway in *Trichosporon cutaneum* modified for methyl-substituted metabolites. *J Bacteriol* 163: 1126–1135.

Prakash O, R Lal. 2006. Description of *Sphingobium fulginis* sp. nov., a phenanthrene-degrading bacterium from a fly ash dumping site, and reclassification of *Sphingomonas cloacae* as *Sphingobium cloacae* com. nov. *Int J Syst Evolut Microbiol* 56: 2147–2152.

Prieto MA, A Perez-Aranda, JL Garcia. 1993. Characterization of an *Escherichia coli* aromatic hydroxylase with a broad substrate range. *J Bacteriol* 175: 2162–2167.

Prieto MA, JL Garcia. 1994. Molecular characterization of 4-hydroxyphenylacetate 3-hydroxylase of *Escherichia coli*. A two-component enzyme. *J Biol Chem* 269: 22823–22829.

Pucci MI, JA Thanassi, H-T Ho, PJ Falk, TJ Dougherty. 1995. *Staphylococcus haemolyticus* contains two D-glutamic acid biosynthetic activities, a glutamate racemase and a D-amino acid transferase. *J Bacteriol* 177: 336–342.

Ragsdale SW. 1991. Enzymology of the acetyl-CoA pathway for CO_2 fixation. *Crit Revs Biochem Mol Biology* 26: 261–300.

Rainey FA, S Klatte, RM Kroppenstedt, E Stackebrandt. 1995. *Dietzia*, a new genus including *Dietzia maris* comb. nov., formerly *Rhodococcus maris. Int J Syst Evol Microbiol* 45: 32–36.

Ramamoorthi R, ME Lidstrom. 1995. Transcriptional analysis of *pqqD* and study of the regulation of pyrroloquinoline quinone biosynthesis in *Methylobacterium extorquens* AM1. *J Bacteriol* 177: 206–211.

Ramana ChV, Ch Sasikala, K Arunasri, PA Kumar, TNR Srinivas, S Shivaji, P Gupta, T Süling, JF Imhoff. 2006. *Rubrivivax benzoatilyticus* nov. sp., an aromatic, hydrocarbon-degrading purple betaproteobacterium. *Int J Syst Evol Microbiol* 56: 2157–2163.

Ramette A, JJ LiPuma, JM Tiedje. 2005. Species abundance and diversity in *Burkholderia cepacia* complex in the environment. *Appl Environ Microbiol* 71: 1193–1201.

Ramos-Vera WH, IA Berg, G Fuchs. 2009. Autotrophic carbon dioxide assimilation in *Thermoproteales* revisited. *J Bacteriol* 191: 4286–4297.

Ramos-Vera WH, V Labonté, M Weiss, J Pauly, G Fuchs. 2010. Regulation of autotrophic CO_2 fixation in the archaeon *Thermoproteus neutrophilus. J Bacteriol* 192: 5329–5340.

Rao JR, JE Cooper. 1994. Rhizobia catabolize *nod* gene-inducing flavonoids via C-ring fission mechanisms. *J Bacteriol* 176: 5409–5413.

Rao JR, ND Sharma, JTG Hamilton, DR Boyd, JE Cooper. 1991. Biotransformation of the pentahydroxy flavone quercitin by *Rhizobium loti* and *Bradyrhizobium* strains Lotus. *Appl Environ Microbiol* 57: 1563–1565.

Rao M, TL Streur, FE Aldwell, GM Cook. 2001. Intracellular pH regulation by *Mycobacterium smegmatis* and *Mycobacterium bovis* BCG. *Microbiology (UK)* 147: 1017–1024.

Rasche ME, MR Hyman, DJ Arp. 1991. Factors limiting aliphatic chlorocarbon degradation by *Nitrosomonas europaea*: Cometabolic inactivation of ammonia monooxygenase and substrate specificity. *Appl Environ Microbiol* 57: 2986–2994.

Rawlings DE, DB Johnson. 2007. The microbiology of biomining: Development and optimization of mineral-oxidizing microbial consortia. *Microbiology (UK)* 153: 314–324.

Reeves RA, MD Gibbs, DD MOrris, KR Gritthiths, DJ Saul, PL Bergquist. 2000. Sequencing and expression of additional xylanases from the hyperthermophile *Thermotoga maritima. Appl Environ Microbiol* 66: 1532–1537.

Rehndorf J, CL Zimmer, UT Bornscheuer. 2009. Cloning, expression, characterization, and biocatalytic investigation of the 4-hydroxyacetophenone monooxygenase from *Pseudomonas putida* JD1. *Appl Environ Microbiol* 75: 3106–3114.

Reinhold-Hurek B, T Hurek. 2000. Reassessment of the taxonomic structure of the diazotrophic genus *Azoarcus sensu lato* and description of three new genera and new species, *Azovibrio restrictus* gen. nov., sp. nov., *Azospira oryzae* gen. nov., sp. nov. and *Azonexus fungiphilus* gen. nov., sp. nov. *Int J Syst Evolut Microbiol* 50: 649–649.

Rhee J-K, D-G Ahn, Y-G Kim, J-W Oh. 2005. New thermophilic and thermostable esterase with sequence similarity to the hormone-sensitive lipase family cloned from a metagenomic library. *Appl Environ Microbiol* 71: 817–825.

Rice CP, HC Sikka. 1973. Uptake and metabolism of DDT by six species of marine algae. *J Agric Food Chem* 21: 148–152.

Rinaudo G, S Orenga, MP Fernandez, H Meugnier, R Bardin. 1991. DNA homologies among members of the genus *Azorhizobium* and other stem- and root-nodulating bacteria isolated from the tropical legume *Sesbania rostrata*. *Int J Syst Evol Microbiol* 41: 114–120.

Rittenberg SC. 1972. The obligate autotroph—the demise of a concept. *Antonie van Leeuwenhoek* 38: 457–478.

Rivkina EM, EI Friedmann, CP McKay, DA Gilichinsky. 2000. Metabolic activity of permafrost bacteria below the freezing point. *Appl Environ Microbiol* 66: 3230–3233.

Robrock KR, M Coelhan, DL Sedlak, L Alvarez-Cohen. 2009. Aerobic biotransformation of polybrominated diphenyl ethers (PBDEs) by bacterial isolates. *Environ Sci Technol* 43: 5705–5711.

Rohdish F, A Wiese, R Feicht, H Simon, A Bacher. 2001. Enoate reductases of *Clostridia*. Cloning, sequencing, and expression. *J Biol Chem* 276: 5779–5787.

Rohwerder T, U Breuer, D Benndorf, U Lechner, RH Müller. 2006. The alkyl *tert*-butyl ether intermediate 2-hydroxyisobutyrate is degraded via a novel cobalamin-dependent mutase pathway. *Appl Environ Microbiol* 72: 4128–4135.

Romanov V, RP Hausinger. 1996. NADPH-dependent reductive *ortho* dehalogenation of 2,4-dichlorobenzoic acid in *Corynebacterium sepedonicum* KZ-4 and coryneform bacterium strain NTB-1 via 2,4-dichlorobenzoyl coenzyme A. *J Bacteriol* 178: 2656–2661.

Romesser JA, DP O'Keefe. 1986. Induction of cytochrome P-450-dependent sulfonylurea metabolism in *Streptomyces griseolus*. *Biochem Biophys Res Comm* 140: 650–659.

Romine MF et al. 1999. Complete sequence of a 184-kilobase catabolic plasmid from *Sphingomonas aromaticivorans*. *J Bacteriol* 181: 1585–1602.

Rontani J-F, MJ Gilewicz, VD Micgotey, TL Zheng, PC Bonin, J-C Bertrand. 1997. Aerobic and anaerobic metabolism of 6,10,14-trimethylpentadecan-2-one by a denitrifying bacterium isolated from marine sediments. *Appl Environ Microbiol* 63: 636–643.

Rosner BM, FA Rainey, RM Kroppenstedt, B Schink. 1997. Acetylene degradation by new isolates of aerobic bacteria and comparison of acetylene hydratase enzymes. *FEMS Microbiol Lett* 148: 175–180.

Rosłoniec KZ, MH Wilbrink, JK Capyk, WH Mohn, M Ostendorf, R van der Geize, L Dijkhuizen, LD Eltis. 2009. Cytochrome P450 125 (CPY125) catalyses C26-hydroxylation to initiate sterol side-chain degradation in *Rhodococcus jostii* RHA1. *Mol Microbiol* 74: 1031–1043.

Rott B, S Nitz, F Korte. 1979. Microbial decomposition of sodium pentachlorophenolate. *J Agric Food Chem* 27: 306–310.

Rouf MA, RF Lomprey. 1968. Degradation of uric acid by certain aerobic bacteria. *J Bacteriol* 96: 617–622.

Rousseau B, L Dostal, JPN Rosazza. 1997. Biotransformations of tocopherols by *Streptomyces catenulae*. *Lipids* 32: 79–84.

Roy R, S Mukund, GJ Schut, DM Dunn, R Weiss, MWW Adams. 1999. Purification and molecular characterization of the tungsten-containing formaldehyde ferredoxin oxidoreductase from the hyperthermophilic archaeon *Pyrococcus furiosus*: The third of a putative five-member tungstoenzyme family. *J Bacteriol* 181: 1171–1180.

Rudnick G, RH Abeles. 1975. Reaction mechanism and structure of the active site of proline racemase. *Biochemistry* 14: 4515–4522.

Sakai, K, A Nakazawa, K Kondo, and H Ohta. 1985. Microbial hydrogenation of nitroolefins. *Agric Biol Chem* 49: 2231–2236.

Sakai Y, H Takahashi, Y Wakasa, T Kotani, H Yurimoto, N Miyachi, PP Van Veldhoven, N Kato. 2004. Role of α-methacyl coenzyme A racemase in the degradation of methyl-branched alkanes by *Mycobacterium* sp. strain P101. *J Bacteriol* 186: 7214–7220.

Salka I, V Moulisivá, G Jost, K Jürgens, M Labrenz. 2008. Abundance, depth distribution, and composition of aerobic bacteriochlorophyll *a*-producing bacteria in four basins of the Central Baltic Sea. *Appl Environ Microbiol* 74: 4398–4404.

Sanford RA, JR Cole, JM Tiedje. 2002. Characterization and description of *Anaeromyxobacter dehalogenans* gen. nov., sp. nov., an aryl-respiring facultative anaerobic myxobacterium. *Appl Environ Microbiol* 68: 893–900.

Sangwan P, S Kovac, KER Davis, M Sait, PH Janssen. 2005. Detection and cultivation of soil verrucomicrobia. *Appl Environ Microbiol* 71: 8402–8410.

Sangwan P, X Chen, P Hugenholtz, PH Janssen. 2004. *Chthoniobacter flavus* gen. nov., sp. nov., the first pure-culture representative of subdivision two, *Spartobacteria* classis nov., of the phylum *Verruco-microbia*. *Appl Environ Microbiol* 70: 5875–5881.

Santini JM, LI Sly, RD Schnagl, JM Macy. 2000. A new chemolithoautotrophic arsenite-oxidizing bacterium isolated from a gold mine: Phylogenetic, physiological, and preliminary biochemical studies. *Appl Environ Microbiol* 66: 92–97.

Sapra R, MFJM Verhagen, MWW Adams. 2000. Purification and characterization of a membrane-bound hydrogenase from the hyperthermophilic archaeon *Pyrococcus furiosus*. *J Bacteriol* 182: 3423–3428.

Sarkanen S, RA Razal, T Piccariello, E Yamamoto, NG Lewis. 1991. Lignin peroxidase: Toward a clarification of its role *in vivo*. *J Biol Chem* 266: 3636–3643.

Sarma PM, D Bhattacharya, S Krishnan, B Lal. 2004. Degradation of polycyclic aromatic hydrocarbons by a newly discovered enteric bacterium *Leclercia adecaroxylata*. *Appl Environ Microbiol* 70: 3163–3166.

Sattley WM, MT Madigan. 2006. Isolation, characterization, and ecology of cold-active, chemolithotrophic, sulfur-oxidizing bacteria from perennially ice-covered Lake Fryxell, Antarctica. *Appl Environ Microbiol* 72: 5562–5568.

Sauer K, U Harms, RK Thauer. 1997. Methanol: Coenzyme M methyltransferase from *Methanosarcina barkeri*. Purification, properties and encoding genes of the corrinoid protein MT1. *Eur J Biochem* 243: 670–677.

Savolainen K, P Bhaumik, W Schmitz, TJ Kotti, E Conzelmann, RK Wierenga, JK Hiltunen. 2005. α-methylacyl-CoA racemase from *Mycobacterium tuberculosis*. Mutational and structural characterization of the active site and the fold. *J Biol Chem* 280: 12611–12620.

Sayama M, M Inoue, M-A Mori, Y Maruyama, H Kozuka. 1992. Bacterial metabolism of 2,6-dinitrotoluene with *Salmonella typhimurium* and mutagenicity of the metabolites of 2,6- dinitrotoluene and related compounds. *Xenobiotica* 22: 633–640.

Schaeffer JK, KD Goodwin, IR McDonald, JC Murrell, RS Oremland. 2002. *Leisingera methylohalidivorans* gen. nov., sp. nov., a marine methylotroph that grows on methyl bromide. *Int J Syst Evol Microbiol* 52: 851–859.

Schauer F, K Henning, H Pscheidl, RM Wittich, P Fortnagel, H Wilkes, V Sinnwell, W Francke. 1995. Biotransformation of diphenyl ether by the yeast *Trichosporon beigelii* SBUG 765. *Biodegradation* 6: 173–180.

Schauder R, F Widdel, G Fuchs. 1987. Carbon assimilation pathways in sulfate-reducing bacteria II. Enzymes of a reductive citric acid cycle in the autotrophic *Desulfobacter hydrogenophilus*. *Arch Microbiol* 148: 218–225.

Schink B, N Pfennig. 1982. *Propionigenium modestum* gen. nov., sp. nov., a new strictly anaerobic, nonsporing bacterium growing on succinate. *Arch Microbiol* 133: 209–216.

Schirmer A, MA Rude, X Li, E Popova, SB del Cardayre. 2010. Microbial biosynthesis of akanes. *Science* 329: 559–562.

Schnell S, B Schink. 1991. Anaerobic aniline degradation via reductive deamination of 4-aminobenzoyl-CoA in *Desulfobacterium anilini*. *Arch Microbiol* 155: 183–190.

Schleinitz KM, S Kleinsteuber, T Vallaeys, W Babel. 2004. Localization and characterization of two novel genes encoding stereospecific dioxygenases catalyzing 2(2,4-dichlorophenoxy)propionate cleavage in *Delftia acidovorans* MC1. *Appl Environ Microbiol* 70: 5357–5365.

Schleinitz KM, S Schmeling, N Jehmlich, M van Bergen, H Harms, S Kleinsteuber, C Vogt, G Fuchs. 2009. Phenol degradation in the strictly anaerobic iron-reducing bacterium *Geobacter metallireducens* GS-15. *Appl Environ Microbiol* 73: 3912–3919.

Schleper C, G Puehler, I Holz, A Gambacorta, D Janekovic, U Santarius, H-P Klenk, W Zillig. 1995. *Picrophilus* gen. nov., fam. nov.: A novel, heterotrophic, thermoacidophilic, genus and family comprising archaea capable of growth around pH 0. *J Bacteriol* 177: 7050–7059.

Schneider J, R Grosser, K Jayasimhulu, W Xue, D Warshawsky. 1996. Degradation of pyrene, benz[*a*]anthracene, and benzo[*a*]pyrene by *Mycobacterium* sp. strain RGHII-135, isolated from a former coal gasification site. *Appl Environ Microbiol* 62: 13–19.

Schoenborn L, PS Yates, BE Grinton, P Hugenholtz, PH Janssen. 2004. Liquid serial dilution is inferior to solid media for isolation of cultures representative of the phylum-level diversity of soil bacteria. *Appl Environ Microbiol* 70: 4363–4366.

Schühle K, G Fuchs. 2004. Phenylphosphate carboxylase: A new C–C lyase involved in anaerobic phenol metabolism in *Thauera aromatica*. *J Bacteriol* 186: 4556–4567.

Schultz JE, PF Weaver. 1982. Fermentation and anaerobic respiration by *Rhodospirillum rubrum* and *Rhodopseudomonas capsulata*. *J Bacteriol* 149: 181–190.

Schut F, E J de Vries, JC Gottschal, BR Robertson, W Harder, RA Prins, DK Button. 1993. Isolation of typical marine bacteria by dilution culture: Growth, maintenance, and characteristics of isolates under laboratory conditions. *Appl Environ Microbiol* 59: 2150–2160.

Segerer A, A Neuner, JK Kristjansson, KO Stetter. 1986. *Acidianus infernus* gen. nov., sp. nov., and *Acidianus brierleyi* comb. nov.: Facultatively aerobic, extremely acidophilic thermophilic sulfur-metabolizing archaebacteria. *Int J Syst Evol Microbiol* 36: 559–564.

Seifritz C, SL Daniel, A Gobner, HL Drake. 1993. Nitrate as a preferred electron sink for the acetogen *Clostridium thermoaceticum*. *J Bacteriol* 175: 8008–8013.

Seigle-Murandi FM, SMA Krivobok, RL Steiman, J-LA Benoit-Guyod, G-A Thiault. 1991. Biphenyl oxide hydroxylation by *Cunninghamella echinulata*. *J Agric Food Chem* 39: 428–430.

Seigle-Murandi F, R Steiman, F Chapella, C Luu Duc. 1986. 5-Hydroxylation of benzimidazole by Micromycetes II. Optimization of production with *Absidia spinosa*. *Appl Microbiol Biotechnol* 25: 8–13.

Seitz H-J, H Cypionka. 1986. Chemolithotrophic growth of *Desulfovibrio desulfuricans* with hydrogen coupled to ammonification with nitrate or nitrite. *Arch Microbiol* 146: 63–67.

Semple KT, RB Cain. 1996. Biodegradation of phenols by the alga *Ochromonas danica*. *Appl Environ Microbiol* 62: 1265–1273.

Servent D, C Ducrorq, Y Henry, A Guissani, M Lenfant. 1991. Nitroglycerin metabolism by *Phanerochaete chrysosporium*: Evidence for nitric oxide and nitrite formation. *Biochim Biophys Acta* 1074: 320–325.

Sessitsch A et al. 2005. *Burkholdetia phytofirmans* sp. nov., a novel plant-associated bacterium woth plant-beneficial properties. *Int J Syst Evol Microbiol* 55: 1187–1192.

Seto M, K Kimbura, M Shimura, T Hatta, M Fukuda, K Yano. 1995a. A novel transformation of polychlori-nated biphenyls by *Rhodococcus* sp. strain RHA1. *Appl Environ Microbiol* 61: 3353–3358.

Seto M, E Masai, M Ida, T Hatta, K Kimbara, M Fukuda, K Yano. 1995b. Multiple polychlorinated biphenyl transformation systems in the Gram-positive bacterium *Rhodococcus* sp. strain RHA1. *Appl Environ Microbiol* 61: 4510–4513.

Seubert W, U Remberger. 1963. Untersuchungen über den bakteriellen Abbau van Isoprenoiden II. Die Rolle der Kohlensäure. *Biochem Z* 338: 245–264.

Sharma PK, PL McCarty. 1996. Isolation and characterization of a facultatively aerobic bacterium that reduc-tively dehalogenates tetrachloroethene to *cis*-dichloroethene. *Appl Environ Microbiol* 62: 761–765.

Sharma P, V Raina, R Kumari, S Malhotra, C Dogra, H Kumari, H-P E Kohler, H-R Buser, C Holliger, R Lal. 2006. Haloalkane dehalogenase LinB is responsible for β- and δ-hexachlorocyclohexane transformation in *Sphingobium indicum* B90A. *Appl Environ Microbiol* 72: 5720–5727.

Sheng X, X Chen, L He. 2008. Characteristics of an endophytic pyrene-degrading bacterium *Enterobacter* sp. 12J1 from *Allium macrostemon*. *Int Biodet Biodeg* 62: 88–95.

Shiba H, T Kawasumi, Y Igarshi, T Kodada, Y Minoda. 1985. The CO_2 assimilation via the reductive tricarbox-ylic acid cycle in an obligately autotrophic, hydrogen-oxidizing bacterium, *Hydrogenobacter thermophi-lus*. *Arch Microbiol* 141: 198–203.

Shimizu S, S Jareonkitmongol, H Kawashima, K Akimoto, H Yamada. 1991. Production of a novel ω1-eicosapentaenoic acid by *Mortierella alpina* 1S-4 grown on 1-hexadecene. *Arch Microbiol* 156: 163–166.

Siegele DA, R Kolter. 1992. Life after log. *J Bacteriol* 174: 345–348.

Sietmann R, E Hammer, M Specht, CE Cerniglia, F Schauer. 2001. Novel ring cleavage products in the biotrans-formation of biphenyl by the yeast *Trichosporon mucoides*. *Appl Environ Microbiol* 67: 4158–4165.

Sikkema J, JAM de Bont. 1993. Metabolism of tetralin (1,2,3,4-tetrahydronaphthalene) in *Corynebacterium* sp. strain C125. *Appl Environ Microbiol* 59: 567–572.

Simon HM, CE Jahn, LT Bergerud, MK Sliwinski, PJ Weimer, DK Willis, RM Goodman. 2005. Cultivation of mesophilic soil crenarchaeotes in enrichment cultures from plant roots. *Appl Environ Microbiol* 71: 4751–4760.

Simoni S, S Klinke, C Zipper, W Angst, H-P E Kohler. 1996. Enantioselective metabolism of chiral 3-phenylbutyric acid, an intermediate of linear alkylbenzene degradation, by *Rhodococcus rhodochrous*. *Appl Environ Microbiol* 62: 749–755.

Singh A, R Lal. 2009. Sphingobium ummariense sp. nov., a hexachlorocyclohexane (HCH)-degrading bacterium, isolated from HCH-contaminated soil. *Int J Syst Evolut Microbiol* 59: 162–166.

Singh BK, A Walker, JAW Morgan, DJ Wright. 2004. Biodegradation of chloropyrifos by *Enterobacter* strain B-14 and its use in bioremediation of contaminated soils. *Appl Environ Microbiol* 70: 4855–4863.

Skidmore ML, JM Focht, MJ Sharp. 2000. Microbial life beneath a high arctic glacier. *Appl Environ Microbiol* 66: 3214–3220.

Sluis MK, RA Larsen, JG Krum, R Anderson, WW Metcalf, SA Ensign. 2002. Biochemical, molecular, and genetic analyses of the acetone carboxylases from *Xanthobacter autotrophicus* strain Py2 and *Rhodobacter capsulatus* strain B10. *J Bacteriol* 184: 2969–2977.

Sluis MK, FJ Small, JR Allen, SA Ensign. 1996. Involvement of an ATP-dependent carboxylase in a CO_2-dependent pathway of acetone metabolism by *Xanthobacter* strain Py2. *J Bacteriol* 178: 4020–4026.

Smith DJ, MA Patrauchan, C Florizone, LD Eltis, WW Mohn. 2008. Distinct roles for two CYP226 family cytochromes P450 in abietane diterpenoid catabolism by *Burkholderia xenovorans* LB400. *J Bacteriol* 190: 1575–1583.

Smith RV, JP Rosazza. 1983. Microbial models of mammalian metabolism. *J Nat Prod* 46: 79–91.

Soeder CJ, E Hegewald, H Kneifel. 1987. Green microalgae can use naphthalenesulfonic acids as sources of sulfur. *Arch Microbiol* 148: 260–263.

Sohn JH, KK Kwon, J-H Kang, H-B Jung, S-J Kim. 2004. *Novosphingobium pentaromaticivorans* sp. nov., a high-molecular-mass polycyclic aromatic hydrocarbon-degrading bacterium isolated from estuarine sediment. *Int J Syst Evolut Microbiol* 54: 1483–14587.

Song B, MM Häggblom, J Zhou, JM Tiedje, NJ Palleroni. 1999. Taxonomic characterization of denitrifying bacteria that degrade aromatic compounds and description of *Azoarcus toluvorans* sp. nov. and *Azoarcus toluclasticus* sp. nov. *Int J Syst Bacteriol* 49: 1129–1140.

Song B, NJ Palleroni, LJ Kerkhof, MM Häggblom. 2001. Characterization of halobenzoate-degrading, denitrifying, *Azoarcus* and *Thauera* isolates and description of *Thauera chlorobenzoica* sp. nov. *Int J Syst Evol Microbiol* 51: 589–602.

Sorokin DY, JG Kuenen. 2005. Haloalkaliphilic sulfur-oxidizing bacteria in soda lakes. *FEMS Microbiol Revs* 29: 685–701.

Sorokin DY, AM Lysenko, LL Mityuishina, TP Tourova, BE Jones, FA Rainey, LA Robertson, GJ Kuenen. 2001. *Thioalkalimicrobium aerophilum* gen. nov., sp. nov and *Thioalkalimicrobium sibericum* sp. nov., and *Thioalkalivibrio versutus* gen. nov., sp. nov., *Thioalkalivibrio nitratis* sp. nov. and *Thioalkalivibrio denitrificans* sp. nov., novel alkaliphilic and obligately chemolithoautotrophic sulfur-oxidizing bacteria from soda lakes. *Int J Syst Evol Microbiol* 51: 565–580.

Spiess T, F Desiere, P Fischer, JC Spain, H-J Knackmuss, H Lenke. 1998. A new 4-nitrotoluene degradation pathway in a *Mycobacterium* strain. *Appl Environ Microbiol* 64: 446–452.

Spring S, B Merkhoffer, N Weiss, RM Kroppenstedt, H Hippe, E Stackebrandt. 2003. Characterization of novel psychrophilic clostridia from an Antarctic microbial mat: Description of *Clostridium frigoris* sp. nov., *Clostridium lacusfryxellense* sp. nov., *Clostridium bowmanii* sp. nov. and *Clostridium psychrophilum* sp. nov. and reclassification of *Clostridium laramiense* as *Clostridium estertheticum* subsp. *laramiense* subsp. nov. *Int J Syst Evol Microbiol* 53: 1019–1029.

Springer N, W Ludwig, B Philipp, B Schink. 1998. *Azoarcus anaerobius* sp. nov., a resorcinol-degrading, strictly anaerobic, denitrifying bacterium. *Int J Syst Bacteriol* 48: 953–956.

Stadtman TC, P Elliot. 1957. Studies on the enzymatic reduction of amino acids II. Purification and properties of a D-proline reductase and a proline racemase from *Clostridium sticklandii*. *J Biol* 983997.

Stenklo K, HD Thorell, H Bergius, R Aasa, T Nilsson. 2001. Chlorite dismutase from *Ideonella dechloratans*. *J Biol Inorg Chem* 6: 601–607.

Stekhanova TN, AV Mardanov, EY Bezsudnova, VM Gumerov, NK Ravin, KG Skryabin, VO Popov. 2010. Characterization of a thermostable short-chain alcohol dehydrogenase from the hyperthermophilic archaeon *Thermococcus sibiricus*. *Appl Environ Microbiol* 76: 4096–4098.

Steven B, G Briggs, CP McKay, WH Pollard, CW Greer, LG Whyte. 2007. Characterization of the microbial diversity in a permafrost sample from the Canadian high Arctic using culture-dependent and culture-independent methods. *FEMS Microbiol Ecol* 59: 513–533.

Stoecker K, B Bendiger, B Schöning, PH Nielsen, JP Nielsen, C Baranyi, ER Toenshoff, H Daims, M Wagner. 2006. Cohn's *Crenothrix* is a filamentous methane oxidizer with an unusual methane monooxygenase. *Proc Natl Acad Sci USA* 103: 2363–2367.

Suar M et al. 2005. Enantioselective transformation of α-hexachlorocyclohexane by the dehydrochlorinases LinA1 and LinA2 from the soil bacterium *Sphingomonas paucimobilis* B90A. *Appl Environ Microbiol* 71: 8514–8518.

Subramanian V, CS Vaidyanathan. 1984. Anthranilate hydroxylase from *Aspergillus niger*: New type of NADPH-linked nonheme iron monooxygenase. *J Bacteriol* 160: 651–655.

Sugimoto M, M Tanaba, M Hataya, S Enokibara, JA Duine, F Kawai. 2001. The first step in polyethylene glycol degradation by sphingomonads proceeds via a flavoprotein alcohol dehydrogenase containing flavin adenine dinucleotide. *J Bacteriol* 183: 6694–6698.

Sun W, R Sierra-Alvarez, L Milner, JA Field. 2010. Anaerobic oxidation of arsenite linked to chlorate reduction. *Appl Environ Microbiol* 76: 6804–6811.

Sutherland JB, AL Selby, JP Freeman, FE Evans, CE Cerniglia. 1991. Metabolism of phenanthrene by *Phanerochaete chrysosporium*. *Appl Environ Microbiol* 57: 3310–3316.

Sutherland JB, DL Crawford, AL Pometto. 1983. Metabolism of cinnamic, *p*-coumaric, and ferulic acids by *Streptomyces setonii*. *Can J Microbiol* 29: 1253–1257.

Sutherland JB, JP Freeman, AL Selby, PP Fu, DW Miller, CE Cerniglia. 1990. Stereoselective formation of a K-region dihydrodiol from phenanthrene by *Streptomyces flavovirens*. *Arch Microbiol* 154: 260–266.

Suzuki T. 1978. Enzymatic methylation of pentachlorophenol and its related compounds by cell-free extracts of *Mycobacterium* sp isolated from soil. *J Pesticide Sci* 3: 441–443.

Suzuki S, A Hiraishi. 2007. *Novosphingobium naphthalenivorans* sp. nov., a naphthalene-degrading bacterium isolated from polychlorinated-dioxin-contaminated environments. *J Gen Appl Microbiol* 53: 221–228.

Swain A, KV Waterhouse, WA Venables, AG Cally, SE Lowe. 1991. Biochemical studies of morpholine catabolism by an environmental mycobacterium. *Appl Microbiol Biotechnol* 35: 110–114.

Swingley WD et al. 2007. The complete genome sequence of *Roseobacter denitrificans* reveals a mixotrophic rather than photosynthetic metabolism. *J Bacteriol* 189: 683–690.

Sylvestre M, R Massé, F Messier, J Fauteux, J-G Bisaillon, R Beaudet. 1982. Bacterial nitration of 4-chlorobiphenyl. *Appl Environ Microbiol* 44: 871–877.

Szewzyk R, N Pfennig. 1987. Complete oxidation of catechol by the strictly anaerobic sulfate-reducing *Desulfobacterium catecholicum* sp. nov. *Arch Microbiol* 147: 163–168.

Takahashi H, B Kimura, M Yoshikawa, T Fujii. 2003. Cloning and sequencing of the histidine decarboxylase genes of Gram-negative, histamine-producing bacteria and their application in detection and identification of these organisms in fish. *Appl Environ Microbiol* 69: 2568–2579.

Takai K, DP Moser, TC Onstott, N Spoelstra, SM Pfiffner, A Dohnalkova, JK Fredrikson. 2001. *Alkaliphilus transvaalensis* gen. nov., sp. nov., an extremely alkaliphilic bacterium isolated from a deep South Africa gold mine. *Int Syst Evol Microbiol* 51: 1245–1256.

Takeuchi M, T Sakane, M Yanagi, K Yamasato, K Hamana, A Yokota. 1995. Taxonomic study of bacteria isolated from plants: Proposal of *Sphingomonas rosa* sp. nov., *Sphingomonas pruni* sp. nov., *Sphingomonas asaccharolytica* sp. nov., and *Sphingomonas mali* sp. nov. *Int J Syst Bacteriol* 45: 334–341.

Takeuchi M, K Hatano, I Sedlácek, Z Pácova. 2002. *Rhodococcus jostii* sp. nov., isolated from a medieval grave. *Int J Syst Evol Microbiol* 52: 409–413.

Takeuchi M, K Hamana, A Hiraishi. 2001. Proposal of the genus *Sphingomonas sensu stricto* and three new genera, *Sphingobium*, *Novosphingobium* and *Sphingopyxis*, on the basis of phylogenetic and chemotaxonomic analyses. *Int J Syst Evol Microbiol* 51: 1405–1417.

Tallant TC, L Paul, JA Krzycki. 2001. The MtsA subunit of the methylthiol: Coenzyme M methyltransferase of *Methanosarcina barkeri* catalyzes both half-reactions of corrin-dependent dimethylsulfide: Coenzyme M methyl transfer. *J Biol Chem* 276: 4485–4493.

Tanaka A, M Ueda. 1993. Assimilation of alkanes by yeasts: Functions and biogenesis of peroxisomes. *Mycol Res* 98: 1025–1044.

Tasaki M, Y Kamagata, K Nakamura, E Mikami. 1991. Isolation and characterization of a thermophilic benzoate-degrading, sulfate-reducing bacterium, *Desulfotomaculum thermobenzoicum* sp. nov. *Arch Microbiol* 155: 348–352.

Tauber MM, A Cavaco-Paulo, K-H Robra, GM Gübitz. 2000. Nitrile hydratase and amidase from *Rhodococcus rhodochrous* hydrolyze acrylic fibers and granular polyacrylonitrile. *Appl Environ Microbiol* 66: 1634–1638.

Taylor SL, JE Stratton, JA Nordlee. 1989. Histamine poisoning (scombroid fish poisoning): An allergy-like intoxication. *Clin Toxicol* 27: 225–240.

Tett VA, AJ Willetts, HM Lappin-Scott. 1994. Enantioselective degradation of the herbicide mecoprop [2-methyl-4-chlorophenoxypropionic acid] by mixed and pure bacterial cultures *FEMS Microbiol Ecol* 14: 191–200.

Thauer RK. 2007. A fifth pathway of carbon fixation. *Science* 318: 1732–1733.

Thompson KT, FH Croker, HL Fredrickson. 2005. Mineralization of the cyclic nitramine explosive hexahydro-1,3,5-trinitro-1,3,5-triazine by *Gordonia* and *Williamsia* spp. *Appl Environ Microbiol* 71: 8265–8272.

Tolli J, GM King. 2005. Diversity and structure of bacterial chemolithotrophic communities in pine forest and groecosystem soils. *Appl Environ Microbiol* 71: 8411–8418.

Toney N, T Adakambi, S Toney, M Yakrus, WR Butler. 2010. Revival and emended description of "*Mycobacterium paraffinicum*" Davis, Chase and Raymond 1956 as *Mycobacterium paraffinicum* sp. nov., nom. rev. *Int J Syst Bacteriol* 60: 2307–2313.

Toyama T, N Momotani, Y Ogata, Y Miyamori, D Inoue, K Sei, K Mori, S Kikuchi, M Ike. 2010. Isolation and characterization of 4-*tert*-butylphenol-utilizing *Sphingobium fulginis* strains from *Phragmites australis* rhizosphere sediment. *Appl Environ Microbiol* 76: 6733–6740.

Toraya T, T Oka, M Ando, M Yamanishi, H Nishihara. 2004. Novel pathway for utilization of cycloropanecarboxylate by *Rhodococcus rhodochrous*. *Appl Environ Microbiol* 70: 224–228.

Trower MK, FS Sariaslani, DP O'Keefe. 1989. Purification and characterization of a soybean flour-induced cytochrome P-450 from *Streptomyces griseus*. *J Bacteriol* 171: 1781–1787.

Tsou AY, SC Ransom, JA Gerlt, DD Buechter, PC Babbitt, GL Kenyon. 1990. Mandelate pathway of *Pseudomonas putida*: Sequence relationships involving mandelate racemase, (*S*)-mandelate dehydrogenase, and benzoylformate decarboxylase and expression of benzoylformate decarboxylase in *Escherichia coli*. *Biochemistry* 29: 9856–9862.

Tuffin IM, P de Groot, SM Deane, DE Rawlings. 2005. An unusual Tn*21*-like transposon containing an unusual *ars* operon is present in highly arsenic-resistant strains of the biomining bacterium *Acidithiobacillus caldus*. *Microbiology (UK)* 151: 3027–3039.

Tuffin IM, SB Hector, SM Deane, DE Rawlings. 2006. Resistance determinants of a highly arsenic-resistant strain of *Leptospirllum ferriphilum* isolated from a commercial biooxidation tank. *Appl Environ Microbiol* 72: 2247–2253.

Tung HC, NE Bramall, PB Price. 2005. Microbial origin of excess methane in glacial ice and implications for life on Mars. *Proc Natl Acad Sci USA* 102: 18292–18296.

Uffen RL. 1973. Growth properties of *Rhodospirillum rubrum* mutants and fermentation of pyruvate in anaerobic dark conditions. *J Bacteriol* 116: 874–884.

Utkin I, C Woese, J Wiegel. 1994. Isolation and characterization of *Desulfitobacterium dehalogenans* gen. nov, sp. nov., an anaerobic bacterium which reductively dechlorinates chlorophenolic compounds. *Int J Syst Bacteriol* 44: 612–619.

Uyttebroek M, S Vermeir, P Wattiau, A Ryngaert, D Springael. 2007. Characterization of cultures enriched from acidic polycyclic aromatic hydrocarbon-contaminated soil for growth on pyrene at low pH. *Appl Environ Microbiol* 73: 3159–3164.

Vadas A, HG Monbouquette, E Johnson, I Schröder. 1999. Identification and characterization of a novel ferric reductase from the hyperthermophilic archaeon *Archaeoglobus fulgidus*. *J Biol Chem* 274: 36715–36721.

van Aken B, CM Peres, SL Doty, JM Yoon, JL Schnoor. 2004. *Methylobacterium populi* sp. nov., a novel aetobic, pink-pigmented, facultatively methylotrophic, methane-utilizing bacterium isolated from poplar trees (*Populus deltoides* × *nigra* DN34). *Int J Syst Evol Microbiol* 54: 1191–1196.

Van Baalen C, R O'Donnell. 1983. Isolation and growth of psychrophilic diatoms from the ice-edge in the Bering Sea. *J Gen Microbiol* 129: 1019–1023.

Valli K, BJ Brock, DK Joshi, MH Gold. 1992a. Degradation of 2,4-dinitrotoluene by the lignin-degrading fungus *Phanerochaete chrysosporium*. *Appl Environ Microbiol* 58: 221–228.

Valli K, H Wariishi, MH Gold. 1992b. Degradation of 2,7-dichlorodibenzo-*p*-dioxin by the lignin-degrading basidiomycete *Phanerochaete chrysosporium*. *J Bacteriol* 174: 2131–2137.

Valli K, MH Gold. 1991. Degradation of 2,4-dichlorophenol by the lignin-degrading fungus *Phanerochaete chrysosporium*. *J Bacteriol* 173: 345–352.

Vandamme P, T Coenye. 2004. Taxonomy of the genus *Cupriavidus*: A tale of lost and found. *Int J Syst Evol Microbiol* 54: 2285–2289.

van der Oost J, WGB Voorhorst, SWM Kengen, ACM Heerling, V Wittenhorst, Y Gueguen, WM de Vos. 2001. Genetic and biochemical characterization of a short-chain alcohol dehydrogenase from the hyperthermo-philic archaeon *Pyrococcus furiosus*. *Eur J Biochem* 268: 3062–3068.

van Ginkel CG, GB Rikken, AGM Kroon, SWM Kengen. 1996. Purification and characterization of chlorite dismutase: A novel oxygen-generating enzyme. *Arch Microbiol* 166: 321–326.

van Ginkel CG, HG J Welten, JAM de Bont. 1987. Oxidation of gaseous and volatile hydrocarbons by selected alkene-utilizing bacteria. *Appl Environ Microbiol* 53: 2903–2907.

van Herwijnen R, D Springael, P Slot, HAJ Govers, JR Parsons. 2003. Degradation of anthracene by *Mycobacterium* sp. strain LB501T proceeds via a novel pathway, through *o*-phthalic acid. *Appl Environ Microbiol* 69: 186–190.

van Hylckama Vlieg JET, H Leemhuis, JHL Spelberg, DB Janssen. 2000. Characterization of the gene cluster involved in isoprene metabolism in *Rhodococcus* sp. strain AD45. *J Bacteriol* 182: 1956–1963.

van Leeuwen JA, BC Nicholson, G Levay, KP Hayes, DE Mulcahy. 1997. Transformation of free tetrachlo-roguaiacol to bound compounds by fungi isolated from Lake Bonney, south-eastern Australia. *Mar Fresh Water Res* 48: 551–557.

Vannelli T, AB Hooper. 1992. Oxidation of nitrapyrin to 6-chloropicolinic acid by the ammonia-oxidizing bacterium *Nitrosomonas europaea*. *Appl Environ Microbiol* 58: 2321–2325.

Vannelli T, M Logan, DM Arciero, AB Hooper. 1990. Degradation of halogenated aliphatic compounds by the ammonia-oxidizing bacterium *Nitrosomonas europaea*. *Appl Environ Microbiol* 56: 1169–1171.

Vercellone-Smith P, DS Herson. 1997. Toluene elicits a carbon starvation response in *Pseudomonas putida* mt-2 containing the TOL plasmid pWWO. *Appl Environ Microbiol* 63: 1925–1932.

Verhagen FJM, HJ Swarts, JBPA Wijnberg, JA Field. 1998. Biotransformation of the major fungal metabolite 3,5-dichloro-*p*-anisyl alcohol under anaerobic conditions and its role in formation of bis(3,5-dichloro-4-hydroxyphenyl)methane. *Appl Environ Microbiol* 64: 3225–3231.

Véron M, L Le Minor. 1975. Nutrition et taxonomie des Enterobacteriaceae et bactéries voisines III Caractéres nutritionnels et différenciation des groupes taxonomiques. *Ann Microbiol (Inst Pasteur)* 126B: 125–147.

Vestal JR, JJ Perry. 1969. Divergent pathways for propane and propionate utilization by a soil isolate. *J Bacteriol* 99: 216–221.

Vilcáez J, K Suto, C Inoue. 2008. Bioleaching of chalcopyrite with thermophiles: Temperature–pH–ORP dependence. *Int J Mineral Process* 88: 37–44.

Vishnivetskaya TA, MA Petrova, J Urbance, M Ponder, CL Moyer, DA Gilichinsky, JM Tiedje. 2006. Bacterial community in ancient Siberian permafrost as characterized by culture and culture-independent methods. *Astrobiol* 6: 400–414.

Wackett LP, GA Brusseau, SR Householder, RS Hanson. 1989. Survey of microbial oxygenases: Trichloroethylene degradation by propane-oxidizing bacteria. *Appl Environ Microbiol* 55: 2960–2964.

Wahman DG, LE Katz, GE Speitel. 2005. Cometabolism of trihalomethanes by *Nitrosomonas europaea*. *Appl Environ Microbiol* 71: 7980–7986.

Ward DE, SWM Kengen, J van der Oost, WM de Vos. 2000. Purification and characterization of the alanine aminotransferase from the hyperthermophilic archaeon *Pyrococcus furiosus* and it role in alanine pro-duction. *J Bacteriol* 182: 2559–2566.

Walker CB et al. 2010. *Nitrosopumilis maritimus* genome reveals unique mechanisms for nitrification and autotrophy in globally distributed marine crenarchaea. *Proc Natl Acad Sci USA* 107: 8818–8823.

Walker N. 1973. Metabolism of chlorophenols by *Rhodotorula glutinis*. *Soil Biol Biochem* 5: 525–530.

Walker JRL, BG Taylor. 1983. Metabolism of phloroglucinol by *Fusarium solani*. *Arch Microbiol* 134: 123–126.

Wallrabenstein C, E Hauschild, B Schink. 1995. *Syntrophobacter pfennigii* sp. nov., new syntrophically propionate-oxidizing anaerobe growing in pure culture with propionate and sulfate. *Arch Microbiol* 164: 346–352.

Wang Y, PCK Lau, DK Button. 1996. A marine oligobacterium harboring genes known to be part of aromatic hydrocarbon degradation pathways of soil pseudomonads. *Appl Environ Microbiol* 62: 2169–2173.

Ward BB. 1987. Kinetic studies on ammonia and methane oxidation by *Nitrosococcus oceanus*. *Arch Microbiol* 147: 126–133.

Ward NL et al. 2009. Three genomes from the phylum *Acidobacteria* provide insight into the lifestyles of these microorganisms in soils. *Appl Environ Microbiol* 75: 2046–2056.

Warshawsky D, T Cody, M Radike, R Reilman, B Schumann, K LaDow, J Schneider. 1995. Biotransformation of benzo[*a*]pyrene and other polycyclic aromatic hydrocarbons and heterocyclic analogues by several green algae and other algal species under gold and white light. *Chem-Biol Interact* 97: 131–148.

Warshawsky D, M Radike, K Jayasimhulu, T Cody. 1988. Metabolism of benzo(a)pyrene by a dioxygenase system of the freshwater green alga *Selenastrum capricornutum*. *Biochem Biophys Res Comm* 152: 540–544.

Wasserman SA, CT Walsh, D Botstein. 1983. Two alanine racemase genes in *Salmonella typhimurium* that differ in structure and function. *J Bacteriol* 153: 1439–1450.

Wayne LG et al. 1991. Fourth report of the cooperative, open-ended study of slowly growing mycobacteria by the international working group on mycobacterial taxonomy. *Int J Syst Bacteriol* 41: 463–472.

Weber CF, GM King. 2007. Physiological, ecological, and phylogenetic characterization of *Stappia*, a marine CO-oxidizing bacterial genus. *Appl Environ Microbiol* 73: 1266–1276.

Webster G, V Jain, MR Davey, C Gough, J Vasse, j Dénarié, EV Cocking. 1998. The flavonoid naringenin stimulates the intercellular colonization of wheat roots by *Azorhizobium caulinodans*. *Plant Cell Environ* 21: 373–383.

Wedemeyer G. 1967. Dechlorination of 1,1,1-trichloro-2,2-bis[p-chlorophenyl]ethane by *Aerobacter aerogenes*. *Appl Microbiol* 15: 569–574.

West PA, GC Okpowasili, PR Brayton, DJ Grimes, RR Colwell. 1984. Numerical taxonomy of phenathrene-degrading bacteria isolated from Chesapeake Bay. *Appl Environ Microbiol* 48: 988–993.

Wetzstein H-G, N Schmeer, W Karl. 1997. Degradation of the fluoroquinolone enrofloxacin by the brown-rot fungus *Gleophyllum striatum*: Identification of metabolites. *Appl Environ Microbiol* 63: 4272–4281.

Whittenbury R, DP Kelly. 1977. Autotrophy: A conceptual phoenix. *Symp Soc Gen Microbiol* 27: 121–149.

Wiberg K, E Brorström-Lundén, TF Bidleman, P Haglund. 2001. Concentrations and fluxes of hexachlorocyclohexanes and chiral composition of α-HCH in environmental samples from the southern Baltic Sea. *Environ Sci Technol* 35: 4739–4746.

Wiberg K, E Brorström-Lundén, TF Bidleman, P Haglund. 2001. Concentrations and fluxes of hexachloro-cyclohexanes and chiral composition of α-HCH in environmental samples from the southern Baltic Sea. *Environ Sci Technol* 35: 4739–4746.

Wieser M, B Wagner, J Eberspächer, F Lingens. 1997. Purification and characterization of 2,4,6-trichlorophenol-4-monooxygenase, a dehalogenating enzyme from *Azotobacter* sp. strain GP1. *J Bacteriol* 179: 202–208.

Willumsen P, U Karlson, E Stackebrandt, RM Kroppenstedt. 2001. *Mycobacterium frederiksbergense* sp. nov., a novel polycyclic aromatic hydrocarbon-degrading *Mycobacterium* species. *Int J Syst Evol Microbiol* 51: 1715–1722.

Willison JC. 2004. Isolation and characterization of a novel sphingomonad capable of growth with chrysene as sole carbon and energy source. *FEMS Microbiol Lett* 241: 143–150.

Willumsen PA, JK Nielsen, U Karlson. 2001. Degradation of phenanthrene-analogue azaarenes by *Mycobacterium gilvum*. *Appl Microbiol Biotechnol* 56: 539–544.

Winters K, PL Parker, C Van Baalen. 1969. Hydrocarbons of blue-green algae: Geochemical significance. *Science* 163: 467–468.

Wong CM, MJ Dilworth, AR Glenn. 1994. Cloning and sequencing show that 4-hydroxybenzoate hydroxylase *Poba* is required for uptake of 4-hydroxybenzoate in *Rhizobium leguminosarum*. *Microbiology (UK)* 140: 2775–2786.

Wood HG, RL Stjernholm. 1962. Assimilation of carbon dioxide by heterotrophic organisms, pp. 41–117. In *The Bacteria* (Eds. IG Gulsalus and RY Stanier), Volume III, Academic Press, New York.

Wright GE, MT Madigan. 1991. Photocatabolism of aromatic compounds by the photo*trophic* purple bacterium *Rhodomicrobium vannielii*. *Appl Environ Microbiol* 57: 2069–2073.

Wu ML, S de Vries, RA van Alen, MK Butler, HJM op den Camp, JT Keltjens, MSM Jetten, M Strous. 2011. Physiological role of the respiratory quinol oxidase in the anaerobeic nitrite-reducing methanotroph "*Candidatus* Methylomirabilis oxyfera." *Microbiology (UK)* 157: 890–898.

Wuchter C, B Abbas, MJL Coolen, L Herfort, J van Bleiswijk, P Timmers, M Strous et al. 2006. Archaeal nitrification in the ocean. *Proc Natl Acad Sci USA* 103: 12317–12322.

Xie C-H, A Yokota. 2006. *Sphingomonas azotifigens* sp. nov., a nitrogen-fixing bacterium isolated from the roots of *Oryza sativa*. *Int J Syst Evol Microbiol* 56: 889–893.

Xu Y, Y Nogi, C Kato, Z Liang, H-J Rüger, D De Kegel, N Glansdorff. 2003. *Moritella profunda* sp. nov. and *Moritella abyssi* sp. nov., two psychropiezophilic organisms isolated from deep Atlantic sediments. *Int J Syst Evol Microbiol* 53: 533–538.

Yabuuchi E, H Yamamoto, S Terakubo, N Okamura, T Naka, N Fujiwara, K Kobayashi, Y Kosako, A Hiraichi. 2001. Proposal of *Sphingomonas wittichii* sp. nov. for strain RW1T, known as a dibenzo-*p*-dioxin metabolizer. *Int J Syst Evol Microbiol* 51: 281–292.

Yadav JS, RE Wallace, CA Reddy. 1995. Mineralization of mono- and dichlorobenzenes and simultaneous degradation of chloro- and methyl-substituted benzenes by the white-rot fungus *Phanerochaete chrysosporium*. *Appl Environ Microbiol* 61: 677–680.

Yakimov MM, PN Golyshin, S Lang, ERB Moore, W-R Abraham, H Lünsdork, KN Timmis. 1998. *Alcanivorax borkumensis* gen. nov., sp. nov., a new hydrocarbon-degrading and surfactant-producing marine bacterium. *Int J Syst Bacteriol* 48: 339–348.

Yang S-J, H-S Lee, C-S Park, Y-R Kim, T-W Moon, K-W Park. 2004. Enzymatic analysis of an amylolytic enzyme from the hyperthermophilic archaeon *Pyrococcus furiosus* reveals its novel catalytic properties as both an α-amylase and a cyclodextrin-hydrolyzing enzyme. *Appl Environ Microbiol* 70: 5988–5995.

Ye D, MA Siddiqui, AE Maccubbin, S Kumar, HC Sikka. 1996. Degradation of polynuclear aromatic hydrocarbons by *Sphingomonas paucimobilis*. *Environ Sci Technol* 30: 136–142.

Ying X, K Ma. 2011. Characterization of a zinc-containing alcohol dehydrogenase with stereoselectivity from a hyperthermophilic archaeon *Thermococcus guaymasensis*. *J Bacteriol* 193: 3009–3019.

Yoon J-H, JJ Lee, SS Kang, M Takeuchi, YK Shin, ST Lee, KH Kang, YH Park. 2000. *Gordonia nitida* sp. nov., a bacterium that degrades 3-ethylpyridine and 3-methylpyridine. *Int J Syst Evol Microbiol* 50: 1202–1210.

Yorifuji T, K Ogata. 1971. Arginine racemase of *Pseudomonas graveolens* I. Purification, crystallization, and properties. *J Biol Chem* 246: 5085–5092.

Yoshinaga DH, HA Frank. 1982. Histamine-producing bacteria in decomposing skipjack tuna (*Katsuwonus pelamis*). *Appl Environ Microbiol* 44: 447–452.

Yoshioka T, JA Krauser, FP Guengerich. 2002. Tetrachloroethylene oxide: Hydrolytic products and reactions with phosphate and lysine. *Chem Res Toxicol* 15: 1096–1105.

Youngblood WW, M Blumer, RL Guillard, F Fiore. 1971. Saturated and unsaturated hydrocarbons in marine benthic algae. *Mar Biol* 8: 190–210.

Yurkov VV, JT Beatty. 1998. Aerobic anoxygenic phototrophic bacteria. *Microbiol Mol Biol Revs* 62: 695–724.

Yumoto I. 2007. Environmental and taxonomic biodiversities of Gram-positive alkaliphiles, pp. 295–310. In *Physiology and Biochemistry of Extremophiles* (Eds. C Gerday, N Glansdorf), ASM Press, Washington, D.C.

Yumoto I, A Nakamura, H Iwata, K Kojima, K Kusumoto, Y Nodasaka, H Matsutama. 2002. *Dietzia psychralcaliphila* sp. nov., a novel facultatively psychrophilic alkaliphile that grows on hydrocarbons. *Int J Syst Evol Microbiol* 52: 85–90.

Yumoto I, K Hirota, Y Nodasaka, Y Yokota, T Horoshino, K Nakajima. 2004. *Alkalibacterium psychrotolerans* sp. nov., a psychrotolerant obligate alkaliphile that reduces an indigo dye. *Int Syst Evol Microbiol* 54: 2379–2383.

Yun J, S Kang, S Park, H Yoon, M-J Kim, S Heu, S Ryu. 2004. Characterization of a novel amylolytic enzyme encoded by a gene from a soil-derived metagenomic library. *Appl Environ Microbiol* 70: 7229–7234.

Zaborina O, M Latus, J Eberspächer, LA Golovleva, F Lingens. 1995. Purification and characterization of 6-chlorohydroquinol 1,2-dioxygenase from *Streptomyces rochei* 303: Comparison with an analogous enzyme from *Azotobacter* sp. strain GP1. *J Bacteriol* 177: 229–234.

Zaitsev GM, JS Uotila, IV Tsitko, AG Lobanok, MS Salkinoja-Salonen. 1995. Utilization of halogenated benzenes, phenols, and benzoates by *Rhodococcus opacus* GM-14. *Appl Environ Microbiol* 61: 4191–4201.

Zarilla KA, JJ Perry. 1984. *Thermoleophilum album*, gen. nov. and sp. nov., a bacterium obligate for thermophily and *n*-alkane substrates. *Arch Microbiol* 137: 286–290.

Zarzycki J, A Schlichting, N Strychlalsky, M Müller, BE Alber, G Fuchs. 2008. Mesaconyl-coenzyme A hydratase, a new enzyme of two central carbon metabolic pathways in bacteria. *J Bacteriol* 190: 1366–1374.

Zhang H, S Hanada, T Shigematsu, K Shibuya, Y Kamagata, T Kanawa, R Kurane. 2000. *Burkholderia kururiensis* sp. nov. a trichloroethylene (TCE)- degrading bacterium isolated from, an aquifer polluted with TCE. *Int J Syst Evol Microbiol* 50: 743–749.

Zhang L, G Xie. 2006. Dioversiry and distribution of *Burkholderia cepacia* complex in the rhizosphere of rice and maize. *FEMS Microbiol Lett* 266: 231–235.

Zhao J-S, A Halasz, L Paquet, C Beaulieu, J Hawari. 2002. Biodegradation of hexahydro-1,3,5-trinitro-1,3,5-triazine and its mononitroso derivative hexahydro-1-nitroso-3,5-dinitro-1,3,5-triazine by *Klebsiella pneumoniae* strain SCZ isolated from an anaerobic sludge. *Appl Environ Microbiol* 68: 5336–5341.

Zhilina TN, GA Zavarzin, FA Rainey, VV Kevbrin, NA Kostrikina, AM Lysenko. 1996. *Spirochaeta alkalika* sp. nov., *Spirochaeta africana* sp. nov., and *Spirochaeta asiatica* sp. nov., alkaliphilic anaerobes from the continental soda lakes in Central Asia and the East African Rift. *Int J Syst Bacteriol* 46: 305–312.

Zhilina TN, GA Zavarzin, FA Rainey, EN Pikuta, GA Osipov, NA Kostrimina. 1997. *Desulfonatronvibrio hydrogenovorans* gen. nov., sp. nov., an alkaliphilic sulfate-reducing bacterium. *Int J Syst Bacteriol* 47: 144–149.

Zhou J, MR Fries, JC Chee-Sanford, JM Tiedje. 1995. Phylogenetic analyses of a new group of denitrifiers capable of anaerobic growth on toluene and description of *Azoarcus tolulyticus* sp. nov. *Int J Syst Bacteriol* 45: 500–506.

Zikmundová M, K Drandarov, L Bigler, M Hesse, C Werner. 2002. Biotransformation of 2-benzoxazolinone and 2-hydroxy-1,4-benzoxazin-3-one by endophytic fungi isolated from *Aphelandra tetragona*. *Appl Environ Microbiol* 68: 4863–4870.

Zipper C, K Nickel, W Angst, H-P E Kohler. 1996. Complete microbial degradation of both enantiomers of the chiral herbicide Mecoprop [(*R*,*S*)-2-(4-chloro-2-methylphenoxy)]propionic acid in an enantioselective manner by *Sphingomonas herbicidovorans* sp. nov. *Appl Environ Microbiol* 62: 4318–4322.

Zylstra GJ, E Kim. 1997. Aromatic hydrocarbon degradation by *Sphingomonas yanoikuyae* B1. *J Indust Microbiol Biotechnol* 19: 408–414.

Part 2: Reactions Mediated by Other Biota

2.11 Aquatic and Terrestrial Biota

2.11.1 Introduction

Very few xenobiotics remain unaltered in the environment for any length of time after their release. Although metabolism by microorganisms has already been discussed, brief comments are made here on the metabolism of contaminants by higher biota—particularly fish. Metabolites such as conjugates may enter the aquatic environment where their ultimate fate is determined by microbial reactions. Metabolism by higher organisms is relevant to the mechanisms whereby organisms detoxify deleterious contaminants or, conversely, induce the synthesis of toxic metabolites. Some illustrations of these are briefly summarized below:

1. The kinetics and products of metabolism critically influence the nature and the concentrations of the xenobiotic and its transformation products to which the cells are exposed. Increasing evidence from different sources has shown that the effective toxicant may indeed be a metabolite synthesized from the compound originally supplied and not the xenobiotic itself. It is important to appreciate that, as with toxicity, the extent of metabolism will generally depend on the nature and position of substituents on aromatic rings as well as on their number. For example, although 2,3,4- and 3,4,5-trichloroaniline were *N*-acetylated in guppy, this did not occur with 2,4,5-trichloroaniline (de Wolf et al. 1993). Metabolism may also be an important determinant of genotoxic effects, estrogenic activity, and teratogenicity.

2. The classic example is that of Prontosil (Figure 2.12) in which the compound is active against bacterial infection in animals although inactive against the bacteria in pure culture. The toxicity in animals is the result of reduction to the sulfanilamide (4-aminobenzenesulfonamide) that competitively blocks the incorporation of 4-aminobenzoate into the vitamin folic acid.

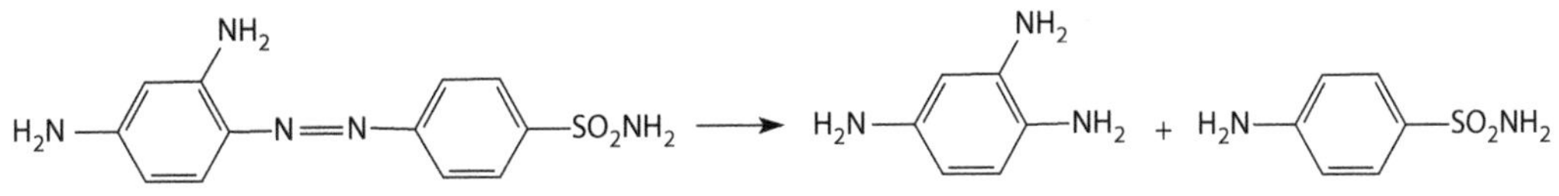

FIGURE 2.12 Metabolism of prontosil in mammals.

3. The mechanism of fluoroacetate toxicity in mammals has been extensively examined and was originally thought to involve simply initial synthesis of fluorocitrate that inhibits aconitase and thereby the functioning of the TCA cycle (Peters 1952). Walsh (1982) has extensively reinvestigated the problem, and revealed both the complexity of the mechanism of inhibition and the stereospecificity of the formation of fluorocitrate from fluoroacetate.

4. Considerable attention has been directed to the synthesis of the epoxides and dihydrodiol epoxides of polycyclic aromatic hydrocarbons mediated by the action of cytochrome P450 systems, and their role in inducing carcinogenesis in fish (Varanasi et al. 1987; de Maagd and Vethaak 1998). Tumors observed in feral fish exposed to PAHs may plausibly—though not necessarily—be the result of this transformation. Even though an apparently causal relationship between exposure of fish to PAHs and disease may have been established (Malins et al. 1985, 1987), caution should be exercised due to the possibility that other—and unknown—substances may have induced carcinogenesis. It is also important to appreciate that other compounds may induce induction of the metabolic system for PAHs. For example, exposure of rainbow trout to PCBs increases the effectiveness of liver enzymes to transform benzo[*a*]pyrene to carcinogenic intermediates (Egaas and Varanasi 1982).

5. The carbamate insecticide aldicarb (Figure 2.13) that exerts its effect by inactivating acetylcholinesterase is metabolized by a flavin monooxygenase from rainbow trout to the sulfoxide, which is a more effective inhibitor (Schlenk and Buhler 1991).

6. Preexposure to the organophosphate diazinon at exposures half the LC_{50} values increased the LC_{50} value by a factor of about five for guppy (*Poecilia reticulata*), but had no effect on the value for zebra fish (*Brachydanio rerio*). This was consistent with the observation that during preexposure of guppy there was a marked inhibition in the synthesis of the toxic metabolites diazoxon and pyrimidinol, whereas this did not occur with zebra fish in which the toxicity was mediated primarily by the parent compound (Keizer et al. 1993).

7. Pyrene is metabolized by the fungus *Crinipellis stipitaria* to 1-hydroxypyrene, and this has a spectrum of toxic effects substantially greater than those of pyrene: these include cytotoxicity to HeLa S3 cells, toxicity to a number of bacteria and to the nematode *Caenorhabditis elegans* (Lambert et al. 1995).

At the other extreme, if metabolism of the xenobiotic by the organism does not occur at all—or at insignificant rates—after exposure, the compound will be persistent in the organism, and may, therefore, be consumed by predators. This is relevant to biomagnification.

Most of the reactions carried out by fish and higher aquatic organisms are relatively limited transformation reactions in which the skeletal structure of the contaminants remains intact. The following three widely distributed reactions are of greatest significance:

1. Cytochrome P450–type monooxygenase systems, which have a generally low substrate specificity, are widely distributed in the species of fish used for toxicity testing (Funari et al. 1987).

FIGURE 2.13 Metabolism of aldicarb by rainbow trout.

2. Glutathione *S*-transferases (Nimmo 1987; Donnarumma et al. 1988), which are important in the metabolism of highly reactive compounds containing electrophilic groups such as epoxides and aromatic rings with several strongly electron-attracting substituents such as halogen, cyano, or nitro groups.

3. Conjugation of polar groups such as amines, carboxylic acids, and phenolic hydroxyl groups produce water-soluble compounds that are excreted and these reactions, therefore, function as a detoxification mechanism.

2.12 Metabolism by Fish

The biotransformation of xenobiotics in many higher organisms is mediated by the cytochrome P450 monooxygenase system and the complexity of factors that regulate the synthesis of this in fish has been reviewed (Andersson and Förlin 1992). The metabolic potential of fish may appear restricted compared with that of microorganisms, but it may have been considerably underestimated. For example, metolachlor (2-chloro-*N*-[2-ethyl-6-methylphenyl]-*N*-[2-methoxy-1-methylethyl]acetamide) is metabolized by bluegill sunfish (*Lepomis macrochirus*) by reactions involving initially *O*-demethylation and hydroxylation (Cruz et al. 1993) (Figure 2.14). These are comparable to the reactions carried out by an actinomycete (Krause et al. 1985), and the benzylic hydroxylation is analogous to that involved in the biotransformation of the structurally similar alachlor by the fungus *Cunninghamella elegans* (Pothuluri et al. 1993). An extensive compilation of the transformation of xenobiotics by fish has been given (Sijm and Opperhuizen 1989), and only a few examples of these reactions are summarized here as illustration:

1. *N*-Dealkylation of dinitramine to 1,3-diamino-2,4-dinitro-6-trifluoromethylbenzene (Olson et al. 1977) by carp (*Cyprinus carpio*) (Figure 2.15).

2. *O*-Demethylation of pentachloroanisole in rainbow trout (*Salmo gairdneri* ≡ *Oncorhynchus mykiss*) (Glickman et al. 1977), and of chlorinated veratroles by zebra fish (Neilson et al. 1989).

3. Acetylation of 3-amino ethylbenzoate in rainbow trout (Hunn et al. 1968).

4. Displacement of the nitro group in pentachloronitrobenzene by hydroxyl and thiol groups (Bahig et al. 1981) (Figure 2.16) in golden orfs (*Idus idus*).

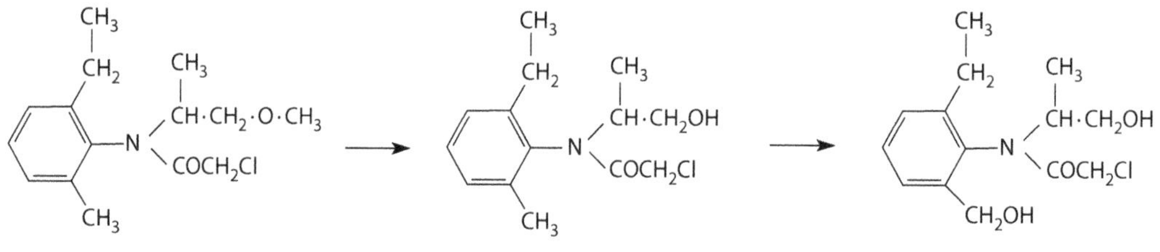

FIGURE 2.14 Metabolism of metolachlor by bluegill sunfish.

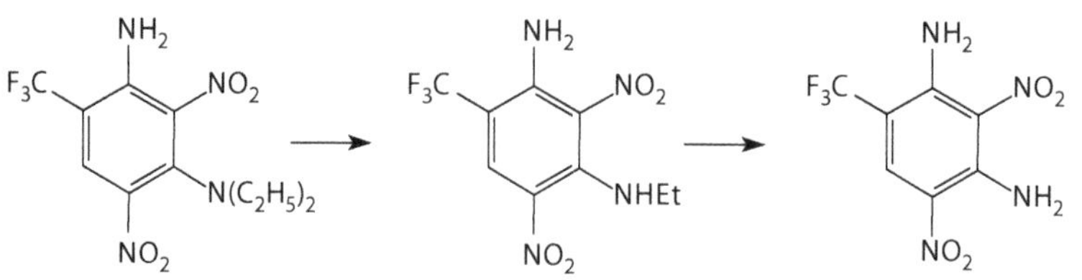

FIGURE 2.15 Metabolism of dinitramine by carp.

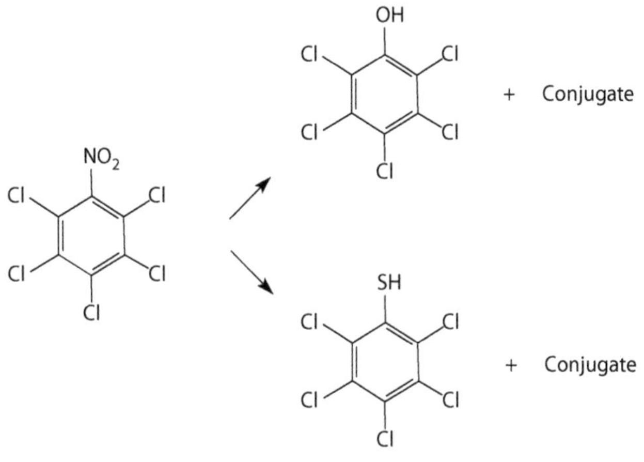

FIGURE 2.16 Metabolism of pentachloronitrobenzene by golden orfs.

5. Oxidation of a number of PAHs has been demonstrated in a variety of fish. A review directed to metabolism and the role of PAH metabolites in inducing tumorigenesis has been given (de Maagd and Vethaak 1998), and only two examples are given here:

   a. Coho salmon (*Oncorhynchus kisutch*) metabolized naphthalene to a number of compounds consistent with oxidation to the epoxide, hydrolysis to the dihydrodiol, and dehydration of the *trans*-dihydrodiol to naphth-1-ol, or by rearrangement of the epoxide (Figure 2.17) (Collier et al. 1978).

   b. For the carcinogen benzo[*a*]pyrene, a wider range of metabolites has been identified in southern flounder (*Paralichthys lethostigma*), including the 4,5-, 7,8-, and 9,10-diols, the 1,6-, 3,6-, and 6,12-quinones in addition to the 1-, 3-, and 9-benzopyreneols (Figure 2.18) (Little et al. 1984).

6. *N*-Hydroxylation of aniline and 4-chloroaniline by rainbow trout to hydroxylamines that could plausibly account for the subchronic toxicity of the original compounds (Dady et al. 1991).

Initially formed polar metabolites such as phenols and amines may be conjugated to water-soluble terminal metabolites that are excreted into the medium and function as an effective mechanism of detoxification. For example, conjugates of pentachlorophenol and pentachlorothiophenol were the major metabolites from pentachloronitrobenzene. Although the naphthalene dihydrodiol was the major metabolite produced from naphthalene, the further transformation product naphth-1-ol was also isolated as the sulfate, glucuronate, and glucose conjugates. Diverse conjugation reactions have been described and include the following:

1. Phenolic compounds with the formation of glucuronides, sulfates, or glucosides as already noted.

2. Reaction of carboxylic acids with the amino groups of glycine (Huang and Collins 1962) or taurine (Figure 2.19) (James and Bend 1976) to form the amides.

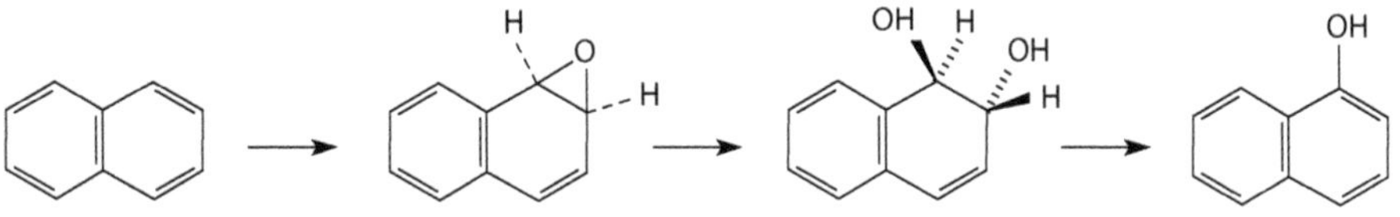

FIGURE 2.17 Metabolism of naphthalene by coho salmon.

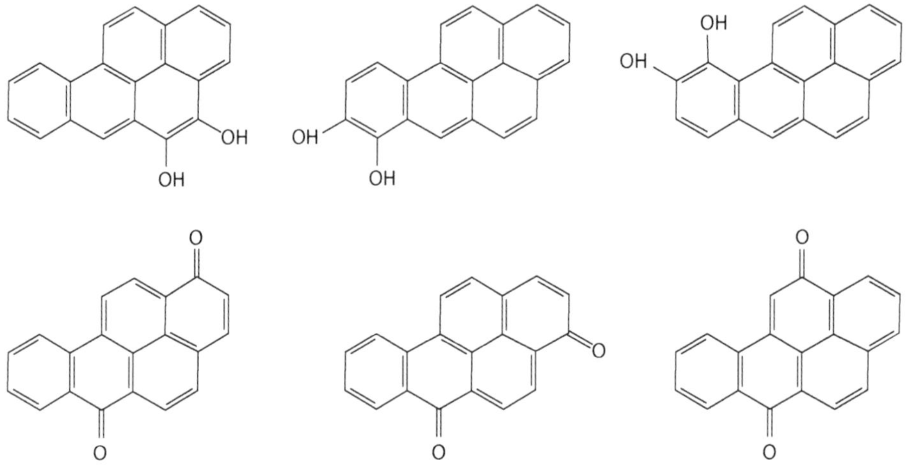

FIGURE 2.18   Metabolism of benzo[*a*]pyrene by southern flounder.

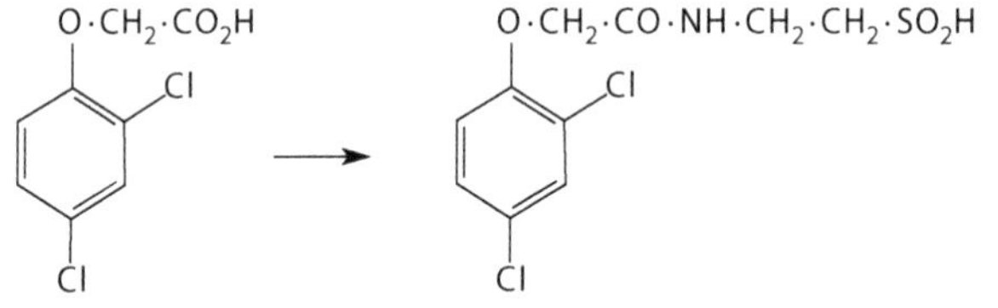

FIGURE 2.19   Conjugation of carboxylic acids with amino acids.

3. Reaction between glutathione and reactive chloro compounds such as 1-chloro-2,4-dinitro-benzene (Niimi et al. 1989), or the chloroacetamide group in demethylated metolachlor (Cruz et al. 1993).

Important investigations have been directed to persistent halogenated aromatic compounds and different mechanisms for their metabolism have been found:

1. Hydroxylated PCBs have been found in a laboratory study using rainbow trout (*Oncorhynchus mykiss*), and were similar to those previously observed (Campbell et al. 2003) in wild lake trout (*Salvelinus namaycush*). Although it was not possible to associate the metabolites unambiguously with their precursor PCB congener, those with neighboring hydrogen atoms in the *ortho/para* positions were most probably transformed, putatively by CYP 2B-type isoforms of cytochrome P450 rather than by CYP 1A enzymes (Buckman et al. 2006).

2. Debromination has been examined in rainbow trout (*Oncorhynchus mykiss*) and carp (*Cyrprinus carpio*) that were fed with a diet containing decabromodiphenyl ether (BDE 209). In both of them, products from debromination were found in whole fish, whole fish homogenates, organs, and microsomal preparations. In trout, debromination produced primarily the octa and nano congeners, whereas penta to octa congeners were found in carp. This was confirmed in microsomal preparations of carp that transformed up to 63% of DBE 209 with the formation of hexa to octa congeners, whereas in trout only 23% of DBE 209 was transformed to octa and nano congeners (Stapleton et al. 2006). It was suggested that the debromination could be analogous to the deiodination of thyroxine to the more active triiodothyronine by loss of iodine from the position *ortho* to the hydroxyl group.

2.13 Metabolism by Other Organisms

2.13.1 Mussels

1. Mussels do not generally carry out more than the limited reactions of oxidation and conjugation, and in the common mussel, *Mytilus edulis*, variations between summer and winter levels for both cytochrome P450 and NADPH-independent 7-ethoxycoumarin *O*-deethylase have been found (Kirchin et al. 1992). Levels of cytochrome P450 and the rates of metabolism of PAHs were apparently low compared with those found in fish (Livingstone and Farrar 1984). An investigation using subcellular extracts of the digestive glands from the mussel *M. galloprovincialis* showed that although the formation of diols and phenols from benzo[*a*]pyrene was dependent on NADPH, the quinones that were the major metabolites were produced in the absence of NADPH apparently by radical-mediated reactions involving lipid peroxidase systems (Michel et al. 1992).

2. A reaction presumably mediated by glutathione *S*-transferase is the replacement of the 4-chloro substituent in octachlorostyrene in the blue mussel (*Mytilus edulis*) by a thiomethyl group (Figure 2.20) (Bauer et al. 1989). A similar reaction of glutathione with arene oxides produced by aquatic mammals from PCBs and DDE results ultimately in the production of methyl sulfones (Bergman et al. 1994; Janák et al. 1998). In the arctic food chain—arctic cod (*Boreogadus saida*), ringed seal (*Phoca hispidus*), and polar bear (*Ursus maritima*)—it has been shown that levels of the dimethyl sulfones of DDE and PCB were low in cod, and that levels in polar bear were the combined result of bioaccumulation from seals and endogenous metabolism (Letcher et al. 1998).

The results of these investigations suggest that caution should be exercised in interpreting not only the results of toxicity assays in which such organisms are employed but also data accumulated in monitoring studies that may not have taken into account the existence of metabolites.

2.13.2 Insects

Resistance of house flies (*Musca domestica*) to DDT was attributed to its transformation to the nontoxic DDE, and the enzyme that carries the dehydrochlorination has been characterized in DDT-resistant flies (Lipke and Kearns 1959a,b). The herbicide alachlor is transformed by chironomid larvae by *O*-demethylation followed by loss of the chloroacetyl group to produce 2,6-diethylaniline (Figure 2.21) (Wei and Vossbrinck 1992).

2.13.3 Invertebrates

The metabolism of xenobiotics by both terrestrial and sediment-dwelling biota has been studied, and provides illustrations of the importance of uptake by food or by sorbed sediment. Some examples of metabolism by terrestrial biota include the following.

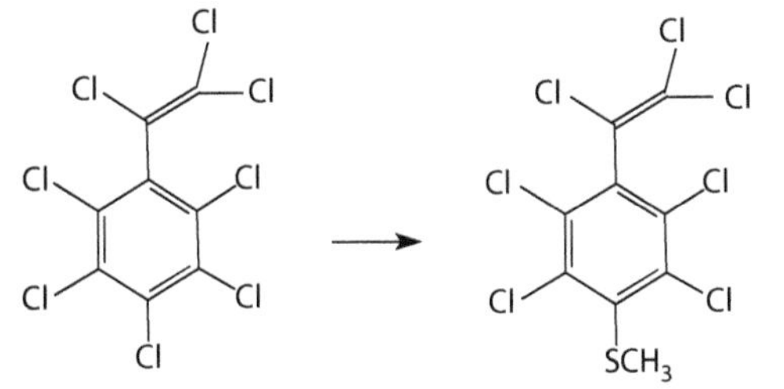

FIGURE 2.20 Metabolism of octachlorostyrene by the blue mussel.

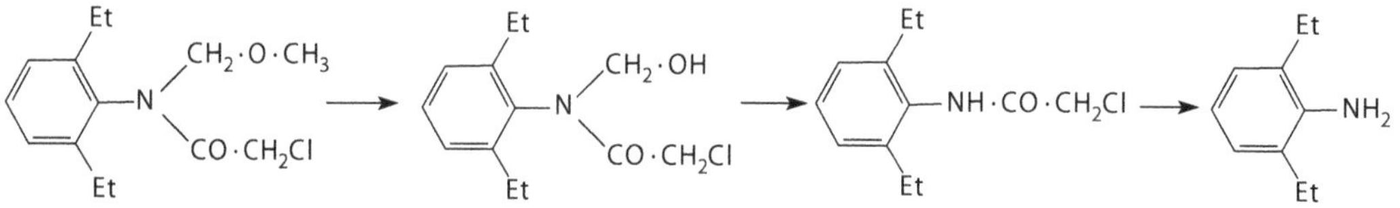

FIGURE 2.21 Metabolism of alachlor by chironomid larvae.

2.13.3.1 Isopods

The uptake and elimination of benzo[a]pyrene by the terrestrial isopod *Porcellio scaber* have been investigated (van Brummelen and van Straalen 1996), and 1-hydroxypyrene was identified among the metabolites of pyrene in this organism (Stroomberg et al. 1996).

2.13.3.2 Oligochaetes

Both the (+)- and (–)-enantiomers of limonene were transformed by larvae of the cutworm *Spodoptera litura* (Miyazawa et al. 1998). For both of them, the reactions involved are (a) dihydroxylation between C_8 and C_9 and (b) oxidation of the C-1 methyl group to carboxyl. These transformations were not dependent on the intestinal microflora in contrast to the transformation of α-terpinene to *p*-mentha-1,3-dien-7-ol and *p*-cymene whose formation could be attributed to the intestinal flora.

2.13.3.3 Polychaetes

Polychaete worms belonging to the genera *Nereis* and *Scolecolepides* have extensive metabolic potential. *Nereis virens* is able to metabolize PCBs (McElroy and Means 1988) and a number of PAHs (McElroy 1990), while *N. diversicolor* and *Scolecolepides viridis* are able to metabolize benzo[a]pyrene (Driscoll and McElroy 1996). It is worth noting that apart from excretion of the toxicant, polar and much more water-soluble metabolites such as the glycosides formed from pyrene by *Porcellio* sp. (Larsen et al. 1998) may be mobile in the interstitial water of the sediment phase.

It has been shown that although the marine terebellid polychaete *Amphitrite ornata* produced no detectable halogenated metabolites, it synthesizes a dehalogenase that is able to carry out oxidative dehalogenation of halogenated phenols, with fluoro, chloro, or bromo substituents (Chen et al. 1996). One of the enzymes (DHB 1) has been purified, and consists of two identical subunits (M_r 15,530) each containing heme and histidine as the proximal Fe ligand. In the presence of H_2O_2, 2,4,6-tribromophenol is oxidized to 2,6-dibromo-benzo-1,4-quinone (LaCount et al. 2000).

2.13.4 Other Organisms

Other groups of biota are able to bring about transformation of structurally diverse compounds and limited investigations have revealed the metabolic potential of taxonomically diverse eukaryotic organisms:

1. The apochlorotic alga (protozoan) *Prototheca zopfii* is able to degrade aliphatic hydrocarbons (Walker and Pore 1978; Koenig and Ward 1983).
2. *Tetrahymena thermophila* transforms pentachloronitrobenzene to the corresponding aniline and pentachlorothioanisole (Figure 2.22) (Murphy et al. 1982).
3. *Daphnia magna* has been shown to bring about dechlorination and limited oxidation of heptachlor (Figure 2.23) (Feroz et al. 1990).

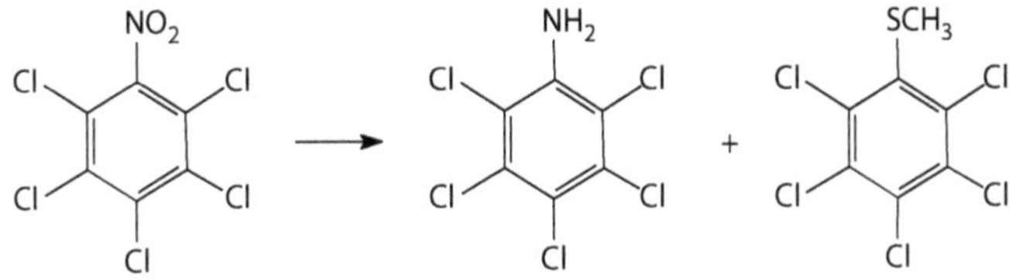

FIGURE 2.22 Biotransformation of pentachloronitrobenzene by *Tetrahymena thermophila*.

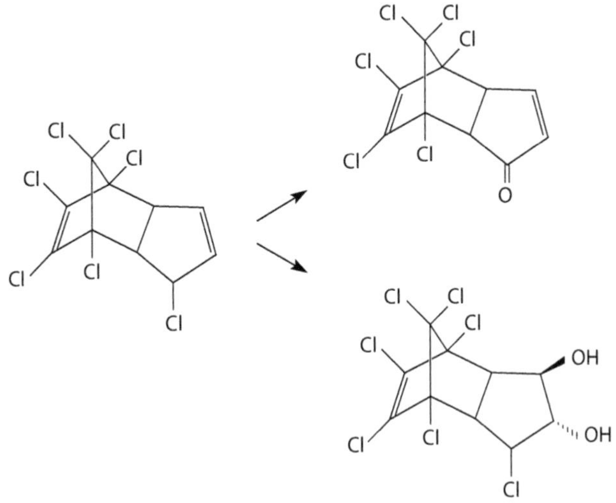

FIGURE 2.23 Biotransformation of heptachlor by *Daphnia magna*.

2.14 References

Andersson T, L Förlin. 1992. Regulation of the cytochrome P450 enzyme system in fish. *Aquat Toxicol* 24: 1–20.

Bahig ME, A Kraus, W Klein, F Korte. 1981. Metabolism of pentachloronitrobenzene-^{14}C quintozene in fish. *Chemosphere* 10: 319–322.

Bauer I, K Weber, W Ernst. 1989. Metabolism of octachlorostyrene in the blue mussel (*Mytilus edulis*). *Chemosphere* 18: 1573–1579.

Bergman Å, RJ Norstrom, K Haraguchi, H Kuroki, P Béland. 1994. PCB and DDE methyl sulfones in mammals from Canada and Sweden. *Environ Toxicol Chem* 13: 121–128.

Bhadra R, DG Wayment, JB Hughes, V Shanks. 1999. Confirmation of conjugation processes during TNT metabolism by axenic plant roots. *Environ Sci Technol* 33: 446–452.

Buckman AH, CS Wong, EA Chow, SB Brown, KR Solomon, AT Fisk. 2006. Biotransformation of polychlorinated biphenyls (PCBs) and bioformation of hydroxylated PCBs in fish. *Aquat Toxicol* 78: 176–185.

Burken JG, JL Schnoor. 1997. Uptake and metabolism of atrazine by poplar trees. *Environ Sci Technol* 31: 1399–1406.

Campbell LM, DCG Muir, DM Whittle, S Backus, RJ Nostrom, AT Fisk. 2003. Hydroxylated PCBs and other chlorinated phenolic compounds in lake trout (*Salvelinus namaycush*) blood plasma from the Great Lakes region. *Environ Sci Technol* 37: 1720–1725.

Chen TP, SA Woodin, DE Lincoln, CR Lovell. 1996. An unusual dehalogenating peroxidase from the marine terebellid polychaete *Amphitrite ornata*. *J Biol Chem* 271: 4609–4612.

Collier TK, LC Thomas, DC Malins. 1978. Influence of environmental temperature on disposition of dietary naphthalene in coho salmon (*Oncorhynchus kisutch*): isolation and identification of individual metabolites. *Comp Biochem Physiol* 61C: 23–28.

Cruz SM, MN Scott, AK Merritt. 1993. Metabolism of [^{14}C]metolachlor in blueguill sunfish. *J Agric Food Chem* 41: 662–668.

Cruz LM, E Maltempi-de Souza, OB Weber, JI Baldani, J Döbereiner, O Pedrosa. 2001. 16S ribosomal DNA characterization of nitrogen-fixing bacteria isolated from banana (*Musa* spp.) and pineapple (*Ananas comosus* (L.) Merril). *Appl Environ Microbiol* 67: 2375–2379.

Dady JM, SP Bradbury, AD Hoffman, MM Voit, DL Olson. 1991. Hepatic microsomal *N*-hydroxylation of aniline and 4-chloroaniline by rainbow trout (*Oncorhyncus mykiss*). *Xenobiotica* 21: 1605–1620.

de Maagd P G-J, AD Vethaak. 1998. Biotransformation of PAHs and their carcinogenic effects in fish. *Handbook Environ Chem* 3J: 265–309.

de Wolf W, W Seinen, LM Hermens. 1993. Biotransformation and toxicokinetics of trichloroanilines in fish in relation to their hydrophobicity. *Arch Environ Contam Toxicol* 25: 110–117.

Donnarumma L, G de Angelis, F Gramenzi, L Vittozzi. 1988. Xenobiotic metabolizing enzyme systems in test fish. III. Comparative studies of liver cytosolic glutathione *S*-transferases. *Ecotoxicol Environ Saf* 16: 180–186.

Driscoll SK, AE McElroy. 1996. Bioaccumulation and metabolism of benzo[*a*]pyrene in three species of polychaete worms. *Environ Toxicol Chem* 15: 1401–1410.

Egaas E, U Varanasi. 1982. Effects of polychlorinated biphenyls and environmental temperature on *in vitro* formation of benzo[*a*]pyrene metabolites by liver of trout (*Salmo gairdneri*). *Biochem Pharmacol* 31: 561–566.

Feroz M, AA Podowski, MAQ Khan. 1990. Oxidative dehydrochlorination of heptachlor by *Daphnia magna*. *Pest Biochem Physiol* 36: 101–105.

Funari E, A Zoppinki, A Verdina, G de Angelis, L Vittozzi. 1987. Xenobiotic metabolizing enzyme systems in test fish. I. Comparative studies of liver microsomal monooxygenases. *Ecotoxicol Environ Saf* 13: 24–31.

Garrison AW, VA Nzengung, JK Avanta, JJ Ellington, WJ Jones, D Rennels, NL Wolfe. 2000. Phytodegradation of *p,p′*-DDT and the enantiomers of *o,p′*-DDT. *Environ Sci Technol* 34: 1663–1670.

Glickman AH, CN Statham, A Wu, JJ Lech. 1977. Studies on the uptake, metabolism, and disposition of pentachlorophenol and pentachloroanisole in rainbow trout. *Toxicol Appl Pharmacol* 41: 649–658.

Höhl H-U, W Barz. 1995. Metabolism of the insecticide phoxim in plants and cell suspension cultures of soybean. *J Agric Food Chem* 43: 1052–1056.

Huang KC, SF Collins. 1962. Conjugation and excretion of aminobenzoic acid isomers in marine fishes. *J Cell Comp Physiol* 60: 49–52.

Hunn JB, RA Schoettger, WA Willford. 1968. Turnover and urinary excretion of free and acetylated MS 222 by rainbow trout, *Salmo gairdneri*. *J Fish Res Bd Can* 25: 215–231.

James MO, JR Bend. 1976. Taurine conjugation of 2,4-dichlorophenoxyacetic acid and phenylacetic acid in two marine species. *Xenobiotica* 6: 393–398.

Janák K, G Becker, A Colmsjö, C Östman, M Athanasiadou, K Valters, Å Bergman. 1998. Methyl sulfonyl polychlorinated biphenyls and 2,2-bis(4-chlorophenyl)-1,1-dichloroethene in gray seal tissues determined by gas chromatography with electron capture detection and atomic emission detection. *Environ Toxicol Chem* 17: 1046–1055.

Just CL, JL Schnoor. 2004. Phytophotolysis of hexahydro-1,3,5-trinitro-1,3,5-triazine (RDX) in leaves of reed canary-grass. *Environ Sci Technol* 38: 290–295.

Keizer J, G d'Agostino, R Nagel, F Gramenzi, L Vittozzi. 1993. Comparative diazinon toxicity in guppy and zebra fish: Different role of oxidative metabolism. *Environ Toxicol Chem* 12: 1243–1250.

Kirchin MA, A Wiseman, DR Livingstone. 1992. Seasonal and sex variation in the mixed-function oxygenase system of digestive gland microsomes of the common mussel, *Mytilus edulis* L. *Comp Biochem Physiol* 101C: 81–91.

Koenig DW, HB Ward. 1983. *Prototheca zopfii* Krüger strain UMK-13 growth on acetate or *n*-alkanes. *Appl Environ Microbiol* 45: 333–336.

Krause A, WG Hancock, RD Minard, AJ Freyer, RC Honeycutt, HM LeBaron, DL Paulson, SY Liu, JM Bollag. 1985. Microbial transformation of the herbicide metolachlor by a soil actinomycete. *J Agric Food Chem* 33: 584–589.

LaCount MW, E Zhang, YP Chen, K Han, MM Whitton, DE Lincoln, SA Woodin, L Lebioda. 2000. The crystal structure and amino acid sequence of dehaloperoxidase from *Amphitrite ornata* indicate common ancestry with globins. *J Biol Chem* 275: 18712–18716.

Lambert M, S Kremer, H Anke. 1995. Antimicrobial, phytotoxic, nematicidal, cytotoxic, and mutagenic activities of 1-hydroxypyrene, the initial metabolite in pyrene metabolism by the basidiomycete *Crinipellis stipitaria. Bull Environ Contam Toxicol* 55: 251–257.

Larsen OFA, IS Kozin, AM Rija, GJ Stroomberg, JA de Knecht, NH Velthorst, C Gooijer. 1998. Direct identification of pyrene metabolites in organs of the isopod *Porcello scaber* by fluorescence line narrowing spectroscopy. *Anal Chem* 70: 1182–1185.

Letcher RJ, RJ Norstrom, DCG Muir. 1998. Biotransformation versus bioaccumulation: Sources of methyl sulfone PCB and 4,4′-DDE metabolites in polar bear food chain. *Environ Sci Technol* 32: 1656–1661.

Lipke H, CW Kearns. 1959a. DDT dehydrochlorinase I. Isolation, chemical properties, and spectrophotometric assay. *J Biol Chem* 234: 2123–2128.

Lipke H, CW Kearns. 1959b. DDT dehydrochlorinase II. Substrate and cofactor specificity. *J Biol Chem* 234: 2129–2132.

Little PJ, MO James, JB Pritchard, JR Bend. 1984. Benzo(*a*)pyrene metabolism in hepatic microsomes from feral and 3-methylcholanthrene-treated southern flounder, *Paralichthys lethostigma. J Environ Pathol Toxicol Oncol* 5: 309–320.

Livingstone DR, SV Farrar. 1984. Tissue and subcellular distribution of enzyme activities of mixed-function oxygenase and benzo[*a*]pyrene metabolism in the common mussel *Mytilis edulis* L. *Sci Tot Environ* 39: 209–235.

Malins DC, BB McCain, DW Brown, MS Myers, MM Krahn, S-L Chan. 1987. Toxic chemicals, including aromatic and chlorinated hydrocarbons and their derivatives, and liver lesions in white croaker (*Genyonemus lineatus*) from the vicinity of Los Angeles. *Environ Sci Technol* 21: 765–770.

Malins DC, MM Krahn, MS Myers, LD Rhodes, DW Brown, CA Krone, BB McCain, S-L Chan. 1985. Toxic chemicals in sediments and biota from a creosote-polluted harbor: Relationships with hepatic neoplasms and other hepatic lesions in English sole (*Parophrys vetulus*). *Carcinogenesis* 6: 1463–1469.

McElroy AE. 1990. Polycyclic aromatic hydrocarbon metabolism in the polychaete *Nereis virens. Aquat Toxicol* 18: 35–50.

McElroy AE, JC Means. 1988. Uptake, metabolism, and depuration of PCBs by the polychaete *Nereis virens. Aquat Toxicol* 11: 416–417.

Michel XR, PM Cassand, DG Ribera, J-F Narbonne. 1992. Metabolism and mutagenic activation of benzo(a) pyrene by subcellular fractions from mussel (*Mytilus galloprovincialis*) digestive gland and sea bass (*Discenthrarcus labrax*) liver. *Comp Biochem Physiol* 103C: 43–51.

Miyazawa M, T Wada, H Kameoka. 1998. Biotransformation of (+) and (–) limonene by the larvae of common cutworm (*Spodoptera litura*). *J Agric Food Chem* 46: 300–303.

Murphy SE, A Drotar, R Fall. 1982. Biotransformation of the fungicide pentachloronitrobenzene by *Tetrahymena thermophila. Chemosphere* 11: 33–39.

Neilson AH, H Blanck, L Förlin, L Landner, P Pärt, A Rosemarin, M Söderström. 1989. Advanced hazard assessment of 4,5,6-trichloroguaiacol in the Swedish environment, pp. 329–374. In *Chemicals in the Aquatic Environment* (Ed. L Landner), Springer, Berlin.

Newman LA, SE Strand, N Choe, J Duffy, G Ekuan, M Ruiszai, BB Shurtleff, J Wilmoth, MP Gordon. 1997. Uptake and biotransformation of trichloroethylene by hybrid poplars. *Environ Sci Technol* 31: 1062–1067.

Niimi AJ, HB Lee, GP Kissoon. 1989. Octanol/water partition coefficients and bioconcentration factors of chloronitrobenzenes in rainbow trout (*Salmo gairdneri*) *Environ Toxicol Chem* 8: 817–823.

Nimmo IA. 1987. The glutathione *S*-transferases of fish. *Fish Physiol Biochem* 3: 163–172.

Olson LE, JL Allen, JW Hogan. 1977. Biotransformation and elimination of the herbicide dinitramine in carp. *J Agric Food Chem* 25: 554–556.

Pavlostathis SG, KK Comstock, ME Jacobson, FM Saunders. 1998. Transformation of 2,4,6-trinitrotoluene by the aquatic plant *Myriophyllum spicatum. Environ Toxicol Chem* 17: 2266–2273.

Peters R. 1952. Lethal synthesis. *Proc Roy Soc (London)* B 139: 143–170.

Pothuluri JV, JP Freeman, FE Evans, TB Moorman, CE Cerniglia. 1993. Metabolism of alachlor by the fungus *Cunninghamella elegans. J Agric Food Chem* 41: 483–488.

Roy S, O Hänninen. 1994. Pentachlorophenol: Uptake/elimination kinetics and metabolism in an aquatic plant *Eichhornia crassipes. Environ Toxicol Chem* 13: 763–773.

Sandermann H. 1994. Higher plant metabolism of xenobiotics: The "green liver" concept. *Pharmacogenetics* 4: 225–241.

Schlenk D, DR Buhler. 1991. Role of flavin-containing monooxygenase in the *in vitro* biotransformation of aldicarb in rainbow trout (*Oncorhyncus mykiss*). *Xenobiotica* 21: 1583–1589.

Shang TQ, SL Doty, AM Wilson, WN Howald, MP Goprdon. 2001. Trichloroethylene oxidative metabolism in plants: The trichloroethanol pathway. *Phytochemistry* 58: 1055–1065.

Sijm DTHM, A Opperhuizen. 1989. Biotransformation of organic chemicals by fish: Enzyme activities and reactions. *Handbook Environ Chem* 2E: 164–235.

Stapleton HM, B Brazil, RD Holbrook, CL Mitchelmore, R Benedict, A Konstatinov, D Potter. 2006. *In vivo* and *in vitro* debromination of decabromodiphenmyl ether (BDE 209) by juvenile rainbow trout and common carp. *Environ Sci Technol* 40: 4653–4658.

Stroomberg GJ, C Reuther, I Konin, TC van Brummelen, CAM van Gestel, C Gooijer, WP Cofino. 1996. Formation of pyrene metabolites by the terrestrial isopod *Porcello scaber. Chemosphere* 33: 1905–1914.

Ugrekhelidze D, F Korte, G Kvesitadze. 1997. Uptake and transformation of benzene and toluene by plant leaves. *Ecotoxicol Environ Saf* 37: 24–29.

van Aken B, JM Yoon, CL Just, JL Schnoor. 2004. Metabolism and mineralization of hexahydro-1,3,5-trinitro-1,3,5-triazine inside poplar tissues (*Populus deltoides × nigra* DN-34). *Environ Sci Technol* 38: 4572–4579.

van Brummelen TC, NM van Straalen. 1996. Uptake and elimination of benzo[*a*]pyrene in the terrestrial isopod *Porcello scaber. Arch Environ Contam Toxicol* 31: 277–285.

Varanasi U, JE Stein, M Nishimoto, WL Reichert, TK Collier. 1987. Chemical carcinogenesis in feral fish: Uptake, activation, and detoxication of organic xenobiotics. *Environ Health Perspect* 71: 155–170.

Walker JD, RS Pore. 1978. Growth of *Prototheca* isolates on *n*-hexadecane and mixed-hydrocarbon substrate. *Appl Environ Microbiol* 35: 694–697.

Walsh C. 1982. Fluorinated substrate analogs: Routes of metabolism and selective toxicity. *Adv Enzymol* 55: 187–288.

Wei LY, CR Vossbrinck. 1992. Degradation of alachlor in chironomid larvae (Diptera: Chironomidae). *J Agric Food Chem* 40: 1695–1699.

Part 3: Plants and Their Microbial Interactions

2.15 Introduction

Plants fulfill a number of widely different functions and bacteria can play a range of primary and secondary roles.

1. A primary role in the metabolism of contaminants derived from the atmosphere through the leaves, or the terrestrial environment through the root system.

2. Plant waxes play a critical role in restricting water loss and are significant in accumulating lipophilic contaminants from the atmosphere.

3. A secondary role includes the excretion of metabolites that serve as growth substrates for bacteria in the rhizosphere—including those capable of fixing nitrogen and functioning as biocontrol agents against pathogens.

4. They can produce metabolites that fulfill an important role in defeating plant pathogenic fungi: resistance to these pathogens is induced by fungal metabolism of these protective plant metabolites.

5. Plant–microbe interactions include the regulation and positive control of the development and function of bacterial and fungal plant pathogens.

6. Physically, the spread of root systems and excretion of plant metabolites improves soil aeration and quality.

On the basis of these issues, this chapter attempts to bring together the wide range of plant–microbe interactions and their several roles. The discussion illustrates the role of plant metabolites in mediating resistance to fungal infection, extends the number of nitrogen-fixing bacteria, the significance of methylotrophs in the plant and soil environment bacterial, and the role of bacterial metabolites in combating plant infection.

2.16 Primary Roles of Plants

2.16.1 Plant Wax

Waxes of higher plants and insects are typically esters of long-chain fatty acids esterified with primary alcohols derived from them by reduction. The biosynthesis of nonisoprenoid hydrocarbons has a history beginning with the interest in the synthesis of waxes and related compounds. In plants, cuticular wax plays an important role in protection against desiccation, and they have been employed for passive sampling of atmospheric contaminants (Jensen et al. 1992; Hellström et al. 2004). Valuable reviews of plant waxes provide cardinal details (Post-Beitenmiller 1996; Kunst and Samuels 2003).

Although fatty acids and their related alcohols and aldehydes are commonly found in most plants, the distribution of alkanes is more restricted: high concentrations are found in *Allium porrum, Arabinopsis thaliana, Brassica oleracea, Arachis hypogea,* and *Medicago sativa,* whereas they are virtually absent in *Hordeum vulgare* and *Zea mays* (Post-Beitenmiller 1996). The study of hydrocarbon biosynthesis in plants was initiated with the studies of Chibnall and his coworkers using Brussels sprout (*Brassica oleracea*). They demonstrated the occurrence of the hydrocarbon $C_{29}H_{60}$ and the possible intermediates nonacos-15-one and nonaco-15-ol which could be consistent with a mechanism involving head-to-head coupling of a C_{15} carboxylic acid and decarboxylation (Sahai and Chibnall 1932).

Substantial studies on wax biosynthesis have used *Arabinopsis* that has a well-developed genetic system (Aarts et al. 1995; Millar et al. 1999; Fiebig et al. 2000). The leaves of the wild type contained principally C_{29} alkanes with lesser amounts of C_{31} and C_{33} alkanes, and the stems mainly C_{29} alkanes (Jenks et al. 1995). The C_{18} carboxylic acids which are the major component of membrane lipids provide the basic structural unit for both plant wax and hydrocarbon synthesis. Briefly, the coenzyme A esters of the fatty acid and malonate take part in fatty acid elongation by a sequence of regulated reactions that have been termed elongases: addition of C_2 units from malonyl-CoA, reduction to β-hydroxyacyl-CoA, dehydration to enoyl-CoA and reduction to produce C_{30} fatty acid CoA esters. After the synthesis of the C_{28} and C_{30} fatty acids by elongation, the biosynthetic pathway diverges into two segments:

1. In the *decarbonylation* pathway, the carboxylic acids are reduced to aldehydes that produce alkanes by loss of CO, followed by oxidation at C_{12} to secondary alkanols.
2. In the *acyl reduction* pathway, the carboxylic acids are sequentially reduced to aldehydes and primary alkanols: esterification of these with the carboxylic acids produces the aliphatic wax components (Kunst and Samuels 2003).

Details of the decarbonylation have emerged from a study of hydrocarbon synthesis in young leaves of *Pisum sativum*, and a particulate aldehyde decarbonylase was partly purified. This had a major protein of 67 kDa, was inhibited by O_2, required phospholipid for activity, and was able to convert octadecanal to heptadecane with the production of carbon monoxide that was trapped as the $RhCl[C_6H_5)_3)P]_3$ complex (Schneider-Belhaddad and Kolattukudy 2000). Hydrocarbons are not, however, synthesized by head-to-head condensation that has been shown to take place in bacteria.

2.16.2 Metabolism

The metabolism of contaminants by plants, including aquatic representatives, is important for several reasons:

1. The metabolites may have deleterious effects on biota at other trophic levels.
2. The metabolites may be translocated from the leaves into the root system and, after partition into interstitial water in the soil, may exert toxic effects on other aquatic and terrestrial biota.

3. There has been interest in the use of plants for bioremediation and this requires a full appreciation of their metabolic potential, including the production of stable end-metabolites.

Detoxification and metabolism by higher plants has been reviewed (Sandermann 1994), and some
illustrative examples are given.

1. Hybrid poplars (*Populus deltoides* × *nigra*) have attracted considerable interest since they
   are able to transport and metabolize diverse xenobiotics—even though they do not normally
   degrade them.

   a. Trichloroethene was metabolized to trichloroethanol and trichloroacetate (Newman et al.
      1997, 1999).

   b. Atrazine was metabolized by reactions involving dealkylation and hydrolytic dechlorina-
      tion to yield 2-hydroxy-4,6-diamino-1,3,5-triazine (Burken and Schnoor 1997).

   c. Tissue cultures of poplar metabolized the explosive hexahydro-1,3,5-trinitro-1,3,5-triazine
      (RDX) by partial reduction to the 1-nitroso- and 1,3-dinitroso derivatives, and in the light,
      these were further metabolized to formaldehyde, methanol, and CO_2 (van Aken et al. 2004b).

2. The metabolism of a range of xenobiotics has been examined in other plants and includes
   aquatic representatives.

   a. Pentachlorophenol is metabolized by the aquatic plant *Eichhornia crassipes* to a number
      of metabolites, including di-, tri-, and tetrachlorocatechol, 2,3,5-tri-, and tetrachloro-
      hydroquinone, pentachloroanisole, and tetrachloroveratrole (Roy and Hänninen 1994). The
      phenolic products should be compared with those produced during the photochemical, and
      the initial, stages in the microbiological metabolism of pentachlorophenol, followed by
      O-methylation (Figure 2.24).

   b. Quite complex transformations may be mediated, and the metabolism of phoxim by plant
      organs and cell suspension of soybean (*Glycine max*) are given as an example (Höhl and
      Barz 1995) (Figure 2.25).

   c. The uptake and biotransformation of benzene from soil and from the atmosphere have
      been studied in a number of plants. For example, it was shown that in leaves of spinach
      (*Spinacia oleracea*), the label from ^{14}C-benzene was found in muconic, fumaric, succinic,
      malic, and oxalic acids, as well as in specific amino acids. In addition, an enzyme prepara-
      tion in the presence of NADH or NADPH produced phenol (Ugrekhelidze et al. 1997).

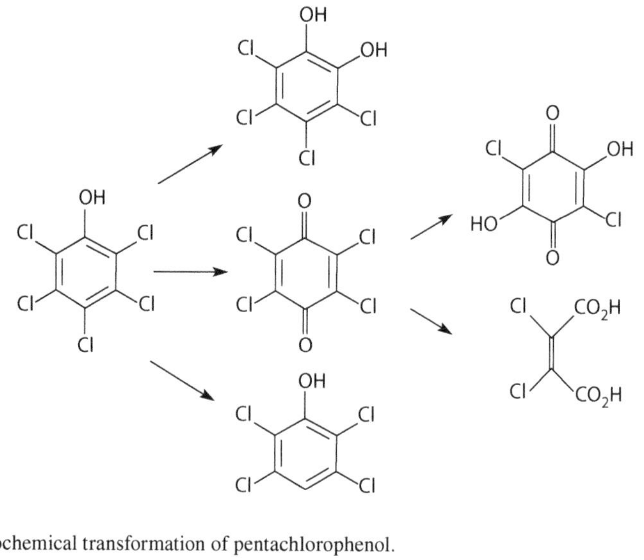

FIGURE 2.24 Photochemical transformation of pentachlorophenol.

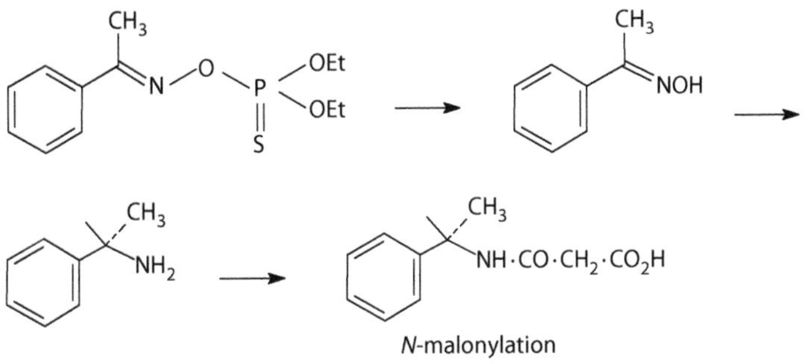

FIGURE 2.25 Metabolism of phoxim by soybean (*Glycine max*).

    d. 2,4,6-Trinitrotoluene was reduced by the aquatic plant *Myriophyllum spicatum* to amino-dinitrotoluenes (Pavlostathis et al. 1998), and in axenic root cultures of *Catharanthus roseus*, the initial metabolites 2-amino-4,6-dinitrotoluene and 4-amino-2,6-dinitrotoluene were conjugated—probably with C_6 units (Bhadra et al. 1999). There are, therefore, several important unresolved issues: these include accumulation in plant tissues (Hughes et al. 1997) and the phytotoxicity of amino metabolites if they were excreted.

3. Reed Canary Grass (*Phalaris arundinacea*) was grown in liquid culture and exposed to hexahydro-1,3,5-trinitro-1,3,5-triazine (RDX) that was metabolized to the potentially toxic 4-nitro-2,4-diazabutanal (Just and Schnoor 2004). This metabolite is also produced from RDX by strains of *Rhodococcus* sp., and from the homologous octahydro-1,3,5,7-tetranitro-1,3,5,7-tetrazocine (HMX) by *Phanerochaete chrysosporium*.

4. Plant cultures have also been examined for their ability to transform contaminants. Generally, only limited changes take place such as hydroxylation and the formation of conjugates with phenolic groups.

    a. The transformation of pentachlorophenol by cultures of wheat (*Triticum aestivum*) produced tetrachlorocatechol as its glucuronide (Schäfer and Sandermann 1988).

    b. Cultures of wheat (*Triticum aestivum*), tomato (*Lycopersicon esculentum*), lettuce (*Lactuca sativa*) and a rose variety, carried out rather limited transformations of fluoranthene with plant-specific formation of 1-, 3-, and 8-hydroxy compounds that were conjugated (Kolb and Harms 2000).

    c. Transformation of PCBs was examined in plant cultures (Wilken et al. 1995). The rates of transformation were generally greater for congeners with lower degrees of substitution, and neither pentachlorobiphenyl (PCB 101) nor hexachlorobiphenyl (PCB 153) was metabolized. The cultures of soybean (*Glycine max*), white clover (*Trifolium repens*), and some of the grasses were most effective. A range of conjugated hydroxylated products was formed from 2-chlorobiphenyl, including 3-chloro-2-hydroxybiphenyl, that involved an NIH shift, while 2,5,2′,5′-tetrachlorobiphenyl (PCB 52) produced four monohydroxylated metabolites (Wilken et al. 1995).

5. a. The metabolism of PCB congeners was studied in callus cultures of *Nicotiana tabacum* exposed in darkness to a range of dichloro-PCBs (2,2′, 2,3′, 2,3′, 2,4′, 2,3, and 2,4) (Rezek et al. 2008). Hydroxylated PCBs were observed, and some were *O*-methylated presumptively as a result of *O*-methyltransferase activity.

    b. Root tissue cultures of black nightshade *Solanum nigrum* were precultivated before exposure to PCBs for 14 days, and the following were examined: 12 dichloro congeners, 7 trichloro congeners, 5 tetrachloro congeners, and one pentachloro PCB. Hydroxylation was observed only for the dichloro congeners and the 2,2′,4-, 2,2′,5- and 2,2′,6-trichloro congeners, while the 2,5-dichloro congener was hydroxylated at the 2-, 3-, and 4- positions in the nonchlorinated ring (Rezek et al. 2007).

6. The transformation of 4-*n*-nonyl[U-^{14}C]phenol was examined in a range of plant cultures. Metabolism took place in the alkyl side chain with the formation of mono- and dihydroxylated products, with a substantial fraction of the label occurring in bound residues, including starch, protein, lignin, and hemicellulose (Bokern and Harms 1997).

2.17 Secondary Role of Plants

2.17.1 Introduction

General comments attempt to address the role of low-molecular-mass products exuded by plants. Bacteria growing at the expense of these compounds may be advantageous both to the plants in their neighborhood (rhizosphere and rhizoplane) and for their possible role in the biodegradation of contaminants in the rhizosphere and neighboring areas. Organic compounds that are exuded from plant roots mediate the colonization of the rhizosphere of many crop plants by fluorescent pseudomonads. Although the structures of these compounds have not always been identified, it has been demonstrated that higher plants excrete a range of phenolic compounds (Fletcher et al. 1995; Hedge and Fletcher 1996). Some of these metabolites include naringin, catechin, and myricitin (Donnelly et al. 1994). Although carbohydrates may predominate, malate and succinate are of primary importance, whereas amino acids are generally of lesser significance except for chemotaxis that is a requisite for virulence of the wilt pathogen *Ralstonia solanacearum* (Yao and Allen 2006). It has been shown that organic sulfur compounds—putatively sulfonates and sulfate esters—are important for the survival of plants in the rhizosphere (Mirleau et al. 2005). In addition, mesophilic archaea belonging to the *Crenarchaeota* have been characterized in extracts of tomato plants by a combination of enrichment and molecular methods (Simon et al. 2005). There may also be an indirect effect of plants. For example, mineralization of PAHs was examined in microcosms prepared with freshwater sediments and planted with the reed *Phragmites australis* (Jouanneau et al. 2005). Inoculation with the pyrene-degrading *Mycobacterium* sp. strain 6PY1 resulted in a greater degree of mineralization than in the absence of plants, and this was accompanied by the release of phenanthrene-4,5-dicarboxylic acid that is an established metabolite. The positive effect of the plants could be attributed to the increased access of oxygen to the sediment and, therefore, direct stimulation of bacterial growth.

2.17.2 Degradative Enzymes

1. The degradation of 2-chlorobenzoate was supported—presumptively involving enzymatic reaction—by exudates from Dahurian wild rye (*Elymus dauricus*) (Siciliano et al. 1998). It was also shown that concentrations of 2-chlorobenzoate aged in soil for 2 years could be reduced by the indigenous microflora (Siciliano and Germida 1999). Experiments have been carried out on the growth of Dahurian wild rye in the presence of bacterial inoculants that did not adversely affect the mineralization of atrazine under various conditions (Burken et al. 1996). Although in the absence of plants of poplar hybrids (*Populus deltoides nigra* DN34), root exudate resulted in only a slight stimulation of the mineralization of ^{14}C-labeled atrazine, addition of crushed roots provided a greater positive effect.

2. The extradiol catechol fission enzyme from *Terrabacter* sp. strain DBf63 was introduced into *Arabidopsis thaliana* using *Agrobacterium tumefaciens*-mediated transformation, and the haloalkane dehalogenase from *Rhodococcus* sp. strain m15-3 into *Nicotiana tabacum*. Seedlings were introduced into hydroponic media and incubated either in darkness or under illumination. It was shown that enzymatic activities from plants expressing the apoplasm-targeted enzymes exceeded those for the cytoplasm-targeted enzymes. This is consistent with transport between the apoplast and the roots, and it was suggested on the basis of these rather preliminary experiments that this methodology could be effective for degrading contaminants in the aquatic environment (Uchida et al. 2005).

3. Mammalian P450 2E1 was introduced into tobacco plants that were exposed to trichloroethene in hydroponic medium for 5 days. Trichloroethene epoxide was produced initially, and was

rearranged to trichloroacetaldehyde that was then reduced to trichloroethanol. This was found in samples of leaves, stems, and roots, and was absent in the control plants. Trichloroethanol was subsequently transported to the leaves where it was apparently metabolized (Doty et al. 2000).

4. *Pseudomonas fluorescens* strain F113 is an important biocontrol strain for sugar beet, and a transposon that contained the genes from *Burkholderia* sp. strain LB400 for the degradation of biphenyl was inserted into the chromosome (Brazil et al. 1995). The resulting hybrid F113pcb was stable under nonselective conditions, and presented a possible candidate for bioremediation, although the level of expression of the *bph* genes was low. Improvement was accomplished by cloning the *bhp* operon under control of the *nod* system in *Sinorhizobium meliloti* to produce *P. fluorescens* strain F113L::1180 that grew faster with biphenyl and was able to transform effectively a range of PCB congeners as great as, or greater than, the original *Burkholderia* sp. strain LB400 (Villacieros et al. 2005).

5. It is well established that one of the pathways for the aerobic degradation of trichloroethene is monooxygenation by the toluene *ortho* monooxygenase (Nelson et al. 1987). Transposon integration was used to insert the gene from *Burkholderia cepacia* strain G4 into the chromosome of *Pseudomonas fluorescens* strain 2–79. The strain was added to the germinated wheat seedling before planting in soil microcosms (Yee et al. 1998). The average degradation using this strain was 63%, and root colonization was comparable to that of the parent strain lacking the monooxygenase genes.

6. *Ralstonia eutropha* (*Alcaligenes eutrophus*) strain NH9 is able to degrade 3-chlorobenzene by the "modified ortho" pathway. The *cbnA* gene that encodes 3-chlorocatechol 1,2-dioxygenase was introduced into rice plants (*Oryza sativa* ssp. *japonica*) under the control of a virus 35S promoter. 3-Chlorocatechol induced dioxygenase activity in the callus of the plants, and leaf tissues oxidized 3-chlorocatechol with the production of 2-chloromuconate (Shimizu et al. 2002). It was, therefore, shown that it is possible to produce transgenic plants with the capability of degrading chlorinated aromatic compounds.

2.17.3 The Role of Plant Exudates in Degradation

The interaction of plants and bacteria in the root system plays an important role in the remediation of contaminants, and may involve the use of plant exudates to stimulate the growth of degradative bacteria as well as the penetration of the soil by plant roots (Kuiper et al. 2004).

1. Metabolites that included naringin, catechin, and myricitin (Donnelly et al. 1994) were able to induce the growth and metabolism of bacteria that degrade a range of PCB congeners (Donnelly et al. 1994). It has also been shown that enzymes for the biodegradation of some PCB congeners could be induced by a number of terpenoids (Gilbert and Crowley 1997) that may originate from higher plants. There is, therefore, a plausible relation between metabolites excreted by higher plants and utilization of them to support bacterial degradation of xenobiotics in the rhizosphere. In addition, some coumarin, flavone, flavanol, and flavanone plant metabolites were able to support the growth of anaerobic bacteria during dechlorination of chlorocatechols (Allard et al. 1992).

2. It has been shown that the fine roots of mulberry (*Morus* sp.) die at the end of the growing season, and that this coincides with the accumulation of flavones (morusin, morusinol, and kuwanon) that have C_5 substituents at C-3 and C-8. These flavones were able to support the growth of *Burkholderia* sp. strain LB400 that is effective in the degradation of a range of PCB congeners. It was, therefore, suggested that the growth of fine roots and the production of phenolic metabolites should be taken into account in evaluating remediation in the rhizosphere (Leigh et al. 2002).

3. *Pseudomonas putida* strain PCL 1444 is an efficient colonizer of the roots of *Lolium multiflorum* that produces a highly branched root system, and it was evaluated for its ability to utilize root exudates and the ability to degrade naphthalene (Kuiper et al. 2001). The exudate from

seeds, seedlings, and roots contained a range of low-molecular-weight fatty acids, including citrate, malate, and succinate, and carbohydrates, particularly fructose, glucose, and arabinose, that could support the growth of the strain. Although concentrations of propionate were low in the exudate, propionate allowed a high rate of expression of the genes for naphthalene degradation. These results illustrated the importance of root colonization by bacteria that used plant exudates for growth (Kuiper et al. 2002).

4. Soil samples from a site in the Czech Republic that had been contaminated with PCBs and was planted with several kinds of mature trees, including ash, birch, pine, willow, and black locust, were taken from the rhizosphere and the root zone, and used to enumerate the numbers of PCB-degrading bacteria using published procedures (Sylvestre 1980; Bedard et al. 1986). Their numbers in the root zones of Austrian pine (*Pinus nigra*) and goat willow (*Salix caprea*) exceeded those from other sites, and the majority of isolates were identified on the basis of 16S rRNA gene sequences as members of the genus *Rhodococcus*, one of which had a spectrum of PCB degradation similar to that of *Burkholderia xenovorans* LB400. Gram-negative strains were isolated only from the root zone, and belonged to the genera *Luteibacter* and *Pseudomonas*. The type strain of *Lutiebacter rhizovicinum* that was isolated from the rhizosphere of spring barley displayed protease (skimmed milk) activity, and the hydrolysis of starch and gelatin that may plausibly be relevant in providing carbon sources (Johansen et al. 2005). The data were interpreted to support the value of rhizoremediation (Leigh et al. 2006).

5. Soil contaminated with PCBs had been planted with horseradish, and samples from the rhizosphere were used to isolate single colonies that could degrade biphenyl. These were examined by MALDI-TOF and a MALDI Biotyper, and a range of bacteria was identified to the genus level, and several also to the species level. Identifications included species of the genera *Aeromonas*, *Arthrobacter*, *Microbacterium*, *Rhizobium*, *Rhodococcus*, and *Serratia* that are represented among strains that are capable of degrading PCB (Uhlik et al. 2011).

6. Bacteria associated with the roots of plants have important potential for the biodegradation of biogenic methane and CO. Examples include methanotrophs associated with aquatic vegetation (King 1994), and the oxidation of CO by bacteria associated with freshwater macrophytes (Rich and King 1998). An important example of considerable environmental significance is the generation of methane in rice paddies where the plant roots provide the carbon source for methanogenesis (Lu et al. 2005).

7. The degradation of trichloroethene by *Rhodococcus* sp. strain L4 was induced by citral, limonene, and cumene (Suttinun et al. 2009), although continued degradation required the periodic addition of cumene and limonene (Suttinun et al. 2010).

2.18 Plant Metabolites as Antagonists

2.18.1 Introduction

Plants synthesize metabolites of dazzling complexity, and some may play a protective role. Two examples are given as introduction:

1. Species of *Crotalaria* synthesize pyrrolizidine alkaloids, and larvae of the moth *Utetheisa* secrete them as protection against predation by spiders and beetles, and parasitism by the wasp *Trichogramma ostriniae* (Bezzerides et al. 2004).

2. Nitroalkanes occur in a range of leguminous plants where they may play a role as toxicants against grazing predators. The metabolism of 2-nitropropane has been examined in extracts from pea seedlings that produced acetone, nitrite, and H_2O_2, and the enzyme was highly specific for this substrate (Little 1957). The metabolism of 3-nitropropionic acid was examined in

extracts from the leaves of the legume *Hippocrepis comosa* that contains high concentrations of this metabolite (Hipkin et al. 1999).

Plants produce a plethora of secondary metabolites that can serve as substrates for bacteria that are capable of degrading xenobiotics and, therefore, play a role in biotransformation and phytoremediation (Singer et al. 2003). These secondary metabolites can offer protection against fungal infection, and they have been divided into two main groups: members of Group I (phytoanticipins) are constitutive and pre-formed in the plants, whereas the synthesis of those of Group II (phytoalexins) is induced at the site only in response to infection.

The function of plant secondary metabolites and their role in combating fungal infection are complex, and only a brief sketch of the main issues is given. Emphasis is placed on the metabolic reactions of pathogenic fungi that mediate resistance to the plant metabolites which function as part of the plant defense system. It should be noted, however, that there may not always be a simple correlation between fungal metabolism and pathogenicity.

A review on fungal resistance to plant metabolites with extensive references has been given (Morrissey and Osbourn (1999), and attention is drawn to reviews (Brooks and Watson 1985, 1991) that deal extensively with the chemical structures of phytoalexins, and the metabolic network of diterpenoid biosynthesis in cereal crops (Peters 2006).

Conversely, some bacteria are able to synthesize compounds that serve to protect plants from fungal infection, and are discussed later in this chapter as biocontrol agents.

2.19 Group 1

2.19.1 Saponins

Saponins are secondary plant metabolites that are found in major food crops, and some are able to provide protection against fungal infection. Avenacins are pentacyclic triterpenoids with a branched trisaccharide at the β-C_3-position that occur in the roots of oat (*Avena* spp.), and they may protect oat roots from fungal attack. Avenacin A-1 and A-2 have hydroxyl groups at C_{23}, whereas in the avenacins B-1 and B-2, this is replaced by hydrogen (Crombie et al. 1986a). Substantial studies have used the root-infecting fungus *Gaeumannomyces graminis* var. *avenae* (Gga) that is insensitive to avenacin A-1, in contrast to var. *tritici* (Ggt) that causes "take all" disease in wheat and barley and, although unable to infect oat species, it is sensitive to avenacin A-1 (Crombie et al. 1986b). The hydrolase avenacinase from Gga is able to remove both β-1,2- and β-1,4-terminal glucose residues to produce aglycones that display less toxicity to fungi (Crombie et al. 1986b). The specific role of avenacinase was confirmed using avenacinase-minus mutants of *Gaeumannomyces graminis* that were unable to infect the saponin-containing oat, although this was retained for wheat that does not contain saponins (Bowyer et al. 1995). In addition, saponin-deficient mutants of the diploid oat *Avena strigosa* were considerably more susceptible to infection with *G. graminis* var. *tritici* than the wild type (Papadoupoulou et al. 1999).

2.19.2 Steroidal Glycoalkaloids

Steroidal glycoalkaloids are found in solanaceous plants, including potato and tomato. They have different branched-glycosides at C_3: α-tomatine with a trisaccharide is found in tomato and α-chaconine with a tetrasaccharide in potato. Fungal pathogens, including *Septoria lycopersici*, *Botrytis cinerea*, and *Fusarium oxysporum* f. sp. *Lycopersici*, are able to detoxify α-tomatine by hydrolysis at different positions of the tetrasaccharide at C_3 (Quidde et al. 1998), although the aglycone tomatidine and lycotetraose have important physiological effects in the suppression of the defense response in suspension-cultured tomato cells (Ito et al. 2004). The physicochemical and immunological properties of avenacinase from *Gaeumannomyces graminis* and tomatinase from *Septoria lycopersici*, both of which bring about the analogous hydrolysis of the terminal β-1,2- and β-1,4-terminal glucose residues of α-tomatine, are quite similar—though not identical (Osbourn et al. 1995). Several lines of evidence have supported the pathogenic role of α-tomatine. Gene disruption of the β$_2$-tomatinase gene in the tomato leaf fungal pathogen *Septoria lycopersici* resulted in the loss of both enzyme activity and toler-ance of α-tomatine (Sandrock and Vanetten 2001). Further evidence using the tomato pathogen *Fusarium oxysporum* f. sp. *lycopersici* that can degrade α-tomatine to the aglycone tomatidine with loss of the entire oligosaccharide unit at C$_3$ (Lairini et al. 1996) showed that tomatinase is required for the full display of virulence in this fungus (Pareja-Jaime et al. 2008). It was also shown that several different tomatinase activities were present. On the other hand, metabolism of α-tomatine by *Gibberella pulicaris* not only removed the C$_3$ lycotetraose by hydrolysis, but also brought about hydroxylation at the axial C$_7$ and formation of the Δ^5-7α-hydroxy-tomatidenol (Weltring et al. 1998). Bacteria are also important plant pathogens and some also produce tomatinase. *Clavibacter michiganensis* is a Gram-positive bacterium that causes bacterial wilt of tomato, and tomatinase is a phytoanticipin that is pro-duced by tomato as a defense against infection (Kaup et al. 2005). Although *Streptomyces scabies* 87-22 is the causal agent of common scab in potato and possesses a functional tomatinase, this is not impor-tant for the pathogenicity on tomato although it may fulfill other functions (Seipke and Loria 2008).

2.19.3 Cyclic Hydroxamates

Cyclic hydroxamates related to 4-hydroxybenz-1,4-oxazin-3-ones can serve as defense chemicals in the Gramineae, including rye, maize, and wheat, though not in rice, barley, or oats (Niemeyer 1988). Maize (*Zea mays*) contains a number of cyclic hydroxamates of which 2,4-dihydroxybenzoxazine-3-one and the 7-methoxy derivative are dominant in plant tissue in the form of the 2-glycosides that are spontane-ously transformed to benzoxaxolin-3-ones by a β-glucosidase on disruption of plant tissue. The benzoxaxolin-3-ones play an important role in protection against fungal infection by *Fusarium subglu-tinans* that is a common and important pathogen of maize. It was able to degrade these benzoxazolin-3-ones to the nontoxic *N*-(2-hydroxyphenyl) malonates and thereby disable the normal protection mechanism of the plant against fungal toxicity (Vilich et al. 1999). This metabolism is also mediated by a number of species of *Fusarium* (Glenn et al. 2001), including *Fusarium verticillioides* (*Fusarium moniliforme*) (Richardson and Bacon 1995; Yue et al. 1998), that is generally a symptomless endophyte of maize but is highly undesirable on maize that is to be used for human consumption. In addition, these secondary plant metabolites can effect *in vitro* the interactions among endophytes in maize (Sauders and Kohn 2008).

2.19.4 Acetoxyalkenes and Acetoxyalkadienes

Acetoxyalkenes and acetoxyalkadienes are long-chain esters of unsaturated fatty acids that occur in avocado and mango. Although the regulation of their levels is complex, their concentrations in avo-cado decrease during ripening—more rapidly in disease-susceptible cultivars. In unripe fruit, their level is enhanced by various insults, including exposure to the fungus *Colletotrichum gloeosporioides* that doubled the concentration of 2-hydroxy-4-oxo-eneicosa-12,15-dienol acetate after infection (Prusky et al. 1990), that has, therefore, been implicated in interactions with avocado (Prusky et al. 1991b). Two long-chain alkanol acetates have been isolated and considered in quiescent infection of avocado by this fungus—2,4-dihydroxy-*n*-heptadec-16-enol and 2-hydroxy-4-oxoeneicosa-12,15-di-enol as their acetates (Prusky et al. 1991a). The level of the monoene decreased during ripening with the appearance of disease symptoms from the fungus, while levels of the diene decreased. Several lines of evidence support the role of lipoxygenase in modifying levels of the diene and its role in fungal infection (Prusky et al. 1985): (a) the development of *Colletotrichum gloeosporioides* on disks of avo-cado was reduced by a factor of ca. 10 by the linoleate inhibitor 5,1,8,11,14-eicosatetraenoate, (b) epi-catechin (*cis*-3-hydroxyflavanol) was a natural inhibitor of avocado dioxygenase, and during expression of fungal infection, levels of this decreased during 11 days when the first appearance of fungal infec-tion was observed. Collectively, these results suggest that lipoxygenase inhibitors result in increased resistance (Prusky et al. 1991b) and that these long-chain alcohol esters contribute to the resistance of unripe fruit to fungal attack.

2.20 Group II Phytoalexins

2.20.1 Cyanogenic Plants

1. Millet (*Sorghum bicolor*) is one of some 200 plant species that produce cyanogenic glycosides, and is unusual in producing the 3-deoxyanthocyanidins apigenidin and luteolinidin. These flavanols inhibit the growth of phytopathogenic fungi *in vitro*, and are synthesized (phytoanticipin) within a host that will subsequently be penetrated by the fungus *Colletotrichum graminicola* (Snyder and Nicholson 1990). Luteolinidin accumulated to levels of 0.24–0.91 ng/cell and apigenidin to levels of 0.48–1.20 ng/cell, and it was shown that these levels were sufficient to inhibit the fungus at the site of infection (Snyder et al. 1991). Cyanogenic glycosides produce cyanide in response to tissue damage or infection, and it has been suggested that attack by pathogenic fungi on these plants is limited by cyanide hydratase produced by the fungi. This is supported by the isolation and characterization of the gene for this enzyme from the sorghum pathogen *Gloeocercospora sorghi* (Wang and VanEtten 1992).

2. A study using birdsfoot trefoil (*Lotus corniculus*) and the pathogen *Stemphyllium loti* revealed that cyanogenesis in the plant was associated with pathogenesis: HCN was released as a result of infection, there was a correlation between the severity of the symptoms and the amount of HCN that was released, and *S. loti* was markedly tolerant of HCN (Miller and Higgins 1970). On the other hand, detoxification of HCN by formamide hydro-lyase to produce formamide was found not only in fungi that were pathogens of cyanogenic plants, but in pathogens for noncyanogenic plants (Fry and Evans 1977).

2.20.2 Rice

Rice blast caused by the fungus *Magneporthe grisea* (originally *Pyricularia oryzae*) is a serious problem in rice (*Oryza sativa*) that produces a number of metabolites, including the flavanone sakuranetin, and the diterpenoids momilactones A and B, oryzalexins C to F, and S, and phytocassanes D to E that share biosynthetic pathways (Peters 2006). Sakuranetin and momilactone A were observed to serve as phytoalexins against infection with avirulent strains of the rice pathogen *Magneporthe grisea*, and levels of these and oryzalexin S that had the greatest antifungal activity were produced in greater amounts in resistant cultivars compared with susceptible cultivars (Dillon et al. 1997). Momilactones A and B are phytoalexins since they display antifungal activity against, and appear in rice leaves only after infection (Cartwright et al. 1981). Cartwright et al. also demonstrated that momilactone synthesis could be induced by UV-irradiation, and this procedure has been extensively employed in later studies. The analogous induction was also found for the oryzalexins B and C (Akatsuka et al. 1985; Kono et al. 1985), and in addition, the rice phytoalexins were localized at blast disease infections. Phytocassanes A to E have been shown to be induced at the edges of necrotic lesions and in the central zone, and increased from healthy to slightly withered leaves (Umemura et al. 2003). Sakuranetin biosynthesis is induced by infection against *M. grisea*, and it was proposed that it might present a defense (Kodama et al. 1992). In addition to the synthesis of phytoalexins, there is an important role for elicitors that are produced by the pathogens. Synthesis of the phytoalexins phytocassanes and momilactones was induced by spraying rice leaves with the glycosphingolipid elicitors cerebroside A and C, and disease severity from rice blast fungus was reduced by a factor of 5–6 after treatment with cerebroside (Umemura et al. 2000). Collectively, all these diterpenoid plant metabolites serve as phytoalexins produced in response to fungal infection.

2.20.3 Grapevine

Stilbenes occur in a number of herbaceous and woody plants. Resveratrol (*trans*-3,5,4′-trihydroxystilbene) occurs in grapevines (*Vitis vinifera*) where it has been associated with the flavor of wine, and is the most prevalent stilbene along with pterostilbene (3′,5′-dimethoxy-4-hydroxystilbene).

Inhibition of the mycelial growth of *Botrytis cinerea*, the causal agent of gray mold, was greater for pterostilbene than for resveratrol, although the former is quantitatively a minor component (Adrian et al. 1997). Resveratrol was found in noninfected fruit close to the necrotic area caused by *Botrytis cinerea*, and under favorable climatic conditions, lesions were observed in spite of the increased production of resveratrol (Jeandet et al. 1995). In addition, a complex picture emerged on the level of phytoalexins and resistance to the fungus among different strains of *Vitis* spp. (Dercks and Creasy 1989a,b), although the synthesis of resveratrol, and α- and ε-viniferin is found in lesions cased by *B. cinerea* in both *Vitis vinifera* and *Vitis riparia* (Langcake 1981). It is worth noting that these stress metabolites are produced even in response to downy mildew (*Plasmopara viticola*) and was greater in resistance than in susceptible strains (Langcake 1981). It was tentatively suggested that the degradation of stilbenes and loss of activity were associated with the polyphenol oxidase laccase (Sbagi et al. 1996), and the pathogenicity of *Botrytis cinerea* was associated with the ability to degrade the stilbenes resveratrol and pterostilbene—and conversely. The phytoalexins resveratrol and pterostilbene were also produced in berries of *Vitis vinifera* following infection by *Rhizopus stolonifer* that is prevalent in Israel (Sarig et al. 1997). The genes for the enzyme required for the synthesis of resveratrol have been transferred by *Agrobacterium tumefaciens* to tomato (*Lycopersicon esculentum*), and were integrated into the genome. The accumulation of *trans*-resveratrol was observed after fungal infection by *Phytophthora infestans* and resulted in increased resistance of the transgenic tomato to this fungus (Thomzik et al. 1997).

2.20.4 Potato

Fungal infection by *Gibberella pulicaris* is a major cause of dry rot in potato tubers and elicits the synthesis of the sequiterpenoids rishitin and lubilin that have been identified after fugal infection, although the analysis of these metabolites and their direct role in pathogenesis has been examined only in a restricted range of strains of the fungus. It was shown that there was a direct correlation between the ability of selected strains of *Gibberella pulicaris* to metabolize rishitin and the expression of virulence on tubers of *Solanum tuberosum* (Desjardins and Gardner 1989, 1991). Details of the detoxification by *Gibberella pulicaris* have been examined for both rishitin and lubimin—rishitin by oxidation to 11,12-epoxyrishitin (Gardner et al. 1994), lubimin by reduction to isolubimin that is still toxic, though it can be transformed to the nontoxic cyclodehydroisolubimin by the formation of a furan ring at C_4 with the C_{15} hydroxymethyl group (Gardner et al. 1988; Desjardins et al. 1989).

2.20.5 Crucifers

Considerable attention has been directed to the phytoalexins in *Arabinopsis thaliana* and to species of *Brassica*, in both of which there is a complex relation between the fungal metabolism of phytoalexins and plant resistance to attack by fungi, insects, and herbivores. Attention is drawn only to a few salient features.

2.20.5.1 Arabinopsis thaliana

Camalexin (thiazolyl-3-indole) is the only phytoalexin produced by *Arabidopsis thaliana* when challenged by bacterial or fungal pathogens, and it has a wide variety of biological effects. In addition, the resistance of the wild crucifers *Camelina sativa* and *Capsella bursa-pastoris* to fungal infection has been associated with the synthesis of camalexin. Camalexin is synthesized in *Arabidopsis* from tryptophan via indole-3-acetaldoxime and subsequent reaction with cysteine (Glawischnig et al. 2004). Its induction is, however, signaled by reactive oxygen species, and is upregulated at the site of pathogen infection (Glawischnig 2007). Tissue damage of plants or attack by pathogens results in the hydrolysis of glucosinolates by a specialized β-thioglucoside glucohydrolase (myrosinase). A range of metabolites was identified in various genotypes of *Arabinopsis thaliana* after challenge with the powdery mildew fungus *Blumeria gramimis hordei* to which it is, however, immune. These metabolites included indolyl-3-methylamine, camalexin, indolyl-3-methylglucosinolate, and in particular 4-methoxyindolyl-3-methylglucosinolate. This metabolite could be activated for antifungal defense by the activity of an atypical myrosinase, and it was proposed that cyclic generation and detoxification of glucosinates were involved (Bednarek et al. 2009). In another study, callose formation at the site of fungal contact could be induced by 4-methoxyindolyl-3-methylglucosinolate, and it was shown that two genes are necessary for both glucosinolate activation and callose formation (Clay et al. 2009). Both of these investigations provide dual mechanisms for the response to plant pathogens and the avoidance of damage by insects and herbivores. Although camalexin was not metabolized by the blackleg fungus *Leprosphaeria maculans* (*Phoma lingam*), it can be transformed by the phytopathogen *Rhizoctonia solani* to nontoxic 5-hydroxy-camalexin derivatives (Pedras and Khan 2000).

2.20.5.2 Brassica *spp.*

A wide range of sulfur-containing indole derivatives that include *brassinin*, and its close relatives *spirobrassinin* and *cyclobrassinin*, occur in the genus *Brassica* that comprises important food crops, including cabbage, broccoli, and cauliflower. Their role in combating fungal infection by the "blackleg" fungus *Leptosphaeria maculans* = *Phoma lingam* has been extensively documented in a review (Pedras et al. 2000). Among the interesting features are differences in metabolism by virulent and avirulent strains of the fungus: (a) The metabolism of brassinin by virulent strains results in conversion of the *N*-thioester to the *S*-oxide and then to the indole-3-carboxaldehyde, whereas avirulent strains carry out acylation or methylation (Pedras and Taylor 1993). (b) Brassicanal A is metabolized by virulent strains of the fungus by the successive formation of the sulfone and reduction to 3-methylindole-2-sulfone, though only slowly by an avirulent strain (Pedras and Khan 1996). (c) Cyclobrassinin is metabolized by the virulent fungus to dioxibrassinin by fission of the tiazine ring and by an avirulant strain to the thiazolindole brassilexin (Pedras and Okanga 1999). Although camalexin is not metabolized by either type of *L. maculans*, the rate of detoxication of brassinin in cultures was increased in its presence (Pedras et al. 2005).

2.20.6 Legumes

The phytoalexins kievitone and phaseollidin that are produced by legumes are isoflavonoids. In *Phaseolus vulgaris*, kievitone (5,7-2′,4′-tetrahydroxy-8-allylchromone) can be detoxified by *Fusarium solani* f.s. *phaseoli* by hydroxylation of the C_8 allyl group, and phaseollinisoflavan (Zhang and Smith 1983). Pea (*Pisum sativum*) produces pisatin and chickpea (*Cicer arietinum*) produces maackiain—both of which contain a 5,6-methylendioxy-2,3-dihydrobenzofuran fused to 7-hydroxy-3,4-chroman. *Nectria haematococca* is a pathogen of pea *Pisum sativum* that is capable of de-*O*-methylating pisatin to 6(a)-hydroxy-maackiain, and virulence was determined by the level of demethylase activity (Kistler and Vanetten 1984). Maackiain and medicarpin are phytoalexins of chickpea (*Cicer arietinum*) that are structurally similar to pisatin with opposite configurations at C_{6a} and C_{11a}. Metabolism of maackiain by *N. haematococca* involved three reactions: (a) hydroxylation at C_{6a}, (b) hydroxylation at C_{1a} to produce a 1a-hydroxydienone, and (c) production of isoflavanones by scission of the furan ring and oxidation at C_{11a} (Lucy et al. 1988) that were less toxic (Miao and Vanetten 1992). Strains of the bean pathogen *Fusarium solani phaseoli* and *Nectria haematococca* were examined to establish whether or not there was a correlation between the production of kievitone hydrate and the elicitation of disease on beans *Phaseolus vulgaris* (Smith et al. 1982). The relation was, however, highly variable; only three strains of 17 were highly virulent and these were capable of synthesizing the hydrate in cell-free cultures.

2.21 Roles of Bacteria Including Biocontrol Agents

Considerable effort has been devoted to the application of bacteria as agents for the biocontrol of plant pathogens—mainly fungi. This is motivated by the desire to diminish the application of agrochemical and, in addition, improve soil quality. Bacteria in the rhizosphere are able to fulfill a number of quite different functions: direct stimulation of plant growth, and the production by strains of *Pseudomonas* of secondary metabolites, including 2,4-diacetylphloroglucinol, phenazine-1-carboxylate, and pyrrolnitrin that are toxic to seed- and root-rotting plant pathogenic fungi. There are useful reviews of agents for the control of soil pathogens: (a) detailed discussions of the roles of fluorescent pseudomonads (Haas and Défago 2005), (b) application of strains of *Pseudomonas* sp. against soil-borne pathogens with a valuable account of the strains that have achieved prominence (Weller 2007), and (c) the mechanisms of biocontrol (Compant et al. 2005a; Kim et al. 2011).

2.21.1 Volatile Metabolites

In addition to these metabolites, microorganisms produce a variety of volatile secondary metabolites. Their potential significance is illustrated by a few examples:

1. Cyanide is produced by a few bacteria, mainly fluorescent pseudomonads and *Chromobacterium violaceum*, and by the hydrolysis of cyanogenic glycosides, including amygdalin and linamarin. *Pseudomonas fluorescens* CHA0 suppresses black rot of tobacco caused by the fungus *Thielaviopsis basicola*, and was able to produce 80 μM HCN after 2 days growth on tobacco root pieces. Its antifungal activity was assayed in an artificial Fe-rich soil system where cyanide biosynthesis would be encouraged, and it was shown that HCN was an important—though not necessarily the only—factor for fungal suppression, without having adverse affects on the plants (Voisard et al. 1989). Cyanide produced by *Pseudomonas fluorescens* CHA0 and *P. chlororaphis*, and to a much lesser degree by *Serratia plymuthica* was deleterious to the growth of *Arabidopsis thaliana* (Blom et al. 2011).

2. Bacteria isolated from the stems, root tips, or stubble of Canola (*Brassica camprestris/napa*) were examined for the production of volatile metabolites and their ability to inhibit sclerotia production and mycelial growth of *Sclerotinia scleretiorum* (Fernando et al. 2005). Activity was observed for *n*-decanal, *n*-nonanal (*Pseudomonas aurantiaca*, *P. chlororaphis*), benzothiazole (*P. fluorescens*), and dimethyltrisulfide (*P. chlororaphis*).

3. It has been shown that *Bacillus amyloliquifaciens* IN937a and *B. subtilis* GB03 were active in airborne signaling of growth promotion for *Arabidopsis thaliana* seedlings (Ryu et al. 2004). This was attributed to 2,3-butanediol and 3-hydroxybutan-2-one (acetoin) since analysis of volatiles from these strains showed the presence of these compounds as well as other metabolites that included 3-methylbutanol, 3-methylbutanol methyl ether, and isoprene (Farag et al. 2006).

4. A plethora of volatile metabolites is produced by fungi. Attention is directed only to those produced by species of the endophytic fungus *Muscodor* on account of their adverse effect on a range of plant fungal pathogens. The principal metabolites produced by *Muscodor albus* were 3-methylbutanol and the corresponding acetate that together made up almost half of the metabolites, of which the esters as a group were the most lethal (Strobel et al. 2001). *M. crispans* also produced a range of metabolites, including methyl 2-methylpropionate, that were effective against a range of fungal pathogens and the bacterium *Xanthomonas axinopodis* pv. *citri* (Mitchell et al. 2010).

2.21.2 Plant–Bacteria Associations

The role of quinolinate phosphoribosyl transferase requires a brief aside on the biosynthesis of NAD. In *Escherichia coli* and *Salmonella typhimurium*, this involves a series of reactions: an FAD-oxidase encoded by *nadB* on aspartate produces the imine that reacts with acetone dihydroxyacetone phosphate using quinolinate synthase encoded by *nadA* to produce quinolinic acid (pyridine-2,3-dicarboxylate). This then reacts with phosphoribosyl diphosphate to form nicotinic acid mononucleotide catalyzed by quinolinate phosphoribosyltransferase encoded by *nadC* (Flachmann et al. 1988).

The role of quinolinate phosphoribosyl transferase in promoting potato root growth by *Burkholderia* sp. strain PsJN was assessed on the basis of several lines of evidence. A mutant H41 that was produced by transposon mutagenesis had lost its plant-growth activity and several lines of evidence suggested the mechanism involved: (i) it was shown that the lost gene exhibited high homology with the *nadC* gene encoding quinolinate phosphoribosyl transferase in *Ralstonia solanacearum*; (ii) expression of the gene in *Escherichia coli* gave a protein that catalyzed the synthesis of nicotinic acid mononucleotide; and (iii) the growth-promoting activity of the mutant could be restored by adding nicotinic mononucleotide to the culture medium. It was shown that the *nadA* gene was cotranscribed with *nadC* as an operon in the wild-type strain.

2.22 Plant Endophytes

This is a brief introduction with a few illustrative examples from a vastly expanding field of current interest. Endophytic bacteria may colonize plant tissues without bringing about adverse effects on the plant, in contrast to epiphytes that remain in the external environment. They can carry out a number of functions which are positive for the plant host: these include growth promotion, suppression of pathogens, provision of assimilable nitrogen by nitrogen fixation, and solubilization of phosphate. In addition, some of them are able to degrade xenobiotics and have been proposed for phytoremediation programs (Weyens et al. 2009). A valuable summary has been given of the endophytic bacteria in plants that include α-, β-, and γ-*Proteobacteria* as well as some *Actinobacteria* (Rosenblueth and Martínez-Romero 2006). A survey of endophytes in crop and prairie plants in the Western USA revealed the frequency on sorghum (*Sorghum bicolor*) and corn (*Zea mays*) of about equal numbers of Gram-positive and Gram-negative bacteria (Zinniel et al. 2002). A number of studies have used poplar (*Populus deltoides*), and a revealing examination of the root rhizosphere and the root endophytes showed that whereas the first was dominated by *Acidobacteria* and *Alphaproteobacteria*, the second consisted of *Gammaproteobacteria* and *Alphaproteobacteria* (Gottel et al. 2011).

Illustrative examples are given while further details occur in later sections that deal with specific organisms and their function.

2.22.1 Positive Effects on Plants

1. Endophytic colonization of *Vitis vinifera* by the plant growth promoting *Burkholderia* sp. strain PsJN induced a local host defense reaction that spread systemically to the aerial parts by transpiration (Compant et al. 2005b).

2. Endophytic strains of *Enterobacter* sp. strain 638, *Stenotrophomomas maltophilia* strain R-551-3 and *Serratia proteamaculans* strain 568 were assessed for their effect on growth of poplar cuttings. The *Enterobacter* strain had the greatest positive effect, even though it produced only low levels of indole-3-acetic acid and was unable to metabolize 1-aminocyclo-propane-1-carboxylate, and γ-aminobutyrate—in contrast to the greater metabolic ability of *Serratia proteamaculans* that was less effective in promoting growth (Taghavi et al. 2009).

3. Rhizosphere soil from plants of coffee plant (*Coffea arabica*) was used to inoculate LGI medium and several pure strains of bacteria were recovered: they were able to reduce acetylene and were assigned to the genus *Acetobacter*. RFLP analysis of *Eco*RI DNA digests revealed four hybridization patterns, and the strains showed only low DNA–DNA homologies with the reference strain of *Acetobacter diazotrophicus* and were clearly different species (Jiménez-Salgado et al. 1997).

4. Strains of *Herbaspirillum* were isolated from rice and identified using 16s rDNA. The *gfp* gene encoding green fluorescent protein was introduced into strain B501 from wild rice *Oryza officinalis*, and after inoculation of seeds it colonized the shoots and seed of aseptically grown seedlings—though not for cultivated rice *Oryza sativa*. Acetylene reduction and $^{15}N_2$ incorporation in seedlings confirmed that this strain of *Herbaspirillum* was compatible with wild rice (Elbeltagy et al. 2001).

5. Populations of cultivable endophytic bacteria were isolated from seedlings and the pulp of fruits of the cardon cactus (*Pachycereus pringlei*). Based on partial 16s rRNA sequencing, the bacteria belonged to the genera *Bacillus* (SENDO 6), *Klebsiella* (SENDO 1 and 2), *Pseudomonas* sp. (SENDO 2) and *Acinetobacter* sp (SENDO1), albeit with only low similarity to established species. A range of volatile acids dominated by gluconic acid was produced by all strains, and comparable amounts of succinate by *Bacillus* SENDO 6 and *Klebsiella* SENDO 1. Mineral

solubilization of P, K, Fe and Mg was observed after 28 days incubation (Puente et al. 2009a), and inoculation of seeds in pulverized rock with these endophytes allowed growth for at least 1 y without fertilization (Puente et al. 2009b).

2.22.2 Biodegradation of Xenobiotics

1. Endophytes associated with several prairie plants were isolated and the dominant species of *Pseudomonas* were responsible for the degradation of alkanes, whereas the increased degradation of PAHs was associated with *Brevundimonas* sp. and *Pseudomonas rhodesiae* (Phillips et al. 2008).

2. An endophyte of *Allium macrostemon* growing in PAH-contaminated soil was identified as *Enterobacter* sp. and it was able to degrade pyrene in addition to producing indole-3-acetic acid and the ability to solubilize inorganic phosphate enabled it to colonize the tissue and rhizosphere (Sheng et al. 2008).

3. Hybrid poplars (*Populus deltoides* × *nigra*) have attracted considerable interest since they are able to transport and metabolize diverse xenobiotics while their endophytes may be able to degrade them:

    i. *Methylobacterium populi* isolated from tissue cultures of poplar (*Populus deltoides* × *nigra* DN34) (van Aken et al. 2004a) was able to utilize methylamine and, uniquely in this genus, methane as a growth substrate. The latter claim seems to merit verification with details of the physiology and biochemistry of the organism (Dedysh et al. 2004).

    ii. A species of *Enterobacter* closely related to *Ent. asburiae* isolated from the hybrid DN177 was able to reduce levels of 72 μM trichloroethene to 30 μM with the release of 127 μM chloride without the need for the addition of inducers such as phenol or toluene (Kang et al. 2012).

4. Bacteria were isolated from the bulk soil, the rhizosphere, and the root interior of plants at sites contaminated with hydrocarbons, and nitroarenes related to TNT manufacture. Gene probes were used to determine the prevalence of genes for alkane degradation (*alkB*), and naphthalene dioxygenase (*ndoB*). The prevalence of *alkB* in the bacteria from the root interior exceeded that for the bulk soil and the rhizosphere, whereas the opposite was true for *ndoB*, although addition of oil brought about a relative increase of *ndoB* in the endophytes. The prevalence of genes for 2-nitrotoluene reductase (*ntdAa*) and nitrotoluene monooxygenase (*ntnM*) was greater for bacteria from the root interior although values for *ntnM* were rather low (Siciliano et al. 2001).

5. The plasmid pTOM of *Burkholderia cepacia* G4 was introduced into *B. cepacia* strain L.S 2.4 that is a natural endophyte of yellow lupin (*Lupinus luteus*), and the resulting strain VM 1330 was used to inoculate plants of yellow lupin. These plants were tolerant of toluene up to a concentration of 1 g/L in hydroponic cultivation, and up to 500 mg/L in nonsterile soil (Barac et al. 2004).

6. Cuttings of hybrid poplar (*Populus trichocarpa* × *deltoides*) were inoculated with a strain of *Burkholderia cepacia* VM 1468 carrying genes for the constitutive expression of toluene degradation. Compared with the uninoculated control, plant growth increased both in the absence and in the presence of toluene and, although the introduced strains did not establish themselves in the endophytic bacterial community, gene transfer to other bacteria with the Tol[+] phenotype took place in the roots and stems (Taghavi et al. 2005).

2.22.3 Nitrogen-Fixing Bacteria

Associations have been observed, including nonsymbiotic diazotrophs, that may contribute to the nitrogen nutrition of plants. It is important to appreciate that the effective function of nitrogen-fixing species such as *Azospirillum* sp. cannot always be attributed solely to their role in nitrogen fixation. Other important factors that are discussed later include the role of polyamines such as putrescine and cadaverine that have a well-established structural role, and the synthesis of indole-3-acetic acid (IAA) and phosphate solubilization in the soil (Mehnaz and Lazarovits 2006). Another mechanism has been proposed for *Burkholderia* sp. strain PsJN that can stimulate growth of potato nodal explants in tissue culture. while the *nadB* gene was located downstream of the *nadA-nadC* operon.

1. Species of *Azospirillum* and *Herbaspirillum* can penetrate the roots of cereal crops and are able to fix nitrogen under microaerophilic conditions. Species of nitrogen-fixing *Azospirillum* have been isolated from the roots and rhizosphere of grasses such as *Digitaria*, and *Herbaspirillum seropedicae* from the roots of sorghum, maize, and rice (Baldani et al. 1986). *Herbaspirillum rubrisubalbicans* and *Burkholderia brasiliensis* and *B. tropicalis* have been isolated from leaf, stem, and root samples of banana (*Musa* spp.) and pineapple (*Ananas comosus*) (Cruz et al. 2001).

2. Nitrogen-fixing species (diazotrophs) of *Burkholderia* have been isolated from the rhizosphere and rhizoplane mainly, though not exclusively, from maize and sugarcane, and these are summarized in Table 2.11. It has been established that *Burkholderia vietnamiensis* is able to fix nitrogen (Gillis et al. 1995; Estrada-de los Santos et al. 2001) and can have a positive effect on plant growth: inoculation of rice seedlings with strain TVV75 significantly increased the final yield of grain (Trân Van et al. 2000) and inoculation with strain MG43 improved the yield of sugarcane plantlets (Govindarajan et al. 2006). Several species of diazotrophic species of *Burkholderia* were found in the rhizosphere and rhizoplane of tomatoes, including *B. unamae*, *B. xenovorans*, and *B. tropica* (Caballero-Mellado et al. 2007). Several additional features included the following: *B. unamae* and *B. xenovorans* displayed 1-aminocyclopropane-1-carboxylate deaminase activity, high concentration of unidentified hydroxamate siderophores, and *B. unamae* was able to grow at the expense of phenol and benzene, like the established ability of the type strain of *B. kururiensis* (Zhang et al. 2000).

3. Diazotrophic strains of the genera *Azoarcus*, *Azovibrio*, and *Azospira* have been isolated from the roots of the alkali- and salt-tolerant *Leptochloa fusca* from the Punjab and *Azonexus fungiphilus* from fungal sclerotium from rice soil, although these organisms have a rather restricted range of growth substrates (Reinhold-Hurek and Hurek 2000).

4. It has clearly emerged that genomovars of *Pseudomonas stutzeri* associated with the roots of rice and as an endophyte, are able to fix nitrogen under microaerophilic conditions (Desnoues et al. 2003; Yan et al. 2008). Some strains had previously been assigned to *Alcaligenes faecalis*.

5. *Gluconacetobacter (Acetobacter) diazotrophicus* is an endophyte that has been isolated from the tissues and roots of sugarcane plants, and the ability to fix nitrogen could contribute to the nitrogen requirement of the plants (Caballero-Mellado et al. 1995). This has been verified using sugarcane plants inoculated with wild-type and Nif⁻ mutants: (i) whereas the former was beneficial under conditions of nitrogen limitation when ^{15}N incorporation was greater in the roots, and (ii) when nitrogen was not limiting; however, the Nif⁻ mutants were able to enhance growth putatively due to the provision of a growth-promoting factor (Sevilla et al. 2001).

6. Surface-sterilized roots and stems of rice were used to isolate bacteria that were identified as *Serratia marcescens* by 16S rRNA gene analysis, although their number on the roots was three logs less than the number of heterotrophs. It was shown by electron microscopy that they were

TABLE 2.11

Diazotrophic Species of *Burkholderia* Isolated from the Rhizosphere, Rhizoplane, and Roots

Species	Plant Source	Reference
tropica	Maize, sugarcane	Reis et al. (2004)
unamae	Maize, sugarcane	Caballero-Mellado et al. (2004)
mimosarum	Mimosa root nodules	Chen et al. (2006)
unamae, tropica	Maize, sugarcane	Perin et al. (2006)
unamae, tropica	Tomato	Caballero-Mellado et al. (2007)
phymatum	Banana/pineapple	Weber et al. (1999)
vietnamiensis	Rice	Gillis et al. (1995)
	Coffee	Estrada-de los Santos et al. (2001)

established as endophytes, and that they displayed nitrogenase activity, though only in media containing a low level of combined nitrogen as yeast extract (Gyaneshwar et al. 2001). *Gluconacetobacter azotocaptans* has been isolated from soil adhering to corn roots (Mehnaz et al. 2006), although under greenhouse conditions for corn, nitrogenase activity, synthesis of indole-3-acetic acid, and phosphate solubilization was less for both *G. azocaptans* and *G. diazotrophicus* than for *Azospirillum lipoferum* and *Azosp. brasilense* (Mehnaz and Lazarovits 2006).

7. Bacteria isolated from the surface-sterilized roots *of Casuarina equisetifolia* were able to fix nitrogen, and analysis of the 16S rRNA genes revealed that they were actinomycetes. The low level of DNA–DNA hybridization with authentic species of *Frankia* showed, however, that they were different from *Frankia*. The *nifH* sequences were closely similar to those of *Frankia*, and analysis of the complete 16S rRNA sequence showed that they were most closely related to actinobacteria (Valdés et al. 2005). In a wider context, nonpathogenic actinobacteria belonging to the genera *Streptomyces* and *Nocardioides* have been isolated from surface-sterilized roots of wheat (*Triticum aestivum*) (Coombs and Franco 2003).

8. *Agrobacterium tumefaciens* is the classical agent of crown gall and *Agrobacterium* strains of biotype 1 are the cause of root mat in hydroponic cucumber and tomato plants (Weller and Stead 2002). It has emerged, however, that free-living strains of *Agrobacterium tumefaciens* were able to fix nitrogen under anoxic conditions (Kanvinde and Sastry 1990).

2.22.4 Solubilization of Phosphate

1. Phosphate in soil may exist as inositol hexakisphosphates (Turner et al. 2002) and under conditions of phosphorus limitation, extracellular phytase from *Bacillus amyloliquefaciens* (Makarewicz et al. 2006) may make a positive contribution to the enhancement of plant growth (Idriss et al. 2002).

2. An indole-3-acetic acid–overproducing strain of *Sinorhizobium meliloti* that showed increased ability to fix nitrogen was also able to mobilize phosphate from insoluble phosphate rock. This was associated with upregulation of the *pho* operon genes and increase in the concentrations of malate, succinate, and fumarate excreted into the medium (Bianco and Defez 2010). In addition, *Medicago truncatula* nodulated with this strain under phosphorus-deficient conditions released higher concentration of 2-hydroxyglutaric acid.

2.22.5 Bacterial Synthesis of Plant Hormones

Bacteria can influence plant development in several ways, and the synthesis of plant hormones may be significant for plants when they are produced by bacteria in the rhizome or rhizoplane. Substantial attention has been directed to the plant hormones: indole-3-acetate (auxin), cytokinins (adenine alkylated at N_6 and their ribosides), the C_{19} and C_{20} diterpenoid gibberellins, and ethylene (Baca and Elmerich 2003). These are discussed in later paragraphs, and probably several contribute simultaneously. An attempt has been made to dissect this interaction for only a few of them—auxins, acetoin, and phenazines (Kim et al. 2011).

The biosynthesis of cytokinins and auxins can be briefly summarized:

1. Bacterial synthesis of cytokinins takes place by two pathways: direct isopentylation of AMP carried out by dimethylallyl transferase and turnover of tRNA that has been shown to produce low levels of the active isomer *trans*-zeatin in species *of Methylobacterium*, including *M. extorquens* (Koenig et al. 2002).

2. Auxins are produced by several pathways in bacteria: the indole-3-acetamide pathway is used by *Pseudomonas savastanoi*, *Erwinia herbicola*, and species of *Agrobacterium* where it is involved in pathogenesis, whereas in *Erwinia herbicola* pv. *gypsophilae*, both the indole-3-acetamide and the indole-3-pyruvate pathways are used (Lambrecht et al. 2000).

Illustrative examples include the following:

1. Indole-3-acetate (IAA) is produced by several bacteria isolated from the rhizosphere, and notably high concentrations produced by two Enterobacteriaceae inhibited the growth of sugar beet seedlings (Loper and Schroth 1986). IAA was produced by epiphytic bacteria, mainly *Erwinia herbicola* (*Pantoea agglomerans*) on pear trees and was associated with the development of fruit russet (Lindow et al. 1998). Indeed, IAA production was observed in most pathovars of *Pseudomonas syringae* (Glickmann et al. 1998).

2. Cytokinin production has been demonstrated, and was particularly significant in *Agrobacterium tumefaciens* strain T37, *Pseudomonas syringae* pv. *savastanoi* strains 1006, TK1050, from olive, *Pseudomonas solanacearum* strain 248 from tobacco, and strain 258 from potato (Akiyoshi et al. 1987). *Paenibacillus polymyxa* synthesized isopentenyl-adenine in the late stationary phase (Timmusk et al. 1999), and zeatin, ribosylzeatin, and ribosyl-1″-methylzeatin were produced by *Pseudomonas savastanoi* (MacDonald et al. 1986). Cytokinins are essential for the production of galls by *Erwinia herbicola* pv. *gypsophilae* (Lichter et al. 1995).

3. Gibberellin that is produced by *Gibberella fujikoroi* is also produced by several bacteria. These include *Azospirillum brasilense* (Janzen et al. 1992), *Azospirillum lipoferum* (Bottini et al. 1989), and both auxins and gibberellins GA1 and GA3 by *Gluconacetobacter* (*Acetobacter*) *diazotrophicus* and *Herbaspirillum seropedicae* (Bastián et al. 1998). It is worth noting that all these bacteria are diazotrophs.

4. The rhizosphere of the salt-marsh plants cordgrass (*Spartina alterniflora*) and black needle rush (*Juncus roemerianus*) harbors diazotrophs whose gene sequences were not analogous to those for established sequences (Lovell et al. 2000). Indole-3-acetic acid was produced by vibrios isolated from these plants and, although synthesis of IAA was observed for all vibrio species, it was greatest for strains of *Vibrio parahaemolyticus*, *V. natriegens*, and *V. pacinii* (Gutierrez et al. 2009).

5. In medium supplemented with tryptophan, *Rhizobium* sp. strain NGR234 synthesized a range of indoles, including indole-3-pyruvate, indole-3-acetamide, and indole-3-acetaldehyde that are established precursors of indole-3-acetic acid (IAA). The synthesis of IAA was stimulated 100-fold by growth with succinate in the presence of the flavonoid daidzein (7,4′-dihydroxyisoflavone) and also by luteolin (5,7,3′,4′-tetrahydroxy-flavone) (Theunis et al. 2004).

In addition to these bacteria, considerable attention has been given to methylotrophs. These are widely distributed on plant surfaces and in the rhizosphere and rhizoplane, and they have important metabolic features in the context of plant growth that include the following examples:

1. Nodulation of species of the legumes *Crotalaria* by *Methylobacterium nodulans* that is able to fix nitrogen symbiotically (Jourand et al. 2004).

2. The presence and expression of the genes for the synthesis of cytokinins by *Methylobacterium mesophilicum* and *Methylovorus mays* (Ivanova et al. 2000).

3. *Methylovorus mays* that was isolated from the maize phyllosphere is able to synthesize both cytokinins and auxins (Doronina et al. 2005), and of auxins from exogenous tryptophan by a wider range of methylotrophs and methanotrophs (Doronina et al. 2002).

Some additional comments on auxins are merited:

1. It has been established that the synthesis of auxins may exert a range of effects, both positive and negative:

   a. The development of plant root system from seeds treated with auxin from *Pseudomonas putida* (Patten and Glick 2002).

   b. In *Pseudomonas savastanoi*, the genes for auxin synthesis are required for the development of galls or knots on olives and oleander (Comai and Kosuge 1982), and in *Erwinia herbicola* pv. *gypsophilae*, for galls on *Gypsophila paniculata* (Clark et al. 1993). The oxidative decarboxylation is carried out by tryptophan 2-monooxygenase whose regulation and synthesis in *P. savastanoi* have been examined (Hutcheson and Kosuge 1985).

   c. *Bacillus amyloliquefaciens* exerts a positive effect on plant growth that has been attributed to auxin synthesis (Idriss et al. 2002).

   d. Gram-positive *Rhodococcus fascines* produces auxin that causes diverse adverse symptoms, including leaf deformation, growth inhibition, leafy galls, and witches' brooms (Vandeputte et al. 2005).

2. In the wider context of auxin activity, the analogue phenylacetic acid has been identified as a natural auxin in shoots of higher plants where its concentration exceeded that of indole-3-acetic acid (Wightman and Lighty 1982). It is produced in bacteria by *Streptomyces humidus* where it was significantly active against the fungal pathogens *Pythium ultimum*, *Phytophthora capsicum*, and *Rhizoctonia solani* (Hwang et al. 2001), and by *Azospirillum brasilense* that colonizes the rhizosphere where it has a positive effect in fixing nitrogen and had an antifungal effect against *Alternaria brassicicola* and *Fusarium oxysporum* f. sp. *matthiolae* (Somers et al. 2005).

2.22.6 Ethylene and ACC Deaminase

Ethylene has a historically established biological activity in plants where it functions as a growth regulator at low concentrations. The roles of ethylene are complex and its biosynthesis is different for higher plants and bacteria.

1. Ethylene is an important plant metabolite since it inhibits cell division and is produced during the ripening of fruit (Burg 1973). Its synthesis in plant tissue from *S*-adenosylmethionine involved production of the pyridoxal-5′-phosphate conjugate of 1-aminocyclopropane-1-carboxylate (ACC) that underwent oxidation (Adams and Yang 1979). In contrast, the synthesis of ethylene by a few bacteria involves elimination from 2-oxo-4-thiobutyric acid (Mansouri and Bunch 1989).

2. 1-Aminocyclopropane-1-carboxylate (ACC) that is the precursor of ethylene in plants is, however, important for a number of other reasons. It is able to serve as a nitrogen source for the growth of bacteria (Sheehy et al. 1991); the deaminase that is a key enzyme in plant promotion by bacteria has been purified from *Pseudomonas putida* strain UW4 (Hontzeas et al. 2004a) and the genes have been found in a range of bacteria, including pseudomonads, and strains of *Rhodococcus* and *Variovorax* (Hontzeas et al. 2005). The deamination activity may modulate ethylene production by the plants by diverting the substrate to deamination and the production of 2-oxobutyrate, particularly under conditions of stress (Glick et al. 2007).

3. The deaminase plays a role in promoting plant growth by *Enterobacter cloacae* (*Pseudomonas putida*) strain UW4 (Li and Glick 2001; Hontzeas et al. 2004b), and has been identified in bacteria that included *Pseudomonas brassicacearum* from the rhizoplane of pea (*Pisum sativum*) and Indian mustard (*Brassica juncea*). Several of these isolates stimulated root elongation of germinating seedlings of *B. juncea* and *B. napus* var. *oleifera* (Belimov et al. 2001). *Pseudomonas brassicacearum* strain Am3 is able to utilize ACC by the activity of a deaminase and stimulated root development of tomato (*Lycopersicon esculentum*) at low concentrations (Belimov et al. 2007). It has been shown that ACC deaminase is distributed in a range of species of *Burkholderia* from the rhizosphere and rhizoplane of tomato (*Lycopersicon esculentum*)—including nitrogen-fixing strains—where it has been suggested that ACC deaminase could play a positive role in the growth and development of tomatoes (Onofre-Lemus et al. 2009). On the other hand, ethylene is able to inhibit the nodulation of legumes by rhizobia, and ACC deaminase promotes nodulation by *Rhizobium leguminosarum* bv. *viciae* (Ma et al. 2003) by modulating the synthesis of ethylene by the plants. This is consistent with the effect of rhizobitoxine that is a metabolite of *Bradyrhizobium elkanii* which inhibits the biosynthesis of ethylene and enhances nodulation of the forage plant siratro (*Macroptilium atropurpureum*) by *Bradyrhizobium elkanii* (Yuhashi et al. 2000).

2.22.7 Polyamines

Cations of the polyamines putrescine (1,4-diaminobutane), cadaverine (1,5-diamino-pentane), and the related spermine and spermidine play important structural roles in membranes, proteins, and nucleic acids. Although substantial investigations have analyzed their roles in eukaryotic systems, including mammals (Pegg 1986), they are important also in plants and bacteria. They should, therefore, be taken into consideration in addition to conventional plant hormones.

1. In seedlings of soybean (*Glycine max*), concentrations of spermidine and spermine exceeded that of cadaverine except in hypocotyls and shoots where putrescine was dominant (Gamarnik and Frydman 1991). This was examined in detail using α-difluoromethylornithine that is an irreversible inhibitor of ornithine decarboxylase in mammalian systems (Pegg 1986); this resulted in a rapid decrease in cadaverine and simultaneous morphological changes in the seedling that could be ameliorated by the addition of cadaverine. There was a temporal relation between the increase in cadaverine and the activity of lysine decarboxylase that was, therefore, the precursor of cadaverine.

2. The synthesis of cadaverine in cultures of *Azospirillum brasilense* was examined in a chemically defined medium by decarboxylation of lysine, and cadaverine promoted the growth of rice (*Oryza sativa*) grown hydroponically (Cassán et al. 2009).

3. In addition to the synthesis of the phytohormone indole-3-acetic acid and its ability to fix nitrogen, *Azospirillum* sp. also synthesized the polyamine putrescine and lower concentrations of spermidine and spermine when NH_4^+ was present in the medium, whereas the concentration of indole-3-acetic acid was greater in the basal medium (Thuler et al. 2003).

2.23 Microorganisms with Activity as Biocontrol Agents

There has been increasing interest in the application of these biocontrol agents instead of chemical agents, and the mechanisms of their activity have been established for some of them. They may be applied to the roots or leaves of plants, or applied as surface coatings to seeds before planting. Although a structurally wide range of secondary metabolites is synthesized, the important aspect of their biosynthesis is not discussed. It is important to emphasize that taxonomic rigor is not attempted in the following paragraphs, and the epithets are those given by the authors.

2.23.1 General Mechanisms

1. Bacteria must be present on the roots of the plants, be able to produce microcolonies, and synthesize inhibitory compounds. Examples include the production against root infection by fungal pathogens of (i) the antimicrobial 2,4-diacetylphloroglucinol by *Pseudomonas fluorescens* CHA0 (Bottiglieri and Keel 2006), (ii) pyoluteorin by *Pseudomonas fluorescens* Pf-5 (Brodhagen et al. 2004), and (iii) phenazine-1-carboxamide by *Pseudomonas chlororaphis* PCL 1391 (Chin-A-Woeng et al. 2000). The combination of *Pseudomonas fluorescens* CHA0 and *P. fluorescens* Q2-87 each of which produces 2,4-diacetyl-phloroglucinol resulted in reciprocal stimulation of the expression of 2,4-diacetyl-phloroglucinol (Maurhofer et al. 2004). Conversely, *Fusarium oxysporum* can induce resistance to 2,4-diacetylphloroglucinol by deacylation to the less toxic 2-acetylphlorglucinol and phloroglucinol (Schouten et al. 2004).

2. The production of phenazine-1-carboxamide by *Pseudomonas chlororaphis* PCL 1391 is a requirement for its function in controlling tomato rot caused by *Fusarium oxysporum*, and quorum sensing mediated by *N*-hexanoyl-L-homoserine lactone in the medium (Chin-A-Woeng et al. 2001) regulates its synthesis. Its production in this strain is promoted by a number of conditions including the presence of amino acids, especially phenylalanine and tryptophan

(Tjeerd van Rij et al. 2004). This is consistent with the biosynthesis of aromatic amino acids and phenazines in branches of the chorismate pathway, whereas both the *N*-9 and the *N*-10 heteroatoms originate from glutamine (Mavrodi et al. 1998, 2001).

3. *Pseudomonas chlororaphis* strain 30-84 is able to control the take-all infection of wheat by *Gaeumannomyces graminis* var. *tritici*, and the primary agent of control resides in the production of phenazines. The primary agents are phenazine-1-carboxylate (phzA) and 2-hydroxyphenazine-1-carboxylate (phzB) which are produced from chorismate followed by hydroxylation. Construction of mutants that either produced only phzA or overproduced phzB showed that the loss of phzB adversely affected the production of biofilm attachment and dispersal, as well as a significant reduction in the elimination of the pathogen (Maddula et al. 2008).

4. Degradation of fungal toxins may contribute to the protection of the plants. For example, fusaric acid produced by *Fusarium oxysporum* can be degraded by *Burkholderia* spp. (Utsumi et al. 1991). In addition, fusaric acid produced by the fungus may have other adverse effects:

   a. In the production of phenazine-1-carboxamide that is required for the suppression of tomato rot by *Pseudomonas chlororaphis* PCL1391 (Tjeerd van Rij et al. 2005)

   b. In the biosynthesis of 2,4-diacetylphloroglucinol in *Pseudomonas fluorescens* CHA0 that is important for biocontrol in the wheat rhizosphere (Notz et al. 2002)

5. It has been shown that in *Pseudomonas fluorescens* CHA0, which is an antagonist of the plant pathogenic nematode *Meloidogyne incognita*, the extracellular protease encoded by the gene *aprA* was involved either directly or indirectly (Siddiqui et al. 2005). In the biocontrol agent *Pseudomonas fluorescens* CHA0, gluconate production had several effects that were elucidated using in-frame mutants, Δ*gcd* lacked glucose dehydrogenase, and Δ*gad* lacked gluconate dehydrogenase. Several effects were observed: (a) the Δ*gad* mutant resulted in acid production and the solubilization of phosphate minerals, (b) the production of gluconate inhibited the production of pyoluteorin and partly of 2,4-diacetyl-phloroglucinol, (c) the severity of disease produced by the pathogen *Gaeumannomyces graminis* var. *tritici* was reduced in the Δ*gcd* mutant (de Werra et al. 2009).

6. Phase variation regulates the production of flagellae, surface lipoproteins, secondary metabolites, and hydrolytic enzymes, including proteases, and is important in the colonization of the rhizosphere by biocontrol agents as a result of increased motility. This has been examined in *Pseudomonas fluorescens* strain F113 during the colonization of alfalfa (*Medicago sativa*). Colony variation was implicated in its effectiveness (Sánchez-Contreras et al. 2002), and is mediated by the site-specific recombinases Sss and XerD (Martínez-Granero et al. 2005, 2006). An analogous phenomenon occurred in *Pseudomonas* sp. strain PCL1171 in which mutations in the genes *gacA* and *gacS* were shown to be responsible (van den Broek et al. 2005).

In the following paragraphs, an outline is given of a selection of bacteria some of which have achieved attention as biocontrol agents against fungal infection.

2.23.1.1 Burkholderia cepacia sensu lato

Although *Burkholderia cepacia* is the causal agent of soft-rot in onion that is mediated by the synthesis of a plasmid-encoded polygalacturonase (Gonzalez et al. 1997), several strains have been considered for biocontrol agents. Some illustrations are given, and additional examples are available (Burkhead et al. 1994; Cartwright et al. 1995; Hebbar et al. 1998). No attempt is made to discuss taxonomy and identification within the *Burkholderia cepacia* complex. *Burkholderia ambifaria* (Coenye et al. 2001) is a member of this complex and has achieved importance as a biocontrol agent. However, since *B. cepacia* has also been isolated from patients with cystic fibrosis, caution would have to be exercised in its application.

1. *Burkholderia* (*Pseudomonas*) *cepacia* strains J82rif and J51rif that had been isolated from the rhizospheres of barley and sorghum were able to mitigate the deleterious effect of the sunflower wilt fungus *Sclerotiorum sclerotorium* in a field experiment (McLoughlin et al. 1992).

2. After application to seeds of pea and maize, *Burkholderia cepacia* strain AMMD suppressed seed rot and damping-off by *Pythium* spp. In addition, the yield of peas grown in soil infested with *Aphanomyces euteiches* was increased (Bowers and Parke 1993).

3. Application of *Burkholderia cepacia* strain PHQM100 to seeds of corn reduced seed rot inflicted by *Pythium* spp., and in soils infested by *Fusarium* spp., it reduced root and mesocotyl necrosis (Hebbar et al. 1998).

4. Coating seeds of radish with *Burkholderia* (*Pseudomonas*) *cepacia* RB425 and RB3292 suppressed colonization by *Rhizoctonia solani* AG4 (Homma and Suzui 1989). In this study, the concentrations of the pseudanes (HMQ and NMQ) exceeded that of pyrollnitrin, while SEM showed that the effective strains multiplied in the spermatosphere and rhizosphere of the seedlings.

2.23.1.2 Pseudomonas *spp.*

Species of *Pseudomonas* synthesize a range of secondary metabolites, including 2,4-diacetylphloroglucinol and pyoluteorin. Some are effective biocontrol agents, and examples are given in the following paragraphs:

1. Studies have been directed to *Pseudomonas putida* strain KT2440 that has attracted interest both as a biocontrol agent and for the degradation of contaminants. It was able to maintain high cell densities when it was introduced as a coating of seeds during growth of agricultural crops (Molina et al. 2000). This organism is able to utilize a number of amino acids, including proline, lysine, and glutamate for growth, and attention has been directed to their metabolism. The enzymes in the *davDt* operon for proline degradation are induced by corn root exudates (Vílchez et al. 2000). There are several pathways for the degradation of lysine (Revelles et al. 2005), and the operon that is involved in its metabolism is induced by δ-aminovalerate which is one of the early degradation products (Revelles et al. 2004). Particular attention has been directed to 1-aminocyclopropane-1-carboxylate (ACC) that is a key intermediate in the biosynthesis of ethene by plants (Adams and Yang 1979), and the role of the bacterial deaminase in promoting the growth of agricultural plants.

    a. It has been shown that *Pseudomonas putida* strain GR12-2 exhibits ACC deaminase activity and that this organism promoted the root development of canola (*Brassica campestris*) (Glick et al. 1997). The specific role of ACC was inferred from the fact that mutants unable to utilized ACC and lacking the deaminase did not exhibit this stimulation (Glick et al. 1994).

    b. ACC deaminase has been demonstrated in a range of bacteria isolated from the rhizosphere of *Pisum sativum* and *Brassica juncea* (Belimov et al. 2001). It has been proposed on the basis of several lines of evidence that the stimulation by ACC deaminase is accomplished in several steps initiated by extrusion of ACC from the plant that is then removed through bacterial degradation mediated by ACC deaminase. This resulted in decreased ethene production by the plant and, thereby, limitation of growth inhibition. Bacteria in the rhizosphere are, therefore, able to play an important role in seed and plant development. ACC deaminase is distributed in a range of species of *Burkholderia* from the rhizosphere and rhizoplane of tomato (*Lycopersicon esculentum*), including nitrogen-fixing strains and it has been suggested that ACC deaminase could play a positive role in the growth and development of tomatoes (Onofre-Lemus et al. 2009).

2. *Pseudomonas putida* strain P9 is an endophyte of potato (*Solanum tuberosum*) that is able to suppress the pathogen *Phytophthora infestans*. For the study of its influence on bacteria indigenous to the potato phytosphere, a rifampin-resistant derivative P9R was used (Andreote et al. 2009). Roots of potato cultivars were dipped in bacterial suspensions before planting, and it was shown that their presence competed with several endophytic strains in the potato phytosphere, including *Pseudomonas syringae* and *P. azotoformans*, that may putatively fix nitrogen.

3. *Pseudomonas fluorescens* strain Pf-5 was isolated from the rhizosphere of cotton seedlings and produced the antibiotic pyrrolnitrin that was an antagonist for the pathogen *Rhizoctonia solani* (Howell and Stipanovic 1979). Most significantly, it produced a group of 16-member macrolides that are rhizoxin analogues, and two of them displayed differential toxicity to the important pathogens *Botrytis cinerea* and *Phytophthora ramorum* as well as the typical toxicity to rice that induced thickened and shortened root morphology (Loper et al. 2008).

4. *Pseudomonas chlororaphis* has been applied as seed coating for barley as a protection against the fungal pathogen *Drechslera teres*. Strain MA 342 has been evaluated in field experiments (Johnsson et al. 1998) and is in use commercially (Hökeberg et al. 1997).

2.23.1.3 Lysobacter *spp.*

There has been considerable interest in these since the mechanisms of their function as biocontrol agents have been established. *Lysobacter* is a genus of gliding, flexing cells of chemoorganotrophic aerobic organisms that are characteristically able to degrade chitin (Christensen and Cook 1978). *Lysobacter enzymogenes* strain C3 (that has also been designated as *Stenotrophomonas maltophilia*) is a biological control agent effective against a range of fungal pathogens. Although the greatest attention has been devoted to strain C3, several other strains, including 3.1Tb against *Pythium* spp., and strain N4-7 against *Magnaporthe poae* have been examined (references in Kobayashi et al. 2005). It has been shown that strain C3 possessed enzymatic activities that have been attributed to the activities of several extracellular hydrolytic enzymes, and a group of heat-stable antifungal factors (HSAFs) which consists of a tricylic structure linked to a polyketide derived from ornithine (Yu et al. 2007).

2.23.1.3.1 Heat-Stable Antifungal Factors

A mutant (5E4) of strain C3 was isolated that was defective in hydrolytic activity as well as inhibitory activity toward *Bipolaris sorokiniana* and *Pythium ultimum*. It was shown that transposon mutagenesis had inserted a *clp* gene homologue, and chromosomal insertion of the *clp* gene into 5E4 produced a complemented strain P1 with restored hydrolytic activity, although the inhibitory activity on leaf area against *B. sorokiana* on tall fescue (*Festuca arundinacea*) was significantly less than that on seedling emergence on *Pythium* damping-off of sugar beet seedlings. A number of important biological effects were observed using purified HSAF (Li et al. 2008):

1. The purified complex exhibited inhibitory activity toward a range of fungal pathogens, including *Bipolaris sorokiana*, *Fusarium graminearum*, *Rhizoctonia solani*, and *Sclerotinia sclerotiorum*.

2. (b) Inhibition of spore germination and hyphal growth were observed in *B. sorokiana*, *Fusarium graminearum*, and *Pythium ultimum*, and the adverse effect on hyphal growth and appressorium on the surface of leaves of tall fescue (*Festuca arundinacea*) was clearly shown.

3. Overall, there was a clear correlation between concentrations of HSAF and the area of diseased leaf area in turf, although mutants lacking the HSAF of the parent strain did not differ in their ability to control *Fusarium* head blight in wheat caused by *Fusarium graminearum*.

An organism designated *Stenotrophomonas* sp. strain SB-K88 produced a series of xanthobaccins whose structures closely resembled the tricyclic HSAF that has already been discussed. It was first characterized as a rhizobacterium of sugar beet that suppressed damping-off disease (Nakayama et al. 1999), and direct application to sugar beet seeds suppressed damping-off in soil that was infested with *Pythium* spp. An examination has also been carried out using this strain against damping-off by *Aphanomyces cochlioides* (Islam et al. 2005).

2.23.1.3.2 Hydrolytic Activities

In addition to HCAFs, *Lysobacter* spp. produce a number of hydrolytic and lipolytic enzymes, and the complexity and probable existence of multiple determinants in pathogenesis is suggested by the following:

1. The role of chitinase production has been examined in *Stenotrophomonas maltophilia* strain C3, and fractions of that displayed activity were separated and the chitinolytic fractions showed antifungal activity against *Bipolaris sorokiniana* in *Festuca arundinacea* (Zhang and Yuen 2000). On the other hand, although chitinase-deficient mutants exhibited reduced antifungal activity, they did not totally lose biocontrol activity.

2. *Stenotrophomonas maltophilia* strain 34A has been suggested as a biocontrol agent against *Magnaporthe poae* in Kentucky bluegrass (*Poa pratensis*), and it has been shown to produce chitinase that is a plausible antifungal component (Kobayashi et al. 2002). The single role of this is, however, compromised by the fact that a mutant devoid of chitinolytic activity displayed weak fungal suppression, and that the progression of the disease in controlled chambers was only slightly reduced.

3. *Lysobacter enzymogenes* strain N4-7 produces multiple β-1,3-glucanase activities, and the genes *gluA*, *gluB*, and *gluC* have been identified (Palumbo et al. 2003).

Although these may possibly be the active biocontrol components, it was suggested that more probably these were contributory rather then singly effective.

2.23.1.4 Streptomyces *spp.*

Species of *Streptomyces* are important pathogens that produce scab in root and tuber crops. The pathogens are produced by several species of *Streptomyces*, and it has been shown by analysis of mutants that the active agent in *Streptomyces acidiscabies* requires the synthesis of an *N*-methylated peptide that belongs to a group of thaxtomins (Healy et al. 2000). The pathogenic agents are peptides in which the indole ring is nitrated at C_4, and this is carried out by a nitric oxide synthase whose gene has been characterized in *Streptomyces turgidiscabies* (Kers et al. 2004). Several species have also been examined as biocontrol agents, and some examples are given as illustration. It is worth noting that in spite of the attention devoted to *Streptomyces*, attention should also be directed to the genus *Micromonospora* (El-Tarabily and Sivasithamparam 2006).

1. The severity of damping-off disease in sugar beet (*Beta vulgaris*) caused by *Sclerotium rolfsii* was significantly reduced by treatment with biomass and culture filtrates of *Streptomyces* spp. isolated from Moroccan soils (Errakhi et al. 2007).

2. The exponential phase culture filtrate of strain SR14 of *Streptomyces hygroscopicus* that produced extracellular chitinase and β-1,3-glucanase contained antagonists against *Colletotrichium gloeosporiodes* and *Sclerotium rolfsii*, although antifungal activity in the stationary cultures was due to an unknown compound (Prapagdee et al. 2008).

3. Root-rot in grapevine caused by *Fusarium oxysporum* is a serious disease, and antagonism by a strain of *Streptomyces alni* was demonstrated by malformation and lysis of the fungus (Ziedan et al. 2010).

4. A formulation consisting of actively growing *Streptomyces* sp. Di-944, alginate beads, and talcum powder effectively suppressed damping-off by *Rhizoctonia solani* in plug transplants of tomato (*Lycopersicon esculentum*), and the formulation was also effective when applied as a seed coating (Sabaratnam and Traquair 2002).

5. a. The biocontrol agent *Streptomyces lydicus* WYEC108 is effective against *Pythium* (Yuan and Crawford 1995) and produces an extracellular chitinase that is putatively responsible for its antifungal activity (Mahdevan and Crawford 1996).

   b. The root-colonizing *Streptomyces lydicus* WYEC109 was able to establish an interaction with *Pisum sativum* that had a number of positive effects, including root nodulation (Tokala et al. 2002). It was suggested that this could be of major importance to pea, and possibly other legumes.

6. Phenylacetic acid and its sodium were isolated and purified from cultures of *Streptomyces humidus* strain S5-55. They had MIC values (μg/mL) of 50, 10, and 50 toward *Phytophthora capsici, Pythium ultimum,* and *Rhizoctonia solani,* although *in vivo,* the effect on disease severity compared with Metalaxl was lower at concentrations below 1000 μg/L. There was no significant effect *in vitro* on species of *Fusarium* (Hwang et al. 2001).

2.23.1.5 Serratia marcescens

A strain of *Serratia marcescens* that occurred as an epiphyte on the aquatic plant *Rhyncholacis pedicillata* produced a chlorinated macrocyclic lactone oocydin A $C_{23}H_{31}O_8Cl$. This possessed activity against several oomycetes (water molds), including *Phytophthora parasitica* that is the agent of citrus root rot and *P. cinnamomi* that causes forest disease as well as damage to avocado (Strobel et al. 1999). It displayed no activity, however, against pathogenic fungi imperfecti, ascomycetes, or basidiomycetes.

2.23.2 Biocontrol Agents against Bacterial Infection

Bacteria can play an antagonistic role not only against fungal infection but also against bacterial infection, and some have been employed as biocontrol agents, including *Pantoea agglomerans* that has been commercialized (Dew et al. 1997). Extreme care should be exercised in the taxonomic designation of these genera of the facultative anaerobes *Brenneria, Erwinia,* and *Pantoea* that are included among the Enterobacteriaceae. This is briefly summarized, without any attempt to be taxonomically rigorous:

Erwinia: Species of *Erwinia,* for example, *E. amylovorans, E. carotovora, E. chrysanthemi, E. herbicola,* and *E. pyrifoliae* are important and well-established plant pathogens.

Brenneria: Species of *Brenneria* include tree pathogens for willow (*B. salicis*), oak (*B. quercina*), and walnut (*B. nigrifluens*).

Pantoea: There has been some confusion in the herbicola-agglomerans group, and the latter group of organisms have been variously assigned to the genera *Enterobacter* and *Pantoea* that includes the biocontrol strain *Pantoea vagans. P. stewartii* subsp. *stewartii* is an important agent of corn blight, and *Pantoea anantis* that is a pathogen for a wide range of crops and trees has antifungal and antibacterial properties which have the potential for biocontrol. *Pantoea agglomerans* has attracted the greatest attention as a biocontrol agent and is illustrated in the following examples:

1. A strain that is an epiphyte on apple blossom was active against fire blight caused by *Erwinia amylovorans* that is a serious pathogen for apple and pear (Vanneste et al. 1992; Kearns and Mahanty 1998). It produced two antibiotics pantocin A and B, whose effect has been demonstrated *in vitro* (Wright et al. 2001), and the structures of these unusual peptides have been determined (Sutton and Clardy 2000; Jin et al. 2003).

2. A strain is able to afford control of infection by *Pseudomonas syringae* pv. *syringae* that is the agent of basal kernel blight in barley (Braun-Kiewnick et al. 2000).

3. Control of fire blight in hawthorn (*Crataegus monogyna*) has been achieved (Wilson et al. 1990).

2.24 Siderophores in Plants: Roles of Iron

Iron is required for the synthesis of enzymes in all biota. However, its excess, due to the synthesis of peroxy compounds, is highly undesirable and its concentration must be controlled, for example, by chelation of iron by siderophores. The role of these in determining the virulence of human pathogens, including *Yersinia pestis*, and in circumventing the host defenses against *Bacillus anthracis* has been demonstrated (Miethke and Marahiel 2007). The significance of siderophores in plant infection has been less clearly shown, although they may play a critical role in the functioning of pathogenic epiphytic bacteria that may derive their Fe from the plants. The range of Fe-chelators in plant pathogens includes salicylic acid and yersiniabactin into which it is incorporated; achromobactin; pyoverdin; the peptides syringomycin, a cyclic nonapeptide linked through serine to 3-hydroxydodecanoate, and syringopeptin, a cyclic octapeptide linked to a linear peptide; and to 3-hydroxydodecanoate through 2,3-dehydroaminobutyrate in the peptide.

In spite of extensive investigation, particularly with strains of *Pseudomonas syringae* (Jones et al. 2007; Braun et al. 2010; Wensing et al. 2010), the role of these chelators in fungal pathogenesis has not been finally resolved. Some brief comments on this important issue are, however, given.

1. *Pseudomonas syringae* pv. *syringae* strain B301D is able to synthesize syringopeptin and syringomycin that induce necrosis, and their role was examined in mutants unable to synthesize syringopeptin (*sypA*) and syringomycin (*syrB 1*) (Scholz-Schroeder et al. 2001). When assayed against immature sweet cherry, the greatest reduction in virulence was observed in the double mutants that exceeded that displayed for single mutants. Although the role of iron chelators was clear, only incomplete protection resulted from their absence.

2. *Pseudomonas syringae* pv. *tomato* DC 3000 has three high-affinity iron-scavenging systems, yersiniabactin, pyoverdin, and citrate, that were operational under conditions of iron limitation. Although growth of a triple mutant that was defective in these was restricted under iron-limited conditions, it was fully pathogenic. Several suggestions for this are given (Jones and Wildermuth 2011).

3. *Dickeya dadantii* (*Erwinium chrysanthemi* = *Pectobacterium chrysanthemi*) is the agent of soft-rot in a range of plants associated with the production of plant cell wall-degrading pectinases, cellulases, and proteases. Virulence also requires the production of both the high-affinity siderophores chrysobactin and achromobactin, while mutants of these compromise the full virulence to the African violet (*Saintpaulia ionantha*) (Franza et al. 2005). It is worth noting that the epiphyte *Pseudomonas syringae* pv. *syringae* B728a produced both siderophores pyoverdin and achromobactin under conditions of iron limitation (Berti and Thomas 2009).

4. The extracellular nonribosomal peptide coprogen contains hydroxamates, two of which form a hexadentate complex with Fe. The synthetase of this (*NPS6*) is a determinant of the virulence of *Cochliobolus heterostrophus* for maize (*Zea mays*). Deletion of orthologues of *NPS6* in the pathogen of rice (*Cochliobolus miyabeanus*) reduced the virulence from 2.3 to 0.73 mm, in *Alternaria brassicicola*, the pathogen for *Arabidopsis thaliana*, from 3.78 to 1.48 mm, and also in the wheat pathogen (*Fusarium graminearum*). In addition, it was shown that the application of exogenous Fe enhanced the virulence of the Δ*nps6* strain in these fungi: for *C. miyabeanus* by ferric citrate from 1.68 to 2.75 mm, and for ferric EDTA from 0.27 to 0.81 mm. It was proposed that the role of these extracellular siderophores was to supply Fe to plants (Oide et al. 2006).

5. *Xanthomonas oryzae* pv. *oryzae* is the causal agent of bacterial blight in rice, and the role of iron has been examined. FeoB is required for uptake of ferrous iron and growth of the pathogen in the plant and the requirement for iron is regulated by an *xss* gene cluster for biosynthesis of siderophore and *xsuA* for utilization. It has been shown that mutation of the *feoB* gene results in a decrease in virulence of the pathogen on rice leaves and overproduction of siderophores to compensate for loss of direct iron uptake (Pandey and Sonti 2010).

2.25 The Role of Mycorrhizal and Other Fungi

Mycorrhizal fungi play an important role in the nutrition and survival of higher plants, and the importance of their associated bacteria has been recognized. In a study of the bacterial communities in the mycorrhizospheres of *Pinus sylvestris–Suillus bovinus* and *P. sylvestris–Paxillus involutis*, the occurrence of both Gram-negative and Gram-positive bacteria was recognized (Timonen et al. 1998). A wide range included taxa such as *Burkholderia cepacia* and *Paenibacillus* sp. in which strains with established biodegradative capability have been recognized. The relevance of rhizosphere bacteria was illustrated by the demonstration of a biofilm community at the interface of the petroleum-contaminated soil/mycorrhizosphere. Strains of *Pseudomonas fluorescens* biovars were isolated and these were able to grow at the expense of 1,3-dimethyl benzene and 3-methylbenzoate, brought about fission of catechol, and harbored the plasmid-borne genes *xylE* and *xylM* (Sarand et al. 1998).

Although considerable attention has been directed to the synthesis of auxins by plants and bacteria in the context of their interaction, auxins are also synthesized by fungi and some brief comments on fungi are justified.

1. The synthesis of indole-3-acetic acid has been examined in a range of mycorrhizal fungi and particularly high concentrations were established in the growth medium of *Suillus bovinus*, *Pisolithus tinctorius*, and *Laccaria laccata* (Ek et al. 1983).

2. The biosynthesis of IAA by the pine ectomycorrhizal fungus *Pisolethus tinctorius* was established, and it was shown that in the presence of L-tryptophan, there was a positive growth response in seedlings of Douglas fir (*Pseudotsuga menziesii*) and that the mycelium formed ectomycorrhizae on the seedlings (Frankenberger and Poth 1987).

3. *Ustilago esculenta* incites the formation of galls and disturbs seed production in the aquatic perennial grass (*Zizania latifolia*), and was able to synthesize indole-3-acetic acid that was inhibited by the addition of thiamine to the growth medium (Chung and Tzeng 2004).

4. *Colletotrichum gloeosporioides* f. sp. *aeschynomene* is the agent of anthracnose disease on a weed that infests the fields of rice and soybean in North America and has been registered as a biocontrol agent. It produces indole-3-acetic acid from tryptophan in culture, and it has been shown that this was produced by the indole-3-pyruvate pathway and by the major indole-3-acetamide pathway (Robinson et al. 1998).

5. *Aciculosporium take* is the causal agent of witches' broom in bamboo, and the role of auxins has been suggested. Cultures of the fungus in L-tryptophan-supplemented medium produced indole-3-acetic acid that was synthesized by the indole-3-pyruvate pathway (Tanaka et al. 2002).

In a more general context, it is worth noting the role of auxins produced by yeasts in the presence of L-tryptophan. *Saccharomyces cerevisiae* has been shown to respond to IAA by inducing pseudohyphal growth. IAA also induced hyphal growth in the human pathogen *Candida albicans*, and thus may function as a secondary metabolite signal that regulates virulence traits such as hyphal transition in pathogenic fungi (Rao et al. 2010).

2.25.1 Fungi and Yeasts as Control Agents

2.25.1.1 Trichoderma *spp.*

Attention is directed to reviews (Howell 2003; Benítez et al. 2004; Harman et al. 2004) which cover mechanisms, and only a few aspects are briefly summarized. It is clear that several mechanisms may be involved, including the role of hydrolytic enzymes, antibiotics, and phytoalexins among the species aggregate *Trichoderma harzianum*.

1. Mycelial inoculum of *Trichoderma harzianum* was used to treat seeds of cucumber (*Cucumis sativus*) and growth was studied in a hydroponic system for up to 21 days. The treated roots were penetrated by the fungus and, after 48 h there was a transient increase of chitinase activity in the roots and leaves, and a marginal increase in the peroxidase activity in the roots with a much greater increase in the leaves after 48 h (Yedidia et al. 1999).

2. A suspension of *Trichoderma asperellum* in the plant growth medium was prepared and used to inoculate seedlings of cucumber (*Cucumis sativus*) (Yedidia et al. 2003). A suspension of the

pathogen *Pseudomonas syringae* pv. *lachrymans* was applied to the leaves of plants, and it was shown that during 120 h after application there was a reduction in the areas of necrosis. Leaves from 11- and 12-day plants were ground and extracted to obtain an extract part of which was hydrolyzed and used for analysis of the components. The phenolic extract after hydrolysis was used to assess the antimicrobial spectrum and it was shown that this was dose dependent.

3. Treatment of the seeds of pepper (*Capsicum annuum*) with *Trichoderma harzianum* reduced the development of necrosis in the stems of the plants induced by the pathogen *Phytophthora capsisi*, and this was also found when plant roots were drenched with spores. Six days after inoculation, the concentration of the terpenoid capsidiol in the stems of inoculated plants was seven times greater than in the untreated plants, which suggested its putative role as a phytoalexin (Ahmed et al. 2000).

4. The metabolites hartzionalide and 6-*n*-pentylpyr-2-one were examined for their effect on the growth of seedlings of tomato (*Lycopersicon esculentum*) and canola (*Brassica napa*) that were treated with the metabolites before exposure to *Botrytis cinerea* or *Leptosphaeris maculans*. This was effective in inhibiting infection and the pyrone was more effective (Vinale et al. 2008).

There are several limitations to the application of biocontrol agents, and two of them have been addressed with *Trichoderma* spp.: the use of a saline-tolerant strain and of a transgenic strain.

1. There is an important application of biocontrol agents where crops are irrigated with saline water. A study examined the potential value of strains of *Trichoderma* spp. isolated from the Mediterranean sponge *Psammocinia* sp. that were halotolerant (Gal-Hemed et al. 2011). Soluble metabolites secreted into agar plates of strain OY 3807 were effective in inhibiting the growth of *Alternaria alternata*, *Rhizoctonia solani*, and, to a lesser extent, *Botrytis cinerea*, and this inhibition of infection was also observed with volatiles from this strain. An analogous effect under greenhouse conditions against *R. solani* was also observed, and a range of volatile metabolites was observed that included the terpenoids α-bergamotene and β-sesquiphellandrene.

2. A transgenic strain SJ3-4 was prepared from *Trichoderma atroviride* (*harzianum*) P1 by introducing the gene *goxA* that encodes glucose oxidase in *Aspergillus niger* into the parent that does not possess this activity (Brunner et al. 2005). Plate assays with P1 and SJ3-4 revealed inhibition of *Rhizoctonia solani* and *Pythium ultimum* by the latter. The effect of the introduction was examined in bean (*Phaseolus vulgaris*) seedlings coated with strains P1 or SJ3-4 and planted in soil deliberately infested with *Rhizoctonia solani*: no growth occurred in the unprotected control 3 weeks after planting, and the growth of the SJ3-4 strain exceeded that of the parent P1. Plants whose seeds had been coated with P1 or SJ3-4 were evaluated for the development of lesions induced by *Botrytis cinerea* applied to the leaves and there was a clear advantage of the SJ3-4 treatment to the seeds that was repeated with *Rhizoctonia solani*.

2.25.1.2 Yeasts

There has also been interest in yeasts as biocontrol agents, and only a few brief illustrations of the broad types of mechanisms are given.

1. *Candida oleophila* has been used for the control of postharvest decay of fruit, and application of *Candida oleophila* to surface wounds of grapefruit (*Citrus paradisi*) elicited resistance to infection by *Penicillium digitatum* at the infected sites. Analysis of treated disks revealed the synthesis of ethene, phenylalanine-ammonia lyase activity, and phytoalexins that is consistent with the role of these biochemical events in biocontrol activity (Droby et al. 2002). The yeast was able to synthesize exo-β-1,3-glucanase and chitinase in the early stages of growth and this was stimulated by the presence of cell wall fragments of the pathogen for grapefruit *Penicillium digitatum* and secreted into the physically wounded fruit (Bar-Shimon et al. 2004).

2. An early study established that *Saccharomyces cerevisiae* could produce a variety of volatile compounds which inhibited the growth and sporulation of *Aspergillus niger* and the germination of seeds of *Lepidium sativum* (Glen and Hutchinson 1969). Cultures of *Candida intermedia* strain C410 grown on a complex medium produced a range of aliphatic metabolites among which 3-methyl-butan-1-ol was abundant. All of them inhibited conidial germination and mycelial growth of *Botrytis cinerea* while they reduced the severity of *Botrytis* fruit rot on strawberry (Huang et al. 2011).

3. Killer toxins are produced by a range of yeasts of which the K1 from *Saccharomyces cerevisiae* has been characterized. Killer toxin from *Pichia membranifaciens* (PMKT) has been described that disrupts plasma membrane electrochemical gradients, and has been suggested for control of gray mold disease by *Botrytis cinerea* of grapevine (*Vitis vinifera*) (Santos and Marquina 2004). Another killer toxin PMK2 has been isolated from the same yeast, and its activity has been assessed against the spoilage yeast *Brettanomyces bruxellensis*: it differed from that of PMKT that was active against *Saccharomyces cerevisiae*, though not for *B. bruxellensis* (Santos et al. 2009).

4. In a wider context, it is noted that milky disease in the crab *Portunus trituberculatus* is caused by the marine yeast *Metschnikowia bicuspidate*, and several killer yeasts that produce toxin against this have been isolated: *Pichia anomala* with an optimum for production of toxin at pH 4.5 and 6% NaCl (Wang et al. 2007); *Kluyveromyces siamensis* from a mangrove with high activity at pH 4.0 even in the absence of NaCl (Buzdar et al. 2011); *Mrakia frigida* from an Antarctic marine sediment with maximal activity at 4.5 with a salinity of 3% NaCl (Hua et al. 2010).

2.26 Conclusions

Important conclusions can be drawn collectively from the results of all these studies:

- Microbial degradation of contaminants in the root system of higher plants can be significant.
- Higher plants play a cardinal role in the provision of organic carbon and of exudates.
- Plants produce metabolites that may deter attack by fungal pathogens.
- Bacteria are important as biocontrol agents against a number of plant pathogens, including both bacteria and fungi.
- Fungi in biocontrol.

2.27 References

Aarts MGM, CJ Keijzer, WJ Stiekema, A Pereira. 1995. Molecular characterization of the *CER1* gene of *Arabidopsis* involved in epicuticular wax biosynthesis and pollen fertility. *The Plant Cell* 7: 2115–2127.

Adams DO, SF Yang. 1979. Ethylene biosynthesis: Identification of 1-aminocyclopropane-1-carboxylic acid as an intermediate in the conversion of methionione to ethylene. *Proc Natl Acad Sci USA* 76: 170–174.

Adrian M, P Jeandet, J Veneau, LA Weston, R Bessis. 1997. Biological activity of resveratrol, a stilbenic compound from grapevines, against *Botrytis cinerea*, the causal agent for gray mold. *J Chem Ecol* 23: 1689–1702.

Ahmed AS, CP Sánchez, ME Candela. 2000. Evaluation of induction of systemic resistance in pepper plants (*Capsicum annuum*) to *Phytophthora capsici* using *Trichoderma harzianum* and its relation with capsidiol accumulation. *Eur J Plant Pathol* 106: 817–824.

Akatsuka T, O Kodama, H Kato, Y Kono, S Takeuchi. 1985. Novel phytoalexins (oryzalexins A, B, and C) from rice blast leaves infected with *Pyricularia oryzae*. Part I: Isolation, characterization and biological activities of oryzalexins. *Agric Biol Chem* 49: 1689–1694.

Akiyoshi DE, DA Regier, MP Gordon. 1987. Cytokinin production by *Agrobacterium* and *Pseudomonas* spp. *J Bacteriol* 169: 4242–4248.

Allard A-S, P-Å Hynning, M Remberger, AH Neilson. 1992. Role of sulfate concentration in dechlorination of 3,4,5-trichlorocatechol by stable enrichment cultures grown with coumarin and flavanone glycones and aglycones. *Appl Environ Microbiol* 58: 961–968.

Andreote FD, WL de Araúo, JL de Azevedo, JD van Elsas, UN da Rocha, LS van Overbeek. 2009. Endophytic colonization of potato (*Solanum tuberosum* L.) by a novel competent bacterial endophyte, *Pseudomonas putida* strain P9, and its effect on associated bacterial communities. *Appl Environ Microbiol* 75: 3396–43406.

Aprill W, RC Sims. 1990. Evaluation of the use of prairie grasses for stimulating polycyclic aromatic hydrocarbon treatment in soil. *Chemosphere* 20: 253–265.

Baca BE, C Elmerich. 2003. Microbial production of plant hormones Chapter 6, pp. 1–31. In *Associative and Endophytic Nitrogen-Fixing Bacteria and Cyanobacterial Associations* (Eds. C Elmerich, WE Newton), Kluwer Academic Publishers. The Netherlands.

Baldani JI, VLD Baldani, L Seldin, J Döbereiner. 1986. Characterization of *Herbaspirillum seropedicae* gen. nov., sp. nov., a root-associated nitrogen-fixing bacterium. *Int J Syst Bacteriol* 36: 86–93.

Barac T, S Taghavi, B Borremans, A Provoost, L Oeyen, JV Colpaert, J Vangronsveld, D van der Lelie. 2004. Engineered endophytic bacteria improve phytoremediation of water-soluble, volatile, organic pollutants. *Nat Biotechnol* 22: 583–588.

Bar-Shimon M, H Yehuda, L Cohen, B Weiss, A Kobeshnikov, A Daus, M Goldway, M Wisniewski, S Droby. 2004. Characterization of extracellular lytic enzymes produced by the yeast biocontrol agent *Candida oleophila. Curr Genet* 45: 140–148.

Bastián F, A Cohen, P Piccolli, V Luna, R Bottini, R Baralfi, R Bottini. 1998. Production of indole-3-acetic acid and gibberellins A1 and A3 by *Acetobacter diazotrophicus* and *Herbaspirillum seropedicae* in chemically-define culture media. *Plant Growth Regul* 24: 7–11.

Bedard DL, R Unterman, LH Bopp, MJ Brennan, ML Haberl, C Johnson. 1986. Rapid assay for screening and characterizing microorganisms for the ability to degrade polychlorinated biphenyls. *Appl Environ Microbiol* 51: 761–768.

Bednarek P and 11 coauthors. 2009. A glucosinolate metabolism pathway in living plant cells mediates broad-spectrum antifungal defense. *Science* 323: 101–106.

Belimov AA et al. 2001. Characterization of plant growth promoting rhizobacteria isolated from polluted soils and containing 1-aminocyclopropane-1-carboxylate deaminase. *Can J Microbiol* 47: 642–652.

Belimov AA, IC Dodd, VI Safronova, N Hontzeas, WJ Davies. 2007. *Pseudomonas brassicacearum* strain Am3 containing 1-aminocyclopropane-1-carboxylate deaminase can show both pathogenic and growth-promoting properties in its interaction with tomato. *J Exp Bot* 58: 1485–1495.

Benítez T, AM Rincón, MC Limón, AC Codón. 2004. Biocontrol mechanisms of *Trichoderma* strains. *Int Microbiol* 7: 249–260.

Berti AD, MG Thomas. 2009. Analysis of achromobactin biosynthesis by *Pseudomonas syringae* pv *syringae* B728a. *J Bacteriol* 191: 4594–4604.

Bezzerides A, T-H Yong, J Bezzerides, J Husseini, J Ladau, M Eisner, T Eisner. 2004. Plant-derived pyrrolizidine alkaloid protects eggs of a moth (*Utetheisa ornatrix*) against a parasitoid wasp (*Trichogramma ostriniae*) *Proc Natl Acad Sci USA* 101: 9029–9032.

Bhadra R, DG Wayment, JB Hughes, JV Shanks. 1999. Confirmation of conjugation processes during TNT metabolism by axenic plant roots. *Environ Sci Technol* 33: 446–452.

Bianco C, R Defez. 2010. Improvement of phosphate solubilization and *Medicago* plant yield by an indole-3-acetic acid-overproducing strain of *Sinorhizobium meliloti. Appl Environ Microbiol* 76: 4626–4632.

Blom D, C Fabbri, L Eberl, L Weisskopf. 2011. Volatile-mediated killing of *Arabidopsis thaliana* by bacteria is mainly due to hydrogen cyanide. *Appl Environ Microbiol* 77: 1000–1008.

Bokern M, HH Harms. 1997. Toxicity and metabolism of 4-*n*-nonylphenol in cell suspension cultures of different plant species. *Environ Sci Technol* 31: 1849–1854.

Bottiglieri M, C Keel. 2006. Characterization of PhlG, a hydrolase that specifically degrades the antifungal compound diacetylphloroglucinol in the biocontrol agent *Pseudomonas fluorescens* CHA0. *Appl Environ Microbiol* 72: 418–427.

Bottini R, M Fulchieri, D Pearce, RP Pharis. 1989. Identification of gibberellins A1, A3, and iso-A3 in cultures of *Azospirillum lipoferum. Plant Physiol* 90: 45–47.

Bowers JH, JL Parke. 1993. Epidemiology of *Pythium* damping-off and *Aphenomyces* root rot of peas after seed treatment with bacterial agents for biological control. *Phytopathol* 83: 1466–1473.

Bowyer P, BR Clarke, P Lunness, MJ Daniels, AE Osbourn. 1995. Host range of plant pathogenic fungus determined by a saponin detoxifying enzyme. *Science* 267: 371–374.

Braun SD, J Hofmann, A Wensing, H Weingart, MS Ullrich, D Spiteller, B Völksch. 2010. *In vitro* antibiosis by *Pseudomonas syringae* Pss22d, acting against the bacterial blight pathogen of soybean plants, does not influence *in planta* biocontrol. *J Phytopathol* 158: 288–295.

Braun-Kiewnick A, BJ Jacobsen, DC Sands. 2000. Biological control of *Pseudomonas syringae* pv. *syringae*, the causal agent of basal kernel blight of barley, by antagonistic *Pantoea agglomerans*. *Phytopath* 90: 368–375.

Brazil GM, L Kenefick, M Callanan, A Haro, V de Lorenzo, DN Dowling, F O'Gara. 1995. Construction of a rhizosphere pseudomonad with potential to degrade polychlorinated biphenyls and detection of *bph* gene expression in the rhizosphere. *Appl Environ Microbiol* 61: 1946–1952.

Brodhagen M, MD Henkels, JE Loper. 2004. Positive autoregulation and signaling properties of pyoluteorin, an antibiotic produced by the biological control organism *Pseudomonas fluorescens* Pf-5. *Appl Environ Microbiol* 70: 1758–1766.

Brooks CLW, DG Watson. 1985. Phytoalexins. *Nat Prod Rep* 2: 427–459.

Brooks CLW, DG Watson. 1991. Terpenoid phytoalexins. *Nat Prod Rep* 8: 367–389.

Brunner K, S Zeilinger, R Cilento, SL Woo, M Lorito, CP Kubicek, RL Mach. 2005. Improvement of the fungal biocontrol agent *Trichoderma atroviride* to enhance both antagonism and induction of plant systemic disease resistance. *Appl Environ Microbiol* 71: 3959–3965.

Burg SP. 1973. Ethylene in plant growth. *Proc Natl Acad Sci USA* 70: 591–597.

Burken JG, JL Schnoor. 1996. Phytoremediation: Plant uptake of atrazine and role of root exudates. *J Environ Eng* 122: 958–963.

Burken JG, JL Schnoor. 1997. Uptake and metabolism of atrazine by poplar trees. *Environ Sci Technol* 31: 1399–1406.

Burkhead KD, DA Schisler, PJ Slinger. 1994. Pyrrolnitrin production by biological control agent *Pseudomonas cepacia* B2372 in culture and in colonized wounds of potatoes. *Appl Environ Microbiol* 60: 2031–2039.

Buzdar MA, Z Chi, Q Wang, MX Hua, ZM Chi. 2011. Production, purification, and characterization of a novel toxin from *Kluyveromyces siamensis* a pathogenic yeast in crab. *Appl Microbiol Biotechnol* 91: 1–9.

Caballero-Mellado J, LE Fuentes-Ramirez, VM Reis, E Martinez-Romero. 1995. Genetic structure of *Acetrobacter diazotropohicus* and identification of a new genetically distant group. *Appl Environ Microbiol* 61: 3008–3013.

Caballero-Mellado J, L Martínez-Aguilar, G Paredes-Valdez, P Estrada-de los Santos. 2004. *Burkholderia unamae* sp. nov., an N_2-fixing rhizospheric and endophytic species. *Int J Syst Evol Microbiol* 54: 1165–1172.

Caballero-Mellado J, J Onofre-Lemus, P Estrada-de los Santos, L Martínez-Aguilar. 2007. The tomato rhizosphere, an environment rich in nitrogen-fixing *Burkholderia* species with capabilities of interest for agriculture and bioremediation. *Appl Environ Microbiol* 73: 5308–5319.

Cartwright DK, WS Chilton, DM Benson. 1995. Pyrrolnitrin and phenazine production by *Pseudomonas cepacia* strain 5.5B, a control agent of *Rhizoctonia solani*. *Appl Microbiol Biotechnol* 43: 211–216.

Cartwright DW, P Langcake, RJ Pryce, DP Leworthy, JP Ride. 1981. Isolation and characterization of two phytoalexins from rice as momilactones A and B. *Phytochemistry* 20: 535–537.

Cassán F, S Maiale, O Masciarelli, A Vidal, V Luna, O Ruiz. 2009. Cadaverine production by *Azospirillum brasilense* and its possible role in plant growth promotion and osmotic stress mitigation. *Eur J Soil Biol* 45: 12–19.

Chen W-M et al. 2006. *Burkholderia mimosarum* sp. nov., isolated from root nodules of *Mimosa* spp. from Taiwan and South America. *Int J Syst Evol Microbiol* 56: 1847–1851.

Chibnall AC, SH Piper. 1934. The metabolism of plant and insect waxes. *Biochem J* 28: 2209–2219.

Chin-A-Woeng TFC, D van den Broek, G de Voer, KMGK van der Drift, S Tuinman, JE Thomas-Oates, BJJ Lugtenberg, GV Bloemberg. 2001. Phenazine-1-carboxamide production in the biocontrol strain *Pseudomonas chlororaphis* PC L 1391 is regulated by multiple factors secreted into the growth medium. *Mol Plant-Microbe Interact* 14: 869–879.

Chin-A-Woeng TFC, GV Bloemberg, IHM Mulders, LC Dekkers, BJJ Lugtenberg. 2000. Root colonization by phenazine-1-carboxamide-producing bacterium *Pseudomonas chlororaphis* PCL 1391 is essential for biocontrol of tomato foot and root rot. *Mol Plant-Microbe Interact* 13: 1340–1345.

Christensen P, FD Cook. 1978. *Lysobacter*, a new genus of nonfruiting, gliding bacteria with a high base ratio. *Int J Syst Bacteriol* 28: 367–393.

Chung KR, DD Tzeng. 2004. Biosynthesis of indole-3-acetic acid by the gall-inducing fungus *Ustilago esculenta*. *J Biol Sci* 4: 744–750.

Clark E, S Manulis, Y Ophir, I Barash, Y Gafni. 1993. Cloning and characterization of iaaM and iaaH from *Erwinia herbicola* pathovar gypsophilae. *Phytopathol* 83: 234–240.

Clay NK, AM Adio, C Denoux, G Jander, FM Ausubel. 2009. Glucosinolate metabolites required for an *Arabidopsis* innate immune response. *Science* 323: 95–101.

Coenye T, E Mahenthiralingam, D Henry, JL LiPuma, S Laevens, M Gillis, DP Speert, P Vandamme. 2001. *Burkholderia ambifaria* sp. nov., a novel member of the *Burkholderia cepacia* complex including biocontrol and cystic fibrosis-related isolates. *Int J Sys Evol Microbiol* 51: 1481–1490.

Comai L, T Kosuge. 1982. Cloning characterization of iaaM, a virulence determinant of *Pseudomonas savastanoi. J Bacteriol* 149: 40–46.

Compant S, B Reiter, A Sessitsch, J Nowak, C Clément, EA Barka. 2005b. Endophytic colonization of *Vitis vinifera* L. by plant growth-promoting bacterium *Burkholderia* sp. strain PsJN. *Appl Environ Microbiol* 71: 1685–1693.

Compant S, B Duffy, J Nowak, C Clément, EA Barka. 2005a. Use of plant growth-promoting bacteria for biocontrol of plant diseases: Principles, mechanisms of action and future prospects. *Appl Environ Microbiol* 71: 4951–4859.

Coombs JT, CMM Franco. 2003. Isolation and identification of actinobacteria from surface-sterilized wheat roots. *Appl Environ Microbiol* 69: 5603–5608.

Crombie WML, L Crombie, JB Green, JA Lucas. 1986a. Pathogenicity of "take-all" fungus to oats: Its relationship to the concentration and detoxification of the four avenacins *Phytochemistry* 25: 2075–2083.

Crombie L, WML Crombie, DA Whiting. 1986b. Structures of the oat resistance factors to "take-all" disease, avenacins A-1, A-2, B-1 and B-2 and their companion substances. *J Chem Soc Perkin Trans* 1: 1917–1922.

Dedysh SN, PF Dunfield, YA Trotsenko. 2004. Methane utilization by *Methylobacterium* species: New evidence but still no proof for an old controversy. *Int J Syst Evol Micobiol* 54: 1919–1920.

Desnoues N, M Lin, X Guo, L Ma, R Carreño-Lopez, C Elmerich. 2003. Nitrogen fixation genetics and regulation in a *Pseudomonas stutzeri* strain associated with rice. *Microbiology (UK)* 149: 2251–2262.

Dercks W, LL Creasy. 1989a. The significance of stilbene phytoalexins in the *Plasmopara viticola*–grapevine interaction. *Physiol Mol Plant Pathol* 34: 189–202.

Dercks W, LL Creasy. 1989b. The significance of stilbene phytoalexins in the *Plasmopara viticola*–grapevine interaction. *Physiol Mol Plant Pathol* 34: 2903–2913.

Desjardins AE, HW Gardner. 1989. Genetic analysis in *Gibberella pulicaris*: Rishitin tolerance, rishitin metabolism and virulence on potato tubers. *Mol Plant-Microbe Interact* 2: 26–34.

Desjardins AE, HW Gardner. 1991. Virulence of *Gibberella pulicaris* on potato tubers and its relation to a gene for rishitin metabolism. *Phytopathology* 81: 429–435.

Desjardins AE, HW Gardner, RD Plattner. 1989. Detoxification of the potato phytoalexin lubimin by *Gibberella pulicaris. Phytochemistry* 28: 431–437.

Dew DW, EN Lawson, JL Broadhurst. 1997. The BIOX^R process for biooxidation of gold-bearing ores or concentrates, pp. 45–79. In *Biomining: Theory, Microbes and Industrial Processes* (Ed. DE Rawlings), Springer-Verlag, Berlin.

de Werra P, M Péchy-Tarr, C Keel, M Maurhofer. 2009. Role of gluconic acid production in the regulation of biocontrol traits of *Pseudomonas fluorescens* CHA0. *Appl Environ Microbiol* 75: 4162–4174.

Dillon VM, J Overton, RJ Grayer, JB Harborne. 1997. Difference in phytoalexin response among rice cultivars of different resistance to blast. *Phytochemistry* 44: 599–603.

Donnelly PK, RS Hedge, JS Fletcher. 1994. Growth of PCB-degrading bacteria on compounds from photosynthetic plants. *Chemosphere* 28: 981–988.

Doronina NV, EG Ivanova, YA Trotsenko. 2002. New evidence for the ability of methylobacteria and methanotrophs to synthesize auxins. *Microbiology* 71: 116–118.

Doronina NV, EG Ivanova, YA Trotsenko. 2005. Phylogenetic position and emended description of the genus *Methylovorus. Int J Syst Evol Microbiol* 55: 903–906.

Doty SL, TW Shang, AM Wilson, J Tangen, AD Westergreen, LA Newman, SE Strand, MP Gordon. 2000. Enhanced metabolism of halogenated hydrocarbons in transgenic plants containing mammalian cytochrome P450 2E1. *Proc Natl Acad Sci USA* 97: 6287–6291.

Droby S, V Vinokur, B Weiss, L Cohen, A Daus, EE Goldschmidt, R Porat. 2002. Induction of resistance to *Penicillium digitatum* in grapefruit by the yeast biocontrol agent *Candida oleophila. Phytopath* 92: 393–399.

Ek M, PO Ljungquist, E Stenström. 1983. Indole-3-acetic acid production by mycorrhizal fungi determined by gas chromatography-mass spectrometry. *New Phytol* 94: 401–407.

Elbeltagy A, K Nishioka, T Sato, J Suzuki, B Ye, T Hamada, T Isawa, H Mitsui, K Minamisawa. 2001. Endophytic colonization and in planta nitrogen fixation by a *Herbaspirillum* sp. isolated from wild rice species. *Appl Environ Microbiol* 67: 5285–5293.

El-Tarabily KA, K Sivasithamparam. 2006. Non-streptomycete actinomycetes as biocontrol agents of soil-borne fungal plant pathogens and as plant growth promoters. *Soil Biol Biochem* 38: 1505–1520.

Errakhi R, F Bouteau, A Lebrihi, M Barakate. 2007. Evidences of biological control capacities of *Streptomyces* spp. against *Sclerotium rolfsii* responsible for damping-off disease in sugar beet (*Beta vulgaris* L). *World J Microbiol Biotechnol* 23: 1503–1509.

Estrada-de Los Santos P, R Bustillos-Cristales, J Caballero-Mellado. 2001. *Burkholderia*, a genus rich in plant-associated nitrogen fixers with wide envrionmental and geographical distribution. *Appl Environ Microbiol* 67: 2790–2798.

Farag MA, CM Ryu, LW Sumner, PW Paré. 2006. GC-MS SPME profiling of rhizobacterial volatiles reveals prospective inducers of growth promotion and induced systemic resistance in plants. *Phytochemistry* 67: 2262–2268.

Fernando WGD, R Ramarathnam, AS Krishnamoorthy, SC Savchuk. 2005. Identification and use of potential bacterial organic antifungal volatiles in biocontrol. *Soil Biol Biochem* 37: 955–964.

Fiebig A, JA Mayfield, NL Miley, S Chau, RL Fischer, D Preuss. 2000. Alterations in *CER6*, a gene identical to *CUT1*, differentially affects long-chain lipid content on the surface of pollen and stems. *The Plant Cell* 12: 2001–2008.

Flachmann R, N Kunz, J Seifert, M Gütlich, F-J Wientjes, A Läufer, HG Gassen. 1988. Molecular biology of pyridine nucelotide biosynthesis in *Escherichia coli*. Cloning and characterization of quinolinate synthetic genes *nadA* and *nadB*. *Eur J Biochem* 175: 221–228.

Fletcher JS, RS Hedge. 1995. Release of phenols by perennial plant roots and their potential importance in bioremediation. *Chemosphere* 31: 3009–3016.

Frankenberger WT, M Poth. 1987. Biosynthesis of indole-3-acetic acid by the pine ectomycorrhizal fungus *Pisolithus tinctorius*. *Appl Environ Microbiol* 53: 2908–2913.

Franza T, B Mahe, D Expert. 2005. *Erwinia chrysanthemi* requires a second iron transport route dependent on the siderochrome achromobactin for extracellular growth and plant infection. *Mol Microbiol* 55: 261–275.

Fry WE, PH Evans. 1977. Association of formamide hydro-lyase with fungal pathogenicity to cyanogenic plants. *Phytopathology* 67: 1001–1006.

Gal-Hemed I, L Atanasova, M Komon-Zelazowska, IS Druzhinina, A Viterbo, O Yarden. 2011. Marine isolates of *Trichoderma* spp. as potential halotolerant agents of biological control for arid-zone agriculture. *Appl Environ Microbiol* 77: 5100–5109.

Gamarnik A, RB Frydman. 1991. Cadaverine, an essential diamine for the normal root development of germinating soybean (*Glycine max*) seeds. *Plant Physiol* 97: 778–784.

Gardner HW, AE Desjardins, SP McCormick, D Weisleder. 1994. Detoxification of the potato phytoalexin rishitin by *Gibberella pulicaris*. *Phytochemistry* 37: 1001–1005.

Gardner HW, AE Desjardins, D Weisleder, RD Plattner. 1988. Biotransformation of the potato phytoalexin, lubimin by *Gibberella pulicaris*. Identification of major products. *Biochim Biophys Acta* 966: 347–356.

Gilbert ES, DE Crowley. 1997. Plant compounds that induce polychlorinated biphenyl degradation by *Arthrobacter* sp. strain B1B. *Appl Environ Microbiol* 63: 1933–1938.

Gillis M, V Trân Van, R Bardin, M Goor, P Hebbar, A Willems, P Segers, K Kersters, T Heulin, MP Fernandez. 1995. Polyphasic taxonomy of the genus *Burkholderia* leading to an emended description of the genus and proposition of *Burkholderia vietnamensis* sp. nov. for N₂-fixing isolates from rice in Vietnam. *Int J Syst Bacteriol* 45: 274–289.

Glawischnig E. 2007. Camalexin. *Phytochemistry* 68: 401–406.

Glawischnig E, BG Hansen, CE Olson, BA Halkier. 2004. Camalexin is synthesized from indol-3-acetaldoxime, a key branching point between primary and secondary metabolism in *Arabidopsis*. *Proc Natl Acad Sci USA* 101: 8245–8250.

Glen AT, SA Hutchinson. 1969. Some biological effects of volatile metabolites from cultures of *Saccharomyces cerevisiae* Meyen ex Hansen. *J Gen Microbiol* 55: 19–27.

Glenn AE, DM Hinton, IE Yates, CW Bacon. 2001. Detoxification of corn antimicrobial compounds as the basis for isolating *Fusarium verticillioides* and some other *Fusarium* species from corn. *Appl Environ Microbiol* 67: 2973–2981.

Glick BR, Z Cheng, J Czarny, J Duan. 2007. Promotion of plant growth by ACC deaminase-producing soil bacteria. *Eur J Plant Pathol* 119: 329–339.

Glick BG, CB Jacobson, MML Schwarze, JJ Pasternak. 1994. 1-aminocyclopropane-1-carboxylic acid deaminase mutants of the plant growth promoting rhizobacterium *Pseudomonas putida* GR12-2 do not stimulate canola root elongation. *Can J Microbiol* 40: 911–915.

Glick BR, C Liu, S Ghosh, EB Dumbroff. 1997. Early development of canola seedlings in the presence of the plant growth-promoting rhizobacterium *Pseudomonas putida* GR12-2. *Soil Biol Biochem* 29: 1233–1239.

Glickmann E, L Gardan, S Jacquet, S Hussain, M Elasri, A Petit, Y Dessaux. 1998. Auxin production is a common feature of most pathovars of *Pseudomonas syringae*. *Mol Plant-Microbe Interact* 11: 156–162.

Gonzalez CF, EA Pettit, VA Valadez, EM Provin. 1997. Mobilization, cloning, and sequence determination of plasmid-encoded polygalacturonase from a phytopathogenic *Burkholderia* (*Pseudomonas*) *cepacia*. *Mol Plant-Microbe Interact* 10: 840–851.

Gottel NR et al. 2011. Distinct communities within the endosphere and rhizosphere of *Populus deltoides* roots across contrasting soil types. *Appl Environ Microbiol* 77: 5934–5944.

Govindarajan M, J Balandreau, M Ramachandran, R Gopalakrishnan, L Cunthipuram. 2006. Improved yield of micropropagated sugarcane following inoculation by endophytic *Burkholderia vietnamiensis*. *Plant Soil* 280: 239–342.

Gutierrez CK, GY Maysui, DE LIncoln, CR Lowell. 2009. Production of the phytohormone indole-3-acetic acid by estuarine species of the genus *Vibrio*. *Appl Environ Microbiol* 75: 2253–2258.

Gyaneshwar P, EK James, N Mathan, PM Reddy, B Reinhold-Hurek, JK Ladha. 2001. Endophytic colonization of rice by a diazotrophic strain of *Serratia marcescens*. *J Bacteriol* 183: 2634–2645.

Haas D, G Défago. 2005. Biological control of soil-borne pathogens by fluorescent pseudomonads. *Nature Revs Microbiol* 3: 307–319.

Harman GE, CR Howell, A Viterbo, I Chet, M Lorita. 2004. *Trichoderma* species: Opportunistic, avirulent plant symbionts. *Nature Revs Microbiol* 2: 43–56.

Healy FG, M Wach, SB Krasnoff, DM Gibson, R Loria. 2000. The txtAB genes of the plant pathogen *Streptomyces acidiscabies* encode a peptide synthetase required for phytotoxin thaxtomin A production and pathogenicity. *Mol Microbiol* 38: 794–804.

Hebbar KP, MH Martel, T Heulin. 1998. Suppression of pre- and postemergence damping-off in corn by *Burkholderia cepacia*. *Eur J Plant Pathol* 104: 29–36.

Hedge RS, JS Fletcher. 1996. Influence of plant growth stage and season on the release of root phenolics by mulberry as related to development of phytoremediation technology. *Chemosphere* 32: 2471–2479.

Hellström A, H Kylin, WMJ Strachan, S Jensen. 2004. Distribution of some organochlorine compounds in pine needles from Central and Northern Europe. *Environ Pollut* 128: 29–48.

Hipkin CR, MA Salem, D Simpson, SJ Wainwright. 1999. 3-nitropropionic acid oxidase from horseshoe vetch (*Hippocrepis comosa*): a novel plant enzyme. *Biochem J* 340: 491–495.

Höhl H-U, W Barz. 1995. Metabolism of the insecticide phoxim in plants and cell suspension cultures of soybean. *J Agric Food Chem* 43: 1052–1056.

Hökeberg M, B Gerhardson, L Johnsson. 1997. Biological control of cereal seed-born diseases by seed bacterization with greenhouse-selected bacteria. *Eur J Plant Pathol* 103: 25–33.

Homma Y, T Suzui. 1989. Role of antibiotic production in suppression of radish damping-off by seed bacterization with *Pseudomonas cepacia*. *Ann Phytopathol Soc Japan* 55: 643–652.

Hontzeas N, AO Richardson, A Belimov, V Safronova, MM Abu-Omar, BR Glick. 2005. Evidence for horizontal transfer of 1-aminocyclopropane-1-carboxylate deaminase genes. *Appl Environ Microbiol* 71: 7556–7558.

Hontzeas N, SS Saleh, BR Glick. 2004b. Changes in gene expression in canol roots by ACC-deaminase containing plant growth-promoting bacteria. *Mol Plant-Microbe Interact* 17: 865–871.

Hontzeas N, J Zoidakis, BR Glick, MM Abu-Omar. 2004a. Expression and characterization of 1-aminocyclopropane-1-carboxylate deaminase from the rhizobacterium *Pseudomonas putida* UW4: A key enzyme in bacterial plant growrh promotion. *Biochim Biophys Acta* 1703: 11–19.

Howell CR. 2003. Mechanisms employed by *Trichoderma* species in the biological control of plant disease: The history and evolution of current concepts. *Plant Dis* 87: 1–10.

Howell CR, RD Stipanovic. 1979. Control of *Rhizoctonia solani* on cotton seedings with *Pseudomonas fluorescens* and with an antibiotic produced by the bacterium. *Phytopathol* 69: 480–482.

Hua MX, Z Chi, GL Liu, MA Buzdar, Zm Chi. 2010. production of a novel and cold-active killer toxin by *Mrakia frigida* 2E00797 isolated from sea sediment in Antarctica. *Extremophiles* 14: 515–521.

Huang R, GQ Li, J Zhang, L Yang, HJ Che, DH Jiang, HC Huang. 2011. Control of postharvest Botrytis fruit rot of strawberry by volatile organic compounds of *Candida intermedia*. *Phytopathol* 101: 859–869.

Hughes JB, J Shanka, M Vanderford, J Lauritzsen, R Bhadra. 1997. Transformation of TNT by aquatic plants and plant tissues cultures. *Environ Sci Technol* 31: 266–271.

Hutcheson SW, T Kosuge. 1985. Regulation of 3-indolacetic acid production in *Pseudomonas savastanoi*. Purification and properties of trypyptophan 2-monooxygenase. *J Biol Chem* 260: 6281–6287.

Hwang BK, SW Lim, BS Kim, JY Lee, SS Moon. 2001. Isolation and *in vivo* and *in vitro* antifungal activity of phenylacetic acid and sodium phenylacetate from *Streptomyces humidus*. *Appl Environ Microbiol* 67: 3739–3745.

Idriss EE, O Makarewicz, A Farouk, K Rosner, R Greiner, H Bochow, T Richter, R Borriss. 2002. Extracellular phytase activity of *Bacillus amyloliquefaciens* FZB45 contributes to its plant-growth-promoting effect. *Microbiology (UK)* 148: 2097–2109.

Islam MT, Y Hashidoko, A Deora, T Ito, S Tahara. 2005. Suppression of damping-off disease in host plants by the rhizoplane bacterium *Lysobacter* sp. strain SB-K88 is linked to plant colonization and antibiosis against soilborne peronosporomycetes. *Appl Environ Microbiol* 71: 3786–3796.

Ito S-I, T Eto, S Tanaka, N Yamauchi, H Takahara, T Ikeda. 2004. Tomatidine and lycotetraose, hydrolysis products of α-tomatine by *Fusarium oxysporum* tomatinase, suppress induced defense responses in tomato cells. *FEBS Lett* 571: 31–34.

Ivanova EG, NV Doronina, AO Shepelyakovskaya, AG Laman, FA Brovko, YA Trotsenko. 2000. Facultative and obligate aerobic methylobacteria synthesize cytokinins. *Microbiology* 69: 646–651.

Janzen RA, SB Rood, JF Dormaar, WB McGill. 1992. *Azospirillum brasilense* produces gibberellin in pure culture on chemically-defined medium and in co-culture on straw. *Soil Biol Biochem* 24: 1061–1064.

Jeandet P, R Bessis, M Sbaghi, P Meunier. 1995. Production of the phytoalexin resveratrol by grapes as a response to *Botrytis* attack under natural conditions. *J Phytopathol* 143: 135–139.

Jenks MA, HA Tuttle, SD Eigenbroide, KA Feldmann. 1995. Leaf epicuticular waxes of the *Eceriferum* mutants in *Arabidopsis*. *Plant Physiol* 108: 369–377.

Jensen S, G Eriksson, H Kylin, WMJ Strachan. 1992. Atmospheric pollution by persistent organic compounds: Monitoring with pine needles. *Chemosphere* 24: 229–245.

Jiménez-Salgado T, LE Fuentes-Ramirez, A Tapia-Hernandez, MA Mascuara-Esparza, E Martínez-Romera, J Caballero-Melado. 1997. *Coffea arabica* L. a new host plant for *Acetobacter diazotrophicus*, and isolation of other nitrogen-fixing acetobacteria. *Appl Environ Microbiol* 63: 3676–3683.

Jin M, L Liu, SA Wright, SV Beer, J Clardy. 2003. Structural and functional analysis of pantocin A: An antibiotic from *Pantoea agglomerans* discovered by heterologous expression of cloned genes. *Angew Chem Int Ed Engl* 42: 2898–2901.

Johansen JE, SJ Binnerup, N Kroer, L Mølbak. 2005. *Luteibacter rhizovicinus* gen. nov., sp nov., a yellow-pigmented gammaproteobacterium isolated from the rhizosphere of barley (*Hordeum vulgarew* L.) *Int J Syst Evol Microbiol* 55: 2285–2291.

Johnsson L, M Hökeberg, B Gerhardson. 1998. Performance of the *Pseudomonas chlororaphis* biocontrol agent MA 342 against cereal seed-borne diseases in field experiments. *Eur J Plant Pathol* 104: 710–711.

Jones AM, MC Wildermuth. 2011. The phytopathogen *Pseudomonas syringae* pv. *tomato* DC3000 has three high-affinity iron-scavenging systems functional under iron limitation conditions but dispensible for pathogenesis. *J Bacteriol* 193: 2767–2775.

Jones AM, SE LIndow, MC Wildermuth. 2007. Salicylic acid, yersiniabactin, and pyoverdin production by the model phytopathogen *Pseudomonas syringae* p. *tomato* DC3000: Synthesis, regulation, and impact on tomato and *Arabidopsis* host plants. *J Bacteriol* 189: 6773–6786.

Jouanneau Y, JC Willison, C Meyer, S Krivobok, N Chevron, J-L Bescombes, G Blake. 2005. Stimulation of pyrene mineralization in freshwater sediments by bacterial and plant augmentation. *Environ Sci Technol* 39: 5729–5735.

Jourand P, E Giraud, G Béna, A Sy, A Willems, M Gillis, B Dreyfus, P de Lajudie. 2004. *Methylobacterium nodulans* sp. nov., for a group of aerobic, facultatively methylotrophic, legume root-nodule-forming and nitrogen fixing bacteria. *Int J Syst Evol Microbiol* 54: 2269–2273.

Just CL, JL Schnoor. 2004. Phytophotolysis of hexahydro-1,3,5-trinitro-1,3,5-triazine (RDX) in leaves of reed canary-grass. *Environ Sci Technol* 38: 290–295.

Kang JW, Z Khan, SL Doty. 2012. Biodegradation of trichloroethylene by an endophyte of hybrid poplar. *Appl Environ Microbiol* 78: 3504–3507.

Kanvinde L, GKR Sastry. 1990. *Agrobacterium tumefaciens* is a diazotrophic bacterium. *Appl Environ Microbiol* 56: 2087–2092.

Kaup O, I Gräfen, E-M Zellermann, R Eichenlaub, K-H Gartemann. 2005. Identification of a tomatinase in the tomato-pathogenic actinomycete *Clavibacter michiganensis* subsp. michiganensis NCPPB382. *Mol Plant-Microbe Interact* 18: 1090–1098.

Kearns LP, HK Mahanty. 1998. Antibiotic production by *Erwinia herbicola* Eh1087: its role in inhibition of *Erwinia amylovora* and partial characterization of antibiotic biosynthesis genes. *Appl Environ Microbiol* 64: 1837–1844.

Kers JA, MJ Wach, SB Krasnoff, J Widom, KD Cameron, RA Bukhalid, DM Gibson, BR Crane, R Loria. 2004. Nitration of a peptide phytotoxin by bacterial nitric oxide synthase. *Nature* 429: 79–82.

Kim YC, J Leveau, BBMcS Gardener, EA Pierson, LS Pierson, C-M Ryu. 2011. The multifactorial basis for plant health promotion by plant-associated bacteria. *Appl Environ Microbiol* 77: 1548–1555.

King GM. 1994. Association of methanotrophs with the roots and rhizomes of aquatic vegetation. *Appl Environ Microbiol* 60: 3220–3227.

Kistler HC, HD Vanetten. 1984. Regulation of pisatin demethylation in *Nectria haematococca* and its influence on pisatin tolerance and virulence. *J Gen Microbiol* 130: 2605–2613.

Kobayashi DY, RM Reedy, JA Bick, PV Oudemans. 2002. Characterization of a chitinase gene from *Stenotrophomonas maltophilia* strain 34S1 and its involvement in biological control. *Appl Environ Microbiol* 68: 1047–1054.

Kobayashi DY, RM Reedy, JD Palumbo, J-M Zhou, GY Yuen. 2005. A *clp* gene homologue belonging to the Crp gene family globally regulates lytic enzyme production, antimicrobial activity, and biological control activity expressed by *Lysobacter enzymogenes* strain C3. *Appl Environ Microbiol* 71: 261–269.

Kodama O, J Miyakawa, T Akatsuka, S Kiyosawa. 1992. Sakurnetin, a flavanone phytoalexin from ultra-violet-irradiated rice leaves. *Agric Biol Chem* 52: 3807–3809.

Koenig RL, RO Morris, JC Polacco. 2002. tRNA is the source of low-level *trans*-zeatin production in *Methylobacterium* spp. *J Bacteriol* 184: 1832–1842.

Kolb M, H Harms. 2000. Metabolism of fluoranthene in different plant cell cultures and intact plants. *Environ Toxicol Chem* 19: 1304–1310.

Kono Y, S Takeuchi, O Kodama, H Sedido, T Akatsuka. 1985. Novel phytoalexins (oryzalexins A, B and C) isolated from rice blast leaves infected with *Pyricularia oryza*e. Part II: Structural studies of oryzalexins. *Agric Biol Chem* 49: 1695–1701.

Kuiper I, GV Bloemberg, BJJ Lugtenberg. 2001. Selection of a plant-bacterium pair as a novel tool for rhizostimulation of polycyclic aromatic hydrocarbon-degrading bacteria. *Mol Plant-Microbe Interact* 14: 1197–1205.

Kuiper I, LV Kravchenko, GV Bloemberg, BJJ Lugtenberg. 2002. *Pseudomonas putida* strain PC L1444, selected for efficient root colonization and naphthalene degradation, effectively utilizes root exudate components. *Mol Plant-Microbe Interact* 15: 734–741.

Kuiper I, EL Lagendijk, GV Bloemberg, BJJ Lugtenberg. 2004. Rhizoremediation: A beneficial plant-microbe interaction. *Mol Plant-Microbe Interact* 17: 6–15.

Kunst L, AL Samuels. 2003. Biosynthesis and secretion of plant cuticular wax. *Prog Lipid Res* 42: 51–80.

Lairini K, A Perez-Espinoza, M Pineda, M Ruiz-Rubio. 1996. Purification and characterization of tomatinase from *Fusarium oxysporium* f. sp. *lycopersici*. *Appl Environ Microbiol* 62: 1604–1609.

Lambrecht M, Y Okon, AV Broek, J Vanderleyden. 2000. Indole-3-acetic acid: A reciprocal signalling molecule in bacteria-plant interactions. *Trends Microbiol* 8: 298–300.

Langcake P. 1981. Disease resistance of *Vitis* spp. and the production of the stress metabolites resveratrol, ε-viniferin, α-vineferin and pterostilbene. *Physiol Plant Pathol* 18: 213–226.

Leigh MB, JS Fletcher, X Fu, FJ Schmitz. 2002. Root turnover: An important source of microbial substrates in rhizosphere remediation of recalcitrant metabolites. *Environ Sci Technol* 36: 1579–1583.

Leigh MB, P Prouzová, M Macková, T Macek, DP Nagle, JS Fletcher. 2006. Polychlorinated biphenyl (PCB)-degrading bacteria associated with trees in a PCB-contaminated site. *Appl Environ Microbiol* 72: 2331–2342.

Li J, BR Glick. 2001. Transcriptional regulation of the *Enterobacter cloacae* UW4 1-aminocyclopropane-1-carboxylate (ACC) deaminase gene (acdS). *Can J Microbiol* 47(9): 359–367.

Li S, CC Jochum, F Yu, K Zaleta-Rivers, L Du, SD Harris, GY Yuen. 2008. An antibiotic complex from *Lysobacter enzymogenes* strain C3: Antimicrobial activity and role in plant disease control. *Phytopathology* 98: 695–701.

Lichter A, I Barash, L Valinsky, A Manulis. 1995. The genes involved in cytokinin biosynthesis in *Erwinia herbicola* pv. *gypsophilae*: Characterization and role in gall formation. *J Bacteriol* 177: 4457–4465.

Lindow SE, C Desurmont, R Elkins, G McGourty, E Clark, MT Brandl. 1998. Occurrence of indole-3-acetic acid-producing bacteria on pear trees and their association with fruit russet. *Phytopathol* 88: 1149–1147.

Little HN. 1957. The oxidation of 2-nitropropane by extracts of pea plants. *J Biol Chem* 229: 231–238.

Loper JE, MD Henkels, BT Shaffer, FA Valeriote, H Gross. 2008. Isolation and identification of rhizoxin analogues fron *Pseudomonas fluorescens* Pf-5 using a genomic mining strategy. *Appl Environ Microbiol* 74: 3085–3083.

Loper JE, MN Schroth. 1986. Influence of bacterial sources of indole-3-acetic acid on root elongation of sugar beet. *Phytopathol* 76: 386–389.

Lovell CR, YM Piceno, JM Quatro, CE Bagwell. 2000. Molecular analysis of diazotroph diversity in the rhizosphere of the smooth cordgrass *Spartina alterniflora*. *Appl Environ Microbiol* 66: 3814–2822.

Lu Y, T Lueders, MW Friedrich, R Conrad. 2005. Detecting active methanogenic populations on rice roots using stable isotope probing. *Environ Microbiol* 7: 326–336.

Lucy MC, PS Mattews, HD Vanetten. 1988. Metabolic detoxification of the phytoalexins maackiain and medicarpin by *Nectria haematococca* field isolates: Relationship to virulence on chickpea. *Physiol Mol Plant Pathol* 33: 187–199.

Ma W, FC Guinel, BR GLick. 2003. *Rhizobium leguminosarum* biovar *viciae* 1-amino-cyclopropane-1-carboxylate deaminase promotes nodulation of pea plants. *Appl Environ Microbiol* 69: 4396–4402.

MacDonald EMS, GK Powell, DA Regier, NL Glass, F Roberto, T Kosuge, RO Morris. 1986. Secretion of zeatin, ribosylzeatin, and ribosyl-1″-methylzeatin by *Pseudomonas savastanoi*. *Plant Physiol* 82: 742–747.

Maddula, VSRK, EA Pierson, LS Pierson. 2008. Altering the ratio of phenazines in *Pseudomonas chlororaphis* (*aureofaciens*) strain 30–84: Effects on biofilm formation and pathogen inhibition. *J Bacteriol* 190: 2759–2766.

Mansouri S, AW Bunch. 1989. Bacterial synthesis from 2-oxo-4-thiobutyric acid and from methionine. *J Gen Microbiol* 135: 2819–2827.

Mahadevan B, DL Crawford. 1996. Purification of chitinase from the biocontrol agent *Streptomyces lydicus* WYEC108. *Enzyme Microb Technol* 20: 489–493.

Makarewicz O, S Dubrac, T Masdek, R Borriss. 2006. Dual role of the PhoP ~ P response regulator: *Bacillus amyloliquefaciens* FZB45 phytase gene transcription is directed by positive and negative interactions with the *phyC* promoter. *J Bacteriol* 188: 6953–6965.

Martínez-Granero F, S Capdevila, M Sánchez-Contreras, M Martín, R Rivilla. 2005. Two site-specific recombinases are implicated in phenotypic variation and competitive rhizosphere colonization in *Pseudomonas fluorescens*. *Microbiology (UK)* 151: 975–983.

Martínez-Granero F, R Rivilla, M Martín. 2006. Rhizosphere selection of highly motile phenotypic variants of *Pseudomonas fluorescens* with enhanced competitive colonization ability. *Appl Environ Microbiol* 72: 3429–3434.

Maurhofer M, E Baehler, R Nitz, V Martinez, C Keel. 2004. Cross talk between 2,4-diacetyl-phloroglucinol-producing biocontrol pseudomonads on wheat roots. *Appl Environ Microbiol* 70: 1990–1998.

Mavrodi DV, RF Bonall, SMK Delaney, MJ Soule, G Phillips, LS Thomashow. 2001. Functional analysis of genes for biosynthesis of pyocyanin and phenazine-1-carboxamide from *Pseudomonas aeruginosa* PAO1. *J Bacteriol* 183: 6454–6465.

Mavrodi DV, VN Ksenzenko, RF Bonsall, RJ Cook, AM Borodin, LS Thosashow. 1998. A seven-gene locus for synthesis of phenazine-1-carboxylic acid by *Pseudomonas fluorescens* 2–79. *J Bacteriol* 180: 2541–2548.

McLoughlin TJ, JP Quinn, A Bettermann, R Bookland. 1992. *Pseudomonas cepacia* suppression of sunflower wilt fungus and role of antifungal compounds in controlling disease. *Appl Environ Microbiol* 58: 1760–1763.

Mehnaz S, G Lazarovits. 2006. Inoculation effects of *Pseudomonas putida*, *Gluconacetobacter azocaptans*, and *Azospirillum lipoferum* on corn plant growth under greenhouse conditions. *Microb Ecol* 51: 326–335.

Mehnaz S, B Weselowski, H Lazarovits. 2006. Isolation and identification of *Gluconobacter azocaptans* from corn rhizosphere. *Syst Appl Microbiol* 29: 496–501.

Miao VPW, HD Vanetten. 1992. Genetic analysis of the role of phytoalexin detoxification in virulence of the fungus *Nectria haematococca* on chickpea (*Cicer arietinum*). *Appl Environ Microbiol* 58: 809–814.

Miethke M, MA Marahiel. 2007. Siderophore-based iron acquisition and pathogen control. *Microbiol Mol Biol Revs* 71: 413–451.

Miller AA, S Clemens, S Zachgo, EM Giblin,DC Taylor, L Kunst. 1999. *CUT1*, and *Arabidopsis* gene required for cuticular wax biosynthesis and pollen fertility, encodes a very-long-chain fatty acid condensing enzyme. *The Plant Cell* 11: 825–838.

Miller RL, VJ Higgins. 1970. Association of cyanide with infection of birdsfoot trefoil by *Stemphyllium loti*. *Phytopathology* 60: 104–110.

Mirleau P, R Wogelius, A Smith, MA Kertesz. 2005. Importance of organosulfur utilization for survival of *Pseudomonas putida* in soil and rhizosphere. *Appl Environ Microbiol* 71: 6571–6577.

Mitchell AM, GA Strobel, E Moore, R Robison, J Sears. 2010. Volatile antimicrobials from *Muscodor crispans*, a novel endophtic fungus. *Microbiology (UK)* 156: 270–277.

Molina L, C Ramos, E Duque, MC Ronchel, JM García, L Wyke, JL Ramos. 2000. Survival of *Pseudomonas putida* KT2440 in soil and in the rhizosphere of plants under greenhouse and environmental conditions. *Soil Biol Biochem* 32: 315–321.

Morrissey JP, AE Osbourn. 1999. Fungal resistance to plant antibiotics as a mechanism of pathogenesis. *Microbiol Mol Biol Revs* 63: 708–724.

Nakayama T, Y Homma, Y Hashidoko, J Mizutani, S Tahara. 1999. Possible role of xanthobaccins produced by *Stenotrophomonas* sp. strain SB-K88 in suppression of sugar beet damping-off disease. *Appl Environ Microbiol* 65: 4334–4339.

Nelson MJK, SO Montgomery, WR Mahaffey, PH Prichard. 1987. Biodegradation of trichloroethylene and involvement of an aromatic biodegradative pathway. *Appl Environ Microbiol* 53: 949–954.

Newman LA et al. 1999. Remediation of trichloroethylene in an artificial aquifer with trees: A controlled field study. *Environ Sci Technol* 33: 2257–2265.

Newman LA, SE Strand, N Choe, J Duffy, G Ekuan, M Ruiszai, BB Shurtleff, J Wilmoth, MP Gordon. 1997. Uptake and biotransformation of trichloroethylene by hybrid poplars. *Environ Sci Technol* 31: 1062–1067.

Niemeyer HM. 1988. Hydroxamic acids (4-hydroxy-1,4-benzoxazin-3-ones), defence chemicals in the Gramineae. *Phytochemistry* 27: 3349–3358.

Notz R, M Maurhofer, H Dubach, D Haas, G Défago. 2002. Fusaric acid-producing strains of *Fusarium oxysporum* alter 2,4-diacetylphloroglucinol biosynthetic gene expression in *Pseudomonas fluorescens* CHA0 and in the rhizosphere of wheat. *Appl Environ Microbiol* 68: 2229–2235.

Oide S, W Moeder, S Krasnoff, D Gibson, H Haas, K Yoshioka, BG Turgeon. 2006. *NPS6*, encoding a nonribosomal peptide synthetase involved in siderophore-mediated iron metabolism, is a conserved virulence determinant of plant pathogenic ascomycetes. *Plant Cell* 18: 2836–2953.

Onofre-Lemus J, I Hernández-Lucas, L Girard, J Caballero-Mellado. 2009. ACC (1-amincyclopropane-1-carboxylate) deaminase activity, a widespread trait in *Burkholderia* species, and its growth-promoting effect on tomato plants. *Appl Environ Microbiol* 6581–6590.

Osbourn A, P Bowyer, P Lunness, B Clarke, M Daniels. 1995. Fungal pathogens of oat roots and tomato leaves employ closely related enzymes to detoxify different host plant pathogens. *Mol Plant-Microbe Interact* 8: 971–978.

Palumbo JD, RF Sullivan, DY Kobayashi. 2003. Molecular characterization and expression in *Escherichia coli* of three β-1,3-glucanase genes from *Lysobacter enzymogenes* strain N4-7. *J Bacteriol* 185: 4362–4370.

Papadoupoulou K, RE Melton, M Leggett, MJ Daniels, AE Osbourn. 1999. Compromised disease resistance in saponin-deficient plants. *Proc Natl Acad Sci USA* 96: 12923–12928.

Pareja-Jaime Y, MIG Roncero, MC Ruiz-Roldán. 2008. Tomatinase from *Fusarium oxysporum*f. sp. *lycopersici* is required for full virulence on tomato plants. *Mol Plant-Microbe Interact* 21: 728–736.

Pandey A, RV Sonti. 2010. Role of the FeoB protein and siderophore in promoting virulence of *Xanthomonas oryzae* pv. *oryzae* on rice. *J Bacteriol* 192: 3187–3203.

Patten CL, BR Glick. 2002. Role of *Pseudomonas putida* indoleacetic acid in development of the host plant root system. *Appl Environ Microbiol* 68: 3795–3801.

Pavlostathis SG, KK Comstock, ME Jacobson, FM Saunders. 1998. Transformation of 2,4,6-trinitrotoluene by the aquatic plant *Myriophyllum spicatum*. *Environ Toxicol Chem* 17: 2266–2273.

Pedras MC, AQ Khan. 1996. Biotransformation of the *Brassica* phytoalexin brassicanal A by the blackleg fungus. *J Agric Food Chem* 44: 303–3407.

Pedras MS, AQ Khan. 2000. Biotransformation of the phytoalexin camalexin by the phytopathogen *Rhizoctonia solani*. *Phytochemistry* 53: 59–69.

Pedras MS, M Jha, OG Okeola. 2005. Camalexin induces detoxification of the phytoalexin brassinin in the plant pathogen *Leptosphaeria maculans*. *Phytochemistry* 66: 2609–2616.

Pedras MSC, FI Okanga. 1999. Strategies of cruciferous pathogenic fungi: Detoxification of the phytoalexin cyclobrassin by mimicry. *J Agric Food Chem* 47: 1196–1202.

Pedras MSC, FI Okanga, IL Zaharia, AQ Khan. 2000. Phytoalexins from crucifers: Synthesis, biosynthesis, and biotransformation. *Phytochemistry* 53: 161–176.

Pedras MS, JL Taylor. 1993. metabolism of the phytoalexin brassinin by the "blackleg" fungus. *J Nat Prod* 56: 731–728.

Pegg AE. 1986. Recent advances in the biochemistry of polyamines in eukaryotes. *Biochem J* 234: 249–262.

Perin L, L Martínez-Aguilar, R Castro-González, P Estrada-de los Santos, T Cabellos-Avelar, HV Guedes, VM Reis, J Caballero-Mallado. 2006. Diazotrophic *Burkholderia* species associated with field-grown maize and sugarcane. *Appl Environ Microbiol* 72: 3103–3110.

Peters RJ. 2006. Uncovering the complex metabolic network underlying diterpenoid phytoalexin biosynthesis in rice and other cereal crop plants. *Phytochemistry* 67: 2307–2317.

Phillips LA, JJ Germida, RE Farrerll, CW Greer. 2008. Hydrocarbon degradation potential and activity of endophytic bacteria associated with prairie plants. *Soil Biol Biochem* 40: 3054–3064.

Post-Beitenmiller D. 1996. Biochemistry and molecular biology of wax production in plants. *Annu Rev Plant Physiol Plant Mol Biol* 47: 405–430.

Prapagdee B, C Kuekulvong, S Mongkolsuk. 2008. Antifungal potential of extracellular metabolite produced by *Streptomyces hygroscopicus* against phytopathogenic fungi. *Int J Biol Sci* 4: 330–338.

Prusky D, I Kobiler, Y Fishman, JJ Sims, SL Midland, NT Keen. 1991a. Identification of an antifungal compound in unripe avocado fruits and its possible involvement in the quiescent infections of *Colleotrichum gloeosporioides*. *J Phytopath* 132: 319–327.

Prusky D, I Kobiler, B Jacoby, JJ Sims, SL Maitland. 1985. Inhibitors of avocado lipoxygenase; their possible relationship with the latency of *Colletotrichum gloeosporoides*. *Physiol Mol Plant Pathol* 27: 269–279.

Prusky DI, L Karni, I Kobiler, RA Plumpley. 1990. Induction of the antifiugal diene in unripe avocado fruits: Effect of inoculation with *Colleotrichum gloeosporioides*. *Physiol Mol Plant Path* 37: 425–435.

Prusky D, RA Plumley, I Kobiler. 1991b. The relationship between antifungal diene levels and fungal inhibition during quiescent infection of unripe avocado fruits with *Colletotrichum gloeosporioides*. *Plant Pathol* 40: 45–52.

Puente ME, C Y Li, Y Bashan. 2009a. Rock-degrading endophytic bacteria in cacti. *Environ Exp Bot* 66: 389–401.

Puente ME, C Y Li, Y Bashan. 2009b. Endophytic bacteria in cacti seeds can improve the development of cactus seedlings. *Environ Exp Bot* 66: 402–408.

Quidde T, AE Osbourn, P Tudzynski. 1998. Detoxificatgion of α-tomatine by *Botrytis cinerea*. *Physiol Mol Plant Pathol* 52: 151–165.

Rao RP, A Hunter, O Kashpur, J Normanly. 2010. Aberrant synthesis of indole-3-acetic acid in *Saccharomyces cerevisiae* triggers morphogenic transition, a virulence trait of pathogenic fungi. *Genetics* 185: 211–220.

Reinhold-Hurek B, T Hurek. 2000. Reassessment of the taxonomic structure of the diazotophic genus *Azoarcus sensu lato* and description of three new genera and new species, *Azovibrio restrictus* gen. nov., sp. nov., *Azospira oryzae* gen. nov., sp. nov. and *Azonexus fungiphilus* gen nov., sp. nov. *Int J Syst Evolut Microbiol* 50: 649–649.

Reis VM et al. 2004. *Burkholderia tropica* sp. nov., a novel nitrogen-fixing, plant-associated bacterium. *Int J Syst Evol Microbiol* 54: 2155–2162.

Revelles O, M Espinosa-Urgel, T Fuhrer, U Sauer, JL Ramos. 2005. Multiple and interconnected pathways for l-lysine catabolism in *Pseudomonas putida* KT2440. *J Bacteriol* 187: 7500–7510.

Revelles O, M Espinosa-Urgel, S Molin, JL Ramos. 2004. The *davDT* operon of *Pseudomonas putida*, involved in lysine catabolism, is induced in response to the pathway intermediate δ-aminovaleric acid. *J Bacteriol* 186: 3439–3446.

Rezek, T Macek, M Mackova, J Triska. 2007. Plant metabolites of polychlorinated biphenyls in hairy root culture of black nightshade *Solanum nigrum* SNC-90. *Chemosphere* 69: 1221–1227.

Rezek, T Macek, M Mackova, J Triska, K Ruzickova. 2008. Hydroxy-PCBs, methoxy-PCBs and hydroxy-methoxy-PCBs: Metabolites of polychlorinated biphenyls formed *in vitro* by tobacco cells. *Environ Sci Technol* 42: 5746–5751.

Rich JJ, GM King. 1998. Carbon monoxide oxidation by bacteria associated with the roots of freshwater macrophytes. *Appl Environ Microbiol* 64: 4939–4943.

Richardson MD, CW Bacon. 1995. Catabolism of 6-methoxy-benzoxazolinone and 2- benzoxazolinone by *Fusarium moniliforme*. *Mycologia* 87: 510–517.

Robinson M, J Riov, A Sharon. 1998. Indole-3-acetic acid biosynthesis in *Colletotrichum gloeosporioides* f. sp. *aeschynomene. Appl Environ Microbiol* 64: 5030–5031.

Rosenblueth M, E Martínez-Romero. 2006. Bacterial endophytes and their interactions with hosts. *Mol Plant-Microbe Inter* 19: 827–837.

Roy S, O Hänninen. 1994. Pentachlorophenol: Uptake/elimination kinetics and metabolism in an aquatic plant *Eichhornia crassipes. Environ Toxicol Chem* 13: 763–773.

Ryu C-M, MA Farag, C-H Hu, MS Reddy, JW Kloepper, PW Paré. 2004. Bacterial volatiles induce systemic resistance in *Arabidopis. Plant Physiol* 134: 1017–1026.

Sabaratnam S, JA Traquair. 2002. Formulation of a *Streptomyces* biocontrol agent for the suppression of *Rhizoctonia* damping-off in tomato transplants. *Biol Control* 23: 245–253.

Sahai PN, AC Chibnall. 1932. Wax metabolism in the leaves of Brussels sprouts. *Biochem J* 26: 403–412.

Sánchez-Contreras M, M Martín, M Villacieros, F O'Gara, I Bonilla, R Rivalla. 2002. Phenotropic selection and phase variation occur during alfalfa root colonization by *Pseudomonas fluorescens* F113. *J Bacteriol* 184: 1587–1596.

Sandermann H. 1994. Higher plant metabolism of xenobiotics: The "green liver" concept. *Pharmacogenetics* 4: 225–241.

Sandrock RW, HD Vanetten. 2001. The relevance of tomatinase activity in pathogens of tomato; disruption of the b_2-tomatinase gene in *Colletotrichum coccodes* and *Septoria lycopersici* and heterologous expression of the *Septoria lycopersici* b_2-tomatinase in *Nectria haematococca*, a pathogen of tomato fruit. *Physiol Mol Plant Pathol* 58: 159–171.

Santos A, D Marquina. 2004. Killer toxin of *Pichia membranifaciens* and its possible use as a biocontrol agent against grey mould disease of grapevine. *Microbiology UK* 150: 2527 –2534.

Santos A, M San Mauro, E Bravo, D Marquina. 2009. PMKT2, a new killer toxin from *Pichia membranifaciens*, and its promising biotechnological properties for control of the spoilage yeast *Brettanomyces bruxellensis. Microbiology UK* 155: 624–634.

Sarand I, S Timonen, E-L Nurmiaho-Lassila, T Koivula, K Haatela, M Romantschuk, R Sen. 1998. Microbial biofilms and catabolic plasmid harbouring degradative fluorescent pseudomonads in Scots pine mycorrhizospheres developed on petroleum contaminated soil. *FEMS Microbiol Ecol* 27: 115–126.

Sarig P, Y Zutkhi, A Monjauze, N Lisker, R BenArie. 1997. Phytoalexin elicitation in grape berries and their susceptibility to *Rhizopus stolonifer. Physiol Mol Plant Pathol* 50: 337–347.

Saunders M, Lm Kohn. 2008. Host-synthesized secondary compounds influence the *in vitro* interactions between fungal endophytes of maize. *Appl Environ Microbiol* 74: 136–142.

Sbagi M, P Jeandet, R Beiss, P Leroux. 1996. Degradation of stilbene-like phytoalexins in relationship to the pathogenicity of *Botrytis cinerea* to grapevine. *Plant Pathol* 45: 139–144.

Schäfer W, H Sandermann. 1988. Metabolism of pentachlorophenol in cell suspension cultures of wheat (*Triticum aestivum* L.). Tetrachlorocatechol as a primary product. *J Agric Food Chem* 36: 370–377.

Schneider-Belhadded F, P Kolattukudy. 2000. Solubilization, partial purification, and characterization of a fatty aldehyde decarbonylase from a higher plant, *Pisum sativum. Arch Biochem Biophys* 377: 341–349.

Scholz-Schroeder BK, ML Hutchison, I Grgurina, DC Gross. 2001. The contribution of syringopeptin and syringomycin to virulence of *Pseudomonas syringae* pv. *syringae* strain B301D on the basis of *sypA* and *syrB1* biosynthesis mutants. *Mol Plant-Microbe Interact* 14: 336–348.

Schouten A, G van den Berg, C Edel-Hermann, C Steinberg, N Gautheron, C Alabouvette, CH de Vos, P Lemanceau, JM Raaijmakers. 2004. Defense responses of *Fusarium oxysporum* to 2,4-diacetylphloroglucinol, a broad-spectrum antibiotic produced by *Pseudomonas fluorescens. Mol Plant-Microbe Interact* 17: 1201–1211.

Seipke RF, R Loria. 2008. *Streptomyces scabies* 87-22 possesses a functional tomatinase. *J Bacteriol* 190: 7684–7692.

Sevilla M, RH Burris, N Gunapala, C Kennedy. 2001. Comparison of benefit to sugarcane plant growth and $^{15}N_2$ incorporation following inoculation of sterile plants with *Acetobacter diazotrophicus* wild-type and Nif⁻ mutant strains. *Mol Plant-Microbe Interact* 14: 358–366.

Sheng X, X Chen, L He. 2008. Characteristics of an endophytic pyrene-degrading bacterium of *Enterobacter* sp. 12J1 from *Allium macrostemon* Bunge. *Int Biodet Biodeg* 62: 88–95.

Sheehy RE, M Honma, M Yameda, T Sasaki, B Martineau, WR Hiatt. 1991. Isolation, sequence, and expression in *Escherichia coli* of the *Pseudomonas* sp. strain ACP gene encoding 1-aminocyclopropane-1-carboxylate deaminase. *J Bacteriol* 173: 5260–5265.

Shimizu M, T Kimura, T Koyama, K Suzuki, N Ogawa, K Miyashita, K Dakka, K Ohmiya. 2002. Molecular breeding of transgenic rice plants expressing a bacterial chlorocatechol dioxygenase gene. *Appl Environ Microbiol* 68: 4061–4066.

Siciliano SD, JJ Germida. 1999. Enhanced phytoremediation of chlorobenzoates in rhizosphere soil. *Soil Biol Biochem* 31: 299–305.

Siciliano SD, H Holdie, JJ Germida. 1998. Enzymatic activity in root exudates of Dahurian wild rye (*Elymus daurica*) that degrades 2-chlorobenzoic acid. *J Agric Food Chem* 46: 5–7.

Siciliano SD et al. 2001. Selection of specific endophytic bacterial genotypes by plants in response to soil contamination. *Appl Environ Microbiol* 67: 2469–2475.

Siddiqui IA, D Haas, S Heb. 2005. Extracellular protease of *Pseudomonas fluorescens* CHA0, a biocontrol factor with activity against the root-knot nematode *Meloidogyne incognita*. *Appl Environ Microbiol* 71: 5646–5649.

Simon HM, CE Jahn, LT Bergerud, MK Sliwinski, PJ Weimer, DK Willis, RM Goodman. 2005. Cultivation of mesophilic soil crenarchaeotes in enrichment cultures from plant roots. *Appl Environ Microbiol* 71: 4751–4760.

Singer AC, DE Crowly, IP Thompson. 2003. Secondary plant metabolites in phytoremediation and biotranformation. *Trends Biotechnol* 21: 123–130.

Smith DA, JM Harrer, TE Cleveland. 1982. Relation between production of extracellular kievitone hydratase by isolates of *Fusarium* and their pathogenicity on *Phaseolus vulgaris*. *Phytopathol* 72: 1319–1323.

Snyder BA, RL Nicholson. 1990. Synthesis of phytoalexins in sorghum as a site-specific response to fungal ingress. *Science* 97: 1637–1639.

Snyder BA, B Leite, J Hipkind, LG Butler, RL Nicholson. 1991. Accumulation of sorghum phytoalexins by *Colletotrichum graminicola* at the infection site. *Physiol Mol Plant Pathol* 39: 463–470.

Somers E, D Ptacek, P Gysegom, M Srinivasan, V Vanderleyden. 2005. *Azospirillum brasilense* produces the auxin-like phenylacetic acid by using the key enzyme for indole-3-acetic acid biosynthesis. *Appl Environ Microbiol* 71: 1803–1810.

Strobel GA, E Dirkse, J Sears, C Markworth. 2001. Volatile antimicrobials from *Muscodor albus*, a novel endophytic fungus. *Microbiology (UK)* 147: 2943–2950.

Strobel G, J-Y Li, F Sugawara, H Koshino, J Harper, WM Hess. 1999. Oocydin A, a chlorinated macrocylic with potent anti-oomycete activity from *Serratia marcescens*. *Microbiology (UK)* 145: 3557–3564.

Suttinun O, R Müller, E Luepromchai. 2009. Trichloroethylene cometabolic degradation by *Rhodococcus* sp. L4 with plant essential oils. *Biodegradation* 20: 281–291.

Suttinun O, R Müller, E Luepromchai. 2010. Cometabolic degradation of trichloroethene by *Rhodococcus* sp. strain L4 immobilized on plant materials rich in essential oils. *Appl Environ Microbiol* 76: 4684–4690.

Sutton AE, J Clardy. 2000. The total synthesis of pantocin B. *Org Lett* 2: 319–321.

Sylvestre M. 1980. Isolation method for bacterial isolates capable of growth on *p*-chlorobiphenyl. *Appl Environ Microbiol* 39: 1223–1224.

Taghavi S, T Barac, B Greenberg, B Borremans, J Vangronsveld, D van der Lelie. 2005. Horizontal gene transfer to endogenous endophytic bacteria from poplar trees improves phytoremediation of toluene. *Appl Environ Microbiol* 71: 8500–8505.

Taghavi S, C Garafola, S Monchy, L Newman, A Hoffman, N Weyens, T Barac, J Vangronsveld, D van der Lelie. 2009. Genome survey and characterization of endophytic bacteria exhibiting a beneficial effect on growth and development of poplar trees. *Appl Environ Microbiol* 75: 748–757.

Tanaka E, C Tanaka, A Ishihara, Y Kuwahara, M Tsuda. 2002. Indole-3-acetic acid biosynthesis in *Aciculosporium take*, a causal agent of witches' broom of bamboo. *J Gen Plant Pathol* 69: 1–6.

Theunis M, H Kobayashi, WJ Broughton, E Prinsen. 2004. Flavonoids, NodD1, NodD2, and *Nod*-Box NB15 modulate expression of the y4wEFG locus that is required for indole-3-acetic acid synthesis in *Rhizobium* sp. strain NGR234. *Mol Plant-Microbe Interact* 17: 1153–1161.

Thompson PL, LA Ramer, JL Schnoor. 1998. Uptake and transformation of TNT by hybrid poplar trees. *Environ Sci Technol* 32: 975–980.

Thomzik JE, K Stenzel, R Stocker, PH Schreier, R Hain, DJ Stahl. 1997. Synthesis of a grapevine phytoalexin in transgenic tomatoes (*Lycopersicum esculentum* Mill.) conditions resistance against *Phytophthora infestans*. *Physiol Mol Plant Pathol* 51: 265–278.

Thuler DS, EIS Floh, W Handro, HR Barbosa. 2003. Plant growth regulators and amino acids released by *Azospirillum* sp. in chemically defined media. *Lett Appl Microbiol* 37: 174–178.

Timmusk S, B Nicander, U Granhall, E Tillberg. 1999. Cytokinin production by *Paenibacillus polymyxa*. *Soil Biol Biochem* 31: 1847–1852.

Timonen S, KS Jörgensen, K Haahtela, R Sen. 1998. Bacterial community structure at defined locations of *Pinus sylvestris–Suillus bovinus* and *Pinus sylvestris–Paxillus involutus* mycorrhizospheres in dry pine forest humus and nursery peat. *Can J Microbiol* 44: 499–513.

Tjeerd van Rij E, G Girard, BJJ Lugtenberg, GV Bloemberg. 2005. Influence of fusaric acid on phenazine-1-carboxamide synthesis and gene expression of *Pseudomonas chlororaphis* strain PCL1391. *Microbiology (UK)*: 151: 2805–2814.

Tjeerd van Rij E, M Wesselink, TFC Chin-A-Woeng, GV Bloemberg, BJJ Lugtenberg. 2004. Influence of environmental conditions on the production of phenazine-1-carboxamide by *Pseudomonas chlororaphis* PCL 1391. *Mol Plant-Microbe Interact* 17: 557–566.

Tokala RK, JL Strap, CM Jung, DL Crawford, MH Salove, LE Deobald, JF Bailey, MJ Morra. 2002. Novel plant-microbe rhizosphere interaction involving *Streptomyces lydicus* WYEC108 and the pea plant (*Pisum sativum*). *Appl Environ Microbiol* 68: 2161–2171.

Trân Van V, O Berge, SN Ke, J Balandreau, T Heulin. 2000. Repeated beneficial effects of rice inoculation with a strain of *Burkholderia vietnamiensis* on early and late yield components in low fertility sulphate acid soils of Vietnam. *Plant Soil* 218: 273–284.

Turner BL, MJ Papházy, PM Hatgarth, ID McKelvie. 2002. Inositol phosphates in the environment. *Philos Trans Roy Soc London* B 357: 449–469.

Uchida E, T Ouchi, Y Suzuki, T Yoshida, H Habe, I Yamaguchi, T Omori, H Nojiri. 2005. Secretion of bacterial xenobiotic-degrading enzymes from transgenic plants by an apoplastic expression system: Applicability for phytoremediation. *Environ Sci Technol* 39: 7671–7677.

Ugrekhelidze D, F Korte, G Kvesitadze. 1997. Uptake and transformation of benzene and toluene by plant leaves. *Ecotoxicol Environ Saf* 37: 24–29.

Uhlik O, M Strejcek, P Junkova, M Sanda, M Hroludova, C Vlcek, M Mackova, T Macek. 2011. Matrix-assisted laser desorption ionization (MALDI)-time of flight mass spectrometry- and MALDI biotyper-based identification of cultured biphenyl-metabolizing bacteria from contaminated horseradish rhizosphere soil. *Appl Environ Microbiol* 77: 6858–6866.

Umemura K, N Ogawa, M Shimura, J Koga, H Usami, T Kono. 2003. Possible role of phytocassane, rice phytoalexin, in disease resistance again to the blast fungus *Magnaporthe grisea*. *Biosci Biotechnol Biochem* 67: 899–902.

Umemura K, N Ogawa, T Yamauchi, M Ikata, M Shimura, J Koga. 2000. Cerebroside elicitors found in divers phytopathogens activate defence responses in rice plants. *Plant Cell Physiol* 41: 676–683.

Utsumi R et al. 1991. Molecular cloning and characterization of the fusaric acid-resistance gene from *Pseudomonas cepacia*. *Agric Biol Chem* 55: 1913–1918.

Valdés M, N-O Pérez, P Estrada-de los Santos, J Caballero-Mellado, JJ Peña-Cabriales, P Normand, AM Hirsch. 2005. Non-*Frankia* actinomycetes isolated from surface-sterilized roots of *Casuarina equisetifolia* fix nitrogen. *Appl Environ Microbiol* 71: 460–466.

van Aken B, CM Peres, SL Doty, YM Yoon, JL Schnoor. 2004a. *Methylobacterium populi* sp. nov., a novel aerobic, pink-pigmented, facultatively methylotrophic, methane-utilizing bacterium isolated from poplar trees (*Populus deltoides × nigra* DN34). *Int J Syst Evol Microbiol* 54: 1191–1196.

van Aken B, JM Yoon, CL Just, JL Schnoor. 2004b. Metabolism and mineralization of hexahydro-1,3,5-trinitro-1,3,5-triazine inside poplar tissues (*Populus deltoides × nigra* DN-34). *Environ Sci Technol* 38: 4572–4579.

van den Broek D, TFC Chin-A-Wong, GV Bloemberg, BJJ Lugtenberg. 2005. Molecular nature of spontaneous modifications in *gacS* which cause colony phase variation in *Pseudomonas* sp. strain PCL1171. *J Bacteriol* 187: 593–600.

Vandeputte O, S Öden, A Mol, D Vereecke, K Goethals, M El Jaziri, E Prinsen. 2005. Biosynthesis of auxin by the Gram-positive phytopathogen *Rhodococcus fascians* is controlled by compounds specific to infected tissues. *Appl Environ Microbiol* 71: 1169–1177.

Vanneste JL, J Yu, SV Beer. 1992. Role of antibiotic production by *Erwinia herbicola* Eh351 in biological control of *Erwinia amylovora*. *J Bacteriol* 174: 2785–2796.

Vílchez S, L Molina, C Ramos, JL Ramos. 2000. Proline catabolism by *Pseudomonas putida*: Cloning, characterization, and expression of the *put* genes in the presence of root exudates. *J Bacteriol* 182: 91–99.

Vilich V, B Löhndorf, RA Silora, A Friebe. 1999. Metabolism of benzoxazolinone allelochemicals of *Zea mays* by *Fusarium subglutinans. Mycol Res* 103: 1529–1532.

Villacieros M et al. 2005. Polychlorinated biphenyl rhizoremediation by *Pseudomonas fluorescens* F113 derivatives, using a *Sinorhizobium meliloti nod* system to drive *bph* gene expresssion. *Appl Environ Microbiol* 71: 2687–2694.

Vinale F, K Sivasithamparam, EL Ghisalberti, R Marra, MJ Barbetti, H Li, SL Woo, M Lorito. 2008. A novel role for *Trichoderma* secondary metabolites in the interactions with plants. *Physiol Mol Plant Pathol* 72: 80–86.

Voisard C, C Keel, D Haas, G Défago. 1989. Cyanide production by *Pseudomonas fluorescens* helps suppress black root rot of tobacco under gnotobiotic conditions. *EMBO Journal* 8: 351–358.

Wang X, Z Xhi, L Yue, J Li, L Wu. 2007. A marine killer yeast against the pathogenic yeast strain in crab (*Portunus trituberculatus*) and an optimization of the toxic production. *Microbiol Res* 162: 77–85.

Weyens N, D van der Lelie, S Taghavi, J Vangronsveld. 2009. Phytoremediation: Plant-endophyte partnerships take the challenge. *Curr Opin Biotechnol* 20: 248–254.

Walton BT, TA Anderson. 1990. Microbial degradation of trichloroethylene in the rhizosphere: Potential application to biological remediation of waste sites. *Appl Environ Microbiol* 56: 1012–1016.

Wang K, K Conn, G Lazarovits. 2006. Involvement of quinolinate phosphoribosyl transferase in promotion of potato growth by a *Burkholderia* strain. *Appl Environ Microbiol* 72: 760–768.

Wang P, HD VanEtten. 1992. Cloning and properties of a cyanide hydratase gene from the phytopathogenic fungus *Gloeocercospora sorghi. Biochem Biophys Res Comm* 187: 1048–1054.

Weber OB, VLD Baldani, KRS Teixeira, G Kirchhof, JI Baldani, J Döbereiner. 1999. Isolation and characterization of diazotrophic bacteria from banana and pineapple plants. *Plant Soil* 210: 103–113.

Weller DM. 2007. *Pseudomonas* biocontrol agents of soilborne pathogens: Looking back on 30 years. *Phytopath* 97: 250–256.

Weller SA, DE Stead. 2002. Detection of root mat associated *Agrobacterium* strains from plant material and other samples by post-enrichment TaqMan PCR. *J Appl Microbiol* 92: 118–126.

Weltring K-M, J Wessels, GF Pauli. 1998. Metabolism of the tomato saponin α-tomatine by *Gibberella pulicaris. Phytochemistry* 48: 1321–1328.

Wensing A, SD Braun, P Büttner, D Expert, B Völksch, MS Ullrich, H Weinart. 2010. Impact of siderophore production by *Pseudomonas syringae* pv. *syringae* 22d/93 on epiphytic fitness and biocontrol activity against *Pseudomonas syringae* pv. *glycinea* 1a/96. *Appl Environ Microbiol* 76: 2704–2711.

Widdowson MA, S Shearer, RK Andersen, JT Novak. 2005. Remediation of polycyclic aromatic hydrocarbon compounds in groundwater using poplar trees. *Environ Sci Technol* 39: 1598–1605.

Wightman Frank, DL Lighty. 1982. Identification of phenylacetic acid as a natural auxin in the shoots of higher plants. *Physiologia Plantarum* 55: 17–24.

Wilken A, C Bock, M Bokern, H Harms. 1995. Metabolism of PCB congeners in plant cell cultures. *Environ Toxicol Chem* 14: 2017–2022.

Wilson M, HAS Epton, DC Sigee. 1990. Biological control of fire blight of hawthorn (*Crataegus monogyna*) with *Erwinia herbicola* under protected conditions. *Plant Path* 39: 301–308.

Wright SAI, CH Zumhoff, L Schneider, SV Beer. 2001. *Pantoea agglomerans* strain EH318 produces two antibiotics that inhibit *Erwinia amylovora* in vitro. *Appl Environ Microbiol* 67: 284–292.

Yan Y et al. 2008. Nitrogen fixation island and rhizosphere competence traits in the genome of root-associated *Pseudomonas stutzeri* A1501. *Proc Natl Scad Sci USA* 105: 7564–7569.

Yao J, C Allen. 2006. Chemotaxis is required for virulence and competitive fitness of the bacterial wilt pathogen *Ralstonia solanacearum. J Bacteriol* 188: 3697–3708.

Yedidia I, N Benhamou, I Chet. 1999. Induction of defense responses in cucumber plants (*Cucumis sativus* L.) by the biocontrol agent *Trichoderma harzianum. Appl Environ Microbiol* 65: 1061–1070.

Yedidia I, M Shoresh, Z Kerem, N Benhanou, Y Kapulnik, I Chet. 2003. Concomitant induction of systemic resistance to *Pseudomonas syringae* pv. *lachrymans* in cucumber by *Trichoderma asperellum* (T-203) and accumulation of phytoalexins. *Appl Environ Microbiol* 69: 7343–7353.

Yee DC, JA Maynard, TK Wood. 1998. Rhizoremediation of trichloroethylene by a recombinant, root-colonizing *Pseudomonas fluorescens* strain expressing *ortho*-monooxygenase constitutively. *Appl Environ Microbiol* 64: 112–118.

Yuan WM, DL Crawford. 1995. Characterization of *Streptomyces lydicus* as a potential biocontrol agent against fungal root and seed rots. *Appl Environ Microbiol* 61: 3119–3128.

Yu F, K Zaleta-Rivera, X Zhu, J Huffman, JC Millet, SD Harris, G Yuen, X-C LI, L Du. 2007. Structure and biosynthesis of heat-stable antifungal factor (HSAF), a broad-spectrum antimycotic with a novel mode of action. *Antimicrobial Agents Chemother* 51: 64–72.

Yue Q, CW Bacon, MD Richardson. 1998. Biotransformation of 2-benzoxazolinone and 6-methoxy-benzoxazolinone by *Fusarium moniliforme*. *Phytochemistry* 48: 451–454.

Yuhashi KI, N Ichikawam H Ezura, S Akao, Y Minakawa, N Nukui, T Yasuta, K Minamisawa. 2000. Rhizobitoxine production by *Bradyrhizobium elkanii* enhances nodulation and competitiveness on *Macroptilium atropurpureum*. *Appl Environ Microbiol* 66: 2658–2663.

Zamani M, AS Tehrani, M Ahmadzadeh, V Hosseininaveh, Y Mostofy. 2009. Control of *Penicillium digitatum* on orange fruit combining *Pantoea agglomerans* with hot sodium dicarbonate dipping. *J Plant Pathol* 91: 437–442.

Zhang H, S Hanada, T Shigematsu, K Shibuya, Y Kamagata, T Kanawa, R Kurane. 2000. *Burkholderia kururiensis* sp. nov. a trichloroethylene (TCE)- degrading bacterium isolated from,an aquifer polluted with TCE. *Int J Syst Evol Microbiol* 50: 743–749.

Zhang Y, DA Smith. 1983. Concurrent metabolism of the phytoalexins phaseollin, kievitone and phaseollinisoflavan by *Fusarium solani* f.sp. *phaseoli*. *Physiol Plant Pathol* 23: 89–100.

Zhang Z, GY Yuen. 2000. The role of chitinase production by *Stenotrophomonas maltophilia* strain C3 in biological control of *Bipolaris sorokiniana*. *Phytopathology* 90: 384–389.

Ziedan E-S H, ES Farrag, RA El-Mohamedy, MA Abd Alla. 2010. *Streptomyces alni* as a biocontrol agent to root rot of grapevine and increasing their efficiency by biofertilizers inocula. *Arch Phytopath Plant Prot* 43: 634–646.

Zinniel DK, P Lambrecht, NB Harris, Z Feng, D Kuczmarski, P Higley, CA Ishimura, A Arfunakumari, RG Barletta, AK Vidaver. 2002. Isolation and characterization of endophytic colonizing bacteria from agronomic crops and prairie plants. *Appl Environ Microbiol* 68: 2198–2208.

3

Mechanisms

Part 1: Oxidation, Dehydrogenation, and Reduction

Oxidation of organic compounds can be carried out by insertion of one or both atoms of oxygen, or by the oxygen from water. Electron acceptors other than oxygen may be used, generally under anaerobic conditions: they include a range of oxyanions and some metal cations and oxyanions. Dehydrogenation is used here to encompass both desaturation, and reactions in which an atom of oxygen from water is used. Reductases carry out reactions ranging from the simple reduction of nitrate to ammonia to the complex reactions involved in anaerobic phenol dehydroxylation and aromatic ring reduction. Examples of all these are given as illustration.

3.1 Introduction

The division of the sections in this chapter is pragmatic rather than rigid: some reactions that are carried out by dehydrogenases are widely understood to include those in which an oxygen atom from water is incorporated into the substrate and should, therefore, more properly be designated oxidoreductases. The oxygen is derived from either one or both atoms of dioxygen, or from water.

1. Monooxygenation—Introduction of one atom of O_2
2. Dioxygenation—Introduction of both atoms of O_2
3. Oxidases, peroxidases, haloperoxidases
4. Oxygen derived from water—oxidoreductase
5. Dehydrogenation
   a. Removal of H_2—desaturation
   b. Dehydrogenation—including incorporation of oxygen derived from water

These reactions are summarized in the following examples:

Monooxygenation $R-H + O_2 + NAD(P)H + H^+ \rightarrow R-OH + H_2O + NAD(P)$

Dioxygenation $\quad R-CH=CH-R' + NADH + O_2 + H^+ \rightarrow R-CH(OH)-CH(OH)R' + NAD^+$

Oxidase $\quad\quad\quad R-CH_2-NH_2 + H_2O + O_2 \rightarrow R-CHO + H_2O_2 + NH_3$

Oxidoreductase $\quad R-H + H_2O + Mo^{VI}=S \rightarrow R-OH + Mo^{IV}-SH + H^+$

Dehydrogenase 1. $R-CH_2-CH_2-CO-SCoA + FAD \rightarrow R-CH=CH-CO-SCoA + FADH_2$

$\quad\quad\quad\quad\quad$ 2. $C\equiv O + H_2O \rightarrow CO_2 + 2H^+$

$\quad\quad\quad\quad\quad\quad HSO_3^- + H_2O \rightarrow SO_4^{2-} + 3H^+$

$\quad\quad\quad\quad\quad\quad (CH_3)_3N + H_2O \rightarrow (CH_3)_2NH + CH_2O.$

3.2 Monooxygenation

3.2.1 Monooxygenases

Diiron soluble hydroxylases
   Methane, phenols, monocyclic arenes
Membrane-integral Fe hydroxylases
   Alkane (*alkBAC*) operon
Flavoprotein monooxygenases
   Alkene epoxidation
   Cycloalkanone monooxygenation
Flavoproteins introducing oxygen at site adjacent to existing aromatic hydroxyl groups
   4-hydroxybenzoate hydroxylase
   Salicylate hydroxylase
   Anthranilate hydroxylase
Hydrocarbon hydroxylases introducing oxygen into nonoxygenated rings
   Toluene, phenanthrene, pyrene monooxygenases
Hydrocarbon epoxidases
   Fungal and yeast biotransformations of PAHs: production of phenols by NIH shift
Cytochrome P450 hydroxylation

3.2.2 Monooxygenation

Monooxygenases belong to a family of enzymes that introduce one atom of dioxygen into a wide range of substrates. These include aliphatic hydrocarbons and ketones, arene hydrocarbons, phenols, and hydroxybenzoates. An outline comparison of the activities of monooxygenases (Figure 3.1a,b) and dioxygenases (Figure 3.1c,d) is given.

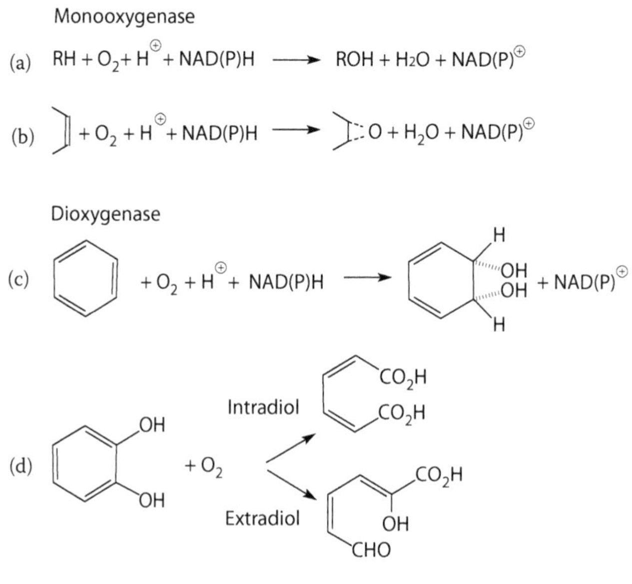

FIGURE 3.1   Examples of (a,b) monooxygenase and (c,d) dioxygenase activities.

3.2.3 Hydroxylation of Alkanes

It is appropriate to introduce this section with an account of the introduction of oxygen into alkanes. As will appear, this is more complex than one might have imagined, and it is necessary to divide this into several parts.

3.2.3.1 Soluble Diiron Monooxygenases

This group of enzymes carries out a range of monooxygenations (hydroxylations) that are involved in the degradation of substrates ranging from methane to phenols and monocyclic arenes. The oxidation of the simplest alkane methane is carried out by methylotrophs, which may be obligate or facultative. Methane monooxygenase that catalyzes the introduction of oxygen exists in both a soluble (sMMO) and a particulate (pMMO) form, of which the former has been more extensively studied.

The paradigm of the soluble diiron monooxygenases is methane monooxygenase (sMMO) which consists of a dimeric μ-hydroxobridged binuclear Fe hydroxylase, an NADH oxidoreductase with FAD/[2Fe–2S] centers, and an effector. Examples include the enzymes from *Methylococcus capsulatus* (Davydov et al. 1999), *Methylosinus trichosporium* (Fox et al. 1989), *Methylocystis* sp. (Grosse et al. 1999), and *Methylomonas* sp. (Shigematsu et al. 1999). Similar enzymes include the soluble propane monooxygenase (sPMO) and butane monooxygenase (sBMO) in *Thauera* (*Pseudomonas*) *butanivorans* (Sluis et al. 2002; Dubbels et al. 2007). The unique alkene monooxygenase of *Rhodococcus corallinus* is rather similar (Gallagher et al. 1997, 1998).

The structure of the hydroxylase in the soluble enzyme and the mechanism of its action involving the Fe^{III}–O–Fe^{III} at the active site are given in a review (Lipscomb 1994), and further details have been added on the basis EPR spectra (Davydov et al. 1999). On the other hand, the particulate enzyme contains copper (Nguyen et al. 1994), or both copper and iron (Zahn and DiSpirito 1996), and the complex genetics that govern the regulation of the soluble enzyme in *Methylococcus capsulatus* have been delineated (Csáki et al. 2003). The concentration of copper determines the catalytic activity of the enzyme (Sontoh and Semrau 1998), and the prevalence of strains exhibiting the soluble form of the enzyme that are obtained by enrichment in copper-limiting media (McDonald et al. 2006).

It is important to appreciate that diiron clusters are not restricted to methane monooxygenase, but are encountered as components of the multi-component monooxygenases for toluene/xylene and phenol that are discussed later.

3.2.3.2 Membrane-Integral Monooxygenases

This group of monooxygenases differs from the first group in a number of ways. They are membrane-associated, and catalyze essentially only one reaction: the hydroxylation of alkanes at both the primary, secondary, and terminal positions—although epoxidation of alkenes has also been demonstrated. They have been most extensively studied in the plasmid OCT encoding octanol hydroxylation in *Pseudomonas putida* (*P. oleovorans*) GPO1 that contains an operon *alkBAC*. Hydroxylation is carried out by three components, an integral membrane alkane hydroxylase encoded by *alkB*, a soluble NADH-rubredoxin reductase encoded by a*lkT* (Eggink et al. 1990) and a soluble rubredoxin encoded by *alkG* that transfers the necessary electrons from NADH. The alkane hydroxylase has been identified and sequenced in *Ps. putida* (*oleovorans*) (Kok et al. 1989), while the components of the ω-hydroxylase in this strain are analogous to those involved in α-hydroxylation (McKenna and Coon 1970). The ω-hydroxylase has been reported to require phospholipid for catalytic activity (Ruettinger et al. 1974). The hydroxylase is a representative of a large nonheme diiron cluster that has been characterized in the ω-hydroxylase by Mössbauer spectroscopy (Shanklin et al. 1997), while *alkC* encodes the dehydrogenase for the resulting alkanol (van Beilen et al. 1992).

The range of substrates is wide ranging from C_3 to $>C_{34}$ including branched-chain alkanes: alkane hydroxylation has been studied in both Gram-negative organisms including strains of *Pseudomonas*, *Acinetobacter*, and *Alcanivorax;* and the Gram-positive strains of *Rhodococcus*, *Mycobacterium*, and

Gordonia (*Corynebacterium*). In addition, biotechnological interest has prompted effort to produce strains for the terminal hydroxylation of short-chain alkanes.

Some additional comments are added:

1. Although alkane hydroxylase activity for higher alkanes is distributed among a number of Gram-negative bacteria, sequence analysis of proteins shows that these differ widely even among pseudomonads (Smits et al. 2002);
2. Alkane hydroxylases belong to a family of nonheme iron oxygenases: there is some structural similarity between the nucleotide sequence of the integral-membrane alkane hydroxylase and the subunits of the monooxygenase encoded by *xylA* and *xylM* in the TOL plasmid that are involved in hydroxylation of the methyl groups in toluene and xylene in *Ps. putida* PaW1 (Suzuki et al. 1991).

3.2.4 Flavoprotein Monooxygenases

There are several classes of flavoprotein monooxygenases. A summary of these and typical reactions that are carried out is given in Table 3.1 (van Berkel et al. 2006). Details of the enzymes are provided later.

3.2.4.1 Epoxidation of Alkenes

The first step established in the aerobic degradation of alkenes is generally epoxidation and involves the critical participation of coenzyme M (Krishnakumar et al. 2008). In *Xanthobacter* sp. strain Py2, the soluble alkene monooxygenase consists of four components: an NADH reductase that provides the reductant for activation of O_2, a Rieske-type ferredoxin to mediate electron transfer, an oxygenase, and a small protein that is putatively analogous to component B of soluble methane monooxygenases (Small and Ensign 1997). Epoxidation is then followed by several alternatives. In one of them, the epoxides may undergo carboxylation: the enzyme complex has been purified from *Xanthobacter* sp. strain Py2, and includes an NADPH: disulfide oxidoreductase and three carboxylation components (Allen and Ensign

TABLE 3.1

Bacterial Monooxygenases

Group	Substrates of Monooxygenases	Selected References
A One gene: Monooxygenase	4-Hydroxybenzoate 3-monooxygenase[a]	Entsch and van Berkel (1995)
B One gene: Monooxygenase	Cyclohexanone monooxygenase[a]	Donoghue et al. (1976)
	4-Hydroxyacetophenone monooxygenase	Tanner and Hopper (2000) Kamerbeek et al. (2001)
C Two genes: monooxygenase, reductase	Alkanesulfonate monooxygenase	Higgins et al. (1996) Eichhorn et al. (1999)
	Nitrilotriacetate monooxygenase	Uetz et al. (1992) Xu et al. (1997)
	Dibenzothiophene monooxygenase	Lei and Tu (1996)
D Two genes: monooxygenase, reductase	4-Hydroxyphenylacetate 3-monooxygenase[a]	Prieto and Garcia (1994)
	4-Nitrophenol monooxygenase	Kadiyla and Spain (1998)
	4-chlorophenol monooxygenase	Gisi and Xun (2003)
E Two genes: monooxygenase, reductase	Styrene monooxygenase[a]	Hartmans et al. (1990)
F Two genes: monooxygenase, reductase	Tryptophan 7-halogenase[a]	Yeh et al. (2005)

[a] The prototype enzymes.

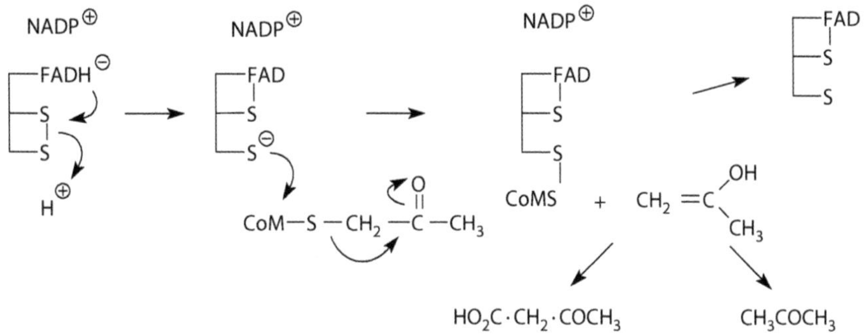

FIGURE 3.2 Epoxidation of propene epoxide followed by carboxylation.

1997; Allen et al. 1999) (Figure 3.2). The isoprene monooxygenase from *Rhodococcus* sp. strain AD45 is analogous (van Hylckama et al. 2000), and consists of an $(\alpha\beta\gamma)_2$ hydroxylase, FAD reductase, a Rieske ferredoxin, and an effector. Alkene monooxygenase from this strain is closely related to the aromatic monooxygenases, and is able to hydroxylate benzene, toluene, and phenol (Zhou et al. 1999), while the alkane hydroxylase from *Pseudomonas oleovorans* is able to carry out both hydroxylation and epoxidation (May and Abbott 1973; Ruettinger et al. 1977).

3.2.4.2 Cycloalkanone Monooxygenation

These flavoprotein 1,2-monooxygenases are used for the insertion of an atom of oxygen into the ring to produce cyclic lactones that is the first step in the degradation of both cyclopentanone and cyclohexanone before hydrolysis and ring fission (Figure 3.3). There are two types of monooxygenase enzymes, one that is FAD–NADPH-dependent, whereas the other is FMN–NADH-dependent. The crystal structure of the enzyme from *Thermobifida fusca* with phenyl acetone has been established (Malito et al. 2004), and revealed that the initially formed 4a-hydroperoxide reacted with the carbonyl group of the ketone followed by an intramolecular rearrangement to the lactone. The FAD-containing cyclohexanone mono-oxygenase has been purified and characterized from *Nocardia glomerula* CL1, and *Acinetobacter* sp. NCIB 9871 (Donoghue et al. 1976). There has been renewed interest in these reactions in view of their biotechnological importance (Iwaki et al. 2002). Cycloalkanone monooxygenase for substrates with more than seven carbon atoms is, however, different (Kostichka et al. 2001). These monooxygenases are also involved in the degradation of cyclic terpenoids, and the enzyme from *Rhodococcus erythropolis* strain DCL14 is able to catalyze three different insertions of oxygen in *R. erythropolis* (van der Werf 2000). Analogous FAD-containing monooxygenases are involved as the initial reactions in the degradation of acetophenone and chloroacetophenone (Higson and Fucht 1990), and 4-hydroxyacetophenone by *Ps. putida* JD1 (Tanner and Hopper 2000) and *Ps. fluorescens* ACB (Kamerbeek et al. 2001).

3.2.4.3 Thioamide Monooxygenation

Thioamides represent a range of second-tier drugs against *Mycobacterium tuberculosis*. They include ethionamide, thiacetazone (TAC) and isoxyl (ISO) (Dover et al. 2007), and all of them require activation

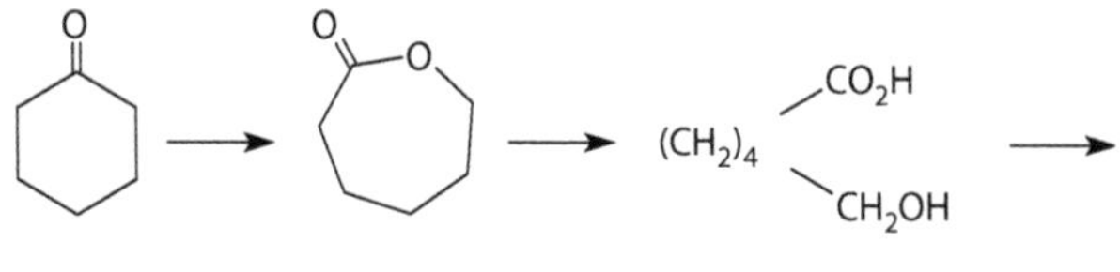

FIGURE 3.3 Monooxygenation of cyclohexanone.

before interacting with their cellular targets (Baulard et al. 2000; DeBarber et al. 2000). For ethionamide this involves monooxygenation, and it has been shown that activation is carried out by a flavoprotein monooxygenase EtaA encoded by a gene that has been characterized. The *S*-oxide was produced from ethionamide followed by 2-ethyl-4-hydroxymethylpyridine (Vannelli et al. 2002). Sequence analysis of the monooxygenase EtaA showed that it belonged to the family of Baeyer–Villiger monooxygenases (BVMOs) and, consistent with this, the enzyme from *M. tuberculosis* was able to bring about typical monooxygenation of ketones to esters or lactones (Fraaije et al. 2004). To avoid intrusion from spontaneous modification brought about during the extraction/purification process, high resolution magic angle spinning-NMR has been used to follow ethionamide activation *in vivo* after entering the mycobacterial cell (Hanoulle et al. 2006). It was shown that the activated drug ETH* is held intracellularly—in contrast to the other metabolite 2-ethyl-4-hydroxymethylpyridine that is extracellular. Although the *S*-epoxide that has been consistently isolated by previous workers was bound to membranes, it may not, however, represent the *in vivo* structure of the activated drug (Hanoulle et al. 2006).

These monooxygenases should be contrasted with the dioxygenases that bring about oxygenation on nitrogen and sulfur which are discussed in the next section, and do not involve structural modification.

3.2.5 Other Monooxygenases

1. *Alkansulfonate Monooxygenase:* The metabolism of alkanesulfonates is carried out by monooxygenation with the production of the corresponding aldehyde and sulfite. The activity in *Escherichia coli* is mediated by a two-component monooxygenase that consists of a hydroxylase and an oxidoreductase which have been purified (Eichhorn et al. 1999). The genes encoding the monooxygenase in *Methylosulfuromonas* designated *msmABCD* are clustered on the chromosome, and the enzyme from *Methylosulfuromonas methylovora* strain M2 is a multi-component monooxygenase whose two-component hydroxylase MsmA and MsmB have amino acid sequences similar to those of dioxygenases, although the electron transfer units McmC and MsmD were typical of monooxygenases (De Marco et al. 1999). In strain TR3 of this organism, the genes encoding methanesulfonate monooxygenase are duplicated (Baxter et al. 2002).

2. *Nitronate Monooxygenase:* The degradation of nitroalkanes has been examined in a range of biota including: the yeast *Hansenula mrakii*, the fungi *Neurospora crassa*, *Fusarium oxysporum* and *Podospora anserina*, the leguminous plants *Pisum sativum* and *Hippocrepis comosa*, and the bacteria *Streptomyces ansochromogenes* and *Pseudomonas* sp. These are discussed in Chapter 11, and only brief comments on the enzymology are given here.

In summary, ketones are produced from secondary nitrates or aldehydes from primary nitrates, nitrite is the initial product, the enzyme acts preferentially on the acid-form of the substrate, and the enzyme is a flavoprotein containing FAD or FMN. The enzyme has been variously described as an oxidase on the basis of H_2O_2 production, a dioxygenase in which both oxygen atoms are incorporated separately into two molecules of the substrate, and a flavoprotein monooxygenase that acts on the acid-nitro form (nitronate). The cardinal investigations are briefly outlined.

The Role of Flavins: Reaction at N-5: The adduct of nitroethane with FAD showed the presence of a nitrobutyl group at N-5 formed by reaction with the nitroethane anion (Gadda et al. 1997). This was confirmed by x-ray crystallography, during turnover in reaction with nitroethane (Nagpal et al. 2006) and, in the presence of cyanide as a trapping agent (Héroux et al. 2009). In summary, the anion of nitroethane reacted with the N-5 of FAD to produce an adduct that formed an iminium cation after loss of nitrite (Valley et al. 2005). This could then react: (a) with a further nitroethane anion to produce the nitrobutyl group at N5, (b) by hydrolysis to the aldehyde, or (c) by trapping with cyanide.

The oxidant: Important details of the reaction have been added using the *ncd-2* gene from *Neurospora crassa* that was cloned and expressed in *E. coli* (Francis et al. 2005). It was shown that a flavosemiquinone was involved in the oxidation of *aci*-nitroalkanes (nitronates) and a mechanism was outlined: The nitronate anion reacted with the flavin to produce a semiquinone and the nitroalkane radical: this reacted with O_2 to form the HO_2 radical and produced the peroxide of the nitroalkane that collapsed to the ketone (Francis et al. 2005). This has been supported by examination of the crystal structure of 2-nitropropane from *Pseudomonas aeruginosa* that revealed two important features: His^{152} was the catalytic base that formed the nitroalkane anion and the proximity of the nitronate to the N-5 of FAD (Ha et al. 2006).

On the basis of the collective evidence, the designation of the enzyme as a nitronate monooxygenase is most appropriate (Gadda and Francis 2010).

3. The degradation of dimethylsulfone by facultative methylotrophs including *Hyphomicrobium sulfonivorans* proceeds by successive reduction by membrane-associated reductases to dimethyl sulfide (Borodina et al. 2002). This degraded by monooxygenation in the type strain to methylsulfide and formaldehyde by a $FMNH_2$-dependent monooxygenase (Boden et al. 2011). The enzyme was specific for methyl sulfides and was unable to accept alkanesulfonate, nitrilotriacetate or dibenzothiophene whose monoxygenses belong to group C above.

4. *Lysine Monooxygenase:* Lysine monooxygenase catalyzes the first step in the degradation of L-lysine by *Pseudomonas fluorescens* with the formation of 5-amino-*n*-valeramide and CO_2. The enzyme has been purified from *Ps. fluorescens* and consists of four equal subunits and four FAD prosthetic groups (Flashner and Massey 1974a). It is relevant to note that the enzyme plays a dual role, and also catalyzes a reaction that is characteristic of amino acid oxidases: the conversion of L-ornithine and O_2 to 2-oxo-5-aminovalerate, ammonia, and H_2O_2 (Nakazawa et al. 1972; Flashner and Massey 1974a), and the regulatory properties of the monooxygenase have been elucidated (Flashner and Massey 1974b).

5. *Oxosteroid 9α-Hydroxylase:* The 9α-hydroxylation of steroids is an essential step in their degradation. This is accomplished in *R. rhodochrous* DSM 43269 and *M. tuberculosis* by a two-component monooxygenase that consists of a ferredoxin reductase (KshB) and a terminal oxygenase (KshA) (Capyk et al. 2009: Petrusma et al. 2009). The reductase requires NADH, uses flavin adenine dinucleotide as cofactor and the [Fe–S] cluster $[2Fe–2S]Cys_4$, while a Rieske-type [Fe–S] cluster $[2Fe–2S]Cys_2His_2$ and nonheme Fe^{2+} are required for the activity of the oxygenase. *Rhodococcus rhodochrous* DSM 43279 possesses several homologues of *kshA* (*kshA1* to *kshA5*) and these have a differential spectrum of activity for steroids: for example, whereas KahA5 was active for a broad range of 3-keto steroids including 11β-hydrocortisone, KshA1 was more limited although it had a high activity for 23,24-bis-norcholesta-4-ene-22-oate and the analogous 1,4-diene (Petrusma et al. 2011).

6. *Heme Oxygenase:* Heme is a derivative of porphyrin **IX** that occurs in mammals, insects, yeasts, plants, algae and bacteria, and fulfills a multiplicity of functions including oxygen transport and respiration. The concentration of free heme must, however, be controlled to avoid the production potentially damaging reactive oxygen species. Degradation of heme is initiated by a heme oxygenase that carries out a hydroxylation, even though dioxygen is involved later in fission of the tetrapyrrole ring. The mechanism has been extensively examined in mammals for the oxygenases HO-1 and HO-2, and for a number of bacterial heme oxygenases that differ in detail. A variant of the mechanism of heme oxygenation has, however, been observed in the blood-sucking insect *Rhodnius prolixus*. The reaction is initiated by reaction at the vinyl positions with the thiol groups of glu–cys–gly and exceptional hydroxylation at the γ-position, followed by hydrolysis of the glycine in the side chain to yield free carboxylic acids leaving cysteine bound at the α- (or β-) positions of the vinyl side chains (Paiva-Silva et al. 2006).

Several bacteria are able to use heme as a source of iron that becomes available after oxidative fission of the tetrapyrrole ring. This produces linear biliverdins, carbon monoxide from the meso positions, and the release of Fe. This enzyme has been examined in several pathogenic bacteria, although it should be noted that *Bradyrhizobium japonicum* strain H 10 is able to use heme as a source of iron. Bioinformatics analysis revealed the presence of two genes *hmuQ* and *hmuD* that were identified as encoding heme oxygenase (Puri and O'Brian 2006), and it was shown that the HmuQ enzyme was capable of producing biliverdin from heme degradation.

The following attempts only a brief summary of the reactions that are involved. Degradation is initiated by O_2 and 2-electron reduction, and the protonated peroxide intermediate $[Fe^{3+}–O-OH]$ carries out hydroxylation at a mesoheme position (generally the α-meso) to produce α-hydroxyheme that is the first oxygenated intermediate. This is then converted by O_2 and a single electron to verdoheme with loss of CO, followed by 4-electron reduction of verdoheme and reaction with O_2 to produce the linear biliverdins—generally biliverdin **IX**α with loss of Fe^{2+} (Yoshida and Migita 2000)

$$— Fe^{III} — \leftrightarrow — Fe^{II} — \leftrightarrow — Fe–O=O — \rightarrow — Fe^{III}–O–OH— \rightarrow — Fe^{III}—OH$$

Heme oxygenase occurs in the opportunistic pathogenic yeast *Candida albicans* where its synthesis was induced by hemoglobin and by iron deprivation (Santos et al. 2003), and is regulated by hemoglobin (Pendrak et al. 2004). Phycobiliproteins are accessory pigments in cyanobacteria, red algae, and crypto-monads, and heme oxygenase is required to transform heme to biliverdin IXα that is converted by a reductase to 3Z-phycocyanobilin before conjugation to produce the holo-α phycocyanin (Tooley et al. 2001; Alvey et al. 2011).

A number of bacterial pathogens are able to use heme (or hemoglobin) as sources of iron: they include the Gram-positive *Corynebacterium diphtheriae*, *Bacillus anthracis*, *Staphylococcus aureus* and *M. tuberculosis* (references in Allen and Schmitt 2011), and the Gram-negative *Neisseria meningitides*, *Escherichia coli*, and *Ps. aeruginosa*.

Several groups of heme oxygenases have been found in bacteria, and it has been shown that the HmuO oxygenase of *Corynebacterium diphtheriae* involves both the Asp[136] and an adjacent water molecule at the active site (Matsui et al. 2005). Heme oxygenases have been characterized in several bacteria and the sequence homologies have been examined.

1. In *Corynebacterium diphtheriae* the heme oxygenase that encodes the *hmuO* gene was shown to have a mass of 24 kDa and showed extensive sequence similarity to the human oxygenase (Schmitt 1997; Wilks and Schmitt 1998).

2. The heme oxygenase from *Neisseria meningitidis* has been purified and in the presence of ascorbate the heme–HemO complex was converted into ferric bilirubin IXα and carbon mon-oxide. Homologies with several other isolates of the genus were identified and the sequence homology among the *Neisseria* enzymes was high, although that with other heme oxygenases was low (Zhu et al. 2000).

3. In *Staphylococcus aureus* it has been shown that *isdG* and *isdI* encode cytoplasmic proteins, and that fission of the porphyrin ring took place in the presence of NADPH cytochrome P450 reductase (Skaar et al. 2004). In addition, the crystal structures of IsdG and IsdI have been analyzed, and on this basis it was suggested that there was an analogy with the monooxygenase ActVA used for the oxidation of polyketides in *Streptomyces* (Wu et al. 2005).

4. The reaction in *Ps. aeruginosa* is different. Heme is an essential source of Fe, and there are two different heme oxygenases: PigA that is part of an operon is induced only under conditions of Fe limitation and hydroxylates at the β-meso position to produce biliverdins **IX**β (C_3, C_7-propionate) and at the γ-meso position to produce biliverdin **IX**δ (C_{13}, C_{17}-propionate), whereas BphO is a conventional oxygenase that produced biliverdin **IX**α (C_8, C_{12}-propionate) after hydroxylation at the α-meso position (Ratliff et al. 2001; Wegele et al. 2004). It was shown that reduced ferredox-ins were effective as electron donors yielding ferric bilirubin as the final product (Wegele et al. 2004).

5. Heme oxygenase is an important component of the enterohemorrhagic *E. coli* O157:H7, and the x-ray structure of the enzyme has been determined and revealed that exposure of the meso-α carbon to oxygen was a result of the orientation of the propionic acid side chains of the pyrrole rings (Suits et al. 2006).

6. *M. tuberculosis* is able to use heme as a source of iron, and the degradation of heme in *M. tuber-culosis* was carried out by MhuD that shared sequence homology with the IsdG and IsdI of *Staphylococcus aureus*, while x-ray analysis showed that it contained a diheme complex (Chim et al. 2010).

7. Purines can serve as sources of nitrogen for some strains of *Klebsiella pneumoniae*, and it has been shown that the hydroxylation of urate at C_5 to 5-hydroxyisourate is carried out by a monooxygenase that is dependent on a flavin cofactor (FAD): the enzyme HpxO is, therefore, a bound flavin hydroperoxide (O'Leary et al. 2009).

3.2.6 Aromatic Hydroxylases

These mediate the introduction of hydroxyl groups into alkyl arene side chains, into the ring of monocyclic arenes, and at the *ortho* position of substrates already containing a phenolic group. The paradigm for these hydroxylases is the FAD-dependent 4-hydroxybenzoate hydroxylase that produces 3,4-dihydroxybenzoate. This is further exemplified by the following examples that illustrate the range of substrates. Hydroxylases are also involved in the degradation of phenols, and phenols carrying chloro and nitro substituents.

3.2.6.1 Monooxygenation of Aromatic Alkyl Side Chains

Although the bacterial degradation of aromatic compounds is usually initiated by dioxygenation, monooxygenases may be involved additionally, or alternatively, for the introduction of oxygen into the ring. An alternative pathway for the degradation of alkyl benzenes involves oxidation of methyl substituents to carboxylates (Figure 3.4). This is also used for the degradation of 4-nitrotoluene by *Pseudomonas* sp. strain TW3, and the *ntnWCMAB** genes that encode the enzymes that convert the substrate to 4-nitrobenzoate are similar to those in the upper pathway of the TOL plasmid (Harayama et al. 1989): an alcohol dehydrogenase, benzaldehyde dehydrogenase, a two-component monooxygenase and part of a benzyl alcohol dehydrogenase (James and Williams 1998). In this strain, however, the enzymes are chromosomal, and the benzyl alcohol dehydrogenase ntnB* differs from the corresponding xylB protein.

The oxidation by strains of *Ps. putida* of the methyl group in compounds containing a hydroxyl group in the *para* position is, however, carried out by a different mechanism. The initial step is a dehydrogenation to a quinone-methide followed by hydration (hydroxylation) to the benzyl alcohol (Hopper 1976) (Figure 3.5). The reaction with 4-ethylphenol is partially stereospecific (McIntire et al. 1984), and the enzymes that catalyze the first two steps are flavocytochromes (McIntire et al. 1985). The role of hydroxylation in the degradation of azaarenes is discussed in the section on oxidoreductases (hydroxylases).

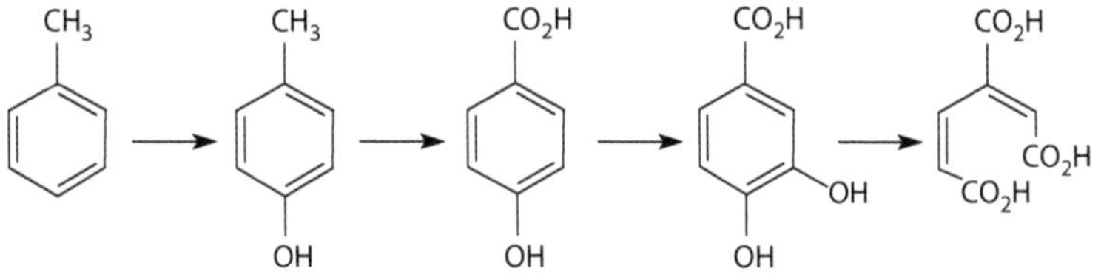

FIGURE 3.4 Degradation of toluene by side-chain oxidation.

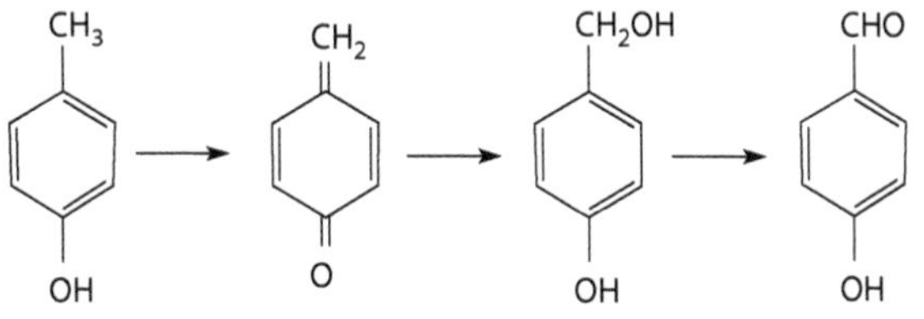

FIGURE 3.5 Degradation of 4-methylphenol by hydroxylation.

3.2.6.2 Ring Hydroxylation of Aromatic Hydrocarbons

3.2.6.2.1 Hydroxylation of Arenes

1. The four-component alkene/aromatic monooxygenases includes *Pseudomonas mendocina* sp. KR1 (Whited and Gibson 1991a,b; Pikus et al. 1996; Hemmi et al. 2001) that carries out the hydroxylation of toluene to 4-hydroxytoluene (Tmo). This consists of a $(\alpha\beta\gamma)_2$ hydroxylase, an NADH reductase, an effector and a Rieske-type ferredoxin. Related hydroxylases include toluene-2-hydroxylase in *Burkholderia cepacia* strain G4 that consists of three components, FAD- [2Fe–2S] cluster that functions as an oxidoreductase, an effector and a $\alpha_2\beta_2\gamma_2$ containing 5–6 Fe atoms that are probably two diiron centers (Newman and Wackett 1995). These are analogous to the toluene-3-monooxygenase (Tbu) in *Ralstonia pickettii* PKO1 that has been shown to be a toluene-4-monooxygenase (Fishman et al. 2004); the toluene/xylene monooxygenase (Tou) in *Pseudomonas stutzeri* strain OX1 (Bertoni et al. 1998); and the phenol hydroxylase (Phl) in *Ralstonia eutropha* strain JMP134 (Ayoubi et al. 1998). It has been shown that the structure of the oxidized diiron center of the monooxygenase in *Ps. stutzeri* OX1 is similar to that of the soluble methane hydroxylase (Sazinsky et al. 2004) while, in a wider context, it is worth noting the structure of the diiron center in *Streptomyces thioluteus* that brings about the successive monooxygenation of 4-aminobenzoate to 4-nitrobenzoate that is required for the biosynthesis of the antibiotic aureothin (Choi et al. 2008), and of the epoxidase in *Geobacter metallireducens* which initiates the anaerobic degradation of benzoyl-coenzyme A (Rather et al. 2011).

   On the other hand, the enzyme that oxidizes toluene-4-sulfonate to the benzyl alcohol in the first step of its degradation is a monooxygenase that consisted of only two components: an [Fe–S] flavoprotein that serves as a reductase and an oxygenase (Locher et al. 1991b).

2. Toluene/*o*-xylene monooxygenase in *Ps. stutzeri* strain OX1 was able to carry out successive monooxygenation of *o*-xylene, and the gene cluster has been analyzed (Bertoni et al. 1998). The subunits of this monooxygenase have been purified and reconstituted to the active enzyme complex (Cafaro et al. 2002), and two different monooxygenases have been demonstrated in this organism: a phenol hydroxylase, and a toluene/*o*-xylene monooxygenase (Cafaro et al. 2004). The toluene-4-monooxygenase of *Ps. mendocina* KR1 and toluene-3-monooxygenase of *Ralstonia pickettii* PKO1 can hydroxylate benzene, toluene, and *o*-xylene (Tao et al. 2004; Vardar and Wood 2004). It has been shown, however, that the monooxygenase is, in fact a *para*-hydroxylating enzyme (Fishman et al. 2004).

3. *Xanthobacter* sp. strain Py2 was isolated by enrichment on propene that is metabolized by initial metabolism to the epoxide. The monooxygenase that is closely related to aromatic monooxygenases is, however, able to hydroxylate benzene to phenol before degradation, and toluene to a mixture of 2, 3, and 4-methylphenols that are not further metabolized (Zhou et al. 1999).

3.2.6.2.2 Concurrent Synthesis of Monooxygenase and Dioxygenase

1. Several strains of *Pseudomonas* sp. that were induced for toluene dioxygenase activity catalyzed the enantiomeric monooxygenation of indan to indan-1-ol and indene to inden-1-ol and *cis* indan-1,2-diol (Wackett et al. 1988), while purified naphthalene dioxygenase from a strain of *Pseudomonas* sp. catalyzed the enantiomeric monooxygenation of indan to indan-1-ol and the dehydrogenation of indan to indene (Figure 3.6) (Gibson et al. 1995).

2. By cloning genes for benzene/toluene degradative enzymes in *Pseudomonas* (*Burkholderia*) sp. strain JS150, it has been found that this strain also carries genes for a monooxygenase in addition to those for the dioxygenase (Johnson and Olsen 1995). Initial products from the metabolism of toluene are, therefore, 3-methyl catechol produced by 2,3-dioxygenation, 4-Methylcatechol by 4-monooxygenation, and both 3- and 4-methylcatechols by 2-monooxygenation (Johnson and Olsen 1997).

3. *Sphingomonas yanoikuyae* (*Beijerinckia* sp.) strain B1 metabolizes biphenyl by initial dioxygenation followed by dehydrogenation to 2,3-dihydroxybiphenyl. Cells of a mutant (strain

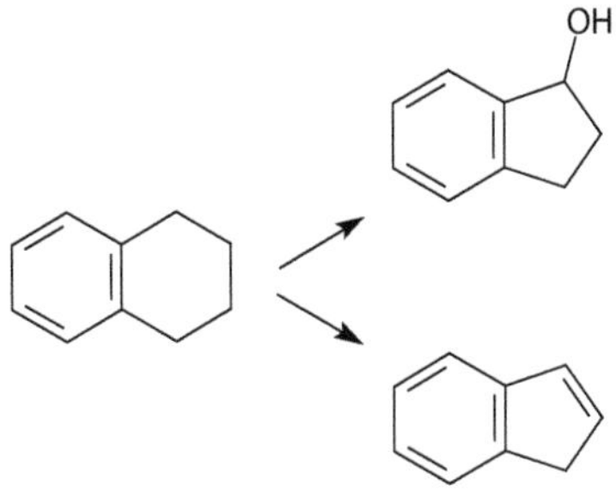

FIGURE 3.6   Transformation of indane to 3-hydroxyindane and indene.

B8/36) lacking *cis*-biphenyl dihydrodiol dehydrogenase were induced with 1,3-dimethylbenzene, and could transform dihydronaphthalene by three reactions:

a.   Monooxygenation to (+)-(*R*)-2-hydroxy-1,2-dihydronaphthalene

b.   Dioxygenation to (+)-(1*R*,2*S*)-*cis*-naphthalene dihydrodiol

c.   Dehydrogenation to naphthalene followed by dioxygenation to (+)-(1*R*,2*S*)-*cis*-naphthalene dihydrodiol (Eaton et al. 1996)

4. The degradation of pyrene by a *Mycobacterium* sp. involves both dioxygenase and monooxygenase activities (Heitkamp et al. 1988).

5. Arene oxides can be intermediates in the bacterial transformation of aromatic compounds and initiate rearrangements (NIH shifts) (Dalton et al. 1981; Cerniglia et al. 1984; Adriaens 1994). The formation of arene oxides may plausibly account for the formation of nitro-substituted products during degradation of aromatic compounds when nitrate is present in the medium (Sylvestre et al. 1982; Omori et al. 1992).

6. Toluene monooxygenases provide alternatives to dioxygenation, and the *ortho*-monooxygenase in *Burkholderia* (*Pseudomonas*) *cepacia* G4 (Newman and Wackett 1995) has been shown to be carried on a plasmid TOM (Shields et al. 1995).

3.2.6.3 Monooxygenation of Phenols and Related Compounds

Monooxygenases generally initiate the degradation of phenols and their ethers, and hydroxybenzoates and hydroxyphenylacetates. It is convenient to separate the monooxygenation of unsubstituted phenols from those carrying chlorine or nitro substituents.

3.2.6.3.1 Phenols Lacking Halogen or Nitro Substituents

1. The degradation of phenol is generally initiated by hydroxylation using a single-component NAD(P)H-dependent monooxygenase which, in *Pseudomonas* sp. EST 1001 is encoded by *pheA* (Nurk et al. 1991a,b; Powlowski and Shingler 1994b). The phenol hydroxylase from *Bacillus thermoglucosidasius* strain A7 is, however, a two-component hydroxylase consisting of an oxygenase (PheA1) and a flavin reductase (PheA2) (Kirchner et al. 2003).

2. The hydroxylation of phenol and 3,4-dimethylphenol by *Pseudomonas* sp. strain CF600 is unusually carried out by a multicomponent enzyme whose genes occur in a plasmid, and have been termed *dmpKLMNOP* (Shingler et al. 1992; Powlowski and Shingler 1994a). The three components consist of a reductase DmpP containing FAD and [2Fe–2S], an activator DmpM and the three subunits of the hydroxylase DmpLNO that contain the diiron center that has been characterized by Mössbauer and EPR studies (Cadieuz et al. 2002). The degradation of 3- and 4-alkylphenols in *Pseudomonas* sp. strain KL28 (Jeong et al. 2003) involves an analogous multi-component hydroxylation encoded by *dmpKLMNOP*.

3. On the other hand, the degradation of 2,5-dimethylphenol by *Pseudomonas alcaligenes* strain P25X involves oxidation of one of the methyl groups to a carboxylate and hydroxylation by a 3-hydroxybenzoate 6-hydroxylase to form a 1,4-dihydroxybenzenecarboxylate before fission by gentisate 1,2-dioxygenase. This strain can, in addition, produce two 3-hydroxybenzoate 6-hydroxylases, although one is strictly inducible by 3-hydroxybenzoate itself, whereas the other is used for the degradation of the carboxylate from 2,5-dimethylphenol (Gao et al. 2005).

4. The degradation of 4-methylphenol by pseudomonads is initiated by conversion into 4-hydroxy-benzyl alcohol by a flavocytochrome by dehydrogenation to a quinone-methide, following by hydroxylation using the oxygen atom of H_2O (Cunane et al. 2000).

5. The degradation of 2-hydroxybiphenyl by *Pseudomonas azelaica* HBP1 is initiated by 2-hydroxybiphenyl 3-monooxygenase (Suske et al. 1999).

6. The degradation of carbaryl (1-naphthyl-*N*-methylcarbamate) by *Pseudomonas* sp. strain C4 proceeds by sequential formation of naphth-1-ol, 1,2-dihydroxynaphthalene and gentisate (Swetha et al. 2007). The hydroxylase putatively contains FAD, and was able to use both NADH and, preferably, NADPH as electron donors, but showed only low activity toward naphth-2-ol and naphthalene.

3.2.6.3.2 Chlorophenols

There are two reactions that may be distinguished: (a) formation of a catechol that undergoes ring fission directly with subsequent loss of chlorine from the products of fission, and (b) hydroxylation with concomitant loss of chlorine that is generally involved for substrates with three or more chlorine substituents. Further details are given in Chapter 9, Part 2. In the degradation of 2,4,6-trichlorophenol by *Ralstonia eutropha* (Louie et al. 2002) and *Burkholderia cepacia* (Gisi and Xun 2003) they are $FADH_2$-dependent (Gisi and Xun 2003).

1. The hydroxylase that converts 2,4-dichlorophenol into 3,5-dichlorocatechol (Figure 3.7a) before ring fission has been purified from a strain of *Acinetobacter* sp. (Beadle and Smith 1982), and from *Alcaligenes eutrophus* JMP 134 (Don et al. 1985; Perkins et al. 1990). The reductant is NADPH, the enzyme is a flavoprotein containing FAD, and in the presence of compounds that are not substrates, NADPH and O_2 are consumed with the production of H_2O_2.

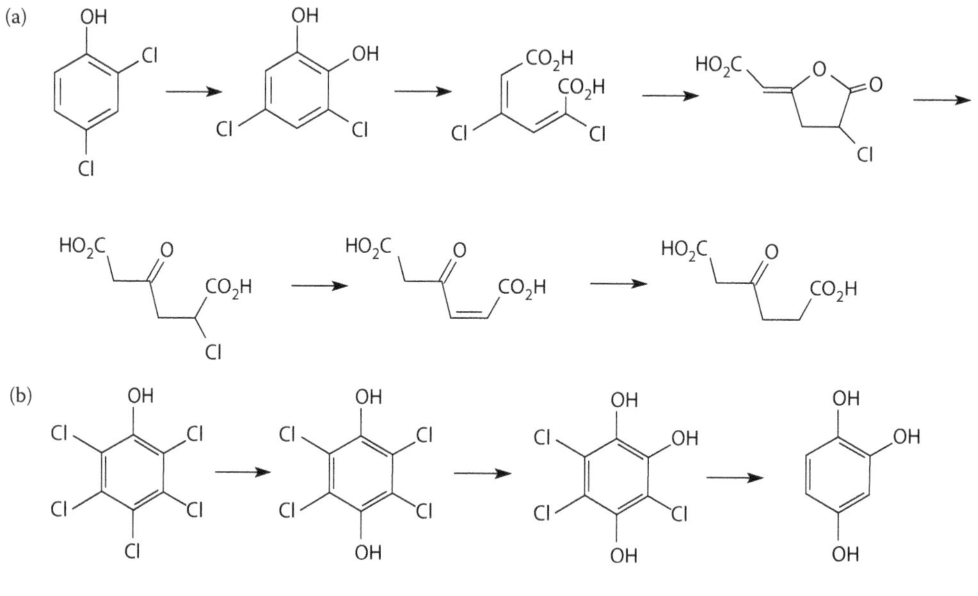

FIGURE 3.7 Degradation of (a) 3,5-dichlorophenol and (b) pentachlorophenol.

2. A range of mechanisms operates for the degradation of chlorophenols with three or more substituents.

a. The mooxygenation of pentachlorophenol is the first step in the degradation of pentachlorophenol by *Sphingobium chlorophenolicum* ATCC 39732 (*Flavobacterium* sp.) and involves introduction of oxygen at C_4 to produce tetrachlorbenzoquinone that is subsequently reduced by a reductase PcpD (Dai et al. 2003)—that was originally suggested to be a pentachlorophenol hydroxylase. The enzyme is a flavin-containing monooxygenase (Xun and Orser 1991) that can accept a number of halogenated phenols including 2,4,6-triiodophenol (Xun et al. 1992). The subsequent steps have been delineated (Cai and Xun 2002) (Figure 3.7b).

b. An NADH-requiring chlorophenol monooxygenase from *Burkholderia cepacia* AC 1100 successively dehalogenates 2,4,5-trichlorophenol to 2,5-dichloro-hydroquinone and 5-chloro-1,2,4-trihydroxybenzene (Xun 1996). It is a two-component enzyme, component A contains FAD and an NADH reductase, and component B was a heterodimer containing 2.9 mol of Fe and 2.1 mol of labile sulfide (Xun and Wagnon 1995).

c. The enzyme from *Azotobacter* sp. strain GP1 that catalyzes the formation of 2,6-dichlorohydroquinone from 2,4,6-trichlorophenol is also a monooxygenase that requires NADH, FAD, and O_2 (Wieser et al. 1997). The enzyme is able to accept other chlorophenols with the consumption of NADH including 2,4-, 2,6-, 3,4-dichloro-, 2,4,5-trichloro-, and 2,3,4,5- and 2,3,4,6-tetrachlorophenol, and in the absence of a substrate results in unproductive formation of H_2O_2.

d. The monooxygenase involved in the degradation of 2,4,6-trichlorophenol to 6-chloro-2-hydroxyhydroquinone by *Ralstonia eutropha* strain JMP134 combines monooxygenation with hydrolytic activity, and it was shown that oxygen is introduced from both O_2 and H_2O (Xun and Webster 2004).

3.2.6.3.3 Nitrophenols

There are two distinct pathways involved in the degradation of 4-nitrophenol.

1. In a strain of *Moraxella* sp., monooxygenation carried out monooxygenation to benzoquinone with the loss of nitrite, followed by reduction to benzohydroquinone before ring fission (Spain and Gibson 1991) (Figure 3.8a). The monooxygenase was characterized as a membrane-bound flavoprotein. These activities have been characterized in a strain of *Pseudomonas* sp. WBC-3 as a single-component FAD-dependent monooxygenase and an FMN-NADPH-dependent quinone reductase (Zhang et al. 2009).

2. An alternative pathway in *Bacillus sphaericus* strain JS 905 involved hydroxylation to 4-nitrocatechol followed by loss of nitrite to produce 1,2,4-trihydroxybenzene before ring fission to 3-oxoadipate (Figure 3.8b). A single two-component monooxygenase and a flavoprotein reductase were partially purified (Kadiyala and Spain 1998). Although this pathway was also observed with

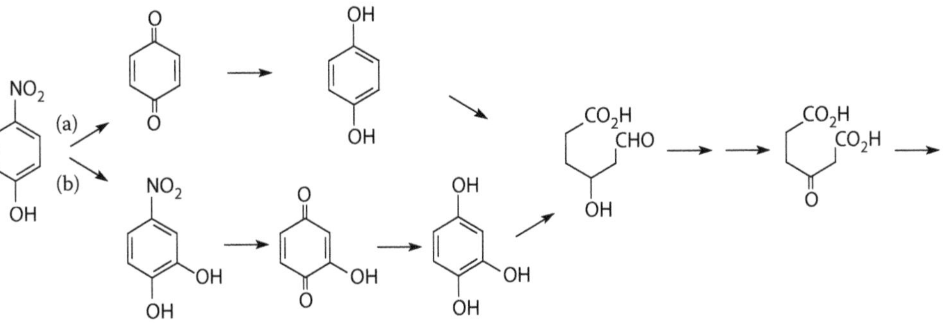

FIGURE 3.8 Degradation of 4-nitrophenol. (a) Oxidative loss of nitrite to benzoquinone; (b) hydroxylation to 4-nitrocatechol.

Arthrobacter sp. strain JS443, it has been shown to occur in Gram-negative bacteria, including *Rhodococcus* sp. strain PN1 (Takeo et al. 2008), involving a two-component hydroxylase.

3.2.6.4 Other Hydroxylations/Monooxygenations

1. Phenylalanine is hydroxylated to tyrosine and then sequentially to 4-hydroxyphenylpyruvate and, by dioxygenation and rearrangement to 2,5-dihydroxyphenylpyruvate (Figure 3.9) (Arias-Barrau et al. 2004). Hydroxylation involves 6,7-dimethyltetrahydrobiopterin that is converted into 4a-carbinolamine (Song et al. 1999). Copper is not a component of the active enzyme, although there is some disagreement on whether or not Fe is involved in the reaction for the hydroxylase from *Chromobacterium violaceum* (Chen and Frey 1998).

2. The enzyme in *Rhodococcus* sp. strain IGTS8 that brings about successive oxidation of dibenzothiophene to the sulfoxide and the sulfone is a flavin mononucleotide-dependent monooxygenase that carries out both reactions by sequential incorporation of a single atom of oxygen from O_2 (Lei and Tu 1996).

3. The transformations of endosulfan with elimination of sulfite are carried out by flavin-dependent monooxygenation in *Mycobacterium* sp. (Sutherland et al. 2002) and *Arthrobacter* sp. (Weir et al. 2006).

4. (a) There are two mechanism which operate in the demethylation of aromatic *O*-methyl ethers, one by a tetrahydrofolate-dependent demethylase in *Sphingomonas paucimobilis* SYK-6 (Abe et al. 2005), the other by monooxygenation. The conversion of vanillin to 3,4-dimethoxybenzoate is carried out in *Pseudomonas* sp. strain HR199 by the demethylases encoded by *van A* and *van B* (Priefert et al. 1997) and in *Comamonas testosteroni* by IvaA and Van A that are Class IA phthalate family oxygenases (Providenti et al. 2006). The conversion of 2-methoxy-3,6-dichlorobenzoate to 3,6-dichlorosalicylate in *Pseudomonas maltophilia* strain DI-6 is carried out by a three-component *O*-demethylase which has been characterized. The individual components were: (i) reductases encoded by *ddmA1* and *ddmA2* with molecular masses of 43.7 and 43.9 kDa, (ii) a [2Fe-2S] ferredoxin encoded by *ddmB* with a mass of 11.4 kDa, and (iii) an oxygenase with a mass of 37.3 kDa encoded by *ddmC* and homologous to members of the phthalate dioxygenase family that function as monooxygenases (Herman et al. 2005). (b) The ability to grow at the expense of methylated purines is dependent on demethylation. This has been examined in *Pseudomonas putida* CBB5 that is able to use caffeine as a sole source of carbon and energy, and to *N*-demethylate all related methylxanthines (Yu et al. 2009). The reductases NdmA and NdmB are independent Rieske non-heme iron monooxygenases that carry out *N*-demethylation of caffeine to theobromine, and theobromine to 7-methylxanthine. These are dependent on a Reiske reductase NdmD that contains one Rieske [2Fe-2S] cluster, one plant-like [2Fe-2S] cluster, and one FMN. The final conversion of 7-methylxanthne to xanthine by NdmC completes the pathway (Summers et al. 2012).

3.2.7 Hydroxylation of Hydroxybenzoates and Related Compounds

These are widely distributed metabolites from the monooxygenation of alkylated phenols, and dioxygenation of arenes with one to four rings. It is convenient first to summarize briefly the spectrum of

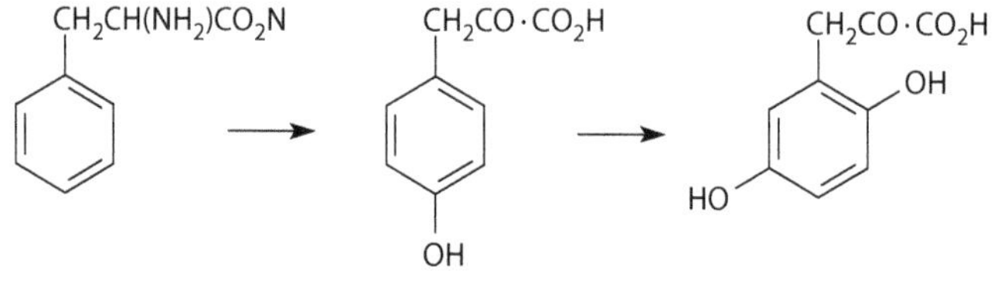

FIGURE 3.9 Hydroxylation of phenylalanine followed by rearrangement to 2,5-dihydroxyphenylpyruvate.

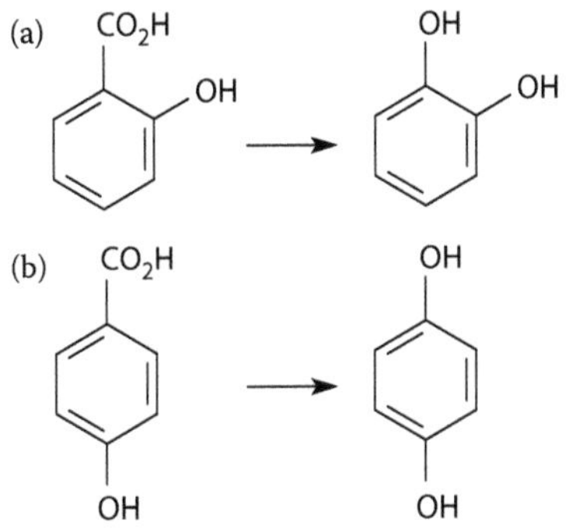

FIGURE 3.10 Monoxygenation of (a) salicylate and (b) 4-hydroxybenzoate.

substrates that undergo monooxygenation (hydroxylation) and the products formed: the mechanisms for specific substrates are given later.

1. Hydroxybenzoates can undergo hydroxylation with or without concomitant loss of CO_2. For example, salicylate → catechol + CO_2 (salicylate-1-hydroxylase) (Figure 3.10a) (White-Stevens et al. 1972) and 4-hydroxybenzoate → 1,4-dihydroxybenzene + CO_2 [4-hydroxybenzoate 1-hydroxylase (decarboxylating)] in *Candida parapsilosis* (Figure 3.10b) (Eppink et al. 1997).
2. Hydroxylation without elimination of CO_2 is illustrated by the examples in Table 3.2.

Degradation of the products then requires fission of the catechols (or 2,5-dihydroxybenzoates) that are produced. All three fission pathways have been observed for 3,4-dihydroxybenzoate:

1. Intradiol 3,4-dioxygenation in *Ps. putida* and the 3-oxoadipate pathway (Figure 3.12)
2. Extradiol 4,5-dioxygenation in the *Pseudomonas acidovorans* group (Figure 3.13a)
3. Extradiol 2,3-dioxygenation in species of *Bacillus* (Figure 3.13b)

TABLE 3.2

Hydroxylation of Hydroxybenzoates, Anthranilate, Naphthoate, and Quinol-2-one

Substrate	Product	Enzyme	Organism	Figure	Reference
Salicylate	2,5-Dihydroxybenzoate	Salicylate-5-hydroxylase	*Ralstonia* sp.	3.11a	Zhou et al. (2002)
Anthranilate	5-Hydroxyanthranilate + 2,5-Dihydroxybenzoate	Anthranilate-5-hydroxylase	*Nocardia opaca*	3.11b	Cain (1968)
	2,3-Dihydroxybenzoate		*Trichosporon cutaneum*	3.11c	Powlowski et al. (1987)
3-Hydroxybenzoate	2,5-Dihydroxybenzoate	3-Hydroxybenzoate-6- hydroxylase	*Ps. alcaligenes*		Gao et al. (2005)
4-Hydroxybenzoate	3,4-Dihydroxybenzoate	4-Hydroxybenzoate-3-hydroxylase	*Ps. aeruginosa*		Entsch and Ballou (1989); Entsch and van Berkel (1995)
3-Aminobenzoate	5-Aminosalicylate	3-Aminobenzoate-6-hydroxylase		3.11d	Russ et al. (1994)
Naphth-2-oate	1-Hydroxynaphth-2-oate[a]		*Burkholderia* sp.		Morawski et al. (1997)
2-Hydroxy-quinoline	2,8-Dihydroxyquinoline		*Ps. putida*	3.11e	Rosche et al. (1995, 1997)

[a] Or by dioxygenation and elimination.

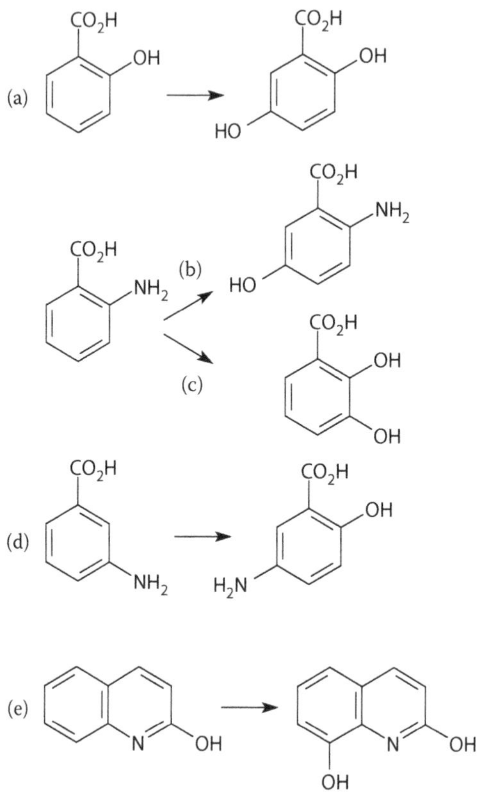

FIGURE 3.11 Monooxygenation of (a) salicylate, (b,c) 2-aminobenzoate, (d) 3-aminobenzoate, and (e) 2-hydroxyquinoline.

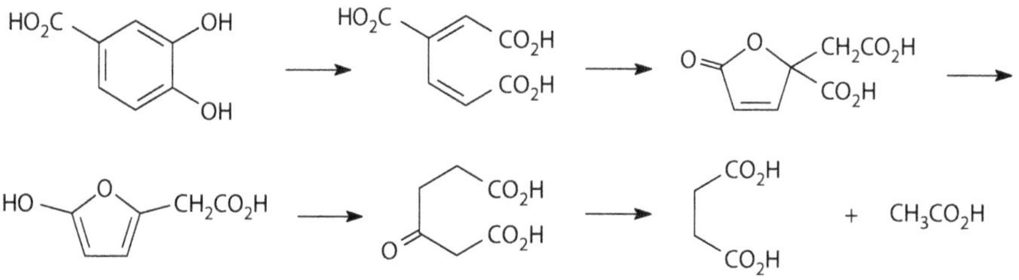

FIGURE 3.12 The β-oxoadipate pathway.

When 1,4-dihydroxy compounds are produced by monooxygenation, these are degraded by the gentisate pathway (Figure 3.14) mediated by gentisate dioxygenase (Wergath et al. 1998).

3.2.7.1 Mechanisms of Hydroxylation

1. The metabolism of 4-hydroxybenzoate involves conversion into 3,4-dihydroxybenzoate by a hydroxylase that has been purified and characterized from a strain of *Ps. fluorescens* (Howell et al. 1972), and from *Ps. aeruginosa* strain PAO1 (Entsch and Ballou 1989; Entsch and van Berkel 1995). The enzyme is a flavoprotein containing FAD and requires NADPH for activity. On the other hand, the degradation of 4-hydroxybenzoate by *Haloarcula* sp. D1 proceeds via 2,5-dihydroxybenzoate (gentisate) that involves an NIH shift (Fairley et al. 2002).

2. Naphthalene is metabolized by *Polaromonas naphthalenivorans* CJ2 by the gentisate pathway. Efforts failed, however, to find evidence for the expected salicylate 5-hydroxylase, but established instead the presence of 3-hydroxybenzoate-6-hydroxylase (Park et al. 2007) that is also

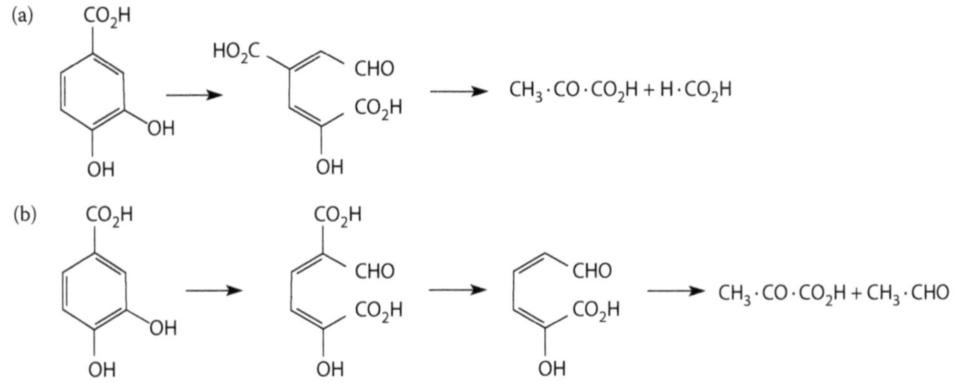

FIGURE 3.13 Biodegradation of 3,4-dihydroxybenzoate mediated by (a) 4,5-dioxygenase in *Pseudomonas acidovorans* and (b) by 2,3-dioxygenase in *Bacillus macerans*.

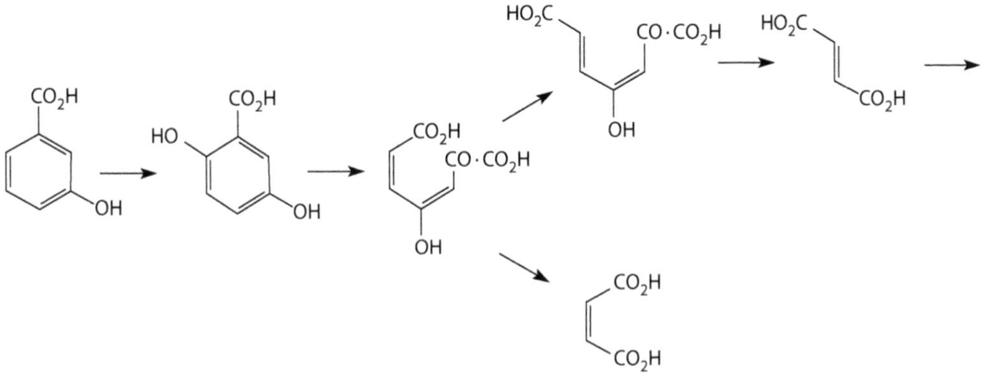

FIGURE 3.14 The gentisate pathway.

involved in the degradation of 2,5-dimethylphenol by the gentisate pathway in *Ps. alcaligenes* NCIMB 9867 (Gao et al. 2005).

3. a. Monooxygenases in a strain of *Ps. putida* (Arunachalam et al. 1992) and in *E. coli* strain W (Prieto and Garcia 1994) introduce an oxygen atom into 4-hydroxyphenylacetate to produce 3,4-dihydroxyphenylacetate. They were reported to be flavoprotein monooxygenases that require a further protein component (coupling protein) for activity: in the absence of this protein, oxidation of NADH produces H_2O_2 (Arunachalam et al. 1992). This phenomenon has also been observed with other oxygenases, including salicylate hydroxylase, 2,4-dichlorophenol hydroxylase, and 2,4,6-trichlorophenol monooxygenase. A further examination using a strain of *E. coli* revealed that the monooxygenase is unusual in the requirement for $FMNH_2$: the hydroxylation is carried out by HpaB coupled with a reductase that reduces $NADH^+$ (Xun and Sandvik 2000).

   b. The hydroxylase in *Acinetobacter baumannii* that carries out hydroxylation of 4-hydroxyphenylacetate to 3,4-dihydroxyphenylacetate is a two-component enzyme (Chaiyen et al. 2001).

4. a. Salicylate is an intermediate in the bacterial metabolism of PAHs including naphthalene and phenanthrene, and its degradation involves oxidative decarboxylation to catechol. The hydroxylase (monooxygenase) has been extensively studied (references in White-Stevens and Kamin 1972) and, in the presence of an analogue that does not serve as a substrate, NADH is oxidized with the production of H_2O_2 (White-Stevens and Kamin 1972). This "uncoupling" is characteristic of flavoenzymes and is exemplified also by the chlorophenol hydroxylase from an *Azotobacter* sp. that is noted later.

b. The salicylate 1-hydroxylase from *Sphingomonas* sp. strain CHY-1 is, unusually, a three-component Fe–S protein complex (Jouanneau et al. 2007). The oxygenase contained one [2Fe–2S] cluster and one mononuclear Fe per α subunit. The enzyme produced catechol from salicylate and 2-aminophenol from anthranilate, but was inactive with naphth-1-ol-2-carboxylate and produced H_2O_2.

c. In *Pseudomonas stutzeri* strain AN10, there are two inducible salicylate hydroxylases encoded by *nahG* and *nahW*, both of which display broad substrate specificities (Bosch et al. 1999).

5. The metabolism of 4-hydroxyphenylacetate that can be used to support the growth of *Ps. acidovorans* has been examined. Degradation was initiated by 1-hydroxylation with the formation of 2,5-dihydroxyphenylacetate involving an NIH shift (Hareland et al. 1975). In contrast to analogous NIH shifts in dioxygenation that are discussed later, the side chain is incorporated unchanged. An analogous hydroxylation of homovanillate putatively produced 3,6-dihydroxy-2-methoxyphenylacetate before ring fission by 1,2-dioxygenation (Allison et al. 1995).

6. The 4-methoxybenzoate monooxygenase from *Ps. putida* shows low substrate specificity. Although it introduces only a single atom of oxygen into 3-hydroxy-, and 4-hydroxybenzoate, it accomplishes the conversion of 4-vinylbenzoate into the corresponding side-chain diol (Wende et al. 1989).

7. Salicylate hydroxylase has been purified and characterized from the yeast *Trichosporon cutaneum*. It was monomeric with a molecular mass of 45,000, and contained FAD. It was capable of hydroxylating a range of *ortho*-substituted benzoates including all dihydroxybenzoates, 4- and 5-aminosalicylate and, less effectively 4- and 5-chlorosalicylate (Sze and Dagley 1984).

3.2.7.2 Cytochrome P450 Hydroxylation

This large family of heme enzymes carries out hydroxylation by a completely different mechanism, and is represented in both prokaryotes and eukaryotes, as well as in mammals. Cytochrome P450s are thiolate-ligated heme enzymes that use dioxygen and the formal reducing equivalents of H_2 to carry out the introduction of oxygen into a wide range of compounds. Extensive reviews (Sono et al. 1996; Ortiz de Montelano and De Vos 2002; Meunier et al. 2004) provide examples that illustrate both the mechanisms and the wide range of reactions that are catalyzed by cytochrome P450s, and the crystal structures of two important cytochrome P450s, P450$_{terp}$ (Hasemann et al. 1994) and 450$_{cam}$ (Poulos et al. 1987) have been determined. The key intermediate in oxidation is a heme iron bearing a single oxygen atom $Fe^{IV}=O$. Although this short-lived intermediate does not accumulate during turnover, it has been prepared by the rapid *m*-chloroperbenzoic acid oxidation of the ferric CYP119 protein from *Sulfolobus acidocaldarius*, and has been extensively characterized after ≈35 ms by UV/visible, Mössbauer and EPR spectroscopy. The electronic structure of the oxidation product was described as an $S = 1$.

Fe^{IV}oxo unit exchanged coupled with an $S = \frac{1}{2}$ ligand-bawsed radical, and the rate of oxidation of C_{12} carboxylate exceeded that of the C_6 carboxylate (Rittle and Green 2010). The structures of mammalian cytochrome P450 and the reactions that they catalyze have been summarized (Vaz et al. 1998):

$$Fe^{3+} + 2\varepsilon + O_2 \rightarrow Fe^{III}O_2^-(peroxo) + H^+ \Leftrightarrow Fe^{III}O_2(hydroperoxo) \rightarrow Fe^{IV} = O(oxenoid)$$

Specific Reactions

Peroxo	$Fe^{III}O_2^-$	$\Leftrightarrow$ Aldehyde deformylation: cyclohexane carboxaldehyde $\rightarrow$ cyclohexene
Hydroperoxo	$Fe^{III}O_2$	$\Leftrightarrow$ Epoxidation of alkenes
Oxenoid	$Fe^{IV}=O$	$\Leftrightarrow$ Epoxidation, hydroxylation of alkenes

3.2.7.2.1 Prokaryotic Organisms

The functioning of cytochrome P450 involves both oxygenation and reduction reactions (Tyson et al. 1972), and this is consistent with its role in both hydroxylation and reductive transformations. This activity is widely distributed among both Gram-negative and Gram-positive organisms, and mediates a number of important degradations and transformations including hydroxylations at quaternary carbon atoms. Cytochrome P450 systems in actinomyces have been reviewed (O'Keefe and Harder 1991), and the systematic nomenclature of some important bacterial cytochrome P450 systems has been given (Munro and Lindsay 1996). Bacterial cytochrome P450 enzymes are involved in the transformation of a wide range of substrates including the following illustrations.

3.2.7.2.1.1 Alkanes and Cycloalkanes: Including Terpenoids and Steroids

1. It has been shown that some alkane-degrading bacteria lack the membrane-integral iron hydroxylating enzymes that have already been discussed. This was first demonstrated for cell extracts of *Gordonia* (*Corynebacterium*) sp. strain 7E1C (Cardini and Jurtshuk 1968), and was extended to strains of *Acinetobacter calcoaceticus* (Asperger et al. 1981). Since then this has been shown to operate in a range of Gram-negative bacteria (van Beilen et al. 2006), and the cytochrome P450 enzyme that is designated CYP153 has been characterized in *Acinetobacter* sp. strain EB104 (Maier et al. 2001). The cytochrome P450 CYP153 A6 in *Mycobacterium* sp. strain HXN-1500 is also able to hydroxylate alkanes from C_6 to C_{11} including 2-methyloctane (Funhoff et al. 2006).

2. The hydroxylation of cyclohexane by a strain of *Xanthobacterium* sp. (Trickett et al. 1990). In cell extracts, a range of other substrates were oxidized including cyclopentane, pinane, and toluene (Warburton et al. 1990).

3. The initial hydroxylation in the degradation of terpenes: the ring methylene group of camphor by *Ps. putida* (Katagiri et al. 1968; Tyson et al. 1972; Koga et al. 1986), and the isopropylidene methyl group of linalool by a strain of *Ps. putida* (Ullah et al. 1990).

4. When the 5- and 6-positions of camphor are blocked by substituents, hydroxylation at other positions may take place. For example, the quaternary methyl group of 5,5-difluorocamphor is hydroxylated to the 9-hydroxymethyl compound (Figure 3.15a) (Eble and Dawson 1984).

5. Adamantane (A) and adamantan-4-one (B) were specifically hydroxylated at the quaternary C_1 by bacterial cytochrome $P450_{cam}$ to produce C and D. On the other hand, the eukaryotic cytochrome $P450_{LM2}$ formed in addition to the C_2 compound from adamantane, and both

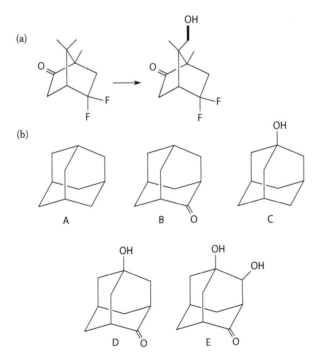

FIGURE 3.15 Hydroxylation of (a) 5,5-difluorocamphor, (b) adamantane (A) and adamantan-4-one (B).

5-hydroxyadamantan-1-one (D) and the 4-*anti*-hydroxyadamantan-1-one (E) from adamantan-4-one (Figure 3.15b) (White et al. 1984).

6. Hydroxylation of progesterone and closely related compounds at the 15-β position is carried out by cell extracts of *Bacillus megaterium* (Berg et al. 1976).

7. Hydroxylation of vitamin D_3 derivatives at C_{25} and $C_1(\alpha)$ has been carried out by cytochrome P450 transformation using strains of *Streptomyces* sp. (Sasaki et al. 1991).

8. The degradation of cholesterol *Rhodococcus jostii* RHA1 is initiated by hydroxylation at C_{26} by cytochrome P450 (CYP125) that has been purified (Rosłoniec et al. 2009).

3.2.7.2.1.2 Arenes

1. The transformation of benzene, toluene, naphthalene, biphenyl, and benzo[*a*]pyrene to the corresponding phenols (Trower et al. 1989) by *Streptomyces griseus*, and of phenanthrene by *Streptomyces flavovirens* to (-)*trans*-[9*S*,10*S*]-9,10-dihydrodihydroxyphenanthrene with minor amounts of 9-hydroxyphenanthrene are carried out by cytochrome P450 hydroxylases (Sutherland et al. 1990).

2. In addition to dioxygenases, *Mycobacterium vanbaalenii* PYR-1 possesses three cytochrome P450 monooxygenase genes whose sequence was determined. The proteins resembled those of CYP151, CYP150, and CPY 51 in mycobacteria, and whole-cell monooxygenation was observed with pyrene and 7-methylbenz[*a*]anthracene. It was shown that genes for these cytochrome P450s in addition to the PAH dioxygenase genes *nidA* and *nidB* were widely distributed in PAH-degrading mycobacteria (Brezna et al. 2006).

3. A mutant of cytochrome P450$_{cam}$ (CYP101) from *Ps. putida* was constructed (Jones et al. 2001), and was able to carry out the hydroxylation (monooxygenation) of polychlorinated arenes (Chen et al. 2002).

4. The bacterial cytochrome P450$_{cam}$ is able to bring about the stereoselective epoxidation of *cis*-methylstyrene to the 1*S*,2*R* epoxide (Ortiz de Montellano et al. 1991).

5. The degradation of bisphenol-A by *Sphingomonas* sp. strain AO1 is initiated by hydroxylation to intermediates that undergo fission to 4-hydroxyacetophenone and 4-hydroxybenzoate. The components have been purified, and they comprise cytochrome P450, ferredoxin reductase, and ferredoxin (Sasaki et al. 2005).

6. The first step in the aerobic degradation of dehydroabietic acid by *Pseudomonas abietaniphila* strain BMKE-9 is hydroxylation at C_7 (Smith et al. 2004). A study with *Burkholderia xenovorans* provided additional details: abietic acid was oxidized by cytochrome P450 (CYP226) DitU to the 7,8-epoxide, and dehydroabietic acid by DitQ to 7-hydroxydehydroabietic acid (Smith et al. 2008).

3.2.7.2.1.3 Aliphatic Compounds: Thiols, Ethers, Nitramines

1. The transformation of sulfonylureas by *Streptomyces griseolus* grown in a complex medium containing glucose in which the methyl group of the heterocyclic moieties is hydroxylated, and for some substrates, subsequently oxidized to the carboxylic acid (Romesser and O'Keefe 1986; O'Keefe et al. 1988).

2. The dealkylation of *S*-ethyl dipropylthiocarbamate and atrazine by a strain of *Rhodococcus* sp. (Nagy et al. 1995a,b), and of 2-ethoxyphenol and 4-methoxybenzoate by *Rhodococcus rhodochrous* (Karlson et al. 1993). In *R. erythropolis* NI86/21, however, the *thcF* gene involved in the degradation of thiocarbamate herbicides is a nonheme haloperoxidase that does not occur in other strains of rhodococci that can degrade thiocarbamates (de Schrijver et al. 1997): this specificity occurs also for the cytochrome P450 system encoded by the *thcRBCD* genes (Nagy et al. 1995a,b).

3. The oxidation of *tert*-butyl methyl ether to *tert*-butanol (Steffan et al. 1997) which is also mediated by the cytochrome P450 from camphor-grown *Ps. putida* CAM, although not by that from *R. rhodochrous* strain 116.

4. Initiation of the degradation of morpholine by *Mycobacterium* strain RP1 (Poupin et al. 1998), and also of pyrrolidine and piperidine.

5. The cytochrome P450 system XplA and XplB carry out the degradation of hexahydro-1,3,5-trinitro-1,3,5-triazine (RDX) in strains of *Rhodococcus* sp. (Seth-Smith et al. 2008). XplA was expressed in *E. coli*, purified, and crystallized. X-ray data showed that the heme domain was fused to the flavodoxin redox partner XplB (Sabbadin et al. 2009).

6. The reductive dehalogenation of polyhalogenated methanes (Castro et al. 1985) and polyhalogenated ethanes (Li and Wackett 1993) by *Ps. putida* strain PpG786.

3.2.7.2.2 Eukaryotic Organisms

Cytochrome P450 hydroxylation activity is well established in eukaryotic yeasts and some fungi, and fungal cytochrome P450 has been used as a surrogate and as a model for mammalian transformation of xenobiotics (Ferris et al. 1976; Smith and Rosazza 1983).

3.2.7.2.2.1 Degradation of Alkanes

1. Hydroxylation of yeasts, typically with chain lengths >4 has been found primarily in species of *Candida* and *Endomycopsis* (references in Käppeli 1986). Although the enzyme designated cytochrome $P450_{aO}$ is induced in organisms including species of *Candida* and *Endomycopsis* during growth with alkanes and, therefore, differs from the enzyme ($P450_{14DH}$), it is synthesized during the growth of saccharophilic yeasts under conditions of low oxygen concentration.

2. Hydroxylation is carried out within microsomes that contain cytochrome P450 and NADPH-cytochrome *c* reductase. *Candida tropicalis* is also able to convert alkanes and alkanoates to α,ω-dicarboxylates (Craft et al. 2003) that has been attributed to the activity of CYP 52 and has been characterized (Eschenfeldt et al. 2003). In addition, the monooxygenase in *Candida maltosa* is able to carry out the conversion of hexadecane in a cascade of reactions by bis-terminal hydroxylation and dehydrogenation to the α,ω-dicarboxylate (Scheller et al. 1998).

3. It is important to note that most eukaryotic organisms such as yeasts and fungi contain peroxisomes, and the degradation of long-chain alkanes by yeasts is carried out in two separate organelles: hydroxylation to alkanols in microsomes (Käppeli 1986), and oxidation to the alkanoic acid CoA-esters that are further metabolized in peroxisomes. Peroxisomes are induced during growth with alkanes and are multipurpose organelles. Some of the salient features that differentiate the pathway from that used for bacterial degradation of alkanoates include the following (Tanaka and Ueda 1993).

4. In *Candida* sp., degradation of the CoA-alkanoic esters to the alkenoic acid esters is catalyzed by an acyl-CoA oxidase and results in the production of H_2O_2 that is converted into O_2 by catalase activity. The enzyme from *C. tropicalis* contains FAD (Jiang and Thorpe 1983), and in *C. lipolytica* carries out a stereospecific *anti*-elimination of hydrogen (Kawaguchi et al. 1980).

5. A bifunctional enol-CoA hydratase and 3-hydroxyacetyl-CoA dehydrogenase are used in the degradation of CoA-alkenoic esters to the 3-oxo acid. This is then degraded to acetyl-CoA and the lower alkanoate ester by 3-oxoacetyl CoA thiolase and acetyl-CoA thiolase activities.

6. With the exception of the acetyl-CoA thiolase, all of these enzymes are located exclusively in the peroxisomes, whereas the enzymes that are involved in lipid synthesis are found in the microsomes and the mitochondrion.

3.2.7.2.2.2 Transformation of Arenes

1. The fungal transformation of arenes including polycyclic representatives is mediated by cytochrome P450 monooxygenation that has been established for *Cunninghamella bainieri* (Ferris et al. 1976). The initial product is an epoxide that may then undergo different reactions: rearrangement to a phenol with a concomitant NIH shift, or by the activity of an epoxyhydrolase to a *trans*-1,2-dihydrodiol. This was illustrated for the metabolism of naphthalene by *Cunninghamella elegans* (Cerniglia et al. 1983).

2. The cytochrome P450 aryl hydroxylase of *Saccharomyces cerevisiae* transforms benzo[*a*]pyrene to the 3- and 9-hydroxy compounds and the 7,8-dihydrodiol (King et al. 1984).

3.2.7.2.2.3 Transformation of Steroids and Related Compounds

1. The hydroxylation of steroids at various positions is carried out by a range of fungi, and probably involves the cytochrome P450 system (Breskvar and Hudnik-Plevnik 1977). Examples are given in Chapter 7, Part 2 dealing with steroids.

2. The removal of angular methyl groups is important in the transformation of steroids and related compounds. In these reactions the methyl group is oxidized to the aldehyde before fission in which the carbonyl group oxygen is retained in formate (or acetate), and one oxygen atom from dioxygen is incorporated into the hydroxyl group. A group of formally analogous reactions using $^{18}O_2$ has confirmed the mechanisms. The range of reactions is illustrated by the following.

   a. The oxidative removal of the 14α-methyl group in lanosterol by *S. cerevisiae* with concomitant production of the Δ^{14-15} bond (Aoyama et al. 1984, 1987) and loss of formate (Figure 3.16a) is plausibly facilitated by formation of the conjugated 8–9,14–15-diene. The enzyme designated cytochrome $P450_{14DM}$ is induced under "semi-anaerobic" conditions during growth with glucose, and is different from both the $P450_{aO}$ enzyme, and from $P450_{22DS}$ that brings about Δ^{22}-dehydrogenation (Hata et al. 1983). The mechanism has been examined in a constructed strain containing the 14α-demethylase from *Candida albicans* and involves reaction of the carbonyl group with Fe^{III}–O–OH (Shyadehi et al. 1996).

   b. The conversion of androgens to estrogens involves removal of the C_{10}-methyl group and aromatization of the A ring of $\Delta^{4,5}$-3-oxosterols. This is accomplished by oxidation to the C_{10}-methyl group to the aldehyde, followed by loss of formate and aromatization of the ring to a 3-hydroxy-desmethylsterol (Figure 3.16b) (Stevenson et al. 1988).

   c. Cytochrome $P450_{17a}$ carries out comparable reactions for removal of the C_{17}-side chain of pregnenolone. Two reactions have been described both of which involved loss of acetate—17α-hydroxylation and formation of the 17-oxo compound, and direct formation of the $\Delta^{16,17}$-ene (Figure 3.16c) (Akhtar et al. 1994).

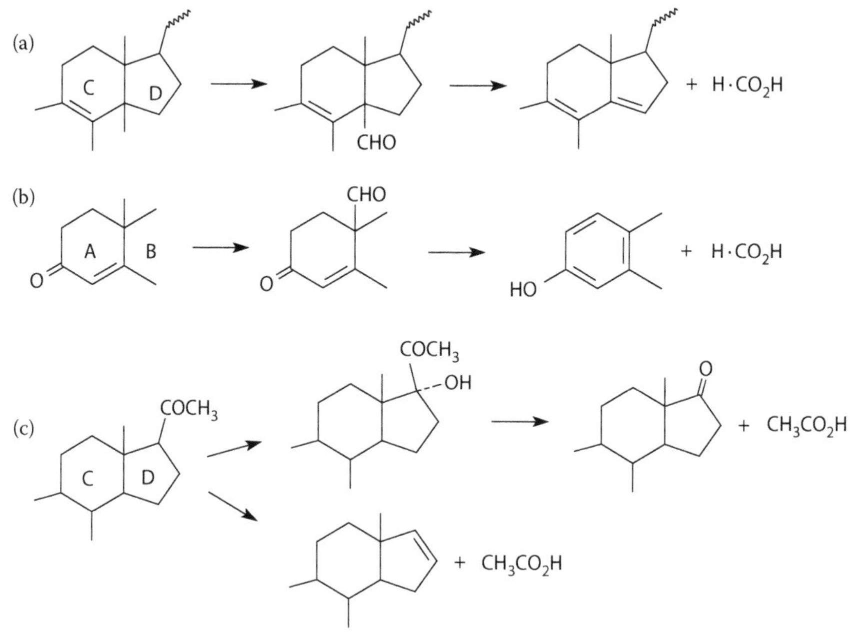

FIGURE 3.16 Cytochrome P450 monooxygenation of (a) lanosterol, (b) aromatization of $\Delta^{4,5}$-3-oxosterols, and (c) pregnenolone.

3.3 References

Abe T, E Masai, K Miyauchi, Y Katayama, M Fukuda. 2005. A tetrahydrofolate-dependent *O*-demethylase, LigM, is crucial for catabolism of vanillate and syringate in *Sphingomonas paucimobilis* SYK-6. *J Bacteriol* 187: 2030–2037.

Adriaens P. 1994. Evidence for chlorine migration during oxidation of 2-chlorobiphenyl by a type II methanotroph. *Appl Environ Microbiol* 60: 1658–1662.

Akhtar M, DL Corina, SL Miller, AZ Shyadehi. 1994. Incorporation of label from $^{18}O_2$ into acetate during side-chain cleavage catalyzed by cytochrome P-450$_{17\alpha}$ (17α-hydroxylase-17,20-lyase). *J Chem Soc Perkin Trans* 1: 263–267.

Allen CE, MP Schmitt. 2011. Novel hemin binding domains in the *Corynebacterium diphtheriae* HtaA protein interact with hemoglobin and are critical to heme iron utilization by HtaA. *J Bacteriol* 193: 5374–5385.

Allen JR, DD Clark, JG Krumn, SA Ensign. 1999. A role for coenzyme M (2-mercaptoethanesulfonic acid) in a bacterial pathway of aliphatic epoxide carboxylation. *Proc Natl Acad Sci USA* 96: 8432–8437.

Allen JR, SA Ensign. 1997. Purification to homogeneity and reconstitution of the individual components of the epoxide carboxylase multiprotein enzyme complex from *Xanthobacter* strain Py2. *J Biol Chem* 272: 32121–32128.

Allison N, JE Turner, R Wait. 1995. Degradation of homovanillate by a strain of *Variovorax paradoxus* via ring hydroxylation. *FEMS Microbiol Lett* 134: 213–219.

Aoyama Y, Y Yoshida, R Sato. 1984. Yeast cytochrome P-450 catalyzing lanosterol 14α-demethylation. II. Lanosterol metabolism by purified P-450$_{14DM}$ and by intact microsomes. *J Biol Chem* 259: 1661–1666.

Aoyama Y, Y Yoshida, Y Sonoda, Y Sato. 1987. Metabolism of 32-hydroxy-24,25-dihydrolanosterol by purified cytochrome P-450$_{14DM}$ from yeast. Evidence for contribution of the cytochrome to whole process of lanosterol 14α-demethylation. *J Biol Chem* 262: 1239–1243.

Arias-Barrau E, ER Olivera, JM Luengo, C Fernández, B Gálan, JL Garcia, E Díaz, B Minambres. 2004. The homogentisate pathway: A central catabolic pathway involved in the degradation of L-phenylalanine, L-tyrosine, and 3-hydroxyphenylacetate in *Pseudomonas putida*. *J Bacteriol* 186: 5062–5077.

Arunachalam U, V Massey, CS Vaidyanathan. 1992. *p*-Hydroxyphenylacetate-3-hydroxylase. A two-protein component enzyme. *J Biol Chem* 267: 25848–25855.

Asperger O, A Naumann, H-P Kleber. 1981. Occurrence of cytochrome P-450 in *Acinetobacter* strains after growth on *n*-hexadecane. *FEMS Microbiol Lett* 11: 309–312.

Ayoubi PJ, AR Harker. 1998. Whole-cell kinetics of trichloroethylene degradation by phenol hydroxylase in *Ralstonia eutropha* JMP134 derivative. *Appl Environ Microbiol* 64: 4353–4356.

Baulard AR, JC Betts, J Engohang-Ndong, S Quan, RA McAdam, PJ Brennard, C Locht, GS Besra. 2000. Activation of the pro-drug ethionamide is regulated in mycobacteria. *J Biol Chem* 275: 28326–28331.

Baxter NJ, J Scanlan, P De Marco, AP Wood, JC Murrell. 2002. Duplicate copies of genes encoding methanesulfonate monooxygenase in *Marinosulfonomonas methylotrophica* strain TR3 and detection of methansulfonate utilizers in the environment. *Appl Environ Microbiol* 68: 289–296.

Beadle TA, ARW Smith. 1982. The purification and properties of 2,4-dichlorophenol hydroxylase from a strain of *Acinetobacter* species. *Eur J Biochem* 123: 323–332.

Berg A, JÅ Gustafsson, M Ingelman-Sundberg, K Carlström. 1976. Characterization of cytochrome P-450-dependent steroid hydroxylase present in *Bacillus megaterium*. *J Biol Chem* 251: 2831–2838.

Bertoni G, M Martino, E Galli, P Barbieri. 1998. Analysis of the gene cluster encoding toluene/*o*-xylene monooxygenase from *Pseudomonas stutzeri* OX1. *Appl Environ Microbiol* 64: 3626–3632.

Boden R, E Borodina, AP Wood, DP Kelly, JC Murrell, H Schäfer. 2011. Purification and characterization of dimethylsulfide monooxygenase from *Hyphomicrobium sulfonivorans*. *J Bacteriol* 193: 1250–1258.

Borodina E, DP Kelly, P Schumann, FA Rainey, NL Ward-Rainey, AP Wood. 2002. Enzymes of dimethylsulfone metabolism and the phylogenetic characterization of the facultative methylotrophs *Arthrobacter sulfonivorans* sp. nov., *Arthrobacter methylotrophus* sp. nov., and *Hyphomicrobium sulfonivorans* sp. nov. *Arch Microbiol* 77: 173–183.

Bosch R, ERB Moore, E García-Valdés, DH Pieper. 1999. NahW, a novel, inducible salicylate hydroxylase involved in mineralization of naphthalene by *Pseudomonas stutzeri* AN10. *J Bacteriol* 181: 2315–2322.

Breskvar K, T Hudnik-Plevnik. 1977. A possible role of cytochrome P-450 in hydroxylation of progesterone by *Rhizopus nigricans*. *Biochem Biophys Res Commun* 74: 1192–1198.

Brezna B, O Kweon, RI Stingley, JP Freeman, AA Khan, B Polek, RC Jones, CE Cerniglia. 2006. Molecular characterization of cytochrome P450 genes in the polycyclic aromatic hydrocarbon degrading *Mycobacterium vanbaalenii* PYR-1. *Appl Microbiol Biotechnol* 71: 522–532.

Cadieuz E, V Vrajmasu, C Achim, J Powlowski, E Mäunck. 2002. Biochemical, Mössbauer, and EPR studies of the diiron cluster of phenol hydroxylase from *Pseudomonas* sp. strain CF 600. *Biochemistry* 41: 10680–10691.

Cafaro V, R Scognamiglio, A Viggiani, V Izzo, I Passaro, E Notomista, PD Piaz et al. 2002. Expression and purification of the recombinant subunits of toluene/*o*-xylene monooxygenase and reconstitution of the active complex. *Eur J Biochem* 269: 5689–5699.

Cafaro V, V Izzo, R Scognamiglio, E Notomista, P Capasso, A Casbarra, P Pucci, A Di Donato. 2004. Phenol hydroxylase and toluene/*o*-xylene monooxygenase from *Pseudomonas stutzeri* OX1: Interplay between two enzymes. *Appl Environ Microbiol* 70: 2211–2219.

Cai M, L Xun. 2002. Organization and regulation of pentachlorophenol-degrading genes in *Sphingobium chlorophenolicum* ATCC 39723. *J Bacteriol* 184: 4672–4680.

Cain RB. 1968. Anthranilic acid metabolism by microorganisms. Formation of 5-hydroxyanthranilate as an intermediate in anthranilate metabolism by *Nocardia opaca*. *Anthonie van Leewenhoek* 34: 417–432.

Capyk JK, I d'Angelo, NC Strynadya, LD Eltis. 2009. Characterization of 3-ketosteroid 9α-hydroxylase, a Rieske oxygenase in the cholesterol degradation pathway of *Mycobacterium tuberculosis*. *J Biol Chem* 284: 9937–9946.

Cardini G, P Jurtshuk. 1968. Cytochrome P-450 involvement in the oxidation of *n*-octane by cell-free extracts of *Corynebacterium* sp. strain 7E1C. *J Biol Chem* 243: 6070–6072.

Castro CE, RS Wade, NO Belser. 1985. Biodehalogenation reactions of cytochrome P-450 with polyhalomethanes. *Biochemistry* 24: 204–210.

Cerniglia CE, JP Freeman, FE Evans. 1984. Evidence for an arene oxide-NIH shift pathway in the transformation of naphthalene to 1-naphthol by *Bacillus cereus*. *Arch Microbiol* 138: 283–286.

Cerniglia CE, JR Althaus, FE Evans, JP Freeman, RS Mitchum, SK Yang. 1983. Stereochemistry and evidence for an arene oxide-NIH shift pathway in the fungal metabolism of naphthalene. *Chem-Biol Interactions* 44: 119–132.

Chaiyen P, C Suadee, P Wilairat. 2001. A novel two-protein component flavoprotein hydroxylase *p*-hydroxyphenylacetate hydroxylase from *Acinetobacter baumannii*. *Eur J Biochem* 268: 5550–5561.

Chen D, PA Frey. 1998. Phenylalanine hydroxylase from *Chromobacterium violaceum*. Uncoupled oxidation of tetrahydropterin and the role of iron in hydroxylation *J Biol Chem* 273: 25594–25601.

Chen X, A Christopher, JP Jones, SG Bell, Q Guo, F Xu, Z Rao, L-K Wong. 2002. Crystal structure of the F87W/Y96F/V247L mutant of cytochrome P-450cam with 1,3,5-trichlorobenzene bound and further protein engineering for the oxidation of pentachlorobenzene and hexachlorobenzene. *J Biol Chem* 277: 37519–37526.

Choi YS, H Zhang, JS Brunzelle, SK Nair, H Zhao. 2008. *In vitro* reconstitution and crystal structure of *p*-aminobenzoate *N*-oxygenase (AurF) involved in aureothin biosynrthesis. *Proc Natl Acad USA* 105: 6858–6863.

Craft DL, KM Madduri, M Eshoo, CR Wilson. 2003. Identification and characterization of the *CYP52* family of *Candida tropicalis* ATCC 20336, important for the conversion of fatty acids and alkanes to α,ω-dicarboxylic acids. *Appl Environ Microbiol* 69: 5983–5991.

Csáki R, L Bodrossy, J Klem, JC Murrell, KL Kovács. 2003. Genes involved in the copper-dependent regulation of soluble methane monooxygenase of *Methylococcus capsulatus* (Bath): Cloning, sequencing and mutational analysis. *Microbiology (UK)* 149: 1785–1795.

Cunane LM, Z-W Chen, N Shamala, FS Mattews, CN Cronin, WS McIntire. 2000. Structures of the flavocytochrome *p*-cresol methylhydroxylase and its substrate complex: Gated substrate entry and proton relays support the proposed catalytic mechanism. *J Mol Biol* 295: 357–374.

Dai M, JB Rogers, JR Warner, SD Copley. 2003. A previously unrecognized step in pentachlorophenol degradation in *Sphingobium chlorophenolicum* is catalyzed by tetrachlorobenzoquinone reductase (PcpD). *J Bacteriol* 185: 302–310.

Dalton H, BT Golding, BW Waters, R Higgins, JA Taylor. 1981. Oxidation of cyclopropane, methylcyclopropane, and arenes with the mono-oxygenase systems from *Methylococcus capsulatus*. *J Chem Soc Chem Commun* 482–483.

Davydov R, AM Valentine, S Komar-Panicucci, BM Hoffman, SJ Lippard. 1999. An EPR study of the dinuclear iron site in the soluble methane monooxygenase from *Methylococcus capsulatus* (Bath) reduced by one electron at 77 K: The effects of component interactions and the binding of small molecules to the diiron (III) center. *Biochemistry* 38: 4188–4197.

DeBarber AE, K Mdluli, M Bosman, L-G Bekker, CE Barry. 2000. Ethionamide activation and sensitivity in multidrug-resistant *Mycobacterium tuberculosis*. *Proc Natl Acad Sci USA* 97: 9677–9682.

De Marco P, P Moradas-Ferreira, TP Higgins, I McDonald, EM Kenna, JC Murrell. 1999. Molecular analysis of a novel methanesulfonic acid monooxygenase from the methylotroph *Methylosulfonomonas methylovora*. *J Bacteriol* 181: 2244–2251.

de Schrijver A, I Nagy, G Schoofs, P Proost, J Vanderleyden, K-H van Pée, R De Mot. 1997. Thiocarbamate herbicide-inducible non-heme haloperoxidase of *Rhodococcus erythropolis* NI86/21. *Appl Environ Microbiol* 62: 1811–1816.

Don RH, AJ Wightman, HH Knackmuss, KN Timmis. 1985. Transposon mutagenesis and cloning analysis of the pathways of degradation of 2,4-dichlorophenoxyacetic acid and 3-chlorobenzoate in *Alcaligenes eutrophus* JMP134 (pJP4). *J Bacteriol* 161: 85–90.

Donoghue NA, DB Norris, PW Trudgill. 1976. The purification and properties of cyclohexanone oxygenase from *Nocardia globerula* CL1 and *Acinetobacter* NCIB 9871. *Eur J Biochem* 63: 175–192.

Dover LG, A Alahari, P Gratraud, JM Gobes, V Bhowruth, RC Reynolds, GS Besra, L Kremer. 2007. EthA, a common activator of thiocarbamide-containing drugs acting on different mycobacterial targets. *Antimicrob Agents Chemother* 51: 1055–1063.

Dubbels BL, LA Sayavedra-Soto, DJ Arp. 2007. Butane monooxygenase of *Pseudomonas butanovora*: Purification and biochemical characterization of a terminal-alkane hydroxylating diiron monooxygenase. *Microbiology (UK)* 153: 1808–1816.

Eaton SL, SM Resnick, DT Gibson. 1996. Initial reactions in the oxidation of 1,2-dihydronaphthalene by *Sphingomonas yanoikuyae* strains. *Appl Environ Microbiol* 62: 4388–4394.

Eble KS, JH Dawson. 1984. Novel reactivity of cytochrome P-450 CAM methyl hydroxylation of 5,5-difluoro-camphor. *J Biol Chem* 259: 14389–14393.

Eggink G, H Engel, G Vriend, P Terpstra, B Withold. 1990. Rubredoxin reductase of *Pseudomonas oleovorans*. Structural relationship to other flavoprotein oxidoreductases based on one NAD and two FAD fingerprints. *J Mol Biol* 212: 135–142.

Eichhorn E, JR van der Ploeg, T Leisinger. 1999. Characterization of a two-component alkanesulfonate monooxygenase from *Escherichia coli*. *J Biol Chem* 274: 26639–26646.

Entsch B, DP Ballou. 1989. Purification, properties, and oxygen reactivity of *p*-hydroxybenzoate hydroxylase from *Pseudomonas aeruginosa*. *Biochim Biophys Acta* 999: 253–260.

Entsch B, WJH van Berkel. 1995. Structure and mechanism of *para*-hydroxybenzoate hydroxylase. *FASEB J* 9: 476–483.

Eppink MHM, SA Boeren, J Vervoort, WJH van Berkel. 1997. Purification and properties of 4-hydroxybenzoate 1-hydroxylase (decarboxylating), a novel flavin adenine dinucleotide-dependent monooxygenase from *Candida parapsilosis* CBS604. *J Bacteriol* 179: 6680–6687.

Eschenfeldt WH, Y Zhang, H Samaha, L Stols, LD Eirich, CR Wilson, MI Donnelly. 2003. Transformation of fatty acids catalyzed by cytochrome P450 monooxygenase enzymes of *Candida tropicalis*. *Appl Environ Microbiol* 69: 5992–5999.

Fairley DJ, DR Boyd, ND Sharma, CCR Allen, P Morgan, MJ Larkin. 2002. Aerobic metabolism of 4-hydroxybenzoic acid in *Archaea* via an unusual pathway involving an intramolecular migration (NIH shift). *Appl Environ Microbiol* 68: 6246–6255.

Ferris JP, LH MacDonald, MA Patrie, MA Martin. 1976. Aryl hydrocarbon hydroxylase in the fungus *Cunninghamella bainieri*: Evidence for the presence of cytochrome P-450. *Arch Biochem Biophys* 175: 443–452.

Fishman A, Y Tao, TK Wood. 2004. Toluene 3-monooxygenase of *Ralstonia pickettii* PKO1 is a *para*-hydroxylating enzyme. *J Bacteriol* 186: 3117–3123.

Flashner MIS, V Massey. 1974a. Purification and properties of L-lysine monooxygenase from *Pseudomonas fluorescens*. *J Biol Chem* 249: 2579–2586.

Flashner MIS, V Massey. 1974b. Regulatory properties of the flavoprotein L-lysine monooxygenase. *J Biol Chem* 249: 2587–2592.

Fox BG. WA Froland, JE Dege, JD Lipscomb. 1989. Methane monooxygenase from *Methylosinus trichosporium* OB3b. *J Biol Chem* 264: 10023–10033.

Fraaije MW, NM Kammerbeek, AJ Heidkamp, R Fortin, DB Janssen. 2004. The prodrug activator EtaA from *Mycobacterium tuberculosis* is a Baeyer–Villiger monooxygenase. *J Biol Chem* 279: 3354–3360.

Francis K, B Russell, G Gadda. 2005. Involvement of a flavosemiquinone in the enzymatic oxidation of nitroalkanes catalyzed by 2-nitropropane dioxygenase. *J Biol Chem* 280: 5195–5204.

Funhoff EG, E Bauer, I García-Rubio, B Witholt, JB van Beilen. 2006. CYP153A6, a soluble P450 oxygenase catalyzing terminal-alkane hydroxylation. *J Bacteriol* 188: 5220–5227.

Gadda G, RD Edmondson, DH Russell, PF Fitzpatrick. 1997. Identification of the naturally occurring flavin of nitroalkane oxidase from *Fusarium oxysporum* as a 5-nitrobutyl-FAD and conversion of the enzyme to the active FAD-containing form. *J Biol Chem* 272: 5563–5570.

Gadda G, K Francis. 2010. Nitronate monooxygenase, a model for anionic flavin semiquinone intermediates in oxidative catalysis. *Arch Biochem Biophys* 493: 53–61.

Gallagher SC, R Cammack, H Dalton. 1997. Alkene monooxygenase from *Nocardia corallina* B-276 is a member of the class of dinuclear proteins capable of stereospecific epoxygenation reactions. *Eur J Biochem* 247: 635–641.

Gallagher, SC, A George, H Dalton. 1998. Sequence-alignment modelling and molecular docking studies of the epoxygenase components of alkene monooxygenase from *Nocardia corallina* B-276. *Eur J Biochem* 254: 480–489.

Gao X, CL Tan, CC Yeo, CL Poh. 2005. Molecular and biochemical characterization of the *xlnD*-encoded 3-hydroxybenzoate 6-hydroxylase involved in the degradation of 2,5-xylenol via the gentisate pathway in *Pseudomonas alcaligenes* NCIMB 9867. *J Bacteriol* 187: 7696–7702.

Gibson DT, SM Resnick, K Lee, JM Brand, DS Torok, LP Wackett, MJ Schocken, BE Haigler. 1995. Desaturation, dioxygenation, and monooxygenation reactions catalyzed by naphthalene dioxygenase from *Pseudomonas* sp. strain 9816-4. *J Bacteriol* 177: 2615–2621.

Gisi MR, L Xun. 2003. Characterization of chlorophenol 4-monooxygenase (TfdD) and NADH:flavin adenine dinucleotide oxidoreductase (TftC) of *Burkholderia cepacia* AC1100. *J Bacteriol* 185: 2786–2792.

Grosse S, L Laramee, K-D Wendlandt, IR McDonald, CB Miguez, H-P Kleber. 1999. Purification and characterization of the soluble methane monooxygenase of the type II methanotrophic bacterium *Methylocystis* sp. strain WI 14. *Appl Environ Microbiol* 65: 3929–3935.

Ha JY, JY Min, SK Lee, HS Kim, DJ Kim, DJ Kim, KH Kim et al. 2006. Crystal structure of 2-nitropropane dioxygenase complexed with FMN and substrate. Identification of the catalytic base. *J Biol Chem* 281: 18660–18667.

Hanoulle X, J-M Wieruszeski, P Rousselot-Pailley, I Landrieu, C Locht, G Lippens, AR Baulard. 2006. Selective intracellular accumulation of the major metabolite issued from the activation of the prodrug ethionamide in mycobacteria. *J Antimicrob Chemother* 58: 768–772.

Harayama S, M Rekik, M Wubbolts, K Rose, RA Leppik, KN Timmis. 1989. Characterization of five genes in the upper-pathway operon of TOL plasmid pPWW0 from *Pseudomonas putida* and identification of the gene products. *J Bacteriol* 171: 5048–5055.

Hareland WA, RL Crawford, PJ Chapman, S Dagley. 1975. Metabolic function and properties of 4-hydroxy-phenylacetic acid 1-hydrolase from *Pseudomonas acidovorans*. *J Bacteriol* 121: 272–285.

Hartmans S, MJ van der Werf, JA de Bont. 1990. Bacterial degradation of styrene involving a novel flavin adenine dinucleotide-dependent monooxygenase. *Appl Environ. Microbiol* 56: 1347–1351.

Hasan SA, MIM Ferreira, MJ Koetsier, MI Arif, DB Janssen. 2011. Complete biodegradation of 4-fluorocinnamic acid by a consortium comprising *Arthrobacter* sp. strain G1 and *Ralstonia* sp. strain H1. *Appl Environ Microbiol* 77: 572–579.

Hasemann CA, KG Ravichandran, JA Peterson, J Deisenhofer. 1994. Crystal structure and refinement of cytochrome P450$_{terp}$ at 2.3 Å resolution. *J Mol Biol* 236: 1169–1185.

Hata S, T Nishino, H Katsuki, Y Aoyama, Y Yoshida. 1983. Two species of cytochrome P-450 involved in ergosterol biosynthesis in yeast. *Biochem Biophys Res Commun* 116: 162–166.

Heitkamp MA, JP Freeman, DW Miller, CE Cerniglia. 1988. Pyrene degradation by a *Mycobacterium* sp.: Identification of ring oxidation and ring fission products. *Appl Environ Microbiol* 54: 2556–2565.

Hemmi H, JM Studts, YK Chae, J Song, JL Markley, BG Fox. 2001. Solution structure of the toluene 4-monooxygenase effector protein (T4moD). *Biochemistry* 40: 3512–3524.

Herman PL, M Behrens, S Chakraborty, BM Chrastil, J Barycki, DP Weeks. 2005. A three-component dicamba *O*-demethylase from *Pseudomonas maltophilia*, strain DI-6. Gene isolation, characterization, and heterologous expression. *J Biol Chem* 280: 24759–24767.

Héroux A, DM Bozinovski, MP Valley, PF Fitzpatrick, AM Orville. 2009. Crystal structure of intermediate in the nitroalkane oxidase reaction. *Biochemistry* 48: 3407–3416.

Higgins TP, M Davey, T Trickett, DP Kelly, JC Murrell. 1996. Metabolism of methanesulfonic acid involves a multicomponent monooxygenase system. *Microbiology (UK)* 142: 251–260.

Higson FK, DD Fucht. 1990. Bacterial degradation of ring-chlorinated acetophenones. *Appl Environ Microbiol* 56: 3678–3685.

Hopper DJ. 1976. The hydroxylation of *p*-cresol and its conversion to *p*-hydroxybenzaldehyde in *Pseudomonas putida. Biochem Biophys Res Commun* 69: 462–468.

Howell JG, T Spector, V Massey. 1972. Purification and properties of *p*-hydroxybenzoate hydroxylase from *Pseudomonas fluorescens. J Biol Chem* 247: 4340–4350.

Iwaki H, Y Hosegawa, S Wang, MM Kayser, PCK Lau. 2002. Cloning and characterization of a gene cluster involved in cyclopentanol metabolism in *Comamonas* sp. strain NCIMB 9872 and biotransformations effected by *Escherichia coli*-expressed cyclopentanone 1,2-monooxygenase. *Appl Environ Microbiol* 68: 5671–5684.

James KD, PA Williams. 1998. *ntn* genes determining the early steps in the divergent catabolism of 4-nitrotoluene and toluene in *Pseudomonas* sp. strain TW3. *J Bacteriol* 180: 2043–2049.

Jeong JJ, JH Kim, C-K Kim, I Hwang, K Lee. 2003. 3- and 4-Alkylphenol degradation pathway in *Pseudomonas* sp. strain KL28: Genetic organization of the *lap* gene cluster and substrate specificities of phenol hydroxylase and catechol 2,3-dioxygenase. *Microbiology (UK)* 149: 3265–3277.

Jiang Z-Y, C Thorpe. 1983. Acyl-CoA oxidase from *Candida tropicalis. Biochemistry* 22: 3752–3758.

Johnson GR, RH Olsen. 1995. Nucleotide sequence analysis of genes encoding a toluene/benzene-2-monooxygenase from *Pseudomonas* sp. strain JS150. *Appl Environ Microbiol* 61: 3336–3346.

Johnson GR, RH Olsen. 1997. Multiple pathways for toluene degradation in *Burkholderia* sp. strain JS150. *Appl Environ Microbiol* 63: 4047–4052.

Jones JP, EJ O'Hare, L- L Wong. 2001. Oxidation of polychlorinated benzenes by genetically engineered CYP 101 (cytochrome P450$_{cam}$). *Eur J Biochem* 268: 1460–1467.

Jouanneau Y, J Micoud, C Meyer. 2007. Purification and characterization of a three-component salicylate 1-hydroxylase from *Sphingomonas* sp. strain CHY-1. *Appl Environ Microbiol* 73: 7515–7521.

Kadiyla V, JC Spain. 1998. A two-component monooxygenase that catalyzes both the hydroxylation of p-nitrophenol and the oxidative release of nitrite from 4-nitrophenol in *Bacillus sphaericus. Appl Environ Microbiol* 64: 2479–2484.

Kamerbeek NM, MJH Moonen, JGM van der Ven, WJH van Berkel, MW Fraaije, DB Janssen. 2001. 4-Hydroxyacetophenone monooxygenase from *Pseudomonas fluorescens* ACB. A novel flavoprotein catalyzing Baeyer–Villiger oxidation of aromatic compounds. *Eur J Biochem* 268: 2547–2557.

Käppeli O. 1986. Cytochromes P-450 of yeasts. *Microbiol Revs* 50: 244–258.

Karlson U, DF Dwyer, SW Hooper, ERB Moore, KN Timmis, LD Eltis. 1993. Two independently regulated cytochromes P-450 in a *Rhodococcus rhodochrous* strain that degrades 2-ethoxyphenol and 4-methoxybenzoate. *J Bacteriol* 175: 1467–1474.

Katagiri M, BN Ganguli, IC Gunsalus. 1968. A soluble cytochrome P-450 functional in methylene hydroxylation. *J Biol Chem* 243: 3543–3546.

Kawaguchi A, S Tsubotani, Y Seyama, T Yamakawa, T Osumi, T Hashimoto, Y Kikuchi, M Ando, S Okuda. 1980. Stereochemistry of dehydrogenation catalyzed by acyl-CoA oxidase. *J Biochem* 88: 1481–1486.

King DJ, MR Azari, A Wiseman. 1984. Studies on the properties of highly purified cytochrome P-448 and its dependent activity benzo[*a*]pyrene hydroxylase, from *Saccharomyces cerevisiae. Xenobiotica* 14: 187–206.

King DJ, MR Azari, A Wiseman. 1984. Studies on the properties of highly purified cytochrome P-448 and its dependent activity benzo[*a*]pyrene hydroxylase, from *Saccharomyces cerevisiae. Xenobiotica* 14: 187–206.

Kirchner U, AH Westphal, R Müller, WJH van Berkel. 2003. Phenol hydroxylase from *Bacillus thermoglucosidasius* A7, a two-component monooxygenase with a dual role for FAD. *J Biol Chem* 278: 47545–47553.

Koga H, H Aramaki, E Yamaguchi, K Takeuchi, T Horiuchi, IC Gunsdalus. 1986. *camR*, a negative regulator locus of the cytochrome P-450$_{cam}$ hydroxylase operon. *J Bacteriol* 166: 1089–1095.

Kok M, R Oldenhuis, MPG van der Linden, P Raatjes, K Kingma, PH van Lelyveld, B Witholt. 1989. The *Pseudomonas oleovorans* alkane hydroxylase gene. Sequence and expression. *J Biol Chem* 264: 5435–5441.

Kostichka K, SM Thomas, KJ Gibson, V Nagarajan, Q Cheng. 2001. Cloning and characterization of a gene cluster for cyclododecanone oxidation in *Rhodococcus ruber* SC1. *J Bacteriol* 183: 6478–6486.

Krishnakumar AM, D Sliwa, JA Endrizzi, ES Boyd, SA Ensign, JW Peters. 2008. Getting a handle on the role of coenzyme *M* in alkene metabolism. *Microbiol Mol Biol Revs* 72: 445–456.

Lei B, S-C Tu. 1996. Gene overexpression, purification, and identification of a desulfurization enzyme from *Rhodococcus* sp. strain IGTS8 as a sulfide/sulfoxide monooxygenase. *J Bacteriol* 178: 5699–5705.

Li S, LP Wackett. 1993. Reductive dehalogenation by cytochrome $P450_{cam}$: Substrate binding and catalysis. *Biochemistry* 32: 9355–9361.

Lipscomb JD. 1994. Biochemistry of the soluble methane monooxygenase. *Annu Rev Microbiol* 48: 371–399.

Locher HH, T Leisinger, AM Cook. 1991b. 4-Toluene sulfonate methyl monooxygenase from *Comamonas testosteroni* T-2: Purification and some properties of the oxygenase component. *J Bacteriol* 173:3741–3748.

Louie TM, CM Webster, L Xun. 2002. Genetic and biochemical characterization of a 2,4,6-trichlorophenol degradation pathway in *Ralstonia eutropha* JMP134. *J Bacteriol* 184: 3492–3500.

Maier T, H-H Förster, O Asperger, U Hahn. 2001. Molecular characterization of the 56-kDa CYP153 from *Acinetobacter* sp. EB104. *Biochem Biophys Res Commun* 286: 652–658.

Malito E, A Slfieri, MW Fraaije, A Mattevi. 2004. Crystal structure of a Baeyer–Villiger monooxygenase. *Proc Natl Acad Sci USA* 101: 13157–13162.

Matsui T, M Furukawa, M Unno, T Tomita, M Ikeda-Saito. 2005. Roles of distal Asp in heme oxygenase from *Corynebacterium diphtheriae*, HmuO. A water-driven oxygen activation system. *J Biol Chem* 280: 2981–2989.

May SW, BJ Abbott. 1973. Enzymatic epoxidation II. Comparison between the epoxidation and hydroxylation reactions catalyzed by the ω-hydroxylation system of *Pseudomonas oleovorans*. *J Biol Chem* 248: 1725–1730.

McDonald IR, CB Miguez, G Rogge, D Bourque, KD Wendlandt, D Groleau, JC Murrell. 2006. Diversity of soluble methane monooxygenase-containing methanotrophs isolated from polluted environments. *FEMS Microbiol Lett* 255: 225–232.

McIntire W, DJ Hopper, JC Craig, ET Everhart, RV Webster, MJ Causer, TP Singer. 1984. Stereochemistry of 1-(4′-hydroxyphenyl)ethanol produced by hydroxylation of 4-ethylphenol by *p*-cresol methylhydroxylase. *Biochem J* 224: 617–621.

McIntire W, DJ Hopper, TP Singer. 1985. *p*-Cresol methylhydroxylase. Assay and general properties. *Biochem J* 228: 325–335.

McKenna EJ, MJ Coon. 1970. Enzymatic ω-oxidation IV. Purification and properties of the ω-hydroxylase of *Pseudomonas oleovorans*. *J Biol Chem* 245: 3882–3889.

Meunier B, SP de Visser, S Shalk. 2004. Mechanisms of oxidation reactions catalyzed by cytochrome P3450 enzymes. *Chem Rev* 194: 3947–3980.

Morawski B, RW Eaton, JT Rossiter, S Guoping, H Griengl, DW Ribbons. 1997. 2-naphthoate catabolic pathway in *Burkholderia* strain JT 1500. *J Bacteriol* 179: 115–121.

Munro AW, JG Lindsay. 1996. Bacterial cytochromes P-450. *Mol Microbiol* 20: 1115–1125.

Nagpal A, MP Valley, PF Fitzpatrick, AM Orville. 2006. Crystal structures of nitroalkane oxidase: Insights into the reaction mechanism from a covalent complex of the flavoenzymes trapped during turnover. *Biochemistry* 45: 1138–1150.

Nagy I, F Compernolle, K Ghys, J Vanderleyden, R De Mot. 1995a. A single cytochrome P-450 system is involved in degradation of the herbicides EPTC (*S*-ethyl dipropylthiocarbamate) and atrazine by *Rhodococcus* sp. N186/21. *Appl Environ Microbiol* 61: 2056–2060.

Nagy I, G Schoofs, F Compermolle, P Proost, J Vanderleyden, R De Mot. 1995b. Degradation of the thiocarbamate herbicide EPTC *S*-ethyl dipropylcarbamoylthioate and biosafening by *Rhodococcus* sp. strain N186/21 involve an inducible cytochrome P-450 system and aldehyde dehydrogenase. *J Bacteriol* 177: 676–687.

Nakazawa, K Hori, O Hayaishi. 1972. Studies on monooxygenases V. Manifestation of amino acid oxidase activity by L-lysine monooxygenase. *J Biol Chem* 247: 3439–3444.

Newman LM, LP Wackett. 1995. Purification and characterization of toluene 2-monooxygenase from *Burkholderia cepacia* G4. *Biochemistry* 34: 14066–14076.

Nguyen H-HT, AK Shiemka, SJ Jacobs, BJ Hales, ME Lidstrom, S I Chan. 1994. The nature of the copper ions in the membranes containing the particulate methane monooxygenase from *Methylococcus capsulatus* (Bath). *J Biol Chem* 269: 14995–15005.

Nurk A, L Kasak, M Kivisaar. 1991a. Sequence of the gene (*pheA*) encoding phenol monooxygenase from *Pseudomonas* sp. EST 1001: Expression in *Escherichia coli* and *Pseudomonas putida*. *Gene* 102: 13–18.

Nurk A, L Kasak, M Kivisaar. 1991b. Sequence of the gene (*pheA*) encoding phenol monooxygenase from a two-component monooxygenase with a dual role for FAD. *J Biol Chem* 278: 47545–47553.

O'Keefe DP, JA Romesser, KJ Leto. 1988. Identification of constitutive and herbicide inducible cytochromes P-450 in *Streptomyces griseolus*. *Arch Microbiol* 149: 406–412.

O'Keefe DP, PA Harder. 1991. Occurrence and biological function of cytochrome P-450 monoxygenase in the actinomycetes. *Mol Microbiol* 5: 2099–2105.

O'Leary SE, KA Hicks, SE Ealick, TP Begley. 2009. Biochemical characterization of the HpxO enzyme from *Klebsiella pneumoniae*, a novel FAD-dependent urate oxidase. *Biochemistry* 48: 3033–3035.

Omori T, L Monna, Y Saiki, T Kodama. 1992. Desulfurization of dibenzothiophene by *Corynebacterium* sp. strain SY1. *Appl Environ Microbiol* 58: 911–915.

Ortiz de Montellano PR, JA Freutel, JR Collins, DL Camper, GH Loew. 1991. Theoretical and experimental analysis of the absolute configuration of *cis*-β-methylstyrene epoxidation by cytochrome P-450$_{cam}$. *J Am Chem Soc* 113: 3195–3196.

Ortiz de Montelano PR, JJ De Vos. 2002. Oxidizing species in the mechanism of cytochrome P450. *Nat Prod Rep* 19: 477–493.

Park M, Y Jeon, HH Jang, H-S Ro, W Park, EL Madsen, CO Jeon. 2007. Molecular and biochemical characterization of 3-hydroxybenzoate 6-hydroxylase from *Polaromonas naphthalenivorans* CJ2. *Appl Environ Microbiol* 73: 5146–5152.

Perkins EJ, MP Gordon, O Caceres, PF Lurquin. 1990. Organization and sequence analysis of the 2,4-dichlorophenol hydroxylase and dichlorocatechol oxidative operons of plasmid pJP4. *J Bacteriol* 172: 2351–2359.

Petrusma M, L Dijkhuizen, R van der Geize. 2009. *Rhodococcus rhodochrous* DSM 43269 3-ketosteroid 9α-hydroxylase, a two-component iron–sulfur-containing monooxygenase with subtle steroid substrate specificity. *Appl Environ Microbiol* 75: 5300–5307.

Petrusma M, G Hessels, L Dijkhuizen, R van der Geize. 2011. Multiplicity of 3-ketosteroid-9α-hydroxylase enzymes in *Rhodococcus rhodochrous* DSM43269 for specific degradation of different classes of steroids. *J Bacteriol* 193: 3931–3940.

Pikus JD, LM Studts, C Achim, KE Kaufmann, E Münck, RJ Steffan, K McClay, BG Fox. 1996. Recombinant toluene-4-monooxygenase: Catalytic and Mössbauer studies of the purified diiron and Rieske components of a four-protein complex. *Biochemistry* 35: 9106–9119.

Poulos TL, BC Finzel, AJ Howard. 1987. High resolution structure of cytochrome P450$_{cam}$. *J Mol Biol* 195: 687–700.

Poupin P, N Truffaut, B Combourieu, M Sancelelme, H Veschambre, AM Delort. 1998. Degradation of morpholine by an environmental *Mycobacterium* strain involves a cytochrome P-450. *Appl Environ Microbiol* 64: 159–165.

Powlowski J, S Dagley, V Massey, DP Ballou. 1987. Properties of anthranilate hydroxylase (deaminating), a flavoprotein from *Trichosporon cutaneum*. *J Biol Chem* 262: 69–74.

Powlowski J, V Shingler. 1994a. Genetics and biochemistry of phenol degradation by *Pseudomonas* sp. CF600. *Biodegradation* 5: 219–236.

Powlowski J, V Shingler. 1994b. Genetics and biochemistry of phenol degradation by *Pseudomonas aeruginosa* POAO1c. *J Bacteriol* 172: 4624–4630.

Priefert H, J Rabenhorst, A Steinbüchel. 1997. Molecular characterization of genes of *Pseudomonas* sp. strain HR199 involved in bioconversion of vanillin to protocatechuate. *J Bacteriol* 179: 2595–2607.

Prieto MA, JL Garcia. 1994. Molecular characterization of 4-hydroxyphenylacetate 3-hydroxylase of *Escherichia coli*. A two-component enzyme. *J Biol Chem* 269: 22823–22829.

Providenti MA, JM O'Brien, J Ruff, AM Cook, IB Lambert. 2006. Metabolism of isovanillate, vanillate, and veratrate by *Comamonas testosteroni*. *J Bacteriol* 188: 3862–3869.

Rather LJ, T Weinert, U Demmer, W Ismael, G Fuchs, U Elmer. 2011. Structure and mechanism of the diiron benzoyl-coenzyme A epoxide BoxB. *J Biol Chem* M 111.236893.

Rittle J, MT Green. 2010. Cytochrome P450 compound I: Capture, characterization, and C–H bond activation kinetics. *Science* 330: 933–936.

Romesser JA, DP O'Keefe. 1986. Induction of cytochrome P-450-dependent sulfonylurea metabolism in *Streptomyces griseolus*. *Biochem Biophys Res Commun* 140: 650–659.

Rosłoniec KZ, MH Wilbrink, JK Capyk, WH Mohn, M Ostendorf, R van der Geize, L Dijkhuizen, LD Eltis. 2009. Cytochrome P450 125 (CPY125) catalyses C26-hydroxylation to initiate sterol side-chain degradation in *Rhodococcus jostii* RHA1. *Mol Microbiol* 74: 1031–1043.

Rosche B, B Tshisuaka, B Hauer, F Lingens, S Fetzner. 1997. 2-Oxo-1,2-dihydroquinoline 8-monooxygenase: Phylogenetic relationship to other multicomponent nonheme iron oxygenases. *J Bacteriol* 179: 3549–3554.

Rosche B, B Tshisuaka, S Fetzner F Lingens. 1995. 2-Oxo-1,2-dihydroquinoline 8-monooxygenase, a two-component enzyme system from *Pseudomonas putida* 86. *J Biol Chem* 270: 17836–17842.

Ruettinger RT, GR Griffith, MJ Coon. 1977. Characteristics of the ω-hydroxylase of *Pseudomonas oleovorans* as a non-heme iron protein. *Arch Biochem Biophys* 183: 528–537.

Ruettinger, RT, ST Olson, RF Boyer, MJ Coon. 1974. Identification of the ω-hydroxylase of *Pseudomonas oleovorans* as a nonheme iron protein requiring phospholipid for catalytic activity. *Biochem Biophys Res Commun* 57: 1011–1017.

Russ R, C Müller, H-J Knackmuss, A Stolz. 1994. Aerobic biodegradation of 3-aminobenzoate by Gram-negative bacteria involves intermediate formation of 5-aminosalicylate as ring-cleavage substrate. *FEMS Microbiol Lett* 122: 137–144.

Sabbadin F, R Jackson, K Haider, G Tampi, JP Turkenburg, S Hart, NC Bruce, G Grogan. 2009. The 1.5-Å structure of the XplA-heme, an unusual cytochrome P450 heme domain that catalyzes reductive biotransformation of royal demolition explosive. *J Biol Chem* 284: 28467–28475.

Sasaki J, A Mikami, K Mizoue, S Omura. 1991. Transformation of 25- and 1α-hydroxyvitamin D_3 to 1α, 25-dihydroxyvitamin D_3 by using *Streptomyces* sp. strains. *Appl Environ Microbiol* 57: 2841–2846.

Sasaki M, A Akahira, L-i Oshiman, T Tsuchid, Y Matsumura. 2005. Purification of cytochrome P450 and ferredoxin involved in bisphenol A degradation from *Sphingomonas* sp. stain AO1. *Appl Environ Microbiol* 71 8024–8030.

Sazinsky MH, J Bard, A Di Donato, SJ Lippard. 2004. Crystal structure of the toluene-*o*-xylene monooxygenase hydroxylase from *Pseudomonas stutzeri* OX1. *J Biol Chem* 279: 30600–30610.

Scheller U, T Zimmer, D Becher, F Schauer, W-H Schunk. 1998. Oxygenation cascade in conversion of *n*-alkanes to α,ω-dioic acids catalyzed by cytochrome P450 52A3. *J Biol Chem* 273: 43528–43534.

Seth-Smith HMB, J Edwards, SJ Rosser, DA Rathbone, NC Bruce. 2008. The explosive-degrading cytochrome P450 system is highly conserved among strains of *Rhodococcus* spp. *Appl Environ Microbiol* 74: 4550–4552.

Shanklin J, C Achim, H Schmidt, BG Fox, E Münck. 1997. Mössbauer studies of alkane ω-hydroxylase: Evidence for a diiron cluster in an integral-membrane enzyme. *Proc Natl Acad Sci USA* 94: 2981–2986.

Shields MS, MJ Reagin, RR Gerger, R Campbell, C Soverville. 1995. TOM, a new aromatic degradative plasmid from *Burkholderia* (*Pseudomonas*) *cepacia* G4. *Appl Environ Microbiol* 61: 1352–1356.

Shigematsu T, S Hanada, M Eguchi, Y Kamagata, T Kanagawam R Kurane. 1999. Soluble methane monooxygenase gene clusters from trichloroethylene-degrading *Methylomonas* sp. strains and detection of methanotrophs during *in situ* bioremediation. *Appl Environ Microbiol* 65: 5198–5206.

Shingler V, J Powlowski, U Marklund. 1992. Nucleotide sequence and functional analysis of the complete phenol/3,4-dimethylphenol catabolic pathway of *Pseudomonas* sp. strain CF600. *J Bacteriol* 174: 711–724.

Shyadehi AZ, DC Lamb, SL Kelly, DE Kelly, W-H Schunck, JN Wright, D Corina, M Akhtar. 1996. The mechanism of the acyl-carbon bond cleavage reaction catalyzed by recombinant sterol 14α-demethylase of *Candida albicans* (Other names are: Lanosterol 14α-demethylase, P-450$_{14DM}$, and CYP51). *J Biol Chem* 271: 12445–12450.

Sluis MK, LA Sayavedra-Soto, DJ Arp. 2002. Molecular analysis of the soluble butane monooxygenase from *Pseudomonas butanovora*. *Microbiology (UK)* 148: 3617–3629.

Small FJ, SA Ensign. 1997. Alkene monooxygenase from *Xanthobacter* strain Py2. Purification and characterization of a four-component system central to the bacterial metabolism of aliphatic alkenes. *J Biol Chem* 272: 24913–24920.

Smith DJ, MA Patrauchan, C Florizone, LD Eltis, WW Mohn. 2008. Distinct roles for two CYP226 family cytochromes P450 in abietane diterpenoid catabolism by *Burkholderia xenovorans* LB400. *J Bacteriol* 190: 1575–1583.

Smith DJ, VJJ Martin, WH Mohn. 2004. A cytochrome P450 involved in the metabolism of abietane diterpenoids by *Pseudomonas abietaniphila* BMKE-9. *J Bacteriol* 186: 3631–3639.

Smith RV, JP Rosazza. 1983. Microbial models of mammalian metabolism. *J Nat Prod* 46: 79–91.

Smits THM, SB Balada, B Witholt, JB van Beilen. 2002. Functional analysis of alkane hydroxylases from Gram-negative and Gram-positive bacteria. *J Bacteriol* 184: 1733–1742.

Song J, T Xia, RA Jensen. 1999. PhhB, a *Pseudomonas aeruginosa* homolog of mammalian pterin 4a-carbinolamine dehydratase/DCcoH, does not regulate expression of phenylalanine hydroxylase at the transcriptional level. *J Bacteriol* 181: 2789–2796.

Sono M, MP Roach, ED Coulter, JH Dawson. 1996. Heme-containing oxygenases. *Chem Rev* 96: 2841–2887.

Sontoh S, JD Semrau. 1998. Methane and trichloroethylene degradation by *Methylosinus trichosporium* OB3b expressing particulate methane monooxygenase. *Appl Environ Microbiol* 64: 1106–1114.

Spain JC, DT Gibson. 1991. Pathway for biodegradation of *p*-nitrophenol in a *Moraxella* sp. *Appl Environ Microbiol* 57: 812–819.

Steffan RJ, K McClay, S Vainberg, CW Condee, D Zhang. 1997. Biodegradation of the gasoline oxygenates methyl *tert*-butyl ether, ethyl *tert*-butyl ether, and amyl *tert*-butyl ether by propane-oxidizing bacteria. *Appl Environ Microbiol* 63: 4216–4222.

Stevenson DE, JN Wright, M Akhtar. 1988. Mechanistic consideration of P-450 dependent enzyme reactions: Studies on oestriol biosynthesis. *J Chem Soc Perkin Trans I*: 2043–2052.

Summers RM, TM Louie, C-L Yu, L Gakhar, KC Louie, M Subramanian. 2012. Novel, highly specific *N*-demethylase enable bacteria to live on caffeine and related purine alkaloids. *J Bacteriol* 194: 2041–2049.

Suske WA, WJH van Berkel, H-P Kohler. 1999. Catalytic mechanism of 2-hydroxybiphenyl 3-monooxygenase a flavoprotein from *Pseudomonas azelaica* HBP1. *J Biol Chem* 274: 33355–33365.

Sutherland JB, JP Freeman, AL Selby, PP Fu, DW Miller, CE Cerniglia. 1990. Stereoselective formation of a K-region dihydrodiol from phenanthrene by *Streptomyces flavovirens*. *Arch Microbiol* 154: 260–266.

Sutherland TD, I Horne, RJ Russell, JG Oakeshott. 2002. Gene cloning and molecular characterization of a two-enzyme system catalyzing the oxidative detoxification of β-endosulfan. *Appl Environ Microbiol* 68: 6237–6245.

Suzuki M, T Hayakawa, JP Shaw, M Rekik, S Harayama. 1991. Primary structure of xylene monooxygenase: Similarities to and differences from the alkane hydroxylation system. *J Bacteriol* 173: 1690–1695.

Swetha VP, A Basu, PS Phale. 2007. Purification and characterization of 1-naphthol-2-hydroxylase from carbaryl-degrading *Pseudomonas* strain C4. *J Bacteriol* 189: 2660–2666.

Sylvestre M, R Massé, F Messier, J Fauteux, J-G Bisaillon, R Beaudet. 1982. Bacterial nitration of 4-chlorobiphenyl. *Appl Environ Microbiol* 44: 871–877.

Sze IS-Y, S Dagley. 1984. Properties of salicylate hydroxylase and hydroxyquinol 1,2-dioxygenase purified from *Trichosporon cutaneum*. *J Bacteriol* 159: 353–359.

Takeda H, S Yamamoto, Y Kojima, O Hazaishi. 1969. Studies on monooxygenases I. General properties of crystalline L-lysine monooxygenase. *J Biol Chem* 244: 2935–2941.

Takeo M, M Murakami, S Niihara, K Yamamoto, M Nishimura, D Kato, S Negoro. 2008. Mechanism of 4-nitrophenol oxidation in *Rhodococcus* sp. strain PN1: Characterization of the two component 4-nitrophenol hydroxylase and regulation of its expression. *J Bacteriol* 190: 7367–7374.

Tanaka A, M Ueda. 1993. Assimilation of alkanes by yeasts: Functions and biogenesis of peroxisomes. *Mycol Res* 98: 1025–1044.

Tanner A, DJ Hopper. 2000. Conversion of 4-hydroxyacetophenone into 4-phenyl acetate by a flavin adenine dinucleotide-containing Baeyer–Villiger-type monooxygenase. *J Bacteriol* 182: 6565–6569.

Tao Y, A Fishman, WE Bentley, TK Wood. 2004. Oxidation of benzene to phenol, catechol, and 1,2,3-trihydroxybenzene by toluene 4-monooxygenase of *Pseudomonas mendocina* KR 1 and toluene 3-monooxygenase of *Ralstonia pickettii* PKO1. *Appl Environ Microbiol* 70: 3814–3820.

Tormos JR, AB Taylor, SC Daubner, PJ Harts, PF Fitzpatrick. 2010. Identification of a hypothetical protein from *Podospora anserina* as a nitroalkane oxidase. *Biochemistry* 49: 5035–5041.

Trickett JM, EJ Hammonds, TL Worrall, MK Trower, M Griffin. 1990. Characterization of cyclohexane hydroxylase; a three-component enzyme system from a cyclohexane-grown *Xanthobacter* sp. *FEMS Microbiol Lett* 82: 329–334.

Trower MK, FS Sariaslani, DP O'Keefe. 1989. Purification and characterization of a soybean flour-induced cytochrome P-450 from *Streptomyces griseus*. *J Bacteriol* 171: 1781–1787.

Tyson CA, JD Lipscomb, IC Gunsalus. 1972. The roles of putidaredoxin and P450$_{cam}$ in methylene hydroxylation. *J Biol Chem* 247: 5777–5781.

Uetz T, R Schneider, R Snozzi, T Egli. 1992. Purification and characterization of a two-component monooxygenase that hydroxylates nitrilotriacetate from "*Chelatobacte*" strain ATCC 29600. *J Bacteriol* 174: 1179–1188.

Ullah AJH, RI Murray, PK Bhattacharyya, GC Wagner, IC Gunsalus. 1990. Protein components of a cytochrome P-450 linalool 8-methyl hydroxylase. *J Biol Chem* 265: 1345–1351.

Valley MP, SE Tichy, PF Fitzpatrick. 2005. Establishing the kinetic competency of the cationic imine intermediate in nitroalkane oxidase. *J Am Chem Soc* 127: 2062–2066.

van Beilen JB, EG Funhoff, A van Loon, A Just, L Kaysser, M Bouza, R Holtackers, M Röthlisberger, Z Li, B Witholt. 2006. Cytochrome P450 alkane hydroxylases of the CYP153 family are common in alkane-degrading eubacteria lacking integral membrane alkane hydroxylases. *Appl Environ Microbiol* 72: 59–65.

van Beilen JR, D Pennings, B Witholt. 1992. Topology of the membrane-bound alkane hydroxylase of *Pseudomonas oleovorans. J Biol Chem* 267: 9194–9201.

van Berkel WJH, NM Kamerbeek, MW Fraaije. 2006. Flavoprotein monooxygenases, a diverse class of oxidative biocatalysts. *J Biotechnol* 124: 670–689.

van der Werf M. 2000. Purification and characterization of a Baeyer–Villiger mono-oxygenase from *Rhodococcus erythropolis* strain DCL14 involved in three different monocyclic degradation pathways. *Biochem J* 347: 693–701.

van Hylckama Vlieg JET, H Leemhuis, JHL Spelberg, DB Janssen. 2000. Characterization of the gene cluster involved in isoprene metabolism in *Rhodococcus* sp. strain AD45. *J Bacteriol* 182: 1956–1963.

Vardar G, TK Wood. 2004. Protein engineering of toluene-*o*-xylene monooxygenase from *Pseudomonas stutzeri* OX1 for synthesizing 4-methylresorcinol, methylhydroquinone, and pyrogallol. *Appl Environ Microbiol* 70: 3253–3262.

Vannelli TA, A Dykman, PR Ortiz de Montellano. 2002. The antituberculosis drug ethionamide is activated by a flavin monooxygenase. *J Biol Chem* 277: 12824–12829.

Vaz ADN, DF McGinnity, MJ Coon. 1998. Epoxidation of olefins by cytochrome P450: Evidence from site specific mutagenesis for hydroperoxo-iron as an electrophilc oxidant. *Proc Natl Acad Sci USA* 95: 3555–3560.

Wackett, LP, LD Kwart, DT Gibson. 1988. Benzylic monooxygenation catalyzed by toluene dioxygenase from *Pseudomonas putida. Biochemistry* 27: 1360–1367.

Warburton EJ, AM Magor, MK Trower, M Griffin. 1990. Characterization of cyclohexane hydroxylase; involvement of a cytochrome P-450 system from a cyclohexane-grown X*anthobacter* sp. *FEMS Microbiol Lett* 66: 5–10.

Wegele R, R Tasler, Y Zeng, M Riveras, N Frankenberg-Dinkel. 2004. The heme oxygenase(s)-phytochrome system of *Pseudomonas aeruginosa. J Biol Chem* 279: 45791–45802.

Weir KM, TD Sutherland, I Horne, RJ Russell. JG Oakshott. 2006. A singe monooxygtenase, Ese, is involved in the metabolism of the organochlorines endosulfan and endosulfate in an *Arthrobacter* sp. *Appl Environ Microbiol* 72: 3524–3530.

Wende R, F-H Bernhardt, K Pfleger. 1989. Substrate-modulated reactions of putidamonooxin the nature of the active oxygen species formed and its reaction mechanism. *Eur J Biochem* 81: 189–197.

Wergath J, H-A Arfmann, DH Pieper, KN Timmis, R-M Wittich. 1998. Biochemical and genetic analysis of a gentisate 1,2-dioxygenase from *Sphingomonas* sp. strain RW 5. *J Bacteriol* 180: 4171–4176.

White RE, MB McCarthy, KD Egeberg, SG Sligar. 1984. Regioselectivity in the cytochromes P-450: Control by protein constraints and by chemical reactivities. *Arch Biochem Biophys* 228: 493–502.

Whited GM, DT Gibson. 1991a. Toluene-4-monooxygenase, a three-component enzyme system that catalyzes the oxidation of toluene to *p*-cresol in *Pseudomonas mendocina* KR1. *J Bacteriol* 173: 3010–3016.

Whited GM, DT Gibson. 1991b. Separation and partial characterization of the enzymes of the toluene-4-monooxygenase catabolic pathway in *Pseudomonas mendocina* KR1. *J Bacteriol* 173: 3017–3020.

White-Stevens RH, H Kamin, QH Gibson. 1972. Studies of a flavoprotein, salicylate hydroxylase II Enzyme mechanism. *J Biol Chem* 247: 2371–2381.

White-Stevens, RH, H Kamin. 1972. Studies of a flavoprotein, salicylate hydroxylase. I. Preparation, properties, and the uncoupling of oxygen reduction from hydroxylation. *J Biol Chem* 247: 2358–2370.

Wieser M, B Wagner, J Eberspächer, F Lingens. 1997. Purification and characterization of 2,4,6-trichlorophenol-4-monooxygenase, a dehalogenating enzyme from *Azotobacter* sp. strain GP1. *J Bacteriol* 179: 202–208.

Xu Y, MW Mortimer, TS Fisher, ML Kahn FJ Brockman, L Xun. 1997. Cloning, sequencing, and analysis of gene cluster from *Chelatobacter heintzii* ATCC 29600 encoding nitrolotriacetate monooxygenase and NADH:flavin mononucleotide oxidoreductase. *J Bacteriol* 179: 1112–1116.

Xun L. 1996. Purification and characterization of chlorophenol 4-monooxygenase from *Burkholderia cepacia* AC1100. *J Bacteriol* 178: 2645–2649.

Xun L, CM Webster. 2004. A monooxygenase catalyzes sequential dechlorinations of 2,4,6-trichlorophnel by oxidative and hydrolytic reactions. *J Biol Chem* 279: 6696–6700.

Xun L, CS Orser. 1991. Purification and properties of pentachlorophenol hydroxylase, a flavoprotein from *Flavobacterium* sp. strain ATCC 39723. *J Bacteriol* 173: 4447–4453.

Xun L, E Topp, CS Orser. 1992. Diverse substrate range of a *Flavobacterium* pentachlorophenol hydroxylase and reaction stoichiometries. *J Bacteriol* 174: 2898–2902.

Xun L, ER Sandvik. 2000. Characterization of 4-hydroxyphenylacetate 3-hydroxylase (HpaB) of *Escherichia coli* as a reduced flavin adenine dinucleotide-utilizing monooxygenase. *Appl Environ Microbiol* 66: 481–486.

Xun L, KB Wagnon. 1995. Purification and properties of component B of 2,4,5-trichlorophenoxyacetate oxygenase from *Pseudomonas cepacia* AC1100. *Appl Environ Microbiol* 61: 3499–3502.

Yeh E, S Garneau, CT Walsh. 2005. Robust *in vitro* activity of RebF and RebH, a two-component reductase/halogenase, generating 7-chlorotryptophan during rebeccamycin biosynthesis. *Proc Natl Acad Sci USA* 102: 3960–3965.

Yu CL, TM Louie, R Summers, Y Kale, S Gopishetty, M Subramanian. 2009. Two distinct pathways for metabolism of theophylline and caffeine are coexpressed in *Pseudomonas putida* CBB5. *J Bacteriol* 191: 4624–4632.

Zahn JA, AA DiSpirito. 1996. Membrane associated methane monooxygenase from *Methylococcus capsulatus* (Bath). *J Bacteriol* 178: 1018–1029.

Zhang J-J, H Liu, Y Xiao, X-E Zhang, N-Y Zhou. 2009. Identification and characterization of catabolic *para*-nitrophenol 4-monooxygenase and *para*-benzoquinone reductase from *Pseudomonas* sp. strain WBC-3. *J Bacteriol* 191: 2703–2710.

Zhou N-Y, A Jenkins, CKN Chan, KW Chion, DJ Leak. 1999. The alkene monooxygenase from *Xanthobacter* strain Py2 is closely related to aromatic monooxygenases and catalyzes aromatic monooxygenation of benzene, toluene, and phenol. *Appl Environ Microbiol* 65: 1589–1595.

Zhou N-Y, J Al-Dulayymi, MS Baird, PA Williams. 2002. Salicylate 5-hydroxylase from *Ralstonia* sp. strain U2: A monooxygenase with close relationships to and shared electron transport proteins with naphthalene dioxygenase. *J Bacteriol* 184: 1547–1555.

3.4 Dioxygenation

3.4.1 Dioxygenases in Dihydroxylation of Arenes

The aerobic degradation of aromatic compounds involves the introduction of oxygen into the rings as a prelude to ring fission. The most common initial reaction is dioxygenation in which both atoms of oxygen are introduced to produce *cis*-dihydrodiols. Catechols are formed from these, either by dehydrogenation or by elimination of substituents before further dioxygenation and fission of the ring. The biochemistry, enzymology, and genetics have been enunciated in great detail. There are three classes of bacterial dioxygenases that are summarized in Figure 3.17. The following overview attempts to bring together the reactions for many of the compounds that are discussed later in the appropriate parts of Chapters 8 and 9.

3.4.2 Dihydroxylation Enzymes

3.4.2.1 Group I

The dioxygenases that catalyze the initial introduction of molecular oxygen into aromatic hydrocarbons have been extensively studied, and are multicomponent enzymes that generally carry out three

Classification of bacterial dioxygenases
Class I: Two component enzyme, e.g., benzoate, 4-sulfobenzoate
    1. Reductase—flavin + [2Fe–2S] redox center
    2. Oxygenase
Class II: Three component enzyme, e.g., benzene and toluene
    1. Reductase—flavoprotein
    2. Ferredoxin—Rieske type
    3. Oxygenase
Class III: Three component type, e.g., naphthalene
    1. Reductase—flavin + [2Fe–2S] redox center
    2. [2Fe–2S] ferredoxin
    3. Oxygenase
Further division may be made according to the FMN or FAD flavin requirement, and the nature of the [2Fe–2S] protein

FIGURE 3.17 Classes of bacterial dioxygenases.

functions—(a) reduction of NAD by a reductase, (b) electron transport by a ferredoxin and transfer of electrons onto the substrate simultaneously with (c) the introduction of oxygen by a terminal oxygenase. The paradigm dihydroxylating enzyme is naphthalene dioxygenase from *Pseudomonas* sp. NCIB 9816-4. This consists of a 36 kDa reductase which contains FAD, a plant-type [2Fe–2S] cluster, a 14 kDa ferredoxin electron transfer protein with a [2Fe–2S] Rieske cluster, and the 210 kDa $\alpha_3\beta_3$ oxygenase component that contains the Rieske and mononuclear iron centers that are essential for catalysis (Wolfe et al. 2001). The crystal structure has been determined (Kauppi et al. 1998) and a route of electron transfer between adjacent α subunits was suggested: the α subunit A containing the [2Fe–2S] cluster bound to His[104], the α subunit B with His[213] and His [208] bound at the active site with Asp[205] serving as a linker with the histidines in the A and B α subunits. Additional comments illustrate the flexibility of some arene dioxygenases.

1. *Sphingomonas yanoikuyae* (*Beijerinckia* sp.) strain B1 metabolizes biphenyl by initial dioxygenation followed by dehydrogenation to 2,3-dihydroxybiphenyl (Gibson et al. 1973). *Sphingobium yanoikuyae* strain B1 is able to grow at the expense of naphthalene, phenanthrene, anthracene, and biphenyl that is initiated by dioxygenation to *cis*-1,2-dihydrodiols before dehydrogenation and ring fission. The dioxygenase from *Pseudomonas* sp. NCIB is additionally capable of carrying out monooxygenation, and desaturation (Gibson et al. 1995). It has been shown in strain B8/36 that the initially formed dihydrodiols may undergo further dioxygenation to *bis-cis*-dihydrodiols (*cis*-tetraols). Chrysene formed 3,4,8,9-*bis-cis*-dihydrodiol, and *bis-cis*-dihydroxylation at the analogous positions has been demonstrated for benzo[*c*] phenanthridine and benzo[*b*]naphtho[2,1-*d*]thiophene (Boyd et al. 1999).

2. Dihydroxylation of carbazole at the 1,9a-angular positions is carried out by a dioxygenase that has been examined in *Janthinobacterium* sp. strain J3. It consisted of a terminal oxygenase (CARDO-O) component that had the unusual homotrimeric structure α_3 that contained one Rieske [2Fe–2S] cluster and one Fe(II) active site. The electron transport components ferredoxin (CARDO-F) and ferredoxin reductase (CARDO-R) (Nojiri et al. 2005), and details of the electron transport complex have been analyzed (Ashikawa et al. 2006). CARDO-O is noteworthy regarding the range of dioxygenations carried out by the carbazole angular dioxygenase from *Pseudomonas* strain CA10 (Nojiri et al. 1999). These include carbazole to 2′-aminobiphenyl-2.3-diol produced by elimination after dioxygenation, and analogously dibenzo[1,4]dioxin to 2,2′,3-trihydroxydiphenyl ether, dibenzofuran to 2,2′,3-trihydroxybiphenyl, xanthene to 2,2′,3-trihydroxydiphenylmethane and phenoxanthiin to 2,2′,3-trihydroxydiphenyl sulfide (Nojiri et al. 1999).

3. In the quinol-2-one monooxygenation by *Ps. putida* strain 86 that hydroxylates C_8, the Rieske [2Fe–2S] center plays a key role by removing the barrier of access of O_2 to the mononuclear Fe(II) site posed by binding to the amide group (Martins et al. 2005).

There are, however, a number of two-component dioxygenases that consist of only a reductase and an oxygenase. These contain a flavin cofactor and a [2Fe–2S] center, but lack the ferredoxin component that is otherwise required for aromatic hydrocarbon dioxygenation. Examples of these two-component dioxygenases are given in Table 3.3.

The main classes of bacterial dioxygenases have been outlined in Figure 3.2. Generally the *cis*-dihydrodiols are dehydrogenated to the corresponding catechol. In some, however, dehydrogenation is not required and elimination of CO_2H, NH_3, Hal⁻, NO_2^-, and SO_3^{2-}, groups—or exceptionally an OH group produces catechols directly. These are assigned to Group II.

3.4.2.2 Group II

In these reactions, dehydrogenation is not required, and formation of the catechol is attained by spontaneous elimination of CO_2, NH_3, NO_2^-, SO_3^{2-}, or halide (Figure 3.18). Illustrative examples include:

1. a. Amine groups in the conversion of aniline to catechol (Bachofer and Lingens 1975; Fukumori and Saint 1997).

TABLE 3.3

Two-Component Arene Dioxygenases

Substrate	Organism	Reference
4-Sulfobenzoate-3,4-dioxygenase	*Comamonas testosteroni* T-2	Locher et al. (1991a)
2-Aminobenzoate-1,2-dioxygenase	*Alcaligenes* sp. O-1	Mampel et al. (1999)
o-phthalate 4,5-dioxygenase	*Ps. cepacia*	Batie et al. (1987)
Benzoate 1,2-dioxygenase	*Ps. arvilla* C-1	Yamaguchi and Fujisawa (1982)
3-Chlorobenzoate 3,4-dioxygenase	*Alcaligenes* sp. strain BR 60	Nakatsu et al. (1995)
4-Chlorophenylacetate-3,4-dioxygenase	*Ps. cepacia* CBS 3	Schweizer et al. (1987)
2-Halobenzoate 1,2-dioxygenase[a]	*Ps. cepacia* 2 CBS	Fetzner et al. (1992)
Quinol-2-one 8-monooxygenase	*Ps. putida* strain 86	Martins et al. (2005)

[a] The corresponding dioxygenase from *Ps. aeruginosa* 142 is, however, a three-component enzyme (Romanov and Hausinger 1994).

b. Both amine and carboxyl groups in the conversion of anthranilic acid to catechol by anthranilate dioxygenase (Taniuchi et al. 1964; Eby et al. 2001; Chang et al. 2003).

c. 2-Aminobenzenesulfonate to catechol-3-sulfonate and ammonia (Mampel et al. 1999). However, there are also anthranilate monooxygenases that have already been noted.

2. a. Carboxyl groups in the conversion of benzoate to catechol by benzoate dioxygenase (Neidle et al. 1991).

b. Decarboxylation with concomitant dehydrogenation may occur during the degradation of phthalates. The degradation of 4-methyl-*o*-phthalate by *Ps. fluorescens* strain JT701 produced 4-methyl-2,3-dihydroxybenzoate (Ribbons et al. 1984), terephthalate (*p*-phthalate) produced 3,4-dihydroxybenzoate, and isophthalate (1,3-dibenzoate) produced 3,4-dihydroxybenzoate in *Comamonas* sp. (Fukuhara et al. 2010). It is noted that concomitant decarboxylation and dehydrogenation is not, however, a component of all pathways.

3. a. Both chloride and carboxyl during the degradation of 2-chlorobenzoate by *Pseudomonas cepacia* (Fetzner et al. 1992), and 2-chlorobenzoate and 2,4-dichlorobenzoate by *Ps. aeruginosa* 142 (Romanov and Hausinger 1994).

b. Chloride during the conversion of 1,2,4,5-tetrachlorobenzene to 3,4,6-trichlorocatechol by *Pseudomonas* sp. strain PS14 (Sander et al. 1991).

c. Chloride during the degradation of 3,4-dichlorobenzoate to 3,4-dihydroxy-5-chlorobenzoate and 3,5-dicarboxypyrone by *Alcaligenes* sp. strain BR 6024, although dehydrogenation is required for conversion of 3-chlorobenzoate-4,5-dihydrodiol (Nakatsu et al. 1997).

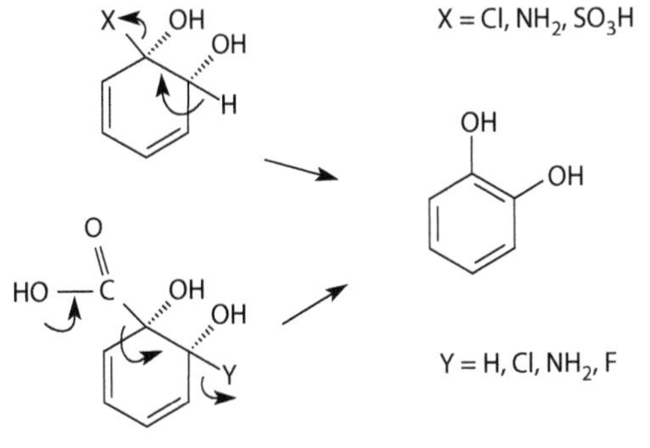

FIGURE 3.18 Dioxygenation with concomitant loss of substituents.

4. Fluoride during the conversion of 3-fluorotoluene to 3-methylcatechol (Renganathan 1989), and of both fluoride and carboxyl during the degradation of 2-fluorobenzoate (Engesser et al. 1980).

5. Nitrite during conversion of 2-nitrotoluene to catechol (An et al. 1994), 2,4-dinitrotoluene to 4-nitrocatechol (Spanggord et al. 1991), and 1,3-dinitrobenzene to 4-nitrocatechol (Dickel and Knackmuss 1991).

6. Sulfite during the conversion of naphthalene-1-sulfonate to 1,2-dihydroxynaphthalene (Kuhm et al. 1991), and 4-carboxybenzenesulfonate to 3,4-dihydroxybenzoate (4-carboxycatechol) (Locher et al. 1991a).

Phenols may exceptionally be formed by spontaneous loss of H_2O, and this is apparently favored by the presence of the electron-attracting substituents in 4-nitrotoluene, pyridine, quinoline and indole.

1. Degradation of *o*-xylene by *Rhodococcus* sp. DK17 is initiated by formation of an unstable *o*-xylene *cis*-3,4-dihydrodiol. This may be dehydrogenated to 3,4-dimethylcatechol followed by extradiol ring fission, or undergo dehydration to produce 2,3- and 3,4-dimethylphenol (Figure 3.19a) (Kim et al. 2004).

2. 4-Methylpyridine → 3-hydroxy-4-methylpyridine by dioxygenation and spontaneous dehydration (Figure 3.19b) (Sakamoto et al. 2001).

3. 4-Nitrotoluene → 2-methyl-5-nitrophenol+3-methyl-6-nitrocatechol (Figure 3.19c) (Robertson et al. 1992).

4. Quinoline → 3-hydroxyquinoline (Boyd et al. 1987).

5. Quinol-2-one → 8-hydroxyquinol-2-one (Rosche et al. 1995).

6. Degradation of carbazole by dioxygenation (Figure 3.19d) (Gieg et al. 1996).

3.4.3 Other Reactions Catalyzed by Dioxygenases

Arene hydrocarbon dioxygenases are capable of carrying out a number of reactions in addition to the introduction of both atoms of oxygen into the substrate. Illustrative examples of monooxygenation carried out by dioxygenases include the following.

1. *Hydroxylation*: Although naphthalene dioxygenase is enantiomer-specific in producing the (+)-*cis*-(1R,2S)-dihydrodiol, it also possesses both dehydrogenase and monooxygenase activities (Gibson et al. 1995). This duality is also displayed by toluene dioxygenase so that monooxygenase and dioxygenase activities are not exclusive.

   a. Naphthalene dioxygenase from *Pseudomonas* sp. strain NCIB 9816-4 has monooxygenase activities:

      i. For indene dioxygenation to *cis*-(1R,2S)-indandiol, monooxygenation to 1S-indenol, and in addition dehydrogenation of indan to indene (Lee et al. 1997).

      ii. Toluene is oxidized by purified naphthalene dioxygenase to benzyl alcohol and benzaldehyde, and ethylbenzene to (S)-1-phenylethanol and acetophenone. Whereas the initial reactions involve monooxygenation, oxidation to the aldehyde and ketone are dioxygen-dependent (Lee and Gibson 1996).

   b. Toluene dioxygenase from *Ps. putida* strain F1 displays several metabolic features:

      i. This dioxygenase and the analogous chlorobenzene dioxygenase from *Burkholderia* sp. strain PS12 are involved in degradation of these substrates by dioxygenation. In addition, they can carry out monooxygenation of 2- and 3-chlorotoluene by side chain oxidation to produce the corresponding benzyl alcohols that were only slowly oxidized further (Lehning et al. 1997).

      ii. It is able to carry out hydroxylation of phenol and 2,5-dichlorophenol (Spain et al. 1989).

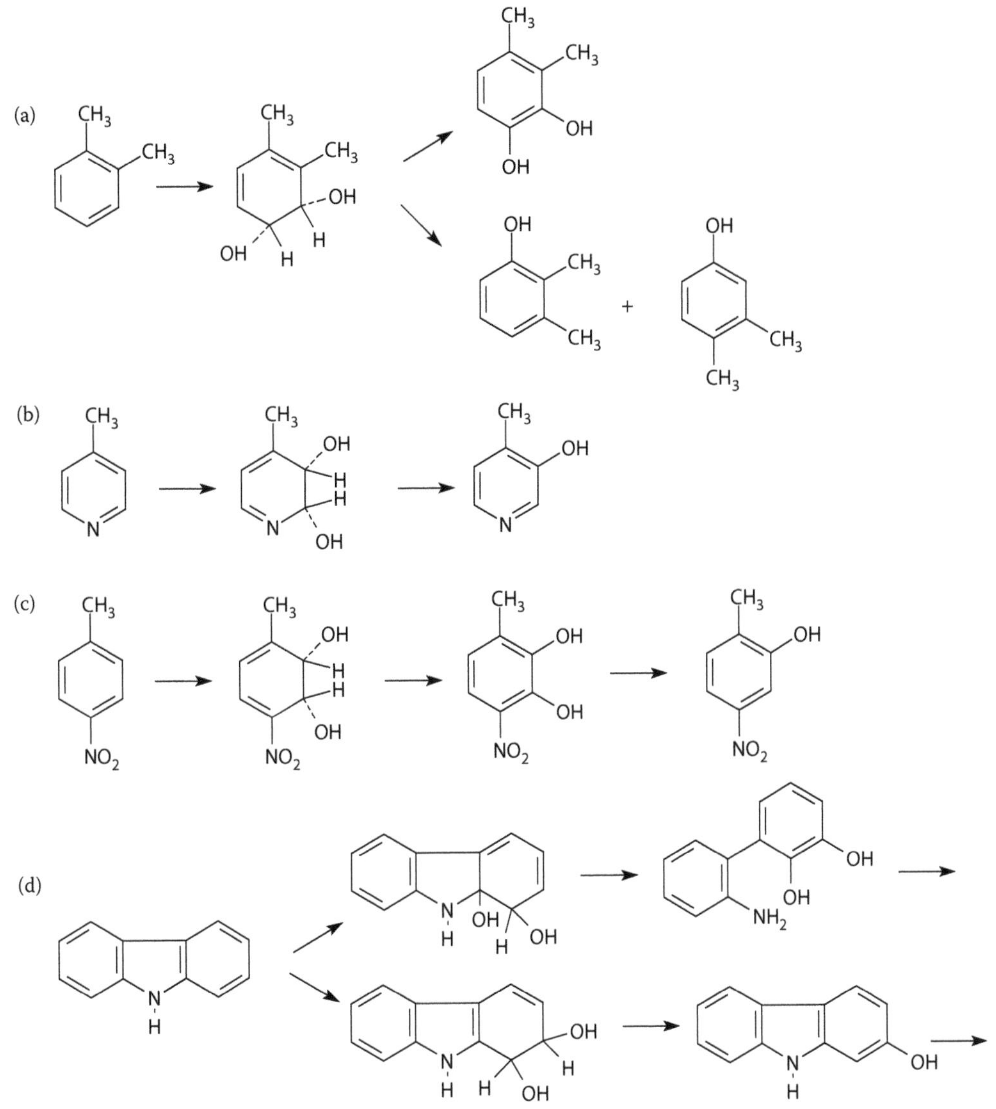

FIGURE 3.19 (a) Degradation of 1,2-dimethylbenzene by dioxygenation, and elimination with formation of dimethylphe-nols, (b) dioxygenation of 4-methylpyridine followed by elimination to 3-hydroxy-4-methylpyridine, (c) dioxygenation of 4-nitrotoluene followed by dehydrogenation or elimination, and (d) degradation of carbazole by dioxygenation followed by elimination.

2. Naphthalene dioxygenase from *Ps. putida* strain F1 is able to oxidize a number of halogenated ethenes, propenes and butenes, and *cis*-hept-2-ene and *cis*-oct-2-ene (Lange and Wackett 1997). Alkenes with halogen and methyl substituents at double bonds form allyl alcohols, whereas those with only alkyl or chloromethyl groups form diols.

There are some metabolic similarities between naphthalene dioxygenase and 2,4-dinitrotoluene dioxygenase from *Burkholderia* sp. strain DNT, and it has been suggested that these enzymes may have a common ancestor (Suen et al. 1996). Protein sequences among biphenyl dioxygenases may be similar or identical in spite of distinct differences in the range of PCB congeners that are attacked (Kimura et al. 1997; Mondello et al. 1997). The roles of the α- and β-subunits of the terminal dioxygenase (ISP) have been examined in more detail with divergent conclusions.

1. Using enzymes from hybrid naphthalene dioxygenase and 2,4-dinitrotoluene dioxygenase genes introduced into *E. coli* (Parales et al. 1998). Although the rates were different for the wild-type and hybrid enzymes, the products and the enantiomeric specificities were the same for each. It was shown that whereas the β-subunit of the dioxygenase was necessary for activity, it was the large α-subunit containing a Rieske [2Fe–2S] center that determined substrate specificity.

2. Chimeras were constructed from the α- and β-subunits of the terminal dioxygenase proteins ISP$_{BPH}$ (Hurtubise et al. 1998) from two strains of PCB-degrading strains, *Comamonas testosteroni* strain B-356 and *Pseudomonas* sp. strain LB400. The enzymes were purified, and it was shown from the substrate specificities of the purified enzymes that the structures of both subunits influenced the specificities to various substrates.

3. The function of the β-subunit that contains no detectable prosthetic groups is not fully understood, although the Tod C2 subunit is needed to obtain catalytic activity of the α-subunit (Tod C1) in *Ps. putida* F1 (Jiang et al. 1999).

It is clear from the preceding comments that there is no absolute distinction between the oxygenase activities mediated by dioxygenases. This is even less clear for heteroarenes than it is for carbocyclic compounds. An illustrative example is provided by *Ps. putida* strain 86 in which 8-hydroxy-quinol-2-one is produced from quinol-2-one (Rosche et al. 1997). The enzyme system consists of a reductase that transfers electrons from NADH and contains FAD and a [2Fe–2S] ferredoxin, and a high molecular weight oxygenase consisting of six identical subunits and six Rieske [2Fe–2S] clusters. It was suggested that this complex belonged to class IB oxygenases that include benzoate-1,2-dioxygenase. It is, therefore, possible that this apparent monooxygenase is, in fact, a dioxygenase introducing oxygen at the 8,8a positions followed by elimination (Rosche et al. 1995). The formation of 1-hydroxynaphthalene-2-carboxylate from naphthalene-2-carboxylate was apparently not via (1*R*,2*S*)-*cis*-1,2-dihydro-1,2-dihydroxynaphthalene-2-carboxylate (Morawski et al. 1997). This could have taken place either by monooxygenation that is supported by the activity of naphthalene dioxygenase in which both activities are catalyzed by the same enzyme (Gibson et al. 1995), or by an elimination that has already been noted.

Exceptionally, an acyl chloride that is spontaneously hydrolyzed to the carboxylic acid may be formed.

1. *Ps. putida* strain GJ31 is able to grow with both toluene and chlorobenzene. Degradation of chlorobenzene proceeds by formation of the dihydrodiol and dehydrogenation to 3-chlorocatechol. This is degraded by a catechol 2,3-dioxygenase (extradiol or proximal) to produce the acyl chloride that is hydrolyzed to 2-hydroxymuconate (Figure 3.20a) (Mars et al. 1997; Kaschabek et al. 1998).

2. The degradation of γ-hexachlorocyclohexane by *Sphingomonas paucimobilis* UT 6 takes place by several steps that result in the production of 2-chlorohydroquinone (Miyauchi et al. 2002; Endo et al. 2005). This is degraded by dioxygenation to an acyl chloride that is hydrolyzed to maleylacetate (Figure 3.20b) (Miyauchi et al. 1999).

3. *Sphingomonas chlorophenolica* ATCC 39723 degrades pentachlorophenol in a series of steps to 2,6-dichlorohydroquinone that undergoes dioxygenation to an acyl chloride that is hydrolyzed to 2-chloromaleylacetate (Figure 3.20c) (Ohtsubo et al. 1999; Xun et al. 1999).

3.4.3.1 Ring-Fission Dioxygenases

After dioxygenation and dehydrogenation (or elimination) with the formation of catechols, ring fission is mediated by a different group of dioxygenases. The mechanisms of ring fission depend on both the organism and the substrate. For example, whereas intradiol fission is used for catechol and 3,4-dihydroxybenzoate by *Ps. putida*, extradiol fission is used for both substrates by *Ps. acidovorans* and *Ps. testosteroni*. On the other hand, for species of *Azotobacter*, although extradiol fission is used for catechol (Sala-Trepat and Evans 1971), intradiol fission is used for 3,4-dihyroxybenzoate (Hardisson et al. 1969). In addition, the extradiol fission pathway for catechol has two branches after ring fission—one involving a dehydrogenase and tautomerase followed by decarboxylation, the other a hydrolase with the loss of formate (Figure 3.21) (Sala-Trepat and Evans 1971).

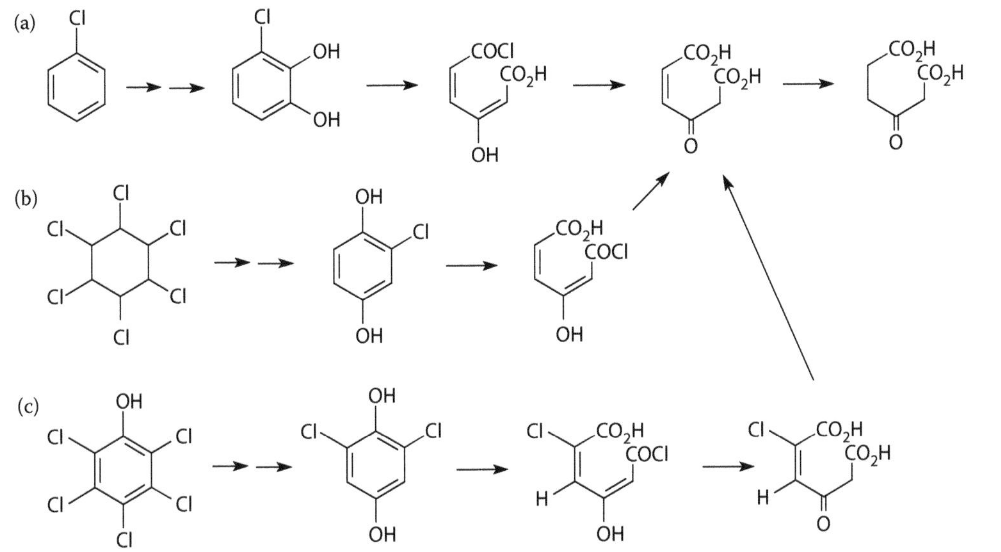

FIGURE 3.20 Degradation involving dioxygenation via acyl chlorides of (a) chlorobenzene, (b) γ- hexachlorocyclohexane, and (c) pentachlorophenol.

A great deal of attention has been devoted to the mechanisms of *intradiol* fission by the Fe(III) enzyme, and of *extradiol* fission by the Fe(II) enzymes even though these share little structural similarity (Vaillancourt et al. 2006). This review provides an extensive bibliography that includes several publications on the crystal structures of dioxygenases.

1. The paradigm *intradiol* dioxygenase (3,4-dihydroxybenzoate 3:4-dioxygenase) from *Ps. aeruginosa* has a trigonal bipyramidal structure in which the Fe(III) is coordinated to two tyrosines, two histidines and one hydroxyl group (Ohlendorf et al. 1994), and the reaction is initiated by binding of catechol to the Fe(III) (Figure 3.22) (Que and Ho 1996). The structure of the 1,2,4-trihydroxybenzene 1:2-dioxygenase in *Nocardioides simplex* strain E3 is essentially similar, except for significant differences in the tertiary structure (Ferraroni et al. 2005).

2. The *extradiol* dioxygenase (2,3-dihydroxybiphenyl 1:2 dioxygenase) from *Pseudomonas* sp. strain KKS102 has a square pyramidal structure with two histidines, one glutamate and two water

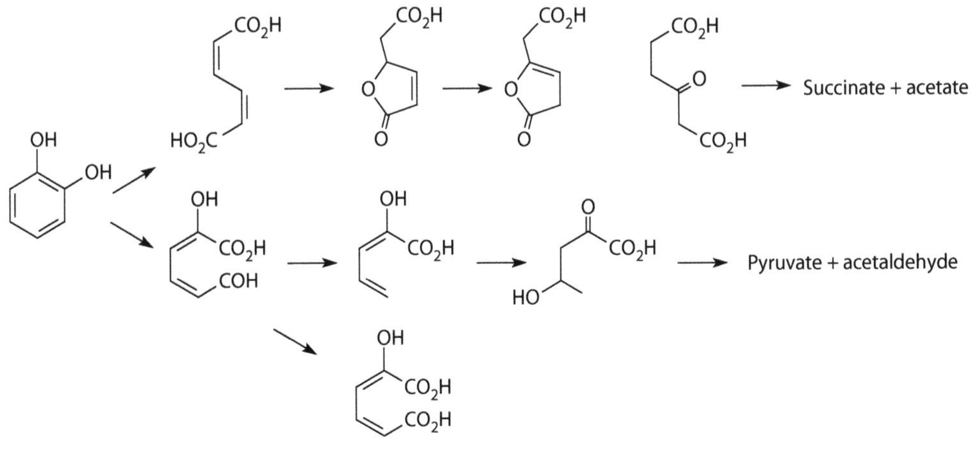

FIGURE 3.21 Branching of extradiol fission pathway after ring fission.

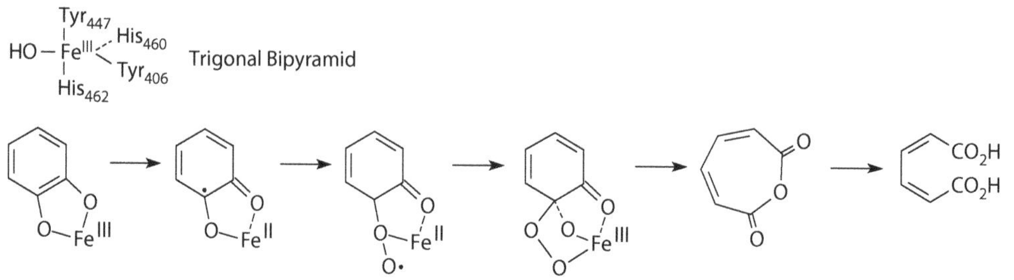

FIGURE 3.22 Mechanism of intradiol catechol fission.

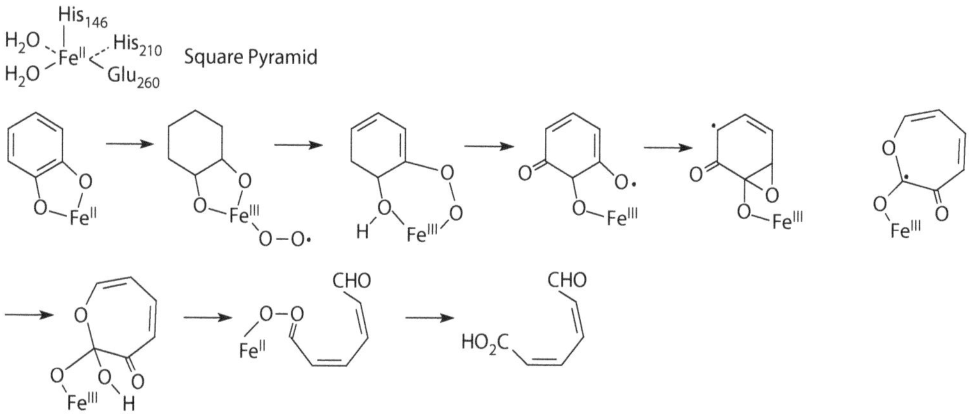

FIGURE 3.23 Oxepin mechanism of extradiol catechol fission.

molecules coordinated to the Fe(II) (Senda et al. 1996). Investigations of the extradiol enzyme have included spectroscopic examination, determination of the crystal structure of enzyme–substrate complexes, and the application of density functional theory (references in Siegbahn and Haeffner 2004). For the *extradiol* enzyme, there is essential agreement that the initial reaction is the binding of dioxygen to the catechol-chelated Fe; subsequent reactions involve fission of the catechol ring with the incorporation of one atom of oxygen into the ring to form an oxapin followed by ring opening of the lactone (Figure 3.23) (Siegbahn and Haeffner 2004).

3.4.3.1.1 Group I

3.4.3.1.1.1 Catechols and 2-Aminophenols There are three possibilities, each of which has been realized (Figure 3.24a–c) and the choice among them depends both on the substituents and on the organism.

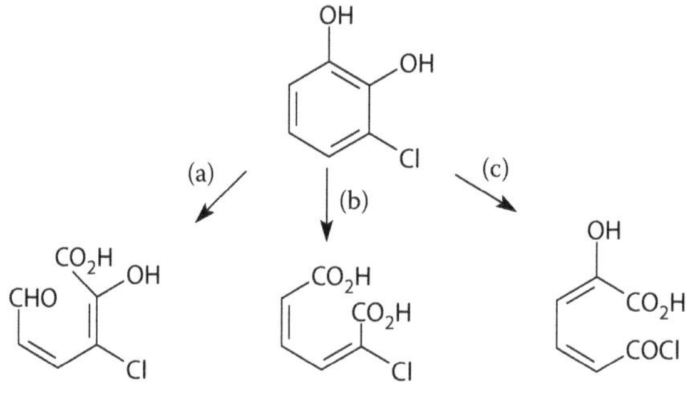

FIGURE 3.24 Alternative dioxygenation pathways for 3-substituted catechols (a) extradiol, (b) intradiol, and (c) distal.

1. Intradiol (or *ortho*) fission by catechol 1:2 dioxygenase breaks the C–C bond between the atoms bearing the phenolic groups:
   a. Catechol-1,2-dioxygenase contains Fe(III), and lacks heme and iron–sulfur components. The protein consists of α- and β-subunits with masses of 30 and 32 in the combinations (α,β), or (α,α), or (β,β), and the molecular mass of the enzyme is 63 kDa.
   b. 3,4-Dihydroxybenzoate-3,4-dioxygenase is an Fe(III) protein and consists of an aggregate of α- and β-subunits with masses of 22.2 and 26.6 in an $(α,β)_{12}$ structure with a molecular mass of 587.

2. Extradiol (or *meta*) fission by catechol dioxygenases are Fe(II)-enzymes that break the C–C bond between one of the hydroxyl groups and the adjacent nonhydroxylated carbon atom. They have been divided into three major classes (Table 3.4) (Spence et al. 1996; Peng et al. 1998). All possible fission mechanisms for vicinal dihydroxybenzoates have been described in various bacteria and include the following illustrations:
   a. *Catechol-1:2-dioxygenation* takes place in the degradation of 2,3-dihydroxyphenylpropionate by *E. coli* (Spence et al. 1996; Díaz et al. 1998);
   b. *Catechol-2:3-dioxygenation* has been demonstrated in *E. coli* C for the degradation of 3,4-dihydroxyphenylacetate (Roper and Cooper 1990), and in *Paenibacillus (Bacillus macerans)* for the degradation of 3,4-dihydroxybenzoate (Kasai et al. 2009);
   c. *Catechol-4:5-dioxygenation* has been characterized in *Sphingomonas paucimobilis* SYK-6 for the degradation of 3,4-dihydroxybenzoate as an alternative to the more widespread *ortho* catechol- 3:4-fission (Noda et al. 1990; Masai et al. 2000).

3. Distal fission: Although substituted catechols generally undergo extradiol fission, this may produce toxic metabolites, and that can be circumvented by regioselective distal-dioxygenation. Although this is rather uncommon, it has been observed in several important degradations.
   i. 1,6-dioxygenation in the metabolism of 2-hydroxyaniline by *Pseudomonas pseudoalcaligenes* JS45 and *Comamonas* strain JS 765 (Davis et al. 1999; Wu et al. 2005b) and by *Pseudomonas* sp. AP-3 (Takenaka et al. 1997; Wu et al. 2005a);

TABLE 3.4

Classes of Extradiol Dioxygenases

Class	Organism	Substrate for Fission by Dioxygenation	Reference
I	*Rhodococcus globerulus* P6	2,3-Dihydroxybiphenyl–(1:2-dioxygenase)	Asturias et al. (1994)
II	*Burkholderia cepacia* LB 400	2,3-Dihydroxybiphenyl–(1:2-dioxygenase)	Haddock and Gibson (1995)
	Ps. pseudoalcaligenes KF 707	2,3-Dihydroxybiphenyl–(1:2-dioxygenase)	Furukara and Arimura (1987)
	Pseudomonas sp. KKS 102	2,3-Dihydroxybiphenyl–(1:2-dioxygenase)	Kimbara et al. (1989)
III	*Sphingomonas paucimobilis* SYK 6	3,4-Dihydroxybenzoate–(4:5-dioxygenase)	Noda et al. (1990); Masai et al. (2000)
	S. paucimobilis SYK 6	2,2′,3-Trihydroxy-3′-methoxybiphenyl-5,5′-dicarboxylate	Peng et al. (1998)
	Pseudomonas sp. CA 10	2′-Amino-2,3-dihydroxybiphenyl–(1:2-dioxygenase)	Sato et al. (1997)
	E. coli C	3,4-Dihydroxyphenylacetate–(2:3-dioxygenase)	Roper and Cooper (1990)
	E. coli	2,3-Dihydroxyphenylpropionate–(1:2-dioxygenase)	Spence et al. (1996); Díaz et al. (1998)
	Alcaligenes eutrophus JMP 222	Catechol 2:3-dioxygenase 1	Kabisch and Fortnagel (1990)

ii. 3,4-dioxygenation of 3-hydroxyanthranilate (4-amino-3-hydroxybenzoate) in *Bordetella* sp. strain 10d (Orii et al. 2006) and in *Ralstoniametallireducens* (Zhang et al. 2005).

These are representatives of ring-fission dioxygenases that are characterized by the presence of a cupin fold and are illustrated in the table (Fetzner 2012).

Substrate	Cupin	Metal		Reference
Gentisate 1,2-dioxygenase	Bicupinhomotetramer	Fe^{2+}		Harpel et al. 1990
Salicylate 1,2-dioxygenase	Bicupinhomotetramer	Fe^{2+}		Hintner et al. 2004
1-hydroxy-2-naphthoate 1,2-dioxygenase	Bicupinhomotetramer	Fe^{2+}		Iwabuchi and Harayama (1998)
3-hydroxyanthanilate 3,4-dioxygenase	Monocupinhomodimer	Fe^{2+}		Zhang et al. 2005
4-amino-3-hydroxybenzoate 2,3-dioxygenase	Monocupinhomodimer	Fe^{2+}		Orii et al. 2006
Flavanol 2,4-dioxygenase	Bicupinhomodimer	Cu^{2+}	(*Aspergillus japonicus*)	Steiner et al. 2002
	Bicupinhomodimer	Mn^{2+}	(*Bacillus subtilis*)	Schaab et al. 2006
	Monocupin homodimer	Ni^{2+}	(*Streptomyces* sp. FLA)	Merkens et al. 2008
	Bicupin homodimer	Cu^{2+}	(*Aspergillus japonicus*)	Steiner et al. 2002

b. 3-Chlorocatechol by *Sphingomonas xenophaga* BN6 (Riegert et al. 2001), and 3-fluoro-benzoate by strain FLB300 (that belongs to the α-2 subclass and is probably a species of *Agrobacterium*) (Engesser et al. 1990; Schreiber et al. 1980).

c. i. The metabolism of 2,3-difluorohydroquinone and 2,5-difluorohydroquinone by *Ps. fluorescens* ACB is carried out by a distal dioxygenase with fission of the $C_1 - C_6$ bond to produce 2,3-difluoro-4-hydroxymuconic semialdehyde and 2,5-difluoro-4-hydroxymuconic semialdehyde (Moonen et al. 2008a).

ii. The aerobic degradation of 4-hydroxyacetophenone by this organism (*Pseudomonas fluorescens* ACB) is initiated by a Baeyer–Villiger oxidation to 4-hydroxyphenylacetate that is hydrolyzed to hydroquinone. The further metabolism is carried out by dioxygenation, ring scission to 4-hydroxymuconic semialdehyde and conversion to 3-oxoadipate (Moonen et al. 2008a). The nonheme iron dioxygenase involved has been shown to belong to a novel sub-class of this family (Moonen et al. 2008b).

4. The choice of pathway is complicated by a number of factors. These include (a) the production of acyl chloride metabolites that inhibit the enzyme producing them, for example, by *extradiol* fission of 3-chloro- and 4-chlorocatechols (Francisco et al. 2001), and the consequent use of intradiol dioxygenation and (b) the formation of terminal metabolites, for example, by *intradiol* fission of 4-methylcatechol (Marín et al. 2010). In both of these, alternatives termed the "modified" *ortho* pathway may be used.

5. Tryptophan 2,3-dioxygenase that mediates the conversion of tryptophan to *N*-formylkynurenine, contains noncovalently bound heme, and is active in the Fe(II) form. Oxygenation is initiated at C_3 with the subsequent formation of a dioxetane that decomposes to the products (Figure 3.25b) (Leeds et al. 1993). The reaction may formally be considered analogous to catechol extradiol fission dioxygenation (Figure 3.25a).

6. In addition, there are some fission dioxygenases that are dependent on a range of other divalent metals.

Copper This is an unusual component of dioxygenases but is contained in the 2,3-dioxygenase from *Aspergillus japonicus* that brings about fission of the C ring of quercetin with formation of the depside of 3,4-dihydroxybenzoate and 2,4,6-trihydroxybenzoate, and the equally unusual metabolite carbon monoxide (Steiner et al. 2002). The enzyme from *Aspergillus niger* DSM 821 was a glycoprotein, and the EPR spectrum of the dioxygenase displayed hyperfine lines at $g_{parallel} = 2.293$ mT and $g_{perp} = 1.3$ mT that are typical of a non-blue type

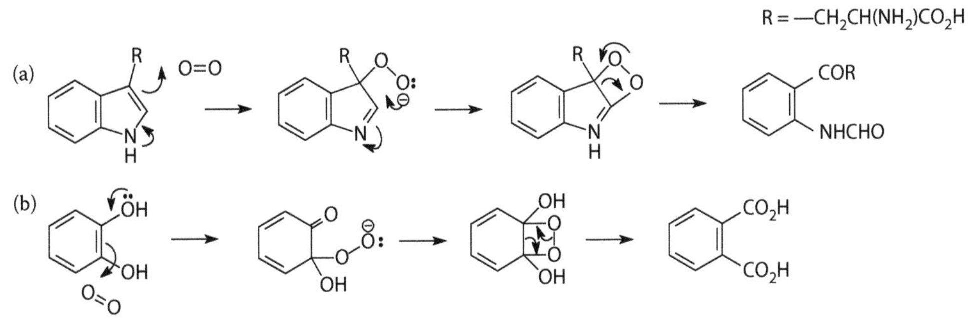

$$R = -CH_2CH(NH_2)CO_2H$$

FIGURE 3.25   Dioxygenation of (a) catechol and (b) tryptophan.

2 Cu^{II} protein (Hund et al. 1999). On the other hand, although copper can be bound quite tightly to tryptophan 2,3-dioxygenase, the evidence suggests that it probably does not play a role in the catalytic activity of the enzyme (Ishimura et al. 1980).

Magnesium   3,4-Dihydroxyphenylacetate extradiol dioxygenase from *K. pneumoniae* strain M5a1 contains magnesium, and is not activated by Fe^{2+} (Gibello et al. 1994).

Manganese

1. The 3,4-dihydroxyphenylacetate-2,3-dioxygenase from *Arthrobacter globiformis* (Boldt et al. 1995) and the 2,3-dihydroxybiphenyl-1,2-dioxygenase from *Bacillus brevis* strain JFG8 (Que et al. 1991) are Mn(II) enzymes, and are neither activated by Fe(II) nor rapidly inhibited by H_2O_2.

2. The 2,3-dihydroxybiphenyldioxygenase from *Bacillus* sp. strain JF8 that mediated fission of the biphenyl ring is unusual in being Mn(II)-dependent, and differs in structure from those of *Burkholderia* sp. strain LB400 and *Ps. paucimobilis* strain KF707 (Hatta et al. 2003).

3. Bacterial quercetinases vary considerably in their metal requirements. The enzyme in *Bacillus subtilis* had a preference for Mn for which the turnover is ≈ 40 times higher than that of the Fe-containing enzyme, and similar to that of the Cu-enzyme in *Aspergillus*. Mn^{2+} was incorporated nearly stoichiometrically in the two cuprin motives, and the EPR spectrum showed that the two Mn^{2+} ions occurred in an octahedral configuration (Schaab et al. 2006).

Nickel or Cobalt

1. The methionine salvage pathway has been elucidated in *K. pneumoniae* (Wray and Abeles 1995), and includes an intermediate enediol that undergoes dioxygenation. A single protein can carry out two oxygenations—the Fe-dependent enzyme carries out fission of the double bond, whereas the other Ni-dependent enzyme leads to loss of CO and formation of formate (Dai et al. 1999) (Figure 3.29). These reactions are formally analogous to the intradiol and extradiol fission of catechols, and the 2,3-dioxygenases involving flavanols and quinolones.

2. Quercetinase 2,3-dioxygenase activity from *Streptomyces* sp. strain FLA was high when the medium was supplemented with Ni^{2+} or Co^{2+}—though not with Fe^{2+}, Cu^{2+}, or Zn^{2+}. The ESR spectrum was consistent with the presence of high spin $S^{3/2}$ Co^{2+}, and the enzyme had a preference for Ni or Co although the K_m values suggested that the Ni enzyme would be more effective under physiological conditions (Merkens et al. 2008).

3.4.3.1.2 Group II

3.4.3.1.2.1 Hydroxybenzoates   The second large group of ring-fission dioxygenases are used for the degradation of *ortho*-dihydroxybenzoates including, 2,3-, 3,4-, and 4,5-dihydroxybenzoate. As for simple catechols, fission may be either intradiol or extradiol, and the choice between them depends on the organism.

Whereas the 3,4-dihydroxybenzoate dioxygenase in *Ps. putida* mediates intradiol fission to a muconic acid by an Fe(III) enzyme (Bull and Ballou 1981), the 3,4-dihydroxybenzoate dioxygenase in *Pseudomonas testosteroni* mediates extradiol fission to a muconic semialdehyde by an Fe(II) enzyme (Arciero et al. 1983).

3.4.3.1.3 Group III

3.4.3.1.3.1 Other Dioxygenases   This includes a heterogeneous group used for the degradation of substrates including gentisate, salicylate, and 1-hydroxynaphthalene-2-carboxylate by pathways that do not involve catechols

1. 2,5-dihydroxybenzoate (gentisate) 1,2-dioxygenase brings about ring fission between the carboxyl and hydroxyl groups, and produces maleylpyruvate without loss of the carboxyl group (Figure 3.26/3.14) (Harpel and Lipscomb 1990). It belongs to a different class of ring-cleaving dioxygenases, since the protein sequences of the gene product from *Sphingomonas* sp. strain RW5 revealed little or no similarities to those of either intradiol or extradiol dioxygenases (Wergath et al. 1998; Hintner et al. 2001). In addition, gentisate 1,2-dioxygenase occurs only in *Acinetobacter oleivorans* strain DR1, and is lacking in other acinetobacters (Jung et al. 2011). By contrast, the degradation of the analogous salicylate proceeds by hydroxylation to catechol with loss of carboxyl, followed by ring fission of the catechol.

2. There are some dioxygenases that carry out fission of the aromatic ring although they do not have vicinal hydroxyl groups.

   a. The dioxygenase from *Nocardioides* sp. strain KP7 brings about ring fission of 1-hydroxynaphthalene-2-carboxylate to 2′-carboxybenzalpyruvate, and is different from the dioxygenases that catalyze ring fission of catechols (Iwabuchi and Harayama 1998).

   b. The metabolism of 5-aminosalicylate by *Pseudomonas* sp. strain BN9 is initiated by dioxygenation concomitant with ring fission to produce 4-amino-6-oxo-hepta-2,4-diendioate which undergoes deamination to fumarylpyruvate. Although the reaction is similar to that carried out by gentisate 1,2-dioxygenase, the enzyme is distinct (Stolz et al. 1992).

   c. The dioxygenase from *Pseudaminobacter salicylatoxidans* is able to carry out direct ring fission of salicylates having a range of substituents including chlorine and bromine without the formation of intermediate catechols (Hintner et al. 2004). The deduced amino acid sequence of the dioxygenase encoded a protein to which the highest degree of similarity was found in a presumptive gentisate 1,2-dioxygenase. Whereas the ring fission products from 3-halosalicylates underwent decarboxylation to 6-halo-2-oxo-3,5-hexa-dienoates, the corresponding products from 5-halosalicylates formed lactones that were hydrolyzed to 4-hydroxy-6-oxo-hepta-2,4-diendioates.

   d. The metabolism of 5-nitroanthanilic acid by *Bradyrhizobium* sp. strain JS329 is initiated by deamination mediated by an aminohydrolase to produce 5-nitrosalicylate that undergoes

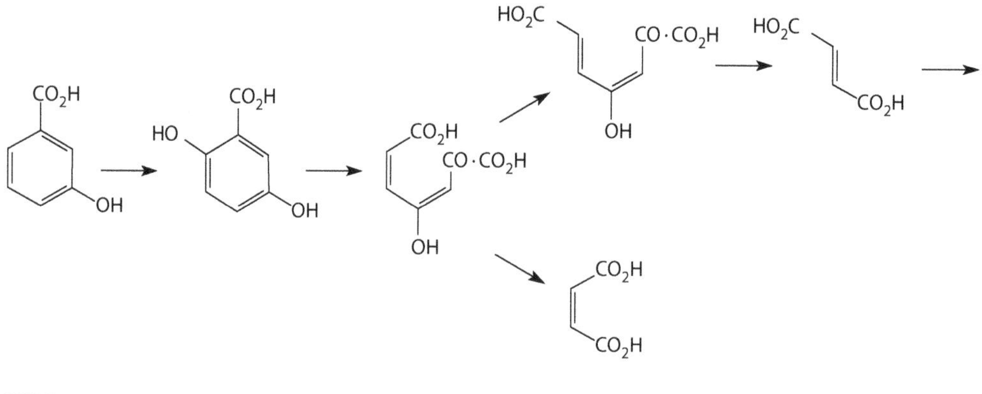

FIGURE 3.26   The gentisate pathway.

dioxygenation with simultaneous fission to 4-nitro-6-oxo-hepta-2,4-diendioate, from which nitrite was lost to form methylpyruvate (Qu and Spain 2011).

e. Ring fission of 3-hydroxyquinol-4-one is carried out by an unusual dioxygenation that is independent of metal or organic cofactors (Steiner et al. 2010), and incorporates both atoms of oxygen at C_2 and C_4, with the formation of formyl anthranilate and carbon monoxide (Fischer et al. 1999). It was proposed that these belonged to the α/β hydrolase-fold super-family functionally related to serine hydroxylases and a plausible mechanism was put forward (Fischer et al. 1999).

3. The dioxygenase that brings about the degradation of 2-methyl-3-hydroxypyridine-5-carboxylate to α-(*N*-acetylaminomethylene)succinic acid (Sparrow et al. 1969). This is discussed in greater detail in Chapter 10.

4. The dioxygenase that catalyzes the degradation of 2,5-dihydroxypyridine to maleamate and formate (Gauthier and Rittenberg 1971).

3.4.3.1.4 Group IV

There is a large heterogeneous group of dioxygenases that do not bear any similarity to the foregoing dioxygenases.

One important group is dependent on Fe^{2+} and 2-oxoglutarate: one atom from O_2 is incorporated into the resulting succinate concomitant with decarboxylation, while the other is incorporated into the substrate by formal hydroxylation.

1. The degradation of 2,4-dichlorophenoxyacetate by *Alcaligenes eutrophus* strain JMP134 involves initial formation of 2,4-dichlorophenol and glyoxylate. This is accomplished by a dioxygenase that couples this reaction to the conversion of 2-oxoglutarate to succinate and CO_2 (Fukumori and Hausinger 1993a (Figure 3.27a). The enzyme has been purified and characterized, and is able to accept other phenoxyacetates including 2,4,5-trichlorophenoxyacetate and 2-chloro-4-methylphenoxyacetates (Fukumori and Hausinger 1993b). In *Burkholderia* spp that also degraded 2,4-dichlorophenoxyacetate by this pathway, the sequence of the chromosomal gene *tfdA* was only 77.2% homologous to the plasmid-borne gene of *A. eutrophus* strain JMP 134. On the other hand, it shared a 99.5% identity with the chromosomal gene from another

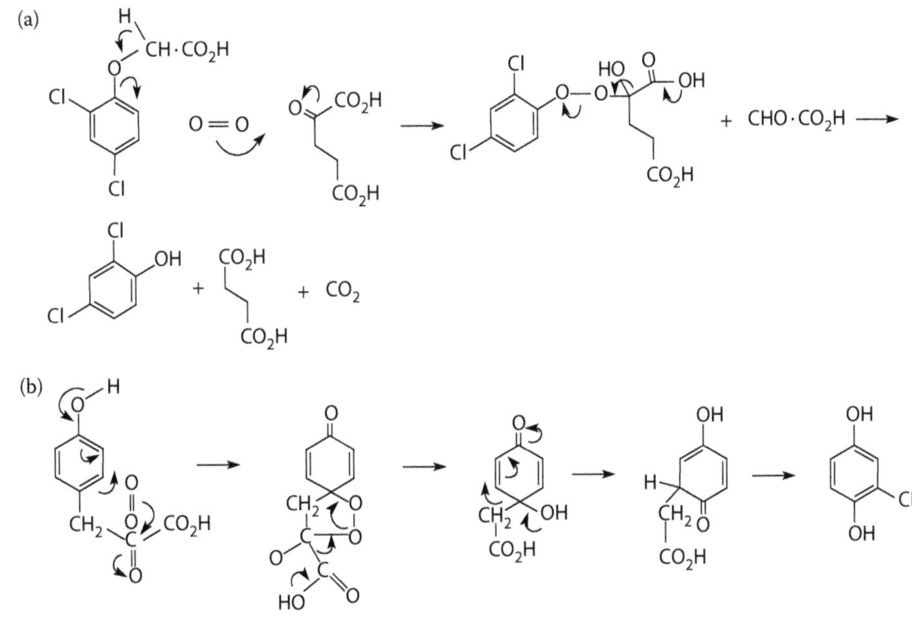

FIGURE 3.27 Dioxygenation of (a) 2,4-dichlorophenoxyacetate and (b) 4-hydroxyphenylpyruvate.

strain of *Burkholderia* sp. from a geographically distinct area (Matheson et al. 1996). In *Sphingomonas herbicidovorans*, the 2-oxoglutarate-dependent dioxygenase converted 4-chloro-2-methylphenoxypropionate (mecoprop) into 4-chloro-2-methylphenol, pyruvate and succinate each of which were formed with the incorporation of one atom of O_2 (Nickel et al. 1997). The enantioselective α-oxoglutarate-dependent dioxygenases have been purified and characterized from *Sphingomonas herbicidovorans* MH (Müller et al. 2006) and from *Delftia acidovorans* (Schleinitz et al. 2004).

2. A formally analogous reaction mediated by a dioxygenase is involved in the intramolecular dioxygenation of 4-hydroxyphenylpyruvate to 2,5-dihydroxyphenylacetate (Lindblad et al. 1970), in which both the 2-hydroxyl and the carboxylate oxygen atoms are derived from O_2 (Figure 3.27b). The degradation of 4-hydroxyphenylpyruvate formed from tyrosine by *Pseudomonas* sp. strain PJ 874 grown at the expense of tyrosine is mediated by 4-hydroxyphenylpyruvate dioxygenase that has been purified (Lindstedt et al. 1977; Johnson-Winters et al. 2003), and the primary structure of the pseudomonas enzyme determined (Rüetschi et al. 1992). The metabolism of L-tyrosine may proceed by transamination, dioxygenation to 4-phenylpyruvate followed by either ring fission, or oxidation to 2-carboxymethyl 1,4-benzoquinone and polymerization (Denoya et al. 1994).

3. a. The use of 2-aminoethanesulfonate (taurine) as a sulfur source by *E. coli* under sulfate limitation takes place by hydroxylation at the position adjacent to the sulfonate group with the production of aminoacetaldehyde and sulfite, and is mediated by a 2-oxoglutarate-dependent dioxygenase (Eichhorn et al. 1999; Ryle et al. 2003).

   b. The utilization of linear alkyl sulfate esters (C_4–C_{12}) as a sulfur source by *Ps. putida* strain S-313 under sulfate limitation takes place by an analogous reaction to produce the corresponding aldehyde and sulfate. This enzyme is, however, able to use a wider range of 2-oxoacids (Kahnert and Kertesz 2000).

4. The hydroxylation of amino acids can be accomplished by the 2-oxoacid Fe^{2+}-dependent dioxygenases.

   a. The dioxygenase from *Streptomyces griseovidirus* is able to hydroxylate L-proline to produce *trans*-4-hydroxy-L-proline that is a component of the antibiotic etamycin (Lawrence et al. 1996).

   b. The enzyme from *Bacillus thuringiensis* has been extensively examined for the hydroxylation of a rangeof amino acids that includes C_4-hydroxylation of L-isoleucine, C_5-hydroxylation of L-norleucine,sulfoxidation of L-methionine and L-ethionine (Hibi et al. 2011).

5. The formal hydroxylation of ectoine—a compatible solute in *Salibacillus salexigens*—to ectoine 5-hydroxylase is carried out by dioxygenation (Bursy et al. 2007).

6. Hypophosphite dioxygenase in *Ps. stutzeri* (White and Metcalf 2002) belongs to the class of 2-oxoglutarate-dependent dioxygenases and catalyzes the reaction:

$$H_2P(=O)-OH \rightarrow HP(=O)(OH)_2 \rightarrow P(=O)(OH)_3$$

Another group of reactions that are carried out by dioxygenation includes the following.

1. The degradation of 3-hydroxy-4-oxoquinaldine by *Arthrobacter ilicis* Rü61a and of 3-hydroxyquinol-4-one by *Ps. putida* 33/1 is mediated by a 2:4-dioxygenase that produced *N*-acetyl- or *N*-formylanthranilate and CO (Figure 3.28a) (Fischer et al. 1999). This dioxygenase is unusual in the lack of cofactor requirements and a mechanism involving the substrate dianion has been proposed (Frerichs-Deeken et al. 2004). The crystal structures of the enzymes have been analyzed and elucidated details of the mechanism (Steiner et al. 2010). The independence of metal or organic cofactors is reminiscent of the monooxygenase in *Streptomyces coelicolor* that catalyzes the conversion of 6-deoxydihydrokalafungin to the 6-dihydro compound

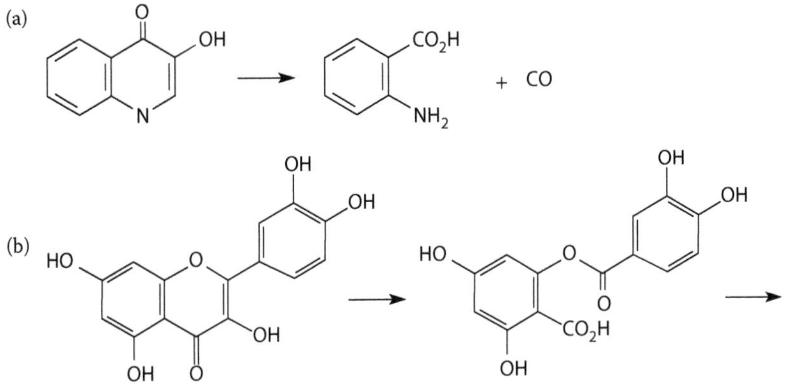

FIGURE 3.28 Dioxygenation of (a) 3-hydroxyquinol to anthranilate with loss of carbon monoxide, (b) quercitin to a depside with loss of carbon monoxide.

and that is also independent of the cofactors generally associated with the activation of oxygen (Sciara et al. 2003).

2. The dioxygenation of quercetin by *Aspergillus flavus* produced the depside 2-(3,4-dihydroxy-benzoyloxy)-4,6-dihydroxybenzoate and CO (Figure 3.28b) (Krishnamurty and Simpson 1970) and is mediated by a uniquely copper-dependent dioxygenase (quercetinase) that has been characterized by EPR (Kooter et al. 2002) and by x-ray crystallography (Steiner et al. 2002). Quercetinase in *Streptomyces* sp. strain FLA has, however, a requirement for Ni or Co and the K_m values suggest that the nickel enzyme would be more effective under physiological conditions (Merkens et al. 2008).

3. In the methionine salvage pathway that has been elucidated in *K. pneumoniae* (Wray and Abeles 1995), a key intermediate is an enediol that undergoes dioxygenation. A single protein can carry out two oxygenations—the enzyme that is Fe-dependent carries out fission of the double bond, whereas the other that is Ni-dependent leads to loss of CO and formation of formate (Dai et al 1999) (Figure 3.29). These reactions are formally analogous to the intradiol and extradiol fission of catechols.

4. Dioxygenation of 2-chlorohydroquinone to produce 3-hydroxymuconate in *Sphingomonas paucimobilis* UT26 (Miyauchi et al. 1999; Endo et al. 2005). The product of dioxygenation is an acyl chloride that is hydrolyzed to a hydroxymuconate to avoid toxifying the dioxygenase.

5. In the oxidation of *n*-propane to *n*-propanol by *Nocardia paraffinicum* (*R. rhodochrous*) ATCC 21198: the ratio of hydrocarbon to oxygen consumed was 2:1 and suggests that the reaction of two molecules of propane and one of dioxygen produced two molecules of alkanol (Babu and Brown 1984). This reaction is formally comparable to the oxidation of 2-nitropropane to acetone by flavoenzymes in *Hansenula mrakii* (Kido et al. 1978b) and *Neurospora crassa* (Kido et al. 1978a; Gorlatova et al. 1998).

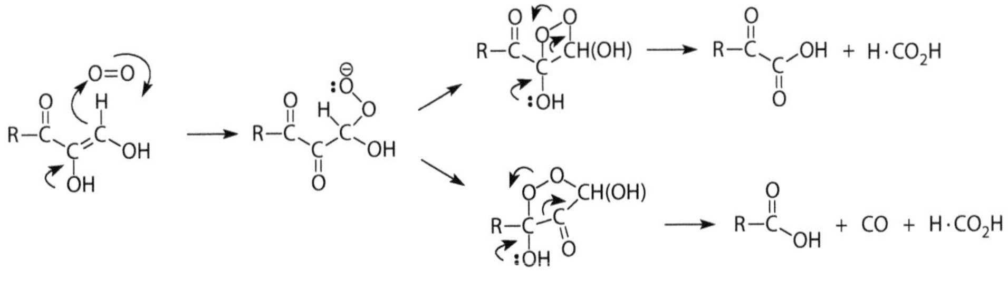

FIGURE 3.29 Dioxygenation of enediol to formate + CO in the methionine salvage pathway.

6. The polyisoprenoid oxygenase (Braaz et al. 2004) in *Xanthomonas* sp. is extracellular and functions as a heme-dependent oxygenase with the production of 12-oxo-4,8-dimethyltrideca-4,8-diene-1-al (Braaz et al. 2004). The enzyme did not display peroxidase activity, it containied two *c*-type hemes and EPR showed that the iron at the heme centers was low-spin Fe(III) (Schmitt et al. 2010).

7. There are a few dioxygenases that carry out dioxygenation of nitrogen or sulfur substituents without the fission of structural bonds.

   a. The oxygenation of a substituted 2-chloroaniline to the corresponding nitro compound $Ar-NH_2 \rightarrow Ar-NO_2$ (pyrrolonitrin) that is a metabolite of *Ps. fluorescens* is carried out by an enzyme that contains a Rieske [Fe–S] cluster and a mononuclear Fe center, although it is able to oxygenate only a very limited range of anilines (Lee et al. 2005). This should be compared with the diiron monooxygenase involved in the biosynthesis of nitrobenzoate by successive oxygenation of 4-aminobenzoate (Choi et al. 2008).

   b. Cysteine dioxygenase is involved in an alternative pathway for the degradation of cysteine to cysteine sulfinate $-SH \rightarrow -SO_2$ followed by scission to pyruvate and sulfite, or decarboxylation to hypotaurine. This was initially found in *B, subtilis*, *B. cereus* and *Streptomyces coelicolor*, although the activity is distributed among a range of bacteria (Dominy et al. 2006), as well as in mammalian systems where the enzyme has been shown to involve a mononuclear Fe center (Simmons et al. 2006).

   c. The degradation of 3,3-thiodipropionate by *Variovorax paradoxus* is initiated by scission to 3-hydroxypropionate and 3-mercaptopropionate that undergoes dioxygenation to 3-sulfinopropionate. After conversion to the coenzyme A ester, this was degraded to sulfite and propionyl-CoA, and the dioxygenase that is homologous to cysteine dioxygenase has been characterized (Bruland et al. 2009).

   d. Aerobic organisms that use elementary S_8 as source of energy metabolize this to sulfite $S + O_2 + H_2O \rightarrow SO_3^{2-} + 2H^+$ and employ an oxygenase, while CO_2 is assimilated as a source of carbon. The oxidizing enzyme has been termed sulfur dioxygenase, and the reaction appears to be initiated with a thiol to produce reactive sulfanes that are the true substrates (Rohwerder and Sand 2003). The nature of the thiols has not, however, been established although glutathione has been used as a surrogate (Rohwerder and Sand 2008). There is also some confusion over the nomenclature of several organisms.

      i. Sulfur oxygenase has been purified from *Sulfolobus brierleyi* grown in a medium with S_8 and supplemented with yeast extract. It oxidized sulfur to sulfite that was shown by incorporation of $^{18}O_2$-enriched oxygen (Emmel et al. 1986). The enzyme had a mass of 560 kDa and showed optimal activity at 65°C and pH 7.

      ii. On the basis of similar molecular mass and physiological properties, the same enzyme was produced by *Desufurolobus ambivalens* with the unexpected ability to produce sulfide, and was termed sulfur oxygenase reductase (Kletzin 1989). This is consistent with the activities of species of *Acidianus* that were designated for three species of *Sulfolobus*. They are facultative aerobes with the ability to grow aerobically as lithotrophs and anaerobically with S_8 by reduction with hydrogen (Segerer et al. 1986).

There are dioxygenases from eukaryotic organisms that are involved in important reactions.

1. The oxidation of 2-nitropropane by the yeast *Hansenula mrakii* is carried out by a flavoenzyme and produces two moles of acetone from 2 moles of substrate and one molecule of O_2 that is activated by conversion to superoxide (Kido et al. 1978b) (Figure 3.30). A similar enzyme for which 2-nitropropane is the optimal substrate has been purified and characterized from the heterothallic ascomycete *Neurospora crassa* (Gorlatova et al. 1998).

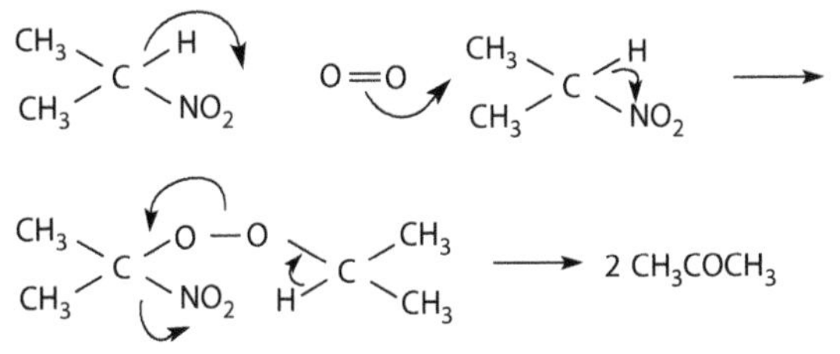

FIGURE 3.30 Dioxygenation of 2-nitropropane.

2. 1,2,4-Trihydroxybenzene is an intermediate in the degradation by *Phanerochaete chrysosporium* of vanillate, 2,4-dichlorophenol, 2,4-dinitrophenol, and 2,7-dichlorodibenzo[1,4]dioxin, and its degradation is mediated by a dioxygenase that carries out intradiol ring fission (Rieble et al. 1994).

3. The degradation of rutin by *Aspergillus flavus* has already been noted, and proceeds by hydrolysis to the aglycone followed by degradation to a depside with release of the unusual metabolite carbon monoxide (Figure 3.28b) and is accomplished by dioxygenation (Krishnamurty and Simpson 1970). In addition, this dioxygenase is unusual in containing Cu in place of the more common Fe (Steiner et al. 2002).

4. The yeast *Rhodotorula glutinis* is able to carry out unusual 2-oxoglutarate-dependent dioxygenations:

   a. Conversion of deoxyuridine by 1′-hydroxylation into uracil and ribonolactone (Stubbe 1985);

   b. Thymine-7-hydroxylase catalyzed the oxidation of the 5-methyl group to 5-hydroxymethyl uracil (Wondrack et al. 1978).

3.5 References

Alvey RA, A Biswas, WM Schluchter, DA Bryant. 2011. Effect of modified phycobilins biosynthesis in the cyanobacterium *Synechococcus* sp. strain PCC 7002. *J Bacteriol* 193: 1663–1671.

An D, DT Gibson, JC Spain. 1994. Oxidative release of nitrite from 2-nitrotoluene by a three component enzyme system from *Pseudomonas* sp. strain JS42. *J Bacteriol* 176: 7462–7467.

Arciero DM, JD Lipscomb, BH Hiynh, TA Kent, E Münck. 1983. EPR and Mössbauer studies of protocatechuate 4,5-dioxygenase. Characterization of a new Fe^{2+} environment. *J Biol Chem* 258: 14981–14991.

Ashikawa Y, Z Fujimoto, H Noguchi, H Habe, T Omori, H Yamane, H Nojiri. 2006. Electron transfer complex formation between oxygenase and ferredoxin components in Rieske nonheme iron oxygenase system. *Structure* 14: 1779–1789.

Asturias JA, LD Eltis, M Bruchna, KN Timmis. 1994. Analysis of three 2,3-dihydroxybiphenyl-1,2-dioxygenase genes found in *Rhodococcus globerulus* P6. Identification of a new family of extradiol dioxygenases. *J Biol Chem* 269: 7807–7815.

Babu JP, LR Brown. 1984. New type of oxygenase involved in the metabolism of propane and isobutane. *Appl Environ Microbiol* 48: 260–264.

Bachofer R, F Lingens. 1975. Conversion of aniline into pyrocatechol by a *Nocardia* sp.: Incorporation of oxygen-18. *FEBS Lett* 50: 288–290.

Batie CJ, E LaHaie, DP Ballou. 1987. Purification and characterization of phthalate oxygenase and phthalate oxygenase reductase from *Pseudomonas cepacia*. *J Biol Chem* 262: 1510–1518.

Boldt YR, MJ Sadowsky, LBM Ellis, L Que, LP Wackett. 1995. A manganese-dependent dioxygenase from *Arthrobacter globiformis* CM-2 belongs to the major extradiol dioxygenase family. *J Bacteriol* 177: 1225–1232.

Boyd DR, ND Sharma, F Hempenstall, MA Kennedy, JF Malone, CCR Allen. 1999. *bis-cis*-Dihydrodiols: A new class of metabolites resulting from biphenyl dioxygenase-catalyzed sequential asymmetric *cis*-dihydroxylation of polycyclic arenes and heteroarenes. *J Org Chem* 64: 4005–4011.

Boyd DR, RAS McMordie, HP Porter, H Dalton, RO Jenkins, OW Howarth. 1987. Metabolism of bicyclic aza-arenes by *Pseudomonas* to yield vicinal *cis*-dihydrodiols and phenols. *J Chem Soc Chem Commun* 1722–1724.

Braaz R, P Fischer, D Jendrossek. 2004. Novel type of heme-dependent oxygenase catalyzes oxidative cleavage of rubber (poly-*cis*-1,4-isoprene). *Appl Environ Microbiol* 70: 7388–7395.

Bruland N, JH Wübbeler, A Steinbüchel. 2009. 3-Mercaptopropionate dioxygenase, a cycteine dioxygenase homologue, catalyzed the initial step of 3-mercaptopropionate catabolism in the 3,3-thiodipropionic acid-degrading bacterium *Variovorax paradoxus*. *J Biol Chem* 284: 660–672.

Bull C, DP Ballou. 1981. Purification and properties of protocatechuate 3,4-dioxygenase from *Pseudomonas putida*. A new iron to subunit stoichiometry. *J Biol Chem* 256: 12673–12680.

Bursy J, AJ Oierik, N Pica, E Bremer. 2007. Osmotically induced synthesis of the compatible solute hydroxyec-toine is mediated by an evolutionarily conserved ectoine hydroxylase. *J Biol Chem* 282: 31147–31155.

Chang H-K, P Mohsen, GJ Zylstra. 2003. Characterization and regulation of the genes for a novel anthranilate 1,2-dioxygenase from *Burkholderia cepacia* DBO1. *J Bacteriol* 185: 5871–5881.

Chim N, A Iniguez, TW Nguyen, CW Goulding. 2010. Unusual diheme conformation of the heme-degrading protein from *Mycobacterium tuberculosis*. *J Mol Biol* 395: 595–608.

Choi YS, H Zhang, JS Brunzelle, SK Nair, H Zhao. 2008. *In vitro* reconstitution and crystal structure of *p*-aminobenzoate *N*-oxygenase (AurF) involved in aureothin biosynthesis. *Proc Natl Acad USA* 105: 6858–6863.

Cooper RA, MA Skinner. 1980. Catabolism of 3- and 4-hydroxyphenylacetate by the 3,4-dihydroxyphenylacetate pathway in *Escherichia coli*. *J Bacteriol* 143: 302–306.

Dai Y, PV Wensink, RH Abeles. 1999. One protein, two enzymes. *J Biol Chem* 274: 1193–1199.

Davis JK, Z He, CC Somerville, JC Spain. 1999. Genetic and biochemical comparison of 2-aminophenol 1,6-dioxygenase of *Pseudomonas pseudoalcaligenes* JS45 to *meta*-cleavage dioxygenases: Divergent evolution of 2-aminophenol *meta*-cleavage pathway. *Arch Microbiol* 172: 330–339.

Denoya CD, DD Skinner, MR Morgenstern. 1994. A *Streptomyces avermitilis* gene encoding a 4-hydroxyphe-nylpyruvic acid dioxygenase-like protein that directs the production of homogentisic acid and an ochro-notic pigment in *Escherichia coli*. *J Bacteriol* 176: 5312–5319.

Díaz E, A Fernández, JL Garcia. 1998. Characterization of the *hca* cluster encoding the dioxygenolytic pathway for initial catabolism of 3-phenylpropionic acid in *Escherichia coli* K-12. *J Bacteriol* 180: 2915–2923.

Dickel OD, H-J Knackmuss. 1991. Catabolism of 1,3-dinitrobenzene by *Rhodococcus* sp. QT-1 *Arch Microbiol* 157: 76–79.

Dominy JE, CR Simmons, PA Karplus, AM Gehring, MH Stipanuk. 2006. Identification and characterization of bacterial cysteine dioxygenases: A new route of cysteine degradation for eubacteria. *J Bacteriol.* 188: 5561–5569.

Eby DM, ZM Beharry, ED Coulter, DM Kurtz, EL Neidle. 2001. Characterization and evolution of anthranilate 1,2-dioxygenase from *Acinetobacter* sp. strain ADP1. *J Bacteriol* 183: 109–118.

Eichhorn E, JR van der Ploeg, T Leisinger. 1999. Characterization of a two-component alkanesulfonate mono-oxygenase from *Escherichia coli*. *J Biol Chem* 274: 26639–26646.

Emmel T, W Sand, WA König, E Bock. 1986. Evidence for the existence of sulfur oxygenase in *Sulfolobus brierleyi*. *J Gen Microbiol* 132: 3415–3420.

Endo R, M Kamakura, K Miyauchi, M Fukuda, Y Ohtsubo, M Tsuda, Y Nagata. 2005. Identification and char-acterization of genes involved in the downstream degradation pathway of γ-hexachlorocyclohexane in *Sphingomonas paucimobilis* UT26. *J Bacteriol* 187: 847–853.

Engesser KH, E Schmidt, H-J Knackmuss. 1980. Adaptation of *Alcaligenes eutrophus* B9 and *Pseudomonas* sp. B13 to 2-fluorobenzoate as growth substrate. *Appl Environ Microbiol* 39: 68–73.

Engesser KH, G Auling, J Busse, H-J Knackmuss. 1990. 3-Fluorobenzoate enriched bacterial strain FLB 300 degrades benzoate and all three isomeric monofluorobenzoates. *Arch Microbiol* 153: 193–199.

Ferraroni M, J Seifert, VM Travkin, M Thiel, S Kaschabek, A Scozzafava, L Gorleva, M Schlömann, F Briganti. 2005. Crystal structure of the hydroxyquinol 1,2-dioxygenase from *Nocardiodides simplex* 3E, a key enzyme involved in polychlorinated aromatics biodegradation. *J Biol Chem* 280: 21144–21154.

Fetzner, S, R Müller, F Lingens. 1992. Purification and some properties of 2-halobenzoate 1,2-dioxygenase, a two-component enzyme system from *Pseudomonas cepacia* 2CBS. *J Bacteriol* 174: 279–290.

Fetzner S. 2012. Ring-cleaving dioxygenases with cupin fold. *Appl Environ Microbiol* 78: 2505–2514.

Fischer F, S Künne, S Fetzner. 1999. Bacterial 2,4-dioxygenases: New members of the α/β hydrolase-fold superfamily of enzymes functionally related to serine hydrolases. *J Bacteriol* 181: 5725–5733.

Francisco PB, N Ogawa, K Suzuki, K Miyashita. 2001. The chlorobenzoate dioxygenase genes of *Burkholderia* sp. strain NK8 involved in the catabolism of chlorobenzoates. *Microbiology (UK)* 147: 121–133.

Frerichs-Deeken U, K Ranguelova, R Kappl, J Hüttermann, S Fetzner. 2004. Dioxygenases without requirement for cofactors and their chemical model reaction: Compulsory ternary complex mechanism of 1*H*-3-hydroxy-4-oxoquinaldine 2,4-dioxygenase involving general base catalysis by histidine 251 and singe-electron oxidation of the substrate dianion. *Biochemistry* 43: 14485–14499.

Fukuhara Y, K Inakazu, N Kodama, N Kamimura, D Kasai, Y Katayama, M Fukuda, E Masai. 2010. Characterization of the isophthalate degradation genes of *Comamonas* sp. strain E6. *Appl Environ Microbiol* 76: 519–527.

Fukumori F, CP Saint. 1997. Nucleotide sequences and regulatory analysis of genes involved in conversion of aniline to catechol in *Pseudomonas putida* UCC22 (pTDN1). *J Bacteriol* 179: 399–408.

Fukumori F, RP Hausinger. 1993a. *Alcaligenes eutrophus* JMP 134 2,4-chlorophenoxyacetate "monooxygenase" is an α-ketoglutarate-dependent dioxygenase. *J Bacteriol* 175: 2083–2086.

Fukumori F, RP Hausinger. 1993b. Purification and characterization of 2,4-dichlorophenoxyacetate/α-ketoglutarate dioxygenase. *J Biol Chem* 268: 24311–24317.

Furukawa K, N Arimura. 1987. Purification and properties of 2,3-dihydroxybiphenyl dioxygenase from polychlorinated biphenyl-degrading *Pseudomonas pseudoalcaligenes* and *Pseudomonas aeruginosa* carrying the cloned *bphC* gene. *J Bacteriol* 169: 924–927.

Gauthier JJ, SC Rittenberg. 1971. The metabolism of nicotinic acid I. Purification and properties of 2,5-dihydroxypyridine oxygenase from *Pseudomonas putida* N-9. *J Biol Chem* 246: 3737–3742.

Gibello A, E Ferrer, M Martín, A Garroido-Pertierra. 1994. 3,4-Dihydroxyphenylacetate 2,3-dioxygenase from *Klebsiella pneumoniae*, a Mg^{2+} containing dioxygenase involved in aromatic catabolism. *Biochem J* 301: 145–150.

Gibson DT, RL Roberts, LC Wells, VM Kopbal. 1973. Oxidation of biphenyl by a *Baijerinckia* species. *Biochem Biophys Res Commun* 50: 211–219.

Gibson DT, SM Resnick, K Lee, JM Brand, DS Torok, LP Wackett, MJ Schocken, BE Haigler. 1995. Desaturation, dioxygenation, and monooxygenation reactions catalyzed by naphthalene dioxygenase from *Pseudomonas* sp. strain 9816-4. *J Bacteriol* 177: 2615–2621.

Gieg LM, A Otter, PM Fodorak. 1996. Carbazole degradation by *Pseudomonas* sp. LD2: Metabolic characteristics and identification of some metabolites. *Environ Sci Technol* 30: 575–585.

Gorlatova N, M Tchorzewski, T Kurihara, K Soda, N Esaki. 1998. Purification, characterization, and mechanism of a flavin mononucleotide-dependent 2-nitropropane dioxygenase from *Neurospora crassa*. *Appl Environ Microbiol* 64: 1029–1033.

Haddock JD, DT Gibson. 1995. Purification and characterization of the oxygenase component of biphenyl 2,3-dioxygenase from *Pseudomonas* sp. strain LB400. *J Bacteriol* 177: 5834–5839.

Hardisson C, JM Sala-Trapat, RY Stanier. 1969. Pathways for the oxidation of aromatic compounds by *Azotobacter*. *J Gen Microbiol* 59: 1–11.

Harpel MR, JD Lipscomb. 1990. Gentisate 1,2-dioxygenase from *Pseudomonas*. Purification, characterization, and comparison of the enzymes from *Pseudomonas testosteroni* and *Pseudomonas acidovorans*. *J Biol Chem* 265: 6301–6311.

Hatta T, G Muklerjee-Dhar, J Damborsky, H Kiyohara. 2003. Characterization of a novel thermostable Mn(II)-dependent 2,3-dihydroxybiphenyl 1,2-dioxygenase from polychlorinated biphenyl- and naphthalene-degrading *Bacillus* sp. JF8. *J Biol Chem* 278: 21483–21492.

Hibi M, T Kawashima, T Kodera, SV Smirnov, PM Sokolov, M Sugiyama, S Shimizu, K Yokozeki, J Ogawa. 2011. Characterization of *Bacillus thuringiensis* L-isoleucine dioxygenase for production of useful amino acids. *Appl Environ Microbiol* 77: 6926–6930.

Hintner J-P, C Lechner, U Riegert, AE Kuhm, T Storm, T Reemtsma, A Stolz. 2001. Direct ring fission of salicylate by a salicylate 1,2-dioxygenase activity from *Pseudaminobacter salicyloxidans*. *J Bacteriol* 183: 6936–6942.

Hintner J-P, T Reemtsma, A Stolz. 2004. Biochemical and molecular characterization of a ring fission dioxygenase with the ability to oxidize (substituted) salicylate(s) from *Pseudaminobacter salicylatoxidans*. *J Biol Chem* 279: 37250–37260.

Hund H-K, J Breuer, F Lingens, J Hüttermann, R Kappl, S Fetzner. 1999. Flavonol 2,4-dioxygenase from *Aspergillus niger* DSM 821, a type 2 Cu^{II}-containing glycoprotein. *Eur J Biochem* 263: 871–878.

Hurtubise Y, D Barriault, M Sylvestre. 1998. Involvement of the terminal oxygenase β subunit in the biphenyl dioxygenase reactivity pattern towards chlorobiphenyls. *J Bacteriol* 180: 5828–5835.

Ishimura Y, R Makino, R Ueno, K Sakaguchi, FO Brady, P Feigelson, P Aisen, O Hayaishi. 1980. Copper is not essential for the catalytic activity of L-tryptophan 2,3-dioxygenase. *J Biol Chem* 255: 3835–3837.

Iwabuchi T, S Harayama. 1998. Biochemical and molecular characterization of 1-hydroxy-2-naphthoate dioxygenase from *Nocardioides* sp. KP7. *J. Biol. Chem.* 273: 8332–8336.

Jiang H, RE Parales, DT Gibson. 1999. The α subunit of toluene dioxygenase from *Pseudomonas putida* F1 can accept electrons from reduced ferredoxin$_{TOL}$ but is catalytically inactive in the absence of the β subunit. *Appl Environ Microbiol* 65: 315–318.

Johnson-Winters K, VM Purpero, M Kavana, T Nelson, GR Moran. 2003. (4-Hydroxyphenyl)pyruvate dioxygenase from *Streptomyces avermitilis*: The basis for ordered substrate addition. *Biochemistry* 42: 2072–2080.

Jung J, EL Madsen, CO Jeon, W Park. 2011. Comparative genome analysis of *Acinetobacter oleivorans* DR1 to determine strain-specific genomic regions and gentisate biodegradation. *Appl Environ Microbiol* 77: 7418–7424.

Kabisch M, P Fortnagel. 1990. Nucleotide sequences of metapyrocatechase 1 (catechol 2,3-dioxygenase 1) gene *mpc1* from *Alcaligenes eutrophus* JMP222. *Nucleic Acids Res* 18: 3504–3506.

Kahnert A. MA Kertesz. 2000. Characterization of a sulfur-regulated oxygenative alkylsulfatase from *Pseudomonas putida* S-313. *J Biol Chem* 275: 31661–31667.

Kasai D, T Fujinami, T Abe, K Mase, Y Katayama, M Fukuda, E Masai. 2009. Uncovering the protocatechuate 2,3-cleavage pathway genes. *J Bacteriol* 191: 6758–6768.

Kaschabek SR, T Kasberg, D Müller, AE Mars, DB Janssen, W Reineke. 1998. Degradation of chloroaromatics: Purification and characterization of a novel type of chlorocatechol 2,3-dioxygenase of *Pseudomonas putida* GJ31. *J Bacteriol* 180: 296–302.

Kauppi B, K Lee, E Carredano, RE Parales, DT Gibson, H Eklund, S Ramaswamy. 1998. Structure of an aromatic ring-hydroxylating dioxygenase—Naphthalene 1,2-dioxygenase. *Structure* 6: 571–586.

Kido T, K Hashizume, K Soda. 1978a. Purification and properties of nitroalkane oxidase from *Fusarium oxysporium. J Bacteriol* 133: 53–58.

Kido T, K Sida, K Asada. 1978b. Properties of 2-nitropropane dioxygenase of *Hansenula mrakii. J Biol Chem* 253: 226–232.

Kim D, J-C Chae, GJ Zylstra, Y-S Kim, MH Nam, YM Kim, E Kim. 2004. Identification of a novel dioxygenase involved in metabolism of *o*-xylene, toluene, and ethylbenzene by *Rhodococcus* sp. strain DK17. *Appl Environ Microbiol* 70: 7086–7092.

Kimbara K, T Hashímoto, M Fukuda, T Koana, M Takagi, M Oishi, K Yano. 1989. Cloning and sequencing of two tandem genes involved in degradation of 2,3-dihydroxybiphenyl to benzoate in the polychlorinated biphenyl-degrading *Pseudomonas* sp. strain KKS102. *J Bacteriol* 171: 2740–2747.

Kimura N, A Nishi, M Goto, K Furukawa. 1997. Functional analyses of a variety of chimeric dioxygenases constructed from two biphenyl dioxygenases that are similar structurally but different functionally. *J. Bacteriol* 179: 3936–3943.

Kletzin A. 1989. Coupled enzymatic production of sulfite, thiosulfate, and hydrogen sulfide from sulfur: Purification and properties of a sulfur oxygenase reductase from the facultatively anaerobic archaebacterium *Desulfurolobus ambivalens. J Bacteriol* 171: 1638–1643.

Kooter IM, RA Steiner, BW Dijkstra, PI van Noort, MR Egmond, M Huber. 2002. EPR characterization of the mononuclear Cu-containing *Aspergillus japonicus* quercetin 2,3-dioxygenase reveals dramatic changes upon anaerobic binding of substrates. *Eur J Biochem* 269: 2971–2979.

Krishnamurty HG, FJ Simpson. 1970. Degradation of rutin by *Aspergillus flavus*. Studies with oxygen 18 on the action of a dioxygenase on quercetin. *J Biol Chem* 245: 1467–1471.

Kuhm AE, A Stolz, K-L Ngai, H-J Knackmuss. 1991. Purification and characterization of a 1,2-dihydroxynaphthalene dioxygenase from a bacterium that degrades naphthalenesulfonic acids. *J Bacteriol* 173: 3795–3802.

Lange CC, LP Wackett. 1997. Oxidation of aliphatic olefins by toluene dioxygenase: Enzyme rates and product identification. *J. Bacteriol.* 179: 3858–3865.

Lawrence CC, WJ Sobey, RA Field, JE Baldwin, CJ Schofield. 1996. Purification and initial characterization of proline 4-hydroxylase from *Streptomyces griseoviridus* P8648: A 2-oxoacid, ferrous-dependent dioxygenase involved in etamycin biosynthesis. *Biochem J* 313: 185–192.

Lee K, DT Gibson. 1996. Toluene and ethylbenzene oxidation by purified naphthalene dioxygenase from *Pseudomonas* sp. strain NCIB 9816-4. *Appl Environ Microbiol* 62: 3101–3106.

Lee K, SM Resnick, DT Gibson. 1997. Stereospecific oxidation of (*R*)- and (*S*)-1-indanol by naphthalene dioxygenase from *Pseudomonas* sp. strain NCIB 9816-4. *Appl Environ Microbiol* 63: 2067–2070.

Lee J, M Simurdiak, H Zhao. 2005. Reconstitution and characterization of aminopyrrolnitrin oxygenase, a Rieske *N*-oxygenase that catalyzes unusual arylamine oxidation. *J Biol Chem* 280: 36719–36727.

Leeds JM, PJ Brown, GM McGeehan, FK Brown, JS Wiseman. 1993. Isotope effects and alternative substrate reactivities for tryptophan 2,3 dioxygenase. *J Biol Chem* 268: 17781–17786.

Lehning A, U Fgock, R-M Wittich, KN Timmis, DH Pieper. 1997. Metabolism of chlorotoluenes by *Burkholderia* sp. strain PS12 and toluene dioxygenase of *Pseudomonas putida* F1: evidence for mono-oxygenation by toluene and chlorobenzene dioxygenases. *Appl Environ Microbiol* 63: 1974–1979.

Lendenmann U, JC Spain. 1996. 2-Aminophenol 1,6-dioxygenase: A novel aromatic ring cleavage enzyme purified from *Pseudomonas pseudoalcaligenes* JS 45. *J Bacteriol* 178: 6227–6232.

Lindblad B, G Lindstedt, S Lindsted. 1970. The mechanism of enzymatic formation of homogentisate from *p*-hydroxyphenylpyruvate. *J Am Chem Soc* 92: 7446–7449.

Lindstedt S, B Odelhög, M Rundgren. 1977. Purification and properties of 4-hydroxyphenylpyruvate dioxygenase from *Pseudomonas* sp. P.J. 874. *Biochemistry* 16: 3369–3377.

Locher HH, T Leisinger, AM Cook. 1991a. 4-Sulfobenzoate 3,4-dioxygenase. Purification and properties of a desulfonative two-component system from *Comamonas testosteroni* T-2. *Biochem J* 274: 833–842.

Mampel J, J Ruff, F Junker, AM Cook. 1999. The oxygenase component of the 2-aminobenzenesulfonate dioxygenase system from *Alcaligenes* sp. strain O-1. *Microbiology (UK)* 145: 3255–3264.

Marín M, D Pérez-Pantoja, R Donoso, V Wray, B González, DH Pieper. 2010. Modified 3-oxoadipate pathway for the biodegradation of methylaromatics in *Pseudomonas reinekei* MT1. *J Bacteriol* 192: 1543–1552.

Martins BM, T Svetlitchnaia, H Dobbek. 2005. 2-oxoquinoline 8-monoxygenase component: active site modulation by Rieske-[2Fe-2S] center oxidation/reduction. *Structure* 13: 665–675.

Mars AE, T Kasberg, SR Kaschabek, MH van Agteren, DB Janssen, W Reineke. 1997. Microbial degradation of chloroaromatics: Use of the *meta*-cleavage pathway for mineralization of chlorobenzene. *J Bacteriol* 179: 4540–4537.

Martins BM, T Svetlitchnaia, H Dobbek. 2004. 2-Oxoquinoline 8-monooxygenase oxygenase component: Active site modulation by Rieske-[2Fe–2S] center oxidation/reduction. *Structure* 13: 817–824.

Masai E, K Momose, H Hara, S Nishikawa, Y Katayama, M Fukuda. 2000. Genetic and biochemical characterization of 4-carboxy-2-hydroxymuconate-6-semialdehyde dehydrogenase and its role in the protocatechuate 4,5-cleavage pathway in *Sphingomonas paucimobilis* SYK-6. *J Bacteriol* 182: 6651–6658.

Matheson VG, LJ, Forney, Y Suwa, CH Nakatsu, AJ Sextone, WE Holben. 1996. Evidence for acquisition in nature of a chromosomal 2,4-dichlorophenoxyacetic acid/α-ketoglutarate dioxygenase gene by different *Burkholderia* spp. *Appl Environ Microbiol* 62: 2457–2463.

Matsui T, M Furukawa, M Unno, T Tomita, M Ikeda-Saito. 2005. Roles of distal asp in heme oxygenase from *Corynebacterium diphtheriae*, HmuO. *J Biol Chem* 280: 2981–2989.

Merkens H, R Kappl, RP Jakob, FX Schmid, S Fetzner. 2008. Quecetinase QueD of *Streptomyces* sp. FLA, a monocupin dioxygenase with a preference for nickel or cobalt. *Biochemistry* 47: 475–487.

Miyauchi K, Y Adachi, Y Nagata, M Takagi. 1999. Cloning and sequencing of a novel *meta*-cleavage dioxygenase gene whose product is involved in degradation of γ-hexachlorocyclohexane in *Sphingomonas paucimobilis*. *J Bacteriol* 181: 6712–6719.

Miyauchi K, H-S Lee, M Fukuda, M Takagi, Y Nagata. 2002. Cloning and characterization of *linR*, involved in regulation of the downstream pathway for γ-hexachlorocyclohexane degradation in *Sphingomonas paucimobilis* UT26. *Appl Environ Microbiol* 68: 1803–1807.

Mondello FJ, MP Turcich, JH Lobos, BD Erickson. 1997. Identification and modification of biphenyl dioxygenase sequences that determine the specificity of polychlorinated biphenyl degradation. *Appl Environ Microbiol* 63: 3096–3103.

Moonen MJH, NM Kamerbeek, AH Westphal, SA Boeren, DB Janssen, MW Fraaije, WJH van Berkel. 2008b. Elucidation of the 4-hydroxyacetophenone catabolic pathway in *Pseudomonas fluorescens* ACB. *J Bacteriol* 190: 5190–5198.

Moonen MJH, SA Synowsky, WAM van der Berg, AH Westphal, AJR Heck, RHH van den Heuvel, MW Fraaije, WJH van Berkel. 2008a. Hydroquinone dioxygenase from *Pseudomonas fluorescens* ACB: A novel member of the family of nonheme-iron (II)-dependent dioxygenases. *J Bacteriol* 190: 5199–5209.

Morawski B, RW Eaton, JT Rossiter, S Guoping, H Griengl, DW Ribbons. 1997. 2-Naphthoate catabolic pathway in *Burkholderia* strain JT 1500. *J Bacteriol* 179: 115–121.

Müller TA, T Fleischmann, JR van der Meer, H-PE Kohler. 2006. Purification and characterization of two enantioselective α-ketoglutarate-dependent dioxygenases, RdpA and SdpA, from *Sphingomonas herbicidovorans*. *Appl Environ Microbiol* 72: 4853–4861.

Nakatsu CH, M Providenti, RC Wyndham. 1997. The *cis*-diol dehydrogenase *cba* gene of Tn*5271* is required for growth on 3-chlorobenzoate but not 3,4-dichlorobenzoate. *Gene* 196: 209–218.

Nakatsu CH, NA Straus, RC Wyndham. 1995. The nucleotide sequence of the Tn *5271* 3-chlorobenzoate 3,4-dioxygenase genes (*cbaAB*) unites the class 1A oxygenases in a single linkage. *Microbiology (UK)* 141: 485–595.

Neidle EL, C Hartnett, LN Ornmston, A Bairoch, M Rekin, S Harayama. 1991. Nucleotide sequences of the *Acinetobacter calcoaceticus benABC* genes for benzoate 1,2-dioxygenase reveal evolutionary relationship among multicomponent oxygenases. *J Bacteriol* 173: 5385–5395.

Nickel K, MJ-F Suter, H-PE Kohler. 1997. Involvement of two α-ketoglutarate-dependent dioxygenases in enantioselective degradation of (*R*)- and (*S*)-mecoprop by *Sphingomonas herbicidovorans* MH. *Appl Environ Microbiol* 63: 6674–6679.

Noda Y, S Nishikawa, K-I Shiozuka, H Kadokura, H Nakajima, K Yoda, Y Katayama, N Morohoshi, T Haraguchi, M Yamasaki. 1990. Molecular cloning of the protocatechuate 4,5-dioxygenase genes of *Pseudomonas paucimobilis*. *J Bacteriol* 172: 2704–2709.

Nojiri H, J-W Nam, M Kosaka, K-I Morii, T Takemura, K Furihata, H Yamane, T Omori. 1999. Diverse oxygenations catalyzed by carbazole 1,9a-dioxygenase from *Pseudomonas* sp. strain CA 10. *J Bacteriol* 181: 3105–3113.

Nojiri H et al. 2005. Structure of the terminal oxygenase component of angular dioxygenase, carbazole 1,9a-dioxygenase. *J Mol Biol* 351: 355–370.

Ohlendorf DH, AM Orville, JD Lipscomb. 1994. Structure of protocatechuate 3,4-dioxygenase from *Pseudomonas aeruginosa* at 2.15 Å resolution. *J Mol Biol* 244: 586–608.

Ohtsubo Y, K Miyauchi, K Kanda, T Hatta, H Kiyohara, T Senda, Y Nagata, Y Mitsui, M Takagi. 1999. PcpA, which is involved in the degradation of pentachlorophenol in *Sphingomonas chlorophenolica* ATCC 39723, is a novel type of ring-cleavage dioxygenase. *FEBS Lett* 459: 395–398.

Orii C, S Takanaka, S Murakami, K Aoki. 2006. Metabolism of 4-amino-3-hydroxybenzoic acid by *Bordetella* sp. strain 10d: A different modified meta-cleavage pathway for 2-aminophenols. *Biosci Biotechnol* 70: 2653–2661.

Paiva-Silva GO, C Cruz-Olibeira, ES Nakaysu, CM Maya-Monteiro, BC Dunkov, H Masuda, IC Almeida, PL Oliveira. 2006. A heme-degradation pathway in a blood-sucking insect. *Proc Natl Acad USA* 103: 8030–8035.

Parales RE, MD Emig, NA Lynch, DT Gibson. 1998. Substrate specificities of hybrid naphthalene and 2,4-dinitrotoluene dioxygenase enzyme systems. *J Bacteriol* 180: 2337–2344.

Pendrak ML, MP Chao, SS Yan, DD Roberts. 2004. Heme oxygenase in *Candida albicans* is regulated by hemoglobin and is necessary for metabolism of exogenous heme and hemoglobin to α-biliverdin. *J Biol Chem* 279: 3426–3433.

Peng X, T Egashira, K Hanashiro, E Masai, S Nishikawa, Y Katayama, K Kimbara, M Fukuda. 1998. Cloning of a *Sphingomonas paucimobilis* SYK-6 gene encoding a novel oxygenase that cleaves lignin-related biphenyl and characterization of the enzyme. *Appl Environ Microbiol* 64: 2520–2527.

Puri S, MR O'Brian. 2006. the *hmuQ* and *hmuD* genes from *Bradyrhizobium japonicum* encode heme-degrading enzymes. *J Bacteriol* 188: 6476–6482.

Qu Y, JC Spain. 2011. Molecular and biochemical characterization of the 5-nitroanthranilic acid degradation pathway in *Bradyrhizobium* sp. strain JS 329. *J Bacteriol* 193: 3057–3063.

Que L, RHN Ho. 1996. Dioxygen activation of enzymes with mononuclear non-heme-iron active sites. *Chem Rev* 96: 2607–2624.

Que L, J Widom, RL Crawford. 1981. 3,4-Dihydroxyphenylacetate 2,3-dioxygenase: A manganese(II) dioxygenase from *Bacillus brevis*. *J Biol Chem* 256: 10941–10944.

Ratliff M, W Zhu, R Deshmukh, A Wilks, I Stojilkovic. 2001. Homologues of *Neisseria* heme oxygenase in Gram-negative bacteria: Degradation of heme by the product of the *pigA* gene of *Pseudomonas aeruginosa*. *J Bacteriol* 183: 6394–6403.

Renganathan V. 1989. Possible involvement of toluene-2,3-dioxygenase in defluorination of 3-fluoro-substituted benzenes by toluene-degrading *Pseudomonas* sp. strain T-12. *Appl Environ Microbiol* 55: 330–334.

Ribbons DW, P Keyser, DA Kunz, BF Taylor, RW Eaton, BN Anderson. 1984. Microbial degradation of phthalates pp. 371–395. In *Microbial Degradation of Organic Compounds* (Ed. DT Gibson), Marcel Dekker, New York.

Rieble S, DK Joshi, MH Gold. 1994. Purification and characterization of a 1,2,4-trihydroxybenzene 1,2-dioxygenase from the basidiomycete *Phanerochaete chrysosporium*. *J Bacteriol* 176: 4838–4844.

Riegert U, S Bürger, A Stolz. 2001. Altering catalytic properties of 3-chlorocatechol-oxidizing extradiol dioxygenase from *Sphingomonas xenophaga* BN6 by random mutagenesis. *J Bacteriol* 183: 2322–2330.

Robertson JB, JC Spain, JD Haddock, DT Gibson. 1992. Oxidation of nitrotoluenes by toluene dioxygenase: Evidence for a monooxygenase reaction. *Appl Environ Microbiol* 58: 2643–2648.

Rohwerder T, W Sand. 2003. The sulfane sulfur of persulfides is the actual substrate of the sulfur-oxidizing enzymes from *Acidithiobacillus* and *Acidiphilium* sp. *Microbiology UK* 149: 1699–1709.

Rohwerder T, W Sand. 2008. Properties of thiols required for sulfur dioxygenase activity at acidic pH. *J Sulfur Chem* 29: 293–301.

Romanov V, RP Hausinger. 1994. *Pseudomonas aeruginosa* 142 uses a three-component *ortho*-halobenzoate 1,2-dioxygenase for metabolism of 2,4-dichloro- and 2-chlorobenzoate. *J Bacteriol* 176: 3368–3374.

Roper DI, RA Cooper. 1990. Subcloning and nucleotide sequence of the 3,4-dihydroxyphenylacetate (homo-protocatechuate) 2,3-dioxygenase gene from *Escherichia coli* C. *FEBS Lett* 275: 53–57.

Rosche B, B Tshisuaka, S Fetzner F Lingens. 1995. 2-Oxo-1,2-dihydroquinoline 8-monooxygenase, a two-component enzyme system from *Pseudomonas putida* 86. *J Biol Chem* 270: 17836–17842.

Rosche B, B Tshisuaka, B Hauer, F Lingens, S Fetzner. 1997. 2-Oxo-1,2-dihydroquinoline 8-monooxygenase: Phylogenetic relationship to other multicomponent nonheme iron oxygenases. *J Bacteriol* 179: 3549–3554.

Rüetschi U, B Odelhög, S Lindstedt, J Barros-Söderling, B Persson, H Jörnvall. 1992. Characterization of 4-hydroxyphenylpyruvate dioxygenase. Primary structure of the *Pseudomonas* enzyme. *Eur J Biochem* 205: 459–466.

Ryle MJ, KD Koehntorp, A Liu, L Que, RP Hausinger. 2003. Interconversion of two oxidized forms of taurine/α-ketoglutarate dioxygenase, a non-heme iron hydroxylase: Evidence for bicarnbonate binding. *Proc Natl Acad Sci USA* 100: 3790–3795.

Sakamoto T, JM Joern, A Arisawa, FM Arnold. 2001. Laboratory evolution of toluene dioxygenase to accept 4-picoline as a substrate. *Appl Environ Microbiol* 67: 3882–3887.

Sala-Trapat J, WC Evans. 1971. The *meta* cleavage of catechol by *Azotobacter* species. *Eur J Biochem* 20: 400–413.

Sala-Trapat J, K Murray, PA Williams. 1972. The metabolic divergence in the *meta* cleavage of catechols by *Pseudomonas putida* NCIB 10015. Physiological significance and evolutionary implications. *Eur J Biochem* 28: 347–356.

Sander P, R-M Wittich, P Fortnagel, H Wilkes, W Francke. 1991. Degradation of 1,2,4-trichloro- and 1,2,4,5-tetrachlorobenzene by *Pseudomonas* strains. *Appl Environ Microbiol* 57: 1430–1440.

Santos R, N Buisson, S Knight, A Dancis, J-M Camadro, W Lesuisse. 2003. Hemin uptake and use as an iron source by *Candida albicans*: Role of *CaHMX-1*-encoded haem oxygenase. *Microbiology (UK)* 149: 579–588.

Sato SI, N Ouchiyama, T Kimura, H Noijiri, H Yamane, T Omori. 1997. Cloning of genes involved in carbazole degradation of *Pseudomonas* sp. strain CA10: Nucleotide sequences of genes and characterization of *meta*-cleavage enzymes and hydrolase. *J Bacteriol* 179: 4841–4849.

Schaab MR, BM Barney, WA Francisco. 2006. Kinetic and spectoscopic studies on the quercetin 2,3-dioxygenase from *Bacillus subtilis*. *Biochemistry* 45: 1009–1016.

Schleinitz KM, S Kleinsteuber, T Vallaeys, W Babel. 2004. Localization and characterization of two novel genes encoding stereospecific dioxygenases catalyzing 2(2,4-dichlorophenoxy)propionate cleavage in *Delftia acidovorans* MC1. *Appl Environ Microbiol* 70: 5357–5365.

Schmitt MP. 1997. Utilization of host iron sources by *Corynebacterium diphtheriae*: Identification of a gene whose product is homologous to eukaryotic heme oxygenases and is required for acquisition of iron from heme and hemoglobin. *J Bacteriol* 179: 838–845.

Schmitt G, G Seiffert, PMH Kroneck, R Braaz, D Jendrossek. 2010. Spectroscopic properties of the rubber dioxygenase RoxA from *Xanthonbacter* sp., a new type of diheme dioxygenase. *Microbiology UK* 156: 2537–2548.

Schreiber A, M Hellwig, E Dorn, W Reineke, H-J Knackmuss. 1980. Critical reactions in fluorobenzoic acid degradation by *Pseudomonas* sp. B13. *Appl Environ Microbiol* 39: 58–67.

Schweizer D, A Markus, M Seez, HH Ruf, F Lingens. 1987. Purification and some properties of component B of the 4-chlorophenylacetate 3,4-dioxygenase from *Pseudomonas* sp. strain CBS3. *J Biol Chem* 262: 9340–9346.

Sciara G, SG Kendrew, AE Miele, NG Marsh, L Federíci, F Malatesta, G Schimperna, C Savino, B Vallone. 2003. The structure of ActVA-Orf6, a novel type of monooxygenase involved in actinorhodin biosynthesis. *EMBO J* 22: 205–215.

Segerer A, A Neuner, JK Kristjansson, KO Stetter. 1986. *Acidianus infernus* gen. nov., sp. nov., and *Acidianus brierleyi* comb. nov.: Facultatively aerobic, extremely acidophilic thermophilic sulfur-metabolizing archaebacteria. *Int J Syst Evol Microbiol* 36: 559–564.

Senda T, K Sugiyama, H Narita, T Yamamoto, K Kimbara, M Fukuda, M Sato, K Yano, Y Mitsui. 1996. Three dimensional structures of free form and two substrate complexes of an extradiol ring-cleavage type dioxygenase, the BphC enzyme from *Pseudomonas* sp. strain KKS102. *J Mol Biol* 255: 735–752.

Siegbahn PEM, F Haeffner. 2004. Mechanism for catechol ring-cleavage by non-heme iron extradiol dioxygenases. *J Am Chem Soc* 126: 8919–8932.

Simmons CR, Q Liu, Q Huang, Q Hao, TP Begley, PA Karplus, MH Stipanuk. 2006. Crystal structure of mammalian cysteine dioxygenase. A novel mononuclear iron center for cysteine thiol oxidation. *J Biol Chem* 281: 18723–18733.

Skaar ER, AH Gaspar, O Schneewind. 2004. IsdG and IsdI, heme-degrading enzymes in the cytoplasm of *Staphylococcus aureus*. *J Biol Chem* 279: 436–443.

Spain JC, GJ Zylstra, CK Blake, DT Gibson. 1989. Monohydroxylation of phenol and 2,5-dichlorophenol by toluene dioxygenase in *Pseudomonas putida* F1. *Appl Environ Microbiol* 55: 2648–2652.

Spanggord RJ, JC Spain, SF Nishino, KE Mortelmans. 1991. Biodegradation of 2,4-dinitrotoluene by a *Pseudomonas* sp. *Appl Environ Microbiol* 57: 3200–3205.

Sparrow LG, PPK Ho, TK Sundaram, D Zach, EJ Nyns, EE Snell. 1969. The bacterial oxidation of vitamin B_6 VII. Purification, properties, and mechanism of action of an oxygenase which cleaves the 3-hydroxypyridine ring. *J Biol Chem* 244: 2590–2660.

Spence EL, M Kawamukai, J Sanvoisin, H Braven, TDH Bugg. 1996. Catechol dioxygenases from *Escherichia coli* (MhpB) and *Alcaligenes eutrophus* (MpcI): Sequence analysis and biochemical properties of a third family of extradiol dioxygenases. *J Bacteriol* 178: 5249–5246.

Steiner RA, HJ Janssen, P Roversi, AJ Oakley, S Fetzner. 2010. Structural basis for cofactor—Independent dioxygenation of *N*-heteroaromatic compounds at the α/β-hydrolase fold. *Proc Natl Acad Sci USA* 107: 657–662.

Steiner RA, KH Kalk, BW Dijkstra. 2002. Anaerobic enzyme substrate structures provide insight into the reaction mechanism of the copper-dependent quercetin 2,3-dioxygenase. *Proc Natl Acad USA* 99: 16625–16630.

Stolz A, B Nörtemann, H-J Knackmuss. 1992. Bacterial metabolism of 5-aminosalicylic acid. Initial ring cleavage. *Biochem J* 282: 675–680.

Stubbe J. 1985. Identification of two α-ketoglutarate-dependent dioxygenases in extracts of *Rhodotorula glutinis* catalyzing deoxyuridine hydroxylation. *J Biol Chem* 260: 9972–9975.

Suen W-C, BE Haigler, JC Spain. 1996. 2,4-Dinitrotoluene dioxygenase from *Burkholderia* sp. strain DNT: Similarity to naphthalene dioxygenase. *J Bacteriol* 178: 4926–4934.

Suits MDL, N Jaffer, Z Jia. 2006. Structure of the *Escherichia coli* O157:H7 heme oxygenase ChuS in complex with heme and enzymatic inactivation by mutation of the heme-coordinating residue His-193. *J Biol Chem* 281: 36776–36782.

Takenaka S, S Murakami, R Shinke, K Hatakeyma, H Yukawa, K Aoki. 1997. Novel genes encoding 2-aminophenol 1,6-dioxygenase from *Pseudomonas* species AP-3 growing on 2-aminophenol and catalytic properties of the purified enzyme. *J Biol Chem* 272: 14727–14739.

Taniuchi H, M Hatanaka, S Kuno, O Hayashi, M Nakazima, N Kurihara. 1964. Enzymatic formation of catechol from anthranilic acid. *J Biol Chem* 239: 2204–2211.

Tooley AJ, YA Cai, AN Glazer. 2001. Biosynthesis of a fluorescent cyanobacterial C-phycocyanin holo-α subunit in a heterologous host. *Proc Natl Acad USA* 98: 10560–10565.

Vaillancourt FH, JT Bolin, LD Eltis. 2006. The ins and outs of ring-cleaving dioxygenases. *Crit Revs Biochem Mol Biol* 41: 241–267.

Wegele R, R Tasler, Y Zeng, M Rovera, N Frankenberg-Dinkel. 2004. Heme oxygenase(s)-phytochrome system of *Pseudomonas aeruginosa*. *J Biol Chem* 279: 45791–45802.

Wergath J, H-A Arfmann, DH Pieper, KN Timmis, R-M Wittich. 1998. Biochemical and genetic analysis of a gentisate 1,2-dioxygenase from *Sphingomonas* sp. strain RW 5. *J Bacteriol* 180: 4171–4176.

White AK, WW Metcalf. 2002. Isolation and biochemical characterization of hypophosphite/2-oxoglutarate dioxygenase. *J Biol Chem* 277: 38262–38271.

Wilks A, MP Schmitt. 1998. Expression and characterization of a heme oxygenase (Hmu O) from *Corynebacterium diphtheriae. J Biol Chem* 273: 837–841.

Wolfe MD, JV Parales, DT Gibson, JD Lipscomb. 2001. Single turnover chemistry and regulation of O_2 activation by the oxygenase component of naphthalene 1,2-dioxygenase. *J Biol Chem* 276: 1945–1953.

Wondrack LM, C-A Hsu, MT Abbott. 1978. Thymine-7-hydroxylase and pyrimidine deoxyribonucleoside 2′-hydroxylase activities in *Rhodotorula glutinis. J Biol Chem* 253: 6511–6515.

Wray JW, RH Abeles. 1995. The methionine salvage pathway in *Klebsiella pneumoniae* and rat liver. *J Biol Chem* 270: 3147–3153.

Wu R, EP Skaar, R Zhang, G Joachimiak, P Goprnicki, O Schneewind, A Joachimiak. 2005a. *Staphylococcus aureus* IsdG and IsdI, heme-degrading enzymes with structural similarity to monooxygenases. *J Biol Chem* 280: 2840–2846.

Wu J-F, C-W Sun, C-Y Jiang Z-P Liu, S-J Liu. 2005b. A novel 2-aminophenol 1,6-dioxygenase involved in the degradation of *p*-chloronitrobenzene by *Comamonas* strain CNB-1: Purification, properties, genetic cloning and expression in *Escherichia coli. Arch Microbiol* 183: 1–8.

Xun L, J Bohuslavek, M Cai. 1999. Characterization of 2,6-dichloro-*p*-hydroquinone 1,2-dioxygenase (PcpA) of *Sphingomonas chlorophenolica* ATCC 39723. *Biochem Biophys Res Commun* 266: 322–325.

Yamaguchi M, H Fujisawa. 1982. Subunit structure of oxygenase components in benzoate 1,2-dioxygenase system from *Pseudomonas arvilla* C-1. *J Biol Chem* 257: 12497–12502.

Yoshida T, CT Migita. 2000. Mechanism of heme degradation by heme oxygenase. *J Inorg Biochem* 82: 33–41.

Zhang Y, KL Colabroy, TP Begley, SE Ealick. 2005. Structural studies on 3-hydroxyanthranilate-3,4-dioxygenase: the catalytic mechanism of a complex oxidation involved in NAD biosynthesis. *J Biol Chem* 44: 7632–7643.

Zhu W, A Wilks, I Stojiljkovic. 2000. Degradation of heme in Gram-negative bacteria: The product of the *hemO* gene of *Neisseria* is a heme oxygenase. *J Bacteriol* 182: 6783–6790.

3.6 Oxidases, Peroxidases, and Haloperoxidases

3.6.1 Oxidases

These are produced by both prokaryotes and eukaryotes, and catalyze a number of important reactions. They are flavoproteins that produce potentially destructive H_2O_2 that is generally removed by the activity of catalase or peroxidase. These reactions in which O_2 is reduced to H_2O_2 should be compared with oxidoreductases (dehydrogenases) and oxygenases. Reactions catalyzed by oxidases are formally outlined in the following examples:

$$R–CH_2–NH_2 + H_2O + O_2 \rightarrow R–CHO + NH_3 + H_2O_2$$

$$R–CH(NH_2)–CO_2H + H_2O + O_2 \rightarrow R–CO–CO_2H + NH_3 + H_2O_2$$

$$R–CH_2–CH_2–CO_2H + O_2 \rightarrow R–CH=CH–CO_2H + H_2O_2$$

There are two groups of NADH oxidases that produce either H_2O_2 or H_2O by 2ε reduction: $2NADH + O_2 \rightarrow 2NAD + H_2O_2$ or by 4ε reduction $4NADH + O_2 \rightarrow 4NAD + 2H_2O$. Both types have been encountered in anaerobes, where they may play an additional role in protection against oxygen toxicity. This aspect is noted briefly in Chapter 4 in the context of oxygen stress in anaerobes. Arsenite oxidase that catalyzes the reaction $AS^{III}O_2^- + 2H_2O \rightarrow As^VO_4^{3-} + 4H^+ + 2ε$ will be discussed later in the context of alternative electron acceptors.

1. Primary amines are widely used as a nitrogen source by bacteria, and the first step in their degradation involves formation of the corresponding aldehyde:

$$R–CH_2–NH_2 + NADH \rightarrow R–CHO + NAD + NH_3$$

There are several mechanisms whereby this may be accomplished. This may be carried out by dehydrogenation in a pyridoxal 5′-phosphate-mediated reaction, for example in *Ps. putida* (ATCC 12633) and *Ps. aeruginosa* (ATCC 17933). The alternative of using an amine oxidase has been demonstrated for several organisms.

Bacteria

a. The oxidative deamination of biogenic amines such as tyramine, octopamine, norepinephrine, and dopamine is carried out by oxidases. These membrane-bound enzymes have been found in several *Enterobacteriaceae* including *K. pneumoniae*, *Serratia marcescens*, and *Proteus rettgeri*, and in *Ps. aeruginosa* (Murooka et al. 1979). This has been confirmed for *Klebsiella oxytoca* (ATCC 8724) and *E. coli* (ATCC 9637) (Hacisalihoglu et al. 1997), and for the catabolism of 2-phenylethylamine in *E. coli* K12 (Parrott et al. 1987), and methylamine in *Arthrobacter* sp. strain P1 (Zhang et al. 1993).

   The amine oxidases function by oxidation of the amine to the aldehyde concomitant with the reduction of the cofactor 2,4,5-trihydroxyphenylalanine quinone (TPQ)—that is generated by posttranslational modification of tyrosine—to an aminoquinol which, in the form of a Cu(I) radical reacts with O_2 to form H_2O_2, Cu(II) and the imine in the last step in the reaction (Wilmot et al. 1999):

$$R–CH_2–NH_2 + O_2 + H_2O \rightarrow R–CHO + H_2O_2 + NH_3$$

   The copper-quinoprotein amine oxidases are composed of two copper ions and one, or two, molecules of the cofactor 2,4,5-trihydroxyphenylalanine quinone (TPQ). They have been examined for histamine oxidase in *Arthrobacter globiformis* (Choi et al. 1995) and tryptamine oxidase in *E. coli* (Steinebach et al. 1996). Details of the overall reaction have been studied using EPR and rapid stopped-flow spectrophotometry to reveal the three reactions involving the quinone/hydroquinone/semiquinone that are coupled to the reduction and oxidation of Cu^{2+}/Cu^+, and dehydrogenation of the amine to the aldehyde (Steinebach et al. 1996).

b. An alternative flavoprotein oxidase is, however, used for the deamination of tyramine by *Sarcina lutea* (Kumagai et al. 1969), putrescine by *Rhodococcus erythropolis* (van Hellemond et al. 2008), and by the enzyme in *Ps. fluorescens* that is capable of the dual oxidase/monooxygenase function for production of the oxoacid and H_2O_2 from L-lysine (Nakazawa et al. 1972; Flashner and Massey 1974a).

Yeasts

The copper amine oxidase in the yeast *Hansenula polymorpha* grown with diethylamine as the nitrogen source also uses 2,4,5-trihydroxyphenylalanine quinone (TPQ) as the cofactor (Mu et al. 1992; Cai and Klingman 1994). The crystal structure has been explored in which there are two stages: reaction of a tyrosine residue with H_2O + Cu(II) + $2O_2 \rightarrow$ TPQ + H_2O_2 + Cu(I) + · OH followed by sequential reactions of the quinone and the alkylamine to produce the aldehyde and conversion of the bound tyrosine to a 2,4-dihydroxy-3-amino residue, followed by reaction with O_2 and H_2O to form ammonia and H_2O_2 (Johnson et al. 2007).

2. *Amino acid oxidases*

a. Glycine oxidase is a homotetrameric flavoenzyme that contains one molecule of noncovalently bound FAD. It is able to carry out several reactions with the production of ammonia (or primary amines) and the ketoacids pyruvate or glyoxylate (Job et al. 2002).

b. Some amino acid oxidases contain FAD covalently linked through an 8α-*S*-cysteine at Cys[315] or Cys[308]. In the presence of tetrahydrofolate, sarcosine oxidase converts sarcosine (*N*-methylglycine) to glycine, 5,10-methylene-tetrahydrofolate and H_2O_2. In two strains of *Corynebacterium* sp. (U-96 and P-1), the oxidase had a heterotetrameric structure, and was unusual since it contained both one mole of covalently bound and one mole of noncovalently bound FAD: the noncovalent FAD accepted electrons from sarcosine that were transferred to the covalent FAD where oxygen was reduced to H_2O_2 (Chlumsky et al. 1995). The x-ray structure of the monomeric sarcosine oxidase from *Bacillus* sp. strain B-0618 has been solved and clearly defined the details of this mechanism (Trickey et al. 1999). An analogous enzyme NikD that is involved in the synthesis of nikkomycin carried out a 4-electron oxidation (aromatization) of piperideine-2-carboxylate to pyridine-2-carboxylate. The crystal structure has made it possible to establish details of the mechanism: the Δ^2 enamine substrate is oxidized by FAD to the $\Delta^{1,6;2,3}$ diene that is in equilibrium with the $\Delta^{2,3;5,6}$ diene and isomerized to the $\Delta^{2,3;5,6}$-diene that is oxidized by FAD to pyridine-2-carboxylate (Carrell et al. 2007). This has been explored by kinetic and spectroscopic examination (Bruckner and Jorns 2009), and an alternative interpretation based on stopped-flow spectroscopy, proposed that the reduced flavin served as an acid–base catalyst (Kommoju et al. 2009).

 c. Tryptophan 2-monooxygenase oxidizes L-tryptophan to indol-3-acetamide as the first step in the biosynthesis of indole-3-acetic acid by *Pseudomonas savastanoi* (Comai and Kosuge 1982) and *Erwinia herbicola* pv *gypsophilae* (Clark et al. 1993). It belongs to a group of enzymes that carry out the oxidative decarboxylation of amino acids, and it has been purified from *Ps. savastanoi* where it is obligatory for virulence (Hutcheson and Kosuge 1985). Amino acids may undergo FAD-dependent reactions that have been described as oxidases, and it has been shown that the mechanism involves hydride transfer (Ralph et al. 2007). The reactions are initiated by the reaction: $FADH^- + H_2N^+ = C(R) - CO_2^-$ that can then undergo two reactions: (i) $\rightarrow H_2O_2 + NH_3 + R - CO - CO_2^-$ and (ii) $\rightarrow CO_2 + H_2O + R - CO - NH_2$. Extensive study has been given to the first reaction and the second play a key role in the alternative synthesis of indole-3-acetic acid by production of the amide from tryptophan followed by hydrolysis. It has been shown on the basis of extensive study of deuterium and ^{15}N kinetic isotope effects that the critical reaction is the hydride transfer (Ralph et al. 2006). It is noted that the enzyme has also been termed an oxidoreductase:decarboxylating.

 d. D-amino acid oxidase has been isolated from a number of yeasts, and the nucleotide sequence of the enzyme from *Rhodotorula gracilis* ATCC 26217 has been established (Alonso et al. 1998). The gene could be overexpressed in *E. coli*, and levels of the enzyme were greater under conditions of low aeration: the enzyme isolated from the recombinant organisms was apparently the apoenzyme since maximum activity required the presence of FAD.

3. The aerobic degradation of nicotine by *Arthrobacter nicotinovorans* pAO1 produces *N*-methylpyrrolidine as the first metabolite by dehydrogenation. This is then hydroxylated at the benzylic carbon atom by an FAD-containing oxidase (Dai et al. 1968), and the γ-*N*-methylaminobutyrate that is produced by fission of the *N*-methylpyrolidine ring is demethylated by an oxidase to 4-aminobutyrate (Chiribau et al. 2004).

4. The monooxygenase that mediates the hydroxylation of a number of bicyclic azaarenes has been termed an oxidase—quinaldine 4-oxidase—since oxygen functions effectively as an electron acceptor (Stephan et al. 1996), and is, therefore, distinguished from the more usual dehydrogenase/oxidoreductase in which the oxygen originates in water.

5. *Hyphomicrobium* sp. strain EG is able to grow at the expense of dimethyl sulfide or dimethyl sulfoxide and initially produces methanethiol that is then further oxidized to formaldehyde, sulfide, and H_2O_2 by an oxidase that has been purified (Suylen et al. 1987).

6. The oxidation of cholesterol to cholest-4-ene-3-one is carried out by an oxidase in several bacteria. This activity has been found in *Brevibacterium sterolicum* and *Streptomyces* sp. strain SA-COO (Ohta et al. 1991), and the extracellular enzyme that has been purified from *Pseudomonas* sp. strain ST-200 (Doukyu and Aono 1998) has a preference for 3β-hydroxy compounds. Sequencing of the genome of *Gordonia cholesterolivorans* revealed the presence of two class I oxidases: ChoOx-1 that is not transcribed during growth with cholesterol and ChoOx-2 that appeared to be similar to other cholesterol oxidases that have been described (Drzyzga et al. 2011).

7. In *Corynebacterium cyclohexanicum*, degradation of cyclohexane was initiated by sequential hydroxylation and dehydrogenation to 4-oxocyclohexane-1-carboxylate. This then underwent

stepwise aromatization to 4-hydroxybenzoate mediated by enzymes that were termed desaturases: desaturase I was an oxidase that carried out dehydrogenation to cyclohex-2-ene-4-oxocar-
boxylate with the formation of H_2O_2, while the unstable desaturase II produced 4-hydroxybenzoate
putatively via cyclohexane-2,5-diene-4-oxocarboxylate (Kaneda et al. 1993).

8. The conversion of 2-hydroxyglutarate to 2-oxoglutarate with the formation of H_2O_2 is carried
   out in *E. coli* by an oxidase encoded by YgaF that possesses noncovalently bound FAD (Kallili
   et al. 2008).

9. Vanillyl-alcohol oxidase that brings about oxidation to vanillin is mediated by a flavoprotein
   that is reduced with production of a quinonemethine that reacts with H_2O to form the aldehyde.
   The reduced flavoprotein is then reoxidized to produce H_2O_2 (Fraaije and van Berkel 1997).
   The enzyme is also able to accept a range of 4-alkylphenols. Two reactions occurred, and their
   ratio depended on the alkyl substituent: for example, whereas 4-hydroxyphenylethane produced
   4-hydroxyacetophenone (76%) and 4-hydroxyphenylethene (24%), only the corresponding
   alkenes were produced from substrates with C_4–C_7 alkyl groups (van den Heuvel et al. 1998).
   The reaction is formally analogous to the bacterial reaction catalyzed by 4-ethylphenol methylene hydroxylase that proceeds by dehydrogenation followed by hydroxylation with H_2O
   (Hopper and Cottrell 2003).

10. Yeasts belonging to the genera *Candida*, *Endomycopsis*, and *Yarrowia* are able to degrade
    alkanes. The initial hydroxylation is carried out by cytochrome P450 that is found in the micro-
    some (references in Käppeli 1986). The degradation of the resulting alkanoic acids is carried
    out by enzymes that are contained in peroxisomes, and are induced during growth with alkanes.
    Acyl-CoA oxidase carries out the first step in the degradation of the alkanoic acid CoA-esters,
    and although this is formally a dehydrogenase, acyl-CoA dehydrogenase activity is absent
    (Kawamoto et al. 1978; Tanaka and Ueda 1993). The three subsequent enzymatic activities are
    apparently contained in a single protein. Acyl-CoA oxidases are produced for the degradation
    of long-chain alkanes and carboxylates, and several genes have been found to be present, one
    in *S. cerevisiae*, two in *Candida maltosa*, three in *Candida tropicalis*, and five in *Yarrowia
    lipolytica* (references in Wang et al. 1999). The isoenzymes in *Yarrowia lipolytica* have differ-
    ent substrate specificities, Aox3 is specific for C_4–C_{10}, and Aox2 for C_{10}–C_{16} (Wang et al. 1999).

11. A critical step in the synthesis of protoporphyrin IX that is the immediate precursor of heme
    and chlorophyll/bacteriochlorophyll involves decarboxylation to vinyl groups of the propionic
    acid side chains in rings A and B of uroporphyrinogen III. This can be accomplished in *E. coli*
    by two different reactions: an oxygen-dependent Mn-oxidase HemF (Breckau et al. 2003), and
    an oxygen-independent mechanism by HemN that involves a free radical generated from
    S-adenosyl-L-methionine and an [4Fe–4S] cluster which has been confirmed from the x-ray
    structure (Layer et al. 2003) and characterized by the complex pattern of splitting in the EPR
    spectrum (Layer et al. 2006).

The biosynthesis of porphyrin and corrin enzymes including siroheme, cobalamin, and coenzyme F_{420}
in archaea such as methanogens (Storbeck et al. 2010) and sulfate-reducing bacteria (Ishida et al. 1998;
Lobo et al. 2009) are synthesized by a different pathway in which uroporphyrinogen III is the branch
point before methylation to precorrin-2 from which cobalamin and sirohydrochlorin are synthesized.
These organisms seem to lack the enzymes already noted for completing the biosynthesis of heme by
oxidative decarboxylation of the propionic acid groups in rings A and B to vinyl groups.

3.6.2 Peroxidases

Extracellular H_2O_2 is required for the activity of peroxidases in white-rot fungi, and this can be produced
by several fungal reactions.

1. Glyoxal oxidase is produced from *Phanerochaete chrysosporium* under high concentrations of
   oxygen, is stimulated by Cu^{2+}, and oxidizes a range of substrates with the production of H_2O_2

from O_2. Substrates include methylglyoxal, glyoxylic acid, and glycolaldehyde, and the pure enzyme requires activation by lignin peroxidase: under these conditions, in the presence of catalytic amounts of H_2O_2, pyruvate and veratraldehyde are produced respectively from methylglyoxal and 3,4-dimethoxybenzyl alcohol (veratryl alcohol) (the lignin peroxidase substrate) (Kersten 1990).

2. An aryl-alcohol oxidase produced optimally under carbon limitation from *Bjerkandera adusta* oxidized a number of benzyl alcohols including 4-methoxybenzyl alcohol, 3,4-dimethoxybenzyl alcohol (veratryl alcohol) and 4-hydroxy-3-methoxybenzyl alcohol, with the production of H_2O_2 from O_2: monosaccharides were not oxidized (Muheim et al. 1990). An aryl-alcohol oxidase from *Pleurotus eryngii* is a flavoprotein with range of substrates comparable to that from *B. adusta* (Guillén et al. 1992).

3. Peroxide may also be produced by the manganese-peroxide dependent oxidation of glycolate and oxalate that are synthesized by *Ceriporiopsis subvermispora* (Urzúa et al. 1998).

Two distinct extracellular enzymes are produced by white-rot fungi—lignin peroxidase (LiP), and manganese peroxidase (MnP). These are produced under specific growth conditions that include carbon, nitrogen or sulfur limitation, manganese concentration, and increased oxygen concentration. They are involved in the degradation of lignin and in the biotransformation of xenobiotics. These enzymes require extracellular H_2O_2 that is produced by a number of these organisms as a result of oxidase activity that has already been noted. Details of the relevant reactions have emerged from a study of the degradation of a model lignin substrate by *Phanerochaete chrysosporium* (Hammel et al. 1994). 1-(3,4-Dimethoxyphenyl)-2-phenoxypropane-1,3-diol was metabolized in the presence of H_2O_2 to glycolaldehyde that was identified by ^{13}C-NMR: the oxidase activity of glyoxal oxidase that was synthesized simultaneously with lignin peroxidase produced oxalate and 3 mol H_2O_2 that could then be recycled.

In the degradation of lignin and model compounds, lignin peroxidase functions by generating cation radicals from aromatic rings. This results in fission of the alkyl chain between C_1 and C_2 (Kirk and Farrell 1987), or—for substrates such as PAHs—may be followed by nonenzymatic nucleophilic reactions of the cation radical (Hammel et al. 1986; Haemmerli et al. 1986). Manganese peroxidase oxidizes Mn(II) to Mn(III) which is the active oxidant, and in the presence of H_2O_2 is capable of oxidizing a number of PAHs (Bogan and Lamar 1996) and mineralizing substituted aromatic compounds (Hofrichter et al. 1998). Both activities may bring about oxidation of PAHs in reactions that are mimicked by oxidation with manganese (III) acetate (Cremonesi et al. 1989) or electrochemical oxidation (Jeftic and Adams 1970).

3.6.3 Haloperoxidases

These are of primary significance in the biosynthesis of organohalogen compounds (Neilson 2003) that are distributed among mammals, marine biota, bacteria, and fungi.

These enzymes catalyze the reaction:

$$S-H + H_2O_2 + HalH \rightarrow S-Hal + 2H_2O$$

where Hal may be chloride, bromide, or iodide (references in Neidleman and Geigert 1986). Haloperoxidases have been isolated and purified from a number of organisms in which they mediate biosynthetic reactions. There are several structural groups of haloperoxidases that are summarized in Table 3.5.

1. The chloroperoxidase from *Caldaromyces fumago* has been isolated in pure form. It is a glycoprotein containing ferroprotoporphyrin IX, and displays, in addition both peroxidase and catalase activities. In the absence of organic substrates it catalyzes the formation of Cl_2 and Br_2 from chloride and bromide, respectively (Morris and Hager 1966; Libby et al. 1982).

2. Bromoperoxidase has been isolated from *Pseudomonas aureofaciens* ATCC 15926, also displays peroxidase and catalase activities, and contains ferriprotoporphyrin IX (van Pée and Lingens 1985b). Four different bromoperoxidases have been isolated from *Streptomyces griseus* (Zeiner et al. 1988). Only one of them, however, contains ferriprotoporphyrin IX, and

TABLE 3.5

Types of Haloperoxidases

Type of Enzyme	Group	Organism	Halogen	Reference
Heme	Bacterium	*Streptomyces phaeochromogenes*	Br	van Pée and Lingens (1985a)
	Rhodophyta	*Cystoclonium purpureum*	Br	Pedersén (1976)
	Chlorophyta	*Penicillus capitatus*	Br	Baden and Corbett (1980); Manthey and Hager (1981)
	Fungus	*Caldariomyces fumago*	Cl	Morris and Hager (1966)
Heme + f Flavin	Polychaete	*Notomastus lobatus*	Cl	Chen et al. (1991)
Vanadium	Phaeophyta	*Ascophyllum nodosum*	Br	Vilter (1984)
	Phaeophyta	*Ecklonia stolonifera*	Br	Hara and Sakurai (1998)
	Phaeophyta	*Laminaria digitata*	Br	Jordan and Vilter (1991)
	Phaeophyta	*Laminaria saccharina*	Br	De Boer et al. (1986)
	Phaeophyta	*Macrocystis pyrifera*	Br	Soedjak and Butler (1990)
	Rhodophyta	*Corallina pilulifera*	Br	Krenn et al. (1989); Itoh et al. (1988)
	Fungus	*Curvularia inequalis*	Cl	Simons et al. (1995)
Nonheme/Flavin	Bacterium	*Ps. fluorescens*	Cl	Keller et al. (2000)
Nonheme/nonflavin	Bacterium	*Ps. putida*	Br	Itoh et al. (1994)
Nonheme/nonmetal	Bacterium	*Pseudomonas pyrrocinia*	Cl	Wiesner et al. (1988)
		Streptomyces aureofaciens	Br	Weng et al. (1991)
		Acinetobacter calcoaceticus	Br	Kataoka et al. (2000)

FIGURE 3.31　Degradation of fluorene.

displays peroxidase and catalase activities. This illustrates that there are different groups of enzymes one of which lacks heme prosthetic groups.

Chloroperoxidase activity has also been found among degradative enzymes.

1. An inducible enzyme in the bacterium *R. erythropolis* NI86/21 that is involved in the degradation of thiocarbamate herbicides is a nonheme haloperoxidase that does not occur in other strains of rhodococci that can degrade thiocarbamates (de Schrijver et al. 1997).

2. *Acinetobacter calcoaceticus* strain F46 was enriched with fluorene that is degraded by dioxygenation, loss of C_3, decarboxylation and Baeyer–Villiger oxygenation to 3,4-dihydrocoumarin (Figure 3.31). The hydrolase for this was able to brominate monochlorodimedone in the presence of H_2O_2 and 3,4-dihydrocoumarin, or acetate, or butyrate (Kataoka et al. 2000). It was proposed that acyl peroxides be produced from an acylated serine site on the enzyme by the action of H_2O_2 and oxidized bromide to the active brominating agent BrO⁻. This is analogous to the mechanism proposed earlier (Picard et al. 1997).

3. The gene encoding the esterase from *Ps. fluorescens* was expressed in *E. coli*, and the enzyme displayed both hydrolytic and bromoperoxidase activity (Pelletier and Altenbuchner 1995).

3.7　References

Alonso J, JL Barredo, B Díez, E Mellado, F Salto, JL García, E Cortés. 1998. D-amino-acid oxidase gene from *Rhodotorula gracilis* (*Rhodosporidium toruloides*) ATCC 26217. *Microbiology (UK)* 144: 1095–1101.

Baden DG, MD Corbett. 1980. Bromoperoxidases from *Penicillus capitatus, Penicillus lamourouxii* and *Rhipocephalus phoenix. Biochem J* 187: 205–211.

Bogan BW, RT Lamar. 1996. Polycyclic aromatic hydrocarbon-degrading capabilities of *Phanerochaete laevis* HHB-1625 and its extracellular lignolytic enzymes. *Appl Environ Microbiol* 62: 1597–1603.

Bruckner RC, MS Jorns. 2009. Spectral and kinetic characterization of intermediates in the aromatization reaction catalyzed by NikD, an unusual amino acid oxidase. *Biochemistry* 48: 4455–4465.

Breckau D, E Mahlitz, A Sauerwald, G Layer, D Jahn. 2003. Oxygen-dependent coproporphyrinogen III oxidase (HemF) from *Escherichia coli* is stimulated by manganese. *J Biol Chem* 278: 46625–46631.

Cai D, JP Klingman. 1994. Evidence of a self-catalytic mechanism of 2,4,5-trihydroxyphenylalanine quinone biogenesis in yeast copper amine oxidase. *J Biol Chem* 269: 32039.

Carrell CJ, RC Bruckner, D Venci, G Zhao, MS Jorns, FS Mathews. 2007. NikD an unusual amino acid oxidase essential for nikkomycin biosynthesis: Structures of closed and open forms at 1.15 and 1.90 Å resolution. *Structure* 2007(15): 928–941.

Chen YP, DE Lincol, SA Woodin, CR Lovell. 1991. Purification and properties of a unique flavin-containing chloroperoxidase from the capitellid polychaete *Notomastus lobatus. J Biol Chem* 266:23909–23915.

Chiribau CB, C Sandu, M Fraaije, E Schiltz, R Bradsch. 2004. A novel γ-*N*-methylaminobutyrate demethylating oxidase involved in catabolism of the tobacco alkaloid nicotine by *Arthrobacter nicotinovorans* pAO1. *Eur J Biochem* 271: 4677–4684.

Chlumsky LJ, L Zhang, MS Jorns. 1995. Sequence analysis of sarcosine oxidase and nearby genes reveals homologies with key enzymes of folate one-carbon metabolism. *J Biol Chem* 270: 18252–18259.

Choi Y-H, R Matsuzaki, T Fukui, E Shimizu, T Yorifigi, H Sato, Y Ozaki, K Tanizawa. 1995. Copper/topa quinone-containing histamine oxidase from *Arthrobacter globiformis*. Molecular cloning and sequencing, overproduction of precursor enzyme, and generation of topa quinone cofactor. *J Biol Chem* 270: 4812–4720.

Clark E, S Manulis, Y Ophir, I Barash, Y Gafni. 1993. Cloning and characterization of *iaaM* and *iaaH* from *Erwinia herbicola* pathovar *gypsophilae. Phytopathol* 83: 234–240.

Comai L, T Kosuge. 1982. Cloning and characterization of iaaM, a virulence determinant of *Pseudomonas savastanoi. J Bacteriol* 149: 40–46.

Cremonesi P, EL Cavalieri, EG Rogan. 1989. One-electron oxidation of 6-substituted benzo(a)pyrenes by manganese acetate. A model for metabolic activation. *J Org Chem* 54: 3561–3570.

Dai VD, K Decker, H Sund. 1968. Purification and properties of L-6-hydroxynicotine oxidase. *Eur J Biochem* 4: 95–102.

de Boer E, MGM Tromp, H Plat, BE Krenn, R Wever. 1986. Vanadium (V) as an essential element for haloperoxidase activity in marine brown algae: Purification and characterization of a vanadium (V)-containing bromoperoxidase from *Laminaria saccharina. Biochim Biophys Acta* 872: 104–115.

de Schrijver A, I Nagy, G Schoofs, P Proost, J Vanderleyden, K-H van Pée, R de Mot. 1997. Thiocarbamate herbicide-inducible nonheme haloperoxidase of *Rhodococcus erythropolis* NI86/21. *Appl Environ Microbiol* 63: 1811–1916.

Doukyu N, R Aono. 1998. Purification of extracellular cholesterol oxidase with high activity in the presence of organic solvents from *Pseudomonas* sp. strain ST-200. *Appl Environ Microbiol* 64: 1929–1932.

Drzyzga O, LF de las Heras, V Morales, JM Navarro Llorens, J Perera. 2011. Cholesterol degradation by *Gordonia cholesterolivorans. Appl Environ Microbiol* 77: 4802–4810.

Fraaije MW, WJH van Berkel. 1997. Catalytic mechanism of the oxidative demethylation of 4-(methoxymethyl)phenol by vanillyl-alcohol oxidase. Evidence for formation of a *p*-quinone intermediate. *J Biol Chem* 272: 18111–18116.

Guillén F, AT Martinez, MJ Martinez. 1992. Substrate specificity and properties of the aryl-alcohol oxidase from the lignolytic fungus *Pleurotus eryngii. Eur J Biochem* 209: 603–611.

Hacisalihoglu A, JA Jongejan, JA Duine. 1997. Distribution of amine oxidases and amine dehydrogenases in bacteria grown on primary amines and characterization of the amine oxidase from *Klebsiella oxytoca. Microbiology (UK)* 143: 505–512.

Haemmerli D, MSA Leisola, D Sanglard, A Fiechter. 1986. Oxidation of benzo[a]pyrene by extracellular ligninases of *Phanerochaete chrysosporium. J Biol Chem* 261: 6900–6903.

Hammel KE, B Kalyanaraman, TK Kirk. 1986. Oxidation of polycyclic aromatic hydrocarbons and dibenzo(*p*) dioxins by *Phanerochaete chrysosporium* ligninase. *J Biol Chem* 261: 16948–16952.

Hammel KE, MD Mozuch, KA Jensen, PJ Kersten. 1994. H_2O_2 recycling during oxidation of the arylglycerol β-aryl ether lignin structure by lignin peroxidase and glyoxal oxidase. *Biochemistry* 33: 13349–13354.

Hara I, T Sakurai. 1998. Isolation and characterization of vanadium bromoperoxidase from a marine macroalga *Ecklonia stolonifera. J Inorg Chem* 72: 23–28.

Hofrichter M, K Scheibner, I Schneegaβ, W Fritsche. 1998. Enzymatic combusion of aromatic and aliphatic compounds by manganese peroxidase from *Nematoloma frowardii. Appl Environ Microbiol* 64: 399–404.

Hopper DJ, L Cottrell. 2003. Alkylphenol biotransformations catalyzed by 4-ethylphenol methylenehydroxylase. *Appl Environ Microbiol* 69: 3650–3652.

Hutcheson SW, T Kosuge. 1985. Regulation of 3-indolacetic acid production in *Pseudomonas savastanoi*. Purification and properties of tryptophan 2-monooxygenase. *J Biol Chem* 260: 6281–6287.

Ishida T, L Yu, H Akutsu, K Ozawa, S Kawanishi, A Seto, T Inubushi, S Sano. 1998. A primitive pathway of porphyrin biosynthesis and enzymology in *Desulfurovibrio vulgaris. Proc Natl Acad Sci USA* 95: 4853–4858.

Itoh N, N Morinaga, T Kouzai. 1994. Purification and characterization of a novel metal-containing nonheme bromoperoxidase from *Pseudomonas putida. Biochim Biophys Acta* 1207: 208–216.

Itoh N, AKM Quamrul Hasan, Y Izumi, H Yamada. 1988. Substrate specificity, regiospecificity and stereospecificity of halogenation reactions catalyzed by non-heme-type bromoperoxidase of *Corallina pilulifera. Eur J Biochem* 172: 477–484.

Jeftic L, RN Adams. 1970. Electrochemical oxidation pathways of benzo(a)pyrene. *J Am Chem Soc* 92: 1332–1337.

Job V, GL Marcone, MS Pilone, L Pollegioni. 2002. Glycine oxidase from *Bacillus subtilis*. Characterization of a new flavoprotein. *J Biol Chem* 277: 6985–6993.

Johnson BJ, J Cohen, RW Welford, A R Pearson, K Schulten, JP Klinman, CM Wilmot. 2007. Exploring molecular oxygen pathways in *Hansenula polymorpha* copper-containing amine oxidase. *J Biol Chem* 282: 17767–17776.

Jordan P, H Vilter. 1991. Extraction of proteins from material rich in anionic mucilages: Partition and fractionation of vanadate-dependent bromoperoxidases from the brown algae *Laminaria digitata* and *L. saccharina* in aqueous polymer two-phase systems. *Biochim Biophys Acta* 1073: 98–106.

Kallili E, SB Mulooney, RP Hausinger. 2008. Identification of *Escherichia coli* YgaF as an L-2-hydroxyglutarate oxidase. *J Bacteriol* 190: 3793–3798.

Kaneda T, H Obata, M Tokumoto. 1993. Aromatization of 4-oxocyclohexanecarboxylic acid to 4-hydroxybenzoic acid by two distinct desaturases from *Corynebacterium cyclohexanicum. Eur J Biochem* 218: 997–1003.

Kataoka M, K Honda, S Shimizu. 2000. 3,4-Dihydrocoumarin hydrolase with haloperoxidase activity from *Acinetobacter calcoaceticus. Eur J Biochem* 267: 3–10.

Kawamoto S, C Nozaki, A Tanaka, S Fukui. 1978. Fatty acid −β-oxidation system in microbodies of *n*-alkane-grown *Candida tropicalis. Eur J Biochem* 83: 609–613.

Keller S, T Wage, K Hohaus, M Hölzer, E Eichjorn, K-H van Pée. 2000. Purification and partial characterization of tryptophan 7-halogenase (PrnA) from *Pseudomonas fluorescens. Angew Chem Int Ed* 39: 2300–2302.

Kersten PJ. 1990. Glyoxal oxidase of *Phanerochaete chrysosporium*: Its characterization and activation by lignin peroxidase. *Proc Natl Acad Sci USA* 87: 2936–2940.

Kirk TK, RL Farrell. 1987. Enzymatic "combustion": The microbial degradation of lignin. *Annu Rev Microbiol* 41: 465–505.

Kommoju P-R, RC Bruckner, P Ferreira, MS Jorns. 2009. Probing the role of active site residues in NikD, an unusual amino acid oxidase that catalyzes an aromatization reaction important in nikkomycin biosynthesis. *Biochemistry* 48: 6951–6962.

Krenn BE, Y Izumi, H Yamada, R Wever. 1989. A comparison of different (vanadium) peroxidases: The bromoperoxidase of *Corallina pilulifera* is also a vanadium enzyme. *Biochim Biophys Acta* 998:63–68.

Kumagai H, H Matsui, K Ogata, H Yamada. 1969. Properties of crystalline tyramine oxidase from *Sarcina lutea. Biochim Biophys Acta* 171: 1–8.

Käppeli O . 1986. Cytochromes P-450 of yeasts. *Microbiol Revs* 50: 244–258.

Layer G, J Moser, DW Heinz, D Jahn, W-D Schubert. 2003. Crystal structure of coproporphyrigogen III oxidase reveals cofactor geometry of radical SAM enzyme. *EMBO J* 22: 6214–6224.

Layer G et al. 2006. The substrate radical of *Escherichia coli* oxygen-independent coproporphyrinogen III oxidase HemN. *J Biol Chem* 281: 15727–15734.

Libby RD, JA Thomas, LW Kaiser, LP Hager. 1982. Chloroperoxidase halogenation reactions. *J Biol Chem* 257: 5030–5037.

Lobo SAL, A Brindley, MJ Warren, LM Saraiva. 2009. Functional characterization of the early steps of tetrapyrrole biosynthesis and modification in *Desulfovibrio vulgaris Hildenborough. Biochem J* 420: 317–325.

Manthey JA, LP Hager. 1981. Purification and properties of bromoperoxidase from *Penicillus capitatus*. *J Biol Chem* 256:11232–11238.

Morris DR, LP Hager. 1966. Chloroperoxidase I Isolation and properties of the crystalline protein. *J Biol Chem* 241: 1763–1768.

Mu D, SM Janes, AJ Smith, DE Brown, D M Dooley, JP Kinman. 1992. Tyrosine codon corresponds to topa quinone at the active site of copper amine oxidases. *J Biol Chem* 267: 7979–7982.

Muheim A, R Waldner, MSA Leisola, A Fiechter. 1990. An extracellular aryl-alcohol oxidase from the white-rot fungus *Bjerkandera adusta*. *Enzyme Microbiol Technol* 12: 204–209.

Murooka Y, N Doi, T Harada. 1979. Distribution of membrane-bound monoamine oxidase in bacteria. *Appl Environ Microbiol* 38: 565–569.

Nakazawa, K Hori, O Hayaishi. 1972. Studies on monooxygenases V. Manifestation of amino acid oxidase activity by L-lysine monooxygenase. *J Biol Chem* 247: 3439–3444.

Neidleman SL, J Geigert. 1986. *Biohalogenation: Principles, Basic Roles and Applications*. Ellis Horwood, Chichester, England.

Neilson AH. 2003. Biological effects and biosynthesis of brominated metabolites. *Handbook Environ Chem* 3R: 75–204.

Ohta T, K Fujishoro, K Yamaguchi, Y Tamura, K Aisaka, T Uwajima, M Hasegawa. 1991. Sequence of gene *choB* encoding cholesterol oxidase of *Brevibacterium sterolicum*: Comparison with *choA* of *Streptomyces* sp. SA-COO. *Gene* 103: 93–96.

Parrott S, S Jones, RA Cooper. 1987. 2-Phenylethylamine catabolism by *Escherichia coli* K12. *J Gen Microbiol* 133: 347–351.

Pelletier I, J Altenbuchner. 1995. A bacterial esterase is homologous with non-heme haloperoxidases and displays brominating activity. *Microbiology (UK)* 141: 459–468.

Pedersén M. 1976. A brominating and hydroxylating peroxidase from the red alga *Cystoclonium purpureum*. *Physiol Plant* 37: 6–11.

Picard M, J Gross, E Lübbert, S Tölzer, S Krauss, K-H van Pée. 1997. Metal-free haloperoxidases as unusual hydrolases: Activation of H_2O_2 by the formation of peracetic acid. *Angew Chem Int Ed* 36: 1196–1199.

Ralph EC, MA Anderson, WW Cleland, PF Fitzpatrick. 2006. Mechanistic studies of the flavoenzyme tryptophan 2-monooxygenase: Deuterium and ^{15}N kinetic isotope effects on alanine oxidation by an L-amino acid oxidase. *Biochemistry* 45: 15844–15852.

Ralph EC, JS Hirsch, MA Anderson, WW Cleland, DA Singleton, PF Fitzpatrick. 2007. Insights into the mechanism of flavoprotein-catalyzed amine oxidation from nitrogen isotope effects on the reaction of *N*-methyltryptophan oxidase. *Biochemistry* 46: 7655–7664.

Simons BH, P Barnett, EGM Vollenbroek, HL Dekker, AO Muijsers A Messerschmidt, R Wever. 1995. Primary structure and characterization of the vanadium chloroperoxidase from the fungus *Curvularia inaequalis*. *Eur J Biochem* 229: 566–574.

Soedjak HS, A Butler. 1990. Characterization of vanadium bromoperoxidase from *Macrocystis* and *Fucus*: Reactivity of bromoperoxidase towards acyl and alkyl peroxides and bromination of amines. *Biochemistry* 29: 7974–7981.

Steinebach V, S de Vries, JA Duine. 1996. Intermediates in the catalytic cycle of copper-quinoprotein amine oxidase from *Escherichia coli*. *J Biol Chem* 271: 5580–5588.

Stephan I, B Tshisuaka, S Fetzner, F Lingens. 1996. Quinaldine 4-oxidase from *Arthrobacter* sp. Rü61a, a versatile procaryotic molybdenum-containing hydroxylase active towards N-containing heterocyclic compounds and aromatic aldehydes. *Eur J Biochem* 236: 155–162.

Storbeck S, S Rolfes, E Raux-Deery, MJ Warren, D Jahn, G Layer. 2010. A novel pathway for the biosynthesis of heme in Archaea: Genome-based bioinformatic predictions and experimental evidence. *Archaea* 2010: 175050.

Suylen GMH, PJ Large, JP van Dijken, JG Kuenen. 1987. Methyl mercaptan oxidase, a key enzyme in the metabolism of methylated sulphur compounds by *Hyphomicrobium* EG. *J Gen Microbiol* 133: 2989–2997.

Tanaka A, M Ueda. 1993. Assimilation of alkanes by yeasts: functions and biogenesis of peroxisomes. *Mycol Res* 98: 1025–1044.

Trickey P, MA Wagner, MS Jorns, FS Mathews. 1999. Monomeric sarcosine oxidase: Structure of a covalently flavinylated amine oxidizing enzyme. *Structure* 7: 331–345.

Urzúa U, PJ Kersten, R Vicuna. 1998. Manganese peroxidase-dependent oxidation of glycolic and oxalic acids synthesized by *Ceriporiopsis subvermispora* produces extracellular hydrogen peroxide. *Appl Environ Microbiol* 64: 68–73.

van den Heuvel RH, MW Fraaije, C Laane, WJH van Berkel. 1998. Regio- and stereospecific-conversion of 4-alkylphenols by the covalent flavoprotein vanillyl-alcohol oxidase. *J Bacteriol* 180: 5646–5651.

van Hellemond EW, M van Dijk, DPHM Heuts, DB Janssen, MW Fraaije. 2008. Discovery and characterization of a putrescine oxidase from *Rhodococcus erythropolis*. *Appl Microbiol Biotechnol* 78: 455–463.

van Pée K-H, F Lingens. 1985a. Purification and molecular catalytic properties of bromoperoxidase from *Streptomyces phaeochromogenes*. *J Gen Microbiol* 131: 1911–1916.

van Pée, K-H, F Lingens. 1985b. Purification of bromoperoxidase from *Pseudomonas aureofaciens*. *J Bacteriol* 161: 1171–1175.

Vilter H. 1984. Peroxidase from *Phaeophyceae*: A vanadium (V)-dependent peroxidase from *Ascophyllum nodosum*. *Phytochem* 23: 1387–1390.

Wang HJ, M-T Le Dall, Y Waché, C Laroche, J-M Belin, C Gaillardin, J-M Nicaud. 1999. Evaluation of acyl coenzyme A oxidase (Aox) isoenzyme function in the *n*-alkane-assimilating yeast *Yarrowia lipolytica*. *J Bacteriol* 181: 5140–5148.

Weng M, O Pfeifer, S Kraus, F Lingens, K-H van Pée. 1991. Purification, characterization and comparison of two non-heme bromoperoxidases from *Streptomyces aureofaciens*. *J Gen Microbiol* 137: 2539–2546.

Wiesner W, K-H van Pée, F Lingens. 1988. Purification and characterization of a novel non-heme chloroperoxidase from *Pseudomonas pyrrocinia*. *J Biol Chem* 263: 13725–13732.

Wilmot CM, J Hajdu, MJ McPherson, PF Knowles, SEV Phillips. 1999. Visualization of dioxygen bound to copper during enzyme catalysis. *Science* 286: 1724–1728.

Zeiner R, K-H van Pée, F Lingens. 1988. Purification and partial characterization of multiple bromoperoxidases from *Streptomyces griseus*. *J Gen Microbiol* 134: 3141–3149.

Zhang X, JH Fuller, WS McIntire. 1993. Cloning, sequencing, expression, and regulation of the structural gene for the copper/Topa quinone-containing methylamine oxidase from *Arthrobacter* strain P1, a Gram-positive facultative methylotroph. *J Bacteriol* 175: 5617–5627.

3.8 Incorporation of Oxygen from Water: Hydrolases, Oxidoreductases, and Hydratases

This section covers the incorporation of oxygen from water by different mechanisms—hydrolases, oxidoreductases, and hydratases. It should be recognized that some reactions in which the oxygen is derived from water are discussed in the section dealing with dehydrogenation.

3.8.1 Hydrolases

These carry out the hydrolysis of esters of carboxylic, sulfuric, and phosphoric acids that are described in Chapter 11, Part 1. There are two important issues that are worth noting here: degradation of nitrate esters is carried out by reductive reactions rather than by hydrolysis, and the class of sulfatases that necessitates posttranslational modification.

3.8.1.1 Sulfatases

Sulfatases are responsible for the hydrolysis of a wide range of substrates and they can display a high degree of specificity. This is illustrated for alkyl sulfates in Chapter 11, for the bile-acid sulfates by *Clostridium* sp. strain S1 where the conformation and position of the sulfate are critical (Huijghebaert and Eyssen 1982), and in *Flavobacterium heparinum* that has been extensively examined for several enzymes including heparinases and the chondroitin sulfatases (Tkalec et al. 2000), and heparin/heparan sulfatase (Myette et al. 2009).

There are several metabolically distinct groups of alkyl sulfatases: Group I enzymes require posttranslational modification of serine or cysteine residues to α-formyl-glycine, Group II comprises the Fe(II)-2-oxoglutarate-dependent enzymes that is used for the degradation of C_4–C_{12} alkyl sulfates by *Ps. putida* S-313 (Kahnert and Kertesz 2000), and group III enzymes have a binuclear Zn cluster for the degradation of dodecyl sulfate (Hagelueken et al. 2006).

Under aerobic conditions posttranslational modification is generally carried out by an oxygenase FGE (Carlson et al. 2008), though details among prokaryotes depend on the organism: for the arylsulfatase in *E. coli* cysteine—though not serine—was modified, whereas in *K. pneumoniae* modification of a serine was involved (Dierks et al. 1998; Miech et al. 1998). In aerobes such as pseudomonads the hydrated formyl group executes nucleophilic attack on sulfur that is shown by the crystal structure of the enzyme from *Ps. aeruginosa* strain PAS (Boltes et al. 2001). In anaerobes on the other hand, the modification is carried out by SME enzymes which are Fe–S enzymes using *S*-adenosylmethionine (Berteau et al. 2006; Benjdia et al. 2007, 2010).

3.8.1.2 *Amidases*

The hydrolysis of amides and related compounds is discussed in Chapter 7, Part 1, and in the context of β-lactam antibiotic resistance in Chapter 4, Part 3. It is sufficient to note here about the existence of two hydrolytic mechanisms, one involving nucleophilic attack at the β-lactam carbonyl by a serine residue, and the other one or two Zn atoms coordinated to three histidines and one aspartate.

The hydrolysis of cyclic amides displays an interesting dichotomy since it may be either ATP-dependent or ATP-independent.

3.8.1.2.1 *ATP Dependent*

a. The hydrolysis of 5-oxo-L-proline to glutamate by *Pseudomonas putida* is carried out by an ATP-dependent reaction. The enzyme activity consists of two enzymes that were purified, neither of which functions alone, and component A required K^+ for optimal activity (Seddon et al. 1984).

b. The enzyme that catalyzed the hydrolysis of *N*-methylhydantoin to *N*-carbamoylsarcosine from *Pseudomonas putida* strain 77 had a molecular mass of 300 kDa and consisted of a tetramer with two small subunits (70 kDa) and two large subunits (80 kDa). ATP was required for activity as well as both divalent cations Mg, Mn, Co, and monovalent cations K, Rb, Cs (Ogawa et al. 1995).

c. It has been suggested that hydrolysis of 2-oxoidoleacetate produced from indoleacetate may belong to the same group (Ebenau-Jehle et al. 2012).

The mechanism of these reactions may plausibly involve phosphorylation of the enolate form of the amide by ATP with the formation of ADP, followed by reaction of the phosphoenamine cation with water and elimination of phosphate.

3.8.1.2.2 *ATP Independent*

The 5-oxoprolinase that has been characterized from *Alcaligenes faecalis* N-38A had a molecular mass of 47 kDa and had no requirement for either divalent or monovalent cations (Nishimura et al. 1999).

3.8.2 **Oxidoreductases**

Both aerobic and anaerobic bacteria degrade azaarenes by hydroxylation in which the oxygen originates from water. This distinguishes them from the activities of monooxygenases and oxidases. They comprise a redox reaction that is mediated by oxidoreductases (dehydrogenases), although the physiological electron acceptors have not been generally established. These oxidoreductases contain a molybdenum cofactor pyranopterin with a 1,2-dithiolene to which the molybdenum is attached in a distorted pyramidal geometry with an M=O at the apex; they are generally conjugated in the active form with cytosine or guanosine diphosphates. The enzymatic reactions also involve several [2Fe–2S] clusters and FAD, while in addition, the formation of active xanthine dehydrogenase in purinolytic clostridia requires a labile cofactor containing selenium. The paradigm reaction is the hydroxylation of hypoxanthine to xanthine:

$$\text{Hypoxanthine} + H_2O + NAD^+ \rightarrow \text{Xanthine} + NADH + H^+$$

A great deal of attention has been devoted to this hydroxylation, and details of the various mechanisms have been covered in a review (Hille 2005). In addition, it is worth pointing out that the enzyme which mediates the hydroxylation of a number of bicyclic azaarenes has been termed an oxidase—quinaldine 4-oxidase—since oxygen functions effectively as an electron acceptor (Stephan et al. 1996).

3.8.2.1 Five-Membered Heteroarenes

The aerobic degradation of the 2-carboxylates of furan, pyrrole, and thiophene is initiated by hydroxyl-ation before fission of the rings. Although details of the enzymes are limited, it was suggested on the basis of tungstate inhibition and [185]W [tungstate] labeling that the degradation of 2-furoyl-Coenzyme A involves a molybdenum-dependent dehydrogenase (Koenig and Andreesen 1990).

3.8.2.2 Pyridine

The aerobic degradation of pyridines with carboxyl and hydroxyl substituents generally involves hydroxyl-ation (dehydrogenation). The dehydrogenases for the degradation of pyridine-2-carboxylate by *Arthrobacter picolinophilus* (Siegmund et al. 1990), and for two steps in the degradation of nicotine by *Arthrobacter nicotinovorans (oxidans)* (Freudenberg et al 1988; Baitsch et al. 2001) contain molybdenum, flavin and nonheme iron–sulfur. The nicotinate dehydrogenase from *Clostridium barkeri* contains, in addition, a labile selenium cofactor (Dilworth 1982) that is coordinated with molybdenum (Gladyshev et al. 1994).

3.8.2.3 Quinoline

The quinoline enzymes have been examined in much greater detail than those for pyridines, and occur in both Gram-negative and Gram-positive organisms. The oxidoreductases have been purified from a number of organisms that degrade quinoline. These include *Ps. putida* strain 86 (Bauder et al. 1990), *Comamonas testosteroni* strain 63 (Schach et al. 1995), *Rhodococcus* sp. strain B1 (Peschke and Lingens 1991), and *Agrobacterium* sp. strain 1B that degrades quinoline-4-carboxylate (Bauer and Lingens 1992). They have a molecular mass of 300–360 kDa and contain per molecule, 8 atoms of Fe, 8 atoms of acid-labile S, 2 atoms of Mo, and 2 molecules of FAD. The organic component of the pterin molybdenum cofactor is generally molybdopterin cytosine dinucleotide (Hetterich et al. 1991; Schach et al. 1995). The crystal structure of the enzyme from *Ps. putida* strain 86 showed that the Mo was coordinated with two ene-dithiolate sulfur atoms, two oxo-ligands and an equatorial catalytically critical sulfide-ligand (Bonin et al. 2004). The oxidoreductase that catalyzes the hydroxylation of 2-methylquinoline is able, in addi-tion, to accept a range of other six-membered azaarenes (Stephan et al. 1996).

3.8.2.4 Isoquinoline

The degradation of isoquinoline by *Pseudomonas diminuta* strain 7 is initiated by an oxidoreductase that contains [2Fe–2S] centers and the cofactor molybdopterin cytosine dinucleotide (Lehmann et al. 1994).

3.8.2.5 Purines

Xanthine dehydrogenase that mediates the conversion of hypoxanthine to xanthine and uric acid has been studied extensively since it is readily available from cow's milk. It has also been studied (Leimkühler et al. 2004) in the anaerobic phototroph *Rhodobacter capsulatus*, and the crystal structures of both enzymes have been solved. Xanthine dehydrogenase is a complex flavoprotein containing Mo, FAD, and [2Fe–2S] redox centers, and the reactions may be rationalized (Hille and Sprecher 1987):

$$R\text{–}H + H_2O + Mo^{VI}\text{=}S \rightarrow ROH + Mo^{IV}\text{-}SH + H^+$$

Some additional details are worth noting.

1. Xanthine dehydrogenase from the anaerobic *Clostridium purinolyticum*, *Cl. acidi-urici*, and *Cl. cylindrosporum* contain a labile selenium cofactor.
2. In *Clostridium purinolyticum*, purine hydroxylase—that is a separate enzyme from xanthine dehydrogenase—hydroxylates purine to hypoxanthine (6-hydroxypurine) and xanthine (2,6-dihydroxypurine), which is then further hydroxylated to uric acid (2,6,8-trihydroxypurine) by xanthine dehydrogenase (Self 2002). Like the xanthine dehydrogenases, it contains a sele-nium cofactor.

3. The degradation of purines has been examined in *K. oxytoca*, and is initiated by the oxidation of 6-hydroxypurine (hypoxanthine) and 2,6-dihydroxypurine (xanthine) to 2,4,6-trihydroxypurine (uric acid). This is mediated, however, by a dioxygenase/reductase system whose genes have been identified: they encode a two-component Rieske nonheme iron, aromatizing hydroxylase (HpxD and HpxE) (Pope et al. 2009). In addition, genetic analysis of *K. pneumoniae* (de la Riva et al. 2008) failed to find evidence for the existence of typical molybdenum-containing enzymes such as xanthine dehydrogenase. This is, therefore, not involved in the introduction of oxygen into the purine ring as a requisite to degradation of both *K. pneumoniae* and *K. oxytoca*.

3.8.3 Hydratases

These enzymes catalyze the addition of the elements of water to carbon–carbon double bonds (C=C), carbon–carbon triple bonds (C≡C), carbon–nitrogen double bonds (C=N), or carbon–nitrogen triple bonds (C≡N). These reactions are completely different from oxidoreductases since no redox reactions are involved. Illustrative examples include the following:

C=C

1. Hydration of nonactivated C=C bonds is uncommon, although it is part of the biosynthetic pathway even in nonphototrophic bacteria including *Brevundimonas* sp. strain SD212 (Nishida et al. 2005), and *Deinococcus radiodurans* (Sun et al. 2009). The stereospecific hydration of oleate at the C_{10} position in a pseudomonad produced Δ^{10}-*trans*-octadecenoate from oleic acid and 10-hydroxypalmitic acid from palmitoleic acid (Niehaus et al. 1970). This has been extended to a number of organisms including *Candida lipolytica*, *Mycobacterium fortuitum*, *Nocardia aurantia*, and species of *Pseudomonas*, and some were able to dehydrogenate this further to the 10-ketone (El-Sharkawy et al. 1992). *Nocardia cholesterolicum* was able to hydroxylate several *cis*-alkenoates including hexadec-9-enoate, and tetradec-9-enoate—though not octadec-6-enoate or *trans*-alkenoates (Koritala et al. 1989). The genes for the hydration of lycopene to 1-hydroxy and 1,1'-dihydroxylycopene in *Rubrivivax gelatinosus* and *Thiocapsa roseopersicina* have been expressed in *E. coli* (Hiseni et al. 2011), and the hydroxylating enzyme has been purified and characterized from *Rubrivivax gelatinosa* (Steiger et al. 2003) and from *Elizabethkingia* (*Chrysobacterium*) *meningoseptica* (Bevers et al. 2009).

2. By contrast, hydration of C=C bonds activated by carbonyl groups is exemplified in a number of reactions including the hydration of *trans*-alk-2-enoates in the degradation of alkanes and alkanoates that are discussed in greater detail in Chapter 7, Part 1. Some additional examples include the following.

   a. The dechlorination of 2-chloroacrylate is carried out by *Pseudomonas* sp. strain YL by initial hydration by a FAD ($FADH_2$) enzyme followed by loss of chloride to produce pyruvate (Mowafy et al. 2010).

   b. The transformation of L-carnityl-CoA to γ-butyrobetainyl-CoA in *E. coli* is carried out by a hydratase encoded by CaiD (Elssner et al. 2001).

   c. The degradation of 4-alkylphenols in *Ps. putida* is initiated by dehydrogenation to a quinone methide followed by hydroxylation to the 4-alkylbenzyl alcohol (Hopper and Cottrell 2003). The degradation of lupanine involves a quinocytochrome *c* hydroxylation of the -CH₂- group adjacent to the ternary N (Hopper et al. 1991), and a mechanism involving dehydrogenation followed by hydroxylation was put forward. The prosthetic group of the flavocytochrome that is involved has been shown to be 8a-(*O*-tyrosyl)flavin adenine dinucleotide (McIntire et al. 1981).

   d. The degradation of tetralin is initiated by dioxygenation and dehydrogenation to 1,2-dihydroxy-5,6,7,8-tetrahydronaphthalene followed by ring fission. The product is a vinylogous 1,3-dioxocarboxylate that is hydrolyzed to deca-2-oxo-3-ene-1,10-dioate followed by hydration and fission to pyruvate and pimelic semialdehyde (Figure 3.32) (Hernáez et al. 2002).

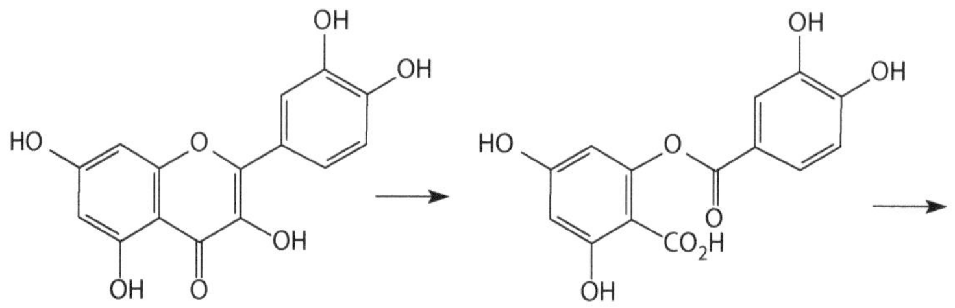

FIGURE 3.32 Aerobic degradation of tetralin.

e. In addition to the oxidative pathway via (4-hydroxy-3-methoxybenzoyl)acetyl-SCoA for the degradation of feruloyl-SCoA to vanillin, the alternative pathway by hydration followed by the loss of acetate and formation of vanillin from the β-hydroxycarboxyl-CoA has been found in *Ps. fluorescens* (Gasson et al. 1998).

f. The degradation of phenanthrene by *Nocardioides* sp. strain KP7 is initiated by dioxygenation to 1-hydroxynaphthalene-2-carboxylate that is further dioxygenated to 2′carboxybenzalpyruvate before hydration and aldolase fission to 2-carboxybenzaldehyde (Iwabuchi and Harayama 1998).

g. The anaerobic degradation of cyclohexanol involves the hydration of cyclohex-2-enone followed by dehydrogenation to cyclohexan-1,3-dione and hydrolytic fission to 5-oxocaproate (Dangel et al. 1989).

h. Initial steps in the anaerobic degradation of benzoate consist of the formation of benzoyl-CoA that is reduced to either a cyclohexa-1,5-diene-1-carboxyl-CoA or cyclohex-1-ene-1-carboxyl-CoA. This is hydrated as the next step before scission of the ring, and the hydration pathway is used both by facultative (Boll et al. 2000) and by obligate anaerobes (Peters et al. 2007).

$$C \equiv C$$

Acetylene hydratase from the anaerobe *Pelobacter acetylenicus* is a tungsten–iron–sulfur enzyme that catalyzes the addition of the elements of water to acetylene (Figure 3.33a) (Meckenstock et al. 1999). Its structure has been determined and it contains two molybdopterin guanine dinucleotides to which the tungsten is bound through the dithiolene, and in addition a cubane [4Fe–4S] cluster (Seiffert et al. 2007). Details have been added to elucidate the mechanism by producing active-site variants that could be heterologously expressed in *E. coli*. These revealed the determinative roles of Asp[13] that is close to W(IV), Lys[48] that couples the [4Fe–4S] cluster to the W site, and Ile[142] that is part of the ring in the substrate channel that accommodates acetylene (tenBrink et al. 2011).

$$R–C \equiv N \quad \text{and} \quad R–N \equiv C$$

1. Nitriles may be degraded to carboxylic acids either directly by the activity of a nitrilase in *Bacillus* sp. strain OxB-1 and *Pseudomonas syringae* B728a, or undergo hydration to amides followed by hydrolysis in *Ps. chlororaphis* B23 (Oinuma et al. 2003). *R. rhodochrous* strain J1 is able to synthesize two nitrile hydratases with low or high molecular mass: these are selectively induced by specific substrates, and the first contains cobalt that is essential for the formation of the catalytically active enzyme (Komeda et al. 1996).

2. The hydratase in *Ps. putida* N19–2 converts isonitriles to *N*-substituted formamides, and its activity is dependent on the presence of Cys-101 (Figure 3.33b) (Goda et al. 2002).

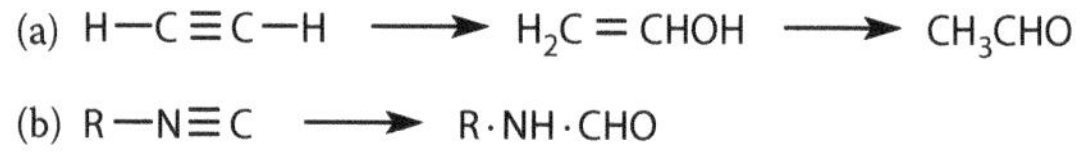

(a) $H–C \equiv C–H \longrightarrow H_2C = CHOH \longrightarrow CH_3CHO$

(b) $R–N \equiv C \longrightarrow R \cdot NH \cdot CHO$

FIGURE 3.33 (a) Hydration of acetylene to acetaldehyde, (b) transformation of isonitrile to *N*-alkylformamide.

3.9 References

Baitsch D, C Sandu, R Brandsch, GL Igloi. 2001. Gene cluster on pAO1 of *Arthrobacter nicotinovorans* involved in degradation of the plant alkaloid nicotine: Cloning, purification, and characterization of 2,6-dihydroxypyridine 3-hydrolase. *J Bacteriol* 183: 5262–5267.

Bauder R, B Tshisuaka, F Lingens. 1990. Microbial metabolism of quinoline and related compounds VII. Quinoline-4-carboxylic acid from *Pseudomonas putida*: A molybdenum-containing enzyme. *Biol Chem Hoppe-Seyler* 370: 1183–1189.

Bauer G, F Lingens. 1992. Microbial metabolism of quinoline and related compounds XV. Quinoline-4-carboxylic acid oxidoreductase from *Agrobacterum* spec. 1B: A molybdenum-containing enzyme. *Biol Chem Hoppe-Seyler* 370: 1183–1189.

Benjdia A, J Leprince, A Guillot, H Vaudry, S Rabot, O Berteau. 2007. Anaerobic sulfatase-maturing enzymes: Radical SAM enzymes able to catalyze *in vitro* sulfatase post-translational modification. *J Am Chem Soc* 129: 3462–3463.

Benjdia A, S Subramanian, J Lepince, H Vaudry, MK Johnson, O Berteau. 2010. Anaerobic sulfatase-maturing enzyme—A mechanistic link with glycyl radical-activating enzymes. *FEBS J* 277: 1906–1920.

Berteau O, A Guillot, A Benjdia, S Rabot. 2006. A new type of bacterial sulfatase reveals a novel maturation pathway in prokaryotes. *J Biol Chem* 281: 22464–22470.

Bevers LE, MWH Pinkse, PDEM Verhaert, WR Hagen. 2009. Oleate hydratase catalyzes the hydration of a non activated carbon-carbon bond. *J Bacteriol* 191: 5010–5012.

Bogan BW, RT Lamar. 1996. Polycyclic aromatic hydrocarbon-degrading capabilities of *Phanerochaete laevis* HHB-1625 and its extracellular lignolytic enzymes. *Appl Environ Microbiol* 62: 1597–1603.

Boll M, D Laempe, W Eisenreich, A Bacher, T Mittelnerger, J Heinze, G Fuchs. 2000. Nonaromatic products from anoxic conversion of benzoyl-CoA with benzoyl-CoA reductase and cyclohexa-1,5-diene-1-carbonyl-CoA hydratase. *J Biol Chem* 275: 21889–21895.

Boltes I, H Czapinska, A Kahnert, R van Bülow, T Dierks, B Schmidt, K von Figure, MA Kertesz, I Usón. 2001. 1.3 Å structure of arylsulfatase from *Pseudomonas aeruginosa* establishes the catalytic mechanism of sulfate ester cleavage in the sulfatase family. *Structure* 9: 483–491.

Bonin I, BM Martins, V Puranov, S Fetzer, R HUber, H Dobbek. 2004. Active site geometry and substrate ecognition of the molybdenum hydroxylase quinoline 2-oxidoreductase. *Structure* 12: 1425–1435.

Carlson BL, ER Ballister, E Skordalakes, DS King, MA Breidenbach, SA Gilmore, JM Berger, CR Bertozzi. 2008. Function and structure of a prokaryotic formylglycine-generating enzyme. *J Biol Chem* 283: 20117–20125.

Dangel W, A Tschech, G Fuchs. 1989. Enzyme reactions involved in anaerobic cyclohexanol metabolism by a denitrifying *Pseudomonas* species. *Arch Microbiol* 152: 273–279.

de la Riva L, J Badia, J Aguilar, RA Bender, L Baldoma. 2008. The *hpx* genetic system for hypoxanthine assimilation as a nitrogen source in *Klebsiella pneumoniae*: Gene organization and transcriptional regulation. *J Bacteriol* 190: 7892–7903.

Dierks T, C Miech, J Hummerjohann, B Schmidt. 1998. Posttranslational formation of formylglycine in prokaryotic sulfatases by modification of either cysteine or serine. *J Biol Chem* 273: 25560–25564.

Dilworth GL. 1982. Properties of the selenium-containing moiety of nicotinic acid hydroxylase from *Clostridium barkeri*. *Arch Biochem Biophys* 219: 30–38.

Ebenau-Jehle C, M Thomas, G Scharf, D Kockelhorn, B Knapp, K Schühle, J Heider, G Fuchs. 2012. Anaerobic metabolism of indoleacetate. *J Bacteriol* 194: 2894–2903.

Elssner T, C Engemann, K Baumgart, H-P Kleber. 2001. Involvement of coenzyme A esters and two new enzymes, an enoyl-CoA hydratase and a CoA-transferase, in the hydration of cronobetaine to L-carnitine by *Escherichia coli*. *Biochemistry* 40: 11140–11148.

El-Sharkawy A, W Yang, L Dostal, JPN Rosazza. 1992. Microbial oxidation of oleic acid. *Appl Environ Microbiol* 58: 2116–2122.

Freudenberg W, König K, Andressen JR. 1988. Nicotine dehydrogenase from *Arthrobacter oxidans:* A molybdenum-containing hydroxylase. *FEMS Microbiol Lett* 52: 13–18.

Gasson MJ, Y Kitamura, WR McLauchlan, A Narbad, AJ Parr, ELH Parsons, J Payne, MJC Rhodes, NJ Walton. 1998. Metabolism of ferulic acid to vanillin. *J Biol Chem* 273: 4163–4170.

Gladyshev VN, SV Khangulov, TC Stadtman. 1994. Nicotinic acid hydroxylase from *Clostridium barkeri*: Electron paramagnetic resonance studies show that selenium is coordinated with molybdenum in the catalytically active selenium-dependent enzyme. *Proc Natl Acad Sci USA* 91: 232–236.

Goda M, Y Hasiimoto, M Takase, S Herai, Y Iwahara, H Higashibata, M Kobayashi. 2002. Isonitrile hydratase from *Pseudomonas putida* N-19–2. *J Biol Chem* 277: 45860–45865.

Hagelueken G, TM Adams, L Wiehlmann, U Widow, H Kolmar, B Tümmler, DW Heinz, W-D Schubert. 2006. The crystal structure of SdsA1, an alkylsulfatase from *Pseudomonas aeruginosa* defines a third class of sulfatases. *Proc Natl Acad Sci USA* 103: 7631–7636.

Hernáez MJ, B Floriano, JJ Ríos, E Santero. 2002. Identification of a hydratase and a class II aldolase involved in biodegradation of the organic solvent tetralin. *Appl Environ Microbiol* 68: 4841–4846.

Hetterich D, B Peschke, B Tshisuaka, F Lingens. 1991. Microbial metabolism of quinoline and related compounds. X. The molybdopterin cofactors of quinoline oxidoreductases from *Pseudomonas putida* 86 and *Rhodococcus* spec. B1 and of xanthine dehydrogenase from *Pseudomonas putida* 86. *Biol Chem Hoppe-Seyler* 372: 513–517.

Hille R. 2005. Molybdenum-containing hydroxylases. *Arch Biochem Biophys* 433: 107–116.

Hille R, H Sprecher. 1987. On the mechanism of action of xanthine oxidase. *J Biol Chem* 262: 10914–10917.

Hiseni A, IWCE Arends, LG Otten. 2011. Biochemical characterization of the carotenoid 1,2-hydratases (CrtC) from *Rubrivivax gelatinosus* and *Thiocapsa roseopersicina*. *Appl Microbiol Biotechnol* 91: 1029–1036.

Hopper DJ, J Rogozinski, M Toczko. 1991. Lupanine hydroxylase, a quinocytochrome *c* from an alkaloid-degrading *Pseudomonas* sp. *Biochem J* 279: 105–109.

Hopper DJ, L Cottrell. 2003. Alkylphenol biotransformations catalyzed by 4-ethylphenol methylenehydroxylase. *Appl Environ Microbiol* 69: 3650–3652.

Huijghebaert SM, HJ Eyssen. 1982. Specificity of bile salt sulfatase activity from *Clostridium* sp. strain S_1. *Appl Environ Microbiol* 44: 1030–1034.

Iwabuchi T, S Harayama. 1998. Biochemical and genetic characterization of trans-2′-carboxybenzalpyruvate hydratase-aldolase from a phenanthrene-degrading *Nocardioides* strain. *J Bacteriol* 180: 945–949.

Kahnert A. MA Kertesz. 2000. Characterization of a sulfur-regulated oxygenative alkylsulfatase from *Pseudomonas putida* S-313. *J Biol Chem* 275: 31661–31667.

Koenig K, JR Andreesen. 1990. Xanthine dehydrogenase and 2-furoyl-coenzyme A dehydrogenase from *Pseudomonas putida* Fu1: Two molybdenum-containing dehydrogenases of novel structural composition. *J Bacteriol* 172: 5999–6009.

Komeda H, M Kobayashi, S Shimizu. 1996. A novel gene cluster including the *Rhodococcus rhodochrous* J1 *nhlBA* genes encoding a low molecular mass nitrile hydratase (L-NHase) induced by its reaction product. *J Biol Chem* 271: 15796–15902.

Koritala S, L Hosie, CT Hou, CW Hesseltine, MO Bagby. 1989. Microbial conversion of oleic acid to 10-hydroxystearic acid. *Appl Microbiol Biotechnol* 32: 299–304.

Lehmann M, B Tshisuaka, S Fetzner, P Röger, F Lingens. 1994. Purification and characterization of isoquinoline 1-oxidoreductase from *Pseudomonas diminuta* 7, a molybdenum-containing hydroxylase. *J Biol Chem* 269: 11254–11260.

Leimkühler S, AL Stockert, K Igarashi, T Nishino, R Hille. 2004. The role of active site glutamate residues in catalysis of *Rhodobacter capsulatus* xanthine dehydrogenase. *J Biol Chem* 279: 40437–40444.

McIntire W, DE Edmonson, DJ Hopper, TP Singer. 1981. 8a-(*O*-tyrosyl)flavin adenine dinucleotide, the prosthetic group of bacterial *p*-cresol methylhydroxylase. *Biochemistry* 20: 3068–3075.

Meckenstock RU, R Krieger, S Ensign, PMH Kroneck, B Schink. 1999. Acetylene hydratase of *Pelobacter acetylenicus*. Molecular and spectroscopic properties of the tungsten iron–sulfur enzyme. *Eur J Biochem* 264: 176–182.

Miech C, T Dierks, T Selmer, K von Figura, B Schmidt. 1998. Arylsulfatase from *Klebsiella pneumoniae* carries a formylglycine generated from a serine. *J Biol Chem* 283: 4835–4837.

Mowafy AM, T Kurihara, A Kurata, T Uemura, N Esaki. 2010. 2-Haloacrylate hydratase, a new class of flavoenzyme that catalyzses the addition of water to the substrate for dehalogenation. *Appl Environ Microbiol* 76: 6032–6137.

Myette JR, V Soundarajan, Z Shriver, R Raman, R Sasisekharan. 2009. Heparin/heparin 6-*O*-sulfatase from *Flavobacterium heparinum*. Integrated structural and biochemical investigation of enzyme active site and substrate specificity. *J Biol Chem* 284: 35177–35188.

Niehaus WG, A Kisic, A Torkelson, DJ Bednarczyk, GJ Schroepfer. 1970. Stereospecific hydration of the Δ^9 double bond of oleic acid. *J Biol Chem* 245: 3790–3797.

Nishida Y, K Adachi, J Kasai, Y Shizuri, K Shindo, A Sawabe, S Komemushi, W Miki, N Misawa. 2005. Elucidation of a carotenoid biosynthesis gene cluster encoding a novel 2,2′-β-hydroxylase from *Brevundimonas* sp. strain SD2b12 and combinatorial biosynthesis of new or rare xanthophylls. *Appl Environ Microbiol* 71: 4286–4296.

Nishimura A, Y Ozaki, H Oyama, T Shin, S Murao. 1999. Purification and characterization of a novel 5-oxoprolinase (without ATP-hydrolyzing activity) from *Alcaligenes faecalis* N-38A. *Appl Environ Microbiol* 65: 712–717.

Ogawa J, JM Kim, W Nirdnoy, Y Amano, H Yamada, S Shimizu. 1995. Purification and characterization of an ATP-independent amidohydrolase *N*-methylhydantoin amidohydrolase, from *Pseudomonas putida* 77. *Eur J Biochem* 229: 284–290.

Oinuma K-I, Y Hashimoto, K Konishi, M Goda, T Noguchi, H Higashibata, M Kobayashi. 2003. Novel aldoxime dehydratase involved in carbon–nitrogen triple bond synthesis of *Pseudomonas chlororaphis* B23. *J Biol Chem* 278: 29600–29608.

Peschke B, F Lingens. 1991. Microbial metabolism of quinoline and related compounds XII. Isolation and characterization of the quinoline oxidoreductase from *Rhodococcus* spec. B1 compared with the quinoline oxidoreductase from *Pseudomonas putida* 86. *Biol Chem Hoppe-Seyler* 370: 1081–1088.

Peters F, Y Shinoda, MJ McInerney, M Boll. 2007. Cyclohexa-1,5-diene-1-carbonyl-coenzyme A (CoA) hydratases of *Geobacter metallireducens* and *Syntrophus aciditrophicus*: Evidence for a common benzoyl-CoA degradation pathway in facultative and strict anaerobes. *J Bacteriol* 189: 1055–1060.

Pope SD, L-L Chen, V Stewart. 2009. Purine utilization by *Klebsiella oxytoca* M5a1: Genes for ring-oxidizing and -opening enzymes. *J Bacteriol* 191: 1006–1017.

Schach S, B Tshisuaka, S Fetzner, F Lingens F. 1995. Quinoline 2-oxidoreductase and 2-oxo-1,2-dihydroquinoline 5,6-dioxygenase from *Comamonas testosteroni* 63. The first two enzymes in quinoline and 3-methylquinoline degradation. *Eur J Biochem* 232: 536–544.

Seddon AP, L Li, A Meister. 1984. Resolution of 5-oxo-L-prolinase into a 5-oxo-L-proline-dependent ATPase and a coupling protein. *J Biol Chem* 259: 8091–8094.

Seiffert GB, GM Ullmann, A Messerschmidt, B Schink, PMH Kroneck. 2007. Structure of the non-redox-active tungsten/[4Fe:4S] enzyme acetylene hydratase. *Proc Natl Acad Sci USA* 104: 3073–3077.

Self WT. 2002. Regulation of purine hydroxylase and xanthine dehydrogenase from *Clostridium purinolyticum* in response to purines, selenium and molybdenum *J Bacteriol* 184: 2039–2044.

Siegmund I, Koenig K, Andreesen JR. 1990. Molybdenum involvement in aerobic degradation of picolinic acid by *Arthrobacter picolinophilus*. *FEMS Microbiol Lett* 67: 281–284.

Steiger S, A Mazetm G Sandemann. 2003. Heterologous expression, purification, and enzymatic characterization of the acyclic carotenoid 1,2-hydratase from *Rubrivivax gelatinosus*. *Arch Biochem Biophys* 414: 51–58.

Stephan I, B Tshisuaka, S Fetzner, F Lingens. 1996. Quinaldine 4-oxidase from *Arthrobacter* sp. Rü61a, a versatile procaryotic molybdenum-containing hydroxylase active towards N-containing heterocyclic compounds and aromatic aldehydes. *Eur J Biochem* 236: 155–162.

Stubbe J, WA van der Donk. 1998. Protein radicals in enzyme catalysis. *Chem Revs* 98: 705–762.

Sun Z, S Shen, C Wang, H Wang, Y Hu, J Jiao, T Ma, B Tiang, Y Hua. 2009. A novel carotenoid 1,2-hydratase (CruF) from two species of the non-photosynthetic bacterium *Deinococcus*. *Microbiology UK* 155: 2775–2783.

tenBrink F, B Schink, PMH Kroneck. 2011. Exploring the active site of the tungsten, iron–sulfur enzyme acetylene hydratase. *J Bacteriol* 193: 1229–1236.

Tkalec AL, D Fink, F Blain, G Zhang-Sun, M Laliberte, DC Bennett, K Gu, JJF Zimmermann, H Su. 2000. Isolation and expression in *Escherichia coli* of *cslA* and *cslab*, genes coding for the chondroitin sulfate-degrading enzymes chondroitinase AC and chondroitinase B, respectively, from *Flavobacterium heparinum*. *Appl Environ Microbiol* 66: 29–35.

van Pée K-H, F Lingens. 1985a. Purification and molecular catalytic properties of bromoperoxidase from *Streptomyces phaeochromogenes*. *J Gen Microbiol* 131: 1911–1916.

3.10 Electron Acceptors Other than Oxygen

3.10.1 Inorganic

3.10.1.1 Nitrate and Related Compounds

3.10.1.1.1 Primary Role

Nitrate can fulfill several functions: (i) assimilatory as a nitrogen source after reduction, (ii) dissimilatory in reduction, and (iii) respiration in which the reduction is coupled to the synthesis of ATP. The structural genes for assimilatory reduction are encoded by *Nas* genes and contain FAD, [4Fe–4S], and *bis*-molybdenum guanine dinucleotide (MGD); in dissimilation, the FAD is replaced by cytochrome *b* or *c* (Moreno-Vivián et al. 1999). The membrane-bound nitrate reductase that is involved in proton translocation and energy production is encoded by the genes *NarGH*, and the enzyme contains molybdopterin guanine dinucleotide, [Fe–S] clusters, and diheme cytochrome b_{556}. The periplasmic nitrate reductase encoded by *NapABC* is primarily involved in dissimilation.

The degradation of organic compounds with nitrate in the absence of oxygen—denitrification or nitrate dissimilation—has been known for a long time, and has been used as a valuable diagnostic character in bacterial classification. The products are either dinitrogen or nitrous oxide, and the reaction is generally inhibited by oxygen, so that it occurs to a significant extent only under anoxic conditions. For example, although it has been reported that *Thiosphaera pantotropha* (*Paracoccus denitrificans*) is capable of both denitrification and nitrification under aerobic conditions (Robertson and Kuenen 1984), it has been shown that the suite of enzymes necessary for denitrification is not expressed constitutively. In addition, the rates of denitrification under aerobic conditions were very much slower than under anaerobic conditions (Moir et al. 1995).

Denitrification involves the sequential formation of nitrite, nitric oxide, and nitrous oxide that are discussed in a later section. Diverse aspects of nitric oxide have attracted attention: (a) chemical oxidation of biogenic nitric oxide to N_{ox} in the context of increased ozone formation (Stohl et al. 1996), and (b) the physiological role in mammalian systems (Feldman et al. 1993; Stuehr et al. 2004), in parasitic infections (James 1995), in the inhibition of bacterial respiration (Nagata et al. 1998), and in bacterial signaling (Zumft 2002). Nitric oxide may be produced—microbiologically in widely different reactions, and various forms of nitrogen oxides are chemically and biologically reactive.

Renewed interest has been focused on degradation of xenobiotics under anaerobic conditions in the presence of nitrate—partly motivated by the extent of leaching of nitrate fertilizer from agricultural land into groundwater. In studies with such organisms, a clear distinction should be made between degradation of the substrate under three conditions that may or may not be biochemically equivalent: (a) aerobic conditions, (b) anaerobic conditions in the presence of nitrate, and (c) fermentation by strictly anaerobic conditions in the absence of any electron acceptor. A few examples are given to illustrate the diversity of degradations that have been observed with facultatively anaerobic organisms using nitrate as the electron acceptor:

1. The degradation of carbon tetrachloride to CO_2 by a *Pseudomonas* sp. (Criddle et al. 1990), although a substantial part of the label was retained in nonvolatile water-soluble residues (Lewis and Crawford 1995). The nature of this was revealed by isolation of adducts with cysteine and *N,N'*-dimethylethylenediamine in which intermediates formally equivalent to $COCl_2$ and $CSCl_2$ were trapped, presumably formed by reaction of the substrate with water and a thiol, respectively. Further consideration of these reactions is given in Chapter 7, Part 2.

2. The nonstoichiometric production of trichloromethane from tetrachloromethane by *Shewanella putrefaciens* (Picardal et al. 1993).

3. The degradation of benzoate (Taylor and Heeb 1972; Williams and Evans 1975; Ziegler et al. 1987) and *o*-phthalate (Afring and Taylor 1981; Nozawa and Maruyama 1988).

4. Attention has been given to denitrifying organisms that could degrade toluene and were referred to the genera *Thauera* and *Azoarcus* (Anders et al. 1995). A critical examination of toluene-degrading strains using a wide range of taxonomic characters was carried out, and descriptions

of *Azoarcus tolulyticus, Azoarcus toluclasticus*, and *Azoarcus toluvorans* were provided (Song et al. 1999).

5. A study of the degradation of halobenzoates under denitrifying conditions led to assignment of strains to *Thauera chlorobenzoica* (Song et al. 2001).

6. *Azoarcus anaerobius* is a strictly anaerobic denitrifying organism that was able to carry out the degradation of resorcinol (1,3-dihydroxybenzene)—though not catechol or hydroquinone—and 16S rRNA sequence analysis showed it to be highly similar to *Azoarcus evansii* and *Azoarcus tolulyticus* (Springer et al. 1998).

7. The degradation of alkyl benzenes (Altenschmidt and Fuchs 1991; Evans et al. 1991a,b; Hutchins 1991). In these studies, some of the organisms referred to the genus *Pseudomonas* have been transferred to the genus *Thauera* (Anders et al. 1995).

8. The degradation of pristane in microcosms and in enrichment cultures (Bregnard et al. 1997).

9. The mineralization of cholesterol by an organism related to *Rhodocyclus, Thauera*, and *Azoarcus* (Harder and Probian 1997).

It is also important to appreciate that an organism that can degrade a given substrate under conditions of nitrate dissimilation may not necessarily display this potential under aerobic conditions. For example, a strain of *Pseudomonas* sp. could be grown with vanillate under anaerobic conditions in the presence of nitrate but was unable to grow under aerobic conditions with vanillate. In contrast, cells grown anaerobically with nitrate and vanillate were able to oxidize vanillate under both aerobic and anaerobic conditions. The cells were also able to demethylate a much wider spectrum of aromatic methoxy compounds under anaerobic conditions than under aerobic conditions (Taylor 1983). Such subtleties should be clearly appreciated and taken into consideration in evaluating the degradative potential of comparable organisms under different physiological conditions.

During degradation in the presence of nitrate this may be reduced, and it has been observed that nitro groups may be introduced into the substrate. Details of this have been discussed in Chapter 2.

Denitrification involves the sequential formation of nitrite, nitric oxide, and nitrous oxide, and several aspects of nitric oxide have attracted attention: (a) chemical oxidation of biogenic nitric oxide to N_{ox} in the context of increased ozone formation (Stohl et al. 1996) and (b) the physiological role in mammalian systems (Feldman et al. 1993; Stuehr et al. 2004), in parasitic infections (James 1995).

3.10.1.1.2 Secondary Roles

Oxides of nitrogen are implicated in a number of diverse reactions, and some examples are given.

1. *Nitric oxide*

   a. Nitric oxide may be produced—microbiologically in widely different reactions. Oxidation of L-arginine by a strain of *Nocardia* sp. produced nitric oxide and L-citrulline (Chen and Rosazza 1995). The enzyme (nitric oxide synthase) carried out two distinct reactions: (i) *N*-hydroxylation of the guanidine of L-arginine catalyzed by an enzyme analogous to cytochrome P450 and (ii) a one-electron oxidation of N^ω-hydroxy-L-arginine to NO and L-citrulline. Evidence has been presented for the occurrence of nitric oxide synthase in *B.subtilis* (Adak et al. 2002a), and *Deinococcus radiodurans* (Adak et al. 2002b), and the mechanism has been reviewed (Stuehr et al. 2004).

      Unusually, NO is produced during the metabolism of glycerol trinitrate by *Phanerochaete chrysosporium* (Servent et al. 1991).

   b. The methanotroph *Methylomirabilis oxyfera* is able to oxidize methane anaerobically using oxygen produced from nitrite. This is reduced to nitric oxide followed by dismutation to N_2 and O_2 (Ettwig et al. 2010). The major part of the oxygen is used for methane activation and oxidation by the route established for aerobic methanotrophs and involves a membrane-bound *bo*-type terminal oxidase (Wu et al. 2011). This reaction seems formally analogous to the formation of oxygen and chloride from chlorite by chlorite dismutase (van Ginkel et al. 1996; Stenklo et al. 2001).

2. *Nitrite as electron acceptor*

Anammox bacteria play an important role in the release of N_2. They carry out the anaerobic reaction between NH_4^+ and NO_2^- : $NH_4^+ + NO_2^- \rightarrow N_2 + 2H_2O$, although a fraction of the nitrite is generally oxidized to nitrate (Schmid et al. 2005). Several groups of bacteria have been implicated, and all of them belong to the phylum Planctomyces, although several quite distinct organisms may be involved. Although their activity was first confirmed in a fluidized wastewater reactor treating an ammonia-rich effluent (van de Graaf et al. 1996), it has since been shown that these organisms are widely distributed geographically in both marine environments and in some terrestrial systems. Specific primers were used to reveal the presence of these organisms in sediments belonging to the provisional genera *Scalindua*, *Jettenia*, *Brocadia*, and *Kuenenia* (Penton et al. 2006). Whereas *Scalindua* is distributed widely in the oceans (references in Hu et al. 2011), species of *Jettenia* and *Brocadia* were found by enrichment of a nitrogen-loaded peat soil (Hu et al. 2011).

3. *Nitrite and NO⁺ as reactants*

Nitrite and NO⁺ produced from it are chemically reactive compounds that can react with nucleophilic substrates. Illustrative examples include the following.

a. Bacteria grown in the presence of 4-chloroaniline and nitrate produced 4,4′-dichloroazobenzene, and further examination using *E. coli* revealed the production of both the azo compound and the corresponding triazine, in addition to the unexpected 3,3′,4,4′-tetrachlorobiphenyl (Corke et al. 1979). These products may plausibly be the product of initial formation of diazonium compounds produced by reaction of the amines with nitrite in a chemical reaction.

b. *E. coli* grown with tryptophan under anaerobic conditions using nitrate as electron acceptor produced a range of di- and trimeric indoles with C=O or C=NOH at C_3, that are formed by reaction of indole produced by tryptophanase (Vederas et al. 1978) and nitrite (Kwon and Weiss 2009). This is the basis of the classical "cholera red" reaction for *Vibrio cholerae*.

4. *Nitration*

The presence of nitrate in the culture medium may bring about nitration of the substrate putatively through the synthesis of NO_2^+ or, NO⁺ followed by oxidation.

a. The formation of nitro-containing metabolites during the degradation of 4-chlorobiphenyl by strain B-206 (Sylvestre et al. 1982).

b. *Corynebacterium* sp. that utilizes dibenzothiophene as a sulfur source produced 2-hydroxybiphenyl, and subsequently nitrated this to form two hydroxynitrobiphenyls (Omori et al. 1992).

c. The transformation of α-tocopherol by *Streptomyces catenulae* produced 5-nitro-tocopherol in addition to quinones (Rousseau et al. 1997).

d. The transformation of 2-hydroxybenz[1,4]oxazin-3-one with *Gliocladium cibotii* produced the intermediate 2-hydroxyacetanilide that was used to produce nitro derivatives (Zikmundová et al. 2002).

e. Nitric oxide synthase from *Streptomyces* sp. that are causal agents of potato scab disease carried out nitration to produce 4-nitrothaxtomin A (Kers et al. 2004).

3.10.1.2 Sulfate and Related Compounds

3.10.1.2.1 Sulfate

Sulfate can be either assimilated as sulfur source or dissimilated by anaerobic bacteria as an electron acceptor when it is reduced to sulfide: details of the reduction are given in a later section. Both reactions involve the formation of AMP anhydrides (APS) catalyzed by ATP sulfurylase. In dissimilation, this is reduced successively to sulfite and sulfide. In assimilation, however, the ester is phosphorylated by ATP and an APS kinase to the anhydride, which after additional phosphorylation at C3′ (PAPS), undergoes further reduction by PAPS reductase encoded by the gene *cysH* (Bick et al. 2000). The assimilatory sulfite reductase from *E. coli* is a complex enzyme containing two different polypeptides. The α-chain coded by the *cysJ* gene in *E. coli* binds one of the flavins (FAD or FMN), while the β-chain coded by the *cysI* gene binds one $[Fe_4S_4]$ cluster and one siroheme that is the site of sulfite reduction (Eschenbrenner et al. 1995). Siroheme is an iron tetrahydroporphyrin with eight carboxylate groups and two angular methyl groups at the site of the acetic acid groups in the reduced A and B rings (Murphy et al. 1973).

3.10.1.2.2 Sulfite

1. *Desulfitibacter alkalitolerans* is an anaerobe that is able to use sulfite—though not sulfate—as electron acceptor for growth with formate, lactate, pyruvate, and choline as carbon and energy sources (Nielsen et al. 2006).

2. Under anaerobic conditions sulfite is able to function as a terminal electron acceptor in *Shewanella oneidensis* MR-1 during growth with lactate: reduction to sulfide required menaquinones and the reductase is carried out by an octaheme cytochrome *c* SirA (Shirodkar et al. 2011).

3.10.1.2.3 Thiosulfate

Thiosulfate is able to adjust its metabolism to a range of conditions: (a) aerobically as a source of energy, (b) anaerobic respiration in the absence of alternatives such as oxygen and nitrate, (c) anaerobically as an electron acceptor, (d) as an energy source by dismutation when a suitable carbon source is available. For the sake of simplicity, all these are considered here, although they are accomplished by different groups of organisms.

1. A number of bacteria can use thiosulfate as energy source for chemolithotrophic growth using oxygen as electron acceptor while the electrons released are used for the assimilation of CO_2. This has been extensively examined in the facultative lithoautotrophs *Paracoccus pantotrophus*, *P. versutus*, and *Starkeya novella* in which thiosulfate is oxidized to sulfate: $O_3S-S^{2-} + 2O_2 + H_2O \rightarrow 2SO_4^{2-} + 2H^+$. The thiosulfate oxidation pathway Sox is carried out by a multienzyme complex comprising four essential proteins: SoxAX cytochromes are heterodimeric proteins with two or three *c*-type hemes, SoxYZ is a carrier protein, SoxB contains two atoms of manganese and SoxCD is a sulfur dehydrogenase is an $\alpha_2\beta_2$ heterotetramer in which SoxC is the molybdenum cofactor, and SoxD is a diheme *c*-type cytochrome (Friedrich et al. 2001). It has been shown that in addition to two active site heme groups, SoxAX contains a mononuclear Cu^{II} site having a distorted tetragonal structure (Kappler et al. 2008). In addition to these organisms, strains of *Bradyrhizobium japonicum* and the slow-growing *Agromonas oligotrophica* (Masuda et al. 2010) are able to grow lithotrophically with thiosulfate, and in *B. japonicum* strain USDA 110, it was shown that, as for *Para. pantotrophus*, 1 mol of thiosulfate produced 2 mol sulfate with the consumption of 2 mol oxygen, and that CO_2 assimilation was thioulfate dependent. Additional examples of other genera that can utilize thiosulfate and tetrathionate are given in Chapter 11 in the context of elemental sulfur metabolism.

   Anaerobic photolithotrophs can use sulfide to carry out the reduction of CO_2 with the production of elemental sulfur, and most species of *Chlorobium* can, in addition, use thiosulfate as electron donor for anoxygenic photosynthesis. In *Chlorobium tepidum* three proteins SoxYZ, Sox B, and SoxAX-CT1020 are required for thiosulfate reduction of cytochrome cyt c_{554} and are analogues of the enzymes used by aerobic chemolithotrophs (Ogawa et al. 2008).

2. Anaerobic respiration can be carried out using several sulfur oxyanions. Whereas sulfite reduction is carried out by a soluble enzyme in *Salmonella enterica* serovar *typhimurium* consisting of three genes *asrA*, *asrB*, and *asrC* (Huang and Barrett 1991). The thiosulfate reductase is mediated by a membrane-bound system, encoded by *phsA, phsB, and phsC*, although it seems that thiosulfate does not make a major contribution to energy conservation during growth with a range of substrates (Heinzinger et al. 1995). All species of *Shewanella* are able to reduce thiosulfate to sulfide (Venkateswaran et al. 1999), and sulfite, thiosulfate, tetrathionate, and elemental sulfur were able to support anaerobic growth of *Shewanella oneidensis*. It has been shown that a gene homologous to that *phsA* gene in *Salmonella enterica* serovar *typhimurium*

is required for anaerobic respiration (Burns and DiChristina 2009). There were similarities between the PsrA, PsrB, and PsrC homologues of *Shewanella* sp., *Salmonella typhimurium* LT2, and *Wolinella succinogenes.*

3. Members of the genus *Desulfatibacillum* of anaerobic sulfate-reducing bacteria are able to use both sulfate and thiosulfate as electron acceptors for the oxidation of alkanes (C_{13}–C_{18}) by *Desulf. aliphaticivorans* (Cravo-Laureau et al. 2004a) and alkenes (C_8–C_{23}) by *Desulf. alkenivorans* (Cravo-Laureau et al. 2004b). Under anaerobic conditions, *Archaeoglobus fulgidus* was able to use thiosulfate as terminal electron acceptor during the degradation of (a) C_4–C_{18} alkanoates and (b) $C_{12:1}$ to $C_{21:1}$ terminal alkenes with the production of sulfide (Khelifi et al. 2010).

4. Thiosulfate may supply the energy for anaerobic growth by dismutation:

$S_2O_3^{2-} + H_2O \rightarrow SO_4^{2-} + HS^- + H^+$. This has been demonstrated for the Gram-negative *Desulfovibrio sulfodismutans* that required acetate as a carbon source (Bak and Pfennig 1987), and for *Desulfocapsa sulfoexigens* that was also able to use elemental sulfur when a sulfide scavenger such as ferrihydrite was present (Finster et al. 1998). The Gram-positive *Desulfotomaculum thermobenzoicum* also illustrated this dismutation in which acetate enhanced both growth and dismutation (Jackson and McInerney 2000).

3.10.1.2.4 Tetrathionate

Tetrathionate is able to serve as electron acceptor for the anaerobic growth of several *Enterobacteriaceae*. The metabolism of tetrathionate may take place with the formation of thiosulfate, sulfite, and sulfide in two stages: tetrathionate reductase produces thiosulfate, while thiosulfate reductase then forms sulfite and sulfide. Neither of these activities occurs in the presence of the preferred electron acceptors oxygen or nitrate. Both activities are present in a number of *Enterobacteriaceae* including species of *Salmonella*, *Citrobacter* and *Proteus* (references in Barrett and Clark 1987), for example, in *Proteus mirabilis*, tetrathionate reductase and thiosulfate reductase were purified as a single component from the cytoplasmic membrane (Oltmann et al. 1974). Tetrathionate, that is a component of some media for the enrichment of salmonellas, is able to support the anaerobic growth of *Salmonella* sp. using glycerol or acetate, and is the *only* electron acceptor that is able to support anaerobic growth of *S. enterica* at the expense of ethanolamine or propan-1,2-diol when endogenous vitamin B_{12} is available (Price-Carter et al. 2001). Tetrathionate reductase in *Salmonella enterica* serovar *typhimurium* comprises three components: TtrA contains a molybdopterin guanine dinucleotide and an [4Fe–4S] cluster, TtrB binds four [4Fe–4S] clusters that are involved in electron transfer from TtrC to TtcA, and TtrC is an integral membrane protein that contains a quinol-oxidizing site. TtrS and TtrR are part of a two-component regulatory system that is required for transcription of the *ttrBSA* operon (Hensel et al. 1999).

3.10.1.3 Selenate and Arsenate

Reductions in which energy is produced during reduction and supports growth of the cells should be carefully distinguished from the situation in which selenate is gratuitously reduced during aerobic growth (Maiers et al. 1988). For example, the membrane-bound selenate reductase from *Enterobacter cloacae* SLD1a-1 is not able to function as electron acceptor under anaerobic conditions (Ridley et al. 2006). Reductases also function in the detoxification of arsenate and selenate, and the three groups of arsenate reductase have been reviewed (Rosen 2002). The arsenate reductase gene *ArsC* both in *Staphylococcus aureus* (Ji and Silver 1992) and in *E. coli* (Liu et al. 1995) is a determinant of arsenate resistance, and the arsenite that is produced is extruded from the cell by the ArsA–ArsB anion-translocating ATPase (Walmsley et al. 1999). Whereas the reductase from *E. coli* is coupled to the glutathione and glutaredoxin system, that from *S. aureus* requires thioredoxin and thioredoxin reductase. The arsenate reductase from *B. subtilis* shows structural similarity at the active site to that of low-molecular-weight tyrosine phosphatase and this enabled a mechanism for its activity to be proposed (Bennett et al. 2001). These reductases are discussed further in Section 3.13 of this chapter.

1. Selenate and arsenate can serve as terminal electron acceptors, and strains of facultatively anaerobic bacteria have been isolated that are able to use them during growth with lactate. These include a strain designated SES-3 (Laverman et al. 1995), and *Bacillus arsenicoselenatis* and *Bacillus selenitrificans* (Blum et al. 1998). In these strains, utilization of acetate is unusual, but acetate is used specifically by *Chrysiogenes arsenatis* (Macy et al. 1996). The reductase from this organism contains Mo, Fe, and acid-labile S (Krafft and Macy 1998). A facultatively anaerobic organism *Thauera selenatis* is able to use both selenate and selenite as electron acceptors for anaerobic growth (Macy et al. 1993). The selenate reductase that produces selenite has been shown to contain molybdenum, iron, acid-labile sulfur, and cytochrome *b*. It is located in the periplasmic space, and is specific for selenate and nitrate, whereas neither chlorate nor sulfate was reduced at significant rates (Schröder et al. 1997). The soluble periplasmic selenate reductase in *Thauera selenatis* has been examined to reveal several features. X-band EPR suggested the presence of Mo(V), heme *b*, and $[3Fe–4S]^{1+}$, and $[4Fe–4S]^{1+}$ clusters that were involved in electron transfer to the active site of the periplasmic reductase under turnover conditions. Collectively, the analysis showed that the reductase belonged to the Type II molybdoenzymes (Dridge et al. 2007). Reduction of selenite to Se^0 can be accomplished by the periplasmic nitrite reductase that also occurs in this strain (DeMoll-Decker and Macy 1993). An organism assigned to *Desulfotomaculum auripigmentum* was able to grow with lactate and could use both arsenate and sulfate as terminal electron acceptors (Newman et al. 1997). Several strains of *Desulfitobacterium* can use arsenate, generally nitrate, and exceptionally selenate as electron acceptors during growth with lactate or pyruvate as electron donors (Niggemyer et al. 2001).

2. A strain of *Shewanella* ANA-3 can couple the reduction of As(V) to As(III) to the oxidation of lactate to acetate (Saltikov et al. 2003). Although this strain contains the *ars* operon, which confers resistance to As(III), the gene *arsB* that encodes the efflux system and *arsC* for the reductase, although advantageous, are not obligatory. Genetic analysis showed the presence of an additional gene cluster that encoded a respiratory As(V) reductase. The gene *arrA* is predicted to encode a molybdenum cofactor, and *arrB* to encode an Fe–S cluster, and are predicted to be similar to those of the DMSO reductase family (Saltikov and Newman 2003). Therefore, although there are two available mechanisms for arsenate respiration, they differ in their expression: the *ars* system is expressed under both aerobic and anaerobic conditions, whereas the *arr* system is expressed only anaerobically and is repressed by oxygen or nitrate (Saltikov et al. 2005). *Shewanella* sp. strain CN-32 is able to use arsenate as terminal electron acceptor that depends both on the gene *arrA* and an additional gene cymA encoding a membrane-associated tetraheme *c*-type cytochrome that is induced anaerobically by fumarate and arsenate (Murphy and Saltikov 2007).

3. *Sulfurospirillum barnesii* is able to use arsenate and selenate, and *Sulfurospirillum arsenophilum* arsenate—though not selenate—as electron acceptors during microaerophilic growth using lactate, pyruvate, fumarate, and formate as electron donors (Stolz et al. 1999).

4. a. An anaerobic bacterium strain Y5 that was closely related to the genus *Desulfosporosinus* on the basis of 16S rDNA phylogeny was able to use arsenate, nitrate, sulfate, and thiosulfate as electron acceptors, although in metabolic capacity it did not resemble either *D. auripigmentum* or *D. meridiei*. It was notable for its ability to use a range of electron donors that included not only succinate and lactate but, rather unusually, syringate, ferulate, phenol, benzoate, and toluene (Liu et al. 2004).

   b. A strictly anaerobic halophile *Halarsenatibacter silvermannii* strain SLAS-1 was able to use arsenate and sulfur (S°) as electron acceptors and a limited range of carbohydrates and pyruvate as electron donors. Arsenate was reduced to arsenite and lactate was oxidized to acetate + CO_2, or sulfur to sulfide. The organism was also able to grow with sulfide as electron donor and arsenate as electron acceptor (Blum et al. 2009).

3.10.1.4 Arsenite

Arsenite is produced by the activity of arsenate reductases, and its extrusion from the cell represents a component of detoxification. There are two mechanisms: one by carrier-mediated efflux and the other by an arsenite-translocating ATPase (Dey and Rosen 1995). Arsenite can also be oxidized to arsenate under different conditions. This includes heterotrophs containing the *aoxAB*, *asoAB* genes encoding the oxidase, and lithotrophs (autotrophs) and contain the *aroAB* gene (Rhine et al. 2007).

1. For aerobic heterotrophs such as *Alcaligenes faecalis*, the bacteria serve only as detoxifying agents by the oxidation of arsenite to arsenate (Phillips and Taylor 1976). The crystal structure of arsenite oxidase has been determined in this organism. It contained two subunits, the larger containing Mo bound to two pyranopterin cofactors and a [3Fe–4S] cluster, and the smaller with a Rieske-type [2Fe–2S] cluster (Ellis et al. 2001). It is worth noting that in spite of the name the oxidase carried out the reaction:

$$As^{III}O_2^- + 2H_2O \rightarrow As^V O_4^{3-} + 4H^+ + 2e$$

   During heterotrophic growth of *Hydrogenophaga* sp. NT-14 oxidation of arsenite took place, and characterization of arsenite oxidase revealed basic features including the existence of two subunits in an $\alpha3\beta3$ configuration, that the enzyme contained Mo and Fe in cofactors, and that it shared similarities with the arsenite oxidase from *Alcaligenes faecalis* (vanden Hoven and Santini 2004). Strains of *Herminiimonas* that have been described as ultramicrobacteria have been isolated from unusual sources that include a deep Greenland ice-core for *Her. glaciei* (Loveland-Curtze et al. 2009) and a highly metal cation-, arsenite-, and arsenate-resistant strain ULPAs1 of *Her. arsenicoxydans* (Muller et al. 2006) from a sludge sample heavily contaminated with arsenic. This is a heterotroph albeit with a limited range of growth substrates and is resistant to arsenite which can be oxidized to arsenate. It can also reduce arsenate to arsenite mediated by several arsenate reductases—although neither arsenate nor selenate is able to support anaerobic growth with acetate or lactate. It, therefore, adds another example of bacteria able to metabolize both arsenite by oxidation and arsenate by reduction.

2. a. A strain HR13 of *Thermus* was a facultative anaerobe isolated from an arsenic-rich geothermal habitat. Under aerobic conditions, it was able to oxidize arsenite to arsenate, and under anaerobic conditions, growth at the expense of lactate could be carried out using arsenate as the electron acceptor (Gihring and Banfield 2001).

   b. Some unusual features were observed in the regulation of arsenite oxidation in *Agrobacterium tumefaciens* (Kashyap et al. 2006). This strain exhibited a complex control of the regulation of arsenite oxidation by arsenite oxidase *aoxAB*. In addition, it contained a gene *arsC* for arsenate oxidation that is not normally expressed, and it was shown that a mutant in strain 5A that had lost the arsenite oxidation phenotype was capable of reducing arsenate to arsenite.

   c. Biofilms from a hot spring contained *Ectothiorhodospira* as the dominant organism that was able to carry out the following reactions (Hoeft et al. 2010):

      i. Anoxic oxidation of As(III) in the light and reduction of As(V) in the dark;

      ii. Acetate was oxidized to CO_2 in the light, but not in darkness;

      iii. Dark fixation of HCO_3^- linked to As(V);

      iv. Functional genes for arsenate respiratory reductase *arrA* were detected though not for arsenite oxidase *aoxB*.

   These add important examples of the cycling of arsenite between oxidation levels by the same organism. In a wider context the *aoxB* genes encode the catalytic subunit of arsenite oxidase, and it has been shown that AoxB phylogeny of this gene could usefully serve as a functional marker of aerobic arsenite oxidizers that are distributed among α-, β-, and γ-Proteobacteria (Quéméneur et al. 2008).

3. Chemolithotrophic organisms obtain energy from the oxidation of arsenite to arsenate and use CO_2/HCO_3^- as carbon source. An organism with this capability was isolated from arsenopyrite (FeAsS) after enrichment with arsenite, and on the basis of 16S rDNA sequence analysis, it was assigned to α-Proteobacteria including *Agrobacterium* and *Rhizobium* (Santini et al. 2000). A facultative chemolithotroph *Thiomonas arsenivorans* isolated from gold mine samples in France, was able to use arsenite, sulfur, thiosulfate and Fe(III) for autotrophic growth, and organic substrates including pyruvate, succinate, aspartate, glutamate, sucrose and raffinose for heterotrophic growth (Battaglia-Brunet et al. 2006).

4. a. A facultative autotrophic lithotrophic strain *Alkalilimnicola ehrlichii* MLHE-1 was able to oxidize arsenite under anaerobic conditions to arsenate using nitrate as electron acceptor (Oremland et al. 2002; Hoeft et al. 2007). [The type genus is *Alcalilimnicola* gen. nov. with the species *halodurans* sp. nov. (Yakimov et al. 2001)]. Oxidation was carried out by a novel arsenite oxidase encoded by *arxA* in place of the normally used arsenite oxidase encoded by *aoxB* in a wide range of bacteria including strains of *Rhizobium* sp., *Agrobacterium tumefaciens*, *Xanthobacter autotrophicus*, *Alcaligenes faecalis*, and *Burkholderia multivorans* (Zargar et al. 2010).

b. Anaerobic oxidation of arsenite has also been linked to chlorate reduction by strains identified by their 16S rRNAs genes as *Dechloromonas*, *Azospira* of which pure cultures ECC1-pb1 and ECC1-pb2, respectively, were isolated and, under autrophic conditions *Stenotrophomonas*. The advantage of chlorate is that chloride is the final product, since the oxygen produced by chlorite dismutase is rapidly consumed. Only the subunit *aroA* of arsenite oxidase could be demonstrated in these strains (Sun et al. 2010). The bacterial oxidation of arsenite produced from the leaching of arsenopyrite is discussed in Chapter 2.

3.10.1.5 Chlorate and Perchlorate

Perchlorate has become virtually ubiquitous in the United States as a contaminant in water supplies and, in view of a range of its adverse effects, regulation by EPA has, therefore, been implemented. Interest has, therefore, centered on procedures for its elimination from wastewater by treatment in bioreactors (Logan et al. 2001). Although perchlorate is a component of rocket fuels, it may be produced in the atmosphere by ozonization of chloride (Dasgupta et al. 2005), and it has been found in several hot-desert samples (Bao and Gu 2004) that are presumably nonanthropogenic and could clearly be distinguished on the basis of the anomaly in the ^{17}O signature, and in the Dry Valleys of Antarctica above the ice table, where it was associated with nitrate deposition (Kounaves et al. 2010).

Oxygen isotopes have been used to distinguish natural and anthropogenic sources of perchlorate using the signatures of O_2 formed by pyrolysis of the samples. In this application, two important factors had to be taken into consideration: (i) removal of nitrate that would compromise values for O_2 and (ii) to make a correction from the determined values of ^{17}O that are generally obtained from values for $^{18}O=^{17}O \approx 0.52 \times \delta^{18}O$. This was used to calculate the ^{17}O anomaly $\Delta^{17}O=^{17}O - 0.52 \times \delta^{18}O$ (Bao and Gu 2004). Values for man-made perchlorate that is produced by electrolysis of chlorate and, therefore, contains oxygen from water had $\delta^{17}O$ values ranging from −9.1 o/oo to −10.5 o/oo, and of the anomaly $\Delta^{17}O$ from −0.06 to −0.20 compared with values of the anomaly from the Atacama Desert of +4.2 to +9.6 o/oo, although this did not finally resolve the origin of the perchlorate in the Atacama Desert.

Chlorate and perchlorate can serve as electron acceptors under anaerobic conditions (Coates et al. 1999; Thorell et al. 2003), and chlorate reductase has been found both in organisms such as *Proteus mirabilis* that can reduce chlorate—although it is unable to couple this to growth—and in chlorate-respiring organisms. Only *Dechloromonas agitata* (Achenbach et al. 2001) and *Dechloromonas hortensis* (Wolterink et al. 2005) were able to use nitrate as electron acceptor as well as chlorate and oxygen. There are two cardinal enzymatic activities required for chlorate respiration: (a) the sequential reduction of perchlorate and chlorate to chlorite $ClO_4^- \rightarrow ClO_3^- \rightarrow ClO_2^-$ and (b) dismutation of the intermediate chlorite: $ClO_3^- \rightarrow ClO_2^- \rightarrow Cl^- + O_2$ that may be used to support oxygenic growth. The genomic island for perchlorate reductase has been examined for the *Betaproteobacterium*:

Dechloromonas aromatica strain RCB, *Azospira suillum* strain PS and *Dechloromonas agitata* strain CKB, and the single *Alphaproteobacterium Magnetospirillum bellicus* strain VDY, and all of them contained genes for chlorite dismutase (*cld*), perchlorate reductase (*pcrABCD*), and *c*-type cytochromes (Mlynek et al. 2011b).

Chlorate has been shown to support the growth of an anaerobic community growing at the expense of acetate (Malmqvist et al. 1991), and a pure culture designated *Ideonella dechloratans* has been isolated (Malmqvist et al. 1994). A number of other organisms can use chlorate as electron acceptors during anaerobic growth, and organisms capable of dissimilatory (per)chlorate reduction are both ubiquitous and diverse: on the basis of 16S r RNA they are distinct from the previously described *Ideonella dechloratans* and have been assigned to species of *Dechloromonas* (Achenbach et al. 2001). These include *Azospira oryzae* GR-1 (*Dechlorosoma suillum*) (Wolterink et al. 2005) that is able to carry out the sequential reduction of perchlorate to chlorate, and further to chloride at the expense of acetate (Rikken et al. 1996). On the other hand, *Pseudomonas chloritidismutans* (Wolterink et al. 2002) is clearly different from these, and is similar to *Ps. stutzeri*—except for its ability of using chlorate as electron acceptor. It is able to use chlorate—though not perchlorate or nitrate as electron acceptors using acetate and propionate as sources of carbon. Although some isolates are capable of high rates of growth using both perchlorate and chlorate, some are able to use only the latter (Logan et al. 2001).

Chlorate and perchlorate reductases have been characterized from a number of different organisms.

1. Chlorate reductase in *Azospira oryzae* strain GR-1 (Kengen et al. 1999) was found in the periplasm. It is oxygen-sensitive, and contains molybdenum and [3Fe–4S] and [4Fe–4S] clusters.

2. A different cytoplasmic reductase has been described in *Ps. chloritidismutans* (Wolterink et al. 2003).

3. Chlorate reductase from *I. dechloratans* has been characterized (Thorell et al. 2003), and consists of three subunits: it contains heme *b* and resembles enzymes belonging to the molybdopterin DMSO reductase class II family.

4. Perchlorate reductase from *Dechloromonas agitata* is different from that of *I. dechloratus*, although the *pcrAB* gene products are similar to α- and β-subunits of other reductases including nitrate, selenate and chlorate reductases, and dimethylsulfide dehydrogenase. The *pcrC* gene product was similar to a *c*-type cytochrome, and the *pcrD* product similar to the molybdenum-containing proteins of the DMSO reductase family (Bender et al. 2005).

The chlorite dismutase that generates oxygen from chlorite produced by reduction of chlorate has been purified and characterized from *Dechloromonas hortensis* strain GR-1 (van Ginkel et al. 1996) and from *Ideonella dechloratans* (Stenklo et al. 2001). The enzyme from *Dechloromonas agitata* has been suggested as a metabolic probe (Bender et al. 2002), and primers for *cld* gene sequences that encode perchlorate reductase from a range of (per)chlorate reducing bacteria (Bender et al. 2004) have been proposed for environmental screening of putative perchlorate-reducing bacteria. It has been shown that chlorite dismutase occurs in the nitrite oxidizer *Nitrobacter winogradskyi*, is catalytically active and the crystal structure of the enzyme revealed it to be, unusually, a dimeric enzyme (Mlynek et al. 2011a).

Whereas chlorite dismutase in *Dechloromonas*, *Dechlorosoma*, *Ideonella*, and *Ps. chloritidismutans* form a cluster in Lineage I, the enzyme from *Nitrobacter winogradskyi* belonged to Lineage II together with enzymes that have not hitherto been shown to possess this facility. Genomic analysis of the nitrite-oxidizing *Nitrospira defluvii* revealed the presence of a gene analogous to that for chlorite dismutase, and after heterologous expression in *E. coli*, the product was able to carry out dismutation of chlorite and contained heme (Maixner et al. 2008).

A range of substrates can be used for growth at the expense of chlorate reduction and includes the following.

1. *Dechloromonas agitata*, *Dechloromonas hortensis*, *Ps. chloritidismutans*, and *Azospira oryzae* can utilize acetate and propionate as electron acceptors. *Dechloromonas agitata* is also able to use butyrate, succinate, and malate (Achenbach et al. 2001), *Ps. chloritidismutans* can

use glucose, maltose, and gluconate (Wolterink et al. 2002), and *Azospira oryzae* is able to use caprionate, malate, and succinate (Rikken et al. 1996).

2. *Magnetospirillum bellicus* strain VDY and strain WD that were not magnetotactic were able to use O_2, perchlorate, chlorate, and nitrate as electron acceptors, and acetate, propionate butyrate and valerate as electron donors (Thrash et al. 2010).

3. A strain identified as *Moorella perchloratireducens* was able to use a range of substrates including carbon monoxide, methanol, pyruvate, and some carbohydrates (Balk et al. 2008). Perchlorate reductase and chlorite dismutase were found in cell extracts, the strain displayed considerable tolerance of oxygen and contained a membrane-bound cytochrome *bd* oxidase that is also found in the strict anaerobe *Moorella thermoacetica* (Das et al. 2005).

4. The degradation of monocyclic arenes has been described for *Dechloromonas* sp. strain RCB (Chakraborty et al. 2005), and the pathway of degradation of benzene putatively involved hydroxylation followed by carboxylation of the resulting phenol and dehydroxylation (Chakraborty and Coates 2005). Only very low incorporation of ^{18}O from $H_2^{18}O$ was, however, observed and no evidence was presented for occurrence of the intermediate 4-hydroxybenzoate. On the other hand, the degradation of benzene by *Alicycliphilus denitrificans* was carried out by successive hydroxylation to catechol followed by ring fission (Balk et al. 2008). This was consistent with the formation of oxygen by the activity of chlorite dismutase, and the detection of a dioxygenase gene (BC-BMOa) and a gene for catechol-2,3-fission dioxygenase (BC-C23O).

Only *Ps. chloritidismutans* and *Dechloromonas hortensis* were able to reduce both chlorate and bromate (Wolterink et al. 2002, 2005). On the other hand, *Pseudomonas* sp. strain SCT isolated from a marine sediment, and similar to *Ps. stutzeri* was able to reduce iodate to iodide in sediment slurries supplemented, in decreasing order of effectiveness with malate > succinate > acetate > citrate (Amachi et al. 2007). This strain was able to grow anaerobically using iodate as electron acceptor and malate as electron donor, and during growth the iodate was quantitatively reduced to iodide.

3.10.1.6 *V(V), Mn(IV), Fe(III), Co(III), Tc(VII), and U(VI)*

There has been considerable interest in the remediation of sites contaminated with U(VI) and Tc(VII) by conversion of these soluble forms into less soluble reduced states that may be precipitated. In assessing results involving the reduction of U(VI) and Tc(VII), it is important to take into consideration (a) intermediate levels of reduction that may be critical and (b) the formation of complexes, for example, with carbonate that may determine the final products and their association with cells or particles. The discussion is assembled by organism, since some of these are able to reduce several oxidants.

3.10.1.6.1 *Shewanella spp*

Strains of *Shewanella putrefaciens* (*Alteromonas putrefaciens*) are widely distributed in environmental samples and are generally considered as aerobic organisms with the capability of reducing thiosulfate to sulfide in complex media. They are also able, however, to grow anaerobically using Fe(III) as electron acceptor, and to oxidize formate, lactate, or pyruvate. These substrates cannot, however, be completely oxidized to CO_2 since the acetate produced from the C_2 compounds is not further metabolized. Mn[IV] may function analogously to Fe(III) (Lovley et al. 1989). The bioenergetics of the system has been examined in cells of another strain of this organism grown anaerobically with lactate using either fumarate or nitrate as electron acceptors. Respiration-linked proton translocation in response to Mn(IV), fumarate, or oxygen was clearly demonstrated (Myers and Nealson 1990). Levels of Fe(III) reductase, nitrate reductase, and nitrite reductase are elevated by growth under microaerophilic conditions, and the organism probably possesses three reductase systems, each of which apparently consists of low- or high-rate components (DiChristina 1992). Reductase activity is located on the outer membranes of *Shewanella putrefaciens* and has a requirement for cytochrome *c* (Myers and Myers 1997) and menaquinones (Saffarini et al. 2002). Another organism with hitherto unknown taxonomic affinity has been isolated (Caccavo et al. 1992), and is able to couple the oxidation of lactate to the reduction of Fe(III), Mn(IV), and U(VI). Under anaerobic conditions, *Shewanella oneidensis* MR1 can couple the reduction of V(V) to V(IV) to the oxidation of lactate, formate, and pyruvate (Carpentier et al. 2003; Myers et al. 2004). Respiration and growth can be carried out by reduction of vanadate: proton translocation across the cytoplasmic membrane takes place during the reduction and is abolished by CCCP, HOQNO, and antimycin (Carpentier et al. 2005). *Shewanella oneidensis* MR-1 is able to respire using Co(III)EDTA⁻ with lactate as carbon source but only when $MgSO_4$ is added to remove the toxicity caused by the reduction product Co(II)EDTA²⁻ (Hau et al. 2008).

Under anaerobic conditions with H_2 as electron donor—when growth did not take place—*S. putrefaciens* reduced Tc(VII) to Tc(IV) that was associated with the cell or with Tc(IV) complexes in the presence of carbonate (Wildung et al. 2000).

3.10.1.6.2 *Geobacter spp*

A strictly anaerobic organism designated GS-15 and now assigned to the taxon *Geobacter metallireducens* (Lovley et al. 1993) is able to use Fe(III) as electron acceptor under anaerobic conditions for the oxidation of a number of substrates including acetate (Lovley and Lonergan 1990), and toluene and phenols (Champine and Goodwin 1991). This organism is also able to oxidize acetate by reduction of Mn(IV) and U(VI) (Lovley et al. 1993), and additional comments are given later in this chapter.

Organisms belonging to the genus *Geobacter* are widely distributed in anaerobic environments in which Fe(III) occurs, and members of the *Geobacteraceae* were enriched at a site contaminated with U(VI) (Anderson et al. 2003). All species in the family are able to use Fe(III) as electron acceptor and acetate as electron donor, generally propionate and benzoate and, for *G. metallireducens* and *G. grbiciae*, additionally toluene using anthraquinone-2,6-disulfonate (AQDS) as an electron acceptor (Coates et al. 2001a). *Geoglobus ahangeri* was able to oxidize acetate, butyrate, valerate, and stearate by using poorly crystalline ferric oxide as electron acceptor, but was unable to use sulfur, sulfate, thiosulfate, nitrate, nitrite, or Mn(IV) as electron acceptors (Kashevi et al. 2002). The phylogenetically distinct *Geovibrio ferrireducens* strain PAL-1 is able to use a wide range of organic compounds as electron donors including acetate, propionate, succinate, and proline (Caccavo et al. 1996). It has been shown in *Geobacter sulfurreducens* that although U(VI) [UO_2^{2+}] is reduced to insoluble U(IV), this occurs by initial reduction to U(V) followed by disproportionation (Renshaw et al. 2005). Therefore, it was suggested that caution be exercised in the use of microbial remediation for transuranic elements whose oxidation levels differ from those of uranium. There are other limitations to their application for bioremediation of uranium-contaminated sites:

1. In long-term experiments, U(IV), which was initially produced by microbial reduction of U(VI), could be reoxidized putatively using residual Fe(III) or Mn(IV) even under anaerobic conditions (Wan et al. 2005).
2. Uranium that is associated with sediments is resistant to reduction (Ortiz-Bernad et al. 2004a).

It has been shown that in the presence of nitrate and an electron donor such as acetate, both U(VI) and Tc(VII) can be reduced (Istok et al. 2004). Important conclusions were: (a) whereas reduction of Tc(VII) occurred concurrently with nitrate reduction, reduction of U(VI) took place only after Fe(II) was detected, (b) reoxidation of U(IV) took place in the presence of high—though not with low—concentrations of nitrate.

The reductase in *Geobacter sulfurreducens* is located in the outer membrane, and a soluble Fe(III) reductase has been characterized from cells grown anaerobically with acetate as electron donor and Fe(III) citrate or fumarate as electron acceptor (Kaufmann and Lovley 2001). The enzyme contained Fe, acid-labile S, and FAD. An extracellular *c*-type cytochrome is distributed in the membranes, the periplasm, and the medium, and functions as a reductase for electron transfer to insoluble iron hydroxides, sulfur, or manganese dioxide (Seeliger et al. 1998). The molecular mass of 9.6 kDa suggested its similarity to cytochromes from sulfate-reducing bacteria, which was consistent with the fact that it contained three hemes although there is conflicting evidence on the role of this component (Lloyd et al. 1999). In contrast, Fe(III) reduction in the hyperthermophilic archaeon *Pyrobaculum islandicum* and in *Pelobacter carbinolicus* (Lovley et al. 1995) appear not to involve cytochromes (Childers and Lovley 2001). Although the hyperthermophilic sulfate-reducing archaeon *Archaeoglobus fulgidus* contains an

Fe(III)–EDTA reductase, this shows sequence similarity to the family of NAD(P)H:FMN oxidoreductases (Vadas et al. 1999).

Cell suspensions of *Geobacter sulfurreducens* can couple the oxidation of hydrogen to the reduction of Tc(VII) to insoluble Tc(IV). An indirect mechanism involving Fe(II) was also observed, and was substantially increased in the presence of the redox mediator AQDS (Lloyd et al. 2000). *Geobacter metallireducens* is also able to grow by vanadate respiration supported by acetate, and it has been suggested that this could provide a new strategy for removing vanadate from groundwater (Ortiz-Bernad et al. 2004b).

3.10.1.6.3 Other Organisms

1. Sulfate-reducing bacteria: A number of sulfate-reducing anaerobic bacteria can oxidize S^0 to sulfate at the expense of Mn(IV) (Lovley and Phillips 1994), and a strain of *Desulfovibrio desulfuricans* was able to use hydrogen to couple the reduction of Tc(VII) to an insoluble, more reduced form that was precipitated on the periphery of the cells (Lloyd et al. 1999). Reduction of U(VI) complexes with aliphatic carboxylates was examined using *Desulfovibrio desulfuricans*: the acetate complex was more rapidly reduced than those of dicarboxylates such as malonate, oxalate, and citrate, and the complex with 4,5-dihydroxybenzene-1,3-disulfonate was readily accessible (Ganesh et al. 1997). In the presence of H_2, *Desulfovibrio fructosovorans* was able to reduce Tc(VII) to soluble Tc(V) or putatively to an insoluble precipitate of Tc(IV), and this was mediated by the Ni–Fe hydrogenase in the organism (De Luca et al. 2001).

2. *Deinococcus radiodurans* strain R1: Anaerobic cultures grown with lactate could reduce chromate, although reduction was increased by the presence of electron transfer to AQDS. Reduction of U(VI) and Tc(VII) could also be accomplished in the presence of the electron transfer agent (Fredrickson et al. 2000).

3. *Pyrobaculum aerophilum*: Reduction of Cr(VI) to Cr(III), U(VI) to insoluble U(IV), and Tc(VII) to insoluble Tc(IV or V) was accomplished with this hyperthermophilic organism in the presence of H_2 (Kashefi and Lovley 2000).

4. *Cellulomonas* spp: Under anaerobic conditions, several strains could carry out reduction of Cr(VI) and U(VI) at the expense of lactate (Sani et al. 2002).

5. *Desulfitobacterium metallireducens*: Growth occurred with lactate using Fe(III) and MnO_2 as electron acceptors, and Cr(VI) is also reduced (Finneran et al. 2002).

6. *Anaeromyxobacter dehalogenans*: This organism that is able to use ortho-substituted phenols for chlororespiration is able to use various forms of Fe(III) (He and Sanford 2003).

All these transformations illustrate important processes for the cycling of organic carbon in sediments where Fe(III) has been precipitated, and it seems likely that comparable geochemical cycles involving manganese (Lovley and Phillips 1988) will also achieve greater prominence (Lovley 1991; Nealson and Myers 1992). It has also been suggested that such organisms could be used for the immobilization of soluble U(VI) in wastewater containing both U(VI) and organic compounds by conversion into insoluble U(IV) (Lovley et al. 1991). The reductases that produce the insoluble Cr(III) from chromate are noted in Part 3 of this chapter.

3.10.2 Organic Acceptors

3.10.2.1 Dehalorespiration

Dehalorespiration in which dehalogenation is coupled to the synthesis of ATP has been demonstrated in a number of bacteria and for a range of chlorinated substrates: this is illustrated in greater detail in Chapter 7, Part 3, and Chapter 9, Parts 1 and 2. Bacteria able to carry out dehalorespiration belong to a wide range of phyla—*Dehalococcoides* to the *Chloroflexi*; *Sulfurospirillum* to the ε-proteobacteria; *Desulfovibrio*, *Desulfobacterium*, *Desulfuromonas*, *Desulfomonile*, and *Geobacter* to the δ-proteobacteria; *Dehalobacter* and *Desulfitobacterium* to the *Firmicutes*. These genera include some that possess a wider range of energy production and are not dependent on dehalorespiration including, *Sulfurospirillum*, *Desulfovibrio*, *Desulfuromonas*, *Geobacter*, and *Desulfitobacterium* (Hiraishi 2008).

Sulfurospirillum (*Dehalospirillum*) *multivorans*, *Dehalobacter restrictus*, *Desulfuromonas chloro-ethenica*, and strains of *Desulfitobacterium* including *Desulf. hafniense* (*frappieri*), and *Desulf. chloro-respirans*) can carry out reductive dechlorination of tetrachloroethene (references in Holliger et al. 1999). *Dehalococcoides ethenogenes* 195 is unusual for several reasons: it is capable of reductively dehaloge-nating tetrachloroethene completely to ethene (Maymó-Gatell et al. 1999; Magnuson et al. 2000) that is discussed further in Chapter 7, and *Dehalococcoides* sp. strain CBDB1 is able to use a range of haloge-nated arenes including chlorobenzenes (Jayachandran et al. 2003) and chlorophenols (Adrian et al. 2007) as electron acceptors. *Desulfitobacterium hafniense* (*frappieri*) has a wide spectrum of substrates that can be partially dechlorinated, including chlorinated phenols, chlorocatechols, chloroanilines, penta-chloronitrobenzene, and pentachloropyridine (Dennie et al. 1998).

3.10.2.1.1 Chloroethenes

The electron donors were generally H_2 or pyruvate, or for *Dehalococcoides ethenogenes* methanol. The chloroethene reductases contain cobalamin and [Fe–S] clusters, and the range of substrates for trichloro-ethene reductase from *Dehalococcoides ethenogenes* includes tribromoethene and bromoethene (Magnuson et al. 2000). In addition, the TCE reductive dehalogenase from this organism is able to debrominate sub-strates containing C_2, C_3, and C_4 or C_5 carbon atoms albeit in decreasing ease: an illustrative reaction is the debromination of tribromoethene to dibromoethenes, vinyl bromide and ethene. Further examples include debrominations carried out by methanogenic bacteria—ethene from 1,2-dibromoethane and ethyne from 1,2-dibromoethene (Belay and Daniels 1987). Details of these dehalogenations have emerged from studies with methanogens. The formation of ethene from 1,2-dichloroethane with hydrogen as electron donor has been demonstrated in cell extracts of *Methanobacterium thermoautotrophicum* DH, and in *Methanosarcina barkeri* has been shown to involve cobalamin and F_{430} using Ti (III) as reductant (Holliger et al. 1992).

3.10.2.1.2 Chlorobenzoates

The dechlorination of 3-chlorobenzoate by *Desulfomonile tiedjei* was the first example of anaerobic dechlorination by a pure culture (DeWeerd et al. 1990) with the concomitant generation of ATP (Dolfing 1990). The dehalogenase from the cytoplasmic membrane of this taxon strain DVCB-1 has been purified and characterized (Ni et al. 1995).

3.10.2.1.3 Chlorophenols

Chlorophenols have also been shown to couple the reductive dechlorination of chlorophenols to the oxi-dation of organic substrates under anaerobic conditions.

1. *Desulfovibrio dechloracetivorans* was able to dechlorinate 2-chlorophenol and 2,6-dichloro-phenol to phenol using a range of substrates: these included acetate, lactate, propionate, and alanine for growth, and sulfate, nitrate, and fumarate as electron acceptors in addition to *ortho*-chlorophenols (Sun et al. 2000).

2. A spore-forming strain of *Desulfitobacterium chlororespirans* coupled the dechlorination of 3-chloro-4-hydroxybenzoate to the oxidation of lactate to acetate, pyruvate, or formate (Sanford et al. 1996). Whereas 2,4,6-trichlorophenol and 2,4,6-tribromophenol supported growth with production of 4-chlorophenol and 4-bromophenol, neither 2-bromophenol nor 2-iodophenol was able to do so. The membrane-bound dehalogenase contains cobalamin, iron and acid-labile sulfur, and is apparently specific for *ortho*-substituted phenols (Krasotkina et al. 2001).

3. *Desulfitobacterium hafniense* (*frappieri*) strain PCP-1 was able to bring about successive dechlorination of pentachlorophenol to 3-chlorophenol (Bouchard et al. 1996). This strain has in addition, a much wider spectrum of substrates for dehalogenation (Dennie et al. 1998). It possesses three reductive dehalogenases, specific respectively for 2,4,6-trichlorophenol (Boyer et al. 2003), 3,5-dichlorophenol (Thibodeau et al. 2004) and pentachlorophenol (Bisaillon et al. 2010).

4. The anaerobic dechlorination of 3-chloro-4-hydroxybenzoate by *Desulfitobacterium chloro-respirans* (Löffler et al. 1996) and of 3-chloro-4-hydroxy-phenylacetate by *Desulfitobacterium hafniense* (Christiansen et al. 1998) has been demonstrated, and a growing culture *Desulfitobacterium hafniense* strain DCB2 was also able to use vanillate or syringate as electron donors for the dechlorination of 3-chloro-4-hydroxyphenylacetate to 4-hydroxyphenylacetate (Neumann et al. 2004).

5. The facultatively anaerobic myxobacter *Anaeromyxobacter dehalogenans* was able to couple the oxidation of acetate, succinate, and formate to the dechlorination of 2-chlorophenol, 2,4-, 2,5-, and 2,6-dichlorophenol (Sanford et al. 2002).

3.10.2.1.4 Chlorinated Arenes

1. Polychlorinated benzenes have been shown to support chlororespiration by *Dehalococcoides* sp. strain CBDB1 using H_2 as electron donor (Hölscher et al. 2003). Hexachlorobenzene was successively dechlorinated to pentachloro-, 1,2,4,5-tetrachloro-, 1,2,4-trichloro-, and 1,4-dichlorobenzene (Jayachandran et al. 2003).

2. An anaerobic ultramicrobacterium strain DF-1 was able to couple the dechlorination of 2,3,4,5-tetrachlorobiphenyl to 2,3,5-trichlorobiphenyl using formate or H_2 as electron donors that could not be replaced by acetate and other carboxylates (May et al. 2008). Dechlorination was, however, dependent on the presence of cultures of *Desulfovibrio* spp or of sterilized extracts.

3.10.2.2 Dimethylsulfoxide

Dimethylsulfoxide (DMSO) can serve as terminal electron acceptor for several organisms under anoxic conditions. The crystal structure of DMSO reductase has been analyzed in *Rhodobacter sphaeroides*, *Rhodobacter capsulatus*, and *E. coli* and, for completeness, a note is added on *Halobacterium* sp. Further comments are given in Part 3 in the context of molybdenum.

1. *Rhodobacter sphaeroides* is a highly versatile purple nonsulfur facultative phototroph that is able to use DMSO as terminal electron acceptor for anaerobic respiration in the absence of light. It contains two molybdopterin guanine dinucleotides and *at* one of the molybdenum redox center, two single electrons are transferred from cytochrome C556 to the substrate dimethyl sulfoxide to generate dimethyl sulfide and (with two protons) water (Schneider et al. 1996). The genes *dorC, dorD, dorB*, and *dorR* that is a response regulator have been identified and sequenced (Shaw et al. 1999).

2. *E. coli* is able to use dimethylsulfoxide as an electron acceptor for anaerobic growth, and dimethylsulfoxide reductase is an integral trimeric membrane enzyme: DmsA is the catalytic subunit containing two molybdopterin guanine dinucleotides, DmsB is an electron transferring units with four [4Fe–4S] clusters and DmsC is a menaquinol oxidizing subunit with eight transmembrane helices (Geijer and Weiner 2004).

3. Genomic analysis of anaerobic respiration has been examined in the archaeon *Halobacterium* sp. strain NRC-1, and an operon *dmsREABCD* in which *dms R* was a regulatory protein, and *dmsEABC* contained the molybdopterin oxidoreductase (Müller and DasSarma 2005).

3.10.2.3 Fumarate as Electron Acceptor

Fumarate reductases (quinol:fumarate reductases, QFR) are involved in anaerobic respiration and are part of the electron transport chain. This is coupled by an electrochemical proton gradient to phosphorylation of ADP with inorganic phosphate catalyzed by ATP synthase. Succinate dehydrogenases (succinate:quinone reductase, SQR) catalyze the opposite oxidation of succinate that is part of the TCA cycle and the aerobic respiratory chain.

3.10.2.3.1 Wolinella succinogenes

Fumarate respiration in *Wolinella succinogenes* has been extensively examined and is the subject of a review (Kröger et al. 2002). The reduction of fumarate to succinate can be coupled with formate dehydrogenase in *Wolinella succinogenes* in the following reactions:

$$\text{Formate} + \text{menaquinone} + \text{H+} \rightarrow CO_2 + \text{menaquinol (formate dehydrogenase)}$$
$$\text{Fumarate} + \text{menaquinol} \rightarrow \text{succinate} + \text{menaquinone (Fumarate reductase).}$$

Fumarate reduction by H_2 of formate as electron donors is catalyzed by the transport chain in the membranes and is coupled to the generation of an electrochemical proton potential Δp. The electron transport chain consists of fumarate reductase, menaquinone, hydrogenase, or formate dehydrogenase, Δp is generated by MK reduction with formate or H_2. Fumarate reductase consists of a catalytic subunit FrdA with conserved flavin adenine dinucleotide and dicarboxylic acid binding residues, a subunit FrdB with the [2Fe–2S], [4Fe–4S], and [3Fe–4S] clusters that interpose FAD with the proximal heme b_p and the distal heme b_d, and the subunit FrdC that carries the membrane anchor unit (Lancaster et al. 1999). The formate dehydrogenase has three subunits of which FdhA contains Mo, and FdhC contains cytochrome *b*. A diagram of the electron transport system and the interaction of the enzymes between the cytoplasm and the periplasm have been provided in a review (Kröger et al. 1992).

3.10.2.3.2 Escherichia coli

Succinate-ubiquinone oxidoreductase (SQR) is part of the TCA cycle and menaquinone-fumarate oxidoreductase (QFR) is used for anaerobic fumarate respiration (Maklashina et al. 1998), and both of them are structurally related membrane-bound enzyme complexes. The fumarate reductase consists of four subunits: a covalently linked FAD in the FrdA subunit, three [Fme-S] clusters in the FrdB subunits that connect with a proximate quinone, and a membrane anchor in the FrdC and FrdD subunits. Electron transfer proceeds successively from FAD to the [2Fe–2S], [4Fe–4S], and [3Fe–4S] clusters, and to the proximate quinone that is separated by 27 Å from a distal quinone (Iverson et al. 1999). Fumarate reductase and succinate dehydrogenase are separate enzymes that use different quinones, although both complexes contain a catalytic domain composed of a subunit with a covalently bound flavin cofactor, the dicarboxylic acid binding site, and an iron–sulfur subunit with three [Fe–S] clusters. Heme of b_{556} is located at the SQR site adjacent to ubiquinone analogous to the position of menaquinone in QFR (Cecchini et al. 2002).

3.10.2.3.3 Geobacter sulfurreducens

The genome contained a heterotrimeric fumarate reductase that was homologous to the fumarate reductase of *W. succinogenes* and a succinate dehydrogenase homologous to that of *B. subtilis*. Examination of mutants that lacked fumarate reductase activity and were unable to grow with fumarate as terminal electron acceptor and lacked also succinate dehydrogenase and was unable to grow with acetate as electron donor and Fe(III) as electron acceptor. It was concluded that the same enzyme was used for both fumarate reduction and succinate dehydrogenation. There was apparently an energetic cost in using the enzyme for oxidation of succinate (Butler et al. 2006).

3.10.2.3.4 Shewanella oneidensis (putrefaciens)

This organism has a complex network of electron transfers that are used for respiration under anaerobic conditions, using fumarate, nitrate, dimethyl sulfoxide, and Fe(III) and Mn(IV) as electron acceptors. CymA is a cytoplasmic membrane-bound tetraheme *c*-type cytochrome that is the principal hydroquinone dehydrogenase (Myers and Myers 1997; Cordova et al. 2011) whereas microbial reduction of insoluble ferric oxide periplasmic electron transfer involves fumarate reductase FccA and the decaheme *c*-type cytochrome MtrA (Schuetz et al. 2009).

In addition to these, fumarate is involved in a few anaerobic degradations.

1. *Desulfitobacterium hafniense* is able to grow anaerobically with fumarate as electron acceptor using the reductant provided by de-*O*-methylation of vanillin and syringate to 3,4-dihydroxy- and 3,4,5-trihydroxybenzoate (Neumann et al. 2004).

2. The degradation of toluene by denitrifying bacteria has been extensively investigated. It involves the direct formation of benzylsuccinate by the reaction of toluene with fumarate and has been delineated in several bacteria including *Thauera aromatica* (Biegert et al. 1996), a denitrifying strain designated *Pseudomonas* sp. strain T2 (Beller and Spormann 1997a), and a sulfate-reducing bacterium strain PRTOL1 (Beller and Spormann 1997b): this is apparently a widespread reaction. These are mediated by free radicals that are discussed in Section 3.22 of this chapter and exemplified in Chapter 8.

3.10.2.4 Caffeate as Electron Acceptor

The anaerobic acetogen *Acetobacterium woodii* was able to couple the reduction of 3,4-dihydroxy-cinnamate (caffeate) to 3,4-dihydroxypropionate with substrate oxidation (Dilling et al. 2007). Caffeate is activated to caffeyl-coenzyme A by an enzyme encoded by CarB that is induced by 4-hydroxycinnamates including, not only caffeate but also 4-hydroxycinnamate, 4-hydroxy-3-methoxycinnamate, and 4-hydroxy-3,5-dimethoxycinnamate, although activity was less for these other substrates (Hess et al. 2011). The regulation of alternative substrates has been examined in *Acetobacterium woodii*, and caffeate-induced cells were able to reduce ferulate (3-methoxy-4-hydroxycinnamate) in the absence of an external reductant, so that the enzymes for oxidation of the methyl group were induced. Reduction of caffeate with ATP synthesis could also be accomplished with H_2 in this organism (Imkamp et al. 2007).

3.10.2.5 Carnitine as Electron Acceptor

Carnitine is transformed by *Escherichia coli* by dehydration to crotonobetaine and reduction to γ-butyrobetaine (Eichler et al. 1994) that is consistent with the use of carnitine and crotonobetaine as electron acceptors during anaerobic growth of *E. coli* (Walt and Kahn 2002). In *Salmonella typhimurium* LT_2, it was shown that anaerobic growth in a complex medium was stimulated by the reduction to γ-butyrobetaine of cronobetaine that was produced by dehydration of L-carnitine (Seim et al. 1982). The metabolism of carnitine is discussed in detail in Chapter 7.

3.10.2.6 Alkane Sulfonates as Terminal Electron Acceptor

A few aliphatic sulfonates such as 2-hydroxyethylsulfonate, alanine-3-sulfonate, and acetaldehyde-2-sulfonate are able to serve as sulfur sources and electron acceptors during anaerobic growth of some sulfate-reducing bacteria when lactate is supplied as the carbon source (Lie et al. 1996). Several sulfite-reducing species of *Desulfitobacterium* are able to use 2-hydroxyethanesulfonate as terminal electron acceptor producing acetate and sulfide (Lie et al. 1999). It is relevant to note that species of *Desulfitobacterium* are also able to use chlorinated organic compounds and arsenate as terminal electron acceptors.

3.10.2.7 Nitroalkanes as Electron Acceptor

An anaerobic organism isolated from the rumen by enrichment for the metabolism of the toxic aglycone of miserotoxin (3-nitro-1-propyl-β-D-glucopyramnoside) was assigned to *Denitrobacterium detoxificans* (Anderson et al. 2000). It was able to use a number of nitroalkanes including 3-nitropropanol, 3-nitropropionate, nitroethane, and 2-nitrobutane as electron acceptors during growth with lactate that was converted into acetate.

3.10.2.8 Humic Substances in Redox Reactions

Humic acids and the corresponding fulvic acid are complex polymeric substances whose structures remain incompletely resolved. They represent major forms of organic matter in soils and sediments, and it is accepted that the structures of humic acid contain oxygenated arenes including quinones that can function as electron acceptors, while reduced humic acid may carry out reductions. Although oxidized anthraquinone-2,6-disulfonate (AQDS) and the hydroquinone (H_2AQDS) have been used as an artificial electron acceptor and reductant as surrogates for humic acids, they differ importantly in redox potential as well as in aqueous solubility.

Bacteria are able to reduce Fe(III), AQDS, and oxidized humic acids (Hum^{Ox}), and they belong to several genera including strains of *Geobacter* and *Desulfitobacterium*. Conversely, other bacteria can make use of Hum^{Red} or H_2AQDS to carry out reductions. There are some qualifications that should be taken into consideration: (a) commercial samples of humic acids do not have the same structure as natural humic acids, and (b) the redox potential of anthraquinone-2,6-disulfonate that is frequently used as a surrogate does not have a redox potential comparable to that of humic acids.

1. The frequency of humic-reducing bacteria has been assessed using anthraquinone-2,6-disulfonate (AQDS) to mimic humic material as electron acceptor, with acetate as electron donor. The organisms that were isolated belonged to the *Geobacteraceae* and could, therefore, be significant as humic acid reducers (Coates et al. 1998). Specifically, the anaerobic *Geobacter metallireducens* and *G. grbiciae* could oxidize toluene and acetate using Fe(III) or anthraquinone-2,6-disulfonate (AQDS) as electron acceptors (Coates et al. 2001a).

2. The microbial degradation of contaminants under anaerobic conditions using humic acids (Hum^{ox}) as electron acceptors has been demonstrated. These included the oxidations: (a) chloroethene and 1,2-dichloroethene to CO_2 that was confirmed using ^{14}C-labeled substrates (Bradley et al. 1998), and (b) toluene to CO_2 with AQDA or humic acid as electron acceptors (Cervantes et al. 2001), (c) transformation of hexahydro-1,3,5-trinitro-1,3,5-triazine (RDX) using *Geobacter metallireducens* and humic material with AQDS as electron shuttle (Kwon and Finneran 2006), and degradation of RDX and HMX by *Clostridium* sp. strain EDB2 in the presence of Fe(III), humic acid and AQDS in which both direct interaction and electron-shuttling were involved (Bushan et al. 2006).

3. Bacteria were isolated using reduced anthraquinone-2,6-disulfonate (H_2AQDS) as electron donor and nitrate as electron acceptor (Coates et al. 2002) The organisms belonged to the α,β,γ and δ subdivision of the *Proteobacteria*, and were able to couple the oxidation of H_2AQDS to the reduction of nitrate with acetate as the carbon source. In addition a number of C_2 and C_3 substrates could be used including propionate, butyrate, fumarate, lactate, citrate, and pyruvate.

   Dechloromonas strain JJ was isolated using anthraquinone-2,6-disulfonate as electron donor and nitrate as electron acceptor, and was able to oxidize benzene anaerobically using nitrate as electron acceptor (Coates et al. 2001b).

4. The role of humic acid as electron acceptor has been examined in the strict anaerobes *Desulfitobacterium* strain PCE-1, *Desulfovibrio* strain G11 and *Methanospirillum hungatei* strain JF (Cervantes et al. 2002). They were able to oxidize H_2 using humic acid (hum^{ox}) or AQDS as electron acceptors, and lactate could be oxidized to acetate by the first two using hum^{red}.

5. The presence of humic acid altered the fermentation pattern of several anaerobic bacteria (Benz et al. 1998). For *Propionibacterium freudenreichii* using lactate as substrate, the ratio of propionate to acetate that was produced diminished from 1.8 to 1.0 in the presence of humic acid, while during the fermentation of glucose by *Enterococcus cecorum* there was an increase in lactate and a decrease in acetate. This study underlined the significance in soils and sediments of iron-reducing and fermenting bacteria in electron transfer via humic acids.

6. *Geobacter metallireducens*, *G. sulphurreducens*, and *Geothrix fermentans* are capable of reducing Fe(III) and oxidized humic acids (Hum^{ox}), and conversely they could oxidize reduced humic acid (Hum^{Red}) or H_2AQDS using nitrate or fumarate as electron acceptors (Lovley et al. 1999). It was shown that in anaerobic bicarbonate medium, *Wolinella succinogenes* was able to oxidize H_2AQDS using nitrate, fumarate, arsenate or selenate as electron acceptors. In the presence of H_2AQDS, *G. metallireducens* and *Wollinella* were able to oxidize acetate as electron donor and fumarate as electron acceptor.

7. *Desulfitobacterium metallireducens* is notable for the range of reactions that it can accomplish: (a) dehalorespiration with 3-chloro-4-hydroxyphenylacetate, trichloroethene, and tetrachloroethene,

(b) oxidation of lactate to acetate $+ CO_2$ at the expense of AQDS reduction, (c) oxidation of lactate using Fe(III) citrate, Mn (IV), S^0, and Hum^{ox} as electron acceptors (Finneran et al. 2002).

8. The anaerobe *Carboxydothermus ferrireducens* is able to carry out a range of reactions: these include (a) dehydrogenation of carbon monoxide to CO_2 and H_2 using water as electron donor, (b) using AQDS as electron acceptor it can carry out the oxidation of CO, H_2, formate, lactate, and glycerol (Henstra and Stams 2004), and (c) in the presence of Fe(III) as electron acceptor it can grow lithoautotrophically with H_2 as electron donor with CO_2 as the sole carbon source (Slobodkin et al. 2006).

9. In a wider context, redox mediators have been implicated in the anaerobic reduction of azo dyes by several bacteria including *Sphingomonas xenophaga* BN6 and *E. coli* strain K-12 (Rau et al. 2002), and the specific role of 4-aminonaphtho-1,2-quinone and 4-ethanolaminonaphtho-1,2-quinone that are produced during the degradation of naphthalene-2-sulfonate by *Sphingomonas xenophaga* BN6 (Keck et al. 2002).

3.11 References

Achenbach LA, U Michaelidou, RA Bruce, J Fryman, JD Coates. 2001. *Dechloromonas agitata* gen. nov., sp. nov. and *Dechloromonas suillum* gen. nov., sp. nov., two novel environmentally dominant (per)chlorate-reducing bacteria and their phylogenetic position. *Int J Syst Evol Microbiol* 51: 527–533.

Adak S, KS Aulak, DJ Stuehr. 2002a. Direct evidence for nitric oxide production by a nitric oxide synthase-like protein from *Bacillus subtilis*. *J Biol Chem* 277: 16167–16171.

Adak S, AM Bilwes, K Panda, D Hosfield, KS Aulak, JF McDonald, JA Tainer, ED Getzoff, BR Crane, DJ Stuehr. 2002b. Cloning, expression, and characterization of a nitric oxide synthase protein from *Dienococcus radiodurans*. *Proc Natl Acad USA* 99: 107–112.

Adrian L, SK Hansen, JM Fung, H Gölrisch, SH Zinder. 2007. Growth of *Dehaloccoides* strains with chlorophenols as electron acceptors. *Environ Sci Technol* 41: 2318–2323.

Afring RP, BF Taylor. 1981. Aerobic and anaerobic catabolism of phthalic acid by a nitrate-respiring bacterium. *Arch Microbiol* 130: 101–104.

Altenschmidt U, G Fuchs. 1991. Anaerobic degradation of toluene in denitrifying *Pseudomonas* sp.: Indication for toluene methylhydroxylation and benzoyl-CoA as central aromatic intermediate. *Arch Microbiol* 156: 152–158.

Amachi S, N Kawaguchi, Y Muramatsu, S Tsuchiya, Y Watanabe, H Shinoyama, T Fujii. 2007. Dissimilatory iodate reduction by marine *Pseudomonas* sp. strain SCT. *Appl Environ Microbiol* 73: 5725–5730.

Anders H-J, A Kaetzke, P Kämpfer, W Ludwig, G Fuchs. 1995. Taxonomic position of aromatic-degrading denitrifying pseudomonad strains K 172 and KB 740 and their description as new members of the genera *Thauera*, as *Thauera aromatica* sp. nov., and *Azoarcus*, as *Azoarcus evansii* sp. nov., respectively, members of the beta subclass of the Proteobacteria. *Int J Syst Bacteriol* 45: 327–333.

Anderson RC, MA Rasmussen, NS Jensen, MJ Allison. 2000. *Denitrobacterium detoxificans* gen. nov. sp. nov., a ruminal bacterium that respires on nitro compounds. *Int J Syst Evol Microbiol* 50: 633–638.

Anderson RT et al. 2003. Stimulating the in site activity of *Geobacter* species to remove uranium from the groundwater of a uranium-contaminated aquifer. *Appl Environ Microbiol* 69: 5884–5891.

Bak F, N Pfennig. 1987. Chemolithotrophic growth of *Desulfovibrio sulfodismutans* sp. nov. by disproportionation of inorganic sulfur compounds. *Arch Microbiol* 147: 184–189.

Beller HR, AM Spormann. 1997a. Anaerobic activation of toluene and *o*-xylene by addition to fumarate in denitrifying strain T. *J Bacteriol* 179: 670–676.

Balk M, T van Gelder, SA Weelink, AJM Stams. 2008. (Per)chlorate reduction by the thermophilic bacterium *Moorella perchloratireducens* sp. nov., isolated from underground gas storage. *Appl Environ Microbiol* 74: 403–409.

Bao H, B Gu. 2004. Natural perchlorate has a unique oxygen isotope signature. *Environ Sci Technol* 38: 5073–5077.

Barrett EL, MA Clark. 1987. Tetrathionate reduction and production of hydrogen sulfide from thiosulfate. *Microbiol Revs* 51: 192–205.

Battaglia-Brunet F, C Joulian, F Garrido, M-C Dictor, D Morin, K Coupland, DB Johnson, KB Hallberg, P Baranger. 2006. Oxidation of arsenite by *Thiomonas* strains and characterization of *Thiomonas arsenivorans* sp. nov. *Antonie van Leeuwenhoek* 89: 99–108.

Belay N, L Daniels. 1987. Production of ethane, ethylene and acetylene from halogenated hydrocarbons by methanogenic bacteria. *Appl Environ Microbiol* 53: 1604–1610.

Beller HR, AM Spormann. 1997b. Benzylsuccinate formation as a means of anaerobic toluene activation by sulfate-reducing strain PRTOL1. *Appl Environ Microbiol* 63: 3729–3731.

Bender KS, SM O'Connor, R Chakraborty, JD Coates, LA Achenbach. 2002. Sequencing and transcriptional analysis of the chlorite dismutase gene of *Dechloromonas agitata* and its use as a metabolic probe. *Appl Environ Microbiol* 68: 4820–4826.

Bender KS, MR Rice, WH Fugate, JD Coates, LA Achenbach. 2004. Metabolic primers for detection of (per) chlorate-reducing bacteria in the environment and phylogenetic analysis of *cld* gene sequences. *Appl Environ Microbiol* 70: 5651–5658.

Bender KS, C Shang, R Chakraborty, SM Belchik, JD Coates, LA Achenbach. 2005. Identification, characterization, and classification of genes encoding perchlorate reductase. *J Bacteriol* 187: 5090–5096.

Bennett MS, Z Guan, M Laurberg, X-D Su. 2001. *Bacillus subtilis* arsenate reductase is structurally and functionally similar to low molecular weight protein tyrosine phosphatases. *Proc Natl Acad Sci USA* 98: 13577–13582.

Benz M, B Schink, A Brune. 1998. Humic acid reduction by *Propionibacterium freudenreichii* and other fermentative bacteria. *Appl Environ Microbiol* 64: 4507–4512.

Bick JA, JJ Dennis, GJ Zylstra, J Nowack T Leustek. 2000. Identification of a new class of 5′ adenylsulfate (APS) reductases from sulfate-assimilating bacteria. *J Bacteriol* 182: 135–142.

Biegert T, G Fuchs, J Heider. 1996. Evidence that anaerobic oxidation of toluene in the denitrifying bacterium *Thauera aromatica* is initiated by formation of benzylsuccinate from toluene and fumarate. *Eur J Biochem* 238: 661–668.

Bisaillon A, R Beaudet, F Lépine, E Déziel, R Villemur. 2010. Identification and characterization of a novel CprA reductive dehalogenase specific for highly chlorinated phenols from *Desulfitobacterium hafniense* strain PCP-1. *Appl Environ Microbiol* 76: 7536–7754.

Blum J, AB Bindi, J Buzzelli, JF Stolz, RS Oremland. 1998. *Bacillus arsenicoselenatis* sp. nov., and *Bacillus selenitrificans* sp. nov.: Two haloalkaliphiles from Mono Lake, California that respire oxyanions of selenium and arsenic. *Arch Microbiol* 171: 19–30.

Blum, JS, S Han, B Lanoil, C Saltikov, B Witte, FR Tabita, S Langley, TJ Beveridge, L Jahnke, RS Oremland. 2009. Ecophysiology of "*Halarsenatibacter silvermanii*" strain SLAS-1^T, gen. nov., sp. nov., a facultative chemoautotrophic arsenate respirer from salt-saturated Searles Lake, California. *Appl Environ Microbiol* 75: 1950–1960.

Bouchard B, R Beaudet, R Villemur, G MacSween, F Lépine, J-G Bisaillon. 1996. Isolation and characterization of *Desulfitobacterium frappieri* sp. nov., an anaerobic bacterium which reductively dechlorinates pentachloro-phenol to 3-chlorophenol. *Int J Syst Bacteriol* 46: 1010–1015.

Boyer A, R-P Bélanger, M Saucier, R Villemur, F Lépine, P Juteau, R Baudet. 2003. Purification, cloning and sequencing of an enzyme mediating the reductive dechlorination of 2,4,6-trichlorophenol from *Desulfitobacterium frappieri* PCP-1. *Biochem J* 373: 297–303.

Bradley PM, FH Chapelle, DR Lovley. 1998. Humic acids as electron acceptors for anaerobic microbial oxidation of vinyl chloride and dichloroethene. *Appl Environ Microbiol* 68: 3102–3103.

Bregnard TP-A, A Häner, P Höhener, J Zeyer. 1997. Anaerobic degradation of pristane in nitrate-reducing microcosms and enrichment cultures. *Appl Environ Microbiol* 63: 2077–2081.

Butler JE, RH Glaven, A Esteve Núñez, C Núñez, ES Shelobolina, DR Bond, DR Lovley. 2006. Genetic characterization of a single bifunctional enzyme for fumarate reduction and succinate oxidation in *Geobacter sulfurreducens* and engineering of fumarate reduction in *Geobacter metallireducens*. *J Bacteriol* 188:450–455.

Burns JL, TJ DiChristina. 2009. Anaerobic respiration of elemental sulfur and thiosulfate by *Shewanella oneidensis* MR-1 requires *psr*A, a homologue of the *phs*A gene of *Salmonella enterica* serovar *typhimurium* LT2. *Appl Environ Microbiol* 75: 5209–5217.

Bushan A, A Halasz, J Hawari. 2006. Effect of iron (III), humic acids and anthraquinone-2,6-disulfonate on biodegradation of cyclic nitaramines by *Clostridium* sp. EDB2. *J Appl Microbiol* 100: 555–563.

Caccavo F, JD Coates, R Rossello-Mora, W Ludwig, KH Schleifer, DR Lovley, MJ McInerney. 1996. *Geovibrio ferrireducens*, a phylogenetically distinct dissimilatory Fe(III)-reducing bacterium. *Arch Microbiol* 165: 370–376.

Caccavo F, RP Blakemore, DR Lovley. 1992. A hydrogen-oxidizing, Fe(III)-reducing microorganism from the Great Bay estuary, New Hampshire. *Appl Environ Microbiol* 58: 3211–3216.

Carpentier W, K Sandra, I De Smet, A Brigé, L De Smet, J Van Beeuman. 2003. Microbial reduction and precipitation of vanadium by *Shewanella oneidensis*. *Appl Environ Microbiol* 69: 3636–3639.

Carpentier W, L De Smet, J Van Beeuman, A Brigé. 2005. Respiration and growth of *Shewanella oneidensis* MR-1 using vanadate as the sole electron acceptor. *J Bacteriol* 187: 3293–3301.

Cecchini G, I Schhröder, RP Gunsalus, E Maklashina. 2002. Succinate dehydrogenase and fumarate reductase from *Escherichia coli*. *Biochim Biophys Acta* 1553: 140–157.

Cervantes FJ, FAM de Bok, T Duong-Dac, AJM Stams, G Lettinga, JA Field. 2002. Reduction of humic substances by halorespiring, sulphate-reducing and methanogenic microorganisms. *Environ Microbiol* 4: 51–57.

Cervantes FJ, W Dijksma, T Duong-Dac, A Ivanova, G Lettinga, JA Field. 2001. Anaerobic mineralization of toluene by enriched sediments with quinones and humus as terminal electron acceptors. *Appl Environ Microbiol* 67: 4471–4478.

Chakraborty R, JD Coates. 2005. Hydroxylation and carboxylation—Two crucial steps of anaerobic benzene degradation by *Dechloromonas* strain RCB. *Appl Environ Microbiol* 71: 5427–5432.

Chakraborty R, SM O'Connor, E Chan, JD Coates. 2005. Anaerobic degradation of benzene, toluene, ethylbenzene, and xylene compounds by *Dechloromonas* strain RCB. *Appl Environ Microbiol* 71: 8649–8655.

Champine JE, S Goodwin. 1991. Acetate catabolism in the dissimilatory iron-reducing isolate GS-15. *J Bacteriol* 173: 2704–2706.

Chen Y, JPN Rosazza. 1995. Purification and characterization of nitric oxide synthase NOSNoc from a *Nocardia* sp. *J Bacteriol* 177: 5122–5128.

Childers SE and DR Lovley. 2001. Differences in Fe(III) reduction in the hyperthermophilic archaeon, *Pyrobaculum islandicum*, versus mesophilic Fe(III)-reducing bacteria. *FEMS Microbiol Lett* 195: 253–258.

Christiansen N, BK Ahring, G Wohlfarth, G Diekert. 1998. Purification and characterization of the 3-chloro-4-hydroxyphenylacetate reductive dehalogenase of *Desulfitobacterium hafniense*. *FEBS Lett* 436: 159–162.

Coates JD, VK Bhupathiraju, LA Achenbach, MJ McInerney, DR Lovely. 2001a. *Geobacter hydrogenophilus*, *Geobacter chapellei* and *Geobacter grbiciae*, three new, strictly anaerobic, dissimilatory Fe(III) reducers. *Int J Syst Evolut Microbiol* 51: 581–588.

Coates JD, R Chakraborty, JG Lack, SM O'Connor, KA Cole, KS Bender, LA Achebach. 2001b. Anaerobic benzene oxidation coupled to nitrate reduction in pure cultures by two strains of *Dechloromonas*. *Nature* 411: 1039–1043.

Coates JD, KA Cole, R Chakraborty, SM O'Connor, LA Achenbach. 2002. Diversity and ubiquity of bacteria capable of utilizing humic substances as electron donors for anaerobic respiration. *Appl Environ Microbiol* 68: 2445–2452.

Coates, JD, DJ Ellis, EL Blunt-Harris, CV Gaw, EE Roden, DR Lovley. 1998. Recovery of humic-reducing bacteria from a diversity of environments. *Appl Environ Microbiol* 64: 1504–1509.

Coates JD, U Michaelidou, RA Bruce, SM O'Connor, JN Crespi, LA Achenbach. 1999. Ubiquity and diversity of dissimilatory (per)chlorate-reducing bacteria. *Appl Environ Microbiol* 65: 5234–5241.

Cordova CD, MFR Schicklberger, Y Yum AM Spormann. 2011. Partial functional replacement of CymA by SirCD in *Shewanella oneidensis* MR-1. *J Bacteriol* 193: 2312–2321.

Corke CT, NJ Bunce, A-L Beaumont, RL Merrick. 1979. Diazonium cations as intermediates in the microbial transformations of chloroanilines to chlorinated biphenyls, azo compounds, and triazines. *J Agric Food Chem* 27: 644–646.

Cravo-Laureau C, R Matheron, J-L Cayol, C Joulian, A Hirschler-Réa. 2004a. *Desulfatibacillum aliphaticivorans* gen. nov., sp. nov., an *n*-alkane and *n*-alkene-degrading sulfate-reducing bacterium. *Int J Syst Evol Microbiol* 54: 77–83.

Cravo-Laureau C, R Matheron, C Joulian, J-L Cayol, A Hirschler-Réa. 2004b. *Desulfatibacillum alkenivorans* sp. nov., a novel *n*-alkene-degrading sulfate-reducing bacterium, and emended description of the genus *Desulfatibacillum*. *Int J Syst Evol Microbiol* 54: 1639–1642.

Criddle CS, JT DeWitt, D Grbic-Galic, PL McCarty. 1990. Transformation of carbon tetrachloride by *Pseudomonas* sp strain KC under denitrifying conditions. *Appl Environ Microbiol* 56: 3240–3246.

Das A, R Silaghi-Dumitrescu, LG Ljungdahl, MD Kurz. 2005. Cytochrome *bd* oxidase, oxidative stress, and dioxygen tolerance of the strictly anaerobic bacterium *Moorella thermoacetica*. *J Bacteriol* 187: 2020–2029.

Dasgupta PK, PK Martinelango, WA Jackson, K Tian, RW Tock, S Rajagopalan. 2005. The origin of naturally occurring perchlorate: The role of atmospheric processes. *Environ Sci Technol* 39: 1569–1575.

De Luca G, P de Philip, Z Dermoun, M Rouseet, A Verméglio. 2001. Reduction of Technetium (VII) by *Desulfovibrio fructosovorans* is mediated by the nickel–iron hydrogenase. *Appl Environ Microbiol* 67: 468–475.

DeMoll-Decker H, JM Macy. 1993. The periplasmic nitrite reductase of *Thauera selenatis* may catalyze the reduction of selenite to elementary selenium. *Arch Microbiol* 160: 241–247.

Dennie D, I Gladu, F Lépine, R Villemur, J-G Bisaillon, R Beaudet. 1998. Spectrum of the reductive dehalogenation activity of *Desulfitobacterium frappieri* PCP-1. *Appl Environ Microbiol* 64: 4603–4606.

DeWeerd KA, L Mandelco, RS Tanner, CR Woese, JM Suflita. 1990. *Desulfomonile tiedjei* gen. nov. and sp. nov., a novel anaerobic, dehalogenating, sulfate-reducing bacterium. *Arch Microbiol* 154: 23–30.

Dey S, BP Rosen. 1995. Dual mode of energy coupling by the oxyanion-translocating ArsB protein. *J Bacteriol* 177: 385–389.

DiChristina TJ. 1992. Effects of nitrate and nitrite on dissimilatory iron reduction by *Shewanella putrefaciens* 200. *J Bacteriol* 174: 1891–1896.

Dilling S, F Imkamp, S Schmidt, V Müller. 2007. Regulation of caffeate respiration in the acetogenic bacterium *Acetobacterium woodii*. *Appl Environ Microbiol* 73: 3630–3636.

Dolfing J. 1990. Reductive dechlorination of 3-chlorobenzoate is coupled to ATP production and growth in an anaerobic bacterium, strain DCB-1. *Arch Microbiol* 153: 264–266.

Dridge E J, CA Watts, BJN Jepson, K Line, JM Santini, DJ Richardson, CS Butler. 2007. Investigation of the redox centres of periplasmic selenate reductase from *Thauera selenatis* by EPR spectroscopy. *Biochem J* 408: 19–28.

Eichler K, W-H Schunck, H-P Kleber, M-A Mandrand-Berthelot. 1994. Cloning, nucleotide sequence, and expression of the *Escherichia coli* gene encoding carnitine dehydratase. *J Bacteriol* 176: 2970–2973.

Ellis PJ, T Conrads, R Hille, P Kuhn. 2001. Crystal structure of the 100 kDa arsenite oxidase from *Alcaligenes faecalis* in two crystal forms at 1.64Å and 2.0 3Å. *Structure* 9: 125–132.

Eschenbrenner M, F Covès, M Fontecave. 1995. The flavin reductase activity of the flavoprotein component of sulfite reductase from *Escherichia coli*. *J Biol Chem* 270: 20550–20555.

Ettwig KF et al. 2010. Nitrite-driven anaerobic methane oxidation by oxygenic bacteria. *Nature* 464: 543–548.

Evans PJ, DT Mang, KS Kim, LY Young. 1991a. Anaerobic degradation of toluene by a denitrifying bacterium. *Appl Environ Microbiol* 57: 1139–1145.

Evans PJ, DT Mang, LY Young. 1991b. Degradation of toluene and m-xylene and transformation of *o*-xylene by denitrifying enrichment cultures. *Appl Environ Microbiol* 57: 450–454.

Feldman PL, OW Griffith, DJ Stuehr. 1993. The surprising life of nitric oxide. *Chem Eng News* 71(51): 26–39.

Finneran KT, HM Forbuch, CV Gaw VanPraagh, DK Lovley. 2002. *Desulphitobacterium metallireducens* sp. nov., an anaerobic bacterium that couples growth with the reduction of humic acids as well as chlorinated compounds. *Int J Syst Bacteriol* 52: 1929–1935.

Finster K, W Liesack, B Thamdrup. 1998. Elemental sulfur and thiosulfate disproportionation by *Desulfocapsa sulfoexigens* sp. nov., a new anaerobic bacterium isolated from marine surface sediment. *Appl Environ Microbiol* 64: 119–125.

Francis CA, AY Obraztsova, BM Tebo. 2000. Dissimilatory metal reduction by the facultative anaerobe *Pantoea agglomerans*. *Appl Environ Microbiol* 66: 543–548.

Fredrickson JK, HM Kostandarithes, SW Li, AE Plymale, MJ Daly. 2000. Reduction of Fe(III), Cr(VI), U(VI), and Tc(VII) by *Deinococcus radiodurans* R1. *Appl Environ Microbiol* 66: 2006–2011.

Friedrich CG, D Rother, F Bardischewsky, A Quentmeier, J Fischer. 2001. Oxidation of reduced inorganic sulfur compounds by bacteria: Emergence of a common mechanism? *Appl Environ Microbiol* 67: 2873–2882.

Ganesh R, KG Robinson, GD Reed, G Sayler. 1997. Reduction of hexavalent uranium from organic complexes by sulfate- and iron-reducing bacteria. *Appl Environ Microbiol* 63: 4385–4391.

Geijer P, JH Weiner. 2004. Glutamate 87 is important for menaquinol binding *in DmdC* of the DMSO reductase (*DmsABC*) from *Escherichia coli*. *Biochim Biophys Acta* 1660: 66–74.

Gihring TM, JF Banfield. 2001. Arsenite oxidation and arsenate respiration by a new *Thermus* isolate. *FEMS Microbiol Lett* 204: 335–340.

Harder J, C Probian. 1997. Anaerobic mineralization of cholesterol by a novel type of denitrifying bacterium. *Arch Microbiol* 167: 269–274.

Hau HH, A Gilbert, D Coursolle, JA Gralnick. 2008. Mechanism and consequences of anaerobic respiration of cobalt by *Shewanella oneidensis* strain MR-1. *Appl Environ Microbiol* 74: 6880–6886.

He Q, RA Sanford. 2003. Characterization of Fe(III) reduction by chlororespiring *Anaeromyxobacter dehalogenans*. *Appl Environ Microbiol* 69: 2712–2718.

Heinzinger NK, SY Fujimoto, MA Clark, MS Moreno, EL Barrett. 1995. Sequence analysis of the *phs* operon in *Salmonella typhimurium* and the contribution of thiosulfate reduction to anaerobic energy metabolism. *J Bacteriol* 117: 2813–2830.

Hensel M, AP Hinsley, T Nikolaus, G Sawers, BC Berks. 1999. The genetic basis of tetrathionate respiration in *Salmonella typhimurium*. *Mol Microbiol* 32: 275–287.

Henstra AM, AJM Stams. 2004. Novel physiological features of *Carboxydothermus hydrogenoformans* and *Thermoterrabacterium ferrireducens*. *Appl Environ Microbiol* 70: 7236–7240.

Hess V, S Vitt, V Müller. 2011. A caffeyl-coenzyme A synthetase initiates caffeate activation prior to caffeate reduction in the acetogenic bacterium *Acetobacterium woodii*. *J Bacteriol* 193: 971–978.

Hiraishi A. 2008. Biodiversity of dehalorespiring bacteria with special emphasis on polychlorinated biphenyl/dioxin dechlorinators. *Microbes Environ* 23: 1–12.

Hoeft SE, JS Blum, JF Stolz, FR Tabita, B Witte, GM King, JM Santini, RS Oremland. 2007. *Alkalilimnicola ehrlichii* sp. nov., a novel arsenite-oxidizing haloalkaliphilic gammaproteobacterium capable of chemoautotrophic or heterotrophic growth with nitrate or oxygen as the electrom acceptor. *Int J Syst Evol Microbiol* 57: 504–512.

Hoeft SE, TR Kulp, S Han, B Lanoil, RS Oremland. 2010. Coupled arsenotrophy in a hot spring photosynthetic biofilm at Mono Lake, California. *Appl Environ Microbiol* 76: 4633–4639.

Holliger C, G Schraa, E Stuperich, AJM Stams, AJB Zehnder. 1992. Evidence for the involvement of corrinoids and factor F_{430} in the reductive dechlorination of 1,2-dichloroethane by *Methanosarcina barkeri*. *J Bacteriol* 174: 4427–4434.

Holliger C, G Wohlfarth, G Diekert. 1999. Reductive dechlorination in the energy metabolism of anaerobic bacteria. *FEMS Microbiol Revs* 22: 383–398.

Hölscher T, H Görisch, L Adrian. 2003. Reductive dehalogenation of chlorobenzene congeners in cell extracts of *Dehalococcoides* sp. strain CBDB1. *Appl Environ Microbiol* 69: 2999–3001.

Hu BL, D Rush, E van der Biezen, P Zheng, M van Mullekom, S Schouten, JS Sinninghe Damsté, AJP Smolders, MSM Jetten, B Kartal. 2011. New anaerobic, ammonium-oxidizing community enriched from peat soil. *Appl Environ Microbiol* 77: 966–971.

Huang CJ, EL Barrett. 1991. Sequence analysis and expression of the *Salmonella typhimurium asr* operon encoding production of hydrogen sulfide from sulfite. *J Bacteriol* 183: 1544–1553.

Hutchins SR. 1991. Biodegradation of monoaromatic hydrocarbons by aquifer microorganisms using oxygen, nitrate or nitrous oxide as the terminal electron acceptors. *Appl Environ Microbiol* 57: 2403–2407.

Imkamp F, E Biegel, E Jayamani, W Buckel, V Müller. 2007. Dissection of the caffeate respiratory chain in the acetogen *Acetobacterium woodii*: Identification of an Rnf-type NADH dehydrogenase as a potential coupling site. *J Bacteriol* 189: 8145–8153.

Istok JD, JM Senko, LR Krumholz, D Watson, MA Bogle, A Peacock, Y-J Chang, DC White. 2004. *In situ* bioremediation of technetium and uranium in a nitrate-contaminated aquifer. *Environ Sci Technol* 38: 468–475.

Iverson TM, C Luna-Chavez, G Cecchini, DC Rees. 1999. Structure of the *Escherichia coli* fumarate reductase respiratory complex. *Science* 284: 1961–1966.

Jackson BE, MJ McInerney. 2000. Thiosulfate disproportionation by *Desulfotomaculum thermobenzoicum*. *Appl Environ Microbiol* 66: 3650–3653.

James SL. 1995. Role of nitric oxide in parasitic infections. *Microbiol Revs* 59: 533–547.

Jayachandran G, H Görisch, L Adrian. 2003. Dehalorespiration with hexachlorobenzene and pentachlorobenzene by *Dehaloccoides* sp. strain CBDB1. *Arch Microbiol* 180: 411–416.

Ji G, S Silver. 1992. Reduction of arsenate to arsenite by the ArsC protein of the arsenic resistance operon of *Staphylococcus aureus* plasmid pI258. *Proc Natl Acad Sci USA* 89:9474–9478.

Kappler U, PV Bernhardt, J Kilmartin, MJ Riley, J Techner, KJ McKenzie, GR Hanson. 2008. SoxAX cytochromes, a new type of heme copper protein involved in bacterial energy generation from sulfur compounds. *J Biol Chem* 283: 22206–22214.

Kashefi K, DR Lovley. 2000. Reduction of Fe(III), Mn(IV), and toxic metals at 100°C by *Pyrobaculum islandicum*. *Appl Environ Microbiol* 66: 1050–1056.

Kashevi K, JM Tor, DE HOlmes, CVG Van Praagh, A-L Reysenbach, DR Lovley. 2002. *Geoglobus ahangari* gen. nov., sp. nov., a novel hyperthermophilic archaeon capable of oxidizing organic acids and growing autotrophically on hydrogen with Fe(III) serving as sole electron acceptor. *Int J Syst Evol Microbiol* 52: 719–728.

Kashyap DR, LM Botero, WL Franck, DJ Hassett, TR McDermott. 2006. Complex regulation of arsenite oxidation in *Agrobacterium tumefaciens*. *J Bacteriol* 188: 6028–6093.

Kaufmann F, DR Lovley. 2001. Isolation and characterization of a soluble NADPH-dependent Fe(III) reductase from *Geobacter sulfurreducens*. *J Bacteriol* 183: 4468–4476.

Keck A, J Rau, T Reemtsma, R Mattes, A Stolz, J Klein. 2002. Identification of quinoid redox mediators that are formed during the degradation of naphthalene-2-sulfonate by *Sphingomonas xenophaga* BN6. *Appl Environ Microbiol* 68: 4341–4349.

Kengen SWM, GB Rikken, WR Hagen, CG van Ginkel, ALM Stams. 1999. Purification and characterization of (per)chlorate reductase from the chlorate-respiring strain GR-1. *J Bacteriol* 181: 6706–6711.

Kers JA, MJ Wach, SB Krasnoff, J Widom, KD Cameron, RA Bukhalid, DM Gibson, BR Crane, R Loria. 2004. Nitration of a peptide phytotoxin by bacterial nitric oxide synthase. *Nature* 429: 79–82.

Khelifi N, V Grossi, M Hamdi, A Dolla, J-L Tholozan, B Ollivier, A Hirshler-Réa. 2010. Anaerobic oxidation of fatty acids and alkenes by the hyperthermophilic sulfate-reducing archaeon *Archaeoglobus fulgidus*. *Appl Environ Microbiol* 76: 3057–3060.

Kounaves SP et al. 2010. Discovery of natural perchlorate in the Antarctic Dry Valleys and its global implications. *Environ Sci Technol* 44: 2360–2364.

Krafft T, JM Macy. 1998. Purification and characterization of the respiratory arsenate reductase of *Chrysiogenes arsenatis*. *Eur J Biochem* 255: 647–653.

Krasotkina J, T Walters, KA Maruya, SW Ragsdale. 2001. Characterization of the B_{12}- and iron–sulfur-containing reductive dehalogenase from *Desulfitobacterium chlororespirans*. *J Biol Chem* 276: 40991–40997.

Kröger A, S Biel, R Gross, G Unden, CRD Lancaster. 2002. Fumarate respiration of *Wolinella succinogenes*: Enzymology, energetics and coupling mechanism. *Biochim Biophys Acta* 1553: 23–38.

Kröger A, V Geisler, E Lemma, F Theis, R Lenger. 1992. Bacterial fumarate respiration. *Arch Microbiol* 158: 311–314.

Kwon MJ, KT Finneran. 2006. Microbially mediated biodegradation of hexahydro-1,3,5-trinitro-1,3,5-triazine by extracellular electron shuttling compounds. *Appl Environ Microbiol* 72: 5933–5941.

Kwon Y-M, B Weiss. 2009. Production of 3-nitrosoindole derivatives by *Escherichia coli* during anaerobic growth. *J Bacteriol* 191: 5369–5376.

Lancaster CRD, A Kröger, M Auer, H Michel. 1999. Structure of fumarate reductase from *Wolinella succinogenes* at 2.2 Å. *Nature* 402: 377–385.

Laverman AM, JS Blum, JK Schaefer, EJP Phillips, DR Lovley, RS Oremland. 1995. Growth of strain SES-3 with arsenate and other diverse electron acceptors. *Appl Environ Microbiol* 61: 3556–3561.

Lewis TA, RL Crawford. 1995. Transformation of carbon tetrachloride via sulfur and oxygen substitution by *Pseudomonas* sp. strain JC. *J Bacteriol* 177: 2204–2208.

Lie TJ, T Pitta, ER Leadbetter, W Godchaux, JR Leadbetter. 1996. Sulfonates: Novel electron acceptors in anaerobic respiration. *Arch Microbiol* 166: 204–210.

Lie TJ, W Godchaux, JR Leadbetter. 1999. Sulfonates as terminal electron acceptors for growth of sulfite-reducing bacteria (*Desulfitobacterium* spp) and sulfate-reducing bacteria: Effects of inhibitors of sulfidogenesis. *Appl Environ Microbiol* 65: 4611–4617.

Liu A, E Garcia-Dominguez, ED Rhine, LY Young. 2004. A novel arsenate respiring isolate that can utilize aromatic substrates. *FEMS Microbiol Ecol* 48: 323–332.

Liu J, TB Gladysheva, L Lee, BP Rosen. 1995. Identification of an essential cysteinyl residue in the ArsC arsenate reductase plasmid R 773. *Biochemistry* 34: 13472–13476.

Lloyd JR, J Ridley, T Khizniak, NN Lyalikova, LE Macaskie. 1999. Reduction of technetium by *Desulfovibrio desulfuricans*: Biocatalyst characterization and use in a flowthrough bioreactor. *Appl Environ Microbiol* 65: 2691–2696.

Lloyd JR, VA Sole, CVH van Praagh, DR Lovley. 2000. Direct and Fe(II)-mediated reduction of technetium by Fe(III)-reducing bacteria. *Appl Environ Microbiol* 66: 3743–3749.

Löffler FE, RA Sanford, JM Tiedje. 1996. Initial characterization of a reductive dehalogenase from *Desulfitobacterium chlororespirans* Co23. *Appl Environ Microbiol* 62: 3809–3813.

Logan BE, H Zhang, P Mulvaney, MG Milner, IM Head, RF Unz. 2001. Kinetics of perchlorate- and chlorate-respiring bacteria. *Appl Environ Microbiol* 67: 2499–2506.

Lorenzen JP, A Kröger, G Unden. 1993. Regulation of anaerobic respiratory pathways in *Wolinellla succinogenes* by the presence of electron acceptors. *Arch Microbiol* 159: 477–483.

Loveland-Curtze J, VI Miteva, JE Brenchley. 2009. *Herminiiomonas glaciei* sp. nov., a novel ultramicrobacterium from 3041 m deep Greenland glacial ice. *Int J S yst Evol Microbiol* 59: 1272–1277.

Lovley DR. 1991. Dissimilatory Fe(III) and Mn(IV) reduction. *Microbiol Rev* 55: 259–287.

Lovley DR, JL Fraga, JD Coates, EL Blunt-Harris. 1999. Humics as an electron donor for anaerobic respiration. *Environ Microbiol* 1: 89–98.

Lovley DR, SJ Giovannoni, DC White, JE Champine, EJP Phillips, YA Gorby, S Goodwin. 1993. *Geobacter metallireducens* gen. nov. sp. nov., a microorganism capable of coupling the complete oxidation of organic compounds to the reduction of iron and other metals. *Arch Microbiol* 159: 336–344.

Lovley DR, DJ Lonergan. 1990. Anaerobic oxidation of toluene, phenol, and *p*-cresol by the dissimilatory iron-reducing organism, GS-15. *Appl Environ Microbiol* 56: 1858–1864.

Lovley DR, EJP Phillips. 1988. Novel mode of microbial energy metabolism: Organic carbon oxidation coupled to dissimilatory reduction of iron or manganese. *Appl Environ Microbiol* 54: 1472–1480.

Lovley DR, EJP Phillips. 1994. Novel processes for anaerobic sulfate production from elemental sulfur by sulfate-reducing bacteria. *Appl Environ Microbiol* 60: 2394–2399.

Lovley DR, EJP Phillips, DJ Lonergan. 1989. Hydrogen and formate oxidation coupled to dissimilatory reduction of iron or manganese by *Alteromonas putrefaciens*. *Appl Environ Microbiol* 55: 700–706.

Lovley DR, EJP Phillips, DJ Lonergan, PK Widman. 1995. Fe(III) and So reduction by *Pelobacter carbinolicus*. *Appl Environ Microbiol* 61: 2132–2138.

Lovley DR, EJP Phillips, Y A Gorby, ER Landa. 1991. Microbial reduction of uranium. *Nature* 350: 413–416.

Macy JM, K Nuna, KD Hagen, DR Dixon, PJ Harbour, M Cahill, LI Sly. 1996. *Chrysiogenes arsenatis* gen. nov., sp. nov., a new arsenate-respiring bacterium isolated from gold mine wastewater. *Int J Syst Bacteriol* 46: 1153–1157.

Macy JM, S Rech, G Auling, M Dorsch, E Stackebrandt, LI Sly. 1993. *Thauera selenatis* gen. nov., sp. nov., a member of the beta subclass of *Proteobacteria* with a novel type of anaerobic respiration. *Int J Syst Bacteriol* 43: 135–142.

Magnuson JK, MF Romine, DR Burris, MT Kingsley. 2000. Trichlorethene reductive dehalogenase from *Dehalococcoides ethenogenes*: Sequence of tceA and substrate characterization. *Appl Environ Microbiol* 66: 5141–5147.

Maiers, DT, PL Wichlacz, DL Thompson, DF Bruhn. 1988. Selenate reduction by bacteria from a selenium-rich environment. *Appl Environ Microbiol* 54: 2591–2593.

Maixner F, M Wagner, S Lücker, E Pelletier, S Schmitz-Esser, K Hace, E Spieck, R Konrat, D Le Paslier, H Daims. 2008. Environmental genomics reveals a functional chlorite dismutase in the nitrite-oxidizing bacterium `Candidatus Nitrospira defluvii´. *Environ Microbiol* 10: 3043–3056.

Maklashina E, DA Berthold, G Cecchini. 1998. Anaerobic expression of *Escherichia coli* succinate dehydrogenase: Functional replacement of fumarate reductase in the respiratory chain during anaerobic growth. *J Bacteriol* 180: 5989–5996.

Malmqvist Å, T Welander, L Gunnarsson. 1991. Anaerobic growth of microorganisms with chlorate as an electron acceptor. *Appl Environ Microbiol* 57: 2229–2232.

Malmqvist, Å, T Welander, E Moore, A Ternström, G Molin, I-M Stenström. 1994. *Ideonella dechloratans* gen. nov., sp. nov., a new bacterium capable of growing anaerobically with chlorate as electron acceptor. *System Appl Microbiol* 17: 58–64.

Masuda S, S Eda, S Ikeda, H Mitsui, K Minamisawa. 2010. Thiosulfate-dependent chemolithoautotrophic growth of *Bradyrhizobium japonicum*. *Appl Environ Microbiol* 76: 2401–2409.

May HD, GS Miller, BV Kjellerup, KR Sowers. 2008. Dehalorespiration with polychlorinated biphenyls by an anaerobic ultramicrobacterium. *Appl Environ Microbiol* 73: 2089–2094.

Maymó-Gatell X, T Anguish, SH Zinder. 1999. Reductive dechlorination of chlorinated ethenes and 1,2-dichloroethane by "*Dehalococcoides ethenogenes*" 195. *Appl Environ Microbiol* 65: 3108–3113.

McEwan AG, JP Ridge, CA McDevitt, P Hugenholtz. 2002. the DMSO reductase family of microbial molybdenum enzymes; molecular properties and the role in the dissimilatory reduction of toxic elements. *Geomicrobiol J* 19: 3–21.

Mlynek RA, A Engelbrecktson, IC Clark, HK Carlson, K Byrne-Bailey, JD Coates. 2011b. Identification of a perchlorate reduction genomic island with novel regulatory and metabolic genes. *Appl Environ Microbiol* 77: 7401–7404.

Mlynek G et al. 2011a. Unexpected diversity of chlorite dismutasea: A catalytically efficient dimeric enzymes fom *Nitrobacter winogradskyi. J Bacteriol* 193: 2408–2417.

Moir, JWB, DJ Richardson, SJ Ferguson. 1995. The expression of redox proteins of denitrification in *Thiosphaera pantropha* grown with oxygen, nitrate, and nitrous oxide. *Arch Microbiol* 164: 43–49.

Moreno-Vivián C, P Cabello, M Martínez-Luque, R Blasco, F Castillo. 1999. Prokaryotic nitrate reduction: Molecular properties and functional distinction among bacterial nitrate reductases. *J Bacteriol* 181: 6573–6584.

Muller D, DD Simeonova, P Riegel, S Mangenot, S Koechler, D Lièvrement, PN Bertin, M-C Lett. 2006. *Herminiimonas arsenicoxydans* sp. nov., a metalloresistant bacterium. *J Syst Evol Microbiol* 56:1765–1769.

Müller JA, S DasSarma. 2005. Genomic analysis of anaerobic respiration in the archaeon *Halobacterium* sp. strain NRC-1: Dimethylsulfoxide and trimethylamine *N*-oxide as terminal electron acceptors. *J Bacteriol* 187: 1659–1667.

Murphy JN, CW Saltikov. 2007. The cymeA gene, encoding a tetraheme c-type cytochrome, is required for arsenate respiration in *Shewanella species. J Bacteriol* 189: 2283–2290.

Murphy MJ, LM Siegel, H Kamin. 1973. Reduced nicotinamide adenine dinucleotide phosphate-sulfite reductase of enterobacteria II. Identification of a new class of heme prosthetic group: An iron–tetrahydroporphyrin (isobacteriochlorin type) with eight carboxylic acid groups. *J Biol Chem* 248: 2801–2814.

Myers CR, JD Myers. 1997. Cloning and sequence of *cymA*, a gene encoding a tetraheme cytochrome *c* required for reduction of iron (III), fumarate, and nitrate by *Shewanella putrefaciens* MR-1. *J Bacteriol* 179: 1143–1152.

Myers CR, KH Nealson. 1990. Respiration-linked proton translocation coupled to anaerobic reduction of manganese(IV) and iron (III) in *Shewanella putrefaciens. J Bacteriol* 172: 6232–6238.

Myers JM, WE Antholine, CR Myers. 2004. Vanadium(V) reduction by *Shewanella oneidensis* MR-1 requires menaquinone and cytochromes from the cytoplasmic and outer membranes. *Appl Environ Microbiol* 70: 1405–1412.

Nagata K, H Yu, M Nishikawa, M Kashiba, A Nakamura, EF Sato, T Tamura, M Inoue. 1998. *Helicobacter pylori* generates superoxide radicals and modulates nitric oxide metabolism. *J Biol Chem* 273: 14071–14073.

Nealson KH, CR Myers. 1992. Microbial reduction of manganese and iron: New approaches to carbon cycling. *Appl Environ Microbiol* 58: 439–443.

Neumann A, T Engelmann, R Schmitz, Y Greiser, A Orthaus, G Diekert. 2004. Phenyl methyl ethers: Novel electron donors for respiratory growth of *Desulfitobacterium hafniense* and *Desulfitobacterium* sp. strain PCE-S. *Arch Microbiol* 181: 245–249.

Newman DK, EK Kennedy, JD Coates, D Ahmann, DJ Ellis, DR Lovley, FMM Morel. 1997. Dissimilatory arsenate and sulfate reduction in *Desulfotomaculum auripigmentum* sp. nov. *Arch Microbiol* 168:380–388.

Ni S, JK Fredrickson, L Xun. 1995. Purification and characterization of a novel 3-chlorobenzoate-reductive dehalogenase from the cytoplasmic membrane of *Desulfomonile tiedje* DVCB-1. *J Bacteriol* 177: 5135–5139.

Nielsen MB, KU Kjeldsen, K Ingvirsen. 2006. *Desulfitibacter alkalitolerans* gen. nov., sp. nov., an anaerobic, alkalitolerant, sulfite-reducing bacterium isolated from a district heating plant. *Int J Syst Evol Microbiol* 56: 2831–2836.

Niggemyer A, S Spring, E Stackebrandt, RF Rosenzweig. 2001. Isolation and characterization of a novel As(V)-reducing bacterium: Implication for arsenic mobilization and the genus *Desulfitobacterium. Appl Environ Microbiol* 67: 5568–5580.

Nozawa T, Y Maruyama. 1988. Anaerobic metabolism of phthalate and other aromatic compounds by a denitrifying bacterium. *J Bacteriol* 170: 5778–5784.

Ogawa T, T Furusawa, R Nomura, D Seo, N Hosoya-Matsuda, H Sakurai, K Inoue. 2008. SoxAX binding protein, a novel component of the thiosulfate-oxidizing multienzyme system in the green sulfur bacterium *Chlorobium tepidum. J Bacteriol* 190: 6097–6110.

Oltmann LF, GS Schoenmaker, AH Stouthamer. 1974. Solubilization and purification of a cytoplasmic membrane bound enzyme catalyzing tetrathionate and thiosulphate reduction in *Proteus mirabilis. Arch Microbiol* 98: 19–30.

Omori T, L Monna, Y Saiki, T Kodama. 1992. Desulfurization of dibenzothiophene by *Corynebacterium* sp. strain SY1. *Appl Environ Microbiol* 58: 911–915.

Oremland RS, SE Hoeft, JM Santini, N Bano, RA Hollibaugh, JT Hollibaugh. 2002. Anaerobic oxidation of arsenite in Mono Lake water and by a facultative, arsenite-oxidizing chemoauthotroph strain MLHE-1. *Appl Environ Microbiol* 68: 4795–4802.

Ortiz-Bernad I, RT Anderson, HA Vrionis, DR Lovley. 2004a. Resistance of solid-phase U(VI) to microbial reduction during *in situ* bioremediation of uranium-contaminated groundwater. *Appl Environ Microbiol* 70: 7558–7560.

Ortiz-Bernad I, RT Anderson, HA Vrionis, DR Lovley. 2004b. Vanadium respiration by *Geobacter metallireducens*: Novel strategy for *in situ* removal of vanadium from groundwater. *Appl Environ Microbiol* 70: 3091–3095.

Penton CR, AH Devol, JM Tiedje. 2006. Molecular evidence for the broad distribution of anaerobic ammonium-oxidizing bacteria in freshwater and marine sediments. *Appl Environ Microbiol* 72: 6829–6832.

Phillips SE, ML Taylor. 1976. Oxidation of arsenite to arsenate by *Alcaligenes faecalis*. *Appl Environ Microbiol* 32: 392–399.

Picardal FW, RG Arnold, H Couch, AM Little, ME Smith. 1993. Involvement of cytochromes in the anaerobic biotransformation of tetrachloromethane by *Shewanella putrefaciens* 200. *Appl Environ Microbiol* 59: 3763–3770.

Price-Carter M, J Tingey, TA Bobik, JR Roth. 2001. The alternative electron acceptor tetrathionate supports B_{12}-dependent anaerobic growth of *Salmonella enterica* serovar *typhimurium* on ethanolamine or 1,2-propandiol. *J Bacteriol* 183: 2463–2475.

Quéméneur M, A Heinrich-Salmeron, D Muller, D Lièvremont, M Jauzein, PN Bertin, F Garrido, C Joulian. 2008. Diversity surveys and evolutionary relationships of *aoxB* genes in aerobic arsenite-oxidizing bacteria. *Appl Environ Microbiol* 74: 4567–4573.

Rau J, H-J Knackmuss, A Stolz. 2002. Effects of different quinoid redox mediators on the anaerobic reduction of azo dyes by bacteria. *Environ Sci Technol* 36: 1497–1504.

Renshaw JC, LJC Butchins, FR Livens, I May, JR Lloyd. 2005. Bioreduction of uranium: Environmental implications of a pentavalent intermediate. *Environ Sci Technol* 39: 5657–5660.

Rhine ED, SMN Chadhain, GJ Zylstra, LY Young. 2007. The arsenite oxidase genes (*aroAB*) in novel chemoautotrophic arsenits oxidizers. *Biochem Biophys Res Commun* 354: 662–667.

Ridley H, CA Watts, DJ Richardson, CS Butler. 2006. Resolution of distinct membrane-bound enzymes from *Enterobacter cloacae* SKLD1a-1 that are responsible for selective reduction of nitrate and selenate anions. *Appl Environ Microbiol* 72: 5173–5180.

Rikken GB, AGM Croon, CG van Ginkel. 1996. Transformation of perchlorate into chloride by a newly isolated bacterium: Reduction and dismutation. *Appl Microbiol Biotechnol* 45: 420–426.

Robertson LA, JG Kuenen. 1984. Aerobic denitrification: A controversy revived. *Arch Microbiol* 139: 351–354.

Rosen BP. 2002. Biochemistry of arsenic detoxification. *FEBS Lett* 529: 86–92.

Rousseau B, L Dostal, JPN Rosazza. 1997. Biotransformations of tocopherols by *Streptomyces catenulae*. *Lipids* 32: 79–84.

Saffarini DA, SL Blumerman, KJ Mansoorabadi. 2002. Role of menaquinones in Fe(III) reduction by membrane fractions of *Shewanella putrefaciens*. *J Bacteriol* 184: 846–848.

Saltikov CW, A Cifuentes, K Venkateswaran, DK Newman. 2003. The *ars* detoxification system is advantageous but not required for As(V) respiration by the genetically tractable *Shewanella* species strain ANA-3. *Appl Environ Microbiol* 69: 2800–2809.

Saltikov CW, DK Newman. 2003. Genetic identification of a respiratory arsenate reductase. *Proc Natl Acad Sci USA* 100: 10983–10988.

Saltikov CW, RA Wildman, DK Newman. 2005. Expression dynamics of arsenic respiration and detoxification in *Shewanella* sp. strain ANA-3. *J Bacteriol* 187: 7390–7396.

Sanford RA, JR Cole, FE Löffler, JM Tiedje. 1996. Characterization of *Desulfitobacterium chlororespirans* sp. nov., which grows by coupling the oxidation of lactate to the reductive dechlorination of 3-chloro-4-hydroxybenzoate. *Appl Environ Microbiol* 62: 3800–3808.

Sanford RA, JR Cole, JM Tiedje. 2002. Characterization and description of *Anaeromyxobacter chlororespirans* gen., nov., sp. nov., an aryl-halorespiring facultative anaerobic myxobacterium. *Appl Environ Microbiol* 68: 893–900.

Sani R, B Peyton, W Smith, W Apel, J Petersen. 2002. Dissimilatory reduction of Cr(VI) and U(VI) by *Cellulomonas* isolates. *Appl Microbiol Biotechnol* 60: 192–199.

Santini JM, LI Sly, RD Schnagl, JM Macy. 2000. A new chemolithoautotrophic arsenite-oxidizing bacterium isolated from a gold mine: Phylogenetic, physiological, and preliminary biochemical studiea. *Appl Environ Microbiol* 66: 92–97.

Schmid MC and 16 coauthors. 2005. Biomarkers for *in situ* detection of anaerobic ammonium-oxidizing (Anammox) bacteria. *Appl Environ Microbiol* 71: 1677–1684.

Schneider F, J Löwe, R Huber, H Schindelin. C Kisker, J Knäblein. 1996. Crystal structure of dimethyl sulfoxide reductase from *Rhodobacter capsulatus* at 1.88 A resolution. *J Mol Biol* 263: 53–69.

Schröder I, S Rech, T Krafft, JM Macey. 1997. Purification and characterization of the selenate reductase from *Thauera selenatis*. *J Biol Chem* 272: 23765–23768.

Schuetz B, M Schicklberger, J Kuermann, AM Spormann, J Gescher. 2009. Periplasmic electron transfer via the *c*-type cytochromes MtrA and FccA of *Shewanella oneidensis* MR-1. *Appl Environ Microbiol* 75: 7789–7796.

Seeliger S, R Cord-Ruwisch B Schink. 1998. A periplasmic and extracellular c-type cytochrome of *Geobacter sulfurreducens* acts as a ferric iron reductase and as an electron carried to other acceptors or to partner bacteria. *J Bacteriol* 180: 3686–3691.

Seim H, H Löster, R Claus, H-P Kleber, E Strack. 1982. Stimulation of the anaerobic growth of *Salmonella typhimurium* by reduction of L-carnitine, carnitine derivatives and structure-related trimethylammonium compounds. *Arch Microbiol* 132: 91–95.

Servent D, C Ducrorq, Y Henry, A Guissani, M Lenfant. 1991. Nitroglycerin metabolism by *Phanerochaete chrysosporium*: Evidence for nitric oxide and nitrite formation. *Biochim Biophys Acta* 1074: 320–325.

Shaw AL, S Leimkuhler, W Klipp, GR Hanson, AG McEwan. 1999. Mutational analysis of the dimethylsulfoxide respiratory (*dor*) operon of *Rhodobacter capsulatus*. *Microbiology UK* 145: 1409–1420.

Shen H, PH Pritchard, GW Sewell. 1996. Microbial reduction of CrVI during anaerobic degradation of benzoate. *Environ Sci Technol* 30: 1667–1674.

Shirodkar S, S Reed, M Romine, D Saffarini. 2011. The octahaem SirA catalyses dissimilatory sulfite reduction in *Shewanella oneidensis MR-1*. *Environ Microbiol* 13: 108–115.

Slobodkin AI, TG Sokolova, AM Lysenko, J Wiegel. 2006. Reclassification of *Thermoterrabacterium ferrireducens* as *Carboxydothermus ferrireducens* comb. nov., and emended description of the genus *Carboxydothermus*. *Int J Syst Evol Microbiol* 56: 2349–2351.

Song B, MM Häggblom, J Zhou, JM Tiedje, NJ Palleroni. 1999. Taxonomic characterization of denitrifying bacteria that degrade aromatic compounds and description of *Azoarcus toluvorans* sp. nov. and *Azoarcus toluclasticus* sp. nov. *Int J Syst Bacteriol* 49: 1129–1140.

Song B, NJ Palleroni, LJ Kerkhof, MM Häggblom. 2001. Characterization of halobenzoate-degrading, denitrifying, *Azoarcus* and *Thauera* isolates and description of *Thauera chlorobenzoica* sp. nov. *Int J Syst Evol Microbiol* 51: 589–602.

Springer N, W Ludwig, B Philipp, B Schink. 1998. *Azoarcus anaerobius* sp. nov., a resorcinol-degrading, strictly anaerobic, denitrifying bacterium. *Int J Syst Bacteriol* 48: 953–956.

Stenklo K, HD Thorell, H Bergius, R Aaasa, T Nilsson. 2001. Chlorite dismutase from *Ideonella dechloratans*. *J Biol Inorg Chem* 6: 601–608.

Stohl A, E Williams, G Wotawa, H Kromp-Kolb. 1996. A European inventory of soil nitric oxide emissions and the effect of these emissions on the photochemical formation of ozone. *Atmos Environ* 30: 3741–3755.

Stolz JF, DJ Ellis, JS Blum, D Ahmann, DR Lovley, RS Oremland. 1999. *Sulfurospirillum barnesii* sp. nov. and *Sulfurospirillum arsenophilum* sp. nov., new members of the *Sulfurosopirillum* clade of the ε-Proteobacteria. *Int J Syst Bacteriol* 49: 1177–1180.

Stuehr DJ, J Santolini, Z-Q Wang, C-C Wei, S Adak. 2004. Update on mechanism and catalytic regulation in the NO synthases. *J Biol Chem* 279: 36167–36170.

Sun B, JR Cole, RA Sanford, JM Tiedje. 2000. Isolation and characterization of *Desulfovibrio dechloroacetivorans* sp. nov., a marine dechlorinating bacterium growing by coupling the oxidation of acetate to the reductive dechlorination of 2-chlorophenol. *Appl Environ Microbiol* 66: 2408–2413.

Sun W, R Sierra-Alvarez, L Milner, JA Field. 2010. Anaerobic oxidation of arsenite linked to chlorate reduction. *Appl Environ Microbiol* 76: 6804–6811.

Sylvestre M, R Massé, F Messier, J Fauteux, J-G Bisaillon, R Beaudet. 1982. Bacterial nitration of 4-chlorobiphenyl. *Appl Environ Microbiol* 44: 871–877.

Taylor BF. 1983. Aerobic and anaerobic catabolism of vanillic acid and some other methoxy-aromatic compounds by *Pseudomonas* sp strain PN-1. *Appl Environ Microbiol* 46: 1286–1292.

Taylor BF, MJ Heeb. 1972. The anaerobic degradation of aromatic compounds by a denitrifying bacterium. Radioisotope and mutant studies. *Arch Microbiol* 83: 165–171.

Thibodeau J, A Gauthier, M Duguay, R Villemur, F Lépine, P Juteau, R Beaudet. 2004. Purification, cloning, and sequencing of a 3,5-dichlorophenol reductive dehalogenase from *Desulfitobacterium frappieri* PCP-1. *Appl Environ Microbiol* 70: 4532–4537.

Thorell HD, K Stenklo, J Karlsson, T Nilsson. 2003. A gene cluster for chlorate metabolism in *Ideonella dechloratans*. *Appl Environ Microbiol* 69: 5585–5592.

Thrash JC, S Ahmadi, T Torok, JD Coates. 2010. *Magnetospirillum bellicus* sp. nov., a novel dissimilatory perchlorate-reducing alphaproteobacterium isolated from a bioelectrical reactor. *Appl Environ Microbiol* 76: 4730–4737.

Vadas A, HG Monbouquette, E Johnson, I Schröder. 1999. Identification and characterization of a novel ferric reductase from the hyperthermophilic archaeon *Archaeoglobus fulgidus*. *J Biol Chem* 274: 36715–36721.

van de Graaf AA, P de Bruijn, LA Robertson, MSM Jetten, JG Kuenen. 1996. Autotrophic growth of anaerobic ammonium-oxidizing micro-organisms in a fluidized bed reactor. *Microbiology (UK)* 142: 2187–2196.

van Ginkel CG, GB Rikken, AGM Kroon, SWM Kengen. 1996. Purification and characterization of chlorite dismutase: A novel oxygen-generating enzyme. *Arch Microbiol* 166: 321–326.

vanden Hoven, RN, JM Santini. 2004. Arsenite oxidation by the heterotroph *Hydrogenophaga* sp. str. NT14: The arsenite oxidase and its physiological electron acceptor. *Biochim Biophys Acta* (BBA) *Bioenergetics* 1656: 148–155.

Venkateswaran K et al. 1999. Polyphasic taxonomy of the genus *Shewanella* and description of *Shewanella oneidensis* sp. nov. *Int J Syst Bacteriol* 49: 705–724.

Vederas JC, E Schleicher, M-D Tsai, HG Floss. 1978. Stereochemistry and mechanism of reactions catalyzed by tryptophanase from *Escherichia coli*. *J Biol Chem* 253: 5350–5354.

Walmsley AR, T Zhou, MI Boirges-Walmsley, BP Rosen. 1999. The ATPase mechanism of ArsA, the catalytic subunit of the arsenite pump. *J Biol Chem* 274: 16153–16161.

Walt A, ML Kahn. 2002. The *fixA* and *fixB* genes are necessary for anaerobic carnitine reduction in *Escherichia coli*. *J Bacteriol* 184: 4044–4047.

Wan J, TK Tokunaga, E Brodie, Z Wang, Z Zheng, D Herman, TC Hazen, MK Firestone, SR Sutton. 2005. Reoxidation of bioreduced uranium under reducing conditions. *Environ Sci Technol* 39: 6162–6169.

Weelink SAB, NCG Tan, H ten Broeke, C van den Kieboom, W van Doesburg, AAM Lagenhoff, J Gerritse, H Junca, AJM Stams. 2008. Isolation and characterization of *Alicycliphilus denitrificans* strain BC, as the electron acceptor. *Appl Environ Microbiol* 74: 6672–6681.

Weiner, JH, DP MacIsaac, RE Bishop, PT Bilous. 1988. Purification and properties of *Escherichia coli* dimethyl sulfoxide reductase, an iron–sulfur molybdoenzyme with broad substrate specificity. *J Bacteriol* 170: 1505–1510.

Wildung RE, YA Gorby, KM Krupka, NJ Hess, SW LI, AE Plymale, JP McKinley, JK Fredrickson. 2000. Effect of electron donor and solution chemistry on products of dissimilatory reduction of Technetium by *Shewanella putrefaciens*. *Appl Environ Microbiol* 66: 2451–2460.

Williams RJ, WC Evans. 1975. The metabolism of benzoate by *Moraxella* species through anaerobic nitrate respiration. *Biochem J* 148: 1–10.

Wolterink A, S Kim, M Muusse, IS Kim, PJM Roholl, CG van Ginkel, AJM Stams, SWM Kengen. 2005. *Dechloromonas hortensis* sp. nov., and strain ASK-1, two novel (per)chlorate-reducing bacteria, and taxonomic description of strain GR-1. *Int J Syst Evol Microbiol* 55: 2063–2068.

Wolterink AF, AB Jonker, SW Kengen, AJ Stams. 2002. *Pseudomonas chloritidismutans* sp. nov., a non-denitrifying, chlorate-reducing bacterium. *Int J Syst Evol Microbiol* 52: 2183–2190.

Wolterink AFWM, E Schiltz, P-L Hagedoorn, WR Hagen, SWM Kengen, AJM Stams. 2003. Characterization of the chlorate reductase from *Pseudomonas chloritidismutans J Bacteriol* 185: 3210–3213.

Wu ML, S de Vries, RA van Alen, MK Butler, HJM op den Camp, JT Keltjens, MSM Jetten, M Strous. 2011. Physiological role of the respiratory quinol oxidase in the anaerobeic nitrite-reducing methanotroph "*Candidatus Methylomirabilis oxyfera*". *Microbiology (UK)* 157: 890–898.

Yakimov MM, L Giuliano, TN Chernikova, G Gentile, W-R Abrfaham, H Lünsdorf, KN Timmis, PN Golyshin. 2001. *Alcalilimnicola halodurans* gen.nov., sp. nov, an alkaliphilic, moderately halophilic and extremely halotolerant bacterium, isolated from sediments of soda-depositing Lake Natron, East Afrca Rift Valley. *Int J Syst Evol Microbiol* 51: 2133–2143.

Zargar K, S Hoeft, R Oremland, CW Saltikov. 2010. Identification of a novel arsenite oxidase gene, *arxA*, in the haloalkaliphilic, arsenite-oxidizing bacterium *Alkalilimnicola ehrlichii* strain MLHE-1. *J Bacteriol* 192: 3755–3762.

Ziegler K, K Braun, A Böckler, G Fuchs. 1987. Studies on the anaerobic degradation of benzoic acid and 2-aminobenzoic acid by a denitrifying *Pseudomonas* strain. *Arch Microbiol* 149: 62–69.

Zikmundová M, K Drandarov, L Bigler, M Hesse, C Werner. 2002. Biotransformation of 2-benzoxazolinone and 2-hydroxy-1,4-benzoxazin-3-one by endophytic fungi isolated from *Aphelandra tetragona*. *Appl Environ Microbiol* 68: 4863–4870.

Zumft. 2002. Nitric oxide signaling and NO dependent transcriptional control in bacterial denitrification by members of the FNR-CRP regulator family. *J Mol Microbiol Biotechnol* 4: 277–286.

3.12 Dehydrogenation

This section is divided into two parts: the first dealing with dehydrogenation in the strict meaning of loss of hydrogen≡desaturation, and the second with reactions that are widely termed dehydrogenases although, in these an oxygen atom from water is incorporated into the substrate. Important examples have already been included in the section dealing with the incorporation of the oxygen from water into the substrate where they have been designated oxidoreductases, although the physiological electron acceptors have not been generally established.

This is an eclectic selection from many reactions, and important dehydrogenases that are not discussed include those involved in intermediate metabolism including those of simple aliphatic substrates and carbohydrates.

3.12.1 Loss of H$_2$: Desaturation

Dehydrogenation is a widely distributed reaction in the degradation of organic substrates—under both aerobic and anaerobic conditions. A few typical examples may serve as illustration:

Alkanol/alkanone

Alkan-1-oate → alk-*trans*-2-ene-1-oate

Arene *cis*-1,2-dihydrodiol → 1,2-dihydroxyarene

Steroid ring A: 3-oxo-4-ene → 3-oxo-1,4-diene

3.12.1.1 Short-Chain Alkanol Dehydrogenases

The short-chain alkanol dehydrogenases/reductases (Jörnvall et al. 1995) that have been extensively studied involve NAD(P)H and the amino acid residues Lys[156] and Tyr[152] of the enzyme. There are several groups of primary alkanol dehydrogenases: Zn-containing enzymes, metal-free enzymes, Fe-containing/activated enzymes, and an unusual enzyme containing both Zn and Fe in *Pyrococcus furiosus* (Ma and Adams 1999). Two metal-free enzymes have been described from *Geobacillus thermodenitrificans* strain NG80-2 that is unusual in the range of primary alkanols from C$_1$ to C$_{30}$ that can be accepted (Liu et al. 2009). Specific dehydrogenases for secondary alkanols have been described (Hou et al. 1981), and exemplified by the formation of methyl ketones from propan-2-ol and butan-2-ol in a range of Gram-negative pseudomonads and Gram-positive species of *Arthrobacter*, *Mycobacterium* (*Rhodococcus*) and *Nocardia* (Hou et al. 1983).

1. A complex situation occurs with arylglycerol-β-ethers that are partial substructures of lignin. This structure contains two asymmetric carbon atoms and, therefore, two diastereoisomers, each of which consists of two enantiomers. The degradation of this by *Sphingobium* sp. strain SYK-6, therefore, involves the activity of four dehydrogenases that belong to the short-chain dehydrogenase/reductase families (Figure 3.34) (Sato et al. 2009).

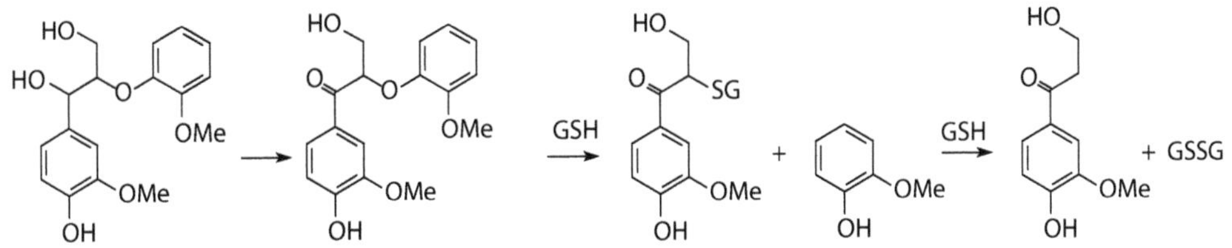

FIGURE 3.34 Dehydrogenations in the degradation of arylglycerol-β-ethers. (Adapted from Sato Y et al. 2009. *Appl Environ Microbiol* 75: 5195–5201.)

2. Dehydrogenations play key roles in the initial reactions used for the aerobic degradation of steroids, and the alkanol/alkanone reaction has been studied particularly for secondary hydroxysteroids, for example, dehydrogenation using NADPH of 3β-hydroxyl to the 3-oxo, and 17β-hydroxyl → 17-oxo; 3α-hydroxyl → 3-oxo; (Marcus and Talalay 1955). Since oxygen is not involved dehydrogenation can also take place anaerobically, for example, the 3α-hydroxy group in cholate by the anaerobe genera *Eubacterium* and *Clostridium* (Wells and Hylemon 2000).

   The degradation of cholate is initiated by the same reactions that are used for steroids with the production of the 3-oxo-1,4-diene. Further reaction requires inversion of the 12α-hydroxyl group to 12β. In *Comamonas testosteroni* TA441, it has been established that this is carried out in two stages: (i) dehydrogenation to the 12-ketone catalyzed by SteA followed by (ii) reduction to the 12β-hydroxy by SteB, both of which belong to the short-chain dehydrogenase/reductase superfamily (Horinouchi et al. 2008).

3. In addition, there are essentially different alternative mechanisms for the dehydrogenation of primary alkanols and alkylamines. These use quinone cofactors as electron acceptors in place of NAD(P)$^+$: pyrroloquinoline quinone (PQQ) that is not covalently bound, and tryptophan tryptophanyl quinone (TTQ), and 2,4,5-trihydroxyphenylalanine quinone (TPQ) both of which are formed by posttranslational modification of amino acids in peptides. It is worth noting that quinoprotein cofactors are not limited to prokaryotes, and they also occur in eukaryotes (Klinman 1996).

4. Methanol dehydrogenase is probably the most thoroughly investigated quinoprotein. In *Methylobacterium extorquens* pyrroloquinoline quinone (PQQ) and Ca^{2+} are used (Ramamoorthi and Lidstrom 1995). Its structure has been determined by x-ray analysis (Xia et al. 1992), and it belongs to one of the groups of alkanol dehydrogenases. Mechanisms for the reaction include nucleophilic reaction at C$_5$ of the quinone involving Ca^{2+} and Asp-303 with formation of a hemiacetal or hydride transfer to C$_5$ of the quinone. The other group of dehydrogenases is exemplified by ethanol dehydrogenase in *Comamonas testosteroni* that contains covalently bound heme in addition to PQQ (de Jong et al. 1995). Dehydrogenation involves free radicals and application of density functional theory has been used to suggest details of the intermediates (Kay et al. 2006). The crystal structure has been determined (Oubrie et al. 2002), and the genes in the *qhedh* operon are organized differently from those in methylotrophic bacteria (Stoorvogel et al. 1996). Whereas this dehydrogenase is soluble and monomeric, a different group of quinohemoprotein dehydrogenases is membrane-associated and contains several subunits. Different substrates may elicit a range of dehydrogenases:

   a. *Ps. putida* strain HK5 produced three soluble PQQ-dependent dehydrogenases that had different activities toward primary and secondary alkanols, and vicinal diols (Toyama et al. 1995);

   b. *Thauera (Pseudomonas) butanovora* expressed a PQQ NAD$^+$-independent primary alkanol dehydrogenase containing heme *c* that was mainly responsible for the dehydrogenation of butanol produced by oxidation of *n*-butane. In addition, NAD$^+$-dependent secondary alkanol dehydrogenases were produced by growth on butan-2-ol (Vangnai and Arp 2001).

3.12.1.2 Degradation of Primary Amines

There are several mechanisms for the dehydrogenation of primary amines to the corresponding aldehydes: (a) pyridoxal-dependent transamination, (b) dehydrogenation by quinoproteins, and oxidase activity that is discussed in the section on oxidases.

1. The pyridoxal reaction has been fully described, and in brief involves formation of an imine between pyridoxal 5′-phosphate and the amine, followed by hydrolysis with the production of pyridoxamine 5′-phosphate and the aldehyde.

2. In quinoprotein dehydrogenation of amines by bacteria, the quinoprotein involved is tryptophan tryptophylquinone (TTQ)—the alternative to PQQ that is used for methanol dehydrogenation in methylotrophs. TTQ is produced within the dehydrogenase complex by posttranslational modification of tryptophan by hydroxylation and dimerization mediated by MauG that is a diheme enzyme whose x-ray structure has been resolved from *Paracoccus denitrificans* (Jensen et al. 2010). This dehydrogenation has been demonstrated for methylamine in *Methylobacterium extorquens* (McIntire et al. 1991), in *Paracoccus denitrificans* (Chen et al. 1994), and in *Paracoccus (Thiobacillus) versutus* (Villieux et al. 1989), and for primary aromatic amines in *Alcaligenes faecalis* (Govindaraj et al. 1994). The crystal structure of the quinoprotein from *Ps. putida* has been determined (Satoh et al. 2002): it is composed of three subunits, the α-subunit with a diheme cytochrome *c*, a β-subunit, and a γ-subunit with three thioether bridges linked to the tryptophylquinone. Mass spectrometry after chemical and enzymatic cleavage revealed the presence of thioether bridges involving a covalent bond to the C_7 of the tryptophanyl quinone, and links with cysteine between the acidic side-chain residues Asp and Glu (Vandenberghe et al. 2001). Dehydrogenation is initiated by reaction of the amine with the quinone to form an *ortho*-hydroxyaniline that is hydrated at the R–CH=NH$^+$– to produce a carbinolamine which forms the aldehyde and an amine (Roujeinikova et al. 2006, 2007). In *Alcaligenes faecalis*, the redox proteins comprising phenylethylamine dehydrogenase, azurin and cytochrome *c* were located in the periplasm (Zhu et al 1999), and the crystal structure of the electron transfer complex between the aromatic amine dehydrogenase in *Alcaligenes faecalis* and the type I copper of the cupredoxin azurin has been determined (Sukumar et al. 2006) and was compared with the analogous amicyanin complex of methylamine. There are a number of additional features that are worth comment.

   a. In *Paracoccus denitrificans*, there are two amine quinoprotein dehydrogenases, one of which functions with methylamine and another that is strongly induced by *n*-butylamine (Takagi et al. 1999). Electrons funneled from the successive dehydrogenation of methylamine, formaldehyde and formate, and from the amicyanin electron carrier are channeled into the ba_3-, cbb_3—and aa_3-type oxidases that reduce O_2 to H_2O (Otten et al. 2001).

   b. Blue copper proteins mediate electron transport from the reduction of the amines to the respiratory cytochrome *c*-550—amicyanin for methylamine dehydrogenase in *Paracoccus denitrificans* (Husain and Davidson 1985; Husain et al. 1986) and azurin for aromatic amines in *Alcaligenes faecalis* (Edwards et al. 1995). The crystal structure of the electron transfer complex in *Ps. denitrificans* consisting of methylamine dehydrogenase, amicyanin and cytochrome c_{551i} has been determined to provide details of the interactions (Chen et al. 1994), and the analogous structure for the dehydrogenase, 2-phenylethylamine, azurin complex in *Alcaligenes faecalis* (Sukumar et al. 2006).

3.12.1.3 Pyrroloquinoline Quinone (PQQ) Dehydrogenation

PQQ-dependent dehydrogenases are also involved in a range of aerobic degradations involving alkanols:

1. The dehydrogenase (QuiA) that catalyzes the first step in the aerobic metabolism of quinate and shikimate by *Acinetobacter calcoaceticus* is a PQQ-dependent enzyme (van Kleef and Duine 1988; Elsemore and Ornston 1994).

2. The degradation of tetrahydrofurfuryl alcohol by *Ralstonia eutropha* is initiated by a PQQ-dependent dehydrogenase that is also able to accept C_4, C_5, and C_6 secondary alkanols as well as butan-1,3- diol and butan-1,4-diol (Zarnt et al. 1997).

3. The degradation of lupanine involves a quinocytochrome *c* hydroxylation of the $-CH_2-$ group adjacent to the ternary N (Hopper et al. 1991) and a mechanism involving dehydrogenation followed by hydroxylation was put forward.

3.12.1.4 Acyl-CoA Dehydrogenases

Acyl-CoA dehydrogenases are involved in the degradation of alkanoates in fatty acid metabolism, and in the metabolism of alkanoates and amino acids. They represent a large family of flavoproteins and the mechanism of their action involves converted scission of both C–H bonds and transfer of the electrons to the respiratory chain (Ghisla and Thorpe 2004). In addition, they are intermediates in the synthesis of poly-3-hydroxyalkanoates. These are reserve materials produced by the intermolecular esterification of (*R*)-3-hydroxyalkanoates, and are found in a wide range of organisms. This is discussed in detail in the section dealing with alkanoates. It is worth emphasizing the difference between these enzymes and fatty acid desaturases that are found in lower eukaryotes, plants, animals, and some bacteria. Studies have been directed to *B. subtilis* where the acyl-lipid desaturase introduces a double bound exclusively at C_5 (Altabe et al. 2003). These are Fe-dependent integral-membrane proteins that utilize molecular oxygen and reductant from the electron transport chain. It has been proposed that the structure consisted of six transmembrane domains, and contained histidine motifs that are involved in Fe ligation (Diaz et al. 2002), and it is remarkable that similar histidine-rich clusters are found in the AlkB alkane hydrolases in pseudomonads that are discussed in Chapter 7.

3.12.1.5 Dehydrogenation of 3-Oxosteroids

Dehydrogenation of 3-oxosteroids with the introduction of a double bond at C_1 or $C_{4(5)}$ is the prelude to the degradation of steroid ring A, for example: 3-oxosteroid $\rightarrow$ Δ1,4-dien-3-one (Plesiat et al. 1991; Florin et al. 1996; van der Geize et al. 2000). It is worth noting that the aerobic degradation of the bile acid cholate by *Pseudomonas* sp. strain Chol1 (Birkenmeier et al. 2007), and that both the anoxic degradation of cholesterol by *Sterolibacterium denitrificans* (Chiang et al. 2008b) and the aerobic degradation by *Rhodococcus* sp. strain RHA1 (van der Geize et al. 2007), involve parallel reactions.

3.12.1.6 Dehydrogenation of Arene cis-1,2-Dihydrodiols

This is discussed in Chapters 8 and 9, and only a few illustrations are given here. The arene *cis*-1,2-dihydrodiol dehydrogenases play a cardinal role in the degradation of arenes including PAHs in pseudomonads and sphingomonads, since the resulting catechols provide the entry for ring fission by 1,2-(*intradiol*) of 2,3-(*extradiol*) dioxygenations. Illustrative examples include the dehydrogenases for toluene *cis*-2,3-dihydrodiol in *Ps. putida* (Rogers and Gibson 1977) that has a molecular weight much lower than that the analogous enzyme for benzene *cis*-1,2-dihydrodiol (Axcell and Geary 1973), and naphthalene *cis*-1,2-dihydrodiol from *Ps. putida* (Patel and Gibson 1974).

3.12.1.7 Aromatization of Cyclohexane Carboxylate

The degradation of cyclohexane carboxylate in *Corynebacterium cyclohexanicum* proceeds by sequential hydroxylation and dehydrogenation to 4-oxocyclohexane-1-carboxylate. This then undergoes stepwise aromatization to 4-hydroxybenzoate mediated by enzymes that were termed desaturases: desaturase I carried out dehydrogenation to cyclohex-2-ene-4-oxocarboxylate and hydrogen peroxide, and desaturase II the formation of 4-hydroxybenzoate putatively via cyclohexane-2,5-diene-4-oxocarboxylate (Kaneda et al. 1993).

3.12.1.8 Formate Dehydrogenase: Distribution and Mechanism

This enzyme is distributed among yeasts, aerobic bacteria, anaerobic bacteria, and archaea. They carry out the reaction: $H \cdot CO_2H \rightarrow CO_2 + 2H^+ + 2e$. The following significant aspects are briefly noted.

1. *Aerobic bacteria*

    a. In methylotrophs, formate dehydrogenase generally supplies electrons to methane mono-oxygenase and nitrogenase, and generates NADH. The enzyme has been partly purified from *Methylosinus trichosporium* OB3b (Yoch et al. 1990; Jollie and Lipscomb 1991), and the holoenzyme contained Fe, acid-labile sulfide as [Fe–S] centers, and Mo that were confirmed in the EPR spectra.

    b. In the facultative methylotroph *Methylobacterium extorquens*, formate dehydrogenase catalyzes the final steps in formaldehyde metabolism by the tetrahydrofolate and tetrahydromethanopterin pathways (Vorholt et al. 2000), while three formate dehydrogenases have been identified in a whole-genome study although they are not required for growth on methanol (Chistoserdova et al. 2004). Formate dehydrogenase has been characterized and contained per mole enzyme: ≈5 mol nonheme iron and acid-labile sulfur, 0.6 mol noncovalently bound FMN and ≈2 mol tungsten (Laukel et al. 2003).

    c. In chemolithotrophic bacteria such as *Ralstonia eutropha* that can be grown on H_2 or formate, FDH plays a central role. There are two types of enzyme in this organism, one a soluble NAD-linked and the other directly coupled to the respiratory chain. The soluble enzyme contains molybdenum or tungsten cofactors and [Fe–S] centers (Friedebold and Bowien 1993), and is part of a *fdsGBACD* operon (Oh and Bowien 1998).

    d. Formate dehydrogenase has been characterized from *Pseudomonas oxalaticus* that was grown with formate as the principal source of carbon with the addition of pyruvate. The enzyme was a complex flavoprotein dimer containing FMN, nonheme Fe and acid-labile sulfur (Müller et al. 1978). Formate dehydrogenase produces electrons that are channeled to NAD^+, and it was pointed out that this organism contains also an insoluble formate oxidase.

    e. In the facultative anaerobe *E. coli*, membrane-bound FDH plays a versatile metabolic role. It exists in a separable complex with formate-hydrogen lyase, and the FDH consisted of a single polypeptide with a molecular weight of 80,000 (Axley et al. 1990). It contains a molybdopterin cofactor, an Fe–S center and selenocysteine as part of a polypeptide chain that was shown by EPR to be bound to the molybdenum (Gladyshev et al. 1994). The structure of formate dehydrogenase H in *E. coli* has been determined by x-ray analysis. It contained two molybdopterin guanine dinucleotide cofactors and an $[Fe_4–S_4]$ cluster. In addition, it contained molybdenum that was directly bound to selenium and both molybdenum cofactors (Boyington et al. 1997). A detailed structural analysis confirmed these features, revealed details of the structure and the interaction of the three subunits, and provided a mechanism for the coupling through the menaquinone pool with the nitrate reductase (Jormakka et al. 2002).

2. *Anaerobic bacteria*

    a. Acetogenic bacteria synthesize acetate from CO_2, and in *Moorella (Clostridium) thermoaceticum*, the first step in the reduction to the methyl group of acetate involves an NADP-dependent formate dehydrogenase. The purified enzyme from this organism consists of two subunits, with molecular weights of 96,000 (α-subunit) and 76,000 (β-subunit), and contains molar ratios: 2 tungsten, 2 selenium, 36 iron, and 40 sulfur, with the selenium occurring in the two α-subunits (Yamamoto et al. 1983).

    b. Formate is metabolized using fumarate as terminal electron acceptor by *Wolinella succinogenes* in combined reactions involving formate dehydrogenase and fumarate reductase. The electrons are channeled through a cytochrome *b* to menaquinone and the membrane-bound formate reductase after solubilization contained (mol/mol): 1 molybdenum, 19 nonheme iron and 18 acid-labile sulfur in two identical subunits with molecular weights of 110,000 (Kröger et al. 1979).

3. *Archaea*

Most methanotrophic archaea contain formate dehydrogenase, and some brief comments are given. Two unusual features are the role of the coenzymes M (2-marcapto-ethanesulfonate) and coenzyme F_{420} (8-hydroxy-5-deazaflavin) in these organisms.

a. The formate dehydrogenase from *Methanobacterium formicicum* has been purified, and in molar ratios contained 1 flavin adenine dinucleotide, 1 molybdenum, 2 zinc, 21–24 iron, and 25–29 inorganic sulfur. The structures of the molybdenum and the Fe/S center was examined by EPR to reveal the fine structure (Barber et al. 1983) while denaturation of the enzyme released the molybdopterin cofactor (Schauer and Ferry 1986). Formate hydrogen lyase activity could be reconstituted with three components from this strain: coenzyme-F_{420}-reducing hydrogenase, coenzyme F_{420}-reducing formate dehydrogenase and coenzyme F_{420} (Baron and Ferry 1989).

b. There are two formate dehydrogenases in *Methanococcus vannielii*, one containing molybdenum, iron, and acid-labile sulfur, while the other is a high molecular weight complex composed of selenoprotein and molybdenum iron–sulfur protein subunits (Jones and Stadtman 1981). The formate-$NADP^+$ oxidoreductase system was composed of formate dehydrogenase, coenzyme F_{420} cofactor and a 5-deazaflavin-dependent $NADP^+$ reductase (Jones and Stadtman 1980).

c. *Methanobacterium ruminantium* possessed a formate dehydrogenase linked to coenzyme F_{420}, and a formate hydrogen lyase system was obtained by combining the formate dehydrogenase with the F_{420}-dependent hydrogenase (Tzeng et al. 1975).

3.12.1.9 Anaerobic Conditions

Dehydrogenation plays a central role in the degradation and transformation of a range of substrates under anaerobic conditions. Only some representative examples are given.

1. Dehydrogenation in ring A in steroids and bile acids plays an important part in the initial steps in the anoxic degradation of cholesterol by *Sterolibacterium denitrificans* (Chiang et al. 2008b). The initial steps in the metabolism of cholesterol under denitrifying conditions (anoxic) by *Sterolibacterium denitrificans* involve dehydrogenation with the formation of the Δ-1,4-diene-3-one (Chiang et al. 2008b), and the flavoprotein dehydrogenase has been partly purified (Chiang et al. 2008a). These reactions closely parallel those under aerobic conditions by *Rhodococcus* strain RHA1 (van der Geize et al. 2007).

2. Dehydrogenation is frequently encountered in the fermentation of organic substrates. The following examples illustrate various dehydrogenations in clostridia.

   a. The fermentation of succinate to acetate and butyrate by *Clostridium kluyveri* takes place by successive dehydrogenations of succinyl-CoA to malonate semialdehyde and 4-hydroxybutyrate (Wolff et al. 1993). Essentially the same initial reactions are involved in the fermentation of 4-aminobutyrate by *Clostridium aminobutyricum* (Gerhardt et al. 2000).

   b. The initial reactions in the fermentation of L-lysine (Yorifugi et al. 1977) and L-ornithine (Tsuda and Friedman 1970) are carried out by free-radical mediated mutases with the production of 3,5-diaminohexanoate and 2,4-diaminopentanoate respectively. These are then dehydrogenated to 3-oxo-5-aminohexanoate and 2-amino-4-oxopentanoate before scission.

3. Dehydrogenation plays an important role in the anaerobic transformation of steroids. On account of its clinical importance, a good deal of attention has been devoted to the 7α-dehydroxylation of cholate in anaerobes, especially species of *Clostridium* and *Eubacterium*. In *Clostridium* sp. strain TO-931, the dehydroxylation involves a complex series of reactions whose genes form an operon (Wells and Hylemon 2000). Dehydrogenations include the following part reactions:

Dehydrogenation of 3α-hydroxy (cholate) → 3-oxo; 3-oxo → 3-oxo-4-ene

Reduction of 3-oxo-4-ene → 3-oxo-5α (3-oxo-deoxycholate)

Dehydrogenation of 3-oxo-5α → 3α-hydroxy.

Elimination of the 7α-hydroxyl is accomplished by dehydration of 3-oxo-4-ene → 3-oxo-Δ-4,6-diene.

3.12.1.10 Desaturases

Although naphthalene dioxygenase in *Pseudomonas* sp. strain 9816-4 represents the paradigm arene dioxygenase, the plurality of desaturation, dioxygenation and monooxygenation of indan has been demonstrated in this strain (Gibson et al. 1995). In a specific context, desaturases play essential roles in the biosynthesis of unsaturated fatty acids, and in the modification of fatty acids in response to low temperatures which necessitates an increase in the fluidity of the membranes. This has been discussed in several reviews (Mansilla et al. 2004; Zhang and Rock 2009) and is noted again in Chapter 4. The synthesis of unsaturated fatty acids plants, bacteria, and yeasts may be carried out by an anaerobic pathway, or by the activity of desaturases that carry out regioselective introduction of a double bond at an inactivated position in acyl-CoA or acyl lipid substrates by reaction with oxygen. This increases the fluidity of the membranes and enables them to function at lower temperatures. In plants these desaturases are soluble enzymes, whereas in bacteria and yeasts they are membrane-bound enzymes with an integral diiron center that is analogous to that in methane monooxygenase, ribonucleotide reductase, and xylene monooxygenase.

1. *E. coli* illustrates the anaerobic pathway for the synthesis of Δ^9 palmitoyl-S-ACP and Δ^{11} *cis*-vaccenoyl-*S*-ACP when the latter is increased in phospholipids with lowering temperature. The biosynthesis is initiated from 3-hydroxydecanoyl-S-ACP by FabA to produce Δ^2 *trans*-2-decanoyl-*S*-ACP, Δ^3 *cis*-decanoyl-*S*-ACP that undergoes chain elongation by FabB to Δ^9-palmitoyl-*S*-ACP and, at low temperature by FabF to Δ^{11}-*cis*-vaccenoyl-*S*-ACP (Mansilla et al. 2004).

2. The mechanism of membrane lipoid fluidity has been extensively examined in *B. subtilis* where cold shock induces the synthesis of a Δ^5 acyl lipid desaturase (Altabe et al. 2003). The regulation of this has been explored in the two-stage signal transduction system that consists of a sensor (histidine kinase) DesK and a response regulator DesR that binds to the promoter region which it controls (Aguilar et al. 2001). Changes in the environment result in phosphorylation of the sensor followed by transphosphorylation into the response regulator and a model for the function of DesR has been given (Cybulski et al. 2004).

3. *Ps. aeruginosa* possesses three pathways (Cronan 2006) to regulate membrane lipid fluidity that includes the anaerobic pathway for synthesis of unsaturated fatty acids that has been discussed for *E. coli*. The two additional aerobic pathways have been examined (Zhu et al. 2006). The anaerobic pathway is used by used by FabA to introduce the Δ^9 and Δ^{11} double bonds into thioesters that are incorporated into phospholipids, while the phospholipid desaturase DesA, and for exogenous C_{18} and C_{16} fatty acids DesB then introduced double bond at Δ^9. A mechanism for the control of the desaturases was proposed.

4. Carotenoids are required by anaerobic phototrophic bacteria to carry out a number of functions including absorption of light energy, and formation and stabilization of the pigment–protein complex. The biosynthetic pathway for lycopene has been examined for *Rubrivivax gelatinosus* in a strain of *E. coli* that can produce phytoene (Harada et al. 2001). The first C_{40} carotene Phytoene is successively desaturated by the desaturase CrtI to phytofluene, ζ-carotene and neurosporene to lycopene with 11 conjugated double bonds and hence, by CrtC, CrtD, CrtF, CrtC, CrtD, CrtF to spirilloxanthin with 13 conjugated double bonds.

5. *Deinococcus radiodurans* is a red-pigmented nonphotosynthetic bacterium with an established tolerance of radiation, oxidants, and desiccation. Deinoxanthin is an important carotenoid in this organism whose biosynthesis is carried out in stages from phytoene and lycopene: (a) cyclization by CrtLm to γ-carotene, (b) 3′,4′-desaturation by CrtD, simultaneous hydroxylation at C_2 by CrtG and dehydrogenation by CrtO at C_4, and (c) CrtC hydration at C_1 to 2-hydroxy-4-oxocarotene (deinoxanthin) (Tian et al. 2008).

6. Integral-membrane desaturases have been explored in fungi and yeasts, and a few examples are given as illustration. The *OLE1* gene in *S. cerevisiae* encodes a Δ^9 fatty acid desaturase (Stukey et al. 1990), and it has been shown that, analogous to the soluble desaturase in the plastids of higher plants the integral membrane desaturase exists as a dimer (Lou and Shanklin 2010). The fatty acid desaturase in *Saccharomyces kluyveri* is a Δ^{12} desaturase and, exceptionally was able to synthesize hexadecadienoic acid in addition to linoleic acid (Watanabe et al. 2004). There is considerable interest in the synthesis of arachidonic acid ($C_{20} \Delta^5, \Delta^8, \Delta^{11}, \Delta^{14}$) that has motivated interest in the fungus *Mortierella alpina* of the Δ^5 fatty acid desaturase (Michaelson et al. 1998; Knutzon et al. 1998) and the Δ^{12} fatty acid desaturase (Sakuradani et al. 1999).

3.12.2 Incorporation of Oxygen from Water

In the reactions that have been discussed above, only the removal/addition of hydrogen atoms is involved, in contrast to another important group in which an oxygen atom from water is incorporated into the substrate that is now discussed. Examples of dehydrogenation in which an oxygen atom from water is incorporated into the substrate include the following quite different types of reaction:

1. The dehydrogenation of trimethylamine to dimethylamine and formaldehyde in a bacterium strain W3A1 involves incorporation of the oxygen from H_2O: $(CH_3)_3 N \rightarrow (CH_3)_2NH + CH_2O$. The enzyme contains two unequal subunits, one containing an [4Fe–4S] and the other an unusual covalently bound coenzyme that has been shown to be FMN with a cysteine bound through the sulfur to position 6 of the FMN (Steenkamp et al. 1978).

2. The dehydrogenation of dimethyl sulfide to dimethyl sulfoxide has been examined in *Rhodovulum sulphidophilum* where this is used for photolithotrophic, though it is not necessary for chemotrophic growth (McDevitt et al. 2002). The enzyme, therefore, carries out the reverse reaction of DMSO reductase. It is a heterotrimer, the genes *ddhA, ddhB*, and *ddhC* form an operon, and it has been shown that DdhA is a type II cofactor of the DMSO reductase family while DdhB contains four putative [Fe–S] clusters and a *b*-type cytochrome. During photoautotrophic growth electrons from the dehydrogenation of dimethyl sulfide are channeled though cytochrome c_2 to the P870 photosystem. The dehydrogenase contains *bis*-molybdopterin guanine dinucleotide and belongs to the Type II dimethylsulfoxide reductase family along with the selenate reductase from *Thauera selenatis* and the ethylbenzene dehydrogenase in *Aromatoleum aromaticum*.

3. The conversion of ethylbenzene to (*S*)-1-phenylethanol in the facultatively anaerobic *Aromatoleum aromaticum* (*Azoarcus* sp.) strain EbN1 (Kniemeyer and Heider 2001) is carried out by a soluble dehydrogenase, and the crystal structure has revealed details of the mechanism (Kloer et al. 2006). The enzyme is a heterotrimer in which the α subunit contains the catalytic center with molybdenum ligated to two molybdopterin-guanine dinucleotides, and the β subunit contains four [Fe–S] clusters that serve as electron carriers to the γ subunit that contains *b*-heme.

4. Sulfite dehydrogenases are encountered in two different groups of bacteria, and they apparently belong to different mechanistic classes.

   a. In chemolithotrophs such as those belonging to the genus *Thiobacillus*, sulfite dehydrogenases play a key role in energy transduction. They have been purified from *Paracoccus denitrificans* in which two c-type cytochromes and a flavoprotein are required for sulfite oxidation (Wodara et al. 1997). The oxidoreductase (dehydrogenase)—that has been described previously (Toghrol and Sutherland 1983) as a sulfite oxidase—has been characterized in *Thiobacillus novellus* grown with thiosulfate as a heterodimer consisting of subunits containing a molybdopterin cofactor and a monoheme *c*-type cytochrome respectively (Kappler et al. 2000). Details of the intramolecular electron transfer have been elucidated from the crystal structure (Kappler and Bailey 2005).

   b. During the degradation of sulfonates by chemotrophs, the sulfonate is generally released as sulfite, and observed as sulfate. The dehydrogenase from *Comamonas acidovorans* P53 has been termed an "atypical" sulfite dehydrogenase that required ferricyanide as electron acceptor in cell-free systems (Reichenbecher et al. 1999), and is clearly different from a sulfite oxidase that is oxygen dependent. This enzyme has been characterized in *Cupriavidus necator* strain H 16 and *Delftia acidovorans* strain SPH-1 (Denger et al. 2008).

3.12.3 Carbon Monoxide Dehydrogenase

These Ni-containing enzymes belong to two essentially different groups: one is monofunctional, and catalyzes the reaction:

$$CO + H_2O \rightarrow CO_2 + 2H^+ + 2e$$

whereas the other is bifunctional and catalyzes two reactions.

1. The synthesis of acetyl-CoA that is generally termed acetyl-CoA synthesis/carbon monoxide dehydrogenase (ACS/CODH):

$$[CH_3-Co^{3+}FeSP]^{2+} + CO + HS-CoA \rightarrow CH_3-CO-SCoA + Co^+FeSP + H^+$$

   where CoFeSP is the corrinoid – iron – sulfur protein with cobalamin in one subunit and a $[Fe_4-S_4]$ cluster in the other;

2. The decarbonylation of acetyl-CoA where H_4SPT is tetramethylsarcinapterin.

$$CH_3-CO-SCoA + H_4SPT + H_2O \rightarrow CH_3-H_4SPT + CO_2 + HS - CoA + 2H^+ + 2e$$

These enzymes have been divided into four classes:

Class I in autotrophic methanogens synthesize acetyl-CoA from CO_2 and H_2, whereas Class II are used for acetoclastic methanogenesis. Both of them consist of five subunits α, β, γ, δ, ε, of which the γ-subunit contains the $[Fe_4-S_4]$ cluster and the δ-subunit contains Co. Class III enzymes are used by homoacetogens and consist of only four subunits α, β, γ, δ, and the γ- and δ-subunits comprise the CoFeSP protein. Class IV enzymes are monofunctional that catalyze CO/CO_2 redox reactions but not the synthesis or degradation of acetate (Lindahl 2002).

1. *Aerobic bacteria*

   Carbon monoxide dehydrogenase catalyzes the reaction $CO + H_2O \rightarrow CO_2 + 2H^+ + 2e$, and is found in aerobic bacteria that can use carbon monoxide as a source of carbon and energy. In aerobic organisms, the CO_2 is assimilated by the ribulose bisphosphate pathway, while the electrons are channeled into the electron transport pathway using CO-resistant enzymes. For anaerobes a structurally diverse range of enzymes have been characterized, and these include the enzyme complex used by methanogens for acetate assimilation.

   a. Carbon monoxide dehydrogenase has been characterized from a number of aerobic bacteria including *Hydrogenophaga pseudoflava* (Kang and Kim 1999), *Bradyrhizobium japonicum* (Lorite et al. 2000), and *Mycobacterium* (*Acinetobacter*) sp. strain DSM 3803 (Park et al. 2003). From the several analyses, the following values for metal content are typical: per mole 2 atoms Mo, 8 atoms Fe, 8 atoms acid-labile S, and 2 mol flavin. The molybdenum exists as molybdopterin cytosine dinucleotide (Schübel et al. 1995), and protein-bound selenium has been demonstrated (Gremer et al. 2000; Meyer and Rajagopalan 1984).

   b. Analysis of the structure of the enzyme from *Oligotropha carboxidovorans* yielded important structural differences from an earlier result, including the fact that the presence of selenium could not be confirmed. The molybdenum in the molybdopterin was linked to a copper atom though a sulfur atom Cu–S–Mo(=O)OH and established a new class of binuclear molybdopterins (Dobbek et al. 2002).

2. *Anaerobic bacteria*

Carbon monoxide dehydrogenase plays a central role in acetogenesis and is a component of the acetyl-CoA decarbonylase synthase complex in anaerobic acetogens. It has been characterized from several anaerobes.

a. In the acetogen *Clostridium thermoaceticum* the enzyme existed in two different subunits with molecular weights of 78,000 (α-subunit) and 71,000 (β-subunit). The dimer with greater activity contained per mol 2 nickel, 1 zinc, 11 iron, and 14 acid-labile sulfur (Ragsdale et al. 1983). EPR was used to show that a previously described paramagnetic complex between nickel and carbon monoxide dehydrogenase also contains the iron (Ragsdale et al. 1985).

b. The enzyme from *Rhodospirillum rubrum* that can use carbon monoxide as a source of carbon and energy has been characterized, and was found in two forms differing in metal content: the more active had a metal content per monomer of 7 Fe, 6S, 0.6 Ni, and 0.4 Zn. The Ni was EPR silent, and the enzyme was antigenically distinct from the enzyme from *Clostridium thermoaceticum* that has already been described (Bonam and Ludden 1987). Although the application of x-ray absorption spectroscopy suggested that the nickel was not involved in the core of the $NiFe_3S_4$ cubane cluster (Tan et al. 1992), an x-ray analysis (Drennan et al. 2001) revealed that there were two metal clusters, one of which contained the $[NiFe_3S_4]$ cluster.

c. In the anaerobe *Carboxydothermus hydrogenoformans*, two functional dehydrogenases CODH I and CODH II were present, and both catalyzed the same reaction although it was suggested that they mediated different functions—CODH I in energy transduction, and CODH II in anabolic functions. Both forms contained ca. 1.5 atoms of Ni, 22 atoms of Fe and 25 atoms of acid-labile S, and although the Ni was EPR silent, evidence suggested the existence of [4Fe–4S] clusters (Svetlitchnyi et al. 2001). An x-ray analysis has been carried out that revealed the complexity of the structure (Dobbek et al. 2001). It contained five metal clusters including [4Fe–4S] clusters, and the active site consisted of an asymmetric [Ni–4Fe–5S] cluster in which the carbon monoxide was bound to nickel.

d. In the acetogen *Moorella (Clostridium) thermoacetica* carbon monoxide dehydrogenase plays a key role in the formation of acetate, in which carbon monoxide provides the carboxyl group of acetate. There are five components involved in the conversion of pyruvate and tetrahydrofolate to acetyl-CoA (Drake et al. 1981). Carbon monoxide was implicated in fraction F_3 that contained high levels of carbon monoxide dehydrogenase and a methyl acceptor that were required for the synthesis of acetyl-CoA from carbon monoxide and tetrahydrofolate. On the basis of these results a pathway for the synthesis of acetyl-CoA from carbon monoxide or $CO_2 + H_2$ could be outlined (Doukov et al. 2002). There are several clusters in the enzyme where two reactions are carried out:

  i. In the C-cluster: $CO_2 + 2H^+ + 2e \rightarrow CO + H_2O$, while the B- and D-clusters take part in electron transfer;

  ii. In the A-cluster of the ACDS:

$$CH_3-Co^{III}-CoFeSP + CO + HSCoA \rightarrow CH_3-CO-SCoA + Co^I-CoFeSP + H^+$$

The ACDS subunit consists of a [4Fe–4S] cubane cluster with sulfur bridging one of the Fe atoms and Cu that is coordinated to two sulfur atoms to which the nickel is coordinated. A mechanism proposed that CO was bound to the Cu, followed by transfer of the CH_3-group from Ni to provide $-COCH_3$ bound to Cu, followed by reaction with HSCoA to give the product $CH_3-CO-SCoA$ (Doukov et al. 2002).

3. *Archaea*

Acetate is used for methanogenesis in archaea of the genera *Methanosarcina*, *Methanosaeta* (*Methanothrix*), and *Methanococcus*. Enormous effort has been expended to elucidate the details of the metabolism in acetotrophic methanogens, mainly belonging to species of

Methanosarcina—M. barkeri, *M. thermophila*, and *M. acetivorans*; and *Methanosaeta (Methanothrix) soehngenii*. The reactions are carried out by a complex termed the acetyl-CoA decarbonylase synthase enzyme (ACDS) that has an estimated molecular weight of 1,600,000 (Grahame 1991). The enzyme activities of the five fractions that have been obtained by fractionation of partial enzymolysis have been resolved (Grahame and DeMoll 1996). Two of the partial reactions carried out by the complex are:

CO: acceptor oxidoreductase (carbon monoxide dehydrogenase)

$$CO_2 + 2H^+ + 2e \quad \Leftrightarrow \quad CO + H_2O$$

Acetyl-CoA synthase reaction:

$$CO_2 + 2\ Fd_{red} + 2H^+ + CoA + CH_3–H_4SPt \rightarrow acetyl\text{-}CoA + H_4SPt + 2\ Fd_{ox} + H_2O$$

where H_4SPt refers to tetrahydrosarcinapterin and $CH_3–H_4SPt$ to N^6-methyltetrahydro sarcinopterin.

3.13 References

Aguilar PS, AM Hernandez-Arriaga, LE Cymbulski, C Erazo, D de Mendoza. 2001. Molecular basis of thermosensing: A two component signal transduction thermometer in *Bacillus subtilis*. *EMBO J* 20: 1681–1691.

Altabe SG, P Aguilar, GM Caballero, D de Mendoza. 2003. The *Bacillus subtilis* acyl lipid desaturase is a Δ5 desaturase. *J Bacteriol* 185: 3228–3231.

Axcell BC, PJ Geary. 1973. The metabolism of benzene by bacteria. Purification and some properties of the enzyme *cis*-1,2-dihydroxycyclohexa-3,5-diene (nicotinamide adenine dinucleotide) oxidoreductase (*cis*-benzene glycol dehydrogenase). *Biochem J* 136: 927–934.

Axley MJ, DA Grahame, TC Stadtman. 1990. *Escherichia coli* formate-hydrogen lyase. Purification and properties of the selenium-dependent formate dehydrogenase component. *J Biol Chem* 265: 18213–18218.

Baitsch D, C Sandu, R Brandsch, GL Igloi. 2001. Gene cluster on pAO1 of *Arthrobacter nicotinovorans* involved in degradation of the plant alkaloid nicotine: Cloning, purification, and characterization of 2,6-dihydroxypyridine 3-hydrolase. *J Bacteriol* 183: 5262–5267.

Barber MJ, LM Siegel, NL Schauer, HD May, JG Ferry. 1983. Formate dehydrogenase from *Methanobacterium formicicum*. Electron paramagnetic resonance spectroscopy of the molybdenum and iron–sulfur centers. *J Biol Chem* 258: 10839–10845.

Baron F, JG Ferry. 1989. Purification and properties of the membrane-associated coenzyme F_{420}-reducing hydrogenase from *Methanobacterium formicicum*. *J Bacteriol* 171: 3854–3859.

Bauder R, B Tshisuaka, F Lingens. 1990. Microbial metabolism of quinoline and related compounds VII. Quinoline-4-carboxylic acid from *Pseudomonas putida*: A molybdenum-containing enzyme. *Biol Chem Hoppe-Seyler* 370: 1183–1189.

Bauer G, F Lingens. 1992. Microbial metabolism of quinoline and related compounds XV. Quinoline-4-carboxylic acid oxidoreductase from *Agrobacterum* spec. 1B: A molybdenum-containing enzyme. *Biol Chem Hoppe-Seyler* 370: 1183–1189.

Birkenmeier A, J Holert, H Erdbrink, HM Moeller, A Friemel, R Schoenberger, MJ-G Suter, J Klebsberger, B Philipp. 2007. Biochemical and genetic investigation of initial reactions in aerobic degradation of the bile acid cholate in *Pseudomonas* sp. strain Chol1. *J Bacteriol* 189: 7165–7173.

Bonam D, PW Ludden. 1987. Purification and characterization of carbon monoxide dehydrogenase, a nickel, zinc, iron–sulfur protein, from *Rhodospirillum rubrum*. *J Biol Chem* 262: 2980–2987.

Boyington JC, VN Gladyshev, SV Khangulov, TC Stadtman, PD Sun. 1997. Crystal structure of formate dehydrogenase H: Catalysis involving Mo, molybdopterin, selenocysteine, and an Fe_4S_4 cluster. *Science* 275: 1305–1308.

Chen L, RCE Durley, FS Matthews, VL Davidson. 1994. Structure of an electron transfer complex: Methylamine dehydrogenase, amicyanin, and cytochrome c_{551i}. *Science* 264: 86–90.

Chiang Y-R, W Ismail, S Gallien, D Heinz, A Van Dorsselaer, G Fuchs. 2008a. Cholest-4-ene-3-one-Δ¹-dehydrogenase, a flavoprotein catalyzing the second step in anoxic cholesterol metabolism. *Appl Environ Microbiol* 74: 107–113.

Chiang Y-R, W Ismail, D Heinz, C Schaeffer, A Van Dorsselaer, G Fuchs. 2008b. Study of anoxic and oxic cholesterol metabolism by *Sterolibacterium denitrificans*. *J Bacteriol* 190: 905–914.

Chistoserdova L, M Laukel, J-C Portais, JA Vorholt, ME Lidstrom. 2004. Multiple formate dehydrogenase enzymes in the facultative methylotroph *Methylobacterium extorquens* AM1 are dispensable for growth on methanol. *J Bacteriol* 186: 22–28.

Choi Y-H, R Matsuzaki, T Fukui, E Shimizu, T Yorifuji, H Sato, Y Ozaki, K Tanizawa. 1995. Copper/TOPA quinone-containing histamine oxidase from *Arthrobacter globiformis* Molecular cloning and sequencing, overproduction of precursor enzyme, and generation of TOPA quinone cofactor. *J Biol Chem* 270: 4712–4720.

Cronan JE. 2006. A bacterium that has three pathways to regulate membrane lipid fluidity. *Mol Microbiol* 60: 256–259.

Cybulski LE, D Albanesi, MC Mansilla, S Altabe, PS Aguilar, D de Mendoza. 2002. Mechanism of membrane fluidity optimization: Isothermal control of the *Bacillus subtilis* acyl-lipid desaturase. *Mol Microbiol* 45: 1379–1388.

Cybulski LE, G del Solar, PO Craig, M Espinosa, D de Mendoza. 2004. *Bacillus subtilis* DesR functions as a phosphorylation-activated switch to control membrane lipid fluidity. *J Biol Chem* 279: 39340–39347.

de Jong GAH, J Caldeira, J Sun, JA Jongejan, S de Vries, TM Loehr, I Moura, JJG Moura, JA Duine. 1995. Characterization of the interaction between PQQ and heme *c* in the quinohemoprotein ethanol dehydrogenase from *Comamonas testosteroni*. *Biochemistry* 34: 9451–9458.

de la Riva L, J Badia, J Aguilar, RA Bender, L Baldoma. 2008. The *hpx* genetic system for hypoxanthine assimilation as a nitrogen source in *Klebsiella pneumoniae*: Gene organization and transcriptional regulation. *J Bacteriol* 190: 7892–7903.

Denger K, S Weinitschke, THM Smits, D Schleheck, AM Cook. 2008. Bacterial sulfite dehydrogenase in organotrophic metabolism: Separation and identification in *Cupriavidus necator* H16 and in *Delftia acidovorans* SPH-1. *Microbiology (UK)* 154: 256–263.

Diaz AR, MC Mansilla, AJ Vila, D de Mendoza. 2002. Membrane topology of the acyl-lipid desaturase from *Bacillus subtilis*. *J Biol Chem* 277: 48099–48106.

Dilworth GL. 1982. Properties of the selenium-containing moiety of nicotinic acid hydroxylase from *Clostridium barkeri*. *Arch Biochem Biophys* 219: 30–38.

Dobbek H, L Gremer, R Kiefersauer, R Huber, O Meyer. 2002. Catalysis at a dinuclear [CuSMo(=O)OH] cluster in a CO dehydrogenase resolved at 1.1 Å resolution. *Proc Natl Acad Sci USA* 99: 15971–15976.

Dobbek H, V Svetlitchnyi, L Gremer, R Huber, O Meyer. 2001. Crystal structure of a carbon monoxide dehydrogenase reveals a [Ni–4Fe–5S] cluster. *Science* 293: 1281–1285.

Dodson RM, RD Muir. 1961a. Microbiological transformations. VI. The microbiological aromatization of steroids. *J Am Chem Soc* 83: 4627–4631.

Dodson RM, RD Muir. 1961b. Microbiological transformations. VII. The hydroxylation of steroids at 9. *J Am Chem Soc* 83: 4631–4635.

Doukov TI, TM Iverson, J Seravalli, SW Ragsdale, CI Drennan. 2002. A Ni–Fe–Cu center in a bifunctional carbon monoxide dehydrogenase/acetyl-CoA synthase. *Science* 298: 567–572.

Drake HL, S-I Hu, HG Wood. 1981. Purification of five components from *Clostridium thermoaceticum* which catalyze synthesis of acetate from pyruvate and methyltetrahydrofolate. *J Biol Chem* 256: 11137–11144.

Drennan CL, J Heo, MD Sintchak, E Schreiter, PW Ludden. 2001. Life on carbon monoxide: X-ray structure of *Rhodospirillum rubrum* Ni–Fe–S carbon monoxide dehydrogenase. *Proc Natl Acad Sci USA* 98: 11973–11978.

Drenth J, WGJ Hol. 1989. Structure of quinoprotein methylamine dehydrogenase at 2.25 Å resolution. *EMBO J* 8: 2171–2178.

Edwards SL, VL Davidson, Y-L Hyun, PT Wingfield. 1995. Spectroscopic evidence for a common electron transfer pathway for two tryptophan tryptophylquinone enzymes. *J Biol Chem* 270: 4293–4298.

Elsemore DA, LN Ornston. 1994. The *pca-pob* supraoperonic cluster of *Acinetobacter calcoaceticus* contains *quiA*, the structural gene for quinate-shikimate dehydrogenase. *J Bacteriol* 176: 7659–7666.

Florin C, T Köhler, M Grandguillot, P Plesiat. 1996. *Comamonas testosteroni* 3-oxosteroid-$\Delta^{4(5a)}$-dehydrogenase: Gene and protein characterization. *J Bacteriol* 178: 3322–3330.

Freudenberg W, K König, JR Andressen. 1988. Nicotine dehydrogenase from *Arthrobacter oxidans:* A molybdenum-containing hydroxylase. *FEMS Microbiol Lett* 52: 13–18.

Friedebold J, B Bowien. 1993. Physiological and biochemical characterization of the soluble formate dehydro-genase, a molybdoenzyme from *Alcaligenes eutrophus*. *J Bacteriol* 175: 4719–4728.

Gerhardt A, I Cinkaya, D Linder, G Huisman, W Buckel. 2000. Fermentation of 4-aminobutyrate by *Clostridium aminobutyricum*: Cloning of two genes involved in the formation and dehydration of 4-hydroxybutyryl-CoA. *Arch Microbiol* 174: 189–199.

Ghisla S, C Thorpe. 2004. Acyl-CoA dehydrogenases. A mechanistic overview. *Eur J Biochem* 271: 494–508.

Gibson DT, SM Resnick, K Lee, JM Brand, DS Torok, LP Wackett, MJ Schocken, BE Haigler. 1995. Desaturation, dioxygenation, and monooxygenation reactions catalyzed by naphthalene dioxygenase from *Pseudomonas* sp. strain 9816-4. *J Bacteriol* 177: 2615–2621.

Gladyshev VN, SV Khangulov, MJ Axley, TC Stadtman. 1994. Coordination of selenium to molybdenum in formate dehydrogenase H from *Escherichia coli*. *Proc Natl Acad Sci USA* 91: 7708–7711.

Govindaraj S, E Eisenstein, LH Jones, J Sanders-Loehr, AY Chistoserdov, VL Davidson, SL Edwards. 1994. Aromatic amine dehydrogenase, a second tryptophan tryptophylquinone enzyme. *J Bacteriol* 176: 2922–2929.

Grahame DA. 1991. Catalysis of acetyl-CoA cleavage and tetrahydrosarcinapterin methylation by a carbon monoxide dehydrogenase-corrinoid enzyme complex. *J Biol Chem* 266: 22227–22233.

Grahame DA, E DeMoll. 1996. Partial reactions catalyzed by protein components of the acetyl-CoA decarbon-ylase synthase enzyme complex from *Methanosarcina barkeri*. *J Biol Chem* 271: 8352–8358.

Gremer L, S Kellner, H Dobbek, R Huber, O Meyer. 2000. Binding of flavin adenine dinucleotide to molybdenum-containing carbon monoxide dehydrogenase from *Oligotropha carboxidovorans*. Structure and functional analysis of carbon monoxide dehydrogenase species in which the native flavoprotein has been replaced by its recombinant counterpart produced in *Escherichia coli*. *J Biol Chem* 275: 1864–1872.

Harada J, KVP Nagashima, S Takaichi, N Misawa, K Matsuura, K Shimada. 2001. Phytoene desaturase, CrtI of the purple photosynthetic bacterium *Rubrivivax gelatinosus*, produces both neurosporene and lyco-pene. *Plant Cell Physiol* 42: 1112–1118.

Hetterich D, B Peschke, B Tshisuaka, F Lingens. 1991. Microbial metabolism of quinoline and related com-pounds. X. The molybdopterin cofactors of quinoline oxidoreductases from *Pseudomonas putida* 86 and *Rhodococcus* spec. B1 and of xanthine dehydrogenase from *Pseudomonas putida* 86. *Biol Chem Hoppe-Seyler* 372: 513–517.

Hille R. 2005. Molybdenum-containing hydroxylases. *Arch Biochem Biophys* 433: 107–116.

Hille R, H Sprecher. 1987. On the mechanism of action of xanthine oxidase. *J Biol Chem* 262: 10914–10917.

Hopper DJ, J Rogozinski, M Toczko. 1991. Lupanine hydroxylase, a quinocytochrome *c* from an alkaloid-degrading *Pseudomonas* sp. *Biochem J* 279: 105–109.

Horinouchi M, T Hayashi, H Koshino, M Malon, T Yamamoto, T Kudo. 2008. Identification of genes involved in inversion of stereochemistry of a C-12 hydroxyl group in the catabolism of cholic acid by *Comamonas testosteroni* TA441. *J Bacteriol* 190: 5545–5554.

Hou CT, N Barnabe, I Marczak. 1981. Stereospecificity and other properties of a secondary alcohol-specific, alcohol dehydrogenase. *Eur J Biochem* 119: 359–364.

Hou CT, R Patel, AI Laskin, N Barnabe, I Barist. 1983. Production of methyl ketones from secondary alcohols by cell suspensions of C_2 to C_4 *n*-alkane-grown bacteria. *Appl Environ Microbiol* 46: 178–184.

Husain M, VL Davidson. 1985. An inducible periplasmic blue copper protein from *Paracoccus denitrificans*. Purification, properties and physiological role. *J Biol Chem* 260: 14626–14629.

Husain M, VL Davidson. AJ Smith. 1986. Properties of *Paracoccus denitrificans* amicyanin. *Biochemistry* 25: 2431–2436.

Jensen LMR, R Sanishvili, VL Davidson, CM Wilmot. 2010. In crystallo posttranslational modification within a MauG/pre-methylamine dehydrogenase complex. *Science* 327: 1392–1394.

Jollie DR, JD Lipscomb. 1991. Formate dehydrogenase from *Methylosinus trichosporium* OB3b. Purification and spectroscopic characterization of the cofactors. *J Biol Chem* 266: 21853–21863.

Jones JB, TC Stadtman. 1980. Reconstitution of a formate-NADP+ oxidoreductase from formate dehydroge-nase and a 5-deazaflavin-linked NADP+ reductase isolated from *Methanococcus vannielii*. *J Biol Chem* 255: 1049–1053.

Jones JB, TC Stadtman. 1981. Selenium-dependent and selenium-independent formate dehydrogenases of *Methanococcus vannielii*. *J Biol Chem* 256: 656–663.

Jormakka, S Törnroth, B Byrne, S Iwata. 2002. Molecular basis of proton motive force generation: Structure of formate dehydrogenase-N. *Science* 295: 1863–1968.

Jörnvall H, B Persson, M Krook, S Atrian, R Gonzàlez-Duarte, J Jeffery, D Ghosh. 1995. Short-chain dehydrogenases/reductases (SDR). *Biochemistry* 34: 6003–6013.

Kaneda T, H Obata, M Tokumoto. 1993. Aromatization of 4-oxocyclohexanecarboxylic acid to 4-hydroxybenzoic acid by two distinctive desaturases from Corynebacterium cyclohexanicum. Properties of two desaturases. *Eur J Biochem* 218: 997–1003.

Kang BS, YM Kim. 1999. Cloning and molecular characterization of the genes for carbon monoxide dehydrogenase and localization of molybdopterin, flavin adenine dinucleotide, and iron–sulfur centers in the enzyme of *Hydrogenophaga pseudoflava*. *J Bacteriol* 181: 5581–5590.

Kappler U, S Bailey. 2005. Molecular basis of intramolecular electron transfer in sulfite-oxidizing enzymes is revealed by high resolution structure of a heterodimeric complex of the catalytic molybdopterin subunit and a c-type cytochrome subunit. *J Biol Chem* 280: 24999–25007.

Kappler U, B Bennett, J Rethmeier, G Schwarz, R Deutzmann, AG McEwan, C Dahl. 2000. Sulfite: Cytochrome c oxidoreductase from *Thiobacillus novellus*. Purification, characterization, and molecular biology of a heterodimeric member of the sulfite oxidase family. *J Biol Chem* 275: 13202–13212.

Kay CWM, B Mennenga, H Görisch, R Bittl. 2006. Structure of the pyrroloquinoline quinone radical in quinoprotein ethanol dehydrogenase. *J Biol Chem* 281: 1470–1476.

Klinman JP. 1996. New quinocofactors in eukaryotes. *J Biol Chem* 271: 27189–27192.

Kloer DP, C Hagel, J Heider, GE Schulz. 2006. Crystal structure of ethylbenzene dehydrogenase from *Aromatoleum aromaticum*. *Structure* 14: 1377–1388.

Kniemeyer O, J Heider. 2001. Ethylbenzene dehydrogenase, a novel hydrocarbon-oxidizing molybdenum/iron–sulfur/heme enzyme. *J Biol Chem* 276: 21381–21386.

Knutzon DS, JM Thurmond, Y-S Huang, S Chaudhary, EG Bobik, GM Chan, SJ Kirchner, P Mukerji. 1998. Identification of Δ5-desaturase from *Mortierella alpina* by heterologous expression in bake's yeast and canola. *J Biol Chem* 273: 29360–29366.

Koenig K, JR Andreesen. 1990. Xanthine dehydrogenase and 2-furoyl-coenzyme A dehydrogenase from *Pseudomonas putida* Fu1: Two molybdenum-containing dehydrogenases of novel structural composition. *J Bacteriol* 172: 5999–6009.

Korotkova N, L Chistoserdova, V Kuksa, ME Lidstrom. 2002. Poly-β-hydroxybutyrate biosynthesis in the facultative methylotroph *Methylobacterium extorquens* AM1. *J Bacteriol* 184: 1750–1758.

Kröger A, E Winkler, A Innerhofer, H Hackenberg, H Schägger. 1979. The formate dehydrogenase involved in electron transport from formate to fumarate in *Vibrio succinogenes*. *Eur J Biochem* 94: 465–475.

Laukel M, L Chistoserdova, ME Lidstrom, JA Vorholt. 2003. The tungsten-containing formate dehydrogenase from *Methylobacterium extorquens*: Purification and properties. *Eur J Biochem* 270: 325–333.

Lehmann M, Tshisuaka B, Fetzner S, Röger P, Lingens F. 1994. Purification and characterization of isoquinoline 1-oxidoreductase from *Pseudomonas diminuta* 7, a molybdenum-containing hydroxylase. *J Biol Chem* 269: 11254–11260.

Leimkühler S, AL Stockert, K Igarashi, T Nishino, R Hille. 2004. The role of active site glutamate residues in catalysis of *Rhodobacter capsulatus* xanthine dehydrogenase. *J Biol Chem* 279: 40437–40444.

Lindahl PA. 2002. The Ni-containing carbon monoxide dehydrogenase family: Light at the end of the tunnel? *Biochemistry* 41: 2097–2105.

Liu X, Y Dong, J Zhang, A Zhang, L Wang, L Feng. 2009. Two novel metal-independent long-chain alkyl alcohol dehydrogenases from *Geobacillus thermodenitrificans* NG80-2. *Microbiology (UK)* 155: 2078–2085.

Lorite ML, J Tachil, J Sanjuán, O Meyer, E Bedmar. 2000. Carbon monoxide dehydrogenase activity in *Bradyrhizobium japonicum*. *Appl Environ Microbiol* 66: 1871–1876.

Lou Y, J Shanklin. 2010. Evidence that the yeast desaturase Ole1p exists as a dimer *in vivo*. *J Biol Chem* 285: 19384–19280.

Ma K, NW Adams. 1999. An unusual oxygen-sensitive, iron- and zinc-containing alcohol dehydrogenase from the hyperthermophilic archaeon *Pyrococcus furiosus*. *J Bacteriol* 181: 1163–1170.

Mansilla MC, LE Cybulski, D Albanesi, D de Mendoza. 2004. Control of membrane lipid fluidity by molecular thermosensors. *J Bacteriol* 186: 5581–6688.

Marcus PI, P Talalay. 1955. Induction and purification of α- and β-hydroxysteroid dehydrogenases. *J Biol Chem* 230: 661–674.

McDevitt CA, P Hugenholtz, GR Hanson, AG McEwan. 2002. Molecular analysis of dimethyl sulfide dehydrogenase from *Rhodovulum sulfidophilum*: Its place in the dimethyl sulfoxide reductase family of microbial molybdopterin enzymes. *Mol Microbiol* 44: 1575–1587.

McIntire WS, DE Wemmer, A Chistoserdov, ME Lidstrom. 1991. A new cofactor in a prokaryotic enzyme: Tryptophan trypotophylquinone as the redox prosthetic group in methylamine dehydrogenase. *Science* 252: 817–824.

Meyer O, KV Rajagopalan. 1984. Selenite binding to carbon monoxide oxidase from *Pseudomonas carboxydovorans*. *J Biol Chem* 259: 5612–5617.

Michaelson LV, CM Lazarus, G Griffiths, JA Napier, AK Stobart. 1998. isolation of a Δ^5-fatty acid desaturase gene from *Mortierella alpina*. *J Biol Chem* 273: 19055–19059.

Müller U, P Willnow, U Ruschig, T Höpner. 1978. Formate dehydrogenase from *Pseudomonas oxalaticus*. *Eur J Biochem* 83: 485–498.

Oh J-I, B Bowien. 1998. Structural analysis of the *fds* operon encoding the NAD+-linked formate dehydrogenase of *Ralstonia eutropha*. *J Biol Chem* 273: 26349–26360.

Otten MF, J van der Oost, WNM Reijnders, HV Westerhoff, B Ludwig, RJM Van Spanning. 2001. Cytochromes c_{550}, c_{552}, and c_1 in the electron transport network of *Paracoccus denitrificans*: Redundant or subtly different in function? *J Bacteriol* 183: 7017–7026.

Oubrie A, HJ Rozeboom, KH Kalk, EG Huizinga, BW Dijkstra. 2002. Crystal structure of quinohemoprotein alcohol dehydrogenase from *Comamonas testosteroni*. *J Biol Chem* 277: 3727–3732.

Park SW, EH Hwang, H Park, JA Kim, J Heo, KH Lee, T Song et al. 2003. Growth of mycobacteria on carbon monoxide and methanol. *J Bacteriol* 185:142–147.

Patel TR, DT Gibson. 1974. Purification and properties of (+)-*cis*-naphthalene dihydrodiol dehydrogenase. *J Bacteriol* 119: 879–888.

Peschke B, F Lingens. 1991. Microbial metabolism of quinoline and related compounds XII. Isolation and characterization of the quinoline oxidoreductase from *Rhodococcus* spec. B1 compared with the quinoline oxidoreductase from *Pseudomonas putida* 86. *Biol Chem Hoppe-Seyler* 370: 1081–1108.

Plesiat P, M Grandguillot, S Harayama, S Vragar, Y Michel-Briand. 1991. Cloning, sequencing, and expression of the *Pseudomonas testosteroni* gene encoding 3-oxosteroid Δ^1-dehydrogenase. *J Bacteriol* 173: 7219–7227.

Pope SD, L-L Chen, V Stewart. 2009. Purine utilization by *Klebsiella oxytoca* M5a1: Genes for ring-oxidizing and -opening enzymes. *J Bacteriol* 191: 1006–1017.

Ragsdale SW, JE Clark, LG Ljungdahl, LL Lundie, HL Drake. 1983. Properties of purified carbon monoxide dehydrogenase from *Clostridium thermoaceticum*, a nickel, iron–sulfur protein. *J Biol Chem* 258: 2364–2369.

Ragsdale SW, HG Wood, WE Antholine. 1985. Evidence that an iron–nickel–carbon complex is formed by reaction of CO with the CO dehydrogenase from *Clostridium thermoaceticum*. *Proc Natl Acad Sci USA* 82: 6811–6814.

Ramamoorthi R, ME Lidstrom. 1995. Transcriptional analysis of *pqqD* and study of the regulation of pyrroloquinoline quinone biosynthesis in *Methylobacterium extorquens* AM1. *J Bacteriol* 177: 206–211.

Reichenbecher W, DP Kelly, JC Murrell. 1999. Desulfonation of propanesulfonic acid by *Comamonas acidovorans* strain P53: Evidence for an alkanesulfonate sulfonatase and an atypical sulfite dehydrogenase. *Arch Microbiol* 172: 387–392.

Rogers JE, DT Gibson. 1977. Purification and properties of *cis*-toluene dihydrodiol dehydrogenase from *Pseudomonas putida*. *J Bacteriol* 130: 1117–1124.

Roujeinikova A, P Hothi, L Masgrau, MJ Sutcliffe, NS Scrutton, D Leys. 2007. New insights into the reductive half-reaction mechanism of aromatic amine dehydrogenase revealed by reaction with carbinolamine substrates. *J Biol Chem* 282: 23766–23777.

Roujeinikova A, NS Scrutton, D Leys. 2006. Atomic level insight into the oxidative half-reaction of aromatic amine dehydrogenase. *J Biol Chem* 281: 40264–40272.

Sakuradani E, M Kobayashi, T Ashikari, S Shimizu. 1999. Identification of $\Delta12$-fatty acid desaturase from archidonic acid-producing *Mortieralla* fungus by heterologous expression in the yeast *Saccharomyces cerevisiae* and the fungus *Aspergillus oryzae*. *Eur J Biochem* 261: 812–820

Sato Y, H Moriuchi, S Hishiyama, Y Otsuka, K Oshima, D Kasai, M Nakamura et al. 2009. Identification of three alcohol dehydrogenase genes involved in the stereospecific catabolism of arylglycerol-β-aryl ether by *Sphingobium* sp. strain SYK-6. *Appl Environ Microbiol* 75: 5195–5201.

Satoh A et al. 2002. Crystal structure of quinoprotein amine dehydrogenase from *Pseudomonas putida*. Identification of a novel quinone cofactor encaged by multiple thioether cross-bridges. *J Biol Chem* 277: 2830–2834.

Schach S, B Tshisuaka, S Fetzner, F Lingens F. 1995. Quinoline 2-oxidoreductase and 2-oxo-1,2-dihydroquinoline 5,6-dioxygenase from *Comamonas testosteroni* 63. The first two enzymes in quinoline and 3-methylquinoline degradation. *Eur J Biochem* 232: 536–544.

Schauer NL, JG Ferry. 1986. Composition of the coenzyme F_{420}-dependent formate dehydrogenase from *Methanobacterium formicicum*. *J Bacteriol* 165: 405–411.

Schübel U, M Kraut, G Mörsdorf, O Mayer. 1995. Molecular characterization of the gene cluster *coxMSL* encoding the molybdenum-containing carbon monoxide dehydrogenase of *Oligotropha carboxidovorans*. *J Bacteriol* 177: 2197–2203.

Self WT. 2002. Regulation of purine hydroxylase and xanthine dehydrogenase from *Clostridium purinolyticum* in response to purines, selenium and molybdenum. *J Bacteriol* 184: 2039–2044.

Siegmund I, K Koenig, JR Andreesen. 1990. Molybdenum involvement in aerobic degradation of picolinic acid by *Arthrobacter picolinophilus FEMS Microbiol Lett* 67: 281–284.

Steenkamp DJ, W McIntire, WC Kenney. 1978. Structure of the covalently bound coenzyme of trimethylamine dehydrogenase. Evidence for a 6-substituted flavin. *J Biol Chem* 253: 2818–2824.

Steinbach V, S de Vries, JA Duine. 1996. Intermediates in the catalytic cycle of copper-quinoprotein amine oxidase from *Escherichia coli*. *J Biol Chem* 271: 5580–5588.

Stephan I, B Tshisuaka, S Fetzner, F Lingens. 1996. Quinaldine 4-oxidase from *Arthrobacter* sp. Rü61a, a versatile procaryotic molybdenum-containing hydroxylase active towards N-containing heterocyclic compounds and aromatic aldehydes. *Eur J Biochem* 236: 155–162.

Stoorvogel J, DE Kraayveld, CA van Sluis, JA Jongejan, S de Vries, JA Duine. 1996. Characterization of the gene encoding quinohemoprotein ethanol dehydrogenase of *Comamonas testosteroni*. *Eur J Biochem* 235: 690–698.

Stukey JE, VM McDonough, CE Martin. 1990. The *OLE1* gene of *Saccharomyces cerevisiae* encodes the Δ9 fatty acid desaturase and can be functionally replaced by the rat stearoyl-CoA desaturase gene. *J Biol Chem* 265: 20144–20149.

Sukumar N, Z-w Chen, D Ferrari, A Merli, GL Rossi, HD Bellamy, A Chistoserdov, VL Davidson, FS Mattews. 2006. Crystal structure of an electron transfer complex between aromatic amine dehydrogenase and azurin from *Alcaligenes faecalis*. 45: 13500–13510.

Takagi K, M Torimura, k Kawaguchi, K Kano, T Ikeda. 1999. Biochemical and electrochemical characterization of quinohemoprotein amine dehydrogenase from *Paracoccus denitrificans*. *Biochemistry* 38: 6035–6942.

Tan GO, SA Ensign, S Ciurli, MJ Scott, B Hedman, RH Holm, PW Ludden, ZR Korszun, PJ Stephens, KO Hodgson. 1992. On the structure of the nickel/iron/sulfur center of the carbon monoxide dehydrogenase from *Rhodospirillum rubrum*: An x-ray absorption spectroscopy study. *Proc Natl Acad Sci USA* 89: 4427–4431.

Tian B, Z Sun, Z Xu, S Shen, H Wang, YHua. 2008. Carotenoid 3′,4′-desaturase is involved in carotenoid biosynthesis in the radioresistant bacterium *Deinococcus radiodurans*. *Microbiology (UK)* 154: 3697–3706.

Toghrol F, WM Sutherland. 1983. Purification of *Thiobacillus novellus* sulfite oxidase. *J Biol Chem* 258: 6762–6766.

Toyama H, A Fujii, K Matsushita, E Shinagawa, M Ameyama, O Adachi. 1995. Three distinct quinoprotein alcohol dehydrogenases are expressed when *Pseudomonas putida* is grown on different alcohols. *J Bacteriol* 177: 2442–2450.

Tsuda Y, HC Friedman. 1970. Ornithine metabolism by *Clostridium sticklandii*. Oxidation of ornithine to 2-amino-4-ketopentanoic acid via 2,4-diaminopentanoic acid: Participation of B_{12} coenzyme, pyridoxal phosphate and pyridine nucleotide. *J Biol Chem* 245: 889–898.

Tzeng SF, MP Bryant, RS Wolfe. 1975. Factor 420-dependent pyridine nucleotide-linked formate metabolism of *Methanobacterium ruminantium*. *J Bacteriol* 121: 192–196.

van der Geize R, GI Hessels, M Nienhuis-Kuiper, L Dijkhuizen. 2008. Characterization of a second *Rhodococcus erythropolis* SQ1 3-ketosteroid 9α-hydroxylase activity comprising a terminal oxygenase homologue, KshA2 active with oxygenase–reductase component KshB. *Appl Environ Microbiol* 74: 7197–7203.

van der Geize R, GI Hessels, R van Gerwen, JW Vrijbloed, P van der Meijden, L Dijkhuizen. 2000. Targeted disruption of the *kstD* gene encoding a 3-ketosteroid $Δ^1$-dehydrogenase isoenzyme of *Rhodococcus erythropolis* strain SQ1. *Appl Environ Microbiol* 66: 2029–2036.

van der Geize R, K Yam, T Heuser, MH Wilbrink, H Hara, MC Anderton, E Sim et al. 2007. A gene cluster encoding cholesterol catabolism in a soil actinomycete provides insight into *Mycobacterium tuberculosis* survival in macrophages. *Proc Natl Acad Sci USA* 104: 1947–1952.

van Kleef MAG, JA Duine. 1988. Bacterial NAD(P)-independent quinate dehydrogenase is a quinoprotein. *Arch Microbiol* 150: 32–36.

Vandenberghe I, J-K Kim, B Devreese, A Hacisalihoglu, H Iwabuki, T Okajima, S Kuroda et al. 2001. The covalent structure of the small subunit from *Pseudomonas putida* amine dehydrogenase reveals the presence of three novel types of internal cross-linkages, all involving cysteine in a thioether bond. *J Biol Chem* 276: 42923–42931.

Vangnai AS, DJ Arp. 2001. An inducible 1-butanol dehydrogenase, a quinoprotein, is involved in the oxidation of butane by "*Pseudomonas*" *butanovora*. *Microbiology (UK)* 147: 745–756.

Vellieux FMD, F Huitema, H Groendijk, KH Kalk, JF Jzn, JA Longejan, JA Duine, K Petratos, J Drenthm, WGJ Hol. 1989. Structure of quinoprotein methylamine dehydrogenase at 2.25 Å resolution. *EMBO J* 8: 2171–2178.

Vorholt JA, CJ Marx, ME Lidstrom, RK Thauer. 2000. Novel formaldehyde-activating enzyme in *Methylobacterium extorquens* AM1 required for growth on methanol. *J Bacteriol* 182: 6645–6650.

Watanabe K, T Oura, H Sakai, S Kajimara. 2004. Yeast Δ12 fatty acid desaturase: Gene cloning, expression, and function. *Biosci Biotechnol Biochem* 68: 721–727.

Wells JF, PB Hylemon. 2000. Identification and characterization of a bile acid 7α-dehydroxylation operon in *Clostridium* sp. strain TO-931, a highly active 7α-dehydroxylating strain isolated from human feces. *Appl Environ Microbiol* 66: 1107–1113.

Wodara C, F Bardischewsky, CG Friedrich. 1997. Cloning and characterization of sulfite dehydrogenase, two *c*-type cytochromes, and a flavoprotein of *Paracoccus denitrificans* GB17: Essential role of sulfite dehydrogenase in lithotrophic sulfur oxidation. *J Bacteriol* 179: 5014–5023.

Wolff RA, GW Urben, SM O'Herrin, WR Kenealy. 1993. Dehydrogenases involved in the conversion of succinate to 4-hydroxybutanoate by *Clostridium kluyveri*. *Appl Environ Microbiol* 59: 1876–1882.

Xia Z-x, W-w Dai, J-p Xiong, Z-p Hao, VL Davidson, S White, FS Mattews. 1992. The three-dimensional structures of methanol dehydrogenase from two methylotrophic bacteria at 2.6 Å resolution. *J Biol Chem* 267: 22289–22297.

Yamamoto I, T Saiki, S-M Liu, LG Ljungdahl. 1983. Purification and properties of NADP-dependent formate dehydrogenase from *Clostridium thermoaceticum*, a tungsten–selenium–iron protein. *J Biol Chem* 258: 1826–1832.

Yoch DC, Y-P Chen, MG Hardin. 1990. Formate dehydrogenase from the methane oxidizer *Methylosinus trichosporium* OB3b. *J Bacteriol* 172: 4456–4463.

Yorifugi T, I-M Jeng, HA Barker. 1977. Purification and properties of 3-keto-5-aminohexanoate cleavage from a lysine-fermenting *Clostridium*. *J Biol Chem* 252: 20–31.

Zarnt G, T Schrader, JR Andreesen. 1997. Degradation of tetrahydrofurfuryl alcohol by *Ralstonia eutropha* is initiated by an inducible pyrroloquinone-dependent alcohol dehydrogenase. *Appl Environ Microbiol* 63: 4891–4898.

Zhang Y-M, CO Rock. 2009. Transcriptional regulation in bacterial membrane lipid synthesis. *J Lipid Res* April(Suppl): S115–S119.

Zhu K, P-H Choi, HP Schweizer, CO Rock, Y-M Zhang. 2006. Two possible pathways for formation of unsaturated fatty acids in *Pseudomonas aeruginosa*. *Mol Microbiol* 60: 260–273.

Zhu Z, D Sun, VL Davidson. 1999. Localization of periplasmic redox proteins of *Alcaligenes faecalis* by a modified general method for fractionating Gram-negative bacteria. *J Bacteriol* 181: 6540–6542.

3.14 Reductases and Related Enzymes

Reduction is an important reaction under both aerobic and anaerobic conditions. The enzymes involved in the reduction of oxyanions such as nitrate, sulfate, chlorate, selenate and arsenate, and metal cations are discussed in Chapter 4, and in a part of this chapter that deals with metalloenzymes. The reductases that are components of the aromatic dioxygenases and that play a key role in the aerobic degradation of aromatic hydrocarbons by bacteria are noted parenthetically in Chapters 8 and 9. The reductases involved in anaerobic degradation of benzoates are, however, discussed here, and are noted again in Chapter 8.

3.14.1 Ribonucleotide Reductase

These are essential enzymes responsible for the reduction of the nucleotide triphosphates to the corresponding deoxynucleotide triphosphates. There are three main groups (Nordlund and Reichard 2006):

those that are dependent on oxygen, those for which its presence is immaterial and those for which it is inimical. They differ in the radical that is obligatory for the activity of all of them.

> *Group I*: Oxygen is required for the generation of a tyrosyl radical, and shuttled to cysteine to produce a thiyl radical. There are three subclasses (Sjöberg 2010) that differ in their metal components: Class 1a Fe(III)/Fe(III); Class 1b Mn(III)/Mn(III); Class 1c Mn(IV)/Fe(III). The crystal structure of the Class 1b enzyme from *E. coli* has been used to clarify the role of the FMN cofactor, the mechanism of activation to a Mn(III)$_2$ – Y˙ radical, and the role of peroxide (Boal et al. 2010). Details of the Class 1b enzymes are given in later paragraphs dealing with manganese.

> *Group II*: The thiyl radical is generated directly via a deoxyadenosyl radical;

> *Group III*: A glycyl radical is produced by the activity of a redox center [4Fe–4S] that together with *S*-adenosylmethionine and reduced flavodoxin: this then generates a thiyl radical.

These groups are not, however, mutually exclusive since the anaerobe *Bacteroides fragilis* contains an aerobic-type I reductase (Smalley et al. 2002), *E. coli* may express type I or, under microaerophilic or anaerobic conditions type III (Torrens et al. 2007), while in *Ps. aeruginosa* and *Ps. stutzeri* active enzymes from class I and II were simultaneously present—though not class III although the genes *nrdA*, *nrdJ*, and *nrdD* for three classes were found (Jordan et al. 1999).

3.14.2 Nitroarene Reductase

Nitroarene reductases catalyze the reduction of nitroarenes. They are flavoprotein (FMN) enzymes (Blehert et al. 1999) of two types:

> *Type I*: These are O_2-insensitive 6-electron reductases and catalyze the sequential reduction of nitroarenes to nitroso, hydroxylamino, and amino arenes. They are encoded in *E.coli* by *nfsA* for the major enzyme and *nfsB* for the minor (Rau and Stolz 2003), and are also important in establishing resistance to nitrofuran drugs (Koziarz et al. 1998).

> *Type II*: These are O_2-sensitive 2-electron reductases and catalyze the reduction to $ArNO_2$. radicals that react with O_2 to produce superoxide (Bryant and DeLuca 1991; Bryant et al. 1991).

The reductive activation of nitrofuran and nitroimidazole prodrugs is discussed in Chapter 4.

3.14.3 "Old Yellow Enzyme"

This is a family of flavoenzymes that are able to carry out a range of reductions of double bonds activated by carbonyl or nitro groups (Williams and Bruce 2002). Illustrative examples include the following.

3.14.3.1 Nitrate Ester Reductase

The degradation of aliphatic nitrate esters produces nitrite and the corresponding alkanol by a reductive rather than a hydrolytic mechanism. The enzymes that mediate reduction of glycerol trinitrate by *Agrobacterium radiobacter* (Snape et al. 1997), pentaerythritol tetranitrate by *Enterobacter cloacae* strain PB2 (French et al. 1996; Khan et al. 2004), and flavoprotein reductases from *Ps. fluorescens* strain I-C (Blehert et al. 1999) have been characterized. They belong to a group of flavoproteins that are related to "old yellow enzyme" with varying substrate specificities, and can reduce pentaerythritol tetranitrate, 1-nitrocyclohexene, cyclohex-2-enone, *N*-ethylmaleamide, and morphinone (Williams et al. 2004). It has been shown that hydrogen transfer in the reaction with NADH in pentaerythritol tetranitrate and morphinone reductase involves hydrogen-tunneling in the reaction between the reduced flavin and the substrate (Basran et al. 2003).

3.14.3.2 Enones (α,β-Unsaturated Ketones) and Related Reductases

A group of flavoenzymes reductases mediate the reduction of nitrate esters and C=C double bonds generally activated with carbonyl or nitro functions. The structures of the substrates vary widely, and some—though not all—are related to "old yellow enzyme" (references in Faber 1997; Williams and Bruce 2002).

1. The reduction of cyclohex-2-ene-1-one to cyclohexanone is highly specific in *Ps. syringae* (Rohde et al. 1999), whereas the reductases from *S. cerevisiae* are able to accept a wide range of activated aliphatic enones (Wanner and Tressel 1998).

2. The related "old yellow enzyme" and pentaerythritol tetranitrate reductase can carry out reduction of androsta-$\Delta^{1,4}$-3,17-dione to androsta-Δ^4-3,17-dione (Vaz et al. 1995) and prednisone to pregna-Δ^4-17α,20-diol-3,11,20-trione (Barna et al. 2001).

3. "Old yellow enzyme" is also able to bring about the unusual dismutations of several conjugated cyclohexenones and cyclodecenones (Vaz et al. 1995) that are discussed in Chapter 7, Part 2.

4. The first step in the metabolism of the isoflavanone daidazin to equol by *Lactococcus* sp. strain 20-92 involved reduction of the double bond in ring B, and it was proposed that the enzyme could be classified as belonging to the family of "old yellow enzymes" (Shimada et al. 2010).

5. A reductase in *E. coli* can reduce *N*-ethylmaleamide to *N*-ethylsuccinimide (Miura et al. 1997).

6. Morphinone can be reduced by *Ps. putida* M10 to hydromorphone using an enzyme of which one of the subunits contains FMN (French and Bruce 1994).

7. It has been shown that an NADPH-linked reduction of α,β-unsaturated nitro compounds may also be accomplished by "old yellow enzyme" via the *aci*-nitro form (Meah and Massey 2000). The substrates include nitrocyclohexene, nitrostyrene and nitrovinylthiophene. This reaction is formally analogous to the reduction and dismutation of cyclic enones by the same enzyme (Vaz et al. 1995).

8. The 2-haloacrylate reductase from *Burkholderia* sp. strain WS does not directly contribute to dehalogenation: 2-haloacrylate is converted to (*S*)-2-halopropionate that is then dehalogenated by a dehalogenase (Kurata et al. 2005). Dehalogenation is accomplished by the activity of a reduced flavin (FAD) enzyme that carries out hydration followed by loss of chloride to produce pyruvate (Mowafy et al. 2010).

9. An enzyme that catalyzes the reduction of Δ^1-piperidein-2-carboxylate to piperidine-2-carboxylate (L-pipecolate) in the catabolism of D-lysine by *Ps. putida* ATCC12633 is an NADPH-dependent representative of a large family of reductases distributed among bacteria and archaea (Muramatsu et al. 2005). It also catalyzes the reduction of Δ^1-pyrrolidine-2-carboxylate to L-proline.

3.14.3.3 Polynitroarenes

In addition to the reduction of the nitro group in nitroarenes to amines and hydroxylamines, 2,4,6-trinitrophenol and 2,4,6-trinitrobenzene can undergo reduction of the ring to Meisenheimer-type hydride complexes. These are discussed in Chapter 9, and only salient features are noted here.

1. Pentaerythritol tetranitrate, glycerol trinitrate, morphinone, and *N*-ethylmaleamide reductases have variable preference for their potential substrates (Williams et al. 2004): for example, all of them were able to carry out partial reduction of the nitro groups of 2,4,6-trinitrotolune, whereas ring reduction was carried out only by pentaerythritol tetranitrate reductase.

2. Analysis of hydride transferases has been carried out (van Dillewijn et al. 2008b), and the activities of homologues of the flavoproteins in *Ps. putida* revealed the variability of XenA

to XenF: although all of them had type I hydride transferase activity to produce hydroxylamines, Xen B had type II hydride transferase activity that produced Meisenheimer-type hydride complexes. Products from the reaction between the hydroxylamines and the Meisenheimer-type hydride complexes to produce *N*-hydroxydiphenylamines have been discussed (van Dillewijn et al. 2008a) in Chapter 9. Analysis of amino acid sequences of members of the OYE family showed that these could be divided into two clades: Clade 1 included most of the reductases while Clade 2 comprised XenA, XenD, and XenE in *Ps. putida* KT2440.

3.14.4 Anaerobic Reductions

1. The reductases from *Clostridium tyrobutyricum* and *Cl. kluyveri* can reduce both aliphatic enoates and cinnamates to the dihydro compounds (Bühler and Simon 1982). The reductase in *Cl. tyrobutyricum* contains iron, labile sulfur, and both FAD and FMN (Kuno et al. 1985), while ESR measurements suggest the presence of a semiquinone radical and Mössbauer spectra the presence of an [4Fe–4S] cluster (Caldeira et al. 1996). The amino acid sequences of these enzymes show similarity to a group of oxidoreductases and dehydrogenases, including the dehydrogenase for dimethylamine in *Hyphomicrobium* X and trimethylamine in *Methylotrophus methylophilus* (Rohdich et al. 2001).

2. Important reductions are involved in the anaerobic transformations of steroid and flavanoids.

   a. Reduction of a steroid 3-oxo-4,6-dienone to the 3-oxo compound involved a complex sequence of reactions in a *Clostridium* sp. (Wells and Hylemon 2000) that is discussed in Chapter 7, Part 2.

   b. Reduction of the B-ring of flavanoids has been found in *Clostridium orbiscindens* (Schoefer et al. 2003) and in *Eubacterium ramulus* (Braune et al. 2001), and these are discussed in Chapter 10, Part 2.

3. The reduction of amino acids is important in the degradation of more complex substrates including amino acids and purines.

   a. The reductive segment (Figure 3.35) of the pathway for the degradation of L-ornithine by *Clostridium sticklandii* involves the formation of D-proline that is reduced to the terminal metabolite 5-aminovalerate. The D-specific reductase has been purified from *Clostridium sticklandii* (Stadtman and Elliot 1957), and consists of three subunits of which one containing a pyruvoyl group is used to bind the proline, while selenocysteine carries out a nucleophilic reaction with release of the product (Kabisch et al. 1999). Analysis of the genome of pathogen *Clostridium difficile* revealed the presence of two selenium-dependent reductases—glycine reductase and D-proline reductase: the latter had a complex of PrdA and PrdB, and was specific for D-proline (Jackson et al. 2006).

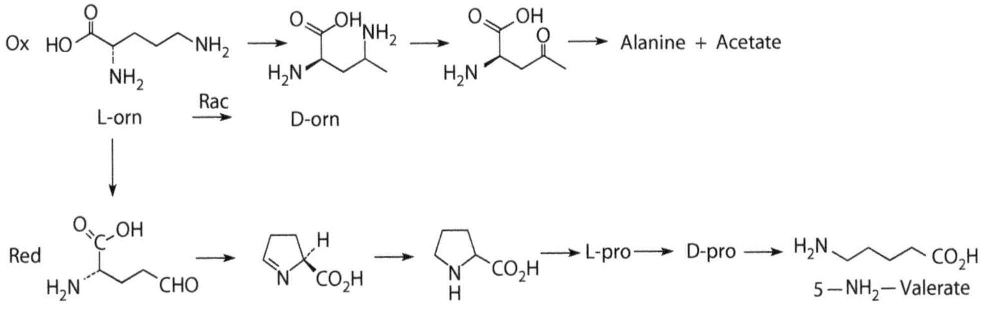

FIGURE 3.35 Oxidative and reductive segments of the pathway for the degradation of L-ornithine by *Clostridium sticklandii*.

b. The anaerobic degradation of glycine is involved in several reactions in which glycine reductase plays an important role, since its reduction to acetate and ammonia is coupled to the synthesis of ATP from inorganic phosphate (P_i).

    i. The fermentation of glycine alone by *Eubacterium acidaminophilum* (Andreesen 1994);

    ii. It is important in the Stickland reaction between pairs of amino acids such as glycine and several other amino acids;

    iii. Glycine is produced during the anaerobic degradation of purines where it is produced from the penultimate intermediates (Dürre and Andreesen 1982).

    Glycine reductase has been purified from *Clostridium sticklandii* (Seto and Stadtman 1976). The reductase consists of a complex of three proteins, a selenoprotein A, a carbonyl group protein B, and a sulfhydryl protein C; the selenium protein serves as a carrier for the C_2 substrate (Stadtman and Davis 1991). A plausible mechanism for the reaction has been outlined (Arkowitz and Abeles 1991) and specifically includes the interaction of the three proteins in the complex (Wagner et al. 1999). Three groups of interacting reactions are involved: (a) reaction between glycine and the carbonyl group of the enzyme, (b) reaction with the Se of the selenoprotein and sulfide of the protein to form a thioester, and (c) production of acetyl phosphate by reaction with inorganic phosphate.

3.14.5 Azo Reductase

Reduction of the azo group in dyes and colorants is the key reaction in their decolorization. Although an azo reductase mediates the reduction of azo groups to amines, the metabolic situation is quite complex. The enzyme apparently is synthesized under both aerobic (Blümel et al. 2002) and anaerobic conditions (Rau et al. 2002; Rau and Stolz 2003), and the anaerobic reduction is facilitated by the presence of quinonoid mediators. The enzyme from *Staphylococcus aureus* is tetrameric with four noncovalently bound FMN, and requires NADPH as reductant for activity (Chen et al. 2005).

3.14.6 Aldehyde Oxidoreductases

There are quite different reactions that have been carried out by aldehyde oxidoreductases.

Aerobic reduction of aryl and alkyl carboxylates to the corresponding aldehydes The reaction involves formation of an acyl-AMP intermediate by reaction of the carboxylic acid with ATP; NADPH then reduces this to the aldehyde (Li and Rosazza 1998; He et al. 2004). The oxidoreductase from *Nocardia* sp. is able to accept a range of substituted benzoic acids, naphthoic acids and a few heterocyclic carboxylic acids (Li and Rosazza 1997).

Phenylacetaldehyde reductase This is involved in the degradation of styrene and is able, in addition, to accept long-chain aliphatic aldehydes and ketones, and halogenated acetophenones (Itoh et al. 1997).

2-Oxoglutarate anaerobic oxidoreductases There are several of these, for example in *Thauera aromatica* (Dörner and Boll 2002), and *Azoarcus evansii* (Ebenau-Jehle et al. 2003), and their role in the metabolism of arene carboxylates is discussed in Chapter 8, Part 3.

Hydroxylation of isoquinoline This oxidoreductase from *Pseudomonas diminuta* strain 7 that carries out at C_2, is a molybdenum enzyme containing [Fe–S] centers and is comparable to the aldehyde oxidoreductase from *Desulfovibrio gigas* (Lehmann et al. 1994).

3.14.7 F_{420}-Dependent (Desazatetrahydrofolate) Reductases

These play a key role both in methanogenesis and in the degradation of phenols that carry several nitro groups and are discussed further in Chapter 9, Part 5. Although these are typically found in methanogens, they have been encountered in a number of other bacteria and archaea.

1. A group of enzymes are involved in the final steps of (a) the synthesis of methane and (b) the generation of electrochemical energy from the resulting product.

   a. Methyl-coenzyme M reductase (component C) brings about the final step in the biosynthesis of methane: $CH_3–S–CoM + HS–HPT \rightarrow CH_4 + CoM–S–S–HPT$ where HS–CoM is 2-mercapto-ethansulfonate and HS–HPT is $N–$(7methylmercaptoyl)threonine-O^3-phosphate. It is carried out by a complex enzyme system involving FAD, cobalamin, and N-(7-methylmercaptoyl)threonine-O^3-phosphate (component B) (Ellermann et al. 1989) that has been characterized from the crystal structure in *Methanobacterium thermoautotrophicum* (Ermler et al. 1997) that confirmed the presence of two molecules of the nickel-containing coenzyme F_{430} (Ellefson et al. 1982). In some strains of this organism there are two methyl-CoM reductases (Rospert et al. 1990).

   b. The heterodisulfide reductase play a key role in the production of an electrochemical proton potential, and has been characterized in several methanogens, that can utilize a number of reductants including H_2, reduced ferredoxin, and reduced F_{420}. The complex has been purified from *Methanobacterium thermoautotrophicum*, and contained FAD, Ni, nonheme Fe and acid-labile sulfur (Setzke et al. 1994). In *Methanosarcina barkeri* the reductase contains heme and Fe/S, though not, however, FAD (Künkel et al. 1997).

   c. The acetotrophic *Methanococcus thermoautotrophicum* contains in addition a thiol: fumarate reductase in which the reduction of fumarate to succinate is driven by the simultaneous oxidation of HS-coenzyme M and HS-HPT to the heterodisulfide (Bobik et al. 1989). The reductase consists of two subunits of which TfrA contains the active site for fumarate reduction and TfrB for CoM–S–S–HPT reduction.

2. Degradation of 2,4,6-trinitrophenol and 2,4-dinitrophenol involves a Meisenheimer hydride complex (Vorbeck et al. 1998; Behrend and Heesche-Wagner 1999; Ebert et al. 1999; Heiss et al. 2002; Hofmann et al. 2004).

3. The sulfite reductase from the hyperthermophilic methanogen *Methanocaldococcus jannashii* is able to reduce the otherwise toxic sulfite to sulfide that is required for growth. In contrast to most organisms that use nicotinamides and cytochromes as electron carriers, this organism uses a coenzyme F_{420}-dependent reductase (Johnson and Mukhopadhyay 2005).

4. Glucose-6-phosphate is normally dehydrogenated to 6-phosphogluconate using an NADP-dependent dehydrogenase. In *Mycobacterium smegmatis* strain mc²6, however, the reaction is mediated by coenzyme F_{420}, and neither NAD or NADP were effective (Purwantini and Daniels 1996).

3.14.8 Anaerobic Reduction of Benzoyl-CoA

Reductases play a cardinal role in the anaerobic degradation of diverse aromatic compounds. They carry out the following: (a) reduction of benzoyl-CoA that is a prelude to the fission of the ring (Harwood et al. 1999) and is discussed in Chapter 8, Part 3, (b) the dehydroxylation of phenols that is discussed in Chapter 8, Part 4, and (c) the degradation of phloroglucinol via dihydrophloroglucinol to acetate and butyrate (Haddock and Ferry 1989, 1993) that is discussed in Chapter 8, Part 4.

Benzoate occupies a central position in the anaerobic degradation of a range of aromatic compounds: these include alkylbenzene, L-phenylalanine, phenylglyoxylate, phenylacetate (Schneider et al. 1997), phenol and 4-hydroxybenzoate. The overall pathways have been summarized in a review (Harwood et al. 1999), and only the reaction involving reduction of benzoate is discussed here (Möbitz and Boll 2002). In *Thauera aromatica* and the obligate anaerobes *Geobacter metallireducens* and *Syntrophus aciditrophicus*, cyclohexa-1,5-diene carboxyl-CoA is produced, whereas the more highly reduced 1-ene carboxyl-CoA is produced from *Rhodopseudomonas palustris*. In addition, the mechanisms for metabolism of the benzoyl-CoA differs for facultative and obligate anaerobes.

Benzoate is converted by a ligase to benzoyl-CoA that is reduced to nonaromatic compounds before ring scission. Different mechanisms are used for facultative anaerobes (*Thauera aromatica, Azoarcus evansii, Rhodopseudomonas palustris*): in these, benzoyl-CoA reductase carries out the two-electron transfer from a reduced ferredoxin to the aromatic ring and is coupled to stoichiometric hydrolysis of ATP. The enzyme from *Thauera aromatica* K172 contained 10.8 mol Fe and 10.5 mol acid-labile S, and—with some uncertainty 0.3 mol flavin per mole, and required MgATP and a reductant such as Ti(III) (Boll and Fuchs 1995). EPR and Mössbauer studies using ^{57}Fe-enriched enzyme from *Thauera aromatica* excluded [2Fe–2S] and [3Fe–4S] clusters and revealed the presence of three [4Fe–4S] clusters that were partly converted into [3Fe–4S] clusters by oxygen inactivation of the enzyme (Boll et al. 2000). The enzyme was a tetramer of nonidentical subunits: A, mass 30 kDa and D, mass 49 kDa each containing an ATP-binding site, and together 1 cubane [4Fe–4S]$^{+1/+2}$; B, mass 48 kDa and C, mass 44 kDa with each containing 1 cubane [4Fe–4S]$^{+1/+2}$ in which the iron is bound to cysteine. 2-oxoglutarate: ferredoxin oxidoreductase in *Thauera aromatica* that supplies reduced ferredoxin contained per mol protein 8.3 mol Fe, 7.2 mol acid-labile sulfur, and 1.6 mol thiamine diphosphate (Dörner and Boll 2002).

In the obligate anaerobes (*Geobacter metallireducens, Desulfoccus multivorans, Syntrophus aciditrophicus*) the reductases are ATP-independent, and in *Desulfoccus multivorans* the ATP-independent benzoyl-CoA reductase contained Fe/S, Mo or W and Se (Peters et al. 2004; Wischgoll et al. 2005). This has been examined in *Geobacter metallireducens*, and it has been shown that the enzyme which catalyzed the reverse reaction was a 185-kDa enzyme with subunits of 73-kDa (BamB) and 21-kDa (BamC) that would be consistent with an $\alpha_2\beta_2$ structure. The $\alpha\beta$ unit contained 0.9 W, 15 Fe, and 12.7 acid-labile S, and EPR indicated the presence of one [3Fe–4S]$^{0/+1}$ and three [4Fe–4S]$^{+1/+2}$ clusters and, in the oxidized form the presence of W^v (Kung et al. 2009).

3.14.9 Anaerobic Dehydroxylation

The anaerobic degradation of aromatic substrates with a hydroxyl group such as phenols and hydroxybenzoates is dependent on dehydroxylation. For hydroxybenzoates the first steps are the synthesis of the benzoyl-CoA ester followed by reduction of the ring catalyzed by 4-hydroxybenzoyl-CoA reductase from *Thauera aromatica*, and an analogous enzyme from *R. palustris* (Gibson et al. 1997). The dehydroxylase consists of three subunits and contains molybdenum, a flavin nucleotide, both [4Fe–4S] and [2Fe–2S] clusters, and FAD (Breese and Fuchs 1998). It had a mass of 260 kDa and contained three subunits: α mass 75 contains molybdpterin cytosine dinucleotide, β mass 35 contains [2Fe–2S]$^{+1/+2}$ one each of type I and II, γ mass 17 contains [4Fe–4S]$^{+1/+2}$ and FAD. The redox centers have been examined by EPR and, by addition of ^{57}Fe to the medium, Mössbauer spectroscopy (Boll et al. 2001). A free-radical mechanism involving the phenyl radical ring was proposed with involvement of Mo(IV), Mo(V) and Mo(VI). Electron transfer takes place Fd$_{red}$ → [4Fe–4S] → FAD → [2Fe–2S] → Mo. The reaction proceeds in stages that involve binding of the substrate with Mo^{IV}, generation of Mo^V and a radical anion, protonation of the radical anion at C$_4$, followed by scission to produce benzoyl-CoA and the Mo^{VI} center that is reduced to Mo^{IV} (Boll et al. 2001).

It is worth noting that whereas in *Th. aromatica* the metabolism of 4-hydroxybenzoyl-CoA, dehydroxylation precedes ring reduction, for 3-hydroxybenzoate ring reduction takes place first (Laempe et al. 2001). In *Sporotomaculatum hydroxybenzoicum*, however, dehydroxylation precedes ring reduction (Müller and Schink 2000). In the degradation of 3,4-dihydroxybenzoyl-CoA by *Th. aromatica*, dehydroxylation at the 4-position precedes ring reduction (Philipp et al. 2002).

3.14.10 Reduction of Azaarene Rings

The aerobic degradation of a number of azaarenes involves reduction of the rings at some stage and is discussed in Chapter 10, Part 1. Illustrative examples include the degradation of pyridines (3-alkylpyridine, pyridoxal), and pyrimidines (catalyzed by dihydropyrimidine dehydrogenases). Reductions are involved in both the aerobic and the anaerobic degradation of uracil and orotic acid.

3.14.11 Reductive Dehalogenation

Reductive dehalogenation is found among anaerobes that carry out dehalorespiration, and in a restricted range of aerobic bacteria and fungi. In dehalorespiration, the dehalogenation is coupled to the synthesis of ATP and has been described in pure culture for a range of genera—*Desulfomonile, Desulfitobacterium, Desulfovibrio, Dehalobacter, Dehaloccoides,* and *Sulfurospirillum (Dehalospirillum).* The substrates range from chlorinated ethenes to chlorophenols, chlorobenzoates (including 3-chloro-4-hydroxyben-zoates), and chloroarenes: these are, most often, only partially dechlorinated. These examples are discussed in greater detail in Chapter 7, Part 1, and Chapter 9, Parts 1 and 2, while dehalorespiration is also discussed in Chapter 3 in the context of electron acceptors other than oxygen.

In several organisms the reductase is associated with membranes, and the enzyme has been solubilized using Triton X (Table 3.6). The reductases belong to three groups: Group 1 having a corrinoid and 2 [Fe–S] clusters (Table 3.7); Group 2 with a corrinoid, but lacking an [Fe–S] cluster that is found in *Desulfitobacterium frappieri* (Boyer et al. 2003); Group 3 with a heme protein containing two subunits that is found in *Desulfomonile tiedjei* (Ni et al. 1995).

Attention is drawn to the following additional details.

1. Two distinct enzyme systems for the dechlorination of tetrachloroethene and chlorophenols are present in *Desulfitobacterium* sp. strain PCE1 (van de Pas et al. 2001).
2. The dechlorinase from *Desulfitobacterium* sp. strain Y51 is specific for tetrachloro- and tri-chloroethenes, although it is able to dechlorinate hexachloro-, pentachloro-, and tetrachloroeth-anes (Suyama et al. 2002).
3. The complete dechlorination of tetrachloroethene to ethene has been accomplished hitherto only by *Dehalococcoides ethenogenes* (Magnuson et al. 1998). Other strains with reductive competence for chloroethenes are noted in Chapter 7, and the anaerobic dechlorination of

TABLE 3.6

Membrane-Bound Dehalogenases

Organism	Substrate	Reference
Desulfitobacterium frappieri	2,4,6-Trichlorophenol	Boyer et al. (2003)
≡*Desulfitobacterium hafniense*	3,5-Dichlorophenol	Thibodeau et al. (2004)
	Pentachlorophenol	Bisaillon et al. (2010)
Desulfitobacterium hafniense	3-Chloro-4-hydroxy-phenylacetate	Christiansen et al. (1998)
Desulfitobacterium dehalogenans	3-Chloro-4-hydroxy-phenylacetate	van de Pas et al. (1999)
Desulfitobacterium chlororespirans	3-Chloro-4-hydroxy-benzoate	Krasotkina et al. (2001); Löffler et al. (1996)
Desulfitobacterium sp. strain PCE1	Tetrachloroethene	van de Pas et al. (2001)
Desulfitobacterium sp. strain Y51	Tetrachloroethene	Suyama et al. (2002)
Desulfitobacterium sp. strain PCE-S	Tetrachloroethene	Miller et al. (1998)
Dehalococcoides ethenogenes	Tetrachloroethene	Magnuson et al. (1998)
Dehaloccoides sp.	Polychlorinatedbenzene	Hölscher et al. (2003)
Desulfomonile tiedjei DCB-1	3-Chlorobenzoate	Ni et al. (1995)

TABLE 3.7

Group 1 Dehalogenases: Corrinoid + 2 [Fe–S] Clusters

Organism	Substrate	Reference
Desulfitobacterium hafniense	3-Chloro-4-hydroxyphenylacetate	Christiansen et al. (1998)
Desulfitobacterium chlororespirans	3-Chloro-4-hydroxybenzoate	Krasotkina et al. (2001)
Desulfitobacterium frappieri	3,5-Dichlorophenol	Thibodeau et al. (2004)
Desulfitobacterium dehalogenans	*ortho*-Chlorophenols	van de Pas et al. (1999)
Dehalobacter restrictus	Tetrachloroethene	Schumacher et al. (1997)
Desulfitobacterium sp. strain PCE-S	Tetrachloroethene	Miller et al. (1998)
Desulfitobacterium PCE1	Tetrachloroethene	van de Pas et al. (2001)
Sulfurospirillum multivorans	Tetrachloroethene	Neumann et al. (1996)
Dehalococcoides spp	Tetrachloroethene	Magnuson et al. (1998)

PCB congeners has been examined using *Dehalococcoides* sp. strain CBDB 1 (Adrian et al. 2009) that was able to grow using highly chlorinated benzenes as electron acceptors (Jayachandran et al. 2004).

4. *Desulfitobacterium hafniense (frappieri)* is able to produce three dehalogenases, specific for 3,5-dichlorophenol (Thibodeau et al. 2004), for 2,4,6-trichlorophenol (Boyer et al. 2003), and for pentachlorophenol (Bisaillon et al. 2010).

5. In *Desulfomonile tiedjei*, reductases for 3-chlorobenzoate and tetrachloroethene are inducible (Cole et al. 1995), and after growth with 3-chlorobenzoate, this organism is able to dechlorinate polychlorinated phenols including pentachlorophenol (Mohn and Kennedy 1992).

6. *Anaeromyxobacter dehalogenans* is able to dechlorinate *ortho*-chlorophenols using acetate as electron donor (Cole et al. 1994; Sanford et al. 2002).

Reductive dechlorination has also been observed in aerobic degradations of halogenated substrates.

1. In the degradation of pentachlorophenol by *Mycobacterium chlorophenolicum*, tetrachlorohydroquinone is formed initially. This is successively dechlorinated to 1,2,4-trihydroxybenzene before ring fission (Apajalahti and Salkinoja-Salonen 1987a,b). The dechlorination is carried out by a glutathione *S*-transferase that has been purified (Xun et al. 1992; Orser et al. 1993). The degradation by *Sphingobium (Flavobacterium) chlorophenolicum* is analogous except that 2,6-dichlorohydroquinone is produced from tetrachlorohydroquinone, and may undergo fission directly (Cai and Xun 2002; Dai et al. 2003).

2. The degradation of γ-hexachlorocyclohexane by *Sphingomonas paucimobilis* strain UT26 results in the production of 2,5-dichlorohydroquinone that is dechlorinated to hydroquinone in a glutathione-dependent reaction (Miyauchi et al. 1998).

3. The degradation of 3-chlorobiphenyl by *Burkholderia xenovorans* involves the production of 3-chloro-2-hydroxy-6-oxo-6-phenol-2,4-dienoate that may inhibit degradation, but it can be reductively dechlorinated in this strain by a glutathione *S*-transferase (Miyauchi et al. 1998; Fortin et al. 2006).

Reductive dechlorination has been observed in a few other degradations that are, apparently not glutathione-dependent. It is worth noting that the transformation of 1,2,3,5-tetrachloro-4,6-dicyano-benzene to 2,3,5-trichloro-4–6-dicanophenol by *Pseudomonas* sp. CTN-3 is a purely hydrolytic—and not a reductive—reaction that is independent of ATP and CoA (Wang et al. 2010).

1. The degradation of 3,5-dichlorophenol by *Ralstonia (Alcaligenes) eutropha* takes place by dioxygenation and intradiol fission followed by lactonization with the loss of chloride, and lactone hydrolysis to 2-chloromaleylacetate. Maleylacetate reductase is able to dechlorinate this to maleylacetate and further to 3-oxoadipate that can enter the TCA cycle. This is explored further in Chapter 9. The reductase did not contain a flavin, and NADH was preferred to NADPH as reductant for both reactions (Seibert et al. 1993).

2. The degradation of 2,4-dichlorobenzoate by *Corynebacterium sepedonicum* is initiated by formation of the benzoyl-CoA ester that is dechlorinated to 4-chlorobenzoyl-CoA in an NADPH-dependent reaction (Romanov and Hausinger 1996).

3. The degradation of pentachlorophenol by the white-rot fungus *Phanerochaete chrysosporium* involves formation of tetrachlorohydroquinone that undergoes successive reductive dechlorination to trichlorohydroquinone, 2,5-dichlorohydroquinone and 2-chlorohydroquinone (Reddy and Gold 2000).

3.14.12 Metal Cations and Oxyanions

Although some metal cations and oxyanions can serve as electron acceptors for growth under anaerobic conditions, reduction may also take place gratuitously, and reductases have been characterized from a number of organisms. Reduction has been implicated in resistance to metal cations, and metalloid oxyanions that are discussed in Chapter 4. Hydrogenase that is putatively the simplest reductase is discussed in Part 3 of this chapter dealing with enzymes containing nickel and/or iron.

1. Soluble reductases at an optimum temperature of 80°C have been described from (a) *Ps. putida* that reduces chromate to insoluble Cr(III) (Park et al. 2000), and (b) *Archaeoglobus fulgidus* that can reduce Fe(III)-EDTA (Vadas et al. 1999). A membrane-bound chromate reductase has been purified from *Enterobacter cloacae* (Wang et al. 1990).

2. Dimeric flavoprotein chromate reductases have been purified from *Ps. putida* (ChrR) and *E. coli* (YieF). The former produces a semiquinone and transiently reactive oxygen species, whereas the latter is an obligate four-electron reductant. One-electron reduction of Cr(VI) to Cr(V) has, however, been observed as an intermediate in the reduction by the NAD(P)H-dependent reductase of *Pseudomonas ambigua* strain G-1 (Suzuki et al. 1992).

3. It has been shown that the nitroreductases from *E. coli* and *Vibrio harveyi* are homologous to the chromate reductase from *Ps. ambigua* (Kwak et al. 2003).

4. Reductase activity is located on the outer membranes of *Shewanella putrefaciens* and has a requirement for cytochrome c (Myers and Myers 1997) and menaquinones (Saffarini et al. 2002).

5. The reductase in *Geobacter sulfurreducens* is located in the outer membrane, and a soluble Fe(III) reductase has been characterized from cells grown anaerobically with acetate as electron donor and ferric citrate or fumarate as electron acceptor (Kaufmann and Lovley 2001). The enzyme contained Fe, acid-labile S, and FAD. An extracellular c-type cytochrome is distributed in the membranes, the periplasm and the medium, and functions as a reductase for electron transfer to insoluble iron hydroxides, sulfur or manganese dioxide (Seeliger et al. 1998).

3.14.13 Nonmetal Oxyanions and Elemental Nonmetals

3.14.13.1 Sulfur Oxyanions and S^0

Reduction of sulfate to sulfide is carried out in a series of steps all of which can be used to support microbial growth in a range of organisms that includes anaerobes, aerobes, in addition to facultative representatives: $(S^{VI}) \rightarrow SO_3^{2-} (S^{IV}) \rightarrow S^0 \rightarrow S^{2-} (S^{-II})$, and for some of them in the reverse direction. Sulfate can serve both an assimilatory role to provide sulfur for the synthesis of amino acids, or as an electron acceptor and energy source under anaerobic conditions. Elementary sulfur is able to serve as a source of energy for aerobic chemolithotrophs and can be reduced to sulfide by hyperthermophilic organisms that are discussed in Chapter 11.

The 8-ε reduction of sulfate to sulfide takes place in steps: (a) reaction of sulfate with ATP to APS catalyzed by ATP-sulfurylase, (b) reduction of APS by APS reductase to produce sulfite and AMP, (c) reduction of sulfite to sulfide. Details from the x-ray crystal structure of adenylsulfate reductase from *Archaeoglobus fulgidus* revealed the role of the flavin by reaction of N_5 with the sulfur atom and transfer of electrons for the reduction from [4Fe–4S] clusters (Fritz et al. 2002b), and further details of the reduction were provided by EPR studies (Fritz et al. 2002a). Additional details have been added (Schnell et al. 2005) from the siroheme-Fe_4S_4-dependent sulfite reductase NirA from *M. tuberculosis*.

The reduction of sulfite to sulfide can be carried out under different conditions.

1. In sulfate-reducing bacteria, the initially formed sulfite is reduced to sulfide and the reductase consists of an $\alpha_2\beta_2$ structure that contains siroheme, nonheme iron, and acid-labile sulfur. The reductase has been characterized and sequenced in *A. fulgidus* and contains a siroheme-[4Fe–4S] complex that contains two sirohemes and six [Fe–S] clusters per molecule (Dahl et al. 1993).

2. Under anaerobic conditions sulfite is able to function as a terminal electron acceptor in *Shewanella oneidensis* MR-1 during growth with lactate: reduction to sulfide required mena-quinones and the reductase was carried out by an octaheme cytochrome c SirA (Shirodkar et al. 2011). This mechanism clearly differs from the siroheme-[4Fe–4S] that has been described.

3.14.13.1.1 Elementary Sulfur S^0

This is metabolized in several reactions, and only a few examples are given as illustration.

1. It can serve in the microaerophilic *Sulfurospirillum deleyianum* as an electron acceptor with the production of sulfide: during growth, H_2, formate, lactate, and pyruvate can serve as electron donors and acetate as a source of carbon (Finster et al. 1997).

2. *Pyrococcus furiosus* is an anaerobic hyperthermophile that is able to carry out fermentation both in the presence and in the absence of elementary sulfur. The oxidoreductase has been shown to be a homodimeric flavoprotein that has a coenzyme A-dependent NAD(P)H sulfur oxidoreductase (Schut et al. 2007).

3. Sulfur can be used as an electron donor in a number of sulfate-reducing bacteria and strains of *Desulfuromonas acetexigens* were able to use elemental sulfur or polysulfide and used exclusively acetate as carbon source (Finster et al. 1994).

4. The novel production of sulfate from elemental S^0 has been observed in several strains of sulfate-reducing bacteria, specifically when Mn(IV) was provided as electron acceptor (Lovley and Phillips 1994).

3.14.13.2 Nitrogen Oxyanions and Dinitrogen

Denitrification involves the successive 5-ε reduction of nitrate to nitrite, nitric oxide, nitrous oxide, and nitrogen in which the formation of an N–N bond occurs in the last two stages. For nitrate this includes the following: reductions of the oxyanions, and of N_2 to NH_3:

$$NO_3^- \, (N^V) \rightarrow NO_2^- \, (N^{III}) \rightarrow NO \, (N^{II}) \rightarrow N_2O \, (N^I) \rightarrow N_2(N^0) \rightarrow NH_3 \, (N^{-III})$$

Nitrate can fulfill several functions: assimilation as a source of nitrogen for growth, as terminal electron acceptor, and for dissimilation, and these are covered in another section of this chapter. Nitrate reductase has been discussed in a review (Moreno-Vivián et al. 1999), and only nitrite reductase is briefly noted here since it is discussed in an extensive review (Zumft 1997). The reduction of dinitrogen (N_2) to ammonia is not discussed except to provide some examples in Chapter 2 in the context of plant nutrition.

The metabolism of nitrite can take place by a 1-ε reduction to nitric oxide, by a 2-e reduction of nitrous oxide to N_2, or by a 6-ε reduction to ammonia when nitrite is used as a source of nitrogen. Denitrification produces NO, N_2O, and N_2, and there are two groups of dissimilatory nitrite reductases, one containing cd_1 heme as the prosthetic group and the other containing copper: examples of the first group include *Ralstonia eutropha* and *Ps. stutzeri*, and of the second *Alcaligenes xylosoxidans*, and *Pseudomonas chlororaphis*. There are in addition, two copper centers: type I that colors the enzyme blue or green, and type II that does not contribute to the visible spectrum.

3.14.13.2.1 Nitrite Reductase

1. An x-ray crystallographic study of the nitrite reductase from *Alcaligenes faecalis* showed that copper occupied a tetrahedral site with three histidines and a water molecule, and a model was proposed in which the water molecule was replaced by nitrite followed by reduction to a Cu^+ complex from which NO was released (Murphy et al. 1997).

2. The nirA gene that encodes the blue dissimilatory nitrite reductase in *Alcaligenes xylosoxidans* has been cloned and expressed in *E. coli* (Prudêncio et al. 1999). The enzyme was trimeric and

on the basis of EPR and metal analysis had only type I copper center with low nitrite reductase activity, while the enzyme obtained by incubation of the periplasmic fraction with Cu^{2+} before purification was trimeric, had high reductase activity and had an EPR spectrum characteristic of type I and type II copper centers.

3. Analysis of the reduction potentials of both copper centers type I and type II was carried out for the blue nitrite reductase isolated from the soluble extract of *Ps. chlororaphis*. The catalytic reactions are: reduction of the type I center, oxidation of the type I center by the type II center, and reoxidation of the type II center Cu by nitrite and, on the basis of EPR spectroscopy it was proposed that in the presence of nitrite, electron transfer between the type I copper and type II copper was the most rapid (Pinho et al. 2004).

4. The heme nitrite reductases consist of two subunits each containing one covalently bound *c*-heme and one d_1-heme. In *Ps. aeruginosa* electrons enter at the c-heme domain and catalysis takes place at the level of the d_1-heme, and a mechanism has been proposed to enable rapid release from intermediates of the potentially inhibiting nitric oxide that is produced by reduction (Cutruzzolà et al. 2001).

3.14.13.2.2 Nitric Oxide Reductase

This carries out the 2-e reduction of nitric oxide to nitrous oxide that is the penultimate step in denitrification. Several aspects are noteworthy:

1. The membrane-associated binuclear enzyme consists of the subunits NorB and NorC. Heme *c* is bound to Norco and serves for electron entry, and NorB contains heme *b* and a five coordinate heme b_3 which form the dinuclear reduction site together with a nonheme iron atom. Although resonance Raman scattering in the enzyme from *Paracoccus denitrificans* has suggested a mechanism involving splitting of the Fe–O–Fe between the tris-histidine nonheme Fe (III) and the heme that is coordinated to an additional histidine (Pinakoulaki et al. 2002), the details remain unresolved in spite of a biomimetic model of nitric oxide reductase (Collman et al. 2008).

2. It has been shown that in some fungi the nitric oxide reductase that produces nitrous oxide is carried out by a cytochrome P450nor that can use using NADH directly as electron donor. This has been observed in the yeasts *Fusarium oxysporum* (Nakahara et al. 1993), *Trichosporon cutaneum* (Tsuruta et al. 1998), and *Cylindrocarpon tonkinense* in which two isoforms of the enzyme have been characterized (Usuda et al. 1995).

3. An unusual phenomenon termed codenitrification has been observed in the fungus *Fusarium oxysporum* (Tanimoto et al. 1992; Su et al. 2004) in which only one atom of nitrous oxide is produced from the nitric oxide substrate, whereas the other comes from a nitrogen source in the medium.

4. Codenitrification has been observed only in a few bacteria: *Streptomyces antibiotics* was able to produce N_2 and some N_2O from nitrate (Kumon et al. 2002). By using ^{15}N-labeled nitrate it emerged that the N_2 consisted of [15]N[14]N in addition to [15]N[15]N. Initial reduction of NO_3^- to NO_2^- took place, and this was termed codenitrification since one of the atoms of N_2O and N_2 was derived from a source other than the ^{15}N nitrate that was supplied.

3.14.13.2.3 Nitrous Oxide Reductase

The dissimilation of nitrous oxide to dinitrogen completes the pathway from nitrite reduction. The enzyme extracted from *Ps. stutzeri* has been extensively examined by EPR and x-ray crystallography. The enzyme contains eight copper atoms in two identical subunits and occurs in various forms and colors, and the EPR spectrum showed an unpaired electron distributed over two copper atoms (Antholine et al. 1992). Electrons enter the reductase at the CuA site, and multi-wavelength anomalous dispersion of the enzyme from *Pseudomonas nautica* revealed that the catalytic center CuZ consists of four copper ions coordinated to seven histidine residues and S^{2-}, and that the substrate N_2O binds to a single copper ion (Brown et al. 2000).

3.14.13.2.4 Nitrite Ammonification

Respiratory nitrite ammonification takes place by the reaction: $NO_2^- + 8H^+ + 6\varepsilon \rightarrow NH_4^+ + 2H_2O$ and has been found in several anaerobes that are discussed in a review which discusses their enzymology and bioenergetics (Simon 2002). Only some examples are given as illustration.

1. Membrane preparations from *Desulfovibrio gigas* grown in a lactate–sulfate medium were examined: H_2 oxidation was coupled to the reduction of nitrite and hydroxylamine coupled to the synthesis of ATP and the production of ammonia (Barton et al. 1983). The nitrite reductase in *Desulfovibrio desulfuricans* is membrane associated and has been examined by EPR and Mössbauer spectroscopy that revealed six ferric heme groups (Costa et al. 1990).

2. *Wolinella succinogenes* was able to carry out phosphorylation during growth with formate as electron donor and the reduction of nitrite to ammonia (Bokranz et al. 1983). Growth with succinate, formate as electron donor and nitrate as terminal electron acceptor and produced two nitrite reductases, although only the membrane-bound enzyme was formed during growth with fumarate: it had a molecular mass of 63 kDa and contained four heme c groups (Schröder et al. 1985).

3. During growth of *E. coli* K-12 in the presence of nitrite and formate, genes for the reduction to ammonia occur in an operon containing seven genes nrfA to nrfG (Hussain et al. 1994) and the periplasmic cytochromes c_{552} and c_{550} encoded by *nrfA* and *napB* are synthesized coordinately with the membrane-associated cytochromes NrfB and NapC (Eaves et al. 1998). The NrfA nitrite complex is a decaheme homodimer and the reductase that NrfB that provides electrons is also a decaheme dimer and a 20-heme $[NrfB]_2[NrfA)]_2$ is transiently formed (Clarke et al. 2004), and the complete pathway includes a $[4Fe-4S]$ ferredoxin encoded by nfrC and an integral membrane quinol dehydrogenase encoded by nfrD.

3.14.13.2.5 Dinitrogenase

Dinitrogenase is induced under nitrogen-limiting conditions by a range of bacteria under anaerobic or microaerophilic conditions. Although the paradigm reaction is the ATP-dependent reduction of dinitrogen to ammonia, a range of other substrates are accepted including acetylene that is reduced to ethylene and has been widely used as a surrogate, as well as a number of novel substrates (Seefeldt et al. 1995; Rasche and Seefeldt 1997). The vanadium enzyme from *Azotobacter vinelandii* has been his-tagged and purified to homogeneity and it was shown to have an $\alpha_2\beta_2\delta_4$ heterooctameric structure and contained a P-cluster that was apparently different from that of the MoFe protein (Lee et al. 2009). Differences between the MoFe and VFe enzymes are noted in Chapter 7, including the two mechanisms for the reduction of H^+ to H_2 only one of which was suppressed by CO.

Other important reductions mediated by nitrogenase and nitrogenase-like enzymes have emerged.

1. Reduction of carbon monoxide to ethene, ethane, propene and propane can be carried out by the Mo-nitrogenase of *Azotobacter vinelandii* (Hu et al. 2011) and, additionally to methane when α-valine[70] was replaced by alanine or glycine in the MoFe protein (Yang et al. 2011). This is discussed in Chapter 7.

2. The synthesis of chlorophyll and bacteriochlorophyll in algae and phototrophic bacteria involves stereospecific reduction of ring D of protochlorophyllide by a dark-operative oxidoreductase. The reductase from *Chlorobium tepidum* has been heterologously produced and purified in *E. coli*, and is an ATP-dependent enzyme that is mechanistically similar to the reduction of dinitrogen to ammonia—although there were important differences since this enzyme lacked the FeMoco- and P-clusters of nitrogenase (Bröcker et al. 2008). The synthesis of bacteriochlorophyll in

Rhodobacter capsulatus is carried out by two reductions of protochlorophyllide, a dark-operative reduction of the D ring to chlorophyllide *a*, followed by a further ATP-dependent reduction of ring B, both of which carried out by an enzyme similar to nitrogenase (Nomata et al. 2006).

3.14.13.3 Selenate Reductase

There are several groups of selenate reductase, although only some are able to support anaerobic respiration using acetate or lactate.

1. The selenate reductase in *Thauera selenatis* is a soluble periplasmic protein consisting of three subunits, a catalytic subunit containing molybdenum cofactor (Mo(V)), a subunit containing [Fe–S] clusters, one [3Fe–4S] and three [4Fe–4S], and one containing heme *b* (Schröder et al. 1997). It is able to reduce selenate—though neither arsenate, nitrate nor sulfate.

2. *Bacillus selenatarsenatis* strain SF-1 is able to reduce selenate to selenite, arsenate to arsenite and nitrate to nitrite during anaerobic respiration using lactate as a carbon source. An operon *srdBC* has been characterized in which SrdA shows the molybdenum—containing catalytic subunits, SrdB is four-cluster protein of molybdopterin oxidoreductase that takes part in electron transfer between quinones and the catalytic subunit that contains four 4Fe–4S clusters while SrdC is probably a transmembrane anchor protein (Kuroda et al. 2011).

3. The selenate reductase from *Enterobacter cloacae* SLDa-1 is membrane-bound containing three subunits (Ridley et al. 2006). Selenate reductase is able to reduce chlorate and bromate but not nitrate, contains Mo, heme and nonheme iron, and consists of three subunits in an $\alpha_3\beta_3\gamma_3$ configuration with a *b*-type cytochrome in the active complex. It functions only under aerobic conditions reducing selenate to Se^0, but is not able to serve as an electron acceptor for anaerobic growth. There are separate nitrate and selenate reductases, both of which are membrane bound.

3.14.13.4 Chlorate and Perchlorate Reductases

Chlorate and perchlorate reductases have been characterized from a number of different organisms.

1. Chlorate reductase in *Azospira oryzae* strain GR-1 (Kengen et al. 1999) that is found in the periplasm is oxygen-sensitive, and contains molybdenum and [3Fe–4S] and [4Fe–4S] clusters.

2. A different cytoplasmic reductase has been described in *Pseudomonas chloritidismutans* (Wolterink et al. 2003).

3. Chlorate reductase from *I. dechloratans* consists of three subunits: it contains heme *b* and resembles enzymes belonging to the molybdopterin DMSO reductase class II family (Thorell et al. 2003).

4. Perchlorate reductase from *Dechloromonas agitata* is different from that of *I. dechloratus*, although the *pcrAB* gene products are similar to the α- and β-subunits of other reductases including nitrate, selenate and chlorate reductases. The *pcrC* gene product was similar to a *c*-type cytochrome, and the *pcrD* product similar to the molybdenum-containing proteins of the DMSO reductase family (Bender et al. 2005).

3.15 References

Adrian L, V Dudková, K Demnerová, DL Bedard. 2009. "*Dehalococcoides*" sp. strain CBDB1 extensively dechlorinates the commercial polychlorinated biphenyl mixture Arochlor 1260. *Appl Environ Microbiol* 75: 4516–4524.

Andreesen JR. 1994. Glycine metabolism in anaerobes. *Antonie van Leeuwenhoek* 66: 223–237.

Antholine WE, DHW Kastrau, GCM Stevens, G Buse, WG Zumft, PMH Kroneck. 1992. A comparative EPR investigation of the multicopper proteins nitrous-oxide reductase and cytochrome *c* oxidase. *Eur J Biochem* 209: 875–881.

Apajalahti JHA, MS Salkinoja-Salonen. 1987a. Dechlorination and para-hydroxylation of polychlorinated phenols by *Rhodococcus chlorophenolicus*. *J Bacteriol* 169: 675–681.

Apajalahti JHA, MS Salkinoja-Salonen. 1987b. Complete dechlorination of tetrachlorohydroquinone by cell extracts of pentachlorophenol-induced *Rhodococcus chlorophenolicus*. *J Bacteriol* 169: 5125–5130.

Arkowitz RA, RH Abeles. 1991. Mechanism of action of clostridial glycine reductase: Isolation and characterization of a covalent acetyl enzyme intermediate. *Biochemistry* 30: 4090–4097.

Barker HA. 1961. Fermentations of nitrogenous organic compounds pp. 151–207. In *The Bacteria*, Vol. 2 (Eds IC Gunsalus, RY Stanier). Academic Press, New York.

Barna TM, H Khan, NC Bruce, I Barsukov, NS Scrutton, PCE Moody. 2001. Crystal structure of pentaerythritol tetranitrate reductase: "Flipped" binding geometries for steroid substrates in different redox states of the enzyme. *J Mol Biol* 310: 433–447.

Barton LL, J LeGall, JM Odom, HD Peck. 1983. Energy coupling to nitrite respiration in the sulfate-reducing bacterium *Desulfovibrio gigas*. *J Bacteriol* 153: 867–871.

Basran J, RJ Harris, MJ Sutcliffe, NS Scrutton. 2003. H-tunneling in the multiple H-transfers of the catalytic cycle of morphinone reductase and in the reductive half-reaction of the homologous pentaerythritol tetranitrate reductase. *J Biol Chem* 278: 43973–43982.

Behrend C, K Heesche-Wagner. 1999. Formation of hydride-Meisenheimer complexes of picric acid (2,4,6-trinitrophenol) and 2,4-dinitrophenol during mineralization of picric acid by *Nocardioides* sp. strain CB 22-2. *Appl Environ Microbiol* 65: 1372–1377.

Bender KS, C Shang, R Chakraborty, SM Belchik, JD Coates, LA Achenbach. 2005. Identification, characterization, and classification of genes encoding perchlorate reductase. *J Bacteriol* 187: 5090–5096.

Bisaillon A, R Beaudet, F Lépine, E Déziel, R Villemur. 2010. Identification and characterization of a novel CprA reductive dehalogenase specific for highly chlorinated phenols from *Desulfitobacterium hafniense* strain PCP-1. *Appl Environ Microbiol* 76: 7536–7540.

Blehert DS, BG Fox, GH Chambliss. 1999. Cloning and sequence analysis of two *Pseudomonas* flavoprotein xenobiotic reductases. *J Bacteriol* 181: 6254–6263.

Blümel S, H-J Knackmuss, A Stolz. 2002. Molecular cloning and characterization of the gene coding for the aerobic azoreductase from *Xenophilus azovorans* KF46F. *Appl Environ Microbiol* 68: 3948–3955.

Boal AK, JA Cotruvo, JoAnne Stubbe, AC Rosenweig. 2010. Structural basis for activation of Class I b ribonucleotide reductase. *Science* 329: 1526–1530.

Bobik TA, RS Wolfe. 1989. An unusual thiol-driven fumarate reductase in *Methanobacterium* with the production of the heterodisulphide of coenzyme M and *N*-(7methylmercaptoyl)threonine-O^3-phosphate. *J Biol Chem* 264: 18714–18718.

Bokranz M, J Katz, I Schröder, AM Robertson, A Kröger. 1983. Energy metabolism and biosynthesis of *Vibrio succinogenes* growing with nitrate as terminal electron acceptor. *Arch Microbiol* 135: 36–41.

Boll M, G Fuchs. 1995. Benzoyl-coenzyme A reductase (dearomatizing), a key enzymes in anaerobic metabolism. ATP dependence of the reaction, purification and some properties of the enzyme from Thauera aromatica strain K172. *Eur J Biochem* 234: 921–933.

Boll M, SJP Albracht, G Fuchs. 1997. Benzoyl-CoA reductase (dearomatizing), a key enzyme of anaerobic aromatic metabolism. A study of adenosine triphosphate, activity, ATP stoichiometry of the reaction and EPR properties of the enzyme. *Eur J Biochem* 244: 840–851.

Boll M, G Fuchs, C Meier, A Trautwein, A El Kasma, SW Ragsdale, G Buchanan, DJ Lowe. 2001. Redox centers of 4-hydroxybenzoyl-CoA reductase, a member of the xanthine oxidase family of molybdenum-containing enzymes. *J Biol Chem* 276: 47853–47862.

Boll M, G Fuchs, C Meier, AX Trautwein, DJ Lowe. 2000. EPR and Mössbauer studies of benzoyl-CoA reductase. *J Biol Chem* 275: 31857–31868.

Boyer A, R-P Bélanger, M Saucier, R Villemur, F Lépine, P Juteau, R Baudet. 2003. Purification, cloning and sequencing of an enzyme mediating the reductive dechlorination of 2,4,6-trichlorophenol from *Desulfitobacterium frappieri* PCP-1. *Biochem J* 373: 297–303.

Braune A, M Gütschow, W Engst, M Blaut. 2001. Degradation of quercetin and luteolin by *Eubacterium ramulus*. *Appl Environ Microbiol* 67: 558–5567.

Breese K, G Fuchs. 1998. 4-hydroxybenzoyl-CoA reductase (dehydroxylating) from the denitrifying bacterium *Thauera aromatica*. Prosthetic groups, electron donor, and genes of a member of the molybdenum–flavin–iron–sulfur proteins. *Eur J Biochem* 251: 916–923.

Bröcker MJ, S Virus, S Ganskow, P Heathcote, DW Heinz, W-D Schubert, D Jahn, J Moser. 2008. ATP-driven reduction by dark-operative protochlorophyllide oxidoreductase from *Chlorobium tepidum* mechanistically resembles nitrogenase catalysis. *J Biol Chem* 283: 10559–10567.

Brown K et al. 2000. A novel type of catalytic copper cluster in nitrous oxide reductase. *Nature Struct Mol Biol* 7: 191–195.

Bryant C, M DeLuca. 1991. Purification and characterization of an oxygen-insensitive NAD(P)H nitroreductase from *Enterobacter cloacae*. *J Biol Chem* 266: 4119–4125.

Bryant C, L Hubbard, WD McElroy. 1991. Cloning, nucleotide sequence, and expression of nitroreductase gene from *Enterobacter cloacae*. *J Biol Chem* 266: 4126–4130s.

Bühler M, H Simon. 1982. On the kinetics and mechanism of enoate reductase. *Hoppe-Seylers Z physiol Chemie* 363: 609–625.

Cai M, L Xun. 2002. Organization and regulation of pentachlorophenol-degrading genes in *Sphingobium chlorophenolicum* ATCC 39723. *J Bacteriol* 184: 4672–4680.

Caldeira J, R Feicht, H White, M Teixeira, JJG Moura, H Simon, I Moura. 1996. EPR and Mössbauer spectroscopic studies on enoate reductase. *J Biol Chem* 271: 18743–18748.

Chen H, SL Hopper, CE Cerniglia. 2005. Biochemical and molecular characterization of an azoreductase from *Staphylococcus aureus*, a tetrameric NADPH-dependent flavoprotein. *Microbiology (UK)* 151: 1433–1441.

Christiansen N, BK Ahring, G Wohlfarth, G Diekert. 1998. Purification and characterization of the 3-chloro-4-hydroxyphenylacetate reductive dehalogenase of *Desulfitobacterium hafniense*. *FEBS Lett* 436: 159–1632.

Clarke TA, V Dennison, HE Seward, B Burlat, JA Cole, AM Hemmings, DJ Richardson. 2004. Purification and spectroscopic characterization of *Escherichia coli* NrfB, a decaheme homodimer that transfers electrons to the decaheme periplasmic nitrite reductase complex. *J Biol Chem* 279: 41333–41339.

Cole JR, AL Cascarelli, WW Mohn, JM Tiedje. 1994. Isolation and characterization of a novel bacterium growing via reductive dehalogenation of 2-chlorophenol. *Appl Environ Microbiol* 60: 3536–3542.

Cole JC, BZ Fathepure, JM Tiedje. 1995. Tetrachloroethene and 3-chlorobenzoate dechlorination activities are co-induced in *Desulfomonile tiedjei* DCB-1. *Biodegradation* 6: 167–172.

Collman JP, Y Yang, A Dey, RA Decréau, S Ghosh, T Ohta, EI Solomon. 2008. A functional nitric oxide reductase model. *Proc Natl Acad Sci USA* 105: 15660–15665.

Costa C, JJG Moura, I Moura, MY Liu, HD Peck, J LeGall, Y Wang, BH Huynh. 1990. Hexaheme nitrite reductase from *Desulfovibrio desulfuricans*. *J Biol Chem* 265: 14382–14387.

Cutruzzolà F, K Brown, EK Wilson, A Bellelli, M Arese, M Tregoni, C Cambillaua, M Brunori. 2001. The nitrite reductase from *Pseudomonas aeruginosa*: Essential role of two active-site histidines in the catalytic and structural properties. *Proc Natl Acad Sci USA* 98: 2232–2237.

Dahl C, NM Kredich, R Deutzmann, HG Trüper. 1993. Dissimilatory sulfite reductase from *Archaeoglobus fulgidus*: Physico-chemical properties of the enzyme and cloning, sequencing and analysis of the reductase genes. *J Gen Microbiol* 139: 1817–1828.

Dai M, JB Rogers, JR Warner, SD Copley. 2003. A previously unrecognized step in pentachlorophenol degradation in *Sphingobium chlorophenolicum* is catalyzed by tetrachlorobenzoquinone reductase (PcpD). *J Bacteriol* 185: 302–310.

de Jong RM, W Brugman, GJ Poelarends, CP Witman, BW Dijkstra. 2004. The X-ray structure of *trans*-3-chloroacrylic acid dehalogenase reveals a novel hydration mechanism in the tautomerase superfamily. *J Biol Chem* 279: 11546–11552.

de Wever H, JR Cole, MR Fettig, DA Hogan, JM Tiedje. 2000. Reductive dehalogenation of trichloroacetic acid by *Trichlorobacter thiogenes* gen. nov., sp. nov. *Appl Environ Microbiol* 66: 2297–2301.

DeWeerd KA, L Mandelco, RS Tanner, CW Woese, JM Suflita. 1990. *Desulfomonile tiedjei* gen. nov. and sp. nov., a novel anaerobic dehalogenating, sulfate-reducing bacterium. *Arch Microbiol* 154: 23–30.

Dietrich W, O Klimmek. 2002. The function of methyl-menaquinone-6 and polysulfide reductase membrane anchor (PsrC) in polysulfide respiration of *Wolinella succinogenes*. *Eur J Biochem* 269: 1086–1095.

Dörner E, M Boll. 2002. Properties of 2-oxoglutarate: Ferredoxin oxidoreductase from *Thauera aromatica* and it role in enzymatic reduction of the aromatic ring. *J Bacteriol* 184: 3975–3983.

Dürre P, JR Andreesen. 1982. Purine and glycine metabolism by purinolytic clostridia. *J Bacteriol* 154: 192–199.

Eaves DJ, J Grove, W Staudenmann, P James, RK Poole, SA White, I Griffiths, JA Cole. 1998. Involvement of products of the *nrfEFG* genes in the covalent attachment of haem c to a novel cycteine-lysine motif in the cytochrome c_{552} nitrite reductase from *Escherichia coli. Mol Microbiol* 28: 205–216.

Ebenau-Jehle, M Boll, G Fuchs. 2003. 2-oxoglutarate: $NADP^+$ oxidoreductase in *Azoarcus evansii*: Properties and function in electron transfer reactions in aromatic ring reduction. *J Bacteriol* 185: 6119–6129.

Ebert S, P-G Rieger, H-J Knackmuss. 1999. Function of coenzyme F_{420} in aerobic catabolism of 2,4,6-trinitrophenol and 2,4-dinitrophenol by *Nocardioides simplex* FJ2–1A. *J Bacteriol* 181: 2669–2674.

Ellefson WL, WB Whitman, RS Wolf. 1982. Nickel-containing factor F_{430}: chromophore of the methylreductase of *Methanobacterium. Proc Natl Acad Sci USA* 79: 3707–3710.

Ellermann J, R Hedderich, R Böcher, RK Thauer. 1988. The final step in methane formation. Investigations with highly purified methyl-CoM reductase (component C) from *Methanobacterium thermoautotrophicum* (strain Marburg). *Eur J Biochem* 172: 669–677.

Ermler U, W Grabarse, S Shima, M Goubeaud, RK Thauer. 1997. Crystal structure of methyl-coenzyme M reductase: The key enzyme of biological methane formation. *Science* 278: 1457–1462.

Faber K. 1997. *Biotransformations in Organic Chemistry*, 3rd edition. Springer-Verlag, Berlin.

Finster K, F Bak, N Pfennig. 1994. *Desulfuromonas acetexigens* sp. nov., a dissimilatory sulfur-reducing eubacterium from anoxic freshwater sediments. *Arch Microbiol* 161: 328–332.

Finster K, W Liesack, BJ Tindall. 1997. *Sulfurospirillum arcachonense* sp. nov., a new microaerophilic sulfur-reducing bacterium. *Int J Syst Bacteriol* 47: 1212–1217.

Fortin PD, GP Horsman, HM Yang, LD Eltis. 2006. A glutathione *S*-transferase catalyzes the dehalogenation of inhibitory metabolites of polychlorinated biphenyls. *J Bacteriol* 188: 4424–4430.

French CE, NC Bruce. 1994. Purification and characterization of morphinone reductase from *Pseudomonas putida* M10. *Biochem J* 301: 97–103.

French CE, S Nicklin, NC Bruce. 1996. Sequence and properties of pentaerythritol tetranitrate reductase from *Enterobacter cloacae* PB2. *J Bacteriol* 178: 6623–6627.

Fritz G, T Büchert, PMH Kroneck. 2002b. The function of the [4Fe–4S] clusters and FAD in bacterial and archaeal adenylsulfate reductases. *J Biol Chem* 277: 26066–26072.

Fritz G, A Roth, A Schiffer, T Bücher, G Bourenkov, HD Bartunik, H Huber, KO Stetter, PMH Kroneck, U Ermler. 2002a. Structure of adenylsulfate reductase from the hyperthermophilic *Archaeoglobus fulgidus* at 1.6 Å resolution. *Proc Natl Acad Sci USA* 99: 1836–1841.

Gibson J, M Dispensam CS Harwood. 1997. 4-hydroxybenzoyl coenzyme A reductase (dehydroxylating) is required for anaerobic degradation of 4-hydroxybenzoate by *Rhodopseudomonas palustris* and shares features with molybdenum-containing hydroxylases. *J Bacteriol* 179: 634–642.

Haddock JD, JG Ferry. 1989. Purification and properties of phloroglucinol reductase from *Eubacterium oxidoreducens. J Biol Chem* 264: 4423–4427.

Haddock JD, JG Ferry. 1993. Initial steps in the anaerobic degradation of 3,4,5-trihydroxybenzoate by *Eubacterium oxidoreducens*: characterization of mutants and role of 1,2,3,5-tetrahydroxybenzene. *J Bacteriol* 175: 669–673.

Harwood CS, G Burchardt, H Herrmann, G Fuchs. 1999. Anaerobic metabolism of aromatic compounds via the benzoyl-CoA pathway. *FEMS Microbiol Rev* 22: 439–458.

He A, T Li, L Daniels, I Fotheringham, JPN Rosazza. 2004. *Nocardia* sp. carboxylic acid reductase: cloning, expression, and characterization of a new aldehyde oxidoreductase family. *Appl Environ Microbiol* 70: 1874–1881.

Heiss G, KW Hofmann, N Trachtmann, DM Walters, P Rouvière, H-J Knackmuss. 2002. *npd* gene functions of *Rhodococcus (opacus) erythropolis* HL-PM-1 in the initial steps of 2,4,6-trinitrophenol degradation. *Microbiology (UK)* 148: 799–806.

Hofmann KW, H-J Knackmuss, G Heiss. 2004. Nitrite elimination and hydrolytic ring cleavage in 2,4,6-trinitrophenol (picric acid) degradation. *Appl Environ Microbiol* 70: 2854–2860.

Hölscher T, H Görisch, L Adrian. 2003. Reductive dehalogenation of chlorobenzene congeners in cell extracts of *Dehalococcoides* sp. strain CBDB1. *Appl Environ Microbiol* 69: 2999–3001.

Hu Y, CC Lee, MW Ribbe. 2011. Extending the carbon chain: Hydrocarbon formation catalyzed by Vanadium/Molybdenum nitrogenases. *Science* 333: 753–755.

Hussain H, J Grove, L Griffiths, S Busby, J Cole. 1994. A seven-gene operon essential for formate-dependent nitrite reduction to ammonia by enteric bacteria. *Mol Microbiol* 12: 153–163.

Itoh N, R Morihama, J Wang, K Okada, N Mizuguchi. 1997. Purification and characterization of phenylacetaldehyde reductase from a styrene-assimilating *Corynebacterium* strain, ST-10. *Appl Environ Microbiol* 63: 3783–3788.

Jackson S, M Calos, A Myers, WT Self. 2006. Analysis of proline reduction in the nosocomial pathogen *Clostridium difficile*. *J Bacteriol* 188: 8487–8495.

Jayachandran G, H Görisch, L Adrian. 2004. Studies on hydrogenase activity and chlorobenzene respiration in *Dahalococcoides* sp. strain CBDB1. *Arch Microbiol* 182: 498–504.

Johnson EF, B Mukhopadhyay. 2005. A new type of sulfite reductase, a novel coenzyme F_{420}-dependent enzyme, from the methanarchaeon *Methanocaldocccus jannaschii*. *J Biol Chem* 280: 38776–38786.

Jordan A, E Torrents, I Sala, U Hellman, I Gilbert, P Reichard. 1999. Ribonucleotide reduction in *Pseudomonas* species: Simultaneous presence of active enzymes from different classes. *J Bacteriol* 181: 3974–3980.

Kabisch UC, A Gräntzdörffer, A Schierhorn, KP Rüecknagel, JR Andreesen, A Pich. 1999. Identification of D-proline reductase from *Clostridium sticklandii* as a selenoenzyme and indications for a catalytically active pyruvoyl group derived from a cysteine residue by cleavage of a proprotein. *J Biol Chem* 274: 8445–8454.

Kaschabek SR, W Reineke. 1995. Maleylacetate reductase of *Pseudomonas* sp. strain B13: Specificity of substrate conversion and halide elimination. *J Bacteriol* 177: 320–325.

Kaufmann F, DR Lovley. 2001. Isolation and characterization of a soluble NADPH-dependent Fe(III) reductase from *Geobacter Sulfurreducens*. *J Bacteriol* 183: 4468–4478.

Kengen SWM, GB Rikken, WR Hagen, CG van Ginkel, ALM Stams. 1999. Purification and characterization of (per)chlorate reductase from the chlorate-respiring strain GR-1. *J Bacteriol* 181: 6706–6711.

Khan H, T Barna, RJ Harris, NC Bruce, I Barsukov, AW Munro, PCE Moody, NS Scrutton. 2004. Atomic resolution structures and solution behavior of enzyme-substrate complexes of *Enterobacter cloacae* PB2 pentaerythritol tetranitrate reductase. *J Biol Chem* 279: 30563–30572.

Körtner C, F Lauterbach, D Tripier, G Unden, A Kröger. 1990. *Wolinella succinogenes* fumarate reductase contains a dihaem cytochrome *b*. *Mol Microbiol* 4: 855–860.

Koziarz JWP, J Veall, N Sandhu, P Kumar, B Hoecher, IB Lambert. 1998. Oxygen-insensitive nitroreductases: analysis of the roles of *nfsA* and *nfsB* in development of resistance to 5-nitrofuran derivatives in *Escherichia coli*. *J Bacteriol* 180: 5529–5539.

Krafft T, M Bokranz, O Klimmek, I Schröder, F Fahrenholz, E Kojro, A Kröger. 1992. Cloning and nucleotide sequence of the *psfA* gene *of Wolinella succinogenes* polysulfide reductase. *Eur J Biochem* 206: 503–510.

Krasotkina J, T Walters, KA Maruya, SW Ragsdale. 2001. Characterization of the B_{12}- and iron–sulfur-containing reductive dehalogenase from *Desulfitobacterium chlororespirans*. *J Biol Chem* 276: 40991–40997.

Kröger A, V Geisler, E Lemma, F Theis, R Lenger. 1992. Bacterial fumarate respiration. *Arch Microbiol* 158: 311–314.

Kumon Y, Y Sasaki, I Kato, N Talaya, H Shoun, T Beppu. 2002. Codenitrification and denitrification are dual metabolic pathways through which dinitrogen evolves from nitrate in *Streptomyces antibioticus*. *J Bacteriol* 184: 2963–2968.

Künkel A, M Vaupel, S Heim, RK Thauer, R Hedderich. 1997. Heterodisulfide reductase from methanol-grown cells of *Methanosarcina barkeri* is not a flavoenzyme. *Eur J Biochem* 244: 226–234.

Kung JW, C Löffler, K Dörner, D Heintz, S Gallien, A Van Dorsselaer, T Friedrich, M Boll. 2009. Identification and characterization of the tungsten-containing class of benzoyl-coenzyme A reductases. *Proc Natl Acad Sci USA* 106: 17687–17692.

Kuno S, A Bacher, H Simon. 1985. Structure of enoate reductase from a *Clostridium tyrobutyricum* (*C. spec.* La1). *Biol Chem Hoppe-Seyler* 366: 463–472.

Kurata A, T Kurihara, H Kamachi, N Esaki. 2005. 2-haloacrylate reductase: A novel enzyme of the medium-chain dehydrogenase/reductase superfamily that catalyzes the reduction of carbon-carbon double bond of unsaturated organohalogen compounds. *J Biol Chem* 280: 20286–20291.

Kuroda M, M Yamashita, E Miwa, K Imao, N Fujimoto, H Ono, K Nagano, K Sei, M Ike. 2011. molecular cloning and characterization of the *srdBCA* operon, encoding the respiratory selenate reductase complex from the selenate-reducing bacterium *Bacillus selenatarsenatis*. *J Bacteriol* 193: 2141–2148.

Kwak YH, DS Lee, HB Kim. 2003. *Vibrio harveyi* nitroreductase is also a chromate reductase. *Appl Environ Microbiol* 69: 4390–4395.

Laempe D, M Jahn, K Breese, H Schägger, G Fuchs. 2001. Anaerobic metabolism of 3-hydroxybenzoate by the denitrifying bacterium *Thauera aromatica*. *J Bacteriol* 183: 968–979.

Lee CC, Y Hu, MW Ribbe. 2009. Unique features of the nitrogenase VFe protein from *Azotobacter vinelandii*. *Proc Natl Acad Sci USA* 106: 929–9214.

Lehmann M, B Tshisuaka, S Fetzner, P Röger, F Lingens. 1994. Purification and characterization of isoquino-line 1-oxidoreductase from *Pseudomonas diminuta* 7, a molybdenum-containing hydroxylase. *J Biol Chem* 269: 11254–11260.

Li T, JPN Rosazza. 1997. Purification, characterization, and properties of an aryl aldehyde oxidoreductase from *Nocardia* sp. strain NRRL 5646. *J Bacteriol* 179: 3482–3487.

Li T, JPN Rosazza. 1998. NMR identification of an acyl-adenylate intermediate in the aryl-aldehyde oxidore-ductase catalyzed reaction. *J Biol Chem* 273: 34230–34233.

Lovley DR, EJP Phillips. 1994. Novel process for anaerobic sulfate production from elemental sulfur by sul-fate-reducing bacteria. *Appl Environ Microbiol* 60: 2394–2399.

Löffler FE, RA Sanford, JM Tiedjei. 1996. Initial characterization of a reductive dehalogenase from *Desulfitobacterium chlororespirans* Co23. *Appl Environ Microbiol* 62: 3809–3813.

Luijten MLGC, J de Weert, H Smidt, HTS Boschker, WM de Vos, G Schraa, AJM Stams. 2003. description of *Sulfurospirillum halorespirans* sp. nov., an anaerobic, tetrachloroethene-respiring bacterium, and transfer of *Dehalospirillum multivorans* to the genus *Sulfurospirillum* as *Sulfurospirillum multivorans* comb. nov. *Int J Syst Evol Microbiol* 53: 787–793.

Magnuson JK, RV Stern, JM Gossett, SH Zinder, DR Burris. 1998. Reductive dechlorination of tetrachloroeth-ene to ethene by a two-component enzyme pathway. *Appl Environ Microbiol* 64: 1270–1275.

Meah Y, V Massey. 2000. Old yellow enzyme: stepwise reduction of nitroolefins and catalysis of acid-nitro tautomerization. *Proc Natl Acad Sci USA* 97: 10733–10738.

Miller E, G Wohlfarth, G Diekert. 1998. Purification and characterization of the tetrachloroethene reductive dehalogenase of strain PCE-S. *Arch Microbiol* 169: 497–502.

Miura K, Y Tomioka, H Suzuki, M Yonezawa, T Hishinuma, M Mizugaki. 1997. Molecular cloning of the *nemA* gene encoding *N*-ethylmaleimide reductase from *Escherichia coli*. *Biol Pharm Bull* 20: 110–112.

Miyauchi K, S-K Suh, Y Nagata, M Takagi. 1998. Cloning and sequencing of a 2,5-dichlorohydroquinone reductive dehalogenase whose product is involved in degradation of γ-hexachlorocyclohexane by *Sphingomonas paucimobilis*. *J Bacteriol* 180: 1354–1359iii.

Miyauchi K, Y Adachi, Y Nagata, M Takagi. 1999. Cloning and sequencing of a novel *meta*-cleavage dioxygen-ase gene whose product is involved in degradation of γ-hexachlorocyclohexane in *Sphingomonas pauci-mobilis*. *J Bacteriol* 181: 6712–6719.

Möbitz H, M Boll. 2002. A Birch-like mechanism in enzymatic benzoyl-CoA reduction: A kinetic study of substrate analogues combined with an *ab initio* model. *Biochemistry* 41: 1752–1758.

Mohn WW, KJ Kennedy. 1992. Reductive dehalogenation of chlorophenols by *Desulfomonile tiedjei* DCB-1. *Appl Environ Microbiol* 58: 1367–1370.

Moreno-Vivián C, P Cabello, M Martínez-Ljuque, R Blasco, F Castillo. 1999. Prokaryotic nitrate reduction: Molecular properties and functional distinction among bacterial nitrate reductases. *J Bacteriol* 181: 6573–6584.

Mowafy AM, T Kurihara, A Kurata, T Uemura, N Esaki. 2010. 2-haloacrylate hydratase, a new class of flavo-enzyme that catalyzses the addition of water to the substrate for dehalogenation. *Appl Environ Microbiol* 76: 6032–6137.

Müller JA, B Schink. 2000. Initial steps in the fermentation of 3-hydroxybenzoate by *Sporotomaculum hydroxy-benzoicum*. *Arch Microbiol* 173: 288–295.

Muramatsu H, H Mihara, R Kakutani, M Yasuda, M Ueda, T Kurihara, N Esaki. 2005. The putative malate/lactate dehydrogenase from *Pseudomonas putida* is an NADPH-dependent Δ^1-piperideine-2-carboxylate/Δ^1-pyrroline-2-carboxylate reductase involved in the catabolism of D-lysine and D-proline. *J Biol Chem* 280: 5329–5335.

Murphy MEP, S Turley, ET Adman. 1997. Structure of nitrite bound to copper-containing nitrite reductase from *Alcaligenes faecalis*. *J Biol Chem* 272: 28455–28460.

Myers CD, JD Myers. 1997. Cloning and sequence of *cymA*, a gene encoding a tetraheme cytochrome *c* required for reduction of iron (III), fumarate, and nitrate by *Shewanella putrefaciens* MR-1. *J Bacteriol* 179: 1143–1152.

Nakahara K, T Tanimoto, K-I Hatano, K Usuda, H Shoun. 1993. Cytochrome P-450 55A1 (P-450dNIR) acts as nitric oxide reductase employing NADH as a direct electron donor. *J Biol Chem* 268: 8350–8355.

Neumann A, G Wohlfarth, G Diekert. 1996. Purification and characterization of tetrachloroethene reductive dehalogenase from *Dehalospirillum multivorans*. *J Biol Chem* 271: 16515–16519.

Ni S, JK Fredrickson, L Xun. 1995. Purification and characterization of a novel-3-chlorobenzoate-reductive dehalogenase from the cytoplasmic membrane of *Desulfomonile tiedjei* DCB-1. *J Bacteriol* 177: 5135–5139.

Nomata J, T Mizoguchi, H Tamiaki, Y Fujita. 2006. A second nitrogenase-like enzyme for bacteriochlorophyll biosynthesis. Reconstitution of chlorophyllide *a* reductase with purified X-protein (BchX) and YZ protein (BachY-BchZ) from *Rhodobacter capsulatus*. *J Biol Chem* 281: 15021–15028.

Nordlund P, P Reichard. 2006. Ribonucleotide reductases. *Annu Rev Biochem* 75: 681–706.

Orser CS, J Dutton, C Lange, P Jablonsli, L Xun, M Hargis. 1993. Characterization of a *Flavobacterium* glutathione *S*-transferase gene involved in reductive dechlorination. *J Bacteriol* 175: 2640–2644.

Park CH, M Keyhan, B Wielinga, S Fendorf, A Matin. 2000. Purification to homogeneity and characterization of a novel *Pseudomonas putida* chromate reductase. *Appl Environ Microbiol* 66: 1788–1795.

Peters F, M Rother, M Boll. 2004. Selenocysteine-containing proteins in anaerobic benzoate metabolism of *Desulfococcus multivorans*. *J Bacteriol* 186: 2156–2163.

Philipp B, D Kemmler, J Hellstern N Gorny, A Caballero, B Schink. 2002. Anaerobic degradation of protocatechuate (3,4-dihydroxybenzoate) by *Thauera aromatica* strain AR-1. *FEMS Microbiol Lett* 212: 139–143.

Pinakoulaki E, S Gemeinhardt, M Saraste, C Varotsis. 2002. Nitric-oxide reductase. Structure and properties of the catalytic site from resonance Raman scattering. *J Biol Chem* 277: 23407–23413.

Pinho D, S Besson, CD Brondino, B de Castro, I Moura. 2004. Copper-containing nitrite reductase from *Pseudomonas chlororaphis* DSM 50135. Evidence for modulation of the rate of intramolecular electron transfer through nitrite binding to the type 2 copper center. *Eur J Biochem* 271: 2361–2369.

Prudêncio M, RR Eady, G Sawers. 1999. The blue copper-containing nitrite reductase from *Alcaligenes xylosoxidans*: cloning of the nirA gene and characterization of the recombinant enzyme. *J Bacteriol* 181: 2323–2329.

Purwantini E, L Daniels. 1996. Purification of a novel coenzyme F_{420}-dependent glucose-6-phosphate dehydrogenase from *Mycobacterium smegmatis*. *J Bacteriol* 178: 2861–2866.

Rasche ME, LC Seefeldt. 1997. Reduction of thiocyanate, cyanate, and carbon disulfide by nitrogenase: kinetic characterization and EPR spectroscopic analysis. *Biochemistry* 36: 8574–8585.

Rau J, H-J Knackmuss, A Stolz. 2002. Effects of different quinoid redox mediators on the anaerobic reduction of azo dyes by bacteria. *Environ Sci Technol* 36: 1497–1504.

Rau J, A Stolz. 2003. Oxygen-insensitive nitroreductases NfsA and NfsB of *Escherichia coli* function under anaerobic conditions as lawsone-dependent azo reductases. *Appl Environ Microbiol* 69: 3448–3455.

Reddy GVB, MH Gold. 2000. Degradation of pentachlorophenol by *Phanerochaete chrysosporium*: Intermediates and reactions involved. *Microbiology (UK)* 146: 405–413.

Ridley H, CA Watts, DJ Richardson, CS Butler. 2006. Resolution of distinct membrane-bound enzymes from *Enterobacter cloacae* SKLD1a-1 that are responsible for selective reduction of nitrate and selenate anions. *Appl Environ Microbiol* 72: 5173–5180.

Rohde BH, R Schmid, MS Ullrich. 1999. Thermoregulated expression and characterization of an NAD(P)H-dependent 2-cyclohexen-1-one reductase in the plant pathogenic bacterium *Pseudomonas syringae* pv. glycinea. *J Bacteriol* 181: 814–822.

Rohdich F, A Wiese, R Feiucht, H Simnon, A Bacher. 2001. Enoate reductases of *Clostridia*. Cloning, sequencing, and expression. *J Biol Chem* 276: 5779–5787.

Romanov V, RP Hausinger. 1996. NADPH-dependent reductive *ortho* dehalogenation of 2,4-dichlorobenzoic acid in *Corynebacterium sepedonicum* KZ-4 and coryneform bacterium strain NTB-1 via 2,4-dichlorobenzoyl coenzyme A. *J Bacteriol* 178: 2656–2661.

Rospert S, D Linder, J Ellermann, RK Thauer. 1990. Two genetically distinct methyl-coenzyme M reductases in *Methanobacterium thermoautotrophicum* strain Marburg and DH. *Eur J Biochem* 194: 871–877.

Saffarini DA, SL Blumerman, KJ Mansoorabadi. 2002. Role of menaquinones in Fe(III) reduction by membrane fractions of *Shewanella putrefaciens*. *J Bacteriol* 184: 846–848.

Sanford RA, JR Cole, JM Tiedje. 2002. Characterization and description of *Anaeromyxobacter dehalogenans* gen., nov., sp. nov., an aryl-halorespiring facultative anaerobic myxobacterium. *Appl Environ Microbiol* 68: 893–900.

Schneider S, M El-Said Mohamed. 1997. Anaerobic metabolism of L-phenylalanine via benzoyl-CoA in the denitrifying bacterium *Thauera aromatica. Arch Microbiol* 168: 310–320.

Schnell R, T Sandalova, U Hellman, Y Lindqvist, G Schneider. 2005. Sirohem- and [Fe_4-S_4]-dependent NirA from *Mycobacterium tuberculosis* is a sulfite reductase with a covalent cys-tyr bond in the active site. *J Biol Chem* 280: 27319–27328.

Schoefer L, R Mohan, A Schwiertz, A Braune, M Blaut. 2003. Anaerobic degradation of flavonoids by *Clostridium orbiscindens. Appl Environ Microbiol* 69: 5849–5854.

Scholz-Muramatsu, A Neumann, M Meßmer, E Moore, G Diekert. 1995. isolation and characterization of *Dehalospirillum multivorans* gen. nov., sp. nov., a tetrachloroethene-utilizing, strictly anaerobic bacterium. *Arch Microbiol* 163: 48–56.

Schröder I, S Rech, T Krafft, JM Macey. 1997. Purification and characterization of the selenate reductase from *Thauera selenatis. J Biol Chem* 272: 23765–23768.

Schröder I, AM Robertson, M Bokranz, G Unen, R Böcher, A Kröger. 1985. The membranous nitrite reductase involved in the electron transport of *Wolinella succinogenes. Arch Microbiol* 140: 380–386.

Schumacher W, C Holliger, AJB Zehnder, WR Hagen. 1997. Redox chemistry of cobalamin and iron–sulfur cofactors in the tetrachloroethene reductase of *Dahalobacter restrictus. FEBS Lett* 409: 421–425.

Schut GJ, SL Bridger, MWW Adams. 2007. Insights into the metabolism of elemental sulfur by the hyperthermophilic archaeon *Pyrococcus furiosus*: Characterization of a coenzyme A-dependent NAD(P)H sulfur oxidoreductase. *J Bacteriol* 189: 4431–4441.

Seefeldt LC, ME Rasche, SA Ensign. 1995. Carbonyl sulfide and carbon dioxide as new substrates, and carbon disulfide as a new inhibitor, of nitrogenase. *Biochemistry* 34: 5382–5389.

Seeliger S, R Cord-Ruwisch, B Schink. 1998. A periplasmic and extracellular *c*-type cytochrome of *Geobacter sulfurreducens* acts as a ferric iron reductase and as a electron carrier to other acceptors or to partner bacteria. *J Bacteriol* 180: 3686–3691.

Seibert V, K Stadler-Fritsche, M Schlomann. 1993. Purification and characterization of maleylacetate reductase from *Alcaligenes eutrophus* JMP134 (pJP4). *J Bacteriol* 175: 6745–6754.

Seto B, TC Stadtman. 1976. Purification and properties of proline reductase from *Clostridium sticklandii. J Biol Chem* 251: 2435–2439.

Setzke E, R Hedderich, S Heiden, RK Thauer. 1994. H_2: Heterodisulfide oxidoreductase complex from *Methanobacterium thermoautotrophicum*. Composition and properties. *Eur J Biochem* 220: 139–148.

Shimada Y, S Yasuda, M Takahashi, T Hayashi, N Miyazawam, I Sato, Y Abiru, S Uchiyama, H Hishigaki. 2010. Cloning and expression of a novel NADP(H)-dependent daidzein reductase, an enzyme involved in the metabolism of daidzein, from equol-producing *Lactococcus* strain 20–92. *Appl Environ Microbiol* 76: 5892–5901.

Shirodkar S, S Reed, M Romine, D Saffarini. 2011. The octahaem SirA catalyses dissimilatory sulfite reduction in *Shewanella oneidensis* MR-1. *Environ Microbiol* 13: 108–115.

Simon J. 2002. Enzymology and bioenergetics of respiratory nitrite ammonification. *FEMS Microbiol Revs* 26: 285–309.

Sjöberg B-M. 2010. A never-ending story. *Science* 329: 1475–1476.

Smalley D, ER Rocha, CJ Smith. 2002. Aerobic-type ribonucleotide reductase in the anaerobe *Bacteroides fragilis. J Bacteriol* 184: 895–903.

Snape JR, NA Walkley, AP Morby, S Nicklin, GF White. 1997. Purification, properties, and sequence of glycerol trinitrate reductase from *Agrobacterium radiobacter. J Bacteriol* 179: 7796–7802.

Stadtman TC, JN Davis. 1991. Glycine reductase protein C. Properties and characterization of its role in the reductive cleavage of Se-carboxymethyl-selenoprotein A. *J Biol Chem* 266: 22147–22153.

Stadtman TC, P Elliot. 1957. Studies on the enzymatic reduction of amino acids II. Purification and properties of a D-proline reductase and a proline racemase from *Clostridium sticklandii. J Biol Chem* 228: 983–997.

Su F, N Takaya, H Shoun. 2004. Nitrous oxide-forming codenitrification catalyzed by cytochrome P450nor. *Biosci Biotechnol Biochem* 68: 473–475.

Sun B, JR Cole, RA Sanford, JM Tiedje. 2000. Isolation and characterization of *Desulfovibrio dechloracetivorans* sp. nov., a marine dechlorinating bacterium growing by coupling the oxidation of acetate to the reductive dechlorination of 2-chlorophenol. *Appl Environ Microbiol* 66: 2408–2413.

Suyama A, M Yamashita, S Yoshino, K Furukawa. 2002. Molecular characterization of the PceS reductive dehalogenase of *Desulfitobacterium* sp. strain Y51. *J Bacteriol* 184: 3419–3425.

Suzuki T, N Miyata, H Horitsu, K Kawai, K Takamizawa, Y Tai, M Okazaki. 1992. NAD(P)H-dependent chromium (VI) reductase of *Pseudomonas ambigua* G-1: A Cr(V) intermediate is formed during the reduction of Cr(VI) to Cr(III). *J Bacteriol* 174: 5340–5345.

Tanimoto T, K-I Hatano, D-h Kim, H Uchiyama, H Shoun. 1992. Co-denitrification by the denitrifying system of the fungus *Fusarium oxysporum*. *FEMS Microbiol Lett* 93: 177–180.

Thibodeau J, A Gauthier, M Duguay, R Villemur, F Lépine, P Juteau, R Beaudet. 2004. Purification, cloning, and sequencing of a 3,5-dichlorophenol reductive dehalogenase from *Desulfitobacterium frappieri* PCP- 1. *Appl Environ Microbiol* 70: 4532–4537.

Thorell HD, K Stenklo, J Karlsson, T Nilsson. 2003. A gene cluster for chlorate metabolism in *Ideonella dechloratans*. *Appl Environ Microbiol* 69: 5585–5592.

Torrens E, I Grinberg, B Gorovitz-Harris, H Lundström, I Borovok, y Aharonowitz, B-M Sjöberg, G Cohen. 2007. NrdR controls differential expression of the *Escherichia coli* ribonucleotide reductase genes. *J Bacteriol* 189: 5012–5021.

Tsuruta S, N Takaya, L Zhang, H Shoun, K Kimura, M Hamamoto, T Nakase. 1998. Denitrification by yeasts and occurrence of cytochrome P450nor in *Trichosporon cutaneum*. *FEMS Microbiol Lett* 168: 105–110.

Usuda K, N Toritsuka, Y Matsuo, H Shoun. 1995. Denitrification by the fungus *Cylindrocarpon tonkinense*: Anaerobic cell growth and two isoenzyme forms of cytochrome P-450nor. *Appl Environ Microbiol* 61: 883–889.

Vadas A, HG Monbouquette, E Johnson, I Schröder. 1999. Identification and characterization of a novel ferric reductase from the hyperthermophilic archaeon *Archaeoglobus fulgidus*. *J Biol Chem* 274: 36715–36721.

van de Pas BA, J Gerritse, WM de Vos, G Schraa, AJM Stams. 2001. Two distinct enzyme systems are responsible for tetrachloroethene and chlorophenol reductive dehalogenation in *Desulfitobacterium* strain PCE1. *Arch Microbiol* 176: 165–169.

van de Pas BA, H Smidt, WR Hagen, J van der Oost, G Schraa, AJM Stams, WM de Vos. 1999. Purification and molecular characterization of *ortho*-chlorophenol reductive dehalogenase, a key enzyme of halorespiration in *Desulfitobacterium dehalogenans*. *J Biol Chem* 274: 20287–20292.

van Dillewijn P, R-M Wittich, A Caballero, J-L Ramos. 2008a. Type II hydride transferases from different microorganisms yield nitrite and diarylamines from polynitroaromatic compounds. *Appl Environ Microbiol* 74: 6820–6823.

van Dillewijn P, R-M Wittich, A Caballero, J-L Ramos. 2008b. Subfunctionality of hydride transferases of the old yellow enzyme family of flavoproteins of *Pseudomonas putida*. *Appl Environ Microbiol* 73: 6703–6708.

Vaz ADN, S Chakraborty, V Massey. 1995. Old yellow enzyme: Aromatization of cyclic enones and the mechanism of a novel dismutation reaction. *Biochemistry* 34: 4246–4256.

Vorbeck C, H Lenke, P Fischer, JC Spain, H-J Knackmuss. 1998. Initial reductive reactions in aerobic microbial metabolism of 2,4,6-trinitrotoluene. *Appl Environ Microbiol* 64: 246–252.

Wagner M, D Sonntag, R Grimm, A Pich, C Eckerskorn, B Söhling, JR Andreesen. 1999. Substrate-specific selenoprotein B of glycine reductase from *Eubacterium acidaminophilum*. Biochemical and molecular analysis. *Eur J Biochem* 260: 38–49.

Wang G, R Li, S Li, J Jiang. 2010. A novel hydrolytic dehalogenase for the chlorinated aromatic compound chlorothalonil. *J Bacteriol* 192: 2737–2745.

Wang P, T Mori, K Toda, H Ohtake. 1990. Membrane-associated chromate reductase activity from *Enterobacter cloacae*. *J Bacteriol* 172: 1670–1672.

Wanner P, R Tressel. 1998. Purification and characterization of two enone reductases from *Saccharomyces cerevisiae*. *Eur J Biochem* 255: 271–278.

Wells JF, PB Hylemon. 2000. Identification and characterization of a bile acid 7α-dehydroxylation operon in *Clostridium* sp. strain TO-931, a highly active 7α-dehydroxylating strain isolated from human feces. *Appl Environ Microbiol* 66: 1107–1113.

Williams RE, NV Bruce. 2002. "New uses for an old enzyme"–the old ellow enzyme family of flavoenzymes. Microbiology (UK) 148: 1607–1614.

Williams RE, DA Rathbone, NS Scrutton, NC Bruce. 2004. Biotransformation of explosives by the Old Yellow Enzyme family of flavoproteins. *Appl Environ Microbiol* 70: 3566–3574.

Wischgoll S, D Heintz, F Peters, A Erxleben, E Sarnighausen, R Reski, A Van Dorsselaer, M Boll. 2005. Gene clusters involved in anaerobic benzoate degradation of *Geobacter metallireducens. Mol Microbiol* 58: 1238–1252.

Wolterink AFWM, E Schiltz, P-L Hagedoorn, WR Hagen, SWM Kengen, AJM Stams. 2003. Characterization of the chlorate reductase from *Pseudomonas chloritidismutans. J Bacteriol* 185: 3210–3213.

Xun L, E Topp, CS Orser. 1992. Purification and characterization of a tetrachloro-*p*-hydroquinone reductive dehalogenase from a *Flavobacterium* sp. *J Bacteriol* 174: 8003–8007.

Yang Z-Y, DR Dean, LC Seefeldt. 2011. Molybdenum nitrogenase catalyzes the reduction and coupling of CO to form hydrocarbons. *J Biol Chem* 286: 19714–19723.

Zumft WG. 1997. Cell biology and m*olecular basis of denitrification. Microbiol Mol Biol Revs* 61: 533–616.

Part 2: Non-Redox Reactions

This section deals with a range of non-redox reactions and is an attempt to bring together reactions that are distributed among the other chapters, and to provide some mechanistic details.

3.16 Pyridoxal-5′-Phosphate (PLP)-Dependent Reactions

The initial reaction is a condensation between the aldehyde and an amino group in the substrate. These accomplish a wide range of reactions, some representatives of which are given as illustration.

3.16.1 Transaminases

Transamination between an amino acid and a 2-oxocarboxylate represents the paradigm reaction involving pyridoxal 5′-phosphate, and provides the basic mechanism for the reactions that are described below, including the use of L-amino acids as sources of carbon and energy.

3.16.2 Aminomutases

Pyridoxal 5′-phosphate is required for the activity of the aminomutases that play a cardinal role in the fermentation of amino acids by clostridia, for example, lysine 2,3-aminomutase (Figure 3.36a) (Moss and Frey 1990), D-lysine 5,6-aminomutase (Chang and Frey 2000) and D-ornithine 5,4-aminomutase (Somack and Costilow 1973).

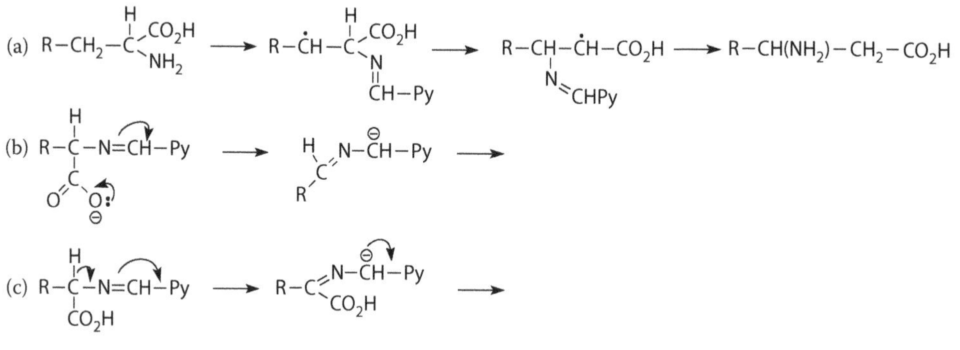

FIGURE 3.36 PLP-dependent reactions: (a) lysine 2,3-aminomutase, (b) decarboxylation of histidine, and (c) racemization of amino acids.

3.16.3 Decarboxylases

Histidine decarboxylases from the Gram-negative *Enterobacteriaceae* including *Klebsiella planticola* and *Enterobacter aerogenes* are pyridoxal 5′-phosphate dependent (Kamath et al. 1991) (Figure 3.36b). On the other hand, the inducible histidine decarboxylase from Gram-positive bacteria such as *Lactobacillus buchneri* that has been studied intensively belongs to the pyruvoyl group of enzymes (Martín et al. 2005).

3.16.4 Racemases

Although amino acids normally have the L-configuration, those with the D-configuration are required for: (a) cell-wall peptidoglycan biosynthesis, (b) synthesis of the poly-γ-D-glutamate capsule in *Bacillus anthracis*, and (c) the synthesis of a range of peptide antibiotics in *Bacillus brevis* including the gramicidins and the tyrocidins.

The synthesis of these is generally carried out by either of two reactions, both of which require pyridoxal-5′-phosphate: racemization where protonation of the R–C=N– provides the possibility of isomerization (Figure 3.36c), or by a D-transaminase from the appropriate 2-oxoacid precursor and D-alanine. Alanine racemase is widely distributed in bacteria. Racemase activity has also been demonstrated for other amino acids including threonine from *Ps. putida* (Lim et al. 1993), arginine from *Pseudomonas graveolens* (Yorifuji and Ogata 1971) and ornithine from *Clostridium sticklandii* (Chen et al. 2000)—in all of which pyridoxal phosphate is bound to the enzyme. A number of bacteria possess two racemases, for glutamate in *Bacillus sphaericus* (Fotheringham et al. 1998) and *Bacillus anthracis* (Dodd et al. 2007), and for alanine in *Salmonella typhimurium* (Wasserman et al. 1983), while *Staphylococcus haemolyticus* is able to synthesize D-glutamate both by the activity of a racemase and by a D-amino acid transaminase (Pucci et al. 1995). The degradation of amino acids as sources of carbon and energy has generally been studied with the L-isomers: use of the D-isomer may, therefore, be dependent on the existence of a racemase although different pathways may be used for the D- and L-isomers, for example, lysine in pseudomonads (Muramatsu et al. 2005).

Not all racemases depend on the activity of pyridoxal-5′-phosphate. These include those, for example, proline in *Clostridium* sp. (Rudnick and Abeles 1975), 2-hydroxyproline in *Pseudomonas striata* (Adams and Noton 1964), aspartate in *Streptococcus thermophilus* (Yamauchi et al. 1992), and gutamate in *Lactobacillus fermenti* (Gallo and Knowles 1993). An alternative reaction is involved for some racemases which are cofactor independent: they involve a direct 1:1 proton transfer mechanism involving pairs of amino acids, for example, lysine/histidine, two lysines, two cysteines or two aspartates, followed by protonation of the delocalized anion. Examples include proline racemase (Figure 3.37a) (Rudnick and Abeles 1975) and mandelate racemase from *Ps. putida* (Figure 3.37b) (Hasson et al. 1998) that are formally analogous to the reactions in Figure 3.36c and a for the analogous PLP-dependent reactions. The involvement of two cysteines has been suggested for the non-cofactor racemases for glutamate in *Lactobacillus fermenti* (Gallo and Knowles 1993), and this is supported by the crystal structure of aspartate racemase that has been determined in *Pyrococcus horikoshii* (Liu et al. 2002).

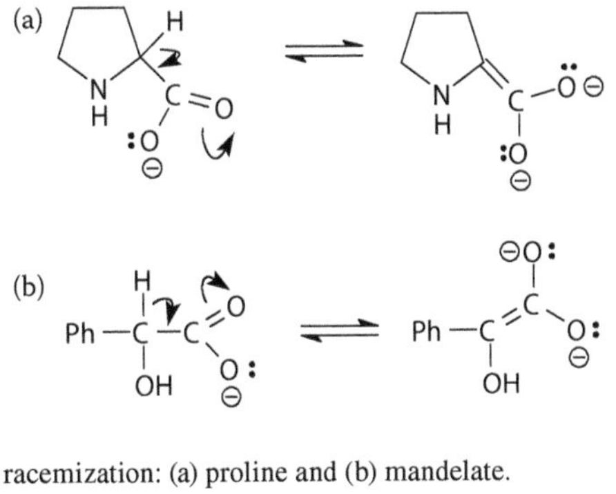

FIGURE 3.37 PLP-independent racemization: (a) proline and (b) mandelate.

3.16.5 Enzyme Inactivation

Effort has been directed to compounds that function as suicide substrates in which the initial reactions parallel those involved in reactions of the relevant amino acid especially to alanine that serves as a component of bacterial cell walls. In the following examples, the initial reaction is an α,β-elimination— that should be distinguished from the γ-lyases that are discussed in the following paragraph. For alanine racemase, its inactivation has been achieved with both the D- and L-isomers of β-fluoro- and β-chloroalanine that partition between α,β-elimination to pyruvate, halide and ammonia, and inactivation (Figure 3.38) (Wang and Walsh 1978). An analogous reaction for the deactivation of L-aspartate β-decarboxylase also involves β-chloroalanine in which the rapid α,β-elimination is accompanied by a slower irreversible enzyme inactivation (Tate et al. 1969). For 3-halovinyl glycine, an initial α,β-elimination of halide produces an allenic intermediate that partitions between reversible and irreversible reactions in which the latter involves irreversible alkylation of tyrosine (Figure 3.39) (Thornberry et al. 1991). In a variant mechanism, glutamate decarboxylase is inactivated by serine O-sulfate bound to PLP: aminoacrylate produced by an α,β-elimination reacts with pyridoxal phosphate by transimination to form a product that can be rapidly converted at high pH to pyruvyl pyridoxal (Likos et al. 1982). This mechanism is supported also by the results of studies with the inactivation of alanine racemase in *Salmonella typhimurium* (Badet et al. 1984) and the broad specificity amino acid racemase in *Ps. striata* (Roise et al. 1984).

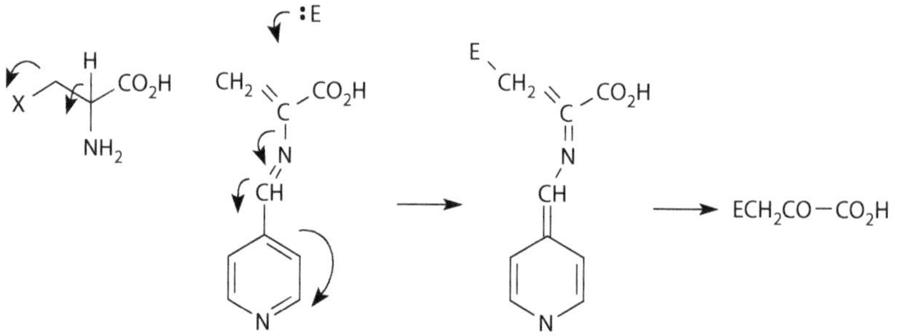

FIGURE 3.38 Inactivation of alanine racemase by β-haloalanine. (Adapted from Wang E, C Walsh. 1978. *Biochemistry* 17: 1313–1321.)

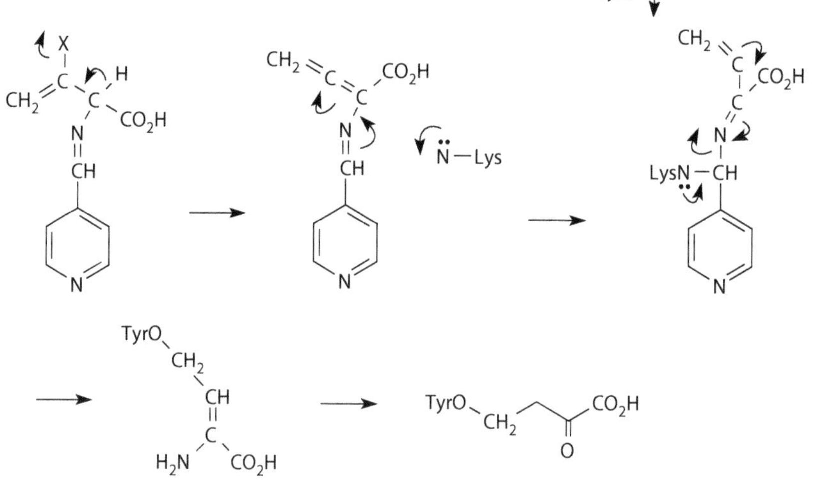

FIGURE 3.39 Inactivation of L-aspartate decarboxylase by 3-halovinylglycine. (Adapted from Thornberry NA et al. 1991. *J Biol Chem* 266: 21657–21665.)

Experiments analogous to those already noted, enzyme inactivation have been directed to methionine γ-lyase. For both L-propargylglycine (Johnston et al. 1979) and L-2-amino-4-chloropent-4-enoate (Esaki et al. 1989), the initial elimination to an allene is followed by reaction of the enzyme at C_3 with resulting inactivation of the enzyme.

3.16.6 Lyases

The following PLP-dependent reactions may involve α,β-eliminations that have been noted in the preceding paragraph, and those that may profitably be considered γ-scissions.

1. The conversion of L-selenocysteine to L-alanine and Se° is carried out by a β-lyase in *Citrobacter freundii* and contains PLP as a cofactor (Chocat et al. 1985). It is found in a limited range of bacteria, although it is absent in yeasts and fungi (Chocat et al. 1983).

$$HSe-CH_2-CH(NH_2)CO_2H \rightarrow CH_3-CH(NH_2)CO_2H + Se°$$

2. The γ-scission of cysteine sulfinate to alanine and sulfite is carried out gratuitously by aspartate β-decarboxylase (Soda et al. 1964), and by an enzyme with a wider range of substrate that has been designated cysteine sulfinate desulfinase (Mihara et al. 1997).

$$HS-CH_2-CH(NH_2)-CO_2H \rightarrow HO_2C-CH(NH_2)-CH_2-SO_2H$$
$$\rightarrow CH_3-CH(NH_2)-CO_2H + HSO_3^-$$

3. The degradation of cysteate (2-amino-3-sulfopropionate) by *Paracoccus pantotrophus* takes place by the formation of 3-sulfopyruvate that is reduced to 3-sulfolactate; this undergoes γ-scission to pyruvate and sulfite by a sulfo-lyase (Rein et al. 2005).

$$HSO_3-CH_2-CH(NH_2)-CO_2H \rightarrow HSO_3-CH_2-CO$$
$$-CO_2H + NH_3 \rightarrow HSO_3-CH_2-CH(OH)-CO_2H \rightarrow CH_3$$
$$-CO-CO_2H + HSO_3^- \rightarrow CH_3-CO-CO_2H + NH_3 + HSO_3^-$$

On the other hand, in *Silicibacter pomeroyi* L-cysteate sulpholyase brings about the conversion of L-cysteate (2-amino-3-sulfopropionate) to sulfite, ammonia, and pyruvate in a pyridoxal 5'-phosphate-dependent reaction (Denger et al. 2006).

4. L-Methionine γ-lyase produces methanethiol, ammonia and 2-oxobutyrate from L-methionine by an α,γ-elimination, and has been characterized from *Brevibacterium linens* strain BL2 (Dias and Weimer 1998a,b): this reaction is probably major source of volatile organosulfur compounds in this organism (Amarita et al. 2004).

$$CH_3S-CH_2-CH_2-CH(NH_2)CO_2H \rightarrow CH_3-SH + HO_2C-CO-CH_2-CH_3$$

5. Cystathionine γ-lyase catalyzes the α,γ-elimination of L-cystathionine to produce L-cysteine, 2-oxobutyrate and ammonia. It has been purified from *Lactococcus lactis* subsp. *cremoris* SK11 (Bruinenberg et al. 1997), and is widely distributed in actinomycetes (Nagasawa et al. 1984).

6. Threonine aldolase carries out the fission of threonine to glycine and acetaldehyde, and has been characterized for L-threonine in several species of *Pseudomonas* (Bell and Turner 1977) and for D-threonine in *Arthrobacter* sp. (Kataoka et al. 1997). For both of them, the reaction is dependent on pyridoxal-5'-phosphate.

7. a. There are some formal α,β-eliminations of a few amino acids that involve pyridoxal-5'-phosphate: these include serine deaminase to produce aminoacrylate before isomerization to pyruvate and ammonia, and threonine deamination to 2-oxobutyrate that is the key

intermediate in the synthesis of isoleucine. They should be contrasted with α-eliminations in the cofactor-independent racemization of mandelate and some amino acids that have already been noted, although all of them are facilitated by electron delocalization of the amino acid α-anion.

b. Tryptophanase is a pyridoxal-dependent enzyme that carries out α,β-elimination of tryptophan to indole, pyruvate and ammonia (Vederas et al. 1978). Growth of *E. coli* under anaerobic conditions using nitrate as electron acceptor produced a range of di- and trimeric indoles with C=O or C=NOH at C_3 that are formed by reaction of indole produced by tryptophanase and nitrite (Kwon and Weiss 2009). This is the basis of the classical "cholera red" reaction for *Vibrio cholerae*.

The biosynthesis of selenocysteine is unique in its biosynthesis on cognate transfer RNA. This has been studied in *E. coli* (Forchhammer and Böck 1991), a phosphoseryl-tRNA kinase has been described (Carlson et al. 2004) and the role of phosphoserine has been identified in eukaryotes and archaea (Yuan et al. 2006). A catalytic mechanism involving pyridoxal 5′-phosphate and serine has been outlined for the eukaryotic selenocysteine synthase (Ganichkin et al. 2008), although there are minor differences in the mechanism for bacteria, eukaryotes, and archaea. Details of the reactions have been elucidated in a study of the crystal structure of the complex containing the synthase, phosphoserine, and selenophosphate (Figure 3.40) (Palioura et al. 2009). Pyridoxal 5′-phosphate plays an active role by binding to the δ-amino group of lysine followed by transamination with the amino group of tRNA serine phosphate, followed by reaction with selenophosphate and release of $tRNA–CO_2–CH(NH_2)–CH_2Se$. It has been proposed that in methanogenic archaea including *Methanothermobacter thermoautotrophicus* and *Methanopyrus kandleri* which lack cysteinyl transfer RNA synthetase, an analogous pathway is used involving *O*-phosphoseryl-tRNA for the synthesis of cysteine (Sauerwald et al. 2005).

Summary of reactions:

1. Reaction of PLP with δ-amino group of lysine
2. Transamination with amino group of tRNA serine phosphate
3. Rearrangement to produce—$CH=NH^+–C(=CH_2)–CO_2–tRNA$
4. Reaction with selenophosphate to form $–CH=NH^+–CH(CH_2–Se)–CO_2–tRNA$
5. Reaction with the δ-amino group of lysine and PLP with release of $tRNA–CO_2–CH(NH_2)–CH_2Se$

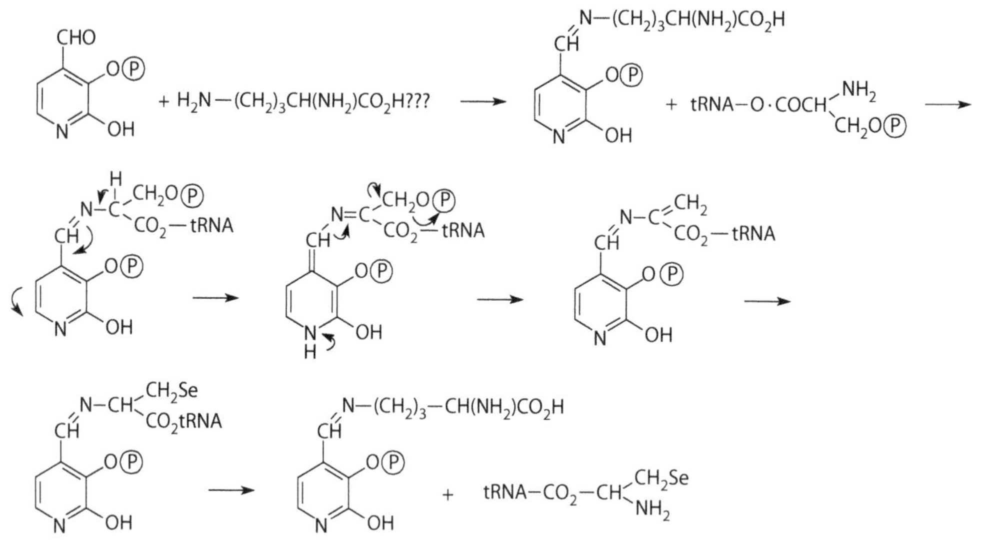

FIGURE 3.40 Synthesis of selenocysteine. (Adapted from Palioura S et al. 2009. *Science* 325: 321–325.)

3.17 References

Adams E, IL Noton. 1964. Purification and properties of inducible hydroxyproline 2-epimerase from *Pseudomonas*. *J Biol Chem* 239: 1525–1535.

Amarita F, M Yvon, M Nardi, E Chambellon, J Delettre, P Bonnarme. 2004. Identification and functional analysis of the gene encoding methionine-γ-lyase in *Brevibacterium linens*. *Appl Environ Microbiol* 70: 7348–7354.

Badet B, D Roise, CT Walsh. 1984. Inactivation of the *dadB Salmonella typhimurium* alanine racemase by D and L isomers of β-substituted alanines: Kinetics, stoichiometry, active site peptide sequencing, and reaction mechanism. *Biochemistry* 23: 5188–5194.

Bell SC, JM Turner. 1977. Bacterial catabolism of threonine. Threonine degradation initiated by L-threonine acetaldehyde-lyase (aldolase). *Biochem J* 166: 209–216.

Bruinenberg PG, G de Roo, GY Limsowtin. 1997. Purification and characterization of cystathionine γ-lyase from *Lactococcus linens* subsp. *cremoris* SK11: Possible role in flavor compound formation during cheese maturation. *Appl Environ Microbiol* 63: 561–566.

Carlson BA, X-M Xu, GV Kryukov, M Rao, MJ Berry, VN Gladyshev. 2004. Identification and characterization of phosphoseryl-tRNA[Ser]Sec kinase. *Proc Natl Acad Sci USA* 101: 12848–12853.

Chang CH, PA Frey. 2000. Cloning, sequencing, heterologous expression, purification, and characterization of adenosylcobalamin-dependent D-lysine 5,6-aminomutase from *Clostridium sticklandii*. *J Biol Chem* 275: 106–114.

Chen H-P, C-F Lin, Y-J Lee, S-S Tsay, S-H Wu. 2000. Purification and properties of ornithine racemase from *Clostridium sticklandii*. *J Bacteriol* 182: 2052–2054.

Chocat P, N Esaki, T Nakamura, H Tanaka, K Soda. 1983. Microbial distribution of selenocysteine lyase. *J Bacteriol* 156: 455–457.

Chocat P, N Esaki, K Tanizawa, K Nakamura, H Tanaka, K Soda. 1985. Purification and characterization of selenocysteine β-lyase from *Citrobacter freundii*. *J Bacteriol* 163: 669–676.

Denger K, THM Smits, AM Cook. 2006. L-cysteate sulfo-lyase, a widespread pyridoxal 5′-phosphate-coupled desulphonative enzyme purified from *Silicibacter pomeroyi* DSS-3. *Biochem J* 394: 657–664.

Dias B, B Weimer. 1998a. Purification and characterization of L-methionine γ-lyase from *Brevibacterium linens* BL2. *Appl Environ Microbiol* 64: 3327–3331.

Dias B, B Weimer. 1998b. Conversion of methionine to thiols by lactococci, lactobacilli, and brevibacteria. *Appl Environ Microbiol* 64: 3320–3326.

Dodd D, JG Reese, CR Louer, JD Ballard, MA Spies, SR Blanke. 2007. Functional comparison of the two *Bacillus anthracis* glutamate racemases. *J Bacteriol* 189: 5265–5275.

Esaki N, H Takada, M Moriguchi, S-i Hatanaka, H Tanaka, K Soda. 1989. Mechanism-based inactivation of L-methionine γ-lyase by L-2-amino-4-chloro-4-pentenoate. *Biochemistry* 28: 2111–2116.

Forchhammer K, A Böck. 1991. Selenocysteine synthase from *Escherichia coli*. *J Biol Chem* 266: 6324–6328.

Fotheringham IG, SA Bledig, PP Taylor. 1998. Characterization of the genes encoding D-amino acid transaminase and glutamate racemase, two D-glutamate biosynthetic enzymes of *Bacillus sphaericus* ATCC 10208. *J Bacteriol* 180: 4319–4323.

Gallo KA, JR Knowles. 1993. Purification, cloning, and cofactor independence of glutamate racemase from *Lactobacillus*. *Biochemistry* 32: 3981–3990.

Ganichkin OM, X-M Xu, BA Carlson, H Mix, DL Hatfield, VN Gladyshev. 2008. Structure and catalytic mechanism of eukaryotic selenocysteine synthase. *J Biol Chem* 283: 5849–5865.

Hasson MS et al.. 1998. Evolution of an enzyme active site: the structure of a new crystal form of muconate lactonizing enzyme compared with mandelate racemase and enolase. *Proc Natl Acad Sci USA* 95: 10396–10401

Johnston M, R Raines, M Chang, N Esaki, K Soda, C Walsh. 1979. Mechanistic studies on reactions of bacterial methionine γ-lyase with olefinic amino acids. *Biochemistry* 20: 4325–4333.

Kamath AV, GL Vaaler, EE Snell. 1991. Pyridoxal phosphate-dependent histidine decarboxylases. Cloning, sequencing, and expression of genes from *Klebsiella planticola* and *Enterobacter aerogenes* and properties of the overexpressed enzymes. *J Biol Chem* 266: 9432–9437.

Kataoka M, M Ikemi, T Morikawa, T Miyoshi, K Nisji, M Wada, H Yamada, S Shimizu. 1997. Isolation and characterization of D-threonine aldolase, a pyridoxal-5′phosphate-dependent enzyme from *Arthrobacter* sp. DK-38. *Eur J Biochem* 248: 385–393.

Kwon Y-M, B Weiss. 2009. Production of 3-nitrosoindole derivatives by *Escherichia coli* during anaerobic growth. *J Bacteriol* 191: 5369–5376.

Likos JL, H Ueno, RW Feldhaus, DE Metzler. 1982. A novel reaction of the coenzyme of glutamate decarboxylase with L-serine *O*-sulfate. *Biochemistry* 21: 4377–4386.

Lim Y-H, K Yokoigawa, N Esaki, K Soda. 1993. A new amino acid racemase with threonine α-epimerase activity from *Pseudomonas putida*: Purification and characterization. *J Bacteriol* 175: 4213–4217.

Liu L, K Iwata, A Kita, Y Kawarabayasi, M Yohda, K Miki. 2002. Crystal structure of aspartate racemase from *Pyrococcus horikoshi* OT3 and its implication for molecular mechanism of PLP-independent racemization. *J Mol Biol* 319: 479–489.

Martín MC, M Fernández, DM Linares, MA Alvarez. 2005. Sequencing, characterization and transcriptional analysis of the histidine decarboxylase operon on *Lactobacillus buchneri*. *Microbiology (UK)* 151: 1219–1229.

Mihara H, T Kurihara, T Yoshimura, K Soda, N Esaki. 1997. Cysteine sulfinate desulfinase, a NIFS-like protein of *Escherichia coli* with selenocysteine lyase and cysteine desulfurase activities. *J Biol Chem* 272: 22417–22424.

Moss ML, PA Frey. 1990. Activation of lysine 2,3-aminomutase by *S*-adenosylmethionine. *J Biol Chem* 265: 18112–18115.

Muramatsu H, H Mihara, R Kakutani, M Yasuda, M Ueda, T Kurihara, N Esaki. 2005. The putative malate/lactate dehydrogenase from *Pseudomonas putida* is an NADPH-dependent D^1-piperideine-2-carboxylate/D^1-pyrroline-2-carboxylate reductase involved in the catabolism of L-lysine and D-proline. *J Biol Chem* 280: 5329–5335.

Nagasawa T, H Kanaki, H Yamada. 1984. Cystathionine γ-lyase of *Streptomyces phaeochromogenes*. The occurrence of cystathionine γ-lyase in filamentous bacteria and its purification and characterization. *J Biol Chem* 259: 10393–10403.

Palioura S, RL Sherrer, TA Steitz, D Söll, M Simonovíc. 2009. The human SepSecS-tRNA[sec] complex reveals the mechanism of selenocysteine formation. *Science* 325: 321–325.

Pucci MI, JA Thanassi, H-T Ho, PJ Falk, TJ Dougherty. 1995. *Staphylococcus haemolyticus* contains two D-glutamic acid biosynthetic activities, a glutamate racemase and a D-amino acid transferase. *J Bacteriol* 177: 336–342.

Rein U, R Gueta, K Denger, J Ruff, K Hollemeyer, AM Cook. 2005. Dissimilation of cysteate via 3-sulfolactate sulfo-lyase and a sulfate exported in *Paracoccus pantotrophus* NKNCYSA. *Microbiology (UK)* 151: 737–747.

Roise D, K Soda, T Yagi, CT Walsh. 1984. Inactivation of the *Pseudomonas striata* broad specificity amino acid racemase by D and L isomers of β-substituted alanines: kinetics, stoichiometry, active site peptide, and mechanistic studies. *Biochemistry* 23: 5195–5201.

Rudnick G, RH Abeles. 1975. Reaction mechanism and structure of the active site of proline racemase. *Biochemistry* 14: 4515–4522.

Rudnick G, RH Abeles. 1975. Reaction mechanism and structure of the active site of proline racemase. *Biochemistry* 14: 4515–4522.

Sauerwald, W Zhu, TA Major, H Roy, S Palioura, D Jahn, WB Whitman, JR Yates, M Ibba, D Söll. 2005. RNA-dependent cysteine biosynthesis in archaea. *Science* 307: 1969–1972.

Soda K, A Novogrodsky, A Meister. 1964. Enzymatic desulfination of cysteine sulfinic acid. *Biochemistry* 3: 1450–1454.

Somack R, RN Costilow. 1973. Purification and properties of a pyridoxal phosphate and coenzyme B$_{12}$-dependent D-ornithine 5,4-aminomutase. *Biochemistry* 12: 2597–2604.

Tate SS, NM Relyea, A Meister. 1969. Interaction of L-aspartate β-decarboxylase with β-chloro-L-alanine. β-elimination reaction and active-site labeling. *Biochemistry* 8: 5016–5021.

Thornberry NA, HG Bull, D Taub, KE Wilson, G Giménez-Gallego, A Rosegay, DS Sodermann, AA Patchett. 1991. Mechanism-based inactivation of alanine racemase by 3-halovinylglycines. *J Biol Chem* 266: 21657–21665.

Vederas JC, E Schleicher, M-D Tsai, HG Floss. 1978. Stereochemistry and mechanism of reactions catalyzed by tryptophanase from *Escherichia coli*. *J Biol Chem* 253: 5350–5354.

Wang E, C Walsh. 1978. Suicide substrates for the alanine racemase of *Escherichia coli* B. *Biochemistry* 17: 1313–1321.

Wasserman SA, CT Walsh, D Botstein. 1983. Two alanine racemase genes in *Salmonella typhimurium* that differ in structure and function. *J Bacteriol* 153: 1439–1450.

Yamauchi T, SY Choi, H Okada, M Yohda, H Kumagai, N Esaki, K Soda. 1992. Properties of aspartate race-mase, a pyridoxal 5′-phosphate-independent amino acid racemase. *J Biol Chem* 267: 18361–18364.

Yorifuji T, K Ogata. 1971. Arginine racemase of *Pseudomonas graveolens* I. Purification, crystallization, and properties. *J Biol Chem* 246: 5085–5092.

Yuan J, S Palioura, JC Salazar, D Su, P O'Donoghue, MJ Hohn, AM Cardoso, WB Whitman, D Söll. 2006. RNA-dependent conversion of phosphoserine forms selenocysteine in eukaryotes and archaea. *Proc Natl Acad Sci USA* 103: 18923–18927.

3.18 Glutathione-Dependent Reactions

This section attempts to summarize the miscellany of glutathione-dependent reactions that are discussed more fully in later chapters.

These reactions can be classified into the following classes: nucleophilic displacement, reductive displacement of halide, and isomerization.

1. Glutathione-dependent formaldehyde metabolism involves reaction between reduced glutathione and formaldehyde to produce $HO-CH_2$-GSH nonenzymatically, followed by an NAD- and gluta-thione-dependent dehydrogenase (GD-FALDH) to *S*-formylglutathione that is oxidized to formate and CO_2. The GD-FALDH has been characterized from *Paracoccus denitrificans* for which it is necessary for methylotrophic growth on methanol, methylamine, and choline (Ras et al. 1995).

2. The metabolism of methylglyoxal by *E. coli* is carried out by glyoxylase I with reduced glutathione to produce *S*-lactoylglutathione and glyoxylase II to D-lactate (Ko et al. 2005).

3. The degradation of dichloromethane by *Methylophilus* sp. strain DM11 is initiated by formation of an *S*-chloromethyl glutathione conjugate that undergoes nonenzymatic hydrolysis to *S*-hydroxymethyl GSH, and then to formaldehyde (Bader and Leisinger 1994) (Figure 3.41a).

4. The metabolism of isoprene epoxide (1,2-epoxy-2-methylbut-3-ene) is initiated by nucleophilic reaction with glutathione *S*-transferase in *Rhodococcus* sp. strain AD45 (van Hylckama Vlieg et al. 1999) (Figure 3.41b). One mechanism for antibiotic resistance to the oxirane fosfomycin is mediated by an analogous reaction with glutathione *S*-transferase to produce an adduct at C_1 (Arca et al. 1990) that is noted in Chapter 4 (Fillgrove et al. 2003).

5. a. The glutathione-dependent (Orser et al. 1993) degradation of pentachlorophenol by *Sphingomonas chlorophenolica* (*Flavobacterium* sp.) is initiated by formation of tetrachlorohydroquinone that undergoes dehydrogenation to the quinone followed by glutathione *S*-transferase successive reductive dechlorination of tetrachlorohydroquinone to 2,3,6-trichloro- and 2,6-dichlorohydroquinone (Figure 3.42a) (Warner et al. 2005). The dechlorination of 2,5-dichlorohydroquinone by LinD in the degradation of lindane is an analogous reaction, although the enzyme has only low identity with the *Sphingomonas* enzyme and maleylacetoacetate isomerase.

   b. The degradation of 3-chlorobiphenyl by *Burkholderia xenovorans* LB400 may produce the inhibitory 3-chloro-2-hydroxy-6-oxo-6-phenyl-2,4-dienoate by

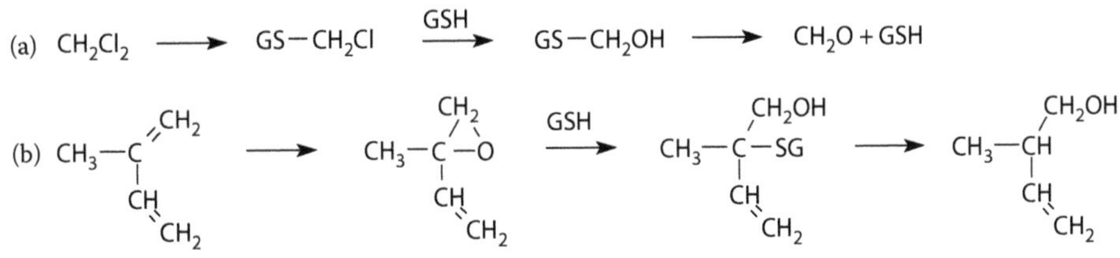

FIGURE 3.41 (a) Degradation of isoprene. (Adapted from Arca P, C Hardisson, JE Suárez. 1990. *Antimicrob Agents Chemother* 34: 844–848.) (b) Degradation of dichloromethane. (Adapted from Bader R, T Leisinger. 1994. *J Bacteriol* 176: 3466–3473.)

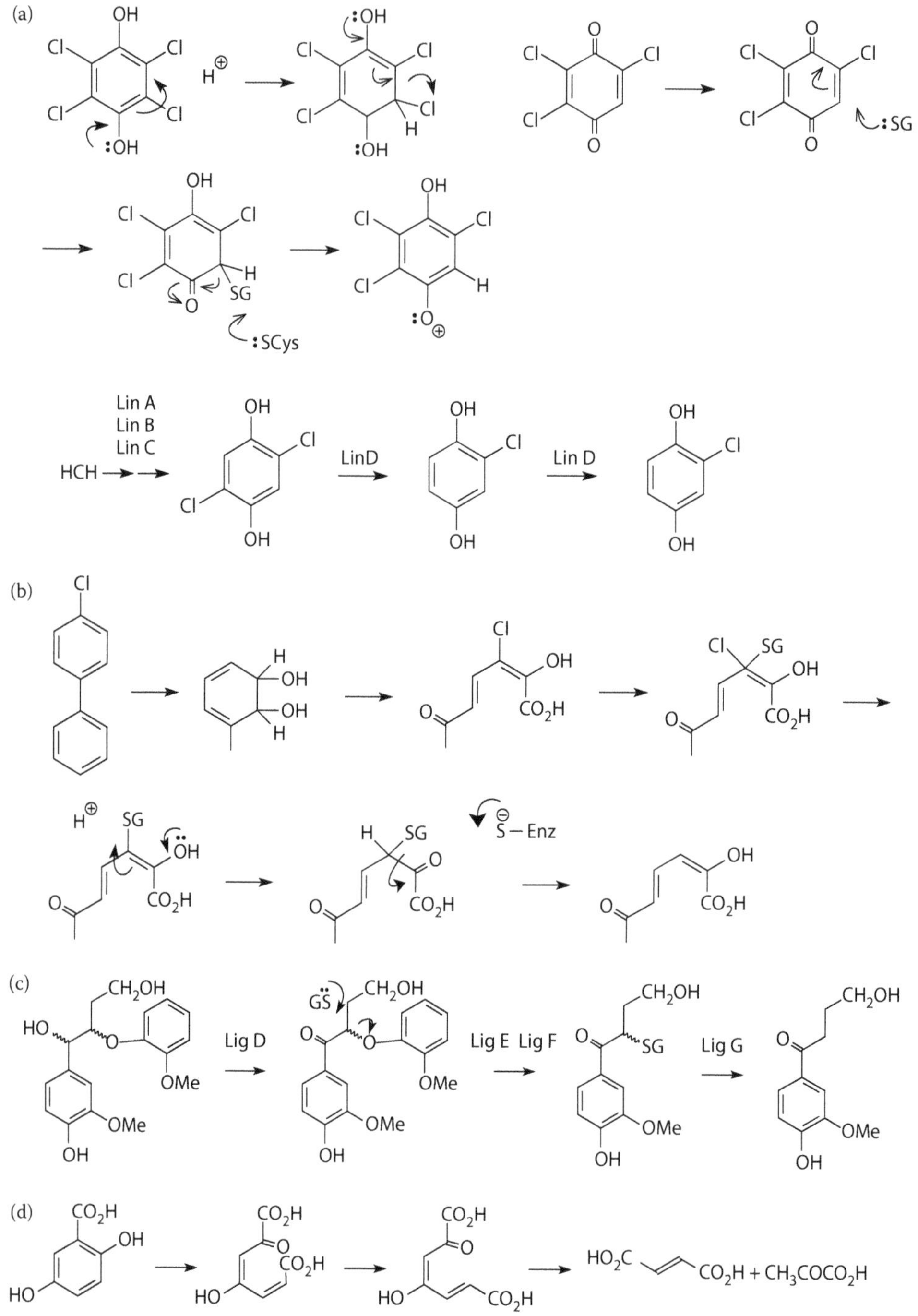

FIGURE 3.42 Glutathione-dependent reactions. (a) Degradation of pentachlorophenol. (Adapted from Warner JR, SL Lawson, SD Copley. 2005. *Biochemistry* 44: 10360–10368.) (b) Dechlorination in the degradation of 3-chlorobiphenyl. (Adapted from Fortin PD et al. 2006. *J Bacteriol* 188: 4424–4430.) (c) Degradation of β-aryl ether. (Adapted from Masai E et al. 2003. *J Bacteriol* 185: 1768–1775.) (d) *cis–trans* isomerization in the degradation of gentisate. (Adapted from Zhou N-Y, SL Fuenmayor, PA Williams. 2001. *J Bacteriol* 183: 700–708.)

TABLE 3.8

Occurrence of Glutathione *S*-Transferase in Gram-Negative PAH-Degrading Bacteria

Strain	Origin	Substrate	Reference
Sphingomonas yanoikuyae B1	Texas	Benzo[*a*]pyrene/ Benz[*a*]anthracene	Jerina et al. (1984)
Sphingomonas paucimobilis EPA 505	Florida	Fluoranthene	Mueller et al. (1990)
Sphingomonas aromaticivorans F199	South Carolina	PAH	Balkwill et al. (1997)
Burkholderia xenovorans LB 400	USA	Biphenyl/PCB	Bopp (1986)
Pseudomonas pseudoalcaligenes KF707	Japan	Biphenyl/PCB	Furukawa and Miyazaki (1989)

dioxygenation, dehydrogenation and *meta*-ring fission. The chlorine is then reductively lost by a glutathione *S*-transferase encoded by BphK that is also able to accept 5-chloro- and 3,9,11-trichloro-2-hydroxy-6-oxo-6-phenylhexa-2,4-dienoate (Fortin et al. 2006) (Figure 3.42b).

6. Degradation of the β-aryl ether by *Sphingomonas paucimobilis* SYK-6 involves reduction of the oxo group by LigD, followed by dehydrogenation of the benzylic hydroxyl group, displacement of guaiacol by a glutathione-dependent etherase (LigE, LigF) and reduction to β-hydroxypropio-vanillone mediated by a glutathione lyase (Lig G) (Figure 3.42c) (Masai et al. 2003).

7. The degradation of 3-hydroxybenzoate by the gentisate pathway involves ring fission to maleyl-pyruvate that is either hydrolyzed to pyruvate and maleate, or is isomerized by a glutathione-dependent isomerase in *Ralstonia* sp. strain U2 to fumarylpyruvate before hydrolysis to pyruvate and fumarate (Figure 3.42d) (Zhou et al. 2001). The enzyme from *Vibrio* 01 that catalyzes the analogous isomerization of *cis*-maleylacetone to *trans*-maleylacetone has been purified (Seltzer 1973), and a mechanism for the reaction proposed (Seltzer and Lin 1979). On the other hand, it should be noted that GSH-independent maleylpyruvate isomerase has been determined in some Gram-positive bacteria including species of *Bacillus* (Crawford 1975) and *Corynebacterium glutamicum* (Shen et al. 2005). The isomerization of *cis*-maleylacetone to *trans*-maleylacetone has achieved prominence in the light of clinical manifestations resulting from genetic errors in the mammalian metabolism of phenylalanine and tyrosine (references in Fernández-Cañón and Peñalva 1998).

8. Genome analysis revealed the presence of the glutathione *S*-transferase gene in a number of PAH-degrading bacteria (Lloyd-Jones and Lau 1997) although the reactions involved were not defined. Strains ANT17, ANT23 were from the Ross Sea, Antarctica; strains RP003, RP006, and WP01 from Rotowaro, New Zealand. The designations of the other strains, their PAH substrates and the references to their sources are given in Table 3.8.

9. It has been shown in *Lactobacillus lactis* that glutathione can serve to mitigate both acid (Zhang et al. 2007) and osmotic (Zhang et al. 2010) stress.

10. A formally analogous reaction is involved in the transformation of alkene epoxides. This involves nucleophilic attack on the epoxide with the formation of alkanols. From epoxypropane both the *R*- and *S*-alkanols are produced by reaction with coenzyme M, and they can be specifically dehydrogenated to ketones (Sliwa et al. 2010). These can then be reductively carboxylated in the presence of CO_2 to yield acetoacetate.

3.19 References

Arca P, C Hardisson, JE Suárez. 1990. Purification of a glutathione *S*-transferase that mediates fosfomycin resistance in bacteria. *Antimicrob Agents Chemother* 34: 844–848.

Bader R, T Leisinger. 1994. Isolation and characterization of the *Methylophilus* sp. strain D M11 gene encoding dichloromethane dehalogenase/glutathione *S*-transferase. *J Bacteriol* 176: 3466–3473.

Balkwill DL et al. 1997. Taxonomic study of aromatic-degrading bacteria from deep-terrestrial-subsurface sediments and description of *Sphingomonas aromaticivorans* sp. nov., *Sphingomonas subterranea* sp. nov., and *Sphingomonas stygia* sp. nov. *Int J Syst Bacteriol* 47: 191–201.

Bopp LH. 1986. Degradation of highly chlorinated PCBs by *Pseudomonas* strain LB400. *J Indust Microbiol* 1: 23–29.

Crawford RL. 1975. Degradation of 3-hydroxybenzoate by bacteria of the genus *Bacillus. Appl Microbiol* 30: 439–444.

Fernández-Cañón JM, MA Peñalva. 1998. Characterization of a fungal maleylacetoacetate isomerase and identification of its human homologue. *J Biol Chem* 273: 329–337.

Fillgrove KL, S Pakhomova, ME Newcomer, RN Armstrong. 2003. Mechanistic diversity of fosfomycin resistance in pathogenic microorganisms. *J Am Chem Soc* 125: 15730–15731.

Fortin PD, GP Horsman, HM Yang, LD Eltis. 2006. A glutathione *S*-transferase catalyzes the dehalogenation of inhibitory metabolites of polychlorinated biphenyls. *J Bacteriol* 188: 4424–4430.

Furukawa K, T Miyazaki. 1989. Cloning of a gene cluster encoding biphenyl and chlorobiphenyl degradation in *Pseudomonas pseudoalcaligenes. J Bacteriol* 166: 392–398.

Gibson DT, DL Cruden, JD Haddock, GJ Zylstra, JM Brande. 1993. Oxidation of polychlorinated biphenyls by *Pseudomonas* sp. strain LB400 and *Pseudomonas pseudoalcaligenes* KF707. *J Bacteriol* 175: 4561–4564.

Jerina DM, PJ van Bladeren, H Yagi, DT Gibson, V Mahadevan, AS Neese, M Koreeda, ND Sharma, DR Boyd. 1984. Synthesis and absolute configuration of the bacterial *cis*-1,2-, *cis*-8,9-, and *cis*-10, 11-dihydrodiol metabolites of benz[*a*]anthacene by a strain of *Beijerinckia. J Org Chem* 49: 3621–3628.

Ko J, I Kim, S Yoo, B Min, K Kim, C Park. 2005. Conversion of methylglyoxal to acetol by *Escherichia coli* aldo-keto reductases. *J Bacteriol* 187: 5782–5789.

Lloyd-Jones G, PCK Lau. 1997. Glutathione *S*-transferase-encoding gene as a potential probe for environmental bacterial isolates capable of degrading polycyclic aromatic hydrocarbons. *Appl Environ Microbiol* 63: 3286–3290.

Masai E, A Ichimura, Y Sato, K Miyauchi, Y Katayma, M Fukuda. 2003. Roles of the enantioselective glutathione *S*-transferases in cleavage of β-aryl ether. *J Bacteriol* 185: 1768–1775.

Mueller JG, Chapman PJ, Blattman BO, Pritchard PH. 1990. Isolation and characterization of a fluoranthene-utilizing strain of *Pseudomonas paucimobilis. Appl Environ Microbiol* 56: 1079–1086.

Orser CD, J Dutton, C Lange, P Jablonski, L Xun, M Hargis. 1993. Characterization of a *Flavobacterium* glutathione *S*-transferase gene involved in reductive dechlorination. *J Bacteriol* 175: 2640–2644.

Ras J, PV van Ophem, WNM Reijnders, RJM van Spanning, JA Duine, AH Stiuthamer, N Harms. 1995. Isolation, sequencing, and mutagenesis of the gene encoding NAD- and glutathione-dependent formaldehyde dehydrogenase (GD-FALDH) from *Paracoccus denitrificans*, in which GD-FALDH is essential for methylotrophic growth. *J Bacteriol* 177: 247–251.

Seltzer S. 1973. Purification and properties of maleylacetone *cis-trans* isomerase from *Vibrio* 01. *J Biol Chem* 248: 215–222.

Seltzer S, M Lin. 1979. Maleylacetone *cis-trans*-isomerase. Mechanism of the interaction of coenzyme glutathione and substrate maleylacetone in the presence and absence of enzyme. *J Am Chem Soc* 101: 3091–3097.

Shen X-H, C-Y Jiang, Y Huang, Z-P Liu, S-J Liu. 2005. Functional identification of novel genes involved in the glutathione-independent gentisate pathway in *Corynebacterium glutamicum. Appl Environ Microbiol* 71: 3442–3452.

Sliwa DA, AM Krishnakumar, JW Peters, SA Ensign. 2010. Molecular basis for enantioselectivity in the (*R*)- and (*S*)-hydroxypropylthioethylsulfonate dehydrogenases, a unique pair of stereoselective short-chain dehydrogenases/ reductases involved in aliphatic epoxide carboxylation. *Biochemistry* 49: 3487–3498.

van Hylckama Vlieg JET, J Kingma, W Kruizinga, DB Janssen. 1999. Purification of a glutathione *S*-transferase and a glutathione conjugate-specific dehydrogenase involved in isoprene metabolism in *Rhodococcus* sp. strain AD45. *J Bacteriol* 181: 2094–2101.

Warner JR, SL Lawson, SD Copley. 2005. A mechanistic investigation of the thiol–disulfide exchange step in the reductive dehalogenation catalyzed by tetrachlorohydroquinone dehalogenase. *Biochemistry* 44: 10360–10368.

Zhang Y, RY Fu, J Hugenholtz, Y Li, J Chen. 2007. Glutathione protects *Lactococcus lactis* against acid stress. *Appl Environ Microbiol* 73: 5268–5275.

Zhang Y, Y Zhang, Y Zhu, A Mao, Y Li. 2010. Proteomic analyses to reveal the protective role of glutathione in resistance of *Lactococcus lactis* to osmotic stress. *Appl Environ Microbiol* 76: 3177–3186.

Zhou N-Y, SL Fuenmayor, PA Williams. 2001. *nag* genes of *Ralstonia* (formerly *Pseudomonas*) sp. strain U2 encoding enzymes for gentisate catabolism. *J Bacteriol.* 183: 700–708.

3.20 Corrinoid-Dependent Reactions

Corrinoids belong to a group of macrocycles in which four conjugated pyrrole rings are coordinated with metals: Fe in heme, Mg in chlorophyll, Co in vitamin B_{12}, and Ni in coenzyme F_{430}. The vitamin B_{12} coenzyme corrins that are found only in anaerobes vary, however, in the other ligands. Briefly, these belong to two axial groups: (i) -CN or -Me and (ii) a range of azaarenes including 5,6-dimethylbenziminazole, adenine, 2-methyladenine, and N_6-methylaminoadenine (references in Hoffmann et al. 2000). The reactions in which they are involved given in detail later. Some involve free radicals that are schematically shown in Figure 3.43a–c: (a) Dioldehydratase (X=OH) and ethanolamine ammonia lyase (X=NH_2) (b) methylmalonate mutase, (c) L-lysine 2,3-aminomutase.

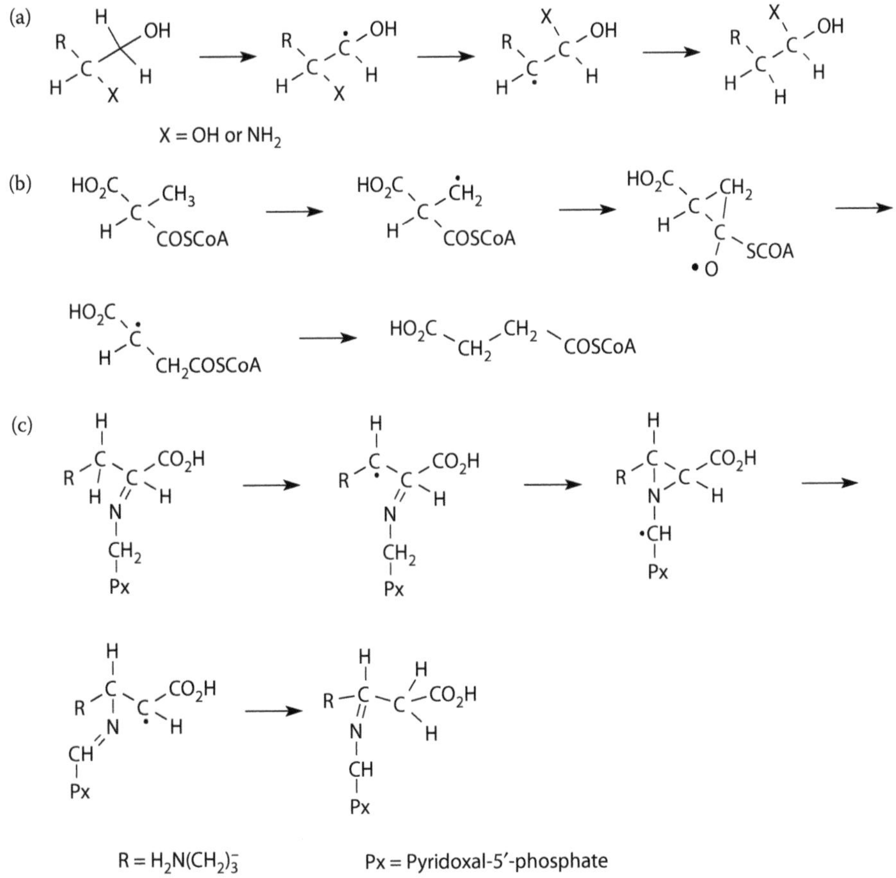

FIGURE 3.43 Coenzyme B_{12}-mediated reactions: (a) dioldehydratase, (b) methylmalonate mutase, and (c) lysine 2,3-aminomutase.

This section attempts to bring together the range of corrin-dependent reactions that are discussed in later chapters. These reactions are generally divided into three groups, and all of them involve radical reactions involving the $AdoCH_2$ radical, formed either from coenzyme B_{12} (AdoCbl) or from *S*-adenosylmethionine:

Class I: Reversible reactions—for example, methylmalonyl-CoA mutase, L-glutamate mutase

Class II: Irreversible reactions—for example, diol dehydratase, ethanolamine ammonia-lyase

Class III: Migrations—for example, L-lysine 2,3-aminomutase, ornithine 5,4-aminomutase, glutamate 2,3-aminomutase

Methylmalonyl-CoA mutase (Methylmalonate → succinate): This free-radical reaction is widely distributed in both aerobic and anaerobic bacteria. Several mechanisms have been proposed, and on the basis of molecular orbital calculations, an intramolecular reaction is supported (Smith et al. 1999).

Ethylmalonate-CoA mutase (Ethylmalonate → methylsuccinate): Growth of organisms on acetate that lack isocitrate lyase of the glyoxylate cycle use the ethylmalonate pathway: crotonyl-CoA is produced from acetate followed by reduction to (*2S*)-ethylmalonyl-CoA and the combined activity of ethylmalonate mutase and malonyl-CoA epimerase to produce (*2S*)-methylsuccinyl-CoA: the mutase in *Rhodobacter sphaeroides* is coenzyme B_{12}-dependent and is highly specific for ethylmalonyl-CoA, whereas the epimerase is less exacting in its substrate requirement (Erb et al. 2008).

Isobutyryl-CoA mutase (Isobutyrate → butyrate): The interconversion isobutyryl-CoA and butyryl-CoA in *Streptomyces cinnamonensis* is involved in polyketide biosynthesis (Brendelberger et al. 1988), and, although the reaction involved is formally analogous to that catalyzed by methylmalonate mutase, the enzymes are distinct (Ratnatilleke et al. 1999). The reverse reaction is involved in the fermentation of glutarate in which butyryl-CoA produced isobutyryl-CoA by the activity of a mutase (Matthies and Schink 1992).

2-hydroxyisobutyryl-CoA mutase (2-hydroxybutyrate → 3-hydroxybutyrate): This has been adduced from one of the pathways for the degradation of methyl *tert*-butyl ether (MTBE) in which 2-hydroxyisobutyryl-CoA is converted into the 3-hydroxybutyrate (Rohwerder et al. 2006).

3.20.1 Diol Dehydratase

The degradation under anaerobic conditions of 1:2-diols including ethanediol, propane-1,2-diol, and glycerol with the production of acetaldehyde, propionaldehyde, and 3-hydroxypropionaldehyde respectively, is mediated by a coenzyme-B_{12} (adenosylcobalamine AdoCbl)-dependent diol dehydratase. This has been established in *Enterobacteriaceae* including *K. pneumoniae* (Toraya et al. 1979) and *Citrobacter freundii* (Seyfried et al. 1996). In *Salmonella enterica* the toxicity of the propionaldehyde produced is restricted by its protection within a micro compartment (metabolosome) (Sampson and Bobik 2008). This also occurs in *Lactobacillus reuteri* in which the propionaldehyde undergoes disproportionation to propanol and propionate (Sriramulu et al. 2008).

An analogous reaction is catalyzed by glycerol dehydratase that produces 2-hydroxypropionate under anaerobic conditions in *Enterobacteriaceae* including *K. pneumoniae* (Toraya et al. 1979) and *Citrobacter freundii* (Seyfried et al. 1996).

There are, however, important exceptions to coenzyme B_{12}-dependent diol dehydratases. This has been established in *Clostridium butyricum* (Raynaud et al. 2003), and *Clostridium glycolycum* (Hartmannis and Stadtman 1986) where the activity is bound to membranes although it has been solubilized (Hartmannis and Stadtman 1987).

3.20.2 Ethanolamine–Ammonia Lyase

A formally analogous reaction with the formation of acetaldehyde and ammonia is involved in the degradation of ethanolamine by *Salmonella enterica* as a source of carbon and nitrogen. This takes place only under anaerobic conditions when adenosylcobalamin (AdoCbl) is synthesized, or under aerobic conditions when this is generated from exogenous vitamin B_{12} (Roof and Roth 1989). An unusual situation emerges with *Paracoccus denitrificans* that is able to synthesize cobalamin under both aerobic and anaerobic conditions, and can, therefore, utilize ethanolamine under both conditions (Shearer et al. 1999). Ethanolamine ammonia-lyase from *Clostridium* sp. has been shown to generate an EPR spectrum resulting from production of the $CH_3-CH(NH_2)-CHOH^*$ radical from 2-aminopropanol (Babior et al. 1974), and the reaction is formalized in Figure 3.43a.

3.20.3 Amino Acid Mutase

This group of enzymes plays a cardinal role in the anaerobic degradation of amino acids, most widely in species of *Clostridium*. The degradations involve migrations that involve free radicals produced by reaction with AdoCbl produced from coenzyme B_{12} or from *S*-adenosylmethione, although for some of the amino acids, an alternative involves dehydration of the hydroxyacids produced by reduction of the ketones formed by deamination (Kim et al. 2004).

1. L-Leucine 2,3-aminomutase: The fermentation of leucine by *Clostridium sporogenes* is initiated by a coenzyme B_{12}-dependent leucine 2,3-mutase to produce 3-amino-4-methylpentanoate (Poston 1976).

2. L-Glutamate mutase: One of the reactions involved in the degradation of glutamate by *Clostridium tetanomorphum* is a free radical reaction catalyzed by a coenzyme B_{12}-dependent glutamate mutase with the production of β-methylasparate (Holloway and Marsh 1994; Bothe et al. 1998). The alternative pathway involving 2-hydroxyacids was first established in *Acidaminococcus fermentans* (Buckel and Barker 1974).

3. L-Lysine 2,3-aminomutase and lysine 5,6-aminomutase: The 2,3-mutase in *Clostridium* mediates the migration of the amino group to β-lysine but the radical is produced from *S*-adenosylmethionine by reduction with an [Fe–S] cluster, and not from coenzyme B_{12} (Moss and Frey 1990). Further metabolism is accomplished with D-lysine 5,6-aminomutase (Chang and Frey 2000) with production of 3,5-diaminohexanoate before scission with acetyl-CoA (Yorifugi et al. 1977). The activity of both mutases requires pyridoxal-5′phosphate.

4. L-Ornithine 5,4-mutase: The fermentation of L-ornithine by *Clostridium sticklandii* takes place in two stages corresponding to the oxidation and reduction of a typical Stickland reaction. The oxidative part is initiated by ornithine racemase (Chen et al. 2000) followed by the activity of a coenzyme B_{12}-dependent 5,4-mutase (Somack and Costilow 1973; Chen et al. 2001) with the formation of 2,4-diaminopentanoate that is dehydrogenated to 2-amino-4-oxo-pentanoate before thiolytic scission to acetate and alanine (Jeng et al. 1974). On the other hand, the reductive segment does not involve amino acid mutases.

3.20.4 2-Methyleneglutarate Mutase

The fermentation of nicotinate by *Clostridium barkeri* is initiated by hydroxylation at C_6 followed by elimination that results in the production of 2-methyleneglutarate. This undergoes rearrangement to 3-methylitaconate followed by isomerization to *cis*-1,2-dimethylmaleate that is fissioned to propionate and pyruvate (Kung and Stadtman 1971; Kung and Tsai 1971).

3.20.5 Metabolism of Chloromethane and Dichloromethane

Corrin-mediated aerobic metabolism of chloromethane (Coulter et al. 1999; Vannelli et al. 1999) is different from the glutathione-mediated degradation of dichloromethane. This pathway is found in strains of *Hyphomicrobium* and in methylotrophs such as *Methylobacterium chloromethanicum* in which a corrin mediates the first step in a series of reactions that produces formaldehyde (Studer et al. 2002). A comparable pathway is used for the anaerobic degradation of chloromethane by the homoacetogen

Acetobacterium dehalogenans (Messmer et al. 1996), and dichloromethane by *Dehalobacterium formicoaceticum* (Mägli et al. 1998).

3.20.6 O-Demethylation

Methoxylated aromatic acid have used for the selective isolation of *Acetobacterium woodii* that ferments these to acetate (Bache and Pfennig 1981). The demethylation of aromatic methyl ethers by *Moorella* (*Clostridium*) *thermoacetica* has been shown to require the operation of a demethylase, tetrahydrofolate as a methyl receptor followed by reaction with the corrinoid/Fe–S protein and with the Ni/Fe–S carbon monoxide dehydrogenase which catalyzes the final step in the synthesis of acetyl-CoA (El Kasmi et al. 1994; Naidu and Ragsdale 2001). The *O*-demethylation of phenyl methyl ethers in other acetogens including *Acetobacterium dehalogenans* (Schilhabel et al. 2009) involves two methyltransferases and a corrin protein at several oxidation states that transfer the methyl group to tetrahydrofolate. This bears formal resemblance to the metabolism of chloromethane, while methyl transfer to coenzyme M (2-marcapto-ethanesulfonate) from methanol, dimethylamine, and trimethylamine (Paul et al. 2000), and dimethylsulfide (Tallant et al. 2001) in the methylotrophic methanogen *Methanosarcina barkeri* are also corrin-dependent.

As an alternative, methanethiol and dimethylsulfide may be produced from methoxybenzoates in the presence of sulfide from the medium (Bak et al. 1992). In a homoacetogen strain TMBS 4, inhibition of the reaction by propyl iodide and reactivation by light was consistent with a role for corrinoid that was tentatively identified as 5-hydroxybenziminazolyl cobamide (Kreft and Schink 1993). In the homoacetogen *Holophaga foetida*, the methyl groups of methoxybenzoates can participate in a range of reactions: (a) successive methylation of sulfide to methyl sulfide and dimethylsulfide, (b) formation of acetate mediated by carbon-monoxide dehydrogenase (Kreft and Schink 1993). A mechanism involving a corrinoid and tetramethylfolate has been proposed (Kreft and Schink 1994).

3.21 References

Adrian L, J Rahneführer, J Gobom, T Hölscher. 2007. Identification of a chlorobenzene reductive dehalogenase in *Dehalococcoides* sp. strain CBDB1. *Appl Environ Microbiol* 73: 7717–7724.

Babior BM, TH Moss, WH Orme-Johnson, H Beinert. 1974. The mechanism of action of ethanolamine ammonia-lyase, a B_{12}-dependent enzyme. The participation of paramagnetic species in the catalytic deamination of 2-aminopropanol. *J Biol Chem* 249: 4537–4544.

Bache R, N Pfennig. 1981. Selective isolation of Acetobacterium woodii on methoxylated aromatic acids and determination of growth yields *Arch Microbiol* 130: 255–261.

Bak F, K Finster, T Rothfuß. 1992. Formation of dimethyl sulfide and methanethiol from methoxylated aromatic compounds and inorganic sulfide by newly isolated anaerobic bacteria. *Arch Microbiol* 157: 529–534.

Bothe H, DJ Darley, SPJ Albracht, GJ Gerfen, BT Golding, W Buckel. 1998. Identification of the 4-glutamyl radical as an intermediate in the carbon skeleton rearrangement catalyzed by coenzyme B_{12}-dependent glutamate mutase from *Clostridium cochlearium*. *Biochemistry* 37: 4105–4113.

Brendelberger G, J Rétey, DM Ashworth, K Reynolds, F Willenbrock, JA Robinson. 1988. The enzymatic interconversion of *n*-butyryl(dethia)-coenzyme A: A coenzyme-B_{12}-dependent carbon skeleton rearrangement. *Angew Chem Int Ed Engl* 27: 1089–1090.

Buckel W, HA Barker. 1974. Two pathways of glutamate fermentation by anaerobic bacteria. *J Bacteriol* 117: 1248–1260.

Chang CH, PA Frey. 2000. Cloning, sequencing, heterologous expression, purification, and characterization of adenosylcobalamin-dependent D-lysine 5,6-aminomutase from *Clostridium sticklandii*. *J Biol Chem* 275: 106–114.

Chen HP, CF Lin, YJ Lee, SS Tsay, SH Wu. 2000. Purification and properties of ornithine racemase from *Clostridium sticklandii*. *J Bacteriol* 182: 2052–2054.

Chen HP, SH Wu, YL Lin, CM Chen, SS Tsay. 2001. Cloning, sequencing, heterologous expression, purification, and characterization of adenosylcobalamin-dependent D-ornithine aminomutase from *Clostridium sticklandii. J Biol Chem* 276: 44744–44750.

Christiansen N, BK Ahring, G Wohlfarth, G Diekert. 1998. Purification and characterization of the 3-chloro-4-hydroxy-phenylacetate reductive dehalogenase of *Desulfitobacterium hafniense. FEBS Lett* 436: 159–162.

Coulter C, JTG Hamilton, WC McRoberts, L Kulakov, MJ Larkin, DB Harper. 1999. Halomethane:bisulfide/halide ion methyltransferase, an unusual corrinoid enzyme of environmental significance isolated from an aerobic methylotroph using chloromethane as the sole carbon source. *Appl Environ Microbiol* 65: 4301–4312.

El Kasmi A, S Rajasekharan, SW Ragsdale. 1994. Anaerobic pathway for conversion of the methyl group of aromatic methyl ethers to acetic acid by *Clostridium thermoaceticum. Biochemistry* 33: 11217–11224.

Erb TJ, J Rétey, G Fuchs, BE Alber. 2008. Ethylmalonyl-CoA mutase from *Rhodobacter sphaeroides* defines a new subclade of coenzyme B_{12}-dependent acyl-CoA mutases. *J Biol Chem* 283: 32283–32293.

Hartmannis MGN, TC Stadtman. 1986. Diol metabolism and diol dehydratase in *Clostridium glycolicum. Arch Biochem Biophys* 245: 144–152.

Hartmannis MGN, TC Stadtman. 1987. Solubilization of a membrane-bound diol dehydratase with retention of EPR g = 2.02 signal by using 2-(N-cyclohexylamino)ethanesulfonic acid buffer. *Proc Natl Acad Sci USA* 84: 76–79.

Hiraishi A. 2008. Biodiversity of dehalorespiring bacteria with special emphasis on polychlorinated biphenyl/dioxin dechlorinators. *Microbes Environ* 23: 1–12.

Hoffmann B, M Oberhuber, E Stupperich, H Bothe, W Buckel, R Conrat, B Kräutler. 2000. Native corrinoids from *Clostridium cochlearium* are adeninylcobamides: Spectroscopic analysis and identification of pseudovitamin B_{12} and factor A. *J Bacteriol* 182: 4773–4782.

Holliger C, G Wohlfarth, G Diekert. 1999. Reductive dechlorination in the energy metabolism of anaerobic bacteria. *FEMS Microbiol Revs* 22: 383–398.

Holloway DE, ENG Marsh. 1994. Adenosylcobalamin-dependent glutamate mutase from *Clostridium tetanomorphum. J Biol Chem* 269: 20425–20430.

Jeng IM, R Somack, HA Barker. 1974. Ornithine degradation in *Clostridium sticklandii*; pyridoxal phosphate and coenzyme A dependent thiolytic cleavage of 2-amino-4-ketopentanoate to alanine and acetyl coenzyme A. *Biochemistry* 13: 2898–2903.

Kim, J, M Hetzel, CD Boiangiu, W Buckel. 2004. Dehydration of (R)-2-hydroxyacyl-CoA to enoyl-CoA in the fermentation of α-amino acids by anaerobic bacteria. *FEMS Microbiol Revs* 28: 455–468.

Krasotkina J, T Walters, KA Maruya, SW Ragsdale. 2001. Characterization of the B_{12}- and iron–sulfur-containing reductive dehalogenase from *Desulfitobacterium chlororespirans. J Biol Chem* 276: 40991–40997.

Kreft J-U, B Schink. 1993. Demethylation and degradation of phenylmethyl ethers by the sulfide-methylating homoacetogenic bacterium strain TMBS4. *Arch Microbiol* 159: 308–315.

Kreft J-U, B Schink. 1994. O-demethylation by the homoacetogenic anaerobe Holophaga foetida studied by a new photometric methylation assay using electrochemically produced cob(I)alamin. *Eur J Biochem* 226: 945–951.

Kung H-F, TC Stadtman. 1971. Nicotinic acid metabolism. VI. Purification and properties of α-methylene glutarate mutase (B_{12}-dependent) and methylitaconate isomerase. *J Biol Chem* 246: 3378–3388.

Kung H-F, L Tsai. 1971. Nicotinic acid metabolism. VII. Mechanisms of action of clostridial α-methylene glutarate mutase (B_{12}-dependent) and methylitaconate isomerase. *J Biol Chem* 246: 6436–6443.

Magnuson, JK, MF Romine, DR Burris. MT Kingsley. 2000. Trichloroethene reductive dehalogenase from *Dehalococcoides ethenogenes*: Sequence of tceA and substrate characterization. *Appl Environ Microbiol* 66: 4628–4638.

Maillard J, W Schumacher, F Vazquez, C Regeard, WR Hagen, C Holliger. 2003. Characterization of the corrinoid iron–sulfur protein tetrachloroethene reductive dehalogenase of *Dehalobacter restrictus. Appl Environ Microbiol* 69: 4628–4638.

Matthies C, B Schink. 1992. Fermentative degradation of glutarate via decarboxylation by newly isolated strictly anaerobic bacteria. *Arch Microbiol* 157: 290–296.

Messmer M, S Reinhardt, G Wohlfarth, G Diekert. 1996. Studies on methyl chloride dehalogenase and O-demethylase in cell extracts of the homoacetogen strain MC based on a newly developed coupled enzyme assay. *Arch Microbiol* 165: 383–387.

Moss M, PA Frey. 1990. The role of S-adenosylmethionone in the lysine 2,3-aminomutase reaction. *J Biol Chem* 262: 14859–14862.

Müller JA, BM Rosner, G von Abendroth, G Meshulam-Simon, PL McCarthy, AM Spormann. 2004. Molecular identification of the catabolic vinyl chloride reductase from *Dehalococcoides* sp. strain VS and its environmental distribution. *Appl Environ Microbiol* 70: 4880–4888.

Mägli A, M Messmer, T Leisinger. 1998. Metabolism of dichloromethane by the strict anaerobe *Dehalobacterium formicoaceticum*. *Appl Environ Microbiol* 64: 646–650

Naidu S, SW Ragsdale. 2001. Characterization of a three-component vanillate O-demethylase from *Moorella thermoacetica*. J Bacteriol 183: 3276–3281.

Neuman A, G Wohlfarth, G Diekert. 1996. Purification and characterization of tetrachloroethene reductive dehalogenase from *Dehalospirillum multivorans*. *J Biol Chem* 271: 16515–16519.

Paul L, DJ Ferguson, JA Krzycki. 2000. The trimethylamine methyltransferase gene and multiple dimethylamine methyltransferase genes of *Methanosarcina barkeri* contain in-frame and read-through amber codons. *J Bacteriol* 182: 2520–2529.

Poston JM. 1976. Leucine 2,3-aminomutase, an enzyme of leucine catabolism. *J Biol Chem* 251: 1859–1863.

Ratnatilleke A, JW Vrijbloed, JA Robinson. 1999. Cloning and sequencing of he coenzyme B12-binding domain of isobutyryl-CoA mutase from *Streptomyces cinnamonensis*, reconstitution of mutase activity and characterization of the recombinant enzyme produced in *Escherichia coli*. *J Biol Chem* 274: 31679–31685.

Raynaud C, P Sarçabal, I Meynial-Salles, C Croux, P Soucaille. 2003. Molecular characterization of the 1,3-propandiol (1,3-PD) operon of *Clostridium butyricum*. *Proc Natl Acad USA* 100: 5010–5015.

Rohwerder T, U Breuer, D Benndorf, U Lechner, RH Müller. 2006. The alkyl tert-butyl ether intermediate 2-hydroxybutyrate is degraded via a novel cobalamin-dependent mutase pathway. *Appl Environ Microbiol* 72: 1428–1435.

Roof DM, JR Roth. 1989. Functions required for vitamin B_{12}-dependent ethanolamine utilization in *Salmonella typhimurium*. *J Bacteriol* 171: 3316–3323.

Sampson EM, TA Bobik. 2008. Microcompartments for B_{12}-dependent 1,2-propanediol degradation provide protection from DNA and cellular damage by a reactive metabolic intermediate. *J Bacteriol* 190: 2966–2971.

Schilhabel A, S Studenik, M Vödisch, S Kreher, B Schlott, AY Pierik, G Diekert. 2009. The ether-cleaving methyltransferase system of the strict anaerobe *Acetobacterium dehalogenans*: Analysis and expression of the encoding genes. *J Bacteriol* 191: 588–599.

Seyfried M, R Daniel, G Gottschalk. 1996. Cloning, sequencing, and overexpression of the genes encoding coenzyme B_{12}-dependent glycerol dehydratase of *Citrobacter freundii*. *J Bacteriol* 178: 5793–5796.

Shearer N, AP Hinsley, RJM van Spanning, S Spiro. 1999. Anaerobic growth of *Paracoccus denitrificans* requires cobalamin: Characterization of cobK and cobJ genes. *J Bacteriol* 181: 6907–6913.

Smidt H, AD L Akkermans, J van der Oost, WM de Vos. 2000. Halorespiring bacteria—Molecular characterization and detection. *Enzyme Microb Technol* 27: 812–820.

Smith DM, BT Golding, L Radom. 1999. Understanding the mechanism of B_{12}-dependent methylmalonyl-CoA mutase: Partial proton transfer in action. *J Am Chem Soc* 121: 9388–9399.

Somack R, RN Costilow. 1973. Purification and properties of a pyridoxal phosphate and coenzyme B_{12} dependent D-ornithine 5,4-aminomutase. *Biochemistry* 12: 2597–2604.

Sriramulu DD, M Liang, D Hernandez-Romero, E Raux-Deery, H Lünsdorf, JB Parsons, MJ Warren, MB Prentice. 2008. *Lactobacillus reuteri* DSM 20016 produces cobalamin-dependent diol dehydratase in metabolosomes and metabolizes 1,2-propanediol by disproportionation. *J Bacteriol* 190: 4559–4567.

Studer A, C McAnulla, R Büchele, T Leisinger, S Vuilleumier. 2002. Chloromethane-induced genes define a third C_1 utilization pathway in *Methylobacterium chloromethanicum* CM4. *J Bacteriol* 184: 3476–3484.

Tallant TC, L Paul, JA Krzycki. 2001. The MtsA subunit of the methylthiol:coenzyme M methyltransferase of *Methanosarcina barkeri* catalyzes both half-reactions of corrin-dependent dimethylsulfide:coenzyme M methyl transfer. *J Biol Chem* 276: 4485–4493.

Terzenbach DP, M Blaut. 1994. Transformation of tetrachloroethylene to trichloroethylene by homoacetogenic bacteria. *FEMS Microbiol Lett* 123: 213–218.

Toraya T, S Honda, S Fukui. 1979. Fermentation of 1,2-propanediol and 1,2-ethanediol by some genera of Enterobacteriaceae involving coenzyme B_{12}-dependent diol dehydratase. *J Bacteriol* 139: 39–47.

van de Pas B, H Smidt, WR Hagen, J van der Oost G Schraa, AJM Stams, WM de Vos. 1999. Purification and molecular characterization of ortho-chlorophenol reductive dehalogenase, a key enzyme of halorespiration in *Desulfitobacterium dehalogenans*. *J Biol Chem* 274: 20287–20292.

Vannelli T, M Messmer, A Studer, S Vuilleumier, T Leisinger. 1999. A corrinoid-dependent catabolic pathway for growth of a *Methylobacterium* strain with chloromethane. *Proc Natl Acad Sci USA* 96: 4615–4620.

Yorifugi Y, I-M Jeng, HA Barker. 1977. Purification and properties of 3-keto-5-aminohexanoate cleavage enzyme from a lysine-fermenting *Clostridium*. *J Biol Chem* 252: 20–31.

3.22 Free-Radical Reactions

3.22.1 Introduction

As in synthetic organic chemistry, there has been belated—and increasing—interest in free-radical reactions in microbiology. They encompass a wide range of anaerobic reactions that include coenzyme B_{12}-dependent mutases, dehydrations involving nonactivated positions, reductions and dehydroxylation of aromatic rings, and reactions involving glycyl radicals. Most of them have been characterized by EPR or Mössbauer spectroscopy that has been described in Chapter 6, Part 2. Their ESR spectra are, therefore, frequently used for comparison. Attention is drawn to a number of reviews (Boll 2005; Buckel et al. 2005; Frey 2001; Kim et al. 2004; Stubbe and van der Donk 1998). These include coenzyme B_{12}-catalyzed reactions in the anaerobic degradation of amino acids that have already been discussed.

3.22.2 Radical S-Adenosylmethionine Proteins

In free radical S-adenosylmethionine (SAM) reactions the $[4Fe-4S]^{1+}$ cluster is produced by one-electron reduction and brings about fission of S-adenosylmethionine to produce the reactive 5′-deoxyadenosine radical. These radical enzymes are important and they accomplish a range of activities both catabolic and anabolic (Booker and Grove 2010). Only a few of these will be briefly illustrated here beginning with the canonical reactions pyruvate formate-lyase and ribonucleotide reductase.

3.22.3 Pyruvate Formate-Lyase

The reaction carried out by pyruvate formate-lyase (Becker and Kabsch 2002) is carried out in two steps: (1) E + pyruvate → acetyl-E + formate, (2) Acetyl-E + CoA → E + acetyl-CoA. This conversion of pyruvate to formate and acetyl CoA is a key enzyme in the anaerobic degradation of carbohydrates in some *Enterobacteriaceae*. Using an enzyme selectively ^{13}C-labeled with glycine, it was shown by EPR that the reaction involves production of a free radical at C_2 of glycine (Wagner et al. 1992). This was confirmed by destruction of the radical with O_2, and determination of part of the structure of the small protein that contained an oxalyl residue originating from Gly^{734}. The reaction is initiated by activation (Frey et al. 1994) that generates the glycyl radical from Gly^{734} (Wagner et al. 1992). This then reacts with Cys^{418} and Cys^{419} to produce the reactive thiyl radical. ENDOR (electron-nuclear double resonance) spectroscopy and EPR have been used to reveal the mechanism in which the [4Fe–4S] cluster reacts with S-adenosylmethionine that is bound to Fe by the carboxylate and to S by the $CH_3–S^+–$ to produce the active $Ado–CH_2•$ radical for reaction with the glycine in the enzyme (Walsby et al. 2002). This has been confirmed using Mössbauer spectroscopy (Krebs et al. 2002), and structural details of glycyl radical formation in the pyruvate formate-lyase reaction have been added (Vey et al. 2008).

3.22.4 Ribonucleotide Reductase Class III

These are essential enzymes responsible for the reduction of the nucleotide triphosphates to the corresponding deoxynucleotide triphosphates. There are three classes of ribonucleotide reductase in all of which radical reactions are involved (Nordlund and Reichard 2006) that have been summarized in Section 3.13 of this chapter. Briefly, Class I enzymes are oxygen dependent and generate a tyrosyl radical, Class II enzymes occur under both aerobic and anaerobic conditions and use coenzyme B_{12} for the production of radical, and Class III enzymes occur under anaerobic conditions and generate a glycyl radical from

S-adenosylmethionine and a [4Fe–4S]$^{1+}$ produced by reduction of a [4Fe–4S]$^{2+}$ cluster that have been characterized by Mössbauer and EPR spectroscopy, respectively. Attention is directed here to the group III enzyme which has undergone exhaustive study so that only salient issues can be summarized.

The reductase has been extensively examined in *E. coli* where it consists of a dimeric protein α on which the ribonucleotides are bound, and a protein β whose function is to produce the catalytically essential glycyl radical at Gly^{681} on protein α. The small activating protein β contains [4Fe–4S]$^{2+}$ and [4Fe–4S]$^{1+}$ clusters, and formation of the glycyl radical requires the concerted action of (a) a reducing system NADPH, flavodoxin, and NADPH:flavodoxin reductase, (b) the 17.5 kDa [Fe–S] activating protein to catalyze reductive fission of *S*-adenosylmethionine, (c) thioredoxin and NADPH:thioredoxin oxidoreductase (Padovani et al. 2001). In contrast to the other ribonucleotide reductases, the electrons required are produced from formate, and it has been pointed out that there are important similarities between Class III ribonucleotide reductase and the pyruvate formate-lyase activation systems.

Radical *S*-adenosylmethionine enzymes are important in other reactions although only a few of these will be illustrated here where the [4Fe–4S]$^{1+}$ cluster functions in producing the 5′-deoxyadenosyl radical from *S*-adenosylmethionine.

1. The maturation of group I sulfatases is dependent on the conversion of cysteine or serine residues in the enzyme to α-glycyl formyl. Although this may involve an oxygenase in eukaryotes, the reaction in *K. pneumoniae* does not depend on dioxygen and the maturase AtsB is a member of the radical *S*-adenosylmethionine superfamily. This contains several [4Fe–4S] clusters and the reaction involves nucleophilic attack of the hydrated geminal diol at the sulfur atom of the alkyl sulfate (Grove et al. 2008). Maturation in anaerobes is carried by reactions that involve several [4Fe–4S] clusters which also belong to the radical SAM super-family (Benjdia et al. 2010).

2. The maturation of several hydrogenases such as the H-cluster in [Fe–Fe] hydrogenase and the mononuclear [Fe] hydrogenase involve [Fe–S] clusters which are discussed in greater detail in the section on hydrogenase. The HydE and HydF in *Chlamydomonas reinhardtii* are involved in the synthesis of active hydrogenase (Posewitz et al. 2004), and in *Thermotoga maritima*, the HydE and HydG proteins are involved in the maturation of the Fe-only hydrogenase; in addition, the HydE protein contains a second [4Fe–4S] cluster, and HydG at least one [4Fe–4S] cluster while, after dithionite reduction both enzymes are able to fission *S*-adenosylmethionine (Rubach et al. 2005).

3. a. 2-Deoxystreptamine is a component of the antibiotic butirosin synthesized by *Bacillus circulans*, and is produced from D-glucose-6-phosphate that is converted into 2-deoxy-*scyllo*-inosose. This is followed by successive reductive amination to 2-deoxy-*scyllo*-inosamine and dehydrogenation at C_1 to the ketone by BtrN and further reductive amination by BtrS to 2-deoxystreptamine. The reaction catalyzed by BtrN is mediated by a radical SAM involving the 5′deoxyadenosy 5′radical and abstraction of hydrogen to form a C1-radical that has been characterized by EPR (Yokoyama et al. 2008) and that showed the presence of a second [Fe–S] cluster (Grove et al. 2010).

   b. D-Desosamine (3-(*N,N*-dimethylamine)-3,4,6-trideoxyhexose is a key component of several antibiotics produced by *Streptomyces venezulae*. It is synthesized from thymidine diphosphate-D-glucose in sequence of reactions: dioxygenation of C_6-OH to C_6-methyl, dehydrogenation to the C_4-ketone, and reductive amination to the C_4-amino sugar. This undergoes a reaction catalyzed by a radical SAM reaction to the C_3-keto sugar catalyzed by DesI and presumptively takes place by formation of the radical at C_3 followed by loss of NH_3 to produce the enol of the C_3-keto sugar (Szu et al. 2009).

4. The degradation of L-lysine by *Clostridium subterminale* is initiated by the activity of L-α-lysine mutase to produce L-β-lysine. In clostridia, these mutases are generally dependent on coenzyme B$_{12}$, but it has been shown that in this strain an SEM/[4Fe–4S] cluster is involved (Lepore et al. 2005).

5. The decarboxylation of vinyl groups in the propionic acid side-chains in rings A and B of uroporphyrinogen III, is a critical step in the biosynthesis of protoporphyrin IX that is the immediate precursor of heme and chlorophyll/bacteriochlorophyll. In *E. coli* this can be accomplished by two different reactions: an oxygen-dependent Mn-oxidase HemF (Breckau et al. 2003), and an

oxygen-independent mechanism by HemN that involves a free radical generated from *S*-adenosyl-L-methionine and an [4Fe–4S] cluster. Details were provided by the crystal x-ray structure (Layer et al. 2003) and by the complex pattern of splitting in the EPR spectrum (Layer et al. 2006).

The following illustrate the range of free-radical reactions in addition to those already discussed for coenzyme B_{12}-mediated reactions.

3.22.5 Glycine Free Radicals

A number of reactions involve the glycine free radicals, and these play cardinal roles in the anaerobic metabolism of hydrocarbons. The glycine radical occurs on an sp^2 carbon atom, is close to the C-terminal end of the protein, and is stabilized by electron delocalization over adjacent peptides. It is transformed by reaction with cysteine to a thiyl cys˙ radical. The enzymes are activated by a $[4Fe–4S]^{1+/2+}$ cubane in which the iron is coordinated to *S*-adenosyl-methionine.

3.22.6 Anaerobic Degradation of Toluene

The anaerobic degradation of toluene is carried out by benzylsuccinate synthase in *Azoarcus* sp. strain T and in *Thauera aromatica*, and in both a radical reaction results in the stereospecific production of (*R*)-2-benzylsuccinate by reaction with fumarate. The radical has been characterized in *Azoarcus* sp. T, and on the basis of similarities between the EPR signal with that of the glycyl radicals of pyruvate formate-lyase and anaerobic ribonucleotide reductase, it was proposed that the synthase radical is located on a glycine residue (Krieger et al. 2001). The enzyme in *Thauera aromatica* consists of four subunits of 94(α), 90(α′), 12(β), and 10 kDa(γ), and the native enzyme was, therefore, described as $\alpha_2\beta_2\gamma_2$ (Leuthner et al. 1998), and it was possible to describe the dioxygen-mediated reaction at the radical center by reaction with the O_2 diradical. The further reaction involves the activity of a Co-A transferase followed by β-oxidation to benzoyl-CoA. The substrates include not only toluene and xylene, but also cresols. The different specificities of the enzymes from *Azoarcus* sp. strain T and *Th. aromatica* have been explored, and the EPR spectra were closely similar to those of typical glycyl radicals (Verfürth et al. 2004).

3.22.7 Anaerobic Degradation of *n*-Alkanes

The degradation of *n*-alkanes has been examined in the denitrifying bacterium strain HxN1 and is carried out by reaction of fumarate at C_2 to produce (1-methypentyl)succinate that was confirmed by labeling (Rabus et al. 2001). On the basis of the EPR spectrum, the reaction probably involves a glycyl radical analogous to that involved in the activity of benzylsuccinate synthase. On the basis of partial amino acid sequences, it was proposed that the gene products MasD, MasC, and MasE were analogous to the subunits of BssC that is involved in the degradation of toluene by benzylsuccinate synthase (Grundmann et al. 2008).

3.22.8 Anaerobic Decarboxylation of 4-Hydroxyphenylacetate

The enzyme from *Clostridium difficile* brings about the decarboxylation of 4-hydroxyphenylacetate to produce *p*-cresol Selmer and Andrei (2001). The subunit of the enzyme has been examined, and revealed the presence of a small subunit HpdC that is essential for catalytic activity (Andrei et al. 2004).

3.22.9 Anaerobic Dehydration of Nonactivated Carboxyl CoA-Esters

Free-radical reactions are implicated in the dehydration of carboxy-CoA metabolites that are produced during anaerobic degradation of α-amino acids by deamination followed by reduction. These radical-mediated dehydrations involve loss of hydrogen from nonactivated positions (Smith et al. 2003; Scott et al. 2004; Kim et al. 2004) (Figure 3.44), and are discussed in greater detail in Chapter 7.

The fermentation of glutamate by *Acidaminococcus fermentans*, *Clostridium symbiosum*, and *Fusobacterium nucleatum* involves the oxidation to 2-oxoglutarate followed by reduction to

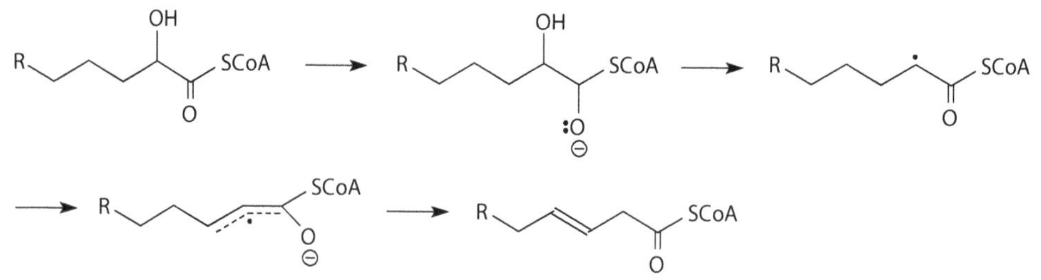

FIGURE 3.44 Dehydration at nonactivated positions.

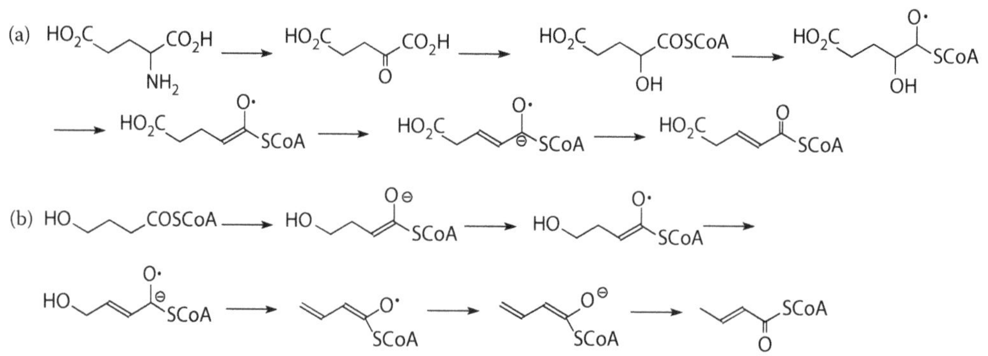

FIGURE 3.45 (a) Dehydration of 2-hydroxyglutarate and (b) dehydration of 4-hydroxysuccinyl-CoA.

(R)-2-hydroxyglutarate. The CoA ester is transformed by dehydration to (E)-glutaconyl-CoA that may undergo decarboxylation to crotonyl-CoA followed by disproportion to acetate, butyrate and H_2. Dehydration involves loss of hydrogen from a nonactivated position (Figure 3.45a). In *A. fermentans*, there are two components, a homodimeric activator A that has one [4Fe–4S]$^{+1/2+}$ and the heterodimeric dehydratase D that contains on [4Fe–4S]$^{2+}$ and reduced FMN (Hans et al. 2002). In component A of *A. fermentans* each subunit is spanned by the cubane cluster that is reduced by flavodoxin hydroquinone, and two ADP-binding sites that are activated by 2 ATP. Component D from *Cl. symbiosum* contains a second [4Fe–4S]$^{2+}$ cluster and Mössbauer spectroscopy showed a symmetric cubane [Fe–S] cluster (Hans et al. 1999). Analogous dehydratases have been characterized in *Cl. sporogenes* that produces cinnamoyl-CoA from (R)-phenyl lactate (Dickert et al. 2002), and in *Cl. difficile* that dehydrates (R)-2-hydroxyisocaproyl-CoA followed by reduction to isocaproyl-CoA that is involved in the fermentation of L-leucine (Kim et al. 2005).

A formally analogous dehydration from a nonactivated position is involved in the dehydration of 4-hydroxybutyryl-CoA to crotonyl-CoA in the fermentation of 4-aminobutyrate via succinate semialdehyde by *Clostridium aminobutyricum* (Gerhardt et al. 2000). The dehydratase is a tetramer with up to one [4Fe–4S]$^{2+}$ cluster that was detected by ESR after photoreduction and one noncovalently bound FAD per 54 kDa subunit (Müh et al. 1996, 1997). The radical-mediated mechanism has been explored on the basis of the crystal structure to reveal the roles of the quinone, semiquinone anion, and neutral semiquinone (Figure 3.45b) (Martins et al. 2004).

3.23 References

Andrei PI, AJ Pierik, S Zauner, LC Andrei-Selmer, T Selmer. 2004. Subunit composition of the glycyl radical enzyme *p*-hydroxyphenylacetate decarboxylase. A small subunit, HpdC, is essential for catalytic activity. *Eur J Biochem* 1–6.

Becker A, W Kabsch. 2002. X-ray structure of pyruvate formate-lyase in complex with pyruvate and CoA. *J Biol Chem* 277: 40036–40042.

Benjdia A, S Subramanian, J Lepince, H Vaudry, MK Johnson, O Berteau. 2010. Anaerobic sulfatase-maturing enzyme—A mechanistic link with glycyl radical-activating enzymes. *FEBS J* 277: 1906–1920.

Boll M. 2005. Key enzymes in the anaerobic aromatic metabolism catalyzing Birch-like reductions. *Biochim Biophys Acta* 1707: 34–50.

Booker SJ, TL Grove. 2010. Mechanistic and functional versatility of radical SAM enzymes. *F 1000 Biol Repts* 2: 52, doi: 10.3410/B2-52.

Breckau D, E Mahlitz, A Sauerwald, G Layer, D Jahn. 2003. Oxygen-dependent coproporphyrinogen III oxidase (HemF) from *Escherichia coli* is stimulated by manganese. *J Biol Chem* 278: 46625–46631.

Buckel W, BM Martins, A Messerschmidt, BT Golding. 2005. Radical-mediated dehydration in anaerobic bacteria. *Biol Chem* 386: 951–959.

Dickert S, AJ Pierik, W Buckel. 2002. Molecular characterization of phenyllactate dehydratase and its initiator from *Clostridium sporogenes*. *Mol Microbiol* 44: 49–60.

Frey M, M Rothe, AFV Wagner, J Knappe. 1994. Adenosylmethionine-dependent synthesis of the glycyl radical in pyruvate formate-lyase by abstraction of the glycine C-2 pro-*S* hydrogen atom. *J Biol Chem* 269: 12432–12437.

Frey PA. 2001. Radical mechanisms of enzymic catalysis. *Annu Rev Biochem* 70: 121–148.

Gerhardt A, I Çinkaya, D Linder, G Huisman, W Buckel. 2000. Fermentation of 4-aminobutyrate by *Clostridium aminobutyricum*: Cloning of two genes involved in the formation and dehydration of 4-hydroxybutyryl-CoA. *Arch Microbiol* 174: 189–199.

Grove TL, JH Ahlum, P Sharma, C Krebs, SJ Booker. 2010. A consensus mechanism for radical SAM-dependent dehydrogenation? *Biochemistry* 49: 3783–3785.

Grove TL, K-H Lee, J St Clair, C Krebs, SJ Booker. 2008. *In vitro* characterization of AtsB, a radical SAM formylglycine-generating enzyme that contains three [4Fe–4S] clusters. *Biochemistry* 47: 7523–7538.

Grundmann O, A Behrends, R Rabus, J Amann, T Halder, J Heider, F Widdel. 2008. Genes encoding the candidate enzyme for anaerobic activation of *n*-alkanes in the denitrifying bacterium strain HxN1. *Environ Microbiol* 10: 376–385.

Hans M, J Sievers, U Müller, E Bill, JA Vorholt, D Linder, W Buckel. 1999. 2-hydroxyglutaryl-CoA dehydratase from *Clostridium symbiosum*. *Eur J Biochem* 265: 404–414.

Hans M, E Bill, I Cirpus, AJ Pierik, M Hetzel, D Alber, W Buckel. 2002. Adenosine triphosphate-induced electron transfer in 2-hydroxyglutaryl-CoA dehydratase from *Acidaminococcus fermentans*. *Biochemistry* 41: 5873–5882.

Kim J, D Darley, W Buckel. 2005. 2-hydroxyisocaproyl-CoA dehydratase and its activator from *Clostridium difficile*. *FEBS J* 272: 550–561.

Kim J, M Hetzel, CD Boiangiu, W Buckel. 2004. Dehydration of (*R*)-2-hydroxyacyl-CoA to enoyl-CoA in the fermentation of α-amino acids by anaerobic bacteria. *FEMS MIcrobiol Revs* 28: 455–468.

Krebs C, WE Broderick, TF Henshaw, JB Broderick, BH Huynh. 2002. Coordination of adenosylmethionine to a unique iron site of the [4Fe–4S] of pyruvate formate-lyase activating enzyme: A Mössbauer spectroscopic study. *J Am Chem Soc* 124: 912–913.

Krieger CJ, W Roseboom, SPJ Albracht, AM Spormann. 2001. A stable organic free radical in anaerobic benzylsuccinate synthase of *Azoarcus* sp. strain T. *J Biol Chem* 276: 12924–12927.

Layer G, J Moser, DW Heinz, D Jahn, W-D Schubert. 2003. Crystal structure of coproporphyrinogen III oxidase reveals cofactor geometry of radical SAM enzyme. *EMBO J* 22: 6214–6224.

Layer G et al. 2006. The substrate radical of Escherichia coli oxygen-independent coproporphyrinogen III oxidase HemN. *J Biol Chem* 281: 15727–15734.

Lepore BW, FJ Ruzicka, PA Frey, D Ringe. 2005. The x-ray crystal structure of lysine-2,3-aminomutase from *Clostridium subterminale*. *Proc Natl Acad Sci USA* 102: 13819–13824.

Leuthner B, C Leutwein, H Schulz, P Hörth, W Haehnel, E Schiltz, H Schägger, J Heider. 1998. Biochemical and genetic characterisation of benzylsuccinate synthase from *Thauera aromatica*: A new glycyl-radical catalysing the first step in anaerobic toluene degradation. *Mol Microbiol* 28: 515–628.

Martins BM, H Dobbek, I Çinkaya, W Buckel, A Messerschmidt. 2004. Crystal structure of 4-hydroxybutyryl-CoA dehydratase: Radical catalysis involving a [4Fe–4S] cluster and flavin. *Proc Natl Acad Sci USA* 101: 15645–15649.

Müh U, I Çinkaya, SPJ Albracht, W Buckel. 1996. 4-hydroxybutyryl-CoA dehydratase from *Clostridium aminobutyricum*: Characterization of FAD and iron–sulfur clusters involved in an overall non-redox reaction. *Biochemistry* 35: 11710–11718.

Müh U, W Buckel, E Bill. 1997. Mössbauer study of 4-hydroxybutyryl-CoA dehydratase. Probing the role of an iron–sulfur cluster in an overall non-redox reaction. *Eur J Biochem* 248: 380–384.

Nordlund P, P Reichard. 2006. Ribonucleotide reductases. *Annu Rev Biochem* 75: 681–706.

Padovani D, E Mulliez, M Fontecave. 2001. Activation of Class III ribonucleoide reductase by thioredoxin. *J Biol Chem* 276: 9587–9589.

Posewitz MC, PW King, SL Smolinski, L Zhang, M Seibert, ML Ghirardi. 2004. Discovery of two novel radical *S*-adenosylmethionine proteins required for the assembly of an active [Fe] hydrogenase. *J Biol Chem* 279: 25711–25720.

Rabus R, H Wilkes, A Behrends, A Armstroff, T Fischer, AJ Pierik, F Widdel. 2001. Anaerobic initial reaction of *n*-alkanes in a denitrifying bacterium: Evidence for (1-methylpentyl) succinate as initial product and for involvement of an organic radical in *n*-hexane metabolism. *J Bacteriol* 183: 1707–1715.

Rubach JR, X Brazzolotto, J Gaillard, M Fontecave. 2005. Biochemical characterization of the HydE and HydG iron-only hydrogenase maturation enzymes from *Thermotoga maritima*. *FEBS Lett* 579: 579: 5055–5060.

Scott R, U Näser, P Friedrich, T Selmer, W Buckel, BT Golding. 2004. Stereochemistry of hydrogen removal from the `unactivated´ C-3 position of 4-hydroxybutyryl-CoA catalyzed by 4-hydroxybutyryl-CoA dehydratase. *Chem Comm* 1210–1211.

Selmer T, PI Andrei. 2001. *p*-Hydroxyphenylacetate decarboxylase from *Clostridium difficile*. A novel glycyl radical enzyme catalysing the formation of *p*-cresol. *Eur J Biochem* 268: 1363–1372.

Smith DM, W Buckel, H Zipse. 2003. Deprotonation of enoxy radicals: Theoretical validation of a 50-year-old mechanistic proposal. *Angew Chem Int Ed* 42: 1867–1870.

Stubbe J, WA van der Donk. 1998. Protein radicals in enzyme catalysis. *Chem Rev* 98: 705–762.

Szu P-H, MW Ruszczcky, S-h Choi, F Yan, H-w Liu. 2009. Characterization and mechanistic studies of DesII: A radical S-adenosyl-L-methionine enzymes involved in the biosynthesis of TDP-D-desosamine. *J Am Chem Soc* 131: 14030–14042.

Tamarit J, E Mulliez, C Meier, A Trautwein, M Fontecave. 2008. The anaerobic ribonucleotide reductase from *Escherichia coli*. The small protein is an activating enzyme containing a [4Fe–4S]$^{2+}$ center. *J Biol Chem* 274: 31291–31296.

Verfürth K, AJ Pierik, C Leutwein, S Zorn, J Heider. 2004. Substrate specificities and electron paramagnetic resonance properties of benzylsuccinate synthesis in anaerobic toluene and *m*-xylene metabolism. *Arch Microbiol* 181: 155–162.

Vey JL, J Yang, M Li, WE Broderick, JB Broderick, CL Drennan. 2008. Structural basis for glycyl radical formation by pyruvate formate-lyase activating enzyme. *Proc Natl Acad Sci USA* 105: 16137–16141.

Wagner AFV, M Frey, FA Neugebauer, W Schäfer, J Knappe. 1992. The free radical in pyruvate formate-lyase is located on glycine-734. *Proc Natl Acad Sci USA* 89: 996–1000.

Walsby CJ, W Hong, WE Broderick, J Cheek, S Ortillo, JB Broderick, BM Hoffman. 2002. Electron-nuclear double resonance spectroscopic evidence that *S*-adenosylmethionine binds in contact with the catalytically active [4Fe–4S]$^+$ cluster of pyruvate formate-lyase activating enzyme. *J Am Chem Soc* 124: 3143–3151.

Yokoyama K, D Ohmori, F Kuda, T Eguchi. 2008. Mechanistic study on the reaction of a radical SAM dehydrogenase BtrN by electron paramagnetic resonance spectroscopy. *Biochemistry* 47: 8950–8960.

3.24 Coenzyme A: Ligase, Transferase, Synthetase

Although coenzyme A esters are involved in a range of reactions that are not discussed here, a few merit brief mention:

1. Acyl-CoA dehydrogenases (Ghisla and Thorpe 2004) that are involved in the β-oxidation pathway for the degradation of *n*-alkanoates and amino acids, and the synthesis of poly(3-hydroxyalkanoates) are discussed in Chapter 7.

2. CO_2 assimilation by the 3-hydroxypropionate pathway in anaerobes (Teufel et al. 2009), and the dicarboxylate/4-hydroxybutyrate cycle in *Thermoproteus neutrophilus* (Ramos-Vera et al. 2009) that are discussed in Chapter 2.

Carboxylic acids are involved in a wide range of reactions. In many of them—particularly for anaerobic reactions—they must be transformed into CoA esters to retain the reactivity of the carbonyl group that is lost in formation of the carboxylate anion. There are three quite different reactions that can accomplish this: coenzyme A ligase, coenzyme A transferase, and coenzyme A synthetase.

3.24.1 Coenzyme A Ligase

The anaerobic degradation of benzoate and related substrates is initiated by formation of the coenzyme A esters by a CoA ligase that is dependent on ATP. This activates the reduction of the aromatic ring as well as later stages in the degradation of the reduced acyclic products. A selection of those that have been characterized is given in Table 3.9.

A coenzyme A ligase is used in the aerobic degradation of several substrates.

1. Some organisms have evolved a single benzoate-CoA ligase that is able to function under both aerobic and anaerobic conditions, for example, in *Thauera aromatica* (Schühle et al. 2003), and in strains of *Magnetospirillum* (Kawaguchi et al. 2006), while the aerobic degradation of benzoate by *Azoarcus evansii* is also initiated by formation of the coenzyme-A ester (Gescher et al. 2002).
2. Coenzyme A esters are involved in the aerobic degradation of several aromatic carboxylates that are summarized in Table 3.10.
3. The activity of a ligase carries out formation of the coenzyme A to initiate the degradation of pimelate:

TABLE 3.9

Anaerobic Activation of Carboxylic Acids by Formation of Coenzyme-A Esters

Substrate	Organism	Selected Reference
Benzoate	*Thauera aromatica*	Schühle et al. (2003)
	Rhodopseudomonas palustris	Egland et al. (1995)
	Azoarcus sp.	López-Barragán et al. (2004)
	Desulfococcus multivorans	Peters et al. (2004)
	Magnetospirillum sp. strain TS-6	Kawaguchi et al. (2006)
3-Hydroxybenzoate	*Thauera aromatica*	Laempe et al. (2001)
	Azoarcus sp.	Wöhlbrand et al. (2008)
	Sporotomaculum hydroxybenzoicum	Müller and Schink (2000)
4-Hydroxybenzoate	*Thauera aromatica*	Biegert et al. (1993)
	Rhodopseudomonas palustris	Gibson et al. (1997)
4-Hydroxycinnamate	*Rhodopseudomonas palustris*	Pan et al. (2008)
4-Hydroxyphenylacetate	*Thauera aromatica*	Mohamed et al. (1993)
	Azoarcus evansii	Hirsch et al. (1998)

TABLE 3.10

Selection of Aerobic Aromatic Degradations Initiated by Synthesis of Coenzyme A Esters

Substrate	Organism	Reference
Phenylacetate	*Pseudomonas putida* strain U	Martínez-Blanco et al. (1990)
	Pseudomonas sp. strain Y2	del Peso-Santos et al. (2006)
	E. coli K-12	Ismail et al. (2003)
	Azoarcus evansii	Mohamed (2000)
Salicylate	*Rhodococcus* sp. strain B4	Grund et al. (1992)
	Streptomyces sp. strain WA 46	Ishiyama et al. (2004)
Chlorobenzoate	*Pseudomonas* sp. strain CBS3	Löffler et al. (1991, 1992); Chang and Dunaway-Mariano (1996)
4-Fluorocinnamate	*Arthrobacter* sp. strain G1	Hasan et al. (2011)

a. In the denitrifying organism LP1, degradation of the coenzyme A ester was carried out in a series of steps with the production of glutaryl-CoA that was decarboxylated to crotonyl-CoA (Härtel et al. 1993; Gallus and Schink 1994; Harrison and Harwood 2005).

b. Aerobic degradation of pimelate produced from 1,2,3,4-tetrahydronaphthalene by *Sphingomonas macrogolitabida* TFA was carried out by β-oxidation of pimelyl-CoA to produce pyruvoyl-CoA and glutaryl-CoA that was degraded further to CO_2, crotonyl-CoA and acetyl-CoA (López-Sanchez et al. 2010).

4. Phenylacetate is produced as an intermediate in the aerobic degradation of substrates including styrene by *Pseudomonas* sp. strain Y2 (del Peso-Santos et al. 2006), and *n*-phenyl-alkanoates by *Ps. putida* strain U (Olivera et al. 2001). Its degradation is initiated by formation of the coenzyme A ester that is carried out by a ligase that has been isolated and characterized from *Ps. putida* strain U (Martínez-Blanco et al. 1990).

5. Degradation of steroid side chains of β-sitosterol and campestrol by *R. rhodochrous* DSM43269 has been delineated by analysis gene inactivation in mutant strain RG32 (Wilbrink et al. 2011). The gene Fad19 was identified as encoding a steroid-coenzyme A ligase that was involved in the degradation which takes place by a reaction involving carboxylation which is discussed in Chapter 7 (Fujimoto et al. 1982a,b), and is analogous to that involved in the degradation of geraniol.

3.24.2 Coenzyme A Transferase

Coenzyme A-transferase catalyzes the reversible transfer of CoA between carboxylic acids and is employed in a wide range of anaerobic reactions:

$$R_2-CO-O^- + R_1-CO-SCoA \leftrightarrow R_2-CO-SCoA + R_1-CO-O^-$$

for example: succinyl–CoA + 3-oxoacid → succinate + 3-oxoacyl-CoA. These are widely used in the anaerobic metabolism of amino acids by the 2-hydroxyacyl pathway which involves a free-radical dehydratase when the coenzyme A ester is activated by ATP and Mg^{2+}.

There are three families of these coenzyme A transferases (Heider 2001) that include the following illustrative examples which avoid the consumption of ATP by coenzyme A ligases. It is worth noting, however, that there does not seem to be a direct correlation between the type of reaction and the family to which the enzyme belongs. For example (Kim et al. 2006),

1. Propionyl CoA and glutaconyl CoA transferases belong to family I, whereas phenyllactyl CoA and 2-hydroxyisocaproyl CoA transferases belong to family III.

2. Among family III CoA transferases there is a rather wide range of similarities in amino acid sequences between 2-hydroxyisocaproyl CoA and other CoA transferases: phenyllactyl CoA transferase 45%, 24% to benzylsuccinyl CoA transferase, 25% to oxalyl CoA transferase, and 25%, and carnitine CoA transferase.

Family I
This typically involves short-chain alkanoates, β-oxoacids and glutaconate. Extensive studies have been devoted to the succinyl-CoA:3-oxoacid coenzyme A transferase: purification of the enzyme from pig heart (White and Jencks 1976a), kinetics of the reaction (Hersh and Jencks 1967) that supported the existence of two half-reactions, properties of the γ-glutamyl intermediate after reduction with ^{3}H-labeled sodium borohydride and hydrolysis to α-amino-δ-hydroxyvalerate (Solomon and Jencks 1969), and specificity of substituted acetates (White and Jencks 1976b). Two key observations are the existence of a two-stage reaction, and a reaction that may involve the terminal γ-carboxylic acid of glutamate. Examples of this family include the following:

Succinyl-CoA:acetoacetyl-CoA transferase *Helicobacter pylori* (Corthésy-Theulaz et al. 1997)
Succinyl-CoA:3-oxoadipyl-CoA transferase *Ps. putida* (Parales and Harwood 1992)

Succinyl-CoA:3-oxoadipyl-CoA transferase *Pseudomonas* sp. strain B13 (Kaschabek et al. 2002)

Succinyl-CoA:3-oxoacid-CoA transferase (White and Jencks 1976a,b)

Succinyl-CoA:acetyl-CoA transferase *Acetobacter aceti* (Mullins et al. 2008)

Acetyl-CoA:gluconyl-CoA transferase *Acidaminococcus fermentans* (Buckel et al. 1981; Selmer and Buckel 1999)

Acetyl-CoA:4-hydroxybutyryl-CoA transferase *Clostridium aminobutyricum* (Scherf and Buckel 1991)

Butyryl-CoA:acetoacetyl-CoA transferase *Clostridium* sp. strain SB4 (Barker et al. 1978)

Further examples include a range of dehydrations of amino acids from nonactivated positions that generally involve free radical reactions (Scott et al. 2004; Smith et al. 2003; Kim et al. 2004).

- (R)-Phenyllactyl-CoA to (E)-cinnamate by *Clostridium sporogenes* (Dickert et al. 2000, 2002)
- 4-Hydroxybutyryl-CoA to crotonyl-CoA by *Clostridium aminobutyricum* (Müh et al. 1996; Gerhardt et al. 2000; Martins et al. 2004)
- –(R)-2-Hydroxyglutaryl-CoA to (E)-glutaconyl-CoA by *Acidaminococcus fermentans* (Schweiger et al. 1987; Müller and Buckel 1995; Hans et al. 2002) by *Clostridium symbiosum* (Hans et al. 1999)
- (R)-2-Hydroxyisocaproyl-CoA to 2-isocaproyl-CoA by *Clostridium difficile* (Kim et al. 2004a, 2006)
- 5-Aminovaleryl-CoA by *Clostridium aminovalericum* (Barker et al. 1987)

Family II

This small family includes citrate lyase in *Klebsiella aerogenes*, citramalate lyase in *Clostridium tetanomorphum* and succinyl-CoA:D-citramalate CoA transferase in *Chloroflexus aurantiacus*.

1. In *Clostridium tetanomorphum* S-citramalate lyase brings about fission to pyruvate and acetate, and this reaction is analogous to citrate lyase in *K. aerogenes*. For the former, the initial reaction was a transferase that carried out acyl exchange:

   S-citramalate + acetyl-CoA → S-citramalyl-CoA + acetate,

   and the pantothenate-requiring enzyme has been purified (Buckel and Bobi 1976).

2. In *Chloroflexus aurantiacus* there are two interacting cycles: (a) carboxylation of acetyl-CoA to malonyl-CoA followed by several steps with the formation of L-malyl-CoA that is incorporated into the second cycle in which D-citramalate is converted to the Co-A ester by the activity of succinyl-CoA: D-citramalate CoA transferase before fission to pyruvate and acetyl-CoA that is carboxylated to malonyl-CoA (Friedmann et al. 2006).

Family III

This group includes important examples of anaerobic metabolism, in particular benzylsuccinate-CoA transferase that is involved in degradations that use the fumarate pathway.

- Cinnamoyl-CoA:(R)-phenyllactate-CoA transferase *Clostridium sporogenes* (Dickert et al. 2000)
- Formyl-CoA:oxalate-CoA transferase *Oxalobacter formigenes* (Jonsson et al. 2004)
- Succinyl-CoA:benzylsuccinate-CoA transferase *Thauera aromatica* (Leutwein and Heider 2001)
- Carnitine-CoA transferase (Elssner et al. 2001; Stenmark et al. 2004)

3.24.3 Coenzyme A Synthetase

The degradation of substrates or intermediate metabolites can be dependent on the activity of coenzyme A synthetases.

1. The degradation of furan-2-oate by *Pseudomonas* F2 is initiated by formation of the coenzyme A ester that undergoes hydroxylation to 5-hydroxy-2-furoyl-CoA and tautomerization to 5-oxo-Δ^2-dihydro-2-furanoyl-CoA before hydrolysis (Trudgill 1969).

2. The degradation of 4-methylcatechol by *R. rhodochrous* N75 involves intradiol fission and isomerization to 3-methylmuconolactone. This is converted to 3-methylmuconolactone-coenzyme A before isomerization to the Δ^3-ene and hydrolysis to 4-methyl-3-oxo-adipyl-CoA (Cha et al. 1998). It should be noted that, in *Pseudomonas reinekei* the hydrolysis of 3-methylmuconolactone occurs directly although the product is converted into 4-methyl-3-oxoadipyl-CoA before hydrolysis (Marín et al. 2010), putatively mediated by the analogous succinyl-CoA:3-oxoadipyl-CoA transferase that has already been noted for *Pseudomonas* spp (Parales and Harwood 1992; Kaschabek et al. 2002).

3. Caffeate (3,4-dihydroxycinnamate) is able to couple its reduction to the synthesis of ATP, and it has been shown that this is initiated by formation of caffeyl-CoA by a coenzyme A synthetase that is produced by CarB that was expressed in *E. coli* and purified. The purified enzyme was able to accept also 4-hydroxycinnamate and 3-methoxy-4-hydroxycinnamate (Hess et al. 2011).

4. The aerobic degradation of 3-sulfinopropionate produced from 3,3'-dithiodipropionate by dehydrogenation and dioxygenation necessitates conversion into the coenzyme A ester. This is carried out by a succinyl coenzyme A synthetase that has been purified from this organism and characterized by MS (Schürmann et al. 2011).

3.24.4 Mechanisms

These enzymes may function by using an aspartate or glutamate in the enzyme to form the aspartyl-CoA or γ-glutamyl-CoA with the CoA donor (Solomon and Jencks 1969; Selmer and Buckel 1999).

1. For formyl-CoA:oxalate transferase, it has been proposed that the reaction is initiated by reaction of Asp[169] with formyl-S-CoA followed by reaction of a mixed anhydride with acetyl-CoA to produce the product (Jonsson et al. 2004).

2. For glutaconate CoA-transferase, in *Acidaminococcus fermentans* there are three partial reactions: (i) reaction of the γ-carboxyl of Glu[54] with acetyl-CoA to form an anhydride with intramolecular production of Glu[54]-COSCoA, (ii) reaction with gluconate to form the anhydride Glu[54]-CO-O-CO-gluconate, (iii) reaction with ⁻SCoA to produce gluconate-SCoA (Solomon and Jencks 1969; Mack and Buckel 1995; Selmer and Buckel 1999).

3. Alanine is fermented by *Clostridium propionicum* by the α-hydroxyacyl pathway that involves activation of (*R*)-lactate to (*R*)-lactoyl-CoA and the transferase has been characterized (Selmer et al. 2002). Identification of the active site was carried out by borohydride reduction or reaction with hydroxylamine followed by carboxylmethylation, protease digestion and analysis by MALDI-TOF-MS that revealed glutamate 324 at the active site.

3.25 References

Barker HA, I-M Jeng, N Neff, JM Robertson, FK Tam, S Hosaka. 1978. Butyryl-CoA:acetoacetate CoA transferase from a lysine-fermenting *Clostridium*. *J Biol Chem* 253: 1219–1225.

Barker HA, L D'Ari, J Kahn. 1987. Enzymatic reactions in the degradation of 5-aminovalerate by *Clostridium aminovalericum*. *J Biol Chem* 262: 8994–9003.

Biegert T, U Altenschmidt, C Eckerskorn, G Fuchs. 1993. Enzymes of anaerobic metabolism of phenolic compounds. 4-Hydroxybenzoate-CoA ligase from a denitrifying *Pseudomonas* species. *Eur J Biochem* 213: 555–561.

Buckel W, A Bobi. 1976. The enzyme complex citramalate lyase from *Clostridium tetanomorphum*. *Eur J Biochem* 64: 255–262.

Buckel W, U Dorn, R Semmler. 1981. Glutaconate CoA-transferase from *Acidaminococcus fermentans*. *Eur J Biochem* 118: 315–321.

Cha C-J, RB Cain, NC Bruce. 1998. The modified β-ketoadipate pathway in *Rhodococcus rhodochrous* N75: Enzymology of 3-methylmuconolactone metabolism. *J Bacteriol* 180: 6668–6673.

Chang K-H, D Dunaway-Mariano. 1996. Determination of the chemical pathway for 4-chlorobenzoate: Coenzyme A ligase catalysis. *Biochemistry* 35: 13478–13484.

Corthésy-Theulaz IE, GE Bergonzelli, H Henry, D Bachmann, DF Schorderet, AL Blum, LN Ornston. 1997. Cloning and characterization of *Helicobacter pylori* succinyl CoA:acetoacetate Co-A transferase, a novel prokaryotic member of the CoA transferase family. *J Biol Chem* 272: 25659–25667.

del Peso-Santos T, D Bartolomé-Martín, C Fernández, S Alonso, JL Gardía, E Díaz, V Shingler, J Perera. 2006. Coregulation by phenylacetyl-coenzyme A-responsive PaaX integrates control of the upper and lower pathways for catabolism of styrene by *Pseudomonas* sp. strain Y2. *J Bacteriol* 188: 4812–4821.

Dickert S, AJ Pierik, D Linder, W Buckel. 2000. The involvement of coenzyme A esters in the dehydration of (*R*)-phenyllactate to (*E*)-cinnamate by *Clostridium sporogenes*. *Eur J Biochem* 267: 3874–3884.

Dickert S, AJ Pierik, W Buckel. 2002. Molecular characterization of phenyllactate dehydratase and its initiator from *Clostridium sporogenes*. *Mol Microbiol* 44: 49–60.

Egland PG, J Gibson, CS Harwood. 1995. Benzoate-coenzyme A ligase, encoded *badA*, is one of three ligases to catalyze benzoyl-coenzyme A formation during anaerobic growth of *Rhodopseudomonas palustris* on benzoate. *J Bacteriol* 177: 6545–6551.

Elssner T, C Engemann, K Baumgart, H-P Kleber. 2001. Involvement of coenzyme A esters and two new enzymes, an enoyl-CoA hydratase and a CoA-transferase, in the hydration of cronobetaine to L-carnitine by *Escherichia coli*. *Biochemistry* 40: 11140–11148.

Fujimoto Y, C-S Chen, Z Szelecky, D DiTullio, CJ Sih. 1982a. Microbial degradation of the phytosterol side chain. 1. Enzymatic conversion of 3-oxo-24-ethyl-4-en-26-oic acid into 3-oxochol-4-en-24-oic acid and androst-4-ene-3,17-dione. *J Am Chem Soc* 104: 4718–4720.

Fujimoto Y, C-S Chen, AS Gopalan, CJ Sih. 1982b. Microbial degradation of the phytosterol side chain. 2. Incorporation of $NaH^{14}CO_3$ into the C-28 position. *J Am Chem Soc* 104: 4720–4722.

Friedmann S, BE Alber, G Fuchs. 2006. Properties of succinyl-coenzyme A: D-citramalate coenzyme A transferase and its role in the autotrophic 3-hydroxypropionate cycle of *Chloroflexus aurianticus*. *J Bacteriol* 188: 6460–6468.

Gallus C, B Schink. 1994. Anaerobic degradation of pimelate by newly isolated denitrifying bacteria. *Microbiology (UK)* 140: 409–416.

Gescher J, A Zaar, M Mohamed, H Schägger, G Fuchs. 2002. Genes coding for a new pathway of aerobic benzoate metabolism in *Azoarcus evansii*. *J Bacteriol* 184: 6301–6315.

Ghisla S, C Thorpe. 2004. Acyl-CoA dehydrogenases. A mechanistic overview. *Eur J Biochem* 271: 494–508.

Gibson J, M Dispensa, CS Harwood. 1997. 4-Hydroxybenzoyl coenzyme A reductase dehydroxylating is required for anaerobic degradation of 4-hydrozybenzoate by *Rhodopseudomonas palustris* and shares features with molybdenum-containing hydroxylases. *J Bacteriol* 179: 634–642.

Grund E, B Denecke, R Eichenlaub. 1992. Naphthalene degradation via salicylate and gentisate by *Rhodococcus* sp. strain B4. *Appl Environ Microbiol* 58: 1874–1877.

Harrison FH, CS Harwood. 2005. The *pimFABCDE* operon from *Rhodopseudomonas palustris* mediates dicarboxylic acid degradation and participates in anaerobic benzoate degradation. *Microbiology (UK)* 151: 727–736.

Hasan SA, MIM Ferreira, MJ Koetsier, MI Arif, DB Janssen. 2011. Complete biodegradation of 4-fluorocinnamic acid by a consortium comprising *Arthrobacter* sp. strain G1 and *Ralstonia* sp. strain H1. *Appl Environ Microbiol* 77: 572–579.

Härtel U, E Eckel, J Koch, G Fuchs, D, Linder, W Buckel. 1993. Purification of glutaryl-CoA dehydrogenase from *Pseudomonas* sp., an enzyme involved in the anaerobic degradation of benzoate. *Arch Microbiol* 159: 174–181.

Heider J. 2001. A new family of CoA-transferases. *FEBS Lett* 509: 345–349.

Hersh KB, WP Jencks. 1967. Coenzyme A transferase. Kinetics and exchange reactions. *J Biol Chem* 242: 3458–3480.

Hess V, S Vitt, V Müller. 2011. A caffeyl-coenzyme A synthetase initiates caffeate activation prior to caffeate reduction in the acetogenic bacterium *Acetobacterium woodii. J Bacteriol* 193: 971–978.

Hirsch W, H Schägger, G Fuchs. 1998. Phenylglyoxalate: NAD+ oxidoreductase (CoA benzoylating), a new enzyme of anaerobic phenylalanine metabolism in the denitrifying bacterium *Azoarcus evansii. Eur J Biochem* 251: 907–915.

Ishiyama D, D Vujaklija, J Davies. 2004. Novel pathway of salicylate degradation by *Streptomyces* sp. strain WA46. *Appl Environ Microbiol* 70: 1297–1306.

Ismail W, M El-Said Mohamed, BL Wanner, KA Datsenko, W Eisenreich, F Rohdich, A Bacher, G Fuchs. 2003. Functional genomics by NMR spectroscopy. Phenylacetate catabolism in *Escherichia coli. Eur J Biochem* 270: 3047–3054.

Jonsson S, S Ricagno, Y Lindqvist, NGJ Richards. 2004. Kinetic and mechanistic characterization of the for-myl-CoA transferase from *Oxalobacter formigenes. J Biol Chem* 279: 36003–36012.

Kaschabek SR, B Kuhn, D Müller, E Schmidt, W Reineke. 2002. Degradation of aromatics and chloroaromat-ics by *Pseudomonas* sp. strain B13: Purification and characterization of 3-oxoadipate:succinyl-coenzyme A (CoA) transferase and 3-oxoadipyl-CoA thiolase. *J Bacteriol* 184: 207–215.

Kawaguchi K, Y Shinoda, H Yurimoto, Y Sakai, N Kato. 2006. Purification and characterization of benzoate-CoA ligase from *Magnetospirillum* sp. strain TS-6 capable of aerobic and anaerobic degradation of aro-matic compounds. *FEMS Microbiol Lett* 257: 208–213.

Kim J, D Darley, T Selmer, W Buckel. 2006. Characterization of (*R*)-2-hydroxyisocaproate dehydrogenase and a family III coenzyme A transferase involved in reduction of L-leucine to isocaproate by *Clostridium difficile. Appl Environ Microbiol* 72: 6062–6069.

Kim J, D Darley, W Buckel. 2004. 2-hydroxyisocaproyl-CoA dehydratase and its activator from *Clostridium difficile. FEBS J* 272: 550–561.

Laempe D, M Jahn, K Breese, H Schägger, G Fuchs. 2001. Anaerobic metabolism of 3-hydroxybenzoate by the denitrifying bacterium *Thauera aromatica. J Bacteriol* 183: 968–979.

Leutwein C, J Heider. 2001. Succinyl-CoA:(*R*)-benzylsuccinate CoA-transferase: An enzyme of the anaerobic toluene catabolic pathway in denitrifying bacteria. *J Bacteriol* 183: 4288–4295.

Löffler F, R Müller, F Lingens. 1991. Dehalogenation of 4-chlorobenzoate by 4-chlorobenzoate dehalogenase from *Pseudomonas* sp. CBS3: An ATP/coenzyme A dependent reaction. *Biochem Biophys Res Commun* 176: 1106–1111.

Löffler F, R Müller, F Lingens. 1992. Purification and properties of 4-halobenzoate-coenzyme A ligase from a *Pseudomonas* sp. CBS3. *Biol Chem Hoppe-Seyler* 373: 1001–1007.

López-Barragán MJ, M Carmona, MT Zamarro, B Thiele, M Boll, G Fuchs, JL Garciá, E Díaz. 2004. The *bzd* gene cluster, coding for anaerobic benzoate metabolism, in *Azoarcus* sp. strain CIB. *J Bacteriol* 186: 5762–5774.

López-Sanchez A, B Floriano, E Andújar, MJ Hernáez, E Santero. 2010. Tetralin-induced and ThnR-regulated aldehyde dehydrogenase and β-oxidation genes in *Sphingomonas macrogolitabida* strain TFA. *Appl Environ Microbiol* 76: 110–118.

Mack M, W Buckel. 1995. Identification of glutamate b54 as the covalent-catalytic residue in the active site of glutaconate CoA-transferase from *Acidaminococcus fermentans. FEBS Lett* 357: 145–148

Marín M, D Pérez-Pantoja, R Donoso, V Wray, B González, DH Pieper. 2010. Modified 3-oxoadipate pathway for the biodegradation of methylaromatics in *Pseudomonas reinekei* MT1. *J Bacteriol* 192: 1543–1552.

Martínez-Blanco H, A Reglero, L Rodriguez-Aparicio, JM Luengo. 1990. Purification and biochemical char-acterization of phenylacetyl-CoA ligase from *Pseudomonas putida. J Biol Chem* 265: 7084–7090.

Mohamed ME-S. 2000. Biochemical and molecular characterization of phenyacetate-coenzyme A ligase, an enzyme catalyzing the first step in the aerobic metabolism of phenylacetic acid in *Azoarcus evansii. J Bacteriol* 182: 286–294.

Mohamed ME-S, B Seyfried, A Tschech, G Fuchs. 1993. Anaerobic oxidation of phenylacetate and 4-hydroxy-phenylacetate to benzoyl-coenzyme A and CO_2 in denitrifying *Pseudomonas* sp. *Arch Microbiol* 159:563–573.

Müh U, I Çinkaya, SPJ Albracht, W Buckel. 1996. 4-hydroxybutyryl-CoA dehydratase from *Clostridium aminobutyricum*: Characterization of FAD and iron–sulfur clusters involved in an overall non-redox reaction. *Biochemistry* 35: 11710–11718.

Müller U, W Buckel. 1995. Activation of (R)-2-hydroxyglutaryl-CoA dehydratase from *Acidaminococcus fer-mentans. Eur J Biochem* 230: 698–704.

Müller JA, B Schink. 2000. Initial steps in the fermentation of 3-hydroxybenzoate by *Sporotomaculum hydroxybenzoicum. Arch Microbiol* 173: 288–295.

Mullins EA, JA Francois, TJ Kappock. 2008. A specialized citric acid cycle requiring succinyl-coenzyme A (CoA):acetate CoA-transferase (AarC) confers acetic acid resistance on the acidophile *Acetobacter aceti. J Bacteriol* 190: 4933–4940.

Olivera ER, D Carnicero, B García, B Miñambres, MA Moreno, L Cañedo. 2001. Two different pathways are involved in the oxidation of *n*-alkalnoic acid and *n*-phenylalkanoic acids in *Pseudomonas putida* U: Genetic studies and biotechnological applications. *Mol Microbiol* 39: 863–874.

Pan C, Y Oda, PK Lankford, B Zhang, NF Samatova, DA Pelletier, CS Harwood, RL Hettich. 2008. Characterization of anaerobic catabolism of *p*-coumarate in *Rhodopseudomonas palustris* by integrating transcriptomics and quantitative proteomics. *Mol. Cell Proteomics* 7: 938–948.

Parales RE, CS Harwood. 1992. Characterization of the genes encoding β-ketoadipate:succinyl-coenzyme A transferase in *Pseudomonas putida. J Bacteriol* 174: 4657–4666.

Ramos-Vera WH, IA Berg, G Fuchs. 2009. Autotrophic carbon dioxide assimilation in *Thermoproteales* revisited. *J Bacteriol* 191: 4286–4297.

Scherf U, W Buckel. 1991. Purification and properties of 4-hydroxybutyrate coenzyme A transferase from *Clostridium aminobutyricum. Appl Environ Microbiol* 57: 2699–2702.

Schneider S, MEl-Said Mohamed, G Fuchs. 1997. Anaerobic metabolism of L-phenylalanine via benzoyl-CoA in the denitrifying bacterium *Thauera aromatica. Arch Microbiol* 168: 310–320.

Schühle K, J Gescher, U Feil, M Paul, M Jahn, H Schägger, G Fuchs. 2003. Benzoate-coenzyme A ligase from *Thauera aromatica*: an enzyme acting in anaerobic and aerobic pathways. *J Bacteriol* 185: 4920–4929.

Schürmann M, JH Wübbeler, J Grote, A Steinbüchel. 2011. Novel reaction of succinyl coenzyme A (succinyl-CoA) synthetase: Activation of 3-sulfinopropionate to 3-sulfinopropionyl-CoA in *Advenella mimigardefordensis* strain DPN7 during degradation of 3,3′-dithiodipropionic acid. *J Bacteriol* 193: 3078–3089.

Schweiger G, R Dutscho, W Buckel. 1987. Purification of 2-hydroxyglutaryl-CoA dehydratase from *Acidaminococcus fermentans*. An iron-sulfur protein. *Eur J Biochem* 169: 441–448.

Scott R, U Näser, P Friedrich, T Selmer, W Buckel, BT Golding. 2004. Stereochemistry of hydrogen removal from the "unactivated" C-3 position of 4-hydroxybutyryl-CoA catalyzed by 4-hydroxybutyryl-CoA dehydratase. *Chem Comm* 1210–1211.

Selmer T, A Willanzheimer, M Hetzel. 2002. Propionate CoA-transferase from *Clostridium propionicum*. Cloning of the gene and identification of glutamate at the active site. *Eur J Biochem* 269: 372–380.

Selmer T, W Buckel. 1999. Oxygen exchange between acetate and the catalytic glutamate residue in glutaconate CoA.-transferase from *Acidaminococcus fermentans. J Biol Chem* 274: 20772–20778.

Smith DM, W Buckel, H Zipse. 2003. Deprotonation of enoxy radicals: Theoretical validation of a 50-year-old mechanistic proposal. *Angew Chem Int Ed* 42: 1867–1870.

Solomon F, WP Jencks. 1969. Identification of an enzyme-γ-glutamyl coenzyme A intermediate from coenzyme A transferase. *J Biol Chem* 244: 1079–1081.

Stenmark P, D Gurumu, P Nordlund. 2004. Crystal structure of CaiB, a type III CoA transferase in carnitine metabolism. *Biochemistry* 43: 13996–14003.

Teufel R, JW Kung, D Kockelhorn, BE Alber, G Fuchs. 2009. 3-hydroxypropionyl-coenzyme A dehydratase, and acryloyl-coenzyme A reductase, enzymes of the autotrophic 3-hydroxypropionate/4-hydroxybutyrate cycle in the *Sulfolobales. J Bacteriol* 191: 4572–4581.

Trudgill PW. 1969. The metabolism of 2-furoic acid by *Pseudomonas* F2. *Biochem J* 113: 577–587.

White H, WP Jencks. 1976a. Properties of succinyl-CoA:3-ketoacid coenzyme A transferase. *J Biol Chem* 251: 1708–1711.

White H, WP Jencks. 1976b. Mechanism and specificity of succinyl-CoA:3-ketoacid coenzyme A transferase. *J Biol Chem* 251: 1688–1699.

Wilbrink MH, M Petrusma, L Dijkhuisen, R van der Geize. 2011. Fad19 of *Rhodococcus rhodochrous* DSM43269, a steroid-coenzyme A ligase essential for degradation of C-14 branched sterol side chains. *Appl Environ Microbiol* 77: 4455–4464.

Wöhlbrand L, H Wilkes, T Halder, R Rabus. 2008. Anaerobic degradation of *p*-ethylphenol by "*Aromatoleum aromaticum*" strain EbN1: pathway, regulation, and involved proteins. *J Bacteriol* 190: 5699–5709.

Part 3: Metalloenzymes

3.26 Enzymes Containing Manganese, Iron, Cobalt, Copper, Zinc, Molybdenum, Tungsten, and Vanadium

In this section, attention is drawn briefly to metals which play a key role in the function of enzymes that catalyze important reactions. Whereas iron is the most widely distributed metal component of enzymes and is followed in frequency by zinc and molybdenum, a number of important enzymes contain nickel, copper, manganese, tungsten, or vanadium.

3.26.1 Manganese

Iron and manganese exist at several oxidation levels and both play important roles in redox systems and in oxidative stress. Manganese is the catalytic center of several enzymes, is able to replace iron in some of them, and is involved in spore development in *Bacillus*. Transport systems for manganese have been explored and, since manganese can be toxic, mechanisms are required for homeostasis including efflux pumps while some of them are members of the ATP-dependent ABC transporter system (Kehres et al. 2003; Waters et al. 2011).

Although metabolism by Fe-cofactored enzymes is well established, problems may arise during extreme iron limitation, or oxidative stress in which iron is the target in Fenton-like reactions. This may be circumvented by replacement of Fe with Mn: for example, ribulose-5-phosphate 3-epimerase in *E. coli* is rapidly damaged by H_2O_2 which can be alleviated by manganese when mononuclear Fe enzymes can be adapted to a Mn-centred metabolism (Sobota and Imlay 2011). The crystal structure of the Class I b enzyme from *E. coli* has been used to clarify the role of the FMN cofactor, the mechanism of activation to a $Mn(III)_2$–Y• radical, and the role of peroxide (Boal et al. 2010).

The role of manganese has been explicitly examined in the context of biodegradation, where it is essential for the growth of the purple nonsulfur anaerobic phototrophs *Rhodospirillum rubrum* and *Rhodopseudomonas capsulata* during growth with N_2—though not with glutamate (Yoch 1979). Illustrative examples of Mn-containing enzymes are provided.

1. Extradiol dioxygenases containing Mn in place of Fe
   a. 3,4-Dihydroxyphenylacetate-2,3-dioxygenase from *Arthrobacter globiformis* (Boldt et al. 1995) and from *Bacillus brevis* (Que et al. 1981) are Mn(II) enzymes, and are neither activated by Fe(II) nor rapidly inhibited by H_2O_2.
   b. The 2,3-dihydroxybiphenyl dioxygenase from *Bacillus* sp. strain JF8 that mediates fission of the biphenyl ring is unusual in being Mn(II)-dependent, and differs in structure from the analogous enzymes in *Burkholderia* sp. strain LB400 and *Pseudomonas paucimobilis* strain KF707 (Hatta et al. 2003).
   c. Bacterial quercetin 2,3-dioxygenases (quercetinases) vary considerably in their metal requirements. The enzyme in *B. subtilis* had a preference for Mn for which the turnover is ≈40 times higher than that of the Fe-containing enzyme, and was similar to that of the Cu-enzyme in *Aspergillus*. Mn^{2+} was incorporated nearly stoichiometrically in the two cuprin motives, and the EPR spectrum showed that the two Mn^{2+} ions occurred in an octahedral configuration (Schaab et al. 2006).
2. The aerobic and anaerobic denitrifying degradation of acetone is initiated by ATP-dependent carboxylation to acetoacetate. The involvement of manganese has been examined in photoheterotrophically grown *Rhodobacter capsulatus* strain B10, and the presence of Mn verified was from the X-band EPR spectrum (Boyd et al. 2004). The carboxylase of acetophenone that is produced from ethylbenzene by *Aromatoleum aromaticum* is, however, a zinc-containing enzyme although it is inhibited by Zn^{2+} (Jobst et al. 2010).

3. Ribonucleotide reductase
   a. The metal-requirements of ribonucleotide reductase (RNR) have already been noted. The protein in the small subunit of *E. coli* normally contains two antiferromagnetically coupled oxo-bridged ferric ions. The apoenzyme produced after metal chelation can, however, react with manganese to produce a protein with two binuclear manganese clusters that have been characterized by EPR and x-ray crystallography (Atta et al. 1992). It has been shown that Class I b RNR in *E. coli* is active with both diferric-tyrosyl and dimanganic-tyrosyl cofactors (Cotruvo and Stubbe 2010), and the Class I b RNR in *E. coli* under conditions of iron-limitation produced a Mn_2^{III}-Y• radical cofactor *in vivo* (Cotruvo and Stubbe 2011).

   b. Less attention has been devoted to ribonucleotide reductase in Gram-positive bacteria, and *Streptomyces* spp contained both oxygen-dependent Class Ia and oxygen-independent and Class II enzymes that function at different stages of the growth cycle (Borovok et al. 2002). Ribonucleotide reductase has been purified and characterized in *Corynebacterium ammoniagenes* and was composed of two protein subunits, a nucleotide binding B1 and the catalytic B2 both of which were necessary for activity. Although the purified enzyme did not require additional metal cations, the electronic spectrum after removal of protein with trichloroacetic acid suggested the presence of an oxo-bridged binuclear Mn(III) which was confirmed by growth in the presence of $^{54}MnCl_2$ when the protein B2 was specifically labeled (Willing et al. 1988). The transcriptional regulation of this I b RNR has an atypical cluster of genes in the *nrdEF* region and expression of *nrdHIE* and *nrdF* were induced by manganese (Torrents et al. 2003).

   c. The anaerobic Class III ribonucleotide reductase in *Lactococcus lactis* contained genes *nrdD* and *nrdG* both of which were necessary for growth. The proteins have been purified and required *S*-adenosylmethione, reduced flavodoxin or reduced deazaflavin, and dithiothreitol, and EPR revealed the presence of a $[4Fe-4S]^+$ cluster in reduced NrdG, and a glycine radical in activated NrdD (Torrents et al. 2000).

4. The metabolism of 3,5-dichlorocatechol by *Ralstonia (Alcaligenes) eutropha* JMP 134 is carried out by intradiol fission to 2,4-dichloro-*cis*-*cis*-muconate that undergoes cyclization with loss of chloride to 4-carboxymethylene-3-chlorobut-2-ene-4-olide and thence to 3-chloromaleylacetate. The dichloromuconate cycloisomerase has been purified and consists of six to eight identical subunits, and required Mn^{2+} and thiol groups for activity (Kuhm et al. 1990).

5. Superoxide oxidoreductases may contain manganese and in *Streptomyces thermoautotrophicus* it unusually functions as the reductant in St2 to transfer electrons from CO dehydrogenase to St1 nitrogenase (Ribbe et al. 1997).

6. The aspartyl dipeptidase DapE in *Salmonella enterica* serovar *Typhimurium* can bind two divalent cations, Zn^{2+} in the high-affinity site and Mn^{2+} in the low-affinity site (Broder and Miller 2003).

7. Several peroxidases are produced by fungi, including manganese peroxidase. This plays an essential role in the metabolic capability of the white-rot fungus *Phanaerochaete chrysosporium*. Two groups of peroxidases are produced during secondary metabolism—lignin peroxidases and manganese-dependent peroxidases. Both are synthesized when only low levels of Mn(II) are present in the growth medium, whereas high concentrations of Mn result in repression of the synthesis of the lignin peroxidases and an enhanced synthesis of manganese-dependent peroxidases (Bonnarme and Jeffries 1990; Brown et al. 1990). Experiments with a nitrogen-deregulated mutant have shown that nitrogen-regulation of both these groups of peroxidases is independent of Mn(II)-regulation (Van der Woude 1993).

3.26.2 Iron

Cells require Fe that is a component of Class I ribonucleotide reductases, cytochromes and of enzymes that carry out degradation by oxygenation. Its concentration within the cell must, however, be controlled to avoid the production of deleterious hydroxyl radicals from molecular oxygen (Touati et al. 1995) or hydrogen peroxide (McCormick et al. 1998). Under Fe limitation, the transport of Fe into the cell is controlled by specific Fe(III) siderophores that contain catechol, α-hydroxycarboxylate, and hydroxamate groups in which the Fe occurs as a hexadentate octahedral complex. These are synthesized within the cell and excreted into the medium. They are bound by specific outer membrane receptors, and cross the potential of the cytosolic membrane using the energy-transducing system TonB-ExcB-ExbD. In Gram-positive bacteria, however, ABC permeases are used (Andrews et al. 2003; Wandersman and Delepelaire 2004). In addition to hemin, it is worth drawing attention to the Fe siroheme in which the methyl and vinyl groups are oxidized to carboxylates to produce an octacarboxylate.

Some pathogenic microorganisms are able to use heme as a source of iron. This requires a sequence of proteins to transport external heme into the cell where it can be processed by heme oxygenase. In brief, the transport process is different for the Gram-negative cells, for example, of *Ps. aeruginosa*, *Serratia marcescens*, and *Yersinia enterocolitica* that have been most extensively examined. For these organisms, heme acquisition involves proteins encoded by the heme acquisition systems operon *has-RADEF* that is controlled by Fur (Ferric uptake regulator). On the other hand, for Gram-positive cells such as *Corynebacterium diphtheriae* and *Staphylococcus aureus* proteins encoded by the iron-regulated surface determinant operon *isdABCDEF* are required. Details of both of these mechanisms are discussed in a review (Tong and Guo 2009). The mechanisms for release of Fe from heme have already been discussed in Part 1 of this chapter.

Iron is involved in the functioning of a wide range of enzymes that include the following.

1. Superoxide dismutase that catalyzes the production of H_2O_2 from the superoxide radical O_2^- generally contains either Mn and Fe, or Cu and Zn.

2. A range of cytochromes containing heme are key components in the energy metabolism of both aerobes and anaerobes, and of cytochrome P450 hydroxylase.

3. The paradigm of the soluble diiron monooxygenases is methane monooxygenase (sMMO) which consists of a dimeric m-hydroxobridged binuclear Fe hydroxylase, an NADH oxidoreductase with FAD/[2Fe–2S] centers, and an effector (Davydov et al. 1999).

4. Aromatic ring fission dioxygenases contain Fe, in the form of Fe(III) for intradiol dioxygenases and Fe(II) for extradiol dioxygenases (Siegbahn and Haeffner 2004). In addition there is another group of dioxygenases that are dependent on Fe^{2+} and 2-oxocarboxylates. The activities of these and their distribution are discussed in the section on dioxygenation.

5. Although the active sites of hydrogenases may contain both Fe and Ni, only Fe, or lack an [Fe–S] cluster, these are, for convenience discussed in the later section dealing with Ni hydrogenases.

6. a. The Fe–Fe nitrogenase system of *Azotobacter vinelandii* has been purified from a strain with an *nifHDK* deletion (Chisnell et al. 1988) and was composed of two components: the dinitrogenase complex and the reductant. The dinitrogenase with subunits α M_r 58, 000 and β M_r 50,000 was assembled into two configurations, $α_2β_2$ and $α_1β_1$. The former with M_r of 216,000 contained 24 Fe and 18 acid-labile sulfide, and concentrations of Mo (0.126), $V < 0.010$, $W < 0.057$, and $Cr < 0.006$ g atom per mole protein were extremely low. The reductase consisted of two identical proteins γ M_r 32,500 with 4 Fe and 4 acid-labile sulfide.

   b. The protein components of the iron-only nitrogenase were isolated from *Rhodobacter capsulatus* and two proteins were produced with which the highest specific activities were obtained with a 40:1 molar ratio. The significant catalytic property was the high activity for H_2-production and the pH-dependent production of ethane from acetylene (Schneider et al. 1997).

3.26.3 Cobalt

1. Cobalt is coordinated to the reduced polypyrrolic rings of vitamin B_{12} that is synthesized by anaerobic bacteria, and some of the reactions in which it plays a key role in anaerobic reactions are discussed in Chapter 3, Part 2, and Chapter 7, Part 1. The vitamin B_{12} coenzyme corrins that are found only in anaerobes vary, however, in the other ligands. Briefly, these belong to two axial

groups: (i) -CN or -Me and (ii) a range of azaarenes including 5,6-dimethylbenziminazole, adenine, 2-methyladenine and N_6-methylaminoadenine (references in Hoffmann et al. 2000).

2. The activity of quercetin 2,3-dioxygenase (quercetinase) from *Streptomyces* sp. strain FLA was high when the medium was supplemented with Ni^{2+} or Co^{2+}—though not with Fe^{2+}, Cu^{2+}, or Zn^{2+}. The ESR spectrum was consistent with the presence of high spin $S^{3/2}$ Co^{2+}, although the enzyme had a preference for Ni or Co and the K_m values suggested that the Ni enzyme would be more effective under physiological conditions (Merkens et al. 2008).

3. The amidohydrolase in *Gulosibacter molinativorax* that brings about hydrolysis of molinate to azepane-1-carboxylate and ethanthiol contains cobalt rather than zinc that is found in most amidohydrolases, although the enzyme shares the greatest sequence identity to phenylurea hydrolases (Duarte et al. 2011).

3.26.4 Nickel

Nickel exists in the tunicate *Trididemnum solidum* as the nickel complex of a modified chlorin (Bible et al. 1988), and is a component of a wide range of enzymes. Urease is the classic example of a nickel-containing enzyme, and several enzymes contain both nickel and iron. Details of enzymes that contain nickel have been provided in a review (Mulrooney and Hausinger 2003), so that only brief summaries will be provided.

1. a. Urease catalyzes the two-stage hydrolysis of urea to ammonia and bicarbonate and is involved in the pathogenesis of reactive arthritis following infection with *Yersinia enterocolitica*, in urinary tract infection, and in exacerbating the function of *Helicobacter pylori* by increasing the pH. The enzyme is found in several *Enterobacteriaceae*—including *E. coli*, and species of *Proteus* and *Providencia*—in which it is induced by urea, although in *K. aerogenes* its synthesis is regulated by nitrogen limitation. The crystal structure of urease from *K. aerogenes* has been determined (Jabri et al. 1995). It contains two nickel centers, each of which is coordinated with four histidines His[134, 136, 246, 272], Aspartate[360], and the oxygen atoms of an 5-*N*-carbamoyl lysine[217]. In *K. aerogenes*, urease is encoded by the gene cluster *ureDABCEFG* of which the four accessory proteins UreD, UreE, UreF, and UreG are required to form the metallocenter, and UreE functions as a chaperone to deliver Ni^{2+} to the nascent active site (Carter and Hausinger 2010).

   b. The virulence of *Helicobacter pylori* is dependent on a membrane-bound [Ni–Fe] hydrogenase and a urease, and it has been shown that the assembly of the proteins in these enzymes is dependent on maturation by HypA and HypB analogous to those that are discussed in Section 3.16.5. HypA was able to bind Ni^{2+} while HypB displayed GTPase activity, and the HypA and HypB proteins were able to form a heterodimeric complex in solution (Mehta et al. 2003).

2. Glyoxalase I is involved in the detoxification of methylglyoxal (Figure 3.46) that is produced during unregulated metabolism of carbohydrates. It is a widely distributed and is typically a Zn-containing enzyme. The enzyme from *E. coli* is, however, activated by Ni (Clugston et al. 1998), and the basis of metal ion activation has been examined (He et al. 2000).

3. Superoxide dismutase is important for the detoxification of the superoxide radical (O_2^-) by reacting with protons to produce H_2O_2: $O_2^- + 2H^+ \rightarrow O_2 + H_2O_2$. Although the enzyme generally contains Mn, or Fe, or Cu and Zn, the enzyme from *Streptomyces seoulensis* contains Ni(III) (Wuerges et al. 2004).

4. Exceptionally in *E. coli*, there are two enzymatic activities aci-reductone dioxygenase (enediol dioxygenase) that are responsible for the salvage of methionine, although they are encoded by the same gene. Whereas one enzyme that is dependent on Fe, produces an oxoacid and formate

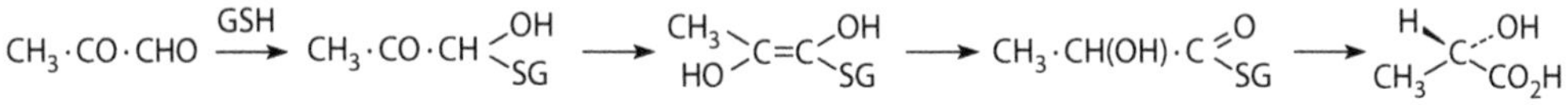

$$CH_3 \cdot CO \cdot CHO \xrightarrow{GSH} CH_3 \cdot CO \cdot CH \underset{SG}{\overset{OH}{<}} \longrightarrow \underset{HO}{\overset{CH_3}{>}} C=C \underset{SG}{\overset{OH}{<}} \longrightarrow CH_3 \cdot CH(OH) \cdot C \underset{SG}{\overset{O}{\lessgtr}} \longrightarrow \underset{CH_3}{\overset{H}{>}} C \underset{CO_2H}{\overset{OH}{<}}$$

FIGURE 3.46 Degradation of methylglyoxal by glyoxylase I.

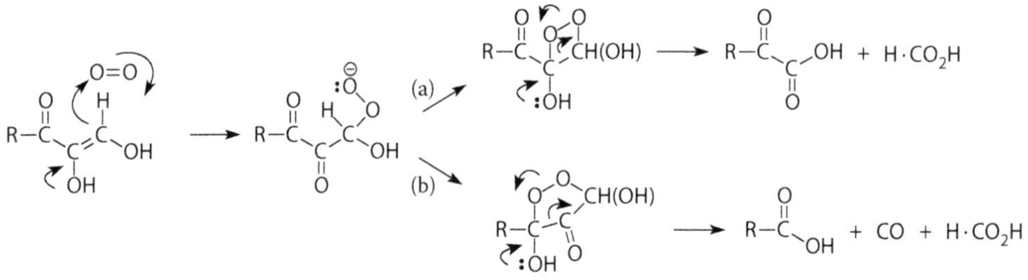

FIGURE 3.47 Alternative dioxygenations of an enediol.

(Figure 3.47a), the other that is nickel-dependent produces a carboxylic acid, formate and CO (Figure 3.47b) (Dai et al. 1999).

5. Whereas quercetinase in *Aspergillus niger* DSM 821 that brings about fission of the C-ring of quercetin is a copper-containing glycoprotein (Hund et al. 1999), the enzyme of *Streptomyces* sp. strain FLA has a preference for Ni or Co, with a possible preference for Ni under physiological conditions (Merkens et al. 2008).

6. Methyl coenzyme M reductase plays a key role in the production of methane in archaea. It catalyzes the reduction of methyl-coenzyme M with coenzyme B to produce methane and the heterodisulfide (Figure 3.48). The enzyme is an $\alpha_2\beta_2\gamma_2$ hexamer, embedded between two molecules of the nickel-porphinoid F_{430}, and the reaction sequence has been determined (Ermler et al. 1997).

7. The acetyl-coenzyme A decarbonylase synthase complex contains five polypeptide subunits and in acetate-degrading methanotrophs such as *Methanosarcina barkeri* and *M. thermophila* catalyzes the formation of methane and CO_2 from acetyl-CoA:

$$CH_3 \cdot CO \cdot S \cdot CoA + Co(I)\text{-FeS-protein} \Leftrightarrow CH_3\text{-Co(III) FeS protein} + CO + CoA.$$

The methyl group is subsequently transferred to a tetrahydropterin and coenzyme M. The β-subunit contains Ni and an Fe/S center, and a $Ni_2[4Fe-4S]$ arrangement at the active site has been proposed (Gencic and Grahame 2003).

8. Carbon monoxide dehydrogenase can participate in different reactions. Both aerobic and anaerobic organisms can oxidize CO using the reaction $CO + H_2O \rightarrow CO_2 + H_2$. Whereas the former uses a Mo–[2Fe–2S]–FAD enzyme, the enzyme from the latter contains Fe and Ni. Carbon monoxide dehydrogenase occurs in several anaerobic bacteria including the homoacetate-fermenting *Clostridium thermoaceticum* (Drake et al. 1980) and the acetate-utilizing methanogen *Methanosarcina thermophila* (Lu et al. 1994). It has been suggested that in *Carboxydothermus hydrogenoformans* there are two forms of the enzyme, both of which contain Ni, one involved in energy generation and one in biosynthetic reactions (Svetlitchnyi et al. 2001). *Carboxydothermus hydrogenoformans* contains several [4Fe–4S] metal clusters, and the active site contains an asymmetrical [Ni–4Fe–5S] cluster in which three of the Fe atoms and the Ni atom are coordinated with cysteine (Dobbek et al. 2001). On the other hand, a [Ni–3Fe–4S] cluster was found in *Rhodospirillum rubrum* (Drennan et al. 2001).

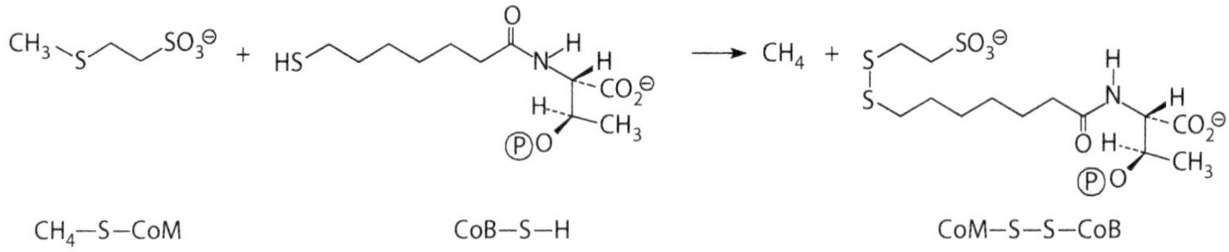

FIGURE 3.48 Biosynthesis of methane from *S*-methylcoenzyme M by methyl coenzyme M reductase.

3.26.5 Hydrogenase

A great deal of attention has been devoted to this enzyme in the context of biological hydrogen generation from available organic substrates. The enzyme catalyzes the reversible reaction $H_2 \Leftrightarrow 2H + 2\varepsilon$ but this simplicity is not reflected in the enzymology. An attempt is made to extract the salient features of the enzymes and their maturation, with illustrations from organisms that include bacteria and archaea. Although only one group contains nickel, an attempt is made to discuss briefly all of them, including the Fe-only enzymes. Reviews have covered salient aspects (Vignais et al. 2001; Calusinska et al. 2010).

1. The enzyme contains either [FeFe], [FeNi], or lacks the [Fe–S] cluster, and is found in a wide range of microorganisms:

   - The aerobes *Rhizobium leguminosarum* (Manyani et al. 2005), *Bradyrhizobium japonicum* (Olson and Maier 2000), the carbon monoxide-utilizing *Oligotropha carboxidovorans* (Santiago and Meyer 1997), and the oxygen-tolerant chemolithotrophs such as *Ralstonia eutropha* (Ludwig et al. 2009)
   - The facultative anaerobe *E. coli*
   - The microaerophilic *Helicobacter pylori* (Olson and Maier 2002)
   - Anaerobes including species of *Desulfovibrio* and *Clostridium*
   - Anaerobic phototrophs including *Rhodobacter capsulatus* (Vignais et al. 2000) and *Rhodospirillum rubrum* (Fox et al. 1996)
   - Hyperthermophilic archaea including *Thermotoga maritima*, *Thermococcus kodakaraensis*, and *Pyrococcus furiosus*
   - Methanogenic archaea including *Methanothermobacter marburgensis* (Lyon et al. 2004) and *Methanocaldococcus jannaschii* (Pilak et al. 2006)

   Some additional features are worth noting

   a. A number of organisms possess several hydrogenases: two in *Pyrococcus furiosus* (van Haaster et al. 2008), four in *E. coli* (Blokesch et al. 2004), and six in *Desulfovibrio vulgaris* Hildenborough (Caffrey et al. 2007). Clostridia are particularly rich in hydrogenases: the [Fe–Fe] enzymes contain trimeric, dimeric, and monomeric domains all of which occur in *Clostridium botulinum* A3 strain Loch Maree and the noncellulose-degrading *Cl. thermocellum*. In addition these two organisms are among the few clostridia that contain a [Ni-Fe] enzyme.

   b. Oxygenic phototrophs can develop hydrogenases under appropriate conditions.

      i. The unicellular *Synechococcus* sp. PCC 6301 (*Anacystis nidulans*) possesses both an uptake and a bidirectional hydrogenase, and the genome harbors the structural genes *hoxUYH* encoding the three subunits of the bidirectional hydrogenase (Boison et al. 1996). In filamentous cyanobacteria hydrogenase activity is carried out by nitrogenase, and in *Anabaena variabilis* ATCC 29413, the uptake hydrogenase genes hupSL were induced under nitrogen-fixing conditions, and it reoxidized the hydrogen produced in an oxyhydrogen reaction (Happe et al. 2000).

      ii. Hydrogenase activity can be induced in *Chlamydomonas reinhardtii* after anaerobic adaptation or under sulfur deprivation which showed higher expression rates of the [FeFe]-hydrogenase (HydA1) (Kamp et al. 2008). Preliminary analysis by EPR spectroscopy showed characteristic axial EPR signals of the CO inhibited forms that are typical for the H_{ox}-CO state of the active site from [Fe–Fe] hydrogenases.

2. Briefly, there are three types of hydrogenases distinguished by the presence of (a) Ni and Fe, [Ni–Fe], (b) only Fe, [Fe–Fe] generally with a binuclear iron center in anaerobic organisms, and (c) unrelated [Fe–S] cluster-free in methanogenic archaea, including the H_2-forming N^5,N^{10}-methylenetetrahydromethanopterin (H_4MPT) dehydrogenase that catalyzes the reaction $CH_2 = H_4MPT^+ \rightarrow CH\equiv H_4MPT^+ + H_2$ from *Methanobacterium thermoautotrophicum* (Zirngibl

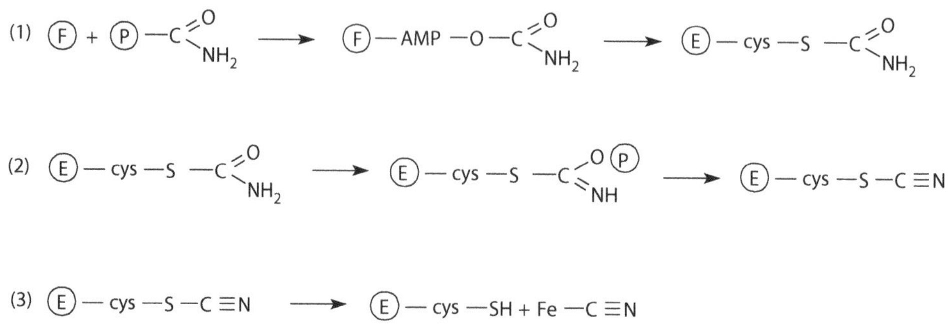

(1)   (E) + (P) —C⟨O / NH₂   ⟶   (F)—AMP —O —C⟨O / NH₂   ⟶   (E) — cys —S —C⟨O / NH₂

(2)   (E) — cys —S —C⟨O / NH₂   ⟶   (E) — cys —S —C⟨O (P) / NH   ⟶   (E) — cys —S —C ≡N

(3)   (E) — cys —S —C ≡N   ⟶   (E) — cys —SH + Fe —C ≡N

FIGURE 3.49   Synthesis of Fe–C≡N ligand.

et al. 1992), *Methanococcus thermolithotrophicus* (Hartmann et al. 1996), and *Methanopyrus kandleri* (Ma et al. 1991).

The [Fe–Fe] hydrogenases are typically involved in the evolution of H_2, whereas the Ni–Fe enzymes are used for the oxidation of H_2. Both the [Ni–Fe] and [Fe–Fe] hydrogenases have nonprotein CO and CN ligands at the Fe sites, and this process has been described as maturation. In [Ni–Fe] hydrogenases, in which Fe is coordinated with one CO and two CN ligands (Pierik et al. 1999), synthesis of the latter involve two proteins HypF and HypE that carry out the consecutive reactions (Figure 3.49) (Reissmann et al. 2003). The analogous active site in the [Fe–Fe] hydrogenase of the hyperthermophilic *Thermotoga maritima* contains CO and CN ligands at both Fe atoms that are linked by an additional CO and sulfur ligands one of which contains the [4Fe–4S] cubane cluster. In this organism, the maturation protein HydF is able to bind GTP and catalyze GTP hydrolysis, and contains a [4Fe–4S] cluster with an unusual EPR signal, which suggested that only three cysteines were, coordinated (Brazzolotto et al. 2006).

The Fe-only hydrogenase from *Clostridium pasteurianum* is monomeric containing 20 g atoms of Fe per mole of protein arranged in five [Fe–S] clusters. Its structure has been resolved by x-ray analysis (Peters et al. 1998), and the active site cluster consists of six atoms of Fe in a cubane [4Fe–4S] subcluster covalently bridged by cysteinate to a [2Fe] subcluster.

3. a. The crystal structure of the cofactor of the [NiFe] hydrogenase has been examined in *Desulfovibrio gigas* (Volbeda et al. 1996). The Ni is coordinated to the protein by four cysteine groups of which two are shared by the Fe which in addition is coordinated to two CN and one CO group, and this has become the paradigm for [NiFe] hydrogenases (Pierik et al. 1999; Reissmann et al. 2003). There are additional features of the hydrogenases in *Desulfovibrio*. Exceptionally, in *Desulfovibrio baculatus*, the active site contained [NiFeSe] (He et al. 1989; Wang et al. 1992). In *Desulfovibrio vulgaris* Hildenborough, there are four periplasmic hydrogenases that carry out the oxidation of H_2: a soluble [Fe–Fe] enzyme, two membrane [Fe–Ni] isoenzymes, and one membrane [NiFeSe] enzyme (Caffrey et al. 2007). The presence of Se in the medium upregulated the synthesis of the [NiFeSe] enzyme and brought about repression of the synthesis of the [Fe–Fe] and [Ni–Fe] enzymes (Valente et al. 2006).

   b. In the chemolithotroph *Ralstonia eutropha*, the membrane-associated hydrogenase consists of the large subunit containing the [NiFe] active site and a small subunit with one [3Fe–4S] and two [4Fe–4S] clusters. FTIR and EPR examination showed that [NiFe] states were analogous to those of anaerobic hydrogenases with the exception of the absence of the oxygen-activated Ni_u-A state (Saggu et al. 2009). The synthesis of the cofactor in this organism is mediated by six Hyp proteins while the auxiliary proteins HoxL and HoxV are concerned with the assembly of the $Fe(CN^-)_2CO$ (Ludwig et al. 2009).

   c. Nickel is required for the anaerobic growth of *E. coli*, and there are four [Ni–Fe] hydrogenases whose genes occur in operons (references in Blokesch et al. 2004). The product of the *hypB* gene that is required for Ni incorporation into the large subunit of hydrogenase 3 is a novel guanine nucleotide-binding protein (Maier et al. 1993). The enzymes consist of a large subunit that contains the reaction center with Fe and Ni, while the small subunit contains several [Fe–S] clusters. These hydrogenases fulfill a range of functions. Two of the operons *hya* and *hyb* are induced under anaerobic conditions and allow the cells to utilize hydrogen to produce H$^+$ and electrons that are used as a source of energy. The third, *hyc* is part of the formate hydrogen lyase that converts formate into CO$_2$ and hydrogen, while the fourth, *hyf* is part of a different formate hydrogen lyase system. In addition, synthesis of the active NiFe enzyme requires the cooperation of several proteins in the *hyp* operon that are involved in maturation: HypG and HypF in the maturation of hydrogenase 1 and 2, and HypC and HypA in the maturation of hydrogenase 3 (Hube et al. 2002). One of them, HybF involves carbamoyl phosphate (Paschos et al. 2002) and has been reported to bind zinc at a site apparently different from the Ni-binding site (Blokesch et al. 2004), while HypE and HypF interact to produce a stable tetramer (Rangarjan et al. 2008). The maturation of the hydrogenase system involves the proteins HydE, HydF, and HydG, and the H cluster that is required for the biosynthesis and insertion of the catalytic [Fe–S] cluster (King et al. 2006).

d. *Thermotoga maritima* is able to ferment carbohydrates with the production of H$_2$ and the degradation of glucose produces both NADH and reduced ferredoxin, both of which are required for production of H$_2$ (Shut and Adams 2009). The [Fe–Fe] hydrogenase in this organism is complex (Verhagen et al. 1999): it is trimeric, the α-subunit that contains 2 [2Fe–2S] and 3 [4Fe–4S] is involved in the oxidation of H$^+$, the β-subunit with 3 [4Fe–4S] and 1 [2Fe–S] in the FMN-coupled oxidation of NADH and the γ-subunit with [2Fe–2S] in the oxidation of reduced ferredoxin. It was proposed (Shut and Adams 2009) that ferredoxin and NADH were used synergistically, and that trimeric bifurcating hydrogenases occurred in other anaerobic bacteria including *Thermoanaerobacter tengcongensis* that is also able to ferment carbohydrates with the evolution of H$_2$. *Thermo. tengcongensis* possesses two hydrogenases, one a ferredoxin-dependent [NiFe] membrane-bound enzyme and the other an NADH-dependent heterotetrameric [Fe–Fe] hydrogenase (Soboh et al. 2004).

Considerable effort has been devoted to elucidating the complex process of maturation of the [Fe–Fe] hydrogenase in *Thermotoga maritima* that depends on the expression of HydE, HydF, and HydG.

i. The maturation protein in HydF in *Thermotoga maritima* was able to bind GTP, catalyze GTP hydrolysis, and contains an [4Fe–4S] cluster with an unusual EPR signal which suggested that only three cysteines were coordinated (Brazzolotto et al. 2006).

ii. The x-ray structure of HydE has been determined (Nicolet et al. 2008) to elucidate the role of the *S*-adenosylmethionine radical in the synthesis of dithiomethylamine and the function of HydE and HydG that are conserved in a wide range of [Fe–Fe] hydrogenases (Posewitz et al. 2004).

e. *Pyrococcus furiosus* is able to ferment carbohydrates with the formation of H$_2$, and contains several hydrogenases, a heterotetrameric cytoplasmic soluble enzyme I, and a multimeric transmembrane enzyme II. The soluble enzyme has been characterized as a [Ni–Fe] enzyme (van Haaster et al. 2008), and the reduced membrane-bound after solubilization produced an EPR signal typical of the Ni center in mesophilic enzymes (Sapra et al. 2000). The soluble enzyme was able to catalyze both the production of H$_2$ and the reduction of S^0 to H$_2$S, and an additional hydrogenase of this type has been characterized (Ma et al. 2000). It was also a [Ni–Fe] tetrameric enzyme that contained flavin adenine dinucleotide, ferredoxin was not efficiently used by either enzyme, and several [4Fe–4S] and [2Fe–2S] clusters were predicted on the basis of amino acid sequences.

f. In hyperthermophilic archaea, hydrogenases are also involved in the reduction of S^0 to H$_2$S (Ma et al. 2000; Laska et al. 2003) so that both hydrogenase and sulfur reductase activities are required. These have been purified from solubilized membrane fractions of *Acidianus ambivalens*: the [Fe–Ni] hydrogenase was characterized, and a number of genes in both the sulfur reductase and hydrogenase were required for maturation of the [Ni–Fe] hydrogenase (Laska et al. 2003). The sulfur reductase belonged to the large family of dimethyl sulfoxide molybdenum reductases.

g. Hydrogenase plays a central role in the metabolism of *Thermococcus kodakarensis* that assimilates organic carbon by the reduction of elementary sulfur or with the generation of H_2. Three operons containing the [Ni–Fe] hydrogenase were examined: Hyh (cytosolic [Ni–Fe] hydrogenase, Mbh (membrane-bound [Ni–Fe] hydrogenase, and Mbx, and their function was determined using mutants. The Hyh-deficient mutant grew well under both sulfide-producing and H_2-evolving conditions; the Mbh-deficient mutant did not grow with evolution of H_2; whereas the Mbx-deficient mutant grew more slowly only under sulfide-producing conditions. This enabled the construction of the reactions that took place during growth with pyruvate lacking S^0, and with amino acids in the presence of S^0 (Kanai et al. 2011).

h. The hydrogenase in archaeal methanogens belongs to the iron–sulfur cluster-free enzymes. The crystal structure of the enzyme from *Methanocaldococcus jannaschii* revealed the nature of the iron-containing cofactor that is coordinated to two CO molecules, one sulfur atom, and a pyridone that is linked via a phosphodiester to a guanosine base (Pilak et al. 2006). The hydrogenase catalyzes the reversible reduction of N^5,N^{10}-methenyltetrahydromethanopterin with H_2 to the methyltetrahydromethanopterin that is an intermediate in methanogenesis, and H^+.

3.26.6 Copper

Resistance to Cu^{2+} has been explored since it is able to produce reactive oxygen species with consequent cellular damage. At the same time, copper is necessary for the functioning of a number of cellular enzymes, and homeostasis must be maintained by regulation of its concentration.

1. Methane monooxygenase may exist either in a soluble (sMMO) form that has been more extensively studied, or a particulate (pMMO) form that is found in most methanotrophs (Hanson and Hanson 1996). This contains Cu whose concentration determines the catalytic activity of the enzyme (Sontoh and Semrau 1998), and enrichments using limiting concentrations of Cu result in the prevalence of strains with the soluble form of the enzyme (McDonald et al. 2006). Proteomic analysis of *Methylococcus capsulatus* (Bath) under a range of copper-to-biomass ratios revealed the role of Cu in methane oxidation, and switching of soluble to particulate MMO (Kao et al. 2004). There are several factors that have complicated the determination of the structure and reactivity of this membrane-associated enzyme (pMMO): these have been clearly summarized (Choi et al. 2003). With the availability of an enzyme preparation of high activity (Yu et al. 2003a, important aspects have been resolved: the pMMO-detergent complex had a mass of 220 kDa, with subunits of 45 kDa (PmoB), 27 kDa (PmoA) and 23 kDa (PmoC), contained 13.6 mol Cu per mol pMMO, was essentially free of both Fe and Zn, and was produced maximally with a copper concentration in the medium of 30 μM. In addition, hydroxylation of both ethane and butane by pMMO takes place with retention of configuration. A mechanism has been proposed involving reaction with an oxene orthogonal to the two QAWCu centers (Yu et al. 2003b). Some additional issues on the reactivity and specificity of the pMMO enzyme are given in Chapter 7, Part 1.

2. a. Although dioxygenases generally contain Fe(II) or Fe(III), copper replaces Fe in the dioxygenase from *Aspergillus japonicus* that brings about fission of the C-ring of quercetin with formation of the depside 2-(3,4-dihydroxybenzoyloxy)-4,6-dihydroxybenzoate, and the mechanism has been examined by both EPR (Kooter et al. 2002) and x-ray crystallography (Steiner et al. 2002).

b. *Acinetobacter* sp. strain M-1 has an unusual range of substrates including both alkanes and alkenes and the purified enzyme contains FAD and requires Cu^{2+} for its activity (Maeng et al. 1996).

3. The carbon monoxide dehydrogenase of the aerobe *Oligotropha carboxidovorans* contains both Cu and Mo in the form of a cluster in which the Mo is bound to the thiol groups of molybdopterin cytosine nucleotide, and the Cu to cysteine residue in the form of a Cu–S–Mo(=O)OH cluster (Dobbek et al. 2002).

4. Glyoxal oxidase is required to produce the H_2O_2 required for the activity of lignin peroxidase, carries out the reaction $RCHO + O_2 + H_2O \rightarrow RCO_2H + H_2O_2$ and requires Cu for full activity. It is a radical-copper oxidase closely resembling galactose oxidase (Whittaker et al. 1996).

5. Copper is a component of the oxalate decarboxylase in *B. subtilis* that converts oxalate into formate and CO_2 (Tanner et al. 2001).

6. The deamination of primary amines such as phenylethylamine by *E. coli* (Cooper et al. 1992), *K. oxytoca* (Hacisalihoglu et al. 1997), and *Hansenula polymorpha* grown with diethylamine as nitrogen source (Mu et al. 1992) is carried out by an oxidase. This contains copper and 2,4,5-trihydroxyphenylalanine quinone (Topaquinone, TPQ) that is produced from tyrosine by dioxygenation. TPQ is reduced to an aminoquinol which, in the form of a Cu(I) radical reacts with O_2 to form H_2O_2, Cu(II) and the imine. The mechanism has been elucidated (Wilmot et al. 1999), and involves formation of a Schiff base followed by hydrolysis, in reactions that are formally analogous to those involved in pyridoxal-mediated transamination.

7. Superoxide dismutases may contain a range of metals: Mn, Fe, or both Cu and Zn, and representatives of all these are found in prokaryotes. The nickel enzyme is noted later.

8. In *Lactococcus lactis*, copper, silver and cadmium—though not zinc—are able to induce the synthesis of oxygen-insensitive nitroreductase that is able to redress oxidative stress including that from the presence of nitroaromatic compounds (Mermod et al. 2010).

9. One group of dissimilatory nitrite reductases catalyzes the reduction of nitrite to nitric oxide and is discussed in a previous part of this chapter. These contain copper: type I is trimeric and is responsible for the blue color of the enzyme (Prudêncio et al. 1999), whereas type II does not contribute to the visible spectrum. A mechanism for the reaction has been proposed (Murphy et al. 1997) and redox titration suggested electron transfer between type I and type II (Pinho et al. 2004).

10. Nitrous oxide reductase is another component of dissimilatory nitrate reduction and has been explored in *Ps. stutzeri*. The enzyme contains 8 copper atoms in two identical subunits and occurs in various forms and colors, and the EPR spectrum showed an unpaired electron distributed over two mixed-valence copper atoms (Antholine et al. 1992). Further examination suggested that electrons enter the reductase at the CuA site, and multi-wavelength anomalous dispersion of the enzyme from *Ps. nautica* revealed that the catalytic center CuZ consists of four copper ions coordinated to seven histidine residues and S^{2-}, and that the substrate N_2O binds to a single copper ion (Brown et al. 2000).

11. Multicopper enzymes occur in plants, fungi, and humans. Tyrosinase that is among the most studied catalyzes both the *ortho*-hydroxylation of phenols and a two-electron oxidation to *o*-quinones. It has been characterized from the actinomycete *Streptomyces glaucescens* and the fungus *Neurospora crassa*, in both of which they are monomeric proteins with two copper atoms (references in Solomon et al. 1996).

3.26.7 Zinc

There are two large groups of enzymes that contain zinc. In one, the active sites of hydrolytic enzymes contain Zn, and these enzymes belong to the amidohydrolase superfamily with a number of subtypes (Seibert and Raushel 2005). Alkyl transferases are another important group of zinc-containing enzymes. The following illustrate examples of each.

Group 1: Hydrolases

1. The phosphotriesterase in *Ps. diminuta* can hydrolyze a range of phosphate triesters including organophosphate agrochemicals, and is representative of zinc enzymes with a binuclear metal center. The structure has been determined by x-ray analysis of the Cd analogue, and clearly revealed the coordination of the both Cd atoms with histidines, and a carbamoylated lysine and a water molecule that bridged the two Cd atoms (Benning et al. 1995, 2001).

2. The oxidative degradation of uracil by *R. erythropolis* JCM 3132 is initiated by dehydrogenation to barbiturate. This is followed by hydrolysis to ureidomalonate by barbiturase that has been characterized and shown to belong to a novel group of hydrolases (Soong et al. 2002).

3. The active sites of α-amino-β-carboxymuconate-ε-semialdehyde decarboxylase (ACMSD) that is an intermediate in one of the pathways in the degradation of 2-nitrobenzoate (Muraki et al. 2003) and of kynurenine, and adenosine deaminase contain zinc coordinated with histidine (Martynowski et al. 2006). It has been suggested (Liu and Zhang 2006) that these decarboxylases be designated ACMSD-related since none of them are amidohydrolases and should include several other decarboxylases and hydratases: 5-carboxyvanillate decarboxylase (Peng et al. 2005), γ-resorcylate decarboxylase (Yoshida et al. 2004), and 4-oxalomesaconate hydratase (Hara et al. 2000).

4. Several schemes have been proposed for the classification of β-lactamases (Bush et al. 1995). They belong to the superfamily of amidohydrolases, and employ either a serine-70 residue (Groups A, C, and D) or a zinc atom (Group B) at the active site to carry out nucleophilic attack at the lactamase carbonyl group: this results in hydrolysis and thereby resistance to β-lactam antibiotics. The metallo-β-lactams have been divided into three groups B1, B2, and B3 (Galleni et al. 2001) that may contain a single Zn atom, or two Zn atoms. Subclass B1 contains most of the metallo-β-lactamases, and the subclass B2 metallo-β-lactamase from *Aeromonas hydrophila* contains only a single Zn atom (Garau et al. 2005). The subclass B3 enzyme from *Stenotrophus maltophilia* contains two Zn atoms that are weakly bound, one with a tetrahedral structure composed of three histidines, the other with a distorted trigonal bipyramidal structure with two histidines (Nauton et al. 2008). In the subclass B3 lactamase from *Legionella (Fluoribacter) gormanii*, both Zn atoms are bridged by a water molecule that carries out nucleophilic attack on the β-lactam carbonyl group (García-Sáez et al. 2003). Although the Zn atom of the metallo-β-lactamase in *Elizabethkingia meningoseptgica* may be replaced by Co or Cd, Zn is essential for stabilizing the anion produced by hydrolytic fission of the β-lactam before protonation (Lisa et al. 2010).

5. Pyrazinamide is a front-line drug for the treatment of tuberculosis that must be activated by hydrolysis to the carboxylic acid by a nicotinamidase/pyrazinamidase. The crystal structure of the enzyme has been determined: in *Pyrococcus horokoshii*, the enzyme contained Zn coordinated with one aspartate and two histidines (Du et al. 2001), whereas in *M. tuberculosis* it contained equal amounts of Mn^{2+} and Fe^{2+} (Zhang et al. 2008), whereas a later study revealed the presence at the active site only if Fe^{2+} coordinated with one aspartate and three histidines (Petrella et al. 2010).

6. The crystal structure of an alkylsulfatase from *Ps. aeruginosa* has been determined (Hagelueken et al. 2006). This defined a third class of sulfatases, and contained two zinc atoms in a binuclear structure that constituted a distinct member of the metallo-β-lactam family.

Group 2: Alkyl transferases

Another important group of zinc-containing enzymes are the alkyl transferases that are involved in a range of quite different reactions. Illustrative examples of these are briefly summarized.

1. The methanogen *Methanosarcina barkeri* is able to use a range of C_1 substrates, methanol, and methylamines. Methanogenesis involves two sets of reactions:

a. Catalysis by the transferase MT1:

$$CH_3-X + [Co^{1+}]corrinoid\ protein + H^+ \rightarrow H-X + [CH_3-Co^{3+}]corrinoid\ protein$$

b. Catalysis by the transferase MT2:

$$HS-CoM + [CH_3-Co^{3+}]corrinoid\ protein \rightarrow CH_3-S-CoM + [Co^{1+}]corrinoid\ protein + H^+$$

It has been shown that in this methanogen a zinc–thiolate is an intermediate in the reaction catalyzed by MT2 (Gencic et al. 2001).

2. The microbial metabolism of short-chain alkenes is initiated by monooxygenation to the epoxide. Further reaction is catalyzed by the zinc-containing enzyme epoxyalkane: CoM transferase (Krum et al. 2002) that carries out the reaction: HS–CoM + epoxypropane → 2-hydroxypropyl–CoM before further metabolism (Krishnakumar et al. 2008). Evidence including the ESR spectrum of the Co^{2+}-substituted enzyme provided additional insight into the Co-thiolate interaction that is activated for arrack on the epoxide substrate (Boyd and Ensign 2005).

3. The synthesis of methionine in *E. coli* is carried out by transfer of the methyl group in 5-methyl-tetrahydropteroylpolyglutamate to L-homocysteine. The cobalamin-independent synthase contains zinc that is essential for catalytic activity, and sequence alignment has shown that His^{641}, Cys^{643}, and Cys^{726} are the only conserved residues. X-ray absorption spectroscopy has been used to examine the bonding of selenohomocysteine to both the cobalamin independent (MetE) and the cobalamin-dependent (MetH) enzymes in both of which the Se and, by analogy the S of homocysteine were directly bound to zinc (Peariso et al. 2005).

Group 3: Alkanol dehydrogenases

These generally contain zinc, and *S. cerevisiae* has three Zn-containing enzymes with differing catalytic constants toward ethanol: these enzymes are also effective for primary alcohols up to C_{10} and, with lower activity for propan-2-ol (Leskovac et al. 2002). Considerable attention has been directed to thermophilic and hyperthermophilic organisms (Radianingtyas and Wright 2003), and the alcohol dehydrogenase from *Pyrococcus furiosus* is remarkable since it contains both Zn and Fe whose presence was detected by EPR (Ma and Adams 1999). It is important to note, however, that Fe may serve only in activating the enzyme without being structurally present at the active site. Additional comments on these enzymes in hyperthermophiles may be found in Chapter 2.

3.26.8 Molybdenum

Molybdopterin oxidoreductases

In the presence of Mo, nitrogenase 1 is produced: this consists of dinitrogen reductase—the Fe protein (*nifH*) and the catalytic dinitrogenase Mo–Fe protein (*nifDK*)—that are organized in the *nifHDK* operon. Mo is a component of the molybdenum cofactor (molybdopterin) that consists of a dihydropteridine fused to a dihydropyran ring containing an enedithiol: this is coordinated to the molybdenum and $-CH_2OPO_3^{2-}$ group. In prokaryotes, the cofactor is formed by coupling to a nucleoside phosphate (generally cytidine, or guanidine) forming a pyrophosphate (Llamas et al. 2004) (Figure 3.50). The biosynthesis of molybdopterin is catalyzed by at least six gene products, and is initiated from guanosine triphosphate by guanosine hydrolase I that provides the pteridine nucleus, while the C_5 of ribose and the purine C_8 provide the dihydropyran ring.

Three groups of enzymes in azotobacters (Rüttimann-Johnson et al. 2003) and phototrophs (Oda et al. 2005) are able to reduce dinitrogen to ammonia: (i) the *nifHDK* enzyme containing molybdenum and iron, (ii) the *vnfHDK* containing a vanadium and iron in *Azotobacter chroococcum* (Eady et al. 1987), and (iii) *anfHDK* an iron-only enzyme in *Azotobacter vinelandii* (Chisnell et al. 1988) and in *Rhodobacter capsulatus* (Schneider et al. 1997).

Molybdenum is a component of one group of nitrogenases, while a later paragraph on vanadium draws attention to two other nitrogenase enzymes including the Fe only enzyme that has already been noted. In

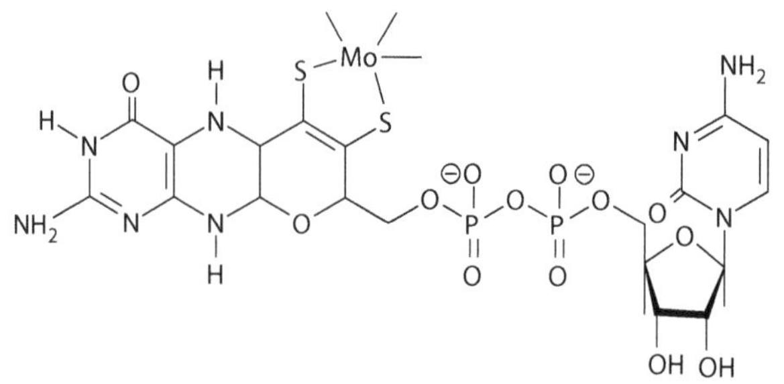

FIGURE 3.50   Structure of molybdopterin conjugated with cytidine diphosphate.

brief, the MoFe nitrogenase *nifHDK* consists of the Fe protein encoded by *nifH* that is a homodimer bridged by an intersubunit [4Fe–4S] cluster and carries out electron donation to the MoFe protein, the MoFe protein encoded by *nifDK* is an $\alpha_2\beta_2$ tetramer containing [8Fe–7S] clusters that shuttle electrons to the FeMo cofactors which are [Mo–7Fe–9S] homocitrate clusters where the reduction takes place (references in Hamilton et al. 2011).

There are four families of enzymes that contain the molybdenum cofactor (Moco): DMSO reductase, xanthine oxidase, sulfite oxidase, and aldehyde ferrredoxin oxidoreductase. DMSO reductase belongs to a large group of enzymes that carry out a wide range of reactions including the function of electron acceptors. They have been divided into three clusters Type I, II, and III (McDevitt et al. 2002), and some examples are given in Table 3.11. Details of the enzymes are given in parts of this chapter dealing with alternative electron acceptors and reductases, respectively.

In addition, there a number of tungsten-containing enzymes with structures analogous to molybdopterin in which W (VI) replaces Mo(VI) (Hille 2002). Molybdopterin is a structural part of the important oxidoreductases involved in the aerobic degradation of azaarenes. Using a bioinformatics screen, a molybdenum-containing oxidoreductase has also been found in the majority of Gram-negative bacteria (Loschi et al. 2004). It contains one molybdopterin, and the soluble subunit designated YedY functions as a reductase for trimethylamine-*N*-oxide, and some sulfoxides. Molybdenum oxidoreductases are widely used for the introduction of the oxygen atom from H_2O into heteroarenes, especially azaarenes including pyridine, quinoline, pyrimidine, and purine: representative examples are given as illustration.

3.26.9 Heteroarenes

The degradation of the five-membered heteroarenes furan-2-carboxylate, and pyrrole-2-carboxylate is initiated by the corresponding hydroxylases (Koenig and Andreesen 1990, 1991; Hormann and Andreesen 1991).

TABLE 3.11

Types of DMSO Reductase

Type	Subunit	Enzyme	Organism
Type I	$\alpha\beta$ and $\alpha\beta\gamma$	Assimilatory nitrate reductase	*K. pneumoniae*
		Periplasmic nitrate reductase	*E. coli*
		Formate dehydrogenase complex	*Wolinella succinogenes*
		Thiosulfate reductase/polysulfide reductase	*Salmonella typhimurium*
Type II	$\alpha\beta\gamma$	Selenate reductase	*Thauera selenatis*
		Ethylbenzene dehydrogenase	*Aromatoleum aromaticum*
Type III	α	Trimethylamine *N*-oxide reductase	*Shewanella putrefaciens*
		Dimethylsulfoxide reductase	*Rhodobacter capsulatus*

3.26.10 Pyridine

The degradation of several pyridine carboxylates is initiated by hydroxylation. In *Arthrobacter pico-linophilus* pyridine-2-carboxylate is hydroxylated to 3,6-dihydroxypyridine-2-carboxylate (as for 2-hydroxypyridine) and this is carried out by a particulate enzyme that introduces oxygen from H_2O (Tate and Ensign 1974), has a molecular mass of 130 kDa, and contains Mo (Siegmund et al. 1990). Pyridine-4-carboxylate is hydroxylated by *Mycobacterium* sp. strain INA1 to the 2,6-dihydroxy-4-carboxylate. Two different hydroxylation enzymes were involved, and were apparently Mo-dependent (Kretzer and Andreesen 1991). The degradation of nicotine by *Arthrobacter nicotinovorans* involves two dehydrogenations by enzymes containing Mo, Fe–S, and FAD (Baitsch et al. 2001).

3.26.11 Quinoline

Experiments with $H_2^{18}O$ using *Ps. putida* strain 86 (Bauder et al. 1990) showed that the oxygen incorporated into quinol-2-one originates from water. The oxidoreductases have been purified from a number of organisms that degrade quinoline including *Ps. putida* (Bauder et al. 1990), *Rhodococcus* sp. strain B1 (Peschke and Lingens 1991), and *Comamonas testosteroni* (Schach et al. 1995), and from *Agrobacterium* sp. strain 1B that degrades quinoline-4-carboxylate (Bauer and Lingens 1992). They have a molecular mass of 300–360 kDa, and contain per molecule, 8 atoms of Fe, 8 atoms of acid-labile S, 2 atoms of Mo, and 2 molecules of FAD. The organic component of the pterin molybdenum cofactor is generally molybdopterin cytosine dinucleotide (Hetterich et al. 1991; Schach et al. 1995).

The metabolism of 2-methylquinoline in *Arthrobacter* sp. strain Rü 61a is comparable, with introduction of oxygen at C_4 (Hund et al. 1990). The enzymes (oxidoreductases) that introduce oxygen into the pyridine rings in *Rhodococcus* sp. strain B1, *Arthrobacter* sp. strain Rü61a, and *Ps. putida* strain 86 are virtually identical and, like those already noted have molecular masses of 300–320 kDa and contain Mo, Fe, FAD, and acid-labile sulfur (De Beyer and Lingens 1993). The enzymes from *Comamonas testosteroni* for hydroxylation of quinoline to 2-hydroxylation (quinoline 2-oxidoreductase), and the dioxygenase responsible for the introduction of oxygen in the benzene ring (2-oxo-1,2-dihydroquinoline 5,6-dioxygenase) have been described (Schach et al. 1995).

Isoquinoline: The degradation of isoquinoline by *Alcaligenes faecalis* strain Pa and *Ps. diminuta* strain 7 (Röger et al. 1990, 1995) is mediated by an oxidoreductase that produces 1,2-dihydroisoquinoline-1-one, followed by ring fission with the production of *o*-phthalate and oxidation to 3,4-dihydroxybenzoate (Figure 3.51). The oxidoreductase has been purified and, like most typical azaarene oxidoreductases contained per mol, 0.85 g atoms Mo, 3.9 g atoms of Fe and acid-labile S (Lehmann et al. 1994).

Ring fission of 3-hydroxyquinol-4-one is carried out by an unusual dioxygenation that is independent of metal or organic cofactors (Steiner et al. 2010), and incorporates both atoms of oxygen at C_2 and C_4, with formation of formyl anthranilate and carbon monoxide (Fischer et al. 1999). It was proposed that these belonged to the α/β hydrolase-fold superfamily functionally related to serine hydroxylases and a plausible mechanism was put forward (Fischer et al. 1999).

3.26.12 Reductases

1. The dehydroxylase in *Thauera aromatica* that brings about dehydroxylation of 4-hydroxybenzoyl-CoA to benzoyl-CoA is a molybdenum-flavin-iron–sulfur enzyme (Breese and Fuchs 1998).

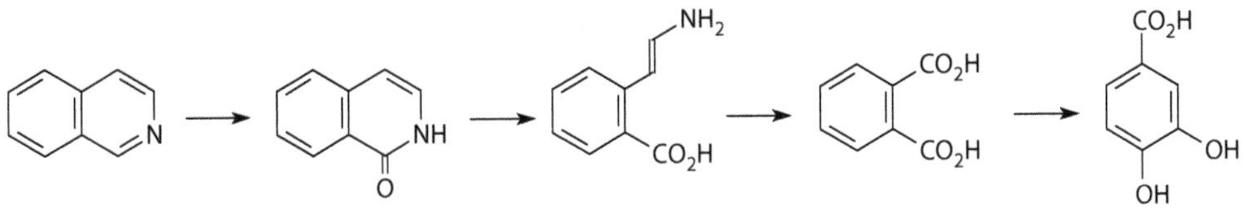

FIGURE 3.51 Degradation of isoquinoline initiated by an oxidoreductase.

2. The DMSO reductase family contains molybdenum generally in the form of *bis* (molybdopterin guanine nucleotide) (references in McEwan et al. 2002). In addition to DMSO reductase, they include TMAO reductase, chlorate reductase, selenate reductase, and arsenate reductase that are able to couple the reduction to the generation of energy (dissimilatory reduction). They catalyze the following reactions that have already been noted in Section 3.1 of this chapter:

$Me_2SO \rightarrow Me_2S$ (dimethyl sulfoxide reductase);

$Me_3NO \rightarrow Me_3N$ (trimethylamine *N*-oxide reductase);

$HCO_2H \rightarrow CO_2 + H_2$ (formate dehydrogenase);

$ClO_4^- \rightarrow ClO_3^- \rightarrow ClO^-$ (perchlorate reductase);

$SeO_4^{2-} \rightarrow SeO_3^{2-}$ (selenate reductase);

$AsO_2^- \rightarrow AsO_4^{3-}$ (arsenite oxidase)

The chlorate reductase has been characterized in strain GR-1 where it was found in the periplasm, is oxygen-sensitive, and contains molybdenum, and both [3Fe–4S] and [4Fe–4S] clusters (Kengen et al. 1999). The arsenate reductase from *Chrysiogenes arsenatis* contains Mo, Fe, and acid-labile S (Krafft and Macy 1998), and the reductase from *Thauera selenatis* that is specific for selenate, is located in the periplasmic space, and contains Mo, Fe, acid-labile S, and cytochrome *b* (Schroeder et al. 1997). On the other hand, the membrane-bound selenate reductase in *Enterobacter cloacae* SLD1a-1 that also contains Mo and Fe is distinct from nitrate reductase, and cannot function as an electron acceptor under anaerobic conditions (Ridley et al. 2006).

In the anaerobic photoheterotroph *Rhodobacter* (*Rhodopseudomonas*) *spheroides*, a molybdenum enzyme in which the metal is coordinated by two equivalents of a pyranopterin cofactor is involved in sequential oxidations and reductions involving Mo(VI), Mo(V), and Mo(IV) (Cobb et al. 2005).

3.26.13 Tungsten

Tungsten-containing enzymes have been found to mediate a variety of reactions carried out by both aerobic and anaerobic bacteria. Their structures may plausibly be assumed to be analogous to the molybdopterins.

1. *Acetylene hydratase* catalyzes the hydration of acetylene in a nonredox reaction in the anaerobe *Pelobacter acetylenicus*. The enzyme that is a monomer with a molecular mass of 83,210 is a tungsten–iron–sulfur enzyme that resembles molybdopterin in which W replaces Mo and contains a [4Fe–4S] cluster (Meckenstock et al. 1999). Important details of the active site have been added by analysis of site-directed mutagenesis and expression of these in *E. coli* (tenBrink et al. 2011).

2. *Oxidoreductases*—A number of these have been characterized and include the following:

   a. Aldehyde oxidoreductase from *Desulfovibrio gigas* (Hensgens et al. 1995).

   b. An aldehyde dehydrogenase from *Desulfovibrio simplex* is stimulated by tungsten and oxidizes aliphatic and aromatic aldehydes using flavins (Zelner and Jargon 1997).

   c. Aldehyde oxidoreductase from *Clostridium thermoaceticum* (Strobl et al. 1992).

   d. In the obligate anaerobes (*Geobacter metallireducens*, *Desulfoccus multivorans*, *Syntrophus aciditrophicus*) the reductases are ATP-independent, and the obligate anaerobe *Desulfoccus multivorans* benzoyl-CoA reductase was independent of ATP and contained Fe/S, Mo or W and Se (Peters et al. 2004; Wischgoll et al. 2005). This reductase has been examined in *Geobacter metallireducens*, and it has been shown that the enzyme which catalyzed the reverse reaction was a 185-kDa enzyme with subunits of 73-kDa (BamB) and 21-kDa (BamC) that would be consistent with an $\alpha_2\beta_2$ structure. The $\alpha\beta$ unit contained 0.9 W, 15 Fe, and 12.7 acid-labile S, and EPR indicated the presence of one $[3Fe–4S]^{0/+1}$ and three $[4Fe–4S]^{+1/+2}$ clusters and, in the oxidized form the presence of W^V (Kung et al. 2009). In addition, the BamBC components of this obligately anaerobic bacterium bore no similarities to the established ATP-dependent reductases from facultatively anaerobic bacteria.

   e. Several tungsten-containing oxidoreductases have attracted attention in the biochemistry of the extreme thermophile *Pyrococcus furiosus*: (i) aldehyde ferredoxin oxidoreductase (Mukund and Adams 1991), (ii) glyceraldehyde ferredoxin oxidoreductase (Mukund and Adams 1995), (iii) formaldehyde ferredoxin oxidoreductase (Mukund and Adams 1993; Roy et al. 1999), (iv) an aldehyde oxidoreductase with a broad substrate specificity that probably involves ferredoxin (Bevers et al. 2005), (v) an enzyme that may be involved in the reduction of S^0 (Roy and Adams 2002).

3. *Carboxylic acid reductase*—Anaerobic reduction of aryl and alkyl carboxylates to the corresponding aldehydes has been demonstrated in a few tungsten-containing reductases and dehydrogenases. In *Clostridium formicoaceticum* the enzyme catalyzes the reduction of propionate, butyrate, benzoate to the corresponding aldehydes $RCO_2H \rightarrow RCHO$. The enzyme contains, per dimer, 11 iron, 16 acid-labile sulfur, 1.4 tungsten + molybdopterin (White et al. 1991). In addition, a Mo oxidoreductase has been described from *Cl. formicoaceticum* (White et al. 1993).

4. Formate dehydrogenase

   a. Formate dehydrogenase from *Moorella (Clostridium) thermoaceticum* catalyzes the reduction of CO_2 to formate and contains W, Se, Fe, and S (Yamamoto et al. 1983), and the FDH-II enzyme from *Eubacterium acidamophilum* that uses formate as electron donor during amino acid fermentations contains W, Se, Fe (Graentzdoerffer et al. 2003). The formate dehydrogenase involved in the oxidation of formate to CO_2 in *E. coli* is, however, a Mo, Se, (Fe_4S_4) enzyme (Boyington et al. 1997).

   b. The different roles of molybdate and tungstate in formate dehydrogenase have been examined in sulfate-reducing bacteria.

       i. There are different forms of formate dehydrogenase in *Desulfuricans alaskensis* and these contain either tungsten (W-FDH), or either molybdenum or tungsten (Mo/W-FDH). Both enzymes contain one α and one β subunit, and in medium with higher concentrations of tungstate only the W-FDH was formed whereas after growth in molybdate-containing medium both W-FDH and Mo/W-FDH were formed (Mota et al. 2011).

      ii. In *Desulfovibrio vulgaris* lactate, formate and H_2 can serve as electron donor, and there are three formate dehydrogenases. The tungsten enzyme that is dominant after growth in medium with tungstate is a dimeric FdhAB, whereas the molybdenum enzyme after growth in medium with molybdate is the trimeric $FdhABC_3$. FdhAB can incorporate both metals and has a higher catalytic activity (da Silva et al. 2011).

   c. NAD-dependent formate dehydrogenase was isolated from cells of *Methylobacterium extorquens* strain AM1 grown with methanol and purified under oxic conditions. It contained mol per mol enzyme: ≈5 nonheme iron and acid-labile sulfur, 0.6 noncovalently bound FMN, and ≈1.8 tungsten (Laukel et al. 2003).

   d. Formate dehydrogenases FDH-1 and FDH-2 (CO_2-reductases) were isolated from *Syntrophobacter fumaroxidans* grown axenically with fumarate and purified under strictly anaerobic conditions. Both reductases contained tungsten and selenium, whereas molybdenum was not detected. On the basis of EPR spectroscopy, FDH-1 contained four [2Fe–2S] clusters and paramagnetically coupled [4Fe–4S] clusters, while FDH-2 contained two [4Fe–4S] clusters (de Bok et al. 2003).

5. Hydrogenase *Thermotoga maritima* grows at the expense of carbohydrates producing lactate, acetate, CO_2, and H_2. Although production of the hydrogenase and its *in vitro* activity is greatly *stimulated* by the presence of tungstate, the purified enzyme contains only Fe and acid-labile S, and tungsten is absent (Juszczak et al. 1991). This is also true for pyruvate: ferredoxin oxidoreductase from the same organism.

3.26.14 Vanadium

1. The vanadium nitrogenase has been characterized in a strain of *Azotobacter chroococcum* (Eady et al. 1987). The major protein contained subunits with M_r of 50,000 and 55,000 in equal amounts, and the complete enzyme with M_r of 210,000 contained per molecule, 2 atoms V, 23 atoms Fe and 20 atoms acid-labile sulfide. The vanadium enzyme from *Azotobacter vinelandii* has been his-tagged and purified to homogeneity and it was shown to have an $\alpha_2\beta_2\delta_4$ heterooctameric structure and contains a P-cluster that is apparently different from that of the MoFe protein (Lee et al. 2009). The reductase protein encoded by *vnfH* is analogous to the corresponding protein *nifH* in the molybdenum enzyme. This is supported by genetic evidence that a strain of *Azotobacter vinelandii* in which the genes for both the molybdenum and vanadium nitrogenases were deleted was able to fix nitrogen, growth was inhibited by Mo and V and was limited by Fe (Pau et al. 1989). There are metabolic differences between the MoFe and VFe enzymes that are noted in Chapter 7, including the two mechanisms for the reduction of H^+ to H_2 only one of which was suppressed by CO in the VFe enzyme, and the catenation products from CO (Hu et al. 2011). The enzymes lacking molybdenum, and containing instead vanadium or iron are in fact widely distributed in the environment (Betancourt et al. 2008).

2. Some haloperoxidases contain vanadium, and a review of vanadium peroxidases has been given (Butler 1998). The structure of the vanadium enzyme in the terrestrial fungus *Curvularia inaequalis* has been determined by x-ray analysis (Messerschmidt et al. 1997), and the apochloroperoxidase possesses in addition phosphatase activity that can be rationalized on the basis of the isomorphism of phosphate and vanadate (Renirie et al. 2000).

3.27 References

Adams MWW. 1990. The structure and mechanism of iron-hydrogenases. *Biochim Biophys Acta* 1020: 115–145.

Andrews SC, AK Robinson, F Rodriguez-Quinones. 2003. Bacterial iron homeostasis. *FEMS Microbiol Rev* 27: 215–237.

Antholine WE, DHW Kastrau, GCM Stevens, G Buse, WG Zumft, PMH Kroneck. 1992. A comparative EPR investigation of the multicopper proteins nitrous-oxide reductase and cytochrome *c* oxidase. *Eur J Biochem* 209: 875–881.

Atta M, P Nordlund, A Åberg, H Eklund, M Fontecave. 1992. Substitution of manganese for iron in ribonucleotide reductase from *Escherichia coli*. *J Biol Chem* 267: 20682–20688.

Baitsch D, C Sandu, R Brandsch, GL Igloi. 2001. Gene cluster on pAO1 of *Arthrobacter nicotinovorans* involved in degradation of the plant alkaloid nicotine: Cloning, purification, and characterization of 2,6-dihydroxypyridine 3-hydrolase. *J Bacteriol* 183: 5262–5267.

Bauder R, B Tshisuaka, F Lingens. 1990. Microbial metabolism of quinoline and related compounds VII. Quinoline oxidoreductase from *Pseudomonas putida*: A molybdenum-containing enzyme. *Biol Chem Hoppe-Seyler* 371: 1137–1144.

Bauer G, F Lingens. 1992. Microbial metabolism of quinoline and related compounds XV. Quinoline-4-carboxylic acid oxidoreductase from *Agrobacterium* spec. 1B: A molybdenum-containing enzyme. *Biol Chem Hoppe-Seyler* 370: 1183–1189.

Benning MM, JM Kuo, FM Raushel, HM Holden. 1995. Three-dimensional structure of the binuclear metal center of phosphotriesterase. *Biochemistry* 34: 7973–7978.

Benning MM, H Shim, FM Raushel, HM Holden. 2001. High resolution x-ray structures of different metal-substituted forms of phosphotriesterase from *Pseudomonas diminuta*. *Biochemistry* 40: 2712–2722.

Betancourt DA, TM Loveless, JW Brown, PE Bishop. 2008. Characterization of diazotrophs containing Mo-independent nitrogenases, isolated from diverse natural environments. *Appl Environ Microbiol* 74: 3471–3480.

Bevers LE, E Bol, P-L Hagedoorn, WR Hagen. 2005. WOR5, a novel tungsten-containing aldehyde oxidoreductase from *Pyrococcus furiosus* with a broad substrate specificity. *J Bacteriol* 187: 7056–7071.

Bible KC, M Buytendorp, PD Zierath, KL Rinehart. 1988. Tunichlorin: a nickel chlorin isolated from the Caribbean tunicate *Trididemnum solidum*. *Proc Natl Acad Sci USA* 85: 4582–4586.

Blokesch M, M Rohrmoser, S Rode, A Böck. 2004. HybF, a zinc-containing protein involved in NiFe hydrogenase maturation. *J Bacteriol* 186: 2603–2611.

Boal AK, JA Cotruvo, JoAnne Stubbe, AC Rosenweig. 2010. Structural basis for activation of Class 1b ribonucleotide reductase. *Science* 329: 1526–1530.

Boison G, O Schmitz, L Mikheeva, S Shestakov, H Bothe. 1996. Cloning, molecular analysis and insertional mutagenesis of the bidirectional hydrogenase genes from the cyanobacterium *Anacystis nidulans*. *FEBS Lett* 394: 153–158.

Boldt YR, MJ Sadowsky, LBM Ellis, L Que, LP Wackett. 1995. A manganese-dependent dioxygenase from *Arthrobacter globiformis* CM-2 belongs to the major extradiol dioxygenase family. *J Bacteriol* 177: 1225–1232.

Bonnarme P, TW Jeffries. 1990. Mn(II) regulation of lignin peroxidases and manganese-dependent peroxidases from lignin-degrading white-rot fungi. *Appl Environ Microbiol* 56: 210–217.

Borovok I, R Kreisberg-Zakarin, M Yanko, R Schreiber, M Myslovati, F Åslund, A Holmgren, G Cohen, Y Aharonowitz. 2002. *Streptomyces* spp. contain class Ia and class II ribonucleotide reductases: expression analysis of the genes in vegetative growth. *Microbiology (UK)* 148: 391–404.

Boyd JM, SA Elsworth, SA Ensign. 2004. Bacterial acetone carboxylase is a manganese-dependent metalloenzyme. *J Biol Chem* 46644–46651.

Boyd JM, SA Ensign. 2005. Evidence for a metal–thiolate intermediate in alkyl group transfer from epoxpropane to coenzyme M and cooperative metal binding in epoxyalkane: CoM transferase. *Biochemistry* 44: 13151–13163.

Boyington JC, VN Gladyshev, SV Khangulov, TC Stadtman, PD Sun. 1997. Crystal structure of formate dehydrogenase H: catalysis involving Mo, molybdopterin, selenocysteine, and an Fe_4S_4 cluster. *Science* 275: 1305–1308.

Brazzolotto X, JK Rubach, J Gaillard, S Gambrelli, M Atta, M Fontecave. 2006. The (Fe–Fe]-hydrogenase maturation protein HydF from *Thermotoga maritima* is a GTPase with an iron–sulfur cluster. *J Biol Chem* 279 281: 769–774.

Breese K, G Fuchs. 1998. 4-hydroxybenzoyl-CoA reductase (dehydroxylating) from the denitrifying bacterium *Thauera aromatica*: prosthetic groups, electron donor, and genes of a member of the molybdenum–flavin–iron–sulfur proteins. *Eur J Biochem* 251: 916–923.

Broder DH, CG Miller. 2003. DapE can function as an aspartyl peptidase in the presence of Mn^{2+}. *J Bacteriol* 185: 4748–4754.

Brown JA, JK Glenn, MH Gold. 1990. Manganese regulates expression of manganese peroxidase by *Phanerochaete chrysosporium*. *J Bacteriol* 172: 3125–3130.

Brown K et al.. 2000. A novel type of catalytic copper cluster in nitrous oxide reductase. *Nature Struct Mol Biol* 7: 191–195.

Bush K, GA Jacob, AA Medicos. 1995. A functional classification for β-lactamases and its correlation with molecular structure. *Antimicrob Agents Chemother* 39: 1211–1223.

Butler A. 1998. Vanadium peroxidases. *Curr Opinion Chem Biol* 2: 279–285.

Caffrey SM, H-S Park, JK Voordouw, Z He, J Zhou, G Voordouw. 2007. Function of periplasmic hydrogenases in the sulfate-reducing bacterium *Desulfovibrio vulgaris* Hildenborough. *J Bacteriol* 189: 6159–6167.

Calusinska M, T Happe, B Joris, A Wilmotte. 2010. The surprising diversity of clostridial hydrogenases: A comparative genomic perspective. *Microbiology UK* 156: 1575–1588.

Carter EL, RP Hausinger. 2010. Characterization of the *Klebsiella aerogenes* urease accessory protein UreD in fusion with the maltose binding protein. *J Bacteriol* 192: 2294–2304.

Chisnell JR, R Premakumar, PE Bishop. 1988. Purification of a second alternative nitrogenase from *nifHDK* deletion strain of *Azotobacter vinelandii*. *J Bacteriol* 170: 27–33.

Choi D-W, RC Kunz, ES Boyd, JD Semrau, WE Antholine, J-I Han, JA Zahn, JM Boyd, AM de la Mora, AA DiSpirito. 2003. The membrane-associated methane monooxygenase (pMMO) and pMMO-NADH: Quinone oxidoreductase complex from *Methylococcus capsulatus* Bath. *J Bacteriol* 185: 5755–5764.

Clugston SL, JFJ Barnard, R Kinach, D Miedema, R Ruman, E Daub, JF Honek. 1998. Overproduction and characterization of a dimeric non-zinc glyoxalase I from *Escherichia coli*: Evidence for optimal activation by nickel ions. *Biochemistry* 37: 8754–8763.

Cobb N, T Conrads, R Hille. 2005. Mechanistic studies of *Rhodobacter spheroides* Me$_2$SO reductase. *J Biol Chem* 280: 11007–11017.

Colabroy KL, TP Begley. 2005. Tryptophan catabolism: Identification and characterization of a new degradative pathway. *J Bacteriol* 187: 7866–7869.

Cooper RA, PF Knowles, DE Brown, MA McGuirl, DM Dooley. 1992. Evidence for copper and 3,4,6-trihydroxyphenylalanine quinone cofactor as an amine oxidase from Gram-negative *Escherichia coli* K-12. *Biochem J* 288: 337–340.

Cotruvo JA, J Stubbe. 2010. An active dimanganese (III)-tyrosyl radical cofactor in *Escherichia coli* Class 1b ribonucleotide reductase. *Biochemistry* 49: 1297–1309.

Cotruvo JA, J Stubbe. 2011. *Escherichia coli* Class 1b ribonucleotide reductase contains a dimanganese (III)-tyrosyl radical cofactor in vivo. *Biochemistry* 50: 1672–1681.

Dai Y, PV Wensink, RH Abeles. 1999. One protein, two enzymes. *J Biol Chem* 274: 1193–1199.

da Silva SM, C Pimentel, PMA Valente, C Rodrigues-Pousada IAC Pereira. 2011. Tungsten and molybdenum regulation of formate dehydrogenase expression in *Desulfovibrio vulgaris* Hildenborough. *J Bacteriol* 193: 2909–2916.

Davydov R, AM Valentine, S Komar-Panicucci, BM Hoffman, SJ Lippard. 1999. An EPR study of the dinuclear iron site in the soluble methane monooxygenase from *Methylococcus capsulatus* (Bath) reduced by one electron at 77 K: The effects of component interactions and the binding of small molecules to the diiron (III) center. *Biochemistry* 38: 4188–4197.

de Bok FAM, P-L Hagedoorn, PJ Silva, WR Hagen, E Schiltz, K Fritsche, AJM Stams. 2003. Two W-containing formate dehydrogenases (CO$_2$-reductases) involved in syntrophic propionate oxidation by *Syntrophobacter fumaroxidans*. *Eur J Biochem* 270: 2476–2485.

De Beyer A, F Lingens. 1993. Microbial metabolism of quinoline and related compounds XVI. Quinaldine oxidoreductase from *Arthrobacter* spec. Rü 61a: a molybdenum-containing enzyme catalysing the hydroxylation at C-4 of the heterocycle. *Biol Chem Hoppe-Seyler* 374: 101–120.

Dobbek H, L Gremer, R Kiefersauer, R Huber, O Meyer. 2002. Catalysis at a dinuclear [CuSMo(=O)OH] cluster in a CO dehydrogenase resolved at 1.1-Å resolution. *Proc Natl Acad Sci USA* 99: 15971–15976.

Dobbek H, V Svetlichnyi, L Gremer, R Huber, O Meyer. 2001. Crystal structure of a carbon monoxide dehydrogenase reveals a (Ni–4Fe–5S] cluster. *Science* 293: 1281–1285.

Drake HL. S-I Hu, HG Wood. 1980. Purification of carbon monoxide dehydrogenase, a nickel enzyme from *Clostridium thermoaceticum*. *J Biol Chem* 255: 7174–7180.

Drennan CL, J Heo, MD Sintchak, E Schreiter, PW Ludden. 2001. Life on carbon monoxide: X-ray structure of *Rhodospirillum rubrum* Ni–Fe–S carbon monoxide dehydrogenase. *Proc Natl Acad Sci USA* 98: 11973–11978.

Du X, WR Wang, R Kim, H Yakota, H Nguyen, SH Kim. 2001. Crystal structure and mechanism of catalysis of a pyrazinamidase from *Pyrococcus horokoshii*. *Biochemistry* 40: 14166–14172.

Duarte M, F Ferreira-da-Silva, H Lünsdorf, H Junca, L Gales, DH Pieper, OC Nunes. 2011. *Gulosibacter molinativorax* ON4[T] molinate hydrolase, a novel cobalt-dependent amidohydrolase. *J Bacteriol* 193: 5810–5816.

Eady R, RL Robson, TH Richardson, RW Miller, M Hawkins. 1987. The vanadium nitrogenase of *Azotobacter chroococcum*. *Biochem J* 244: 197–207.

Ermler U, W Grabarse, S Shima, M Goubeaud, RK Thauer. 1997. Crystal structure of methyl-coenzyme M reductase: The key enzyme of biological methane formation. *Science* 278: 1457–1462.

Fischer F, S Künne, S Fetzner. 1999. Bacterial 2,4-dioxygenases: New members of the α,β-hydrolase-fold superfamily of enzymes functionally related to serine hydrolases. *J Bacteriol* 181: 5725–5733.

Fox JD, RK Kerby, GP Roberts, PW Ludden. 1996. Characterization of the CO-induced, CO-tolerant hydrogenase from *Rhodospirillum rubrum* and the enzyme encoding the large subunit of the enzyme *J Bacteriol* 178: 1515–1524.

Galleni M, J Lamotte-Brasseur, GM Rossolini, J Spencer, O Dideberg, J-M Frère et al. 2001. Standard numbering scheme for class B β-lactamases. *Antimicrob Agents Chemother* 45: 660–663.

Garau G, C Bebrone, C Anne, M Galleni, J-M Frère, O Dideberg. 2005. A metallo-β-lactamase enzyme in action: Crystal structures of the monozinc carbapenemase CphA and its complex with biapenem. *J Mol Biol* 345: 785–795.

García-Sáez I, PS Mercuri, C Papamicael, R Kahn, JM Frère, M Galleni, GM Rossolini, O Dideberg. 2003. Three dimensional structure of FEZ-1, a monomeric subclass B3 metallo-β-lactamase from *Fluoribacter gormanii*, in native form and in complex with D-captopril. *J Mol Biol* 325: 650–660.

Gencic S, DA Grahame. 2003. Nickel in subunit *b* of the acetyl-CoA decarboxylase/synthase multienzyme complex in methanogens. *J Biol Chem* 278: 6101–6110.

Gencic S, GM LeClerc, N Gorlatova, K Pearison, JE Penner-Hahn, DA Grahame. 2001. Zinc–thiolate intermediate in catalysis of methyl group transfer in *Methanosarcina barkeri*. *Biochemistry* 40: 13068–13078.

Graentzdoerffer A, D Rauh, A Pich, JR Andreesen. 2003. Molecular and biochemical characterization of two tungsten- and selenium-containing formate dehydrogenases from *Eubacterium acidamophilum* that are associated with components of an iron-only hydogenase. *Arch Microbiol* 179: 116–130.

Gutzke G, B Fischer, RR Mendel, G Schwarz. 2001. Thiocarboxylation of molybdopterin synthase provides evidence for the mechanism of dithiolene formation in metal-binding pterins. *J Biol Chem* 276: 36268–36274.

Hacisalihoglu A, JA Jongejan, JA Duine. 1997. Distribution of amine oxidases and amine dehydrogenases in bacteria grown on primary amines and characterization of the amine oxidase from *Klebsiella oxytoca*. *Microbiology (UK)* 143: 505–512.

Hagelueken G, TM Adams, L Wiehlmann, U Widow, H KOlmar, B Tümmler, DW Heinz, W-D Schubert. 2006. The crystal structure of SdsA1, an alkylsulfatase from *Pseudomonas aeruginosa* defines a third class of sulfatases. *Proc Natl Acad Sci USA* 103: 7631–7636.

Hamilton TL, M Ludwig, R Dixon, ES Boyd, PC Dos Santos, JC Setubal, DA Bryant, DR Dean, JW Peters. 2011. Transcriptional profiling of nitrogen fixation in *Azotobacter vinelandii*. *J Bacteriol* 193: 4477–4486.

Hanson RS, TE Hanson. 1996. Methanotrophic bacteria. *Microbiol Rev* 60: 439–471.

Happe T, K Schütz, H Böhme. 2000. Transcriptional and mutational analysis of the uptake hydrogenase of the filamentous cyanobacterium *Anabaena variabilis* ATCC 29413. *J Bacteriol* 182: 1624–1631.

Hara H, E Masai, Y Katayama, M Fukuda. 2000. The 4-oxalomesaconate hydratase gene, involved in the protocatechuate 4,5-cleavage pathway, is essential to vanillate and syringate degradation in *Sphingomonas paucimobilis* SYK-6. *J Bacteriol.* 182: 6950–6957.

Hartmann GC, AR Klein, M Linder, RK Thauer. 1996. Purification, properties and primary structure of H_2-forming N^5,N^{10}-methylenetetrahydromethanopterin dehydrogenase from *Methanococcus thermolithotrophicus*. *Arch Microbiol* 165: 187–193.

Hatta T, G Muklerjee-Dhar, J Damborsky, H Kiyohara. 2003. Characterization of a novel thermostable Mn(II)-dependent 2,3-dihydroxybiphenyl 1,2-dioxygenase from polychlorinated biphenyl- and naphthalene-degrading *Bacillus* sp. JF8. *J Biol Chem* 278: 21483–21492.

He MM, SL Clugston, JF Honek, BW Mattews. 2000. Determination of the structure of *Escherichia coli* glyoxalase I suggests a structural basis for differential metal activation. *Biochemistry* 39: 8719–8727.

He SH, M Teixeira, J LeGall, DS Patil, I Moura, JJG Moira, DV DerVartanian, BH Huynh, HD Peck. 1989. EPR studies with [77]Se-enriched (NiFeSe) hydrogenase of *Desulfovibrio baculatus*. *J Biol Chem* 264: 2678–2682.

Hensgens CMH, WR Hagen, TA Hansen. 1995. Purification and characterization of a benzylviologen-linked, tungsten-containing aldehyde oxidoreductase from *Desulfovibrio gigas*. *J Bacteriol* 177: 6195–6200.

Hetterich D, B Peschke, B Tshisuaka, F Lingens. 1991. Microbial metabolism of quinoline and related compounds. X. The molybdopterin cofactors of quinoline oxidoreductases from *Pseudomonas putida* 86 and *Rhodococcus* spec. B1 and of xanthine dehydrogenase from *Pseudomonas putida* 86. *Biol Chem Hoppe-Seyler* 372: 513–517.

Hille R. 2002. Molybdenum and tungsten in biology. *Trends in Biochem Sci* 27: 360–367.

Hoffmann B, M Oberhuber, E Stupperich, H Bothe, W Buckel, R Conrat, B Kräutler. 2000. Native corrinoids from *Clostridium cochlearium* are adeninylcobamides: Spectroscopic analysis and identification of pseudovitamin B_{12} and factor A. *J Bacteriol* 182: 4773–4782.

Hormann K, JR Andreesen. 1991. A flavin-dependent oxygenase reaction initiates the degradation of pyrrole-2-carboxylate in *Arthrobacter* strain Py1 (DSM 6386). *Arch Microbiol* 157: 43–48.

Hu Y, CC Lee, MW Ribbe. 2011. Extending the carbon chain: Hydrocarbon formation catalyzed by Vanadium/Molybdenum nitrogenases. *Science* 333: 753–755.

Hube M, M Blokesh, A Böck. 2002. Network of hydrogenase maturation in *Escherichia coli*: Role of accessory proteins HypA and HybF. *J Bacteriol* 184: 3879–3885.

Hund H-K, J Breuer, F Lingens, J Hüttermann, R Kappl, S Fetzner. 1999. Flavonol 2,4-dioxygenase from *Aspergillus niger* DSM 821, a type 2 Cu^{II}-containing glycoprotein. *Eur J Biochem* 263: 871–878.

Hund HK, A de Beyer, F Lingens. 1990. Microbial metabolism of quinoline and related compounds. VI. Degradation of quinaldine by *Arthrobacter* sp. *Biol Chem Hoppe-Seyler* 371: 1005–1008.

Jabri E, MB Carr, RP Hausinger, PA Karplus. 1995. The crystal structure of urease from *Klebsiella aerogenes*. *Science* 268: 998–1004.

Jobst B, K Schühle, U Linne, J Heider. 2010. ATP-dependent carboxylation of acetophenone by a novel type of carboxylase. *J Bacteriol* 192: 1387–1394.

Juszczak A, S Aono, MWW Adams. 1991. The extremely thermophilic eubacterium *Thermotoga maritima*, contains a novel iron-hydrogenase whose cellular activity is dependent upon tungsten. *J Biol Chem* 266: 13834–13841.

Kamp C, A Silakov, M Winkler, WSJ Reijerse, W Lubitz, T Happe. 2008. Isolation and first EPR characterization of the [FeFe]-hydrogenases from green algae. *Biochim Biophys Acta* 1777: 410–416.

Kanai T, R Matsuoka, H beppu, A Nakajka, Y Okada, H Atomi, T Imanaka. 2011. Distinct physiological roles of the three [NiFe]-hydrogenase orthologues in the hyperthermophilic archaeon *Thermococcus kodakarensis*. *J Bacteriol* 193: 3109–3116.

Kao W-C, Y-R Chen, EC Yi, H Lee, Q Tian, K-M Wu, S-F Tsai et al. 2004. Quantitative proteomic analysis of metabolic regulation by copper ions in *Methylococcus capsulatus* (Bath). *J Biol Chem* 279: 51554–52560.

Kehres DG, ME Maguire. 2003. Emerging themes in manganese transport, biochemistry, and pathogenesis in bacteria. *FEMS Microbiol Revs* 27: 263–290.

Kengen SWM, GB Rikken, WR Hagen, CG van Ginkel, ALM Stams. 1999. Purification and characterization of (per)chlorate reductase from the chlorate-respiring strain GR-1. *J Bacteriol* 181: 6706–6711.

King PW, MC Posewitz, ML Ghirardi, M Seibert. 2006. Functional studies of (FeFe] hydrogenase maturation in an *Escherichia coli* biosynthetic system. *J Bacteriol* 188: 2163–2172.

Koenig K, JR Andreesen. 1990. Xanthine dehydrogenase and 2-furoyl-coenzyme A dehydrogenase from *Pseudomonas putida* Fu1: two molybdenum-containing dehydrogenases of novel structural composition. *J Bacteriol* 172: 5999–6009.

Koenig K, JR Andreesen. 1991. Aerobic and anaerobic degradation of furan-3-carboxylate by *Paracoccus denitrificans* strain MK33. *Arch Microbiol* 157: 70–75.

Kooter IM, RA Steiner, BW Dijkstra, PI van Noort, MR Egmond, M Huber. 2002. EPR characterization of the mononuclear Cu-containing *Aspergillus japonicus* quercetin 2,3-dioxygenase reveals dramatic changes upon anaerobic binding of substrates. *Eur J Biochem* 269: 2971–2979.

Krafft T, JM Macy. 1998. Purification and characterization of the respiratory arsenate reductase of *Chrysiogenes arsenatis*. *Eur J Biochem* 255: 647–653.

Kretzer A, JR Andreesen. 1991. A new pathway for isonicotinate degradation by *Mycobacterium* sp. INA1. *J Gen Microbiol* 137: 1073–1080.

Krishnakumar AM, D Sliwa, JA Endrizzi, ES Boyd, SA Ensign, JW Peters. 2008. Getting a handle on the role of coenzyme M in alkene metabolism. *Microbiol Mol Biol Revs* 72: 445–456.

Krum JG, H Ellsworth, RR Sargeant, G Rich, SA Ensign. 2002. Kinetic and microcalorimetric analysis of substrate and cofactor interactions in epoxyalkene CoM:transferase, a zinc-dependent epoxidase. *Biochemistry* 41: 5005–5014.

Kuhm AE, M Schlömann, H-J Knackmuss, DH Pieper. 1990. Purification and characterization of dichloromuconate cycloisomerase from *Alcaligenes eutropha* JMP 134. *Biochem J* 266: 877–883.

Kung JW, C Löffler, K Dörner, D Heintz, S Gallien, A Van Dorsselaer, T Friedrich, M Boll. 2009. Identification and characterization of the tungsten-containing class of benzoyl-coenzyme A reductases. *Proc Natl Acad Sci USA* 106: 17687–17692.

Laska S, F Lottspeicht, A Kletzin. 2003. Membrane-bound hydrogenase and sulfur reductase of the hyperthermophilic and acidophilic archaeon *Acidianus ambivalens*. *Microbiology (UK)* 149: 2357–2371.

Laukel M, L Chistoserdova, ME Lidstrom, JA Vorholt. 2003. The tungsten-containing formate dehydrogenase from *Methylobacterium extorquens* AM1: purification and properties. *Eur J Biochem* 270: 324–333.

Lee CC, Y Hu, MW Ribbe. 2009. Unique features of the nitrogenase VFe protein from *Aztobacter vinelandii*. *Proc Natl Acad Sci USA* 106: 929–9214.

Lehmann M, B Tshisuaka, S Fetzner, P Röger, F Lingens. 1994. Purification and characterization of isoquinoline 1-oxidoreductase from *Pseudomonas diminuta* 7, a molybdenum-containing hydroxylase. *J Biol Chem* 269: 11254–11260.

Leskovac V, S Trivíc, D Pericin. 2002. The three zinc-containing alcohol dehydrogenases from bake's yeast, *Saccharomyces cerevisiae. FEMS Yeast Res* 2: 481–494.

Lisa M-N, L Hemmingsen, AJ Vila. 2010. Catalytic role of the metal ion in the metallo-β-lactamase GOB. *J Biol Chem* 285: 4570–4577.

Liu A, H Zhang. 2006. Transition metal-catalyzed nonoxidative decarboxylation reactions. *Biochemistry* 45: 104007–10411.

Llamas A, RR Mendel, G Schwarz. 2004. Synthesis of adenylated molybdopterin. An essential step for molybdenum insertion. *J Biol Chem* 279: 55241–55246.

Loschi L, SJ Brokx, TL Hills, G Zhang, MG Bertero, AL Lovering, JH Weiner, NCJ Strynadka. 2004. Structural and biochemical identification of a novel bacterial oxidoreductase. *J Biol Chem* 279: 50391–50400.

Lu W-P, PE Jablonski, M Rasche, JG Ferry, SW Ragsdale. 1994. Characterization of the metal centers of the Ni/Fe–S component of the carbon-monoxide dehydrogenase enzyme complex of *Methanosarcina thermophila. J Biol Chem* 269: 9736–9742.

Ludwig M, T Schubert, I Zebger, N Wisitruangsakul, M Saggu, A Strack, O Lenz, P Hildebrandt. B Friedrich. 2009. Concerted action of two novel auxiliary proteins in assembly of the active site in a membrane-bound (NiFe] hydrogenase. *J Biol Chem* 284: 2159–2168.

Lyon EJ, S shima, G Buur,an, S Chowdhuri, A Batschauer, K Steinbach, RK Thauer. 2004. UV-A/blue-light inactivation of the "metal-free" hydrogenase (Hmd) from methanogenic archaea: The enzyme contains functional iron after all. *Eur J Biochem* 271: 195–204.

Ma K, MWW Adams. 1999. An unusual oxygen-sensitive, iron- and zinc-containing alcohol dehydrogenase from the hyperthermophilic archaeon *Pyrococcus furiosus. J Bacteriol* 181: 1163–1170.

Ma K, R Weiss, MWW Adams. 2000. Characterization of hydrogenase II from the hyperthermophilic archaeon *Pyrococcus furiosus* and assessment of its role in sulfur reduction. *J Bacteriol* 182: 1864–1871.

Ma K, C Zirngibl, D Linder, KO Stetter, RK Thauer. 1991. $N5,N10$-methylenetetrahydro-methanopterin dehydrogenase (H_2-forming) from the extreme thermophile *Methanopyrus kandleri. Arch Microbiol* 156: 43–48.

Maeng JH, Y Sakai, Y Tani, N Kato. 1996. Isolation and characterization of a novel oxygenase that catalyzes the first step of *n*-alkyl oxidation in *Acinetobacter* sp. strain M-1. *J Bacteriol* 178: 3695–3700.

Maier T, A Jacobi, M Sauter, A Böck. 1993. The product of the *hypB* gene, which is required for nickel incorporation into hydrogenases, is a novel guanine nucleotide-binding protein. *J Bacteriol* 175: 630–635.

Manyani H, L Rey, JM Palacios, J Imperial, T Ruiz-Argüeso. 2005. Gene products of the *hupGHIK* operon are involved in maturation of the iron–sulfur subunit of the (NiFe] hydrogenase from *Rhizobium leguminosarum* bv. viciae. *J Bacteriol* 187: 7018–7026.

Martynowski D, Y Eyobo, T Li, K Yang, A Liu, H Zhang. 2006. Crystal structure of α-amino-β-carboxymuconate-ε-semialdehyde decarboxylase: Insight into the active site and catalytic mechanism of a novel decarboxylation reaction. *Biochemistry* 45: 10412–10421.

McCormick ML, GR Buettner, BE Britigan. 1998. Endogenous superoxide dismutase levels regulate iron-dependent hydroxyl radical formation in *Escherichia* coli exposed to hydrogen peroxide. *J Bacteriol* 189: 622–625.

McDevitt CA, P Hugenholtz, GR Hanson, AG McEwan. 2002. Molecular analysis of dimethyl sulfide dehydrogenase from *Rhodovulum sulfidophilum*: Its place in the dimethyl sulfoxide reductase family of microbial molybdopterin enzymes. *Mol Microbiol* 44: 1575–1587.

McDonald IR, CB Miguez, G Rogge, D Bourque, KD Wendlandt, D Groleau, JC Murrell. 2006. Diversity of soluble methane monooxygenase-containing methanotrophs isolated from polluted environments. *FEMS Microbiol Lett* 255: 225–232.

McEwan AG, JP Ridge, CA McDevitt, P Hugenholtz. 2002. the DMSO reductase family of microbial molybdenum enzymes; molecular properties and the role in the dissimilatory reduction of toxic elements. *Geomicrobiol J* 19: 3–21.

Meckenstock RU, R Krieger, S Ensign, PMH Kroneck, B Schink. 1999. Acetylene hydratase of *Pelobacter acetylenicus*. Molecular and spectroscopic properties of the tungsten iron–sulfur enzyme. *Eur J Biochem* 264: 176–182.

Mehta N, JW Olson, RJ Maier. 2003. Characterization of *Helicobacter pylori* nickel metabolism accessory proteins needed for maturation of both urease and hydrogenase. *J Bacteriol* 185: 726–734.

Merkens H, R Kappl, RP Jakob, FX Schmid, S Fetzner. 2008. Quecetinase QueD of *Streptomyces* sp. FLA, a monocupin dioxygenase with a preference for nickel or cobalt. *Biochemistry* 47: 12185–12196.

Mermod M, F Mourlane, S Waltersperger, AE Oberholzer, U Baumann, M Solioz. 2010. Structure and function of CinD (YtjD) of *Lactococcus lactis*, a copper-induced nitroreductase involved in defense against oxidative stress. *J Bacteriol* 192: 4172–4180.

Messerschmidt A, L Prade, R Wever. 1997. Implications for the catalytic mechanism of the vanadium-containing enzyme chloroperoxodase from the fungus *Curvularia inaequalis* by X-ray structures of the native and peroxide form. *Biol Chem* 378: 309–315.

Mobley HLT, MD Island, RP Hausinger. 1995. Molecular biology of microbial ureases. *Microbiol Revs* 59: 451–480.

Mota CS, O Valette, PJ González, CD Brondino, JJG Moura, I Moura, A Dolla, MG Rivas. 2011. Effects of molybdate and tungstate on expression levels and biochemical characteristics of formate dehydrogenases produced by *Desulfovibrio alaskensis* NCIB 13491. *J Bacteriol* 193: 2917–2923.

Mu D, SM Janes, AJ Smith, DE Brown, DM Dooley, JP Klinman. 1992. Tyrosine codon corresponds to topa quinone at the active site of copper amine oxidases. *J Biol Chem* 267: 7979–7982.

Mukund S, MWW Adams. 1991. The novel tungsten–iron–sulfur protein of the hyperthermophilic archaebacterium, *Pyrococcus furiosus*, is an aldehyde ferredoxin oxidoreductase *J Biol Chem* 266: 14208–14216.

Mukund S, MWW Adams. 1993. Characterization of a novel tungsten-containing formaldehyde oxidoreductase from the extremely thermophilic archaeon, *Thermococcus litoralis*. *J Biol Chem* 268: 13592–13600.

Mukund S, MWW Adams. 1995. Glyceraldehyde-3-phosphate ferredoxin oxidoreductase, a novel tungsten-containing enzyme with a potential glycolytic role in the hyperthermophilic archaeon, *Pyrococcus furiosus*. *J Biol Chem* 270: 8389–8392.

Mulrooney SB, RP Hausinger. 2003. Nickel uptake and utilization by microorganisms. *FEMS Microbiol Revs* 27: 239–261.

Muraki T, M Taki, Y Hasegawa, H Iwaki, PCK Lau. 2003. Prokaryotic homologues of the eukaryotic 3-hydroxyanthranilate 3,4-dioxygenase and 2-amino-carboxymuconate-6-semialdehyde decarboxylase in the 2-nitrobenzoate degradation pathway of *Pseudomonas fluorescens* strain KU-7. *Appl Environ Microbiol* 69: 1564–1572.

Murphy MEP, S Turley, ET Adman. 1997. Structure of nitrite bound to copper-containing nitrite reductase from *Alcaligenes faecalis*. *J Biol Chem* 272: 28455–28460.

Nauton L, R Kahn, G Garau, JF Hernandez, O Dideberg. 2008. Structural insights into the design of inhibitors for the L1 metallo-β-lactamase from *Stenotrophomomas maltophilia*. *J Mol Biol* 375: 257–269.

Nicolet Y, JK Rubach, MC Posewitz, P Amara, C Mathevon, M Atta, M Fontecave, JC Fontecilla-Camps. 2008. X-ray structure of the (Fe–Fe]-hydrogenase maturase HydE from *Thermotoga maritima*. *J Biol Chem* 283: 18861–18872.

Nordlund P, P Reichard. 2006. Ribonucleotide reductases. *Annu Rev Biochem* 75: 681–706.

Oda Y, SK Samanta, FE Rey, L Wu, X Liu, T Yan, J Zhou, CS Harwood. 2005. Functional genomic analysis of three nitrogenase isoenzymes in the photosynthetic bacterium *Rhodopseudomonas palustris*. *J Bacteriol* 187: 7784–7794.

Olson JW, RJ Maier. 2000. Dual roles of *Bradyrhizobium japonicum* nickelin protein in nickel storage and GTP-dependent Ni mobilization. *J Bacteriol* 182: 1702–1705.

Olson JW, RJ Maier. 2002. Molecular hydrogen as an energy source for *Helicobacter pylori*. *Science* 298: 1788–1790.

Paschos A, A Bauer, A Zimmermann, E Zehelein, A Böck. 2002. HypF, a carbamoyl phosphate enzyme involved in (NiFe] hydrogenase maturation. *J Biol Chem* 277: 49945–49951.

Pau RN, LA Mitchenall, RL Robson. 1989. Genetic evidence for an *Azotobacter vinelandii* nitrogenase lacking molybdenum and vanadium. *J Bacteriol* 171: 124–129.

Peariso K, ZS Zhou, AE Smith, RG Mattews, JE Penner-Hahn. 2005. Characterization of the zinc sites in cobalamin-independent and cobalamin-dependent methionine synthase using zinc and selenium X-ray absorption spectroscopy. *Biochemistry* 40: 987–993.

Peng X, E Masai, D Kasai, K Miyauchi, Y Katayama, M Fukuda. 2005. A second 5-carboxyvanillate decarboxylase gene and its role in lignin-related catabolism in *Sphingomonas paucimobilis* SYK-6. *Appl Environ Microbiol* 71: 5014–5021.

Peschke B, F Lingens. 1991. Microbial metabolism of quinoline and related compounds XII. Isolation and characterization of the quinoline oxidoreductase from *Rhodococcus* spec. B1 compared with the quinoline oxidoreductase from *Pseudomonas putida* 86. *Biol Chem Hoppe-Seyler* 370: 1081–1088.

Peters JW, WN Lanzilotta, BJ Lemon, LC Seefeldt. 1998. X-ray crystal structure of the Fe-only hydrogenase (Cpl) from *Clostridium pasteurianum* to 1.8 angstrom resolution. *Science* 282: 1853–1858.

Peters F, M Rother, M Boll. 2004. Selenocysteine-containing proteins in anaerobic benzoate metabolism of *Desulfococcus multivorans*. *J Bacteriol* 186: 2156–2163.

Petrella S, N Gelus-Ziental, A Maudry, C Laurans, R Boudjelloul, W Sougakoff. 2010. Crystal structure of the pyrazinamidase of *Mycobacterium tuberculosis*: Insight into natural and acquired resistance to pyrazinamide. *PloS ONE* 6: 15785.

Pierik AJ, W Roseboom, RP Happe, KA Bagley, SPJ Albracht. 1999. Carbon monoxide and cyanide as intrinsic ligands to iron in the active site of (NiFe]-hydrogenases. *J Biol Chem* 274: 3331–3337.

Pilak O, B Mamat, S Vogt, CH Hagemeier, RK Thauer, S Shima, C Vonrhein, E Warkentin, U Ermler. 2006. The crystal structure of the apoenzyme of the iron–sulfur cluster-free hydrogenase. *J Mol Biol* 358: 798–809.

Pinho D, S Besson, CD Brondino, B de Castro, I Moura. 2004. Copper-containing nitrite reductase from *Pseudomonas chlororaphis* DSM 50135. Evidence for modulation of the rate of intramolecular electron transfer through nitrite binding to the type 2 copper center. *Eur J Biochem* 271: 2361–2369.

Posewitz MC, PW King, SL Smolinski, L Zhang, M Seibert, MK Ghirardi. 2004. Discovery of two novel radical *S*-adenosylmethionine proteins required for the assembly of an active (Fe] hydrogenase. *J Biol Chem* 279: 25711–25720.

Prudêncio M, RR Eady, G Sawers. 1999. The blue copper-containing nitrite reductase from *Alcaligenes xylosoxidans*: Cloning of the *nirA* gene and characterization of the recombinant enzyme. *J Bacteriol* 181: 2323–2329.

Que L, J Widom, RL Crawford. 1981. 3,4-dihydroxyphenylacetate 2,3-dioxygenase: A manganese(II) dioxygenase from *Bacillus brevis*. *J Biol Chem* 256: 10941–10944.

Radianingtyas H, PC Wright. 2003. Alcohol dehydrogenases from thermophilic and hyperthermophilic archaea and bacteria. *FEMS Microbiol Rev* 27: 593–616.

Rangarjan ES, A Asinas, A Proteau, C Munger, J Baardsnes, P Iannuzzi, A Matte, M Cygler. 2008. Structure of (NiFe] hydrogenase maturation protein HypE from *Escherichia coli* and its interaction with HypF. *J Bacteriol* 190: 1447–1458.

Reissmann S, E Hochleitner, H Wang, A Paschos, F Lottspeich, RS Glass, A Böck. 2003. Taming of a poison: Biosynthesis of the NiFe-hydrogenase cyanide ligands. *Science* 299: 1067–1070.

Renirie R, W Hemrika, R Wever. 2000. Peroxidase and phosphatase activity of active-site mutants of vanadium chloroperoxidase from the fungus *Curvularia inaequalis*. *J Biol Chem* 275: 11650–11657.

Ribbe M, D Gadkari, O Meyer. 1997. N$_2$ fixation by *Streptomyces thermoautotrophicum* involves a molybdenum-dinitrogenase and a manganese-superoxide oxidoreductase that couple N$_2$ reduction to the oxidation of superoxide produced from O$_2$ by a molybdenum-CO dehydrogenase. *J Biol Chem* 272: 26627–26633.

Richard DJ, G Sawers, F Sargent, L McWalter, DH Boxer. 1999. Transcriptional regulation in response to oxygen and nitrate of the operons encoding the (NiFe] hydrogenases 1 and 2 of *Escherichia coli*. *Microbiology (UK)* 145: 2903–2912.

Ridley H, CA Watts, DJ Richardson, CS Butler. 2006. Resolution of distinct membrane-bound enzymes from *Enterobacter cloacae* SKLD1a-1 that are responsible for selective reduction of nitrate and selenate anions. *Appl Environ Microbiol* 72: 5173–5180.

Röger P, G Bär, F Lingens. 1995. Two novel metabolites in the degradation pathway of isoquinoline by *Pseudomonas diminuta* 7. *FEMS Microbiol Lett* 129: 281–286.

Röger P, A Erben, F Lingens. 1990. Microbial metabolism of quinoline and related compounds IV. Degradation of isoquinoline by *Alcaligenes faecalis* Pa and *Pseudomonas diminuta* 7. *Biol Chem Hoppe-Seyler* 370: 1183–1189.

Roy R, MWW Adams. 2002. Characterization of a fourth tungsten-containing enzyme from the hyperthermophilic *Pyrococcus furiosus*. *J Bacteriol* 184: 6952–6956.

Roy R, S Mukund, GJ Schut, DM Dunn, R Weiss, MWW Adams. 1999. Purification and molecular characterization of the tungsten-containing formaldehyde ferredoxin oxidoreductase from the hyperthermophilic archaeon *Pyrococcus furiosus*: The third of a putative five-member tungstoenzyme family. *J Bacteriol* 181: 1171–1180.

Rüttimann-Johnson C, LM Rubio, DR Dean, PW Ludden. 2003. VnfY is required for full activity of the vanadium-containing dinitrogenase in *Azotobacter vinelandii. J Bacteriol* 185: 2383–2386.

Saggu M, I Zebger, M Ludwig, O Lenz, B Friedrich, P Hildebrandt, F Lenzian. 2009. Spectroscopic insights into the oxygen-tolerant membrane-associated (NiFe] hydrogenase of *Ralstonia eutropha* H 16. *J Biol Chem* 284: 16264–16276.

Santiago B, O Meyer. 1997. Purification and molecular characterization of the H_2 uptake membrane-bound NiFe-hydrogenase from the carboxidotrophic bacterium *Oligotropha carboxidovorans. J Bacteriol* 179: 6053–6060.

Sapra R, MFJM Verhagen, MWW Adams. 2000. Purification and characterization of a membrane-bound hydrogenase from the hyperthermophilic archaeon *Pyrococcus furiosus. J Bacteriol* 182: 3423–3428.

Schaab MR, BM Barney, WA Francisco. 2006. Kinetic and spectoscopic studies on the quercetin 2,3-dioxygenase from *Bacillus subtilis. Biochemistry* 45: 1009–1016.

Schach S, B Tshisuaka, S Fetzner F Lingens. 1995. Quinoline 2-oxidoreductase and 2-oxo-1,2-dihydroquinoline 5,6-dioxygenase from *Comamonas testosteroni* 63. The first two enzymes in quinoline and 3-methylquinoline degradation. *Eur J Biochem* 232: 536–544.

Schneider K, U Gollan, M Dröttboom, S Selsmeier-Voigt, A Müller. 1997. Comparative biochemical characterization of the iron-only nitrogenase and the molybdenum nitrogenase from *Rhodobacter capsulatus. Eur J Biochem* 244: 789–800.

Schroeder I, S Rech, T Krafft, JM Macey. 1997. Purification and characterization of the selenate reductase from *Thauera selenatis. J Biol Chem* 272: 23765–23768.

Seibert CM, FM Raushel. 2005. Structural and catalytic diversity within the amidohydrolase superfamily. *Biochemistry* 44: 6383–6391.

Shut GJ, MWW Adams. 2009. The iron-hydrogenase of *Thermotoga maritima* utilizes ferredoxin and NADH synergistically: A new perspective on anaerobic hydrogen production. *J Bacteriol* 191: 4451–4457.

Siegbahn PEM, F Haeffner. 2004. Mechanism for catechol ring-cleavage by non-heme iron extradiol dioxygenases. *J Am Chem Soc* 126: 8919–8932.

Siegmund I, K Koenig, JR Andreesen. 1990. Molybdenum involvement in aerobic degradation of picolinic acid by *Arthrobacter picolinophilus FEMS Microbiol Lett* 67: 281–284.

Soboh B, D Linder, R Hedderich. 2004. A multisubunit membrane-bound (NiFe] hydrogenase and an NADH-dependent Fe-only hydrogenase in the fermenting bacterium *Thermoanaerobacter tengcongensis. Microbiology (UK)* 150: 2451–2463.

Sobota JM, JA Imlay. 2011. Iron enzyme ribulose-5-phosphate 3-epimerase in *Escherichia coli* is rapidly damaged by hydrogen peroxide but can be protected by manganese. *Proc Natl Acad Sci USA* 108: 5402–5407.

Solomon EI, UM Sundaram, TE Machonkin. 1996. Multicopper oxidases and oxygenases. *Chem Rev* 96: 2563–2605.

Sontoh S, JD Semrau. 1998. Methane and trichloroethylene degradation by *Methylosinus trichosporium* OB3b expressing particulate methane monooxygenase. *Appl Environ Microbiol* 64: 1106–1114.

Soong C-L, J Ogawa, E Sakuradani, S Shimizu. 2002. Barbiturase, a novel zinc-containing amidohydrolase involved in oxidative pyrimidine metabolism. *J Biol Chem* 277: 7051–7058.

Steiner RA, HJ Janssen, P Roversi, AJ Oakley, S Fetzner. 2010. Structural basis for cofactor-independent dioxygenation of *N*-heteroaromatic compounds at the α/β-hydrolase fold. *Proc Natl Acad Sci USA* 107: 657–662.

Steiner RA, KH Kalk, BW Dijkstra. 2002. Anaerobic enzyme substrate structures provide insight into the reaction mechanism of the copper-dependent quercetin 2,3-dioxygenase *Proc Natl Acad USA* 99: 16625–16630.

Strobl G, R Feicht, H White, F Lottspeich, H Simon. 1992. The tungsten-containing aldehyde oxidoreductase from *Clostridium thermoaceticum* and its complex with a viologen-accepting NADPH oxidoreductase. *Biol Chem Hoppe-Seyler* 373: 123–132.

Svetlitchnyi V, C Peschel, G Acker, O Meyer. 2001. Two membrane-associated NiFeS-carbon monoxide dehydrogenases from the anaerobic carbon-monoxide-utilizing eubacterium *Carboxydothermus hydrogenoformans. J Bacteriol* 183: 5134–5144.

Tallant TC, L Paul, JA Krzycki. 2001. The MtsA subunit of the methylthiol: Coenzyme M methyltransferase of *Methanosarcina barkeri* catalyzes both half-reactions of corrinoid-dependent dimethylsulfide: Coenzyme M methyl transfer. *J Biol Chem* 276: 4485–4493.

Tanner A, L Bowater, SA Fairhurst, S Bornemann. 2001. Oxalate decarboxylase requires manganese and dioxygen for activity. *J Biol Chem* 276: 43627–43634.

Tate RL, Ensign JC. 1974. Picolinic acid hydroxylase of *Arthrobacter picolinophilus*. *Can J Microbiol* 20: 695–702.

tenBrink F, B Schink, PMH Kroneck. 2011. Exploring the active site of the tungsten, iron–sulfur enzyme acetylene hydratase. *J Bacteriol* 193: 1229–1236.

Tong Y, M Guo. 2009. Bacterial heme-transport proteins and their heme-coordination modes. *Arch Biochem Biophys* 481: 1–15.

Torrents E, G Buist, A Liu, J Kok, I Gibert, A Gräslund, P Reichert. 2000. The anaerobic (Class III) ribonucleotide reductase from *Lactococcus lactis*. Catalytic properties and allosteric regulation of the pure enzyme system. *J Biol Chem* 275: 2463–2471.

Torrents E, I Roca, I Gibert. 2003. *Corynebacterium ammoniagenes* class Ib ribonucleotide reductase: Transcriptional regulation of an atypical genomic organization in the *nrd* cluster. *Microbiology (UK)* 149: 1011–1020.

Touati D, M Jacques, B Tardat, L Bouchard, S Despied. 1995. Lethal oxidative damage and mutagenesis are generated by iron *fur* mutants of *Escherichia coli*: Protective role of superoxide dismutase. *J Bacteriol* 177: 2305–2314.

Valente FMA, CC Almeida, I Pacheco, J Carita, LM Saraiva, IAS Pereira. 2006. Selenium is involved in regulation of periplasmic hydrogenase gene expression in *Desulfovibrio vulgaris* Hildenborough. *J Bacteriol* 188: 3228–3235.

van der Woude MW, K Boominanathan, CA Reddy. 1993. Nitrogen regulation of lignin peroxidase and manganese-dependent peroxidase production is independent of carbon and manganese regulation in *Phanerochaete chrysosporium*. *Arch Microbiol* 160: 1–4.

van Haaster DJ, PJ Silva, P-L Hagedoorn, JA Jongejan, WR Hagen. 2008. Reinvestigation of the steady-state kinetics and physiological function of the soluble NiFe-hydrogenase I of *Pyrococcus furiosus*. *J Bacteriol* 190: 1584–1587.

Verhagen NFJM, T O'Rourke, MWW Adams. 1999. The hyperthermophilic bacterium, *Thermotoga maritima*, contains an unusually complex iron-hydrogenase: Amino acid sequence analyses versus biochemical characterization. *Biochim Biophys Acta* 1412: 212–229.

Vignais PM, B Billoud, J Meyer. 2001. Classification and phylogeny of hydrogenases. *FEMS Microbiol Revs* 25: 4355–501.

Vignais PM, B Dimon, NA Zorin, M Tomiyama, A Colbeau. 2000. Characterization of the hydrogen-deuterium exchange activity of the energy-transducing HupSL hydrogenase and H_2-signilling HupUV hydrogenase in *Rhodobacter capsulatus J Bacteriol* 182: 5997–6994.

Volbeda A, E Garcin, C Piras, AL de Lacey, VM Fernandez, E Claude, EC Hatchikan, M Frey, JC Fontecilla-Camps. 1996. Structure of the (NiFe] hydrogenase active site: Evidence for biologically uncommon Fe ligands. *J Am Chem Soc* 118: 12989–12996.

Wandersman C, P Delepelaire. 2004. Bacterial iron sources: From siderophores to hemophores. *Annu Rev Microbiol* 58: 611–647.

Wang C-P, R Franco, JJG Moura, I Moura, EP Day. 1991. The nickel site in active *Desulfovibrio baculatus* (NiFeSe] hydrogenase is diamagnetic. *J Biol Chem* 267: 7378–7380.

Wang C-P, R Franco, JJG Moura, I Moura, EP Day. 1992. The nickel site in active *Desulfovibrio baculatus* (NiFeS] hydrogenase is diamagnetic. Multifield saturation magnetization measurement of the spin state of Ni(II). *J Biol Chem* 267: 7378–7380.

Waters LS, M Sandoval, G Storz. 2011. The *Escherichia coli* MntR miniregulon includes genes encoding a small protein and an efflux pump required for manganese homeostasis. *J Bacteriol* 193: 5887–5897.

White H, R Feicht, C Huber, F Lottspeich, H Simon. 1991. Purification and some properties of the tungsten-containing carboxylic acid reductase from *Clostridium formicoaceticum*. *Biol Chem Hoppe-Seyler* 372: 999–1005.

White H, C Huber, R Feicht, H Simon. 1993. On a reversible molybdenum-containing aldehyde oxidoreductase from *Clostridium formicoaceticum*. *Arch Microbiol* 159: 244–249.

Whittaker MM, PJ Kersten, N Nakamura, J Sanders-Loehr, ES Schweizer, JW Whittaker. 1996. Glyoxal oxidase from *Phanerochaete chrysosporium* is a new radical copper oxidase. *J Biol Chem* 271: 681–687.

Willing A, H Follmann, G Auling. 1988. Ribonucleotide reductase of *Brevibacterium ammoniagenes* is a manganese enzyme. *Eur J Biochem* 170: 603–611.

Wilmot CM, J Hajdu, MJ Mc Pherson, PF Knowles, SEV Phillips. 1999. Visualization of dioxygen bound to copper during enzyme catalysis. *Science* 286: 1724–1728.

Wischgoll S, D Heintz, F Peters, A Erxleben, E Sarnighausen, R Reski, A Van Dorsselaer, M Boll. 2005. Gene clusters involved in anaerobic benzoate degradation of *Geobacter metallireducens*. *Mol Microbiol* 58: 1238–1252.

Wuerges J, J-W Lee, Y-I Yim, H-S Yim, S-O Kang, KD Carugo. 2004. Crystal structure of nickel-containing superoxide dismutase reveals another type of active site. *Proc Natl Acad Sci USA* 101: 8569–8574.

Yamamoto I, T Saiki, S-M Liu, LG Ljungdahl. 1983. Purification and properties of NADP-dependent formate dehydrogenase from *Clostridium thermoaceticum*, a tungsten-selenium-iron protein. *J Biol Chem* 258: 1826–1832.

Yoch DC. 1979. Manganese, an essential trace element for N_2 fixation by *Rhodospirilllum rubrum* and *Rhodopseudomonas capsulata*: role in nitrogenase regulation *J Bacteriol* 140: 987–995.

Yoshida M, N Fukuhara, T Oikawa. 2004. Thermophilic, reversible γ-resorcylate decarboxylase from *Rhizobium* sp. strain MTP-10005: purification, molecular characterization, and expression. *J Bacteriol* 186: 6855–6863.

Yu S S-F, K H-C Chen, M Y-H Tseng, Y-S Wang, C-F Tseng, Y-J Chen, D-S Huang, SI Chan. 2003a. Production of high-quality particulate methane monooxygenase in high yields from *Methylococcus capsulatus* (Bath) with a hollow-fiber membrane bioreactor. *J Bacteriol* 185: 5915–5924.

Yu S S-F, L-Y Wu, K H.-C Chen, W-I Luo, D-S Huang, SI Chan. 2003b. The stereospecific hydroxylation of [2,2-^{2}H$_2$]butane and chiral dideuteriobutanes by the particulate methane monooxygenase from *Methylococcus capsulatus* (Bath). *J Biol Chem* 278: 40658–40669.

Zellner G, A Jargon. 1997. Evidence for a tungsten-stimulated aldehyde dehydrogenase activity of *Desulfovibrio simplex* that oxidizes aliphatic and aromatic aldehydes *Arch Microbiol* 168: 480–485.

Zhang H, J-Y Deng, L-J Bi, Y-F Zhou, Z-P Zhang, C-G Zhang, Y Zhang, X-E Zhang. 2008. Characterization of *Mycobacterium tuberculosis* nicotinamidase/pyrazinamidase. *FEBS J* 275: 753–762g.

Zirngibl C, W van Dongen, B Schwörer, R Von Bünau, M Richter, A Klein, RK Thauer. 1992. H$_2$-forming methylenetetrahydromethanopterin dehydrogenase without iron–sulfur clusters in methanogenic archaea. *Eur J Biochem* 208: 511–520.

4

Interactions

Part 1: Environmental Stress

A book (Gerday and Glansdorff 2007) has been devoted to what have conveniently been termed extremophiles. It deals with organisms subject to a range of extreme physical environments—temperature, salinity, pH, and pressure—and contains an extensive bibliography of the relevant organisms. The journal *Extremophiles* has also been devoted to them. In the following account, emphasis is placed on metabolic aspects of such organisms.

4.1 Adaptation to Stress

Bacteria may be subjected to stress in their natural environment. This may be physical such as temperature, acidity, osmotic pressure as a result of salinity, or chemical such as exposure to hydrocarbons, antibiotics, metal cations, and metalloid oxyanions. A wide range of responses has been revealed, while the important issue of toxic metabolites synthesized by the organisms themselves is discussed later in this chapter.

4.2 Physical Stress

4.2.1 Low Temperature: Psychrophiles

This section is restricted to metabolic aspects since general issues have been covered in Chapter 2, and the role of desaturases in membrane fluidity is briefly noted in Chapter 3. The role of temperature may be of importance both in laboratory studies and for mixed cultures of organisms in the natural environment. Temperature may effect important changes both in the composition of the microbial flora and on the rates of reaction. Greatest attention has hitherto been directed to hydrocarbons and PCBs. Some illustrative examples are provided of important issues that are especially relevant to degradation in extreme environments:

1. Experiments using a fluidized-bed reactor showed that removal of chlorophenols could be accomplished by organisms adapted to a temperature of 5–7°C (Järvinen et al. 1994).

2. An anaerobic sediment sample was incubated with 2,3,4,6-tetrachlorobiphenyl at various temperatures between 4°C and 66°C (Wu et al. 1997a). The main products were 2,4,6- and 2,3,6-trichlorobiphenyl, and 2,6-dichlorobiphenyl. The first was produced maximally and discontinuously at 12°C and 34°C, the second maximally at 18°C, and the third was dominant from 25°C to 30°C. Dechlorination was not observed above 37°C. In a further comparable study with Arochlor 1260, hexa- to nonachlorinated congeners were dechlorinated with a corresponding increase in tri- and tetrachloro congeners, and four dechlorination patterns associated with different temperature ranges could be distinguished (Wu et al. 1997b).

3. Sediment samples from a contaminated site were spiked with Arochlor 1242 and incubated at 4°C for several months (Williams and May 1997). Degradation by aerobic organisms in the

upper layers of the sediment—but not in those at >15 mm from the surface—occurred with the selective production of di- and trichlorobiphenyls. Some congeners were not found, including 2,6- and 4,4-dichlorobiphenyls, and a range of trichlorobiphenyls.

4. Enrichment of arctic soils from Northwest Territories, Canada with biphenyl yielded organisms that were assigned to the genus *Pseudomonas* (Master and Mohn 1998). Rates of removal of individual congeners of Arochlor 1242 were examined at 7°C and compared with those for the mesophilic *Burkholderia cepacia* strain LB400. The spectrum of rates for all congeners was similar for the arctic strains and, for some of the trichlorinated congeners was considerably greater than those obtained for *Burkholderia cepacia* strain LB400. Further investigation suggested that the dioxygenase from the cold-tolerant strain *Pseudomonas* sp. Cam-1 is not, however, cold adapted (Master et al. 2008).

5. Soil samples from a putatively contaminated site on Ellesmere Island (82°N, 62°W) contained bacteria belonging to the genera *Pseudomonas* and *Sphingomonas* that were psychrotolerant. They belonged to two mutually exclusive groups, one able to grow with abietanes and *n*-dodecane, and the other with pimaranes, benzoate, and toluene. Bacteria that are able to degrade resin acids were not found in pristine tundra soils, and the isolation of the resin acid–degrading bacteria presented a conundrum (Yu et al. 2000).

6. Mineralization of ^{14}C-ring-labeled toluene was examined in contaminated and uncontaminated samples from an aquifer in Alaska, and one in South Carolina (Bradley and Chapelle 1995). A number of important facts emerged:

   a. Rates for organisms from South Carolina were greater for the contaminated sample than for the uncontaminated sample, and for the uncontaminated sample were zero at 5°C.

   b. Rates for organisms from the contaminated site in Alaska were highly sensitive to temperature, and showed a distinct maximum at 20°C that was not apparent for those from the uncontaminated site. On the other hand, rates for both groups of organisms were virtually identical at 5°C and 35°C.

7. The aerobic degradation of hydrocarbons has been examined in many cold-temperature regions—Arctic (Whyte et al. 1996; Yu et al. 2000), Alpine (Margesin et al. 2003), and Antarctic (Ruberto et al. 2003; Aislabie et al. 2004)—and in both terrestrial and marine environments. A number of psychrotrophic organisms that were isolated from various sites in northern Canada were capable of mineralizing naphthalene, toluene, and linear dodecane and hexadecane at 5°C. Although some of the positive strains possessed genes homologous with those required for metabolism of the substrates by established pathways—*nahB* for naphthalene, and *xylE* and *todC1* for toluene—others showed only low homologies and may have possessed novel catabolic genes (Whyte et al. 1997). Two organisms assigned to the genus *Pseudomonas* degraded both alkanes (C_5 to C_{12}), toluene, and naphthalene at both 5°C and 25°C (Whyte et al. 1998). It was shown by PCR and DNA sequence analysis that the plasmid-borne catabolic genes were comparable to those for the Alk pathway in *P. oleovorans* and the nah pathway in *P. putida* G7.

8. Anaerobic psychrophilic sulfate-reducing bacteria isolated using lactate as substrate were assigned as to the genera *Desulfofrigus*, *Desulfofaba*, and *Desulfotalea*, and could grow at temperatures down to −1.8°C. Most strains were able to use Fe(III)citrate as electron acceptors as well as sulfate, and a range of C_1 to C_4 carboxylates as electron donors (Knoblauch et al. 1999).

4.2.2 Alterations in Membrane Structure

1. An organism *Psychrobacter arcticus* strain 273-4 was isolated from continuously frozen permafrost in Kolyma, and examination revealed a number of salient features. It was capable of growth in the range −10°C to +28°C (Bakermans et al. 2006), and at temperatures below 0°C synthesized increasing amounts of unsaturated membrane fatty acids (Ponder et al. 2005). A detailed analysis (Bergholz et al. 2009) showed that at low temperatures, the organism down-regulated

genes for energy and carbon substrate metabolism, and increased the expression of genes for maintaining the structures of cell membranes, cell walls, and the mobility of nucleic acids.

2. Alterations in membrane structure are widely associated with temperature stress.

   a. It has been shown in the archaeon *Methanococcoides burtonii* that the degree of unsaturated lipids in cells grown at 4°C was higher than in cells grown at 23°C, and that the degree of unsaturation is specific for each class of phospholipid (Nichols et al. 2004).

   b. *Colwellia psychrerythraea, C. demingae, C. hornerase*, and *C. rossensis* which had been isolated from Antarctic sea-ice were psychrophilic facultatively anaerobic organisms that had high concentrations of docosahexaenoic acid $C_{22:6\omega3}$ that is associated with tolerance of low temperatures (Bowman et al. 1998).

   c. A range of investigations with *Bacillus subtilis* has been devoted to alterations in the structure of membranes, when cultures are transferred from 37°C to 20°C—that has been termed cold shock. *Bacillus subtilis* has developed a number of strategies to deal with these temperature variations.

      i. Unsaturated fatty acids are not components during growth at 37°C, and unsaturation is induced by a fatty acid desaturase encoded by the gene *des* to maintain fluidity in the membranes (Aguilar et al. 1998), and it has been shown that this specifically involves a $\Delta5$ desaturase (Altabe et al. 2003).

      ii. The fatty acid profile is characterized by a high content of branched-chain fatty acids, and it has been shown that tolerance of a temperature shift from 37°C to 15°C was associated with an increase in the amounts of the C_{15} and C_{17} anteiso-fatty acids, and that an isoleucine-dependent switch was able to alter the fatty acid branching pattern for adaptation to the lower temperature (Klein et al. 1999).

      iii. The activity of the $\Delta5$ desaturase for the synthesis of unsaturated fatty acids requires oxygen. Low-temperature adaptation is, however, also observed under anaerobic conditions in the absence of oxygen and, although the desaturase was still active, it was suggested that this adaptation is mediated by the synthesis of an increased ratio of anteiso/iso branched-chain fatty acids (Beranová et al. 2010).

   d. The enzymes in cyanobacteria that introduce the double bonds $\Delta6$, $\Delta9$, $\Delta12$, $\Delta15$ into C_{18} fatty acids are acyl-lipid desaturases play a role in cold tolerance. They belong to four groups of cyanobacteria that include the most extensively studied unicellular strains of *Synechococcus* and *Synechocystis* (Murata and Wada 1995), and a comparative genomic analysis that included the marine species of *Prochlorococcus* and filamentous N_2-fixing strains has been provided (Chi et al. 2008).

3. The activities of betaines as compatible solutes in protection against high osmolarity are discussed in the next section. In addition, it has been shown that glycine betaine, γ-butyrobetaine, carnitine, and proline betaine could serve as protectants of *Bacillus subtilis* strain 168 against chill stress, in which each of the glycine betaine transporters were involved. Stress imposed by osmolarity (1.2 M NaCl), heat (52°C), and cold stress (15°C) was also observed with the same group of betaines, although proline, ectoine, and glutamate were not able to affect the resistance to cold stress (Hoffmann and Bremer 2011).

4. Analysis of a promoter library from *Bacillus cereus* revealed a number of differences during growth at 10°C compared with 37°C. Among these, a gene encoding a lipase and a transcriptional regulator were identified. In addition, a mutant deficient in the lipase-encoding gene displayed reduced growth at low temperatures (Brillard et al. 2010).

4.2.3 Osmotic Stress

Bacteria subjected to greater than normal saline environments have developed a number of strategies to combat water efflux by the synthesis of organic osmolytes (compatible solutes). These neutral molecules are water-soluble neutral compounds that are either synthesized by the cell or accumulated from the medium by transport systems. They range considerably in complexity, for example, trehalose in *Escherichia coli* (Purvis et al. 2005), 1,1′-diglyceryl phosphate in the hyperthermophile *Archaeoglopbus fulgidus* under saline stress (Lamosa et al. 2000), and glycine betaine in members of the Enterobacteriaceae (Le Rudulier and Bouillard 1983) and in *Bacillus subtilis* (Kappes et al. 1996). Mannosylglycerate and mannosylglyceramide function in the thermophile *Rhodothermus marinus* under thermal stress (Borges et al. 2004), and ectoine (2-methyl-5-hydroxy-1,4,5,6-tetrahydropyrimidine-5-carboxylate) is widely distributed, for example, in species of *Salibacillus* (Bursy et al. 2007), and in *Vibrio cholera*, it plays a role in its persistence in saline environments when seawater and freshwater mix (Pflughoeft et al. 2003). It is worth noting that ectoine is also synthesized as a thermoprotectant in *Chromohalobacter salexigens* (García-Estepa et al. 2006), and in response to either salt or temperature (39°C) stress in *Streptomyces coelicolor* (Bursy et al. 2008), while the synthesis of ectoine in *Virgibacillus pantotheticus* is also brought about by reduction in the growth temperature (Kuhlmann et al. 2008). Small peptides such as N-acetylglutaminylglutamine amide (NGGA) play an important role in the tolerance of osmotic stress in *Pseudomonas aeruginosa* that is a significant opportunistic pathogen in the clinical environment (D'Souza-Ault et al. 1993). Some additional comments are worth adding.

1. In several strains of rhizobia, glycine betaine serves primarily as a source of energy (Boncompagni et al. 1999). In *Sinorhizobium meliloti*, choline is transported into the cell by the activity of several transport systems (Pocard et al. 1989), and glycine betaine is synthesized by the activities of choline oxidase and choline aldehyde dehydrogenase, while the specific enzymes involved in its degradation by successive demethylation to glycine and then to serine and pyruvate are reduced when the cells are subjected to inhibitory osmotic stress (Smith et al. 1988). This takes place analogously in *Staphylococcus aureus* where choline is transported into the cell and metabolized to glycine betaine that accumulates (Kaenjak et al. 1993). In Enterobacteriaceae, however, choline and glycine betaine serve only an osmoregulatory role, since they are unable to serve as sources of carbon or nitrogen (Le Redulier and Bouillard 1983).

2. A good deal of attention has been devoted to the salinity and low temperature tolerance of *Listeria monocytogenes* in the context of food-borne disease. The tolerance of *L. monocytogenes* to osmotic and low temperature stress is mitigated by glycine betaine (Ko et al. 1994), and since *L. monocytogenes* is unable to synthesize osmolytes, it is, therefore, dependent on uptake from the medium. Glycine betaine and carnitine transport systems are driven by ATP-dependent mechanisms, and three transport systems are known to operate in *L. monocytogenes*—BetL and Gbu for glycine betaine transport, and OpuC for carnitine (Angelidis and Smith 2003a) while they are also an adaptation to chill stress (Angelidis and Smith 2003b).

3. Carnitine is transformed in *Escherichia coli* by dehydration to crotonobetaine and then by reduction to γ-butyrobetaine (Eichler et al. 1994) which is consistent with the use of carnitine and crotonobetaine as electron acceptors during anaerobic growth of *Escherichia coli* (Walt and Kahn 2002). Carnitine functions as a compatible solute in *L. monocytogenes* (Angelidis and Smith 2003a,b), and the uptake of carnitine, crotonobetaine and γ-butyrobetaine by the ABC transport system in *Bacillus subtilis* takes place as a response to high osmolarity (Kappes and Bremer 1998).

4. Glutathione is able to protect *Lactobacillus lactis* against osmotic stress (Zhang et al. 2010), and proteomic analysis demonstrated a number of alterations, including upregulation of several glycolytic enzymes.

4.2.4 Acid Stress

Many bacteria normally grow at neutral pH, although adverse conditions of acidity may be encountered in their natural environment, while acid conditions may be produced by substrate fermentation. The role of acidophiles in mineral processing and the significance of moderately acidophilic Acidobacteria have been noted in Chapter 2. Some illustrative examples, primarily of clinical significance, are given in the following discussion.

4.2.5 Gram-Negative Bacteria

1. Enterobacteriaceae can adapt themselves to conditions of high acidity by the synthesis of amino acid decarboxylases that function as proton antiporters. Examples include lysine decarboxylase CadA and the antiporter CadB in *Salmonella typhimurium* (Park et al. 1996); glutamate decarboxylases GadA and GadB and the glutamate/γ-aminobutyrate antiporter GadC (Hersh et al. 1996); arginine decarboxylase AdiA and the arginine/agmatine antiporter AdiC (Iyer et al. 2003) that are able to protect *Escherichia coli* against conditions of high acidity. The last of these also is involved in the acid resistance of *Salmonella enterica* serovar Typhimurium under anaerobic conditions (Kieboom and Abee 2006). The challenge of low pH may be mitigated by decarboxylation of amino acids that may also result in the generation of a protomotive force, for example, tyrosine decarboxylation to tyramine in *Enterococcus faecium* (Pereira et al. 2009). Additional factors have been implicated in acid resistance in *Escherichia coli*: (a) CO_2 increased tolerance during adaptation at pH 5.5 to a greater extent than during challenge at pH 2.5 and is consistent with the significance of decarboxylation mechanisms (Sun et al. 2005) and (b) there is a requirement for metabolic ATP produced from inosine phosphate by PurA and PurB, or by Adk from AMP (Sun et al. 2011).

2. a. Considerable attention has been directed to the clinically important issue of acid tolerance of *Vibrio cholerae* that is mediated by decarboxylation, and it has been established that the genes encoding lysine decarboxylase *cadA*, and the lysine/cadaverine antiporter *cadB* exist in an operon (Merrell and Camilli 2000). The virulence cascade in *Vibrio cholerae* is determined by a number of signals that depend upon the transcription factors AphA and AphB. AphB is a LysR-type activator that initiates expression of the virulence cascade, and a number of genes are activated by AphB, including *cadA*, *cadB*, and *cah*, that also encodes carbonic anhydrase. It has been shown that AphB activates the expression of these genes in response to low pH and at neutral pH under anaerobic conditions (Kovacikova et al. 2010). AphA plays an additional role since it functions as a repressor of an operon involved in the production of acetoin which counteracts lethal acidification by directing pyruvate produced from glucose into neutral rather than acidic products (Kovacikova et al. 2005).

   b. The synthesis of lysine decarboxylase in *Vibrio vulnificus* is upregulated not only by acid stress but also by superoxide stress (Kim et al. 2006), while under conditions of extreme phosphate limitation, lysine decarboxylation offers protection of *Escherichia coli* against acid produced by fermentation (Moreau 2007).

3. Urease is a virulent factor of the human pathogen *Helicobacter pylori* that must be able to withstand fluctuating pH in its natural environment. This organism is able to withstand pH 1 in the presence of urea, and it has been proposed that the cytoplasmic pH was a factor in the activation of urease at low pH (Stingl et al. 2002). In addition, the periplasmic α-carbonic anhydrase is essential for acid acclimation (Marcus et al. 2005).

4. The acetic acid-producing *Acetobacter aceti* is resistant to high levels of ethanol that is the growth substrate and its product acetic acid. It has been shown that although this organism possesses a compete citric acid cycle, the production of succinate from succinyl-CoA by succinyl-CoA synthase is replaced by succinyl-CoA:acetate CoA transferase (Mullins et al. 2008).

4.2.6 Gram-Positive Bacteria

1. In Gram-positive bacteria, several mechanisms contribute to acid tolerance (Cotter and Hill 2003). Glutamate decarboxylase plays a key role since decarboxylation results in the consumption of a proton and the export of γ-aminobutyrate by an antiporter. It has been proposed that this would bring about ATP generation (Higuchi et al. 1997) analogous to the decarboxylation in *O. formigenes* (Berthold et al. 2005), where oxalate and formate form a one-to-one antiport system that involves the consumption of an internal proton during decarboxylation, and serves as an indirect proton pump to generate ATP by decarboxylative phosphorylation (Dimroth and Schink 1998). In *Lactococcus lactis*, the glutamate decarboxylase genes *gad* and *gad* were

maximally expressed at low pH and the onset of stationary phase, while in *Listeria monocyto-genes* the homologues *gad A* and *gad B* have been characterized in acid tolerance: *gad* is cotranscribed with *gad* which encodes a potential glutamate/γ-aminobutyrate antiporter and its expression is upregulated by mild acid conditions in contrast to *gad A* that is only weakly regulated (Cotter et al. 2001). The situation is, however, more complex, since it has been shown that the production of γ-aminobutyrate can be uncoupled to its efflux (Karat et al. 2010). An alternative protective mechanism in *Lactococcus lactis* involves glutathione (Zhang et al. 2007).

2. Mycobacteria are found both in clinical specimens and in soil samples, and they vary markedly in their tolerance of acidity.

   a. *Mycobacterium tuberculosis* resides in the phagocyte vacuole of host macrophages where the pH lies between 6.2 and 4.5 (Vandal et al. 2009), and its growth in a defined medium is highly restricted by pH, and increased levels of Mg^{2+} were required for growth at pH 6.0. Other mycobacteria, including *Mycob. fortuitum*, *Mycob. scrofulaceum*, and *Mycob. Chelonae*, were not similarly affected (Piddington et al. 2000). By contrast, *Mycob. smegmatis* and *Mycob. bovis* were able to grow at pH values of 4.5 and 5.0 respectively (Rao et al. 2001). At an external pH of 5.0, the ΔpH of *Mycob. smegmatis* was dissipated by the protonophore carbonyl cyanide *m*-chlorophenylhydrazone (CCCP), and by dicyclohexyl-carbodiimide (DCCD) that inhibits the proton translocating F_1F_0-ATPase. It was, therefore, concluded that the permeability of the cytoplasmic membrane to protons played a key role in maintaining the internal pH at near neutral.

   b. Enrichments with pyrene of soils from an acidic environment contaminated with PAHs from a gas manufacturing site were carried out, and growth of the enrichment cultures was optimal at pH ≤ 7 (Uyttebroek et al. 2007). The cultures were dominated by a single organism that was identified by DGGE fingerprinting and sequencing analysis of 16S rRNA as a slow-growing mycobacterium closely related to *Mycobacterium montefiorense* that has been characterized and is related to *Mycobacterium triplex* (Levi et al. 2003). On the other hand, analogous enrichment at pH 7 from a pH 7 soil at the same site yielded a culture dominated by *Pseudomonas putida* with a mycobacterium closely related to the fast-growing *Mycobacterium aurum*.

3. *Acid tolerance in Streptococci*: Streptococci are agents of dental caries that is mediated by the acidic conditions produced by carbohydrate fermentation. A spectrum of mechanisms has been proposed for their acid tolerance.

   a. Increase in pH by the fermentation of malate to lactate and CO_2 (Sheng and Marquis 2007), and by the production of ammonia, CO_2 and ATP from the metabolism of agmatine encoded by the *aguBDAC* operon (Griswold et al. 2004, 2006).

   b. Upregulation of proton-pump F_1-F_0 ATPase for removal of protons (Kuhnert et al. 2003, 2004).

   c. Alteration of the relative proportion of monounsaturated fatty acids $C_{18:1}$ and $C_{20:1}$ in the absence of de novo protein synthesis by the FabM hydratase/isomerase (Marrakchi et al. 2002) that is obligatory for acid tolerance (Fozo et al. 2004a,b).

   d. Upregulation of the genes for biosynthesis of branched-chain amino acids as a result of the diversion of pyruvate to lactate and branched-chain amino acids, and decreased formation of formic acid (Len et al. 2004).

   e. Upregulation of the amino acid biosynthetic genes *ilvC* and *ilvE* while loss of *ivlE* exhibited a defect in F_1-F_0 ATPase activity and a decrease in acid tolerance (Santiago et al. 2012).

4.2.7 Extremophiles

This is discussed in a wider context in Chapter 2 to which reference may be made. Methane oxidation has been described for several organisms of the phylum Verrucomicrobia and all of them were extremophiles. They exhibited some unusual features, including the mechanisms for degradation of methane.

a. An organism (V4) that grew optimally at pH 2.0 to 2.5 was isolated from a thermal area in New Zealand (Dunfield et al. 2007). It grew robustly only on C_1 substrates (methane and methanol), contained the genes required for the operation of the Benson–Calvin cycle and for those encoding particulate methane monooxygenase that belonged to a new group of these enzymes.

b. An organism (SolV) was isolated at pH 2.0 and 50°C from the Solfatara volcano mudpot (Pol et al. 2007), and the organism was assigned to *Acidimethylosilex fumarolicum*. Particulate methane monooxygenase activity was measured and sequencing of the DNA suggested that the organism used a combination of the serine, tetrahydrofolate, and ribulose-1,5-bisphosphate pathways for assimilation of carbon.

c. An extremophile that carried out oxidation at 55°C and pH 2 lacked intracytoplasmic membrane, and most remarkably, lacked genes for methane monooxygenase (Islam et al. 2008).

4.2.8 Tolerance of Oxygen

4.2.8.1 Oxygen in Anaerobic Bacteria

It is well established from laboratory practice that anaerobic microorganisms differ widely in their oxygen tolerance, ranging from the strictly anaerobic methanogens to the generally more tolerant clostridia.

There are, however, important exceptions: *Clostridium acetobutylicum* and *Cl. aminovalericum* can grow under microaerophilic conditions and consume oxygen, and the anaerobic *Geobacter sulfurreducens* is able to use oxygen as terminal electron acceptor and can grow at oxygen concentrations up to 10% (Lin et al. 2004). In addition, some anaerobes are able to use nitrate as electron acceptors. An obligately anaerobic organism uses nitrate as electron acceptor during the degradation of resorcinol (Gorny et al. 1992), and the normally strictly anaerobic sulfate-reducing organisms *Desulfobulbus propionicus* and *Desulfovibrio desulfuricans* may grow by reducing nitrate or nitrite to ammonia using hydrogen as electron donor (Seitz and Cypionka 1986). A range of mechanisms for oxygen tolerance has been found among anaerobes, and a few representative examples are provided as illustration:

1. Among bacteria considered strict anaerobes, cytochrome *bd*-oxidases have been found, and these enable the organisms to tolerate and reduce low levels of oxygen. Such cytochromes have been found in *Bacteroides fragilis* (Baughn and Malamy 2004), *Desulfovibrio gigas* (Lemos et al. 2001), and *Moorella thermoacetica* (Das et al. 2005).

2. a. *Desulfovibrio vulgaris* Hillsborough is capable of growth at oxygen concentrations of 0.24–0.48 µM (Johnson et al. 1997). Although some strains known to respond to oxidative stress remained unaltered, the upregulation of the predicted tetraheme cytochrome c_3 complex was observed in others (Mukhopadhyay et al. 2007).

   b. Considerable effort had been directed to the oxygen tolerance of *Desulfovibrio gigas* (Fareleira et al. 2003), and it has been proposed that in *Desulfovibrio vulgaris*, protection against oxygen is accomplished by using two redox systems (Lumppio et al. 2001), rubredoxin oxidoreductase (Rbo) to convert O_2^- to H_2O_2 followed by the activity of rubrerythrin (Rbr) to produce H_2O, and that these enzymes interact in the periplasm with redox enzymes and superoxide dismutase.

3. Several investigations have examined the specific role of NADH oxidase in protecting anaerobic bacteria from the deleterious effects of oxygen. Two different reactions have been found that differ in the products of their oxidase activity.

   a. In *Clostridium aminovalericum*, the purified NADH oxidase contained FAD as a cofactor and produced H_2O as the final product (Kawasaki et al. 2004). The gene that encoded NADH oxidase was upregulated in *Cl. aminovalericum* when microaerophilic conditions were attained, and a number of NAD(P)H-dependent enzyme activities were induced in *Cl. acetobutylicum* (Kawasaki et al. 2005).

   b. *Archaeoglobus fulgidus* is able to grow by sulfate reduction at the expense of a number of substrates, or lithoautotrophically with H_2, CO_2 and thiosulfate. There are several NADH oxidases, of which NoxA-1 and NoxB-1 are FAD-containing enzymes, and both of them produce H_2O rather than H_2O_2 (Kengen et al. 2003).

c. *Thermotoga maritima* is a hyperthermophilic bacterium that can utilize carbohydrates with the production of H_2, CO_2, acetate, and lactate. The NADH oxidase that was a flavo-protein has been purified and catalyzed the reduction of O_2 to H_2O_2 (Yang and Ma 2007).

d. Different species of *Bifidobacterium* display either oxygen sensitivity or microaerophilic growth. *Bifid. bifidum* is oxygen sensitive, and the NADH oxidase in this species that was purified contained both FAM and FAD, and catalyzed two different reactions: the forma-tion of H_2O_2 from O_2 and the dehydrogenation of dihydroorotate (Kawasaki et al. 2009).

4. A different situation is illustrated by enzymatic activities that are, on the one hand dependent on oxygen, but on the other contain enzymes that are deactivated by oxygen. Classic examples are provided by nitrogen fixation in aerobes such as *Azotobacter*, facultative anaerobes such as *Klebsiella* and oxygenic cyanobacteria. The chemolithotrophs *Ralstonia eutropha* and *R. metallidurans* obtain energy by oxidizing H_2 using O_2 as the terminal acceptor. Although hydrogenases are generally inhibited by O_2, in this organism, levels of inhibitory O_2 were sub-stantially greater than those widely encountered in anaerobes (Ludwig et al. 2009). In addition, it was shown by FTIR and EPR that states of the [NiFe] hydrogenase were analogous to those of typical anaerobic hydrogenases—with the important exception of the absence of the oxygen-activated Ni_u-A state (Saggu et al. 2009).

4.2.8.2 Oxygen in Aerobic and Facultatively Anaerobic Organisms

1. Facultatively anaerobic organisms can employ either fermentative or oxidative modes for the metabolism of appropriate substrates such as carbohydrates, while others can use alternative electron acceptors such as nitrate, chlorate, or tetrathionate in the absence of oxygen. Such organisms have, therefore, two metabolic options, even though they may often be mutually exclusive. An illustrative example of the metabolic flexibility of facultatively anaerobic organ-isms is provided by the type strains of all the species of the enteric genus *Citrobacter*. These organisms were able to degrade a number of amino acids, including glutamate, using either respiratory or fermentative metabolism, and they could rapidly switch between these alterna-tives (Gerritse and Gottschal 1993b). Under anaerobic conditions, the initial steps of glutamate degradation involved the formation of 3-methylaspartate, mesaconate, and citramalate, which are typical of clostridial fermentations. Although the oxygen concentration in natural environ-ments may be highly variable, organisms with such a high degree of metabolic flexibility may reasonably be presumed to be at an advantage.

2. Oxygen toxicity under aerobic conditions may result from the synthesis of toxic compounds, including superoxide, hydrogen peroxide, or hydroxyl radicals. Further details have emerged from observations that the Fe isoenzyme of superoxidase dismutase is synthesized by *Escherichia coli* under anaerobic conditions, in contrast to the Mn isoenzyme that is produced only under aerobic conditions. The synthesis of the former facilitates the transition from anaer-obic to aerobic conditions by destroying the superoxide radical generated after exposure to oxygen (Kargalioglu and Imlay 1994). This flexibility is important in environments where fluc-tuating oxygen concentrations prevail. Studies with mutants of *Pseudomonas aeruginosa* in which the synthesis of Mn superoxide dismutase and Fe superoxide dismutase were impaired showed that growth in a complex medium or in a defined glucose medium was adversely affected by the latter but only insignificantly by the former (Hassett et al. 1995). Concentrations of Fe must be regulated, since excess of Fe within the cell would result in the production of deleterious hydroxyl radicals from molecular oxygen (Touati et al. 1995). The transport of Fe into the cell is regulated by Fe(III) high-affinity siderophores which are synthesized within the cell and excreted into the medium before transport of the complexes into the cell by specific membrane transporters (Wandersman and Delepelaire 2004). In this way, homeostasis is achieved.

3. There is substantial evidence that organisms that are strictly dependent on the aerobic metabolism of substrates for growth and replication may nonetheless accomplish biodegradations and biotransformations under conditions of low oxygen concentration. Indeed, such conditions may inadvertently prevail in laboratory experiments using dense cell suspensions; it is important to appreciate that the growth of these aerobic organisms may be limited by the availability of oxygen. A number of examples are given below to illustrate the environmental role of putatively aerobic organisms under conditions of low oxygen concentration:

a. The rate of biodegradation of hexadecane in a marine enrichment culture was unaffected until oxygen concentrations were lower than 1% saturation (Michaelsen et al. 1992).

b. It has been shown that the degradation of pyrene by a strain of *Mycobacterium* sp. can take place at low oxygen concentrations (Fritzsche 1994).

c. *Alcaligenes* sp. strain L6 that was obtained after enrichment with 3-chlorobenzoate in an atmosphere of only 2% oxygen in the gas phase possessed both 2,5- and 3,4-dihydroxybenzote dioxygenase activities although it lacked catechol dioxygenase activity that would be required for fission of the aromatic ring (Krooneman et al. 1996). The affinity for the substrate of cells grown under oxygen limitation was three times greater than was observed under excess oxygen, and exceeded values for other bacteria growing with benzoate or 2,5-dichlorobenzoate. This strain that metabolized 3-chlorobenzoate by the gentisate pathway was, under low oxygen concentrations, able to compete successfully with a *Pseudomonas* sp. strain A that used the catechol pathway (Krooneman et al. 1998).

d. Further details emerged from the results of experiments that compared the kinetic properties of catechol-2,3-dioxygenases from toluene-degrading pseudomonads isolated from aquifer sands or groundwater (strains W 31 and CFS 215) with other strains (*P. putida* F1). This revealed a number of significant differences that may plausibly be associated with the low oxygen concentrations pertaining in the environment from which the strains were isolated. For strains that degraded toluene via *cis*-toluene-2,3-dihydrodiol and 3-methylcatechol, lower values of K_m were observed for O_2 and 3-methylcatechol, whereas higher values of V_{max} for O_2 and 3-methylcatechol were found (Kukor and Olsen 1996) (Table 4.1).

e. The growth of *Pseudomonas putida* strains KT2442 and mt-2 on aromatic carboxylates decreased with O_2 concentrations <10 μM, and under these conditions, strain KT2442 excreted catechol during growth with benzoate, or 3,4-dihydroxybenzoate from 4-hydroxybenzoate (Arras et al. 1998). In these experiments, the rate of oxygenation rather than the pO_2 was the limiting factor, since the K_m for catechol 1,2-dioxygenase was 20 μM O_2.

f. The synthesis of cytochrome P450 in yeasts that assimilate long-chain alkanes has been examined in facultatively anaerobic species of *Candida*. In *Candida guilliermondii*, levels of cytochrome P450 that were induced during growth with hexadecane—though not with glucose—increased at dissolved O_2 concentrations <20% (Mauersberger et al. 1980). For cells of *Candida tropicalis* grown with hexadecane, levels of cytochrome P450 increased when the oxygen concentration was <2 kPa (14% dissolved O_2). The levels of alcohol dehydrogenase, aldehyde dehydrogenase, catalase, and cytochrome *c* were not, however, influenced (Gmünder et al. 1981).

TABLE 4.1

Values of K_m and V_{max} for O_2 and 3-Methylcatechol in Strains F1, W31, and CFS 215

	K_m Substrate		V_{max} Substrate	
Strain	**O_2**	**3-Methylcatechol**	**O_2**	**3-Methylcatechol**
F1	9.7	16.9	4.3	17
W31	2.0	0.3	293	125
CFS 215	0.9	0.5	400	180

4. Issues of particular relevance in the context of bioremediation are illustrated by the following.

   a. The increased rate of cell death in an established naphthalene-degrading *Pseudomonas putida* G7 brought about by the substrate (naphthalene) under conditions of oxygen (or combined oxygen and nitrogen) limitation (Ahn et al. 1998).

   b. The effect of oxygen concentration on the degradation of toluene has been examined in *Pseudomonas putida* mt-2 (Martínez-Lavanchy et al. 2010). Expression of the catabolic gene *xylM* in the upper degradation pathway and *xylE* in the lower pathway decreased during oxygen depletion but returned to normal after a period of anoxia.

5. The growth of *Bacillus subtilis* may take place under a variety of conditions: (a) aerobic conditions, (b) using nitrate as electron acceptor, and (c) fermentative conditions with glucose provided pyruvate is available as an electron acceptor since the organism lacks pyruvate formate hydrogen lyase (Nakano and Zuber 1998).

6. Facultative strains may be able to regulate the degradation pathway to the availability of oxygen. An illustration of this is provided by *Thauera* sp. strain DNT-1 that is able to degrade toluene under aerobic conditions mediated by a dioxygenase, and under denitrifying conditions in the absence of oxygen by the anaerobic benzylsuccinate pathway (Shinoda et al. 2004). Whereas the *tod* genes were induced under aerobic conditions, the *bss* genes were induced under both aerobic and anaerobic conditions.

It should also be appreciated that the oxygen concentration may determine the outcome of a reaction, or even the possibility of either oxidative or reductive reactions:

1. The degradation of pentachlorophenol by *Mycobacterium chlorophenolicum* (*Rhodococcus chlorophenolicus*) proceeds by the initial formation of tetrachlorohydroquinone. Whereas its formation is oxygen dependent, its subsequent degradation can be accomplished in the absence of oxygen (Apajalahti and Salkinoja-Salonen 1987).

2. *Pseudomonas putida* G786 containing the CAM plasmid carried our reductive dehalogenation of a number of haloalkanes (Hur et al. 1994) and when a plasmid containing genes encoding toluene dioxygenase was incorporated, both reduction and oxidation were observed (Wackett et al. 1994).

3. Biotransformation as opposed to biodegradation may, in fact, be favored by limited oxygen concentration. A good example is provided by the synthesis of $7\alpha,12\beta$-dihydroxy-1,4-androstadien-3,17-dione from cholic acid by a strain of *Pseudomonas* sp. (Smith and Park 1984), in which oxygen limitation restricts the rate of C_9 α-hydroxylation that initiates degradation.

4. Oxidation of the trichloromethyl group in 2-chloro-6-trichloromethylpyridine to the corresponding carboxylic acid by *Nitrosomonas europaea* occurs at high oxygen concentrations during cooxidation of ammonia or hydrazine (Figure 4.1a). In contrast, at low oxygen concentrations in the presence of hydrazine, reductive dechlorination to 2-chloro-6-dichloromethyl-pyridine occurs (Vannelli and Hooper 1992) (Figure 4.1b).

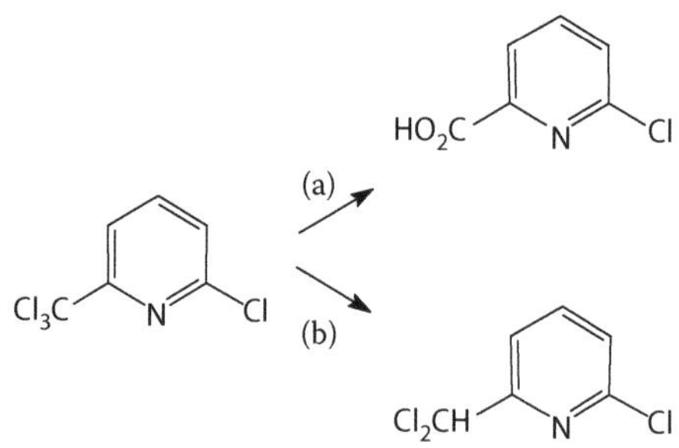

FIGURE 4.1 Biotransformation of the trichloromethyl group of 2-chloro-5-trichloromethyl pyridine by (a) oxidation, (b) dechlorination.

4.2.9 Redox Potential

For strictly anaerobic bacteria, it is usually not sufficient that the medium is anoxic. It must also be poised at the correct redox potential. It is for this reason that prereduced media are used. This is accomplished by the addition of, for example, sulfide, dithionite, or titanium(III) citrate, and the media generally contain a redox indicator such as resazurin. Examples have been given earlier of reactions carried out by aerobic organisms at low oxygen concentrations, and the outcome of reactions carried out by such organisms may be influenced by the redox potential. For example, cells of the aerobic *Pseudomonas cepacia* carried out the degradation of tetrachloromethane only when a negative potential was maintained, and the maximum rate occurred at a potential of ca.–150 mV (Jin and Englande 1997).

4.3 References

Aguilar PS, JE Cronan, D de Mendoza, 1998. A *Bacillus subtilis* gene induced by cold chock encodes a membrane phospholipid desaturase. *J Bacteriol* 180: 2194–2200.

Ahn I-S, WC Ghiorse, LW Lion, ML Shuler, 1998. Growth kinetics of *Pseudomonas putida* G7 on naphthalene and toxicity during nutrient deprivation. *Biotechnol Bioeng* 59: 587–594.

Aislabie JM, MR Balks, JM Foght, EJ Waterhouse, 2004. Hydrocarbon spills on Antarctic soils: Effects and management. *Environ Sci Technol* 38: 1265–1274.

Altabe SG, P Aguilar, GM Caballero, D de Mendoza, 2003. The *Bacillus subtilis* lipid desaturase is a $\Delta 5$ desaturase. *J Bacteriol* 185: 3228–3231.

Angelidis AS, GM Smith, 2003a. Role of glycine betaine and carnitine transporters in adaptation of *Listeria monocytogenes* to chill stress in defined medium. *Appl Environ Microbiol* 69: 7492–7498.

Angelidis AS, GM Smith, 2003b. Three transporters mediate uptake of glycine betaine and carnitine by *Listeria monocytogenes* in response to hyperosmotic stress. *Appl Environ Microbiol* 69: 1013–1022.

Apajalahti JH, MS Salkinoja-Salonen, 1987. Dechlorination and *para*-hydroxylation of polychlorinated phenols by *Rhodococcus chlorophenolicus*. *J Bacteriol* 169: 675–681.

Arras T, J Schirawski, G Unden, 1998. Availability of O_2 as a substrate in the cytoplasm of bacteria under aerobic and microaerobic conditions. *J Bacteriol* 180: 2133–2136.

Bakermans C, HL Ayala-del-Rio, MA Pander, T Vishnovbetskaya, D Gilichinsky, MF Thomashow, JM Tiedje, 2006. *Psychrobacter cryohalolentis* sp. nov., and *Psychrobacter arcticus* sp. nov., isolated from Siberian permafrost. *Int J Syst Evol Microbiol* 56: 1285–1291.

Baughn AD, MN Malamy, 2004. The strict anaerobe *Bacteroides fragilis* grown in and benefits from nanomolar concentrations of oxygen. *Nature* 427: 441–444.

Bergholz PW, C Bakermans, JM Tiedje, 2009. *Psychrobacter arcticus* 273-4 uses resource efficiency and molecular motion adaptations for subzero temperature growth. *J Bacteriol* 191: 2340–2352.

Beranová J, MC Mansilla, D de Mendoza, D Elhottová, I Konopásek, 2010. Differences in cold adaptation of *Bacillus subtilis* under anaerobic and aerobic conditions. *J Bacteriol* 192: 4164–4171.

Berthold CL, P Moussatche, NGJ Richards, Y Lindqvist, 2005. Structural basis for activation of the thiamin diphosphate-dependent enzyme oxalyl-CoA decarboxylase by adenosine diphosphate. *J Biol Chem* 280: 41645–41654.

Boncompagni E, M Østerås, M-C Pogg, Dl Rudulier, 1999. Occurrence of choline and glycine betaine uptake and metabolism in the family *Rhizobiaceae* and their roles in osmoprotection. *Appl Environ Microbiol* 65: 2072–2077.

Borges MN, JD Marugg, N Empadinhasm MS da Costa, H Santos, 2004. Specialized roles of the two pathways for the synthesis of mannosylglycerate in osmoadaptation and thermoadaptation of *Rhodothermus marinus*. *J Biol Chem* 279: 9892–9898.

Bowman JP, JJ Gosink, SA McCammon, TE Lewis, DS Nichols, PD Nichols, JH Skerratt, JT Staley, TA McMeekin, 1998. *Colwellia demingae* sp. nov., *Colwellia hornerae* sp. nov., *Colwellia rossensis*, sp nov. and *Colwellia psychrotropica* sp. nov.: Psychrophilic Antarctic species with the ability to synthesize docosahexaenoic acid (22:6ω3). *Int J Syst Bacteriol* 48: 1171–1180.

Bradley PM, FH Chapelle, 1995. Rapid toluene mineralization by aquifer microorganisms at Adak, Alaska: Implications for intrinsic bioremediation in cold environments. *Environ Sci Technol* 29: 2778–2781.

Brillard J, I Jéhanno, C Dargaignaratz, I Barbosa, C Ginies, F Carlin, S Fedhila, C Nguyen-the, V Broussolla, V Sanchis, 2010. Identification of *Bacillus cereus* genes specifically expressed during growth at low temperatures. *Appl Environ Microbiol* 76: 2562–2473.

Bursy J, AJ Pierik, E Bremer, 2007. Osmotically induced synthesis of the compatible solute hydroxyectoine is mediated by an evolutionary conserved ectoine hydroxylase. *J Biol Chem* 282: 3147–31155.

Bursy J, AU Kuhlmann, M Pittelkow, H Hartmann, M Jebberm AJ Pierik, E Bremer, 2008. Synthesis and uptake of the compatible solutes ectoine and 5-hydroxyectoine by *Streptomyces coelicolor* A3(2) in response to salt and heat stresses. *Appl Environ Microbiol* 74: 7286–7296.

Chi X, Q Yang, F Zhao, S Qin, Y Yang, J Shen, H Lin, 2008. Comparative analysis of fatty acid desaturases in cyanobacterial genomes. *Comp Funct Genomics* doi: 10.1155/2008/284508.

Cotter PD, C Hill, 2003. Surviving the acid test: Responses of Gram-positive bacteria to low pH. *Microbiol Mol Biol Revs* 67: 429–453.

Cotter PD, CG Gahan, C Hill, 2001. A glutamate decarboxylase system protects *Listeria monocytogenes* in gastric fluid. *Mol Microbiol* 40: 465–475.

D'Souza-Ault MR, LT Smith, GM Smith, 1993. Roles of *N*-acetylglutaminylglutamine amide and glycine in adaptation of *Pseudomonas aeruginosa* to osmotic stress. *Appl Environ Microbiol* 59: 473–478.

Das A, R Silaghi-Dumitrescu, LG Ljungdahl, DM Kurtz, 2005. Cytochrome *bd* oxidase, oxidative stress, and dioxygen tolerance of the strictly anaerobic bacterium *Moorella thermoacetica*. *J Bacteriol* 187: 2020–2029.

Dedysh SN, C Knief, PF Dunfield, 2005. *Methylocella* species are facultatively methanotrophic. *J Bacteriol* 187: 4665–4767.

Dimroth P, B Schink. 1998. Energy conservation in the decarboxylation of dicarboxylic acids by fermenting bacteria. *Arch Microbiol* 170: 69–77.

Dunfield PF and 18 coauthors, 2007. Methane oxidation by an extremely acidophilic bacterium of the phylum Verrumicrobia. *Nature* 450: 879–882.

Eichler K, F Bourgis, A Buchet, H-P Kleber, M-A Mandrand-Berthelot, 1994. Molecular characterization of the *cai* operon necessary for carnitine metabolism in *Escherichia coli*. *Mol Microbiol* 13: 775–786.

Fareleira P, BS Santos, C António, P Moradas-Ferreira, J LeGall, AV Xavier, H Santos, 2003. Response of a strict anaerobe to oxygen: Survival strategies in *Desulfovibrio gigas*. *Microbiology (UK*. 149: 1513–1522.

Fozo EM, RG Quivey, 2004a. Shifts in the membrane fatty acid profile of *Streptococcus mutans* enhance survival in acidic environments. *Appl Environ Microbiol* 70: 929–936.

Fozo EM, RG Quivey, 2004b. The *fabM* gene product of *Streptococcus mutans* is responsible for the synthesis of mono unsaturated fatty acids and is necessary for survival at low pH. *J Bacteriol* 186: 4152–4158.

Fritzsche C, 1994. Degradation of pyrene at low defined oxygen concentrations by *Mycobacterium* sp. *Appl Environ Microbiol* 60: 1687–1689.

García-Estepa R, M Argandoña, M Reina-Bueno, N Capote, F Iglesias-Guerra, JJ Nieto, C Vargas, 2006. The ectD gene which is involved in the synthesis of the compatible solute hydroxyectoine, is essential for thermoprotection of the halophilic bacterium *Chromohalobacter salexigens*. *J Bacteriol* 188: 3774–3784.

Gerday C, N Glansdorff (Eds.), 2007. *Physiology and Biochemistry of Extremophiles*. ASM Press, Washington, D.C.

Gerritse J, JC Gottschal, 1993b. Oxic and anoxic growth of a new *Citrobacter* species on amino acids. *Arch Microbiol* 160: 51–61.

Griswold AR, YY Chen, RA Burne, 2004. Analysis of an agmatine deiminase gene cluster in *Streptococcus mutans* UA159. *J Bacteriol* 186: 1902–1904.

Griswold AR, M Lee-Jameson, RA Burne, 2006. Regulation and physiological significance of the agmatine deiminase system of *Streptoccus mutans*. *J Bacteriol* 188: 834–841.

Gmünder FK, O Käppeli, A Fiechter, 1981. Chemostat studies on the hexadecane assimilation by the yeast *Candida tropicalis*. *Eur J Appl Microbiol Biotechnol* 12: 135–142.

Gorny N, G Wahl, A Brune, B Schink, 1992. A strictly anaerobic nitrate-reducing organism growing with resorcinol and other aromatic compounds. *Arch Microbiol* 158: 48–53.

Hassett DJ, HP Schweitzer, DE Ohman, 1995. *Pseudomonas aeruginosa sodA* and *sodB* mutants defective in manganese- and iron-cofactored superoxide dismutase activity demonstrate the importance of the iron-cofactored form in aerobic metabolism. *J Bacteriol* 177: 6330–6337.

Hersh BM, FT Farooq, DN Barstad, DL Blankenhorn, JL Slonczewski, 1996. A glutamate-dependent acid resistance gene in *Escherichia coli*. *J Bacteriol* 178: 3978–3981.

Higuchi T, H Hayashi, K Abe, 1997. Exchange of glutamate and γ-aminobutyrate in a *Lactobacillus* strain. *J Bacteriol* 179: 3362–3364.

Hoffmann T, E Bremer, 2011. Protection of *Bacillus subtilis* against cold stress via compatible-solute acquisition. *J Bacteriol* 193: 1552–1562.

Hur H-G, M Sadowsky, LP Wackett, 1994. Metabolism of chlorofluorocarbons and polybrominated compounds by *Pseudomonas putida* G786phG-2 via an engineered metabolic pathway. *Appl Environ Microbiol* 60: 4148–4154.

Islam T, S Jensen, LJ Reigstas, Ø Larsen, N-K Birekeland, 2008. Methane oxidation at 55°C and pH 2 by a thermoacidophilic bacterium belonging to the Verrucomicrobia phylum. *Proc Natl Acad Sci USA* 105: 300–304.

Iyer R, C Williams, C Miller, 2003. Arginine-agmatine antiporter in extreme acid resistance in *Escherichia coli*. *J Bacteriol* 185: 6556–6561.

Järvinen KT, Melin, ES, KA Puhakka, 1994. High-rate bioremediation of chlorophenol-contaminated groundwater at low temperatures. *Environ Sci Technol* 28: 2387–2392.

Jin G, AJ Englande, 1997. Biodegradation kinetics of carbon tetrachloride by *Pseudomonas cepacia* under varying oxidation-reduction potential conditions. *Water Environ Res* 69: 1094–1099.

Johnson MS, IB Zhulin, M-E R Gapuzan, BL Taylor, 1997. Oxygen-dependent growth of the obligate anaerobe *Desulfovibrio vulgaris* Hildenborough. *J Bacteriol* 179: 5598–5601.

Kaenjak A, JE Graham, BJ Wilkinson, 1993. Choline transport activity in *Staphylococcus aureus* induced by osmotic stress and low phosphate concentrations. *J Bacteriol* 175: 2400–2406.

Kappes RM, B Kekpf, E Bremer, 1996. Three transport systems for the osmoprotectant glycine betaine in *Bacillus subtilis*: Characterization of OpuD. *J Bacteriol* 178: 5071–5079.

Kappes RM, E Bremer, 1998. Response of *Bacillus subtilis* to high osmolarity: Uptake of carnitine, cronobetaine, and γ-butyrobetaine via the ABC transport system OpuC. *Microbiology (UK)* 144: 83–90.

Karatzas K-A, O Brennan, S Heavin, J Mossissey, CP O'Byrne, 2010. Intracellular accumulation of high levels of γ-aminobutyrate by *Listeria monocytogenes* 10403S in response to low pH: Uncoupling of γ-aminobutyrate synthesis from efflux in a chemically defined medium. *Appl Environ Microbiol* 76: 3529–3537.

Kargalioglu Y, JA Imlay, 1994. Importance of anaerobic superoxide dismutase synthesis in facilitating outgrowth of *Escherichia coli* upon entry into an aerobic habitat. *J Bacteriol* 176: 7653–7658.

Kawasaki S, J Ishikura, D Shiba, T Nishino, Y Niimura, 2004. Purification and characterization of an H_2O-forming oxidase from *Clostridium aminovalericum*: Existence of an oxygen-detoxifying enzyme in an obligate anaerobic bacterium. *Arch Microbiol* 181: 324–330.

Kawasaki S, T Satoh, M Todoroki, Y Niimura, 2009. *b*-Type dihydroorotate dehydrogenase is purified as an H_2O_2-forming NADH oxidase from *Bifidobacterium bifidum*. *Appl Environ Microbiol* 75: 629–636.

Kawasaki S, Y Watamura, M Ono, T Watanabe, K Takeda, Y Niimura, 2005. Adaptive responses to oxygen stress in obligatory anaerobes *Clostridium acetobutylicum* and *Clostridium aminovalericum*. *Appl Environ Microbiol* 71: 8442–8450.

Kengen SWM, J van der Oost, WM de Vos, 2003. Molecular characterization of H_2O_2-forming NADH oxidases from *Archaeoglobus fulgidus*. *Eur J Biochem* 270: 2885–2894.

Kieboom J, T Abee, 2006. Arginine-dependent acid resistance in *Salmonella enterica* serovar Typhimurium. *J Bacteriol* 188: 5650–5653.

Kim J-S, SH Choi, JK Lee, 2006. Lysine decarboxylase expression by *Vibrio vulnificus* is induced by SoxR in response to superoxide stress. *J Bacteriol* 188: 8586–8592.

Klein W, MHW Weber, MA Marahiel, 1999. Cold shock response of *Bacillus subtilis*: Isoleucine-dependent switch in the fatty acid branching pattern for membrane adaptation to low temperature. *J Bacteriol* 181: 5341–5349.

Knoblauch C, K Sahm, BB Jörgesen, 1999. Psychrophilic sulfate-reducing bacteria isolated from permanently cold arctic marine sediments: Description *of Desulfofrigus oceanense* gen. nov., sp. nov., *Desulfofrigus fragile* sp. nov., *Desulfofaba gelida* gen. nov., sp. nov., *Desulfotalea psychrophila* gen. nov., sp. nov., and *Desulfotalea arctica* sp. nov. *Int J Syst Bacteriol* 49: 1631–1643.

Ko R, LT Smith, GM Smith, 1994. Glycine betaine confers enhanced osmotolerance and cryotolerance on *Listeria monocytogenes*. *J Bacteriol* 176: 426–431.

Kovacikova G, W Lin, K Skorupski, 2005. Dual regulation of genes involved in acetoin biosynthesis and motility/biofilm formation by the virulence activator AphA and the acetate-responsive LysR-type regulator AlsR in *Vibrio cholerae*. *Mol Microbiol* 57: 420–433.

Kovacikova G, W Lin, K Skorupski, 2010. The LysR-type virulence activator AphB regulates the expression of genes in *Vibrio cholerae* in response to low pH and anaerobiosis. *J Bacteriol* 192: 4181–4191.

Krooneman J, ERB Moore, JCL van Velzen, RA Prins, LJ Forney, JC Gottschal, 1998. Competition for oxygen and 3-chlorobenzoate between two aerobic bacteria using different degradation pathways. *FEMS Microbiol Ecol* 26: 171–179.

Krooneman J, EBA Wieringa, ERB Moore, J Gerritse, RA Prins, JC Gottschal, 1996. Isolation of *Alcaligenes* sp. strain L6 at low oxygen concentrations and degradation of 3-chlorobenzoate via a pathway not involving (chloro)catechols. *Appl Environ Microbiol* 62: 2427–2434.

Kuhlmann AU, J Bursy, S Gimpel, T Hoffmann, E Bremer, 2008. Synthesis of the compatible solute ectoine in *Virgibacillus pantothenticus* is triggered by high salinity and low growth temperature. *Appl Environ Microbiol* 74: 4560–5463.

Kuhnert WL, RG Quivey, 2003. Genetic and biochemical characterization of the F-.ATPase operon from *Streptococcus sanguis*. *J Bacteriol* 185: 1525–1533.

Kuhnert WL, G Zheng, RC Faustoferri, RG Quivey, 2004. The F-ATPase operon promoter of *Streptococcus mutans* is transcriptionally regulated in response to external pH. *J Bacteriol* 186: 8524–8528.

Kukor JJ, RH Olsen, 1996. Catechol 2,3-dioxygenases functional in oxygen-limited hypoxic environments. *Appl Environ Microbiol* 62: 1728–1740.

Lamosa P, A Burke, R Peist, R Huber, M-Y Liu, G Silva, C Rodrigues-Pousada, J LeGall, C Maycock, H Santos, 2000. Thermostabilization of proteins by diglycerol phosphate, a new compatible solute from the hyperthermophile *Archaeoglobus fulgidus*. *Appl Environ Microbiol* 66: 1974–1979.

Len ACL, DWS Harty, NA Jacques. 2004. Proteome analysis of *Streptococcus mutans* metabolic phenotype during acid tolerance. *Microbiology UK* 150: 1363–1366.

Le Rudulier D, L Bouillard, 1983. Glycine betaine, an osmotic effector in *Klebsiella pneumoniae* and other members of the Enterobacteriaceae. *App Environ Microbiol* 46: 152–159.

Lemos RS, CM Gomes, M Santana, J LeGall, AV Xavier, M Teixeira, 2001. The "strict anaerobe" *Desulfovibrio gigas* contains a membrane-bound oxygen-reducing respiratory chain. *FEBS Lett* 496: 40–43.

Levi MH, J Bartell, L Gandolfo, SC Mole, SF Costa, LM Weiss, JK Johnson, G Osterhout, LH Herbst, 2003. Characterization of *Mycobacterium montefiorense* sp. nov., a novel pathogenic *Mycobacterium* from moray eels that is related to *Mycobacterium triplex*. *J Clin Microbiol* 41: 2147–2152.

Lin WC, MV Coppi, DR Lovley, 2004. *Geobacter sulfurreducens* can grow with oxygen as terminal electron acceptor. *Appl Environ Microbiol* 70: 2525–2528.

Ludwig M, JA Cracknell, KA Vincent, FA Armstrong, O Lenz, 2009. Oxygen-tolerant H_2 oxidation by membrane-bound [NiFe] hydrogenases of *Ralstonia* species. Coping with low levels of H_2 in air. *J Biol Chem* 284: 465–477.

Lumppio HL, NV Shenvi, AO Summers, G Voordouw, DM Kurtz, 2001. Rubrerythrin and rubredoxin oxidoreductases in *Desulfovibrio vulgaris*: A novel oxidative stress protection system. *J Bacteriol* 183: 101–108.

Marcus EA, AP Moshfegh, G Sachs, DR Scott, 2005. The periplasmic α-carbonic anhydrase activity of *Helicobacter pylori* is essential for acid acclimation. *J Bacteriol* 187: 729–738.

Margesin R, D Labbé, F Schinner, CW Greer, LG Whyte, 2003. Characterization of hydrocarbon-degrading microbial populations in contaminated and pristine alpine soils. *Appl Environ Microbiol* 69: 3085–3092.

Marrakchi H, K-H Choi, CO Rock, 2002. A new mechanism for anaerobic unsaturated fatty acid formation in *Streptococcus pneumoniae*. *J Biol Chem* 277: 44809–44816.

Martínez-Lavanchy PM, C Müller, I Nijenhuis, U Kappelmeyer, M Buffing, K McPherson, HJ Heipieper, 2010. High stability and fast recovery of expression of the TOL plasmid-carried toluene catabolism genes of *Pseudomonas putida* mt-2 under conditions of oxygen limitation and oscillation. *Appl Environ Microbiol* 76: 6715–6723.

Master ER, NYR Agar, L Gómez-Gil, JB Powlowski, WW Mohn, LD Eltis, 2008. Biphenyl dioxygenase from an arctic isolate is not cold adapted. *Appl Environ Microbiol* 74: 3908–3911.

Master EM, WW Mohn, 1998. Psychrotolerant bacteria isolated from arctic soil that degrade polychlorinated biphenyls at low temperatures. *Appl Environ Microbiol* 64: 4823–4829.

Mauersberger S, RN Matyashova, H-G Müller, AB Losinov, 1980. Influence of the growth substrate and the oxygen concentration in the medium on the cytochrome P-450 content in *Candida guilliermondii*. *Eur J Appl Microbiol Biotechnol* 9: 285–294.

Merrell DS, A Camilli, 2000. Regulation of *Vibrio cholerae* genes required for acid tolerance by a member of the "ToxR-like" family of transcriptional regulators. *J Bacteriol* 182: 5342–5350.

Michaelsen M, R Hulsch, T Höpner, L Berthe-Corti, 1992. Hexadecane mineralization in oxygen-controlled sediment-seawater cultivations with autochthonous microorganisms. *Appl Environ Microbiol* 58: 3072–3077.

Miura-Fraboni J, S Englard, 1983. Quantitative aspects of γ-butyrobetaine and D- and L-carnitine utilization by growing cell cultures of *Acinetobacter calcoaceticus* and *Pseudomonas putida*. *FEMS Microbiol Lett* 18: 113–116.

Moreau P, 2007. The lysine decarboxylase CadA protects *Escherichia coli* starved of phosphate against fermentation acids. *J Bacteriol* 189: 2249–2261.

Mukhopadhyay A et al., 2007. Cell-wide responses to low-oxygen exposure in *Desulfovibrio vulgaris* Hildenborough. *J Bacteriol* 189: 5996–6010.

Mullins EA, JA Francois, TJ Kappock, 2008. A specialized citric acid cycle requiring succinyl-coenzyme A (CoA):acetate CoA-transferase (AarC) confers acetic acid resistance on the acidophile *Acetobacter aceti*. *J Bacteriol* 190: 4933–4940.

Murata N, H Wada, 1995. Acyl-lipid desaturases and their importance in the tolerance and acclimitization to cold of cyanobacteria. *Biochem J* 308: 1–8.

Nakano NM, P Zuber, 1998. Anaerobic growth of a "strict aerobe" (*Bacillus subtilis*). *Annu Rev Microbiol* 52: 165–190.

Nichols DS, MR Miller, NW Davies, A Goodchild, M Raferty, R Cavicchioli, 2004. Cold adaptation in the Antarctic archaeon *Methanococcoides burtonii* involves membrane lipid unsaturation. *J Bacteriol* 186: 8508–8515.

Park Y-K, B Bearson, SH Bang, LS Bang, JW Foster, 1996. Internal pH crisis, lysine decarboxylase and the acid tolerance response of *Salmonella typhimurium*. *Mol Microbiol* 20: 605–611.

Pereira CI, D Matos, MV San Romão, MT Narreto-Crespo, 2009. Dual role for the tyrosine decarboxylation pathway in *Enterococcus faecium* E17: Response to an acid challenge and generation of a proton motive force. *Appl Environ Microbiol* 75: 345–352.

Pflughoeft KJ, K Kierek, PO Watnick, 2003. Role of ectoine in *Vibrio cholerae* osmoadaptation. *Appl Environ Microbiol* 69: 5959–5927.

Piddington, DL, A Kashkouli, NA Buchmeier, 2000. Growth of *Mycobacterium tuberculosis* in a defined medium is very restricted by acid pH and Mg^{2+} levels. *Infect Immun* 68: 4518–4522.

Pocard J-A, T Bernard, LT Smith, D Le Rudulier, 1989. Characterization of three choline transport activities in *Rhizobium meliloti*. *J Bacteriol* 171: 531–537.

Pol A, K Heijmans, HR Harhangi, D Tedesco, MSM Jetten, HJM Op den Camp, 2007. Methanotrophy below pH 1 by a new Verrucomicrobia species. *Nature* 450: 874–878.

Ponder MA, SK Gilmour, PW Bergholz, CA Mindock, R Hollingsworth, MF Thomashow, JM Tiedje, 2005. Characterization of potential stress responses in ancient Siberian permafrost psychroactive bacteria. *FEMS Microbiol Ecol* 53: 103–115.

Purvis JE, LP Yomano, LO Ingram, 2005. Enhanced trehalose production improves growth of *Escherichia coli* under osmotic stress. *Appl Environ Microbiol* 71: 3761–3769.

Rao M, TL Streur, FE Aldwell, GM Cook, 2001. Intracellular pH regulation by *Mycobacterium smegmatis* and *Mycobacterium bovis* BCG. *Microbiology (UK)* 147: 1017–1024.

Ruan Z-S, V Anantharam, IT Crawford, SV Ambudkar, SY Rhee, MJ Allison, PC Maloney, 1992. Identification, purification, and reconstitution of OxlT, the oxalate:formate antiport protein of *Oxalobacter formigenes*. *J Biol Chem* 267: 10537–10543.

Ruberto L, SC Vazquez, WP MacCormack, 2003. Effectiveness of the natural flora, biostimulation and bioaugmentation on the bioremediation of a hydrocarbon contaminated Antarctic soil. *Int Biodet Biodeg* 52: 115–125.

Saggu M, I Zebger, M Ludwig, O Lenz, B Friedrich, P Hildebrandt, F Lenzian, 2009. Spectroscopic insights into the oxygen-tolerant membrane-associated [NiFe] hydrogenase of *Ralstonia eutropha* H 16. *J Biol Chem* 284: 16264–16276.

Santiago B, M MacGilvray, RC Faustoferri, RG Quivey, 2012. The branched-chain amino acid aminotransferase encoded by *ilvE* is involved in acid tolerance in *Streptococcus mutans*. *J Bacteriol* 194: 2010–2019.

Seitz H-J, H Cypionka, 1986. Chemolithotrophic growth of *Desulfovibrio desulfuricans* with hydrogen coupled to ammonification with nitrate or nitrite. *Arch Microbiol* 146: 63–67.

Sheng J, RE Marquis, 2007. Malolactic fermentation by *Streptococcus mutans. FEMS Microbiol Lett* 272: 196–201.

Shinoda Y, Y Sakai, H Uenishi, Y Uchihashi, A Hiraishi, H Yukawa, H Yurimoto, N Kato, 2004. Aerobic and anaerobic toluene degradation by a newly isolated denitrifying bacterium, *Thauera* sp. strain DNT-1. *Appl Environ Microbiol* 70: 1385–1392.

Smith LT, J-A Pocard, T Bernard, D Le Rudulier, 1988. Osmotic control of glycine betaine biosynthesis and degradation in *Rhizobium meliloti. J Bacteriol* 170: 3142–3149.

Smith MG, RJ Park, 1984. Effect of restricted aeration on catabolism of cholic acid by two *Pseudomonas* species. *Appl Environ Microbiol* 48: 108–113.

Stingl K, E-M Uhlemann, R Schmid, K Altendorf, EP Bakker, 2002. Energetics of *Helicobacter pylori* and its implication for the mechanism of urease-dependent acid tolerance at pH 1. *J Bacteriol* 184: 3053–3060.

Sun L, T Fukamachi, H Saito, H Kobayashi, 2005. CO_2 increases acidic resistance in *Escherichia coli. Lett Appl Microbiol* 40: 397–400.

Sun L, T Fukamachi, H Saito, H Kobayashi, 2011. ATP requirement for acidic resistance in *Escherichia coli. J Bacteriol* 193: 3072–3077.

Touati D, M Jacques, B Tardat, L Bouchard, S Despied, 1995. Lethal oxidative damage and mutagenesis are generated by iron Δ*fur* mutants of *Escherichia coli:* Protective role of superoxide dismutase. *J Bacteriol* 177: 2305–2314.

Uyttebroek M, S Vermeir, P Wattiau, A Ryngaert, D Springael, 2007. Characterization of cultures enriched from acidic polycyclic aromatic hydrocarbon-contaminated soil for growth on pyrene at low pH. *Appl Environ Microbiol* 73: 3159–3164.

Vandal OH, CF Nathan, S Ehrt, 2009. Acid resistance in *Mycobacterium tuberculosis. J Bacteriol* 191: 4714–4721.

Vannelli T, AB Hooper, 1992. Oxidation of nitrapyrin to 6-chloropicolinic acid by the ammonia-oxidizing bacterium *Nitrosomonas europaea. Appl Environ Microbiol* 58: 2321–2325.

Wackett LP, MJ Sadowsky, LM Newman, H-G Hur, S Li, 1994. Metabolism of polyhalogenated compounds by a genetically engineered bacterium. *Nature* 368: 627–629.

Walt A, ML Kahn, 2002. The *fixA* and *fixB* genes are necessary for anaerobic carnitine reduction in *Escherichia coli. J Bacteriol* 184: 4044–4047.

Wandersman C, P Delepelaire, 2004. Bacterial iron sources: From siderophores to hemophores. *Annu Rev Microbiol* 58: 611–647.

Whyte LG, L Bourbonnière, CW Greer, 1997. Biodegradation of petroleum hydrocarbons by psychotrophic *Pseudomonas* strains possessing both alkane (*alk.* and naphthalene (*nah*) catabolic pathways. *Appl Environ Microbiol* 63: 3719–3723.

Whyte LG, CW Greer, WE Innis, 1996. Assessment of the biodegradation potential of psychrotrophic microorganisms. *Can J Microbiol* 42: 99–106.

Whyte LG, J Hawari, E Zhou, L Mournonnière WE Inniss, CW Greer, 1998. Biodegradation of variable-chain length alkanes at low temperatures by a psychotrophic *Rhodococcus* sp. *Appl Environ Microbiol* 64: 2578–2584.

Williams WA, RJ May, 1997. Low-temperature microbial aerobic degradation of polychlorinated biphenyls in sediment. *Environ Sci Technol* 31: 3491–3496.

Wu Q, DL Bedard, J Wiegel, 1997a. Effect of incubation temperature on the route of microbial reductive dechlorination of 2,3,4,6-tetrachlorobiphenyl in polychlorinated biphenyl (PCB)-contaminated and PCB-free freshwater sediments. *Appl Environ Microbiol* 63: 2836–2843.

Wu Q, DL Bedard, J Wiegel, 1997b. Temperature determines the pattern of anaerobic microbial dechlorination of Arochlor 1260 primed by 2,3,4,6-tetrachlorobiphenyl in Woods Pond sediment. *Appl Environ Microbiol* 63: 4818–4825.

Yang X, K Ma, 2007. Characterization of an exceedingly active NADH oxidase from the anaerobic hyperthermophilic bacterium *Thermotoga maritima. J Bacteriol* 189: 3312–3317.

Yu Z, GR Stewart, W Mohn, 2000. Apparent contradiction: Psychrotolerant bacteria from hydrocarbon-contaminated arctic tundra soils that degrade diterpenoids synthesized by trees. *Appl Environ Microbiol* 66: 5148–5154.

Zhang Y, RY Fu, J Hugenholtz, Y Li, J Chen, 2007. Glutathione protects *Lactococcus lactis* against acid stress. *Appl Environ Microbiol* 73: 5268–5275.

Zhang Y, Y Zhang, Y Zhu, A Mao, Y Li, 2010. Proteomic analyses to reveal the protective role of glutathione in resistance of *Lactococcus lactis* to osmotic stress. *Appl Environ Microbiol* 76: 3177–3186.

4.4 Chemical Stress

4.4.1 Hydrocarbon Tolerance

4.4.1.1 Gram-Negative Bacteria

A number of pseudomonads have been shown to tolerate high concentrations of aromatic hydrocarbons.

a. *Pseudomonas putida* strain Idaho was able to grow in a two-phase system containing *p*-xylene from 5% to 50% (v/v), and in a complex medium with a range of solvents, including *n*-alkanes, and heptan-1-ol with log $P_{oct} \leq 2.2$ (Cruden et al. 1992). During growth with *p*-xylene, the outer cell membrane became convoluted and the cytoplasmic membrane formed vesicles. This strain was solvent-tolerant to a greater extent than *P. putida* strains F1 and Mt-2 and *P. mendocina*.

b. Another strain of *P. putida* DOT-T1 was able to grow in the presence of 90% (v/v) toluene, and with a range not only of aromatic hydrocarbons, including toluene, styrene, ethylbenzene, but also of long-chain *n*-alkanes (Ramos et al. 1995). The advantage of this tolerance was taken by using the hydroxylase (*tmoABCDEF*) from this strain to produce 4-hydroxytoluene that could then undergo successive side-chain oxidation to 4-hydroxybenzoate (Ramos-González et al. 2003) and by using a strain lacking the *xylE* gene on a plasmid to produce 3-methylcatechol from *m*-xylene in an alkanol/water two-phase system (Rojas et al. 2004).

4.4.1.2 Alteration of Membrane Structure

Bacterial degradation of highly lipophilic aromatic compounds, including phenol and hydrocarbons such as toluene, necessitates their resistance to potentially adverse effects, and a number of mechanisms whereby this is overcome have been considered (Sikkema et al. 1995). This may be accomplished by alteration of the structure of their lipid membranes, although it should be emphasized that none of the mechanisms that have been proposed are entirely conclusive, or exclusive.

It is important to distinguish between short-term and long-term response. In *Pseudomonas putida* DOT-T1, which is a solvent-tolerant strain, the short-term response has been associated with the transformation of *cis*-9,10-methylene hexadecanoic acid to 9-*cis*-hexadecenoate and then to the *trans*-isomer (Ramos et al. 1997). *Pseudomonas putida* strain S12 is able to grow with concentrations of toluene up to 50% and growth with acetate in the presence of 1% toluene was accompanied by the replacement of *cis*-unsaturated fatty acids with the *trans*-isomers. This was maintained during several generations even in the absence of toluene (Weber et al. 1994). Enzymes that bring about the *cis/trans* isomerization of unsaturated fatty acids (Cti) have been described in *Pseudomonas oleovorans* GPo12 that has been cured of the plasmid determining the degradation of octanol (Pedrotta and Witholt 1999). Although greater activity could be shown in the presence of crude membrane preparations from *Pseudomonas putida* DOT-T1 in the presence of alcohols, the gene *cti* for this enzyme was present even in solvent-sensitive strains (Junker and Ramos 1999). Therefore, although the presence of this enzyme is important in alleviating short-term damage to solvents, it cannot be the primary cause of solvent resistance in this strain. It has been pointed out that details of how the organisms are harvested and the samples prepared could jeopardize the results, and that the *cis–trans* isomerization is competitive with the formation of cyclopropane fatty acids from the *cis*-isomers (Härtig et al. 2005). The synthesis of classes of phospholipid (phosphatidylethanolamine, phosphatidyl-glycerol, cardiolipin, phosphatidic acid, and phosphatidylserine) was examined in two strains of *Pseudomonas putida*, neither of which could degrade *o*-xylene and one of which was tolerant to *o*-xylene. For all strains, there were alterations in the concentration of total phospholipid, with a decrease in the sensitive strain and an increase in the tolerant strain. The fatty acid composition of the lipids was also altered in the tolerant strain, with a general decrease in *cis*-unsaturated and increase in *trans*-unsaturated fatty acids (this has already been noted). There was a higher rate of phospholipid biosynthesis in the tolerant strain and this suggests that the tolerant strain was able to repair membranes damaged by *o*-xylene more effectively than the sensitive strain (Pinkart and White 1997).

4.4.2 Efflux Pumps

4.4.2.1 *Gram-Negative Bacteria*

An alternative—and possibly complementary—mechanism that is highly attractive is the existence of an efflux pump. Multidrug efflux pumps play a critical role in antibiotic resistance and it has been shown that three of them MexA-MexB-OprM, MexC-MexD-OprJ, and MexE-MexF-OprN are able to contribute to tolerance of *n*-hexane and toluene by *Pseudomonas aeruginosa* PAO1 as well as antibiotic resistance (Li et al. 1998). Evidence has been provided for an energy-dependent export system in a toluene-resistant strain of *Pseudomonas putida* S-12 (Isken and de Bont 1996). The genes for this system have been cloned and their nucleotide sequence determined (Kieboom et al. 1998). Proteins coded by the three genes *srpA*, *srpB*, and *srpC* have extensive similarity to those for proton-dependent multidrug efflux systems, which are discussed in the section dealing with antibiotic resistance. In addition, the genes could be transferred to another strain of *P. putida* with the development of solvent resistance. Consistent with the role of the MexA-MexB-OprM operon that is expressed in *Pseudomonas aeruginosa* under normal growth conditions and that mediates antibiotic resistance, these genes also contribute to tolerance to *n*-hexane and *p*-xylene—though not toluene—in this strain (Li et al. 1998). As with antibiotic resistance, tolerance is abolished in the presence of the protonophore CCCP. In the toluene-resistant *Pseudomonas putida* DOT-T1E, three efflux pumps have been identified. The TtgABC and TtgGHI pumps extruded toluene, *m*-xylene, ethylbenzene, and styrene, whereas TtgDEF extruded only toluene and styrene (Rojas et al. 2001). In addition, mutants lacking the TtgABC pump displayed reduced resistance to nalidixic acid, chloramphenicol, and tetracycline, whereas a mutant lacking TtgGHI was as sensitive as the parent strain. An efflux system has been characterized in a strain of *Pseudomonas fluorescens* cLP6a (Hearn et al. 2003). This system was selective for extrusion of phenanthrene, anthracene, and fluoranthene, whereas it was inactive for toluene and naphthalene.

4.4.2.2 *Gram-Positive Bacteria*

Hydrocarbon tolerance has also been found in Gram-positive rhodococci. Tolerance to high concentrations of benzene has been demonstrated in a strain of *Rhodococcus* that is, in addition, tolerant of pH in the range 2–10 (Paje et al. 1997). For *Rhodococcus opacus*, resistance to benzene, toluene, phenol, and chlorobenzene was accompanied by an increase in the synthesis of 10-methyl branched fatty acids at the putative expense of their unsaturated fatty acids (Tsitko et al. 1999).

A number of factors may, therefore, be involved in the tolerance of bacteria to aromatic hydrocarbons, although greatest attention has hitherto been centered on Gram-negative organisms.

4.4.3 Tolerance of Phenols

1. It has been proposed that the phenol-degrading *Pseudomonas putida* strain P8 is protected from phenol toxicity by conversion of *cis-* → *trans*-unsaturated fatty acids (Heipieper et al. 1992), and this seems to be a widespread defense mechanism.

2. There are two putative mechanisms whereby bacteria become resistant to toxic polychlorinated phenols. These are illustrated by the following examples:

    a. Two strains of *Sphingomonas* sp. that could degrade pentachlorophenol maintained their levels of ATP even in the presence of high concentrations of pentachlorophenol. Analysis of the lipids using ^{31}P NMR showed that this could be attributed to the increased levels of cardiolipin (Lohmeier-Vogel et al. 2001).

    b. Bacterial *O*- and *S*-methylation has been demonstrated in a range of halogenated phenols and thiophenols (Neilson et al. 1988), thiophenols (Drotar and Fall 1985), and methanethiol and methaneselenol (Drotar et al. 1987). Methylation has been putatively attributed to reaction with *S*-adenosylmethionine.

4.4.4 Antibiotic Resistance

Since the genes are often carried on transmissible plasmids, resistance to a range of antibiotics is of increasing concern in clinical practice. This has been confirmed from the results of an investigation of strains of *Streptomyces* isolated from a range of urban, agricultural, and forest soils that were isolated on a standard medium. They displayed resistance to antibiotics that is well established in clinical practice, and resistance was found among all classes of antibiotics, while a disturbing number of isolates were resistant to between 5 and 10 of them (D'Costa et al. 2006). In addition, antibiotic tolerance has been associated with the growth of microbial films that are characteristic of several chronic microbial infections. Bacteria in these infections, display extreme tolerance toward antibiotics and the nutrient starvation that occurs in the periphery of biofilms as a result of reduced substrate diffusion (Nguyen et al. 2011) activates the stringent response (SR). This is mediated by starvation for carbon, amino acids, or iron, and induces *relA* and *spoT* gene products that results in the synthesis of the alarmone guanosine 3′, 5′-bisdiphosphate (ppGpp) (Trexler et al. 2008).

There are several types of mechanism that confer resistance, including enzymatic covalent modification of the antibiotic, effective efflux systems, and induction of a cellular enzyme that is resistant to the antibiotic. This has produced an extensive literature on drug resistance and only a selection of these is given as illustration.

4.5 Enzymatic Covalent Modification of Antibiotic: Range of Reactions

1. The β-lactam antibiotics function by inhibiting bacterial D,D-transpeptidases that are responsible for the final step in peptidoglycan cross-linking. These antibiotics include penicillins, cephalosporins, and carbapenems that are employed against both Gram-negative and Gram-positive bacteria. Drug resistance is, however, a serious and developing problem for many organisms of clinical significance. Resistance to these is generally mediated by a group of β-lactamases that bring about hydrolysis of the antibiotics before they reach the desired site of infection. Several schemes have been proposed for the classification of these β-lactamases (Bush et al. 1995) that has been updated (Bush and Jacoby 2010). They belong to a large family of hydrolases that employ at the active site either a serine[70] (Groups A, C, D) or in the metallo-β-lactamase one or two zinc atoms (Group B1, B2, B3) (Galleni et al. 2001) that are involved in the hydrolysis of the important group of carbapenems. Both groups carry out nucleophilic attack on the lactamase carbonyl group which results in hydrolysis, loss of activity, and thereby resistance to β-lactam antibiotics. As a result of this activity, many pathogenic bacteria in clinical use have become resistant to the β-lactam antibiotics. This has been only partly circumvented by the development of synthetic penicillins and cephalosporins whose structural differences make them less susceptible to the action of β-lactamases: these include substituents at C_6 and in cephalosporins at C_3 and C_6. There has been particular concern over the emergence of carbapenem-hydrolyzing lactamases (Rasmussen and Bush 1997) that belong both to the serine[70] group in *Enterobacter cloacae* and *Serratia marcescens* (Group 2f) and the metallo-β-lactamases in *Aeromonas hydrophila* (subclass B2) (Garau et al. 2005). Carbapenem-resistant strains of *Pseudomonas aeruginosa* that produce metallo-β-lactamase present a serious problem for patients especially those with impaired defense systems (Senda et al. 1996).

2. *O*-Acetylation of chloramphenicol (Shaw and Leslie 1991) and zwittermicin A (Stohl et al. 1999). For chloramphenicol, the reaction is complex involving three reactions: (i) acetylation at C_3, (ii) nonenzymatic acetyl migration to C_1, followed by (iii) acetylation at C_3 to the 1,3-diacetyl compound.

3. *O*-Phosphorylation of (a) chloramphenicol (Mosher et al. 1995) and spectinomycin that lead to resistance in *Legionella pneumophila* (Thompson et al. 1998), and (b) erythromycin at the 2′-position (Noguchi et al. 1995).

4. *O*-Glucosylation of telithromycin at the 2′-position in *Streptomyces* sp. strain Ja#7 (D'Costa et al. 2006).

5. The macrolide erythromycin inhibits protein synthesis and resistance is induced by N^6-dimethylation of adenine within the 23S rRNA, which results in reduced affinity of ribosomes for antibiotics related to erythromcin (Skinner et al. 1983).

6. Monooxygenation of tetracycline by TetX (Yang et al. 2004).

7. Fosfomycin (1R,2S)-epoxypropylphosphonic acid) is a broad-spectrum antibiotic that has attracted attention for the treatment of multidrug-resistant Gram-negative bacteria (Falagas et al. 2010; Wishino et al. 2010). Resistance to the drug is mediated by three enzymes that depend on the reactivity of the oxirane to nucleophilic attack:

    a. Glutathione S-transferase (Arca et al. 1990) (FosA) has one mononuclear manganese center on the basis of EPR, and is related to glyoxylase I and the extradiol dioxygenases (Bernat et al. 1997).

    b. (FosB) is a metallothiol transfer related to (Fos A), though with a preference for Mg^{2+} as cofactor and cysteine as nucleophile (Cao et al. 2001): both (a) and (b) form adducts by reaction at C_1 of the oxirane ring (Fillgrove et al. 2003).

    c. Hydrolase (FosX) catalyzes hydration at C_1 with inversion of configuration involving Mn^{2+} activation of the oxirane (Fillgrove et al. 2007).

4.5.1 Antituberculosis Drugs

These have one common and unusual feature—they require activation to achieve their therapeutic effect, and this has clear implications for the mechanisms of drug resistance.

4.5.1.1 Isonicotinic Hydrazide

Isonicotinic acid hydrazide (INH) is a frontline drug against *Mycobacterium tuberculosis* and the *M. tuberculosis* complex—though not against *Mycobacterium leprae*—and resistance to it illustrates the converse from the foregoing examples. INH is a prodrug that is converted into the active form, and resistance can be achieved by mutation of the gene KatG that encodes a catalase–peroxidase (Bertrand et al. 2004). Details of the mechanism of activation are complex, and several possibilities for resistance have been considered: these include failure to activate the drug, and interference with the synthesis of mycolic acids.

1. Activation by KatG catalase/peroxidase (Heym et al. 1993) produces metabolites that include a putative isonicotinic acid radical that reacts with NADH to form a competitive inhibitor (Rozwarski et al. 1998). This is supported by the mechanism of enzyme-catalyzed activation in peroxidases (Pierattelli et al. 2004), whereas activation of ethionamide takes place by S-monooxygenation (DeBarber et al. 2000) that is noted later. Failure to activate KatG was, therefore, suggested as leading to INH resistance, and is supported by the results of a study of sequencing *katG* mutants from a range of INH-resistant clinical isolates of *Mycobacterium tuberculosis*. These included assays for INH oxidase, peroxidase, and catalase, both high-level and low-level oxidase activities were observed, and there was a clear relation between levels of oxidase activity and levels of INH resistance (Ando et al. 2010).

2. The role of NO is supported by several lines of evidence:

    a. Its formation has been shown by electron paramagnetic spin resonance using ^{15}N-labeled INH, and CPTIO that is a specific scavenger of NO provided protection against INH activity in liquid cultures of *Mycobacterium tuberculosis* (Timmins et al. 2004).

    b. Formation of nitrotyrosine by nitric oxide carriers during oxidation of INH (van Zyl and van der Walt 1994).

    c. Incorporation into ciprofloxacin of an ester with a terminal $-O-NO_2$ group that can release NO stimulates its activity against *Mycobacterium tuberculosis* (Ciccone et al. 2003). An alternative—and not necessarily conflicting view—is the formation of a free radical at a tyrosine center that was trapped using added NO to form 3-nitrotyrosine (Zhao et al. 2004).

3. An alternative mechanism attributes resistance to inhibition of 2-*trans*-enoyl-ACP reductase that is linked to impaired synthesis of mycolic acids that are principal components of the cell wall. There is substantial evidence to support the view that resistance to INH is associated with the inhA gene that encodes an NADH-dependent *trans*-enoyl acyl carrier protein reductase which is part of the FAS II fatty acid synthase using the preferred primers C16 to C24 enoyl-ACP to produce meromycolate by methylmalonate elongation. A study using the INH-sensitive *Mycobacterium aurum* A+ showed that mycolic acid biosynthesis was inhibited by INH and, on the basis of several observations, the 24:1 *cis*-5 elongase was implicated as the primary target of INH (Wheeler and Anderson 1996). NIH inhibits mycolic acid synthesis in *Mycobacterium smegmatis* and the role of the enoyl-ACP reductase was examined using temperature-sensitive mutants. Exposure to the nonpermissive temperature of 420°C resulted in inactivation of the enzyme, while the level of C26 fatty acids increased concomitantly with the decrease in the level of C16:0 fatty acids. In addition, these cells in which mycolic acid synthesis was inhibited underwent lysis (Vilchèze et al. 2000).

4. Some other drugs that target tuberculosis are prodrugs which require activation by reduction to achieve their therapeutic effect. Examples include:

   a. The tricyclic benzo-1,3-thiazin-4-ones that inhibit the formation of cell wall arabinans require reduction of the 8-nitro group to the nitroso group which then reacts with a cysteine residue in the active site of the decaprenylphosphoryl-β-D-ribose 2′-epimerase (Christophe et al. 2009; Crellin et al. 2011; Trefzer et al. 2012).

   b. Bicyclic 2-nitroimidazoles based on the lead compound PA-824 have been shown to be active even against non-replicating cells and require reduction to the active des-nitro metabolite that is produced together with nitric oxide whose effect can be reversed by specific NO inhibitors (Singh et al. 2008).

4.5.1.2 Pyrazinamide

Pyrazinamide is a frontline drug for the treatment of tuberculosis and is an important component of tuberculosis therapy since it kills dormant bacilli in the acidic environment of macrophages where the pH lies between 6.2 and 4.5 (Vandal et al. 2009). It requires activation by hydrolysis that is carried out by a single enzyme with nicotinamide and pyrazinamidase activities. The amidase encoded by the gene *pncA* has been characterized from naturally pyrazinamide-resistant *Mycobacterium avium*, and was able to confer susceptibility on pyrazinamide-resistant *Myco. tuberculosis* (Sun et al. 1997). A line-probe assay has been developed to detect mutations in loci that result in loss of pyrazinamidase activity and consequent resistance: it was shown that there was complete coincidence in sensitivity and specificity with the results for conventional pyrazinamide assay (Sekiguchi et al. 2007). This is not, however, the only important mechanism, and the combination of an acidic pH and a deficient efflux system for the active pyrazinoic acid contribute to the unique susceptibility of *M. tuberculosis* to pyrazinamide (Zhang et al. 1999). Mutation in the gene *pncA* has been demonstrated to be significant in strains of *M. tuberculosis* that were resistant to pyrazinamide (Juréen et al. 2008), although other mutations have been reported.

4.5.1.3 Thioamides Including Ethionamide

The second-tier antituberculosis drug ethionamide is also a prodrug that is activated by an FAD-containing monooxygenase (EtaA) to the *S*-oxide before interacting with its cellular target (DeBarber et al. 2000), and then to the amide (Vannelli et al. 2002). Since this differs from the mechanism for resistance to INH, these INH-resistant strains can retain their susceptibility to ethionamide. The initial reaction is carried out by EtaA by a Baeyer–Villiger monooxygenase that functions typically with other substrates such as ketones and 4-thiomethyltoluene (Fraaije et al. 2004), as well as the second-tier thioamides prodrugs thiacetazone (TAC) and isoxyl (ISO) (Dover et al. 2007). It has been shown that mutations in the gene encoding EtaA confer resistance not only to ethionamide but also to the thiosemicarbazone (thiacetazone TAC) and the diphenylthiourea (isoxyl ISO) (DeBarber et al. 2000).

4.5.2 β-Lactam Inhibitors: Isoniazid Susceptibility

1. As a complement to the development of structurally new antibiotics, attention has been directed to the use of β-lactam inhibitors that have been used in combination with β-lactam antibiotics against Gram-negative pathogens. Inhibitors include clavulanate, tazobactam, and sulbactam that themselves contain β-lactam rings. The principle is that these mimics would acylate the active site of the antibiotics and then proceed to the formation of covalent acyclic linear products that would inhibit the antibiotics. They would thereby inhibit hydrolysis of the antibiotics with the concomitant development of resistance before they reached their target. This is illustrated for clavanulate inhibition in which reaction with the carbonyl group of the β-lactam ring with the formation of an acyl-enzyme is followed by ring opening to a *trans*-enamine intermediate (Padayatti et al. 2005) that may then undergo either formation of cross-linked intermediates or decarboxylation and further degradation (Figure 4.2) (Therrien and Levesque 2000). Clavulanate-resistant variants of TEM-1 and SHV-1 β-lactamase have, however, emerged and efforts have been made to study the mechanism of their action with a view to the development of novel β-lactamase inhibitors (Sulton et al. 2005).

2. This concept has also been applied to drug-resistant *Mycobacterium tuberculosis*, since there has been an alarming increase in strains of *Mycobacterium tuberculosis* that are resistant to frontline drugs, including isonicotinic hydrazide, in addition to second-line drugs. *Mycobacterium tuberculosis* is not generally sensitive to β-lactams, since it synthesizes a constitutive β-lactamase that is the major source of resistance to β-lactam antibiotics, so that little attention has been paid to their possible application to infection by *Mycobacterium tuberculosis*. For both of the above reasons, the combination of β-lactam antibiotics with the lactamase inhibitor clavulanate has, therefore, been examined with promising results (Cynamon et al. 1983; Segura et al. 1998; Voladri et al. 1998). The crystal structure of the β-lactamase has been determined for *Mycobacterium tuberculosis* (Wang et al. 2006), and the mechanism of inhibition of the β-lactamase BlaC in *Mycobacterium tuberculosis* has been examined for the three inhibitors containing β-lactam rings: the sulfoxides tazobactam, sulbactam, and clavulanate. On the basis of the structure of the lactamase *blaC* and data on the slow rates of hydrolysis of the carbanem meropenem, an examination of resistance in the combination with clavulanate— that is an inhibitor of β-lactamase—was carried out (Hugonnet et al. 2009). The combination was effective both as shown by values of MIC and importantly in lowering the survival under anaerobic conditions that mimic the persistent state. This involves upregulation of genes encoding an L,D-transpeptidase with the formation of $3 \rightarrow 3$ cross-links (Lavollay et al. 2008). Values of the combined MIC for *Mycobacterium tuberculosis* and for a resistant strain were 0.32 and

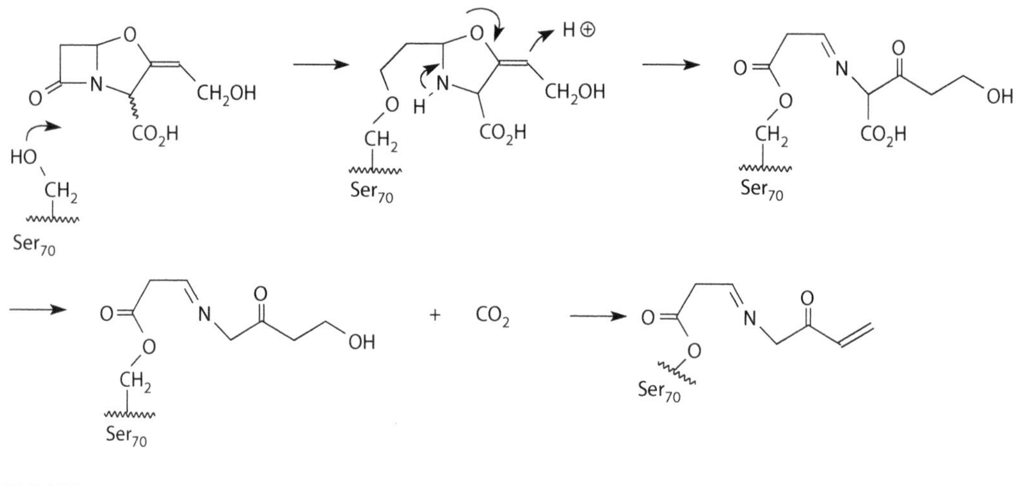

FIGURE 4.2 Mechanism of clavulanate inhibition. (Adapted from Sulton D et al. 2005. *J Biol Chem* 42: 35528–35536.)

0.94, respectively, and for a range of resistant strains between 1.25 and 0.32 against values of >10 for the amoxicillin combination and 5.0 for the ampicillin combination. For strains subjected to anaerobic conditions, there was a significant loss of survival that was pronounced for all the resistant strains after 2 weeks incubation with meropenem and clavulanate.

4.5.3 Nitrofurans and Nitroimidazoles

Nitrofurans such as nitrofurantoin have been used against urinary tract infections primarily caused by *Escherichia coli*, while nitroimidazoles have seen diverse application to human infection: with the protists *Giardia lamblia*, *Entamoeba histolytica*, and *Trichomonas vaginalis*, the bacteria *Helicobacter pylori* and *Bacteroides fragilis*, and, most recently, *Mycobacterium tuberculosis*. 2-Nitroimidazole was active against replicating cells of *Mycobacterium bovis* BCG and *Mycobacterium tuberculosis* H37Ra, more effectively than nitrofurantoin. In human monocytic cell line THP-1, it reduced viability and, after long-term exposure, induced sterility (Khan et al. 2008), although the compound was ineffective in the Wayne *in vitro* cell model for the study of nonreplicating cells under oxygen limitation (Wayne and Hayes 1996).

Nitrofurans and nitroimidazoles are prodrugs that require activation by reduction to reactive free radicals, and are effective only in microaerophilic protists and a few bacteria. Metronidazole (1-[2-hydroxyethyl]-2-methyl-5-nitroimidazole) is an important drug against the microaerophilic *Guardia lamblia*, *Entamoeba histolytica*, and *Trichomonas vaginalis*, and activation is achieved by an O_2-insensitive reductase (ntr) that is homologous to the metronidazole (rdx A), although they also possess a reductase nim that is homologous to the metronidazole-resistance gene in *Bacteroides fragilis* (Pal et al. 2009). In *Trichomonas vaginalis* that has been extensively investigated, metronitazole enters the hydrogenosome by diffusion and is reduced by ferredoxin to its cytotoxic form in which the drug functions as an alternative electron acceptor for hydrogenase (Rasoloson et al. 2002). Activity involves reduction of the nitro group to a free radical that has been demonstrated by EPR in the hydrogenosome of *Trichomonas vaginalis* using electrons released by the activity of pyruvate:oxidoreductase followed by transfer of ferredoxin to the drug. An alternative source has been established in which malate is converted to pyruvate and CO_2 by malic enzyme and the electrons that are released reduce NAD^+ and ferredoxin (Hrdý et al. 2005).

Nitroheteroarenes have been used as drugs against *Helicobacter pylori* (Sisson et al. 2002) in which metronidazole is reduced by the product encoded by the O_2-insensitive reductase rdxA or the NADH flavin oxidoreductase frxA, and resistance to the drug is induced by premature stop codons. Metronidazole has been used against infection by the anaerobe *Bacteroides fragilis*, and resistance has been associated with the reductases nim A, B, C, D which occur on small mobilizable plasmids and on the chromosome (Trinh and Reysset 1996). The same principles apply to nitrofuran drugs. Nitrofurantoin is used against urinary tract infections, and the first step in acquisition of resistance to furazolidone in *Escherichia coli* is mutation of nfsA that encodes an oxygen-insensitive nitro reductase, while the increase in resistance is a result of a second-step mutation of nfsB (Whiteway et al. 1998). Reduction results in the production of nitro-anion free radicals that interact with protein and DNA.

There is continuing interest in the attempt to find new drugs that are effective against *Mycobacterium tuberculosis*. Among these, the nitroimidazo-oxazine (PA-824) has elicited particular interest. It is a prodrug that requires reductive activation that is accomplished by hydride reaction at C_3 which is activated to nucleophilic attack by the C_2 nitro group to produce an intermediate anion: from this three reactions take place: (i) loss of nitrite to produce the des-nitro imidazole, (ii) formation of 2-hydroxyimidazole and H–N–O that is the source of the active NO, and (iii) reduction to the hydroxylamine (Singh et al. 2008). The hydrolytic reaction (ii) is analogous to that involved in the degradation of 2-nitroimidazole by *Mycobacterium* sp. JS 330 (Qu and Spain 2011). Resistance to PA-824 is most commonly mediated by loss of a specific glucose-6-phosphate dehydrogenase (FGD1) or its deazaflavin cofactor F_{420}, which together provide electrons for the reductive activation of this class of molecules, although additional accessory proteins that directly interact with the nitroimidazole are required (Manjunatha et al. 2006).

It is worth noting that compared with the reductive that operates for other nitroarenes, the resistance to chloramphenicol is accomplished by *O*-acetylation (Shaw and Leslie 1991), and the hydroxylamine and nitroso metabolites formed by reduction of the nitro group had no antibiotic effect (Corbett and Chipko 1978).

4.6 Efflux Systems

An alternative to most of these mechanisms is the existence of efficient efflux systems, so that toxic concentrations of the drug are not achieved. There are three major families of proton-dependent multi-drug efflux systems: (1) the major facilitator superfamily, (2) the small multidrug resistance family, and (3) the resistance/nodulation/cell division family (Paulsen et al. 1996). It should be emphasized that several of these systems are involved not with antibiotic efflux but with, for example, acriflavine, chlorhexi-dine, and crystal violet. An attempt is made only to outline a few salient features of the resistance/nodulation/cell division family that mediates antibiotic efflux, and these are given in Table 4.2 (Nikaido 1996). They consist of a transporter, a linker, and an outer membrane channel.

4.6.1 Antifungal Agents

This is not discussed in detail since mechanisms of resistance have been carefully reviewed (Ghannoum and Rice 1999) where it was pointed out that resistance has not been generally associated with modifica-tion of the structure of the drug but rather by specific inhibition of key enzymes.

Only a brief account is given to illustrate the different modes of action of two of them, the 1,2,4-triazoles and 5-fluorocytosine.

1. The 1,2,4-triazoles have been widely used and their target is the synthesis of ergosterol that is the dominant component of fungal cell membranes. They achieve this by blocking steps in its synthesis from lanosterol, and resistance is generally associated with modification of the target enzymes, for example, the epoxidation of squalene (Terbinafine), or 14α-demethylase (Fluconazole). Resistance of *Candida albicans* to the azole antifungal agent fluconazole dem-onstrated, however, the simultaneous occurrence of several types of mechanism for resistance (Perea et al. 2001):

   a. Levels in expression of the genes encoding lanosterol 14α-demethylase (*ERG11*) both by overexpression that necessitates higher intracellular concentration of the drug and by point mutations that result in diminished affinity for the agent.

   b. The existence of multidrug efflux transporters, including both *MDR1* and *CDR*.

   The increasing resistance of *Mycosphaerella graminicola* toward *Septaria* leaf blotch and the decline in the effectiveness of azole fungicides has been attributed to increase in mutations of CYP51 that is responsible for the synthesis of sterol 14α-demethylase (Cool et al. 2011).

2. Many types of yeast are inhibited by 5-fluorocytosine and its mode of action is quite different from the triazoles since it targets both RNA and DNA synthesis. Briefly, after uptake of the drug into the cell by a permease, 5-fluorouracil is produced by deamination followed by phos-phorylation to 5-fluorouridine diphosphate that is either phosphorylated further to 5-fluorouri-dine triphosphate which interferes with RNA synthesis or deoxygenated to 5-fluorodeoxyuridine phosphate that interferes with thymidine synthase and consequent DNA synthesis. A number of factors have been associated with resistance: these include impaired permease activity, disruption of the deaminase, uracil phosphoribosyltransferase, and the pyrophorylase (Jund

TABLE 4.2

Summary of Resistance/Nodulation/Cell Division Family

Transporter	Linker	Outer Membrane Channel	Organism	Substrates
AcrB	AcrA	TolC	*E. coli*	TC, CA, FQ, ERY, NOV, RIF, β-LAC
MexB	MexA	OprM	*P. aeruginosa*	TC, CP, FQ, β-LAC
MtrD	MtrC	MtrE	*Neisseria gonorrhoeae*	TC, CP, β-LAC, RIF

Source: Adapted from Nikaido H, 1996. *J Bacteriol* 178: 5853–5859.

and Lacoute 1970, Whelan and Kerridge 1984). On account of widespread resistance, the drug is seldom used alone but in combination with amphoterocin B and the triazine fluconazole.

4.6.2 Resistance to Metal Cations, Metal Oxyanions, and Metalloid Oxyanions

There is general concern over the toxicity of a number of metal cations and metalloid oxyanions. As for antibiotics, the genes for resistance are often plasmid-borne. There are several mechanisms that may operate—reduction, methylation, efflux, and the synthesis of metal-binding metallothioneins. Tolerance to Cu(II), Zn(II), Cd(II), and Ni(II) is widespread among acidophilic organisms such as the bacteria *Acidiphilum*, *Acidicella*, and *Acidithiobacillus ferroxidans* and the archaea *Ferroplasma*, and *Metallosphaera sedula* (Dopson et al. 2003). The following illustrate aspects of these resistance mechanisms.

4.6.2.1 Reduction

A mechanism of resistance may be mediated by reduction to less toxic forms, for example, Hg^{2+} to Hg^0, CrO_4^{2-} to Cr^{3+}, U(VI) to U(IV), SeO_3^{2-} to Se^0, TeO_3^{2-} to Te^0, and AsO_4^{3-} to AsO_3^{3-}. Extrusion of the reduced form may also be involved in, for example, Se^0, Te^0, and AsO_3^{2-} so that synthesis of the more toxic AsO_3^{2-} is circumvented.

4.6.2.1.1 Mercury

Mercuric reductase is a key enzyme in detoxifying inorganic mercury (Hg^{2+}) by reducing it to the non-toxic Hg^0. The reductase in a range of rapid-growing mycobacteria from clinical sources increased from 20- to >100-fold after exposure to $HgCl_2$ (Steingrube et al. 1991). The reductases from the plasmid-carrying *Escherichia coli* strain J53-1(R831) (Schottel 1978) and from *Pseudomonas aeruginosa* carrying the plasmid pVS1 (Fox and Walsh 1982) have been purified and both contained FAD. The enzyme from the latter was shown to contain two thiol groups that are released after reduction with NADPH, and was considered to be similar to lipoamide dehydrogenases.

The degradation of phenylmercuric acetate to benzene, methylmercuric chloride to methane, and ethylmercuric chloride to ethane and Hg^{2+} is apparently carried out by different enzymes from the plasmid-carrying *Escherichia coli* strain K12 (R831) (Schottel 1978) and *Pseudomonas* sp. Resistance to organic mercury compounds has also been found in clinical isolates of nontuberculous, rapidly growing mycobacteria (Steingrube et al. 1991) and can present a challenge in the clinical environment.

4.6.2.1.2 Chromium

Although reduction of chromate Cr^{VI} to Cr^{III} has been observed in a number of bacteria, these are not necessarily associated with chromate resistance. For example, reduction of chromate has been observed with cytochrome c_3 in *Desulfovibrio vulgaris* (Lovley and Phillips 1994), soluble chromate reductase has been purified from *Pseudomonas putida* (Park et al. 2000), and a membrane-bound reductase has been purified from *Enterobacter cloacae* (Wang et al. 1990). The flavoprotein reductases from *Pseudomonas putida* (ChrR) and *Escherichia coli* (YieF) have been purified and can reduce Cr^{VI} to Cr^{III} (Ackerley et al. 2004). Whereas ChrR generated a semiquinone and reactive oxygen species, YieR yielded no semiquinone, and is apparently an obligate four-electron reductant. It could, therefore, present a suitable enzyme for bioremediation. A chromium-resistant organism assigned to *Delftia* sp. was able to reduce Cr(VI) to Cr(III) in liquid medium and had a soluble chromate reductase (Morel et al. 2011). In addition, it produced indole-3-acetic acid and had nitrogenase activity.

4.6.2.1.3 Arsenic and Antimony

a. A number of redox systems are involved in microbial reactions that confer resistance to inorganic arsenic (Silver and Phung 2005). The arsenate reductase gene *ArsC* both in *Staphylococcus aureus* (Ji and Silver 1992) and in *Escherichia coli* (Liu et al. 1995) is a determinant of arsenate resistance, and the arsenite that is produced is extruded from the cell by the ArsA–ArsB anion-translocating ATPase (Walmsley et al. 1999). In contrast, ArsB catalyzes arsenite extrusion coupled to electrochemical energy, and transport of Sb(III) into *E. coli* is catalyzed by the ArsB carrier protein (Meng et al. 2004).

b. Another mechanism for detoxification of As(III) has been found in the legume symbiont *Sinorhizobium meliloti*. In the operon that is involved in As(III) detoxification, the *arsb* gene is replaced by *aqpS* that encodes an aquaglyceroporin. AqpS, therefore, confers resistance to arsenate by facilitated extrusion of arsenite produced by ArsC-catalyzed reduction of arsenate (Yang et al. 2005). In contrast, it has been proposed that in *Escherichia coli* the polyol transporter GlpF accumulates both arsenite and the analogous antimonite.

c. High resistance to both arsenite (12 mM) and arsenate (500 mM) has been found in *Corynebacterium glutamicum*. These values exceeded those for virtually all other bacteria except *Rhodococcus fascians*, and a complex array of genetic determinants were involved (Ordóñez et al. 2005). There were two operons on the chromosome *ars1* and *ars2* for resistance to arsenite both of which contain genes encoding an arsenite permease *arsB*, an arsenate reductase *arsC*, and a regulatory protein *arsR*, while additional genes for arsenite permease *arsB3* and arsenate reductase *arsC4* were also found in the chromosome.

d. The acidophile *Ferroplasma acidarmanus* strain Fer1 is highly resistant to arsenate which does not diminish growth at 133 mM, though more sensitive to arsenite (Baker-Austin et al. 2007). The genome sequence (Gihring et al. 2003) includes a novel operon that includes *arsR* which encodes a transcriptional regulator and *arsB* that encodes a transmembrane pump, though the gene *arsC* that encodes the arsenate reductase which is generally implicated in arsenate resistance was lacking, so that alternative hitherto unestablished mechanisms for arsenate resistance must operate.

Arsenite is also an intermediate in the fungal biomethylation of arsenic (Bentley and Chasteen 2002) and oxidation to the less toxic arsenate can be accomplished by heterotrophic bacteria, including *Alcaligenes faecalis*. Exceptionally, arsenite can serve as electron donor for chemolithotrophic growth of an organism designated NT-26 (Santini et al. 2000), and both selenate and arsenate can be involved in dissimilation reactions as alternative electron acceptors that are discussed in Chapter 3.

In a wider context, reduction of Sb(V) in the native drug to Sb(III) is required for the activity of preferred drugs for the treatment of leishmaniasis, and is supported by the following evidence. The sequence for the arsenate reductase ScAcr2p in *Saccharomyces cerevisiae* was used to clone the homolog LmACR2 in *Leishmania major*, and it was shown that the purified enzyme was able to reduce both As(V) and Sb(V). In addition, transfect ion of *L. infantum* with *LmACR2* increased the sensitivity of amastigotes to the Sb(V) drug Pentostam (Zhou et al. 2004).

4.6.2.1.4 Selenium and Tellurium

Whereas selenium is a necessary trace element particularly for anaerobic bacteria, tellurium is not known to play an essential role. Both selenite (SeO_3^{2-}) and tellurite (TeO_3^{2-}) can be toxic, and in *Escherichia coli*, selenite is able to induce the genes for Mn and Fe superoxide dismutases that may be protective under aerobic conditions (Bébien et al. 2002). Resistance to selenite and tellurite can be developed in a range of bacteria. A number of controversial mechanisms have been proposed for the reduction of SeO_3^{2-} to Se^0, and the involvement of glutathione has been suggested on the plausible basis of a chemical analogy (Kessi and Hanselmann 2004). In resistant bacteria related to the genus *Pseudoalteromonas* (Rathgeber et al. 2002) and in *Shewanella oneidensis* (Klonowska et al. 2005), the selenite that has been reduced to Se^0 can be extruded from the cells, whereas the Te^0 that is produced from tellurite is located within the cells. As for the mechanism of reduction, several reasons have been adduced for resistance to selenite and tellurite. In *Escherichia coli*, tellurite resistance has been associated with the activity of nitrate reductase (Avazéri et al. 1997; Sabaty et al. 2001). The *ars* operon on the plasmid R773, which functions as an anion-translocating ATPase for arsenite, arsenate, and antimonite, also provides moderate resistance to tellurite (Turner et al. 1992). In the Gram-positive *Geobacillus stearothermophilus*, which is naturally resistant to tellurite, transfer of the *iscS* gene that encodes cysteine desulfurase into

E. coli confers tellurite resistance (Tantaleán et al. 2003). Detoxification of tellurite has also been attributed to methyltransferase activity involving *S*-adenosylmethionine (Liu et al. 2000) or thiopurine methyltransferase (Cournoyer et al. 1998), even when the expected methylated products were not observed (Liu et al. 2000). Selenate can be reduced to Se^0 via selenite in *Enterobacter cloacae* SLD1a-1 (Losi and Frankenberger 1997), although it cannot be used as an electron acceptor for anaerobic growth (Ridley et al. 2006). Selenate can also serve as an electron acceptor, which is discussed in Part 2 of this chapter. Although in *Rhodobacter sphaeroides* (Van Fleet-Stalder et al. 2000) and in the metal-resistant *Ralstonia metallidurans* CH34 (Sarret et al. 2005), both reduction of selenite to Se^0 and formation of organic selenium may take place, reduction appears to be the dominant reaction.

4.6.2.2 Methylation

Methylation of both metals and metalloids has been observed for both fungi and bacteria. These metabolites may, however, be toxic to higher biota as a result of their volatility. The Minamata syndrome represents the classic example of the toxicity of forms of methylated Hg in man, even though the formation of $Hg(CH_3)_2$ was probably the result of both biotic and abiotic reactions.

Transmethylation is important not only in the biosynthesis of cellular components but also in detoxification. Although the resulting metabolites are less toxic to the cell, they are often more lipophilic and may have serious adverse effects on other biota. Toxic volatile arsenic compounds produced from inorganic arsenic were among the first examples, and methylmercury is a well-established toxicant both to wildlife and to humans. Transmethylation is also involved in the degradation of methyl halides by the corrinoid pathway.

Alkylated compounds were previously of industrial importance, for example, tetraethyl lead as an antiknock in gasoline engines and butyltin compounds as biocides. Biogeochemical cycles must take into account biogenic and anthropogenic inputs as well as biogenic and abiotic degradation and transformation. Monitoring has revealed the presence in the environment of a wide variety of methylated metals and metalloids. In the following summary, however, attention is directed only to results that conclusively demonstrate methylation under controlled laboratory conditions. Reference should be made to substantive reviews with extensive bibliographies dealing with methylation of metals and metalloids (Bentley and Chasteen 2002; Thayer 2002; Chasteen and Bentley 2003).

A cardinal issue is the species of the metal or metalloid that is examined. Metals such as mercury or tin are methylated from cationic Hg^{2+} or Sn^{4+} whereas the metalloids are transformed from the oxyanions of As, Sb, Se, or Te. The classical Challenger mechanism that involves sequential reductions and methylations is well established, at least for fungal methylation of the oxyanions of As (Bentley and Chasteen 2002), and Se—and is assumed to be—for Te (Chasteen and Bentley 2003). Methylation may take place under aerobic conditions for fungi or under anaerobic conditions for bacteria.

The formation of methylated derivatives may occur in metals and metalloids belonging to groups 15 and 16, and a few of group 14 of the periodic table, and these have been discussed in a critical review (Thayer 2002). Although bacteria can carry out several of the methylations, fungal methylation is probably most widely distributed, for example, of As and Sb (Tamaki and Frankenberger 1992; Bentley and Chasteen 2002), and Se and Te (Chasteen and Bentley 2003). Although the mechanism has not been established for all methylations, methyltransferases generally use either *S*-adenosylmethionine or methylcobalamin as methyl donors. In contrast, methylation of selenite in *Escherichia coli* is carried out by the activity of thiopurine methyltransferase (Ranjard et al. 2003).

Methylation of mercury seems, however, to be carried out only by bacteria, although the mechanisms for methylation under anaerobic conditions have not been finally resolved (Bentley and Chasteen 2002). Sulfate-reducing bacteria are generally presumed to be the primary source of $HgCH_3$ under anaerobic conditions (Choi et al. 1994; King et al. 2000), and generally use methylcobalamin as methyl donor. Methylation may also take place by reactions, which do not involve methylcobalamin (Ekstrom et al. 2003). It has been shown that *Geobacter* strain CLFeRB was able to carry out methylation of Hg^{2+} at a rate comparable to that of *Desulfobulbus propionicus* strain Ipr3 (Fleming et al. 2006). Under anaerobic conditions, methanogenic archaea use pathways that probably involve methylcobalamin for the methylation of arsenate to dimethylarsine (McBride and Wolfe 1971); arsenate to mono-, di-, and trimethylarsines; $SbCl_3$

to trimethylstibine; and $Bi(NO_3)_3$ to trimethylbismuth (Michalke et al. 2000). Conversely, among the organic arsenic compounds that were evaluated, only methylarsine appreciably inhibited acetoclastic methanogenesis (Sierra-Alvarez et al. 2004).

Under aerobic conditions, *S*-adenosylmethionine is the methyl donor for methylation of methanethiol and methaneselenol (Drotar et al. 1987), and probably for the bacterial methylation of halogenated phenols and thiophenols (Neilson et al. 1988). It is also the probable methyl donor for fungal methylation of the oxyanions of As and Sb (Bentley and Chasteen 2002).

Although the pathway has not been established, relatively high yields of trimethyltin from inorganic tin have been observed in yeast concomitant with the degradation of butyltin compounds (Errécalde et al. 1995). Exceptionally, methionine transferase may carry out the methylation of Hg in *Neurospora crassa* (Landner 1971) and thiopurine methyltransferase may carry out the methylation of inorganic Se in *Escherichia coli* (Ranjard et al. 2003).

Although their source has not been identified, it has been shown that volatile compounds from landfills contain carbonyls of Mo and W in addition to the known hydrides and methylated derivatives of As, Se, Sn, Sb, Te, Hg, Pb, and Bi (Feldmann and Cullen 1997).

4.6.2.3 Efflux Systems

As for antibiotics, resistance to toxic cations may be mediated by the existence of efflux systems that have been discussed in reviews (Silver and Phung 1996; Mergeay et al. 2003; Nies 2003). A great deal of attention has been directed to *Ralstonia metallidurans* that has masqueraded under several names— *Alcaligenes eutropha*, *Ralstonia eutropha*, and *Achromobacter xylosoxidans*—and details of the mechanisms have been presented in a review (Mergeay et al. 2003). In this brief summary of efflux systems that alleviate heavy metal toxicity, the convenient system of classification used by Nies (2003) will be adopted. The range of metals in this section is restricted to resistance to the divalent cations of Zn, Cd, Co, and Ni, and the monovalent Cu and Ag, and occurs in a range of bacteria and a few yeasts. Mechanisms for resistance to mercury have already been discussed, and chromate resistance in *Pseudomonas aeruginosa* is achieved by an efflux system that is driven by the membrane potential (Alvarez et al. 1999).

a. *The resistance–nodulation–cell division (RND) family of proteins.* In *Ralstonia metallidurans*, the CzcA protein mediates resistance to Zn^{2+}, Cd^{2+}, Co^{2+}, and Ni^{2+} in two megaplasmids, one associated (pMOL30) with resistance to Cd^{2+}, Zn^{2+}, and Co^{2+} and the other (pMOL28) with increased resistance to Co^{2+} and Ni^{2+}. In *Escherichia coli*, resistance to copper is mediated by a periplasmic protein (CusF).

b. *Cation-diffusion facilitators (CDF).* The primary substrate for the protein is Zn^{2+}, although other divalent cations, including Ni^{2+} and Co^{2+}, may also be effective. Transport is driven by gradients of several types—concentration, chemiosmotic $\Delta\psi$ or ΔpH, or potassium, and most studies have examined *R. metallidurans* in which the encoding genes are chromosomal, although species of *Saccharomyces* have also been included.

c. *P-type ATPases.* These function both as uptake and as efflux mediators, and one family is involved in the specific efflux of monovalent Cu^+ and Ag^+, but not Cu^{2+}, and has been examined in Gram-positive bacteria and the archaeon *Archaeoglobus fulgidus*. Another family that is involved in efflux of $Zn^{2+}/Cd^{2+}/Pb^{2+}$ has been found in Gram-negative bacteria and yeast.

4.6.2.4 Polyphosphates

There is some evidence for the role of phosphate produced by hydrolysis of polyphosphates in metal tolerance and remediation. Polyphosphates are accumulated by many bacteria and serve as an energy source and for chelating metal cations. Their biosynthesis is mediated by a kinase *ppk* that catalyzes their formation from the terminal phosphate of ATP, while an exopolyphosphatase *ppx* brings about their hydrolysis to inorganic phosphate. It has been suggested that the extruded phosphate could bind metal cations and, therefore, lead to the development of resistance to the toxic cation. For example, the thermophilic archaeon *Sulfolobus metallicus* was highly resistant to Cu^{2+} during the concomitant decrease in the level of polyphosphate, increase in the level of exopolyphosphatase, and the efflux of phosphate (Remonsellez et al. 2006). Overexpression of the polyphosphate kinase gene in *Pseudomonas aeruginosa* induced accumulation of polyphosphate that could be mobilized under conditions of carbon starvation with the release of phosphate into the medium. Phosphate was utilized to remove UO_2^{2+} from solution that was initially precipitated on the cell walls (Renninger et al. 2004).

4.6.2.5 Metallothionein

Metallothioneins are proteins that are rich in cysteine residues, and the genes encoding them are found in eukaryotes and a few prokaryotes. Although they have been found in a wide range of biota, their function has not been finally resolved, and different roles have been assigned to them (Palmiter 1998). To demonstrate their function in metal resistance, it is necessary to show not only their presence but also that the genes encoding them are amplified after exposure to the metal. Resistance has been explored extensively for copper that can produce reactive oxygen species with consequent cellular damage. At the same time, copper is necessary for the functioning of several cellular enzymes. Homeostasis is maintained by regulation.

Although levels of copper in prokaryotes can be maintained by P1-type ATPases in combination with a regulatory protein, for example, in *Pseudomonas putida* (Adaikkalam and Swarup 2002), tolerance may also be mediated by metallothioneins. They have been found in a few cyanobacteria within the genus *Synechococcus* (Turner and Robinson 1992), and the structure in strain Tx-20 has been elucidated (Olafson et al. 1988). It has been shown in a culture of *Synechococcus* PCC 6301 adapted to Cd^{2+} that amplification of the *smtAB* genes took place, and that multiple copies of them were found after exposure to Cd^{2+} to which the cells are resistant (Gupta et al. 1992). Genes homologous to these have also been found in Cd-resistant strains of sulfate-reducing bacteria. Metal-translocating ATPases were, however, also found in two of the strains (Naz et al. 2005), so that several mechanisms of resistance could be involved.

Substantial studies have been directed to yeasts. Several studies have examined the resistance to Cu^{2+} and Ag^+ in the yeast *Candida*. In *C. albicans*, genes encoding a P1-type ATPase, and one encoding a metallothionein were cloned, and their expression studied in response to Cu^{2+}. Transcription of only the former increased with increasing levels of Cu^{2+}, so that this was putatively the primary determinant of resistance (Riggle and Kumamoto 2000). A similar study showed that extracellular Cu^{2+} induced the genes for a metallothionein (CaCUP1) and a P-type ATPase (CaCRP1). The former was responsible for the unusual resistance of *C. albicans* to copper, whereas the latter was essential for survival under conditions of low copper concentration (Weissman et al. 2000). In *Candida glabrata*, whereas glutathione-related $(\gamma EC)_n F$ peptides are induced by Cd^{2+}, metallothioneins encoded by the genes MT-1 and MT-2 are induced by Cu^{2+}. In cells induced by increasing concentrations of Cu, chromosomal amplification of the MT-II took place, whereas the MT-I gene remained as a single copy and was shown to map on different chromosomes (Mehra et al. 1990). Although protection from copper toxicity in *Saccharomyces cerevisiae* involves a metallothionein encoded by *CUP1*, there is an additional element encoded by *CRS5* that is regulated by both Cu^{2+} and oxidative stress and possesses a metallothionein-like amino acid sequence (Culotta et al. 1994).

It is, therefore, clear that care should be exercised in assigning metallothioneins to a cardinal role in conferring resistance to metals.

4.7 References

Ackerley DF, CF Gonzalez, CH Park, R Blake, M Keyhan, A Matin, 2004. Chromate-reducing properties of soluble flavoproteins from *Pseudomonas putida* and *Escherichia coli*. *Appl Environ Microbiol* 70: 873–882.

Adaikkalam V, S Swarup, 2002. Molecular characterization of an operon, *cueAR*, encoding a putative P1-type ATPase and a MerR-type regulatory protein involved in copper homeostasis in *Pseudomonas putida*. *Microbiology (UK)* 148: 2857–2867.

Alvarez AH, R Moreno-Sánchez, C Cervantes, 1999. Chromate efflux by means of the ChrA chromate resistance protein from *Pseudomonas aeruginosa*. *J Bacteriol* 181: 7398–7400.

Ando H, Y Kondo, T Suetake, E Toyota, S Kato, T Mori, T Kirikae, 2010. Identification of *katG* mutations associated with high-level isoniazid resistance in *Mycobacterium tuberculosis*. *Antimicrob Agents Chemother* 54: 1793–1799.

Arca P, C Hardisson, JE Suárez, 1990. Purification of a glutathione *S*-transferase that mediates fosfomycin resistance in bacteria. *Antimicrob Agents Chemother* 34: 844–848.

Avazéri C, R Turner, J Pommier, J Weiner, GG Giordano, A Verméglio, 1997. Tellurite resistance activity of nitrate reductase is responsible for the basal resistance of *Escherichia coli* to tellurite. *Microbiology (UK)* 143: 1181–1189.

Baker-Austin C, M Dobson, M Wexler, RG Sawers, A Stemmler, BP Rosen, PL Bond, 2007. Extreme arsenic resistance by the acidophilic archaeon *"Ferroplasma acidarmanus"* Fr1. *Extremophiles* 11: 425–434.

Bébien M, G Lagniel, J Garin, D Touati, A Verméglio, J Labarre, 2002. Involvement of superoxide dismutases in the response of *Escherichia coli* to selenium oxides. *J Bacteriol* 184: 1556–1564.

Bentley R, TG Chasteen, 2002. Microbial methylation of metalloids: Arsenic, antimony and bismuth. *Microbiol Mol Biol Rev* 66: 250–271.

Bernat BA, T Laughlin, RN Armstrong, 1997. Fosfomycin resistance protein (FosA) is a manganese metalloglutathione transferase related to glyoxylase and the extradiol dioxygenases. *Biochemistry* 36: 3050–3055.

Bertrand T, NAJ Eady, JN Jones, Jesmin, JM Nagy, B Jamart-Grégoire, EL Raven, KA Brown, 2004. Crystal structure of *Mycobacterium tuberculosis* catalase-peroxidase. *J Biol Chem* 279: 38991–38999.

Bush K, G Jacoby, AA Medeiros, 1995. A functional classification for β-lactamases and its correlation with molecular structure. *Antimicrob Agents Chemother* 39: 1211–1233.

Bush K, GA Jacoby, 2010. Updated functional classification of β-lactamases. *Antimicrob Agents Chemother* 54: 969–976.

Cao M, BA Bernat, Z Wang, RN Armstrong, JD Helmann, 2001. FosB, a cysteine-dependent fosfomycin resistance protein under the control σ^W, an extracyctoplasmic-function σ factor in *Bacillus subtilis*. *J Bacteriol* 183: 2380–2383.

Chasteen TG, R Bentley, 2003. Biomethylation of selenium and tellurium: Microorganisms and plants. *Chem Rev* 103: 1–25.

Choi S-C, TT Chase, R Bartha, 1994. Metabolic pathways leading to mercury methylation in *Desulfovibrio desulfuricans* LS. *Appl Environ Microbiol* 60: 4072–4077.

Christophe T et al. 2009. High content screening identifies decaprenyl-phosphoribose 2′ epimerase as a target for intracellular antimycobacterial inhibitors. *PLoS Pathogens* 5: e10000645, doi:10.1371/journal.ppat.1000645.

Ciccone R, F Mariani, A Cavone, T Persichini, G Venturini, E Ongini, V Volizzi, M Colasanti, 2003. Inhibitory effect of NO-releasing ciprofloxacin (NCX 976) on *Mycobacterium tuberculosis* survival. *Antimicrob Agents Chemother* 47: 2299–2302.

Cool HJ, JGL Mullins, BA Fraaije, JE Parker, DE Kelly, JA Lucas, SL Kelly, 2011. Impact of recently emerged sterol 14α-demethylase (CYP51) variants of Mycosphaerella graminicola on azole fungicide activity. *Appl Environ Microbiol* 77: 3830–3837.

Corbett MD, BR Chipko, 1978. Synthesis and antibiotic properties of chloramphenicol reduction products. *Appl Environ Microbiol* 13: 193–198.

Cournoyer B, S Watanabe, A Vivian, 1998. A tellurite-resistance genetic determinant from pathogenic pseudomonads encodes a thiopurine methyltransferase: Evidence of a widely conserved family of methyltransferases. *Biochim Biophys Acta* 1397: 161–168.

Crellin PK, R Brammananth, RL Coppel, 2011. Decaprenylphosphoryl-β-D-ribose 2′-epimerase, the target of benzothiazinones and dinitrobenzamides, is an essential enzyme in *Mycobacterium smegmatis*. *PLoS One* 6: e16869.

Cruden DI, JH Wolfram, RD Rogers, DT Gibson, 1992. Physiological properties of a *Pseudomonas* strain which grows with *p*-xylene in a two-phase (organic-aqueous) medium. *Appl Environ Microbiol* 58: 2723–2629.

Culotta VC, WR Howards, XF Liu, 1994. *CRS5* encodes a metallothionein-like protein in *Saccharomyces cerevisiae*. *J Biol Chem* 269: 25295–25302.

Cynamon MH, GS Palmer, 1983. *In vitro* activity of amoxicillin in combination with clavulanic acid against *Mycobacterium tuberculosis*. *Antimicrob Agents Chemother* 24: 429–431.

D'Costa VM, KM McGrann, DW Hughes, GD Wright, 2006. Sampling the antibiotic resistome. *Science* 311: 374–377.

DeBarber AE, K Mdluli, M Bosman, L-G Bekker, CE Barry, 2000. Ethionamide activation and sensitivity in multidrug-resistant *Mycobacterium tuberculosis*. *Proc Natl Acad Sci USA* 97: 9677–9682.

Dopson M, C Baker-Austin, PR Koppineedi, PL Bond, 2003. Growth in sulfidic mineral environments: Metal resistance mechanisms in acidophilic micro-organisms. *Microbiology (UK)* 149: 1959–1970.

Dover LG, A Alahari, P Gratraud, JM Gobes, V Bhowruth, RC Reynolds, GS Besra, L Kremer, 2007. EthA, a common activator of thiocarbamide-containing drugs acting on different mycobacterial targets. *Antimicrob Agents Chemother* 51: 1055–1963.

Drotar A-M, R Fall, 1985. Methylation of xenobiotic thiols by *Euglena gracilis*: Characterization of a cytoplasmic thiol methyltransferase. *Plant Cell Physiol* 26: 847–854.

Drotar A-M, LR Fall, EA Mishalanie, JE Tavernier, R Fall, 1987. Enzymatic methylation of sulfide, selenide, and organic thiols by *Tetrahymena thermophila*. *Appl Environ Microbiol* 53: 2111–2118.

Ekstrom EB, FM Morel, JM Benoit, 2003. Mercury methylation independent of the acetyl-coenzyme A pathway in sulfate-reducing bacteria. *Appl Environ Microbiol* 69: 5414–5422.

Errécalde O, M Astruc, G Maury, R Pinel, 1995. Biotransformation of butyltin compounds using pure strains of microorganisms. *Appl Organomet Chem* 9: 23–28.

Falagas M, AC, Kastoris, AM Lapaskelis, DE Karageorgopoulos, 2010. Fosfomycin for the treatment of multidrug resistant, including extended-spectrum β-lactamase producing, Enterobacteriaceae infections: A systematic review. *Lancet Infect Dis* 10: 43–50.

Feldmann J, WR Cullen, 1997. Occurrence of volatile transition metal compounds in landfill has: Synthesis of molybdenum and tungsten carbonyls in the environment. *Environ Sci Technol* 31: 2125–2129.

Fillgrove KL, S Pakhomova, ME Newcomer, RN Armstrong, 2003. Mechanistic diversity of fosfomycin resistance in pathogenic microorganisms. *J Am Chem Soc* 125: 15730–15731.

Fillgrove KL, S Pakhomova, MR Schaab, ME Newcomer, RN Armstrong, 2007. Structure and mechanism of the genetically encoded fosfomycin resistance protein, FosX, from *Listeria monocytogenes*. *Biochemistry* 46: 8110–8120.

Fleming EJ, EE Mack, PG Green, DC Nelson, 2006. Mercury methylation from unexpected sources: Molybdate-inhibited freshwater sediments and an iron-reducing bacterium. *Appl Environ Microbiol* 72: 457–464.

Fox B, CT Walsh, 1982. Mercuric reductase. Purification and characterization of a transposon-encoded flavoprotein containing an oxidation–reduction active disulfide. *J Biol Chem* 257: 2498–2503.

Fraaije MW, NM Kammerbeek, AJ Heidkamp, R Fortin, DB Janssen, 2004. The prodrug activator EtaA from *Mycobacterium tuberculosis* is a Baeyer–Villiger monooxygenase. *J Biol Chem* 279: 3354–3360.

Galleni M, J Lamotte-Brasseur, GM Rossolini, J Spencer, O Dideberg, J-M Frère et al., 2001. Standard numbering scheme for class B β-lactamases. *Antimicrob Agents Chemother* 45: 660–663.

Garau G, C Bebrone, C Anne, M Galleni, J-M Frère, O Dideberg, 2005. A metallo-β-lactamase enzyme in action: Crystal structures of the monozinc carbapenemase CphA and its complex with biapenem. *J Mol Biol* 345: 785–795.

Ghannoum MA, LB Rice, 1999. Antifungal agents: Mode of action, mechanisms of resistance, and correlation of these mechanisms with bacterial resistance. *Clin Microbiol Rev* 12: 501–517.

Gihring TM, PL Bond, SC Peters, JF Banfield, 2003. Arsenic resistance in the archaeon "*Ferroplasma acidarmanus*": New insights into thhe structure and evoluation of th *ars* genes. *Extremophiles* 7: 123–130.

Gupta A, BA Whitton, AP Morby, JW Huckle, NJ Robinson, 1992. Amplification and rearrangement of a prokaryotic metallothionein locus *smt* in *Synechococcus* PCC 6301 selected for tolerance to cadmium. *Proc Roy Soc London Ser B* 248: 273–281.

Härtig C, N Loffhagen, H Harms, 2005. Formation of *trans* fatty acids is not involved in growth-linked membrane adaptation of *Pseudomonas putida*. *Appl Environ Microbiol* 71: 1915–1922.

Heipieper HR, R Diefenbach, H Keweloh, 1992. Conversion of *cis*-unsaturated fatty acids to *trans*, a possible mechanism for the protection of phenol-degrading *Pseudomonas putida* P8 from substrate toxicity. *Appl Environ Microbiol* 58: 195–205.

Hearn EM, JJ Dennis, MR Gray, JJ Foght, 2003. Identification and characterization of the *emhABC* efflux system for polycyclic aromatic hydrocarbons in *Pseudomonas fluorescens* cLP6a. *J Bacteriol* 185: 6233–6240.

Heym B, Y Zhang, S Poulet, D Young, ST Cole, 1993. Characterization of the *katF* gene encoding a catalase-peroxidase required for the isoniazid susceptibility of *Mycobacterium tuberculosis*. *J Bacteriol* 185: 4255–4259.

Hrdý I, R Cammack, P Stopka, J Kulda, J Tachezy, 2005. Alternative pathway of metronidazole activation in Trichomonas vaginalis hydogenosomes. *Antimicrob Agents Chemother* 49: 5033–5036.

Hugonnet J-E, LW Tremblay, HI Boschoff, CE Barry, JS Blanchard, 2009. Meropenem-clavulanate is effective against extensively drug-resistant *Mycobacterium tuberculosis. Science* 323: 1215–1218.

Isken S, JAM de Bont, 1996. Active efflux of toluene in a solvent-resistant bacterium. *J Bacteriol* 178: 6056–6058.

Ji G, S Silver, 1992. Reduction of arsenate to arsenite by the ArsC protein of the arsenic resistance operon of *Staphylococcus aureus* plasmid pI258. *Proc Natl Acad Sci USA* 89: 9474–9478.

Jund R, F Lacoute, 1970. Genetic and physiological aspects of resistance to 5-fluoropyrimidines in Saccharomyces cerevisiae. *J Bacteriol* 102: 707–715.

Junker F, JL Ramos, 1999. Involvement of the *cis/trans* isomerase Cti in solvent resistance of *Pseudomonas putida* DOT-T1E. *J Bacteriol* 181: 5693–5700.

Juréen P, J Werngren, J-C Toro, S Hoffner, 2008. Pyrazinamide resistance and *pncA* gene mutations in *Mycobacterium tuberculosis. Antimicrob Agents Chemother* 52: 1852–1854.

Kessi J, KW Hanselmann, 2004. Similarities between the abiotic reduction of selenite with glutathione and the dissimilatory reaction mediated by *Rhodospirillum rubrum* and *Escherichia coli. J Biol Chem* 279: 50662–50669.

Khan A, S Sarkar, D Sarker, 2008. Bactericidal activity of 2-nitroimidazole against the active replicating stage of *Mycobacterium bovis* BCG and *Mycobacterium tuberculosis* with intracellular efficiency in THP-1 macrophages. *Int J Antimicrob Agents* 32: 40–45.

Kieboom, J, JJ Dennis, JAM de Bont, GJ Zylstra, 1998. Identification and molecular characterization of an efflux pump involved in *Pseudomonas putida* S12 solvent tolerance. *J Biol Chem* 273: 85–91.

King JK, JE Kostka, ME Frischer, FM Saunders, 2000. Sulfate-reducing bacteria methylate mercury at variable rates in pure culture and in marine sediments. *Appl Environ Microbiol* 66: 2430–2437.

Klonowska A, T Heulin, A Vermeglio, 2005. Selenite and tellurite reduction by *Shewanella oneidensis. Appl Environ Microbiol* 71: 5607–5609.

Landner L, 1971. Biochemical model for biological methylation of mercury suggested from methylation studies in-vivo with *Neurospora crassa. Nature* 230: 452–454.

Lavollay M, M Arthur, M Fourgeaud, L Dubost, A Marie, N Veziris, D Blanot, L Gutmann, J-L Mainardi, 2008. The peptidoglycan of stationary-phase *Mycobacterium tuberculosis* predominantly contains cross-links generated by L,D-transpeptidation. *J Bacteriol* 190: 4360–4366.

Li X-Z, L Zhang, K Poole, 1998. Role of multidrug efflux systems of *Pseudomonas aeruginosa* in organic solvent tolerance. *J Bacteriol* 180: 2987–2991.

Liu J, TB Gladysheva, L Lee, BP Rosen, 1995. Identification of an essential cysteinyl residue in the ArsC arsenate reductase plasmid R 773. *Biochemistry* 34: 13472–13476.

Liu M, RJ Turner, TL Winstone, A Saetre, M Dyllick-Brenzinger, G Jickling, LW Tari, JH Weiner, DE Taylor, 2000. *Escherichia coli* TehB requires *S*-adenosylmethionine as a cofactor to mediate tellurite resistance. *J Bacteriol* 182: 6509–6513.

Lohmeier-Vogel EM, KT Leung, H Lee, JT Trevors HJ Vogel, 2001. Phosphorus-31 nuclear magnetic resonance study of the effect of pentachlorophenol on the physiologies of PCP-degrading microorganisms. *Appl Environ Microbiol* 67: 3549–3556.

Losi ME, WT Frankenberger, 1997. Reduction of selenium oxyanions by *Enterobacter cloacae* SLD1a-1: Isolation and growth of the bacterium and its expulsion of selenium particles. *Appl Environ Microbiol* 63: 3079–3084.

Lovley DR, EJP Phillips, 1994. Novel processes for anaerobic sulfate production from elemental sulfur by sulfate-reducing bacteria. *Appl Environ Microbiol* 60: 2394–2399.

Manjunatha UH, H Boshoff, CS Dowd, L Zhang, TJ Albert, JE Norton, L Daniels, T Dick, SS Pang, CE Barry, 2006. Identification of a nitroimidazo-oxazine-specific protein involved in PA-824 resistance in *Mycobacterium tuberculosis. Proc Natl Acad Sci USA 103: 431–436.*

McBride BC, RS Wolfe, 1971. Biosynthesis of dimethylarsine by a methanobacterium. *Biochemistry* 10: 4312–4317.

Mehra RK, JR Garey, DR Winge, 1990. Selective and tandem amplification of a member of the metallothionein gene family in *Candida glabrata. J Biol Chem* 265: 6369–6375.

Meng Y-L, Z Liu, BP Rosen, 2004. As(III) and Sb(III) uptake by GlpF and efflux by ArsB in *Escherichia coli. J Biol Chem* 279: 18334–18341.

Mergeay M, S Monchy, T Vallaeys, V Auquier, A Benotmane, P Bertin, S Taghavi, J Dunn, D van der Lelie, R Wattiez, 2003. *Ralstonia metallidurans*, a bacterium specifically adapted to toxic metals: Towards a catalogue of metal-responsive genes. *FEMS Microbiol Rev* 27: 385–410.

Michalke K, EB Wickenheiser, M Mehring, AV Hirner, R Hensel, 2000. Production of volatile derivatives of metal(loid)s by microflora involved in anaerobic digestion of sewage sludge. *Appl Environ Microbiol* 55: 2791–2705.

Morel MA, MC Ubalde, V Brana, S Castro-Sowinski, 2011. *Delftia* sp. JD2: a potential Cr(VI)-reducing agent with plant growth-promoting activity. *Arch Microbiol* 193: 63–68.

Mosher RH, DJ Camp, K Yang, MP Brown, WV Shaw, LC Vining, 1995. Inactivation of chloramphenicol by O-phosphorylation. *J Biol Chem* 270: 27000–27006.

Naz N, HK Young, N Ahmed, GM Gadd, 2005. Cadmium accumulation and DNA homology with metal resistance genes in sulfate-reducing bacteria. *Appl Environ Microbiol* 71: 4610–4618.

Nguyen D et al. 2011. Active starvation responses mediate antibiotic tolerance in biofilms and nutrient-limited bacteria. *Science* 334: 982–986.

Neilson AH, C Lindgren, P-Å Hynning, M Remberger, 1988. Methylation of halogenated phenols and thiophenols by cell extracts of Gram-positive and Gram-negative bacteria. *Appl Environ Microbiol* 54: 524–530.

Nies DH, 2003. Efflux-mediated heavy metal resistance in prokaryotes. *FEMS Micobiol Rev* 27: 313–339.

Nikaido H, 1996. Multidrug efflux pumps in Gram-negative bacteria. *J Bacteriol* 178: 5853–5859.

Noguchi N, A Emura, H Matsuyama, K O'Hara, M Sasatsu, M Kono, 1995. Nucleotide sequence and characterization of erythromycin resistance determinant that encodes macrolide 2-phosphotransferase I in *Escherichia coli*. *Antimicrob Agents Chemother* 39: 2359–2363.

Olafson RW, WD McCubbin, CM Kay, 1988. Primary- and secondary-structural analysis of a unique prokaryotic metallothionein from a *Synechococcus* sp. cyanobacterium. *Biochem J* 251: 691–699.

Ordóñez E, M Letek, N Valbuena, JA Gil, LK Mateos, 2005. Analysis of genes involved in arsenic resistance in *Corynebacterium glutamicum* ATCC 13032. *Appl Environ Microbiol* 71: 6206–6215.

Padayatti PS, MS Helfand, MA Totir, MP Carey, PR Carey, RA Bonomo, F van den Akker, 2005. High resolution crystal structures of the *trans*-enamine intermediates formed by sulbactam and clavulanic acid and E166a SHV-1 β-lactamase. *J Biol Chem* 280: 34900–34907.

Paje MLF, BA Neilan, I Couperwhite, 1997. A *Rhodococcus* species that thrives on medium saturated with liquid benzene. *Microbiology (UK)* 143: 2975–2981.

Pal D, S Banerjee, J Cui, A Schwartz, SK Ghosh, J Samuelson, 2009. Giardia, Entamoeba, and Trichomonas enzymes activate metronidazole (nitroreductases) and inactivate metronidazole (nitroimidazole reductases). *Antimicrob Agents Chemother* 53: 458–464.

Palmiter RD, 1998. The elusive function of metallothioneins. *Proc Natl Acad Sci USA* 95: 8428–8430.

Park CH, M Keyhan, B Wielinga, S Fendorf, A Matin, 2000. Purification to homogeneity and characterization of a novel *Pseudomonas putida* chromate reductase. *Appl Environ Microbiol* 66: 1788–1795.

Paulsen IT, MH Brown, RA Skurray, 1996. Proton-dependent multidrug efflux systems. *Microbiol Rev* 60: 575–608.

Pedrotta V, B Witholt, 1999. Isolation and characterization of the *cis/trans*-unsaturated fatty acid isomerase of *Pseudomonas oleovorans* Gpo12. *J Bacteriol* 181: 3256–3261.

Perea S, JL López-Ribot, WR Kirkpatrick, RK McAtee, RA Santillán, M Martínez, D Calabrese, D Sanglard, TF Patterson, 2001. Prevalence of molecular mechanisms of resistance to azole antifungal agents in *Candida albicans* strains displaying high-level fluconazole resistance isolated from human immunodeficiency virus-infected patients. *Antibicrob Agents Chemother* 45: 2676–2684.

Pierattelli R, L Banci, NA Eady, J Bodiguel, JN Jones, PCE Moody, EL Raven, B Jamart-Grégoire, KA Brown, 2004. Enzyme-catalyzed mechanism of isoniazide activation in Class I and Class III peroxidases. *J Biol Chem* 279: 39000–39009.

Pinkart HC, DC White, 1997. Phospholipid biosynthesis and solvent tolerance in *Pseudomonas putida* strains. *J Bacteriol* 179: 4219–4226.

Qu Y, JC Spain, 2011. Catabolic pathway for 2-nitromidazole involves a novel nitrohydrolase that also confers drug resistance. *Environ Microbiol* 13: 1010–1017.

Ramos-González M-I, A Ben-Bassat, M-J Campos, JL Ramos, 2003. Genetic engineering of a highly solvent-tolerant *Pseudomonas putida* strain for biotransformation of toluene to *p*-hydroxybenzoate. *Appl Environ Microbiol* 69: 5120–5127.

Ramos JL, E Duque, M-J Huertas, A Haïdour, 1995. Isolation and expansion of the catabolic potential of a *Pseudomonas putida* strain able to grow in the presence of high concentrations of aromatic hydrocarbons. *J Bacteriol* 177: 3911–3916.

Ramos JL, E Duque, J-J Rodríguez-Herva, P Godoy, A Haïdour, F Reyes, A Fernández-Barrero, 1997. Mechanisms for solvent tolerance in bacteria. *J Biol Chem* 272: 3887–3890.

Ranjard L, S Nazaret, B Cournoyer, 2003. Freshwater bacteria can methylate selenium through the thiopurine methyltransferase pathway. *Appl Environ Microbiol* 69: 3784–3790.

Rasmussen BA, K Bush, 1997. Carbapenem-hydrolyzing β-lactamases. *Antimicrob Agents Chemother* 41: 223–232.

Rasoloson D, S Vanacova, E Tomkova, J Razga, I Hrdý, J Tachezy, J Kulda, 2002. Mechanisms of *in vitro* development of resistance to metronidazole in Trichomonas vaginalis. *Microbiology (UK)* 148: 2467–2477.

Rathgeber C, N Yurkova, E Stackebrandt, JT Beatty, V Yurkov, 2002. Isolation of tellurite and selenite-resistant bacteria from hydrothermal vents of the Juan de Fuca Ridge in the Pacific Ocean. *Appl Environ Microbiol* 68: 4613–4622.

Remonsellez F, A Orell, CA Jerez, 2006. Copper tolerance of the thermoacidophilic archaeon *Sulfolobus metallicus*: Possible role of polyphosphate metabolism. *Microbiology (UK)* 152: 59–66.

Renninger N, R Knopp, H Nitsche, DS Clark, JD Keasling, 2004. Uranyl precipitation by *Pseudomonas aeruginosa* via controlled polyphosphate metabolism. *Appl Environ Microbiol* 70: 7404–7412.

RH, DJ Camp, K Yang, MP Brown, WV Shaw, LC Vining, 1995. Inactivation of chloramphenicol by O-phosphorylation. *J Biol Chem* 270: 27000–27006.

Ridley H, CA Watts, DJ Richardson, CS Butler, 2006. Resolution of distinct membrane-bound enzymes from *Enterobacter cloacae* SKLD1a-1 that are responsible for selective reduction of nitrate and selenate anions. *Appl Environ Microbiol* 72: 5173–5180.

Riggle PJ, CA Kumamoto, 2000. Role of a *Candida albicans* P1-type ATPase in resistance to copper and silver toxicity. *J Bacteriol* 182: 4899–4905.

Rojas A, E Duque, G Mosqueda, G Golden, A Hurtado, JL Ramos, A Segura, 2001. Three efflux pumps are required to provide efficient tolerance toluene in *Pseudomonas putida* DOT-T1E. *J Bacteriol* 183: 3967–3973.

Rojas A, E Duque, A Schmid, A Hurtado, J-L Ramos, A Segura, 2004. Biotransformation in double-phase systems: Physiological responses of *Pseudomonas putida* DOT.T1E to a double phase made of aliphatic alcohols and biosynthesis of substituted catechols. *Appl Environ Microbiol* 70: 3637–3643.

Rozwarski DA, GA Grant, DHR Barton, WR Jacobs, JC Sacchettini, 1998. Modification of the NADH of the isoniazid target (InhA) from *Mycobacterium tuberculosis*. *Science* 279: 98–102.

Sabaty M, C Avazeri, D Pignol, A Vermeglio, 2001. Characterization of the reduction of selenate and tellurite by nitrate reductases. *Appl Environ Microbiol* 67: 5122–5126.

Santini JM, LI Sly, RD Schnagl, JM Macy, 2000. A new chemolithoautotrophic arsenite-oxidizing bacterium isolated from a gold mine: Phylogenetic, physiological and preliminary biochemical studies. *Appl Environ Microbiol* 66: 92–97.

Sarret G, L Avoscan, M Carrière, R Collins, N Geoffroy, F Carrot, J Covès, B Gouget, 2005. Chemical forms of selenium in the metal-resistant bacterium *Ralstonia metallireducens* CH34 exposed to selenite and selenate. *Appl Environ Microbiol* 71: 2331–2337.

Schottel JL, 1978. The mercuric and organomercurial detoxifying enzymes from a plasmid-bearing strain of *Escherichia coli*. *J Biol Chem* 253: 4341–4349.

Segura C, M Salvadó, I Collado, J Chaves, A Coira, 1998. Contribution of β-lactamases to β-lactam susceptibilities of susceptible and multi-drug resistant *Mycobacterium tuberculosis* clinical isolates. *Antimicrob Agents Chemother* 42: 1524–1526.

Sekiguchi J et al., 2007. Development and evaluation of a line-probe assay for rapid identification of *pncA* mutations in pyrazinamide-resistant *Mycobacterium tuberculosis*. *J Clin Microbiol* 45: 2802–2807.

Senda K, Y Arakawa, K Nakashima, H Ito, S Ichiyama, K Shimokata, N Kato, M Ohta, 1996. Multifocal outbreaks of metallo-β-lactamase-producing *Pseudomonas aeruginosa* resistant to broad-spectrum β-lactams, including carbapenems. *Antimicrob Agents Chemother* 40: 349–353.

Shaw WV, AGW Leslie, 1991. Chloramphenicol acetyltransferase. *Annu Rev Biophys Chem* 20: 363–386.

Sierra-Alvarez R, I Cortinas, U Venal, JA Field, 2004. Methanogenesis inhibited by arsenic compounds. *Appl Environ Microbiol* 70: 5688–5691.

Sikkema J, JAM de Bont, B Poolman, 1995. Mechanisms of membrane toxicity of hydrocarbons. *Microbiol Rev* 59: 201–222.

Silver S, LT Phung, 1996. Bacterial heavy metal resistance: New surprizes. *Annu Rev Microbiol* 50: 753–789.

Silver S, LT Phung, 2005. Genes and enzymes in bacterial oxidation and reduction of inorganic arsenic. *Appl Environ Microbiol* 71: 599–608.

Singh R et al. 2008. PA-824 Kills nonreplicating *Mycobacterium tuberculosis* by intracellular NO release. *Science* 322: 1392–1395.

Sisson G, A Goodwin, A Raudonikiene, NJ Hughes, AK Mukhopadhyay, DE Berg, PS Hoffman, 2002. Enzymes associated with reductive activation and action of nitazoxanide, nitrofurans, and metronidazole in *Helicobacter pylori*. *Antimicrob Agents Chemother* 46: 2116–2123.

Skinner R, E Cundliffe, FJ Schmidt, 1983) Site of action of a ribosomal RNA methylase responsible for resistance to erythromycin and other antibiotics. *J Biol Chem* 258: 12702–12706.

Steingrube VA, RJ Wallace, LC Steele, Y Pang, 1991. Mercuric reductase activity and evidence of broad-spectrum mercury resistance among clinical isolates of rapidly growing mycobacteria. *Antimicrob Agents Chemother* 35: 819–823.

Stohl EA, SF Brady, J Clardy, J Handelsman, 1999. ZmaR, a novel and widespread antibiotic resistance determinant that acetylates zwittermycin. *J Bacteriol* 181: 5455–5460.

Sulton D, D Pagan-Rodriguez, X Zhou, Y Liu, AM Hujer, CR Bethel, MS Helfand et al., 2005. Clavulanic acid inactivation of SMV-1 and the inhibitor-resistant S130G SMV-1 β-lactamase. Insights into the mechanism of inhibition. *J Biol Chem* 42: 35528–35536.

Sun Z, A Scorpio, Y Zhang, 1997. The *pncA* gene from naturally pyrazinamide-resistant *Mycobacterium avium* encodes pyrazinamidase and confers pyrazinamide susceptibility to resistant *M. tuberculosis* complex organisms. *Microbiology UK* 143: 3367–3373.

Tamaki S, WT Frankenberger, 1992. Environmental biochemistry of arsenic. *Rev Environ Contam Toxicol* 124: 79–110.

Tantaleán JC, MA Araya. CP Saavedra, DE Fuementes, JM Pérez, IL Calderón, P Youderian, CC Vásquez, 2003. The *Geobacillus stearothermophilus* V *iscS* gene, encoding cysteine desulfurase, confers resistance to potassium tellurite in *Escherichia coli*. *J Bacteriol* 185: 5831–5837.

Thayer JS, 2002. Biological methylation of less-studied elements. *Appl Organometal Chem* 16: 677–691.

Therrien C, RC Levesque, 2000. Molecular basis of antibiotic resistance and β-lactamase inhibition by mechanism-based inactivators: Perspectives and future directions. *FEMS Microbiol Revs* 24: 251–262.

Thompson PR, DW Hughes, NP Cianciotto, GD Wright, 1998. Spectinomycin kinase from *Legionella pneumoniae*. *J Biol Chem* 273: 14788–14795.

Timmins GS, S Master, F Rusnak, V Deretic, 2004. Nitric oxide generated from isoniazid activation by KatG: Source of nitric oxide and activity against tuberculosis. *Antimicrob Agents Chemother* 48: 3006–3009.

Trefzer C et al. 2012. Benzothiazinones are suicide inhibitors of mycobacterial decaprenylphosphoryl-β-D-ribofuranose 2′-oxidase DprE1. *J Amer Chem Soc* 134: 912–915.

Trexler MF, SM Summers, H-T Nguyen, VM Zacharia, GA Hightower, JT Smith, T Conway, 2008. The global, ppGpp-mediated stringent response to amino-acid starvation in *Escherichia coli*. *Mol Microbiol* 68: 1128–1148.

Trinh S, Reysset, 1996. Detection by PCR of the nim genes encoding 5-nitroiminidazole resistance in *Bacteroides* spp. *J Clin Microbiol* 34: 2078–2084.

Tsitko IV, GM Zaitsev, AG Lobanok, MS Salkinoja-Salonen, 1999. Effect of aromatic compounds on cellular fatty acid composition of *Rhodococcus opacus*. *Appl Environ Microbiol* 65: 853–855.

Turner RJ, Y Hou, JH Weiner, DE Taylor, 1992. The arsenical APTase efflux pump mediates tellurite resistance. *J Bacteriol* 174: 3092–3094.

Turner JS, NJ Robinson, 1992. Cyanobacterial metallothioneins: Biochemistry and molecular genetics. *J Ind Microbiol* 14: 119–125.

van Fleet-Stalder V, TG Chasteen, IJ Pickering, GN George, RC Prince, 2000. Fate of selenate and selenite metabolized by *Rhodobacter sphaeroides*. *Appl Environ Microbiol* 66: 4849–4853.

Van Zyl JM, BJ van der Walt, 1994. Apparent hydroxyl radical generation without transition metal catalysts and tyrosine nitration during oxidation of the anti-tubercular drug, isonicotinic acid hydrazide. *Biochem Pharmacol* 48: 2033–2042.

Vannelli TA, A Dykman, PR Ortiz de Montellano, 2002. The antituberculosis drug ethionamide is activated by a flavin monooxygenase. *J Biol Chem* 277: 12824–12829.

Vilchèze C, HR Morison, TR Leisured, H Iwamoto, M Koop, JC Sachitting, WR Jacobs, 2000. Inactivation of the *inhA*-encoded fatty acid synthase II (FASII) enoyl-acyl carrier protein reductase induces accumulation of the FASI end products and cell lysis of *Mycobacterium smegmatis*. *J Bacteriol* 2000 182: 4059–4067.

Voladri RKR, DL Lakey, SH Hennigan, BE Menzies, KM Edwards, DS Kernodle, 1998. Recombinant expression and characterization of the major β-lactamase of *Mycobacterium tuberculosis*. *Antimicrob Agents Chemother* 42: 1375–1381.

Wachino J-I, K Yamane, S Suzuki, K Kimura, Y Arakawa, 2010. Prevalence of Fosfomycin resistance among CTX-M-producing *Escherichia coli* clinical isolates in Japan and identification of novel plasmid-mediated fosfomycin-modifying enzymes. *Antimicrob Agents Chemother* 54: 3061–3064.

Walmsley AR, T Zhou, MI Boirges-Walmsley, BP Rosen, 1999. The ATPase mechanism of ArsA, the catalytic subunit of the arsenite pump. *J Biol Chem* 274: 16153–16161.

Wang F, C Cassidy, JC Sacchetti, 2006. Crystal structure and activity studies of the *Mycobacterium tuberculosis* β-lactamase reveal its critical role in resistance to β-lactam antibiotics. *Antimicrob Agents Chemother* 50: 2762–2771.

Wang P, T Mori, K Toda, H Ohtake, 1990. Membrane-associated chromate reductase activity from *Enterobacter cloacae*. *J Bacteriol* 172: 1670–1672.

Wayne LG, LG Hayes, 1996. An *in vitro* model for sequential study of shiftdown of *Mycobacterium tuberculosis* through two stages of non-replicating persistence. *Infect Immun* 64: 2062–2069.

Weber FJ, S Isken, JAM de Bont, 1994. *Cis/trans* isomerization of fatty acids as a defence mechanism of *Pseudomonas putida* strains to toxic concentrations of toluene. *Microbiology (UK)* 140: 2013–2017.

Weissman Z, I Berdicevsky, BZ Cavari, D Kornotzer, 2000. The high copper tolerance of *Candida albicans* is mediated by a P-type ATPase. *Proc Natl Acad Sci USA* 97: 3520–3525.

Wheeler PR, PM Anderson, 1996. Determination of the primary target for isoniazid in mycobacterial mycolic acid biosynthesis with *Mycobacterium aurum* A+. *Biochem J* 318: 451–457.

Whelan WL, D Kerridge, 1984. Decreased activity of UMP pyphosphorylase associated with resistance to 5-fluorocytosine in *Candida albicans*. *Antimicrob Agents Chemotherap* 41: 1482–1487.

Whiteway J, P Koziarz, J Veall, N Sandhu, P Kumer, B Hoecher, IB Lambert, 1998. Oxygen-insensitive nitroreductases: Analysis of the roles of nfsA and nfsB in development of resistance to 5-nitrofuran derivatives in *Escherchia coli*. *J Bacteriol* 180: 5529–5539.

Wishino J-I, K Yamane, S Suzuki, K Kimura, Y Arakawa, 2010. Prevalence of fosfomycin resistance among CTX-M-producing *Escherichia coli* clinical isolates in Japan and identification of novel plasmid-mediated fosfomycin-modifying enzymes. *Antimicrob Agents Chemother* 54: 3061–3064.

Yang H-C, J Cheng, TM Finan, BP Rosen, H Bhattacharjee, 2005. Novel pathway for arsenic detoxification in the legume symbiont *Sinorhizobium meliloti*. *J Bacteriol* 187: 6991–6997.

Yang W, IF Moore, KP Koteva, DC Bareich, DW Hughe, GD Wright, 2004. TetX is a flavin-dependent monooxygenase conferring resistance to tetracycline antibiotics. *J Biol Chem* 279: 52346–52352.

Zhang Y, A Scorpio, H Nikaido, Z Sun, 1999. Role of acid pH and deficient efflux of pyrazinoic acid in unique susceptibility of *Mycobacterium tuberculosis* to pyrazinamide. *J Bacteriol* 181: 2044–2049.

Zhao X, S Girotto, S Yu, RS Magliozzo, 2004. Evidence for radical formation at Tyr-353 in *Mycobacterium tuberculosis* catalase-peroxidase (KatG) *J Biol Chem* 279: 7606–7612.

Zhou Y, N Messier, M Ouellette, BP Rosen, R Mukhopadhyay, 2004. *Leishmania major* LmACR2 is a pentavalent antimony reductase that confers sensitivity to the drug Pentostam. *J Biol Chem* 279: 37445–37451.

Part 2: Metabolic Interactions

4.8 Single Substrates: Several Organisms

Cultures of a single microorganism may occur naturally only in circumstances where extreme selection pressure operates. Generally, however, many different organisms with diverse metabolic potential will exist side by side. Therefore, metabolic interactions are probably the rule rather than the exception in natural ecosystems and biological treatment systems. Both the nature and the tightness of the association may vary widely, and the degradation of a single compound may necessitate the cooperation of two (or more) organisms. Some well-defined interactions and the different mechanisms underlying the cooperation are illustrated by the following examples:

1. One of the organisms fulfills the need for a growth requirement by the other, for example, vitamin requirements of one organism that is provided by the other. Examples are provided by biotin in cocultures of *Methylocystis* sp. and *Xanthobacter* sp. (Lidstrom-O'Connor et al. 1983), and thiamin in cocultures of *Pseudomonas aeruginosa* and an undefined *Pseudomonas* sp. that degraded the phosphonate herbicide glyphosate (Moore et al. 1983).

2. One organism may be able to carry out only a single step in biodegradation. Many examples among aerobic organisms have been provided (Reanney et al. 1983; Slater and Lovatt 1984) so that this is probably widespread. The following examples are illustrative.

   a. The degradation of parathion was carried out by a mixed culture of *Pseudomonas stutzeri* and *P. aeruginosa* (Daughton and Hsieh 1977) in which the 4-nitrophenol initially formed by the former is metabolized by the latter (Figure 4.3).

   b. The degradation of 4-chloroacetophenone was accomplished by a mixed culture of an *Arthrobacter* sp. and a *Micrococcus* sp. The first organism was able to carry out all the degradative steps except the conversion to 4-chlorocatechol of the intermediate 4-chlorophenol that is toxic to this organism (Havel and Reineke 1993).

   c. A consortium of a *Flavobacterium* sp. and a *Pseudomonas* sp. could carry out the aerobic degradation of polyethylene glycol. The latter was required for the degradation of the glycolate produced by the former (Kawai and Yamanaka 1986).

$$HO-(CH_2-CH_2O)_n-CH_2-CH_2-O-CH_2-CH_2OH \rightarrow HO-(CH_2-CH_2O)_n-CH_2-CH_2$$
$$-O-CH_2-CO_2H \rightarrow HO-(CH_2-CH_2O)_n \cdot CH_2-CH_2-OH \rightarrow HO-CH_2-CO_2H$$

3. Two organisms are required to maintain the redox balance. Among anaerobic bacteria, hydrogen transfer is important since the redox balance must be maintained, and the hydrogen concentration in mixed cultures may be critical. Interspecies hydrogen transfer has been demonstrated especially among populations of rumen bacteria containing methanogens where the concentration of hydrogen must be limited for effective functioning of the consortia. Illustrative examples have been summarized (Wolin 1982), and some additional comments on degradative reactions that are dependent on hydrogen transfer mediated by one of the cooperating organisms are added here.

Stable metabolic associations generally between pairs of anaerobic bacteria have been termed syntrophs, and these are effective in degrading a number of aliphatic carboxylic acids or benzoate

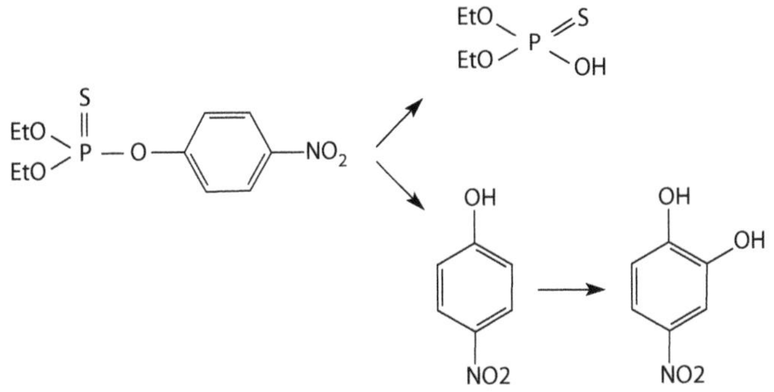

FIGURE 4.3 Degradation of parathion by a mixed culture of two pseudomonads.

under anaerobic conditions. These reactions have been discussed in reviews (Schink 1991, 1997; Lowe et al. 1993) that provide lucid accounts of the role of syntrophs in the degradation of complex organic matter. Two examples are given here to illustrate the experimental intricacy of the problems besetting the study of syntrophic metabolism under anaerobic conditions, and further details are given in Chapter 7.

1. Oxidation under anaerobic conditions of long-chain aliphatic carboxylic acids was established in syntrophic cultures of *Syntrophospora* (*Clostridium bryantii*), *Desulfovibrio* sp. (Stieb and Schink 1985); *Syntrophomonas sapovorans*, *Methanospirillum hungatei* (Roy et al. 1986); and *Syntrophomonas wolfei* in coculture with H_2-utilizing anaerobic bacteria (McInerney et al. 1981). The role of the syntroph was to metabolize the reducing equivalents produced by oxidation of the carboxylic acids. *S. wolfei* was subsequently, however, adapted to grow with crotonate in pure culture (Beaty and McInerney 1987), and this procedure was also used for *Syntrophospora* (*Clostridium*) *bryantii*. 16S rRNA sequence analysis was then used to show the close relationship of these two organisms and to assign them to a new genus *Syntrophospora* (Zhao et al. 1990). Anaerobic oxidation of carboxylic acids with chain lengths of up to C_{10} has, however, been demonstrated in pure cultures of species of *Desulfonema* (Widdel et al. 1983), and aliphatic hydrocarbons may be completely oxidized to CO_2 by a sulfate-reducing bacterium (Aeckersberg et al. 1991, 1998).

2. There has been considerable interest in the anaerobic degradation of propionate that is a fermentation product of many complex substrates, and syntrophic associations of acetogenic and methanogenic bacteria have been obtained. During the metabolism of propionate in a syntrophic culture (Houwen et al. 1991), the methanogens serve to remove hydrogen produced during the oxidation of propionate to acetate. The growth of syntrophic propionate-oxidizing bacteria in the absence of methanogens has, however, been accomplished using fumarate as the sole substrate (Plugge et al. 1993). Fumarate played a central role in the metabolism of this organism since it is produced from propionate via methylmalonate and succinate, and is metabolized by the acetyl-CoA cleavage pathway via malate, oxalacetate, and pyruvate.

Pure cultures of organisms that can oxidize propionate either in the presence of a methanogen or using sulfate as electron acceptor have, however, been obtained that include both *Syntrophobacter wolinii* and *Syntrophobacter pfenigii* (Wallrabenstein et al. 1995). The interaction of two organisms, therefore, is clearly not obligatory for the ability to degrade these carboxylic acids underanaerobic conditions.

4.9 Cometabolism and Related Phenomena

In natural ecosystems, it is indeed seldom that either a pure culture or a single substrate exists. In general, several substrates will be present, and these will include compounds of widely varying susceptibility to microbial degradation. The phenomenon where degradation occurs in the presence of two substrates has been termed "cooxidation" or, less specifically, "cometabolism" or "concurrent metabolism." Unfortunately, the term "cometabolism" (Horvath 1972) has been used in different and conflicting ways. Since the prefix "co" implies "together," it should not, therefore, be used when only a single substrate is present. The term "biotransformation" is unambiguous, and seems appropriate and entirely adequate. Detailed discussions have been presented (Dalton and Stirling 1982), and some of the conflicting aspects have been briefly summarized (Neilson et al. 1985). An illustration is provided by the following in which the mechanisms for the simultaneous degradation of both substrates that were used for growth were established. The degradation of atrazine by *Pseudomonas* sp. strain ADP is initiated by three hydrolases AtzA, AtzB, and AtzC to produce 2,4,6-trihydroxytriazine (cyanuric acid) (Cheng et al. 2005; Shapir et al. 2005), and the degradation requires the presence of a carbon source such as succinate. This strain can also utilize phenol as a carbon substrate by oxidation to catechol and degradation by the intradiol pathway (Neumann et al. 2004). The growth rate using ammonia or cyanuric acid as sources of nitrogen and phenol as carbon source were comparable, although the rate for growth with atrazine and phenol was reduced by about one-third. During growth with increasing concentrations of phenol, the ratio *trans/cis*-unsaturated fatty acids declined in contrast to that for atrazine where only a minor change was observed at the highest concentration of atrazine. This is consistent with the response to stress by phenol that has been observed for *Pseudomonas putida* strain P8 (Heidpieper et al. 1992).

Whereas probably most aerobic bacteria that have been used in experiments on biodegradation were obtained by elective growth using the chosen substrate as sole source of both carbon and energy, anaerobic bacteria are often more fastidious in their nutritional demands. The addition of nutrient supplements, such as yeast extract or ruminant fluid, may be needed to stimulate or maintain growth. Although the metabolic conclusions from such laboratory experiments are unambiguous, care should be exercised in uncritically extrapolating the results to natural ecosystems that are unlikely to provide such a nutritious environment. In such experiments, unequivocal results may often advantageously be obtained by using suspensions of washed cells.

Cometabolism merits, however, careful analysis since important metabolic principles underlie most of the experiments, even though confusion may have arisen as a result of ambiguous terminology. An attempt is, therefore, made to ignore semantic implications and to adopt a broad perspective in discussing this environmentally important issue. A pragmatic point of view has been adopted, and the following issues attempt to illustrate the kinds of experiments, which have been carried out under various conditions.

4.10 Induction of Catabolic Enzymes

4.10.1 Preexposure to an Analog Substrate

Organisms may be obtained after elective enrichment with a given substrate but are subsequently shown to be unable to use the substrate for growth, although they are able to accomplish its incomplete metabolism. These situations should rather be termed biotransformation. A typical example is the partial oxidation of 2,3,6-trichlorobenzoate to 3,5-dichlorocatechol (Figure 4.4) (Horvath 1971). The catabolic enzymes may, however, be induced by preexposure to an analog substrate. Cells may be grown with this before exposure to the xenobiotic. Although the xenobiotic is extensively degraded, it cannot be used alone to support the growth of cells. When both substrates are simultaneously present, the term cometabolism is appropriate, but this term is quite unjustifiable for the situation in which cells grown with a given inducing substrate are then used for studying the metabolism of a single xenobiotic. The following examples are given as illustration:

a. Several strains of bacteria grown with nonchlorinated substrates such as phenol, naphth-2-ol, or naphthalene were able to oxidize 4-chlorophenol to 4-chlorocatechol, or 3-chlorobenzoate to 3-chloro-5,6-dihydroxybenzoate (Figure 4.5) even though they were unable to use them for growth (Spokes and Walker 1974).

b. The oxidation of nitrobenzene to 3-nitrocatechol by strains of *Pseudomonas* sp. grown with toluene or chlorobenzene (Haigler and Spain 1991).

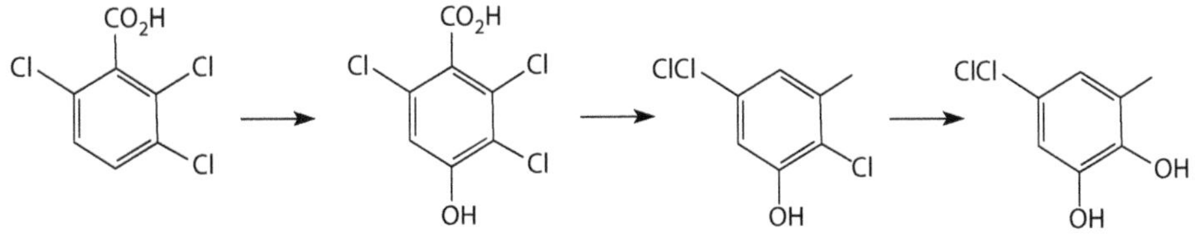

FIGURE 4.4 Biotransformation of 2,3,6-trichlorobenzoate.

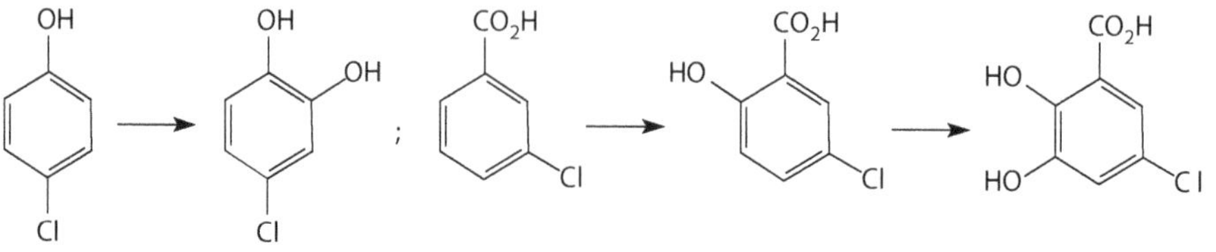

FIGURE 4.5   Biotransformation of 2-chlorophenol and 3-chlorobenzoate.

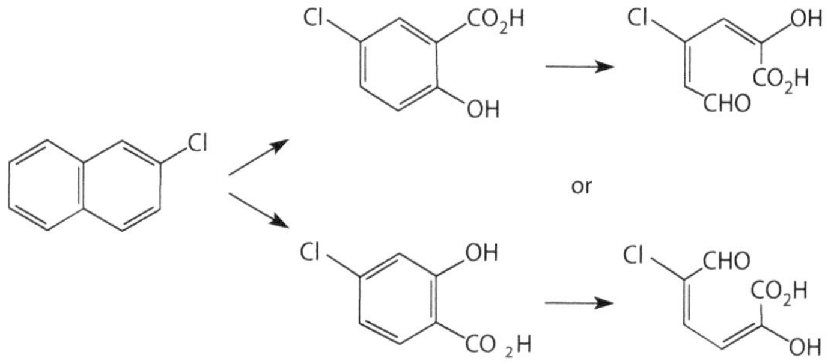

FIGURE 4.6   Biotransformation of 2-chloronaphthalene.

c. The oxidation of 2-chloronaphthalene to chloro-2-hydroxy-6-oxohexa-2,4-dienoic acids (Figure 4.6) by cells in which biotransformation of the substrate was induced by growth with naphthalene (Morris and Barnsley 1982).

d. The degradation of polychlorinated biphenyls has been established in a number of organisms enriched with biphenyl as carbon source, for example, *Pseudomonas* sp. strain LB400 (Bopp 1986), *Alcaligenes eutrophus* strain H850 (Bedard et al. 1987), and *P. pseudoalcaligenes* (Furukawa and Arimura 1987).

e. The use of brominated biphenyls has been examined to induce anaerobic dechlorination of highly chlorinated biphenyls, including the hepta-, hexa-, and pentachloro congeners (Bedard et al. 1998). Di- and tribromo congeners were the most effective and were themselves reduced to biphenyl.

f. The oxidation of methylbenzothiophenes by cells grown with 1-methylnaphthalene (Saflic et al. 1992).

g. The mineralization of benz[*a*]anthracene, chrysene, and benzo[*a*]pyrene by organisms isolated by enrichment with and grown at the expense of phenanthrene (Aitken et al. 1998).

Comparable situations could be encountered during experiments on bioremediation in which it may be experimentally expedient to grow cells on a suitable analog (Klecka and Maier 1988), or to introduce the organisms into the contaminated site (Harkness et al. 1993). There are, however, inherent dangers in this procedure. For example, cells able to degrade 5-chlorosalicylate (Crawford et al. 1979), 2,6-dichlorotoluene (Vandenbergh et al. 1981), and pentachlorophenol (Stanlake and Finn 1982) were unable to degrade the nonhalogenated analogs. Further, the degradation of 4-nitrobenzoate was inhibited by benzoate even though the strain could use either substrate separately (Haller and Finn 1978).

4.10.2 Enzyme Induction by Growth on Structurally Unrelated Compounds

Enzymes necessary for the metabolism of a substrate may be induced by growth on structurally unrelated compounds. In the examples used for illustration, monooxygenases play a cardinal role as a result of the versatility of methane monooxygenase, while monooxygenases that may be involved in toluene degradation are discussed in Chapter 3, Part 1 and Chapter 8, Part 1.

1. *Trichloroethene and aromatic compounds.* A striking example is the degradation of trichloroethene by bacteria during growth with aromatic substrates whose degradation is mediated by monooxygenation (hydroxylation), for example, phenol, or dioxygenation when monooxygenase activity may also be present, for example, toluene. This is illustrated in the following examples of trichloroethene degradation: growth with phenol or toluene by *Pseudomonas* sp. (Folsom et al. 1990); phenol by *Burkholderia kururiensis* (Zhang et al. 2000) and strains *Variovorax* (Futamata et al. 2005); toluene by *Pseudomonas putida* F1 (Wackett and Gibson 1988); and isopropylbenzene by *Pseudomonas* sp. strain JR1 (Pflugmacher et al. 1996). The use of phenol or toluene has attracted attention for the bioremediation of sites contaminated with chloroethenes (Hopkins and McCarty 1995). Conversely, toluene degradation is induced (a) by trichloroethene in a strain of *P. putida* (Heald and Jenkins 1994) and (b) by trichloroethene, pentane, and hexane in *P. mendocina* (McClay et al. 1995)—though not in *Burkholderia* [*Pseudomonas*] *cepacia* or *P. putida* strain F1. This metabolic versatility is consistent with the different pathways that are followed in the degradation of toluene (Figure 4.7):

   a. *P. putida* F1 by the classical toluene dioxygenase system (Zylstra et al. 1989)

   b. *B. cepacia* G4 by monooxygenation to 2-methylphenol (Shields et al. 1989)

   c. *P. mendocina* KR by monooxygenation to 4-methylphenol (Whited and Gibson 1991)

   d. *P. pickettii* PKO1 by monooxygenation to 4-methylphenol (Olsen et al. 1994)

   The last example was mediated by a monooxygenase that can be induced by benzene, toluene, and ethylbenzene, and also by xylenes and styrene. A plausibly analogous situation exists for strains of *Pseudomonas* sp. and *Rhodococcus erythropolis* that were obtained by enrichment with isopropylbenzene, and that could be shown to oxidize trichloroethene (Dabrock et al. 1992). In addition, one of the pseudomonads could oxidize 1,1-dichloroethene, vinyl chloride, trichloroethane, and 1,2-dichloroethane.

2. *Methane, butane, and chloroform.* Cells of *Methylosinus trichosporium* grown with methane and of *Pseudomonas butanovora* and *Mycobacterium vaccae* grown with butane were able to partially degrade chloroform (Hamamura et al. 1997). Again, this may be the result of the induction of monooxygenase activity.

3. Strain G4/PR1 of *Burkholderia cepacia* in which the synthesis of toluene-2-monooxygenase is constitutive is able to degrade a number of ethers, including diethyl ether and *n*-butyl methyl ether, but not *t*-butyl methyl ether (Hur et al. 1997).

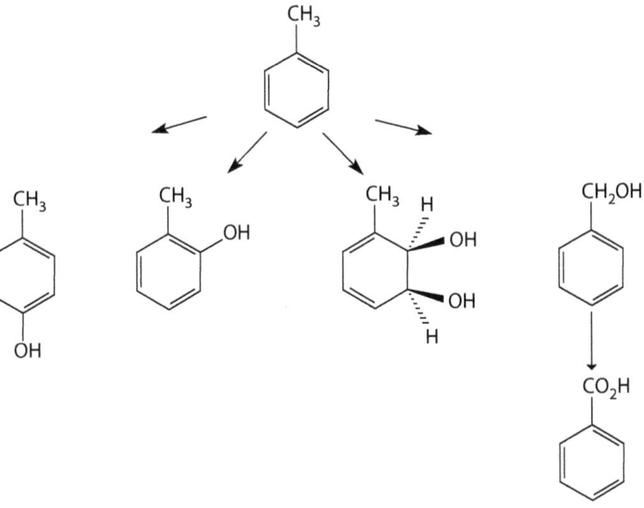

FIGURE 4.7 Pathways for the biotransformation of toluene.

4. A good example is provided by DDE, which is the first metabolite in the conventional degradation pathway of DDT and is apparently persistent in the environment. Pure cultures of aerobic and anaerobic bacteria that were able to degrade 1,1-dichloroethene and 4,4′-dichlorobiphenyl, which were considered to represent important structural features of DDE, were, however, unable to degrade DDE even during incubation with dense cell suspensions (Megharaj et al. 1997). Cell extracts of the aerobic organisms were also ineffective, and it was, therefore, concluded that recalcitrance lay in the structure of 1,1-diphenyl-2,2-dichloroethene since 1,1-diphenylethene could be used as the sole substrate for the growth of styrene-degrading strains of *Rhodococcus* sp. (Megharaj et al. 1998). In contrast, cells of *Pseudomonas acidovorans* strain M3GY during growth with biphenyl have been shown to degrade DDE with the fission of one ring and production of 4-chlorobenzoate (Hay and Focht 1998).

This situation may be of widespread occurrence and further examples of its existence will be facilitated by insight into the mechanisms of pathways for biodegradation.

4.11 Role of Readily Degraded Substrates

Results from experiments on biodegradation in which readily degraded substrates such as glucose are added have probably restricted relevance to natural ecosystems in which such substrates exist in negligible concentration. However, readily degraded substrates in addition to those less readily degradable undoubtedly occur in biological waste treatment systems. In these circumstances, at least three broadly different metabolic situations may exist:

1. The presence of glucose may suppress or lower degradation of a recalcitrant compound:
   a. Strains of *Pseudomonas pickettii* degrade 2,4,6-trichlorophenol, and its degradation is induced by several other chlorophenols, but is repressed in the presence of glucose or succinate (Kiyohara et al. 1992).
   b. The presence of glucose decreased the rate of degradation of phenol in a natural lake-water community, though the rate was increased by arginine (Rubin and Alexander 1983).

2. The presence of glucose may, however, enhance the degradation of a recalcitrant compound. Several different metabolic situations may be discerned, each probably representing a different mechanism for the stimulation:
   a. Experiments in which degradation of fluorobenzoates by a mixed bacterial flora was enhanced by the presence of glucose might plausibly be attributed to an increase in the cell density of the appropriate organism(s) (Horvath and Flathman 1976). A comparable conclusion could also be drawn from the data for the degradation of 2,4-dichlorophenoxyacetate and *O,O*-dimethyl-*O*-[3-methyl-4-nitrophenyl] phosphorothioate in cyclone fermentors (Liu et al. 1981).
   b. The presence of readily degraded substrates such as glutamate, succinate, or glucose had a stimulatory effect on the degradation of pentachlorophenol by a *Flavobacterium* sp., and could be attributed to ameliorating the toxic effects of pentachlorophenol (Topp et al. 1988). These additional substrates also enhanced the ability of natural communities to degrade a number of xenobiotics (Shimp and Pfaender 1985a).
   c. The presence of glucose facilitated the anaerobic dechlorination of pentachlorophenol and may plausibly be attributed to the increased level of reducing equivalents (Henriksen et al. 1991). A comparable phenomenon is the enhanced dechlorination of tetrachloroethene in anaerobic microcosms by the addition of carboxylic acids, including lactate, propionate, butyrate, and crotonate (Gibson and Sewell 1992).

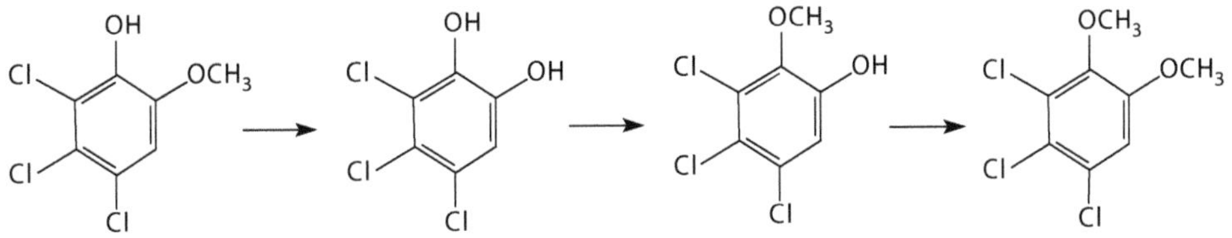

FIGURE 4.8 Biotransformation of 4,5,6-trichloroguaiacol by *Rhodococcus* sp. during growth with vanillate.

3. The xenobiotic may be degraded in preference to glucose, which is not a universal growth substrate. This situation is encountered in a phenol-utilizing strain of the yeast *Trichosporon cutaneum* that possesses a partially constitutive catechol 1,2-dioxygenase (Shoda and Udaka 1980), and illustrates the importance of regulatory mechanisms in determining the degradation of xenobiotics. Constitutive synthesis of the appropriate enzyme systems may indeed be of determinative significance in many natural ecosystems, and merits investigation.

In some circumstances, therefore, the presence of readily degraded substrates may clearly facilitate the degradation of recalcitrant xenobiotics, although a generally valid mechanism for these positive effects has not emerged. Whereas the addition of metabolizable analogs may increase the overall rates of degradation (Klecka and Maier 1988), it should be emphasized that the presence of readily degraded substrates in enrichments would generally be expected to be counterselective to the development of organisms that could degrade a specific xenobiotic. In addition, the observed enhancements summarized earlier were generally observed during relatively short time intervals.

Simultaneous metabolism of two structurally related substrates in which only one of them serves as growth substrate during biotransformation of the other may exist. A simple example is the *O*-demethylation of 4,5,6-trichloroguaiacol to 3,4,5-trichlorocatechol followed by successive *O*-methylation during the growth of a strain of *Rhodococcus* sp. at the expense of vanillate (Figure 4.8) (Allard et al. 1985). The results of experiments with mixtures of benzoate and 2,5-dichlorobenzoate using variants of a strain of *Pseudomonas aeruginosa* grown in chemostat cultures have revealed the following important aspects of environmental significance (van der Woude et al. 1995):

a. Variants that formed stable cultures could be obtained by growth limitation with both benzoate and 2,5-dichlorobenzoate.

b. One of these variants was capable of 2,5-dichlorobenzoate-limited growth at an oxygen concentration of 11 μM, although the presence of benzoate increased the residual concentration of 2,5-dichlorobenzoate from 0.05 to 1.27 μM.

A number of factors may be responsible for these observations, including the oxygen gradient within the cell and the oxygen concentration required for synthesis of the degradative oxygenases. It was concluded that under the low oxygen tensions that might exist in natural ecosystems, the presence of the more readily degraded benzoate necessitated high oxygen affinity for organisms to achieve complete degradation of the 2,5-dichlorobenzoate. In this case, it is relevant to note that the organism was originally isolated after enrichment at high oxygen concentrations. The issue of oxygen concentration is discussed in a wider context later in this chapter.

4.12 Association of Bacteria with Particulate Material: "Free" and "Bound" Substrates

Xenobiotics exist not only in the "free" state but also in association with organic and mineral components of particles in the water mass, and the soil and sediment phases. This association is a central determinant of the persistence of xenobiotics in the environment, since the extent to which the reactions are reversible is generally unknown. Such residues may, therefore, be inaccessible to microbial attack and apparently persistent. This is a critical factor in determining the effectiveness of bioremediation (Harkness et al. 1993). Although the most persuasive evidence for the significance of reduced bioavailability comes from data on the persistence of agrochemicals in terrestrial systems (Calderbank 1989), the principles can be translated with modification to aquatic and sediment phases that contain organic matter that resembles structurally that of soils.

4.12.1 Biological Mechanisms for Association with Organic Components of Soil and Sediment

These may involve both the original compound and its metabolites produced by biological reactions. This mechanism has wide implications, although it has been most extensively documented in the terrestrial environment.

1. Naphth-1-ol is an established fungal metabolite of naphthalene and may play a role in the association of naphthalene with humic material (Burgos et al. 1996).

2. ^{13}C-labeled metabolites of 9-[^{13}C])-anthracene, including 2-hydroxyanthracene-3-carboxylate and phthalate, were not extractable with acetone or dichloromethane, but could be recovered after alkaline hydrolysis (Richnow et al. 1998).

3. The results of experiments with ^{14}C-labeled pyrene added to a pristine forest soil illustrated a number of important issues (Guthrie and Pfaender 1998):

   a. Extensive mineralization took place only in samples amended with a pyrene-degrading microbial community.

   b. Compared with an azide-treated control, there was a substantially greater nonextractable fraction of the label in soils containing either the natural or the introduced microflora.

   c. Metabolites that could be released by acid and base extraction remained in the soil, even after incubation for 270 days.

4. The metabolism of ^{14}C-labeled BTX has been examined in soil cultures, and a mass balance constructed after 4 weeks of aerobic incubation (Tsao et al. 1998). Mineralization of all substrates was ca. 70% but ca. 20% of the label in toluene and ca. 30% in *o*-xylene were found in humus. It was suggested that alkylated catechol metabolites were responsible for this association.

5. The mechanism for the interaction of cyprodinil (4-cyclopropyl-6-methyl-2-phenylaminopyrimidine) with soil organic matter has been examined. The association with soil organic carbon was biologically mediated, and it was shown that this increased during incubation for up to 180 days (Dec et al. 1997a). After 169 days of incubation, the fractions obtained by methanol extraction, and the humic acid and fulvic acid fractions after alkali extraction were examined by ^{13}C NMR (Dec et al. 1997b). Both the phenyl and the pyrimidine rings were associated with humic material, though only partly in the form of intact cyprodinil.

6. Considerable attention has been directed to enzymatic reactions mediated by fungal oxidoreductase enzymes such as phenol oxidase, peroxidase, and laccase. These systems have been used to copolymerize structurally diverse xenobiotics to lignin-like structures, and include substituted anilines (Bollag et al. 1983) and benzo[*a*]pyrene quinone (Trenck and Sandermann 1981). One substantial advantage of using these model systems is that it is possible to isolate the products of the reactions and determine their chemical structures. Some examples are given below to illustrate the different substrates involved and the types of products that may be produced:

   a. Incubation of pentachlorophenol with a crude supernatant from *Phanerochaete chrysosporium* in the presence of a lignin precursor (ferulic acid) and H_2O_2 produced a high-molecular-mass polymer (Rüttimann-Johnson and Lamar 1996). It was suggested that this could mimic the association of pentachlorophenol with humic material and the formation of heteropolymers between pentachlorophenol and lignin monomers.

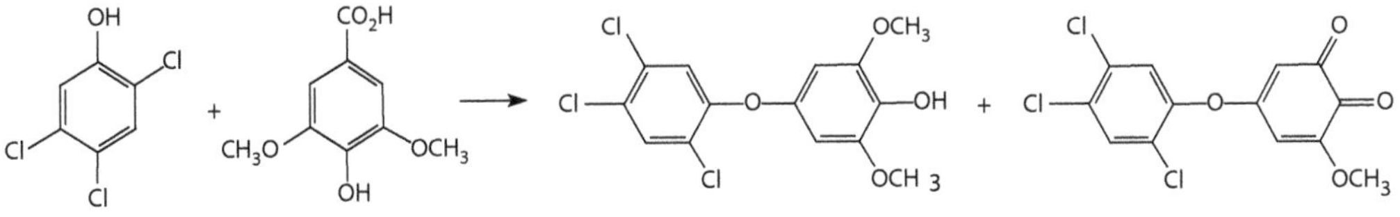

FIGURE 4.9 Reaction between 2,4,5-trichlorophenol and syringic acid catalyzed by laccase.

b. The reaction between halogenated phenols and syringic acid in the presence of laccase from the fungus *Rhizoctonia praticola* resulted in the formation of a series of diphenyl ethers. One ring originated from the chlorophenol, together with 1,2-quinonoid products resulting from partial *O*-demethylation and oxidation (Bollag and Liu 1985) (Figure 4.9). Comparable reactions have also been postulated to occur between 2,4-dichlorophenol and fulvic acid (Sarkar et al. 1988).

c. It has been shown that oligomerization of 4-chloroaniline mediated by oxidoreductases may produce 4,4′-dichloroazobenzene and 4-chloro-4′-aminodiphenyl as well as trimers and tetramers (Simmons et al. 1987). A study with guaiacol and 4-chloroaniline using a number of oxidoreductases demonstrated the synthesis of oligomeric quinone imines together with compounds resulting from the reaction of the aniline with diphenoquinones produced from guaiacol (Simmons et al. 1989) (Figure 4.10).

d. Direct evidence for the existence of covalent bonding between 2,4-dichlorophenol and peat humic acid in the presence of horseradish peroxidase has been provided from the results of an NMR study using 2,6-[^{13}C]-2,4-dichlorophenol (Hatcher et al. 1993). In the absence of suitable model compounds, interpretation of the results was based on estimated chemical shifts for a range of plausible structures. The most important contributions came from those with an ester linkage with the phenol group, and covalent bonds between carbon atoms of the humic acid and C_4 (with loss of chlorine) and C_6 of the chlorophenol.

e. Laccase-catalyzed reactions between bentazon (3-isopropyl-*H*-2,1,3-benzothiadiazine-4(3*H*)-one 2,2-dioxide) and various humic acid monomeric components have been studied, and the products from reactions with catechol examined in detail using both ^{1}H and ^{13}C NMR (Kim et al. 1997). Products with masses of 348 and 586 mu were isolated, and these were assigned to products formed by reactions between the nitrogen atom of bendazon and the 1,2-quinone produced by the laccase.

f. Coniferyl alcohol—the monomeric precursor of lignin—was copolymerized by peroxidase and H_2O_2 in the presence of ^{15}N aniline and 3,4-dichloroaniline in various ratios, and the

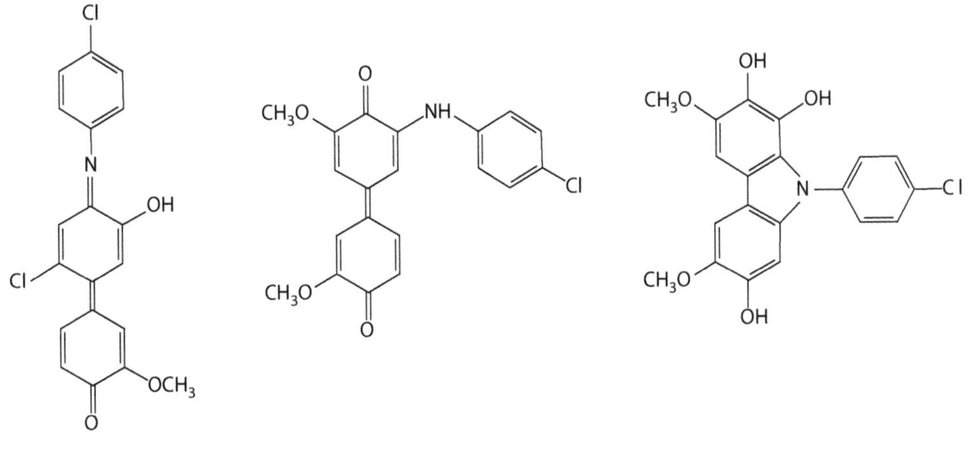

FIGURE 4.10 Products from the enzymatic copolymerization of guaiacol and 4-chloroaniline.

products were examined by ^{1}H, ^{13}C, and ^{15}N NMR (Lange et al. 1998). The conjugates were formed by reaction at the benzylic carbon atom of the coniferyl alcohol polymer. Although the anilines could be recovered by acid hydrolysis, it was pointed out that this could result from the high molar ratio of anilines used for copolymerization.

There is, therefore, extensive evidence that may be used to rationalize the occurrence of "bound" residues in soils, and this phenomenon is of particular significance for agrochemicals. Such processes influence not only their recovery by chemical procedures but also their biological effect and their biodegradability (Calderbank 1989). The extent to which these principles are applicable to aquatic systems appears to have been established less frequently, though it is plausible that comparable mechanisms exist in the environment.

4.12.2 Aging

This process should be considered in the light of the preceding comments on association. Many experiments on the recoverability, persistence, and toxicity of xenobiotics have used spiked samples that do not take into account the cardinal issue of alterations in the contaminant that have taken place after deposition. This is termed aging, and should be evaluated critically in determining persistence. Some examples are given below as illustration for both terrestrial and aquatic systems:

1. Experiments have examined the effect of aging on biodegradability using sterilized samples of soils that were spiked and aged under laboratory conditions. These have shown that the rates and extent of degradation of phenanthrene and 4-nitrophenol by an added strain of *Pseudomonas* sp. decreased markedly with prolonged aging (Hatzinger and Alexander 1995).

2. The sorption–desorption of PAHs has been extensively investigated, and the role of desorption in determining their biodegradability in aged sediments has been widely accepted (references in Carmichael et al. 1997). A definitive study using ^{14}C-phenanthrene and ^{14}C-chrysene showed that in contaminated soils, their rates of mineralization were much lower than the rates of desorption from spiked sediments. In contrast, for aged substrates, desorption rates were essentially comparable to rates of mineralization. This suggested that the indigenous microflora may have become adapted to the low substrate concentrations available by desorption (Carmichael et al. 1997).

3. Suspensions of 2,4-dichlorophenoxyacetate sorbed onto sterile soils were completely protected from degradation by either free or sorbed bacteria, and degradation of the substrate required access of the bacteria to the free compound in solution (Ogram et al. 1985). Rates of degradation in soil with a high organic content were lower than for one with a low organic content (Greer and Shelton 1992), and this adds additional support for the significance of desorption of the xenobiotic in determining its biodegradability.

4. ^{14}C-labeled 2,4-dichlorophenol bound to synthetic or natural humic acids or polymerized by H_2O_2 and peroxidase was mineralized to CO_2 only to a limited extent (<10%), and the greater part remained bound to the polymers (Dec et al. 1990).

5. ^{14}C-labeled 3,4-dichloroaniline–lignin conjugates were degraded to $^{14}CO_2$ by *Phanerochaete chrysosporium* as effectively as the free compound (Arjmand and Sandermann 1985), and it was, therefore, concluded that these "bound" residues were not persistent in the environment. This may, however, represent a special case for the following reasons: (a) although this organism is able to degrade lignin, the relevance of such organisms in most aquatic environments is possibly marginal, and (b) the lignin peroxidases implicated in lignin degradation are generally extracellular so that soluble substrates are probably not necessary.

6. The presence of humic acids had a detrimental effect on the degradation of substituted phenols by a microbial community after lengthy adaptation to the humic acids, and was not alleviated by the addition of inorganic nutrients (Shimp and Pfaender 1985b). The diminished number of organisms with degradative capability was responsible for the reduced degree of degradation,

so that the predominant effect was probably the toxicity of the humic acids even toward adapted microorganisms.

7. In short-term experiments with carbofuran (2,3-dihydro-2,2-dimethyl-7-benzofuranyl-*N*-methylcarbamate), degradation was accomplished by organisms in an enrichment culture obtained from soils with a low carbon content where sorption of the substrate is low, though it was essentially absent in cultures obtained from soils with a high organic matter (Singh and Sethunathan 1992).

8. Experiments examined a chlorocatechol-contaminated sediment, and interstitial water prepared from it. These showed that the concentrations of total 3,4,5-tri- and tetrachlorocatechols (i.e., including the fraction that is released only after alkaline extraction) were apparently unaltered during prolonged incubation even after addition of cultures with established dechlorinating capability for the soluble chlorocatechols (Allard et al. 1994).

Fewer controlled experiments have been carried out for purely aquatic systems. Montmorillonite complexes with benzylamine at concentrations <200 µg/L decreased the extent of mineralization in lakewater samples, although a similar effect was not noted with benzoate (Subba-Rao and Alexander 1982). Even in apparently simple systems, general conclusions cannot, therefore, be drawn even for two structurally similar aromatic compounds, both of which are readily degradable under normal circumstances in the dissolved state.

All the preceding investigations have been concerned with polar compounds for which plausible mechanisms for their association with organic components of water, soil, and sediment may be more readily conceptualized. To provide a wider perspective, examples are given below for neutral compounds:

1. The aerobic mineralization of α-hexachloro[*aaaaee*]cyclohexane by endemic bacteria in the soil is limited by the rate of its desorption and by intraparticle mass transfer (Rijnaarts et al. 1990).

2. Whereas degradation of the readily extractable toluene in spiked soil by *Pseudomonas putida* was rapidly accomplished, there was a residue that was degraded much more slowly at a rate that was apparently dependent on its desorption (Robinson et al. 1990).

3. The extent of bioremediation of sediments contaminated with PCBs appears to be limited by the association of a significant fraction with organic components of the sediment phase (Harkness et al. 1993).

4. Immobilization of neutral xenobiotics in soils by quaternary ammonium cations has been established, and its significance on the bioavailability of naphthalene to bacteria has been examined. Bioavailability was determined by the rates of desorption, and these differed between a strain of *Pseudomonas putida* and one of *Alcaligenes* sp. (Crocker et al. 1995).

5. An important development has been the isolation of bacteria that were able to degrade phenanthrene that was sorbed to humic acid material (Vacca et al. 2005). Enrichment was carried out with PAH-contaminated soils using phenanthrene sorbed to commercial humic acid. Only the strains isolated from this enrichment were able to carry out degradation of ^{14}C-labeled phenanthrene, and this exceeded by factors of 4–9 the amount estimated to be available from the aqueous phase alone. It was suggested that specially adapted bacteria might interact specifically with naturally occurring colloidal material.

The results of these experiments in both aquatic systems and terrestrial systems may profitably be viewed against the extensive evidence for the persistence of agrochemicals in the terrestrial environment. Considerable effort has been directed to the issue of bound residues of agrochemicals (Calderbank 1989) and to its significance in determining both their biological effects and their persistence. This is now fully accepted in contemporary thinking. At the same time, it should be appreciated that from an economic point of view, enhanced rates of degradation of agrochemicals in the terrestrial environment may be highly undesirable (Racke and Coats 1990). Nonetheless, care should be exercised in making generalizations. For example, whereas it has been established that soil microorganisms may significantly increase the evolution of $^{14}CO_2$ from ^{14}C parathion (Racke and Lichtenstein 1985), this was not observed with chlorpyrifos (Racke et al. 1990).

The analysis of these diverse observations clearly demonstrates that the persistence of a xenobiotic in the aquatic or in the terrestrial environment may be significantly greater when it is bound either to inorganic minerals or to any of a range of complex natural polymers such as humic and fulvic acids. One of the key issues is the rate of desorption of a xenobiotic from the matrices. This may depend critically on the mechanism of the association, and the extent of bioavailability may also depend on the organisms. For example, (a) soil-sorbed naphthalene was degraded at markedly different rates by two naphthalene-degrading organisms (Guerin and Boyd 1992), and (b) at low substrate concentrations, 2,4-dichlorophen-oxyacetate was degraded at different rates by the two strains that were examined (Greer and Shelton 1992). Although a number of important unresolved issues remain, it is clear that the degree of bioavailability of xenobiotics in natural systems introduces an important additional uncertainty in extrapolating to processes and rates in natural ecosystems the results of studies on biodegradation and biotransformation of "free" xenobiotics in laboratory experiments.

This is important not only in field investigations. Even in laboratory experiments on the metabolism of xenobiotics, problems of association between the substrate and the microbial cells may occur. If this were not quantitatively evaluated or eliminated, the results and interpretation of such experiments would be seriously compromised.

Increasing attention has been directed to the degradation of xenobiotics in aquifers, and it has been shown that most of the relevant bacteria are associated with fine particles rather than existing as free entities. It is, therefore, important to include such material in laboratory experiments using unenriched communities that attempt to simulate natural conditions (Holm et al. 1992). The interdependence of surfactant sorption and biodegradability is supported by the results of laboratory experiments. For example, *Pseudomonas* sp. strain DES1 that was able to degrade sodium dodecyl sulfate attached to sediment particles did so more effectively than organisms that were unable to degrade analogous (nondegradable) substrates (Marchesi et al. 1994). Although this was attributed to the effect of the metabolite dodecan-1-ol, further study has revealed a more complex situation (Owen et al. 1998) as a result of differences in the responses of different strains and toward different surfactants. Bacterial associations with particulate material should also be evaluated in the context of the bioavailability of the substrate and in bioremediation strategies.

4.13 Substrate Concentration, Transport into Cells, and Toxicity

Procedures used for growth of organisms to be used in studies of biodegradation may be pragmatically divided into major groups:

1. Complex media for organisms that have undetermined and complex nutritional requirements, including some fungi and anaerobic bacteria.

2. Defined mineral media containing only the substance whose biodegradation is being examined.

3. Use of analogs of the substrate, for example, biphenyl that has been widely used for isolating PCB-degrading organisms.

4. Substrates that are related to those naturally available to the degrading organism but unrelated to that of the contaminant have been exploited occasionally, and have been discussed in a review (Singer et al. 2003).

   a. The use of a range of coumarin (esculetin) and flavanone (quercetin, naringenin) glycones and aglycones, and a flavane (catechin) (Figure 4.11) as substrates for the anaerobic dechlorination of chlorocatechols (Allard et al. 1991, 1992).

   b. The growth of established PCB-degrading bacteria by growth on a range of flavones, flavanols, and flavanones (catechin, naringenin, and myricitin) (Figure 4.11) (Donnelly et al. 1994), and the induction by terpenoids, including carvone and limonene (Figure 4.11) of the degradation of PCB congeners (Gilbert and Crowley 1997).

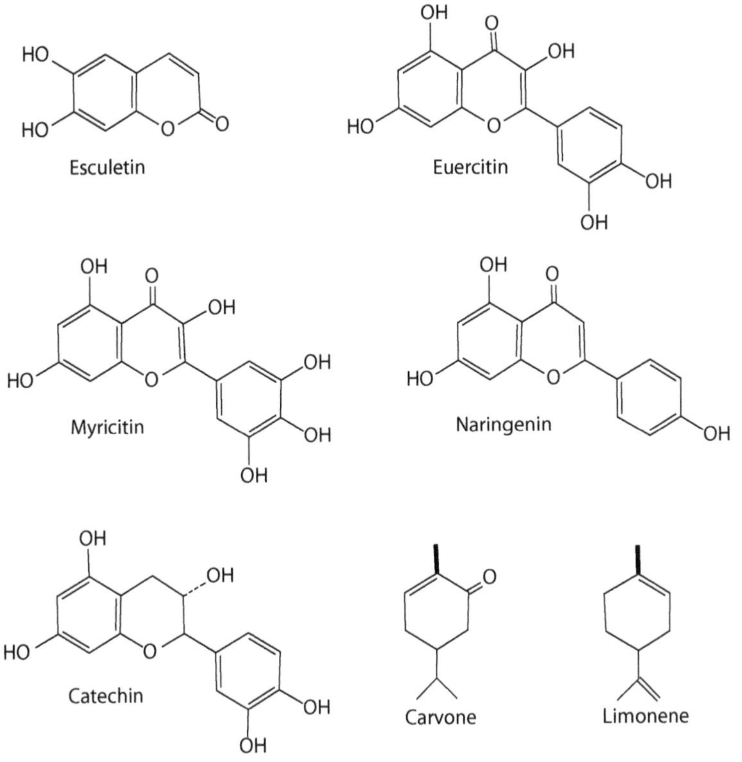

FIGURE 4.11 Structures of natural products used as substrates or inducers.

c. The differential enantioselectivity of carvone, cymene, and biphenyl on the degradation of the atropisomeric PCBs 45, 84, 91, and 95 with *ortho*-chlorinated congeners was examined using both Gram-positive and Gram-negative PCB-degrading strains: carvone and cymene-grown cultures showed the greatest extent of biotransformation (Singer et al. 2002).

d. The degradation of trichloroethene by *Rhodococcus* sp. strain L4 was induced by citral, limonene, and cumene (Suttinun et al. 2009), although continued degradation required the periodic addition of cumene or limonene (Suttinun et al. 2010).

4.13.1 Utilization of Low Substrate Concentrations

Most laboratory experiments on biodegradation and biotransformation have been carried out using relatively high concentrations of the appropriate substrates, even though these may be far in excess of those that are likely to be encountered in natural ecosystems (Subba-Rao et al. 1982; Alexander 1985). This limitation is particularly severe in conventional tests for biodegradability, and seriously restricts the degree to which the results of these experiments are environmentally relevant. The investigations by Nyholm et al. (1992) have addressed this important issue in the marine environment by using ^{14}C-labeled substrates. Except in the immediate neighborhood of industrial discharge, low concentrations of xenobiotics are almost certainly the rule rather than the exception in natural ecosystems, and a number of significant experimental observations should be taken into consideration.

Investigations of the flora of natural waters have revealed the presence of bacteria able to grow with extremely low substrate concentrations, which have been termed oligotrophs (Poindexter 1981). It is, of course, well established that organisms such as *Aeromonas hydrophila* (van der Kooij et al. 1980), *Pseudomonas aeruginosa* (van der Kooij et al. 1982), and a species of *Spirillum* (van der Kooij and Hijnen 1984) may proliferate in natural waters supplemented with low concentrations of additional organic carbon. It has been suggested that there exist bacteria specially adapted to such conditions, although doubt has been expressed on the absolute distinction between eutrophic and oligotrophic bacteria (Martin and MacLeod 1984). This distinction is blurred by the fact that after prolonged nutrient starvation under laboratory conditions, initially oligotrophic marine bacteria can be isolated on media containing high substrate concentrations (Schut et al. 1993); these organisms are, therefore, facultatively oligotrophic. These authors point out that obligate oligotrophy may be determined by life history rather than by invariant physiological characteristics. One critical issue seems to be the effectiveness and regulation of substrate transport into the cells (Schut et al. 1995). The metabolism of series of 1,2,4,5-tetrachloro-, 1,2,4-trichloro-, and three isomeric dichlorobenzenes was examined in *Burkholderia* sp. strain PS14. At nanomolar concentrations in liquid culture, all of them were rapidly metabolized from initial concentrations of 500 nM to below their detection limits, whereas in soil microcosms tetrachlorobenzene at 65 ppb or trichlorobenzene at 54 ppb was metabolized only after 72 h (Rapp and Timmis 1999).

An important development that improved the range of bacteria that could be isolated from soil samples in pure culture has been described by Janssen et al. (2002). They used a medium containing 80 mg/L nutrient broth for isolation and succeeded in obtaining representatives of the divisions Actinobacteria, Acidobacteria, and Verrucomicrobia, which were identified on the basis of matching 16S rRNA gene sequences. An impressively high rate of isolation was achieved, although the reasons for success have not been finally resolved; these could include the lengthy incubation, the use of gellan gum in place of agar for solid media, or the addition of $CaCl_2$. This procedure made it possible to isolate the first pure culture of *Chthoniobacter flavus* (Sangwan et al. 2004). Additional experiments using the V55 medium with a range of carbon sources at 0.5 mM concentration and incubation times up to 2 months resulted in extension of the range of bacteria belonging to the phyla Acidobacteria, Actinobacteria, and Proteobacteria (Davis et al. 2005).

4.13.2 Existence of Threshold Concentrations

This has received only fragmentary attention but is relevant when bacteria are exposed to extremely low concentrations of a xenobiotic—of the order of nanograms per liter or less. For example, although the rates of biodegradation of phenol, benzoate, benzylamine, 4-nitrophenol, and di(2-ethylhexyl)phthalate in natural lake water were linear over a wide range of substrate concentrations between nanograms per liter and micrograms per liter (Rubin et al. 1982), it has been shown that the rates of degradation of 2,4-dichlorophenoxyacetate at concentrations of the order of micrograms per liter were extremely low (Boethling and Alexander 1979). This observation has subsequently been extended to a greater range of compounds (Hoover et al. 1986). These and other data could be interpreted as supporting the concept of a threshold concentration below which growth and degradation does not take place—or occurs at insignificant rates. Although the reasons for the existence of such threshold concentrations have not been entirely resolved, a number of plausible hypotheses may be put forward:

1. The substrate concentrations may be too low for effective transport into the cells.
2. There may be a limiting substrate concentration required for induction of the appropriate catabolic enzyme; at low substrate concentrations the necessary enzymes would simply not be synthesized, and this could be the determining factor in some circumstances (Janke 1987).

Two contrasting results are instructive. Experiments with chlorinated benzenes in which the effect of substrate concentration was examined in batch cultures and in recirculating fermentors showed that although substrates could be degraded completely in the former, a residual concentration of the substrate persisted in the latter (Van der Meer et al. 1992). However, experiments using *Burkholderia* sp. strain PS14 failed to detect residual concentrations after mineralization of 1,2,4,5-tetra- and 1,2,4-trichlorobenzene at concentrations >0.5 nM (Rapp and Timmis 1999).

All these observations emphasize that tests for biodegradability carried out at high substrate concentrations may not adequately predict the rates of degradation occurring in natural ecosystems where only low concentrations of xenobiotics are encountered (Alexander 1985). This phenomenon is, therefore, of enormous environmental importance, since it would imply the possible extreme persistence of low concentrations in natural ecosystems. The further exploration of this phenomenon is probably only limited in practice by the access to analytical methods for measuring sufficiently accurate substrate concentrations at the level of nanograms per liter or lower. Most studies have been carried out using ^{14}C-labeled substrates, which has restricted the range of accessible compounds, although the use of ^{13}C-labeled substrates and application of NMR could extend the range of possibilities.

4.13.3 Strategies Used by Cells for Substrates with Low or Negligible Water Solubility

This issue is critical to the design of laboratory experiments on biodegradation and biotransformation. Since many contaminants such as alkanes, PAHs, and polyhalogenated compounds, including PCBs, belong to this class of compounds, water solubility may present a serious obstacle in attempts at the bioremediation of contaminated terrestrial systems. Indigenous organisms may have been able to develop their own strategies to circumvent low water solubility of their substrates, and a few examples of mechanisms are given below as illustration:

1. They may produce extracellular enzymes, which attack the substrate without the need for transport into the cell, for example, cellulase, DNAse, or gelatinase.
2. They may synthesize surface-active emulsifying compounds during growth. This problem has been extensively investigated on account of its commercial application (Gerson and Zajic 1979), and details of the structural aspects of biosurfactants elaborated for the degradation of water-insoluble substrates have been given (Hommel 1994). Some of the key conclusions from a wide range of studies are briefly summarized:
   a. Glycolipids consisting of long-chain carboxylic acids and rhamnose (Itoh and Suzuki 1972; Rendell et al. 1990) or trehalose (Suzuki et al. 1969; Singer et al. 1990) have been isolated during the growth of a number of different bacteria on *n*-alkanes. The rhamnolipid surfactant produced by a strain of *Pseudomonas* sp. was effective in enhancing degradation of octadecane (Zhang and Miller 1992), though the concentration used in these experiments was rather high (300 mg/L) to encourage its practical application. A model that included the effect of rhamnolipids on solubilization, biodegradation, and bioavailability within surfactant micelles has been presented to rationalize the data from batch studies on the dissolution, bioavailability, and biodegradation of phenanthrene (Zhang et al. 1997).
   b. A polyanionic heteropolysaccharide (emulsan) is produced during the growth of a strain of *Acinetobacter calcoaceticus* with hydrocarbon mixtures, and the high-molecular-weight polymer is necessary for emulsifying activity (Shoham and Rosenberg 1983). The value of emulsan for treating oil spills seems, however, equivocal in the light of results that demonstrate reduced biodegradation in its presence (Foght et al. 1989).
   c. *Rhodococcus* sp. synthesized a glycolipid during growth with *n*-alkanes and *n*-alkanols though not with carboxylic acids, triglycerides, or carbohydrates, and its formation was favored by nitrogen limitation (Singer and Finnerty 1990).
   d. *Pseudomonas maltophilia* produced an extracellular surfactant during growth with naphth-1-oate, and displayed much greater emulsifying activity toward aromatic hydrocarbons with one or two rings than toward aliphatic hydrocarbons (Phale et al. 1995).
   e. A strain of *Pseudomonas* sp. produced a surfactant in the presence of high concentrations of glucose or mannitol, and naphthalene or phenanthrene (Déziel et al. 1996).
   f. The synthesis of an emulsifying agent produced by *Candida lipolytica* is inducible during growth with a number of *n*-alkanes, but it is not synthesized during growth with glucose (Cirigliano and Carman 1984).

Different mechanisms have, therefore, clearly emerged and it seems premature to draw general conclusions, especially in the application of synthetic and natural surfactants to bioremediation, which is discussed in greater detail in Chapter 14. It is important to note, however, that the production of biosurfactants may not be the only mechanism for facilitating the uptake of substrates with low water solubility. For a strain of *Rhodococcus* sp. that did not produce surfactants, the rates of degradation of pyrene dissolved in water or in the nondegradable 2,2,4,4,6,8,8-heptamethyl-nonane exceeded those predicted for physicochemical transfer from the solvent to the aqueous phase. They could, however, be accounted for on the basis of uptake both from the phase interface and from the aqueous solution (Bouchez et al. 1997).

4.13.4 Transport Mechanisms

The mechanisms whereby carbohydrates, carboxylic acids, and glycerol are transported across the bacterial cell membrane prior to metabolism have been elucidated in great detail. Attention should, however, be drawn to an investigation of the regulation of the transport of amino acids and glucose into a facultatively oligotrophic marine bacterium that revealed the lack of specificity of the constitutive system for amino acids, and the interaction of the regulatory systems for amino acids and glucose (Schut et al. 1995). It should also be noted that different transport systems for C_4 dicarboxylic acids into *Escherichia coli* operate under aerobic and anaerobic conditions (Six et al. 1994) so that uptake is dependent on the physiological state of the cells.

   In contrast, less effort has apparently been directed to the transport of xenobiotics, and there is an intrinsic difficulty that in contrast to organisms that utilize carbohydrates or amino acids, suitable mutants defective in the metabolism of the substrate may not be available. This limitation makes it impossible to determine directly whether active transport is involved. Although the genes encoding permeases have been described quite frequently, details of their mechanisms have been less well documented:

1. It has been suggested that active transport systems for benzoate (Thayer and Wheelis 1982) and for mandelate (Higgins and Mandelstam 1972) are involved. In *Rhizobium leguminosarum*, 4-hydroxybenzoate hydroxylase activity is required for the uptake of 4-hydroxybenzoate (Wong et al. 1994), while in *Pseudomonas putida*, a gene cluster *pcaRKF* is involved not only in the transport of 4-hydroxybenzoate into the cells but also their chemotactic response to the substrate and its degradation by ring hydroxylation (Harwood et al. 1994; Nichols and Harwood 1997). The situation for phenylacetate transport into *Pseudomonas putida* U is apparently different (Schleissner et al. 1994), and this is consistent with the fact that the pathway for degradation of phenylacetate by this strain is distinct from that used for 4-hydroxyphenylacetate.

2. A detailed investigation by Groenewegen et al. (1990) has examined the uptake of 4-chlorobenzoate by a coryneform bacterium that was able to degrade this compound. The uptake was inducible and occurred in cells grown with 4-chlorobenzoate but not with glucose. A proton motive force (Δp)–driven mechanism was almost certainly involved, and uptake could not take place under anaerobic conditions unless an electron acceptor such as nitrate was present.

3. The transport of toluene-4-sulfonate into *Comamonas testosteroni* has been examined (Locher et al. 1993), and rapid uptake required growth of the cells with toluene-4-sulfonate or 4-methylbenzoate. From the results of experiments with various inhibitors, it was concluded that a toluenesulfonate anion/proton symport system operates rather than transport driven by a difference in electrical potential ($\Delta \psi$), and uptake could not take place under anaerobic conditions unless an electron acceptor such as nitrate was present.

4. The uptake of benzoate was examined in two strains of *Alcaligenes denitrificans*. The transport system was inducible, carrier-mediated, energy-dependent, and involved a proton symport system. In contrast, the uptake of 2,4-dichlorobenzoate by one of the strains was constitutive, displayed no saturation kinetics, and appeared to occur by passive diffusion (Miguez et al. 1995). The uptake of 2,4-dichlorophenoxyacetate has been studied in *Ralstonia eutropha* JMP134(pJP4) in which the degradative genes are plasmid-borne. Uptake was inducible by the substrate, did not occur with fructose-grown cells, and was inhibited by cyanide that prevents development of a protomotive force, and by the protophore carbonylcyanide-3-chlorophenylhydrazone (Leveau et al. 1998). The protein involved was encoded by an open reading frame on the plasmid designated *tfdK*.

5. *Sphingobium* (*Sphingomonas*) *herbicidovorans* strain MH is able to grow at the expense of both enantiomers of 2-(2,4-dichlorophenoxy)propionate with a preference for the *S*-enantiomer, and the uptake of both enantiomers, and of 2,4-dichlorophenoxyacetate is inducible. Although the ATPase inhibitor *N,N'*-dicyclohexylcarbodiimide (DCCD) had only slight effect on intracellular levels of ATP, uptake was inhibited by the protophore carbonylcyanide-3-chlorophenylhydrazone and by nigericin that dissipates ΔpH in the presence of high concentration of K+, but not by valinomycin. It was suggested that uptake is driven by a protomotive force (Δ*p*), and that ΔpH rather than Δψ is the determinant of uptake (Zipper et al. 1998).

6. In *Oxalobacter formigenes*, oxalate and its decarboxylation product formate form a one-to-one antiport system, which involves the consumption of an internal proton during decarboxylation, and serves as a proton pump to generate ATP by decarboxylative phosphorylation (Anantharam et al. 1989).

7. The transport of EDTA into a bacterial strain capable of its degradation has been examined (Witschel et al. 1997). Inhibition was observed with DCCD (ATPase inhibitor), nigericin (dissipates ΔpH), but not valinomycin (dissipates Δψ), and was dependent on the stability constant of metal–EDTA complexes.

8. A graphic example of the significance of effective transport is provided by an aerobic reductase from *Xenophilus azovorans*. This was expressed in *Escherichia coli*, and was able to reduce a range of important sulfonated colorants, even though whole cells were unable to do so (Blümel et al. 2002).

These results illustrate both the potential complexity and the significance of transport systems in bacteria. This has hitherto been a rather neglected aspect of the degradation of xenobiotics, and extension to other organisms and to a wider range of xenobiotics is clearly merited.

4.14 Preexposure: Pristine and Contaminated Environments

Although experimental aspects of the elective enrichment procedure are discussed in Chapter 5, the question of its existence and significance in natural populations already exposed to xenobiotics is addressed here. It is important to distinguish between induction (or derepression) of catabolic enzymes, and selection for a specific phenotype. The former is a relatively rapid response, so that exposure of samples from uncontaminated areas to xenobiotics for a period of weeks or months would be expected to result in the selection of organisms with degradative potential rather than merely be the result of low rates of enzyme induction. Caution should, however, be exercised in establishing a correlation between exposure to xenobiotics and the existence of organisms with the relevant degradative capacity. The following diverse examples are given as illustration:

1. It has been shown (Kamagata et al. 1997) that bacteria isolated from a pristine site with no established contamination were capable of degrading 2,4-dichlorophenoxyacetate (2,4-D) and differed from those traditionally isolated from contaminated sites. The new isolates grew slowly and although one of them could be assigned to the genus *Variovorax* and carried the *tfdA* gene, the other five did not and had no 2,4-D-specific 2-oxoglutarate-dependent dioxygenase activity.

2. These results are consistent with previous evidence (Fulthorpe et al. 1996) from a study of pristine soils that although populations existed that could degrade both 3-chlorobenzoate and 2,4-dichlorophenoxyacetate, isolation of the latter strains was generally unsuccessful by the methods that were used. These results should be viewed against the general comments on oligotrophs and bradytrophs (slow-growing organisms) that have already been discussed.

3. Soils from putatively pristine areas in Southwest Australia; South Africa; California, USA; Central Chile; Saskatchewan Canada; and Russia were enriched with 3-chlorobenzoate and assayed for mineralization of the substrate (Fulthorpe et al. 1998). Genetic procedures were

used to show that 91% of the genotypes were unique to the sites from which the organisms were isolated. These results suggest that the genotypes were endemic and were not the result of global dispersion of single genotype.

4. Comparison has been made of the degradability in organisms isolated from a contaminated and a pristine system in Canada. It was shown that the established genotypes for the degradation of 3-chlorobenzoate, that is, *clc*, *cba*, and *fcb* that encode enzymes for 3-chlorocatechol 2,3-dioxygenase (Frantz and Chakrabarty 1987), 3,4-(4,5)-dioxygenase (Nakatsu and Wyndham 1993), and the hydrolytic pathway for 4-chlorobenzoate (Chang et al. 1992) were present in strains from the contaminated site. However, these were absent in strains isolated from the pristine site. Based on substrate utilization patterns, these traits were distributed among phenotypically distinct groups (Peel and Wyndham 1999).

5. Degradation of contaminants may occur with bacteria that have been isolated from pristine environments without established exposure to the contaminants, and exhibit no dependence on substrate concentration. For example, organisms from a previously unexposed forest soil were able to degrade 2,4,6-trichlorophenol at concentrations up to 5000 ppm, and terminal restriction fragment length polymorphism analysis revealed that at concentrations up to 500 ppm, the bacterial community was unaltered (Sánchez et al. 2004).

The classic experiments by Tattersfield (1928) clearly showed that the biocidal effect of naphthalene decreased during successive application, and that this was due to the biodegradation of naphthalene. Since then, many results have accumulated that support the view that preexposure increases the number of organisms capable of degrading a given xenobiotic, though fewer attempts have been made to quantify the number of organisms involved. There is convincing evidence that exposure to unusual substrates in laboratory experiments elicits the synthesis of genes for their degradation (Mortlock 1982). Increasing support for the view that exposure to xenobiotics increases the probability of mutations that are favorable to the degradation of these substrates has also been found (Hall 1990; Thomas et al. 1992). The following examples attempt to illustrate the spectrum of responses that have been observed:

1. Rates of mineralization of the more readily degraded PAHs such as naphthalene and phenanthrene were greater in samples from PAH-contaminated areas than in those from pristine sediments, although it is significant that even in the former, the rates for benz[*a*]anthracene and benzo[*a*]pyrene were extremely low (Herbes and Schwall 1978). Examination of the sequences of 16S rRNA and of naphthalene dioxygenase [Fe–S] protein genes (*nahAc*) of nine strains of bacteria capable of degrading naphthalene and isolated from the same site showed that whereas in seven strains the dioxygenase differed by as much as 7.9%, they had a single *nahAc* allele. All strains contained plasmids of different sizes that contained the gene for naphthalene degradation, and it was suggested that horizontal transfer of plasmids might play a role in the adaptation of microbial communities to xenobiotics (Herrick et al. 1997).

2. Experiments using marine sediment slurries have examined the effect of preexposure to various aromatic hydrocarbons on the rate of subsequent degradation of the same, or other hydrocarbons. The results clearly illustrated the complexity of the selection process: for example, whereas preexposure to benzene, naphthalene, anthracene, or phenanthrene enhanced the rate of mineralization of naphthalene, similar preexposure to naphthalene stimulated the degradation of phenanthrene but had no effect on that of anthracene (Bauer and Capone 1988).

3. In experiments using soil samples from a pristine aquifer exposed in the laboratory to a range of compounds, the following widely diverse responses were observed (Aelion et al. 1987, 1989):

   a. The bacterial population was apparently already adapted to some of the compounds such as phenol, 4-chlorophenol, and 1,2-dibromoethane at the start of the experiment, and these substrates were, therefore, rapidly degraded.

   b. No adaptation was found for chlorobenzene or 1,2,4-trichlorobenzene, and only slight mineralization was observed.

  c. A linear increase in the rate of degradation with increasing length of exposure was noted for some substrates such as aniline, which are generally regarded as readily degradable.

  d. True adaptation by selection of the appropriate organisms was observed only for 4-nitrophenol.

4. Systematic studies on the degradation of 4-nitrophenol (Spain et al. 1984) showed that the rates of adaptation in a natural system were comparable to those observed in a laboratory test system and were associated with an increase in the number of degrading organisms by up to 1000-fold.

5. It has been consistently observed that a wide range of agrochemicals applied successively to the same plots are increasingly readily degraded, presumably due to enrichment of the appropriate degradative microorganisms (Racke and Coats 1990). Examples include compounds as diverse as naphthalene (Gray and Thornton 1928), γ-hexachloro[*aaaeee*]-cyclo-hexane (Wada et al. 1989), and triazines (Cook and Hütter 1981).

6. The dehalogenation of polychlorinated or polybrominated biphenyls was more rapid in cultures using inocula prepared from sediments contaminated with the chlorinated or brominated biphenyl, respectively (Morris et al. 1992).

7. The enrichment cultures from soils treated with the urea herbicide linuron for more than 10 years showed specific degradation of linuron and the related bromuron, though not the urea herbicides lacking the methoxy group. Samples from untreated soil did not display this behavior (El-Fantroussi 2000). The use of reverse transcription-PCR and denaturing gradient gel electrophoresis showed that a bacterial consortium was required for the complete degradation of linuron. *Variovorax* sp. strain WDL1, *Delftia acidovorans* WDL34, and *Pseudomonas* sp. strain WDL5 could be isolated directly from the enrichment, and *Hyphomicrobium sulfonivorans* WDL6 and *Comamonas testosteroni* strain WDL7 from subenrichments (Dejonghe et al. 2003). Only *Variovorax* sp. WDL1 was, however, able to use linuron as the sole source of carbon and energy.

8. Long-term laboratory enrichments have been shown to yield cultures that were able to degrade initially recalcitrant compounds. Several examples have been given by Slater, who was a pioneer in this area (Slater and Lovatt 1984). An additional illustration is provided by an organism that was able to grow at the expense of 4,4′-dicarboxyazobenzene. This had been isolated during prolonged continuous cultivation (≈500 days) under nonsterile conditions using the more complex naphthalene analogs (Kulla 1981). Pure cultures of *Xenophilus azovorans* KF46 and *Pigmentiphaga kullae* were then isolated from the adapted culture (references in Blümel et al. 2002).

It, therefore, seems premature to draw general conclusions on the influence of preexposure on the biodegradability of structurally different substrates.

These results are particularly relevant to bioremediation, and suggest that organisms originally isolated from sites either contaminated naturally (Fredrickson et al. 1991) or as a result of industrial activity (Grosser et al. 1991) may be particularly suitable. The discussion hitherto has been devoted to the issue of selection, but for the sake of completeness it should be noted that the alternative approach of deliberately adding an inducer has also been examined. Salicylate is an inducer of the enzymes for degradation of naphthalene. Its addition to soil has been shown to result in a modest increase in the number of organisms degrading naphthalene (Ogunseitan et al. 1991), and in the mineralization of some 4- and 5-ring PAHs (Chen and Aitken 1999). However, since naphthalene and phenanthrene are readily degradable compounds, the wider evaluation of this interesting idea has to be explored before its application to practical situations can be justified. Substrates whose structures are unrelated to the contaminant have been noted for PCB degradation: (a) growth on a range of flavones, flavanols, and flavanones (catechin, naringenin, and myricitin) (Donnelly et al. 1994) and (b) induction by terpenoids, including carvone and limonene (Gilbert and Crowley 1997).

This discussion may also be viewed against the background of studies on genetic transfer of catabolic activity toward a given xenobiotic. Considerable care should, however, be exercised in the interpretation even of data that seem supportive of this process, since selection and enrichment from a small population of organisms initially present must be excluded. One interesting investigation on the biodegradation of aniline revealed the existence of two genotypes differing in their tolerance to the substrate. Although the dominant organism that was originally present assimilated aniline at micromolar concentrations, it was inhibited at higher concentrations. A mutant could, however, be isolated from a population of several hundred cells or by continuous culture, and this organism tolerated millimolar concentrations. Populations of the two organisms in a natural system were apparently regulated by the prevailing concentration of aniline (Wyndham 1986).

Although great advances have been made using the technology of molecular biology to determine the components of natural populations without the limitations of specific enrichment, it is hardly possible to escape the conclusion that our understanding of the processes regulating the population dynamics of microorganisms in natural systems is still limited. Details of procedures for investigating natural populations in the context of bioremediation are addressed in Chapter 13.

4.15 Rates of Metabolic Reaction

4.15.1 Kinetic Aspects

The rates at which xenobiotics are degraded or transformed are of cardinal importance, since it is upon their quantification that a given compound can, in the final analysis be designated persistent or otherwise (Battersby 1990). It should be appreciated at the outset that even if acceptable rates of degradation are observed in laboratory experiments, the final assessment of persistence depends upon the demonstration that the compound is indeed degraded under natural conditions. Ultimately, this is of primary environmental significance. It has already been pointed out that persistence is determined not only by rates of biotic and abiotic degradation but also by the accessibility of the substrate that may be associated with organic or inorganic components in the water mass, or in the soil or sediment phases. There are a number of additional issues that should be addressed in the discussions of rates:

1. Even if rate constants can be measured in laboratory experiments, these must be normalized to the number of microbial cells. This may pose only minor problems with an axenic culture in the laboratory, and has been consistently carried out in investigations where well-defined kinetics prevailed (Allard et al. 1987). However, this becomes a major problem in natural situations: how many of the organisms are metabolically active in accomplishing the given reaction? Use of specific DNA probes has been used for detection of genes coding for heavy-metal resistance (Diels and Mergeay 1990), and for the detection of pathogens (Samadapour et al. 1990). This procedure has, however, been less extensively applied to organisms of catabolic significance (Sayler et al. 1985; Holben et al. 1988), although the values obtained for 2,4-dichlorophenoxy-acetic-degrading populations agreed well with those using conventional most-probable-number methods (Holben et al. 1992).

2. Although relevant models have been assembled (Simkins and Alexander 1984) and their application evaluated (Simkins et al. 1986), microbial reactions may not follow well-defined kinetics. They may, for example, exhibit multiphase kinetics that has been illustrated during the transformation of methyl parathion by a *Flavobacterium* sp. (Lewis et al. 1985). System I was a high-affinity, low-capacity system, whereas system II was the opposite. In experiments with mixed microbial cultures on the degradation of phenol, 4-chlorophenol, 4-methylphenol, acetone, and methanol, multiphase kinetics were encountered. Failure to take this into account would have resulted in serious errors in evaluating the rates of degradation (Hwang et al. 1989). The degradation of 1,2,4-trichlorobenzene by *Burkholderia* sp. strain PS14 at low substrate concentrations showed a first-order relation between the specific rate of transformation and the substrate concentration. However, at higher concentration, this was replaced by a second-order one (Rapp 2001). The results of these studies illustrate only

one of the several factors that may invalidate predictions that fail to take into account multi-phase kinetics.

3. Attempts have been made to apply the structure–activity concept (Hansch and Leo 1995) to environmental problems, and this has been successfully applied to the rates of hydrolysis of carbamate pesticides (Wolfe et al. 1978), and of esters of chlorinated carboxylic acids (Paris et al. 1984). This has been extended to correlating rates of biotransformation with the structure of the substrates and has been illustrated with a number of single-stage reactions. Clearly, this approach can be refined with the increased understanding of the structure and function of the relevant degradative enzymes. Some examples illustrate the application of this procedure:

   a. Rates of bacterial hydroxylation of substituted phenols to catechols by *Pseudomonas putida* correlated well with the van der Waals radii of the substituents (Paris et al. 1982). This was also demonstrated for the biotransformation of anilines to catechols both by this strain and by a natural population of bacteria (Paris and Wolfe 1987).

   b. Rates of hydrolysis of substituted aromatic amides by the bacterial population of pond water correlated well with the infrared C·O stretching frequencies of the substrates (Steen and Collette 1989).

   c. Rates of anaerobic dechlorination of aromatic hydrocarbons (Peijnenburg et al. 1992), and of the hydrolysis of aromatic nitriles under anaerobic conditions (Peijnenburg et al. 1993) have been correlation with a number of parameters, including Hammett σ-constants, inductive parameters, and evaluations of the soil/water partitioning.

For a number of reasons, there are some important limitations to the extension of this principle. Biodegradation—as opposed to biotransformation—of complex molecules necessarily involves a number of sequential reactions each of whose rates may be determined by complex regulatory mechanisms. For novel compounds containing structural entities that have not been previously investigated, the level of prediction is necessarily limited by lack of the relevant data. Too Olympian a view of the problem of rates should not, however, be adopted. An overly critical attitude should not be allowed to pervade the discussions—provided that the limitations of the procedures that are used are clearly appreciated and set forth. In view of the great practical importance of quantitative estimates of persistence to microbial attack, any procedure—even if it provides merely orders of magnitude—should not be neglected.

4.16 Metabolic Aspects: Nutrients

In natural ecosystems, microbial growth and metabolism may be limited by the concentrations of inorganic nutrients such as nitrogen, phosphorus, or even iron. Systematic investigation of these limitations on the biodegradation of xenobiotics has seldom been carried out except when the substrates contain nitrogen or phosphorus in a form that is used for growth of the cells. In growth media containing nitrate, unusual metabolites containing nitro groups have been found. These have been discussed in Chapter 2.

The role of Cu, Ni, Mo, and W are discussed in the context of metalloenzymes in Chapter 3. Only two examples are noted here of the importance of nutrient limitation in determining the realization of biodegradation:

1. Expression of lignin peroxidases in *Phanerochaete chrysosporium* is induced by nitrogen limitation, and by the concentration of Mn(II) in the medium (Perez and Jeffries 1990).

2. Under conditions of selenium starvation, *Clostridium purinilyticum* degrades uric acid by an unusual pathway involving cleavage of the iminazole ring to produce 5,6-diaminouracil, which is then degraded to formate, acetate, glycine, and CO_2 (Dürre and Andreesen 1982).

In experiments using field samples containing natural assemblies of microorganisms, at least two effects of supplementation with nitrogen or phosphorus have been encountered:

1. A decreased lag phase was observed before transformation of 4-methylphenol (Lewis et al. 1986).

2. The size of the bacterial population increased, although there was no effect on the second-order rates of transformation of a number of compounds, including phenol, and the agrochemicals propyl 2,4-dichlorophenoxyacetate, methyl parathion and methoxychlor (Paris and Rogers 1986).

Limitation in the concentrations of these inorganic nutrients does not, therefore, appear to have had a dramatic effect on the persistence of the relatively few compounds that have been examined systematically. However, it has been suggested that competition between organisms for inorganic phosphorus may account for differences in biodegradability when several carbon sources are present (Steffensen and Alexander 1995).

Consistent with the preceding comments on the metabolism of xenobiotics in the presence of additional carbon substrates, the result of deliberate addition of organic carbon may be quite complex and will not be addressed in detail. Two examples on rates of mineralization are given as illustration in which addition of glucose apparently elicited two different responses. It should, however, be emphasized that since the concentration of readily degradable substrates in natural aquatic systems will generally be extremely low, the environmental relevance of such observations will inevitably be restricted:

1. The rate of mineralization of phenol by the flora of natural lake water decreased (Rubin and Alexander 1983).

2. The rate of mineralization of 4-nitrophenol was enhanced in lake water inoculated with a *Corynebacterium* sp. when rates of mineralization were low (Zaidi et al. 1988).

4.17 Regulation and Toxic Metabolites

4.17.1 Regulation

Whereas the following discussion is directed primarily to the role of metabolites, environmental factors may also be important. Some examples including the role of oxygen concentration have already been given as illustration.

Classic studies were devoted to the regulation of the enzymes for conversion of catechol and 3,4-dihydroxybenzoate (protocatechuate) to 3-oxoadipate by *Pseudomonas putida* (Ornston 1966), the 3-oxoadipate pathway in *Moraxella calcoacetica* (Cánovas and Stanier 1967), and the mandelate pathway in *Pseudomonas aeruginosa* (Rosenberg 1971). In these organisms, catechol is degraded by intradiol fission catalyzed by 1,2-dioxygenase, and the induction patterns are briefly summarized. Synthesis of the enzymes for the primary oxidations is induced by the growth substrates, whereas synthesis of those for later transformations is induced by their products—either by *cis,cis*-muconate or by 3-oxoadipate. However, for *Alcaligenes eutrophus* that metabolizes phenol and 4-methylphenol by extradiol fission catalyzed by catechol 2,3-dioxygenase, all six enzymes that result in the production of 4-hydroxy-2-oxovalerate are inducible by the initial growth substrates (Hughes and Bayly 1983). The induction patterns may, however, be quite complex and depend strongly on the growth substrate. For example, *Pseudomonas paucimobilis* (strain Q1) that was isolated by enrichment with biphenyl is capable of utilizing biphenyl, xylene and toluene, salicylate, and benzoate (Furukawa et al. 1983). All the enzymes for degradation of the first four of these that are degraded by 2,3-dioxygenase ring fission were induced by growth on these substrates, whereas growth with benzoate induced only the 1,2-dioxygenase rather than the 2,3-dioxygenase. It is instructive to note that whereas the enzymes of the 3-oxoadipate pathway are generally inducible, this is not always the case: those in slow-growing species of *Bradyrhizobium* spp. are constitutive (Parke and Ornston 1986). Except for aromatic hydrocarbons and related compounds, somewhat fewer studies have been directed to the genetics and the regulation of the enzymes for the degradation of xenobiotics. The following illustrative examples have received extensive study:

a. *Pseudomonas putida* strain F1 that degrades toluene with the methyl group intact (Finette et al. 1984).

b. For the *nah* operon in *Pseudomonas putida* strain G1064 involved in the degradation of naphthalene to salicylate (Eaton and Chapman 1992), the enzymes are generally induced by growth with salicylate (Austen and Dunn 1980).

c. The initial steps in the degradation of biphenyls by *Pseudomonas* sp. strain LB400 (Mondello 1989), by *Pseudomonas pseudoalcaligenes* strain KF707 (Taira et al. 1992), and by *Rhodococcus globerulus* (Asturias and Timmis 1993). This is discussed in greater detail in Chapter 9, Part 2.

Further details of the pathways for the degradation of PAHs are given in Chapter 8, Part 1, and in reviews that are cited there. It seems that most of the degradative enzymes are inducible, and this is consistent with the fact that most strains have been isolated after specific enrichment with the xenobiotic. Synthesis of catechol 1,2-dioxygenase in the yeast *Trichosporon cutaneum* is, however, partially constitutive (Shoda and Udaka 1980). For biotransformation, however, there are sporadic examples of the constitutive synthesis of enzymes. For example, the system carrying out the *O*-methylation of halogenated phenolic compounds was apparently constitutive (Neilson et al. 1988), and is consistent with the isolation of the strains by enrichment with C_1 compounds structurally unrelated to the halogenated substrates. The *O*-methylation reaction may function primarily as a detoxification system, so that in this case constitutive synthesis of the enzyme would clearly be advantageous to the survival of the cells.

Since it is only seldom that organisms are exposed to a single substrate, attention has been directed to the metabolism of mixtures of benzoate and aliphatic carboxylic acids that may be the products of degradation. In *Ralstonia eutropha* (*Alcaligenes eutrophus*), diauxic growth was exhibited when both benzoate and acetate were supplied as substrates, though not when benzoate and succinate were used (Ampe et al. 1997). When both of these were supplied, the growth rate was increased, and under these circumstances there was putatively a more optimal distribution of metabolites among the various anabolic and catabolic pathways. In contrast, the simultaneous presence of acetate and benzoate would result in the production of energy that would exceed the requirements of the cell. However, during growth with both acetate and phenol, synthesis of both phenol hydroxylase and the extradiol dioxygenase involved in the metabolism of catechol decreased, with the result that benzoate blocked diauxic growth with phenol (Ampe et al. 1998).

An interesting and ecologically relevant observation has been made on the induction of the metabolism of PCBs by *Arthrobacter* sp. strain B1B (Gilbert and Crowley 1997). Cells were grown in a mineral medium with fructose and carvone (50 mg/L). Effective degradation of a number of congeners of Arochlor 1242 was induced by carvone that could not, however, be used as a growth substrate and was toxic at high concentrations (>500 mg/L). Other structurally related compounds, including limonene, *p*-cymene, and isoprene, were also effective, and such results may be relevant to bioremediation programs for PCBs. It has also been shown by protein sequencing in *Comamonas testosteroni* that testosterone induces both the synthesis of enzymes that degrade steroids and those required for the degradation of aromatic compounds (Möbus et al. 1997).

The induction of the monooxygenases for the degradation of trichloroethene by aromatic substrates and vice versa is discussed in Chapter 8, Part 1.

4.17.2 Toxic or Inhibitory Metabolites

There are several examples in which metabolites that toxify the organism responsible for their synthesis are produced. The classic example is fluoroacetate (Peters 1952), which enters the TCA cycle and is thereby converted into fluorocitrate. This effectively inhibits aconitase—the enzyme involved in the next metabolic step—so that cell metabolism itself is inhibited with the resulting death of the cell. Walsh (1982) has extensively reinvestigated the problem and revealed both the complexity of the mechanism of inhibition and the stereospecificity of the formation of fluorocitrate from fluoroacetate. It should be noted, however, that bacteria able to degrade fluoroacetate to fluoride exist so that some organisms have developed the capability for overcoming this toxicity (Meyer et al. 1990).

The significance of toxic metabolites is important in diverse metabolic situations: (a) when a pathway results in the synthesis of a toxic or inhibitory metabolite, and (b) when pathways for the metabolism of two (or more) analogous substrates supplied simultaneously are incompatible due to the production of a toxic metabolite by one of the substrates. A number of examples are provided to illustrate these possibilities that have achieved considerable attention in the context of the biodegradation of chlorinated aromatic compounds (further discussion is given in Chapter 9, Part 1):

1. Biotransformation of chlorobenzene by *Pseudomonas putida* grown with toluene or benzene resulted in the formation of 3-chlorocatechol. This inhibited further metabolism by catechol 2,3-dioxygenase, so that its presence resulted in the formation of catechols even from benzene and toluene (Gibson et al. 1968; Klecka and Gibson 1981; Bartels et al. 1984). The same situation has emerged in the degradation of 3-chlorobenzoate: the inhibition that would result from the inhibitory effect of 3-chlorocatechol on catechol 2,3-dioxygenase is lifted by the synthesis of catechol 1,2-dioxygenase (Reinecke et al. 1982). This is indeed a general strategy, and inactivation of extradiol dioxygenases—reversible or otherwise—by other substituted catechols has been reported:

    a. 2,3-Dihydroxybiphenyl 1,2-dioxygenase from a biphenyl-degrading strain of *Pseudomonas* sp. by 4-phenyl catechol that is not a substrate for the dioxygenase (Lloyd-Jones et al. 1995).

    b. Catechol 2,3-dioxygenase by 4-ethyl catechol that is a metabolite of 4-ethylbenzoate (Ramos et al. 1987).

    c. Catechol 2,3-dioxygenase by 3-vinylcatechol produced from the dioxygenation of styrene by *Pseudomonas putida* F1 (George et al. 2011).

    Although this inactivation may present a limitation in the degradative potential of the relevant strains, *Pseudomonas putida* strain GJ31 degrades chlorobenzene via 3-chlorocatechol and extradiol fission (Kaschabek et al. 1998). This is accomplished by a chlorocatechol 2,3-dioxygenase that hydrolyses the initially formed *cis,cis*-hydroxy muconic acid chloride to 2-hydroxymuconate. Thereby the irreversible reaction of the acid chloride with nucleophiles or the formation of pyr-2-one-6-carboxylate as a terminal metabolite is avoided. The catechol 2,3-dioxygenase from this strain encoded by *cbzE* is plasmid-borne, and is capable of metabolizing both 3-chlorocatechol and 3-methylcatechol. It belongs to the 2.C subfamily of type 1 extradiol dioxygenases (Mars et al. 1999). The alternative extradiol fission of 3-chlorocatechol may take place between the 1 and 6 positions (distal fission), and this has been shown for the 2,3-dihydroxybiphenyl 1,2-dioxygenase from the naphthalene sulfonate degrading *Sphingomonas* sp. strain BN6 (Riegert et al. 1998).

    The converse situation occurs during the degradation of 4-chlorobenzoate in which the synthesis of catechol 1,2-dioxygenase circumvents the production of chloroacetaldehyde by the action of catechol 2,3-dioxygenase (Reinecke et al. 1982). It has been shown that the synthesis of protoanemonin may take place during the degradation of 4-chlorobenzoate, which is formed by ring fission of 4-chlorobiphenyl (Blasco et al. 1995), and that this may inhibit the growth of PCB-degrading organisms in the soil (Blasco et al. 1997). Although protoanemonin is produced from chloro-*cis-cis*-muconates, its formation may be circumvented by the action of chloromuconate cycloisomerases to form dienelactones, and by conversion by dienelactone hydrolase into *cis*-acetylacrylate (Brückmann et al. 1998) (Figure 4.12).

2. Ring fission during the degradation of chloroaromatic compounds is generally catalyzed by catechol-1,2-dioxygenases, whereas for the corresponding methyl compound a catechol-2,3-dioxygenase is involved. These pathways are generally incompatible due to the inactivation of catechol-2,3-oxygenase by 3-chlorocatechol. Simultaneous degradation of chloro- and methyl-substituted aromatic compounds has, however, been shown to occur in some strains (Taeger 1988; Pettigrew et al. 1991), while a strain of *Ralstonia* sp. strain JS705 carries genes for both chlorocatechol degradation and toluene dioxygenation (Van der Meer et al. 1998). Although

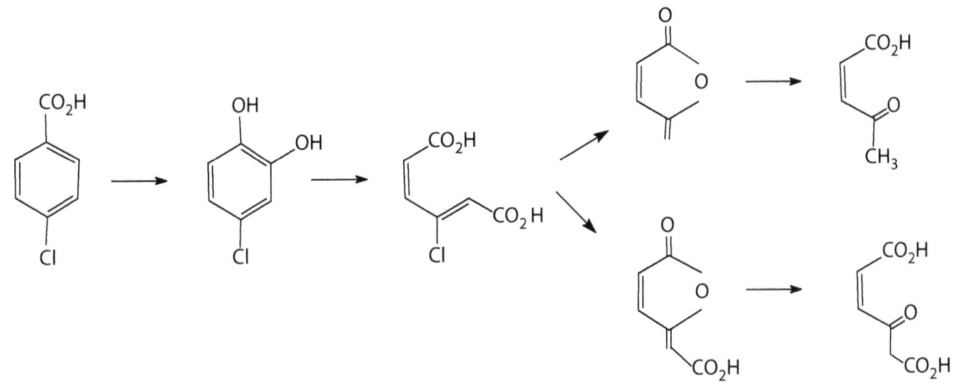

FIGURE 4.12 Biodegradation of 4-chlorobenzoate.

Comamonas testosteroni strain JH5 degraded 4-chlorophenol and 2-methylphenol simultaneously, 4-chlorophenol and 4-methylphenol were degraded sequentially. This involved a catechol-2,3-dioxygenase for both 4-methylcatechol and 4-chlorocatechol that was induced by growth with 4-chlorophenol or 4-methylphenol (Hollender et al. 1994). *Phanerochaete chrysosporium* is able to degrade simultaneously chlorobenzene and toluene (Yadav et al. 1995). This phenomenon is highly relevant to the biological treatment of industrial effluents, since most of these consist of complex mixtures of substrates, and to bioremediation.

3. Total degradation of PCBs necessitates degradation of the chlorobenzoates produced by dioxygenation, dehydrogenation, and ring fission. Metabolism of chlorobenzoate may produce chlorocatechols and then muconic acids. It has been shown that in *Pseudomonas testosteroni* strain B-356, the metabolites from 3-chlorobenzoate strongly inhibited the activity of 2,3-dihydroxybiphenyl 1,2-dioxygenase, and, therefore, the degradation of the original PCB substrates (Sondossi et al. 1992). This inhibition is reminiscent of the inhibition of catechol 2,3-dioxygenase by 3-chlorocatechol, which has already been noted.

4. Simultaneous metabolism of even closely related substrates may be restricted by the synthesis of inhibitory metabolites. For example, cells of an *Acinetobacter* sp. grown with 4-chlorobenzoate could dehalogenate 3,4-dichlorobenzoate even though the organism cannot use this as the sole growth substrate, and the dichlorobenzoate inhibited both its own metabolism and that of the growth substrate 4-chlorobenzoate (Adriaens and Focht 1991a,b).

5. A strain of *Pseudomonas* sp. grew with a wide range of aromatic compounds, including phenol, benzoate, benzene, toluene, naphthalene, and chlorobenzene (Haigler et al. 1992), and mixtures of these substrates were degraded in continuous cultures without evidence for accumulation of metabolites. In this case, degradation depended on the presence of a nonspecific toluene dioxygenase, and of induction of enzymes for the intradiol (*ortho*) fission, extradiol (*meta*) fission, and the "modified intradiol fission (*ortho*) pathway" for the degradation of the catechols. The results implied the existence of a metabolic situation in addition to that normally encompassed by the terms biodegradation and biotransformation—enzymes induced by the presence of one substrate facilitated the degradation of another substrate that is not normally used as the sole source of carbon and energy. These results have obvious implications for implementing bioremediation programs.

6. There may be several reasons why an analog substrate cannot be metabolized by an organism. This is well illustrated by 4-ethylbenzoate that could not be used by strains that degraded 3- and 4-methylbenzoate (Ramos et al. 1987). There were two reasons:

    a. The metabolite 4-ethylcatechol inactivated catechol 2,3-dioxygenase that is obligatory for its degradation.

    b. 4-Ethylcatechol, in contrast to 3- and 4-methylcatechol, does not activate the xylS protein whose gene is the positive regulator of the promoter of the TOL plasmid extradiol fission pathway. These limitations could, however, be overcome by the construction of mutant strains.

7. The degradation of trichloroethene by methylotrophic bacteria involves the epoxide as intermediate (Little et al. 1988). Further transformation of this may produce CO that can toxify the bacterium, both by competition for reductant and by enzyme inhibition (Henry and Grbic-Galic 1991). The inhibitory effect of CO may, however, be effectively overcome by adding a reductant such as formate.

8. The growth of a strain of *Alcaligenes* sp. that could degrade 2-aminobenzenesulfonate was inhibited by equimolar concentrations of 3-methylbenzoate, although an exconjugant after insertion of the TOL catabolic genes was able to carry out sequential degradation of both substrates (Jahnke et al. 1993).

4.18 Catabolic Plasmids

Plasmids may be defined as fragments of DNA that replicate outside the bacterial chromosome and they are important in a number of different contexts:

1. As carriers of antibiotic resistance (Chapter 3, Part 4): the emergence of antibiotic-resistant strains has had serious repercussions in the application of antibiotic therapy, and has seriously increased the danger of nosocomial infections.

2. The presence of unusual carbohydrate fermentation patterns (particularly for lactose), and the ability to use citrate among Enterobacteriaceae has hindered, and sometimes jeopardized, the identification of pathogenic strains, including *Salmonella typhi*.

3. Resistance to heavy metals, including Hg, may be mediated by plasmid-borne genes. Further discussion of resistance to metal cations and metalloid anions is given in Chapter 3, Part 4.

4. Genes coding for the catabolism of a large number of diverse xenobiotics are often carried on plasmids. This is highly germane to the present discussion, and some examples to illustrate the range of substrates are given in Table 4.3.

Degradative plasmids for substrates, including PAHs and chlorophenolic compounds, are widely distributed among strains of *Sphingomonas*, and a large number of them were megaplasmids, although some degradative genes were also chromosomal (Basta et al. 2004) (Table 4.4). Plasmid transfer may occur within the genus *Sphingomonas* and closely related genera, but apparently infrequently to other genera. In addition, many bacteria harbor plasmids with no hitherto established function.

Genes for specific substrates may be borne either on plasmids or chromosomally:

1. In the degradation of toluene, the plasmid-borne genes encode enzymes for the oxidation to benzoate, whereas those that are chromosomally borne encode ring 2,3-dioxygenation.

2. The degradation of trichloroethene by *Alcaligenes eutrophus* JMP134 is carried out either by a chromosomal phenol-dependent pathway or by the plasmid-borne 2,4-dichlorophenoxyacetate pathway (Harker and Kim 1990).

3. The degradation of 2,4-dichlorophenoxyacetate is initiated by a 2-oxoglutarate-dependent dioxygenase that results in the production of 2,4-dichlorophenol. The genes are generally plasmid-borne, although chromosomally borne genes have been identified in *Burkholderia* sp. strain RASC (Suwa et al. 1996). The transfer of both the plasmid-borne and chromosomally borne *tfdA* gene may contribute to the dissemination of capability to degrade 2,4-dichlorophenoxyacetate (Matheson et al. 1996; McGowan et al. 1998).

Probably the greatest interest in catabolic plasmids stems from the possibility of constructing strains with increased metabolic potential toward xenobiotics, and from the potential application of such strains

TABLE 4.3

Examples of Plasmids Carrying Genes Coding Enzymes for the Biodegradation or Biotransformation of Xenobiotics

Hydrocarbons and Related Compounds	Reference
Octane	Chakrabarty et al. (1973)
Benzene	Tan and Mason (1990)
Toluene and xylene	Williams and Worsey (1976)
Naphthalene	Connors and Barnsley (1982)
Naphthalene, phenanthrene, and anthracene	Sanseverino et al. (1993)
Dibenzothiophene	Monticello et al. (1985)
Terpenoids	
Camphor	Rheinwald et al. (1973)
Citronellol, geraniol	Vandenbergh et al. (1983)
Linalool	Vandenbergh et al. (1986)
Halogenated Compounds	
Bromoxynil	Stalker and McBride (1987)
Chloroalkanoates	Kawasaki et al. (1981); Hardman et al. (1986)
Chlorophenoxyacetates	Don and Pemberton (1985); Chaudhry and Huang (1988)
Dichloromethane	Gälli and Leisinger (1988)
Chlorobenzenes	Van der Meer et al. (1991)
Chlorobenzoates	Chatterjee et al. (1981)
Chlorobiphenyls	Furukawa and Chakrabarty (1982); Shields et al. (1985)
Chloridazon	Kreiss et al. (1981)
γ-Hexachlorocyclohexane	Miyazaki et al. (2006)
Other Structural Groups	
6-Aminohexanoate cyclic dimer	Negoro et al. (1980); Kanagawa et al. (1989)
Parathion	Serdar et al. (1982); Mulbry et al. (1986)
Nicotine	Brandsch et al. (1982)
Cinnamic acid	Andreoni and Bestetti (1986)
Ferulic acid	Andreoni and Bestetti (1986)
S-ethyl-*N,N*-dipropylthiocarbamate	Tam et al. (1987)
Aniline	Anson and Mackinnon (1984); McClure and Venables (1987); Karns and Eaton (1997)
2-Aminobenzenesulfonate	Jahnke et al. (1990)
1,1′-Dimethyl-4,4′-bipyridinium dichloride	Salleh and Pemberton (1993)
Arylsulfonates	Junker and Cook (1997)

to waste treatment systems. Considerable effort has been devoted to aromatic chlorinated compounds, and conspicuous success has been achieved in overcoming problems with, for example, the synthesis of toxic metabolites such as 3-chlorocatechol during the degradation of chlorobenzenes by natural strains. There are several important issues that should, however, be addressed before considering the application of such artificially constructed strains to biological treatment systems or bioremediation programs:

1. Competition with endemic strains, which may eventually outnumber and eliminate the introduced strains.

2. Genetic instability if selection pressure is removed as a result of fluctuations in the loading of the xenobiotic.

3. Concern over the discharge of such strains into the environment. Attempts have, therefore, been made to incorporate safeguards so that the strains are unlikely to survive in competition with natural strains in the ecosystem.

TABLE 4.4

Large Plasmids in Species of *Sphingomonas* and Their Substrates

Species	Substrate	Reference
Yanoikuyae	Tol, Bip, Nap, Anth, Phen	Yabuuchi et al. (1990)
Herbicidovorans	2,4-dichlorophenoxyacetate	Zipper et al. (1996)
Chlorophenolica	Pentachlorophenol	Nohynek et al. (1995)
Subarctica	2,3,4,6-tetrachlorophenol	Nohynek et al. (1996)
Macrogoltabidus	Polyethyleneglycol	Takeuchi et al. (2001)
Subterranea	Nap, Fle	Balkwill et al. (1997)
Aromaticivorans	Phen, Anth, Flu,	Shi et al. (2001)
Stygia	Dibenzothiophene	Balkwill et al. (1997)

Source: Basta T et al. 2004. *J Bacteriol* 186: 3862–3872.

Note: Tol = toluene; Bip = biphenyl; Nap = naphthalene; Anth = anthracene; Phen = phenanthrene; Fle = fluorene; Flu = fluoranthene.

Plasmid transmission and the stability of plasmids in natural ecosystems have received considerable attention, but caution should be exercised in drawing general conclusions on the basis of the sometimes fragmentary evidence from laboratory experiments. Some important principles are illustrated by the following:

1. A study with a strain of plasmid-borne antibiotic-resistant *Escherichia coli* indicated that the strain did not transmit these plasmids to indigenous strains after introduction into the terrestrial environment (Devanas et al. 1986).

2. It has been shown (Smith et al. 1978) that in enteric bacteria carrying thermosensitive plasmids coding for the utilization of citrate and for resistance to antibiotics, rates of transmission were negligible at 37°C but appreciable at 23°C—a temperature more closely approaching that which prevails in natural ecosystems.

3. There is evidence that bacterial populations may retain a small number of organisms carrying the relevant degradative plasmids even in the absence of selective pressure exerted by the presence of a xenobiotic. For example, strains of *Pseudomonas putida* in which the degradation of toluene is mediated by genes on the nonconjugative TOL plasmid maintain a small population of cells carrying the plasmid even in the absence of toluene (Keshavarz et al. 1985). It has also been shown (Duetz et al. 1994) that cultures of *Pseudomonas putida* grown with growth-limiting concentrations of succinate express TOL catabolic genes that are responsible for degradation by both the upper and lower pathways. This was observed in response to *o*-xylene that is not metabolized, although not during nonlimiting growth with succinate.

4. Strains of indigenous groundwater bacteria and a strain of *Pseudomonas putida* carrying a TOL plasmid were introduced into microcosms prepared from a putatively pristine aquifer. DNA-specific probes were used to monitor the numbers of organisms carrying the genotypes and it was found that the stability of genotypes for the degradation of toluene was maintained in the absence of selective pressure over the 8-week period of the experiment (Jain et al. 1987).

5. A strain of *Alcaligenes* sp. carrying a plasmid bearing the genes for the degradation of 3-chlorobenzoate was introduced into a freshwater microcosm system, and a specific DNA probe was used to enumerate the organisms bearing this gene (Fulthorpe and Wyndham 1989). A number of generally important results were obtained:

   a. The presence of 3-chlorobenzoate was needed to maintain the catabolic genotype

   b. The number of probe-positive organism often greatly outnumbered that of the original organism determined by plate counts

   c. The nature of these "additional" probe-positive organisms was not established

6. Detailed analysis of bacteria isolated from the sediment of a coal-tar-contaminated site demonstrated horizontal transfer of genes involved in naphthalene degradation, and that identical alleles NahAc were shared among seven taxonomically diverse hosts (Herrick et al. 1997). Analysis of restriction fragment length polymorphism (RFPL) revealed the existence of two catabolic plasmids in 12 of the isolates, and that these were closely related both to each other and to the plasmid pDGT1 of *Pseudomonas putida* strain NCIB 9816-4. The plasmids were stable in this environment, and it was, therefore, suggested that these plasmids were involved in the development of the naphthalene-degrading community as a result of the selective pressure exerted by the coal tar contamination (Stuart-Keil et al. 1998).

Clearly, therefore, further investigation is required before generalizations can be made on the cardinal issues of the stability of plasmids in natural ecosystems, the extent to which these plasmids are transmissible, and the stability of the genotypes in the absence of selective pressure. At least some of the apparently conflicting views may be attributed to the different organisms that have been used, including their nutritional demands, compounded by the widely varying environments in which their stability have been examined (Sobecky et al. 1992). Currently, the greatest volume of research is being devoted to primarily genetic aspects of these problems, particularly in biotechnology.

4.19 References

Adriaens P, DD Focht. 1991a. Evidence for inhibitory substrate interactions during cometabolism of 3,4-dichlorobenzoate by *Acinetobacter* sp. strain 4-CB1. *FEMS Microbiol Ecol* 85: 293–300.

Adriaens P, DD Focht. 1991b. Cometabolism of 3,4-dichlorobenzoate by *Acinetobacter* sp. strain 4-CB1. *Appl Environ Microbiol* 57: 173–179.

Aeckersberg F, F Bak, F Widdel. 1991. Anaerobic oxidation of saturated hydrocarbons to CO_2 by a new type of sulfate-reducing bacterium. *Arch Microbiol* 156: 5–14.

Aeckersberg F, FA Rainey, F Widdel. 1998. Growth, natural relationships, cellular fatty acids and metabolic adaptation of sulfate-reducing bacteria that utilize long-chain alkanes under anoxic conditions. *Arch Microbiol* 170: 361–369.

Aelion CM, DC Dobbins, FK Pfaender. 1989. Adaptation of aquifer microbial communities to the biodegradation of xenobiotic compounds: Influence of substrate concentration and preexposure. *Environ Toxicol Chem* 8: 75–86.

Aelion CM, CM Swindoll, FK Pfaender. 1987. Adaptation to and biodegradation of xenobiotic compounds by microbial communities from a pristine aquifer. *Appl Environ Microbiol* 53: 2212–2217.

Ahn I-S, WC Ghiorse, LW Lion, ML Shuler. 1998. Growth kinetics of *Pseudomonas putida* G7 on naphthalene and toxicity during nutrient deprivation. *Biotechnol Bioeng* 59: 587–594.

Aislabie JM, MR Balks, JM Foght, EJ Waterhouse. 2004. Hydrocarbon spills on Antarctic soils: Effects and management. *Environ Sci Technol* 38: 1265–1274.

Aitken MD, WT Stringfellow, RD Nagel, C Kazuga, S-H Chen. 1998. Characteristics of phenanthrene-degrading bacteria isolated from soils contaminated with polycyclic aromatic hydrocarbons. *Can J Microbiol* 44: 743–752.

Alexander M. 1985. Biodegradation of organic chemicals. *Environ Sci Technol* 18: 106–111.

Allard A-S, P-Å Hynning, C Lindgren, M Remberger, AH Neilson. 1991. Dechlorination of chlorocatechols by stable enrichment cultures of anaerobic bacteria. *Appl Environ Microbiol* 57: 7784.

Allard A-S, P-Å Hynning, M Remberger, AH Neilson. 1992. Role of sulfate concentration in dechlorination of 3,4,5-trichlorocatechol by stable enrichment cultures grown with coumarin and flavanone glycones and aglycones. *Appl Environ Microbiol* 58: 961–968.

Allard A-S, P-Å Hynning, M Remberger, AH Neilson. 1994. Bioavailability of chlorocatechols in naturally contaminated sediment samples and of chloroguaiacols covalently bound to C_2-guaiacyl residues. *Appl Environ Microbiol* 60: 777–784.

Allard A-S, M Remberger, AH Neilson. 1985. Bacterial *O*-methylation of chloroguaiacols: Effect of substrate concentration, cell density and growth conditions. *Appl Environ Microbiol* 49: 279–288.

Allard A-S, M Remberger, AH Neilson. 1987. Bacterial *O*-methylation of halogen-substituted phenols. *Appl Environ Microbiol* 53: 839–845.

Ampe F, D Léonard, ND Lindley. 1998. Repression of phenol catabolism by organic acids in *Ralstonia eutropha*. *Appl Environ Microbiol* 64: 1–6.

Ampe F, J-L Uribelarrea, GMF Aragao, ND Lindley. 1997. Benzoate degradation via the *ortho* pathway in *Alcaligenes eutrophus* is perturbed by succinate. *Appl Environ Microbiol* 63: 2765–2770.

Anantharam V, MJ Allison, PC Maloney. 1989. Oxalate: Formate exchange. *J Biol Chem* 264: 7244–7250.

Andreoni V, G Bestetti. 1986. Comparative analysis of different *Pseudomonas* strains that degrade cinnamic acid. *Appl Environ Microbiol* 52: 930–934.

Anson JG, G Mackinnon. 1984. Novel plasmid involved in aniline degradation. *Appl Environ Microbiol* 48: 868–869.

Apajalahti JH, MS Salkinoja-Salonen. 1987. Dechlorination and *para*-hydroxylation of polychlorinated phenols by *Rhodococcus chlorophenolicus*. *J Bacteriol* 169: 675–681.

Arjmand M, H Sandermann. 1985. Mineralization of chloroaniline/lignin conjugates and of free chloroanilines by the white rot fungus *Phanerochaete chrysosporium*. *J Agric Food Chem* 33: 1055–1060.

Arras T, J Schirawski, G Unden. 1998. Availability of O_2 as a substrate in the cytoplasm of bacteria under aerobic and microaerobic conditions. *J Bacteriol* 180: 2133–2136.

Asturias JA, KN Timmis. 1993. Three different 2,3-dihydroxybiphenyl-1,2-dioxygenase genes in the Gram-positive polychlorobiphenyl-degrading bacterium *Rhodococcus globerulus* P6. *J Bacteriol* 175: 4631–4640.

Austen RA, NW Dunn. 1980. Regulation of the plasmid-specified naphthalene catabolic pathway of *Pseudomonas putida*. *J Gen Microbiol* 117: 521–528.

Balkwill DL et al., 1997. Taxonomic study of aromatic-degrading bacteria from deep-terrestrial-subsurface sediments and description of *Sphingomonas aromaticivorans* sp. nov., *Sphingomonas subterranea* sp. nov., and *Sphingomonas stygia* sp. nov. *Int J Syst Bacteriol* 47: 191–201.

Bartels I, H-J Knackmuss, W Reineke. 1984. Suicide inactivation of catechol 2,3-dioxygenase from *Pseudomonas putida* mt-2 by 3-halocatechols. *Appl Environ Microbiol* 47: 500–505.

Basta T, A Keck, J Klein, A Stolz. 2004. Detection and characterization of conjugative plasmids in xenobiotic-degrading *Sphingomonas* strains. *J Bacteriol* 186: 3862–3872.

Battersby NS. 1990. A review of biodegradation kinetics in the aquatic environment. *Chemosphere* 21: 1243–1284.

Bauer JE, DG Capone. 1988. Effects of co-occurring aromatic hydrocarbons on degradation of individual polycyclic aromatic hydrocarbons in marine sediment slurries. *Appl Environ Microbiol* 54: 1649–1655.

Baughn AD, MN Malamy. 2004. The strict anaerobe *Bacteriodes fragilis* grown in and benefits from nanomolar concentrations of oxygen. *Nature* 427: 441–444.

Beaty PS, MJ McInerney. 1987. Growth of *Syntrophomonas wolfei* in pure culture on crotonate. *Arch Microbiol* 147: 389–393.

Bedard DL, H van Dort, KA Deweerd. 1998. Brominated biphenyls prime extensive microbial reductive dehalogenation of Arochlor 1260 in Housatonic River sediment. *Appl Environ Microbiol* 64: 1786–1795.

Bedard DL, ML Haberl, RJ May, MJ Brennan. 1987. Evidence for novel mechanisms of polychlorinated biphenyl metabolism in *Alcaligenes eutrophus* H 850. *Appl Environ Microbiol* 53: 1103–1112.

Blasco R, M Mallavarapu, R-M Wittich, KN Timmis, DH Pieper. 1997. Evidence that formation of protoanemonin from metabolites of 4-chlorobiphenyl degradation negatively affects the survival of 4-chlorobiphenyl-cometabolizing microorganisms. *Appl Environ Microbiol* 63: 434.

Blasco R, R-M Wittich, M Mallavarapu, KN Timmis, DH Pieper. 1995. From xenobiotic to antibiotic, formation of protoanemonin from 4-chlorocatechol by enzymes of the 3-oxoadipate pathway. *J Biol Chem* 270: 29229–29235.

Blümel S, H-J Knackmuss, A Stolz. 2002. Molecular cloning and characterization of the gene coding for the aerobic azoreductase from *Xenophilus azovorans* KF46F. *Appl Environ Microbiol* 68: 3948–3955.

Boethling RS, M Alexander. 1979. Effect of concentration of organic chemicals on their biodegradation by natural microbial communities. *Appl Environ Microbiol* 37: 1211–1216.

Bollag J-M, S-Y Liu. 1985. Copolymerization of halogenated phenols and syringic acid. *Pest Biochem Physiol* 23: 261–272.

Bollag J-M, RD Minard, S-Y Liu. 1983. Cross-linkage between anilines and phenolic humus constituents. *Environ Sci Technol* 17: 72–80.

Bopp LH. 1986. Degradation of highly chlorinated PCBs by *Pseudomonas* strain LB400. *J Ind Microbiol* 1: 23–29.

Bouchez M, D Blanchet, J-P Vandecasteele. 1997. An interfacial uptake mechanism for the degradation of pyrene by a *Rhodococcus* strain. *Microbiology (UK)* 143: 1087–1093.

Bradley PM, FH Chapelle. 1995. Rapid toluene mineralization by aquifer microorganisms at Adak, Alaska: Implications for intrinsic bioremediation in cold environments. *Environ Sci Technol* 29: 2778–2781.

Brandsch R, AE Hinkkanen, K Decker. 1982. Plasmid-mediated nicotine degradation in *Arthrobacter oxidans*. *Arch Microbiol* 132: 26–30.

Brückmann M, R Blasco, KH Timmis, DH Pieper. 1998. Detoxification of protoanemonin by dienelactone hydrolase. *Appl Environ Microbiol* 64: 400–402.

Burgos WD, JT Novak, DF Berry. 1996. Reversible sorption and irreversible binding of naphthalene and α-naphthol to soil: Elucidation of processes. *Environ Sci Technol* 30: 1205–1211.

Calderbank A. 1989. The occurrence and significance of bound pesticide residues in soil. *Rev Environ Contam Toxicol* 108: 1–110.

Cánovas JL, RY Stanier. 1967. Regulation of the enzymes of the β-ketoadipate pathway in *Moraxella calcoacetica*. *Eur J Biochem* 1: 289–300.

Carmichael LM, RF Christman, FK Pfaender. 1997. Desorption and mineralization kinetics of phenanthrene and chysene in contaminated soils. *Environ Sci Technol* 31: 126–132.

Chakrabarty AM, G Chou, IC Gunsalus. 1973. Genetic regulation of octane dissimilation plasmid in *Pseudomonas*. *Proc Natl Acad Sci USA* 70: 1137–1140.

Chang KH, P-H Liang, W Beck, JD Scholten, D Dunaway-Mariano. 1992. Isolation and characterization of the three polypeptide components of 4-chlorobenzoate dehalogenase from *Pseudomonas* sp. strain CBS-3. *Biochemistry* 31: 5605–5610.

Chatterjee DK, ST Kellogg, S Hamada, AM Chakrabarty. 1981. Plasmid specifying total degradation of 3-chlorobenzoate by a modified *ortho* pathway. *J Bacteriol* 146: 639–646.

Chaudhry GS, GH Huang. 1988. Isolation and characterization of a new plasmid from a *Flavobacterium* sp. which carries the genes for degradation of 2,4-dichlorophenoxyacetate. *J Bacteriol* 170: 3897–3902.

Chen S-W, MD Aitken. 1999. Salicylate stimulates the degradation of high-molecular weight polycyclic aromatic hydrocarbons by *Pseudomonas saccharophila* P15. *Environ Sci Technol* 33: 435–439.

Cheng G, N Shapir, MJ Sadowsky, LP Wackett. 2005. Allophanate hydrolase, not urease functions in bacterial cyanuric acid metabolism. *Appl Environ Microbiol* 71: 4437–4445.

Cirigliano MC, GM Carman. 1984. Isolation of a bioemulsifier from *Candida lipolytica*. *Appl Environ Microbiol* 48: 747–750.

Connors MA, EA Barnsley. 1982. Naphthalene plasmids in pseudomonads. *J Bacteriol* 149: 1096–1101.

Cook AM, R Hütter. 1981. Degradation of *s*-triazines: A critical view of biodegradation, pp. 237–249. In *Microbial Degradation of Xenobiotics and Recalcitrant Compounds* (Eds. T Leisinger, AM Cook, R Hütter, J Nüesch), Academic Press, London.

Crawford RL, PE Olson, TD Frick. 1979. Catabolism of 5-chlorosalicylate by a *Bacillus* isolated from the Mississippi River. *Appl Environ Microbiol* 38: 379–384.

Crocker FH, WA Guerin, SA Boyd. 1995. Bioavailability of naphthalene sorbed to cationic surfactant-modified smectite clay. *Environ Sci Technol* 29: 2953–2958.

Dabrock B, J Riedel, J Bertram, G Gottschalk. 1992. Isopropylbenzene (cumene)—A new substrate for the isolation of trichloroethene-degrading bacteria. *Arch Microbiol* 158: 9–13.

Dalton H, DI Stirling. 1982. Co-metabolism. *Phil Trans Roy Soc London* B 297: 481–496.

Das A, R Silaghi-Dumitrescu, LG Ljungdahl, DM Kurtz. 2005. Cytochrome *bd* oxidase, oxidative stress, and dioxygen tolerance of the strictly anaerobic bacterium *Moorella thermoacetica*. *J Bacteriol* 187: 2020–2029.

Daughton CG, DPH Hsieh. 1977. Parathion utilization by bacterial symbionts in a chemostat. *Appl Environ Microbiol* 34: 175–184.

Davis KER, SJ Joseph, PH Janssen. 2005. Effects of growth medium, imoculum size, and incubation time on culturability and isolation of soil bacteria. *Appl Environ Microbiol* 71: 826–835.

Dec J, K Haider, A Benesi, V Rangaswamy, A Schäffer, E Fernandes, J-M Bollag. 1997a. Formation of soil-bound residues of cyprodinil and their plant uptake. *J Agric Food Chem* 45: 514–520.

Dec J, K Haider, A Benesi, V Rangaswamy, A Schäffer, U Plücken, J-M Bollag. 1997b. Analysis of soil-bound residues of ^{13}C-labelled fungicide cyprodinil by NMR spectroscopy. *Environ Sci Technol* 31: 1128–1135.

Dec J, KL Shuttleworth, J-M Bollag. 1990. Microbial release of 2,4-dichlorophenol bound to humic acid or incorporated during humification. *J Environ Qual* 19: 546–551.

Dejonghe W, E Berteloot, J Goris, N Boon, K Crul, S Maertens, M Höfte, P de Vos, W Verstraete, EM Top. 2003. Synergistic degradation of linuron by a bacterial consortium and isolation of a single linuron-degrading *Variovorax* strain. *Appl Environ Microbiol* 69: 1532–1542.

Devanas MA, D Rafaeli-Eshkol, G Stotsky. 1986. Survival of plasmid-containg strains of *Escherichia coli* in soil: Effect of plasmid size and nutrients on survival of hosts and maintenance of plasmids. *Curr Microbiol* 13: 269–277.

Déziel É, G Paquette, R Villemur, F Lépine, J-G Bisaillon. 1996. Biosurfactant production by a soil *Pseudomonas* strain growing on polycyclic aromatic hydrocarbons. *Appl Environ Microbiol* 62: 1908–1912.

Diels L, M Mergeay. 1990. DNA probe-mediated detection of resistant bacteria from soils highly polluted by heavy metals. *Appl Environ Microbiol* 56: 1485–1491.

Don RH, JM Pemberton. 1985. Genetic and physical map of the 2,4-dichlorophenoxyacetic acid-degradative plasmid pJP4. *J Bacteriol* 161: 466–468.

Donnelly PK, RS Hegde, JS Fletcher. 1994. Growth of PCB-degrading bacteria on compounds from photosynthetic plants. *Chemosphere* 28: 981–988.

Duetz WA, S Marqués, C de Jong, JL Ramos, J G van Andel. 1994. Inducibility of the TOL catabolic pathway in *Pseudomonas putida* (pWWO) growing on succinate in continuous culture: Evidence of carbon catabolite repression control. *J Bacteriol* 176: 2354–2361.

Dürre P, JR Andreesen. 1982. Anaerobic degradation of uric acid via pyrimidine derivatives by selenium-starved cells of *Clostridium purinolyticum*. *Arch Microbiol* 131: 255–260.

Eaton RW, PJ Chapman. 1992. Bacterial metabolism of naphthalene: Construction and use of recombinant bacteria to study ring cleavage of 1,2-dihydroxynaphthalene and subsequent reactions. *J Bacteriol* 174: 7542–7554.

El-Fantroussi S. 2000. Enrichment and molecular characterization of bacterial culture that degrades methoxy-methyl urea herbicides and their aniline derivatives. *Appl Environ Microbiol* 66: 5110–5115.

Eschbach M, K Schreiber, K Trunk, J Buer, D Jahn, M Schobert. 2004. Long-term anaerobic survival of the opportunistic pathogen *Pseudomonas aeruginosa* via pyruvate fermentation. *J Bacteriol* 186: 4596–4604.

Finette BA, V Subramanian, DT Gibson. 1984. Isolation and characterization of *Pseudomonas putida* PpF1 mutants defective in the toluene dioxygenase enzyme system. *J Bacteriol* 160: 1003–1009.

Foght JM, DL Gutnick, DWS Westlake. 1989. Effect of emulsan on biodegradation of crude oil by pure and mixed bacterial cultures. *Appl Environ Microbiol* 55: 36–42.

Folsom BR, PJ Chapman, PH Pritchard. 1990. Phenol and trichloroethylene degradation by *Pseudomonas cepacia* G4: Kinetics and interactions between substrates. *Acinetobacter* sp. strain 4-CB1. *Appl Environ Microbiol* 56: 1279–1285.

Frantz B, AM Chakrabarty. 1987. Organization and nucleotide sequence determination of a gene cluster involved in 3-chlorocatechol degradation. *Proc Natl Acad Sci USA* 84: 4460–4464.

Fredrickson JK, FJ Brockman, DJ Workmnan, SW Li, TO Stevens. 1991. Isolation and characterization of a subsurface bacterium capable of growth on toluene, naphthalene, and other aromatic compounds. *Appl Environ Microbiol* 57: 796–803.

Fritzsche C. 1994. Degradation of pyrene at low defined oxygen concentrations by *Mycobacterium* sp. *Appl Environ Microbiol* 60: 1687–1689.

Fulthorpe RR, AN Rhodes, JM Tiedje. 1996. Pristine soils mineralize 3-chlorobenzoate and 2,4-dichlorophenoxyacetate via different microbial populations. *Appl Environ Microbiol* 62: 1159–1166.

Fulthorpe RR, AN Rhodes, JM Tiedje. 1998. High levels of endemicity of 3-chlorobenzoate-degrading soil bacteria. *Appl Environ Microbiol* 64: 1620–1627.

Fulthorpe RR, RC Wyndham. 1989. Survival and activity of a 3-chlorobenzoate-catabolic genotype in a natural system. *Appl Environ Microbiol* 55: 1584–1590.

Furukawa K, N Arimura. 1987. Purification and properties of 2,3-dihydroxybiphenyl dioxygenase from poly-chlorinated biphenyl-degrading *Pseudomonas pseudoalcaligenes* and *Pseudomonas aeruginosa* carrying the cloned *bphC* gene. *J Bacteriol* 169: 924–927.

Furukawa K, AM Chakrabarty. 1982. Involvement of plasmids in total degradation of chlorinated biphenyls. *Appl Environ Microbiol* 44: 619–626.

Furukawa K, JR Simon, AM Chakrabarty. 1983. Common induction and regulation of biphenyl, xylene/toluene, and salicylate catabolism in *Pseudomonas paucimobilis. J Bacteriol* 154: 1356–1362.

Futamata H, Y Nagano, K Watanabe, A Hitaishi. 2005. Unique kinetic properties of phenol-degrading *Variovorax* strains responsible for efficient trichloroethylene degradation in a chemostat enrichment culture. *Appl Enviiron Microbiol* 71: 904–911.

Gälli R, T Leisinger. 1988. Plasmid analysis and cloning of the dichloromethane-utilizing genes of *Methylobacterium* sp. DM4. *J Gen Microbiol* 134: 943–952.

George KW, J Kagle, L Junker, A Risen, AG Hay. 2011. Growth of *Pseudomonas putida* F1 on styrene requires increased catechol-2,3-dioxygenase activity, not a new hydrolase. *Microbiology (UK)* 157: 89–98.

Gerritse J, JC Gottschal. 1993a. Two-membered mixed cultures of methanogenic and aerobic bacteria in O_2-limited chemostats. *J Gen Microbiol* 139: 1853–1860.

Gerritse J, JC Gottschal. 1993b. Oxic and anoxic growth of a new *Citrobacter* species on amino acids. *Arch Microbiol* 160: 51–61.

Gibson DT, JR Koch, CL Schuld, RE Kallio. 1968. Oxidative degradation of aromatic hydrocarbons by microorganisms. II. Metabolism of halogenated aromatic hydrocarbons. *Biochemistry* 7: 3795–3802.

Gibson SA, GW Sewell. 1992. Stimulation of reductive dechlorination of tetrachloroethene in anaerobic aquifer microcosms by addition of short-chain organic acids or alcohols. *Appl Environ Microbiol* 58: 1392–1393.

Gerson DF, JE Zajic. 1979. Comparison of surfactant production from kerosene by four species of *Corynebacterium. Antonie van Leeuwenhoek* 45: 81–94.

Gilbert ES, DE Crowley. 1997. Plant compounds that induce polychlorinated biphenyl degradation by *Arthrobacter* sp. strain B1B. *Appl Environ Microbiol* 63: 1933–1938.

Gmünder FK, O Käppeli, A Fiechter. 1981. Chemostat studies on the hexadecane assimilation by the yeast *Candida tropicalis. Eur J Appl Microbiol Biotechnol* 12: 135–142.

Gorny N, G Wahl, A Brune, B Schink. 1992. A strictly anaerobic nitrate-reducing bacterium growing with resorcinol and other aromatic compounds. *Arch Microbiol* 158: 48–53.

Gray PHH, HG Thornton. 1928. Soil bacteria that decompose certain aromatic compounds. *Centralbl Bakteriol Parasitenkd Infektionskr* (2 Abt) 73: 74–96.

Greer LE, DR Shelton. 1992. Effect of inoculant strain and organic matter content on kinetics of 2,4-dichlorophenoxyacetic acid degradation in soil. *Appl Environ Microbiol* 58: 1459–1465.

Groenewegen, PEJ, AJM Diessen, WM Konings, JAM de Bont. 1990. Energy-dependent uptake of 4-chlorobenzoate in the coryneform bacterium NTM-1. *J Bacteriol* 172: 419–423.

Grosser RJ, D Warshawsky, JR Vestal. 1991. Indigenous and enhanced mineralization of pyrene, benzo[*a*] pyene, and carbazole in soils. *Appl Environ Microbiol* 57: 3462–3469.

Guerin WF, SA Boyd. 1992. Differential bioavailability of soil-sorbed naphthalene by two bacterial species. *Appl Environ Microbiol* 58: 1142–1152.

Guthrie EA, FK Pfaender. 1998. Reduced pyrene bioavailability in microbially active soils. *Environ Sci Technol* 32: 501–508.

Haigler BE, CA Pettigrew, JC Spain. 1992. Biodegradation of mixtures of substituted benzenes by *Pseudomonas* sp. strain JS150. *Appl Environ Microbiol* 58: 2237–2244.

Haigler BE, JC Spain. 1991. Biotransformation of nitrobenzene by bacteria containing toluene degradative pathways. *Appl Environ Microbiol* 57: 3156–3162.

Hall BG. 1990. Spontaneous point mutations that occur more often when advantageous than when neutral. *Genetics* 126: 5–16.

Haller HD, RK Finn. 1978. Kinetics of biodegradation of *p*-nitrobenzoate and inhibition by benzoate in a pseudomonad. *Appl Environ Microbiol* 35: 890–896.

Hamamura N, C Page, T Long, L Semprini, DH Arp. 1997. Chloroform cometabolism by butane-grown CF8, *Pseudomonas butanovora*, and *Mycobacterium vaccae* JOB5 and methane-grown *Methylosinus trichosporium* OB3b. *Appl Environ Microbiol* 63: 3607–3613.

Hansch C, A Leo. 1995. *Exploring QSAR: Fundamentals and Applications in Chemistry and Biology*. American Chemical Society, Washington, DC.

Hardman DJ, PC Gowland, JH Slater. 1986. Large plasmids from soil bacteria enriched on halogenated alkanoic acids. *Appl Environ Microbiol* 51: 44–51.

Harker AR, Y Kim. 1990. Trichloroethylene degradation by two independent aromatic-degrading pathways in *Alcaligenes eutrophus* JMP134. *Appl Environ Microbiol* 56: 1179–1181.

Harkness MR, JB McDermott, DA Abramowicz, JJ Salvo, WP Flanagan, ML Stephens, FJ Mondello et al., 1993. *In situ* stimulation of aerobic PCB biodegradation in Hudson River sediments. *Science* 259: 503–507.

Harwood CS, NN Nichols, M-K Kim, JJ Ditty, RE Parales. 1994. Identification of the *pcaRKF* gene cluster from *Pseudomonas putida*: Involvement in chemotaxis, biodegradation, and transport of 4-hydroxybenzoate. *J Bacteriol* 176: 6479–6488.

Hassett DJ, HP Schweitzer, DE Ohman. 1995. *Pseudomonas aeruginosa sodA* and *sodB* mutants defective in manganese- and iron-cofactored superoxide dismutase activity demonstrate the importance of the iron-cofactored form in aerobic metabolism. *J Bacteriol* 177: 6330–6337.

Hatcher PG, JM Bortiatynski, RD Minard, J Dec, J-M Bollag. 1993. Use of high-resolution ^{13}C NMR to examine the enzymatic covalent binding of ^{13}C-labeled 2,4-dichlorophenol to humic substances. *Environ Sci Technol* 27: 2098–2103.

Hatzinger PB, M Alexander. 1995. Effect of aging of chemicals in soil on their biodegradability and extractability. *Environ Sci Technol* 29: 537–545.

Havel J, W Reineke. 1993. Microbial degradation of chlorinated acetophenones. *Appl Environ Microbiol* 59: 2706–2712.

Hay AG, DD Focht. 1998. Cometabolism of 1,1-dichloro-2,2-bis(4-chlorophenyl)ethylene by *Pseudomomas acidovorans* M3GY grown on biphenyl. *Appl Environ Microbiol* 64: 2141–2146.

Heald S, RO Jenkins. 1994. Trichloroethylene removal and oxidation toxicity mediated by toluene dioxygenase of *Pseudomonas putida*. *Appl Environ Microbiol* 60: 4634–4637.

Heidpieper HR, R Diefenbach, H Keweloh. 1992. Conversion of *cis*-unsaturated fatty acids to *trans*, a possible mechanism for the protection of phenol-degrading *Pseudomonas putida* P8 from substrate toxicity. *Appl Environ Microbiol* 58: 195–205.

Henriksen HV, S Larsen, BK Ahring. 1991. Anaerobic degradation of PCP and phenol in fixed-film reactors: The influence of an additional substrate. *Water Sci Technol* 24: 431–436.

Henry SM, D Grbic-Galic. 1991. Inhibition of trichloroethylene oxidation by the transformation intermediate carbon monoxide. *Appl Environ Microbiol* 57: 1770–1776.

Herbes SE, LR Schwall. 1978. Microbial transformation of polycyclic aromatic hydrocarbons in pristine and petroleum-contaminated sediments. *Appl Environ Microbiol* 35: 306–316.

Herrick JB, KG Stuart-Keil, WC Ghiorse, EL Madsen. 1997. Natural horizontal transfer of a naphthalene dioxygenase gene between bacteria native to a coal tar-contaminated field site. *Appl Environ Microbiol* 63: 2330–2337.

Higgins SJ, J Mandelstam. 1972. Evidence for induced synthesis of an active transport factor for mandelate in *Pseudomonas putida*. *Biochem J* 126: 917–922.

Holben WE, BM Schroeter, VGM Calabrese, RH Olsen, JK Kukor, VO Biederbeck, AE Smith, JM Tiedje. 1992. Gene probe analysis of soil microbial populations selected by amendment with 2,4-dichlorophenoxyacetic acid. *Appl Environ Microbiol* 58: 3941–3948.

Hollender J, W Dott, J Hopp. 1994. Regulation of chloro- and methylphenol degradation in *Comamonas testosteroni* JH5. *Appl Environ Microbiol* 60: 2330–2338.

Holben WE, JK Jansson, BK Chelm, JM Tiedje. 1988. DNA probe method for the detection of specific microorganisms in the soil bacterial community. *Appl Environ Microbiol* 54: 703–711.

Holm PE, PH Nielsen, H-J Albrechtsen, TH Christensen. 1992. Importance of unattached bacteria and bacteria attached to sediment in determining potentials for degradation of xenobiotic organic contaminants in an aerobic aquifer. *Appl Environ Microbiol* 58: 3020–3026.

Hommel RK. 1994. Formation and function of biosurfactants for degradation of water-insoluble substrates, pp. 63–87. In *Biochemistry of Microbial Degradation* (Ed. C Rattledge), Kluwer Academic Publishers, Dordrecht, The Netherlands.

Hoover DG, GE Borgonovi, SH Jones, M Alexander. 1986. Anomalies in mineralization of low concentrations of organic compounds in lake water and sewage. *Appl Environ Microbiol* 51: 226–232.

Hopkins GD, PL McCarty. 1995. Field evaluation of *in situ* aerobic cometabolism of trichloroethylene and three dichloroethylene isomers using phenol and toluene as primary substrates. *Environ Sci Technol* 29: 1628–1637.

Horvath RS. 1971. Cometabolism of the herbicide 2,3,6-trichlorobenzoate. *J Agric Food Chem* 19: 291–293.

Horvath RS. 1972. Microbial co-metabolism and the degradation of organic compounds in nature. *Bacteriol Rev* 36: 146–155.

Horvath RS, P Flathman. 1976. Co-metabolism of fluorobenzoates by natural microbial populations. *Appl Environ Microbiol* 31: 889–891.

Houwen FP, C Dijkema, AJM Stams, AJB Zehnder. 1991. Propionate metabolism in anaerobic bacteria; determination of carboxylation reactions with ^{13}C-NMR spectroscopy. *Biochim Biophys Acta* 1056: 126–132.

Hughes EJL, RC Bayly. 1983. Control of catechol *meta*-cleavage pathway in *Alcaligenes eutrophus*. *J Bacteriol* 154: 1363–1370.

Hur H-G, LM Newman, LP Wackett, MJ Sadowsky. 1997. Toluene 2-monooxygenase-dependent growth of *Burkholderia cepacia* G4/PR1 on diethyl ether. *Appl Environ Microbiol* 63: 1606–1609.

Hur H-G, M Sadowsky, LP Wackett. 1994. Metabolism of chlorofluorocarbons and polybrominated compounds by *Pseudomonas putida* G786pHG-2 via an engineered metabolic pathway. *Appl Environ Microbiol* 60: 4148–4154.

Hwang H-M, RE Hodson, DL Lewis. 1989. Microbial degradation kinetics of toxic organic chemicals over a wide range of concentrations in natural aquatic systems. *Environ Toxicol Chem* 8: 65–74.

Itoh S, T Suzuki. 1972. Effect of rhamnolipids on growth of *Pseudomonas aeruginosa* mutant deficient in *n*-paraffin-utilizing ability. *Agric Biol Chem* 36: 2233–2235.

Jahnke M, T El-Banna, R Klintworth, G Auling. 1990. Mineralization of orthanilic acid is a plasmid-associated trait in *Alcaligenes* sp. O-1. *J Gen Microbiol* 136: 2241–2249.

Jahnke M, F Lehmann, A Schoebel, G Auling. 1993. Transposition of the TOL catabolic genes Tn*46521* into the degradative plasmid pSAH of *Alcaligenes* sp. O-1 ensures simultaneous mineralization of sulpho- and methyl-substituted aromatics. *J Gen Microbiol* 139: 1959–1966.

Jain RK, GS Sayler, JT Wilson, L Houston, D Pacia. 1987. Maintenance and stability of introduced genotypes in groundwater aquifer material. *Appl Environ Microbiol* 53: 996–1002.

Janke D. 1987. Use of salicylate to estimate the threshold inducer level for *de novo* synthesis of the phenol-degrading enzymes in *Pseudomonas putida* strain H. *J Basic Microbiol* 27: 83–89.

Janssen PH, PS Yates, BE Grinton, PM Taylor, M Sait. 2002. Improved cultivability of soil bacteria and isolation in pure culture of novel members of the divisions Acidobacteria, Actinobacteria, Proteobacteria, and Verrucomicrobia. *Appl Environ Microbiol* 68: 2391–2396.

Järvinen KT, ES Melin, KA Puhakka. 1994. High-rate bioremediation of chlorophenol-contaminated groundwater at low temperatures. *Environ Sci Technol* 28: 2387–2392.

Jin G, AJ Englande. 1997. Biodegradation kinetics of carbon tetrachloride by *Pseudomonas cepacia* under varying oxzidation-reduction potential conditions. *Water Environ Res* 69: 1094–1099.

Kukor JJ, RH Olsen. 1996. Catechol 2,3-dioxygenases functional in oxygen-limited hypoxic environments. *Appl Environ Microbiol* 62: 1728–1740.

Johnson MS, IB Zhulin, M-ER Gapuzan, BL Taylor. 1997. Oxygen-dependent growth of the obligate anaerobe *Desulfovibrio vulgaris* Hildenborough. *J Bacteriol* 179: 5598–5601.

Junker F, AM Cook. 1997. Conjugative plasmids and the degradation of arylsulfonates in *Comamonas testosteroni*. *Appl Environ Microbiol* 63: 2403–2410.

Kamagata Y, RR Fulthorpe, K Tamura, H Takami, LJ Forney, JM Tiedje. 1997. Pristine environments harbor a new group of oligotrophic 2,4-dichlorophenoxyacetic acid-degrading bacteria. *Appl Environ Microbiol* 63: 2266–2272.

Kanagawa K, S Negoro, N Takada, H Okada. 1989. Plasmid dependence of *Pseudomonas* sp. strain NK87 enzymes that degrade 6-aminohexanoate-cyclic dimer. *J Bacteriol* 171: 3181–3186.

Kargalioglu Y, JA Imlay. 1994. Importance of anaerobic superoxide dismutase synthesis in facilitating outgrowth of *Escherichia coli* upon entry into an aerobic habitat. *J Bacteriol* 176: 7653–7658.

Karns JS, RW Eaton. 1997. Genes encoding *s*-triazine degradation are plasmid-borne in *Klebsiella pneumoniae* strain 99. *J Agric Food Chem* 45: 1017–1022.

Kaschabek SR, T Kasberg, D Müller, AE Mars, DB Janssen, W Reineke. 1998. Degradation of chloroaromatics: Purification and characterization of a novel type of chlorocatechol 2,3-dioxygenase of *Pseudomonas putida* GJ31. *J Bacteriol* 180: 296–302.

Kawai F, H Yamanaka. 1986. Biodegradation of polyethyelene glycol by symbiotic mixed culture (obligate mutualism). *Arch Microbiol* 146: 125–129.

Kawasaki H, H Yahara, K Tonomura. 1981. Isolation and characterization of plasmid pUO1 mediating dehalogenation of haloacetate and mercury resistance in *Moraxella* sp. B. *Agric Biol Chem* 45: 1477–1481.

Keshavarz T, MD Lilly, PC Clarke. 1985. Stability of a catabolic plasmid in continuous culture. *J Gen Microbiol* 131: 1193–1203.

Kim J-E, E Fernandes, J-M Bollag. 1997. Enzymatic coupling of the herbicide bentazon with humus monomers and characterization of reaction products. *Environ Sci Technol* 31: 2392–2398.

Kiyohara H, T Hatta, Y Ogawa, T Kakuda, H Yokoyama, N Takizawa. 1992. Isolation of *Pseudomonas pickettii* strains that degrade 2,4,6-trichlorophenol and their dechlorination of chlorophenols. *Appl Environ Microbiol* 58: 1276–1283.

Klecka GM, DT Gibson. 1981. Inhibition of catechol 2,3-dioxygenase from *Pseudomonas putida* by 3-chlorocatechol. *Appl Environ Microbiol* 41: 1159–1165.

Klecka GM, WJ Maier. 1988. Kinetics of microbial growth on mixtures of pentachlorophenol and chlorinated aromatic compounds. *Biotechnol Bioeng* 31: 328–335.

Knoblauch C, K Sahm, BB Jörgesen. 1999. Psychrophilic sulfate-reducing bacteria isolated from permanantly cold arctic marine sediments: Descríption of *Desulfofrigus oceanense* gen. nov., sp. nov., *Desulfofrigus fragile* sp. nov., *Desulfofaba gelida* gen. nov., sp. nov., *Desulfotalea psychrophila* gen. nov., sp. nov., and *Desulfotalea arctica* sp. nov. *Int J Syst Bacteriol* 49: 1631–1643.

Kreiss M, J Eberspächer, F Lingens. 1981. Detection and characterization of plasmids in Chloridazon and Antipyrin degrading bacteria. *Zbl Bakteriol Hyg, I Abt Orig* C2: 45–60.

Krooneman J, ERB Moore, JCL van Velzen, RA Prins, LJ Forney, JC Gottschal. 1998. Competition for oxygen and 3-chlorobenzoate between two aerobic bacteria using different degradation pathways. *FEMS Microbiol Ecol* 26: 171–179.

Krooneman J, EBA Wieringa, ERB Moore, J Gerritse, RA Prins, JC Gottschal. 1996. Isolation of *Alcaligenes* sp. strain L6 at low oxygen concentrations and degradation of 3-chlorobenzoate via a pathway not involving (chloro)catechols. *Appl Environ Microbiol* 62: 2427–2434.

Kulla HG. 1981. Aerobic bacterial degradation of azo dyes, pp. 387–399. In *Microbial Degradation of Xenobiotics and Recalcitrant Compounds* (Eds. T Leisinger, AM Cook, R Hütter, J Nüesch), FEMA Symposium 12. Academic Press, London.

Lange BM, N Hertkorn, H Sandermann. 1998. Chloroaniline/lignin conjugates as model system for nonextractable pesticide residues in crop plants. *Environ Sci Technol* 32: 2113–2118.

Lemos RS, CM Gomes, M Santana, J LeGall, AV Xavier, M Teixeira. 2001. The 'strict anaerobe' *Desulfovibrio gigas* contains a membrane-bound oxygen-reducing respiratory chain. *FEBS Lett* 496: 40–43.

Leveau JHL, AJB Zehnder, JR van der Meer. 1998. The *tfdK* gene product facilitates uptake of 2,4-dichlorophenoxyacetate by *Ralstonia eutropha* JMP134(pJP4). *J Bacteriol* 180: 2237–2243.

Lewis DL, RE Hodson, LF Freeman. 1985. Multiphase kinetics for transformation of methyl parathion by *Flavobacterium* species. *Appl Environ Microbiol* 50: 553–557.

Lewis DL, HP Kollig, RE Hodson. 1986. Nutrient limitation and adaptation of microbial populations to chemical transformations. *Appl Environ Microbiol* 51: 598–603.

Lidstrom-O'Connor ME, GL Fulton, AE Wopat. 1983. '*Methylobacterium ethanolicum*': A syntrophic association of two methylotrophic bacteria. *J Gen Microbiol* 129: 3139–3148.

Little CD, AV Palumbo, SE Herbes, ME Lidström, RL Tyndall, PJ Gilmour. 1988. Trichloroethylene biodegradation by a methane-oxidizing bacterium. *Appl Environ Microbiol* 54: 951–956.

Liu D, WMJ Strachan, K Thomson, K Kwasniewska. 1981. Determination of the biodegradability of organic compounds. *Environ Sci Technol* 15: 788–793.

Lloyd-Jones G, RC Ogden, PA Williams. 1995. Inactivation of 2,3-dihydroxybiphenyl 1,2-dioxygenase from *Pseudomonas* sp. strain CB406 by 3,4-dihydroxybiphenyl (4-phenyl catechol). *Biodegradation* 6: 11–17.

Locher HH, B Poolman, AM Cook, WN Konings. 1993. Uptake of 4-toluene sulfonate by *Comamonas testosteroni* T-2. *J Bacteriol* 175: 1075–1080.

Lowe SE, MK Jain, JG Zeikus. 1993. Biology, ecology, and biotechnological applications of anaerobic bacteria adapted to environmental stresses in temperature, pH, salinity, or substrates. *Microbiol Rev* 57: 451–509.

Marchesi JR, SA Owen, GF White, WA House, NJ Russell. 1994. SDS-degrading bacteria attach to riverine sediment in response to the surfactant or its primary degradation product dodecan-1-ol. *Microbiology (UK)* 140: 2999–3006.

Margesin R, D Labbé, F Schinner, CW Greer, LG Whyte. 2003. Characterization of hydrocarbon-degrading microbial populations in contaminated and pristine alpine soils. *Appl Environ Microbiol* 69: 3085–3092.

Mars AE, J Kingma, SR Kaschabek, W Reinke, DB Janssen. 1999. Conversion of 3-chlorocatechol by various catechol 2,3-dioxygenases and sequence analysis of the chlorocatechol dioxygense region of *Pseudomonas putida* GJ31. *J Bacteriol* 181: 1309–1318.

Martin P, RA MacLeod. 1984. Observations on the distinction between oligotrophic and eutrophic marine bacteria. *Appl Environ Microbiol* 47: 1017–1022.

Master EM, WW Mohn. 1998. Psychrotolerant bacteria isolated from arctic soil that degrade polychlorinated biphenyls at low temperatures. *Appl Environ Microbiol* 64: 4823–4829.

Matheson VG, LJ Forney, Y Suwa, CH Nakatsu, AJ Sextone, WE Holben. 1996. Evidence for acquisition in nature of a chromosomal 2,4-dichlorophenoxyacetic acid/α-ketoglutarate dioxygenase gene by different *Burckholderia* spp. *Appl Environ Microbiol* 62: 2457–2463.

Mauersberger S, RN Matyashova, H-G Müller, AB Losinov. 1980. Influence of the growth substrate and the oxygen concentration in the medium on the cytochrome P-450 content in *Candida guilliermondii. Eur J Appl Microbiol Biotechnol* 9: 285–294.

McClay K, SH Streger, RJ Steffan. 1995. Induction of toluene oxidation in *Pseudomonas mendocina* KR1 and *Pseudomonas* sp. strain ENVPC5 by chlorinated solvents and alkanes. *Appl Environ Microbiol* 61: 3479–3481.

McClure NC, WA Venables. 1987. pTDN1, a catabolic plasmid involved in aromatic amine catabolism in *Pseudomonas putida* mt-2. *J Gen Microbiol* 133: 2073–2077.

McGowan C, R Fulthorpe, A Wright, JM Tiedje. 1998. Evidence for interspecies gene transfer in the evolution of 2,4-dichlorophenoxyacetic acid degraders. *Appl Environ Microbiol* 64: 4089–4092.

McInerney MJ, MP Bryant, RB Hespell, JW Costerton. 1981. *Syntrophomonas wolfei* gen. nov., sp. nov., an anaerobic, syntrophic, fatty acid-oxidizing bacterium. *Appl Environ Microbiol* 41: 1029–1039.

Megharaj M, S Hartmans, K-H Engesser, H Thiele. 1998. Recalcitrance of 1,1-dichloro-2,2-bis (*p*-chlorophenyl)ethylene to degradation by pure cultures of 1,1-diphenylethylene-degrading aerobic bacteria. *Appl Microbiol Biotechnol* 49: 337–342.

Megharaj M, A Jovcic, HL Boul, JH Thiele. 1997. Recalcitrance of 1,1-dichloro-2,2-bis(*p*-chlorophenyl) ethylene (DDE) to cometabolic degradation by pure cultures of aerobic and anaerobic bacteria. *Arch Environ Contam Toxicol* 33: 141–146.

Meyer JJM, N Grobbelaar, PL Steyn. 1990. Fluoroacetate-metabolizing pseudomonad isolated from *Dichapetalum cymosum. Appl Environ Microbiol* 56: 2152–2155.

Michaelsen M, R Hulsch, T Höpner, L Berthe-Corti. 1992. Hexadecane mineralization in oxygen-controlled sediment-seawater cultivations with autochthonous microorganisms. *Appl Environ Microbiol* 58: 3072–3077.

Miguez CB, CW Greer, JM Ingram, RA MacLeod. 1995. Uptake of benzoic acid and chloro-substituted benzoic acids by *Alcaligenes denitrificans* BRI 3010 and BRI 6011. *Appl Environ Microbiol* 61: 4152–4159.

Miyazaki R, Y Sato, M Ito, Y Ohtsubo, Y Nagata, M Tsuda. 2006. Complete nucleotide sequence of an exogenously isolated plasmid, pLB1, involved in γ-hexachlorocyclohexane degradation. *Appl Environ Microbiol* 72: 6923–6933.

Möbus E, M Jahn, R Schmid, D Jahn, E Maser. 1997. Testosterone-regulated expression of enzymes involved in steroid and aromatic hydrocarbon catabolism in *Comamonas testosteroni. J Bacteriol* 179: 5951–5955.

Mondello FJ. 1989. Cloning and expression in *Escherichia coli* of *Pseudomonas* strain LB400 genes encoding polychlorinated biphenyl degradation. *J Bacteriol* 171: 1725–1732.

Monticello DJ, D Bakker, WR Finnerty. 1985. Plasmid-mediated degradation of dibenzothiophene by *Pseudomonas species. Appl Environ Microbiol* 49: 756–760.

Moore JK, HD Braymer, AD Larson. 1983. Isolation of a *Pseudomonas* sp. which utilizes the phosphonate herbicide glyphosate. *Appl Environ Microbiol* 46: 316–320.

Morris CM, EA Barnsley. 1982. The cometabolism of 1- and 2-chloronaphthalene by pseudomonads. *Can J Microbiol* 28: 73–79.

Morris PJ, JF Quensen III, JM Tiedje, SA Boyd. 1992. Reductive debromination of the commercial polybrominated biphenyl mixture Firemaster BP6 by anaerobic microorganisms from sediments. *Appl Environ Microbiol* 58: 3249–3256.

Mortlock RP. 1982. Metabolic acquisition through laboratory selection. *Annu Rev Microbiol* 34: 37–66.

Mulbry WW, JS Karns, PC Kearney, JO Nelson, CS McDaniel, JR Wild. 1986. Identification of a plasmid-borne parathion hydrolase gene from *Flavobacterium* sp. by Southern hybridization with *opd* from *Pseudomonas diminuta*. *Appl Environ Microbiol* 51: 926–930.

Nakano NM, P Zuber. 1998. Anaerobic growth of a "strict aerobe" (*Bacillus subtilis*). *Annu Rev Microbiol* 52: 165–190.

Nakatsu CH, RC Wyndham. 1993. Cloning and expression of the transposable chlorobenzoate-3,4-dioxygenase genes of *Alcaligenes* sp. strain BR60. *Appl Environ Microbiol* 59: 3625–3633.

Negoro S, H Shinagawa, A Naata, S Kinoshita, T Hatozaki, H Okada. 1980. Plasmid control of 6-aminohexanoic acid cyclic dimer degradation enzymes of *Flavobacterium* sp. KI72. *J Bacteriol* 143: 238–345.

Neilson AH, A-S Allard, M Remberger. 1985. Biodegradation and transformation of recalcitrant compounds. *Handb Environ Chem* 2C: 29–86.

Neilson AH, C Lindgren, P-Å Hynning, M Remberger. 1988. Methylation of halogenated phenols and thiophenols by cell extracts of Gram-positive and Gram-negative bacteria. *Appl Environ Microbiol* 54: 524–530.

Neumann G, R Teras, L Monson, M Kivisaar, F Schauer, HJ Heipieper. 2004. Simultaneous degradation of atrazine and phenol by *Pseudomonas* sp. strain ADP: Effects of toxicity and adaptation. *Appl Environ Microbiol* 70: 1907–1912.

Nichols NN, CS Harwood. 1997. PcaK, a high-affinity permease for the aromatic compounds 4-hydroxybenzoate and protocatechuate from *Pseudomonas putida*. *J Bacteriol* 179: 5056–5061.

Nohynek JL, E-L Nurmiaho-Lassila, EL Suhonen, H-J Busse, M Mohammadi, M Hantula, F Raney, M Salkinoja-Salonen. 1996. Description of chlorophenol-degrading *Pseudomonas* sp. strains KF1, KF3, and NKF1 as a new species of the genus *Sphingomonas*, *Sphingomonas subarctica* sp. nov. *Int J Syst Bacteriol* 46: 1042–1055.

Nohynek JL, EL Suhonen, E-L Nurmiaho-Lassila, M Hantula, M Salkinoja-Salonen. 1995. Description of four pentachlorophenol-degrading bacterial strains as *Sphingomonas chlorophenolica* sp. nov. *Syst Appl Microbiol* 18: 527–538.

Nyholm N, A Damborg, P Lindgaard-Jörgensen. 1992. A comparative study of test methods for assessment of the biodegradability of chemicals in seawater—Screening tests and simulation tests. *Ecotoxicol Environ Saf* 23: 173–190.

Ogram AV, RE Jessup, LT Ou, PSC Rao. 1985. Effects of sorption on biological degradation rates of (2,4-dichlorophenoxy)acetic acid in soils. *Appl Environ Microbiol* 49: 582–587.

Ogunseitan OA, IL Deklgado, Y-L Tsai, BH Olson. 1991. Effect of 2-hydroxybenzoate on the maintenance of naphthalene-degrading pseudomonads in seeded and unseeded soils. *Appl Environ Microbiol* 57: 2873–2879.

Olsen RH, JJ Kukor, B Kaphammer. 1994. A novel toluene-3-monooxygenase pathway cloned from *Pseudomonas pickettii* PKO1. *J Bacteriol* 176: 3749–3756.

Ornston LN. 1966. The conversion of catechol and protocatechuate to β-ketoadipate by *Pseudomonas putida*. IV. Regulation. *J Biol Chem* 241: 3800–3810.

Owen SA, NJ Russell, WA House, GF White. 1998. Re-evaluation of the hypothesis that biodegradable surfactants stimulate surface attachment of competent bacteria. *Microbiology (UK)* 143: 3649–3659.

Paris DF, JE Rogers, 1986. Kinetic concepts for measuring microbial rate constants: Effects of nutrients on rate constants. *Appl Environ Microbiol* 51: 221–225.

Paris DF, NL Wolfe. 1987. Relationship between properties of a series of anilines and their transformation by bacteria. *Appl Environ Microbiol* 53: 911–916.

Paris DF, NL Wolfe, WC Steen. 1982. Structure–activity relationships in microbial transformation of phenols. *Appl Environ Microbiol* 44: 153–158.

Paris DF, NL Wolfe, WC Steen. 1984. Microbial transformation of esters of chlorinated carboxylic acids. *Appl Environ Microbiol* 47: 7–11.

Parke D, LN Ornston. 1986. Enzymes of the β-ketoadipate pathway are indicible in *Rhizobium* and *Agrobacterium* spp. and constitutive in *Bradyrhizobium* spp. *J Bacteriol* 165: 288–292.

Peel MC, RC Wyndham. 1999. Selection of *clc*, *cba*, and *fbc* chlorobenzoate-catabolic genotypes from groundwater and surface waters adjacent to the Hyde Park, Niagara Falls, chemical landfill. *Appl Environ Microbiol* 65: 1627–1635.

Peijnenburg WJGM, KGM de Beer, HA den Hollander, MHL Stegeman, H Verboom. 1993. Kinetics, products, mechanisms and QSARs for the hydrolytic transformation of aromatic nitriles in anaerobic sediment slurries. *Environ Toxicol Chem* 12: 1149–1161.

Peijnenburg WJGM, MJ 't Hart, HA den Hollander, D van de Meent, HH Verboom, NL Wolfe. 1992. QSARs for predicting reductive transformation constants of halogenated aromatic hydrocarbons in anoxic sediment systems. *Environ Toxicol Chem* 11: 301–314.

Perez J, TW Jeffries. 1990. Mineralization of ^{14}C-ring-labelled synthetic lignin correlates with the production of lignin peroxidase, not of manganese peroxidase or laccase. *Appl Environ Microbiol* 56: 1806–1812.

Peters R. 1952. Lethal synthesis. *Proc Roy Soc (Lond)* B 139: 143–170.

Pettigrew CA, BE Haigler JC Spain. 1991. Simultaneous biodegradation of chlorobenzene and toluene by a *Pseudomonas* strain. *Appl Environ Microbiol* 57: 157–162.

Pflugmacher U, B Averhoff, G Gottschalk. 1996. Cloning, sequencing, and expression of isopropylbenzene degradation genes from *Pseudomonas* sp, strain JR1: Identification of isopropylbenzene dioxygenase that mediates trichloroethene oxidation. *Appl Environ Microbiol* 62: 3967–3977.

Phale PS, HS Savithri, NA Rao, CS Vaidyanathan. 1995. Production of biosurfactant "Biosur-Pm" by *Pseudomonas maltophilia* CSV89: Characterization and role in hydrocarbon uptake. *Arch Microbiol* 163: 424–431.

Plugge CM, C Dijkema, AJM Stams. 1993. Acetyl-CoA cleavage patyhways in a syntrophic propionate oxidizing bacterium growing on fumarate in the absence of methanogens. *FEMS Microbiol Lett* 110: 71–76.

Poindexter JS. 1981. Oligotrophy. Fast and famine existence. *Adv Microb Ecol* 5: 63–89.

Racke KD, JR Coats. 1990. *Enhanced Biodegradation of Pesticides in the Environment.* American Chemical Society Symposium Series 426. American Chemical Society, Washington, DC.

Racke KD, DA Laskowski, MR Schultz. 1990. Resistance of chloropyrifos to enhanced biodegradation in soil. *J Agric Food Chem* 38: 1430–1436.

Racke KD, EP Lichtenstein. 1985. Effects of soil microorganisms on the release of bound ^{14}C residues from soils previously treated with [^{14}C]parathion. *J Agric Food Chem* 33: 938–943.

Ramos JL, A Wasserfallen, K Rose, KN Timmis. 1987. Redesigning metabolic routes: Manipulation of TOL plasmid pathway for catabolism of alkylbenzoates. *Science* 235: 593–596.

Rapp P. 2001. Multiphase kinetics of transformation of 1,2,4-trichlorobenzene at nano- and micromolar concentrations by *Burkholderia* sp. strain PS14. *Appl Environ Microbiol* 67: 3496–3500.

Rapp P, KN Timmis. 1999. Degradation of chlorobenzenes at nanomolar concentrations by *Burkholderia* sp. strain PS14 in liquid cultures and in soil. *Appl Environ Microbiol* 65: 2547–2552.

Reanney DC, PC Gowland, JH Slater. 1983. Genetic interactions among microbial communities. *Symp Soc Gen Microbiol* 34: 379–421.

Reinecke W, DJ Jeenes, PA Williams, H-J Knackmuss. 1982. TOL plasmid pWWO in constructed halobenzoate-degrading *Pseudomonas* strains: Prevention of meta pathway. *J Bacteriol* 150: 195–201.

Rendell NB, GW Taylor, M Somerville, H Todd, R Wilson, PJ Cole. 1990. Characterization of *Pseudomonas* rhamnolipids. *Biochim Biophys Acta* 1045: 189–193.

Rheinwald JG, AM Chakrabarty, C Gunsalus. 1973. A transmissible plasmid controlling camphor oxidation in *Pseudomonas putida. Proc Natl Acad Sci USA* 70: 885–889.

Richnow HH, A Eschenbach, B Mahro, R Seifert, P Wehrung, P Albrecht, W Michaelis. 1998. The use of ^{13}C-labelled polycyclic aromatic hydrocarbons for the analysis of their transformation in soil. *Chemosphere* 36: 2211–2224.

Riegert U, G Heiss, P Fischer, A Stolz. 1998. Distal cleavage of 3-chlorocatechol by an extradiol dioxygenase to 3-chloro-2-hydroxymuconic semialdehyde. *J Bacteriol* 180: 2849–2853.

Rijnaarts HHM, A Bachmann, JC Jumelet, AJB Zehnder. 1990. Effect of desorption and intraparticle mass transfer on the aerobic biomineralization of alpha-hexachlorocyclohexane in a contaminated calcareous soil. *Environ Sci Technol* 24: 1349–1354.

Robinson KG, WS Farmer, JT Novak. 1990. Availability of sorbed toluene in soils for biodegradation by acclimated bacteria. *Water Res* 24: 345–350.

Rosenberg S. 1971. Regulation of the mandelate pathway in *Pseudomonas aeruginosa. J Bacteriol* 108: 1257–1269.

Roy F, E Samain, HC Dubourguier, G Albagnac. 1986. *Synthrophomonas* (sic) *sapovorans* sp. nov., a new obligately proton reducing anaerobe oxidizing saturated and unsaturated long chain fatty acids. *Arch Microbiol* 145: 142–147.

Ruberto L, SC Vazquez, WP MacCormack. 2003. Effectiveness of the natural flora, biostimulation and bioaugmentation on the bioremediation of a hydrocarbon contaminated Antarctic soil. *Int Biodet Biodeg* 52: 115–125.

Rubin HE, M Alexander. 1983. Effect of nutrients on the rates of mineralization of trace concentrations of phenol and *p*-nitrophenol. *Environ Sci Technol* 17: 104–107.

Rubin HE, RV Subba-Rao, M Alexander. 1982. Rates of mineralization of trace concentrations of aromatic compounds in lake water and sewage samples. *Appl Environ Microbiol* 43: 1133–1138.

Rüttimann-Johnson C, RT Lamar. 1996. Polymerization of pentachlorophenol and ferulic acid by fungal extracellular lignin-degrading enzymes. *Appl Environ Microbiol* 62: 3890–3893.

Sabra W, E-J Kim, A-P Zeng. 2002. Physiological responses of *Pseudomonas aeruginosa* PAO1 to oxidative stress in controlled microaerophilic and aerobic cultures. *Microbiology (UK)* 148: 3195–3202.

Saflic S, PM Fedorak, JT Andersson. 1992. Diones, sulfoxides, and sulfones from the aerobic cometabolism of methylbenzothiophenes by *Pseudomonas* strain BT1. *Environ Sci Technol* 26: 1759–1764.

Salleh MA, JM Pemberton. 1993. Cloning of a DNA region of a *Pseudomonas* plasmid that codes for detoxification of the herbicide paraquat. *Curr Microbiol* 27: 63–67.

Samadapour M, J Liston, JE Ongerth, PI Tarr. 1990. Evaluation of DNA probes for detection of Shiga-like-toxin-producing *Escherichia coli* in food and calf fecal samples. *Appl Environ Microbiol* 56: 1212–1215.

Sánchez MA, M Vásquez, B González. 2004. A previously unexposed forest soil microbial community degrades high levels of the pollutant 2,4,6-trichlorophenol. *Appl Environ Microbiol* 70: 7567–7570.

Sangwan P, X Chen, P Hugenholtz, PH Janssen. 2004. *Chthoniobacter flavus* gen. nov., sp. nov., the first pure-culture representative of subdivision two, *Spartobacteria* classis nov., of the phylum Verrucomicrobia. *Appl Environ Microbiol* 70: 5875–5881.

Sanseverino J, BM Applegate, JMH King, GS Sayler. 1993. Plasmid-mediated mineralization of naphthalene, phenanthrene, and anthracene. *Appl Environ Microbiol* 59: 1931–1937.

Sarkar JM, RL Malcolm, J-M Bollag. 1988. Enzymatic coupling of 2,4-dichlorophenol to stream fulvic acid in the presence of oxidoreductases. *Soil Sci Am J* 52: 688–694.

Sayler GS, MS Shields, ET Tedford, A Breen, SW Hooper, KM Sirotkin, W Davis. 1985. Application of DNA–DNA colony hybridization to the detection of catabolic genotypes in environmental samples. *Appl Environ Microbiol* 49: 1295–1303.

Schink B. 1991. Syntrophism among prokaryotes, pp. 276–299. In *The Prokaryotes* (Eds. A Balows, HG Trüper, M Dworkin, W Harder, K-H Schleifer), Springer, Heidelberg.

Schink B. 1997. Energetics of syntrophic cooperation in methanogenic degradation. *Microbiol Mol Biol Rev* 61: 262–280.

Schleissner C, ER Olivera, M Fernández-Valverde, M Luengo. 1994. Aerobic catabolism of phenylacetic acid in *Pseudomonas putida* U: Biochemical characterization of a specific phenylacetic acid transport system and formal demonstration that phenylacetyl-coenzyme A is a catabolic intermediate. *J Bacteriol* 176: 7667–7676.

Schreiber K, N Boes, M Eschbach, L Jaensch, J Wehland, T Bjarnsholt, M Givskov, M Hentzer, M Schobert. 2006. Anaerobic survival of *Pseudomonas aeruginosa* by pyruvate fermentation requires an Usp-type stress protein. *J Bacteriol* 188: 659–668.

Schut F, EJ de Vries, JC Gottschal, BR Robertson, W Harder, RA Prins, DK Button. 1993. Isolation of typical marine bacteria by dilution culture: Growth, maintenance, and characteristics of isolates under laboratory conditions. *Appl Environ Microbiol* 59: 2150–2160.

Schut F, M Jansen, TMP Gomes, JC Gottschal, W Harder, RA Prins. 1995. Substrate uptake and utilization by a marine ultramicrobacterium. *Microbiology (UK)* 141: 351–361.

Seitz H-J, H Cypionka. 1986. Chemolithotrophic growth of *Desulfovibrio desulfuricans* with hydrogen coupled to ammonification with nitrate or nitrite. *Arch Microbiol* 146: 63–67.

Serdar CM, DT Gibson, DM Munnecke, JH Lancaster. 1982. Plasmid involvement in parathion hydrolysis in *Pseudomonas diminuta*. *Appl Environ Microbiol* 44: 246–249.

Shapir N, MJ Sadowsky, LP Wackett. 2005. Purification and characterization of allophanate hydrolase/AtzF) from *Pseudomonas* sp. strain ADP. *J Bacteriol* 187: 3731–3738.

Shi T, JK Fredrickson, DL Balkwill. 2001. Biodegradation of polycyclic aromatic hydrocarbons by *Sphingomonas* strains isolated from terrestrial subsurface. *J Ind Microbiol Biotechnol* 26: 283–289.

Shields MS, SW Hooper, GS Sayler. 1985. Plasmid-mediated mineralization of 4-chlorobiphenyl. *J Bacteriol* 163: 882–889.

Shields MS, SO Montgomery, PJ Chapman, SM Cuskey, PH Pritchard. 1989. Novel pathway of toluene catabolism in the trichloroethylene-degrading bacterium G4. *Appl Environ Microbiol* 55: 1624–1629.

Shimp RJ, FK Pfaender. 1985a. Influence of easily degradable naturally occurring carbon substrates on biodegradation of monosubstituted phenols by aquatic bacteria. *Appl Environ Microbiol* 49: 394–401.

Shimp J, FK Pfaender. 1985b. Influence of naturally occurring humic acids on biodegradation of monosubstituted phenols by aquatic bacteria. *Appl Environ Microbiol* 49: 402–407.

Shinoda Y, Y Sakai, H Uenishi, Y Uchihashi, A Hiraishi, H Yukawa, H Yurimoto, N Kato. 2004. Aerobic and anaerobic toluene degradation by a newly isolated denitrifying bacterium, *Thauera* sp. strain DNT-1. *Appl Environ Microbiol* 70: 1385–1392.

Shoda M, S Udaka. 1980. Preferential utilizatiion of phenol rather than glucose by *Trichosporon cutaneum* possessing a partiallty constitutive catechol 1,2-oxygenase. *Appl Environ Microbiol* 39: 1129–1133.

Shoham Y, E Rosenberg. 1983. Enzymatic depolymerization of emulsan. *J Bacteriol* 156: 161–167.

Simkins S, M Alexander. 1984. Models for mineralization kinetics with the variables of substrate concentration and population density. *Appl Environ Microbiol* 47: 1299–1306.

Simkins S, R Mukherjee, M Alexander. 1986. Two approaches to modeling kinetics of biodegradation by growing cells and application of a two-compartment model for mineralization kinetics in sewage. *Appl Environ Microbiol* 51: 1153–1160.

Simmons KE, RD Minard, J-M Bollag. 1987. Oligomerization of 4-chloroaniline by oxidoreductases. *Environ Sci Technol* 21: 999–1003.

Simmons KE, RD Minard, J-M Bollag. 1989. Oxidative co-oligomerization of guaiacol and 4-chloroaniline. *Environ Sci Technol* 23: 115–121.

Singer AC, CS Wong, DE Crowley. 2002. Differential enantioselective transformation of atropisomeric polychlorinated biphenyls by multiple strains with different inducing compounds. *Appl Environ Microbiol* 68: 5756–5759.

Singer AC, DE Crowley, IP Thompson. 2003. Secondary plant metabolites in phytoremediation and biotransformation. *Trends Biotechnol* 21: 123–130.

Singer MEV, WR Finnerty. 1990. Physiology of biosurfactant synthesis by *Rhodococcus* species H13-A. *Can J Microbiol* 36: 741–745.

Singer MEV, WR Finnerty, A Tunelid. 1990. Physical and chemical properties of a biosurfactant synthesized by *Rhodococcus* sp. H13-A. *Can J Microbiol* 36: 746–750.

Singh N, N Sethunathan. 1992. Degradation of soil-sorbed carbofuran by an enrichment culture from carbofuran-retreated *Azolla* plot. *J Agric Food Chem* 40: 1062–1065.

Six S, SC Andrews, G Unden, JR Guest. 1994. *Escherichia coli* possesses two homologous anaerobic C_4-dicarboxylate membrane transporters (*Dcua* and *Dcub*) distinct from the aerobic dicarboxylate transport system (*Dct*). *J Bacteriol* 176: 6470–6478.

Slater JH, D Lovatt. 1984. Biodegradation and the significance of microbial communities, pp. 439–485. In *Microbial Degradation of Organic Compounds* (Ed. DT Gibson), Marcel Dekker Inc, New York.

Smith HW, Z Parsell, P Green. 1978. Thermosensitive antibiotic resistance plasmids in enterobacteria. *J Gen Microbiol* 109: 37–47.

Smith MG, RJ Park. 1984. Effect of restricted aeration on catabolism of cholic acid by two *Pseudomonas* species. *Appl Environ Microbiol* 48: 108–113.

Sobecky PA, MA Schell, MA Moran, RE Hodson. 1992. Adaptation of model genetically engineered microorganisms to lake water: Growth rate enhancements and plasmid loss. *Appl Environ Microbiol* 58: 3630–3637.

Sondossi M, M Sylvestre, D Ahmad. 1992. Effects of chlorobenzoate transformation on the *Pseudomonas testosteroni* biphenyl and chlorobiphenyl degradation pathway. *Appl Environ Microbiol* 58: 485–495.

Spain JC, PA van Veld, CA Monti, PH Pritchard, CR Cripe. 1984. Comparison of *p*-nitrophenol biodegradation in field and laboratory test systems. *Appl Environ Microbiol* 48: 944–950.

Spokes JR, N Walker. 1974. Chlorophenol and chlorobenzoic acid co-metabolism by different genera of soil bacteria. *Arch Microbiol* 96: 125–134.

Stalker DM, KE McBride. 1987. Cloning and expresssion in *Escherichia coli* of a *Klebsiella ozaenae* plasmid-borne gene encoding a nitrilase specific for the herbicide Bromoxynil. *J Bacteriol* 169: 955–960.

Stanlake GJ, RK Finn. 1982. Isolation and characterization of a pentachlorophenol-degrading bacterium. *Appl Environ Microbiol* 44: 1421–1427.

Steen WC, TW Collette. 1989. Microbial degradation of seven amides by suspended bacterial populations. *Appl Environ Microbiol* 55: 2545–2549.

Steffensen WS, M Alexander. 1995. Role of competition for inorganic nutrients in the biodegradation of mixtures of substrates. *Appl Environ Microbiol* 61: 2859–2862.

Stieb M, B Schink. 1985. Anaerobic oxidation of fatty acids by *Clostridium bryantii* sp. nov., a sporeforming, obligately syntrophic bacterium. *Arch Microbiol* 140: 387–390.

Stuart-Keil KG, AM Hohnstock, KP Drees, JB Herrick, EL Madsen. 1998. Plasmids responsible for horizontal transfer of naphthalene catabolism genes between bacteria at a coal tar-contaminated site are homologous to pDTG1 from *Pseudomonas putida* NCIB 9816-4. *Appl Environ Microbiol* 64: 3633–3640.

Subba-Rao RV, M Alexander. 1982. Effect of sorption on mineralization of low concentrations of aromatic compounds in lake water samples. *Appl Environ Microbiol* 44: 659–668.

Subba-Rao RV, HE Rubin, M Alexander. 1982. Kinetics and extent of mineralization of organic chemicals at trace levels in freshwater and sewage. *Appl Environ Microbiol* 43: 1139–1150.

Suttinun O, R Müller, E Luepromchai. 2009. Trichloroethylene cometabolic degradation by *Rhodococcus* sp. L4 with plant essential oils. *Biodegradation* 20: 281–291.

Suttinun O, R Müller, E Luepromchai. 2010. Cometabolic degradation of trichloroethene by *Rhodococcus* sp. strain L4 immobilized on plant materials rich in essential oils. *Appl Environ Microbiol* 76: 4684–4690.

Suwa Y, WE Olben, LJ Forney. 1996. Characterization of chromosomally encoded 2,4-dichlorophenoxyacetic acid-α-ketoglutarate dioxygenase from *Burkholderia* sp. strain RASC. *Appl Environ Microbiol* 62: 2464–2469.

Suzuki T, K Tanaka, I Matsubara, S Kinoshita. 1969. Trehalose lipid and alpha-branched-beta-hydroxy fatty acid formed by bacteria grown on *n*-alkanes. *Agric Biol Chem* 33: 1619–1627.

Taeger K, H-J Knackmuss, E Schmidt. 1988. Biodegradability of mixtures of chloro- and methyl- substituted aromatics: Simultaneous degradation of 3-chlorobenzoate and 3-methylbenzoate. *Appl Microbiol Biotechnol* 28: 603–608.

Taira K, J Hirose, S Hayashida, K Furukawa. 1992. Analysis of *bph* operon from the polychlorinated biphenyl-degrading strain of *Pseudomonas pseudoalcaligenes* KF707. *J Biol Chem* 267: 4844–4853.

Takeuchi M, K Hamana, A Hiraishi. 2001. Proposal of the genus *Sphingomonas sensu stricto* and three new genera, *Sphingobium*, *Novosphingobium* and *Sphingopyxis*, on the basis of phylogenetic and chemotaxonomic analyses. *Int J Syst Evol Microbiol* 51: 1405–1417.

Tam AC, RM Behki, SU Khan. 1987. Isolation and characterization of an *S*-ethyl-*N*,*N*-dipropylthiocarbamate-degrading *Arthrobacter* strain and evidence for plasmid-associated *S*-ethyl-*N*,*N*-dipropylthiocarbamate degradation. *Appl Environ Microbiol* 53: 1088–1093.

Tan H-M, JR Mason. 1990. Cloning and expression of the plasmid-encoded benzene 2,3,4,6-tetrachlorobiphenyl dioxygenase genes from *Pseudomonas putida* ML2. *FEMS Microbiol Lett* 72: 259–264.

Tattersfield F. 1928. The decomposition of naphthalene in the soil, and the effect upon its insecticidal action. *Ann Appl Biol* 15: 57–80.

Thayer JR, ML Wheelis. 1982. Active transport of benzoate in *Pseudomonas putida*. *J Gen Microbiol* 128: 1749–1753.

Thomas AW, J Lewington, S Hope, AW Topping, AJ Weightman, JH Slater. 1992. Environmentally directed mutations in the dehalogenase system of *Pseudomonas putida* strain PP3. *Arch Microbiol* 158: 176–182.

Topp E, RL Crawford, RS Hanson. 1988. Influence of readily metabolizable carbon on pentachlorophenol metabolism by a pentachlorophenol-degrading *Flavobacterium* sp. *Appl Environ Microbiol* 54: 2452–2459.

Touati D, M Jacques, B Tardat, L Bouchard, S Despied. 1995. Lethal oxidative damage and mutagenesis are generated by iron Δ*fur* mutants of *Escherichia coli*: Protective role of superoxide dismutase. *J Bacteriol* 177: 2305–2314.

Trenck Tvd, H Sandermann. 1981. Incorporation of benzo[*a*]pyrene quinones into lignin. *FEBS Lett* 125: 72–76.

Tsao C-W, H-G Song, R Bartha, 1998. Metabolism of benzene, toluene, and xylene hydrocarbons in soil. *Appl Environ Microbiol* 64: 4924–4929.

Vacca DJ, WF Bleam, WJ Hickey. 2005. Isolation of soil bacteria adapted to degrade humic acid-sorbed phenanthrene. *Appl Environ Microbiol* 71: 3797–3805.

van der Kooij D, WAM Hijnen. 1984. Substrate utilization by an oxalate-consuming *Spirillum* species in relation to its growth in ozonated water. *Appl Environ Microbiol* 47: 551–559.

van der Kooij D, JP Oranje, WAM Hijnen. 1982. Growth of *Pseudomonad aeruguinosa* in tap water in relation to utilization of substrates at concentrations of few micrograms per liter. *Appl Environ Microbiol* 44: 1086–1095.

van der Kooij D, A Visser, JP Oranje. 1980. Growth of *Aeromonas hydrophila* at low concentrations of substrates added to tap water. *Appl Environ Microbiol* 39: 1198–1204.

van der Meer JR, TNP Bosma, WP de Bruin, H Harms, C Holliger, HHM Rijnaarts, ME Tros, G Schraa, AJB Zehnder. 1992. Versatility of soil column experiments to study biodegradation of halogenated compounds under environmental conditions. *Biodegradation* 3: 265–284.

van der Meer JR, ARW van Neerven, EJ de Vries, WM de Vos, AJB Zehnder. 1991. Cloning and characterization of plasmid-encoding genes for the degradation of 1,2-dichloro-, 1,4-dichloro-, and 1,2,4-trichlorobenzene of *Pseudomonas* sp. strain P51. *J Bacteriol* 173: 6–15.

van der Meer JR, C Werlen, SF Nishino, JC Spain. 1998. Evolution of a pathway for chlorobenzene metabolism leads to natural attenuation in contaminated groundwater. *Appl Environ Microbiol* 64: 4185–4193.

van der Woude BJ, JC Gottschal, RA Prins. 1995. Degradation of 2,5-dichlorobenzoic acid by *Pseudomonas aeruginosa* JB2 at low oxygen tensions. *Biodegradation* 6: 39–46.

Vandenbergh PA, RL Cole. 1986. Plasmid involvement in linalool metabolism by *Pseudomonas fluorescens*. *Appl Environ Microbiol* 52: 939–940.

Vandenbergh PA, RH Olsen, JE Colaruotolo. 1981. Isolation and genetic characterization of bacteria that degrade chloroaromatic compounds. *Appl Environ Microbiol* 42: 737–739.

Vandenbergh PA, AM Wright, 1983. Plasmid involvement in acyclic isoprenoid metabolism by *Pseudomonas putida*. *Appl Environ Microbiol* 45: 1953–1955.

Vannelli T, AB Hooper. 1992. Oxidation of nitrapyrin to 6-chloropicolinic acid by the ammonia-oxidizing bacterium *Nitrosomonas europaea*. *Appl Environ Microbiol* 58: 2321–2325.

Wackett LP, DT Gibson. 1988. Degradation of trichloroethylene by toluene dioxygenase in whole-cell studies with *Pseudomonas putida* F1. *Appl Environ Microbiol* 54: 1703–1708.

Wackett LP, MJ Sadowsky, LM Newman, H-G Hur, S Li. 1994. Metabolism of polyhalogenated compounds by a genetically engineered bacterium. *Nature* 368: 627–629.

Wada H, K Sendoo, Y Takai. 1989. Rapid degradation of γ-HCH in the upland soil after multiple application. *Soil Sci Plant Nutr* 35: 71–77.

Wallrabenstein C, E Hauschild, B Schink. 1995. *Syntrophobacter pfennigii* sp. nov., new syntrophically propionate-oxidizing anaerobe growing in pure culture with propionate and sulfate. *Arch Microbiol* 164: 346–352.

Walsh C. 1982. Fluorinated substrate analogs: Routes of metabolism and selective toxicity. *Adv Enzymol* 55: 187–288.

Wandersman C, P Delepelaire. 2004. Bacterial iron sources: From siderophores to hemophores. *Annu Rev Microbiol* 58: 611–647.

Whited GM, DT Gibson. 1991. Separation and partial characterization of the enzymes of the toluene-4-monooxygenase catabolic pathway in *Pseudomonas mendocina* KR1. *J Bacteriol* 173: 3017–3020.

Whyte LG, CW Greer, WE Innis. 1996. Assessment of the biodegradation potential of psychrotrophic microorganisms. *Can J Microbiol* 42: 99–106.

Whyte LG, J Hawari, E Zhou, L Mournonnière, WE Inniss, CW Greer. 1998. Biodegradation of variable-chain length alkanes at low temperatures by a psychotrophic *Rhodococcus* sp. *Appl Environ Microbiol* 64: 2578–2584.

Widdel F, G-W Kohring, F Mayer. 1983. Studies on dissimilatory sulfate-reducing bacteria that decompose fatty acids. *Arch Microbiol* 134: 286–294.

Williams WA, RJ May. 1997. Low-temperature microbial aerobic degradation of polychlorinated biphenyls in sediment. *Environ Sci Technol* 31: 3491–3496.

Williams PA, MJ Worsey. 1976. Ubiquity of plasmids in coding for toluene and xylene metabolism in soil bacteria: Evidence for the existence of new TOL plasmids. *J Bacteriol* 125: 818–828.

Witschel M, S Nagel, T Egli. 1997. Identification and characterization of the two-enzyme system catalyzing the oxidation of EDTA in the EDTA-degrading bacterial strain DSM 9103. *J Bacteriol* 179: 6937–6943.

Wolfe NL, RG Zepp, DF Paris. 1978. Use of structure–reactivity relationships to estimate hydrolytic persistence of carbamate pesticides. *Water Res* 12: 561–563.

Wolin MJ. 1982. Hydrogen transfer in microbial communities, pp. 323–356. In *Microbial Interactions and Communities* (Eds. AT Bull, JH Slater), Vol. 1, Academic Press, New York.

Wong CM, MJ Dilworth, AR Glenn. 1994. Cloning and sequencing show that 4-hydroxybenzoate hydroxylase *Poba* is required for uptake of 4-hydroxybenzoate in *Rhizobium leguminosarum. Microbiology (UK)* 140: 2775–2786.

Wu Q, DL Bedard, J Wiegel. 1997a. Effect of incubation temperature on the route of microbial reductive dechlorination of 2,3,4,6-tetrachlorobiphenyl in polychlorinated biphenyl (PCB)-contaminated and PCB-free freshwater sediments. *Appl Environ Microbiol* 63: 2836–2843.

Wu Q, DL Bedard, J Wiegel. 1997b. Temperature determines the pattern of anaerobic microbial dechlorination of Arochlor 1260 primed by 2,3,4,6-tetrachlorobiphenyl in Woods Pond sediment. *Appl Environ Microbiol* 63: 4818–4825.

Wyndham RC. 1986. Evolved aniline catabolism in *Acinetobacter calcoaceticus* during continuous culture of river water. *Appl Environ Microbiol* 51: 781–789.

Yabuuchi E, I Yano, H Oyaizu, Y Hashimoto, T Ezaki, H Yamamoto. 1990. Proposals of *Sphingomonas paucimobilis* gen. nov. and comb. nov., *Sphingomonas parapaucimobilis* sp. nov., *Sphingomonas yanoikuyae* sp. nov., *Sphingomonas adhaesiva* sp. nov., *Sphingomonas capsulata* comb. nov., and two genospecies of the genus *Sphingomonas. Microbiol Immunol* 34: 99–119.

Yadav JS, RE Wallace, CA Reddy. 1995. Mineralization of mono- and dichlorobenzenes and simultaneous degradation of chloro- and methyl-substituted benzenes by the white-rot fungus *Phanerochaete chrysosporium Appl Environ Microbiol* 61: 677–690.

Yu Z, GR Stewart, W Mohn. 2000. Apparent contradiction: Psychrotolerant bacteria from hydrocarbon-contaminated arctic tundra soils that degrade diterpenoids synthesized by trees. *Appl Environ Microbiol* 66: 5148–5154.

Zaidi BR, Y Murakami, M Alexander. 1988. Factors limiting success of inoculation to enhance biodegradation of low concentrations of organic chemicals. *Environ Sci Technol* 22: 1419–1425.

Zhang H, S Hanada, T Shigematsu, K Shibuya, Y Kamagata, T Kanawa, R Kurane. 2000. *Burkholderia kururiensis* sp. nov. a trichloroethylene (TCE)-degrading bacterium isolated from an aquifer polluted with TCE. *Int J Syst Evol Microbiol* 50: 743–749.

Zhang Y, RM Miller. 1992. Enhanced octadecane dispersion and biodegradation by a *Pseudomonas* rhamnolipid surfactant (biosurfactant). *Appl Environ Microbiol* 58: 3276–3282.

Zhang Y, WJ Maier, RM Miller. 1997. Effect of rhamnolipids on the dissolution, bioavailability and biodegradation of phenanthrene. *Environ Sci Technol* 31: 2211–2217.

Zhao H, D Yang, CR Woese, MP Bryant. 1990. Assignment of *Clostridium bryantii* to *Syntrophospora bryantii* gen. nov., comb. nov. on the basis of a 16S rRNA sequence analysis of its crotonate-grown pure culture. *Int J Syst Bacteriol* 40: 40–44.

Zipper C, M Bunk, AJB Zehnder, H-PE Kohler. 1998. Enantioselective uptake and degradation of the chiral herbicide dichloroprop [(*RS*)-2-(2,4-dichlorophenoxy)propionic acid] by *Sphingomonas herbicidovorans* MH. *J Bacteriol* 180: 3368–3374.

Zipper C, K Nickel, W Angst, H-P E Kohler. 1996. Complete microbial degradation of both enantiomers of the chiral herbicide Mecoprop [(*R,S*)-2-(4-chloro-2-methylphenoxy)]propionic acid in an enantioselective manner by *Sphingomonas herbicidovorans* sp. nov. *Appl Environ Microbiol* 62: 4318–4322.

Zylstra GJ, LP Wackett, DT Gibson. 1989. Trichloroethylene degradation by *Escherichia coli* containing the cloned *Pseudomonas putida* F1 toluene dioxygenase genes. *Appl Environ Microbiol* 55: 3162–3166.

Section II

Experimental Procedures

5

Experimental Procedures

Part 1: General

5.1 Introduction

A general overview of the processes that determine the fate and persistence of xenobiotics in the environment has been presented in Chapters 1 through 4, but it is important to anchor these to the experimental methods on which the conclusions from such investigations must ultimately be based. This is the aim of the present chapter. The application of isotopes and physical methods for structure determination are given in the next chapter.

5.2 Abiotic Reactions

Although these reactions have already been illustrated, a few brief comments on experimental aspects should be given. For chemical reactions, few details are needed since their design follows those used in traditional preparative organic chemistry integrated with measurements of rates by methods that are widely used in physical organic chemistry. Two issues are worth noting and should be taken into account:

1. Substrate concentrations that are environmentally realistic are much lower than those often used in purely chemical investigations.
2. Environmental temperatures are highly variable, and range from low summer water temperatures in northern latitudes to high temperatures on the surface of soils during periods of high insolation.

Experiments on photochemical reactions and transformations have been carried out under a number of different conditions:

1. Reactions in liquid medium can be carried out by illumination with radiation of the relevant wavelengths generally in quartz vessels of various configurations.
2. Gas-phase reactions have been carried out in 160 mL quartz vessels, and the products analyzed online by mass spectrometry (Brubaker and Hites 1998). Hydroxyl radicals were produced by photolysis of ozone in the presence of water:

$$O_3 + h\nu \rightarrow O(^1D) + O_2; \qquad O(^1D) + H_2O \rightarrow 2OH$$

3. Reactions that simulate tropospheric conditions have been carried out in Teflon bags with volumes of ca. 6 m^3 fitted with sampling ports for introduction of reactants and substrates, and removal of samples for analysis. Substrates can be added in the gas phase or as aerosols that form a surface film. The primary reactants are the hydroxyl and nitrate radicals, and ozone. These must be prepared before use by reactions (a) through (c).

a. Hydroxyl radicals by photolysis of methyl nitrite:

$$CH_3-ONO + h\nu \rightarrow CH_3O + NO$$

$$CH_3O + O_2 \rightarrow CH_2O + HO_2$$

$$HO_2 + NO \rightarrow OH + NO_2$$

NO is added to inhibit photolysis of NO_2 that would produce O_3 and NO_3 radicals:

$$NO + h\nu \rightarrow NO + O(^3P)$$

$$O(^3P) + O_2 \rightarrow O_3$$

$$O_3 + NO_2 \rightarrow NO_3 + O_2$$

b. Nitrate radicals are prepared by photolysis of dinitrogen pentoxide in the dark:

$$N_2O_5 \rightarrow NO_3 + NO_2$$

c. Ozone generated by a corona discharge in O_2.

Products from the reactions are collected on Tenax cartridges, and the analytes desorbed by heating, or on polyurethane form plugs from which the analytes can be recovered by elution with a suitable organic solvent.

4. Outdoor reactors (EUROPHORE) that use atmospheric illumination have been constructed in Valencia, Spain using hemispherical chambers with a volume of 170 m^3 covered with ethene–propene foil that is suitably transparent at the appropriate wavelengths. It is possible to carry out reactions throughout the day, and by using differential optical spectroscopy with a path length of 326.8 m and Fourier transform spectroscopy with a path length of 128 m that is built into the system. As a result, possible artifacts from sampling or treatment for analysis can be minimized. Two examples are given that are illustrated using toluene that is an important atmospheric contaminant

a. In one experiment (Klotz et al. 1998) on the reaction of toluene, which is a major atmospheric contaminant, a range of concentrations of NO_x was used that are observed in the lower atmosphere, and ozone concentrations at the beginning of the reactions were low. It was, therefore, possible to provide an unequivocal description of the photochemical reactions that were involved.

b. In another experiment (Alvarez et al. 2007), concentrations of NO_x resembled those found in the atmosphere, and concentrations of ozone were low at the beginning of the experiment. The objective was to determine the yields of the products from ring-scission of benzene and toluene, and complement the results of Berndt and Boge (2006). The major reaction involved OH radicals that initiated two reactions: (a) removal of a hydrogen atom from the methyl group to form benzaldehyde and (b) formation of a toluene-hydroxyl adduct (methyl-hydroxy-cyclohexadienyl radical) that reacted with oxygen to produce *o*-, *m*-, and *p*-cresol that reacted further with nitrate radicals. Glyoxal, methylglyoxal, 4-ketopent-2-enal, and butenedialdehyde (*cis* and *trans*) were produced from toluene, and glyoxal, butenedialdehyde (*cis* and *trans*) from benzene.

5.3 Microbial Reactions

The study of microbial metabolism is a relatively young discipline, not more than 100 years old, like organic chemistry on which it critically depends. And the cardinal experimental procedures for isolation of microorganisms for studies on the metabolism of xenobiotics remain those of elective enrichment pioneered by Winogradsky, Beijerinck, Kluyver, van Niel, and their successors, backed up by the use of pure cultures using procedures developed by Koch in his classic investigations on anthrax. The major innovation has been the development of generally applicable procedures for the isolation of strictly anaerobic organisms that were introduced by Hungate (1969). There have, however, been major developments in methods for the elucidation of metabolic pathways that are discussed in later sections:

1. Availability of isotopically labeled compounds—in particular ^{14}C, ^{13}C, $^{18}O_2$, $^{18}OH_2$, and ^{15}N.
2. Application of genetic procedures and molecular biology.
3. Application of modern analytical procedures and physical methods for structure determination. In studies of microbial metabolism, the advantages resulting from the requirement for only extremely small quantities of material needed for gas- and liquid-chromatographic quantification, coupled to mass spectrometric identification, can hardly be overestimated.

There has, at the same time, been an increasing realization of the role of populations and the intrinsic limitations in enrichment. These are noted later in this chapter, and procedures for their examination are discussed in the second part of this chapter.

In the following sections, an attempt is made to provide a critical outline of experimental aspects of investigations directed to biodegradation and biotransformation, with particular emphasis on outstanding issues to which sufficient attention has not always been paid, and which have not, therefore, received ultimate resolution. Before proceeding further, it is desirable to define clearly some operational terms:

Mineralization is conventionally used for the aerobic degradation of a compound to CO_2 and H_2O.

Ready biodegradability refers to the situation in which the test compound is totally degraded (under aerobic conditions to CO_2, H_2O, etc.) within the time span of a standardized test usually lasting 5, 7, or 28 days.

Inherent biodegradability is applied when the compound *may* be degraded, though not under the standard conditions generally used: their degradation may require, for example, preexposure to the xenobiotic.

Recalcitrance is a valuable concept (Alexander 1975) that has been applied to compounds that have not been demonstrably degraded under the conditions used for their examination.

Biotransformation is applied to situations in which, even though degradation is not achieved, minor structural modifications of the test compound have occurred.

Rigid boundaries between these terms should not, however, be drawn, since all of them are operational rather than absolute.

5.4 Storage of Samples

In both field investigations and laboratory experiments, it may be necessary to store samples before examination for biodegradation of the contaminants. For laboratory experiments there is little choice except freezing. There has been conflicting advice for soil samples. Although the ISO Standard ISO/TC/190/SC 2 does not recommend freezing, for samples from three horizons of the unsaturated zone of a sandy soil, it has been shown that rates of degradation by the endogenous organisms of both 4-chloro-2-methylphenoxyacetic acid (MCPA) and metribuzin (4-amino-6-*tert*-butyl-3-methylthio-1,2,4-triazin-5-(4*H*)-one) were unaffected by freezing the samples (Mortensen and Jacobsen 2004). In contrast, storage under conditions of drying followed by addition of water to the original content produced substantial changes: there were significant differences in the population of archaea and bacteria, and degradation of metalaxyl-M and lufenuron were delayed by factors up to 6 (Pesaro et al. 2004).

5.5 Determination of Ready Biodegradability

Because of the central role that estimates of biodegradability play in environmental impact assessments, a great deal of effort has been devoted to developing standardized test procedures (Gerike and Fischer 1981). In spite of this, conventional tests for biodegradability under aerobic conditions retain some questionable, or even undesirable, features from an environmental point of view. Attention is, therefore, drawn to two valuable critiques of widely used procedures (Howard and Banerjee 1984; Battersby 1990). A methodology based on continuous measurement of CO_2 evolution by measuring conductivity provides continuous measurements in an essentially closed system (Strotmann et al. 2004). This makes it possible to evaluate substrates that are poorly soluble in water or are volatile. Comparison with other widely used systems showed that it was both accurate and reliable. Some of the important issues in the design of such tests are, therefore, only briefly summarized here.

5.5.1 The Inoculum

For assessment of biodegradation in freshwater systems that have been most extensively examined, the inoculum is often taken from municipal sewage treatment plants and is, therefore, dominated by microorganisms that have been subjected to selection primarily for their ability to use readily degraded substrates. These organisms are clearly valuable in evaluating the persistence of compounds that might be poorly degraded in such treatment systems, and might, therefore, be discharged unaltered into the environment. They are, however, not necessarily equally suited to investigations on the degradability of possibly recalcitrant substances in natural ecosystems that are dominated by microorganisms adapted to a different environment. In addition, attention is drawn to organisms that have been isolated from pristine environments (Kitagawa et al. 2002; Sánchez et al. 2004; Hennessee et al. 2009).

5.5.2 Concentration of the Substrate

In conventional assay systems, substrate concentrations are generally used at levels appropriate to measurements of the uptake of oxygen or the evolution of carbon dioxide, and these concentrations greatly exceed those likely to be encountered in receiving waters; attention has been already been drawn to this issue in this chapter and in Chapter 4. Application to poorly water-soluble (De Morsier et al. 1987) or to volatile substrates may be difficult or even impossible due to the need for the high substrate concentrations. In addition, high substrate concentrations may be toxic to the test organism and thereby provide false-negative results. Use of ^{14}C-labeled substrates and measurement of the evolution of [14]CO_2 enables very much lower substrate concentrations to be used, and this is particularly important if the contribution of marine oligotrophic bacteria is to be evaluated. This has been applied to populations of marine bacteria using ^{14}C-labeled acetate, aniline, 4-nitrophenol, desmethyl- and methylparathion, and 4-chloroaniline (Nyholm et al. 1992). The wider application of the use of isotopes is limited primarily by the availability of labeled substrates, though an increasingly wider range of industrially important compounds, including agrochemicals, is becoming commercially available.

5.5.3 Endpoints

For aerobic degradation, uptake of oxygen or the evolution of carbon dioxide is most widely used. Use of the concentration of dissolved organic carbon may present technical problems when particulate matter is present, though analysis of dissolved inorganic carbon in a closed system has been advocated (Birch and Fletcher 1991), and may simultaneously overcome problems with poorly soluble or volatile compounds.

For anaerobic degradation, advantage has been taken of methane production (Battersby and Wilson 1989; Birch et al. 1989). Whereas this may be valuable in the context of municipal sewage treatment plants, it is more questionable whether this is generally a valid parameter in investigations concerned with anaerobic degradation in natural ecosystems in view of the extensive evidence for anaerobic degradation by nonmethanogenic bacteria such as sulfate-reducing anaerobes.

5.6 Design of Experiments on Inherent Biodegradability

Probably most investigations have been carried out in conventional batch cultures, but attention should be drawn to an attractive and flexible procedure using a cyclone fermentor (Liu et al. 1981).

5.6.1 Metabolic Limitations

In the most widely applied procedures, the test system is restricted in flexibility by the salinity and pH requirements of the test organisms, but probably the most serious limitation of these test systems is that no account is generally taken of biotransformation reactions, nor is identification of their products routinely attempted. Examples where this has been carried out include the following:

1. In a study of the degradation of sodium dodecyltriethoxy sulfate under mixed-culture die-away conditions using acclimated cultures (Griffiths et al. 1986), the metabolites were identified and the kinetics of their synthesis compared with the degradation pathways elucidated in investigations using pure cultures (Hales et al. 1982, 1986).
2. The biodegradation of branched-chain alkanol ethoxyethylates was carried out by the standard OECD confirmatory tests and the metabolites fractionated after solid-phase extraction. The structures of the metabolites were determined by electrospray mass spectrometry and this made it possible to derive a scheme for the partial degradation of the compounds (Di Corcia et al. 1998).

These procedures could advantageously receive general application.

5.6.2 Application to Marine Systems

In view of the substantial quantities of xenobiotics that enter the marine environment, surprisingly limited effort has been directed to this problem. The degradation of several structurally diverse substrates, including nitrilotriacetate, 3-methylphenol, and some chlorobenzenes, was evaluated from the rates of incorporation of ^{14}C-labeled substrates into biomass and production of $^{14}CO_2$. These were used to evaluate the differences between freshwater, estuarine, and marine environments and revealed the difficulty of correlating rates with characteristics of the microbial community (Bartholomew and Pfaender 1983). Recent studies have used both dissolved organic carbon and oxygen uptake as parameters (Nyholm and Kristensen 1992), or analysis for specific ^{14}C-labeled compounds at low substrate concentration, which are proposed as "simulation tests" (Nyholm et al. 1992). Both these tests used the indigenous organisms present in seawater and thereby provided a valuable degree of relevance even though they inevitably encountered the variability in the nutritional status—particularly for organic carbon in seawater. Attention is drawn to oligotrophic marine ultramicrobacteria that are of undoubted importance in oceanic systems. There are, however, important aspects of their isolation that should be appreciated, including the extent to which their physiology may be altered during maintenance under severe carbon limitation in the laboratory (Schut et al. 1993). Biodegradation of xenobiotics by these organisms has not attracted the attention it merits, and it is hazardous to extrapolate results from freshwater or brackish-water systems to marine systems since all the factors noted above coupled with the low temperatures that characterize ocean water introduce complexities that remain to be systematically investigated. Attention is drawn to the extensive metabolic potential of anaerobic sulfate-reducing bacteria isolated from marine muds and sediments. Examples are provided throughout in Chapters 7 through 9.

5.6.3 Isolation and Elective Enrichment

It is only seldom that it has been possible to obtain bacteria with a desired metabolic capability directly from natural habitats. Almost always, large numbers of other organisms are present, so that some form of selection or enrichment is generally adopted before metabolic studies are attempted. Use of antibiotics or even more drastic procedures using alkali or hypochlorite have been used only infrequently except for the isolation of pathogenic bacteria such as *Mycobacterium tuberculosis*. Valuable results have been obtained from experiments using metabolically stable mixed cultures, which overcome to a large extent the problems of repeatability. Probably most metabolic studies on xenobiotics have, however, been carried with pure cultures of organisms. With the possible exception of anaerobic bacteria, and in particular methanogens, only a limited range of these organisms have been isolated from samples of municipal sewage sludge. Most have been obtained after elective enrichment of natural samples of water, sediments, or soils. This methodology was developed by Beijerinck and Winogradsky, and has been extensively exploited in the pioneering investigations carried out by the Delft school and their successors over many years. In the present context, one of its particularly attractive features is the inherent degree of environmental realism introduced by its application, and the flexibility whereby virtually any environmental condition can be reproduced must be considered as one of the most attractive features of this procedure. There is extensive evidence for the existence of enrichment in the natural environment and a number of examples illustrating its operation have been given in Chapter 4. Possibly the last word should be left to one of the pioneers of its application:

> But once an elective culture method for a particular microbe is available, it may safely be concluded that this organism will also be found in nature under conditions corresponding in detail to those of the culture, and that it will carry out the same transformations. (Van Niel 1955)

Enrichment may, however, introduce a bias, not only in the organisms but also in their degradative genes (Marchesi and Weightman 2003). In addition, there are many organisms that have not hitherto been cultivated (Stevenson et al. 2004), and these almost certainly play an important role in degradation. Natural enrichment certainly takes place in environments that have been chronically exposed to contaminants and is an important mechanism for the loss of agrochemicals. Degradation of contaminants may, however, be observed in bacteria that have been isolated from pristine environments with no established exposure to the contaminants (e.g., Kitagawa et al. 2002; Sánchez et al. 2004; Hennessee et al. 2009). Procedures that have been used to investigate the role of natural populations without isolation of the relevant organisms are discussed later in this chapter.

5.6.4 General Procedures

In its simplest form, the procedure consists of the following steps:

1. Elective enrichment of the microorganisms in an environmental sample by growth at the expense of a single compound serving as the sole source of carbon and energy.
2. Successive transfer into fresh medium after growth has occurred.
3. Isolation of the appropriate organisms, generally by plating on solid media or by dilution.

Some of the experimental details are briefly described in the following paragraphs. The three successive stages used in isolating the desired organisms are outlined first, followed by a more extensive discussion of media.

1. An appropriate mineral medium supplemented with the organic compound that is to be studied is inoculated with a sample of water, soil, or sediment. In studies of the environmental fate of a xenobiotic in a specific ecosystem, samples are generally taken from the area putatively contaminated with the given compound so that a degree of environmental relevance is automatically incorporated. Attention has, in addition, been directed to pristine environments, and the issues of adaptation or preexposure have already been discussed.

   If the test compound is to serve as a source of sulfur, nitrogen, or phosphorus, these elements will necessarily be omitted from the medium, and an appropriate carbon source supplied either by the xenobiotic under investigation or by another substrate that is added. In such experiments, glassware must be scrupulously cleaned to remove interfering traces of, for example, detergents since these may contain residues of all of these nutrients and could, therefore, compromise the outcome of the experiment. The metabolism of phosphonates and sulfonates, and of sulfate and phosphate esters, may in addition, be inducible only in the absence of inorganic sources of phosphorus or sulfur. Although aromatic sulfonates have been used as sources of sulfur, the pathway for their aerobic degradation differs from that when they are used as sources of carbon and energy. The following are given as illustration of enrichments designed to obtain organisms using organic compounds as sources of nitrogen, sulfur, or phosphorus—although they are not necessarily able to use the test substrates as carbon sources:

a. Nitrophenols as nitrogen source using succinate as carbon source (Bruhn et al. 1987)

b. 2-Chloro-4,6-diamino-1,3,5-triazine as nitrogen source and lactate as carbon source (Grossenbacher et al. 1984), or 2-chloro-4-aminoethyl-6-amino-1,3,5-triazine as nitrogen source and glycerol as carbon source (Cook and Hütter 1984)

c. Arylsulfonic acids as sulfur source and succinate or glycerol as carbon sources (Zürrer et al. 1987)

d. 2-(1-Methylethyl)amino-4-hydroxy-6-methylthio-1,3,5-triazine as sulfur source and glucose as carbon source (Cook and Hütter 1982)

e. Glyphosate (*N*-phosphonomethylglycine) as phosphorus source and glucose as carbon source (Talbot et al. 1984)

These procedures may clearly result in the dominance of organisms that carry out only biotransformation of the xenobiotic, although the biodegradation of many of these compounds has also been demonstrated using the same or other organisms.

2. The cultures are then incubated under relevant conditions of temperature, pH, and oxygen concentration, and after growth has occurred, successive transfer to fresh medium is carried out; interfering particulate matter will thereby be removed by dilution, and a culture suitable for isolation may be obtained. Incubation is generally carried out in the dark, though for the isolation of phototrophic organisms, illumination at suitable wavelength and intensity must obviously be supplied. There are no rigid rules on how many transfers should be carried out, but especially for anaerobic organisms, sufficient time should elapse between transfers to allow growth of these often slow-growing organisms. Enrichment may, therefore, take up to a year or even longer.

3. After metabolically stable cultures have been obtained, pure cultures of the relevant organisms may then be obtained by any of several procedures:

a. By preparing serial dilution of the culture in a suitable buffer medium and spreading portions onto solid media (agar plates or roll tubes for anaerobic bacteria) containing the organic compound as source of carbon, sulfur, nitrogen, or phosphorus. The plates (or tubes) are then incubated under appropriate conditions. After growth has taken place, single colonies are then selected and pure cultures obtained by repeated restreaking on the original defined medium. Use of complex media or substrate analogs may introduce serious ambiguities since overgrowth of unwanted, rapidly growing organisms may occur. Considerable difficulty may be experienced when "spreading" organisms conceal the desired organism and this may make the isolation of single colonies a tedious procedure.

b. Serial dilution may be carried out in a defined liquid growth medium and the dilutions incubated under suitable conditions; successive transfers are then made from the highest liquid dilution showing growth. This experimentally tedious procedure may indeed be obligatory for organisms such as *Thermomicrobium fosteri* (a name no longer accepted by Zarilla and Perry 1984), which are unable to produce colonies on agar plates (Phillips and Perry 1976). It has been quite extensively used for anaerobic bacteria in which the liquid medium is replaced by a soft agar medium. A cyclic procedure involving an additional plating step was used to reduce the complexity of the original population and to isolate an organism utilizing high concentrations of 2,4,6-trichlorophenol (Maltseva and Oriel 1997).

c. Mechanical methods may be used for the isolation of filamentous organisms. For example, washing on the surface of membrane filters, or micromanipulation on the surface of agar medium (Skerman 1968) has occasionally been employed, and an ingenious procedure, which uses an electron microscope grid for preliminary removal of other organisms, has been described for use under anaerobic conditions (Widdel 1983).

d. A technique using floating filters has been used for the isolation of organisms from enrichment cultures: the culture is filtered through a membrane of suitable porosity and the filter is floated on the surface of a mineral medium and exposed to the substrate in the gas phase (de Bruyn et al. 1990). Details of its application are given later.

Lack of repeatability of the results of metabolic studies using laboratory strains that have been maintained by repeated transfer for long periods under nonselective conditions may be encountered. These strains may no longer retain their original metabolic capabilities, and this may be particularly prevalent when the strains carry catabolic plasmids that may have been lost under nonselective conditions. For these reasons, strains should be maintained in the presence of a cryoprotectant such as glycerol at low temperatures ($-70°C$ or in liquid N_2) as soon as possible after isolation. Freeze-drying is also widely adopted, and is recommended.

In some cases, difficulty may be experienced in isolating pure cultures with the desired metabolic capability, and the mixed enrichment cultures or consortia must be used for further studies. For example, although an enrichment culture effectively degraded atrazine (2-chloro-4-ethylamine-6-isopropylamine-1,3,5-triazine), none of the 200 pure cultures isolated from this were able to use the substrate as nitrogen source (Mandelbaum et al. 1993). Two (or more) organisms may cooperate in the degradation of the substrate. Considerable effort may then be required to determine the appropriate combination of organisms. For example, an enrichment using 4-chloroacetophenone yielded eight pure strains although none of these could degrade the substrate in pure culture. All the pair combinations were then analyzed, and this revealed that the degradation of the substrate was accomplished by strains of an *Arthrobacter* sp. and a *Micrococcus* sp. These were then used to elucidate the metabolic pathway (Havel and Reineke 1993). Attention is drawn to the fact that, in this case, isolation of the pure strains from the enrichment culture was carried out using a complex medium under nonselective conditions.

These batch procedures for enrichment and successive transfer may be replaced by the use of continuous culture. This may be particularly attractive when the test compound is toxic, when it is poorly soluble in water, or where the investigations are directed to substrate concentrations so low that clearly visible growth is not to be expected. These problems remain, however, for subsequent isolation of the relevant organisms. One considerable problem in long-term use arises from growth in the tubing of the pump system that is used to administer the medium and should be renewed periodically.

5.6.5 Basal Media

The choice of appropriate basal media is of cardinal importance and a number of important practical considerations should be taken into account.

5.6.5.1 Mineral Media

A plethora of basal media for the growth of freshwater organisms has been formulated. These may, however, differ significantly, particularly in the concentrations of phosphate, while for anaerobic bacteria the inclusion of bicarbonate and a suitable reductant is a standard practice. Numerous examples of suitable media have been collected in *The Prokaryotes* (e.g., Balows et al. 1992). Clearly, if the organic substrate is to serve as a source of nitrogen, sulfur, or phosphorus, these elements must be omitted from the basal medium. Otherwise, these inorganic nutrient requirements will generally be supplied by the following: sulfur (generally as sulfate except for organisms such as chlorobia, which require reduced sulfur as S^{2-}); nitrogen (generally as ammonium or nitrate except for N_2-fixing organisms); phosphorus (as phosphate); and Mg^{2+}, Ca^{2+}, and lower concentrations of Na^+ and K^+. For marine organisms, the basal medium is constructed to resemble natural seawater in the concentrations of Na^+, K^+, Mg^{2+}, Ca^{2+}, Cl^-, and SO_4^{2-}, while glycerophosphate may be used as phosphorus to avoid problems with precipitation. A wide range of different formulations have been used (Neilson 1980; Taylor et al. 1981), though the precise composition does not seem to be critical. The inclusion of nitrate may lead to various complications, which have already been discussed in Chapter 2. In addition, the nitrogen status of the growth medium determines the levels of lignin peroxidases and manganese-dependent peroxidases that are synthesized in *Phanerochaete chrysosporium*. The role of Mn concentration is noted later and in Chapter 3, Part 5.

It is important to underscore the fact that carbon dioxide is required for the growth not only of photo-trophic and lithotrophic organisms, but also of heterotrophs since some of these have an obligate requirement of carbon dioxide. It is worth noting that some pathogenic bacteria such as *Streptococcus pneumoniae* use carbonic anhydrase to minimize loss of internal CO_2 since HCO_3^- is unable to diffuse across membranes (Burghout et al. 2010), while in *Helicobacter pylori* the periplasmic α-carbonic anhy-drase is essential for acid acclimation (Marcus et al. 2005). Illustrative examples of heterotrophic growth where CO_2 is required include the following:

1. Anaerobic bacteria such as the acetogens, methanogens, and the propionic bacteria—the medium is generally supplemented with bicarbonate
2. Aerobic bacteria include the following:
   a. Degradation of propane (MacMichael and Brown 1987) and the branched-chain hydrocar-bon 2,6-dimethyloct-2-ene (Fall et al. 1979)
   b. Metabolism of epoxides by *Xanthobacter* sp. strain Py2 (Ensign et al. 1998)
   c. Oxidation of carbon monoxide (Meyer and Schlegel 1983)
   d. Degradation of acetone by both aerobic and anaerobic bacteria

5.6.5.2 *Trace Elements*

Many enzymes and coenzymes contain metals (Chapter 3, Part 5), and these elements—Zn, Cu, Fe, Mn, Co, Ni, Mo, and W—are generally provided at low concentrations. These are provided as the cation salts, for Mo as molybdate and for W as tungstate. A large number of different formulations of trace elements have been published. The A4 formulation (Arnon 1938) supplemented with Co^{2+} and MoO_4^{2-} has been widely used, or one of the SL series of formulations developed by Pfennig and his coworkers particularly for the cultivation of anaerobic bacteria. SL 9 is typical of the later series of formulations (Tschech and Pfennig 1984). Somewhat conflicting views exist on the possibly deleterious effects resulting from the incorporation of complexing agents, particularly ETDA, so that their concentration should probably be kept to a minimum.

Anaerobic bacteria are more fastidious in their trace-metal requirements, and selenium as selenite and tungsten as tungstate are routinely added (Tschech and Pfennig 1984). Selenium is required for the syn-thesis of active enzymes by a number of anaerobes, in some of them as a labile cofactor: xanthine dehy-drogenase in purine-fermenting *Clostridium acidiurici* and *Cl. cylindrosporum* (Wagner and Andreesen 1979); purine hydroxylase in *Cl. purinolyticum* (Self 2002); nicotinate hydroxylase in *Cl. barkeri* (Gladyshev et al. 1994); glycine fermentation in *Cl. purinolyticum* (Durre and Andreesen 1982); the glycine reductase complex in clostridium (Cone et al. 1977); formate dehydrogenase in a number of organisms, including methanogens (Jones and Stadtman 1981); and *Clostridium thermoaceticum* (Yamamoto et al. 1983). Addition of selenite enhanced the growth of *Bradyrhizobium japonicum* with H_2 and it was shown that selenium was incorporated into the hydrogenase (Boursier et al. 1988).

Tungsten is required for the synthesis of a range of enzymes necessary for the growth of various methanogens (Winter et al. 1984; Zellner et al. 1987), an anaerobic cellulose-degrading bacterium (Taya et al. 1985), and for the synthesis of various enzymes in anaerobic hyperthermophilic bacteria. Important examples of tungsten requirement that are dealt with in later sections include the following:

1. The carboxylic acid reductase in acetogenic clostridia such as *Clostridium thermoaceticum* (White et al. 1989; Strobl et al. 1992).
2. The benzylviologen-linked aldehyde oxidoreductase in *Desulfovibrio gigas* grown with ethanol (Hensgens et al. 1995) and the corresponding enzyme in *Desulfovibrio simplex* (Zellner and

Jargon 1997). For the latter, it was suggested that the flavins FMN or FAD were the natural cofactors.

3. The acetylene hydratase of *Pelobacter acetylenicus* (Rosner and Schink 1995).

4. Tungsten may be incorporated into some proteins of purinolytic clostridia (Wagner and Andreesen 1987), and into formylmethanofuran dehydrogenase in *Methanobacterium thermoautotrophicum* (Bertram et al. 1994).

5. Ferredoxin reductases in *Pyrococcus furiosus*, including aldehyde ferredoxin reductase, glyceraldehyde-3-phosphate ferredoxin oxidoreductase, and formaldehyde ferredoxin reductase (Roy et al. 1999).

It should be noted, however, that whereas the production and activity of some enzymes in *Thermotoga maritima* was stimulated by tungstate, the purified enzymes apparently did not contain tungsten. For the sake of completeness, it is worth noting the role of vanadium in the synthesis of alternative nitrogenase systems by several aerobic bacteria, particularly those within the genus *Azotobacter* (Fallik et al. 1991); low yields of hydrazine are produced, and this suggests a different affinity of the nitrogenase for the N_2 substrate (Dilworth and Eady 1991).

5.6.5.3 Control of pH

Although media are often buffered by the inclusion of phosphate, excessive concentrations should be avoided since precipitation of insoluble Ca and Mg phosphates may occur during sterilization by autoclaving. A number of organic buffers have been used effectively in several applications. In studies employing organic phosphorus compounds as sources of phosphorus, phosphate buffer has been replaced by, for example, HEPES (Cook et al. 1978), and MOPS has been incorporated into media for growth of Enterobacteriaceae (Neidhardt et al. 1974) and for moderately acidophilic *Acidobacteria* (Eichorst et al. 2007, 2011). For media requiring low pH, MES has been used in a medium with an extremely low phosphate concentration (Angle et al. 1991), and TRIS has been incorporated into media for growth of marine bacteria (Taylor et al. 1981). For all of these, it should be established that metabolic complications do not arise as a result of the ability of the organisms to use the buffer as sources of nitrogen or sulfur. It is also possible for the buffer to react with metabolic intermediates, and this is illustrated by the isolation of a compound produced by reaction of the carbon atom (formally CO) of CCl_4 with HEPES (Lewis and Crawford 1995). Metal chelating agents such as NTA or ETDA may be used, although their concentrations should be kept to a minimum in view of their potential toxicity: NTA must obviously be omitted in studies of utilization of organic nitrogen as N-source since both NTA and its metal complexes are apparently quite readily degraded (Firestone and Tiedje 1975).

5.6.5.4 Vitamins

Vitamins such as thiamin, biotin, and vitamin B_{12} are often added. Once again, the requirements of anaerobes are somewhat greater, and a more extensive range of vitamins that includes pantothenate, folate, and nicotinate is generally employed. In some cases, additions of low concentrations of peptones, yeast extract, casamino acids or rumen fluid may be used, though in higher concentrations, metabolic ambiguities may be introduced since these compounds may serve as additional carbon sources.

5.6.6 Sterilization

Mineral–basal media may be sterilized by autoclaving, but for almost all organic compounds that are used as sources of C, N, S, or P, it is probably better to prepare concentrated stock solutions and sterilize these by filtration, generally using 0.2 μm cellulose nitrate or cellulose acetate filters. The same applies to solutions of vitamins, and to solutions of bicarbonate and sulfide that are components of many media used for anaerobic bacteria.

5.6.7 Metal Concentration in Metabolism

These requirements should be assessed in the context of metalloenzymes, which are discussed in Chapter 3, Part 3.

5.6.7.1 Iron

Although Fe is required as a trace element, its uptake is critically regulated since excess leads to the generation of toxic hydroxyl radicals, and complex interactions involving Fe(II) and Fe(III) exist within the cell (Touati et al. 1995). The role of Fe(III)-complexing siderophores in maintaining homeostasis has been noted in Chapter 3, Part 5. Details of the role of Fe and its relation to the generation of toxic hydroxyl radicals have been explored by analysis of a strain of *Escherichia coli* and a mutant strain lacking both Fe and Mn superoxide dismutase. The mutant strain showed a marked increase in the hydroxyl radical after exposure to H_2O_2 (McCormick et al. 1998): preincubation with an Fe chelator inhibited this difference, and redox-active Fe defined as EPR-detectable ascorbyl radicals was greater in the mutant than in the wild-type strain. Iron may also play a subtler role in determining the biodegradability of a substrate that forms complexes with Fe. Two examples are used as illustration of probably different underlying reasons for this result:

1. A strain of *Agrobacterium* sp. was able to degrade ferric EDTA, though not the free compound (Lauff et al. 1990). This may be due to either the adverse effect of free EDTA on the cells or the inability of the cells to transport the free compound. The former is supported by the established sensitivity (Wilkinson 1968) of some Gram-negative organisms to ETDA, and the increased surface permeability in enteric organisms exposed to EDTA (Leive 1968). These results are consistent with the requirement for high concentrations of Ca^{2+} in the enrichment medium used for isolating a mixed culture capable of degrading EDTA (Nörtemann 1992). In contrast, although the degradative enzymes have been purified and characterized from cells of a pure culture grown with Mg ETDA, the enzyme complex was unable to use Fe EDTA as a substrate (Witschel et al. 1997). This is consistent with the results using an unenriched culture and ¹⁴C-labeled EDTA (Allard et al. 1996). In the aquatic environment with low concentrations of other complexing cations, degradation of EDTA—and probably DTPA—is accomplished primarily by photolysis of the Fe complex (Kari and Giger 1995).
2. A strain of *Pseudomonas fluorescens* biovar II is able to degrade citrate whose metabolism requires access to the hydroxyl group. This group is, however, implicated in the tridentate ligand with Fe(II) so that this complex is resistant to degradation in contrast to the bidentate ligand with Fe(III) that has a free hydroxyl group and is readily degraded (Francis and Dodge 1993). Similarly, the bidentate complexes containing Fe(III), Ni, and Zn are readily degraded by *Pseudomonas fluorescens*, in contrast to the tridentate complexes containing Cd, Cu, and U that are not degraded (Joshi-Tope and Francis 1995).

These results may be viewed in the wider context of interactions between potential ligands of multifunctional xenobiotics and metal cations in aquatic environments and the subtle effects of the oxidation level of cations such as Fe. The Fe status of a bacterial culture has an important influence on the synthesis of the redox systems of the cell since many of the electron transport proteins contain Fe. This is not generally evaluated systematically, although the degradation of tetrachloromethane by a strain of *Pseudomonas* sp. under denitrifying conditions clearly illustrated the adverse effect of Fe on the biotransformation of the substrate (Lewis and Crawford 1993; Tatara et al. 1993). This possibility should, therefore, be taken into account in the application of such organisms to bioremediation programs.

5.6.7.2 Manganese

The role of manganese concentration has seldom been explicitly examined in the context of biodegradation. A few extradiol dioxygenases contain Mn in place of Fe, and it is essential for the growth of the purple nonsulfur anaerobic phototrophs *Rhodospirillum rubrum* and *Rhodopseudomonas capsulata* during growth with N_2, though not with glutamate (Yoch 1979). It plays an essential role in the metabolic capability of the white-rot fungus *Phanaerochaete chrysosporium*. This organism produces two groups of peroxidases during secondary metabolism—lignin peroxidases and manganese-dependent peroxidases. Both are synthesized when only low levels of Mn (II) are present in the growth medium, whereas high concentrations of Mn result in repression of the synthesis of the lignin peroxidases and an enhanced synthesis of manganese-dependent peroxidases (Bonnarme and Jeffries 1990; Brown et al. 1990). Experiments with a nitrogen-deregulated mutant have shown that N regulation of both these groups of peroxidases is independent of Mn(II) regulation (Van der Woude et al. 1993). The unusual manganese-dependent dioxygenase (Hatta et al. 2003) would also necessitate adequate concentrations.

5.6.7.3 Copper

In *Methylosinus trichosporium* OB3b, which expresses a particulate monooxygenase, the concentration of copper plays a significant role in the kinetic parameters for the consumption of both methane and trichloroethene (Sontoh and Semrau 1998). For methane, V_{max} decreased from 300 to 82 when the concentration of Cu increased from 2.5 to 20 μM, and K_s decreased from 62 to 8.3 under these conditions. For trichloroethene, V_{max} and K_s were unmeasurable at Cu concentrations of 2.5 μM even in the presence of formate, but were 4.1 and 7.9 for concentrations of 20 μM in the presence of formate. In addition, enrichments using limiting concentrations of Cu result in the prevalence of strains with the soluble form of the enzyme (McDonald et al. 2006). Although copper is an unusual component of dioxygenases, it is a component of the dioxygenase from *Aspergillus japonicus*, which brings about fission of the C ring of quercitin with the formation of carbon monoxide (Steiner et al. 2002). It is also a component of oxalate decarboxylase in *Bacillus subtilis*, which converts oxalate into formate and CO_2 (Tanner et al. 2001).

5.6.7.4 Other Metals

Heavy-metal cations and oxyanions are generally toxic to bacteria although resistance may be induced by various mechanisms after exposure. Attention is drawn to an unusual example in which Al^{3+} may be significant, since the catechol 1,2-dioxygenase and 3,4-dihydroxybenzoate (protocatechuate) 3,4-dioxygenase that are involved in the metabolism of benzoate by strains of *Rhizobium trifolii* are highly sensitive to inhibition by Al^{3+} (Chen et al. 1985).

5.6.8 Redox Potential of Media

Cultivation of strictly anaerobic organisms requires not only that the medium be oxygen-free, but also that the redox potential of the medium be compatible with that required by the organisms. This may be accomplished by addition of reducing agents such as sulfide, dithionite, titanium(III) citrate, or titanium(III) nitrilotriacetate. Any of these may, however, be toxic so that only low concentrations should be employed. Attention has been drawn to the fact that titanium(III) citrate–reduced medium may be inhibitory to bacteria during initial isolation (Wachenheim and Hespell 1984).

5.7 Organic Substrates

Although organic substrates such as carboxylic acids are thermally stable and may be sterilized with the basal media, many others, including, for example, carbohydrates, esters, or amides, are more safely prepared as concentrated stock solutions, sterilized by filtration through 0.22 μm filters and added to the sterile basal medium.

5.7.1 Aqueous Solubility and Toxicity

A problem arises when organic substrates are poorly soluble in water, and this may be exacerbated by toxicity when substrates are too toxic to be added in the free state at concentrations sufficient for growth.

For example, the toxicity of long-chain aliphatic compounds with low water solubility has been examined in yeasts (Gill and Ratledge 1972). For substrates with poor water solubility, several methods have been used which introduce the substrates at low concentration and may at the same time overcome potential toxicity. Illustrative examples have included a range of strategies:

1. Application of toluene in an inert hydrophobic carrier (Rabus et al. 1993).
2. Use of the water-soluble though toxic hexadecyltrimethylammonium chloride adsorbed on silica (Van Ginkel et al. 1992).
3. A gas-phase delivery system using tetrachloroethene sorbed on beads of Tenax for continuous supply to an anaerobic *Desulfuromonas* sp. strain BB1 (Brennan and Sanford 2002).
4. Dibutyl phthalate as a cosolvent for the toxic cyclohexane epoxide has been examined, though in this case, the essential problem was the susceptibility of the substrate to hydrolysis to the diol that was more readily degradable (Carter and Leak 1995).
5. Bio-Sep beads amended with ^{13}C-labeled monocyclic arenes to assess *in situ* biodegradation potential (Geyer et al. 2005).
6. In the degradation of 4-chloroacetophenone (Havel and Reineke 1993), the 4-chlorophenol produced was toxic to one of the components of the consortium, and low concentrations of the toxic intermediate were maintained by adding gelatin to the medium: this procedure facilitated the growth of one of the components.

5.7.2 Volatility

For gaseous substrates or those with sufficient volatility, the substrate may be supplied in the gas phase in desiccators containing the appropriate mineral medium. It must be borne in mind that limitation of oxygen or CO_2 for organisms which require CO_2 supplementation should not occur. It may be convenient to supply toxic volatile substrates such as hydrogen cyanide in the gas phase (Harris and Knowles 1983). Gaseous or highly volatile substrates may be overcome by the use of enclosed systems such as desiccators, sealed ampoules, or by bottles capped with Teflon-lined crimp caps and inversion during incubation. Attention is drawn, however, to the permeability to organic compounds of many types of rubber sealing. This procedure has been employed for 4-chloroacetophenone (Havel and Reineke 1993), since the 4-chlorophenol produced was toxic to one of the components of the consortium. In this example, low concentrations of the toxic intermediate could also be maintained by adding gelatin to the medium, and this procedure facilitated the growth of one of the components. It should be noted that, particularly in sealed systems, it is important to satisfy the obligate requirement of many organisms for CO_2. The result of enrichments may sometimes depend upon whether the substrate is applied in the vapor phase or in the aqueous phase; for example, enrichments with α-pinene in the vapor phase yielded predominantly Gram-positive organisms in contrast to the Gram-negative organisms obtained when the substrate was added to the liquid medium. This may plausibly be related to the greater sensitivity of Gram-positive organisms that are exposed to only low substrate concentrations in the vapor phase (Griffiths et al. 1987).

A technique using floating filters has been used for the isolation organisms from enrichment cultures. It seems particularly suitable for the isolation of autotrophs when conventional methods may have resulted in overgrowth of heterotrophs, or when the substrate is volatile, or volatile and potentially toxic. The culture is filtered through a membrane of suitable porosity and the filter is floated on the surface of a mineral medium during exposure to the substrate (de Bruyn et al. 1990).

1. This was used for the enumeration and isolation of the iron-oxidizing autotroph *Thiobacillus ferrooxidans*—specifically using sterile Nucleopore polycarbonate 0.2 μm filters (de Bruyn et al. 1990).
2. This was successfully used for the isolation of a sulfide-oxidizing chemolithotroph from a reactor when conventional methods of isolation had failed. The filters were exposed in desiccators

containing an atmosphere enriched with CO_2 and supplemented with controlled concentrations of hydrogen sulfide at a sublethal concentration (Visser et al. 1997).

3. It has been used for isolating a species belonging to the phylum *Verrucomicrobia* that was able to oxidize methane at pH < 1 (Pol et al. 2007).

5.7.3 Substrate Concentration

Particularly for soil bacteria, there has been a considerable gap between the numbers obtained by cultivation on solid media and those estimated, for example, by most-probable-number methods in liquid culture, or by comparative analysis of 16S rRNA or from genes derived from nucleic acid directly extracted from the soil. The number of these "unculturable" bacteria has remained distressingly high, with the result that that many groups of bacteria remained essentially inaccessible while their metabolic potential was unrealized. For many purposes it is, therefore, highly desirable to resolve this substantial discrepancy, and increasingly successful experiments have been devoted to obtaining pure cultures of these organisms in order to reduce this number. A number of procedures for isolation of these previously "uncultured" or "underestimated" organisms have been employed that include the following:

1. Addition of anthraquinone disulfonate as a protectant against exogenous peroxides, inclusion of acyl homoserine lactones and incubation at increased concentrations of CO_2 (Stevenson et al. 2004)

2. Encapsulation of microorganisms in gel micro droplets and separation of these by chromatography and flow cytometry (Zengler et al. 2002): this technique has been applied successfully to both marine and soil organisms

Traditional bacteriological media contain relatively high concentrations of organic growth substrates, and successful isolations of these "unculturable" organisms have used defined mineral media, or complex media, with unusually low substrate concentrations. These methods have been used to isolate moderately acidophilic acidobacteria that were able to degrade plant polymers, including xylan and starch, for example, *Terriglobus* strain KBS 63 (Eichorst et al. 2007), strains of *Edaphobacter* (Koch et al. 2008), and *Granulicella* (Pankratov and Dedysh 2010). Further details are given in the following examples:

1. This procedure was successfully used to obtain *Chthoniobacter flavus* in pure culture (Sangwan et al. 2004), and used a dilute nutrient broth medium solidified with gellan gum in place of agar (Janssen et al. 2002), and this has since been extended (Sangwan et al. 2005).

2. In further experiments (Joseph et al. 2003), a mixture of monosaccharides and uronic acids at 0.5 mM concentrations was used, and resulted in an impressive addition to the range of cultivated bacteria. These included members of subdivisions of the phyla Acidobacteria, Actinobacteria, and Verrucomicrobia, and suggest that many of these bacteria should no longer be considered uncultivable.

3. A strain of *Gemmatimonas aurantiaca* was isolated from a laboratory wastewater treatment system using a semicomplex medium containing low concentrations of glucose, polypeptone, glutamate, and yeast extract (Zhang et al. 2003).

4. Soils from land cultivated for corn, soybean, and wheat, and from unplowed grassland were used to isolate *Acidobacteria* using medium buffered to pH 5.5–6.5. Either readily oxidized substrates such as glucose or plant polymers both at concentrations of 50 mg/L were used as substrates and the medium was solidified with gellan gum. Two strains KBS 83 and KBS 93 were obtained and their nutritional range and their 16S rRNA were determined. In addition to the utilization of a range of carbohydrates, ferulate, and syringate, these strains were able to degrade xylan, cellulose, pectin, and methyl cellulose (Eichorst et al. 2011).

5. It is worth pointing out that, contrary to expectation, it has been shown that isolation on solid medium using low concentrations of monomeric as well as polymeric substrates was more effective for members of *Acidobacteria*, *Verrucomicrobia*, and *Gemmatimonadetes* than the use of liquid serial dilution (Schoenborn et al. 2004).

Considerable attention has been directed to isolating dominant marine bacteria, including those from the open ocean, in which members of the SAR 11 clade dominate (Morris et al. 2002). A successful procedure for isolating marine bacteria belonging to several clades has been developed, and was dependent on several important features: use of acid-washed micro-titer plates, use of seawater medium containing only extremely low concentrations of nutrients (DOC ca. 100 μM), cultivation of only a few cells, long-term incubation for several weeks, and dilution-to-extinction. This has been exemplified by isolation of SAR 11 (α-subclass), OM 43 (β-subclass), SAR 92 (γ-subclass), OM 60 (γ-subclass) (Connon and Giovannini 2002), and subsequently SAR 11 (Song et al. 2009).

Attention has been drawn in Chapter 2 to aerobic methanotrophic cultures of the phylum Verrucomicrobia that were obtained under the highly selective conditions of high temperature and low pH (Dunfield et al. 2007; Pol et al. 2007; Islam et al. 2008). It is appropriate to note also the isolation of methanotrophs from *Sphagnum* spp. in acidic peat bogs (pH 3.8–4.3) in the Netherlands (Kip et al. 2011). Whereas acidophilic α-proteobacteria were isolated using a conventional medium, use of a medium with substantially lowered concentrations of phosphate potassium and magnesium, and increased concentrations of aluminum and silicon made possible the isolation of members of the γ-proteobacteria.

5.7.4 Preparation of Solid Media

Although solid media have been prepared from silica gels, these have not been widely used. Attention has already been directed to the use of gellan gum in place of agar (Janssen et al. 2002; Joseph et al. 2003; Eichorst et al. 2011). When agar is used for preparing solid medium it should be of the highest quality, and as free as possible from alternative carbon sources. It is generally preferable to autoclave agar separately from the mineral base; both are prepared at double the final concentration and mixed after autoclaving (Stanier et al. 1966). The problem of preparing plates for testing the metabolic capacity of substrates that are only poorly soluble is a serious one for which no universal solutions are available. Some of the techniques that have been advocated include the following:

1. Liquid hydrocarbons have been adsorbed on silica powder and dispersed in the agar medium. The silica may be autoclaved (Baruah et al. 1967), though this may be avoided by sterilizing the silica by heating, and carrying out the sorption and removal of solvents such as acetone or dichloromethane under sterile conditions.

2. Solutions of the substrate have been prepared, for example, in acetone or diethyl ether, and added to or spread over the surface of agar plates either before or after inoculation (Sylvestre 1980; Shiaris and Cooney 1983).

3. For compounds that are sufficiently volatile such as benzene, toluene, or naphthalene, the substrate may be contained in a tube placed above the level of the medium (Claus and Walker 1964), or on the lid of a Petri dish (Söhngen 1913). For benzene and toluene, this also obviates problems with toxicity since the organisms are exposed to only low concentrations of the substrate.

4. A solution of the substrate in ethanol may be mixed with the bacterial suspension in agarose and poured over agar plates of the base medium (Bogardt and Hemmingsen 1992). This is the general procedure used with top agar in the Ames test (Maron and Ames 1983), although dimethyl sulfoxide is generally used as the water-miscible solvent.

5. *Chthoniobacter flavus* has been obtained in pure culture (Sangwan et al. 2004), which was made possible by the use of a dilute nutrient broth medium solidified with gellan gum in place of agar (Janssen et al. 2002), and this has since been extended (Sangwan et al. 2005). It has been shown that, contrary to expectation, isolation on solid medium was more effective for members of *Acidobacteria*, *Verrucomicrobia*, and *Gemmatimonadetes* than the use of liquid serial dilution (Schoenborn et al. 2004).

5.7.4.1 Growth at the Expense of Alternative Substrates

It may be found that after enrichment, growth does not occur on plates prepared with the compound that showed satisfactory growth in liquid medium. An alternative that is worth examining is that of attempting to grow the organisms with a potential metabolite, though it should be kept in mind that organisms may be unable to utilize compounds that are clearly established metabolites of the desired substrate. Examples include the inability of fluorescent pseudomonads, which degrade aromatic compounds via *cis,cis*-muconate to use this as a substrate (Robert-Gero et al. 1969) or of alkane-degrading bacteria to grow with the corresponding carboxylic acids which are the first metabolites in alkane degradation (Zarilla and Perry 1984). Plausible reasons could be the lack of an effective transport system for the metabolite or the failure of the metabolite to induce the enzymes necessary for its production. For example, whereas salicylate normally induces the enzymes required for the degradation of naphthalene, this is apparently not the case for a naphthalene-degrading strain of *Rhodococcus* sp. (Grund et al. 1992).

Use of complex media for isolating organisms after elective enrichment is, in contrast, a potentially hazardous procedure. Media that are routinely used for nonmetabolic studies in clinical laboratories generally contain high concentrations of peptones, casamino acids, yeast extract, or carbohydrates. These may provide alternative carbon sources, and their use may, therefore, result in only mild selection pressure for the emergence of the desired organisms: overgrowth by undesired microorganisms may then take place all too readily.

5.8 Techniques for Anaerobic Bacteria

Increased attention has been directed to the growth and isolation of anaerobic bacteria. In addition to the nutritional requirements noted above, their general requirement for CO_2 should be taken into consideration. Broadly, three types of experimental procedures have been used:

1. Anaerobe jars containing a catalyst for the reaction between oxygen and hydrogen that is either added to or generated within the system. These systems have significant limitations in the kinds of experiment that can be carried out, since at some stage exposure to air cannot be avoided. In addition, many workers have experienced the unreliability of these systems, and they are not suitable for work with highly oxygen-sensitive organisms, including methanogens.

2. The classical technique (Hungate 1969) has been successfully used over many years, and incorporates a number of features designed to minimize exposure to oxygen. This procedure enables incubation to be carried out under a variety of gas atmospheres that include CO_2 and H_2, and is designed to produce a redox potential in media that is suitable for growth. Roll-tubes have been used instead of Petri dishes, and thereby a strictly anoxic environment may be maintained during manipulation. A modification using serum bottles has been introduced (Miller and Wolin 1974), and roll-tubes have also been successfully used for isolation of anaerobic phycomycetous fungi from rumen fluid (Joblin 1981).

3. Anaerobe chambers of varying design have achieved increasing popularity since they enable standard manipulations to be carried out under anoxic conditions. These systems maintain a gas atmosphere of N_2, H_2, and CO_2 (generally 90:5:5), and include a heated catalyst for the maintenance of anaerobic conditions. They may employ either a glove-box design, or free access through wristbands, and enable quite sophisticated experiments to be carried out and cultures maintained over lengthy periods. These can, therefore, be unequivocally recommended although their maintenance costs should not be underestimated.

5.9 Design of Experiments on Biodegradation and Biotransformation

There are several essentially different kinds of experiments that may be carried out. Two are laboratory-based and two are field investigations:

1. Laboratory experiments using pure cultures or stable consortia
2. Laboratory experiments using communities in microcosms simulating natural systems

3. Field experiments in model ecosystems—pools or steams
4. Large-scale field experiments under natural conditions

It should be clearly appreciated that the objectives of these various investigations are rather different. The first two aim at elucidating the basic facts of metabolism, the products formed, and the kinetics of their synthesis. Studies using pure cultures may ultimately be directed to studying more sophisticated aspects of the regulation and genetics of biodegradation. In contrast, the last two procedures are designed to obtain data of greater environmental relevance and may profitably—and even necessarily—draw upon the results obtained using the first two procedures. While the degree of environmental realism increases from Procedures 1 through 4, so do the experimental difficulties and the interpretative ambiguities. All the procedures have clear advantages for specific objectives and are complementary. Indeed, it is desirable to use several of them to provide a broader perspective on the biodegradability of the substrate.

There are significant differences in the control experiments that are possible in each of these systems. Before the quantifier *bio-* can be applied, the possibility of abiotic alteration of the substrate during incubation must be eliminated or taken into consideration. Only the first design lends itself readily to this control. For experiments using cell suspensions, the obvious controls are incubation of the substrate in the absence of cells or using autoclaved cultures. Care should be exercised in the interpretation of the results, however, since some reactions may apparently be catalyzed by cell components in purely chemical reactions. The question may then legitimately be raised whether or not these are biochemically mediated. Two examples are given as illustration of apparently chemically mediated reactions, which have been referred to in Chapter 1:

1. Dechlorination reactions of organochlorine compounds that involve corrinoids or porphyrins.
2. Reduction of aromatic nitro compounds by sulfide catalyzed by extracellular compounds excreted into the medium during growth of *Streptomyces griseoflavus* (Glaus et al. 1992).

In experiments where relatively small volumes of sediment suspensions are employed, autoclaving may significantly alter the structure of the sediment as well as introducing possibly severe analytical difficulties. In such circumstances, there are few alternatives to incubation in the presence of toxic agents such as NaN_3, which has been used at a concentration of 2 g/L. There remains, of course, the possibility that azide-resistant strains could emerge during prolonged incubation, and the possible occurrence of reactions between the substrate and azide must also be taken into consideration.

Only controls using inhibitors of microbial growth are possible for microcosm experiments, and these may be impractical for outdoor systems, which, therefore, combine and may fail to discriminate abiotic and biotic reactions.

5.10 Pure Cultures and Stable Consortia

Different kinds of experimental procedures have been used and these should be evaluated against the background in which several organisms or several substrates are simultaneously present. There are no essential differences in the design of experiments using pure cultures and those employing metabolically stable consortia. It should be emphasized, however, that even in the latter, the experiments should be carried out under aseptic conditions; otherwise, interpretation of the results may be compromised by adventitious organisms.

5.11 Cell Growth at the Expense of the Xenobiotic

In the simplest case, growth of the organism that has been isolated may be studied using the test substance that fulfills the nutritional requirement of the organism as the sole source of carbon, sulfur, nitrogen, or phosphorus. For a compound that is used as the sole source of carbon and energy, the endpoints could be growth and conversion of the substance into CO_2 under aerobic conditions or growth under anaerobic conditions accompanied by, for example, production of methane or sulfide from sulfate. In some studies, however, only diminution in the concentration of the initial substrate has been demonstrated, and this alone clearly does not constitute evidence for biodegradation. Biotransformation is equally possible, and should be taken into consideration. Ideally, use may be made of radiolabeled substrates followed by identification of the labeled products. ^{14}C, ^{35}S, ^{36}Cl, and ^{31}P have been used, though the relevant labeled products may not always be available commercially, and the required synthetic expertise may not be available in all laboratories. Further comments on the use of these isotopes together with the application of the nonradioactive isotopes ^{13}C and ^{19}F are given in Chapter 6, Part 1. It should be noted that CO_2 is incorporated not only into phototrophs and chemolithotrophs, but also into heterotrophic organisms. A review (Ensign et al. 1998) provides a brief summary of the role of CO_2 in the metabolism of epoxides by *Xanthobacter* sp. strain Py2, and of acetone by both aerobic and anaerobic bacteria.

In later chapters, considerable weight is given to the environmental significance of biotransformation and the synthesis of toxic metabolites. It is particularly desirable, therefore, to direct effort to the identification of such metabolites. This may present a substantially greater challenge than that of quantifying the original substrate for several reasons:

1. The structure of the metabolite will often be unknown, and must be predicted from knowledge of putative degradation pathways and confirmed by chemical analysis, generally including mass spectrometry and increasingly by NMR.
2. The metabolite will frequently be more polar than the initial substrate, so that specific procedures for extraction and analysis must generally be developed, and the pure compound must be available for quantification.
3. The metabolite may be transient with unknown kinetics of its formation and further degradation.

In practice, there is only one really satisfactory solution: the kinetics of the transformation must be followed. The justification for this substantial increase in effort is the dividend resulting in the form of a description of the metabolic pathway, including the synthesis of possibly inhibitory metabolites. An important dividend is that it may be possible to make generalizations on the degradation of other xenobiotics—structurally related or otherwise.

5.12 Stable Enrichment Cultures

Some investigations have used metabolically stable enrichment cultures that are particle free to study biotransformation. These are preferable to the use of sediment slurries even though these may be environmentally realistic, and there are several important reasons:

They avoid the ambiguities resulting from the presence of xenobiotics in the original soil or sediment.

They eliminate association of the added xenobiotic with particulate matter.

They increase the accuracy of analytical procedures.

They overcome the unresolved function of organic components in the sediment.

They make it possible to carry out reproducible experiments under clearly defined conditions.

Particularly for anaerobes, however, it is often difficult to obtain pure cultures, and sediment or sludge slurries have frequently been used. Examples of studies that have used stable enrichment cultures that are free of sediment by successive transfer in defined media during extensive periods of time, include the following:

1. A series of experiments on the anaerobic dechlorination of chlorocatechols (references in Allard et al. 1994)
2. The anaerobic dechlorination of 2,3,5,6-tetrachlorobiphenyl (Cutter et al. 1998)
3. The anaerobic dechlorination of Arochlor 1260 (Bedard et al. 2006)

5.13 Use of Dense Cell Suspensions

Dense cell suspensions have traditionally been used for experiments on the respiration of microbial cells at the expense of organic substrates, and they are equally applicable to experiments on biodegradation and biotransformation. Cells are grown in a suitable medium generally containing the test compound (or an alternative growth substrate) collected by centrifugation, washed in a buffer solution to remove remaining concentrations of the growth substrate and its metabolites, and resuspended in fresh medium before further exposure to the xenobiotic. For aerobic organisms, there are generally few experimental difficulties with three important exceptions:

1. For organisms that grow poorly in liquid medium, it may be difficult to obtain sufficient quantities of cells; cultures may then be grown on the surface of agar plates and the cells removed by scraping.
2. Organisms that have fastidious nutritional requirements may require undefined growth additives such as peptones, yeast extract, rumen fluid, or serum. Subsequent exposure to the xenobiotic may then be used to induce synthesis of the relevant catabolic enzymes. For example, the chlorophenol-degrading bacterium *Mycobacterium chlorophenolicum* (*Rhodococcus chlorophenolicus*) has been grown in media containing yeast extract or rhamnose, while exposure to pentachlorophenol was used to induce the enzymes required for the degradation of a wide range of chlorophenols (Apajalahti and Salkinoja-Salonen 1986).
3. Organisms such as actinomycetes may not produce well-suspended growth and shaking, for example, in baffled flasks or in flasks with coiled-wire inserts may be advantageous in partially overcoming this problem.

For anaerobic bacteria, the same principles apply, except that additional attention must be directed to preparing the cell suspensions. Use of an anaerobic chamber in which cultures can be transferred to tightly capped centrifuge tubes is virtually obligatory, and addition of an anaerobic indicator should be used to ensure that subsequent entrance of oxygen does not take place inadvertently. On account of the inserts, screw-capped Oak Ridge tubes are convenient for centrifugation.

5.14 Use of Immobilized Cells

Cells can be immobilized on a number of suitable matrices, confined in a reactor and the medium containing the test substrate circulated continuously. Although this methodology has been motivated by interest in biotechnology and in bioremediation technology, it is clearly applicable to laboratory experiments on biodegradation and biotransformation that could readily be carried out under sterile conditions. This procedure has been used, for example, to study the biodegradation of 4-nitrophenol (Heitkamp et al. 1990), pentachlorophenol (O'Reilly and Crawford 1989), and 6-methylquinoline (Rothenburger and Atlas 1993). A few additional comments are inserted:

1. It is possible to carry out experiments with immobilized cells in essentially nonaqueous media (Rothenburger and Atlas 1993), and this could prove an attractive strategy for compounds with limited water solubility, provided that solvents can be found that are compatible with the solubility of the substrate and the sensitivity of the cells to organic solvents.
2. Encapsulated cells have been successfully used for the commercial biosynthesis of a number of valuable compounds such as amino acids, and this technique could readily be adapted to investigations on biodegradation. This methodology offers the advantage that the metabolic activity of the cells can be maintained over long periods of time so that a high degree of reproducibility is guaranteed, and the stability of such systems may be particularly attractive in studies of recalcitrant compounds.

3. With appropriate experimental modifications, the various procedures could be adapted to study biodegradation under anaerobic conditions. For example, sparging with air could be replaced by the use of a gas mixture containing appropriate concentrations of CO_2 and H_2 in an inert gas such as N_2.

5.15 Application of Continuous Culture Procedures

These have been particularly valuable in studies using low concentrations of xenobiotics and for the isolation of consortia that have been used in elucidating metabolic interactions between the various microbial components. In many cases, consortia containing several organisms are obtained even though only a few of their members are actively involved in the metabolism of the xenobiotic. It is possible that the low substrate concentrations that have been used in these experiments favor selection for organisms that are able to take advantage of the lysis products from cells that do not play a direct role in the degradation of the xenobiotic. Three examples suffice as illustration:

1. Enrichment in a two-stage chemostat with parathion (*O,O*-diethyl-*O*-(4-nitrophenyl)-phosphorothioate) as the sole source of carbon and sulfur resulted in a community that was stable for several years (Daughton and Hsieh 1977). It should be noted that on account of the toxicity of parathion to the culture, only low substrate concentrations could be used, and this methodology is ideally adapted to such situations. Degradation was accomplished by two organisms, *Pseudomonas stutzeri* and *P. aeruginosa*, whereas the third organism in the stable community had no defined function; *P. stutzeri* functioned only in ester hydrolysis, which is the first step in the degradation of parathion. This was the first demonstration of degradation of parathion by a metabolically defined microbial consortium, though degradation by a culture consisting of nine organisms had already been demonstrated (Munnecke and Hsieh 1976).

2. Chemostat enrichment was carried out with a mixture of linear alkylbenzene sulfonates as the sole sources of carbon and sulfur at a concentration of 10 mg/L, and resulted in the development of a four-component consortium (Jiménez et al. 1991). Three of the organisms were apparently necessary to accomplish this apparently straightforward degradation, though the isolation procedure that used a complex medium with glucose as carbon source is not entirely unequivocal. A similar situation arose with hexadecyltrimethylammonium chloride from which three strains that could grow with the substrate were obtained, again after streaking on yeast–glucose medium (Van Ginkel et al. 1992).

3. Chemostat enrichment with 2-chloropropionamide yielded a community of at least six organisms: one of these, a *Mycoplana* sp., carried out hydrolysis of the amide, while various other components used the resulting free acid for growth. An interesting observation was that after prolonged incubation at a dilution rate of 0.01/h, a single strain of *Pseudomonas* sp. capable of growth solely on 2-chloropropionamide as carbon source could be isolated (Reanney et al. 1983).

There are some general conclusions that may be drawn from the results of these experiments:

1. Relatively simple reactions were involved in the degradations, and these can be expected to be available to single organisms.
2. The first stages for two of these degradations were straightforward hydrolytic reactions.
3. In all the examples, organisms with undefined metabolic functions were present and probably fulfilled an important role in providing complex organic substrates in the form of cell lysis products or nutritional requirements.

Reaction sequences used for the degradation of xenobiotics in natural systems may, therefore, be more complex than might plausibly be predicted on the basis of studies with pure cultures using relatively high substrate concentrations.

Attention should be drawn to experiments in which solutions of the substrate in a suitable mineral medium are percolated through soil that is used as the source of inoculum. This is one of the classical procedures of soil microbiology and has been exploited to advantage in studies on the degradation of a range of chlorinated contaminants in groundwater (Van der Meer et al. 1992). Apart from the fact that this mimics closely the natural situation and incorporates the features inherent in any enrichment methodology, this procedure offers a degree of flexibility that enables systematic exploration of the following:

1. The effect of varying redox conditions since by altering the gas-phase experiments can readily be carried out under aerobic, microaerophilic, or anaerobic conditions.
2. The effect of substrate concentration and the important issue of the existence or otherwise of threshold concentrations below which degradation is not effectively accomplished.
3. The influence of sorption/desorption on biodegradation, which has been discussed in a wider context in Chapter 4.

Apart from its application to the specific problem of groundwater contamination, this procedure offers a potentially valuable procedure for simulating bioremediation of contaminated soils.

5.16 Simultaneous Presence of Two Substrates

Analogous to the fact that pure cultures of microorganisms seldom occur in natural ecosystems, it is very rare for a single organic substrate to exist in appreciable concentrations. The relevant microorganisms under natural situations are, therefore, exposed simultaneously to several compounds and this situation can be simulated in laboratory experiments. Although the term "cometabolism" has been used extensively, it has been applied to conflicting metabolic situations and the pragmatic term "concurrent metabolism" (Neilson et al. 1985) offers an attractive alternative when more than one substrate is present. Three examples are used to illustrate the application of this procedure to experiments in which the pathways for the biotransformation of different xenobiotics have been established:

1. Experiments have used cells with a metabolic capability that may plausibly be predicted as relevant to that of the xenobiotic. For example, although elective enrichment failed to yield organisms that were able to grow at the expense of dibenzo[1,4]dioxin, its metabolism could be studied using a strain of *Pseudomonas* sp. that was capable of growth with naphthalene (Klecka and Gibson 1979). Cells were grown with salicylate (1 g/L) in the presence of dibenzo[1,4] dioxin (0.5 g/L), and two metabolites of the latter were isolated: dibenzo[1,4]dioxin-*cis*-1,2-dihydrodiol and 2-hydroxydibenzo[1,4]dioxin. The former is consistent with the established dioxygenation of naphthalene and the role of salicylate as coordinate inducer of the relevant enzymes for conversion of naphthalene into salicylate.
2. An environmentally relevant situation may be simulated by the growth of an organism with a single substrate at a relatively high concentration with simultaneous exposure to a structurally unrelated xenobiotic present at a significantly lower concentration. A series of investigations has used growth substrates at concentrations of 200 mg/L and xenobiotic concentrations of 100 µg/L; it may reasonably be assumed that growth with compounds at the latter concentration was negligible in these experiments. For example, during growth of a stable anaerobic enrichment culture with 3,4,5-trimethoxybenzoate, 4,5,6-trichloroguaiacol was transformed into 3,4,5-trichlorocatechol that was further dechlorinated to 3,5-dichlorocatechol (Neilson et al. 1987).
3. There has been considerable discussion on the mineralization of DDT, and in particular the biodegradation of the intermediate DDE. Cells of *Pseudomonas acidovorans* strain M3GY have been shown, however, during growth with biphenyl to degrade DDE with the fission of one ring and production of 4-chlorobenzoate (Hay and Focht 1998).

5.17 Use of Unenriched Cultures: Undefined Natural Consortia

Laboratory experiments using natural consortia under defined conditions have particular value from several points of view. They are of direct environmental relevance, and their use minimizes the ambiguities in extrapolation from the results of studies with pure cultures. They provide valuable verification of the results of studies with pure cultures and make it possible to evaluate the extent to which the results of such studies may justifiably be extended to the natural environment. It should be appreciated, however, that in some cases the habitats from which the inoculum was taken might already have been exposed to xenobiotics so that "natural" enrichment (preexposure) could already have taken place. This has been discussed in Chapter 4.

Extensive studies on the effect of substrate concentration and on the bioavailability of the substrate to the appropriate microorganisms have employed samples of natural lake water supplemented with suitable nutrients. There are few additional details that need to be added since the experimental methods are straightforward and present no particular difficulties. Considerable use has also been made of a comparable methodology to determine the fate of agrochemicals in the terrestrial environment.

Because of the difficulty in obtaining pure cultures of anaerobic bacteria, use has been made of anaerobic sediment slurries in laboratory experiments. In some of these, although no enrichment was deliberately incorporated, experiments were carried out over long periods of time in the presence of contaminated sediments and adaptation of the natural flora to the xenobiotic during exposure in the laboratory might, therefore, have taken place. The design of these experiments may also inevitably result in interpretative difficulties. A few illustrations are provided:

1. Although the results of experiments on the dechlorination of pentachlorophenol (Bryant et al. 1991) enabled elucidation of the pathways to be elucidated, this study also revealed one of the limitations in the use of such procedures. Detailed interpretation of the kinetics of pentachlorophenol degradation using dichlorophenol-adapted cultures was equivocal due to carryover of phenol from the sediment slurries.

2. The biodegradation of acenaphthene and naphthalene under denitrifying conditions was examined in soil–water slurries (Mihelcic and Luthy 1988), though in this case only analyses for the concentrations of the initial substrates were carried out.

In both of these examples, growth of the degradative organisms was supported at least partly by organic components of the soil and sediment, so that the physiological state of the cells could not be precisely defined. It would, therefore, be desirable to avoid ambiguity by using cultures in which the sediment is no longer present. This approach is illustrated by extensive investigations on the anaerobic dechlorination of chlorocatechols (Allard et al. 1991, 1994), and 2,3,5,6-tetrachlorobiphenyl (Cutter et al. 1998).

5.18 Microcosm Experiments

Microcosms are laboratory systems generally consisting of tanks such as fish aquaria containing natural sediment and water or soil. In those that have been most extensively evaluated for aquatic systems, continuous flow systems are used. In all of them, continuous measurement of $^{14}CO_2$ evolved from ^{14}C-labeled substrates may be incorporated, and recovery of both volatile and nonvolatile metabolites is possible so that a material balance may be constructed (Huckins et al. 1984). It should be pointed out that the term microcosm has also been used to cover much smaller-scale experiments that have been carried out in flasks or bottles under anaerobic conditions (Edwards et al. 1992), and to systems for evaluating the effect of toxicants on biota. Some examples are given to illustrate different facets of the application of microcosms to study various aspects of biodegradation.

1. Biodegradation of *t*-butylphenyl diphenyl phosphate was examined using sediments either from an uncontaminated site or from one having a history of chronic exposure to agricultural

chemicals (Heitkamp et al. 1986). Mineralization was very much more extensive in the latter case, but was inhibited by substrate concentrations exceeding 0.1 mg/L. Low concentrations of diphenyl phosphate, 4-*t*-butylphenol, and phenol indicated the occurrence of esterase activity, while the recovery of triphenyl phosphate suggested dealkylation by an unestablished pathway. A comparable study of naphthalene biodegradation (Heitkamp et al. 1987) found more rapid degradation when sediments chronically exposed to petroleum hydrocarbons were used, and isolation of *cis*-1,2-dihydronaphthalene-1, 2-diol, 1- and 2-naphthol, salicylate, and catechol confirmed the pathway established for the degradation of naphthalene. The results of both investigations illustrate the potential for a more extensive application of the procedure and at the same time the significance of preexposure to the xenobiotic.

2. One of the key issues in bioremediation is the survival of the organisms deliberately introduced into the contaminated system. A microcosm prepared from a pristine ecosystem was inoculated with a strain of *Mycobacterium* sp. that had a wide capacity for degrading PAHs, and this organism was used to study the degradation of 2-methylnaphthalene, phenanthrene, pyrene, and benzo[*a*]pyrene (Heitkamp and Cerniglia 1989). The test strain survived in the system with or without exposure to PAHs, but the addition of organic nutrients was detrimental to its maintenance. Clearly, an almost unlimited range of parameters could be varied to enable a realistic evaluation of the effectiveness of bioremediation in natural circumstances.

3. Concern has been expressed on the potential hazards from discharge into the environment of organisms carrying catabolic genes on plasmids. A number of investigations (Jain et al. 1987; Fulthorpe and Wyndham 1989; Sobecky et al. 1992) have used a set of microcosms to determine the conditions needed to preserve the genotype and its stability. Once again the advantage of the technique is the ease of incorporating important variables that may be difficult to analyze in natural systems.

4. A sediment–water system was used to study the partition and the degradation of ^{14}C-labeled 4-nitrophenol and 3,4-dichloroaniline (Heim et al. 1994). The results clearly illustrated the importance of water-to-sediment partitioning, and that a substantial fraction of the substrates existed in the form of nonextractable residues.

5. A study using resuspended river sediment (Marchesi et al. 1991) illustrated the important interdependence of substrate attachment to particulate matter and its biodegradability. Addition of sodium dodecyl sulfate that is degradable resulted in a relative increase in the number of particle-associated bacteria, whereas this was not observed with the nondegradable analogs such as sodium tetradecyl sulfate or sodium dodecane sulfonate.

6. A series of soil microcosms were used to study the biodegradation and bioavailability of pyrene during long-term incubation. The nonextractable fraction of ^{14}C-labeled pyrene that had been introduced into pristine soil and incubated with or without the addition of azide was substantially greater in the latter (Guthrie and Pfaender 1998). It was also shown that microbial activity produced a number of unidentified polar metabolites that might plausibly be involved in the association.

Truly field experiments on microbial reactions are extremely difficult to carry out, but a series of microcosm experiments on the substrates that may support anaerobic sulfate reduction quite closely approached this ideal situation (Parkes et al. 1989). This investigation used inhibition of sulfate reduction by molybdate to study the increase in the levels of a wide range of organic substrates endogenous in the sediments that were used. These substrates included a range of alkanoic acids and amino acids and, at the same time, considerably increased the range of organic substrates able to support sulfate reduction.

Both *in situ* microcosms and laboratory systems were used to compare and evaluate first-order rates of degradation for a range of mixed substrates, including aromatic hydrocarbons and phenolic compounds (Nielsen et al. 1996). The observed rates were comparable, although no systematic differences were observed with the exception of 2,6-dichlorophenol, which was not degraded in the laboratory system.

5.19 Experiments in Models of Natural Aquatic Systems

It is extremely difficult to carry out field investigations in natural ecosystems with the rigor necessary to unravel metabolic intricacies, although such experiments have been successfully carried out in investigations aimed at determining the fate and persistence of agrochemicals in the terrestrial environment, and in the context of bioremediation. In general, however, simplified systems have been developed. These attempt to simulate critical segments of natural ecosystems in a clearly defined way. Outdoor model systems have been used and two examples are used to illustrate the kinds of data that can be assembled and the range of conclusions—and their limitations—that may be drawn from such experiments. Not only purely microbiological determinants of persistence may be revealed, but also important data on the distribution and fate of the xenobiotic may be acquired.

1. Studies in an artificial stream system were designed to provide confirmation in a field situation of the results from laboratory experiments that had demonstrated the biodegradability of pentachlorophenol. Pentachlorophenol was added continuously to the system for 88 days and its degradation followed (Pignatello et al. 1983, 1985). The results confirmed that pentachlorophenol was indeed degraded by the natural populations of microorganisms and, in addition, drew attention to the significance of both sediments and surfaces in the partitioning of pentachlorophenol between the phases within the system.

2. 4,5,6-Trichloroguaiacol was added continuously during several months to mesocosm systems simulating the Baltic Sea littoral zone. Samples of water, sediment, and biota, including algae, were removed periodically for analysis both of the original substance and of metabolites identified previously in extensive laboratory experiments (Neilson et al. 1989). A complex of metabolic transformations of 4,5,6-trichloroguaiacol was identified, including *O*-methylation to 3,4,5-trichloroveratrole, *O*-demethylation to 3,4,5-trichlorocatechol, and partial dechlorination to a dichlorocatechol, and these metabolites were distributed among the various matrices in the system. Of particular significance was the fact that a material balance unequivocally demonstrated the role of the sediment phase as a sink for both the original substrate and the metabolites, so that a number of interrelated factors determined the fate of the initial substrate.

3. Mesocosms placed in shallow Finnish lakes were used to evaluate changes brought about by extended incubation of biologically treated bleachery effluent from mills that used chloride dioxide. The mesocosms had a volume of ca. 2 m^3 and were constructed of translucent polyether or black polyethene to simulate dark reactions. The experiments were carried out at ambient temperatures throughout the year, and sum parameters were used to trace the fate of the organically bound chlorine. In view of previous studies on the molecular mass distribution of effluents (Jokela and Salkinoja-Salonen 1992), this was measured as an additional marker. The important features were that (a) sedimentation occurred exclusively within the water mass within the mesocosm, (b) the atmospheric input could be estimated from control mesocosms, and (c) the microbial flora included both indigenous organisms in lake water and those carried over from the treatment process (Saski et al. 1996a, 1996b). There were a number of important general conclusions:

   • The atmospheric input of organically bound halogen was negligible compared with that of the effluent.

   • There was >50% loss of organohalogen in the water phase and this included compounds with masses of both <500 and >500.

   • The tetrahydrofuran-extractable organic chlorine in the *de novo* sediment had a molecular mass (average 1400) that was much higher than that from the water-phase extracts (average 360). There was no evidence for selective sorption of higher-molecular-mass components although transformation of lower-molecular-mass components to higher-molecular-mass hydrophobic compounds cannot be excluded.

4. A mesocosm was constructed to study the dynamics of populations of bacteria in an aquifer contaminated with BTEX (principally benzene). Water from the site was contained in permeable

tubes placed within perforated polypropylene cylinders at sites within the aquifer. Periodic samples were removed and analyzed to characterize the bacterial community using denaturing gel electrophoresis, analysis of PCR-amplified 16S rRNA genes, and BTEX degradative genes (Hendricks et al. 2005). The results showed that a stable bacterial community developed downstream after access to bacteria from the contaminated site, and that the community was characterized by genotypes similar to *xylM/xylE1*.

5.20 Evaluation of Degradation Using Metabolites

Comparable experiments in natural aquatic ecosystems are generally difficult to design (Madsen 1991), although some examples of what may be accomplished are given as illustration, and are applicable when there is sufficient knowledge about the degradative pathways of the xenobiotic:

1. Analysis of chlorobenzoates in sediments, which had been contaminated with PCBs, was used to demonstrate that the lower PCB congeners that had initially been produced by anaerobic dechlorination were subsequently degraded under aerobic conditions. The chlorobenzoates were transient metabolites and their concentrations were extremely low since bacteria that could successfully degrade them were present in the sediment samples (Flanagan and May 1993).
2. The bacterial aerobic degradation of pyrene is initiated by the formation of *cis*-pyrene-4,5-dihydrodiol. Analysis for this metabolite was used to demonstrate the biodegradability of pyrene in an environment in which there was continuous input of the substrate, when it was not possible to use any diminution in its concentration as evidence for biodegradation (Li et al. 1996). The corresponding metabolite from naphthalene—*cis*-naphthalene-1,2-dihydrodiol—has been used to demonstrate biodegradation of naphthalene both in site-derived enrichment cultures and in leachate from the contaminated site (Wilson and Madsen 1996).
3. It has been shown that pure cultures of bacteria under anaerobic denitrifying conditions may produce benzylsuccinate as a metabolite of toluene (Evans et al. 1992; Beller et al. 1996; Migaud et al. 1996). Demonstration of this and the corresponding methylbenzyl succinates from xylenes has been used to demonstrate metabolism of toluene and xylene in an anaerobic aquifer (Beller et al. 1995; Beller 2002).

However, care must be exercised in the interpretation of results that show the presence of putative metabolites. An illustrative example is provided by a study of the biodegradation of a range of PAHs in compost-amended, unamended, and sterilized soil (Wischmann and Steinhart 1997):

1. Neither dihydrodiols formed by bacterial dioxygenation nor phenols from fungal monooxygenation followed by rearrangement or hydrolysis and elimination were found.
2. In contrast, plausible oxidation products of anthracene, acenaphthylene, fluorene, and benz[*a*]anthracene—anthracene-9,10-quinone, acenaphthene-9,10-dione, fluorene-9-one, and benz[*a*]anthracene-7,12-quinone—were found transiently in compost-amended soil. It was shown, however, that these were formed even in sterile controls by undetermined abiotic reactions.

These results clearly illustrate the care that must be exercised in interpreting the occurrence of PAH oxidation products as evidence of biodegradation.

On account of the potential health hazard, the application of radiolabeled substrates, which have been the cornerstone of metabolic experiments, is not generally acceptable. Examples of experiments using stable isotopes have, however, been carried out to determine biodegradation under field conditions and to establish the source of contaminants. The application of stable isotope fractionation is discussed in Chapter 6, Part 1, and a single illustration of the use of deuterium-labeled substrates is given here. Fully deuterated benzene, toluene, 1,4-dimethylbenzene, and naphthalene were used to determine their dissemination and biodegradation in an aquifer plume with bromide as an inert marker (Thierrin et al. 1995).

Analysis of samples was readily accomplished by GC-MS, and after taking into account loss by sorption and dispersion, the half-lives of the substrates were calculated. At the oxic upper surface of the plume, rapid degradation occurred and continued at a slower rate into the anoxic zone. Although benzene was the most persistent substrate, there is evidence that it is degradable under anaerobic methanogenic, sulfate-reducing, Fe(III)-reducing (Kazumi et al. 1997), and nitrate-reducing (Burland and Edwards 1999) conditions.

5.21 Experimental Problems: Water Solubility, Volatility, Sampling, and Association of the Substrate with Microbial Cells

Although these issues have already been discussed, they deserve a few additional comments. For freely water-soluble substrates that have low volatility, there are few difficulties in carrying out the appropriate experiments described above. There is, however, increasing interest in xenobiotics, including polycyclic aromatic hydrocarbons (PAHs) and highly chlorinated compounds, including, for example, PCBs, which have only low water solubility. In addition, attention has been focused on volatile chlorinated aliphatic compounds such as the chloroethenes, dichloromethane, and carbon tetrachloride. For these and other volatile substrates, attention is drawn to the floating filter method that has already been illustrated.

1. Whereas suspensions of poorly water-soluble substrates can be used for experiments on the identification of metabolites, these methods are not suitable for kinetic experiments that necessitate the quantification of substrate concentrations. In such cases, the whole sample must be sacrificed at each sampling time, and care must be taken to ensure that substrate concentrations in each incubation vessel are as far as possible equal. In addition, the whole sample must be extracted for analysis since representative aliquots cannot be removed. Experiments are initiated by preparing solutions of the substrates in suitable solvents such as acetone, diethyl ether, methanol, or ethyl acetate, sterilizing the solutions by filtration, and dispensing appropriate volumes into each incubation vessel. The solvent is then removed in a stream of sterile N_2 and the cell suspension added. As an alternative, solutions in any of these solvents or in others that are much less volatile, such as dimethyl sulfoxide and dimethyl formamide, have been added directly to media after sterilization by filtration. However, since the remaining concentration of the solvents may not be negligible, care must be taken to ensure that these are neither toxic nor compromise the results of the metabolic experiments. When the test substrate is a solid, it may be preferable to prepare saturated solutions in the basal medium, remove undissolved substrate by filtration through glass-fiber filters, and sterilize the solution by filtration before dispensing and adding cell suspensions. An attractive procedure has used PAHs sorbed onto sheets of hydrophobic membranes and incubated with samples of contaminated soils (Bastiaens et al. 2000). After rinsing the sheets to remove particulate material, the membranes were placed on agar mineral medium with or without addition of PAHs. After incubation, cultures were then made from the cell mass that had developed.

2. Investigation of substrates with appreciable volatility—and many organic compounds, including solids having significant vapor pressures at ambient temperatures—presents a greater experimental challenge, especially if experiments are to be conducted over any length of time. Incubation vessels such as tubes or bottles may be closed with rubber stoppers or with teflon-lined crimp caps fitted with rubber seals, and these are particularly convenient for withdrawing samples using syringes. Even with Teflon seals, however, these cap inserts may be permeable to compounds with appreciable vapor pressure, and sorption of the test substrate may also occur to a significant extent. Both of these factors result in controls that display undesirable diminution in the concentration of the test substrate. Good illustrative examples are provided by the results of a study with endosulfans and related compounds (Guerin and Kennedy 1992), and on the metabolism of 4,4′-dichlorobiphenyl by *Phanerochaete chrysosporium* (Dietrich et al. 1995). These results support the importance even in laboratory experiments of taking

into account gas/liquid partitioning. Completely sealed glass ampoules may be used for less volatile compounds—though clearly not for highly volatile compounds—and subsampling cannot be carried out. For aerobic organisms, one serious problem with all these closed systems is that of oxygen limitation. The volume of the vessels should, therefore, be very much greater than that of the liquid phase, and conclusions from the experiments should recognize that microaerophilic conditions will almost certainly prevail during prolonged incubation. Such limitations are clearly not relevant for anaerobic organisms.

3. Two procedures were compared for the isolation of bacteria able to degrade poorly soluble polycyclic aromatic hydrocarbons by enrichment of a contaminated sample (Bastiaens et al. 2000). Conventional enrichment was carried out in which solid PAHs were added to a mineral medium containing the sediment and incubated before removal of the solid material and enrichment of the aqueous phase, followed by isolation on the mineral medium supplemented with the appropriate PAH. In the alternative, the soil suspension was incubated in the presence of a PAH-sorbing membrane spiked with the appropriate PAH followed by rinsing bacteria from the filter and purification as above to obtain single colonies. It was established that whereas the liquid-phase procedure yielded primarily sphingomonads, the filter method produced mycobacteria: the single mycobacterium from the liquid culture degraded phenanthrene, fluoranthene, and pyrene but differed from the analogous strain obtained by the filter procedure. The results were rationalized on the basis of PAH adhesion to the filters and the negative charge carried by the PAHs.

4. It is well established that associations occur between xenobiotics and biota, and these may introduce ambiguities into the interpretation of the results of metabolic experiments using microorganisms. Illustrative examples are given for the association of xenobiotics with microorganisms:

   a. Chloroguaiacols with Gram-positive bacteria (Allard et al. 1985)

   b. PCB congeners with the hyphae of *Phanerochaete chrysosporium* (Dietrich et al. 1995)

   c. 2,4,6-Trichlorophenol with *Bacillus subtilis* (Daughney and Fein 1998)

Analytical procedures for these substrates should, therefore, be designed to take into account such associations. If they are not quantitatively evaluated or eliminated, the results of such experiments may be seriously compromised.

5.22 References

Alexander M. 1975. Environmental and microbiological problems arising from recalcitrant molecules. *Microbiol Ecol* 2: 17–27.

Allard A-S, P-Å Hynning, M Remberger, AH Neilson. 1994. Bioavailability of chlorocatechols in naturally contaminated sediment samples and of chloroguaiacols covalently bound to C_2-guaiacyl residues. *Appl Environ Microbiol* 60: 777–784.

Allard A-S, C Lindgren, P-Å Hynning, M Remberger, AH Neilson. 1991. Dechlorination of chlorocatechols by stable enrichment cultures of anaerobic bacteria. *Appl Environ Microbiol* 57: 77–84.

Allard A-S, M Remberger, AH Neilson. 1985. Bacterial *O*-methylation of chloroguaiacols: Effect of substrate concentration, cell density and growth conditions. *Appl Environ Microbiol* 49: 279–288.

Allard A-S, L Renberg, AH Neilson. 1996. Absence of $^{14}CO_2$ evolution from ^{14}C-labelled EDTA and DTPA and the sediment/water partition ratio using a sediment sample. *Chemosphere* 33: 577–583.

Alvarez EG, J Viidanoja, A Munoz, K Wirtz, J Hjorth. 2007. Experimental confirmation of the dicarbonyl route in the photo-oxidation of toluene and benzene. *Environ Sci Technol* 41: 8362–8369

Angle JS, SP McGrath, RL Chaney. 1991. New culture medium containing ionic concentrations of nutrients similar to concentrations found in soil solution. *Appl Environ Microbiol* 57: 3674–3676.

Apajalahti JHA, MS Salkinoja-Salonen. 1986. Degradation of polychlorinated phenols by *Rhodococcus chlorophenolicus*. *Appl Microbiol Biotechnol* 25: 62–67.

Arnon DI. 1938. Microelements in culture-solution experiments with higher plants. *Am J Bot* 25: 322–325.

Balows A, HG Trüper, M Dworkin, W Harder, K-H Schleifer. 1992. (Eds.) *The Prokaryotes*. Springer-Verlag, Heidelberg.

Bartholomew GW, FK Pfaender. 1983. Influence of spatial and temporal variations on organic pollutant biodegradation rates in an estuarine environment. *Appl Environ Microbiol* 45: 103–109.

Baruah JN, Y Alroy, RI Mateles. 1967. Incorporation of liquid hydrocarbons into agar media. *Appl Microbiol* 15: 961.

Bastiaens L, D Springael, P Wattiau, H Harms, R deWachter, H Verachtert, L Diels. 2000. Isolation of adherent polycyclic aromatic hydrocarbon (PAH)-degrading bacteria using PAH-sorbing carriers. *Appl Environ Microbiol* 66: 1834–1843.

Battersby NS. 1990. A review of biodegradation kinetics in the aquatic environment. *Chemosphere* 21: 1243–1284.

Battersby NS, V Wilson. 1989. Survey of the anaerobic biodegradation potential of organic chemicals in digesting sludge. *Appl Environ Microbiol* 55: 433–439.

Bedard DL, JJ Bailey, BL Reiss, G van S Jerzak. 2006. Development and characterization of stable sediment-free anaerobic bacterial enrichment cultures that dechlorinate Arochlor 1260. *Appl Environ Microbiol* 72: 2460–2470.

Beller HR. 2002. Analysis of benzylsuccinates in groundwater by liquid chromatography/tandem mass spectrometry and its use for monitoring *in situ* BTEX biodegradation. *Environ Sci Technol* 36: 2724–2728.

Beller HR, W-S Ding, M Reinhard. 1995. Byproducts of anaerobic alkylbenzene metabolism useful as indicators of *in situ* bioremediation. *Environ Sci Technol* 29: 2864–2870.

Beller HR, AM Spormann, PK Sharma, JR Cole, M Reinhard. 1996. Isolation and characterization of a novel toluene-degrading sulfate-reducing bacterium. *Appl Environ Microbiol* 62: 1188–1196.

Berndt T, O Boge. 2006. Formation of phenol and carbonyls from the atmospheric reaction of OH carbonyls with benzene. *Phys Chem Chem Phys* 8: 1205–1214.

Bertram PA, RA Schmitz, D Linder, RK Thauer. 1994. Tungsten can substitute for molybdate in sustaining growth of *Methanobacterium thermoautotrophicum*: Identification and characterization of a tungsten isoenzyme of formylmethanofuran dehydrogenase. *Arch Microbiol* 161: 220–228.

Birch RR, C Biver, R Campagna, WE Gledhill, U Pagga. 1989. Screening chemicals for anaerobic degradability. *Chemosphere* 19: 1527–1550.

Birch RR, RJ Fletcher. 1991. The application of dissolved inorganic carbon measurements to the study of aerobic biodegradability. *Chemosphere* 23: 507–524.

Bogardt AH, BB Hemmingsen. 1992. Enumeration of phenanthrene-degrading bacteria by an overlay technique and its use in evaluation of petroleum-contaminated sites. *Appl Environ Microbiol* 58: 2579–2582.

Boursier P, FJ Hanus, H Papen, MM Becker, SA Russell, HJ Evans. 1988. Selenium increases hydrogenase expression in autotrophically cultured *Bradyrhizobium japonicum* and is a constituent of the purified enzyme. *J Bacteriol*. 170: 594–5600.

Bonnarme P, TW Jeffries. 1990. Mn(II) regulation of lignin peroxidases and manganese-dependent peroxidases from lignin-degrading white-rot fungi. *Appl Environ Microbiol* 56: 210–217.

Brennan RA, RA Sanford. 2002. Continuous steady-state method using Tenax for delivering tetrachloroethene to chloro-respiring bacteria. *Appl Environ Microbiol* 68: 1464–1467.

Brown JA, JK Glenn, MH Gold. 1990. Manganese regulates expression of manganese peroxidase by *Phanerochaete chrysosporium*. *J Bacteriol* 172: 3125–3130.

Brubaker WW, RA Hites. 1998. OH reaction kinetics of gas-phase α- and β-hexachlorocyclohexane and hexachlorobenzene. *Environ Sci Technol* 32: 766–769.

Bruhn C, H Lenks, H-J Knackmuss. 1987. Nitrosubstituted aromatic compounds as nitrogen source for bacteria. *Appl Environ Microbiol* 53: 208–210.

Bruns A, H Cypionka, J Overmann. 2002. Cyclic AMP and acyl homoserine lactones increase cultivation efficiency of heterotrophic bacteria from the Central Baltic Sea. *Appl Environ Microbiol* 68: 3978–3987.

Bryant FO, DD Hale, JE Rogers. 1991. Regiospecific dechlorination of pentachlorphenol by dichlorophenol-adapted microorganisms in freshwater anaerobic sediment slurries. *Appl Environ Microbiol* 57: 2293–2301.

Burghout P, LE Cron, H Gradstedt, B Quintero, E Simonetti, JJE Bijlsma, HJ Bootsma, PWM Hermans. 2010. Carbonic anhydrase is essential for *Streptococcus pneumoniae* growth in environmental ambient air. *J Bacteriol* 192: 4054–4062.

Burland SM, EA Edwards. 1999. Anaerobic benzene biodegradation linked to nitrate reduction. *Appl Environ Microbiol* 65: 529–533.

Carter SF, DJ Leak. 1995. The isolation and characterisation of a carbocyclic epoxide-degrading. *Corynebacterium* sp. *Biocatalysis Biotrans* 13: 111–129.

Chen YP, AR Glenn, MJ Dilworth. 1985. Aromatic metabolism in *Rhizobium trifolii*—Catechol 1,2-dioxygenase. *Arch Microbiol* 141: 225–228.

Claus D, N Walker. 1964. The decomposition of toluene by soil bacteria. *J Gen Microbiol* 36: 107–122.

Cone JE, RM del Rio, TC Stadtman. 1977. Clostridial glycine reductase complex. Purification and characterization of the selenoprotein component. *J Biol Chem* 252: 5337–5344.

Connon SA, SJ Giovannini. 2002. High-throughput methods for culturing microorganisms in very-low nutrient medium yield divers new marine isolates. *Appl Environ Microbiol* 68: 3878–3885.

Cook AM, CG Daughon, M Alexander. 1978. Phosphonate utilization by bacteria. *J Bacteriol* 133: 85–90.

Cook AM, R Hütter. 1982. Ametryne and prometryne as sulfur sources for bacteria. *Appl Environ Microbiol* 43: 781–786.

Cook AM, R Hütter. 1984. Deethylsimazine: Bacterial dechlorination, deamination and complete degradation. *J Agric Food Chem* 32: 581–585.

Corke CT, NJ Bunce, A-L Beaumont, RL Merrick. 1979. Diazonium cations as intermediates in the microbial transformations of chloroanilines to chlorinated biphenyls, azo compounds and triazenes. *J Agric Food Chem* 27: 644–646.

Cutter L, KR Sowers, HD May. 1998. Microbial dechlorination of 2,3,5,6-tetrachlorobiphenyl under anaerobic conditions in the absence of soil or sediment. *Appl Environ Microbiol* 64: 2966–2969.

Daughney CJ, JB Fein. 1998. Sorption of 2,4,6-trichlorophenol by *Bacillus subtilis*. *Environ Sci Technol* 32: 749–752.

Daughton CG, DPH Hsieh. 1977. Parathion utilization by bacterial symbionts in a chemostat. *Appl Environ Microbiol* 34: 175–184.

de Bruyn JC, FC Boogerd, P Bos, JG Kuenen. 1990. Floating filters, a novel technique for isolation and enumeration of fastidious, acidophilic, iron-oxidizing autotrophic bacteria. *Appl Environ Microbiol* 56: 2891–2894.

De Morsier A, J Blok, P Gerike, L Reynolds, H Wellens, WJ Bontinck. 1987. Biodegradation tests for poorly-soluble compounds. *Chemosphere* 16: 269–277.

Di Corcia A, C Crescenzi, A Marcomini, R Samperi. 1998. Liquid-chromatography-electrospray-mass spectrometry as a valuable tool for characterizing biodegradation intermediates of branched alcohol ethoxylate surfactants. *Environ Sci Technol* 32: 711–718.

Dietrich D, WJ Hickey, R Lamar. 1995. degradation of 4,4′-dichlorobiphenyl, 3,3′4,4′-tetrachlorobiphenyl, and 2,2′,4,4′,5,5′-hexachlorobiphenyl by the white rot fungus *Phanerochaete chrysosporium*. *Appl Environ Microbiol* 61: 3904–3909.

Dilworth MJ, RR Eady. 1991. Hydrazine is a product of dinitrogen reduction by the vanadium-nitrogenase from *Azotobacter chroococcum*. *Biochem J* 277: 465–468.

Dunfield PF and 18 co-authors. 2007. Methane oxidation by an extremely acidophilic bacterium of the phylum Verrumicrobia. *Nature* 450: 879–882.

Durre P, JR Andreesen. 1982. Selenium-dependent growth and glycine fermentation by *Clostridium purinolyticum*. *J Gen Microbiol* 128: 1457–1466.

Edwards EA, LE Williams, M Reinhard, D Grbic-Galic. 1992. Anaerobic degradation of toluene and xylene by aquifer microorganisms under sulfate-reducing conditions. *Appl Environ Microbiol* 58: 794–800.

Eichorst SA, JA Breznak, TM Schmidt. 2007. Isolation and characterization of newly isolated soil bacteria that define *Terriglobus* gen. nov., in the phylum *Acidobacteria*. *Appl Environ Microbiol* 73: 2708–2717.

Eichorst SA, CR Kuske, TM Schmidt. 2011. Influence of plant polymers on the distribution and cultivation of bacteria in the phylum *Acidobacteria*. *Appl Environ Microbiol* 77: 586–596.

Ensign SA, FJ Smakk, JR Allen, MK Sluis. 1998. New roles for CO_2 in the microbial metabolism of aliphatic epoxides and ketones. *Arch Microbiol* 169: 179–187.

Evans PJ, W Ling, B Goldschmidt, ER Ritter, LY Young. 1992. Metabolites formed during anaerobic transformation of toluene and *o*-xylene and their proposed relationship to the initial steps of toluene mineralization. *Appl Environ Microbiol* 58: 496–501.

Fall RR, JI Brown, TL Schaeffer. 1979. Enzyme recruitment allows the biodegradation of recalcitrant branched hydrocarbons by *Pseudomonas citronellolis*. *Appl Environ Microbiol* 38: 715–722.

Fallik E, Y-K Chan, RL Robson. 1991. Detection of alternative nitrogenases in aerobic Gram-negative nitrogen-fixing bacteria. *J Bacteriol* 173: 365–371.

Ferrari BC, SJ Binnerup, M Gillings. 2005. Microcolony cultivation on a soil substrates membrane system selects for previously uncultured soil bacteria. *Appl Environ Microbiol* 71: 8714–8720.

Firestone MK, JM Tiedje. 1975. Biodegradation of metal-nitrilotriacetate complexes by a *Pseudomonas* species: Mechanism of reaction. *Appl Microbiol* 29: 758–764.

Flanagan WP, RJ May. 1993. Metabolite detection as evidence for naturally occurring aerobic PCB degradation in Hudson River sediments. *Environ Sci Technol* 27: 2207–2212.

Francis AJ, CJ Dodge. 1993. Influence of complex structure on the biodegradation of iron-citrate complexes. *Appl Environ Microbiol* 59: 109–113.

Fulthorpe RR, RC Wyndham. 1989. Survival and activity of a 3-chlorobenzoate-catabolic genotype in a natural system. *Appl Environ Microbiol* 55: 1584–1590.

Gerike P, WK Fischer. 1981. A correlation study of biodegradability determinations with various chemicals in various tests. II. Additional results and conclusions. *Ecotoxicol Environ Saf* 5: 45–55.

Geyer R, AD Peacock, A Miltner, H-H Richnow, DC White, KL Sublette, M Kästner. 2005. *In situ* assessment of biodegradation potential using biotraps amended with ^{13}C-labeled benzene or toluene. *Environ Sci Technol* 39: 4983–4989.

Gill CO, C Ratledge. 1972. Toxicity of *n*-alkanes, *n*-alkenes, *n*-alkan-1-ols and *n*-alkyl-1-bromides towards yeasts. *J Gen Microbiol* 72: 165–172.

Gladyshev VN, SV Khangulov, TC Stadtman. 1994. Nicotinic acid hydroxylase from *Clostridium barkeri*: Electron paramagnetic resonance studies show that selenium is coordinated with molybdenum in the active site. *Proc Natl Acad Sci USA* 91: 232–236.

Glaus MA, CG Heijman, RP Schwarzenbach, J Zeyer. 1992. Reduction of nitroaromatic compounds mediated by *Streptomyces* sp. exudates. *Appl Environ Microbiol* 58: 1945–1951.

Griffiths ET, SG Hales, NJ Russell, GK Watson, GF White. 1986. Metabolite production during biodegradation of the surfactant sodium dodecyltriethoxy sulphate under mixed-culture die-away conditions. *J Gen Microbiol* 132: 963–972.

Griffiths ET, SM Bociek, PC Harries, R Jeffcoat, DJ Sissons, PW Trudgill. 1987. Bacterial metabolism of α-pinene: Pathway from α-pinene oxide to acyclic metabolites in *Nocardia* sp strain P183. *J Bacteriol* 169: 4972–4979.

Grossenbacher H, C Horn, AM Cook, R Hütter. 1984. 2-chloro-4-amino-1,3,5-triazine-6(5*H*)-one: A new inter-mediate in the biodegradation of chlorinated s-triazines. *Appl Environ Microbiol* 48: 451–453.

Grund E, B Denecke, R Eichenlaub. 1992. Naphthalene degradation via salicylate and gentisate by *Rhodococcus* sp strain B4. *Appl Environ Microbiol* 58: 1874–1877.

Guerin TF, IR Kennedy. 1992. Distribution and dissipation of endosulfan and related cyclodienes in sterile aqueous systems: Implication for studies on biodegradation. *J Agric Food Chem* 40: 2315–2323.

Guthrie EA, FK Pfaender. 1998. Reduced pyrene bioavailability in microbially active soils. *Environ Sci Technol* 32: 501–508.

Hales SG, GK Watson, KS Dodson, GF White. 1986. A Comparative study of the biodegradation of the surfac-tant sodium dodecyltriethoxy sulphate by four detergent-degrading bacteria. *J Gen Microbiol* 132: 953–961.

Hales SG, KS Dodgson, GF White, N Jones, GK Watson. 1982. Initial stages in the biodegradation of the surfactant sodium dodecyltriethoxy sulfate by *Pseudomonas* sp. strain DES1. *Appl Environ Microbiol* 44: 790–800.

Harris R, CJ Knowles. 1983. Isolation and growth of a *Pseudomonas* species that utilizes cyanide as a source of nitrogen. *J Gen Microbiol* 129: 1005–1011.

Harwood CS, NN Nichols, M-K Kim, JJ Ditty, RE Parales. 1994. Identification of the *pcaRKF* gene cluster from *Pseudomonas putida*: Involvement in chemotaxis, biodegradation, and transport of 4-hydroxybenzoate. *J Bacteriol* 176: 6479–6488.

Hatta T, G Muklerjee-Dhar, J Damborsky, H Kiyohara. 2003. Characterization of a novel thermostable Mn(II)-dependent 2,3-dihydroxybiphenyl 1,2-dioxygenase from polychlorinated biphenyl- and naphthalene-degrading *Bacillus* sp. JF8. *J Biol Chem* 278: 21483–21492.

Havel J, W Reineke. 1993. Microbial degradation of chlorinated acetophenones. *Appl Environ Microbiol* 59: 2706–2712.

Hay AG, DD Focht. 1998. Cometabolism of 1,1-dichloro-2,2-bis(4-chlorophenyl)ethylene by *Pseudomomas acidovorans* M3GY grown on biphenyl. *Appl Environ Microbiol* 64: 2141–2146.

Heim K, I Schuphan, B Schmidt. 1994. Behaviour of [^{14}C]-4-nitrophenol and [^{14}C]-3,4-dichloroaniline in lab sediment-water systems. 1. Metabolic fate and partitioning of radioactivity. *Environ Toxicol Chem* 13: 879–888.

Heitkamp MA, CE Cerniglia. 1989. Polycyclic aromatic hydrocarbon degradation by a *Mycobacterium* sp. in microcosms containing sediment and water from a pristine ecosystem. *Appl Environ Microbiol* 55: 1968–1973.

Heitkamp MA, JP Freeman, CE Cerniglia. 1986. Biodegradation of *tert*-butylphenyl diphenyl phosphate. *Appl Environ Microbiol* 51: 316–322.

Heitkamp MA, V Camel, TJ Reuter, WJ Adams. 1990. Biodegradation of *p*-nitrophenol in an aqueous waste stream by immobilized bacteria. *Appl Environ Microbiol* 56: 2967–2973.

Heitkamp MA, JP Freeman, CE Cerniglia. 1987. Naphthalene biodegradation in environmental microcosms: Estimates of degradation rates and characterization of metabolites. *Appl Environ Microbiol* 53: 129–136.

Hendricks B et al. 2005. Dynamics of an oligotrophic bacterial aquifer community during contact with a groundwater plume contaminated with benzene, toluene, ethylbenzene, and xylenes: An *in situ* mesocosm study. *Appl Environ Microbiol* 71: 3815–3823.

Hennessee CT, J-S Seo, AM Alvarez, QX Li. 2009. Polycyclic aromatic hydrocarbon-degrading species isolated from Hawaiian soils: *Mycobacterium crocinum*, sp. nov., *Mycobacterium pallens* sp. nov., *Mycobacterium rutilum* sp. nov., *Mycobacterium rufum* sp. nov. and *Mycobacterium aromaticivorans* sp. nov. *Int J Syst Evol Microbiol* 59: 378-387.

Hensgens CMH, WR Hagen, TA Hansen. 1995. Purification and characterization of a benzylviologen-linked, tungsten-containing aldehyde oxidoreductase from *Desulfovibrio gigas*. *J Bacteriol* 177: 6195–6200.

Howard PH, S Banerjee. 1984. Interpreting results from biodegradability tests of chemicals in water and soil. *Environ Toxicol Chem* 3: 551–562.

Huckins JN, JD Petty, MA Heitkamp. 1984. Modular containers for microcosm and process model studies on the fate and effects of aquatic contaminants. *Chemosphere* 13: 1329–1341.

Hungate R E. 1969. A roll tube method for cultivation of strict anaerobes, Vol 3B, pp. 117–132. In *Methods in Microbiology* (Eds. Norris and DW Ribbons), Academic Press, New York.

Islam T, S Jensen, LJ Reigstas, Ø Larsen, N-K Birekeland. 2008. Methane oxidation at 55°C and pH 2 by a thermoacidophilic bacterium belonging to the *Verrucomicrobia* phylum. *Proc Natl Acad Sci USA* 105: 300-304

Jain RK, GS Sayler, JT Wilson, L Houston, D Pacia. 1987. Maintenance and stability of introduced genotypes in groundwater aquifer material. *Appl Environ Microbiol* 53: 996–1002.

Janssen PH, PS Yates, BE Grinton, PM Taylor, M Sait. 2002. Improved culturability of soil bacteria and isolation in pure culture of novel members of the divisions *Acidobacteria*, *Actinobacteria*, *Proteobacteria*, and *Verrucomicrobia*. *Appl Environ Microbiol* 68: 2391–2396.

Jiménez L, A Breen, N Thomas, TW Federle, GS Sayler. 1991. Mineralization of linear alkylbenzene sulfonate by a four-member aerobic bacterial consortium. *Appl Environ Microbiol* 57: 1566–1569.

Joblin KN. 1981. Isolation, enumeration, and maintenance of rumen anaerobic fungi in roll tubes. *Appl Environ Microbiol* 42: 1119–1122.

Jokela JK, M Salkinoja-Salonen. 1992. Molecular weight distributions of organic halogens in bleached kraft pulp mill effluents. *Environ Sci Technol* 26: 1190–1197.

Jones JB, TC Stadtman. 1981. Selenium-dependent and selenium-independent formate dehydrogenase of *Methanococcus vannielii*. Separation of the two forms and characterization of the purified selenium-independent form. *J Biol Chem* 256: 656–663.

Joseph SJ, P Hugenholtz, P Sangwan, CA Osborne, PH Janssen. 2003. Laboratory cultivation of widespread and previously uncultured soil bacteria. *Appl Environ Microbiol* 69: 7210–7215.

Joshi-Tope G, AJ Francis. 1995. Mechanisms of biodegradation of metal-citrate complexes by *Pseudomonas fluorescens*. *J Bacteriol* 177: 1989–1993.

Kari FG, W Giger. 1995. Modelling the photochemical degradation of ethylenediaminetetraacetate in the River Glatt. *Environ Sci Technol* 29: 2814–2827.

Kazumi J, ME Caldwell, JM Suflita, DR Lovley, LY Young. 1997. Anaerobic degradation of benzene in diverse anoxic environments. *Environ Sci Technol* 31: 813–818.

Kip N et al. 2011. Detection, isolation, and characterization of acidophilic methanotrophs from *Sphagnum* mosses. *Appl Environ Microbiol* 77: 56453–5654.

Kitagawa W, S Takami, K Miyauchi, E Masai, Y Kamagata, JM Tiedje, M Fukuda. 2002. Novel 2,4-dichlorophenoxyacetic acid degradation genes from oligotrophic *Bradyrhizobium* sp. strain HW 13 isolated from a pristine environment. *J Bacteriol* 184: 509–518.

Klecka GM, DT Gibson. 1979. Metabolism of dibenzo[1,4]dioxan by a *Pseudomonas* species. *Biochem J* 180: 639–645.

Klotz B et al. 1998. Atmospheric oxidation of toluene in a large-volume outdoor photoreactor: *In situ* determination of ring-retaining product yields. *J Phys Chem A* 102: 10289–10299.

Koch IH, F Gich, PF Dunfield, J Overmann. 2008. *Edaphobacter modestus* gen. nov., sp. nov., and *Edaphobacter aggregans* sp. nov., two novel acidobacteria isolated from alpine and forest soils. *Int J Syst Bacteriol* 58: 1114–1122.

Lauff JJ, DB Steele, LA Coogan, JM Breitfeller. 1990. Degradation of the ferric chelate of EDTA by a pure culture of an *Agrobacterium* sp. *Appl Environ Microbiol* 56: 3346–3353.

Leive L. 1968. Studies on the permeability change produced in coliform bacteria by ethylenediaminetetra-acetate. *J Biol Chem* 243: 2373–2380.

Lewis TA, RL Crawford. 1993. Physiological factors affecting carbon tetrachloride dehalogenation by the denitrifying bacterium *Pseudomonas* sp. strain KC. *Appl Environ Microbiol* 59: 1635–1641.

Lewis TA, RL Crawford. 1995. Transformation of carbon tetrachloride via sulfur and oxygen substitution by *Pseudomonas* sp. strain KC. *J Bacteriol* 177: 2204–2208.

Li X-F, X-C Le, CD Simpson, WR Cullen, KJ Reimer. 1996. Bacterial transformation of pyrene in a marine environment. *Environ Sci Technol* 30: 1115–1119.

Liu D, WMJ Strachan, K Thomson, K Kwasniewska. 1981. Determination of the biodegradability of organic compounds. *Environ Sci Technol* 15: 788–793.

MacMichael GJ, LR Brown. 1987. Role of carbon dioxide in catabolism of propane by "*Nocardia paraffini-cum*" (*Rhodococcus rhodochrous*). *Appl Environ Microbiol* 53: 65–69.

Madsen EL. 1991. Determining *in situ* biodegradation. Facts and challenges. *Environ Sci Technol* 25: 1663–1673.

Maltseva O, P Oriel. 1997. Monitoring of an alkaline 2,4,6-trichlorophenol-degrading enrichment culture by DNA fingerprinting methods and isolation of the responsible organism, haloalkaliphilic *Nocardioides* sp. strain M6. *Appl Environ Microbiol* 63: 4145–4149.

Mandelbaum RT, LP Wackett, DL Allan. 1993. Mineralization of the *s*-triazine ring of atrazine by stable bacterial mixed cultures. *Appl Environ Microbiol* 59: 1695–1701.

Marchesi JR, NJ Russel, GF White, WA House. 1991. Effects of surfactant adsorption and biodegradability on the distribution of bacteria between sediments and water in a freshwater microcosm. *Appl Environ Microbiol* 57: 2507–2513.

Marchesi JR, AJ Weightman. 2003. Comparing the dehalogenase gene pool in cultivated α-halocarboxylic acid-degrading bacteria with the environmental matagene pool. *Appl Environ Microbiol* 69: 4375–4382.

Maron DM, BN Ames. 1983. Revised methods for the *Salmonella* mutagenicity test. *Mutat Res* 113: 173–215.

Marcus EA, AP Moshfegh, G Sachs, DR Scott. 2005. The periplasmic α-carbonic anhydrase activity of *Helicobacter pylori* is essential for acid acclimatiom. *J Bacteriol* 187: 729–738.

McCormick ML, GR Buettner, BE Britigan. 1998. Endogenous superoxide dismutase levels regulate iron-dependent hydroxyl radical formation in *Escherichia coli* exposed to hydrogen peroxide. *J Bacteriol* 180: 622–625.

McDonald IR, CB Miguez, G Rogge, D Bourque, KD Wendlandt, D Groleau, JC Murrell. 2006. Diversity of soluble methane monooxygenase-containing methanotrophs isolated from polluted environments. *FEMS Microbiol Lett* 255: 225–232.

Meyer O, HG Schlegel. 1983. Biology of aerobic carbon monoxide-oxidizing bacteria. *Annu Rev Microbiol* 37: 277–310.

Migaud ME, JC Chee-Sandford, JM Tiedje, JW Frost. 1996. Benzylfumaric, benzylmaleic, and Z- and E-phenylitaconic acids: Synthesis, characterization, and correlation with a metabolite generated by *Azoarcus tolulyticus* Tol-4 during anaerobic toluene degradation. *Appl Environ Microbiol* 62: 974–978.

Mihelcic JR, RG Luthy. 1988. Microbial degradation of acenaphthene and naphthalene under denitrification conditions in soil–water systems. *Appl Environ Microbiol* 54: 1188–1198.

Miller TL, MJ Wolin. 1974. A serum bottle modification of the Hungate technique for cultivating obligate anaerobes. *Appl Microbiol* 27: 985–987.

Morris RM et al. 2002. SAR11 clade dominates ocean surface bacterioplankton communities. *Nature* 420 806–810.

Mortensen SKL, CS Jacobsen. 2004. Influence of frozen storage on herbicide degradation capacity in surface and subsurface sandy soils. *Environ Sci Technol* 38: 6625–6632.

Munnecke DM, DPH Hsieh. 1976. Pathways of microbial metabolism of parathion. *Appl Environ Microbiol* 31: 63–69.

Neidhardt FC, PL Bloch, DF Smith. 1974. Culture medium for enterobacteria. *J Bacteriol* 119: 736–747.

Neilson AH. 1980. Isolation and characterization of bacteria from the Swedish west coast. *J Appl Bacteriol* 49: 215–223.

Neilson AH, A-S Allard, C Lindgren, M Remberger. 1987. Transformations of chloroguaiacols, chloroveratroles and chlorocatechols by stable consortia of anaerobic bacteria. *Appl Environ Microbiol* 53: 2511–2519.

Neilson AH, A-S Allard, M Remberger. 1985. Biodegradation and transformation of recalcitrant compounds. *Handb Enviorn Chem* 2C: 29–86.

Neilson AH, H Blanck, L Förlin, L Landner, P Pärt, A Rosemarin, M Söderström. 1989. Advanced hazard assessment of 4,5,6-trichloroguaiacol in the Swedish environment, pp. 329–374. In *Chemicals in the Aquatic Environment* (Ed. L Landner), Springer, Berlin.

Nielsen PH, PL Bjerg, P Nielsen, P Smith, TH Christensen. 1996. *In situ* and laboratory determined first-order degradation rate constants of specific organic compounds in an aerobic aquifer. *Environ Sci Technol* 30: 31–37.

Nörtemann B. 1992. Total degradation of EDTA by mixed cultures and a bacterial isolate. *Appl Environ Microbiol* 58: 671–676.

Nyholm N, A Damborg, P Lindgaard-Jörgensen. 1992. A comparative study of test methods for assessment of the biodegradability of chemicals in seawater—Screening tests and simulation tests. *Ecotoxicol Environ Saf* 23: 173–190.

Nyholm N, P Kristensen. 1992. Screening methods for assessment of biodegradability of chemicals in seawater—Results from a ring test. *Ecotoxicol Environ Saf* 23: 161–172.

O'Reilly KT, RL Crawford. 1989. Degradation of pentachlorophenol by polyurethane-immobilized *Flavobacterium* cells. *Appl Environ Microbiol* 55: 2113–2118.

Omori T, L Monna, Y Saiki, T Kodama. 1992. Desulfurization of dibenzothiophene by *Corynebacterium* sp. strain SY1. *Appl Environ Microbiol* 58: 911–915.

Pankratov TA, SN Dedysh. 2010. *Granulicella paludicola* gen. nov., sp. nov., *G. pectinovorans* sp. nov., *G. aggregans* sp. nov., and *G. rosea* sp. nov., novel acidophilic polymer-degrading acidobacteria from *Sphagnum* peat bogs. *Int J Syst Evol Microbiol* 60: 2951–2959.

Parkes RJ, GR Gibson, I Mueller-Harvey, WJ Buckingham, RA Herbert. 1989. Determination of the substrates for sulphate-reducing bacteria within marine and estuarine sediments with different rates of sulphate reduction. *J Gen Microbiol* 135: 175–187.

Pesaro M, G Nicollier, J Zeyer, F Widmer. 2004. Impact of soil drying-rewetting stress on microbial communities and activities of two crop protection products. *Appl Environ Microbiol* 70: 2577–2587.

Phillips WE, JJ Perry. 1976. *Thermomicrobium fosteri* sp. nov., a hydrocarbon-utilizing obligate thermophile. *Int J Syst Bacteriol* 26: 220–225.

Pignatello JJ, LK Johnson, MM Martinson, RE Carlson, RL Crawford. 1985. Response of the microflora in outdoor experimental streams to pentachlorophenol: Compartmental contributions. *Appl Environ Microbiol* 50: 127–132.

Pignatello JJ, MM Martinson, JG Steiert, RE Carlson, RL Crawford. 1983. Biodegradation and photolysis of pentachlorophenol in artificial freshwater streams. *Appl Environ Microbiol* 46: 1024–1031.

Pol A, K Heijmans, HR Harhangi, D Tedesco, MSM Jetten, HJM Op den Camp. 2007. Methanotrophy below pH 1 by a new Verrucomicrobia species. *Nature* 450: 874–878.

Rabus R, R Nordhaus, W Ludwig, F Widdel. 1993. Complete oxidation of toluene under strictly anoxic conditions by a new sulfate-reducing bacterium. *Appl Environ Microbiol* 59: 1444–1451.

Reanney DC, PC Gowland, JH Slater. 1983. Genetic interactions among microbial communities. *Symp Soc Gen Microbiol* 34: 379–421.

Robert-Gero M, M Poiret, RY Stanier. 1969. The function of the beta-ketoadipate pathway in *Pseudomonas acidovorans*. *J Gen Microbiol* 57: 207–214.

Rosner BM, B Schink. 1995. Purification and characterization of acetylene hydratase of *Pelobacter acetyleni-cus*, a tungsten iron-sulfur protein. *J Bacteriol* 177: 5767–5772.

Rothenburger S, RM Atlas. 1993. Hydroxylation and biodegradation of 6-methylquinoline by pseudomonads in aqueous and nonaqueous immobilized-cell bioreactors. *Appl Environ Microbiol* 59: 2139–2144.

Roy R, S Mukund, GJ Schut, DM Dunn, R Weiss, MWW Adams. 1999. Purification and molecular characterization of the tungsten-containing formaldehyde ferredoxin oxidoreductase from the hyperthermophilic archaeon *Pyrococcus furiosus*: The third of a putative five-member tungstoenzyme family. *J Bacteriol* 181: 1171–1180.

Sánchez MA, M Vásquez, B González. 2004. A previously unexposed forest soil microbial community degrades high levels of the pollutant 2,4,6-trichlorophenol. *Appl Environ Microbiol* 70: 7567–7570.

Sangwan P, X Chen, P Hugenholtz, PH Janssen. 2004. *Chthoniobacter flavus* gen. nov., sp. nov., the first pure-culture representative of subdivision two, *Spartobacteria* classis nov., of the phylum *Verrucomicrobia*. *Appl Environ Microbiol* 70: 5875–5881.

Sangwan P, X Chen, P Hugenholtz, PH Janssen. 2004. *Chthoniobacter flavus* gen. nov., sp. nov., the first pure-culture representative of subdivision two, *Spartobacteria* classis nov., of the phylum *Verrucomicrobia*. *Appl Environ Microbiol* 70: 5875–5881.

Saski EK, JK Jokela, MS Salkinoja-Salonen. 1996a. Biodegradability of different size classes of bleached kraft mill effluent organic halogens during wastewater treatment and in lake environments, pp. 179–193. In *Environmental Fate and Effects of Pulp and Paper Mill Effluents* (Eds. MR Servos, KR Munkittrick, JH Carey, and GJ van der Kraak), St Lucie Press, Delray Beach, FL.

Saski EK, A Vähätalo, K Salonen, MS Salkinoja-Salonen. 1996b. Mesocosm simulation on sediment for-mation induced by biologically treated bleached kraft pulp mill wastewater in freshwater recipients, pp. 261–270. In *Environmental Fate and Effects of Pulp and Paper Mill Effluents* (Eds. MR Servos, KR Munlittrick, JH Carey, and GJ van der Kraak), St Lucie Press, Delray Beach, FL.

Schoenborn L, PS Yates, BE Grinton, P Hugenholtz, PH Janssen. 2004. Liquid serial dilution is inferior to solid media for isolation of cultures representative of the phylum-level diversity of soil bacteria. *Appl Environ Microbiol* 70: 4363–4366.

Schut F, E J de Vries, JC Gottschal, BR Robertson, W Harder, RA Prins, DK Button. 1993. Isolation of typical marine bacteria by dilution culture: Growth, maintenance, and characteristics of isolates under laboratory conditions. *Appl Environ Microbiol* 59: 2150–2160.

Self WT. 2002. Regulation of purine hydroxylase and xanthine dehydrogenase from *Clostridium purinolyticum* in response to purines, selenium and molybdenum. *J Bacteriol* 184: 2039–2044.

Shiaris MP, JJ Cooney. 1983. Replica plating method for estimating phenanthrene-utilizing and phenanthrene-cometabolizing microorganisms. *Appl Environ Microbiol* 45: 706–710.

Skerman VBD. 1968. A new type of micromanipulator and microforge. *J Gen Microbiol* 54: 287–297.

Sobecky PA, MA Schell, MA Moran, RE Hodson. 1992. Adaptation of model genetically engineered microorganisms to lake water: Growth rate enhancements and plasmid loss. *Appl Environ Microbiol* 58: 3630–3637.

Söhngen NL. 1913. Benzin, Petroleum, Paraffinöl und Paraffin als Kohlenstoff- und Energiequelle für Mikroben. *Centr Bakteriol Parisitenkd Infektionskr Zweite Abt* 37: 595–609.

Song J, H-M Oh, J-C Cho. 2009. Improved culturability of SAR 11 strains in dilution-to-extinction culturing from the East Sea, Western Pacific Ocean. *FEMS Microbiol Lett* 295: 141–147.

Sontoh S, JD Semrau. 1998. Methane and trichloroethylene degradation by *Methylosinus trichosporium* OB3b expressing particulate methane monooxygenase. *Appl Environ Microbiol* 64: 1106–1114.

Steiner RA, KH Kalk, BW Dijkstra. 2002. Anaerobic enzyme substrate structures provide insight into the reac-tion mechanism of the copper-dependent quercitin 2,3-dioxygenase. *Proc Natl Acad USA* 99: 16625–16630.

Stanier RY, NJ Palleroni, M Doudoroff. 1966. The aerobic pseudomonads: A taxonomic study. *J Gen Microbiol* 43: 159–271.

Stevenson BS, SA Eichorst, JT Wertz, TM Schmidt, JA Breznak. 2004. New strategies for cultivation and detection of previously uncultured microbes. *Appl Environ Microbiol* 70: 4748–4755.

Strobl G, R Feicht, H White, F Lottspeich, H Simon. 1992. The tungsten-containing aldehyde oxidoreductase from *Clostridium thermoaceticum* and its complex with a viologen-accepting NADPH oxidoreductase. *Biol Chem Hoppe-Seyler* 373: 123–132.

Strotmann U, P Reuschenbach, H Schwarz, U Pagga. 2004. Development and evaluation of an online CO_2 evolution test and a multicomponent biodegradation test system. *Appl Environ Microbiol* 70: 4621–4628.

Sylvestre M. 1980. Isolation method for bacterial isolates capable of growth on *p*-chlorobiphenyl. *Appl Environ Microbiol* 39: 1223–1224.

Talbot HW, LM Johnson, DM Munnecke. 1984. Glyphosate utilization by *Pseudomonas* sp and *Alkaligenes* sp isolated from environmental sources. *Curr Microbiol* 10: 255–260.

Tanner A, L Bowater, SA Fairhurst, S Bornemann. 2001. Oxalate decarboxylase requires manganese and dioxygen for activity. *J Biol Chem* 276: 43627–43634.

Tatara GM, MJ Dybas, CS Criddle. 1993. Effect of medium and trace metals on kinetics of carbon tetrachloride transformation by *Pseudomonas* sp. strain KC. *Appl Environ Microbiol* 59: 2126–2131.

Taya M, H Hinoki, T Kobayashi. 1985. Tungsten requirement of an extremely thermophilic cellulolytic anaerobe (strain NA 10). *Agric Biol Chem* 49: 2513–2515.

Taylor BF, RW Curry, EF Corcoran. 1981. Potential for biodegradation of phthalic acid esters in marine regions. *Appl Environ Microbiol* 42: 590–595.

Thierrin J, GB Davis, C Barber. 1995. A ground-water tracer test with deuterated compounds for monitoring *in situ* biodegradation and retardation of aromatic hydrocarbons. *Ground Water* 33: 469–475.

Touati D, M Jacques, B Tardat, L Bouchard, S Despied. 1995. Lethal oxidative damage and mutagenesis are generated by iron Δ*fur* mutants of *Escherichia coli*: Protective role of superoxide dismutase. *J Bacteriol* 177: 2305–2314.

Tschech A, N Pfennig. 1984. Growth yield increase linked to caffeate reduction in *Acetobacterium woodii*. *Arch Microbiol* 137: 163–167.

Van der Meer JR, TNP Bosma, WP de Bruin, H Harms, C Holliger, HHM Rijnaarts, ME Tros, G Schraa, AJB Zehnder. 1992. Versatility of soil column experiments to study biodegradation of halogenated compounds under environmental conditions. *Biodegradation* 3: 265–284.

Van der Woude MW, K Boominathan, CA Reddy. 1993. Nitrogen regulation of lignin peroxidase and manganese-dependent peroxidase production is independent of carbon and manganese regulation in *Phanerochaete chrysosporium*. *Arch Microbiol* 160: 1–4.

Van Ginkel CG, JB van Dijl, AGM Kroon. 1992. Metabolism of hexadecyltrimethylammonium chloride in *Pseudomonas* strain B1. *Appl Environ Microbiol* 58: 3083–3087.

Van Niel CB. 1955. Natural selection in the microbial world. *J Gen Microbiol* 13: 201–217.

Visser JM, GC Stefess, LA Robertson, JG Kuenen. 1997. *Thiobacillus* sp. W5, the dominant autolithotroph oxidizing sulfide to sulfur in a reactor for aerobic treatment of sulfide waters. *Antonie van Leeuwenhoek* 72: 127–134.

Wachenheim DE, RE Hespell. 1984. Inhibitory effects of titanium(III) citrate on enumeration of bacteria from rumen contents. *Appl Environ Microbiol* 48: 444–445.

Wagner R, JR Andreesen. 1979. Selenium requirement for active xanthine dehydrgenase from *Clostridium acidiurici* and *Clostridium cylindrosporum*. *Arch Microbiol* 121: 255–260.

Wagner R, JR Andreesen. 1987. Accumulation and incorporation of [185]W-tungsten into proteins of *Clostridium acidiurici* and *Clostridium cylindrosporum*. *Arch Microbiol* 147: 295–299.

White H, G Strobl, R Feicht, H Simon. 1989. Carboxylic acid reductase: A new tungsten enzyme catalyses the reduction of non-activated carboxylic acids to aldehydes. *Eur J Biochem* 184: 89–96.

Widdel F. 1983. Methods for enrichment and pure culture isolation of filamentous gliding sulfate-reducing bacteria. *Arch Microbiol* 134: 282–285.

Wilkinson SG. 1968. Studies on the cell walls of pseudomonas species resistant to ethylenediaminetetra-acetic acid. *J Gen Microbiol* 54: 195–213.

Wilson MS, EL Madsen. 1996. Field extraction of a transient intermediary metabolite indicative of real time *in situ* naphthalene biodegradation. *Environ Sci Technol* 30: 2099–2103.

Winter J, C Lerp, H-P Zabel, F X Wildenauer, H König, F Schindler. 1984. *Methanobacterium wolfei*, sp. nov., a new tungsten-requiring, thermophilic, autotrophic methanogen. *Syst Appl Microbiol* 5: 457–466.

Wischmann H, H Steinhart. 1997. The formation of PAH oxidation products in soils and soil/compost mixtures. *Chemosphere* 35: 1681–1698.

Witschel M, S Nagel, T Egli. 1997. Identification and characterization of the two-enzyme system catalyzing the oxidation of EDTA in the EDTA-degrading bacterial strain DSM 9103. *J Bacteriol* 179: 6937–6943.

Yamamoto I, T Saiki, SM Liu, LG Ljungdahl. 1983. Purification and properties of NADP-dependent formate dehydrogenase from *Clostridium thermoaceticum*, a tungsten-selenium-iron protein. *J Biol Chem* 258: 1826–1832.

Yoch DC. 1979. Manganese, an essential trace element for N_2 fixation by *Rhodospirilllum rubrum* and *Rhodopseudomonas capsulata*: Role in nitrogenase regulation. *J Bacteriol* 140: 987–995.

Zarilla KA, JJ Perry. 1984. *Thermoleophilum album* gen. nov., and sp. nov., a bacterium obligate for thermophily and *n*-alkane substrates. *Arch Microbiol* 137: 286–290.

Zellner G, C Alten, E Stackebrandt, EC de Macario, J Winter. 1987. Isolation and characterization of *Methanocorpusculum parvum*, gen. nov., sp. nov., a new tungsten-requiring, coccoid methanogen. *Arch Microbiol* 147: 13–20.

Zellner G, A Jargon. 1997. Evidence for a tungsten-stimulated aldehyde dehydrogenase activity of *Desulfovibrio simplex* that oxidizes aliphatic and aromatic aldehydes with flavins as coenzymes. *Arch Microbiol* 168: 480–485.

Zengler K, G Toledo, M Rappé, J Elkins, EJ Mathur, JM Short, M Keller. 2002. Cultivating the uncultured. *Proc Natl Acad Sci USA* 99: 15681–15686.

Zhang H, Y Sekiguchi, S Hanada, P Hugenholtz, H Kim, Y Kamagata, K Nakamura. 2003. *Gemmatimonas aurantiaca* gen. nov., sp. nov., a Gram-negative, aerobic, polyphosphate-accumulating micro-organism, the first cultured representative of the new bacterial phylum *Gemmatimonadetes* phyl. nov. *Int J Syst Evol Microbiol* 53: 1155–1163.

Zürrer D, AM Cook, T Leisinger. 1987. Microbial desulfonation of substituted naphthalenesulfonic acids and benzenesulfonic acids. *Appl Environ Microbiol* 53: 1459–1463.

Part 2: Study of Microbial Populations

5.23 Introduction

Pathways for the degradation of synthetic substrates have been elucidated in laboratory experiments using cultures of bacteria that have been obtained by elective enrichment with a given, or related, substrate. This may, however, introduce a bias, not only in the organisms but also in the degradative genes that they contain (Marchesi and Weightman 2003). In addition, although there has been impressive success in isolating previously "uncultivable" bacteria (Joseph et al. 2003), there are many organisms in natural habitats that have not hitherto been cultivated (e.g., Stevenson et al. 2004): these may play an important role in degradation. There has, therefore, been greater appreciation of the role of populations and increasing interest in procedures for their analysis. Natural enrichment certainly takes place in environments that have been chronically exposed to contaminants, and is an important mechanism for the loss of agrochemicals. At the same time, bacteria that have been isolated from pristine environments with no established exposure to the contaminants have also been shown to degrade, for example, 2,4-dichlorophenoxyacetic acid (Kitagawa et al. 2002), 2,4,6-trichlorophenol (Sánchez et al. 2004), and PAHs (Hennessee et al. 2009).

A number of procedures have been used for the analysis of natural populations, including some which are directed to specific metabolic activity, and a variety of methods developed in molecular biology have been used. In general, these presuppose knowledge of the mechanisms for degradation of the contaminants, and this is equally true for methods using stable isotope analysis. In spite of the caution that been directed to the limitations of pure culture methods, it is the use of these that has built up the background for these studies which are not dependent on isolation of specific organisms. Application of the PCR (polymerase chain reaction) to specific degradative organisms has been reviewed (Steffan and Atlas 1991).

A different procedure has used intact cells embedded in a solid organic matrix: ionization of proteins is achieved with a UV laser followed by separation of the ions on the basis of molecular mass and charge, and analysis for identification using a library of established spectral data. Typically, whole-cell mass spectrometry is carried out by MALDI-TOF followed by analysis of data using a MALDI-TOF Biotyper that has been commercially developed. This has been extensively applied to important groups of clinical organisms, and has been successfully used for clinical and environmental strains of *Staphylococcus* (Carbonnelle et al. 2007; Dubois et al. 2010), *Listeria* (Barbuddhe et al. 2008), and *Campylobacter* (Mandrell et al. 2005).

Analysis of populations has revealed a number of cardinal aspects of microbial ecology:

1. The complexity of bacterial ecosystems
2. The number of organisms of hitherto undetermined taxonomy, and the existence of organisms that have not been cultivated
3. The difficulty of assigning metabolic roles to all the genomic sequences that have been revealed

5.24 Analysis of Degradative Populations

There has been substantial interest in identifying bacteria in environmental samples with a view to determining the components of bacterial communities and understanding their role in biodegradation and biotransformation. This has been possible by using molecular procedures, including the use of appropriate genetic probes to determine the existence of specific metabolic activity, and of analysis of 16S rRNA and 16S rDNA to assess the similarity of isolates to other taxa. A valuable overview of the principles and application of a range of molecular procedures has been given (Power et al. 1998). Potential applications include

1. Comparison of the degradation potential of natural populations with that of organisms isolated by enrichment
2. Prospecting for novel enzyme activity by analysis of soil metagenomes
3. Analysis of natural degradation may usefully be combined with the use of stable isotope enrichment factors that are discussed in Chapter 6

Some examples are given to illustrate the range and application of these procedures. These include dioxygenation of aromatic compounds, alkane degradation, dechlorination, and the discovery of hydrolytic enzymes for natural polymers.

5.25 Procedures Directed to Populations for the Degradation of Specific Contaminants

5.25.1 Hydrocarbons

1. Analysis of the occurrence and frequency of genes encoding the dioxygenases for aromatic hydrocarbons have been used to evaluate the potential of a site for biodegradation.
   a. An evaluation of the effect of plants on the remediation of a PAH-contaminated combined mineralization of ^{13}C-labeled naphthalene in the bulk soil and in the rhizosphere, with the occurrence of genes specific for dioxygenation of naphthalene (*ndoB*), alkane monooxygenation (*alkB*), and catechol 2,3-dioxygenase (*xylE*) (Siciliano et al. 2003). The results showed that even aged contaminants were susceptible to plant-mediated degradation in the rhizosphere.
   b. A study directed to the analysis of aromatic hydrocarbon degradation used real-time PCR amplification of oxygenase genes. The primer sets were identified for the large subunits (a) of genes for oxygenases, including dioxygenases for naphthalene, biphenyl, and toluene, and monooxygenases for xylene, phenol, and ring-monooxygenases for toluene (Baldwin et al. 2003).

c. The abundance of genes similar to the a-subunit of the *nagAc* naphthalene dioxygenase gene in *Ralstonia* sp. strain U2 that encodes the degradation of naphthalene to salicylate was analyzed by real-time PCR in samples from a coal-tar-contaminated site (Dionisi et al. 2004). There was a strong positive correlation between the number of gene copies and the concentration of naphthalene at the site, and a lower correlation with concentrations of phenanthrene, and it was concluded that active biodegradation was occurring.

d. Isolates from a site contaminated with BTEX were analyzed to assess the diversity of dioxygenases in the a-subunit of the toluene/biphenyl subfamily and the ability to grow with benzene, toluene, and ethylbenzene. All strains could degrade benzene, but only a few of them could degrade toluene and ethylbenzene. There was a strong correlation between sequence type and substrate utilization, and sequence analysis showed that dioxygenases belonging to the isopropylbenzene branch dominated (Witzig et al. 2006).

2. The reverse sample genome probe procedure has been used to monitor sulfate-reducing bacteria in oil-field samples (Voordouw et al. 1991). It was extended to include 16 heterotrophic bacteria (Telang et al. 1997), and applied to evaluating the effect of nitrate on an oil-field bacterial community. Denatured genomic DNAs were used in a reverse sample genomic probe procedure to examine the effect of toluene and dicyclopentadiene on the community structure of bacteria in a contaminated soil. Hybridization of total-community DNA isolated from soil exposed to toluene showed enrichment of strains that were able to metabolize toluene. At the same time, DNA from the soil contaminated with dicyclopentadiene indicated enrichment of organisms that were able to form products by oxygenation with masses of 146 and 148 that corresponded to derivatives with epoxide or ketone groups. On the other hand, no mineralization of dicyclopentadiene was observed (Shen et al. 1998).

3. The archaeal population in anaerobic oil-contaminated water at a crude oil storage site was examined using PCR amplification of fragments of 16S rDNA using a range of primers. Although two of the sequences were closely related to *Methanosaeta concilii* and *Methanovorans hollandica*, other sequences could not be identified, including the most abundant (KuA12) that accounted for ≈50% of the total archaeal rDNA copies that were detected. Species of novel archaea were enriched at the site and were putatively active methanogens (Watanabe et al. 2002).

4. A stable mixed culture that could degrade benzene was established, and the specific role of sulfate-reducing bacteria was inferred from the inhibition of degradation in the presence of molybdate. The consortium was characterized from the sequence of small subunit rRNA genes, some of which could be assigned to the family *Desulfobacteriaceae* (Phelps et al. 1998).

5.25.2 Trichloroethene

1. Populations of bacteria had been enriched by adding lactate to enrich the indigenous organisms for *in situ* dechlorination of trichloroethene to ethene. The consortia in the groundwater source were assessed using PCR-amplification of 16S rRNA genes. In a clone library, a homoacetogen dominated the archaea, with a clone affiliated with the acetoclastic methanogen *Methanosaeta concilii*, while proteobacteria similar to *Geobacter (Trichlorobacter) thiogenes* and *Sulfurospirillum multivorans* were also found. A mixed culture that dechlorinated trichloroethene to ethene was enriched from the groundwater, and this was compared with that in the groundwater by terminal restriction fragment length polymorphism. Although the enrichment community was less diverse than that from the groundwater, its archaeal structure was similar. It was concluded that acetate produced from fermentation of lactate was the source of methane by the acetoclastic pathway (Macbeth et al. 2004).

2. The use of toluene to induce oxygenation of haloalkanes has been discussed in Chapter 7, Part 3, and probes for toluene-2-monooxygenase have been used to evaluate the potential number of trichloroethene-degrading organisms in an aquifer (Fries et al. 1997). In this study, repetitive extragenic palindromic PCR (REP-PCR) (de Bruijn 1992) of isolates was used to classify their metabolic capability.

3. To assess the population of organisms capable of degrading trichloroethene at a contaminated site, samples were examined from a contiguous, but pristine site. Attention was directed to methane- and ammonia-oxidizing bacteria, and included both free-living and attached communities. Primer sets included genes for particulate methane monooxygenase and ammonia monooxygenase. Whole-genome amplification was used for metagenomic DNA from each of the communities at the sampling sites. Bacteria related to *Methylocystis* dominated both the free-living and attached communities, and gene sequences associated with trichloroethene metabolism were observed for both of them (Erwin et al. 2005).

5.25.3 Phenol

1. Analysis of the populations of two phenol-degrading bacteria *Pseudomonas putida* BH and *Comamonas* sp. strain E6 introduced into an activated sludge system was carried out by extraction of DNA, PCR amplification of the *gyrB* gene fragments using strain-specific primers, and quantification after electrophoresis by densitometry. Appropriate dilutions of the extracted DNA were used to maintain linearity between the measured intensity and population density. The initial inocula of ca. 10^8 cells/mL fell to ca. 10^4 cell/mL after 10 days and thereafter remained steady for a further 18 days (Watanabe et al. 1998a).

2. A study of the community in a phenol-degrading population used amplification of the gene encoding the largest subunit of phenol hydroxylase, followed by analysis using temperature gradient gel electrophoresis. Bacteria were also isolated by three procedures: (i) direct plating on a complex medium or a mineral medium supplemented with phenol, (ii) by enrichment in batch culture, and (iii) by chemostat enrichment. The dominant population could be isolated only by direct plating or by chemostat enrichment. One of the dominant organisms that contained a novel large phenol hydroxylase subunit was closely related to *Variovorax paradoxus*, and it was proposed that this was the principal organism for degradation of phenol in the community (Watanabe et al. 1998b).

3. A site at the Agricultural Experimental Station (Ithaca, NY) was treated in microcosms with ^{13}C-labeled glucose, phenol, caffeine, and naphthalene. Levels of ^{13}CO$_2$ were measured to assess utilization of the substrates, and the populations analyzed by separating the ^{13}C-labeled DNA by density centrifugation, followed by PCR amplification and sequencing of 16S rRNA (Padmanabhan et al. 2003). Populations contained relatives to a range of bacteria that depended on the substrate. Only relatives of *Acinetobacter* were found in all the samples, and for caffeine, only *Pantoea*.

4. In a further study (DeRito et al. 2005) at the same site, attention was concentrated on phenol using respiration of ^{13}C-labeled phenol and stable isotope probes of soil DNA. It was shown that the whole community could utilize products from the degradation of phenol. The distribution of labeled carbon was determined by using different regimes for the addition of substrate, and populations could be distinguished on the basis of their diversity. The greatest diversity was found in organisms from soil that had not been enriched with phenol, while members of the Gram-positive genera *Kocuria* and *Staphylococcus* dominated the phenol-degrading population. Members of the genus *Pseudomonas* dominated those that were metabolically versatile and utilized the carbon from other organisms.

5.25.4 Chlorophenol

Soil samples were enriched in medium containing 2,4,6-trichlorophenol and subjected to a procedure to reduce the complexity of the mixed flora. This involved cyclic serial dilution, plating, and growth in liquid medium. Repetitive extragenic palindromic PCR (REP-PCR) and amplified ribosomal DNA restriction analysis (ARDRA) were used to monitor the flora during the cycles. From the fourth cycle, four organisms were isolated. Although three of them were unable to utilize 2,4,6-trichlorophenol, an additional slow-growing organism that was able to do so was isolated, and tentatively assigned to the genus *Nocardioides* (Maltseva and Oriel 1997).

5.25.5 Chlorobenzoate

The occurrence of bacteria that can mineralize 3-chlorobenzoate has been examined in soil samples from widely separated regions in five continents (Fulthorpe et al. 1998). The genotypes of the isolates were examined by two procedures: (a) repetitive extragenic palindromic PCR genomic fingerprints (REP-PCR) (de Bruyn et al. 1990) and (b) analysis of restriction digests of the 16S rRNA (amplified ribosomal DNA restriction analysis, ARDRA) (Weisberg et al. 1991). The results showed that each genotype was generally (91%) restricted to the site from which the samples were collected, and that, therefore, the genotypes were not derived by global dispersion.

5.25.6 Phenylurea Herbicides

1. DNA was extracted from soil samples that had been treated with three phenylurea herbicides during 10 years, and from an untreated control. The PCR products using two primers were analyzed by denaturing gradient gel electrophoresis (DGGE), and the patterns used to assess the quantitative similarities of the bands. PCR using two different primers followed by DGGE was used to obtain 16S rDNA for sequencing (El Fantroussi et al. 1999). The microbial diversity determined from the gel profiles had decreased in the treated soils, and the sequencing revealed that the organisms that were most affected belonged to uncultivated taxa. Enrichment cultures showed that dichlorinated linuron was more readily degraded in cultures using the treated soils than from those using the untreated soils. DGGE analysis and sequencing showed that one of the components that was found only in these enrichments showed a 95% similarity to *Variovorax* sp. that was also found in enrichments using 2,4-dichlorophenoxyacetate (Kamagata et al. 1997). The effect of the herbicide applications had, therefore, affected both the composition of the bacteria flora and the metabolic capabilities of its components.

2. Analysis of a bacterial population that could degrade the chlorinated phenylurea herbicide linuron revealed the presence of several strains of bacteria which were identified on the basis of 16S rRNA gene sequences, isolation of the strains, and determination of their metabolic role (Dejonghe et al. 2003). Five strains were identified of which the *Variovorax* strain WDL1 was able to use linuron as a source of carbon, nitrogen, and energy. Among the other strains, *Hyphomicrobium sulfonivorans* was able to degrade *N,O*-dimethylhydroxylamine, while degradation of linuron was stimulated by the presence of these other strains.

5.25.7 Dehalogenation of Chloroalkanoates

Dehalogenases from bacteria enriched from activated sludge for the degradation of 2,2-dichloropropionate were compared with those from the environmental metagene pool. Although the dehalogenases found in pure cultures dominated the enrichment culture, they were only a minor part of the community used for enrichment. Analysis of dehalogenase genes found in pure cultures with those from the metagenome pool from which they were isolated revealed the substantial bias introduced by culturing both in the bacteria and in their degradative genes (Marchesi and Weightman 2003).

5.25.8 PCBs

1. Metabolic inhibitors were used in an elegant set of experiments designed to enrich for the organisms that carried out the *ortho* dechlorination of 2,3,5,6-tetrachlorobiphenyl (Holoman et al. 1998), and the community structure was followed by analysis of total community genes for 16S rRNA. It was shown that the diversity of the community could be reduced in mineral medium by addition of inhibitors for methanogens (2-bromoethanesulfonic acid), and for *Clostridium* spp. (vancomycin) without eliminating dechlorination that was inhibited by addition of molybdate that inhibited sulfate reduction. The bacteria that actively carried out *ortho* dechlorination belonged to three groups: the d group, the low G + C Gram-positive group, and the *Thermotogales* subgroup that had not hitherto been implicated in anaerobic dechlorination.

2. Based on experience with clinical bacteria, a different approach has been used. MALDI-TOF and a MALDI-TOF Biotyper have been applied to the analysis of strains of single colonies of biphenyl-utilizing organisms that were isolated from the horseradish rhizosphere at a PCB-contaminated site. A range of bacteria was identified to the genus level, and several also to the species level. Identifications included species of the genera *Aeromonas, Arthrobacter, Microbacterium, Rhizobium, Rhodococcus,* and *Serratia* all of which are represented among strains that have been shown to be able to degrade PCBs.

5.26 Application to Populations of Specific Groups of Organisms

There has been considerable interest in the anaerobic metabolism of methane in the large reservoirs that lie beneath the seafloor, since little of this reaches the oxic conditions in the water column. Consortia of archaea that have so far resisted isolation, and sulfate-reducing bacteria have been implicated (Orphan et al. 2002).

1. A hydrothermal vent in the Gulf of California produces hydrocarbons that support the growth of sulfate-reducing bacteria. PCR-amplified genes for sulfite reductase (*dsrAB*) and 16S rRNA were analyzed in sediment cores. Groups related to *Desulfobacter* that could utilize acetate represented a major group, while organisms affiliated with members of the genus *Desulfotomaculum,* and organisms in the δ proteobacteria were also found (Dhillon et al. 2003).

2. Cores from the same vent were examined by 16S rRNA sequencing complemented with isotopic analysis of lipids. Uncultured groups associated with anaerobic methane oxidation ANME-1 and ANME-2 were identified, while $\delta^{13}C$ values for archaeal lipids (archaeol, *sn*-2-hydroxyarcheol, and diphytanediol) indicated their origin in methanotrophic archaea (Teske et al. 2002). Analysis of the sequences for methyl coenzyme M reductase A (*mcrA*) in ANME-1 and ANME-2 suggested its potential catalytic activity (Hallam et al. 2003). It was suggested that anaerobic methane oxidation could occur by reverse methanogenesis since many—though not all—of the genes could be identified in both pathways (Hallam et al. 2004).

5.27 Nonspecific Examination of Natural Populations

1. Core samples from three redox zones at depths of 6.1, 7.3, and 9.2 m in an aquifer contaminated with hyd\rocarbons and chlorinated were examined (Dojka et al. 1998). Small subunit rRNA genes from core DNA were amplified directly by PCR with bacteria- or archaea-specific primers and cloned, and the clones screened by restriction fragment length polymorphisms (RFLP). Analysis of the sequences showed the existence of the following types of sequence:

   a. 10 sequences with no known taxonomic divisions

   b. 21 sequences with no cultured representatives

   c. 63 sequences with recognized divisions

   The sequences were classified into seven major groups, and it was shown that the two most abundant sequence types could be correlated with sequences of species of *Methanosaeta* and *Syntrophus*. It was proposed that hydrocarbon degradation proceeded by oxidation of the hydrocarbons, fermentation of the resulting carboxylic acids by organisms with sequences related to *Syntrophus* sp., followed by acetoclastic methanogenesis by organisms related in sequence to *Methanosaeta* sp.

2. Small-subunit rRNA genes were amplified directly by PCR in DNA from a sediment sample in Yellowstone National Park and cloned: universally conserved, or bacteria-specific rDNA primers were used (Hugenholtz et al. 1998). Restriction fragment length polymorphism was used to classify rDNA fragments. Most of the sequences were representatives of established bacterial

division, although 30% were not so related. Database matches suggested the presence of organisms closely related to: (a) *Thermodesulfovibrio yellowstonii* that would be able to carry out anaerobic sulfate reduction, (b) the organotrophs *Thermus* sp. and *Dictyoglomus thermophilum*, and (c) the hydrogen-oxidizing *Calderiobacterium hydrogenophilum*. In contrast to previous perceptions, however, members of bacteria dominated over archaea in this sediment, so that the ecological boundaries between bacteria and archaea were less clearly defined.

3. The organisms in a high-salinity evaporation pond were examined by plating on solid media and prolonged incubation, and by analysis of a library of PCR-amplified 16S rRNA genes (Burns et al. 2004a). The isolates were related to species of several genera, including *Haloferax*, *Halorubrum*, and *Natromonas*. The first of these was not, however, represented in the gene library, and it was suggested that this could be due to its ability to form colonies even though it was not a dominant group. A major group identified in the library was the SHOW square organisms of Walsby that have been isolated and cultivated (Bolhuis et al. 2004; Burns et al. 2004b).

5.28 References

Baldwin BR, CH Nakatsu, L Lies. 2003. Detection and enumeration of aromatic oxygenase genes by multiplex and real-time PCR. *Appl Environ Microbiol* 69: 3350–3358.

Barbuddhe SB, T Maier, G Schwarz, M Kostrzewa, H Hof, E Domann, T Chakraborty, T Hain. 2008. Rapid identification and typing of *Listeria* species by matrix-assisted laser desorption ionization–time of flight mass spectrometry. *Appl Environ Microbiol* 74: 5402–5407.

Bolhuis H, EM te Poele, F Rodríguez-Valera. 2004. Isolation and cultivation of Walsby's square archaeon. *Environ Microbiol* 6: 1287–1291.

Burns DG, HM Camakaris, PH Janssen, ML Dyall-Smith. 2004a. Combined use of cultivation-dependent and cultivation-independent methods indicates that members of most haloarchaeal groups in an Australian crystallizer pond are cultivable. *Appl Environ Microbiol* 70: 5258–5265.

Burns DG, HM Camakaris, PH Janssen, ML Dyall-Smith. 2004b. Cultivation of Walsby's square archaeon. *FEMS Microbiol Lett* 238: 469–473.

Carbonnelle E, J-L Beretti, S Cottyn, G Quesne, P Berche, X Nassif, A Ferroni. 2007. Rapid identification of staphylococci isolated in clinical microbiology laboratories by matrix-assisted laser desorption ionization–time of flight mass spectrometry. *J Clin Microbiol* 45: 2156–2161.

de Bruyn JC, FC Boogerd, P Bos, JG Kuenen. 1990. Floating filters, a novel technique for isolation and enumeration of fastidious, acidophilic, iron-oxidizing autotrophic bacteria. *Appl Environ Microbiol* 56: 2891–2894.

Dejonghe W, E Berteloot, J Goris, N Boon, K Crul, S Maertens, M Höfte, P de Vos, W Verstraete, EM Top. 2003. Synergistic degradation of linuron by a bacterial consortium and isolation of a single linuron-degrading *Variovorax* strain. *Appl Environ Microbiol* 69: 1532–1542.

DeRito CM, GM Pumphrey, EL Madsen. 2005. Use of field-based stable isotope probing to identify adapted populations and track carbon flow through a phenol-degrading soil microbial community. *Appl Environ Microbiol* 71: 7858–7865.

Dhillon A, A Teske, J Dillon, DA Stahl, ML Sogin. 2003. Molecular characterization of sulfate-reducing bacteria in the Guaymas Basin. *Appl Environ Microbiol* 69: 2765–2772.

Dionisi HM, CS Chewning, KH Morgan, F-M Menn, JP Easter, GS Sayler. 2004. Abundance of dioxygenase genes similar to *Ralstonia* sp. strain U2 *nagAc* is correlated with naphthalene concentrations in coal tar-contaminated freshwater sediments. *Appl Environ Microbiol* 70: 3988–3995.

Dojka MA, P Hugenholtz, SK Haack, NR Pace. 1998. Microbial diversity in a hydrocarbon- and chlorinated-solvent-contaminated aquifer undergoing intrinsic bioremediation. *Appl Environ Microbiol* 64: 3869–3877.

Dubois D, D Leyssene, JP Chacornac, M Kostrzewa, PO Schmit, R Talon, R Bonnet, J Delmas. 2010. Identification of a variety of *Staphylococcus* species by matrix-assisted laser desorption ionization–time of flight mass spectrometry. *J Clin Microbiol* 48: 941–945.

El Fantroussi S, L Verschuere, W Verstraete, EM Top. 1999. Effect of phenylurea herbicides on soil microbial communities estimated by analysis of 16S rRNA gene fingerprints and community-level physiological profiles *Appl Environ Microbiol* 65: 982–988.

Erwin DP, IK Erickson, ME Delwiche, FS Colwell, JL Strap, RL Crawford. 2005. Diversity of oxygenase genes from methane- and ammonia-oxidizing bacteria in the Eastern Snake River plain aquifer *Appl Environ Microbiol* 71: 2016–2925.

Fries MR, LJ Forney, JM Tiedje. 1997. Phenol- and toluene-degrading microbial populations from an aquifer in which successful trichloroethene cometabolism occurred. *Appl Environ Microbiol* 63: 1523–1530.

Fulthorpe RR, AN Rhodes, JM Tiedje. 1998. High levels of endomicity of 3-chlorobenzoate-degrading soil bacteria. *Appl Environ Microbiol* 64: 1620–1627.

Hallam, PR Girguis, CM Preston, PM Richardson, EF DeLong. 2003. Identification of methyl coenzyme M reductase A (*merA*) genes associated with methane-oxidizing archaea. *Appl Environ Microbiol* 69: 5483–5491.

Hallam SJ, N Putnam, CM Preston, JC Detter, D Rokhsar, PM Richardson, ED DeLong. 2004. Reverse methanogenesis: Testing the hypothesis with environmental genomics. *Science* 305: 1457–1462.

Hennessee CT, J-S Seo, AM Alvarez, QX Li. 2009. Polycyclic aromatic hydrocarbon-degrading species isolated from Hawaiian soils: *Mycobacterium crocinum*, sp. nov., *Mycobacterium pallens* sp. nov., *Mycobacterium rutilum* sp. nov., *Mycobacterium rufum* sp. nov. and *Mycobacterium aromaticivorans* sp. nov. *Int J Syst Evol Microbiol* 59: 378–387.

Holoman TRP, MA Elberson, LA Cutter, HD May, KR Sowers. 1998. Characterization of a defined 2,3,5,6-tetrachlorobiphenyl-*ortho*-dechlorinating microbial community by comparative sequence analysis of genes coding for 16S rRNA. *Appl Environ Microbiol* 64: 3359–3367.

Hugenholtz P, C Pitulle, KL Hershberger, NR Pace. 1998. Novel division level bacterial diversity in a Yellowstone hot spring. *J Bacteriol* 180: 366–376.

Joseph SJ, P Hugenholtz, P Sangwan, CA Osborne, PH Janssen. 2003. Laboratory cultivation of widespread and previously uncultured soil bacteria. *Appl Environ Microbiol* 69: 7210–7215.

Kamagata Y, RR Fulthorpe, K Tamura, H Takami, LJ Forney, JM Tiedje. 1997. Pristine environments harbor a new group of oligotrophic 2,4-dichlorophenoxyacetic acid-degrading bacteria. *Appl Environ Microbiol* 63: 2266–2272.

Kitagawa W, S Takami, K Miyauchi, E Masai, Y Kamagata, JM Tiedje, M Fukuda. 2002. Novel 2,4-dichlorophenoxyacetic acid degradation genes from oligotrophic *Bradyrhizobium* sp. strain HW 13 isolated from a pristine environment *J Bacteriol* 184: 509–518.

Macbeth TW, DE Cummings, S Spring, LM Petzke, KS Sorenson. 2004. Molecular characterization of a dechlorinating community resulting from *in situ* biostimulation in a trichloroethene-contaminated fractured basalt aquifer and comparison to a derivative laboratory culture *Appl Environ Microbiol* 70: 7329–7341.

Maltseva O, P Oriel. 1997. Monitoring of an alkaline 2,4,6-trichlorophenol-degrading enrichment culture by DNA fingerprinting methods and isolation of the responsible organism, haloalkaliphilic *Nocardioides* sp. strain M6. *Appl Environ Microbiol* 63: 4145–4149.

Mandrell RE, LA Harden, A Bates, WG Miller, WF Haddon, CK Fagerquist. 2005. Speciation of *Campylobacter coli*, C. *jejueni*, C. *helveticus*, *C Lari*, *C sputorum*, and C. *upsaliensis* by matrix-assisted laser desorption ionization–time of flight mass spectrometry. *Appl Environ Microbiol* 71: 6292–6307.

Marchesi JR, AJ Weightman. 2003. Comparing the dehalogenase gene pool in cultivated α-halocarboxylic acid-degrading bacteria with the environmental metagene pool. *Appl Environ Microbiol* 69: 4375–4382.

Orphan VL, CH House, K-U Hinrichs, KD McKeegan, EF DeLong. 2002. Multiple archaeal groups mediate methane oxidation in anoxic cold seep sediments. *Proc Natl Acad USA* 99: 7663–7668.

Padmanabhan P, S Padmanabhan, C DeRito, A Gray, D Gannon, JR Snape, DC Tsai, W Park, C Jeon, EL Madsen. 2003. respiration of ^{13}C-labeled substrates added to soil in the field and subsequent 16S rRNA gene analysis of ^{13}C-labeled soil DNA. *Appl Environ Microbiol* 69: 1614–1622.

Phelps CD, LJ Kerkhof, LY Young. 1998. Molecular characterization of sulfate-reducing consortium which mineralizes benzene. *FEMS Microbiol Ecol* 27: 269–279.

Power M, JR van der Meer, R Tchelet, T Egli, R Eggen. 1998. Molecular-based methods can contribute to assessment of toxicological risks and bioremediation strategies. *J Microbiol Methods* 32: 1078–1119.

Sánchez, M Vásquez, B González. 2004. A previously unexposed forest soil microbial community degrades high levels of the pollutant 2,4,6-trichlorophenol. *Appl Environ Microbiol* 70: 7567–7570.

Shen Y, LG Stehmeier, G Voordouw. 1998. Identification of hydrocarbon-degrading bacteria in soil by reverse sample genome probing *Appl Environ Microbiol* 63: 637–645.

Siciliano SD, JJ Germida, K Banks, CW Greer. 2003. Changes in microbial community composition and function during a polyaromatic hydrocarbon phytoremediation field trial. *Appl Environ Microbiol* 69: 483–489.

Steffan RJ, RM Atlas. 1991. Polymerase chain reactions, applications in environmental microbiology. *Annu Rev Microbiol* 45: 137–161.

Stevenson BS, SA Eichorst, JT Wertz, TM Schmidt, JA Breznak. 2004. New strategies for cultivation and detection of previously uncultivated microbes. *Appl Environ Microbiol* 70: 4748–4755.

Telang AJ, S Ebert, LM Focht, DW Westlake, GE Jenneman, D Gevertz, G Voordouw. 1997. Effect of nitrate injection on the microbial community in an oil field monitored by reverse sample genome probing. *Appl Environ Microbiol* 63: 1785–1793.

Teske A, K-U Hinrichs, V Edgcomb, A de Vera Gomez, D Lysela, SP Sulva, ML Sogin, HW Jannach. 2002. Microbial diversity of hydrothermal sediments in the Guaymas Basin: Evidence for anaerobic methanotrophic communities. *Appl Environ Microbiol* 68: 1994–2007.

Uhlik O, M Strejcek, P Junkova, M Sanda, M Hroludova, C Vlcek, M Mackova, T Macek. 2011. Matrix-assisted laser desorption ionization (MALDI)–time of flight mass spectrometry- and MALDI biotyper-based identification of cultured biphenyl-metabolizing bacteria from contaminated horseradish rhizosphere soil. *Appl Environ Microbiol* 77: 6858–6866.

Voordouw G, JK Voordouw, RR Karkhoff-Schweiser, PM Fedorak, DWS Westlake. 1991. Reverse sample genome probing, a new technique for identification of bacteria in environmental samples by DNA hybridization, and its application to the identification of sulfate-reducing bacteria in oil field samples. *Appl Environ Microbiol* 57: 3070–3078.

Watanabe K, M Teramoto, H Futamata, S Harayama. 1998b. Molecular detection, isolation, and physiological characterization of functionally dominant phenol-degrading bacteria in activated sludge *Appl Environ Microbiol* 64: 4396–4402.

Watanabe K, S Yamamoto, S Hino, S Harayama. 1998a. Population dynamics of phenol-degrading bacteria in activated sludge determined by *gyrB*-targeted quantitative PCR. *Appl Environ Microbiol* 64: 1203–1209.

Watanabe K, Y Kodama, N Hamamura, N Kaku. 2002. Diversity, abundance, and activity of archaeal populations in oil-contaminated groundwater accumulated at the bottom of an underground crude oil storage cavity *Appl Environ Microbiol* 68: 3899–3907.

Weisberg WG, SM Barns, DA Pelletier, DJ Lane. 1991. 16S ribosomal DNA amplification for phylogenetic study. *J Bacteriol* 173: 697–703.

Witzig R, H Junca, H-J Hecht, DH Pieper. 2006. Assessment of toluene/biphenyl dioxygenase gene diversity in benzene-polluted soils: Links between benzene degradation and genes similar to those encoding isopropylbenzene dioxygenases. *Appl Environ Microbiol* 72: 3504–3514.

6

Procedures for Elucidation of Metabolic Pathways

6.1 Introduction

The sections in this chapter deal with procedures for investigating the pathways used in biodegradation and biotransformation. They cover briefly several applications of isotopes, and of nondestructive physical methods for structure determination that include NMR, EPR, and x-ray crystallographic analysis.

Part 1: Application of Natural and Synthetic Isotopes

The application of substrates isotopically labeled in specific positions makes it possible to follow the fate of individual atoms during the microbial degradation of xenobiotics. Under optimal conditions, both the kinetics of the degradation and the formation of metabolites may be followed—ideally, when samples of the labeled metabolites are available. Many of the classical studies on the microbial metabolism of carbohydrates, carboxylic acids, and amino acids used radioactive ^{14}C-labeled substrates and specific chemical degradation of the metabolites to determine the position of the label. The method is indeed obligatory for distinguishing between degradative pathways when the same products are produced from the substrate by different pathways. A suitable example is provided by the β-methylaspartate and hydroxyglutarate pathways for fermentations of glutamate, both of which produce butyrate but which could clearly be distinguished by the use of [4-^{14}C]-glutamate (Buckel and Barker 1974).

6.2 Carbon (^{14}C and ^{13}C)

Conventional use has been made of the radioisotope ^{14}C, and details need hardly be given here. Illustrative examples include the elucidation of pathways for the anaerobic degradation of amino acids (Chapter 7, Part 1) and purines (Chapter 10, Part 1). Some applications have used ^{13}C with high-resolution Fourier transform NMR in whole-cell suspensions, and this is equally applicable to molecules containing the natural ^{19}F or the synthetic ^{31}P nuclei. As noted later, major advances in NMR have made it possible to use natural levels of ^{13}C.

Illustrative examples of the application of ^{13}C-labeled substrates to biodegradation include the following:

1. The degradation of $^{13}CCl_4$ by *Pseudomonas* sp. strain KC involved the formation of intermediate $COCl_2$ that was trapped as a HEPES complex, and by the reaction with cysteine (Lewis and Crawford 1995). Further details of the pathway that is mediated by the metabolite pyridine-dithiocarboxylic acid have been elucidated (Lewis et al. 2001).

2. ^{13}C[bicarbonate] and NMR were used to demonstrate that the first product in the metabolism of propene epoxide is acetoacetate, which is then reduced to β-hydroxybutyrate (Allen and Ensign 1996).

3. ^{13}C[bicarbonate] and mass spectrometry were used to demonstrate the formation of carboxylic acids during the sulfidogenic mineralization of naphthalene and phenanthrene (Zhang and Young 1997).

4. 9[^{13}C]-anthracene was used to study its degradation in soil, and the formation of labeled metabolites that could be released only after alkaline hydrolysis (Richnow et al. 1998). It was possible to construct a carbon balance during the 599-d incubation, and to distinguish metabolically formed phthalate from indigenous phthalate in the soil.

5. Traps with Bio-Sep beads amended with [^{13}C$_6$]-benzene and [^{13}C]-toluene were used to assess biodegradation in an aquifer (Geyer et al. 2005). Beads were lyophilized after exposure, lipids were extracted with chloroform–methanol, and the fatty acids and δ^{13}C values analyzed. High enrichment of ^{13}C was observed in several fatty acids, which showed that the label from the substrates had been incorporated. In addition, there were differences in the abundance of the fatty acids in beads amended with benzene or toluene that suggested the existence of different microbial degradative populations.

6. The degradation of ^{13}C$_6$-benzene was studied in anaerobic enrichment cultures when phenol, benzoate, and toluene were detected, and the kinetics of their formation studied (Ulrich et al. 2005).

6.3 Sulfur (^{35}S) and Chlorine (^{36}Cl)

Quite extensive use of ^{35}S has been made in studies on the degradation of alkyl sulfonates (Hales et al. 1986), and was used to study the source of sulfur in the production of methanethiol from methionine, cysteine and sulfate in *Ferroplasma acidarmanus* (Baumler et al. 2007). On the other hand, ^{36}Cl has achieved only limited application on account of technical difficulties resulting from the low specific activities and the synthetic inaccessibility of appropriately labeled substrates. One of the few examples of its application to the degradation of xenobiotics is provided by a study of the anaerobic dechlorination of hexachlorocyclohexane isomers (Jagnow et al. 1977), the results of which are discussed in Chapter 7, Part 3.

6.4 Hydrogen (^{2}H) and Oxygen (^{18}O)

Although the radioactive isotope ^{3}H has been extensively used for studies on the uptake of xenobiotics into whole cells, the intrusion of exchange reactions and the large isotope effect renders this isotope rather less straightforward for metabolic studies. Both deuterium ^{2}H-labeled substrates, and oxygen ^{18}O$_2$ and ^{18}OH$_2$ have, however, been extensively used in metabolic studies, since essentially pure labeled compounds are readily available and mass spectrometer facilities have become an essential part of structural determination.

Illustrative examples of the use of ^{2}H in different applications include the following:

1. Deuterium labeling has been invaluable in studying rearrangements involving protons. For example, it has been used to reveal the operation of the NIH shift during metabolism of [4-^{2}H] ethylbenzene by the monooxygenase system from *Methylococcus capsulatus* (Dalton et al. 1981) of [2,2′,3,3′,5,5′,6,6′-^{2}H]biphenyl by *Cunninghamella echinulata* (Smith et al. 1981), and of [1-^{2}H]naphthalene and [2-^{2}H]naphthalene by *Oscillatoria* sp. (Narro et al. 1992).

2. The conversion of long-chain alkanoate CoA esters into the alkenoate-CoA esters by acyl-CoA oxidase involves an *anti* elimination reaction. The stereochemistry of the reaction in *Candida lipolytica* was established using stearoyl-CoA labeled with ^{2}H at the 2(*R*)-, 3(*R*)-, and 3(*S*)-positions (Kawaguchi et al. 1980).

3. ^{2}H-labeled substrates have been used to determine details of the dehydrogenation of *cis*-dihydrodiols produced by dioxygenases from aromatic substrates (Morawski et al. 1997), and it was possible to demonstrate the specificity of hydrogen transfer from the dihydrodiol substrates to NAD.

4. The fumarate pathway for the metabolism of alkanes under sulfate-reducing conditions was examined using fully deuterated hexadecane (Callaghan et al. 2006). Identification of the metabolites including labeled methylpentadecylsuccinate and 4-methyloctadecanoate was used to determine details of the pathway that involved a rearrangement during the initial reaction.

5. The enantiomerization of phenoxyalkanoic acids containing a chiral side chain has been studied in soil using ^2H_2O (Buser and Müller 1997). It was shown that there was an equilibrium between the *R*- and *S*-enantiomers of 2-(4-chloro-2-methylphenoxy)propionic acid (MCPP) and 2-(2,4-dichlorophenoxy)propionic acid (DCPP) with an equilibrium constant favoring the herbicidally active *R*-enantiomer. The exchange reactions proceeded with both retention and inversion of configuration at the chiral sites.

6. ^2H-labeled substrates have been used to determine the dissipation and degradation of aromatic hydrocarbons in a contaminated aquifer plume (Thierrin et al. 1995). Its application was particularly appropriate since the site was already contaminated with the substrates. With suitable precautions, this procedure seems capable of extension to determining the presence—though not the complete structure—of metabolites, provided that the possibility of exchange reactions were taken into account.

7. The fate of toluene and *o*-xylene in an aquifer contaminated with BTEX was examined by injecting toluene-d_8 and *o*-xylene-d_{10} followed by quantification of the label in benzyl succinate and 2-methylbenzene succinate (Reusser et al. 2002) that are established metabolites on the anaerobic pathway for the degradation of toluene and *o*-xylene.

8. Styrene-degrading bacteria from full-scale and experimental biofilters were exposed to $[^2H_8]$ styrene, and analysis of fatty acids was used to distinguish the bacterial flora of the two systems (Alexandrino et al. 2001).

9. Biodegradation of toluene was evaluated at a site that was contaminated with benzene and toluene from the earlier production of benzene. Anaerobic degradation was supported with sulfate as the electron acceptor, and ring-deuterated (d_5) and fully deuterated (d_8) toluene were injected into the anoxic zone. Biodegradation was assessed by several procedures such as (a) isotope fractionation using the Raleigh equation combined with laboratory-determined isotope fractionation and (b) analysis of changes in the concentration of the deuterated toluene relative to bromide that was used as a marker. Both procedures yielded comparable results, and either of them was able to provide a record of toluene biodegradation. Degradation was confirmed by demonstration of benzylsuccinate-d_5 and an initial increase in the ratio of deuterium to hydrogen in the groundwater (Fischer et al. 2006).

Although the application of ^{18}O has been less frequent, it has been used effectively to determine the source of oxygen and the number of oxygen atoms incorporated during metabolism of xenobiotics under both aerobic and anaerobic conditions. Some typical examples are given below:

1. In a classical study, it was shown that during bacterial oxidation of benzene to catechol, both atoms of oxygen came from $^{18}O_2$ (Gibson et al. 1970). This initiated the appreciation of the role of dioxygenases in the degradation of aromatic xenobiotics, and many examples are given in Chapter 8, Parts 1 and 2.

2. During the biodegradation of 2,4-dinitrotoluene by a strain of *Pseudomonas* sp., two atoms of oxygen were incorporated from $^{18}O_2$ during the formation of 4-methyl-5-nitrocatechol by dioxygenation with loss of nitrite (Spanggord et al. 1991).

3. The degradation of nicotine has been examined extensively in *Arthrobacter nicotinovorans* (*oxydans*). In strain P34, the first metabolite was 6-hydroxynicotine, and experiments with $^{18}O_2$ and $H_2^{18}O$ showed that the oxygen in the hydroxyl group was derived from water (Hochstein and Dalton 1965).

4. Experiments with $H_2^{18}O$ using *Pseudomonas putida* strain 86 (Bauder et al. 1990) showed that the oxygen incorporated into quinol-2-one is derived from water.

5. During the degradation of 2-chloroacetophenone by a strain of *Alcaligenes* sp., one atom of $^{18}O_2$ was incorporated into 2-chlorophenol formed from the 2-chlorophenyl acetate that was initially formed by Baeyer–Villiger monooxygenation (Higson and Focht 1990).

6. Benzene and toluene were anaerobically hydroxylated to phenol and 4-hydroxytoluene, and experiments with $H_2^{18}O$ showed that the oxygen atoms had originated from water (Vogel and Grbic-Galic 1986).

7. The mechanism of the cytochrome $P450_{14DM}$ conversion of a C_{14} methyl sterol into formate and the $\Delta^{14,15}$ was studied using $^{18}O_2$, and it could be shown that the hydroxyl oxygen atom in the formate that was produced contained one atom of ^{18}O (Shyadehi et al. 1996).

In experiments involving the use of $^{18}O_2$ and $H_2^{18}O$, care should be taken to exclude chemical exchange reactions involving potentially labile C–H or C–O bonds, since these reactions could seriously compromise the conclusions. A good illustration of the pitfalls in such investigations is shown by a study on the dechlorination of pentachlorophenol by the dehalogenase from a strain of *Arthrobacter* sp. The initial reaction in the degradation of pentachlorophenol is mediated by a pentachlorophenol dehalogenase that produces tetrachloro-1,4-dihydroxy-benzene. Experiments using the enzyme showed that ^{18}O is incorporated into this metabolite only after incubation with $H_2^{18}O$ and not with $^{18}O_2$. It has been shown that the labeling occurs as a result of exchange between the initially formed unlabeled metabolite and $H_2^{18}O$. An unambiguous elucidation of the mechanism of the reaction was not, therefore, possible, since even if ^{18}O had been incorporated during the reaction with $^{18}O_2$, exchange with the excess $H_2^{16}O$ in the medium would have yielded an unlabeled product (Schenk et al. 1990).

6.5 Other Isotopes

Studies of biodegradation have made more limited use of metal isotopes:

1. The aerobic degradation of the 2-carboxylates of furan, pyrrole, and thiophene is initiated by hydroxylation before fission of the rings. Although details of the enzymes are limited, it was suggested on the basis of tungstate inhibition and ^{185}W [tungstate] labeling that the degradation of 2-furoyl-coenzyme A involves a molybdenum-dependent dehydrogenase (Koenig and Andreesen 1990).

2. Use of Mössbauer spectra generally requires the addition of ^{57}Fe to the growth medium. This procedure has received extensive application to studying the environment of Fe in enzymes, and illustrative examples include the following, and further details of its application in EPR are given later.

   a. EPR analysis and Mössbauer spectra of the enoate reductase in *Clostridium tyrobutyricum* revealed the presence of a semiquinone and an [4Fe–4S] cluster (Caldeira et al. 1996).

   b. Mössbauer spectra of the sulfate reductase of *Bacillus subtilis* showed the presence of a [4Fe–4S] cluster in the isolated enzyme, and that this was degraded to a [2Fe–2S] cluster. It was suggested that the interconversion of these clusters functioned in suppressing the formation of sulfite under aerobic conditions (Berndt et al. 2004).

3. The purified ribonucleotide reductase from *Corynebacterium ammoniagenes* did not require additional metal cations but after removal of protein with trichloroacetate, the electronic and EPR spectra suggested the presence of an oxo-bridged binuclear Mn(III) which was confirmed by growth in the presence of $^{54}MnCl_2$ when the catalytic B2 protein was specifically labeled (Willing et al. 1988).

4. The anaerobic degradation of nicotinate by *Clostridium barkeri* is initiated by nicotinate hydroxylase that introduces oxygen from water at C_6. The active enzyme contains Se, and it was

shown that in cultures grown with ^{77}Se the signal of Mo(V) in the EPR spectrum was split without affecting the Fe signals of the iron–sulfur cluster (Gladyshev et al. 1994). The selenium was, therefore, coordinated with the molybdopterin cofactor.

6.6 Isotope Effects and Stable Isotope Fractionation

These procedures have had two principal applications: analysis of degradation and transformation in natural populations, and the determination of degradation pathways. Only a brief outline of experimental aspects is provided. Depending on the objective, results from these experiments have been reported in a number of ways, and in comparing results it is important to take this into consideration. This aspect is, therefore, discussed here.

Elements such as C, N, O, S, and Cl that are components of many organic compounds exist naturally as mixtures of stable isotopes. The ratios of these in a compound reflect the different rates of reaction at isotopically labeled positions, and, therefore, reflect the fractionation—biotic or abiotic—by which it was synthesized or to which the compound has been subjected. Techniques have been developed whereby the ratios $^{13}C/^{12}C$ ($\delta^{13}C$), $^{15}N/^{14}N$ ($\delta^{15}N$), $^{18}O/^{16}O$ ($\delta^{18}O$), $^{34}S/^{32}S$ ($\delta^{34}S$), and $^{37}Cl/^{35}Cl$ ($\delta^{37}Cl$) can be accurately measured by mass spectrometry. The differences are expressed as per mille (‰) deviations with respect to a standard:

$$\delta = 1000 \times (\text{ratio for sample/ratio for standard}) - 1$$

Standard samples are used for calibration, and detailed procedures for C and O have been given by Craig (1957); these include details of the standards, the appropriate correction factors that should be applied, and mass spectrometric techniques. The standard for carbon is the Vienna Pee Dee Belemnite (VPDM), and for oxygen the Vienna Standard Mean Ocean Water (VMOW). Secondary standards have included graphite and atmospheric oxygen that have been related to the primary standards. Atmospheric air has been used for nitrogen and ocean water chloride for chlorine. These values may differ slightly from those derived from the relative isotope contributions given in compilations of the elements.

Determination of $\delta^{13}C$ has been used extensively in geochronology, and $\delta^{18}O$ in glaciology and climatology, and these techniques have also been used in determining biosynthetic pathways of, for example, lipids, and the various processes whereby CO_2 is incorporated into and disseminated in biota (Hayes 1993). They have been used increasingly in environmental research. Some examples are given to illustrate the potential of this methodology, using analysis of individual compounds to establish their sources, or alterations during transformation or dissemination. Extensive applications have been made to the biodegradation of chlorinated ethenes, monocyclic aromatic hydrocarbons, polychlorinated biphenyls, and methyl *tert*-butyl methyl ether. They have been directed to anaerobic reactions since these conditions are widespread in contaminated groundwater systems (Chapter 14, Part 7). The results have been reported in a number of ways, and valuable comments on pitfalls that may be encountered when heavy isotopes are used have been given (Hunkeler 2002).

Results have been expressed in a number of ways. In the Raleigh model that has been used extensively, the fractionation factor α is given by $R/R_0 = f^{(\alpha-1)}$ when the fraction of the remaining substrate is f and where R is the isotopic composition of the substrate during degradation and R_0 is the initial value. The enrichment factor ε where $\varepsilon = 1000(\alpha - 1)$ has also been used. There are certain conditions that must be fulfilled for the Raleigh model to be applicable:

1. The concentrations of the heavy isotope must be low, which is true for experiments using natural levels of abundance, or the reaction must be near the initial starting point.
2. The model is strictly applicable only when applied to a single-step reaction, and not to a sequence of reactions.

Kinetic isotope effects have also been given in terms of ε; for reactions at a single carbon atom: $^{12}k/^{13}k = 1/(1 + \varepsilon/1000)$, where ^{12}k and ^{13}k are rates for ^{12}C and ^{13}C substrates.

In the application of the Raleigh equation to contaminated sites, complications due to the heterogeneity of the system have been addressed (Abe and Hunkeler 2006). Detailed analysis has revealed systematic problems that could result in underestimation of first-order rates, and procedures for evaluating these were provided.

6.7 References

Abe Y, D Hunkeler. 2006. Does the Raleigh equation apply to evaluate field isotope data in contaminant hydrogeology? *Environ Sci Technol* 40: 1588–1596.

Aggarwal PK, ME Fuller, MM Gurgas, JF Manning, MA Dillon. 1997. Use of stable oxygen and carbon isotope analyses for monitoring the pathways and rates of intrinsic and enhanced *in situ* biodegradation. *Environ Sci Technol* 31: 590–596.

Alexandrina M, C Knife, A Lip ski. 2001. Stable-isotope based labeling of styrene-degrading microorganisms in boilers. *Appl Environ Microbiol* 67: 4796–4804.

Allen JR, SA Ensign. 1996 Carboxylation of epoxides to β-keto acids in cell extracts of *Xanthobacter* strain Py2. *J Bacteriol* 178: 1469–1472.

Bauder R, B Tshisuaka, F Lingens. 1990 Microbial metabolism of quinoline and related compounds VII Quinoline oxidoreductase from *Pseudomonas putida*: A molybdenum-containing enzyme. *Biol Chem Hoppe-Seyler* 371: 1137–1144.

Baumler DJ, K-F Hung, KC Jeong, CW Kaster. 2007. Production of methanethiol and volatile sulfur compounds by the archaeon "*Ferroplasma acidarmanus*". *Extremophiles* 11: 841–851.

Berndt C, CH Lillig, M Wollenberg, E Bill, MC Mansilla, D de Mendoza, A Seidler, JD Schwenn. 2004. Characterization and reconstitution of a 4Fe–4S adenyl sulfate/phosphoadenyl sulfate reductase from *Bacillus subtilis*. *J Biol Chem* 279: 7850–7855.

Buckel W, HA Barker. 1974. Two pathways of glutamate fermentation by anaerobic bacteria. *J Bacteriol* 117: 1248–1260.

Buser H-R, MD Müller. 1997. Conversion reactions of various phenoxyalkanoic acid herbicides in soil 2 Elucidation of the enantiomerization process of chiral phenoxy acids from incubation in a D_2O/soil system. *Environ Sci Technol* 31: 1960–1967.

Caldeira J, R Feicht, H White, M Teixeira, JJG Mourat, H Simon, I Moura. 1996. EPR and Mössbauer spectroscopic studies on enoate reductase. *J Biol Chem* 271: 18743–18748.

Callaghan AV, LM Gieg, KG Kropp, JM Suflita, LY Young. 2006. Comparison of mechanisms of alkane metabolism under sulfate-reducing conditions among two bacterial isolates and a bacterial consortium. *Appl Environ Microbiol* 72: 4274–4282.

Craig H. 1957. Isotopic standards for carbon and oxygen and correction factors for mass-spectrometric analysis of carbon dioxide. *Geochim Cosmochim Acta* 12: 133–149.

Dalton H, BT Golding, BW Waters, R Higgins, JA Taylor. 1981. Oxidations of cyclopropane, methylcyclopropane, and arenes with the mono-oxygenase system from *Methylococcus capsulatus*. *J Chem Soc Chem Comm* 482–483.

Fischer A, J Bauer, RU Meckenstock, W Stichler, C Griebler, P Maloszewski, M Kästner, HH Richnow. 2006. A multitracer test proving the reliability of Rayleigh equation-based approach for assessing biodegradation in a BTEX contaminated aquifer. *Environ Sci Technol* 40: 4245–4252.

Geyer R, AD Peacock, A Miltner, H-H Richnow, DC White, KL Sublette, M Kästner. 2005. *In situ* assessment of biodegradation potential using biotraps amended with ^{13}C-labeled benzene or toluene. *Environ Sci Technol* 39: 4983–4989.

Gibson DT, GE Cardini, FC Maseles, RE Kallio. 1970. Incorporation of oxygen-18 into benzene by *Pseudomonas putida*. *Biochemistry* 9: 1631–1635.

Gladyshev VN, SV Khangulov, TC Stadtmann. 1994. Nicotinic acid hydrolase from *Clostridium barkeri*: Electron paramagnetic studies show that selenium is coordinated with molybdenum in the catalytically active selenium-dependent enzyme. *Proc Natl Acad USA* 91: 232–236.

Hales SG, GK Watson, KS Dodgson, GF White. 1986. A comparative study of the biodegradation of the surfactant sodium dodecyltriethoxy sulphate by four detergent-degrading bacteria. *J Gen Microbiol* 132: 953–961.

Hayes JM. 1993. Factors controlling ^{13}C contents of sedimentary organic compounds: Principles and evidence. *Mar Geol* 113: 111–125.

Higson FK, DD Focht. 1990. Bacterial degradation of ring-chlorinated acetophenones. *Appl Environ Microbiol* 56: 3678–3685.

Hochstein LI, BP Dalton. 1965. The hydroxylation of nicotine: The origin of the hydroxyl oxygen. *Biochem Biophys Res Commun* 21: 644–648.

Holt BD, NC Sturchio, TA Abrajano, LJ Heraty. 1997. Conversion of chlorinated volatile organic compounds to carbon dioxide and methyl chloride for isotopic analysis of carbon and chlorine. *Anal Chem* 69: 2727–2733.

Hunkeler D .2002. Quantification of isotope fractionation in experiments with deuterium-labeled substrate. *Appl Environ Microbiol* 68: 5205–5207.

Jagnow G, K Haider, P-C Ellwardt. 1977. Anaerobic dechlorination and degradation of hexachlorocyclohexane by anaerobic and facultatively anaerobic bacteria. *Arch Microbiol* 115: 285–292.

Kawaguchi A, S Tsubotani, Y Seyama, T Yamakawa, T Osumi, T Hashimoto, Y Kikuchi, M Ando, S Okuda. 1980. Stereochemistry of dehydrogenation catalyzed by acyl-CoA oxidase. *J Biochem* 88: 1481–1486.

Koenig K, JR Andreesen. 1990. Xanthine dehydrogenase and 2-furoyl-coenzyme A dehydrogenase from *Pseudomonas putida* Fu1: Two molybdenum-containing dehydrogenases of novel structural composition. *J Bacteriol* 172: 5999–6009.

Kucklick JR, HR Harvey, PH Ostrom, NE Ostrom, JE Baker. 1996. Organochlorine dynamics in the pelagic food web of Lake Baikal. *Anal Chim Acta* 15: 1388–1400.

Lewis TA, RL Crawford. 1995. Transformation of carbon tetrachloride via sulfur and oxygen substitution by *Pseudomonas* sp. strain KC. *J Bacteriol* 177: 2204–2208.

Lewis TA, A Paszczynski, SW Gordon-Wylie, S Jeedigunta, C-H Lee, RL Crawford. 2001. Carbon tetrachloride dechlorination by the bacterial transition metal chelator pyridine,2,6-bis(thiocarboxylic acid). *Environ Sci Technol* 35: 552–559.

Merritt DA, KH Freeman, MP Ricci, SA Studley, JM Hayes. 1995. Performance and optimization of a combustion interface for isotope ratio monitoring gas chromatography/mass spectrometry. *Anal Chem* 67: 2461–2473.

Morawski B, G Casy, C Illaszewicz, H Griengl, DW Ribbons. 1997. Stereochemical course of two arene-*cis*-diol dehydrogenases specifically induced in *Pseudomonas putida*. *J Bacteriol* 179: 4023–4029.

Narro ML, CE Cerniglia, C Van Baalen, DT Gibson. 1992. Evidence for an NIH shift in oxidation of naphthalene by the marine cyanobacterium *Oscillatoria* sp. strain JCM. *Appl Environ Microbiol* 58: 1360–1363.

Numata M, N Nakamura, H Koshikawa, Y Terashima. 2002. Chlorine isotope fractionation during reductive dechlorination of chlorinated ethenes by anaerobic bacteria. *Environ Sci Technol* 36: 4389–4394.

Reusser DE, JD Istok, HR Beller, JA Field. 2002. *In situ* transformation of deuterated toluene and xylene to benzylsuccinic acid analogues in BTEX-contaminated aquifers. *Environ Sci Technol* 36: 4127–4134.

Richnow HH, A Eschenbach, B Mahro, R Seifert, P Wehrung, P Albrecht, W Michaelis. 1998. The use of ^{13}C-labelled polycyclic aromatic hydrocarbons for the analysis of their transformation in soil. *Chemosphere* 36: 2211–2224.

Saurer M, I Robertson, R Siegwolf, M Leuenberger. 1998. Oxygen isotope analysis of cellulose: An interlaboratory comparison. *Anal Chem* 70: 2074–2080.

Schenk T, R Müller, F Lingens. 1990. Mechanism of enzymatic dehalogenation of pentachlorophenol by *Arthrobacter* sp. strain ATCC 33790. *J Bacteriol* 172: 7272–7274.

Shyadehi AZ, DC Lamb, SL Kelly, DE Kelly, W-H Schunck, JN Wright, D Corina, M Akhtar. 1996. The mechanism of the acyl-carbon bond cleavage reaction catalyzed by recombinant sterol 14α-demethylase of *Candida albicans* (other names are: lanosterol 14α-demethylase, P-450$_{14DM}$, and CYP51). *J Biol Chem* 271: 12445–12450.

Smith RV, PJ Davis, AM Clark, SK Prasatik. 1981. Mechanism of hydroxlation of biphenyl by *Cunninghamella echinulata*. *Biochem J* 196: 369–371.

Spanggord RJ, JC Spain, SF Nshino, KE Mortelmans. 1991. Biodegradation of 2,4-dinitrotoluene by a *Pseudomonas* sp. *Appl Environ Microbiol* 57: 3200–3205.

Thierrin J, GB Davis, C Barber. 1995. A ground-water tracer test with deuterated compounds for monitoring *in situ* biodegradation and retardation of aromatic hydrocarbons. *Ground Water* 33: 469–475.

Ulrich AC, HR Beller, EA Edwards. 2005. Metabolites detected during biodegradation of ^{13}C$_6$-benzene in nitrate-reducing and methanogenic enrichment cultures. *Environ Sci Technol* 39: 6681–6691.

Vogel TM, D Grbic-Galic. 1986. Incorporation of water into toluene and benzene during anaerobic fermentative transformation. *Appl Environ Microbiol* 52: 200–202.

Westaway KC, T Koerner, Y-R Fang, J Rudzinski, P Paneth. 1998. A new method of determining chlorine kinetic isotope effects. *Anal Chem* 70: 3548–3552.

Zhang X, LY Young. 1997. Carboxylation as an initial reaction in the anaerobic metabolism of naphthalene and phenanthrene by sulfidogenic consortia. *Appl Environ Microbiol* 63: 4759–4764.

Part 2: Stable Isotope Probes and Stable Isotope Fractionation

The general application of isotopes has been discussed in Part 1 of this chapter. The following discussion is devoted to two applications of stable isotopes—stable isotope probes and stable isotope enrichment. Both depend on the differential rates of reaction of molecules containing different isotopes. This application has proliferated, and only some representative examples are given as illustration. These procedures have had two principal applications: analysis of degradation and transformation in natural populations, and the determination of degradation pathways. Illustrative examples are given here but only an outline of experimental aspects is attempted. Depending on the objective, results from these experiments have been reported in a number of ways, and in comparing results it is important to take this into consideration. These aspects are, therefore, discussed first.

6.8 Experimental Procedures

The analyte must be converted into a volatile compound suitable for mass spectrometric analysis. Procedures for C, N, and O follow those developed for conventional organic microanalysis—oxidation of organic C to CO_2, reduction of organic N to N_2, and conversion of O_2 to CO or CO_2. In most procedures, cryogenic purification of the products is carried out before mass spectrometry, and both off-line and online procedures have been developed.

Carbon—Determination of $\delta^{13}C$ has been carried out by GC separation of the analytes followed by high-temperature oxidation. Although oxidation with CuO at 850°C has been used, the use of NiO at 1050°C supplemented by O_2 is preferable, and it is important to remove H_2O before isotopic analysis (Merritt et al. 1995).

Nitrogen—For $\delta^{15}N$, the sample is heated at 850°C with an excess of Cu (to reduce NO_x to N_2) and CuO, followed by cryogenic purification of the N_2 (Kucklick et al. 1996).

Oxygen—For the determination of $\delta^{18}O$, the compound is pyrolyzed with graphite embedded with platinum wire at 800°C (Aggarwal et al. 1997) or glassy carbon at 1080°C (Saurer et al. 1998) for conversion to CO_2 and CO, respectively. For oxyanions such as nitrate and chlorate, the formation of O_2 by pyrolysis is measured directly.

Chlorine—For the determination of $\delta^{37}Cl$ in organic compounds, several methods have been used:

  a. Conversion into CuCl and then into CH_3Cl by reaction with CH_3I at 300°C (Holt et al. 1997)

  b. Reduction to chloride with sodium biphenyl followed by conversion to CsCl and determination by thermal ionization mass spectrometry (Numata et al. 2002)

  c. Conversion to AgCl followed by ionization by xenon atoms and determination of isotopes using an FAB–isotope ratio mass spectrometer (Westaway et al. 1998)

Sulfur—In experiments with anaerobic sulfate-reducing bacteria, the sulfur exists as sulfate from the residual substrate and sulfide produced by microbial reduction. Dissolved sulfate was precipitated as $BaSO_4$ and sulfide as ZnS and subsequent conversion to Ag_2S. $BaSO_4$ and Ag_2S were oxidized to SO_2 that was used for isotope ratio analysis of $^{34}S/^{32}S$ (Bolliger et al. 2001; Knöller et al. 2006).

6.9 Expression of Results

Results have been expressed in a number of ways. In the Raleigh model that has been extensively used, the fractionation factor α is given by $R/R_o = f^{(\alpha-1)}$ when the fraction of remaining substrate is f and where R is the isotopic composition of the substrate during degradation and R_o is the initial value. The enrichment factor e, where $\varepsilon = 1000(\alpha - 1)$, has also been used. There are certain conditions that must be fulfilled for the Raleigh model to be applicable:

1. The concentrations of the heavy isotope must be low, which is true for experiments using natural levels of abundance, or the reaction must be near the initial starting point.
2. The model is strictly applicable only when applied to a single-step reaction, and not to a sequence of reactions.

Kinetic isotope effects have also been given in terms of ε; for reactions at a single carbon atom, $^{12}k/^{13}k = 1/(1 + \varepsilon/1000)$, where ^{12}k and ^{13}k are respective rates for ^{12}C and ^{13}C substrates.

In the application of the Raleigh equation to contaminated sites, complications due to the heterogeneity of the system have been addressed (Abe and Hunkeler 2006). Detailed analysis has revealed systematic problems that could result in underestimation of first-order rates, and procedures for evaluating these were provided.

6.10 Stable Isotope Probes

This procedure makes it possible to relate the function of microorganism in the environment to specific organisms or enzymes. Samples are treated with the appropriate isotope-labeled compound and identification of the organisms carried out by extraction of DNA fractionation by CsCl density centrifugation, and analysis of the fractions by PCR and 16S rRNA probes. There are inherent problems in the interpretation of the results of stable isotope probing. The concentration of the isotope probe may be substantially higher than that to which the cells are normally exposed. During the lengthy exposure that may be necessary, cross-feeding of the substrate and its metabolites among organisms may take place.

6.11 Application to Processes

6.11.1 Methanotrophy

1. The population of putative methanotrophs in a peat-soil microcosm was examined by long-term incubation with $^{13}CH_4$ followed by separation of DNA by gradient centrifugation. This was used as a template for PCR. The amplified products of 16S rRNA genes and those for genes that encode enzymes established for methane oxidation were analyzed (Morris et al. 2002). Although sequences related to type I and type II methanotrophs were identified, the existence of clones related to the β-subclass of *Proteobacteria* indicated the presence of other groups.
2. In an analogous study, the assimilation of $^{13}CH_3OH$ or $^{13}CH_4$ was examined (Radajewski et al. 2002). Analysis showed the presence of bacteria belonging to the genera *Methylocella*, *Methylocystis*, and *Methylocapsa*, while DNA from the microcosms that was used as a template for PCR with ammonia oxygenase-specific primers showed sequences belonging to *Nitrosomomas* and *Nitrosospira*. A number of other sequences were observed although their function was unresolved.
3. The methylotrophic community in the soil from rice paddies was examined by a number of methods including ^{13}C-labeling with methanol followed by incubation under aerobic conditions and analysis of ^{13}C-labeled rRNA by measurement of the density in a cesium trifluoroacetate gradient. Sequence analysis of labeled rRNA revealed the presence of bacteria belonging to the Methylobacteriaceae, although with increasing length of incubation the Methylophilaceae dominated. In addition, labeled nucleic acids were observed in fungi and protozoans, and it was concluded that the ^{13}C accumulated in the methylotrophs had been transferred during incubation to these organisms (Lueders et al. 2004a).

4. Stable isotope probing of methanotrophic bacteria from soil in the Canadian Arctic at Eureka in Ellesmere Island was carried out (Martineau et al. 2010). In spite of the high latitude, the site was not pristine since a hydrocarbon spill had occurred some years before sampling. Several methods were used including stable isotope $^{13}CH_4$ probing of DNA, DNA purification by CsCl density gradient centrifugation to separate enriched fractions, and denaturing gradient gel electrophoresis. Sequencing the 16S rRNA and particulate methane monooxygenase (*pmoA*) genes showed that only a low diversity of methanogens was present, and that they belonged to the type I methanotrophs *Methylobacter* and *Methylosarcina*. These bacteria were not, however, dominant and their metabolic activity was low at 4°C even when nutrients were added. The availability of nitrogen and phosphorus, and the low average yearly temperature were clearly limiting factors.

6.11.2 Methanogenesis

It has been established that methane is produced on rice roots by reduction of CO_2. This was examined in rice roots using a combination of 16S rRNA sequencing and density gradient fractionation of ^{13}C-labeled DNA after incubation with $^{13}CO_2$. The major groups of archaea detected were *Methanosarcinaceae* that decreased with time to be replaced by the hitherto uncultured Rice Cluster I, although the former subsequently dominated (Lu et al. 2005).

6.11.3 Nitrogen Fixation

Core samples collected from a field that had lain fallow for 20 years were enriched in an atmosphere of 20% O_2 and 80% $^{15}N_2$. DNA was extracted and purified, and subjected to CsCl density gradient centrifugation followed by additional centrifugation of enriched fractions with bis-benzimide. Fractions were subjected to quantitative PCR. Classification of the 16S rRNA genes showed the presence of a large number of sequences. Those for which the number of sequences recovered from the $^{15}N_2$ library exceeded the number on the control group belonged to three groups: *Rhodoplanes*, unclassified *Betaproteobacteria*, and unclassified actinobacteria, none of which have been associated with nitrogen fixation (Buckley et al. 2007).

6.12 Application to Biodegradation

6.12.1 Methanol

A consortium that was grown anaerobically with methanol and nitrate as the electron acceptor was used to characterize the principal components (Ginige et al. 2004). ^{13}C-methanol was used to label DNA, and extracted samples of [^{13}C]DNA and [^{12}C]DNA were used in a full-cycle rRNA analysis. The dominant 16S rRNA gene phylotype in the [^{13}C]DNA clone library was closely related to those of *Methylobacillus* and *Methylophilus*, and provided evidence of their significance in denitrification. Oligonucleotide probes designed for fluorescence studies did not show any correlation with the abundance of the genera *Hyphomicrobium* or *Paracoccus*.

6.12.2 Propionate

The syntrophic population of anaerobic organisms that carry out the oxidation of propionate was examined using [^{13}C]propionate. Analysis of the "heavy" ^{13}C-labeled rRNA by terminal restriction length polymorphism fingerprinting and sequence analysis of the resolved populations showed the presence of the genera *Syntrophobacter*, *Smithella*, and *Pelotomaculum* whose degradation of propionate had already been confirmed in other laboratory studies. Among the archaea, species of *Methanobacterium*, *Methanosarcina*, and members of the Rice Cluster I were found (Lueders et al. 2004b).

6.12.3 C₁, C₂, C₃-Alkane Oxidation

Sediment from a marine hydrocarbon seep was incubated with ^{13}C-methane, ^{13}C-ethane, and ^{13}C-propane. DNA from the ^{13}C-labeled samples was prepared and fractions from CsCl density gradient centrifugation were used to carry out terminal restriction fragment length polymorphism analysis and to construct 16S rRNA clone libraries. Briefly, the results showed that whereas the major consumers of ^{13}C-methane were affiliated with *Methylophilaceae* or *Methylophaga*, and *Methylococcaceae* made only a minor contribution, the last ones were the primary consumers for ^{13}C-ethane oxidation. For ^{13}C-propane, on the other hand, substrate consumption was attributed mainly to unclassified *Gammaproteobacteria* (Redmond et al. 2010).

6.12.4 Benzene

1. Gasoline-contaminated groundwater containing BTEX from an aquifer was supplemented with [^{13}C]benzene and an electron acceptor. Analysis of the ^{13}C-RNA fraction revealed the presence of a phylotype related to the genus *Azoarcus* only when nitrate was used as supplement. This was confirmed by denaturing gradient gel electrophoresis (DGGE) of 16S rRNA gene fragments. Isolation on a nonselective medium and screening of the colonies by DGG resulted in the isolation of two strains that were able to degrade benzene, toluene, and *m*-xylene—although not *o*- or *p*-xylene. It was suggested that the success of this procedure lay in combining stable isotope probing with DGGE screening of strains isolated on nonselective medium (Kasai et al. 2006).

2. A sulfidogenic enrichment culture that degraded benzene was used to add further details to an earlier study that had been shown to contain a number of sulfate-reducing bacteria (Oka et al. 2008). Cultures were grown with ^{13}C-benzene and DNA was separated by density gradient centrifugation. Dialyzed samples were used for 16S rRNA gene PCR and terminal restriction fragment length polymorphism analysis. During incubation, the significance was revealed of a 270 base-pair peak that incorporated the greater part of the ^{13}C. Its sequence suggested a close relationship to a clone SB-21 that had been found to be dominant in other sulfate-reducing benzene-degrading enrichments.

3. A stable anaerobic enrichment culture with benzene that used amorphous ferric oxide as sole electron acceptor was used to identify the principal organisms (Kunapuli et al. 2007). Cultures were grown with ^{13}C-benzene and ^{12}C-benzene, and used for isolation of DNA by density gradient centrifugation and analyzed by terminal restriction fragment length polymorphism fingerprinting. The recovered DNA was used to recover 16S rRNA gene fragments that were amplified and used for analysis. Two fragments with base-pairs of 289 and 162 were dominant during the incubation up to 120 d. The first was proposed as the organism responsible for primary reaction and was clustered among clostridia of the *Peptococcaceae*, while the latter that was presumed to be an auxiliary organism clustered among *Desulfobulbaceae*.

6.12.5 Phenol

A phenol-degrading community was examined using ^{13}C phenol followed by analysis of the stable-isotope-labeled RNA by equilibrium density centrifugation, and complemented with reverse-transcription PCR and denaturing gradient gel electrophoresis. The results suggested the dominance of a species of *Thauera* in addition to organisms conventionally associated with phenol degradation (Manefield et al. 2002).

6.12.6 Naphthalene

Application to a site contaminated with coal-tar waste and containing aromatic hydrocarbons was examined using a number of procedures. These included the use of ^{13}C-labeled naphthalene followed by fractionation of labeled DNA, and construction of a library of cloned bacterial 16S rRNA. These were complemented with conventional assays for respiration, and isolation of bacteria by serial dilution and plating on a mineral medium with exposure to naphthalene. A strain CJ2 that closely resembled *Polaromonas vacuolata* was isolated from the contaminated sediment, and was active *in situ*, even though the naphthalene dioxygenase gene was not related to that of existing strains of the genus (Jeon et al. 2003).

6.12.7 Biphenyl

These investigations have been carried out with samples from PCB-contaminated soil and sediment.

1. This was carried out to compare the organisms in the rhizosphere of horseradish that was contaminated with PCBs with those in the bulk soil (Uhlik et al. 2009). Cultures were grown with uniformly labeled [^{13}C]biphenyl and DNA stable isotope probing combined with terminal restriction fragment length polymorphism was used to characterize the rhizosphere and bulk soil samples. The rhizosphere PCB-contaminated soil was dominated by a species of *Hydrogenophaga*, whereas the bulk soil was dominated initially by *Paenibacillus* that was accompanied after 14 days by a range of organisms including *Stenotrophomonas* and *Methylophilus*. In short-term incubation, the amino acid sequence of the biphenyl dioxygenase α-subunit closely resembled that of *Pseudomonas alcaligenes* strain B-357.

2. A sediment sample that had been historically contaminated with Arochlor 1248 was used (Sul et al. 2009). Using uniformly labeled [^{13}C]biphenyl, a 16S rRNA clone library was established after incubation for 14 days with [^{13}C]biphenyl. The dominant organisms belonged to the genera *Achromobacter* and *Pseudomonas*, and a library built from PCR amplification of genes for aromatic ring dihydroxylating dioxygenases showed the presence of sequences similar to those of *Comamonas testosteroni* B-356 and *Rhodococcus* sp. RHA1. In addition, a cosmid library revealed the presence of functional genes from an uncultivated organism that could degrade PCB.

3. Soil from the root zone of Austrian pine (*Pinus nigra*) in an area in the Czech Republic that had been highly contaminated with PCBs (15 mg/kg) soil was used to establish soil microcosms. Samples were taken at intervals of 14 d for conventional plating using biphenyl in the vapor phase, and for stable isotope probing (SIP) using uniformly labeled ^{13}C biphenyl. DNA was extracted from isolates that were positive for growth on biphenyl and this was analyzed by density gradient centrifugation, while the microbial community was profiled by terminal restriction fragment length polymorphism (Leigh et al. 2007). ^{13}C-labeled CO_2 was evolved successively in the microcosms over 14 d, although only a few genes putatively involved in the degradation of biphenyl and monocyclic arenes were recovered in ^{13}C-DNA. A wide range of bacteria was found that included species of *Sphingomonas* with established PCB degradation, *Pseudonocardia* with established degradation of monocyclic arene carboxylates, and *Nocardia* with established phenanthrene degradation: this is consistent with the degradation of biphenyl and its further degradation. Only a single strain of *Arthrobacter* was identified using both SIP and plate cultivation, and although species of *Rhodococcus* were strongly represented by cultivation, they were not detected by SIP. The consequences of these findings in suggesting interpretation and limitations of the procedures were discussed.

6.13 Stable Isotope Fractionation

This has attracted considerable attention in evaluating the effectiveness of remediation either by natural populations or those enriched by specific additions. Isotopic analysis of the contaminant supplemented with that of established metabolites makes it possible to determine whether degradation and transformation has taken place within the impacted area. It may also be possible to determine the rates at which this has taken place and, sometimes degradation pathways. Results from these experiments have been reported in a number of ways depending on the objective. Illustrative examples include the following:

6.13.1 Application of $\delta^{13}C$

6.13.1.1 Chlorinated Aromatic Hydrocarbons

1. Values of $\delta^{13}C$ have been measured for a number of PCB congeners and applied to a number of commercial PCB mixtures. Both the number and the position of the chlorine substituents affected the depletion of ^{13}C, and this reflected the manufacturing procedures that involved kinetic isotope effects as well as the source of the biphenyl starting material (Jarman et al. 1998). It was suggested that this could be applied to determine the source of PCBs in the environment.

2. An anaerobic bacterial enrichment culture was used to examine the dechlorination of 2,3,4,5-tetrachlorobiphenyl that produced exclusively 2,3,5-trichlorobiphenyl. Although there was no alteration in the values of $\delta^{13}C$, compound-specific analysis of Arochlor 1268 showed that there was a trend for decreasing ^{13}C abundance with increasing content of chlorine. This is consistent with dechlorination of the congeners with more chlorine substituents. It was suggested that reductive dechlorination of PCBs would produce congeners with more depleted values of $\delta^{13}C$ compared with unweathered Arochlors (Drenzer et al. 2001).

3. Values of $\delta^{13}C$ were measured for the degradation of trichlorobenzenes (Griebler et al. 2004). Values for aerobic degradation by *Pseudomonas* sp. strain P51 that carried out degradation by dioxygenation were not significant. In contrast, the isotope enrichment factor (e) for anaerobic dechlorination by *Dehalococcoides* sp. strain CDB1 that produced 1,3-dichlorobenzene from 1,2,3-trichlorobenzene was −3.4‰, and for 1,2,4-trichlorobenzene that produced 1,4-dichlorobenzene was −3.2‰.

6.13.1.2 Methyl Tert-Butyl Ether

Enrichment factors have been used in studies with methyl *tert*-butyl ether that is a fuel additive and is a widespread contaminant in subsurface aquatic systems.

1. In laboratory enrichments with MTBE under both methanogenic and sulfate-reducing conditions, carbon enrichment of -14.4 ± 0.7‰ was found (Somsamak et al. 2006), and in a methanogenic enrichment -14 ± 4.5‰ for *tert*-amyl methyl ether (Somsamak et al. 2005).

2. Both $\delta^{13}C$ and δ^2H were used in an examination of several contaminated groundwater sites (Kuder et al. 2005). These isotopes were extensively fractionated during anaerobic degradation with stable enrichment factors of −13‰ for carbon and −16‰ for hydrogen. These values were used to predict the extent of biodegradation at the sites using the Raleigh model for degradation.

3. In a field investigation of groundwater where there were several sources of MTBE, values of $\delta^{13}C$ ranged from −26‰ to +40‰, and δ^2H from −73.1‰ to +60.3‰ that were consistent with degradation along the plume (Zwank et al. 2005). At the same time, the carbon composition of *tert*-butanol that is produced by demethylation was constant, which showed that this was not further degraded. Comparison with previous measurements and important aspects of the methodology were discussed.

6.13.1.3 Aromatic Hydrocarbons

Investigations have examined the application to monocyclic aromatic hydrocarbons and phenols under both aerobic and anaerobic conditions.

6.13.1.3.1 Aerobic Conditions

1. Two strains for which the mechanism of biodegradation of benzene had not, however, been established were examined for carbon and hydrogen isotope fractionation (Hunkeler et al. 2001). Values for *Acinetobacter* sp. and *Burkholderia* sp. were respectively −1.46‰ and −3.53‰ for carbon and −12.8‰ and −11.2‰ for hydrogen. Values of $\delta^{13}C$ and δ^2H for benzene differed from various sources, and it was pointed out that these differences should be taken into consideration.

2. Fractionation factors (ε) were measured for the aerobic degradation of aromatic hydrocarbons using several strains of bacteria that use different mechanisms for its initiation (Morasch et al. 2002). For strains of *Pseudomonas putida* strain F1 and naphthalene (strain BCIB 9816) that degrade toluene by dioxygenation, values for toluene were not significant. On the other hand, for strains that initiated the degradation of alkylbenzenes by monooxygenation, values for *Pseudomonas putida* strain mt-2 were −3.3‰ for toluene, −1.7‰ for *m*-xylene, and −2.3‰ for *p*-xylene and, for *Ralstonia pickettii* strain PJO1 −1.1‰ for toluene.

6.13.1.3.2 Anaerobic Conditions

1. Under laboratory methanogenic and sulfidogenic conditions, only low values of enrichment factors were observed, although ^{13}C enrichment of toluene was found at late stages of transformation (Ahad et al. 2000).

2. In a continuous-flow headspace system, use was made of d^2H analysis during methanogenic degradation of toluene (Ward et al. 2000). Enrichment values exceeding −60‰ were found, and these are about 10 times greater than the values reported for $\delta^{13}C$. It was suggested that combined use of both $\delta^{13}C$ and δ^2H would be attractive for assessing bioremediation at contaminated sites.

3. Four different conditions for the degradation of toluene were examined in cultures that used different electron acceptors (oxygen, nitrate, ferric iron, or sulfate), and different pathways for degradation (Meckenstock et al. 1999). The fractionation factors for *Pseudomonas putida* (1.0026), *Thauera aromatica* (1.0017), *Geobacter metallireducens* (1.0018), and a sulfate-reducing organism (1.0017) were in the same range for all the organisms.

4. Enrichment factors for both carbon and hydrogen were measured during anaerobic degradation of benzene using different electron acceptors (Mancini et al. 2003). There were no significant differences in the enrichment factors among the cultures, with values for $\delta^{13}C$ from −1.9‰ to −3.6‰, and much larger values for δ^2H (−29‰ to −79‰). This would be consistent with different mechanisms for degradation involving C–H bond fission.

5. Enrichment factors during the anaerobic degradation of *o*-xylene, *m*-xylene, *m*-cresol, and *p*-cresol by pure cultures of sulfate-reducing bacteria that use the fumarate addition reaction ranged from −1.5‰ to −3.9‰ (Morasch et al. 2004). It was, therefore, proposed that this could be applied to evaluating *in situ* bioremediation of contaminants that use the fumarate pathway for biodegradation.

6. A sulfate enrichment culture prepared from a contaminated site gave enrichment factors (ε) of −1.1 for naphthalene and −0.9 for 2-methylnaphthalene (Griebler et al. 2004). These values combined with literature values from analogous laboratory experiments were used to quantify degradation of toluene, xylenes, and naphthalene at the site. Additional evidence for degradation of BTEX was derived from analyses of established metabolites produced by anaerobic degradation.

7. Direct evidence for the biodegradation of benzene and toluene in a contaminated aquifer was lacking, and an alternative strategy was examined. Bio-Sep beads were maintained in tubes and [^{13}C] benzene or [^{13}C]toluene were sorbed on to the surface. Analysis of $\delta^{13}C$ in fatty acids extracted from lipids showed enrichments up to 13,500‰ for benzene and toluene, and significant differences between values of individual fatty acids for the benzene and toluene amendments. It was, therefore, concluded that both benzene and toluene in the aquifer were being degraded, and in addition that different groups of organisms were responsible (Geyer et al. 2005).

8. a. Sulfur fractionation was examined during degradation of toluene by a sulfate-reducing *Desulfobacula toluolica*, a culture PRTOL1, and an enrichment culture. Enrichment factors differed only slightly on the sulfate concentration ranged from $-19.8‰$ for the enrichment to $-47‰$ for PRTOL1 but there was no significant correlation between the ε-values and the initial sulfate concentration (Bolliger et al. 2001).

   b. Examination of sulfur enrichment was examined using the PRTOL1 and the enrichment culture for a range of substrates, including acetate, benzoate, naphthalene, and 1,3,4-trimethylbenzene (Kleikemper et al. 2004). Enrichment factors ($^{34}\varepsilon$) for PRTOL1 ranged from $-30‰$ (benzoate) to $34.5‰$ for the enrichment culture 3,4,5-(1,3,5-trimethylbenzene) and there was no relation between these values and the rates of cell-specific sulfate reduction.

   c. Oxygen and sulfur fractionation was examined during degradation of toluene by a sulfate-reducing *Desulfobacula toluolica* and an enrichment culture from an aquifer contaminated with BTEX. Experiments with the pure strain yielded values of $^{34}\varepsilon$ using $FeSO_4$ that ranged from $-17.5‰$ to $-43.1‰$ while the value using Na_2SO_4 was $-34.2‰$, whereas highly variable values for $^{18}\varepsilon$ were found (Knöller et al. 2006).

6.13.1.4 Chloroalkanes and Chloroalkenes

6.13.1.4.1 Aerobic Conditions

1. For C_1 and C_2 chloroalkanes and chloroalkenes, $\delta^{13}C$ values ranged from -25.58 (trichloroethane) to -58.77 (CH_3Cl) and $\delta^{37}Cl$ values from -2.86 (trichloroethane) to $+1.56$ (CH_2Cl_2). Although for analysis of environmental samples the method has the disadvantage that water must first be removed from the samples, it has been used to determine the distribution of trichloroethene in a contaminated aquifer (Sturchio et al. 1998).

2. During the aerobic degradation of chloroethene (vinyl chloride) by strains of *Mycobacteria* and *Nocardioides*, enrichment factors (ε) lay within the range $-8.2 \pm 0.1‰$ to $-7.0 \pm 0.3‰$ (Chartrand et al. 2005). These values were lower than those for the anaerobic degradation of chloroethene, and values of the kinetic isotope effect were 1.10 ± 0.001 for aerobic degradation, compared with the value of 1.03 ± 0.007 for anaerobic degradation. These are consistent with the pathways in which the aerobic reaction involves the C=C bond compared with the higher mass of the C–Cl bond in the anaerobic reaction.

3. Experiments have been carried out with 1,2-dichloroethane for which the pathways of degradation have been established. For degradation by *Pseudomonas* sp. strain DCA1 in which degradation is initiated by monooxygenation, the enrichment factor was $-3.0‰$, whereas for organisms including *Xanthobacter autotrophicus* in which a dehalogenase initiates degradation values of $-32.3‰$ were found (Hirschorn et al. 2004). This procedure, therefore, differentiated the mechanisms by which degradation took place, and would be applicable to field studies.

4. Carbon isotope fractionation was examined during the aerobic degradation of trichloroethene by *Burkholderia cepacia* strain G4 that possesses toluene monooxygenase activity (Barth et al. 2002). There were substantial differences in values of isotope shifts during degradation, from $-57‰$ to $-17‰$ and, when the data were corrected to correspond to the same amount of substrate reduction, the Releigh enrichment factor was -18.2.

6.13.1.4.2 Anaerobic Conditions

1. A methanogenic enrichment culture was used to determine $\delta^{13}C$ enrichment factors during the dechlorination of trichloroethene (TCE), *cis*-1,2-dichloroethene (*cis*-DCE), and chloroethene (VC) (Bloom et al. 2000). Using the Raleigh model, enrichment factors (ε) were $-6.6‰$ and $2.5‰$ for TCE, $-14.1‰$ and $-16.1‰$ for *cis*-DCE, and $-26.6‰$ and $-21.5‰$ for VC. The large isotope effects that were observed suggested the value of this procedure for evaluating anaerobic dechlorination of chlorinated ethenes.

2. Anaerobic microbial consortia from different sources were used to determine enrichment factors for trichloroethene, *cis*-dichloroethene, and chloroethene (vinyl chloride) during dechlorination (Slater et al. 2001). For one of them (KB-1) that was able to bring about reductive dechlorination using methanol as electron donor, Raleigh enrichment factors values were −13.8, −20.4 and −22.4‰, whereas for tetrachloroethene for which the Raleigh model could not be applied values varied among the consortia. It was suggested that this procedure could provide a valuable indication of dechlorination in spite of the rather wide spread of the enrichment factors that were observed.

3. A site at which earlier attempts at remediation of trichloroethene did not proceed beyond the production of *cis*-dichloroethene was augmented with the KB-1 culture that had already been examined in microcosms. The isotope levels of *cis*-dichloroethene and chloroethene were increased at the final sampling, although that of the continuous input of trichloroethene was essentially constant (Chartrand et al. 2005).

4. A comparison was been made of ^{13}C fractionation during the dechlorination of tetrachloroethene by *Sulfurospirillum multivorans* and *Desulfitobacterium* sp. strain PCE-S in laboratory experiments (Nijenhuis et al. 2005). Isotope fractionation in growing cultures was 1.0052 for *Desulfitobacterium* sp., and only 1.00042 for *Sulfurospirillum multivorans*, while fractionation was greater in crude cell extracts from both strains. It was concluded that caution should, therefore, be exercised in applying fractionation factors to the evaluation of *in situ* bioremediation.

Although isotopes of oxygen, nitrogen, sulfur, and chlorine have been much less extensively used, illustrations include the following.

6.13.1.5 Application of $\delta^{18}O/\delta^{17}O$

1. The values for $\delta^{13}C$ and $\delta^{18}O$ were examined during degradation of diesel oil by a mixed culture under aerobic conditions. Oxygen in the gas phase of closed samples was analyzed by conversion to CO_2 after cryogenic separation. The values for oxygen were particularly valuable in correlating the production of carbon dioxide with the loss in substrate concentration (Aggarwal et al. 1997), and it was, therefore, suggested that this methodology could be used to provide rates of *in situ* biodegradation.

2. Oxygen isotopes have been used to distinguish natural and anthropogenic sources of perchlorate using the signatures of O_2 formed by pyrolysis of the samples. In this application, two important factors had to be taken into consideration: (i) removal of nitrate that would compromise values for O_2 and (ii) to make a correction from the determined values of ^{17}O that are generally obtained from values for $^{18}O = {}^{17}O \approx 0.52 \times \delta^{18}O$. This was used to calculate the ^{17}O anomaly: $\Delta^{17}O = {}^{17}O - 0.52 \times \delta^{18}O$ (Bao and Gu 2004). Values for man-made perchlorate that is produced by electrolysis of chlorate and, therefore, contains oxygen from water had $\delta^{17}O$ values ranging from −9.1‰ to −10.5‰, and of the anomaly $\Delta^{17}O$ from −0.06 to −0.20 compared with values of the anomaly from the Atacama Desert of +4.2 to +9.6‰, although this did not finally resolve the origin of the perchlorate in the Atacama Desert.

6.13.1.6 Application of $\delta^{15}N$

Examples are given to illustrate the application to different mechanisms.

1. Abiotic reduction of nitroarenes was carried out by Fe(II) sorbed on goethite or 8-hydroxynaphtho-1,4-quinone reduced with H_2S. Reduction proceeded successively to the nitroso, hydroxylamine, and amine (Hartenbach et al. 2006). Enrichment factors for nitrogen using Fe(II)/goethite ranged from −29‰ for 4-chloronitrobenzene to −32‰ for 2-methylnitrobenzene, and were independent of the reductant and consistent with the fission of one N−O bond.

2. The metabolism of nitrobenzene was examined in two strains of bacteria—*Comamonas* sp. strain JS765 that degrades the substrate by dioxygenation and *Pseudomonas pseudoalcaligenes* strain JS45 that carried out only partial reduction (Hofstetter et al. 2008). For *Comamonas* strain JS765, the enrichment factors were −3.9‰ for carbon with a kinetic isotope effect of 1.0241 that is consistent with the reaction involving two of the six carbon atoms including a correction for the total number of carbon atoms, and a value for nitrogen of 0.75‰ with a kinetic isotope effect of only 1.0008 that is consistent with the fact that the fission of the $O_2N–C$ bond, that is, loss of nitrite is not the rate-determining event. The enrichment factor for the *Pseudoalcaligenes* strain was −0.57‰ for carbon and −26.6‰ for nitrogen since there is no N–C bond fission, with values comparable to those for abiotic reduction of nitrobenzene.

3. The metabolism of RDX was examined under aerobic and anaerobic conditions (Bernstein et al. 2008). Compound-specific isotope analysis was carried out for reactions in batch cultures, and the enrichment factor for $\delta^{15}N$ was −2.1‰ under aerobic conditions and −5.0‰ under anaerobic conditions. The method also allowed the simultaneous determination of values for $\delta^{18}O$ that were −1.7‰ and −5.3‰ under aerobic and anaerobic conditions.

6.13.1.7 Application of $\delta^{34}S$

1. Determinations of $\delta^{34}S$ have been extensively used in studies on the sulfur cycle, including reactions involving microbial anaerobic reduction of sulfate and thiosulfate (Smock et al. 1998).

2. Isotope fractionation during sulfate reduction by the hyperthermophilic *Archaeoglobus fulgidus* varied with the concentration of sulfate, and it was suggested that different pathways were operative at concentrations >0.6 mM or <0.3 mM (Habicht et al. 2005).

3. Enrichment factors with a range of anaerobic sulfate-reducing bacteria under optimal conditions yielded fractionation factors ranging from −2‰ to −42‰ based on the isotopic composition of sulfate (Detmers et al. 2001). Although values for organisms that completely oxidized the substrate (benzoate, acetate or butyrate) were generally greater than for those that carried out only the incomplete oxidation of lactate, the physiology and metabolic competence of the organisms clearly determined the extent of fractionation.

6.13.1.8 Application of $\delta^{37}Cl$

1. Isotope ratios for ^{13}C and ^{37}Cl have been measured and applied to chloroalkanes and chloroalkenes (Holt et al. 1997). For the C_1 and C_2 compounds, $\delta^{13}C$ values ranged from 25.58 to −58.77, and $\delta^{37}Cl$ values from −2.86 to +1.56. It was suggested that the method could be used to study the fate and distribution of such compounds. The method has, however, the disadvantage that water must first be removed from the sample.

2. Isotope ratios for ^{13}C and ^{37}Cl were measured for the aerobic degradation of dichloromethane by a methanotroph MC8b (Heraty et al. 1999). Values of the fractionation factor (α) were 0.9586 for carbon and 0.9962 for chlorine, and kinetic isotope effects were 1.0424 for carbon and 1.0038 for chlorine.

3. Three anaerobic dechlorinating consortia were used to examine fractionation during dechlorination of tetrachloroethene and trichloroethene to *cis*-dichloroethene (Numata et al. 2002). Fractionation factors (α) for the first reaction ranged from 0.987 to 0.991 for the three consortia, and for the second reaction were 0.9944 for all consortia. Some important limitations were pointed out: (i) the chlorinated ethenes were not separated so that the isotopic composition represented the totality of tetrachloroethene, trichloroethene, and *cis*-dichloroethene and (ii) the chlorine isotopic composition of the trichloroethene that was used differed from the expected sum value for the chlorine atoms.

4. Kinetic isotope effects for the dechlorination by *Xanthobacter autotrophicus* strain GJ10 were 1.0045 for 1,2-dichloroethane and 1.0066 for 1-chlorobutane (Lewandowicz et al. 2001).

5. Carbon and chlorine isotope enrichment factors were examined during the aerobic oxidation of vinyl chlorine by *Nocardioides* sp. strain JS614 and *cis*-1,2-dichloroethene by strain JS666, and anaerobic dechlorination using a commercially available consortium containing strains of *Dehalococcoides*. For oxidation, ε values for carbon were −7.2 and −8.5 with values for chlorine of only −0.3 that is compatible with epoxidation and the absence of C−Cl bond fission. Values for dechlorination were −25.2 and −18.5 for carbon and −1.8 and −1.5 for chlorine. The values for chlorine were unexpectedly low, and the authors suggested that during the rate-limiting step, fission of carbon−chlorine bonds did not take place (Abe et al. 2009).

Collectively, these results illustrate the value of this procedure for determining the occurrence of biodegradation and biotransformation in natural environments, and their application to assessing the effectiveness of bioremediation. A number of important limitations should be addressed.

1. Values may be strictly dependent on the pathway used for degradation, and this must be taken into consideration.

2. For aromatic hydrocarbons and chloroethenes, large differences exist between reactions carried out under aerobic or anaerobic conditions, and between organisms carrying out anaerobic dechlorination. These should be carefully taken into account.

3. For monocyclic aromatic hydrocarbons under aerobic conditions, values may depend on the mechanism of degradation, and the use of values for both ^{13}C and ^{2}H can provide valuable information. Combination with analyses of metabolites established from laboratory investigations provides a valuable complement.

4. When mixtures of chlorinated alkanes are examined, it is essential to consider the range of possible mechanisms that could produce the products, since these may not necessarily reflect values for conventional sources (Hunkeler et al. 2005).

6.14 References

Abe Y, R Aravena, J Zopfi, O Shouaker-Stash, E Cox, JD Roberts, D Hunkeler. 2009. Carbon and chlorine isotope fractionation during aerobic oxidation and reductive dechlorination of vinyl chlorine and *cis*-1,2-dichloroethene. *Environ Sci Technol* 43: 101–107.

Abe Y, D Hunkeler. 2006. Does the Raleigh equation apply to evaluate field isotope data in contaminant hydrogeology? *Environ Sci Technol* 40: 1588–1596.

Aggarwal PK, ME Fuller, MM Gurgas, JF Manning, MA Dillon. 1997. Use of stable oxygen and carbon isotope analyses for monitoring the pathways and rates of intrinsic and enhanced *in situ* biodegradation. *Environ Sci Technol* 31: 590–596.

Ahad JME, BS Lollar, EA Edwards, GF Slater, BE Sleep. 2000. Carbon isotope fractionation during anaerobic biodegradation of toluene: Implications for intrinsic bioremediation. *Environ Sci Technol* 34: 892–896.

Bao H, B Gu. 2004. Natural perchlorate has a unique oxygen isotope signature. *Environ Sci Technol* 38: 5073–5077.

Barth JAC, G Slater, C Schüth, M Bill, A Downey, M Larkin, RM Kalin. 2002. Carbon isotope fractionation during aerobic biodegradation of trichloroethene by *Burkholderia cepacia* G4: A tool to map degradation mechanisms. *Appl Environ Microbiol* 68: 1728–1734.

Bernstein A, Z Ronen, E Adar, R Nativ, H Lowag, W Stichler, RU Meckenstock. 2008. Compound-specific isotope analysis of RDX and stable isotope fractionation during aerobic and anaerobic biodegradation. *Environ Sci Technol* 42: 7772–7777.

Bloom Y, R Aravena, D Hunkler, E Edwards, SK Frape. 2000. Carbon isotope fractionation during microbial dechlorination of trichloroethene, *cis*-1,2-dichloroethene and vinyl chloride: implication for assessment of natural attenuation. *Environ Sci Technol* 34: 2768–2772.

Bolliger C, MH Schroth, SM Bernasconi, J Kleikemper, J Zeyer. 2001. Sulfur isotope fractionation during microbial sulfate reduction by toluene-degrading bacteria. *Geochim Cosmochim Acta* 65: 3289–3298.

Buckley DH, V Huangyutitham, S-F Hsu, TA Nelson. 2007. Stable isotope probing with $^{15}N_2$ reveals novel noncultivated diazotrophs in soil. *Appl Environ Microbiol* 73: 3196–3204.

Chartrand MMG, A Waller, TE Mattes, M Elsner, G Lacrampe-Couloume, JM Gossett, EA Edwards, BS Lollar. 2005. Carbon isotope fractionation during aerobic vinyl chloride degradation. *Environ Sci Technol* 39: 1064–1070.

Detmers J, V Brüchert, KS Habicht, J Kuever. 2001. Diversity of sulfur isotope fractionation by sulfate-reducing prokaryotes. *Appl Environ Microbiol* 67: 888–894.

Drenzer NJ, TI Eglinton, CO Wirsen, HD May, Q Wu, KR Sowers, CM Reddy. 2001. The absence and application of stable carbon isotope fractionation during the reductive dechlorination of polychlorinated biphenyls. *Environ Sci Technol* 35: 3310–3313.

Geyer R, AD Peacock, A Miltner, H-H Richnow, DC White, KL Sublette, M Kästner. 2005. *In situ* assessment of biodegradation potential using biotraps amended with ^{13}C-labeled benzene or toluene. *Environ Sci Technol* 39: 4983–4989.

Ginige MP, P Hugenholtz, H Daims, M Wagner, J Keller, LL Blackall. 2004. Use of stable-isotope probing, full-cycle rRNA analysis and fluorescence in situ hybridization-microautoradiography so study a methanol-fed denitrifying microbial community. *Appl Environ Microbiol* 70: 588–596.

Griebler C, M Safinowski, A Vieth, HH Richnow, RU Meckenstock. 2004. Combined application of stable carbon isotope analysis and specific metabolites determination for assessing *in situ* degradation of aromatic hydrocarbons in a tar oil-contaminated aquifer. *Environ Sci Technol* 38: 617–631.

Habicht KS, L Salling, B Thamdrup, DE Canfield. 2005. Effect of low sulfate concentrations on lactate oxidation and isotope fractionation during sulfate reduction by *Archaeoglobus fulgidus* strain Z. *Appl Environ Microbiol* 71: 3770–3777.

Hartenbach A, TB Hofstetter, M Berg, J Bolotin, RP Schwarzenbach. 2006. Using nitrogen isotope fractionation to assess abiotic reduction of nitroaromatic compounds. *Environ Sci Technol* 40: 7710–7716.

Hayes JM. 1993. Factors controlling ^{13}C contents of sedimentary organic compounds: Principles and evidence. *Mar Geol* 113: 111–125.

Heraty LJ, ME Fuller, L Huang, T Abrajano, NC Sturchio. 1999. Isotope fractionation of carbon and chlorine by microbial degradation of dichloromethane. *Org Geochem* 30: 793–799.

Hirschorn SK, MJ Dinglasan, M Elsner, SA Mancini, G Lacrampe-Couloume, EA Edwards, BS Lollar. 2004. Pathway dependent isotopic fractionation during aerobic biodegradation of 1,2-dichloroethane. *Environ Sci Technol* 38: 4775–4781.

Hofstetter TB, JC Spain, SF Nishino, J Bolotin, RP Schwarzenbach. 2008. Identifying competing aerobic nitrobenzene biodegradation pathways by compound-specific isotope analysis. *Environ Sci Technol* 42: 4764–4770.

Holt BD, NC Sturchio, TA Abrajano, LJ Heraty. 1997. Conversion of chlorinated volatile organic compounds to carbon dioxide and methyl chloride for isotopic analysis of carbon and chlorine. *Anal Chem* 69: 2727–2733.

Hunkeler D. 2002. Quantification of isotope fractionation in experiments with deuterium-labeled substrate. *Appl Environ Microbiol* 68: 5205–5207.

Hunkeler D, N Andersen, R Aravena, SM Bernasconi, BJ Butler. 2001. Hydrogen and carbon isotope fractionation during aerobic biodegradation of benzene. *Environ Sci Technol* 35: 3462–3467.

Hunkeler D, R Aravena, K Berry-Spark, E Cox. 2005. Assessment of degradation pathways in an aquifer with mixed chlorinated hydrocarbon contamination using stable isotope analysis. *Environ Sci Technol* 39: 5975–5981.

Jarman WM, A Hilkert, CE Bacon, JW Collister, K Ballschmiter, RW Risebrough. 1998. Compound-specific carbon isotopic analysis of Arochlors, Clophens, Kanechlors, and Phenoclors. *Environ Sci Technol* 32: 833–836.

Jeon CO, W Park, P Padmanabhan, C DeRito, JR Snape, EL Madsen. 2003. Discovery of a bacterium, with distinctive dioxygenase, that is responsible for *in situ* biodegradation in contaminated sediment *Proc Natl Acad Sci USA* 100: 13591–13596.

Kasai Y, Y Takahata, M Manefield, K Watanabe. 2006. RNA-based stable isotope probing and isolation of anaerobic benzene-degrading bacteria from gasoline-contaminated groundwater. *Appl Environ Microbiol* 72: 3586–3592.

Kleikemper J, MH Schroth, SM Bernasconi, B Brunner, J Zeyer. 2004. Sulfur isotope fractionation during growth of sulfate-reducing bacteria on various carbon sources. *Geochim Cosmochim Acta* 68: 4891–4904.

Knöller K, C Vogt, H-H Richnow, SM Weise. 2006. Sulfur and oxygen isotope fractionation during benzene, ethyl benzene, and xylene degradation by sulfate-reducing bacteria. *Environ Sci Technol* 40: 3779–3885.

Kucklick JR, HR Harvey, PH Ostrom, NE Ostrom, JE Baker. 1996. Organochlorine dynamics in the pelagic food web of Lake Baikal. *Anal Chim Acta* 15: 1388–1400.

Kuder T, JT Wilson, P Kaiser, R Kolhatkar, P Philp, J Allen. 2005. Enrichment of stable carbon and hydrogen isotopes during anaerobic biodegradation of MTBE: Microcosm and field evidence. *Environ Sci Technol* 39: 213–220.

Kunapuli U, T Lueders, RU Meckenstock. 2007. The use of stable isotope probing to identify key iron-reducing microorganisms involved in anaerobic benzene degradation. *ISME J* 1: 643–653.

Leigh MB, VH Pellizari, O Uhlík, R Sutka, J Rodrigues, NE Ostrom, J Zhou., JM Tiedje. 2007. Biphenyl-utilizing bacteria and their functional genes in a pine root zone contaminated with polychlorinated biphenyls (PCBs). *ISME J* 1: 134–148.

Lewandowicz A, J Rudzinski, L Tronstad, M Widersten, P Ryberg, O Matsson, P Paneth. 2001. Chlorine isotope effects on the haloalkane dehalogenase reaction. *J Am Chem Soc* 123: 4550–4555.

Lu Y, T Lueders, MW Friedrich, R Conrad. 2005. Detecting active methanogenic populations on rice roots using stable isotope probing. *Environ Microbiol* 7: 326–336.

Lueders T, B Pommerenke, MW Friedrich. 2004b. Stable-isotope probing of microorganisms thriving at thermodynamic limits: Syntrophic propionate oxidation in flooded soil. *Appl Environ Microbiol* 70: 5778–5786.

Lueders T, B Wagner, P Claus, MW Friedrich. 2004a. Stable isotope probing of rRNA and DNA reveals a dynamic methylotroph community and trophic interactions with fungi and protozoa in oxic rice field soil. *Environ Microbiol* 6: 60–72.

Mancini SA, AC Ulrich, G Lacrampe-Couloume, B Sleep, EA Edwards, BS Lollar. 2003. Carbon and hydrogen isotope fractionation during anaerobic biodegradation of benzene. *Appl Environ Microbiol* 69: 191–198.

Manefield M, AS Whiteley, RI Griffiths, MJ Bailey. 2002. RNA stable isotope probing, a novel means of linking microbial community function to phylogeny *Appl Environ Microbiol* 68: 5367–5373.

Martineau C, LG Whyte, CW Greer. 2010. Stable isotope probing analysis of the diversity and activity of methanotrophic bacteria in soils from the Canadian High Arctic. *Appl Environ Microbiol* 76: 5773–5784.

Meckenstock RU, B Morasch, R Warthmann, B Schink, E Annweiler, W Michaelis, HH Richnow. 1999. $^{13}C/^{12}C$ isotope fractionation of aromatic hydrocarbons during microbial degradation. *Environ Microbiol* 1: 409–425.

Merritt DA, KH Freeman, MP Ricci, SA Studley, JM Hayes. 1995. Performance and optimization of a combustion interface for isotope ratio monitoring gas chromatography/mass spectrometry. *Anal Chem* 67: 2461–2473.

Morasch B, HH Richnow, B Schink, A Vieth, RU Meckenstock. 2002. Carbon and hydrogen stable isotope fractionation during aerobic bacterial degradation of aromatic hydrocarbons. *Appl Environ Microbiol* 68: 5191–5194.

Morasch B, HH Richnow, A Vieth, B Schink, RU Meckenstock. 2004. Stable isotope fractionation caused by glycyl radical enzymes during bacterial degradation of aromatic compounds. *Appl Environ Microbiol* 70: 2935–2940.

Morris SA, S Radajewski, TW Willison, JC Murrell. 2002. Identification of the functionally active methanotroph population in a peat soil microcosm by stable-isotope probing *Appl Environ Microbiol* 68: 1446–1453.

Nijenhuis I, J Andert, K Beck, M Kästner, G Diekert, H-H Richow. 2005. Stable isotope fractionation of tetrachloroethene during reductive dechlorination by *Sulfospirillum multivorans* and *Desulfitobacterium* sp. strain PCE-S and abiotic reactions with cyanocobalamin *Appl Environ Microbiol* 71: 3413–3419.

Numata M, N Nakamura, H Koshikawa, Y Terashima. 2002. Chlorine isotope fractionation during reductive dechlorination of chlorinated ethenes by anaerobic bacteria. *Environ Sci Technol* 36: 4389–4394.

Oka AR, CD Phelps, LM McGuinness, A Mumford, LY Young, LJ Kerkhof. 2008. Identification of critical members in a sulfidogenic benzene-degrading consortium by DNA stable isotope probing. *Appl Environ Microbiol* 74: 6476–6480.

Radajewski S, G Webster, DS Reay, SA Morris, P Ineson, DB Nedwell, JI Prosser, JC Murrell. 2002. Identification of active methylotroph populations in an acidic forest soil by stable-isotope probing. *Microbiology (UK)* 148: 2331–2341.

Redmond MC, DL Valentine, AL Sessions. 2010. Identification of novel methane-, ethane-, and propane-oxidizing bacteria at marine hydrocarbon seeps by stable isotope probing. *Appl Environ Microbiol* 76: 6412–6422.

Saurer M, I Robertson, R Siegwolf, M Leuenberger. 1998. Oxygen isotope analysis of cellulose: An interlaboratory comparison. *Anal Chem* 70: 2074–2080.

Slater GF, BS Lollar, BE Sleep, EA Edwards. 2001. Variability in carbon isotope fractionation during biodegradation of chlorinated ethenes: Implications for field applications. *Environ Sci Technol* 35: 901–907.

Smock AM, ME Böttcher, H Cypionka. 1998. Fractionation of sulfur isotopes during thiosulfate reduction by *Desulfovibrio desulfuricans. Arch Microbiol* 169: 460–463.

Somsamak P, HH Richnow, MM Häggblom. 2005. Carbon isotope fractionation during anaerobic biotransformation of methyl *tert*-butyl ether and *tert*-amyl methyl ether. *Environ Sci Technol* 39: 103–109.

Somsamak P, HH Richnow, MM Häggblom. 2006. Carbon isotope fractionation during anaerobic degradation of methyl *tert*-butyl ether under sulfate-reducing and methanogenic conditions. *Appl Environ Microbiol* 72: 1157–1163.

Sturchio NC, JL Clausen, IJ Heraty, L Huang, BD Holt, TA Abrajano. 1998. Chlorine isotope investigation of natural attenuation of trichloroethene in an aerobic aquifer. *Environ Sci Technol* 32: 3037–3042.

Sul WS, J Park, JF Quensen, JLM Rodrigues, L Seliger, TV Tsoi, GJ Zylstra, JM Tiedje. 2009. DNA-stable isotope probing integrated with metagenomics for retrieval of biphenyl dioxygenase genes from polychlorinated biphenyl-contaminated river sediment. *Appl Environ Microbiol* 75: 5501–5506.

Uhlik O, K Jeena, M Mackova, C Vlcek, M Hroudova, K Demnerova, V Paces, T Macek. 2009. Biphenylmetabolizing bacteria in the rhizosphere of horseradish and bulk soil contaminated by polychlorinated biphenyls as revealed by stable isotope proving. *Appl Environ Microbiol* 75: 6371–6477.

Ward JAM, JME Ahad, G Lacrampe-Couloume, GF Slater, EA Edwards, BS Lollar. 2000. Hydrogen isotope fractionation of toluene: Potential for direct verification of bioremediation. *Environ Sci Technol* 34: 4577–4581.

Westaway KC, T Koerner, Y-R Fang, J Rudzinski, P Paneth. 1998. A new method of determining chlorine kinetic isotope effects. *Anal Chem* 70: 3548–3552.

Willing A, H Follmann, G Auling. 1988. Ribonucleotide reductase of *Brevibacterium ammoniagenes* is a manganese enzyme. *Eur J Biochem* 170: 603–611.

Zwank L, M Berg, M Elsner, TC Schmidt, RP Schwarzenbach, SB Haderlien. 2005. New evaluation scheme for two-dimensional isotope analysis to decipher biodegradation processes: Application to groundwater contamination by MTBE. *Environ Sci Technol* 39: 1018–1029.

Part 3: Physical Methods of Structure Determination

Considerable attention has been directed to physical methods of structure determination that have become increasingly sophisticated. This is not the place to discuss these developments, and the following brief sections attempt only to provide examples that attempt to illustrate and highlight their application to studies of biodegradation and biotransformation. In many cases, a combination of several methods has been valuably employed. In most applications, the sample can be recovered after analysis.

6.15 Nuclear Magnetic Resonance

NMR in its various forms has been increasingly used. This is largely the result of technical developments that are briefly summarized in the following paragraphs:

1. During the years since its introduction, there have been substantial increases in the magnetic field strengths that are attainable. Greater sensitivity and resolution are thereby achieved with modern instruments, so that sample size has become less significant—provided that the sample is of adequate purity. Application to naturally occurring levels of ^{13}C has become routine, and

examination of compounds with ^{15}N, [17]O, and ^{19}F has been reported. NMR has been increasingly used for *in vivo* investigations of microbial degradation, since samples containing cell suspensions and the ^{13}C-labeled substrates can be examined directly, and the kinetics of the appropriate resonance signals monitored. A ^{13}C NMR method has been developed to determine the transformation and degradation of chloroacetonitrile and related compounds in various samples of soil and water (Castro et al. 1996).

2. The size of the sample required has been reduced by a number of technical developments including micro inverse probes and micro cells (references in Martin et al. 1998), and has been reduced even further using a newly developed 1.7 mm submicro inverse-detection gradient probe (Martin et al. 1998). Combined use of inverse-detection probes with solenoid microcoils has also been developed to reduce sample volumes for ^{13}C NMR (Subramanian and Webb 1998).

3. Multidimensional spectra have been increasingly used, as well as techniques including DEPT (distortionless enhancement by polarization transfer), COSY (correlated spectroscopy), and ROESY (rotating-frame Overhauser enhancement spectroscopy).

4. Mass spectra are often carried out after HPLC separation of components, and when soft ionization techniques are used, they provide the molecular mass, but relatively little structural information. The combined application of an interface to NMR and to MS would provide a powerful structural tool, and for some mass spectrometry methods that provide limited structural information this is a particularly attractive combination. Identification by NMR provides detailed structural information, and a system for the analysis of peptides involves initial HPLC followed by splitting of the fractions for subsequent MS and NMR analysis in a flow-through system (Holt et al. 1997). Application to environmental samples after fractionation is attractive, and a configuration has been developed for combining capillary zone electrophoresis, capillary HPLC and capillary electrochromatography with NMR including use in stopped-flow experiments (Pusecker et al. 1998). Direct coupling to LC systems has been achieved, and application to whole cells has been possible in a number of applications. Both the continuous-flow and stop-flow modes have proved valuable, and a combination of continuous-flow HPLC interfaced with ^{1}H NMR and HPLC interfaced to thermospray mass spectrometry has been used to identify the products from the photolysis of 2,4,6-trinitrotoluene (Godejohann et al. 1998). The application of NMR analysis coupled to high-performance liquid chromatography and supercritical fluid procedures has been made possible by a radical change in the design of the NMR instrument (Albert 2002), and further technical developments are likely to offer a wider range of applications. The procedure has been valuable for identifying urinary metabolites of pharmaceuticals and has used both continuous-flow and stopped-flow-methods (Sidelmann et al. 1997).

There are many advantages of NMR, including the fact that it is nondestructive. It may be carried out not only in extracts after purification by conventional means, but also directly in culture supernatants, and even during their synthesis in the spectrometer tubes. These may be equipped with gas inlets which make it possible to carry out studies under virtually any metabolic conditions. In addition, there are no experimental restrictions in handling radioactive material, although for studies of carbon metabolism, access to ^{13}C-labeled substrates is advantageous. Increasingly, however, many of these are no longer less accessible than their ^{14}C analogs. On the other hand, the relatively low sensitivity of the method may preclude identification of metabolites which are formed transiently only in low concentration.

6.15.1 Hydrogen [1]H

This has received widespread application, and a single example has been chosen as representative. The pathway for the degradation of morpholine by *Mycobacterium aurum* MO1 and *Mycobacterium* strain RP1 was examined using whole cells, and this confirmed its identity to the one that had been proposed earlier for *M. chelonae* (Combourieu et al. 1998; Poupin et al. 1998).

6.15.2 Carbon ^{13}C

For reasons given in the introductory paragraphs, NMR has become a standard procedure for structure determination and the elucidation of metabolic pathways. Only a few general comments supplemented by some illustrations are, therefore, given.

Interpretative difficulties may arise from the inherent design of NMR experiments that might necessitate the use of high substrate concentrations due to the relatively low sensitivity of the procedure. A single illustrative example is given of the occurrence of the artifacts that may be encountered. During a study of the metabolism of mandelate by *Pseudomonas putida* (Halpin et al. 1981), benzyl alcohol was unexpectedly identified when experiments were carried out at high substrate concentration (50 mM). This was, however, subsequently shown to be due to the action of a nonspecific alcohol dehydrogenase under the anaerobic conditions prevailing at the high substrate concentration used for the identification of the metabolites (Collins and Hegeman 1984).

A single example is given to illustrate both the strengths and the limitations of NMR. During the metabolism of 2-^{13}C-acetate in methylotrophic strains of *Pseudomonas* sp., it was shown that the substrate was converted into α,α-trehalose in isocitrate lyase-negative strains, although not in one which synthesized this enzyme. In addition, an unknown compound was revealed by the *in vitro* experiments, but was not present in perchloric acid extracts of the cells. Possibly more disturbing, however, was the fact that the analysis with a strain during growth with ^{13}C-methanol did not reveal the presence of intermediates known to be part of the serine pathway that functions in this organism (Narbad et al. 1989).

^{13}C NMR using whole cells has been applied to the study of a number of metabolic reactions involving small molecules that include the following:

1. Metabolism of acetate and methanol by *Pseudomonas* sp. (Narbad et al. 1989)
2. Combined use of ^1H and ^{13}C NMR was used in a powerful combination to deduce the degradation pathways of chloroethene (vinyl chloride) (Castro et al. 1992a,b)
3. Incorporation of $^{13}CO_2$ into poly-β-hydroxybutyrate by a strain of *Xanthomonas* sp. that metabolizes propene or its oxide (Small and Ensign 1995; Allen and Ensign 1996)
4. Degradation of acetonitrile by *Methylosinus trichosporium* (Castro et al. 1996)
5. Metabolism of $^{13}CCl_4$ by *Pseudomonas* sp. strain KC (Lewis and Crawford 1995; Lewis et al. 2001)

Examples of more complex metabolic pathways include the following:

1. Detection of glycolate and 2-(2-aminoethoxy)acetate as intermediates in the degradation of morpholine by *Mycobacterium aurum* strain MO1 (Combourieu et al. 1998).
2. The degradation of phenylacetate had remained enigmatic for several years, and details of the pathway used by *Escherichia coli* were elucidated using the ^{13}C-labeled substrate. Using the full complement of NMR technology, the structures of critical intermediates were determined and provided details of an unusual dioxygenation pathway (Ismail et al. 2003).
3. The degradation of acenaphthene is initiated by benzylic monooxygenation, and the pathway was determined using [1-^{13}C]acenaphthene by isolation of intermediate metabolites (Selifonov et al. 1998). Importantly, the method proved applicable even when only limited biotransformation of the substrates had taken place by partial oxidation.

The procedure is particularly suited to the study of anaerobic transformations since there are no problems resulting from problems with oxygen limitation. Two illustrative examples are given.

1. Application of ^{13}C NMR was used to elucidate the intricate relations of fumarate, succinate, propionate, and acetate metabolism in a syntrophic organism, both in the presence (Houwen et al. 1991) and in the absence of methanogens (Plugge et al. 1993; Stams et al. 1993).

2. The complex pathway for the anaerobic degradation of propionate by *Smithella propionica* and *Methanospirillum hungatei* involves reaction of two molecules of propionate followed by rearrangement to 3-oxohexanoate. The details were elucidated using ^{13}C-propionate labeled at C_1, C_2, C_3, or at both C_2 and C_3 (de Bok et al. 2001).

6.15.3 Nitrogen ^{15}N

Apart from its use in studies of nitrogen fixation, ^{15}N-labeled substrates have been used to assess the association of contaminants with organic matter in soil, and in synthesis and metabolism.

6.15.3.1 Association of Xenobiotics with Soil Organic Material

1. ^{15}N NMR has been used in conjunction with ^{13}C NMR in studies on reactions of [^{15}N] hydroxylamine with fulvic and humic substances (Thorn et al. 1992).
2. The availability of ^{15}N aniline has made possible a direct study of the reactions of aniline with humic and fulvic acids (Thorn et al. 1996), and the detection of resonances attributed to anilinoquinone, imines and *N*-heterocyclic compounds are fully consistent with reactions involving quinone and ketone groups.
3. The ^{15}N-labeled amines produced by partial and total reduction of the nitro groups in 2,4,6-trinitrotoluene reacted with carbonyl groups (quinones and ketones) in humic acid to produce a range of products (Thorn and Kennedy 2002).

6.15.3.2 Synthesis and Metabolism

1. The metabolism of methylamine by *Pseudomonas* sp. strain MA was studied using combined ^{13}C and ^{15}N NMR (Jones and Bellion 1991). This organism uses a complex pathway for the degradation of methylamine involving *N*-methylglutamate and γ-glutamylmethylamide, and these could be identified at different rates of oxygenation.
2. The degradation of benzothiazole was examined in two strains of *Rhodococcus*. The structure of the metabolite as 2,6-dihydroxybenzothiazole was determined by long-range ^{1}H-^{15}N gradient heteronuclear multiple-bond correlation without ^{15}N enrichment (Besse et al. 2001).
3. The metabolism of 2,4,6,8,10,12-hexanitro-2,4,6,8,10,12-hexaazaisowurtzitane (CL-20) labeled in the ring atoms with ^{15}N was examined using salicylate 1-monooxygenase from *Pseudomonas* sp. strain FA1, and it was possible to propose a pathway (Bhushan et al. 2004).
4. The biosynthesis of the plant pathogen thaxtomin by *Streptomyces turgidiscabies* was examined using L-arginine-guanidino-$^{15}N_2$ and the ^{15}N NMR spectrum showed a single resonance at 375 ppm that could be assigned to the 4-nitro group on indole, whereas when $^{15}NH_4NO_3$ was used, the ^{15}N NMR spectrum showed three resonances at 375 ppm, 132 ppm (indole N), and 125 and 114 ppm (*N*-methyl amide N) (Kers et al. 2004).

6.15.4 Oxygen ^{17}O

Application of ^{17}O NMR has been made to a restricted range of chlorinated aromatic compounds (Kolehmainen et al. 1992), and has been used to establish the source of oxygen in the metabolites produced from ^{17}O acetate and $^{17}O_2$ by *Aspergillus melleus* (Staunton and Sutkowski 1991).

6.15.5 Fluorine ^{19}F

Application of NMR to fluorine-containing molecules is particularly attractive since naturally occurring fluorine is monoisotopic, and since the range of chemical shifts in fluorine compounds is very much greater than the proton shifts for hydrogen-containing compounds. Although only a few examples are given as illustration, there is a vast potential for the application of ^{19}F NMR to metabolic studies of fluorine compounds. The principles and applications have been given in a substantive review (Stanley 2002).

1. Substantial attention has been devoted to the metabolism of 5-fluorouracil and related compounds. For example, ^{19}F NMR was used successfully both in cell extracts and in whole mycelia to elucidate anabolic reactions involving pyrimidine nucleotides and degradation to α-fluoro-β-alanine in the fungus *Nectria haematococca* (Parisot et al. 1989, 1991).

2. Fluoroacetate and 4-fluorothreonine are synthesized from fluoride by *Streptomyces cattleya*, and analysis of supernatants was used to elucidate the details of their biosynthesis. They were apparently synthesized by independent routes, and it was suggested that what is formally glycolate could be their precursor (Reid et al. 1995).

3. Oxidative decarboxylation of hydroxybenzoates by the yeast *Candida parapsilosis* is catalyzed by a flavin monooxygenase that is able to use a range of fluorinated hydroxybenzoates that were examined by ^{19}F NMR (Eppink et al. 1997).

4. ^{19}F NMR was used to study the effect of pH on the hydroxylation of 3-fluorophenol by the hydroxylase from *Trichosporon cutaneum*, and revealed that the ratio of products as well as the yield was pH dependent (Peelen et al. 1993).

5. The metabolism of 3,5- and 2,5-difluorobenzoate was studied in a mutant of *Pseudomonas putida* PpJT103 that is unable to aromatize the 1,2-dihydrodiol of benzoate. ^{19}F NMR was used to establish that in both 3,5- and 2,5-difluorobenzoate dioxygenation took place at C_1 and C_2, and that for the latter, the dihydrodiol lost fluoride to produce 4-fluorocatechol (Cass et al. 1987).

6. Dioxygenation of 2,3-difluoro- and 2,5-difluorohydroquinone was examined by ^{19}F NMR and demonstrated the occurrence of 1:6 distal fission with the formation of products tentatively identified as difluoro-4-hydroxymuconic semialdehydes (Moonen et al. 2008).

7. The metabolism of fluorophenols by phenol hydroxylase from *Trichosporium cutaneum*, catechol 1,2-dioxygenase from *Pseudomonas arvilla* strain C-1, and by the fungus *Exophilia jeanselmei* has been examined, and detailed NMR data were given for the ring fission fluoromuconates (Boersma et al. 1998).

8. A ^{19}F NMR study used several species of *Rhodococcus*, and a range of monofluoro-, difluoro-, and trifluorophenols. Both ring hydroxylation and hydrolytic defluorination were observed, followed by intradiol ring fission (Bondar et al. 1998).

This methodology is worthy of wider exploitation in the study of other groups of organofluorine compounds that are of industrial importance as agrochemicals and pharmaceuticals and fluorinated compounds such as polyfluorinated octanoates and sulfonates that have awakened increasing environmental interest.

6.15.6 Phosphorus ^{31}P

There has been only limited interest in the catabolism of organophosphorus compounds, although considerable attention has been directed to anabolic reactions. One example is given to illustrate the strengths and limitations of the technique. ^{31}P was used to examine the effect of ethanol on the metabolism of glucose by *Zymomonas mobilis*. Whereas the sensitivity was sufficient to establish changes in nucleoside triphosphates during *in vivo* experiments, details of the changes in the various phosphorylated metabolites necessitated the use of perchloric acid cell extracts (Strohhäcker et al. 1993). Illustrative examples include the following:

1. Nucleotide pools and transmembrane potential in bacteria after exposure to pentachlorophenol were investigated using ^{31}P NMR. Differences were used to differentiate *Escherichia coli* which does not degrade this substrate and a *Flavobacterium* sp. which is able to do so (Steiert et al. 1988).

2. Two strains of *Sphingomonas* sp. that could degrade pentachlorophenol maintained their levels of ATP even in the presence of high concentrations of pentachlorophenol. Analysis of the lipids using ^{31}P NMR showed that this could be attributed to the increased levels of cardiolipin (Lohmeier-Vogel et al. 2001).

Escherichia coli has a unique a bacterial alkaline phosphatase (BAP) that is able to catalyze the oxidation of phosphite to phosphate and molecular H_2, and ^{31}P NMR was used to demonstrate that this took place without the intervention of other phosphorus compounds (Yang and Metcalf 2004).

6.15.7 Silicon ^{29}Si

^{29}Si NMR of hexamethyldisiloxane has been examined using a 750 MHz ^{1}H resonance frequency, and a number of technical issues were discussed (Knight and Kinrade 1999). The use of chromium^{III} acetylacetonate was used to reduce the ^{29}Si relaxation time with a DEPT-45 pulse sequence. This indicated the potential application to the metabolism of organosilicon compounds.

6.16 References

Albert K. 2002. 1 LC-NMR: Theory and experiment, pp. 1–22, In: Albert K (Ed.) *On-Line LC-NMR and Related Techniques*. Wiley and Sons, Ltd., Chichester, U.K.

Allen JR, SA Ensign. 1996. Carboxylation of epoxides to β-keto acids in cell extracts of *Xanthobacter* strain Py2. *J Bacteriol* 178: 1469–1472.

Besse P, B Combourieu, G Boyse, M Sancelme, H de Wever, A-M Delort. 2001. Long-range ^{1}H-^{15}N heteronuclear shift correlation at natural abundance: A tool to study benzothiazole biodegradation by two *Rhodococcus* strains. *Appl Environ Microbiol* 67: 1412–1417.

Bhushan B, A Halasz, J Spain, J Hawari. 2004. Initial reaction(s) in biotransformation of CL-20 is catalyzed by salicylate 1-monooxygenase from *Pseudomonas* sp strain ATCC 29352. *Appl Environ Microbiol* 70: 4040–4047.

Boersma MG, TY Dinarieva, WJ Middelhoven, WJH van Berkel, J Doran, J Vervoort, IM-CM Rietjens. 1998. ^{19}F nuclear magnetic resonance as a tool to investigate microbial degradation of fluorophenols to fluorocatechols and fluoromuconates. *Appl Environ Microbiol* 64: 1256–1263.

Bondar VS, MG Boersma, EL Golovlev, J Vervoort, WJH van Berkel, ZI Finkelstein, IP Solyanikova, LA Golovleva, IMCM Rietjens. 1998. ^{19}F NMR study on the biodegradation of fluorophenols by various *Rhodococcus species*. *Biodegradation* 9: 475–486.

Boyd JM, H Ellsworth, SA Ensign. 2004. Bacterial acetone carboxylase is a manganese-dependent metalloenzyme. *J Biol Chem* 279: 46644–46651.

Cass AEG, DW Ribbons, JT Rossiter, SR Williams. 1987. Biotransformation of aromatic compounds. Monitoring fluorinated analogues by NMR. *FEBS Lett* 220: 353–347.

Castro CE, DM Riebeth, NO Belser. 1992a. Biodehalogenation: The metabolism of vinyl chloride by *Methylosinus trichosporium* OB-3b. A sequential oxidative and reductive pathway through chloroethylene oxide. *Environ Toxicol Chem* 11: 749–755.

Castro CE, RS Wade, DM Riebeth, EW Bartnicki, NO Belser. 1992b. Biodehalogenation: Rapid metabolism of vinyl chloride by a soil *Pseudomonas* sp. Direct hydrolysis of a vinyl C-Cl bond. *Environ Toxicol Chem* 11: 757–764.

Castro CE, SK O'Shea, W Wang, EW Bartnicki. 1996. Biodehalogenation: Oxidative and hydrolytic pathways in the transformations of acetonitrile, chloroacetonitrile, chloroacetic acid, and chloroacetamide by *Methylosinus trichosporium* OB-3b. *Environ Sci Technol* 30: 1180–1184.

Collins J, G Hegeman. 1984. Benzyl alcohol metabolism by *Pseudomonas putida*: A paradox resolved. *Arch Microbiol* 138: 153–160.

Combourieu B, P Besse, M Sancelme, H Veschambre, AM Delort, P Poupin, N Truffaut. 1998. Morpholine degradation pathway of *Mycobacterium aurum* MO1: Direct evidence of intermediates by *in situ* ^{1}H nuclear magnetic resonance. *Appl Environ Microbiol* 64: 153–158.

de Bok F, AAJ Stams, C Dijkema, DR Boone. 2001. Pathway of propionate oxidation by a syntrophic culture of *Smithella propionica* and *Methanospirillum hungatei*. *Appl Environ Microbiol* 67: 1800–1804.

Eppink MHM, SA Boeren, J Vervoort, WJH van Berkel. 1997. Purification and properties of 4-hydroxybenzo-ate 1-hydroxylase (decarboxylating), a novel flavin adenine dinucleotide-dependent monooxygenase from *Candida parapilosis* CBS604. *J Bacteriol* 179: 6680–6687.

Godejohann M, M Astratov, A Preiss, K Levsen, C Mügge. 1998. Application of continuous-flow HPLC-proton-nuclear magnetic resonance spectroscopy and HPLC-thermospray spectroscopy for the structural elucidation of phototransformation products of 2,4,6-trinitrotoluene. *Anal Chem* 70: 4104–4110.

Halpin RA, GD Hegeman, GL Kenyon. 1981. Carbon-13 nuclear magnetic resonance studies of mandelate metabolism in whole bacterial cells and in isolated, *in vivo* cross-linked enzyme complexes. *Biochemistry* 20: 1525–1533.

Holt BD, NC Sturchio, TA Abrajano, LJ Heraty. 1997. Conversion of chlorinated volatile organic compounds to carbon dioxide and methyl chloride for isotopic analysis of carbon and chlorine *Anal Chem* 69: 2727–2733.

Houwen FP, C Dijkema, AJM Stams, AJB Zehnder. 1991. Propionate metabolism in anaerobic bacteria; determination of carboxylation reactions with ^{13}C-NMR spectroscopy. *Biochim Biophys Acta* 1056: 126–132.

Ismail W, M El-Said Mohamed, BL Wanner, KA Datsenko, W Eisenreich, F Rohdich, A Bacher, G Fuchs. 2003. Functional genomics by NMR spectroscopy Phenylacetate catabolism in *Escherichia coli*. *Eur J Biochem* 270: 3047–3054.

Jiang Z-H, DS Argyropoulos, A Granata. 1995. Correlation analysis of ^{31}P NMR chemical shifts with substituent effects of phenols. *Magn Reson Chem* 33: 375–382.

Jones JG, E Bellion. 1991. *In vivo* ^{13}C and ^{15}N NMR studies of methylamine metabolism in *Pseudomonas* species MA. *J Biol Chem* 266: 11705–11713.

Kers JA, MJ Wach, SB Krasnoff, J Widom, KD Cameron, RA Bukhalid, DM Gibson, BR Crane, R Loria. 2004. Nitration of a peptide phytotoxin by bacterial nitric oxide synthase. *Nature* 429: 79–82.

Knight CTG, SD Kinrade. 1999. Silicon-29 nuclear magnetic resonance spectroscopy detection limits. *Anal Chem* 71: 265–267.

Kolehmainen E, K Laihia, J Knuutinen, J Hyötyläinen. 1992. ^{1}H, ^{13}C and ^{17}O NMR study of chlorovanillins and some related compounds. *Magn Reson Chem* 30: 253–258.

Lewis TA, RL Crawford. 1995. Transformation of carbon tetrachloride via sulfur and oxygen substitution by *Pseudomonas* sp. strain KC. *J Bacteriol* 177: 2204–2208.

Lewis TA, A Paszczynski, SW Gordon-Wylie, S Jeedigunta, C-H Lee, RL Crawford. 2001. Carbon tetrachloride dechlorination by the bacterial transition metal chelator pyridine-2,6-bis(thiocarboxylic acid). *Environ Sci Technol* 35: 552–559.

Lohmeier-Vogel EM, KT Leung, H Lee, JT Trevors HJ Vogel. 2001. Phosphorus-31 nuclear magnetic resonance study of the effect of pentachlorophenol on the physiologies of PCP-degrading microorganisms. *Appl Environ Microbiol* 67: 3549–3556.

Martin GE, JE Guido, RH Robins, MHM Sharaf, PL Schiff, AN Tackie. 1998. Submicro inverse-detection gradient NMR: A powerful new way of conducting structure elucidation studies with <005 µmol samples. *J Nat Prod* 61: 555–559.

Moonen MJH, SA Synowsky, WAM van der Berg, AH Westphal, AJR Heck, RHH van den Heuvel, MW Fraaije, WJH van Berkel. 2008. Hydroquinone dioxygenase from *Pseudomonas fluorescens* ACB: A novel member of the family of nonheme-iron (II)-dependent dioxygenases. *J Bacteriol* 190: 5199–5209.

Narbad A, MJ Hewlins, A G Callely. 1989. ^{13}C-NMR studies of acetate and methanol metabolism by methylotrophic *Pseudomonas* strains. *J Gen Microbiol* 135: 1469–1477.

Parisot D, MC Malet-Martino, P Crasnier, R Martino. 1989. ^{19}F nuclear magnetic resonance analysis of 5-fluorouracil metabolism in wild-type and 5-fluorouracil-resistant *Nectria haematococca*. *Appl Environ Microbiol* 55: 2474–2479.

Parisot D, MC Malet-Martino, R Martino, P Crasnier. 1991. ^{19}F nuclear magnetic resonance analysis of 5-fluorouracil metabolism in four differently pigmented strains of *Nectria haematococca*. *Appl Environ Microbiol* 57: 3605–3612.

Peelen S, IMCM Rietjens, WJH van Berkel, WAT van Workum, J Vervoort. 1993. ^{19}F-NMR study on the pH-dependent regioselectivity and rate of the *ortho*-hydroxylation of 3-fluorophenol by phenol hydroxylase from *Trichosporon cutaneum*. *Eur J Biochem* 218: 345–353.

Plugge CM, C Dijkema, AJM Stams. 1993. Acetyl-CoA cleavage pathway in a syntrophic propionate oxidizing bacterium growing on fumarate in the absence of methanogens. *FEMS Microbiol Lett* 110: 71–76.

Poupin P, N Truffaut, B Combourieu, M Sancelelme, H Veschambre, AM Delort. 1998. Degradation of morpholine by an environmental *Mycobacterium* strain involves a cytochrome P-450. *Appl Environ Microbiol* 64: 159–165.

Pusecker K, J Schewitz, P Gfrörer, L-H Tseng, K Albert, E Bayer. 1998. On-line coupling of capillary electrochromatography, capillary electrophoresis, and capillary HPLC with nuclear magnetic resonance spectroscopy. *Anal Chem* 70: 3280–3285.

Reid KA, JTG Hamilton, RD Bowden, D O'Hagan, L Dasaradhi, MR Amin, DB Harper. 1995. Biosynthesis of fluorinated secondary metabolites by *Streptomyces cattleya*. *Microbiology (UK)* 141: 1385–1393.

Riedel A, S Fetzner, M Rampp, F Lingens, U Liebl, J-L Zimmermann, W Nitschke. 1995. EPR, electron spin echo envelope modulation, and electron nuclear double resonance studies of the 2Fe-2S centers of the 2-halobenzoate 1,2-dioxygenase from *Burkholderia* (*Pseudomonas*) *cepacia* 2CBS. *J Biol Chem* 270: 30869–30873.

Selifonov SA, PJ Chapman, SB Akkerman, JE Gurst, JM Bortiatynski, MA Nanny, PG Hatcher. 1998. Use of ^{13}C nuclear magnetic resonance to assess fossil fuel biodegradation: Fate of [1-^{13}C]acenaphthene in creosote polycyclic aromatic compound mixtures degraded by bacteria. *Appl Environ Microbiol* 64: 1447–1453.

Sidelmann U, M Braumann, M Hofmann, M Spraul, JC Lindon, JK Nicholson, SH Hansen. 1997. Directly coupled 800 MHz HPLC-NMR spectroscopy of urine and its application to the identification of the major phase II metabolites of tolfenamic acid. *Anal Chem* 69: 607–612.

Small FJ, SA Ensign. 1995. Carbon dioxide fixation in the metabolism of propylene and propylene oxide by *Xanthobacter* strain Py2. *J Bacteriol* 177: 6170–6175.

Stams AJM, JB van Dijk, C Dijkema, CM Plugge. 1993. Growth of syntrophic propionate-oxidizing bacteria with fumarate in the absence of methanogenic bacteria. *Appl Environ Microbiol* 59: 1114–1119.

Stanley PD. 2002. Principles and topical applications of ^{19}F NMR spectrometry. *Handbook Environ Chem* 3N: 1–56.

Staunton J, AC Sutkowski. 1991. ^{17}O NMR in biosynthetic studies: Aspyrone, asperolactone isoasperoloactone, metabolites of *Aspergillus melleus*. *J Chem Soc Chem Commun* 1106–1108.

Steiert JG, WJ Thoma, K Ugurbil, RL Crawford. 1988. ^{31}P nuclear magnetic resonance studies of effects of some chlorophenols on *Escherichia coli* and a pentachlorophenol-degrading bacterium. *J Bacteriol* 170: 4954–4957.

Steiner RA, KH Kalk, BW Dijkstra. 2002. Anaerobic enzyme substrate structures provide insight into the reaction mechanism of the copper-dependent quercitin 2,3-dioxygenase *Proc Natl Acad USA* 99: 16625–16630.

Strohhäcker J, AA de Graaf, SM Schoberth, RM Wittig, H Sahm. 1993. ^{31}P nuclear magnetic resonance studies of ethanol inhibition in *Zymomonas mobilis*. *Arch Microbiol* 159: 484–490.

Subramanian R, AG Webb. 1998. Design of solenoidal microcoils for high-resolution ^{13}C NMR spectroscopy. *Anal Chem* 70: 2454–2458.

Thorn KA, JB Arterburn, MA Mikita. 1992. ^{15}N and ^{13}C NMR investigation of hydroxylamine-derivatized humic substances. *Environ Sci Technol* 26: 107–116.

Thorn KA, KR Kennedy. 2002. ^{15}N NMR investigation of the covalent binding of reduced TNT amines to soil humic acid, model compounds, and lignocellulose. *Environ Sci Technol* 36: 3787–3796.

Thorn KA, PJ Pettigrew, WS Goldenberg, EJ Weber. 1996. Covalent binding of aniline to humic substances. 2. ^{15}N NMR studies of nucleophilic addition reactions. *Environ Sci Technol* 30: 1764–1775.

Unkefer CJ, RE London. 1984. *In vivo* studies of pyridine nucleotide metabolism in *Escherichia coli* and *Saccharomyces cerevisiae* by carbon-13 NMR spectroscopy. *J Biol Chem* 259: 2311–2320.

Yang K, WW Metcalf. 2004. A new activity for an old enzyme: *Escherichia coli* bacterial alkaline phosphatase is a phosphite-dependent hydrogenase. *Proc Natl Acad Sci USA* 101: 7919–7924.

6.17 Electron Paramagnetic Resonance

There are two rather different applications of electron paramagnetic resonance (EPR): (a) the identification of the metal and its valence state at the active site of enzymes to provide clues to the mechanism and (b) the identification of free radical intermediates. It is instructive to compare the EPR results for dioxygenases with their emphasis on the nature of the metal with the x-ray results that provide details of the surrounding ligands and provide a basis for the mechanism of these reactions. This summary presents merely a selection of reactions in which EPR or Mössbauer spectroscopy have been used to elucidate reactions many of which have been discussed in greater detail in other chapters of this book.

6.17.1 Radical Reactions

1. The degradation of CCl_4 carried out by the Cu complex of the bacterial metabolite pyridine-1,6-bis(thiocarboxylic acid) produced a number of products. The reaction was initiated by the formation of the CCl_3 radical that was characterized from the reaction with $^{13}CCl_4$ by EPR using the spin-trap α-phenyl-*t*-butyl nitrone as a triplet of doublet of doublets (Lewis et al. 2001).

2. Pyruvate formate lyase catalyzes the conversion of pyruvate to formate and acetyl-CoA and is a key enzyme in the anaerobic degradation of carbohydrates in some *Enterobacteriaceae*. Using an enzyme selectively ^{13}C-labeled with glycine, it was shown by EPR that the reaction involves production of a free radical at C_2 of glycine (Wagner et al. 1992). This was confirmed by destruction of the radical with O_2, and determination of part of the structure of the small protein that contained an oxalyl residue originating from Gly^{734}.

3. The significance of radical *S*-adenosylmethionine reactions has been discussed in Chapter 3, and it is sufficient to provide only a few illustrations of the application of EPR.

   a. Dehydrogenation of 2-deoxy-scyllo-inosamine to the C_1 ketone is required in the synthesis of the antibiotic butisorin by *Bacillus circulans*. This is mediated by the enzyme BtrN, and the reaction has been followed by EPR that was used to reveal details of the reaction including the roles of 5′-deoxyadenosine. This was used to propose a reaction mechanism (Yokoyama et al. 2008).

   b. Class III sulfatases that catalyze the hydrolysis of aliphatic sulfates require posttranslational modification to carry out essential modification of cystein or serine residues to C-α formylglycine. The reaction has been studied by EPR which showed the presence of several [4Fe–4S] clusters that were involved in the production and function of the 4′-deoxyademosyl radical (Benjdia et al. 2010).

4. The dehydration of 4-hydroxybutyryl-CoA to crotonyl-CoA by *Clostridium aminobutyricum* involves the fission of a nonactivated C–H bond mediated by FAD and an [4Fe–4S] cluster (Müh et al. 1996), and proceeds via enoxy and dienoxy radicals. After reduction with dithionite, the FAD semiquinone was characterized by EPR, and further details were obtained by the analysis of the x-band electron-nuclear double resonance (ENDOR) spectroscopy (Çinkaya et al. 1997). The ^{57}Fe-enriched enzyme Mössbauer spectroscopy was used to define the [4Fe–4S]$^{2+}$ cluster in the enzyme (Müh et al. 1997).

5. The anaerobic degradation of toluene by *Azoarcus* sp. strain T involved the reaction of the benzyl radical to fumarate with the production of benzylsuccinate and was mediated by a free radical that was identified by EPR associated with benzylsuccinate synthase (Krieger et al. 2001). The enzyme was also active with *o*-, *m*-, and *p*-xylene and *o*-, *m*-, and *p*-cresol, and the doublet splitting in the EPR spectrum was consistent with the presence of a glycyl radical (Verfürth et al. 2004) that has also been observed in the anaerobic degradation of hexane by *Azoarcus* strain HxN1 with the formation of 1-methylpentylsuccinate (Rabus et al. 2001).

6. The metabolism of 2-aminopropanol to propionaldehyde and ammonia takes place under anaerobic conditions when adenosylcobalamin (AdoCbl) is synthesized, or under aerobic conditions when this is generated from exogenous cobalamin (Roof and Roth 1989). The ESR spectrum consisted of a broad signal and an asymmetrical doublet whose width was increased in the 1-[^2H]$_2$ and 1-[^{13}C] analogs. This is consistent with the production of the $CH_3–CH(NH_2)–CHOH^*$ radical at C_1 (Babior et al. 1974).

7. The anaerobic degradation of benzoyl-CoA by *Thauera aromatica* strain K172 involved several [Fe–S] clusters and an ATP-driven reduction to nonaromatic products before further

degradation (Boll et al. 1997). The nature of the redox centers was examined by ESR and Mössbauer spectroscopy of the ^{57}Fe-enriched reductase. The Mössbauer spectra of the oxidized reductase excluded the presence of [2Fe–2S] and [3Fe–4S] clusters, so that only cubane [4Fe–4S] clusters were involved in the dearomatization reaction (Boll et al. 2000).

8. Nitroimidazoles are prodrugs that necessitate activation for their activity. This is carried out by reduction of the nitro groups to free radical intermediates that are the active species in damaging protein and DNA. The generation of these radicals has been demonstrated from metronidazole (1-[2-hydroxyethyl]-2-methyl-5-nitroimidazole) using electrons from pyruvate: oxidoreductase (Rasoloson et al. 2002) or malic enzyme (Hrdý et al. 2005).

6.17.2 Hydroxylation

1. Groups of hydroxylases contain diiron at the active sites.

   a. The soluble methane monooxygenase from *Methylococcus capsulatus* comprises a diiron Fe(III) hydroxylase, a reductase, and a coupling protein B. Since the structure in the native state is EPR silent, it has been examined by EPR in the reduced state that contains a mixed-valence Fe(II)–Fe(III) center. This revealed the structure of the diiron complex that contains histidine and glutamate ligands, and led to a proposal for the structure, including the role of protein B in altering the oxidation state of the redox center (Davydov et al. 1999).

   b. The hydroxylation of phenol and 3,4-dimethylphenol by *Pseudomonas* sp. strain CF600 is unusually carried out by a multicomponent enzyme whose genes occur in a plasmid, and have been designated *dmpKLMNOP* (Shingler et al. 1992). The hydroxylase consists of three components: a reductase DmpP containing FAD and [2Fe–2S], an activator DmpM and the three subunits of the hydroxylase DmpLNO complex. Analysis of ^{57}Fe-enriched hydroxylase showed that the active site contained two types of diiron clusters, and suggested that the two Fe sites in cluster I were bridged by an oxo group and those in cluster II were hydroxo-bridged (Cadieux et al. 2002).

   c. The plasmid-borne genes of toluene/*o*-xylene hydroxylase in *Pseudomonas* sp, strain OX1 enable this organism to degrade toluene and *o*-xylene (Bertoni et al. 1998). Examination by ^{57}Fe Mössbauer spectroscopy (the reaction is also accomplished with phenol) has revealed details of the mechanism that putatively involves an Fe(III)peroxo adduct and the formation of H_2O_2 from an intermediate (Murray et al. 2007).

2. *Pseudomonas oleovorans* is able to carry out ω-hydroxylation of alkanes. The alkane hydroxylase AlkB is an integral-membrane oxygenase and after expression in *Escherichia coli* it has been examined by Mössbauer spectroscopy (Shanklin et al. 1997). These revealed that it contains four Fe environments, two associated with a dinuclear iron cluster, and two mononuclear species, one Fe(II) S_A and the other with a high-spin Fe(III) S_B that were present in variable amounts. In the native state of the enzyme, the cluster contained an antiferromagnetically coupled pair of high-spin Fe(III) that have different coordination.

3. The anaerobic degradation of nicotinic acid by *Clostridium barkeri* is initiated by hydroxylation to produce 6-hydroxynicotinate in a selenium-requiring reaction. The EPR spectrum of Mo(V) was altered in the ^{77}Se-enriched enzyme, and changed dramatically in the ^{95}Mo-enriched hydroxylase. The results were interpreted to show that Se is coordinated with molybdenum at the active site (Gladyshev et al. 1994).

6.17.3 Dioxygenation

1. a. The degradation of 3,4-dihydroxyphenylacetate by *Bacillus brevis* takes place by extradiol fission, although the enzyme was atypically not activated by Fe(II), not inhibited by

cyanide, and not inactivated by H_2O_2. Analysis by x-ray fluorescence and atomic absorption revealed the presence of manganese, and the ESR spectra were consistent with this unusual presence of Mn(II) (Que et al. 1981).

    b. The 2,3-dihydroxybiphenyl 1,2-dioxygenase in *Bacillus* sp. strain JF8 that has been studied for its ability to degrade PCBs contains manganese instead of iron. The presence of Mn was confirmed by EPR for which signals at $g = 2.02$ and $g = 4.06$ having the sixfold splitting characteristic of Mn(II) were observed (Hatta et al. 2003).

2. *Pseudomonas testosteroni* degrades 3,4-dihydroxybenzoate (protocatechuate) by extradiol fission to produce α-hydroxy-γ-carboxymuconate semialdehyde. The enzyme has been characterized by Mössbauer spectroscopy using ^{57}Fe-enriched enzyme and by ESR in the complexes with nitric oxide that suggest the value of this application for the study of the ESR-silent active site in extradiol dioxygenases (Arciero et al. 1983).

3. The degradation of 2-halobenzoate by *Burkholderia cepacia* 2CBS involves dioxygenation with concomitant decarboxylation and loss of halide. After reduction with dithionite the EPR spectrum showed the presence of two different [Fe–S] clusters, a ferredoxin-type and a Rieske-type (Riedel et al. 1995).

4. The dioxygenation of 4-hydroxyphenylpyruvate to 2,5-dihydroxyphenylacetate (homogentisate) and CO_2 is carried out with rearrangement in *Pseudomonas* sp. strain PJ 874. The EPR spectrum consisted of a single signal for high-spin Fe(III) with an intensity for only a single atom of Fe. These results were interpreted to support the presence of an iron-tyrosinate protein (Bradley et al. 1986).

5. The dioxygenation of quercitin by *Aspergillus japonicus* is carried out exceptionally by a Cu-containing dioxygenase, and its EPR has been examined (Kooter et al. 2002). At pH 6, the spectrum consists of two EPR species that are replaced by a single signal after anaerobic addition of quercitin. The EPR data were consistent with initial binding of the Cu^{2+} to the C_3 and C_4 oxygen functions, followed by radical formation with reduction of the Cu^{2+} to Cu^+. Reaction with oxygen at C_2 and C_4 is then followed by elimination of CO and formation of the product.

6.17.4 Dehydrogenation

1. Alcohol dehydrogenases generally contain Zn at the active site, and the alcohol dehydrogenase from *Pyrococcus furiosus* is, therefore, remarkable since it contains both Zn and Fe, and the presence of low-spin Fe(II) was detected by EPR (Ma and Adams 1999). It is important to note, however, that Fe may serve only in activating the enzyme without being structurally present at the active site.

2. The dehydrogenation of primary amines in *Escherichia coli* is carried out using a copper-quinoprotein oxidase that contains two copper ions, and one or two molecules of the cofactor topaquinone 2,4,5-trihydroxyphenylalanine quinone (TPQ). Details of the overall reaction have been studied using EPR of stable intermediates and rapid stopped-flow spectrophotometry to reveal the three primary reactions involving the quinone/hydroquinone/semiquinone that is coupled to the reduction, and oxidation of Cu^{2+}/Cu^+, and dehydrogenation of the amine to the aldehyde: the TPQH radical with Cu^+ reacts with O_2 to produce H_2O_2 and TPQ/Cu^{2+} (Steinebach et al. 1996).

3. Formate can be used for methanogenesis in *Methanobacterium formicicum* (Schauer and Perry 1986), and the functioning of formate dehydrogenase is dependent on coenzyme F_{420}. The formate dehydrogenase has been examined by ESR, and although the oxidized enzyme was ESR-silent, reduction by dithionite produced two paramagnetic signals that could also be found in whole cells anaerobically reduced with formate. Use of partially reduced ^{95}Mo-enriched enzyme displayed additional splitting that is indicative of the Mo center in the dehydrogenase. A summary was provided of the redox midpoint potentials for the Mo center in formate dehydrogenases (Barber et al. 1983).

4. Obligate and facultative methylotrophic bacteria contain an NAD^+-dependent formate dehydrogenases that catalyze the oxidation of formate to CO_2 with concomitant reduction of NAD^+.
The purified enzyme has been prepared from *Methylosinus trichosporium* OB3b and contained
flavin, iron, molybdenum, and inorganic sulfide. Reduction of the enzyme with formate,
NADH, or dithionite produced an EPR spectrum in which five centers could be resolved: four
of them were characteristic of [Fe–S] clusters and the fifth displayed hyperfine splitting in
^{95}Mo-enriched enzyme (Barber et al. 1983).

5. Carbon monoxide dehydrogenases have been described for both aerobic and anaerobic
CO-utilizing organisms.

   a. *Oligotropha carboxidovorans* oxidizes carbon monoxide using a complex consisting of a
   molybdenum subunit with molybdopterin cytosine dinucleotide that contains the active
   site, a flavoprotein subunit, and an iron–sulfur protein subunit. ESR was used to distinguish the Mo/FAD, and the [2Fe–2S] centers, using preparations in which the flavoprotein
   was dissociated by detergent (Gremer et al. 2000).

   b. *Carboxydothermus hydrogenoformans* metabolizes carbon monoxide anaerobically by the
   reaction $CO + H_2O \rightarrow CO_2 + 2H^+ + 2e$ that involves NiFeS proteins. Two membrane-associated carbon monoxide dehydrogenases have been isolated and examined by EPR
   (Svetlitchnyi et al. 2001). Although the Ni was EPR-silent, the dehydrogenases COD 1 and
   COD 2 have EPR spectra indicative of [4Fe–4S] clusters both in the native form, and after
   reduction by CO or dithionite. It was suggested that COD 1 was involved in energy generation, whereas COD 2 served catabolic functions.

   c. *Methanotrix* and *Methanosarcina* are archaea that can utilize acetate, and both contain carbon monoxide dehydrogenase. The enzyme in *Methanothrix soehngenii* has been examined
   by EPR (Jetten et al. 1991). When purified aerobically, the signal at $g = 2.014$ was characteristic of a $[3Fe–4S]^{1+}$ cluster, whereas the correspondingly purified enzyme under anaerobic
   conditions had complex EPR spectra typical of $[4Fe–4S]^{1+}$ clusters. Acetyl-CoA/CO exchange
   activity could be demonstrated under anaerobic conditions but not under aerobic conditions.
   These results were used make a proposal for the mechanism of activity of the enzyme.

6.17.5 Hydrogenase

1. The hydrogenase of *Desulfovibrio baculatus* is unusual in containing Ni, Se, and nonheme Fe.
The periplasmic enzyme was purified after growth in ^{77}Se-enriched medium, and EPR spectra
of the reduced enzymes showed that the unpaired electron was shared by the Ni and Se atoms,
and that the Se serves as a ligand to the nickel redox center (He et al. 1989).

2. The [NiFe] hydrogenase of *Ralstonia eutropha* H16 is unusual in its tolerance to oxygen. It consists of a large subunit containing the active site and a small subunit coordinating one [3Fe–4S]
and two [4Fe–4S] clusters. The solubilized dimer in its oxidized state had an EPR spectrum
which showed split signals of the $[3Fe–4S]^+$ cluster and the signal for Ni(III). On reduction with
H_2, the signal of the [3Fe–4S] was replaced by a complex spectrum from $[4Fe–4S]^+$ clusters. In
membrane preparations that included the *b*-type cytochrome, analysis by EPR and FTIR was
unremarkable except for the significant absence of the O_2-inactivated Ni_u-A state that seems to
be an attribute of oxygen-tolerant [NiFe] hydrogenases (Saggu et al. 2009).

6.17.6 Reductase

a. The soluble periplasmic selenate reductase in *Thauera selenatis* has been examined to reveal
several features. X-band EPR suggested the presence of Mo(V), heme *b*, and $[3Fe–4S]^{1+}$, and
$[4Fe–4S]^{1+}$ clusters that were involved in electron transfer to the active site of the periplasmic
reductase under turnover conditions. Collectively, the analysis showed that the reductase
belonged to the Type II molybdoenzymes (Dridge et al. 2007).

b. Benzoyl-coenzyme A reductase plays a central role in the dearomatization of benzoyl-CoA by both facultative and anaerobic bacteria. Whereas the first are dependent on ATP, the second are independent. The ATP-independent reductase has been examined for the obligate anaerobe *Geobacter metallireducens*, and EPR was used to determine the presence of one [3Fe–4S] and three [4Fe–4S] clusters at different oxidation levels, as well as a signal $g_z = 2.013$ in the oxidized state attributed to a rhombic W(V) species (Kung et al. 2009).

6.17.7 Acetone Carboxylase

The conversion of acetone to acetoacetate at the expense of ATP in *Rhodobacter capsulatus* is carried out by a carboxylase. Acetone-grown cells had an EPR signal centered at $g = 2$ that could be associated with the enzyme since this was absent in malate-grown cells and was attributed to the presence of Mn (Boyd et al. 2004).

6.17.8 Hydroxyl Radicals: Role in Toxicity of H_2O_2 to Bacteria

In a study directed to the analysis of the role of Fe and the generation of H_2O_2 in *Escherichia coli* (McCormick et al. 1998), hydroxyl radicals were specifically trapped by reaction with ethanol to give the α-hydroxyethyl radical. This formed a stable adduct with α-(4-pyridyl-1-oxide)-*N*-*t*-butyl nitroxide that was not formed either by superoxide or hydroxyl radicals. The role of redox-reactive iron used EPR to analyze the EPR-detectable ascorbyl radicals.

6.18 References

Arciero DM, JD Lipscomb, BH Hiynh, TA Kent, E Münck. 1983. EPR and Mössbauer studies of protocatechuate 4,5-dioxygenase. Characterization of a new Fe^{2+} environment. *J Biol Chem* 258: 14981–14991.

Babior BM, TH Moss, WH Orme-Johnson, H Beinert. 1974. The mechanism of action of ethanolamine ammonia-lyase, a B_{12}-dependent enzyme. The participation of paramagnetic species in the catalytic deamination of 2-aminopropanol. *J Biol Chem* 249: 4537–4544.

Barber MJ, LM Siegel, NL Schauer, HD May, JG Ferry. 1983. Formate dehydogenase from *Methanobacterium formicum*. Electron paramagnetic resonance spectroscopy of the molybdenum and iron-sulfur centers. *J Biol Chem* 258: 10839–10845.

Benjdia A, S Subramanian, J Leprince, H Vaudry, MK Johnson, O Berteau. 2010. Anaerobic sulfatase-meturing enzyme—a mechanistic link with glycyl radical-activating enzymes? *FEBS J* 277: 1906–1920.

Bertoni G, M Martino, E Galli, P Barbieri. 1998. Analysis of the gene cluster encoding toluene/*o*-xylene monooxygenase from *Pseudomonas stutzeri* OX1. *Appl Environ Microbiol* 64: 3626–3632.

Boll M, G Fuchs, C Meier, A Trautwein, A El Kasmi, SW Ragsdale, G Buchanan, DJ Lowe. 2001. Redox centers of 4-hydroxybenzoyl-CoA reductase, a member of the xanthine oxidase family of Molybdenum-containing enzymes. *J Biol Chem* 276: 47853–47862.

Boll M, G Fuchs, C Meier, AX Trautwein, DJ Lowe. 2000. EPR and Mössbauer studies of benzoyl-CoA reductase. *J Biol Chem* 275: 31857–31868.

Boll M, SJP Albracht, G Fuchs. 1997. Benzoyl-CoA reductase (dearomatizing), a key enzyme of anaerobic aromatic metabolism. A study of adenosinetriphosphate, activity, ATP stoichiometry of the reaction and EPR properties of the enzyme. *Eur J Biochem* 244: 840–851.

Boyd JM, H Ellsworth, SA Ensign. 2004. Bacterial acetone carboxylase is a manganese-dependent metalloenzyme. *J Biol Chem* 279: 46644–46651.

Bradley FC, S Lindstedt, JD Lipscomb, L Que, AL Roe, M Rundgren. 1986. 4-hydroxyphenylpyruvate dioxygenase is an iron-tyrosinate protein. *J Biol Chem* 261: 11693–11696.

Cadieux E, V Vrajmasu, C Achim, J Powlowski, E Mäunck. 2002. Biochemical, Mössbauer, and EPR studies of the diiron cluster of phenol hydroxylase from *Pseudomonas* sp. strain CF 600. *Biochemistry* 41: 10680–10691.

Çinkaya I, W Buckel, M Medina, C Gomez-Moreno, R Cammick. 1997. Electron-nuclear double resonance spectroscopy investigation of 4-hydroxybutyryl-CoA dehydratase from *Clostridium aminobutyricum*: Comparison with other flavin radical enzymes. *Biol Chem* 378: 843–849.

Davydov R, AM Valentine, S Komar-Panicucci, BM Hoffman, SJ Lippard. 1999. An EPR study of the dinuclear iron site in the soluble methane monooxygenase from *Methylococcus capsulatus* (Bath) reduced by one electron at 77 K: The effects of component interactions and the binding of small molecules to the diiron (III) center. *Biochemistry* 38: 4188–4197.

Dridge EJ, CA Watts, BJN Jepson, K Line, JM Santini, DJ Richardson, CS Butler. 2007. Investigation of the redox centres of periplasmic selenate reductase from *Thauera selenatis* by EPR spectroscopy. *Biochem J* 408: 19–28.

Gladyshev VN, SV Khangulov, TC Stadtman. 1994. Nicotinic acid hydroxylase from *Clostridium barkeri*: Electron paramagnetic resonance studies show that selenium is coordinated with molybdenum in the catalytically active selenium-dependent enzyme. *Proc Natl Acad Sci USA* 91: 232–236.

Gremer L, S Kellner, H Dobbek, R Huber, O Meyer. 2000. Binding of flavin adenine dinucleotide to molybdenum-containing carbon monoxide dehydrogenase from *Oligotropha carboxidovorans*. Structural and functional analysis of a carbon monoxide dehydrogenase species in which the native flavoprotein has been replaced by its counterpart produced in *Escherichia coli*. *J Biol Chem* 275: 1864–1872.

Hatta T, G Muklerjee-Dhar, J Damborsky, H Kiyohara. 2003. Characterization of a novel thermostable Mn(II)-dependent 2,3-dihydroxybiphenyl 1,2-dioxygenase from polychlorinated biphenyl- and naphthalene-degrading *Bacillus* sp. JF8. *J Biol Chem* 278: 21483–21492.

He SH, M Teixeira, J LeGall, DS Patil, I Moura, JJG Moira, DV DerVartanian, BH Huynh, HD Peck. 1989. EPR studies with [77]Se-enriched (NiFeSe) hydrogenase of *Desulfovibrio baculatus*. *J Biol Chem* 264: 2678–2682.

Hrdý I, R Cammack, P Stopka, J Kulda, J Tachezy. 2005. Alternative pathway of metronidazole activation in *Trichomonas vaginalis* hydogenosomes. *Antimicrob Agents Chemother* 49: 5033–5036.

Jetten MSM, WR Hagen, AJ Pierik, AJM Stams, AJB Zehnder. 1991. Paramagnetic centers and acetyl-coenzyme A/CO exchange activity of carbon monoxide dehydrogenase from *Methanothrix soehngenii*. *Eur J Biochem* 195: 385–391.

Jablonski PE, W-P Lu, SW Ragsdale, JG Ferry. 1993. Characterization of the metal centers of the corrinoid/iron-sulfur component of the CO dehydrogenase enzyme complex from *Methanosarcina thermophila* by EPR spectroscopy and spectroelectrochemistry. *J Biol Chem* 268: 325–329.

Jollie DR, JD Lipscomb. 1991. Formate dehydrogenase from *Methylosinus trichosporium* OB3b. Purification and spectroscopic characterization of the cofactors. *J Biol Chem* 266: 21853–21863.

Kooter IM, RA Steiner, BW Dijkstra, PI van Noort, MR Egmond, M Huber. 2002. EPR characterization of the mononuclear Cu-containing *Aspergillus japonicus* quercitin 2,3-dioxygenase reveals dramatic changes upon anaerobic binding of substrates. *Eur J Biochem* 269: 2971–2979.

Krieger SJ, W Roseboom, SPJ Albracht, AM Spormann. 2001. A stable organic radical in anaerobic benzylsuccinate synthase of *Azoarcus* sp. strain T. *J Biol Chem* 276: 12924–12927.

Kung JW, C Löffler, K Dörner, D Heintz, S Gallien, A Van Dorsselaer, T Friedrich, M Boll. 2009. Identification and characterization of the tungsten-containing class of benzoyl-coenzyme A reductases. *Proc Natl Acad Sci USA* 106: 17687–17692.

Leuthner B, C Leutwein, H Schulz, P Hörth, W Haehnel, E Schiltz, H Schägger, J Heider. 1998. Biochemical and genetic characterisation of benzylsuccinate synthase from *Thauera aromatica*: A new glycyl-radical catalysing the first step in anaerobic toluene degradation. *Mol Microbiol* 28: 515–628.

Lewis TA, A Paszczynski, SA Gordon-Wylie, S Jeedigunta, CH Lee, RL Crawford. 2001. Carbon tetrachloride dechlorination by the bacterial transition metal chelator pyridine-2,6-bis (thiocarboxylic acid). *Environ Sci Technol* 35: 552–559.

Ma K, MWW Adams. 1999. An unusual oxygen-sensitive, iron- and zinc-containing alcohol dehydrogenase from the hyperthermophilic archaeon *Pyrococcus furiosus*. *J Bacteriol* 181: 1163–1170.

McCormick ML, GR Buettner, BE Britigan. 1998. Endogenous superoxide dismutase levels regulate iron-dependent hydroxyl radical formation in *Escherichia coli* exposed to hydrogen peroxide. *J Bacteriol* 180: 622–625.

Müh U, W Buckel, E Bill. 1997. Mössbauer study of 4-hydroxybutyryl-CoA dehydratase. Probing the role of an iron-sulfur cluster in an overall non-redox reaction. *Eur J Biochem* 248: 380–384.

Müh U, I Çinkaya, SPJ Albracht, W Buckel. 1996. 4-hydroxybutyryl-CoA dehydratase from *Clostridium aminobutyricum*: Characterization of FAD and iron–sulfur clusters involved in an overall non-redox reaction. *Biochemistry* 35: 11710–11718.

Murray LJ, SG Naik, DO Ortillo, R Garcia-Serres, JK Lee, BH Huynh, SJ Lippard. 2007. Characterization of the arene-oxidizing intermediate in ToMOH as a diiron(III) species. *J Amer Chem Soc* 129: 14500–14510.

Que L, J Widom, RL Crawford. 1981. 3,4-dihydroxyphenylacetate 2,3-dioxygase. A manganese (II) dioxygenase from *Bacillus brevis*. *J Biol Chem* 256: 10941–10944.

Rabus R, H Wilkes, A Behrends, A Armstroff, T Fischer, AJ Pierik, F Widdel. 2001. Anaerobic initial reaction of *n*-alkanes in a denitrifying bacterium: Evidence for (1-methylpentyl) succinate as initial product and for involvement of an organic radical in *n*-hexane metabolism. *J Bacteriol* 183: 1707–1715.

Rasoloson D, S Vanacova, E Tomkova, J Razga, I Hrdý, J Tachezy, J Kulda. 2002. Mechanisms of in vitro development of resistance to metronidazole in *Trichomonas vaginalis*. *Microbiology UK* 148: 2467–2477.

Riedel A, S Fetzer, M Rampp, F Lingens, U Liebl, J-L Zimmermann, W Nitschke. 1995. EPR, electron spin echo envelope modulation, and electron nuclear double resonance studies of the 2Fe–2S centers of the 2-halobenzoate 1,2-dioxygenase from *Burkholderia* (*Pseudomonas*) *cepacia* 2CBS. *J Biol Chem* 270: 30869–31873.

Roof DM, JR Roth. 1989. Functions required for vitamin B_{12}-dependent ethanolamine utilization in *Salmonella typhimurium*. *J Bacteriol* 171: 3316–3323.

Saggu M, I Zebger, M Ludwig, O Lenz, B Friedrich, P Hildebrandt, F Lenzian. 2009. Spectroscopic insights into the oxygen-tolerant membrane-associated [NiFe] hydrogenase of *Ralstonia eutropha* H 16. *J Biol Chem* 284: 16264–16276.

Schauer NL, JG Ferry. 1986. Composition of the coenzyme F_{420}-dependent formate dehydrogenase from *Methanobacterium formicicum*. *J Bacteriol* 165: 405–411.

Shanklin J, C Achim, H Schmidt, BG Fox, E Münck. 1997. Mössbauer studies of alkane ω-hydroxylase: Evidence for a diiron cluster in an integral-membrane enzyme. *Proc Natl Acad Sci USA* 94: 2981–2986.

Shingler, V, J Powlowski, U Marklund. 1992. Nucleotide sequence and functional analysis of the complete phenol/3,4-dimethylphenol catabolic pathway of *Pseudomonas* sp. strain CF600. *J Bacteriol* 174: 711–724.

Steinebach V, S de Vries, JA Duine. 1996. Intermediates in the catalytic cycle of copper-quinoprotein amine oxidase from *Escherichia coli*. *J Biol Chem* 271: 5580–5588.

Svetlitchnyi V. C Peschel, G Acker, O Meyer. 2001. Two membrane-associated NiFeS-carbon monoxide dehydrogenases from the anaerobic carbon-monoxide-utilizing eubacterium *Carboxydothermus hydrogenoformans*. *J Bacteriol* 183: 5134–5144.

Verfürth K, AJ Pierik, C Leutwein, S Zorn, J Heider. 2004. Substrate specificities and electron paramagnetic resonance properties of benzylsuccinate synthesis in anaerobic toluene and *m*-xylene metabolism. *Arch Microbiol* 181: 155–162.

Wagner AFV, M Frey, FA Neugebauer, W Schäfer. 1992. The free radical in pyruvate formate-lyase is located on glycine-734. *Proc Natl Acad USA* 89: 996–1000.

Wolfe MD, JV Parales, DT Gibson, JD Lipscomb. 2001. Single turnover chemistry and regulation of O_2 activation by the oxygenase component of naphthalene 1,2-dioxygenase. *J Biol Chem* 276: 1845–1953.

Yokoyama K, D Ohmori, F Kudo, T Eguchi. 2008. Mechanistic study on the reaction of a radical SAM dehydrogenase BtrN by electron paramagnetic resonance spectroscopy. *Biochemistry* 47: 8950–8960.

Part 4: X-Ray Crystallographic Analysis

The application of x-ray analysis to enzymes has been facilitated by a number of technical developments: amplification of degradative genes and expression in *Escherichia coli* has provided adequate quantities of the enzymes, procedures have been developed for obtaining crystals of proteins suitable for crystallographic analysis, synchrotron sources of irradiation have become widely accessible, and the necessary computational machinery has been developed. A range of illustrations has been chosen in which attention has been focused on the structures, while details of the reactions are examined at greater length in other chapters.

6.19 Acetylene Hydratase

The anaerobic hydration of acetylene to acetaldehyde by the *Pelobacter acetylenicus* involves tungsten coordinated to the bis-thiolane of the bis-molybdopterin guanine dinucleotide and a cubane [4Fe–4S] cluster. The structure has been elucidated to binding pocket for acetylene that shows water bound by the tungsten and an adjacent aspartate hydroxyl group whose pK_a is reduced by the adjacent [4Fe–4S] cluster (Seiffert et al. 2007).

6.20 Triesterase

The phosphotriesterase in *Pseudomonas diminuta* can hydrolyze a range of phosphate triesters, including organophosphate agrochemicals, and is representative of binuclear enzymes at the active center at which a number of divalent cations are catalytically active. The structure has been determined by x-ray analysis of the Cd analog, and clearly revealed the coordination of Cd 1 with two histidines and aspartate, and Cd 2 with two histidines. A carbamoylated lysine and a water molecule bridged the two Cd atoms (Benning et al. 1995, 2001).

6.21 Dehydrogenases

1. *Formate dehydrogenase*
   a. Formate dehydrogenase H in *Escherichia coli* catalyzes the oxidation of formate to CO_2, and is a complex consisting of a selenocysteine ligand to Mo coordinated to each of two molybdopterin guanosine dinucleotide cofactors, and a cubane [4Fe–4S] that takes place in the redox reactions. On the basis of the structure, the role of selenocysteine in proton abstraction was suggested (Boyington et al. 1997).
   b. The structure of the membrane formate hydrogenase N that participates in nitrate respiration in *Escherichia coli* has been determined (Jormakka et al. 2002). It contained redox centers including molybdopterin guanosine dinucleotides, [4Fe–4S] clusters, two heme *b* groups, and a menoquinone analog. The structure was used to delineate the proton motive force generated by the redox loop (Jormakka et al. 2002).

2. *Carbon monoxide dehydrogenase*   Carbon monoxide dehydrogenases are able to catalyze the reaction $CO + H_2O \rightarrow CO_2 + 2H^+ + 2\varepsilon$ in both aerobic and anaerobic bacteria, and in the bifunctional carbon monoxide dehydrogenase/acetyl-CoA synthase.
   a. After oxidative washing with H_2O_2, the native enzyme from the aerobe *Oligotropha carboxidovorans* displayed an ESR signal for a single copper atom, and multiple wavelength anomalous dispersion used to determine the structure of the active site. This contained Mo and Cu in which S linked the two sites, and the Mo that was bound to the pyran dithiolene of molybdopterin cytosine dinucleotide, also contained a hydroxo ligand. On the basis of these observations, a catalytic cycle was proposed in which the oxidized state containing Mo(VI)–S–Cu–S– is reduced by CO to a Mo(IV) intermediate with insertion of CO before reacting with H_2O and loss of CO_2 to produce a reduced state before loss of 2e to produce the initial state (Dobbek et al. 2002).
   b. The dehydrogenase of the anaerobe *Carboxydothermus hydrogenoformans* was examined by multiple wavelength anomalous dispersion. It contained five metal clusters including [4Fe–4S] clusters whose spatial arrangement provided a plausible electron transfer path, and an active site that contained both Ni and Fe as an asymmetric cluster [Ni–4Fe–4S] cluster to which CO was bound to the Ni that carried four sulfur ligands (Dobbek et al. 2001).

c. Details of the acetyl-CoA decarboxylase/synthase and the roles of the subunits have been described for the enzyme complex in *Methanosarcina thermophila* (Gencic and Grahame 2003). Crystallographic analysis of the complex from *Moorella thermoacetica* revealed an assembly in which a cubane [4Fe–4S] cluster was linked through cysteine S to Cu that had two cysteine sulfur ligands that were shared with the Ni. On the basis of these structural observations, a putative mechanistic cycle for enzymatic activity was proposed (Doukov et al. 2002).

3. *Quinohemoprotein dehydrogenase*   Alternative mechanisms for the dehydrogenation of primary alkanols and alkylamines use quinone cofactors as electron acceptors in place of NAD(P)$^+$.

a. The structure of the quinohemoprotein alcohol dehydrogenase in *Comamonas testosteroni* has been determined and consists of two domains. The N-terminal binds one molecule of pyrroloquinoline quinone cofactor and one Ca^{2+} ion, and the C-terminal is a *c*-type cytochrome with his and met ligands to Fe. On the basis of this, a mechanism was proposed for electron transfer between the redox centers that is facilitated by a disulfide bond between cysteines (Oubrie et al. 2002).

b. The structure of the quinohemoprotein amine dehydrogenase in *Pseudomonas putida* consists of three nonidentical subunits. The α-subunit contains a di-heme cytochrome *c*, a β-subunit that contains part of the active site, and a γ-subunit that contains a cross-linked cysteine tryptophyl quinone, and additional complex cross-linkage involving cysteine attached to tryptophan C$_4$ and cys-to-asp/glu thioethers (Satoh et al. 2002).

6.22 Quinoprotein Amine Oxidase

The deamination of primary amines such as phenylethylamine by *Escherichia coli* and *Klebsiella oxytoca* is carried out by an oxidase. This contains copper and 2,4,5-trihydroxyphenylalanine quinone (topaquinone, TPQ) that is produced from tyrosine by dioxygenation. The mechanism that consists of two coupled reactions has been elucidated on the basis of the crystal structure in *Escherichia coli* (Wilmot et al. 1999). In the reductive part, TPQ reacts with the amine to form a Schiff base that undergoes hydrolysis to the aldehyde, followed by a redox reaction of the aminoquinol and Cu^{2+} to Cu$^+$ and a semiquinone that reacts with O$_2$ to form H$_2$O$_2$ (Wilmot et al. 1999).

6.23 Dehalogenases

Dehalogenation was been examined in both alkyl and alkenyl halides that represent different types of reaction.

1. The degradation of 1,2-dichloroethane is initiated in *Xanthobacter autotrophicus* by hydrolysis to 2-chloroethanol. The enzyme responsible for dehalogenase activity that has been purified from *Xanthobacter autotrophicus* strain GJ10 consists of a single polypeptide chain with a molecular mass of 36 kDa. Details of the mechanism have been explored using an ingenious method of producing crystal at different stages of the reaction (Verschuren et al. 1993). The overall reaction involves a catalytic triad at the active site: asp$_{124}$ binds to one of the carbon atoms, and hydrolysis with inversion is accomplished by cooperation of Asp[260] and His[289] with a molecule of water bound to Glu[56].

2. The dehalogenase from *Pseudomonas pavonaseae* that is able to transform the vinyl chloride *trans*-3-chloroacrylate—albeit activated by a terminal carboxyl group—to malonate semialdehyde has been examined to determine details of the reaction. This involves the formation of the chlorohydrin through participation of α-Arg[8], α-Arg[11], β-Pro[1], and α-Glu[52] at the active site (de Jong et al. 2004).

6.24 Atypical Dehydratases

Anaerobic degradations may involve dehydrations from nonactivated positions. These occur, for example, in the anaerobic degradation of 2-hydroxyglutaryl-CoA by *Acidaminococcus fermentans* and of 4-hydroxybutyryl-CoA by *Clostridium aminobutyricum*. Details of these reactions are provided in parts of Chapter 3 and in Chapter 7.

6.24.1 2-Hydroxyglutaryl-CoA Dehydratase

The anaerobic degradation of glutarate by *Acidaminococcus fermentans* is initiated by reduction to 2-hydroxyglutarate. The 2-hydroxyglutaryl-CoA undergoes dehydration from a nonactivated position to produce (*E*)-glutaconyl-CoA catalyzed by a hydratase that consists of two components, component A contains a bridging [4Fe–4S] cluster and uses the hydrolysis of ATP to deliver an electron to the reductase component D. The crystal structure of the reductase from *Acidaminococcus fermentans* was used to provide details of the reaction and the functional similarity to the ATP-dependent nitrogenase was pointed out (Locher et al. 2001).

6.24.2 4-Hydroxybutyryl-CoA Dehydratase

The fermentation of succinate to butyrate by *Clostridium kluyveri* proceeds via the formation of 4-hydroxybutyrate with the production of crotonate that is reduced to butyrate. 4-Aminobutyrate is produced by decarboxylation of glutamate and is fermented by *Clostridium aminobutyricum* to succinate semialdehyde and 4-hydroxybutyrate. Dehydration of the 4-hydroxybutyryl-CoA to yield crotonyl-CoA involves loss of a hydrogen atom from a nonactivated position, and involves FAD and an [Fe–S] cluster (Müh et al. 1996). The crystal structure of 4-hydroxybutyryl-CoA dehydratase has been provided to support a free radical mechanism involving reaction of an FAD semiquinone anion with an enoxy radical followed by the formation of a dienoxy radical and hydration (Martins et al. 2004, 2005).

6.25 Baeyer–Villiger Monooxygenase

Baeyer–Villiger monooxygenases catalyze the introduction of an oxygen atom into ketones to produce esters. The crystal structure of phenylacetone monooxygenase from *Thermobifida fusca* was been determined to reveal details of the reaction. The enzyme complex with FAD and NADPH reacts with O_2 to produce the 5a-peroxide that carries out a nucleophilic attack on the ketone, and the structure revealed the critical position of Arg[337] to stabilize the intermediates (Malito et al. 2004). This intermediate then undergoes rearrangement to the lactone.

6.26 Arsenite Oxidase

This enzyme carries out the oxidation of arsenite to arsenate $As^{III}O_2^- + 2H_2O \rightarrow As^VO_4^3 + 4H^+ + 2\epsilon$ and does not produce H_2O_2. The crystal structure of arsenite oxidase from *Alcaligenes faecalis* was determined using a combination of procedures. It consisted of two subunits, the larger contained Mo bound to two pyranopterin cofactors and a [3Fe–4S] cluster, and the smaller that had a Rieske-type [2Fe–2S] cluster (Ellis et al. 2001). A pathway for electron transport was outlined: from the large subunit N_5 of one pyranopterin through the [3Fe–4S] cluster and Ser[99] to the small cluster His[62] in the [2Fe–2S] cluster of the Rieske subunit, and then via His[81] to azurin or cytochrome *c*.

6.27 Methyl Coenzyme M Reductase

Methyl coenzyme reductase plays a cardinal role in the production of methane by the reaction with 7-thioheptanoyl-threoninephosphate (coenzyme B) and formation of the disulfide. The structure from *Methanobacterium thermoautotrophicum* has been determined to show that in its inactive state two molecules of the Ni porphinoid coenzyme F_{430} are imbedded between the subunits to produce two active sites (Ermler et al. 1997). A stepwise mechanism was proposed involving the forming a Ni(III) followed by successive formation of Ni(II), and in the subterminal reaction Ni(I) and methane.

6.28 Urease

The crystal structure of urease from *Klebsiella aerogenes* has been determined (Jabri et al. 1995). It contains an $(\alpha\beta)_8$ barrel domain that contains two nickel centers, each of which is coordinated with four histidines His[134, 136, 246, 272], Asp[360], and the oxygen atoms of an 5-*N*-carbamoyl lysine[217]. Attention was drawn to the high structural similarity with the structure of the mammalian Zn-containing adenosine, and it is worth noting that histidines have also been shown to be the ligands in metalloenzyme β-lactamases (Baldwin et al. 1978).

6.29 Hydrogenase

There are two groups of metal-containing hydrogenases, and structures of representative [Ni–Fe] and [Fe–Fe] hydrogenases have been examined.

1. The crystal structure of the [Fe–Fe] hydrogenase in *Clostridium pasteurianum* has been solved by x-ray multiwavelength anomalous dispersion (Peters et al. 1998). The active site consisted of a cubane [4Fe–4S] cluster linked through the sulfur of cysteine to a [2Fe] subcluster in which one CN and one CO are bound to each octahedral Fe atom that are linked to each other by one shared CO and two sulfur atoms. The structure was used as a model for the proton reduction.

2. The structure of the [Ni–Fe] hydrogenase in *Desulfovibrio gigas* is the prototype for this group of hydrogenases. The structure confirmed the presence of both Ni and Fe, and the presence of a bridging structure between the Fe and Ni centers. The structure of the subunit contains one [3Fe–4S] and two [4Fe–4S] clusters (Volbeda et al. 1995), and further details have added the nature of the ligands to Ni and Fe: Ni with two S atoms of cysteines and Fe with one CO and two CN ligands. In addition, the two metals are linked by shared sulfur atoms of cysteine (Volbeda et al. 1996).

6.30 β-Lactamases

The hydrolysis of β-lactams in penicillins, cephalosporins, and carbapenem is brought by hydrolytic fission of the β-lactam ring, and is the immediate cause of resistance to these antibiotics, and has given rise to the development of β-lactamase inhibitors. There are two broad types of enzymes depending on the structures at the active site.

6.30.1 Serine Residues at the Active Site: Classes A, C, and D

1. Hydrolysis of the inhibitors sulbactam and clavulanate is initiated by reaction with a serine[70] residue that carries out nucleophilic attack on the sulbactam carbonyl with the formation of

the acyl enzyme. The structures of deacylation-deficient lactamases were used to determine further reactions involving the *trans*-enamine that is critical for transient inactivation (Padayatti et al. 2005).

2. β-Lactam antibiotics are generally ineffective against infection by *Mycobacterium tuberculosis*. Cloned plasmids of the lactamase blaC were introduced into and amplified in *Escherichia coli*. The crystal structure displayed an overall similarity to those of other class A lactamases, except for amino acid substitutions that are consistent with the broad specificity of the enzyme, low penicillinase activity, and resistance to clavulanate, although this has been effectively used in combination with β-lactam antibiotics (Wang et al. 2006).

6.30.2 Metallo-β-Lactamases: Groups B1, B2, B3

These contain either one or two zinc atoms at the active site(s). X-ray analysis has been used to determine the structure of the carbapenemase in *Aeromonas hydrophila* and to propose a mechanism for β-lactam hydrolysis. This involved the association of the zinc with the C_3-carboxyl group and, transiently, the nitrogen atom of the β-lactam (Garau et al. 2005). The structure of the lactamase from *Legionella gormanii* that contains two atoms of zinc revealed the role of the participating zinc, and the coordination of one Zn with three histidines, and the other Zn with two histidines and two aspartates (García-Sáez et al. 2003). In view of the interest in the design of β-lactamase inhibitors, the structure of the enzyme from *Stenotrophomonas maltophilia* was determined (Nauton et al. 2008). This revealed the binding of the inhibitors with both zinc atoms through the sulfur atoms, and the triazole with the sulfur atom of one zinc and the nitrogen of the triazole with the other.

6.31 Alkylsulfatase

There are several groups of alkylsulfates including the Fe(II) 2-oxoglutarate-dependent enzymes, and a distinct group containing a binuclear Zn cluster enzymes has been characterized. The crystal structure of an alkylsulfatase from *Pseudomonas aeruginosa* was determined (Hagelueken et al. 2006), and defined a third class of sulfatases containing two zinc atoms in a binuclear structure that constituted a distinct member of the metallo-β-lactam family.

6.32 Dioxygenases

These enzymes initiate the aerobic degradation of a range of PAHs. Catechols are produced by dioxygenation of the aromatic substrates followed by dehydrogenation. Their further metabolism is carried out by ring-fission dioxygenases that are of two types: intradiol enzymes for fission of the C–C bond between the catechol hydroxyl groups, and extradiol for fission between one of them and the adjacent C–C bond. The crystal structures of important representatives of each of them have been determined.

6.32.1 Arene Dihydroxylases

1. a. Naphthalene 1,2-dioxygenase is the paradigm for aromatic ring dihydroxylation and brings about the dioxygenation of naphthalene to naphthalene 1,2-dihydrodiol. The crystal structure of naphthalene 1,2-dioxygenase from *Pseudomonas* sp. NCIB 9816-4 has been determined (Kauppi et al. 1998). The Fe at the active site was bound to His[208], His[213], a bidentate Asp[362], and a water molecule in a distorted octahedral pyramidal geometry lacking one ligand. The structure made it possible to suggest a route of electron transfer between adjacent α subunits: the α subunit A containing the [2Fe–2S] cluster bound to His[104], the α

subunit B with His^{213} and His^{208} bound at the active site with Asp^{205} serving as a linker with the histidines in the A and B α subunits.

b. *Sphingobium yanoikuyae* B1 was isolated by its ability to degrade biphenyl, and the dioxygenase has been extensively examined. It consists of a set of multiple terminal Rieske nonheme iron oxygenases and an electron donor of a reductase and a ferredoxin. The crystal structure showed the general similarity to the dioxygenase from NCIB 9816-4 that has already been discussed, but revealed that the entrance to the active site was wider and that this facilitated the acceptance of PAHs with four or more rings: chrysene, benzo[*a*]pyrene and benz[*a*]anthracene (Ferraro et al. 2007).

2. The structure of the hydroxylase from *Pseudomonas putida* strain 86 that carries out hydroxylation of quinol-2-one to 8-hydroxyquinolone has been examined and revealed the critical role of the Rieske [2Fe–2S] center whose reduction increases the accessibility of O_2 to the active site from the amide group (Martins et al. 2005).

3. The degradation of carbazole by *Pseudomonas putida* strain 86 is initiated by dioxygenation at the angular positions. The dioxygenase comprised a terminal oxygenase and the electron transfer components ferredoxin and ferredoxin reductase. Details of the crystal structure have made possible a description of electron transfer from the ferredoxin reductase to the Rieske oxygenase cluster (Ashikawa et al. 2006).

6.32.2 Ring-Fission Dioxygenases

1. The structure of 3,4-dihydroxybenzoate 3,4-dioxygenase in *Pseudomonas aeruginosa* that carries out *intradiol*-fission has been determined (Ohlendorf et al. 1994). The coordination of the Fe(III) forms a trigonal bipyramid with tyrosine and histidine as axial ligands, and the catechol is bound to produce a complex that accepts O_2: subsequent reactions produce a diketooxapine that is fissioned to the final muconate.

2. The structure of the *extradiol* ring-fission enzyme for 3-chlorobiphenyl from *Pseudomonas* sp. strain KKS102 has been determined (Senda et al. 1996). The Fe(II) is coordinated with two histidines and a glutamate in a square pyramidal structure. Oxygen reacts with the Fe complex at the active site to produce one –Fe–O–O– that is bound to C_3 of the catechol, followed by fission to the 6-(4-chlorophenyl)-6-oxo-2-hydroxycarboxylate.

3. The structure of the unusual copper-containing dioxygenase in *Aspergillus japonicus* bound to its natural substrates quercitin and kaemferol made it possible to provide a mechanism for its activity (Steiner et al. 2002). The copper atom is coordinated to three histidines and one water molecule that is displaced to form a monodentate ligand with the hydroxyl group at C_3 of the substrates. Oxygen reacts to produce an endoperoxide between C_3 and the carbonyl at C_4 that loses CO to form the product, while the intervention of a radical intermediate was proposed to overcome the spin-forbidden reaction with the triplet ground state of O_2.

6.33 Superoxide Dismutase

Superoxide dismutases catalyze the production of H_2O_2 from the superoxide radical O_2^-, and generally contain either Mn and Fe, or Cu and Zn. Exceptionally, however, the enzyme may contain Ni (Wuerges et al. 2004). This has been demonstrated in the enzyme from *Streptomyces seoulensis* in which the Ni(III) at the catalytic site was coordinated with the amino group of histidine, the amide group of cysteine, two thiolate groups of cysteine, and the imidazolate group of histidine that was lost in the reduced state of the enzyme (Wuerges et al. 2004). The authors found that in addition, Ni occurs in a number of *Actinomyces*.

6.34 Formyl-CoA Transferase

This plays a key role in the anaerobic metabolism of oxalate by *Oxalobacter formigenes* in which oxalate is metabolized to formate and CO_2 by the coupled reaction of formyl-CoA transferase and oxalyl-CoA decarboxylase. The transferase is initiated by the reaction with Asp[169] to form a mixed anhydride that reacts with oxalate to produce aspartyl oxalate whose structure was resolved crystallographically (Jonsson et al. 2004), and made possible a complete description of the reaction.

6.35 PLP-Independent Racemases

1. The x-ray structures of mandelate racemase (MR) and muconate lactonizing enzyme (MLE) have been examined (Hasson et al. 1998). The structure of mandelate racemase showed the involvement of Lys[166] and His[297] assisted by Asp[270] and the carboxylate anion stabilized by both association with Lys[164] and a cation. It was suggested that in both of them the reactions involved a 1:1 proton transfer to produce $R–C(R')=C(O^-)_2$ where $R'=OH$ for MR or H for MLE, followed by reversible reprotonation to either enantiomer.

2. Although racemization of amino acids is normally dependent on pyridoxal-5′-phosphate, an alternative mechanism that had been explored with mandelate racemase involved a two-base direct 1:1 proton transfer. The crystal structure of aspartate racemase from the archaeon *Pyrococcus horikoshii* OT3 provided support for this mechanism: the enzyme consisted of a dimeric structure whose inter-subunits are held by a Cys[73]–Cys[73′] disulfide and, between the two domains the Cys[82] and Cys[194] that carry out the two-base racemization that are strictly conserved (Liu et al. 2002).

6.36 References

Ashikawa Y, Z Fujimoto, H Noguchi, H Habe, T Omori, H Yamane, H Nojiri. 2006. Electron transfer complex formation between oxygenase and ferredoxin components in Rieske nonheme iron oxygenase system. *Structure* 14: 1779–1789.

Baldwin GS, A Glades, AH Hill, BE Smith, SAG Wally, EP Abraham. 1978. Histidine residues as zinc ligands in β-lactamases II. *Biochem J* 175: 441–447.

Benning MM, H Shim, FM Raushel, HM Holden. 2001. High resolution X-ray structures of different metal-substituted forms of phosphotriesterase from *Pseudomonas diminuta*. *Biochemistry* 40: 2712–2722.

Benning MM, JM Kuo, FM Raushel, HM Holden. 1995. Three-dimensional structure of the binuclear metal center of phosphotriesterase. *Biochemistry* 34: 7973–7978.

Boyington JC, VN Gladyshev, SV Khangulov, TC Stadtman, PD Sun. 1997. Crystal structure of formate dehydrogenase H: Catalysis involving Mo, molybdopterin, selenocysteine, and an Fe_4S_4 cluster. *Science* 275: 1305–1308.

de Jong RM, W Brugman, GJ Poelarends, CP Witman, BW Dijkstra. 2004. The X-ray structure of *trans*-3-chloroacrylic acid dehalogenase reveals a novel hydration mechanism in the tautomerase superfamily. *J Biol Chem* 279: 11546–11552.

Dobbek H, L Gremer, R Kiefersauer, R Huber, O Meyer. 2002. Catalysis at a dinuclear [CuSMo(=O)OH] cluster in a CO dehydrogenase resolved at 1.1 Å resolution. *Proc Natl Acad Sci USA* 99: 15971–15976.

Dobbek H, V Svetlitchnyi, L Gremer, R Huber, O Meyer. 2001. Crystal structure of a carbon monoxide dehydrogenase reveals a [Ni-4Fe-5S] cluster. *Science* 293: 1281–1285.

Doukov TI, TM Iverson, J Seravalli, SW Ragsdale, CI Drennan. 2002. A Ni-Fe-Cu center in a bifunctional carbon monoxide dehydrogenase/acetyl-CoA synthase. *Science* 298: 567–572.

Ellis PJ, T Conrads, R Hille, P Kuhn. 2001. Crystal structure of the 100 kDa arsenite oxidase from Alcaligenes faecalis in two crystal forms at 1.64 Å and 2.03 Å. *Structure* 9: 125–132.

Ermler U, W Grabarse, S Shima, M Goubeaud, RK Thauer. 1997. Crystal structure of methyl-coenzyme M reductase: The key enzyme of biological methane formation. *Science* 278: 1457–1462.

Ferraro DJ, EN Brown, C-L Yu, RE Parales, DT Gibson, S Ramaswamy. 2007. Structural investigation of the ferredoxin and terminal oxygenase components of the biphenyl 2,3-dioxygenase from *Sphingobium yanoikuyae* B1. *BMC Struct Biol* 7: 10.

Garau G, C Bebrone, C Anne, M Galleni, J-M Frère, O Dideberg. 2005. A metallo-β-lactamase enzyme in action: Crystal structures of the monozinc carbapenemase CphA and its complex with biapenem. *J Mol Biol* 345: 785–795.

García-Sáez I, PS Mercuri, C Papamicael, R Kahn, JM Frère, M Galleni, GM Rossolini, O Dideberg. 2003. Three dimensional structure of FEZ-1, a monomeric subclass B3 metallo-β-lactamase from *Fluoribacter gormanii*, in native form and in complex with D-captopril. *J Mol Biol* 325: 650–660.

Gencic S, DA Grahame. 2003. Nickel in subunit *b* of the acetyl-CoA decarbonylase/synthase multienzyme complex in methanogens. Catalytic properties and evidence for a binuclear Ni-Ni site. *J Biol Chem* 278: 6101–6110.

Hagelueken G, TM Adams, L Wiehlmann, U Widow, H Kolmar, B Tümmler, DW Heinz, W-D Schubert. 2006. The crystal structure of SdsA1, an alkylsulfatase from *Pseudomonas aeruginosa* defines a third class of sulfatases. *Proc Natl Acad Sci USA* 103: 7631–7636.

Hasson MS, I Schlichting, J Moulai, K Taylor, W Barrett, GL Kenyon, PC Babbitt, JA Gerlt, GA Petsko, D Ringe. 1998. Evolution of an enzyme active site: The structure of a new crystal form of muconate lactonizing enzyme compared with mandelate racemase and enolase. *Proc Natl Acad Sci USA* 95: 10396–10401.

Jabri E, MB Carr, RP Hausinger, PA Karplus. 1995. The crystal structure of urease from *Klebsiella aerogenes*. *Science* 268: 998–1004.

Jonsson S, S Ricagno, Y Lindqvist, NGJ Richards. 2004. Kinetic and mechanistic characterization of the formyl-CoA transferase from *Oxalobacter formigenes*. *J Biol Chem* 279: 36003–36012.

Jormakka M, S Törnroth, B Byrne, S Iwata. 2002. Molecular basis of proton motive force generation: Structure of formate dehydrogenase-N. *Science* 295: 1863–1968.

Kauppi B, K Lee, E Carredano, RE Parales, DT Gibson, H Eklund, S Ramaswamy. 1998. Structure of an aromatic ring-hydroxylating dioxygenase—naphthalene 1,2-dioxygenase. *Structure* 6: 571–586.

Li Y-F, Y Hata, T Fujii, T Hisano, M Nishihara, T Kurihara, N Esaki. 1998. Crystal structures of reaction intermediates of L-2-haloacid dehalogenase and implications for the reaction mechanism. *J Biol Chem* 273: 15035–15044.

Liu L, K Iwata, A Kita, Y Kawarabayasi, M Yohda, K Miki. 2002. Crystal structure of aspartate racemase from *Pyrococcus horikoshii* OT3 and its implications for molecular mechanism of PLP-independent racemization. *J Mol Biol* 319: 479–489.

Locher KP, M Hans, AP Yeh, B Schmid, W Buckel, DC Rees. 2001. Crystal structure of the *Acidaminococcus fermentans* 2-hydroxyglutaryl-CoA dehydratase component A. *J Mol Biol* 307: 297–308.

Malito E, A Alfieri, MW Fraaije, A Mattevi. 2004. Crystal structure of a Baeyer-Villiger monooxygenase. *Proc Natl Acad Sci USA* 101: 13157–13162.

Martins BM, H Dobbek, I Çinkaya, W Buckel, A Messerschmidt. 2004. Crystal structure of 4-hydroxybutyryl-CoA dehydratase: Radical catalysis involving a [4Fe–4S] cluster and flavin. *Proc Natl Acad Sci USA* 101: 15645–15649.

Martins BM, T Svetlitchnaia, H Dobbek. 2005. 2-oxoquinoline 8-monooxygenase oxygenase component: Active site modification by Rieske-[2Fe-2S] center oxidation/reduction. *Structure* 13: 817–824.

Müh U, I Çinkaya, SPJ Albracht, W Buckel. 1996. 4-hydroxybutyryl-CoA dehydratase from *Clostridium aminobutyricum*: Characterization of FAD and iron – sulfur clusters involved in an overall non-redox reaction. *Biochemistry* 35: 11710–11718.

Nauton L, R Kahn, G Garau, JF Hernandez, O Dideberg. 2008. Structural insights into the design of inhibitors for the L1 metallo-β-lactamase from *Stenotrophomomas maltophilia*. *J Mol Biol* 375: 257–269.

Nicolet Y, JK Rubach, MC Posewitz, P Amara, C Mathevon, M Atta, M Fontecave, JC Fontecilla-Camps. 2008. X-ray structure of the [Fe-Fe]-hydrogenase maturase HydE from *Thermotoga maritima*. *J Biol Chem* 283: 18861–18872.

Ohlendorf DH, AM Orville, JD Lipscomb. 1994. Structure of protocatechuate 3,4-dioxygenase from *Pseudomonas aeruginosa* at 2.15 Å resolution. *J Mol Biol* 244: 586–608.

Oubrie A, HJ Rozeboom, KH Kalk, EG Huizinga, BW Dijkstra. 2002. Crystal structure of quinohemoprotein alcohol dehydrogenase from *Comamonas testosteroni. J Biol Chem* 277: 3727–3732.

Padayatti PS, MS Helfand, MA Totir, MP Carey, PR Carey, RA Bonomo, F van den Akker. 2005. High resolution crystal structures of the *trans*-enamine intermediates formed by sulbactam and clavulanic acid and E166a SHV-1 β-lactamase. *J Biol Chem* 280: 34900–34907.

Peters JW, WN Lanzilotta, BJ Lemon, LC Seefeldt. 1998. X-ray crystal structure of the Fe-only hydrogenase (Cpl) from *Clostridium pasteurianum* to 1.8 angstrom resolution. *Science* 282: 1853–1858.

Satoh A, and 13 co-authors. 2002. Crystal structure of quinoprotein amine dehydrogenase from *Pseudomonas putida*. Identification of a novel quinone cofactor encaged by multiple thioether cross-bridges. *J Biol Chem* 277: 2830–2834.

Seiffert GB, GM Ullmann, A Messerschmidt, B Schink, PMH Kroneck. 2007. Structure of the non-redox-active tungsten/[4Fe:4S] enzyme acetylene hydratase. *Proc Natl Acad Sci USA* 104: 3073–3077.

Senda T, K Sugiyama, H Narita, T Yamamoto, K Kimbara, M Fukuda, M Sato, K Yano, Y Mitsui. 1996. Three dimensional structures of free form and two substrate complexes of an extradiol ring-cleavage type dioxygenase, the BphC enzyme from *Pseudomonas* sp. strain KKS102. *J Mol Biol* 255: 735–752.

Steiner RA, KH Kalk, BW Dijkstra. 2002. Anaerobic enzyme-substrate structures provide insight into the reaction mechanism of the copper-dependent quercitin 2,3-dioxygenase *Proc Natl Acad USA* 99: 16625–16630.

Verschuren KHG, F Seljée, HJ Rozeboom, KH Kalk, BW Dijkstra. 1993. Crystallographic analysis of the catalytic mechanism of haloalkane dehalogenase. *Nature* 363: 693–698.

Volbeda A, M-H Charon, C Piras, EC Hatchikian, M Frey, JC Fontecilla-Camps. 1995. Crystal structure of the nickel-iron hydrogenase from *Desulfovibrio gigas. Nature* 373: 580–587.

Volbeda A, E Garcin, C Piras, AL de Lacey, VM Fernandez, E Claude, EC Hatchikan, M Frey, JC Fontecilla-Camps. 1996. Structure of the [NiFe] hydrogenase active site: Evidence for biologically uncommon Fe ligands. *J Am Chem Soc* 118: 12989–12996.

Wang F, C Cassidy, JC Sacchetti. 2006. Crystal structure and activity studies of the *Mycobacterium tuberculosis* β-lactamase reveal its critical role in resistance to β-lactam antibiotics. *Antimicrob Agents Chemother* 50: 2762–2771.

Wilmot CM, J Hajdu, MJ McPherson, PF Knowles, SEV Phillips. 1999. Visualization of dioxygen bound to copper during enzyme catalysis. *Science* 286: 1724–1728.

Wuerges J, J-W Lee, Y-I Yim, H-S Yim, S-O Kang, KD Carugo. 2004. Crystal structure of nickel-containing superoxide dismutase reveals another type of active site. *Proc Natl Acad Sci USA* 101: 8569–8574.

Section III

Pathways and Mechanisms of Degradation and Transformation

7

Aliphatic Compounds

7.1 Alkanes

Petroleum hydrocarbons are used as automotive fuels and as monomers for the production of a range of plastics. They provide the basis of the petrochemical industry, while the halogenated derivatives that have diverse application are discussed in later sections of this chapter. There is an enormous literature on the microbial degradation of alkanes: this has been motivated by applications as diverse as the utilization of methane and methanol for the production of single-cell protein, and to combating oil spills. The number and range of microorganisms is impressive, and includes many different taxa of bacteria, yeasts, and fungi. Bacteria include several thermophilic species of *Geobacillus* (Nazina et al. 2001) and *Thermoleophilum* (Yakimov et al. 2003a); the marine organisms *Marinobacter hydrocarbonoclasticus* (Gauthier et al. 1992), *Alcanivorax borkumensis* (Yakimov et al. 1998), *Thalassolituus oleivorans* (Yakimov et al. 2004), *Oleispira antarctica* (Yakimov et al. 2003b); and *Oceanobacter*-related bacteria (Teramoto et al. 2009).

7.1.1 Aerobic Conditions

7.1.1.1 Utilization of Methane

The simplest alkane is methane. In outline, the pathway used by bacteria for aerobic degradation is straightforward and involves three stages:

1. Oxidation to methanol catalyzed by methane monooxygenase (MMO).
2. Conversion of methanol to formaldehyde by methanol dehydrogenase. A complex array of genes is involved in this oxidation, and the dehydrogenase contains pyrroloquinoline quinone (PQQ) as a cofactor (references in Ramamoorthi and Lidstrom 1995). Details of its function differ, however, from that of methylamine dehydrogenase that also contains a quinoprotein— tryptophan tryptophylquinone (TTQ).
3. Further degradation to formate may involve the tetrahydrofolate or tetrahydromethanopterin pathways (Marx et al. 2003) that are analogous to those used for methyl chloride and methyl bromide.

In addition, the organisms must be capable of synthesizing cell material from the substrate, so that some fraction of the C_1 metabolites must also be assimilated. Several distinct pathways for this have been described, but these are merely summarized here since a comprehensive and elegant presentation of the details has been given (Anthony 1982):

1. The ribulose 1,5-bisphosphate pathway for the assimilation of CO_2 that is identical to the Benson–Calvin cycle used by photosynthetic organisms
2. The ribulose monophosphate cycle for the incorporation of formaldehyde
3. The serine pathway for the assimilation of formaldehyde

Methylotrophs using methane as substrate belong to type I or type II strains that differ in their membrane structure. In addition, MMO may exist in both a particulate MMO (pMMO) and a soluble MMO (sMMO) form that has been more extensively studied. The particulate form is found in almost all methanotrophs, while sMMO has been established in *Methylococcus* sp. *Methylosinus* sp. *Methylocyctis* sp., *Methylomonas* sp., and *Methylomicrobium* sp.: references to these are given in Chapter 3, Section 3.2 dealing with monooxygenases. The particulate pMMO contains Cu, and enrichments using limiting concentrations of Cu, therefore, resulted in the prevalence of strains with the soluble form of the enzyme (McDonald et al. 2006). The enzymes differ markedly in their substrate specificity: sMMO is able to hydroxylate a range of *n*-alkanes, *n*-alkenes, cycloalkanes, and aromatic substrates including benzene and styrene, whereas pMMO is restricted to *n*-alkanes and *n*-alkenes with ≤5 carbon atoms, and is unreactive toward aromatic substrates. In addition, sMMO hydroxylates at both C_1 and C_2 in contrast to pMMO that hydroxylates predominantly at C_2 in *Methylosinus trichosporium* OB3b (Burrows et al. 1984).

The soluble MMO from both type I and type II methanotrophs is a three-component system. It comprises: (1) a nonheme iron hydroxylase containing an oxo-bridged binuclear Fe cluster, (2) a metal-free protein component without redox cofactors, and (3) an NADH reductase containing FAD and a [2Fe–2S] cluster (Fox et al. 1989). The structure of the soluble hydroxylase and the mechanism of its action involving the $Fe^{III} – O – Fe^{III}$ at the active site have been given in a review (Lipscomb 1994), and additional details added from the results of EPR spectroscopy (Davydov et al. 1999). Substantial studies have been devoted to the metabolic versatility of the sMMO of *Methylococcus capsulatus* that include the terminal and sub-terminal hydroxylation of *n*-alkanes, the epoxidation of alkenes, and further examples that are given in Figure 7.1 (Colby et al. 1977; Patel et al. 1982). It is worth noting the difference between the metabolic activity of this strain and that of other methylotrophs, and it is appropriate to drawn attention to the similarity of this activity to the monooxygenase involved in the oxidation of ammonia, and the broad substrate specificity of cyclohexane oxygenase that is been noted in Chapter 2, Figure 2.10.

On the other hand, the particulate enzyme contains copper, and the concentration of copper determines the catalytic activity of the enzyme (Sontoh and Semrau 1998). The substrate specificity is noted above, and hydroxylation of both ethane and butane by pMMO takes place with retention of

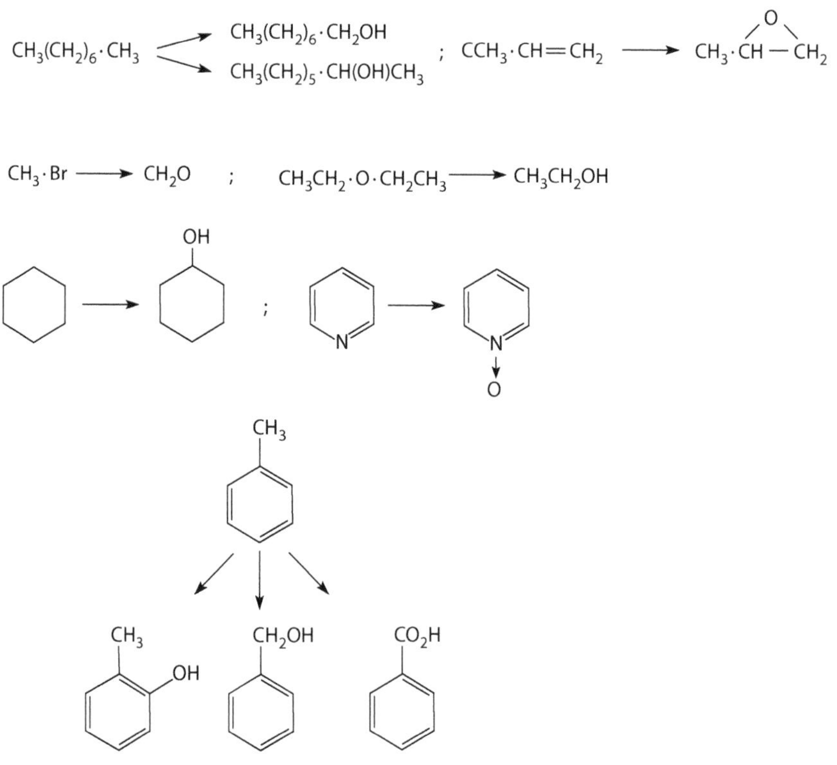

FIGURE 7.1 Examples of the reactions catalyzed by methane monooxygenase.

configuration. A mechanism has been proposed involving reaction with an oxene orthogonal to the two Cu centers (Yu et al. 2003).

Although the majority of methylotrophs prefer oxidized C_1 substrates, methanotrophs are able to grow at the expense of methane, and some additionally with methane or methylamine. In addition to these, attention is directed to a number of methanotrophic extremophiles that are discussed in Chapter 2. Notable features among these are the following:

1. Species of *Methylocella* appear to possess only the soluble form of MMO, and lack the intra-cytoplasmic membrane system to which particulate MMO is generally bound (Dedysh et al. 2005).

2. Methane oxidation has been described for several organisms of the *Verrucomicrobia* phylum all of which were extremophiles.

   a. An organism (V4) grew robustly only on C_1 substrates (methane and methanol), contained the genes required for the operation of the Benson–Calvin cycle and for those encoding pMMO that belonged to a new group of these enzymes (Dunfield et al. 2007).

   b. An organism (SolV) assigned to *Acidimethylosilex fumarolicum* displayed pMMO activity, and sequencing of the DNA suggested that the organism used a combination of the serine, tetrahydrofolate, and ribulose-1,5-d-bisphosphate pathways for assimilation of carbon (Pol et al. 2007).

   c. An extremophile that carried out oxidation at 55°C and pH 2, lacked intracytoplasmic membrane, and most remarkably, lacked genes for MMO (Islam et al. 2008).

7.1.1.2 Utilization of Higher Alkanes

7.1.1.2.1 Introduction

Bacteria and some yeasts are able to carry out the degradation of higher alkanes by hydroxylation, followed by sequential dehydrogenation to alkane carboxylates. The alkanes range from propane to those with 30 or more carbon atoms, and include normal and branched-chained alkanes. Even-numbered alkanes are degraded to acetate while odd-numbered ones produce propionate and acetate. Hydroxylation of even-numbered alkanes may take place at both the terminal α- and ω-positions, or in the β-position. Both Gram-negative and Gram-positive bacteria, and several yeasts of the genera *Candida* and *Endomycopsis* have been examined.

7.1.1.2.2 Terminal Hydroxylation

Degradation of *n*-alkanes proceeds by terminal hydroxylation followed by successive dehydrogenation to the aldehyde and carboxylate, β-oxidation, and fission with the production of acetate (Figure 7.2). For sub-terminal hydroxylation, degradation proceeds by dehydrogenation to the ketone (Figure 7.3) from which several modes of degradation are possible.

$$R \cdot CH_2 \cdot CH_2 \cdot CH_2 \cdot CH_3 \longrightarrow R \cdot CH_2 \cdot CH_2 \cdot CH_2 \cdot CH_2OH \longrightarrow R \cdot CH_2 \cdot CH_2 \cdot CH_2 \cdot CHO$$
$$\longrightarrow R \cdot CH_2 \cdot CH_2 \cdot CH_2 \cdot CO_2H \longrightarrow R \cdot CH_2 \cdot CO_2H + CH_3 \cdot CO_2H \longrightarrow$$

FIGURE 7.2 Terminal oxidation of alkanes.

$$RCH_2CH_2CH_3 \longrightarrow RCH_2CH(OH)CH_3 \longrightarrow RCH_2COCH_3 \longrightarrow$$
$$RCH_2O \cdot COCH_3 \longrightarrow RCH_2OH \longrightarrow RCO_2H$$

FIGURE 7.3 Subterminal oxidation of alkanes.

The metabolism of acetate is unexceptional except for organisms that lack isocitrate lyase which is an essential component of the glyoxylate cycle: for these organisms the ethylmalonate pathway (Figure 7.4) is operative (Erb et al. 2007). Propionate is produced from the degradation of odd-numbered alkanes, and the side chain and A/B ring of cholesterol by *Rhodococcus* sp. strain RHA1 (van der Geize et al. 2007). Propionate can be metabolized by several mechanisms (Figure 7.5) including the methylmalonate (Figure 7.3a), the acrylate (Figure 7.3b), and the 2-methylcitrate pathways (Figure 7.3c). The latter is initiated by reaction of propionate with oxalacetate followed by formation of 2-methylcitrate from which succinate is produced with loss of pyruvate (Figure 7.6). This reaction has been established in enteric bacteria (Textor et al. 1997; Horswill and Escalante-Semerena 1999), in *Ralstonia eutropha* (Brämer and Steinbüchel 2001), and in *Corynebacterium glutamicum* (Claes et al. 2002).

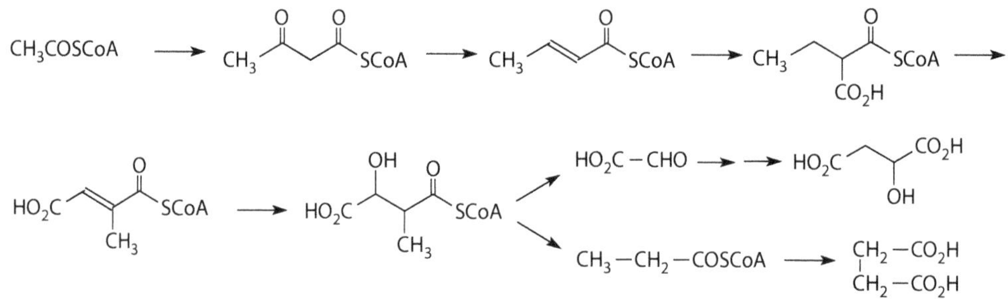

FIGURE 7.4   Ethylmalonate pathway. (Adapted from Erb TJ et al. 2007. *Proc Natl Acad USA* 104: 10631–10636.)

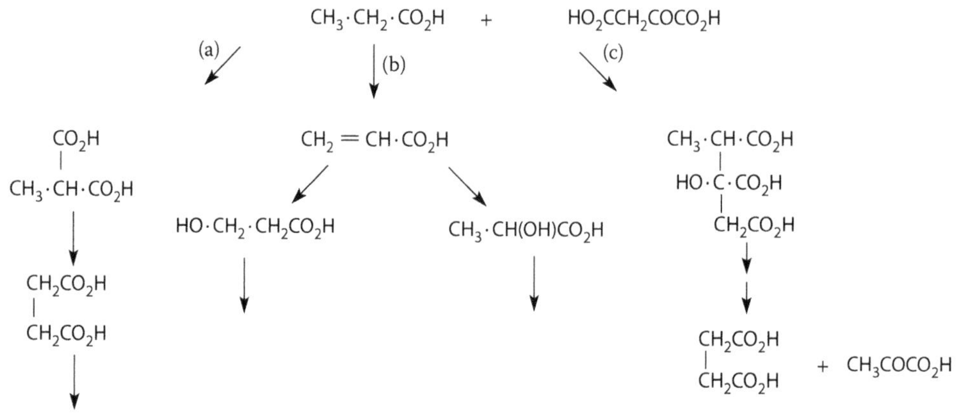

FIGURE 7.5   Alternative pathways for the aerobic degradation of propionate.

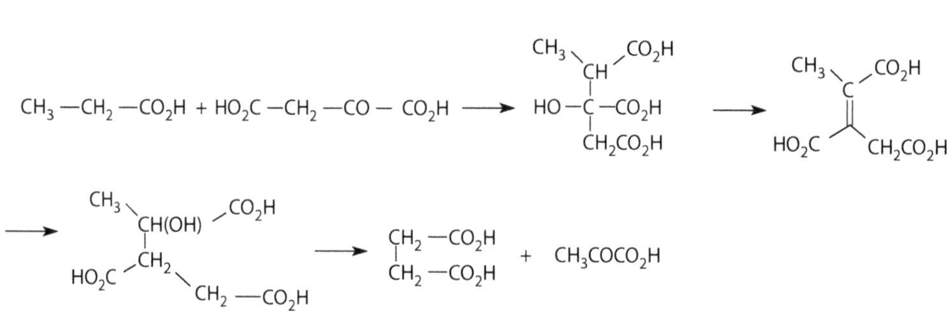

FIGURE 7.6   Methylcitrate pathway.

Subterminal oxidation to ketones may be followed by Baeyer–Villiger monooxygenation to lactones followed by hydrolysis with the loss of a C_2 unit. Additional pathways involving carboxylation are given in the section dealing with alkanones.

7.1.1.2.3 Bis-Terminal Hydroxylation

For long-chain alkanes, bis-terminal hydroxylation may produce the α,ω-dicarboxylates after dehydrogenation of the alkanols. In one of the pathways for the degradation of pristane (Figure 7.7), this is initiated by α,ω-oxidation with sequential β-oxidation (McKenna and Kallio 1971; Pirnik et al. 1974), and in the transformation of squalene by *Corynebacterium* sp. (Seo et al. 1983), and the putative degradation of squalane by species of *Mycobacterium* (Berekaa and Steinbüchel 2000). The production of α,ω-carboxylates has attracted attention, and this may depend on the critical chain-length of the alkane. For example, although *Gordonia* (*Corynebacterium*) sp. strain 7E1C produces α,ω-dodecandoic acid during growth with dodecane, the analogous α,ω-hexadecanedioc acid is not produced from hexadecane (Broadway et al. 1993).

7.1.1.2.4 Chain Branching

Branching at some positions of an alkane may prevent normal β-oxidation of the initially formed carboxylates. This is clearly shown by the observation that, although a number of strains could utilize octane as carbon source, they were unable to use 3,6-, 2,7-, and 2,6-dimethyloctane (Schaeffer et al. 1979). On the other, 2,2-dimethylheptane, 2,2-dimethylhexane, and 2,2-dimethylpentane can be used as substrates for growth by strains of *Achromobacter* (Catelani et al. 1977), degradation is terminated with the formation of 2,2-dimethylpropionate (pivalate). A strain of *Rhodococcus* TMP2 was able to degrade

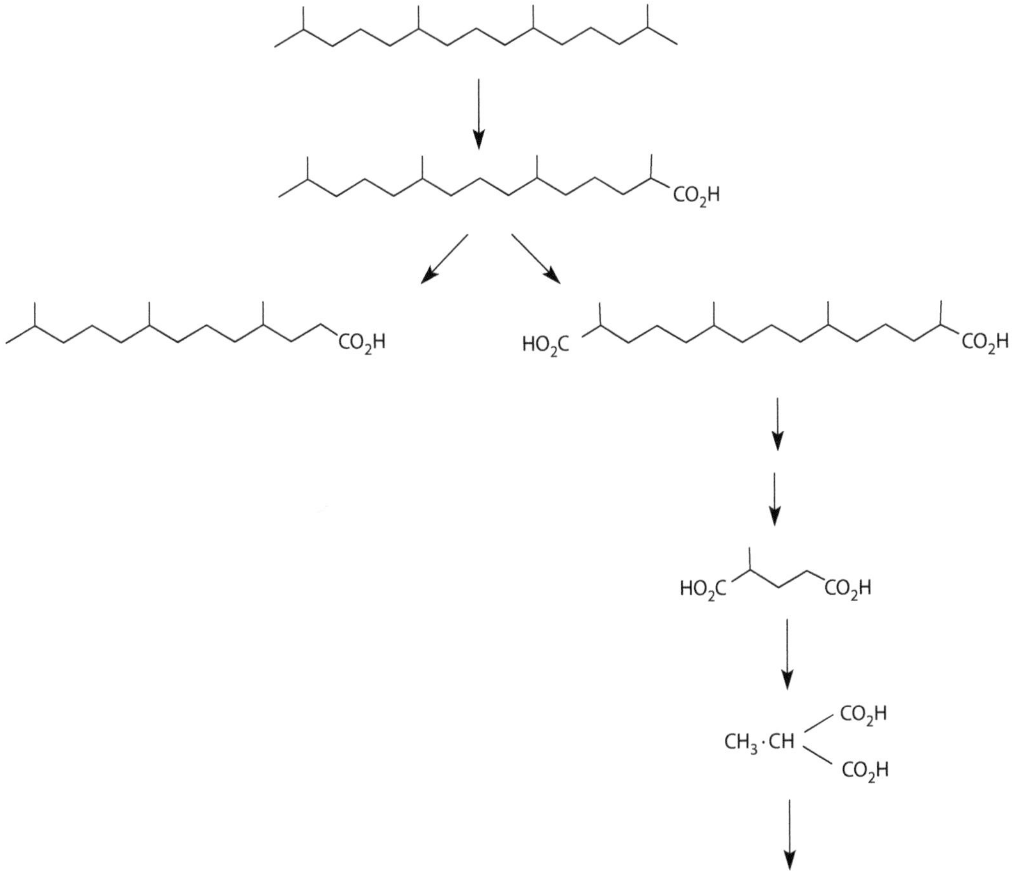

FIGURE 7.7 Pathways for the biodegradation of pristane.

pristane at 20°C, though not at 30°C. The key genes encoding enzymes for alkane degradation AlkB were identified (Takei et al. 2008).

When several methyl groups occur in branched-chain alkanes, the complex stereochemistry of the alkane may present an obstacle to degradation, and necessitate racemization at the centers with the methyl groups. This is encountered, for example, in the degradation of pristane, phytane, and squalane in *Mycobacterium* sp. strain P101. A recombinant α-methyacyl coenzyme A racemase Mcr has been purified from this strain and catalyzed the partial conversion of (*R*)-2-methylpentadecanoyl-CoA to the corresponding (*S*)-enantiomer (Sakai et al. 2004). A β-oxidation pathway in this strain was proposed to illustrate the several stereochemical configurations—and complications—that may be encountered in the degradation of pristane. In the rubber-degrading *Gordonia polyisoprenivorans* it was shown that the analogous *mcr* gene was expressed during degradation of poly(*cis*-1,4-isoprene) (Arenskötter et al. 2004).

The existence of chain branching may clearly present an obstacle to degradation by preventing β-oxidation, although this can be circumvented by a carboxylation pathway. This is outlined for the degradation of a branched-chain alkane (Figure 7.8): after dehydrogenation of the alkanol, metabolism of the carboxylates involved the coenzyme A esters. The critical step is carboxylation of the methyl group at C_3 followed by hydration of the unsaturated coenzyme A ester, oxidation, and loss of -CH_2CO_2H that is followed by hydrolytic fission of the β-ketoester. The cardinal details were initially elucidated for the degradation of the acyclic terpenoids citronellol and geraniol by *Pseudomonas citronellolis* (Seubert 1960), and the cardinal role of CO_2 was established (Seubert and Remberger 1963; Seubert et al. 1963). The alkanols were dehydrogenated to the carboxylates, and the oxidation of geraniol was dependent on molybdenum whereas the oxidation of citronellol was not (Höschle and Jendrossek 2004). This was followed by dehydrogenation to the 2-enecarboxyl-CoA esters. The further pathway involved two rather unusual enzyme activities for degradation of geraniol:

1. Carboxylation of the C_2 methyl group where the pathways converge with the formation of isohexenyl glutaconyl-CoA (Díaz-Pérez et al. 2004; Höschle et al. 2005)
2. Scission of 3-hydroxy-3-isohexenylglutaryl-CoA (Figure 7.9)

Additional details were provided by showing that the key degradative enzyme geranyl-CoA carboxylase was strongly induced by growth with citronellol in *Ps. citronellolis*, *Ps. Mendocina*, and *Ps. aeruginosa* (Cantwell et al. 1978), and the biotin-dependent enzyme has been purified and characterized from *Ps. citronellolis* (Seubert et al. 1963; Fall and Hector 1977). Indeed, this organism possesses several acyl-CoA carboxylases, including those for acetyl-, propionyl-, 3-methylcrotonyl-, and geraniol-CoA after growth on the relevant sources of carbon (Hector and Fall 1976).

This pathway is analogous to the carboxylation used for the degradation of both D- and L-leucine, and 3-methylcrotonate in *Pseudomonas aeruginosa* (Höschle et al. 2005). It is involved in the degradation of

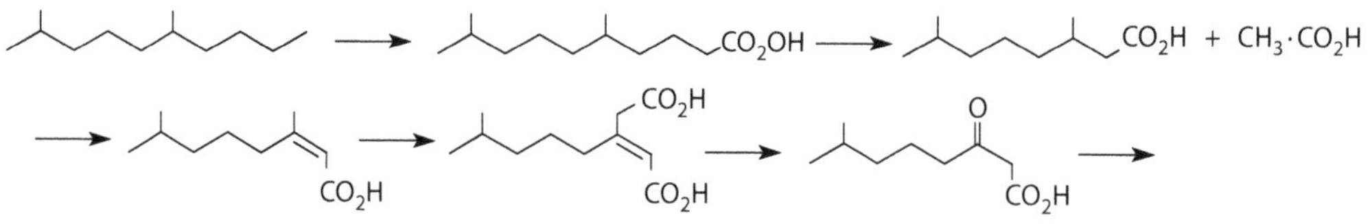

FIGURE 7.8   Degradation of a branched-chain alkane.

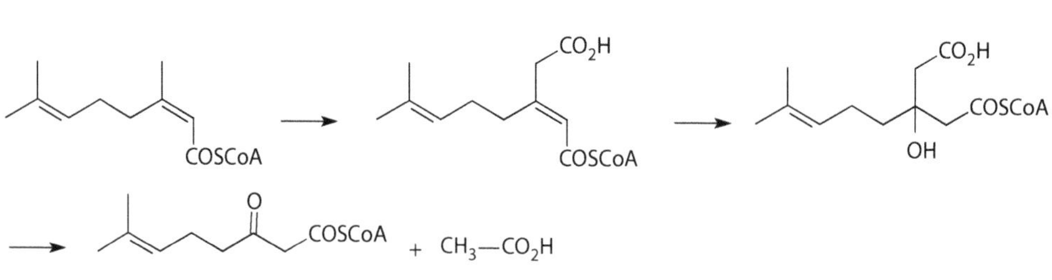

FIGURE 7.9   Scission of 3-hydroxy-3-isohexenylglutaryl-CoA.

6,10,14-trimethylpentadecan-2-one by *Marinobacter* sp. strain CAB (Rontani et al. 1997), and presumptively, in the degradation of squalane by *Mycobacterium* sp. (Berekaa and Steinbüchel 2000).

7.1.1.2.5 Quaternary-Substituted Substrates

An additional hindrance occurs for substrates with quaternary carbon atoms. These can be produced as terminal products, for example, 2,2-dimethylpropionate from the aerobic degradation of *tert*-butylbenzene and 2,2-dimethylheptane (Catelani et al. 1977). Both pivalate (Probian et al. 2003) and dimethylmalonate (Kniemeyer et al. 1999) can, however, be degraded under denitrifying anaerobic conditions (Figure 7.10). This may putatively involve a corrin-mediated rearrangement of pivalyl-CoA to 2-methylbutyryl-CoA (Rohwerder et al. 2006; Rohwerder and Müller 2008), analogous to the metabolism of *tert*-butanol that is initiated by oxidation to 2-hydroxyisobutyrate followed by an adenosylcobalamin-dependent rearrangement to 3-hydroxybutyrate (Figure 7.11) (Rohwerder et al. 2006; Müller et al. 2008). There are further examples of these reactions, including important substrates containing phenolic rings that provide entry to degradation by oxidation and rearrangement.

1. Rearrangement of bisphenol-A is analogous to that involved in the metabolism of 2,2-dimethylpropionate, and produces 4-hydroxyacetophenone and 4-hydroxybenzoate (Lobos et al. 1992; Spivack et al. 1994) after fission of the hydroxylated intermediates. These are produced in *Sphingomonas* sp. strain AO1 by a hydroxylase whose components have been purified, and consist of cytochrome P450, ferredoxin reductase, and ferredoxin (Sasaki et al. 2005) (Figure 7.12). An alternative mechanism produced hydroquinone and a number of metabolites arising from an intermediate cation produced from the side chain (Kolvenbach et al. 2007).

2. Although the degradation of alkylphenols can be initiated by oxidation of the alkyl groups to carboxylic acids, the degradation of nonylphenol isomers with quaternary side chains in strains of *Sphingomonas* sp. displays unusual features (Corvini et al. 2004; Gabriel et al. 2005a,b). This may be described formally as oxygenation at the position *para* to the phenol group, followed by either of two rearrangements: (a) migration of the alkyl group to the C-3 position of the phenolic ring with formation of a hydroquinone (Figure 7.13a), or (b) insertion of the oxygen from the hydroxyl group between the alkyl group and the phenolic ring followed by oxygenation with loss of the tertiary alkanol and formation of benzoquinone (Figure 7.13b). An alternative mechanism that differs slightly has also been proposed (Porter and Hay 2007).

7.1.1.2.6 Distribution of Hydroxylation

Alkane degradation has been observed in the following bacteria: these include the Gram-negative *Pseudomonas putida*; *Ps. aeruginosa*; *Acinetobacter* sp.; *Hydrocarboniphaga effusa* (Palleroni et al.

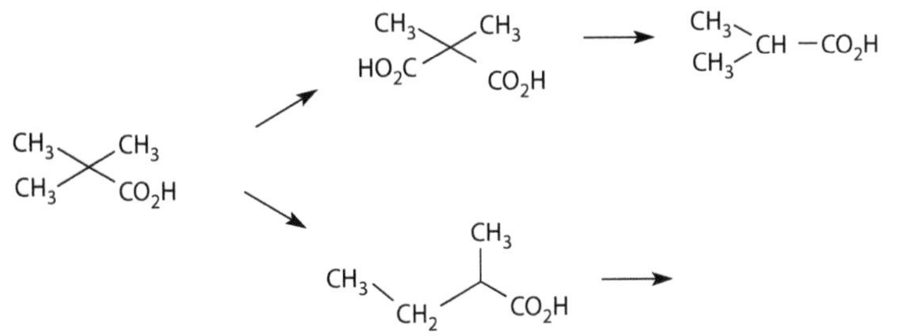

FIGURE 7.10 Degradation of 2,2-dimethylpropionate. (Adapted from Probian C, A Wülfing, J Harder. 2003. *Appl Environ Microbiol* 69: 1866–1870; Ratnatilleke A, JW Vrijbloed, JA Robinson. 1999. *J Biol Chem* 274: 31679–31685.)

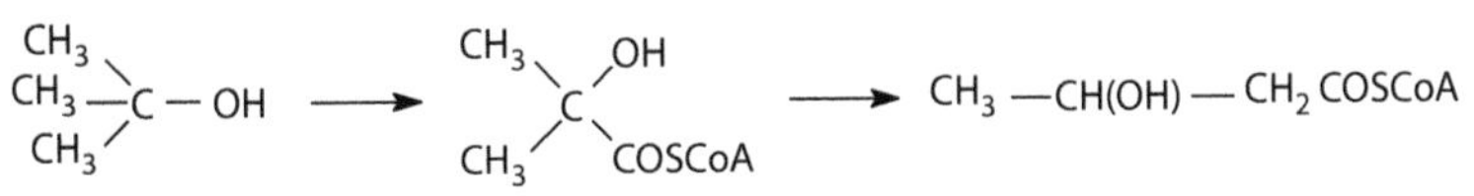

FIGURE 7.11 Rearrangement mechanism in degradation.

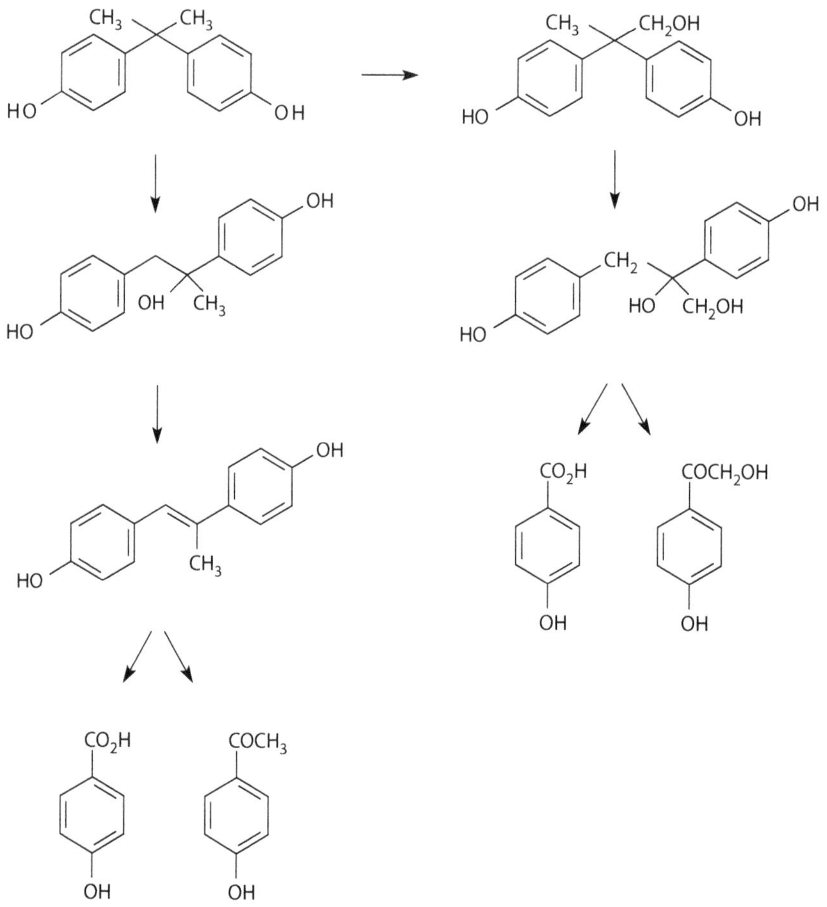

FIGURE 7.12 Degradation of bisphenol-A. (Adapted from Spivack J, TK Leib, JH Lobos. 1994. *J Biol Chem* 269: 7323–7329.)

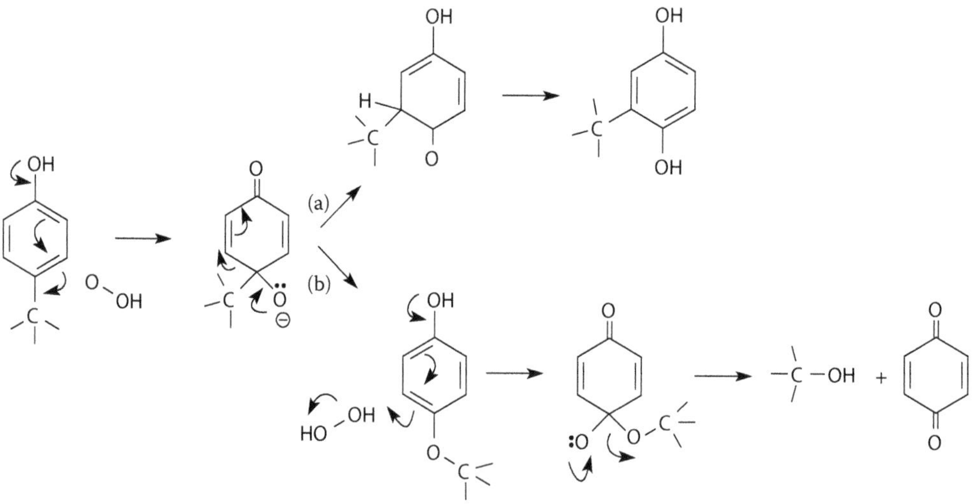

FIGURE 7.13 Degradation of nonylphenol. (a) Migration from C3 to produce 2-alkylhydroquinone, (b) oxygenation at C4 to produce benzoquinone.

2004); the marine organisms *A. borkumensis* (Yakimov et al. 1998), *T. oleivorans* (Yakimov et al. 2004) and *O. antarctica* (Yakimov et al. 2003b); and the Gram-positive *Rhodococcus erythropolis*; *Rh. rhodochrous*; *Mycobacterium* sp., and *Geobacillus subterraneus*, and *Geobacillus uzenensis* (Nazina et al. 2001). Studies using oil-contaminated seawater showed the prevalence of *Alcanivorax* in the degradation of the branched-chain pristane and phytane (Hara et al. 2003), whereas the degradation of C_{12} to C_{32} *n*-alkanes was carried out by *T. oleivorans* (McKew et al. 2007).

Gordonia sp. strain SoCg is able to grow on *n*-alkanes from C_{12} to C_{36}, and terminal hydroxylation has been demonstrated for C_{16} (Piccolo et al. 2011).

7.1.1.3 Yeasts

7.1.1.3.1 C_1 Degradation

Yeasts are not generally able to grow at the expense of methane, although many of them can utilize methanol. This is noted in another section.

7.1.1.3.2 $C_{>4}$ Degradation

There has been considerable interest in alkane degradation by yeasts that have been assigned to several genera such as *Candida*, *Endomycopsis*, and *Yarrowia*. They are able to degrade alkanes with chain lengths >4 (Käppeli 1986; Tanaka and Ueda 1993): examples include C_{10} to C_{14} *n*-alkanes and C_{12} to C_{14} *n*-alk-1-enes by *Candida tropicalis* (Gill and Ratledge 1972), C_{10} to C_{18} *n*-alkanes by *Yarrowia lipolytica* (Mauersberger et al. 2001), and C_{10} and C_{16} by *Trichosporon adeninovorans* (Middelhoven et al. 1984). The alkane hydroxylase of *C. tropicalis* is located within the microsomes that contain cytochrome P450 and NADPH-cytochrome *c* reductase (references in Käppeli 1986). The degradation of the resulting alkanoates is, however, carried out in peroxisomes that contain the β-oxidation enzymes—alkanoate oxidase, enoyl-CoA hydratase, and 3-hydroxyacyl-CoA dehydrogenase. *C. tropicalis* is also able to convert alkanes and alkanoates to α,ω-dicarboxylates (Craft et al. 2003) that may be excreted into the medium, and this has been attributed to the activity of *CYP 52* that has been characterized (Eschenfeldt et al. 2003). In yeasts here are several cytochrome P450s that catalyze the primary hydroxylation of *n*-alkanes, and in *Y. lipolytica* there are multiple genes encoding cytochrome P450s that display different substrate selectivity (Iida et al. 2000). In addition, the monooxygenase in *Candida maltosa* is able to carry out the metabolism of hexadecane in a cascade of reactions by bis-terminal hydroxylation and dehydrogenation to the α,ω-dicarboxylate (Scheller et al. 1998), in addition to the monohydroxylation to the terminal carboxylate that is fed into the peroxisomes.

7.1.1.4 Range of Alkane Substrates for Bacteria

It is convenient to discuss the range of substrates into the following groups: C_1–C_4; C_5–C_{16}; and >C_{16} (van Beilen and Funhoff 2007). These are not, however, reciprocally exclusive and there is considerable overlap between the metabolic activities of Gram-negative and Gram-positive organisms. It is worth noting that genomic analysis of the methylotroph *Methylibium petroleiphilum* revealed genes that encoded the oxidation not only of C_5 to C_{12} *n*-alkanes, but also monocyclic arenes (Kane et al. 2007).

In the following sections there is some inevitable overlap in the range of substrates.

7.1.1.4.1 C_2 to C_5

Monooxygenases that can carry out hydroxylation of butane have been found in the Gram-negative *Thauera (Pseudomonas) butanovora* (Hamamura et al. 1999) and *Ps. putida* GPo1 (Johnson and Hyman 2006). In *Thauera butanovora*, the broad-substrate soluble diiron hydroxylase (sBMO) is also able to accept up to C_9 alkanes and, like the sMMO in *M. trichosporium* OB3b (Fox et al. 1989) consists of three proteins: a hydroxylase with three subunits, an oxidoreductase, and an effector (Sluis et al. 2002; Dubbels et al. 2007). Whereas the hydroxylase carried out primarily terminal hydroxylation, it was also able to accept isobutane and isopentane (Dubbels et al. 2007). An important metabolic limitation has been observed for the degradation of odd-numbered alkanes: propionate produced by hydroxylation and dehydrogenation is a potent repressor of the monooxygenase (Doughty et al. 2006).

The metabolism of propane and butane by *Mycobacterium vaccae* strain JOB-5 exhibits significant structural differences: propane is hydroxylated at C_2 (Vestal and Perry 1969) followed by dehydrogenation to acetone and fission to acetate and a C_1 fragment that is incorporated into the C_1 metabolite pool (Coleman and Perry 1984), whereas in this strain, butane undergoes terminal hydroxylation (Phillips and Perry 1974). The hydroxylation of propane at C_2 has also been observed for *Gordonia* sp. strain TY-5 that is also able to grow with alkanes from C_{13} to C_{22} (Kotani et al. 2003). Attention has been directed to the evolution of butane oxidation (Koch et al. 2009) by manipulation of the terminal hydroxylase AlkB from *Ps. putida* GPo1 and the class II soluble CYP 153A5 from *Mycobacterium* sp. strain HXN-1500 both of which normally have substrate ranges of C_6–C_{16}.

7.1.1.4.2 C_8 to C_{16}

This range of *n*-alkanes are degraded optimally both by the Gram-negative *Ps. putida* GPo1; *Ps. aeruginosa* PAO1; *A. borkumensis* (van Beilen et al. 2004); and the Gram-positive *Rh. rhodochrous*; and *Mycobacterium* strain HXN 1500 (van Beilen et al. 2002a). Indeed, the thermophilic Gram-negative species of *Theromleophilum* were restricted to growth at the expense of C_{12} to C_{19} *n*-alkanes and were reported to be unable to use C_{12} to C_{19} alk-1-enes, C_{12} to C_{18} *n*-alkanols, or C_{13} to C_{19} *n*-ketones (Zarilla and Perry 1984). A later study (Yakimov et al. 2003a) showed, however, that the cell-wall morphology revealed the ultrastructure typical of Gram-positive bacteria, and it was suggested that these organisms belonged to a subclass of *Actinobacteria*. A number of Gram-negative marine bacteria have limited metabolic capacity except for alkanes: the growth of *Thalassolitus oleivorans* was restricted to C_7 to C_{20} alkanes and their oxygenated relatives (Yakimov et al. 2004), and the marine *O. antarctica* was also restricted to C_{10} to C_{18} alkanes (Yakimov et al. 2003b).

7.1.1.4.3 $C_{>16}$

1. The Gram-negative marine *M. hydrocarbonoclasticus* is able to use C_{14}, C_{16}, and C_{20} *n*-alkanes as well as pristane (Gauthier et al. 1992) and *Pseudomonas fluorescens* contains a second hydroxylase for oxidation of C_{18} to C_{28} alkanes (Smits et al. 2002). The hydroxylase from the Gram-negative *Acinetobacter* strain M-1 is able to accept alkanes up to C_{30}, although maximally C_{12} (Maeng et al. 1996), and alkanes from C_{10} to C_{40} by *Acinetobacter* sp. strain DSM 17874 (Throne-Holst et al. 2007).

2. Some Gram-positive strains of *Rh. erythropolis* can grow with *n*-alkanes with chain lengths from C_{20} and C_{32} (van Beilen et al. 2002a). Analysis of the genome and proteome of strain NG890-2 of the facultative anaerobe *Geobacillus thermodenitrificans* revealed the presence of enzymes required for the degradation of *n*-hexadecane, and an alkane monooxygenase that was able to accept C_{15} to C_{36} *n*-alkanes was characterized (Feng et al. 2007). Specifically, it did not have enzymes either homologous to AlkB that is required for many Gram-negative *n*-alkane-degrading bacteria or to established monooxygenases. A strain of *Gordonia* SoCg is able to grow with alkanes from C_{12} to C_{36} (Piccolo et al. 2011).

3. A number of strains of the Gram-positive *Dietzia* sp. are able to grow at the expense of long-chain *n*-alkanes:

    a. *Dietzia maris* with C_6 to C_{17}, C_{19}, and C_{23} *n*-alkanes (Rainey et al. 1995)

    b. The facultatively psychrophilic alkaliphile *Dietzia psychralcaliphila* with decanes, eicosane, and tetracosane—though not with pristane (Yumoto et al. 2002)

    c. *Dietzia* sp. strain DQ 12-45-1b with C_6 to C_{40} *n*-alkanes. This strain contained the alkane integral-membrane monooxygenase (AlkB) fused to rubredoxin (Rd) encoded by *alkW1* and *alkW2* of which the first was active for the degradation of C_8 to C_{32} *n*-alkanes (Nie et al. 2011).

4. Degradation of pristane 2,6,10,12-tetramethylhexadecane by *Mycobacterium ratisbonense* (Silva et al. 2007) was initiated by hydroxylation at terminal positions followed by dehydrogenation to 3,7,11,15-tetramethylhexadecanoic acid and 2,6,10,14-tetramethylhexadecanoic acid, or by oxidation to 2-ethyl-6,10,14-trimethylpentadecanoic acid. These may undergo either

β-oxidation to C_2-C_3 units or intra-esterification to wax esters. The 3,7,11,15-tetramethylhexa-decan-2-one was produced by subterminal oxidation.

7.1.1.4.4 Mechanisms

The aerobic degradation of alkanes is initiated by the introduction of oxygen that can be accomplished by different mechanisms: monooxygenation, hydroxylation, and cytochrome P450 monooxygenation.

1. Although the metabolism of methane by MMOs has already been discussed, it is important to appreciate that the degradation of propane and butane may also be carried out by monooxygenases (PMO and BMO). These activities have been found in both the Gram-positive *M. vaccae* strain JOB-5 and the Gram-negative *Thauera (Pseudomonas) butanovora* (Hamamura et al. 1999).

2. Considerable study has been devoted to the OCT plasmid in *Ps. putida (oleovorans)* GPo1 that provided much of the basic details of the enzymology, including that of ω-hydroxylation. Briefly, the genes that encode the enzymes for hydroxylation and sequential dehydrogenation of the carboxylate are carried in an operon (*alkBAC*). Hydroxylation is carried out by three components, an integral membrane diiron alkane hydroxylase encoded by *alkB*, while *alkC* encodes the dehydrogenase for the resulting alkanol (van Beilen et al. 1992). The activity of the hydroxylase requires cooperation with rubredoxin 2 encoded by *alkG* (Kok et al. 1989b) and rubredoxin reductase encoded by *alkT* (Eggink et al. 1990) that transfer the necessary electrons from NADH. The alkane hydroxylase has been identified and sequenced in *Ps. putida (oleovorans)* (Kok et al. 1989a), while the components of the ω-hydroxylase in this strain are analogous to those involved in α-hydroxylation (McKenna and Coon 1970). The ω-hydroxylase has been reported to require phospholipid for catalytic activity (Ruettinger et al. 1974), and has been characterized as a diiron cluster in an integral-membrane by Mössbauer spectroscopy (Shanklin et al. 1997). Rubredoxin has been characterized in *Pseudomonas oleovorans*, and may contain one or two atoms of iron (Lode and Coon 1971; Kok et al. 1989b).

There are a number of additional details that should be added.

1. The identity of sequences of AlkB hydroxylase among homologues from *Ps. putida* GPo1, *Ps. aeruginosa* PAO1, *Ps. fluorescens* CHAO and *A. borkumensis* was rather low (Smits et al. 2002).

2. Some strains have multiple hydroxylases. These include *Ps. aeruginosa* strains PAO1 and RR1 (Marín et al. 2003), *Rhodococcus* sp. strains Q 15 and NRRL B-16531 (Whyte et al. 2002), and *A. borkumensis* (van Beilen et al. 2004). *Acinetobacter* sp. strain M-1 has two hydroxylases whose regulation is dependent on the chain-length of the alkane C_{16} to C_{22} for *alkMb* and $C_{<22}$ for *alkMa*, while both share the same rubredoxin and rubredoxin reductase (Tani et al. 2001).

3. In *Ps. putida* ppG1 that carries the OCT plasmid, there is duplication of some of the loci. Whereas those for alkane hydroxylation (*alkA, alkB,* and *alkC*) and for alkanol dehydrogenation (*alcO*) occur on the plasmid, those for *alcA* and *alcB*, and for aldehyde dehydrogenation (*aldA, aldB*) occur in the chromosome (Grund et al. 1975).

4. The *alk genes* that are involved in alkane hydroxylation in the OCT plasmid of *Ps. putida* GPo1 can be induced by dicyclopropane (Grund et al. 1975), and this facility has been applied to the oxidation of methyl *t*-butyl ether (MTBE) (Smith and Hyman 2004).

5. The alkane hydroxylase from *Acinetobacter* sp. strain ADP1 is encoded by *alkM* that is quite different from *alkb* and whose genes are not linked to the rubredoxin genes (Ratajczak et al. 1998).

6. It has been shown that there are two groups of rubredoxin reductases (AlkG1 and AlkG2), both of which generally occur in strains that can degrade alkanes (van Beilen et al. 2002b).

7. For alkane degradation by *Nocardioides* strain CF 8, the hydroxylation of C_2 to C_4 alkanes is carried out by a Cu-monooxygenase, whereas for C_6 to C_{16} alkanes a binuclear Fe integral-membrane hydroxylase is employed (Hamamura et al. 2001).

8. The three-component ω-hydroxylation system of *Ps. oleovorans* is able, in addition, to carry out epoxidation of octa-1,7-diene to 7,8-epoxy-1-octene and 1,2,7,8-diepoxyoctane (May and Abbott 1973).

7.1.1.4.5 Regulation of Hydroxylation

The regulation of the genes encoding alkane hydroxylation is complex and only a brief outline is given here. The *alkBFGHJKL* operon encodes genes for the hydroxylation of alkanes and successive dehydrogenation to alkanoates. Induction of the enzymes is regulated by the AlkS protein that activates expression of the promotor *PalkB*, while catabolic repression lowers expression of the two AlkS-activated promotors *PalkB* and *PalkS2* (Yuste and Rojo 2001). Catabolic repression control (Crc) has been implicated in repression of genes involved in the nitrogen metabolism of strains of *Pseudomonas*, and is effective in *Ps. putida* GPo1. Although Crc is important in catabolic repression when cells are grown in a complex medium (Luria-Bertani broth) (Yuste and Rojo 2001), repression with cells grown with organic acids such as succinate or lactate is dependent on the activity of cytochrome *o* ubiquinol oxidase (Dimamarca et al. 2002). Both activities are, therefore, generally involved in catabolite repression by substrates other than alkanes. It is worth noting that the alkane hydroxylase gene of *Burkholderia cepacia* strain RR10 is under carbon catabolite repression to an extent much greater than for other alkane degraders (Marín et al. 2001).

7.1.1.4.6 Cytochrome P450

The hydroxylation of *n*-octane in *Gordonia* (*Corynebacterium*) sp. strain 7E1C was resolved into a flavoprotein and a cytochrome P450 (Cardini and Jurtshuk 1970). A number of strains with the capacity for alkane hydroxylation do not contain the integral Fe-membrane hydroxylase, and hydroxylation is then carried out by a cytochrome P450 that has been designated CYP 153 (van Beilen et al. 2006). The reductant from NADH is provided by ferredoxin and ferredoxin reductase. Whereas in *Acinetobacter calcoaceticus* strain 69-V, cytochromes *b* and *o* were present during growth with *n*-hexadecane, cytochrome P450 was found in other strains growing at the expense of the same substrate (Asperger et al. 1981), and the cytochrome P450 that is designated CYP 153 has been characterized in *Acinetobacter* sp. strain EB104 (Maier et al. 2001). The cytochrome P450 CYP153 A6 in *Mycobacterium* sp. strain HXN-1500 is able to hydroxylate alkanes from C_6 to C_{11} including 2-methyloctane (Funhoff et al. 2006), while proteomic analysis of *A. borkumensis* suggests the presence of a cytochrome P450 monooxygenase in addition to *alkB1* (Sabirova et al. 2006).

Sub-terminal ω-oxidation of C_{12} to C_{18} long-chain carboxylic acids, amides and alcohols—though the esters or the corresponding alkanes—at the Δ^1, Δ^2, and Δ^3 positions has been observed with a soluble cytochrome $P450_{BM-3}$ system from *Bacillus megaterium* (Miura and Fulco 1975). This enzyme is, exceptionally, inducible with barbiturates, and the fact that the enzyme is unusual in having a single 119 kDa protein coupling NADH reduction to oxygenation (Narhi and Fulco 1986, 1987).

7.1.1.4.7 Other Mechanisms

Occasionally, dioxygenation has been postulated.

1. In the hydroxylation of *n*-propane by *Nocardia paraffinicum* (*Rh. rhodochrous*), the ratio of alkane to oxygen consumed was 2:1, and this is consistent with an intermolecular reaction involving two molecules of propane and one molecule of oxygen to produce two molecules of alkanol (Figure 7.14a) (Babu and Brown 1984). This reaction is formally analogous to the oxidation of 2-nitropropane to acetone by flavoenzymes in the *Hansenula mrakii* (Kido et al. 1978) and *Neurospora crassa* (Gorlatova et al. 1998) whose details are discussed in Chapter 11.

2. The involvement of a hydroperoxide in alkane degradation was proposed many years ago and appears to operate in *Acinetobacter* sp. strain M-1 that has an unusual range of substrates

(a) $2CH_3-CH_2-CH_3 + O_2 \longrightarrow 2CH_3CH(OH)CH_3 \longrightarrow 2CH_3-CO-CH_3$

(b) $R-CH_2-CH_3 + O_2 \longrightarrow R-CH_2-CH_2-OOH \longrightarrow R-CH_2-CO-OOH \longrightarrow R-CH_2-CHO \longrightarrow R-CH_2-CO_2H$

FIGURE 7.14 Metabolism of n-propane by (a) dioxygenation to acetone, (b) hydroperoxidation to ethane carboxaldehyde.

including both alkanes and alkenes (Maeng et al. 1996). The purified enzyme contained FAD and required Cu^{2+} for its activity but, unusually did not require NAD(P)H (Figure 7.14b).

7.1.2 Anaerobic Conditions

7.1.2.1 Methane

It is 30 years since the anaerobic oxidation of methane was reported by Zehnder and Brock (1980), and there has been renewed interest in this in the light of interest in the fate of methane and the possibility that anaerobic sediments where it is produced by methanogenesis represent an important sink. It is worth noting the generation of methane in the roots of rice plants where the roots provide the source of carbon for methanogenesis (Lu et al. 2005). Although the organisms responsible have not hitherto been isolated in pure culture, several possibilities have been put forward.

1. In the marine environment, the anaerobic oxidation of methane probably may involve consortia of archaea with sulfate-reducing bacteria (Teske et al. 2002; Schouten et al. 2003; Girguis et al. 2005), and may possibly occur by reverse methanogenesis (Hallam et al. 2004). This is supported by the results of simulation experiments using methane and sulfate in the presence of the methane-producing archaeon ANME-2 and the cluster of *Desulfosarcina/Desulfococcus* (Nauhaus et al. 2007). Microscopic examination showed that the archaea formed a central cluster with peripheral layers of bacteria, and lipid analysis showed the presence of those attributed to ANME-2 archaea. In another study (Inagaki et al. 2004), genes for both anaerobic methane oxidation and methanogenesis were found together with those for aerobic methane oxidation at the surface of a single site in Japan.

2. An alternative reaction has been suggested for freshwater habitats involving nitrite that is not present at effective concentrations in the marine environment (Ettwig et al. 2008). In long-term enrichment, methane was oxidized at the expense of the reduction of nitrite to molecular nitrogen. Significantly, after a period exceeding one year, there was no inhibition of methane oxidation by bromoethane sulfonate that is a specific inhibitor of the cardinal gene in archaea. The reaction was described by the reaction $3CH_4 + 8HO-NO \rightarrow 3CO_2 + 4N_2 + 10H_2O$, and quantitative PCR revealed enrichment of bacteria belonging to the NC10 phylum (Ettwig et al. 2009).

3. Anaerobic methane oxidation has been observed in the freshwater oligotrophic Lake Constance. Oxidation of $^{14}CH_4$ to $^{14}CO_2$ was carried out by organisms in the NC10 phylum that was verified by 16S rRNA amplicons from DNA extracts at various sites (Deutzmann and Schink 2011).

On the other hand, while the biochemistry of methanogenesis is fully understood, it is appropriate to note the roles of two specialized coenzymes that occur in methanogens:

1. Coenzyme F_{420} a 5-deaza-8-hydroxypterin with a glutamylglutamate side chain that is involved in the activity of hydrogenase, formate dehydrogenase, and $NADP^+$ reductase

2. The pterin methanopterin that functions as a C_1-carrier and is used also by methylotrophic bacteria

It is worth noting that (a) the anaerobic methanogen *Methanomethylovorans hollandica* is able to grow with both dimethyl sulfide and methanethiol in addition to methanol and methylamines (Lomans et al. 1999), and (b) that in *Methanosarcina*, a number of C_1 substrates are able to take part in methanogenesis: these include methanol (Sauer et al. 1997), dimethyl sulfide (Tallant et al. 1997, 2001), and di- and trimethylamines (Paul et al. 2000). In all of them, the key reaction is the methyl transfer of these C_1 units to coenzyme M (2-mercapto-ethanesulfonate).

7.1.2.2 Higher Alkanes

There is evidence for the anaerobic degradation of alkanes to CO_2 plausibly under conditions of sulfate-reduction. In experiments with sediment slurries from contaminated marine areas, $^{14}CO_2$ was recovered from ^{14}C-hexadecane (Coates et al. 1997), and was inhibited by molybdate that is consistent with the involvement of sulfate reduction, and ^{14}CO$_2$ was produced from ^{14}C[14,15]octacosane (C$_{28}$H$_{58}$) under sulfate-reducing conditions (Caldwell et al. 1998). Aliphatic hydrocarbons may be completely oxidized to CO$_2$ by a sulfate-reducing bacterium (Aeckersberg et al. 1998), and further examples are given later.

Different mechanisms have been elucidated for the anaerobic degradation of higher alkanes, and both of them occurred simultaneously in a sulfate-reducing consortium (Callaghan et al. 2006). Anaerobic growth at the expense of hexadecane without evidence of the pathway has been reported for the archaeon *Thermococcus sibiricus* (Mardanov et al. 2009). Two broad types of mechanism have been resolved, without associating the physiology of the organisms with the mechanism.

7.1.2.3   Reaction with Fumarate

1. Reaction with fumarate at the subterminal position with the production of (1-methylpentyl)suc-cinate was initiated by a free-radical reaction in a denitrifying organism during *n*-hexane metabolism (Rabus et al. 2001; Wilkes et al. 2002) (Figure 7.15). It has been suggested (Grundmann et al. 2008) on the basis of genetic evidence that in strain HxN1 the activation by MasDEC is analogous to that by benzylsuccinate synthase BssABC during the degradation of toluene.

2. This reaction has been observed for a sulfate-reducing organism that degraded hexadecane (Callaghan et al. 2006): this was followed by a rearrangement analogous to that of succinate → malonate, with the sequential loss of acetate. Analogous reactions have been described for a sulfate-reducing *Desulfoglaeba alkanexedens* (Davidova et al. 2006), and the sulfate-reducing *Desulfatibacillum aliphaticivorans* CV2803 that could oxidize C$_{13}$ to C$_{18}$ alkanes (Cravo-Laureau et al. 2005). An analogous pathway is used for the degradation of ethyl-cyclopentane by a sulfate-reducing enrichment culture that is discussed in Section 7.3 of this chapter.

7.1.2.4   Carboxylation

An alternative reaction was initiated by carboxylation at C$_3$ and has been found in the sulfate-reducing *Desulfococcus oleovorans* strain Hxd3 (So et al. 2003). This was followed by loss of the terminal C$_2$ fragment to produce a terminal carboxylate with one less carbon atoms, and further degradation of the resulting terminal carboxylate by β-oxidation (Figure 7.16). This pathway has also been proposed for the degradation of hexadecane by a component of a nitrate-reducing enrichment culture. In addition, deute-rium-labeled hexadecane produced a range of long-chain carboxylates, while C-methylation took place at C$_{10}$ (Callaghan et al. 2009).

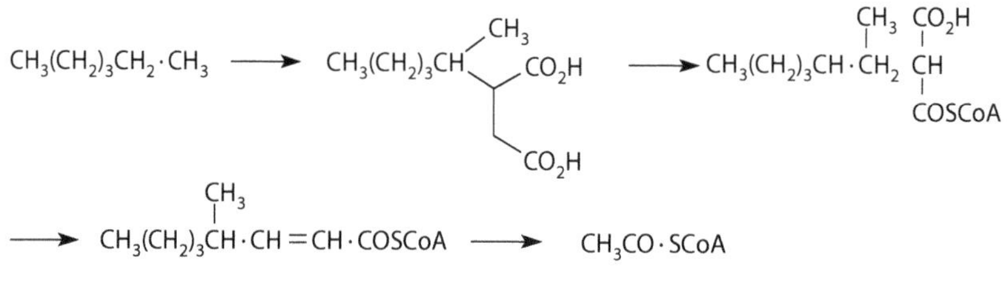

FIGURE 7.15   Anaerobic degradation of *n*-alkanes by the fumarate pathway.

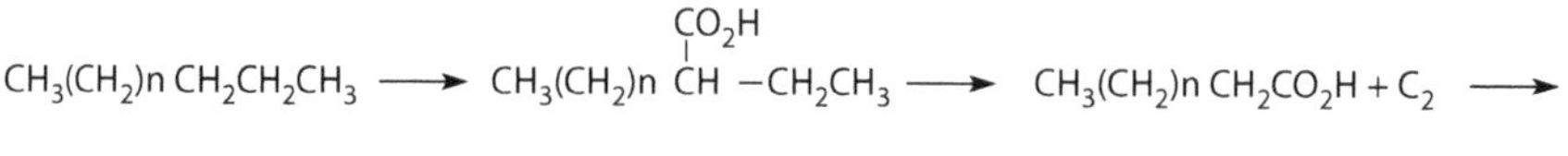

FIGURE 7.16   Anaerobic degradation of *n*-alkanes by carboxylation.

7.2 References

Aeckersberg F, F Rainey, F Widdel. 1998. Growth, natural relationships, cellular fatty acids and metabolic adaptation of sulfate-reducing bacteria that utilize long-chain alkanes under anoxic conditions. *Arch Microbiol* 170: 361–369.

Anthony C. 1982. *The Biochemistry of Methylotrophs.* Academic Press, London.

Arenskötter M, D Bröker, Steinbüchel A. 2004. Biology of the metabolically diverse genus *Gordonia. Appl Environ Microbiol* 70: 3195–3204.

Asperger O, A Naumann, H-P Kleber. 1981. Occurrence of cytochrome P-450 in *Acinetobacter* strains after growth on *n*-hexadecane. *FEMS Microbiol Lett* 11: 309–312.

Babu JP, LR Brown. 1984. New type of oxygenase involved in the metabolism of propane and isobutane. *Appl Environ Microbiol* 48: 260–264.

Berekaa MM, A Steinbüchel. 2000. Microbial degradation of multiply branched alkane 2,6,10,15,19,23-hexamethyltetracosane (squalane) by *Mycobacterium fortuitum* and *Mycobacterium ratisbonense. Appl Environ Microbiol* 66: 4462–4467.

Brämer CO, A Steinbüchel. 2001. The methylcitrate acid pathway in *Ralstonia eutropha*: New genes identified involved in propionate metabolism. *Microbiology (UK)* 147: 2203–2214.

Broadway NM, FM Dickinson, C Ratledge. 1993. The enzymology of dicarboxylic acid formation by *Corynebacterium* sp. strain 7E1C grown on *n*-alkanes. *J Gen Microbiol* 139: 1337–1344.

Burrows KJ, A Cornish, D Scott, IJ Higgins. 1984. Substrate specificities of the soluble and particulate methane mono-oxygenases of *Methylosinus trichosporium* OB3b. *J Gen Microbiol* 130: 3327–3333.

Caldwell ME, RM Garrett, RC Prince, JM Suflita. 1998. Anaerobic biodegradation of long-chain *n*-alkanes under sulfate-reducing conditions. *Environ Sci Technol* 32: 2191–2195.

Callaghan AV, LM Gieg, KG Kropp, JM Suflita, LY Young. 2006. Comparison of mechanisms of alkane metabolism under sulfate-reducing conditions among two bacterial isolates and a bacterial consortium. *Appl Environ Microbiol* 72: 4274–4282.

Callaghan AV, M Tierney, CD Phelps, LY Young. 2009. Anaerobic biodegradation of *n*-hexadecane by a nitrate-reducing consortium. *Appl Environ Microbiol* 75: 1339–1344.

Cantwell SG, EP Lau, DS Watt, RR Fall. 1978. Biodegradation of acyclic isoprenoids by *Pseudomonas* species. *J Bacteriol* 135: 324–333.

Cardini G, P Jurtshuk. 1970. The enzymatic hydroxylation of *n*-octane by *Corynebacterium* sp. strain 7E1C. *J Biol Chem* 245: 2789–2796.

Catelani D, A Colombi, C Sorlini, V Trecani. 1977. Metabolism of quaternary carbon compounds: 2,2-dimethylheptane and *tert*-butylbenzene. *Appl Environ Microbiol* 34: 351–354.

Claes WA, A Pühler, J Kalinowski. 2002. Identification of two *prpDBC* gene clusters in *Corynebacterium glutamicum* and their involvement in propionate degradation via the 2-methylcitrate cycle. *J Bacteriol* 184: 2728–2739.

Coates JD, J Woodward, J Allen, P Philip, DR Lovley. 1997. Anaerobic degradation of polycyclic aromatic hydrocarbons and alkanes in petroleum-contaminated marine harbor sediments. *Appl Environ Microbiol* 63: 3589–3593.

Colby J, DI Stirling, H Dalton. 1977. The soluble methane mono-oxygenase of *Methylococcus capsulatus* (Bath). Its ability to oxygenate *n*-alkanes, *n*-alkenes, ethers, and alicyclic, aromatic and heterocyclic compounds. *Biochem J* 165: 395–402.

Coleman JP, JJ Perry. 1984. Fate of the C_1 product of propane dissimilation in *Mycobacterium vaccae. J Bacteriol* 160: 1163–1164.

Corvini PFX, RJW Meesters, A Schäffer, HF Schröder, R Vinken, J Hollender. 2004. Degradation of a nonylphenol single isomer by *Sphingomonas* sp. strain TTNP3 leads to a hydroxylation-induced migration product. *Appl Environ Microbiol* 70: 6897–6900.

Craft DL, KM Madduri, M Eshoo, CR Wilson. 2003. Identification and characterization of the *CYP52* family of *Candida tropicalis* ATCC 20336, important for the conversion of fatty acids and alkanes to α,ω-dicarboxylic acids. *Appl Environ Microbiol* 69: 5983–5991.

Cravo-Laureau C, V Grossi, D Raphel, R Matherton, A Hirschler-Réa. 2005. Anaerobic *n*-alkane metabolism by a sulfate-reducing bacterium *Desulfatibacillum aliphaticivorans* strain CV2803[T]. *Appl Environ Microbiol* 71: 3458–3467.

Davidova IA, KE Duncan, OK Choi, JM Suflita. 2006. *Desulfoglaeba alkanexedens* gen. nov., sp. nov., an *n*-alkane-degrading, sulfate-reducing bacterium. *Int J Syst Evol Microbiol* 56: 2737–2742.

Davydov R, AM Valentine, S Komar-Panicucci, BM Hoffman, SJ Lippard. 1999. An EPR study of the binuclear iron site in the soluble methane monooxygenase from *Methylococcus capsulatus* (Bath) reduced by one electron at 77 K: The effects of component interactions and the binding of small molecules to the diiron (III) center. *Biochemistry* 38: 4188–4197.

Dedysh SN, C Knief, PF Dunfield. 2005. *Methylocella* species are facultatively methanotrophic. *J Bacteriol* 187: 4665–47670.

Deutzmann JS, B Schink. 2011. Anaerobic oxidation of methane in sediments of Lake Constance, an oligotrophic freshwater lake. *Appl Environ Microbiol* 77: 4429–4436.

Díaz-Pérez AL, AN Zavala-Hernández, C Cervantes, J Campos-García. 2004. The *gnyRDBHAL* cluster is involved in acyclic isoprenoid degradation in *Pseudomonas aeruginosa*. *Appl Environ Microbiol* 70: 5102–5110.

Dimamarca MA, A Ruiz-Manzano, F Rojo. 2002. Inactivation of cytochrome *o* ubiquitol oxidase relieves catabolic repression of the *Pseudomonas putida* GPo1 alkane degradation pathway. *J Bacteriol* 184: 3785–3793.

Doughty DM, LA Sayavedra-Soto, DJ Arp, PJ Bottomley. 2006. Product repression of alkane monooxygenase expression in *Pseudomonas butanovora*. *J Bacteriol* 188: 2586–2592.

Dubbels BL, LA Sayavedra-Soto, DJ Arp. 2007. Butane monooxygenase of "*Pseudomonas butanovora*": purification and biochemical characterization of a terminal-alkane hydroxylating diiron monooxygenase. *Microbiology (UK)* 153: 1808–1816.

Dunfield PF et al. 2007. Methane oxidation by an extremely acidophilic bacterium of the phylum *Verrumicrobia*. *Nature* 450: 879–882.

Eggink G, H Engel, G Vriend, P Terpstra, B Witholt. 1990. Rubredoxin reductase of *Pseudomonas oleovorans*. Structural relationship to other flavoprotein oxidoreductases based on one NAD and two FAD fingerprints. *J Mol Biol* 212: 135–142.

Erb TJ, IA Berg, V Brecht, M Müller, G Fuchs, BE Alber. 2007. Synthesis of C_5-dicarboxylic acids from C_2-units involving crotonyl-CoA carboxylase/reductase: The ethylmalonate pathway. *Proc Natl Acad USA* 104: 10631–10636.

Eschenfeldt WH, Y Zhang, H Samaha, L Stols, LD Eirich, CR Wilson, MI Donnelly. 2003. Transformation of fatty acids catalyzed by cytochrome P450 monooxygenase enzymes of *Candida tropicalis*. *Appl Environ Microbiol* 69: 5992–5999.

Ettwig KF, S Shima, KT van de Pas-Schoonen, J Kahnt, MH Medema, HJM op den Camp, MSM Jetten, M Strous. 2008. Denitrifying bacteria anaerobically oxidize methane in the absence of *Archaea*. *Environ Microbiol* 10: 3164–3173.

Ettwig KF, T van Alen, KT van de Pas-Schoonen, MSM Jetten, M Strous. 2009. Enrichment and molecular detection of denitrifying methanotrophic bacteria of the NC10 phylum. *Appl Environ Microbiol* 75: 3656–3662.

Fall RR, ML Hector. 1977. Acyl-coenzyme A carboxylases. Homologous 3-methylcrotonyl-CoA and geranyl-CoA carboxylases from *Pseudomonas citronellolis*. *Biochemistry* 16: 4000–4005.

Feng L et al. 2007. Genome and proteome of long-chain alkane degrading *Geobacillus thermodenitrificans* BG80–2 isolated from a deep-subsurface oil reservoir. *Proc Natl Acad Sci USA* 104: 5602–5607.

Fox BG, WA Froland, JE Dege, JD Lipscomb. 1989. Methane monooxygenase from *Methylosinus trichosporium* OB3b. Purification and properties of a three-component system with a high specific activity from a type II methanotroph. *J Biol Chem* 264: 10023–10033.

Funhoff EG, E Bauer, I García-Rubio, B Witholt, JB van Beilen. 2006. CYP153A6, a soluble P450 oxygenase catalyzing terminal-alkane hydroxylation. *J Bacteriol* 188: 5220–5227.

Gabriel FLP, A Heidlberger, D Rentsch, W Giger, K Guenther, H-PE Kohler. 2005b. A novel metabolic pathway for degradation of 4-nonylphenol environmental contaminants by *Sphingomonas xenophaga* Bayram. *J Biol Chem* 280: 15526–15533.

Gabriel FLP, W Giger, K Guenther, H-PE Kohler. 2005a. Differential degradation of nonylphenol isomers by *Sphingomonas xenophaga* Bayram. *Appl Environ Microbiol* 71: 1123–1129.

Gauthier MJ, B Lafay, R Christen, L Fernandez, M Acquaviva, P Bonin, J-C Bertrand. 1992. *Marinobacter hydrocarbonoclasticus* gen. nov., sp. nov., a new extremely halotolerant, hydrocarbon-degrading marine bacterium. *Int J Syst Bacteriol* 42: 568–576.

Gill CO, C Ratledge. 1972. Toxicity of *n*-alkanes, *n*-alkenes, *n*-alkan-1-ols and *n*-alkyl-1-bromides towards yeasts. *J Gen Microbiol* 72: 165–172.

Girguis PR, AE Cozen, EF Delong. 2005. Growth and population dynamics of anaerobic methane-oxidizing archaea and sulfate-reducing bacteria in a continuous-flow reactor. *Appl Environ Microbiol* 71: 3725–3733.

Gorlatova N, M Tchorzewski, T Kurihara, K Soda, N Esaki. 1998. Purification, characterization, and mechanism of a flavin mononucleotide-dependent 2-nitropropane dioxygenase from *Neurospora crassa*. *Appl Environ Microbiol* 64: 1029–1033.

Grund A, J Shapiro, M Fennewald, P Bacha, J Leahy, K Markbreiter, M Nieder, M Toepfer. 1975. Regulation of alkane oxidation in *Pseudomonas putida*. *J Bacteriol* 123: 546–556.

Grundmann O, A Behrends, R Rabus, J Amann, T Halder, J Heider, F Widdel. 2008. Genes encoding the candidate enzyme for anaerobic activation of *n*-alkanes in the denitrifying bacterium strain HxN1. *Environ Microbiol* 10: 376–385.

Hallam SJ, N Putnam, CM Preston, JC Detter, D Rokhsar, PM Richardson, EF DeLong. 2004. Reverse methanogenesis: Testing the hypothesis with environmental genomics. *Science* 305: 1457–1462.

Hamamura N, RT Storfa, L Semprini, DJ Arp. 1999. Diversity of butane monooxygenases among butane-grown bacteria. *Appl Environ Microbiol* 65: 4586–4593.

Hamamura N, CM Yeager, DJ Arp. 2001. Two distinct monooxygenases for alkane oxidation in *Nocardioides* sp. strain CF8. *Appl Environ Microbiol* 67: 4992–4998.

Hara A, K Syutsubo, S Harayama. 2003. *Alcanivorax* which prevails in oil-contaminated seawater exhibits broad substrate specificity for alkane degradation. *Environ Microbiol* 5: 746–753.

Hector ML, RR Fall. 1976. Multiple acyl-coenzyme A carboxylases in *Pseudomonas citronellolis*. *Biochemistry* 15: 3465–3472.

Horswill AR, JC Escalante-Semerena. 1999. *Salmonella typhmurium* LT2 catabolizes propionate by the 2-methylcitric acid cycle. *J Bacteriol* 181: 5615–5623.

Höschle B, V Gnau, D Jendrossek. 2005. Methylcrotonyl-CoA and geranyl-CoA carboxylases are involved in leucine/isovalerate utilization (Liu) and acyclic terpene utilization (Atu), and are encoded by *liuB/liuS* and *atuC/atuF*, in *Pseudomonas aeruginosa*. *Microbiology (UK)* 151: 3649–3656.

Höschle B, D Jendrossek. 2004. Utilization of geraniol is dependent on molybdenum in *Pseudomonas aeruginosa*: Evidence for different metabolic routes for oxidation of geraniol and citronellol. *Microbiology (UK)* 151: 2277–2283.

Iida T, T Sumita, A Ohta, M Takagi. 2000. The cytochrome P450ALK multigene family of an *n*-alkane-assimilating yeast, *Yarrowia lipolytica*: Cloning and characterization of genes coding for new *CYP52* family members. *Yeast* 16: 1077–1087.

Inagaki F et al. 2004. Characterization of C1-metabolizing prokaryotic communities in methane seep habitat at the Kurishima Knoll, Southern Tyuku Arc, by analyzing *pmoA*, *mmoX*, *mxaF*, *mcrA*, and 16S rRNA genes. *Appl Environ Microbiol* 70: 7445–7455.

Islam T, S Jensen, LJ Reigstas, Ø Larsen, N-K Birekeland. 2008. Methane oxidation at 55°C and pH 2 by a thermoacidophilic bacterium belonging to the *Verrucomicrobia* phylum. *Proc Natl Acad Sci USA* 105: 300–304.

Johnson EL, MR Hyman. 2006. Propane and *n*-butane oxidation by *Pseudomonas putida* GPo1. *Appl Environ Microbiol* 72: 950–952.

Kane SR et al. 2007. Whole-genome analysis of the methyl *tert*-butyl ether-degrading beta-proteobacterium *Methylibium petroleiphilum* PM1. *J Bacteriol* 178: 1931–1945.

Käppeli O. 1986. Cytochromes P-450 of yeasts. *Microbiol Rev* 50: 244–258.

Khelifi N, V Grossi, M Hamdi, A Dolla, J-L Tholozan, B Ollivier, A Hirshler-Réa. 2010. Anaerobic oxidation of fatty acids and alkenes by the hyperthermophilic sulfate-reducing archaeon *Archaeoglobus fulgidus*. *Appl Environ Microbiol* 76: 3057–3060.

Kido T, K Sida, K Asada. 1978. Properties of 2-nitropropane dioxygenase of *Hansenula mrakii*. *J Biol Chem* 253: 226–232.

Kniemeyer O, C Probian, R Rosselló-Mora, J Harder. 1999. Anaerobic mineralization of quaternary carbon atoms; isolation of denitrifying bacteria on dimethylmalonate. *Appl Environ Microbiol* 65: 3319–3324.

Koch DJ, MM Chen, JB van Beilen, FH Arnold. 2009. *In vivo* evolution of butane oxidation by terminal alkane hydroxylases AlkB and CPY153A6. *Appl Environ Microbiol* 75: 337–344.

Kok M, R Oldenhuis, MPG Van der Linden, CHC Meulenberg, J Kingma, B Witholt. 1989b. The *Pseudomonas oleovorans alkBAC* operon encodes two structurally related rubredoxins and an aldehyde dehydrogenase. *J Biol Chem* 264: 5442–5451.

Kok M, R Oldenhuis, MPG Van der Linden, P Raatjes, K Kingma, PH Van Lelyveld, B Witholt. 1989a. The *Pseudomonas oleovorans* alkane hydroxylase gene. Sequence and expression. *J Biol Chem* 264: 5435–5441.

Kolvenbach B, N Schlaich, Z Raoui, J Prell, S Zühlke, A Schäffer, FP Guengerich, PFX Cordini. 2007. Degradation pathway of bisphenol A: Does *ipso* substitution apply to phenols containing a quaternary α-carbon structure in the *para* position? *Appl Environ Microbiol* 73: 4776–4784.

Kotani T, T Yamamoto, H Yurimoto, Y Sakai, N Kato. 2003. Propane monooxygenase and NAD$^+$-dependent secondary alcohol dehydrogenase in propane metabolism by *Gordonia* sp. strain TY-5. *J Bacteriol* 185: 7120–7128.

Lipscomb JD. 1994. Biochemistry of the soluble methane monooxygenase. *Annu Rev Microbiol* 48: 371–399.

Lobos JH, TK Leib, T-M Su. 1992. Biodegradation of bisphenol A and other bisphenols by a Gram-negative aerobic bacterium. *Appl Environ Microbiol* 58: 1823–1831.

Lode ET, MJ Coon. 1971. Enzymatic ω-oxidation V. Forms of *Pseudomonas oleovorans* rubredoxin containing one or two iron atoms: Structure and function in ω-hydroxylation. *J Biol Chem* 246: 791–802.

Lomans BP, R Maas, R Luderer, HJP op den Camp, A Pol, C van der Drift, GD Vogels. 1999. Isolation and characterization of *Methanomethylovorans hollandica* gen. nov., sp. nov., isolated from freshwater sediment, a methylotrophic methanogen able to grow with dimethyl sulfide and methanethiol. *Appl Environ Microbiol* 65: 3641–3650.

Lu Y, T Lueders, MW Friedrich, R Conrad. 2005. Detecting active methanogenic populations on rice roots using stable isotope probing. *Environ Microbiol* 7: 326–336.

Maeng JH, Y Sakai, Y Tani, N Kato. 1996. Isolation and characterization of a novel oxygenase that catalyzes the first step of *n*-alkyl oxidation in *Acinetobacter* sp. strain M-1. *J Bacteriol* 178: 3695–3700.

Maier T, H-H Förster, O Asperger, U Hahn. 2001. Molecular characterization of the 56-kDa CYP153 from *Acinetobacter* sp. EB104. *Biochem Biophys Res Comm* 286: 652–658.

Mardanov AV, NV Ravin, VA Svetlichnyi, AV Beletsky, ML Miroshnichenko, EA Bonch-Osmolovskaya, KG Skryabin. 2009. Metabolic versatility and indigenous origin of the archaeon *Thermococcus sibiricus* from a Siberian oil reservoir, as revealed by genome analysis. *Appl Environ Microbiol* 75: 4580–4588.

Marín MM, THM Smits, JB van Beilen, F Rojo. 2001. The alkane hydroxylase gene of *Burkholderia cepacia* RR10 is under catabolite repression control. *J Bacteriol* 183: 4202–4209.

Marín MM, L Yuste, F Rojo. 2003. Differential expression of the components of the two alkane hydroxylases from *Pseudomonas aeruginosa*. *J Bacteriol* 185: 3232–3237.

Marx CJ, L Chistoserdova, ME Lidstrom. 2003. Formaldehyde-detoxifying role of the tetrahydromethanopterin-linked pathway in *Methylobacterium extorquens* AM1. *J Bacteriol* 185: 7160–7168.

Mauersberger S, H-J Wang, C Gaillardin, G Barth, J-M Nicaud. 2001. Insertional mutagenesis in the *n*-alkane-assimilating yeast *Yarrowia lipolytica*: Generation of tagged mutations in genes involved in hydrophobic substrate utilization. *J Bacteriol* 183: 5102–5109.

May SW, BJ Abbott. 1973. Enzymatic epoxidation II. Comparison between the epoxidation and hydroxylation reactions catalyzed by the ω-hydroxylation system of *Pseudomonas oleovorans*. *J Biol Chem* 248: 1725–1730.

McDonald IR, CB Miguez, G Rogge, D Bourque, KD Wendlandt, D Groleau, JC Murrell. 2006. Diversity of soluble methane monooxygenase-containing methanotrophs isolated from polluted environments. *FEMS Microbiol Lett* 255: 225–232.

McKenna EJ, MJ Coon. 1970. Enzymatic ω-oxidation IV. Purification and properties of the ω-hydroxylase of *Pseudomonas oleovorans*. *J Biol Chem* 245: 3882–3889.

McKenna EJ, RE Kallio. 1971. Microbial metabolism of the isoprenoid alkane pristane. *Proc Natl Acad Sci USA* 68: 1552–1554.

McKew BA, F Coulon, AM Osborn, KN Timmis, TJ McGenity. 2007. Determining the identity and roles of oil-metabolizing marine bacteria from the Thames estuary, UK. *Environ Microbiol* 9: 165–176.

Middelhoven WJ, MC Hoogkamer-Te Niet, NJ Kreger-Van Rij. 1984. *Trichosporon adeninovorans* sp. nov., a yeast species utilizing adenine, xanthine, uric acid, putrescine and primary *n*-alkylamines as the sole source of carbon, nitrogen and energy. *Antonie van Leeuwenhoek* 50: 369–378.

Miura Y, AJ Fulco. 1975. ω-1, ω-2 and ω-3 hydroxylation of long-chain fatty acids, amides and alcohols by a soluble enzyme system from *Bacillus megaterium*. *Biochim Biophys Acta* 388: 305–317.

Müller RH, T Rohwerder, H Harms. 2008. Degradation of fuel oxygenates and their main intermediates by *Aquincola tertiaricarbonis* L108. *Microbiology (UK)* 154: 1414–1421.

Narhi LO, AJ Fulco. 1986. Characterization of a catalytically self-sufficient 119,000-Dalton cytochrome P-450 monooxygenase induced by barbiturates in *Bacillus megaterium*. *J Biol Chem* 261: 7160–7169.

Narhi LO, AJ Fulco. 1987. Identification and characterization of two functional domains in cytochrome P-450 $_{BM-3}$, a catalytically self-sufficient monooxygenase induced by barbiturates in *Bacillus megaterium*. *J Biol Chem* 262: 6683–6690.

Nauhaus K, M Albrecht, M Elvert, A Boetius, F Widdel. 2007. *In vitro* cell growth of marine archaeal-bacterial consortia during anaerobic oxidation of methane with sulfate. *Environ Microbiol* 9: 187–196.

Nazina TN et al. 2001. Taxonomic study of aerobic thermophilic bacilli: Descriptions of *Geobacillus subterraneus* gen. nov., sp. nov. and *Geobacillus uzenensis* sp. nov. from petroleum reservoirs and transfer of *Bacillus stearothermophilus*, *Bacillus thermocatenulatus*, *Bacillus thermoleovorans*, *Bacillus kaustophilus*, *Bacillus thermoglucosidasius*, and *Bacillus thermodenitrificans* to *Geobacillus* as the new combinations *G. stearothermophilus*, *G. thermocatenulatus*, *G. thermoleovorans*, *G. kaustophilus*, *G. thermoglucosidasius*, and *G. thermodenitrificans*. *Int J Syst Evol Microbiol* 51: 433–446.

Nie Y, J Liang, Y-Q Tang, X-L Wu. 2011. Two novel alkane hydroxylase-rubredoxin fusion genes isolated from a *Dietzia* bacterium and the functions of fused rubredoxin domains in long-chain *n*-alkane degradation. *Appl Environ Microbiol* 77: 7279–7288.

Palleroni NJ, AM Port, H-K Chang. GJ Zylstra. 2004. *Hydrocarboniphaga effusa* gen. nov., sp. nov., a novel member of the γ-proteobacteria active in alkane and aromatic hydrocarbon degradation. *Int J Syst Evol Microbiol* 54: 1203–1207.

Patel RN, CT Hou, AI Laskin, A Felix. 1982. Microbial oxidation of hydrocarbons: Properties of a soluble methane monooxygenase from a facultative methane-utilizing organism, *Methylobacterium* sp. strain CRL-26. *Appl Environ Microbiol* 44: 1130–1137.

Paul L, DK Ferguson, JA Krzycki. 2000. The trimethylamine methyltransferase gene and multiple dimethylamine methyltransferase genes of *Methanosarcina barkeri* contain in-frame and read-through amber codons. *J Bacteriol* 182: 2520–2529.

Phillips WE, JJ Perry. 1974. Metabolism of *n*-butane and 2-butanone by *Mycobacterium vaccae*. *J Bacteriol* 120: 987–989.

Piccolo LL, C De Pasquale, R Fodale, AM Puglia, P Quatrini. 2011. Involvement of an alkane hydroxylase system of *Gordonia* sp. strain SoCg in degradation of solid *n*-alkanes. *Appl Environ Microbiol* 77: 1204–1213.

Pirnik MP, RM Atlas, R Bartha. 1974. Hydrocarbon metabolism by *Brevibacterium erythrogenes*: Normal and branched alkanes. *J Bacteriol* 119: 868–878.

Pol A, K Heijmans, HR Harhangi, D Tedesco, MSM Jetten, HJM Op den Camp. 2007. Methanotrophy below pH 1 by a new *Verrucomicrobia* species. *Nature* 450: 874–878.

Porter AW, AG Hay. 2007. Identification of *opdA*, a gene involved in biodegradation of the endocrine disruptor octylphenol. *Appl Environ Microbiol* 73: 7373–7379.

Probian C, A Wülfing, J Harder. 2003. Anaerobic mineralization of quaternary carbon atoms: Isolation of denitrifying bacteria on pivalic acid (2,2-dimethylpropionic acid). *Appl Environ Microbiol* 69: 1866–1870.

Rabus R, H Wilkes, A Behrends, A Armstroff, T Fischer, AJ Pierik, F Widdel. 2001. Anaerobic initial reaction of *n*-alkanes in a denitrifying bacterium: Evidence for (1-methylpentyl) succinate as initial product and for involvement of an organic radical in *n*-hexane metabolism. *J Bacteriol* 183: 1707–1715.

Rainey FA, S Klatte, RM Kroppenstedt, E Stackebrandt. 1995. *Dietzia*, a new genus including *Dietzia maris* comb. nov., formerly *Rhodococcus maris*. *Int J Syst Evol Microbiol* 45: 32–36.

Ramamoorthi R, ME Lidstrom. 1995. Transcriptional analysis of *pqqD* and study of the regulation of pyrroloquinoline quinone biosynthesis in *Methylobacterium extorquens* AM1. *J Bacteriol* 177: 206–211.

Ratajczak A, W Geißdörfer, W Hillen. 1998. Alkane hydroxylase from *Acinetobacter* sp. strain ADP1 is encoded by *alkM* and belongs to a new family of bacterial integral-membrane hydrocarbon hydroxylases. *Appl Environ Microbiol* 64: 1175–1179.

Ratnatilleke A, JW Vrijbloed, JA Robinson. 1999. Cloning and sequencing of the coenzyme B12-binding domain of isobutyryl-CoA mutase from *Streptomyces cinnamonensis*. Reconstitution of mutase activity and characterization of the recombinant enzyme produced in *Escherichia coli*. *J Biol Chem* 274: 31679–31685.

Rohwerder T, U Breuer, D Benndorf, U Lechner, RH Müller. 2006. The alkyl *tert*-butyl ether intermediate 2-hydroxyisobutyrate is degraded via a novel cobalamin-dependent mutase pathway. *Appl Environ Microbiol* 72: 4128–4135.

Rohwerder T, RH Müller. 2008. New bacterial cobalamin-dependent CoA-carbonyl mutases involved in degradation pathways, pp. 81–98. In *Vitamin B: New Research* (Ed CM Elliot); Nova Science Publishers Inc, Hauppauge, NY.

Rontani J-F, MJ Gilewicz, VD Micgotey, TL Zheng, PC Bonin, J-C Bertrand. 1997. Aerobic and anaerobic metabolism of 6,10,14-trimethylpentadecan-2-one by a denitrifying bacterium isolated from marine sediments. *Appl Environ Microbiol* 63: 636–643.

Ruettinger, RT, ST Olson, RF Boyer, MJ Coon. 1974. Identification of the ω-hydroxylase of *Pseudomonas oleovorans* as a nonheme iron protein requiring phospholipid for catalytic activity. *Biochem Biophys Res Commun* 57: 1011–1017.

Sabirova JS, M Ferrer, D Regenhardt, KN Timmis, PN Golyshin. 2006. Proteomic insights into metabolic adaptations of *Alcalivorax borkumensis* induced by alkane utilization. *J Bacteriol* 188: 3763–3773.

Sakai Y, H Takahashi, Y Wakasa, T Kotani, H Yurimoto, N Miyachi, PP Van Veldhoven, N Kato. 2004. Role of α-methacyl coenzyme A racemase in the degradation of methyl-branched alkanes by *Mycobacterium* sp. strain P101. *J Bacteriol* 186: 7214–7220.

Sasaki M, A Akahira, L-i Oshiman, T Tsuchid, Y Matsumura. 2005. Purification of cytochrome P450 and ferredoxin involved in bisphenol A degradation from *Sphingomonas* sp. strain AO1. *Appl Environ Microbiol* 71: 8024–8030.

Sauer K, U Harms, RK Thauer. 1997. Methanol: Coenzyme M methyltransferase from *Methanosarcina barkeri*. Purification, properties and encoding genes of the corrinoid protein MT1. *Eur J Biochem* 243: 670–677.

Schaeffer TL, SG Cantwell, JL Brown, DS Watt, RR Fall. 1979. Microbial growth on hydrocarbons: Terminal branching inhibits biodegradation. *Appl Environ Microbiol* 38: 742–746.

Scheller U, T Zimmer, D Becher, F Schauer, W-H Schunk. 1998. Oxygenation cascade in conversion of *n*-alkanes to α,ω-dioic acids catalyzed by cytochrome P450 52A3. *J Biol Chem* 273: 43528–32534.

Schouten S, SG Wakeham, EC Hopmans, JSS Damsté. 2003. Biogeochemical evidence that thermophilic archaea mediate the anaerobic oxidation of methane. *Appl Environ Microbiol* 69: 1680–1686.

Seo CW, Y Yamada, N Takada, H Okada. 1983. Microbial transformation of squalene: Terminal methyl group oxidation by *Corynebacterium* sp. *Appl Environ Microbiol* 45: 522–525.

Seubert W. 1960. Degradation of isoprenoid compounds by microorganisms I. Isolation and characterization of an isoprenoid-degrading bacterium, *Pseudomonas citronellolis* n. sp. *J Bacteriol* 79: 426–434.

Seubert WS, E Fass, U Remberger. 1963. Untersuchungen über den bakteriellen Abbau van Isoprenoiden III. Reinignung und Eigenschaften de Geranylcarboxylase. *Biochem Z* 338: 265–275.

Seubert W, U Remberger. 1963. Untersuchungen über den bakteriellen Abbau van Isoprenoiden II. Die Rolle der Kohlensäure. *Biochem Z* 338: 245–264.

Shanklin J, C Achim, H Schmidt, BG Fox, E Münck. 1997. Mössbauer studies of alkane ω-hydroxylase: Evidence for a diiron cluster in an integral-membrane enzyme. *Proc Natl Acad Sci USA* 94: 2981–2986.

Silva RA, V Gross, HM Alvarez. 2007. Biodegradation of phytane (2,6,10,12-tetramethylhexadecane) and accumulation of related isoprenoid wax esters by *Mycobacterium ratisbonense* strain SD4 under nitrogen-starved conditions. *FEMS Microbiol Lett* 272: 220–228.

Sluis MK, LA Sayavedra-Soto, DJ Arp. 2002. Molecular analysis of the soluble butane monooxygenase from `Pseudomonas butanovora´. *Microbiology (UK)*148: 3617–3629.

Smits THM, SB Balada, B Witholt, JB van Beilen. 2002. Functional analysis of alkane hydroxylases from Gram-negative and Gram-positive bacteria. *J Bacteriol* 184: 1733–1742 .

Smith CA, MR Hyman. 2004. Oxidation of methyl *tert*-butyl ether by alkane hydroxylase in dicyclopropane-induced and *n*-octane-grown *Pseudomonas putida* GPo1. *Appl Environ Microbiol* 70: 4544–4550.

So CM, CD Phelps, LY Young. 2003. Anaerobic transformation of alkanes to fatty acids by a sulfate-reducing bacterium strain Hxd3. *Appl Environ Microbiol* 69: 3892–3900.

Sontoh S, JD Semrau. 1998. Methane and trichloroethylene degradation by *Methylosinus trichosporium* OB3b expressing particulate methane monooxygenase. *Appl Environ Microbiol* 64: 1106–1114.

Spivack J, TK Leib, JH Lobos. 1994. Novel pathway for bacterial metabolism of bisphenol A. Rearrangements and stilbene cleavage in bisphenol A metabolism. *J Biol Chem* 269: 7323–7329.

Takei D, K Washio, M Morikawa. 2008. Identification of alkane hydroxylase genes in *Rhodococcus* sp. strain TMP2 that degrades a branched alkane. *Biotechnol Lett* 30: 1447–1452 .

Tallant TC, JA Krzycki. 1997. Methylthiol: Coenzyme M methyltransferase from *Methanosarcina barkeri*, an enzyme of methanogenesis from dimethylsulfide and methylmercaptopropionate. *J Bacteriol* 179: 6902–6911.

Tallant TC, L Paul, JA Krzycki. 2001. The MtsA subunit of the methylthiol: Coenzyme M methyltransferase of *Methanosarcina barkeri* catalyzes both half-reactions of corrinoid-dependent dimethylsulfide: Coenzyme M methyl transfer. *J Biol Chem* 276: 4485–4493.

Tanaka A, M Ueda. 1993. Assimilation of alkanes by yeasts: Functions and biogenesis of peroxisomes. *Mycol Res* 98: 1025–1044.

Tani A, T Ishige, Y Sakai, N Kato. 2001. Gene structures and regulation of the alkane hydroxylase complex in *Acinetobacter* sp. strain M-1. *J Bacteriol* 183: 1819–1823.

Teramoto M, M Suzuki, F Okazaki, A Hatmanti, S Harayama. 2009. *Oceanobacter*-related bacteria are important for the degradation of petroleum aliphatic hydrocarbons in the tropical marine environment. *Microbiology (UK)* 155: 3362–3370.

Teske A, K-U Hinrichs, V Edgcomb, A de Vera Gomez, D Kysela, SP Sylva, ML Sogin, HW Jannasch. 2002. Microbial diversity of hydrothermal sediments in the Guaymas Basin: Evidence for anaerobic methanotrophic communities. *Appl Environ Microbiol* 68: 1994–2007.

Textor S, VF Wendich, AA De Graaf, U Müller, MI Linder, D Linder, W Buckel. 1997. propionate oxidation in *Escherichia coli*: evidence for operation of a methylcitrate cycle in bacteria. *Arch Microbiol* 168: 428–436.

Throne-Holst M, A Wentzel, TE Ellingsen, H-K Kotlar, SB Zotchev. 2007. Identification of novel genes involved in long-chain *n*-alkane degradation by *Acinetobacter* sp. strain DSM 17874. *Appl Environ Microbiol* 73: 3317–3332.

van Beilen JB, EG Funhoff. 2007. Alkane hydroxylases in microbial alkane degradation. *Appl Microbiol Biotechnol* 74: 13–21.

van Beilen JB, EG Funhoff, A van Loon, A Just, L Kaysser, M Bouza, R Holtackers, M Röthlisberger, Z Li, B Witholt. 2006. Cytochrome P450 alkane hydroxylases of the CYP153 family are common in alkane-degrading eubacteria lacking integral membrane alkane hydroxylases. *Appl Environ Microbiol* 72: 59–65.

van Beilen JB, M Neuenschwander, THM Smits, C Roth, SB Balada, B Witholt. 2002b. Rubredoxins involved in alkane oxidation. *J Bacteriol* 184: 1722–1732.

van Beilen JB, MM Marín, THM Smits, M Röthlisberger, AG Franchini, B Witholt, F Rojo. 2004. Characterization of two alkane hydroxylase genes from the marine hydrocarbonoclastic bacterium *Alcanivorax borkumensis*. *Environ Microbiol* 6: 264–273.

van Beilen JR, D Pennings, B Witholt. 1992. Topology of the membrane-bound alkane hydroxylase of *Pseudomonas oleovorans*. *J Biol Chem* 267: 9194–2201.

van Beilen THM Smits, LG Whyte, S Schorcht, M Röthlisberger, T Plaggmeier, K-H Engesser, B Witholt. 2002a. Alkane hydroxylase homologues in Gram-positive strains. *Environ Microbiol* 4: 676–682.

van der Geize R, K Yam, T Heuser, MH Wilbrink, H Hara, MC Anderton, E Sim, L Dijkhuizen, JE Davies, WW Mohn, LD Eltis. 2007. A gene cluster encoding cholesterol catabolism in a soil actinomycete provides insight into *Mycobacterium tuberculosis* survival in macrophages. *Proc Natl Acad Sci USA* 104: 1947–1952.

Vestal JR, JJ Perry. 1969. Divergent pathways for propane and propionate utilization by a soil isolate. *J Bacteriol* 99: 216–221.

Whyte LG, THM Smits, D Labbé, B Witholt, CW Greer, JB van Beilen. 2002. Gene cloning and characterization of multiple alkane hydroxylase systems in *Rhodococcus* strains Q15 and NRRL B-16531. *Appl Environ Microbiol* 68: 5933–5942.

Wilkes H, R Rabus, T Fischer, A Armstroff, A Behrends, F Widdel. 2002. Anaerobic degradation of *n*-hexane in a denitrifying bacterium: Further degradation of the initial intermediate (1-methylpentyl)succinate via a C-skeleton rearrangement. *Arch Microbiol* 177: 235–243.

Yakimov MM, L Giuliano, R Denaro, E Crisafi, TN Chernikova, W-R Abraham, H Lünsdorf, KN Timmis, PN Golyshin. 2004. *Thalassolituus oleivirans* gen. nov., sp. nov., a novel marine bacterium that obligately utilizes hydrocarbons. *Int J Syst Evol Microbiol* 54: 141–148.

Yakimov MM, L Giuliano, G Gentile, E Crisafi, TN Chernikova, W-R Abraham, H Lünsdorf, KN Timmis, PN Golyshin. 2003b. *Oleispira antarctica* gen. nov., sp. nov., a novel hydrocarbonoclastic marine bacterium isolated from Antarctic coastal sea water. *Int J Syst Evol Microbiol* 53: 779–785.

Yakimov MM, PN Golyshin, S Lang, ERB Moore, W-R Abraham, H Lünsdorf, KN Timmis. 1998. *Alcanivorax borkumensis* gen. nov., sp. nov., a new hydrocarbon-degrading and surfactant-producing marine bacterium. *Int J Syst Evol Microbiol* 48: 339–348.

Yakimov MM, H Lünsdorf, PN Golyshin. 2003a. *Thermoleophilum album* and *Thermoleophilum minutum* are culturable reresentatives of group 2 of the *Rubrobacteridae* (*Actinobacteria*). *Int J Syst Evol Microbiol* 53: 377–380.

Yu S S-F, L-Y Wu, K H.-C Chen, W-I Luo, D-S Huang, SI Chan. 2003. The stereospecific hydroxylation of [2,2-^{2}H$_2$]butane and chiral dideuteriobutanes by the particulate methane monooxygenase from *Methylococcus capsulatus* (Bath). *J Biol Chem* 278: 40658–40669.

Yumoto I, A Nakamura, H Iwata, K Kojima, K Kusumoto, Y Nodasaka, H Matsutama. 2002. *Dietzia psychral-caliphila* sp. nov., a novel facultatively psychrophilic alkaliphile that grows on hydrocarbons. *Int J Syst Evol Microbiol* 52: 85–90.

Yuste L, F Rojo. 2001. Role of the *crc* gene in catabolic repression of the *Pseudomonas putida* GPo1 alkane degradation pathway. *J Bacteriol* 183: 6197–6206.

Zarilla KA, JJ Perry. 1984. *Thermoleophilum album*, gen. nov. and sp. nov., a bacterium obligate for thermophily and *n*-alkane substrates. *Arch Microbiol* 137: 286–290.

Zehnder AJB, TD Brock. 1980. Anaerobic methane oxidation—Occurrence and ecology. *Appl Environ Microbiol* 39: 194–204.

7.3 Cycloalkanes: Including Terpenoids and Steroids

The aerobic degradation of cycloalkanes has been examined in both monocyclic and polycyclic substrates that include diterpenoids, steroids, and bile acids. In all of them, monooxygenation is the first step, and this is sometimes accomplished by cytochrome P450 systems. Reviews of the degradation of alicyclic compounds including monoterpenes have been given by one of the pioneers (Trudgill 1978, 1984, 1994), and these should be consulted for further details. Only a brief outline of significant aspects is, therefore, necessary here.

7.3.1 Monocyclic Cycloalkanes and Their Derivatives

The first steps in the degradation of cycloalkanes are carried out by cycloalkane monooxygenase hydroxylation followed by dehydrogenation to the cycloalkanone, and are formally analogous to those used for the degradation of linear alkanes. It was only after considerable effort, however, that pure strains of microorganisms were isolated that could grow with cycloalkanes or their simple derivatives. The degradation of cyclohexane has been examined in detail (Stirling et al. 1977; Trower et al. 1985). There are five steps in the degradation:

1. Hydroxylation of the ring by monooxygenation
2. Dehydrogenation to the ketone
3. Insertion of one atom of oxygen into the ring in a reaction formally similar to the Baeyer–Villiger persulfate oxidation
4. Hydrolysis of the lactone
5. Oxidation to an α,ω-dicarboxylate

The pathway is illustrated for cyclohexane (Figure 7.17) (Stirling et al. 1977), and a comparable one operates also for cyclopentanol (Griffin and Trudgill 1972). The enantiomeric specificity of this oxygen-insertion reaction has been examined in a strain of camphor-degrading *Ps. putida* (Jones et al. 1993). Cyclohexanone monooxygenase that introduces oxygen into the ring has been purified and characterized from *Nocardia globerula* CL1 and *Acinetobacter* NCIB 9871 (Donoghue et al. 1976), and both cyclohexanone monooxygenase (Branchaud and Walsh 1985) and cyclopentanone monooxygenase exhibit a substrate versatility (Iwaki et al. 2002) that is reminiscent of that of methane monooxygenase and has attracted interest in them as biocatalysts.

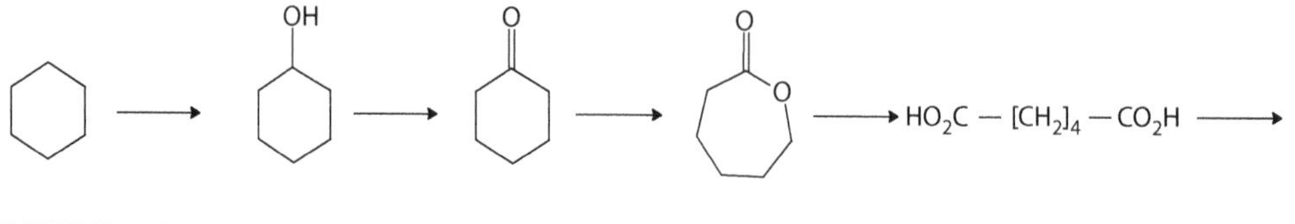

FIGURE 7.17 Biodegradation of cyclohexane.

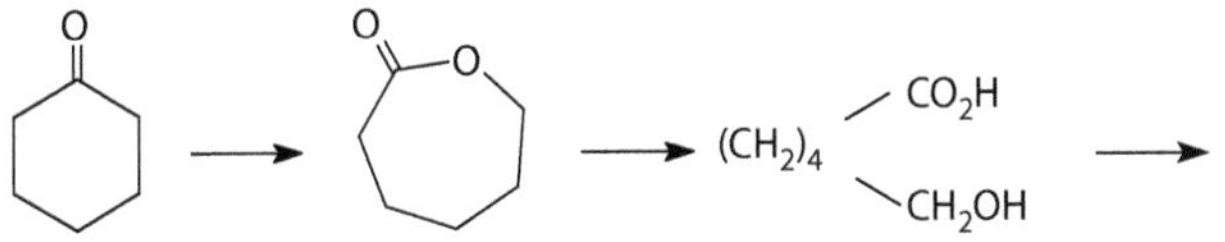

FIGURE 7.18 Biodegradation of cycloalkanones.

7.3.2 Cycloalkanones

Flavoprotein 1,2-monooxygenases are used for the insertion of an atom of oxygen into the ring that is the first step in the degradation of both cyclopentanone and cyclohexanone before hydrolysis and ring fission (Figure 7.18). There are two types, one is FAD-NADPH-dependent, whereas the other is FMN-NADH-dependent. The crystal structure of the enzyme from *Thermobifida fusca* with phenyl acetone has been established (Malito et al. 2004), and revealed that the initially formed 5a-hydroperoxide reacted with the carbonyl group of the ketone. This then underwent intramolecular rearrangement to the lactone. The FAD-containing cyclohexanone monooxygenase has been purified and characterized from *Nocardia glomerula* CL1 and from *Acinetobacter* sp. NCIB 9871 (Donoghue et al 1976). There has been renewed interest in these reactions in view of their biotechnological relevance (Iwaki et al. 2002):

a. Cycloalkanone monooxygenase for substrates with more than seven carbon atoms is, however, different. Cyclododecanone monooxygenase from *Rhodococcus ruber* SC1 is different from those already mentioned, and is active only towards substrates with more than seven carbon atoms (Kostichka et al. 2001).

b. The enzyme derived from *Pseudomonas* sp. strain HI-70 is able to oxidize a wide range of substrates including C_{12} to C_{15} ketones, C_5 and C_6 ketones with methyl substituents, and some bicyclic ketones including decalones (Iwaki et al. 2006).

7.3.3 Inositols (Hexahydroxycyclohexanes)

Although these are not contaminants in any sense, since they are produced from a wide range of biota, it seems appropriate to make a few brief comments on their structures as introduction. The most widespread inositol isomers have the configurations: *myo*-C_1-axial, D-*chiro*- C_1 and C_6-axial, *neo*-C_1 and C_4-axial, *scyllo*- all equatorial. These isomers occur naturally in a range of structures and environments some of which are given as illustration:

1. *myo*-inositol, *scyllo*-inositol and D-*chiro*-inositol occur in plants and animals.

2. *myo*-inositol is a component of lipids such as phosphatidylinositol that consists of *myo*-inositol-1-phosphate linked to the 3-position of glycerol bisesterified with fatty acids, and is an important lipid, both as a key membrane constituent and as a participant in essential metabolic processes in all plants and animals.

3. *myo*-inositol phosphorylated with one, three, and six phosphates are important in plants as storage material for phosphorus, and *scyllo*-, D-*chiro*-, and *neo*-inositol hexakisphosphates have been found to occur in soils (Turner et al. 2002). Their enzymatic dephosphorylation by extracellular phytase that is induced by phosphate limitation in *Bacillus amyloliquefaciens* is one factor that contributes to plant growth (Makarewicz et al. 2006).

4. 1-Deoxy-1-amino-*scyllo*-inositol and L-3-*O*-methyl-*scyllo*-inosamine are important rhizopines (Saint et al. 1993; Murphy et al. 1995) that are essential for the establishment of nodules in *Sinorhizobium meliloti* (Kohler et al. 2010) and *Rhizobium leguminosarum* (Fry et al. 2001)— and only in strains capable of their catabolism that is, therefore, of cardinal importance.

5. Antibiotics contain a wide range of cyclitols that include *myo*-inositol, D-*chiro*-inositol, and 1-amino-1-deoxy-*myo*-inositol and 1-amino-1-deoxy-*scyllo*-inositol (Haskell 1984).

The degradation of *myo*-inositol takes place by a mechanism quite different from the aromatization pathway that is discussed later for cyclohexane carboxylates. The metabolic pathway was elucidated in pioneering

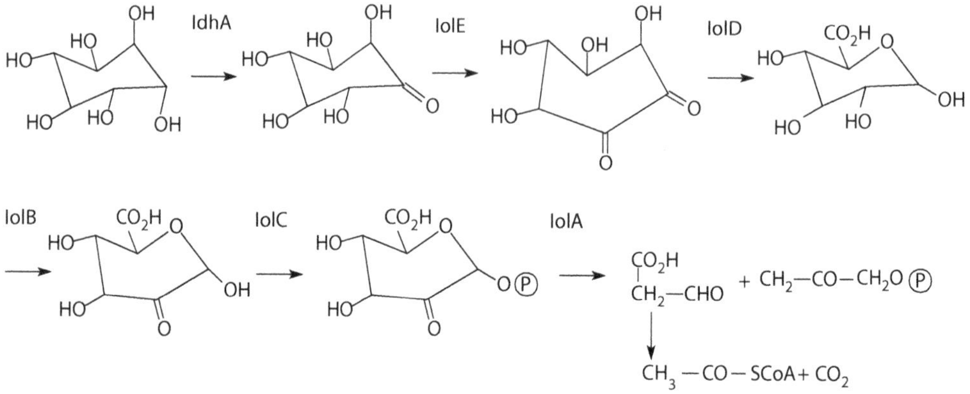

FIGURE 7.19   Biodegradation of *myo*-inositol.

studies of *myo*-inositol degradation using *Aerobacter aerogenes* (*Klebsiella mobilis*) (Anderson and Magasanik 1971a,b), and has been extended to the enzymology of the enzymes for *myo*-inositol catabolism in the *iolABCDEFGHIJ* operon of *Bacillus subtilis* (Yoshida et al. 2008). The enzymes sequentially carried out dehydrogenation (*iolG*), dehydration (*iolE*), hydrolysis (*iolD*), isomerization (*iolB*), phosphorylation (*iolC*), an aldolase (*iolJ*). In *Rhizobium leguminosarum* and *S. meliloti*, inositol plays a cardinal role for competitiveness in root nodulation. The genetics and regulation of the inositol pathway have been analyzed in *S. meliloti*, and degradation was observed not only for *myo*-inositol, but also for *scyllo*-inositol and D-*chiro*-inositol (Kohler et al. 2010). The critical catabolic reaction was hydrolytic scission of the 3/4,5-trihydroxycyclohexane-1-2-dione that was produced by dehydrogenation of inositol followed by dehydration. This was followed by conversion of the C_6 unit to C_3 malonate semialdehyde that produced acetyl-CoA and 1,3-dihydroxyacetone phosphate that was incorporated into the glycolytic cycle (Kohler et al. 2010) (Figure 7.19). The regulation of inositol catabolism in *S. meliloti* is complex, and the catabolic genes occur in three clusters that comprise an operon. Although the pathway is analogous to that for *B. subtilis* different gene designations are used for the enzymes that involve successively dehydrogenation (IdhA), dehydration (IolE), fission of the cyclic 1,2-diketone (IolD), dehydrogenation (IolB), phosphorylation (IolC), and fission of malonate semialdehyde (IolA) (Kohler et al. 2011). It is worth noting that the degradation of inositol has also been described in bacteria that are not determinants of nodulation in the family *Rhizobiaceae*. Although the catabolic pathways are similar in both Gram-positive and Gram-negative organisms, there are some differences that are noted briefly:

1. *Clostridium perfringens* had two genes in the operon that encode the dehydrogenase for the first step in the degradation (Kawsar et al. 2004).
2. *C. glutamicum* contained two clusters of genes of which cluster I contained the enzymes for degradation (Krings et al. 2006).
3. *Lactobacillus casei* specifically strain BL23 had two pathways for the metabolism of malonate semialdehyde: IolA in a reaction with coenzyme A to produce acetyl CoA, and IolK in which decarboxylation produced acetaldehyde (Yebra et al. 2007).
4. *Salmonella enterica* had a complex transcriptional organization and regulation (Kröger and Fuchs 2009).

   The degradation of the rhizopines *scyllo*-inosamine and L-3-*O*-methyl-*scyllo*-inosamine has been examined in *Rh. leguminosarum* (Bahar et al. 1998), and the *mocABCR* operon has been established for *Rhizobium meliloti* L5-30 which is necessary for the degradation of rhizopine. *mocABC* and *mocR* are separated and, on the basis of homology, some suggestions for the degradation were proposed (Rossbach et al. 1994; Saint et al. 1993). It is significant that their degradation is dependent on the existence of a functional pathway for *myo*-inositol catabolism (Galbraith et al. 1998). In support of this it has been shown that the catabolism of these *scyllo*-inosamines is carried out by 3-*O*-demethylation mediated by an oxygenase-ferredoxin to produce *scyllo*-inositol (Bahar et al. 1998). In addition, it is plausible that degradation of *scyllo*-inosamine occurs by the reverse of the activities involved in the synthesis of *scyllo*-inosamine from *myo*-inositol in *Streptomyces* sp.—*myo*-inositol dehydrogenase and transamination with glutamine (Walker 1995). In support of this, an L-glutamine: *scyllo*-inosinose transaminase has been characterized in a streptomycin-producing strain of *Streptomyces* (Ahlert et al. 1997).

7.3.4 Cycloalkane Carboxylates

1. Cyclopropane carboxylic acid is degraded via 3-hydroxybutyrate by both the bacterium *Rh. rhodochrous* (Toraya et al. 2004) (Figure 7.20) and by fungi (Schiller and Chung 1970), although the mechanism for ring fission seems to have been undetermined.

2. The degradation of 11,12-methyleneoctadecanoate is carried out by *Tetrahymena pyriformis* by a modified β-oxidation pathway with formation of propionate and acetate (Figure 7.21) (Tipton and Al-Shathir 1974).

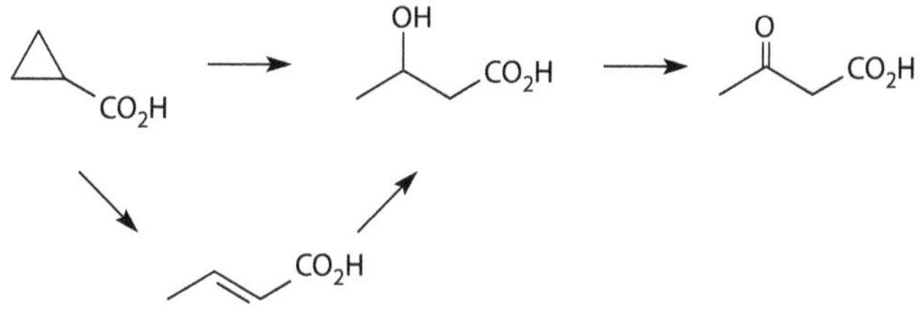

FIGURE 7.20 Aerobic degradation of cyclopropanecarboxylate. (Adapted from Toraya T et al. 2004. *Appl Environ Microbiol* 70: 224–228.)

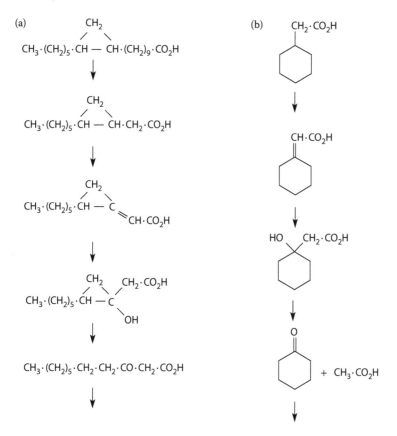

FIGURE 7.21 Degradation (a) 11,12-methyleneoctadecanoate and (b) cyclohexylacetate.

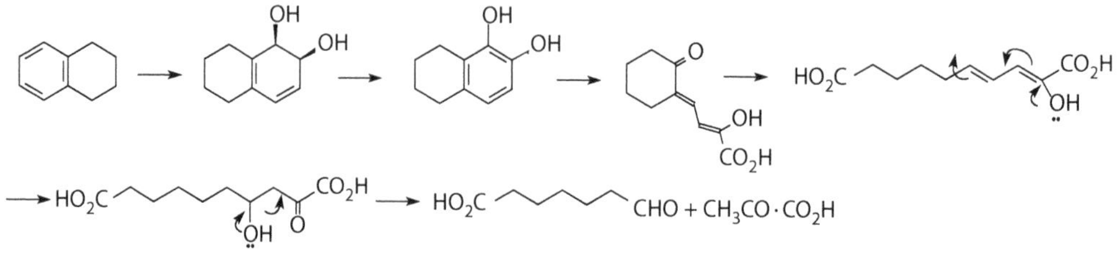

FIGURE 7.22 Aerobic degradation of tetralin. (Adapted from Hernáez MJ et al. 2002. *Appl Environ Microbiol* 68: 4841–4846.)

3. Cyclohexylacetate is degraded to cyclohexanone by elimination of the side chain by dehydogenation followed by hydroxylation at the ring junction (Ougham and Trudgill 1982).

An alternative to monooxygenation is realized in the hydration of the substituted cyclohexanone derived from the fission product of 1,2-dihydroxynaphthalene during the degradation of 1,2,3,4-tetrahydronaphthalene (tetralin) (Hernáez et al. 2002; López-Sánchez et al. 2010) (Figure 7.22).

7.3.4.1 The Aromatization Pathway

A less common pathway for the degradation of cyclohexanecarboxylate has been found in which the ring is dehydrogenated to 4-hydroxybenzoate before fission (Figure 7.23) (Blakley 1974; Taylor and Trudgill 1978). In *Corynebacterium cyclohexanicum* degradation was initiated by sequential hydroxylation and dehydrogenation to 4-oxocyclohexane-1-carboxylate. This then underwent stepwise aromatization to 4-hydroxybenzoate mediated by enzymes termed desaturases: desaturase I carried out dehydrogenation by an oxidase to cyclohex-2-ene-4-oxocarboxylate with the formation of H_2O_2, while desaturase II produced 4-hydroxybenzoate putatively via cyclohexane-2,5-diene-4-oxocarboxylate (Kaneda et al. 1993).

The degradation of polyhydroxylated cyclohexane carboxylates such as quinate and shikimate also involves aromatic intermediates (Ingledew et al. 1971). In *A. calcoaceticus*, the degradation of quinate and shikimate proceeds to the production of 3,4-dihydroxybenzoate (Elsemore and Ornston 1994). This is initiated by dehydrogenation to 3-dehydroquinate by *quiA*, that is dehydrated by *quiB* to dehydroshikimate that is common to the pathways for both substrates. Dehydroshikimate is then dehydrated by *quiC* to 3,4-dihydroxybenzoate (Figure 7.24) that is then degraded by ring fission. It is worth noting that the dehydrogenase encoded by QuiA in *A. calcoaceticus* is a member of the membrane pyrolloquinoline quinone (PQQ)-dependent dehydrogenases (van Kleef and Duine 1988). Although the quinate/shikimate pathway in *C. glutamicum* is formally analogous, QsuD being equivalent to QuiA and QsuB to QuiC, the amino acid sequences have no similarity and are remotely similar to the corresponding fungal enzymes QutBm, QutE, and QutC (Teramoto et al. 2009). The interrelation between the anaerobic metabolism cyclohexane carboxylates and their benzenoid analogues may be seen in the pathways for the degradation of cyclohexane carboxylate by *Rhodopseudomonas palustris*. This takes place by slightly different reactions involving the activity of a ligase encoded by *badA* to form the coenzyme ester, followed by a dehydrogenase to produce cyclohex-1-ene-1-carboxyl-CoA, which is then fed into the pathway used for the anaerobic degradation of benzoyl-CoA (Egland et al. 1995).

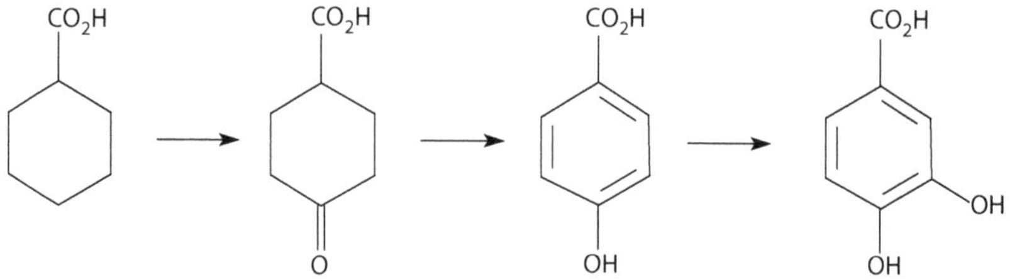

FIGURE 7.23 Alternative pathway for the biodegradation of a cyclohexane carboxylate.

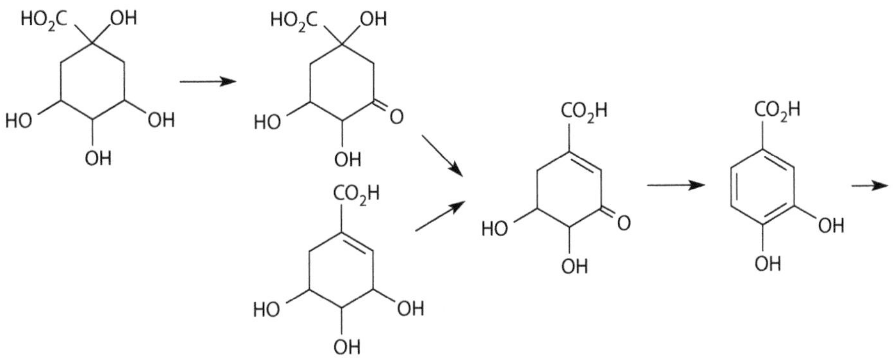

FIGURE 7.24 Biodegradation of quinate and shikimate.

The interrelation between the metabolism cyclohexane carboxylates and their benzenoid analogues may be seen in the pathways for the anaerobic degradation of cyclohexane carboxylate by *Rh. palustris*. This takes place by the action of a ligase (AliA) to form the coenzyme ester, followed by a dehydrogenase (BadJ) to produce cyclohex-1-ene-1-carboxyl-CoA, which is then fed into the pathway used for the anaerobic degradation of benzoyl-CoA (Egland and Harwood 1999) that is discussed in Chapter 8, Parts 1 and 3.

7.3.5 Reactions Catalyzed by "Old Yellow Enzyme"

These reactions are discussed in detail in Chapter 3 in the context of reductions, and is briefly mentioned here.

1. An unusual reaction was been observed in the reaction of "old yellow enzyme" with α,β-unsaturated ketones. A dismutation took place under aerobic or anaerobic conditions with the formation from cyclohex-1-oxo-2-ene of the corresponding phenol and cyclohexanone, and an analogous reaction from representative cyclodec-3-oxo-4-enes–putatively by hydride-ion transfer (Vaz et al. 1995).

2. Reduction of the double bond in α,β-unsaturated ketones has been observed, and the enone reductases from *Saccharomyces cerevisiae* have been purified and characterized. They are able to carry out reduction of the C=C bonds in aliphatic aldehydes and ketones, and ring double bonds in cyclohexenones (Wanner and Tressel 1998).

3. Reductions of steroid 1,4-diene-3-ones can be mediated by the related old yellow enzyme and pentaerythritol tetranitrate reductase, for example, androsta-$\Delta^{1,4}$-3,17-dione to androsta-Δ^4-3,17-dione (Vaz et al. 1995) and prednisone to pregna-Δ^4-3,11,20-trione-17α, 20-diol (Barna et al. 2001), respectively.

7.3.5.1 Microbial Hydroxylation

Numerous single-step transformations—generally hydroxylations, oxidations of alcohols to ketones or dehydrogenations—of both terpenes and sterols have been accomplished using microorganisms, especially fungi. This interest has been motivated by the great interest of the pharmaceutical industry in the products (Smith et al. 1988) and in fungal metabolism as a model for higher organisms (Smith and Rosazza 1983). Reactions catalyzed by cytochrome P450-mediated in both prokaryotes and eukaryotes have been discussed in Chapter 3, Part 1.

7.3.6 Anaerobic Reactions

Anaerobic degradation of cycloalkanes has seldom been reported. The pathway used for the degradation of ethylcyclopentane by a sulfate-reducing enrichment was analogous to the fumarate pathway used for *n*-alkanes (Section 7.1), with the formation of 3-ethyl-cyclopentane-carboxylate followed by ring fission to 3-ethylpentan-1,5-dicarboxylate (Rios-Harnandez et al. 2003). Only a few examples of the degradation of cycloalkanols under anoxic denitrifying conditions have been described.

1. A denitrifying pseudomonad (Dangel et al. 1988) was able to grow with cyclohexanol from which cyclohexane-1,3-dione was produced via cyclohex-2-enone, and 3-hydroxycyclohexanone before ring fission to 5-oxocaproate (Dangel et al. 1989).

2. A strain of *Azoarcus* was able to grow at the expense of cyclohexanecarboxylate, *cis*- and *trans*-cyclohexane-1,2-diols, 2-hydroxycyclohexane, and cylohexane-1,2-dione, as well as a range of aliphatic carboxylates including butyrate, succinate, caproate, pimerate, and adipate (Harder 1997).

   This has been examined in *Azoarcus* sp. strain 22L in that produced 6-oxohexanoate in a reaction catalyzed by an enzyme which was purified as a homodimer with a molecular mass of 105 kDa, and contained thiamin diphosphate, FAD, and Mg^{2+} in the monomer (Steinbach et al. 2011). The reaction is formally analogous to the C–C fission of 2-oxocarboxylates by the thiamin diphosphate enzyme (Tittman 2009).

3. *Alicycliphilus denitrificans* was able to use as substrates cyclohexanol, cyclohexanone, cyclohex-2-enone, and a range of carboxylates including adipate, pimelate, 2-oxoglutarate, 5-oxocaproate, and pyruvate (Mechichi et al. 2003). This is consistent with the proposed pathway supported by the activities of cyclohexanol dehydrogenase, cyclohexanone dehydrogenase, cyclohex-2-enone hydratase/3-hydroxycyclohexanone dehydrogenase, and cyclohexan-1,3-dione hydrolase for the degradation under anaerobic conditions (Dangel et al. 1989).

7.3.6.1 Monoterpenoids

Analogous Baeyer–Villiger oxidations are also employed in the degradation of alicyclic compounds containing several rings such as terpenes and sterols. The degradation of the monoterpenoid camphor has attracted attention since the initial hydroxylation in a plasmid-bearing pseudomonad is carried out by a cytochrome P450 enzyme (designated P450cam) (Schlichting et al. 2000). This enzyme displays a wide versatility that has already been noted in the wider context of monooxygenation in Chapter 3. In *Ps. putida* ATCC 17453 carrying the cam plasmid, degradation involves an initial cytochrome P450 hydroxylation at C_5 followed by oxidation, and the introduction of an oxygen atom adjacent to the quaternary methyl group (Figure 7.25) (Ougham et al. 1983). In an organism designated *Rh. rhodochrous* strain NCIMB 9784, hydroxylation occurs, however, at C_6, followed by ring fission of the 1,3-diketone (Figure 7.26) (Chapman et al. 1966). The hydrolase that carries out fission of the cyclic 1,3-diketone has been

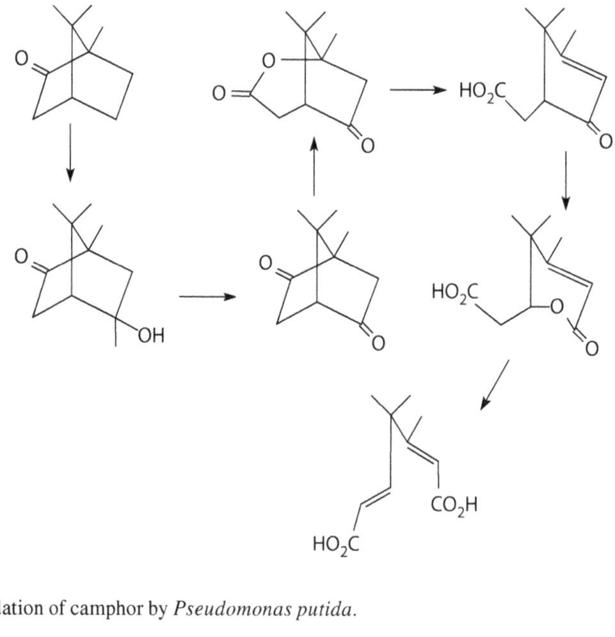

FIGURE 7.25 Degradation of camphor by *Pseudomonas putida*.

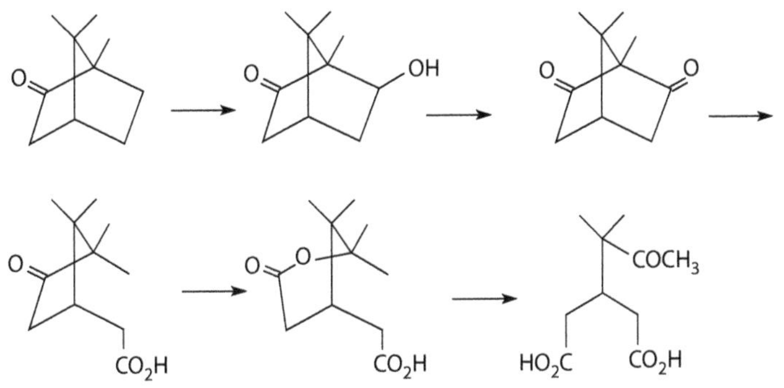

FIGURE 7.26   Degradation of camphor by *Rhodococcus rhodochrous.*

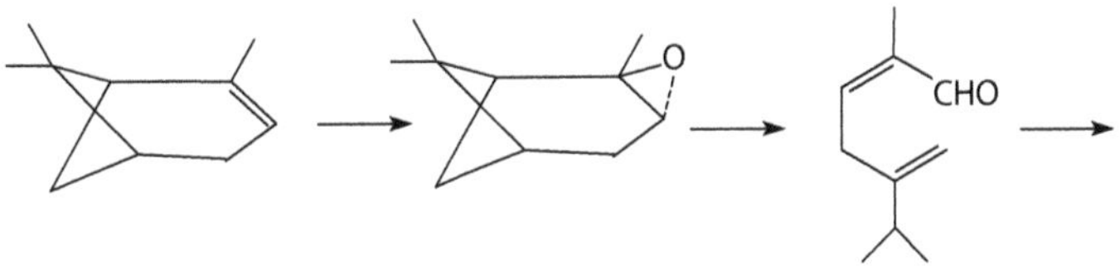

FIGURE 7.27   Degradation of α-pinene.

characterized and was comparable with those carried out the enoyl-CoA hydratase superfamily (Grogan et al. 2001; Whittingham et al. 2003).

Anaerobic growth of *Ps. citronellolis* using nitrate as electron acceptor occurred with the primary alkanols 3,7-dimethyoctan-1-ol and citronellol—though not with geraniol that was, however, produced from linalool—putatively by the activity of a mutase (Harder and Probian 1995). Although the metabolism of monoterpenes was examined anaerobically in denitrifying strains of *Alcaligenes defragrans* without defining the mechanism (Heyen and Harder 2000), the earlier suggestion of the involvement of a mutase has been confirmed in cells of *Castellaniella* (*Alcaligenes*) *defragrans* grown grown anoxically with linalool (Brodkorb et al. 2010). A single bifunctional enzyme was synthesized with a molecular mass of 40 kDa that catalyzed the reversible dehydration of linalool to myrcene and the allylic isomerization of linalool to geraniol.

The biodegradation of cyclic monoterpenes has been investigated under both aerobic and denitrifying conditions (Foss and Harder 1998), and may involve key reactions other than the Baeyer–Villiger type ring cleavage of ketones (Trudgill 1994). For example, in the degradation of α-pinene, although some strains of *Pseudomonas* sp. degrade this by rearrangement to limonene, oxidation, and β-oxidation, in others the initial reaction was formation of the epoxide that produced a carbonium ion at C_1 followed by sequential ring fission of the cyclobutane and cyclohexane rings to produce 2-methyl-5-isopropylhexa-2,5-dienal (Best et al. 1987; Griffiths et al. 1987a,b) (Figure 7.27). Bacterial hydroxylation by cytochrome P450 systems is well established in actinomyces (O'Keefe and Harder 1991). Stereospecific hydroxylation of α-ionone—though not β-ionone—has been observed with strains of *Streptomyces* sp. (Lutz-Wahl et al. 1998), and the racemic substrate is hydroxylated to the (3R,6R)—and (3R,6S)—hydroxy-α-ionones.

7.3.6.2 Diterpenoids and Triterpenoids

The hydroxylation of diterpenoids by fungi has been examined, and includes the following examples.

1. *Cunninghamella elegans* of a synthetic diterpenoid butenolide at the 5α- and both 7α- and 7β-positions, as well as on the isopropyl side chain (Milanova et al. 1994)

2. *C. elegans* of sclareol at the 2α-, 3β-, 18-, and 19-positions (Abraham 1994)

3. Stemodin by *Cephalosporium aphidicola* at the 7α- 7β, 8β-, 18-, and 19-positions (Hanson et al. 1994)

It is worth noting that in these examples, hydroxylation took place not only in ring A and ring B, but also at the quaternary methyl groups of C_{18} and C_{19}. This provides a plausible mechanism for the transformation of the triterpenoids lanosterol and cycloartenol by *Mycobacterium* sp. NRRL B-3805 where androsta-4,8(14)diene-3,17-dione was produced by successive hydroxylations at C_{18} with loss of formaldehyde from the 3-oxo-18-hydroxy intermediate, and analogous hydroxylation at C_{14} and migration of the double bond at $\Delta^{8,9}$ to $\Delta^{8,14}$ (Wang et al. 1995).

The degradation of the diterpenoid abietane derivatives abietic and dehydroabietic acid has been examined in different strains of bacteria.

1. In *Pseudomonas abietaniphila* strain BKME-9, it is degraded by initial cytochrome P450 hydroxylation at C_7 (Smith et al. 2004), oxidation to the 7-oxo compound followed by dioxygenation to the 7-oxo-11,12-diol that undergoes *meta*-fission (Section 3.2) (Martin and Mohn 1999, 2000). Further degradation then takes place by dioxygenation of ring C and ring fission (Figure 7.28a) (Biellmann et al. 1973b) or by an alternative (Figure 7.28b) (Biellmann et al. 1973a).

2. Details of the roles of the complementary roles of cytochrome P450 (CYP226) oxidations in the degradation of abietanes by *Burkholderia xenovorans* LB400 are illustrated by the following (Smith et al. 2008):

   a. The degradation of abietic acid proceeds by oxidation to the 7,8-epoxide by ditU, followed by rearrangement to the 7-oxo compound and dehydrogenation to 7-oxodehydroabietic acid

   b. Dehydroabietic acid is oxidized by ditQ to the 7-hydroxydehydroabietic acid followed by oxidation to the common intermediate 7-oxodehydroabietic acid that is then degraded by dihydroxylation at Δ^{10-11} followed by ring fission

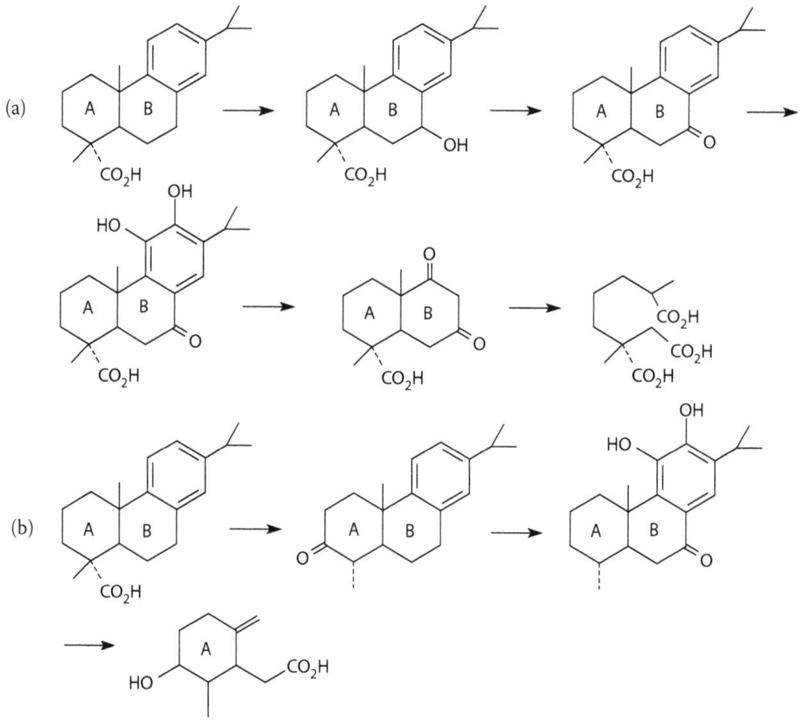

FIGURE 7.28 Cytochrome P450-transformation of abietane. (a) Successive hydroxylation and dioxygenation of ring C, (b) oxidation of ring A and fission of rings B and C.

7.3.6.3 Other Polycyclic Substrates

The following illustrate the variety of hydroxylations of structurally diverse polycyclic substrates.

1. Patchoulol is transformed by *Botrytis cinerea* to a number of products principally to those involving hydroxylation at the C_5 and C_7 quaternary atoms (Aleu et al. 1999) (Figure 7.29).
2. As illustration of the plethora of reactions that may occur is afforded by the transformation of caryophyllene oxide by *B. cinerea*. Although most of the reactions were hydroxylations or epoxidations, two involved transannular reactions: (a) between the C_4-epoxide oxygen and C_7 and (b) between the C_4-epoxide and C_{13} with formation of a caryolane (Figure 7.30) (Duran et al. 1999).
3. The degradation of atropine has been examined in *Pseudomonas* sp. strain AT3 and produces tropine as the initial metabolite. The degradation of this proceeds by oxidative loss of the *N*-methyl group, and elimination of ammonia to form 6-hydroxy-cyclohepta-1,4-dione, followed by fission of the 1,3-diketone to produce 4,6-dioxoheptanoate (Figure 7.31) (Bartholomew et al. 1996). Although the enzymology was unresolved, loss of ammonia presumably occurs either by successive hydroxylation at the tertiary carbon atoms adjacent to the -NH group, or by successive dehydrogenations.

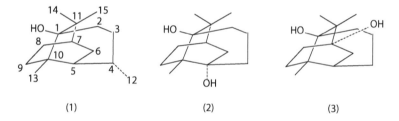

FIGURE 7.29 Hydroxylation of patchoulol.

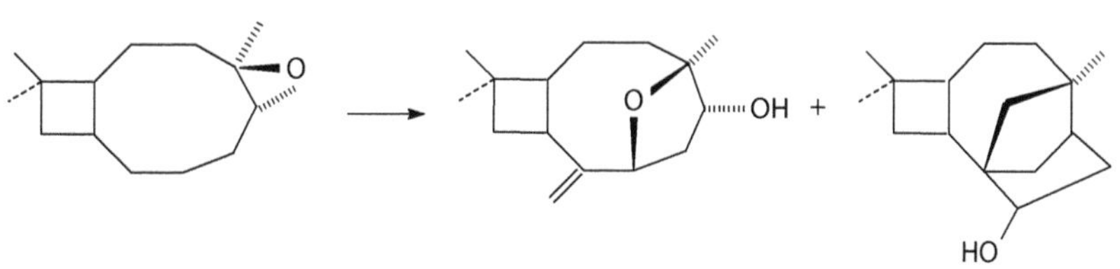

FIGURE 7.30 Transformation of caryophyllene oxide by *Botrytis cinerea*.

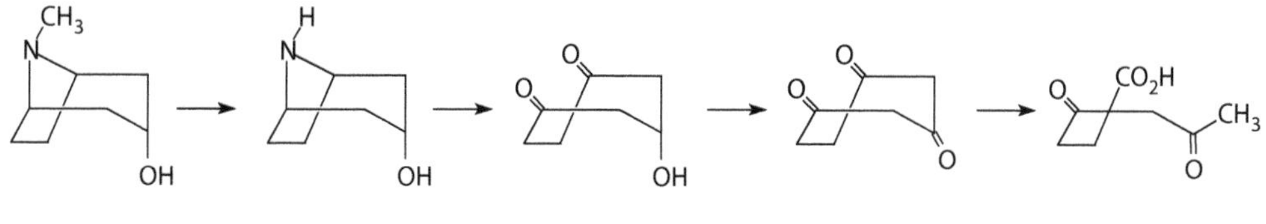

FIGURE 7.31 Degradation of tropine.

7.3.6.4 Steroids

This section deals with the transformation and degradation of three structurally similar compounds: steroids, bile acids, and steroid hormones with aromatic A rings.

7.3.6.4.1 Transformation by Hydroxylation

Microorganisms, particularly fungi, are able to hydroxylate sterols selectively at the 11α-, 12β-, 15-, 17-, 19-, and 21-positions (references in Hudlický 1990), although the product even from the same substrate will depend on the organism.

1. *Calonectria decora*, *Rhizopus nigricans*, and *Aspergillus ochraceus* produce the $12\beta,15\alpha$-; $11\alpha,16\beta$-; and $6\beta,11\alpha$-diols from 3-oxo-5α-androstane (Bird et al. 1980).
2. The biotransformation of pregna-4,17(20)-*cis*-diene-3,16-dione by *Aspergillus niger* produced metabolites with hydroxyl groups at the 7β,- or 7β- and 15β- positions, whereas *C. aphidicola* produces metabolites with hydroxyl groups at the 11α-, or 11β- and 15β- positions (Atta-ur-Rahman et al. 1998).
3. Hydroxylation of vitamin D_3 derivatives at C_{25} and $C_1(\alpha)$ has been carried out by the cytochrome P450 strains of *Streptomyces* sp. (Sasaki et al. 1991).
4. *Penicillium lilacinum* transformed testosterone successively to androst-4-ene-3,17-dione and testololactone (Prairie and Talalay 1963): the oxygen atom is unusually introduced into ring D at the quaternary position between C_{13} and C_{17}.

An important development has been the need for unusual products that may be sterol metabolites. Studies have been directed to the biosynthesis, for example, of the otherwise rare and inaccessible derivatives of progesterone hydroxylated at the 6-, 9-, 14-, or 15-positions that could be accomplished by incubating progesterone in a complex medium with the fungus *Apiocrea chrysosperma* (Smith et al. 1988). The hydroxylation of progesterone and closely related compounds at the 15-β position has been observed in cell extracts of *B. megaterium* (Berg et al. 1976).

7.3.6.4.2 Other Aerobic Transformations

Attention has been directed to the reactions catalyzed by cytochrome P450s that bring about important reactions leading to the loss of angular methyl groups at C_{10} (with concomitant aromatization of ring A) and C_{14}, and the C_{17} -$COCH_3$ side chain. These reactions are discussed further in Chapter 3, Part 1. The transformations of steroids and their precursor lanosterol have been extensively examined as sources of estrone and related compounds. In all of them, the essential reactions involve loss of the angular methyl groups, and cytochrome P450-mediated reactions have already been noted in Chapter 3, Part 1. In *Nocardia* sp., demethylation of steroids is accomplished by sequential formation of the 1,4-diene-3-ones, hydroxylation at C_9, and a dienone–phenol rearrangement (Figure 7.32a) (Sih et al. 1966). In a lanosterol derivative, *Mycobacterium* sp. (NRRL-B3805) carries out analogous reactions for the loss of the C_{29}, C_{30}, and C_{28} methyl groups by hydroxylation (Figure 7.32b) (Wang et al. 1995). Plausible reactions for the formation of formaldehyde from the quaternary methyl groups are outlined (Figure 7.32c, although formate is produced in cytochrome P450-mediated reactions.

7.3.6.4.3 Aerobic Degradation

There is a substantial literature on the microbial degradation of steroids accumulated during more than 50 years, and only a selected summary of the key reactions that have been established is given here. In the earlier studies, a range of bacteria were used including *Pseudomonas* sp. Searle B-20-184, *Arthrobacter* sp. strain Searle B22-9, and *Nocardia* sp. ATCC 13259, while in later studies *Comamonas (Pseudomonas) testosteroni* has been extensively investigated. There has been increased interest in strains of *Mycobacterium smegmatis* and *Mycob. phlei* prepared by insertion mutagenesis, and directed to partial degradation in plant steroids such as sitosterol with a C_{10} side chain (Andor et al. 2006).

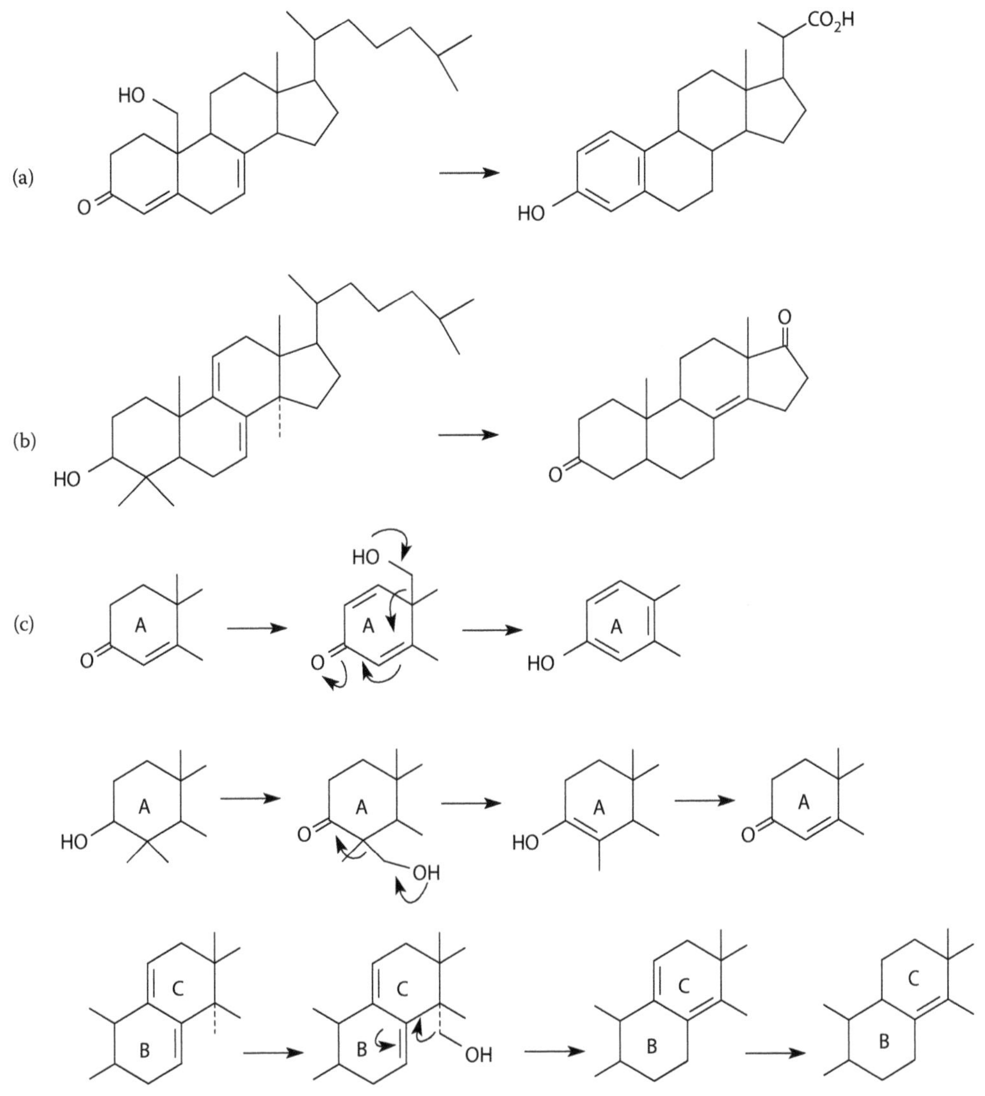

FIGURE 7.32 Demethylation of angular methyl groups (a) in steroids and related compounds, (b) in lanosterol derivative, and (c) reactions producing CH$_2$O.

The degradation of steroids can pragmatically be divided into sequential reactions:

1. Dehydrogenation by NADP of 3β-hydroxyl → 3-oxo; 17β-hydroxyl → 17-oxo; 3α-hydroxyl →
2. 3-Oxo by NADH+ (Marcus and Talalay 1955)
3. Dehydrogenation of 3-oxo → 1,4-diene-3-one (Plesiat et al. 1991; Florin et al. 1996; van der Geize et al. 2000)
4. Hydroxylation at C$_9$ that provides a critical entry into the degradation of rings A and B (Dodson and Muir 1961a,b; van der Geize et al. 2002, 2008)
5. Spontaneous dienone/phenol rearrangement to the 9,10-secosteroid
6. *Ortho*-hydroxylation of the phenolic ring to the 3,4-diol (Sih et al. 1966; Leppik 1989)
7. Fission of the catechol (Gibson et al. 1966; Leppik 1989; Horinouchi et al. 2003)
8. Degradation of the side chains

It should be noted, however, that degradation may not necessarily proceed in this order.

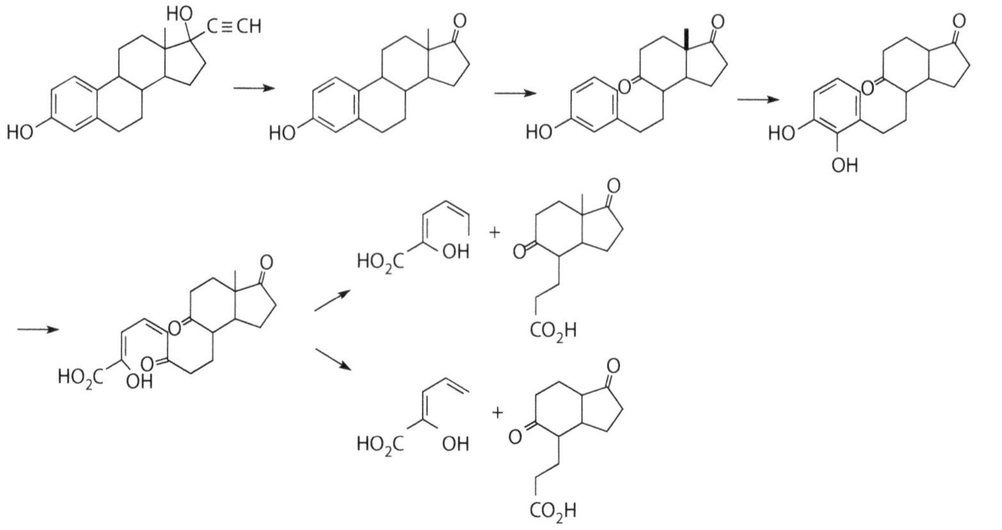

FIGURE 7.33   Aerobic degradation of 17α-ethynylestradiol.

The enzymology of most of these reactions has been elucidated, and this has enabled the assembly of the complete pathway for the degradation of testosterone in *Comamonas testosteroni* (Horinouchi et al. 2005), cholesterol by *Rhodococcus* sp. strain RHA1 (van der Geize et al. 2007), and 17α-ethynylestradiol by *Sphingobacterium* sp. strain JCR5 that shares the critical C_9 oxygenation and dioxygenation of ring A to produce a catechol before ring fission (Figure 7.33) (Haiyan et al. 2007). It is worth noting that the aerobic degradation of the bile acid cholate by *Pseudomonas* sp. strain Chol1 (Birkenmeier et al. 2007), and the anoxic degradation of cholesterol by *Sterolibacterium denitrificans* (Chiang et al. 2008a, 2010) involve the same initial reactions.

The degradation of rings A and B in steroids and bile acids that carry oxygen substitution at C_3 (3-oxo-4-ene and 3-hydroxy-5-ene) is initiated by a number of dehydrogenations that result in the formation of 3-oxo-1,4-dienes. The critical reaction that provides entry to aerobic degradation is hydroxylation at C_9 (Dodson and Muir 1961a,b; Leppik 1989). The 9α-hydroxylase has been characterized from *Rh. erythropolis* SQ1 (van der Geize et al. 2008), *Rh. rhodochrous* DSM 43269 (Petrusma et al. 2009), and from *Mycobacterium tuberculosis* (Capyk et al. 2009). It is a two-component iron–sulfur monooxygenase that consists of a terminal oxygenase KshA containing [2Fe-2S] Cys_2His_2 and the ferredoxin reductant KshB containing FAD and [2Fe-2S] Cys_4. The enzyme is also able to use testosterone (androst-3-one- 4-ene-17β-ol), and progesterone (preg-4-ene-3,17-dione). Dehydrogenation of the 9α-hydroxyl to 9-oxo takes place with concomitant aromatization of the A-ring by a dienone–phenol reaction that results in fission of the B-ring at C_{9-10}. This series of reactions has been exemplified in bacteria that degrade cholesterol, testosterone, cholate, and 17α-ethynylestradiol. The degradation of ring A is accomplished by hydroxylation and extradiol fission of the catechol. These reactions are illustrated for testosterone that carries no substituent at C_{17} and is replaced by a keto group (Figure 7.34) (Horinouchi et al. 2003, 2005). Steroid hormones including estradiol and estrone differ in having an aromatic ring A and lack a side chain at C_{17} that is replaced by a hydroxyl or keto group.

The degradation of cholate is initiated by the same reactions that are used for steroids with the production of the 3-oxo-1,4-diene. Further reaction requires the inversion of the 12α-hydroxyl group to 12-β. In *C. testosteroni* TA441, it has been established that this is carried out in two stages: (i) dehydrogenation to the 12-ketone catalyzed by SteA followed by (ii) reduction to the 12β-hydroxyl by SteB, both of which belong to the short-chain dehydrogenase/reductase superfamily (Horinouchi et al. 2008).

Different reactions have been found for the degradation of the side chains of steroids and related compounds.

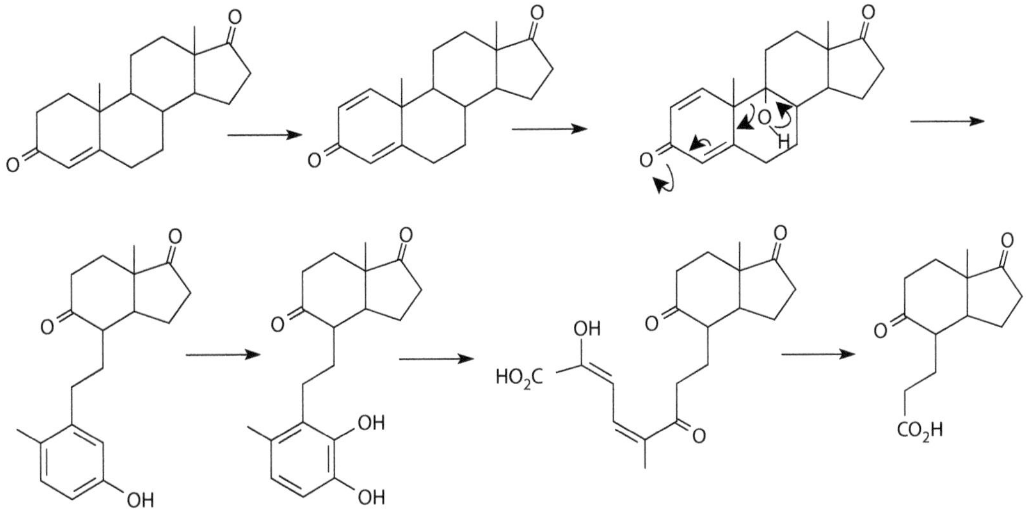

FIGURE 7.34 Aerobic degradation of testosterone. (Adapted from Horinouchi M et al. 2003. *Appl Environ Microbiol* 69: 4421–4430.)

1. The C_{17}–$COCH_3$ side chain of pregnane is degraded by *Cylindrocarpum radicicola* in two stages to the C_{17} ketone: Baeyer–Villiger oxygenation to the C_{17} acetate followed by the sequential activity of an oxygenase and a dehydrogenase (Rahim and Sih 1966).

2. The degradation of the side chain of cholesterol by *Rhodococcus jostii* strain RHA1 is initiated by cytochrome P450 (CYP 235) oxidation of C_{26} methyl group to the C_{26}-carboxylate (Rosłoniec et al. 2009). Further details that have already been described involved successive β-oxidations to produce the 17-oxo compound with the loss of two propionate groups and one acetate group (Figure 7.35) (Sih et al. 1968a,b; Fujimoto et al. 1982a,b; van der Geize et al. 2007). It should be noted that among strains that degrade cholesterol, differences may exist in the order of (a) side-chain degradation, (b) oxidations initiated by 9-α-hydroxylation with subsequent degradation of rings A and B, and (c) C_{19} hydroxylation.

3. Degradation of the side chain of phytosterol is more complicated since it contains a C_{24} ethyl group in addition to the C_{26} methyl group. This is accomplished by a reaction that involves carboxylation (Figure 7.36) (Fujimoto et al. 1982a,b) which is analogous to that involved in the degradation of geraniol, and requires activation of the carboxyl group by a coenzyme A ligase that has been identified as Fad19 in *Rh. rhodochrous* (Wilbrink et al. 2011).

Although steroids and bile acids (cholates) differ in the substitution at C_{17}, their degradations have in common the production of C_{17} ketones by a series of β-oxidations from the following substrates: cholesterol (van der Geize et al. 2007), cholate (Birkenmeier et al. 2007). Exceptionally, hydroxylation of the tertiary C_{25} in cholesterol takes place in *S. denitrificans* under denitrifying conditions (Chiang et al. 2007).

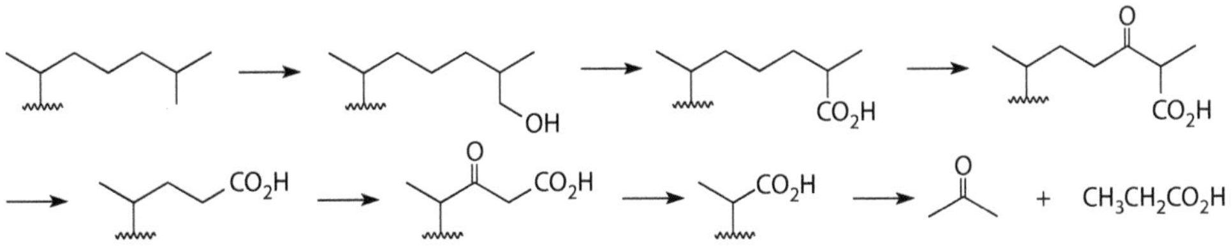

FIGURE 7.35 Aerobic degradation of cholesterol mediated by CYP 235.

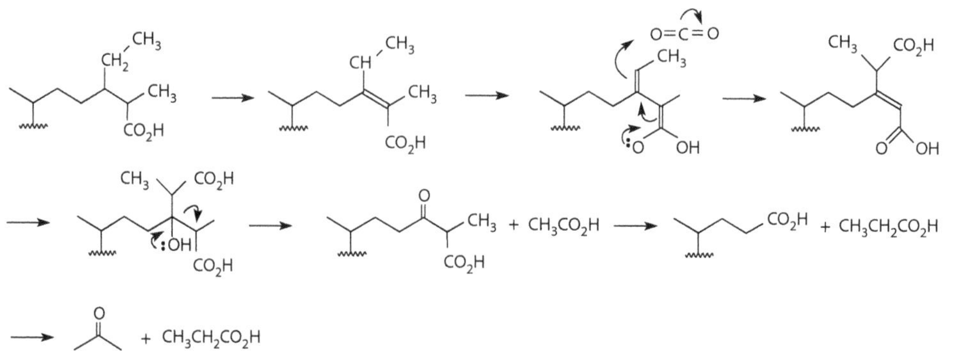

FIGURE 7.36 Aerobic degradation of phytosterol.

7.3.6.4.4 Anaerobic Transformation

Anaerobic reactions have been examined quite extensively in the context of the intestinal metabolism of bile acids, and a number of reactions that are otherwise quite unusual have been observed, most frequently in organisms belonging to the genera *Eubacterium* or *Clostridium*. Illustrative examples include the following:

1. Reduction of the $\Delta^{4,5}$-diene with production of 5β-reduced compounds, although in *Eubacterium coprostanoligenes* this is carried out in stages (Ren et al. 1996):

   3-β-hydroxy-5-ene   (cholesterol) → 3-one-5-ene → 4-ene-3-one → 3-one → 3-α-hydroxy (coprostanol)

2. Reductive dehydroxylation of 7α-hydroxy bile acids (Masuda et al. 1984) has been extensively studied since the deoxycholate is potentially toxic. Both the genetics and the complex of pathways have been elucidated for cholic acid in *Clostridium* sp. strain TO-931(Wells and Hylemon 2000). These reactions involve a complex sequence of reactions within an operon: (a) successive dehydrogenations by BaiB and BiaA to 3-oxo-4-enecholate, (b) dehydration by BaiE to 3-oxo-4,6-dienecholate, and (c) successive reduction to the 3-oxodeoxycholate and 3-hydroxydeoxycholate which are illustrated (Figure 7.37) (Wells and Hylemon 2000).

3. In addition, dehydroxylation of 16α-hydroxy and 21-hydroxy corticosteroids has been observed with fecal bacteria (Bokkenheuser et al. 1980).

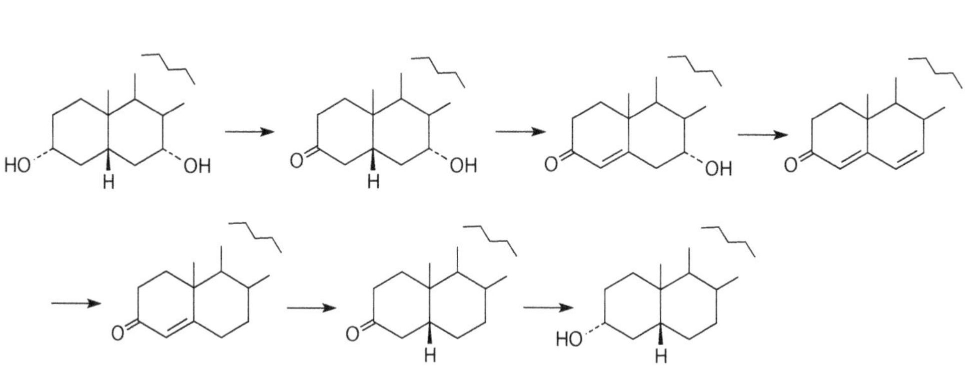

FIGURE 7.37 Anaerobic biotransformation of 3α,7α-cholic acid. (Adapted from Wells JF, PB Hylemon. 2000. *Appl Environ Microbiol* 66: 1107–1113.)

7.3.6.4.5 Anaerobic Degradation

Degradation of steroids has been demonstrated under anoxic conditions using nitrate as electron acceptor:

a. Estradiol and estrone—though not cholesterol or androst-4-ene-3,17-dione—by a strain of *Denitratisoma oestradiolicum*. Growth occurred with a range of C_1 to C_6 carboxylates and, significantly isobutyrate (Fahrbach et al. 2006).

b. Testosterone and estradiol by *Steroidobacter denitrificans* that has a rather restricted nutritional spectrum, being able to grow only with acetate and glutamate, or aerobically only with testosterone (Fahrbach et al. 2008). The metabolism is initiated—as for the aerobic metabolism—by successive dehydrogenations to the 1,4-dienone which undergoes hydration followed by reduction to androstan-1,3,17-triol (Leu et al. 2011).

Details of some of the reactions involved have emerged for the degradation of cholesterol by *S. denitrificans* under denitrifying conditions (Chiang et al. 2007). This was initiated by dehydrogenations to cholesta-1,4-diene-3-one from ring A that were analogous to those employed under aerobic conditions by *Rhodococcus* sp. strain RHA1 (van der Geize et al. 2007). This was, however, accompanied by the exceptional hydroxylation at the tertiary C_{25} (Chiang et al. 2007).

7.4 References

Abraham W-R. 1994. Microbial hydroxylation of sclareol. *Phytochemistry* 36: 1421–1424.

Ahlert J, J Distler, K Mansouri, W Piepersberg. 1997. Identification of *stsC*, the gene encoding the L-glutamine: *Scyllo*-inosose aminotransferase from streptomycin-producing Streptomyces. *Arch Microbiol* 168: 102–113.

Aleu J, JR Hanson, RH Galán, IG Collado. 1999. Biotransformation of the fungistatic *sesquiterpenoid patchoulol* by *Botrytis cinerea*. *J Nat Prod* 62: 437–440.

Anderson WA, B Magasanik. 1971a. The pathway of *myo*-inositol degradation by *Aerobacter aerogenes*: Identification of the intermediate 2-deoxy-5-keto-D-gluconic acid. *J Biol Chem* 246: 5653–5661.

Anderson WA, B Magasanik. 1971b. The pathway of *myo*-inositol degradation in *Aerobacter aerogenes*: Conversion of 2-deoxy-5-keto-D-gluconic acid to glycolytic intermediates. *J Biol Chem* 246: 5662–5675.

Andor A, A Jekkel, DA Hopwood, F Jeanplong, E Ilkôy, A Kónya, I Kurucz, G Ambrus. 2006. Generation of usefully insertionally blocked sterol degradation pathway mutants of fast-growing mycobacteria and cloning, characterization, and expression of the terminal oxygenase of the 3-oxosteroid 9α-hydroxylase in *Mycobacterium smegmatis* Mc2155. *Appl Environ Microbiol* 72: 6554–6559.

Atta-ur-Rahman, MI Choudhary, F Shaheen, M Ashraf, S Jahan. 1998. Microbial transformation of hypolipemic *E*-guggulsterone. *J Nat Prod* 61: 428–431.

Bahar M, J De Majnik, M Wexler, J Fry, PS Poole, PJ Murphy. 1998. A model for the catabolism of rhizopine in *Rhizobium leguminosarum* involves a ferredoxin oxygenase complex and the inositol degradative pathway. *Mol Plant Microbe Interact* 11: 1057–1068.

Barna TM, H Khan, NC Bruce, I Barsukov, NS Scrutton, PCE Moody. 2001. Crystal structure of pentaerythritol tetranitrate reductase: "flipped" binding geometries for steroid substrates in different redox states of the enzyme. *J Mol Biol* 310: 433–447.

Bartholomew BA, MJ Smith, PW Trudgill, DJ Hopper. 1996. Atropine metabolism by *Pseudomonas* sp. strain AT3: evidence for nortropine as an intermediate in atropine breakdown and reactions leading to succinate. *Appl Environ Microbiol* 62: 3245–3250.

Berg A, JÅ Gustafsson, M Ingelman-Sundberg, K Carlström. 1976. Characterization of a cytochrome P-450-dependent steroid hydroxylase present in *Bacillus megaterium*. *J Biol Chem* 251: 2831–2838.

Best DJ, NC Floyd, A Magalhaes, A Burfield, PM Rhodes. 1987. Initial steps in the degradation of *alpha*-pinene by *Pseudomonas fluorescens* NCIMB 11671. *Biocatalysis* 1: 147–159.

Biellmann JF, G Branlant, M Gero-Robert, M Poiret. 1973a. Dégradation bactérienne de l'acide dehydroabiétique par *Flavobacterium resinovorum*. *Tetrahedron* 29: 1227–1236.

Biellmann JF, G Branlant, M Gero-Robert, M Poiret. 1973b. Dégradation bactérienne de l´acide dehydroabiétique par un *Pseudomonas* et un *Alcaligenes. Tetrahedron* 29: 1237–1241.

Bird TG, PM Fredricks, ERH Jones, GD Meakins. 1980. Microbiological hydroxylations. Part 23. Hydroxylations of fluoro-5α-androstanones by the fungi *Calonectria decora*, *Rhizopus nigricans*, and *Aspsrgillus ohraceus. J Chem Soc Perkin* I: 750–755.

Birkenmeier A, J Holert, H Erdbrink, H M Moeller, A Friemel, R Schoenberger, M J-G Suter, J Klebsberger, B Philipp. 2007. Biochemical and genetic investigation of initial reactions in aerobic degradation of the bile acid cholate in *Pseudomonas* sp. strain Chol1. *J Bacteriol* 189: 7165–7173.

Blakley ER. 1974. The microbial degradation of cyclohexanecarboxylic acid: a pathway involving aromatization to form *p*-hydroxybenzoic acid. *Can J Microbiol* 20: 1297–1306.

Bokkenheuser VD, J Winter, S ÓRourke, AE Ritchie. 1980. isolation and characterization of fecal bacteria capable of 16α-dehydroxylating corticoids. *Appl Environ Microbiol* 40: 803–808.

Branchaud BP, CT Walsh. 1985. Functional group diversity in enzymatic oxygenation reactions catalyzed by bacterial flavin-containing cyclohexanone oxygenase. *J Amer Chem Soc* 107: 2153–2161.

Brodkorb D, M Gottschall, R Marmulla, F Lüddeke, J Harder. 2010. Linalool dehydratase-isomerase, a bifunctional enzyme in the anaerobic degradation of monoterpenes. *J Biol Chem* 285:30436–30442.

Capyk JK, I d'Angelo, NC Strynadya, LD Eltis. 2009. Characterization of 3-oxosteroid 9α-hydroxylase, a Rieske oxygenase in the cholesterol degradation pathway of *Mycobacterium tuberculosis. J Biol Chem* 284: 9937–9946.

Chapman PJ, G Meerman, JC Gunsalus, R Srinivasan, KL Rinehart. 1966. A new acyclic metabolite in camphor oxidation. *J Am Chem Soc* 88: 618–619.

Chiang Y-R, W Ismail, S Gallien, D Heinz, A Van Dorsselaer, G Fuchs. 2008b. Cholest-4-ene-3-one-Δ^1-dehydrogenase, a flavoprotein catalyzing the second step in anoxic cholesterol metabolism. *Appl Environ Microbiol* 74: 107–113.

Chiang Y-R, W Ismail, D Heinz, C Schaeffer, A Van Dorsselaer, G Fuchs. 2008a. Study of anoxic and oxic cholesterol metabolism by *Sterolibacterium denitrificans. J Bacteriol* 190: 905–914.

Chiang Y-R, W Ismail, M Müller, G Fuchs. 2007. Initial steps in the anoxic metabolism of cholesterol by the denitrifying *Sterolibacterium denitrificans. J Biol Chem* 282: 13240–13249.

Chiang Y-R, J-Y Fang, W Ismail, P-H Wang. 2010. Initial steps in anoxic testosterone degradation by *Steroidobacter denitrificans. Microbiology (UK)* 156: 2253–2259.

Dangel W, A Tschech, G Fuchs. 1988. Anaerobic metabolism of cyclohexanol by denitrifying bacteria. *Arch Microbiol* 150: 358–362.

Dangel W, A Tschech, G Fuchs. 1989. Enzyme reactions involved in anaerobic cyclohexanol metabolism by a denitrifying *Pseudomonas* species. *Arch Microbiol* 152: 273–279.

Dodson RM, RD Muir. 1961a. Microbiological transformations. VI. The microbiological aromatization of steroids. *J Am Chem Soc* 83: 4627–4631.

Dodson RM, RD Muir. 1961b. Microbiological transformations. VII. The hydroxylation of steroids at C-9. *J Am Chem Soc* 83: 4631–4635.

Donoghue NA, DB Norris, PW Trudgill. 1976. The purification and properties of cyclohexanone oxygenase from *Nocardia globurula* CL1 and *Acinetobacter* NCIB 9871. *Eur J Biochem* 63: 175–192.

Duran R, E Corrales, R Hernández-Galán, IG Collado. 1999. Biotransformation of caryophyllene oxide by *Botrytis cinerea. J Nat Prod* 62: 41–33.

Egland PG, J Gibson, CS Harwood. 1995. Benzoate-coenzyme A ligase, encoded *badA*, is one of three ligases to catalyze benzoyl-coenzyme A formation during anaerobic growth of *Rhodopseudomonas palustris* on benzoate. *J Bacteriol* 177: 6545–6551.

Egland PG, CS Harwood. 1999. BadR, a new MarR family member regulates anaerobic benzoate degradation by *Rhodopseudomonas palustris* in concert with AadR, an Fnr family member. *J Bacteriol* 181: 2102–2109.

Eichhorn E, JR van der Ploeg, T Leisinger. 1999. Characterization of a two-component alkane sulfonate monooxygenase from *Escherichia coli. J Biol Chem* 274: 26639–26646.

Elsemore DA, LN Ornston. 1994. The *pca-pob* supraoperonic cluster of *Acinetobacter calcoaceticus* contains *quiA*, the structural gene for quinate-shikimate dehydrogenase. *J Bacteriol* 176: 7659–7666.

Fahrbach M, J Kuever, R Meinke, P Kämpfer, J Hollender. 2006. *Denitratisoma oestradiolicum* gen. nov., sp. nov., a 17β-oestradiol-degrading, denitrifying betaproteobacterium. *Int J Syst Evol Microbiol* 56: 1547–1552.

Fahrbach M, J Kuever, M Remesch, BE Huber, P Kämpfer, W Dott, J Hollender. 2008. *Steroidobacter denitrificans* gen. nov., sp. nov., a steroidal hormone-degrading gammaproteobacterium. *Int J Syst Evol Microbiol* 58: 2215–2223.

Florin C, T Köhler, M Grandguillot, P Plesiat. 1996. *Comamonas testosteroni* 3-oxosteroid-$\Delta^{4(5a)}$—dehydrogenase: Gene and protein characterization. *J Bacteriol* 178: 3322–3330.

Foss S, J Harder. 1998. *Thauera linaloolentis* sp. nov. and *Thauera terpenica* sp. nov., isolated on oxygen-containing monoterpenes (linalool, menthol, and eucalyptol) and nitrate. *Syst Appl Microbiol* 21: 365–373.

Fry J, M Wood, PS Poole. 2001. Investigation of *myo*-inositol catabolism in *Rhizobium leguminosarum* bv. *viciae* and its effect on nodulation competitiveness. *Mol Plant Microbe Interact* 14: 1016–1025.

Fujimoto Y, C-S Chen, AS Gopalan, CJ Sih. 1982b. Microbial degradation of the phytosterol side chain. 2. Incorporation of $NaH^{14}CO_3$ into the C-28 position. *J Am Chem Soc* 104: 4720–4722.

Fujimoto Y, C-S Chen, Z Szelecky, D DiTullio, CJ Sih. 1982a. Microbial degradation of the phytosterol side chain. 1. Enzymatic conversion of 3-oxo-24-ethyl-4-en-26-oic acid into 3-oxochol-4-en-24-oic acid and androst-4-ene-3,17-dione. *J Am Chem Soc* 104: 4718–4720.

Galbraith MP, SF Feng, J Borneman, EW Triplett, FJ de Bruijn. 1998. A functional *myo*-inositol catabolism pathway is essential for rhizopine utilization by *Sinorhizobium meliloti*. *Microbiology (UK)* 144: 2915–2924.

Gibson DT, KC Wang, CS Sih, H Whitlock. 1966. Mechanisms of steroid oxidation by microorganisms. IX. On the mechanisms of ring A cleavage in the degradation of 9,10-seco-steroids by microorganisms. *J Biol Chem* 241: 551–559.

Griffin M, PW Trudgill. 1972. The metabolism of cyclopentanol by *Pseudomonas* NCIB 9872. *Biochem J* 129: 595–603.

Griffiths ET, SM Bociek, PC Harries, R Jeffcoat, DJ Sissons, PW Trudgill. 1987a. Bacterial metabolism of α-pinene: pathway from α-pinene oxide to acyclic metabolites in *Nocardia* sp. strain P18.3. *J Bacteriol* 169: 4972–4979.

Griffiths ET, PC Harries, R Jeffcoat, PW Trudgill. 1987b. Purification and properties of α-pinene oxide lyase from *Nocardia* sp., strain P18.3. *J Bacteriol* 169: 4980–4983.

Grogan G, GA Roberts, D Bougioukou, NJ Turner, SL Flitsch. 2001. The desymmetrization of bicyclic β-dioxones by an enzymatic retro-claisen reaction. *J Biol Chem* 276: 12565–12572.

Haiyan R, J Shulan, N ud din Ahmad, W Dao, C Chenwu. 2007. Degradation characteristics and metabolic pathway of 17α-ethynylestradiol by *Sphingobacterium* sp. JCR5. *Chemosphere* 66: 340–346.

Hanson JR, PB Reese, JA Takahashi, MR Wilson. 1994. Biotransformation of some stemodane diterpenoids by *Cephalosporium aphidicola*. *Phytochem* 36: 1391–1393.

Harder J. 1997. Anaerobic degradation of cyclohexane-1,2-diol. *Arch Microbiol* 168: 199–204.

Harder J, C Probian. 1995. Microbial degradation of monoterpenes in the absence of molecular oxygen. *Appl Environ Microbiol* 61:3804–3808.

Haskell T. 1984. Cyclitol antibiotics pp. 9–32. In *Handbook of Microbiology*, 2nd Edition (Eds AI Laskin and HA Lechevalier), CRC Press, Boca Raton, Florida.

Hernáez MJ, B Floriano, JJ Ríos, E Santero. 2002. Identification of a hydratase and a class II aldolase involved in biodegradation of the organic solvent tetralin. *Appl Environ Microbiol* 68: 4841–4846.

Heyen U, J Harder. 2000. Geranic acid formation, an initial reaction of anaerobic monoterpene metabolism in denitrifying *Alcaligenes defragrans*. *Appl Environ Microbiol* 66:3004–3009.

Horinouchi M, T Hayashi, T Yamamoto, T Kudo. 2003. A new bacterial steroid degradation gene cluster in *Comamonas testosteroni* TA441 which consists of aromatic-compound degradation genes for seco-steroids and 3-oxosteroid dehydrogenase genes. *Appl Environ Microbiol* 69: 4421–4430.

Horinouchi M, T Hayashi, H Koshino, T Kurita, T Kudo. 2005. Identification of 9,17-dioxo-1,2,3,4,10,19-hexanorandrostan-5-oic acid, 4-hydroxy-2-oxohexanoic acid, and 2-hydroxyhexa-2,4-dienoic acid and related enzymes involved in testosterone degradation in *Comamonas testosteroni* TA441. *Appl Environ Microbiol* 71: 5275–5281.

Horinouchi M, T Hayashi, H Koshino, M Malon, T Yamamoto, T Kudo. 2008. Identification of genes involved in inversion of stereochemistry of a C-12 hydroxyl group in the catabolism of cholic acid by *Comamonas testosteroni* TA441. *J Bacteriol* 190: 5545–5554.

Hudlický M. 1990. *Oxidations in Organic Chemistry*. ACS Monograph 186. American Chemical Society, Washington, DC.

Ingledew WM, MEF Tresguerres, JL Cánovas. 1971. Regulation of the enzymes of the hydroaromatic pathway in *Acinetobacter calco-aceticus*. *J Gen Microbiol* 68: 273–282.

Iwaki H, S Wang, S Grosse, H Bergeron, A Nagahashi, J Lertvorachon, J Yang, Y Konishi, Y Hasegawa, PCK Lau. 2006. Pseudomonad cyclopentadecanone monooxygenase displaying an uncommon spectrum of Baeyer-Villiger oxidations of cyclic ketones. *Appl Environ Microbiol* 72: 2707–2720.

Iwaki H, Y Hosegawa, S Wang, MM Kayser, PCK Lau. 2002. Cloning and characterization of a gene cluster involved in cyclopentanol metabolism in *Comamonas* sp. strain NCIMB 9872 and biotransformations effected by *Escherichia coli*-expressed cyclopentanone 1,2-monooxygenase. *Appl Environ Microbiol* 68: 5671–5684.

Jones KH, RT Smith, PW Trudgill. 1993. Dioxocamphane enantiomer-specific `Baeyer-Villiger´ monooxygenases from camphor-grown *Pseudomonas putida* ATCC 17453. *J Gen Microbiol* 139: 797–805.

Kaneda T, H Obata, M Tokumoto. 1993. Aromatization of 4-oxocyclohexanecarboxylic acid to 4-hydroxybenzoic acid by two distinct desaturases from *Corynebacterium cyclohexanicum*. *Eur J Biochem* 218: 997–1003.

Kawsar HL, K Ohtani, K Okumura, H Hayashi, T Shimizu. 2004. Organization and transcription regulation of *myo*-inositol operon in *Clostridium perfringens*. *FEMS Microbiol Lett* 235: 289–295.

Kohler PRA, E-L Choong, S Rossbach. 2011. The RpiR-like repressor IolR regulates inositol catabolism in *Sinorhizobium melitoti*. *J Bacteriol* 193: 5155–5163.

Kohler PRA, JY Zheng, E Schoffers, S Rossbach. 2010. Inositol catabolism, a key pathway in *Sinorhizobium meliloti* for competitive host nodulation. *Appl Environ Microbiol* 76: 7972–7980.

Kostichka K, SM Thomas, KJ Gibson, V Nagarajan, Q Cheng. 2001. Cloning and characterization of a gene cluster for cyclododecanone oxidation in *Rhodococcus ruber* SC1. *J Bacteriol* 183: 6478–6486.

Krings E, K Krumbach, B Bathe, R Kelle, VF Wendisch, J Sahm, L Eggeling. 2006. Characterization of *myo*-inositol utilization by *Corynebacterium glutamicum*: The stimulon, identification of transporters, and influence on L-lysine formation. *J Bacteriol* 188: 8054–8061.

Kröger C, TM Fuchs. 2009. Characterization of the *myo*-inositol utilization island of *Salmonella enterica* serovar Typhimurium. *J Bacteriol* 191: 545–554.

Leppik RA. 1989. Steroid catechol degradation: Disecoandrostane intermediates accumulated by Pseudomonas transposon mutant strains. *J Gen Microbiol* 135: 1979–1988.

Leu Y-L, P-H Wang, M-S Shiao, W Ismail, Y-R Chiang. 2011. A novel testosterone catabolic pathway in bacteria. *J Bacteriol* 193: 4447–4455.

López-Sánchez A, B Floriano, E Andújar, MJ Henáez, E Santero. 2010. Tetralin-induced and ThnR-regulated aldehyde dehydrogenase and β-oxidation genes in *Sphingomonas macrogolitabida* strain TFA. *Appl Environ Microbiol* 76: 110–118.

Lutz-Wahl S, P Fischer, C Schmidt-Dannert, W Wohlleben, B Hauer, RD Schmid. 1998. Stereo- and regioselective hydroxylation of α-ionone by *Streptomyces* strains. *Appl Environ Microbiol* 64: 3878–3881.

Makarewicz O, S Dubrac, T Masdek, R Borriss. 2006. Dual role of the PhoP ~ P response regulator: *Bacillus amyloliquefaciens* FZB45 phytase gene transcription is directed by positive and negative interactions with the *phyC* promoter. *J Bacteriol* 188: 6953–6965.

Malito E, A Slfieri, MW Fraaije, A Mattevi. 2004. Crystal structure of a Baeyer-Villiger monooxygenase. *Proc Natl Acad Sci USA* 101: 13157–13162.

Marcus PI, P Talalay. 1955. Induction and purification of α- and β-hydroxysteroid dehydrogenases. *J Biol Chem* 230:661–674.

Martin VJJ, WW Mohn. 1999. A novel aromatic ring-hydroxylating dioxygenase from the diterpenoid-degrading *Pseudomonas abietaniphila* BKME-9. *J Bacteriol* 181: 2675–2682.

Martin VJJ, WW Mohn. 2000. Genetic investigation of the catabolic pathway for degradation of abietane diterpenoids by *Pseudomonas abietaniphila* BKME-9. *J Bacteriol* 182: 3784–3793.

Masuda N, H Oda, S Hirano, M Masuda, H Tanaka. 1984. 7α-dehydroxylation of bile acids by resting cells of a *Eubacterium lentum*-like intestinal microbe, strain c-25. *Appl Environ Microbiol* 47: 735–739.

Mechichi T, E Stackebrandt, G Fuchs. 2003. *Alicycliphilus denitrificans* gen. nov., sp. nov., a cyclohexanol-degrading, nitrate-reducing β-proteobacterium. *Int J Syst Evol Microbiol* 53: 147–152.

Milanova R, M Moore, Y Hirai. 1994. Hydroxylation of synthetic abietane diterpenes by *Aspergillus* and *Cunninghamella* species: Novel route to the family of diterpenes isolated from *Triptergium wilfordii*. *J Nat Prod* 57: 882–889.

Möbus E, M Jahn, R Schmid, D Jahn, E Maser. 1997. Testosterone-related expression of enzymes involved in steroid and aromatic hydrocarbon metabolism. *J Bacteriol* 179: 5951–5955.

Murphy PJ, W Wexler, W Grzemski, JP Pao, D Gordon. 1995. Rhizopines—their role in symbiosis and competition. *Soil Biol Biochem* 27: 525–529.

O'Keefe DP, PA Harder. 1991. Occurrence and biological function of cytochrome P-450 monooxygenase in the actinomycetes. *Mol Microbiol* 5: 2099–2105.

Ougham HJ, DG Taylor, PW Trudgill. 1983. Camphor revisited: involvement of a unique monooxygenase in metabolism of 2-oxo-Δ^3-4,5,5-trimethylcyclopentenylacetic acid by *Pseudomonas putida*. *J Bacteriol* 153: 140–152.

Ougham HJ, PW Trudgill. 1982. Metabolism of cyclohexaneacetic acid and cyclohexanebutyric acid by *Arthrobacter* sp. strain CA1. *J Bacteriol* 150: 1172–1182.

Petrusma M, L Dijkhuizen, R van der Geize. 2009. *Rhodococcus rhodochrous* DSM 43269 3-ketosteroid 9α-hydroxylase, a two-component iron-sulfur-containing monooxygenase with subtle steroid substrate specificity. *Appl Environ Microbiol* 75: 5300–5307.

Plesiat P, M Grandguillot, S Harayama, S Vragar, Y Michel-Briand. 1991. Cloning, sequencing, and expression of the *Pseudomonas testosteroni* gene encoding 3-oxosteroid Δ^1-dehydrogenase. *J Bacteriol* 173: 7219–7227.

Prairie RL, P Talalay. 1963. Enzymatic formation of testololactone. *Biochemistry* 2: 203–208.

Rahim MA, CJ Sih. 1966. Mechanisms of steroid oxidation by microorganisms XI. Enzymatic cleavage of the pregnane side chain. *J Biol Chem* 241: 3615–3523.

Ren D, L Li, JW Young, DC Beitz. 1996. Mechanism of cholesterol reduction to coprostanol by *Eubacterium coprostanoligens* ATCC 51222. *Steroids* 61: 33–40.

Rios-Harnandez LA, LM Gieg, JM Suflita. 2003. Biodegradation of an alicyclic hydrocarbon by a sulfate-reducing enrichment from a gas condensate-contaminated aquifer. *Appl Environ Microbiol* 69: 434–443.

Rosłoniec KZ, MH Wilbrink, JK Capyk, WH Mohn, M Ostendorf, R van der Geize, L Dijkhuizen, LD Eltis. 2009. Cytochrome P450 125 (CPY125) catalyses C26-hydroxylation to initiate sterol side-chain degradation in *Rhodococcus jostii* RHA1. *Mol Microbiol* 74: 1031–1043.

Rossbach S, DA Kulpa, U Rossbach, FJ de Bruijn. 1994. Molecular and genetic characterization of the (*mocABCR*) genes of *Rhizobium meliloti* L5–30 *Mol Gen Genet* 245: 11–24.

Saint CP, M Wexler, PJ Murphy, J Tempe, ME Tate. 1993. Characterization of genes for synthesis and catabolism of a new rhizopine induced in nodules by *Rhizobium meliloti* Pm220–3: extension of the rhizopine concept. *J Bacteriol* 175 . 5205–5215.

Sasaki J, A Mikami, K Mizoue, S Omura. 1991. Transformation of 25- and 1α-hydroxyvitamin D_3 to 1α,25-dihydroxyvitamin D_3 by using *Streptomyces* sp. strains. *Appl Environ Microbiol* 57: 2841–2846.

Schiller JG, AE Chung. 1970. Mechanism of cyclopropane ring cleavage in cyclopropanecarboxylic acid. *J Biol Chem* 245: 6553–6557 .

Schlichting I, J Berendzen, K Chu, AM Stock, SA Maves, DE Benson, RM Sweet, D Ringe, GA Petsko, SG Sligar. 2000. The catalytic pathway of cytochrome P450cam at atomic resolution. *Science* 287: 1615–1622.

Sih CJ, SS Lee, YY Tsong, WC Wang. 1966. Mechanisms of steroid oxidation by microorganisms VIII. 3,4-dihydroxy-9,10-secoandrosta-1,3,5(10)-triene-9–17-dione, an intermediate in the microbiological degradation of ring A of androst-4-ene-3,17 dione. *J Biol Chem* 241: 540–550.

Sih CJ, HH Tai, YY Tsong, SS Lee, RG Coombe. 1968b. Mechanisms of steroid oxidation by microorganisms. XIV. Pathway of cholesterol side-chain degradation. *Biochemistry* 7: 808–818.

Sih CJ, KC Wang, HH Tai. 1968a. Mechanisms of steroid oxidation by microorganisms. XIII. C_{22} acid intermediates in the degradation of the cholesterol side chain. *Biochemistry* 7: 796–807.

Smith KE, S Latif, DN Kirk, KA White. 1988. Microbial transformations of steroids—I. Rare transformations of progesterone by *Apiocrea chrysosperma*. *J Steroid Biochem* 31: 83–89.

Smith DJ, VJJ Martin, WW Mohn. 2004. A cytochrome P450 involved in the metabolism of abietane diterpenoids by *Pseudomonas abietaniphila* BKME-9. *J Bacteriol* 186: 3631–3639.

Smith DJ, MA Patrauchan, C Florizone, LD Eltis, WW Mohn. 2008. Distinct roles for two CYP226 family cytochromes P450 in abietane diterpenoid catabolism by *Burkholderia xenovorans* LB400. *J Bacteriol* 190:1575–1583.

Smith RV, JP Rosazza. 1983. Microbial models of mammalian metabolism. *J Nat Prod* 46: 79–91.

Steinbach AK, S Fraas, J harder, A Tabbert, H Brinkmann, A Mayer, U Ermler, PMH Kroneck. 2011. Cyclohexane-1,2-dione hydrolase from denitrifying *Azoarcus* sp strain 22Lin, a novel member of the thiamin diphosphate enzyme family. *J Bacteriol* 193:6760–6769.

Stirling LA, RJ Watkinson, IJ Higgins. 1977. Microbial metabolism of alicyclic hydrocarbons: Isolation and properties of a cyclohexane-degrading bacterium. *J Gen Microbiol* 99: 119–125.

Taylor DG, PW Trudgill. 1978. Metabolism of cyclohexane carboxylic acid by *Alcaligenes* strain W1. *J Bacteriol* 134: 401–411.

Teramoto H, M Inui, H Yulawa. 2009. Regulation of expression of genes involved in quinate and shikimate utilization in *Corynebacterium glutamicum*. *Appl Environ Microbiol* 75: 3461–3468.

Tipton CL, NM Al-Shathir. 1974. The metabolism of cyclopropane fatty acids by *Tetrahymena pyriformis*. *J Biol Chem* 249: 886–889.

Tittmann K. 2009. Reaction mechanisms of thiamin diphosphate enzymes: Redox reactions. *FEBS J* 276:2454–2468.

Toraya T, T Oka, M Ando, M Yamanishi, H Nishihara. 2004. Novel pathway for utilization of cyclopropanecarboxylate by *Rhodococcus rhodochrous*. *Appl Environ Microbiol* 70: 224–228.

Trower MK, RM Buckland, R Higgins, M Griffin. 1985. Isolation and characterization of a cyclohexanemetabolizing *Xanthobacter* sp. *Appl Environ Microbiol* 49: 1282–1289.

Trudgill PW. 1978. Microbial degradation of alicyclic hydrocarbons. pp. 47–84. In *Developments in Biodegradation of hydrocarbons-1* (Ed RJ Watkinson). Applied Science Publishers Ltd, London.

Trudgill PW. 1984. Microbial degradation of the alicyclic ring: Structural relationships and metabolic pathways. pp. 131–180. In *Microbial degradation of organic compounds* (Ed DT Gibson). Marcel Dekker Inc, New York.

Trudgill PW. 1994. Microbial metabolism and transformation of selected monoterpenes. pp. 33–61. In *Biochemistry of Microbial Degradation* (Ed C Ratledge). Kluwer Academic Publishers, Dordrecht, The Netherlands.

Turner BL, MJ Papházy, PM Hatgarth, ID McKelvie. 2002. Inositol phosphates in the environment. *Philos Trans R Soc Lond B Biol Sci* 357: 449–469.

van der Geize R, GI Hessels, L Dijkhuizen. 2002. Molecular and functional characterization of the kstD2 gene of *Rhodococcus erythropolis* SQ1 encoding a second 3-ketosteroid Delta(1)-dehydrogenase isoenzyme. *Microbiology* 148: 3285–3292.

van der Geize R, GI Hessels, M Nienhuis-Kuiper, L Dijkhuizen. 2008. Characterization of a second *Rhodococcus erythropolis* SQ1 3-oxosteroid 9α-hydroxylase activity comprising a terminal oxygenase homologue, KshA2 active with oxygenase-reductase component KshB. *Appl Environ Microbiol* 74: 7197–7203.

van der Geize R, GI Hessels, R van Gerwen, JW Vrijbloed, P van der Meijden, L Dijkhuizen. 2000. Targeted disruption of the *kstD* gene encoding a 3-oxosteroid Δ1-dehydrogenase isoenzyme of *Rhodococcus erythropolis* strain SQ1. *Appl Environ Microbiol* 66: 2029–2036.

van der Geize R, K Yam, T Heuser, MH Wilbrink, H Hara, MC Anderton, E Sim, L Dijkhuizen, JE Davies, WW Mohn, LD Eltis. 2007. A gene cluster encoding cholesterol catabolism in a soil actinomycete provides insight into *Mycobacterium tuberculosis* survival in macrophages. *Proc Natl Acad Sci USA* 104: 1947–1952.

van Kleef MAG, JA Duine. 1988. Bacterial NAD(P)-independent quinate dehydrogenase is a quinoprotein. *Arch Microbiol* 150: 32–36.

Vaz ADN, S Chakraborty, V Massey. 1995. Old yellow enzyme: Aromatization of cyclic enones and the mechanism of a novel dismutation reaction. *Biochemistry* 34: 4246–4256.

Walker JB. 1995. Enzymatic synthesis of aminocyclitol moieties of aminoglycoside antibiotics from inositol by *Streptomyces* spp.: detection of glutamine—aminocyclitol aminotransferase and diaminocyclitol aminotransferase activities in a spectinomycin producer. *J Bacteriol* 177: 818–822.

Wang KC, B-J You, J-L Jan, S-H Lee. 1995. Microbial transformation of lanosterol derivatives with *Mycobacterium* sp. (NRLL B-3805). *J Nat Prod* 58: 1222–1227.

Wanner P, R Tressel. 1998. Purification and characterization of two enone reductases from *Saccharomyces cerevisiae*. *Eur J Biochem* 255: 271–278.

Wells JF, PB Hylemon. 2000. Identification and characterization of a bile acid 7α-dehydroxylation operon in *Clostridium* sp. strain TO-931, a highly active 7α-dehydroxylating strain isolated from human feces. *Appl Environ Microbiol* 66: 1107–1113.

Whittingham JL, JP Turkenburg, CS Verma, MA Walsh, G Grogan. 2003. The 2-Å crystal structure of 6-oxocamphor hydrolase. *J Biol Chem* 278: 1744–1750.

Wilbrink MH, M Petrusma, L Dijkhuisen, R van der Geize. 2011. Fad19 of *Rhodococcus rhodochrous* DSM43269, a steroid-coenzyme A ligase essential for degradation of C-14 branched sterol side chains. *Appl Environ Microbiol* 77: 4455–4464.

Yebra MJ, M Zúñiga, S Beaufils, G Pérez-Martínez, J Deutscher, V Monedero. 2007. Identification of a gene cluster enabling *Lactobacillus casei* BL23 to utilize *myo*-inositol. *Appl Environ Microbiol* 73: 3850–3858.

Yoshida K, M Yamaguchi, T Morinaga, M Kinehara, M Ikeuchi, H Ashida, Y Fujita. 2008. *myo*-inositol catabolism in *Bacillus subtilis*. *J Biol Chem* 283: 10415–10424.

7.5 Alkenes and Alkynes

7.5.1 Alkenes

7.5.1.1 Aerobic Conditions

Two kinds of investigations have been carried out: (a) growth of microorganisms at the expense of alkenes and (b) biotransformations resulting in the synthesis of epoxides. For example, growth has been demonstrated at the expense of propene and butene (van Ginkel and de Bont 1986). An interesting observation is the pathway for the degradation of intermediate *n*-alkenes produced by an aerobic organism under anaerobic conditions (Parekh et al. 1977). Although the generality of this pathway remains unknown, it is clearly possible that even aerobic bacteria under anoxic conditions might accomplish comparable degradations. Attention should also be drawn to the possibility that intermediate metabolites may be incorporated into biosynthetic pathways: hexadecene is oxidized by the fungus *Mortierella alpina* by β-oxidation (Shimizu et al. 1991), but the lipids contain carboxylic acids with both 18 and 20 carbon atoms including the unusual polyunsaturated acid ω1-5*cis*, 8*cis*,11*cis*,14*cis*,19-eicosapentaenoic acid.

Direct fission of the double bond by oxidation exemplifies the simplest, though less common reaction. Some illustrations are given:

1. The degradation of squalene by *Corynebacterium* (Arthrobacter) *terpenotabidum* (Takeuchi et al. 1999) is initiated by oxidative fission at Δ^{10-11} to *trans*-geranylacetone and further oxidation products including isovalerate and 3,3-dimethylacrylate (Figure 7.38) (Yamada et al. 1975; Ikeguchi et al. 1988). The degradation of squalene by *Marinobacter* sp. strain 2sq31 (Rontani et al. 2002) is analogous and subsequent reactions are carried out by β-oxidation and carboxylation that are comparable with those used for branched alkanes.

2. The degradation of synthetic poly(*cis*-1,4-isoprene) was examined in *Streptomyces coelicolor* strain 1A and produced a range of metabolites formed by oxidation of the terminal double bonds—ketones and carboxylates (Bode et al. 2000). An extracellular enzyme from *Xanthomonas* sp. was able to degrade poly(*cis*-1,4-isoprene) with the production of 12-oxo-4,8-dimethyltrideca-4,8-diene-1-al (Braaz et al. 2004), and functioned as a heme-dependent oxygenase (Braaz et al. 2005).

7.5.1.1.1 Epoxidation

The first established step in the aerobic degradation of alkenes is generally epoxidation. Alkene monooxygenase is closely related to the aromatic monooxygenases, and is able to hydroxylate benzene, toluene, and phenol (Zhou et al. 1999) while the alkane hydroxylase from *Ps. oleovorans* is able to carry out both hydroxylation and epoxidation (Ruettinger et al. 1977). In *Xanthobacter autotrophicus* strain Py2 that can carry out epoxidation of alkenes from C_2 to C_6, the monooxygenase has been characterized, and consists of an oxygenase, a ferredoxin, and a reductase (Small and Ensign 1997). Epoxidation is followed by several alternatives that may be summarized before further discussion: (a) hydrolysis to a 1:2-diol; (b) nucleophilic attack by coenzyme M followed by dehydrogenation; (c) a reductive glutathione-mediated reaction.

In detail, the degradation of epoxides is quite varied, and several different pathways have been observed. Mechanisms of resistance to fosfomycin (1*R*, 2*S*)-epoxypropylphosphonic acid) that

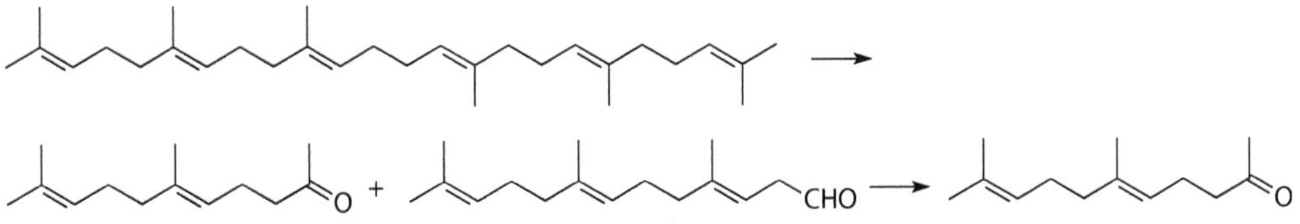

FIGURE 7.38 Degradation of squalene.

involve reactions with the epoxide have been noted in Chapter 4, while additional comments are given below.

1. Degradation of epichlorohydrin (1-chloro-2,3-epoxypropane) may proceed by hydrolysis of the epoxide to 3-chloro-1,2-propanediol that is then converted successively to 3-hydroxy-1,2-epoxypropane (glycidol) followed by hydrolysis to glycerol before degradation (van den Wijngaard et al. 1989). Epoxide hydrolases have been isolated and characterized from bacteria that are able to use epoxides as growth substrates. A *Corynebacterium* sp. is able to grow with alicyclic epoxides, and the sequence of the hydrolase (Misawa et al. 1998) is similar to the enzyme from *Agrobacterium radiobacter* strain AD1 that used epichlorohydrin (1-chloro-2,3-epoxypropane) as growth substrate (Rink et al. 1997). Examination of mutants of this strain prepared by site-directed mutagenesis showed that the mechanism involves nucleophilic attack by Asp107 at the terminal position of the substrate followed by hydrolysis of the resulting ester mediated by His275. Analogy may be noted with the inversion accompanying hydrolysis of 2-haloacids mediated by L-2-haloacid hydrolase. Limonene-1,2-epoxide hydrolases from *Rh. erythropolis* DCL14 that are intermediates in the degradation of both (+)-(4*R*)- and (−)-(4*S*)-limonene (van der Werf et al. 1999) differ from these groups of enzymes, and do not involve the catalytic function of histidine residues (van der Werf et al. 1998).

2. Hydrolysis to the diol followed by dehydration to the aldehyde and oxidation to the carboxylic acid is used by a propene-utilizing species of *Nocardia* (de Bont et al. 1982). Although an ethene-utilizing strain of *Mycobacterium* sp. strain E44 degrades ethane-1,2-diol by this route, the diol is not an intermediate in the metabolism of the epoxide (Wiegant and de Bont 1980).

3. The aldehyde may also be produced directly from the epoxide. This occurs in the metabolism of ethene by *Mycobacterium* sp. strain E44 (Wiegant and de Bont 1980), of styrene by a strain of *Xanthobacter* sp. strain 124X (Hartmans et al. 1989), and by *Corynebacterium* sp. strain ST-5 and AC-5 (Itoh et al. 1997). The reductase in the coryneforms has a low substrate specificity and is able to reduce acetophenone to 3-phenylethan-2-ol with an enantiomeric excess >96%. In *Rh. rhodochrous*, however, styrene is degraded by ring dioxygenation with the vinyl group intact (Warhurst et al. 1994): 2-vinyl-*cis,cis*-muconate is produced by catechol 1,2-dioxygenase as a terminal metabolite, and complete degradation is carried out by catechol 2,3-dioxygenase activity that is also present.

4. *X. autotrophicus* strain Py2 may be grown with propene or propene oxide. On the basis of amino acid sequences, the monooxygenase that produces the epoxide was related to those that catalyze the monooxygenation of benzene and toluene (Zhou et al. 1999). Three reactions are initially involved (Figure 7.39):

   – Metabolism of the (*R*)- and (*S*)-propene epoxides initiated by nucleophilic reaction with coenzyme M in a Zn-dependent transferase (Krum et al. 2002; Boyd and Ensign 2005) to produce the (*R*)- and (*S*)-hydroxypropyl-CoM

   – Enantioselective dehydrogenation to 2-ketopropyl-CoM (Sliwa et al. 2010)

   – Reductive carboxylation to 3-oxobutyrate (acetoacetate) (Krishnakumar et al. 2008)

   The final products depend on whether or not CO_2 is present, and both alternatives are dependent on a pyridine nucleotide-disulfide oxidoreductase (Swaving et al. 1996; Nocek et al. 2002).

   a. In the absence of CO_2, by transformation to acetone that is not further degraded;

   b. In the presence of CO_2 by reductive carboxylation to 3-oxobutyrate (acetoacetate). This is used partly for cell growth, and partly converted into the storage product poly-β-hydroxybutyrate (Small et al. 1995). Kinetic and ^{13}C NMR experiments confirm that acetoacetate is the first product from which β-hydroxybutyrate is formed as a secondary metabolite with acetone as the terminal metabolite (Allen and Ensign 1996). The epoxide carboxylase is a three component enzyme—all three of which are necessary for activity (Allen and Ensign 1997). Coenzyme M (2-mercaptoethanesulfonate) is required (Allen

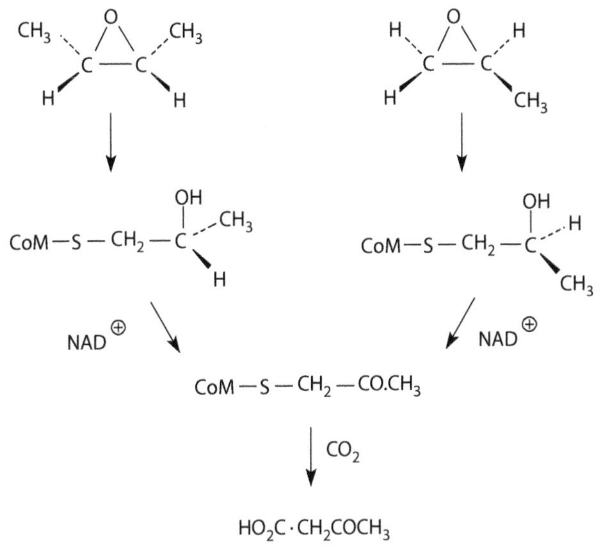

FIGURE 7.39 Reaction of epoxide with coenzyme M.

et al. 1999; Krum and Ensign 2001), component II is a flavin containing NADPH: disulfide oxidoreductase (Nocek et al. 2002), and the interactions involve $NADP^+$, FAD, the disulfide, and the 2-oxopropyl-CoM (Clark et al. 2000). An analogous mechanism operates in the degradation of epichlorohydrin (1-chloro-2,3-epoxypropane) by the same strain (Small et al. 1995);

c. The metabolism of acetone that converges with the metabolism of propene oxides in *X. autotrophicus* strain Py2 is accomplished by an ATP-dependent carboxylase with production of acetoacetate (Sluis et al. 1996, 2002).

5. *Rhodococcus* sp. strain AD45 carried out the transformation of 2-methyl-1,3-butadiene (isoprene) and both *cis*- and *trans*-dichloroethenes to the epoxides (van Hylckama Vlieg et al. 1998). Degradation of the diene (Figure 7.40) takes place by a pathway involving a glutathione *S*-transferase that is able to react with the epoxides and a conjugate-specific dehydrogenase that produces 2-glutathionyl-2-methylbut-3-enoate (van Hylckama Vlieg et al. 1999).

6. The degradation of vinyl chloride and ethene has been examined in *Mycobacterium* sp. strain JS 60 (Coleman and Spain 2003) and in *Nocardioides* sp. strain JS614 (Mattes et al. 2005). For both substrates, the initially formed epoxides underwent reaction with reduced coenzyme M and, after dehydrogenation and formation of the coenzyme A esters, reductive loss of coenzyme M acetate resulted in the production of *S*-acetyl-coenzyme A. The reductive fission is formally analogous to that in the glutathione-mediated reaction.

Epoxides may be formed from alkenes during degradation by *Ps. oleovorans*, although octan-1,2-epoxide is not further transformed, and degradation of oct-1-ene takes place by β-oxidation (May and Abbott 1973;

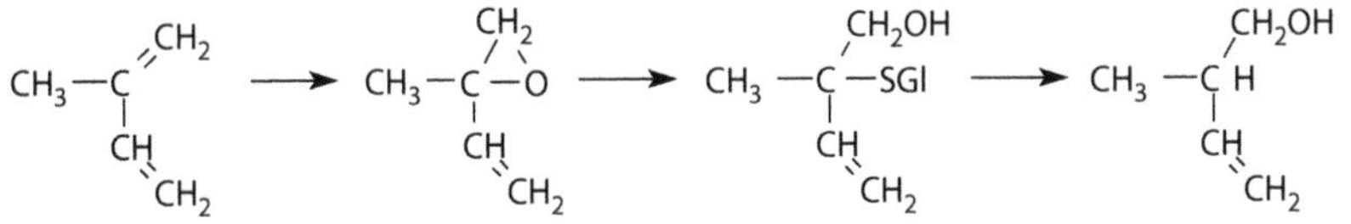

FIGURE 7.40 Glutathione-mediated degradation of epoxides.

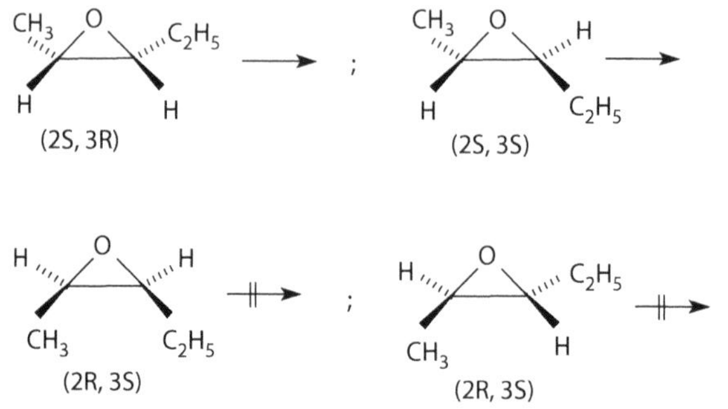

FIGURE 7.41 Biodegradability of enantiomeric of epoxides of *cis*- and *trans*-pent-2-enes.

Abbott and Hou 1973). The β-hydroxylase enzyme is able to carry out either hydroxylation or epoxidation (Ruettinger et al. 1977).

Considerable attention has been directed to the epoxidation of alkenes on account of interest in the epoxides as industrial intermediates. The wide metabolic capability of methane monooxygenase that has already been noted has been applied to the epoxidation of C_2, C_3, and C_4 alkenes (Patel et al. 1982). A large number of propane-utilizing bacteria are also effective in carrying out the epoxidation of alkenes (Hou et al. 1983). Especially valuable is the possibility of using microorganisms for resolving racemic mixtures of epoxides. For example, this has been realized for *cis*- and *trans*-2,3-epoxypentanes using a *Xanthobacter* sp. which is able to degrade only one of the pairs of enantiomers leaving the other intact (Figure 7.41) (Weijers et al. 1988). Bacterial epoxidation of alkenes and fungal enzymatic hydrolysis of epoxides have been reviewed in the context of their application to the synthesis of enantiomerically pure epoxides and their derivatives (Archelas and Furstoss 1997). One of the disadvantages of using bacteria that may carry out undesirable degradation may sometimes be overcome by the use of fungi (Archelas and Furstoss 1992), although the initially formed epoxides are generally hydrolyzed by the fungal epoxide hydrolase activity.

Other aspects of epoxide formation and degradation are worth noting, particularly on account of their biotechnological relevance.

1. Although *Mycobacterium* sp. strain E3 is able to degrade ethene via the epoxide, the epoxide-degrading activity is highly specific for epoxyethane, and degradation requires reductant generated from glycogen or trehalose storage material (de Haan et al. 1993).
2. In *Xanthobacter* sp. strain Py2 both the alkene monooxygenase and the epoxidase are induced by C_2, C_3, and C_4 alkenes, and also by chlorinated alkenes including vinyl chloride, *cis*- and *trans*-dichloroethene, and 1,3-dichloropropene (Ensign 1996).

7.5.1.1.2 Hydration

Hydration of activated double bonds is a key reaction in the metabolism of alkanoates after dehydrogenation to *trans*-alk-2-enoates, and is also widely encountered in the anaerobic metabolism of benzoyl-CoA esters after reduction of the ring. In aerobic organisms, hydration may occasionally serve as an alternative to epoxidation, and is illustrated by the following examples.

1. The hydration of the nonactivated double bond in oleate to 10-hydroxy- and 10-oxostearate has been described for a range of microorganisms (El-Sharkawy et al. 1992), and the hydratase involved has been characterized (Bevers et al. 2009).
2. Hydration is also involved in the degradation of the ring-fission products from dioxygenation and dehydrogenation of tetralin (Figure 7.42) (Hernáez et al. 2000, 2002; López-Sánchez et al. 2010).
3. Although the aerobic degradation of squalene involves oxidation and fission at Δ^{10-11} (Figure 7.43), degradation under anaerobic conditions involves hydration and carboxylation (Rontani et al. 2002).

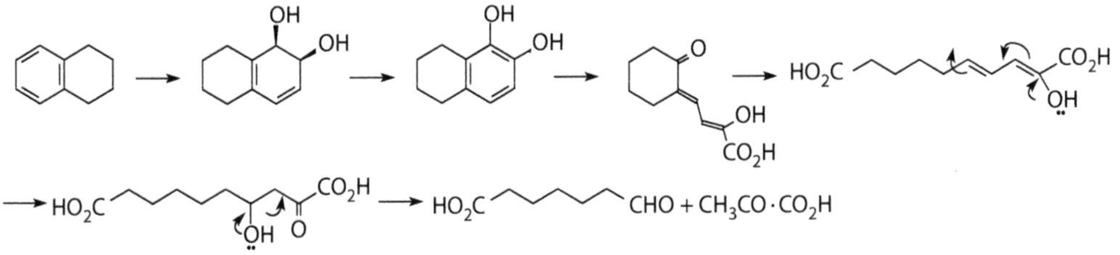

FIGURE 7.42 Aerobic degradation of tetralin by hydration.

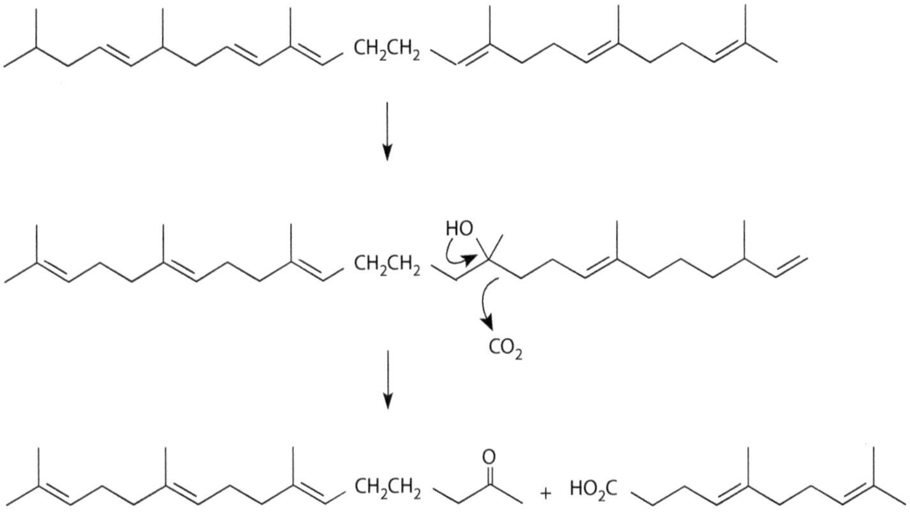

FIGURE 7.43 Aerobic degradation of squalene. (Adapted from Rontani J-F. 2002. *Arch Microbiol* 178: 279–287.)

7.5.1.2 Anaerobic Conditions

Degradation of hex-1-ene has been observed in a methanogenic consortium (Schink 1985a) that converted the substrate to methane, and a plausible pathway involving hydration and oxidation was suggested. A good deal of attention has been directed to the oxidation of long-chain alk-1-enes by sulfate-reducing bacteria. *Desulfatibacillum alkenivorans* was able to use sulfate and thiosulfate as electron acceptors, n-alkanes (C_4 to C_{18}) as electron donors, and was able to oxidize alk-1-enes (C_8 to C_{23}) (Cravo-Laureau et al. 2004). The metabolism of alk-1-enes has been examined in *D. aliphaticivorans* that was also able to degrade alkanes anaerobically, and two mechanisms have been proposed (Grossi et al. 2007). In one, hydroxylation and dehydrogenation took place to produce a terminal carboxylic acid that was degraded either by β-oxidation or transformed into fatty acids. Alternatively, rearrangements took place involving the formation of methyl groups at C_2 and C_4 of the carboxylate, or C_4-ethyl groups. In the presence of thiosulfate, *Archaeoglobus frigidus* was able to oxidize n-alk-1-enes $C_{12:1}$ to $C_{21:1}$ to CO_2 with the production of sulfide, and it was tentatively proposed that this took place by oxidation to carboxylic acids (Khelifi et al. 2010).

7.5.2 Alkynes

The degradation of alkynes has been the subject of sporadic but effective interest during many years so that the pathway has been clearly delineated. It is quite distinct from those used for alkanes and alkenes, and is a reflection of the enhanced nucleophilic character of the alkyne C≡C bond. The initial step is, therefore, hydration of the triple bond followed by ketonization of the initially formed enol. This reaction operates during the degradation of acetylene itself (de Bont and Peck 1980), acetylene carboxylic acids

$$HC\equiv C\cdot CH_2\cdot CH_2OH \longrightarrow HC\equiv C\cdot CH_2\cdot CHO \longrightarrow HC\equiv CH_2\cdot CO_2H$$

$$\longrightarrow CH_3\cdot CO\cdot CH_2\cdot CO_2H \longrightarrow CH_3\cdot CO_2H$$

FIGURE 7.44 Aerobic biodegradation of but-3-ynol.

(Yamada and Jakoby 1959), and more complex alkynes (Figure 7.44) (van den Tweel et al. 1985). It is also appropriate to note that the degradation of acetylene by anaerobic bacteria proceeds in the same pathway (Schink 1985b).

Acetylene was able to support the growth of the anaerobe *Pelobacter acetylenicus* (Schink 1985b), and underwent initial hydration to acetaldehyde followed by dismutation into acetate and ethanol. Although the enzyme was stable in air, it required a strong reductant such as Ti(III) or dithionite for its activity (Rosner and Schink 1995). The hydratase was a tungsten iron–sulfur protein with a [4Fe-4S] cluster in the reduced state and contained a molybdopterin cofactor conjugated with guanosine monophosphate (Meckenstock et al. 1999). Its structure was determined by x-ray crystallography which showed that it contained a bis-molybdopterin guanine dinucleotide to which the tungsten was bound through the dithiolene, and in addition a cubane [4Fe–4S] cluster (Seiffert et al. 2007).

Aerobic enrichment with acetylene as sole source of carbon and energy yielded strains of *Rhodococcus* and *Gordonia*. The activity of cell-free extracts of *Rhodococcus opacus* was dependent on molybdenum and required the presence of the reductant Ti(III) citrate, whereas the hydratase from *Rh. ruber* and *Gordonia* sp. did not require additional reductant. Importantly, there was no cross-reaction between cell extracts of these organisms and the hydratase of the anaerobic *P. acetylenicus* (Rosner et al. 1997).

7.6 References

Abbott BJ, CT Hou. 1973. Oxidation of 1-alkenes to 1,2-epoxides by *Pseudomonas oleovorans*. *Appl Microbiol* 26: 86–91.

Allen JR, DD Clark, JG Krum, SA Ensign. 1999. A role for coenzyme M (2-mercaptoethanesulfonic acid) in a bacterial pathway of aliphatic epoxide carboxylation. *Proc Natl Acad Sci USA* 96: 8432–8437.

Allen JR, SA Ensign. 1996. Carboxylation of epoxides to β-keto acids in cell extracts of *Xanthobacter* strain Py2. *J Bacteriol* 178: 1469–1472.

Allen JR, SA Ensign. 1997. Characterization of three protein components required for functional reconstitution of the epoxide carboxylase multienzyme complex from *Xanthobacter* strain Py2. *J Bacteriol* 179: 3110–3115.

Archelas A, Furstoss. 1992. Synthesis of optically pure pityol—a pheromone of the bark beetle *Pityophthorus pityographus*—using a chemoenzymatic route. *Tetrahedron Lett* 33: 5241–5242.

Archelas A, Furstoss. 1997. Synthesis of enantiopure epoxides through biocatalytic approaches. *Annu Rev Microbiol* 51: 491–525.

Bevers LE, MWH Pinkse, PDEM Verhaert, WR Hagen. 2009. Oleate hydratase catalyzes the hydration of a non activated carbon-carbon bond. *J Bacteriol* 191: 5010–5012.

Bode HG, A Zeeck, K Plückhahn, D Jemdrossek. 2000. Physiological and chemical investigations into microbial degradation of synthetic poly(*cis*-1,4-isoprene). *Appl Environ Microbiol* 66: 3680–3685.

Boyd JM, SA Ensign. 2005. Evidence for a metal–thiolate intermediate in alkyl group transfer from epoxpropane to coenzyme M and cooperative metal binding in epoxyalkane: CoM transferase. *Biochemistry* 44: 13151–13163.

Braaz R, W Armbruster, D Jendrossek. 2005. Heme-dependent rubber oxygenase Roxa of *Xanthomonas* sp. cleaves the carbon backbone of poly(*cis*-1,4-isoprene) by a dioxygenase mechanism. *Appl Environ Microbiol* 71: 2473–2478.

Braaz R, P Fischer, D Jendrossek. 2004. Novel type of heme-dependent oxygenase catalyzes oxidative cleavage of rubber (poly-*cis*-1,4-isoprene). *Appl Environ Microbiol* 70: 7388–7395.

Clark DD, JR Allen, SA Ensign. 2000. Characterization of five catalytic activities associated with the NADPH:2-ketopropyl-coenzyme M [2-(2-ketopropylthio)ethanesulfonate] oxidoreductase/carboxylase of the *Xanthobacter* strain Py2 epoxide carboxylase system. *Biochemistry* 39: 1294–1304.

Coleman NV, JC Spain. 2003. Epoxyalkane: coenzyme M transferase in the ethene and vinyl chloride biodegradation pathways of *Mycobacterium* strain JS60. *J Bacteriol* 185: 5536–5546.

Cravo-Laureau C, R Matheron, C Joulian, J-L Cayol, A Hirschler-Réa. 2004. *Desulfatibacillum alkenivorans* sp. nov., a novel *n*-alkene-degrading sulfate-reducing bacterium, and emended description of the genus *Desulfatibacillum*. *Int J Syst Evol Microbiol* 54: 1639–1642.

de Bont JAM, JP van Dijken, CG van Ginkel. 1982. The metabolism of 1,2-propanediol by the propylene oxide utilizing bacterium *Nocardia* A60. *Biochim Biophys Acta* 714: 465–470.

de Bont JAM, MW Peck. 1980. Metabolism of acetylene by *Rhodococcus* A1. *Arch Microbiol* 127: 99–104.

de Haan A, MR Smith, WGB Voorhorst, JAM de Bont. 1993. Co-factor regeneration in the production of 1,2-epoxypropane by *Mycobacterium* strain E3: The role of storage material. *J Gen Microbiol* 139: 3017–3022.

El-Sharkawy A, W Yang, L Dostal, JPN Rosazza. 1992. Microbial oxidation of oleic acid. *Appl Environ Microbiol* 58: 2116–2122.

Ensign SA. 1996. Aliphatic and chlorinated alkenes and epoxides as inducers of alkene monooxygenase and epoxidase activities in *Xanthobacter* strain Py2. *Appl Environ Microbiol* 62: 61–66.

Grossi V, C Cravo-Laureau, A Méou, D Raphel, F Garzino, A Hirschler-Réa. 2007. Anaerobic 1-alkene metabolism by the alkane- and alkene-degrading sulfate reducer *Desulfatibacillum aliphaticivorans* strain CV2803^T. *Appl Environ Microbiol* 73: 7882–7890.

Hartmans S, JP Smits, MJ van der Werf, F Volkering, JAM de Bont. 1989. Metabolism of styrene oxide and 2-phenylethanol in the styrene-degrading *Xanthobacter* strain 124X. *Appl Environ Microbiol* 55: 2850–2855.

Hernáez MJ, E Andújar, JJ Ríos, SR Kaschabek, W Reineke, E Santero. 2000. Identification of a serine hydrolase which cleaves the alicyclic ring of tetralin. *J Bacteriol* 182: 5488–5453.

Hernáez MJ, B Floriano, JJ Ríos, E Santero. 2002. Identification of a hydratase and a class II aldolase involved in biodegradation of the organic solvent tetralin. *Appl Environ Microbiol* 68:4841–4846.

Hou CT, R Patel, AI Laskin, N Barnabe, I Barist. 1983. Epoxidation of short-chain alkenes by resting-cell suspensions of propane-grown bacteria. *Appl Environ Microbiol* 46: 171–177.

Ikeguchi N, T Nihara, A Kishimoto, T Yamada. 1988. Oxidative pathway from squalene to geranylacetone in *Arthrobacter* sp. strain Y-11. *Appl Environ Microbiol* 54: 381–385.

Itoh N, R Morihama, J Wang, K Okada, N Mizuguchi. 1997. Purification and characterization of phenylacetaldehyde reductase from a styrene-assimilating *Corynebacterium* strain, ST10. *Appl Environ Microbiol* 63: 3783–3788.

Khelifi N, V Grossi, M Hamdi, A Dolla, J-L Tholozan, B Ollivier, A Hirshler-Réa. 2010. Anaerobic oxidation of fatty acids and alkenes by the hyperthermophilic sulfate-reducing archaeon *Archaeoglobus fulgidus*. *Appl Environ Microbiol* 76: 3057–3060.

Krishnakumar AM, D Sliwa, JA Endrizzi, ES Boyd, SA Ensign, JW Peters. 2008. Getting a handle on the role of coenzyme M in alkene metabolism. *Microbiol Mol Biol Revs* 72: 445–456.

Krum JG, SA Ensign. 2001. Evidence that a linear megaplasmid encodes enzymes of aliphatic alkene and epoxide metabolism and coenzyme M (2-mercaptoethanesulfonate) biosynthesis in *Xanthobacter* strain Py2. *J Bacteriol* 183: 2172–2177.

Krum JG, H Ellsworth, RR Sargeant, G Rich, SA Ensign. 2002. Kinetic and microcalorimetric analysis of substrate and cofactor interactions in epoxyalkene CoM:transferase, a zinc-dependent epoxidase. *Biochemistry* 41: 5005–5014.

Mattes TE, NV Coleman, JC Spain, JM Gossett. 2005. Physiological and molecular genetic analysis of vinyl chloride and ethene biodegradation in *Nocardioides* sp. strain JS614. *Archiv Microbiol* 183: 95–106.

May SW, BJ Abbott. 1973. Enzymatic epoxidation. II. Comparison between the epoxidation and hydroxylation reactions catalyzed by the omega-hydroxylation system of *Pseudomonas oleovorans*. *J Biol Chem* 248: 1725–1730.

Meckenstock RU, R Krieger, S Ensign, PMH Kroneck, B Schink. 1999. Acetylene hydratase of *Pelobacter acetylenicus*. Molecular and spectroscopic properties of the tungsten iron-sulfur enzyme. *Eur J Biochem* 264: 176–182.

Misawa E, CKCCK Chion, IV Archer, MP Woodland, N-Y Zhou, SF Carter, DA Widdowson, DJ Leak. 1998. Characterization of a catabolic epoxide hydrolase from a *Corynebacterium* sp. *Eur J Biochem* 253: 173–183.

Nocek B, SB Jang, MS Jeong, DD Clark, SA Ensign, JW Peters. 2002. Structural basis for CO_2 fixation by a novel member of the disulfide oxidoreductase family of enzymes, 2-ketopropyl-coenzyme M oxidoreductase/carboxylase. *Biochemistry* 41: 12907–12913.

Parekh VR, RW Traxler, JM Sobek. 1977. *n*-alkane oxidation enzymes of a pseudomonad. *Appl Environ Microbiol* 33: 881–884.

Patel RN, CT Hou, AI Laskin, A Felix. 1982. Microbial oxidation of hydrocarbons: Properties of a soluble methane monooxygenase from a facultative methane-utilizing organism, *Methylobacterium* sp. strain CRL-26. *Appl. Environ. Microbiol.* 44: 1130–1137.

Rink R, M Fennema, M Smids, U Dehmel, DB Janssen. 1997. Primary structure and catalytic mechanism of the epoxide hydrolase from *Agrobacterium radiobacter* AD1. *J Biol Chem* 272: 14650–14657.

Rontani J-F, A Mouzdahir, V Michotey, P Bonin. 2002. Aerobic and anaerobic metabolism of squalene by a denitrifying bacterium isolated from marine sediment. *Arch Microbiol* 178: 279–287.

Rosner BM, FA Rainey, RM Kroppenstedt, B Schink. 1997. Acetylene degradation by new isolates of aerobic bacteria and comparison of acetylene hydratase enzymes. *FEMS Microbiol Lett* 148: 175–180.

Rosner BM, B Schink. 1995. Purification and characterization of acetylene hydratase of *Pelobacter acetylenicus*, a tungsten iron-sulfur protein. *J Bacteriol* 177: 5767–5772.

Ruettinger RT, GR Griffith, MJ Coon. 1977. Characteristics of the ω-hydroxylase of *Pseudomonas oleovorans* as a non-heme iron protein. *Arch Biochem Biophys* 183: 528–537.

Schink B. 1985a. Degradation of unsaturated hydrocarbons by methanogenic enrichments cultures. *FEMS Microbiol Ecol* 31: 69–77.

Schink B. 1985b. Fermentation of acetylene by an obligate anaerobe, *Pelobacter acetylenicus* sp. nov. *Arch Microbiol* 142: 295–301.

Seiffert GB, GM Ullmann, A Messerschmidt, B Schink, PMH Kroneck. 2007. Structure of the non-redox-active tungsten/[4Fe:4S] enzyme acetylene hydratase. *Proc Natl Acad Sci USA* 104: 3073–3077.

Shimizu S, S Jareonkitmongkol, H Kawashima, K Akimoto, H Yamada. 1991. Production of a novel ω1-eicosapentaenoic acid by *Mortierella alpina* 1S-4 grown on 1-hexadecene. *Arch Microbiol* 156: 63–166.

Sliwa DA, AM Krishnakumar, JW Peters, SA Ensign. 2010. Molecular basis for enantioselectivity in the (R)— and (S)-hydroxypropylthioethylsulfonate dehydrogenases, a unique pair of stereoselective short-chain dehydrogenases/reductases involved in aliphatic epoxide carboxylation. *Biochemistry* 49: 3487–3498.

Sluis MK, RA Larsen, JG Krum, R Anderson, WW Metcalf, SA Ensign. 2002. Biochemical, molecular, and genetic analyses of the acetone carboxylases from *Xanthobacter autotrophicus* strain Py2 and *Rhodobacter capsulatus* strain B10. *J Bacteriol* 184: 2969–1977.

Sluis MK, FJ Small, JR Allen, SA Ensign. 1996. Involvement of an ATP-dependent carboxylase in a CO_2-dependent pathway of acetone metabolism by *Xanthobacter* strain Py2. *J Bacteriol* 178: 4020–4026.

Small FJ, SA Ensign. 1995. Carbon dioxide fixation in the metabolism of propylene and propylene oxide by *Xanthobacter* strain Py2. *J Bacteriol* 177: 6170–6175.

Small FJ, SA Ensign. 1997. Alkene monooxygenase from *Xanthobacter* strain Py2. Puriifcation and characterization of a four-component system central to the bacterial metabolism of aliphatic alkenes. *J Biol Chem* 272: 24913–24920.

Small FJ, JK Tilley, SA Ensign. 1995. Characterization of a new pathway for epichlorohydrin degradation by whole cells of *Xanthobacter* strain Py2. *Appl Environ Microbiol* 61: 1507–1513.

Swaving J, JAM de Bont, A Westphal, A de Kok. 1996. A novel type of pyridine nucleotide-disulfide oxidoreductase is essential for NAD+- and NADPH-dependent degradation of epoxyalkanes by *Xanthobacter* strain Py2. *J Bacteriol* 178: 6644–6646.

Takeuchi M, T Sakane, T Nihira, Y Yamada, K Imai. 1999. *Corynebacterium terpenotabidum* sp. nov., a baterium capable of degrading squalene. *Int J Syst Bacteriol* 49: 223–229.

van den Tweel WJJ, JAM de Bont. 1985. Metabolism of 3-butyl-1-ol by *Pseudomonas* BB1. *J Gen Microbiol* 131: 3155–3162.

van den Wijngaard AJ, DB Janssen, B Withold. 1989. Degradation of epichlorohydrin and halohydrins by bacterial cultures isolated from freshwater sediment. *J Gen Microbiol* 135: 2199–2208.

van der Werf MJ, HJ Swarts, JAM de Bont. 1999. *Rhodococcus erythopolis* DCL14 contains a novel degradation pathway for limonene. *Appl Environ Microbiol* 65: 2092–2102.

van der Werf MJ, KM Overkamp, JAM de Bont. 1998. Limonene-1,2-epoxide hydrolase from *Rhodococcus erythropolis* DCL14 belongs to a novel class of epoxide hydrolases. *J Bacteriol* 180: 5052–5057.

van Ginkel CG, JAM de Bont. 1986. Isolation and characterization of alkene-utilizing *Xanthobacter* spp. *Arch Microbiol* 145: 403–407.

van Hylckama Vlieg JET, J Kingma, W Kruizinga, DB Janssen. 1999. Purification of a glutathione *S*-transferase and a glutathione conjugate-specific dehydrogenase is involved in isoprene metabolism by *Rhodococcus* sp. strain AD 45. *J Bacteriol* 181: 2094–2101.

van Hylckama Vlieg JET, J Kingma, AJ van den Wijngaard, DB Janssen. 1998. A gluathione *S*-transferase with activity towards *cis*-1,2-dichloroepoxyethane is involved in isoprene utilization by *Rhodococcus* strain AD 45. *Appl Environ Microbiol* 64: 2800–2805.

Warhurst AM, KF Clarke, RA Hill, RA Holt, CA Fewson. 1994. Metabolism of styrene by *Rhodococcus rhodochrous* NCIMB 13259. *Appl Environ Microbiol* 60: 1137–1145.

Weijers CAGM, A de Haan, JAM de Bont. 1988. Chiral resolution of 2,3-epoxyalkanes by *Xanthobacter* Py2. *Appl Microbiol Biotechnol* 27: 337–340.

Wiegant WM, JAM de Bont. 1980. A new route for ethylene glycol metabolism in *Mycobacterium* E44. *J Gen Microbiol* 120: 325–331.

Yamada EW, WB Jakoby. 1959. Enzymatic utilization of acetylenic compounds II. Acetylenemonocarboxylic acid hydrase. *J Biol Chem* 234: 941–945.

Yamada Y, H Motoi, S Kinoshita, N Takada, H Okada. 1975. Oxidative degradation of squalene by *Arthrobacter* species. *Appl Microbiol* 29: 400–404.

Zhou N-Y, A Jenkins, CKN Chan, KW Chion, DJ Leak. 1999. The alkene monooxygenase from *Xanthobacter* strain Py2 is closely related to aromatic monooxygenases and catalyzes aromatic monooxygenation of benzene, toluene, and phenol. *Appl Environ Microbiol* 65: 1589–1595.

7.7 Alkanols, Alkanones, Alkanoates, Amides

7.7.1 Alkanols

7.7.1.1 Aerobic Conditions

The aerobic metabolism of methanol—and methylamine—by bacteria and yeasts has been fully described in *The Biochemistry of Methylotrophs* (Anthony 1982), so that only some brief comments are justified. The essential details of the aerobic metabolism of methanol by bacteria have been discussed in dealing with the monooxygenation of methane: in brief, this involves sequential dehydrogenation to formaldehyde and formate. The mechanisms used for the dehydrogenations are discussed in the appropriate section of chapter. For yeasts, the essential difference lies in the xylulose pathway for the assimilation of formaldehyde: this forms part of the dissimilatory dihydroxyacetone pathway that is fully discussed by Anthony (1982). The majority of methanol-utilizing yeasts belong to the genera *Hansenula*, *Candida*, *Pichia*, and *Tolulopsis*.

The initial reactions involved in the bacterial metabolism of higher alkanes ($>C_1$) are formally similar to those used for the metabolism of methane, and the soluble alkanol dehydrogenases also contain the pyrroloquinone (PQQ) (references in Anthony 1982). Terminal alkanols produce alkanoates while subterminal alkanes produce ketones that are subsequently degraded by pathways that are discussed below. Two additional features are worth attention:

1. Details of the enzymology of dehydrogenation may be quite complex. For example, a number of distinct alcohol and fatty acid dehydrogenases have been isolated from an *Acinetobacter* sp. during the metabolism of hexadecane (Singer and Finnerty 1985a,b).

2. Dehydrogenases specific for secondary alkanols have been isolated from methylotrophs (Hou et al. 1981), and illustrated by the formation of methyl ketones from propan-2-ol and butan-2-ol in a range of Gram-negative pseudomonads, and in Gram-positive species of *Arthrobacter*, *Mycobacterium* (*Rhodococcus*), and *Nocardia* (Hou et al. 1983).

The aerobic degradation of *t*-butanol and *t*-pentanol that are the initial metabolites produced by several organisms from the fuel oxygenates methyl *t*-butyl ether (MTBE) and *t*-amyl methyl ether (TAME) has been examined in several bacteria. In *Mycobacterium austroafricanum*, evidence of the relevant genes in MTBE (Ferriera et al. 2006), and in *Aquicola tertiaricarbonis* (Müller et al. 2008) led to a proposed pathway involved oxidation of *t*-butanol to 2-methylpropane-1,2-diol followed by successive dehydrogenation to 2-hydroxyiso-butyraldehyde and isobutyrate (Figure 7.45). The putative degradation of 2-hydroxyisobutyrate may plausibly be mediated by a cobalamin-dependent mutase with the production

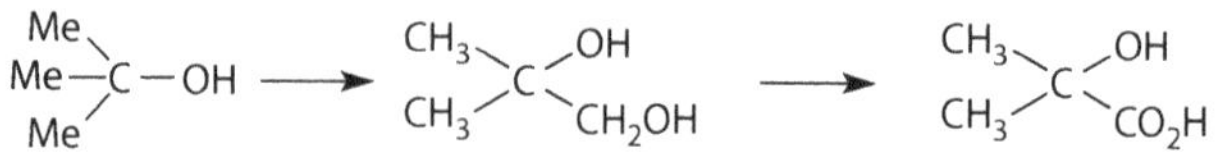

FIGURE 7.45   Aerobic degradation of *t*-butanol.

of 3-hydroxybutyrate (Rohwerder et al. 2006), and the same pathway has been proposed for *Variovorax paradoxus* (Zaitsev et al. 2007). The analogous degradation of *t*-amyl alcohol that can be produced from *t*-amyl methyl ether (TAME) has been resolved from the formation of 2-methylbutal-2,3-diol by the activity of a Rieske non-heme iron monooxygenase MdpJ followed by elimination (isomerization) to 3-methylpent-1-ene-3-ol before dehydrogenation to 3-methylcrotonoic acid; the coenzyme A ester was then carboxylated, dehydrated and underwent scission to acetoacetate and 3-oxovalerate (Schuster et al. 2012).

7.7.1.2 Anaerobic Conditions

Methanol can be metabolized under a range of conditions that are briefly summarized.

1. Methanol is degraded anaerobically by *Butyribacterium methylotrophicum* with the production of acetate, butyrate, and H_2 (Lynd and Zeikus), while in the presence of CO_2, *Eubacterium limosum* converted methanol into acetate and butyrate (Müller et al. 1981).

2. *Thermotoga lettingae* is an anaerobic thermophilic organism that is able to grow at the expense of pyruvate, cellobiose, starch, all the methylamines, and methanol. It is able to grow with acetate in the presence of a methanogen or thiosulfate when alanine and sulfide were produced. In the presence of S^0 or thiosulfate, methanol was converted into CO_2 and alanine. In pure cultures methanol was fermented to acetate, CO_2 and H_2 although this was slower than in the presence of methanogens or electron acceptors such as S^0, thiosulfate, Fe(III) or anthraquinone-2,6-disulfonate (Balk et al. 2002).

3. Anaerobic oxidation of methanol—though not incorporation into cellular material—has been observed in a sulfate-reducing bacterium (Braun and Stolp 1985). Degradation of methanol by the reduction of sulfate has been shown to follow the equation (Nanninga and Gottschal 1987):

$$4CH_3OH + 3SO_4^{2-} \rightarrow 4HCO_3^- + 3HS^- + H^+ + 4H_2O$$

4. Some archaea are able to use a range of C_1 substrates for methanogenesis. For the highly versatile metabolic *Methanosarcina barkeri*, these include methanol (Sauer et al. 1997), dimethyl sulfide (Tallant et al. 2001), and di- and trimethylamines (Paul et al. 2000), CO, and acetate and $CO_2 + H_2$. The methanogen *Methanosphaera stadtmanae* isolated from the human intestine is, however, restricted to the reduction of methanol to methane, and is dependent on acetate as a carbon source. Analysis of the whole genome revealed that several important groups of enzymes that are generally associated with methanogenesis were missing: these were CDS for the synthesis of molybdopterin, and the carbon monoxide dehydrogenase/acetyl-CoA synthase complex (Fricke et al. 2006) that is discussed later.

Ethanol may be converted into methane by *Methanogenium organophilum* (Widdel 1986; Frimmer and Widdel 1989), and oxidation of primary alkanols has been demonstrated in *Acetobacterium carbinolicum* (Eichler and Schink 1984). Secondary alcohols such as propan-2-ol and butan-2-ol may be used as hydrogen donors for methanogenesis with concomitant oxidation to the corresponding ketones (Widdel et al. 1988). An NAD-dependent alcohol dehydrogenase has been purified from *Desulfovibrio gigas* that can oxidize ethanol, and it has been shown that the enzyme does not bear any relation to classical alcohol dehydrogenases (Hensgens et al. 1993).

7.7.1.2.1 Polyols

The metabolism of 1,2-diols has attracted considerable attention, and in particular that of glycerol in view of its ubiquity as a component of lipids. Widely different pathways have been found of which three are given as illustration.

1. *Anaerovibrio glycerini* ferments glycerol to propionate (Schauder and Schink 1989) and *Desulfovibrio carbinolicus* to 3-hydroxypropionate (Nanniga and Gottschal 1987).

2. *Desulfovibrio alcoholovorans* converts glycerol to acetate, and 1,2-propandiol to acetate and propionate (Qatibi et al. 1991).

3. The fermentation of glycerol to propane-1,2-diol, acetate, ethanol, formate, H_2 and CO_2 has been demonstrated in *Paenibacillus macerans* in the absence of an additional electron acceptor (Gupta et al. 2009). Dihydroxyacetone phosphate was converted into propane-1,2-diol via methylglyoxal and hydroxyacetone, while acetyl-CoA and formate were produced from pyruvate. These result in the synthesis of ethanol and, $CO_2 + H_2$. These products, with the addition of succinate by carboxylation have been demonstrated in *Escherichia coli* (Murarka et al. 2008), although glycerol fermentation in several *Enterobacteriaceae* generally produces propan-1,3-diol and pyruvate.

7.7.1.2.1.1 Diol Dehydratase The degradation under anaerobic conditions of 1:2-diols including ethanediol, propane-1,2-diol, and glycerol with the production of acetaldehyde, propionaldehyde, and 3-hydroxypropionaldehyde respectively is mediated by a coenzyme-B_{12} (adenosylcobalamine AdoCbl)-dependent diol- dehydratase (Figure 7.46a). This has been established in *Enterobacteriaceae* including *Klebsiella pneumoniae* (Toraya et al. 1979) and *Citrobacter freundii* (Seyfried et al. 1996). In *Salmonella enterica* the toxicity of the propionaldehyde produced is restricted by its protection within a micro compartment (metabolosome) (Sampson and Bobik 2008). This also occurs in *Lactobacillus reuteri* in which the propionaldehyde undergoes disproportionation to propanol and propionate (Figure 7.46b) (Sriramulu et al. 2008). A comparable reaction is catalyzed by glycerol dehydratase that produces 2-hydroxypropionate under anaerobic conditions in *Enterobacteriaceae* including *Kleb. pneumoniae* (Toraya et al. 1979) and *C. freundii* (Seyfried et al. 1996). This may be analogous to the observation that *S. enterica* was able to conserve acetaldehyde that was produced from ethanolamine by the activity of the diol dehydratase in carboxysome-like organelles (Penrod and Roth 2006).

There are, however, important exceptions to coenzyme B_{12}-dependent diol dehydratases. The genes *dhaB1*, *dhaB2*, and *dhaT* comprise an operon in *Clostridium butyricum*, and their expression in *E. coli*—that does not synthesize coenzyme B_{12}—enabled the synthesis of propan-1,3-diol from glycerol by the combined activities of glycerol dehydratase and glycerol dehydrogenase in the absence of added coenzyme B_{12} (Raynaud et al. 2003). This provides an additional example of coenzyme B_{12}-independent glycerol dehydratases that has also been described in *Clostridium glycolycum*, although that enzyme had no activity with glycerol (Hartmannis and Stadtman 1986). This enzyme is associated with membranes though it has been solubilized and exhibited an EPR signal at $g = 2.02$ (Hartmannis and Stadtman 1987).

7.7.2 Alkanones

7.7.2.1 Formaldehyde

Formaldehyde is the common intermediate in the oxidation of methanol and dehydrogenation of methylamines. Further metabolism to formate may take place by dehydrogenation either with, or without, the intervention of glutathione: the distribution of these activities among methylotrophs has been summarized (Anthony 1982). In the facultative methylotroph *Methylobacterium extorquens*, there are two other pathways, the tetrahydrofolate, and the tetrahydromethanopterin-linked pathways (Marx et al. 2003), that are analogous to those used for the aerobic degradation of methyl

(a) $CH_3 - CH(OH)CH_2CHO \longrightarrow CH_3CH_2CHO$

(b) $CH_3CH_2CHO \longrightarrow CH_3CH_2CO_2H + CH_3CH_2CH_2OH$

FIGURE 7.46 Anaerobic metabolism of 1,2-diols. (a) Coenzyme B12-dependent diol-dehydratase, (b) metabolism of glycerol to propionaldehyde followed by dismutation.

chloride and methyl bromide, while *Burkholderia fungorum* contains three of the four putative formaldehyde oxidation systems (Marx et al. 2004). Formate may be dehydrogenated to CO_2 or, by reversal of the tetrahydrofolate pathway to methylenetetrahydrofolate and incorporation into the serine cycle.

7.7.2.2 Higher Alkanones and Ketones

The aerobic metabolism of acetone by *Xanthobacter* sp. strain Py2 involves at ATP-dependent carboxylation (Sluis et al. 1996), with the direct formation of acetoacetate. The carboxylase has been characterized from *X. autotrophicus* strain Py2, and from the anaerobic phototroph *Rhodobacter capsulatus* strain B10 (Sluis et al. 2002). It has a $\alpha_2\beta_2\gamma_2$ subunit composition, and EPR measurements confirmed the presence of Mn^{2+} (Boyd et al. 2004). The anaerobic degradation of ketones under denitrifying conditions also involves carboxylation followed by hydrolysis (Platen and Schink 1990; Hirschler et al. 1998; Dullius et al. 2011) (Figure 7.47), and the ATP-dependent decarboxylases used for aerobic degradation by *X. autotrophicus* strain Py2 and for the nitrate-reducing strains are similar (Dullius et al. 2011). There are two subsequent pathways in anaerobes: (a) oxidation to CO_2 in denitrifying and sulfate reducing bacteria by the acetyl-CoA-CO dehydrogenase pathway, and (b) an anabolic pathway by a modified tricarboxylate-glyoxylate cycle (Janssen and Schink 1995a).

Higher alkanones may be degraded under aerobic conditions by a sequence involving Baeyer–Villiger monooxygenation to lactones, followed by hydrolysis. The resulting alkanols are then degraded to carboxylic acids by the reactions following terminal hydroxylation that have already been described (Figure 7.2). This is used in one of the pathways used for the degradation of 6,10,14-trimethylpentadecan-2-one by *Marinobacter* sp. strain CAB that involves insertion of an oxygen atom between C_2 and C_3 and subsequent hydrolysis and oxidation with loss of C_2 and C_3 (Rontani et al. 1997).

7.7.2.3 Methylglyoxal

Methylglyoxal is a produced during the unregulated carbohydrate metabolism, and is toxic to the cell. Detoxification in *E. coli* can be achieved by several mechanisms (Ko et al. 2005):

1. Reduction to lactaldehyde (2-hydroxyacetaldehyde)
2. Successive reductions to 2-hydroxyacetone and propane-1,2-diol
3. The activity of glyoxalase I to produce *S*-lactoylglutathione and glyoxylase II to D-lactate (Figure 7.48) (Clugston et al. 1998)

In *Clostridium beijerinckii*, however, detoxification is carried out only by reduction to 1,2-propanediol (Liyanage et al. 2001).

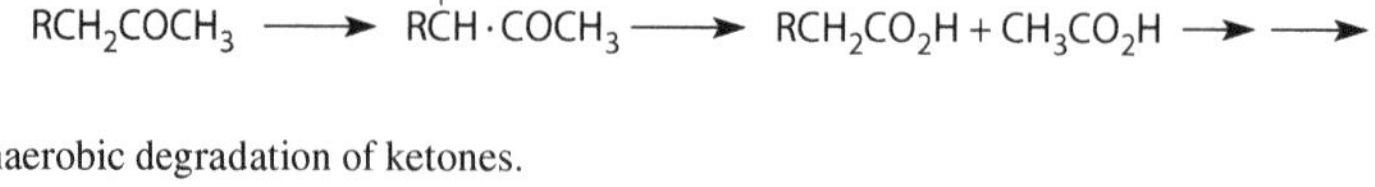

FIGURE 7.47 Anaerobic degradation of ketones.

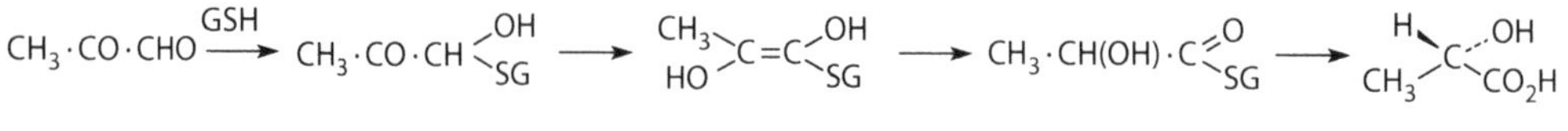

FIGURE 7.48 Conversion of methylglyoxal to D-lactate.

7.7.3 Alkanoates

7.7.3.1 Aerobic Conditions

7.7.3.1.1 C_1: Carbon Monoxide and Formate

The metabolism of carbon monoxide, formate, and oxalate have in common the initiation by dehydrogenases, and the degradation of CO is relevant here in view of the mechanism for its utilization and the metabolism of other C_1 compound including formate.

Carbon monoxide is produced biotically from hemin and amino acids, during fossil-fuel combustion, and in soil. Although normally toxic to aerobic organisms, microorganisms have been isolated that are able to degrade it, and several groups of aerobic bacteria are able to use carbon monoxide as a source of energy and cell carbon (Meyer and Schlegel 1983), while some mycobacteria are able to utilize both carbon monoxide and methanol (Park et al. 2003). CO_2 is formed by a carbon monoxide dehydrogenase and is incorporated into biosynthetic reactions by the ribulose-1,5-bisphosphate pathway (Park et al. 2003).

Formate and oxalate are used by *R. eutropha* for aerobic growth, and formate by bacteria able to grow with H_2 and CO_2, whereas carbon monoxide is used by a much wider range of aerobic organisms including *Oligotropha (Pseudomonas) carboxydovorans*, *Pseudomonas carboxyhydrogena*, *Hydrogenophaga pseudoflava*, *Bradyrhizobium japonicum* and, for dark heterotrophic growth of *Rhodopseudomonas gelatinosa* and *Rhodospirillum rubrum*.

The metabolism of all three of these compounds—carbon monoxide, formate, and oxalate—is initiated by dehydrogenation.

1. In chemolithotrophic bacteria such as *R. eutropha* that can be grown on H_2 or formate, FDH plays a central role. There are two types of enzyme in this organism that catalyze the reaction $H \cdot CO_2H \rightarrow CO_2 + 2H^+ + 2\varepsilon$, one a soluble NAD-linked enzyme, and the other directly coupled to the respiratory chain. The soluble enzyme contains molybdenum or tungsten cofactors and [Fe-S] centers (Friedebold and Bowien 1993), and is part of an *fdsGBACD* operon (Oh and Bowien 1998). CO_2 is assimilated by the reductive pentose phosphate cycle.

2. *Pseudomonas oxalaticus* grown with formate and pyruvate was used for the preparation and characterization of formate dehydrogenase that existed in a complex of two catalytically active species, both of which contained FMN, non-heme Fe and acid-labile sulfide (Müller et al. 1978). Carbon dioxide is assimilated by the Benson–Calvin cycle and ribulose-1,5-bisphosphate has been purified and characterized (Lawlis et al. 1979).

3. Carbon monoxide dehydrogenase catalyzes the reaction $CO + H_2O \rightarrow CO_2 + 2H^+ + 2\varepsilon$ from which the electrons are fed into a CO-resistant respiratory chain. The enzyme has been characterized from a number of organisms including *H. pseudoflava* (Kang and Kim 1999), *Oligotropha carboxydovorans* (Schübel et al. 1995), and *B. japonicum* (Lorite et al. 2000). The enzymes from all of them are essentially similar and are molybdenum–iron–sulfur flavoproteins containing molybdopterin cytosine dinucleotide, and selenocysteine (Gremer et al. 2000).

4. *Streptomyces thermoautotrophicus* is able to grow aerobically with carbon monoxide or H_2 in the presence of CO_2, and is able to reduce dinitrogen, in spite of the normal inhibition of nitrogenase by CO, while in addition this enzyme is able to reduce acetylene to ethylene at only a very low rate. The enzymology is unusual, and it has been proposed that the oxidation of CO to CO_2 results in the production of superoxide that serves as a reductase to the dinitrogenase which has been shown to be a Mo Fe–S enzyme with a molecular mass of ≈ 144 kDa (Ribbe et al. 1997).

7.7.3.1.2 C_2: Glycollate, Glyoxylate, Oxalate

These substrates are related by increasing oxidation levels:

$$HO-CH_2-CO_2H > CHO-CO_2H > HO_2C-CO_2H$$

$$HO_2C-CO_2H \longrightarrow HO_2C-COSCoA \longrightarrow CHO-COSCoA \longrightarrow CHO-CH(OH)-CO_2H \longrightarrow HOCH_2-CH(OH)-CO_2H$$

FIGURE 7.49   Degradation of tartronic acid.

It is convenient to preface the details of their aerobic metabolism by a brief summary of the basic mechanisms involved. The metabolism of methylglyoxal that is produced during unregulated metabolism of carbohydrates has already been noted.

7.7.3.1.2.1 Carboligase Pathway   This is a key reaction in the degradation of all three substrates. In *Ps. oxalaticus* the degradation of oxalate involves the following steps: oxalate is converted to glyoxylate by the reduction of oxalyl-CoA, followed by the condensation of glyoxylate to tartronic semialdehyde mediated by glyoxylic acid carboligase followed by reduction to glycerate that is incorporated into the phosphoglycerate cycle (references in Blackmore and Quayle 1970) (Figure 7.49). Glyoxylate carboligase is not, however, obligatory for oxalate metabolism, and the serine-glyoxyate aminotransferase and hydroxypyruvate reductase pathway has been described for another strain AM2 of *Pseudomonas* (Blackmore and Quayle 1970).

7.7.3.1.2.2 Glycine/Serine/Formate Pathway   This is used for the degradation of oxalate by *Pseudomonas* sp. strains AM1 and AM2, and *Pseudomonas extorquens* in an alternative pathway that involve three interconnecting reactions (Blackmore and Quayle 1970):

1. Oxalate is reduced to glyoxylate by reduction of oxalyl-CoA.
2. Oxalyl-CoA is converted to formyl-CoA by formate dehydrogenase with the loss of CO_2, followed by the incorporation of formate into tetrahydrofolate and reaction with glycine to produce serine.
3. Glyoxylate reacts with serine to produce hydroxypyruvate.

7.7.3.1.2.3 3-Hydroxyaspartate Pathway   The degradation of glycolate in *Micrococcus denitrificans* is initiated by condensation of glyoxylate with glycine to produce 3-hydroxyaspartate that undergoes dehydration and subsequent loss of ammonia to produce oxalacetate that is incorporated into the citric acid cycle (Kornberg and Morris 1965).

$$HO_2C\text{-}CHO + H_2N\text{-}CH_2\text{-}CO_2H \rightarrow HO_2C\text{-}CH(OH)\text{-}CH(NH_2)\text{-}CO_2H \rightarrow$$

$$HO_2C\text{-}CH{=}C(NH_2)CO_2H \rightarrow HO_2C\text{-}CH_2\text{-}CO\text{-}CO_2H$$

7.7.3.1.2.4 Glyoxylate   This is produced during several aerobic degradations:

1. 1,2-Dichloroethane by both the hydrolytic (Janssen et al. 1988) and oxidative (Hage and Hartmans 1999) pathways
2. Allantoate that is a product of purine metabolism (Cusa et al. 1999)
3. Morpholine by *Mycobacterium aurum* strain MO1 (Combourieu et al. 1998)
4. Cyclic ethers by the fungus *Cordyceps sinensis* (Nakamiya et al. 2005)

The pathway for bacterial degradation of glyoxylate by the carboligase pathway (Ornston and Ornston 1969) has already been outlined (Figure 7.50). In the brown-rot fungus *Tyromyces palustris*, however, glyoxylate is dehydrogenated to oxalate by a cytochrome *c* dependent enzyme (Tokimatsu et al. 1998).

7.7.3.1.2.5 Glycolate   The metabolism of glycolate in *E. coli* provided the basic pathway for degradation involving glyoxylate carboligase that was shown to require thiamine pyrophosphate and Mg^{2+} (Krakow et al. 1961), and that was subsequently verified for the degradation of glyoxylate that has already been described.

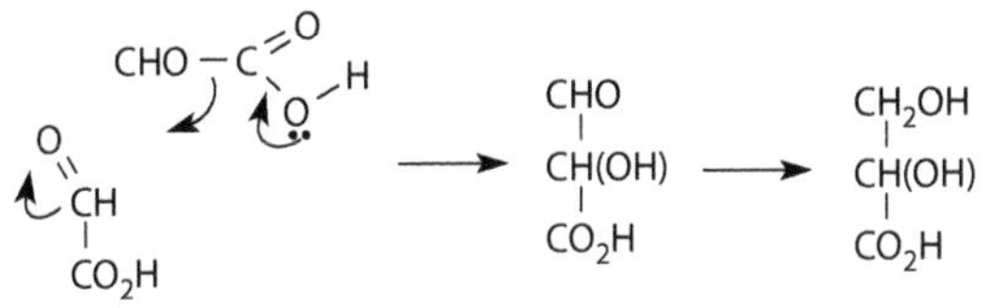

FIGURE 7.50 Aerobic degradation of glyoxylate.

7.7.3.1.2.6 Oxalate This is synthesized from acetyl-CoA and oxalacetate in *Burkholderia glumae* (Li et al. 1999), and in the wood-rotting basidiomycete *Fomitopsis palustris* (Munir et al. 2001). Formate and CO_2 are produced by oxalate decarboxylase, and the enzyme in *B. subtilis* requires manganese for activity (Tanner et al. 2001); it is, however, induced not by oxalate, but by an acidic pH (Tanner and Bornemann 2000).

The degradation of oxalate is apparently widespread, and an ammonia-dependent strain *Ammoniphilus oxalivorans* that was able to grow with high concentrations of ammonium oxalate has been described (Zaitsev et al. 1998). The metabolism of oxalate by the lithotrophic *Ralstonia (Alcaligenes) eutropha* takes place by the reaction $H \cdot CO_2 \cdot CO_2H \rightarrow 2CO_2 + 2H^+ + 2e$. The CO_2 is, however, assimilated not by the ribulose-1,5-bisphosphate cycle used for formate metabolism in this strain, but by dehydrogenation to glyoxylate, decarboxylative dimerization mediated by the activity of glyoxylate carboligase and thence entry into the phosphoglycerate cycle (Friedrich et al. 1979). The degradation of oxalate by the carboligase pathway and the glycine/formate/serine pathways (Blackmore and Quayle 1970) has already been described. *Alcaligenes eutrophus* strain H16 is able to grow autotrophically on formate that undergoes dehydrogenation to CO_2 and assimilation by ribosebisphosphate carboxylase, or with oxalate when this enzyme is not induced and the glycerate pathway with glyoxylate carboligase is used (Friedrich et al. 1979).

7.7.3.1.3 C_2: Acetate

1. The formation of acetate by the aerobic degradation of even-membered alkanes has already been discussed. Since, in addition to oxidation, the pathway for its assimilation must provide cell components, a pathway has been proposed (Kornberg et al. 1959; Kornberg 1966) that involves the functioning of two essential enzymes:

   Isocitrate lyase (ICL) ⇔ Isocitrate → succinate + glyoxylate

   Malate synthase (MS) ⇔ Glyoxylate + acetyl-CoA → malate

2. There are, however, a number of bacteria including the phototroph *Rhodobacter sphaeroides* (Erb et al. 2007), the methylotroph *M. extorquens* (Korotkova et al. 2002a), and the streptomycete *S. coelicolor* (Han and Reynolds 1997) all of which can metabolize acetate, even though they lack isocitrate lyase (ICL) activity. A pathway must circumvent this limitation, and this has been accomplished by the finding of an alternative enzyme in *Rhodobacter capsulatus* (Meister et al. 2005) that carries out the following two reactions catalyzed by L-malyl-CoA/β-methylmalyl-CoA lyase (MCL):

   Acetyl-CoA + glyoxylate ⇔ L-malyl-CoA

   Propionyl-CoA + glyoxylate ⇔ *erythro*-β-methyl-CoA

The genome of the relevant bacteria has been completely sequenced, and the genes encoding *mcl1 and mcl2* were shown to be present in isocitrate lyase-negative strains that could assimilate acetate—*Rh. capsulatus*, *Rhodobacter spheroides*, *Rh. rubrum*, *M. extorquens*, and *S. coelicolor*. Significantly, *mcl1* was absent in the phototroph *Rhodobacter palustris* that is ICL-positive. It is worth noting that MCL has also been shown to operate in the CO_2 assimilation pathway of *Chloroflexus aurianticus* (Herter et al. 2002).

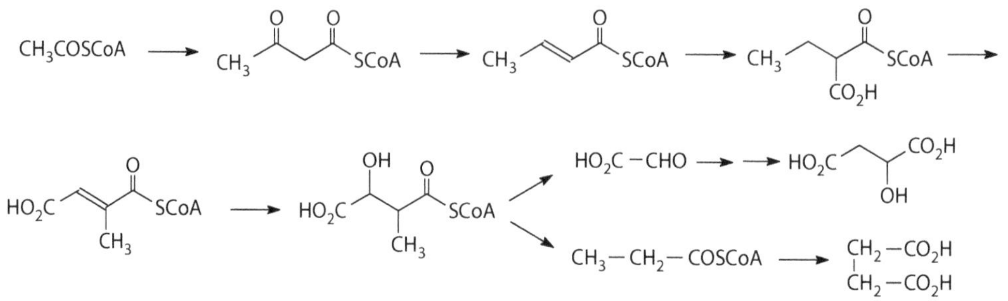

FIGURE 7.51 Ethylmalonate pathway.

3. In organisms that lack isocitrate lyase activity and, therefore, a functional glyoxylate cycle, the ethylmalonate cycle has been proposed for *Rhodobacter sphaeroides* (Figure 7.51) (Erb et al. 2007, 2008). The key enzymes are: crotonyl-CoA carboxylase/reductase and a coenzyme B_{12}-dependent acyl-CoA mutase that carry out the reactions:

$$\text{Crotonyl-CoA} + \text{NADPH} + CO_2 \rightarrow \text{ethylmalonyl-CoA} + NADP^+$$

$$(S)\text{-2-ethylmalonyl-CoA} \rightarrow (S)\text{-2-methylsuccinyl-CoA}$$

This is transformed to β-methylmalyl-CoA that undergoes fission to propionyl-CoA and glyoxylate followed by carboxylation to succinyl-CoA and reaction with acetyl-CoA to produce malate. Use of a combination of ^{13}C-labelling during methylotrophic growth of *M. extorquens* and high-resolution mass spectrometry has suggested that this pathway takes precedence over a pathway involving the synthesis of propionate via isobutyryl-CoA, methacrlyl-CoA, and hydroxybutyryl-CoA (Peyraud et al. 2009).

4. A further alternative has been shown to operate in *Haloarcula marismortui* that does not have the genetic complement to operate the ethylmalonyl pathway. A 3-methylaspartate cycle has been proposed that is supported by the existence of a NADP glutamate dehydrogenase, and a methyl-aspartate ammonia lyase. The key reactions involve the operation of a glutamate mutase to produce 3-methylasparte followed by deamination to mesaconyl-CoA, and hydration to 3-methylmalonyl-CoA that undergoes scission to propionyl-CoA and glyoxylate.

This pathway that requires high concentration of glutamate provides interaction between nitrogen metabolism and carbon assimilation (Khomyakova et al. 2011).

There are additional features that deserve consideration:

1. In *M. extorquens* for growth on C_2 compounds, an additional metabolic route is required to dispose of potentially toxic glyoxylate. This is accomplished by the activity of three enzymes: serine glyoxylate aminotransferase (*sga*), serine hydroxymethyl transferase (*glya*), and glycine decarboxylase (*gcv*) that are required for growth with ethylamine, and accomplish the conversion of glyoxylate into phosphoglycerate (Okubo et al. 2010).

2. The metabolism of acetyl-CoA by *Streptomyces cinnamonensis* involves condensation to 3-oxo-butyryl-CoA followed by dehydrogenation, dehydration, and reduction to butyryl-CoA. This then undergoes coenzyme B_{12}-catalyzed rearrangement to isobutyryl-CoA, followed by carboxylation to methylmalonyl-CoA and further rearrangement to succinyl-CoA (Han and Reynolds 1997).

7.7.3.1.3.1 TCA Cycle   Attention has already been directed to modifications of the TCA cycle including bacteria growing at the expense of acetate that are dependent on the glyoxylate cycle, for those lacking this that use the ethylmalonate cycle, and the 3-methylaspartate cycle. The functioning of the cycle in a reductive mode that is used for the assimilation of CO_2 by a number of anaerobic bacteria—and a few aerobes—has been discussed in Chapter 2. Some additional aspects include the following.

1. The TCA cycle for *Helicobacter pylori* presents several unusual features (Kather et al. 2000):

   a. The conversion of malate to oxalacetate is carried out not by a dehydrogenase with the production of NADH, but by a membrane-bound FAD-dependent malate: quinone oxidoreductase that has also been identified in *C. glutamicum* (Molenaar et al. 1998).

   b. The dehydrogenation of both 2-oxoglutarate and pyruvate are carried out by ferredoxin-dependent enzymes that avoid the production of NADH.

2. The acetic acid-producing *Acetobacter aceti* is resistant to high levels of ethanol that is the growth substrate and its product acetic acid. It has been shown that although this organism possesses a complete TCA cycle, the production of succinate from succinyl-CoA by succinyl-CoA synthase is replaced by succinyl-CoA:acetate CoA transferase with the production of acetyl-CoA (Mullins et al. 2008).

3. Two metabolic limitations have been described in anaerobes: bifurcation of the TCA cycle and mechanisms for the synthesis of oxaloacetate.

   a. *Clostridium acetobutylicum* is capable of growth in a minimal medium with glucose even though the genome lacks homologies of several TCA key cycle enzymes. This has been resolved by showing that a bifurcated cycle operates in which pyruvate fulfils two metabolic functions: (i) formation of acetyl-CoA and further reaction to citrate and thence to succinate (ii) carboxylation to oxalacetate and then via malate and fumarate to succinate (Amador-Noguez et al. 2010).

   b. The anaerobe *Dehalococcoides ethenogenes* grown with hydrogen as reductant and acetate as carbon source is able to carry out complete dehalogenation of some chlorinated ethenes to ethene. Growth with ^{13}C-labeled acetate and CO_2 followed by identification of target genes with quantitative PCR was used to elucidate the carbon metabolic pathway (Tang et al. 2009). The metabolic functions already noted were displayed: bifurcation of the TCA cycle, and assimilation of CO_2 by formation of pyruvate from acetyl-CoA and conversion to oxaloacetate.

7.7.3.1.4 C_3

Propionate is produced by the oxidation of odd-numbered alkanes, and branched-chain alkanes including pristane (Sakai et al. 2004), the side-chain of phytosterol (Sih et al. 1968a,b) and cholesterol (van der Geize et al. 2007), and the degradation of the branched-chain amino acids isoleucine and valine. Propionate can be metabolized by several mechanisms including the methylmalonate, the acrylate, and the 2-methylcitrate pathway. The 2-methylcitrate pathway (Figure 7.52) is initiated by reaction of propionate with oxalacetate followed by formation of 2-methylcitrate from which succinate is produced with the loss of pyruvate. This reaction has been established in enteric bacteria (Textor et al. 1997; Horswill and Escalante-Semerena 1999), in *R. eutropha* (Brämer and Steinbüchel 2001), and in *C. glutamicum* (Claes et al. 2002).

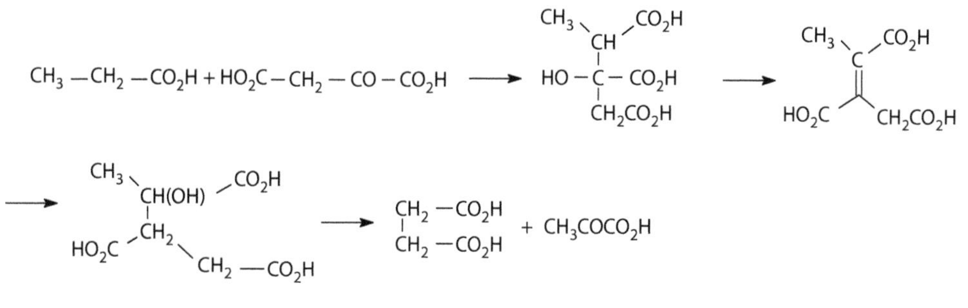

FIGURE 7.52 Methylcitrate pathway.

7.7.3.1.4.1 Higher Carboxylates Polyhydroxyalkanoates: Synthesis The aerobic degradation of alkylcarboxylates has been covered briefly in the discussion on the degradation of alkanes: in brief it involves dehydrogenation of carboxyl-CoA esters to *trans*-enoyl-CoA followed by hydration, dehydrogenation to β-oxocarboxylates and fission by hydrolysis. An alternative is the production of poly-3-hydroxyalkanoates (PHAs) under conditions of nitrogen limitation. This has been examined in *R. eutropha* strain H16 where the genes for the β-oxothiolase (*phaA*), an acetoacetyl reductase (*phaB*), and a polyphydroxyalkanoate synthase *(phaC)* occur on an operon, and it has been shown that the thiolase reaction is rate limiting (Budde et al. 2010). After mobilization that is discussed in the next paragraph, poly-3-hydroxyalkanoates can serve as reserve substrates and as reductants, while in *Azospirillum brasiliensis* the reserve polymer was able to promote protection against stress and increase competitiveness, even though it had no effect on root colonization (Kadouri et al. 2003).

Acyl-CoA dehydrogenases that are involved in the degradation of alkanoates in fatty acid metabolism and in the metabolism of alkanoates and amino acids have already been discussed. They represent a large family of flavoproteins and the mechanism of their action involves concerted scission of both C–H bonds and transfer of the electrons to the respiratory chain (Ghisla and Thorpe 2004). In addition, they are intermediates for the synthesis of poly-3-hydroxyalkanoates (PHAs). These are reserve materials produced by the intermolecular esterification of (*R*)-3-hydroxyalkanoates under conditions of a high C/N ratio. The important difference between the synthesis of poly-3-hydroxybutyrates and poly-3-hydroxyalkanoates has been clearly pointed out (Kessler and Palleroni 2000). These polymers are found in a wide spectrum of organisms: these include the methylotroph *M. extorquens* (Korotkova et al. 2002b), fluorescent pseudomonads (Huisman et al. 1989), the photoheterotroph *Rh. rubrum* during growth with acetate (Stanier et al. 1959), the halophilic archaea *Natrialba aegyptica* strain 56 (Hezayen et al. 2000), and *H. marismortui* (Han et al. 2007).

Studies of this reaction have been motivated by its biotechnological significance (Jendrossek 2009; Andreeßen and Steinbüchel 2010). The following is a simplified summary of the synthesis of PHAs and the organisms involved.

1. The most straightforward situation occurs when these polymers are produced from an enoyl-CoA by an *R*-specific enoyl-CoA hydratase (*phaJ*$_{ac}$), for example, by *Aeromonas caviae* (Fukui et al. 1998; Hisano et al. 2003), whereas the *S*-specific hydratase proceeds to the formation of 3-oxoacyl-CoA and thence by β-oxidation to acetyl-CoA. The homologous genes *phaJ1*$_{Pa}$ and *phaJ2*$_{Pa}$ have been cloned from *Ps. aeruginosa*, and recombinants in *E. coli* displayed substrate specificities for short-chain and medium-chain length enoyl-CoA (Tsuge et al. 1999).

2. As an alternative, acetoacetyl-CoA produced from acetyl-CoA may undergo *R*-specific reduction to (*R*)-3-hydroxyalkyl-CoA: for example, ethylacetoacetate is reduced by *Paracoccus denitrificans* with nitrate under anaerobic conditions to (*R*)-ethyl-3-hydroxybutyrate (Nakashimada et al. 2001).

3. In the facultative methylotroph *M. extorquens*, poly-3-hydroxybutyrate can be produced from no less than three carbon substrates: (i) methanol by interaction of the serine and glyoxylate regeneration pathways, (ii) from succinate when the TCA cycle is involved in the production of acetyl-CoA and thence acetoacetyl-CoA, or (iii) from ethylamine that is dehydrogenated to acetyl-CoA and interaction of the TCA and glyoxylate regeneration pathways (Korotkova and Lidstrom 2001; Korotkova et al. 2002b).

4. PHAs are produced from *Ps. oleovorans* by metabolism of C_6 to C_{12} alkanes. For those alkanes with >7 carbon atoms, the PHA may contain carbon chains with two fewer carbon atoms than the growth substrates (Lageveen et al. 1988). In *Pseudomonas* sp. strain 61-3, both poly(3-hydroxybutyrate) and the copolymer poly(3-hydroxybutyrate-*co*-3-hydroxyalkanoate) can be produced simultaneously (Matsusaki et al. 1998).

5. The fatty acid synthetase from *Ps. putida* CA-3 displays activity not only to long-chain carboxylates (C_6 to C_{12}) but also to those with a distal phenyl group, and was able to convert them to the corresponding polyhydroxyalkanoates (Hume et al. 2009).

6. The PHA poly(3-hydroxyvalerate) is produced by *P. denitrificans* during growth with *n*-pentanol and the production of 3-oxovalerate followed by reduction to D(-) 3-hydroxyvaleryl-CoA.

At the same time, scission of 3-oxovalerate to propionyl-CoA and acetyl-CoA followed by condensation to acetoacetyl-CoA and reduction to D(-) 3-hydroxybutyryl-CoA produced poly(3-hydroxybutyrate) (Yamane et al. 1996).

7. The existence of an epimerase for conversion of the *S*-hydroxyalkanoate to the required *R*-hydroxyalkanoate by epimerization has been reported for *E. coli* (Yang and Elzinga 1993), in spite of putative evidence to the contrary in *Ps. oleovorans* (Fiedler et al. 2002).

8. The synthesis of PHAs by groups of halophilic aerobic (Han et al. 2010) archaea has been reported.

   a. *Haloferax mediterranei* under phosphate limitation with starch as the carbon and energy source (Lillo and Rodriguez-Valera 1990), and additional features have emerged: (i) synthesis of the copolymer poly(3-hydroxybutyrate-*co*-hydroxyvalerate) without the addition of potential carbon sources such as propionate or valerate, and (ii) polymer synthesis myay take place under both nutrient-limited and nutrient-rich media (Lu et al. 2008). Presumably the carbon sources are provided either from the rich medium in which cultures were pre-cultured, or by lesions in the β-oxidation cycle (Ren et al. 2000).

   b. *N. aegyptica* strain 56 was able to produce poly(3-hydroxybutyrate) during growth with *n*-butyric acid and acetate (Hezayen et al. 2000, 2001).

9. The production of poly-(3-hydroxybutyrate) by the thermophile *Chelatococcus* sp. strain MW10 was established at 50°C in cyclic fed-batch fermentation and achieved as high yield (Ibrahim and Steinbüchel 2010).

10. It has been shown that polymers analogous to poly-(3-hydroxybutyrate) can be synthesized from 3-mercaptopropionate or 3,3'-dithiodipropionate by *R. eutropha*. The synthesis of copolymers that contain the basic structural unit

$$[-O-CH(Me)-CH_2-CO-]_a \, [-S-CH_2-CH_2-CO-]_b$$

has been characterized by extensive supporting evidence from NMR, IR and GC-MS (Lütke-Eversloh et al. 2001). Since the physico-chemical properties of these thioester polymers are different from those of poly-(3-hydroxyalkanoates), there has been interest in the synthesis and regulation of such copolymers from *R. eutropha* H16 (Lindenkamp et al. 2010).

11. An alternative mechanism exists in *S. cerevisiae* in which the γ-aminobutyrate shunt can bypass two steps of the TCA cycle with the production of succinate semialdehyde. This may then be either channeled to the TCA cycle by dehydrogenation, or undergo reduction to 4-hydroxybutyrate that can be copolymerized with 3-hydroxybutyrate produced from acetoacetate (Bach et al. 2009).

Polyhydroxyalkanoates: Depolymerization This is introduced by summarizing the results of two classic studies. In *Pseudomonas lemoignei*, the extracellular poly-3-hydroxybutyrate depolymerase was excreted when growth ceased, and produced 3-hydroxybutyrate together with a dimer that could be degraded by an intracellular enzyme (Delafield et al. 1965). The intricacies of depolymerization of poly-3-hydroxybutyrate were clearly displayed in *Rh. rubrum*, and it was suggested that the system had three components, a soluble depolymerase, a heat-stable activator and a 3-hydroxybutyrate dimer hydrolase (Merrick and Doudoroff 1964).

Depolymerization of poly-3-hydroxyalkanoates is determined by a number of factors: (a) the intracellular or native polymer is amorphous and has a surface coating of protein and phospholipid, whereas the extracellular polymer is denatured and takes on a semicrystalline structure, (b) depolymerization may produce not only the monomer, but a series of oligomeric degradation products, (c) there is a considerable heterogeneity among polymerases from different organisms and, indeed some appear to lack the catalytic triad containing serine.

It is important to note some taxonomic changes: the genus *Ralstonia* has undergone a number of changes to *Wautersia* (Vaneechoutte et al. 2004) and to *Cupriavidus* (Vandamme and Coenye 2004), and it has been proposed (Jendrossek 2001) that *Pseudomonas lemoignei* be assigned to *Paucimomas lemoignei*.

Aspects of the hydrolysis of extracellular PHAs are briefly summarized as introduction to the complex aspects of intracellular depolymerization.

Extracellular depolymerases: The extracellular enzyme encoded by the gene *phaZ*$_{Pst}$ in *Pseudomonas stutzeri* had a molecular mass of 57.5 kDa and contained the characteristic, catalytic, linker and substrate-binding domains, although only low sequence homology was shown to other PAH polymerases (Ohura et al. 1999); further, in contrast to several other depolymerases, the product was the monomer 3-hydroxybutyrate, rather than the dimers found in *Comamonas acidovorans* (Kasuya et al. 1997) and *Alcaligenes faecalis* AE122 (Kita et al. 1995). This lack of similarity to other analogous enzymes has also been observed for the oligomer poly-3-hydroxybutyrate hydrolase in *Pseudomonas* sp. strain A1 (Zhang et al. 1997) and the PhaZ7 gene product in *P. lemoignei* (Braaz et al. 2003). The depolymerase in *Ps. fluorescens* GK13 catalyzed the depolymerization of poly-3-hydroxyoctanoic acid, and the soluble enzyme produced mainly the dimeric ester of 3-hydroxyoctanoate, although the enzyme immobilized on polypropylene produced the monomer. The esterase activities for 4-nitrophenyl esters of C_2, C_4, C_6, C_8, C_{10}, C_{12}, and C_{14} chain length were observed in increasing activity (Gangoiti et al. 2010). *Paucimonas lemoignei* is able to synthesize at least six extracellular polyhydroxyalkanoate depolymerases PhaZ1 to PhaZ6, and the hydrolase exhibited a number of unusual characteristics that were previously considered to be characteristic of intracellular depolymerases (Handrick et al. 2001). The purified enzyme hydrolyzed, (i) amorphous granules of (*R*)-poly-3-hydroxybutyrate and poly-3-hydroxyvalerate, (ii) oligomers of 3-hydroxybutyrate with six or more units, and (iii) native polyhydroxybutyrate or polyhydroxyvalerate with the production of pentamers as the main products. Although the structural gene *phaZ7* showed no homology to any other polyhydroxybutyrate depolymerases, it did possess the box-like motif containing serine in the catalytic triad (Braaz et al. 2003).

Intracellular depolymerases: Examination of the depolymerase in *Rh. rubrum* revealed its specific activity toward amorphous short-chain polyhydroxybutyrate, native and artificial PHB and oligomers with 3 or more units. The amino acid sequence showed no similarity to other intracellular depolymerases although there was strong similarity to type II catalytic domains of extracellular PHB, and the catalytic triad containing ser^{42} was identified (Handrick et al. 2004). It was suggested that an intracellular activator was unlikely to play a physiological role.

B. megaterium produced an intracellular PHB depolymerase PhaZ1 that had a high affinity for PHB granules, and was able to carry out their degradation without protease treatment to remove surface proteins, and the enzyme was able to degrade semicrystalline PHB with the production of 3-hydroxybutyrate (Chen et al. 2009). The intracellular depolymerase PhaZ from *Bacillus thuringiensis* showed no similarity to other depolymerases that had been described, and was able to degrade trypsin-activated native PHB granules and artificial amorphous PHB granules. Although PhaZ contained the box-like sequence containing serine, it lacked the signal peptide (Tseng et al. 2006).

P. lemoignei is able to synthesize at least five different extracellular polyhydroxybutyrate depolymerases, and there are technical difficulties in characterizing the intracellular enzyme since the polymer after excretion undergoes a chemical alteration. The situation is complicated by the occurrence of several genes that would putatively enable the organism to use intracellular PHB in the genome of *Ralstonia eutropha* strain H16 that has been extensively studied. It has been shown that PhaZd from this organism degraded artificial PHB granules mainly to oligomers of 3-hydroxybutyrate, and was distributed equally between PHB inclusion bodies and the cytosolic fraction (Abe et al. 2005). *R. eutropha* H16 was able to mobilize intracellular PHB and an oligomer hydrolase PhaZc was cloned and expressed in *E. coli*. Most of the enzyme was present in the cytosolic fraction and, although able to degrade oligomers, native PHB granules and semicrystalline PHB were not degraded (Kobayashi et al. 2005). An analogous enzyme PhaZ2 was able to hydrolyze cyclic PHB oligomers and participate with PhaZ1 was bound exclusively to inclusion bodies whereas PhaZ2 was bound top inclusion bodies and exist as a soluble enzyme (Kobayashi et al. 2003). Of the nine genes that have been found in the genome of this organism, only the product from the PhaZa1 was functional in recombinant *E. coli* strains (Uchino et al. 2008). Most of the gene products that have been described were able to hydrolyze extracellular short-chain PHB, although the five depolymerases PhaZ1 to PhsaZ5 occur in *P. lemoignei*, PhaZ6 was purified from recombinant *E. coli* and was able to accept not only poly-3-hydroxybutyrate but in addition poly-3-hydroxyvalerate and polymers produced after growth on odd-numbered carbon sources (Schöber et al. 2000).

PhaZ depolymerase from *Ps. putida* KT2442 has been purified and biochemically characterized after its expression in *E. coli*. To facilitate these studies a new and very sensitive radioactive method for detecting PHA hydrolysis *in vitro* was developed. PhaZ was an intracellular depolymerase that is located in PHA granules and hydrolyzed specifically medium chain-length PHAs containing aliphatic and aromatic monomers (de Eugenio et al. 2007).

Some additional aspects of the metabolism of high alkanoates are worth noting:

1. An unusual reaction has been discovered in the hydration at the nonactivated double bond of oleate to produce 10-hydroxyoctadecanoate and 10-oxo-octadecanoate by bacteria including *Mycobacterium fortuitum*, *Nocardia aurantia*, some species of *Pseudomonas*, and *Candida lipolytica* (El-Sharkawy et al. 1992). The hydratase has been characterized from *Elizabethkingia meningoseptica* (Bevers et al. 2009).

2. Another important reaction uses 3-methyl-3-hydroxybutanoate as the substrate. Isopentenyl diphosphate is the precursor of squalene and many terpenoids and is produced from mevalonate diphosphate by decarboxylation and elimination of the tertiary hydroxyl group. An analogous reaction has been explored using the enzyme from *S. cerevisiae* when it was shown that under anaerobic conditions the mevalonate diphosphate decarboxylase produced isobutene from 3-hydroxy-3-methylbutyrate in an ATP-dependent reaction (Gogerty and Bobik 2010). A route for the synthesis of 3-hydroxy-3-methylbutyrate has been demonstrated in *Galactomyces reesii* that was able to transform 3-methylbutyrate by dehydrogenation to methylcrotonyl-CoA followed by hydration and hydrolysis (Lee and Rosazza 1998; Dhar et al. 2002). Attention has also been directed to plausible biosynthetic routes to 3-hydroxy-3-methylbutyrate from 3-methylglutaconyl-CoA and methylcrotonyl-CoA.

7.7.3.2 Anaerobic Conditions

7.7.3.2.1 C_1 Compounds: Carbon Monoxide and Formate

It is convenient to summarize the range of C_1 substrates that can be utilized by anaerobic bacteria for the synthesis of acetate from glucose or fructose whose metabolism provides CO, CO_2, and formate by pathways given in a review by one of the pioneers in this area (Ljungdahl 1986).

$$CO_2 \rightarrow H \cdot CO_2H \rightarrow THF\text{--}CHO \rightarrow THF\text{--}CH_3$$

$$THF\text{--}CH_3 + E\text{-}Co \rightarrow E\text{-}Co\text{--}CH_3$$

$$CO_2 + 2H \rightarrow CO + H_2O$$

$$E\text{-}Co\text{--}CH_3 + CO + HS\text{-}CoA \rightarrow CH_3\text{--}CO\text{-}SCoA$$

These involve the cooperation of formate dehydrogenase (FDH), tetrahydrofolate enzymes (THF), carbon monoxide dehydrogenase (CODH), carbon monoxide dehydrogenase/carbon monoxide dehydrogenase disulfide reductase (CODH-S_2 reductase), corrinoid proteins (Co^+, Co^{2+}, and Co^{3+}) and corrinoid methyltransferase.

Substrate	CO	H·CO₂H	CO₂/H₂	CH₃OH/CO₂
Acetobacterium woodii	+	+	+	+
(Moorella)Clostridium thermoacetica	+	+	+	+
Clostridium thermoautotrophicum	+	+	+	+
Sporomusa acidovorans	–	+	+	+
Desulfomaculum orientis	+	+	+	+

7.7.3.2.1.1 Carbon Monoxide   This is metabolized by a range of anaerobes including *Clostridium thermoaceticum*, *Cl. formicoaceticum*, *Carboxydothermus hydrogenoformans*, and the phototrophs *Rh. rubrum*, *Rhodocyclus (Rhodopseudomonas) palustris*, and *Rubrivivax gelatinosus*. They employ dehydrogenases to carry out the reaction $CO + H_2O \rightarrow CO_2 + 2H^+ + 2e$, and the enzymes have been characterized although they differ considerably, and are discussed in greater deal in the section dealing with dehydrogenation. The enzyme from *Cl. thermoaceticum* contains Ni, and ferredoxin and a membrane-bound b-type cytochrome are reduced in the presence of CO (Drake et al. 1980). The purified enzyme consists of two subunits and the dimer produced by dissociation of the monomer contains per mol 2 Ni, 11 Fe, 14 acid-labile S, and 1 Zn (Ragsdale et al. 1983). The authors point out that whereas the enzyme from aerobic bacteria contains Mo, the enzyme from anaerobes contains Ni. The enzyme from *Rh. rubrum* was membrane-bound and existed in two forms, the one having the major activity with 7 Fe, 6 S, 0.6 Ni, and 0.4 Zn, although the nickel was EPR silent (Bonam and Ludden 1987). The enzyme from *Carboxydothermus hydrogenoformans* was also found in two forms with Fe, acid-labile S, and Ni, while, although Ni was EPR silent, EPR suggested the presence of [4Fe–4S] clusters (Svetlitchnyi et al. 2001).

Some unusual features in the growth of the methanogen *Methanosarcina acetivorans* C2A have been uncovered.

1. It is able to grow with carbon monoxide and produced methane only when the cells enter stationary phase, although during growth both acetate and formate were produced (Rother and Metcalf 2004). Proteomic analysis revealed that acetate is produced by reversal of the initial steps for the conversion of acetate to methane, and that a distinct pathway is used for CO_2 reduction to methane (Lessner et al. 2006).
2. During growth with carbon monoxide, cells effectively removed free sulfide from the medium with the resulting production of methanethiol and dimethyl sulfide (Moran et al. 2008).

7.7.3.2.1.2 Reduction and Catenation   The existence of nitrogenases with Mo or V at the active site has been described in Chapter 3. Although carbon monoxide has been regarded as an inhibitor of the Mo-nitrogenase reduction of all substrates except H^+ and, of H^+ reduction by the V-nitrogenase, it has been shown that carbon monoxide can be reduced by both enzymes. Ethene and ethane were produced in addition to lower amounts of propane and propene and, for the V-enzyme much lower concentrations of α-C_4H_8 and n-C_4H_{10}. In addition, methane was found only for the V-nitrogenase (Hu et al. 2011). Another study from different authors revealed that substitution of α-valine[70] by alanine or glycine in the MoFe protein resulted in the activity of the enzyme to reduce carbon monoxide to methane, ethane, ethene, propene, and propane (Yang et al. 2011).

7.7.3.2.1.3 Formate
1. *Methanococcus vannielii* ferments formate to methane and CO_2. The formate-dependent reduction of $NADP^+$ is carried out by coupling formate dehydrogenase with the reduced 8-hydroxy-5-deazaflavin cofactor (coenzyme F_{420}) (Jones and Stadtman 1980). There are two forms of formate dehydrogenase in this strain, one of them contains molybdenum, iron and acid-labile sulfur, and the other comprises a high molecular weight complex containing seleno-protein and molybdenum–iron–sulfur subunits (Jones and Stadtman 1981). Formate can be used for methanogenesis in *Methanobacterium formicicum* (Schauer and Ferry 1986) and *Methanobacterium ruminantium* (Tzeng et al. 1975) and in both organisms formate dehydrogenase is coenzyme F_{420}-dependent.
2. Syntrophic growth at the expense of formate has been demonstrated for syntrophic growth of *Moorella* strain AMP and *Desulfovibrio* strain G11 in the presence of *Methanothermobacter* and *Methanobrevibacter arboriphilus* AZ respectively, neither of which are able to use formate alone while the methanogens can use only H_2 as electron donor (Dolfing et al. 2008).
3. The homoacetogen *Moorella thermoacetica* is able to grow with formate by carrying out the reaction: $4H-CO_2^- + H+ \rightarrow CH_3 + CO_2^- + 2HCO_3^-$ (Drake and Daniel 2004).

7.7.3.2.2 C_2 Compounds: Oxalate, Malonate, Glyoxylate, Glycolate
Although oxalate, malonate, glyoxylate, and glycolate are rather unusual substrates, their degradation by anaerobic bacteria has been described.

7.7.3.2.2.1 Oxalate

1. *Oxalobacter vibrioformis* uses oxalate as sole source of energy and acetate as principal carbon source (Dehning and Schink 1989a). In *O. formigenes*, oxalate is metabolized to formate and CO_2 by the coupled reaction of formyl-CoA transferase (FRC) and oxalyl-CoA decarboxylase followed by thiamin diphosphate-dependent decarboxylation to formate and CO_2 encoded by *oxc*.

$$FRC: HCO-SCoA + HO_2C-CO_2H \rightarrow HCO_2H + HO_2C-CoSCoA$$

$$Oxc: HO_2C-CoSCoA \rightarrow CO_2 + HCO-SCoA$$

Both enzymes have been purified from *O. formigenes*, formyl-CoA transferase by Baetz and Allison (1990) and oxalyl-CoA decarboxylase by Baetz and Allison (1989). X-ray analysis (Jonsson et al. 2004) showed that the class III transferase involved a series of mixed anhydrides: Asp^{169} reacted with formyl-CoA (donor) to produce aspartyl formate followed by reaction with oxalate (acceptor) to form an intermediate aspartyl oxalate from which oxalyl-CoA and formate were produced. The thiamin diphosphate-dependent oxalyl-CoA decarboxylase involves activation by adenosine diphosphate (Berthold et al. 2005). Oxalate and formate form a one-to-one antiport system which involves the consumption of an internal proton during decarboxylation, and serves as an indirect proton pump (Ruan et al. 1992) to generate ATP by decarboxylative phosphorylation. Cell-free extracts contain the enzymes of the biosynthetic glycerate pathway via glyoxylate and tartronic semialdehyde (Cornick and Allison 1996). Oxalyl-CoA decarboxylase has been characterized from several species of anaerobic bifidobacteria including *Bifidobacterium lactis, Bif. animalis, Bif. longum*, and *Bif. adolescens* (Federici et al. 2004). An extensive study using *Bif. animalis* subsp *lactis* showed that oxalate was required for induction of both FRC and Oxc, and that a low pH was obligately required for oxalate degradation (Turroni et al. 2010). It was suggested that this enzyme could be used for the management of oxalate-related kidney disease.

2. The acetogen *Clostridium thermoaceticum* is able to grow at the expense of oxalate with the production of acetate: $4^-O_2C-CO_2^- + 5H_2O + H^+ \rightarrow CH_3CO_2^- + 6HCO_3^- + H_2O$. The essential partial reactions are: (i) formation of oxalyl-CoA and decarboxylation to formyl-CoA, (ii) formate dehydrogenation $H-CO_2^- + H_2O \rightarrow HCO_3^- + 2[H]$, and (iii) $[H] + HCO_3^- \rightarrow$ acetyl-CoA by the Wood pathway (Daniel and Drake 1993).

7.7.3.2.2.2 Malonate

1. In the presence of low concentrations of yeast extract, the strictly anaerobic *Sporomusa malonica* was able to use malonate and succinate as sole sources of carbon, and were decarboxylated to acetate and propionate, while a range of other TCA cycle carboxylates were fermented, generally to acetate (Dehning et al. 1989).

2. *Malonomonas rubra* is a micro-aerotolerant fermenting organism that decarboxylates malonate to acetate (Dehning and Schink 1989b). The organism contains high concentrations of *c*-type cytochromes that are not involved in the metabolism of the substrate, and are presumably remnants of the sulfur-reducing relatives of the organism (Kolb et al. 1998). Biotin-dependent carboxylases couple the decarboxylation to the transport of Na^+ across the cytoplasmic membrane and use the electrochemical potential $\Delta\mu$ Na^+ to mediate the synthesis of ATP (Dimroth and Hilbi 1997).

3. The degradation of malonate to acetate and CO_2 under anaerobic conditions in the presence of yeast extract was carried out by an organism tentatively assigned to *Citrobacter* (Janssen and Harfoot 1990), and this was also established for degradation by *Klebsiella oxytoca* where malonate was activated in cell extracts by the addition of malonyl-CoA (Dehning and Schink 1994).

7.7.3.2.2.3 Glyoxylate/Glycolate

Several anaerobic bacteria have been described that are able to grow with glycolate, and two of them involving the formation of glyoxylate (Friedrich et al. 1996).

1. *Desulfofustis glycolicus* was able to oxidize glycolate and glyoxylate with sulfate as electron acceptor and carried out the complete oxidation of glycolate to CO_2 (Friedrich and Schink 1995b). This organism contained cytochromes of the *b*- and *c*-type as well as menaquinone-5, and a sulfite reductase of the desulforubidin type. The pathway of glycolate degradation could not, however, be identified although several plausible alternatives were explored (Friedrich and Schink 1995b).

2. Glyoxylate was fermented by a pure culture of *Syntrophobotulus glycolicus*. The glyoxylate was disproportionated to glycolate, CO_2, and H_2, and glyoxylate was metabolized by the malate cycle in which glyoxylate reacted with acetyl-CoA to produce malate that was then degraded to pyruvate, CO_2 and H_2. Hydrogen-dependent glyoxylate reduction to glycolate was coupled to the synthesis of ATP (Friedrich and Schink 1995a).

3. Glycolate or fumarate can be fermented by an organism belonging to the family *Lachnospiraceae* to acetate, succinate, and CO_2, without the formation of hydrogen (Janssen and Hugenholtz 2003).

The anaerobic degradation of aliphatic carboxylic acids is of great ecological importance since compounds such as acetate, propionate, or butyrate may be the terminal fermentation products of organisms degrading more complex compounds including carbohydrates, proteins, and lipids, while long-chain acids are produced by the hydrolysis of lipids (Zeikus 1980; Mackie et al. 1991). Degradation of aliphatic carboxylic acids by sulfate-reducing bacteria was traditionally restricted to lactate and its near relative, pyruvate, but recent developments have radically altered the situation and increased the spectrum of compounds which can be oxidized to CO_2 at the expense of sulfate reduction. In the following paragraphs, an attempt will be made to present a brief summary of the anaerobic degradation of the main groups of aliphatic compounds. Studies on the anaerobic degradation of alkanoic acids have been carried out using both pure cultures and syntrophic associations.

7.7.3.2.3 C_2: Acetate

For the anaerobic degradation of acetate, two different reactions may take place: oxidation to CO_2 or dismutation to methane and CO_2. The oxidation of acetate under anaerobic conditions requires the synthesis of the glyoxylate enzymes that are synthesized by *Thiobacillus versutus* growing anaerobically with nitrate, although not during aerobic growth with acetate (Claasen and Zehnder 1986). To some extent, as will emerge, segments of both pathways are at least formally similar, although the mechanisms for anaerobic degradation of these apparently simple compounds are quite subtle. Degradation of acetate and butyrate can be accomplished by *Desulfotomaculum acetoxidans* (Widdel and Pfennig 1981), and of acetate by *Desulfuromonas acetoxidans* (Pfennig and Biebl 1976) and species of *Desulfobacter* (Widdel 1987), while propionate is degraded by species of *Desulfobulbus* (Widdel and Pfennig 1982; Samain et al. 1984). A sulfur-reducing organism with a much wider degradative capability than *Desulfuromonas acetoxidans* has been isolated (Finster and Bak 1993), and this organism is capable of accomplishing the complete oxidation of, for example, propionate, valerate, and succinate.

7.7.3.2.3.1 Acetate Assimilation   The assimilation of acetate under anaerobic conditions can occur by completely different pathways that have been investigated in detail, and their enzymology has been delineated.

1. *The citric acid cycle: Reductive and oxidative modes*
    a. *Reductive mode*

       The assimilation of CO_2 by *Chlorobium limicola (thiosulfatophilum)* was shown many years ago to take place by a reductive citric acid cycle (Evans et al. 1966; Fuchs et al. 1980). It also operates during autotrophic growth of *Desulfobacter hydrogenophilus* in which an ATP-citrate lyase brings about the scission of citrate to acetyl-CoA and oxalacetate (Schauder et al. 1987).

    b. *Oxidative mode*

       Oxidation may take place in the reverse direction by a modification of the classic tricarboxylic acid cycle in which the production of CO_2 is coupled to the synthesis of

NADPH and reduced ferredoxin, and the dehydrogenation of succinate to fumarate is coupled to the synthesis of reduced menaquinone. This is used to carry out oxidation of acetate by *Desulfuromonas acetoxidans* in which S^0 is reduced to S^{2-} (Gebhardt et al. 1985) and, in modified form by *Desulfobacter postgatei* in which an ATP-citrate lyase is used to synthesize citrate from oxalacetate and acetyl-CoA (Brandis-Heep et al. 1983; Möller et al. 1987).

2. *The oxidative acetyl-CoA/carbon monoxide dehydrogenase pathway*

Operation of the oxidative citric acid cycle is not, however, the only mode for oxidation of acetate by sulfate-reducing anaerobes. On the other hand, dissimilation of acetate may take place by reversal of the pathway used by organisms such as *Cl. thermoaceticum* for the synthesis of acetate from CO_2. In the degradation of acetate, the pathway involves a dismutation in which the methyl group is successively oxidized via methyl tetrahydrofolate (THF) to CO_2 while the carbonyl group is oxidized via bound carbon monoxide. The oxidative acetyl-CoA/carbon monoxide dehydrogenase pathway has been demonstrated in several sulfate-reducing anaerobes, and is illustrated for *Desulfobacterium autotrophicum*. The metabolism of lactate by the sulfate-reducing *D. autotrophicum* is initiated by dehydrogenation to pyruvate followed by decarboxylation to acetyl-CoA. This undergoes dismutation in which the methyl group is transferred to methyl-tetrahydromethanopterin (Me-H_4MPt) while the carbonyl group is dehydrogenated by carbon monoxide dehydrogenase to CO_2. The Me-H_4MPt then undergoes successive dehydrogenation: $CH_2=H_4MPt \rightarrow CH\equiv H_4MPt \rightarrow CHO-H_4MPt$ and then to CO_2 (Schauder et al. 1989). This is essentially the reverse of the reactions involved in acetogenesis by organisms such as *Cl. thermoaceticum* and is also used lactate by the sulfate-reducing archaeon *Archaeoglobus fulgidus* (Möller-Zinkhan and Thauer 1990).

3. *Methanogenesis*

An important reaction in anaerobic environments is the conversion of acetate to methane by a group of methanogenic archaea that occur in a variety of environments—terrestrial, marine, and the human intestine. These include *Methanosarcina thermophila*, *M. barkeri*, and *Methanosaeta soehngenii*. Particular emphasis has been directed to those belonging to the genus *Methanosarcina* that are able to use both a number of C_1 substrates, $CO_2 + H_2$, and acetate. The cardinal reaction in methanogenesis from acetate is carried out by a carbon monoxide complex that comprises a Ni/Fe–S component and a corrinoid/Fe–S component. There are differences among these enzymes, including in the subunit composition of CO dehydrogenase: $\alpha_2\beta_2$ in *M. barkeri* and *Methanosaeta soehngenii*, and $\alpha\beta\gamma\delta\epsilon$ in *Methanosarcina thermophila*, and minor differences in the metal composition. In brief, the methyl group is initially converted into methyltetrahydrosarcinopterin (CH_3-H_4SPT corresponding to methyltetrahydrofolate in the acetate oxidations discussed above), before reduction to methane via methyl-coenzyme M and *N*-(7-mercaptoheptanoyl)threonine-O^3-phosphate; the carbonyl group of acetate is oxidized via bound CO to CO_2. The conversion of acetate to methane and CO_2 in *Methanosarcina* is carried out in stages that can be summarized (Jablonski et al. 1993):

a. Activation of acetate to produce acetyl-CoA by acetate kinase and phosphotransacetylase

b. Scission of acetyl-CoA catalyzed by the carbon monoxide dehydrogenase complex

c. Oxidation of the carbonyl-bound CO to CO_2, and transfer of the methyl group to the

d. Co/Fe–S component

e. Transfer of the methyl group successively tetrahydrosarcinapterin and HS–CoM

f. Transmethylation of CH_3-CoM by *N*-(7-mercaptoheptanoyl)threonine-O^3-phosphate and reduction to methane

The effective operation of methanogenesis requires the coordination of a number of reactions. Three of these cardinal reactions are:

$$1 \text{ Acetyl-CoA} + H_4\text{SPt} + H_2O \rightarrow CH_3-H_4\text{SPt} + CO_2 + 2H^+ + \text{CoA-SH}$$

$$2 \, CH_3-H_4SPt + CoM-SH \rightarrow CH_3-SCoM + H_4SPt$$

$$3 \, CH_3-SCoM \rightarrow CH_4$$

The genes carrying the enzymes are arranged in an operon consisting of the following subunits (Gencic and Grahame 2003):

cdhA ≡ CODH 87.7 in the α-subunit

cdhB ≡ CODH 18.6 in the ε-subunit

cdhC ≡ acetyl-CoA decarbonylase synthase in the 52.7 (Ni/Fe) β-subunit

ACDS ORF ≡ accessory protein Ni insertion 28.1

cdhD ≡ corrinoid protein 47.2 in the δ-subunit

cdhE ≡ corrinoid protein 51.2 in the γ-subunit

7.7.3.2.4 C_3: Propionate

Degradation of propionate is an intermediate step in the decomposition of organic matter in anaerobic environments, and growth at the thermodynamic limit is facilitated by syntrophy with methanogenic bacteria. For example, the syntrophic population of anaerobic organisms that carried out the oxidation of propionate in a rice field was examined and showed the presence of the genera *Syntrophobacter*, *Smithella*, and *Pelotomaculum*. Among the archaea species of *Methanobacterium*, *Methanosarcina*, and members of the Rice Cluster I have been found (Lueders et al. 2004). A number of important aspects are noted.

1. Both the synthesis of propionate and its metabolism may take place under anaerobic conditions. In *Desulfobulbus propionicum*, degradation could plausibly take place by reversal of the steps used for its synthesis from acetate (Stams et al. 1984)—carboxylation of propionate to methyl-malonate followed by coenzyme B_{12}-mediated rearrangement to succinate which then enters the citric acid cycle. The converse decarboxylation of succinate to propionate has been observed in *Propionigenium modestum* (Schink and Pfennig 1982), *Sporomusa malonica* (Dehning et al. 1989), *Selenomonas acidaminovorans* (Guangsheng et al. 1992), and *Propionigenium maris* (Janssen and Liesack 1995).

2. Growth of syntrophic propionate-oxidizing bacteria in the absence of methanogens has been accomplished using fumarate as the sole substrate (Plugge et al. 1993). Fumarate plays a central role in the metabolism, since it is produced from propionate via methylmalonate and succinate, while fumarate itself is metabolized by the acetyl-CoA cleavage pathway via malate, oxalacetate, and pyruvate.

3. The pathways and mechanism of interspecies transfer have been examined in syntrophic propionate-oxidizing organisms.

   a. Two pathways for propionate degradation have been proposed. In *Pelotomaculum thermopropionicum* the methylmalonyl-CoA pathway is used, and stringent regulation of fumarase is exerted (Kosaka et al. 2006). In *Smithella propionica* that produces acetate and butyrate, however, the pathway is initiated by condensation of two molecules of propionate by a Claisen reaction (Heath and Rock 2002) involving the α-CH of one with the C=O of the CoA ester of the other. This is followed by rearrangement—putatively by a coenzyme B_{12}-dependent mutase—to 3-oxohexanoate followed by fission to butyrate and acetate (Figure 7.53) (de Bok et al. 2001).

   b. Although it has been generally assumed that interspecies transfer in propionate-oxidizing syntrophs involved H_2, there is increasing evidence that formate transfer is actually involved. Components of the syntrophic culture of *Syntrophobacter fumaroxidans* and *Methanospirillum hungatei* that degrade propionate were separated by Percoll gradient centrifugation. Levels of formate dehydrogenase, hydrogenase and formate-hydrogen lyase in the components were examined, and it was showed that interspecies electron transfer was carried out by primarily via formate (de Bok et al. 2002).

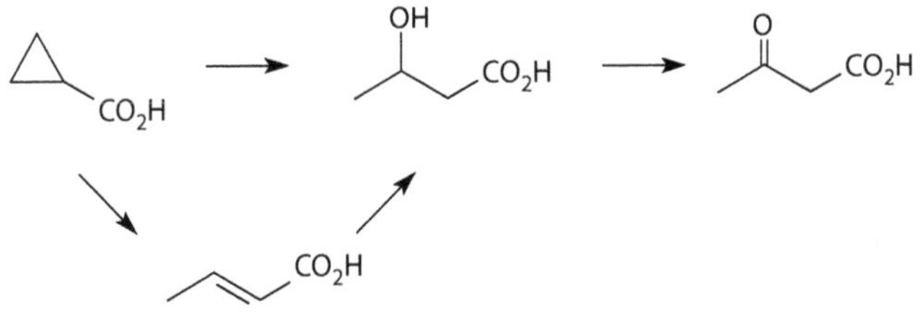

FIGURE 7.53 Anaerobic degradation of propionate by *Smithella propionica*.

7.7.3.2.5 C_4: Succinate, Fumarate, Maleate, Crotonate, and Butyrate

7.7.3.2.5.1 Succinate

1. The fermentation of succinate to acetate and butyrate has been described for *Clostridium kluyveri*. This is carried out by successive dehydrogenation of succinyl-CoA to succinate semialdehyde followed by reduction to 4-hydroxybutyrate. This underwent atypical free-radical-mediated dehydration (Buckel et al. 2005) to but-3-enoate from which the acetate and butyrate were formed (Wolff et al. 1993).

2. The strict anaerobe *Propionigenium modestum* was able to grow by decarboxylating succinate to propionate (Schink and Pfennig 1982) that is carried out by successive formation of succinyl-CoA, methylmalonyl-CoA, and propionyl-CoA (Bott et al. 1997). A group of quite different organisms was isolated with this capability, and one strain Ft1 that was tolerant of oxygen up to 20% was able to disproportionate both fumarate and aspartate to propionate and acetate (Denger and Schink 1990).

3. Succinate can also be dehydrogenated to fumarate using S^0 as electron acceptor in an electron transport chain involving menaquinone (Paulsen et al. 1986).

7.7.3.2.5.2 Fumarate

Fumarate, malate, and maleate have the same redox potential, and fumarate occupies a central position in anaerobic metabolism where it is able to serve either as an electron acceptor or as growth substrate for fermentation. Fumarate can be produced from aspartate by elimination, and this is discussed in later paragraphs dealing with the fermentation of aspartate. Illustrative aspects of fumarate metabolism include the following.

1. Fumarate serves as the terminal electron acceptor in the metabolism of formate in *Wolinella succinogenes* in combined reactions involving formate dehydrogenase and fumarate reductase. The reduction of fumarate to succinate can be coupled to formate dehydrogenase in *W. succinogenes* in the following reactions:

$$\text{Formate} + \text{menaquinone} + \text{H}^+ \rightarrow \text{CO}_2 + \text{menaquinol (Formate dehydrogenase)}$$

$$\text{Fumarate} + \text{menaquinol} \rightarrow \text{succinate} + \text{menaquinone (Fumarate reductase)}$$

Fumarate reductase has three subunits of which FrdA contains FAD, FdrB has a [3Fe–4S] cluster, and FrdC has a cytochrome *b* with two heme groups (Körtner et al. 1990). The formate dehydrogenase also has three subunits of which FdhA contains Mo, and FdhC contains cytochrome *b* (Kröger et al. 1992).

2. The anaerobic de-*O*-methylation of vanillin and syringate can be carried out by strains of *Desulfitobacterium hafniense* using fumarate as electron acceptor: this is reduced to succinate while the methyl groups enter the tetrahydrofolate pathway (Kreher et al. 2008).

3. Fumarate is fermented by *Clostridium formicoaceticum* by dismutation to succinate, acetate, and CO_2 (Dorn et al. 1978):

$$3 \text{ fumarate} + 2\text{H}_2\text{O} \rightarrow 2 \text{ succinate} + 1 \text{ acetate} + \text{CO}_2$$

Malate can also be fermented by this organism, albeit with the production primarily of acetate and CO_2 since the pathways different in detail.

4. *D. gigas* and several strains of *Desulfovibrio desulfuricans* were able to carry out the dismutation of fumarate in sulfate-free medium (Miller and Wakerley 1966). In two strains of the latter, succinate, malate, and oxalalacetate were produced, followed by oxidation to acetate and utilization of the reducing equivalents to reduce fumarate to succinate.

5. Strains of *Propionivibrio dicarboxylaticus* were isolated for enrichment of organisms that were able to ferment maleate. It was shown that they could also ferment fumarate in which 1 mol of fumarate was oxidized to 1 mol acetate and the reducing equivalents that were produced were used to reduce 2 mol fumarate to 2 mol succinate, and by decarboxylation to 2 mol propionate (Tanaka et al. 1990).

7.7.3.2.5.3 Crotonate

1. *Ilyobacter polytrophus* is able to ferment a range of carboxylic acids including crotonate, 3-hydroxybutyrate citrate and pyruvate (Stieb and Schink 1984). Crotonate and 3-hydroxybutyrate were fermented to acetate and butyrate in a ratio of ca. 2:1, citrate and pyruvate produced almost equal amounts of acetate and formate, while fumarate was fermented to acetate, propionate, and formate.

2. *Sproromusa malonica* that was able to use malonate as sole source of carbon and energy, was able to ferment crotonate and 3-hydroxybutyrate to acetate and butyrate with a greater amount of acetate, apparently combining homoacetogenic fermentation with crotonate dismutation (Dehning et al. 1989).

3. A number of syntrophs are able to grow with crotonate alone: these include *Syntrophomonas wolfei* and *Syn. buswellii* (Wallrabenstein and Schink 1994), although as noted below the degradation of butyrate necessitated the complementation with a hydrogen-utilizing organism such as *Methylospirillum hungatei*. The syntrophic anaerobe *Syntrophus aciditrophicus* was able to grow with crotonate as the sole substrate with the formation of acetate and cyclohexene-1-carboxylate (Mouttaki et al. 2007). The latter is produced by a complex series of reactions involving the synthesis of acetoacetate and crotonate followed by carboxylation to glutaconyl-CoA, condensation with acetyl-CoA and reduction to 3-hydroxypimelyl-CoA that underwent cyclization (Figure 7.54). It has been shown that under these conditions, reverse electron transport occurred and that that there were multiple AMP-forming coenzyme A ligases and acyl-coenzyme A synthetases (McInerney et al. 2007).

7.7.3.2.5.4 Butyrate The oxidation of butyrate to acetate by *Syntrophomonas wolfei* requires the presence of the methanotroph *Methanospirillum hungatei* to remove the hydrogen produced by dehydrogenation of butyryl-CoA to crotonate from which the acetate is produced. This is accomplished by a reverse electron transport system to produce NADH for the generation of hydrogen (Müller et al. 2009). The anaerobic oxidation of long-chain carboxylates by syntrophs is noted later.

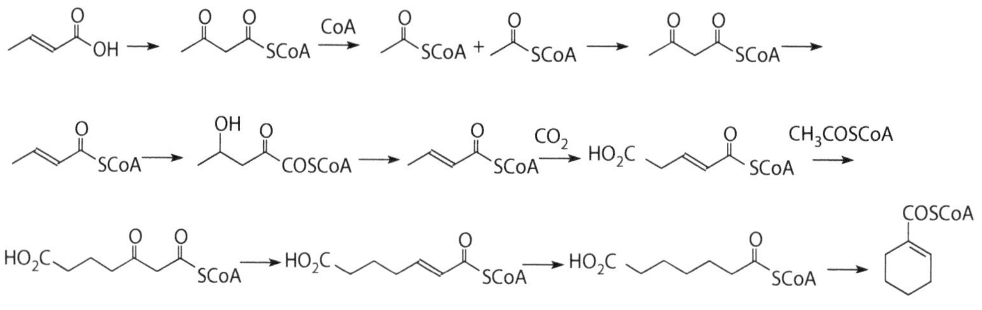

FIGURE 7.54 Biosynthesis of cyclohexene 1-carboxylate from crotonate.

7.7.3.2.5.5 Glutarate Glutarate is fermented via glutaryl-CoA by dehydrogenation followed by decarboxylation and reduction to butyryl-CoA—that may also be produced directly. This may produce butyrate, or by rearrangement catalyzed by a mutase to isobutyryl-CoA and then isobutyrate (Matthies and Schink 1992). Fermentation of glutamate can produce 2-hydroxyglutarate that may be dehydrated by a radical mechanism to glutaconate. This in turn may be decarboxylated to crotonate by an enzyme complex consisting of 4-subunits: carboxyl transferase (α), biotin-containing carrier protein (β), carboxybiotin decarboxylase (γ), membrane anchor (δ).

7.7.3.2.5.6 Long-Chain Carboxylates Long-chain dicarboxylic acids are produced in a number of reactions:

1. The C_{12} dicarboxylate by α,ω-oxidation of dodecane (Broadway et al. 1993)
2. The C_7 dicarboxylate (pimelate) as an intermediate in the anaerobic degradation of benzoate (Harwood et al. 1999)
3. As intermediates in the degradation of cycloalkanones (Kostichka et al. 2001)
4. Atmospheric oxidation of unsaturated fatty acids (Stephanou and Stratigatic 1993)

The degradation of long-chain carboxylic acids is important in the anaerobic metabolism of lipids, and an extensive compilation of the organisms that can accomplish this has been given (Mackie et al. 1991).

There are several situations in which the degradation of long-chain alkanoates has been described.

1. Syntrophic associations. This capability has been demonstrated in several syntrophic bacteria in the presence of hydrogen-utilizing bacteria: for example, β-oxidation of C_4 to C_8, C_5, and C_7 carboxylic acids was carried out by the *Syntrophomonas wolfei* association (McInerney et al. 1981), of C_4 to C_{10}, and C_5 to C_{11} by the *Syntrophospora* (*Clostridium*) *bryantii* syntroph (Stieb and Schink 1985; Zhao et al. 1990), and of C_6 to C_{18} by the syntrophs *Syntrophomonas curva* (Zhang et al. 2004) and *Syntrophomonas sapovorans* (Roy et al. 1986). Acetate and propionate were the respective terminal products from the even- and odd-numbered acids.
2. Pure cultures of several sulfate-reducing bacteria are able to carry out analogous reactions. *Desulfobacterium cetonicum* degraded butyrate to acetate by a typical β-oxidation pathway (Janssen and Schink 1995b), and oxidation of long-chain carboxylates up to C_{10} has been demonstrated in species of *Desulfonema* (Widdel et al. 1983). *Desulfuromonas palmitatis* was able to oxidize acetate, lactate, palmitate, and stearate completely to CO_2 using Fe(III) as the electron acceptor: it was also able to use MnO_2, fumarate, and $S°$ as electron acceptors, though neither sulfate nor thiosulfate could be used (Coates et al. 1995). *Desulforegula conservatrix* was specifically able to oxidize *n*-carboxylates from C_4 to C_{17} using sulfate as electron acceptor (Rees and Patel 2001): even-number substrates were oxidized to acetate and odd-numbered ones to acetate and propionate.
3. Attention has been directed to hyperthermophilic organisms isolated from natural thermally active sites that used other electron acceptors. *Geoglobus ahangeri* was able to oxidize acetate, butyrate, valerate and stearate by using poorly crystalline ferric oxide as electron acceptor, but was unable to use sulfur, sulfate, thiosulfate, nitrate, nitrite, or Mn(IV) (Kashevi et al. 2002). *Archaeoglobus fulgidus* was able to grow at the expense of C_4 to C_{18} alkanoates using thiosulfate as electron acceptor with the production of sulfide (Khelifi et al. 2010).
4. The degradation of pimelate by a denitrifying organism LP1 was initiated by formation of the coenzyme A ester, and was followed by a series of steps with the production of glutaryl-CoA that was decarboxylated to crotonyl-CoA (Härtel et al. 1993; Gallus and Schink 1994; Harrison and Harwood 2005). This produced two molecules of acetate, analogous to the aerobic degradation where the genetics have been explored (Parke et al. 2001) in *Acinetobacter* sp. strain ADP1 that is able to degrade dicarboxylic acids up to C_{14}. In strictly fermenting organisms butyrate and *iso*butyrate were produced (Matthies and Schink 1992). The enzymes for the

degradation of pimelate in *Rh. palustris* are contained in an operon in which the PimA ligase that catalyzes the first step is able to accept an unusually wide range of dicarboxylic acids as substrates (Harrison and Harwood 2005).

In summary, the anaerobic degradation of alkanoic acids may truly be described as ubiquitous and is carried out by organisms of widely different taxonomic affinity, both in pure culture and in syntrophic associations.

7.7.3.2.5.7 Biosynthesis of Long-Chain Alkanes, Alkenes, and Alkanoates The biosynthesis of alkanes and alkanes by phototrophs has been discussed in Chapter 2. In contrast to cyanobacteria where hydrocarbons are produced by decarbonylation of fatty aldehydes (Schirmer et al. 2010), reactions for other bacteria include several other mechanisms.

1. Hydrocarbons produced by *Sarcina lutea* included C_{22} to C_{29} hydrocarbons: these alkanes had branched methyl groups, either *iso* or *anteiso* in both alkyl chains, and unsaturation in C_{25} hydrocarbons that was confirmed by hydrogenation (Tornabene et al. 1967). Identification of the C_{27}, C_{28}, and C_{29} alkanes added important details (Albro and Dittmer 1969a,b) that were substantiated, and a mechanism based on the results of experiments using cell-free extracts was proposed. It was shown that the incorporation of palmitate into the hydrocarbons required coenzyme A, ATP; Mg^{2+}, and either pyridoxal phosphate or pyridoxamine phosphate (Albro and Dittmer 1969c).

2. The biosynthesis of hydrocarbons was examined in *Stenotrophus* (*Pseudomonas*) *maltophilia*. A spectrum of hydrocarbons was produced and was characterized by two features: dominance of mono-, di-, and tri-unsaturated alkenes, and single- and double-branched alkenes. The hydrocarbons occurred in groups with C_{28}, C_{29}, C_{30}, and C_{31} (Suen et al. 1988), and supplementation of the complex growth medium with leucine increased the contributions of the i,i-C_{29} and i-C_{30} alkanes, and with isoleucine the a,i-C_{28} and a,i-C_{30} alkanes.

3. The synthesis of hydrocarbons has been examined in a range of bacteria. A cluster of C_{28} to C_{31} hydrocarbons was produced by *Stenotrophomonas maltophilia*, while a C_{29} hydrocarbon was dominant for *Micrococcus luteus* and species of *Arthrobacter* that included *Arthrobacter chlorophenolicus*, *Arthrobacter oxydans*, and *Arthrobacter nicotianae* (Frias et al. 2009). It was shown by comparison with synthetic standards that the C_{29} hydrocarbon was a mixture of 2,26-dimethyl-heptacosa-13-ene, and 3,25-dimehylheptacosa-13-ene. It was suggested that, in contrast to strains of *Micrococcus*, the biosynthesis of alkenes is not ubiquitous in strains of *Arthrobacter*.

4. A gene cluster from *Micrococcus luteus* was introduced into a fatty acid-overproducing *E. coli* (Beller et al. 2010). The alkenes C_{27} and C_{29} with three double bonds were produced putatively by the reactions: (a) oxidation of the C_{15}-acyl-CoA to a 1,3-dioxoacyl-CoA, (b) condensation with an acyl-CoA ester and elimination of CO_2 to produce a 1,3-diketone. In two series of reduction and dehydration this was then converted to the alkene (Figure 7.55a).

5. A pathway for the synthesis of alkenes by a head-to-head condensation has been examined in *Shewanella oneidensis* strain MR-1, and was shown to be dependent on the *oleABCD* gene cluster. The product 3,6,9,12,15,19,22,25,28-hentriacontanonaene ($C_{31:9\Delta}$) was confirmed by NMR, and reduction to the established *n*-hentriacosane while, in the absence of OleA a ketone was produced. These results are consistent with a mechanism involving Claisen-type condensation between alkanoate-CoA esters followed by (a) loss of HSCoA and CO_2 to the nonaene, or (b) formation of the β-oxoacid followed by loss of CO_2 to the monoketone (Sukovich et al. 2010a). The $C_{31}H_{46}$ alkene was produced by a range of organisms including several strains of *Shewanella* sp., *Geobacter bemidjiensis*, *Colwellia psychrerythraea*, *Planctomyces maris*, $C_{31}H_{58}$ by *Chloroflexus aurantiacus*, $C_{29}H_{58}$ by *Brevibacterium fuscum*, and a range of $C_{28}H_{56}$ to $C_{31}H_{58}$ by *Xanthomonas campestris* (Sukovich et al. 2010b).

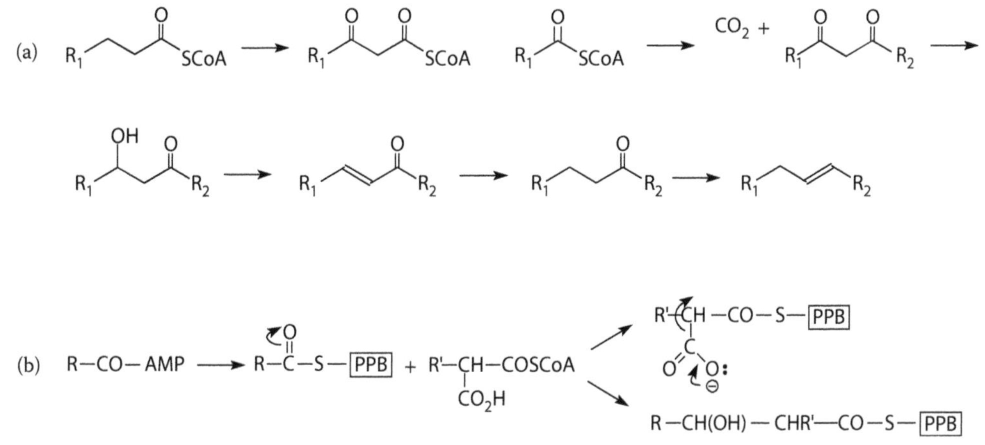

FIGURE 7.55 Biosynthesis of long-chain alkenes. (a) Condensation of acyl-CoA ester with elimination of CO_2 in *Micrococcus luteus*, (b) reaction of *n*-fatty acid ACP thioester with methylmalonyl-CoA in *Mycobacterium tuberculosis*.

6. Strains of *Jeotgalicoccus* produce 18-methylnonadec-1-en (i-C_{21}) and 17-methylnonadec-1-ene (a-C_{20}) as terminal alkenes, and feeding a culture of strain ATCC 8456 with eicosanoate (*n*-C_{20}) increased 20-fold the concentration of nonadec-1-ene (Rude et al. 2011). An enzyme OleT$_{JE}$ with a mass of 50 kDa was assigned to a cytochrome P450 enzyme designated cyp152, and this activity was analyzed in several organisms that were able to produce pentadec-1-ene and β-hydroxypalmitate from palmitate (*n*-C_{16}). A mechanism was proposed for the occurrence of the two reactions mediated by cytochrome P450 OleT$_{JE}$: (i) formation of a radical at C_3 followed by oxidation to the C_3 cation and decarboxylation to the olefin and (ii) hydroxylation at C_3 to produce the 3-hydroxycarboxylate. The authors draw an analogy between OleT$_{JE}$ and P450$_{Rm}$ from *Rhodotorula minuta* that carried out the formation of isobutene from isovalerate, although the range of enzymatic activities is different.

7. a. There are several groups of lipids on the cell surface of *M. tuberculosis* and these complex structures have been extensively investigated since they are important in pathogenesis, and suggest possible routes to the synthesis of inhibitors for this increasingly widespread pathogen. Only a few aspects are briefly noted here: the biosynthesis of the branched-chain mycocerosic acid, phthiocerol, and sulfolipid-1, and the condensation reaction that produces mycolic acids. There are two main groups of methyl-branched fatty acids in *M. tuberculosis*: C_{28} to C_{32} and C_{22} to C_{26}. It is worth noting that the reactions which are involved in their biosynthesis by the reaction of malonyl-*S*-CoA and methylmalonyl-*S*-CoA with an acyl-ACP and concomitant decarboxylation differ from conventional Claisen condensation since the electrons are provided by decarboxylation—and not from an anion by proton loss.

    i. Phthiocerol and phthiodiolone dimycocerosates (DIMs) are complex esters of the branched-chain mycocerosic acids whose biosynthesis is carried out by consecutive chain elongation of *n*-fatty ACP thioester with methylmalonyl-CoA. This has been verified in cell extracts of *M. tuberculosis* var. *bovis* (Rainwater and Kolattukudy 1985) using a C_{12} carboxylate as primer, and the gene has been cloned and sequenced (Mathur and Kolattukudy 1992). The biosynthesis of phthiocerol is initiated by ATP-mediated reaction of a C_{16} to C_{18} *n*-fatty to produce an acyl carrier protein C_{16}-thioester, and the complex reaction is carried out in the synthase modules PpsA to PpsE (Trivedi et al. 2005; Siméone et al. 2010). In PpsA and PpsaB, successive reactions with malonyl-CoA with concomitant decarboxylation produce the 16,18-dihydroxy thioester ACP; further chain elongation then takes place in PpsC by malonyl-CoA and in PpsD with methylmalonyl-CoA to produce the 23-methyl-16,18-dihydroxy-thioester ACP.

This is followed by chain elongation in PpsE with methylmalonyl-CoA to phthiocerol thioester ACP that undergoes esterification with mycocerosic acid to produce the phthiocerol dimycocerosates (DIMs). This is illustrated in Figure 7.55b for a C_{16} carboxylate, where the chain numbering is from the distal carbon atom.

ii. The sulfated glycolipid SL-1 is a trehalose triester of hydroxy hepta- and octa-methyl-phthioceranic acids whose biosynthesis from C_{14} or C_{16}-n-carboxylates takes place by successive reactions with methylmalonate-CoA (Converse et al. 2003; Sirakova et al. 2001) which are analogous to the biosynthesis of mycocerosic acid.

b. The biosynthesis of mycolic acids in *M. tuberculosis* has been discussed in a review (Takayama et al. 2005), and has achieved prominence since some antituberculosis drugs function by inhibiting the synthesis of mycolic acids that are major components of the protective cell layer—in addition to fulfilling structural roles. Mycolic acids are long-chain α-alkyl β-hydroxylated fatty acids: R–CH(OH)–CH(R′)–CO$_2$H where in *M. tuberculosis* R represents the meromycolate chain which contains up to 56 carbon atoms and R′ is a shorter branch with 22–26 carbon atoms. *C. glutamicum* synthesizes structurally simpler C_{32} mycolic acids produced by head-to-head condensation of C_{16} fatty acids and provided a suitable model for the more complex analogues in *M. tuberculosis*. A range of *C. glutamicum* mutants displayed important features: growth of the *accD1* mutant was impaired and this could be recovered by addition of oleate, while *accD2* and *accd3* mutants that encode carboxyltransferases were unable to synthesize mycolic acids, although levels of fatty acids were unaltered (Gande et al. 2004). The crucial step was recognition of the polyketide synthase complex Pks13, since a mutant Δpks13:km did not produce mycolic acids although high levels of C_{16} to C_{18} fatty acids were produced. The final step in the biosynthesis of mycolic acids was the condensation carried out by Pks13 that contained four catalytic domains: PPB (N-terminal phosphopante-thiene binding, KS (ketoacyl synthase), AT (acyl transferase), PPB, and TE (thioesterase) of which the PPBs are involved in the condensation of the two fatty aids to mycolic acids (Portevin et al. 2004). For the mycolic acids in *M. tuberculosis*, the sequence of three reactions is as follows (Takayama et al. 2005):

i. A-meroacyl–S–ACP + ATP → α-meroacyl–AMP
ii. C_{26}–S-CoA + CO$_2$ → 2-carboxyl–C_{26}–S–CoA
iii. Claisen-type condensation

7.7.3.2.5.8 Hydroxyalkanoates

1. The synthesis of poly(3-hydroxyalkanoates) that serve as reserve material under conditions of nitrogen limitation has already been discussed, and the role noted of mevalonate diphosphate decarboxylase that play a key role in the biosynthesis of isopentenyl diphosphate as the precursor of squalene but in the synthesis of isobutene.

2. Hydroxyalkanoates are intermediates in reactions that are involved in the metabolism of a range of substrates: these include citrate, isocitrate, malate in the tricarboxylic acid cycle, and the degradation of propionate via acrylate: these need no discussion here. As an alternative to rearrangements that are outlined above, the degradation of amino acids may take place by transamination with 2-oxoglutarate followed by NADH-dependent reduction to the 2-hydroxy-acids followed by dehydration of their coenzyme A esters with hydrogen at a nonactivated position. These are free-radical reactions (Smith et al. 2003) (Figure 7.56) including the following that are discussed later:

a. 2-Hydroxyglutaryl-CoA in glutamate fermentation by *Acidaminococcus fermentans* (Hans et al. 2002)

b. Phenyllactyl-CoA in the fermentation of phenylalanine by *Clostridium sporogenes* (Dickert et al. 2002): phenylalanine undergoes dismutation—oxidation to phenylacetate, and reduction to phenylpyuvate followed by reduction to phenyllactate

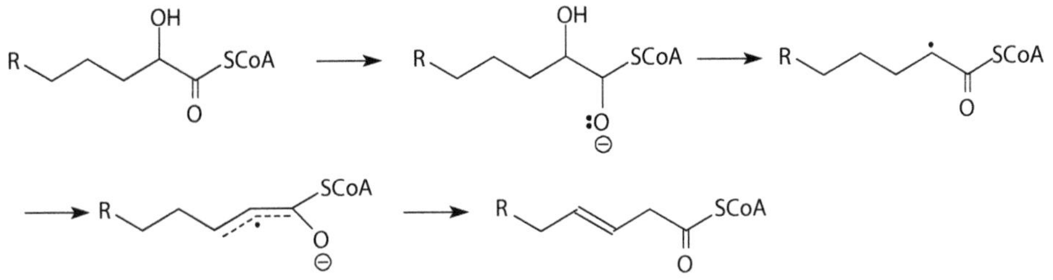

FIGURE 7.56 Anaerobic dehydration of 2-hydroxycarboxyl-CoA.

   c. 4-Hydroxybutyryl-CoA in the fermentation of 4-aminobutyrate by *C. aminobutyricum* (Müh et al. 1996), and the fermentation of succinate by *Clostridium kluyveri* (Söhling and Gottschalk 1996).

3. There are several mechanisms adopted by autotrophic organisms for the assimilation of CO_2.

   a. The 3-hydroxypropionate pathway has been elucidated for the green nonsulfur anaerobic bacterium *Chloroflexus aurantiacus* in which the Benson–Calvin cycle does not operate. In this organism CO_2 assimilation is accomplished by functioning of the 3-hydroxypropionate cycle (Herter et al. 2002). In brief, this operates by carboxylation of acetyl-CoA to malonyl-CoA, followed by sequential reduction by a multifunctional enzyme to 3-hydroxypropionate and propionate whose CoA ester under goes carboxylation to methylmalonyl-CoA and fission to glyoxylate. The metabolism of glyoxylate involves the ethylmalonate cycle (Figure 7.51) (Erb et al. 2007) in place of the classical glyoxylate cycle. Both pathways utilize the novel bifunctional enzyme mesaconyl-CoA hydratase (Zarzycki et al. 2008).

   b. In *Metallosphaera sedula*, a variant of the 3-hydroxypropionate cycle is used. In this, the reduction of 3-hydroxypropionyl-CoA to propionyl-CoA requires the activity of two enzymes in place of the single one in *Chloroflexus aurantiacus*. This involves the activity of a dehydrase to form acryloyl-CoA followed by reduction (Teufel et al. 2009).

   c. A further variant is the dicarboxylate/4-hydroxybutyrate cycle used by *Thermoproteus neutrophilus* (Ramos-Vera et al. 2009). This involves reductive carboxylation of acetyl-CoA to pyruvate, carboxylation of phosphoenol pyruvate to oxalacetate followed by reduction to malate, dehydration to fumarate, and reduction to succinate.

7.7.4 Amides and Related Compounds

7.7.4.1 Amides

Aromatic amides, carbamates, and ureas are components of a number of important agrochemicals, and the first step in their biodegradation is mediated by the activities of amidases, ureases, and carbamylases with the production of amines (Figure 7.57a,b,c). The chloroanilines that are formed from many of them as initial products may, however, be substantially more resistant to further degradation. Application of assays for amidase activity, particularly for pyrazinamidase, has been widely used in the classification of mycobacteria (Wayne et al. 1991): this enzyme is necessary for activation of pyrazinamide in tuberculosis therapy and is discussed in Chapter 4. Sequential hydrolysis of nitriles to amides and carboxylic acids is well established both in aliphatic (Miller and Gray 1982; Nawaz et al. 1992), and aromatic compounds (Harper 1977; McBride et al. 1986). Degradation of the herbicide bromoxynil may, however, take place by the elimination of cyanide from the ring with the initial formation of 2,6-dibromohydroquinone (Topp et al. 1992) (Figure 7.58). There may be a high degree of specificity in the action of these nitrilases and this may have considerable significance in biotechnology. Some examples are given as illustrations:

   1. Racemic 2-(4′-isobutylphenyl)propionitrile is converted by a strain of *Acinetobacter* sp. to *S*-(+)-2-(4′-isobutylphenyl)propionic acid with an optical purity >95% (Yamamoto et al. 1990).

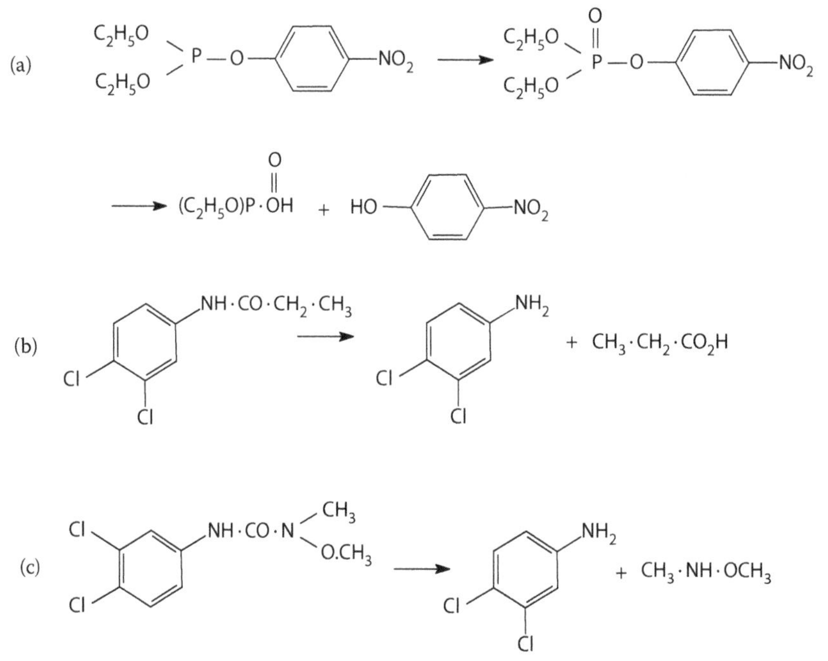

FIGURE 7.57 Hydrolysis of (a) phenylureas, (b) carbamates, and (c) carbamates.

2. The nitrilase from a number of strains of *Pseudomonas* sp. mediated an enantiomerically selective hydrolysis of racemic *O*-acetylmandelonitrile to D-acetylmandelic acid *R* (−)-acetylmandelic acid (Layh et al. 1992).

3. The amidase from *Rhodococcus erythopolis* strain MP50 was used to selectively convert racemic 2-phenylpropionamide to *S*-2-phenylpropiohydroxamate. This was converted into the isocyanate by Lossen rearrangement and then by hydrolysis to *S*(−)-phenylethylamine (Hirrlinger and Stolz 1997).

4. An amidase from *Ochrobactrum anthropi* strain NCIMB 40321 has a wide substrate versatility for L-amides, primarily those with an α-amino group (Sonke et al. 2005), while the condensation product of urea and formaldehyde $H_2N-[CONH.CH_2NH]_n-CONH_2$ is hydrolyzed by another strain of *O. anthropi* (Jahns et al. 1997).

5. A thermostable amidase has been characterized in *Pseudonocardia thermophila* and has a substrate range including aliphatic, aromatic, and amino acid amides (Egorova et al. 2004), and is highly *S*-stereospecific for 2-phenylpropionamide.

6. The cyclic amide β-lactam is a structural component of a number of antibiotics that are widely used in clinical practice. They include derivatives of penicillin, cephalosporin, and carbapenem that are employed against both Gram-negative and Gram-positive bacteria. The activities of β-lactamases that bring about the hydrolysis of β-lactams are highly undesirable, since they

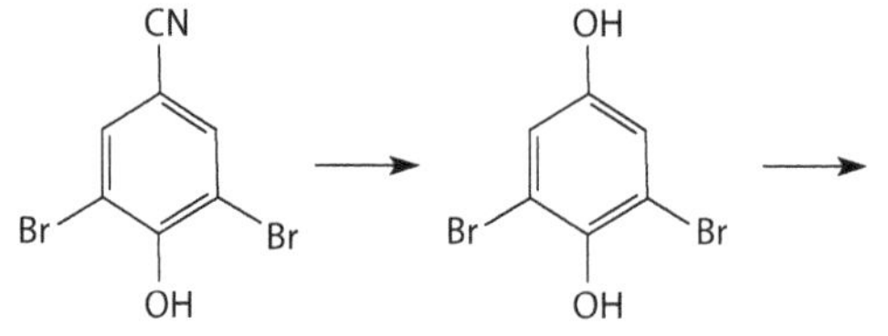

FIGURE 7.58 Metabolism of bromoxynil.

mediate resistance towards a number of important antibiotics that compromise their application. These activities are widely distributed, and have been exacerbated by immoderate application of β-lactam antibiotics and genetic transfer. The β-lactamases belong to a large family of hydrolases, several schemes have been proposed for their classification (Bush et al. 1995), and they may be chromosomally encoded, for example, in *Bacillus licheniformis*, or plasmid-borne, for example, in *Staphylococcus aureus*. The hydrolases employ either a Serine[70] (Groups A, C, D), or one or two zinc atoms (Group B1, B2, B3) at the active site to bring about hydrolysis. There has been particular concern over the emergence of carbapenem-hydrolyzing lactamases that belong both to the serine[70] group in *Enterobacter cloacae* and *Serratia marcescens* (Group 2f) and the metallo-β-lactamases in *Aeromonas hydrophila* (Group 3) (Rasmussen and Bush 1997). Carbapenem-resistant strains of *Ps. aeruginosa* that produce metallo-β-lactamase present a serious problem for patients especially those with impaired defense systems (Senda et al. 1996). Considerable attention has, therefore, been devoted to the application of β-lactamase inhibitors: these aspects are discussed briefly in Chapter 3, Section 3.12 dealing with antibiotic resistance.

7.7.4.2 Nitriles

There are two pathways for the degradation of nitriles. In one, they are hydrolyzed to carboxylic acids directly by the activity of a nitrilase, for example, in *Bacillus* sp. strain OxB-1 and *P. syringae* B728a. In the other, they undergo hydration to amides followed by hydrolysis, for example, in *P. chlororaphis* (Oinuma et al. 2003). Interest has developed in the enantioselective conversion of arylacetonitriles to the carboxylic acids. The difference in selectivity of different bacteria is exemplified by the hydrolysis of (*R,S*)-mandelonitrile to (*R*)-mandelate by *A. faecalis* ATCC 8750, though only a low degree of enantioselectivity was expressed by *Ps. fluorescens* EBC 191 (Kiziak and Stolz 2009). The monomer acrylonitrile occurs in wastewater from the production of polyacrylonitrile (PAN) and is hydrolyzed by bacteria to acrylate by the combined activity of a nitrilase (hydratase) and an amidase. Acrylate is then degraded by hydration to either lactate or β-hydroxypropionate. The nitrilase or amidase is also capable of hydrolyzing the nitrile group in a number of other nitriles (Robertson et al. 2004) including polyacrylonitrile (Tauber et al. 2000).

7.7.4.3 Isonitriles

Isonitriles are widespread metabolites in marine organisms, although little attention has been paid to their degradation. *Ps. putida* strain N-19-2 hydrolyzes nitriles to *N*-substituted formamides (Goda et al. 2001), and the enzyme is active towards cyclohexyl isocyanide and benzyl isocyanide.

7.7.4.4 Sulfonylureas and Thiocarbamates

1. Sulfonylureas are the basis of a large group of herbicides. Cytochrome P450 enzymes in *Streptomyces griseolus* transform the sulfonylureas by hydroxylation (Omer et al. 1990) leaving the −SO$_2$NHCONH—part of the structure unaltered (Harder et al. 1991).
2. Substantial effort has been given to the degradation of thiocarbamates, and the expression of the *thcB* gene encoding cytochrome P450 in strains of *Rhodococcus* is expressed constitutively (Shao and Behki 1996). The degradation of *S*-ethyl *N,N′*-dipropylthiocarbamate (EPTC) has been examined in a number of rhodococci, and in one of them (strain TE1 given as a species of *Arthrobacter* sp.) degradation is associated with a 50.5 megadalton plasmid (Tam et al. 1987). The principal pathway in *Rhodococcus* sp. strain JE1 involves hydroxylation of the propyl group followed by loss of propionaldehyde and degradation of *N*-depropyl EPTC to *S*-ethylformate and propylamine (Figure 7.59) (Dick et al. 1990). Two of the genes induced by EPTC encode a cytochrome P450 system that carries out the initial hydroxylation with formation of the *N*-depropylated product and an aldehyde dehydrogenase that converted the aldehyde produced to the

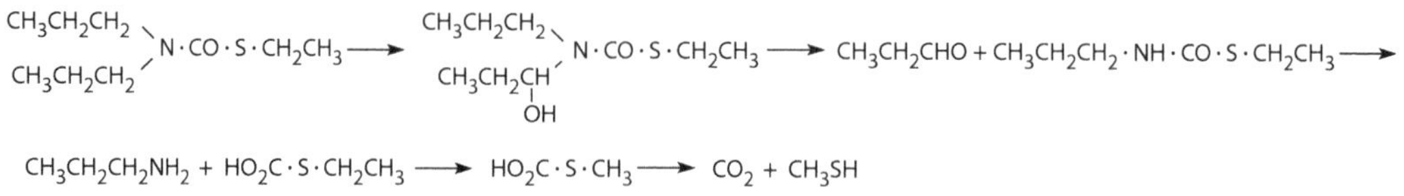

FIGURE 7.59 Aerobic degradation of *S*-ethyl *N,N*′-dipropylthiocarbamate.

corresponding carboxylic acid (Nagy et al. 1995). In addition to these mechanisms, the degradation of thiocarbamates may be carried out in *Rh. erythropolis* NI86/21by a herbicide-inducible nonheme haloperoxidase (De Schrijver et al. 1997).

Although it is moot whether cyanide, and thiocyanate should be included as organic compounds, brief comments on their biodegradation are summarized. The metabolism of the other C_1 substrate carbon monoxide has already been discussed.

7.7.4.5 Cyanide

Cyanide is both anthropogenic and biogenic. It is widely distributed as a component of plant cyanogenic glycosides including amygdalin and linamarin, and is produced by hydrolysis of glycosinolinates that are found in species of *Brassica*. In addition, a range of structurally diverse diterpenoid isonitriles is produced in marine biota including sponges, and in cyanobacterial isonitriles the isonitrile group originates from glycine. Hydrogenases have cyanide as ligands coordinated to Fe, one to each Fe in the Fe-only [Fe–Fe] hydrogenase, and two to Fe in [Ni–Fe] hydrogenases: this is discussed in the section on metals in Chapter 3. Biosynthesis of cyanide in maturation of these enzymes is accomplished from an enzyme-bound cysteine by an ATP-dependent formation of isothiocarbamyl phosphate, followed by successive elimination of phosphate and the enzyme-bound cysteine (Reissmann et al. 2003). Cyanide is a toxic waste from the mining and electroplating industries, and although normally toxic to aerobic organisms, microorganisms have been isolated that are able to degrade it. Cyanide may also be produced by *Ps. fluorescens* and *Ps. aeroginosa*, where it has been shown that the cyanide carbon originates from the C_2 of glycine (Askeland and Morrison 1983).

Two mechanisms have emerged for the degradation of cyanide (Fernandez et al. 2004) and the corresponding nitriles: (a) hydrolysis by nitrilase to the carboxylic acid and ammonia and (b) hydratases that catalyze the formation of the amides. The mechanism of the first reaction in *Ps. fluorescens* NCIMB 11764 is complex, and involves both monooxygenation and hydrolysis. Oxygenation is carried out by reduced pterins and is analogous to the hydroxylation of phenylalanine to tyrosine. The overall reaction involves monooxygenation to cyanate followed by reduction to formamide, and hydrolysis to formate and NH_4^+. An alternative degradation for cyanate has, however, been shown in *E. coli* in which an inducible cyanase brings about a recycling reaction between bicarbonate and cyanate to form formamide that is hydrolyzed to CO_2 and NH_4^+ (Johnson and Anderson 1987). A strain of *Pseudomonas pseudoalcaligenes* isolated by enrichment with cyanide as sole source of nitrogen was able to utilize in addition ferrocyanide, nitroferricyanide, and cuprocyanide, and may involve cyanate (Luque-Almagro et al. 2005).

Noncyanogenic fungi can degrade cyanide to formamide followed by hydrolysis by a hydratase to formate and ammonia (Dumestre et al. 1997). This pathway is used also by some bacteria (Jandhyala et al. 2003) Bacteria also use a number of reactions for the detoxification of cyanide, including monooxygenation to CO_2 and ammonia (Wang et al. 1996).

7.7.4.6 Cyanate

Although the metabolism of cyanate has been examined in a range of bacteria including cyanobacteria, its physiological role has not been finally established although it can serve in recycling CO_2 and bicarbonate. Cyanase mediates the reaction: $CNO- + H^+ + 2H_2O \rightarrow NH_4^+ + HCO_3^-$ and the catalytic activity of the enzyme is unusual in its requirement for bicarbonate where it serves for recycling rather than hydrolysis (Johnson and Anderson 1987). Enzyme activity has been found in *Ps. fluorescens* NCIB 11764 (Kunz and Nagappan 1989), and the gene *cynS* has been sequenced in *E. coli* (Sung et al 1987). A survey of *cynS* sequences revealed its widespread existence in marine cyanobacteria where it may play a role in detoxification or nitrogen metabolism (Kamennaya and Post 2011).

7.7.4.7 Thiocyanate

The degradation of thiocyanate has been examined in the chemolithoautrophic *Paracoccus thiocyanatus* that is able to use this as an energy source (Katayama et al. 1995), and in a range of extreme halophiles that illustrate the range of mechanisms (Sorokin et al. 2001) within the wider context of the microbial sulfur cycle in halophiles (Sorokin et al. 2011). The metabolism of thiocyanate may proceed by alternative pathways:

$$^-S{-}C{\equiv}N + 2H_2O \rightarrow S{=}C{=}O + NH_3 + HO^- - \text{carbonyl sulfide pathway}$$

$$^-S{-}C{\equiv}N + H_2O \rightarrow {}^-O{-}C{\equiv}N + H_2S - \text{cyanate pathway}$$

whereas chemolithotrophic metabolism of thiocyanate by *Thiohalobacter thiocyanaticus* used the cyanate pathway and could utilize thiosulfate and oxidized sulfide, elemental sulfur, and tetrathionate though not trithionate (Sorokin et al. 2010), *Thiohalophilus thiocyanatoxidans* used the carbonyl sulfide pathway (Sorokin et al. 2007).

The enzyme that catalyzes the carbonyl sulfide pathway has been purified from the obligate chemolithotroph *Thiobacillus thioparus* strain THI 115 from cells that used thiocyanate as a source of energy. It had a molecular mass of 126 kDa with subunits α (19 kDa), β (23 kDa), and γ (32 kDa) (Katayama et al. 1992), while the partial similarity to the subunits of nitrile hydrolase in *Pseudomonas chlororaphis* suggested an analogous mechanism (Katayama et al. 1998). The analogous enzyme from *Thiohalophilus thiocyanatoxidans* that is also an obligate chemolithoautroph which uses the carbonyl sulfide pathway had a molecular mass of 140 kDa with subunits of mass 17, 19, and 29 kDa and contained both Co and Fe (Bezsudnova et al. 2007).

7.8 References

Abe T, T Kobayashi, T Saito. 2005. Properties of a novel intracellular poly(3-hydroxybutyrate) depolymerase with high specific activity (PhaZd) in *Wautersia eutropha* H16. *J Bacteriol* 187: 6982–6990.

Aeckersberg F, FA Rainey, F Widdel. 1998. Growth, natural relationships, cellular fatty acids and metabolic adaptation of sulfate-reducing bacteria that utilize long-chain alkanes under anoxic conditions. *Arch Microbiol* 170: 361–369.

Albro PW, JC Dittmer. 1969a. The biochemistry of long-chain, nonisoprenoid hydrocarbons. I. Characterization of the hydrocarbons of *Sarcina lutea* and the isolation of possible intermediates. *Biochemistry* 8: 394–404.

Albro PW, JC Dittmer. 1969b. The biochemistry of long-chain, nonisoprenoid hydrocarbons. II. The incorporation of acetate and the aliphatic chains of isoleucine and valine into fatty acids and hydrocarbons by *Sarcina lutea. Biochemistry* 8: 853–859.

Albro PW, JC Dittmer. 1969c. The biochemistry of long-chain, nonisoprenoid hydrocarbons. IV. Characteristics of synthesis by a cell-free preparation of *Sarcina lutea. Biochemistry* 8: 3317–3324.

Amador-Noguez D, X-J Feng, J Fan, N Roquet, H Rabitz, JR Rabinowitz. 2010. Systems-level metabolic flux profiling elucidates a complete bifurcated tricarboxylic acid cycle in *Clostridium acetobutylicum. J Bacteriol* 192: 4452–4461.

Anantharam V, MJ Allison, PC Maloney. 1989. Oxalate:formate exchange. *J Biol Chem* 264: 7244–7250.

Andreeßen B, A Steinbüchel. 2010. Biosynthesis and biodegradation of 3-hydroxypropionate-containing polyesters. *Appl Environ Microbiol* 76: 4919–4925.

Anthony C. 1982. *The Biochemistry of Methylotrophs.* Academic Press, London.

Askeland RA, SM Morrison. 1983. Cyanide production by *Pseudomonas fluorescens* and *Pseudomonas aeruginosa. Appl Environ Microbiol* 45: 1802–1807.

Bach B, E Meudee, J-P Lepoutre, T Rossignol, B Blondin, S Dequin, C Camarassa. 2009. New insights into γ-aminobutyric acid catabolism: Evidence for γ-hydroxybutyric acid and polyhydroxybutyrate synthesis in *Saccharomyces cerevisiae. Appl Environ Microbiol* 75: 4231–4239.

Baetz AL, MJ Allison. 1989. Purification and characterization of oxalyl-coenzyme A decarboxylase from *Oxalobacter formigenes. J Bacteriol* 171: 2605–2608.

Baetz AL, MJ Allison. 1990. Purification and characterization of formyl-coenzyme A transferase from *Oxalobacter formigenes. J Bacteriol* 172: 3537–3540.

Beller HR, E-B Goh, JD Keasling. 2010. Genes involved in long-chain alkene biosynthesis in *Micrococcus luteus. Appl Environ Microbiol* 76: 1212–1223.

Balk M, J Weijma, AJM Stams. 2002. *Thermotoga lettingae* sp. nov., a novel thermophilic, methanol-degrading bacterium solated from a thermophilic anaerobic reactor. *Int J Syst Evol Microbiol* 52: 1361–1368.

Berthold CL, P Moussatche, NGJ Richards, Y Lindqvist. 2005. Structural basis for activation of the thiamin diphosphate-dependent enzyme oxalyl-CoA decarboxylase by adenosine diphosphate. *J Biol Chem* 280: 41645–41654.

Bevers LE, MWH Pinkse, PDEM Verhaert, WR Hagen. 2009. Oleate hydratase catalyzes the hydration of a non activated carbon-carbon bond. *J Bacteriol* 191: 5010–5012.

Bezsudnova EY, DY Sorokin, TV Tikhova, VO Popov. 2007. Thiocyanate hydrolase, the primary enzyme initiating degradation in the novel obligately chemolithoautotrophic bacterium *Thiohalophilus thiocyanoxidans. Biochim Biophys Acta* 1774:1563–1570.

Blackmore MA, JR Quayle. 1970. Microbial growth on oxalate by a route not involving glyoxylate carboligase. *Biochem J* 118: 53–59.

Bonam D, PW Ludden. 1987. Purification and characterization of carbon monoxide dehydrogenase, a nickel, zinc, iron-sulfur protein, from *Rhodospirillum rubrum. J Biol Chem* 262: 2980–2987.

Bott M, K Pfister, P Burda, O Kalbernmatter, G Woehlke, P Dimroth. 1997. Methylmalonyl-CoA decarboxylase from *Propionigenium modestum.* Cloning and sequencing of the structural genes and purification of the enzyme complex. *Eur J Biochem* 250: 590–599.

Boyd JM, H Ellsworth, SA Ensign. 2004. Bacterial acetone carboxylase is a manganese-dependent metalloenzyme. *J Biol Chem* 279: 46644–46651.

Braaz R, R Handrick, D Jendrossek. 2003. Identification and characterization of the catalytic triad of the alkaliphilic thermotolerant PHA depolymerase PhaZ7 of *Paucimonas lemoignei. FEMS Microbiol Lett* 224: 107–112.

Brämer CO, A Steinbüchel. 2001. The methylcitrate acid pathway in *Ralstonia eutropha*: New genes identified involved in propionate metabolism. *Microbiology (UK)* 147: 2203–2214.

Brandis-Heep A, NA Gebhardt, RK Thauer, F Widdel, N Pfennig. 1983. Anaerobic acetate oxidation to CO_2 by *Desulfobacter postgatei.* 1. Demonstration of all the enzymes required for the operation of the citric acid cycle. *Arch Microbiol* 136: 222–229.

Braun M, H Stolp. 1985. Degradation of methanol by a sulfate reducing bacterium. *Arch Microbiol* 142: 77–80.

Broadway NM, FM Dickinson, C Ratledge. 1993. The enzymology of dicarboxylic acid formation by *Corynebacterium* sp. strain 7E1C grown on *n*-alkanes. *J Gen Microbiol* 139: 1337–1344.

Buckel W, BM Martins, A Messerschmidt, BT Golding. 2005. Radical-mediated dehydration reactions in anaerobic bacteria. *Biol Chem* 386: 951–959.

Budde CF, AE Mahan, J Lu, X Rha, AJ Sinskey. 2010. Roles of multiple acetoacetyl coenzyme A reductases in polyhydroxybutyrate biosynthesis in *Ralstonia eutropha* H16. *J Bacteriol* 192: 5319–5328.

Bush K, GA Jacob, AA Medicos. 1995. A functional classification for β-lactamases and its correlation with molecular structure. *Antimicrob Agents Chemother* 39: 1211–1223.

Chen H-J, S-C Pan, G-C Shaw. 2009. Identification and characterization of a novel intracellular poly(3-hydroxybutyrate) depolymerase from Bacillus megaterium. *Appl Environ Microbiol* 85: 5290–5299.

Claasen PAM, AJB Zehnder. 1986. Isocitrate lyase activity in *Thiobacillus versutus* grown anaerobically on acetate and nitrate. *J Gen Microbiol* 132: 3179–3185.

Clugston SL, JFJ Barnard, R Kinach, D Miedema, R Ruman, E Daub, JF Honek. 1998. Overproduction and characterization of a dimeric non-zinc glyoxylase I from *Escherichia coli*: Evidence for optimal activation by nickel ions. *Biochemistry* 37: 8754–8763.

Coates JD, DJ Lonergan, EJP Philipps, H Jenter, DR Lovley. 1995. *Desulfuromonas palmitatis* sp. nov., a marine dissimilatory Fe(III) reducer that can oxidize long-chain fatty acids. *Arch Microbiol* 164: 406–413.

Combourieu B, P Besse, M Sancelme, H Veschambre, AM Delort, P Poupin, N Truffaut. 1998. Morpholine degradation pathway of *Mycobacterium aurum* MO1: direct evidence of intermediates by *in situ* 1H nuclear magnetic resonance. *Appl Environ Microbiol* 64: 153–158.

Converse SE, JD Mougous, MD Leavell, JA Leary, CR Bertozzi, JS Cox. 2003. MmpL8 is required for sulfolipid-1 biosynthesis and *Mycbacterium tuberculosis* virulence. *Proc Natl Acad Sci USA* 100: 6121–6126.

Cornick NA, MJ Allison. 1996. Anabolic incorporation of oxalate by *Oxalobacter formigenes*. *Appl Environ Microbiol* 62: 3011–3013.

Cusa E, N Obradors, L Baldomá, J Badía, J Aguilar. 1999. Genetic analysis of a chromosomal region containing genes required for assimilation of allantoin nitrogen and linked glyoxylate metabolism in *Escherichia coli*. *J Bacteriol* 181: 7479–7484.

Daniel SL, HL Drake. 1993. Oxalate- and glyoxylate-dependent growth and acetogenesis by *Clostridium thermoaceticum*. *Appl Environ Microbiol* 59: 3062–3069.

de Bok FAM, AJM Stams, C Dijkema, DR Boone. 2001. Pathway of propionate oxidation by a syntrophic culture of *Smithella propionica* and *Methanospirillum hungatei*. *Appl Environ Microbiol* 67: 1800–1804.

de Bok FAM, MLGC Luijten, AJM Stams. 2002. Biochemical evidence for formate transfer in syntrophic propionate-oxidizing cocultures of *Syntrophobacter fumaroxidans* and *Methanospirillum hungatei*. *Appl Environ Microbiol* 68: 4247–4252.

De Eugenio LI, P García, JM Luengo, JM Sanz, J San Román, JL Gardía, MA Prieto. 2007. Biochemical evidence that *phaZ* gene encodes a specific intracellular medium chain length polyhydroxyalkanoate depolymerase in *Pseudomonas putida* KT2442. Characterization of a paradigmatic enzyme. *J Biol Chem* 282: 4951–4962.

de Schrijver A, I Nagy, G Schoofs, P Proost, J Vanderleyden, K-H van Pée, R de Mot. 1997. Thiocarbamate herbicide-inducible nonheme haloperoxidase of *Rhodococcus erythropolis* NI86/21. *Appl Environ Microbiol* 63: 1811–1916.

Dehning I, B Schink. 1989a. Two new species of anaerobic oxalate-fermenting bacteria, *Oxalobacter vibrioformis* sp. nov. and *Clostridium oxalicum* sp. nov. from sediment samples. *Arch Microbiol* 153: 79–84.

Dehning I, B Schink. 1989b. *Malonomonas rubra* gen. nov., sp. nov., a microaerotolerant anaerobic bacterium growing by decarboxylation of malonate. *Arch Microbiol* 151: 427–433.

Dehning I, B Schink. 1994. Anaerobic degradation of malonate via malonyl-CoA by *Sporomusa malonica*, *Klebsiella oxytoca*, and *Rhodobacter capsulatus*. *Antonie van Leeuwenhoek* 66:343–350.

Dehning I, M Stieb, B Schink. 1989. *Sporomusa malonica* sp. nov., a homoacetogenic bacterium growing by decarboxylation of malonate or succinate. *Arch Microbiol* 151: 421–426.

Delafield FP, M Doudoroff, NJ Palleroni, R Contopoulos. 1965. Decomposition of polyhydroxybutyrate by Pseudomonads. *J Bacteriol* 90: 1455–1466.

Denger K, B Schink. 1990. New motile anaerobic bacteria growing by succinate decarboxylation to propionate. *Arch Microbiol* 154: 550–555.

Dennis MW, PE Kolattukudy. 1991. Alkane biosynthesis by decarbonylation of aldehyde catalyzed by a microsomal preparation from *Botryococcus braunii*. *Arch Biochem Biophys* 287: 268–275.

Dhar A, K Dhar, JNP Rosazza. 2002. Purification and characterization of a *Galactomyces reesii* hydratase that converts 3-methylcrotonic acid to 3-hydroxy-3-methylbutyric acid. *J Ind Microbiol Biotechnol* 28: 81–87.

Dick WA, RO Ankumah, G McClung, N Abou-Assaf. 1990. Enhanced degradation of *S*-ethyl *N,N'*-dipropylcarbamothioate in soil and by an isolated soil microorganism pp 98–112. In *Enhanced biodegradation of pesticides in the environment* (Eds KD Racke and JR Coats). American Chemical Society Symposium Series 426, American Chemical Society, Washington DC.

Dullius CH, C.Y Chen, B Schink. 2011. Nitrate-dependent degradation of acetone by *Alicycliphilus* and *Paracoccus* strains and comparison of acetone carboxylase enzymes. *Appl Environ Microbiol* 77: 6821–6825.

Dimroth P, H Hilbi. 1997. Enzymatic and genetic basis for bacterial growth on malonate. *Molec Microbiol* 25: 3–10.

Dolfing J, B Jiang, AM Henstra, AJM Stams, CM Plugge. 2008. Syntrophic growth on formate: A new microbial niche in anoxic environments. *Appl Environ Microbiol* 74: 6126–6131.

Dorn M, JR Andreesen, G Gottschalk. 1978. Fermentation of fumarate and L-malate by *Clostridium formicoaceticum*. *J Bacteriol* 133: 26–32.

Drake HL, SL Daniel. 2004. Physiology of the thermophilic acetogen *Moorella thermoacetica*. *Res Microbiol* 155: 869–883.

Drake HL, SI Hu, HG Wood. 1980. Purification of carbon monoxide dehydrogenase, a nickel enzyme from *Clostridium thermocaceticum*. *J Biol Chem* 255: 7174–7180.

Dumestre A, T Chone, J-M Portal, M Gerard, J Berthelin. 1997. Cyanide degradation under alkaline conditions by a strain of *Fusarium solani* isolated from contaminated soils. *Appl Environ Microbiol* 63: 2729–2734.

Egorova K, H Trautwein, S Verseck, G Antranikian. 2004. Purification and properties of an enantioselective and thermoactive amidase from the thermophilic actinomycete *Pseudonocardia thermophila*. *Appl Microbiol Biotechnol*. 65: 38–45.

Eichler B, B Schink. 1984. Oxidation of primary aliphatic alcohols by *Acetobacterium carbinolicum* sp. nov., a homoacetogenic anaerobe. *Arch Microbiol* 140: 147–152.

El-Sharkawy A, W Yang, L Dostal, JPN Rosazza. 1992. Microbial oxidation of oleic acid. *Appl Environ Microbiol* 58: 2116–2122.

Erb TJ, IA Berg, V Brecht, M Müller, G Fuchs, BE Alber. 2007. Synthesis of C_5-dicarboxylic acids from C_2-units involving crotonyl-CoA carboxylase/reductase: The ethylmalonate pathway. *Proc Natl Acad USA* 104: 10631–10636.

Erb TJ, J Rétey, G Fuchs, BE Alber. 2008. Ethylmalonyl-CoA mutase from *Rhodobacter sphaeroides* defines a new subclade of coenzyme B_{12}-dependent acyl-CoA mutases. *J Biol Chem* 283: 32283–32293.

Evans MCW, BB Buchanan, DI Arnon. 1966. A new ferredoxin-dependent carbon reduction cycle in a photosynthetic bacterium. *Proc Natl Acad Sci USA* 55: 928–933.

Federici F, B Vitali, R Gotti, MR Pasca, S Gobbi, AB Peck, P Brigidi. 2004. Characterization and heterologous expression of the oxalyl coenzyme A decarboxylase gene from *Bifidobacterium lactis*. *Appl Environ Microbiol* 70: 5066–5073.

Ferguson DJ, JA Krzycki. 1997. Reconstitution of trimethylamine-dependent coenzyme M methylation with the trimethylamine corrinoid protein and the isozymes of methyltransferase II from *Methanosarcina barkeri*. *J Bacteriol* 179: 846–852.

Fernandez RF et al. 2004. Enzymatic assimilation of cyanide via pterin-dependent oxygenolytic cleavage to ammonia and formate in *Pseudomonas fluorescens* NCIMB 11764. *Appl Environ Microbiol* 70: 121–128.

Ferriera NL, D Labbé, F Monot, F Fayolle-Guichard, CW Greer. 2006. Genes involved in the methyl *tert*-butyl ether (MTBE) metabolic pathway of *Mycobacterium austroafricanum* IFP 2012. *Microbiology (UK)* 152: 1361–1374.

Fiedler S, S Steinbüchel, BHA Rehm. 2002. The role of the fatty acid β-oxidation multienzyme complex from *Pseudomonas oleovorans* in polyhydroxyalkanoate biosynthesis: Molecular characterization of the *fadBA* operon from *P. oleovorans* and of the enoyl-CoA hydratase genes *phaJ* from *P. oleovorans* and *Pseudomonas putida*. *Arch Microbiol* 178: 149–160.

Finster K, F Bak. 1993. Complete oxidation of propionate, valerate, succinate, and other organic compounds by newly isolated types of marine, anaerobic, mesophilic, Gram-negative sulfur-reducing eubacteria. *Appl Environ Microbiol* 59: 1452–1460.

Frias JA, JE Richman, LP Wackett. 2009. C_{29} olefinic hydrocarbons biosynthesized by *Arthrobacter* species. *Appl Environ Microbiol* 75: 1774–1777.

Fricke WF, H Seedorf, A Henne, M Krüer, H Liesegang, R Hedderich, G Gottschalk, RK Thauer. 2006. The genome sequence of *Methanosphaera stadtmanae* reveals why this intestinal archaeon is restricted to methanol and H_2 for methane formation and ATP synthesis. *J Bacteriol* 188: 642–658.

Friedebold J, B Bowien. 1993. Physiological and biochemical characterization of the soluble formate dehydrogenase, a molybdoenzyme from *Alcaligenes eutrophus*. *J Bacteriol* 175: 4719–4728.

Friedrich CG, B Bowien, B Friedrich. 1979. Formate and oxalate metabolism in *Alcaligenes eutrophus*. *J Gen Microbiol* 115: 185–192.

Friedrich M, B Schink. 1995a. Electron transport phosphorylation driven by glyoxylate respiration with hydrogen as electron donor in membrane vesicles of a glyoxylate-fermenting bacterium. *Arch Microbiol* 163: 268–275.

Friedrich M, B Schink. 1995b. Isolation and characterization of a desulforubidin-containing sulfate-reducing bacterium growing with glycolate. *Arch Microbiol* 164: 271–279.

Friedrich M, N Springer, W Ludwig, B Schink. 1996. Phylogenetic positions of *Desulfofustis glycolicus* gen. nov., sp. nov., and *Syntrophobotulus glycolicus* gen. nov., sp. nov., two new strict anaerobes growing with glycolic acid. *Int J Syst Bacteriol* 46: 1065–1069.

Frimmer U, F Widdel. 1989. Oxidation of ethanol by methanogenic bacteria. Growth experiments and enzymatic studies. *Arch Microbiol* 152: 479–483.

Fuchs G, E Stupperich, G Eden. 1980. Autotrophic CO_2 fixation in *Chlorobium limicola*. Evidence for the operation of a reductive tricarboxylic acid cycle in growing cells. *Arch Microbiol* 128: 64–71.

Fukui T, N Shiomi, Y Doi. 1998. Expression and characterization of (*R*)-specific enoyl-coenzyme A hydratase involved in polyhydroxyalkanoate biosynthesis by *Aeromonas punctata*. *J Bacteriol* 180: 667–673.

Gallus C, B Schink. 1994. Anaerobic degradation of pimelate by newly isolated denitrifying bacteria. *Microbiology (UK)* 140: 409–416.

Gande R, KJC Gibson, AK Brown, K Krumbach, LG Dover, H Sahm, S Shioyama, T Oikawa, GS Besra, L Eggeling. 2004. Acyl-CoA carboxylases (*accD2* and *accD3*), together with a unique polyketide synthase (*Cg-pks*), are key to mycolic acid biosynthesis in *Corynebacterianeae* such as *Corynebacterium glutamicum* and *Mycobacterium tuberculosis. J Biol Chem* 279: 44847–44857.

Gangoiti J, M Santos, MJ Llama, JL Serra. 2010. Production of chiral (*R*)-3-hydroxyoctanoic acid monomers, catalyzed by *Pseudomonas fluorescens* GK13 poly(3-hydroxyoctanoic acid) depolymerase. *Appl Environ Microbiol* 76: 3554–3560.

Gebhardt NA, RK Thauer, D Linder, PM Kaulfers, N Pfennig. 1985. Mechanism of acetate oxidation to CO_2 with elemental sulfur in *Desulfuromonas acetoxidans. Arch Microbiol* 141: 392–398.

Gencic S, DA Grahame. 2003. Nickel in subunit *b* of the acetyl-CoA decarbonylase/synthase multienzyme complex in methanogens. Catalytic properties and evidence for a binuclear Ni-Ni site. *J Biol Chem* 278: 6101–6110.

Ghisla S, C Thorpe. 2004. Acyl-CoA dehydrogenases. A mechanistic overview. *Eur J Biochem* 271: 494–508.

Goda M, Y Hashimoto, S Shimizu, M Kobayashi. 2001. Discovery of a novel enzyme, isonitrile hydratase, involved in nitrogen-carbon triple bond cleavage. *J Biol Chem* 276: 23480–23485.

Gogerty DS, TA Bobik. 2010. Formation of isobutene from 3-hydroxy-3-methylbutyrate by diphosphomevalonate decarboxylase. *Appl Environ Microbiol* 76: 8004–8010.

Gremer L, S Kellner, H Dobbek, R Huber, O Meyer. 2000. Binding of flavin adenine dinucleotide to molybdenum-containing carbon monoxide dehydrogenase from *Oligotropha carboxidovorans*. Structural and functional analysis of a carbon monoxide dehydrogenase species in which the native flavoprotein has been replaced by its counterpart produced in *Escherichia coli. J Biol Chem* 275: 1864–1872.

Guangsheng C, CM Plugge, W Roelofsen, FP Houwen, AJM Stams. 1992. *Selenomonas acidaminovorans* sp. nov., a versatile thermophilic proton-reducing anaerobe able to grow by decarboxylation of succinate to propionate. *Arch Microbiol* 157: 169–175.

Gupta A, A Murarka, P Campbell, R Gonzalez. 2009. Anaerobic fermentation of glycerol in *Paenibacillus macerans*: Metabolic pathways and environmental determinants. *Appl Environ Microbiol* 75: 5871–5883.

Hage JC, S Hartmans. 1999. Monooxygenase-mediated 1,2-dichloroethane degradation by *Pseudomonas* sp. strain DCA1. *Appl Environ Microbiol* 65: 2466–2470.

Han J, J Hou, H Liu, S Cai, B Feng, J Zhou, H Xiang. 2010. Wide distribution among halophilic archaea of a novel polyhydroxyalkanoate synthase subtype with homology to bacterial type III synthases. *Appl Environ Microbiol* 76: 7811–7819.

Han J, Q Lu, L Zhou, J Zhou, H Xiang. 2007. Molecular characterization of the *phaEC*$_{Hm}$ genes, required for biosynthesis of poly(3-hydroxybutyrate) in the extremely halophilic archaeon *Haloarcula marismortui. Appl Environ Microbiol* 73: 6058–6065.

Han L, KA Reynolds. 1997. A novel alternate anaplerotic pathway to the glyoxylate cycle in *Streptomyces. J Bacteriol* 179: 5157–5164.

Handrick R, S Reinhardt, ML Focarete, M Scanfola, G Adamus, M Kowalczuk, D Jendrossek. 2001. A new type of thermoalkaliphilic hydrolase of *Paucimonas lemoignei* with high specificity for amorphous polyesters of short-chain-length hydroxyalkanoic acids. *J Biol Chem* 276: 36215–36224.

Handrick R, S Reinhardt, P Kimmig, D Jendrossek. 2004. The "Intracellular" poly(3-hydoxybutyrate) PHB depolymerase of *Rhodospirillum rubrum* is a periplasm-located protein with specificity for native PHB and with structural similarity to extracellular PHB depolymerases. *J Bacteriol* 186: 7243–7253.

Harder PA, DP ÓKeefe, JA Romesser, KJ Leto, CA Omer. 1991. Isolation and characterization of *Streptomyces griseolus* deletion mutants affected in cytochrome P-450-mediated herbicide metabolism. *Mol Gen Genet* 227: 238–244.

Harper DB. 1977. Microbial metabolism of aromatic nitriles. Enzymology of C-N cleavage by *Nocardia* sp. (*Rhodochrous* group) NCIB 11216. *Biochem J* 165: 309–319.

Harrison FH, CS Harwood. 2005. The *pimFABCDE* operon from *Rhodopseudomonas palustris* mediates dicarboxylic acid degradation and participates in anaerobic benzoate degradation. *Microbiology (UK)* 151: 727–736.

Härtel U, E Eckel, J Koch, G Fuchs, D, Linder, W Buckel. 1993. Purification of glutaryl-CoA dehydrogenase from *Pseudomonas* sp., an enzyme involved in the anaerobic degradation of benzoate. *Arch Microbiol* 159: 174–181.

Hartmannis MGN, TC Stadtman. 1986. Diol metabolism and diol dehydratase in *Clostridium glycolicum. Arch Biochem Biophys* 245: 144–152.

Hartmannis MGN, TC Stadtman. 1987. Solubilization of a membrane-bound diol dehydratase with retention of EPR g = 2.02 signal by using 2-(N-cyclohexylamino)ethanesulfonic acid buffer. *Proc Natl Acad Sci USA* 84: 76–79.

Harwood CS, G Burchardt, H Herrmann, G Fuchs. 1999. Anaerobic metabolism of aromatic compounds via the benzoyl-CoA pathway. *FEMS Microbiol Revs* 22: 439–458.

Heath RJ, CO Rock. 2002. The Claisen condensation in biology. *Nat Prod Rep* 19: 581–596.

Hensgens CMH, J Vonck. J Van Beeumen, EFJ van Bruggen, TA Hansen. 1993. Purification and characterization of an oxygen-labile, NAD-dependent alcohol dehydrogenase from *Desulfovibrio gigas*. *J Bacteriol* 175: 2859–2863.

Herter S, A Busch, G Fuchs. 2002. L-Malyl-CoA/β-methylmalyl-CoA lyase from *Chloroflexus aurantiacus*, a bifunctional enzyme involved in autotrophic CO_2 fixation. *J Bacteriol* 184: 5999–6006.

Hezayen FE, BH Rehm, R Eberhardt, A Steinbuchel. 2000. Polymer production by two newly isolated and extremely halophlilic archaea: Application of a novel corrossive-resistant bioteactor. *Appl Microbiol Biotechnol* 54: 319–325.

Hezayen FE, BH Rehm, R Eberhardt, A Steinbuchel. 2001. Transfer of *Natrialba asiatic* B1T to *Natrialba taiwanensis* sp. nov., and description of *Natrialba aegyptica* sp. nov., a novel, extremely halophilic, aerobic non-pigmented member of the *Archaea* from Egypt that produces extracellular poly(glutamic acid). *Int J Syst Evol Microbiol* 51: 1133–1142.

Hirrlinger B, A Stolz. 1997. Formation of a chiral hydroxamic aid with an amidase from *Rhodococcus erythropolis* MP50 and subsequent chemical Lossen rearrangement to a chiral amine. *Appl Environ Microbiol* 63: 3390–3393.

Hirschler A, J-F Rontani, D Raphel, R Matheron, J-C Bertrand. 1998. Anaerobic degradation of hexadecan-2-one by a microbial enrichment culture under sulfate-reducing conditions. *Appl Environ Microbiol* 64: 1576–1579.

Hisano T, T Tsuge, T Fukui, T Iwata, K Miki. 2003. Crystal structure of the (R)-specific enoyl-CoA hydratase from *Aeromonas caviae* involved in polyhydroxyalkanoate biosynthesis. *J Biol Chem* 278: 617–624.

Horswill AR, JC Escalante-Semerena. 1999. *Salmonella typhmurium* LT2 catabolizes propionate by the 2-methylcitric acid cycle. *J Bacteriol* 181: 5615–5623.

Hou CT, N Barnabe, I Marczak. 1981. Stereospecificity and other properties of a secondary alcohol-specific, alcohol dehydrogenase. *Eur J Biochem* 119: 359–364.

Hou CT, R Patel, AI Laskin, N Barnabe, I Barist. 1983. Production of methyl ketones from secondary alcohols by cell suspensions of C_2 to C_4 n-alkane-grown bacteria. *Appl Environ Microbiol* 46: 178–184.

Hu Y, CC Lee, MW Ribbe. 2011. Extending the carbon chain: Hydrocarbon formation catalyzed by Vanadium/Molybdenum nitrogenases. *Science* 333: 753–755.

Huisman GW, O de Leeuw, G Eggink, B Witholt. 1989. Synthesis of poly-3-hydroxyalkanoates is a common feature of fluorescent pseudomonads. *Appl Environ Microbiol* 55: 1949–1954.

Hume AR, J Nikodinovic-Runic, KE ÓConnor. 2009. FadD from *Pseudomonas putida* CA-3 is a true long-chain fatty acyl coenzyme A synthetase that activates phenylalkanoic and alkanoid acids. *J Bacteriol* 191: 7554–7565.

Ibrahim MHA, A Steinbüchel. 2010. High-cell-density cyclic fed-batch fermentation of a poly(3-hydroxybutyrate)-accumulating thermophile, *Chelatococcus* sp. strain MW10. *Appl Environ Microbiol* 76: 7890–7895.

Jablonski PE, W-P Lu, SW Ragsdale, JG Ferry. 1993. Characterization of the metal centers of the corrinoid/iron-sulfur component of the CO dehydrogenase enzyme complex from *Methanosarcina thermophila* by EPR spectroscopy and spectroelectrochemistry. *J Biol Chem* 268: 325–329.

Jahns T, R Schepp, H Kaltwasser. 1997. Purification and characterization of an enzyme from a strain of *Ochrobactrum anthropi* that degrades condensation products of urea and formaldehyde (ureaform). *Can J Microbiol* 43: 1111–1117.

Jandhyala D, M Berman, PD Myers, BT Sewell, RC Willson, MJ Benedik. 2003. CynD, the cyanide dihydratase from *Bacillus pumilus*: Gene cloning and structural studies. *Appl Environ Microbiol* 69: 4794–4805s.

Janssen DB, J Gerritse, J Brackman, C Kalk, D Jager, B Witholt. 1988. Purification and characterization of a bacterial dehalogenase with activity towards halogenated alkanes, alcohols and ethers. *Eur J Biochem* 171: 67–72.

Janssen PH, W Liesack. 1995. Succinate decarboxylation by *Propionigenium maris* sp. nov., a new anaerobic bacterium from an estuarine sediment. *Arch Microbiol* 164: 29–35.

Janssen PH, CG Harfoot. 1990. Isolation of a *Citrobacter* species able to grow on malonate under strictly anaerobic conditions. *J Gen Microbiol* 136: 1037–1042.

Janssen PH, P Hugenholtz. 2003. Fermentation of glycolate by a pure culture of a strictly anaerobic gram-positive bacterium belonging to the family *Lachnospiraceae. Arch Microbiol* 179: 321–328.

Janssen PH, B Schink. 1995a. Metabolic pathways and energetics of the acetone-oxidizing sulfate-reducing bacterium *Desulfobacterium cetonicum. Arch Microbiol* 163: 188–194.

Janssen PH, B Schink. 1995b. Pathway of butyrate catabolism by *Desulfobacterium cetonicum. J Bacteriol* 177: 3870–3872.

Jendrossek D. 2001. Transfer of [*Pseudomonas*] *lemoignei*, a Gram-negative rod with restricted catabolic capacity, to *Paucimonas* gen. nov. with one species, *Paucimonas lemoignei* comb. nov. *Int J Syst Evol Microbiol* 51: 905–908.

Jendrossek D. 2009. Polyhydroxyalkanoate granules are complex subcellular organelles (carbonosomes). *J Bacteriol* 191: 3195–3202.

Johnson WV, PM Anderson. 1987. Bicarbonate is a recycling substrate for cyanase. *J Biol Chem* 262: 9021–9025.

Jones JB, TC Stadtman. 1980. Reconstitution of a formate-NADP$^+$ oxidoreductase from formate dehydrogenase and a 5-deazaflavin-linked NADP$^+$ reductase isolated from *Methanococcus vannielii. J Biol Chem* 255: 1049–1053.

Jones JB, TC Stadtman. 1981. Selenium-dependent and selenium-independent formate dehydrogenases of *Methanococcus vannielii. J Biol Chem* 256: 656–663.

Jonsson S, S Ricagno, Y Lindqvist, NGJ Richards. 2004. Kinetic and mechanistic characterization of the formyl-CoA transferase from *Oxalobacter formigenes. J Biol Chem* 279: 36003–36012.

Kadouri D, E Jurkevitch, Y Okon. 2003. Involvement of the reserve material poly-ß-hydroxybutyrate in *Azospirillum brasilense* stress endurance and root colonization. *Appl Environ Microbiol* 69:3244—3250.

Kamennaya NA, AE Post. 2011. Characterization of cyanate metabolism in marine *Synechococcus* and *Prochlorococcus* spp. *Appl Environ Microbiol* 77: 291–301.

Kang BS, YM Kim. 1999. Cloning and molecular characterization of the genes for carbon monoxide dehydrogenase and localization of molybdopterin, flavin adenine dinucleotide, and iron-sulfur centers in the enzyme of *Hydrogenophaga pseudoflava. J Bacteriol* 181: 5581–5590.

Kasuya K-I, Y Inoue, T Tanaka, T Akahata, T Iwata, T Fukui, Y Doi. 1997. Biochemical and molecular characterization of the polyhydroxybutyrate depolymerase of *Comamonas acidovorans* YM1609. *Appl Environ Microbiol* 43: 4844–4852.

Katayama Y, A Hiraishi, H Kuraishi. 1995. *Paracoccus thiocyanatus* sp. nov., a new species of thiocyanate-utilizing facultative chemolithotroph, and transfer of *Thiobacillus versutus* to the genus *Paracoccus* as *Paracoccus versutus* comb. nov. with emendation of the genus. *Microbiology (UK)* 141: 1469–1477.

Katayama Y, Y Matsushita, M Kaneko, M Kondo, T Mizuno, H Nyunoya. 1998. Cloning of genes coding for the three subunits of thiocyanate hydrolase of *Thiobacillus thioparus* THI 115 and their evolutionary relationships to nitrile hydratase. *J Bacteriol* 180: 2583–2589.

Katayama Y, Y Narahara, Y Inoue, F Amano, T Kanagawa, H Kuraishi. 1992. A thiocyanate hydrolase of *Thiobacillus thioparus* A novel enzyme catalyzing the formation of carbonyl sulfide from thiocyanate. *J Biol Chem* 267:9170–9175.

Kashevi K, JM Tor, DE HOlmes, CVG Van Praagh, A-L Reysenbach, DR Lovley. 2002. *Geoglobus ahangari* gen. nov., sp. nov., a novel hyperthermophilic archaeon capable of oxidizing organic acids and growing autotrophically on hydrogen with Fe(III) serving as sole electron acceptor. *Int J Syst Evol Microbiol* 52: 719–728.

Kather B, K Stingl, ME van der Rest, K Alterdorf, D Molenaar. 2000. Another unusual type of citric acid cycle enzyme in *Helicobacter pylori*: The malate:quinone oxidoreductase. *J Bacteriol* 182: 3204–3209.

Kessler B, NJ Palleroni. 2000. Taxonomic implications of synthesis of poly-β-hydroxybutyrate and other poly-β-hydroxyalkanoates by aerobic pseudomonads. *Int J Syst Evol Microbiol* 50: 711–713.

Khelifi N, V Grossi, M Hamdi, A Dolla, J-L Tholozan, B Ollivier, A Hirshler-Réa. 2010. Anaerobic oxidation of fatty acids and alkenes by the hyperthermophilic sulfate-reducing archaeon *Archaeoglobus fulgidus. Appl Environ Microbiol* 76: 3057–3060.

Khomyakova M, Ö Bükmez, LK Thomas, TJ Erb, IA Berg. 2011. A methylaspartate cycle in haloarchaea. *Science* 331: 334–337.

Kita K, K Ishimaru, M Teraoka, H Yanase, N Kato. 1995. Properties of (poly-3-hydroxybutyrate) depolymerase from a marine bacterium *Alcaligenes faecalis* AE122. *Appl Environ Microbiol* 61: 1727–1730.

Kiziak C, A Stolz. 2009. Identification of amino acid residues responsible for the enantioselectivity and amide formation capacity of the arylacetonitrilase from *Pseudomonas fluorescens* EBC191. *Appl Environ Microbiol* 75: 5592–5599.

Ko J, I Kim, S Yoo, B Min, K Kim, C Park. 2005. Conversion of methylglyoxal to acetol by *Escherichia coli*. *J Bacteriol* 187: 5782–5789.

Kobayashi T, M Shiraki, T Abe, A Sugiyama, T Saito. 2003. Purification and properties of an intracellular 3-hydroxybutyrate-oligomer hydrolase (PhaZ2) in *Ralstonia eutropha* H16 and its identification as a novel intracellular poly(3-hydroxybutyrate) depolymerase. *J Bacteriol* 185: 3485–3490.

Kobayashi T, K Uchino, T Abe, Y Yamazaki, T Saito. 2005. Novel intracellular 3-hydroxybutyrate-oligomer hydrolase in *Wautersia eutropha* H16. *J Bacteriol* 187: 5129–5135.

Kolb S, S Seeliger, N Springer, W Ludwig, B Schink. 1998. The fermenting bacterium *Malonomonas rubra* is phylogenetically related to sulfur-reducing bacteria and contains a *c*-type cytochrome similar to those of sulfur and sulfate reducers. *System Appl Microbiol* 21: 340–345.

Kornberg HL. 1966. The role and control of the glyxoxylate cycle in *Escherichia coli*. *Biochem J* 99:1–11.

Kornberg HL, JG Morris. 1965. The utilization of glycollate by *Micrococcus denitrificans*: The β-hydroxyaspartate pathway. *Biochem J* 95: 577–586.

Kornberg HL, PJR Phizackerley, RJ Stadler. 1959. Synthesis of cell constituents from acetate by *Escherichia coli*. *Biochem J* 72: 32–33.

Korotkova N, Chistoserdova L, V Kuska, ME Lidstrom. 2002a. Glyoxylate regeneration pathway in the methylotroph *Methylobacterium extorquens* AM1. *J Bacteriol* 184: 1750–1758.

Korotkova N, L Chistoserdova, V Kuksa, ME Lidstrom. 2002b. Poly-β-hydroxybutyrate biosynthesis in the facultative methylotroph *Methylobacterium extorquens* AM1. *J Bacteriol* 184: 1750–1758.

Korotkova N, ME Lidstrom. 2001. A connection between poly-β-hydroxybutyrate biosynthesis and growth on C_1 and C_2 compounds in the methylotroph *Methylobacterium extorquens* AM1. *J Bacteriol* 183:1038–1046.

Kosaka T et al. 2006. Reconstruction and regulation of the central catabolic pathway in the thermophilic propionate-oxidizing syntroph *Pelotomaculum thermopropionicum*. *J Bacteriol* 188: 202–210.

Kostichka K, SM Thomas, KJ Gibson, V Nagarajan, Q Cheng. 2001. Cloning and characterization of a gene cluster for cyclododecanone in *Rhodococcus ruber* SC1. *J Bacteriol* 183: 6478–6486.

Körtner C, F Lauterbach, D Tripier, G Unden, A Kröger. 1990. *Wolinella succinogenes* fumarate reductase contains a dihaem cytochrome *b*. *Mol Microbiol* 4: 855–860.

Krakow G, SS Barkulis, JA Hayashi. 1961. Glyoxylic acid carboligase: An enzyme present in glycolate-grown *Escherichia coli*. *J Bacteriol* 81: 509–518.

Kreher S, A Schilhabel, G Diekert. 2008. Enzymes involved in the anoxic utilization of phenyl methyl ethers by *Desulfitobacterium hafniense* DCB2 and *Desulfitobacterium hafniense* PCE-S. *Arch Microbiol* 190: 489–495.

Kröger A, V Geisler, E Lemma, F Theis, R Lenger. 1992. Bacterial fumarate respiration. *Arch Microbiol* 158: 311–314.

Kunz DA, O Nagappan. 1989. Cyanase-mediated utilization of cyanate in *Pseudomonas fluorescens* NCIB 11764. *Appl Environ Microbiol* 55: 256–258.

Lageveen RG, GW Huisman, H Preusting, P Ketelaar, G Eggink, B Witholt. 1988. Formation of polyesters by *Pseudomonas oleovorans*: Effect of substrates on formation and composition of poly-(R)-3-hydroxyalkanoates and poly-(R)-3-hydroxyalkenoates. *Appl Environ Microbiol* 54: 2924–2932.

Lawlis VB, GLR Gordon, BA McFadden. 1979. Ribulose-1,5-bisphosphate carboxylase/oxygenase from *Pseudomonas oxalaticus*. *J Bacteriol* 139: 287–298.

Layh N, A Stolz, S Förster, F Effenberger, H-J Knackmuss. 1992. Enantioselective hydrolysis of *O*-acetylmandelonitrile to *O*-acetylmandelic acid by bacterial nitrilases. *Arch Microbiol* 158: 405–411.

Lee I-Y, JPN Rosazza. 1998. Enzyme analyses demonstrate that β-methylbutyric acid is converted to β-hydroxy-β-methylbutyric acid via the leucine catabolic pathway by *Galactomyces reessii*. *Arch Microbiol* 169: 257–262.

Lessner DJ, L Li, Q Li, T Rejtar, VP Andreev, M Reichlen, K Hill, JJ Moran, BL Karger, JG Ferry. 2006. An unconventional pathway for reduction of CO_2 to methane in CO-grown *Methanosarcina acetivorans* revealed by proteomics. *Proc Natl Acad Sci USA* 103: 17921–17926.

Li H-Q, I Matsuda, Y Fujise, A Ichiyama. 1999. Short-chain acyl-CoA-dependent production of oxalate from oxaloacetate by *Burkholderia glumae*, a plant pathogen which causes grain rot and seedling rot of rice via oxalate production. *J Biochem* 126: 243–253.

Lillo JG, F Rodriguez-Valera. 1990. Effects of culture conditions on poly (β-hydroxybutyric) acid production by *Haloferax mediterranaei*. *Appl Environ Microbiol* 56: 2517–2521.

Lindenkamp N, K Peplinski, E Volodina, A Ehrenreich, A Steinbüchel. 2010. Impact of multiple β-ketothiolase mutations in *Ralstonia eutropha* H16 on the composition of 3-mercaptopropionic acid-containing copolymers. *Appl Environ Microbiol* 76: 5373–5382.

Liyanage H, S Kashket, M Young, ER Kashket. 2001. *Clostridium beijerinckii* and *Clostridium difficile* detoxify methylglyoxal by a novel mechanism involving glycerol dehydrogenase. *Appl Environ Microbiol* 67: 2004–2010.

Ljungdahl LG. 1986. Autotrophic acetate synthesis. *Annu Rev Microbiol* 40: 415–450.

Lorite MJ, J Tachil, J Sanjuan, O Meyer, EJ Bedmar. 2000. Carbon monoxide dehydrogenase activity in *Bradyrhizobium japonicum*. *Appl Environ Microbiol* 66: 1871–1876.

Lu Q, J Han, L Zhou, J Zhou, H Xiang. 2008. Genetic and biochemical characterization of the poly(3-hydroxybutyrate-*co*-3-hydroxyvalerate) synthase in *Haloferax mediterranei*. *J Bacteriol* 190: 4173–4180.

Lueders T, B Pommerenke, MW Friedrich. 2004. Stable-isotope probing of microorganisms thriving at thermodynamic limits: syntrophic propionate oxidation in flooded soil. *Appl Environ Microbiol* 70: 5778–5786.

Luque-Almagro VM, M-J Huertas, M Martínez-Luque, C Moreno-Vivián, MD Roldán, LJ García-Gil, F Castillo, R Blasco. 2005. Bacterial degradation of cyanide and its metal complexes under alkaline conditions. *Appl Environ Microbiol* 71: 940–947.

Lütke-Eversloh T, K Bergander, H Luftmann, A Steinbüchel. 2001. Identification of a new class of biopolymer: Bacterial synthesis of a sulfur-containing polymer with thioester linkages. *Microbiology (UK)* 147: 11–19.

Lynch JM. 1977. Hydrocarbons. pp. 39–45. In *Handbook of Microbiology 2nd Edition* (Eds AI Laskin, HA Lechevalier). CRC Press, Boca Raton, Florida.

Lynd LH, JG Zeikus. 1983. Metabolism of H2-CO2, methanol, and glucose by *Butyribacterium methylotrophicum*. *J Bacteriol* 153: 1415–1423.

McBride, KE, JW Kenny, DM Stalker. 1986. Metabolism of the herbicide bromoxynil by *Klebsiella pneumoniae* subspecies *ozaenae*. *Appl Environ Microbiol* 52: 325–330.

Mackie RI, BA White, MP Bryant. 1991. Lipid metabolism in anaerobic ecosystems. *Crit Rev Microbiol* 17: 449–479.

Marx CJ, L Chistoserdova, ME Lidstrom. 2003. Formaldehyde-detoxifying role of the tetrahydromethanopterin-linked pathway in *Methylobacterium extorquens* AM1. *J Bacteriol* 185: 7160–7168.

Marx CJ, JA Miller, L Chistoserdova, ME Lidstrom. 2004. Multiple formaldehyde oxidation/detoxification pathways in *Burkholderia fungorum* LB400. *J Bacteriol* 186: 2173–2178.

Mathur M, PE Kolattukudy. 1992. Molecular cloning and sequencing of the gene for mycocerosic acid synthase, a novel fatty acid elongating multifunctional enzyme, from *Mycobacterium tuberculosis* var. *bovis* Bacillus Calmette-Guérin. *J Biol Chem* 267: 19388–19395.

Matsusaki H, S Manij, K Taguchi, M Kato, T Fukui, Y Doi. 1998. Cloning and molecular analysis of the poly(3-hydroxbutyrate) and poly(3-hydroxybutyrate-*co*-3-hydroxyalkanoate) biosynthesis genes in *Pseudomonas* sp. strain 61–3. *J Bacteriol* 180: 6459–6467.

Matthies C, B Schink. 1992. Fermentative degradation of glutarate via decarboxylation by newly isolated strictly anaerobic bacteria. *Arch Microbiol* 157: 290–296.

McInerney MJ, MP Bryant, RB Hespell, JW Costerton. 1981. *Syntrophomonas wolfei* gen. nov. sp. nov., an anaerobic, syntrophic, fatty acid-oxidizing bacterium. *Appl Environ Microbiol* 41: 1029–1039.

McInerney MJ, L Rohlin, H Moputtaki, U Kim, RS Krupp, L Rios-Hernandez, J Sieber, CG Struchtmeyer, A Bhattacharya, JW Campbell, RP Gunsalus. 2007. The genome of *Syntrophus aciditrophicus*: Life at the thermodynamic limit of microbial growth. *Proc Natl Acad USA* 104: 7600–7605.

Meister M, S Saum, BE Alber, G Fuchs. 2005. L-Malyl-Coenzyme A/β-methylmalyl-coenzyme A lyase is involved in acetate assimilation of the isocitrate lyase-negative bacterium *Rhodobacter capsulatus*. *J Bacteriol* 187: 1415–1425.

Merrick JM, M Doudoroff. 1964. Depolymerization of poly-β-hydroxybutyrate by an intracellular enzyme system. *J Bacteriol* 88: 60–71.

Meyer O, HG Schlegel. 1983. Biology of aerobic carbon monoxide-oxidizing bacteria. *Annu Rev Microbiol* 37: 277–310.

Miller JM, DO Gray. 1982. The utilization of nitriles and amides by a *Rhodococcus* species. *J Gen Microbiol* 128: 1803–1809.

Miller JDA, DS Wakerley. 1966. Growth of sulfate-reducing bacteria by fumarate dismutation. *J Gen Microbiol* 43: 101–107.

Molenaar D, ME van der Rest, S Petrović. 1998. Biochemical and genetic characterization of the membrane-associated malate dehydrogenase (acceptor) (EC 1.1.99.16) from *Corynebacterium glutamicum Eur J Biochem* 254: 395–403.

Möller D, R Schauder, G Fuchs, RK Thauer. 1987. Acetate oxidation to CO_2 via a citric acid cycle involving an ATP-citrate lyase: A mechanism for the synthesis of ATP via substrate level phosphorylation in *Desulfobacter postgatei* growing on acetate and sulfate. *Arch Microbiol* 148: 202–207.

Möller-Zinkhan D, RK Thauer. 1990. Anaerobic lactate oxidation to 3 CO_2 by *Archaeoglobus fulgidus* via the carbon monoxide dehydrogenase pathway: Demonstration of the acetyl-CoA carbon-carbon cleavage reaction in cell extracts. *Arch Microbiol* 153: 215–218.

Moran JJ, CH House, JM Vrentas, KH Freeman. 2008. Methyl sulfide production by a novel carbon monoxide metabolism in *Methanosarcina acetivorans*. *Appl Environ Microbiol* 74: 540–542.

Mouttaki H, MA Nanny, MJ McInerney. 2007. Cyclohexane carboxylate and benzoate formation from crotonate in *Syntrophus aciditrophicus*. *Appl Environ Microbiol* 73: 930–938.

Müh U, I Çinkaya, SPJ Albracht, W Buckel. 1996. 4-hydroxybutyryl-CoA dehydratase from *Clostridium aminobutyricum*: Characterization of FAD and iron–sulfur clusters involved in an overall non-redox reaction. *Biochemistry* 35: 11710–11718.

Müller E, K Fahlbusch, R Walter, G Gottschalk. 1981. Formation of *N,N*-imethylglycine, acetic acid, and butyric acid, from betaine by *Eubacterium limosum*. *Appl Environ Microbiol* 42: 439–445.

Müller RH, T Rohwerder, H Harms. 2008. Degradation of fuel oxygenates and their main intermediates by *Aquincola tertiaricarbonis* L108. *Microbiology (UK)* 154:1414–1421.

Müller N, D Schleheck, B Schink. 2009. Involvement of NADH: Acceptor oxidoreductase and butyryl coenzyme A dehydrogenase in reversed electron transport during syntrophic butyrate oxidation in *Syntrophomonas wolfei*. *J Bacteriol* 181: 6167–6177.

Müller U, P Willnow, U Ruschig, T Höpner. 1978. Formate dehydrogenase from *Pseudomonas oxalaticus*. *Eur J Biochem* 83: 485–498.

Mullins EA, JA Francois, TJ Kappock. 2008. A specialized citric acid cycle requiring. succinyl-coenzyme A (CoA):acetate CoA-transferase (AarC) confers acetic acid resistance on the acidophile *Acetobacter aceti*. *J Bacteriol* 190: 4933–4940.

Munir E, JJ Yoon, T Tokimatsu, T Hattori, M Shimada. 2001. A physiological role for oxalic acid biosynthesis in the wood-rotting basidiomycete *Fomitopsis palustris*. *Proc Natl Acad Sci USA* 98: 11126–11130.

Murarka A, Y Dharmadi, SS Yazdani, R Gonzalez. 2008. Fermentative utilization of glycerol by *Escherichia coli* and its implications for the production of fuels and chemicals. *Appl Environ Microbiol* 74: 1124–1135.

Nagy I, G Schoofs, F Compermolle, P Proost, J Vanderleyden, R De Mot. 1995. Degradation of the thiocarbamate herbicide EPTC (*S*-ethyl dipropylcarbamoylthioate) and biosafening by *Rhodococcus* sp. strain N186/21 involve an inducible cytochrome P-450 system and aldehyde dehydrogenase. *J Bacteriol* 177: 676–687.

Nakamiya K, S Hashimoto, H Ito, JS Edmonds, M Morita. 2005. Degradation of 1,4-dioxane and cyclic ethers by an isolated fungus. *Appl Environ Microbiol* 71: 1254–1258.

Nakashimada Y, H Kubota, A Tayayose, T Kakizono, N Nishio. 2001. Asymmetric reduction of ethyl acetoacetate to ethyl (*R*)-3-hydroxybutyrate coupled with nitrate reduction by *Paracoccus denitrificans*. *J Biosci Bioeng* 91: 368–372.

Nanninga HJ, JC Gottschal. 1987. Properties of *Desulfovibrio carbinolicus* sp. nov. and other sulfate-reducing bacteria isolated from an anaerobic-purification plant. *Appl Environ Microbiol* 53: 802–809.

Nawaz MS, TM Heinze, CE Cerniglia. 1992. Metabolism of benzonitrile and butyronitrile by *Klebsiella pneumoniae*. *Appl Environ Microbiol* 58: 27–31.

Oh J-I, B Bowien. 1998. Structural analysis of the *fds* operon encoding the NAD^+-linked formate dehydrogenase of *Ralstonia eutropha*. *J Biol Chem* 273: 26349–26360.

Ohura T, K-I Kasuya, Y Doi. 1999. Cloning and characterization of the polyhydroxybutyrate depolymerase gene of *Pseudomonas stutzeri* and analysis of substrate-binding domains. *Appl Environ Microbiol* 65: 189–197.

Oinuma K-I, Y Hashimoto, K Konishi, M Goda, T Noguchi, H Higashibata, M Kobayashi. 2003. Novel aldoxime dehydratase involved in carbon-nitrogen triple bond synthesis of *Pseudomonas chloroaphis* B23. *J Biol Chem* 278: 29600–29608.

Okubo Y, S Yang, L Chistoserdova, ME Lidstrom. 2010. Alternative route for glyoxylate consumption during growth on two-carbon compounds by *Methylobacterium extorquens* AM1. *J Bacteriol* 192: 1813–1823.

Omer CA, R Lenstra, PJ Little, C Dean, JM Tepperman, KJ Leto, JA Romesser, DP ÓKeefe. 1990. Genes for two herbicide-inducible cytochromes P-450 *from Streptomyces griseolus*. *J Bacteriol* 172: 3335–3345.

Ornston LN, MK Ornston. 1969. Regulation of glyoxylate metabolism in *Escherichia coli* K-12. *J Bacteriol* 98: 1098–1108.

Park M-O. 2005. New pathway for long-chain *n*-alkane synthesis via 1-alcohol in *Vibrio furnissii* M1. *J Bacteriol* 187: 1426–1429.

Park SW, EH Hwang, H Park, JA Kim, J Heo, KH Lee, T Song, E Kim, YT Rao, SW Kim, YM Kim. 2003. Growth of mycobacteria on carbon monoxide and methanol. *J Bacteriol* 185: 142–147.

Parke D, MA Garcia, LN Ornston. 2001. Cloning and genetic characterization of *dca* genes required for β-oxidation of straight-chain dicarboxylic acids in *Acinetobacter* sp. strain ADP1. *Appl Environ Microbiol* 67: 4817–4827.

Paul L, DK Ferguson, JA Krzycki. 2000. The trimethylamine methyltransferase gene and multiple dimethylamine methyltransferase genes of *Methanosarcina barkeri* contain in-frame and read-through amber codons. *J Bacteriol* 182: 2520–2529.

Paulsen J, A Kröger, RK Thauer. 1986. ATP-driven succinate oxidation in the catabolism of *Desulfuromonas acetoxidans*. *Arch Microbiol* 144: 78–83.

Penrod JT, JR Roth. 2006. Conserving a volatile metabolite: A role for carboxysome-like organelles in *Salmonella enterica*. *J Bacteriol* 188: 2865–2874.

Peyraud R, P Kiefer, P Christen, S Massou, J-C Portais, JA Vorholt. 2009. Demonstration of the ethylmalonyl-CoA pathway using ^{13}C metabolomics. *Proc Natl Acad Sci USA* 106: 4846–4851.

Pfennig N, H Biebl. 1976. *Desulfuromonas acetoxidans* gen. nov. and sp. nov., a new anaerobic, sulfur-reducing, acetate-oxidizing bacterium. *Arch Microbiol* 110: 3–12.

Platen H, B Schink. 1990. Enzymes involved in anaerobic degradation of acetone by a denitrifying bacterium. *Biodegradation* 1: 243–251.

Plugge CM, C Dijkema, AJM Stams. 1993. Acetyl-CoA cleavage pathways in a syntrophic propionate oxidizing bacterium growing on fumarate in the absence of methanogens. *FEMS Microbiol Lett* 110: 71–76.

Portevin D, C de Sousa-D´Auria, C Houssin, C Grimaldi, M Chami, M Daffe, C Guilhot. 2004. A polyketide synthase catalyzes the last condensation step of mycolic acid biosynthesis in mycobacteria and related organisms. *Proc Natl Acad USA* 101: 314–319.

Qatibi AI, V Niviére, JL Garcia. 1991. *Desulfovibrio alcoholovorans* sp. nov., a sulfate-reducing bacterium able to grow on glycerol, 1,2- and 1,3-propanediol. *Arch Microbiol* 155: 143–148.

Ragsdale SJ, EW Clark, LG Ljungdahl, LL Lundie, HL Drake. 1983. Properties of purified carbon monoxide dehydrogenase from *Clostridium thermoaceticum* a nickel, iron-sulfur protein. *J Biol Chem* 258: 2364–2369.

Rainwater DL, PE Kolattukudy. 1985. Fatty acid biosynthesis in *Mycobacterium tuberculosis* var. *bovis* Bacillus Calmette-Guérin. Purification and characterization of a novel fatty acid synthase, mycocerosic acid synthase, which elongates *n*-fatty acids with methylmalonyl-CoA. *J Biol Chem* 260: 616–623.

Ramos-Vera WH, IA Berg, G Fuchs. 2009. Autotrophic carbon dioxide assimilation in *Thermoproteales* revisited. *J Bacteriol* 191: 4286–4297.

Rasmussen BA, K Bush. 1997. Carbapenem-hydrolyzing β-lactamases. *Antimicrob Agents Chemother* 41: 223–232.

Raynaud C, P Sarcabal, I Meynial-Salles, C Croux, P Soucaille. 2003. Molecular characterization of the 1,3-propandiol (1,3-PD) operon of *Clostridium butyricum*. *Proc Natl Acad USA* 100: 5010–5015.

Reissmann S, E Hochleitner, H Wang, A Paschos, F Lottspeich, RS Glass, A Böck. 2003. Taming of a poison: Biosynthesis of the NiFe-hydrogenase ligands. *Science* 299: 1067–1070.

Ren Q, N Sierro, B Witholt, B Kessler. 2000. FabG, an NADPH-dependent 3-ketoacyl reductase of *Pseudomonas aeruginosa*, provides precursors for medium-chain-length poly-3-hydroxyalkanoate biosynthesis in *Escherichia coli*. *J Bacteriol* 182: 2978–2981.

Ribbe M, D Gadkari, O Meyer. 1997. N₂ fixation by *Streptomyces thermoautotrophicum* involves a molybde-num-dinitrogenase and a manganese-superoxide oxidoreductase that couple N_2 reduction to the oxidation of superoxide produced from O_2 by a molybdenum-CO dehydrogenase. *J Biol Chem* 272: 26627–26633.

Robertson DE and 20 other authors. 2004. Exploring nitrilase sequence space for enantioselective catalysis. *Appl Environ Microbiol* 70: 2429–2436.

Rohwerder T, U Breuer, D Benndorf, U Lechner, RH Müller. 2006. The alkyl *tert*-butyl ether intermediate 2-hydroxybutyrate is degraded via a novel cobalamin-dependent mutase pathway. *Appl Environ Microbiol* 72: 1428–1435.

Rontani J-F, MJ Gilewicz, VD Micgotey, TL Zheng, PC Bonin, J-C Bertrand. 1997. Aerobic and anaerobic metabolism of 6,10,14-trimethylpentadecan-2-one by a denitrifying bacterium isolated from marine sedi-ments. *Appl Environ Microbiol* 63: 636–643.

Rother M, WW Metcalf. 2004. Anaerobic growth of *Methanosarcina acetivorans* C2A on carbon monoxide: An unusual way of life for a methanogenic archaeon. *Proc Natl Acad Sci USA* 101: 16929–16934.

Roy F, E Samain, HC Dubourguier, G Albagnac. 1986. *Syntrophomonas sapovorans* sp. nov., a new obligately proton reducing anaerobe oxidizing saturated and unsaturated long chain fatty acids. *Arch Microbiol* 145: 142–147.

Ruan Z-S, V Anantharam, IT Crawford, SV Ambudkar, SY Rhee, MY Allison, PC Maloney. 1992. Identification, purification, and reconstitution of OxlT, the oxalate:formate antiport protein of *Oxalobacter formigenes*. *J Biol Chem* 267: 1537–10543.

Rude MA, TS Baron, S Brubaker, M Alibhai, SPD Cardayre, A Schirmer. 2011. Terminal olefin (1-alkene) biosynthesis by a novel P450 fatty acid decarboxylase from *Jeotgalicoccus* species. *Appl Environ Microbiol* 77: 1718–1717.

Sakai Y, H Takahashi, Y Wakasa, T Kotani, H Yurimoto, N Miyachi, PP Van Veldhoven, N Kato. 2004. Role of α-methacyl coenzyme A racemase in the degradation of methyl-branched alkanes by *Mycobacterium* sp. strain P101. *J Bacteriol* 186: 7214–7220.

Samain E, HC Dubourguier, G Albagnac. 1984. Isolation and characterization of *Desulfobulbus elongatus* sp. nov. from a mesophilic industrial digester. *Syst Appl Microbiol* 5: 391–401.

Sampson EM, TA Bobik. 2008. Microcompartments for B₁₂-dependent 1,2-propanediol degradation provide protection from DNA and cellular damage by a reactive metabolic intermediate. *J Bacteriol* 190: 2966–2971.

Sauer K, U Harms, RK Thauer. 1997. Methanol: Coenzyme M methyltransferase from *Methanosarcina barkeri*. Purification, properties and encoding genes of the corrinoid protein MT1. *Eur J Biochem* 243: 670–677.

Seyfried M, R Daniel, G Gottschalk. 1996. Cloning, sequencing and overexpression of the genes encoding coenzyme B12- dependent glycerol dehydratase of *Citrobacter freundii*. *J Bacteriol* 178: 5793–5796.

Schauder R, A Preuβ, M Jetten, G Fuchs. 1989. Oxidative and reductive acetyl-CoA/carbon monoxide dehy-drogenase pathway in *Desulfobacterium autotrophicum* 2. Demonstration of the enzymes of the pathway and comparison of CO dehydrogenase. *Arch Microbiol* 151: 84–89.

Schauder R, B Schink. 1989. *Anaerovibrio glycerini* sp. nov., an anaerobic bacterium fermenting glycerol to propionate, cell matter and hydrogen. *Arch Microbiol* 152: 473–478.

Schauder R, F Widdel, G Fuchs. 1987. Carbon assimilation pathways in sulfate-reducing bacteria II. Enzymes of a reductive citric acid cycle in the autotrophic *Desulfobacter hydrogenophilus*. *Arch Microbiol* 148: 218–225.

Schauer NL, JG Ferry. 1986. Composition of the coenzyme F₄₂₀-dependent formate dehydrogenase from *Methanobacterium formicicum*. *J Bacteriol* 165: 405–411.

Schink B, N Pfennig. 1982. *Propionigenium modestum* gen. nov., sp. nov., a strictly anaerobic non-sporing bacterium growing on succinate. *Arch Microbiol* 133: 209–216.

Schübel U, M Kraut, G Mörsdorf, O Meyer. 1995. Molecular characterization of the gene cluster coxMSL encoding the molybdenum containing carbon monoxide dehydrogenase of *Oligotropha carboxidov-orans*. *J Bacteriol* 177: 2197–2203.

Schirmer A, MA Rude, X Li, E Popova, SB del Cardayre. 2010. Microbial biosynthesis of alkanes. *Science* 329: 559–562.

Schneider-Belhaddad F, P Kolattukudy. 2000. Solubilization, partial purification, and characterization of a fatty aldehyde decarbonylase from a higher plant, *Pisum sativum*. *Arch Biochem Biophys* 377: 341–349.

Schöber U, C Thiel, D Jendrossek. 2000. Poly(3-hydroxyvalerate) depolymerase of *Pseudomonas lemoignei*. *Appl Environ Microbiol* 66: 1385–1392.

Schuster J, F Schäfer, N Hübler, A Brandt, M Rosell, C Härtig, RH Müller, T Rohwerder. 2012. Bacterial degradation of *tert*-amyl alcohol proceeds via hemiterpene 2-methyl-3-buten-2-ol by employing the tertiary alcohol desaturase function of the Rieske nonheme mononuclear iron oxygenase MdpJ. *J Bacteriol* 194: 972–981.

Senda K, Y Arakawa, K Nakashima, H Ito, S Ichiyama, K Shimokata, N Kato, M Ohta. 1996. Multifocal outbreaks of metallo-β-lactamase-producing *Pseudomonas aeruginosa* resistant to broad-spectrum β-lactams, including carbapenems. *Antimicrob Agents Chemother* 40: 349–353.

Shao ZQ, R Behki. 1996. Characterization of the expression of the *thcB* gene, coding for a pesticide-degrading cytochrome P450 in *Rhodococcus* strains. *Appl Environ Microbiol* 62: 403–407.

Sih CJ, HH Tai, YY Tsong, SS Lee, RG Coombe. 1968b. Mechanisms of steroid oxidation by microorganisms. XIV. Pathway of cholesterol side-chain degradation. *Biochemistry* 7: 808–818.

Sih CJ, KC Wang, HH Tai. 1968a. Mechanisms of steroid oxidation by microorganisms. XIII. C_{22} acid intermediates in the degradation of the cholesterol side chain. *Biochemistry* 7: 796–807.

Siméone R, M Léger, P Constant, W Malaga, H Marrakchi, M Daffé, C Guilhot, C Chalut. 2010. Delineation of the roles of FaD22, FaD26 and FaD29 in the biosynthesis of the phthiocerol dimycocerosates and related compounds in *Mycobacterium tuberculosis*. *FEBS J* 277: 2715–2725.

Singer ME, WR Finnerty. 1985a. Fatty aldehyde dehydrogenases in *Acinetobacter* sp. strain HO1-N: Role in hexadecane and hexadecanol metabolism. *J Bacteriol* 164: 1011–1016.

Singer ME, WR Finnerty. 1985b. Alcohol dehydrogenases in *Acinetobacter* sp. strain HO1-N: Role in hexadecane and hexadecanol metabolism. *J Bacteriol* 164: 1017–1024.

Sirakova TD, AK Thirumala. VS Dubey, H Sprecher, PE Kolattukudy. 2001. The *Mycobacterium tuberculosis* *pks2* gene encodes the synthase for the hepta- and octamethyl-branched fatty acids required for sulfolipid synthesis. *J Biol Chem* 276: 16833–16839.

Sluis MK, RA Larsen, JG Krum, R Anderson, WW Metcalf, SA Ensign. 2002. Biochemical, molecular, and genetic analyses of the acetone carboxylases from *Xanthobacter autotrophicus* strain Py2 and *Rhodobacter capsulatus* strain B10. *J Bacteriol* 184: 2969–1977.

Sluis MK, FJ Small, JR Allen, SA Ensign. 1996. Involvement of an ATP-dependent carboxylase in a CO_2-dependent pathway of acetone metabolism by *Xanthobacter* strain Py2. *J Bacteriol* 178: 4020–4026.

Sonke T, S Ernste, RF Tandler, B Kaptein, WPH Peeters, FBJ van Assema, MG Wubbolts, HE Schoemaker. 2005. L-Selective amidase with extremely broad substrate specificity from *Ochrobactrum anthropi* NCIMB 40321. *Appl Environ Microbiol* 71: 7971–7963.

Sriramulu DD, M Liang, D Hernandez-Romero, E Raux-Deery, H Lünsdorf, JB Parsons, M Warren, MB Prentice. 2008. *Lactobacillus reuteri* DSM 20016 produces cobalamin-dependent diol dehydratase in metabolosomes and metabolizes 1,2-propanediol by disproportionation. *J Bacteriol* 190: 4559–4567.

Stams AJM, DR Kremer, K Nicolay, GH Weenk, TA Hansen. 1984. Pathway of propionate formation in *Desulfobulbus propionicus*. *Arch Microbiol* 139: 167–173.

Söhling B, G Gottschalk. 1996. Molecular analysis of the anaerobic succinate degradation pathway in *Clostridium kluyveri*. *J Bacteriol* 178: 871–880.

Sorokin DY, JG Kuenen, G Muyzer. 2011. The microbial sulfur cycle at extremely haloalkaline conditions of soda lakes. *Front Microbiol* doi: 10.3389/fmicb.2011.

Sorokin DY, TP Tourova, A M Lysenko, JG Kuenen. 2001. Microbial thiocyanate utilization under highly alkaline conditions. *Appl Environ Microbiol* 67: 528–538.

Stanier RY, M Doudoroff, R Kunisawa, R Contopulou. 1959. The role of organic substrates in bacterial photosynthesis. *Proc Natl Acad Sci USA* 45: 1246–1260.

Stephanou EG, N Stratigakis. 1993. Keto carboxylic and α,ω-dicarboxylic acids: Photooxidation products of biogenic unsaturated fatty acids present in urban aerosols. *Environ Sci Technol* 27: 1403–1407.

Stieb M, B Schink. 1984. A new 3-hydroxybutyrate fermenting anaerobe, *Ilyobacter polytrophus*, gen. nov. sp. nov., possessing various fermentation pathways. *Arch Microbiol* 140: 139–146.

Stieb M, B Schink. 1985. Anaerobic oxidation of fatty acids by *Clostridium bryantii* sp. nov., a sporeforming, obligately syntrophic bacterium. *Arch Microbiol* 140: 387–390.

Suen Y, GU Holzer, JS Hubbard, TG Tornabene. 1988. Biosynthesis of acyclic methyl branched polyunsaturated hydrocarbons in *Pseudomonas maltophilia*. *J Ind Microbiol* 2: 337–348.

Sukovich DJ, JL Seffernick, JE Richman, JA Gralnick, LP Wackett. 2010b. Widespread head-to-head hydrocarbon biosynthesis in bacteria and role of OleA. *Appl Environ Microbiol* 76: 3850–3862.

Sukovich DJ, JL Seffernick, JE Richman, KA Hunt, JA Gralnick, LP Wackett. 2010a. Structure, function, and insights into biosynthesis of a head-to-head hydrocarbon in *Shewanella oneidensis*. *Appl Environ Microbiol* 76: 3842–3849.

Sung Y-C, PM Anderson, JA Fuchs. 1987. Characterization of high-level expression and sequencing of the *Escherichia coli* K-12 *cynS* gene encoding cyanase. *J Bacteriol* 169: 5224–5230.

Svetlitchnyi V. C Peschel, G Acker, O Meyer. 2001. Two membrane-associated NiFeS-carbon monoxide dehydrogenases from the anaerobic carbon-monoxide-utilizing eubacterium *Carboxydothermus hydrogenoformans*. *J Bacteriol* 183: 5134–5144.

Takayama K, C Wang, GS Besra. 2005. Pathway to synthesis and processing of mycolic acids in *Mycobacterium tuberculosis*. *Clin Microbiol Revs* 18: 81–101.

Tallant TC, JA Krzycki. 1997. Methylthiol: Coenzyme M methyltransferase from *Methanosarcina barkeri*, an enzyme of methanogenesis from dimethylsulfide and methylmercaptopropionate. *J Bacteriol* 179: 6902–6911.

Tallant TC, L Paul, JA Krzycki. 2001. The MtsA subunit of the methylthiol: Coenzyme M methyltransferase of *Methanosarcina barkeri* catalyzes both half-reactions of corrinoid-dependent dimethylsulfide: Coenzyme M methyl transfer. *J Biol Chem* 276: 4485–4493.

Tam AC, RM Behki, SU Khan. 1987. Isolation and characterization of an *S*-ethyl-*N,N*-dipropylthiocarbamate-degrading *Arthrobacter* strain and evidence for plasmid-associated *S*-ethyl-*N,N*-dipropylthiocarbamate degradation. *Appl Environ Microbiol* 53: 1088–1093.

Tanaka K, K Nakamura, E Mikami. 1990. Fermentation of maleate by a gram-negative strictly anaerobic nonspoire former, *Propionivibrio dicarboxylicus* gen. nov., sp. nov. *Arch Microbiol* 154: 323–328.

Tang YJ, S Yi, W-Q Zuang, SH Zinder, JD Keasling, L Alvarez-Cohen. 2009. Investigation of carbon metabolism in *"Dehalococcoides ethenogenes"* strain 195 by use of isotopomer and transcriptomic analysis. *J Bacteriol* 191: 5224–5231.

Tanner A, S Bornemann. 2000. *Bacillius subtilis* YvrK is an acid-induced oxalate decarboxylase. *J Bacteriol* 182: 5271–5273.

Tanner A, L Bowater, SA Fairhurst, S Bornemann. 2001. Oxalate decarboxylase requires manganese and dioxygen for activity. *J Biol Chem* 276: 43627–43634.

Tauber MM, A Cavaco-Paulo, K-H Robra, GM Gübitz. 2000. Nitrile hydratase and amidase from *Rhodococcus rhodochrous* hydrolyze acrylic fibers and granular polyacrylonitrile. *Appl Environ Microbiol* 66: 1634–1638.

Teufel R, JW Kung, D Kockelhorn, BE Alber, G Fuchs. 2009. 3-hydroxypropionyl-coenzyme A dehydratase, and acryloyl-coenzyme A reductase, enzymes of the autotrophic 3-hydroxypropionate/4-hydroxybutyrate cycle in the *Sulfolobales*. *J Bacteriol* 191: 4572–4581.

Textor S, VF Wendich, AA De Graaf, U Müller, MI Linder, D Linder, W Buckel. 1997. Propionate oxidation in *Escherichia coli* evidence for operation of a methylcitrate cycle in bacteria. *Arch Microbiol* 168: 428–436.

Tokimatsu T, Y Nagai, T Hattori, M Shimada. 1998. Purification and characteristics of a novel cytochrome c dependent glyoxylate dehydrogenase from a wood-rotting fungus *Tyromyces palustris*. *FEBS Lett* 437: 117–121.

Topp E, L Xun, CS Orser. 1992. Biodegradation of the herbicide bromoxynil (3,5-dibromo-4-hydroxybenzonitrile) by purified pentachlorophenol hydroxylase and whole cells of *Flavobacterium* sp. strain ATCC 39723 is accompanied by cyanogenesis. *Appl Environ Microbiol* 58: 502–506.

Toraya T, S Honda, S Fukui. 1979. Fermentation of 1,2-Ethanediol by some genera of Enterobacteriaceae involving Coenzyme Bx2-dependent diol dehydratase. *J Bacteriol* 139: 39–47.

Tornabene TG, E Golpi, J Oró. 1967. Identification of fatty acids and aliphatic hydrocarbons in *Sarcina lutea* by gas chromatography and combined gas-chromatography mass spectrometry. *J Bacteriol* 94: 333–343.

Trivedi OA, P Arora, A Vats, MZ Ansari, R Tickkoo, V Sridharan, D Mohany, RS Gokhale. 2005. Dissecting the mechanism and assembly of a complex virulence mycobacterial lipid. *Mol Cell* 17: 631–643.

Tseng C-L, H-J Chen, G-C Shaw. 2006. Identification and characterization of the Bacillus thuringiensis *phaZ* gene encoding new intracellular poly-3-hydroxybutyrate depolymerase. *J Bacteriol* 188: 7592–7599.

Tsuge T, T Fukui, H Matsusaki, S Taguchi, G Kobayashi, A Ishizaki, Y Doi. 1999. Molecular cloning of two (*R*)-specific enoyl-CoA hydratase genes from *Pseudomonas aeruginosa* and their use for polyhydroxyalkanoate synthesis. *FEMS Microbiol Lett* 184: 193–198.

Turroni S, C Bendazzoli, SCF Dipalo, M Candela, B Vitali, R Gotti, P Brigidi. 2010. Oxalate-degrading activity in *Bifidobacterium animalis* subsp. *lactis*: Impact of acidic conditions on the transcriptional levels of the oxalalyl coenzyme A (CoA) decarboxylase and formyl-CoA transferase genes. *Appl Environ Microbiol* 76: 5609–5620.

Tzeng SF, MP Bryant, RS Wolfe. 1975. Factor 420-dependent pyridine nucleotide-linked formate metabolism of *Methanobacterium ruminantium*. *J Bacteriol* 121: 192–196.

Uchino K, T Saito, D Jendrossek. 2008. Poly(3-hydroxybutyrate) (PHB) depolymerase PhaZa1 is involved in mobilization of accumulated PHB in *Ralstonia eutropha* H16. *Appl Environ Microbiol* 74: 1058–1063.

Vandamme P, T Coenye. 2004. Taxonomy of the genus *Cupriavidus*: A tale of lost and found. *Int J Syst Evol Microbiol* 54: 2285–2289.

Vaneechoutte M, P Kämpfer, T De Baere, E Falsen, G Verschraegen. 2004. *Wautersia* gen. nov., a novel genus accommodating the phylogenetic lineage including *Ralstonia eutropha* and related species, and proposal of *Ralstonia [Pseudomonas] syzygii* (Roberts *et al.* 1990) comb. nov. *Int J Syst Evol Microbiol* 54: 317–327.

van der Geize R, K Yam, T Heuser, MH Wilbrink, H Hara, MC Anderton, E Sim, L Dijkhuizen, JE Davies, WW Mohn, LD Eltis. 2007. A gene cluster encoding cholesterol catabolism in a soil actinomycete provides insight into *Mycobacterium tuberculosis* survival in macrophages. *Proc Natl Acad Sci USA* 104: 1947–1952.

Wackett LP, JA Frias, JL Seffernick, DJ Sukovich, SM Cameron. 2007. Genomic and biochemical studies demonstrating the absence of an alkane-producing phenotype in *Vibrio furnissii* M1. *Appl Environ Microbiol* 73: 7192–7198.

Wallrabenstein C, B Schink. 1994. Evidence of reversed electron transport in syntrophic butyrate or benzoate oxidation by *Syntrophomonas wolfei* and *Syntrophomonas buswellii*. *Arch Microbiol* 162: 136–142.

Wang C-S, DA Kunz, BJ Venables. 1996. Incorporation of molecular oxygen and water during enzymatic oxidation of cyanide by *Pseudomonas fluorescens* NCIMB 11764. *Appl Environ Microbiol* 62: 2195–2197.

Wayne LG, et al. 1991. Fourth report of the cooperative, open-ended study of slowly growing mycobacteria by the international working group on mycobacterial taxonomy. *Int J Syst Bacteriol* 41: 463–472.

Widdel F. 1986. Growth of methanogenic bacteria in pure culture with 2-propanol and other alcohols as hydrogen donors. *Appl Environ Microbiol* 51: 1056–1062.

Widdel F. 1987. New types of acetate-oxidizing, sulfate-reducing *Desulfobacter* species, *D. hydrogenophilus* sp. nov., *D. latus* sp. nov., and *D. curvatus* sp. nov. *Arch Microbiol* 148: 286–291.

Widdel F, G-W Kohring, F Mayer. 1983. Studies on dissimilatory sulfate-reducing bacteria that decompose fatty acids. *Arch Microbiol* 134: 286–294.

Widdel F, N Pfennig. 1981. Sporulation and further nutritional characteristics of *Desulfotomaculum acetoxidans* (emend). *Arch Microbiol* 112: 119–122.

Widdel F, N Pfennig. 1982. Studies on dissimilatory sulfate-reducing bacteria that decompose fatty acids. II. Incomplete oxidation of propionate by *Desulfobulbus propionicus* gen. nov., sp. nov. *Arch Microbiol* 131: 360–365.

Widdel F, PE Rouvière, RS Wolfe. 1988. Classification of secondary alcohol-utilizing methanogens including a new thermophilic isolate. *Arch Microbiol* 150: 477–481.

Winters K, PL Parker, C Van Baalen. 1969. Hydrocarbons of blue-green algae: Geochemical significance. *Science* 163: 467–468.

Wolff RA, GW Urben, SM O'Herri, WR Keneal. 1993. Dehydrogenases involved in the conversion of succinate to 4-hydroxybutanoate by *Clostridium kluyveri*. *Appl Environ Microbiol* 59: 1876–1882.

Yamamoto K, Y Ueno, K Otsubo, K Kawakami, K-I Komatsu. 1990. production of *S*-(+)-ibuprofen from a nitrile compound by *Acinetobacter* sp. strain AK 226. *Appl Environ Microbiol* 56: 3125–3129.

Yamane T, X-F Chen, S Ueda. 1996. Growth-associated production of poly(3-hydroxyvalerate) from *n*-pentanol by a methylotrophic bacterium *Paracoccus denitrificans*. *Appl Environ Microbiol* 62: 380–384.

Yang S-Y, M Elzinga. 1993. Association of both enoyl coenzyme A hydratase and 3-hydroxyacyl coenzyme coenzyme A epimerase with an active site in the amino-terminal domain of the multifunctional fatty acid oxidation protein from *Escherichia coli*. *J Biol Chem* 268: 6588–6592.

Yang Z-Y, DR Dean, LC Seefeldt. 2011. Molybdenum nitrogenase catalyzes the reduction and coupling of CO to form hydrocarbons. *J Biol Chem* 286: 19714–19421.

Zaitsev GM, IV Tsitko, FA Rainey, YA Trotsenko, JS Uotila, A Stackebrandt, MS Salkinoja-Salonen. 1998. New aerobic ammonium-dependent obligately oxalotrophic bacteria: Description of *Ammoniphilus oxalaticus* gen. nov., sp. nov. *Int J Sys Bacteriol*. 48: 151–163.

Zaitsev GM, LS Uotila, MM Häggblom. 2007. Biodegradation of methyl *tert*-butyl ether by cold adapted mixed culture and pure bacterial cultures. *Appl Microbiol Biotechnol* 74: 1092–1102.

Zarzycki J, A Schlichting, N Strychalsky, M Müller, BE Alber, G Fuchs. 2008. Mesaconyl-coenzyme A hydratase, a new enzyme of two central carbon metabolic pathways in bacteria. *J Bacteriol* 190: 1366–1374.

Zeikus JG. 1980. Chemical and fuel production by anaerobic bacteria. *Annu Rev Microbiol* 34: 423–464.

Zhang C, X Liu, X Dong. 2004. *Syntrophomonas curvata* sp. nov., an anaerobe that degrades fatty acids in co-culture with methanogens. *Int J Syst Evol Microbiol* 54: 969–973.

Zhang K, M Shiraki, T Saito. 1997. Purification of an extracellular D-(-)-3-hydroxybutyrate oligomer hydrolase from *Pseudomonas* sp. strain A1 and cloning of its gene. *J Bacteriol* 179: 72–77.

Zhao H, D Yang, CR Woese, MP Bryant. 1990. Assignment of *Clostridium bryantii* to *Syntrophospora bryantii* gen. nov., comb. nov. on the basis of a 16S rRNA sequence analysis of its crotonate-grown pure culture. *Int J Syst Bacteriol* 40: 40–44.

7.9 Alkylamines and Amino Acids

7.9.1 Aerobic Conditions

7.9.1.1 *Alkylamines*

7.9.1.1.1 *Primary Amines*

The ability to use methylamines as growth substrates is widespread. The initial reaction in the biodegradation of primary alkylamines is conversion to the aldehyde, and subsequent reactions converge on those for the degradation of primary alkanes and alkanones. There are a number of variations for this apparently straightforward reaction: these include different types of dehydrogenase, and the activity of different amine oxidases.

Primary amines are widely used as a nitrogen source by bacteria, and the first step in their degradation involves formation of the corresponding aldehyde:

$$R\text{-}CH_2\text{-}NH_2 + NADH \rightarrow R\text{-}CHO + NAD + NH_3$$

There are several mechanisms by which this can be accomplished.

1. This may be carried out by dehydrogenation in a pyridoxal 5′-phosphate-mediated reaction: in brief, this involves formation of an imine between pyridoxal 5′-phosphate and the amine, followed by hydrolysis with the production of pyridoxamine 5′-phosphate and the aldehyde. *Methylotenera mobilis* is unique among methylotrophs in its ability to grow at the expense of methylamine that was oxidized by methylamine dehydrogenase and formaldehyde assimilated by the ribulose monophosphate cycle (Kalyuzhnaya et al. 2006). An inducible primary amine dehydrogenase in a strain of *Mycobacterium convolutum* had diverse degradative capability with a broad specificity. It was involved in the degradation of 1-, and 2-aminopropane and 1,3-diaminopropane, and the products were assimilated by the methylmalonate pathway, or by formation of $C_2 + C_1$ fragments (Cerniglia and Perry 1975).

2. Some organisms, however, use an amine oxidase. For example, whereas *Ps. putida* (ATCC 12633) and *Ps. aeruginosa* (ATCC 17933) employ a dehydrogenase, *Klebsiella oxytoca* (ATCC 8724) and *E. coli* (ATCC 9637) use a copper quinoprotein amine oxidase (Hacisalohoglu et al. 1997). The amine oxidase functions by oxidation of the amine to the aldehyde concomitant with the reduction of 2,4,5-trihydroxyphenylalanine quinone (TPQ) to an aminoquinol which, in the form of a Cu(I) radical reacts with O_2 to form H_2O_2 Cu(II) and the imine:

$$R\text{-}CH_2\text{-}NH_2 + O_2 + H_2O \rightarrow R\text{-}CHO + H_2O_2 + NH_3$$

3. An alternative flavoprotein oxidase is, however, used for the deamination of some primary amines: these include tyramine by *Sarcina lutea* (Kumagai et al. 1969), putrescine by *Rh.*

erythropolis (van Hellemond et al. 2008), and *Micrococcus rubens* (DeSa 1972), and by the flavoprotein monooxygenase from L-lysine in *Ps. fluorescens* that is capable of the dual oxidase/monooxygenase function for production of the oxoacid and H_2O_2 (Nakazawa et al. 1972; Flashner and Massey 1974).

4. A further alternative involves quinoproteins. In quinoprotein dehydrogenation, the quinoprotein involved is tryptophan tryptophylquinone (TTQ) that has been established for methylamine in *M. extorquens* (McIntire et al. 1991), in *P. denitrificans* (Chen et al. 1994), and for primary aromatic amines in *A. faecalis* (Govindaraj et al. 1994). This is the alternative to PQQ that is used for methanol dehydrogenation in methylotrophs. The crystal structure of the quinoprotein from *Ps. putida* has been determined (Satoh et al. 2002): it is composed of three subunits, the α-subunit with a diheme cytochrome *c*, a β-subunit, and a γ-subunit with three thioether bridges linked to the tryptophylquinone. Dehydrogenation is initiated by reaction of the amine with the quinone to form an *ortho*-hydroxyarylamine that is hydrated at the R–CH = NH+– to produce a carbinolamine which forms the aldehyde and an amine (Roujeinikova et al. 2006, 2007).

5. An alternative pathway for methylamine in *Pseudomonas* sp. strain MA and *Hyphomicrobium* is initiated by the formation of *N*-methylglutamate with the loss of NH_4^+, and is mediated by an FMN-containing synthetase (Pollock and Hersh 1973). This is dehydrogenated by a flavoprotein to glutamate and formaldehyde (Bamforth and Large 1977), and established pathways then metabolize the formaldehyde. Details of this pathway have been elucidated (Latpova et al. 2010) and it is used for the assimilation of methylamine as a nitrogen source—though not as a carbon source—by the non-methylotrophic *Agrobacterium tumefaciens* (Chen et al. 2010).

6. The transformation of 2-aminopropane to alaninol by a strain of *Pseudomonas* sp. KIE171 involves reaction with glutamate in an ATP-dependent reaction to produce a γ-glutamide. Introduction of a hydroxyl group into the methyl group is followed by hydrolysis to regenerate glutamate, and dehydrogenation to alanine (Wäsch et al. 2002).

7. Species of *Pseudomonas* can utilize the α,γ-diamines putrescine, spermidine and spermine as sources of carbon and nitrogen, while putrescine can be produced from the secondary amine spermidine together with 1,3-diaminopropane (Dasu et al. 2006). Putrescine (1,4-diaminobutane) is an intermediate in the ADC pathway of L-arginine degradation, and can be degraded by two pathways.

   a. In *Ps. aeruginosa* via the analogous intermediates used for primary amines—oxidative deamination to 4-aminobutyraldehyde, followed by dehydrogenation to 4-aminobutyrate and oxidative deamination to succinate semialdehyde (Lu et al. 2002).

   b. In this pathway, the formation of Δ^1-pyrroline by internal condensation of the 4-aminobutyraldehyde is undesirable, and is avoided in *Escherichia coli* K-12 by an ATP-dependent production of γ-glutamyl putrescine by PuuA. This is followed by a sequence of reactions: (a) conversion of the terminal—CH_2NH_2 group to aldehyde by an oxidase PuuB, (b) dehydrogenation to the carboxylate by PuuC, (c) removal of glutamate by PuuD to γ-aminobutyrate, (d) conversion to succinate semialdehyde by PuuE, and (e) to succinate by YneI (Kurihara et al. 2008). The last two are important for the utilization of putrescine as sole carbon source (Kurihara et al. 2010).

8. a. Although yeasts cannot generally use alkylamines as sources of carbon the utilization of alkylamines as sources of nitrogen has been widely demonstrated, the asporogenous yeast *T. adeninovorans* was able to use the primary amines butylamine, pentylamine, hexylamine, and octylamine as well as the purines adenine and xanthine as sole source of carbon, nitrogen, and energy (Middelhoven et al. 1984).

   b. Primary amines are metabolized by the activity of oxidases that produce aldehydes and H_2O_2, and cells of *Candida boidinii* grown on methylamine, dimethylamine, and trimethylamine showed activities for catalase, formaldehyde dehydrogenase *S*-formylglutathione hydolase, formate dehydrogenase, and isocitrate dehydrogenase (Haywood and Large 1981).

c. After growth on glucose with methylamine or butylamine as sources of nitrogen, the activities of the oxidases, formaldehyde dehydrogenase, and isocitrate dehydrogenase could be demonstrated for strains of *C. boidinii*, *Hansenula minuta*, and *Pichia pastoris*. Two significant issues emerged: some strains showed two bands for oxidase activities on gels, and some were able to utilize butylamine, though not methylamine for growth (Green et al. 1982). *Candida utilis* and *Hansenula polymorpha* were able to utilize trimethylamine, dimethylamine, and methylamine consecutively with the production of formaldehyde: amines were not required as inducers of enzyme activity, and the synthesis of amine oxidase paralleled the development of peroxisomes in the cells (Zwart and Harder 1983).

d. *C. boidinii* is able to use spermidine as a nitrogen source though spermine only poorly. Cells grown on spermidine, cadaverine, and putrescine synthesize a polyamine oxidase that produces putrescine from spermidine. This is degraded, however, not by an oxidase but by transamination to 4-aminobutyraldehyde and thence via pyrrole-1-ene to succinaldehyde (Haywood and Large 1984). On the other hand, the benzylamine/putrescine oxidase from *Pichia pastoris* differs importantly from the analogous enzymes from *C. utilis* and specifically was able to catalyze the oxidation of the primary amines spermidine and spermine to the dialdehydes as well as lysine and ornithine (Green et al. 1983).

9. Amino sugars—excluding antibiotics

a. Chitin is a β(1–4)-linked homopolymer of *N*-acetyl-D-glucosamine which is widely distributed in the cells of fungi and marine invertebrates. It is an important source of nitrogen in the biosphere, and hydrolysis to *N*-acetylglucosamine provides a growth substrate for many bacteria including strains of *Vibrio* spp. (Hunt et al. 2008) and *Shewanella* spp. (Yang et al. 2006). Details of the pathway that is organized in the *nagBACD* operon have been explored in *Escherichia coli* which is able to utilize glucosamine and *N*-glucosamine as source of both carbon and nitrogen (Plumbridge and Vimr 1999; Álvarez-Añorve et al. 2005). In summary, transport into the cell is mediated by a phosphotransferase system that produces the 6-phosphates intracellularly that is then degreded successively by deacetylation to glucosamine-6-phosphate by NagA, and deamination and isomerization by NagB to produce ammonium and fructose-6-phosphate which enters the glycolytic pathway.

b. Sialic acids and related compounds derived from 2-oxo-3-deoxy-5-acetamido-D-glycero-D-galacto-nonulosonic acids (*N*-acetylneuraminic acid Neu5Ac) (Vimr et al. 2004) fulfill an important range of activities in eukaryotes including release from complex glycoproteins at sites of bacterial infection. Degradation of sialic acid is restricted to a range of pathogenic bacteria including Gamma-proteobacteria and Firmicutes: for example, the pathogens *Salmonella typhimurium*, *Vibrio cholerae,* and *Yersinia* enterocolitica are able to use sialic acid as sole source of carbon for growth (Almagro-Morena and Boyd 2009). After transport into the cell by NanT, *N*-acetylmannosamine and pyruvate are produced by the lyase (NanA), and degradation is carried out in stages analogous to those used for the degradation of *N*-acetylglucosamine—phosphorylation to the 6-phosphate by NanK, isomerization by NanE to *N*-acetylglucosamine-6-phosphate and successive deacetylation by NagA and deamination by NagB to fructose-6-phosphate which enters the central metabolic pathway. The pathway in *Bacteroides fragilis* differs importantly   after production of *N*-acetylmannosamine by NanL since isomerization is carried out by a novel NanE epimerase and the *N*-acetylglucosamine produced is phosphorylated by an ATP-dependent kinase RokA to the 6-phosphate before conversion to fructose-6-phosphate (Brigham et al. 2009).

7.9.1.1.2 *Ethanolamine: Ammonia Lyase*

A reaction formally analogous to the corrin-dependent rearrangement of 1:2-diols takes place with ethanolamine and the formation of acetaldehyde and ammonia. This is involved in the degradation of ethanolamine by *S. enterica* as source of carbon and nitrogen. This takes place, however, only under anaerobic conditions when adenosylcobalamin (AdoCbl) is synthesized, or under aerobic conditions when this is generated from exogenous vitamin B_{12} (Roof and Roth 1989). It has been suggested—without revealing the mechanism—that *S. enterica* was able to conserve in carboxysome-like organelles acetaldehyde that was produced from the activity of the diol dehydratase (Penrod and Roth 2006). An unusual situation emerges with *P. denitrificans* that is able to synthesize cobalamin under both aerobic and anaerobic conditions, and can, therefore, utilize ethanolamine under both conditions (Shearer et al. 1999). Ethanolamine ammonia-lyase from *Clostridium* sp. has been shown to generate an EPR spectrum resulting from production of the CH_3-$CH(NH_2)$-$CHOH^*$ radical from 2-aminopropanol (Babior et al. 1974).

7.9.1.1.3 N-Alkylamines: Secondary and Tertiary Amines

The mechanisms for the degradation of tri- and dialkylamines depend on the organism. For trimethylamine, this involves de-*N*-methylation with the successive formation of dimethylamine, and methylamine. Demethylation can be accomplished by two mechanisms, dehydrogenation or through formation of the *N*-oxides.

7.9.1.1.3.1 Trimethylamine

1. The dehydrogenation of trimethylamine to dimethylamine and formaldehyde in a bacterial strain W3A1 involves incorporation of the oxygen from H_2O:

$$(CH_3)_3N + H_2O \rightarrow (CH_3)_2NH + CH_2O.$$

   The enzyme contains two unequal subunits, one containing an [4Fe–4S] and the other an unusual covalently bound coenzyme that has been shown to be FMN with a cysteine bound through the sulfur to position 6 of the FMN (Steenkamp et al. 1978b,c) that enabled details of the mechanism that involves an [4Fe–4S] iron–sulfur cluster to be established (Steenkamp et al. 1978a).

2. An alternative mechanism involves formation of the *N*-oxide by a monooxygenase in *Aminobacter (Pseudomonas) aminovorans* (Boulton et al. 1974), followed by a nonoxidative, nonhydrolytic formation of dimethylamine and formaldehyde (Boulton and Large 1977, 1979).

7.9.1.1.3.2 Dimethylamine

Analogous reactions take place for dimethylamine—dehydrogenation to methylamine by *Hyphomicrobium* X by a dehydrogenase that is analogous to the enzyme for trimethylamine (Meiberg and Harder 1978, 1979; Steenkamp 1979), or monooxygenation in *Aminobacter (Pseudomonas) aminovorans* (Brook and Large 1976). It is worth noting that *Hyphomicrobium* X is able to synthesize two enzymes for the metabolism of dimethylamine—dimethylamine monooxygenase at high oxygen concentrations when oxygen is necessary and dimethylamine dehydrogenase under anaerobic conditions (Meiberg et al. 1980).

7.9.1.1.4 Methylamine

The degradation of methylamine in methylotrophs including *P. denitrificans* and *Pa. venustus* that is mediated by tryptophan tryptophylquinone (TTQ) is discussed in the section dealing with dehydrogenation.

7.9.1.1.5 N-Alkylated Amines

The degradation of a range of *N*-methylated amines has been examined, including *N,N'*-dimethylformamide, tetramethylammonium hydroxide, glycine betaine (Me_3N^+-CH_2-CO_2^-), carnitine (Me_3N^+-CH_2-$CH(OH)$-CH_2-CO_2^-), and sarcosine ($MeNH$-CH_2-CO_2H).

1. Species of *Paracoccus* that were able to degrade tetramethylammonium hydroxide have been assigned to the species *Pa. kocurii* (Ohara et al. 1990), and those degrading *N,N'*-dimethylformamide to *Pa. aminophilus* and *Pa. aminovorans* (Urakami et al. 1990). Although *Aminobacter aminovorans*—formerly *Pseudomonas aminovorans* and *Paracoccus aminovorans*—was able to utilize methylamine and trimethylamine, it was unable to use methane, methanol, or dimethylamine (Urakami et al. 1992).

2. In *Sinorhizobium (Rhizobium) meliloti*, glycine betaine is produced from choline by oxidation and dehydrogenation. Successive demethylation produced glycine that, by reaction with tetra-hydrofolate produced serine that underwent dehydration to pyruvate. Significantly for an osmolyte, the specific activities of the enzymes succeeding glycine betaine were reduced under conditions of high osmolarity (Smith et al. 1988).

3. The degradation of dimethylglycine and sarcosine has been examined in species of *Arthrobacter* (Meskys et al. 2001). The degradation of both was catalyzed by an oxidase, and the formaldehyde that was produced from the *N*-methyl groups was processed through 10-formyltetrahydrofolate, while glycine was converted into serine by the activity of serine hydroxymethyltransferase.

4. A number of reactions have been shown to occur during bacterial metabolism of carnitine, and reference should be made to a review (Kleber 1997) for further details. Salient aspects are briefly summarized and attention is drawn to some apparent conflicts.

    a. Under anaerobic conditions, several *Enterobacteriaceae* are able to transform carnitine in the presence of carbon and nitrogen substrates by dehydration to crotonobetaine followed by reduction to γ-butyrobetaine (Eichler et al. 1994a):

$$\text{Me}_3\text{N}^+\text{-CH}_2\text{-CH(OH)-CH}_2\text{-CO}_2^- \rightarrow \text{Me}_3\text{N}^+\text{-CH}_2\text{-CH=H-CO}_2^-$$
$$\rightarrow \text{Me}_3\text{N}^+\text{-CH}_2\text{-CH}_2\text{-CH}_2\text{-CO}_2^-.$$

       In *E. coli*, the enzymes occur in a *caiTABCDE* operon (Eichler et al. 1994b), and it has been shown that *caiD* encodes crotonobetainyl-CoA hydratase and *caiB* carnitine coenzyme A-transferase (Elssner et al. 2001). These findings are compatible with the use of carnitine and crotonobetaine as electron acceptors during anaerobic growth of *E. coli* (Walt and Kahn 2002). Under aerobic conditions, carnitine can also be metabolized by some *Enterobacteriaceae* (Elßner et al. 1999), and the transformation of L-crotonobetaine to L-carnitine has been examined in *Proteus* sp. grown on a complex medium (Engemann et al. 2005). The function of the genes in the operon was elucidated, and it was shown that *caiA* encoded the irreversible reductase for crotobetainyl-CoA, *caiB* the specific CoA-transferase for carnitine, crotonobetaine and γ-butyrobetaine, *caiD* the components of the cronobetaine hydration system, and *caiT* transport proteins for betaines.

    b. In contrast to the reactions already described for anaerobic conditions, some *Enterobacteriaceae* are able to transform L-carnitine to crotonobetaine and γ-butyrobetaine under aerobic conditions in the presence of carbon and nitrogen sources (Elßner et al. 1999). This has been demonstrated for *E. coli* ATCC 25922, and strains of *Proteus mirabilis* and *Prot. vulgaris*, in which L-carnitine dehydratase and cronobetaine reductase were present, although there was no evidence for their presence in *Kleb. pneumoniae*, *Enterobacter cloacae* or *C. freundii*.

    c. A strain of *Ps. putida* grown with D-, L-carnitine degraded only L-carnitine to produce glycine betaine $\text{Me}_3\text{N-CH}_2\text{-CO}_2^-$ or trimethylamine after growth with γ-butyrobetaine (Miura-Fraboni et al. 1982). In *A. calcoaceticus* ATCC 39647, the production of L-carnitine from the racemate involved specific metabolism by of D-carnitine to trimethylamine and malate mediated by a monooxygenase in (Ditullio et al. 1994).

$$\text{D-Me}_3\text{N}^+\text{-CH}_2\text{-CH(OH)-CH}_2\text{-CO}_2^- \rightarrow \text{Me}_3\text{N} + \text{HO}_2\text{C-CH(OH)-CO}_2\text{H}$$

       It was suggested that this could be the result of differential transport systems for the enantiomers.

    d. The metabolism of γ-butyrobetaine has presented some apparent conflicts:

       i. A strain of *Pseudomonas* AK1 was isolated by enrichment with γ-butyrobetaine that was able to serve as the sole source of both carbon and nitrogen. Labeled carnitine was isolated from labeled γ-butyrobetaine, and hydroxylation of γ-butyrobetaine by a

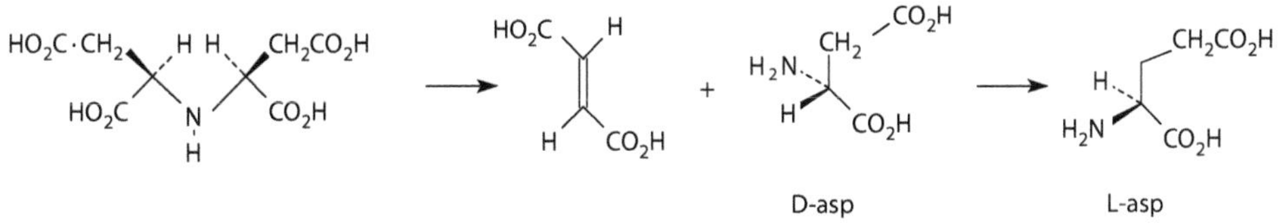

D-asp   L-asp

FIGURE 7.60   Aerobic degradation of iminodisuccinate.

      2-oxoglutarate-dependent enzyme was demonstrated and the enzyme purified (Lindstedt et al. 1977b).

  ii. On the other hand, growth on γ-butyrobetaine and both L- and D-carnitine by *Acinetobacter calcoaceticus* and *Ps. putida* resulted in the formation of trimethylamine (Miura-Fraboni et al. 1982), although the fate of the remaining carbon remained controversial. In that study two apparently conflicting observations were made: 2-oxoglutarate-linked dioxygenase activity was not detected and degradation of γ-butyrobetaine did not involve carnitine as an intermediate.

7.9.1.1.6 *Other* N-*Alkylated Amines: Nitrilotriacetate (NTA), Ethylenediamine Tetraacetate (EDTA)*

The degradation of these and related compounds display some additional features:

1. The degradation of NTA takes place by successive loss of glyoxylate (Cripps and Noble 1973; Firestone and Tiedje 1978) and the monooxygenase system consists of two components, both of which are necessary for hydroxylation (Uetz et al. 1992; Xu et al. 1997).

2. The first step in the degradation of EDTA is carried out by a flavin-dependent monooxygenase which has been purified (Witschel et al. 1997; Payne et al. 1998), and results in the formation of glyoxal and N,N-ethylenediaminediacetate (EDDA). The ultimate formation of ethylenediamine and glyoxylate is, however, carried out by an oxidase (Liu et al. 2001). It may be assumed that an analogous pathway exists for the degradation of triethylenediaminepentaaacetic acid (DTPA).

3. Iminodisuccinate that is a potential replacement for ETDA has two asymmetric centers and, since the C–N lyase that cleaves it to D-aspartate and fumarate is stereospecific, degradation is initiated by the activity of epimerases to the R,S enantiomer. D-Aspartate is then isomerized to L-aspartate that undergoes elimination to fumarate (Figure 7.60) (Cokesa et al. 2004).

7.9.1.1.7 *Morpholine*

The degradation of morpholine has been described in several strains of *Mycobacterium*, including *M. chelonae* (Swain et al. 1991), *M. aurum strain* MO1 (Combourieu et al. 1998), and *Mycobacterium* strain RP1 (Poupin et al. 1998). Initiation of the degradation of the cyclic secondary amines morpholine, piperidine, and pyrrolidine by the strain of *Mycobacterium* RP-1 is carried out by cytochrome P450 hydroxylation (Poupin et al. 1998). The first step in the degradation of the tertiary amine nicotine by *Arthrobacter nicotinovorans* involves hydroxylation to 6-hydroxynicotine that is followed by fission of the N-methylpyrrolidine ring by an oxidase (Dai et al. 1968) and is discussed in Part 1 of Chapter 10.

7.10 Amino Acids

A number of factors complicate the aerobic metabolism of amino acids—different enzymes may be used even for the same amino acid; the enzymes may be inducible or constitutive depending on their function; α-oxoacids may be produced by deamination, or amines by decarboxylation.

    In view of the interest in their application, brief attention is directed to a few polymeric amino acids where the greatest interest has been directed to *Bacillus* spp. For glutamic acid, polymerization may take place to produce either the γ-amide or the α-amide, and correspondingly from lysine to the ε-amide or the α-amide.

1. Poly-γ-glutamate and poly-ε-lysine

   a. *B. subtilis* produces an extracellular glutamylpeptide polymer, whereas the polymer from *Bacillus anthracis* is contained in the capsule: in both of them D-glutamate was the major component (Thorne et al. 1954) and the polymer is also produced by *B. licheniformis* and *B. megaterium* (references in Urushibata et al. 2002). Polymer synthesis in *B. licheniformis* is membrane-mediated and involves activation of glutamate, racemization and polymerization of L-glutamate (Troy 1973). The synthesis is repressed by exogenous glutamate in *B. licheniformis*, though not in *B. subtilis* (Kambourova et al. 2001). A synthase has been purified from *B. subtilis* and in the presence of ATP and Mn^{2+} catalyzes the production of high molecular weight poly-γ-glutamate from L-glutamate through an acylphosphate intermediate (Urushibata et al. 2002). The DL-glutamate served as the best substrate, and use of D- or L-glutamate produced polymers with the respective configuration (Ashiuchi et al. 2004).

   b. Strain 40 of the extreme halophilic archaeon *N. aegyptica* produced poly-(γ-glutamic acid) using casamino acids as substrate (Hezayen et al. 2000, 2001).

   c. The analogous metabolite produced by *Streptomyces albus* from lysine is an ε-L-polyamide consisting of 25–25 units of the monomer with terminal amino and carboxyl groups. Considerable effort has been directed to it biosynthesis that is dependent on a low pH and is regulated by intracellular levels of ATP (Yamanaka et al. 2010).

2. Poly-α-glutamate: An enzyme RimK from *E. coli* K-12 was able to produce poly-α-glutamate by hydrolysis of ATP to ADP and phosphate (Kino et al. 2011). Application of NMR conclusively proved the structure that was clearly differentiated from the γ-polymer, and polymers with a maximum of 46 monomer units was synthesized, while the chain length was dependent on the pH.

7.10.1 Deamination and Decarboxylation of Amino Acids

7.10.1.1 Deamination

Pyridoxal-5′phosphate enzymes are used for oxidative deamination when amino acids are supplied as sources of carbon or nitrogen. An aldimine is initially produced and is isomerized to the corresponding ketimine through loss of the α-proton which is facilitated by a lysine group on the enzyme. The ketimine is then hydrolyzed to pyridoxamine phosphate and an α-oxocarboxylic acid that is incorporated into established catabolic pathways. Transamination using, for example, 2-oxoglutarate as acceptor may be used, while nonoxidative deamination is discussed in the section on amino acid-ammonia lyases. There are a number of important additional features that are illustrated for specific amino acids.

7.10.1.1.1 Lysine

There are several mechanisms for the degradation of lysine by pseudomonads.

1. Details and the enzymology of the degradation of L-lysine in *Pseudomonas putida* have been established (Miller and Rodwell 1971; Chang and Adams 1984; Muramatsu et al. 2005). The δ-aminovalerate pathway for L-lysine is initiated by L-lysine monooxygenase with concomitant decarboxylation to cadaverine that undergoes deamination and dehydrogenation to glutarate (Figure 7.61a). The enzymes for the first four steps have been characterized, and it has been shown that the genes *davBA* encoding L-lysine monooxygenase and 5-aminovaleramide amidase, and *davDT* that encode 5-aminovalerate aminotransferase and glutarate semialdehyde dehydrogenase occur on separate operons (Revelles et al. 2004, 2005). The alternative

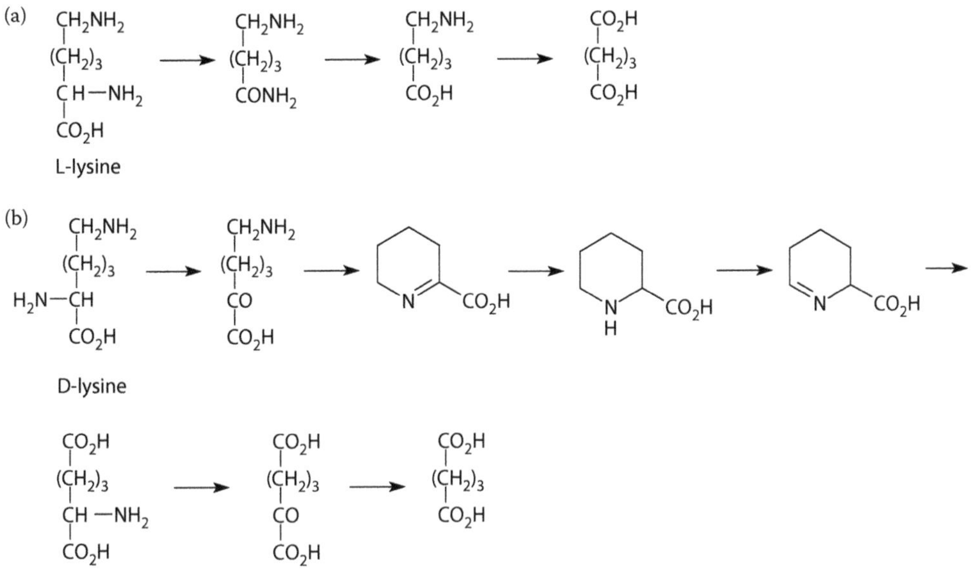

FIGURE 7.61 Degradation of (a) L-lysine and (b) D-lysine.

pipecolate pathway (Figure 7.61b) is used for D-Lysine, and the genetics have been established (Revelles et al. 2007).

L-Lysine monooxygenase is a flavoprotein containing four subunits with four FAD prosthetic groups and has been purified and characterized from *Ps. fluorescens* (Flashner and Massey 1974). The outstanding issue is the physiological interconnection of these pathways that is supported by genetic evidence in *Ps. putida* strain KT2440 (Revelles et al. 2005)—an efficient colonizer of important commercial plants in the rhizosphere (Molina et al. 2000). The critical factor is the extent to which lysine racemase is physiologically active in this interconnection. Although this enzyme has been characterized (Ichihara et al. 1960), is apparently widely distributed in bacteria (Huang and Davisson 1958), and a pyridoxal-5′-phosphate-independent racemase has been isolated from a soil metagenome (Chen et al. 2009), it has not hitherto been unequivocally established that this interconnection actually occurs. There are specific differences in the metabolism of D- and L-lysine that would be required for interconnection of the pathways: (a) the existence of a racemase for L- and D-lysine that has been described (Ichihara et al. 1960), and (b) the channeling of 2-aminoadipate from the D-lysine route to glutarate that is a product of the L-lysine route.

2. L-Lysine cannot be used effectively for growth by *Ps. aeruginosa* and a mutant that could grow with L-lysine lacked two cardinal enzymes normally required for L-lysine degradation—L-lysine monooxygenase that is necessary for operation of the δ-aminovalerate pathway and L-pipecolate dehydrogenase that is a component of the piperid-1-ene-6-carboxylate pathway. This alternative pathway involved decarboxylation of L-lysine to cadaverine followed by dehydrogenation and cyclization to piperid-1-ene, ring opening to 5-aminovalerate and transamination to glutarate semialdehyde (Fothergill and Guest 1977). The decarboxylase pathway has been examined in *Ps. aeruginosa* PAO1 and the pyridoxal-5′-phosphate-dependent L-lysine decarboxylase was characterized as obligatory for growth at the expense of L-lysine (Chou et al. 2010). Examination of the *Pseudomonas* Genome Database showed indeed that the *ldcA* gene that encodes the decarboxylase was quite widely distributed in this genus. The remarkable role of L-arginine was discussed since the *ldcA* promotor was induced by exogenous L-arginine—though not by L-lysine.

3. It is worth noting that the L-lysine monooxygenase in *Ps. fluorescens* shows oxidase activity in the ability to metabolize L-ornithine to 2-oxo-5-aminovalerate (Nakazawa et al. 1972; Flashner and Massey 1974).

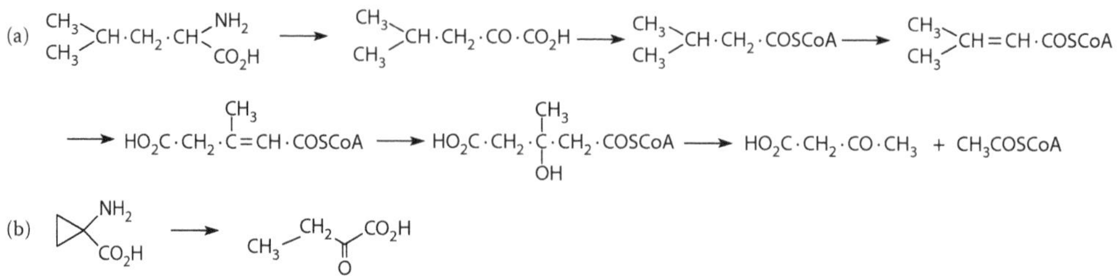

FIGURE 7.62   Degradation of (a) D- and L-leucine and (b) 1-aminocyclopropane-1-carboxylate.

7.10.1.1.2 L- and D-Leucine

These are degraded by *Ps. aeruginosa* in a pathway analogous to that for the branched-chain alkanes and the acyclic terpenoids citronellol and geraniol—formation of isovaleryl-CoA, dehydrogenation to methylcrotonly-CoA that is degraded by terminal carboxylation to 3-methylgutaconyl-CoA before hydroxylation and fission to acetyl-CoA and acetoacetate (Figure 7.62a) (Massey et al. 1974; Höschle et al. 2005).

7.10.1.1.3 1-Aminocyclopropane-1-Carboxylate

This somewhat unusual amino acid is the immediate precursor of the important plant hormone ethene (Adams and Yang 1979) and the bacterial deaminase in the rhizosphere has, therefore, attracted considerable attention (Belimov et al. 2001) that is discussed more fully in Chapter 2. Unlike other amino acids, 1-aminocyclopropane-1-carboxylate does not contain the β-proton that is required for conventional deamination, and its degradation involves concomitant fission of the cyclopropane ring to α-oxobutyrate (Figure 7.62b). The mechanism has been examined, and several possibilities have been put forward that are consistent with the x-ray structure of the enzyme (Karthikeyan et al. 2004) and with isotopic evidence (Walsh et al. 1981).

7.10.1.1.4 L-Arginine

The degradation of arginine can be accomplished by at least four different mechanisms depending on the organism and the growth conditions (Jann et al. 1988; Nakada and Itoh 2002). In pseudomonads, the arginine succinyltransferase (AST) pathway is generally preferred and, in its absence arginine: pyruvate transamination to produce 2-oxoarginine may be used (Yang and Lu 2007). The deiminase ADI pathway enables organisms to grow anaerobically with L-arginine, and includes *Bacillus licheniformis*, *Halobacterium halobium*, *Streptococcus faecalis* and, to a limited extent *Ps. aeruginosa*. The several pathways are summarized:

Arginine deiminase (ADI) pathway (Mercenier et al. 1980; Wauven et al. 1984):

Arginine → citrulline → ornithine + carbamoyl phosphate

Arginine decarboxylase (ADC) pathway (Lu et al. 2002):

Arginine → agmatine → *N*-carbamoylputrescine → putrescine → 4-aminobutyrate

Arginine dehydrogenase (ADH) pathway (Nakada and Itoh 2002):

Arginine → 2-oxoarginine → 4-guanidobutyraldehyde → 4-guanidobutyrate →
4-aminobutyrate + urea

Arginine succinyltransferase (AST) pathway (Schneider et al. 1998):

Arginine → N^2-succinylarginine → N^2-succinylornithine → → N^2-succinylglutamate →
glutamate

Decarboxylation of arginine produces agmatine [(4-aminobutyl)guanidine] that can be metabolized by a deiminase to *N*-carbamoylputrescine. The conversion of this to putrescine can be mediated either by the *aguB* product *N*-carbamoylputrescine amidohydrolase, or by the *ptcA* product putrescine *N*-carbamoylytransferase. These are different enzymes that are employed selectively, *ptcA* in Grampositive and *aguB* in Gram-negative organisms (Landete et al. 2010).

7.10.1.1.5 Threonine

Threonine fulfils both anabolic and catabolic functions. For the latter, the aerobic degradation of L-threonine by strains of *Pseudomonas* is initiated by the activity of an aldolase to produce glycine and acetaldehyde that can be further metabolized for growth (Bell and Turner 1977). An aldolase that is active for D-threonine has been characterized in *Arthrobacter* sp. (Kataoka et al. 1997), and both of these aldolases require pyridoxal-5′-phosphate. In the anabolic function of threonine, it is the precursor of isoleucine whose synthesis is initiated by the activity of threonine dehydratase to produce 2-oxobutyrate followed by a complex sequence of reactions. An alternative reaction has been discovered under anaerobic conditions in a strain of *E. coli* that was unable to synthesize pyruvate formate-lyase. It produced an enzyme that was able to accept 2-oxobutyrate to produce propionyl-CoA and formate and thence propionate (Heßlinger et al. 1998).

7.10.1.1.6 Phenylalanine and Tyrosine

There are several pathways for the microbial metabolism of L-phenylalanine, and it is appropriate to contrast those for bacteria, yeasts and fungi. The pathway for bacterial degradation of aromatic amino acids is generally initiated by oxidative deamination to 2-oxoacids, whereas the activity of phenylalanine ammonia-lyase (PAL) to produce cinnamate is less common in bacteria, although it is widespread in yeasts and fungi.

7.10.1.1.7 Bacteria

1. In the metabolism of phenylalanine and tyrosine by *Streptomyces setonii* the activity of ammonia-lyases followed by hydration and dehydration to produce phenylacetate and 4-hydroxyphenylacetate. These were hydroxylated to 2,5-dihydroxyphenylacetate (homogentisate)—that presumably involved an NIH shift for 4-hydroxyphenylacetate (Pometto and Crawford 1985).

2. Phenylalanine may be deaminated and oxidized to phenylacetate. The degradation by *Ps. putida* strain U and Y2 is carried out by a pathway that is different from that used for the hydroxylated substrates. In brief, it is initiated by formation of the coenzyme A ester that is oxygenated to a 1,2-epoxide that is rearranged to an oxepin. This is hydrated to 3-oxo-5,6-dehydrosuberyl-CoA that undergoes scission to produce acetyl-CoA and succinyl-CoA (Figure 7.63) (Teufel et al. 2010).

3. In *Ps. putida* strain U, the degradation of phenylalanine is initiated by hydroxylation to tyrosine (Arias-Barrau et al. 2004) which circumvents the oxepin pathway that is used for phenylacetate and L-phenylalanine by *Ps. putida* strain Y2 (Teufel et al. 2010). This is followed by transamination with 2-oxoglutarate to 4-hydroxybenzoyl-pyruvate. The degradation of L-phenylalanine and L-tyrosine converge, therefore, at the production of 4-benzoylpyruvate that is dioxygenated to 2,5-dihydroxyphenylacetate (homogentisate)—in which both the 2-hydroxyl

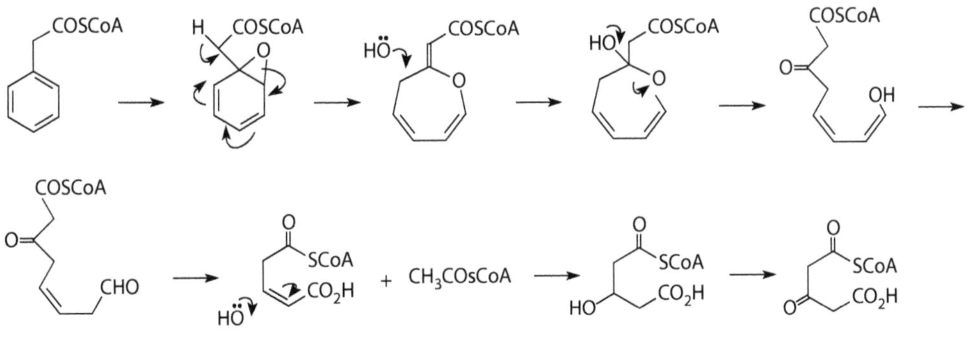

FIGURE 7.63 Oxepin route for degradation of phenylacetate.

and the carboxylate oxygen atoms are derived from O_2 (Lindstedt et al. 1977a; Johnson-Winters et al. 2003). Homogentisate undergoes ring fission to maleylacetoacetate followed by isomerization and hydrolysis to fumarate and acetoacetate (Arias-Barrau et al. 2004). Details of the metabolism of homogentisate metabolism are given in Chapter 8. This pathway involving initial hydroxylation is used by the facultative methylotroph *Nocardia* sp. strain 239 (Boer et al. 1988).

4. Tyrosine phenol lyase that occurs in several facultative anaerobes including *Escherichia coli*, *Citrobacter freundii*, and *Proteus morganii* carries out the pyridoxal-dependent β-elimination of tyrosine to phenol, pyruvate, and ammonia. The reaction that is reversible has attracted biotechnological interest and is initiated by reaction with Lys_{257} followed by formation of a quinonoid intermediate (Milić et al. 2011): this is mechanistically analogous to tryptophanase that is discussed in Chapter 10, Part 10.3.1.1.

7.10.1.1.8 Yeasts

1. In *Sporobolomyces roseus*, the activity of phenylalanine ammonia-lyase (PAL) produced cinnamic acid, followed by oxidation and ring-hydroxylation to 4-hydroxybenzoate that was also produced from tyrosine. Further ring-hydroxylation produced the terminal metabolite 3,4-dihydroxy-benzoate (Moore et al. 1968).

2. In *Rhodotorula glutinis* using $^3H[C_4]$-phenylalanine, analogous reactions produced benzoate although the ring-hydroxylation of this involved an NIH shift to produce $^3H[C_3]$-4-hydroxybenzoate while 3,4-dihydroxybenzoate was not produced (Marusich et al. 1981). It was shown in the yeast *R. graminis* that benzoate hydroxylase was required for the degradation of phenylalanine and mandelate that took place with the formation of 3,4-dihyroxybenzoate which underwent intradiol ring fission and entry into the 3-oxoadipate pathway (Durham et al. 1984).

3. Tyrosine can serve as a growth substrate for the yeast *Trichosporum cutaneum* (Sparnins et al. 1979). Elimination of ammonia produced 4-hydroxycinnamate that was converted to the coenzyme A ester before hydration and fission to acetyl-CoA and 4-hydroxy-benzaldehyde. Subsequent reactions produced 1,2,4-trihydroxybenzene that underwent fission to maleylacetate.

7.10.1.1.9 Fungi

1. Metabolism of phenylalanine by the basidiomycete *Ischnoderma benzoinum* involved two sets of reaction: oxidative deamination to phenylpyruvate followed by oxidation to benzoate and a reductive segment with PAL formation of cinnamate that was reduced to 3-phenylpropanol (Krings et al. 1996).

2. Several transformations of L-phenylalanine by the white-rot fungus *Bjerkandera adusta* are initiated by the activity of phenylalanine ammonia-lyase to produce cinnamate that is followed by a range of reactions (Lapadatescu et al. 2000).

    a. Hydration to 2-hydroxyphenylpropionate and oxidation to phenylpyruvate, followed by decarboxylation to phenylacetaldehyde, dehydrogenation to phenylacetate, and hydroxylation at C_2 to mandelate/benzoylformate

    b. Hydration to 3-hydroxyphenylpropionate followed by dehydrogenation to 3-oxophenyl-propionate and decarboxylation to acetophenone, or hydrolysis to benzoate

    c. Products from these reactions may be hydroxylated at the *para* position to 4-hydroxy-benzoates and *O*-methylation to a series of compounds such as veratric acid

7.10.2 Mechanisms

7.10.2.1 Decarboxylation

1. As an alternative to these mechanisms, decarboxylation is involved in polyamine biosynthesis and in the protection of enteric bacteria from acidic conditions that is discussed in Chapter 4.

Under these conditions, the electrons used to form the ketimine are provided by decarboxylation in place of those from the α-methine group. In addition, decarboxylation is able to provide a protomotive force and hence the generation of ATP from decarboxylation.

a. In *E. coli*—and probably other *Enterobacteriaceae*—amino acids play a role in the resistance to extremely acid conditions through removal of intracellular protons by decarboxylation. Tolerance is mediated by decarboxylase antiporter systems including arginine/agmatine, glutamate/γ-aminobutyrate, and lysine/cadaverine (Iyer et al. 2003). The first of these also is involved in the acid-resistance of *S. enterica* serovar *Typhimurium* under anaerobic conditions (Kieboom and Abee 2006), while decarboxylation of lysine mediates the acid tolerance of both *Vibrio cholerae* and *V. vulnificus* that are noted in Chapter 4.

b. i. In Gram-positive lactobacilli, this is illustrated by the decarboxylation of histidine to histamine by *Lactobacillus buchneri* (Molenaar et al. 1993), and of glutamate to γ-aminobutyrate by *Lactobacillus* sp. strain E1 (Higuchi et al. 1997). In *Listeria monocytogenes* this reaction does not, however, take place in a chemically defined medium, in contrast to a complex medium (Karatzas et al. 2010). It was shown that the antiporter/decarboxylase system (GadT2D2) that has been considered as necessary for acid-tolerance in *L. monocytogenes* was transcribed poorly in the defined medium, whereas *gadD1T1* and *gadD3* were overexpressed at pH 3.5. These results clearly demonstrated that the production of γ-aminobutyrate can be uncoupled from its efflux.

ii. The decarboxylation of tyrosine to tyramine has been examined in *Enterococcus faecalis*, the genetic sequence was determined, and it was shown that the decarboxylase belonged to the class of pyridoxal-phosphate-requiring enzymes (Connil et al. 2002).

iii. The production of putrescine by decarboxylation has been demonstrated to occur in some species of *Staphylococcus*, exceptionally in *S. epidermis* 2015B (Coton et al. 2010) and in virtually all strains of *S. lugdunensis* (Tsoi and Tse 2011).

c. Polyamines have an important function in contributing to the stability of molecular complexes in the cell. The biosynthesis of putrescine generally involves decarboxylation of arginine or ornithine. There are two decarboxylases for arginine and ornithine, both of which are dependent on pyridoxal 5'-phosphate and Mg^{2+} (Refs in Tabor and Tabor 1985), and probably also for lysine (Lemonnier and Lane 1998; Kikuchi et al. 1997). The synthesis of these alternative enzymes is controlled by the conditions under which the cells are grown—aeration at neutral pH, or semi-anaerobic at low pH. Exceptionally, a pyruvoyl-dependent arginine decarboxylase in *Methanococcus jannaschii* functions in polyamine synthesis (Graham et al. 2002). The 3-aminopropyl group in spermine and spermidine is produced by decarboxylation of *S*-adenosylmethionine that is noted in the next paragraph.

2. A distinct enzyme has been found in a number of organisms that carry out the metabolism of amino acids. In this group of bacteria, a pyruvoyl group is covalently bound to the active enzyme that is produced from a proenzyme in a self-maturation process (Toms et al. 2004). The proenzyme contains a serine residue that undergoes rearrangement to an ester followed by conversion to the β-chain of the enzyme and a dehydroalanine residue that forms the *N*-terminal pyruvoyl group of the α-chain. This type of enzyme has been determined for a number of important decarboxylations:

a. Histidine decarboxylation to histamine in *Lactobacillus* including *L. buchneri* (Huynh and Snell 1985; Martín et al. 2005)—that is the most extensively studied in which the histidine/histamine antiport is electrogenic (Molenaar et al. 1993). This is formally analogous to the decarboxylation of glutamate to γ-aminobutyrate in *Lactobacillus* strain E1 (Higuchi et al. 1997).

b. Arginine decarboxylation in *Methanococcus jannaschii* (Graham et al. 2002);

  c. *S*-adenosylmethionine carboxylase in *E. coli* (Markham et al. 1982; Anton and Kutny 1987), *Sulfolobus solfataricus* (Cacciapuoti et al. 1991), *M. jannaschii* (Lu and Markham 2004), and *Thermotoga maritima* (Toms et al. 2004).

S-Adenosylmethionine carboxylase is the source of the propylamine in the polyamines spermine and spermidine. The activity of spermine synthase introduces this into spermidine and spermine that has already been noted. It is worth pointing out that, whereas the inducible histidine decarboxylase from Gram-positive bacteria belongs to the pyruvoyl group, those from the Gram-negative *Enterobacteriaceae* including *Klebsiella planticola* and *Enterobacter aerogenes* are pyridoxal 5′-phosphate dependent (Kamath et al. 1991).

7.10.2.1.1 Oxidases

Amino acids may undergo FAD-dependent reactions that have been described as oxidases. These are initiated by the formation of $FADH^-$ and $H_2N^+{=}C(R){-}CO_2^-$ which can then undergo two reactions, either $R{-}CO{-}CO_2^- + H_2O_2 + NH_3$, or $R{-}CO{-}NH_2 + CO_2 + H_2O$. Extensive study has been given to the first of these enzymes while the second is important in the alternative synthesis of indole-3-acetic acid by production of the amide from tryptophan followed by hydrolysis. It has been shown on the basis of extensive study of deuterium and ^{15}N kinetic isotope effects that hydride transfer is the critical reaction for both (Ralph et al. 2006).

  1. Glycine oxidase from *B. subtilis* is capable of deaminating glycine, sarcosine, *N*-ethylsarcosine to glyoxylate and NH_3 or primary amines (Job et al. 2002).
  2. Sarcosine oxidase converts sarcosine (*N*-methylglycine) to glycine, 5,10-methylene-tetrahydrofolate and H_2O_2. In two strains of *Corynebacterium* sp. U-96 (Suzuki 1981) and P-1 (Chlumsky et al. 1995) it has a heterotetrameric structure, and is unusual since it contains one mole of covalently bound and one mole of noncovalently bound FAD: the noncovalent FAD accepts electrons from sarcosine that are transferred to the covalent FAD where oxygen is reduced to H_2O_2 (Chlumsky et al. 1995).
  3. The plant hormone tryptophan 3-acetic acid (IAA) is synthesized from L-tryptophan in two pathways that depend on the organism. In *Pseudomonas savastanoi*, an FAD-dependent oxidase encoded by *iaaM* produces tryptophan-3-acetamide (Comai and Kosuge 1982) followed by a hydrolase encoded by *iaaH*. On the other hand, in *Erwinia herbicola* (Brandl and Lindow 1996) and *Enterobacter cloacae* (Schütz et al. 2003), L-tryptophan aminotransferase produces indol-3-pyruvate. This undergoes decarboxylation to indole-3-acetaldehyde that is oxidized to tryptophan-3-acetate. These pathways are not, however, exclusive since the oxidase of *Pseudomonas* sp. strain P.501 possessed both deaminating and decarboxylating activities, although it had a rather narrow substrate range—L-phenylalanine, L-tyrosine, L-tryptophan, and L-methionine (Koyama 1982). L-Phenylalanine produced both 3-phenylpyruvate and 3-phenyl-acetamide from L-phenylalanine, whereas L-methionine was mainly oxidized to the α-ketoacid (Koyama 1984). In contrast, the L-amino oxidase from *Bacillus carotarum* had a more extensive range of substrates that included, less effectively some D-amino acids (Brearley et al. 1994).

7.10.2.1.2 Dioxygenation

The 2-oxoacid Fe^{2+}-dependent dioxygenases are able to carry out the hydroxylation of amino acids to products some of which are constituents of antibiotics.

  1. The degradation of 2-aminoethanesulfonate (taurine) by *E. coli* takes place by hydroxylation at the position adjacent to the sulfonate group with the production of aminoacetaldehyde and sulfite (Eichhorn et al. 1999; Ryle et al. 2003).
  2. The dioxygenase from *Streptomyces griseoviridus* is able to hydroxylate L-proline to produce *trans*-4-hydroxy-L-proline that is a component of the antibiotic etamycin (Lawrence et al. 1996).

3. The enzyme from *B. thuringiensis* has been extensively examined for the hydroxylation of a range of amino acids that includes C_4-hydroxylation of L-isoleucine, C_5-hydroxylation of L-norleucine, sulfoxidation of L-methionine and L-ethionine (Hibi et al. 2011).

7.10.2.1.3 Amino Acid Ammonia-Lyase

Amino acid ammonia-lyases have been found in several bacteria where they serve a biosynthetic function. Phenylalanine-ammonia lyase (PAL) catalyzes the nonoxidative deamination of L-phenylalanine to *trans*-cinnamate, and is widespread among plants as a precursor of lignins, flavanoids, and coumarins.

1. The *encP* gene encoding phenylalanine ammonia-lyase has been characterized in "*Streptomyces maritimus*" where it was part of a cluster that carried out the biosynthesis of the antibiotic enterocin (Xiang and Moore 2002). Strains of the insect pathogen *Photorhabdus luminescens* that is a member of the *Enterobacteriaceae* produce the antibiotic 3,5-dihydroxy-4-isopropylstilbene, and the gene *stlA* encoded the analogous PAL (Williams et al. 2005). The crystal structure of the enzyme from the yeast *Rhodosporidium toruloides* showed similarity to the mechanistically analogous histidine ammonia-lyase (Calabrese et al. 2004).
2. The gene for tyrosine ammonia-lyase (TAL) was found in *Rh. capsulatus* and the enzyme after expression in *E. coli* had a much greater activity for tyrosine than phenylalanine (Kyndt et al. 2002).
3. The analogous L-histidine ammonia lyase produces *trans*-urocanate from L-histidine that is the first enzyme in the pathway for the bacterial degradation of L-histidine, and the crystal structure made it possible to propose a mechanism (Schwede et al. 1999).

7.10.3 Anaerobic Conditions

The methanogens *Methanococcoides methylutens* (Sowers and Ferry 1983) and *Methanolobus tindarius* (Konig and Stetter 1982) are able to use methylamines and methanol, and *Methanomethylovorans hollandica* can use a range of methyl substrates including mono-, di-, and trimethylamine as well as dimethyl sulfide and methanethiol (Lomans et al. 1999). The genus *Methanosarcina* contains metabolically versatile species which are able to use the C1 substrates methanol (Sauer et al. 1997), the methylated amines (Paul et al. 2000) and dimethyl sulfide (Tallant et al. 2001) as substrates for methanogenesis. The genomes of *M. barkeri*, *M. mazei*, *M. acetivorans*—though different—display extensive genome rearrangement (Maeder et al. 2006), and *Methanolobus siciliae* has been transferred to this genus as *Methanosarcina siciliae* (Ni et al. 1994).

Methyltransferase proteins MtaB for methanol, MtmB for monomethylamine, MtbB for dimethylamine, and MttB for trimethylamine (Soares et al. 2005) transfer methyl groups from the substrates, and transfer to the cognate corrinoid proteins is carried out by MtaC, MtmC, MtbC, and MttC (Ferguson and Krzycki 1997; Paul et al. 2000). A second transferase carries out demethylation to methyl coenzyme M (2-methylmercaptoethanesulfonate) from the methylated corrinoid proteins (Ferguson et al. 1996, 2000; Ferguson and Krzycki 1997; Harms and Thauer 1996), and methyl-CoM reacts with coenzyme B (7-mercaptoheptanoylthreonine phosphate) that is reduced to methane (Krätzer et al. 2009).

The functional biosynthesis of the methylamine transferase enzymes requires suppression of the UAG amber codon and the incorporation of the amino acid pyrrolysine at this position. This is carried out by pyrrolysyl-tRNA synthetase (PylRS) and the aminoacyl-tRNA synthetase (AARS) whose genes are found in the archaeal family *Methanosarcinaceae* and—remarkably—in *Desulfitobacterium hafniense* (Herring et al. 2007). Pyrrolysine is required for these pathways and constitutes the 22nd amino acid in the genetic code. The genes *pylB*, *pylC,* and *pylD* are required for tRNA-independent pyrrolysine synthesis that involves a single pathway from two molecules of L-lysine: (i) free-radical *S*-adenosyl-methionine isomerization by PylB of L-lysine at C_2 with concomitant methylation to 2*R*, 3*R*-methyl-D-ornithine, (ii) ATP-dependent amidation with Nε-L-lysine by PylC to (3*R*)-3-methyl-D-ornithyl-Nε-L lysine, (iii) dehydrogenation of the terminal Nε-amine to the aldehyde (3*R*)-3-methyl-D-glutamyl-semialdehyde-Nε-L-lysine, and (iv) cyclization to L-pyrrolysine (Gaston et al. 2011).

The corrinoids require reactivation since during transfer of the methyl group in methylamine by MtmB to the corrin CH3-Co(III) MtmC, the oxidation level is reduced to Co(I)-MtmC and Co(II)-MtmC. Reduction is carried out by RamA which has been characterized in *Methanosarcina barkeri* as an iron–sulfur protein with an average molecular mass of 71 kDa containing per mol monomer, 8-mol acid-labile sulfur and 6.5 mol iron that is consistent with two [4Fe–4S] clusters. Purified RamA was able to activate both the trimethylamine:CoM and dimethylamine:CoA methyltransferases (MttB and MtbB) as well as their cognate corrinoid proteins. In addition, multiple *ramA* homologues were shown to occur in several bacterial genomes (Ferguson et al. 2009).

7.10.3.1 Amino Acid Fermentation and Other Reactions

The fermentation of amino acids by clostridia has been extensively examined and is discussed in the succeeding paragraphs. Fermentations by other organism are however important, and the fermentation of aspartate illustrates important differences in its metabolism.

1. In *Campylobacter* sp. the activity of L-aspartate ammonia lyase converts L-aspartate to fumarate which is processed in two metabolic segments: reduction to succinate, and oxidation to succinate and acetate (Laanbroek et al. 1978). In the oxidative segment, fumarate is hydrated to malate and, via oxaloacetate and pyruvate to acetyl-CoA and citrate that is reduced to succinate and CO_2. The reductants that are produced by dehydrogenation are utilized for the reduction of fumarate to succinate in the other segment. This is formally analogous to the fermentation of fumarate by *D. gigas* in which two reactions took place: reduction of fumarate to succinate, and oxidation in which part of the fumarate was hydrated to malate that was oxidized to acetate and CO_2 (Miller and Wakerley 1966).

2. Aspartate and asparagine are the preferred amino acids for fermentation of some species of *Bacteroides*. The fermentation of L-[^{14}C]aspartate by *Bacteroides melananinogenicus* produced labeled succinate as the major product (Wong et al. 1977), and the metabolism of aspartate and asparagine has been examined in *B. intermedius* and *B. gingivalis*. Since aspartase activity was not detected in these organisms, this activity could not account for the production of fumarate and thence succinate. Malate dehydrogenase was found in cell extracts, fumarate reductase in the particulate fraction, and the relevant electron carriers were observed: cytochrome *b* and menaquinone-9 in *B. gingivalis* and cytochrome *c* and menaquinone-11 in *B intermedius*. In the light of these observations, the pathway for aspartate metabolism was postulated to occur by deamination to oxaloacetate from which malate, fumarate and succinate were successively produced (Shah and Williams 1987).

3. The fermentation of aspartate by *Propionibacterium freudenreichii* provides a clear example of cometabolism since the strains that were examined were unable to ferment aspartate except when propionate was present (Rosner and Schink 1990). Metabolism of aspartate occurred by two routes, both initiated by elimination to fumarate with loss of NH_3. Three metabolic sequences were involved: (i) reduction of fumarate to succinate, (ii) hydration of fumarate to malate, dehydrogenation to oxaloacetate from which pyruvate was produced and subsequently acetate, (iii) the CO_2 produced by decarboxylation of oxaloacetate was processed via propionyl-CoA to methylmalonyl-CoA and thence succinate. This was supported by determination of the relevant enzymes, and the reactions could be summarized:

$$3 \text{ aspartate} + \text{propionate} + 2H_2O \rightarrow 3 \text{ succinate} + \text{acetate} + CO_2 + 3NH_3$$

7.10.3.2 Clostridial Fermentations

The fermentation of amino acids by several clostridia involving rearrangements was investigated extensively in classic studies by Stadtman and by Barker, and some examples are given in Table 7.1. There are three groups of reactions involved in the anaerobic metabolism of amino acids. These are

TABLE 7.1

Fermentation of Single Amino Acids by Clostridia

Substrate	Products	Organism
Glutamate	Acetate + butyrate	*Clostridium tetanomorphum*
Lysine	Acetate + butyrate	*Clostridium sticklandii*
Ornithine	Acetate + alanine	*Clostridium sticklandii*
Leucine	Acetate + isobutyrate	*Clostridium sporogenes*

not, however, exclusive and some use coupled redox reactions that are formally analogous to Stickland reactions.

Several fermentations are mediated by free-radical dismutases that involve coenzyme B_{12}: leucine 2,3-aminomutase; glutamate mutase; lysine 2,3 and lysine 5,6 aminomutase; ornithine 5,4-aminomutase (Frey 2001). These are discussed in Chapter 3, and some are illustrated in the following examples.

1. The first step in leucine fermentation by *Clostridium sporogenes* is the formation of β-leucine by an AdoCbl-mediated 2,3-aminomutase to produce 3-amino-4-methylpentanoate (Poston 1976).
2. Glutamate fermentation by *Clostridium tetanomorphum* is initiated by glutamate mutase and involves AdoCbl in a free-radical reaction (Holloway and Marsh 1994; Bothe et al.1998). The product is 3-methylaspartate from which 2-methylmaleate is produced followed by hydration and scission to acetate and pyruvate (Figure 7.64).
3. The fermentation of lysine is twofold and has been summarized (Kreimeyer et al. 2007) (Figure 7.65a,b).

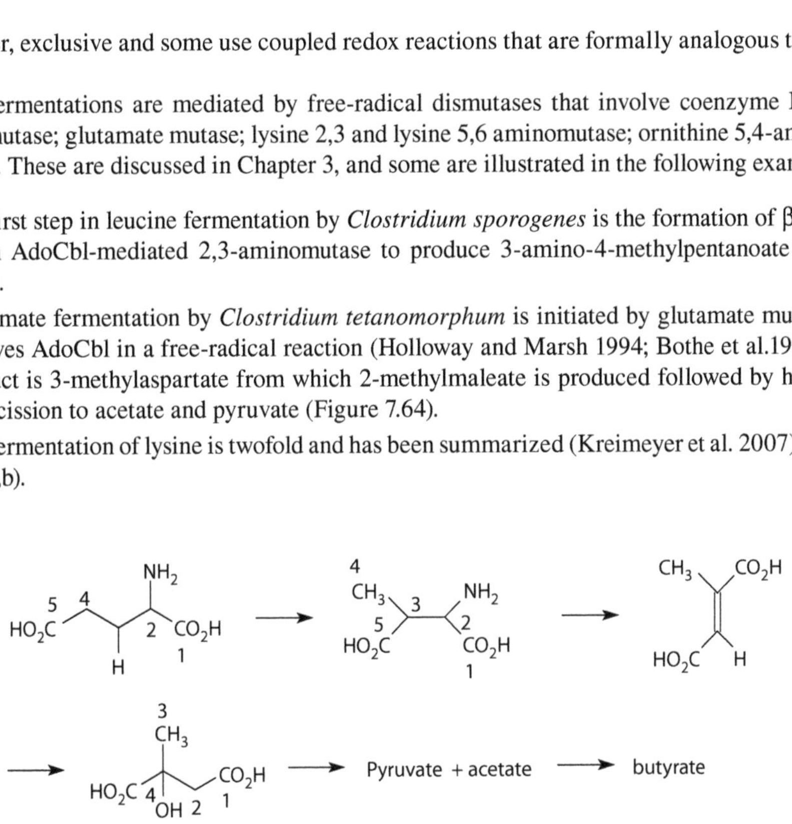

FIGURE 7.64 Fermentation of L-glutamate.

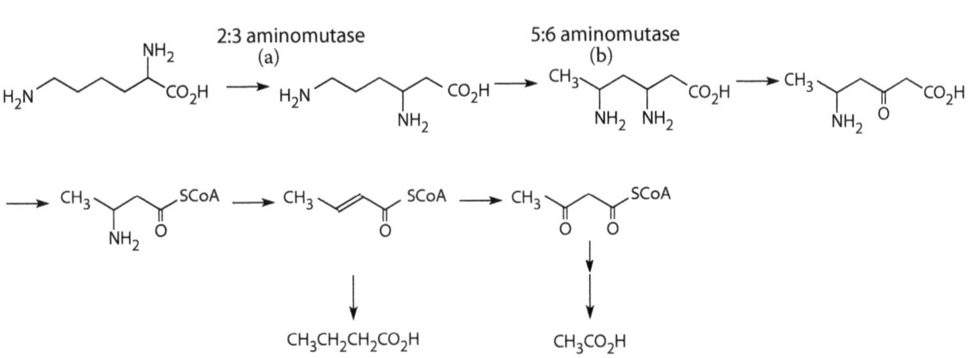

FIGURE 7.65 Fermentation of L-lysine by successive activities of (a) 2,3-aminomutase and (b) 5,6-aminomutase.

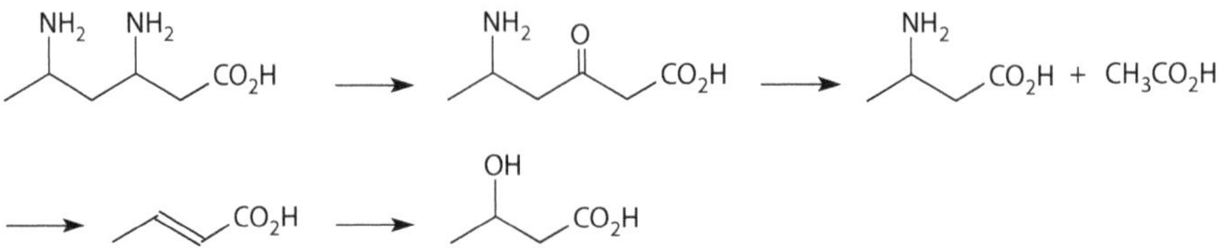

FIGURE 7.66 Metabolism of D,L-erythro-3,5-diaminohexanoate.

    a. The 2,3-mutase in *Clostridium sticklandii* mediates the migration of the amino group to β-lysine (Chirpich et al. 1970; Petrovich et al. 1991) although the radical is produced from *S*-adenosylmethionine by reduction with an [Fe–S] cluster, and not from coenzyme-B_{12} (Moss and Frey 1990);

    b. Further metabolism is accomplished with D-lysine 5,6-aminomutase (Baker et al. 1973; Chang and Frey 2000) with the production of 3,5-diaminohexanoate: both mutases require pyridoxal-5′-phosphate for activity and involve the adenosyl-5′ radical. Dehydrogenation of 3,5-diaminohexanoate produces 3-oxo-5-aminohexanoate that undergoes scission with acetyl-CoA to produce 3-aminobutyryl CoA and acetoacetate (Figure 7.66) (Yorifugi et al. 1977). This fission reaction is also involved in the aerobic degradation of 3,5-diaminohexanoate by *Brevibacterium* sp. (Barker et al. 1980).

7.10.3.3 2-Hydroxyacyl-CoA Dehydratase Pathway

An alternative pathway that is independent of coenzyme B_{12} is initiated by transamination with 2-oxoglutarate followed by NADH-dependent reduction to a 2-hydroxyacyl-CoA (Kim et al. 2004b). It is emphasized that these reactions involve the coenzyme-A esters that are synthesized by transferases which are discussed further in Chapter 3. The acyl-CoA ester undergoes dehydration involving loss of a proton from a nonactivated position in a radical-mediated pathway. The dehydratase requires activation by ATP and Mg^{2+} that has been examined in several organisms, including *Clostridium sporogenes* (Dickert et al. 2002), *Clostridium difficile* (Kim et al. 2004a, 2006), and *Acidamincoccus fermentans* (Hans et al. 2002). A summary of these reactions is given in Table 7.2, and some additional details are given in the following paragraphs.

    1. This alternative is used by glutamate in *Clostridium symbiosum* (Hans et al. 1999) and in *Acidaminococcus fermentans* (Buckel and Barker 1974; Müller and Buckel 1995) (Figure 7.67). The dehydrase for 2-hydroxyglutaryl-CoA in *Acidamincoccus fermentans* has been characterized and consists of two components, one of which (A) contains an [Fe–S] cluster and the active component (D) that contains FMN and [Fe–S] clusters and whose activity requires component A and ATP (Hans et al. 2002).

TABLE 7.2

Fermentation of Amino Acids Using the 2-Hydroxyacyl-CoA Pathway

Substrate	Initial Products	Final Products	Organism	Reference
Alanine[a]	Lactoyl-CoA	Acetate + propionate	*Clostridium propionicum*	Selmer et al. (2002)
Leucine[a]	Hydroxyisocaproyl-CoA	Isocaproate + isovalerate	*Clostridium difficile*	Kim et al. (2004a, 2006)
Phenylalanine[a]	Cinnamoyl-CoA	3-phenylpropionate + phenylacetate	*Clostridium sporogenes*	Dickert et al. (2002)
Glutamate	Hydroxyglutaryl-CoA	Acetate + butyrate	*Acidaminococcus fermentans*	Müller and Buckel (1995)
			Clostridium symbiosum	Hans et al. (1999)

[a] These reactions involve a Strickland oxidation/reduction system.

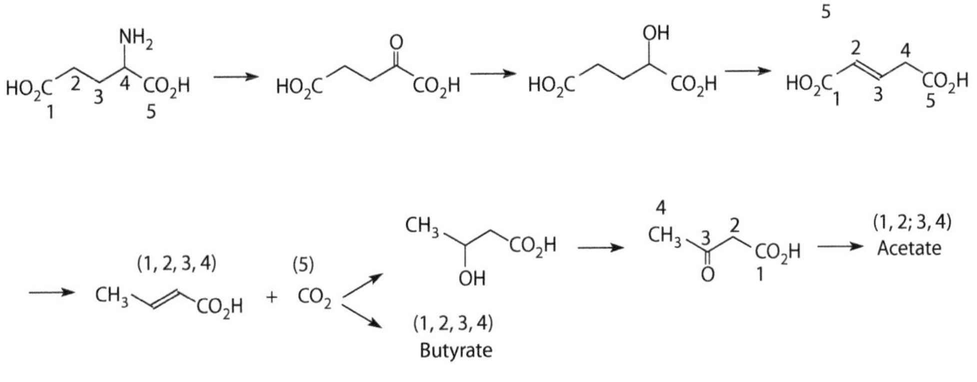

FIGURE 7.67 Alternative fermentation of L-glutamate.

$$H_2N{-}CH_2{-}CH_2{-}CO_2H \longrightarrow CHO{-}CH_2{-}CH_2{-}CO_2H \longrightarrow HOCH_2{-}CH_2{-}CO_2H \longrightarrow CH_3{-}CH{=}CH{-}CO_2H \longrightarrow \text{Butyrate + Acetate}$$

FIGURE 7.68 Fermentation of 4-aminobutyrate.

2. The fermentation of 4-aminobutyryl-CoA by *Clostridium aminobutyricum* is initiated by formation of succinate by transamination with 2-oxoglutarate. This undergoes reduction to 4-hydroxybutyryl-CoA followed by dehydration to crotonyl-CoA, reduction to butyryl-CoA and dismutation to acetate and butyrate (Figure 7.68) (Gerhardt et al. 2000). The dehydratase has been characterized in this organism by EPR, and a mechanism was proposed that involved one-electron transport to activate the β-C–H bond and the role of FAD and a [4Fe–4S] cluster (Müh et al. 1996). The crystal structure has been determined in which the substrate is bound between FAD and $[4Fe{-}4S]^{2+}$ and support a free radical mechanism involving reaction of an FAD semiquinone anion with an enoxy radical followed by formation of a dienoxy radical and hydration (Martins et al. 2004). Analogously, the fermentation of succinate to butyrate by *Clostridium kluyveri* proceeds via formation of 4-hydroxybutyrate with the production of crotonate that is reduced to butyrate (Söhling and Gottschalk 1996).
3. 5-Aminovalerate is produced by the anaerobic degradation of protein hydrolysates, and is degraded by *Clostridium aminovalericum* by the coupled oxidation to acetate and propionate, and the reduction to valerate (Barker et al. 1987). Although the details of the reactions have not been finally clarified, this involves initial formation of 5-hydroxyvaleryl-CoA followed by dehydration steps with sequential formation of pent-4-enoyl-Co-A, penta-2,4-dienoyl-CoA and 2-pent-2-enoyl-CoA—again involving dehydration of the intermediate 5-hydroxyvalerate with loss of hydrogen from a nonactivated position.

7.10.3.4 Redox Systems, the Stickland Reaction

An important dismutation between pairs of amino acids has been studied extensively in *Clostridium sporogenes* (Stickland 1934, 1935a,b), and can be carried out by many other clostridia (Barker 1961). There are two classes of reaction: one involving a pair of different amino acids in which one serves as oxidant and the other as reductant, whereas in the other a single amino acid fulfils both roles. The products from proline and alanine (Stickland 1935a) illustrate the reaction (Figure 7.69) and further examples are given in Table 7.3.

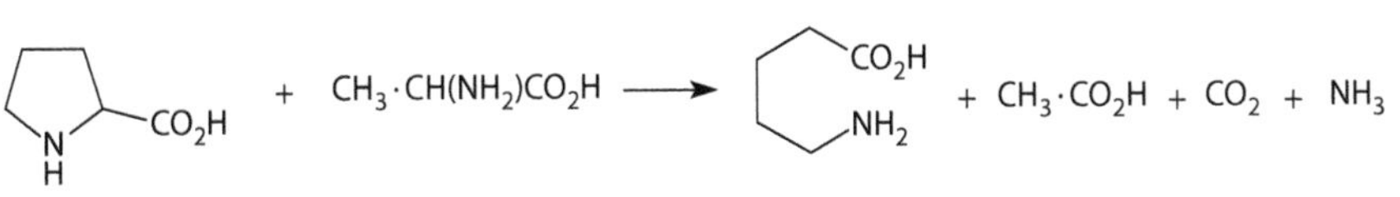

FIGURE 7.69 Stickland reaction between L-proline and L-alanine.

TABLE 7.3

Stickland Oxidations of Clostridia with Proline That Is
Reduced to 5-Aminovalerate

Substrate	Products	Organism
Leucine	Isovalerate	*Clostridium valerianum*
Isoleucine	α-Methylbutyrate	*Clostridium propionicum*
Valine	Isobutyrate	*Clostridium caproicum*
Ornithine	Acetate + alanine	*Clostridium sticklandii*

There are important examples that involve only a single amino acid, and these may additionally involve dehydration with removal of a proton from a nonactivated position by radical reactions. The following illustrate these possibilities.

1. The degradation of leucine by *Clostridium difficile* is initiated by transamination to the keto acid where the pathway diverges: (a) reduction to the 2-hydroxyacid followed by dehydration and reduction to isocaproate and (b) reductive decarboxylation to isovalerate and NADH (Kim et al. 2004a, 2006) (Figure 70a,b).

2. The degradation of ornithine by *Cl. sticklandii* consists of two segments that formally constitute the oxidative and reductive sequences of a typical Stickland reaction (Figure 7.71a,b) (Fonknechten et al. 2009).

   a. In the first stage, L-ornithine is converted by a racemase (Chen et al. 2000) into D-ornithine followed by the activity of a 5,4-mutase (Somack and Costilow 1973a; Chen et al. 2001) to produce 2,4-diaminopentanoate. This undergoes dehydrogenation (Tsuda and Friedmann 1970; Somack and Costilow 1973b) to 2-amino-4-oxopentanoate before CoA-SH scission (Jeng et al. 1974) to alanine and acetyl-CoA.

   b. In the second stage, L-ornithine reacts with 2-oxoglutarate by transamination to produce glutamate semialdehyde that undergoes spontaneous cyclization to Δ¹-pyrrolidine 5-carboxylate, and the subsequent reactions have been characterized in *Cl. sticklandii*. The pyrrolidine undergoes reduction to L-proline followed by epimerization by a racemase to D-proline (Rudnick and Abeles 1975)—that does not require pyridoxal phosphate (Stadtman and Elliot 1957). This is reduced to 5-aminovalerate in a reaction that involves

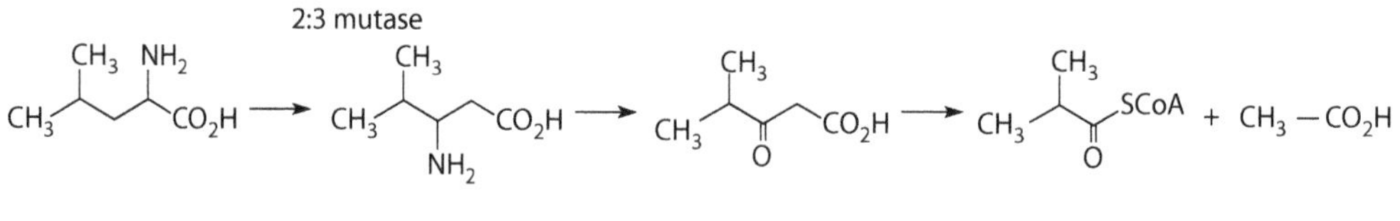

FIGURE 7.70 Fermentation of L-leucine.

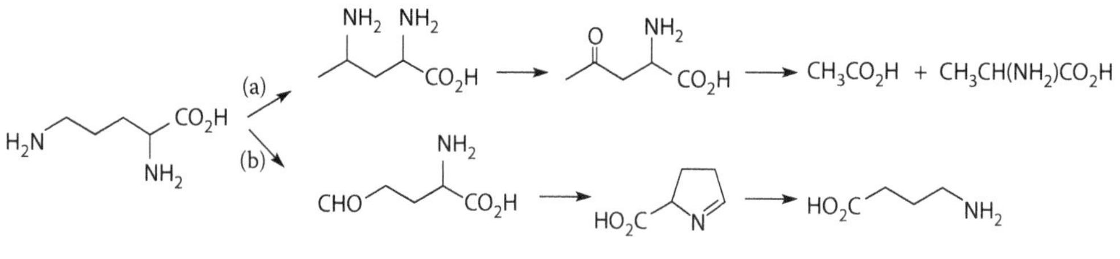

FIGURE 7.71 Fermentation of L-ornithine.

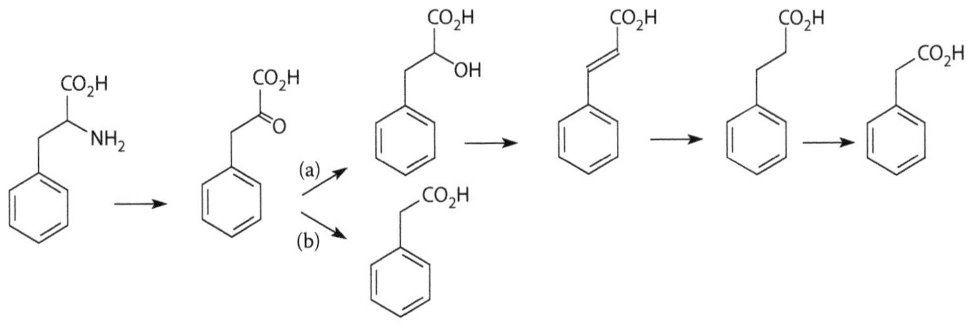

FIGURE 7.72 Fermentation of L-phenylalanine.

interaction between a selenoprotein and D-proline that is covalently bound with pyruvate in a protein (Kabisch et al. 1999; Jackson et al. 2006).

3. Phenylalanine can be fermented both as a single substrate and in a Stickland reaction. In *Clostridium sporogenes* it undergoes transamination to phenylpyruvate when the reaction diverges: (a) reduction followed by dehydration to (*E*)-cinnamate that is reduced to 3-phenyl-propionate and (b) oxidation with decarboxylation to phenylacetate (Figure 7.72a,b) Dickert et al. 2000, 2002).

4. The fermentation of glycine by *Eubacterium acidaminophilum* is a complex redox reaction that has received extensive study (Andreesen 1994) and the mechanism whereby glycine is reduced to acetate by glycine reductase from *Cl. sticklandii* has been elucidated (Arkowitz and Abeles 1989).

   a. Oxidation by glycine decarboxylase/synthase—an enzyme complex of four proteins: $P_1 \equiv P$, $P_2 \equiv H$, $P_3 \equiv L$, and $P_4 \equiv T$—to yield 5,10-methylene tetrahydrofolate that is oxidized to CO_2 by the activities of two dehydrogenases (Freudenberg et al. 1989; Freudenberg and Andreesen 1989).

   b. Reduction to acetyl phosphate and ammonia by glycine reductase (Arkowitz and Abeles 1989), and the production of APT by acetate kinase in a reaction that involves thioredoxin and a selenoprotein (references in Andreesen 1994).

7.10.3.5 Transformation of Aromatic Amino Acids

The terminal products formed by fermentation of aromatic amino acids in clostridia have been described (Elsden et al. 1976). The reactions involve oxidation deamination of the amino acids to the corresponding carboxylates followed by decarboxylation. Examples include:

a. Toluene from phenylalanine (Pons et al. 1984) by *Clostridium aerofoetidum*

b. Toluene from phenylalanine by the facultative anaerobe *Tolumonas auensis* during anaerobic growth with a range of monosaccharides (Fischer-Romero et al. 1996)

c. *p*-Cresol from tyrosine in cell extracts of *Clostridium difficile* (D'Ari and Barker 1985)

d. 3-Methylindole (skatole) from indole-3-acetate by a rumen species of *Lactobacillus* (Yokoyama et al. 1977; Yokoyama and Carlson 1981; Honeyfield and Carlson 1990)

The decarboxylase from *Clostridium difficile* has been purified, and its activity examined toward a range of substituted phenylacetates, 3,4-dihydroxyphenylacetate, and pyridine-4-acetate (Selmer and Andrei 2001). The protein was readily and irreversibly inactivated by oxygen, contained a small sub-unit HpdC that was essential for catalytic activity, and it was proposed on the basis of similarity to pyruvate-formate-like proteins in *E. coli*, that the reaction involved a glycyl radical (Andrei et al. 2004).

7.11 References

Adams DO, SF Yang. 1979. Ethylene biosynthesis: identification of 1-aminocyclopropane-1-carboxylic acid as an intermediate in the conversion of methionine to ethylene. *Proc Natl Acad Sci USA* 76: 170–174.

Almagro-Moreno S, EF Boyd. 2004. Insights into the evolution of sialic acid catabolism among bacteria. *BMC Evolut Biol* 9, doi:10.1186/1471-2148-9-118.

Álvarez-Añorve LI, ML Calcagno, J Plumbridge. 2005. Why does *Escherichia coli* grow more slowly on glucosamine than on *N*-acetylglucosamine? Effects of enzyme levels and allosteric activation of GlcN6P deaminase (NagB) on growth rates. *J Bacteriol* 187: 2974–2982.

Andreesen JR. 1994. Glycine metabolism in anaerobes. *Antonie van Leeuwenhoek* 66:223–237.

Andrei PI, AJ Pierik, S Zauner, LC Andrei-Selmer, T Selmer. 2004. Subunit composition of the glycyl radical enzyme *p*-hydroxyphenylacetate decarboxylase. A small subunit, HpdC, is essential for catalytic activity. *Eur J Biochem* 1–6.

Angelidis AS, GM Smith. 2003. Role of glycine betaine and carnitine transporters in adaptation of *Listeria monocytogenes* to chill stress in defined medium. *Appl Environ Microbiol* 69: 7492–7498.

Anton DL, R Kutny. 1987. *Escherichia coli* S-adenosylmethionine decarboxylase. Subunit structure, reductive amination, and NH$_2$-terminal sequences. *J Biol Chem* 262: 2817–2822.

Arias-Barrau E, ER Olivera, JM Luengo, C Fernández, B Gálan, JL García, E Díaz, B Miñambres. 2004. The homogentisate pathway: a central catabolic pathway involved in the degradation of L-phenylalanine, L-tyrosine, and 3-hydroxyphenylacetate in *Pseudomonas putida*. *J Bacteriol* 186:5062–5077.

Arkowitz RA, RH Abeles. 1989. Identification of acetyl phosphate as the product of clostridial glycine reductase: evidence for an acyl enzyme intermediate. *Biochemistry* 28: 4639–4644.

Ashiuchi M, K Shimanouchi, H Nakamura, T Kamei, K Soda, C Park, M-H Sung, H Misono. 2004. Enzymatic synthesis of high-molecular-mass poly-γ-glutamate and regulation of its stereochemistry. *Appl Environ Microbiol* 70: 4249–4255.

Babior BM, TH Moss, WH Orme-Johnson, H Beinert. 1974. The mechanism of action of ethanolamine ammonia-lyase, a B$_{12}$-dependent enzyme. The participation of paramagnetic species in the catalytic deamination of 2-aminopropanol. *J Biol Chem* 249: 4537–4544.

Baker JJ, C van der Drift, TC Stadtman. 1973. Purification and properties of β-lysine mutase, a pyridoxal phosphate and B$_{12}$ coenzyme dependent enzyme. *Biochemistry* 12: 1054–1063.

Bamforth CW, PJ Large. 1977. Solubilization, partial purification and properties of *N*-methylglutamate dehydrogenase from *Pseudomonas aminovorans*. *Biochem J* 161: 357–370.

Barker HA. 1961. Fermentations of nitrogenous organic compounds pp 151–207. In *The Bacteria*, Volume II (Eds IC Gunsalus, RY Stanier) Academic Press, New York.

Barker HA, JM Kahn, S Chew. 1980. Enzymes involved in 3,5-diaminohexanoate degradation by *Brevibacterium* sp. *J Bacteriol* 143: 1165–1170.

Barker HA, L D'Ari, J Kahn. 1987. Enzymatic reactions in the degradation of 5-aminovalerate by *Clostridium aminovalericum*. *J Biol Chem* 262: 8994–9003.

Belimov AA et al. 2001. Characterization of plant growth promoting rhizobacteria isolated from polluted soils and containing 1-aminocyclopropane-1-carboxylate deaminase. *Can J Microbiol* 47: 642–652.

Bell SC, JM Turner. 1977. Bacterial catabolism of threonine. Threonine degradation initiated by L-threonine acetaldehyde-lyase (aldolase) in species of *Pseudomonas*. *Biochem J* 166: 209–216.

Boer L, W Harder, L Dijkhuizen. 1988. Phenylalanine and tyrosine metabolism in the facultative methylotroph *Nocardia* sp. 239. *Arch Microbiol* 149: 459–465.

Bothe H, DJ Darley, SPJ Albracht, GJ Gerfen, BT Golding, W Buckel. 1998. Identification of the 4-glutamyl radical as an intermediate in the carbon skeleton rearrangement catalyzed by coenzyme B$_{12}$-dependent glutamate mutase from *Clostridium cochlearium*. *Biochemistry* 37: 4105–4113.

Boulton CA, CA Crabbe, PJ Large. 1974. Microbial oxidation of amines. Partial purification of a trimethylamine mono-oxygenase from *Pseudomonas aminovorans* and its role in growth on trimethylamine. *Biochem J* 140: 253–263.

Boulton CA, PJ Large. 1977. Synthesis of certain assimilatory and dissimilatory enzymes during bacterial adaptation to growth on trimethylamine. *J Gen Microbiol* 101: 151–156.

Boulton CA, PJ Large. 1979. Inactivation of trimethylamine *N*-oxide aldolase (demethylase) during preparation of bacterial extracts. *FEMS Microbiol Lett* 5: 159–162.

Brandl MT, SE Lindow. 1996. Cloning and characterization of a locus encoding an indolpyruvate decarboxylase involved in indol-3-acetic acid synthesis *in Erwinia herbicola*. *Appl Environ Microbiol* 62: 4121–4128.

Brearley GM, CP Price, T Atkinson, PM Hammond. 1994. Purification and partial characterization of a broad-range L-amino acid oxidase from *Bacillus carotarum* 2Pfa isolated from soil. *Appl Microbiol Biotechnol* 41: 670–676.

Brigham C, R Caughlan, R Gallegos, MB Dallas, VG Godoy, MH Malamy. 2009. Sialic acid (*N*-acetyl neuraminic acid) utilization by *Bacteroides fragilis* requires a novel *N*-acetyl mannosamine epimerase. *J Bacteriol* 191: 3629–3638.

Brook DF, PJ Large. 1976. A steady-state kinetic study of the reaction catalysed by the secondary-amine mono-oxygenase of *Pseudomonas aminovorans*. *Biochem J* 157: 197–205.

Buckel W, BM Martins, A Messerschmidt, BT Golding. 2005. Radical-mediated dehydration reactions in anaerobic bacteria. *Biol Chem* 386: 951–959.

Buckel W, HA Barker. 1974. Two pathways of glutamate fermentation by anaerobic bacteria. *J Bacteriol* 117: 1248–1260.

Cacciapuoti G, M Porcelli, M de Rosa, A Gambacorta, C Bertoldo, V Zappia. 1991. *S*-adenosylmethionine decarboxylase from the thermophilic archaebacterium *Sulfolobus solfataricus*. Purification, molecular properties and studies on the covalently bound pyruvate. *Eur J Biochem* 199: 395–400.

Calabrese JC, Db Jordan, A Boodhoo, S Sariasliani, T Vannelli. 2004. Crystal structure of phenylalanine ammonia lyase: multiple helix dipoles implicated in catalysis. *Biochemistry* 43: 11403–11416.

Cerniglia CE, JJ Perry. 1975. Metabolism of *n*-propylamine, isopropylamine, and 1,3-propane diamine by *Mycobacterium convolutum*. *J Bacteriol* 124: 285–289.

Chang CH, PA Frey. 2000. Cloning, sequencing, heterologous expression, purification, and characterization of adenosylcobalamin-dependent D-lysine 5,6-aminomutase from *Clostridium sticklandii*. *J Biol Chem* 275: 106–114.

Chang Y-F, E Adams. 1974. D-lysine catabolic pathway in *Pseudomonas putida*; interrelations with L-lysine metabolism. *J Bacteriol* 117: 753–764.

Chen HP, CF Lin, YJ Lee, SS Tsay, SH Wu. 2000. Purification and properties of ornithine racemase from *Clostridium sticklandii*. *J Bacteriol* 182: 2052–2054.

Chen HP, SH Wu, YL Lin, CM Chen, SS Tsay. 2001. Cloning, sequencing, heterologous expression, purification, and characterization of adenosylcobalamin-dependent D-ornithine aminomutase from *Clostridium sticklandii*. *J Biol Chem* 276: 44744–44750.

Chen I-C, W-D Lin, S-K Hsu, V Thiruvengadam, W-H Hsu. 2009. Isolation and characterization of a novel lysine racemase from a soil metagenomic library. *Appl Environ Microbiol* 75: 5161–5166.

Chen L, RCE Durley, FS Matthews, VL Davidson. 1994. Structure of an electron transfer complex: methylamine dehydrogenase, amicyanin, and cytochrome c_{551i}. *Science* 264: 86–90.

Chen Y, KL McAleer, JC Murrell. 2010. Monomethylamine as a nitogen source for a nonmethylotrophic bacterium *Agrobacterium tumefaciens*. *Appl Environ Microbiol* 76: 4102–4104.

Chirpich TP, V Zappia, RN Costilow, HA Barker. 1970. Lysine 2,3-aminomutase. *J Biol Chem* 245: 1778–1789.

Chlumsky LJ, L Zhang, MS Jorns. 1995. Sequence analysis of sarcosine oxidase and nearby genes reveals homologies with key enzymes of folate one-carbon metabolism. *J Biol Chem* 270: 18252–18259.

Chou HT, M Hegazy, C-D Lu. 2010. L-Lysine catabolism is controlled by L-arginine and ArgR in *Pseudomonas aeruginosa* PAO1. *J Bacteriol* 192: 5874–5880.

Cokesa Z, H-J Knackmuss, P-G Rieger. 2004. Biodegradation of all stereoisomers of the EDTA substitute iminodisuccinate by *Agrobacterium tumefaciens* BY6 requires an epimerase and a stereoselective C–N lyase. *Appl Environ Microbiol* 70: 3941–3947 .

Comai L, T Kosuge. 1982. Cloning and characterization of *iaaM*, a virulence determinant of *Pseudomonas savastanoi*. *J Bacteriol* 149:40–46.

Combourieu B, P Besse, M Sancelme, H Veschambre, AM Delort, P Poupin, N Truffaut. 1998. Morpholine degradation pathway of *Mycobacterium aurum* MO1: direct evidence of intermediates by *in situ* 1H nuclear magnetic resonance. *Appl Environ Microbiol* 64: 153–158.

Connil N, Y Le Breton, X Dousset, Y Auffrey, A Rincé. H Prévost. 2002. Identification of the *Enterococcus faecalis* tyrosine decarboxylase operon involved in tyramine production. *Appl Environ Microbiol* 68: 3537–3544.

Coton E, N Mulder, M Coton, S Pochet, H Trip, JS Lolkema. 2010. Origin of the putrescine-producing ability of the coagulase-negative bacterium *Staphylococcus epidermis* 2015B. *Appl Environ Microbiol* 76: 5570–5576.

Cripps RE, AS Noble. 1973. The metabolism of nitrilotriacetate by a pseudomonad. *Biochem J* 136: 1059–1068.

D´Ari L, WA Barker. 1985. *p*-cresol formation by cell-free extracts of *Clostridium difficile. Arch Microbiol* 143: 311–312.

Dai VD, K Decker, H Sund. 1968. Purification and properties of L-6-hydroxynicotine oxidase. *Eur J Biochem* 4: 95–102.

Dasu V V, Y Nakada, M Ohnishi-Kameyama, K Kimura, Y Itoh. 2006. Characterization and a role of *Pseudomonas aeruginosa* spermidine dehydrogenase in polyamine metabolism. *Microbiology (UK)* 152: 2265–2272.

DeSa RJ. 1972. Putrescine oxidase from *Micrococcus rubens*. Purification and properties of the enzyme. *J Biol Chem* 247: 5527–5534.

Dickert S, AJ Pierik, W Buckel. 2002. Molecular characterization of phenyllactate dehydratase and its initiator from *Clostridium sporogenes*. *Mol Microbiol* 44: 49–60.

Dickert S, AJ Pierik, D Linder, W Buckel. 2000. The involvement of coenzyme A esters in the dehydration of (*R*)-phenyllactate to (*E*)-cinnamate by *Clostridium sporogenes*. *Eur J Biochem* 267: 3874–3884.

Ditullio D, D Anderson CS Chen, CJ Sih. 1994. L-Carnitine via enzyme-catalyzed oxidative kinetic resolution. *Bioorg Med Chem* 6:415–420.

Durham DR, CG McNamee, DB Stewart. 1984. Dissimilation of aromatic compounds in *Rhodotorula graminis*: biochemical characterization of pleiotropically negative mutants. *J Bacteriol* 160: 771–777.

Eichhorn E, JR van der Ploeg, T Leisinger. 1999. Characterization of a two-component alkane sulfonate monooxygenase from *Escherichia coli. J Biol Chem* 274: 26639–26646.

Eichler K, W-H Schunck, H-P Kleber, M-A Mandrand-Berthelot. 1994a. Cloning, nucleotide sequence, and expression of the *Escherichia coli* gene encoding carnitine dehydratase. *J Bacteriol* 176: 2970–2973.

Eichler K, F Bourgis, A Buchet, H-P Kleber, MA Mandrand-Berthelot. 1994b. Molecular characterization of the *cai* operon necessary for carnitine metabolism in *Escherichia coli. Mol Microbiol* 13: 775–786.

Elsden SR, MG Hilton, JM Waller. 1976. The end products of the metabolism of aromatic acids by clostridia. *Arch Microbiol* 107: 283–288.

Elßner T, A Preußer, U Wagner, H-P Kleber. 1999. Metabolism of L(-)-carnitine by Enterobacteriaceae under aerobic conditions. *FEMS Microbiol Lett* 174: 295–301.

Elssner T, C Engemann, K Baumgart, H-P Kleber. 2001. Involvement of coenzyme A esters and two new enzymes, an enoyl-CoA hydratase, and a CoA-transferase, in the hydration of crotonobetaine to L-carnitine by *Escherichia coli*. *Biochemistry* 40: 11140–11148.

Engemann C, T Elssner, S Pfeifer, C Krumholz, T Maier, H-P Kleber. 2005. Identification and functional characterization of genes and corresponding enzymes involved in carnitine metabolism of *Proteus* sp. *Arch Microbiol* 183: 176–189.

Ferguson DJ, JA Krzycki, DA Grahame. 1996. Specific roles of methylcobamide:coenzyme M methyltransferase isoenzymes in metabolism of methanol and methylamines in *Methanosarcina barkeri. J Biol Chem* 271: 5189–5194.

Ferguson DJ, JA Krzycki. 1997. Reconstitution of trimethylamine-dependent coenzyme M methylation with the trimethylamine corrinoid protein and the isozymes of methyltransferase II from *Methanosarcina barkeri. J Bacteriol* 179: 846–852.

Ferguson DJ, N Gorlatova, DA Grahame, JA Krzycki. 2000. Reconstitution of dimethylamine: Coenzyme M methyl transfer with a discrete corrinoid protein and two methyltransferases purified from *Methanosarcina barkeri. J Biol Chem* 275: 29053–29060.

Ferguson T, JA Soares, T Lienard, G Gottschalk, JA Krzycki. 2009. RamA, a protein required for reductive activation of corrinoid-dependent methylamine methyltransferase reactions in methanogenic archaea. *J Biol Chem* 284: 2285–2295.

Firestone MK, JM Tiedje. 1978. Pathway of degradation of nitrilotriacetate by a *Pseudomonas* sp. *Appl Environ Microbiol* 35: 955–961.

Fischer-Romero C, BJ Tindall, F Jüttner. 1996. *Tolumonas auensis* gen. nov., sp. nov., a toluene-producing bacterium from anoxic sediments of a freshwater lake. *Int J Syst Bacteriol* 46: 183–188.

Flashner MIS, V Massey. 1974. Purification and properties of L-lysine monooxygenase from *Pseudomonas fluorescens. J Biol Chem* 249: 2579–2586.

Fonknechten N, A Perret, N Perchat, S Tricot, C Lechaplais, D Vallenet, C Vergne, A Zaparucha, D Le Paslier, J Weissenbach, M Salanoubat. 2009. A conserved gene cluster rules anaerobic oxidative degradation of L-ornithine. *J Bacteriol* 191: 3162–3167.

Fothergill JC, JR Guest. 1977. Catabolism of L-lysine by *Pseudomonas aeruginosa*. *J Gen Microbiol* 99: 139–155.

Freudenberg W, D Dietrichs, H Lebertz, JR Andreesen. 1989. isolation of an atypically small lipoamide dehydrogenase involved in the glycine decarboxylase complex from *Eubacterium acidaminophilum*. *J Bacteriol* 171: 1346–1354.

Freudenberg W, JR Andreesen. 1989. Purification and partial characterization of the glycine decarboxylase multienzyme complex from *Eubacterium acidaminophilum*. *J Bacteriol* 171: 2209–2215.

Frey PA. 2001. Radical mechanisms of enzymatic catalysis. *Annu Rev Biochem* 70: 121–148.

Gaston MA, L Zhang, KB Green-Church, JA Krzycki. 2011. The complete biosynthesis of the genetically encoded amino acid pyrrolysine from lysine. *Nature* 471: 647–650.

Gerhardt A, I Çinkaya, D Linder, G Huisman, W Buckel. 2000. Fermentation of 4-aminobutyrate by *Clostridium aminobutyricum*: cloning of two genes involved in the formation and dehydration of 4-hydroxybutyryl-CoA. *Arch Microbiol* 174: 189–199.

Govindaraj S, E Eisenstein, LH Jones, J Sanders-Loehr, AY Chistoserdov, VL Davidson, SL Edwards. 1994. Aromatic amine dehydrogenase, a second tryptophan tryptophylquinone enzyme. *J Bacteriol* 176: 2922–2929.

Green J, GW Haywood, PJ Large. 1982. More than one amine oxidase is involved in the metabolism by yeasts of primary amines supplied as nitrogen source. *J Gen Microbiol* 128: 991–996.

Green J, GW Harwood, PJ Large. 1983. Serological differences between the multiple amine oxidases of yeasts and comparison of the specificities of the purified enzymes from *Candida utilis* and *Pichia pastoris*. *Biochem J* 211: 481–493.

Graham DE, H Xu, RH White. 2002. *Methanococcus jannaschii* uses a pyruvoyl-dependent arginine decarboxylase in polyamine biosynthesis. *J Biol Chem* 277: 23500–23507.

Hacisalihoglu A, JA Jongejan, JA Duine. 1997. Distribution of amine oxidases and amine dehydrogenases in bacteria grown on primary amines and characterization of the amine oxidase from *Klebsiella oxytoca*. *Microbiology (UK)* 143: 505–512.

Hans M, J Sievers, U Müller, E Bill, JA Vorholt, D Linder, W Buckel. 1999. 2-hydroxyglutaryl-CoA dehydratase from *Clostridium symbiosum*. *Eur J Biochem* 265: 404–414.

Hans, M, E Bill, I Cirpus, AJ Pierik, M Hetzel, D Alber, W Buckel. 2002. Adenosine triphosphate-induced electron transfer in 2-hydroxyglutaryl-CoA dehydratase from *Acidaminococcus fermentans*. *Biochemistry* 41: 5873–5882.

Harms U, RK Thauer. 1996. Methylcobalamine coenzyme M methyltransferase enzymes MtaA and MtbA from *Methanosarcina barkeri*. Cloning, sequencing, and differential transcription of the encoding gene, and functional expression of the mtaA gene in *Escherichia coli*. *Eur J Biochem* 235: 653–659.

Haywood GW, PJ Large. 1981. Microbial oxidation of amines. Distribution, purification and properties of two primary-amine oxidases from the yeast *Candida boidinii* grown on amines as sole nitrogen source. *Biochem J* 199: 187–201.

Haywood GW, P Large. 1984. Partial purification of a peroxisomal polyamine oxidase from *Candida boidinii* and its role in growth on spermidine as sole nitrogen source. *J Gen Microbiol* 130: 1123–1136.

Heßlinger C, SA Faithurst, G Sawers. 1998. Novel keto acid formate-lyase and propionate kinase enzymes are components of an anaerobic pathway in *Escherichia coli* that degrades L-threonine to propionate. *Mol Microbiol* 27: 477–492.

Herring S, A Ambrogelly, CR Polycarpo, D Söll. 2007. Recognition of pyrrolysine tRNA by the *Desulfitobacterium hafniense* pyrrolysyl-tRNA synthetase. *Nucleic Acids Res* 35: 1270–1278.

Hezayen FE, BH Rehm, R Eberhardt, A Steinbuchel. 2000. Polymer production by two newly isolated and extremely halophlilic archaea: application of a novel corrossive-resistant bioteactor. *Appl Microbiol Biotechnol* 54: 319–325.

Hezayen FE, BH Rehm, R Eberhardt, A Steinbuchel. 2001. Transfer of *Natrialba asiatic* B1T to *Natrialba taiwanensis* sp. nov., and description of *Natrialba aegyptica* sp. nov., a novel, extremely halophilic, aerobic non-pigmented member of the *Archaea* from Egypt that produces extracellular poly(glutamic acid). *Int J Syst Evol Microbiol* 51: 1133–1142.

Hibi M, T Kawashima, T Kodera, SV Smirnov, PM Sokolov, M Sugiyama, S Shimizu, K Yokozeki, J Ogawa. 2011. Characterization of *Bacillus thuringiensis* L-isoleucine dioxygenase for production of useful amino acids. *Appl Environ Microbiol* 77: 6926–6930.

Higuchi T, H Hayashi, K Abe. 1997. Exchange of glutamate and γ-aminobutyrate in a *Lactobacillus* strain. *J Bacteriol* 179: 3362–3364.

Holloway DE, ENG Marsh. 1994. Adenosylcobalamin-dependent glutamate mutase from *Clostridium tetano-morphum. J Biol Chem* 269: 20425–20430.

Honeyfield DC, JR Carlson. 1990. Assay for the enzymatic conversion of indoleacetic acid to 3-methylindole in a ruminal *Lactobacillus* species. *Appl Environ Microbiol* 56: 724–729.

Höschle B, V Gnau, D Jendrossek. 2005. Methylcrotonyl-CoA and geranyl-CoA carboxylases are involved in leucine/isovalerate utilization (Liu) and acyclic terpene utilization (Atu), and are encoded by *liuB /liuD* and *atuC/atuF*, in *Pseudomonas aeruginosa. Microbiology (UK)* 151: 3649–3656.

Huang HT, JW Davisson. 1958. Distribution of lysine racemase in bacteria. *J Bacteriol* 76: 495–498.

Hunt DE, D Gevers, NM Vahora, MF Polz. 2008. Conservation of the chitin utilization pathway in the *Vibrionaceae. Appl Environ Microbiol* 74: 44–51.

Huynh QK, EE Snell. 1985. Pyruvoyl-dependent histidine decarboxylases. Preparation and amino acid sequences of the β chains of histidine decarboxylase from *Clostridium perfringens* and *Lactobacillus buchneri. J Biol Chem* 260: 2798–2803.

Ichihara A, S Furiya, M Suda. 1960. Metabolism of L-lysine by bacterial enzymes III. Lysine racemase. *J Biochem (Tokyo)* 48: 277–282.

Iyer R, C Williams, C Miller. 2003. Arginine-agmatine antiporter in extreme acid resistance in *Escherichia coli. J Bacteriol* 185: 6556–6561.

Jackson S, M Calos, A Myers, WT Self. 2006. Analysis of proline reduction in the nosocomial pathogen *Clostridium difficile. J Bacteriol* 188: 8387–8495.

Jann A, H Matsumoto, D Haas. 1988. The fourth arginine catabolic pathway of *Pseudomonas aeruginosa. J Gen Microbiol* 134: 1043–1053.

Jeng IM, R Somack, HA Barker. 1974. Ornithine degradation in *Clostridium sticklandii*; pyridoxal phosphate and coenzyme A dependent thiolytic cleavage of 2-amino-4-ketopentanoate to alanine and acetyl coenzyme A. *Biochemistry* 13: 2898–2903.

Job V, GL Marcone, MS PIlone, L Pollegioni. 2002. Glycine oxidase from *Bacillus subtilis*. Characterization of a new flavoprotein. *J Biol Chem* 277: 6985–6993.

Johnson-Winters K, VM Purpero, M Kavana, T Nelson, GR Moran. 2003. (4-hydroxyphenyl)pyruvate dioxygenase from *Streptomyces avermitilis*: the basis for ordered substrate addition. *Biochemistry* 42: 2072–2080.

Kabisch UC, A Gräntzdörffer, A Schierhorn, KP Rüecknagel, JR Andreesen, A Pich. 1999. Identification of D-proline reductase from *Clostridium sticklandii* as a selenoenzyme and indications for a catalytically active pyruvoyl group derived from a cysteine residue by cleavage of a proprotein. *J Biol Chem* 274: 8445–8454.

Kalyuzhnaya M, S Bowerman, J Laram M Lidstrom, L Chistoserdova. 2006. *Methylotenera mobilis* is an obligately methylamine-utilizing organism within the family Methylophilaceae. *Int J Syst Evol Microbiol* 56: 2819–2823.

Kamath AV, GL Vaaler, EE Snell. 1991. Pyridoxal phosphate-dependent histidine decarboxylases. Cloning, sequencing, and expression of genes from *Klebsiella planticola* and *Enterobacter aerogenes* and properties of the overexpressed enzymes. *J Biol Chem* 266: 9432–9437.

Kambourova M, M Tangney, FG Priest. 2001. Regulation of polygutamic acid synthesis by glutamate in *Bacillus licheniformis* and *Bacillus subtilis. Appl Environ Microbiol* 67: 1004–1007.

Karatzas K-A, O Brennan, S Heavin, J Mossissey, CP O'Byrne. 2010. Intracellular accumulation of high levels of γ-aminobutyrate by *Listeria monocytogenes* 10403S in response to low pH: uncoupling of γ-aminobutyrate synthesis from efflux in a chemically defined medium. *Appl Environ Microbiol* 76: 3529–3537.

Karthikeyan S, Q Zhou, Z Zhao, C-L Kao, Z Tao, H Robinson, H-W Liu, H Zhang. 2004. Structural analysis of *Pseudomonas* 1-aminocyclopropane-1-carboxylate deaminase complexes: insight into the mechanism of a unique pyridoxal-5′-phosphate dependent cyclopropane ring-opening reaction. *Biochemistry* 43: 13328–13339.

Kataoka M, M Ikemi, T Morikawa, T Miyoshi, K-I Nishi, M Wada, H Yamada, S Shimizu. 1977. Isolation and characterization of D-threonine aldolase, a pyridoxal-5′-phosphate-dependent enzyme from *Arthrobacter* sp. DK-38 *Eur J Biochem* 248: 385–393.

Kieboom J, T Abee. 2006. Arginine-dependent acid resistance in *Salmonella enterica* serovar Typhimurium. *J Bacteriol* 188: 5650–5653.

Kikuchi Y, H Kojima, T Tanaka, Y Takatsuka, Y Kamio. 1997. Characterization of a second lysine decarboxylase isolated from *Escherichia coli*. *J Bacteriol* 179: 4486–4492.

Kim J, D Darley, W Buckel. 2004a. 2-hydroxyisocaproyl-CoA dehydratase and its activator from *Clostridium difficile*. *FEBS J* 272: 550–561.

Kim, J, M Hetzel, CD Boiangiu, W Buckel. 2004b. Dehydration of (*R*)-2-hydroxyacyl-CoA to enoyl-CoA in the fermentation of α-amino acids by anaerobic bacteria. *FEMS Microbiol Revs* 28: 455–468.

Kim J, D Darley, T Selmer, W Buckel. 2006. Characterization of (*R*)-2-hydroxyisocaproate dehydrogenase and a family III coenzyme A transferase involved in the reduction of L-leucine to isocaproate by *Clostridium difficile*. *Appl Environ Microbiol* 72: 6062–6069.

Kino K, T Arai, Y Arimura. 2011. Poly-α-glutamic acid synthesis using a novel catalytic activity of RimK from *Escherichia coli*. *Appl Environ Microbiol* 77: 2019–2125.

Kleber H-P. 1997. Bacterial carnitine metabolism. *FEMS Microbiol Lett* 147: 1–9.

Konig H, KO Stetter. 1982. Isolation and characterization of *Methanolobus tindarius*, sp. nov., a coccoid methanogen growing only on methanol, and methylamines. *Zentralbl Bakteriol Parasitenkd Infektionskr Hyg Abt 1 Orig Reihe* C 3: 478–490.

Koyama H. 1982. Purification and characterization of a novel L-phenylalanine oxidase (deaminating and decarboxylating) from *Pseudomonas* sp. P-501. *J Biochem (Tokyo)* 92: 1235–1240.

Koyama H. 1984. Oxidation and oxygenation of L-amino acids catalyzed by a L-phenylalanine oxidase (deaminating and decarboxylating) from *Pseudomonas* sp. P-501. *J Biochem (Tokyo)* 96: 421–427.

Krätzer C, P Carini, R Hovey, U Deppenmeier. 2009. Transcriptional profiling of methyltransferase genes during growth of *Methanosarcina mazei* on trimethylamine. *J Bacteriol* 191: 5108–5115.

Kreimeyer A, A Perret, C Lechaplais, D Vallenet, C Médigue, M Salanoubat, J Weissenbach. 2007. Identification of the last unknown genes in the fermentation pathway of lysine. *J Biol Chem* 282: 7191–7197.

Krings U, M Hinz, RG Berger. 1996. Degradation of [^{2}H]phenylalanine by the basidiomycete *Ischnoderma benzoinum*. *J Biotechnol* 51: 123–129.

Kuchta RD, RH Abeles. 1985. Lactate reduction in *Clostridium propionicum*. Purification and properties of lactyl-CoA dehydratase. *J Biol Chem* 260: 13181–13189.

Kumagai H, H Matsui, K Ogata, H Yamada. 1969. Properties of crystalline tyramine oxidase from *Sarcina lutea*. *Biochim Biophys Acta* 171: 1–8.

Kurihara S, A Oda, K Kato, Y Tsuboi, HG Kim, M Oshida, H Kumagai, H Suzuki. 2008. γ-glutamylputrescine synthetase in the putrescine utilization pathway of *Escherichia coli* K-12. *J Biol Chem* 283: 19981–1990.

Kurihara S, K Kato, K Asada, H Kumagai, H Suzuki. 2010. A putrescine-inducible pathway comprising PuuE-YneI in which γ-aminobutyrate is degraded into succinate in *Escherichia coli* K-12. *J Bacteriol* 192: 4582–4591.

Kyndt JA, TE Mayer, MA Cusanovich, JJ van Neeumen. 2002. Characterization of a bacterial tyrosine ammonia lyase, a biosynthetic enzyme for the photoreactive yellow protein. *FEBS Lett* 512: 240–244.

Laanbroek HJ, JT Lambers, WM de Vos, H Veldkamp. 1978. L-Aspartate fermentation by a free-living *Campylobacter* species. *Arch Microbiol* 117: 109–114.

Landete JM, ME Arena, I Pardo, MC Manca de Nadra, S Ferrer. 2010. The role of two families of bacterial enzymes in putrescine synthesis from agmatine via agmatine deiminase. *Int Microbiol* 13: 169–177.

Lapadatescu C, C Giniès, J-L LOl Quéré, P Bonnarme. 2000. Novel scheme for biosynthesis of aryl metabolites from L-phenylalanine in the fungus *Bjerkardera adusta*. *Appl Environ Microbiol* 66: 1517–1522.

Latpova E, S Yang, Y-S Wang, T Wang, TA Chavkin, H Hackett, H Schäfer, MG Kalyuzhnaya. 2010. Genetics of the glutamate-mediated methylamine utilization pathway in the facultative methylotrophic betaproteobacterium *Methyloversatilis universalis* FAM5. *Mol Microbiol* 75: 426–439.

Lawrence CC, WJ Sobey, RA Field, JE Baldwin, CJ Schofield. 1996. Purification and initial characterization of proline 4-hydroxylase from *Streptomyces griseoviridus* P8648: a 2-oxoacid, ferrous-dependent dioxygenase involved in etamycin biosynthesis. *Biochem J* 313: 185–192.

Lemonnier M, D Lane. 1998. Expression of the second lysine decarboxylase gene of *Escherichia coli*. *Microbiology (UK)* 144: 751–761.

Lindstedt S, B Odelhög, M Rundgren. 1977a. Purification and properties of 4-hydroxyphenylpyruvate dioxygenase from *Pseudomonas* sp. P.J. 874. *Biochemistry* 16: 3369–3377.

Lindstedt G, S Lindstedt, I Norin. 1977b. Purification and properties of γ-butyrobetaine hydroxylase from *Pseudomonas* sp. AK1. *Biochemistry* 16: 2181–2188.

Liu Y, TM Louie, J Payne, J Bohuslavek, H Bolton, L Xun. 2001. Identification, purification, and characterization of iminodiacetate oxidase from the EDTA-degrading bacterium BNC-1. *Appl Environ Microbiol* 67:696–701.

Lomans BP, R Maas, R Luderer, HJP op den Camp, A Pol, C van der Drift, GD Vogels. 1999. isolation and characterization of *Methanomethylovorans hollandica* gen. nov. sp. nov., isolated from freshwater sediment, a methylotrophic methanogen able to grow with dimethyl sulfide and methanethiol. *Appl Environ Microbiol* 65: 3641–3650.

Lu CD, Y Itoh, Y Nakada, Y Jiang. 2002. Functional analysis and regulation of the divergent *spuABCDEFGH-spuI* operons for polyamine uptake and utilization in *Pseudomonas aeruginosa* PAO1. *J Bacteriol* 184: 3765–3773.

Lu ZJ, GD Markham. 2004. Catalytic properties of the archaeal S-adenosylmethionine decarboxylase from *Methanococcus jannaschii*. *J Biol Chem* 279: 265–273.

Maeder DL et al. 2006. The *Methanosarcina barkeri genome*: Comparative analysis with *Methanosarcina acetivorans* and *Methanosarcina mazei* reveals extensive rearrangement within methanosarcinal genomes. *J Bacteriol* 188: 7922–7931.

Markham GD, CW Tabor and H Tabor. 1982. S-adenosylmethionine decarboxylase of *Escherichia coli*. Studies on the covalently linked pyruvate required for activity. *J Biol Chem* 257: 12063–12068.

Martín MC, M Fernández, DM Linares, MA Alvarez. 2005. Sequencing, characterization and transcriptional analysis of the histidine decarboxylase operon on *Lactobacillus buchneri*. *Microbiology (UK)* 151: 1219–1229.

Martins BM, H Dobbek, I Çinkaya, W Buckel, A Messerschmidt. 2004. Crystal structure of 4-hydroxybutyryl-CoA dehydratase: radical catalysis involving a [4Fe–4S] cluster and flavin. *Proc Natl Acad Sci USA* 101. 15645–15649.

Marusich WC, RA Jensen, LO Zamir. 1981. Induction of L-phenylalanine ammonia-lyase during utilization of phenylalanine as a carbon or nitrogen source in *Rhodotorula glutinis*. *J Bacteriol* 146: 1013–1019.

Massey LK, RS Conrad, JR Sokatch. 1974. Regulation of leucine catabolism in *Pseudomonas putida*. *J Bacteriol* 118: 112–120.

McIntire WS, DE Wemmer, A Chistoserdov, ME Lidstrom. 1991. A new cofactor in a prokaryotic enzyme: tryptophan trypotophylquinone as the redox prosthetic group in methylamine dehydrogenase. *Science* 252: 817–824.

Meiberg JBM, PM Bruinenberg, W Harder. 1980. Effect of dissolved oxygen tension on the metabolism of methylated amines in *Hyphomicrobium* X in the absence and presence of nitrate: evidence for 'aerobic' denitrification. *J Gen Microbiol* 120: 453–463.

Meiberg JBM, W Harder. 1978. Aerobic and anaerobic metabolism of trimethylamine, dimethylamine and methylamine in *Hyphomicrobium* X. *J Gen Microbiol* 106: 265–276.

Meiberg JBM, W Harder. 1979. Dimethylamine dehydrogenase from *Hyphomicrobium* X: purification and some properties of a new enzyme that oxidizes secondary amines. *J Gen Microbiol* 115: 49–58.

Mercenier A, J-P Simon, CV Wauven, D Haasm, V Stalon. 1980. Regulation of enzyme synthesis in the arginine deiminase pathway of *Pseudomonas aeruginosa*. *J Bacteriol* 144: 159–163.

Meskys R, RJ Harris, V Casaite, J Basran, NS Scrutton. 2001. Organization of the genes involved in dimethylglycine and sarcosine degradation in *Arthrobacter* spp. *Eur J Biochem* 268: 3390–3398.

Middelhoven WJ, MC Hoogkamer-Te Niet, NJ Kreger-Van Rij. 1984. *Trichosporon adeninovorans* sp. nov., a yeast species utilizing adenine, xanthine, uric acid, putrescine and primary *n*-alkylamines as the sole source of carbon, nitrogen and energy. *Antonie van Leeuwenhoek* 50: 369–378.

Miller DL, VW Rodwell. 1971. Metabolism of basic amino acids in *Pseudomonas putida*. Catabolism of lysine by cyclic and acyclic intermediates. *J Biol Chem* 246: 2758–2764.

Miller JDA, DS Wakerley. 1966. Growth of sulfate-reducing bacteria by fumarate dismutation. *J Gen Microbiol* 43: 101–107.

Miura-Fraboni J, H-P Kleber, S Englard. 1982. Assimilation of γ-butyrobetaine, and D- and L-carnitine by resting cell suspensions *of Acinetobacter calcoaceticus* and *Pseudomonas putida*. *Arch Microbiol* 133: 217–221.

Milić D, TV Demidkina, NG Faleev, RS Phillips, D Matković-Calogović, AA Antson. 2011. Crystallographic snapshots of tyrosine phenol-lyase show that substrate strain plays a role in C–C cleavage. *J Amer Chem Soc* 133: 16468–16476.

Molenaar D, JS Bosscher, B ten Brink, AJM Driessen, WN Konengs. 1993. Generation of a proton motive force by histidine decarboxylation and electrogenic histidine/histamine antiport in *Lactobacillus buchneri*. *J Bacteriol* 175: 2864–2870.

Molina L, C Ramos, E Duque, MC Ronchel, JM García, L Lyke, JL Ramos. 2000. Survival of *Pseudomonas putida* KT2440 in soil and in the rhizosphere of plants under greenhouse and environmental conditions. *Soil Biol Biochem* 32: 315–321.

Moore K, PV Subba Rao, GHN Towers. 1968. Degradation of phenylalanine and tyrosine by *Sporobolomyces roseus*. *Biochem J* 106: 507–514.

Moss M, PA Frey. 1990. The role of *S*-adenosylmethionone in the lysine 2,3-aminomutase reaction. *J Biol Chem* 262: 14859–14862.

Müh U, I Çinkaya, SPJ Albracht, W Buckel. 1996. 4-hydroxybutyryl-CoA dehydratase from *Clostridium aminobutyricum*: characterization of FAD and iron–sulfur clusters involved in an overall non-redox reaction. *Biochemistry* 35: 11710–11718.

Müller U, W Buckel. 1995. Activation of (*R*)-2-hydroxyglutaryl-CoA dehydratase from *Acidaminococcus fermentans*. *Eur J Biochem* 230: 698–704.

Muramatsu H, H Mihara, R Kakutani, M Yasuda, M Ueda, T Kurihara, N Esaki. 2005. The putative malate/lactate dehydrogenase from *Pseudomonas putida* is an NADPH-dependent Δ^1-piperideine-2-carboxylate/Δ^1-pyrroline-2-carboxylate reductase involved in the catabolism of L-lysine and D-proline. *J Biol Chem* 280: 5329–5335.

Nakada Y, Y Itoh. 2002. Characterization and regulation of the *gbuA* gene, encoding guanidinobutyrase in the arginine dehydrogenase pathway of *Pseudomonas aeruginosa*. *J Bacteriol* 184: 3377–3384.

Nakazawa T, K Hori, O Hayaishi. 1972. Studies on monooxygenases V. Manifestation of amino acid oxidase activity by L-lysine monooxygenase. *J Biol Chem* 247: 3439–3444.

Ni S, CR Woese, HC Aldrichm, DR Boone. 1994. Transfer of *Methanolobus siciliae* to the genus *Methanosarcina*, naming it *Methanosarcina siciliae*, and emendation of the genus *Methanosarcina*. *Int J Syst Evol Microbiol* 44: 357–359.

Ohara M, Y Katayama, M Tsuzaki, S Nakamoto, H Kuraishi. 1990. *Paracoccus kocurii* sp. nov., a tetramethyl-ammonium-assimilating bacterium. *Int J Syst Bacteriol* 40: 292–296.

Paul L, DJ Ferguson, JA Krzycki. 2000. The trimethylamine methyltransferase gene and multiple dimethyl-amine methyltransferase genes of *Methanosarcina barkeri* contain in-frame and read-through amber codons. *J Bacteriol* 182: 2520–2529.

Payne JW, H Bolton, JA Campbell, L Xun. 1998. Purification and characterization of EDTA monooxygenase from the EDTA-degrading bacterium BNC1. *J Bacteriol* 180: 3823–3827.

Penrod JT, JR Roth. 2006. Conserving a volatile metabolite: a role for carboxysome-like organelles in *Salmonella enterica*. *J Bacteriol* 188: 2865–2874.

Petrovich RM, FJ Ruzicka, GH Reed, PA Frey. 1991. Metal cofactors of lysine 2,3-aminomutase. *J Biol Chem* 266: 7656–7660.

Plumbridge J, E Vimr. 1999. Convergent pathways for utilization of the amino sugars *N*-acetylglucosamine, *N*-acetylmannosamine, and *N*-acetylneuraminic acid by *Escherichia coli*. *J Bacteriol* 181: 47–54.

Pollock RJ, LB Hersh. 1973. *N*-methylglutamate synthetase. The use of flavin mononucleotide in oxidative catalysis. *J Biol Chem* 248: 6724–6733.

Pometto AL, DL Crawford. 1985. L-Phenylalanine and L-tyrosine catabolism by selected *Streptomyces* species. *Appl Environ Microbiol* 49: 727–729.

Pons J-L, A Rimbault, JC Darbord, G Leluan. 1984. Biosynthèse de toluène chez *Clostridium aerofoetidum* souche WS. *Ann Microbiol (Inst Pasteur)* 135 B: 219–222.

Poston JM. 1976. Leucine 2,3-aminomutase, an enzyme of leucine catabolism. *J Biol Chem* 251: 1859–1863.

Poupin P, N Truffaut, B Combourieu, M Sancelelme, H Veschambre, AM Delort. 1998. Degradation of morpholine by an environmental *Mycobacterium* strain involves a cytochrome P-450 *Appl Environ Microbiol* 64: 165–165.

Ralph EC, MA Anderson, WW Cleland, PF Fitzpatrick. 2006. Mechanistic studies of the flavoenzymes trypto-phan 2-monooxygenase: deuterium and ^{15}N kinetic isotope effects on alanine oxidation by an L-amino acid oxidase. *Biochemistry* 45: 15844–15852.

Rees GN, BKC Patel. 2001. *Desulforegula conservatrix* gen. nov., sp. nov., a long-chain fatty acid-oxidizinbg, sulfate-reducing bacterium isolated from sediments of a freshwater lake. *Int J Syst Evolut Microbiol* 51: 1911–1916.

Revelles O, M Espinosa-Urgel, S Molin, JL Ramos. 2004. The *davDT* operon of *Pseudomonas putida*, involved in lysine catabolism is induced in response to the pathway intermediate delta-aminovaleric acid. *J Bacteriol* 186: 3439–3446.

Revelles O, M Espinosa-Urgel, T Fuhrer, U Sauer, JL Ramos. 2005. Multiple and interconnected pathways for L-lysine catabolism in *Pseudomonas putida* KT2440. *J Bacteriol* 187: 7500–7510.

Revelles O, RM Wittich, JL Ramos. 2007. Identification of the initial steps in D-lysine catabolism in *Pseudomonas putida. J Bacteriol* 189: 2787–2792.

Roof DM, JR Roth. 1989. Functions required for vitamin B_{12}-dependent ethanolamine utilization in *Salmonella typhimurium. J Bacteriol* 171: 3316–3323.

Rosner B, B Schink. 1990. Propionate acts as carboxylic group acceptor in aspartate fermentation by *Propionibacterium freudenreichii. Arch Microbiol* 155: 46–51.

Roujeinikova A, NS Scrutton, D Leys. 2006. Atomic level insight into the oxidative half-reaction of aromatic amine dehydrogenase. *J Biol Chem* 281: 40264–40272.

Roujeinikova A, P Hothi, L Masgrau, MJ Sutcliffe, NS Scrutton, D Leys. 2007. New insights into the reductive half-reaction mechanism of aromatic amine dehydrogenase revealed by reaction with carbinolamine substrates. *J Biol Chem* 282: 23766–23777.

Rudnick G, RH Abeles. 1975. Reaction mechanism and structure of the active site of proline racemase. *Biochemistry* 14: 4515–4522.

Ryle MJ, KD Koehntop, A Liu, L Que Jr, RP Hausinger. 2003. Interconversion of two oxidized forms of taurine/α-ketoglutarate dioxygenase, a non-heme iron hydroxylase: Evidence for bicarbonate binding. *PNAS* 100: 3790–3795.

Satoh A et al. 2002. Crystal structure of quinoprotein amine dehydrogenase from *Pseudomonas putida. J Biol Chem* 277: 2830–2834.

Sauer K, U Harms, RK Thauer. 1997. Methanol: coenzyme M methyltransferase from *Methanosarcina barkeri.* Purification, properties and encoding genes of the corrinoid protein MT1. *Eur J Biochem* 243: 670–677.

Schneider BL, AK Kiupakis, LR Reitzer. 1998. Arginine catabolism and the arginine succinyltransferase pathway in *Escherichia coli. J Bacteriol* 180: 4278–4286.

Schwede TF, J Rétey, GE Schulz. 1999. Crystal structure of histidine ammonia-lyase revealed a novel polypeptide modification as the catalytic electrophile. *Biochemistry* 38: 5355–5361.

Schütz A, T Sandalova, S Ricagno, G Hübner, S König, G Schneider. 2003. Crystal structure of thiaminediphosphate-dependent decarboxylase from *Enterobacter cloacae*, an enzyme involved in the biosynthesis of the plant hormone indol-3-acetic acid. *Eur J Biochem* 270: 2312–2312.

Selmer T, PI Andrei. 2001. *p*-hydroxyphenylacetate decarboxylase from *Clostridium difficile.* A novel glycyl radical enzyme catalysing the formation of *p*-cresol. *Eur J Biochem* 268: 1363–1372.

Selmer T, A Willanzheimer, M Hetzel. 2002. Propionate CoA-transferase from *Clostridium propionicum.* Cloning of the gene and identification of glutamate 324 at the active site. *Eur J Biochem* 269: 372–380.

Shah HN, RAD Williams. 1987. Catabolism of aspartate and asparagine by *Bacteroides intermedius* and *Bacteroides gingivalis. Curr Microbiol* 15: 313–318.

Shearer, N, AP Hinsley, RJM van Spanning, S Spiro (19999 Anaerobic growth of *Paracoccus denitrificans* requires cobalamin: characterization of *cobK* and *cobJ* genes. *J Bacteriol* 181: 6907–6913.

Smith LT, J-A Pocard, T Bernard, D Le Rudulier. 1988. Osmotic control of glycine betaine biosynthesis and degradation in *Rhizobium meliloti. J Bacteriol* 170: 3142–3149.

Soares JA, L Zhang, RL Pitsch, NM Kleinholz, RB Jones, JJ Wolff, J Amster, KN Green-Church, JA Krzycki. 2005. The residue mass of L-pyrrolysine in three distinct methylamine methyltransferases. *J Biol Chem* 280: 36962–36969.

Söhling B, G Gottschalk. 1996. Molecular analysis of the anaerobic succinate degradation pathway in *Clostridium kluyveri. J Bacteriol* 178: 871–880.

Somack R, RN Costilow. 1973a. Purification and properties of a pyridoxal phosphate and coenzyme B_{12}–dependent D-ornithine 5,4-aminomutase. *Biochemistry* 12: 2597–2604.

Somack R, RN Costilow. 1973b. 2,4-diaminopentanoic acid C_4 dehydrogenase. *J Biol Chem* 247: 385–388.

Sowers KR, JG Ferry. 1983. Isolation and characterization of a methylotrophic marine methanogen *Methanococcoides methylutens* gen. nov., sp. nov. *Appl Environ Microbiol* 45: 684–690.

Sparnins VL, DG Burbee, S Dagley. 1979. Catabolism of L-tyrosine in *Trichosporon cutaneum. J Bacteriol* 138: 425–430.

Stadtman TC, P Elliot. 1957. Studies on the enzymatic reduction of amino acids II. Purification and properties of a D-proline reductase and a proline racemase from *Clostridium sticklandii. J Biol* 983–997.

Steenkamp DJ. 1979. Identification of the prosthetic groups of dimethylamine dehydrogenase from *Hyphomicrobium* X. *Biochem Biophys Res Comm* 88: 244–259.

Steenkamp DJ, TP Singer, H Beinert. 1978a. Participation of the iron-sulfur cluster and of the covalently bound coenzyme of trimethylamine dehydrogenase in catalysis. *Biochem J* 169: 361–369.

Steenkamp DJ, W McIntire, WC Kenney. 1978b. Structure of the covalently bound coenzyme of trimethylamine dehydrogenase. Evidence for a 6-substituted flavin. *J Biol Chem* 253: 2818–2824.

Steenkamp DJ, WC Kenny, TP Singer. 1978c. A novel type of covalently bound coenzyme in trimethylamine dehydrogenase *J Biol Chem* 253: 2812–2817.

Stickland LH. 1934. Studies in the metabolism of the strict anaerobes (Genus Clostridium) I The chemical reactions by which *Cl. sporogenes* obtains its energy. *Biochem J* 28: 1746–1759.

Stickland LH. 1935a. Studies in the metabolism of the strict anaerobes (Genus Clostridium). II. The reduction of proline by *Cl. sporogenes*. *Biochem J* 29: 288–290.

Stickland LH. 1935b. Studies in the metabolism of the strict anaerobes (Genus Clostridium). III. The oxidation of alanine by *Cl. sporogenes*. IV. The reduction of glycine by *Cl. sporogenes*. *Biochem J* 29: 898.

Suzuki M. 1981. Purification and some properties of sarcosine oxidase from *Corynebacterium* sp. U-96. *J Biochem (Tokyo)* 89: 599–607.

Swain A, KV Waterhouse, WA Venables, AG Callely, SE Lowe. 1991. Biochemical studies of morpholine catabolism by an environmental mycobacterium. *Appl Microbiol Biotechnol* 35: 110–114.

Tabor CW, H Tabor. 1985. Polyamines in microorganisms. *Microbiol Revs* 49: 81–99.

Tallant TC, L Paul, JA Krzycki. 2001. The MtsA subunit of the methylthiol:coenzyme M methyltransferase of *Methanosarcina barkeri* catalyzes both half-reactions of corrin-dependent dimethylsulfide:coenzyme M methyl transfer. *J Biol Chem* 276: 4485–4493.

Teufel R, V Mascaraque, W Ismail, M Voss, J Perrera, W Eisenreich, W Haehnel, G Fuchs. 2010. Bacterial phenylalanine and phenylacetate catabolic pathway revealed. *Proc Natl Acad Sci USA* 107:14390–14395.

Thorne CB, CG Gómez, HE Noyes, RD Housewright. 1954. Production of glutamyl polypeptide by *Bacillus subtilis*. *J Bacteriol* 68: 307–315.

Toms AV, C Kinsland, DE McCloskey, AE Pegg, SE Ealick. 2004. Evolutionary links as revealed by the structure of *Thermotoga maritima* S-adenosylmethionine decarboxylase. *J Biol Chem* 279: 33837–33846.

Troy FA. 1973. Chemistry and biosynthesis of the poly (γ-D-glutamyl) capsule in *Bacillus licheniformis*: properties of the membrane-mediated biosynthetic reaction. *J Biol Chem* 248: 305–315.

Tsoi H-W, H Tse. 2011. *Staphylococcus lugdunensis* is the likely origin of the ornithine decarboxylase operon in *Staphylococcus epidermis* 2015B. *Appl Environ Microbiol* 77: 392–393.

Tsuda Y, HC Friedmann. 1970. Ornithine metabolism by *Clostridium sticklandii*. Oxidation of ornithine to 2-amino-4-ketopentanoic acid via 2,4-diaminopentanoic acid: participation of B_{12} coenzyme, pyridoxal phosphate and pyridine nucleotide. *J Biol Chem* 245: 889–898.

Uetz T, R Schneider, M Snozzi, T Egli. 1992. Purification and characterization of a two-component monooxygenase that hydroxylates nitrilotriacetate from "*Chelatobacter*" strain ATCC 29600. *J Bacteriol* 174: 1179–1188.

Urakami T, H Araki, H Oyanagi, K-I Suzuki, K Komagata. 1990. *Paracoccus aminophilus* sp. nov. and *Paracoccus aminovorans* sp. nov., which utilize *N,N′*-dimethylformamide. *Int J Syst Bacteriol* 40: 287–291.

Urakami T, H Araki, H Oyanagi, K-I Suzuki, K Komagata. 1992. Transfer of *Pseudomonas aminovorans* (den Dooren de Jong 1926) to *Aminobacter* gen. nov. as *Aminobacter aminovorans* comb. nov., and description of *Aminobacter aganoensis* sp. nov. and *Aminobacter niigataensis* sp. nov. *Int J Syst Bacteriol* 42: 84–92.

Urushibata Y, S Tokuyama, Y Tahara. 2002. Characterization of the *Bacillus subtilis yws* gene, involved in γ-polyglutamic acid production. *J Bacteriol* 184: 337–343.

van Hellemond EW, M van Dijk, DPHM Heuts, DB Janssen, MW Fraaije. 2008. Discovery and characterization of a putrescine oxidase from *Rhodococcus erythropolis*. *Appl Microbiol Biotechnol* 78: 455–463.

Vimr ER, KA Kalivoda, EL Deszo, SM Steenbergen. 2004. Diversity of microbial sialic acid metabolism. *Microbiol Mol Biol Revs* 68: 132–153.

Walsh C, RA Pascal, M Johnston, R Raines, D Dikshit, A Krantz, M Honma. 1981. Mechanistic studies on the pyridoxal phosphate enzyme 1-aminocyclopropane-1-carboxylate deaminase from *Pseudomonas* sp. *Biochemistry* 20: 7509–7519.

Walt A, ML Kahn. 2002. The *fixA* and *fixB* genes are necessary for anaerobic carnitine reduction in *Escherichia coli*. *J Bacteriol* 184: 4044–4047.

Wäsch SI de A, JR van der Ploeg, T Maire, A Lebreton, A Kiener, T Leisinger. 2002. Transformation of isopropylamine to L-alaninol by *Pseudomonas* sp. strain KIE171 involves *N*-glutamylated intermediates. *Appl Environ Microbiol* 68: 2368–2375.

Wauven CV, S Piérard, M Kley-Raymann, D Haas. 1984. *Pseudomonas aeruginosa* mutants affected in anaerobic growth on arginine: evidence for a four-gene cluster encoding the arginine deiminase pathway. *J Bacteriol* 160: 928–934.

Williams JS, M Thomas, DJ Clarke. 2005. The gene *stlA* encodes a phenylalanine ammonia-lyase that is involved in the production of a stilbene antibiotic in *Photorhabdus luninescens* TT01, *Microbiology (UK)* 151: 2543–2550.

Wilmot CM, J Haijdu, MJ McPherson, PF Knowles, SEV Phillips. 1999. Visualization of dioxygen bound to copper during enzyme catalysis. *Science* 286: 1714–1728.

Witschel M, S Nagel, T Egli. 1997. Identification and characterization of the two-enzyme system catalyzing the oxidation of EDTA in the EDTA-degrading bacterial strain DSM 9103. *J Bacteriol* 179: 6937–6943.

Wong JC, JK Dyer, JL Tribble. 1977. Fermentation of L-aspartate by a saccharolytic strain of *Bacteroides melaninogenicus. Appl Environ Microbiol* 33: 69–73.

Xiang L, BS Moore. 2002. Inactivation, complementation, and heterologous expression of *encP*, a novel. bacterial phenylalanine ammonia-lyase gene. *J Biol Chem* 277:32505–32509.

Xu Y, MW Mortimer, TS Fisher, ML Kahn FJ Brockman, L Xun. 1997. Cloning, sequencing, and analysis of gene cluster from *Chelatobacter heintzii* ATCC 29600 encoding nitrolotriacetate monooxygenase and NADH:flavin mononucleotide oxidoreductase. *J Bacteriol* 179: 1112–1116.

Yamanaka K, N Kito, Y Imokawa, C Maruyama, T Utagawa, Y Hamano. 2010. Mechanism of ε-poly-L-lysine production and accumulation revealed by identification and analysis of an ε-poly-L-lysine-degrading enzyme. *Appl Environ Microbiol* 76: 5669–5675.

Yang Z, C-D Lu. 2007. Functional genomics enables identification of the genes of the arginine transamination pathway in *Pseudomonas aeruginosa. J Bacteriol* 189: 3945–3953.

Yang C, DA Rodionov, X Li, ON Laikova, MS Gelfand, OP Zagnitko, MF Romine, AY Obraztsova, KH Nealson, AL Osterman. 2006. Comparative genomics and experimental characterization of *N*-acetylglucosamine utilization pathway of *Shewanella oneidensis. J Biol Chem* 281: 29872–29885.

Yokoyama MT, JR Carlson. 1981. Production of skatole and *para*-cresol by a rumen *Lactobacillus* sp. *App Environ MIcrobiol* 41: 71–76.

Yokoyama MT, JR Carlson, LV Holdeman. 1977. Isolation and characteristics of a skatole-producing *Lactobacillu*s sp. from the bovine rumen. *App Environ MIcrobiol* 34: 837–842.

Yorifugi T, I-M Jeng, HA Barker. 1977. Purification and properties of 3-keto-5-aminohexanoate cleavage from a lysine-fermenting *Clostridium. J Biol Chem* 252: 20–31.

Zwart KB, W Harder. 1983. Regulation of the metabolism of some alkylated amines in the yeasts *C. utilis* and *Hansenula polymorpha. J Gen Microbiol* 129: 3157–3169.

7.12 Alkanes, Cycloalkanes, and Related Compounds with Chlorine, Bromine, or Iodine Substituents

Particularly the chlorinated compounds have enjoyed range of applications: vinyl chloride (chloroethene) as monomer for the production of PVC, tetra- and trichloroethenes as solvents for degreasing, and the insecticides 1,1,1-trichloro-2,2-bis(*p*-chlorophenyl)ethane (DDT) and isomers of hexachlorocyclohexane (HCH) (benzene hexachloride). The biodegradation of fluorinated aliphatic compounds is generally different from the outlines that have emerged from investigations on their chlorinated, brominated, and even iodinated analogues. They are, therefore, treated separately in Section 7.14.

7.12.1 Chlorinated, Brominated, and Iodinated Alkanes, Alkenes, and Alkanoates

A range of mechanisms has been found for the biodegradation of halogenated alkanes, alkenes, and alkanoates:

1. Elimination including dehydrohalogenation
2. A corrinoid pathway for C_1 halides, which operates under both aerobic and anaerobic conditions

3. Nucleophilic displacement including hydroxylation, hydrolysis, and glutathione-mediated reactions
4. Reduction in which a halogen atom is replaced by hydrogen including dehalorespiration
5. Cytochrome P450-mediated reactions.

7.12.1.1 Elimination Reactions

In contrast to chemical reactions in which elimination and nucleophilic displacement are alternatives and may occur simultaneously, microbial elimination is less common. This term is also used for the reaction in which, for example, 1,2-dihaloethanes are transformed to ethene, in contrast to dehydrohalogenation in which a haloethene is produced or reductive hydrogenolysis to a haloethane.

Degradation involving elimination is found in several degradations:

1. Several steps in the degradation of DDT by the facultatively anaerobic bacterium *A. aerogenes* involve elimination (Figure 7.73) (Wedemeyer 1967), while a range of products from transformation of the trichloromethyl groups have been isolated (Schwartzbauer et al. 2003). The recovery of the elimination product 1,1-dichloro-2,2-bis(*p*-chlorophenyl)ethene (DDE) from environmental samples long after restriction on the use of DDT suggests a high degree of persistence. DDE can, however, be degraded by ring dioxygenation and extradiol ring fission to 4-chlorobenzoate in cells of *Pseudomonas acidovorans* M3GY grown with biphenyl (Hay and Focht 1998) and *Terrabacter* sp. strain DDE-1 induced with biphenyl (Aislabie et al. 1999). An alternative pathway for the transformation of DDT involves hydroxylation of the ring and displacement of the aromatic ring chlorine atom by hydroxyl (Figure 7.74) (Massé et al. 1989). Bis(*p*-chlorophenyl)acetic acid (DDA) is a polar metabolite that is apparently persistent in the environment (Heberer and Dünnbier 1999). In higher organisms, DDE can be metabolized by formation of sulfones (Letcher et al. 1998).

2. Growth of a range of bacteria including *E. coli*, *B. subtilis*, and *Streptococcus pyogenes* is inhibited by 3-chloro-D-alanine and extracts of the first two of these—inactivate alanine racemase (Manning et al. 1974). The enzyme that brings about degradation by elimination to pyruvate,

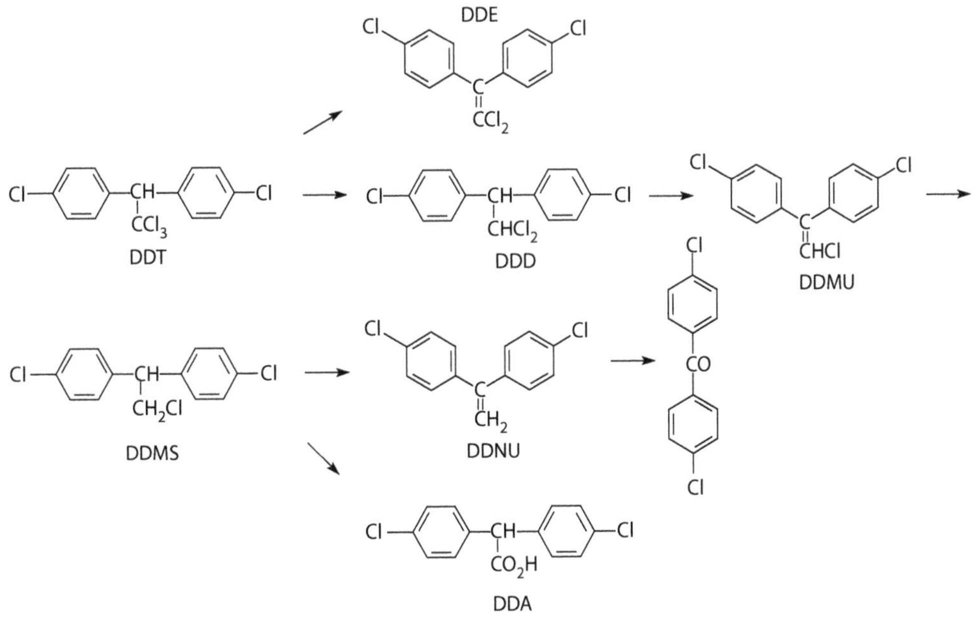

FIGURE 7.73 Biotransformations of DDT.

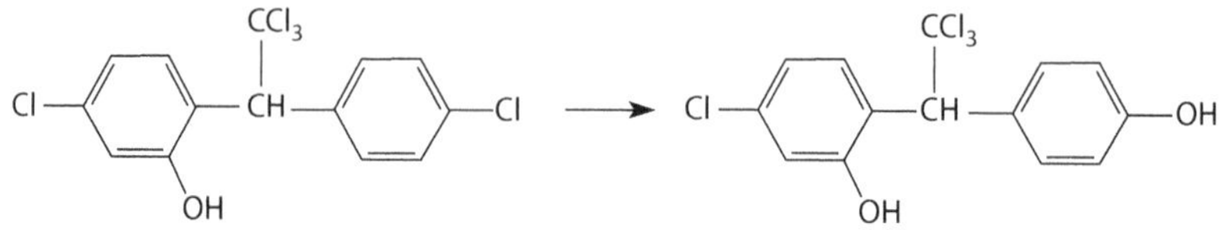

FIGURE 7.74   Alternative pathway for biodegradation of DDT.

ammonia, and chloride has been purified from *Ps. putida* strain CR 1-1 and contains 2 mol pyridoxal 5′-phosphate per mole of enzyme. The enzyme is induced by 3-chloro-D-alanine and also catalyzes the analogous degradation of D-cysteine to pyruvate, sulfide, and ammonia, although this activity could not be detected in medium supplemented with D-cysteine (Nagasawa et al. 1982). There are important differences in the effects of D- and L-3-chloroalanine on membrane transport of proline that are reflections of their metabolism. Whereas in membrane vesicles from *E. coli* B, the L-enantiomer can undergo elimination to pyruvate catalyzed by pyridoxal phosphate, the D-enantiomer brings about irreversible inactivation of membrane transport by dehydrogenation to chloropyruvate (Kaczorowski et al. 1975).

7.12.1.2 Aerobic Conditions

A great deal of effort has been expended in elucidating the details of the degradation of HCH isomers that differ both in their toxicity and their biodegradability. This has been covered in a review that also considers the problems inherent in bioremediation of HCH-contaminated sites (Lal et al. 2010). This has been examined mostly in strains originally classified as *Sphingomonas paucimobilis*, and to a lesser extent in *Rhodobacter lindaniclasticus* (Nalin et al. 1999), though this strain apparently no longer exists. Although many of the reactions are biotransformations in which the ring is retained, evidence for fission of the ring has been obtained for *Sphingobium japonicum* strain UT26 that is noted below.

1. Several bacteria that degrade or transform hexachlorocyclohexane isomers have been described. They vary in selectivity towards the isomers, and the β-isomer that has only equatorial substituents is generally the most recalcitrant. Three strains that have been most studied belong to the genus *Sphingobium*, *Sph. indicum* B90A, *Sph. japonicum* UT26 and *Sph. francese* sp.+ (Pal et al. 2005).

   All have the genes for *linA, linD, linC, and linDER* that are discussed later, but they differ in the ease with which they transform the HCH isomers (Table 7.4) (Sharma et al. 2006). A strain of *Pandoraea* sp. was also able to degrade the γ-isomer (Okeke et al. 2002).

2. There are seven isomers of HCH that differ in their conformations: α-isomer [*aaaaee*]; β-isomer [*eeeeee*]; γ-isomer [*aaaeee*]; δ-isomer [*aeeeee*]; ε-isomer [*aeeaee*]; η-isomer [*aaeaee*]; and φ-isomer [*aeaeee*] (Willett et al. 1998). Their degradation is initiated by dehydrochlorination that is dependent on the stereochemistry of the isomer. For example, the β-isomer

TABLE 7.4

Relative Transformation of Hexachlorocyclohexane Isomers by Strains of *Sphingobium*

Strain	Hexachlorocyclohexane Isomer			
	α	γ	β	δ
B90A	+++	+++	+++	+++
sp+	++	++	++	++
UT26	++	++	+	++

Note: A strain of *Pandoraea* sp. was also able to degrade the γ-isomer (Okeke et al. 2002).

that lacks 1,2-diaxial hydrogen and chlorine groups is not a substrate for the LinA dehydro-halogenase (Nagata et al. 1993) and as a consequence this isomer is less readily degraded. Strain *Sphingobium japonicum* UT26 is, however, capable of converting the persistent β-isomer only into pentachlorocyclohexanol by the activity of the dehalogenase *lin B* (Nagata et al. 2005).

3. In strains of *S. paucimobilis*, most of the genes (*linA-E*) encoding the degradative enzymes are associated with an insertion sequence (IS*6100*) that mediated horizontal gene transfer (Dogra et al. 2004). In another strain that is able to degrade γ-HCH, the enzymes involved in dehydro-chlorination were apparently extracellular (Thomas et al. 1996).

4. a. The first two steps in the biotransformation of HCH involve eliminations (dehydrochlorina-tions), for example, in the degradation of γ-hexachloro(*aaaeee*)cyclohexane from which pentachlorobenzene (Tu 1976) or γ-2,3,4,6-tetrachlorocyclohex-1-ene are formed (Jagnow et al. 1977) (Figure 7.75). The formation of 2,5-dichlorophenol and 2,4,5-trichlorophenol during the aerobic degradation of γ-hexachloro(*aaaeee*)cyclohexane by *Sphingomonas* (*Pseudomonas*) *paucimobilis* putatively involves comparable elimination reactions (Sendoo and Wada 1989). The transformation by *Sphingobium japonicum* (*Pseudomonas paucimo-bilis*) strain UT26 produced 1,2,4-trichlorobenzene as a terminal metabolite by sequential diaxial dehydrochlorinations (Nagasawa et al. 1993a,b). Further details have been given of the stereochemistry of the initial reactions that involve 1,2-*diaxial* dehydrochlorination to pentachlorocyclohexane followed by sequential 1,4-*anti* dehydrochlorinations to 1,2,4,5-tetrachlorocyclohex-2,5-diene and 1,2,4-trichlorobenzene (Figure 7.76) (Trantírek et al. 2001).

   b. In the degradation by *Sphingobium japonicum* strain UT26, the genes leading to 2,5-dichlorohydroquinone have been designated *linA*, *linB*, and *linC*. They encode enzymes that carry out two successive dehydrochlorinations (LinA), two successive dehalogenations

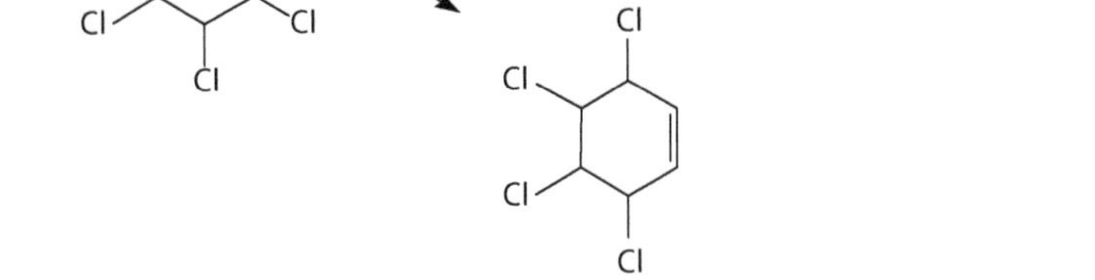

FIGURE 7.75 Pathway for biotransformation of γ-hexachlorocyclohexane.

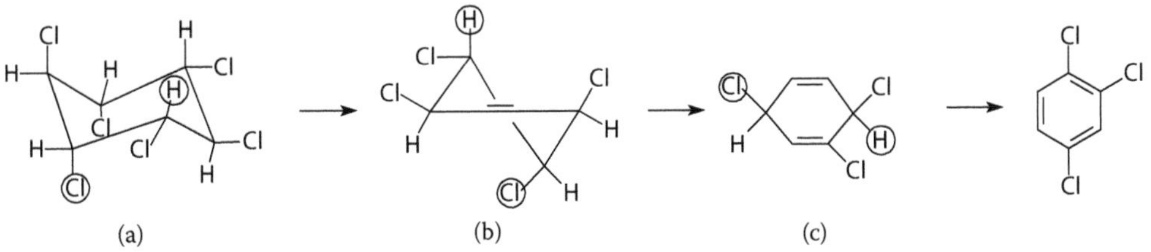

FIGURE 7.76 Stereochemistry of first steps in the biotransformation of γ-hexachlorocyclohexane.

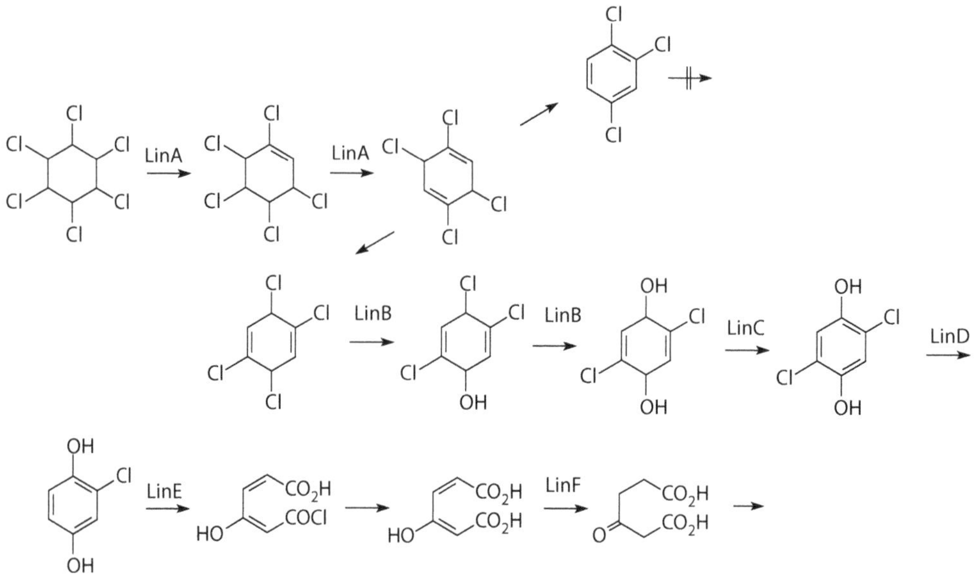

FIGURE 7.77 Aerobic transformation of γ-hexachlorocyclohexane.

(hydrolytic dechlorinations) (LinB), and a dehydrogenase (LinC) (Nagata et al. 1999). Further steps involving *linD*, *linE*, and *linF* in this strain have been described. They comprise dioxygenation of 2-chlorohydroquinone, followed by hydrolysis of the acyl chloride to 3-hydroxymuconate that is mediated by an unusual *meta* fission enzyme encoded by *linE* (Figure 7.77) (Miyauchi et al. 1999; Endo et al. 2005). Since the elimination reactions that are involved are not themselves dependent on the presence of oxygen, they may occur under anaerobic conditions.

5. Whereas strain UT26 is capable of converting the β-isomer of HCH only into pentachlorocyclohexanol by the activity of the dehalogenase *lin B* (Nagata et al. 2005), *Sphingobium indicum* (*Sphingomonas paucimobilis*) strain B90A is able to degrade all four isomers (α-, β-, γ-, and δ-) of HCH mediated by *linA1*, *linA2*, *linB*, *linC*, *linD*, and *linE* (Suar et al. 2004).

6. *Sphingobium indicum* strain B90A is able to transform α-, β-, γ-, and δ-isomers, and extensive effort has been devoted to establishing details of the reactions (Raina et al. 2007, 2008). The cardinal issues are summarized:

   a. Strain B90A contains two copies of *linA* that have important consequences. The corresponding dehydrochlorinases LinA1 and LinA2 are specific for the (+) and (-) enantiomers of α-HCH (Suar et al. 2005), and the presence of both makes possible the degradation of racemic HCH. In strains with only a *single* copy of the enzyme, however, degradation of only one of the enantiomers will take place with consequent enrichment of the nondegradable enantiomer.

   b. The initial hydrolytic dechlorination of the β- and δ-isomers is carried out by LinB (Figure 7.78a,b) while the initial reaction in the γ-isomer is catalyzed by LinA, and subsequent transformations by LinB that produce 2,5- 2,6- and 3,5-dichlorophenol as terminal metabolites (Figure 7.78c).

7.12.1.3 Anaerobic Conditions

Elimination is rather exceptional under anaerobic conditions although there are some examples that involve additionally dehydrogenation.

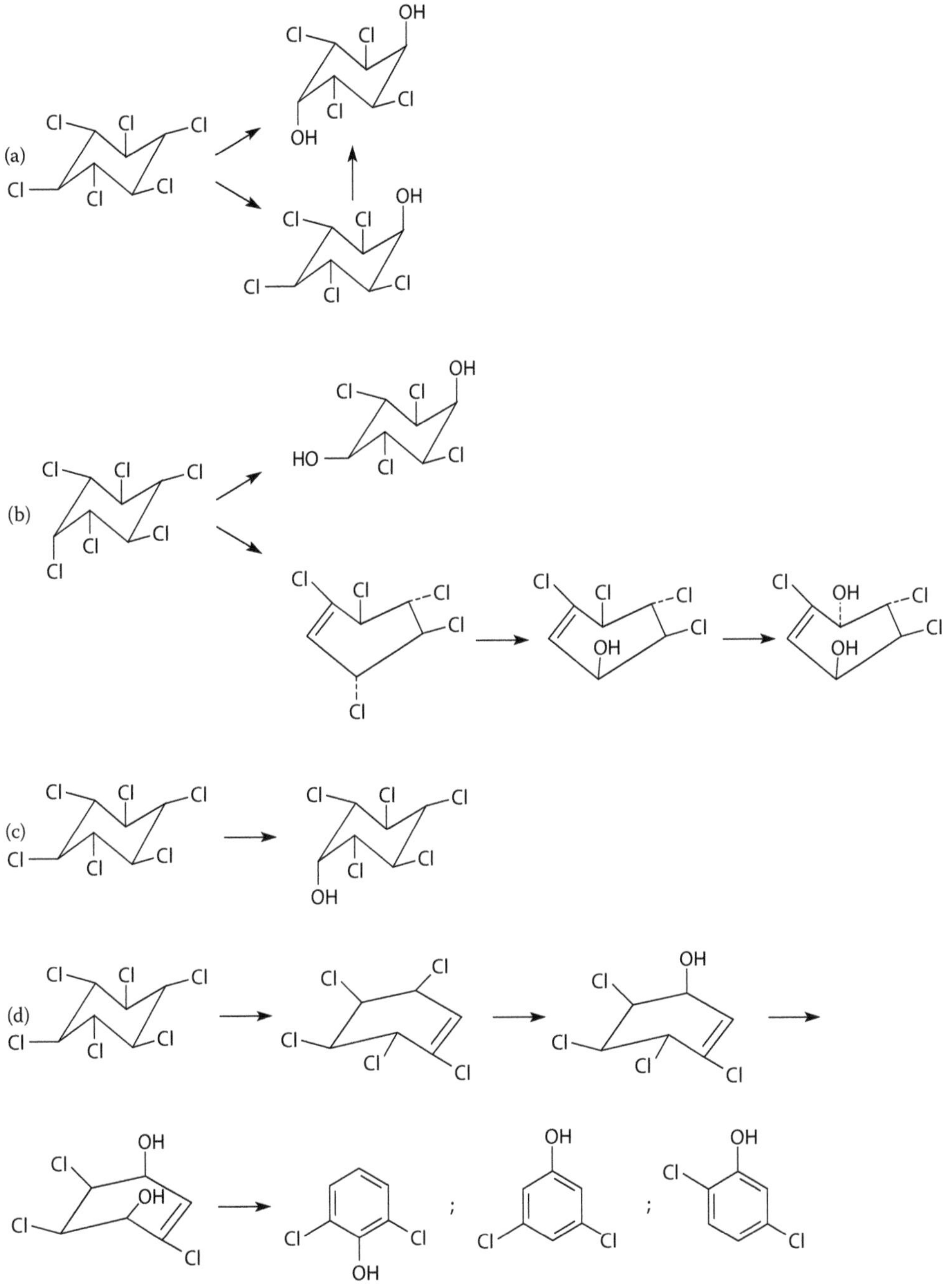

FIGURE 7.78 Hydroxylation of (a) β- and (b) δ-hexachloro-cyclohexane and (c) aerobic transformation of γ- hexachloro-cyclohexane to 2,5- 2,6-, and 3,5-dichlorophenol as terminal metabolites.

1. Elimination, dehydrogenation, and hydrogenolysis may take place during biotransformation of HCH isomers.
   a. A strain of *Clostridium rectum* (Ohisa et al. 1980) converted γ-HCH into 1,2,4-trichloro-benzene, while γ-1,3,4,5,6-pentachlorocyclohexene produced 1,4-dichlorobenzene (Figure 7.79). It was suggested (Ohisa et al. 1982) that this reductive dechlorination and

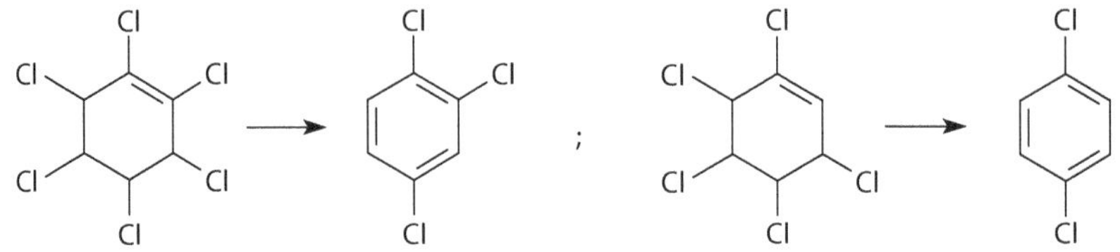

FIGURE 7.79 Anaerobic biotransformation of γ-hexachlorocyclohexene and γ-1,3,4,5,6-pentachlorocyclohexene.

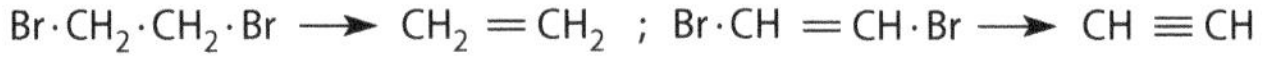

$$Br \cdot CH_2 \cdot CH_2 \cdot Br \longrightarrow CH_2 = CH_2 \; ; \; Br \cdot CH = CH \cdot Br \longrightarrow CH \equiv CH$$

FIGURE 7.80 Metabolism of 1,2-dibromoethane.

elimination was coupled to the synthesis of ATP, and this possibility has been amply confirmed in a number of anaerobic dehalogenations of polychlorinated ethenes and some chlorophenols.

b. During anaerobic incubation of [^{36}Cl]-γ-hexachlorocyclohexane, dechlorination with tetrachlorocyclohexene as an intermediate was demonstrated in *C. butyricum*, *Clostridium pasteurianum*, and *C. freundii* (Jagnow et al. 1977).

c. A methanogenic enrichment culture dehydrochlorinated β-hexachlorocyclohexane to δ-tetrachlorocyclohexene by elimination and hydrogenolysis, with the formation of chlorobenzene and benzene as the stable end product. The α-isomer was dechlorinated at a comparable rate, and the γ- and δ-isomers more slowly (Middeldorp et al. 1996).

2. Analogous reactions have been shown to occur with aliphatic halides:

a. Pure cultures of strictly anaerobic methanogenic bacteria transformed 1,2-dibromoethane to ethene and 1,2-dibromoethene to ethyne (Figure 7.80) (Belay and Daniels 1987).

b. A nonmethanogenic culture produced propene from 1,2-dichloropropane plausibly via the reductive formation of monochloropropanes followed by elimination (Löffler et al. 1997).

c. The dehalorespiring *Desulfitobacterium dichloroeliminans* strain DCA1 carried out exclusive *anti* elimination of vicinal dichloroethane, and all the vicinal dichlorobutanes (de Wildeman et al. 2003).

d. The reductive dehalogenase from *Desulfitobacterium* sp. strain Y51 carried out not only the dehalogenation of tetrachloroethene to trichloroethene and *cis*-1,2-dichloroethene, but also the elimination of a number of polyhalogenated ethanes, for example, hexachloroethane to *cis*-1,2-dichloroethene (Suyama et al. 2002).

7.12.1.4 Corrinoid Pathways

The existence of corrinoids in anaerobic bacteria in substantial concentrations is well established, and their metabolic role in acetogenesis and in methanogenesis has been elucidated. Their involvement in degradation pathways of aerobic organisms is more recent, and it has emerged that their role under these different conditions is similar. These issues are explored in the following paragraphs with a view to illustrating the similar metabolic pathways used by both aerobes and anaerobes.

Corrinoids are involved in aerobic degradation of methyl chloride by the aerobic *Methylobacterium* sp. strain CM4, and also in the degradation of other C_1 compounds by *M. extorquens* (Chistoserdova et al. 1998). Methyl corrins are key components in transmethylation and examples illustrating the similarity of pathways in aerobic and anaerobic metabolism will be summarized. In the following discussion, THF or tetrahydromethanopterin (Figure 7.81) are implicated in the form of their methyl (CH_3), methylene (CH_2), methine (CH), and formyl (CHO) derivatives (Figure 7.82). The formation of a CH_3–Co bond is central to activity, and generally the 5,6-dimethylbenziminazole is replaced by histidine.

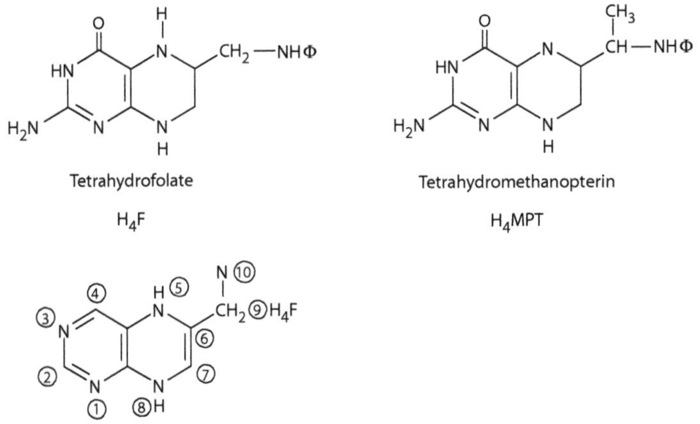

FIGURE 7.81 Partial structures of tetrahydrofolate (H_4F) and tetrahydromethanopterin (H_4MPT).

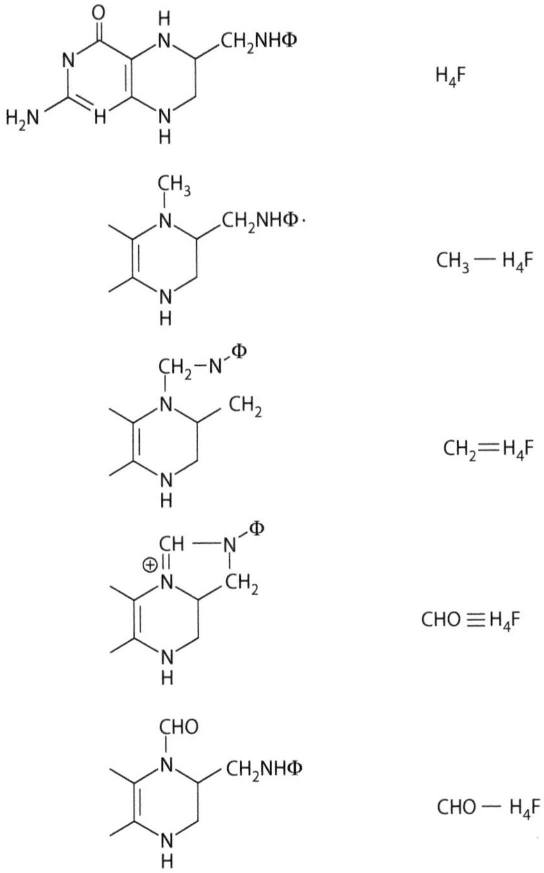

FIGURE 7.82 Dehydrogenation of CH_3-tetrahydrofolate to formyl-tetrahydrofolate.

Strain IMB-1 is able to grow at the expense of methyl bromide (Woodall et al. 2001) and belongs to a group of organisms that can also degrade methyl iodide, but are unable to use formaldehyde or methanol (Schaefer and Oremland 1999). It was postulated that the pathway for chloromethane degradation in this strain was similar to that in *Methylobacterium chloromethanicum* (McAnulla et al. 2001a).

7.12.1.4.1 Aerobic Degradation of Methyl Chloride

Methylotrophic bacteria have been isolated that are able to use methyl chloride aerobically as the sole source of energy and carbon, and bacteria that can utilize methyl chloride are apparently widely distributed and several belong to strains of *Hyphomicrobium* (McAnulla et al. 2001b). The substrate is metabolized to formaldehyde and subsequently oxidized either to formate and CO_2, or incorporated via the serine pathway. A study using strain CC495 that is similar to the strain IMB-1 already noted revealed the complexity of this reaction (Coulter et al. 1999), while details had emerged from a somewhat earlier study of methyl chloride degradation by the aerobic *Methylobacterium* sp. strain CM4 (*Methylobacterium chloromethanicum*). Cobalamin was necessary for growth with methyl chloride, though not for growth with methylamine, and the use of mutants containing a mini*Tn5* insertion and enzyme assays revealed that the mechanism involved initial methyl transfer to a Co(I) corrinoid followed by oxidation via THFs to formyltetrahydrofolate and thence to formate with production of ATP (Figure 7.83) (Vannelli et al. 1999).

7.12.1.4.2 Anaerobic Degradation of Methyl Chloride

The anaerobic methylotrophic homoacetogen *Acetobacterium dehalogenans* is able to grow with methyl chloride and CO_2, and uses a pathway comparable with that noted above for the aerobic degradation: dehydrogenation of the methyl group involving THF and a corrinoid coenzyme. Acetate is simultaneously produced from CO_2 by the activity of CO dehydrogenase and the methyltetrahydrofolate (Figure 7.84) (Meßmer et al. 1993). Some of the gene products are shared with those involved in metabolism of methyl chloride (Vannelli et al. 1999). The methyl transfer reactions and those involved in the subsequent formation of acetate have been explored for the demethylase of this strain

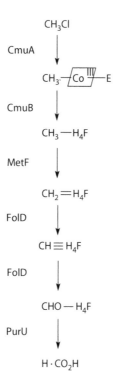

Methylobacterium chloromethanicum

FIGURE 7.83 Degradation of methyl chloride by *Methylobacterium chloromethanicum*. (Redrawn from Vannelli T et al. 1999. *Proc Natl Acad Sci USA* 96: 4615–4620.)

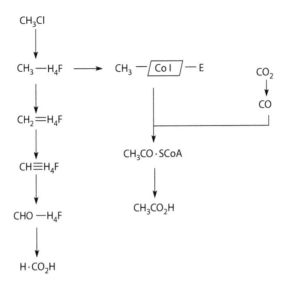

FIGURE 7.84 Degradation of methyl chloride *by Acetobacterium halogenans*. (Redrawn from Meßmer M, G Wohlfarth, G Diekert. 1993. *Arch Microbiol* 160: 383–387; Kaufmann F, G Wohlfarth, G Diekert. 1998. *Eur J Biochem* 253: 706–711.)

(Kaufmann et al. 1998), and resemble closely those for the aerobic metabolism of methyl chloride by aerobic methylotrophs.

7.12.1.4.3 Anaerobic Degradation of Polyhalogenated Methanes

Acetobacterium dehalogenans is able to degrade dichloromethane and the pathway formally resembles that for the anaerobic degradation of methyl chloride. A strain of *Dehalobacterium formicoaceticum* is able to use only dichloromethane as a source of carbon and energy forming formate and acetate (Mägli et al. 1998). The pathway involves initial synthesis of methylenetetrahydrofolate of which two-thirds is degraded to formate with generation of ATP, while the other third is dehydrogenated, transmethylated, and after incorporation of CO forms acetate with production of ATP (Figure 7.85). The formation of [^{13}C] formate, [^{13}C]methanol, and [^{13}CH$_3$]CO$_2$H was elegantly confirmed using a cell suspension and [^{13}C] CH$_2$Cl$_2$. It was suggested that a sodium-independent F$_0$F$_1$-type ATP synthase exists in this organism in addition to generation of ATP from formyltetrahydrofolate.

A strain of *Acetobacterium woodii* strain DSM 1930 dehalogenated tetrachloromethane to dichloromethane as the final chlorinated product, while the carbon atom of [^{14}C]tetrachloromethane was recovered as acetate (39%), CO$_2$ (13%), and pyruvate (10%) (Egli et al. 1988). Since the transformation of tetrachloromethane to chloroform and CO$_2$ is a nonenzymatic corrinoid-dependent reaction (Egli et al. 1990; Hasham and Freedman 1999), it seems safe to assume operation of the acetyl-CoA synthase reaction. The synthesis of acetate that also takes place during the degradation of dichloromethane by *Dehalobacterium formicoaceticum* involves CO$_2$ that originates from the medium (Mägli et al. 1996).

7.12.2 Nucleophilic Substitution: Hydrolytic Reactions of Halogenated Alkanes and Alkanoates

7.12.2.1 Halogenated Alkanes

Displacement of halogen by hydroxyl is a widely distributed reaction in the degradation of haloalkanes and haloalkanoates. Although an apparently simple pathway involving two displacement steps is illustrated (Figure 7.86) (Janssen et al. 1985), it should be emphasized that the enzymology of hydrolytic dehalogenation is quite complex. For example, two different dehalogenases are involved in the dechlorination of 1,2-dichloroethane and chloroacetate (van den Wijngaard et al. 1992). A number of distinct dehalogenases exist, and they may differ significantly in their substrate specificity in respect of chain

CH2Cl2 + H4F
CH2=H4F
CH2= H4F
CH=H4F
CH3—H4F
CHO—H4F
CH3—Co I—E
CO2
CO
H·CO2H
CH3·CO·SCoA
CH3·CO·PO4
CH3CO2H

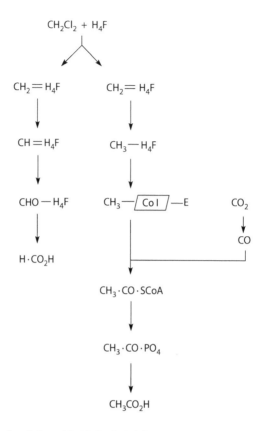

FIGURE 7.85 Degradation of methylene chloride by *Dehalobacterium formicoaceticum*. (Redrawn from Mägli A, M Messmer, T Leisinger. 1998. *Appl Environ Microbiol* 64: 646–650.)

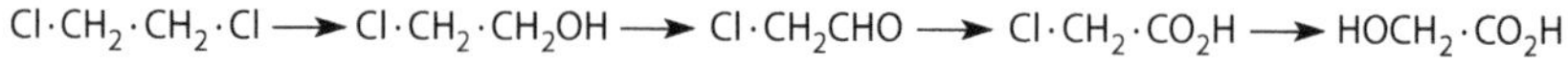

Cl·CH2·CH2·Cl
Cl·CH2·CH2OH
Cl·CH2CHO
Cl·CH2·CO2H
HOCH2·CO2H

FIGURE 7.86 Biodegradation of 1,2-dichloroethane.

length and the influence of halogen atoms at the ω-position (Scholtz et al. 1988; Sallis et al. 1990). The degradation of 1,2-dichloroethane may also be initiated by monooxygenation (Hage and Hartmans 1999), and the pathways for hydrolysis and monooxygenation converge with the production of glyoxylate. The degradation of 2-chloroethylvinyl ether by *Ancylobacter aquaticus* is initiated by a dehalogenase, although fission of the C–O–C bond is nonenzymatic (van den Wijngaard et al. 1993).

Pseudomonas sp. strain ES-2 was able to grow with a range of brominated alkanes that greatly exceeded the range of chlorinated or unsubstituted alkanes, and bromoalkanes with chain lengths of C_6–C_{16} and C_{18} could be utilized (Shochat et al. 1993). A range of chlorinated, brominated, and iodinated alkanes C_4–C_{16} was incubated with resting cells of *Rh. rhodochrous* NCIMB 13064 (Curragh et al. 1994), and dehalogenation was assessed from the concentration of halide produced. The range of substrates is impressive and the yields were approximately equal for chloride and bromide, and greater for iodide. Significant differences between the degradation of chlorine- and bromine-substituted alkanes may, however, exist. For example, although *X. autotrophicus* is able to grow with 1,2-dichloroethane and the dehalogenases can debrominate 1,2-dibromoethane and bromoacetate, these substrates are unable to support growth of the organism. Several reasons have been suggested including the toxicity of bromoacetaldehyde (van der Ploeg et al. 1995). This is consistent with the observations that in this strain, the initially produced 2-chloroethanol is oxidized to the aldehyde by an alkanol dehydrogenase and then to chloroacetate before loss of chloride and mineralization. In contrast,

Mycobacterium sp. strain GP1, which belongs to the group of fast-growing mycobacteria, was able to use 1,2-dibromoethane as a source of carbon and energy. Although a hydrolytic haloalkane dehalogenase produced 2-bromoethanol, this was converted into the epoxide that was used for growth by a pathway that was not established. In this way, production of toxic bromoacetaldehyde was circumvented (Poelarends et al. 1999).

The metabolic capacity of species of mycobacteria including the human pathogen *M. tuberculosis* strain H37Rv is remarkable (Jesenská 2000). Extracts of *M. avium* and *M. smegmatis* were able to dehalogenate a range of halogenated alkanes with chlorine, bromine, and iodine terminal substituents. On the basis of amino acid and DNA sequences, the strain that was used contained three halohydrolases, and the debromination capability of a selected number of other species of mycobacteria is given in Table 7.5. The haloalkane dehalogenase gene from *M. avium* has been cloned and partly characterized (Jesenská et al. 2002), and it has been shown that there are two dehalogenase genes *dmbA* and *dmbB* that are widely distributed in bacteria of the *M. tuberculosis* complex (Jesenská et al. 2005). A dehalogenase from *Agrobacterium tumefaciens* strain C58 had an unusual spectrum of substrates (Hasan et al. 2011): it had only low activity for chlorinated substrates but had a remarkably high activity for brominated and iodinated substrates. High activity was expressed for 1-iodopropane, 1-iodobutane, and 1-iodopentane and for 1,3-dibromopropane and 1-bromohexane. In addition, a high degree of enantioselectivity was exhibited for 2-bromopentane, 2-bromohexane, and ethyl-2-bromopropionate.

Some strains of bacteria are able to use α,ω-dichlorinated long-chain alkanes for growth, and the activity of the hydrolase from *Rh. erythropolis* strain Y2 was high for 1,2-dibromoethane, 1,2-dibromopropane, and the α,ω-dichloroalkanes (Sallis et al. 1990). In contrast, the range of α,ω-dichlorinated alkanes that was used for growth of *Pseudomonas* sp. strain 273 was limited to the C_9 and C_{10} substrates (Wischnak et al. 1998). Dehalogenase activity was demonstrated in a strain of *Acinetobacter* GJ70 that could degrade some α,ω-dichloroalkanes, and 1-bromo- and 1-iodopropane. Although 1,2-dibromoethane could be converted into 2-bromoethanol, this could not be used for growth plausibly due to the toxicity of bromoacetaldehyde that has already been noted, and the inability to use dihydroxyethane as growth substrate (Janssen et al. 1987). In a later study, the enzyme from this strain showed dehalogenase activity towards a wide range of substrates including halogenated alkanes, alkanols, and ethers (Janssen et al. 1988).

A brief summary of the degradation of 1,2-dichloroethane by *X. autotrophicus* strain GJ10 is given, since this strain has been used to delineate all stages of the metabolism and the appropriate enzymes have been demonstrated (Janssen et al. 1987) (Figure 7.87). The activity of a haloalkane dehalogenase initiates degradation and is discussed in this section, while the alkanol dehydrogenase, the aldehyde dehydrogenase, and the haloacetate dehalogenase are discussed subsequently. The enzyme responsible for dehalogenase activity has been purified from *X. autotrophicus* strain GJ10, consists of a single polypeptide chain with a molecular mass of 36 kDa, and was able to dehalogenate chlorinated, and both

TABLE 7.5

Specific Activity (μmol Bromide Produced/mg Protein/min) of Dehalogenase from Selected Species of *Mycobacterium* for 1,2-Dibromoethane

Taxon	Specific Activity	Taxon	Specific Activity
M. bovis BCG MU10	99	*M. avium* MU1	36
M. fortuitum MU8	76	*M. phlei* CCM 5639	22
M. triviale MU3	61	*M. parafortuitum* MU2	22
M. smegmatis CCM4622	49	*M. chelonae*	20

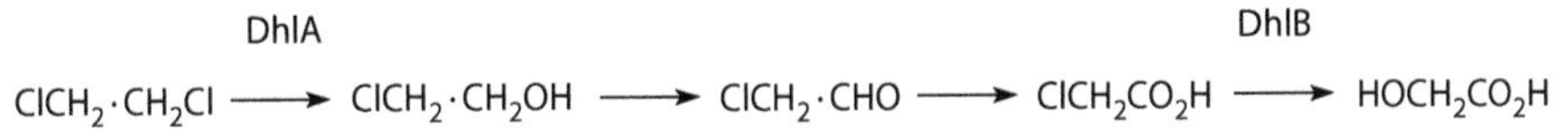

DhlA $\qquad\qquad\qquad\qquad\qquad\qquad\qquad\qquad\qquad\qquad$ DhlB

$ClCH_2 \cdot CH_2Cl \longrightarrow ClCH_2 \cdot CH_2OH \longrightarrow ClCH_2 \cdot CHO \longrightarrow ClCH_2CO_2H \longrightarrow HOCH_2CO_2H$

FIGURE 7.87 Degradation of 1,2-dichloroethane by *Xanthobacter autotrophicus* strain GJ10. (Redrawn from van der Ploeg J et al. 1994. *Appl Environ Microbiol* 60: 1599–1605.)

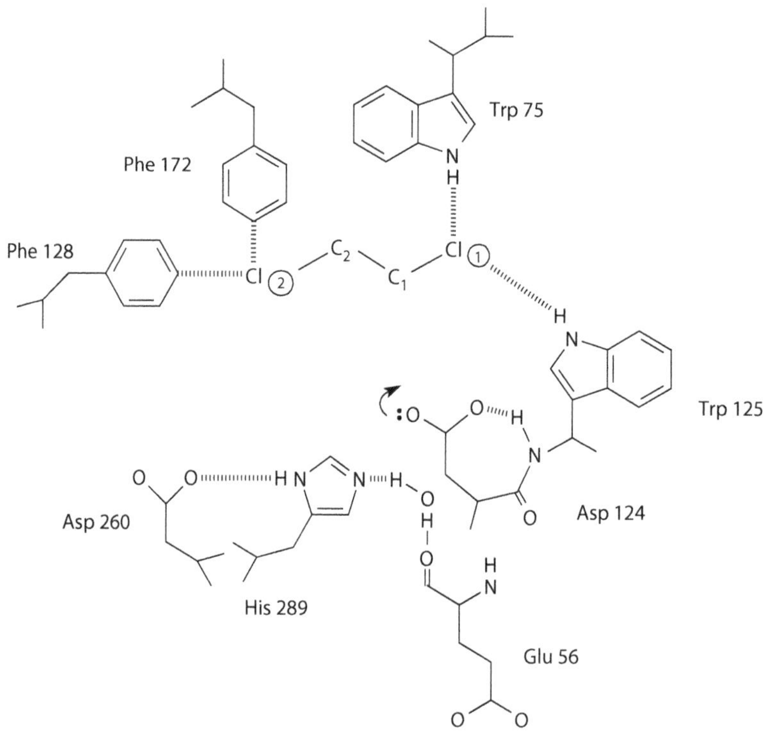

FIGURE 7.88 Catalytic mechanism of 1,2-dichloroethane dehalogenation. (Adapted from Verschuren KHG et al. 1993. *Nature* 363: 693–698.)

brominated and iodinated alkanes (Keuning et al. 1985). Details of the mechanism have been explored using an ingenious method of producing crystal at different stages of the reaction (Verschuren et al. 1993). The overall reaction involves a catalytic triad at the active site: Asp_{124} binds to one of the carbon atoms, and hydrolysis with inversion is accomplished by cooperation of Asp_{260} and His_{289} with a molecule of water bound to Glu_{56} (Figure 7.88).

7.12.2.2 Intermediate Epoxides

The conversion of vicinal haloalkanols into epoxides is of considerable interest and was apparently first recognized in a strain of *Flavobacterium* (Castro and Bartnicki 1968). The *trans*-epoxide from *erythro*-3-bromobutan-2-ol and the *cis*-epoxide from *threo*-3-bromobutan-2-ol were formed stereospecifically, and the epihalohydrins were able to react with halide to produce 2-hydroxy-1,3-dihalobutanes (Bartnicki and Castro 1969). Haloalkanol dehalogenase activity has since been found in a number of different bacteria (Slater et al. 1997). The enantioselective halohydrin hydrogen lyase (alkanol dehalogenase) has been characterized from *Corynebacterium* sp. strain N-1074 (Nakamura et al. 1994a), as well as two different epoxide hydrolases that produce 3-chloropropane-1,2-diol from the same strain (Nakamura et al. 1994b). The haloalkanol dehalogenase has been characterized from *Arthrobacter* sp. strain AD2 that is able to utilize 3-chloro-1,2-propandiol for growth (van den Wijngaard et al. 1991). It has a molecular mass of 29 Da and consists of two equal subunits. The x-ray structure of the epoxide hydrolase from *A. radiobacter* strain AD2 has been carried out and provided details of its mode of action (Nardini et al. 1999). Haloalkanol dehalogenase from this strain is able to form epoxides from a number of vicinal chlorinated and brominated alkanols (van den Wijngaard et al. 1991), and is also able to carry out transhalogenation between epihalohydrins and halide ions.

The activities of both haloalkanol dehalogenase (halohydrin hydrogen lyase) that catalyzes the formation of epoxides from alkanes with vicinal hydroxyl and halogen groups, and epoxide hydrolase that brings about hydrolysis of epoxyalkanes to diols are involved in a number of degradations that involve their sequential operation.

1. In the degradation of epichlorohydrin (3-chloro-1,2-epoxyethane) (van den Wijngaard et al. 1989), the epoxide hydrolase produced 3-chloropropan-1,2-diol that was dehalogenated to glycidol by the dehydrogenase, followed by hydrolysis of the epoxide with the production of glycerol.
2. The degradation of trihalogenated propanes by *A. radiobacter* into which dehalogenases in plasmids from *Rhodococcus* sp. strain m15-3 were incorporated (Bosma et al. 1999) involved a sequence of steps, two of which involve the formation and hydrolysis of epoxides with the ultimate production of glycerol.

7.12.2.3 Halogenated Alkanoates

Halogenated alkanoic acids are produced from the corresponding aldehydes during the degradation of 1,2-dihaloethanes by *X. autotrophicus* (Janssen et al. 1985) and the facultative methanotroph *Ancylobacter aquaticus* (van den Wijngaard et al. 1992), and are degraded by haloacetate dehalogenases. The aerobic degradation of halogenated alkanoic acids has been extensively investigated, and is generally carried out by halohydrolases that have variable specificity for the halogen, its position, and for different enantiomers. The dehalogenases have been grouped according to their reactions with 2-chloropropionate in carrying out hydrolysis with or without inversion—and the effect of sulfhydryl inhibitors (Fetzner and Lingens 1994). The enzymology of 2-haloalkanoate dehalogenases has been discussed in detail (Slater et al. 1997).

The crystal structures of the complexes between the L-2-haloacid dehalogenase from *Pseudomonas* sp. strain YL and 2-chloroalkanoates have been determined (Li et al. 1998). They reveal that the hydrolytic inversion involves the Arg_{41} and Asp_{10} sites as electrophiles and nucleophiles, respectively, followed by interaction of the Asp_{10} ester with Ser_{118}. The enantiomeric specific dehalogenase from *Pseudomonas* sp. strain DL-DEX is able to use *both* enantiomers of 2-haloalkanoic acids as substrates forming products with inversion of the configuration (Nardi-Del et al. 1997). The degradation of 2,2-dichloropropionate involves dehalogenation to pyruvate but even here two different dehalogenases are synthesized (Allison et al. 1983). Di- and trichloroacetate are degradable, though not apparently by the same groups of organisms. Mono- and dichloroacetate were effectively degraded by a strain of "*Pseudomonas dehalogenans*" that had only limited effect on trichloroacetate, whereas conversely, strains of *Arthrobacter* sp. readily degraded trichloroacetate—though not monochloroacetate (Jensen 1960). An organism that was able to grow with trichloroacetate, though not with mono- or dichloroacetate was similar to *A. calcoaceticus* on the basis of 16S rDNA (Yu and Welander 1995).

The anaerobic degradation of halogenated alkanoic acids has, however, been much less exhaustively examined. *Geobacter (Trichlorobacter) thiogenes* was able to transform trichloroacetate to dichloroacetate by coupling the oxidation of acetate to CO_2 with the reduction of sulfur to sulfide that carries out the dechlorination (De Wever et al. 2000).

7.12.2.4 Halogenated Alkenoates

Burkholderia sp. WS is able to grow at the expense of 2-chloroacrylate. This enone is reduced by an NADPH reductase to (*S*)-2-chloropropionate that is then converted by a (*S*)-2-chloroacid dehalogenase into (*R*)-lactate (Kurata et al. 2005). Dehalogenation is accomplished by the activity of a reduced flavin (FAD) enzyme that carries out hydration followed by loss of chloride to produce pyruvate (Mowafy et al. 2010). On the other hand, the degradation of *trans*-3-chloroacrylate by *Pseudomonas pavonaceae* takes place by hydroxylation at the β-carbon followed by loss of chloride from the chlorohydrin to malonate semialdehyde (de Jong et al. 2004). The degradation of 1,3-dichloropropene by *Pseudomonas pavonaceae* (*cichorii*) strain 170 involves a series of steps by which *trans*-3-chloroacrylate is formed (Poelarends et al. 1998). This was degraded by dehalogenation to malonate semialdehyde (Poelarends et al. 2001), and it has been shown that this is accomplished by hydration of *trans*-3-chloroacrylate to an unstable halohydrin that collapses to form malonate semialdehyde (De Jong et al. 2004). The malonate semialdehyde decarboxylase (Poelarends et al. 2003) has been characterized.

7.12.3 Glutathione-Mediated and Other Reactions Involving Nucleophilic Sulfur

Although the reactions described above are formally nucleophilic displacements of the chlorine atoms by hydroxyl groups, a different mechanism clearly operates in the degradation of dichloromethane by *Hyphomicrobium* sp. The enzyme is glutathione-dependent (Stucki et al 1981; Kohler-Staub et al. 1986) and the reaction presumably involves at least two steps.

Glutathione-dependent reactions are also involved in the transformation and detoxification of alkenes including isoprene, and both *cis*- and *trans*-1,2-dichloroethene in *Rhodococcus* sp. strain AD45 (van Hylckama Vlieg et al. 1998). For *cis*-1,2-dichloroethene, the initially formed epoxide is transformed by reaction with glutathione followed by chemical reactions with the final production of glyoxal. Glutathione *S*-transferase and a glutathione dehydrogenase have been characterized in this strain (van Hylckama Vlieg et al. 1999).

The degradation of tetrachloromethane by a strain of *Pseudomonas* sp. presented a number of exceptional features. Although $^{14}CO_2$ was a major product from the metabolism of $^{14}CCl_4$, a substantial part of the label was retained in nonvolatile water-soluble residues (Lewis and Crawford 1995). The nature of these was revealed by the isolation of adducts with cysteine and *N,N'*-dimethylethylenediamine, when the intermediates that are formally equivalent to $COCl_2$ and $CSCl_2$ were trapped—presumably formed by reaction of the substrate with water and a thiol, respectively. Further examination of this strain classified as *Pseudomonas stutzeri* strain KC has illuminated novel details of the mechanism. A spontaneous mutant of this strain that had lost the ability to degrade tetrachloromethane was used to show that production of pyridine-2,6-bis(thiocarboxylic acid) was essential for degradation (Lewis et al. 2000). This metabolite played a key role in the degradation: its copper complex produced trichloromethyl and thiyl radicals, and thence the formation of CO_2, CS_2, and COS (Figure 7.89) (Lewis et al. 2001, 2004).

7.12.3.1 Monooxygenation

7.12.3.1.1 Halogenated Alkanes

A number of halogenated—including polyhalogenated—alkanes are degraded by hydroxylation mediated by MMO.

1. The soluble MMO from *M. capsulatus* (Bath) is able to oxidize chloro- and bromomethane—though not iodomethane—with the presumptive formation of formaldehyde (Colby et al. 1977).

2. The methane degrading *M. trichosporium* OB3b has been shown to degrade both methyl bromide and dibromomethane (Bartnicki and Castro 1994; Streger et al. 1999), and the propane-degrading *M. vaccae* JOB5 can degrade methyl bromide (Streger et al. 1999).

3. A number of haloalkanes including dichloromethane, chloroform, 1,1-dichloroethane, and 1,2-dichloroethane may be degraded by the soluble MMO system of *M. trichosporium* (Oldenhuis et al. 1989).

4. The metabolism of chloroform has been studied in several organisms (Hamamura et al. 1997), and from the inhibitory effect of acetylene it was concluded that a monooxygenase was involved:

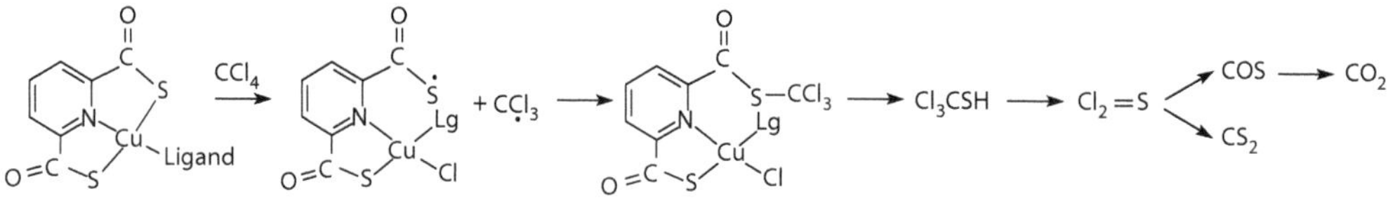

FIGURE 7.89 Degradation of tetrachloromethane (CCl_4) mediated by pyridine-2,6-dithiocarboxylate.

a. Cells of *M. trichosporium* strain OB3b grown with methanol and incubated with formate as electron donor degraded chloroform with the release of 2.1 mol of chloride per mole substrate and at substrate concentrations up to 38.6 μM.

b. Cells of *Pseudomonas butanovora* grown with butane and incubated with butyrate as electron donor degraded chloroform with the release of 1.7 mol chloride per mole substrate and degradation was incomplete even at concentrations of 12.9 μM. Butane inhibited the degradation. This organism also partially degraded other chloroalkanes and chloroalkenes including chloroethene (vinyl chloride), 1,2-*trans*-dichloroethene and trichloroethene.

c. Butane-grown cells of *M. vaccae* strain JOB5 were able to degrade chloroform without addition of an electron donor.

5. The degradation of 1,2-dichloroethane by *Pseudomonas* sp. strain DCA1 was initiated not by hydrolysis, but by monooxygenation with the direct formation of 1,2-dichloroethanol that spontaneously decomposed to chloroacetaldehyde (Hage and Hartmans 1999).

The cometabolism of halogenated methanes has been examined in *Nitrosomonas europaea* and may putatively be mediated by ammonia monooxygenase.

1. Oxidation of methyl fluoride to formaldehyde has been demonstrated (Hyman et al. 1994), and of chloroalkanes at carbon atoms substituted with a single chlorine atom to the corresponding aldehyde (Rasche et al. 1991).

2. Oxidation of a number of chloroalkanes and chloroalkenes including dichloromethane, chloroform, 1,1,2-trichloroethane, and 1,2,2-trichloroethene has been demonstrated (Vannelli et al. 1990). Although the rate of cometabolism of trihalomethanes increased with levels of bromine substitution, so also did the toxicity. Both factors must, therefore, be evaluated (Wahman et al. 2005).

7.12.3.1.2 Halogenated Alkenes

Although degradation of halogenated alkenes by direct displacement of halogen is not expected on purely chemical grounds, this reaction apparently occurs during the degradation of vinyl chloride by a strain of *Pseudomonas* sp. that carries out the direct hydrolysis to acetaldehyde followed by mineralization to CO_2 (Castro et al. 1992a). The degradation of 3-*trans*-3-chloroacrylate, however, involves dehalogenation by hydration to an unstable halohydrin (De Jong et al. 2004).

The regulation of the synthesis of the soluble and particulate monooxygenase enzymes is illustrated by the degradation of trichloroethene by *M. trichosporium* strain OB3b. During copper limitation, the soluble monooxygenase is formed, but not during copper sufficiency when the particulate form is synthesized (Oldenhuis et al. 1989). Epoxidation is the first reaction in the degradation of chlorinated ethenes including chloroethene (vinyl chloride) (Coleman et al. 2002a; Coleman and Spain 2003; Danko et al. 2004) and *cis*-dichloroethene that is a recalcitrant intermediate in the anaerobic dechlorination of tetra- and trichloroethene (Coleman et al. 2002b). The epoxides formed by monooxygenation are toxic to cells and mechanisms must exist for their removal. Although evidence for the existence of epoxide hydrolases is well established in mammalian systems, this has not been so widely observed in bacteria. Some aspects of the pathways are summarized:

1. The degradation of trichloroethene (Figure 7.90) provides a good example of the monooxygenation of haloalkenes (Little et al. 1988). 2,2,2-Trichloroacetaldehyde is produced during oxidation of trichloroethene by several methanotrophs and undergoes a dismutation to form trichloroethanol and trichloroacetate (Newman and Wackett 1991). At least formally, this transformation is analogous to an NIH shift. The particulate enzyme contains copper, or both copper and iron, and the concentration of copper determines the catalytic activity of the enzyme (Sontoh and Semrau 1998).

2. The degradation of 1,2,3-trichloropropane by *A. radiobacter* strain AD1 involves hydrolysis of an intermediate epichlorohydrin (3-chloroprop-1-ene) to the diol (Bosma et al. 1999, 2002). An enzyme from this strain has been modified from the use of epichlorohydrin that is its normal substrate to accept *cis*-1,2-dichloroethene with the release of chloride and the presumptive formation of glyoxal (Rui et al. 2004).

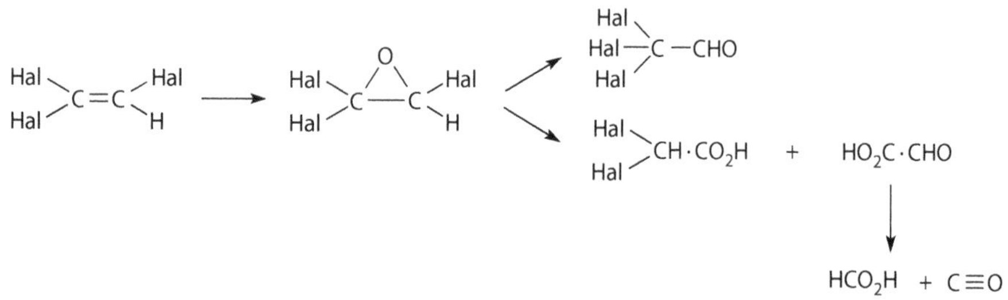

FIGURE 7.90   Generalized degradation of trihaloethenes by methanotrophs. (Redrawn from Little CD et al. 1998. *Appl Environ Microbiol* 54: 951–956.)

3. The aerobic degradation of chloroethene (vinyl chloride) by *M. aurum* strain L1 proceeded by initial formation of an epoxide mediated by an alkene monooxygenase (Hartmans and de Bont 1992). This reaction has also been demonstrated to occur with *M. trichosporium*, even though subsequent reactions were purely chemical (Castro et al. 1992b).

4. The degradation of chloroethene by *Mycobacterium* sp. strain JS60 involved epoxidation followed by reaction with coenzyme M and loss of chloride in formation of the aldehyde. Subsequent steps involved reductive loss of coenzyme M and the production of acetate (Coleman and Spain 2003) and were analogous to those in *Nocardioides* sp. strain JS614 (Mattes et al. 2005). The reductive loss of coenzyme M is formally analogous to that involved in the degradation of isoprene by *Rhodococcus* sp. strain AD45 (van Hylckama Vlieg et al. 2000). The coenzyme M pathway is not limited to Gram-positive organisms and has also been demonstrated in *Ps. putida* strain AJ and *Ochrobactrum* sp. strain TD (Danko et al. 2006).

7.12.3.1.3 Induction of Monooxygenation by Nonhalogenated Substrates

An important observation is the transformation of chlorinated ethenes by monooxygenases induced by a range of "cosubstrates" including phenol, toluene, and methane or propane (references in Kageyama et al. 2005). It is convenient to summarize salient aspects of hydroxylases (monooxygenases) in the metabolism of phenol and toluene which are discussed in greater detail in Chapter 8:

1. Degradation of phenol is normally carried out by hydroxylation (Nurk et al. 1991a; Powlowski and Shingler 1994a,b). This is consistent with the cometabolic monooxygenation of trichloroethene by *Burkholderia kururiensis* that was isolated by enrichment with phenol (Zhang et al. 2000).

2. Hydroxylation of toluene can be carried out by hydroxylation (monooxygenation) (Lee and Gibson 1996), and toluene dioxygenase in *Ps. putida* F1 is able to carry out hydroxylation of phenol (Spain et al. 1989). Although monooxygenation can be involved in the degradation of toluene, there is a complex relation between toluene monooxygenase activity and the degradation of chlorinated hydrocarbons.

Advantage has been taken of these activities for bioremediation of contaminated sites when both aromatic and chlorinated aliphatic contaminants are present. Factors determining the effectiveness of toluene as an inducer of monooxygenase have been examined (Leahy et al. 1996), although strains of *Wautersia* sp. can degrade trichloroethene in the absence of a cosubstrate (Kageyama et al. 2005). The degradation of trichloroethenes has also been examined by strains in which the relevant oxygenase has been induced with a range of cosubstrates. These include: (a) phenol or benzoate (Nelson et al. 1987), phenol (Zhang et al. 2000), (b) toluene (Wackett and Gibson 1988), (c) isopropylbenzene (Pflugmacher et al. 1996), or (d) in a hybrid dioxygenase using elements of the toluene and biphenyl dioxygenase operons (Furukawa et al. 1994). Degradation seems to be limited to trichloroethene and *cis*-1,2-dichloroethene and the toxicity of the substrates is a limiting factor (Wackett and Gibson 1988).

Toluene oxidation is accomplished by *Pseudomonas mendocina* strain KR1, *P. putida* strain F1, *P. picketii* strain PKO1, and *B. cepacia* strain G4 that possess toluene monooxygenase activities. The degradation of chloroform and 1,2-dichloroethane is carried out only by strains with the facility of toluene-4-monooxygenation—*P. mendocina* KR1 and *Pseudomonas* sp. strain ENVPC5 (McClay et al. 1996). In these strains, toluene oxidation could be induced by trichloroethene, which was subsequently degraded, whereas trichloroethene did not induce toluene oxidation in *B. cepacia* strain G4 or *P. putida* strain F1 (McClay et al. 1995). The degradation of trichloroethene by the three components of toluene 2-monooxygenase of *B. cepacia* involves initial formation of the epoxide followed by spontaneous decomposition to carbon monoxide, formate, and glycolate (Newman and Wackett 1997). In contrast to the degradation by MMO or cytochrome P450 monooxygenase, chloral hydrate was not formed.

7.12.3.2 Reductive Reactions

7.12.3.2.1 Cytochrome P450 Reductions

The reductive dehalogenation of polyhalogenated alkanes by cytochrome P450 has been demonstrated. For example, cells of *Ps. putida* G786 containing cytochrome P450cam genes on the cam plasmid were able to carry out the reductive dehalogenation of a number of halogenated alkanes and for some substrates also oxidations. Illustrative examples include the following:

1. Selective reductive debromination of polyhalogenated methanes (Castro et al. 1985; Li and Wackett 1993).
2. Both aerobic and reductive pathways were suggested for the degradation of 1,1,2-trichloroethane (Castro and Belser 1990)—a dominant aerobic pathway to chloroacetate and glyoxylate, and simultaneously a minor reductive reaction, which must also involve an elimination reaction with the formation of chloroethene (Figure 7.91).
3. Under anaerobic conditions, various reactions can be carried out by *Ps. putida* G786pHG-2 and the following are illustrative: (a) trichlorofluoromethane → carbon monoxide; (b) hexachloroethane → tetrachloroethene; (c) 1,1,1-trichloro-2,2,2-trifluoroethane → 1,1-dichloro-2,2-difluoroethene (Hur et al. 1994).
4. When a plasmid containing genes encoding toluene dioxygenase was incorporated into this modified strain of *P. putida*, complete degradations that involved *both* reductive and oxidative steps were accomplished, for example, the degradation of pentachloroethane via trichloroethene to glyoxylate, and of 1,1-dichloro-2,2-difluoroethene to oxalate (Wackett et al. 1994).

7.12.3.2.2 Anaerobic Dehalogenation of Polyhalogenated Alkenes

Considerable effort has been devoted to the anaerobic transformation of polychlorinated C_1 alkanes and C_2 alkenes in view of their extensive use as industrial solvents and their identification as widely distributed groundwater contaminants. Early experiments, which showed that tetrachloroethene was transformed into chloroethene (Vogel and McCarty 1985) (Figure 7.92) aroused concern; though it has now been shown that complete dechlorination can be accomplished by some organisms.

The dechlorination of polychlorinated ethenes by a number of bacteria can be coupled to the synthesis of ATP and has been designated dehalorespiration (Holliger et al. 1999; Drzyzga and Gottschal 2002; Sun et al. 2002): this is discussed more fully in Chapter 3. Dehalorespiration has been demonstrated in *Sulfurospirillum*

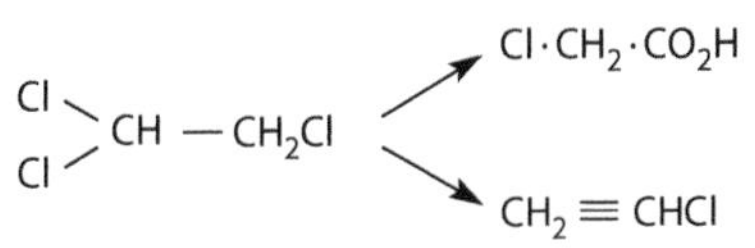

FIGURE 7.91 Biodegradation of trichloroethane by *Ps. putida*.

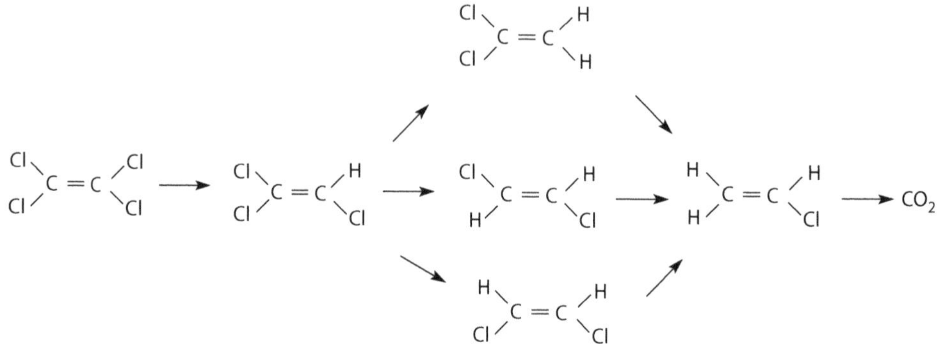

FIGURE 7.92 Anaerobic dechlorination of tetrachloroethene.

(Dehalospirillum) multivorans, Dehalobacter restrictus, Desulfuromonas chloroethenica, and some strains of *Desulfitobacterium* sp. (references in Holliger et al. 1999), and in *D. ethenogenes* that is capable of reductively dehalogenating tetrachloroethene to ethene (Magnuson et al. 2000; Maymó-Gatell et al. 1999). The electron donors were generally H_2 or pyruvate, or methanol for *D. ethenogenes*. The chloroethene reductases contain corrinoid cofactors and [Fe–S] clusters, and the trichloroethene reductase from *D. ethenogenes* is able to debrominate substrates containing two, three, four, or five carbon atoms, albeit with decreasing ease (Magnuson et al. 2000). Further examples include the following dechlorinations that are discussed in more detail later:

1. Polychlorinated ethenes to *cis*-1,2-dichloroethene, chloroethene (vinyl chloride), and ethene or ethane, and of chloroethene (vinyl chloride) to ethene (He et al. 2003; Müller et al. 2004).
2. Tetrachloroethene to *trans*-1,2-dichloroethene (Griffin et al. 2004).
3. Tetra- and trichloroethene at the expense of acetate or pyruvate and H_2, to *cis*-dichloroethene by *Geobacter lovleyi*, which is a relative of *Geobacter thiogenes* that can couple growth from the dechlorination of trichloroacetate to dichloroacetate (Sung et al. 2006a).

Among organisms that can dechlorinate tetrachloroethene, several types of reductase have been found. In *Sulfurospirillum (Dehalospirillum) multivorans*, *Desulfitobacterium* sp. strain PCE-S, *Desulfitobacterium hafniense (frappieri)* strain TCE-1, and *Dehaloccoides ethenogenes*, it consists of a peptide with a single corrinoid cofactor and two [Fe–S] centers. There is considerable interest in the dechlorination of tetrachloroethene, trichloroethene and vinyl chloride, and organisms that are able to do so include the following:

1. A strain of *Sulfurospirillum (Dehalospirillum) multivorans* transformed tetrachloroethene to trichloroethene and *cis*-1,2-dichloroethene (Neumann et al. 1994) using pyruvate as electron donor, and some properties of the dehalogenase have been reported (Neumann et al. 1995). One mol of the dehalogenase contained 1 mol of corrinoid, 9.8 mol of Fe, and 8.0 mol of acid-labile sulfur (Neumann et al. 1996), and the genes have been cloned and sequenced (Neumann et al. 1998). Although comparable values have been reported for the enzyme from *Desulfitobacterium* sp. strain PCE-S (Miller et al. 1997), the N-terminal sequence of the enzyme showed little similarity to that of *Sulfurospirillum multivorans* (Miller et al. 1998). The cardinal role of a corrinoid was proposed for both organisms (Neumann et al. 1996; Miller et al. 1997). A comparable reductive dehalogenation—though only as far as trichloroethene—is carried out by *Desulfomonile tiedjei* in which the reductase is a heme protein that is similar to that involved in dehalogenation of 3-chlorobenzoate (Townsend and Suflita 1996).
2. *D. ethenogenes* is unusual among the dehalorespiring strains in its capability to reductively dehalogenate tetrachloroethene to ethene using methanol as substrate (Magnuson et al. 2000;

Maymó-Gatell et al. 1999). It is also capable, though less readily, of reductively dehalogenating a few halogenated propenes including 3-chloropropene and 1,3-dichloropropene with the formation of propene (Magnuson et al. 2000).

3. The membrane-bound dehalogenase from *Dehalobacter restrictus* contained (per mole of subunit) 1 mol of cobalamin, 0.6 mol cobalt, 7 mol iron, and 6 mol acid-labile sulfur. It carried out dechlorination of tetra- and trichloroethene to *cis*-1,2-dichloroethene, and had a substrate spectrum that included tetrachloromethane, hexachloroethane, and 1,1,1-trichloro-2,2,2-trifluorarethane although the products from these were not apparently identified (Maillard et al. 2003).

4. Two strains of *Desulfuromonas michiganensis* carried out the partial dechlorination of tetrachloroethene to *cis*-1,2-dichloroethene using acetate as the electron donor (Sung et al. 2003).

5. The reduction of tetrachloroethene to *cis*-1,2-dichloroethene by the enteric organism *Pantoea (Enterobacter) agglomerans* may be noted (Sharma and McCarty 1996) as one of the few examples of the ability of *Enterobacteriaceae* to carry out reductive dechlorination.

6. A nonfermentative organism putatively assigned to *Desulfuromonas acetexigens* reduced tetrachloroethene to *cis*-dichloroethene using acetate as electron donor (Krumholz et al. 1996), and a similar species *D. chloroethenica* used both tetra- and trichloroethene as electron acceptors with the production of *cis*-dichloroethene using acetate or pyruvate as electron donors (Krumholz 1997).

7. *Clostridium bifermentans* can dechlorinate tetrachloroethene to *cis*-dichloroethene (Chang et al. 2000; Okeke et al. 2001). The enzyme has been purified and contains a corrinoid cofactor, although the complete nucleotide sequence of the gene encoding the reductase showed no homology with existing dehalogenases. The enzyme was most active with tetrachloroethene and trichloroethene, although lower rates were also observed with other chlorinated ethenes except chloroethene (vinyl chloride).

8. Under methanogenic conditions, a strain of *Methanosarcina* sp. transformed tetrachloroethene to trichloroethene (Fathepure and Boyd 1988). In the presence of suitable electron donors such as methanol, complete reduction of tetrachloroethene to ethene may be achieved in spite of the fact that the dechlorination of vinyl chloride appeared to be the rate-limiting step (Freedman and Gossett 1989).

9. Tetrachloroethene can be dechlorinated to trichloroethene as the sole product by the homoacetogen *Sporomusa ovata* using methanol as the electron donor, and cell extracts of other homoacetogens including *Clostridium formicoaceticum* and *Acetobacterium woodii* were able to carry this out using CO as electron donor (Terzenbach and Blaut 1994).

10. Complete dechlorination of high concentrations of tetrachloroethene in the absence of methanogenesis has been achieved using methanol as electron donor (DiStefano et al. 1991).

There has been substantial interest in the complete dechlorination of polychlorinated ethenes to ethene, since chloroethene (vinyl chloride) is generally an undesirable product of their partial dechlorination. Strains of *Dehalococcoides* sp., and especially *D. ethenogenes* may possibly be unique in being able to carry out the dechlorination of tetrachloroethene to chloroethene and ethene (Magnuson et al. 2000; Maymó-Gatell et al. 1999). These organisms have, therefore, attracted particular attention: it is worth noting, however, that while reductive dechlorination by *D. ethenogenes* strain 195 has been extensively studied, other strains with analogous capability for chlorethenes have been reported: strain VS (Müller et al. 2004), strain FL2 (He et al. 2005), strain BAC1 (Krajmalnik-Brown et al. 2004), strain GT (Sung et al. 2006b).

Additional features are worth of note:

1. It has been shown that the gene *bvcA* that is putatively involved in the reduction of chloroethene to ethene by *Dehalococcoides* sp. strain BAC1 was present in strains that could carry out chloroethene respiration, but was absent in those that could not (Krajmalnik-Brown et al. 2004).

$$BrCH_2 CH_2 Br \longrightarrow CH_2 = CH_2$$

$$BrCH = CHBr \longrightarrow HC \equiv CH$$

$$BrCH_2 CH_2 SO_3 H \longrightarrow CH_2 = CH_2$$

FIGURE 7.93 Reductive debromination by methanogenic bacteria. (Redrawn from Belay N, L Daniels. 1987. *Appl Environ Microbiol* 53: 1604–1610.)

2. The genes for the reductase designated *vcrAB* were detected at sites with established chloro-ethene contamination and it was suggested that this could serve as monitoring probe (Müller et al. 2004).

Further examples include the debromination of 1,2-dibromoethane to ethene, and 1,2-dibromo-ethene to ethyne by methanogenic bacteria (Figure 7.93) (Belay and Daniels 1987). Details of these dehalogenations have emerged from studies with methanogens. The formation of ethene from 1,2-dichloroethane using hydrogen as electron donor has been demonstrated in cell extracts of *Methanobacterium thermoautotrophicum* DH and in *M. barkeri* has been shown to involve cobalamin and F_{430} using Ti(III) as reductant (Holliger et al. 1992). Some additional comments on abiotic reactions are given in Chapter 1.

7.12.3.2.3 Other Substrates

1. Reductive dehalogenation is one of the series of reactions involved in the degradation of γ-HCH (*aaaeee*) (Nagata et al. 1999): (a) initial elimination catalyzed by LinA, (b) hydrolysis by LinB, and (c) glutathione-mediated reductive loss of chloride from 2,5-dichlorohydroquinone cata-lyzed by LinD, which has already been noted (Figure 7.77). The last reaction also occurs during the degradation of pentachlorophenol (Xun et al. 1992).

2. Reductive dechlorination in combination with the elimination of chloride has been demon-strated in a strain of *Clostridium rectum* (Ohisa et al. 1982): γ-hexachlorocyclohexene formed 1,2,4-trichlorobenzene and γ-1,3,4,5,6-pentachlorocyclohexene formed 1,4-dichlorobenzene (Figure 7.94). It was suggested that this reductive dechlorination is coupled to the synthesis of ATP, and this possibility has been clearly demonstrated during the dehalogenation of 3-chloro-benzoate coupled to the oxidation of formate in *Desulfomonile tiedjei* (Mohn and Tiedje 1991). Combined reduction and elimination has also been demonstrated in methanogenic cultures that transform 1,2-dibromoethane to ethene and 1,2-dibromoethene to ethyne (Belay and Daniels 1987).

3. Both the active enzyme, the heat-inactivated enzyme from *Sulfurospirillum (Dehalospirillum) multivorans*, and cyanocobalamin are capable of dehalogenating haloacetates (Neumann et al. 2002), and the rate of abiotic dehalogenation depends on the catalyst that is used.

4. An organism that is able to use methyl chloride as energy source and converting this into ace-tate has been isolated (Traunecker et al. 1991).

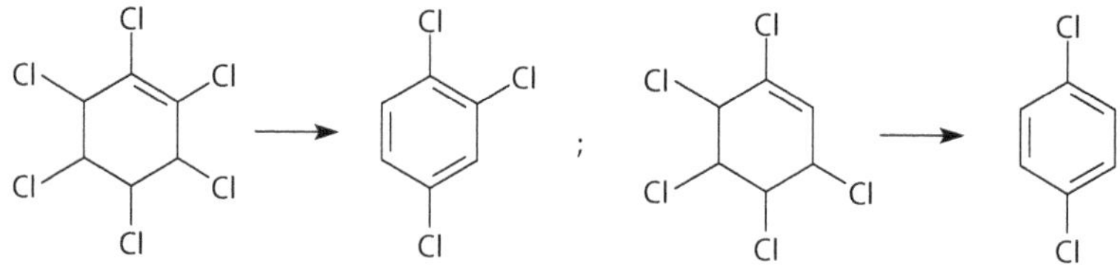

FIGURE 7.94 Biotransformation of γ–hexachlorocyclohexene and γ-1,3,4,5,6-pentachlorocyclohexene.

5. Cultures of a number of anaerobic bacteria are able to dechlorinate tetrachloromethane and *Acetobacterium woodii* formed dichloromethane as the final chlorinated metabolite by successive dechlorination, although CO_2 was also produced by an unknown mechanism (Egli et al. 1988).

6. A strain of *Clostridium* sp. transformed 1,1,1-trichloroethane to 1,1-dichloroethane, and tetrachloromethane successively to trichloromethane and dichloromethane (Gälli and McCarthy 1989).

7. Although dichloromethane is a terminal metabolite in some transformations, an organism assigned to *Dehalobacterium formicoaceticum* is able to use this as a source of carbon and energy (Mägli et al. 1996). Dichloromethane was converted into methylenetetrahydrofolate from which formate is produced by oxidation and acetate by incorporation of CO_2 catalyzed by CO dehydrogenase and acetyl-coenzyme A synthase.

8. Toxaphene is a complex mixture of compounds prepared by chlorinating camphene, and contains several hundred polychlorinated bornanes. After incubation with *Sulfurospirillum (Dehalospirillum) multivorans* only the hexa- (B6-923) and the heptachlorinated (B7-1001) remained. The nonachlorinated congener B9-1679 could be transformed by preferential dechlorination of the *gem*-dichloro groups with formation of the heptachlorinated B7-1001 (Ruppe et al. 2003). Further examination showed that this congener could be dechlorinated to penta- and hexachlorinated bornanes (Ruppe et al. 2004).

7.13 References

Aislabie J, AD Davison, HL Boul, PD Franzmann, DR Jardine, P Karuso. 1999. Isolation of *Terrabacter* sp. strain DDE-1, which metabolizes 1,1-dichloro-2-2-bis(4-chlorophenyl)ethylene when induced with biphenyl. *Appl Environ Microbiol* 65: 5607–5611.

Allison N, AJ Skinner, RA Cooper. 1983. The dehalogenases of a 2,2-dichloropropionate-degrading bacterium. *J Gen Microbiol* 129: 1283–1293.

Bartnicki EW, CE Castro. 1969. Biodehalogenation. The pathway for transhalogenation and the stereochemistry of epoxide formation from halohydrins. *Biochemistry* 8: 4677–4680.

Bartnicki EW, CE Castro. 1994. Biodehalogenation: Rapid oxidative metabolism of mono- and polyhalomethanes by *Methylosinus trichosporium* OB-3b. *Environ Toxicol Chem* 13: 241–245.

Belay N, L Daniels. 1987. Production of ethane, ethylene, and acetylene from halogenated hydrocarbons by methanogenic bacteria. *Appl Environ Microbiol* 53: 1604–1610.

Bosma T, J Damborský, G Stucki, DB Janssen. 2002. Biodegradation of 1,2,3-trichloropropane through directed evolution and heterologous expression of a haloalkane dehalogenase gene. *Appl Environ Microbiol* 68: 3582–3587.

Bosma T, E Kruizinga, EJ de Bruin, GJ Poelarends, DB Janssen. 1999. Utilization of trihalogenated propanes by *Agrobacterium radiobacter* AD1 through heterologous expression of the haloalkane dehalogenase from *Rhododoccus* sp. strain m15-3. *Appl Environ Microbiol* 65: 4575–4581.

Castro CE, EW Bartnicki. 1968. Biodehalogenation. Epoxidation of halohydrins, epoxide opening, and transhalogenation by a *Flavobacterium* sp. *Biochemistry* 7: 3213–3218.

Castro CE, NO Belser. 1990. Biodehalogenation: Oxidative and reductive metabolism of 1,1,2-trichloroethane by *Pseudomonas putida*—biogeneration of vinyl chloride. *Environ Toxicol Chem* 9: 707–714.

Castro CE, DM Riebeth, NO Belser. 1992b. Biodehalogenation: The metabolism of vinyl chloride by *M. trichosporium* OB-3b. A sequential oxidative and reductive pathway through chloroethylene oxide. *Environ Toxicol Chem* 11: 749–755.

Castro CE, RS Wade, NO Belser. 1985. Biodehalogenation reactions of cytochrome P-450 with polyhalomethanes. *Biochemistry* 24: 204–210.

Castro CE, RS Wade, DM Riebeth, EW Bartnicki, NO Belser. 1992a. Biodehalogenation: Rapid metabolism of vinyl chloride by a soil *Pseudomonas* sp. Direct hydrolysis of a vinyl C–Cl bond. *Environ Toxicol Chem* 11: 757–764.

Chang YC, M Hatsu, K Jung, YS Yoo, K Takamizawa. 2000. Isolation and characterization of a tetrachloroethylene dechlorinating bacterium, *Clostridium bifermentans* DPH-1. *J Biosci Bioeng* 89: 489–491.

Chistoserdova L, JA Vorholt, RK Thauer, ME Lidstrom. 1998. C1 transfer enzymes and coenzymes linking methylotrophic bacteria and methanogenic archaea. *Science* 281: 99–102.

Colby J, DI Stirling, H Dalton. 1977. The soluble methane mono-oxygenase of *Methylococcus capsulatus* (Bath). Its ability to oxygenate *n*-alkanes, *n*-alkenes, ethers, and alicyclic, aromatic and heterocyclic compounds. *Biochem J* 165: 395–401.

Coleman NV, JC Spain. 2003. Epoxyalkane:coenzyme M transferase in the ethene and vinyl chloride biodegradation pathways of *Mycobacterium* strain JS60. *J Bacteriol* 185: 5536–5546.

Coleman NV, TE Mattes, JM Gossett, JC Spain. 2002a. Phylogenetic and kinetic diversity of aerobic vinyl chloride-assimilating bacteria from contaminated soils. *Appl Environ Microbiol* 68: 6162–6171.

Coleman NV, TE Mattes, JM Gossett, JC Spain. 2002b. Biodegradation of *cis*-dichloroethene as the sole carbon source by a beta-protobacterium. *Appl Environ Microbiol* 68: 2726–2730.

Coulter C, JTG Hamilton, WC McRoberts, L Kulakov, MJ Larkin, DB Harper. 1999. Halomethane: bisulfite/halide ion methyltransferase, an unusual corrinoid enzyme of environmental significance isolated from an aerobic methylotroph using chloromethane as the sole carbon source. *Appl Environ Microbiol* 65: 4301–4312.

Curragh H, O Flynn, MJ Larkin, TM Stafford, JTG Hamilton, DB Harper. 1994. Haloalkane degradation and assimilation by *Rhodococcus rhodochrous* NCIMB 13064. *Microbiology (UK)* 140: 1433–1442.

Danko AS, M Luo, CE Bagwell, RL Brigmon, DL Freedman. 2004. Involvement of linear plasmids in aerobic biodegradation of vinyl chloride. *Appl Environ Microbiol* 70: 6092–6097.

Danko AS, CA Saski, JP Tomkins, DL Freedman. 2006. Involvement of coenzyme M during aerobic degradation of vinyl chloride and ethene by *Pseudomonas putida* strain AJ and *Ochrobactrum* sp. stain TD. *Appl Environ Microbiol* 72: 3756–3758.

De Jong RM, W Brugman, GJ Poelarends, CP Whitman, BW Dijkstra. 2004. The X-ray structure of *trans*-3-chloroacrylic acid dehalogenase reveals a novel hydration mechanism in the tautomerase superfamily. *J Biol Chem* 279: 11546–11552.

De Wever H, JR Cole, MR Fettig, DA Hogan, JM Tiedje. 2000. Reductive dehalogenation of trichloroacetic acid by *Trichlorobacter thiogenes* gen. nov., sp. nov. *Appl Environ Microbiol* 66: 2297–2301.

de Wildeman S, G Diekert, H van Langenhove, W Verstraete. 2003. Stereoselective microbial dehalorespiration with vicinal dichlorinated alkanes. *Appl Environ Microbiol* 69: 5643–5647.

DiStefano, TD, JM Gossett, SH Zinder. 1991. Reductive dechlorination of high concentrations of tetrachloroethene to ethene by an anaerobic enrichment culture in the absence of methanogenesis. *Appl Environ Microbiol* 57: 2287–2292.

Dogra C, V Raina, R Pal, M Suar, S Lal, K-H Gartemann, C Holliger, JR van der Meer, R Lal. 2004. organization of line genes and IS6100 among different strains of hexachlorocyclohexane-degrading *Sphingomonas paucimobilis*: Evidence for horizontal gene transfer. *J Bacteriol* 186: 2225–2235.

Drzyzga O, JC Gottschal. 2002. Tetrachloroethene dehalorespiration and growth of *Desulfitobacterium frappieri* TCE1 in strict dependence on the activity of *Desulfovibrio fructosivorans*. *Appl Environ Microbiol* 68: 542–549.

Egli T, S Stromeyer, AM Cook, T Leisinger. 1990. Transformation of tetra- and trichloromethane to CO_2 by anaerobic bacteria is a non-enzymic process. *FEMS Microbiol Lett* 68: 207–212.

Egli C, T Tschan, R Scholtz, AM Cook, T Leisinger. 1988. Transformation of tetrachloromethane to dichloromethane and carbon dioxide by *Acetobacterium woodii*. *Appl Environ Microbiol* 54: 2819–2824.

Endo R, M Kamakura, K Miyauchi, M Fukuda, Y Ohtsubo, M Tsuda, Y Nagata. 2005. Identification and characterization of genes involved in the downstream degradation pathway of γ-hexachlorocyclohexane in *Sphingomonas paucimobilis* UT26. *J Bacteriol* 187: 847–853.

Fathepure BZ, SA Boyd. 1988. Dependence of tetrachloroethylene dechlorination on methanogenic substrate consumption by *Methanosarcina* sp. strain DCM. *Appl Environ Microbiol* 54: 2976–2980.

Fetzner S, F Lingens. 1994. Bacterial dehalogenases: Biochemistry, genetics, and biotechnological applications. *Microbiol Rev* 58: 641–685.

Freedman DL, JM Gossett. 1989. Biological reductive dechlorination of tetrachloroethylene and trichloroethylene under methanogenic conditions. *Appl Environ Microbiol* 55: 2144–2151.

Furukawa K, J Hirose, S Hayashida, K Nakamura. 1994. Efficient degradation of trichloroethylene by a hybrid aromatic ring dioxygenase. *J Bacteriol* 176: 2121–2123.

Gälli R, PL McCarthy. 1989. Biotransformation of 1,1,1-trichloroethane, trichloromethane, and tetrachloromethane by a *Clostridium* sp. *Appl Environ Microbiol* 55: 837–844.

Griffin BM, JM Tiedje, FE Löffler. 2004. Anaerobic microbial reductive dechlorination of tetrachloroethene to predominantly *trans*-1,2-dichloroethene. *Environ Sci Technol* 38: 4300–4303.

Gibson DT, SM Resnick, K Lee, JM Brand, DS Torok, LP Wackett, MJ Schocken, BE Haigler. 1995. Desaturation, dioxygenation, and monooxygenation reactions catalysed by naphthalene dioxygenase from *Pseudomonas* sp. strain 9816–4. *J Bacteriol* 177: 2615–2621.

Hage JC, S Hartmans. 1999. Monooxygenase-mediated 1,2-dichloroethane degradation by *Pseudomonas* sp. strain DCA1. *Appl Environ Microbiol* 65: 2466–2470.

Hamamura, N, C Page, T Long, L Semprini, DH Arp. 1997. Chloroform cometabolism by butane-grown CF8, *Pseudomonas butanovora*, and *Mycobacterium vaccae* JOB5 and methane-grown *M. trichosporium* OB3b. *Appl Environ Microbiol* 63: 3607–3613.

Hartmans S, JAM de Bont. 1992. Aerobic vinyl chloride metabolism in *Mycobacterium aurum* L1. *Appl Environ Microbiol* 58: 1220–1226.

Hasan K, A Fortova, T Koudelakova, R Chaloupkova, M Ishituka, Y Nagata, J Damborsky, Z Prokop. 2011. Biochemical characteristics of the novel haloalkane dehalogenase DatA, isolated from the plant pathogen *Agrobacterium tumefaciens* C58. *Appl Environ Microbiol* 77: 1881–1884.

Hasham SA, DL Freedman. 1999. Enhanced biotransformation of carbon tetrachloride by *Acetobacterium woodii* upon addition of hydroxycobalamin and fructose. *Appl Environ Microbiol* 65: 4537–4542.

Hay AG, DD Focht. 1998. Cometabolism of 1,1-dichloro-2-2-bis(4-chlorophenyl)ethylene by *Pseudomonas acidovorans* M3GY grown on biphenyl. *Appl Environ Microbiol* 64: 2141–2146.

He J, KM Ritalahti, MR Aiello, FE Löffler. 2003. Complete detoxification of vinyl chloride by an anaerobic enrichment culture and identification of the reductively dechlorinating population as a *Dehaloccoides* species. *Appl Environ Microbiol* 69: 996–1003.

He J, Y Sung, R Krajmalnik-Brown, KM Ritalahti, FE Löffler. 2005. Isolation and characterization of *Dehalococcoides* sp. strain FL2, a trichloroethene (TCE)—and 1,2-dichloroethene-respiring anaerobe. *Environ Microbiol* 7: 1442–1450.

Heberer T, U Dünnbier. 1999. DDT metabolite bis(chlorophenyl)acetic acid; the neglected environmental contaminant. *Environ Sci Technol* 33: 2346–2351.

Holliger, C, G Schraa, E Stuperich, AJM Stams, AJB Zehnder. 1992. Evidence for the involvement of corrinoids and factor F_{430} in the reductive dechlorination of 1,2-dichloroethane by *M. barkeri. J Bacteriol* 174: 4427–4434.

Holliger C, G Wohlfarth, G Diekert. 1999. Reductive dechlorination in the energy metabolism of anaerobic bacteria. *FEMS Microbiol Rev* 22: 383–398.

Hur H-G, M Sadowsky, LP Wackett. 1994. Metabolism of chlorofluorocarbons and polybrominated compounds by *Pseudomonas putida* G786pHG-2 via an engineered metabolic pathway. *Appl Environ Microbiol* 60: 4148–4154.

Hyman MR, CL Page, DJ Arp. 1994. Oxidation of methyl fluoride and dimethyl ether by ammonia monooxygenase in *Nitrosomonas europaea. Appl Environ Microbiol* 60: 3033–3035.

Jagnow G, K Haider, P-C Ellwardt. 1977. Anaerobic dechlorination and degradation of hexachlorocyclohexane by anaerobic and facultatively anaerobic bacteria. *Arch Microbiol* 115: 285–292.

Janssen DB, J Gerritse, J Brackman, C Kalk, D Jager, B Witholt. 1988. Purification and characterization of a bacterial dehalogenase with activity towards halogenated alkanes, alcohols and ethers. *Eur J Biochem* 171: 67–72.

Janssen DB, D Jager, B Witholt. 1987. Degradation of *n*-haloalkanes and α,ω-dihaloalkanes by wild-type and mutants of *Acinetobacter* sp. strain GJ70. *Appl Environ Microbiol* 53: 561–566.

Janssen DB, A Scheper, L Dijkhuizen, B Witholt. 1985. Degradation of halogenated aliphatic compounds by *Xanthobacter autotrophicus* GJ10. *Appl Environ Microbiol* 49: 673–677.

Jensen HL. 1960. Decomposition of chloroacetates and chloropropionates by bacteria. *Acta Agric Scand* 10: 83–103.

Jesenská A. 2000. Dehalogenation of haloalkanes by *Mycobacterium tuberculosis* H37Rv and other mycobacteria. *Appl Environ Microbiol* 66: 219–222.

Jesenská A, M Bartos, V Czerneková, I Rychlík, I Pavlík, J Damborský. 2002. Cloning and expression of the haloalkane dehalogenase gene *dhmA* from *Mycobacterium avium* N85 and preliminary characterization of DhmA. *Appl Environ Microbiol* 68: 3224–3230.

Jesenská A, M Pavlová, M Strouhal, R Chaloupková, I Tesínská, M Monincová, Z Prokop et al. 2005. Cloning, biochemical properties, and distribution of mycobacterial haloalkane dehalogenases. *Appl Environ Microbiol* 71: 6736–6745.

Kaczorowski G, L Shaw, R Laura, C Walsh. 1975. Active transport in *Escherichia coli* B membrane vesicles. Differential inactivating effects from the enzymatic oxidation of β-chloro-L-alanine and β-chloro-D-alanine. *J Biol Chem* 250: 8921–8930.

Kageyama C, T Ohta, K Hiraoka, M Suzuki, T Okamoto, K Ohishi. 2005. Chlorinated aliphatic hydrocarbon-induced degradation of trichloroethylene by *Wautersia numadzuensis* sp. nov. *Arch Microbiol* 183: 56–65.

Kaufmann F, G Wohlfarth, G Diekert. 1998. *O*-demethylase from *Acetobacterium dehalogenans*. Substrate specificity and function of the participating proteins. *Eur J Biochem* 253: 706–711.

Keuning S, DB Janssen, B Witholt. 1985. Purification and characterization of hydrolytic haloalkane dehalogenase from *Xanthobacter autotrophicus* GJ 10. *J Bacteriol* 163: 635–639.

Kohler-Staub D, S Hartmans, R Gälli, F Suter, T Leisinger. 1986. Evidence for identical dichloromethane dehalogenases in different methylotrophic bacteria. *J Gen Microbiol* 132: 2837–2843.

Krajmalnik-Brown R, T Hölscher, IN Thompson, EM Saunders, KM Ritalahti, FE Löffler. 2004. Genetic identification of a putative vinyl chloride reductase in *Dehalococcoides* sp. strain BAV1. *Appl Environ Microbiol* 70: 6347–6351.

Krumholz LR. 1997. *Desulfuromonas chloroethenica* sp. nov. uses tetrachloroethylene and trichloroethylene as electrom donors. *Int J Syst Bacteriol* 47: 1262–1263.

Krumholz LR, R Sharp, SS Fishbain. 1996. A freshwater anaerobe coupling acetate oxidation to tetrachloroethylene dehalogenation. *Appl Environ Microbiol* 62: 4108–4113.

Kurata A, T Kurihara, H Kamachi, N Esaki. 2005. 2-haloacrylate reductase: A novel enzyme of the medium-chain dehydrogenase/reductase superfamily that catalyzes the reduction of carbon-carbon double bond of unsaturated organohalogen compounds. *J Biol Chem* 280: 20286–20291.

Lal R, et al. 2010. Biochemistry of microbial degradation of hexachlorocyclohexane and prospects for bioremediation. *Microbiol Mol Biol Revs* 74: 58–80.

Leahy JG, AM Byrne, RH Olsen. 1996. Comparison of factors influencing trichloroethylene degradation by toluene-oxidizing bacteria. *Appl Environ Microbiol* 62: 825–833.

Lee K, DT Gibson. 1996. Toluene and ethylbenzene oxidation by purified naphthalene dioxygenase from *Pseudomonas* sp. strain NCIB 9816-4. *Appl Environ Microbiol* 62: 3101–3106.

Letcher RJ, RJ Norstrom, DCG Muir. 1998. Biotransformation versus bioaccumulation: Sources of methyl sulfone PCB and 4,4' – DDE metabolites in the polar bear food chain. *Environ Sci Technol* 32: 1656–1661.

Lewis TA, MS Cortese, JL Sebat, TL Green, C-H Lee, RL Crawford. 2000. A *Pseudomonas stutzeri* gene cluster encoding the biosynthesis of the CCl_4-dechlorination agent pyridine-2,6-bis(thiocarboxylic acid) *Environ Microbiol* 2: 407–416.

Lewis TA, RL Crawford. 1995. Transformation of carbon tetrachloride via sulfur and oxygen substitution by *Pseudomonas* sp. strain KC. *J Bacteriol* 177: 2204–2208.

Lewis TA, L Leach, S Morales, PR Austin, HJ Hartwell, B Kaplan, C Forker, JM Meyer. 2004. Physiological and molecular genetic evaluation of the dechlorination agent pyridine-2,6-bis(monothiocarboxylic acid) (PDTC) as a secondary siderophore of *Pseudomonas*. *Environ Microbiol* 6: 159–169.

Lewis TA, A Paszczynski, SW Gordon-Wylie, S Jeedigunta, C-H Lee, RL Crawford. 2001. Carbon tetrachloride dechlorination by the bacterial transition metal chelator pyridine,2,6-bis(thiocarboxylic acid). *Environ Sci Technol* 35: 552–559.

Li Y-F, Y Hata, T Fujii, T Hisano, M Nishihara, T Kurihara, N Esaki. 1998. Crystal structure of reaction intermediates of L-2-haloacid dehalogenase and implications for the reaction mechanism. *J Biol Chem* 273: 15035–15044.

Li S, LP Wackett. 1993. Reductive dehalogenation by cytochrome $P450_{CAM}$: Substrate binding and catalysis. *Biochemistry* 32: 9355–9361.

Little CD, AV Palumbo, SE Herbes, ME Lidstrom, RL Tyndall, PJ Gilmer. 1988. Trichloroethylene biodegradation by a methane-oxidizing bacterium. *Appl Environ Microbiol* 54: 951–956.

Löffler FE, JE Champine, KM Ritalahti, SJ Sprague, JM Tiedje. 1997. Complete reductive dechlorination of 1,2-dichloropropane by anaerobic bacteria. *Appl Environ Microbiol* 63: 2870–2873.

Mägli A, M Messmer, T Leisinger. 1998. Metabolism of dichloromethane by the strict anaerobe *Dehalobacterium formicoaceticum*. *Appl Environ Microbiol* 64: 646–650.

Mägli A, M Wendt, T Leisinger. 1996. Isolation and characterization of *Dehalobacterium formicoaceticum* gen. nov., sp. nov., a strictly anaerobic bacterium utilizing dichloromethane as source of carbon and energy. *Arch Microbiol* 166: 101–108.

Magnuson JK, MF Romine, DR Burris, MT Kingsley. 2000. Trichlorethene reductive dehalogenase from *Dehalococcoides ethenogenes*: Squence of *tceA* and substrate characterization. *Appl Environ Microbiol* 66: 5141–5147.

Maillard J, W Schumacher, F Vazquez, C Regeard, WR Hagen, C Holliger. 2003. Characterization of the corrinoid iron-sulfur protein tetrachloroethene reductive dehalogenase of *Dehalobacter restrictus*. *Appl Environ Microbiol* 69: 4628–4638.

Manning JM, NE Merrifield, WM Jones, EC Gotschlich. 1974. Inhibition of bacterial growth by β-chloro-D-alanine. *Proc Natl Acad Sci USA* 71: 417–421.

Massé R, D Lalanne, F Mssier, M Sylvestre. 1989. Characterization of new bacterial transformation products of 1,1,1-trichloro-2,2-bis-(4-chlorophenyl)ethane (DDT) by gas chromatography/mass spectrometry. *Biomed Environ Mass Spectrom* 18: 741–752.

Mattes TE, NV Coleman, JC Spain, JM Gossett. 2005. Physiological and molecular genetic analysis of vinyl chloride and ethene biodegradation in *Nocardiodes* sp. strain JS614. *Arch Microbiol* 183: 95–106.

Maymó-Gatell X, T Anguish, SH Zinder. 1999. Reductive dechlorination of chlorinated ethenes and 1,2-dichloroethane by "*Dehalococcoides ethenogenes*" 195. *Appl Environ Microbiol* 65: 3108–3113.

McAnulla C, IR McDonald, JC Murrell. 2001b. Methyl chloride utilising bacteria are ubiquitous in the natural environment. *FEMS Microbiol Lett* 201: 151–155.

McAnulla C, CA Woodall, IR McDonald, A Studer, S Vuilleumier, T Leisinger, JC Murrell. 2001a. Chloromethane utilization gene cluster from *Hyphomicrobium chloromethanicum* strain CM2 and development of functional gene probes to detect halomethane degrading bacteria. *Appl Environ Microbiol* 67: 307–316.

McClay K, BG Fox, RJ Steffan. 1996. Chloroform mineralization by toluene-oxidizing bacteria. *Appl Environ Microbiol* 62: 2716–2732.

McClay K, SH Streger, RJ Steffan. 1995. Induction of toluene oxidation in *Pseudomonas mendocina* KR1 and *Pseudomonas* sp. strain ENVPC5 by chlorinated solvents and alkanes. *Appl Environ Microbiol* 61: 3479–3481.

Meßmer M, G Wohlfarth, G Diekert. 1993. Methyl chloride metabolism of the strictly anaerobic, methyl chloride-utilizing homoacetogen strain MC. *Arch Microbiol* 160: 383–387.

Middeldorp PJM, M Jaspers, AJB Zehnder, G Schraa. 1996. Biotransformation of α,β,γ, and δ-hexachlorocyclohexane under methanogenic conditions. *Environ Sci Technol* 30: 2345–2349.

Miller E, G Wohlfarth, G Diekert. 1997. Comparative studies on tetrachloethene reductive dechlorination mediated by *Desulfitobacterium* sp. strain PCE-S. *Arch Microbiol* 168: 513–519.

Miller E, G Wohlfarth, G Dielkert. 1998. Purification and characterization of the tetrachloroethene reductive dehalogenase of strain PCE-S. *Arch Microbiol* 169: 497–502.

Miyauchi K, Y Adachi, Y Nagata, M Takagi. 1999. Cloning and sequencing of a novel *meta*-cleavage dioxygenase gene whose product is involved in degradation of γ-hexachlorocyclohexane in *Sphingomonas paucimobilis*. *J Bacteriol* 181: 6712–6719.

Mohn WW, JM Tiedje. 1991. Evidence for chemiosmotic coupling of reductive dechlorination and ATP synthesis in *Desulfomonile tiedjei*. *Arch Microbiol* 157: 1–6.

Mowafy AM, T Kurihara, A Kurata, T Uemura, N Esaki. 2010. 2-haloacrylate hydratase, a new class of flavoenzyme that catalyzses the addition of water to the substrate for dehalogenation. *Appl Environ Microbiol* 76: 6032–6137.

Müller JA, BM Rosner, G von Abendroth, G Meshulam-Simon, PL McCarty, AM Spormann. 2004. Molecular identification of the catabolic vinyl chloride reductase from *Dehalococcoides* sp. Strain VS and its environmental distribution. *Appl Environ Microbiol* 70: 4880–4888.

Nagasawa S, R Kikuchi, Y Nagata, M Takagi, M Matsuo. 1993a. Stereochemical analysis of γ-HCH degradation by *Pseudomonas paucimobilis* UT26. *Chemosphere* 26: 1187–1201.

Nagasawa S, R. Kikuchi, Y. Nagata, M. Takagi, M. Matsuo. 1993b. Aerobic mineralization of γ-HCH by *Pseudomonas paucimobilis* UT26. *Chemosphere* 26: 1719–1728.

Nagasawa T, H Ohkishi, B Kawakami, H Yamano, H Hosono, Y Tani, H Yamada. 1982. 3-chloro-D-alanine chloride-lyase (deaminating) of *Pseudomonas putida* CR 1-1. Purification and characterization of a novel enzyme occurring in 3-chloro-D-alanine-resistant pseudomonads. *J Biol Chem* 257: 13749–13756.

Nagata Y, A Futamura, K Miyauchi, M Takagi. 1999. Two different types of dehalogenases LinA and linB, involved in γ-hexachlorocyclohexane degradation in *Sphingomonas paucimobilis* are localized in the periplasmic space without molecular processing. *J Bacteriol* 181: 5409–5413.

Nagata Y, T Nariya, R Ohtomo, M Fukuda, K Yano, M Takagi. 1993. Cloning and sequencing of a dehalogenase gene encoding an enzyme with hydrolase activity involved in the degradation of γ-hexachlorocyclohexane in *Pseudomonas paucimobilis*. *J Bacteriol* 175: 6403–6410.

Nagata Y, Z Prokop, Y Sato, P Jerabek, A Kumar, Y Ohtsubo, M Tsuda, J Damborský. 2005. Degradation of β-hexachlorocyclohexane by haloalkane dehalogenase LinB from *Sphingomonas paucimobilis* UT 26. *Appl Environ Microbiol* 71: 2183–2185.

Nakamura T, T Nagasawa, F Yu, I Watanabe, H Yamada. 1994a. Characterization of a novel enantioselective halohydrin hydrogen-halide lyase. *Appl Environ Microbiol* 60: 1297–1301.

Nakamura T, T Nagasawa, F Yu, I Watanabe, H Yamada. 1994b. Purification and characterization of two epoxide hydrolases from *Corynebacterium* sp. strain N-1074. *Appl Environ Microbiol* 60: 4630–4633.

Nalin R, P Simonet, TM Vogel, P Normand. 1999. *Rhodanobacter lindaniclasticus* gen. nov., sp. nov., a lindane-degrading bacterium. *Int J Syst Bacteriol* 49: 19–23.

Nardi-Del V T, C Kutihara, C Park, N Esaki, K Soda. 1997. Bacterial DL-2-haloacid dehalogenase from *Pseudomonas* sp. strain 113: Gene cloning and structural comparison with D- and L-2-haloacid dehalogenases. *J Bacteriol* 179: 4232–4238.

Nelson MJK, SO Montgomery, WR Mahaffey, PH Prichard. 1987. Biodegradation of trichloroethylene and involvement of an aromatic biodegradative pathway. *Appl Environ Microbiol* 53: 949–954.

Neumann A, H Scholz-Muramatsu, G Diekert. 1994. Tetrachloroethene metabolism of *Dehalospirillum multivorans*. *Arch Microbiol* 162: 295–301.

Neumann A, A Siebert, T Trescher, S Reinhardt, G Wohlfarth, G Dickert. 2002. Tetrachloroethene reductive dehalogenase of *Dehalospirillum multivorans*: Substrate specificity of the native enzyme and its corronoid cofactor. *Arch Microbiol* 177: 420–426.

Newman LM, LP Wackett. 1991. Fate of 2,2,2-trichloroacetaldehyde (chloral hydrate) produced during trichloroethylene oxidation by methanotrophs. *Appl Environ Microbiol* 57: 2399–2402.

Newman LM, LP Wackett. 1997. Trichloroethylene by purified toluene 2-monooxygenase: Products, kinetics, and turnover-dependent inactivation. *J Bacteriol* 179: 90–96.

Neumann A, G Wohlfart G Diekert. 1995. Properties of tetrachloroethene and trichloroethene dehalogenase of *Dehalospirillum multivorans*. *Arch Microbiol* 163: 276–281.

Neumann A, G Wohlfarth, G Diekert. 1996. Purification and characterization of tetrachloroethene dehalogenase from *Dehalospirillum multivorans*. *J Biol Chem* 271: 16515–16519.

Neumann A, G Wohlfarth, G Diekert. 1998. Tetrachloroethene dehalogenase from *Dehalospirillum multivorans*: Cloning, sequencing of the encoding genes, and expression of the *pceA* gene in *Escherichia coli*. *J Bacteriol* 180: 4140–4145.

Nurk A, L Kasak, M Kivisaar. 1991a. Sequence of the gene (*pheA*) encoding phenol monooxygenase from *Pseudomonas* sp. EST 1001: Expression in *Escherichia coli* and *Pseudomonas putida*. *Gene* 102: 13–18.

Ohisa N, N Kurihara, M Nakajima. 1982. ATP synthesis associated with the conversion of hexachlorocyclohexane related compounds. *Arch Microbiol* 131: 330–333.

Ohisa N, M Yamaguchi, N Kurihara. 1980. Lindane degradation by cell-free extracts of *Clostridium rectum*. *Arch Microbiol* 125: 221–225.

Okeke BC, YC Chang, M Hatsu, T Suzuki, K Takamizawa. 2001. Purification, cloning, and sequencing of an enzyme mediating the reductive dechlorination of tetrachloroethene (PCE) from *Clostridium bifermentans* DPH-1. *Can J Microbiol* 47: 448–456.

Okeke BC, T Siddique, MC Arbestain, WT Frankenberger. 2002. Biodegradation of γ-hexachlorocyclohexane (lindane) and α-hexachlorocyclohexane in water and a soil slurry by a *Pandoraea* species. *J Agric Food Chem* 50: 2548–2555.

Oldenhuis R, RLJM Vink, DB Janssen, B Witholt. 1989. Degradation of chlorinated aliphatic hydrocarbons by *M. trichosporium* OB3b expressing soluble methane monooxygenase. *Appl Environ Microbiol* 55: 2819–2816.

Pal R et al. 2005. Hexachlorocyclohexane-degrading bacterial strains *Sphingomonas paucimobilis* B90A, UT26 and Sp+, having similar *lin* genes, represent three distinct species, *Sphingobium indicum* sp. nov., *Sphingobium japonicum* sp. nov., and *Sphingobium francense* sp. nov., and reclassification of [*Sphingomonas*] *chungbukensis* as *Sphingobium chungbukense* comb. nov. *Int J Syst Evol Microbiol* 55: 1965–1972.

Pflugmacher U, B Averhoff, G Gottschalk. 1996. Cloning, sequencing, and expression of isopropylbenzene degradation genes from *Pseudomonas* sp, strain JR1: Identification of isopropylbenzene dioxygenase that mediates trichloroethene oxidation. *Appl Environ Microbiol* 62: 3967–3977.

Poelarends GJ, WH Johnson, AG Murzin, CP Whitman. 2003. Mechanistic characterization of a bacterial malonate semialdehyde decarboxylase. *J Biol Chem* 278: 48674–48683.

Poelarends GJ, R Saunier, DB Janssen. 2001. *trans*-3-chloroacrylic acid dehalogenase from *Pseudomonas pavonaceae* 170 shares structural and mechanistic similarities with 4-oxalocrotonate tautomerase. *J Bacteriol* 183: 4269–4277.

Poelarends GJ, JET van Hylckama Vlieg, JR Marchesi, LM Freitas dos Santos, DB Janssen. 1999. Degradation of 1,2-dibromoethane by *Mycobacterium* sp strain GP1. *J Bacteriol* 181: 2050–2058.

Poelarends GJ, M Wilkens, MJ Larkin, JD van Elsas, DB Janssen. 1998. Degradation of 1,3-dichloropropene by *Pseudomonas cichorii* 170. *Appl Environ Microbiol* 64: 2931–2936.

Powlowski J, V Shingler. 1994a. Genetics and biochemistry of phenol degradation by *Pseudomonas* sp. CF600. *Biodegradation* 5: 219–236.

Powlowski J, V Shingler. 1994b. Genetics and biochemistry of phenol degradation by *Pseudomonas* sp. EST 1001: Expression in *Escherichia coli* and *Pseudomonas putida*. *Gene* 102: 13–18.

Raina V, A Hauser, HR Buser, D Rentsch, P Sharma, R Lal, C Holliger, T Poiger, MD Müller, H-P E Kohler. 2007. Hydroxylated metabolites of β- and δ-hexachlorocyclohexane: Bacterial formation, stereochemical configuration, and occurrence in groundwater at a former production site. *Environ Sci Technol* 41: 4292–4298.

Raina V, D Rentsch, T Geiger, P Sharma, HR Buser, C Holliger, R Lal, H P-E Kohler. 2008. New metabolites in the degradation of α- and γ-hexachlorocyclohexane (HCH): Pentachlorocyclohexenes, and cyclohexendiols by the haloalkane dehalogenase from *Sphingobium indicum* B90A. *J Agric Food Chem* 56: 6594–6603.

Rasche ME, MR Hyman, DJ Arp. 1991. Factors limiting aliphatic chlorocarbon degradation by *Nitrosomonas europaea*: Cometabolic inactivation of ammonia monooxygenase and substrate specificity. *Appl Environ Microbiol* 57: 2986–2994.

Rui L, l Cao, W Chen, KF Reardon, TK Wood. 2004. Active site engineering of the epoxide hydrolase from *A. radiobacter* AD1 to enhance aerobic mineralization of *cis*-1,2-dichloroethylene in cells expressing an evolved toluene *ortho*-monooxygenase. *J Biol Chem* 279: 46810–46817.

Ruppe S, A Neumann, E Braekevalt, GT Tomy, GA Stern, KA Maruya, W Vetter. 2004. Anaerobic transformation of compounds of technical toxaphene. 2. Fate of compounds lacking geminal chlorine atoms. *Environ Toxicol Chem* 23: 591–598.

Ruppe S, A Neumann, W Vetter. 2003. Anaerobic transformation of compounds of technical toxaphene. I. Regiospecific reaction of chlorobornanes with geminal chlorine aroms. *Environ Toxicol Chem* 22: 2614–2621.

Sallis PJ, SJ Armfield, AT Bull, DJ Hardman. 1990. Isolation and characterization of a haloalkane halidohydrolase from *Rhodococcus erythropolis* Y2. *J Gen Microbiol* 136: 115–120.

Schaefer JK, RS Oremland. 1999. Oxidation of methyl halides by the facultative methylotroph strain IMB-1. *Appl Environ Microbiol* 65: 5035–5041.

Scholtz R, F Messi, T Leisinger, AM Cook. 1988. Three dehalogenases and physiological restraints in the biodegradation of haloalkanes by *Arthrobacter* sp. strain HA1. *Appl Environ Microbiol* 54: 3034–3038.

Schwartzbauer J, M Ricking, R Littke. 2003. DDT-related compounds bound to the non-extractable particulate matter in sediments of the Teltow Canal, Germany. *Environ Sci Technol* 37: 488–495.

Sendoo K, H Wada. 1989. Isolation and identification of an aerobic γ-HCH-decomposing bacterium from soil. *Soil Sci Plant Nutr* 35: 79–87.

Sharma PK, PL McCarty. 1996. Isolation and characterization of a facultatively aerobic bacterium that reductively dehalogenates tetrachloroethene to *cis*-dichloroethene. *Appl Environ Microbiol* 62: 761–765.

Sharma P, V Raina, R Kumari, S Malhotra, C Dogra, H Kumari, H-P E Kohler, H-R Buser, C Holliger, R Lal. 2006. Haloalkane dehalogenase LinB is responsible for β- and δ-hexachlorocyclohexane transformation in *Sphingobium indicum* B90A. *Appl Environ Microbiol* 72: 5720–5727.

Shochat E, I Hermoni, Z Cohen, A Abeliovich, S Belkin. 1993. Bromoalkane-degrading *Pseudomonas* strains. *Appl Environ Microbiol* 59: 1403–1409.

Slater JH, AT Bull, DJ Hartman. 1997. Microbial dehalogenation of halogenated alkanoic acids, alcohols and alkanes. *Adv Microbial Physiol* 38: 133–174.

Sontoh S, JD Semrau. 1998. Methane and trichloroethylene degradation by *Methylosinus trichosporium* OB3b expressing particulate methane monooxygenase. *Appl Environ Microbiol* 64: 1106–1114.

Spain JC, GJ Zylstra, CK Blake, DT Gibson. 1989. Monohydroxylation of phenol and 2,5-dichlorophenol by toluene dioxygenase in *Pseudomonas putida* F1. *Appl Environ Microbiol* 55: 2648–2652.

Streger SH, CW Condee, AP Togna, MF Deflaun. 1999. Degradation of halohydrocarbons and brominated compounds by methane- and propane-oxidizing bacteria. *Environ Sci Technol* 33: 4477–4482.

Stucki G, R Gälli, HR Ebersold, T Leisinger. 1981. Dehalogenation of dichloromethane by cell extracts of *Hyphomicrobium* DM2. *Arch Microbiol* 130: 366–371.

Suar M et al. 2005. Enantioselective transformation of α-hexachlorocyclohexane by the dehydrochlorinases LinA1 and LinA2 from the soil bacterium *Sphingomonas paucimobilis* B90A. *Appl Environ Microbiol* 71: 8514–8518.

Suar M, JR van der Meer, K Lawlor, C Holliger, R Lal. 2004. Dynamics of multiple *lin* gene expression in *Sphingomonas paucimobilis* B90A in response to different hexachlorocyclohexane isomers. *Appl Environ Microbiol* 70: 6650–6656.

Sun B, BM Griffin, HL Ayala-delRio, SA Hashsham, JM Tiedje. 2002. Microbial dehalorespiration with 1,1,1-trichloroethane. *Science* 298: 1023–1025.

Sung Y, KE Fletcher, KM Ritalahti, RP Apkarian, N Ramos-Hernández, RA Sanford, NM Mesbah, FE Löffler. 2006a. *Geobacter lovleyi* sp. nov. strain SZ, a novel metal-reducing and tetrachloroethene-dechlorinating bacterium. *Appl Environ Microbiol* 72: 2775–2782.

Sung Y, KM Ritalahti, RP Apkarian, FE Löffler. 2006b. Quantitative PCR confirms purity of strain GT, a novel trichloroethene-to-ethene-respiring *Dehalococcoides* isolate. *Appl Environ Microbiol* 72: 1980–1987.

Sung Y, KM Ritalahti, RA Sanford, JW Urbace, SJ Flynn, JM Tiedje, FA Löffler. 2003. Characterization of two tetrachloroethene-reducing acetate-oxidizing anaerobic bacteria and their description as *Desulfuromonas michiganensis* sp. nov. *Appl Environ Microbiol* 69: 2964–2974.

Suyama A, M Yamashita, S Yoshino, K Furukawa. 2002. Molecular characterization of the PceA reductive dehalogenase of *Desulfitobacterium* sp. strain Y51. *J Bacteriol* 184: 3419–3425.

Terzenbach DP, M Blaut. 1994. Transformation of tetrachloroethylene to trichloroethylene by homoacetogenic bacteria. *FEMS Microbiol Lett* 123: 213–218.

Thomas J-C, F Berger, M Jacquier, D Bernillon, F Baud-Grasset, N Truffaut, P Normand, TM Vogel, P Simonet. 1996. Isolation and characterization of a novel γ-hexachlorocyclohexane-degradingbacterium. *J Bacteriol* 178: 6049–6055.

Townsend GT, JM Suflita. 1996. Characterization of chloroethylene dehalogenation by cell extracts of *Desulfomonile tiedjei* and its relationship to chlorobenzoate dehalogenation. *Appl Environ Microbiol* 62: 2850–2853.

Trantírek L, K Hynková, Y Nagata, A Murzin, A Ansorgová, V Sklenár, J Damborský. 2001. Reaction mechanism and stereochemistry of γ-hexachlorocyclohexane dehydrochlorinase LinA. *J Biol Chem* 276: 7734–7740.

Traunecker J, A Preub, G Diekert. 1991. Isolation and characterization of a methyl chloride utilizing, strictly anaerobic bacterium. *Arch Microbiol* 156: 416–421.

Tu CM. 1976. Utilization and degradation of lindane by soil microorganisms. *Arch Microbiol* 108: 259–263.

van den Wijngaard AJ, DB Janssen, B Withold. 1989. Degradation of epichlorohydrin and halohydrins by bacterial cultures isolated from freshwater sediment. *J Gen Microbiol* 135: 2199–2208.

van den Wijngaard AJ, J Prins, AJAC Smal, DB Janssen. 1993. Degradation of 2-chloroethylvinyl ether by *Ancylobacter aquaticus* AD25 and AD27. *Appl Environ Microbiol* 59: 2777–2783.

van den Wijngaard AJ, PTW Reuvenkamp, DB Janssen. 1991. Purification and characterization of haloalcohol dehalogenase from *Arthrobacter* sp. strain AD2. *J Bacteriol* 173: 124–129.

van den Wijngaard AJ, KWHJ van der Kamp, J van der Ploeg, F Pries, B Kazemier, DB Janssen. 1992. Degradation of 1,2-dichloroethane by *Ancyclobacter aquaticus* and other facultative methylotrophs. *Appl Environ Microbiol* 58: 976–983.

van der Ploeg J, MP Smidt, AS Landa, DB Janssen. 1994. Identification of chloroacetaldehyde dehydrogenase involved in 1,2-dichloroethane degradation. *Appl Environ Microbiol* 60: 1599–1605.

van der Ploeg J, M Willemsen, G van Hall, DB Janssen. 1995. Adaptation of *Xanthobacter autotrophicus* GJ10 to bromoacetate due to activation and mobilization of the haloacetate dehalogenase gene by insertion element IS*1247*. *J Bacteriol* 177: 1348–1356.

van Hylckama Vlieg JET, J Kingma, AJ van den Wijngaard, DB Janssen. 1998. A gluathione *S*-transferase with activity towards *cis*-1,2-dichloroepoxyethane is involved in isoprene utilization by *Rhodococcus* strain AD 45. *Appl Environ Microbiol* 64: 2800–2805.

van Hylckama Vlieg JET, H Leemhuis, JHL Spelberg, DB Janssen. 2000. Characterization of the gene cluster involved in isoprene metabolism in *Rhodococcus* sp. strain AD45. *J Bacteriol* 182: 1956–1963.

van Hylckama Vlieg JET, J Kingma, W Kruizinga, DB Janssen. 1999. Purification of a glutathione *S*-transferase and a glutathione conjugate-specific dehydrogenase is involved in isoprene metabolism by *Rhodococcus* sp. strain AD 45. *J Bacteriol* 181: 2094–2101.

Vannelli T, M Logan, DM Arciero, AB Hooper. 1990. Degradation of halogenated aliphatic compounds by the ammonia-oxidizing bacterium *Nitrosomonas europaea*. *Appl Environ Microbiol* 56: 1169–1171.

Vannelli T, M Messmer, A Studer, S Vuilleumier, T Leisinger. 1999. A corrinoid-dependent catabolic pathway for growth of a *Methylobacterium* strain with chloromethane. *Proc Natl Acad Sci USA* 96: 4615–4620.

Verschuren KHG, F Seljée, HJ Rozeboom, KH Kalk, BW Dijkstra. 1993. Crystallographic analysis of the catalytic mechanism of haloalkane dehalogenase. *Nature* 363: 693–698.

Vogel TM, PL McCarty. 1985. Biotransformation of tetrachloroethylene to trichloroethylene, dichloroethylene, vinyl chloride, and carbon dioxide under methanogenic conditions. *Appl Environ Microbiol* 49: 1080–1083.

Wackett LP, DT Gibson. 1988. Degradation of trichloroethylene by toluene dioxygenase in whole-cell studies with *Pseudomonas putida* F1. *Appl Environ Microbiol* 54: 1703–1708.

Wackett LP, MJ Sadowsky, LM Newman, H-G Hur, S Li. 1994. Metabolism of polyhalogenated compounds by a genetically engineered bacterium. *Nature* 368: 627–629.

Wahman DG, LE Katz, GE Speitel. 2005. Cometabolism of trihalomethanes by *Nitrosomonas europaea*. *Appl Environ Microbiol* 71: 7980–7986.

Wedemeyer G. 1967. Dechlorination of 1,1,1-trichloro-2,2-bis[p-chlorophenyl]ethane by *Aerobacter aerogenes*. I. Metabolic products. *Appl Microbiol* 15: 569–574.

Willett KL, EM Ulrich, RA Hites. 1998. Differential toxicity and environmental fates of hexachlorocyclohexane isomers. *Environ Sci Technol* 32: 2197–2207.

Wischnak C, FE Löffler, J Li, JW Urbance, R Müller. 1998. *Pseudomonas* sp. strain 273, an aerobic α,ω-dichloroalkane-degrading bacterium. *Appl Environ Microbiol* 64: 3507–3511.

Woodall CA, KL Warner, RS Oremland, JC Murrell, IR McDonald. 2001. Identification of methyl halide-utilizing genes in the methyl bromide-utilizing bacterial strain IMB-1 suggests a high degree of conservation of methyl halide-specific genes in Gram-negative bacteria. *Appl Environ Microbiol* 67: 1959–1963.

Xun L, E Topp, CS Orser. 1992. Diverse substrate range of a *Flavobacterium* pentachlorophenol hydroxylase and reaction stoichiometries. *J Bacteriol* 174: 2898–2902.

Yu P, T Welander. 1995. Growth of an aerobic bacterium with trichloroacetic acid as the sole source of energy and carbon. *Appl Microbiol Biotechnol* 42: 769–774.

Zhang H, S Hanada, T Shigematsu, K Shibuya, Y Kamagata, T Kanawa, R Kurane. 2000. *Burkholderia kururiensis* sp. nov. a trichloroethylene (TCE)-degrading bacterium isolated from, an aquifer polluted with TCE. *Int J Syst Evol Microbiol* 50: 743–749.

7.14 Fluorinated Aliphatic Compounds

There has been increase in the application of these compounds since the synthesis of fluorinated alkanes and related compounds in the 1930s. These include fluorinated hydrocarbons that were formerly used as propellants, polymerized tetrafluoroethene, and the polyfluorinated C_4–C_8 carboxylates and sulfonates. All of them are notable for their inertness under normal conditions. Aromatic fluorinated compounds are discussed in Chapter 9, Part 3.

7.14.1 Alkanes and Alkenes

The biodegradation of hydrochlorofluorocarbons and hydrofluorocarbons has attracted considerable attention on account of their presumptive adverse effect on ozone depletion and climate alteration (references in Fabian and Singh 1999). Valuable background on chlorofluorocarbons has been given (Elliott 1994) and on alternatives to them (Rao 1994). The perfluorinated compounds including difluoromethane (HFC-32), trifluoroethane (HFC-143a), tetrafluoroethane (HFC-134a), hexafluoropropane (HFC-236ea), and heptafluoropropane (HFC-227ea) are, however, of potential interest as—environmentally acceptable

TABLE 7.6

Acronyms for Chlorofluorocarbons (CFCs), Hydrochlorofluorocarbons (HCFCs), and Hydrofluorocarbons (HFCs)

Acronym	Structure	Acronym	Structure	Acronym	Structure
CFC-11	CCl_3F	HCFC-123	$CHCl_2–CF_3$	HFC-134a	$CH_2F–CF_3$
CFC-12	CCl_2F_2	HCFC-22	$CHClF_2$	HFC-227ea	$CF_3–CHF–CF_3$
CFC-113	$CCl_2F–CClF_2$	HCFC-141b	$CH_3–CCl_2F$	HFC-152a	$CH_3–CHF_2$
CFC-114	$CClF_2–CClF_2$	HCFC-142b	$CH_3–CClF_2$	HFC-143a	$CH_3–CF_3$
CFC-115	$CClF_3–CF_3$	HCFC-124	$CHClF–CF_3$	HFC-125	$CHF_2–CF_3$
		HCFC-225ca	$CHCl_2–CF_2–CF_3$	HFC-32	CH_2F_2
		HCFC-225cb	$CHFCl–CF_2–CF_3Cl$	HFC-23	CHF_3
				HFC-245ca	$CHF_2–CF_2–CH_2F$

(Harnisch 1999). A list of acronyms is given in Table 7.6 (Midgley and McCulloch 1999). Further details of their degradation are given in a review (Neilson and Allard 2002).

7.14.1.1 Fluorohydrocarbons

Fluoromethane has been used as a selective inhibitor of ammonium oxidation and nitrification-linked synthesis of N_2O in *Nitrosomonas europaea* (Miller et al.1993), while difluoromethane has been proposed as a reversible inhibitor of methanotrophs (Miller et al. 1998).

Monooxygenation is distributed among a variety of bacteria and several have been examined for their potential to degrade fluorinated alkanes:

1. Ammonia monooxygenase in *Nitrosomonas europaea* is able to oxidize fluoromethane to formaldehyde (Hyman et al. 1994).

2. The methanotroph *M. trichosporium* strain OB3b that produces the soluble MMO system consisting of a 40 kDa NADH oxidoreductase, a 245 kDa hydroxylase, and a 16 kDa protein termed component B has a low substrate specificity (Sullivan et al. 1998). It has been shown to metabolize trifluoroethene to glyoxylate, difluoroacetate, and the rearranged product trifluoroacetaldehyde (Fox et al. 1990). The last reaction is analogous to the formation of trichloroacetaldehyde from trichloroethene by the same strain (Oldenhuis et al. 1989).

3. This strain and *M. vaccae* strain JOB 5 that produces propane monooxygenase has been used to examine the degradation of a number of hydrochlorofluorocarbons and hydrofluorocarbons (Streger et al. 1999). It was shown that during complete degradation *by M. trichosporium* and *M. vaccae*, this was accompanied by release of fluoride. It is worth noting, however, that the methylotrophs IMB-1 (Schaefer and Oremland 1999) and CC495 (Coulter et al. 1999) are able to oxidize chloromethane, bromomethane, and iodomethane though not fluoromethane. On the basis of their 16S rRNA sequences, these organisms are related to those classified as *Pseudaminobacter* sp., and more distantly related to the nitrogen-fixing rhizobia (Coulter et al. 1999).

7.14.1.2 Chlorofluorocarbons and Hydrochlorofluorocarbons

In general, halogen is last from organic substrates in the order I > Br > Cl > F. Considerable effort has been directed to degradation and transformation by methylotrophic bacteria. As a generalization, the HCFCs are more readily degraded that the corresponding compounds lack hydrogen. The soluble MMO from *M. trichosporium* OB3b that has a wide substrate spectrum is able to oxidize some hydrochlorofluoroethanes including 1,1,2-trichloro-2-2-fluoroethane and 1,1,2-trifluoroethane, though neither trichlorofluoromethane (CFC-11) nor any hydrochlorofluoroethane with three fluorine substituents on the same carbon atom were oxidized (DeFlaun et al. 1992). Although quantitative loss of fluoride and chloride was shown for dichlorofluoromethane, none of the organic products of oxidation were identified for the others.

Reductive dechlorination of fluorinated substrates has been observed for both methanogens and sulfate-reducing bacteria.

1. The methanogen *M. barkeri* is able to transform trichlorofluoromethane (CFC-11) by successive loss of chlorine to produce chlorofluoromethane (Figure 7.95a) (Krone and Thauer 1992). A similar transformation has been demonstrated in the presence of sulfate and butyrate with a mixed culture containing putatively *Desulfovibrio baarsii* and *Desulfobacter postgatei* (Sonier et al. 1994).

2. In anaerobic microcosms, 1,1,2-trichloro-1,2,2-trifluoroethane (CFC-113) was transformed by successive reductive dechlorination to 1,2-dichloro-1,2,2-trifluoroethane (HCFC-123a), and under methanogenic conditions to 1-chloro-1,2,2-trifluoroethane (HCFC-133) and 1-chloro-1,1,2-trifluoroethane (HCFC-133b) without evidence for the reductive replacement of fluorine (Figure 7.95b) (Lesage et al. 1992).

3. 1,1,1-Trifluoro-2,2-dichloroethane (HCFC-123) was recalcitrant in aerobic soils, but underwent reductive dechlorination anaerobically to produce 1,1,1-trifluoro-2-chloroethane (Oremland et al. 1996).

Cytochrome P450cam is a monooxygenase that can be induced in *Ps. putida* G786 by growth on camphor, and is responsible for the introduction of a hydroxyl group at C_5 as the first step in the degradation of camphor. This enzyme is also able to carry out non-physiological reductive dehalogenation (Castro et al.1985), and it has been shown (Li and Wackett 1993) that the electrons for reduction can be supplied by putidaredoxin. These alternatives are summarized:

$$\text{Monooxygenation} \quad R–H + O_2 + 2H^+ + 2\varepsilon \rightarrow R–OH + H_2O$$

$$\text{Reductive dehalogenation} \quad R–Hal + 2H^+ + 2\varepsilon \rightarrow R–H + H–Hal$$

The production of carbon monoxide from trichlorofluoromethane catalyzed by cytochrome P450cam proceeded through intermediate formation of the dichlorofluorocarbene (Li and Wackett 1993). Other reactions included β-elimination from 1,1,1-trichloro-2,2,2-trifluorethane (Figure 7.95c). *Ps. putida* strain G786 (pGH-2) was constructed to contain both the cytochrome P450cam genes on the cam plasmid and the *tod C1*, *tod C2*, *tod b*, and *tod A* genes of toluene dioxygenase. Toluene dioxygenase was constitutively expressed and cytochrome P450cam after induction by camphor (Wackett et al. 1994). Under anaerobic conditions 1,1,1,2-tetrachloro-2,2-difluoroethane was dehalogenated to 1,1-chloro-2,2-difluoroethene that could be oxidized by the dioxygenase under aerobic conditions to oxalate (Figure 7.96) (Hyman et al. 1994).

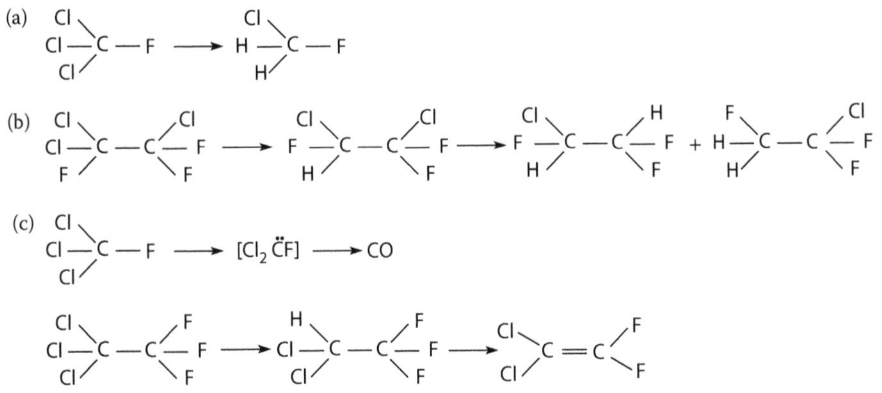

FIGURE 7.95 (a) Metabolism of trichlorofluoromethane by *Methanosarcina barkeri*, (b) transformation of 1,1,2-trichloro-1,2,2-trifluoroethane (CFC-113) under anaerobic conditions, and (c) dehalogenation by cytochrome P450cam of trichlorofluoromethane and 1,1,1-trichloro-2,2,2-trifluoroethane.

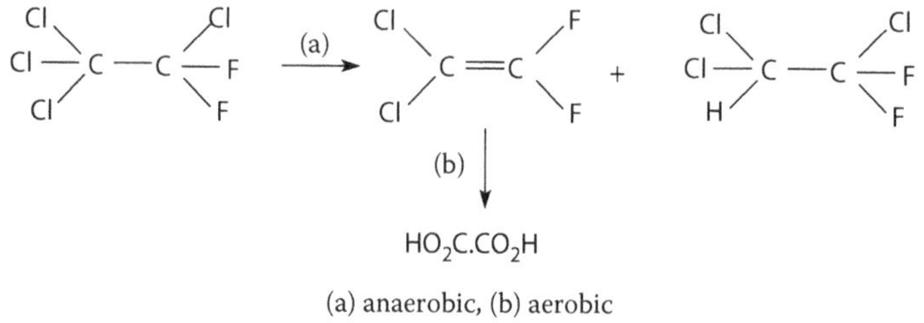

(a) anaerobic, (b) aerobic

FIGURE 7.96 Combined dehalogenation of 1,1,1,2-tetrachloro-2,2-difluoroethane by cytochrome P450cam and oxidation with toluene 2,3-dioxygenase.

7.14.1.3 Carboxylic Acids

Fluorinated aliphatic carboxylic acids are produced by some plants including species of *Dichapetalum* (Gribble 2002), and several mechanisms have been suggested for the biosynthesis of fluorinated metabolites (Harper et al. 2003). The mechanism of fluoroacetate toxicity in mammals has been extensively examined and was originally thought to involve simply initial synthesis of fluorocitrate that inhibits aconitase and thereby functioning the TCA cycle. Walsh has extensively reinvestigated the problem, and revealed both the complexity of the mechanism of inhibition and the stereospecificity of the formation of fluorocitrate from fluoroacetate (Walsh 1982).

7.14.1.3.1 Fluoroacetate

As might be expected, bacteria have been isolated from the plants that produce fluoroacetate, and these include an unidentified *Pseudomonas* sp. (Goldman 1965), a strain of *Burkholderia* (*Pseudomonas*) *cepacia* from *Dichapetalum cymosum* (Meyer et al. 1990), and a strain of *Moraxella* sp. (Kawasaki et al. 1981). In addition, fluoroacetate is an unusual product of microbial metabolism:

1. It is a terminal metabolite formed during the metabolism of 2-fluoro-4-nitrobenzoate by *Nocardia erythropolis* (Figure 7.97) (Cain et al. 1968).
2. It is formed together with 4-fluorothreonine during the late-stage growth of *Streptomyces cattleya* on a defined medium in the presence of fluoride (Reid et al. 1995). It has been shown (Hamilton et al. 1998) that glycine is an effective precursor of both fluoroacetate and 4-fluorothreonine, and that glycine is metabolized via N^5,N^{10}-methylenetetrahydrofolate to serine and thence to pyruvate. However, details of the mechanism for the incorporation of fluoride remain incompletely resolved.

The metabolism of fluoroacetate results in the production of fluoride and glycolate, and the use of $H_2^{18}O$ showed that the oxygen atom was introduced from water by a hydroxylase (Goldman 1966). The

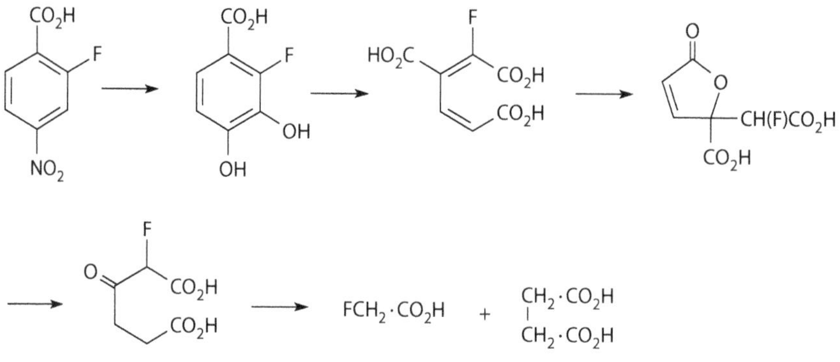

FIGURE 7.97 Degradation of 2-fluoro-4-nitrobenzoate by *Nocardia erythropolis*.

enzyme has been purified from the unidentified pseudomonad (Goldman 1965), and is specific for fluoroacetate: it is inactive toward 2- and 3-fluoropropionate, di- and trifluoroacetates, and fluoroben-zoates. It is rapidly inhibited by 4-chloromercuriphenylsulfonate and slowly by *N*-methylmaleimide, and this suggests the involvement of an active thiol group in the enzyme. Strains of *Moraxella* sp. have also been shown to assimilate fluoroacetate using plasmid-determined dehalogenase activities (Kawasaki et al. 1981). One of these (strain H-1) was active towards both fluoroacetate and chloroac-etate, whereas the other (strain H-2) was active only towards chloroacetate. Further investigation (Au and Walsh 1984) of the haloacetate hydrolase H-1 from *Pseudomonas* sp. strain A used ^{1}H NMR of the (−)-α-methoxyl-α-(trifluoromethyl) phenylacetic acid ester of the phenacyl ester of the glycolate produced from (*S*)-2-[^{2}H$_1$]fluoroacetate. This established that the major metabolite was the (*R*)-enantiomer of 2-[^{2}H$_1$]glycolate and that the reaction proceeded with inversion of the configura-tion at C-2. The mechanism for the hydrolysis of fluoroacetate by *Moraxella* sp. strain B has been examined and confirmed the involvement of Asp$_{105}$ as nucleophile with inversion of the configuration at C$_2$ (Liu et al. 1998).

7.14.1.3.2 Di- and Trifluoroacetate

The soluble MMO from *M. trichosporium* OB3 produced difluoroacetate as one of the main products from the oxidation of trifluoroethene and a low yield of trifluoroacetaldehyde by rearrangement (Fox et al. 1990).

Chlorodifluoroacetic acid has been identified in rain and snow samples (Martin et al. 2000) and may plausibly be an atmospheric degradation product of 1,1,1-trichloro-1,2,2-trifluoroethane. Trifluoroacetate has been found in a wide range of environmental samples (Key et al. 1997), and attention has been directed to its origin, its toxicity, and its recalcitrance. Compared with chloroacetates, however, it is only mildly phytotoxic to algae (Berends et al. 1999) and higher plants (Boutonnet et al. 1999). It is worth noting its formation during the photochemical degradation of 3-trifluoromethyl-4-nitrophenol (Ellis and Mabury 2000). Concern has been expressed over its biodegradability and conflicting results over its recalcitrance under anaerobic conditions have been reported. This has been resolved by the results of a chemostat study using a mixed ethanol-degrading culture (Kim et al. 2000). Clear evidence for anaero-bic degradation was produced on the basis of fluoride release and formation of acetic acid and methane.

7.14.1.4 Perfluoroalkyl Carboxylates and Sulfonates

There has been increasing interest in perfluoroalkyl sulfonates and carboxylates (Key et al. 1997). The sulfonates are valuable surface protectors and surfactants under extreme conditions, and are components of fire-fighting foams (Moody and Field 2000). Perfluorooctane carboxylates and sulfonates have a global distribution in wildlife and humans (Houde et al. 2006). Low levels of perfluoro compounds have been carefully quantified in samples of human sera (Hansen et al. 2001) and determination of perfluori-nated surfactants in surface water samples has been described. Two independent analytical techniques—liquid chromatography/tandem mass spectrometry and ^{19}F NMR were used—and a summary of analytical methods was included (Moody et al. 2001). References to toxicological studies in rats and epidemiological studies in man have been given (Hansen et al. 2001), and the biochemical toxicology of the related perfluorooctanoic acid has been discussed (DePierre 2002). Concern with their apparent persistence has motivated investigations of their potential sources including fluorotelomer alcohols. Attention is drawn to the abiotic degradations that are discussed in Chapter 1.

A strain of *Pseudomonas* sp. strain D2 was used to evaluate the degradation of a range of fluori-nated sulfones (Key et al. 1998). Although the fully fluorinated perfluorooctane sulfonate was resistant to defluorination by an aerobic bacterium, the presence of hydrogen substituents in 1*H*,1*H*,2*H*,2*H*-perfluorooctane sulfonate made it possible for the organism to carry out partial defluo-rination. For growth and defluorination of difluoromethane sulfonate, acetate or glucose were sup-plied as the source of carbon, ammonium as the source of nitrogen, and the substrate as the source of sulfur. It should be noted that perfluoro analogs that lacked hydrogen substituents did not support growth of this strain.

These considerations apply also to the fluorotelomer alcohol $CF_3(CF_2)_7$–CH_2CH_2OH that was degraded in a mixed culture obtained by enrichment with ethanol. Terminal dehydrogenation followed by elimination of fluoride, hydration and further loss of fluoride produced perfluorooctanoate (Dinglasan et al. 2004).

$$CF_3(CF_2)_7CH_2-CH_2OH \rightarrow CF_3(CF_2)_7CH_2CO_2H \rightarrow CF_3(CF_2)_6-CF=CHCO_2H$$
$$\rightarrow\rightarrow CF_3(CF_2)_6COCH_2CO_2H \rightarrow CF_3(CF_2)_6CO_2H$$

An exhaustive study with activated municipal sludge using the telomer alcohol labeled with ^{14}C at the terminal CF_2 established the formation of a range of products that involved fission of several C–F bonds (Wang et al. 2005). Perfluorooctanoate is also formed during simulated atmospheric reactions of the fluorotelomer alcohol (Ellis et al. 2004) and it is suggested that all these reactions are the possible sources of perfluorocarboxylic acids.

7.15 References

Au KG, CT Walsh. 1984. Stereochemical studies on a plasmid-coded fluoroacetate halohydrolase. *Bioorg Chem* 12: 197–295.

Berends AG, JC Bouttonet, CG de Rooij, RS Thompson. 1999. Toxicity of trifluoroacetate to aquatic organisms. *Environ Toxicol Chem* 18: 1053–1059.

Boutonnet JC et al. 1999. Environmental risk assessment of trifluoroacetic acid. *Human Ecol Risk Assess* 5: 59–124.

Cain RB, EK Tranter, JA Darrah. 1968. The utilization of some halogenated aromatic acids by *Nocardia*. Oxidation and metabolism. *Biochem J* 106: 211–227.

Castro CE, RS Wade, NO Belser. 1985. Biodehalogenation reactions of cytochrome P-450 with polyhalomethanes. *Biochemistry* 24: 204–210.

Coulter C, JTG Hamilton, WC McRoberts, L Kulakov, MJ Larkin, DB Harper. 1999. Halomethane: bisulfide/halide ion methyltransferase, an unusual corrinoid enzyme of environmental significance isolated from an aerobic methylotroph using chloromethane as the sole carbon source. *Appl Environ Microbiol* 65: 4301–4312.

DeFlaun ME, BD Ensley, RJ Steffan. 1992. Biological oxidation of hydrochlorofluorocarbons (HCFCs) by a methanotrophic bacterium. *Biotechnology* 10: 1576–1578.

DePierre JW. 2002. Effects on rodents of perfluorofatty acids. *Handbook Environ Chem* 3N: 203–248.

Dinglasan MJA, Y Ye, EA Edwards, SA Mabury. 2004. Fluorotelomer alcohol biodegradation yields poly- and perfluorinated acids. *Environ Sci Technol* 38: 2857–2864.

Elliott AJ. 1994. Chlorofluorocarbons. In *Organofluorine Chemistry Principles and Commercial Applications* pp. 145–157. (Eds RE Banks, BE Smart, and JC Tatlow), Plenum Press, New York.

Ellis DA, JW Martin, AO De Silva, SA Mabury, MD Hurley, MPS Andersen, TJ Wallington. 2004. Degradation of fluorotelomer alcohols: a likely atmospheric source of perfluorinated carboxylic acids. *Environ Sci Technol* 38: 3316–3321.

Ellis DA, SA Mabury. 2000. The aqueous photolysis of TFM and related trifluoromethylphenols. An alternate source of trifluoroacetic acid in the environment. *Environ Sci Technol* 34: 632–637.

Fabian P, ON Singh (Eds). 1999. Reactive halogen compounds in the atmosphere. *Handb Environ Chem* 4E.

Fox BG, JG Borneman, LP Wackett, JD Lipscomb. 1990. Halolkene oxidation by the soluble methane mono-oxygenase from *M. trichosporium* OB3b: mechanistic and environmental implications. *Biochemistry* 29: 6419–6427.

Goldman P. 1965. The enzymatic cleavage of the carbon-fluorine bond in fluoroacetate. *J Biol Chem* 240: 3434–3438.

Goldman P. 1966. Carbon-fluorine bond cleavage II Studies on the mechanism of the defluorination of fluoroacetate. *J Biol Chem* 241: 5557–5559.

Gribble GW. 2002. Naturally occurring organofluorines. *Handbook Environ Chem* 3N: 121–136.

Hamilton JTG, CD Murphy, MR Amin, D O'Hagan, DB Harper. 1998. Exploring the biosynthetic origin of fluoroacetate and 4-fluorothreonine in *Streptomyces cattleya*. *J Chem Soc Perkin Trans* 1: 759–767.

Hansen KJ, LA Clemen, ME Ellefson, HO Johnson. 2001. Compound-specific quantitative characterization of organic fluorochemicals in biological matrices. *Environ Sci Technol* 35: 766–770.

Harnisch J. 1999. Reactive Fluorine Compounds. *Handbook Environ Chem* 4E: 81–111.

Harper DB, D O'Hagan, CD Murphy. 2003. Fluorinated natural products: occurrence and biosynthesis. *Handbook Environ Chem* 3P: 141–169.

Houde M, JW Martin, RJ Letcher, KR Solomon, DCG Muir. 2006. Biological monitoring of perflkuoroalkyl sybstances: a review. *Environ Sci Technol* 40: 3463–3473.

Hyman MR, CL Page, DJ Arp. 1994. Oxidation of methyl fluoride and dimethyl ether by ammonia monooxygenase in *Nitrosomonas europaea*. *Appl Environ Microbiol* 60: 3033–3035.

Kawasaki H, N Tone, K Tonomura. 1981. Plasmid-determined dehalogenation of haloacetates in *Moraxella* species. *Agric Biol Chem* 45: 29–34.

Key BD, RD Howell, CS Criddle. 1997. Fluorinated organics in the biosphere. *Environ Sci Technol* 31: 2445–2454.

Key BD, RD Howell, CS Criddle. 1998. Defluorination of organofluorine sulfur compounds by *Pseudomonas* sp. strain D2. *Environ Sci Technol* 32: 2283–2287.

Kim BR, MT Suidan, TJ Wallington, X Du. 2000. Biodegradability of trifluoroacetic acid. *Environ Eng Sci* 17: 337–342.

Krone UE, RK Thauer. 1992. Dehalogenation of trichlorofluoromethane (CFC-11) by *Methanosarcina barkeri*. *FEMS Microbiol Lett* 90: 201–204.

Lesage S, S Brown, KH Hosler. 1992. Degradation of chlorofluorocarbon-113 under anaerobic conditions. *Chemosphere* 24: 1225–1243.

Li S, LP Wackett. 1993. Reductive dehalogenation by cytochrome P450$_{CAM}$: substrate binding and catalysis. *Biochemistry* 32: 9355–9361.

Liu J-Q, T Kurihara, S Ichiyama, M Miyagi, S Tsunasawa, H Kawasaki, K Soda, N Esaki. 1998. Reaction mechanism of fluoroacetate dehalogenase from *Moraxella* sp. B. *J Biol Chem* 273: 30897–30902.

Martin JW, J Franklin, ML Hanson, KR Solomon, SA Mabury, DA Ellis, BF Scott, and DCG Muir. 2000. Detection of chlorodifluoroacetic acid in precipitation: a possible product of fluorocarbon degradation. *Environ Sci Technol* 34: 274–281.

Meyer JJM, N Grobbelaar, PL Steyn. 1990. Fluoroacetate-metabolizing pseudomonad isolated from *Dichapetalum cymosum*. *Appl Environ Microbiol* 56: 2152–2155.

Midgley PM, A McCulloch. 1999. Properties and applications of industrial halocarbons. *Handbook Environ Chem* 4E: 129–152.

Miller LG, C Sasson, RS Oremland. 1998. Difluoromethane, a new and improved inhibitor of methanotrophy. *Appl Environ Microbiol* 64: 4357–4362.

Miller LG, MD Coutlakis, RS Oremland, BB Ward. 1993. Selective inhibition of ammonium oxidation and nitrification-linked N$_2$O formation by methyl fluoride and dimethyl ether. *Appl Environ Microbiol* 59: 2457–2464.

Moody CA, JA Field. 2000. Perfluorinated surfactants and environmental implications of their use in fire-fighting foams. *Environ Sci Technol* 34: 3864–3870.

Moody CA, WC Kwan, JW Martin, DCG Muir, SA Mabury. 2001. Determination of perfluorinated surfactants in surface water samples by two independent analytical techniques: liquid chromatography/ tandem mass spectrometry and ^{19}F NMR. *Anal Chem* 73: 2200–2206.

Neilson AH, A-S Allard. 2002. Degradation and transformation of organic fluorine compounds. *Handbook Environ Chem* 3N: 138–202.

Oldenhuis R, RLJM Vink, DB Janssen, B Witholt. 1989. Degradation of chlorinated aliphatic hydrocarbons by *Methylosinus trichosporium* OB3b expressing soluble methane monooxygenase. *Appl Environ Microbiol* 55: 2819–2826.

Oremland RS, DJ Lonrergan, CW Culbertson, DR Lovley. 1996. Microbial degradation of hydrochlorofluorocarbons (CHCl$_2$F and CHCl$_2$CF$_3$) in soils and sediments. *Appl Environ Microbiol* 62: 1818–1821.

Rao VNM. 1994. Alternatives to chlorofluorocarbons (CFCs) pp. 159–175. In *Organofluorine Chemistry Principles and Commercial Applications* (Eds RE Banks, BE Smart, and JC Tatlow), Plenum Press, New York.

Reid KA, JTG Hamilton, RD Bowden, DO'Hagan, L Dasaradhi, MR Amin, DB Harper. 1995. Biosynthesis of fluorinated secondary metabolites by *Streptomyces cattleya*. *Microbiology (UK)* 141: 1385–1393.

Schaefer JK, RS Oremland. 1999. Oxidation of methyl halides by the facultative methylotroph strain IMB-1. *Appl Environ Microbiol* 65: 5035–5041.

Sonier DN, NLDuran, GB Smith. 1994. Dechlorination of trichlorofluoromethane (CFC-11) by sulfate-reducing bacteria from an aquifer contaminated with halogenated aliphatic compounds. *Appl Environ Microbiol* 60: 4567–4572.

Streger SH, CW Condee, AP Togna, MF Deflaun. 1999. Degradation of halohydrocarbons and brominated compounds by methane- and propane-oxidizing bacteria. *Environ Sci Technol* 33: 4477–4482.

Sullivan JP, D Dickinson, HA Chase. 1998. Methanotrophs, *Methylosinus trichosporíum* OB3b, sMMO, and their application to bioremediation. *Crit Rev Microbiol* 24: 335–373.

Wackett LP, M J Sadowsky, LM Newman, H-G Hur, S Li. 1994. Metabolism of polyhalogenated compounds by a genetically engineered bacterium. *Nature (London)* 368: 627–629.

Walsh C. 1982. Fluorinated substrate analogs: routes of metabolism and selective toxicity. *Adv Enzymol* 55: 187–288.

Wang N, B Szostek, RC Buck, PW Folsom, LM Sulecki, V Capka, MR Berti, JT Gannon. 2005. Fluorotelomer alcohol biodegradation—direct evidence that perfluorinated carbon chains breakdown. *Environ Sci Technol* 39: 7516–7628.

8

Carbocyclic Aromatic Compounds without Halogen Substituents

Before the advent of the petrochemical industry, aromatic compounds, such as naphthalene, phenol, and pyridine, provided the source of many important industrial chemicals, including dyestuffs, while the monocyclic compounds continue to play an important role as fuels and starting materials.

Part 1: Monocyclic Aromatic Hydrocarbons

8.1 Introduction: Bacteria

The degradation of aromatic compounds, including hydrocarbons and phenols, has attracted interest over many years, for several reasons:

1. They are components of unrefined oil, and there has been serious concern over the hazard associated with their discharge into the marine environment after accidents at sea.
2. A number of the polycyclic representatives have been shown to be human procarcinogens that require metabolic activation.
3. There has been increased concern over air pollution as a result of their presence in the atmosphere from incomplete combustion.
4. The corresponding phenols are significant components of creosote and tar, which have traditionally been used for wood preservation.

Although aerobic growth at the expense of aromatic hydrocarbons has been known for many years (Söhngen 1913; Tausson 1927; Gray and Thornton 1928), it was many years later before details of the ring-fission reactions began to emerge. Two converging lines of investigations have examined them in detail: (a) the degradation of the monocyclic aromatic hydrocarbons benzene, toluene, and the xylenes and (b) the degradation of oxygen-substituted compounds such as benzoate, hydroxybenzoates, and phenols. As a result of this activity, the pathways of degradation and their regulation are now known in considerable detail, and ever-increasing attention has been directed to the degradation of polycyclic aromatic hydrocarbons. Since many of these metabolic sequences recur in the degradation of a wide range of aromatic compounds, a brief sketch of the principal reactions may conveniently be presented here. Reviews that include almost all aspects have been given, for example, by Hopper (1978), Cripps and Watkinson (1978), Ribbons and Eaton (1982), Gibson and Subramanian (1984), Smith (1994), and Neilson and Allard (1998). In addition, developments in regulatory aspects have been presented (Rothmel et al. 1991; van der Meer et al. 1992; Parales and Harwood 1993).

8.2 Monocyclic Arenes

8.2.1 Aerobic Conditions

8.2.1.1 Overview

It is important to appreciate a number of key issues in the degradation of substituted catechols (chloro-catechols are discussed in Chapter 9).

1. For complete degradation of an aromatic hydrocarbon to occur, it is necessary that the products of ring oxidation and fission can be further degraded to molecules that enter anabolic and energy-producing reactions.

2. Essentially, different mechanisms operate in bacteria and fungi, and these differences have important consequences. In bacteria, the initial reaction is carried out by dioxygenation and results in the synthesis of a *cis*-1,2-dihydro-1,2-diol, which is then dehydrogenated to a catechol before ring fission mediated by dioxygenases. In fungi, however, the first reaction is monooxy-genation to an epoxide followed by hydrolysis to a *trans*-1,2-dihydro-1,2-diol and rearrange-ment to a phenol. Ring fission of polycyclic aromatic hydrocarbons does not generally occur in fungi, so that these reactions are essentially biotransformations. These reactions are schemati-cally illustrated in Figure 8.1.

3. Both fungi and some yeasts are able to degrade monocyclic arenes and simpler substituted aromatic compounds such as 3,4-dihydroxybenzoate (Cain et al. 1968) which can be produced as intermediates. The pathways that are used differ, however, from those used by bacteria. Examples are given in which concomitant hydroxylation and decarboxylation are involved:

   a. The degradation of 3,4-dihydroxybenzoate by the yeast *Trichosporon cutaneum* produces 1,2,4-trihydroxybenzene prior to ring fission (Anderson and Dagley 1980). In addition, *Trichosporon cutaneum* is able to degrade 4-methylphenol to 4-methylcatechol by intradiol ring fission and formation of 3-oxoadipate by a pathway that is not normally accessible to bacteria (Powlowski and Dagley 1985; Powlowski et al. 1985).

   b. The pathway used for degradation of vanillate by *Sporotrichium pulverulentum* (Figure 8.2) (Ander et al. 1983).

   c. The oxidative decarboxylation to 1,4-dihydroxybenzenes of a range of 4-hydroxybenzo-ates, including 2,4-dihydroxy- and 3,4-dihydroxybenzoate and their mono-, di-, and tetra-fluorinated analogs, has been examined in *Candida parapsilosis* CBS604 and is carried out by an FAD-dependent monooxygenase (Eppink et al. 1997).

8.2.1.1.1 Methylcatechols

Methylcatechols are produced by dioxygenation of methylarenes, methylbenzoates, and methylsalicy-lates, and by hydroxylation of methyl phenols. In contrast to the general rule that favors extradiol fission of substituted catechols, these undergo intradiol fission.

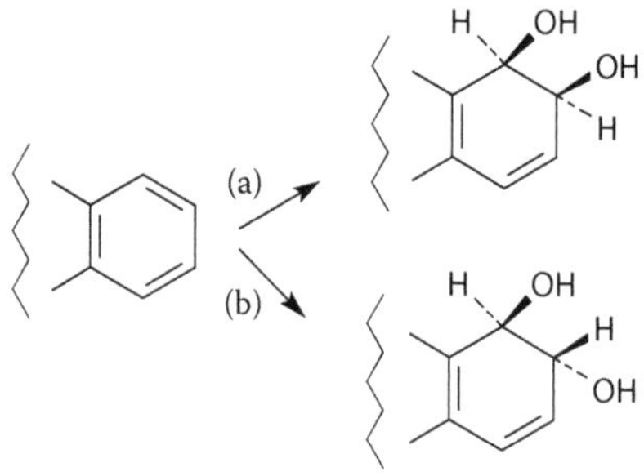

FIGURE 8.1 Alternative pathways for the oxidative metabolism of naphthalene: (a) bacteria and (b) fungi.

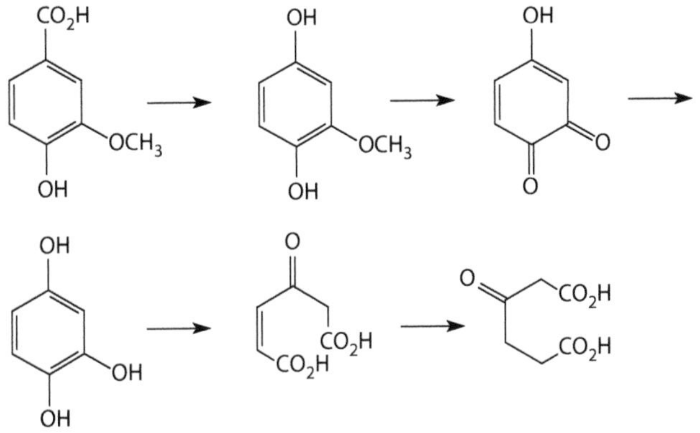

FIGURE 8.2 Biodegradation of vanillic acid by fungi.

1. The degradation of 3-methyl- and 4-methylbenzoate has been examined in *Pseudomonas arvilla* mt-2 and is initiated by dioxygenation with concomitant decarboxylation to 3-methyl- and 4-methylcatechol. These are degraded by extradiol fission followed either by hydrolytic loss of acetate or by decarboxylation (Murray et al. 1972) (Figure 8.3a and b).

2. 4-Methyl- and 5-methylsalicylate can be degraded by *Pseudomonas* sp. strain MT1 with the production of 4-methylcatechol that is metabolized after isomerization of the 4-methylmucono-lactone (Cámara et al. 2007) (Figure 8.4).

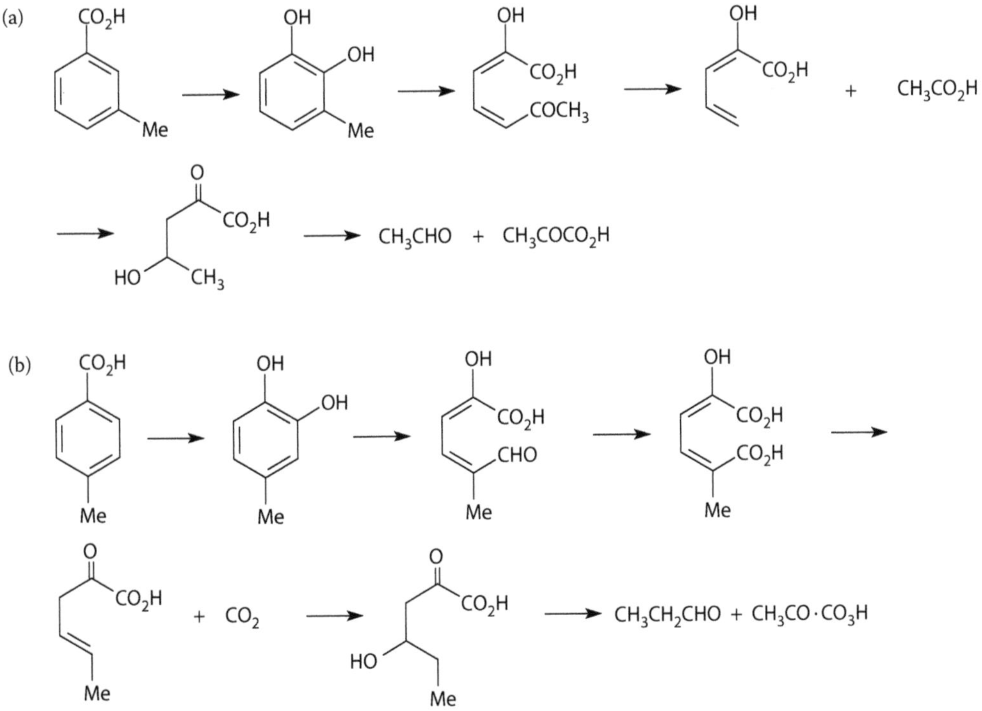

FIGURE 8.3 Degradation of (a) 3-methylbenzoate, (b) 4-methylbenzoate. (Adapted from Marín M et al. 2010. *J Bacteriol* 192: 1543–1552.)

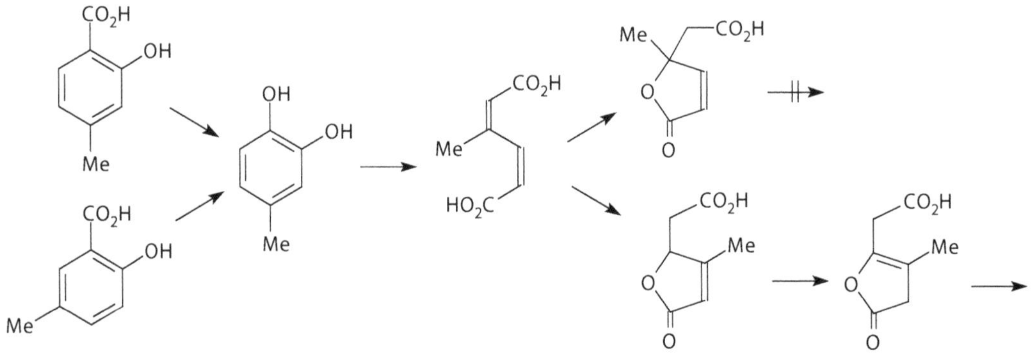

FIGURE 8.4   Degradation of 4-methyl- and 5-methylsalicylate with the production of 4-methylcatechol and 3-methyl-muconate. (Adapted from Cámara B et al. 2007. *J Bacteriol* 189: 1664–1674.)

There are further consequences of the choice of intradiol fission. The presence of a methyl group at the C_4 of a muconolactone may block further metabolism and subsequent entry into the 3-oxoadipate pathway. In *Pseudomonas desmolyticum*, for example, the 4-methylmuconolactone is a terminal metabolite (Figure 8.5a) (Catelani et al. 1971). There are, however, mechanisms that are able to circumvent this block. The degradation of 4-methylcatechol by *Cupriavadus necator* (*Alcaligenes eutrophus*) JMP134 takes place by the formation of the 4-methylmuconolactone followed by isomerization to the 3-methyl isomer, rearrangement to the 3-methyl-but-3-en-1,4-olide, and production of 4-methyl-3-oxoadipate (Figure 8.5b) (Pieper et al. 1985). Isomerization has been examined in both Gram-positive and Gram-negative bacteria, although this is not required in the yeast *Trichosporon cutaneum* or the fungus *Aspergillus niger* where the choice of the cyclization obviates the need for later isomerization of the lactone.

1. In the yeast *Trichosporon cutaneum*, 4-methylphenol is hydroxylated to 4-methylcatechol. Intradiol fission is followed by cyclization to 3-methylmuconolactone that can be isomerized to 3-methyl-but-3-en-1,4-olide and hydrolyzed to 4-methyl-3-oxoadipate (Powlowski and Dagley 1985; Powlowski et al. 1985).

2. In the degradation of 4-methylbenzoate by *Rhodococcus* sp. N75, the 4-methylcatechol that is produced undergoes intradiol fission to the expected 4-methylmuconolactone which undergoes rearrangement to 3-methylmuconolactone that is poised for entry into the 3-oxoadipate pathway. The isomerase has been characterized from this strain, had a mass of 75 kDa and was highly specific for this 1-methybislactone (Figure 8.6) (Bruce et al. 1989). It has also been shown that 3-methylmuconolactone is converted to the coenzyme A ester by a highly specific synthase before hydrolysis to 4-methyl-3-oxoadipyl-CoA (Cha et al. 1998).

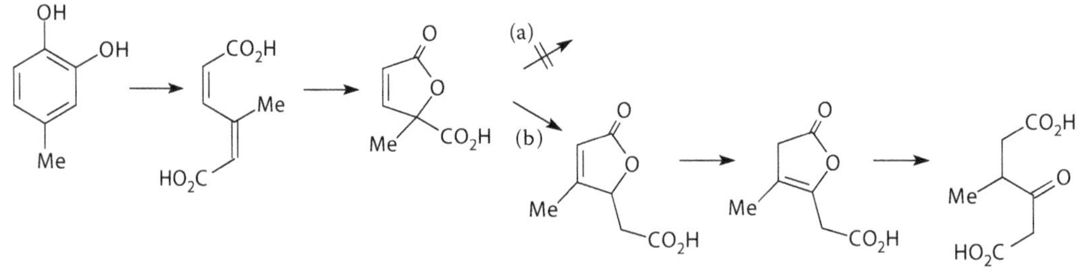

FIGURE 8.5   Metabolism of 4-methylcatechol to (a) 4-methylmuconolactone as terminal metabolite or (b) isomerization and formation of 4-methyl-3-oxoadipate.

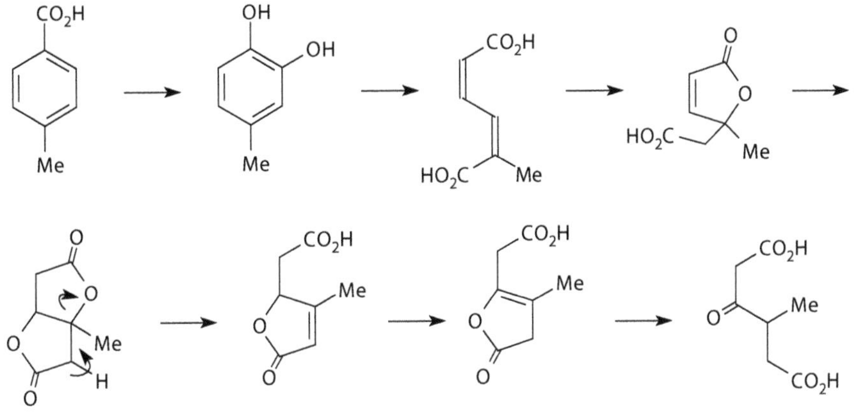

FIGURE 8.6 Activity of specific isomerase for 1-methybislactone.

3. In a study using *Cupriavadus necator* (*Alcaligenes eutrophus*) JMP 134, the formation of the bislactone as intermediate was not favorable, and a more circuitous mechanism was proposed (Pieper et al. 1990). It was shown that although EDTA had no effect, *p*-mercuribenzoate was inhibitory but could be reversed by dithiothreitol that suggested the presence of sulfhydryl groups. The isomerase has been characterized further in the same strain (Erb et al. 1998) and, although this enzyme apparently represented a new type of isomerase, its primary structure was related to those of the muconolactone isomerase from *Acinetobacter*, and strains of *Pseudomonas putida* (Erb et al. 1998).

4. *Pseudomonas reinekei* is able to degrade 4-methylcatechol by intradiol fission: the key enzyme is the isomerase that produces 3-methylmuconolactone from the initially formed 4-methylmuconolactone. The further steps in this strain leading to the formation of 4-methyl-3-oxoadipate and finally acetate and methylsuccinate are unexceptional except for the formation of 4-methyl-3-oxoadipyl-CoA before fission by a thiolase. There are 10 genes that form an operon: *mmlC, mmlL,mmlR,mmlF,mmlG,mmlD,mmlH,mmlI,mmlJ,mmlK* (Marín et al. 2010). Of these, *mmlI* and *mmlK* encode 4-methylmuconolactone methylisomerase and methylmuconolactone isomerase which have been found in other organisms that are able to degrade 4-methylcatechol. This strain is also able to metabolize 4-chloro- and 5-chlorosalicylate that are oxygenated to 4-chlorocatechol before intradiol fission to 3-chloromuconate and cyclization to 4-chloromuconate: this may then undergo decarboxylation with the loss of chloride to produce the terminal metabolite protoanemonin, or to maleylacetate and then to 3-oxoadipate (Figure 8.7). Maleylacetate reductase plays a central role in the metabolism of chlorinated aromatic compounds (Kaschabek and Reineke 1995) where it is successively reduced to 3-oxoadipate that enters the TCA cycle, and the maleylacetate reductase from *Ralstonia* (*Alcaligenes*) *eutropha* has been characterized (Seibert et al. 1993).

Analogous reactions occur in the degradation of 4-methyl- and 5-methylsalicylate by *Pseudomonas* sp. strain MT1 both of which are metabolized to 4-methylcatechol. This undergoes intradiol fission followed by lactonization to 4-methylmuconolactone and isomerization to 3-methylmuconolactone and then to 3-methybut-3-enolide (Figure 8.4) (Cámara et al. 2007).

8.2.1.1.2 Benzene

Benzene is one of a group of related aromatic monocyclic hydrocarbons (BTEX—benzene, toluene, ethylbenzene, and xylene), and since these are water soluble, there has been concern for their dissipation and persistence in groundwater under both aerobic and anaerobic conditions. Although aerobic growth at the expense of benzene was established many years ago, the pathway for its degradation was established only

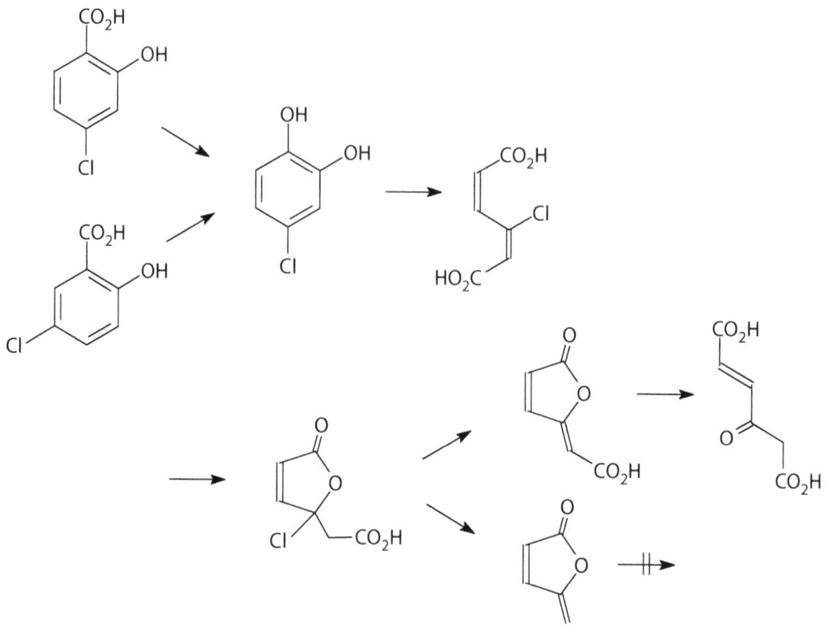

FIGURE 8.7 Degradation of 4-chloro- and 5-chlorosalicylate with cyclization to 4-chloromuconate: Decarboxylation with loss of chloride to produce the terminal metabolite protoanemonin, or to maleylacetate and then to 3-oxoadipate. (Adapted from Marín M et al. 2010. *J Bacteriol* 192: 1543–1552.)

much later. The aerobic degradation of benzene by bacteria is initiated by the formation of a *cis*-dihydrodiol, followed by dehydrogenation to catechol (Gibson et al. 1968). In a strain of *Moraxella* sp., subsequent intradiol fission by catechol-1,2-dioxygenase produced muconate that was degraded by the 3-oxoadipate pathway to succinate and acetate (Högn and Jaenicke 1972). In other organisms, including *Pseudomonas putida*, extradiol ring fission by catechol-2,3-dioxygenase produced 2-hydroxymuconate semialdehyde that was degraded to pyruvate and acetaldehyde.

The formation of the dihydrodiol in a mutant strain of *Pseudomonas putida* involves incorporation of both atoms of $^{18}O_2$ (Gibson et al. 1970). It is mediated by a dioxygenase that has been purified and shown to be a soluble enzyme consisting of three protein components (Axcell and Geary 1975) encoded by the genes *bedA*, *bedB*, and *bedC*. A flavoprotein reductase$_{BED}$ containing FAD accepts electrons from NADH and transfers them to a [2Fe–2S] Rieske ferredoxin$_{BED}$, while the terminal oxygenase is a high-molecular-weight protein consisting of two large subunits containing two [2Fe–2S] Rieske ferredoxins (ISP$_{BED}$) (Crutcher and Geary 1979). The dihydrodiol dehydrogenase that converts the dihydrodiol into catechol is a large enzyme with a molecular weight of 440,000 (Axcell and Geary 1973). In *Pseudomonas putida* ML2, the genes encoding benzene dioxygenase are carried on a 112-kb plasmid (Tan and Mason 1990).

As an alternative, successive monooxygenation of benzene to phenol, catechol, and 1,2,3-trihydroxybenzene may be accomplished by the toluene 4-monooxygenase of *Pseudomonas mendocina* strain JKR1 and the 3-monooxygenase of *Ralstonia* (*Pseudomonas*) *pickettii* strain PKO1 (Tao et al. 2004).

8.2.1.1.3 Toluene

The degradation of toluene has been studied extensively in strains of *Pseudomonas putida*, and details of the three different pathways have been resolved.

1. In strains containing the TOL plasmid, degradation proceeds by successive oxidation of the methyl group to hydroxymethyl, aldehyde, and carboxylate, followed by dioxygenation of

benzoate to catechol and extradiol ring fission (Figure 8.8a) (Keil and Williams 1985). The TOL plasmid also carries the genes for the degradation of 1,3- and 1,4-dimethylbenzenes (Assinder and Williams 1990), and involves comparable oxidations of a single methyl group (Davey and Gibson 1974; Williams and Worsey 1976). These carboxylic acids are then converted into catechols by benzoate dioxygenase and then undergo extradiol ring fission (Williams and Worsey 1976). In *Pseudomonas desmolytica*, *n*-propyl benzene is degraded by alternative pathways, both of which terminate in ring fission: (a) oxidation of the side chain to produce benzoate or (b) dioxygenation and dehydrogenation to 3-*n*-propylcatechol (Jigami et al. 1979).

2. The genes for toluene degradation may also be located on the chromosome, when a different pathway is followed—dioxygenation by toluene 2,3-dioxygenase to (+)-*cis*-(2*R*,3*S*) 2,3-dihydro-2,3-dihydroxytoluene (Ziffer et al. 1977; Boyd et al. 1992), dehydrogenation to 3-methylcatechol followed by extradiol ring fission (Figure 8.8b). The multicomponent dioxygenase enzymes encoded by the genes *todA*, *todB*, *todC1*, and *todC2* are analogous to those involved in benzene degradation: a flavoprotein reductase$_{TOL}$ that accepts electrons from NADH, and a ferredoxin$_{TOL}$ that transfers them to the terminal dioxygenase (ISP$_{TOL}$) which consists of two subunits (Zylstra and Gibson 1989). The dihydrodiol dehydrogenase has, however, a much lower molecular weight (104 kDa) than that for the corresponding benzene dihydrodiol dehydrogenase (Rogers and Gibson 1977).

3. Degradation of toluene may also take place in different species of *Pseudomonas* by initial monooxygenation at the 2-, 3-, or 4-positions. The application of this activity to the degradation of trichloroethene is discussed in Chapter 7. The situation for toluene monooxygenases has been confusing. However, it is now recognized that only the following two initial reactions take place:

   a. Toluene-2-monooxygenase of *Burkholderia cepacia* G4 hydroxylates toluene at the *ortho* position (Shields et al. 1995) and then further to 3-methylcatechol.

   b. The formation of 4-methylphenol by the three-component monooxygenase in *P. mendocina* strain KR1 plausibly involves an arene oxide intermediate since it takes place with an NIH shift of a proton to the 3-position (Whited and Gibson 1991a,b). Further degradation involves oxidation to 4-hydroxybenzoate, hydroxylation to 3,4-dihydroxybenzoate followed by intradiol ring fission (Figure 8.8c).

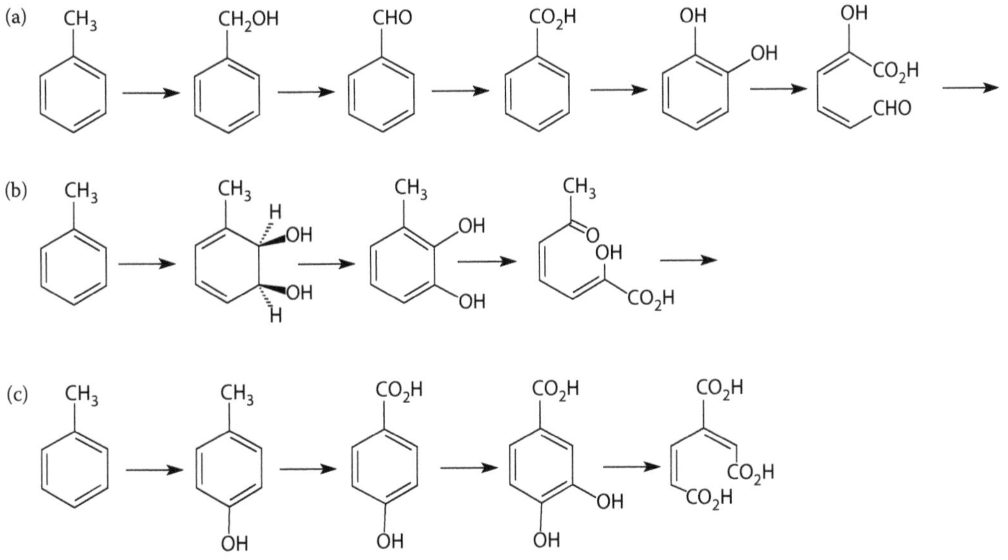

FIGURE 8.8 Aerobic degradation of toluene by (a) side-chain oxidation, (b) dioxygenation of the ring, and (c) monooxygenation.

c. Although the monooxygenase from *Ralstonia pickettii* was originally reported to form 3-hydroxytoluene, it has been shown that 4-hydroxytoluene is produced as the major product (Fishman et al. 2004).

Other examples of the monooxygenation of arenes include the following:

1. The soluble methane monooxygenase from *Methylococcus capsulatus* produced 4-ethylphenol and 1-phenylethanol from ethylbenzene, and both 1- and 2-hydroxynaphthalene from naphthalene (Dalton et al. 1981); the formation of the 4-ethylphenol is accompanied by an NIH shift consistent with the involvement of an intermediate arene oxide.

2. Under conditions of active NH_4^+ oxidation, cells of *Nitrosomonas europaea* can transform ethylbenzene to 4-ethylphenol and acetophenone (Keener and Arp 1994).

3. In cell extracts, the cytochrome P450 system of *Streptomyces griseus* transforms benzene to phenol, toluene to 2-methylphenol, biphenyl to 2- and 4-hydroxybiphenyl, styrene to 4-hydroxy-styrene, and naphthalene to naphth-1-ol (Sariaslani et al. 1989).

8.2.1.1.4 Xylenes

The pathway for the degradation of the xylenes depends critically on the orientation of the methyl groups, and *o*-xylene is considered to be the most recalcitrant, since xylene monooxygenase cannot hydroxylate one of its methyl groups.

1. The degradation of *m*- and *p*-xylene in organisms carrying the TOL plasmid is initiated, as for toluene, by oxidation of one of the methyl groups to *m*- or *p*-toluic acid. These are then degraded to 3- or 4-methylcatechol by dioxygenation and decarboxylation (Davey and Gibson 1974).

2. The degradation of *o*-xylene by chromosomal genes in *Pseudomonas stutzeri* OX1 takes place, however, by successive hydroxylation to 3,4-dimethylcatechol (Bertoni et al. 1998) that is degraded by dioxygenation and ring fission, with the loss of acetate to propionaldehyde and pyruvate.

3. The metabolism of *o*-xylene by *Rhodococcus* strain B3 took place simultaneously by two pathways: (i) oxidation of a methyl group to 2-methylbenzoate, followed by dioxygenation and decarboxylation to 3-methylcatechol and (ii) hydroxylation to 2,3-dimethylphenol followed by further hydroxylation to 3,4-dimethylcatechol (Bickerdike et al. 1997).

4. Degradation of *o*-xylene by *Rhodococcus* sp. DK17 is initiated by dioxygenation to an unstable *o*-xylene *cis*-3,4-dihydrodiol. This may be dehydrogenated to 3,4-dimethylcatechol followed by extradiol ring fission, or undergo dehydration to produce 2,3- and 3,4-dimethylphenol (Kim et al. 2004).

5. *Corynebacterium* sp. strain C125 was able to grow at the expense of *o*-xylene, though not *m*- or *p*-xylene. Degradation took place by dioxygenation and dehydrogenation followed by extradiol fission (Schraa et al. 1987).

6. *Pseudoxanthomonas spadix* BD-a59 was able to degrade benzene, toluene, ethylbenzene as well as, unusually, all xylene isomers in the presence of 150 mg/L yeast extract, or with the addition of an unknown soil component (Kim et al. 2008).

8.2.1.1.5 1,2,4-Trimethylbenzene

Pseudomonas putida strains HS1 and mt-2 were able to degrade toluene, *m*-xylene, *p*-xylene, and 1,2,4-trimethylbenzene (Kunz and Chapman 1981). For 1,2,4-trimethylbenzene, this was carried out by oxidation to 3,4-dimethylbenzoate that was degraded by extradiol fission to 2-hydroxy-6-oxo-5-methyl-heptanoate, and then 4-hydroxy-2-oxovalerate, propionaldehyde, and pyruvate. On the other hand, naphthalene 1,2-dioxygenase in *Pseudomonas aeruginosa* strain PAO1 (pRE 695) was able to oxidize methyl groups by monooxygenation to benzyl alcohols. Substrates included methylnaphthalenes, and 1,2,4-trimethylbenzene that produced 2,5-, 2,4-, and 3,4-dimethylbenzyl alcohols (Selifonov et al. 1996).

8.3 Fungi

The degradation and transformation of monocyclic arenes has been examined in fungi, including a white-rot basidiomycete.

1. Hydroxylation of benzene and toluene was observed in the following examples (Smith and Rosazza 1983):
   a. Benzene to phenol by *Penicillium chrysogenum, Cunninghamella blakesleeana, C. baineri,* and *Gliocladium deliquens*
   b. Toluene to 2- and 4-methylphenol by *Penicillium chrysogenum, Rhizopus stolonifer,* and *C. baineri*

2. a. Benzene, toluene, ethylbenzene, and *o-* and *p*-xylene were degraded by *Phanerochaete chrysosporium* without evidence for inhibitory interactions between them (Yadav and Reddy 1993).
   b. Toluene was able to serve as a growth substrate for the fungus *Cladosporium sphaerospermum.* Oxygen consumption was observed for benzyl alcohol, benzaldehyde, and benzoate, and the activities of the dehydrogenases, as well as catechol dioxygenase, were measured. The pathway was initiated by attack on the methyl group, and was probably analogous to that used by bacteria (Weber et al. 1995).
   c. The growth and oxidation of monocyclic arenes was examined in fungi, including *Cladosporium sphaerospermum,* and species of *Cladophiliophora, Pseudeurotium, Exophilia,* and *Leptodontium.* Toluene was oxidized by all of them and ethylbenzene by *Cladophilia* and *Exophilia,* although none was able to oxidize benzene (Prenafeta-Boldú et al. 2001).
   d. Styrene was transformed by the black yeast *Exophilia jeanselmei* to phenylacetate by a pathway similar to that used by bacteria. The initial monooxygenation was carried out by a cytochrome P450, and phenylacetate was further metabolized to 2-hydroxy- and 2,5- dihydroxyphenylacetate (Cox et al. 1996).

Alkylated benzenes were clearly more readily metabolized than benzene which was generally resistant.

8.3.1 Anaerobic Conditions

8.3.1.1 Introduction

Concern has arisen that leakage of BTEX from gasoline may reach groundwater where its fate may be determined by anaerobic reactions. Although the degradation of benzene can occur anaerobically under sulfate-reducing, denitrifying, methanogenic, and iron (III)-reducing conditions (Mancini et al. 2003), the complete pathway for these reactions has not been established. Details of intermediates have been obtained using enrichment cultures using $^{13}C_6$-benzene: labeled phenol, benzoate, and toluene were detected, and the kinetics of their formation was determined (Ulrich et al. 2005). The anaerobic mineralization of BTEX in Fe-reducing cultures has also been observed (Jahn et al. 2005).

The bacterial degradation of toluene and some xylenes has been examined under a variety of environments, and some of the organisms are able to function under both anaerobic and aerobic conditions. These organisms include anaerobic phototrophs, and facultative anaerobic bacteria that can use chlorate, nitrate, and sulfate as electron acceptors. These studies have led to elucidation of the pathway and mechanism of degradation. It is worth noting that benzoate which is the first product in the degradation of toluene by these facultative organisms can be degraded by obligate anaerobes such as *Geobacter metallireducens,* the syntrophs *Syntrophus gentianae, Syntrophus aciditrophicus,* and the sulfate-reducing *Desulfococcus multivorans* that are discussed in later paragraphs.

8.3.1.1.1 Benzene

The anaerobic oxidation of benzene has been described using a range of electron acceptors in enrichment cultures under different conditions and, although only a few pure cultures have been isolated to date, several putative mechanisms have been put forward.

Nitrate—Members of the *Azoarcus* phylotype were identified by stable isotope probing of groundwater supplemented with ^{13}C-benzene and specifically with nitrate as electron acceptor (Kasai et al. 2006). Two strains of *Dechloromonas* RCB and JJ isolated by enrichment under quite different conditions were used to examine the anaerobic oxidation of benzene: strain RCB used chlorate as electron acceptor and 4-chlorobenzoate as electron donor, and strain JJ used anthraquinone-2,6-disulfonate as electron donor and nitrate as electron acceptor. Both were able to oxidize benzene anaerobically using nitrate as electron acceptor (Coates et al. 2001). Strain RCB was able to utilize not only benzene but also toluene, ethylbenzene, and all isomers of xylene—albeit only partly with *p*-xylene with nitrate as electron acceptor (Chakraborty et al. 2005).

Chlorate—The pathway for benzene degradation by *Dechloromonas aromatica* strain RCB using chlorate as electron acceptor involved hydroxylation to phenol followed by carboxylation and dehydroxylation to benzoate (Chakraborty and Coates 2005). The carboxylation is analogous to that involved in the anaerobic degradation of phenylphosphate by the denitrifying *Thauera aromatica* (Breese and Fuchs 1998; Schühle and Fuchs 2004). *Dechloromonas aromatica* strain RCB was also able to degrade toluene, ethylbenzene, and, partially, xylenes (Chakraborty et al. 2005). Perchlorate could also be used as an electron acceptor for the degradation of benzene, while benzene and toluene could be degraded concurrently.

Sulfate—Use of sulfate as electron acceptor was supported by stable isotope probing of a sulfidogenic consortium (Oka et al. 2008), and by analysis of a sulfate-reducing enrichment culture in which the dominant phylotype was closely related to a clade of Deltaproteobacteria while cell suspensions did not show activity for phenol or toluene (Musat and Widdel 2008).

Ferric iron

1. Anaerobic benzene degradation has been examined in an iron-reducing culture (Laban et al. 2010), and the N-terminal sequence of an enzyme that was specifically found under these conditions showed 43% similarity to the phenylphosphate carboxylase of *Aromatoleum aromaticum* EbN1 and 56% and benzoate-CoA ligase in *Geobacter metallireducens*.

2. The hyperthermophilic *Ferroglobus placidus* was able to oxidize benzene stoichiometrically to CO_2 under anaerobic conditions using Fe(III) as the sole electron acceptor. Microarray comparison of gene transcripts of cells grown with benzene versus acetate showed upregulation of the enzymes for anaerobic degradation of benzoate by ring fission to 3-hydroxypimeloyl-CoA, though not via phenol by phosphorylation. In further support of this pathway, ^{14}C-benzoate was produced transiently from ^{14}C-benzene (Holmes et al. 2011).

8.3.1.1.2 Alkylated Benzenes

The degradation of toluene under denitrifying conditions has been described for strains of *Thauera aromatica* and for *Azoarcus tolulyticus*, *Azoarcus toluclasticus*, and *Azoarcus toluvorans* that are discussed in Chapter 2. The pathways for the degradation of toluene and xylene under denitrifying and sulfate-reducing conditions have been studied most extensively, and they take place by reactions quite different from those used by aerobic bacteria. Some strains are able to regulate the pathway of degradation of toluene to the availability of oxygen—aerobic degradation by dioxygenation or, in the absence of oxygen under denitrifying conditions, by the benzylsuccinate pathway (Shinoda et al. 2004). The anaerobic degradation of toluene results in the production of benzoate that is activated to the coenzyme A ester by an ATP-dependent coenzyme A ligase before reduction and fission of the ring that is discussed in greater detail in the later section dealing with benzoate. Some organisms have evolved a single benzoate-CoA ligase that is able to function under both aerobic and anaerobic conditions, for example, in *Thauera aromatica* (Schühle et al. 2003), and in strains of *Magnetospirillum* (Kawaguchi et al. 2006), while the aerobic degradation of benzoate by *Azoarcus evansii* is also initiated by the formation of the coenzyme A ester (Gescher et al. 2002).

Two anaerobes affiliated with known sulfate-reducing bacteria were isolated from enrichments with crude oil and were able to grow at the expense of a number of alkylated benzenes—strain oXyS1 with toluene, *o*-xylene, and *o*-ethyltoluene and strain mXyS1 with toluene, *m*-xylene, and *m*-ethyltoluene (Harms et al. 1999). The fumarate reaction is employed in the degradation of toluene by *Desulfobacula toluolica* (Rabus and Heider 1998), *o*- and *m*-xylene by a sulfate-reducing strain of *Desulfotomaculum* (Morasch et al. 2004), and for xylene and ethylbenzene under sulfate-reducing conditions (Kniemeyer et al. 2003). The analogous succinates and fumarates that would be produced from dimethyl benzenes have been suggested as markers for the anaerobic degradation of these substrates in an aquifer (Beller et al. 1995; Beller 2002). The fumarate pathway is also used for the degradation of *m*- and *p*-cresol by *Desulfobacterium cetonicum* (Müller et al. 2001), and of 2-methylnaphthalene by a sulfate-reducing enrichment culture (Selesi et al. 2010).

8.3.1.2 Range of Mechanisms

A wide range of mechanisms has been established for the anaerobic degradation of toluene by different organisms.

1. *Dehydrogenation of the methyl group*   Strains of denitrifying bacteria have been shown to degrade toluene in the absence of oxygen using N_2O as electron acceptor (Schocher et al. 1991), and the data are consistent with a pathway involving successive oxidation of the ring methyl group with the formation of benzoate. The details of this pathway involving benzyl alcohol and benzaldehyde have been clearly demonstrated with a strain of *Thauera* (*Pseudomonas*) sp. under denitrifying conditions (Altenschmidt and Fuchs 1992). This pathway is supported by the demonstration of benzyl alcohol dehydrogenase, benzaldehyde dehydrogenase, benzoyl-CoA ligase, and benzoyl-CoA reductase activities in cell extracts (Biegert and Fuchs 1995). The benzyl alcohol dehydrogenase from benzyl alcohol–grown cells was similar in many of its properties to those from the aerobic bacteria *Acinetobacter calcoaceticus* and *Pseudomonas putida* (Biegert et al. 1995).

2. *Condensation reactions*   It has been suggested that the degradation of toluene could proceed by condensation with acetate to form phenylpropionate and benzoate before ring fission (Evans et al. 1992). These are produced as terminal metabolites during anaerobic degradation of toluene by sulfate-reducing enrichment cultures (Beller et al. 1992). A mechanism for the oxidation of the methyl group has been proposed for *Azoarcus tolulyticus*, and involves a condensation reaction with acetyl-CoA followed by dehydrogenation to cinnamoyl-CoA (Migaud et al. 1996). This is then either transformed into benzylsuccinate and benzylfumarate, which are apparently terminal metabolites that have also been isolated from a denitrifying organism (Evans et al. 1992), or further degraded to benzoyl-CoA, which is the substrate for the ring reductase (Figure 8.9). The later reactions are also parts of the fumarate pathway.

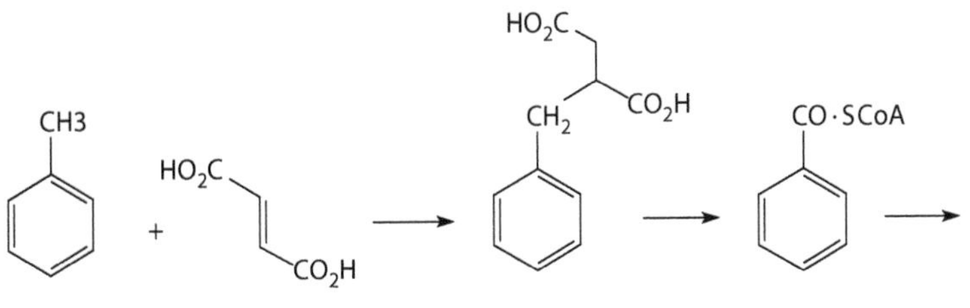

FIGURE 8.9   Anaerobic degradation of toluene via benzylsuccinate.

3. *Reaction with fumarate* An alternative sequence for denitrifying bacteria has been extensively investigated. It involves the direct formation of benzylsuccinate by the reaction of toluene with fumarate and has been delineated in several bacteria, including *Thauera aromatica* (Biegert et al. 1996), a denitrifying strain designated *Pseudomonas* sp. strain T2 (Beller and Spormann 1997a), and a sulfate-reducing bacterium strain PRTOL1 (Beller and Spormann 1997b): this is apparently a widespread reaction, and important additional details have emerged from further studies.

   a. It has been shown in the denitrifying strain T2 that the reaction between toluene and fumarate is stereospecific, yielding (*R*)(+)-benzylsuccinate, and that the proton abstracted from toluene is incorporated into the benzylsuccinate (Figure 8.10) (Beller and Spormann 1998).

   b. The synthesis of benzylsuccinate during the anaerobic degradation of toluene under denitrifying conditions involves activation of toluene to a benzyl radical by a mechanism involving glycyl radical catalysis (Leuthner et al. 1998; Verfürth et al. 2004). The amino acid sequence of the large subunit of the purified enzyme from *Thauera aromatica* showed a high level of homology to glycine radical enzymes and particularly to pyruvate formate lyase. It has been shown (Coschigano et al. 1998) in two mutants of strain T1 that the genes *tutD* and *tutE* involved in the anaerobic degradation of toluene encode proteins with molecular masses of 97.6 and 41.3 kDa, which possess homologies to pyruvate formate lyase and its activating enzyme, respectively. A free radical at Gly[734] is involved in the operation of pyruvate formate lyase (Wagner et al. 1992) and there is, therefore, a formal similarity in at least one step in the two reaction pathways.

   c. The (*R*)-2-benzylsuccinate that is produced is converted to (*R*)-2-benzylsuccinyl-CoA by a class III transferase that has been characterized (Leutwein and Heider 2001). This is necessary for the consecutive reactions that metabolize this (i) to benzoyl-CoA by dehydrogenation to (*E*)-2′-phenylitaconyl-CoA, (ii) hydration to 2-(hydroxymethyl)-succinyl-CoA, and (iii) further dehydrogenation (Leutwein and Heider 2001). Analogous pathways are used by *Thauera aromatica* and *Azoarcus* sp. for the degradation of *m*-xylene under denitrifying conditions (Verfürth et al. 2004). The reactions used by sulfate-reducing bacteria have already been discussed.

4. *Other reactions*

   a. Under denitrifying conditions that have been extensively examined in *Aromatoleum aromaticum* (*Azoarcus* sp.) strain EbN1, whereas the degradation of toluene proceeds via the fumarate pathway (Kube et al. 2004), the degradation of ethylbenzene is quite different. The pathway involves dehydrogenation at the benzylic carbon atom to produce phenylethanol followed by dehydrogenation to acetophenone, carboxylation, and hydrolysis of the β-keto acid (Rabus et al. 2002; Kühner et al. 2005) (Figure 8.11). The dehydrogenase is a

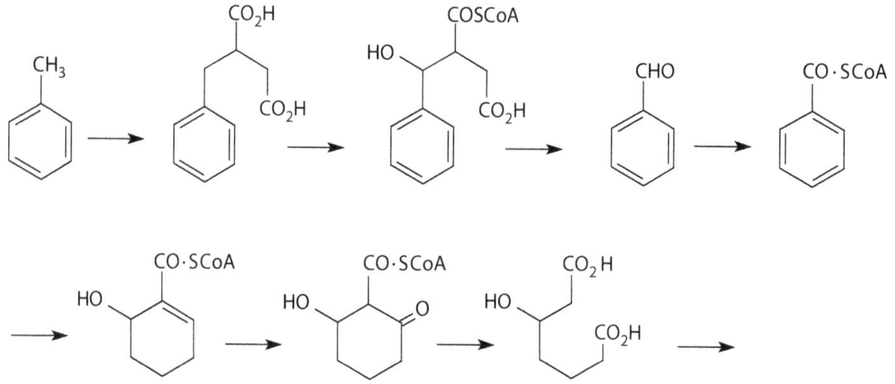

FIGURE 8.10 Anaerobic degradation of toluene.

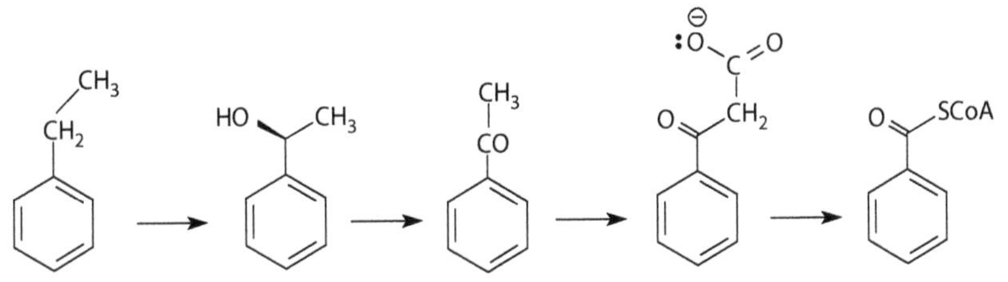

FIGURE 8.11 Anaerobic degradation of ethylbenzene. (Adapted from Rabus R et al. 2002. *Arch Microbiol* 178: 506–516.)

soluble periplasmic trimeric enzyme that contains molybdenum, iron, acid-labile sulfur, and the molybdopterin cofactor (Johnson et al. 2001) whose crystal structure has revealed the electron transport from the molybdopterin to heme via ferredoxins and a [3Fe–4S] cluster (Kloer et al. 2006). The carboxylation that is ATP dependent, plausibly involving the formation of the enol phosphate, and the carboxylase contains Mg rather than Zn (Jobst et al. 2010).

b. Anaerobic hydroxylation of toluene to 4-hydroxytoluene followed by oxidation of the methyl group and dehydroxylation to benzoate has been suggested (Rudolphi et al. 1991). Although the role of these in degradation has not been clearly established, at least two analogies can be suggested: (i) carboxylation and dehydroxylation to 3-methylbenzoate that is produced from 1,3-dimethylbenzene by *Thauera* sp. strain K172 under denitrifying conditions (Biegert and Fuchs 1995) and (ii) oxidation to 4-hydroxybenzoate and further degradation.

c. An alternative pathway for toluene includes side-chain carboxylation to phenyl acetate and oxidation via phenylglyoxylate to benzoate, that is an established pathway for phenylalanine degradation by *Thauera aromatica* (Schneider et al. 1997; Schneider and Fuchs 1998) and for phenyl acetate itself by *Thauera aromatica* (Dangel et al. 1991) and *Azoarcus evansii* (Hirsch et al. 1998).

8.3.1.3 Styrene

This is used in large amounts for the production of polymers, and attention has been directed to the degradation of the volatile monomer that may be discharged into the environment or collected in biofilters. The bacterial degradation and transformation of styrene has attracted considerable attention (Warhurst and Fewson 1994), and several pathways have been described for bacteria.

1. In some bacteria, ring dioxygenation with the production of the *cis*-2,3-dihydrodiol takes place, leaving the vinyl group intact.

   a. In *Rhodococcus rhodochrous* strain NCIMB 13259, 2-vinyl-*cis,cis*-muconate is produced by catechol 1,2-dioxygenase as a terminal metabolite, although complete degradation is possible owing to the existence of catechol 2,3-dioxygenase activity that is also present (Warhurst et al. 1994).

   b. *Pseudomonas putida* F1 is able to degrade styrene by the *tod* operon of the toluene pathway, though it is unable to use it as a growth substrate. It has been shown that catechol 2,3-dioxygenase is inhibited by the 3-vinylcatechol that is produced. This could, however, be overcome through increased dioxygenase activity by overexpression of the dioxygenase on a plasmid in addition to resistance of the enzyme to inactivation by 3-vinylcatechol (George et al. 2011).

2. Naphthalene dioxygenase from *Pseudomonas* sp. strain NCIB 9816-4 produces (*R*)-1-phenyl-1,2-ethanediol by stereospecific dioxygenation of the vinyl group (Lee and Gibson 1996). In a formally analogous reaction, the 4-methoxybenzoate monooxygenase from *Pseudomonas*

putida strain DSM 1868 brings about dioxygenation of 4-vinyl benzoate to 4-(1,2-dihydroxy-ethyl)benzoate (Wende et al. 1989).

3. Monooxygenation of the vinyl group with the formation of the epoxide takes place with a number of bacteria, including *Methylococcus capsulatus* strain Bath (Colby et al. 1977) and *Nitrosomonas europaea* (Keener and Arp 1994, p. 202). The degradation of styrene oxide by *Xanthobacter* sp. strain 124X produced phenylacetaldehyde by the activity of an isomerase (Hartmans et al. 1989), and in an unclassified Gram-positive organism—though not in the 124X strain—the degradation of styrene is accomplished by an FAD-dependent monooxygenase (Hartmans et al. 1990). Further metabolism of phenylacetaldehyde in *Corynebacterium* sp. strain ST-10 is carried out by an aldehyde reductase that is able to accept a wide range of aldehyde substrates (Itoh et al. 1997). In *Pseudomonas* sp. strain Y2, phenylacetaldehyde enters the lower pathway for the degradation of phenylacetate via the coenzyme A ester (del Peso-Santos et al. 2006).

4. The initial steps in the degradation of 2′-methyl-4-methoxystyrene (*trans*-anethole) by a strain of *Arthrobacter aurescens* are comparable to those already noted, except that the epoxide is hydrolyzed to a diol to produce 4-methoxybenzoate (Shimoni et al. 2002).

It is appropriate here to note that styrene is transformed by the black yeast *Exophilia jeanselmei* to phenylacetate by a pathway similar to that of the *Xanthobacter* sp. already noted. The initial monooxygenation was carried out by a cytochrome P450, and phenylacetate was further metabolized to 2-hydroxy- and 2,5-dihydroxyphenylacetate (Cox et al. 1996).

8.3.1.4 Stilbene

Substituted stilbenes are formed during the production of wood pulp, and are components of fluorescent whiting agents. The latter are considered recalcitrant although there are a few examples of their biodegradation. The degradation of stilbenes by oxidative fission of the Ar–C=C–Ar bond has been described for a lignin model compound (Habu et al. 1989b) putatively by dioxygenation (Habu et al. 1989a), and presumptively for α-methyl-4,4′-dihydroxy stilbene that is an intermediate in the degradation of bisphenol-A (Spivack et al. 1994).

8.3.1.5 Synthetic Applications

It has already been shown that biodegradation of many aromatic compounds proceeds by initial dioxygenation to *cis*-dihydrodiols, followed by dehydrogenation and ring fission. The high enantiomeric purity of *cis*-dihydrodiols produced by bacterial dioxygenases has been emphasized. Mutant strains that lack dehydrogenase activity produce only dihydrodiols, and there has been increasing interest in developing the use of these as synthons for the production of novel compounds that would not be readily available by conventional chemical synthesis.

For dihydrodiols derived from substituted benzenes, the key to their significance lies in the availability of two adjacent chiral centers with an established absolute stereochemistry. The dihydrodiol from benzene is, of course, the *meso* compound, although enantiomers produced by subsequent reaction with a chiral reagent are readily separated. There are useful reviews containing numerous applications (Carless 1992; Ribbons et al. 1989), many of which involve, in addition, the use of *cis*-fluoro-, *cis*-chloro-, or *cis*-bromobenzene-2,3-dihydrodiols.

Only a few illustrative syntheses using benzene and toluene *cis*-dihydrodiols are given below:

1. Although the product from the transformation of toluene by mutants of *Pseudomonas putida* lacking dehydrogenase activity is the *cis*-2R,3S dihydrodiol, the *cis*-2S,3R dihydrodiol has been synthesized from 4-iodotoluene by a combination of microbiological and chemical reactions. *P. putida* strain UV4 was used to prepare both enantiomers of the *cis*-dihydrodiol, and iodine was chemically removed using H_2 Pd/C. Incubation of the mixture of enantiomers with

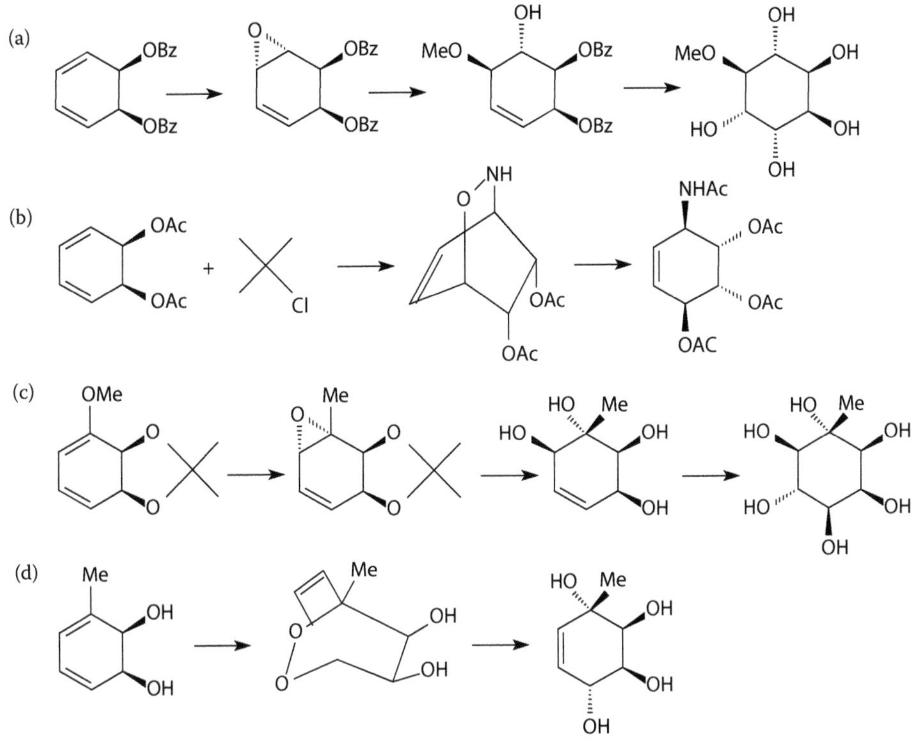

FIGURE 8.12 Examples of chemical syntheses based on cyclohexadiene *cis*-dihydrodiols (a) pinitol, (b) conduramine A, (c) (−)-laminitol, and (d) conduritol analogs.

> *P. putida* NCIMB 8859 selectively degraded the 2*R*,3*S* compound to produce toluene *cis*-2*S*,3*R* dihydrodiol (Allen et al. 1995).

2. Racemic pinitol from benzene *cis*-dihydrodiol benzoate by successive epoxidation and osmylation (Figure 8.12a) (Ley et al. 1987).

3. Conduramine A1 tetraacetate using an activated nitroso-mannose derivative (Figure 8.12b) (Werbitzky et al. 1990).

4. (−)-Laminitol from toluene *cis*-dihydrodiol by successive epoxidations (Figure 8.12c) (Carless and Oak 1991).

5. Analogs of conduritols from toluene *cis*-dihydrodiol by reaction with singlet oxygen followed by scission with thiourea (Figure 8.12d) (Carless et al. 1989).

There is clearly enormous potential using other *cis*-dihydrodiols produced from benzocycloalkenes, or from the numerous dihydrodiols produced from polycyclic carbocyclic and heterocyclic substrates.

8.4 References

Allen CCR, DR Boyd, H Dalton, ND Sharma, I Brannigan, NA Kerley, GN Sheldrake, SC Taylor. 1995. Enantioselective bacterial biotransformation routes to *cis*-diol metabolites of monosubstituted benzenes, naphthalene and benzocycloalkenes of either absolute configuration. *J Chem Soc Chem Commun* 117–118.

Altenschmidt U, G Fuchs. 1992. Anaerobic toluene oxidation to benzylalcohol and benzaldehyde in a denitrifying *Pseudomonas* sp. *J Bacteriol* 174: 4860–4862.

Ander P, K-E Eriksson, H-S Yu. 1983. Vanillic acid metabolism by *Sporotrichium pulverulentum*: Evidence for demethoxylation before ring-cleavage. *Arch Microbiol* 136: 1–6.

Anderson JJ, S Dagley. 1980. Catabolism of aromatic acids in *Trichosporon cutaneum*. *J Bacteriol* 141: 534–543.

Assinder SJ, PA Williams. 1990. The TOL plasmids: Determinants of the catabolism of toluene and the xylenes. *Adv Microb Physiol* 31: 1–69.

Axcell BC, PJ Geary. 1973. The metabolism of benzene by bacteria. Purification and some properties of the enzyme *cis*-1,2-dihydroxycyclohexa-3,5-diene (nicotinamide adenine dinucleotide) oxidoreductase (*cis*-benzene glycol dehydrogenase). *Biochem J* 136: 927–934.

Axcell BC, PJ Geary. 1975. Purification and some properties of a soluble benzene-oxidizing system from a strain of *Pseudomonas*. *Biochem J* 146: 173–183.

Beller HR. 2002. Analysis of benzylsuccinates in groundwater by liquid chromatography/tandem mass spectrometry and its use for monitoring *in situ* BTEX biodegradation. *Environ Sci Technol* 36: 2724–2728.

Beller HR, W-H Ding, M Reinhard. 1995. Byproducts of anaerobic alkylbenzene metabolism useful as indicators of *in situ* bioremediation. *Environ Sci Technol* 29: 2864–2870.

Beller HR, M Reinhard, D Grbic-Galic. 1992. Metabolic by-products of anaerobic toluene degradation by sulfate-reducing enrichment cultures. *Appl Environ Microbiol* 58: 3192–3195.

Beller HR, AM Spormann. 1997a. Anaerobic activation of toluene and *o*-xylene by addition to fumarate in denitrifying strain T. *J Bacteriol* 179: 670–676.

Beller HR, AM Spormann. 1997b. Benzylsuccinate formation as a means of anaerobic toluene activation by sulfate-reducing strain PRTOL1. *Appl Environ Microbiol* 63: 3729–3731.

Beller HR, AM Spormann. 1998. Analysis of the novel benzylsuccinate synthase reaction for anaerobic toluene activation based on structural studies of the product. *J Bacteriol* 180: 5454–5457.

Bertoni G, M Martino, E Galli, P Barbieri. 1998. Analysis of the gene cluster encoding toluene/*o*-xylene mono-oxygenase from *Pseudomonas stutzeri* OX1. *Appl Environ Microbiol* 64: 3626–3632.

Bickerdike SR, RA Holt, GM Stevens. 1997. Evidence for metabolism of *o*-xylene by simultaneous ring and methyl group oxidation in a new soil isolate. *Microbiology (UK)* 143: 2321–2319.

Biegert T, U Altenschmidt, C Eckerskorn, G Fuchs. 1995. Purification and properties of benzyl alcohol dehydrogenase from a denitrifying *Thauera* sp. *Arch Microbiol* 163: 418–423.

Biegert T, G Fuchs. 1995. Anaerobic oxidation of toluene (analogues) to benzoate (analogues) by whole cells and by cell extracts of a denitrifying *Thauera* sp. *Arch Microbiol* 163: 407–417.

Biegert T, G Fuchs, J Heider. 1996. Evidence that anaerobic oxidation of toluene in the denitrifying bacterium *Thauera aromatica* is initiated by formation of benzylsuccinate from toluene and fumarate. *Eur J Biochem* 238: 661–668.

Boyd DR, ND Sharma, R Boyle, RAS McMordie, J Chima, H Dalton. 1992. A ^{1}H NMR method for the determination of enantiomeric excess and absolute configuration of *cis*-dihydrodiol metabolites of polycyclic arenes and heteroarenes. *Tetrahedron Lett* 33: 1241–1244.

Breese K, M Boll, J Alt-Mörbe, H Schäggrer, G Fuchs. 1998. Genes encoding the benzoyl-CoA pathways of anaerobic aromatic metabolism in the bacterium *Thauera aromatica*. *Eur J Biochem* 256: 148–154.

Bruce NC, RB Cain, PH Pieper, K-H Engesser. 1989. Purification and characterization of 4-methylmucolactone methyl-isomerase, a novel enzyme of the modified 3-oxoadipate pathway in nocardioform actinomycetes. *Biochem J* 262: 303–312.

Cain RB, RF Bilton, JA Darrah. 1968. The metabolism of aromatic acids by micro-organisms. Metabolic pathways in the fungi. *Biochem J* 108: 797–828.

Cámara B, P Bielecki, F Kaminski, VM dos Santos, I Plumeier, P Nikodem, DH Pieper. 2007. A gene cluster involved in degradation of substituted salicylates via *ortho*-cleavage in *Pseudomonas* sp. strain MT1 encodes enzymes specifically adapted for transformation of 4-methylcatechol and 3-methylmuconate. *J Bacteriol* 189: 1664–1674.

Carless HAJ. 1992. The use of cyclohexa-3,5-diene-1,2-diols in enantiospecific synthesis. *Tetrahedron Asymmetry* 3: 795–826.

Carless HAJ, Billinge JR, Oak OZ. 1989. Photochemical routes from arenes to inositol intermediates: The photo-oxidation of substituted *cis*-cyclohexane-3,5-diene-1,2-diols. *Tetrahedron Lett* 30: 3113–3116.

Carless HAJ, Oak OZ. 1991. Total synthesis of (–)-laminitol (1D-4*C*-methyl-*myo*-inositol) via microbial oxidation of toluene. *Tetrahedron Lett* 32: 1671–1674.

Catelani D, A Fiecchi, E Galli. 1971. (+)-γ-carboxymethyl-γ-methyl-δ-butenolide. A 1,2-ring fission product of 4-methylcatechol by *Pseudomonas desmolyticum*. *Biochem J* 121: 89–92.

Cha C-J, RB Cain, NC Bruce. 1998. The modified β-ketoadipate pathway in *Rhodococcus rhodochrous* N75: Enzymology of 3-methylmuconolactone metabolism. *J Bacteriol* 180: 6668–6673.

Chakraborty R, JD Coates. 2005. Hydroxylation and carboxylation—Two crucial steps of anaerobic benzene degradation by *Dechloromonas* strain RCB. *Appl Environ Microbiol* 71: 5427–5432.

Chakraborty R, SM O'Connor, E Chan, JD Coates. 2005. Anaerobic degradation of benzene, toluene, ethylbenzene, and xylene compounds by *Dechloromonas* strain RCB. *Appl Environ Microbiol* 71: 8649–8655.

Coates JD, R Chakraborty, JG Lack, SM O'Connor, KA Cole, KS Bender, LA Achebach. 2001. Anaerobic benzene oxidation coupled to nitrate reduction in pure cultures by two strains of *Dechloromonas*. *Nature* 411: 1039–1043.

Colby J, DI Stirling, H Dalton. 1977. The soluble methane mono-oxygenase of *Methylococcus capsulatus* (Bath). Its ability to oxygenate *n*-alkanes, *n*-alkenes, ethers, and alicyclic, aromatic, and heterocyclic compounds. *Biochem J* 165: 395–402.

Coschigano PW, TS Wehrman, LY Young. 1998. Identification and analysis of genes involved in anaerobic toluene metabolism by strain T1: Putative role of a glycine free radical. *Appl Environ Microbiol* 64: 1650–1656.

Cox HHJ, BW Faber, VNM van Heiningen, H Radhoe, HJ Doddema, W Harder. 1996. Styrene metabolism in *Exophilia jeanselmei* and involvement of a cytochrome P-450-dependent styrene monooxygenase. *Appl Environ Microbiol* 62: 1471–1474.

Cripps RE, RJ Watkinson. 1978. Polycyclic hydrocarbons: Metabolism and environmental aspects, pp. 113–134. In *Developments in Biodegradation of Hydrocarbons-1* (Ed. RJ Watkinson), Applied Science Publishers Ltd, London.

Crutcher SE, PJ Geary. 1979. Properties of the iron-sulphur proteins of the benzene dioxygenase system from *Pseudomonas putida*. *Biochem J* 177: 393–400.

Dalton H, BT Gording, BW Watyers, R Higgins, JA Taylor. 1981. Oxidations of cyclopropane, methylcyclopropane, and arenes with the monooxygenase system from *Methylococccus capsulatus*. *J Chem Soc Chem Commun* 482–483.

Dangel W, R Brackmann, A Lack, M Mohamed, J Koch, J Oswald, B Seyfried, A Tschech, G Fuchs. 1991. Differential expression of enzyme activities initiating anoxic metabolism of various aromatic compounds via benzoyl-CoA. *Arch Microbiol* 155: 256–262.

Davey JF, DT Gibson. 1974. Bacterial metabolism of *para*- and *meta*-xylene: Oxidation of a methyl substituent. *J Bacteriol* 119: 923–929.

del Peso-Santos T, D Bartolomé-Martín, C Fernández, S Alonso, JL Gardía, E Díaz, V Shingler, J Perera. 2006. Coregulation by phenylacetyl-coenzyme A-responsive PaaX integrates control of the upper and lower pathways for catabolism of styrene by *Pseudomonas* sp. strain Y2. *J Bacteriol* 188: 4812–4821.

Eppink MHM, SA Boeren, J Vervoort, WJH van Berkel. 1997. Purification and properties of 4-hydroxybenzoate 1-hydroxylase (decarboxylating), a novel flavin adenine dinucleotide-dependent monooxygenase from *Candida parapilosis* CBS604. *J Bacteriol* 179: 6680–6687.

Erb RW, KN Timmis, DH Pieper. 1998. Characterization of a gene cluster from *Ralstonia eutropha* JMP134 encoding metabolism of 4-methylmucomolactone. *Gene* 206: 53–62.

Evans PJ, W Ling, B Goldschmidt, ER Ritter, LY Young. 1992. Metabolites formed during anaerobic transformation of toluene and *o*-xylene and their proposed relationship to the initial steps of toluene mineralization. *Appl Environ Microbiol* 58: 496–501.

Fishman A, Y Tao, TK Wood. 2004. Toluene 3-monooxygenase of *Ralstonia pickettii* PKO1 is a para-hydroxylating enzyme. *J Bacteriol* 186: 3117–3123.

George KW, J Kagle, L Junker, A Risen, AG Hay. 2011. Growth of *Pseudomonas putida* F1 on styrene requires increased catechol-2,3-dioxygenase activity, not a new hydrolase. *Microbiology (UK)* 157: 89–98.

Gescher J, A Zaar, M Mohamed, H Schägger, G Fuchs. 2002. Genes coding a new pathway of aerobic benzoate metabolism in *Azoarcus evansii*. *J Bacteriol* 184: 6301–6315.

Gibson DT, GE Cardini, FC Masales, RE Kallio. 1970. Incorporation of oxygen-18 into benzene by *Pseudomonas putida*. *Biochemistry* 9: 1631–1635.

Gibson DT, JR Koch, RE Kallio. 1968. Oxidative degradation of aromatic hydrocarbons. I. Enzymatic formation of catechol from benzene. *Biochemistry* 9: 2653–2662.

Gibson DT, V Subramanian. 1984. Microbial degradation of aromatic hydrocarbons, pp. 181–252. In *Microbial Degradation of Organic Compounds* (Ed. DT Gibson), Marcel Dekker Inc, New York.

Gray PHH, G Thornton. 1928. Soil bacteria that decompose certain aromatic compounds. *Centralbl Bakteriol Parasitenkd Infektionskr* (2 Abt) 73: 74–96.

Habu N, M Samejima, T Yoshimoto. 1989a. A novel dioxygenase responsible for the Cα-Cβ cleavage of lignin model compounds from *Pseudomonas* sp. TMY1009. *Mokuzai Gakkaishi* 35: 26–29.

Habu N, M Samejima, T Yoshimoto. 1989b. Metabolism of a diarylpropane type lignin model by *Pseudomonas* sp. TMY 1009. *Mokuzai Gakkaishi* 35: 348–355.

Harms G, K Zengle, R Rabus, F Aeckersberg, D Minz, R Rosselló-Mora, F Widdel. 1999. Anaerobic oxidation of *o*-xylene, *m*-xylene, and homologous alkylbenzenes by new types of sulfate-reducing bacteria. *Appl Environ Microbiol* 65: 999–1004.

Hartmans S, JP Smits, MJ van der Werf, F Volkering, JAM de Bont. 1989. Metabolism of styrene oxide and 2-phenylethanol in the styrene-degrading *Xanthobacter* strain 124X. *Appl Environ Microbiol* 55: 2850–2855.

Hartmans S, MJ van der Werf, JAM de Bont. 1990. Bacterial degradation of styrene involving a novel flavin adenine dinucleotide-dependent styrene monooxygenase. *Appl Environ Microbiol* 41: 1045–1054.

Harwood CS, G Burchardt, H Herrmann, G Fuchs. 1999. Anaerobic metabolism of aromatic compounds via the benzoyl-CoA pathway. *FEMS Microbiol Rev* 22: 439–458.

Hirsch W, H Schägger, G Fuchs. 1998. Phenylglyoxalate: NAD$^+$ oxidoreductase (CoA benzoylating), a new enzyme of anaerobic phenylalanine metabolism in the denitrifying bacterium *Azoarcus evansii*. *Eur J Biochem* 251: 907–915.

Högn T, L Jaenicke. 1972. Benzene metabolism of *Moraxella* species. *Eur J Biochem* 30: 369–375.

Holmes DE, C Risso, JA Smith, DR Lovley. 2011. Anaerobic oxidation of benzene by the hyperthermophilic archaeon *Ferroglobus placidus*. *Appl Environ Microbiol* 77: 5926–5933.

Hopper DJ. 1978. Microbial degradation of aromatic hydrocarbons, pp. 85–112. In *Developments in Biodegradation of Hydrocarbons-1* (Ed. RJ Watkinson), Applied Science Publishers Ltd, London.

Itoh N, R Morihama, J Wang, K Okada, N Mizuguchi. 1997. Purification and characterization of phenylacetaldehyde reductase from a styrene-assimilating *Corynebacterium* strain, ST10. *Appl Environ Microbiol* 63: 3783–3788.

Jahn MK, SB Haderlin, RU Meckenstock. 2005. Anaerobic degradation of benzene, toluene, ethylbenzene, and *o*-xylene in sediment-free iron-reducing enrichment cultures. *Appl Environ Microbiol* 71: 3355–3358.

Jigami Y, Y Kawasaki, T Omori, Y Minoda. 1979. Coexistence of different pathways in the metabolism of *n*-propylbenzene by *Pseudomonas* sp. *Appl Environ Microbiol* 38: 783–788.

Jobst B, K Schühle, U Linne, J Heider. 2010. ATP-dependent carboxylation of acetophenone by a novel type of carboxylase. *J Bacteriol* 192: 1387–1394.

Johnson HA, DA Pelletier, AM Spormann. 2001. Isolation and characterization of anaerobic ethylbenzene dehydrogenase, a novel Mo-Fe-S enzyme. *J Bacteriol* 183: 4536–4542.

Kasai Y, Y Takahata, M Manefield, K Watanabe. 2006. RNA-based stable isotope probing and isolation of anaerobic benzene-degrading bacteria from gasoline-contaminated groundwater. *Appl Environ Microbiol* 72: 3586–3592.

Kaschabek SR, W Reineke. 1995. Maleylacetate reductase of *Pseudomonas* sp. strain B13: Specificity of substrate conversion and halide elimination. *J Bacteriol* 177: 320–325.

Kawaguchi K, Y Shinoda, H Yurimoto, Y Sakai, N Kato. 2006. Purification and characterization of benzoate-CoA ligase from *Magnetospirillum* sp. strain TS-6 capable of aerobic and anaerobic degradation of aromatic compounds. *FEMS Microbiol Lett* 257: 208–213.

Keener WK, DJ Arp. 1994. Transformations of aromatic compounds by *Nitrosomonas europaea*. *Appl Environ Microbiol* 60: 1914–1932.

Keil H, PA Williams. 1985. A new class of TOL plasmid deletion mutants in *Pseudomonas putida* MT15 and their reversion by tandem gene amplification. *J Gen Microbiol* 131: 1023–1033.

Kim D, J-C Chae, GJ Zylstra, Y-S Kim, MH Nam, YM Kim, E Kim. 2004. Identification of a novel dioxygenase involved in metabolism of *o*-xylene, toluene, and ethylbenzene by *Rhodococcus* sp. strain DK17. *Appl Environ Microbiol* 70: 7086–7092.

Kim JM, NT Le, BS Chung, JH Park, J-W Bae, EL Madsen, CO Jeon. 2008. Influence of soil components on the biodegradation of benzene, toluene, ethylbenzene, and *o*-, *m*-, and *p*-xylenes by the newly isolated bacterium *Pseudomxanthomonas spadix* BD-a59. *Appl Environ Microbiol* 74: 7313–7320.

Kloer DP, C Hagel, J Heider, GE Schulz. 2006. Crystal structure of ethylbenzene dehydrogenase from *Aromatoleum aromaticum*. *Structure* 14: 1377–1388.

Kniemeyer O, T Fischer, H Wilkes, FO Glöckner, F Widdel. 2003. Anaerobic degradation of ethylbenzene by a new type of marine sulfate-reducing bacterium. *Appl Environ Microbiol* 69: 760–768.

Kube M, J Heider, J Amann, P Hufnagel, S Kühner, A Beck, R Reinhardt, R Rabus. 2004. Genes involved in the anaerobic degradation of toluene in a denitrifying bacterium, strain EbN1. *Arch Microbiol* 181: 182–194.

Kühner S, L Wöhlbrand, I Fritz, W Wruck, C Hultschig, P Hufnagel, M Kube, R Reinhardt, R Rabus. 2005. Substrate-dependent regulation of anaerobic degradation pathways for toluene and ethylbenzene in a denitrifying bacterium strain EbN1. *J Bacteriol* 187: 1493–1503.

Kunz DA, PJ Chapman. 1981. Isolation and characterization of spontaneously occurring TOL plasmid mutants of *Pseudomonas putida* HS1. *J Bacteriol* 146: 952–964.

Laban NA, D Selesi, T Rattei, P Tischler, RU Meckenstock. 2010. Identification of enzymes involved in anaerobic benzene degradation by a strictly iron-reducing enrichment culture. *Environ Microbiol* 12: 2783–2796.

Lee K, DT Gibson. 1996. Stereospecific dihydroxylation of the styrene vinyl group by purified naphthalene dioxygenase from *Pseudomonas* sp. strain NCIB 9816-4. *J Bacteriol* 178: 3353–3356.

Leuthner B, C Leutwein, H Schulz, P Hörth, W Haehnel, E Schiltz, H Schägger, J Heider. 1998. Biochemical and genetic characterisation of benzylsuccinate synthase from *Thauera aromatica*: A new glycyl-radical catalysing the first step in anaerobic toluene degradation. *Mol Microbiol* 28: 515–628.

Leutwein C, J Heider. 2001. Succinyl-CoA:(R)-benzylsuccinate CoA-transferase: An enzyme of the anaerobic toluene catabolic pathway in denitrifying bacteria. *J Bacteriol* 183: 4288–4295.

Ley SV, F Sternfield, S Taylor. 1987. Microbial oxidation in synthesis: A six step preparation of (±)-pinitol from benzene. *Tetrahedron Lett* 28: 225–226.

Mancini SA, AC Ulrich, G Lacrampe-Couloume, B Sleep, EA Edwards, BS Lollar. 2003. Carbon and hydrogen isotopic fractionation during anaerobic biodegradation of benzene. *Appl Environ Microbiol* 69: 191–198.

Marín M, D Pérez-Pantoja, R Donoso, V Wray, B González, DH Pieper. 2010. Modified 3-oxoadipate pathway for the biodegradation of methylaromatics in *Pseudomonas reinekei* MT1. *J Bacteriol* 192: 1543–1552.

Migaud ME, JC Chee-Sandford, JM Tiedje, JW Frost. 1996. Benzylfumaric, benzylmaleic, and Z- and E-phenylitaconic acids: Synthesis, characterization, and correlation with a metabolite generated by *Azoarcus tolulyticus* Tol-4 during anaerobic toluene degradation. *Appl Environ Microbiol* 62: 974–978.

Morasch B, B Schink, CC Tebbe, RU Meckenstock. 2004. Degradation of *o*-xylene and *m*-xylene by a novel sulfate-reducer belonging to the genus *Desulfotomaculum*. *Arch Microbiol* 181: 407–417.

Müller JA, AS Galuschko, A Kappler, B Schink. 2001. Initiation of anaerobic degradation of *p*-cresol by formation of 4-hydroxybenzylsuccinate in *Desulfitobacterium cetonicum*. *J Bacteriol* 183: 752–757.

Murray K, CJ Duggleby, J Sala-Trepat, PA Williams. 1972. The metabolism of benzoate and methylbenzoates via the *meta*-cleavage pathway by *Pseudomonas arvilla* mt-2. *Eur J Biochem* 28: 301–310.

Musat F, F Widdel. 2008. Anaerobic degradation of benzene by a marine sulphate-reducing enrichment culture and cell hybridization of the dominant phylotype. *Environ Microbiol* 10: 10–19.

Neilson AH, A-S Allard. 1998. Microbial metabolism of PAHs and heteroarenes. *Handb Environ Chem* 3J: 1–80.

Oka AR, CD Phelps, L McGuinness, A Mumford, LY Young, LJ Kerkhof. 2008. Identification of critical members in a sulfidogenic benzene-degrading consortium by DNA stable isotope probing. *Appl Environ Microbiol* 74: 6476–6480.

Parales RE, CS Harwood. 1993. Regulation of the *pcaIJ* genes for aromatic acid degradation in *Pseudomonas putida*. *J Bacteriol* 175: 5829–5838.

Pieper DH, K-H Engesser, RH Don, KN Timmis, H-J Knackmuss. 1985. Modified *ortho*-cleavage pathway in *Alcaligenes eutrophus* JMP 134 for the degradation of 4-methylcatechol. *FEMS Microbiol Lett* 29: 63–67.

Pieper DH, K Stadler-Fritzsche, H-J Knackmuss, K-H Engesser, NC Bruce, RB Cain. 1990. Purification and characterization of 4-methylmucolactone methyl-isomerase, a novel enzyme of the modified 3-oxoadipate pathway in the Gram-negative bacterium *Alcaligenes eutrophus* JMP 134. *Biochem J* 271: 529–534.

Powlowski JB, S Dagley. 1985. β-ketoadipate pathway in *Trichosporom cutaneum* modified for methyl-substituted metabolites. *J Bacteriol* 163: 1126–1135.

Powlowski JB, J Ingebrand, S Dagley. 1985. Enzymology of the β-ketoadipate pathway in *Trichosporon cutaneum*. *J Bacteriol* 163: 1135–1141.

Prenafeta-Boldú FX, A Kuhn, DMAM Luykx, H Anke, JW van Groenstijn, JAM de Bont. 2001. Isolation and characterization of fungi growing on volatile aromatic hydrocarbons as their sole carbon and energy source. *Mycol Res* 105: 477–484.

Rabus R, J Heider. 1998. Initial reactions of anaerobic metabolism of alkylbenzenes in denitrifying and sulfate-reducing bacteria. *Arch Microbiol* 170: 377–384.

Rabus R, M Kube, A Beck, F Widdel, R Reinhardt. 2002. Genes involved in the anaerobic degradation of ethylbenzene in a denitrifying bacterium, strain EbN1. *Arch Microbiol* 178: 506–516.

Ribbons DW, RW Eaton. 1982. Chemical transformations of aromatic hydrocarbons that support the growth of microorganisms, pp. 59–84. In *Biodegradation and Detoxification of Environmental Pollutants* (Ed. AM Chakrabarty), CRC Press, Boca Raton, Florida.

Ribbons DW, SJC Taylor, CT Evans, SD Thomas, JT Rossiter, DA Widdowson, DJ Williams. 1989. Biodegradations yield novel intermediates for chemical synthesis, pp. 213–245. In *Biotechnology and Biodegradation* (Eds. D Kamely, A Chakrabarty, GS Omenn), Gulf Publishing Company, Houston, Texas.

Rogers JE, DT Gibson. 1977. Purification and properties of *cis*-toluene dihydrodiol dehydrogenase from *Pseudomonas putida*. *J Bacteriol* 130: 1117–1124.

Rothmel, RK, DL Shinbarger, MR Parsek, TL Aldrich, AM Chakrabarty. 1991. Functional analysis of the *Pseudomonas putida* regulatory protein CatR: Transcriptional studies and determination of the CatR DNA-binding site by hydroxyl-radical footprinting. *J Bacteriol* 173: 4717–4724.

Rudolphi A, A Tschech, G Fuchs G. 1991. Anaerobic degradation of cresols by denitrifying bacteria. *Arch Microbiol* 155: 238–248.

Sala-Trepat JM, WC Evans. 1971. The meta cleavage of catechol by *Azotobacter* species: 4-Oxalocrotonate pathway. *Eur J Biochem* 20: 400–413.

Sariaslani FS, MK Trower, SE Buchloz. 1989. Xenobiotic transformations by *Streptomyces griseus*. *Dev Ind Microbiol* 30: 161–171.

Schneider S, G Fuchs. 1998. Phenylacetyl-CoA: Acceptor oxidoreductase, a new α-oxidizing enzyme that produces phenylglyoxylate. Assay, membrane localization, and differential production in *Thauera aromatica*. *Arch Microbiol* 169: 509–516.

Schneider S, MEl-Said Mohamed, G Fuchs. 1997. Anaerobic metabolism of l-phenylalanine via benzoyl-CoA in the denitrifying bacterium *Thauera aromatica*. *Arch Microbiol* 168: 310–320.

Schocher RJ, B Seyfried, F Vazquez, J Zeyer. 1991. Anaerobic degradation of toluene by pure cultures of denitrifying bacteria. *Arch Microbiol* 157: 7–12.

Schraa G, BM Bethe, ARW Neerven, WJJ Tweel, E Wende, AJB Zehnder. 1987. Degradation of 1,2-dimethyl-benzene by *Corynebacterium* strain C125. *Antonie van Leeuwenhoek* 53: 159–170.

Schühle K, G Fuchs. 2004. Phenylphosphate carboxylase: A new C–C lyase involved in anaerobic phenol metabolism in *Thauera aromatica*. *J Bacteriol* 186: 4556–4567.

Schühle K, J Gescher, U Feil, M Paul, M Jahn, H Schagger, G Fuchs. 2003. Benzoate-coenzyme A ligase from *Thauera aromatica*: An enzyme acting in anaerobic and aerobic pathways. *J Bacteriol* 185: 4920–4929.

Seibert V, K Stadler-Fritsche, M Schlomann. 1993. Purification and characterization of maleylacetate reductase from *Alcaligenes eutrophus* JMP134 (pJP4). *J Bacteriol* 175: 6745–6754.

Selesi D, N Jehmlich, M von Bergen, F Schmidt, T Rattei, P Tischler, T Lueders, RU Meckenstock. 2010. Combined genomic and proteomic approaches identify gene clusters involved in anaerobic 2-methylnaphthalene degradation in the sulfate-reducing enrichment culture N47. *J Bacteriol* 192: 295–306.

Selifonov SA, M Grifoll, RW Eaton, PJ Chapman PJ. 1996. Oxidation of naphthenoaromatic and methyl-substituted aromatic compounds by naphthalene 1,2- dioxygenase. *Appl Environ Microbiol* 62: 507–514.

Shields MS, MJ Reagin, RR Gerger, R Campbell, C Somerville. 1995. TOM, a new aromatic degradative plasmid from *Burkholderia* (*Pseudomonas*) *cepacia* G4. *Appl Environ* Microbiol 61: 1352–1356.

Shimoni E, T Baasov, U Ravid, Y Shoham. 2002. The *trans*-anethole degradation pathway in an *Arthrobacter* sp. *J Biol Chem* 277: 11866–11872.

Shinoda Y, Y Sakai, H Uenishi, Y Uchihashi, A Hiraishi, H Yukawa, H Yurimoto, N Kato. 2004. Aerobic and anaerobic toluene degradation by a newly isolated denitrifying bacterium *Thauera* sp. strain DNT-1. *Appl Environ Microbiol* 70: 1385–1392.

Smith MR. 1994. The physiology of aromatic hydrocarbon degrading bacteria, pp. 347–378. In *Biochemistry of Microbial Degradation* (Ed. C Ratledge), Kluwer Academic Publishers, Dordrecht, The Netherlands.

Smith RV, JP Rosazza. 1983. Microbial models of mammalian metabolism. *J Nat Prod* 46: 79–91.

Söhngen NL. 1913. Benzin, Petroleum, Paraffinöl und Paraffin als Kohlenstoff- und Energiequelle für Mikroben. *Centralbl Bakteriol Parasitenkd Infektionskr* (2 *Abt*) 37: 595–609.

Spivack J, TK Leib, JH Lobos. 1994. Novel pathway for bacterial metabolism of bisphenol A. Rearrangements and stilbene cleavage in bisphenol A metabolism. *J Biol Chem* 269: 7323–7329.

Tan H-M, JR Mason. 1990. Cloning and expression of the plasmid-encoded benzene dioxygenase from *Pseudomonas putida* ML2. *FEMS Microbiol Lett* 72: 259–264.

Tao Y, A Fishman, WE Bentley, TK Wood. 2004. Oxidation of benzene to phenol, catechol, and 1,2,3-trihydroxybenzene by toluene 4-monooxygenase of *Pseudomonas mendocina* KR 1 and toluene 3-monooxygenase of *Ralstonia pickettii* PKO1. *Appl Environ Microbiol* 70: 3814–3820.

Tausson WO. 1927. Naphthalin als Kohlenstoffquelle für Bakterien. *Planta* 4: 214–256.

Ulrich AC, HR Beller, EA Edwards. 2005. Metabolites detected during biodegradation of $^{13}C_6$-benzene in nitrate-reducing and methanogenic enrichment cultures. *Environ Sci Technol* 39: 6681–6691.

van der Meer, WM de Vos, S Harayama, AJB Zehnder. 1992. Molecular mechanisms of genetic adaptation to xenobiotic compounds. *Microbiol Revs* 56: 677–694.

Verfürth K, AJ Pierik, C Leutwein, S Zorn, J Heider. 2004. Substrate specificities and electron paramagnetic resonance properties of benzylsuccinate synthesis in anaerobic toluene and *m*-xylene metabolism. *Arch Microbiol* 181: 155–162.

Wagner AFV, M Frey, FA Neugebauer, W Schäfer, J Knappe. 1992. The free radical in pyruvate formate-lyase is located on glycine-734. *Proc Natl Acad USA* 89: 996–1000.

Warhurst AM, KE Clarke, RA Hill, RA Holt, CA Fewson. 1994. Metabolism of styrene by *Rhodococcus rhodochrous* NCIMB 13259. *Appl Environ Microbiol* 60: 1137–1145.

Warhurst AM, CA Fewson. 1994. Microbial metabolism and biotransformations of styrene. *J Appl Bacteriol* 77: 597–606.

Weber FJ, KC Hage, JAM de Bont. 1995. Growth of the fungus *Cladosporium sphaerospermum* with toluene as the sole carbon and energy source. *Appl Environ Microbiol* 61: 3562–3566.

Wende P, F-H Bernhardt, K Pfleger. 1989. Substrate-modulated reactions of putidamonooxin. The nature of the active oxygen species formed and its reaction mechanism. *Eur J Biochem* 181: 189–197.

Werbitzky O, K Klier, H Felber H. 1990. Asymmetric induction of four chiral cenbters by hetero Diels–Alder reaction of a chiral nitrosodienmophile. *Liebigs Ann Chem* 1990: 267–270.

Whited GM, DT Gibson. 1991a. Separation and partial characterization of the enzymes of the toluene-4-monooxygenase catabolic pathway in *Pseudomonas mendocina* KR1. *J Bacteriol* 173: 3017–3020.

Whited GM, DT Gibson. 1991b. Toluene-4-monooxygenase, a three-component enzyme system that catalyzes the oxidation of toluene to *p*-cresol in *Pseudomonas mendocina* KR1. *J Bacteriol* 173: 3010–3016.

Williams PA, MJ Worsey. 1976. Ubiquity of plasmids in coding for toluene and xylene metabolism in soil bacteria: Evidence for the existence of new TOL plasmids. *J Bacteriol* 125: 818–828.

Yadav JS, CA Reddy. 1993. Degradation of benzene, toluene, ethylbenzene, and xylenes (BTEX) by the lignin-degrading basidiomycete *Phanerochaete chrysosporium*. *Appl Environ Microbiol* 59: 756–762.

Ziffer H, K Kabuta, DT Gibson, VM Kobal, DM Jerina. 1977. The absolute stereochemistry of several *cis* dihydrodiols microbially produced from substituted benzenes. *Tetrahedron* 33: 2491–2496.

Zylstra GJ, DT Gibson. 1989. Toluene degradation by *Pseudomonas putida* F1. Nucleotide sequence of the *todC1C2BADE* genes and their expression in *Escherichia coli*. *J Biol Chem* 264: 14940–14946.

Part 2: Polycyclic Aromatic Hydrocarbons

8.5 Introduction

Aromatic hydrocarbons are components of petroleum and are produced during the gasification of coal, while naturally occurring partially aromatized steroids and terpenoids which occur in sediments and coal have been used as biomarkers (Simoneit 1998). Concern with polycyclic aromatic hydrocarbons has centered on those that display carcinogenicity to mammals after metabolic activation. Since anoxic sediments may be a sink for PAHs, degradation and biodegradation have been examined under both aerobic and anaerobic conditions. Biochemical aspects of the biodegradation of PAHs have been given in a classic review (Gibson and Subramanian 1984), and by Smith (1994), Kanaly and Harayama (2000), and in a short review in the context of bioremediation by Gibson and Parales (2000). Specific aspects of sphingomonads have been presented (Pinyakong et al. 2003; Ferraro et al. 2007). The biodegradation of benzo[*a*]pyrene by bacteria, fungi, and algae has been addressed in a review within the context of remediation (Juhasz and Naidu 2000). Attention is directed to marine organisms that are able of degrading PAHs, and which frequently belong to the genus *Cycloclasticus* (Wang et al. 1996; Geiselbrecht et al. 1998; Kasai et al. 2003), and *Neptunomonas* (Hedlund et al. 1999).

The aerobic degradation of naphthalene was described many years ago (Tausson 1927; Gray and Thornton 1928). As for the degradation of benzene, it was, however, many years later before details of the reaction involved were elucidated. Since then, the degradation of naphthalene, anthracene, phenanthrene, fluoranthene, and pyrene has been described in a number of organisms, that of benzo[*a*] pyrene, benz[*a*]anthracene, and dibenz[*a,h*]anthracene less often, and that of chrysene (Demanèche et al. 2004; Willison 2004) and coronene (Juhasz et al. 1997) only occasionally. It should be realized that growth at the expense of PAHs may require only a contribution from the products of partial degradation, and caution should, therefore, be exercised in the interpretation of experiments using partially labeled substrates (Ye et al. 1996). For example, evolution of $^{14}CO_2$ from ^{14}C-labeled PAHs was reported for a strain of *Acidovorax* that had been enriched with phenanthrene (Singleton et al. 2009). Rather low rates were observed for [U^{14}C] naphthalene, [9-^{14}C] phenanthrene, [5,6,11,12-^{14}C] chrysene, [5,6-^{14}C]benz[*a*]anthacene, and [7-^{14}C] benzo[*a*]pyrene, whereas for [4,5,9,10-^{14}C]pyrene, [3-^{14}C]fluoranthene rates were not significantly above background. Total degradation could not, therefore, be determined. In addition, increases in the expression of the *phnAc* and *phnB* genes in response to these PAHs were found only for naphthalene and phenanthrene. These observations clearly merit careful analysis.

Aerobic degradation is generally initiated by dioxygenation followed by dehydrogenation to catechols that undergo ring fission. For polycyclic arenes, repetition of these reactions is then used for the successive degradation of the other rings, for example, *o*-phthalate is produced from anthracene (van Herwijnen et al. 2003), phenanthrene (Moody et al. 2001), and pyrene (Vila et al. 2001; Krivobok et al. 2003). Exceptionally, elimination rather than dehydrogenation takes place. For example, the dioxygenation of 7,4′-dihydroxyisoflavone by biphenyl-2,3-dioxygenase from *Burkholderia* sp. strain LB400 is followed by elimination to the phenol 7,2′,4′ – trihydroxyisoflavone rather than by dehydrogenation (Seeger et al. 2003).

Substantial attention has been directed to details of the degradation of naphthalene on account of interest in the synthesis of chiral compounds using naphthalene-1,2-dioxygenase, and of biphenyl, since biphenyl-2,3-dioxygenase induced by growth with biphenyl has a broad substrate range, including congeners of PCBs. Genes for the degradation of naphthalene may be carried on plasmids, for example, in pseudomonads (Dunn and Gunsalus 1973; Dunn et al. 1980; Yen et al. 1988), *Novosphingobium aromaticivorans* (Romine et al. 1999), and some species of *Rhodococcus* (Kulakov et al. 2005).

8.6 Aerobic Reactions Carried Out by Bacteria

8.6.1 Naphthalene

The aerobic degradation of naphthalene and its derivatives has been extensively examined, so that the pathway, biochemistry, and genetics are well established. Naphthalene is readily degraded by many bacteria, including Gram-negative pseudomonads, Gram-positive rhodococci, and marine bacteria belonging to the genus *Cycloclasticus* (Geiselbrecht et al. 1998) and *Neptunomonas naphthovorans* (Hedlund et al. 1999).

Both the details of the initial steps and their enzymology have been elucidated. Genes for the degradation of naphthalene and the intermediate metabolite salicylate are generally carried on plasmids in pseudomonads (Dunn and Gunsalus 1973; Dunn et al. 1980; Yen et al. 1988), and in some species of *Rhodococcus* (Kulakov et al. 2005). The overall pathway of degradation is shown (Figure 8.13) and is carried out by a sequence of enzymes. The enzymes for the complete sequence of enzymes involved in

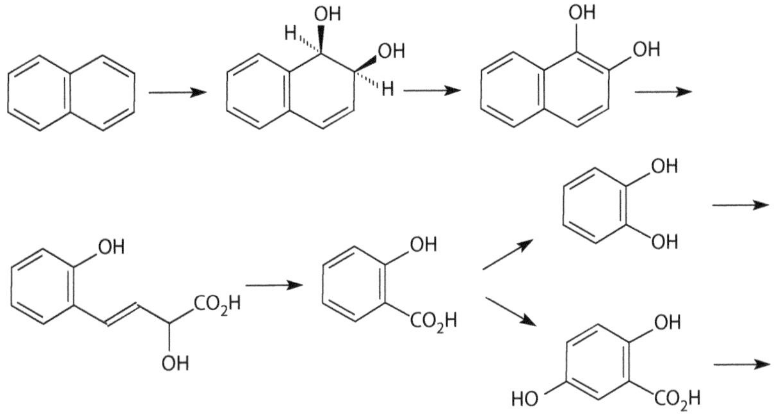

FIGURE 8.13 Degradation of naphthalene.

the oxidative degradation of naphthalene to catechol and for the extradiol (*meta*) fission pathway for the degradation of the catechol are inducible by growth with salicylate in *Pseudomonas putida* (Austen and Dunn 1980). The genes for degradation in *Pseudomonas putida* G7 are organized in two operons, the upper encoding those for the conversion of naphthalene into salicylate, and the lower those for the degradation of salicylate to pyruvate and acetate.

The *nah* genes encoding the enzymes on the plasmid NAH7 for conversion of naphthalene into salicylate have been cloned from *Pseudomonas putida* G1064: *nahA, nahB, nahC, nahD, nahE*, and *nahF* for naphthalene dioxygenase, naphthalene *cis*-dihydrodiol dehydrogenase, 1,2-dihydroxynaphthalene dioxygenase, 2-hydroxychromene-2-carboxylate isomerase, and salicylaldehyde dehydrogenase, respectively (Yen and Serdar 1988). The initial product of naphthalene oxidation in *Pseudomonas putida* by naphthalene dioxygenase is naphthalene (+)-*cis*-(1*R*,2*S*)-1,2-dihydro-1,2-diol (Jeffrey et al. 1975). Naphthalene dioxygenase (Patel and Barnsley 1980; Ensley and Gibson 1983), *cis*-naphthalene dihydrodiol dehydrogenase (Patel and Gibson 1974), and 1,2-dihydroxynaphthalene dioxygenase (Kuhm et al. 1991) have been purified, while further details of the subsequent steps have been added (Eaton and Chapman 1992).

The degradation of naphthalene has also been studied in strains of *Rhodococcus* sp., and two groups of strains have been examined. Differences from pseudomonads have been observed in strain NCIMB 12038, in which the genes are carried on a plasmid: (a) naphthalene is the sole inducer of the whole pathway and (b) the degradation of salicylate proceeds from the activity of salicylate 5-hydroxylase to 2,5-dihydroxybenzoate (gentisate) (Grund et al. 1992; Larkin et al. 1999). In other strains, including strain P200, the genes are chromosomal (Kulakov et al. 2005).

Further details merit brief comment:

1. In some strains of pseudomonads, the degradation of the intermediate catechol produced by the activity of salicylate hydroxylase may proceed either by the extradiol or by the alternative intradiol fission pathway (Barnsley 1976).

2. In *Ralstonia* (*Pseudomonas*) sp. strain U2 (Fuenmayor et al. 1998; Zhou et al. 2002), the alternative gentisate pathway for the degradation of the intermediate salicylate is used.

3. In a strain of *Mycobacterium* sp., both dioxygenation to *cis*-naphthalene-1,2-dihydrodiol and monooxygenation and hydrolysis to *trans*-naphthalene-1,2-dihydrodiol were encountered, with the latter dominating 25-fold (Kelley et al. 1990).

4. Naphthalene dioxygenase from *Pseudomonas* sp. strain 9816-4, that is, the paradigm arene dioxygenase, additionally carried out enantiomeric monooxygenation of indan and dehydrogenation of indene (Gibson et al. 1995), and the stereospecific hydroxylation of (*R*)-1-indanol, (*S*)-1-indanol to *cis*-indan-1,3-diol, and *trans* (1*S*,3*S*)-indane-1,3-diol (Lee et al. 1997); the indantriols

are also formed by further reactions. Essentially, comparable reactions have been observed with *Rhodococcus* sp. strain NCIMB 12038 (Allen et al. 1997).

5. *Novophingobium aromaticivorans* carries catabolic genes for the degradation of aromatic hydrocarbons on a plasmid. It is able to degrade not only naphthalene and biphenyl by dioxygenation but also initiate the degradation of xylenes by monooxygenation of the methyl groups, followed by dioxygenation of the resulting benzoates (Romine et al. 1999).

Naphthalene dioxygenase from *Pseudomonas putida* consists of three proteins: the reductase$_{NAP}$ (Haigler and Gibson 1990b), a ferredoxin$_{NAP}$ (Haigler and Gibson 1990a), and the two-subunit terminal oxygenase ISP$_{NAP}$ (Ensley and Gibson 1983). They are analogous to those involved in the degradation of benzene and toluene, and are encoded by the genes *nadAa*, *nadAb*, and *nadAc/nadAd* (Simon et al. 1993). However, the reductase$_{NAD}$ component of naphthalene dioxygenase contains both FAD and a [2Fe–2S] ferredoxin (Haigler and Gibson 1990a), so that naphthalene is unusual in requiring three redox centers to transfer electrons from NADH to the terminal dioxygenase.

Although naphthalene dioxygenase is enantiomer-specific in producing the (+)-*cis*-(1*R*,2*S*)-dihydrodiol, it also possesses dehydrogenase and monooxygenase activities (Gibson et al. 1995), so that monooxygenase and dioxygenase activities are not exclusive. Exclusive monooxygenation has been observed; for example, the transformation of naphthalene by *Bacillus cereus* yielded naphth-1-ol with retention of the deuterium after incubation with 1[^{2}H]- and 2-[^{2}H]-naphthalene, which is consistent with involvement of an arene oxide and an NIH shift (Cerniglia et al. 1984).

8.6.2 Alkylated Naphthalenes

Marine strains of *Cycloclasticus* were able to use 1- and 2-methyl- and 2,6-dimethylnaphthalene as well as naphthalene and phenanthrene for growth (Geiselbrecht et al. 1998). Methylnaphthalenes are important components of crude oils, and their degradation follows the initial stages used for alkylated benzenes. The degradation of 2,6-dimethylnaphthalene by flavobacteria involves a pathway analogous to that for dimethylbenzenes—successive oxidation of one methyl group to carboxylate, decarboxylative dioxygenation to 1,2-dihydroxy-6-methylnaphthalene, ring fission to 5-methylsalicylate, followed by further degradation by pathways established for naphthalene itself (Barnsley 1988). Degradation of both 1-methyl- and 2-methylnaphthalene was initiated by 7,8-dioxygenation (Mahajan et al. 1994). Whereas 1-methylnaphthalene produced only 3-methylcatechol, 2-methylnaphthalene produced both 4-hydroxymethylcatechol and 4-methylcatechol. The catechols then underwent extradiol fission. The oxidation of 1,5-, 2,6-, 2,7-, and 1,8-dimethylnaphthalene by a recombinant strain of *Pseudomonas aeroginosa* PAO1 involved successive oxidation of only a single methyl group to the monocarboxylates, except for 1,8-dimethylnaphthalene in which both methyl groups were oxidized to the dicarboxylate (Selifonov et al. 1996). The oxidation of a wide range of dimethyl naphthalenes has been examined in *Sphingomonas paucimobilis* strain 2322 (Dutta et al. 1998). Degradative pathways involved successive oxidation to the corresponding carboxylate of a methyl group in one of the rings. Further degradation involved the following alternative pathways:

1. Decarboxylative dioxygenation to a 1,2-dihydroxynaphthalene followed by 1:1a extradiol fission and formation of salicylate
2. Hydroxylation to a 1-hydroxynaphthalene-2-carboxylate and 1:2-intradiol ring fission with the formation of *o*-phthalate

Although tetralin (1,2,3,4-tetrahydronaphthalene) could be considered an alkylated naphthalene, its degradation was initiated by oxidation of the aromatic ring. *Corynebacterium* sp. strain C125 that had been isolated by enrichment with *o*-xylene carried out successive dioxygenation and dehydrogenation to 1,2-dihydroxy-5,6,7,8-tetrahydronaphthalene (Sikkema and de Bont 1993). This was followed by ring fission to a vinylogous 1,3-dioxocarboxylate that was hydrolyzed to deca-2-oxo-3-ene-1,10-dioate followed by hydration and fission to pyruvate and pimelic semialdehyde (Figure 3.32) (Hernáez et al. 2002).

8.6.3 Naphthols and Naphthalene Carboxylates

Less attention has been devoted to the degradation of naphthols and naphthalene carboxylates:

1. a. The degradation of naphth-1-ol has been described (Bollag et al. 1975), and the specificity of oxygen uptake in whole cells—though not in cell-free extracts—has been studied (Larkin 1988). Two strains, *Pseudomonas* sp. strain 12043 and *Rhodococcus* sp. strain NCIB 12038, were less versatile, whereas *Pseudomonas* sp. strain 12042 was more versatile, showing oxygen uptake with both 1- and 2-naphthols, 1,5- and 2,7-dihydroxynaphthalene, and naphthalene itself. The subsequent degradation of the 1,2-dihydroxynaphthalene formed from the naphthols may plausibly be assumed to proceed by ring fission and subsequently established reactions that produce salicylate and then catechol (strain 12042) or gentisate (strains 12043 and 12038).

   b. Naphthalene-1-hydroxy-2-carboxylate is an intermediate in the degradation of phenanthrene, and it has been proposed that as a β-oxoacid, it is decarboxylated to naphth-1-ol followed by pathways leading to salicylate or *o*-phthalate (Samanta et al. 1999).

   c. Naphthalene 1,8-dicarboxylate and naphthalene 1-carboxylate are intermediates in the degradation of acenaphthylene by *Rhizobium* sp. strain CU-A1. They are formed from acenaphthenequinone, and then degraded to salicylate and 2,5-dihydroxybenzoate (Poonthrigpun et al. 2006).

2. Although the degradation of naphthalene-2-carboxylate by *Burkholderia* sp. strain JT 1500 involves the formation of 1-hydroxy naphthalene-2-carboxylate, this is not formed from the expected (1R,2S)-*cis*-1,2-dihydrodiol-2-naphthoate. Possibly, therefore, the reaction is carried out by a monooxygenase, or a dehydration step is involved. Subsequent reactions produced pyruvate and *o*-phthalate that was degraded via 4,5-dihydroxyphthalate (Morawski et al. 1997). Degradation of naphthalene carboxylates formed by oxidation of methyl groups has already been noted.

8.6.4 Biphenyl

An organism strain B1 that could grow and degrade biphenyl was isolated, and it was shown that oxidation produced *cis*-2,3-dihydroxy-1-phenylcyclohexa-4,6-diene (Gibson et al. 1973). The organism was described as a species of *Beijerinckia* although it is now known as *Sphingobium yanoikuyae* strain B1 (Gibson 1999). The degradation of biphenyl has been examined in considerable detail on account of interest in the degradation of PCBs, since virtually all bacteria that degrade PCBs have been isolated after enrichment with biphenyl (Pellizari et al. 1996). Although the degradative pathways of naphthalene and biphenyl are broadly similar, dioxygenation of chlorinated biphenyls by *Pseudomonas* sp. strain LB400 may occur at either the 2,3- or the 4,5-positions (Haddock et al. 1995). A strain of *Pseudomonas paucimobilis*, which also degrades toluene, and 1,3- and 1,4-dimethylbenzene, carries out dioxygenation and dehydrogenation of biphenyl followed by ring fission to benzoate, which is further degraded by intradiol ring fission (Figure 8.14) (Furukawa et al. 1983).

The three-component biphenyl dioxygenase consists of a reductase$_{BPH}$, a ferredoxin$_{BPH}$, and a terminal dioxygenase ISP$_{BPH}$. The dioxygenase contains two subunits and a Rieske-type [2Fe–2S] center that are encoded by the genes *bphA4* (*bphG*), *bphA3* (*bphF*), and *bphA1bph/A2* (*bphA/bphE*) (Erickson and Mondello 1992). The dioxygenase has been purified from *Pseudomonas* sp. strain LB400 (Haddock and Gibson 1995). All the components from *Comamonas testosteroni* strain B-356, which degrades 4-chlorobiphenyl, have been isolated. Purification of the reductase$_{BPH}$ and ferredoxin$_{BPH}$ proteins was made possible by using His-tagged components produced from recombinant strains of *Escherichia coli* (Hurtubise et al. 1995). The reductase$_{BPH}$ contains 1 mol of FAD, the ferredoxin$_{BPH}$ a Rieske [2Fe–2S] center, and together with the terminal oxygenase ISP$_{BPH}$, both biphenyl-2,3-dihydrodiol and biphenyl-3,4-dihydrodiol were produced from biphenyl.

It has been shown that the product of dioxygenation of biphenyl by *Sphingomonas yanoikuyae* strain B1 (*Beijerinckia* sp. B1) is biphenyl *cis*-(2R,3S)-2,3-dihydrodiol (Ziffer et al. 1977). This is then

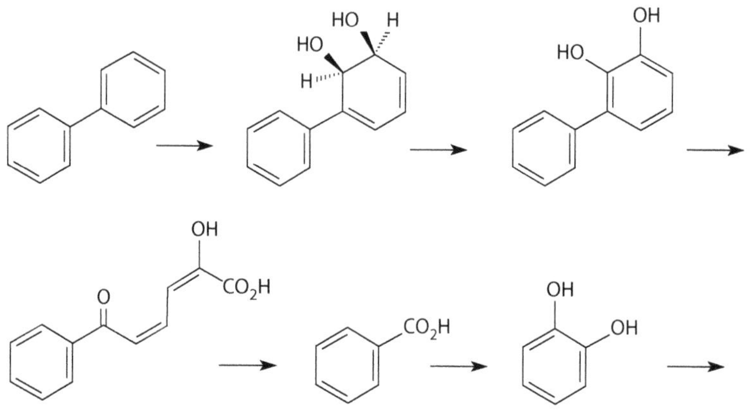

FIGURE 8.14 Degradation of biphenyl.

converted into 2,3-dihydroxybiphenyl by a dehydrogenase encoded by the *bphB* gene followed by ring fission by 2,3-dihydroxybiphenyl 1,2-dioxygenase encoded by the *bphC* gene. The ring fission dioxygenase has been examined in a number of organisms, and although polyclonal antibodies from the enzymes from *P. paucimobilis* strain Q1 and from *P. pseudoalcaligenes* strain KF707 do not cross-react (Taira et al. 1992), there is a high degree of homology between the *bphC gene* in *P. putida* strain OU 83 and those from *P. cepacia* strain LB400 and *P. pseudoalcaligenes* strain KF707 (Khan et al. 1996b). Greater complexity has been observed in Gram-positive strains: in *Rhodococcus globerulus* strain P6, there are three genes encoding 2,3-dihydroxybiphenyl-1,2-dioxygenase (Asturias and Timmis 1993), and in *Rhodococcus erythropolis* strain TA421, there are four (Maeda et al. 1995).

Sphingomonas yanoikuyae (*Beijerinckia* sp.) strain B1 metabolizes biphenyl by initial dioxygenation followed by dehydrogenation to 2,3-dihydroxybiphenyl (Gibson et al. 1973). Cells of a mutant (strain B8/36) lacking *cis*-biphenyl dihydrodiol dehydrogenase were induced with 1,3-dimethyl-benzene, and were able to carry out both monooxygenation and dioxygenation of 1,2-dihydro-naphthalene (Eaton et al. 1996). This is analogous to the duality found in naphthalene dioxygenase (Gibson et al. 1995). *Sphingobium yanoikuyae* strain B1 is able to grow at the expense of naphthalene, phenanthrene, anthracene, and biphenyl that is initiated by dioxygenation to *cis*-1,2-dihydrodiols before dehydrogenation and ring fission. In addition, it has been shown in strain B8/36 that the initially formed dihydrodiols may undergo further dioxygenation to *bis-cis*-dihydrodiols (*cis*-tetraols). Chrysene formed 3,4,8,9-*bis-cis*-dihydrodiol, and *bis*-dihydroxylation at the analogous positions has been demonstrated for benzo[*c*]phenanthridine and benzo[*b*]naphtho[2,1-*d*]thiophene (Boyd et al. 1999).

Although toluene-2,3-dioxygenase, naphthalene-1,2-dioxygenase, and biphenyl-2,3-dioxygenase are broadly similar, an interesting difference has emerged in the products formed from benzocycloheptene by bacterial strains that express these activities (Resnick and Gibson 1996).

8.7 PAHs with Three or More Rings

8.7.1 Introduction

The aerobic degradation of a wide range of PAHs with three or more rings has been described, and the pathways are analogous to those outlined for naphthalene and biphenyl—though sometimes they have been covered in less detail. These reactions have been reviewed (Kanaly and Harayama 2000), and only salient features with some exceptions are briefly noted.

1. Mycobacteria that are able to degrade or metabolize a range of PAHs have been summarized in Chapter 2, and the network of PAH degradations by *Mycobacterium vanbaalenii* that incorporates the reactions given for individual PAHs has been provided (Kweon et al. 2011). The PAH

TABLE 8.1

Bacterial Degradation of PAHs

Substrate	Organism	Reference
Nap, Bi, Phe, Anth	*Sphingobium yanoikuyae*	Gibson (1999)
Nap, Phe, Flu, 3-Me-Chol	*Mycobacterium* sp.	Heitkamp et al. (1988a,b)
Nap, Phe, Anth, Flu	*Alcaligenes denitrificans*	Weissenfels et al. (1991)
Nap, Flu, Pyr	*Enterobacter* sp.	Sarma et al. (2004)
Phe, Pyr, Flu	*Mycobacterium* sp. BB1	Boldrin et al. (1993)
Nap, Phe, Anth, Flu	*Pseudomonas cepacia* F297	Grifoll et al. (1994, 1995)
Nap, Phe, Anth	*Pseudomonas putida* GZ44	Goyal and Zylstra (1996)
Pyr, BaPyr, BaAnth	*Mycobacterium* sp. RJGII-135	Schneider et al. (1996)
Flu, Pyr, BaAnth, DBahAnth	*Burkholderia cepacia*	Juhasz et al. (1997)
Bi, Phe, Anth, Flu, BbFlu, Chr, Pyr	*Pseudomonas paucimobilis*	Mueller et al. (1990)
Flu, Pyr, BaAnth, Chr, BaPyr	*Pseudomonas paucimobilis*	Ye et al. (1996)
Nap, Anth, Phe, Flu, Pyr, BaAnth, BaPyr	*Mycobacterium vanbaalenii* PYR-1	Kim et al. (2006)
Phe, Anth, Flu, Pyr, BaAnth, BaPyr, Chr, BbFlu	*Novosphingobium pentaromaticivorans* US6-1	Sohn et al. (2004)

Notes: Bi, Biphenyl; Nap, Naphthalene; Phe, Phenanthrene; Anth, Anthracene; Fl, Fluoranthene; Pyr, Pyrene; Chr, Chrysene; BaAnth, Benz[*a*]anthracene; BaPyr, Benzo[*a*]pyrene; BbFlu, Benzo[*b*]fluoranthene; DbahAnth, Dibenz[*a,h*]anthracene; 3-Me-Chol, 3-methyl-cholanthrene.

dioxygenases have been characterized (Brezna et al. 2003), while several of them also possess cytochrome P450 activity (Brezna et al. 2006).

2. Sphingomonads have been involved in the degradation of PAHs and include the classic strain of *Sphingobium* (*Sphingomomas*) *yanoikuyae* (Khan et al. 1996a) that was originally assigned to *Beijerinckia* sp. strain B1 (Gibson 1999), *Sphingomonas paucimobilis* strain EPA (Ye et al. 1996), and one that unusually degraded chrysene (Willison 2004; Kweon et al. 2011).

3. The degradation of pyrene by *Leclercia adecarboxylata* (Sarma et al. 2004) and a strain of *Enterobacter* sp. isolated from the roots of *Allium macrostemon* (Sheng et al. 2008) is unusual for Enterobacteriaeae.

For PAHs with four or more rings, dioxygenation to produce *cis*-dihydrodiols may take place at more than one position in the rings: this is discussed later for each PAH. Some illustrative examples of the degradation of PAHs are given in Table 8.1.

The stereospecific dioxygenation of several PAHs by bacteria has been described, and includes the following illustrative examples:

1. Anthracene to *cis*-(1*R*,2*S*)-anthracene-1,2-dihydrodiol (Akhtar et al. 1975)
2. Phenanthrene to *cis*-(1*R*,2*S*)-phenanthrene-1,2-dihydrodiol and *cis*-(3*S*,4*R*)-phenanthrene-3,4-dihydrodiol (Koreeda et al. 1978)
3. Benz[*a*]anthracene to benz[*a*]anthracene-*cis*-(1*R*,2*S*)-1,2-dihydrodiol, *cis*-(8*R*,9*S*)-8,9-dihydrodiol, and *cis*-(10*S*,11*R*)-10,11-dihydrodiol (Jerina et al. 1984)

Further degradation of the dihydrodiols involves dehydrogenation to the catechols followed by ring fission and oxidative degradation of the fission product. However, for higher PAHs, degradation may be more complex. Other issues that intrude include dioxygenation at different positions of the rings, and the alternative of monooxygenation. A great deal of attention has been given to the degradation of higher PAHs, and only selected examples are given as illustration.

8.7.2 Anthracene and Phenanthrene

The degradation of these was described many years ago (Evans et al. 1965), and details have since been added (Figure 8.15a) (van Herwijnen et al. 2003). Phenanthrene is more readily degraded than anthracene,

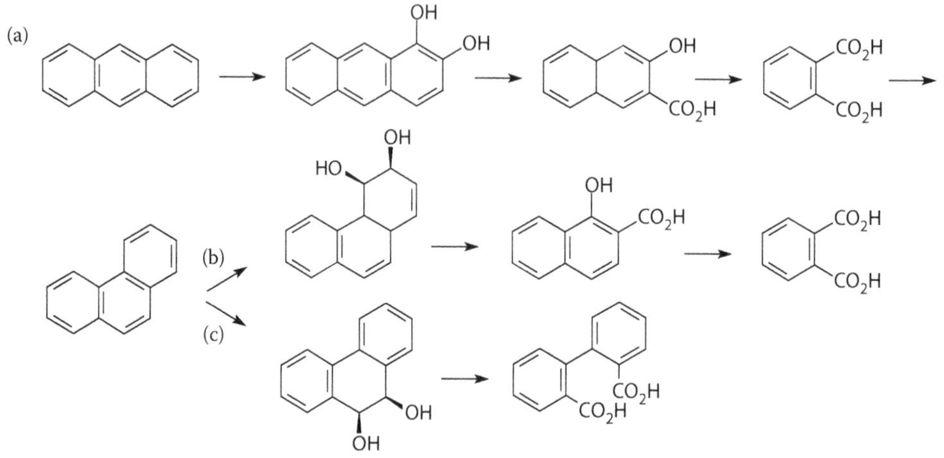

FIGURE 8.15 Degradation of (a) anthracene and (b,c) alternative pathways for phenanthrene.

and both can be degraded to *o*-phthalate. After fission of the peripheral ring in phenanthrene, several pathways have been demonstrated:

1. The naphthalene pathway via salicylate (Evans et al. 1965)
2. The *o*-phthalate pathway followed by fission of 3,4-dihydroxybenzoate (Kiyohara et al. 1976; Kiyohara and Nagao 1978; Barnsley 1983) (Figure 8.15b)
3. Degradation to diphenyl-6,6'-dicarboxylate (Figure 8.15c) (Moody et al. 2001)
4. A pathway that involves naphth-1-ol, which has already been noted (Samanta et al. 1999)

The enzymes involved in the degradation of phenanthrene by *Nocardioides* sp. strain KP7 have been characterized. In this strain, the genes are chromosomal, and *phdA* and *phdB* encoding the α- and β-subunits of the dioxygenase, *phdC* and *phdD* that encode the ferredoxin, and ferredoxin reductase have been cloned and sequenced (Saito et al. 2000). The order of the genes was different from that in pseudomonads, and there was only moderate similarity to the sequence for other dioxygenase subunits. 1-Hydroxy-2-naphthoate dioxygenase (Iwabuchi and Harayama 1998a), *trans*-2'-carboxybenzalpyruvate hydratase-aldolase (Iwabuchi and Harayama 1998b), and 2-carboxybenzaldehyde dehydrogenase, which catalyzes the formation of *o*-phthalate (Iwabuchi and Harayama 1997), have been purified and characterized. The purified dioxygenase was used to determine further details of the degradation pathways (Adachi et al. 1999). The initial product from ring fission is in equilibrium with the lactone formed by reaction between the carboxyl group and the activated double bond of the benzalpyruvate in a reaction formally comparable to the formation of muconolactones during degradation of monocyclic aromatic compounds. An important exception to the generalization that bacterial dioxygenation is used for aromatic hydrocarbons is provided by *Streptomyces flavovirens* in which cytochrome P450 produces (–)-*trans*-[9*S*,10*S*]-9,10-dihydrodihydroxyphenanthrene with minor amounts of 9-hydroxyphenanthrene (Sutherland et al. 1990).

8.7.3 Fluorene

Several pathways have been reported, including oxidation to fluoren-9-one and dioxygenation at the ring junction followed by degradation of the oxygenated ring by extradiol fission to yield phthalate (Grifoll et al. 1994; Trenz et al. 1994). In one variant, an 8-hydroxy benzo[*c*]coumarin is produced (Figure 8.16). Alternatively, dioxygenation may take place to produce 3,4-dihydroxyfluorene followed by extradiol ring fission and further degradation to 3,4-dihydrocoumarin (Grifoll et al. 1992) (Figure 8.17). A strain of *Pseudomonas cepacia* F297 that was able to grow with a range of PAHs, including fluorene, anthracene,

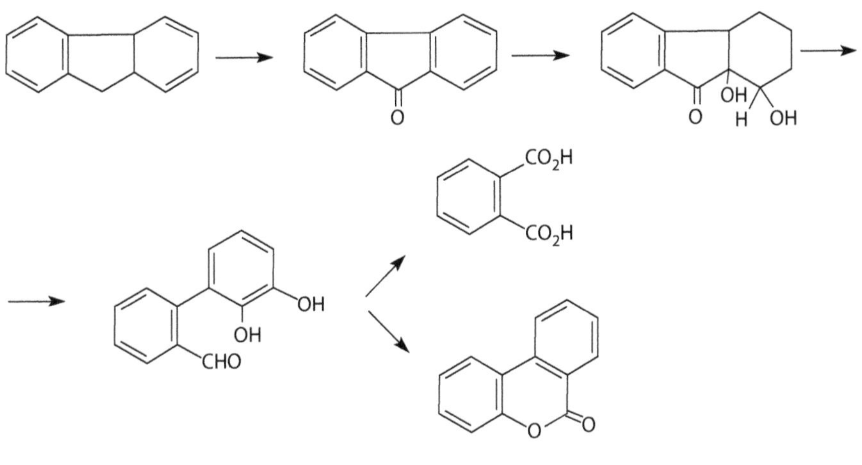

FIGURE 8.16 Pathways for degradation of fluorine.

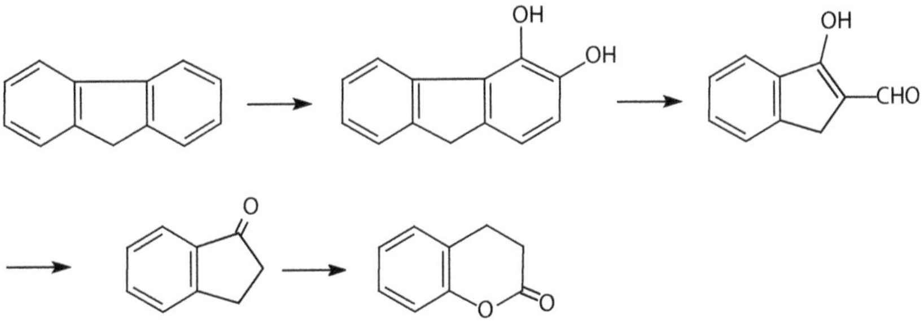

FIGURE 8.17 Degradation of fluorine.

and phenanthrene, degraded fluorene via 3,4-dihydroxyfluorene and extradiol fission to indan-1-one as terminal metabolite (Grifoll et al. 1994).

8.7.4 Fluoranthene

Degradation of fluoranthene and pyrene has been demonstrated in a number of new species of *Mycobacterium* that were isolated from pristine habitats (Hennessee et al. 2009). Degradation by *Mycobacterium vanbaalenii* strain PYR-1 involved dioxygenation at both the 1,2- and 7,8-positions followed by dehydrogenation, and after ring fission fluorene-9-one (Figure 8.18a), or acenaphthene-7-one (Figure 8.18b) (Kelley et al. 1993). Further details have been added by genomic and proteomic analysis of the pathway (Kweon et al. 2007). In a strain of *Alcaligenes denitrificans*, acenaphthene-7-one is further oxidized by a Baeyer–Villiger-type oxidation to 3-hydroxymethyl-3,4-dihydro-benzo[*d,e*]coumarin before further oxidation (Weissenfels et al. 1991) (Figure 8.18c). The unusual transformation products fluoranthene-2,3- and 1,5-quinone have been identified in several strains of bacteria (Kazunga et al. 2001). Both were terminal metabolites and since the 2,3-quinone inhibited the degradation of other PAHs, its formation could have an adverse effect on attempts at bioremediation of PAH-contaminated sites. *Novosphingobium pentaromaticivorans* is able to degrade an unusually wide range of PAHs (Sohn et al. 2004): these included fluoranthene and, unusually, benzo[*b*]fluoranthene.

8.7.5 Benz[a]anthracene

Benz[*a*]anthracene and benzo[*a*]pyrene can be degraded by *Novosphingobium pentaromaticivorans* (Sohn et al. 2004). Transformation by *Sphingomonas yanoikuyae* strain B1 (*Beijerinckia* sp. strain B1)

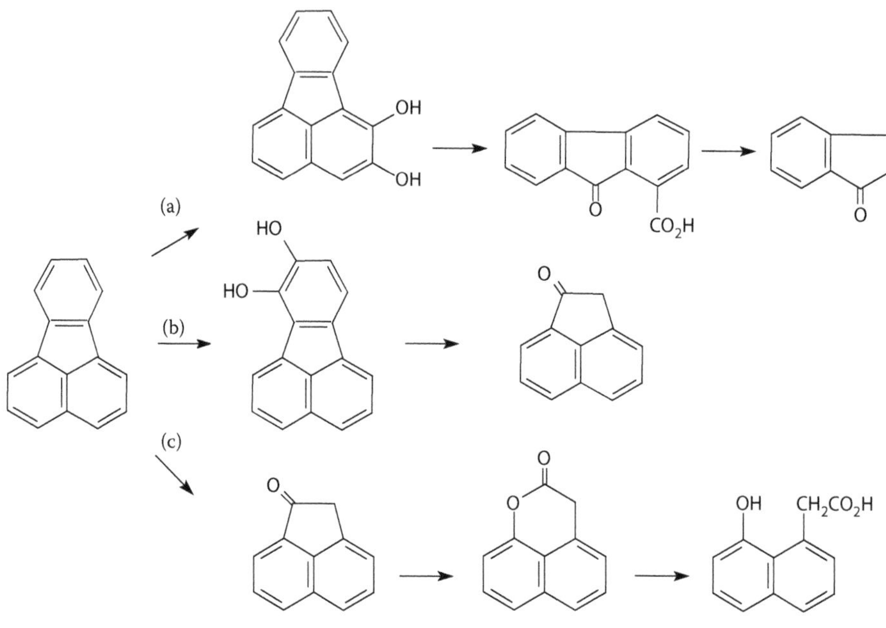

FIGURE 8.18 Degradation of fluoranthene: Dioxygenation followed by ring fission (a) to fluoren-9-one, (b) to acenaphthene-7-one, and (c) direct production of acenaphthene-7-one followed by Baeyer–Villiger oxidation and hydrolysis.

involves dioxygenation at the 1,2-, 8,9-, or 10,11-positions with production of the cis-[1R,2S], cis-[8R,9S], or cis-[10S,11R] dihydrodiols of which the first is the dominant (Jerina et al. 1984). These are then further degraded to 1-hydroxy-2-carboxyanthracene, or the corresponding phenanthrenes (Mahaffey et al. 1988) (Figures 8.19a–c). In contrast, *Mycobacterium* sp. strain RJGII-135 forms the 5,6- and 10,11-dihydrodiols (Schneider et al. 1996) so that four different pathways for degradation are possible. Exceptionally,

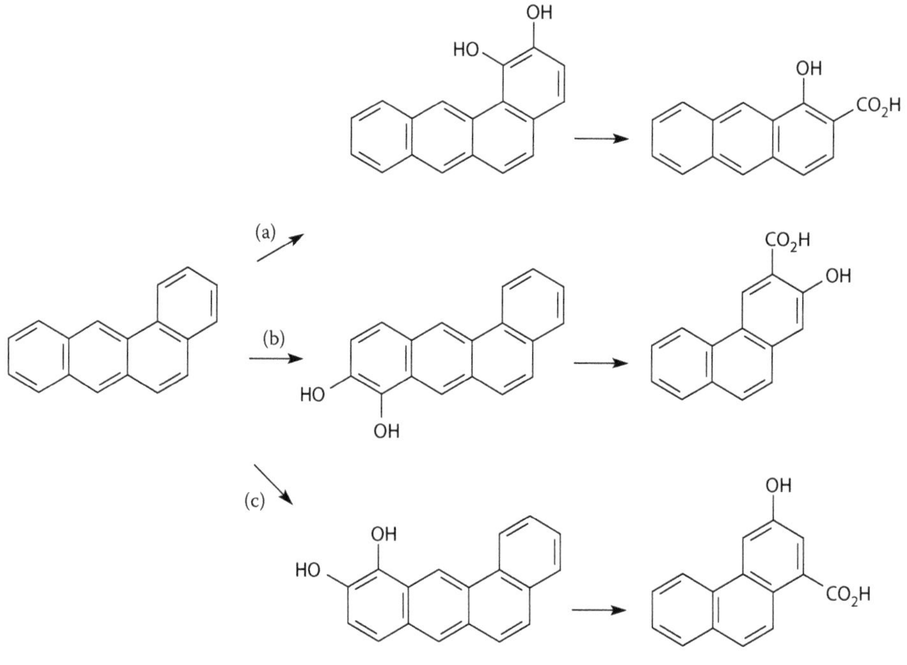

FIGURE 8.19 Dioxygenation of benz[a]anthracene by *Sphingomonas yanoikuyae* strain B1 at positions: (a) 1,2, (b) 8,9, and (c) 10,11.

monooxygenation may occur concomitant with dioxygenation, for example, the transformation of 7,12-dimethylbenz[*a*]anthracene by *Mycobacterium vanbaalenii* PYR-1 takes place by monooxygenation to epoxides followed by hydration to *trans*-dihydrodiols, and also by dioxygenation to *cis*-dihydrodiols (Moody et al. 2003).

8.7.6 Pyrene

Degradation by a *Mycobacterium* sp. involves both monooxygenation and dioxygenation at the 4,5-position with production of the *trans*- and *cis*-dihydrodiols, respectively, which are degraded to phenanthrene-4-carboxylate and ultimately to *o*-phthalate. 4-Hydroxyperinaphthenone was also produced, presumably by initial dioxygenation at the 1,2-positions (Heitkamp et al. 1988a) (Figure 8.20). This is consistent with the structure of the intermediate from the degradation of pyrene by *Rhodococcus* sp. strain UW1 (Walter et al. 1991). Alternatively, degradation by *Mycobacterium* sp. strain AP1 produced both *cis*- and *trans*-4,5-dihydrodiols (Vila et al. 2001). The former was degraded further to phenanthrene-4,5-dicarboxylic acid, phenanthrene-4-carboxylate and phthalate, and 6,6'-dihydroxy-biphenyl-2,2'-dicarboxylate (Figure 8.21), and has been confirmed by proteomic analysis (Kim et al. 2007) in *Mycobacterium vanbaaleni* PYR-1. In this strain, the alternative dioxygenation and dehydrogenation to 1,2-dihydroxypyrene resulted in the formation of a terminal metabolite by *O*-methylation of the catechol. Bacterial formation of pyrene-4,5-quinone has been reported (Kazunga and Aitken 2000). Although the degradation of pyrene by *Mycobacterium* sp. strain A1-PYR was restricted with the formation of several oxygenated metabolites, complete degradation without their formation was accomplished

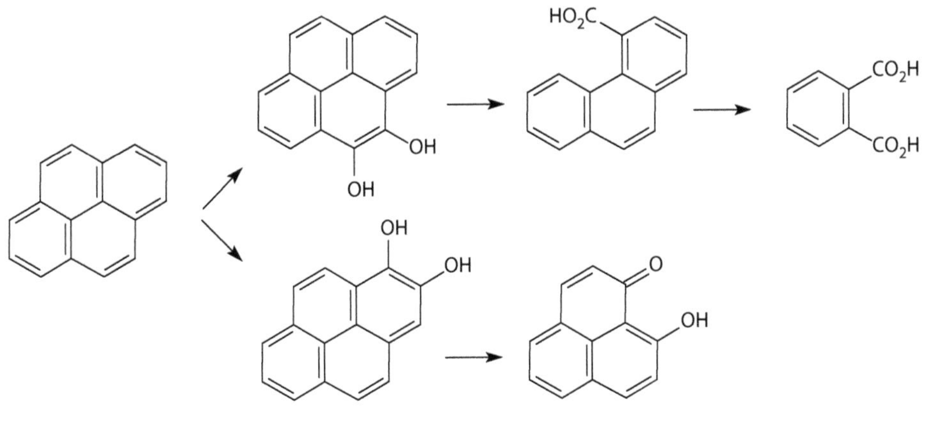

FIGURE 8.20 Degradation of pyrene.

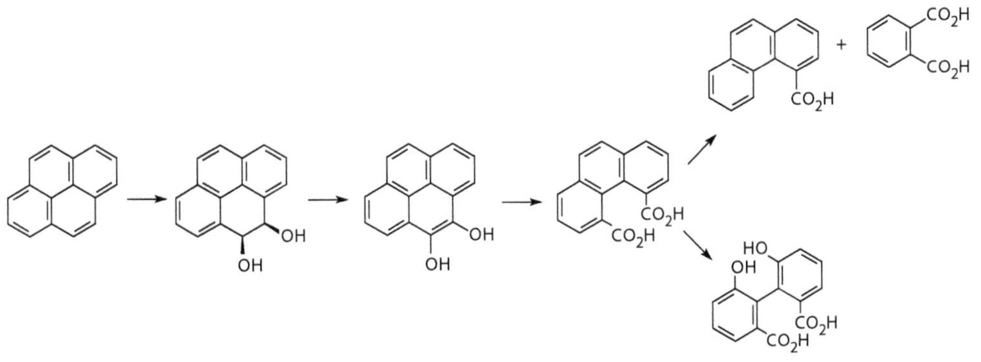

FIGURE 8.21 Degradation of pyrene. (Adapted from Vila J et al. 2001. *Appl Environ Microbiol* 67: 5497–5505.)

when phenanthrene or fluoranthene was also present (Zhong et al. 2006). This is particularly relevant to the degradation of mixtures of PAHs, for example, in bioremediation.

8.7.7 Chrysene

The degradation of chrysene appears to have attracted less attention among strains of mycobacteria and sphingomonads than other PAHs. A strain of *Sphingomonas* sp. strain CHY-1 was isolated by enrichment with chrysene and, based on its 16S rRNA gene sequence, was closely similar to *Burkholderia xenophaga* (Willison 2004). It was able to grow at the expense of naphthalene, phenanthrene and anthracene, and chrysene, and was capable of carrying out the production of dihydrodiols from pyrene-(4,5), fluoranthene-(2,3), benz[*a*]anthracene-(1,2), and benzo[*a*]pyrene-(9,10)—though unable to use these as growth substrates (Demanèche et al. 2004; Jouanneau and Meyer 2006). Two ring-hydroxylating enzymes PhnI and PhnII were induced, and PhnII was capable of hydroxylating 3-methyl salicylate to 3-methylcatechol, and 4-methyl- and 5-methylsalicylate to 4-methylcatechol (Demanèche et al. 2004), and the *cis*-dihydrodiol dehydrogenase has been purified and characterized (Jouanneau and Meyer 2006). The spectrum of PAHs that can be degraded by *Novosphingobium pentaromaticivorans* is remarkable since it is able to degrade pyrene, chrysene, benzo[*a*]pyrene, and benz[*a*]anthracene (Sohn et al. 2004).

8.7.8 Benzo[a]pyrene

Mycobacterium sp. strain RJGH-135 was able to metabolize pyrene, benz[*a*]anthracene, and benzo[*a*] pyrene, and degraded the last by dioxygenation at the 4,5-, 7,8-, and 9,10-positions (Schneider et al. 1996). Chrysene-4,5-dicarboxylate was presumptively formed by intradiol fission of the 4,5-benzo[*a*] pyrene-4,5-diol, while the 7,8- and 9,10-dihydrodiols underwent extradiol fission to 7- and 8-dihydropyrene carboxylates (Figure 8.22). *Mycobacterium vanbaalenii* strain PYR-1 carried out both monooxygenation with the formation of benzo[*a*]pyrene *trans*-11,12-dihydrodiol and dioxygenation to *cis*-dihydrodiols at the 4,5- and 11,12-positions. The former was then degraded further to chrysene-4,5-dicarboxylic acid (Moody et al. 2004). A soil strain of *Rhodanobacter* sp. in a consortium could degrade [7-^{14}C]benzo[*a*]pyrene apparently using metabolites produced by other organisms, since the pure cultures were unable to degrade the substrate (Kanaly et al. 2002). This may be a widespread phenomenon for complex mixtures such as those that were used in that study.

8.8 Anaerobic Degradations Carried Out by Bacteria

The anaerobic degradation of only a few PAHs has been examined using mixed cultures or unenriched samples from widely different habitats. Investigations have often been limited to naphthalene and phenanthrene, and degradation has also been assessed from the diminishing substrate concentration of anthracene, phenanthrene, and pyrene by denitrifying pseudomonads isolated from diverse environments, under both aerobic and anaerobic denitrifying conditions (McNally et al. 1998). Under anaerobic nitrate-reducing conditions, both [ring ^{14}C]toluene and [1-^{14}C]naphthalene were oxidized to ^{14}CO$_2$

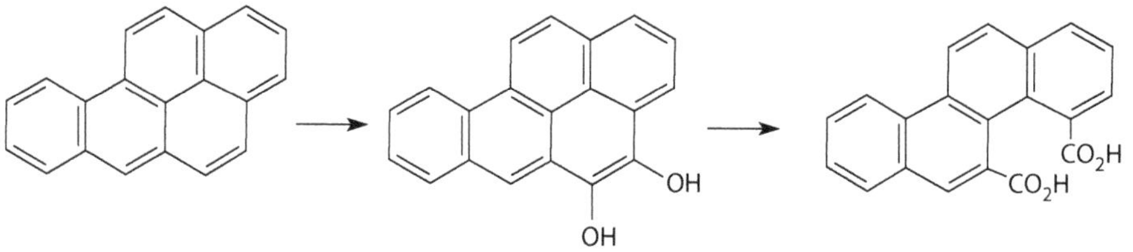

FIGURE 8.22 Products produced by partial degradation of benzo[*a*]pyrene.

(Bregnard et al. 1996), and [U-^{14}C] naphthalene was partly mineralized to $^{14}CO_2$ by pure strains isolated from a consortium (Rockne et al. 2000). Under sulfate-reducing conditions, [1-^{14}C]naphthalene and [9-^{14}C]phenanthrene were oxidized to $^{14}CO_2$ (Coates et al. 1996, 1997). Although carboxylated intermediates of naphthalene and phenanthrene have been identified in a sulfidogenic mixed culture (Zhang and Young 1997), these may not be the primary metabolites for naphthalenes.

Two essentially different pathways have been proposed for naphthalene—methylation and carboxylation. Methylation was the initial reaction in the degradation by a sulfate-reducing culture N47 (Safinowski and Meckenstock 2006), and the metabolites naphthyl-2-methyl succinate and naphthyl-2-methylene succinate were found. Details of the pathway for this strain were constructed by a combination of genomic and proteomic analysis (Selesi et al. 2010), and the pathway is shown for the final production of octahydro-2-naphthoyl-CoA in (Figure 8.23). The genes in the genome identified the following key enzymes:

NmsABC	naphthyl-2-methyl-succinate synthase
BnsEF	naphthyl-2-methyl-succinate CoA transferase
BnsG	naphthyl-2-methyl-succinyl-CoA dehydrogenase
BnsH	naphthyl-2-methylene-CoA hydratase
BnsCD	naphthyl-2-hydroxymethyl-succinyl-CoA dehydrogenase
Bns AB	naphthyl-2-oxomethyl-succinyl-CoA thiolase
NcrABCD	2-naphthoyl-CoA reductase

The gene products were analogous to those that have been found in a range of strains that have been implicated in the anaerobic metabolism of arenes and alkanes: *Aromatoleum aromaticum* EbN1, *Azoarcus* sp. strain HxN1, *Geobacter* sp. strain GS-15, *Thauera aromatica*, *Azoarcus evansii*, *Anaeromyxobacter* sp. strain Fw109-5, *Thermoanaerobacter tengcongensis*, and *Desulfatibacillum alkenivorans* AK-0.

Marine strains (NaphS3 and NaphS6) that were able to carry out naphthalene-dependent sulfate reduction have been isolated from Mediterranean sediment (Musat et al. 2009), and resembled a previously isolated strain (NaphS2) from North Sea sediment (Galushko et al. 1999). All of them could utilize 2-methylnaphthalene and naphth-2-oate, and all of them belonged to the δ-subclass of the Proteobacteria. Naphthalene-grown cells of all these strains produced sulfide from naphthalene—though not from 2-methylnaphthalene—and complete oxidation of naphthalene by strain NaphS3 and NaphS6 at the expense of sulfide was observed (Musat et al. 2009). In addition, labeling experiments using strain NaphS2 grown with a mixture of d_8-naphthalene and unlabeled 2-methylnaphthalene showed that >99.5% of 2-naphthylmethyl-succinate was unlabeled, and not, therefore, derived from naphthalene, whereas 86% naphth-2-oate was labeled and derived from naphthalene. It was, therefore, presumed that different pathways were used for the degradation of naphthalene and 2-methylnaphthalene, and this was supported by the results of denaturing gel electrophoresis where the 2-methylnaphthalene-activating enzyme was absent in cells of NaphS2, NaphS3, and NaphS6 that degraded naphthalene: the initial step in the degradation of naphthalene was not, therefore, 2-methylnaphthalene that is degraded by the fumarate pathway (Selesi et al. 2010), analogous to that used for the anaerobic degradation of toluene (Beller

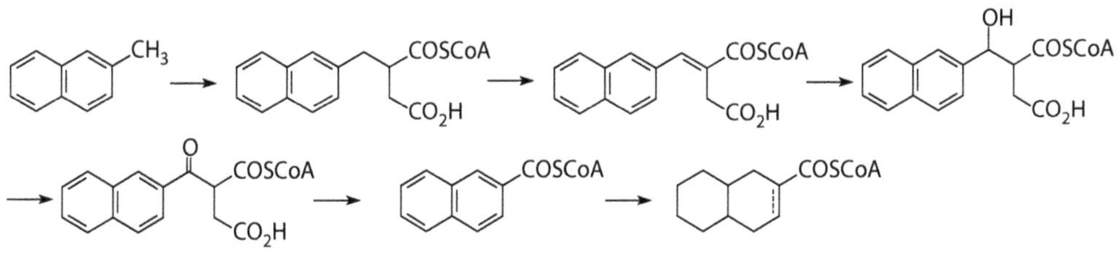

FIGURE 8.23 Anaerobic degradation of 2-methylnaphthalene.

and Spormann 1998). The pathway for naphthalene could alternatively take place by carboxylation that would be consistent with (a) identification of carboxylated intermediates of naphthalene and phenanthrene in a sulfidogenic mixed culture (Zhang and Young 1997) and (b) the demonstration that in another sulfate-reducing consortium reduction of the A-ring and carboxylation of the B-ring of 2-methylnaphthalene with the production of decahydronaphthalene-2-carboxylate was observed (Sullivan et al. 2001).

During co-metabolism with 2-methylnaphthalene, analogous reactions were shown to operate with benzothiophene and benzofuran (Safinowski et al. 2006). It was, therefore, suggested that analysis of polar metabolites could be used to assess the degradation of these arenes in anaerobic groundwater, analogous to the use of benzylsuccinates from monocyclic arenes (BTEX) whose value had already been demonstrated (Beller et al. 1995, Beller 2002).

8.9 Fungal Transformations

In contrast to the degradations carried out by bacteria, eukaryotic organisms such as yeasts and fungi often accomplish only biotransformation. The aryl hydrocarbon monooxygenase in *Cunninghamella baineri* is a cytochrome P450 (Ferris et al. 1976) in which the $[FeO]^{3+}$ complex with the substrate may either carry out direct hydroxylation of the substrate by elimination of a proton, or produce an epoxide or a rearranged (NIH shift) monohydroxy (phenolic) compound (Guengerich 1990). Isotope experiments clearly show that the phenols are generally not produced by either a direct substitution reaction or elimination from the *trans*-dihydrodiols. Epoxide hydrolases produce phenols from the *trans*-dihydrodiols, and can then be conjugated to sulfate or glucuronide esters (Cerniglia et al. 1982b), or *O*-methylated. Monooxygenation of methyl groups to carboxyl groups has also been reported.

Cunninghamella elegans has been used to examine the transformation of a number of polycyclic aromatic hydrocarbons. Reactions are generally confined to oxidation of the rings with the formation of phenols, catechols, and quinones, and ring fission does not generally take place. Different rings may be oxygenated, for example, in 7-methylbenz[*a*]anthracene (Cerniglia et al. 1982c) (Figure 8.24) or oxidation may take place in several rings, for example, in fluoranthene (Pothuluri et al. 1990) (Figure 8.25).

Since fungal metabolism of PAHs has been suggested as a model for mammalian metabolism (Smith and Rosazza 1983), particular attention has been directed to the stereochemistry and absolute configuration of the *trans*-dihydrodiols produced by monooxygenation. There is one very significant difference between the *trans*-dihydrodiols produced by fungi and those from mammalian systems—the absolute configuration of the products. Although *trans*-1,2-dihydroxy-1,2-dihydroanthracene, and *trans*-1,2-dihydroxy-1,2-dihydrophenanthrene are formed from anthracene by *C. elegans*, these dihydrodiols have the *S,S* configuration in contrast to the *R,R* configuration of the metabolites from rat liver microsomes

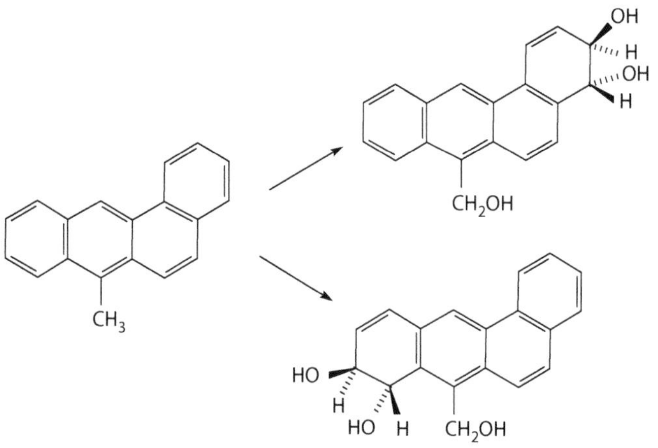

FIGURE 8.24 Biotransformation of 7-methylbenz[*a*]anthracene by *Cunninghamella elegans*.

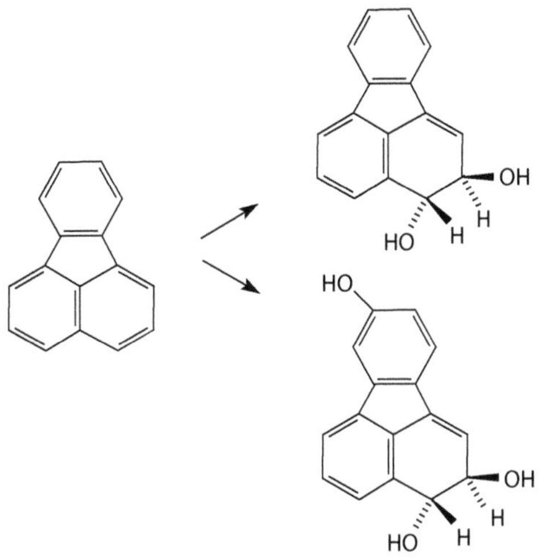

FIGURE 8.25 Alternative pathways for the biotransformation of fluoranthene by *Cunninghamella elegans.*

(Cerniglia and Yang 1984). It has become clear, however, that the situation among a wider range of fungi is much less straightforward. For example, the *trans*-9,10-dihydrodiol produced by *Phanerochaete chrysosporium* was predominantly the 9*S*,10*S* enantiomer, whereas those produced by *C. elegans* and *Syncephalastrum racemosum* were dominated by the 9*R*,10*R* enantiomers (Sutherland et al. 1993). Comparable differences were also observed for the *trans*-1,2-dihydrodiols and *trans*-3,4-dihydrodiols, so that generalizations on the enantiomeric selectivity of these reactions should be viewed with caution.

There are two issues that recur in fungal transformations, and these have been extensively documented:

1. The phenol, which is formed by rearrangement from the initially produced *trans*-dihydrodiol, may be conjugated to form sulfate esters or glucuronides (Cerniglia et al. 1982b; Golbeck et al. 1983; Cerniglia et al. 1986; Lange et al. 1994). The less-common glucosides have also been identified: 1-phenanthreneglucopyranoside is produced from phenanthrene by *C. elegans* (Cerniglia et al. 1989) and 3-(8-hydroxyfluoranthene)-lucopyranoside from fluoranthene by the same organism (Pothuluri et al. 1990). The xylosylation of 4-methylguaiacol and vanillin by the basidiomycete *Coriolus versicolor* (Kondo et al. 1993) represents an even less-common alternative.

2. In reactions involving monooxygenase systems with the formation of intermediate arene oxides, rearrangement of substituents may take place (Figure 8.26a). This is an example of the NIH shift that plays an important role in the metabolism of xenobiotics by mammalian systems (Daly et al. 1972). It has also been observed in fungal (Figure 8.26b) (Faulkner and Woodcock 1965; Smith et al. 1981; Cerniglia et al. 1983) and bacterial systems (Figure 8.26c) (Dalton et al. 1981; Cerniglia et al. 1984a,b; Adriaens 1994), including the marine cyanobacterium *Oscillatoria* sp. (Narro et al. 1992a,b).

The transformation of naphthalene by a large number of fungi has been examined, including representatives of the *Mucorales* such as *Cunninghamella*, *Syncephalastrum*, and *Mucor*, which were the most active. The principal product was naphth-1-ol, with lesser amounts of 2-naphthol, 4-hydroxytetral-1-one, *trans*-naphthalene-1,2-dihydrodiol, and the 1,2- and 1,4-quinones (Cerniglia et al. 1978). The results of the detailed study (Cerniglia et al. 1983) of the transformation of naphthalene by *C. elegans* illustrate most of the basic principles involved in the fungal transformation of all PAHs:

1. The initial product is naphthalene-1,2-epoxide, which is converted into the *trans*-dihydrodiol with the (+)-(1*S*,2*S*) configuration.

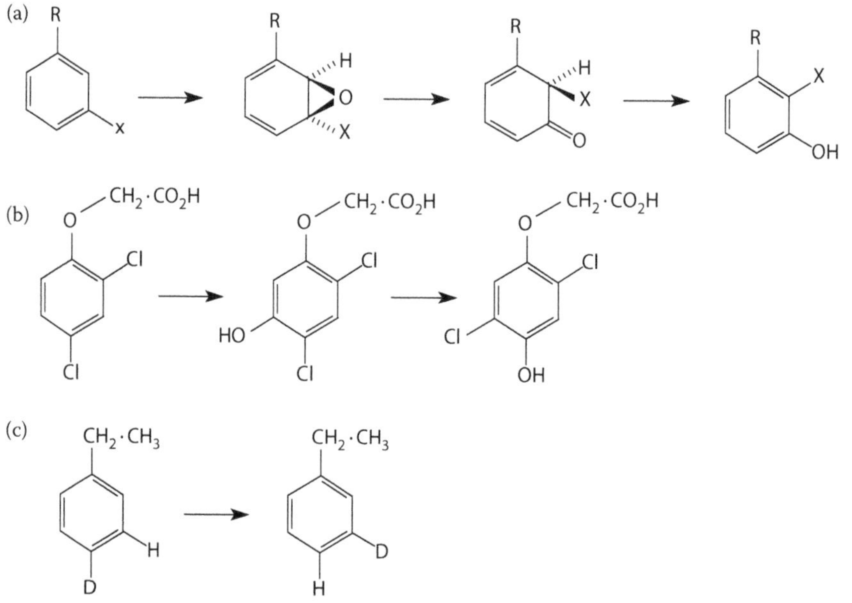

FIGURE 8.26 NIH rearrangement. (a) Bacteria, (b) fungi.

2. The *trans*-dihydrodiol is formed by the introduction of a single atom of $^{18}O_2$ at C_1 while that at C_2 comes from $H_2^{18}O$.

3. The naphth-1-ol formed from [1-^{2}H]-naphthalene retains 78% of the deuterium so that an NIH shift is involved.

Similar experiments with 1-methyl- and 2-methylnaphthalene involved oxidation of the methyl group to hydroxymethyl and carboxyl groups (Cerniglia et al. 1984a,b).

The transformation of biphenyl is analogous to that of naphthalene, with the production of hydroxylated biphenyls as the major metabolites: 4-hydroxybiphenyl from *C. echinulata*, 2-hydroxybiphenyl from *Helicostylum piriforme*, 4,4′-dihydroxybiphenyl from *Aspergillus parasiticus* (Smith et al. 1980; Golbeck et al. 1983), and 2-, 3-, and 4-hydroxybiphenyl, and 4,4′-dihydroxybiphenyl from *C. elegans* (Dodge et al. 1979). During the formation of 4-hydroxybiphenyl from [U-^{2}H]-biphenyl by *C. echinulata*, 20% of the deuterium was retained, which is consistent with the formation of an arene oxide followed by an NIH shift (Smith et al. 1981). Hydroxylation of biphenyl accompanied by ring fission is discussed later.

Higher PAHs may produce a number of products. For example, 7,12-dimethylbenz[*a*]anthracene gives rise to a range of metabolites, including both hydroxymethyl compounds and several *trans*-dihydrodiols. The 7-hydroxymethyl-12-methylbenz[*a*]anthracene-*trans*-5,6-dihydrodiols produced by *C. elegans* and *S. racemosum* were mixtures of the *R,R* and *S,S* enantiomers with the former predominating (McMillan et al. 1987). This is, however, contrary to the products from mammalian systems. From fluoranthene, *C. elegans* produces both the *trans*-2,3-dihydrodiol as well as the 8-hydroxy-3-hydroxyglucoside, and the 8- and 9-hydroxy-3,4-dihydrodiols (Pothuluri et al. 1990).

The metabolism of pyrene and benzo[*a*]pyrene by *C. elegans* is increasingly complex. Pyrene is transformed by hydroxylation at the 1-, 1- and 6-, and 1- and 8-positions, and the bisphenols were glucosylated at the 6- and 8-positions; the 1,6- and 1,8-pyrenequinones were also formed (Cerniglia et al. 1986). The same organism transformed benzo[*a*]pyrene to the *trans*-7,8- and *trans*-9,10-diols, the 3- and 9-hydroxy compounds, and the 1,6- and 3,6-quinones (Cerniglia and Gibson 1979) (Figure 8.27). In addition, *C. elegans* formed sulfate conjugates of polar metabolites, and one of these was identified as the 7β,8α,9α,10β-tetrahydrotetrol formed sequentially from the 7β,8β-epoxide, the 7β,8α-dihydrodiol, and the 7β,8α-dihydrodiol-9α,10α-epoxide (Cerniglia and Gibson 1980a). Tetrahydrotetrols were also

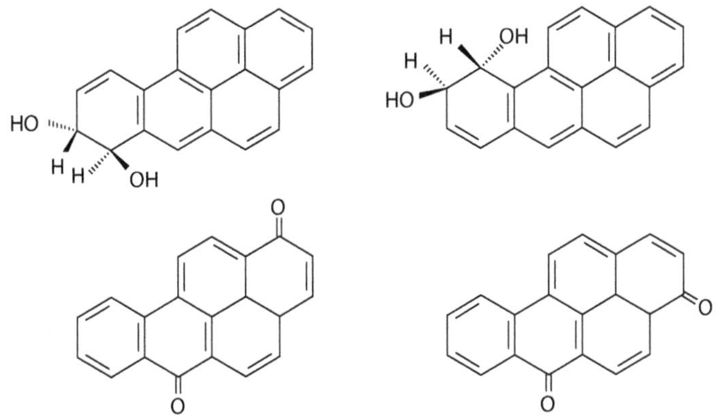

FIGURE 8.27 Products produced by fungal transformation of benzo[*a*]pyrene.

formed by the metabolism of both diastereomers of (+/–)-*trans*-9,10-dihydrodiol, including the 7β,8α,9β,10α- and 7β,8α,9α,10β-compounds; a dihydrodiol epoxide was also isolated that could be hydrolyzed to the former (Cerniglia and Gibson 1980b). The transformation of 3-methylcholanthrene by *C. elegans* was exceptional in that 1-hydroxy-*trans*-9,10-dihydrodiol was formed in only trace amounts, and benzylic oxidations at C$_1$ and C$_2$ dominated (Cerniglia et al. 1982d).

The metabolism of pyrene by the basidiomycete fungus *Crinipellis stipitaria* strain JK 364 illustrates a hitherto largely neglected possibility: biosynthesis from the products of metabolism. This strain can transform pyrene to 1-hydroxypyrene, and *trans*-4,5-dihydro-4,5-dihydroxypyrene. Simultaneously, the secondary metabolite 3-methyl-6,8-dihydroxy*iso*coumarin is produced, which is synthesized from acetate produced by ring fission (Lange et al. 1995). The biosynthesis of the *iso*coumarin was also induced by chrysene, though not by anthracene, phenanthrene, or fluoranthene.

Most of the foregoing studies were conducted with *C. elegans* or *S. racemosum*, but results have been presented showing that a wide range of nonbasidiomycete soil fungi are capable of oxidizing both pyrene and benzo[*a*]pyrene (Launen et al. 1995). Those with high activity included *Penicillium janthinellum* that produced 1-hydroxypyrene as a major metabolite, together with the 1,6- and 1,8-quinones.

8.10 Yeasts and Algae

Hydroxylation of arenes by yeasts and algae is not uncommon, although ring fission is less frequently encountered.

1. The imperfect fungus *Paecilomyces liliacinus* hydroxylated biphenyl to 2,5-, 2,3-, 3,4-, and 4,4′-dihydroxy-biphenyls, and subsequently to 3,4,5- and 3,4,4′-trihydroxybiphenyl. Extradiol ring fission of metabolites with vicinal hydroxyl groups produced muconic acids that were cyclized to 4-arylpyrone-6-carboxylates (Gesell et al. 2001).

2. Hydroxylation of biphenyl has been examined in *Trichosporon cutaneum* and produced a range of products: 2,3- and 3,4-dihydroxybiphenyl, and 3,4,5- and 2,3,4-trihydroxybiphenyl. Ring fission of 2,3-dihydroxyphenyl produced 3-phenyl-pyr-2-one-6-carboxylate and from 3,4-dihydroxybiphenyl 4-phenyl-pyr-2-one-6-carboxylate, and 4-phenyl-2-hydroxy-5-acetate (Sietmann et al. 2001).

3. A cytochrome P450 purified from *Saccharomyces cerevisiae* had benzo[*a*]pyrene hydroxylase activity (King et al. 1984), and metabolized benzo[*a*]pyrene to 3- and 9-hydroxybenzo[*a*]pyrene and benzo[*a*]pyrene-7,8-dihydrodiol (Wiseman and Woods 1979). The transformation of PAHs by *Candida lipolytica* produced predominantly monohydroxylated products: naphth-1-ol from naphthalene, 4-hydroxybiphenyl from biphenyl and 3- and 9-hydroxybenzo[*a*]pyrene from

benzo[*a*]pyrene (Cerniglia and Crow 1981). The transformation of phenanthrene was demonstrated in a number of yeasts isolated from littoral sediments and of these, *Trichosporum penicillatum* was the most active. In contrast, biotransformation of benz[*a*]anthracene by *Candida krusei* and *Rhodotorula minuta* was much slower (MacGillivray and Shiaris 1993).

4. Naphthalene, biphenyl, and phenanthrene were transformed in low yield by oxygenic phototrophs (algae and cyanobacteria) with the formation of several metabolites. A range of algae produced naphth-1-ol from naphthalene (Cerniglia et al. 1980a,b, 1982a), while biphenyl was transformed by a strain of the cyanobacterium *Oscillatoria* sp. to 4-hydroxy-biphenyl (Cerniglia et al. 1984a). A study with the same strain showed that the naphth-1-ol formed from [1-^{2}H]- and [2-^{2}H]-naphthalene retained 68% and 74% of the deuterium, respectively. This shows that the reaction proceeded by the initial formation of naphthalene-1,2-epoxide followed by an NIH shift (Narro et al. 1992a). The transformation of phenanthrene by the unicellular cyanobacterium *Agmenellum quadruplicatum* involved monooxygenation to produce the *trans*-9,10-dihydrodiol and lesser amounts of 3-methoxyphenanthrene. The dihydrodiol which has an enantiomeric excess of the (–)-[9*S*,10*S*] enantiomer was produced by the introduction of a single atom of $^{18}O_2$ (Narro et al. 1992b), and is formally analogous to the transformation of naphthalene to naphth-1-ol by a strain of *Oscillatoria* sp., which has already been noted.

5. The products from the metabolism of benzo[*a*]pyrene by the green alga *Selenastrum capricornutum* strain UTEX 1648 are unexpected, since the *cis*-dihydrodiols at the 11,12-, 7,8-, and 4,5-positions were apparently produced by dioxygenation (Warshawsky et al. 1995b). The light conditions determine both the preferred site of dioxygenation and the relative contribution of the phytotoxic benzo[*a*]pyrene-3,6-quinone (Warshawsky et al. 1995a). An interesting and exceptional example of ring fission by algae is afforded by the degradation of phenol by *Ochromonas danica* via catechol followed by extradiol fission (Semple and Cain 1996).

8.11 White-Rot Fungi

There has been considerable interest in white-rot fungi since they are capable of degrading not only lignin but also a wide range of other substrates, including PAHs. They are discussed separately since the mechanisms whereby they accomplish degradation or transformation of PAHs differ significantly from the fungi already discussed. Although considerable attention has been directed to *Phanerochaete chrysosporium*, the degradation and transformation of PAHs is not limited to this taxon (Ferris et al. 1976). Investigations with *Ph. laevis* and *Pleurotus ostreatus* (Bezalel et al. 1996a) have brought to light significantly different metabolic pathways. It should be pointed out that 10 different strains of *Ph. chrysosporium* have been shown to harbor bacteria, although only a few of these have been identified (Seigle-Murandi et al. 1996). The metabolic consequence of this association is apparently unresolved.

The mechanism and enzymology of transformations by white-rot fungi are complicated for several issues:

1. The synthesis of the lignin and Mn peroxidases is regulated by several factors, including the nitrogen status of the cells, the oxygen concentration, and the concentration of Mn^{2+} in the growth medium.

2. The dependence of enzyme activity on the growth phase of the cells. The results of experiments with whole cells may, therefore, differ from those obtained with pure enzymes or cell extracts.

3. The range of oxygenation mechanisms that has emerged.

A brief summary of these factors includes the following issues:

1. *Peroxidase systems.* There are at least three of these extracellular enzyme systems: (a) lignin peroxidase (LiP), (b) manganese-dependent peroxidase (MnP), and (c) manganese peroxide-dependent lipid peroxidase. The first functions by the formation of a cation radical that undergoes further

reaction, the second by the formation of Mn^{3+} that brings about oxidation of the substrate, and the third probably by proton abstraction followed by the introduction of oxygen. The enzyme system responsible for lignin degradation is expressed during idiophasic growth as a result of nitrogen limitation. The regulation of the synthesis of LiP and MnP is complex, and depends on the Mn^{2+} concentration in the medium (Bonnarme and Jeffries 1990; Brown et al. 1990), although it is not apparently subject to carbon substrate regulation (van der Woude et al. 1993). However, the synthesis of manganese peroxidase is regulated by concentrations of H_2O_2 and O_2 (Li et al. 1995).

2. *Monooxygenases.* Under nonlignolytic conditions, arene monooxygenase and epoxide hydrolase systems may function to produce *trans*-dihydrodiols. Hydrogen abstraction mediated by the lipid peroxidase system may operate, for example, in the formation of fluorene-9-one from fluorene by *Ph. chrysosporium* (Bogan et al. 1996).

3. *Hydroxylases.* Some substrates are activated to cation radicals that react further with H_2O to ultimately produce quinones. This is particularly prevalent in 4- and 5-ring compounds, though important alternatives exist.

4. *Dioxygenase.* Although it is not involved in the transformation of PAHs, it may be noted for the sake of completeness that a catechol intradiol dioxygenase is involved in the fission of 1,2,4-trihydroxybenzene that is formed from a number of aromatic substrates by the basidiomycete *Phanerochaete chrysosporium* (Rieble et al. 1994).

Although a number of white-rot fungi have been examined and shown to degrade PAHs (Field et al. 1992), greatest attention has probably been directed to *Phanerochaete chrysosporium* and *Pleurotus ostreatus*, and to the PAHs anthracene, phenanthrene, pyrene, and benzo[a]pyrene that will be used to illustrate the cardinal principles. A substantial fraction of PAHs may also be sorbed to the biomass—40% for phenanthrene and 22% for benzo[a]pyrene (Barclay et al. 1995). The degree of mineralization of PAHs by white-rot fungi may sometimes be quite low, for example, for *Pleurotus ostreatus*, yields were 3.0%, 0.44%, 0.19%, and 0.19% for phenanthrene, pyrene, fluorene, and benzo[a]pyrene, respectively (Bezalel et al. 1996a).

8.11.1 Anthracene

Degradation by *Ph. chrysosporium* took place by the initial formation of anthra-9,10-quinone followed by ring fission to phthalate (Figure 8.28) (Hammel et al. 1991). This pathway is completely different from that used by both bacteria and fungi, and the fission of the quinone might plausibly involve a Baeyer–Villiger-type insertion of oxygen. In contrast, *P. ostreatus* formed both anthra-9,10-quinone and anthracene *trans*-1,2-dihydrodiols as terminal metabolites, with an *S,S*- to *R,R*- ratio of 58:42 in the latter (Bezalel et al. 1996c).

8.11.2 Phenanthrene

There are different pathways that have been delineated and these are considered separately as follows:

1. The transformation of phenanthrene has been extensively studied and illustrates the operation of alternative pathways that depend on the status of the cells. During cell growth of *Ph. chrysosporium*, the *trans*-3,4- and *trans*-9,10-diols were formed and rearranged to produce 3-, 4-, and 9-phenanthrenols (Figure 8.29a). Since lignin peroxidase activity was not observed, it was suggested that the *trans*-diols were formed by monooxygenase and epoxide hydrolase activities (Sutherland et al. 1991).

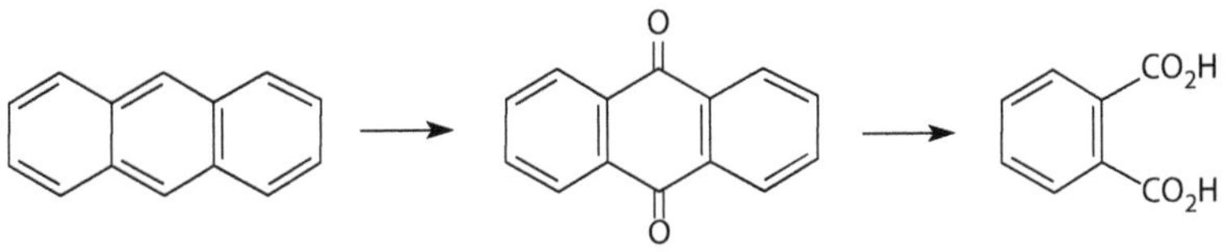

FIGURE 8.28 Degradation of anthracene by *Phanerochaete chrysosporium*.

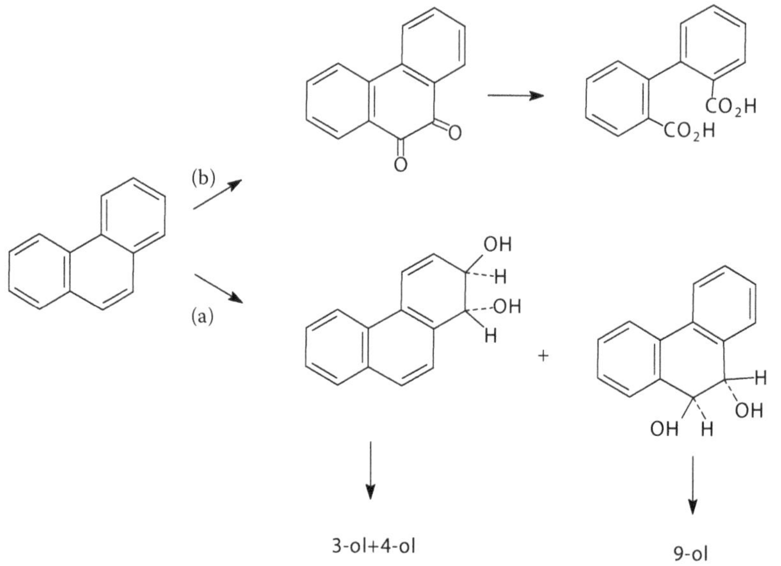

FIGURE 8.29 Pathways for transformation of phenanthrene by *Phanerochaete chrysosporium*. (a) Dihydroxylation to the *trans*-diols followed by formation of phenanthrols, (b) oxidation to the 9,10-quinone followed by ring fission.

2. The pathways for the metabolism of phenanthrene by *Phanerochaete chrysosporium* depend on the conditions. In a complex medium, phenanthrene *trans*-3,4- and *trans*-9,10 dihydrodiols are formed and rearranged to 3- and 4-phenanthrol, and 9-phenanthrol, respectively, (Figure 8.29a) (Sutherland et al. 1991). Under conditions where lignin degradation is induced, oxidation to phenanthrene-9,10-quinone takes place before ring fission to biphenyl-2,2′-dicarboxylate (Figure 8.29b) (Hammel et al. 1992). Cytochrome P450 activation was not apparently involved, and extracellular ligninases were inactive *in vitro*. The same products, with a preponderance of the intermediate *trans*-[*R,R*]-9,10-dihydrodiol, were produced by *P. ostreacus*, and it was suggested that a cytochrome P450 system was involved (Bezalel et al. 1996b, 1997).

3. The transformation of phenanthrene has been examined in *Pleurotus ostreatus* and was carried out in phases (Bezalel et al. 1997). Phase I was mediated by cytosolic and microsomal cytochrome P450 monooxygenases to form the epoxide, followed by the activity of an epoxide hydrolase to produce phenanthrene *trans*-9,10-dihydrodiol. The phase II enzyme glutathione *S*-transferase was found in the cytosolic fraction, whereas the activities of UDP-glucosyltransferase and UDP-glucuronosyltransferase were found in the microsomal fraction in spite of the fact that the expected conjugates were not detected.

4. Biphenyl-2,2′-dicarboxylate was formed by *Ph. chrysosporium* in a reaction mixture with manganese peroxidase, O_2, and unsaturated lipid, and it was suggested that a MnP-mediated lipid peroxidation was involved (Moen and Hammel 1994).

A synthesis of these results has been presented, and evidence presented for the role of alkoxy radicals generated during lipid oxidation (Tatarko and Bumpus 1993; Bogan and Lamar 1995).

8.11.3 Pyrene and Benzo[a]pyrene

Purified ligninase H8 produced by *P. chrysosporium* in stationary cultures oxidized pyrene to pyrene-1,6- and pyrene-1,8-quinones in high yield, and experiments with $H_2^{18}O$ showed that both quinone oxygen atoms originated in water (Figure 8.30). It was suggested that initial one-electron abstraction produced cation radicals at the 1 and 6 or 8-positions (Hammel et al. 1986), whereas in *P. ostreacus* that produced the *trans*-4,5-dihydrodiols with a 63:37 ratio of the *R,R* and *S,S* enantiomers, a cytochrome P450 monooxygenase was involved (Bezalel et al. 1996b). In analogy with the transformation of pyrene, the

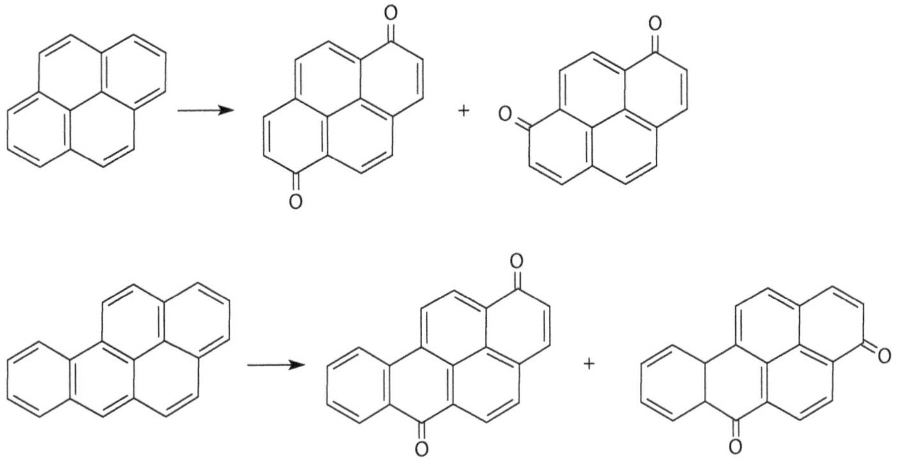

FIGURE 8.30 Quinones produced from pyrene and benzo[*a*]pyrene.

1,6-, 3,6-, and 6,12-quinones were formed from benzo[*a*]pyrene by *Ph. chrysosporium* (Haemmerli et al. 1986). These reactions are analogous to those involved in the electrochemical oxidation of benzo[*a*]pyrene (Jeftic and Adams 1970), and oxidation by Mn(III) acetate (Cremonesi et al. 1989), while substantial evidence supports the role of cation radicals as intermediates in biological oxidations and the tumorogenic properties of some PAHs (Cavalieri and Rogan 1998).

In most of the above illustrations, quinones have been formed and they may be terminal metabolites or only transient intermediates that are produced, for example, during mineralization by the basidiomycete *Stropharia coronilla* (Steffen et al. 2003). Degradation was stimulated by addition of Mn^{2+}, and could be attributed to elevated levels of the lignolytic manganese peroxidase. Experiments with *Ph. laevis* strain HHB-1625 have revealed some additional features (Bogan and Lamar 1996):

1. Under conditions of nitrogen limitation in the presence of Mn^{2+}, manganese peroxidase activity was induced although no lignin peroxidase activity could be demonstrated.

2. Transformation of anthracene, phenanthrene, benz[*a*]anthracene, and benzo[*a*]pyrene was demonstrated without the accumulation of quinones that were, at best, transient metabolites.

3. In comparison with *Ph. chrysosporium*, *Ph. laevis* was more effective in mineralizing [5,6-^{14}C] benz[*a*]anthracene-7-12-quinone.

In summary, it may be stated that for a wide range of PAHs, the specific association of dioxygenation to give *cis*-dihydrodiols by prokaryotes and of monooxygenation to epoxides and thence to *trans*-dihydrodiols by eukaryotes is generally valid. The absolute stereochemistry of the *trans*-dihydrodiols produced by fungi is highly variable, and there are frequently significant differences between the products of fungal monooxygenation and those carried out by rat-liver microsomes. Although quinones are frequently formed as metabolites of the white-rot fungus *Ph. chrysosporium*, these are only transiently formed by *Ph. laevis*, so that care should be exercised in making generalizations on the basis of the results obtained for a single species or strain. The intermediate epoxides may rearrange to produce phenols that are generally conjugated.

8.12 References

Adachi K, T Iwabuchi, H Sano, S Harayama. 1999. Structure of the ring cleavage product of 1-hydroxy-2-naphthoate, an intermediate of the phenanthrene-degradative pathway of *Nocardioides* sp. strain KP7. *J Bacteriol* 181: 757–763.

Adriaens P. 1994. Evidence for chlorine migration during oxidation of 2-chlorobiphenyl by a type II methano-troph. *Appl Environ Microbiol* 60: 1658–1662.

Akhtar MN, DR Boyd, NJ Thompson, M Koreeda, DT Gibson, V Mahadevan, DM Jerina. 1975. Absolute stereochemistry of the dihydroanthracene-*cis*- and -*trans*,1,2-diols from anthracene by mammals and bacteria. *J Chem Soc Perkin* I: 2506–2511.

Allen CC, DR Boyd, MJ Larkin, KA Reid, ND Sharma, K Wilson. 1997. Metabolism of naphthalene, 1-naph-thol, indene, and indole by *Rhodococcus* sp. strain NCIMB 12038. *Appl Environ Microbiol* 63: 151–155.

Annweiler E, W Michaelis, RU Meckenstock. 2002. Identical ring cleavage products during anaerobic degrada-tion of naphthalene, 2-methylnaphthalene, and tetralin indicate a new metabolic pathway. *Appl Environ Microbiol* 68: 853–858.

Asturias JA, KN Timmis. 1993. Three different 2,3-dihydroxybiphenyl-1,2-dioxygenase genes in the Gram-positive polychlorobiphenyl-degrading bacterium *Rhodococcus globerulus* P6. *J Bacteriol* 175: 4631–4640.

Austen RA, NW Dunn. 1980. Regulation of the plasmid-specified naphthalene catabolic pathway of *Pseudomonas putida. J Gen Microbiol* 117: 521–528.

Barclay CD, GF Farquhar, RL Legge. 1995. Biodegradation and sorption of polyaromatic hydrocarbons by *Phanerochaete chrysosporium. Appl Microbiol Biotechnol* 42: 958–963.

Barnsley EA. 1976. Role and regulation of the *ortho* and *meta* pathways of catechol metabolism in pseudomo-nads metabolizing naphthalene and salicylate. *J Bacteriol* 125: 404–408.

Barnsley EA. 1983. Phthalate pathway of phenanthrene metabolism: Formation of 2'-carboxybenzalpyruvate. *J Bacteriol* 154: 113–117.

Barnsley EA. 1988. Metabolism of 2,6-dimethylnaphthalene by flavobacteria. *Appl Environ Microbiol* 54: 428–433.

Beller HR. 2002. Analysis of benzylsuccinates in groundwater by liquid chromatography/tandem mass spec-trometry and its use for monitoring *in situ* BTEX biodegradation. *Environ Sci Technol* 36: 2724–2728.

Beller HR, W-H Ding, M Reinhard. 1995. Byproducts of anaerobic alkylbenzene metabolism useful as indica-tors of *in situ* bioremediation. *Environ Sci Technol* 29: 2864–2870.

Beller HR, AM Spormann. 1998. Analysis of the novel benzyl succinate synthase reaction for anaerobic tolu-ene activation based on structural studies of the product. *J Bacteriol* 180: 5454–5457.

Bezalel L, Y Hadar, CE Cerniglia. 1996a. Mineralization of polycyclic aromatic hydrocarbons by the white-rot fungus *Pleurotus ostreatus. Appl Environ Microbiol* 62: 292–295.

Bezalel L, Y Hadar, CE Cerniglia. 1997. Enzymatic mechanisms involved in phenanthrene degradation by the white-rot fungus *Pleurotus ostreatus. Appl Environ Microbiol* 63: 2495–2501.

Bezalel L, Y Hadar, PP Fu, JP Freeman, CE Cerniglia. 1996b. Metabolism of phenanthrene by the white rot fungus *Pleurotus ostreatus. Appl Environ Microbiol* 62: 2547–2553.

Bezalel L, Y Hadar, PP Fu, JP Freeman, CE Cerniglia. 1996c. Initial oxidation products in the metabolism of pyrene, anthracene, fluorene, and dibenzothiophene by the white rot fungus *Pleurotus ostreatus. Appl Environ Microbiol* 62: 2554–2559.

Bogan BW, RT Lamar. 1995. One-electron oxidation in the degradation of creosote polycyclic aromatic hydro-carbons by *Phanerochaete chrysosporium. Appl Environ Microbiol* 61: 2631–2635.

Bogan BW, RT Lamar. 1996. Polycyclic aromatic hydrocarbon-degrading capabilities of *Phanerochaete laevis* HHB-1625 and its extracellular lignolytic enzymes. *Appl Environ Microbiol* 62: 1597–1603.

Bogan L, RT Lamar, KE Hammel. 1996. Fluorene oxidation *in vivo* by *Phanerochaete chrysosporium* and *in vitro* during manganese peroxidase-dependent lipid peroxidation. *Appl Environ Microbiol* 62: 1788–1792.

Boldrin B, A Tiehm, C Fritzsche. 1993. Degradation of phenanthrene, fluorene, fluoranthene, and pyrene by a *Mycobacterium* sp. *Appl Environ Microbiol* 59: 1927–1930.

Bollag J-M, EJ Czaplicki, RD Minard. 1975. Bacterial metabolism of 1-naphthol. *J Agric Food Chem* 23: 85–90.

Bonnarme P, TW Jeffries. 1990. Mn(II) regulation of lignin peroxidases and manganese-dependent peroxidases from lignin-degrading white-rot fungi. *Appl Environ Microbiol* 56: 210–217.

Boyd DR, ND Sharma, F Hempenstall, MA Kennedy, JF Malone, CCR Allen. 1999. *bis-cis*-Dihydrodiols A new class of metabolites resulting from biphenyl dioxygenase-catalyzed sequential asymmetric *cis*-dihy-droxylation of polycyclic arenes and heteroarenes. *J Org Chem* 64: 4005–4011.

Bregnard TP-A, P Höhener, A Häner, J Zeyer. 1996. Degradation of weathered diesel fuel by microorganisms from a contaminated aquifer in aerobic and anaerobic conditions. *Environ Toxicol Chem* 15: 299–307.

Brezna B, AA Khan, CE Cerniglia. 2003. Molecular characterization of dioxygenases froim polycyclic aromatic hydrocarbon-degrading *Mycobacterium* spp. *FEMS Microbiol Lett* 223: 177–183.

Brezna B, O Kweon, RI Stingley, JP Freeman, AA Khan, B Polek, RC Jones, CE Cerniglia. 2006. Molecular characterization of cytochrome P450 genes in the polycyclic aromatic hydrocarbon degrading *Mycobacterium vanbaalenii* PYR-1. *Appl Microbiol Biotechnol* 71: 522–532.

Brown JA, JK Glenn, MA Gold. 1990. Manganese regulates expression of manganese peroxidase by *Phanerochaete chrysosporium*. *J Bacteriol* 172: 3125–3130.

Cavalieri E, E Rogan. 1998. Mechanisms of tumor initiation by polycyclic aromatic hydrocarbons in mammals. *Handbook Environ Chem* 3J: 82–117.

Cerniglia CE, JR Althus, FE Evans, JP Freeman, RK Mitchum, SK Yang. 1983. Stereochemistry and evidence for an arene oxide-NIH shift pathway in the fungal metabolism of naphthalene. *Chem-Biol Interactions* 44: 119–132.

Cerniglia CE, C van Baalen, DT Gibson. 1980b. Oxidation of biphenyl by the cyanobacterium, *Oscillatoria* sp. strain JCM. *Arch Microbiol* 125: 203–207.

Cerniglia CE, WL Campbell, JP Freeman, FE Evans. 1989. Identification of a novel metabolite in phenanthrene metabolism by the fungus *Cunninghamella elegans*. *Appl Environ Microbiol* 55: 2275–2279.

Cerniglia CE, SA Crow. 1981. Metabolism of aromatic hydrocarbons by yeasts. *Arch Microbiol* 129: 9–13.

Cerniglia CE, RH Dodge, DT Gibson. 1982d. Fungal oxidation of 3-methylcholanthrene: Formation of proximate carcinogenic metabolites of 3-methylcholanthrene. *Chem-Biol Interactions* 38: 161–173.

Cerniglia CE, JP Freeman, FE Evans. 1984a. Evidence for an arene oxide-NIH shift pathway in the transformation of naphthalene to 1-naphthol by *Bacillus cereus*. *Arch Microbiol* 138: 283–286.

Cerniglia CE, JP Freeman, RK Mitchum. 1982b. Glucuronide and sulfate conjugation in the fungal metabolism of aromatic hydrocarbons. *Appl Environ Microbiol* 43: 1070–1075.

Cerniglia CE, PP Fu, SK Yang. 1982c. Metabolism of 7-methylbenz[*a*]anthracene and 7-hydroxymethylbenz[*a*] anthracene by *Cunninghamella elegans*. *Appl. Environ. Microbiol.* 44: 682–689.

Cerniglia CE, DT Gibson. 1979. Oxidation of benzo[*a*]pyrene by the filamentous fungus *Cunninghamella elegans*. *J Biol Chem* 254: 12174–12180.

Cerniglia CE, DT Gibson. 1980a. Fungal oxidation of benzo[*a*]pyrene and (+/−)-*trans*-7,8-dihydroxy-7,8-dihydrobenzo[*a*]pyrene: Evidence for the formation of a benzo[*a*]pyrene 7,8-diol-9,10-epoxide. *J Biol Chem* 255: 5159–5163.

Cerniglia CE, DT Gibson. 1980b. Fungal oxidation of (+/−)-9,10-dihydroxy-9-10-dihydrobenzo[*a*]pyrene: Formation of diastereomeric benzo[*a*]pyrene 9,10-diol-7,8-epoxides. *Proc Natl Acad USA* 77: 4554–4558.

Cerniglia CE, DT Gibson, C van Baalen. 1980a. Oxidation of naphthalene by cyanobacteria and microalgae. *J Gen Microbiol* 116: 495–500.

Cerniglia CE, DT Gibson, C van Baalen. 1982a. Naphthalene metabolism by diatoms isolated from the Kachemak Bay region of Alaska. *J Gen Microbiol* 128: 987–990.

Cerniglia CE, RL Herbert, PJ Szaniszlo, DT Gibson. 1978. Fungal metabolism of naphthalene. *Arch Microbiol* 117: 135–143.

Cerniglia CE, DW Kelley, JP Freeman, DW Miller. 1986. Microbial metabolism of pyrene. *Chem-Biol Interactions* 57: 203–216.

Cerniglia CE, KL Lambert, DW Mille, JP Freeman. 1984b. Transformation of 1- and 2-methylnaphthalene by *Cunninghamella elegans*. *Appl Environ Microbiol* 47: 111–118.

Cerniglia CE, SK Yang. 1984. Stereoselective metabolism of anthracene and phenanthrene by the fungus *Cunninghamella elegans*. *Appl. Environ. Microbiol.* 47: 119–124.

Coates JD, RT Anderson, DR Lovley. 1996. Oxidation of polycyclic aromatic hydrocarbons under sulfate-reducing conditions. *Appl Environ Microbiol* 62: 1099–1101.

Coates JD, J Woodward, J Allen, P Philip, DR Lovley. 1997. Anaerobic degradation of polycyclic aromatic hydrocarbons and alkanes in petroleum-contaminated marine harbor sediments. *Appl Environ Microbiol* 63: 3589–3593.

Cremonesi P, EL Cavalieri, EG Rogan. 1989. One-electron oxidation of 6-substituted benzo[*a*]pyrenes by manganic acetate: A model for metabolic activation. *J Org Chem* 54: 3561–3570.

Dalton H, BT Gording, BW Watyers, R Higgins, JA Taylor. 1981. Oxidations of cyclopropane, methylcyclopropane, and arenes with the monooxygenase system from *Methylococccus capsulatus*. *J Chem Soc Chem Commun* 482–483.

Daly JW, DM Jerina, B Witkop. 1972. Arene oxides and the NIH shift: The metabolism, toxicity and carcinogenicity of aromatic compounds. *Experientia* 28: 1129–1149.

Demanèche S, C Meyer, J Micoud, M Louwagie, JC Willison, Y Jouanneau. 2004. Identification and functional analysis of two aromatic-ring-hydroxylating dioxygenases from a *Sphingomonas* strain that degrades various polycyclic aromatic hydrocarbons. *Appl Environ Microbiol* 70: 6714–6725.

Dodge RH, CE Cerniglia, DT Gibson. 1979. Fungal metabolism of biphenyl. *Biochem J* 178: 223–230.

Dunn NW, HM Dunn, RA Austen. 1980. Evidence for the existence of two catabolic plasmids coding for the degradation of naphthalene. *J Gen Microbiol* 117: 529–533.

Dunn NW, IC Gunsalus. 1973. Transmissible plasmid coding early enzymes of naphthalene oxidation in *Pseudomonas putida. J Bacteriol* 114: 974–979.

Dutta TK, SA Selifonov, IC Gunsalus. 1998. Oxidation of methyl-substituted naphthalenes: Pathways in a versatile *Sphingomomas paucimobilis* strain. *Appl Environ Microbiol* 64: 1884–1889.

Eaton RW, PJ Chapman. 1992. Bacterial metabolism of naphthalene: Construction and use of recombinant bacteria to study ring cleavage of 1,2-dihydroxynaphthalene and subsequent reactions. *J Bacteriol* 174: 7542–7554.

Eaton SL, SM Resnick, DT Gibson. 1996. Initial reactions in the oxidation of 1,2-dihydronaphthalene by *Sphingomonas yanoikuyae* strains. *Appl Environ Microbiol* 62: 4388–4394.

Ensley BD, DT Gibson. 1983. Naphthalene dioxygenase: Purification and properties of a terminal oxygenase component. *J Bacteriol* 155: 505–511.

Erickson BD, FJ Mondello. 1992. Nucleotide sequencing and transcriptional mapping of the genes encoding biphenyl dioxygenase, a multicomponent polychlorinated-biphenyl-degrading enzyme in *Pseudomonas* strain LB400. *J Bacteriol* 174: 2903–2912.

Evans WC, HN Fernley, E Griffiths. 1965. Oxidative metabolism of phenanthrene and anthracene by soil pseudomonads. The ring-fission mechanism. *Biochem J* 95: 819–831.

Faulkner JK, D Woodcock. 1965. Fungal detoxication. Part VII. Metabolism of 2,4-dichlorophenoxyacetic and 4-chloro-2-methylphenoxyacetic acids by *Aspergillus niger. J Chem Soc* 1187–1191.

Ferraro DJ, EN Brown, C-L Yu, RE Parales, DT Gibson, S Ramaswamy. 2007. Structural investigation of the ferredoxin and terminal oxygenase components of the biphenyl 2,3-dioxygenase from *Sphingobium yanoikuyae* B1. *BMC Struct Biol* 7: 10.

Ferris JP, LH MacDonald, MA Patrie, MA Martin. 1976. Aryl hydrocarbon hydroxylase activity in the fungus *Cunninghamella bainieri*: Evidence for the presence of cytochrome P-450. *Arch Biochem Biophys* 175: 443–452.

Field JA, E de Jong, GF Costa GF, JAM de Bont. 1992. Biodegradation of polycyclic aromatic hydrocarbons by new isolates of white-rot fungi. *Appl Environ Microbiol* 58: 2219–2226.

Fuenmayor SL, M Wild, A Boyes, PA Williams. 1998. A gene cluster encoding steps in conversion of naphthalene to gentisate in *Pseudomonas* sp. strain U2. *J Bacteriol* 180: 2522–2530.

Furukawa K, JR Simon, AM Chakrabarty. 1983. Common induction and regulation of biphenyl, xylene/toluene, and salicylate catabolism in *Pseudomonas paucimobilis. J Bacteriol* 154: 1356–1362.

Galushko A, D Minz, B Schink, F Widdel. 1999. Anaerobic degradation of naphthalene by a pure culture of a novel type of marine sulphate-reducing bacterium. *Environ Microbiol* 1: 415–420.

Geiselbrecht AD, BP Hedland, MA Tichi, JT Staley. 1998. Isolation of marine polycyclic aromatic hydrocarbon (PAH)-degrading *Cycloclasticus* strains from the Gulf of Mexico and comparison of their PAH degradation ability with that of Puget Sound *Cycloclasticus* strains. *Appl Environ Microbiol* 64: 4703–4710.

Gesell M, E Hammer, M Specht, W Francke, F Schauer. 2001. Biotransformation of biphenyl by *Paecilomyces lilacinus* and characterization of ring cleavage products. *Appl Environ Microbiol* 67: 1551–1557.

Gibson DT. 1999. *Beijerinckia* sp. strain B1: A strain by any other name. *J Ind Microbiol Biotechnol* 23: 284–293.

Gibson DT, RE Parales. 2000. Aromatic hydrocarbon dioxygenases in environmental biotechnology. *Curr Opinion Biotechnol* 11: 236–243.

Gibson DT, SM Resnick, K Lee, JM Brand, DS Torok, LP Wackett, MJ Schocken, BE Haigler. 1995. Desaturation, dioxygenation, and monooxygenation reactions catalyzed by naphthalene dioxygenase from *Pseudomonas* sp. strain 9816-4. *J Bacteriol* 177: 2615–2621.

Gibson DT, RL Roberts, LC Wells, VM Kopbal. 1973. Oxidation of biphenyl by a *Baijerinckia* species. *Biochem Biophys Res Commun* 50: 211–219.

Gibson DT, V Subramanian. 1984. Microbial degradation of aromatic hydrocarbons, pp. 181–252. In *Microbial Degradation of Organic Compounds* (Ed. DT Gibson), Marcel Dekker Inc., New York.

Golbeck JH, SA Albaugh, R Radmer. 1983. Metabolism of biphenyl by *Aspergillus toxicarius*: Induction of hydroxylating activity and accumulation of water-soluble conjugates. *J Bacteriol* 156: 49–57.

Goyal AK, GJ Zylstra. 1996. Molecular cloning of novel genes for polycyclic aromatic hydrocarbon degradation from *Comamonas testosteroni*. *Appl Environ Microbiol* 62: 230–236.

Gray PHH, HG Thornton. 1928. Soil bacteria that decompose certain aromatic compounds. *Centralbl Bakteriol Parasitenkd Infektionskr Abt II* 73: 74–96.

Grifoll M, M Casellas, JM Bayona, AM Solanas. 1992. Isolation and characterization of a fluorene-degrading bacterium: Identification of ring oxidation and ring fission products. *Appl Environ Microbiol* 58: 2910–2917.

Grifoll M, SA Selifonov, PJ Chapman. 1994. Evidence for a novel pathway in the degradation of fluorene by *Pseudomonas* sp. strain F274. *Appl Environ Microbiol* 60: 2438–2449.

Grifoll M, SA Selifonov, CV Gatlin, PJ Chapman. 1995. Actions of a versatile fluorene-degrading bacterial isolate on polycyclic aromatic compounds. *Appl Environ Microbiol* 61: 3711–3723.

Grund E, B Denecke, R Eichenlaub. 1992. Naphthalene degradation via salicylate and gentisate by *Rhodococcus* sp. strain B4. *Appl Environ Microbiol* 58: 1874–1877.

Guengerich P. 1990. Enzymatic oxidation of xenobiotic chemicals. *Crit Revs Biochem Mol Biol* 25: 97–153.

Haddock JD, DT Gibson. 1995. Purification and characterization of the oxygenase component of biphenyl 2,3-dioxygenase from *Pseudomonas* sp. strain LB400. *J Bacteriol* 177: 5834–5839.

Haddock JD, JR Horton, DT Gibson. 1995. Dihydroxylation and dechlorination of chlorinated biphenyls by purified biphenyl 2,3-dioxygenase from *Pseudomonas* sp. strain LB400. *J Bacteriol* 177: 20–26.

Haemmerli SD, MSA Leisola, D Sanglard, A Fiechter. 1986. Oxidation of benzo[*a*]pyrene by extracellular ligninases of *Phanerochaete chrysosporium*. *J Biol Chem* 261: 6900–6903.

Haigler BE, DT Gibson. 1990a. Purification and properties of ferredoxinNAP, a component of naphthalene dioxygenase from *Pseudomonas* sp. strain NCIB9816. *J Bacteriol* 172: 465–468.

Haigler BE, DT Gibson. 1990b. Purification and properties of NADH-ferredoxinNAP reductase, a component of naphthalene dioxygenase from *Pseudomonas* sp. strain NCIB9816. *J Bacteriol* 172: 457–464.

Hammel KE, WZ Gai, B Green, MA Moen. 1992. Oxidative degradation of phenanthrene by the lignolytic fungus *Phanerochaete chrysposporium*. *Appl Environ Microbiol* 58: 1832–1838.

Hammel KE, B Green, WZ Gai. 1991. Ring fission of anthracene by a eukaryote. *Proc Natl Acad Sci USA* 88: 10605–10608.

Hammel KE, B Kalyanaraman, TK Kirk. 1986. Oxidation of polycyclic aromatic hydrocarbons and dibenzo[*p*]-dioxins by *Phanerochaete chrysosporium*. *J Biol Chem* 261: 16948–16952.

Hedlund BP, AD Geiselbrecht, TJ Bair, JT Staley. 1999. Polycyclic aromatic hydrocarbon degradation by a new marine bacterium *Neptunomonas naphthovorans* gen. nov., sp. nov. *Appl Environ Microbiol* 65: 251–259.

Heitkamp MA, JP Freeman, DW Miller, CE Cerniglia. 1988a. Pyrene degradation by a *Mycobacterium* sp.: Identification of ring oxidation and ring fission products. *Appl Environ Microbiol* 54: 2556–2565.

Heitkamp MA, W Franklin, CE Cerniglia. 1988b. Microbial metabolism of polycyclic aromatic hydrocarbons: Isolation and characterization of a pyrene-degrading bacterium. *Appl Environ Microbiol* 54: 2549–2555.

Hennessee CT, J-S Seo, AM Alvarez, QX Li. 2009. Polycyclic aromatic hydrocarbon-degrading species isolated from Hawaiian soils: *Mycobacterium crocinum*, sp. nov., *Mycobacterium pallens* sp. nov., *Mycobacterium rutilum* sp. nov., *Mycobacterium rufum* sp. nov. and *Mycobacterium aromaticivorans* sp. nov. *Int J Syst Evol Microbiol* 59: 378–387.

Hernáez MJ, B Floriano, JJ Ríos, E Santero. 2002. Identification of a hydratase and a class II aldolase involved in biodegradation of the organic solvent tetralin. *Appl Environ Microbiol* 68: 4841–4846

Hurtubise Y, D Barriault, J Powlowski, M Sylvestre. 1995. Purification and characterization of the *Comamonas testosteroni* B-356 biphenyl dioxygenase components. *J Bacteriol* 177: 6610–6618.

Iwabuchi T, S Harayama. 1997. Biochemical and genetic characterization of 2-carboxybenzaldehyde dehydrogenase, an enzyme involved in phenanthrene degradation by *Nocardioides* sp. strain KP7. *J Bacteriol* 179: 6488–6494.

Iwabuchi T, S Harayama. 1998a. Biochemical and molecular characterization of 1-hydroxy-2-naphthoate dioxygenase from *Nocardioides* sp. KP7. *J Biol Chem* 273: 8332–8336.

Iwabuchi T, S Harayama. 1998b. Biochemical and genetic characterization of *trans*-2'-carboxybenzalpyruvate hydratase-aldolase from a phenanthrene-degrading *Nocardioides* strain. *J Bacteriol* 180: 945–949.

Jeffrey AM, HJC Yeh, DM Jerina, TR Patel, JF Davey, DT Gibson. 1975. Initial reactions in the oxidation of naphthalene by *Pseudomonas putida*. *Biochemistry* 14: 575–584.

Jeftic L, RN Adams. 1970. Electrochemical oxidation pathways of benzo[*a*]pyrene. *J Am Chem Soc* 92: 1332–1337.

Jerina DM, PJ van Bladeren, H Yagi, DT Gibson, V Mahadevan, AS Neese, M Koreeda, ND Sharma, DR Boyd. 1984. Synthesis and absolute configuration of the bacterial *cis*-1,2-, *cis*-8,9-, and *cis*-10, 11-dihydrodiol metabolites of benz[*a*]anthacene by a strain of *Beijerinckia*. *J Org Chem* 49: 3621–3628.

Jouanneau Y, C Meyer. 2006. Purification and characterization of an arene *cis*-dihydrodiol dehydrogenase endowed with broad substrate specificity towards polycyclic aromatic hydrocarbon dihydrodiols. *Appl Environ Microbiol* 72: 4726–4734.

Juhasz AL, ML Britz, GA Stanley. 1997. Degradation of benzo[*a*]pyrene, dibenz[*a,h*]anthracene and coronene by *Burkholderia cepacia*. *Water Sci Technol* 36: 45–51.

Juhasz AL, R Naidu. 2000. Bioremediation of high molecular weight polycyclic aromatic hydrocarbons: A review of the microbial degradation of benzo[*a*]pyrene. *Int Biodet Biodeg* 45: 57–88.

Kanaly RA, S Harayama. 2000. Biodegradation of high molecular weight polycyclic aromatic hydrocarbons. *J Bacteriol* 182: 2059–2067.

Kanaly RA, S Harayama, K Watanabe. 2002. *Rhodanobacter* sp. strain BPC1 in a benzo[*a*]pyrene-mineralizing bacterial consortium. *Appl Environ Microbiol* 68: 5826–5833.

Kasai Y, K Shindo, S Harayama, N Misawa. 2003. Molecular characterization and substrate preference of a polycyclic aromatic hydrocarbon dioxygenase from *Cycloclasticus* sp. strain A5. *Appl Environ Microbiol* 69: 6688–6697.

Kazunga C, MD Aitken. 2000. Products from the incomplete metabolism of pyrene by polycyclic aromatic hydrocarbon-degrading bacteria. *Appl Environ Microbiol* 66: 1917–1922.

Kazunga C, MD Aitken, A Gold, R Sangaiah. 2001. Fluoranthene-2,3- and -1,5-diones are novel products from the bacterial transformation of fluoranthene. *Environ Sci Technol* 35: 917–922.

Kelley I, JP Freeman, CE Cerniglia. 1990. Identification of metabolites from degradation of naphthalene by a *Mycobacterium* sp. *Biodegradation* 1: 283–290.

Kelley I, JP Freeman, FE Evans, CE Cerniglia. 1993. Identification of metabolites from the degradation of fluoranthene by *Mycobacterium* sp. strain PYR-1. *Appl Environ Microbiol* 59: 800–806.

Khan AA, R-F Wang, W-W Cao, W Franklin, CE Cerniglia. 1996a. Reclassification of a polycyclic aromatic hydrocarbon-metabolizing bacterium, *Beijerinckia* sp. strain B1, as *Sphingomonas yanoikuyae* by fatty acids analysis, protein pattern analysis, DNA-DNA hybridization, and 16S ribosomal pattern analysis. *Int J Syst Bacteriol* 46: 466–490.

Khan AA, R-F Wang, MS Nawaz, W-W Cao, CC Cerniglia. 1996b. Purification of 2,3-dihydroxybiphenyl 1,2-dioxygenase from *Pseudomonas putida* OU83 and characterization of the gene (*bphC*). *Appl Environ Microbiol* 62: 1825–1830.

Kim S-J, O Kweon, JP Ftreeman, RC Jones, MD Adjei, J-W Jhoo, Rd Edmonson, CE Cerniglia. 2006. Molecular cloning and expression of genes encoding a novel dioxygenase involved in low- and high-molecular-weight polycyclic aromatic hydrocarbon degradation in *Mycobacterium vanbaalenii* PYR-1. *Appl Environ Microbiol* 72: 1045–1054.

Kim S-J, O Kweon, RC Jones, JP Freeman, RD Edmondson, CE Cerniglia. 2007. Complete and integrated pyrene degradation pathway in *Mycobacterium vanbalenii* PYR-1 based on systems biology. *J Bacteriol* 189: 464–472.

King DJ, MR Azari, A Wiseman. 1984. Studies on the properties of highly purified cytochrome P-448 and its dependent activity benzo[*a*]pyrene hydroxylase, from Saccharomyces cerevisiae. *Xenobiotica* 4: 187–206.

Kiyohara H, K Nagao. 1978. The catabolism of phenanthrene and naphthalene by bacteria. *J Gen Microbiol* 105: 69–75.

Kiyohara H, K Nagao, R Nomi. 1976. Degradation of phenanthrene through *o*-phthalate by an Aeromonas sp. *Agric Biol Chem* 40: 1075–1082.

Kondo R, H Yamagami, K Sakai. 1993. Xylosation of phenolic hydroxyl groups of the monomeric lignin model compounds 4-methylguaiacol and vanillyl alcohol by Coriolus versicolor. *Appl Environ Microbiol* 59: 438–441.

Koreeda M, MN Akhtar, DR Boyd, JD Neill, DT Gibson, DM Jerina. 1978. Absolute stereochemistry of *cis*-1,2-, *trans*-1,2-, and *cis*,3,4-dihydrodiol metabolites of phenanthrene. *J Org Chem* 43: 1023–1027.

Krivobok S, S Kuony, C Meyer, M Louwagie, JC Wilson, Y Jouanneau. 2003. Identification of 51 pyrene-induced proteins in *Mycobacterium* sp. strain 6PY1: Evidence for two ring-hydroxylating dioxygenases. *J Bacteriol* 185: 3828–3841.

Kuhm AE, A Stolz, K-L Ngai, H-J Knackmuss. 1991. Purification and characterization of a 1,2-dihydroxynaphthalene dioxygenase from a bacterium that degrades naphthalenesulfonic acids. *J Bacteriol* 173: 3795–3802.

Kulakov LA, S Chen, CCR Allen, MJ Larkin. 2005. Web-type evolution of *Rhodococcus* gene clusters associated with utilization of naphthalene. *Appl Environ Microbiol* 71: 1754–1764.

Kweon O, S-J Kim, RD Holland, H Chen, D-W Kim, Y Gao, L-R Yu et al. 2011. Polycyclic aromatic hydrocarbon, metabolic network in *Mycobacterium vanbaalenii*. *J Bacteriol* 193: 4326–4337.

Kweon O, S-J Kim, RC Jones, JP Freeman, MD Adjei, RD Edmonson, CE Cerniglia. 2007. A polyomic approach to elucidate the fluoranthene-degradative pathway in *Mycobacterium vanbaalenii* PYR-1. *J Bacteriol* 189: 4635–4647.

Lange B, S Kremer, O Sterner, A Anke. 1994. Pyrene metabolism in *Crinipellis stipitaria*: Identification of *trans*-4,5-dihydro-4,5-dihydroxypyrene and 1-pyrenylsulfate in strain KJ364. *Appl Environ Microbiol* 60: 3602–3607.

Lange B, Kremer S, Sterner O, Anke H. 1995. Induction of secondary metabolism by environmental pollutants: Metabolism of pyrene and formation of 6,8-dihydroxy-3-methylisocoumarin by *Crinipellis stipitaria* JK 364. *Z Naturforsch* 50c: 806–812.

Larkin MJ. 1988. The specificity of 1-naphthol oxygenases from three bacterial isolates, *Pseudomonas* spp. (NCIB 12042 and 12043) and *Rhodococcus* sp. (NCIB 12038) isolated from garden soil. *FEMS Microbiol Lett* 52: 173–176.

Larkin MJ, CCR Allen, LA Kulakov, DA Lipscomb. 1999. Purification and characterization of a novel naphthalene dioxygenase from *Rhodococcus* sp. strain NCIMB 12038. *J Bacteriol* 181: 6200–6204.

Launen L, L Pinto, C Wiebe, E Kiehlmann, M Moore. 1995. The oxidation of pyrene and benzo(*a*)pyrene by nonbasidiomycete soil fungi. *Can J Microbiol* 41: 477–488.

Lee K, SM Resnick, DT Gibson. 1997. Stereospecific oxidation of (*R*)- and (*S*)-1-indanol by napthalene dioxygenase from *Pseudomonas* sp. strain NCIB 9816-4. *Appl Environ Microbiol* 63: 2067–2070.

Li D, M Alic, JA Brown, MH Gold. 1995. Regulation of manganese peroxidase gene transcription by hydrogen peroxide, chemical stress, and molecular oxygen. *Appl Environ Microbiol* 61: 341–345.

MacGillivray AR, MP Shiaris. 1993. Biotransformation of polycyclic aromatic hydrocarbons by yeasts isolated from coastal sediments. *Appl Environ Microbiol* 59: 1613–1618.

Maeda M, S-Y Chung, E Song, T Kudo. 1995. Multiple genes encoding 2,3-dihydroxybiohenyl 1,2-dioxygenase in the Gram-positive polychlorinated biphenyl-degrading bacterium *Rhodococcus erythropolis* TA421, isolated from a termite ecosystem. *Appl Environ Microbiol* 61: 549–555.

Mahaffey WR, DT Gibson, CE Cerniglia. 1988. Bacterial oxidation of chemical carcinogens: Formation of polycyclic aromatic acids from benz[*a*]anthracene. *Appl Environ Microbiol* 54: 2415–2423.

Mahajan MC, PS Phale, CS Vaidyanathan. 1994. Evidence for the involvement of multiple pathways in the biodegradation of 1- and 2-methylnaphthalene by *Pseudomonas putida* CSV86. *Arch Microbiol* 161: 425–433.

McMillan DC, PP Fu, CE Cerniglia. 1987. Stereoselective fungal metabolism of 7,12-dimethylbenz[*a*]anthracene: Identification and enantiomeric resolution of a K-region dihydrodiol. *Appl Environ Microbiol* 53: 2560–2566.

McNally DL, JR Mihelcic, DR Lueking. 1998. Biodegradation of three- and four-ring polycyclic aromatic hydrocarbons under aerobic and denitrifying conditions. *Environ Sci Technol* 32: 2633–2639.

Meckenstock RU, E Annweiler, W Michaelis, HH Richnow, B Schink. 2000. Anaerobic naphthalene degradation by a sulfate-reducing enrichment culture. *Appl Environ Microbiol* 66: 2743–2747.

Moen MA, KE Hammel. 1994. Lipid peroxidation by the manganese peroxidase of *Phanerochaete chrysosporium* is the basis for phenanthrene oxidation by the intact fungus. *Appl Environ Microbiol* 60: 1956–1961.

Moody JD, JP Freeman, DR Doerge, CE Cerniglia. 2001. Degradation of phenanthrene and anthracene by cell suspensions of *Mycobacterium* sp. strain PYR-1. *Appl Environ Microbiol* 67: 1476–1483.

Moody JD, JP Freeman, PP Fu, CE Cerniglia. 2004. Degradation of benzo[*a*]pyrene by *Mycobacterium vanbaalenii* PYR-1. *Appl Environ Microbiol* 70: 340–345.

Moody JD, PP Fu, JP Freeman, CE Cerniglia. 2003. Regio-and stereoselective metabolism of 7,12,dimethyl-benz[*a*]anthracene by Mycobacterium vanbaalenii PYR-1. *Appl Environ Microbiol* 69: 3924–3951.

Morawski B, RW Eaton, JT Rossiter, S Guoping, H Griengl, DW Ribbons. 1997. 2-naphthoate catabolic pathway in *Burkholderia* strain JT 1500. *J Bacteriol* 179: 115–121.

Mueller JG, Chapman PJ, Blattman BO, Pritchard PH. 1990. Isolation and characterization of a fluoranthene-utilizing strain of *Pseudomonas paucimobilis*. *Appl Environ Microbiol* 56: 1079–1086.

Musat F, A Galushko, J Jacob, F Widdel, M Kube, R Reinhardt, H Wilkes, B Schink, R Rabus. 2009. Anaerobic degradation of naphthalene and 2-methylnaphthalene by strains of marine sulphate-reducing bacteria. *Environ Microbiol* 11: 209–219.

Narro ML, CE Cerniglia, C van Baalen, DT Gibson. 1992a. Evidence for an NIH shift in oxidation of naphthalene by the marine cyanobacterium *Oscillatoria* sp. strain JCM. *Appl Environ Microbiol* 58: 1360–1363.

Narro ML, CE Cerniglia, C van Baalen, DT Gibson. 1992b. Metabolism of phenanthrene by the marine cyanobacterium *Agmenellum quadruplicatum* PR-6. *Appl Environ Microbiol* 58: 1351–1359.

Patel TR, DT Gibson. 1974. Purification and properties of (+)-*cis*-naphthalene dihydrodiol dehydrogenase of *Pseudomonas putida*. *J Bacteriol* 119: 879–888.

Patel TR, EA Barnsley. 1980. Naphthalene metabolism by pseudomonads: Purification and properties of 1,2-dihydroxynaphthalene oxygenase. *J Bacteriol* 143: 668–673.

Pellizari VH, S Bezborodnikov, JF Quensen, JM Tiedje. 1996. Evaluation of strains isolated by growth on naphthalene and biphenyl for hybridization of genes to dioxygenase probes and polychlorinated biphenyl-degrading ability. *Appl Environ Microbiol* 62: 2053–2058.

Pinphanichakarn P. 2006. Novel intermediates of acenaphthylene degradation by *Rhizobium* sp. strain CU-A1: Evidence for naphthalene-1-8-dicarboxylic acid metabolism. *Appl Environ Microbiol* 72: 6034–6039.

Pinyakong O, H Habe, T Omori. 2003. The unique aromatic catabolic genes in sphingomonads degrading polycyclic aromatic hydrocarbons. *J Gen Appl Microbiol* 49: 1–19.

Poonthrigpun S, K Pattaragulwanit, S Paengthai, T Kriangkripipat, K Juntongjin, S Thaniyavarn, A Petsom, P Pinphanichakarn. 2006. Novel intermediates of acenaphthylene degradation by *Rhizobium* sp. strain CU-A1: Evidence for naphthalene-1,8-dicarboxylic acid metabolism *Appl Envir Microbiol* 72: 6034–6039.

Pothuluri V, JP Freeman, FE Evans, CE Cerniglia. 1990. Fungal transformation of fluoranthene. *Appl Environ Microbiol* 56: 2974–2983.

Resnick SM, DT Gibson. 1996. Regio- and stereospecific oxidation of fluorene, dibenzofuran, and dibenzothiophene by naphthalene dioxygenase from *Pseudomonas* sp. strain NCIB-4. *Appl Environ Microbiol* 62: 4073–4080.

Rieble S, Joshi DK, Gold MA. 1994. Purification and characterization of a 1,2,4-trihydroxybenzene 1,2-dioxygenase from the basiodiomycete *Phanerochaete chrysosporium*. *J Bacteriol* 176: 4838–4844.

Rockne KJ, JC Chee-Sanford, RA Sanford, BP Hedland, JT Staley, SE Strand. 2000. Anaerobic naphthalene degradation under nitrate-reducing conditions. *Appl Environ Microbiol* 66: 1595–1601.

Romine MF et al. 1999. Complete sequence of a 184-kilobase catabolic plasmid from *Sphingomonas aromaticivorans*. *J Bacteriol* 181: 1585–1602.

Safinowski M, C Griebler, RU Meckenstock. 2006. Anaerobic cometabolic transformation of polycyclic and heterocyclic aromatic hydrocarbons: Evidence from laboratory and field studies. *Environ Sci Technol* 40: 4165–4173.

Safinowski M, RU Meckenstock. 2006. Methylation is the initial reaction in anaerobic naphthalene degradation by a sulfate-reducing enrichment culture. *Environ Microbiol* 8: 347–352.

Saito A, T Iwabuchi, S Harayama. 2000. A novel phenanthrene dioxygenase from *Nocardioides* sp. strain KP7: Expression in *Escherichia coli*. *J Bacteriol* 182: 2134–2141.

Samanta SK, AK Chakraborti, RK Jain. 1999. Degradation of phenanthrene by different bacteria: Evidence for novel transformation sequences involving the formation of 1-naphthol. *Appl Microbiol Biotechnol* 53: 98–107.

Sarma PM, D Bhattacharya, S Krishnan, B Lal. 2004. Degradation of polycyclic aromatic hydrocarbons by a newly discovered enteric bacterium *Leclercia adecaroxylata*. *Appl Environ Microbiol* 70: 3163–3166.

Schneider J, R Grosser, K Jayasimhulu, W Xue, D Warshawsky. 1996. Degradation of pyrene, benz[*a*]anthracene, and benzo[*a*]pyrene by *Mycobacterium* sp. strain RGHII-135, isolated from a former coal gasification site. *Appl Environ Microbiol* 62: 13–19.

Seeger M, M González, B Cámara, L Muñoz, E Ponce, L Mejías, C Mascayano, Y Vásquez, S Sepúlveda-Boza. 2003. Biotransformation of natural and synthetic isoflavanoids by two recombinant microbial enzymes. *Appl Environ Microbiol* 69: 5045–5050.

Seigle-Murandi F, P Guiraud, J Croizé, E Falsen, K-E L Eriksson. 1996. Bacteria are omnipresent on *Phanerochaete chrysosporium* Burdsall. *Appl Environ Microbiol* 62: 2477–2481.

Selesi D, N Jehmlich, M von Bergen, F Schmidt, T Rattei, P Tischler, T Lueders, RU Meckenstock. 2010. Combined genomic and proteomic approaches identify gene clusters involved in anaerobic 2-methyl-naphthalene degradation in the sulfate-reducing enrichment culture N47. *J Bacteriol* 192: 295–306.

Selifonov SA, M Grifoll, RW Eaton, PJ Chapman PJ. 1996. Oxidation of naphthenoaromatic and methyl-substituted aromatic compounds by naphthalene 1,2- dioxygenase. *Appl Environ Microbiol* 62: 507–514.

Semple KT, RB Cain. 1996. Biodegradation of phenols by the alga *Ochromonas danica*. *Appl Environ Microbiol* 62: 1265–1273.

Sheng X, X Chen, L He. 2008. Characteristics of an endophytic pyrene-degrading bacterium *Enterobacter* sp. 12J1 from *Allium macrostemon*. *Int Biodet Biodeg* 62: 88–95.

Sietmann R, E Hammer, M Specht, CE Cerniglia, F Schauer. 2001. Novel ring cleavage products in the biotransformation of biphenyl by the yeast *Trichosporon mucoides*. *Appl Environ Microbiol* 67: 4158–4165.

Sikkema J, JAM de Bont. 1993. Metabolism of tetralin (1,2,3,4-tetrahydronaphthalene) in *Corynebacterium* sp. strain C125. *Appl Environ Microbiol* 59: 567–572.

Simon MJ, TD Osslund, R Saunders, BD Ensley, S Suggs, A Harcourt, W-C Suen, DL Cruden, DT Gibson, GJ Zylstra. 1993. Sequences of genes encoding naphthalene dioxygenase in *Pseudomonas putida* strains G7 and NCIB 9816-4. *Gene* 127: 31–37.

Simoneit BTR. 1998. Biomarker PAHs in the environment. *Handbook Environ Chem* 3I: 175–221.

Singleton DR, LG Ramirez, MD Aitken. 2009. Characterization of a polycyclic aromatic hydrocarbon degradation gene cluster in a phenanthrene-degrading *Acidovorax* strain. *Appl Environ Microbiol* 75: 2613–2620.

Smith MR. 1994. The physiology of aromatic hydrocarbon degrading bacteria, pp. 347–378. In *Biochemistry of Microbial Degradation* (Ed. C Ratledge), Kluwer Academic Publishers, Dordrecht, The Netherlands.

Smith RV, PJ Davis, AM Clark, S Glover-Milton. 1980. Hydroxylations of biphenyl by fungi. *J Appl Bacteriol* 49: 65–73.

Smith RV, PJ Davis, AM Vlark, SK Prasatik. 1981. Mechanism of hydroxylation of biphenyl by *Cunninghamella echinulata*. *Biochem J* 196: 369–371.

Smith RV, JP Rosazza. 1983. Microbial models of mammalian metabolism. *J Nat Prod* 46: 79–91.

Sohn JH, KK Kwon, J-H Kang, H-B Jung, S-J Kim. 2004. *Novosphingobium pentaromaticivorans* sp. nov., a high-molecular-mass polycyclic aromatic hydrocarbon-degrading bacterium isolated from estuarine sediment. *Int J Syst Evolut Microbiol* 54: 1483–14587.

Steffen KT, A Hatakka, M Hofrichter. 2003. Degradation of benzo[*a*]pyrene by the litter-decomposing basidiomycete *Stropharia coronilla*: Role of manganese peroxidase. *Appl Environ Microbiol* 69: 3957–3963.

Sutherland JB, JP Freeman, AL Selby, PP Fu, DW Miller, CE Cerniglia. 1990. Stereoselective formation of a K-region dihydrodiol from phenanthrene by *Streptomyces flavovirens*. *Arch Microbiol* 154: 260–266.

Sutherland JB, PP Fu, SK Yang, LS von Tungeln, RP Casillas, SA Crow, CE Cerniglia. 1993. Enantiomeric composition of the *trans*-dihydrodiols produced from phenanthrene by fungi. *Appl Environ Microbiol* 59: 2145–2149.

Sutherland JB, AL Selby, JP Freeman, FE Evans, CE Cerniglia. 1991. Metabolism of phenanthrene by *Phanerochaete chrysosporium*. *Appl Environ Microbiol* 57: 3310–3316.

Sullivan ER, X Zhang, C Phelps, LY Young. 2001. Anaerobic mineralization of stable-isotope-labeled 2-methylnaphthalene. *Appl Environ Microbiol* 67: 4353–4357.

Taira K, J Hirose, S Hayashida, K Furukawa. 1992. Analysis of *bph* operon from the polychlorinated biphenyl-degrading strain of *Pseudomonas pseudoalcaligenes* KF707. *J Biol Chem* 267: 4844–4853.

Tatarko M, JA Bumpus. 1993. Biodegradation of phenanthrene by *Phanerochaete chrysporium*: On the role of lignin peroxidase. *Lett Appl Microbiol* 17: 20–24.

Tausson WO. 1927. Naphthalin als Kohlenstoffequelle für Bakterien. *Planta* 4: 214–256.

Trenz SP, KH Engesser, P Fischer, H-J Knackmuss. 1994. Degradation of fluorene by *Brevibacterium* sp. strain DPO1361: A novel C–C bond cleavage mechanism via 1,10-dihydro-1,10-dihydroxyfluoren-9-one. *J Bacteriol* 176: 789–795.

van der Woude MW, K Boominanathan, CA Reddy. 1993. Nitrogen regulation of lignin peroxidase and manganese-dependent peroxidase production is independent of carbon and manganese regulation in *Phanerochaete chrysosporium*. *Arch Microbiol* 160: 1–4.

van Herwijnen R, D Springael, P Slot, HAJ Govers, JR Parsons. 2003. Degradation of anthracene by *Mycobacterium* sp. strain LB501T proceeds via a novel pathway, through *o*-phthalic acid. *Appl Environ Microbiol* 69: 186–190.

Vila J, Z López, J Sabaté, C Minguillón, AM Solanasm, M Grifoll. 2001. Identification of a novel metabolite in the degradation of pyrene by *Mycobacterium* sp. strain AP1: Actions of the isolate on two- and three-ring polycyclic aromatic hydrocarbons. *Appl Environ Microbiol* 67: 5497–5505.

Walter U, M Beyer, J Klein, H-J Rehm. 1991. Degradation of pyrene by *Rhodococcus* sp. UW1. *Appl Microbiol Biotechnol* 34: 671–676.

Wang Y, PCK Lau, DK Button. 1996. A marine oligobacterium harboring genes known to be part of aromatic hydrocarbon degradation pathways of soil pseudomonads. *Appl Environ Microbiol* 62: 2169–2173.

Warshawsky D, T Cody, M Radike, R Reilman, B Schujann, K LaDow, J Schneider. 1995b. Biotransformation of benzo[*a*]pyrene and other polycyclic aromatic hydrocarbons and heterocyclic analogues by several green algae and other algal species under gold and white light. *Chem-Biol Interactions* 97: 131–148.

Warshawsky D, M Radike, K Jayasimhulu, T Cody. 1995a. Metabolism of benzo[*a*]pyrene by a dioxygenase enzyme system of the freshwater green alga *Selenstrum capricornutum*. *Biochim Biophys Res Commun* 152: 540–544.

Weissenfels WD, M Beyer, J Klein, HJ Rehm. 1991. Microbial metabolism of fluoranthene: Isolation and identification of fission products. *Appl Microbiol Biotechnol* 34: 528–535.

Willison JC. 2004. Isolation and characterization of a novel sphingomonad capable of growth with chrysene as sole carbon and energy source. *FEMS Microbiol Lett* 241: 143–150.

Wiseman A, LFJ Woods. 1979. Benzo[*a*]pyrene metaboltes formed by the action of yeast cytochrome P-450/P-448. *J Chem Tech Biotechnol* 29: 320–324.

Ye D, MA Siddiqui, AE Maccubbin, S Kumar, HC Sikka. 1996. Degradation of polynuclear aromatic hydrocarbons by *Sphingomonas paucimobilis*. *Environ Sci Technol* 30: 136–142.

Yen K-M, CM Serdar. 1988. Genetics of naphthalene catabolism in pseudomonads. *CRC Crit Rev Microbiol* 15: 247–268.

Zhang X, LY Young. 1997. Carboxylation as an initial reaction in the anaerobic metabolism of naphthalene and phenanthrene by sulfidogenic consortia. *Appl Environ Microbiol* 63: 4759–4764.

Zhong Y, T Luan, H Zhou, C Lan, N Fung, FY Tam. 2006. Metabolite production in degradation of pyrene alone or in mixture with another polycyclic aromatic hydrocarbon by *Mycobacterium* sp. *Environ Toxicol Chem* 25: 2853–2859.

Zhou N-Y, J Al-Dulayymi, MS Baird, PA Williams. 2002. Salicylate 5-hydroxylase from *Ralstonia* sp. strain U2; a monooxygenase with close relationships to and shared electron transport proteins with naphthalene dioxygenase. *J Bacteriol* 184: 1547–1555.

Ziffer H, K Kabuta, DT Gibson, VM Kobal, DM Jerina. 1977. The absolute stereochemistry of several *cis* dihydrodiols microbially produced from substituted benzenes. *Tetrahedron* 33: 2491–2496.

Part 3: Aromatic Carboxylates, Carboxaldehydes, and Related Compounds

8.13 Introduction

Catechols, benzoates, and hydroxylated benzoates are intermediates in the aerobic degradation of aromatic hydrocarbons. The first stage in their degradation is the introduction of the elements of dioxygen into the ring, either both atoms by dioxygenation or only one atom by monooxygenation. The second stage involves fission of the ring by another group of dioxygenases. Details of these oxygenases are given in Chapter 3, Part 1. The anaerobic degradation of benzoate is carried out by different reactions involving hydrogenation of the ring.

8.14 Benzoates

8.14.1 Aerobic Conditions

8.14.1.1 Mechanisms for the Initial Oxygenation

Benzoate is generally degraded to catechol by dioxygenation with concomitant decarboxylation. For benzoates substituted at the *ortho* position with amino or halogen groups, dioxygenation involves the loss of both the carboxyl and amino groups or both the carboxyl and halogen groups. In contrast to arene hydrocarbons, dehydrogenation is not, therefore, required for the formation of catechols. The formation of 1-hydroxynaphthalene-2-carboxylate from naphthalene-2-carboxylate was apparently not via (1R,2S)-*cis*-1,2-dihydro-1,2-dihydroxynaphthalene-2-carboxylate (Morawski et al. 1997). This is supported by the activity of naphthalene dioxygenase in which both activities are catalyzed by the same enzyme (Gibson et al. 1995). The degradation of halogenated benzoates, arene sulfonates, and nitroarenes, by dioxygenation is discussed in Chapter 9, Parts 1, 3 through 5.

8.14.1.2 Benzoate Dioxygenase

In *Acinetobacter calcoaceticus*, the enzyme is chromosomal and consists of a hydroxylase and an electron transport protein that have been designated *benAB* and *benC* (Neidle et al. 1991). The corresponding genes are designated *xylXY* and *xylZ* in the plasmid-encoded toluate dioxygenase in *Pseudomonas putida*, and this dioxygenase accepts a much wider range of substrates.

8.14.1.3 Anthranilate-1,2-Dioxygenase

1. In *Acinetobacter* sp. strain ADP1 (Eby et al. 2001), this enzyme catalyzes the dioxygenation of anthranilate with the loss of both the amino and carboxyl groups. This, like the enzyme from *Pseudomonas putida* and *P. aeruginosa*, is a two-component enzyme consisting of an oxygenase and a reductase, whereas the enzyme from *Burkholderia cepacia* DBO1 consists of three components—a two-subunit oxygenase, a ferredoxin, and a reductase (Chang et al. 2003).

2. Direct ring fission has been found in some substituted anthranilates that is analogous to that displayed by *Pseudaminobacter salicylatoxidans* (Hintner et al. 2004). This is exemplified in the degradation of 5-aminoanthranilate by *Pseudomonas* sp. strain BN9 (Stolz et al. 1992) and 5-nitroanthranilate by *Bradyrhizobium* sp. (Qu and Spain 2011).

Anthranilate may also be hydroxylated to 5-hydroxyanthranilate, and 2,5-dihydroxybenzoate by anthranilate-5-hydroxylase in *Nocardia opaca* (Cain 1968) and in *Ralstonia* sp. strain U2 (Zhou et al. 2002), or to 2,3-dihydroxybenzoate in *Trichosporon cutaneum* (Powlowski et al. 1987) and *Aspergillus niger* (Subramanian and Vaidyanathan 1984).

8.14.1.4 2-Halobenzoate 1,2-Dioxygenases

These are multicomponent enzymes that catalyze dioxygenation with the loss of the halogen and the carboxyl groups, and are discussed in Chapter 9, Part 1.

The application of *cis*-arene dihydrodiols has been noted in Part 1 of this chapter, and it is sufficient to note here the application of a mutant of *Alcaligenes eutrophus* strain B9 that is blocked in the degradation of benzoate (and some halogenated benzoates). This produced the *cis*-1,2-dihydrodiol (Reiner and Hegeman 1971), and has been used as the source of ring B for the synthesis of a range of tetracyclines (Charest et al. 2005). A range of substituted *cis*-dihydrodiols has been produced, and it has been shown that for 3-substituted benzoates, both 3- and 5-substituted *cis*-dihydrodiols were formed (Reineke et al. 1978).

8.15 Hydroxybenzoates and Related Compounds

Phenols and hydroxybenzoates are generally degraded by monooxygenation. Illustrative examples include the following:

1. The bacterial degradation of salicylate can take place by various reactions.

   a. The degradation of salicylate to catechol is initiated by monooxygenation accompanied by decarboxylation (salicylate-1-hydroxylase), and two different and independent salicylate hydroxylases have been found in the naphthalene-degrading *Pseudomonas stutzeri* AN10 (Bosch et al. 1999).

   b. In *Rhodococcus* sp. strain B4, salicylate is initiated by the formation of the coenzyme A esters before hydroxylation by salicylate-5-hydroxylase to 2,5-dihydroxybenzoate (Grund et al. 1992).

   c. An alternative occurs for 5-hydroxy- and 5-aminosalicylate in *Pseudaminobactersalicylatoxidans* where ring fission is accomplished directly (Hintner et al. 2001).

   d. In *Streptomyces* sp. strain WA46, degradation is initiated by the sequential formation of salicyl-AMP and salicyl-CoA before dioxygenation to gentisyl-CoA (Ishiyama et al. 2004).

2. The flavoprotein salicylate hydroxylase in *Trichosporon cutaneum* is able to carry out concomitant hydroxylation and decarboxylation on a range of *ortho*-hydroxybenzoates, including salicylate, 2,5-dihydroxybenzoate (gentisate), and 4-amino- and 5-amino-salicylate (Sze and Dagley 1984). An intradiol ring-fission dioxygenase was characterized and was able to carry out ring fission of 1,2,4-trihydroxybenzene to maleylacetate.

3. The monooxygenase from *Pseudomonas fluorescens* that converts 4-hydroxybenzoate into 3,4-dihydroxybenzoate before ring fission has been characterized (Howell et al. 1972).

4. In *Pseudomonas putida*, the hydroxylation of 4-hydroxyphenylacetate to 3,4-dihydroxy-phenyl-acetate is carried out by an enzyme that consists of a flavoprotein and a coupling factor (Arunachalam et al. 1992). At least in *Escherichia coli*, it has been suggested that the large component is an $FADH_2$-utilizing monooxygenase (Xun and Sandvik 2000; Chaiyen et al. 2001).

5. The first enzyme in the degradation of 2-hydroxybiphenyl by *Pseudomonas azelaica* strain HBP1 is an FAD-dependent monooxygenase that produces 2,3-dihydroxybiphenyl (Suske et al. 1999).

Hydroxylation to 1,4-dihydroxy compounds may also activate the ring to oxidative fission. This is illustrated by the following examples of the gentisate pathway:

1. The degradation of 3-methylphenol, 3-hydroxybenzoate and salicylate can be initiated by hydroxylation to 2,5-dihydroxybenzoate (gentisate). The range of organisms includes species of *Pseudomonas*, and *Bacillus* (Crawford 1975a), and the enteric bacteria *Salmonella typhimurium* (Goetz and Harmuth 1992) and *Klebsiella pneumoniae* (Jones and Cooper 1990). Gentisate dioxygenase carries out fission of the ring to produce pyruvate, and fumarate or maleate (Figure 8.31). Its distribution is noted later.

2. The gentisate pathway is used for the degradation of salicylate produced from naphthalene by a *Rhodococcus* sp. strain B4 (Grund et al. 1992), rather than by the more usual sequence involving simultaneous hydroxylation and decarboxylation of salicylate to catechol.

3. An analogous pathway is used for the degradation of 5-aminosalicylate that is an intermediate in the degradation of 6-aminonaphthalene-2-sulfonate. Direct ring fission of both 5-amino-salicylate and 5-hydroxysalicylate can be accomplished by a salicylate 1,2-dioxygenase in *Pseudaminobacter salicylatoxidans* (Hintner et al. 2001).

4. The gentisate pathway may plausibly be involved in the degradation of benzoate by a denitrifying strain of *Pseudomonas* sp. in which the initial reaction is the formation of 3-hydroxybenzoate (Altenschmidt et al. 1993).

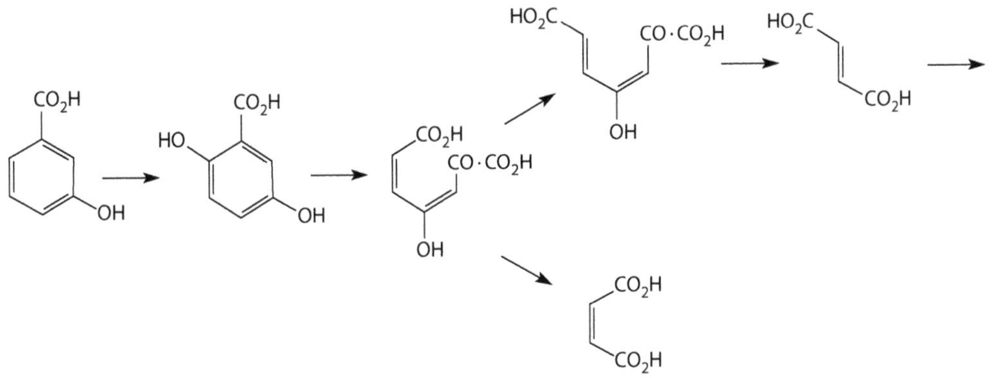

FIGURE 8.31 The gentisate pathway.

Unusual pathways have been found in the bacterial degradation of a number of 4-hydroxy-benzoates and related compounds and in some of them, rearrangements (NIH shifts) are involved:

1. In *Pseudomonas putida*, L-phenylalanine is hydroxylated to tyrosine by a reaction that involves 6,7-dimethyltetrahydrobiopterin, which is converted into 4a-carbinolamine (Song et al. 1999). Tyrosine is then transformed to 2,5-dihydroxyphenylacetate followed by ring fission (Arias-Barrau et al. 2004).

2. Gentisate is formed by a strain of *Bacillus* sp. in an unusual rearrangement from 4-hydroxyben-zoate (Crawford 1976) that is formally analogous to the formation of 2,5-dihydroxyphenylac-etate from 4-hydroxyphenylacetate by *Pseudomonas acidovorans* (Hareland et al. 1975). Similarly, the metabolism of 4-hydroxybenzoate by the archaeon *Haloarcula* sp. strain D1 involves the formation of 2,5-dihydroxybenzoate (Fairley et al. 2002). All these reactions puta-tively involve an NIH shift.

3. The degradation of 4-hydroxyphenylacetate by *Pseudomonas acidovorans* takes place by hydroxylation with the production of 2,5-dihydroxyphenylacetate in a reaction involving an NIH shift. The final products are fumarate and acetate, and the enzyme has been characterized (Hareland et al. 1975). In contrast, the degradation of 3,4-dihydroxyphenylacetate by the same organism takes place by extradiol fission to produce pyruvate and succinate semialdehyde (Sparnins and Dagley 1975).

4. *Pseudomonas* sp. strain P.J. 874 grown with tyrosine carried out dioxygenation of 4-hydroxy-phenylpyruvate to 2,5-dihydroxyphenylacetate accompanied by an NIH shift (Lindstedt et al. 1977). The involvement of a high-spin ferric center coordinated with tyrosine is conclusively revealed in the primary structure of the enzyme (Rüetschi et al. 1992).

5. The chlorophenol-4-hydroxylase from *Burkholderia cepacia* strain AC1100 is able to bring about not only the transformation of 4-hydroxybenzaldehydes to the expected 4-hydroxybenzo-ates but also rearrangement to 2,5-dihydroxybenzaldehydes (Martin et al. 1999) (Figure 8.32).

6. The degradation of nonylphenol isomers with quaternary side chains in strains of *Sphingomonas* sp. displays unusual features that involve rearrangement of the side chain (Corvini et al. 2004; Gabriel et al. 2005a,b). This is discussed in Part 4 of this chapter.

Both fungi and yeasts are able to degrade simpler substituted aromatic compounds such as vanillate (Ander et al. 1983) (Figure 8.33) and 3,4-dihydroxybenzoate (Anderson and Dagley 1981), both of which involve concomitant decarboxylation and hydroxylation. The oxidative decarboxylation of a range of 4-hydroxybenzoates to 1,4-dihydroxy compounds has been examined, and is carried out by an FAD-dependent monooxygenase in *Candida parapsilosis* CBS604 (Eppink et al. 1997). The yeast *Trichosporon cutaneum* is able to metabolize phenol and a number of other aromatic compounds. The pathways by

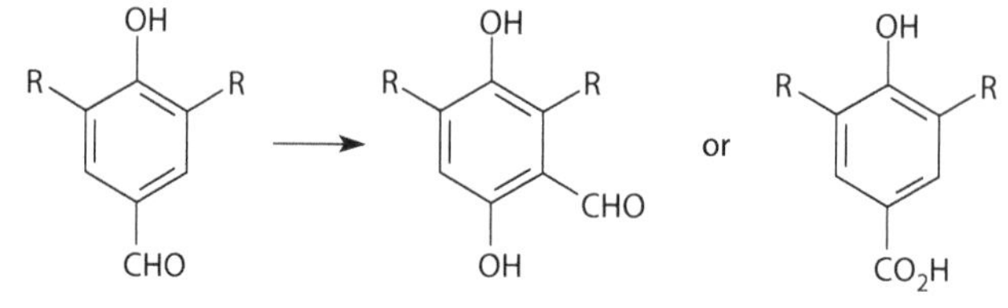

FIGURE 8.32 Aerobic rearrangement of 4-hydroxybenzaldehydes. (Adapted from Martin G et al. 1999. *Eur J Biochem* 261: 533–538.)

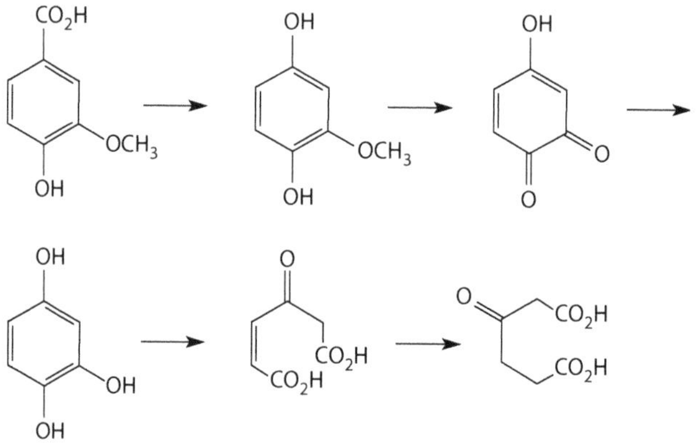

FIGURE 8.33 Biodegradation of vanillic acid by fungi.

which this is accomplished differ, however, from those that are operative in most bacteria. Illustrative examples include the following (Anderson and Dagley 1980):

1. Benzoate is degraded by successive hydroxylation at the 4- and 3-positions followed by concomitant hydroxylation and decarboxylation to benzene-1,2,4-triol, while an analogous pathway is used for 2,5-dihydroxybenzoate. In contrast, 2,3-dihydroxybenzoate is decarboxylated to catechol without hydroxylation.

2. The benzene-1,2,4-triol undergoes intradiol ring fission followed by reduction to 3-oxoadipate and incorporation into the TCA cycle.

3. Analogous hydroxylations are used to initiate the metabolism of phenylacetate and 4-hydroxy-phenyl acetate. Fission of the rings takes place between the *ortho* hydroxyl group and the –CH$_2$–CO$_2$H group to produce acetoacetate, and fumarate or oxalacetate.

8.16 Mechanisms for Fission of Oxygenated Rings

8.16.1 Catechols

In the degradation of the catechols, the next step is the dioxygenation followed by fission of the ring, by either extradiol (2:3) or intradiol (1:2) fission. This is discussed in greater detail in Part 4 dealing with the degradation of phenols. The intradiol and extradiol enzymes are quite specific for their respective substrates, and whereas all of the first group contain Fe^{3+}, those of the latter contain Fe^{2+} (Wolgel et al. 1993). Although the extradiol 2,3-dihydroxybiphenyl 1,2-dioxygenases in *Rhodococcus globerulus* strain P6 are typical in containing Fe (Asturias et al. 1994), the enzyme from *Bacillus* sp. strain JF8 is manganese dependent (Hatta et al. 2003). A few other manganese-dependent ring-fission dioxygenases have been observed for 3,4-dihydroxyphenylacetate in *Arthrobacter globiformis* (Boldt et al. 1995) and *Bacillus brevis* (Que et al. 1981).

After the formation of the 1,2-dihydroxy compounds, ring fission is mediated by 1:2 (intradiol fission) or 2:3 (extradiol and distal fission) dioxygenases. There are, however, important variations in the pathways used by various groups of microorganisms:

1. The pathways and their regulation during the degradation of catechol and 3,4-dihydroxybenzoate in *Pseudomonas putida* have been elucidated in extensive studies (Ornston 1966). In this organism, intradiol ring fission is carried out by a 3,4-dioxygenase to produce 3-oxoadipate (Figure 8.34). The stereochemistry of the reactions after ring fission has been examined in detail (Kozarich 1988), and the regulation and genetics in a range of organisms have been reviewed (Harwood and Parales 1996). In contrast, 3,4,5-trihydroxybenzoate (gallate) is degraded in *P. putida* by extradiol fission with the production of 2 mol of pyruvate (Sparnins and Dagley 1975).

2. In the degradation of 3-hydroxybenzoate, divergent pathways are used by different groups of pseudomonads:

   a. *Pseudomonas testosteroni* uses 4,5-dioxygenase to produce pyruvate and formate from 3,4-dihydroxybenzoate by extradiol fission (Figure 8.35a) (Wheelis et al. 1967). The degradation involves pyrone-4,6-dicarboxylate that was converted by a hydrolase into the open-chain 2-hydroxy-4-carboxymuconic acid that was produced directly by extradiol fission from gallate by the same enzyme (Kersten et al. 1982).

   b. *Pseudomonas acidovorans* produces 2,5-dihydroxybenzoate that is degraded by gentisate 1,2-dioxygenase to fumarate and pyruvate, which has already been noted (Figure 8.26), and two different gentisate dioxygenases have been characterized (Harpel and Lipscomb 1990).

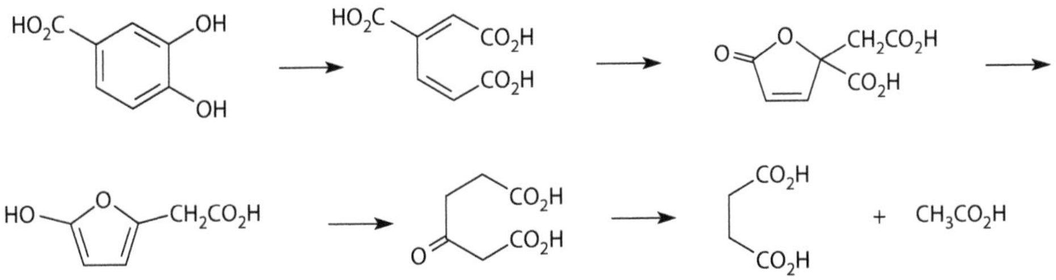

FIGURE 8.34 The β-oxoadipate pathway.

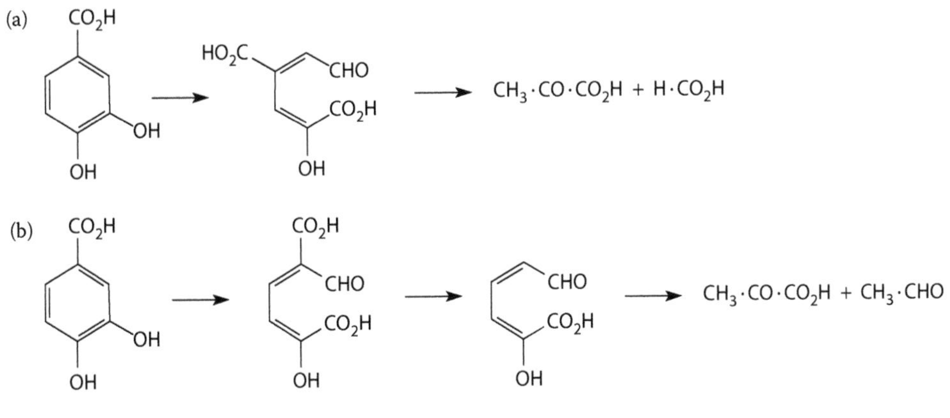

FIGURE 8.35 Biodegradation of 3,4-dihydroxybenzoate mediated by (a) 4,5-dioxygenase in *Pseudomonas testosteroni* and (b) 2,3-dioxygenase in *Bacillus macerans*.

3. The third alternative for ring fission of 3,4-dihydroxybenzoate is exemplified by *Bacillus mac-erans* and *B. circulans* that use a 2,3-dioxygenase to accomplish this (Figure 8.35b) (Crawford 1975a,b, 1976). In addition, a 2,3-dioxygenase is elaborated by Gram-negative bacteria for the degradation of 3,4-dihydroxyphenylacetate (Sparnins et al. 1974), and by Gram-positive bacteria for the degradation of L-tyrosine via 3,4-dihydroxyphenylacetate (Sparnins and Chapman 1976).

4. The enzymes of alternative pathways may be induced in a given strain by growth with different substrates; for example, growth of *Pseudomonas putida* R1 with salicylate induces enzymes of the extradiol fission pathway, whereas growth with benzoate induces those of the intradiol path-way (Chakrabarty 1972). As a broad generalization, the extradiol fission is preferred for the degradation of more complex compounds such as toluene, naphthalene, and biphenyl (Furukawa et al. 1983).

5. For the degradation of methylbenzoates there are two pathways: in one, degradation is initiated by oxidation of the methyl group to carboxylate in which ring fission takes place, while in the other, the ring is fissioned with one or more of the methyl groups intact. This is discussed more fully in Part 4.

8.16.2 Methoxybenzoates and Related Compounds

Studies on the degradation of lignin have attracted interest in the degradation of the monomeric vanillate and syringate, and mechanisms for their de-*O*-methylation are given in Chapter 11, Part 2. For substrates such as vanillate and isovanillate, degradation is initiated by de-*O*-methylation, followed in *Comamonas testosteroni* by extradiol fission (Providenti et al. 2006). 3-*O*-Methylgallate occupies a central position in several degradations such as that of (a) 4-hydroxy-3-methoxymandelate by *Acinetobacter lwoffii* (Sze and Dagley 1987), (b) syringate by *Sphingomonas paucimobilis* SYK-6 (Kasai et al. 2004), and (c) 3,4,5-trimethoxybenzoate by *Pseudomonas putida* (Donnelly and Dagley 1980). In all of them, dioxy-genation carries out extradiol ring fission to an ester that may undergo cyclization to 2-pyrone-4,6-dicar-boxylate with the loss of methanol, and degradation of 4-oxalomesaconitate to pyruvate and oxalacetate (Figure 8.36a–c).

8.16.3 2-Hydroxybenzoate and 1-Hydroxynaphthalene-2-Carboxylate

Salicylate is generally degraded by monooxygenation and decarboxylation mediated by salicylate-1-hydrox-ylase to catechol (White-Stevens and Kamin 1972; White-Stevens et al. 1972), although monooxygenation without decarboxylation can be carried out by salicylate-5-hydroxylase to produce 2,5-dihydroxybenzoate (Zhou et al. 2002). The degradation of naphthalene-2-carboxylate by *Burkholderia* sp. strain JT 1500 involves the formation of 1-hydroxynaphthalene-2-carboxylate rather than initial oxidative decarboxylation. Naphthalene-1,2-dihydrodiol-2-carboxylate is not, however, involved; hence, the reaction is possibly carried out by a monooxygenase, or a dehydration step is involved. Subsequent reactions produced pyruvate and *o*-phthalate, which was degraded via 4,5-dihydroxyphthalate (Morawski et al. 1997).

8.16.4 Fission of 1,4-Dihydroxybenzoates

Gentisate (2,5-dihydroxybenzoate) is produced from a range of substrates and has already been noted. Fission of the ring is carried out by gentisate 1,2-dioxygenase that has been characterized from a range of organisms, including *Moraxella osloensis* (Crawford et al. 1975); *Pseudomonas testosteroni* and *P. acidovorans* (Harpel and Lipscomb 1990); *P. alcaligenes* and *P. putida* (Feng et al. 1999); *Sphingomonas* sp. strain RW 5 (Wergath et al. 1998); and *Klebsiella pneumoniae* (Suárez et al. 1996).

8.16.5 Nonoxidative Decarboxylation of Benzoate and Related Compounds

Although benzoate is generally metabolized by simultaneous dioxygenation and decarboxylation to cat-echol followed by ring fission, nonoxidative decarboxylation of hydroxybenzoates has been observed.

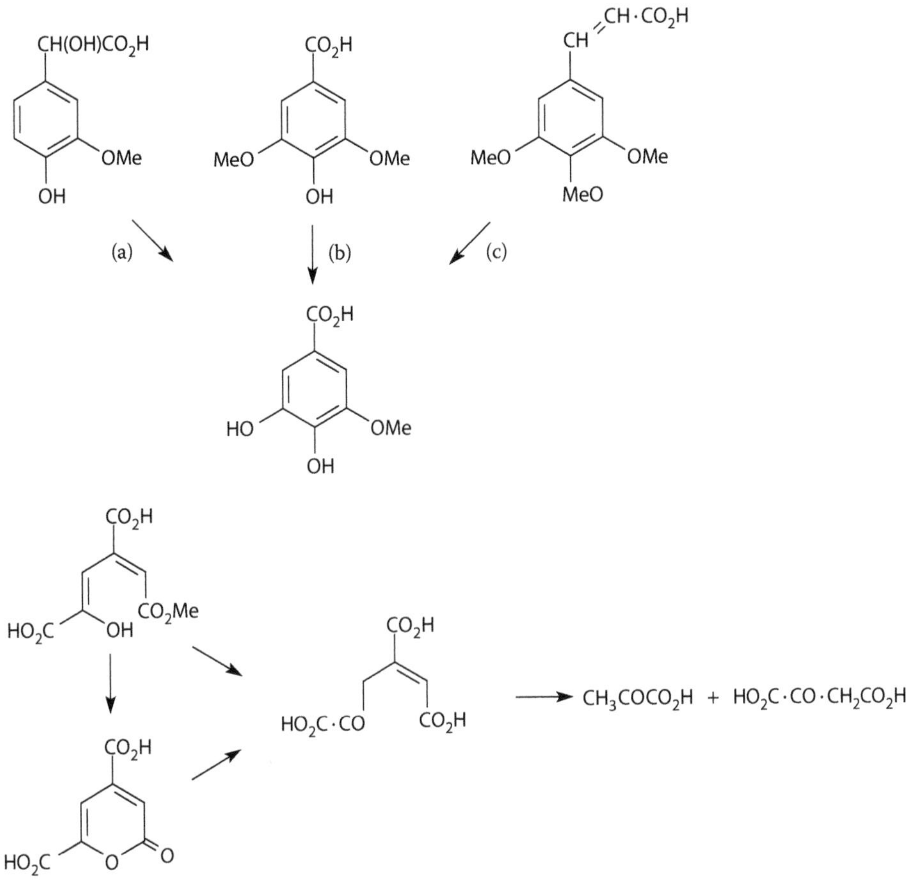

FIGURE 8.36 Degradation of 3-*O*-methylgallate produced from (a) 4-hydroxy-3-methoxymandelate, (b) syringate, and (c) 3,4,5-trimethoxycinnamate.

Important issues include the following: (a) the reaction is carried out both by aerobic and anaerobic bacteria, (b) it may be reversible, (c) some of the enzymes—even from aerobes—are oxygen sensitive, and (d) a high degree of specificity is often observed.

8.16.5.1 Aerobic Bacteria

1. Strains of *Bacillus megaterium* and *Streptomyces* sp. strain 179 transformed vanillate to guaiacol by decarboxylation (Crawford and Olson 1978). The gene cluster encoding decarboxylation in *Streptomyces* sp. strain D7 has been characterized (Chow et al. 1999), and the enzyme has been purified from *Bacillus pumilis* (Degrassi et al. 1995).
2. Decarboxylation of aromatic carboxylic acids has been encountered extensively in facultatively anaerobic *Enterobacteriaceae*. For example, 4-hydroxycinnamic acid is decarboxylated to 4-hydroxystyrene, and ferulic acid (3-methoxy-4-hydroxycinnamic acid) to 4-vinylguaiacol by several strains of *Hafnia alvei* and *H. protea*, and by single strains of *Enterobacter cloacae* and *K. aerogenes* (Figure 8.37) (Lindsay and Priest 1975). The decarboxylase has been characterized from *Pantoae agglomerans*, is specific for 3,4,5-trihydroxybenzoate (gallate), and is oxygen sensitive (Zeida et al. 1998).
3. Decarboxylation is involved in the degradation of a range of aromatic carboxylates.
   a. The degradation of 2,2′-dihydroxy, 3,3′-dimethoxybiphenyl-5,5′-dicarboxylate (5-5′-dehydrodivanillate) by *Sphingomonas paucimobilis* SYK-6 proceeds by partial de-*O*-methylation

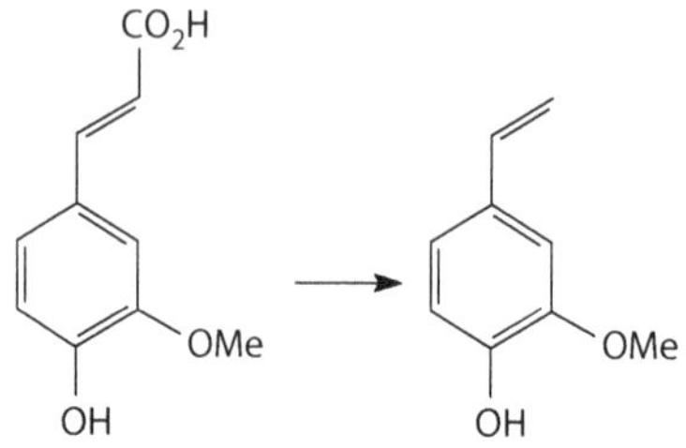

FIGURE 8.37   Decarboxylation of ferulic acid (3-methoxy-4-hydroxycinnamic acid).

followed by extradiol fission of the catechol to 2-hydroxy-3-methoxy-5-carboxybenzoate. Diversion of this into central metabolic pathways involves decarboxylation to vanillate by two separate decarboxylases LigW1 and LigW2 (Peng et al. 2005).

b. Decarboxylation is involved in all the pathways for the degradation of phthalates. The degradation of *o*-phthalate and *p*-phthalate (terephthalate) has been examined in strains of both Gram-negative and Gram-positive bacteria, and the genetics has been established (references in Sasoh et al. 2006). In summary, degradation is accomplished by dioxygenation followed by dehydrogenation and decarboxylation.

   i. The degradation of *o*-phthalate in *Pseudomonas cepacia* is initiated by dioxygenation. A two-component enzyme consisting of a nonheme iron oxygenase and an NADH-dependent oxidoreductase containing FMN and a [2Fe–2S] ferredoxin (Batie et al. 1987) produces the 4,5-dihydrodiol. Degradation is completed by dehydrogenation, decarboxylation (Pujar and Ribbons 1985) to 3,4-dihydroxybenzoate, and ring fission to 2-carboxymuconate. It is worth noting that in contrast, the degradation of 4-methyl-*o*-phthalate by *P. fluorescens* strain JT701 takes place by the formation of the 2,3-dihydrodiol followed by decarboxylation analogous to that of benzoate, with the formation of 4-methyl-2,3-dihydroxybenzoate and extradiol ring fission (Ribbons et al. 1984).

   ii. On the other hand, the pathway for the degradation of *o*-phthalate in *Micrococcus* sp. strain 12B, however, involves the initial formation of the 3,4-dihydrodiol followed by dehydrogenation and decarboxylation to 3,4-dihydroxybenzoate (Eaton and Ribbons 1982).

   iii. The degradation of *p*-phthalate follows an analogous pathway in *Comamonas* sp. strain T-2 via 3,4-dihydroxybenzoate (Schläfli et al. 1994), and that of 5-hydroxyisophthalate via 4,5-dihydroxyisophthalate and decarboxylation to 3,4-dihydroxybenzoate (Elmorsi and Hopper 1979).

   iv. The pathway and the genes have been characterized (Fukuhara et al. 2010) in the degradation of isophthalate (*m*-phthalate) by *Comamonas* sp. strain E6 and involves 4,5-dioxygenation followed by dehydrogenation and decarboxylation to 3,4-dihydroxybenzoate and extradiol 4,5-fission.

      The phthalate dioxygenase from *Ps. cepacia* is also active with pyridine-2,3- and 3,4-dicarboxylates (Batie et al. 1987), and this is substantiated in growth experiments in which different strains were able to metabolize pyridine-2,6-, 2,5- and 2,3-dicarboxylate (Taylor and Amador 1988). It has been shown that there are two regions of DNA that encode the enzymes for *o*-phthalate degradation, with the gene encoding quinolinate phosphoribosyl transferase located between them, and insertional knockout mutants with elevated levels of this enzyme enhanced growth on *o*-phthalate (Chang and Zylstra 1999).

c. The degradation of 2,6-dihydroxybenzoate (γ-resorcylate) by *Rhizobium* sp. strain MTP-10005 is initiated by decarboxylation. The enzyme has been purified and characterized, is specific for γ-resorcylate, and is reversible (Yoshida et al. 2004).

8.16.5.2 Anaerobic Bacteria

1. The reversible decarboxylation of 4-hydroxybenzoate and 3,4-dihydroxybenzoate has been described in *Sedimentibacter* (*Clostridium*) *hydroxybenzoicum* (He and Wiegel 1996), and the oxygen-sensitive enzyme has been purified.

2. The decarboxylase in *Clostridium thermoceticum* produces CO_2 that is essential for growth, and under CO_2 limitation it provides both the methyl and the carboxyl groups of acetate (Hsu et al. 1990).

3. The first reaction in the degradation of gallate by *Eubacterium oxidoreducens* is decarboxylation to pyrogallol that is then rearranged to 1,3,5-trihydroxybenzene and then degraded to acetate and butyrate (Krumholz et al. 1987). For reasons that were not established, the activity of the decarboxylase was low in comparison with the other degradative enzymes.

4. The terminal products formed by fermentation of aromatic amino acids in clostridia have been described (Elsden et al. 1976). These reactions involve oxidative deamination of the amino acids to the α-oxoacids followed by decarboxylation. Examples include

   a. Toluene from phenylalanine (Pons et al. 1984) by *Clostridium aerofoetidum*

   b. Toluene from phenylalanine by the facultative anaerobe *Tolumonas auensis* during anaerobic growth with a range of monosaccharides (Fischer-Romero et al. 1996)

   c. *p*-Cresol from tyrosine in cell extracts of *Clostridium difficile* (D'Ari and Barker 1985)

   d. 3-Methylindole (skatole) from indole-3-acetate by a rumen species of *Lactobacillus* (Yokoyama et al. 1977; Yokoyama and Yokoyama 1981; Honeyfield and Carlson 1990).

The decarboxylase from *Clostridium difficile* has been purified, and its activity examined toward a range of substituted phenylacetates, 3,4-dihydroxyphenylacetate, and pyridine-4-acetate (Selmer and Andrei 2001). The protein was readily and irreversibly inactivated by oxygen, and it was proposed on the basis of similarity to pyruvate formate lyase-like proteins in *Escherichia coli*, that the reaction involved a glycyl radical.

8.16.6 Alternative Pathways for the Degradation of Benzoates and Related Compounds

Unusual reactions have been encountered in the aerobic degradations carried out by *Azoarcus evansii* and *Geobacillus stearothermophilus* (Zaar et al. 2001). The anaerobic degradation of benzoate by *Azoarcus evansii* (Ebenau-Jehle et al. 2003) and *Thauera aromatica* (Dörner and Boll 2002), and of 3-hydroxybenzoate by *Th. aromatica* (Laempe et al. 2001) is discussed later.

1. Aerobic degradation of benzoate in *Azoarcus evansii* takes place via the CoA-ester to produce the CoA-esters of succinate and acetate (Zaar et al. 2001; Gescher et al. 2002) (Figure 8.38). Further examination revealed new features that made possible construction of the complete

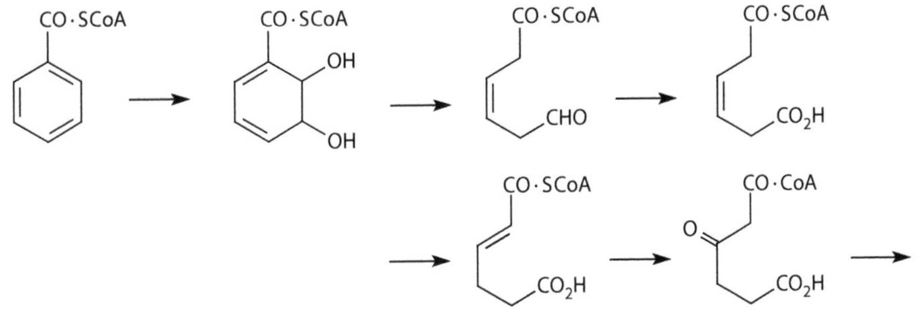

FIGURE 8.38 Alternative aerobic degradation of benzoyl-CoA by *Azoarcus evansii*. (Adapted from Gescher J et al. 2006. *J Bacteriol* 188: 2919–2927.)

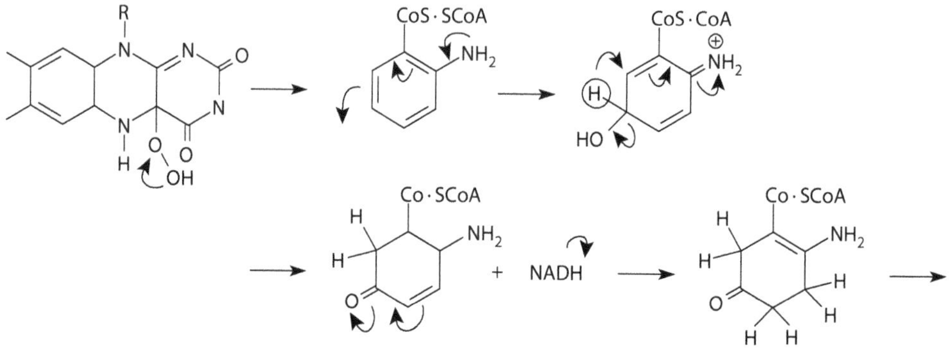

FIGURE 8.39   Aerobic degradation of 2-aminobenzoate by *Azoarcus evansii*. (Adapted from Hartmann S et al. 1999. *Proc Natl Acad USA* 96: 7831–7836; Schühle K et al. 2001. *J Bacteriol* 183: 5268–5278.)

pathway. The initial product of oxygenation that had been characterized as a dihydrodiol using ^{13}C NMR (Ismail et al. 2003) has been revised to an epoxide that is tautomeric with an oxepin (Rather et al. 2010; Teufel et al. 2010). The oxygenase (Zaar et al. 2004) that carries this out is an diiron epoxide BoxB that has been characterized by x-ray crystallography (Rather et al. 2011), and BoxA is NADPH-dependent reductase. BoxC is not dependent on oxygen and produces formate from C_2 (Gescher et al. 2005) and 3,4-dehydroadipyl-CoA semialdehyde that undergoes dehydrogenation by BoxD to 3-dehydro-adipyl-CoA that is then fissioned to acetyl-CoA and succinyl-CoA (Gescher et al. 2006). The antibiotic tropodithietic (TDA) acid is produced by marine bacteria in the *Roseobacter* lineage, and plays a role in the symbiosis of marine bacteria and algae (Geng and Belas 2010; Seyedsayamdost et al. 2011). The biosynthesis of TDA is carried out by a branch of the pathway for the metabolism of phenylacetyl CoA— oxidation to the oxepin, hydrolysis to 3-oxo-5,6-dehydrosuberoyl-CoA and cyclization to tropone before dimethylsulfoniopropionate-mediated thiation to tropodithietic acid (TDA) (Thiel et al. 2010; Berger et al. 2012).

2. *Burkholderia xenovorans* strain LB400 uses several pathways for the degradation of the benzoate produced by degradation of biphenyl. One of them is the classical benzoate dihydroxylation/catechol intradiol fission pathway, while another resembles the pathway that is used by *Azoarcus evansii*. The genes for this pathway occur both in the chromosome when they are expressed during growth with biphenyl, and in a megaplasmid copy where they were detected only in benzoate-grown cells in the transition to stationary phase (Denef et al. 2005).

3. Novel pathways for the aerobic degradation of anthranilate (2-aminobenzoate) were described several years ago in a strain then designated as a *Pseudomonas* sp. (Altenschmidt and Fuchs 1992a,b; Lochmeyer et al. 1992), and now reassigned to *Azoarcus evansii*. The pathway is analogous to that used for benzoate by this strain. This is initiated by the formation of the benzoyl SCoA-ester followed by monooxygenation/reduction to 2-amino-5-oxocyclohex-1-ene that undergoes β-oxidation with a concomitant NIH shift of the hydrogen at C_5 to C_6 (Hartmann et al. 1999; Schühle et al. 2001) (Figure 8.39).

8.17 Aerobic Reduction of Arene Carboxylates

The reduction of aromatic carboxylic acids to the corresponding aldehydes under aerobic conditions is of interest in biotechnology, since the oxidoreductase from *Nocardia* sp. is able to accept a range of substituted benzoic acids, naphthoic acids, and a few heterocyclic carboxylic acids (Li and Rosazza 1997). The reaction involves the formation of an acyl-AMP intermediate by reaction of the carboxylic acid with

ATP; NADPH then reduces this to the aldehyde (Li and Rosazza 1998; He et al. 2004). A comparable reaction for aromatic carboxylates has been demonstrated in *Neurospora crassa* (Gross 1972).

8.18 Arenes with an Oxygenated C₂ or C₃ Side Chain

Several pathways are used for the aerobic degradation of aromatic compounds with an oxygenated C_2 or C_3 side chain. These include acetophenones and reduced compounds that may be oxidized to acetophenones, and compounds, including tropic acid, styrene, and phenylethylamine that can be metabolized to phenylacetate, which has already been discussed.

8.18.1 Mandelate

The mandelate pathway in *Pseudomonas putida* involves successive oxidation to benzoyl formate and benzoate, which is further metabolized via catechol and the 3-oxoadipate pathway (Figure 8.40a) (Hegeman 1966). Both enantiomers of mandelate were degraded through the activity of a mandelate racemase (Hegeman 1966), and the racemase (mdlA) is encoded in an operon that includes the next two enzymes in the pathway—*S*-mandelate dehydrogenase (mdlB) and benzoylformate decarboxylase (mdlC) (Tsou et al. 1990). Mandelate racemase has been the subject of intensive examination and takes place by 1:1 proton transfer (Hasson et al. 1998) that is noted in Chapter 3. A formally comparable pathway is used by a strain of *Alcaligenes* sp. that degrades 4-hydroxyacetophenone to 4-hydroxybenzoyl methanol, which is oxidized in an unusual reaction to 4-hydroxybenzoate and formate. The 4-hydroxybenzoate is then metabolized to 3-oxoadipate via 3,4-dihydroxybenzoate (Figure 8.40b) (Hopper et al. 1985).

8.18.2 Vanillate and Related Substrates

Although the metabolism of vanillate generally involves de-*O*-methylation to 3,4-dihydroxy-benzoate followed by intradiol ring fission, in *Acinetobacter lwoffii*, vanillate is hydroxylated to 3-*O*-methyl

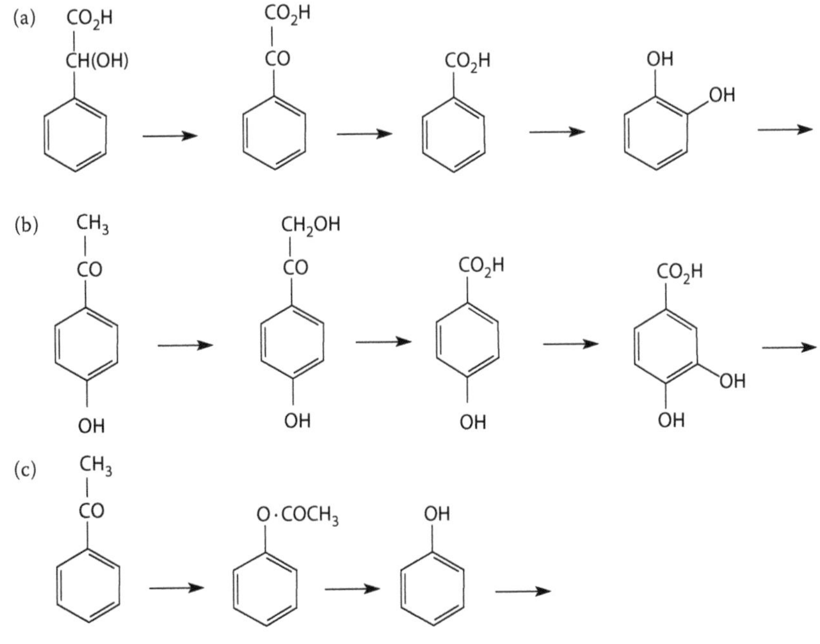

FIGURE 8.40 Degradation of (a) mandelate and (b) 4-hydroxyacetophenone by side-chain oxidation pathways, and (c) acetophenone by Baeyer–Villiger monooxygenation.

gallate, which produces pyruvate and oxalacetate in reactions that have already been noted (Sze and Dagley 1987). The metabolism of ferulate to vanillin by *Pseudomonas fluorescens* strain AN103 is carried out by an enoyl-SCoA hydratase/isomerase rather than by oxidation, and the enzyme belongs to the enoyl-CoA hydratase superfamily (Gasson et al. 1998). A strain of *Pseudomonas* sp. AT3 degraded tropic acid from the hydrolysis of atropine. Phenylacetate was produced by a succession of reactions in the side chain (Long et al. 1997):

$$-CH(CH_2OH)-CO_2H \rightarrow -CH(CHO)-CO_2H \rightarrow -CH_2-CHO + CO_2 \rightarrow -CH_2-CO_2H$$

8.18.3 Acetophenone

Baeyer–Villiger-type oxidations initiate the degradation of acetophenone by strains of *Arthrobacter* sp. and *Nocardia* sp. (Cripps et al. 1978), and of 4-hydroxyacetophenone by *Pseudomonas putida* strain JD1 (Darby et al. 1987). Acetophenone is converted into phenyl acetate, which is hydrolyzed to phenol and then hydroxylated to catechol before ring fission (Figure 8.40c). Similarly, 4-hydroxyacetophenone is oxidized to 4-hydroxyphenyl acetate, which is hydrolyzed to 1,4-dihydroxybenzene before ring fission to 3-oxoadipate. The 4-hydroxy-acetophenone monooxygenase from *Pseudomonas fluorescens* has been purified from recombinant *Escherichia coli*, and was able to carry out Baeyer–Villiger oxidation of a range of substituted acetophenones, bicyclic cyclobutanones, and the enantiomerically specific oxidation of methyl aryl sulfides to sulfoxides (Kamerbeek et al. 2003). The metabolism of chloroacetophenones takes place by analogous monooxygenation to the corresponding chlorophenyl-esters. Some *ortho*-substituted chlorophenols are, however, inhibitory, and only low rates of oxidation have been encountered with di- and trichlorinated acetophenones, so that growth with them is not possible (Higson and Focht 1990). However, degradation of 4-chloroacetophenone has been demonstrated with a mixed culture of an *Arthrobacter* sp. and a *Micrococcus* sp. (Havel and Reineke 1993).

The metabolism of ethylbenzene under denitrifying conditions by *Aromatoleum aromaticum* is initiated by hydroxylation to (*S*)-1-phenylethanol and dehydrogenation to acetophenone. This is then carboxylated by an ATP-dependent enzyme to benzoylacetate that undergoes thiolytic fission to acetyl-CoA and benzoyl-CoA (Jobst et al. 2010). The enzyme contains Zn, although it is inhibited by Zn^{2+} and, therefore, differs from the analogous acetone carboxylase that contains Mg (Boyd et al. 2004).

8.18.4 Phenylacetate and Related Substrates

Some unusual reactions have emerged in the aerobic degradation of phenylalanine, phenylacetate, and 3-phenyl butyrate. Phenylacetate is a central intermediate in the aerobic metabolism of aromatic substrates that include ethylbenzene, styrene, phenylalanine, phenylethylamine, and ω-phenylalkanoates (Luengo et al. 2001). The range of organisms includes Gram-negative *Escherichia coli* K-12 strains (Ferrández et al. 1998), *Pseudomonas putida* strain U (Olivera et al. 1998), *Pseudomonas* sp. strain Y2 (Bartolomé-Martín et al. 2004), and the Gram-positive *Rhodococcus* sp. strain RHA1 that was able to degrade an unusually wide range of aromatic substrates and contained three ring-hydroxylating dioxygenases (Navarro-Llorens et al. 2005). Important comments on the degradation of L-phenylalanine for which several pathways are available are provided in Chapter 7.

The degradation of phenylacetate by the Gram-negative *Pseudomonas putida* strain Y2 and *Escherichia coli* K-12 is carried out by a pathway that is completely different from that used for 3- or 4-hydroxyphenylacetate where degradation is initiated by hydroxylation of the ring to 3,4-dihydroxy- or 1,4-dihydroxy-phenylacetate, both of which are poised to undergo fission of the aromatic ring by dioxygenation. For phenylacetate, on the other hand, degradation is initiated by the formation of the coenzyme A ester by a specific ligase (Martínez-Blanco et al. 1990), and coenzyme A esters are retained throughout the metabolic pathway (Figure 8.41) that is supported by ^{13}C NMR (Ismail et al. 2003). The genes for the degradative enzymes are carried on two operons, *paaABCDEFGHIJK* and a disjoint regulatory operon *paaXY* (Ferrández et al. 2000): *paa ABCDE* encode the oxygenation, *paaGZJ* encode the ring fission and *paaFHJ* encode the final steps. This pathway is also followed by

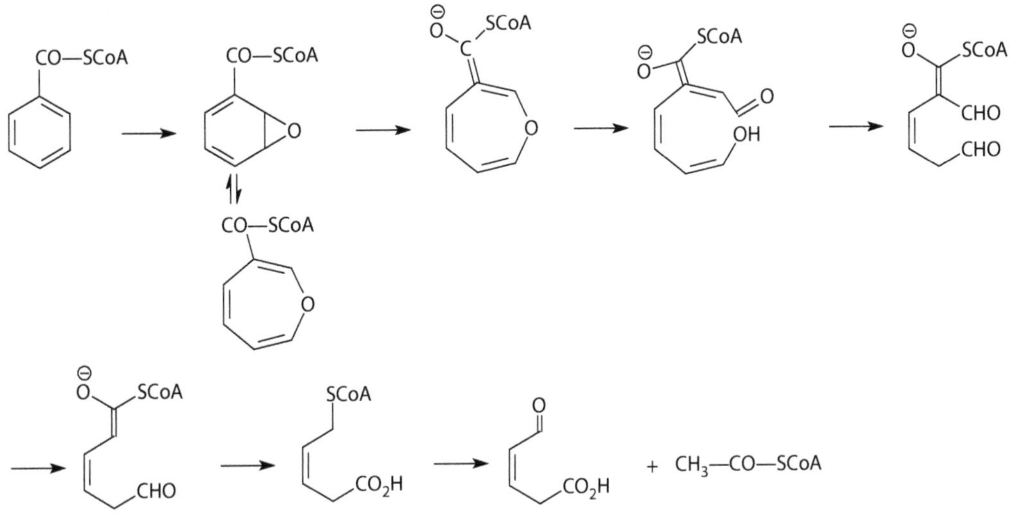

FIGURE 8.41 Aerobic degradation of phenylacetate.

Escherichia coli strain W (Fernández et al. 2006) and *Rhodococcus* sp. strain RHA1 which has an established ability to degrade PCBs and a wide range of monocyclic aromatic compounds (Navarro-Llorens et al. 2005).

In addition, there are two different β-oxidation pathways for the successive degradation of *n*-alkanoates and ω-phenylalkanoates: (a) 8-phenyloctanoyl-CoA to phenylacetyl-CoA involving 3-hydroxy-8-octanoyl-CoA and 3-hydroxyhexanoyl-CoA that can be channeled into poly-3-hydroxyalkanoates, and (b) 9-nonanoyl-CoA to cinnamoyl-CoA (Olivera et al. 2001).

It is worth adding some brief comments on oxepins that have emerged as important in several contexts.

1. *Burkholderia cenocepacia* is a component of the *Burkholderia cepacia* complex, and it has been shown that its pathogenicity toward the nematode *Caenorhabditis elegans* that is used as a surrogate for infection, requires the existence of a functional pathway for the degradation of phenylacetate (Law et al. 2008). By the use of mutants it was shown that the critical gene controlled the oxygenation of the coenzyme A ester.

2. Several models of catechol extradiol fission by dioxygenation have included an oxepin intermediate (Sanvoisin et al. 1995; Shu et al. 1995). This has been supported by a study using density functional theory: the model involved the formation of radicals that included an oxepin fused to an epoxide that was transformed to an oxepin radical before ring fission (Siegbahn and Haeffner 2004).

3. There has been intense interest in the oxidation of benzene in the context of its mammalian carcinogenicity and its oxidation to benzene oxide/oxepin. This may undergo several reactions: (a) ring fission to *trans,trans*-muconaldehyde or to 6-hydroxy-*trans,trans*-2,4-hexadienal or hexadienoic acid, (b) conjugation with glutathione, and (c) reaction with epoxyhydrolase to produce catechol or phenol (Monks et al. 2010).

4. Benzene can be oxidized chemically by dimethyldioxirane to muconaldehyde and, although the 2,3-epoxide could not be identified, the isomeric 4,5-epoxide was a minor product (Bleasdale et al. 1997).

5. A study of benzene oxidation in the atmosphere suggested the production of benzene oxide/oxepin as an intermediate before ring fission, and the 2,3-epoxide was produced by oxidation of the benzene oxide/oxepin with the nitrate radical (Klotz et al. 1997).

8.18.4.1 Hydroxylated Phenylacetates

The degradation of hydroxylated phenylacetates and related compounds such as tyrosine has been described, and are different from that for phenylacetate. These substrates include 3,4-dihydroxyphenyl-acetate by *Escherichia coli* (Cooper and Skinner 1980), and 2,5-dihydroxyphenylacetate produced by rearrangement of 4-hydroxyphenylacetate in *Xanthobacter* sp. strain 124X (van der Tweel et al. 1986). However, these are not intermediates in the degradation of phenylacetate, and an important alternative for this has been described.

8.18.4.2 3-Phenyl Propionate

Although the aerobic degradation of 3-phenylpropionate by *Escherichia coli* proceeded as expected by ring dioxygenation, dehydrogenation, and ring fission to pyruvate, succinate, and acetate (Díaz et al. 1998), the degradation of iboprofen (2-[4-isobutylphenyl]-propionic acid) with a –$CH(CH_3)CO_2H$ side chain was different. The catechol produced by dioxygenation and dehydrogenation was succeeded exceptionally by loss of the side chain (Murdoch and Hay 2005), involving a hitherto unresolved mechanism.

8.18.4.3 Phenyl Butyrate

1. The degradation of 3-phenylbutyrate was examined in a strain of *Pseudomonas* sp. (Sariaslani et al. 1982). The formation of the expected catechol was proved by NMR analysis of its nonenzymatic transformation to a dihydrocoumarin, and characterization of the ring fission product by reaction with NH_3 to produce a pyridine. In addition, however, products that were identified by mass spectrometry included 3-phenylpropionate and 3-phenylacetate. It was tentatively suggested that the first of these was produced by oxidation of the methyl group followed by decarboxylation, although the expected phenylsuccinate could not be utilized.

2. The degradation of (*R*)- and (*S*)-3-phenyl butyrate has been examined in *Rhodococcus rhodochrous* strain PB1. Whereas the (*S*)-enantiomer was dioxygenated to the 2,3-dihydroxy compound that was not metabolized further, the (*R*)-enantiomer was converted into 3-phenylpropionate by a mechanism that was not established. This was followed by dioxygenation and fission of the ring to succinate (Simoni et al. 1996).

8.18.5 Anaerobic Metabolism

Considerable effort has been devoted to the anaerobic degradation of aromatic compounds. It is important to note that several distinct groups of organisms are involved: (a) strictly anaerobic fermentative bacteria, (b) strictly anaerobic photoheterotrophic bacteria, (c) anaerobic sulfate-reducing bacteria, and (d) organisms using nitrate as electron acceptor under anaerobic conditions. Pragmatically, these organisms generally belong to the following groups: methanogenic, sulfidogenic, denitrifying, and phototrophic. Although the greatest attention is given here to studies using pure cultures and investigations in which the relevant enzymes have been characterized, some valuable examples from studies using mixed cultures are provided.

8.18.5.1 Bacteria Using Nitrate Electron Acceptor under Anaerobic Conditions, and Anaerobic Phototrophs

It has become clear that benzoate occupies a central position in the anaerobic degradation of both phenols and alkylated arenes such as toluene and xylenes, and that carboxylation, hydroxylation, and reductive dehydroxylation are important reactions for phenols that are discussed in Part 4 of this chapter. The simplest examples include alkylated benzenes, products from the carboxylation of naphthalene and phenanthrene (Zhang and Young 1997), the decarboxylation of *o*-, *m*-, and *p*-phthalate under denitrifying conditions (Nozawa and Maruyama 1988), and the metabolism of phenols and anilines by carboxylation. Further illustrative examples include the following:

1. Under denitrifying conditions, 2-aminobenzoate is degraded by a *Pseudomonas* sp. to benzoate, which is then reduced to cyclohexene-1-carboxylate (Lochmeyer et al. 1992).

2. The metabolism of cinnamate and ω-phenylalkane carboxylates has been studied in *Rhodopseudomonas palustris* (Elder et al. 1992), and for growth with the higher homologs additional CO_2 was necessary. The key degradative reaction was β-oxidation, for compounds with chain lengths of three, five, and seven carbon atoms, benzoate was formed and further metabolized, but for the even-numbered compounds with four, six, and eight carbon atoms, phenylacetate was a terminal metabolite.

3. The anaerobic metabolism of L-phenylalanine by *Thauera aromatica* under denitrifying conditions involves several steps that result in the formation of benzoyl-CoA: (a) conversion to the CoA-ester by a ligase, (b) transamination to phenylacetyl-CoA, (c) α-oxidation to phenylglyoxylate, and (d) decarboxylation to benzoyl-CoA (Schneider et al. 1997). An analogous pathway is used by *Azoarcus evansii* (Hirsch et al. 1998). The membrane-bound phenylacetyl-CoA: acceptor oxidoreductase that is induced under denitrifying conditions during growth with phenylalanine or phenylacetate has been purified (Schneider and Fuchs 1998). The level of the enzyme was low in cells grown with phenylglyoxylate, the enzyme was insensitive to oxygen, and was absent in cells grown aerobically with phenylacetate.

4. Phenylacetate and 4-hydroxyphenylacetate are oxidized sequentially under anaerobic conditions by a denitrifying strain of *Pseudomonas* sp. to the phenylglyoxylate and benzoate (Mohamed et al. 1993).

8.18.5.2 Benzoate

The degradation of benzoate plays a central role in the degradation of a range of aromatic compounds, including toluene, that has already been discussed. The pathway for the anaerobic degradation of benzoate is entirely different from that used under aerobic conditions, for example, by pseudomonads and, in contrast to the degradation under aerobic conditions (with the exception of that carried out by *Azoarcus evansii*), it invariably involves the initial formation of the CoA-esters. There are broadly two types of reaction, one used by facultative anaerobes that requires ATP, whereas in obligate anaerobes, this is not required.

The pathway is initiated by the activity of an ATP-dependent ligase to form the benzoyl-CoA-ester that is an inducer for both the anaerobic and aerobic pathways, depending on the absence or presence of oxygen (Schühle et al. 2003). The range of organisms that can carry out the anaerobic degradation of benzoate includes facultative anaerobic, phototrophic, and obligately anaerobic bacteria. Although the pathways are essentially similar, they differ in detail between the facultative and obligate anaerobes. Facultative anaerobes include *Thauera aromatica* and *Azoarcus evansii* (β-Proteobacteria), *Rhodopseudomonas palustris*, and *Magnetospirillum* sp. (α-Proteobacteria), and the obligate anaerobes include *Geobacter metallireducens*, and *Desulfococcus multivorans* (δ-Proteobacteria). The enzymology and genetics of these organisms, including *Syntrophus gentianae*, *Syntrophus acidi-trophicus*, and *Desulfococcus multivorans*, have been discussed in an extensive review (Carmona et al. 2009). For all of them, the degradation is initiated by the formation of coenzyme A esters, and this is also used for the anaerobic degradation of 3-hydroxybenzoate, 4-hydroxybenzoate, 4-hydroxyphenylacetate and 4-hydroxycinnamate that are listed in Table 8.2. There are some aspects that are worth noting:

1. Benzoate CoA ligase also initiates the aerobic degradation of benzoate by *Azoarcus evansii* (Gescher et al. 2002, 2006), and under both aerobic and anaerobic conditions by facultatively anaerobic bacteria, including *Thauera aromatica* (Schühle et al. 2003) and *Magnetospirillum* sp. (Kawaguchi et al. 2006).

2. The ligase from *Desulfococcus multivorans* is able to catalyze an unusually broad range of substrates that include the monofluorobenzoates—though not the chlorobenzoates—the aminobenzoates, and among hydroxybenzoates only 3-hydroxybenzoate (Peters et al. 2004).

TABLE 8.2

Anaerobic Activation of Carboxylic Acids by the Formation of Coenzyme A Esters

Substrate	Organism	Selected Reference
Benzoate	*Thauera aromatica*	Schühle et al. (2003)
	Rhodopseudomonas palustris	Egland et al. (1995)
	Azoarcus sp.	López-Barragán et al. (2004)
	Desulfococcus multivorans	Peters et al. (2004)
	Magnetospirillum sp. strain TS-6	Kawaguchi et al. (2006)
3-Hydroxybenzoate	*Thauera aromatica*	Laempe et al. (2001)
	Azoarcus sp.	Wöhlbrand et al. (2008)
	Sporotomaculum hydroxybenzoicum	Müller and Schink (2000)
4-Hydroxybenzoate	*Thauera aromatica*	Biegert et al. (1993)
	Rhodopseudomonas palustris	Gibson et al. (1997)
4-Hydroxycinnamate	*Rhodopseudomonas palustris*	Pan et al. (2008)
4-Hydroxyphenylacetate	*Thauera aromatica*	Mohammed et al. (1993)
	Azoarcus evansii	Hirsch et al. (1998)

The pathway was originally established for the degradation of benzoate under denitrifying and phototrophic conditions. Three cardinal reactions were involved: (1) formation of benzoyl-CoA by a ligase, (2) reduction of the ring (dearomatization) by an ATP-dependent reductase (for facultative anaerobes) with two electrons (to the 1,5-diene), or four electrons (to the 1-ene), and (3) ring scission (references in Harwood et al. 1999; Fuchs 2008; Carmona et al. 2009) (Figure 8.42a) for *Thauera aromatica* and (Figure 8.42b) for *Rhodopseudomonas palustris*. The designation of the enzymes corresponds to those for the genes that are discussed below, and a common pathway is used for both facultative and strict anaerobes (Peters et al. 2007). Differences in benzoyl-CoA reductase (BCR) and glutaryl-CoA dehydrogenase (GDH) are given in Table 8.3 (Wischgoll et al. 2009).

Additional details of the reactions are given in the following paragraphs:

1. Formation of the coenzyme A thioester mediated by an ATP-dependent ligase which has been demonstrated in a number of organisms, including *Pseudomonas* sp. strain K172 (*Thauera* sp. strain K172) (Dangel et al. 1991) and *Rhodopseudomonas palustris* (Egland et al. 1995).

2. A benzoyl-CoA reductase reduces one or more of the double bonds in the aromatic ring (Gibson and Gibson 1992; Koch and Fuchs 1992). In facultative anaerobes such as *Thauera aromatica*, the reductase is ATP dependent, whereas in the obligate anaerobes *Geobacter metallireducens* (Wischgoll et al. 2005) and *Desulfococcus multivorans* (Peters et al. 2004), the reductase is independent of ATP, and is mediated by proteins containing Fe/S, Mo or W, and Se: this is more fully discussed in Chapter 3, Part 7. In *Thauera aromatica*, the reduction is mediated by a 2-oxoglutarate: ferredoxin oxidoreductase (Dörner and Boll 2002), whereas in *Azoarcus evansii*, a 2-oxoglutarate: NADP⁺ oxidoreductase is involved (Ebenau-Jehle et al. 2003). It was used as a biomarker for the anaerobic degradation of aromatic compounds by developing degenerate primers from the established genes from *Thauera aromatica*, *Azoarcus evansii*, and the isolated clones were then assigned to the *bcr* or *bzd* type of the reductase (Song and Ward 2005).

3. The cyclohexa-1,5-dienecarboxylate thioester is hydrated to produce cyclohex-1-ene-6-hydroxycarboxyl-CoA from *Thauera aromatica* (Boll et al. 2000) or 2-hydroxycyclohexane-1-carboxyl-CoA from *Rhodopseudomonas palustris*.

4. The pathways in *T. aromatica* and *Rhps. palustris* also differ slightly in detail.

   a. In *T. aromatica* strain K172, the gene sequence for the enzymes has been determined: *bcr CBAD* for the benzoyl-CoA reductase, *dch* for the dienyl-CoA hydratase, *had* for 6-hydroxycyclohex-1-ene-1-carboxy-CoA dehydrogenase, and they occur in the order *had*, *dch*, *bcrCBAD* (Breese et al. 1998).

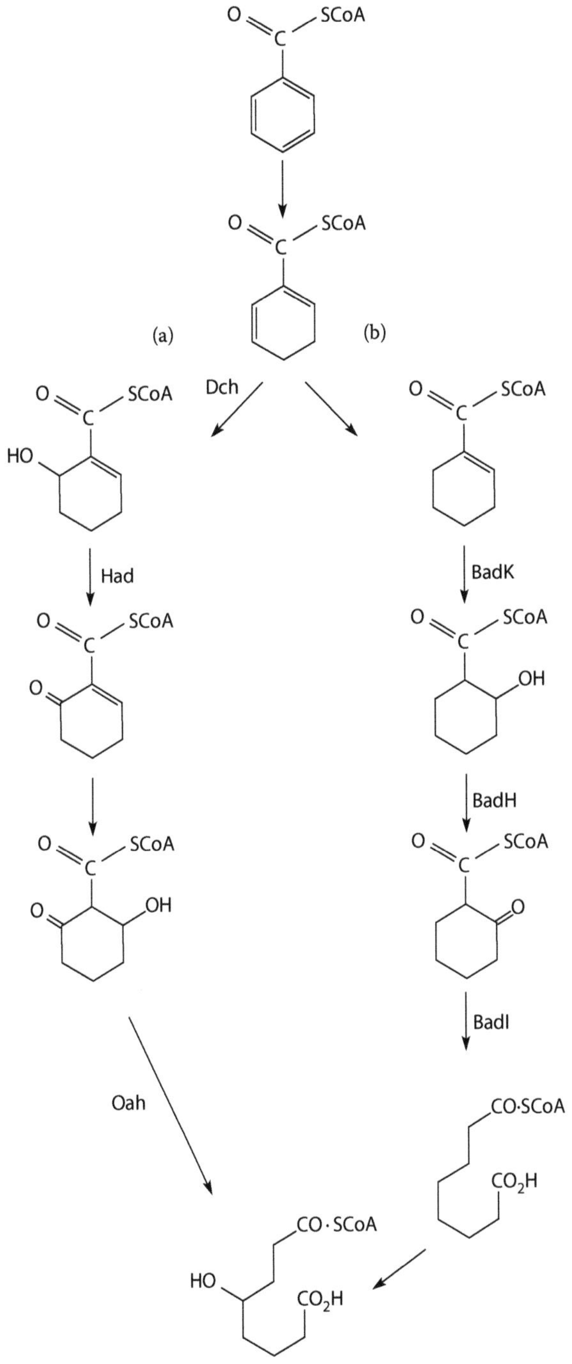

FIGURE 8.42 Anaerobic degradation of benzoate. (a) *Thauera aromatica*, (b) *Rhodopseudomonas palustris*.

b. In *Rhps. palustris*, the genes are *badDEFG* for the benzoyl-CoA reductase, *badK* for the cyclohex-1ene-1-carboxyl-CoA hydratase, and *badH* for the 2-hydroxycyclohexane-1-carboxyl-CoA dehydrogenase, and *badL* for the ketohydrolase. They occur in the order *badK, badL, badH, badD, badE, badF, badG*, the first three transcribing in the opposite direction to the others (Harwood et al. 1999; Egland and Harwood 1999).

TABLE 8.3

Summary of Salient Differences among Groups of Benzoate-Degrading Anaerobic Bacteria

Organism	Type of Reaction	Type of Anaerobe	Type of BCR	Type of GDH
Rhodopseudomonas palustris	Phototrophic	Facultative	ATP dependent	Decarboxylating
Thauera aromatica	Denitrifying	Facultative	ATP dependent	Decarboxylating
Azoarcus sp.	Denitrifying	Facultative	APP dependent	Decarboxylating
Geobacter sp.	Fe(III)-reducing	Obligate	Mo/W/Se-Cys	Decarboxylating
Desulfococcus multivorans	Sulfate-reducing	Obligate	Mo/W/Se-Cys	Nondecarboxylating
Syntrophus aciditrophus	Fermenting	Obligate	Mo/W/Se-Cys	Nondecarboxylating

Note: BCR, benzoyl-CoA reductase; GDH, glutaryl-CoA dehydrogenase.

5. Fission of the ring: In *T. aromatica*, the initial ring fission product is hepta-3-ene-1,7-dicarboxylyl-CoA formed by 3-oxoacyl-CoA hydrolase which is encoded by *oah* that lies immediately to the right, of *had*. The resulting 3-hydroxpimelyl-CoA is degraded by β-oxidation to glutaryl-CoA. With the minor modifications noted above, the same pathway has been demonstrated in *Rhodopseudomonas palustris*, and the ring fission enzyme 2-oxocyclohexanecarboxyl-CoA hydrolase that produces pimelyl-CoA has been purified (Pelletier and Harwood 1998). The gene for the ring-fission enzyme *badL* lies to the left of *badH*.

6. It has been shown in *Rhps. palustris* that the suit of enzymes in the sequence from cyclohexen-1-carboxylate onwards was induced during growth on benzoate but not on succinate (Perotta and Harwood 1994). The further degradation of glutarate takes place by a pathway involving dehydrogenation and decarboxylation to crotonyl-CoA and subsequent formation of acetate (Härtel et al. 1993). After growth of a denitrifying organism with either benzoate or pimelate, both glutaryl-CoA dehydrogenase and glutaconyl-CoA decarboxylase activities were induced—though not the enzymes leading from pimelyl-CoA to glutaryl-CoA: this is consistent with the involvement of 3-hydroxypimelyl-CoA in the degradation of benzoyl-CoA (Gallus and Schink 1994). The initial steps in the degradation of glutarate involving dehydrogenation, decarboxylation, and the β-oxidation pathway have also been demonstrated in other organisms that produce butyrate and *iso*butyrate (Matties and Schink 1992). Details of the various pathways for the anaerobic metabolism of acetate have been reviewed (Thauer et al. 1989).

7. The degradation of benzoate has been examined in obligate anaerobes in which the reduction of the initially produced benzoyl-CoA does not depend on ATP, while there are some other significant differences from the pathways that have been discussed for facultative anaerobes and have been summarized in Table 8.3.

For the obligate anaerobe *Syntrophus gentiane* degradation of benzoate proceeds by the pathway already elucidated for denitrifying and phototrophic bacteria (Figure 8.42a,b)—albeit with significant differences in the enzymology of benzoate activation and the decarboyxlation of glutaconyl-CoA (Schöcke and Schink 1999). In the pathway used by the anaerobes *Geobacter metallireducens* and *Syntrophus aciditrophicus* (Peters et al. 2007), the activity of the hydratases required for the pathway in Figure 8.42b was demonstrated, and was active in benzoate-grown cells. It was, therefore, concluded that the same pathways were functional for the degradation of benzoate in both the facultatively anaerobic *Thauera aromatica* and the obligate anaerobes.

The anaerobic sulfate-reducing *Desulfococcus multivorans* was able to degrade benzoate by a similar pathway, except for details of the mechanism of ring dehydrogenation of benzoyl-CoA: in place of an ATP-dependent [FeS] enzyme molybdenum and selenocysteine proteins were involved (Peters et al. 2004). This has been examined in the obligately anaerobic *Geobacter metallireducens*, and it has been shown that the enzyme which catalyzed the reverse reaction was a 185-kDa enzyme with subunits of 74 kDa (BamB) and 23 kDa (BamC) that would be consistent with an $\alpha_2\beta_2$ structure. The $\alpha\beta$ unit contained 0.9 W, 15 Fe, and 12.7 acid-labile S, as well as Ca 2.1 and Zn 1.2, and EPR indicated the presence of one $[3Fe–4S]^{0/+1}$ and three $[4Fe–4S]^{+1/+2}$ clusters and, in the oxidized form the presence of W^v (Kung et al. 2009).

8. An alternative metabolic pathway took place during the fermentation of benzoate by *Syntrophus aciditrophicus* in the absence of a hydrogen-utilizing bacterium such as *Methanospirillum hungatei*. The benzoate underwent dismutation with the formation of cyclohexane carboxylate by reduction, and of acetate by oxidation to acetate and CO_2 (Elshahed and McInerney 2001). Cyclohexane carboxylate was a terminal metabolite, and the transient formation of cyclohex-1-ene-1-carboxylate, pimelate, and glutarate were consistent with the degradative pathway established for anaerobes and denitrifying bacteria (Peters et al. 2007), and indeed occurred with this organism in the presence of hydrogen-utilizing organisms (Elshahed et al. 2001). The analogous fermentation of 3-hydroxybenzoate by *Sporotomaculum hydroxybenzoicum* produced crotonyl-CoA that functioned as electron acceptor to produce butyrate, acetate, and CO_2 as terminal products (Müller and Schink 2000).

8.18.5.3 Hydroxybenzoates and Related Compounds

There are different pathways for the degradation of hydroxybenzoates:

1. For hydroxybenzoates with hydroxyl groups at the *ortho* or *meta* positions, degradation is initiated by decarboxylation.
   a. For 2,6- and 3,5-dihydroxybenzoates, a denitrifying organism initiates degradation by decarboxylation to 1,3-dihydroxybenzene followed by the reduction of the ring to cylohexan-1,3-dione (Kluge et al. 1990).
   b. For 3,4,5- and 2,4,6-trihydroxybenzoate, *Pelobacter acidigalli* carries out decarboxylation to 1,3,5-trihydroxybenzene followed by reduction of the ring (Brune and Schink 1992).
2. The degradation of 4-hydroxybenzoate by *Thauera aromatica* (Brackmann and Fuchs 1993) and by the anaerobic phototroph *Rhodopseudomonas palustris* (Gibson et al. 1997) is initiated by the formation of the CoA-ester. The next step involves dehydroxylation to benzoate by a reductase that has been characterized in both *Thauera aromatica* (Breese and Fuchs 1998) and *Rh. palustris* (Gibson et al. 1997). The benzoate that is produced then enters the established pathway for reduction of the ring (Heider et al. 1998). The degradation of 3-hydroxybenzoate by *Sporotomaculum hydroxybenzoicum* uses an analogous pathway (Müller and Schink 2000).
3. The fermentation of 3-hydroxybenzoate by *Sporotomaculum hydroxybenzoicum* produces acetate, butyrate, and CO_2, with benzoate as a transient intermediate (Brauman et al. 1998). However, although the degradation of 3-hydroxybenzoate by *Thauera aromatica* begins with the formation of the CoA-ester, this is followed by the reduction of the ring with retention of the original hydroxyl group (Laempe et al. 2001).
4. *Clostridium difficile* is able to decarboxylate 4-hydroxyphenylacetate with the production of *p*-cresol in a glycyl-mediated radical reaction (Selmer and Andrei 2001). Analogously, phenylacetate can be decarboxylated by anaerobic and facultative anaerobic bacteria to produce toluene by *Tolumonas auensis* (Fischer-Romero et al. 1996), *Clostridium aerofoetidum* (Pons et al. 1984) and a strain of *Lactobacillus* (Yokoyama and Yokoyama 1981).

8.19 Aldehydes

The reductive transformation of arene carboxylates to the corresponding aldehydes under aerobic conditions has already been noted. In addition, aromatic aldehydes may undergo both reductive and oxidative reactions, with the possibility of decarboxylation of the carboxylic acid formed:

1. Oxidation of substituted benzaldehydes to benzoates at the expense of sulfate reduction has been demonstrated in strains of *Desulfovibrio* sp., although the carboxylic acids produced were apparently stable to further degradation (Zellner et al. 1990). Vanillin was, however, used as a substrate for growth by a strain of *Desulfotomaculum* sp. and was metabolized via vanillate and catechol (Kuever et al. 1993). Vanillin is transformed by *Clostridium formicoaceticum*

sequentially to vanillate and 3,4-dihydroxybenzoate, while the methyl group is converted into acetate via acetyl coenzyme A (Gößner et al. 1994).

2. Reduction of benzoates to the corresponding benzyl alcohols has been observed in several organisms: (i) cell extracts of *Clostridium formicoaceticum* at the expense of carbon monoxide (Fraisse and Simon 1988), (ii) *Desulfomicrobium escambiense* at the expense of pyruvate that was oxidized to acetate, lactate, and succinate (Sharak Genther et al. 1997).

Oxidoreductases mediate the transfer of electrons between aldehydes and carboxylates. In *Clostridium formicoaceticum*, there are two oxidoreductases. One contains Mo, is reversible, and active toward a range of both aliphatic and aromatic carboxylic acids and aldehydes (White et al. 1993), whereas the other is a W-containing enzyme (White et al. 1991). In addition, there exists a different enzyme that carries out the same reaction in *C. thermoaceticum* (White et al. 1989). Different aldehyde oxidoreductases have been isolated from *Clostridium thermoaceticum* and *Cl. formicoaceticum* (White et al. 1993). Whereas however, both cinnamate and cinnamaldehyde were good substrates for the Mo-containing enzyme from the latter, benzoate was an extremely poor substrate. The W-containing enzyme from *Cl. formicoaceticum*, however, displays high activity toward a greater range of substituted benzoates (White et al. 1991).

8.20 References

Altenschmidt U, G Fuchs. 1992. Novel aerobic 2-aminobenzoate metabolism. *Eur J Biochem* 205: 721–727.

Altenschmidt U, B Oswald, E Steiner, H Herrmann, G Fuchs. 1993. New aerobic benzoate oxidation pathway via benzoyl-coenzyme A and 3-hydroxybenzoyl-coenzyme A in a denitrifying *Pseudomonas* sp. *J Bacteriol* 175: 4851–4858.

Ander P, K-E Eriksson, H-S Yu. 1983. Vanillic acid metabolism by *Sporotrichium pulverulentum*: Evidence for demethoxylation before ring-cleavage. *Arch Microbiol* 136: 1–6.

Anderson JJ, S Dagley. 1980. Catabolism of aromatic acids in *Trichosporon cutaneum*. *J Bacteriol* 141: 534–543.

Anderson JJ, S Dagley. 1981. Catabolism of tryptophan, anthranilate, and 2,3-dihydroxybenzoate in *Trichosporon cutaneum*. *J Bacteriol* 146: 291–297.

Arias-Barrau E, ER Olivera, JM Luengo, C Fernández, B Gálan, JL Garcia, E Díaz, B Miñambres. 2004. The homogentisate pathway: A central catabolic pathway involved in the degradation of L-phenylalanine, L-tyrosine, and 3-hydroxyphenylacetate in *Pseudomonas putida*. *J Bacteriol* 186: 5062–5077.

Arunachalam U, V Massey, CS Vaidyanathan. 1992. *p*-Hydroxyphenacetate-3-hydroxylase. A two-component enzyme. *J Biol Chem* 267: 25848–25855.

Asturias JA, LD Eltis, M Prucha, KN Timmis. 1994. Analysis of three 2,3-dihydroxybiphenyl 1,2-dioxygenases found in *Rhodococcus globerulus* P6. *J Biol Chem* 269: 7807–7815.

Bartolomé-Martín D, E Martínez-García, V Mascaraque, J Rubio. J Perera, S Alonso. 2004. Characterization of a second functional gene cluster for the catabolism of phenylacetic acid in *Pseudomonas* sp. strain Y2. *Gene* 341: 167–179.

Batie CJ, E LaHaie, DP Ballou. 1987. Purification and characterization of phthalate oxygenase and phthalate oxygenase reductase from *Pseudomonas cepacia*. *J Biol Chem* 262: 1510–1518.

Berger M, NL Brock, H Liesegang, M Dogs, I Preuth, M Simon, JS Dickschat, T Brinkhoff. 2012. Genetic analysis of the upper phenylacetate catabolic pathway in the production of tropothietic acid by *Phaeobacter gallaeciensis*. *Appl Environ Microbiol* 78: 3539–3551.

Biegert T, U Altenschmidt, C Eckerskorn, G Fuchs. 1993. Enzymes of anaerobic metabolism of phenolic compounds. 4-Hydroxybenzoate-CoA ligase from a denitrifying *Pseudomonas* species. *Eur J Biochem* 213: 555–561.

Bleasdale C, R Cameron, C Edwards, BT Golding. 1997. Dimethyldioxirane converts benzene oxide/oxepin into (*Z,Z*)-muconaldehyde and *sym*-oxepin oxide: Modeling the metabolism of benzene and its photooxidative degradation. *Chem Res Toxicol* 10: 1314–1318.

Boldt YR, MJ Sadowsky, LBM Ellis, L Que, LP Wackett. 1995. A manganese-dependent dioxygenase from *Arthrobacter globiformis* CM-2 belongs to the major extradiol dioxygenase family. *J Bacteriol* 177: 1225–1232.

Boll M, D Laempe, W Eisenreich, A Bacher, T Mittelnerger, J Heinze, G Fuchs. 2000. Nonaromatic products from anoxic conversion of benzoyl-CoA with benzoyl-CoA reductase and cyclohexa-1,5-diene- 1-carbonyl-CoA hydratase. *J Biol Chem* 275: 21889–21895.

Bosch R, ERB Moore, E García-Valdés, DH Pieper. 1999. NahW, a novel, inducible salicylate hydroxylase involved in mineralization of naphthalene by *Pseudomonas stutzeri* AN10. *J Bacteriol* 181: 2315–2322.

Boyd JM, SA Elsworth, SA Ensign. 2004. Bacterial acetone carboxylase is a manganese-dependent metalloenzyme. *J Biol Chem* 279: 46644–46651.

Brackmann R, G Fuchs. 1993. Enzymes of anaerobic metabolism of phenolic compounds. 4-hydroxybenzoyl-CoA reductase (dehydroxylating) from a denitrifying *Pseudomonas* sp. *Eur J Biochem* 213: 563–571.

Brauman A, JA Müller, J-L Garcia, A Brine, B Schink. 1998. Fermentative degradation of 3-hydroxybenzoate in pure culture by a novel strictly anaerobic bacterium, *Sporotomaculum hydroxybenzoicum* gen. nov., sp. nov. *Int J Syst Bacteriol* 48: 215–221.

Breese K, M Boll, J Alt-Mörbe, H Schäggrer, G Fuchs. 1998. Genes encoding the benzoyl-CoA pathways of anaerobic aromatic metabolism in the bacterium *Thauera aromatica*. *Eur J Biochem* 256: 148–154.

Breese K, G Fuchs. 1998. 4-hydroxybenzoyl-CoA reductase (dehydroxylating) from the denitrifying bacterium *Thauera aromatica*: Prosthetic groups, electron donor, and genes of a member of the molybdenum-flavin-iron-sulfur proteins. *Eur J Biochem* 251: 916–923.

Brune A, B Schink. 1992. Phloroglucinol pathway in the strictly anaerobic *Pelobacter acidigallici* fermentation of trihydroxybenzenes to acetate via triacetic acid. *Arch Microbiol* 157: 417–424.

Cain RB. 1968. Anthranilic acid metabolism by microorganisms. Formation of 5-hydroxyanthranilate as an intermediate in anthranilate metabolism by *Nocardia opaca*. *Anthonie van Leeuwenhoek* 34: 417–432.

Carmona M, MT Zamarro, B Blázquez, G Durante-Rodríguez, JF Juárez, JA Valderrama, MJL Barragán, JL Garciá, E Díaz. 2009. Anaerobic catabolism of aromatic compounds: A genetic and genomic view. *Microbiol Mol Biol Revs* 73: 71–133.

Chaiyen P, C Suadee, P Wilairat. 2001. A novel two-protein component flavoprotein hydroxylase *p*-hydroxyphenylacetate from *Acinetobacter baumannii*. *Eur J Biochem* 268: 5550–5561.

Chakrabarty AM. 1972. Genetic basis of the biodegradation of salicylate in *Pseudomonas*. *J Bacteriol* 112: 815–823.

Chang H-K, P Mohseni, GJ Zylstra. 2003. Characterization and regulation of the genes for a novel anthranilate 1,2-dioxyganase from *Burkholderia cepacia* DBO1. *J Bacteriol* 185: 5871–5881.

Chang H-K, GJ Zylstra. 1999. Role of quinolinate phosphoribosyl transferase in degradation of phthalate by *Burkholderia cepacia* DBO1. *J Bacteriol* 181: 3069–3075.

Charest MG, CD Lerner, JD Brubaker, DR Siegel, AG Myers. 2005. A convenient enantioselective route to structurally diverse 6-deoxytetracyline antibiotics. *Science* 308: 395–398.

Chow KT, MK Pope, J Davies. 1999. Characterization of a vanillic acid non-oxidative decarboxylation gene cluster from *Streptomyces* sp. D7. *Microbiology* (*UK*) 145: 2393–2404.

Cooper RA, MA Skinner. 1980. Catabolism of 3- and 4-hydroxyphenylacetate by the 3,4-dihydroxyphenylacetate pathway in *Escherichia coli*. *J Bacteriol* 143: 302–306.

Corvini PFX, RJW Meesters, A Schäffer, HF Schröder, R Vinken, J Hollender. 2004. Degradation of a nonylphenol single isomer by *Sphingomonas* sp. strain TTNP3 leads to a hydroxylation-induced migration product. *Appl Environ Microbiol* 70: 6897–6900.

Crawford RL. 1975a. Degradation of 3-hydroxybenzoate by bacteria of the genus *Bacillus*. *Appl Microbiol* 30: 439–444.

Crawford RL. 1975b. Novel pathway for degradation of protocatechuic acid in *Bacillus* species. *J Bacteriol* 121: 531–536.

Crawford RL. 1976. Pathways of 4-hydroxybenzoate degradation among species of *Bacillus*. *J Bacteriol* 127: 204–290.

Crawford RL, SW Hutton, PJ Chapman. 1975. Purification and properties of gentisate 1,2-dioxygenase from *Moraxella osloensis*. *J Bacteriol* 121: 794–799.

Crawford RL, PP Olson. 1978. Microbial catabolism of vanillate: Decarboxylation to guaiacol. *Appl Environ Microbiol* 36: 539–543.

Cripps RE, PW Trudgill, JG Whateley. 1978. The metabolism of 1-phenylethanol and acetophenone by *Nocardia* T5 and an *Arthrobacter* species. *Eur J Biochem* 86: 175–186.

D'Ari L, WA Barker. 1985. *p*-cresol formation by cell-free extracts of *Clostridium difficile*. *Arch Microbiol* 143: 311–312.

Dangel W, R Brackmann, A Lack, M Mohamed, J Koch, B Oswald, B Seyfried, A Tschech, G Fuchs. 1991. Differential expression of enzyme activities initiating anoxic metabolism of various aromatic compounds via benzoyl-CoA. *Arch Microbiol* 155: 256–262.

Darby JM, DG Taylor, DJ Hopper. 1987. Hydroquinone as the ring-fission substrate in the catabolism of 4-ethylphenol and 4-hydroxyacetophenone by *Pseudomonas putida* D1. *J Gen Microbiol* 133: 2137–2146.

Degrassi G, PP de Laureto, CV Bruschi. 1995. Purification and characterization of ferulate and *p*-coumarate decarboxylase from *Bacillus pumilis*. *Appl Environ Microbiol* 61: 326–332.

Denef VJ, MA Patrauchan, C Florizone, J Park, TV Tsoi, W Verstraete, JM Tiedje, LD Eltis. 2005. Growth sub- strate- and phase-specific expression of biphenyl, benzoate, and C1 metabolic pathways in *Burkholderia xenovorans* LB400. *J Bacteriol* 187: 7996–8005.

Díaz E, A Ferrández, JL Garcia. 1998. Characterization of the *hca* cluster encoding the dioxygenolytic pathway for initial catabolism of 3-phenylpropionic acid in *Escherichia coli* K-12. *J Bacteriol* 180: 2915–2923.

Donnelly MI, S Dagley. 1980. Production of methanol from aromatic acids by *Pseudomonas putida*. *J Bacteriol* 142: 916–924.

Dörner E, M Boll. 2002. Properties of 2-oxoglutarate: Ferredoxin oxidoreductase from *Thauera aromatica* and its role in enzymatic reduction of the aromatic ring. *J Bacteriol* 184: 3975–3983.

Eaton RW, DW Ribbons. 1982. Metabolism of dibutylphthalate and phthalate by *Micrococcus* sp. strain 12B. *J Bacteriol* 151: 48–57.

Ebenau-Jehle C, M Boll, G Fuchs. 2003. 2-oxoglutarate: NADP⁺ oxidoreductase in *Azoarcus evansii*: Properties and function in electron transfer reactions in aromatic ring reduction. *J Bacteriol* 185: 6119–6129.

Eby DM, ZM Beharry, ED Coulter, DM Kurtz, EL Neidle. 2001. Characterization and evolution of anthranilate 1,2-dioxygenase from *Acinetobacter* sp. strain ADP1. *J Bacteriol* 183: 109–118.

Egland PG, J Gibson, CS Harwood. 1995. Benzoate-coenzyme A ligase, encoded *badA*, is one of three ligases to catalyze benzoyl-coenzyme A formation during anaerobic growth of *Rhodopseudomonas palustris* on benzoate. *J Bacteriol* 177: 6545–6551.

Egland PG, CS Harwood. 1999. BadR, a new MarR family member regulates anaerobic benzoate degradation by *Rhodopseudomonas palustris* in concert with AadR, an Fnr family member. *J. Bacteriol.* 181: 2102–2109.

Elder DJE, P Morgan, DJ Kelly. 1992. Anaerobic degradation of *trans*-cinnamate and ω-phenylalkane carbox- ylic acids by the photosynthetic bacterium *Rhodopseudomonas palustris*: Evidence for a beta-oxidation mechanism. *Arch Microbiol* 157: 148–154.

Elmorsi EA, DJ Hopper. 1979. The catabolism of 5-hydroxyisophthalate by a soil bacterium. *J Gen Microbiol* 111: 145–152.

Elsden SR, MG Hilton, JM Waller. 1976. The end products of the metabolism of aromatic acids by clostridia. *Arch Microbiol* 107: 283–288.

Elshahed MS, VK Bhupathiraju, NO Wofford, MA Nanny, MJ McInerney. 2001. Metabolism of benzoate, cyclohex-1-ene carboxylate, and cyclohexane carboxylate by "*Syntrophus aciditrophicus*" strain SB in syntrophic association with H₂-using microorganisms. *Appl Environ Microbiol* 67: 1728–1738.

Elshahed MS, MJ McInerney. 2001. Benzoate fermentation by the anaerobic bacterium *Syntrophus aciditrophi- cus* in the absence of hydrogen-utilizing microorganisms *Appl Environ Microbiol* 67: 5520–5525.

Eppink MHM, SA Boeren, J Vervoort, WJH van Berkel. 1997. Purification and properties of 4-hydroxybenzo- ate 1-hydroxylase (decarboxylating), a novel flavin adenine dinucleotide-dependent monooxygenase from *Candida parapilosis* CBS604. *J Bacteriol* 179: 6680–6687.

Fairley DJ, DR Boyd, ND Sharma, CCR Allen, P Morgan, MJ Larkin. 2002. Aerobic metabolism of 4-hydroxy- benzoic acid in *Archaea* via an unusual pathway involving an intramolecular migration (NIH shift). *Appl Environ Microbiol* 68: 6246–6255.

Feng Y, HE Khoo, CL Poh. 1999. Purification and characterization of gentisate 1,2-dioxygenases from *Pseudomonas alcaligenes* NCIB 9867 and *Pseudomonas putida* NCIB 9869. *Appl Environ Microbiol* 65: 946–950.

Fernández C, A Ferrández, B Miñambres, E Díaz, JL García. 2006. Genetic characterization of the phenylace- tyl-CoA oxygenase from the aerobic phenylacetic acid degradation pathway of *Escherichia coli*. *Appl Environ Microbiol* 72: 7422–7426.

Ferrández A, JL García, E Díaz. 2000. Transcriptional regulation of the divergent *paa* catabolic operons for phenylacetic acid degradation in *Escherichia coli*. *J Biol Chem* 275: 12214–12222.

Ferrández C, B Miñambres, B García, ER Olivera, JM Luengo, JL García, E Díaz. 1998. Catabolism of phen- ylacetic acid in *Escherichia coli*. Characterization of a new aerobic hybrid pathway. *J Biol Chem* 273: 25974–25986.

Fischer-Romero C, BJ Tindall, F Jüttner. 1996. *Tolumonas auensis* gen. nov., sp. nov., a toluene-producing bacterium from anoxic sediments of a freshwater lake. *Int J Syst Bacteriol* 46: 183–188.

Fraisse L, H Simon. 1988. Observations on the reduction of non-activated carboxylates by *Clostridium formicoaceticum* with carbon monoxide or formate and the influence of various viologens. *Arch Microbiol* 150: 381–386.

Fuchs G. 2008. Anaerobic metabolism of aromatic compounds. *Ann NY Acad Sci* 1125: 82–99.

Fukuhara Y, K Inakazu, N Kodama, N Kamimura, D Kasai, Y Katayama, M Fukuda, E Masai. 2010. Characterization of the isophthalate degradation genes of *Comamonas* sp. strain E6. *Appl Environ Microbiol* 76: 519–527.

Furukawa K, JR Simon, AM Chakrabarty. 1983. Common induction and regulation of biphenyl, xylene/toluene, and salicylate catabolism in *Pseudomonas paucimobilis. J Bacteriol* 154: 1356–1362.

Gabriel FLP, W Giger, K Guenther, H-P E Kohler. 2005b. Differential degradation of nonylphenol isomers by *Sphingomonas xenophaga* Bayram. *Appl Environ Microbiol* 71: 1123–1129.

Gabriel FLP, A Heidlberger, D Rentsch, W Giger, K Guenther, H-PE Kohler. 2005a. A novel metabolic pathway for degradation of 4-nonylphenol environmental contaminants by *Sphingomonas xenophaga* Bayram. *J Biol Chem* 280: 15526–15533.

Gallus C, B Schink. 1994. Anaerobic degradation of pimelate by newly isolated denitrifying bacteria. *Microbiology (UK)* 140: 409–416.

Gasson MJ, Y Kitamura, WR McLaughlan, A Narbad, AJ Parr, ELH Parsons, J Payne, MJC Rhodes, NK Walton. 1998. Metabolism of ferulic acid to vanillin. A bacterial gene of the enoyl-SCoA hydratase/isomerase superfamily encodes an enzyme for the hydration and cleavage of a hydroxycinnamic acid SCoA thioester. *J Biol Chem* 273: 4163–4170.

Geng H, R Belas. 2010. Expression of tropodithietic acid biosynthesis is controlled by a novel inducer. *J Bacteriol* 192: 4377–4387.

Gescher J, W Eisenreich, J Wört, A Bacher, G Fuchs. 2005. Aerobic benzoyl-CoA catabolic pathway in *Azoarcus evansii*: Studies on the non-oxygenolytic ring cleavage enzyme. *Mol Microbiol* 56: 1586–1600.

Gescher J, W Ismail, E Ölgeschläger, W Eisenreich, J Wört, G Fuchs. 2006. Aerobic benzoyl-coenzyme A (CoA) catabolic pathway in *Azoarcus evansii*: Conversion of ring cleavage product by 3,4-dehydroadipyl-CoA semialdehyde dehydrogenase. *J Bacteriol* 188: 2919–2927.

Gescher J, A Zaar, M Mohamed, H Schägger, G Fuchs. 2002. Genes coding a new pathway of aerobic benzoate metabolism in *Azoarcus evansii. J Bacteriol* 184: 6301–6315.

Gibson J, M Dispensa, CS Harwood. 1997. 4-Hydroxybenzoyl coenzyme A reductase dehydroxylating is required for anaerobic degradation of 4-hydrozybenzoate by *Rhodopseudomonas palustris* and shares features with molybdenum-containing hydroxylases. *J Bacteriol* 179: 634–642.

Gibson KJ, J Gibson. 1992. Potential early intermediates in anaerobic benzoate degradation by *Rhodopseudomonas palustris. Appl Environ Microbiol* 58: 696–698.

Gibson DT, SM Resnick, K Lee, JM Brand, DS Torok, LP Wackett, MJ Schocken, BE Haigler. 1995. Desaturation, dioxygenation, and monooxygenation reactions catalyzed by naphthalene dioxygenase from *Pseudomonas* sp. strain 9816-4. *J Bacteriol* 177: 2615–2621.

Goetz FE, L Harmuth. 1992. Gentisate pathway in *Salmonella typhimurium*: Metabolism of *m*-hydroxybenzoate and gentisate. *FEMS Microbiol Lett* 97: 45–50.

Gößner A, SL Daniel, HL Drake. 1994. Acetogenesis coupled to the oxidation of aromatic aldehyde groups. *Arch Microbiol* 161: 126–131.

Gross GG. 1972. Formation and reduction of intermediate acyladenylate by aryl-aldehyde. NADP oxidoreductase from *Neurospora crassa. Eur J Biochem* 31: 585–592.

Grund E, B Denecke, R Eichenlaub. 1992. Naphthalene degradation via salicylate and gentisate by *Rhodococcus* sp. strain B4. *Appl Environ Microbiol* 58: 1874–1877.

Hareland WA, RL Crawford, PJ Chapman, S Dagley. 1975. Metabolic function and properties of 4-hydroxyphenylacetic acid 1-hydrolase from *Pseudomonas acidovorans. J Bacteriol* 121: 272–285.

Harpel MR, JD Lipscomb. 1990. Gentisate 1,2-dioxygenase from *Pseudomonas*. Purification, characterization, and comparison of the enzymes from *Pseudomonas testosteroni* and *Pseudomonas acidovorans. J Biol Chem* 265: 6301–6311.

Härtel U, E Eckel, J Koch, G Fuchs, D, Linder, W Buckel. 1993. Purification of glutaryl-CoA dehydrogenase from *Pseudomonas* sp., an enzyme involved in the anaerobic degradation of benzoate. *Arch Microbiol* 159: 174–181.

Hartmann S, C Hultschig, W Eisenreich, G Fuchs, A Bacher, S Ghisla. 1999. NIH shift in flavin-dependent monooxygenation: Mechanistic studies with 2-aminobenzoyl-CoA monooxygenase/reductase. *Proc Natl Acad USA* 96: 7831–7836.

Harwood CS, G Burchardt, H Herrmann, G Fuchs. 1999. Anaerobic metabolism of aromatic compounds via the benzoyl-CoA pathway. *FEMS Microbiol Rev* 22: 439–458.

Harwood CS, RE Parales. 1996. The β-ketoadipate pathway and the biology of self-identity. *Annu Rev Microbiol* 50: 553–590.

Hasson MS et al. 1998. Evolution of an enzyme active site: The structure of a new crystal form of muconate lactonizing enyme compared with mandelate racemase and enolase. *Proc Natl Acad Sci USA* 95: 10396–10401.

Hatta T, G Mukerjee-Dhar, J Damborsky, H Kiyohara, K Kimbara. 2003. Characterization of a novel thermo-stable Mn(II)-dependent 2,3-dihydroxybiphenyl 1,2-dioxygenase from a polychlorinated biphenyl- and naphthalene-degrading *Bacillus* sp. JF8. *J Biol Chem* 278: 21483–21492.

Havel J, W Reineke. 1993. Microbial degradation of chlorinated acetophenones. *Appl Environ Microbiol* 59: 2706–2712.

He A, T Li, L Daniels, I Fotheringham, JPN Rosazza. 2004. *Nocardia* sp. carboxylic acid reductase: Cloning, expression, and characterization of a new aldehyde oxidoreductase family. *Appl Environ Microbiol* 70: 1874–1881.

He Z, J Wiegel. 1996. Purification and characterization of an oxygen-sensitive, reversible 3,4-dihydroxybenzo-ate decarboxylase from *Clostridium hydroxybenzoicum*. *J Bacteriol* 178: 3539–3543.

Hegeman GD. 1966. Synthesis of the enzymes of the mandelate pathway by *Pseudomonas putida*. I. Synthesis of enzymes of the wild type. *J Bacteriol* 91: 1140–1154.

Heider J et al. 1998. Differential induction of enzymes involved in anaerobic metabolism of aromatic com-pounds in the denitrifying bacterium *Thauera aromatica*. *Arch Microbiol* 170: 120–131.

Higson FK, DD Focht. 1990. Degradation of 2-bromobenzoate by a strain of *Pseudomonas* aeruginosa. *Appl Environ Microbiol* 56: 1615–1619.

Hintner J-P, C Lechner, U Riegert, AE Kuhm, T Storm, T Reemtsma, A Stolz. 2001. Direct ring fission of salicylate by a salicylate 1,2-dioxygenase activity from *Pseudaminobacter salicylatoxidans*. *J Bacteriol* 183: 6936–6942.

Hintner J-P, T Reemtsma, A Stolz. 2004. Biochemical and molecular characterization of a ring fission dioxy-genase with the ability to oxidize (substituted) salicylate(s) from *Pseudaminobacter salicylatoxidans*. *J Biol Chem* 279: 37250–37260

Hirsch W, H Schägger, G Fuchs. 1998. Phenylglyoxalate: NAD$^+$ oxidoreductase (CoA benzoylating), a new enzyme of anaerobic phenylalanine metabolism in the denitrifying bacterium *Azoarcus evansii*. *Eur J Biochem* 251: 907–915.

Honeyfield DC, JR Carlson. 1990. Assay for the enzymatic conversion of indoleacetic acid to 3-methylindole in a ruminal *Lactobacillus* species. *Appl Environ Microbiol* 56: 724–729.

Hopper DJ, HG Jones, EA Elmorisi, ME Rhodes-Roberts. 1985. The catabolism of 4-hydroxyacetophenone by an *Alcaligenes* sp. *J Gen Microbiol* 131: 1807–1814.

Howell JG, T Spector, V Massey. 1972. Purification and properties of *p*-hydroxybenzoate hydroxylase fromn *Pseudomonas fluorescens*. *J Biol Chem* 247: 4340–4350.

Hsu T, MF Lux, HL Drake. 1990. Expression of an aromatic-dependent decarboxylase which provides growth-essential CO$_2$ equivalents for the acetogenic (Wood) pathway of *Clostridium thermoaceticum*. *J Bacteriol* 172: 5901–5907.

Ishiyama D, D Vujaklija, J Davies. 2004. Novel pathway of salicylate degradation by *Streptomyces* sp. strain WA46. *Appl Environ Microbiol* 70: 1297–1306.

Ismail W, M El-Said Mohamed, BL Wanner, KA Datsenko, W Eisenreich, F Rohdich, A Bacher, G Fuchs. 2003. Functional genomics by NMR spectroscopy. Phenylacetate catabolism in *Escherichia coli*. *Eur J Biochem* 270: 3047–3054.

Jobst B, K Schühle, U Linne, J Heider. 2010. ATP-dependent carboxylation of acetophenone by a novel type of carboxylase. *J Bacteriol* 192: 1387–1394

Jones DCN, RA Cooper. 1990. Catabolism of 3-hydroxybenzoate by the gentisate pathway in *Klebsiella pneu-moniae* M5a1. *Arch Microbiol* 154: 489–495.

Kamerbeek NM, AJJ Olsthoorn, MW Fraaije, DB Janssen. 2003. Substrate specificity and enantioselectivity of 4-hydroxyacetophenone monooxygenase. *Appl Environ Microbiol* 69: 419–426.

Kasai K, E Masai, K Miyauchi, Y Katayama, M Fukuda. 2004. Characterization of the 3-*O*-methylgallate dioxygenase gene and evidence of multiple 3-*O*-methylgallate catabolic pathways in *Sphingomonas paucimobilis* SYK-6. *J Bacteriol* 186: 4951–4959.

Kawaguchi K, Y Shinoda, H Yurimoto, Y Sakai, N Kato. 2006. Purification and characterization of benzoate-CoA ligase from *Magnetospirillum* sp. strain TS-6 capable of aerobic and anaerobic degradation of aromatic compounds. *FEMS Microbiol Lett* 257: 208–213.

Kersten PJ, S Dagley, JW Whittaker, DM Arciero, JD Lipscomb. 1982. 2-Pyrone-4,6-dicarboxylic acid, a catabolite of gallic acids in *Pseudomonas* species. *J Bacteriol* 152: 1154–1162.

Klotz B, I Barnes, KH Becker, BT Golding. 1997. Atmospheric chemistry of benzene oxide/oxepin. *J Chem Soc Faraday Trans* 93: 1507–1516.

Kluge C, A Tschech, G Fuchs. 1990. Anaerobic metabolism of resorcylic acids (*m*-dihydroxybenzoates) and resorcinol (1,3-benzenediol) in a fermenting and in a denitrifying bacterium. *Arch Microbiol* 155: 68–74.

Koch J, G Fuchs. 1992. Enzymatic reduction of benzoyl-CoA to alicyclic compounds, a key reaction in anaerobic aromatic metabolism. *Eur J Biochem* 205: 195–202.

Kozarich JW. 1988. Enzyme chemistry and evolution in the β-ketoadipate pathway, pp. 283–302. In *Microbial Metabolism and the Carbon Cycle* (Eds. SR Hagedorn, RS Hanson, and DA Kunz), Harwood Academic Publishers, Chur, Switzerland.

Krumholz LR, RL Crawford, ME Hemling, MP Bryant. 1987. Metabolism of gallate and phloroglucinol in *Eubacterium oxidoreducens* via 3-hydroxy-5-ketohexanoate. *J Bacteriol* 169: 1886–1890.

Kuever J, J Kulmer, S Janssen, U Fischer, K-H Blotevogel. 1993. Isolation and characterization of a new spore-forming sulfate-reducing bacterium growing by complete oxidation of catechol. *Arch Microbiol* 159: 282–288.

Kung JW, C Löffler, K Dörner, D Heintz, S Gallien, A Van Dorsselaer, T Friedrich, M Boll. 2009. Identification and characterization of the tungsten-containing class of benzoyl-coenzyme A reductases. *Proc Natl Acad Sci USA* 106: 17687–17692.

Laempe D, M Jahn, K Breese, H Schägger, G Fuchs. 2001. Anaerobic metabolism of 3-hydroxybenzoate by the denitrifying bacterium *Thauera aromatica*. *J Bacteriol* 183: 968–979.

Law RJ, JNR Hamlin, A Sivro, SJ McCorrister, GA Cardama, ST Cardona. 2008. A functional phenylacetic acid catabolic pathway is required for full pathogenicity of *Burkholderia cenocepacia* in the *Caenorhabditis elegans* host model. *J Bacteriol* 190: 7209–7218.

Li T, JPN Rosazza. 1997. Purification, characterization, and properties of an aryl aldehyde oxidoreductase from *Nocardia* sp. strain NRRL 5646. *J Bacteriol* 179: 3482–3487.

Li T, JPN Rosazza. 1998. NMR identification of an acyl-adenylate intermediate in the aryl-aldehyde oxidoreductase catalyzed reaction. *J Biol Chem* 273: 34230–34233.

Lindsay RF, FG Priest. 1975. Decarboxylation of substituted cinnamic acids by enterobacteria: The influence on beer flavour. *J Appl Bacteriol* 39: 181–187.

Lindstedt S, B Odelhög, M Rundgren. 1977. Purification and properties of 4-hydroxyphenylpyruvate dioxygenase from *Pseudomonas* sp. P.J. 874. *Biochemistry* 16: 3369–3377.

Lochmeyer C, J Koch, G Fuchs. 1992. Anaerobic degradation of 2-aminobenzoic acid (anthranilic acid) via benzoyl-coenzyme A (CoA) and cyclohex-1-enecarboxyl-CoA in a denitrifying bacterium. *J Bacteriol* 174: 3621–3628.

Long MT, BA Bartholomew, MJ Smith, PW Trudgill, DJ Hopper. 1997. Enzymology of oxidation of tropic acid to phenylacetic acid in metabolism of atropine by *Pseudomonas* sp. strain AT3. *J Bacteriol* 179: 1044–1050.

López-Barragán MJ, M Carmona, MT Zamarro, B Thiele, M Boll, G Fuchs, JL Garciá, E Díaz. 2004. The *bzd* gene cluster, coding for anaerobic benzoate metabolism, in *Azoarcus* sp. strain CIB. *J Bacteriol* 186: 5762–5774.

Luengo JM, JL Garciá, ER Olivera. 2001. The phenylacetyl-CoA catabolon: A complex catabolic unit with broad biotechnological applications. *Mol Microbiol* 39: 1434–1442.

Martin G, S Dijols, C Capeillere-Blandin, I Arnaud. 1999. Hydroxylation reaction catalyzed by the *Burkholderia cepacia* AC1100 bacterial strain. Involvement of the chlorophenol-4-monooxygenase. *Eur J Biochem* 261: 533–538.

Martínez-Blanco H, A Reglero, LB Rodriguez-Aparicio, JM Luengo. 1990. Purification and biochemical characterization of phenylacetyl-CoA ligase from *Pseudomonas putida*. *J Biol Chem* 2765: 7084–7090.

Matties C, B Schink. 1992. Fermentative degradation of glutarate via decarboxylation by newly isolated strictly anaerobic bacteria. *Arch Microbiol* 157: 290–296.

Mohamed ME-S, B Seyfried, A Tschech, G Fuchs. 1993. Anaerobic oxidation of phenylacetate and 4-hydroxyphenylacetate to benzoyl-coenzyme A and CO_2 in denitrifying *Pseudomonas* sp. *Arch Microbiol* 159: 563–573.

Monks TJ, M Butterworth, SS Lau. 2010. The fate of benzene oxide. *Chemico Biol Interact* 184: 201–206.

Morawski B, RW Eaton, JT Rossiter, S Guoping, H Griengl, DW Ribbons. 1997. 2-Naphthoate catabolic pathway in *Burkholderia* strain JT 1500. *J Bacteriol* 179: 115–121.

Müller JA, B Schink. 2000. Initial steps in the fermentation of 3-hydroxybenzoate by *Sporotomaculum hydroxybenzoicum*. *Arch Microbiol* 173: 288–295.

Murdoch RW, AG Hay. 2005. Formation of catechols via removal of acid side chains from ibuprofen and related aromatic acids. *Appl Environ Microbiol* 71: 6121–6125.

Navarro-Llorens JM, MA Patrauchan, GR Stewart, JE Davies, LD Eltis, WW Mohn. 2005. Phenylacetate catabolism in *Rhodococcus* sp. strain RHA1: A central pathway for degradation of aromatic compounds. *J Bacteriol* 187: 4497–4504.

Neidle EL, C Harnett, LN Ornston, A Bairoch, M Rekik et al. 1991. Nucleotide sequences of the *Acinetobacter calcoaceticus benABC* genes for benzoate 1,2-dioxygenase reveal evolutionary relationships among multi-component enzymes. *J Bacteriol* 173: 5385–5395.

Nogales J, R Macchi, F Franchi, D Barzaghi, C Fernández, JL García, G Bertoni, E Díaz. 2007. Characterization of the last step of the aerobic phenylacetic acid degradation pathway. *Microbiology (UK)* 153: 357–365.

Nozawa T, Y Maruyama. 1988. Anaerobic metabolism of phthalate and other aromatic compounds by a denitrifying bacterium. *J Bacteriol* 170: 5778–5784.

Olivera ER, D Carnicero, B García, B Miñambres, MA Moreno, L Cañedo, CC DiRusso, F Naharro, JM Luengo. 2001. Two different pathways are involved in the β-oxidation of *n*-alkanoic acid and *n*-phenylalkanoic acids in *Pseudomonas putida* U: Genetic studies and biotechnological applications. *Mol Microbiol* 39: 863–874.

Olivera ER, B Miñambres, B García, C Muniz, MA Moreno, A Ferrández, E Díaz, JL García, JM Luengo. 1998. Molecular characterization of the phenylacetic acid catabolic pathway in *Pseudomonas putida* U: The phenylacetyl-CoA catabolon. *Proc Natl Acad Sci USA* 95: 6419–6425.

Ornston LN. 1966. The conversion of catechol and protocatechuate to beta-ketoadipate by *Pseudomonas putida* IV. Regulation. *J Biol Chem* 241: 3800–3810.

Pan C, Y Oda, PK Lankford, B Zhang, NF Samatova, DA Pelletier, CS Harwood, RL Hettich. 2008. Characterization of anaerobic catabolism of *p*-coumarate in *Rhodopseudomonas palustris* by integrating transcriptomics and quantitative proteomics. *Mol. Cell Proteomics* 7.5: 938–948.

Pelletier DA, CS Harwood. 1998. 2-ketohexanecarboxyl coenzyme A hydrolase, the ring cleavage enzyme required for anaerobic benzoate degradation by *Rhodopseudomonas palustris*. *J Bacteriol* 180: 2330–2336.

Peng X, E Masai, D Kasai, K Miyauchi, Y Katayama, M Fukuda. 2005. A second 5-carboxyvanillate carboxylase gene, *lig W2*, is important for lignin-related biphenyl catabolism in *Sphingomonas paucimobilis* SYK-6. *Appl Environ Microbiol* 71: 5014–5021.

Perrotta JA, CS Harwood. 1994. Anaerobic metabolism of cyclohex-1-ene-1-carboxylate, a proposed intermediate of benzoate degradation by *Rhodopseudomonas palustris*. *Appl Environ Microbiol* 60: 1775–1782.

Peters F, M Rother, M Boll. 2004. Selenocysteine-containing proteins in anaerobic benzoate metabolism of *Desulfococcus multivorans*. *J Bacteriol* 186: 2156–2163.

Peters F, Y Shinoda, MJ McInerney, M Boll. 2007. Cyclohexa-1,5-diene-1-carbonyl-coenzyme A (CoA) hydratases of *Geobacter metallireducens* and *Syntrophus aciditrophicus*: Evidence for a common benzoyl-CoA degradation pathway in facultative and strict anaerobes. *J Bacteriol* 189: 1055–1060.

Pons J-L, A Rimbault, JC Darbord, G Leluan. 1984. Biosynthèse de toluène chez *Clostridium aerofoetidum* souche WS. *Ann Microbiol (Inst Pasteur)* 135 B: 219–222.

Powlowski JB, S Dagley, V Massey, DP Ballou. 1987. Properties of anthranilate hydroxylase (deaminating), a flavoprotein from *Trichosporon cutaneum*. *J Biol Chem* 262: 69–74.

Providenti MA, JM O'Brien, J Ruff, AM Cook, IB Lambert. 2006. Metabolism of isovanillate, vanillate, and veratrate by *Comamonas testosteroni* strain BR6020. *J Bacteriol* 188: 3862–3869.

Pujar BG, DW Ribbons. 1985. Phthalate metabolism in *Pseudomonas florescens* PHK: Purification and properties of 4,5-dihydroxyphthalate decarboxylase. *Appl Environ Microbiol* 49: 374–376.

Qu Y, JC Spain. 2011. Molecular and biochemical characterization of the 5-nitroanthranilic acid degradation pathway in *Bradyrhizobium* sp. strain JS 329. *J Bacteriol* 193: 3057–3063.

Que L, J Widom, RL Crawford. 1981. 3,4-dihydroxyphenylacetate 2,3-dioxygenase: A manganese(II) dioxygenase from *Bacillus brevis*. *J Biol Chem* 256: 10941–10944.

Rather LJ, B Knapp, W Haehnel, G Fuchs. 2010. Coenzyme A-dependent aerobic metabolism of benzoate via epoxide formation. *J Biol Chem* 285: 20615–20624.

Rather LJ, T Weinert, U Demmer, W Ismael, G Fuchs, U Elmer. 2011. Structure and mechanism of the diiron benzoyl-coenzyme A epoxide BoxB. *J Biol Chem* M 111.236893

Reineke W, W Otting, H-J Knackmuss. 1978. *cis*-Dihydrodiols microbially produced from halo- and methylbenzoic acids. *Tetrahedron* 34: 1707–1714.

Reiner AM, GD Hegeman. 1971. Metabolism of benzoic acid by bacteria. Accumulation of (–)-3,5-cyclohexadiene-1,2-diol-1-carboxylic acid by a mutant strain of *Alcaligenes eutrophus*. *Biochemistry* 10: 2530–2536.

Ribbons DW, P Keyser, DA Kunz, BF Taylor, RW Eaton, BN Anderson. 1984. Microbial degradation of phthalates, pp. 371–395. In *Microbial Degradation of Organic Compounds* (Ed. DT Gibson), Marcel Dekker, New York.

Rüetschi U, B Odelhög, S Lindstedt, J Barros-Söderling, B Persson, H Jörnvall. 1992. Characterization of 4-hydroxyphenylpyruvate dioxygenase. Primary structure of the *Pseudomonas* enzyme. *Eur J Biochem* 205: 459–466.

Sanvoisin J, GJ Langley, TDH Bugg. 1995. Mechanism of extradiol catechol dioxygenases: Evidence for a lactone intermediate in the 2,3-dihydroxypropionate 1,2-dioxygenase reaction. *J Am Chem Soc* 117:7836–7837.

Sariaslani FS, JL Sudmeier, DD Foch. 1982. Degradation of 3-phenylbutyric acid by *Pseudomonas* sp. *J Bacteriol* 152: 411–421.

Sasoh M, E Masai, S Ishibashi, H Hara, N Kamimura, K Miyauchi, M Fukuda. 2006. Characterization of the terephthalate degradation genes of *Comamonas* sp. strain E6. *Appl Environ Microbiol* 72: 1825–1832.

Schläfli HR, MA Weiss, T Leisinger, AM Cook. 1994. Terephthalate 1,2-dioxygenase system from *Comamonas testosteroni* T-2: Purification and some properties of the oxygenase component. *J Bacteriol* 176: 6644–6652.

Schneider S, G Fuchs. 1998. Phenylacetyl-CoA: Acceptor oxidoreductase, a new α-oxidizing enzyme that produces phenylglyoxylate. Assay, membrane localization, and differential production in *Thauera aromatica*. *Arch Microbiol* 169: 509–516.

Schneider S, MEl-Said Mohamed, G Fuchs. 1997. Anaerobic metabolism of L-phenylalanine via benzoyl-CoA in the denitrifying bacterium *Thauera aromatica*. *Arch Microbiol* 168: 310–320.

Schöcke L, B Schink. 1999. Energetics and biochemistry of fermentative benzoate degradation by *Syntrophus gentianae*. *Arch Microbiol* 171: 331–337.

Schühle K, J Gescher, U Feil, M Paul, M Jahn, H Schagger, G Fuchs. 2003. Benzoate-coenzyme A ligase from *Thauera aromatica*: An enzyme acting in anaerobic and aerobic pathways. *J Bacteriol* 185: 4920–4929.

Schühle K, M Jahn, S Ghisla, G Fuchs. 2001. Two similar gene clusters coding for enzymes of a new type of aerobic 2-aminobenzoate (anthranilate) metabolism in the bacterium *Azoarcus evansii*. *J Bacteriol* 183: 5268–5278.

Selmer T, PI Andrei. 2001. *p*-hydroxyphenylacetate decarboxylase from *Clostridium difficile*. A novel glycyl radical enzyme catalysing the formation of *p*-cresol. *Eur J Biochem* 268: 1363–1372.

Seyedsayamdost MR, G Carr, R Kolter, J Clardy. 2012. Roseobactericides: Small molecule modulators of an algal-bacterial symbiosis. *J Amer Chem Soc* 133: 18343–18349.

Sharak Genther BR, GT Townsend, BO Blattmann. 1997. Reduction of 3-chlorobenzoate, 3-bromobenzoate, and benzoate to corresponding alcohols by *Desulfomicrobium escambiense*, isolated from a 3-chlorobenzoate-dechlorinating coculture. *Appl Environ Microbiol* 63: 4698–4703.

Shu L, Y-M Chiou, AM Orville, MA Miller, JD Lipscomb, L Que. 1995. X-ray absorption spectroscopic studies of the Fe(II) active site of catechol 2,3-dioxygenase. Implications for the extradiol cleavage mechanism. *Biochemistry* 34: 6649–6659.

Siegbahn PEM, F Haeffner. 2004. Mechanism for catechol ring-cleavage by non-heme iron extradiol dioxygenases. *J Am Chem Soc* 126: 8919–8932.

Simoni S, S Klinke, C Zipper, W Angst, H-P E Kohler. 1996. Enantioselective metabolism of chiral 3-phenylbutyric acid, an intermediate of linear alkylbenzene degradation, by *Rhodococcus rhodochrous*. *Appl Environ Microbiol* 62: 749–755.

Song B, BB Ward. 2005. Genetic diversity of benzoyl coenzyme A reductase genes detected in denitrifying isolates and estuarine sediment communities. *Appl Environ Microbiol* 71: 2036–2045.

Song J, T Xia, RA Jensen. 1999. PhhB, a *Pseudomonas aeruginosa* homolog of mammalian pterin 4a-carbinol-amine dehydratase/DCcoH, does not regulate expression of phenylalanine hydroxylase at the transcriptional level. *J Bacteriol* 181: 2789–2796.

Sparnins VL, PJ Chapman. 1976. Catabolism of L-tyrosine by the homoprotocatechuate pathway in Gram-positive bacteria. *J Bacteriol* 127: 363–366.

Sparnins VL, PJ Chapman, S Dagley. 1974. Bacterial degradation of 4-hydroxyphenylacetic acid and homoprotocatechuic acid. *J Bacteriol* 120: 159–167.

Sparnins VL, S Dagley. 1975. Alternative routes of aromatic catabolism in *Pseudomonas acidovorans* and *Pseudomonas putida*: Gallic acid as a substrate and inhibitor of dioxygenases. *J Bacteriol* 124: 1374–1381.

Stolz A, B Nörtemann, H-J Knackmuss. 1992. Bacterial metabolism of 5-aminosalicylic acid. Initial ring cleavage. *Biochem J* 282: 675–680.

Suárez M, E Ferrer, M Martin. 1996. Purification and biochemical characterization of gentisate 1,2-dioxygenase from *Klebsiella pneumoniae* M5a1. *FEMS Microbiol Lett* 143: 89–95.

Subramanian V, CS Vaidyanathan. 1984. Anthranilate hydroxylase from *Aspergillus niger*: New type of NADPH-linked nonheme iron monooxygenase. *J Bacteriol* 160: 651–655.

Suske WA, WJH van Berkel, H-P Kohler. 1999. Catalytic mechanism of 2-hydroxybiphenyl 3-monooxygenase a flavoprotein from *Pseudomonas azelaica* HBP1. *J Biol Chem* 274: 33355–33365.

Sze I S-Y, S Dagley. 1984. Properties of salicylate hydroxylase and hydroxyquinol 1,2-dioxygenase purified from *Trichosporon cutaneum*. *J Bacteriol* 159: 353–359.

Sze IS-Y, S Dagley. 1987. Degradation of substituted mandelic acids by *meta* fission reactions. *J Bacteriol* 169: 3833–3835.

Sze IS-Y, S Dagley. 1984. Properties of salicylate hydroxylase and hydroxyquinol 1,2-dioxygenase purified from *Trichosporon cutaneum*. *J Bacteriol* 159: 353–359.

Taylor BF, JA Amador. 1988. Metabolism of pyridine compounds by phthalate-degrading bacteria. *Appl Environ Microbiol* 54: 2341–2344.

Teufel R, V Mascaraque, W Ismail, M Voss, J Perrera, W Eisenreich, W Haehnel, G Fuchs. 2010. Bacterial phenylalanine and phenylacetate catabolic pathway revealed. *Proc Natl Acad Sci USA* 107: 14390–14395.

Thauer RK, D Möller-Zinjhan, AM Spormann. 1989. *Biochemistry* of acetate catabolism in anaerobic chemotrophic bacteria. *Annu Rev Microbiol* 43: 43–67.

Thiel V, T Brinkhoff, JS Dickschat, S Wickel, J Gruneberg, I Wagner-Döbler, M Simon, S Schulz. 2010. Identification and biosynthesis of tropone derivatives and sulfur volatiles produced by bacteria of the *Roseobacter* clade. *Org Biomol Chem* 8: 234–246.

Tsou, AY, SC Ransom, JA Gerlt, DD Buechter, PC Babbitt, GL Kenyon. 1990. Mandelate pathway of *Pseudomonas putida*: Sequence relationships involving mandelate racemase, (*S*)-mandelate dehydrogenase, and benzoylformate decarboxylase and expression of benzoylformate decarboxylase in *Escherichia coli*. *Biochemistry* 29: 9856–9862.

van der Tweel WJJ, RJJ Janssens, JAM de Bont. 1986. Degradation of 4-hydroxyphenylacetate by *Xanthobacter* 124X. *Antonie van Leeuwenhoek* 52: 309–318.

Wergath J, H-A Arfmann, DH Pieper, KN Timmis, R-M Wittich. 1998. Biochemical and genetic analysis of a gentisate 1,2-dioxygenase from *Sphingomonas* sp. strain RW 5. *J Bacteriol* 180: 4171–4176.

Wheelis M, NJ Palleroni, RY Stanier. 1967. The metabolism of aromatic acids by *Pseudomonas testosteroni* and *P. acidovorans*. *Arch Mikrobiol* 59: 302–314.

White H, R Feicht, C Huber, F Lottspeich, H Simon. 1991. Purification and some properties of the tungsten-containing carboxylic acid reductase from *Clostridium formicoaceticum*. *Biol Chem Hoppe-Seyler* 372: 999–1005.

White H, C Huber, R Feicht, H Simon. 1993. On a reversible molybdenum-containing aldehyde oxidoreductase from *Clostridium formicoaceticum*. *Arch Microbiol* 159: 244–249.

White H, G Strobl R Feicht, H Simon. 1989. Carboxylic acid reductase: A new tungsten enzyme catalyses the reduction of non-activated carboxylic acids to aldehydes. *Eur J Biochem* 184: 89–96.

White-Stevens RH, H Kamin. 1972. Studies of a flavoprotein, salicylate hydroxylase. I. Preparation, properties, and the uncoupling of oxygen reduction from hydroxylation. *J Biol Chem* 247: 2358–2370.

White-Stevens RH, H Kamin, QH Gibson. 1972. Studies of a flavoprotein, salicylate hydroxylase II Enzyme mechanism. *J Biol Chem* 247: 2371–2381.

Wischgoll S, D Heintz, F Peters, A Erxleben, E Sarnhausen, R Reski, A van Dorsselaer, M Boll. 2005. Gene clusters involved in anaerobic benzoate degradation of *Geobacter metallireducens*. *Mol Microbiol* 58: 1238–1252.

Wischgoll S, M Taubert, F Peters, N Jehmlich, M von Bergen, M Boll. 2009. Decarboxylating and nondecarboxylating glutaryl-coenzyme A dehydrogenases in the aromatic metabolism of obligately anaerobic bacteria. *J Bacteriol* 191: 4401–4409.

Wöhlbrand L, H Wilkes, T Halder, R Rabus. 2008. Anaerobic degradation of *p*-ethylphenol by *Aromatoleum aromaticum* strain EbN1: Pathway, regulation, and involved proteins. *J Bacteriol* 190: 5699–5709.

Wolgel SA, JE Dege, PE Perkins-Olson, CH Juarez-Garcia, RL Crawford, E Münck, JD Lipscomb. 1993. Purification and characterization of protocatechuate 2,3-dioxygenase from *Bacillus macerans*: A new extradiol catecholic dioxygenase. *J Bacteriol* 175: 4414–4426.

Xun L, ER Sandvik. 2000. Characterization of 4-hydroxyphenylacetate 3-hydroxylase (HpaB) *of Escherichia coli* as a reduced flavin adenine dinucleotide-utilizing monooxygenase. *Appl Environ Microbiol* 66: 481–486.

Yokoyama MT, JR Carlson, LV Holdeman. 1977. Isolation and characterization of a skatole-producing *Lactobacillus* sp. from the bovine rumen. *App Environ MIcrobiol* 34: 837–842.

Yokoyama MT, CT Yokoyama. 1981. Production of skatole and *p*- cresol by a ruminal *Lactobacillus* sp. *Appl Environ Microbiol* 41: 71–76.

Yoshida M, N Fukuhara, T Oikawa. 2004. Thermophilic, reversible γ-resorcylate decarboxylase from *Rhizobium* sp. strain MTP-10005: Purification, molecular characterization and expression. *J Bacteriol* 186: 6855–68623.

Zaar A, W Eisenreich, A Bacher, G Fuchs. 2001. A novel pathway of aerobic benzoate catabolism in the bacteria *Azoarcus evansii* and *Bacillus stearothermophilus*. *J Biol Chem* 276: 24997–25004.

Zaar A, J Gescher, W Eisenreich, A Bacher, G Fuchs. 2004. New enzymes involved in aerobic benzoate metabolism in *Azoarcus evansii*. *Mol Microbiol* 54: 223–238.

Zeida M, M Wieser, T Yoshida, T Suigio, T Nagaawa. 1998. Purification and characterization of gallic acid decarboxylase from *Pantoea agglomerans*. *Appl Environ Microbiol* 64: 4743–4747.

Zellner G, H Kneifel, J Winter. 1990. Oxidation of benzaldehydes to benzoic acid derivatives by three *Desulfovibrio* strains. *Appl Environ Microbiol* 56: 2228–2233.

Zhang X, LY Young. 1997. Carboxylation as an initial reaction in the anaerobic metabolism of naphthalene and phenanthrene by sulfidogenic consortia. *Appl Environ Microbiol* 63: 4759–4764.

Zhou N-Y, J Al-Dulayymi, MS Baird, PA Williams. 2002. Salicylate 5-hydroxylase from *Ralstonia* sp. strain U2; a monooxygenase with close relationships to and shared electron transport proteins with naphthalene dioxygenase. *J Bacteriol* 184: 1547–1555.

Part 4: Nonhalogenated Phenols and Anilines

Although phenols and anilines have similar electronic structures, the initial step in their aerobic degradation is different. Whereas the degradation of phenols is initiated by monooxygenation, dioxygenation with loss of the amino group is used for anilines. Catechols are then produced from both of them. The degradation of halogenated phenols and anilines is discussed in Chapter 9, Part 2.

8.21 Phenols

8.21.1 Aerobic Degradation

It is worth noting the role of phenol in inducing monooxygenase activity for the cometabolism of trichloroethene that is given in Chapter 7. The degradation of phenol is initiated by hydroxylation that has been described in a number of organisms in which different hydroxylases have been found.

1. In *Pseudomonas putida* sp. EST 1001, a single component that is an NAD(P)H-dependent monooxygenase is encoded by *pheA* in which it is plasmid-borne and that unusually carries out intradiol fission of the resulting catechol (Nurk et al. 1991a,b; Powlowski and Shingler 1994b).

2. In *Pseudomonas* sp. strain CF600, the hydroxylation of phenol and 3,4-dimethylphenol is unusually carried out by a multicomponent enzyme whose genes occur in a plasmid, and the operon has been designated *dmpKLMNOP* (Shingler et al. 1992; Powlowski and Shingler 1994a). The three components consist of a reductase DmpP containing FAD and [2Fe–2S], an activator DmpM, and the three subunits of the hydroxylase DmpLNO that contain the diiron center.

3. In *Bacillus thermoglucosidasius*, the hydroxylase is a two-component system encoded by *pheA1* and *pheA2*, and hydroxylation requires supplementation by the second component with FAD(NADH) (Kirchner et al. 2003).

4. In *Bacillus thermoleovorans* strain A2, the hydroxylase displayed greatest similarity to the larger component of the two-component 4-hydroxyphenylacetate hydroxylase in *Escherichia coli* strain W (Duffner and Müller 1998).

5. The hydroxylase from *Pseudomonas pickettii* PKO1 was cloned and inserted into *Pseudomonas aeruginosa* PAO1c where all the cresols were also substrates (Kukor and Olsen 1990).

6. *Mycobacterium goodii* strain 12523 and *Myco. smegmatis* strain mc^{2}155 are able to carry out the hydroxylation of phenol to hydroquinone. This is accomplished by a binuclear iron monooxygenase that is part of the *mimABCD* operon which encodes the oxygenase large subunit, a reductase, the oxygenase small subunit, and a coupling protein (Furuya et al. 2011).

Differences have also been observed in the pathways for degradation of catechol that is produced after growth with phenol (monooxygenation) or benzoate (dioxygenation). Catechol 1,2-dioxygenase carries out C–C ring fission between the hydroxyl-bearing atoms, whereas catechol 2,3-dioxygenase involves fission between one of the hydroxyl-bearing atoms and the adjacent carbon atom: these have been termed the intradiol *ortho* or extradiol *meta* fission pathways. There are a number of important complications. For example, in *Pseudomonas putida*, whereas catechol produced from phenol by hydroxylation was degraded by extradiol fission and the enzymes of the pathway were induced by phenol, catechol produced by dioxygenation of benzoate was degraded by intradiol fission and the enzymes were induced by *cis,cis*-muconate (Feist and Hegeman 1969a,b). The analogous distinction also holds for the degradation of phenol and 3-methylbenzoate by *Pseudomonas putida* strain EST1001 (Kivisaar et al. 1989). Additional comments on the choice of these alternative ring-fission mechanisms and the intricacies in their selection by various bacteria for metabolic pathways are given in Chapter 3.

8.22 Alkylated Phenols: Degradation of Methylcatechols

8.22.1 Methylphenols (Cresols and Xylenols)

For alkylated phenols, there are different pathways for degradation, depending on whether or not oxidation of the alkyl group precedes oxygenation and fission of the ring.

Methylcatechols are produced in the degradation of methylphenols by hydroxylation and of methylbenzoates and methylsalicylates by dioxygenation. Important comments on this have already been made in the first part of this chapter, and additional metabolic aspects of their degradation are briefly summarized here.

1. The metabolism of methylphenols can be initiated either by oxidation of a methyl group to carboxylate, or by ring hydroxylation when the methyl groups are retained in the final metabolites.

2. Hydroxylation may introduce an *ortho* or a *para* hydroxyl group. Subsequent ring fission may be carried out by a catechol dioxygenase or, for hydroquinones, by a gentisate dioxygenase.

3. As for chlorocatechols, there are significant differences between the results of intradiol and extradiol fission.

4. There are potential obstacles for entry into the 3-oxoadipate pathway for degradation when a methyl group is situated at C_4 after intradiol fission, although this can be circumvented by isomerization of the methyl group.

8.22.1.1 Initiation by Hydroxylation

Initiation by hydroxylation is widely utilized and is illustrated in the following examples:

1. The degradation of 3-methyl- and 4-methylphenol by *Pseudomonas putida* is initiated by hydroxylation to 3-methyl- and 4-methylcatechol that undergo extradiol fission followed by hydrolysis (Sala-Trapat et al. 1972) (Figure 8.43a and b). This is also used for the degradation of 3-methyl- and 4-methylbenzoate by *Pseudomonas arvilla* mt-2 (Murray et al. 1972).

2. *Pseudomonas* sp. strain KL 28 is able to degrade 4-*n*-alkylphenols with chain lengths of 1–5 carbon atoms. Degradation of both the 3-*n*-alkyl and 4-*n*-alkylphenols is initiated by hydroxylation to 3,4-dihydroxy compounds that are degraded by extradiol fission followed by dehydrogenation to 2-hydroxy-5-*n*-alkylmuconates with the alkyl groups intact (Jeong et al. 2003). The degradation of 4-*n*-alkylphenols by *Pseudomonas* sp. strain KL28 involves hydroxylation to 4-*n*-alkylcatechols followed by extradiol fission, *n*-hexanal is produced from 4-*n*-butylcatechol (Figure 8.44) although 4-*t*-butylphenol cannot be degraded by this strain (Jeong et al. 2003). Whereas the degradation of branched 4-nonylphenol by *Sphingomonas xenophaga* involves migrations that are described later, the degradation of 4-*t*-butylphenol by *Sphingomonas fulginis* strain TIK-1 is different: degradation proceeds analogously to that for 4-*n*-alkylphenols, except that the product of ring fission is *t*-butyl methyl ketone (3,3-dimethylbutan-2-one) (Figure 8.45) that could apparently be further degraded (Toyama et al. 2010). This strain was also able to degrade 4-*t*-pentylphenol and

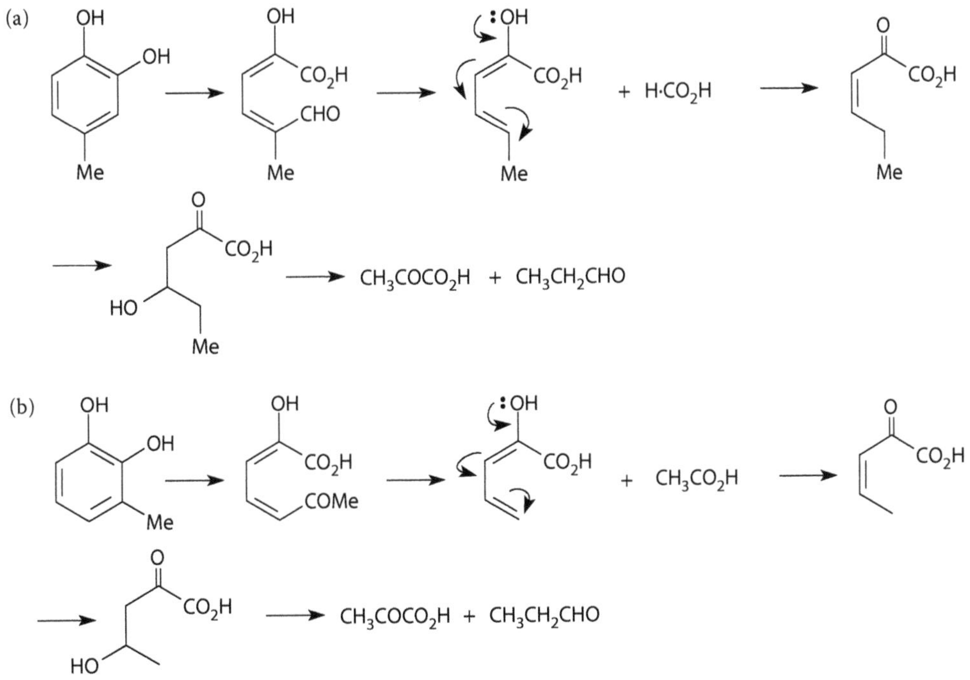

FIGURE 8.43 Aerobic degradation of (a) 3-methyphenol and (b) 4-methylphenol. (Adapted from Sala-Trapat J, K Murray, PA Williams. 1972. *Eur J Biochem* 28: 347–356.)

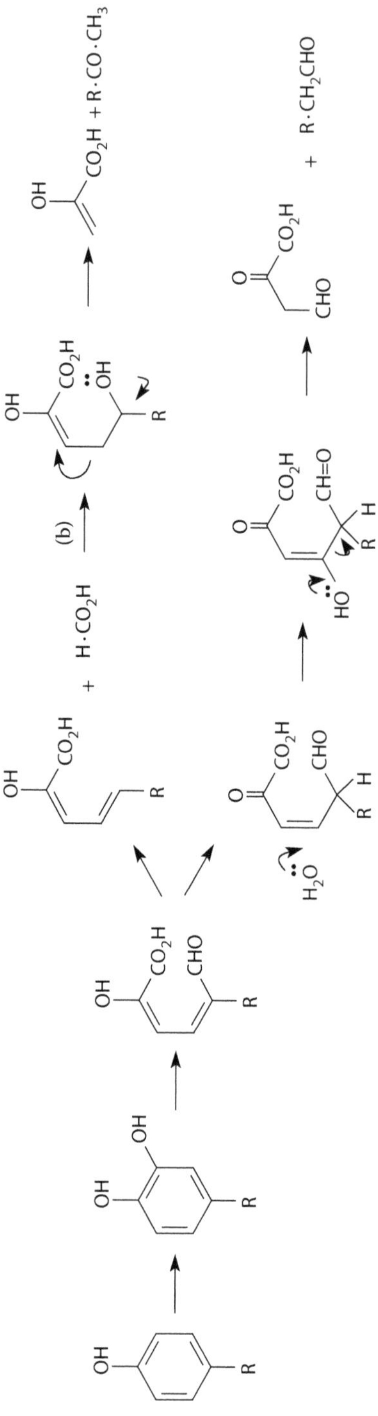

FIGURE 8.44   Aerobic degradation of 4-*n*-butylcatechol.

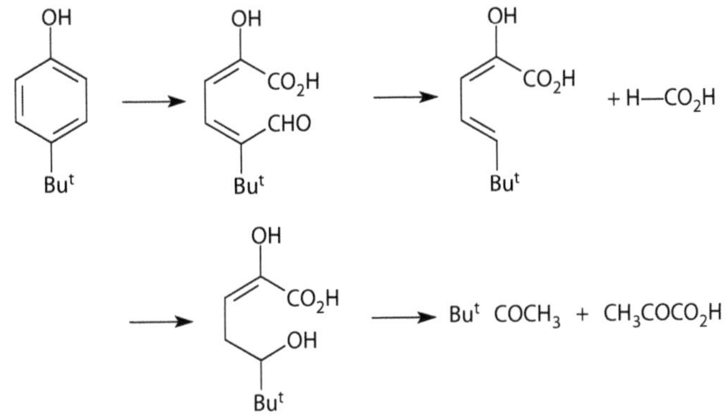

FIGURE 8.45 Degradation of 4-*t*-butylphenol.

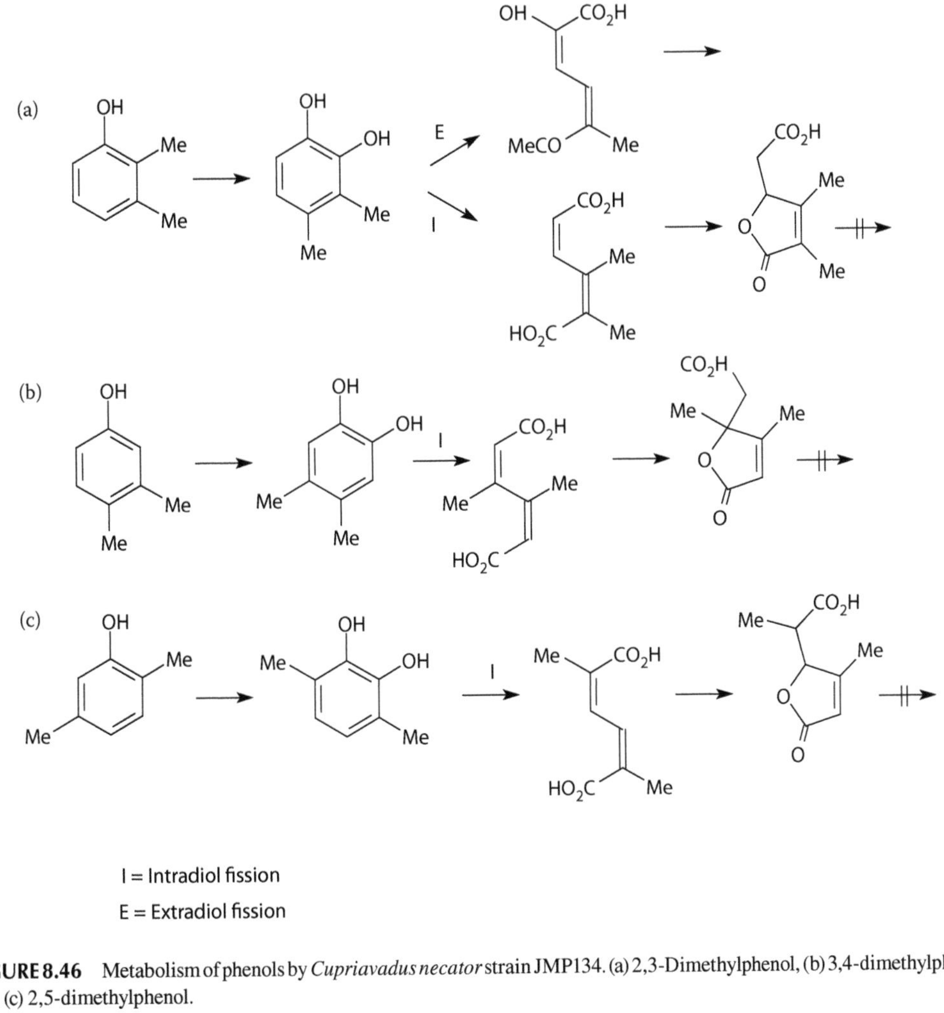

I = Intradiol fission

E = Extradiol fission

FIGURE 8.46 Metabolism of phenols by *Cupriavadus necator* strain JMP134. (a) 2,3-Dimethylphenol, (b) 3,4-dimethylphenol, and (c) 2,5-dimethylphenol.

4-*t*-octylphenol to produce 3,3-dimethylpentan-2-one and 3,3,5,5-tetramethylhexan-2-one, respectively.

3. In *Cupriavadus necator* (*Alcaligenes eutrophus* strain) JMP 134, 2,3-, 3,4-, 2,4-, and 2,5-dimethylphenol can be hydroxylated to 3,4-, 4,5-, 3,5-, and 2,5-dimethycatechol, which differs critically in its further metabolism. Extradiol fission of 3,4-, 4,5-, and 2,5-dimethylcatechol may take place to products that are acceptable for further degradation, whereas the products from extradiol fission after lactonization for 2,3-dimethyl-, 3,4-dimethyl-, and 2,5-dimethylmuconolactones are not metabolized by this strain (Pieper et al. 1995) (Figure 8.46a–c).

8.22.1.2 Intradiol or Extradiol Fission

In the foregoing examples, the catechols were degraded by extradiol fission that is normally preferred for substituted catechols. Intradiol fission may, however, occur in the degradation of 4-methylcatechol, and the methyl group at C_4 of the muconolactone that is produced may block further metabolism and entry into the 3-oxoadipate pathway. 4-Methylmuconolactone is, for example, a terminal metabolite in *Pseudomonas desmolyticum* (Catelani et al. 1971).

Only a brief summary is given of the various complexities encountered in the degradation of 4-methyl-catechol since this is discussed at greater length in the first part of this chapter. In summary, this is observed for *Cupriavadus necator* (*Alcaligenes eutrophus*) JMP134 (Pieper et al. 1990), *Pseudomonas reinekei* MT1 (Marín et al. 2010), *Rhodococcus rhodochrous* N75 (Cha et al. 1998), and the yeast *Trichosporon cutaneum* (Powlowski et al. 1985). This is accomplished for bacteria by an isomerase that transforms the initially produced 4-methylmuconolactone to 3-methylmuconolactone (Figure 8.47), whereas for the yeast this involves the alternative cyclization (Powlowski et al. 1985). It is worth noting that an analogous reaction occurs during the degradation of 4-methyl- and 5-methylsalicylate by *Pseudomonas* sp. strain MT1 when both are metabolized to 4-methylcatechol (Figure 8.48) (Cámara et al. 2007).

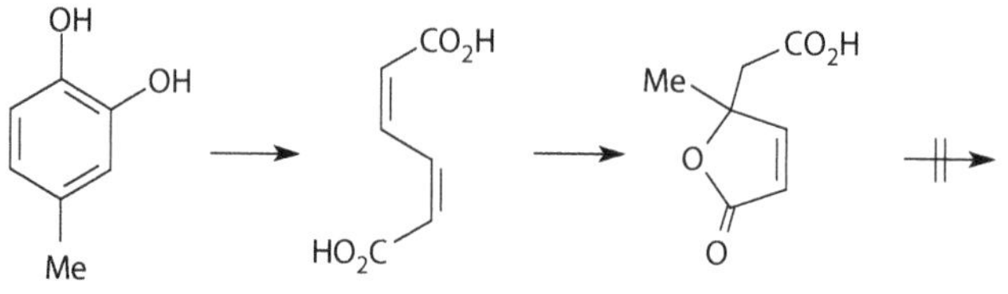

FIGURE 8.47 Specific isomerase for 1-methybislactone.

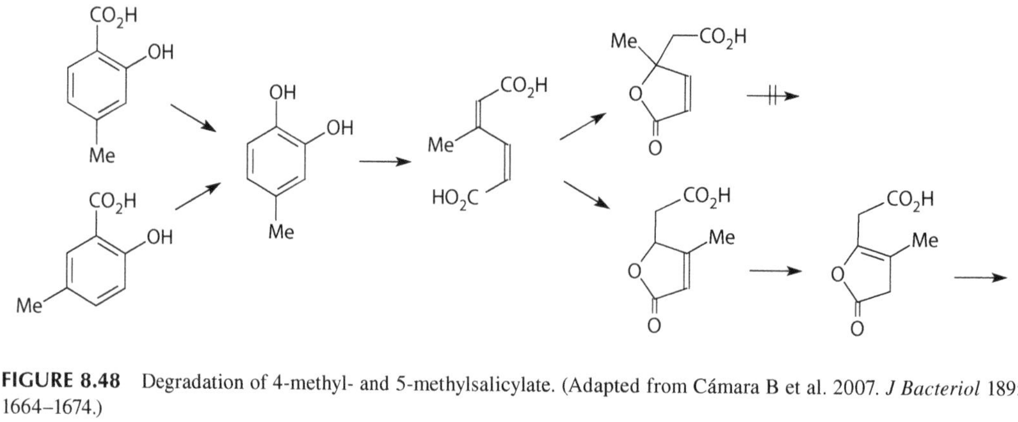

FIGURE 8.48 Degradation of 4-methyl- and 5-methylsalicylate. (Adapted from Cámara B et al. 2007. *J Bacteriol* 189: 1664–1674.)

8.22.1.3 Initiation by Oxidation of Methyl Groups

This is widely used, and there are important differences when several methyl groups are present, and in the mechanisms for Gram-positive organisms.

1. 4-Methyl phenol is oxidized to 4-hydroxybenzoate, which is hydroxylated to 3,4-dihydroxybenzoate and undergoes intradiol ring fission to 3-oxoadipate, whereas for 3-methylphenol, degradation of the 3-hydroxybenzoate involves hydroxylation to 2,5-dihydroxybenzoate (gentisate) that is degraded by gentisate dioxygenase to pyruvate and maleate (Hopper and Chapman 1971).

2. Generally, only one (of several) methyl group is oxidized (Bayly et al. 1988). For example, in the degradation of 2,3,5-trimethyl phenol, two of them are retained in the products from ring fission. The degradation of 2,4-, 2,5-, and 3,5-dimethylphenol has been examined in several strains of *Pseudomonas putida* (Bayly et al. 1988):

   a. In 2,4-dimethylphenol, by successive oxidation of C_2-methyl and C_4-methyl to phenol-2,4-dicarboxylate, catechol-4-carboxylate and ring fission to 3-oxohexane-1,6-dicarboxylate (Figure 8.49a).

   b. In 2,5- and 3,5-dimethylphenol, by oxidation of C_5-methyl, ring hydroxylation to 5-methyl hydroquinone-2-carboxylate and ring fission between C_1 and C_2 by gentisate dioxygenase to produce methylfumarate (Figure 8.49b).

8.22.1.4 Gram-Positive Organisms

Gram-positive organisms have been less extensively examined, but there are two significant examples:

1. The degradation of 2,6-dimethylphenol can be accomplished by *Mycobacterium* sp. strain DM1. This involves successive hydroxylation to 2,6-dimethylhydroquinone and 2,6-dimethyl-3-hydroxyhydroquinone, followed by intradiol ring fission to methylfumarate and, presumptively, propionate (Ewers et al. 1989) (Figure 8.50).

2. 3,4-Dimethylbenzoate and 3,5-dimethylbenzoate can be oxidized by *Rhodococcus rhodochrous* to the corresponding catechols followed by intradiol fission to the muconic acids. Cyclization of these to the 2,3-dimethyl- and 2,4-dimethylmuconloactones does not, however, proceed further (Schmidt et al. 1994). The resistance of the 2,5-dimethylmuconolactone is clearly analogous to that of 4-methylmuconolactone.

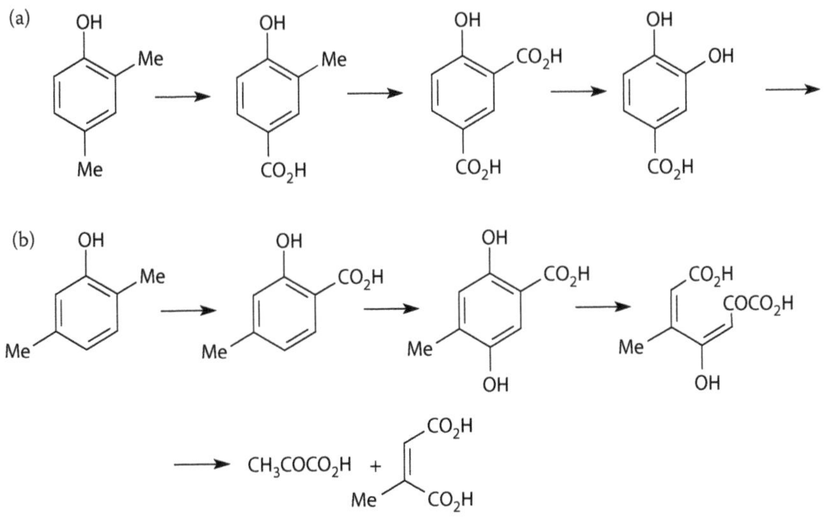

FIGURE 8.49 Degradation of (a) 2,4-dimethylphenol and (b) 2,5-dimethylphenol.

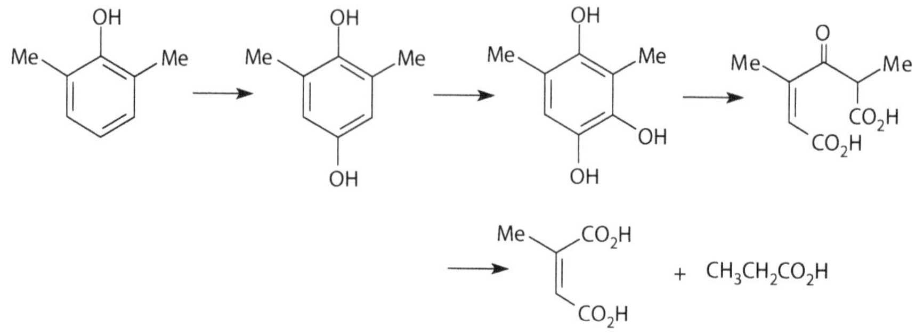

FIGURE 8.50 Degradation of 2,5- and 3,5-dimethylphenol.

8.22.1.5 Terminal Metabolites

1. The recalcitrant dimethylmuconolactones produced by *Alcaligenes eutrophus* strain JMP 134 from dimethylcatechols by extradiol fission have already been noted (Pieper et al. 1995).
2. In *Comamonas testosteroni*, the degradation of 2,4- and 3,4-dimethylphenol and 2,4,6-trimethylphenol (Hollender et al. 1994) occurs by the oxidation of the C_4-methyl groups. The 3,5-dimethyl-4-hydroxybenzoate can, however, be transformed by *Rhodococcus rhodochrous* N75 and *Pseudomonas* HH35 by oxidative decarboxylation to 2,6-dimethylhydroquinone followed by ring fission to 2,6-dimethyl-4-hydroxy-6-oxo-hexa-2,4-dienoate (Figure 8.51) (Cain et al. 1997).

8.22.1.6 Unusual Reactions

A number of less common reactions have been encountered in the degradation of phenols.

1. The degradation of bisphenol-A involves a rearrangement—analogous to those described for compounds with quaternary carbon atoms in Chapter 7, Part 1—followed by oxidative fission of the stilbene produced (Lobos et al. 1992; Spivack et al. 1994) (Figure 8.52).
2. The degradation of 4-ethylphenol and related compounds is initiated not by oxygenation but by dehydrogenation to a quinone methide followed by hydroxylation (Hopper and Cottrell 2003) (Figure 8.53), and the flavocytochrome 4-ethylphenol methylene hydroxylase in *Pseudomonas putida* strain JD1 has been characterized (Reeve et al. 1989).

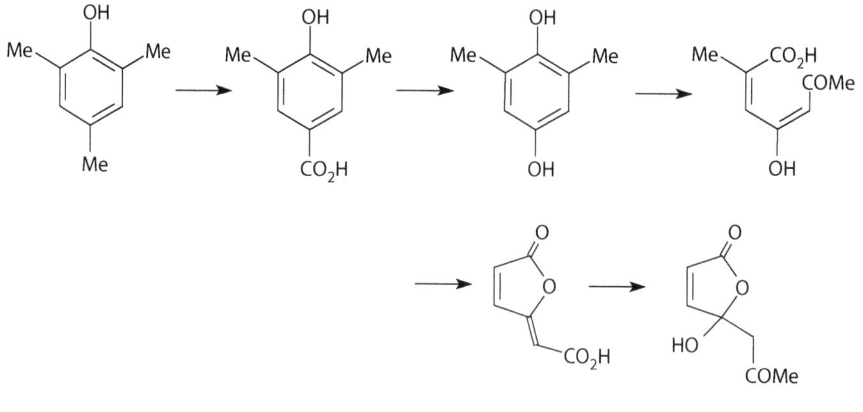

FIGURE 8.51 Degradation of 2,6-dimethylphenol by *Mycobacterium* sp.

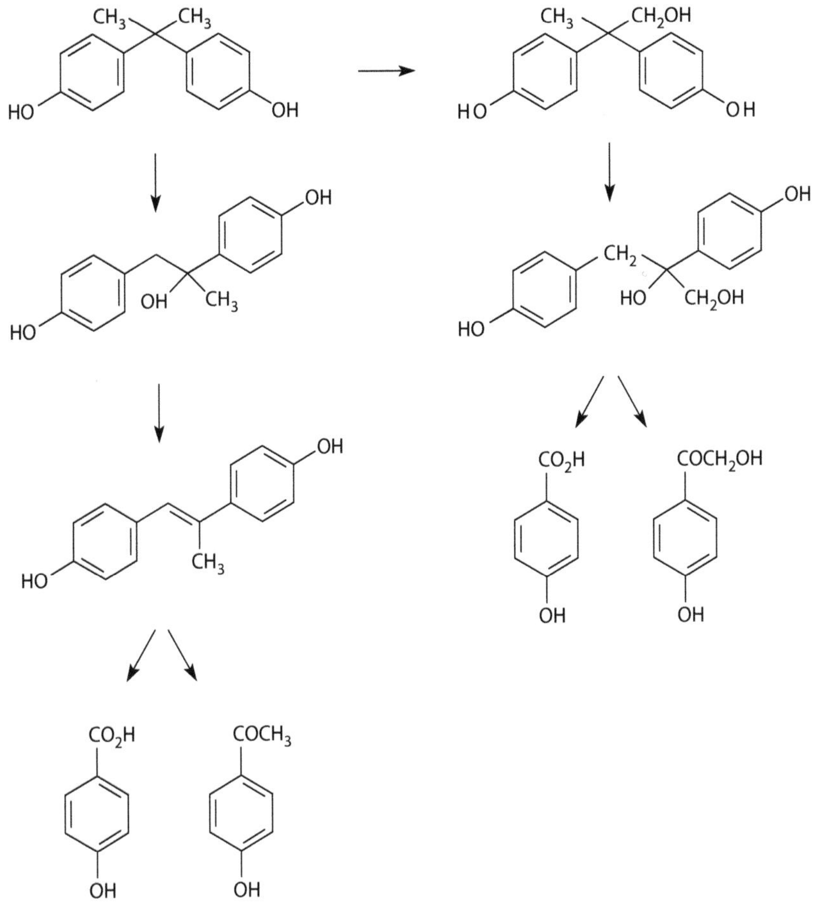

FIGURE 8.52 Degradation of bisphenol-A. (Adapted from Spivack J, TK Leib, JH Lobos. 1994. *J Biol Chem* 269: 7323–7329.)

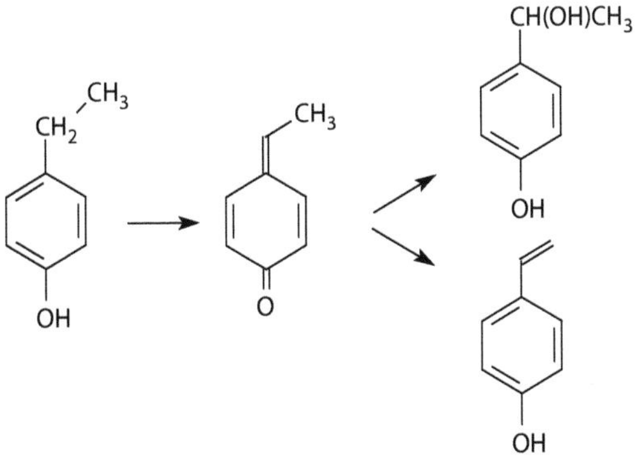

FIGURE 8.53 Aerobic degradation of 4-ethylphenol. (Adapted from Hopper DJ, L Cottrell. 2003. *Appl Environ Microbiol* 69: 3650–3652.)

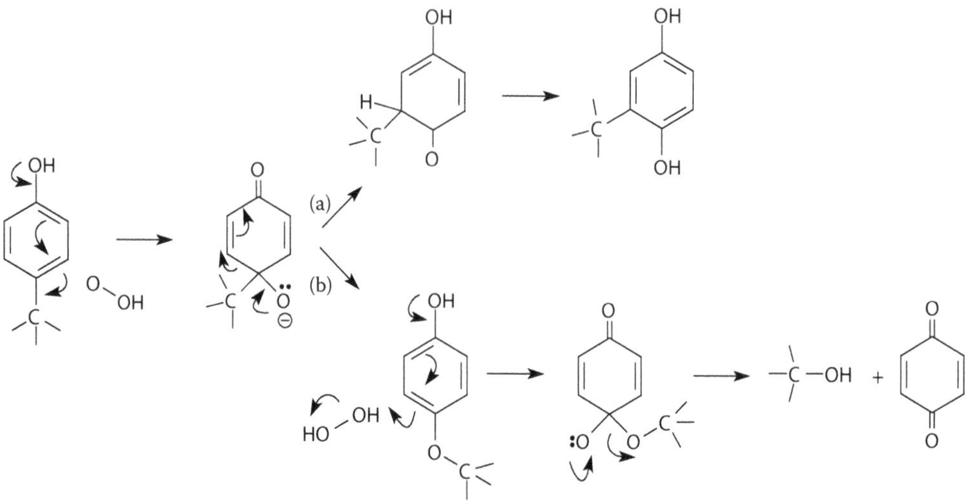

FIGURE 8.54   Degradation of nonylphenol. (a) Migration from C3 to produce 2-alkylhydroquinone, (b) oxygenation at C4 to produce benzoquinone.

3. a.   The degradation of nonylphenol isomers with quaternary side chains in a strain of *Sphingomonas xenophaga* displays unusual features (Corvini et al. 2004; Gabriel et al. 2005a,b). These may be described formally as oxygenation at the position para to the phenol group, followed by either of two rearrangements:

   i.   Migration of the alkyl group to C_3 of the phenolic ring with the formation of an alkylated hydroquinone (Figure 8.54a).

   ii. Rearrangement by insertion of an oxygen atom from the hydroxyl group between the terminal position of the alkyl group and C-4 of the phenolic ring, followed by further oxygenation with the loss of the tertiary alkanol and the formation of benzoquinone (Figure 8.54b).

   b.   Analogous reactions may be presumed to have taken place in the transformation a *tert*-octylphenol by a strain of *Sphingomonas* sp. in which 2,4,4-trimethylpent-2-anol was produced (Tanghe et al. 2000).

8.23 Polyhydric Phenols

The aerobic degradation of polyhydroxybenzenes has been examined less extensively.

1. a.   The degradation of resorcinol can take place by several pathways that are initiated by hydroxylation. 1,2,4-Trihydroxybenzene is degraded by (a) 1:2 dioxygenation and ring fission to maleylacetate and 3-oxoadipate, or by 1:6 dioxygenation and ring fission to acetylpyruvate in strains of *Pseudomonas putida* (Chapman and Ribbons 1976), or by (b) 1:2 dioxygenation and ring fission in *Corynebacterium glutamicum* (Huang et al. 2006). In contrast, degradation by *Azotobacter vinelandii* involved hydroxylation to pyrogallol that was degraded by intradiol fission to 2-hydroxymuconate and pyruvate (Groseclose and Ribbons 1981).

   b.   1,2,4-Trihydroxybenzene is produced by *Trichosporon cutaneum* or is an intermediate in the degradation of several substrates, including the following: the activity of salicylate hydroxylase on 2,5- and 3,4-dihydroxybenzoate (Sze and Dagley 1984), the metabolism of hydroxybenzoates (Anderson and Dagley 1980), and the catabolism of tyrosine (Sparnins et al. 1979). 1,2,4-Trihydroxybenzene can be oxidized to maleylacetate by a ring-fission dioxygenase that has been characterized (Sze et al. 1984).

2. The metabolism of phloroglucinol (1,3,5-trihydroxybenzene) by *Fusarium solani* produced 1,2,3-trihydroxybenzene (pyrogallol) as an unexpected intermediate, which was degraded by intradiol fission to vinylpyruvate, and then to acetaldehyde and pyruvate (Walker and Taylor 1983). Under anaerobic conditions, *Pelobacter massiliensis* uses the transhydroxylation pathway for the degradation of all the trihydroxybenzenes (Brune et al. 1992).

3. 2,4-Diacetylphloroglucinol, which is a secondary metabolite produced by *Pseudomonas fluorescens* CHA0, is the effective biocontrol agent for fungal pathogens. As a 1:3-diketone, it can be hydrolyzed to the monoacetyl compound, which is much less toxic to *Fusarium oxysporum*, and finally to phloroglucinol (Schouten et al. 2004). The hydrolase has been characterized from *Pseudomonas fluorescens* CHA0, is specific for 2,4-diacetylphloroglucinol, and was unreactive against 2-acetyl- and 2,4,6-triacetylphloroglucinol (Bottiglieri and Keel 2006).

8.23.1 Anaerobic Degradation

8.23.1.1 Monohydric and Dihydric Phenols

Briefly, the anaerobic degradation of phenols is dependent on two specific reactions—carboxylation of the ring to hydroxybenzoates and dehydroxylation of these to benzoates. Although details differ, two essentially different mechanisms are involved.

1. a. The anaerobic degradation of phenol by organisms of different taxonomic affiliation involves phosphorylation by phenylphosphate carboxylase (Schühle and Fuchs 2004) followed by carboxylation and dehydroxylation to benzoyl-CoA that is degraded to acetate (Breinig et al. 2000). The enzymes that carry out anaerobic dehydroxylation have been described in Chapter 3, Section 3.13.9. Briefly, the dehydroxylating enzyme from *Thauera aromatica* has been characterized as a molybdenum–flavin–iron–sulfur protein (Breese and Fuchs 1998; Narmandakh et al. 2006), and a similar molybdenum-containing dehydroxylase is also involved in the metabolism of 4-hydroxybenzoyl-CoA by *Rhodopseudomonas palustris* (Gibson et al. 1997). This pathway has been described for several organisms—*Thauera aromatica* (Schühle and Fuchs 2004), *Geobacter metallireducens* GS-15 (Schleinitz et al. 2009), and *Desulfobacterium anilini* (Ahn et al. 2009).

   b. An analogous pathway is used for the degradation of aniline by *Desulfobacterium anilini* (Schnell and Schink 1991), and similar reactions are involved in the degradation of catechol: the production of 3,4-dihydroxybenzoyl-CoA that is dehydroxylated to 3-hydroxybenzoyl-CoA by *Desulfobacterium* sp. strain Cat2 (Gorny and Schink 1994b) and by *Thauera aromatica* (Ding et al. 2008). The anaerobic degradation of 1,4-dihydroxybenzene (hydroquinone) is similar with the formation of 2,5-dihydroxybenzoyl-CoA followed by dehydroxylation (Gorny and Schink 1994a). The anaerobic degradation of 1,3-dihydroxybenzene (resorcinol) by *Azoarcus anaerobius* is, however, completely different (Darley et al. 2007). It is initiated by membrane-associated enzymes with the production of 2-hydroxybenzoquinone that undergoes ring fission to malate and acetate. This strain is unable to degrade catechol or hydroquinone.

2. The anaerobic metabolism of 4-alkylphenols proceeds by two quite different pathways that depend on the organism.

   a. In the strict anaerobe *Geobacter metallireducens*, the degradation of 4-methylphenol is initiated by a membrane-bound enzyme (*p*-cresol methylhydroxylase) that carries out dehydrogenation and hydration to 4-hydroxybenzyl alcohol. This is dehydrogenated to 4-hydroxy-benzoate, esterified to 4-hydroxybenzoyl-CoA and dehydroxylated to benzoyl-CoA (Peters et al. 2007). The catalytically active component of the enzyme complex contains FAD covalently linked to tyrosine (McIntire et al. 1981), and is able to carry out dehydrogenation both of the first step and of 4-hydroxybenzyl alcohol (Johannes et al. 2008).

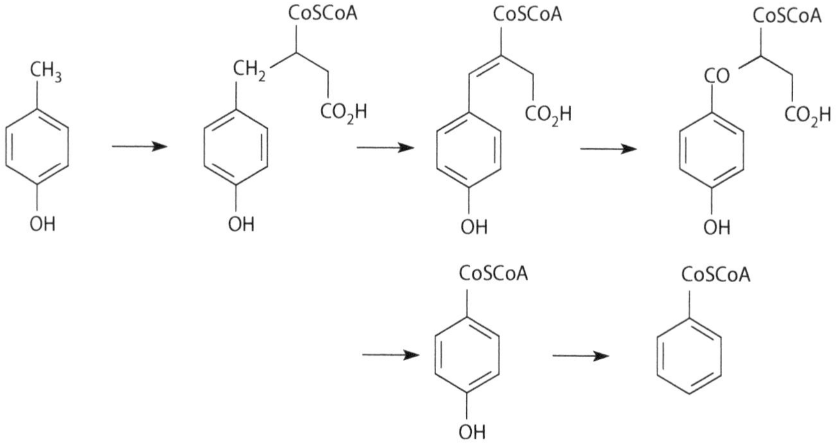

FIGURE 8.55 Anaerobic degradation of 4-methylphenol.

b. An alternative mechanism operates for the degradation of 3- and 4-methylphenol by *Desulfobacterium cetonicum* (Müller et al. 2001) and the analogous 4-ethylphenol by *Aromatoleum aromaticum* strain EbN1 (Wöhlbrand et al. 2008). In these, the degradation is initiated by reaction with fumarate catalyzed by benzylsuccinate synthase in reactions analogous to those already described for the anaerobic degradation of toluene and xylenes (Verfürth et al. 2004). For 3- and 4-methylphenol, this is followed by successive dehydrogenations to 3- or 4-hydroxybenzoyl-CoA and dehydroxylation to benzoyl-CoA (Figure 8.55) (Müller et al. 2001). The degradation 3-methylphenol (*m*-cresol) by *Desulfotomaculum* sp. strain Groll proceeded, however, by hydroxylation of the methyl group to 3-hydroxy-benzyl alcohol, followed by successive dehydrogenation to 3-hydroxybenzaldehyde and 3-hydroxybenzoate, and dehydroxylation to benzoate (Londry et al. 1997). This pathway is also used by a denitrifying organism (Bonting et al. 1995).

8.23.1.2 Polyhydroxybenzoates and Polyhydric Phenols

Gallate (3,4,5-trihydroxybenzoate) can be degraded by *Pelobacter acidigallici* to acetate and CO_2 (Schink and Pfennig 1982), and by *Eubacterium oxidoreducens* in the presence of exogenous H_2 or formate to acetate, butyrate, and CO_2 (Krumholz et al. 1987). The metabolism has been studied in detail, and takes place by an unusual pathway (Figure 8.56) (Krumholtz and Bryant 1988; Brune and Schink 1992).

The formation of phloroglucinol in *P. acidigallici* involves a series of intramolecular hydroxyl transfer reactions with regeneration of 1,2,3,5-tetrahydroxybenzene (Brune and Schink 1990) (Figure 8.57) and in *Eubacterium oxidoreducens* the reaction is catalyzed by an isomerase that contains Fe, acid-labile sulfur, and Mo (Krumholtz and Bryant 1988). Although the reaction is a transhydroxylase, it does not involve hydroxylation by water that is found in many other molybdenum enzymes (Reichenbecher et al. 1996). The

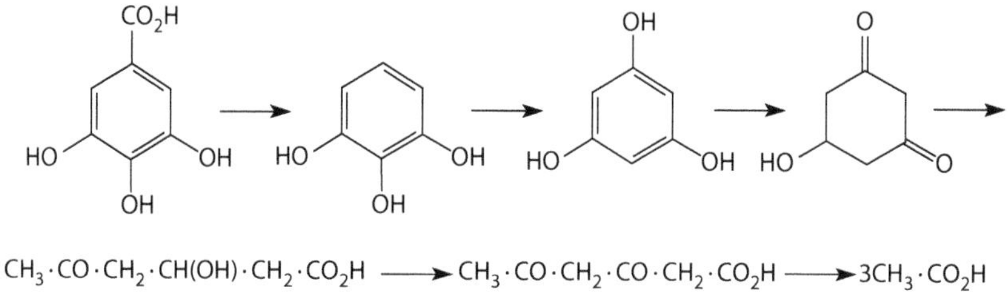

$$CH_3 \cdot CO \cdot CH_2 \cdot CH(OH) \cdot CH_2 \cdot CO_2H \longrightarrow CH_3 \cdot CO \cdot CH_2 \cdot CO \cdot CH_2 \cdot CO_2H \longrightarrow 3CH_3 \cdot CO_2H$$

FIGURE 8.56 Pathway for the biodegradation of 3,4,5-trihydroxybenzoate.

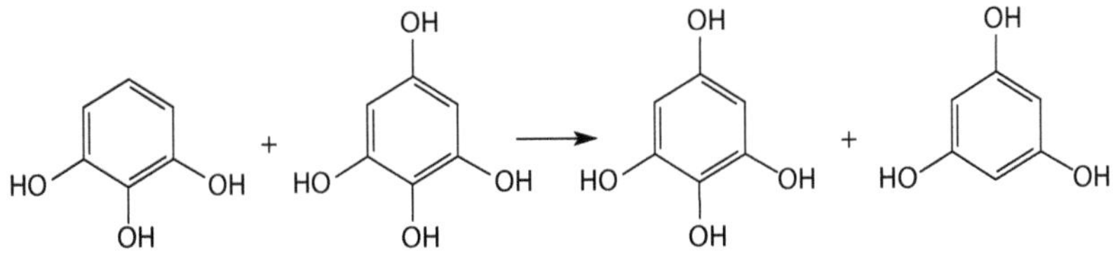

FIGURE 8.57 Intramolecular hydroxyl group transfers in the biodegradation of 1,2,3-trihydroxybenzoate.

crystal structure of the transhydroxylase (Messerschmidt et al. 2004) showed that it was a heterodimer with the active molybdopterin guanine dinucleotide in the α subunit and three [4Fe–4S] clusters in the β subunit. It was proposed that the reaction involved a series of consecutive reactions: (i) oxidation of pyrogallol by Mo(VI) to an *ortho*-quinone with the hydroxyl group coordinated with histidine[144] and Mo(IV), (ii) nucleophilic reaction between this and the anion of 1,2,3,5-tetrahydroxybenzene and tyrosine[404] to produce a diphenyl ether, (iii) the formation of phloroglucinol and regeneration of the catalytic 1,2,3,5-tetrahydroxybenene with oxidation of the Mo(IV) to Mo(VI). An analogous reaction occurs in *E. oxidoreducens* and also involves 1,2,3,5-tetrahydroxybenzene, although details of the pathway may be different (Haddock and Ferry 1993). Subsequent reduction to dihydrophloroglucinol is followed by ring fission to 3-hydroxy-5-oxohexanoate and finally to the formation of butyrate and acetate (Krumholz et al. 1987). This pathway is comparable to that involved in the degradation of resorcinol by species of clostridium (Tschech and Schink 1985), although an alternative pathway involving direct hydrolysis has been observed in denitrifying bacteria (Gorny et al. 1992). *Pelobacter massiliensis* is able to degrade all three trihydroxybenzenes (1,2,3-, 1,3,5-, and 1,2,4-) by the transhydroxylation pathway (Brune et al. 1992), whereas *Desulfovibrio inopinatus* degrades 1,2,4-trihydroxybenzene by a different pathway (Reichenbecher and Schink 1997).

The degradation of phenols has also been examined in mixed cultures, and embodies some of the features that have been observed in pure cultures—in particular the significance of carboxylation.

The following examples illustrate some additional details of these reactions. It is worth noting that metabolism of the methylphenols (cresols) by mixed cultures proceeds by ring carboxylation in contrast to the oxidation pathway that has been demonstrated in pure cultures.

1. Phenol is carboxylated by a defined obligate syntrophic consortium to benzoate that is then degraded to acetate, methane, and CO_2 (Knoll and Winter 1989).
2. 2-Methylphenol is carboxylated by a methanogenic consortium to 4-hydroxy-3-methylbenzoate which was dehydroxylated to 3-methylbenzoate that was the stable endproduct (Bisaillon et al. 1991).

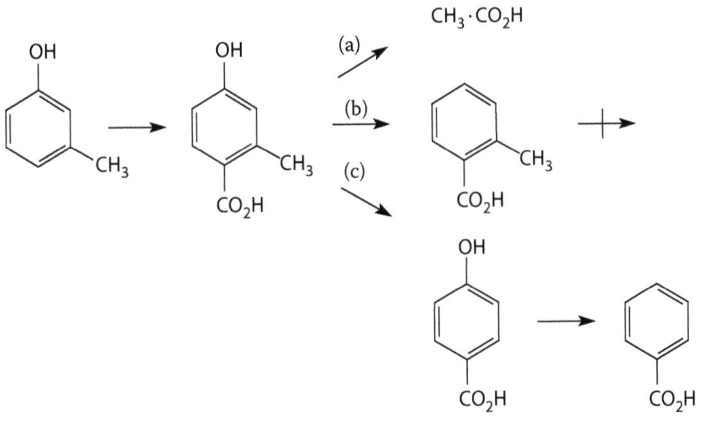

FIGURE 8.58 Anaerobic transformation of 3-methylphenol by carboxylation. (a) Production of methane, (b) dehydroxylation to 2-methylbenzoate, and (c) production of benzoate from 4-hydroxybenzoate.

3. 3-Methylphenol is carboxylated to 2-methyl-4-hydroxybenzoate by a methanogenic enrichment culture before degradation to acetate (Figure 8.58a) (Roberts et al. 1990): ^{14}C-labeled bicarbonate produced carboxyl-labeled acetate, while ^{14}C-methyl labeled 3-methylphenol yielded methyl-labeled acetate. On the other hand, 2-methylbenzoate formed by dehydroxylation of 2-methyl-4-hydroxybenzoate was not further metabolized (Figure 8.58b). A similar reaction occurs with a sulfate-reducing mixed culture (Ramanand and Suflita 1991).

4. An unusual reaction occurred during the degradation of 3-methylphenol by a methanogenic consortium (Londry and Fedorak 1993): although carboxylation to 2-methyl-4-hydroxybenzoate took place as in the preceding example, further metabolism involved loss of the methyl group with the formation of methane before dehydroxylation to benzoate (Figure 8.58c).

It was concluded from these observations that whereas benzoate produced by the carboxylation of phenols can be degraded, dehydroxylation with the formation of substituted benzoates may produce stable terminal metabolites.

8.23.2 Anilines

8.23.2.1 Aerobic

The aerobic degradation of anilines is, in principle, straightforward and involves deamination in a strain of *Nocardia* sp. by a dioxygenase (Bachofer and Lingens 1975), although details of the enzyme have not been fully resolved (Fukumori and Saint 1997). This is followed by ring fission of the resulting catechols by either intradiol or extradiol ring fission. The analogous mechanisms are used for anthranilic acid by deamination and decarboxylation (Chang et al. 2003), and 2-aminobenzenesulfonate to catechol-3-sulfonate with concomitant deamination (Junker et al. 1994). The degradation of 3- and 4-methylanilines by *Pseudomonas putida* mt-2 has been described (McClure and Venables 1986, 1987). It is important to emphasize that since anilines may be incorporated into humic material, their fate is not determined solely by biodegradation. The aerobic degradation of aminobenzenesulfonates that are produced from azobenzenes by reduction and scission is discussed in Chapter 9, Part 4. The degradation of diphenylamine has been examined in *Burkholderia* sp. strain JS667, and is initiated by dioxygenation at the C=C adjacent to the nitrogen atom followed by the formation of aniline and catechol that are then degraded by established pathways (Shin and Spain 2009).

8.23.2.2 Anaerobic

In analogy with phenol, aniline is carboxylated to 4-aminobenzoate followed by reductive deamination to benzoate (Schnell and Schink 1991), which is degraded by pathways that have already been elaborated.

8.24 References

Ahn Y-B, J-C Chae, GJ Zylstra, MM Häggblom. 2009. Degradation of phenol via phenolphosphate and carboxylation to 4-hydroxybenzoate by a newly isolated strain of the sulfate-reducing bacterium *Desulfobacterium anilini. Appl Environ Microbiol* 75: 4248–4253.

Anderson JJ, S Dagley. 1980. Catabolism of aromatic acids in *Trichosporon cutaneum. J Bacteriol* 141: 534–543.

Bachofer R, F Lingens. 1975. Conversion of aniline into pyrocatechol by a *Nocardia* sp.: Incorporation of oxygen-18. *FEBS Lett* 50: 288–290.

Bayly R, R Jain, CL Poh, R Skurry. 1988. Unity and diversity in the degradation of xylenols by *Pseudomonas* spp.: A model for the study of microbial evolution, pp. 359–379. In *Microbial Metabolism and the Carbon Cycle* (Eds. SR Hagedorn, RS Hanson, DA Kunz), Harwood Academic Publishers, Chur, Switzerland.

Bisaillon J-G, F Lépine, R Beaudet, M Sylvestre. 1991. Carboxylation of *o*-cresol by an anaerobic consortium under methanogenic conditions. *Appl Environ Microbiol* 57: 2131–2134.

Bonting CFC, S Schneider, G Schmidtberg, G Fuchs. 1995. Anaerobic degradation of *m*-cresol via methyl oxidation to 3-hydroxybenzoate by a denitrifying bacterium. *Arch Microbiol* 164: 63–69.

Bottiglieri M, C Keel. 2006. Characterization of PhlG, a hydrolase that specifically degrades the antifungal compound diacetylphloroglucinol in the biocontrol agent *Pseudomonas fluorescens* CHA0. *Appl Environ Microbiol* 72: 418–427.

Brackmann R, G Fuchs. 1993. Enzymes of anaerobic metabolism of phenolic compounds 4-hydroxybenzoyl-CoA reductase (dehydroxylating) from a denitrifying *Pseudomonas* sp. *Eur J Biochem* 213: 563–571.

Breese K, G Fuchs. 1998. 4-hydroxybenzoyl-CoA reductase (dehydroxylating) from the denitrifying bacterium *Thauera aromatica*: Prosthetic groups, electron donor, and genes of a member of the molybdenum-flavin-iron-sulfur proteins. *Eur J Biochem* 251: 916–923.

Breinig S, E Schiltz, G Fuchs. 2000. Genes involved in anaerobic metabolism of phenol in the bacterium *Thauera aromatica*. *J Bacteriol* 182: 5849–5863.

Brune A, B Schink. 1990. Pyrogallol-to-phloroglucinol conversion and other hydroxyl-transfer reactions catalyzed by cell extracts of *Pelobacter acidigallici*. *J Bacteriol* 172: 1070–1076.

Brune A, B Schink. 1992. Phloroglucinol pathway in the strictly anaerobic *Pelobacter acidigallici* fermentation of trihydroxybenzenes to acetate via triacetic acid. *Arch Microbiol* 157: 417–424.

Brune A, S Schnell, B Schink. 1992. Sequential transhydroxylations converting hydroxyhydroquinone to phloroglucinol in the strictly anaerobic, fermentative bacterium. *Pelobacter massiliensis. Appl Environ Microbiol* 58: 1861–1868.

Cain RB, P Fortnagel, S Hebenbrock, GW Kirby, HLS McLenaghan, V Rao, S Schmidt. 1997. Biosynthesis of a cyclic tautomer of (3-methylmaleyl(acetone from 4-hydroxy-3,5-dimethylbenzoate by *Pseudomonas* sp. HH35 but not by *Rhodococcus rhodochrous* N75. *Biochem Biophys Res Comm* 238: 197–201.

Cámara B, P Bielecki, F Kaminski, VM dos Santos, I Plumeier, P Nikodem, DH Pieper. 2007. A gene cluster involved in degradation of substituted salicylates via *ortho*-cleavage in *Pseudomonas* sp. strain MT1 encodes enzymes specifically adapted for transformation of 4-methylcatechol and 3-methylmuconate. *J Bacteriol* 189: 1664–1674.

Catelani D, A Fiecchi, E Galli. 1971. (+)-γ-carboxymethyl-γ-methyl-δ-butenolide, a 1,2-ring fission product of 4-methylcatechol by *Pseudomonas desmolyticum*. *Biochem J* 121: 89–92.

Cha C-J, RB Cain, NC Bruce. 1998. The modified β-ketoadipate pathway in *Rhodococcus rhodochrous* N75: Enzymology of 3-methylmuconolactone metabolism. *J Bacteriol* 180: 6668–6673.

Chang H-K, P Mohseni, GJ Zylstra. 2003. Characterization and regulation of the genes for a novel anthanilate 1,2 dioxygenase from *Burkholderia cepacia* DBO 1. *J Bacteriol* 185: 5871–5881.

Chapman PJ, DW Ribbons. 1976. metabolism of resorcinylic compounds by bacteria: Alternative pathways fore resorcinol catabolism in *Pseudomonas putida*. *J Bacteriol* 125: 985–998.

Corvini PFX, RJW Meesters, A Schäffer, HF Schröder, R Vinken, J Hollender. 2004. Degradation of a nonylphenol single isomer by *Sphingomonas* sp. strain TTNP3 leads to a hydroxylation-induced migration product. *Appl Environ Microbiol* 70: 6897–6900.

Darley PI, JA Hellstern, JI Medina-Bellver, S Marqués, B Schink, B Philipp. 2007. Heterologous expression and identification of the genes involved in anaerobic degradation of 1,3-dihydroxybenzene (resorcinol) in *Azoarcus anaerobius*. *J Bacteriol* 189: 3824–3833.

Ding B, S Schmeling, G Fuchs. 2008. Anaerobic metabolism of catechol by the denitrifying bacterium *Thauera aromatica*—a result of promiscuous enzymes and regulators? *J Bacteriol* 190: 1620–1630.

Duffner FM, R Müller. 1998. A novel phenol hydroxylase and catechol 2,3-dioxygenase from the thermophilic *Bacillus thermoleovorans* strain A2: Nucleotide sequence and analysis of the genes. *FEMS Microbiol Lett* 161: 37–45.

Ewers JM, MA Rubio, H-J Knackmuss, D Freier-Schröder. 1989. Bacterial metabolism of 2,5-xylenol. *Appl Environ Microbiol* 55: 2904–2908.

Feist CF, GD Hegeman. 1969a. Phenol and benzoate metabolism by *Pseudomonas putida*: Regulation of tangential pathways. *J Bacteriol* 100: 869–877.

Feist CF, GD Hegeman. 1969b. Regulation of the *meta* cleavage pathway for benzoate oxidation by *Pseudomonas putida*. *J Bacteriol* 100: 1121–1123.

Fukumori F, CP Saint. 1997. Nucleotide sequences and regulational analysis of genes involved in conversion of aniline to catechol in *Pseudomonas putida* UCC22 (pTDN1). *J Bacteriol* 179: 399–408.

Furuya T, S Hirose, H Osanai, H Semba, K Kino. 2011. Identification of the monooxygenase gene clusters responsible for the regioselective oxidation of phenol to hydroquinone in mycobacteria. *Appl Environ Microbiol* 77: 1214–1220.

Gabriel FLP, W Giger, K Guenther, H-P E Kohler. 2005a. Differential degradation of nonylphenol isomers by *Sphingomonas xenophaga* Bayram. *Appl Environ Microbiol* 71: 1123–1129.

Gabriel FLP, A Heidlberger, D Rentsch, W Giger, K Guenther, H-P E Kohler. 2005b. A novel metabolic pathway for degradation of 4-nonylphenol environmental contaminants by *Sphingomonas xenophaga* Bayram. *J Biol Chem* 280: 15526–15533.

Gibson J, M Dispensa, C S Harwood. 1997. 4-hydroxybenzoyl coenzyme A reductase dehydroxylating is required for anaerobic degradation of 4-hydroxybenzoate by *Rhodopseudomonas palustris* and shares features with molybdenum-containing hydroxylases. *J Bacteriol* 179: 634–642.

Gorny N, B Schink. 1994a. Hydroquinone degradation via reductive dehydroxylation of gentisyl-CoA by a strictly anaerobic fermenting bacterium. *Arch Microbiol* 161: 25–32.

Gorny N, B Schink. 1994b. Anaerobic degradation of catechol by *Desulfobacterium* sp. strain cat2 proceeds via carboxylation to protocatechuate. *Appl Environ Microbiol* 60: 3396–3400.

Gorny N, G Wahl, A Brune, B Schink. 1992. A strictly anaerobic nitrate-reducing bacterium growing with resorcinol and other aromatic compounds. *Arch Microbiol* 158: 48–53.

Groseclose EE, DW Ribbons. 1981. Metabolism of resorcinolic compounds by bacteria: A new pathway for resorcinol catabolism in *Azotobacter vinelandii*. *J Bacteriol* 146: 460–466.

Haddock JD, JG Ferry. 1989. Purification and properties of phloroglucinol reductase from *Eubacterium oxidoreducens* G-41. *J Biol Chem* 264: 4423–4427.

Haddock JD, JG Ferry. 1993. Initial steps in the anaerobic degradation of 3,4,5-trihydroxybenzoate by *Eubacterium oxidoreducens*: Characterization of mutants and role of 1,2,3,5-tetrahydroxybenzene. *J Bacteriol* 175: 669–673.

He Z, J Wiegel. 1996. Purification and characterization of an oxygen-sensitive, reversible 3,4-dihydroxybenzoate decarboxylase from *Clostridium hydroxybenzoicum*. *J Bacteriol* 178: 3539–3543.

Hollender J, W Dott, J Hopp. 1994. Regulation of chloro- and methylphenol degradation in *Comamonas testosteroni* JH5. *Appl Environ Microbiol* 60: 2330–2338.

Hopper DJ, PJ Chapman. 1971. Gentisic acid and its 3- and 4-methyl-substituted homologues as intermediates in the bacterial degradation of *m*-cresol, 3,5-xylenol and 2,4-xylenol. *Biochem J* 122: 19–28.

Hopper DJ, L Cottrell. 2003. Alkylphenol biotransformations catalyzed by 4-ethylphenol methylenehydroxylase. *Appl Environ Microbiol* 69: 3650–3652.

Huang Y, K-X Zhao, X-H Shen, MT Chaudhry, C-Y Jiang, S-J Liu. 2006. Genetic characterization of the resorcinol catabolic pathway in *Corynebacterium glutamicum*. *Appl Environ Microbiol* 72: 7238–7245.

Jeong JJ, JH Kim, C-K Kim, I Hwang, K Lee. 2003. 3- and 4-alkylphenol degradation pathway in *Pseudomonas* sp. strain KL28: Genetic organization of the *lap* gene cluster and substrate specificities of phenol hydroxylase and catechol 2,3-dioxygenase. *Microbiology (UK)* 149: 3265–3277.

Johannes J, A Bluschke, N Jemlich, M van Bergen, M Boll. 2008. Purification and characterization of active-site components of the putative *p*-cresol methylhydroxylase membrane complex from *Geobacter metallireducens*. *J Bacteriol* 190: 6493–6500.

Junker F, T Leisinger, AM Cook. 1994. 3-sulfocatechol 2,3-dioxygenase and other dioxygenases (EC 1.13.11.2 and EC 1-14-12) in the degradative pathways of 2-aminobenzenesulfonic, benzenesulphonic and 4-toluenesulfonic acids in *Alcaligenes* sp. strain O-1. *Microbiology (UK)* 140: 1713–1722.

Kirchner U, AH Westphal, R Müller, WJH van Berkel. 2003. Phenol hydroxylase from *Bacillus thermoglucosidasius* A7, a two-component monooxygenase with a dual role for FAD. *J Biol Chem* 278: 47545–47553.

Kivisaar MA, JA Habicht, AL Heinaru. 1989. Degradation of phenol and *m*-toluate in *Pseudomonas* sp. strain EST1001 and its *Pseudomonas putida* transconjugants is determined by a multiplasmid system. *J Bacteriol* 171: 5111–5116.

Knoll G, J. Winter. 1989. Degradation of phenol via carboxylation to benzoate by a defined, obligate syntrophic consortium of anaerobic bacteria. *Appl Microbiol Biotechnol* 30: 318–324.

Krumholz LR, MP Bryant. 1988. Characterization of the pyrogallol-phloroglucinol isomerase of *Eubacterium oxidoreducens*. *J Bacteriol* 170: 2472–2479.

Krumholz LR, RL Crawford, ME Hemling, MP Bryant. 1987. Metabolism of gallate and phloroglucinol in *Eubacterium oxidoreducens* via 3-hydroxy-5-oxohexanoate. *J Bacteriol* 169: 1886–1890.

Kuever J, J Kulmer, S Janssen, U Fischer, K-H Blotevogel. 1993. Isolation and characterization of a new spore-forming sulfate-reducing bacterium growing by complete oxidation of catechol. *Arch Microbiol* 159: 282–288.

Kukor JJ, RH Olsen. 1990. Molecular cloning, characterization, and regulation of a *Pseudomonas pickettii* PKO1 gene encoding phenol hydroxylase and expression of the gene in *Pseudomonas* aeruginosa PAO1c. *J Bacteriol* 172: 4624–4630.

Lobos JH, TK Leib, T-M Su. 1992. Biodegradation of bisphenol A and other bisphenols by a gram-negative aerobic bacterium. *Appl Environ Microbiol* 58: 1823–1831.

Londry KL, PM Fedorak. 1993. Use of fluorinated compounds to detect aromatic metabolites from *m*-cresol in a methanogenic consortium: Evidence for a demethylation reaction. *Appl Environ Microbiol* 59: 2229–2238.

Londry KL, PM Fedorak, JM Suflita. 1997. Anaerobic degradation of *m*-cresol by a sulfate-reducing bacterium. *Appl Environ Microbiol* 63: 3170–3175.

Marín M, D Pérez-Pantoja, R Donoso, V Wray, B González, DH Pieper. 2010. Modified 3-oxoadipate pathway for the biodegradation of methylaromatics in *Pseudomonas reinekei* MT1. *J Bacteriol* 192: 1543–1552.

McClure NC, WA Venables. 1986. Adaptation of *Pseudomonas putida* mt-2 to growth on aromatic amines. *J Gen Microbiol* 132: 2209–2218.

McClure NC, WA Venables. 1987. pTDN1, a catabolic plasmid involved in aromatic amine catabolism in *Pseudomonas putida* mt-2. *J Gen Microbiol* 133: 2073–2077.

McIntire W, DE Edmonson, DJ Hopper, TP Singer. 1981. 8a-(*O*-tyrosyl)flavin adenine dinucleotide, the prosthetic group of bacterial *p*-cresol methylhydroxylase. *Biochemistry* 20: 3068–3075.

Messerschmidt A, H Niessen, D Abt, O Einsle, B Schink, PMH Kroneck. 2004. Crystal structure of pyrogallol-phloroglucinol transhydroxylase, an Mo enzyme capable of intermolecular hydroxyl transfer between phenols. *Proc Natl Acad Sci USA* 101: 11571–11476.

Müller JA, AS Galuschko, A Kappler, B Schink. 2001. Initiation of anaerobic degradation of *p*-cresol by formation of 4-hydroxybenzylsuccinate in *Desulfitobacterium cetonicum*. *J Bacteriol* 183: 752–757.

Murray K, CJ Duggleby, J Sala-Trepat, PA Williams. 1972. The metabolism of benzoate and methylbenzoates via the *meta*-cleavage pathway by *Pseudomonas arvilla* mt-2. *Eur J Biochem* 28: 310–310.

Narmandakh A, N Gad'on, F Drepper, B Knapp, W Haenel, G Fuchs. 2006. Phosphorylation of phenol by phenylphosphate synthase: Role of histidine in catalysis. *J Bacteriol* 188: 7815–7822.

Nurk A, L Kasak, M Kivisaar. 1991a. Sequence of the gene (*pheA*) encoding phenol monooxygenase from *Pseudomonas* sp. EST 1001: Expression in *Escherichia coli* and *Pseudomonas putida*. *Gene* 102: 13–18.

Nurk A, L Kasak, M Kivisaar. 1991b. Sequence of the gene (pheA) encoding phenol monooxygenase from a two-component monooxygenase with a dual role for FAD. *J Biol Chem* 278: 47545–47553.

Peters F, D Heintz, J Johannes, A van Dorsselaer, M Boll. 2007. Genes, enzymes, and regulation of *para*-cresol metabolism in *Geobacter metallireducens*. *J Bacteriol* 189: 4729–4738.

Phillip B, B Schink. 1998. Evidence of two oxidative reaction steps initiating anaerobic degradation of resorcinol (1,3-dihydroxybenzene) by the denitrifying bacterium *Azoarcus anaerobius*. *J Bacteriol.* 180: 3644–3649.

Pieper DH, K-H Engesser, RH Don, KN Timmis, H-J Knackmuss. 1985. Modified *ortho*-cleavage pathway in *Alcaligenes eutrophus* JMP 134 for the degradation of 4-methylcatechol. *FEMS Microbiol Lett* 29: 63–67.

Pieper DH, K Stadler-Fritzsche, H-J Knackmuss, K-H Engesser, NC Bruce, RB Cain. 1990. Purification and characterization of 4-methylmucolactone methyl-isomerase, a novel enzyme of the modified 3-oxoadipate pathway in the Gram-negative bacterium *Alcaligenes eutrophus* JMP 134. *Biochem J* 271: 529–534.

Pieper DH, K Stadler-Fritzsche, H-J Knackmuss, KN Timmis. 1995. Formation of dimethylmuconolactones from dimethylphenols by *Alcaligenes eutrophus* JMP 134. *Appl Environ Microbiol* 61: 2159–2165.

Powlowski JB, J Ingebrand, S Dagley. 1985. Enzymology of the β-ketoadipate pathway in *Trichosporon cutaneum*. *J Bacteriol* 163: 1135–1141.

Powlowski J, V Shingler. 1994a. Genetics and biochemistry of phenol degradation by *Pseudomonas* sp. CF600. *Biodegradation* 5: 219–236.

Powlowski J, V Shingler. 1994b. Genetics and biochemistry of phenol degradation by *Pseudomonas* *Pseudomonas* sp. EST 1001: Expression in *Escherichia coli* and *Pseudomonas putida*. *Gene* 102: 13–18.

Ramanand K, JM Suflita. 1991. Anaerobic degradation of *m*-cresol in anoxic aquifer slurries: Carboxylation reactions in a sulfate-reducing bacterial enrichment. *Appl Environ Microbiol* 57: 1689–1695.

Reeve CD, MA Carver, DJ Hopper. 1989. The purification and characterization of 4-ethylphenol methylene hydroxylase, a flavocytochrome from *Pseudomonas putida* JD1. *Biochem J* 263: 431–437.

Reichenbecher W, A Rüdiger, PMH Kroneck, B Schink. 1996. One molecule of molybdopterin guanine dinucleotide is associated with each subunit of the heterodimeric Mo-Fe-S protein transhydroxylase of *Pelobacter acidigallici* as determined by SDS (PAGE) and mass spectrometry. *Eur J Biochem* 237: 406–413.

Reichenbecher WW, B Schink. 1997. *Desulfovibrio inopinatus* sp nov, a new sulfate-reducing bacterium that degrades hydroxyhydroquinone (1,2,4-trihydroxybenzene). *Arch Microbiol* 168: 338–344.

Roberts DJ, PM Fedorak, SE Hrudey. 1990. CO_2 incorporation and 4-hydroxy-2-methylbenzoic acid formation during anaerobic metabolism of *m*-cresol by a methanogenic consortium. *Appl Environ Microbiol* 56: 472–478.

Sala-Trapat J, K Murray, PA Williams. 1972. The metabolic divergence in the *meta* cleavage of catechols by *Pseudomonas putida* NCIB 10015. Physiological significance and evolutionary implications. *Eur J Biochem* 28: 347–356.

Schink B, N Pfennig. 1982. Fermentation of trihydroxybenzenes by *Pelobacter acidigallici* gen nov., sp nov., a new strictly anaerobic, non-sporeforming bacterium. *Arch Microbiol* 133: 195–201.

Schleinitz KM, S Schmeling, N Jehmlich, M van Bergen, H Harms, S Kleinsteuber, C Vogt, G Fuchs. 2009. Phenol degradation in the strictly anaerobic iron-reducing bacterium *Geobacter metallireducens* GS-15. *Appl Environ Microbiol* 73: 3912–3919.

Schmidt S, RB Cain, GV Rao, GW Kirby. 1994. Isolation and identification of two novel butenolides as products of dimethylbenzoate metabolism by *Rhodococcus rhodochrous* N75. *FEMS Microbiol Lett* 120: 93–98.

Schnell S, F Bak, N Pfennig. 1989. Anaerobic degradation of aniline and dihydroxybenzenes by newly isolated sulfate-reducing bacteria and description of *Desulfobacterium anilini*. *Arch Microbiol* 152: 556–563.

Schnell S, B Schink. 1991. Anaerobic aniline degradation via reductive deamination of 4-aminobenzoyl-CoA in *Desulfobacterium anilini*. *Arch Microbiol* 155: 183–190.

Schouten A, G van den Berg, C Edel-Hermann, C Steinberg, N Gautheron, C Alabouvette, CH de Vos, P Lemanceau, JM Raaijmakers. 2004. Defense responses of *Fusarium oxysporum* to 2,4-diacetylphloroglucinol, a broad-spectrum antibiotic produced by *Pseudomonas fluorescens*. *Mol Plant–Microbe Interact* 17: 1201–1211.

Schühle K, G Fuchs. 2004. Phenylphosphate carboxylase: A new C–C lyase involved in anaerobic phenol metabolism in *Thauera aromatica*. *J Bacteriol* 186: 4556–4567.

Shin KA, JC Spain. 2009. Pathway and evolutionary implictions of diphenylamine biodegradation by *Burkholderia* sp. strain JS667. *Appl Environ Microbiol* 75: 2694–2704.

Shingler, V, J Powlowski, U Marklund. 1992. Nucleotide sequence and functional analysis of the complete phenol/3,4-dimethylphenol catabolic pathway of *Pseudomonas* sp. strain CF600. *J Bacteriol* 174: 711–724.

Sparnins VL, DG Burbee, S Dagley. 1979. Catabolism of L-tyrosine in *Trichosporon cutaneum*. *J Bacteriol* 138: 425–430.

Spivack J, TK Leib, JH Lobos. 1994. Novel pathway for bacterial metabolism of bisphenol A. Rearrangements and stilbene cleavage in bisphenol A metabolism. *J Biol Chem* 269: 7323–7329.

Sze I S-Y, S Dagley. 1984. Properties of salicylate hydroxylase and hydroxyquinol 1,2-dioxygenase purified from *Trichosporon cutaneum*. *J Bacteriol* 159: 353–359.

Szewzyk R, N Pfennig. 1987. Complete oxidation of catechol by the strictly anaerobic sulfate-reducing *Desulfobacterium catecholicum* sp. nov. *Arch Microbiol* 147: 163–168.

Tanghe T, W Dhooge, W Verstraete. 2000. Formation of the metabolic intermediate 2,4,5-trimethyl-2-pentanol during incubation of a *Sphingomonas* sp. strain with the xeno-estrogenic octylphenol. *Biodegradation* 11: 11–19.

Toyama T, N Momotani, Y Ogata, Y Miyamori, D Inoue, K Sei, K Mori, S Kikuchi, M Ike. 2010. Isolation and characterization of 4-*tert*-butylphenol-utilizing *Sphingomonas fulginis* strains from *Phragmites australis* rhizosphere sediment. *Appl Environ Microbiol* 76: 6733–6740.

Tschech A, B Schink. 1985. Fermentative degradation of resorcinol and resorcylic acids. *Arch Microbiol* 143: 52–59.

Verfürth K, AJ Pierik, C Leutwein, S Zorn, J Heider. 2004. Substrate specificities and electron paramagnetic resonance properties of benzylsuccinate synthesis in anaerobic toluene and *m*-xylene metabolism. *Arch Microbiol* 181: 155–162.

Walker JRL, BG Taylor. 1983. Metabolism of phloroglucinol by *Fusarium solani*. *Arch Microbiol* 134: 123–126.

Wöhlbrand L, H Wilkes, T Halder, R Rabus. 2008. Anaerobic degradation of *p*-ethylphenol by *Aromatoleum aromaticum* strain EbN1: Pathway, regulation, and involved proteins. *J Bacteriol* 190: 5699–5709.

9

Halogenated Arenes and Carboxylates with Chlorine, Bromine, or Iodine Substituents[*]

Part 1: Arenes and Carboxylated Arenes with Halogen, Sulfonate, Nitro, and Azo Substituents

9.1 Aerobic Degradation

9.1.1 Introduction

The aerobic degradation of chlorinated arene hydrocarbons and carboxylates, including the important group of PCBs, and chlorobenzoates that are produced from them as metabolites, is generally initiated by dihydroxylation of the rings to dihydrodiols followed by dehydrogenation to catechols. Halide may be lost simultaneously and, for 2-halogenated benzoates both halide and carboxyl. Salient aspects are summarized, and attention is drawn to selected aspects of enzyme inhibition. The aerobic degradation of halogenated phenols takes place, however, by monooxygenation and is discussed in Part 2 of this chapter. It is noted here only for the role of chlorocatechols produced from chlorophenols and chloroanilines.

Emphasis is placed on chlorinated substrates. Reference may be made to a review (Allard and Neilson 2003) for details of their brominated and iodinated analogs, while the degradation of aromatic fluorinated compounds is discussed in Part 3 of this chapter, and in a review (Neilson and Allard 2002).

A brief summary is given of the basic mechanisms involved in the degradation of halogenated arenes. These parallel the pathways for unsubstituted hydrocarbons: formation of a *cis*-dihydrodiol followed by dehydrogenation to the catechol, ring fission to muconic acids or hydroxymuconate semialdehydes, and further metabolism to provide entry to the 3-oxoadipate pathway.

9.1.1.1 Ring Dioxygenation and Dehydrogenation

Dioxygenation initiates the degradation of halogenated arenes and carboxylates with the production of dihydrodiols. The genes encoding the α- and β-subunits of the terminal chlorobenzene oxidase, the ferredoxin, and the reductase have been cloned from *Burkholderia* sp. strain PS14 into *Escherichia coli* (Beil et al. 1997). It was confirmed that whereas metabolism of 1,2- and 1,4-dichlorobenzene and 1,2,4-trichlorobenzene (Figure 9.1), and 1,2,3,4-tetrachlorobenzene produced stable *cis*-dihydrodiols, dechlorination concomitant with dioxygenation took place with 1,2,4,5-tetrachlorobenzene (Figure 9.2) and was confirmed using $^{18}O_2$. *Pseudomonas* sp. strain P51 is able to use 1,2-dichloro-, 1,4-dichloro, and 1,2,4-trichlorobenzene as sources of carbon and energy (van der Meer et al. 1991a). The genes were encoded on two plasmids, and the dioxygenase TcbA and the *cis*-dihydrodiol dehydrogenase TcbB in this

---

[*] These industrial chemicals are widely used as the precursors for a wide range of products, which include agrochemicals, pharmaceutical products, polychlorinated biphenyls (PCBs), and polybrominated diphenyl ether flame retardants. For virtually all of them, serious concern has arisen over their adverse environmental effects including persistence.

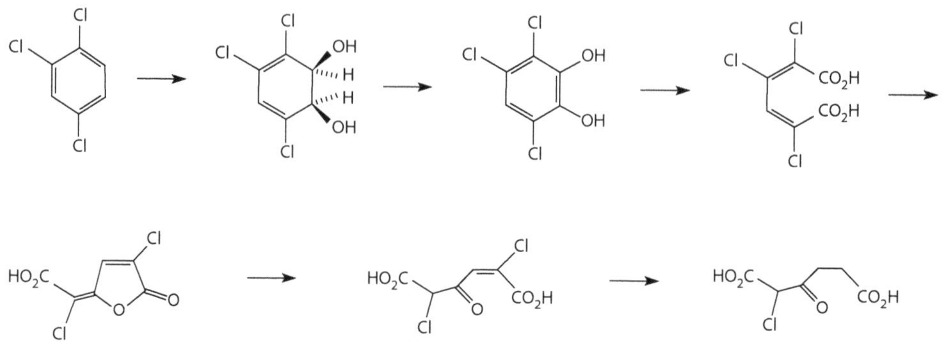

FIGURE 9.1   Metabolism of 1,2,4-trichlorobenzene.

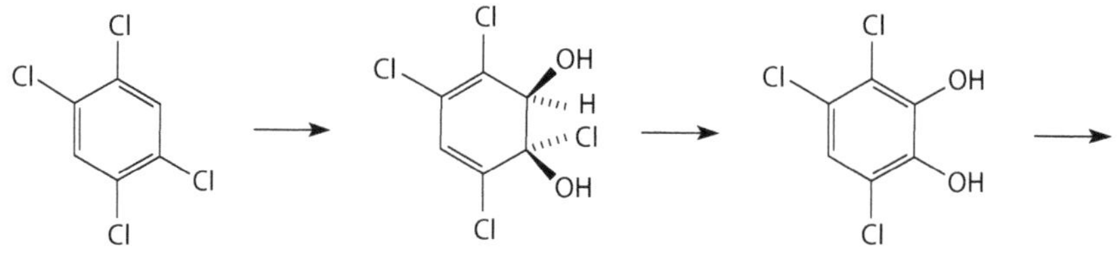

FIGURE 9.2   Metabolism of 1,2,4,5-tetrachlorobenzene.

strain were closely similar to the analogous enzymes TodC and Tod D for the degradation of benzene and toluene (Werlen et al. 1996).

Dioxygenation is generally followed by dehydrogenation, except when loss of halide obviates this, for example, for 2,3-dioxygenation of 1,2,4,5-tetrachlorobenzene (Figure 9.2) (Sander et al. 1991), for 2,3-dioxygenation of chlorinated biphenyls with chlorine at the 2-positions (Seeger et al. 2001), and for 1,2-dioxygenation with decarboxylation of 2-chlorobenzoate (Fetzner et al. 1992). This results in the production of 1,2-dihydroxybenzenes (catechols) that are substrates for ring fission by dioxygenation carried out by a different group of enzymes. Catechols, therefore, occupy a central role in aerobic degradation of halogenated arenes including halogenated phenols that are discussed in Part 2. The importance of both the number and the position of chlorine substituents is illustrated by the results of a study with *Xanthobacter flavus* strain 14p1. The strain grew only with 1,4-dichlorobenzene, which was degraded by dioxygenation, dehydrogenation, and intradiol ring fission, whereas 1,3-dichlorobenzene could not induce dioxygenase activity (Sommer and Görisch 1997).

9.1.1.2 Degradation of Chlorocatechols

The degradation of chlorocatechols involves successive reactions: (1) ring fission by intradiol dioxygenation, (2) cyclization of the *cis–cis*-muconates that are produced to muconolactones with, or without loss of chloride, (3) hydrolysis of diene lactones, and (4) reduction of maleylacetates. These enzymes may display a high degree of specificities: (a) the dioxygenases (Dorn and Knackmuss 1978a), (b) the cycloisomerases (Schmidt and Knackmuss 1980), and (c) the chloromaleylacetate reductases (Kaschabek and Reineke 1995). This is illustrated for the degradation of the catechols produced from 1,3-dichloro-, 1,2,3-trichloro-, 1,2,3,4-tetrachloro, and 1,4-dichlorobenzene (Figure 9.3a–d). In addition, protoanemonin that is a terminal metabolite may be produced from 2-chloro- and 5-chloromuconolactones by decarboxylation and loss of chloride.

The ring fission of substituted catechols can be carried out by 1,2-dioxygenation (intradiol or ortho fission), by 3,4-dioxygenation (extradiol or meta fission) or, less commonly, by 1,6 dioxygenation (distal fission) (Figure 9.4a–c) although this alternative has been demonstrated for 3-chlorocatechol in *Sphingomonas* sp. strain BN6 (Riegert et al. 1998) to produce 2-hydroxy-3-chloromuconic semialdehyde.

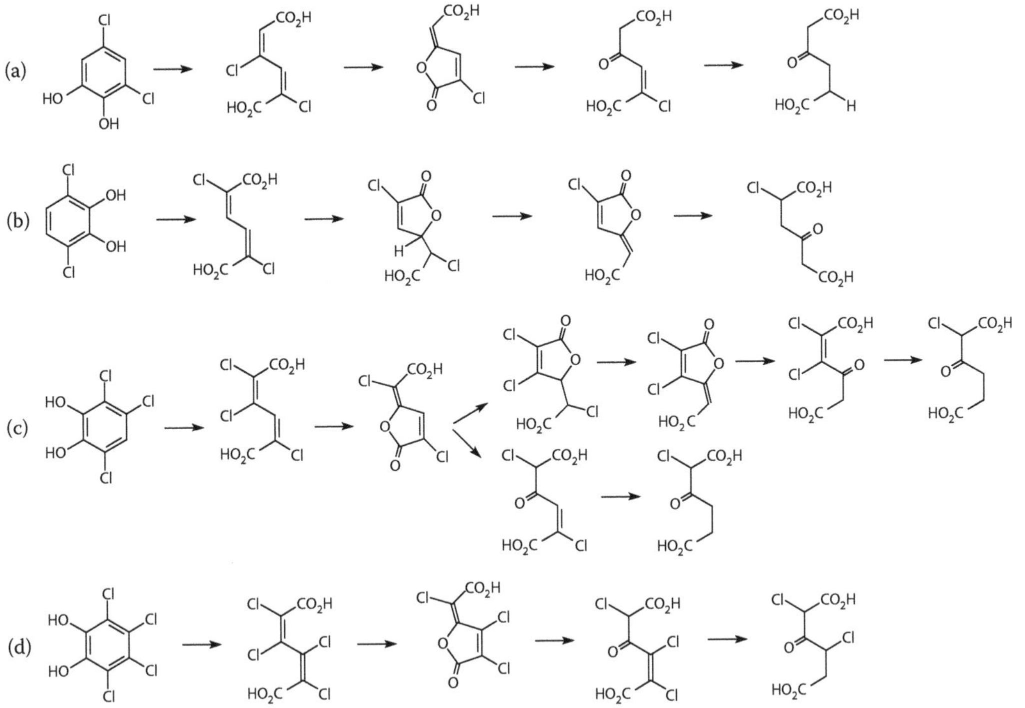

FIGURE 9.3 Degradation of (a) 1,3-dichloro-, (b) 1,4-dichloro-, (c) 1,2,4-trichloro-, and (d) 1,2,3,4-tetrachlorobenzene.

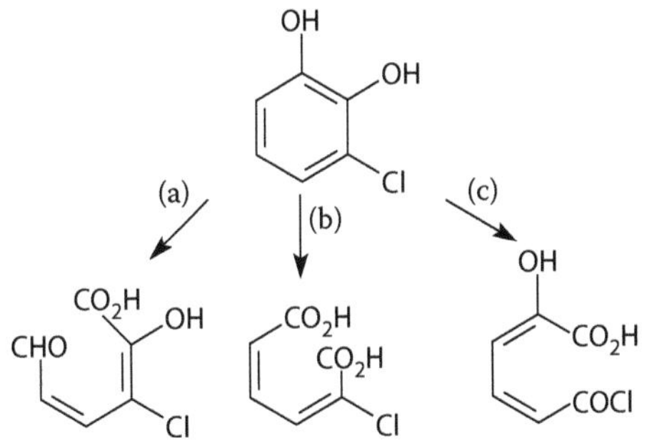

FIGURE 9.4 Alternative dioxygenation pathways for 3-chlorocatechol (a) intradiol, (b) extradiol, and (c) distal, fission.

9.1.1.3 Extradiol Fission

Although this is generally preferred for substituted catechols, for chlorocatechols this may result in the production of a toxic metabolites and the choice of intradiol fission (Figure 9.4a). Extradiol 2,3-dioxygenation of 3-chlorocatechol may produce an acyl chloride that inhibits the enzyme (Figure 9.5) (Bartels et al. 1984; Klečka and Gibson 1981). This may, however, be circumvented when the acyl chloride undergoes further reaction:

1. Hydrolysis by *Pseudomonas putida* GJ 31 to 2-hydroxy-6-chlorocarbonylmuconate followed by the formation of 2-hydroxy-*cis,cis*-muconate, oxalocrotonate, pyruvate, and acetaldehyde (Mars et al. 1997; Kaschabek et al. 1998).
2. Reaction of the extradiol fission product from 3-chloro-4,5-dihydroxybenzoate by *Alcaligenes* sp. strain BR 6024 to produce pyrone-4,6-dicarboxylate that is hydrolyzed to

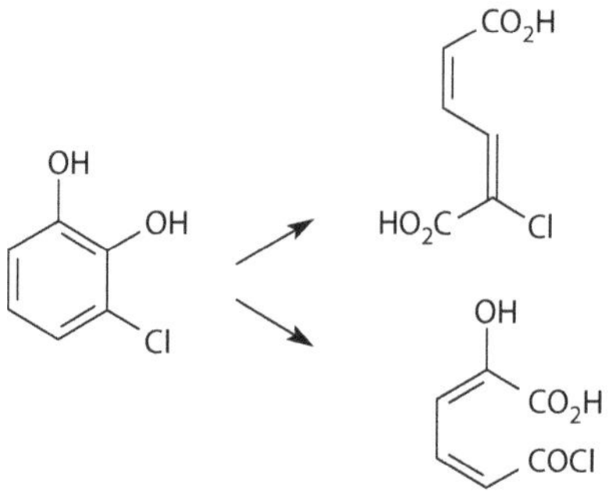

FIGURE 9.5   Extradiol 2,3-dioxygenation of 3-chlorocatechol with production of an acyl chloride.

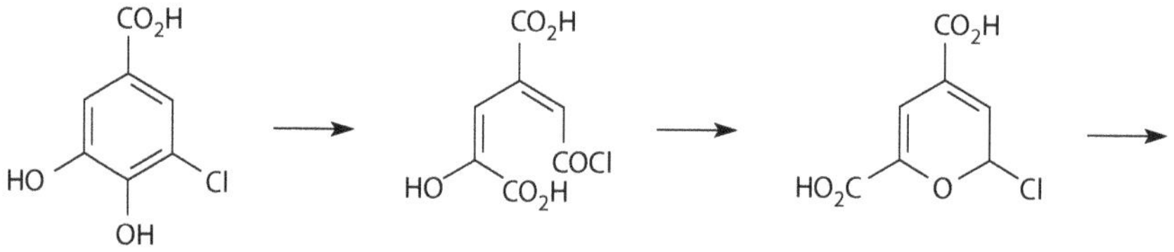

FIGURE 9.6   Metabolism of 3-chloro-4,5-dihydroxybenzoate *Alcaligenes* sp. strain BR 6024 to pyrone-4,6-dicarboxylate.

2-hydroxy-4-carboxymuconate (Nakatsu et al. 1997) (Figure 9.6). This is analogous to the pathway for the degradation of 5-chlorovanillate by *Pseudomonas testosteroni* (Kersten et al. 1985).

3. Spontaneous hydrolysis after chlorohydroquinone 1,2-dioxygenation that is noted in the degradation of chlorophenols with three or more substituents.

9.1.1.4 Intradiol Fission

1. *Type I intradiol 1,2-dioxygenation:* This generally operates for catechol and 3,4-dihydroxybenzoate in a number of bacteria (references in Harwood and Parales 1996), with differences for Azotobacters and fungi.

2. *Type II intradiol 1,2-dioxygenation:* This is used for the degradation of 3-chloro- and 4-chlorocatechol to produce respectively 2-chloro- and 3-chloromuconate (Dorn and Knackmuss 1978a; Schmidt and Knackmuss 1980) (Figure 9.7a,b). The significance of the halogen position in the substrate on the activity of chlorocatechol 1,2-dioxygenase has been analyzed in several organisms: *Pseudomonas* sp. strain B13 (Dorn and Knackmuss 1978b), and *Alcaligenes eutrophus* strain JMP134 (pJP4) (Don et al. 1985) grown with 3-chlorobenzoate; *Pseudomonas* sp. strain P51 grown with 1,2,4-trichlorobenzene (van der Meer et al. 1991a); *Pseudomonas chlororaphis* grown with 1,2,3,4-tetrachlorobenzene (Potrawfke et al. 1998; 2001). Briefly, there were striking differences among them, especially for 3,4-dichloro-, 3,5-dichlor-, and 4,5-dichlorocatechol.

9.1.1.5 Cyclization of Muconates

1. The cycloisomerases that catalyze the formation of 2-chloro-, 3-chloro-, 4-chloro-, or 6-chloromuconolactones from 2-chloro- or 3-chloromuconate play a cardinal role in degradation. Although it has been shown that 2-chloro- and 6-chloromuconolactones may be stable to further reaction (Vollmer et al. 1994), degradation can generally be accomplished, and depends critically on the result of cyclization. The cycloisomerases can bring about 1:4-cyclization and 3:6 cyclization, and these determine the pathway, and the possible synthesis of terminal metabolites. Elimination of chloride from critical positions can produce

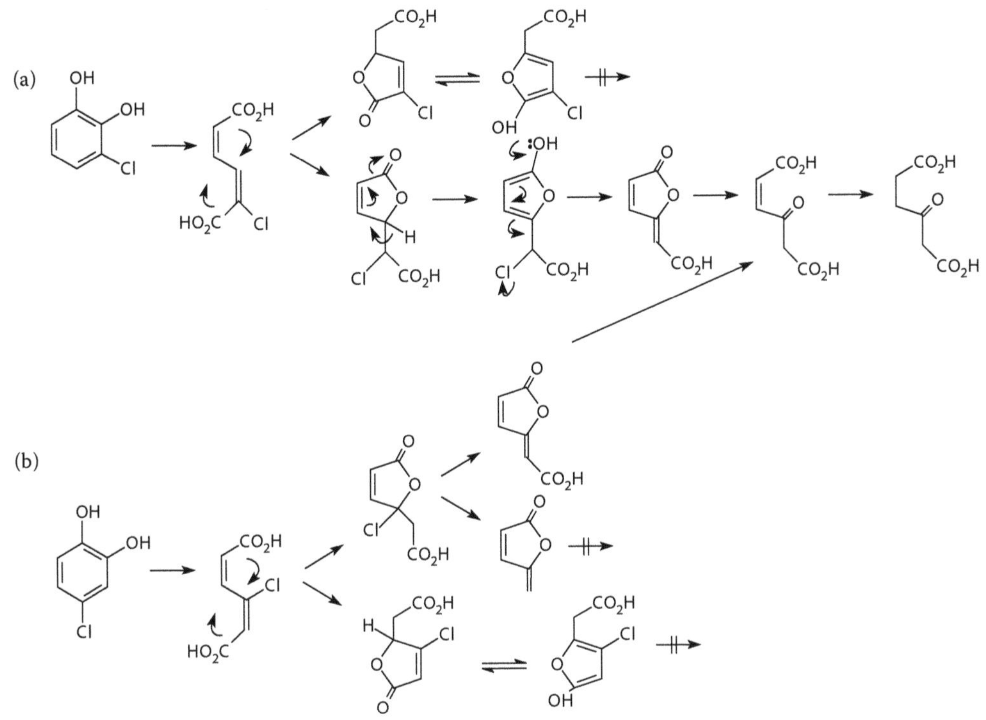

FIGURE 9.7 Degradation of (a) 3-chlorocatechol and (b) 4-chlorocatechol.

dienelactones that are able to enter central metabolic pathways via 3-oxoadipate (Figure 9.7a,b). On the other hand, 2-chloro- and 3-chloromuconolactones may be stable to further reaction. There are differences in the specificity and rates of Gram-negative organisms and the Gram-positive *Rhodococcus opacus* 1CP (Solyanikova et al. 1995; Eulberg et al. 1998; Vollmer et al. 1998).

2. In *Pseudomonas* sp. strain B13, whereas ring fission of 3-chlorocatechol produced 2-chloromuconate followed by cyclization to the *trans*-diene, the analogous reactions with 4-chlorocatechol produced 3-chloromuconate and the *cis*-diene (Schmidt et al. 1980).

   The metabolism of 3-chloro-5-methylcatechol by *Ralstonia eutropha* JMP 134 produced the *cis,cis*-muconate followed by cyclization to 5-chloro-3-methylmuconolactones. Their stereochemistry was deduced from NMR spectra, and it was shown that *syn*-elimination of chloride took place: (4R,5R)-muconolactone produces the *trans*-diene-lactone, while the (4R,5S) muconolactone produced the *cis*-dienelactone (Prucha et al. 1996a,b).

3. Muconolactones with a chlorine substituent at C_4 may undergo either loss of chloride to form the diene that can be processed further, or diverted by decarboxylation and loss of chloride to protoanemonin that was first identified as a product from 4-chlorocatechol, and is generally a terminal metabolite (Blasco et al. 1995) (Figure 9.8). Protoanemonin can also be produced from 2-chloromuconolactone through the activities of muconate cycloisomerase and muconolactone isomerase, and its formation is dependent on the kinetics of competing reactions (Figure 9.9) (Skiba et al. 2002).

9.1.1.6 Reduction of Maleylacetates

Dioxygenation may be followed by loss of chloride to produce a dienelactone followed by hydrolysis and reduction of the resulting maleylacetate by *Pseudomonas* sp. strain B13 (Kaschabek and Reineke 1995) or by *Pseudomonas chlororaphis* (Potrawfke et al. 2001).

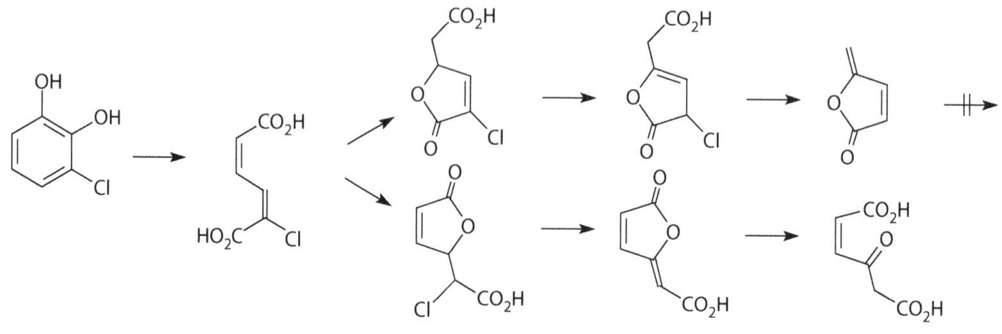

FIGURE 9.8 Metabolism of muconolactones with a chlorine substituent at C_4.

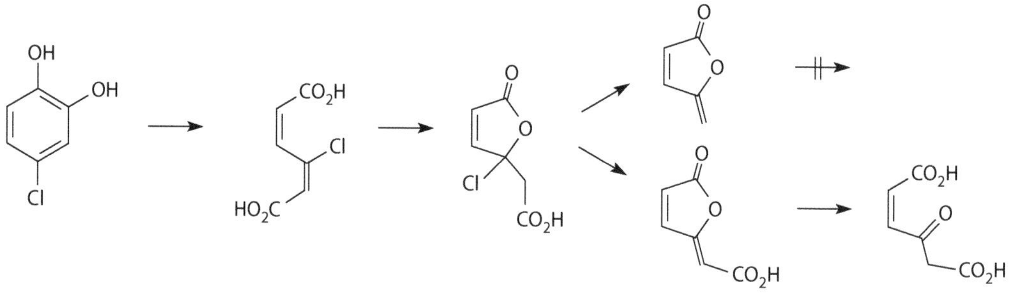

FIGURE 9.9 Production of protoanemonin from 2-chloromuconolactone.

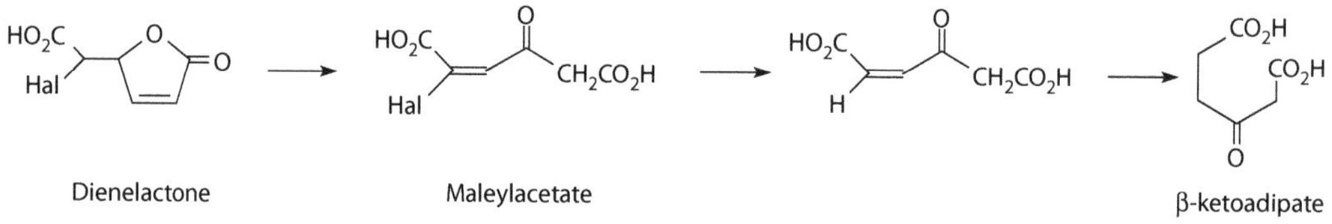

FIGURE 9.10 Degradation of halogenated dienelactones maleylacetate and reduction to 3-oxoadipate. (Redrawn from Kaschabek SR, W Reineke. 1995. *J Bacteriol* 177: 320–325.)

Entry into the 3-oxoadipate pathway requires the reduction of maleylacetates produced by hydrolysis of the corresponding dienelactones. This has been examined in *Pseudomonas* sp. strain B13 grown with 3-chlorobenzoate (Kaschabek and Reineke 1995), and several important issues were resolved (Figure 9.10):

1. For 2-chloromaleylacetate 2 moles of NADH were required in contrast to only 1 mole when there were no substituents at C_2.
2. Elimination of halide occurred for a range of substrates that included 2-chloro-, 2,3-dichloro-, 2,3,5-trichloro-, 2-chloro-3-methyl-, and 2-chloro-5-methylmaleylacetates.
3. Rates were strongly affected by the position of methyl groups: 2-methyl- ≥ 3-methyl- >> 5-methylmaleylacetate.

9.1.2 Monocyclic Chlorinated Arenes

9.1.2.1 Chlorobenzene

The degradation of chlorinated benzenes with one to four substituents has been described and is summarized in Table 9.1. Relevant details have been given in the introduction to the chapter and require no further comment.

TABLE 9.1

Aerobic Degradation of Chlorobenzenes

Substrate	Organism	Reference
1,2-Dichlorbenzene	*Pseudomonas* sp. strain P51	van der Meer et al. (1991b)
1,2-Dichlorbenzene	*Pseudomonas* sp. strain PS12	Sander et al. (1991)
1,4-Dichlorobenzene	*Pseudomonas* sp. strain JS 6	Spain and Nishino (1987)
1,4-Dichlorobenzene	*Xanthobacter flavus* strain 14p1	Sommer and Görisch (1997)
1,2,4-Trichlorobenzene	*Pseudomonas* sp. strain P51	van der Meer et al. (1991b)
1,2,4-Trichlorobenzene	*Pseudomonas* sp. strain PS12	Sander et al. (1991)
1,2,3,4-Tetrachlorobenzene	*Pseudomonas chlororaphis* strain RW71	Potrawfke et al. (1998)
1,2,4,5-Tetrachlorobenzene	*Pseudomonas* sp. PS14	Sander et al. (1991)

9.1.2.2 Hexachlorobenzene

The oxidative degradation of hexachlorobenzene (HCB) has been examined in a strain of *Nocardioides* that is also able to degrade pentachloronitrobenzene (Takagi et al. 2009). Degradation is initiated by formation of pentachlorophenol and proceeds by the pathway already demonstrated—to tetrachlorohydroquinone and 2,6-dichlorohydroquinone. It is worth noting that a mutant of cytochrome $P450_{cam}$ (CYP101) from *Pseudomonas putida* made possible the monooxygenation of chlorinated benzenes with less than three substituents to chlorophenols, with concomitant NIH shifts for 1,3-dichlorobenzene (Jones et al. 2001). Further mutations made it possible to oxidize even pentachlorobenzene and HCB to pentachlorophenol (Chen et al. 2002). *Sphingobium chlorophenolicum* is among the bacteria able to degrade pentachloro-phenol, and protein engineering to introduce the cytochrome $P450_{cam}$ plasmid enabled the mutant to degrade HCB putatively by monooxygenation to pentachlorophenol and then via tetrachloro- and 2,6-dichlorohydroquinone (Yan et al. 2006).

9.1.2.3 Chlorotoluenes

These display metabolic features additional to those for chlorobenzenes. There are two reactions that have been observed in the metabolism of chlorotoluenes: (a) monooxygenation by dioxygenases to benzyl alcohols that are terminal metabolites (Lehning et al. 1997), or (b) dioxygenation to dihydrodiols.

Considerable attention has been directed to the degradation of chlorotoluenes, and a rather complex situation has emerged that impinges on the metabolism of methylcatechols and chlorocatechols. The *tecA* gene that encodes the tetrachlorobenzene dioxygenase and the *tecB* that encodes the dehydrogenase in

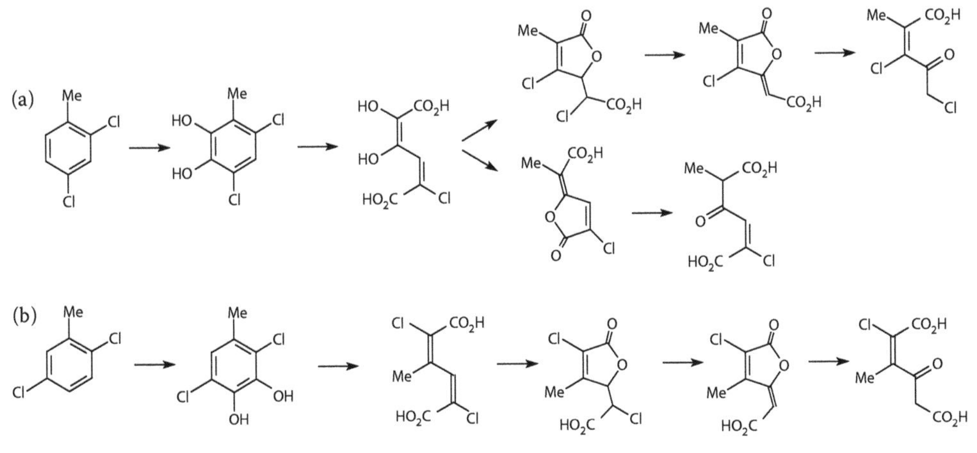

FIGURE 9.11 Catechols from (a) 2,4-dichlorotoluene and (b) 2,5-dichlorotoluene.

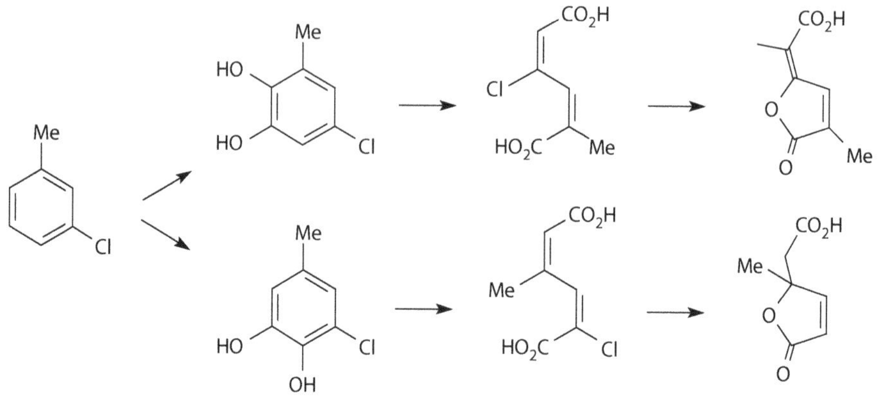

FIGURE 9.12　Metabolism of 3-chlorotoluene: muconolactone resistant to degradation.

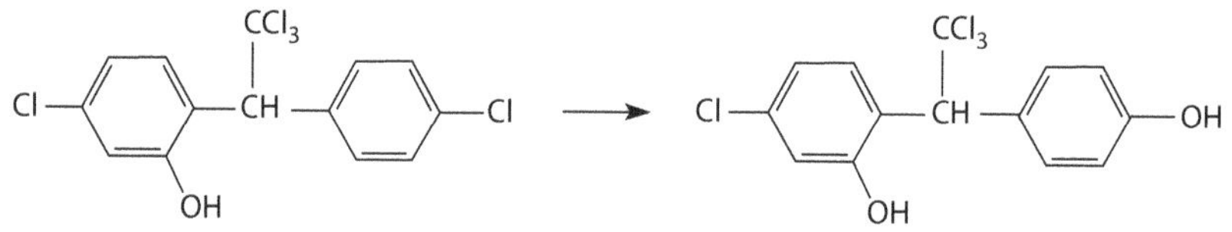

FIGURE 9.13　Alternative pathways for degradation of DDT.

Ralstonia sp. strain PS 12 were able to bring about formation of the corresponding catechols from 2,4-, and 2,5-dichlorotoluene that were, therefore, potentially degradable (Figure 9.11a,b) (Pollmann et al. 2002). On the other hand, 2,3-, 2,6-, and 3,5-dichlorotoluene, and 2,4,5-trichlorotoluene were metabolized to the benzyl alcohols by this strain (Pollmann et al. 2002). The degradation of 3-chlorotoluene can take place by 4,5-dioxygenation and successive formation of 5-chloro-3-methyl-catechol, 4-chloro-2-methylmuconate and 2-methyl-*cis,cis*-dienelactone that can be degraded (Pollmann et al. 2005). On the other hand, the production of 3-chloro-4-methylcatechol by 3-4-dioxygenation, and 2-chloro-4-methyl-muconate results in the synthesis of muconolactones which are resistant to degradation (Figure 9.12).

Although 1,1-dichloro-2,2-bis(*p*-chlorophenyl)ethene (DDE) is appreciably recalcitrant, it can be degraded by dioxygenation and extradiol ring fission to chlorobenzoate by cells of *Pseudomonas acidovorans* M3GY grown with biphenyl (Hay and Focht 1998; Aislabie et al. 1999). Bis(*p*-chlorophenyl)acetic acid (DDA), which is a polar metabolite, is apparently persistent in the environment (Heberer and Dünnbier 1999). Alternatively, degradation of DDT may take place by hydroxylation of the ring and displacement of the aromatic ring chlorine atom by hydroxyl (Figure 9.13) (Massé et al. 1989).

9.2 Anaerobic Conditions

Attention has been directed to the dechlorination of polychlorinated benzenes by strains that use them as an energy source by dehalorespiration. Investigations using *Dahalococcoides* sp. strain CBDB1 have shown its ability to dechlorinate congeners with three or more chlorine substituents (Hölscher et al. 2003). Although there are minor pathways, the major one for HCB was successive reductive dechlorination to pentachlorobenzene, 1,2,4,5-tetrachlorobenzene, 1,2,4-trichlorobenzene, and 1,4-dichlorobenzene (Jayachandran et al. 2003). The electron transport system has been examined by the use of specific inhibitors. Ionophores had no effect on dechlorination, whereas the ATP-synthase inhibitor *N,N′*-dicyclohexylcarbodiimide (DCCD) was strongly inhibitory (Jayachandran et al. 2004).

9.3 Polychlorinated Biphenyls

9.3.1 Aerobic Degradation

Although this is discussed in some detail, both on account of the importance of the degradation of PCBs, which are widely distributed contaminants and since it illustrates a number of important principles, it is, however, a synopsis since reviews have been given elsewhere.

9.3.1.1 Overview

The degradation of PCBs has attracted enormous attention. Impressive investigations have been directed to the Gram-negative *Burkholderia xenovorans* strain LB400 and *Pandorea pnomusa* (*Comamonas testosteroni*) strain B-356, and the Gram-positive *Rhodococcus jostii* strain RHA1 (Seto et al. 1995) and *Paenibacillus* sp. strain KBC101 (Sakai et al. 2005) on account of the wide spectrum of PCBs that is oxidized. There are, however, a number of factors that determine PCB biodegradability that include: (a) both the number and position of the chlorine substituents, (b) the positions at which dioxygenation occurs, (c) whether or not the initially formed dihydrodiol can be dehydrogenated, and (d) the degradability of the metabolites produced by ring fission. Considerable effort has been made to enlarge the range of strains that are susceptible to degradation by genetic manipulation.

Since PCBs are lipophilic contaminants, a range of biota including fish, fish-eating birds, and mammals, can accumulate PCBs. In contrast to the reactions described in this section, PCBs are metabolized by the formation of sulfones in mammals (Letcher et al. 1998), or by monooxygenation with the production of hydroxylated PCBs in fish (Buckman et al. 2006), and cultures of *Nicotiana tabacum* (Rezek et al. 2008) and *Solanum nigrum* (Rezek et al. 2007). The anaerobic dechlorination of these hydroxylated PCB metabolites has been described (Wiegel et al. 1999) in the context of the degradation of halogenated phenols.

9.3.1.2 The Range of Organisms

In this section, the strains that have most frequently been explored are given to minimize confusion, since there have been substantial changes in the nomenclature of some organisms, for example:

Acinetobacter sp. strain P6 → *Rhodococcus globurulus* strain P6 (Asturias and Timmis 1993).

Comamonas (*Pseudomonas*) *testosteroni* B-356 → *Pandoraea pnomenusa* B-356 (Gómez-Gil et al. 2007).

Pseudomonas sp. strain LB400 → *Burkholderia xenovorans* strain LB400 (Goris et al. 2004).

Pseudomonas pseudoalcaligenes KF707 → *Pseudomonas paucimobilis* KF707 (Taira et al. 1992).

A number of organisms able to degrade PCB congeners has been isolated, generally from biphenyl enrichments. These include the following—giving the names used on their isolation:

1. Gram-negative strains—*Alcaligenes eutrophus* strain H850 (Bedard et al. 1987a), *Pseudomonas pseudoalcaligenes* strain KF707 (Furukawa and Miyazaki 1986), *Pseudomonas* sp. strain LB400 (Bopp 1986), *Pseudomonas testosteroni* B-356 (Ahmad et al. 1991), and *Sphingomonas aromaticivorans* F199 (Romine et al. 1999).

2. Gram-positive strains—*Rhodococcus globurulus* strain P6 (Furukawa et al 1979), *Rhodococcus erythropolis* strain TA421 (Maeda et al. 1995), *Rhodococcus jostii* strain RHA1 (Seto et al. 1995), *Bacillus* sp. strain JF8 (Shimura et al. 1999), and *Paenibacillus* sp. strain KBC101 (Sakai et al. 2005).

9.3.1.3 The Diversity of Activity

There are considerable differences in the oxidative activity of these organisms toward PCB congeners.

1. The versatility of *Burkholderia (Pseudomonas)* sp. strain LB400 (Arnett et al. 2000) and *Pandoraea pnomenusa* strain B-356 (Gómez-Gil et al. 2007) are rather similar, and greater than that of *Pseudomonas paucimobilis (pseudoalcaligenes)* strain KF707 (Gibson et al. 1993).

2. The range of congeners transformed by *Rhodococcus jostii* strain RHA1 differs from that of *Burkholderia* sp. strain LB400 and *Pseudomonas paucimobilis* strain KF707 (Seto et al. 1995).

3. *Paenibacillus* sp. strain KBC101 was able not only to degrade a wide range of highly chlorinated PCB congeners but was unusual in degrading the planar 3,4,3′,4′congener.

4. The spectrum of PCB congeners degraded, and the preference for specific substituent orientation differs: strain LB400 has a general preference for *ortho*-substituted congeners, KF 707 for *para* congeners, and B-356 for *meta* congeners.

5. The degradation of *ortho–ortho*-substituted congeners seems to be unusual, with a few exceptions:

   a. An organism SK-4 tentatively assigned to *Alcaligenes* sp. was able to grow at the expense of 2,2′-dichlorobiphenyl with the production of 2-chlorobenzoate and chloride, although it was unable to grow with 2,2′,5- or 2,2′,3-trichlorobiphenyl (Kim and Picardal 2001).

   b. The Gram-positive *Microbacterium* sp. strain B51 was able to oxidize 2,2′- and 4,4′-dichlorobiphenyl, and 2,2′4- and 2,4,4′-trichlorobiphenyls (Rybkina et al. 2003).

9.3.1.4 Outline of Pathway for Degradation

The pathway for degradation of PCBs follows the pattern used for biphenyl (and ethylbenzene)—ring dioxygenation, dehydrogenation, and ring fission encoded by *bphA*, *bphB*, *bphC* (Furukawa and Miyazaki 1986; Ahmad et al. 1991; Taira et al. 1992). This is followed by the activity of enzymes for the degradation of the ring-fission products encoded by *bphD*, *bphE*, *bphF*, *and bphG* that finally produce pyruvate and acetaldehyde.

9.3.1.5 The Multiplicity of Enzymes

Gram-positive strains including *Rhodococcus jostii* strain RHA1 and *Rhodococcus globerulus* strain P6 have several enzymes that are involved in dioxygenation followed by fission of the dihydroxybiphenyls that are produced after dehydrogenation.

1. *Rhodococcus* sp. strain RH1 is able to degrade congeners with one to eight chlorine substituents during cometabolism with biphenyl or ethylbenzene (Seto et al. 1995). The ring-hydroxylation system is complex in this organism, and consists of the following isozymes (Iwasaki et al. 2006) all of which (Table 9.2) are required for activity.

   Further examination revealed the presence of a wide range of isozymes involved in the biphenyl pathway in this strain (Gonçalves et al. 2006).

TABLE 9.2

Isozymes of *Rhodococcus* sp. Strain RH1

Enzyme	Genes		
Dioxygenase, large subunit	*bphA1*	*ebdA1*	*etbA1*
Dioxygenase, small subunit	*bphA2*	*ebdA2*	*etbA2*
Ferredoxin	*bphA3*	*ebdA3*	*etbA3*
Ferredoxin reductase	*bphA4*	*ebdA4*	*etbA4*

2. In *Rhodococcus globerulus* strain P6 there are three 2,3-dihydroxybiphenyl 1,2-extradiol dioxygenases BphC1, BphC2, and BphC3 with different specificities (McKay et al. 2003).

9.3.1.6 Degradation Is Initiated by Dioxygenation

Dioxygenation generally takes place at the 2,3-position, or less commonly, at the 3,4-position, to produce *cis*-2,3-dihydrodiols or *cis*-3,4-dihydrodiols. For example, a strain of *Pseudomonas* sp. that was able to degrade a range of PCB congeners had both 2,3-dioxygenase and 3,4-dioxygenase activity, and four of the open-reading frames were homologous to components of toluene dioxygenase (Erickson and Mondelo 1992). It was not resolved whether a single dioxygenase was able to introduce oxygen at the 2,3- or the 3,4-positions, or whether there were two different enzymes. The dioxygenase has been purified and characterized from several organisms including *Burkholderia* (*Pseudomonas*) sp. strain LB400 (Haddock and Gibson 1995) and *Comomonas testosteroni* strain B-356 (Hurtubise et al. 1995). Dioxygenation required three components—a [2Fe–2S] oxygenase ISP_{BDH}, a [2Fe–2S] ferredoxin (FER_{BPH}), and a flavin reductase (FER_{BPH}). These are encoded by *bphA* (α-subunit), *bphE* (β-subunit), *bphF*, and *bphG*. The iron–sulfur protein consists of a large subunit (α) with a molecular mass of 51–52 kDa and a small subunit (β) with a molecular mass of 22–27 kDa.

Whole-cell studies using oxygen uptake by *Burkholderia* (*Pseudomonas*) sp. strain LB400 had clearly shown the versatility of this organism in the degradation of PCB congeners containing up to four chlorine substituents. The biphenyl 2,3-dioxygenase from this strain has been purified and revealed additional mechanistic details (Haddock and Gibson 1995):

1. The dihydrodiol was produced from all congeners, and as an alternative to dehydrogenation, an *ortho*-chlorine substituent could be eliminated from the 2,3-dihydrodiol to produce the catechol.
2. 4,5-Dihydrodiols were produced from 3-3 (3,3′), (25-2) 2,2′,5, and the (25–25) 2,2′,5,5′congeners, although these are not substrates for the dihydrodiol dehydrogenase.

These results are consistent with the results accumulated over many years, and the general operation of a single dioxygenase enzyme.

9.3.1.7 Specificity

Several factors determine the specificity of strains for the degradation of PCBs, and extensive efforts have been made to increase the range of strains that can be degraded. The specificity reflects differences in the protein sequences in two regions (III and IV) of the BphA large α-subunit of the dioxygenase (Kimura et al. 1997; Mondello et al. 1997), and this offers the possibility of expanding the range of degradable PCBs by alterations of specific amino acids. The purified dioxygenase from *Burkholderia* sp. strain LB400—which is one of the most versatile—has been examined in the presence of electron-transport proteins and cofactors for its reactivity toward congeners (Arnett et al. 2000). Reactivity depended both on the number and position of substituents, and some congeners were recalcitrant, including (24-4) 2,4,4′-trichloro-, (24-24) 2,4,2′,4′-tetrachloro-, and (345-25) 3,4,5,2′5′-pentachlorobiphenyl. Substantial effort has, therefore, been directed to elucidating determinants of the specificity for degradation, and to procedures for increasing the range of substrates that can be accepted. The *ortho*-substituted congeners are particularly recalcitrant.

1. It has been suggested that the substrate specificity of the dioxygenase is determined by the α-subunit of the dioxygenase. In contrast, the substrate specificities of the four chimeras constructed from the respective α- and β-subunits of the terminal dioxygenase ISP_{BPH} of *Pseudomonas* sp. strain LB400 and *Comamonas* (*Pseudomonas*) *testosteroni* strain B-356 were dependent on the presence of both proteins (Hurtubise et al. 1998). In addition, the catalytic activity of hybrid dioxygenases comprising α- and β-subunits from distant biphenyl dioxygenases is not determined specifically by one or other of the subunits (Chebrou et al. 1999). There is, therefore, a complex dependency on the presence of both subunits.

2. Shuffling the genes in a fragment of the *bphA* gene in *Burkholderia* sp. strain LB400, *Comamonas testosteroni* strain B-356, and *Rhodococcus globerulus* strain P6 resulted in variants that had high activity toward the generally persistent 2,6-dichloro- and 2,4,4′-trichlorobiphenyls (Barriault et al. 2002).

3. Site-directed mutants of amino acids that coordinate the catalytic iron center of *Pseudomonas pseudoalcaligenes* strain KF707 were produced and expressed in *E. coli*. One mutant was able to degrade (25-25) 2,5,2′,5′tetrachlorobiphenyl by 3,4-dioxygenation and displayed 2,3- and 3,4-dioxygenase activities for 2,5,2′- and 2,5,4′-trichlorobiphenyls (Suenaga et al. 2002).

4. Efforts have been made to increase the range of substrates accepted by the dioxygenase by mutagenesis of multiple sites in region III (Barriault and Sylvestre 2004). The changes were in the positions at which dioxygenation took place:

   a   2-2 (2,2′)-Dichloro from the 2,3-position with loss of chloride, to the 3,4-dihydrodiol that is a terminal metabolite.

   b. 23-23 (2,2′,3,3′)-Tetrachloro to the 4,5-dihydrodiol without dechlorination.

   c. 2,2′,5,5′-Tetrachloro to the 3,4-dihydrodiol without dechlorination.

5. A suggestion has been made to improve the spectrum of PCBs that are degraded by *Pseudomonas* sp. strain KKS102 by altering the promoter of the *bph* operon (Ohtsubo et al. 2003). By this means, strains were obtained that showed enhanced degradation of tri-, tetra-, and pentachlorobiphenyl.

The next step in degradation is generally dehydrogenation that results in the production of the dihydroxybiphenyl, which is the substrate for ring fission. Alternatively, in the degradation of 2,4′- and 4,4′-dichlorobiphenyl by *Pseudomonas testosteroni* strain B-356 involving 3,4-dioxygenation, this is accomplished by loss of chloride (Ahmad et al. 1991), as in *Burkholderia* (*Pseudomonas*) sp. strain LB400 with congeners carrying *ortho*-chlorine substituents (Haddock and Gibson 1995; Seeger et al. 1999).

9.3.1.8 Ring Fission by 2,3-Dihydroxybiphenyl Dioxygenase

Degradation of the initially formed 2,3-dihydroxybiphenyl is carried out by extradiol fission, and 2,3-dihydroxybiphenyl 1,2-dioxygenase has been purified and characterized from several organisms, including *Pseudomonas pseudoalcaligenes* strain KF707 (Furukawa and Arimura 1987), *Pseudomonas paucimobilis* strain Q1 (Taira et al. 1988), *Burkholderia* (*Pseudomonas*) sp. strain LB400 (Haddock and Gibson 1995), and the Gram-positive *Rhodococcus* sp. strain RHA1 (Hauschild et al. 1996). In *Rhodococcus globerulus* strain P6 (*Acinetobacter* sp. strain P6), there are several nonhomologous 2,3-dihydroxybiphenyl-1,2-dioxygenases with a narrow substrate specificity (Asturias and Timmis 1993), and one of them encoded by *bphC2* was appreciably different from other extradiol dioxygenases (Asturias et al. 1994). The enzyme from *Bacillus* sp. strain JF8 is unusual in being Mn(II)-dependent and differs in structure from those of *Burkholderia* sp. strain LB400 and *Pseudomonas paucimobilis* strain KF707 (Hatta et al. 2003). Detailed investigations have revealed a number of important conclusions that affect the degradability of PCB congeners:

1. In *Burkholderia* sp. strain LB400, which is one of the most versatile, dioxygenation at C-2 and C-3 is preferred, and hydroxylation at atoms bearing chlorine substituents is restricted to the *ortho* positions with elimination of chloride in place of dehydrogenation (Seeger et al. 1999). The preferred site also depends, however, on the substitution pattern of the other ring, and 2′- and 2′,6′-dichlorobiphenyls seriously inhibit the activity of dihydroxybiphenyl dioxygenase (Dai et al. 2002). It had been suggested that substrate inhibition of the dioxygenase was a result of enzyme inactivation, and a general mechanism for inhibition of the dioxygenase by catechols involving formation of the complex between the substrate and the Fe(II) enzyme has been suggested (Vaillancourt et al. 2002). This has been explored further (Fortin et al. 2005) using extradiol dioxygenases from organisms with established ability to degrade PCBs—*Burkholderia* sp. strain LB400, two isoenzymes from *Rhodococcus globerulus* strain P6, and *Sphingomonas* sp. strain RW1. Rates of dioxygenation were measured from the rate of formation of the fission

product 2-hydroxy-6-oxo-6-phenylhexa-2,4-dienoate. Important conclusions emerged: (i) rates for the monochlorinated isomers were lower than those for the nonchlorinated 2,3-dihydrodi-hydroxybenzoate and (ii) among polychlorinated congeners, the rate for the recalcitrant 2′,6′dichloro-dihydroxybiphenyl was the lowest.

2. The next step in degradation involves hydrolysis of 2-hydroxy-6-oxo-6-phenylhexa-2,4-dienoate to benzoate and 2-hydroxy-penta-2,4-dienoate encoded by the hydrolase BphD. A study of the hydrolase in *Burkholderia cepacia* LB 400 has revealed significant details (Seah et al. 2000) (Figure 9.14a):

   a. There are substantial differences in the rates of hydrolysis by the hydrolase.

   b. The 4-chloro compound is hydrolyzed nonenzymatically to the 4-hydroxy compound, which may undergo cleavage to acetophenone and 3-oxoadipate (Figure 9.14b).

   c. The 3-chloro compound is stable to enzymatic hydrolysis and strongly inhibits the hydrolase.

3. The hydrolase is competitively inhibited by 3-chloro- and 4-chloro-2-hydroxy-6-oxo-6-phenyl-hexa-2,4-dienoate (Seah et al. 2000), and a cardinal role has been assigned to a glutathione *S*-transferase BphK that had not hitherto been assigned a specific function. The *bphK* gene does not occur in all bacteria that are recognized as being able to degrade PCBs, and it has been shown in *Burkholderia (Pseudomonas) xenovorans* strain LB400 that the transferase has the ability to dechlorinate 3-chloro-, 5-chloro-, and 3,9,11-trichloro-2-hydroxy-6-oxo-6-phenyl-hexa-2,4-dienoate in reactions involving addition of GS⁻, elimination of Cl⁻, followed by gen-eration of GSSG (Fortin et al. 2006). These observations contribute to understanding the relative recalcitrance of PCB congeners, rationalize the formation of the metabolites, and focus attention on the formation of the 3- and 4-chlorohexadienoates that would be produced from 3- and 4-chlorinated biphenyls by 2-hydroxy-6-oxo-6-phenylhexa-2,4-dienoate hydrolases.

Collectively, these conclusions are supported by the results of a study of 33 congeners using *Acinetobacter* sp. strain P6 (*Rhodococcus globerulus*) (Asturias and Timmis 1993). The extradiol fission

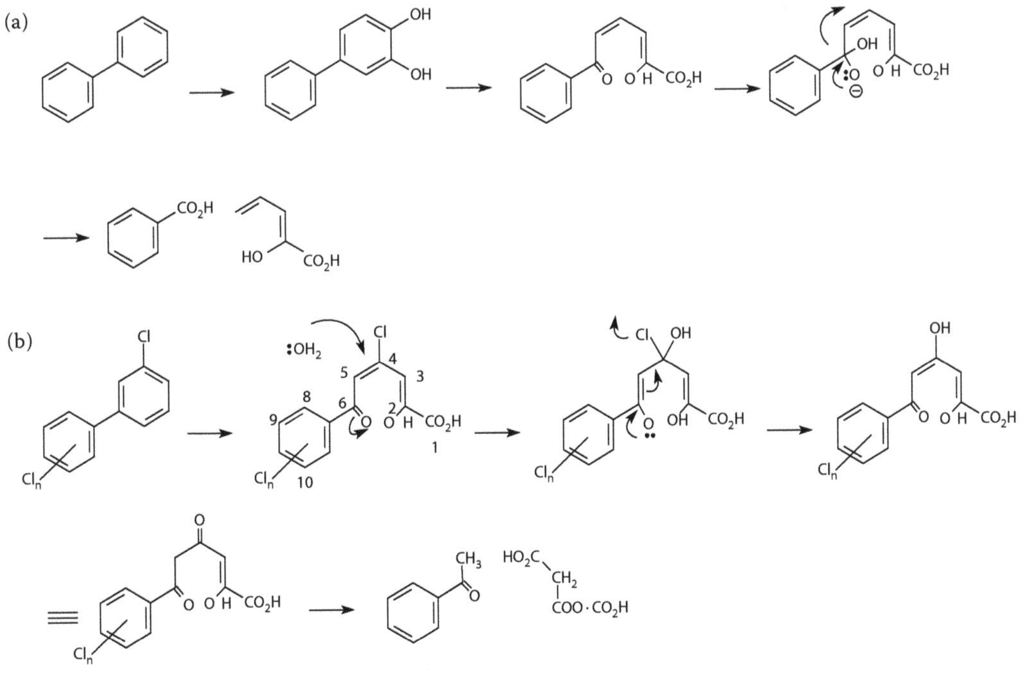

FIGURE 9.14 Aerobic degradation of (a) biphenyl by biphenyl-2,3-dioxygenase and (b) a polychlorinated 4′-chlorobiphenyl.

product greatly exceeded the chlorobenzoates for the 2,5,4'- and 2,4,4'-trichlorobiphenyls (Furukawa et al. 1979). In this strain, however, although the specificities of the hydrolases in *Burkholderia cepacia* LB 400 and *Rhodococcus globerulus* P6 were essentially similar, significant differences have been observed (Seah et al. 2001). In the former, inhibition by 3-chlorohexadienoate exceeded that of 4-chlorohexadienoate, whereas the opposite was found in the latter.

9.3.1.9 Metabolites

9.3.1.9.1 Chlorobenzoates and Catechols

1. Total degradation of PCBs necessitates degradation of the chlorobenzoates produced by the foregoing reactions. It has been suggested that the inability of strains to degrade chlorobenzoates produced from some PCB congeners may be related to the restricted metabolism of chlorobenzoate fission products (Hernandez et al. 1995). The distribution of chlorobenzoates has indeed been suggested as biomarkers for the aerobic degradation of PCBs (Flanagan and May 1993). The inhibitory nature of chlorocatechols that are metabolites of chlorobenzoates has already been noted.

2. Another metabolic limitation involves the formation of protoanemonin (4-methylenebut-2-ene-4-olide) as an intermediate in the degradation of 4-chlorobenzoate formed by partial degradation of 4-chlorobiphenyl. This is formed by the intradiol fission of 4-chlorocatechol to 3-chloro-*cis,cis*-muconate followed by loss of CO_2 and chloride (Blasco et al. 1995) (Figure 9.15). The synthesis of this metabolite adversely affects the survival of organisms that metabolize 4-chlorobiphenyl in soil microcosms, although its formation can be obviated by organisms using a modified pathway that produces the *cis*-dienelactone (Blasco et al. 1997).

9.3.1.10 Acetophenones

Quite different metabolites may also be formed. Although the ultimate products from the degradation of PCBs are generally chlorinated muconic acids, the unusual metabolite 2,4,5-trichloroacetophenone has been isolated (Bedard et al. 1987b) from the degradation of 2,4,5,2',4',5'-hexachlorobiphenyl (Figure 9.16) by *Alcaligenes eutrophus* H850, which has an unusually wide spectrum of degradative activity for PCB congeners. The mechanism for the formation of 2,4,5-trichloroacetophenone may be analogous to the formation of acetophenones by nonenzymatic hydrolysis of the 4'-chlorinated compound. The metabolism of chloroacetophenones takes place by monooxygenation to the corresponding chlorophenylesters. Some *ortho*-substituted chlorophenols are, however, inhibitory, and only low rates of oxidation have been encountered with di- and trichlorinated acetophenones, so that growth with them is not possible (Higson and Focht 1990). However, degradation of 4-chloroacetophenone has been demonstrated with a mixed culture of an *Arthrobacter* sp. and a *Micrococcus* sp. (Havel and Reineke 1993).

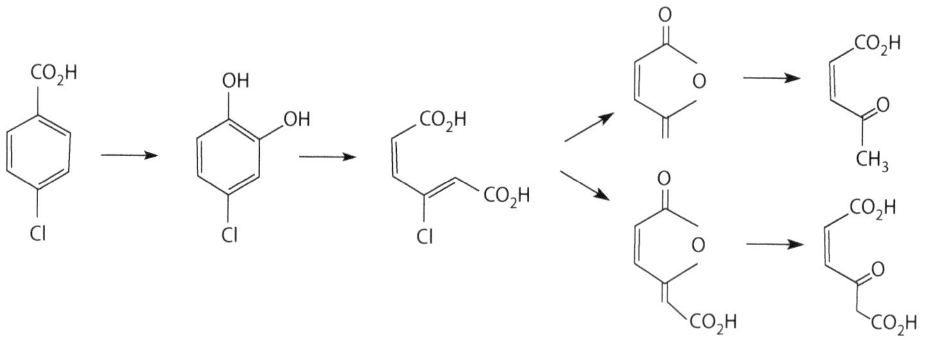

FIGURE 9.15   Formation of protoanemonin from 4-chlorobenzoate.

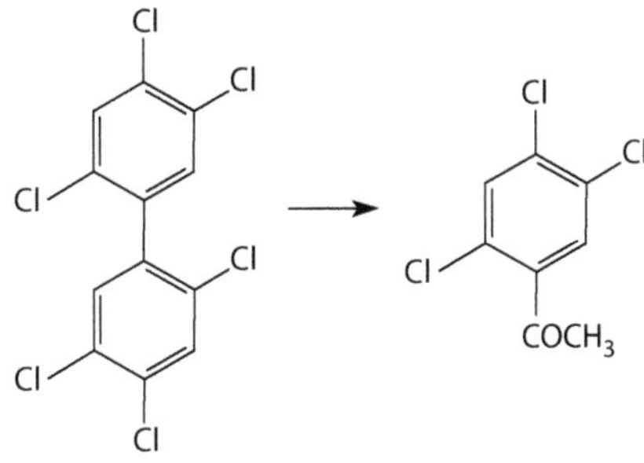

FIGURE 9.16 Biodegradation of 2,4,5,2′,4′,5′-hexachlorobiphenyl.

9.3.1.11 The NIH Shift

A rearrangement (NIH shift) occurred during the transformation of 2-chlorobiphenyl to 2-hydroxy-3-chlorobiphenyl by a methanotroph, and is consistent with the formation of an intermediate arene oxide (Adriaens 1994). The occurrence of such intermediates also offers plausible mechanisms for the formation of nitro-containing metabolites that have been observed in the degradation of 4-chlorobiphenyl in the presence of nitrate (Sylvestre et al. 1982).

9.3.1.12 Induction of PCB Metabolism

Apart from induction of degradative enzymes by growth with biphenyl, nonrelated naturally occurring substrates have been shown to induce the enzymes for PCB degradation.

1. A range of related compounds was examined (Donnelly et al. 1994) for their capacity to support the growth of *Alcaligenes eutrophus* H850, *Pseudomonas putida* LB400, and *Corynebacterium* sp. MB1. For strains H850 and MB1, growth with biphenyl equaled that using a wide range of substrates including naringin, catechin, and myricitin (Figure 9.17). These results suggest that natural plant metabolites are able to mediate the growth of PCB-degrading organisms. In addition, the pattern of metabolism of PCB congeners was identical using biphenyl or the naturally occurring plant metabolites.
2. Cells of *Arthrobacter* sp. strain B1B were grown in a mineral medium with fructose and carvone (50 mg/L). Effective degradation of a number of congeners in Arochlor 1242 was induced by carvone that could not, however, be used as a growth substrate, and was toxic at high concentrations (>500 mg/L). Other structurally related compounds including limonene, *p*-cymene, and isoprene were also effective (Gilbert and Crowley 1997).

There are other changes that are involved in the metabolism of biphenyls and PCBs. A study of the degradation of biphenyl by *Burkholderia xenovorans* LB400 showed that elevated levels of the enzymes involved in the upper Bph pathway were induced by growth with biphenyl, whereas those for benzoate degradation were upregulated by growth on benzoate. Enzymes for the latter involved conversion to benzoyl-CoA and use of the pathway delineated for *Azoarcus evansii* (Denef et al. 2004). Accumulation of inorganic polyphosphate is induced in cells exposed to stress, and has been observed during the growth of *Pseudomonas* sp. strain B4 in a defined medium with biphenyl, and on changing the growth substrate from glucose to biphenyl (Chávez et al. 2004).

9.3.2 Fungal Dehalogenation

Dehalogenation of commercial PCB mixtures has been observed using *Phanerochaete chrysosporium* in different media including low nitrogen, high nitrogen, and complex malt extract (Yadav et al. 1995). Dechlorination of congeners with *ortho*, *meta*, and *para* substituents occurred. Maximal dechlorination

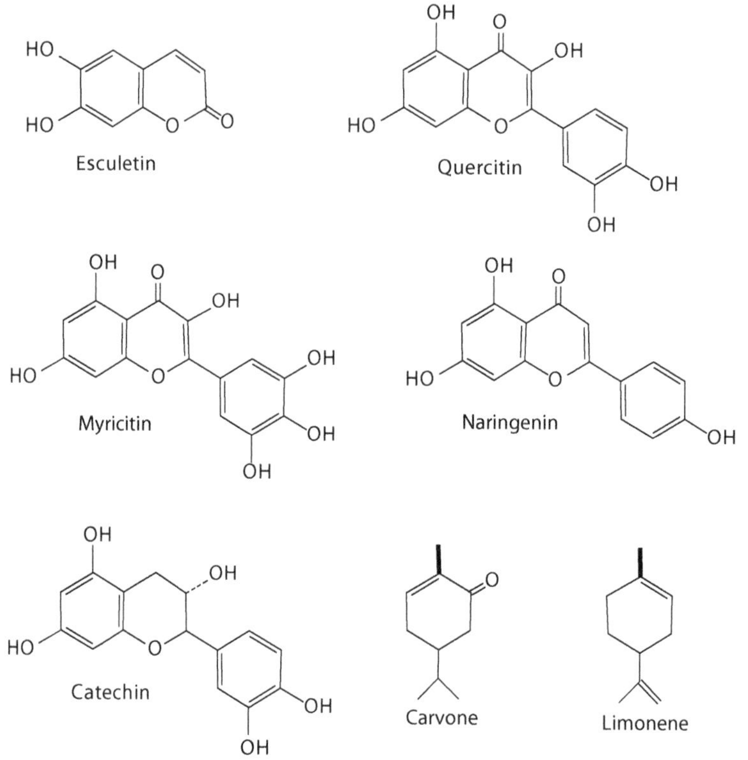

FIGURE 9.17 Natural substrates or inducers.

ranged from 61% for Arochlor 1242 to 18% for Arochlor 1260, and although it was greatest in malt extract medium, it also occurred in low- and high-nitrogen media.

9.3.3 Reductive Dehalogenation

The anaerobic dechlorination of PCBs has been extensively studied both in laboratory microcosms and in field samples from heavily contaminated sites in the United States. A complex pattern of dechlorination has been described that depends on the site and environmental factors such as concentration of organic carbon and sulfate (Bedard and Quensen 1995). In a study where dechlorination was stimulated by addition of brominated biphenyls (Bedard et al. 1998; Wu et al. 1999), three main patterns were found—**N** that removed flanked meta chlorines, **P** that removed para chlorines, and **LP** that removed unflanked para chlorines. The example of the 2,2′,3,4,4′,5,5′-heptachloro (2345-245) congener is given in Figure 9.18. For the heptachloro congener 2,2′,3,3′,4,4′,5 (2345-234), a combination of processes **N**, **P**, and **LP** resulted in the production of the 2,2′-dichloro- and 2,2′,5-trichloro congeners (Bedard et al. 2005). In contrast, *ortho* chlorines were more recalcitrant. An important study using sediment contaminated with Arochlor 1260 was initiated by priming with 2,6-dibromobiphenyl to select for **N** dechlorination. Successive dilution produced sediment-free cultures that grew with acetate, butyrate or pyruvate, and H₂ and resulted in a range of bacteria. These were able to dechlorinate hexa- through nonachlorinated congeners to tri- through pentachlorinated congeners, and dechlorination of heptachloro congeners produced the 2,2′,4′,5 (25-24), 2,2′,4,6′ (24-26) and 2,2′,5,6′(25-26) tetrachloro congeners with retention of the *ortho* substituents (Bedard et al. 2006).

It is important to appreciate important factors that affect the relative effectiveness of dechlorination: (a) temperature (Wu et al. 1997b) that is discussed in greater detail below and (b) the nature of the PCB congener that primes the dehalogenation (Wu et al. 1997a). In contrast to the results discussed above, an enrichment culture supplemented with acetate was able to dechlorinate 2,3,5,6-tetrachlorobiphenyl to

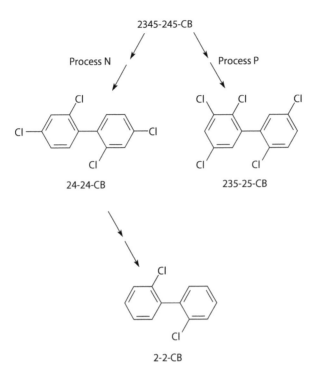

FIGURE 9.18 Anaerobic dechlorination of 2,2′,3,4,4′,5,5′-heptachlorobiphenyl. (Redrawn from Bedard DL, HM van Dort, KA Deweerd. 1998. *Appl Environ Microbiol* 64: 1786–1795.)

3,5-dichlorobiphenyl with the removal of *ortho* chlorines (Holoman et al. 1998), and success is being achieved in defining the organisms responsible. For example, a nonmethanogenic mixed culture containing strain *o*-17 was able to bring about dechlorination of *ortho*-halogenated congers 2,3,5,6-tetrachlorobiphenyl to 3,5-dichlorobiphenyl (May et al. 2006).

Substantial attention has been given to organisms within the *Dehalococcoides* group and the green nonsulfur phylum *Chloroflexi* (Fagervold et al. 2005). Examples include the following:

1. Laboratory studies using enriched cultures specifically dechlorinated PCBs with doubly flanked chlorines (Wu et al. 2002).
2. Cultures containing *Dehalococcoides ethenogenes* carried out sequential dechlorination of 2,3,4,5,6-pentachlorobiphenyl to 2,4,6-trichlorobiphenyl (Fennell et al. 2004).
3. It has been shown that individual species of *Chloroflexi* have different specificities (Fagervold et al. 2005).
4. The anaerobic dechlorination of PCB congeners has been examined using *Dehalococcoides* sp. strain CBDB 1 (Adrian et al. 2009) that was able to grow using highly chlorinated benzenes as electron acceptors (Jayachandran et al. 2003). This strain was able to bring about dechlorination in many of the congeners in Arochlor 1260 including penta- (IUPAC 101), hexa- UPAC, hepta- (IUPAC 170, 174, 177, 183), and octa- (IUPAC 194, 199) congeners. Table 9.3 illustrates the patterns of dechlorination that are consistent with the pattern designated **H** for the dechlorination of Arochlor 1254 in sediments (Brown and Wagner 1990).

9.3.4 The Role of Temperature

This is an important parameter particularly for naturally occurring mixed cultures of organisms in the natural environment: temperature may result in important changes in the compositionof the microbial

TABLE 9.3

Anaerobic Dechlorination of PCB Congeners by *Dehalococcoides* sp. CBDB1

	Congener	Product
Heptachloro	2345-245	235-25 + 25-25
	2345-234	235-24 + 235-23/25-24 > 25-23
Hexachloro	234-245	24-25 > 23-25
	234-234	23-24 > 24-24 > 23-23
	245-245	25-25

Source: Adapted from Adrian L et al. 2009. *Appl Environ Microbiol* 75: 4516–45524.

flora as well as on the rates for different processes. An illustrative exampleof its importance includes the following. An anaerobic sediment sample was incubated with 2,3,4, 6-tetrachlorobiphenyl at various temperatures between 4°C and 66°C (Wu et al. 1997a,b). The main products were 2,4,6- and 2,3,6-trichlorobiphenyl and 2,6-dichlorobiphenyl; the first was produced maximally and discontinuously at 12°C and 34°C, the second maximally at 18°C, and the third was dominant from 25°C to 30°C. Dechlorination was not observed above 37°C.

Collectively, there is, therefore, extensive evidence for the anaerobic dechlorination of PCBs but the extent and specificity of this depends on both the organism and the degree and pattern of substituents in the congener.

9.4 Polybrominated Biphenyls and Diphenylmethanes

Mixed cultures of organisms that were isolated from sediments contaminated with PCBs and polybrominated biphenyls (PBBs) were shown to debrominate PBBs under anaerobic conditions (Morris et al. 1992), and the dominant congener—2,2′,4,4′,5,5′(245-245)-hexabromobiphenyl—could be successively debrominated to 2,2′-dibromobiphenyl. However, in sediments from the most heavily contaminated site containing contaminants in addition to PBBs, very little debromination occurred and the recalcitrance was attributed to the toxicity of the other contaminants (Morris et al. 1993).

Ingenious experiments to which reference has already been made have used addition of specific PCB congeners that are more readily dechlorinated to "prime" dechlorination at specific positions (Bedard and Quensen 1995). They have been extended to the use of dibrominated biphenyls in the presence of malate to stimulate dechlorination of the hexachloro- to nanochlorobiphenyls (Bedard and Quensen 1995). Results from these experiments provide valuable evidence of important differences between anaerobic dechlorination and anaerobic debromination, and the greater facility of the latter.

The use of brominated biphenyls to induce dechlorination of highly chlorinated biphenyls has been examined in detail. Di- and tribromobiphenyls were the most effective in dechlorinating the PCBs including the heptachloro, hexachloro, and pentachloro congeners, and were themselves reduced to biphenyl (Bedard et al. 1998). In addition, 2,6-dibromobiphenyl stimulated the growth of anaerobes that effectively dechlorinated hexa-, hepta-, octa-, and nanochlorobiphenyls over the temperature range 8–30°C (Wu et al. 1999). In anaerobic sediment microcosms, a range of tribrominated biphenyls was successively debrominated to dibromo- and monobromo compounds before complete debromination to biphenyl. Of particular interest, both the 2,2′,4,5 (25-24) and 2,2′,5,5′ (25-25) tetrabromo congeners were debrominated to the 2,2′-dibromo congener that was slowly and completely debrominated to biphenyl within 54 weeks (Bedard and van Dort 1998). The pathways for debromination of 2,2′,4,5′- and 2,2′,5,5′-tetrabromobiphenyls are shown in Figure 9.19.

In comparable microcosm experiments, a number of important features were observed:

1. All the tribrominated congeners were debrominated to products including biphenyl in acclimation times between <1 and 7 weeks.

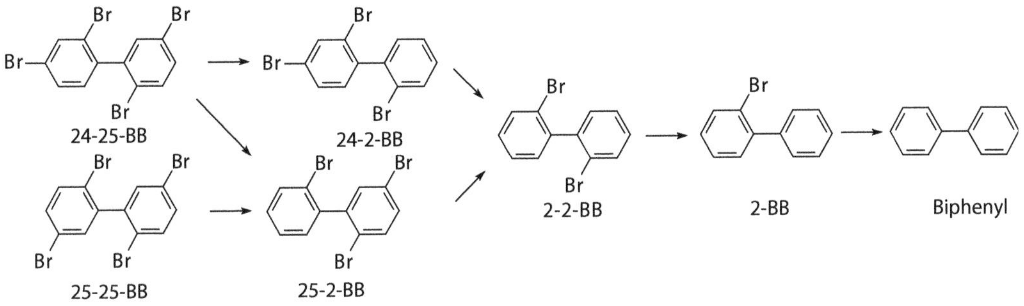

FIGURE 9.19 Anaerobic dechlorination of 2,2′,5,5′-tetrabromobiphenyl. (Redrawn from Bedard DL, HM van Dort. 1998. *Appl Environ Microbiol* 64: 940–947.)

2. For the corresponding chlorinated congeners with comparable acclimation times, the loss of only single chlorine was noted and biphenyl was not observed.

3. The mono- and dichlorinated congeners required long periods of acclimation before the onset of dechlorination.

Other compounds have also been examined as "primers" for the dechlorination of hexachloro to non-achloro PCB congeners (DeWeerd and Bedard 1999): a number of substituted brominated monocyclic aromatic compounds were examined, and 4-bromobenzoate was effective—though less so than 2,6-dibromobiphenyl. In contrast, the chlorobenzoates that are metabolites of aerobic degradation were ineffective. The positive effect of brominated biphenyls in "priming" the anaerobic dechlorination of CBs has also been encountered in the dechlorination of octachlorodibenzo[1,4]dioxin to the 2,3,7,8 congener induced by 2-bromodibenzo[1,4]dioxin in the presence of H_2 (Albrecht et al. 1999).

PBBs are, therefore, not only debrominated microbially under anaerobic conditions, but are also able to induce effective dechlorination of their chlorinated analogs. Debromination may, however, be limited in the presence of other contaminants.

The analogous dehalogenation of halogenated dibenzo[1,4]dioxins and diphenyl ethers is discussed in Chapter 10, and of halogenated dibenzo[1,4]dioxins and dibenzofurans in Chapter 11.

9.4.1 Halogenated Benzoates

Chlorobenzoates may be formed during the initial steps in the aerobic degradation of PCBs, and their further metabolism illustrates a number of pathways. There are several reactions that carry out dehalogenation including dioxygenation, hydrolysis, and reduction.

9.4.1.1 Dioxygenation

The aerobic degradation of benzoate is initiated by dioxygenation with the formation of catechol and concomitant decarboxylation. An analogous pathway is used for halogenated benzoates with *ortho* substituents, which involves dioxygenation with decarboxylation and loss of halide. For example, a broad-spectrum two-component 1,2-dioxygenase is used by *Pseudomonas cepacia* 2CBS for 2-halobenzoates, and a three-component 1,2-dioxygenase for 2-chloro- and 2,4-dichlorobenzoate by *Pseudomonas aeruginosa* strain 142 (Romanov and Hausinger 1994). The alternative 2,3-dioxygenation to give 2,3-dihydroxybenzoate with loss of chloride has been observed in a *Pseudomonas* sp. (Fetzner et al. 1989). For 3- and 4-halogenated benzoates, dioxygenation and dehydrogenation with concomitant decarboxylation without loss of halogen is generally observed. For example, 3-chlorobenzoate may produce 3-chlorocatechol by 1,2-dioxygenation or 4-chlorocatechol by 1,6-dioxygenation, and 4-chlorobenzoate would produce 4-chlorocatechol. The pathway for further metabolism of the chlorocatechols may be

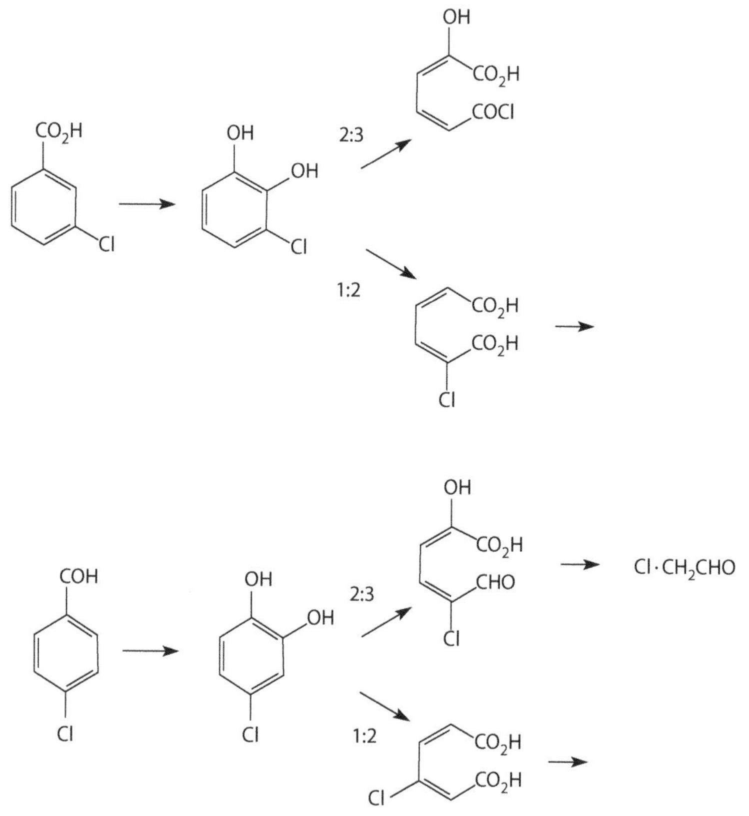

FIGURE 9.20 Ring-cleavage pathways for the biodegradation of chlorocatechols formed from 3- and 4-chlorobenzoates.

critical due to the formation of toxic metabolites by extradiol (2:3) fission—an acyl chloride from 3-chlorocatechol, or chloroacetaldehyde from 4-chlorocatechol (Figure 9.20). The "modified *ortho*" (1:2) fission may exceptionally take place, and is noted again later.

1. A two-component 2-halobenzoate 1,2-dioxygenase has been purified from *Pseudomonas cepacia* strain 2CBS that is able to metabolize 2-fluorobenzoate, 2-chlorobenzoate, 2-bromobenzoate, and 2-iodobenzoate to catechol by concomitant decarboxylation and loss of halide (Fetzner et al. 1992). The inducible 2-halobenzoate 1,2-dioxygenase consisted of two components: an oxygenase A and a reductase B. The oxygenase A is an iron–sulfur protein and consists of nonidentical subunits, α (M_r 52,000) and β (M_r 20,000) in an $\alpha 3 \beta 3$ structure. The reductase B is an iron–sulfur flavoprotein (M_r 37,500) containing FAD. It is broadly similar to benzoate dioxygenase though different from the FMN-containing dioxygenases. It has been shown that the gene cluster *cdbABC* encoding the enzyme is localized on a plasmid (Haak et al. 1995). Significant homology existed with the amino acid sequence of *benABC*, which encodes benzoate dioxygenase in *Acinetobacter calcoaceticus* and to *xylXYZ*, which encodes toluate dioxygenase in *Pseudomonas putida* mt-2.

2. The 2-halobenzoate dioxygenase of *Burkholderia* sp. strain TH2 was active toward a number of 2-halobenzoates, and the predicted amino acid sequences of all the gene products *cbdABC* were highly similar to those of *Pseudomonas* sp. strain 2CBS. It was shown that the effectors of the transcriptional regulatory gene *Cbds* were 2-chloro-, 2-bromo-, 2-iodo-, and 2-methyl-benzoate (Suzuki et al. 2001).

3. *Pseudomonas aeruginosa* strain 142 was isolated from a PCB-degrading consortium and is able to degrade a range of 2-halogenated benzoates (Figure 9.21). The enzyme consists of three

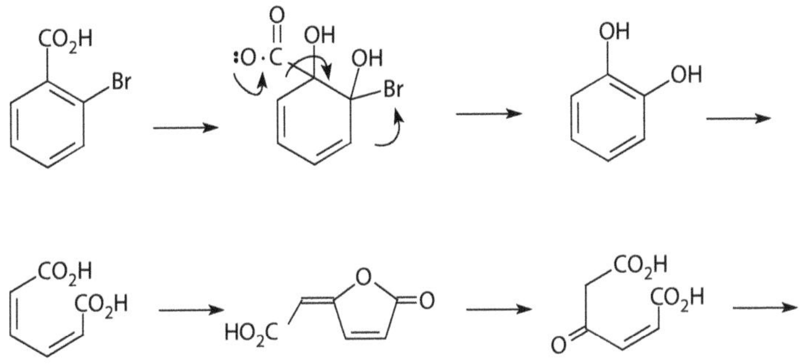

FIGURE 9.21 Degradation of 2-bromobenzoate by *Pseudomonas aeruginosa* strain 142. (Redrawn from Romanov V, RP Hausinger. 1994. *J Bacteriol* 176: 3368–3374.)

components, one of which is a ferredoxin containing a single Rieske-type [2Fe–2S] cluster (Romanov and Hausinger 1994), and the amino acid sequence is similar to those of other three-component dioxygenases containing ferredoxin such as benzene dioxygenase, toluene-2,3-dioxygenase, biphenyl dioxygenase, and naphthalene dioxygenases. This pathway is used for the degradation of both 2,3- and 2,5-dichlorobenzoate (Hickey and Focht 1990), and for a number of other 2-substituted benzoates (Romanov and Hausinger 1994).

4. *Burkholderia* sp. strain NK8 can be grown with benzoate, and both 3- and 4-chlorobenzoate. The genes encoding enzymes for the dioxygenation of benzoates (cbeABCD) and catechols (catA, catBC) have been cloned and analyzed (Francisco et al. 2001). Both 3-chlorocatechol (1:6 dioxygenation) and catechol (1:2 dioxygenation) were formed from 2-chlorobenzoate, and 4-chlorocatechol (1:2 dioxygenation) from both 3- and 4-chlorobenzoate. Degradation of both the 3- and 4-chlorocatechol was accomplished by intradiol fission. The degradation of muconates derived from 3- and 4-chlorocatechol by intradiol fission mediated by the 1,2-dioxygenase is well established (Figure 9.22a–c) and involves the activity of a sequence of enzymes: chloromuconate cycloisomerase, dienelactone hydrolase, and maleylreductase. One of the key steps in degradation occurs after ring fission, and involves maleylreductase. For maleylacetates with a halogen atom at the 2-position, reductive loss of halide takes place using 2 mol of NADH per mol of substrate, whereas only 1 mol of NADH is required for those lacking a halogen substituent (Figure 9.23) (Seibert et al. 1993; Kaschabek and Reineke 1995). For 3,5-dichlorocatechol that would be produced from 3,5-dichlorobenzoate, or 2,4-dichlorophenol, from 2,4-dichlorophenoxy-.acetate, there is, however, some ambiguity over the details of the formation of the final products (succinate and acetate).

5. *Alcaligenes* sp. strain BR6024 degraded 3-chlorobenzoate by dioxygenation to 3-chlorobenzoate-*cis*-4,5-dihydrodiol, dehydrogenation to 5-chloro-3,4-dihydroxybenzoate, and formation of pyr-2-one-2,4-dicarboxylate and 3,4-dihydroxybenzoate (Nakatsu et al. 1997). The *cbaC* gene that encodes 3-chlorobenzoate-*cis*-4,5-dihydrodiol dehydrogenase is not, however, required for the metabolism of 3,4-dichlorobenzoate when spontaneous loss of halide occurs (Nakatsu et al. 1997).

6. For 3-chlorobenzoate, an alternative pathway in *Alcaligenes* sp. strain BR60 may involve 3,4- or 4,5-dioxygenation. Ring fission of the catechols resulted in the production of pyruvate and oxalacetate (Nakatsu and Wyndham 1993).

These considerations are clearly relevant to the degradation of PCBs in which chlorobenzoates are produced during the initial dioxygenation of the upper pathway. The products from metabolism of these, such as 3-chlorocatechol, may inhibit the function of the ring-fission dioxygenase (Sondossi et al. 1992), while the synthesis of protoanemonin from 4-chlorobenzoate may inhibit the survival of organisms that degrade 4-chlorobiophenyl (Blasco et al. 1997).

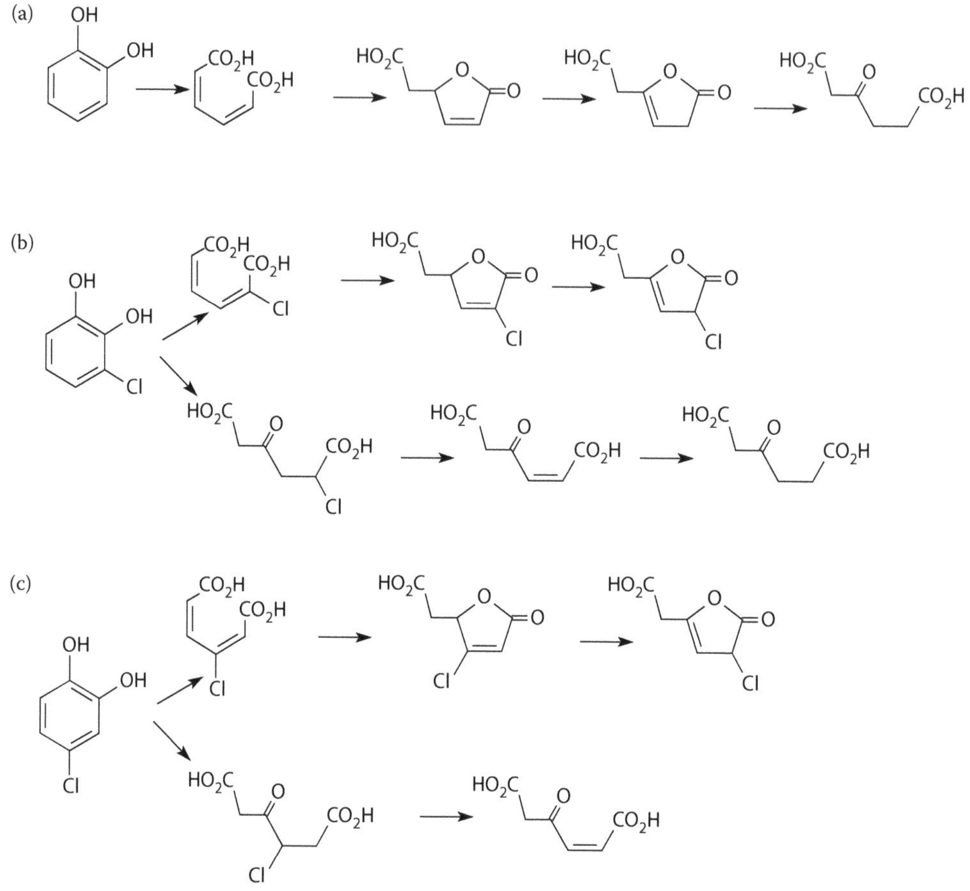

FIGURE 9.22 Degradation of (a) catechol, (b) 3-chlorocatechol, and (c) 4-chlorocatechol.

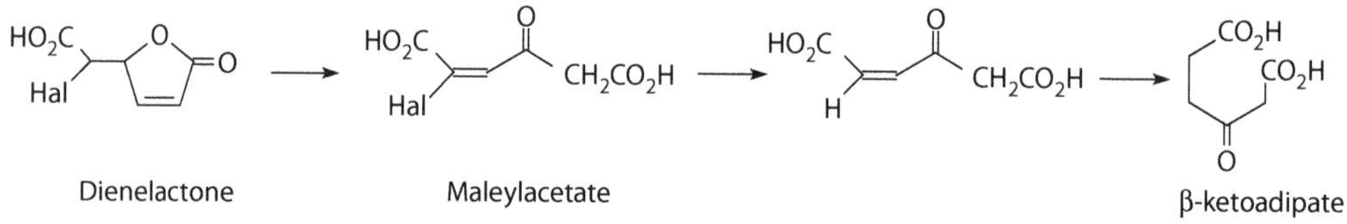

FIGURE 9.23 Degradation of halogenated dienelactones maleylacetate and reduction to 3-oxoadipate. (Redrawn from Kaschabek SR, W Reineke. 1995. *J Bacteriol* 177: 320–325.)

9.4.1.2 Hydrolytic Reactions

9.4.1.2.1 4-Halobenzoates

The metabolism of 4-halobenzoates is different from that of 2-halobenzoates since simultaneous loss of halide and the carboxyl group cannot occur. Instead, hydrolytic loss of halogen is involved as an early step in the degradation of 4-halogenated benzoate by a number of strains. It has been shown that the carboxyl group of strains with hydrolytic dehalogenating activity is generally activated by a coenzyme A ligase to form benzoyl-coenzyme A esters (Chang and Dunaway-Mariano 1996), and the substrate specificity of the 4-chlorobenzoate ligase in *Pseudomonas* sp. strain CBS3 has been described (Löffler et al. 1992). The three polypeptide components of the 4-chlorobenzoate dehalogenase have been isolated and characterized (Chang et al. 1992). Evidence for the occurrence of nucleophilic catalysis in the

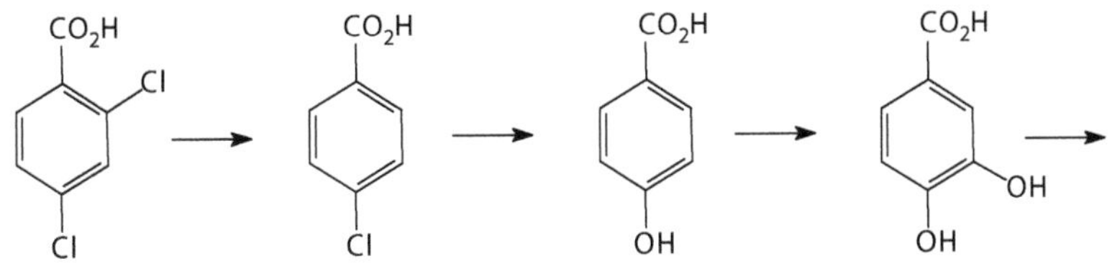

FIGURE 9.24   Reductive step in the biodegradation of 2,4-dichlorobenzoate.

dehalogenation has been provided (Yang et al. 1994), and the kinetic parameters for the analogous dehalogenation of 4-halobenzoyl-CoA by *Acinetobacter* sp. CB1 have been established (Crooks and Copley 1993).

1. Hydrolytic dehalogenation of 4-chlorobenzoate has been shown in cell extracts of *Pseudomonas* sp. strain CBS3 (Löffler et al. 1991), and the resulting 4-hydroxybenzoate is then readily degraded by hydroxylation to 3,4-dihydroxybenzoate followed by ring fission. The dehalogenase consists of three components, one with a molecular weight of 3000 that is unstable, and two stable components with molecular weights of ≈86,000 and 92,000 (Elsner et al. 1991; Chang et al. 1992).

2. A strain of *Acinetobacter* sp. 4-CB1 is able to dehalogenate 4-chlorobenzoate with the formation of 4-hydroxybenzoate after the initial formation of the 4-chlorobenzoyl CoA ester (Copley and Crooks 1992).

3. *Alcaligenes denitrificans* strain NTB-1 is able to use 4-chloro-, 4-bromo-, and 4-iodobenzoates as sole sources of carbon and energy. The pathway involves hydrolytic dehalogenation to 4-hydroxybenzoate followed by hydroxylation to 3,4-dihydroxybenzoate (van den Tweel et al. 1987).

4. A comparable pathway was used by *Arthrobacter* sp. strain TM-1, which was able to grow with 4-chlorobenzoate and 4-bromobenzoate (Marks et al. 1984).

An additional variant that involves both hydrolytic and reductive reactions has been found in the degradation of 2,4-dichlorobenzoate by *Alcaligenes denitrificans* strain NTB-1 (van den Tweel et al. 1987) (Figure 9.24), and via the CoA ester by *Corynebacterium sepedonicum* (Romanov and Hausinger 1996).

9.5 Mechanisms for the Ring Fission of Substituted Catechols

It is convenient to provide a short introduction to the issues encountered in the degradation of chlorinated substrates that produce chlorocatechols as intermediates. An outline of the ring-fission mechanisms has been given in Chapter 3, Part 1, and Chapter 8, Part 3. Three pathways for fission of 3-substituted catechols by dioxygenation are possible—1,2-intradiol (*ortho*), 2,3-extradiol (*meta*), and 1,6- (meta/distal) (Figure 9.25a–c). The rather complex situation may be summarized as follows:

1. Extradiol fission of substituted catechols is generally favored except for 3-chlorocatechol when the enzyme is inhibited by the substrate (Klecka and Gibson 1981; Bartels et al. 1984). Intradiol fission is then favorable using what has been termed a "modified *ortho*" pathway. After growth of *Pseudomonas chlororaphis* strain RW71 with 1,2,3,4-tetrachlorobenzene, the chlorocatechol 1,2-dioxygenase is able to catalyze the oxidation of a wide range of chlorocatechols (Potrawfke et al. 1998). Intradiol fission has been observed even for 4,5-dichlorocatechol, although this is not considered to be normally accepted by chlorocatechol dioxygenases (Potrawfke et al. 2001). An important implication is that the degradation of 3-methylcatechol, for which extradiol fission is preferred, may be incompatible with the simultaneous degradation of 3-alkyl and 3-halogenated catechols.

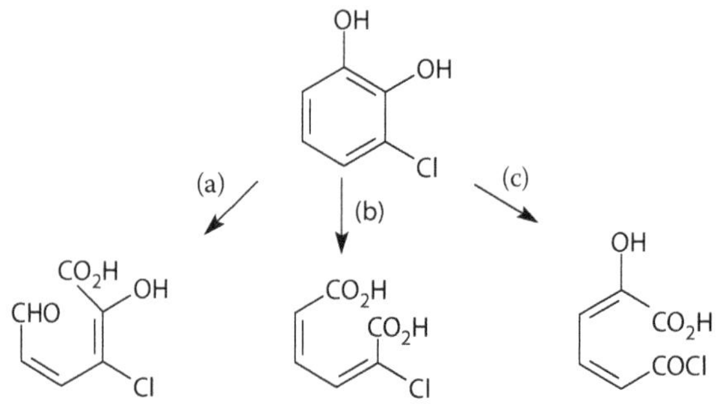

FIGURE 9.25 Ring-cleavage pathways for the biodegradation of 3-substituted catechols: (a) distal, (b) intradiol, and (c) extradiol.

2. *Pseudomonas putida* strain GJ31 degrades chlorobenzene via 3-chlorocatechol and extradiol fission (Kaschabek et al. 1998). This is accomplished by a chlorocatechol 2,3-dioxygenase that hydrolyzes the initially formed *cis,cis*-hydroxymuconacyl chloride to 2-hydroxymuconate. Thereby, the irreversible reaction of the acid chloride with nucleophiles or the formation of pyr-2-one-6-carboxylate as a terminal metabolite is avoided (Mars et al. 1997). The 3-chlorocatechol 2,3-dioxygenase has been characterized, and is clearly different from other catechol extradiol dioxygenases (Kaschabek et al. 1998; Mars et al. 1999). An analogous spontaneous hydrolysis has been observed in the product from 1:2 dioxygenation of 2-chlorohydroquinone, which is a metabolite of γ-hexachlorocyclohexane (Endo et al. 2005).

3. The alternative extradiol fission between C_1 and C_6 (distal fission) of catechol has been observed in *Azotobacter vinelandii* strain 206 (Sala-Trepat and Evans 1971). Although this is rather inefficient due to the low turnover capacity for the enzyme, it has been observed for 3-chlorocatechol in *Sphingomonas xenophaga* (Riegert et al. 1998), and its catalytic properties have been improved by random mutagenesis of *bphC1* that encodes 2,3-dihydroxybiphenyl dioxygenase (Riegert et al. 2001).

Which of the pathways is followed depends, therefore, on both the organism and the substrate that is being metabolized.

In summary, a range of pathways are available for the degradation of halogenated benzoates:

- Dioxygenation with concomitant loss of carboxyl and halogen to produce catechols (2-, 2,3-, and 2,5-halogenated benzoates).
- Dioxygenation with decarboxylation but without loss of halogen to produce chlorocatechols that may be degraded by several ring-fission pathways (3- and 4-chlorobenzoates).
- Dioxygenation with loss of only halogen to produce a dihydroxybenzoate (2-chlorobenzoate).
- Dehalogenation of the CoA ester to hydroxybenzoate (4-halobenzoates).
- Both hydrolytic and reductive elimination of halogen (2,4-dichlorobenzoate).

There is an additional problem that has important implications for the bioremediation of contaminated sites when two substrates such as a chlorinated and an alkylated aromatic compound are present. The extradiol fission pathway is generally preferred for the degradation of alkylbenzenes (Figure 9.26), although this may be incompatible with the degradation of chlorinated aromatic compounds since the 3-chlorocatechol produced inhibits the activity of the catechol-2,3-oxygenase (Klecka and Gibson 1981; Bartels et al. 1984).

This has been overcome in some strains:

1. Mutant strains have successfully reconciled this incompatibility (Taeger et al. 1988; Pettigrew et al. 1991).

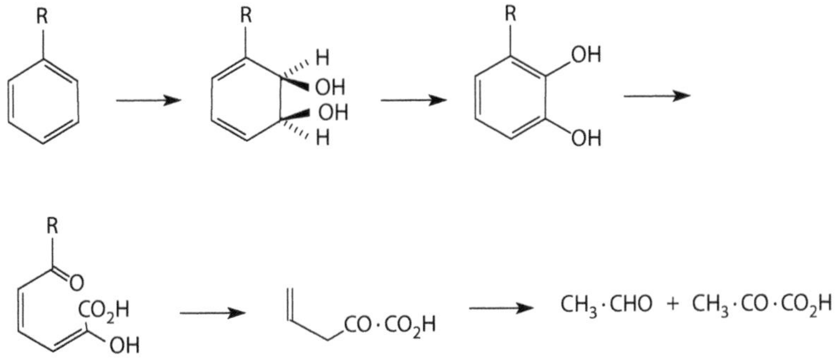

FIGURE 9.26 Biodegradation of alkylbenzenes.

2. 2,3-Dihydroxybiphenyl 1,2-dioxygenase from the naphthalene sulfonate-degrading *Sphingomonas* sp. strain BN6 metabolized 3-chlorocatechol by extradiol fission between the 1- and 6-positions (distal fission) (Riegert et al. 1998).

3. Chlorocatechol 2,3-dioxygenase from *Pseudomonas putida* GJ31 metabolized 3-chlorocatechol with concomitant elimination of chloride to form 2-hydroxymuconate (Kaschabek et al. 1998), while the catechol 2,3-dioxygenase from this strain encoded by *cbzE* is plasmid-borne and is capable of metabolizing both 3-chlorocatechol and 3-methylcatechol (Mars et al. 1999). It belongs to the 2.C subfamily of type 1 extradiol dioxygenases.

4. Attempts have been made to overcome this limitation by using random mutagenesis of the genes of the 2,3-dihydroxybiphenyl dioxygenase from *Sphingomonas xenophaga* (Riegert et al. 2001). This resulted in higher rates of reaction with 3-chlorocatechol and mutants that were able to degrade 3-methylcatechol and 2,3-dihydroxybiphenyl by distal fission.

9.5.1 Reductive Loss of Halogen

9.5.1.1 Aerobic Reactions

There are important reactions carried out by anaerobic bacteria, and they may involve only partial dehalogenation; for example, 2,4,6-chlorobenzoate is successively dechlorinated by enrichment cultures to 2,4-dichlorobenzoate and 4-chlorobenzoate that is the terminal product (Gerritse et al. 1992). Although a number of strains have been isolated by the enrichment of halogenated benzoates under denitrifying conditions, no growth on either 2- or 4-bromobenzoate was observed under either aerobic or denitrifying conditions. Some strains were able to grow with 3-bromobenzoate under denitrifying conditions but not aerobically. On the basis of 16S rRNA sequences, these organisms belonged to *Thauera aromatica*, *Pseudomonas stutzeri*, or *Ochrobactrum anthropi* (Song et al. 2000). Other illustrations of reductive pathways employed by aerobic organisms include the following:

1. *Alcaligenes denitrificans* strain NTB-1 (now designated as a coryneform) that was isolated with 4-chlorobenzoate as substrate was able to grow with 4-bromo- and 4-iodobenzoate and, in addition, on 2,4-dichlorobenzoate. The first step in the pathway for 2,4-dichlorobenzoate was the reductive loss of the *ortho* chlorine substituent, whereas the second step is carried out by a halohydrolase that forms 4-hydroxybenzoate, followed by ring fission (Figure 9.24) (van den Tweel et al. 1987).

2. It has been shown that the reductive dechlorination of this strain and of *Corynebacterium sepedonicum* involved initial synthesis of a coenzyme A thioester and involves NADPH as the reductant. In addition, hydrolytic 4-chlorobenzoyl-CoA dehalogenase, 4-hydroxybenzoate 3-monooxygenase, and 3,4-dihydroxybenzoate 3,4-dioxygenase activities were found and enabled the construction of the degradative pathway (Romanov and Hausinger 1996).

3. The dechlorination of 2,4,5,6-tetrachloroisophthalonitrile (chlorothalonil) by *Pseudomonas* sp. strain CTB-3 took place under either aerobic or anaerobic conditions with replacement of chlorine

para to the −CN group with hydroxyl and retention of both −CN groups (Wang et al. 2010). For this dehalogenation no cofactors such as CoA or NADPH were required, and it was suggested that the enzyme was a hydrolase analogous to members of the metallo-β-lactamase family.

9.5.1.2 Anaerobic Reactions

It has been shown that under anaerobic phototrophic conditions, *Rhodopseudomonas palustris* was able to carry out the conversion of 3-chlorobenzoate to the CoA ester followed by reductive loss of chloride, and ultimately degradation of benzoate to acetyl-CoA and CO_2 (Egland et al. 2001). Although reductive dehalogenation by anaerobic bacteria in which the substrate functions as source of ATP has been most extensively explored for halogenated phenols that are discussed in Part 2 of this chapter, this has been demonstrated also for halogenated benzoates. *Desulfomonile tiedjei* was the first organism isolated in pure culture that was able to carry this out: it is a sulfate-reducing bacterium that is capable of reducing 3-chlorobenzoate to benzoate (DeWeerd et al. 1990) when ATP is synthesized by coupling proton translocation to dechlorination (Dolfing 1990; Mohn and Tiedje 1991). Cells induced by growth with 3-chlorobenzoate were able to partially dechlorinate polychlorinated phenols, specifically at the 3-position, whereas the monochlorophenols were apparently resistant to dechlorination (Mohn and Kennedy 1992). The membrane-bound reductive dehalogenase from *Desulfomonile tiedjei* has been solubilized and purified (Ni et al. 1995). It is distinct from the tetrachlorohydroquinone enzyme from a strain of *Flavobacterium* sp. (Xun et al. 1992c), and it plausibly plays a role in the energy transduction of *Desulfomonile tiedjei*. A membrane-bound cytochrome *c* is coinduced with the activity for reductive dechlorination, and has been purified and shown to be a high-spin diheme cytochrome distinct from previously characterized *c*-type cytochromes (Louie et al. 1997). It has been suggested that a chemiosmotic process may be used to rationalize the coupling of energy production with concomitant dechlorination (Louie and Mohn 1999). The related *Desulfomonile limimaris* is able to dechlorinate several meta-chlorinated benzoates including 2,3,5-trichlorobenzoate (Sun et al. 2001), and the anaerobic dechlorination of chlorobenzoates by *Desulfomonile* is summarized in Table 9.4 (DeWeerd and Suflita 1990; Sun et al. 2001).

   4,5,6,7-Tetrachlorophthalide is a fungicide against rice blast disease. Anaerobic dechlorination by an enrichment culture containing two strains of *Dehalobacter* closest to *Dehalobacter restrictus* carried out loss of one chlorine to produce the 4,5,7 and 4,6,7 isomers, two chlorines to give the 4,5- and 4,6-isomers, and finally 4-chlorophthalide that was the stable terminal product (Yoshida et al. 2009). This is consistent with the recalcitrance of the chlorine atom *para* to the carbonyl group.

9.6 Halogenated Phenylacetates

The degradation of 4-halogenated phenylacetates is carried out by 3:4 dioxygenation with loss of the halogen. *Pseudomonas* sp. strain CBS is able to grow at the expense of 4-chlorophenylacetate, and loss

TABLE 9.4

Anaerobic Dechlorination of Chlorobenzoates by *Desulfomonile* sp.

Substrate	Product(s)
3-Chlorobenzoate	Benzoate
2,5-Dichlorobenzoate	2-Chlorobenzoate
3,4-Dichlorobenzoate	4-Chlorobenzoate
3,5-Dichlorobenzoate	3-Chlorobenzoate + benzoate
2-,3-,4-Iodobenzoate	Benzoate

Source: Adapted from DeWeerd KA, JM Suflita. 1990. *Appl Environ Microbiol* 56: 2999–3005; Sun B, JR Cole, JM Tiedje. 2001. *Int J Syst Evol Microbiol* 51: 365–371.

TABLE 9.5

Substrate Specificity of 4-Chlorophenylacetate 3,4-Dioxygenase in Component A of *Pseudomonas* sp. Strain CBS

Substrate	Relative Activity (%)	Substrate	Relative Activity (%)
4-Fluorophenylacetate	30	4-Bromophenylacetate	102
4-Chlorophenylacetate	100	3-Chlorophenylacetate	10

Source: Adapted from Markus A, D Krekel, F Lingens. 1986. *J Biol Chem* 261: 12883–12888.

of chloride is mediated by a dioxygenase, which produces 3,4-dihydroxyphenylacetate that is degraded by an extradiol catechol dioxygenase (Markus et al. 1986). The dioxygenase consists of two components A and B. Component A has been purified and has a molecular weight of 140,000 consisting of three equal subunits, and contains iron and acid-labile sulfur. The substrate specificity is given in Table 9.5.

Component B is a monomeric reductase with a molecular weight of 35,000 and contains per mol of enzyme, 1 mol of FMN, 2.1 mol of Fe, and 1.7 mol of labile sulfur. After reduction with NADH, the ESR spectrum showed signals that were attributed to a [2Fe–2S] structure and a flavosemiquinone radical (Schweizer et al. 1987). The molecular and kinetic properties of the enzyme are broadly similar to the Class IB reductases of benzoate 1,2-dioxygenase and 4-methoxybenzoate monooxygenase-*O*-demethylase.

9.6.1 Anaerobic Reactions

The anaerobic dechlorination of 3-chloro-4-hydroxybenzoate and 3-chloro-4-hydroxy-phenylacetate has been demonstrated (Table 9.6), and a growing culture *Desulfitobacterium hafniense* strain DCB2 was also able to use vanillate or syringate as electron donors for the dechlorination of 3-chloro-4-hydroxyphenylacetate to 4-hydroxyphenylacetate (Neumann et al. 2004).

9.6.2 Fungal Reactions

Substituted 2,4,6-triiodinated benzoates are incorporated into x-ray contrast agents, and their transformation has been examined in the white-rot fungus *Trametes versicolor* (Rode and Müller 1998). Since these compounds are putatively unable to pass the cell walls of the fungus, it is important that although lignin peroxidase activity was not observed, nonspecific extracellular manganese-dependent peroxidase and laccase activities were found. There was no introduction of oxygen into the ring and the main reactions were successive deiodination to the monoiodinated compound (Figures 9.27 and 9.18).

TABLE 9.6

Anaerobic Dechlorination of 3-Chloro-4-Hydroxybenzoate by *Desulfitobacterium chlororespirans* and of 3-Chloro-4-Hydroxyphenylacetate by *Desulfitobacterium hafniense*, and *Desulfitobacterium metallireducens*

Substrate	Organism	Reference
3-Chloro-4-hydroxybenzoate	*Desulfito chlororespirans*	Löffler et al. (1996)
3-Chloro-4-hydroxyphenylacetate	*Desulfito chlororespirans*	Löffler et al. (1996)
	Desulfito hafniense	Christiansen et al. (1998)
	Desulfito metallireducens	Finneran et al. (2002)

Source: Adapted from Löffler FE, RA Sanford, JM Tiedje. 1996. *Appl Environ Microbiol* 62: 3809–3813; Christiansen N et al. 1998. *FEBS Lett* 436: 1159–1632; Finneran KT et al. 2002. *Int J Syst Bacteriol* 52: 1929–1935.

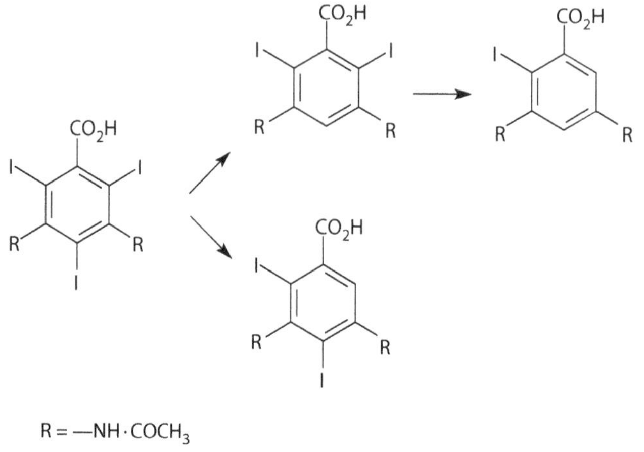

R = —NH·COCH₃

FIGURE 9.27 Transformation of 3,5-diacetamido-2,4,6-triiodobenzoate by *Trametes versicolor*. (Redrawn from Rode U, R Müller. 1998. *Appl Environ Microbiol* 64: 3114–3117.)

9.7 References

Adriaens P. 1994. Evidence for chlorine migration during oxidation of 2-chlorobiphenyl by a type II methanotroph. *Appl Environ Microbiol* 60: 1658–1662.

Adrian L, V Dudková, K Demnerová, DL Bedard. 2009. "*Dehalococcoides*" sp. strain CBDB1 extensively dechlorinates the commercial polychlorinated biphenyl mixture Arochlor 1260. *Appl Environ Microbiol* 75: 4516–4524.

Ahmad D, R Massé, M Sylvestre. 1990. Cloning, physical mapping and expression in *Pseudomonas putida* of 4-chlorobiphenyl transformation genes from *Pseudomonas testosteroni* strain B-356 and their homology to the genomic DNA from other PCB-degrading bacteria. *Gene* 86: 53–61.

Ahmad D, M Sylvestre, M Sondossi. 1991. Subcloning of *bph* genes from *Pseudomonas testosteroni* B-356 in *Pseudomonas putida* and *Escherichia coli*: Evidence for dehalogenation during initial attack on chlorobiphenyls. *Appl Environ Microbiol* 57: 2880–2887.

Aislabie J, AD Davison, HL Boul, PD Franzmann, DR Jardine, P Karuso. 1999. Isolation of *Terrabacter* sp. strain DDE-1, which metabolizes 1,1-dichloro-2-2-bis(4-chlorophenyl)ethylene when induced with biphenyl. *Appl Environ Microbiol* 65: 5607–5611.

Albrecht ID, AL Barkovskii, P Adriaens. 1999. Production and dechlorination of 2,3,7,8-tetrachlorodibenzo-*p*-dioxin in historically-contaminated estuarine sediments. *Environ Sci Technol* 33: 737–744.

Allard A-S, AH Neilson. 2003. Degradation and transformation of organic bromine and iodine compounds: Comparison with their chlorinated analogues. *Handbook Environ Chem* 3R: 1–74.

Arnett CM, JV Parales, JD Haddock. 2000. Influence of chlorine substituents on rates of oxidation of chlorinated biphenyls by the biphenyl dioxygenase of *Burkholderia* sp. strain LB400. *Appl Environ Microbiol* 66: 2928–2933.

Asturias JA, LD Eltis, M Prucha, KN Timmis. 1994. Analysis of three 2,3-dihydroxybiphenyl 1,2-dioxygenases found in *Rhodococcus globerulus* P6. Identification of a new family of extradiol dioxygenases. *J Biol Chem* 269: 7807–7815.

Asturias JA, KN Timmis. 1993. Three different 2,3-dihydroxybiphenyl-1,2-dioxygenase genes in the gram-positive polychlorobiphenyl-degrading bacterium *Rhodococcus globerulus* P6. *J Bacteriol* 175: 4631–4640.

Barriault D, M-M Plante, M Sylvestre. 2002. Family shuffling of a targeted *bphA* region to engineer biphenyl dioxygenase. *J Bacteriol* 184: 3794–3800.

Barriault D, M Sylvestre. 2004. Evolution of the biphenyl dioxygenase BphA from *Burkholderia xenovorans* LB400 by random mutagenesis of multiple sites in Region III. *J Biol Chem* 279: 47480–47488.

Bartels I, H-J Knackmuss, W Reineke. 1984. Suicide inactivation of catechol 2,3-dioxygenase from *Pseudomonas putida* mt-2 by 3-halocatechols. *Appl Environ Microbiol* 47: 500–505.

Bedard DL, JJ Bailey, BL Reiss, G van S Jerzak. 2006. Development and characterization of stable sediment-free anaerobic bacterial enrichment cultures that dechlorinate Arochlor 1260. *Appl Environ Microbiol* 72: 2460–2470.

Bedard DL, ML Haberl, RJ May, MJ Brennan. 1987a. Evidence for novel mechanisms of polychlorinated biphenyl metabolism in *Alcaligenes eutrophus* H 850. *Appl Environ Microbiol* 53: 1103–1112.

Bedard DL, EA Pohl, JJ Bailey, A Murphy. 2005. Characterization of the PCB substrate range of microbial dechlorination process LP. *Environ Sci Technol* 39: 6831–6838.

Bedard DL, JF Quensen III. 1995. Microbial reductive dechlorination of polychlorinated biphenyls. In *Microbial Transformation and Degradation of Toxic Organic Chemicals* (Eds LY Young, CE Cerniglia), pp. 127–216. Wiley-Liss, New York.

Bedard DL, HM van Dort. 1998. Complete reductive dehalogenation of brominated biphenyls by anaerobic microorganisms in sediment. *Appl Environ Microbiol* 64: 940–947.

Bedard DL, HM van Dort, KA Deweerd. 1998. Brominated biphenyls prime extensive microbial reductive dehalogenation of Arochlor 1260 in Housatonic River sediment. *Appl Environ Microbiol* 64: 1786–1795.

Bedard DL, RE Wagner, MJ Brennan, ML Haberl, RJ May, JF Brown Jr. 1987b. Extensive degradation of Arochlors and environmentally transformed polychlorinated biphenyls by *Alcaligenes eutrophus* H 850. *Appl Environ Microbiol* 53: 1094–1102.

Beil S, B Happe, KN Timmis, DH Pieper. 1997. Genetic and biochemical characterization of the broad spectrum chlorobenene dioxygenase from *Burkholderia* sp. strain PS12. Dechlorination of 1,2,4,5-tetrachlorobenzene. *Eur J Biochem* 247: 190–199.

Blasco R, M Mallavarapu, R-M Wittich, KN Timmis, DH Pieper. 1997. Evidence that formation of protoanemonin from metabolites of 4-chlorobiphenyl degradation negatively affects the survival of 4-chlorobiphenyl-cometabolizing microorganisms. *Appl Environ Microbiol* 63: 427–434.

Blasco R, R-M Wittich, M Mallavarapu, KN Timmis, DH Pieper. 1995. From xenobiotic to antibiotic, formation of protoanemonin from 4-chlorocatechol by enzymes of the 3-oxoadipate pathway. *J Biol Chem* 270: 29229–29235.

Bopp LH. 1986. Degradation of highly chlorinated PCBs by *Pseudomonas* strain LB400. *J Indust Microbiol* 1: 23–29.

Brown JF, RE Wagner. 1990. PCB movement, dechlorination, and detoxication in the Acusnet Estuary. *Environ Toxicol Chem* 9: 1215–1233.

Buckman AH, CS Wong, EA Chow, SB Brown, KR Solomon, AT Fisk. 2006. Biotransformation of polychlorinated biphenyls (PCBs) and bioformation of hydroxylated PCBs in fish. *Aquat Toxicol* 78: 176–185.

Chang K-H, D Dunaway-Mariano. 1996. Determination of the chemical pathway for 4-chlorobenzoate: Coenzyme A ligase catalysis. *Biochemistry* 35: 13478–13484.

Chang KH, P-H Liang, W Beck, JD Scholten, D Dunaway-Mariano. 1992. Isolation and characterization of the three polypeptide components of 4-chlorobenzoate dehalogenase from *Pseudomonas* sp. strain CBS-3. *Biochemistry* 31: 5605–5610.

Chávez FP, H Lünsdorf, CA Jerez. 2004. Growth of polychlorinated-biphenyl-degrading bacteria in the presence of biphenyl and chlorobiphenyls generates oxidative stress and massive accumulation of inorganic polyphosphate. *Appl Environ Microbiol* 70: 3064–3072.

Chebrou H, Y Hurtubise, D Barriault, M Sylvestre. 1999. Heterologous expression and characterization of the purified oxygenase component of *Rhodococcus globerulus* P6 biphenyl dioxygenase and of chimeras derived from it. *J Bacteriol* 181: 4805–4811.

Chen X, A Christopher, JP Jones, SG Bell, Q Guo, F Xu, Z Rao, L-L Wong. 2002. Crystal structure of the F87W/ V96F/ V247L mutant of cytochrome P-450cam with 1,3,5-trichlorobenzene bound and further protein engineering for the oxidation of pentachlorobenzene and hexachlorobenzene. *J Biol Chem* 277: 37519–37526.

Christiansen N, BK Ahring, G Wohlfarth, G Diekert. 1998. Purification and characterization of the 3-chloro-4-hydroxyphenylacetate reductive dehalogenase of *Desulfitobacterium hafniense*. *FEBS Lett* 436: 1159–1632.

Copley SD, GP Crooks. 1992. Enzymatic dehalogenation of 4-chlorobenzoyl coenzyme A in *Acinetobacter* sp. strain 4-CB1. *Appl Environ Microbiol* 58: 1385–1387.

Crooks GP, SD Copley. 1993. A surprizing effect of leaving group on the nucleophilic aromatic substitution reactions catalyzed by 4-chlorobenzoyl-CoA dehalogenase. *J Am Chem Soc* 115: 6422–6423.

Dai S, FH Vaillancourt, H Maaroufi, NM Drouin, DB Neau, V Snieckus, JT Bolin, LD Eltis. 2002. Identification and analysis of a bottleneck in PCB biodegradation. *Nat. Struct Biol* 9: 934–939.

Denef VJ, J Park, TV Tsoi, J-M Rouillard, H Zhang, JA Wibbenmeyer, W Verstraete, E Gulari, SA Hasham, JM Tiedje. 2004. Biphenyl and benzoate metabolism in a genomic context: Outlining genome-wide metabolic networks in *Burkholderia xenovorans* LB400. *Appl Environ Microbiol* 70: 4961–4970.

DeWeerd KA, DL Bedard. 1999. Use of halogenated benzoates and other halogenated aromatic compounds to stimulate the microbial dechlorination of PCBs. *Environ Sci Technol* 33: 2057–2063.

DeWeerd KA, L Mandelco, RS Tanner, CR Woese, JM Suflita. 1990. *Desulfomonile tiedjei* gen. nov. and sp. nov., a novel anaerobic, dehalogenating, sulfate-reducing bacterium. *Arch Microbiol* 154: 23–30.

DeWeerd KA, JM Suflita. 1990. Anaerobic aryl reductive dehalogenation of halobenzoates by cell extracts of "*Desulfomonile tiedjei*" *Appl Environ Microbiol* 56: 2999-3005.

Dolfing J. 1990. Reductive dechlorination of 3-chlorobenzoate is coupled to ATP production and growth in an anaerobic bacterium, strain DCB-1. *Arch Microbiol* 153: 264–266.

Don RH, AJ Weightman, HJ Knackmuss, KN Timmis. 1985. Transposon mutagenesis and cloning analysis of the pathways for degradation of 2,4-dichlorophenoxyacetic acid and 3-chlorobenzoate in *Alcaligenes eutrophus* JMPl34(pJP4). *J Bacteriol* 161: 85–90.

Donnelly PK, RS Hegde, JS Fletcher. 1994. Growth of PCB-degrading bacteria on compounds from photosynthetic plants. *Chemosphere* 28: 981–988.

Dorn E, H-J Knackmuss. 1978a. Chemical structure and biodegradability of halogenated aromatic compounds. Substituent effects on 1,2-dioxygenation of catechol. *Biochem J* 174: 85–94.

Dorn E, H-J Knackmuss. 1978b. Chemical structure and biodegradability of halogenated aromatic compounds. Two catechol 1,2-dioxygenases from a 3-chlorobenzoate-grown pseudomonad. *Biochem J* 174: 73–84.

Egland PG, J Gibson, CS Harwood. 2001. Reductive, coenzyme A-mediated pathway for 3-chlorobenzoate degradation in the phototroophic bacterium *Rhodopseudomonas palustris*. *Appl Environ Microbiol* 67: 1396–1399.

Elsner A, F Löffler, K Miyashita, R Müller, F Lingens. 1991. Resolution of 4-chlorobenzoate from *Pseudomonas* sp strain CBS3 into three components. *Appl Environ Microbiol* 57: 324–326.

Endo R, M Kamakura, K Miyauchi, M Fukuda, Y Ohtsubo, M Tsuda, Y Nagata. 2005. Identification and characterization of genes involved in the downstream degradation pathway of γ-hexachlorocyclohexane in *Sphingomonas paucimobilis* UT26. *J Bacteriol* 187: 847–853.

Erickson BD, FJ Mondelo. 1992. Nucleotide sequencing and transcriptional mapping of the genes encoding biphenyl dioxygenase, a multicomponent polychlorinated-biphenyl-degrading enzyme in *Pseudomonas* strain LB 400. *J Bacteriol* 174: 2903–2912.

Eulberg D, EM Kourbatova, LA Golovleva, M Schlömann. 1998. Evolutionary relationship between chlorocatechol catabolic enzymes from *Rhodococcus opacus* 1CP and their counterparts in proteobacteria: Sequence divergence and functional convergence. *J Bacteriol* 180: 1082–1094.

Fagervold SK, JEM Watts, HD May, KR Sowers. 2005. Sequential reductive dechlorination of *meta*-chlorinated polychlorinated biphenyl congeners in sediment microcosms by two different types of *Chloroflexi* phylotypes. *Appl Environ Microbiol* 71: 8085–8090.

Fennell DE, I Nijenhuis, SF Wilson, SH Zinder, MM Häggblom. 2004. *Dehalococcoides ethenogenes* strain 195 reductively dechlorinates diverse chlorinated aromatic pollutants. *Environ Sci Technol* 38: 2075–2081.

Fetzner S, R Müller, F Lingens. 1989. A novel metabolite in the degradation of 2-chlorobenzoate. *Biochem Biophys Res Commun* 161: 700–705.

Fetzner S, R Müller, F Lingens. 1992. Purification and some properties of 2-halobenzoate 1,2-dioxygenase, a two-component enzyme system from *Pseudomonas cepacia* 2CBS. *J Bacteriol* 174: 279–290.

Finneran KT, HM Forbuch, CV VanPraagh, DK Lovley. 2002. *Desulphitobacterium metallireducens* sp. nov., an anaerobic bacterium that couples growth with the reduction of humic acids as well as chlorinated compounds. *Int J Syst Bacteriol* 52: 1929–1935.

Finkelstein ZI, BP Baskunov, MG Boersma, J Vervoort, EL Golovlev, WJH van Berkel, LA Golovleva, IMCM Rietjens. 2000. Identification of fluoropyrogallols as new intermediates in biotransformation of monofluorophenols in *Rhodococcus opacus* 1cp. *Appl Environ Microbiol* 66: 2148–2153.

Flanagan WP, RJ May. 1993. Metabolite detection as evidence for naturally occurring aerobic PCB degradation in Hudson River sediments. *Environ Sci Technol* 27: 2207–2212.

Fortin PD, GP Horsman, HM Yang, LD Eltis. 2006. A glutathione S-transferase catalyzes the dehalogenation of inhibitory metabolites of polychlorinated biphenyls. *J Bacteriol* 188: 4424–4430.

Fortin PD, AT-F Lo, M-A Haro, SR Kaschabek, W Reineke, LF Eltis. 2005. Evolutionarily divergent extradiol dioxygenases possess higher specificities for polychlorinated biphenyl metabolites. *J Bacteriol* 187: 415–421.

Francisco PB, N Ogawa, K Suzuki, K Miyashita. 2001. The chlorobenzoate dioxygenase genes of *Burkholderia* sp. strain NK8 involved in the catabolism of chlorobenzoates. *Microbiology (UK)* 147: 121–133.

Furukawa K, N Arimura. 1987. Purification and properties of 2,3-dihydroxybiphenyl dioxygenase from polychlorinated biphenyl-degrading *Pseudomonas pseudoalcaligenes* and *Pseudomonas aeruginosa* carrying the cloned *bphC* gene. *J Bacteriol* 169: 924–927.

Furukawa K, T Miyazaki. 1986. Cloning of a gene cluster encoding biphenyl and chlorobiphenyl degradation in *Pseudomonas pseudoalcaligenes*. *J Bacteriol* 166: 392–398.

Furukawa K, N Tomizuka, A Kamibayashi. 1979. Effect of chlorine substitution on the bacterial metabolism of various polychlorinated biphenyls. *Appl Environ Microbiol* 38: 301–310.

Gerritse J, BJ van der Woude, JC Gottschal. 1992. Specific removal of chlorine from the *ortho*-position of halogenated benzoic acids by reductive dechlorination in anaerobic enrichment cultures. *FEMS Microbiol Lett* 100: 273–280.

Gibson DT, DL Cruden, JD Haddock, GJ Zylstra JM Brande. 1993. Oxidation of polychlorinated biphenyls by *Pseudomonas* sp. strain LB400 and *Pseudomonas pseudoalcaligenes* KF707. *J Bacteriol* 175: 4561–4564.

Gilbert ES, DE Crowley. 1997. Plant compounds that induce polychlorinated biphenyl degradation by *Arthrobacter* sp. strain B1B. *Appl Environ Microbiol* 63: 1933–1938.

Gómez-Gil L, P Kumar, D Barriault, JT Bolin, M Sylvestre, LD Eltis. 2007. Characterization of biphenyl dioxygenase of *Pandoraea pnomenusa* B-356 as a potent polychlorinated biphenyl-degrading enzyme. *J Bacteriol* 189: 5705–5715.

Gonçalves ER, H Hara, D Miyazawa, JE Davies, LD Eltis, WW Mohn. 2006. Transcriptonic assessment of isoenzymes in the biphenyl pathway of *Rhodococcus* sp. strain RHA1. *Appl Environ Microbiol* 72: 6183–6193.

Goris J, P De Vos, J Caballero-Melado, J Park, E Falsen, JF Quensen, JM Tiedje, P Vandamme. 2004. Classification of the biphenyl- and polychlorinated biphenyl-degrading strain LB400T and relatives as *Burkholderia xenovorans*. *Int J Syst Evol Microbiol* 54: 1677–1681.

Haak B, S Fetzner, F Lingens. 1995. Cloning, nucleotide sequence, and expression of the plasmid-encoded genes for the two-component 2-halobenzoate 1,2-dioxygenase from *Pseudomonas cepacia* 2CBS. *J Bacteriol* 177: 667–675.

Haddock JD, DT Gibson. 1995. Purification and characterization of the oxygenase component of biphenyl 2,3-dioxygenase from *Pseudomonas* sp. strain LB400. *J Bacteriol* 177: 5834–5839.

Haddock JD, JR Horton, DT Gibson. 1995. Dihydroxylation and dechlorination of chlorinated biphenyls by purified biphenyl 2,3-dioxygenase from *Pseudomonas* sp. strain LB400. *J Bacteriol* 177: 20–26.

Harwood CS, RE Parales. 1996. The beta-ketoadipate pathway and the biology of self-identity. *Annu Rev Microbiol* 50: 553–590.

Hatta T, G Muklerjee-Dhar, J Damborsky, H Kiyohara. 2003. Characterization of a novel thermostable Mn(II)-dependent 2,3-dihydroxybiphenyl 1,2-dioxygenase from polychlorinated biphenyl- and naphthalene-degrading *Bacillus* sp. JF8. *J Biol Chem* 278: 21483–21492.

Hauschild JE, E Masai, K Sugiyama, T Hatta, K Kimbara, M Fukuda, K Yano. 1996. Identification of an alternative 2,3-dihydroxybiphenyl 1,2-dioxygenase in *Rhodococcus* sp. strain RHA1 and cloning of the gene. *Appl Environ Microbiol* 62: 2940–2946.

Havel J, W Reineke. 1993. Microbial degradation of chlorinated acetophenones. *Appl Environ Microbiol* 59: 2706–2712.

Hay AG, DD Focht. 1998. Cometabolism of 1,1-dichloro-2-2-bis(4-chlorophenyl)ethylene by *Pseudomonas acidovorans* M3GY grown on biphenyl. *Appl Environ Microbiol* 64: 2141–2146.

Heberer T, U Dünnbier. 1999. DDT metabolite bis(chlorophenyl)acetic acid; the neglected environmental contaminant. *Environ Sci Technol* 33: 2346–2351.

Hernandez BS, JJ Arensdorf, DD Focht. 1995. Catabolic characteristic of biphenyl-utilizing isolates which cometabolize PCBs. *Biodegradation* 6: 75–82.

Hickey WJ, DD Focht. 1990. Degradation of mono-, di-, and trihalogenated benzoic acids by *Pseudomonas aeruginosa* JB2. *Appl Environ Microbiol* 56: 3842–3850.

Higson FK, DD Focht. 1990. Bacterial degradation of ring-chlorinated acetophenones. *Appl Environ Microbiol* 56: 3678–3685.

Holoman TRP, MA Elberson, LA Cutter, HD May, KR Sowers. 1998. Characterization of a defined 2,3,5,6-tetrachlorobiphenyl-*ortho*-dechlorinating microbial community by comparative sequence analysis of genes coding for 16S rRNA *Appl Environ Microbiol* 64: 3359–3367.

Hölscher T, H Görisch, L Adrian. 2003. Reductive dehalogenation of chlorobenzene congeners in cell extracts of *Dehalococcoides* sp. strain CBDB1. *Appl Environ Microbiol* 69: 2999–3001.

Hurtubise Y, D Barriault, M Powlowski, M Sylvestre. 1995. Purification and characterization of the *Comamonas testosteroni* B-356 biphenyl dioxygenase components. *J Bacteriol* 177: 6610–6618.

Hurtubise Y, D Barriault, M Sylvestre. 1998. Involvement of the terminal oxygenase β subunit in the biphenyl dioxygenase reactivity pattern towards chlorobiphenyls. *J Bacteriol* 180: 5828–5838. .

Iwasaki T, K Miyauchi, E Masai, M Furukada. 2006. Multiple-subunit genes of the aromatic-ring-hydroxylating dioxygenase play an active role in biphenyl and polychlorinated biphenyl degradation in *Rhodococcus* sp. strain RHA1. *Appl Environ Microbiol* 72: 5396–5402.

Jayachandran G, H Görisch, L Adrian. 2003. Dehalorespiration with hexachlorobenzene and pentachloroben-zene by *Dehalococcoides* sp. strain CBDB1. *Arch Microbiol* 180: 411–416.

Jayachandran G, H Görisch, L Adrian. 2004. Studies on hydrogenase activity and chlorobenzene respiration in *Dahalococcoides* sp. strain CBDB1. *Arch Microbiol* 182: 498–504.

Jones JP, EJ O'Hare, L-L Wong. 2001. Oxidation of polychlorinated benzenes by genetically engineered CYP101 (cytochrome P450$_{cam}$). *Eur J Biochem* 268: 1460–1467.

Kaschabek SR, T Kasberg, D Müller, AE Mars, DB Janssen, W Reineke. 1998. Degradation of chloroaromat-ics: Purification and characterization of a novel type of chlorocatechol 2,3-dioxygenase of *Pseudomonas putida* GJ31. *J Bacteriol* 180: 296–302.

Kaschabek SR, W Reineke. 1995. Maleylacetate reductase of *Pseudomonas* sp. strain B13: Specificity of sub-strate conversion and halide elimination. *J Bacteriol* 177: 320–325.

Kersten PJ, PJ Chapman, S Dagley. 1985. Enzymatic release of halogens or methanol from some substituted protocatechuic acids. *J Bacteriol* 162: 693–697.

Kim S, F Picardal. 2001. Microbial growth on dichlorobiphenyls chlorinated on both rings as sole carbon and energy sources. *Appl Environ Microbiol* 67: 1953–1955.

Kimura N, A Nishi, M Goto, K Furukawa. 1997. Functional analysis of a variety of chimeric dioxygenases constructed from two biphenyl dioxygenases that are similar structurally but different functionally. *J Bacteriol* 179: 3936–3943.

Klecka GM, DT Gibson. 1981. Inhibition of catechol 2,3-dioxygenase from *Pseudomonas putida* by 3-chlorocatechol. *Appl Environ Microbiol* 41: 1159–1165.

Lehning A, U Fock, R-M Wittich, KN Timmis, DH Pieper. 1997. Metabolism of chlorotoluenes by *Burkholderia* sp. strain PS12 and toluene dioxygenase of *Pseudomonas putida* F1: Evidence for monooxygenation by toluene and chlorobenzene dioxygenases. *Appl Environ Microbiol* 63: 1974–1979.

Letcher RJ, RJ Norstrom, DCG Muir. 1998. Biotransformation versus bioaccumulation: Sources of methyl sulfone PCB and 4,4′-DDE metabolites in polar bear food chain. *Environ Sci Technol* 32: 1656–1661.

Löffler F, R Müller, F Lingens. 1991. Dehalogenation of 4-chlorobenzoate by 4-chlorobenzoate dehalogenase from *Pseudomonas* sp. CBS3: An ATP/coenzyme A dependent reaction. *Biochem Biophys Res Commun* 176: 1106–1111.

Löffler F, R Müller, F Lingens. 1992. Purification and properties of 4-halobenzoate-coenzyme A ligase from a *Pseudomonas* sp. CBS3. *Biol Chem Hoppe-Seyler* 373: 1001–1007.

Löffler FE, RA Sanford, JM Tiedje. 1996. Initial characterization of a reductive dehalogenase from *Desulfitobacterium chlororespirans* Co23. *Appl Environ Microbiol* 62: 3809–3813.

Louie TM, S Ni, L Xun, WW Mohn. 1997. Purification, characterization and gene sequence analysis of a novel cytochrome *c* coinduced with reductive dechlorination activity in *Desulfomonile tiedjei* DCB-1. *Arch Microbiol* 168: 520–527.

Louie TM, WW Mohn. 1999. Evidence for a chemiosmotic model of dehalorespiration in *Desulfomonile tiedjei* DCB-1. *J Bacteriol* 181: 40–46.

Maeda M, S-Y Chung, E Song, T Kudo. 1995. Multiple genes encoding 2,3-dihydroxybiphenyl 1,2-dioxygenase in the gram-positive polychlorinated biphenyl-degrading bacterium *Rhodococcus erythropolis* TA421, isolated from a termite ecosystem. *Appl Environ Microbiol* 61: 549–555.

Marks TS, ARW Smith, AV Quirk. 1984. Degradation of 4-chlorobenzoic acid by *Arthrobacter* sp. *Appl Environ Microbiol* 48: 1020–1025.

Markus A, D Krekel, F Lingens. 1986. Purification and some properties of component A of the 4-chlorophenyl-acetate 3,4-dioxygenase from *Pseudomonas* species strain CBS. *J Biol Chem* 261: 12883–12888.

Mars AE, T Kasberg, SR Kaschabek, MH van Agteren, DB Janssen, W Reineke. 1997. Microbial degradation of chloroaromatics: Use of the *meta*-cleavage pathway for mineralization of chlorobenzene. *J Bacteriol* 179: 4530–4537.

Mars AE, J Kingma, SR Kaschabek, W Reineke, DB Janssen. 1999. Conversion of 3-chlorocatechol by various catechol 2,3-dioxygenases and sequence analysis of the chlorocatechol dioxygenase region of *Pseudomonas putida* GJ31. *J Bacteriol* 181: 1309–1318.

Massé R, D Lalanne, F Mssier, M Sylvestre. 1989. Characterization of new bacterial transformation products of 1,1,1-trichloro-2,2-bis-(4-chlorophenyl)ethane (DDT) by gas chromatography/mass spectrometry. *Biomed Environ Mass Spectrom* 18: 741–752.

May HD, LA Cutter, GS Miller, CE Milliken, JEM Watts, KR Sowers. 2006. Stimulatory and inhibitory effects of organohalides on the dehalogenating activities of PCB-dechlorinating bacterium *o*-17. *Environ Sci Technol* 40: 5704–5709.

McKay DB, M Prucha, W Reineke, KN Timmis, DH Pieper. 2003. Substrate specificity and expression of three 2,3-dihydroxybiphenyl 1,2-dioxygenases from *Rhodococcus globerulus* strain P6. *J Bacteriol* 185: 2944–2951.

Mohn WW, KJ Kennedy. 1992. Reductive dehalogenation of chlorophenols by *Desulfomonile tiedjei* DCB-1. *Appl Environ Microbiol* 58: 1367–1370.

Mohn WW, JM Tiedje. 1991. Evidence for chemiosmotic coupling of reductive dechlorination and ATP synthesis in *Desulfomonile tiedjei*. *Arch Microbiol* 157: 1–6.

Mondello FJ, MP Turcich, JH Lobos, BD Erickson. 1997. Identification and modification of biphenyl dioxygenase sequences that determine the specificity of polychlorinated biphenyl degradation. *Appl Environ Microbiol* 63: 3096–3103.

Morris PJ, JF Quensen III, JM Tiedje, SA Boyd. 1992. Reductive debromination of the commercial polybrominated biphenyl mixture Firemaster BP6 by anaerobic microorganisms from sediments. *Appl Environ Microbiol* 58: 3249–3256.

Morris PJ, JF Quensen III, JM Tiedje, SA Boyd. 1993. An assessment of the reductive debromination of polybrominated biphenyls in the Pine River reservoir. *Environ Sci Technol* 27: 1580–1586.

Nakatsu CH, M Providenti, RC Wyndham. 1997. The *cis*-diol dehydrogenase *cbaC* gene of Tn*5271* is required for growth on 3-chlorobenzoate but not 3,4-dichlorobenzoate. *Gene* 196: 200–218.

Nakatsu CH, RC Wyndham. 1993. Cloning and expression of the transposable chlorobenzoate-3,4-dioxygenase genes of *Alcaligenes* sp. strain BR60. *Appl Environ Microbiol* 59: 3625–3633.

Neilson AH, A-S Allard. 2002. Degradation and transformation of organic fluorine compounds. *Handbook Environ Chem* 3N: 137–202.

Neumann A, T Engelmann, R Schmitz, Y Greiser, A Orthaus, G Diekert. 2004. Phenyl methyl ethers: Novel electron donors for respiratory growth of *Desulfitobacterium hafniense* and *Desulfitobacterium* sp. strain PCE-S. *Arch Microbiol* 181: 245–249.

Ni S, JK Fredrickson, L Xun. 1995. Purification and characterization of a novel 3-chlorobenzoate-reductive dehalogenase from the cytoplasmic membrane of *Desulfomonile tiedje* DVCB-1. *J Bacteriol* 177: 5135–5139.

Ohtsubo Y, M Shimura, M Delawary, K Kimbara, M Takagi, T Kudo, A Ohta, Y Nagata. 2003. Novel approach to the improvement of biphenyl and polychlorinated biphenyl degradation activity: Promoter implantation by homologous recombination. *Appl Environ Microbiol* 69: 146–153.

Pettigrew CA, BE Haigler, JC Spain. 1991. Simultaneous biodegradation of chlorobenzene and toluene by a *Pseudomonas* strain. *Appl Environ Microbiol* 57: 157–162.

Prucha M, A Peterseim, KN Timmis, DH Pieper. 1996a. Muconolactone isomerase of the 3-oxoadipate pathway catalyzes dechlorination of 5-chloro-substituted muconolactones. *Eur J Biochem* 237: 350–356.

Prucha M, V Wray, DH Pieper. 1996b. Metabolism of 5-chlorosubstituted muconolactones. *Eur J Biochem* 237: 357–366.

Pollmann K, S Beil, DH Pieper. 2001. Transformation of chlorinated benzenes and toluenes by *Ralstonia* sp. strain PS12 *tecA* (tetrachlorobenzene dioxygenase) and *tecB* (chlorobenzene dihydrodiol dehydrogenase) gene products. *Appl Environ Microbiol* 67: 4057–4063.

Pollmann K, S Kaschabek, V Wray, W Reineke, DH Pieper. 2002. Metabolism of dichlomethylcatechols as central intermediates in the degradation of dichlorotoluenes by *Ralstonia* sp. strain PS12. *J Bacteriol* 184: 5261–5274.

Pollmann K, V Wray, DH Pieper. 2005. Chloromethylmuconolactones as critical metabolites in the degradation of chloromethylcatechols: Recalcitrance of 2-chlorotoluene. *J Bacteriol* 187: 2332–2340.

Potrawfke T, J Armangaud, R-M Wittich. 2001. Chlorocatechols substituted at positions 4 and 5 are substrates of the broad-spectrum chlorocatechol 1,2-dioxygenase of *Pseudomonas chlororaphis* RW71. *J Bacteriol* 183: 997–1011.

Potrawfke T, KN Timmis, R-M Wittich. 1998. Degradation of 1,2,3,4-tetrachlorobenzene by *Pseudomonas chlororaphis* RW71. *Appl Environ Microbiol* 64: 3798–3806.

Rezek, T Macek, M Mackova, J Triska. 2007. Plant metabolites of polychlorinated biphenyls in hairy root culture of black nightshade *Solanum nigrum* SNC-90. *Chemosphere* 69: 1221–1227.

Rezek, T Macek, M Mackova, J Triska, K Ruzickova. 2008. Hydroxy-PCBs, methoxy-PCBs and hydroxy-methoxy-PCBs: Metabolites of polychlorinated biphenyls formed in vitro by tobacco cells. *Environ Sci Technol* 42: 5746–5751.

Riegert U, S Bürger, A Stolz. 2001. Altering catalytic properties of 3-chlorocatechol-oxidizing extradiol dioxygenase from *Sphingomonas xenophaga* BN6 by random mutagenesis. *J Bacteriol* 183: 2322–2330.

Riegert U, G Heiss, P Fischer, A Stolz. 1998. Distal cleavage of 3-chlorocatechol by an extradiol dioxygenase to 3-chloro-2-hydroxymuconic semialdehyde. *J Bacteriol* 180: 2849–2853.

Rode U, R Müller. 1998. Transformation of the ionic X-ray contrast agent Diatrizoate and related triiodinated benzoates by *Trametes versicolor*. *Appl Environ Microbiol* 64: 3114–3117.

Romanov V, RP Hausinger. 1994. *Pseudomonas aeruginosa* 142 uses a three-component *ortho*-halobenzoate 1,2-dioxygenase for metabolism of 2,4-dichloro- and 2-chlorobenzoate. *J Bacteriol* 176: 3368–3374.

Romanov V, RP Hausinger. 1996. NADPH-dependent reductive *ortho* dehalogenation of 2,4-dichlorobenzoic acid in *Corynebacterium sepedonicum* KZ-4 and coryneform bacterium strain NTB-1 via 2,4-dichloro-benzoyl coenzyme A. *J Bacteriol* 178: 2656–2661.

Romine MF, LC Stillwell, K-K Wong, SJ Thurston, EC Sisk, C Sensen, T Gaasterland, JK Fredrickson, JD Saffer. 1999. Complete sequence of a 184-kilobase catabolic plasmid from *Sphingomonas aromaticivorans* F199. *J Bacteriol* 181: 1585–1602.

Rybkina DO, EG Plotnikova, LV Dorofeeva, YL Mironenko, VA Demakov. 2003. A new aerobic Gram-positive bacterium with a unique ability to degrade *ortho*- and *para*-chlorinated biphenyls. *Microbiology* 72: 672–677 DOI: 10.1023/B: MICI.

Sakai M, S Ezaki, N Suzuki, R Kurane. 2005. Isolation and characterization of a novel polychlorinated biphenyl-degrading bacterium, *Paenibacillus* sp. KBC101. *Appl Microbiol Biotechnol* 68: 111–116.

Sala-Trepat JM, WC Evans. 1971. The meta cleavage of catechol by *Azotobacter* species: 4-oxalocrotonatev pathway. *Eur J Biochem* 20: 400–413.

Sander P, R-M Wittich, P Fortnagel, H Wilkes, W Francke. 1991. Degradation of 1,2,4-trichloro- and 1,2,4,5-tetrachlorobenzene by *Pseudomonas* strains. *Appl Environ Microbiol* 57: 1430–1440.

Schmidt E, H-J Knackmuss. 1980. Chemical structure and biodegradability of halogenated aromatic compounds. Conversion of chlorinated muconic acids into maleoylacetic acid. *Biochem J* 192: 339–347.

Schmidt E, G Remberg, H-J Knackmuss. 1980. Chemical structure and biodegradability of halogenated aromatic compounds. Halogenated muconic acids as intermediates. *Biochem J* 192: 331–337.

Schweizer D, A Markus, M Seez, HH Ruf, F Lingens. 1987. Purification and some properties of component B of the 4-chlorophenylacetate 3,4-dioxygenase from *Pseudomonas* species strain CBs 3. *J Biol Chem* 262: 9340–9346.

Seah SYK, G Labbé, SR Kaschabek, F Reifenrath, W Reineke, LD Eltis. 2001. Comparative specificities of two evolutionarily divergent hydrolases involved in microbial degradation of polychlorinated biphenyls. *J Bacteriol* 183: 1511–1516.

Seah SYK, G Labbé, S Nerdinger, M Johnson, V Snieckus, LD Eltis. 2000. Identification of a serine hydrolase as a key determinant in the microbial degradation of polychlorinated biphenyls. *J Biol Chem* 275: 15701–15708.

Seeger M, B Cámara, B Hofer. 2001. Dehalogenation, denitration, dehydroxylation, and angular attack on substituted biphenyls and related compounds by biphenyl dioxygenase. *J Bacteriol* 183: 3548–3555.

Seeger M, M Zielinski, KN Timmis, B Hofer. 1999. Regiospecificity of dioxygenation of di- to pentachlorobiphenyls and the degradation to chlorobenzoates by the bph-encoded catabolic pathway of *Burkholderia* sp. strain LB400. *Appl Environ Microbiol* 65: 3614–3621.

Seibert V, K Stadler-Fritsche, M Schlomann. 1993. Purification and characterization of maleylacetate reductase from *Alcaligenes eutrophus* JMP134 (pJP4). *J Bacteriol* 175: 6745–6754.

Seto M, E Masai, M Ida, T Hatta, K Kimbara, M Fukuda, K Yano. 1995. Multiple polychlorinated biphenyl transformation systems in the Gram-positive bacterium *Rhodococcus* sp. strain RHA1. *Appl Environ Microbiol* 61: 4510–4513.

Shimura M, G Mukerjee-Dhar, K Kimbara, H Nagato, H Kiyohara, T Hatta. 1999. isolation and characterization of a thermophilic *Bacillus* sp. JF8 capable of degrading polychlorinated biphenyls and naphthalene. *FEMS Microbiol Lett* 178: 87–93.

Skiba A, V Hecht, DH Pieper. 2002. Formation of protoanemonin from 2-chloro-*cis,cis*-muconate by the combined action of muconate cycloisomerase and muconolactone isomerase. *J Bacteriol* 184: 5402–5409.

Solyanikova IP, OV Maltseva, MD Vollmer, LA Golovleva, M Schlömann. 1995. Characterization of muconate and chloromuconate cycloisomerase from *Rhodococcus erythropolis* 1CP: Indications for functionally convergent evolution among bacterial cycloisomerases. *J Bacteriol* 177: 2821–2826.

Sommer C, H Görisch. 1997. Enzymology of the degradation of (di)chlorobenzenes by *Xanthobacter flavus* 14p1. *Arch Microbiol* 167: 384–391.

Sondossi M, M Sylvestre, D Ahmad. 1992. Effects of chlorobenzoate transformation on the *Pseudomonas testosteroni* biphenyl and chlorobiphenyl degradation pathway. *Appl Environ Microbiol* 58: 485–495.

Song B, NJ Palleroni, MM Häggblom. 2000. Isolation and characterization of diverse halobenzoate-degrading denitrifying bacteria from soils and sediments. *Appl Environ Microbiol* 66: 3446–3453.

Spain JC, SF Nishino. 1987. Degradation of 1,4-dichlorobenzene by a *Pseudomonas* sp. *Appl Environ Microbiol* 53: 1010–1019.

Suenaga H, T Watanabe, M Sato, Ngadiman, K Furukawa. 2002. Alteration of regiospecificity in biphenyl dioxygenase by active-site engineering. *J Bacteriol* 184: 3682–3688.

Sun B, JR Cole, JM Tiedje. 2001. *Desulfomonile limimaris* sp. nov., an anaerobic dehalogenating bacterium from marine sediments. *Int J Syst Evol Microbiol* 51: 365–371.

Suzuki K, N Ogawa, K Miyashita. 2001. Expression of 2-halobenzoate dioxygenase genes (*cbdSABC*) involved in the degradation of benzoate and 2-halobenzoate in *Burkholderia* sp. TH2. *Gene* 262: 137–145.

Sylvestre M, R Massé, F Messier, J Fauteux, J-G Bisaillon, R Beaudet. 1982. Bacterial nitration of 4-chlorobiphenyl. *Appl Environ Microbiol* 44: 871–877.

Taeger K, H-J Knackmuss, E Schmidt. 1988. Biodegradability of mixtures of chloro- and methylsubstituted aromatics: Simultaneous degradation of 3-chlorobenzoate and 3-methylbenzoate. *Appl Microbiol Biotechnol* 28: 603–608.

Taira K, N Hayase, N Arimura, S Yamashita, T Miyazaki, K Furukawa. 1988. Cloning and nucleotide sequence of the 2,3-dihydroxybiphenyl dioxygenase gene from the PCB-degrading strain of *Pseudomonas paucimobilis* Q1. *Biochemistry* 27: 3990–3996.

Taira K, J Hirose, S Hayashida, K Furukuwa. 1992. Analysis of bph operon from polychlorinated biphenyl-degrading strain of *Pseudomonas pseudoalcaligenes* KF707. *J Biol Chem* 267: 4844–4853.

Takagi K, A Iwasaki, I Kamei, K Satsuma, Y Yoshioka, N Harada. 2009. Aerobic mineralization of hexachlorobenzene by newly isolated pentachloronitrobenzene-degrading *Nocardioides* sp. strain PD653. *Appl Environ Microbiol* 75: 4452–4458.

Vaillancourt FH, G Labbé, NM Drouin, PD Fortin, LD Eltis. 2002. The mechanism-based inactivation of 2,3-dihydroxybiphenyl 1,2-dioxygenase by catecholic substrates. *J Biol Chem* 277: 2019–2027.

van der Meer JR, ARW van Neerven, EJ de Vries, WM de Vos, AJB Zehnder. 1991b. Cloning and characterization of plasmid-encoded genes for the degradation of 1,2-dichloro-, 1,4-dichloro- and 1,2,4-trichlorobenzene of *Pseudomonas* sp. strain P51. *J Bacteriol* 173: 6–15.

van der Meer JR, AJB Zehnder, WM de Vos. 1991a. Identification of a novel composite transposable element, Tn 5280, carrying chlorobenzene dioxygenase genes of *Pseudomonas* sp. strain P51. *J Bacteriol* 173: 7177–7083.

van den Tweel WJ, JB Kok, JA de Bont. 1987. Reductive dechlorination of 2,4-dichlorobenzoate to 4-chlorobenzoate and hydrolytic dehalogenation of 4-chloro, 4-bromo, and 4-iodobenzoate by *Alcaligenes denitrificans* NTB-1. *Appl Environ Microbiol* 53: 810–815.

Vollmer MK, P Fischer, H-J Knackmuss, M Schlömann. 1994. Inability of muconate cycloisomerases to cause dehalogenation during conversion of 2-chloro-*cis,cis*-muconate. *J Bacteriol* 176: 4366–4375.

Vollmer MD, H Hoier, HJ Hecht, U Schell, J Gröning, A Goldman, M Schlömann. 1998. Substrate specificity of and product formation by muconate cycloisomerases: An analysis of wild-type enzymes and engineered variants. *Appl Environ Microbiol* 64: 3290–3299.

Wang G, R Li, S Li, J Jiang. 2010. A novel hydrolytic dehalogenase for the chlorinated aromatic compound chlorothalonil. *J Bacteriol* 192: 2737–2745.

Werlen C, H-P E Kohler, JR van der Meer. 1996. The broad substrate chlorobenzene dioxygenase and *cis*-chlorobenzene dihydrodiol dehydrogenase of *Pseudomonas* sp. strain P51 are linked evolutionarily to the enzymes for benzene and toluene degradation. *J Biol Chem* 271: 4009–4016.

Wiegel J, X Zhang, Q Wu. 1999. Anaerobic dehalogenation of hydroxylated polychlorinated biphenyls by *Desulfitobacterium dehalogenans*. *Appl Environ Microbiol* 65: 2217–2221.

Wu Q, DL Bedard, J Wiegel. 1997a. Effect of incubation temperature on the route of microbial reductive dechlorination of 2,3,4,6-tetrachlorobiphenyl in polychlorinated biphenyl (PCB)-contaminated and PCB-free freshwater sediments. *Appl Environ Microbiol* 63: 2836–2843.

Wu Q, DL Bedard, J Wiegel. 1997b. Temperature determines the pattern of anaerobic microbial dechlorination of Arochlor 1260 primed by 2,3,4,6-tetrachlorobiphenyl in Woods Pond sediment. *Appl Environ Microbiol* 63: 4818–4825.

Wu Q, DL Bedard, J Wiegel. 1999. 2,6-dibromobiphenyl primes extensive dechlorination of Arochlor 1260 in contaminated sediment at 8–30°C by stimulating growth of PCB-dehalogenating microorganisms. *Environ Sci Technol* 33: 595–602.

Wu Q, JEM Watts, KR Sowers, HD May. 2002. Identification of a bacterium that specifically catalyzes the reductive dechlorination of polychlorinated biphenyls with doubly flanked chlorines. *Appl Environ Microbiol* 68: 807–812.

Xun L, E Topp, CS Orser. 1992c. Purification and characterization of a tetrachloro-*p*-hydroquinone reductive dehalogenase from a *Flavobacterium* sp. *J Bacteriol* 174: 8003–8007.

Yadav JS, JF Quensen, JM Tiedje, CA Reddy. 1995. Degradation of polychlorinated biphenyl mixtures (Arochlors 1242, 1254, 1260) by the white rot fungus *Phanerochaete chrysosporium* as evidenced by congener-specific analysis. *Appl Environ Microbiol* 61: 2560–2565.

Yan D-Z, H Liu, N-Y Zhou. 2006. Conversion of *Sphingobium chlorophenolicum* ATCC 39723 to a hexachlorobenzene degrader by metabolic engineering. *Appl Environ Microbiol* 72: 2283–2286.

Yang G, P-H Liang, D Dunaway-Mariano. 1994. Evidence for nucleophilic catalysis in the aromatic substitution reaction catalyzed by (4-chlorobenzoyl)coenzyme A dehalogenase. *Biochemistry* 33: 8527–8531.

Yoshida N, L Ye, D Baba, A Katayama. 2009. A novel *Dehalobacter* species is involved in extensive 4,5,6,7-tetrachlorophthalide dechlorination. *Appl Environ Microbiol* 75: 2400–2405.

Part 2: Halogenated (Chlorine, Bromine, and Iodine) Phenols and Anilines

9.8 Phenols

9.8.1 Aerobic Conditions

There are several mechanisms for the degradation of chlorophenols. For those with less than three halogen substituents, degradation is initiated by hydroxylation to catechols followed by ring fission during which halogen is lost. For those with three or more substituents, monooxygenation takes place with the loss of halogen to produce chlorinated hydroquinones that undergo ring fission. In addition, chlorophenols may take place in dehalorespiration under anaerobic conditions with reductive loss of halide.

9.8.1.1 Hydroxylation and Ring Fission

Phenol hydroxylase has been characterized for 2,4-dichlorophenol (Beadle and Smith 1982; Perkins et al. 1990), and the pathway is shown in Figure 9.28. Halogenated catechols occupy a central position in the metabolism of these chlorophenols, and they undergo intradiol fission for the same reasons that are

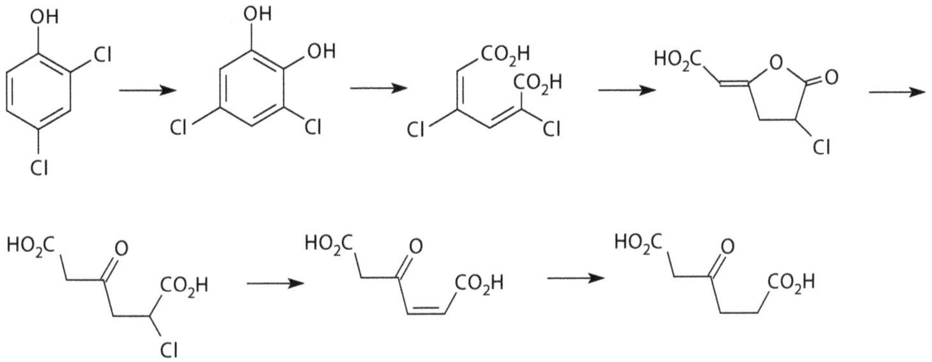

FIGURE 9.28 Aerobic degradation of 3,5-dichlorophenol.

chosen for the degradation of chloroarenes and chlorobenzoates. There are some further aspects of these degradations that merit comment, in particular the stereospecific 2-oxoglutarate dioxygenases and the chlorocatechol dioxygenases.

1. A different 3-chlorocatechol dioxygenase is used for the degradation of 2-chloro- and 2,4-dichlorophenol by *Rhodococcus opacus* ICP, and this involves a *"modified ortho"* ring fission, a second chloromuconate isomerase, a second dienelactone hydrolase and muconolactone isomerase (Moiseeva et al. 2002).

2. The degradation of 2,4-dichlorophenoxyacetate and dichloroprop [(R,S)-2-(2,4-dichloro-phenoxypropanoate)] in *Sphingomonas herbicivorans* is initiated by production of 2,4-dichlorophenol by 2-oxoglutarate dioxygenases to produce chlorophenols (Nickel et al. 1997). Specific *R-* and *S-*2-oxoglutarate dioxygenases have been characterized in *Sphingomonas herbicidivorans* (Müller et al. 2006) that is able to degrade both enantiomers of the racemate (R,S)-4-chloro-2-methylphenoxypropionate (mecoprop) (Zipper et al. 1996), and in *Delftia acidovorans* (Schleinitz et al. 2004). 2,4-Dichlorophenol is then hydroxylated to 3,5-dichlorocatechol (Müller et al. 2004), and there are two different 3,5-dichloro-catechol 1,2-dioxygenases, encoded by dccA1 and dccA2.

9.8.1.2 Monooxygenation

This has been described for several chlorophenols and results in the production of 2,6-dichlorohydroquinone or 2-chlorohydroquinone (Figure 9.29). These are produced from phenols with three or more substituents by initial loss of halide and monooxygenation. For example, 2,6-dichlorohydroquinone is produced from both pentachlorophenol by *Sphingomonas chlorophenolica* (Ohtsubo et al. 1999; Xun et al. 1999), and 2,4,6-trichlorophenol (Li et al. 1991; Xun and Webster 2004), although 2-chloro-6-hydroxyhydroquinone is produced by *Ralstonia eutropha* JMP 134 (Louie et al. 2002). These may then undergo ring fission mediated by 6-chlorohydroxyquinol 1,2-dioxygenases. Further details are given later.

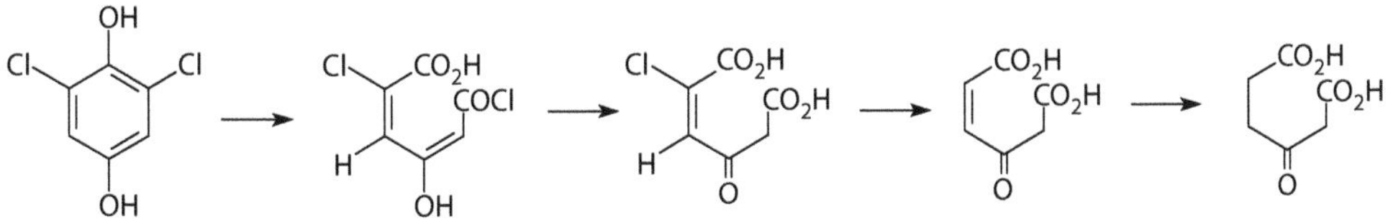

FIGURE 9.29 Degradation of 2,6-dichlorohydroquinone.

9.8.1.3 Chlorohydroquinones

The 2,6-dichlorohydroquinone is degraded by *intradiol* dioxygenation to the acyl chloride followed by spontaneous hydrolysis to 2-chloro-4-hydroxymuconate. The 6-chlorohydroxyquinol 1,2-dioxygenase has been purified from several organisms:

a. The enzyme from *Streptomyces rochei* 303 (Zaborina et al. 1995) had a specificity for 6-chlorohydroxyquinol similar to that of the enzyme from *Azotobacter* GP1 (Wieser et al. 1997), and both enzymes had a preference for the chlorinated quinol rather than the nonchlorinated substrate.

b. The enzyme has also been characterized from the pentachlorophenol-degrading *Sphingomonas chlorophenolica* (Ohtsubo et al. 1999; Xun et al. 1999).

In the degradation of 2,4,5-trichlorophenol produced from the phenoxyacetate by *Burkholderia phenoliruptrix (cepacia)* AC1100, however, chloride is removed from the 2-hydroxy-5-chlorohydroquinone before ring fission (Daubaras et al. 1996).

9.8.1.4 Reductive Displacement of Halide

In addition to hydrolytic replacement of halogen, reductive displacement has been shown to occur during the degradation of a few aromatic compounds, even under aerobic conditions. For example, both hydrolytic and reductive reactions are involved in the degradation of a number of chlorophenols with three or more substituents. Degradation takes place by a pathway that is initiated by monooxygenation accompanied by dechlorination with the formation of hydroquinones. This is followed by formal displacement of chlorine substituents by hydroxyl, or by reduction before ring cleavage. These reactions have been shown to occur in the degradation of a number of polychlorinated phenols, including 2,4,5- and 2,4,6-trichlorophenol, and pentachlorophenol which has attracted particular attention in view of its toxicity.

9.8.1.4.1 Trichlorophenols and Pentachlorophenol

The following examples illustrate initiation by monooxygenation followed by further loss of halogen before ring fission:

1. In the degradation of 2,4,6-trichlorophenol that has been examined in a number of bacteria, monooxygenation plays a key role, and 2,6-dichlorobenzoquinone and 6-chloro-hydroxyquinol have been recognized as intermediates.

   a. The monooxygenase that produced 2,6-dichlorohydroquinone as the initial metabolite has been purified and characterized from *Azotobacter* sp. strain GP1, and is able to accept a number of other polychlorinated phenols (Wieser et al. 1997).

   b. In *Ralstonia eutropha* strain JMP134, both monooxygenation to 2,6-dichloro-hydroquinone and hydrolysis to 2-chloro-6-hydroxybenzoquinone are accomplished by a single enzyme, and this was confirmed by the incorporation of ^{18}O from both $^{18}O_2$ and $H_2^{18}O$ (Xun and Webster 2004). 2-Chloro-6-hydroxybenzoquinone is not, however, a necessary intermediate, since a dioxygenase can bring about ring fission of 2,6-dichlorohydroquinone directly to produce 2-chloromaleylacetate (Figure 9.30a) (Ohtsubo et al. 1999; Xun et al. 1999). Details including the genetics and biochemistry have been given by Louie et al. (2002).

   c. In *Streptomyces rochei* strain 303, 2,6-dichlorohydroquinone and 6-chloro-2-hydroxyhydroquinone are produced, and the dioxygenase that brings about intradiol fission of this to 2-chloromaleylacetate has been purified. The NH_2-terminal amino acid sequence showed a high degree of similarity to the corresponding enzyme from *Azotobacter* sp. strain GP1 (Zaborina et al. 1995).

2. The degradation of 2,4,5-trichlorophenoxyacetate by *Burkholderia* (*Pseudomonas*) *phenoliruptrix (cepacia)* AC1100 (Haugland et al. 1990) is initiated by a 2-oxoglutarate-dependent dioxygenation to 2,4,5-trichlorophenol, which is degraded by initial monooxygenation followed

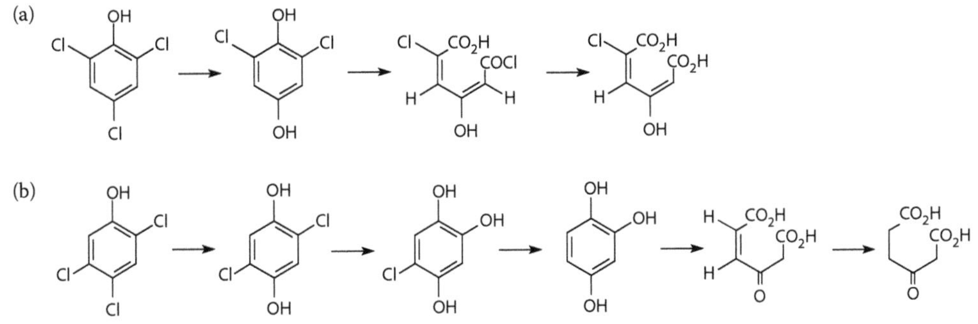

FIGURE 9.30 Degradation of (a) 2,4,6-trichlorophenol and (b) 2,4,5-trichlorophenol.

by reduction to 2,5-dichlorohydroquinone and 5-chloro-2-hydroxy-hydroquinone (Daubaras et al. 1995; Gisi and Xun 2003). This is then dechlorinated to 2-hydroxy-hydroquinone before ring fission to maleylacetate (Figure 9.30b) (Latus et al. 1995; Daubaras et al. 1996; Zaborina et al. 1998). This strain also brings about para-hydroxylation of dichlorophenols, independent of the existence of a chlorine substituent (Tomasi et al. 1995).

3. The degradation of pentachlorophenol has been extensively examined in Gram-positive and Gram-negative genera.

   a. Gram-positive strains of *Mycobacterium* sp. (Suzuki 1983) and *Mycobacterium [Rhodococcus] chlorophenolicum* (Apajalahti and Salkinoja-Salonen 1987b).

   The pathway was delineated for *Mycobacterium chlorophenolicum (Rhodococcus chlorophenolicus)* strain PCP-1, and as for 2,4,6- and 2,4,5-trichlorophenols is initiated by monooxygenation with the formation of tetrachlorohydroquinone. Subsequent steps involve hydrolytic dechlorination followed by three reductive dechlorinations to 1,2,4-trihydroxybenzene, which undergo ring fission (Apajalahti and Salkinoja-Salonen 1987a,b; Uotila et al. 1995) (Figure 9.31a). In *Mycobacterium fortuitum* strain CG-2, the monooxygenase that initiates degradation is membrane associated, whereas the enzyme that carried out hydroxylation and reductive dechlorination to 1,2,4-trihydroxybenzene was soluble (Uotila et al. 1992). Loss of fluoride from pentafluorophenol and bromide from pentabromophenol is also catalyzed by the monooxygenase from *Mycobacterium fortuitum* strain CG-2 (Uotila et al. 1992).

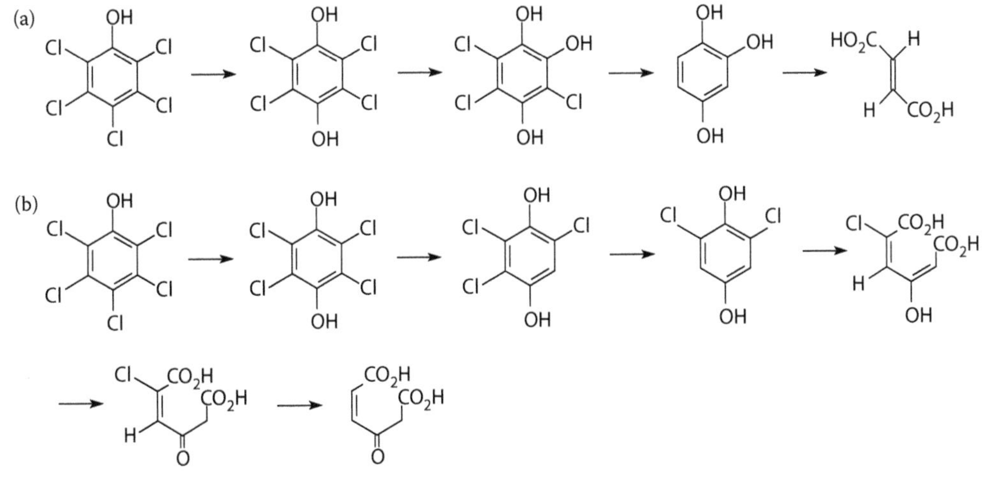

FIGURE 9.31 Degradation of pentachlorophenol by (a) *Mycobacterium chlorophenolicum*, (b) *Sphingobium chlorophenolicum*.

b. Gram-negative sphingomonads (Tiirola et al. 2002b) which include *Sphingobium* (*Sphingo-monas*) *chlorophenolicum* [*Flavobacterium* sp.] (Steiert and Crawford 1986), strains of *Sphingomonas chlorophenolica* (Nohynek et al. 1995), *Sphingomonas chlorophenolica* strain RA2 (Miethling and Karlson 1996), and *Novosphingobium* strain MT1 (Tiirola et al. 2002a).

Degradation by *Sphingobium chlorophenolicum* ATCC 39723 is initiated by monooxygenation (Dai et al. 2003) with the formation of tetrachlorohydroquinone via the corresponding tetrachlorobenzoquinone, and the pathway for its degradation has been described (Cai and Xun 2002). Analysis of the *pcb* gene encoding the monooxygenase in sphingomonads has been carried out (Tiirola et al. 2002b) and suggested natural horizontal gene transfer. The following reactions involve sequential reductive dechlorinations to 2,3,6-trichlorohydroquinone and 2,6-dichloro-hydroquinone mediated by a glutathione *S*-transferase system (Orser et al. 1993). Although a putative chlorohydrolase has been implicated in the degradation of 2,6-dichlorohydroquinone (Chanama and Crawford 1997; Lee and Xun 1997), the substrate can be oxidized with fission of the ring to 2-chloromaleylacetate before reductive dechlorination to maleylacetate (Kaschabek and Reineke 1995) (Figure 9.31b). Comparable pathways are used to bring about dehalogenation of phenols with other *para*-substituted halogen substituents. The pentachlorophenol monooxygenase from *Sphingobium chlorophenolicum* (*Flavobacterium* sp.) ATCC 39723 is also able to hydroxylate a range of other para-substituted phenols with elimination of, for example, chloride, bromide, iodide, cyanide, and nitrite from the 4-position (Xun et al. 1992a–c).

9.9 *O*- and *S*-methylation

In contrast to degradation, bacterial transformation of chlorophenols by *O*-methylation to chloroanisoles (Allard et al. 1987; Häggblom et al. 1989) is noted in view of the potentially adverse environmental effects of these neutral anisoles and veratroles (Neilson et al. 1984). *S*-methylation has been also demonstrated in a range of halogenated thiophenols by bacteria (Neilson et al. 1988) and methanethiol and methaneselenol by *Tetrahymena thermophila* (Drotar et al. 1987).

9.9.1 Dioxygenation

It is worth noting that hydroxylation and dioxygenation are not exclusive, since the toluene dioxygenase from *Pseudomonas putida* F1 can hydroxylate both phenol and 2,5-dichlorophenol (Spain et al. 1989). Chlorosalicylates, like chlorobenzoates are degraded by dioxygenation with decarboxylation to

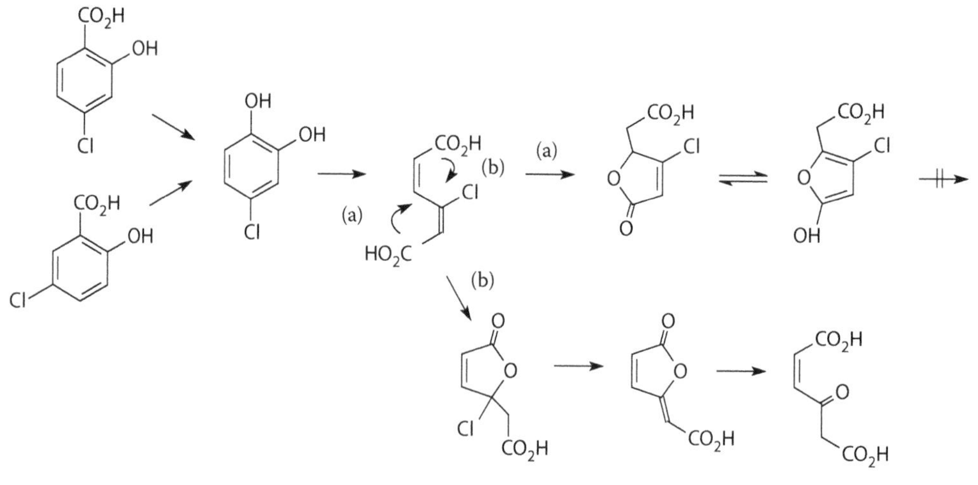

FIGURE 9.32 Loss of chloride to produce the lactone diene and thence 3-oxoadipate. (Adapted from Nikodem P. et al. 2003. *J Bacteriol* 185: 6790–6800.)

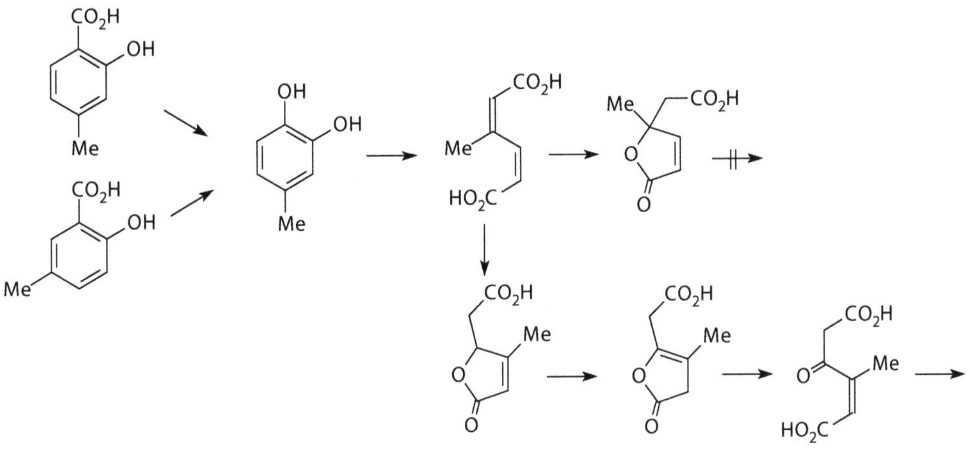

FIGURE 9.33 1:4 Cyclization of 3-methylmuconate to 4-methylmuconolactone. (Adapted from Cámara B et al. 2007. *J Bacteriol* 189: 1664–1674.)

chlorocatechols. The degradation of 4-chloro- and 5-chlorosalicylates by *Pseudomonas reinekei* strain MT1 is initiated by dioxygenation (Nikodem et al. 2003) followed by ring fission to 3-chloromuconate and 3:6-cyclization to 4-chloromuconolactone. There are alternatives for this: either loss of chloride to produce the lactone diene and thence 3-oxoadipate (Figure 9.32), or loss of chloride and CO_2 to protoanemonin that is a terminal product (Nikodem et al. 2003; Cámara et al. 2007). Further analysis revealed the presence of two novel enzymes that were involved in the degradation of 4-chlorocatechol: the gene *ccaA* encoded a chlorocatechol 1,2-dioxygenase, and *ccaB* a chloromuconate cycloisomerase which formed a gene cluster with a *trans*-dienelactone hydrolase (*ccaC*) and maleylacetate reductase (*ccaD*) that enabled this strain to degrade 4- and 5-chlorosalicylate (Cámara et al. 2009). This strain can also metabolize 4-methyl- and 5-methylsalicylate to produce 3-methylmuconate as expected. This, however, undergoes 1:4 cyclization to 4-methylmuconolactone since the alternative 3:6 cyclization is blocked (Figure 9.33).

9.10 Fungi and Yeasts

Although these have been less exhaustively investigated than their bacterial counterparts, the results of these investigations have revealed a number of significant features:

1. The NIH shift (Daly et al. 1972) with translocation of chlorine has been demonstrated during the biotransformation of 2,4-dichlorophenoxyacetate by *Aspergillus niger* (Figure 9.34) (Faulkner and Woodcock 1965). The NIH shift is not restricted to fungi since it has also been demonstrated with protons—though less frequently with other substituents—in prokaryotes.

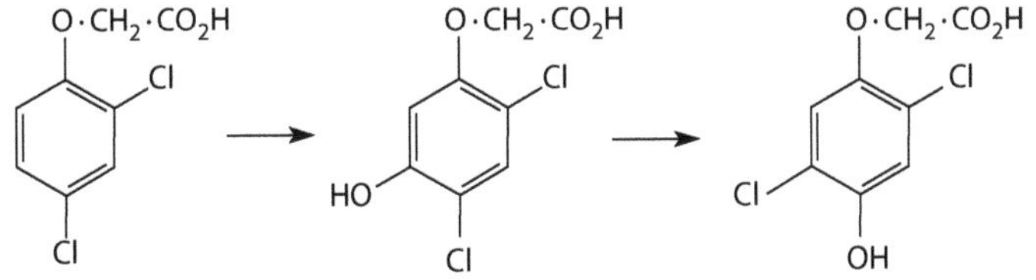

FIGURE 9.34 NIH shift during the metabolism of 2,4-dichlorophenoxyacetate by *Aspergillus niger*.

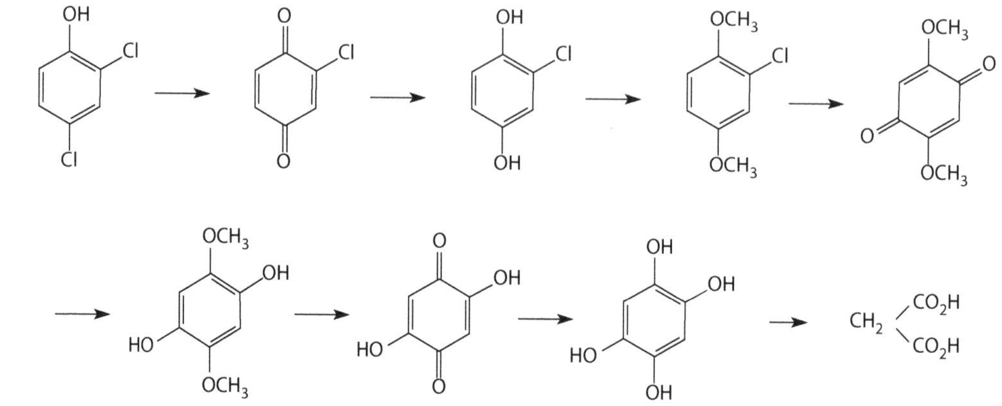

FIGURE 9.35 Biodegradation of 2,4-dichlorophenol by *P. chrysosporium.*

2. The degradation of chlorinated phenols has been examined with the white-rot basidiomycete *P. chrysosporium* under conditions of nitrogen limitation, and apparently involves both lignin peroxidase and manganese-dependent peroxidase activities (Valli and Gold 1991).

    For 2,4-dichlorophenol, the reaction involves a series of oxidations and reductions (Figure 9.35) that are entirely different from the sequence that is employed by bacteria. Essentially similar reactions are involved in the degradation of 2,4,5-trichlorophenol, and this is accomplished more rapidly—possibly due to less interference from polymerization reactions (Joshi and Gold 1993). In addition, 2-chloro-1,4-dimethoxybenzene is produced from 2,4-dichlorophenol, and 2,5-dichloro-1,4-dimethoxybenzene from 2,4,5-trichlorophenol that also produced small amounts of a dimeric product that was tentatively identified as 2,2′-dihydroxy-3,3′,5,5′,6,6′-hexachlorobiphenyl. The degradation of 2,4,6-trichlorophenol by *P. chrysosporium* is analogous to that of 2,4-dichlorophenol, and forms 2-chloro-1,4-dihydroxybenzene that is converted by alternative pathways to 1,2,4-trihydroxybenzene before ring cleavage (Reddy et al. 1998). Parenthetically, a formally comparable sequence of reactions is used for the degradation of 2,4-dinitrotoluene (Valli et al. 1992). The degradation of pentachlorophenol is initiated by formation of tetrachloro-1,4-benzoquinone, and is followed by a series of reductive dechlorinations to produce 1,2,4-trihydroxybenzene that undergoes ring fission (Reddy and Gold 2000).

3. The yeast *Candida maltosa* is capable of assimilating phenol, and phenol-grown cells were able to hydroxylate 2-chlorophenol to 3-chlorocatechol and bring about ring fission to *cis,cis*-2-chloromuconate. 3-Chlorophenol and 4-chlorophenol produced 4-chlorocatechol that was subsequently hydroxylated to 5-chloropyrogallol, which underwent ring fission to 4-carboxy-methylenebut-2-en-4-olide (Polnisch et al. 1992).

9.10.1 Anaerobic Conditions

The persistence of halogenated phenols and anilines in anaerobic environments is determined by the activity of anaerobic dehalogenating bacteria. Extensive effort has, therefore, been devoted to isolating the relevant organisms and an increasing number of strains have been obtained in pure culture. They display different specificities for the position of the halogen; several bring about dechlorination specifically at positions *ortho* to the chlorine, while some of them can use the aromatic substrate as an electron acceptor. The role of reductive dechlorination in energy metabolism (dehalorespiration) has been reviewed (Holliger et al. 1999), and results for chlorophenols and their substrates are summarized in Table 9.7 for *Desulfomonile tiedjei* and Table 9.8 for *Desulfitobacterium hafniense (frappieri)* that are followed by brief comments. Additional details of the reductases are given in Chapter 3.

1. Cells of *Desulfomonile tiedjei* induced by growth with 3-chlorobenzoate were able to partially dechlorinate polychlorinated phenols, specifically at the 3-position, whereas the monochlorophenols

TABLE 9.7

Reductive Dechlorination by *Desulfomonile tiedjei*

Substrate	Product
Pentachlorophenol	2,4,6-Trichlorophenol
2,3,4,6-Tetrachlorophenol	2,4,6-Trichlorophenol
2,3,4-Trichlorophenol	2,4-Dichlorophenol
2,3,5-Trichlorophenol	2,3-Dichlorophenol + 2 chlorophenol
3,4-Dichlorophenol	4-Chlorophenol

Source: Adapted from Mohn WW, KJ Kennedy. 1992. *Appl Environ Microbiol* 58: 1367–1370.

TABLE 9.8

Reductive Dechlorination by *Desulfitobacterium hafniense (frappieri)*

Substrate	Product
Pentachlorophenol	3-Chlorophenol
2,3,4,5-Tetrachlorophenol	3,4,5-Trichloro- → 3,5-dichloro → 3-chlorophenol
2,3,5,6-Tetrachlorophenol	2,3,5-Trichloro- → 3,5-dichloro → 3 chlorophenol
2,3,4-Trichlorophenol	3,4-Dichlorophenol
2,3,5-Trichlorophenol	2,5-Dichlorophenol
3,5-Dichlorophenol	3-Chlorophenol
2,4-Di-chlorophenol	4-Chlorophenol

Source: Adapted from Bouchard B et al. 1996. *Int J Syst Evol Microbiol* 46: 1010–1015; Boyer A et al. 2003. *Biochem J* 373: 297–303.

were apparently resistant to dechlorination (Mohn and Kennedy 1992). Details of the reductase have already been described in Part 1 of this chapter.

2. A sulfite-reducing organism *Desulfitobacterium dehalogenans*, which is unable to carry out dissimilatory reduction of sulfate, dechlorinates 3-chloro-4-hydroxyphenylacetate to 4-hydroxyphenylacetate as a terminal metabolite during growth with pyruvate (Utkin et al. 1994). The specificity of partial dechlorination of *ortho*-chlorinated phenols has been examined (Utkin et al. 1995). The reductive halogenase was characterized, and electron paramagnetic resonance (EPR) analysis showed the presence of one [4Fe–4S] cluster and one cobalamin per monomer (van de Pas et al. 1999). This strain is also able to completely dehalogenate biphenyls with *ortho* chlorine and hydroxyl substituents, such as 3,3′, 5,5′-tetrachloro-4,4′-dihydroxybiphenyl, which is a PCB metabolite from higher organisms (Wiegel et al. 1999). This preference for the presence of an *ortho*-hydroxyl group has been explored in *Desulfitobacterium chlororespirans* and was attributed to its necessity for binding to the substrate (Krasotkina et al. 2001).

3. *Desulfitobacterium hafniense* (*frappieri*) strain PCP-1 was isolated from a methanogenic consortium and could dechlorinate pentachlorophenol to 3-chlorophenol via 2,3,4,5-tetra-, 3,4,5-tri-, and 3,5-dichlorophenol (Bouchard et al. 1996). In addition, it is able to dechlorinate a wide range of polychlorinated aromatic substrates including phenols, catechols, anilines, pentachloronitrobenzenes, and pentachloropyridine (Dennie et al. 1998). There are several membrane-bound reductive dehalogenases have been characterized in this strain, each specific to the substrate: 3,5-dichlorophenol (Thibodeau et al. 2004), 2,4,6-trichlorophenol (Boyer et al. 2003) and pentachlorophenol (Bisaillon et al. 2010). *D. hafniense* strain DCB-2 is able to dechlorinate 3-chloro-4-hydroxyphenylacetate to 4-hydroxyphenylacetate and 2,4,6-trichlorophenol to 4-chlorophenol. The membrane-bound dehalogenase has been characterized and contains per mol of subunit 0.7 mol corrinoid, 12 mol iron, and 13 mol acid-labile sulfur (Christiansen et al. 1998).

4. A facultatively anaerobic organism designated *Anaeromyxobacter dehalogenans* (Sanford et al. 2002) was capable of dechlorinating *ortho*-chlorinated phenols using acetate as electron donor—2-chlorophenol was reduced to phenol and 2,6-dichlorophenol to 2-chlorophenol (Cole et al. 1994). A strain of *Desulfovibrio dechloracetivorans* was also able to couple the dechlorination of *ortho*-substituted chlorophenols to the oxidation of acetate, fumarate, lactate, and propionate (Sun et al. 2000).

5. A spore-forming strain of *Desulfitobacterium chlororespirans* was able to couple the dechlorination of 3-chloro-4-hydroxybenzoate to the oxidation of lactate to acetate, pyruvate, or formate (Sanford et al. 1996). Whereas 2,4,6-trichlorophenol and 2,4,6-tribromophenol supported growth with the production of 4-chlorophenol and 4-bromophenol, neither 2-bromophenol nor 2-iodophenol was able to do this. The membrane-bound dehalogenase contains cobalamin, iron, and acid-labile sulfur, and is apparently specific for *ortho*-substituted phenols (Krasotkina et al. 2001).

6. *Desulfitobacterium chlororespirans* can use *ortho*-substituted phenols as electron acceptors for anaerobic growth, and is able to debrominate 2,6-dibromo-4-cyanophenol (Bromoxynil) and 2,6-dibromo-4-carboxyphenol. In contrast, 2,6-diiodo-4-cyanophenol (Ioxynil) was deiodinated only in the presence of 3-chloro-4-hydroxybenzoate (Cupples et al. 2005).

7. *Desulfovibrio* strain TBP-1 grown with lactate and sulfate as electron acceptor is able to debrominate 2-bromo-, 2,4-dibromo, 2,6-dibromo-, and 2,4,6-tribromophenol (Boyle et al. 1999).

8. *Desulfovibrio dechloracetivorans* is able to use acetate and lactate as electron donors for the dechlorination of 2-chlorophenol and 2,6-dichlorophenol although it was unable to carry this out with 2,3-, 2,4-, and 2,5-dichlorophenol, and 2,4,6-trichlorophenol (Sun et al. 2000).

In addition to these investigations using pure strains, some have used enrichment cultures in various experimental designs that are important to appreciate in interpreting the results. Complete dechlorination of polyhalogenated compounds under anaerobic conditions has not always been observed, and most commonly only partial dehalogenation occurs: all these reactions are, therefore, strictly biotransformations. Illustrative examples include HCB (Figure 9.36a) (Fathepure et al. 1988) and tetrachloroaniline (Figure 9.36b) (Kuhn et al. 1990), although pentachlorophenol can be completely dechlorinated and mineralized (Mikesell and Boyd 1986). The flame-retardant tetrabromobisphenol A was debrominated under both methanogenic and sulfidogenic conditions (Ronen and Abeliovich 2000; Voordeckers et al. 2002).

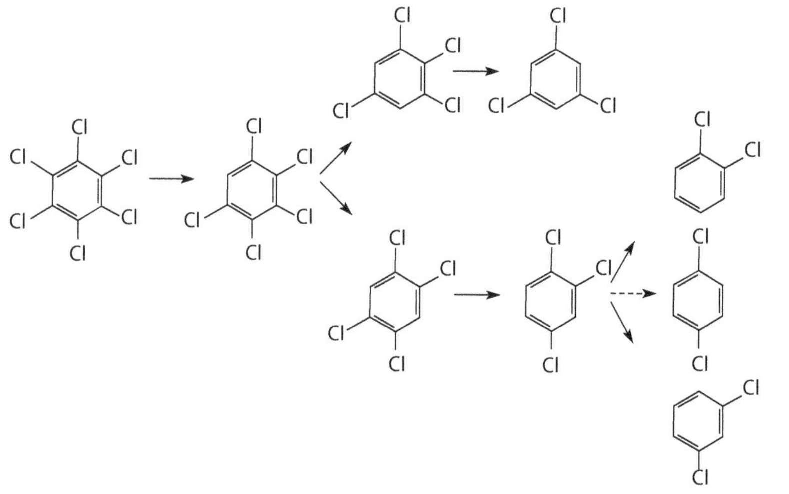

FIGURE 9.36 Partial anaerobic dechlorination of (a) pentachlorophenol and (b) chloroanilines.

9.11 Anilines

9.11.1 Aerobic Conditions

Chlorinated anilines are produced by the hydrolysis of a range of acetanilide, urea, and carbamate herbicides, and are, therefore, widely distributed in agricultural soils. Mechanisms for their loss include not only biodegradation but also the formation of stable complexes with humic material. The range of halogenated anilines that have been examined includes the following:

1. 2-, 3-, and 4-chloroaniline, and 4-fluoroaniline by a *Moraxella* sp. strain G (Zeyer et al. 1985), and 3- and 4-chloroaniline by a strain of *Pseudomonas acidovorans* (Loidl et al. 1990).
2. 2-Methylaniline and 4-chloro-2-methylaniline by *Rhodococcus rhodochrous* strain CTM (Fuchs et al. 1991).
3. 3-Chloro-4-methylaniline by *Burkholderia* (*Pseudomonas*) *cepacia* strain CMA1 (Stockinger et al. 1992).
4. 3,4-Dichloroaniline by *Variovorax* sp. strain WDL1 (Dejonghe et al. 2003).
5. The mineralization of the 2,3-, 2,4-, 2,5- and 3,5-dichloroanilines by a strain of *Bacillus megaterium* IMT21 (Yao et al. 2011). Whereas 3,4- and 3,5-dichloroaniline were acetylated, hydroxylation of 2,3-dichloroaniline occurred although the position of hydroxylation was unresolved (Yao et al. 2011).

Degradation of chloroanilines is most plausibly initiated by dioxygenation with the elimination of NH_4^+, followed by ring fission of the resulting chlorocatechols. The pathway for 4-chloroaniline involves oxidative deamination to 4-chlorocatechol and intradiol ring fission to the same intermediates that are involved in the degradation of halogenated catechols (Zeyer et al. 1985). From the broad specificity of aniline oxygenase, it is plausible to assume that the pathway is the same for all 4-halogenated anilines. The degradation of 3-chloroaniline has been examined in strains of *Comamonas testosteroni* and *Delftia* (*Pseudomonas*) *acidovorans*, and is initiated by dioxygenation with the production of 4-chlorocatechol that could undergo either intradiol or extradiol ring fission (Boon et al. 2001). The herbicide linuron, which is a urea derivative of 3,4-dichloroaniline, can be degraded by *Variovorax* sp. strain WDL1 to 3,4-dichloroaniline before fission of 3,4- or 4,5-dichlorocatechol by an unresolved pathway (Dejonghe et al. 2003).

There are, however, a number of complicating issues that determine the degradability of chloroanilines:

1. The degradation of aniline may be induced by aniline, although both 3- and 4- chloroaniline—which are poor substrates—were able to induce the enzymes for aniline degradation in a strain of *Pseudomonas* sp. (Konopka et al. 1989). This strain was able to degrade aniline in the presence of readily degradable substrates such as lactate.
2. The degradation of 3- and 4-chloroaniline may require the presence of either aniline or glucose (references in Zeyer et al. 1985), while the metabolism of methylanilines required the addition of ethanol as additional carbon source (Fuchs et al. 1991).
3. The degradation of 3-chloro-4-methylaniline by *Pseudomonas cepacia* strain CMA1 involved ring fission of 3-chloro-4-methylcatechol by an intradiol enzyme (Stockinger et al. 1992).
4. The gene *tdnQ* that encodes the oxidative deamination of aniline is carried on a plasmid in strains of *Comamomas testosteroni* and *Delftia acidovorans*. However, it has not been resolved whether this enzyme is involved in the degradation of 3-chloroaniline, because the gene is transcribed in the presence of aniline, but not when only 3-chloroaniline is present (Boon et al. 2001).

A few additional comments are worth noting.

1. The degradation of 5-nitroanthranilic acid by *Bradyrhizobium* sp. strain JS 329 displayed two unusual features: (i) the amine group underwent hydrolytic deamination to 5-nitrosalicylate

and (ii) this was followed by dioxygenation with simultaneous ring fission to produce 4-nitro-6-oxo-hepta-2,4-diendioate from which nitrite was lost to form methylpyruvate (Qu and Spain 2011). This can be compared with the degradation of 5-aminosalicylate that is initiated by dioxygenation and concomitant ring fission, followed by hydrolytic loss of ammonia from the ring fission product (Stolz et al. 1992).

2. Although fungal transformation of anilines by acylation is well established, the degradation of 3,4-dichloroaniline by the white-rot fungus *P. chrysosporium* has been shown to occur at 37°C when oxygen was provided and oligomerization was restricted. Degradation took place by reaction with 2-oxoglutaryl-CoA followed by formation of the succinimide that was then degraded to CO_2 (Sandermann et al. 1998).

9.11.2 Anaerobic Conditions

1. A strain of *Rhodococcus* sp. grown anaerobically in the presence of nitrate carried out an unusual reaction with 3,4-dichloro-, 3,4-difluoro-, and 3-chloro-4-fluoroaniline. In addition to the formation of 3,4-dichloroacetanilide, 1,2-dichlorobenzene and 1,2-difluorobenzene were produced by deamination, and ^{19}F NMR was used to delineate the products (Travkin et al. 2002).

2. The partial dechlorination of chlorinated anilines has been examined in anaerobic slurries, which has already been noted (Kuhn et al. 1990), while *D. hafniense (frappieri)* strain PCP-1 is able to dechlorinate a wide range of polychlorinated aromatic substrates including phenols, catechols, anilines, pentachloronitrobenzene, and pentachloropyridine (Dennie et al. 1998). Both pentachloroaniline and 2,3,5,6-tetrachloroaniline were partially dechlorinated to dichloroanilines, although the orientation of the substituents was unresolved. The kinetics of dechlorination of pentachloroaniline have been described (Tas et al. 2006).

9.12 References

Allard A-S, M Remberger, AH Neilson. 1987. Bacterial *O*-methylation of halpogen-substituted phenols. *Appl Environ Microbiol* 53: 839–845.

Apajalahti JHA, MS Salkinoja-Salonen. 1987a. Dechlorination and para-hydroxylation of polychlorinated phenols by *Rhodococcus chlorophenolicus. J Bacteriol* 169: 675–681.

Apajalahti JHA, MS Salkinoja-Salonen. 1987b. Complete dechlorination of tetrachlorohydroquinone by cell extracts of pentachlorophenol-induced *Rhodococcus chlorophenolicus. J Bacteriol* 169: 5125–5130.

Beadle CA, ARW Smith. 1982. The purification and properties of 2,4-dichlorophenol hydroxylase from a strain of *Acinetobacter* sp. *Eur J Biochem* 123: 323–332.

Bisaillon A, R Beaudet, F Lépine, E Déziel, R Villemur. 2010. Identification and characterization of a novel CprA reductive dehalogenase specific for highly chlorinated phenols from *Desulfitobacterium hafniense* strain PCP-1. *Appl Environ Microbiol* 76: 7536–754.

Boon N, J Goris, P de Vos, W Verstraete, EM Top. 2001. Genetic diversity among 3-chloroaniline- and aniline-degrading strains of *Comomonadaceae. Appl Environ Microbiol* 67: 1107–1115.

Bouchard B, R Beaudet, R Villemur, G McSween, F Lépine, J-G Bisaillon. 1996. Isolation and characterization of *Desulfobacterium frappieri* sp. nov., an anaerobic bacterium which reductively dechlorinates pentachlorophenol to 3-chlorophenol. *Int J Syst Evol Microbiol* 46: 1010–1015.

Boyer A, R P-Bélanger, M Saucier, R Villemur, F Lépine, P Juteau, R Baudet. 2003. Purification, cloning and sequencing of an enzyme mediating the reductive dechlorination of 2,4,6-trichlorophenol from *Desulfitobacterium frappieri* PCP-1. *Biochem J* 373: 297–303.

Boyle AW, CD Phelps, LY Young. 1999. Isolation from estuarine sediments of a *Desulfovibrio* strain which can grow on lactate coupled to the reductive dehalogenation of 2,4,6-tribromophenol. *Appl Environ Microbiol* 65: 1133–1140.

Cai M, L Xun. 2002. Organization and regulation of pentachlorophenol-degrading genes in *Sphingobium chlorophenolicum* ATCC 39723. *J Bacteriol* 184: 4672–4680.

Cámara B, P Bielecki, F Kaminski, VM dos Santos, I Plumeier, P Nikodem, DH Pieper. 2007. A Gene cluster involved in degradation of substituted salicylates via *ortho*-cleavage in *Pseudomonas* sp. strain MT1 encodes enzymes specifically adapted for transformation of 4-methylcatechol and 3-methylmuconate. *J Bacteriol* 189: 1664–1674.

Cámara B, P Nikodem, P Bielecki, R Bobadilla, H Junca, DH Pieper. 2009. Characterization of a gene cluster involved in 4-chlorocatechol degradation by *Pseudomonas reinekei* MT1. *J Bacteriol* 191: 4905–4915.

Chanama S, RL Crawford. 1997. Mutational analysis of *pcpA* and its role in pentachlorophenol degradation by *Sphingomonas* (*Flavobacterium*) *chlorophenolica* ATCC39723. *Appl Environ Microbiol* 63: 4833–4838.

Christiansen N, BK Ahring. 1996. *Desulfitobacterium hafniense* sp. nov., an anaerobic reductively dechlorinating bacterium. *Int Syst Bacteriol* 46: 442–448.

Christiansen N, BK Ahring, G Wohlfarth, G Diekert. 1998. Purification and characterization of the 3-chloro-4-hydroxyphenylacetate reductive dehalogenase of *Desulfitobacterium hafniense*. *FEBS Lett* 436: 159–162.

Cole JR, AL Cascarelli, WW Mohn, JM Tiedje. 1994. Isolation and characterization of a novel bacterium growing via reductive dehalogenation of 2-chlorophenol. *Appl Environ Microbiol* 60: 3536–3542.

Cupples AM, RA Sanford, GK Sims. 2005. Dehalogenation of the herbicides Bromoxynil (3,5-dibromo-4-hydroxybenzonitrile) and Ioxynil (3,5-diiodo-4-hydroxybenzonitrile) by *Desulfitobacterium chlororespirans*. *Appl Environ Microbiol* 71: 3741–3746.

Dai M, JB Rogers, JR Warner, SD Copley. 2003. A previously unrecognized step in pentachlorophenol degradation in *Sphingobium chlorophenolicum* is catalyzed by tetrachlorobenzoquinone reductase (PcpD). *J Bacteriol* 185: 302–310.

Daly J, DM Jerina, B Witkop. 1972. Arene oxides and the NIH shift: The metabolism, toxicity and carcinogenicity of aromatic compounds. *Experientia* 28: 1129–1149.

Daubaras DL, CD Hershberger, K Kitano, AM Chakrabarty. 1995. Sequence analysis of a gene cluster involved in metabolism of 2,4,5-trichlorophenoxyacetic acid by *Burkholderia cepacia* AC1100. *Appl Environ Microbiol* 61: 1279–1289.

Daubaras DL, K Saido, AM Chakrabarty. 1996. Purification of hydroxyquinol 1,2-dioxygenase and maleylacetate reductase: The lower pathway of 2,4,5-trichlorophenoxyacetic acid metabolism by *Burkholderia cepacia* AC1100. *Appl Environ Microbiol* 62: 4276–4279.

Dejonghe W, E Berteloot, J Goris, N Boon, K Crul, S Maertens, M Höfte, P de Vos, W Verstraete, EM Top. 2003. Synergistic degradation of linuron by a bacterial consortium and isolation of a single linuron-degrading *Variovorax* strain. *Appl Environ Microbiol* 69: 1532–1542.

Dennie D, I Gladu, F Lépine, R Villemur, J-G Bisaillon, R Beaudet. 1998. Spectrum of the reductive dehalogenation activity of *Desulfitobacterium frappieri* PCP-1. *Appl Environ Microbiol* 64: 4603–4606.

Drotar A-M, LR Fall, EA Mishalanie, JE Tavernier, R Fall. 1987. Enzymatic methylation of sulfide, selenide, and organic thiols by *Tetrahymena thermophila*. *Appl Environ Microbiol* 53: 2111–2118.

Endo R, M Kamakura, K Miyauchi, M Fukuda, Y Ohtsubo, M Tsuda, Y Nagata. 2005. Identification and characterization of genes involved in the downstream degradation pathway of γ-hexachlorocyclohexane in *Sphingomonas paucimobilis* UT26. *J Bacteriol* 187: 847–853.

Fathepure BZ, JM Tiedje, SA Boyd. 1988. Reductive dechlorination of hexachlorobenzene to tri- and dichlorobenzenes in anaerobic sewage sludge. *Appl Environ Microbiol* 54: 327–330.

Faulkner JK, D Woodcock. 1965. Fungal detoxication. Part VII. Metabolism of 2,4-dichlorophenoxyacetic and 4-chloro-2-methylphenoxyacetic acids by *Aspergillus niger. J Chem Soc* 1187–1191.

Fuchs K, A Schreiner, F Lingens. 1991. Degradation of 2-methylaniline and chlorinated isomers of 2-methylaniline by *Rhodococcus rhodochrous* strain CTM. *J Gen Microbiol* 137: 2033–2039.

Gisi MR, L Xun. 2003. Characterization of chlorophenol 4-monooxygenase (TftD) and NADH: Flavin adenine dinucleotide oxidoreductase (TftC) of *Burkholöderia cepacia* AC100. *J Bacteriol* 185: 2786–2792.

Häggblom MM, D Janke, PJM Middeldorp, MS Salknoja-Salonen. 1989. *O*-methylation of chlorinated phenols in the genus *Rhodococcus*. *Arch Microbiol* 152: 6–9.

Haugland RA, DJ Schlemm, RP Lyons, PR Sferra, AM Chakrabarty. 1990. Degradation of the chlorinated phenoxyacetate herbicides 2,4-dichlorophenoxyacetic acid and 2,4,5-trichlorophenoxyacetic acid by pure and mixed bacterial cultures. *Appl Environ Microbiol* 56: 1357–1362.

Holliger C, G Wohlfarth, G Diekert. 1999. Reductive dechlorination in the energy metabolism of anaerobic bacteria. *FEMS Microbiol Revs* 22: 383–398.

Joshi DK, MH Gold. 1993. Degradation of 2,4,5-trichlorophenol by the lignin-degrading basidiomycete *Phanerochaete chrysosporium*. *Appl Environ Microbiol* 59: 1779–1785.

Kaschabek SR, W Reineke. 1995. Maleylacetate reductase of *Pseudomonas* sp. strain B13: Specificity of substrate conversion and halide elimination. *J Bacteriol* 177: 320–325.

Konopka A, D Knoght, RF Turco. 1989. Characterization of a *Pseudomonas* sp. capable of aniline degradation in the presence of secondary carbon sources. *Appl Environ Microbiol* 55: 385–389.

Krasotkina J, T Walters, KA Maruya, SW Ragsdale. 2001. Characterization of the B_{12}- and iron–sulfur-containing reductive dehalogenase from *Desulfitobacterium chlororespirans*. *J Biol Chem* 276: 40991–40997.

Kuhn EP, GT Townsend, JM Suflita. 1990. Effect of sulfate and organic carbon supplements on reductive dehalogenation of chloroanilines in anaerobic aquifer slurries. *Appl Environ Microbiol* 56: 2630–2637.

Latus M, H-J Seitz, J Eberspächer, F Lingens. 1995. Purification and characterization of hydroxyquinol 1,2-dioxygenase from *Azotobacter* sp. strain GP1. *Appl Environ Microbiol* 61: 2453–2460.

Lee J-Y, L Xun. 1997. Purification and characterization of 2,6-dichloro-*p*-hydroquinone chlorohydrolase from *Flavobacterium* sp. strain ATCC 39723. *J Bacteriol* 179: 1521–1524.

Li D-Y, J Eberspächer, B Wagner, J Kuntzer, F Lingens. 1991. Degradation of 2,4,6-trichlorophenol by *Azotobacter* sp. strain GP1. *Appl Environ Microbiol* 57: 1920–1928.

Löffler FE, RA Sanford, JM Tiedje. 1996. Initial characterization of a reductive dehalogenase from *Desulfitobacterium chlororespirans* Co23. *Appl Environ Microbiol* 62: 3809–3813.

Loidl M, C Hinteregger, G Ditzelmüller, A Ferschl, F Streichsbier. 1990. Degradation of aniline and monochlorinated anilines by soil-borne *Pseudomonas acidovorans* strains. *Arch Microbiol* 155: 56–61.

Louie TM, CM Webster, L Xun. 2002. Genetic and biochemical characterization of a 2,4,6-trichlorophenol degradation pathway in *Ralstonia eutropha* JMP134. *J Bacteriol* 184: 3492–3500.

Miethling R, Karlson U. 1996. Accelerated mineralization of pentachlorophenol in soil upon inoculation with *Mycobacterium chlorophenolicum* PCP1 *and Sphingomonas chlorophenolica* RA2. *Appl Environ Microbiol* 62: 4361–4366.

Mikesell MD, SA Boyd. 1986. Complete reductive dechlorination and mineralization of pentachlorophenol by anaerobic microorganisms. *Appl Environ Microbiol* 52: 861–865.

Miyauchi K, Y Adachi, Y Nagata, M Takagi. 1999. Cloning and sequencing of a novel *meta*-cleavage dioxygenase gene whose product is involved in degradation of γ-hexachlorocyclohexane in *Sphingomonas paucimobilis*. *J Bacteriol* 181: 6712–6719.

Mohn WW, KJ Kennedy. 1992. Reductive dehalogenation of chlorophenols by *Desulfomonile tiedjei* DCB-1. *Appl Environ Microbiol* 58: 1367–1370.

Moiseeva OV, IP Solyanikova, SR Kaschabek, J Gröning, M Thiel, LA Golovleva, M Schlömann. 2002. A new modified *ortho* cleavage pathway of 3-chlorocatechol degradation by *Rhodococcus opacus* 1CP: Genetic and biochemical evidence. *J Bacteriol* 184: 5282–5292.

Müller TA, SM Byrde, C Werlen, JR van der Meer, H-P E Kohler. 2004. Genetic analysis of phenoxyalkanoic acid degradation in *Sphingomonas herbicidovorans*. *Appl Environ Microbiol* 70: 6066–6175.

Müller TA, T Fleischmann, JR van der Meer, H-P E Kohler. 2006. Purification and characterization of two enantioselective α-ketoglutarate-dependent dioxygenases, RdpA and SdpA, from *Sphingomonas herbicidovorans*. *Appl Environ Microbiol* 72: 4853–4861.

Neilson AH, A-S Allard, S Reiland, M Remberger, A Tärnholm, T Viktor, L Landner. 1984. Tri- and tetrachloroveratrole, metabolites produced by bacterial *O*-methylation of tri- and tetrachloroguaiacol: An assessment of their bioconcentration potential and their effects on fish reproduction. *Can J Fish Aquat Sci* 41: 1502–1512.

Neilson AH, C Lindgren, P-Å Hynning, M Remberger. 1988. Methylation of halogenated phenols and thiophenols by cell extracts of Gram-positive and Gram-negative bacteria. *Appl Environ Microbiol* 54: 524–530.

Nickel K, MJ-F Suter, H-PE Kohler. 1997. Involvement of two α-ketoglutarate-dependent dioxygenases in enantioselective degradation of (*R*)- and (*S*)-mecoprop by *Sphingomonas herbicidovorans* MH. *Appl Environ Microbiol* 63: 6674–6679.

Nikodem P, V Hecht, M Schlömann DH Pieper. 2003. New bacterial pathway for 4- and 5-chlorosalicylate degradation via 4-chlorocatechol and maleylacetate in *Pseudomonas* sp. strain MT1. *J Bacteriol* 185: 6790–6800.

Nohynek JL, EL Suhonen, E-L Nurmiaho-Lassila, M Hantula, M Salkinoja-Salonen. 1995. Description of four pentachlorophenol-degrading bacterial strains as *Sphingomonas chlorophenolica* sp. nov. *Syst Appl Microbiol* 18: 527–538.

Ohtsubo Y, K Miyauchi, K Kanda, T Hatta, H Kiyohara, T Senda, Y Nagata, Y Mitsui, M Takagi. 1999. PchA, which is involved in the degradation of pentachlorophenol in *Sphingomonas chlorophenolica* ATCC 39723, is a novel type of ring-cleavage dioxygenase. *FEBS Letters* 459: 395–398.

Orser CS, J Dutton, C Lange, P Jablonsli, L Xun, M Hargis. 1993. Characterization of a *Flavobacterium* glutathione *S*-transferase gene involved in reductive dechlorination. *J Bacteriol* 175: 2640–2644.

Perkins EJ, MP Gordon, O Caceres, PF Lurquin. 1990. Organization and sequence analysis of the 2,4-dichlorophenol hydroxylase and dichlocatechol oxidative operons of plasmid pJP4. *J Bacteriol* 172: 2351–2359.

Polnisch E, H Kneifel, H Franzke, KL Hofmann. 1992. Degradation and dehalogenation of monochlorophenols by the phenol-assimilating yeast *Candida maltosa. Biodegradation* 2: 193–199.

Qu Y, JC Spain. 2011. Molecular and biochemical characterization of the 5-nitroanthranilic acid degradation pathway in *Bradyrhizobium* sp. strain JS 329. *J Bacteriol* 193: 3057–3063.

Reddy GVB, MDS Gelpke, MH Gold. 1998. Degradation of 2,4,6-trichlorophenol by *Phanerochaete chrysosporium*: Involvement of reductive dechlorination. *J Bacteriol* 180: 5159–5164.

Reddy GVB, MH Gold. 2000. Degradation of pentachlorophenol by *Phanerochaete chrysosporium*: Intermediates and reactions involved. *Microbiology (UK)* 146: 405–413.

Ronen Z, A Abeliovich. 2000. Anaerobic–aerobic process for microbial degradation of tetrabromobisphenol-A. *Appl Environ Microbiol* 66: 2372–2377.

Sandermann H, W Heller, N Hertkorn, E Hoque, D Pieper, R Winkler. 1998. A new intermediate in the mineralization of 3,4-dichloroaniline by the white-rot fungus *Phanerochaete chrysosporium. Appl Environ Microbiol* 64: 3305–3312.

Travkin V, BP Baskunov, EL Golovlev, MG Boersma, S Boeren, J Vervoort, WJH van Berkel, IMCM Rietjens, LA Golovleva. 2002. Reductive deamination as a new step in the anaerobic microbial degradation of halogenated anilines. *FEMS Microbiol Lett* 209: 307–312.

Sanford RA, JR Cole, FE Löffler, JM Tiedje. 1996. Characterization of *Desulfitobacterium chlororespirans* sp. nov., which grows by coupling the oxidation of lactate to the reductive dechlorination of 3-chloro-4-hydroxybenzoate. *Appl Environ Microbiol* 62: 3800–3808.

Sanford RA, JR Cole, JM Tiedje. 2002. Characterization and description of *Anaeromyxobacter dehalogenans* gen., nov., sp. nov., an aryl-halorespiring facultative anaerobic myxobacterium. *Appl Environ Microbiol* 68: 893–900.

Schleinitz KM, S Kleinsteuber, T Vallaeys, W Babel. 2004. Localization and characterization of two novel genes encoding stereospecific dioxygenases catalyzing 2(2,4-dichlorophenoxy)propionate cleavage in *Delftia acidovorans* MC1. *Appl Environ Microbiol* 70: 5357–5365.

Spain JC, GJ Zylstra, CK Blake, DT Gibson. 1989. Monohydroxylation of phenol and 2,5-dichlorophenol by toluene dioxygenase in *Pseudomonas putida* F1. *Appl Environ Microbiol* 55: 2648–2652.

Steiert JG, RL Crawford. 1986. Catabolism of pentachlorophenol by a *Flavobacterium* sp. *Biochem Biophys Res Commun* 141: 1421–1427.

Stockinger J, C Hinteregger, M Loidl, A Ferschl, F Streichsbier. 1992. Mineralizarion of 3-chloro-4-methylamiline via an *ortho*-cleavage pathway by *Pseudomonas cepacia* strain CMA1. *Appl Microbiol Biotechnol* 38: 421–428.

Stolz A, B Nörtemann, H-J Knackmuss. 1992. Bacterial metabolism of 5-aminosalicylic acid. Initial ring cleavage. *Biochem J* 282: 675–680 .

Sun B, JR Cole, RA Sanford, JM Tiedje. 2000. Isolation and characterization of *Desulfobvibrio dechloracetivorans* sp. nov., a marine dechlorinating bacterium growing by coupling the oxidation of acetate to the reductive dechlorination of 2-chlorophenol. *Appl Environ Microbiol* 66: 2408–2413.

Suzuki T. 1983. Methylation and hydroxylation of pentachlorophenol by *Mycobacterium* sp. isolated from soil. *J Pesticide Sci* 8: 419–428.

Tas DO, IN Thomson, FE Löffler, SG Pavlostathis. 2006. Kinetics of the microbial reductive dechlorination of pentachloroaniline. *Environ Sci Technol* 40: 4467–4472.

Thibodeau J, A Gauthier, M Duguay, R Villemur, F Lépine, P Juteau, R Beaudet. 2004. Purification, cloning, and sequencing of a 3,5-dichlorophenol reductive dehalogenase from *Desulfitobacterium frappieri* PCP-1. *Appl Environ Microbiol* 70: 4532–4537.

Tiirola MA, MK Männistö, JA Puhakka, MS Kulomaa. 2002a. Isolation and characterization of *Novosphingobium* sp. strain MT1, a dominant polychlorophenol-degrading strain in a groundwater bioremediation system. *Appl Environ Microbiol* 68: 173–180.

Tiirola MA, H Wang, L Paulin, MS Kulomaa. 2002b. Evidence for natural horizontal transfer of the *pcpB* gene in the evolution of polychlorophenol-degrading sphingomonads. *Appl Environ Microbiol* 68: 4495–4501.

Tomasi I, I Artaud, Y Bertheau, D Mansuy. 1995. Metabolism of polychlorinated phenols by *Pseudomonas cepacia* AC1100: Determination of the first two steps and specific inhibitory effect of methimazole. *J Bacteriol* 177: 307–311.

Uotila JS, VH Kitunen, T Coote, T Saastamoinen, MS Salkinoja-Salonen, JHA Apajalahtri. 1995. Metabolism of halohydroquinones in *Rhodococcus chlorophenolicus* PCP-1. *Biodegradation* 6: 119–126.

Uotila JS, VH Kitunen, T Saastamoinen, T Coote, MM Häggblom, MS Salkinoja-Salonen. 1992. Characterization of aromatic dehalogenases of *Mycobacterium fortuitum* CG-2. *J Bacteriol* 174: 5669–5675.

Utkin I, DD Dalton, J Wiegel. 1995. Specificity of reductive dehalogenation of substituted *ortho*chlorophenols by *Desulfitobacterium dehalogenans* JW/IU-DC1. *Appl Environ Microbiol* 61: 346–351.

Utkin I, C Woese, J Wiegel. 1994. Isolation and characterization of *Desulfitobacterium dehalogenans* gen. nov., sp. nov., an anaerobic bacterium which reductively dechlorinates chlorophenolic compounds. *Int J Syst Bacteriol* 44: 612–619.

Valli K, BJ Brock, DK Joshi, MH Gold. 1992. Degradation of 2,4-dinitrotoluene by the lignin-degrading fungus *Phanerochaete chrysosporium*. *Appl Environ Microbiol* 58: 221–228.

Valli K, MH Gold. 1991. Degradation of 2,4-dichlorophenol by the lignin-degrading fungus *Phanerochaete chrysosporium*. *J Bacteriol* 173: 345–352.

van de Pas BA, H Smidt, WR Hagen, J van der Oost, G Schraa, AJM Stams, WM de Vos. 1999. Purification and molecular characterization of *ortho*-chlorophenol reductive dehalogenase, a key enzyme of halorespiration in *Desulfitobacterium dehalogenans*. *J Biol Chem* 274: 20287–20292.

Voordeckers JW, DE Fennell, K Jones, MM Häggblom. 2002. Anaerobic biotransformation of tetrabromobisphenol-A, tetrachlorobisphenol-A, and bisphenol A in estuarine sediments. *Environ Sci Technol* 36: 696–701.

Wiegel J, X Zhang, Q Wu. 1999. Anaerobic dehalogenation of hydroxylated polychlorinated biphenyls by *Desulfitobacterium dehalogenans*. *Appl Environ Microbiol* 65: 2217–2221.

Wieser M, B Wagner, J Eberspächer, F Lingens. 1997. Purification and characterization of 2,4,6-trichlorophenol-4-monooxygenase, a dehalogenating enzyme from *Azotobacter* sp. strain GP1. *J Bacteriol* 179: 202–208.

Xun L, J Bohuslavek, M Cai. 1999. Characterization of 2,6-dichloro-*p*-hydroquinone 1,2-dioxygenase (PcpA) of *Sphingomonas chlorophenolica* ATCC 39723. *Biochem Biophys Res Commun* 266: 322–325.

Xun L, E Topp, CS Orser. 1992a. Confirmation of oxidative dehalogenation of pentachlorophenol by a *Flavobacterium* pentachlorophenol hydroxylase. *J Bacteriol* 174: 5745–5747.

Xun L, E Topp, CS Orser. 1992b. Diverse substrate range of a *Flavobacterium* pentachlorophenol hydroxylase and reaction stoichiometries. *J Bacteriol* 174: 2898–2902.

Xun L, E Topp, CS Orser. 1992c. Purification and characterization of a tetrachloro-*p*-hydroquinone reductive dehalogenase from a *Flavobacterium* sp. *J Bacteriol* 174: 8003–8007.

Xun L, CM Webster. 2004. A monooxygenase catalyzes sequential dechlorination of 2,4,6-trichlorophenol by oxidative and hydrolytic reactions. *J Biol Chem* 279: 6696–6700.

Yao X-F, F Khan, R Pandey, J Pandey, RG Mourant, RK Jain, J-H Guo, RJ Russell, JG Oakenshot, G Pandey. 2011. Degradation of dichloroaniline isomers by a newly isolated strain *Bacillus megaterium* IMT21. *Microbiology (UK)* 157: 721–726.

Zaborina O, DL Daubaras, A Zago, Y Xun, K Saido, T Klem, D Nikolic, AM Chakrabarty. 1998. Novel pathway for conversion of chlorohydroxyquinol to maleylacetate in *Burkholderia cepacia* AC1100. *J Bacteriol* 180: 4667–4675.

Zaborina O, M Latus, J Eberspächer, L Golovleva, F Lingens. 1995. Purification and characterization of 6-chlorohydroquinol 1,2-dioxygenase from *Streptomyces rochei* 303: comparison with an analogous enzyme from *Azotobacter* sp. strain GP1. *J Bacteriol* 177: 229–234.

Zeyer J, A Wasserfallen, KN Timmis. 1985. Microbial mineralization of ring-substituted anilines through an *ortho*-cleavage pathway. *Appl Environ Microbiol* 50: 447–453.

Zipper C, K Nickel, W Angst, H-P E Kohler. 1996. Complete microbial degradation of both enantiomers of the chiral herbicide Mecoprop [(*R*,S)-2-(4-chloro-2-methylphenoxy)]propionic acid in an enantioselective manner by *Sphingomonas herbicidovorans* sp. nov. *Appl Environ Microbiol* 62: 4318–4322.

Part 3: Fluorinated Hydrocarbons, Carboxylates, Phenols, and Anilines

The basic mechanisms for the aerobic degradation of carbocyclic aromatic compounds are well established and initial dioxygenation is now recognized as one of the fundamental mechanisms for activating aromatic rings to aerobic degradation. Many of the mechanisms that operate for chlorinated and brominated arenes apply with relatively minor modification to their fluorinated analogs. Further details are given in a review (Neilson and Allard 2002).

9.13 Fluorinated Aromatic Hydrocarbons

9.13.1 Aerobic Conditions

The transformation of a series of 3-substituted fluorobenzenes has been examined in *Pseudomonas* strain T-2 in which toluene 2,3-dioxygenase was synthesized (Renganathan 1989). Dioxygenation took place for all of them, and was sometimes accompanied by concomitant loss of fluorine (Figure 9.37). Degradation of fluorobenzene by strain F11 within the order *Rhizobiales* took place by dioxygenation at C_3 and C_4 to produce 4-fluorocatechol, or at C_1 and C_2 to produce catechol with elimination of fluoride: 4-Fluorocatechol was then degraded by intradiol fission to 3-oxoadipate (Carvalho et al. 2006). The biphenyl dioxygenase BphA in *Burkholderia* sp. strain LB400, which is able to degrade a wide range of PCBs, was able to transform a number of 2,2′-substituted biphenyls. These included 2,2′-difluorobiphenyl from which a fluorinated 2,3-dihydrodiol was initially formed, and by loss of fluoride this produced 2′-hydroxy-2,3-dihydroxybiphenyl (Seeger et al. 2001). A strain of *Sphingomonas* strain SS33 adapted to the growth of 4,4′-difluorobiphenyl ether produced the transient intermediates 4-fluorophenol and 4-fluorocatechol before loss of fluoride (Schmidt et al. 1993). These may plausibly occur as a result of the activities of phenol hydroxylase and catechol 1,2-dioxygenase since both these enzymes were found in the parent strain SS33 (Schmidt et al. 1992).

9.13.2 Metabolism by Yeasts and Fungi

Whereas the metabolism of aromatic hydrocarbons takes place by dioxygenation, their biotransformation by yeasts and fungi is normally initiated by monooxygenation to the epoxide followed by hydrolysis to

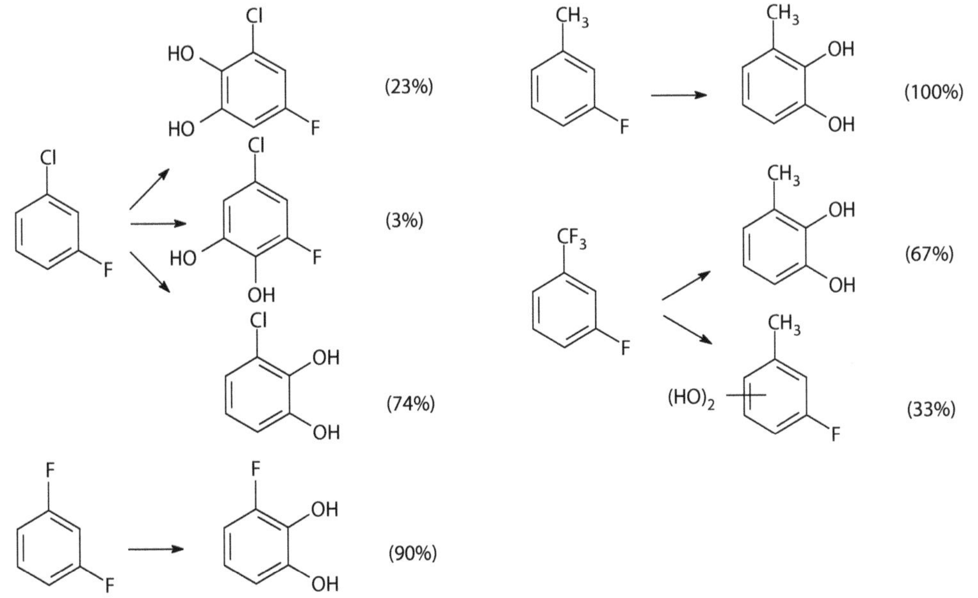

FIGURE 9.37 Transformation of fluorinated substrates by toluene 2,3-dioxygenase and yields (%).

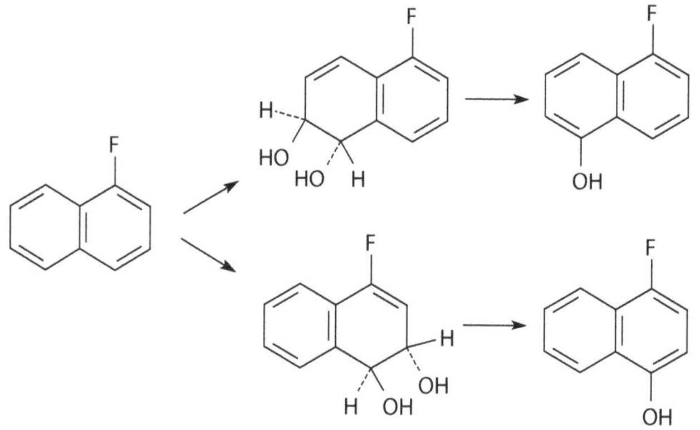

FIGURE 9.38 Transformation of 1-fluoronaphthalene by *Cunninghamella elegans.*

the *trans*-dihydrodiols. Phenols may subsequently be formed either by elimination or by nonenzymatic rearrangement of the epoxide:

1. The transformation of 4-fluorobiphenyl by the ectomycorrhizal fungus *Tylospora fibrilosa* was studied by ^{19}F NMR (Green et al. 1999) and the principal products were 4-fluorobiphen-3′-ol and 4-fluorobiphen-4′-ol.

2. *Cunninghamella elegans* metabolized 1-fluoronaphthalene to a number of products whose synthesis was clearly initiated by the formation of epoxides and *trans*-dihydrodiols (Figure 9.38) (Cerniglia et al. 1984). This illustrates the apparent indifference of the monooxygenase to the presence of fluorine atoms.

9.13.3 Anaerobic Denitrifying Conditions

Some strains originally designated as species of *Pseudomonas* have now been assigned to other genera such as *Thauera* or *Azoarcus* (Anders et al. 1995). There are two major pathways for the anaerobic degradation of toluene, both of which involve formation of benzoate as an intermediate followed by reduction of the benzene ring. One pathway is initiated by reaction with fumarate to produce benzylsuccinate, whereas the other involves dehydrogenation of the methyl group to carboxylate.

Although fluorotoluenes have yielded the corresponding fluorobenzoates, these reactions are merely biotransformations since the fluorobenzoates are terminal metabolites:

1. The *Pseudomonas* sp. strain T oxidized 2-, 3-, and 4-fluorotoluenes to the fluorobenzoates without loss of fluorine, although the yields from 2- and 3-fluorotoluene were low (Seyfried et al. 1994).

2. The toluene-degrading organism *Azoarcus tolulyticus* strain Tol-4 transforms toluene by the fumarate condensation pathway. In the presence of 2-, 3-, or 4-fluorotoluene, the corresponding fluorobenzoates were produced together with the nonfluorinated (*E*-)phenylitaconate (Chee-Sanford et al. 1996). Presumably, the fluorinated analogs were too transient for unambiguous detection.

3. The *Thauera* sp. strain K172 was capable of dehydrogenating a range of fluorinated toluenes to the corresponding benzoates, including 3- and 4-fluorotoluene and, less effectively, 2-fluorotoluene (Biegert and Fuchs 1995).

9.14 Fluorobenzoates

9.14.1 Aerobic Conditions

The metabolism of fluorobenzoates has been examined over many years. Early studies using *Nocardia erythropolis* (Cain et al. 1968) and *Pseudomonas fluorescens* (Hughes 1965) showed that although the rates of whole-cell oxidation of fluorobenzoates were less than for benzoate, they were comparable to, and greater than for, the chlorinated analogs. As for their chlorinated analogs, both dioxygenation and hydrolytic pathways may be involved, and studies have revealed that the different pathways depended on the positions of the fluorine substituents.

1. The rates of dioxygenation of the fluorobenzoates and chlorobenzoates by *Pseudomonas* sp. B13 grown with 3-chlorobenzoate were greater for the 3-halogenated compounds than those for the other isomers—though less than that for benzoate (Reineke and Knackmuss 1978).

2. Whereas 3-fluoro- and 3-chloro-*cis,cis*-muconic acids prepared from *Pseudomonas* sp. B13 by catechol oxidation are cycloisomerized abiotically at pH values between 4 and 5, the corresponding 2-halogenated muconic acids are stable at pH 6 (Schmidt et al. 1980).

3. Loss of halogen may take place before or subsequent to ring fission:

   a. For the *ortho*-fluorinated compounds, dioxygenation may occur with simultaneous decarboxylation and loss of fluoride.

   b. For the 4-fluorinated compounds, loss of fluoride may take place hydrolytically before ring fission.

The pathways for the degradation of 3-halogenated compounds are determined by the inhibitory effect of 3-halogenated catechols, which is discussed in greater detail in Part 1 of this chapter and Chapter 3, Part 1.

The metabolism of 2-fluorobenzoate has been described (Goldman et al. 1967) in a strain of *Pseudomonas* sp. that was able to use this as the sole source of carbon. 3-Fluorocatechol and 2-fluoromuconic acid were identified as metabolites, while studies with $^{18}O_2$ (Milne et al. 1968) established that both atoms of catechol originated from oxygen. Further details of this reaction were provided by a study of the metabolism of *Burkholderia* (*Pseudomonas*) *cepacia* strain 2CBS, which is able to metabolize 2-fluorobenzoate, 2-chlorobenzoate, 2-bromobenzoate, and 2-iodobenzoate to catechol by concomitant decarboxylation and loss of halide (Figure 9.39) (Fetzner et al. 1992). The inducible 2-halobenzoate 1,2-dioxygenase consisted of two components: an oxygenase (A), which is an iron–sulfur protein with nonidentical subunits, α (M_r 52,000) and β (M_r 20,000) in an α3β3 structure, and a reductase (B), which is an iron–sulfur flavoprotein (M_r 37,500) containing FAD. The enzyme is also active for 2,4-difluorobenzoate and less active for 2,6- and 3,4-difluorobenzoate. It is broadly similar to benzoate dioxygenase though different from the FMN-containing dioxygenases. This 2-halobenzoate-1,2-dioxygenase is also different from the three-component enzyme in *Pseudomonas aeruginosa* strain 142 (Romanov and Hausinger 1994), which contains a [2Fe–2S] ferredoxin, and is active toward 2,5-dichlorobenzoate though less toward 2,6-, 3,4-, and 3,5-dichlorobenzoate.

A contrasting pathway was found for the metabolism of 2-fluoro-4-nitrobenzoate by *Nocardia erythropolis* that does not involve loss of fluorine concomitant with decarboxylation (Cain et al. 1968). The pathway (Figure 9.40), therefore, differs from what has subsequently emerged as the principal pathway for metabolism of nonhydroxylated 2-halogenated benzoates.

A bacterial strain FLB300 is able to use 3-fluorobenzoate for growth (Schreiber et al. 1980; Engesser et al. 1990). This strain minimizes the production of toxic 3-fluorocatechol as an intermediate by regioselective distal 1:6 dioxygenation with the formation of 4-fluorocatechol (Figure 9.41). This strategy minimizes the production of the inhibitory 3-fluorocatechol. For maleylacetates that contain a halogen

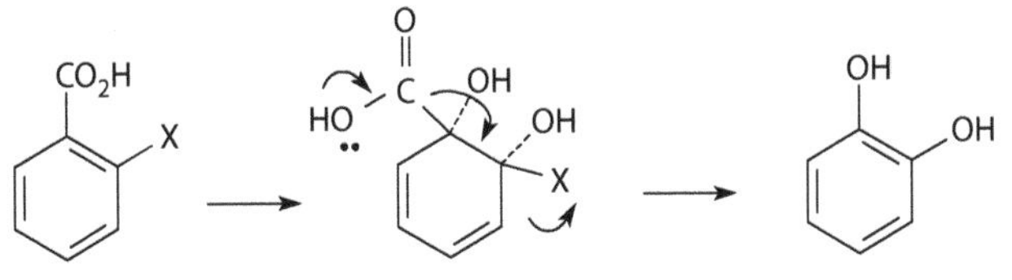

FIGURE 9.39 Transformation of 2-halogenated benzoates to catechol by concomitant decarboxylation and dehalogenation.

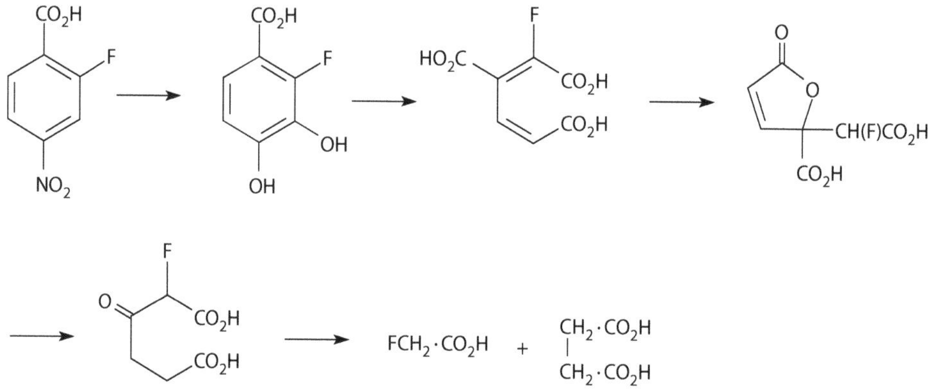

FIGURE 9.40 Degradation of 2-fluoro-4-nitrobenzoate by *Nocardia erythropolis.*

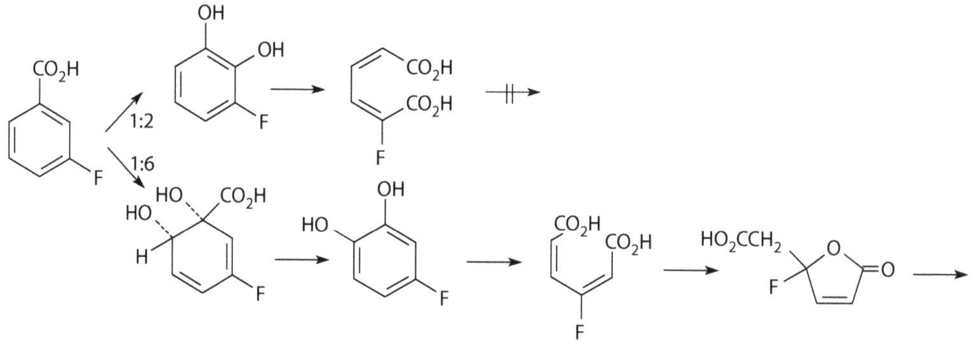

FIGURE 9.41 Degradation of 3-fluorobenzoate by regioselective dioxygenation.

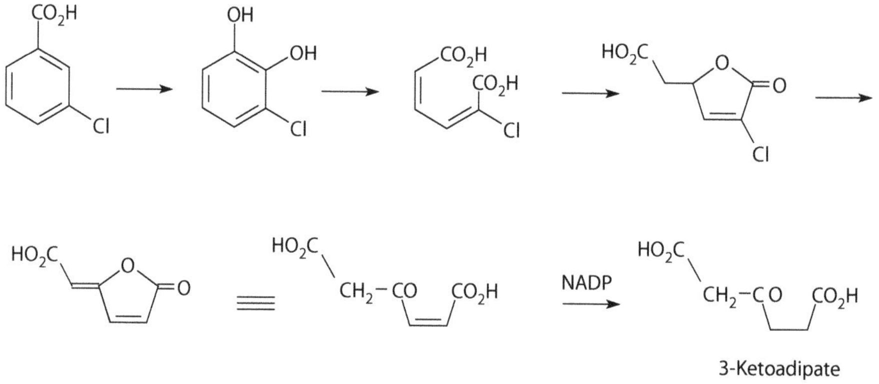

3-Ketoadipate

FIGURE 9.42 Degradation of halogenated maleylacetate to β-ketoadipate in *Pseudomonas* sp. strain B13.

atom (fluorine, chlorine, or bromine) at the 2-position, loss of halogen consumes 1 mol of NADH per mol of substrate, and a further reduction to 3-oxoadipate, an additional mol (Kaschabek and Reineke 1995) (Figure 9.42).

A mutant strain of *Alcaligenes eutrophus* was used (Reiner and Hegeman 1971) to demonstrate that the initial step in the metabolism of 4-fluorobenzoate involves 1:2 dioxygenation to 4-fluoro-1,2-dihydrodihydroxybenzene-1-carboxylate. 4-Fluorobenzoate may be used for growth by several bacteria,

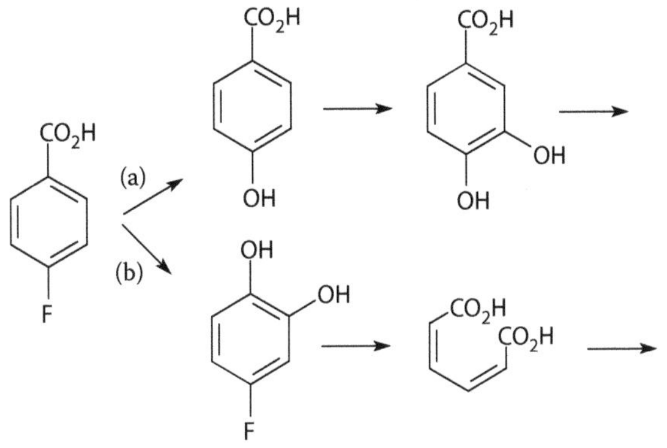

FIGURE 9.43 Alternative pathways for degradation of 4-fluorobenzoate by *Alcaligenes* sp. (a) strain RH 025 and (b) strain RH 021.

and metabolites produced by a strain of *Pseudomonas* sp., which was also able to degrade 4-fluorophenylacetic acid, were identified and used to construct the pathway. Degradation took place by dioxygenation (decarboxylating), intradiol ring fission of the resulting catechol, and loss of fluoride from the muconolactone (Harper and Blakley 1971c). Part of this pathway was confirmed in later studies in which the important role of maleylacetate reductase was determined (Schreiber et al. 1980; Schlömann et al. 1990a,b).

Strains of *Alcaligenes* RH021 and RH022 were able to utilize all the monofluorobenzoates for growth, and two degradative pathways for 4-fluorobenzoate were observed:

1. In strain RH025, a hydrolytic mechanism brought about the loss of fluoride to produce 4-hydroxybenzoate, which was further hydroxylated to 3,4-dihydroxybenzoate before intradiol ring fission (Figure 9.43a).

2. In strain RH021, a benzoate dioxygenase produced 4-fluorocatechol, which was degraded by intradiol fission (Figure 9.43b) (Oltmanns et al. 1989). This pathway is analogous to those proposed for the degradation of 4-chlorobenzoate by a strain of *Arthrobacter* sp. (Marks et al. 1984), for 2,4-dichlorobenzoate by *Alcaligenes denitrificans* strain NTB-1 after the initial reductive loss of the *ortho* chlorine substituent (van der Tweel et al. 1987), and in a coryneform bacterium through an intermediate 2,4-dichlorobenzoyl coenzyme A (Romanov and Hausinger 1996).

9.15 4-Fluorocinnamate

This is degraded aerobically by a consortium of *Arthrobacter* sp. strain G1 and *Ralstonia* sp. strain H1: strain G1 carries out the formation of 4-fluorobenzoate that is degraded by strain H1. The first stage is initiated by the formation of CoA ester that is successively hydrated and dehydrogenated to 4-fluorophenyl-3-oxopropionyl-CoA that undergoes fission by a thiolase. The resulting 4-fluorobenzoyl-CoA is hydrolyzed and dioxygenated to 4-fluorocatechol that is degraded by intradiol fission (Hasan et al. 2011).

9.16 Difluorobenzoates

The metabolism of 2,5- and 3,5-difluorobenzoate by *Pseudomonas putida* strain JT103 has been examined by ^{19}F NMR (Cass et al. 1987). 1:2 Dioxygenation took place with both substrates (Figure 9.44a), and for the 2,5-difluorobenzoate this dominated the alternative distal 1:6 dioxygenation, and resulted in the production of 4-fluorocatechol that was then degraded with loss of the additional fluorine substituent (Figure 9.44b). This pathway was used by *Pseudomonas aeruginosa* JB2 for the

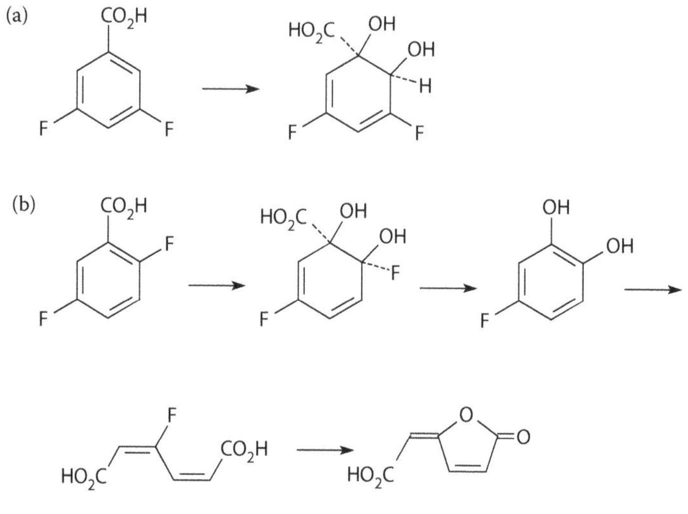

FIGURE 9.44 (a) Transformation of 3,5-difluorobenzoate and (b) degradation of 2,5-difluorobenzoate by *Pseudomonas putida* JT 102.

degradation of the analogous 2,5-dichlorobenzoate, which also degraded 2- and 3-chlorobenzoate (Hickey and Focht 1990).

9.16.1 Degradation under Denitrifying Conditions

The degradation of benzoate under anaerobic denitrifying conditions has been studied in detail and involves the formation of the CoA thioester by a ligase followed by reduction of the ring, hydroxylation, and dehydrogenation before fission. This pathway has been reviewed for both denitrifying and anaerobic phototrophic bacteria (Harwood et al. 1999), although care should be exercised in extrapolating these mechanisms to other groups of organisms.

Degradation of 2-fluorobenzoate under denitrifying conditions has been reported without metabolic details several times during the past 25 years (Taylor et al. 1979; Schennen et al. 1985; Song et al. 2000a). It was shown that by analogy with benzoate itself, benzoyl-coenzyme A synthetase was induced by 2-fluorobenzoate in *Pseudomonas* sp. strain KB740 under anaerobic conditions and also by 3- and 4-fluorobenzoate that were not, however, degraded (Schennen et al. 1985).

A number of bacterial strains isolated by enrichment with halobenzoates have been examined for growth under both aerobic and denitrifying conditions (Song et al. 2000a). A rather complex pattern of assimilation was observed that depended on whether aerobic or anaerobic conditions were employed, and is illustrated in Table 9.8 The fate of fluorine was not resolved in these studies, and different mechanisms apparently operated, and this was confirmed in a study with a strain designated *Thauera aromatica* genomovar *chlorobenzoica* (Song et al. 2000b). Importantly, all the fluorobenzoate isomers were recalcitrant under sulfate-reducing, iron-reducing, and methanogenic conditions (Vargas et al. 2000). This contrasts with the anaerobic degradation of 3-chlorobenzoate, which has been extensively investigated in the sulfate-reducing *Desulfomonile tiedjei* that synthesizes ATP by dehalogenation (Louie and Mohn 1999). This produces benzoate by reductive dehalogenation of 3-chlorobenzoate, 3-bromobenzoate, and 3-iodobenzoate—but *not* 3-fluorobenzoate (DeWeerd and Suflita 1990).

9.16.1.1 Anaerobic Transformation

The syntrophic anaerobe *Syntrophus aciditrophus* is able to grow with crotonate as the sole substrate with the production of acetate and cyclohexanecarboxylate (Mouttaki et al. 2007). The latter is produced by a complex series of reactions involving the synthesis of acetoacetate and crotonate followed by carboxylation to glutaconyl-CoA, condensation with acetyl-CoA and reduction to 3-hydroxypimelyl-CoA that underwent cyclization. The metabolism of fluorobenzoates has been examined, and the following briefly summarizes the rather complex results (Mouttaki et al. 2009).

- During growth with crotonate, 2-hydroxybenzoate served as the electron acceptor and was reduced to cyclohexane carboxylate, while crotonate as electron donor formed acetate.
- Whereas 4-fluorobenzoate had no adverse effect on the growth with crotonate and was not metabolized, 3-fluorobenzoate underwent stoichiometric formation of benzoate and fluoride, albeit in low yield.
- Prolonged incubation with crotonate and 3-fluorobenzoate produced a fluorinated metabolite that was, on the basis of ^{19}F NMR tentatively assigned to 3-fluorocyclohexa-2,6-diene-1 carboxylate or 3-fluorocyclohexa-3,6-diene-1-carboxylate.

9.17 Fluorinated Phenols

9.17.1 Aerobic Conditions

9.17.1.1 Bacterial Metabolism

Three broadly different metabolic routes have been demonstrated for fluorophenols, and parallel those described for the analogous chlorophenols.

9.17.1.1.1 Hydroxylation of the Ring to a Fluorocatechol

1. Whole cells of *Rhodococcus opacus* strain 1cp were used to study the metabolism of 3- and 4-fluorophenol (Finkelstein 2000): both fluorocatechols and fluoropyrogallols were produced (Figure 9.45a), and both isomers produced 5-fluoropyrogallol that was transformed into 2-pyrone-4-fluoro-6-carboxylate (Figure 9.45b).
2. A ^{19}F NMR study used several species of *Rhodococcus*, and a range of monofluoro- difluoro-, and trifluorophenols. As illustrated in Table 9.9 both ring hydroxylation and hydrolytic defluorination were observed, followed by intradiol ring fission (Figure 9.46) (Bondar et al. 1998).

9.17.1.1.2 Monooxygenation

This brings about hydroxylation with concomitant removal of a fluorine substituent and production of a quinone or hydroquinone that then undergoes ring fission.

1. The degradation of 4-fluorophenol by *Arthrobacter* sp. strain IF1 produced benzoquinone by monooxygenation with the loss of fluoride, followed by oxidation to 1,2,4-trihydroxybenzene and intradiol ring fission (Ferreira et al. 2009).
2. The metabolism of pentafluoro-, pentachloro-, and pentabromophenol by *Mycobacterium fortuitum* strain CG-2 was initiated by a monooxygenase that carried out hydroxylation at the *para*

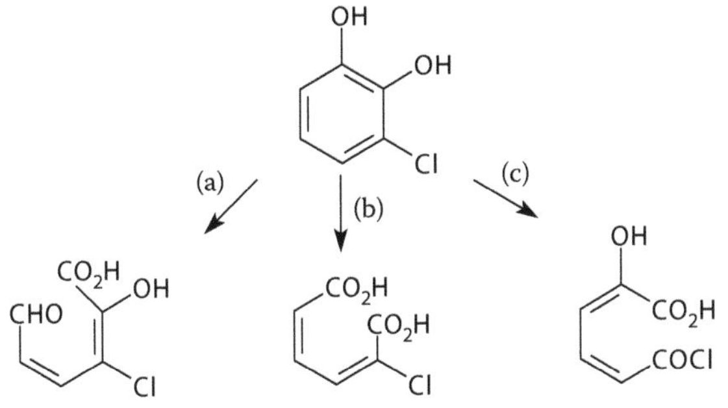

FIGURE 9.45   Transformation of 3- and 4- fluorophenol by *Rhodococcus* sp. strain 1 cp.

TABLE 9.9

Utilization of Fluorobenzoates under Aerobic and Denitrifying Conditions

Substrate Strain	2-Fluorobenzoate		3-Fluorobenzoate		4-Fluorobenzoate	
	Aerobic	Denitrifying	Aerobic	Denitrifying	Aerobic	Denitrifying
2 FB11	+	+	+	−	+	−
4 FB 11	+	+	−	−	+	+
2 FB 6	+	+	−	−	−	−
4 FB 10	−	+	−	−	−	+

Source: Adapted from Song B, NJ Palleroni, MM Häggblom. 2000a. *Appl Environ Microbiol* 66: 3446–3453.

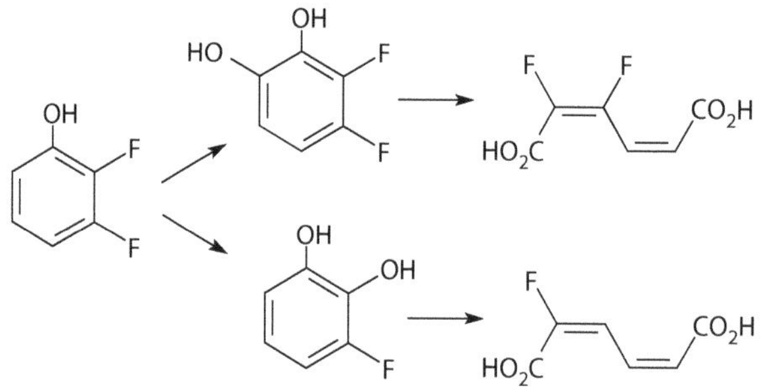

FIGURE 9.46 Transformation of 2,3-difluorophenol by *Rhodococcus opacus* strain 135.

position (Uotila et al. 1992). Cell extracts of *Rhodococcus chlorophenolicus* (*Mycobacterium chlorophenolicum* strain PCP-1) in the presence of reductant transformed tetrafluoro-, tetra-chloro-, and tetrabromohydroquinone to 1,2,4-trihydroxybenzene by reactions that clearly involve both hydrolytic and reductive loss of fluorine (Uotila et al. 1995).

9.17.1.1.3 Dioxygenation

The hydroquinone dioxygenase from *Pseudomonas fluorescens* ACB was able to accept 2,3-difluoro- and 2,5-difluorohydroquinone with ring fission at the distal 1:6-positions and formation of metabolites tentatively identified by NMR as 2,3- and 2,5-difluoro-4-hydroxymuconic semialdehydes (Moonen et al. 2008).

9.17.1.2 Metabolism by Yeasts and Fungi

The metabolism of fluorinated phenols has been examined quite extensively in yeasts:

1. The metabolism of a range of fluorophenols containing up to five fluorine substituents was examined using phenol hydroxylase from *Trichosporon cutaneum* (Peelen et al. 1995). Fluorocatechols were formed, with loss of fluoride for some substrates.

2. ^{19}F NMR was used to examine the metabolites produced by the fungus *Exophiala jeanselmei*. Hydroxylation was observed for 4-fluorophenol, 2,4-difluorophenol, and 2,3,4-trifluorophenol, and the muconates produced from purified phenol hydroxylase and catechol 1,2-dioxygenase are given in Table 9.10 (Boersma et al. 1998).

9.17.2 Anaerobic Conditions

The metabolism of phenols under anaerobic conditions has been examined under denitrifying, sulfate-reducing, Fe (III)-reducing, and anaerobic nonmethanogenic conditions. It is plausible to suggest a common pathway that has been elucidated for denitrifying bacteria. This consists of three steps: (a) activation

TABLE 9.10

Metabolites of Fluorophenols Produced by *Rhodococcus corallinus* Strain 135

Substrate	Metabolite (%)	
2,3-Difluorophenol	3,4-Difluorocatechol (23)	2-Fluoromuconate (24)
		2,3-Difluoromuconate (1)
2,5-Difluorphenol	4-Fluorocatechol (9)	2,5-Difluoromuconate (74)
	3,6-Difluorocatechol (12)	
2,3,5-Trifluoro-phenol	3,5-Difluorocatechol (78)	2,4-Difluoromuconate (1)
	3,4,6-Trifluorocatechol (2)	2,3,5-Trifluoromuconate (5)

Source: Adapted from Bondar VS et al. 1998. *Biodegradation* 9: 475–486.

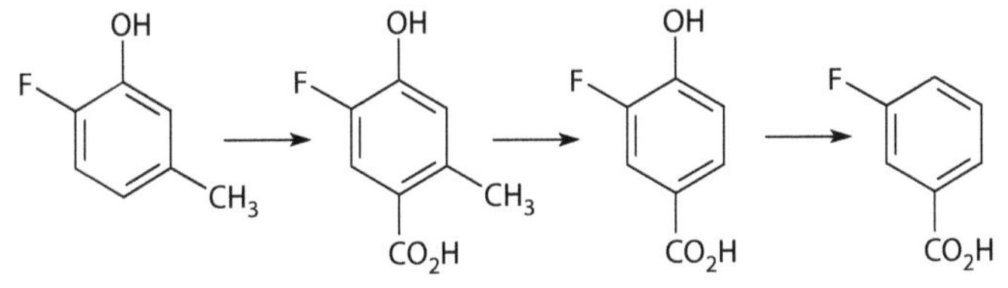

FIGURE 9.47 Anaerobic transformation of 6-fluoro-3-methylphenol.

of phenol by the formation of phenylphosphate, (b) carboxylation at a position para to the hydroxyl group (Breinig et al. 2000), and (c) dehydroxylation of the 4-hydroxybenzoyl CoA to benzoate (Breese and Fuchs 1998). Successive carboxylation and dehydroxylation reactions have been demonstrated with 2- and 3-fluorophenol and resulted in the production of 2- and 3-fluorobenzoate (Genther et al. 1990). Under methanogenic conditions, 6-fluoro-3-methylphenol was degraded to 3-fluorobenzoate via 5-fluoro-4-hydroxy-2-methylbenzoate and 3-fluoro-4-hydroxybenzoate with the apparent loss of a methyl group (Figure 9.47) (Londry and Fedorak 1993).

9.18 Aromatic Trifluoromethyl Compounds

Compounds bearing trifluoromethyl substituents have a wide range of application, and consistent with their chemical stability, the strongly electron-attracting trifluoromethyl group is recalcitrant to defluorination. During the metabolism of aromatic compounds with trifluoromethyl substituents, this group generally remains intact even after ring fission, although the trifluoromethyl group apparently does not inhibit either ring dioxygenation or ring fission. It is the recalcitrance of the ring-fission product that is responsible for the partial degradation of these substrates. The following illustrative examples are given:

1. The *cis*-dihydrodiols have been produced from a number of trifluoromethylbenzoates:
   a. 4-Trifluoromethylbenzoate by *Pseudomonas putida* strain JTIO7 (Figure 9.48a) (DeFrank and Ribbons 1976).
   b. 2-Trifluoromethylbenzoate by *P. aeruginosa* strain 142 (Figure 9.48b) (Selifonov et al. 1995).
   c. 4-Methyl- and 4-iodotrifluorobenzene by *P. putida* strain UV-4 (Boyd et al. 1993).
2. Both trifluoromethylbenzoates and trifluoromethylphenols produced 2-hydroxy-6-oxo-7,7,7-trifluorohepta-2,4-dienoate, which is the immediate ring-fission product. The trifluoromethyl group of the original substrates was retained in the terminal metabolites.
   a. 4-Trifluoromethylbenzoate was cometabolized by 4-isopropylbenzoate-grown cells of *P. putida* strain JT (Figure 9.49a) (Engesser et al. 1988b).

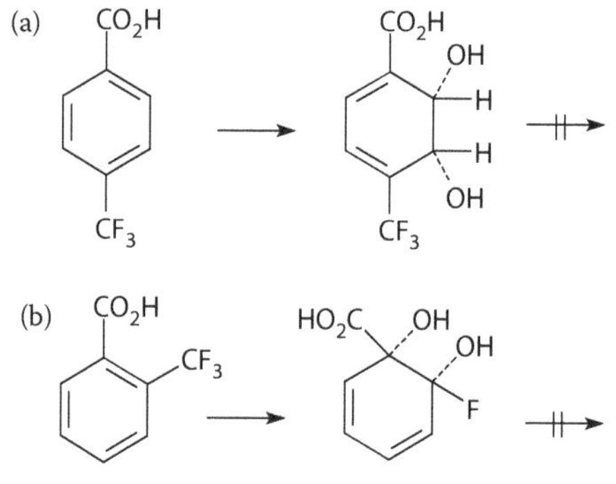

FIGURE 9.48 Dioxygenation of (a) 4-trifluoromethylbenzoate by *Pseudomonas putida* strain JT 107 and (b) 2-trifluoro-methylbenzoate by *Pseudomonas aeruginosa* strain 142.

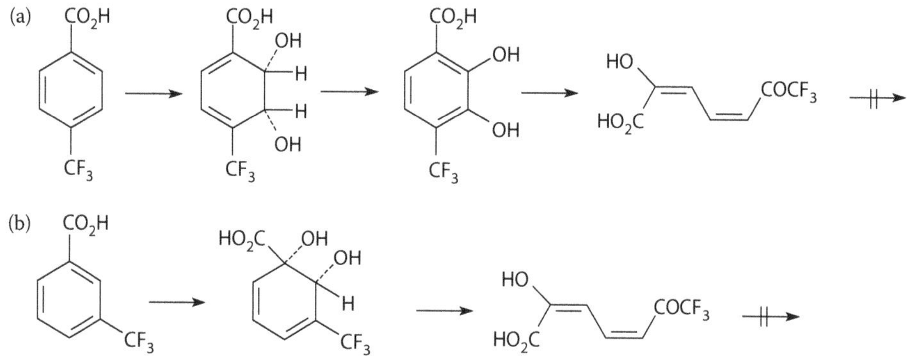

FIGURE 9.49 Metabolism of (a) 4-trifluoromethylbenzoate by *Pseudomonas putida* strain JT grown with 4-isopropyl-benzoate and (b) 3-trifluoromethylbenzoate by *Pseudomonas putida* strain mt-2 grown with 3-methylbenzoate.

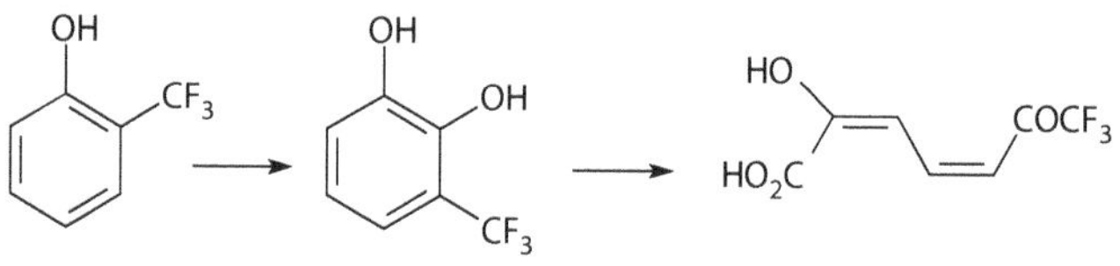

FIGURE 9.50 Transformation of 2-trifluoromethylphenol by *Bacillus thermoleovorans.*

    b. 3-Trifluoromethylbenzoate was cometabolized by 3-methylbenzoate-grown cells of *P. putida* strain mt-2 (Figure 9.49b) (Engesser et al. 1988a).

    c. The hydroxylation of 2-trifluoromethylphenol to 3-trifluoromethylcatechol by *Bacillus thermoleovorans* strain A2 was followed by extradiol ring fission (Figure 9.50)

(Reinscheid et al. 1998), while the transformation of 3-trifluoromethylcatechol Engesser et al. 1990) produced the stable terminal metabolite.

9.19 References

Anders H-J, A Kaetzke, P Kämpfer, W Ludwig, G Fuchs. 1995. Taxonomic position of aromatic-degrading denitrifying pseudomonad strains K 172 and KB 740 and their description as new members of the genera *Thauera*, as *Thauera aromatica* sp. nov., and *Azoarcus*, as *Azoarcus evansii* sp. nov., respectively, members of the beta subclass of the Proteobacteria. *Int J Syst Bacteriol* 45: 327–333.

Biegert T, G Fuchs. 1995. Anaerobic oxidation of toluene (analogues) to benzoate (analogues) by whole cells and by cell extracts of a denitrifying *Thauera* sp. *Arch Microbiol* 163: 407–417.

Boersma MG, TY Dinarieva, WJ Middelhoven, WJH van Berkel, J Doran, J Vervoort, IMCM Rietjens. 1998. ^{19}F nuclear magnetic resonance as a tool to investigate microbial degradation of fluorophenols to fluoro-catechols. *Appl Environ Microbiol* 64: 1256–1263.

Bondar VS, MG Boersma, EL Golovlev, J Vervoort, WJH van Berkel, ZI Finkelstein, IP Solyanikova, LA Golovleva, IMCM Rietjens. 1998. ^{19}F NMR study on the biodegradation of fluorophenols by various *Rhodococcus species*. *Biodegradation* 9: 475–486.

Boyd DR, ND Sharma, MV Hand, MR Groocock, NA Kerley, H Dalton, J Chima, GN Sheldrake. 1993. Stereodirecting substituent effects during enzyme-catalysed synthesis of *cis*-dihydrodiol metabolites of 1,4-disubstituted benzene substrates. *J Chem Soc Chem Commun* 974–976.

Breese K, G Fuchs. 1998. 4-Hydroxybenzoyl-CoA reductase (dehydroxylating) from the denitrifying bacterium *Thauera aromatica*— Prosthetic groups, electron donor, and genes of the molybdenum-flavin-iron-sulfur proteins. *Eur J Biochem* 251: 916–923.

Breinig S, E Schiltz, G Fuchs. 2000. Genes involved in anaerobic metabolism of phenol in the bacterium *Thauera aromatica*. *J Bacteriol* 182: 5849–5863.

Cain RB, EK Tranter, JA Darrah. 1968. The utilization of some halogenated aromatic acids by Nocardia. Oxidation and metabolism. *Biochem J* 106: 211–227.

Carvalho MF, MIM Ferreira, IS Moreira, PML Castro, DB Janssen. 2006. Degradation of fluorobenzene by *Rhizobiales* strain F11 via *ortho* cleavage of 4-fluorocatechol and catechol. *Appl Environ Microbiol* 72: 7413–7417.

Cass AEG, DW Ribbons, JT Rossiter, SR Williams. 1987. Biotransformation of aromatic compounds monitoring fluorinated analogues by NMR. *FEBS Lett* 220: 353–357.

Cerniglia CE, DW Miller, SK Yang, JP Freeman. 1984. Effects of a fluoro substituent on the fungal metabolism of 1-fluoronaphthalene. *Appl Environ Microbiol* 48: 294–300.

Chee-Sanford JC, JW Frost, MR Fries, Z Zhou, JM Tiedje. 1996. Evidence for acetyl coenzyme A and cinnamoyl coenzyme A in the anaerobic toluene mineralization pathway in *Azoarcus tolulyticus* Tol-4. *Appl Environ Microbiol* 62: 964–873.

DeFrank JJ, DW Ribbons. 1976. The p-cymene pathway in *Pseudomonas putida* PL: Isolation of a dihydrodiol accumulated by a mutant. *Biochem Biophys Res Commun* 70: 1129–1135.

DeWeerd KA, JM Suflita. 1990. Anaerobic aryl reductive dehalogenation of halobenzoates by cell extracts of "*Desulfomonile tiedjei.*" *Appl Environ Microbiol* 56: 2999–3005.

Engesser KH, RB Cain, HJ Knackmuss. 1988b. Bacterial metabolism of side chain fluorinated aromatics: Cometabolism of 3-trifluoromethyl (TFM)-benzoate by *Pseudomonas putida* (arvilla) mt-2 and *Rhodococcus rubropertinctus* N657. *Arch Microbiol* 149: 188–197.

Engesser KH, MA Rubio, H-J Knackmuss. 1990. Bacterial metabolism of side-chain-fluorinated aromatics: Unproductive meta cleavage of 3-trifluoromethylcatechol. *Appl Microbiol Biotechnol* 32: 600–608.

Engesser KH, MA Rubio, DW Ribbons. 1988a. Bacterial metabolism of side chain fluorinated aromatics: Cometabolism of 4-trifluoromethyl (TFM)-benzoate by 4-isopropylbenzoate grown *Pseudomonas putida* JT strains. *Arch Microbiol* 149: 198–206.

Ferreira MIM, T Iida, SA Hasan, K Nakamura, MW Fraaije, DB Janssen, T Kudo. 2009. Analysis of two gene clusters involved in the degradation of 4-fluorophenol by *Arthrobacter* sp. strain IF1. *Appl Environ Microbiol* 75: 7767–7773

Fetzner S, R Müller, F Lingens. 1992. Purification and some properties of 2-halobenzoate 1,2-dioxygenase, a two-component enzyme system from *Pseudomonas cepacia* 2CBS. *J Bacteriol* 174: 279–290.

Finkelstein ZI, BP Baskunov, MG Boersma, J Vervoort, EL Golovlev, WJH van Berkel, LAA Gololeva, IMCM Rietjens. 2000. Identification of fluoropyrogallols as new intermediates in biotransformation of mono-fluorophenols in *Rhodococcus opacus* 1 cp. *Appl Environ Microbiol* 66: 2148–2153.

Genther BRS, GT Townsend, PJ Chapman. 1990. Effect of fluorinated analogues of phenol and hydroxybenzo-ates on the anaerobic transformation of phenol to benzoate. *Biodegradation* 1: 65–74.

Goldman P, GWA Milne, MT Pignataro. 1967. Fluorine containing metabolites formed from 2-fluorobenzoic acid by *Pseudomonas* species. *Arch Biochem Biophys* 118: 178–184.

Green NA, AA Meharg, C Till, J Troke, JK Nicholson. 1999. Degradation of 4-fluorobiphenyl by mycorrizal fungi as determined by ^{19}F nuclear magnetic resonance spectroscopy and ^{14}C radiolabelling analysis. *Appl Environ Microbiol* 65: 4021–4027.

Harper DB, ER Blakley. 1971c. The metabolism of *p*-fluorobenzoic acid by a *Pseudomonas* sp. *Can J Microbiol* 17: 1015–1023.

Harwood CS, G Burchardt, H Herrmann, G Fuchs. 1999. Anaerobic metabolism of aromatic compounds via the benzoyl-CoA pathway. *FEMS Microbiol Revs* 22: 439–458.

Hasan SA, MIM Ferreira, MJ Koetsier, MI Arif, DB Janssen. 2011. Complete biodegradation of 4-fluorocinnamic acid by a consortium comprising *Arthrobacter* sp. strain G1 and *Ralstonia* sp. strain H1. *Appl Environ Microbiol* 77: 572–579

Hickey WJ, DD Focht. 1990. Degradation of mono-, di-, and trihalogenated benzoic acids by *Pseudomonas aeruginosa* JB2. *Appl Environ Microbiol* 56: 3842–3850.

Hughes DE. 1965. The metabolism of halogen-substituted benzoic acids by *Pseudomonas fluorescens*. *Biochem J* 96: 181–188.

Kaschabek SR, W Reineke. 1995. Maleylacetate reductase of *Pseudomonas* sp. strain B13: specificity of substrate conversion and halide elimination. *J Bacteriol* 177: 320–325.

Londry KL, PM Fedorak. 1993. Use of fluorinated compounds to detect aromatic metabolites from *m*-cresol in a methanogenic consortium: Evidence for a demethylation reaction. *Appl Environ Microbiol* 59: 2229–2238.

Louie TM, WW Mohn. 1999. Evidence for a chemiosmotic model of dehalorespiration in *Desulfomonile tiedjei* DCB-1. *J Bacteriol* 181: 40–46.

Marks TS, ARW Smith, AV Quirk. 1984. Degradation of 4-chlorobenzoic acid by *Arthrobacter* sp. *Appl Environ Microbiol* 48: 1020–1025.

Milne GWA, P Goldman, JL Holzman. 1968. The metabolism of 2-fluorobenzoic acid. *J Biol Chem* 243: 5374–5376.

Moonen MJH, SA Synowsky, WAM van der Berg, AH Westphal, AJR Heck, RHH van den Heuvel, MW Fraaije, WJH van Berkel. 2008. Hydroquinone dioxygenase from *Pseudomonas fluorescens* ACB: A novel member of the family of nonheme-iron (II)-dependent dioxygenases. *J Bacteriol* 190: 5199–5209

Mouttaki H, MA Nanny, MJ McInerney. 2007. Cyclohexane carboxylate and benzoate formation from crotonate in *Syntrophus aciditrophicus*. *Appl Environ Microbiol* 73: 930–938

Mouttaki H, MA Nanny, MJ McInerney. 2009. Metabolism of hydroxylated and fluorinated benzoates by *Syntrophus aciditrophicus* and detection of a fluorodiene metabolite. *Appl Environ Microbiol* 75: 998–1004

Oltmanns RH, R Müller, MK Otto, F Lingens. 1989. Evidence for a new pathway in the bacterial degradation of 4-fluorobenzoate. *Appl Environ Microbiol* 55: 2499–2504.

Peelen S, IMC Rietjens, MG Boersma, J Vervoort. 1995. Conversion of phenol derivatives to hydroxylated products by phenol hydroxylase from *Trichosporon cutaneum*. *Eur J Biochem* 227: 284–291.

Reineke W, H-J Knackmuss. 1978. Chemical structure and biodegradability of halogenated aromatic compounds. Substituent effects on 1,2-dioxygenation of benzoic acid. *Biochim Biophys Acta* 542: 412–423.

Reiner AM, GD Hegeman. 1971. Metabolism of benzoic acid by bacteria. Accumulation of (-)-3,5-cyclohexadien-1,2-diol,1-carboxylic acid by a mutant strain of *Alcaligenes eutrophus*. *Biochemistry* 10: 2530–2536.

Reinscheid UM, H Zuilhog, R Müller, J Vervoort. 1998. Biological, thermal and photochemical transformation of 2-trifluoromethylphenol. *Biodegradation* 9: 487–499.

Renganathan V. 1989. Possible involvement of toluene-2,3-dioxygenase in defluorination of 3-fluoro-substituted benzenes by toluene-degrading *Pseudomonas* sp. strain T-12. *Appl Environ Microbiol* 55: 330–334.

Romanov V, RP Hausinger. 1994. *Pseudomonas aeruginosa* 142 uses a three-component ortho-halobenzoate 1,2-dioxygenase for metabolism of 2,4-dichloro- and 2-chlorobenzoate. *J Bacteriol* 176: 3368–3374.

Romanov V, RP Hausinger. 1996. NADPH-dependent reductive ortho dehalogenation of 2,4-dichlorobenzoic acid in *Corynebacterium sepedonicum* KZ-4 and coryneform bacterium strain NTB-1 via 2,4-dichlorobenzoyl coenzyme A. *J Bacteriol* 178: 2656–2661.

Schennen U, K Braun, H-J Knackmuss. 1985. Anaerobic degradation of 2-fluorobenzoate by benzoate-degrading, denitrifying bacteria. *J Bacteriol* 161: 321–325.

Schlömann M, P Fischer, E Schmidt, H-J Knackmuss. 1990b. Enzymatic formation, stability, and spontaneous reactions of 4-fluoromuconolactone, a metabolite of the bacterial degradation of 4-fluorobenzoate. *J Bacteriol* 172: 5119–5129.

Schlömann M, E Schmidt, H-J Knackmuss. 1990a. Different types of dienelactone hydrolase in 4-fluorobenzoate-utilizing bacteria. *J Bacteriol* 172: 5112–5118.

Schmidt S, P Fortnagel, R-M Wittich. 1993. Biodegradation and transformation of 4,4′- and 2,4-dihalo-diphenyl ethers by *Sphingomonas* sp. strain SS33. *Appl Environ Microbiol* 58: 3931–3933.

Schmidt E, G Remberg, H-J Knackmuss. 1980. Chemical structure and biodegradability of halogenated aromatic compounds. Halogenated muconic acids as intermediates. *Biochem J* 192: 331–337.

Schmidt S, R-M Wittich, D Erdmann, H Wilkes, W Francke, P Fortnagel. 1992. Biodegradation of diphenyl ether and its monohalogenated derivatives by *Sphingomonas* sp. strain SS3. *Appl Environ Microbiol* 58: 2744–2750.

Schreiber A, M Hellwig, E Dorn, W Reineke, H-J Kmackmuss. 1980. Critical reactions in fluorobenzoic acid degradation by *Pseudomonas* sp. B13. *Appl Environ Microbiol* 39: 58–67.

Seeger M, B Cámara, B Hofer. 2001. Dehalogenation, denitrification, dehydroxylation, and angular attack on substituted biphenyls and related compounds by a biphenyl dioxygenase. *J Bacteriol* 183: 3548–3555.

Selifonov SA, JE Gurst, LP Wackett. 1995. Regioselective dioxygenation of *ortho*-trifluoromethylbenzoate by *Pseudomonas aeruginosa* 142: evidence for 1,2-dioxygenation as a mechanism in *ortho*-halobenzoate dehalogenation. *Biochem Biophys Res Comm* 213: 759–767.

Seyfried B, G Glod, R Schocher, A Tschech, J Zeyer. 1994. Initial reactions in the anaerobic oxidation of toluene and *m*-xylene by denitrifying bacteria. *Appl Environ Microbiol* 60: 4047–4052.

Song B, NJ Palleroni, MM Häggblom. 2000a. Isolation and characterization of diverse halobenzoate-degrading denitrifying bacteria from soils and sediments. *Appl Environ Microbiol* 66: 3446–3453.

Song B, NJ Palleroni, MM Häggblom. 2000b. Description of strain 3CB-1, a genomovar of *Thauera aromatica*, capable of degrading 3-chlorobenzoate coupled to nitrate reduction. *Int J Syst Evol Microbiol* 50: 551–558.

Taylor BF, WL Hearn, S Pincus. 1979. Metabolism of monofluoro- and monochlorobenzoates by a denitrifying bacterium. *Arch Microbiol* 122: 301–306.

Uotila JS, VH Kitunen, T Coote, T Saastamoinen, M Salkinoja-Salonen, JHA Apajalahti. 1995. Metabolism of halohydroquinones in *Rhodococcus chlorophenolicus* PCP-1 *Biodegradation* 6: 119–126.

Uotila JS, VH Kitunen, T Saastamoinen, T Coote, MM Hägblom, M Salkinoja-Salonen. 1992. Characterization of aromatic dehalogenases of *Mycobacterium fortuitum* CG-2. *J Bacteriol* 174: 5669–5675.

van der Tweel WJJ, JB Kok, JAM de Bont. 1987. Reductive dechlorination of 2,4-dichlorobenzoate to 4-chlorobenzoate and hydrolytic dehalogenation of 4-chloro-, 4-bromo-, and 4-iodobenzoate by *Acaligenes denitrificans* NTB-1. *Appl Environ Microbiol* 53: 810–815.

Vargas C, B Song, M Camps, MM Häggblom. 2000. Anaerobic degradation of fluorinated aromatic compounds. *Appl Microbiol Biotechnol* 53: 342–347.

Part 4: Arene Sulfonates

Anthropogenic arene sulfonates are structural elements of many industrially important dyes, pigments, and anionic surfactants, although they are very seldom encountered as naturally occurring metabolites.

Whereas chlorine is only exceptionally removed from aromatic rings during dioxygenation, this generally takes place with carboxyl and sulfonate groups. The pathway for the degradation of aromatic sulfonates has been elucidated in a detailed study (Cain and Farr 1968), and is illustrated for 4-toluenesulfonate in Figure 9.51. The basic reaction is dioxygenation with concomitant elimination of sulfite, which is illustrated for naphthalene-1-sulfonate (Figure 9.52). The dioxygenase has been purified from a naphthalene-2-sulfonate-degrading pseudomonad (Kuhm et al. 1991). The metabolites may, however, be

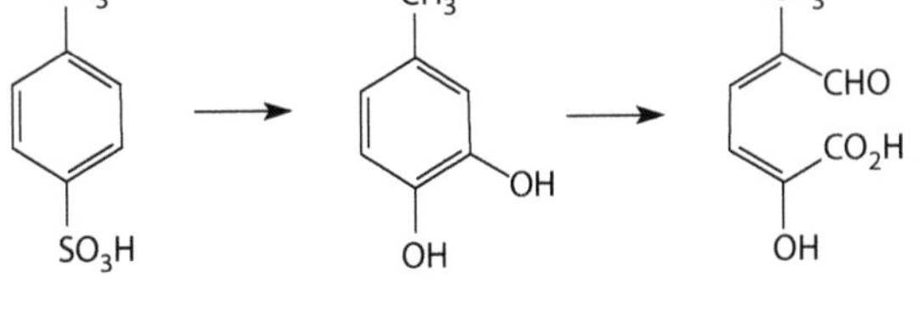

FIGURE 9.51 Biodegradation of 4-toluenesulfonate.

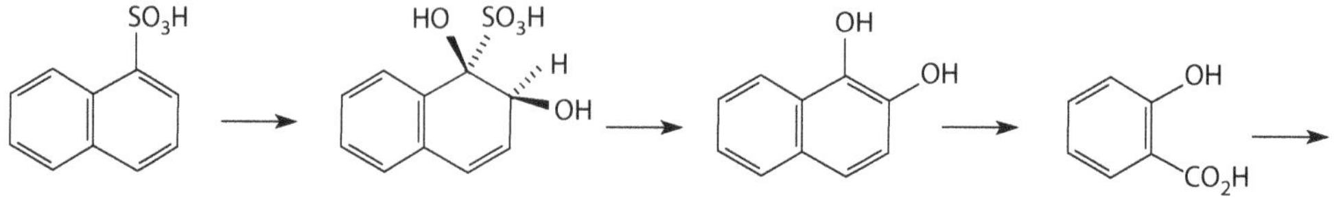

FIGURE 9.52   Biodegradation of naphthalene-1-sulfonate.

sufficiently reactive to react with NH_4^+ in the medium and result in the formation of 5-hydroxyquinoline-2-carboxylate from 5-aminonaphthalene-2-sulfonate (Nörtemann et al. 1993). In contrast, when arene sulfonates serve as sources of sulfur in the absence of sulfate, they are degraded by monooxygenation to the corresponding phenol and sulfite (Kertesz 1999).

There are a number of alternative pathways:

1. a.   Linear alkylbenzene sulfonates are important anionic surfactants whose degradation has attracted enormous attention. Degradation is initiated by the activity of *Parvibaculum laventivorans* that produces 3-(4-sulfophenyl)butyrate by oxidation and *b*-elimination from both congeners. Further degradation by *Comamonas testosteroni* involves successive dehydrogenation, hydration of the alkene carboxylate, and dehydrogenation to the aceto-phenone. This undergoes monooxygenation to 4-sulfophenyl acetate before dioxygenation, intradiol ring fission and loss of sulfite (Schleheck et al. 2010).

   b.   The degradation of 4-toluenesulfonate has been extensively investigated in *Comamonas testosteroni* in which the operon *tsaMBCD* and *tsaR* has been characterized (Junker et al. 1997). Two pathways were proposed: (a) successive monooxygenation of the methyl group to a carboxyl group before elimination of sulfite by dioxygenation (Locher et al. 1991) (Figure 9.53), and (b) a putatively reductive pathway with initial loss of sulfite (Cook et al. 1999).

2. Considerable attention has been directed to the degradation of naphthalenesulfonates, and an additional reaction has emerged for the elimination of sulfite. The pathway for the degradation of naphthalene 1,6- and 2,6-disulfonates involves the expected elimination of the 1- or 2-sulfonate groups, followed by ring fission with the formation of 5-sulfosalicylate. This is then converted into 2,5-dihydroxybenzoate by what may formally be represented as hydroxylation with elimination of sulfite (Figure 9.54) (Wittich et al. 1988). The 1,2-dihydroxynaphthalene dioxygenase from *Pseudomonas* sp. strain (BN6) that degraded naphthalene sulfonates oxidized both 1,2-dihydroxynaphthalene and 2,3-dihydroxybiphenyl (Kuhm et al. 1991), although the same

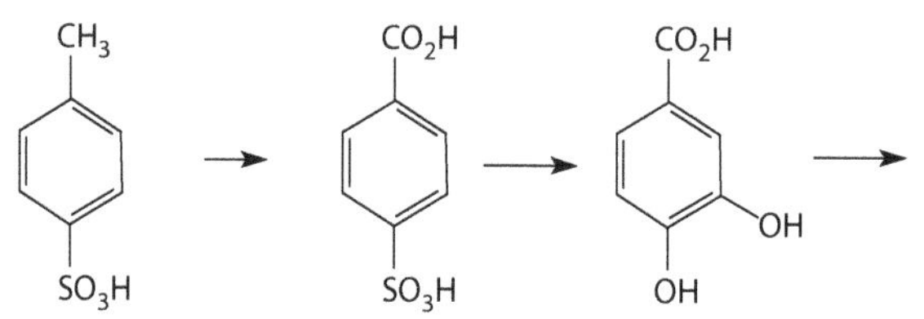

FIGURE 9.53   Alternative pathway for the biodegradation of 4-toluenesulfonate.

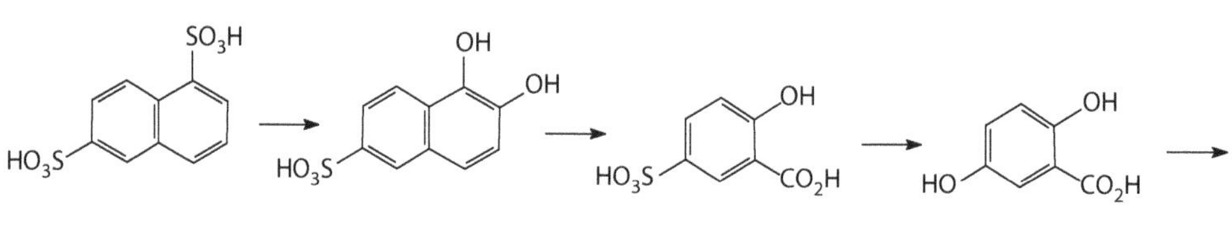

FIGURE 9.54   Biodegradation of naphthalene-1,6-disulfonate.

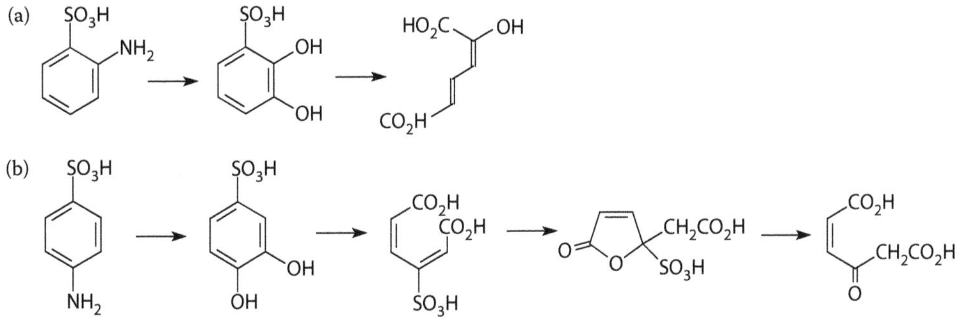

FIGURE 9.55   Degradation of (a) 2-aminobenzenesulfonate and (b) 4-aminobenzenesulfonate.

organism also synthesizes a different 2,3-dihydroxybiphenyl dioxygenase that is not, however, involved in the degradation of naphthalene sulfonates (Heiss et al. 1995).

3. Many azo pigments and dyes contain aromatic rings containing sulfonate substituents. They are degraded by reduction and scission of the resulting products to amines. This is discussed in Part 6 of this chapter. 6-Aminonaphthalene-2-sulfonate is degraded by *Pseudaminobacter salicylatoxidans* via 5-aminosalicylate with subsequent ring fission (Hintner et al. 2001). The degradation of 2-aminobenzenesulfonate is carried out by dioxygenation with concomitant deamination to catechol 3-sulfonate, from which sulfite is eliminated during extradiol ring fission and hydrolysis (Figure 9.55a) (Junker et al. 1994).

4. Sulfite may not necessarily be eliminated before ring fission. The degradation of 4-aminobenzenesulfonate by a mixed culture of *Hydrogenophaga palleroni* and *Agrobacterium radiobacter* produced 4-sulfocatechol from which the sulfonate was eliminated spontaneously as sulfite from the 2-sulfolactone after ring fission (Figure 9.55b) (Hammer et al. 1996).

   Although anaerobic desulfonation of arene sulfonates, in which the sulfonate is used as a source of sulfur, has been described in a *Clostridium* sp., the product was not identified (Denger et al. 1996; Denger and Cook 1997).

9.20 References

Cain RB, DR Farr. 1968. Metabolism of arylsulphonates by micro-organisms. *Biochem J* 106: 859–877.

Cook AM, H Laue, F Junker. 1999. Microbial desulfonation. *FEMS Microbiol Rev* 22: 399–419.

Denger K, AM Cook. 1997. Assimilation of sulfur from alkyl and aryl sulfonates by *Clostridium* spp. *Arch Microbiol* 167: 177–181.

Denger K, MA Kertesz, EM Vock, R Schön, A Mägli, AM Cook. 1996. Anaerobic desulfonation of 4-tolylsulfonate and 2-(4-sulfophenyl)butyrate by a *Clostridium* sp. *Appl Environ Microbiol* 62: 1526–1530.

Hammer A, A Stolz, and H-J Knackmuss. 1996. Purification and characterization of a novel type of proto-catechuate 3,4-dioxygenase with the ability to oxidize 4-sulfocatechol. *Arch Microbiol* 166: 92–100.

Heiss G, A Stolz, AE Kuhm, C Müller, J Klein, J Altenbuchner, H-J Knackmuss. 1995. Characterization of a 2,3-dihydroxybiphenyl dioxygenase from the naphthalenesulfonate-degrading bacterium strain BN6. *J Bacteriol* 177: 5865–5871.

Hintner J-P, C Lechner, U Riegert, AE Kuhm, T Storm, T Reemtsma, A Stolz. 2001. Direct ring fission of salicylate by a salicylate 1,2-dioxygenase activity from *Pseudaminobacter salicylatoxidans*. *J Bacteriol* 183: 6936–6942.

Junker F, R Kiewitz, AM Cook. 1997. Characterization of the *p*-toluenesulfonate operon *tsaMBCD* and *tsaR* in *Comamonas testosteroni* T-2. *J Bacteriol* 179: 919–927.

Junker F, T Leisinger, AM Cook. 1994. 3-sulfocatechol 2,3-dioxygenase and other dioxygenases (EC 1.13.11.2 and EC 1-14-12) in the degradative pathways of 2-aminobenzenesulfonic, benzenesulphonic and 4-toluenesulfonic acids in *Alcaligenes* sp. strain O-1. *Microbiology* (U.K.) 140: 1713–1722.

Kertesz MA. 1999. Riding the sulfur cycle—Metabolism of sulfonates and sulfate esters by Gram-negative bacteria. *FEMS Microbiol Revs* 24: 135–175.

Kuhm AE, A Stolz, K-L Ngai, H-J Knackmuss. 1991. Purification and characterization of a 1,2-dihydroxy-naphthalene dioxygenase from a bacterium that degrades naphthalenesulfonic acids. *J Bacteriol* 173: 3795–3802.

Locher HH, T Leisinger, AM Cook. 1991. 4-Toluene sulfonate methyl-monooxygenase from *Comamonas testosteroni*: Purification and some properties of the oxygenase component. *J Bacteriol* 173: 3741–3748.

Nörtemann B, A Glässer, R Machinek, G Remberg, H-J Knackmuss. 1993. 5-Hydroxyquinoline-2-carboxylic acid, a dead-end metabolite from the bacterial oxidation of 5-aminonaphthalene-2-sulfonic acid. *Appl Environ Microbiol* 59: 1898–1903.

Schleheck D, F van Netzer, T Fleishmann, D Rentsch, T Huhn, AM Cook, H-P Kohler. 2010. The missing link in linear alkylbenzenesulfonate surfactant degradation: 4-sulfoacetophenone as a transient intermediate in the degradation of 3-(4-sulfophenyl)butyrate by *Comamonas testosteroni*. KF-1 *Appl Environ Microbiol* 76: 196–202.

Wittich RM, HG Rast, H-J Knackmuss. 1988. Degradation of naphthalene-2,6- and naphthalene-1,6-disulfonic acid by a *Moraxella* sp. *Appl Environ Microbiol* 54: 1842–1847.

Part 5: Aromatic Compounds with Nitro Substituents

These are readily produced by nitration of aromatic compounds and are important explosives. The amines formed by reduction are able to undergo a number of reactions, and have a wide range of application in the production of agrochemicals, dyestuffs, and pharmaceuticals.

9.21 Nitroarenes

Nitrotoluenes including 2,4,6-trinitrotoluene (TNT) are components of explosives and several nitroarenes including the antibacterial nitrofurans have established mutagenicity (Purohit and Basu 2000*)*. Nitrofuran and nitroimidazole drugs that necessitate reduction to express activity are discussed in Chapter 4. The biodegradation of 2,4,6-trinitrotoluene has been reviewed (Esteve-Núñez et al. 2001), while another review discusses the biodegradation of nitroaromatic compounds and provides a valuable summary of nitroarenes explosives and pesticides, and their natural occurrence as metabolites of *Streptomyces* (Ju and Parales 2010).

Substantial effort has been directed to the degradation of nitroarenes, and to their reduction to amines. Although nitroarene reductases, noted in Chapter 3 are distributed in a range of biota, the products may not necessarily represent intermediates in the degradation pathway. The amines produced may undergo a number of reactions that may contribute to their removal by transformation.

1. *N*-acetylation to neutral acetanilides that are terminal metabolites (Noguera and Freedman 1996).
2. Dimerization of partially reduced intermediates to nitroazoxytoluenes (Bayman and Radkar 1997).
3. Association with organic matter in soil to produce nonextractable residues (Thorn and Kennedy 2002).

Nitroarenes are reactive compounds, and may undergo cyclization to form benzimidazoles when the nitro group is vicinal to an amino or substituted amine group. This has been demonstrated as one of the pathways during the fungal metabolism of dinitramine (Figure 9.56) (Laanio et al. 1973), and in a mixed bacterial culture, which metabolized 2-nitroaniline to 2-methylbenzimidazole (Hallas and Alexander 1983).

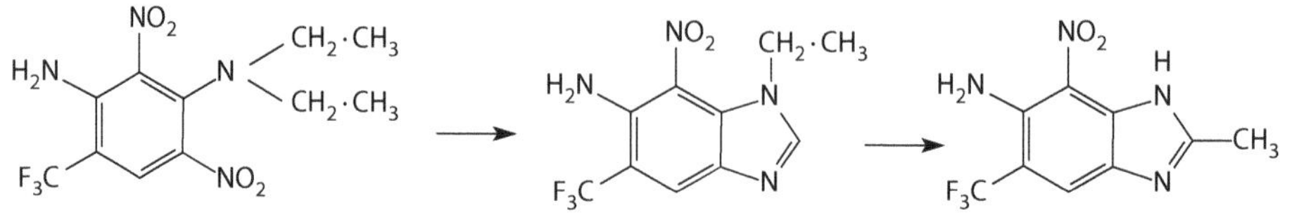

FIGURE 9.56 Biodegradation of dinitramine.

A wide range of mechanisms is involved in the degradation and transformation of aromatic compounds with nitro substituents. These include reduction of the nitro group, reduction of the ring, dioxygenation, monooxygenation, and peroxidation.

9.21.1 Reduction of Nitro Groups

Reduction can take place both of the nitro group and of the aromatic ring: the two reactions are not, however, exclusive and both may occur simultaneously. Reduction of the nitro group can be carried out by specific reductase enzymes and by the flavoprotein oxidoreductase "Old yellow Enzyme" which are discussed in Chapter 3. Examples include the following in which the reduction of the nitro groups to amines may occur under either aerobic or anaerobic conditions.

1. The complete sequence of reduction products was produced from 2,6-dinitrotoluene by *Salmonella typhimurium* strain TA 98 (Sayama et al. 1992)—2-nitroso-6-nitrotoluene, 2-hydroxylamino-6-nitrotoluene, and 2-amino-6-nitrotoluene.

2. Reduction of nitroarenes has been demonstrated in species of *Clostridium* and *Eubacterium*, and was associated with the reduction in the mutagenic activity of 1-nitropyrene, and 1,3- and 1,6-dinitropyrene (Rafii et al. 1991).

3. Anaerobic extracts of *Clostridium acetobutylicum* reduced 2,4,6-trinitrotoluene to 2,4-dihydroxylamino-6-nitrobenzene, which underwent an enzymatic Bamberger-type rearrangement (noted later) to 2-amino-4-hydroxylamino-5-hydroxy-6-nitrotoluene (Hughes et al. 1998). This is especially remarkable since the enzymatic activity was not dependent on the presence of nitroaromatic compounds as growth substrates, and the reduction was mediated by the Fe-only hydrogenase which is typically associated with hydrogen production in clostridia (Watrous et al. 2003). Analogous reactions were found for the carbon monoxide dehydrogenase from *Clostridium thermoaceticum* (Huang et al. 2000).

4. *Clostridium bifermentans* reduced 2,4,6-trinitrotoluene to 2,4,6-triaminotoluene, and a metabolite was formed by reaction of one of the amino groups with methylglyoxal (Lewis et al. 1996).

5. Reduction of the nitro group of nitrodiphenylamines was accomplished by strains of *Desulfovibrio* sp., *Desulfococcus* sp., and *Desulfomicrobium* sp. during growth with lactate or benzoate. During incubations with 2-aminodiphenylamine, phenazine and 4-aminoacridine were formed (Drzyzga et al. 1996). The mechanism whereby the C_{10} of 4-aminoacridine is formed from the growth substrate lactate was not resolved, although it might plausibly be suggested to involve carboxylation—analogous to that of aniline—followed by cyclization and reduction of the 9-hydroxyacridine.

6. Under anoxic conditions, TNT can serve as a terminal electron acceptor (Esteve-Núñez et al. 2000), with utilization of the compound as a source of nitrogen. A number of products were formed by oxidation of the methyl group and loss of nitrite to 4-hydroxybenzoate (Esteve-Núñez and Ramos 1998).

7. The degradation of 2,4,6-trinitrotoluene under aerobic conditions involves the reduction of one or more nitro groups to amino groups before dioxygenation and ring fission to methylnitrocatechols (Fiorella and Spain 1997; Johnson et al. 2001).

8. The transformation of TNT has been studied in several yeasts, and reduction of one nitro group as well as of the ring was observed. *Saccharomyces* sp. strain ZS produced 2-hydroxylamino-3,5-dinitrobenzene and 4-hydroxylamine-2,6-dinitrobenzene that were also produced by *Candida* sp. strain AN-L14 which produced in addition the ring-reduced 3-hydride (Zaripov et al. 2002).

As an alternative, partial reduction to hydroxylamines may be involved in the degradation of nitroarenes.

1. This has been demonstrated for a number of compounds: 4-nitrobenzoate (Groenewegen et al. 1992; Hughes and Williams 2001), 4-nitrotoluene (Spiess et al. 1998), and in one of the

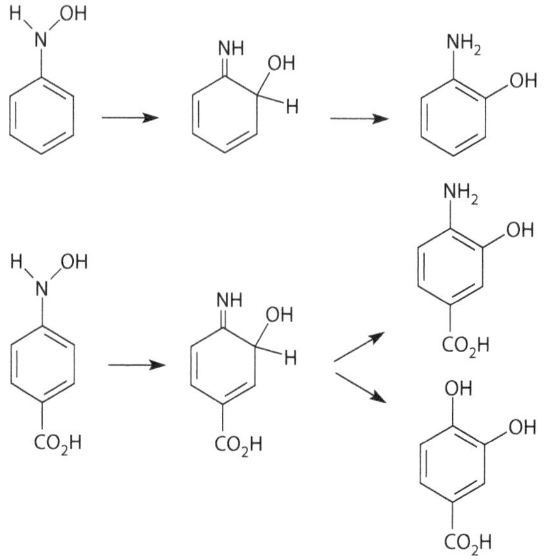

FIGURE 9.57 Aerobic transformation of phenylhydroxylamines. (Adapted from Nadeau LJ, Z He, JC Spain. 2003. *Appl Environ Microbiol* 69: 2786–2793.)

pathways of nitrobenzene (Somerville et al. 1995). The resulting hydroxylamines may then undergo a Bamberger rearrangement by a combination of mutases and lyases to *ortho*-amino-phenols (Nadeau et al. 2003), which are then degraded by oxygenation and ring fission (Figure 9.57). The 3-hydroxylaminophenol mutase from *Ralstonia eutropha* JMP134 grown with 3-nitrophenol as nitrogen source has been purified (Schenzle et al. 1997), and is able to catalyze the rearrangement of a number of substituted aromatic hydroxylamines. The formation of both 2- and 4-hydroxyphenol from hydroxylaminobenzene is formally comparable to the classic Bamberger rearrangement. As an alternative, the intermediate *ortho*-aminophenols may pro-duce 2-aminophenoxazine-3-ones as terminal metabolites (Spiess et al. 1998; Hughes et al. 2002). The phenoxazinone nucleus in the secondary metabolite grixazone B of *Streptomyces griseus* is produced from 3-amino-4-hydroxybenzoate by analogous reactions—oxidation to the quinone imine followed by reaction with *N*-acetylcysteine, and further oxidation and dimer-ization to grixazone with loss of a carboxyl group (Suzuki et al. 2006a). The biosynthesis of the precursor 3-amino-4-hydroxybenzoate is accomplished by the condensation of aspartate 4-semialdehyde and dihydroxyacetone phosphate, followed by a sequence of Mn^{2+}-catalyzed reactions involving dihydroxypiperidine and piperidone intermediates (Suzuki et al. 2006b).

2. The biodegradation of nitrobenzene by a strain of *Pseudomonas pseudoalcaligenes* is initiated by reduction to the phenylhydroxylamine followed by rearrangement to 2-aminophenol (Bamberger rearrangement). This is then degraded by a 2-aminophenol 1,6-dioxygenase (Nishino and Spain 1993; Lendenmann and Spain 1996; Takenaka et al. 1997) to 2-amino-muconic semialdehyde, which is further degraded by dehydrogenation and deamination to 2-hydroxymuconate (4-oxalocrotonate) (Figure 9.58) (He and Spain 1997, 1998). Although nitrosobenzene was not detected as an intermediate, it is a substrate for the nitroreductase,

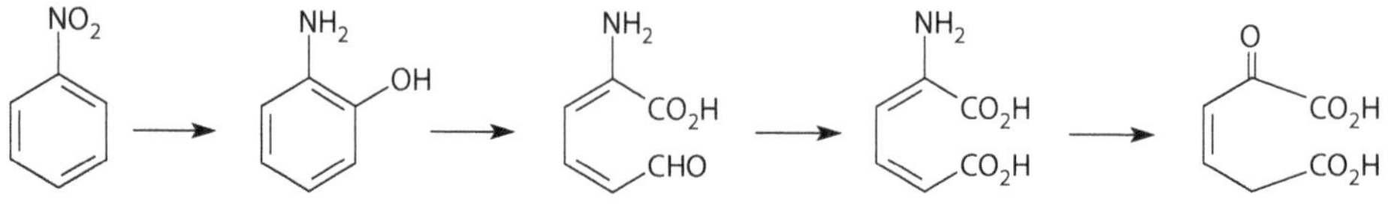

FIGURE 9.58 Degradation of nitrobenzene via hydroxylamine.

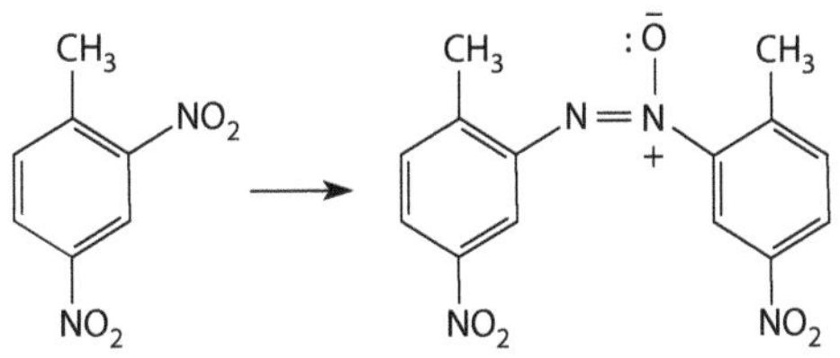

FIGURE 9.59 Metabolism of 2,4-dinitrotoluene by *Mucrosporium* sp.

whereas hydroxylaminobenzene is not further reduced by the enzyme (Somerville et al. 1995). In *Pseudomonas putida* strain HS-12, the gene encoding the hydroxylamino mutase occurs on a plasmid pNB2, whereas the other genes are carried by plasmid pNB1 (Park and Kim 2000). This pathway is also used by a strain of *Mycobacterium* sp. for the degradation of 4-nitrotoluene (Spiess et al. 1998), with the formation of 3-hydroxy-4-aminotoluene as an intermediate before further degradation. Analogous pathways are used for the degradation of 4-nitrochloro-benzene involving reduction, dioxygenation, and deamination of the 2-amino-5-chloromuconate (Wu et al. 2006).

3. Interactions may take place between metabolites at different levels of reduction. This may plausibly account for the dimerization of partially reduced intermediates to nitro-azoxytoluenes (Bayman and Radkar 1997), and the identification of azoxy compounds as biotransformation products of 2,4-dinitrotoluene by the fungus *Mucrosporium* sp. (Figure 9.59) (McCormick et al. 1978).

Other important interactions involving hydroxylamines and amines produced by reduction are discussed later in the context of Meisenheimer-type reactions and reduction of the ring. It is sufficient here to note a few examples in which biotransformation of 2,4,6-trinitrotoluene resulted in the release of nitrite that was used as a source of nitrogen by mechanisms that are described in the later paragraph dealing with Meisenheimer-type complexes.

1. A reaction involving reductive elimination of nitrite has been observed in cultures of a *Pseudomonas* sp. that can use 2,4,6-trinitrotoluene as a source of nitrogen (Duque et al. 1993). The substrate was transformed by successive reductive loss of nitro groups with the formation of toluene. Although toluene cannot be metabolized by this strain, it could be degraded by a transconjugant containing the TOL plasmid from *Pseudomonas putida*.

2. A strain of *E. coli* was able to use 2,4,6-trinitrotoluene as nitrogen source, and cell extracts of cells grown with glucose or glycerol formed nitrite using NADPH or NADH as cofactor (Stenuit et al. 2006).

3. *Pseudomonas putida* strain JLR11 was able to assimilate nitrite from 2,4,6-trinitrotoluene (Caballero et al. 2005).

9.21.2 Dioxygenation

Dioxygenation may result in the elimination of nitrite in reactions that are analogous to the elimination of sulfite from aromatic sulfonates, or halogen from 2-halobenzoates. The structure of nitrobenzoate dioxygenase has been examined for nitrotoluenes in *Comamonas* sp. strain JS7865 to determine which specific amino acids at the active site discriminate ring dioxygenation to catechols with the loss of nitrite from side-chain oxidation to benzyl alcohols of Ju and Parales (2006). As an alternative to dioxygenation, toluene-3- and toluene-4-monooxygenases can transform nitrobenzene to 4- or 3-nitrophenol, with the former dominating (Fishman et al. 2004).

1. The degradation of nitrobenzene by *Comamonas* sp. strain JS765 (Nishino and Spain 1995; Lessner et al. 2002) and of 2-nitrotoluene by *Pseudomonas* sp. strain JS42 (Haigler et al. 1994) is mediated by dioxygenation with the formation of catechol and 3-methylcatechol,

respectively. The enzyme involved in the degradation of 2-nitrotoluene by this strain has been purified and shown to consist of three components—an [Fe–S] protein ISP_{2NT} that serves as the terminal oxygenase, a reductase$_{2NT}$ that may be a flavoprotein, and an electron-transfer protein ferredoxin$_{2NT}$ similar to Rieske-type ferredoxins (An et al. 1994). The dioxygenase involved in the degradation of 2,4-dinitrotoluene by *Burkholderia* (*Pseudomonas*) sp. strain DNT also consists of three components—a terminal oxygenase, an [Fe–S] ferredoxin, and a reductase—and is broadly similar to naphthalene dioxygenase (Suen et al. 1996). Nonetheless, although the 2-nitrotoluene dioxygenase from *Pseudomonas* sp. strain JS42, the 2,4-dinitrotoluene dioxygenase from *B. cepacia*, and naphthalene dioxygenase from *Pseudomonas* sp. strain 9816-4 have comparable dioxygenase activities toward naphthalene, they have widely different specificities toward the isomeric mono-nitrotoluenes and 2,4-dinitrotoluene (Parales et al. 1998). In the same way, nitrocatechols can be formed initially from 1,3-dinitrobenzenes, followed by elimination of nitrite in a second step (Dickel and Knackmuss 1991; Spanggord et al. 1991).

2. The degradation of both 2,4- and 2,6-dinitrotoluene is initiated by dioxygenation with elimination of nitrite and the formation of nitrocatechols. For the 2,4-isomer, loss of an additional nitrite takes place with the formation of 2-hydroxy-5-methylbenzoquinone (Spanggord et al. 1991), whereas for the 2,6-isomer, the catechol undergoes fission and nitrite is lost during the last stage of the sequence (Figure 9.60) (Nishino et al. 2000). The degradation of 2,4-dinitrotoluene by *Burkholderia cepacia* strain R34 involves dioxygenation with elimination of nitrite to 4-methyl-5-nitrocatechol that underwent monooxygenation to 2-hydroxy-5-methylquinone. The 2,4,5-trihydroxytoluene that is formed from this is degraded by extradiol fission to pyruvate and methylmalonate-semialdehyde (Figure 9.61) (Johnson et al. 2002).

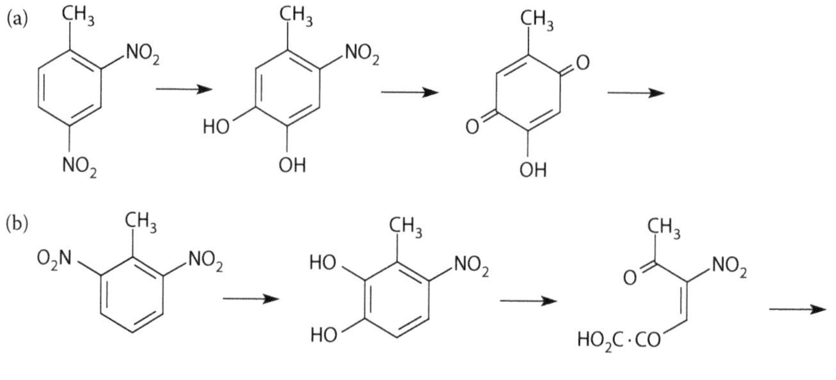

FIGURE 9.60 Degradation of (a) 2,4-dinitrotoluene and (b) 2,6-dinitrotoluene.

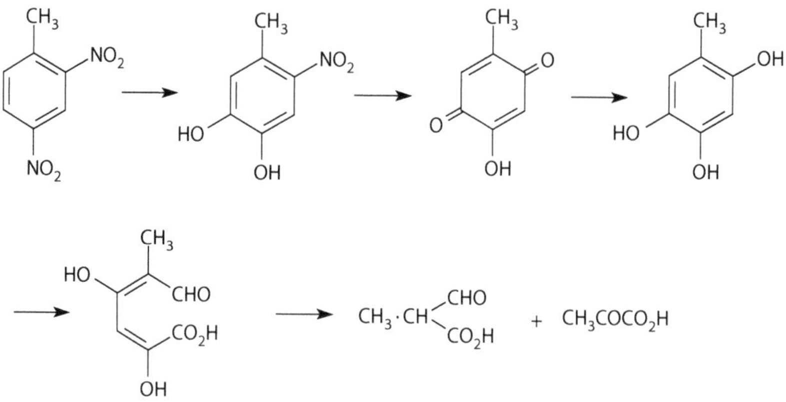

FIGURE 9.61 Biodegradation of 2,4-dinitrotoluene.

3. The degradation of 2,4,6-trinitrotoluene under aerobic conditions is dependent on the reduction of one or more nitro groups to amino groups before dioxygenation (Fiorella and Spain 1997; Johnson et al. 2001). The primary reaction was dioxygenation to yield nitrite and the corresponding aminomethylnitrocatechol whereas monooxygenation to produce the benzyl alcohol was a secondary reaction.

4. Exceptionally, loss of nitrite may take place after dioxygenation and ring fission. The degradation of 5-nitroanthranilic acid by *Bradyrhizobium* sp. strain JS 329 displayed two unusual features: (i) the amine group underwent hydrolytic deamination to 5-nitrosalicylate and (ii) this was followed by dioxygenation with simultaneous ring fission to produce 4-nitro-6-oxo-hepta-2,4-diendioate from which nitrite was lost to form methylpyruvate (Qu and Spain 2011).

9.21.3 Side-Chain Oxidation

The degradation of 4-nitrotoluene by *Pseudomonas* sp. takes place by oxidation to 4-nitrobenzoate via 4-nitrobenzyl alcohol and 4-nitrobenzaldehyde (Haigler and Spain 1993; Rhys-Williams et al. 1993; James and Williams 1998).

9.21.4 Peroxidase Oxidation

The degradation of 2,4-dinitrotoluene by *P. chrysosporium* involves a mechanism completely different from those already outlined, and involves the function of both the manganese-dependent, and the lignin peroxidase systems (Valli et al. 1992). The pathway is shown in Figure 9.62, and is reminiscent of that used by the same organism for the degradation of 2,4-dichlorophenol (Valli and Gold 1991). The second step is catalyzed by the manganese peroxidase system with elimination of methanol, and the subsequent loss of nitrite with partial de-*O*-methylation is carried out by the lipid peroxidase system.

9.22 Nitrobenzoates

The degradation of nitrobenzoates displays the same diversity of reactions that has been found for nitroarenes, and the pathway likewise depends on the position of the nitro group.

1. The transformation of nitrobenzoates by reduction has been demonstrated even though the reduced products were not on the direct pathway for biodegradation (Cartwright and Cain 1959). The degradation of 2-nitrobenzoate is, however, initiated by reduction.

   a. Degradation by *Pseudomonas fluorescens* strain KU-7 is initiated by partial reduction to the hydroxylamine followed by rearrangement to 2-amino-3-hydroxybenzoate

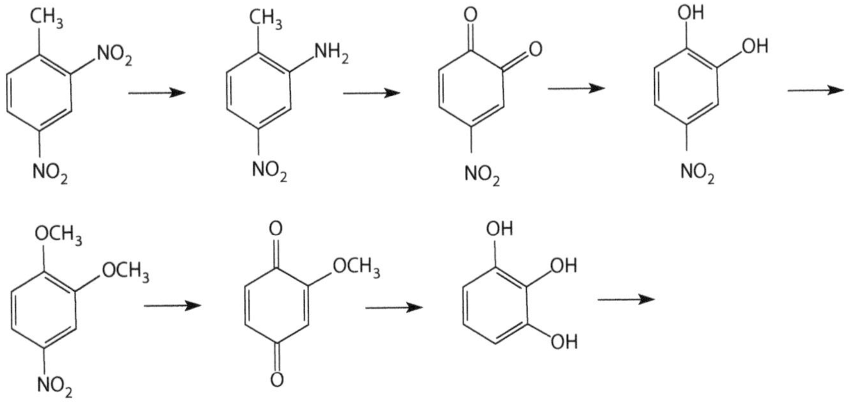

FIGURE 9.62 Biodegradation of 2,4-dinitrotoluene by *P. chrysosporium*.

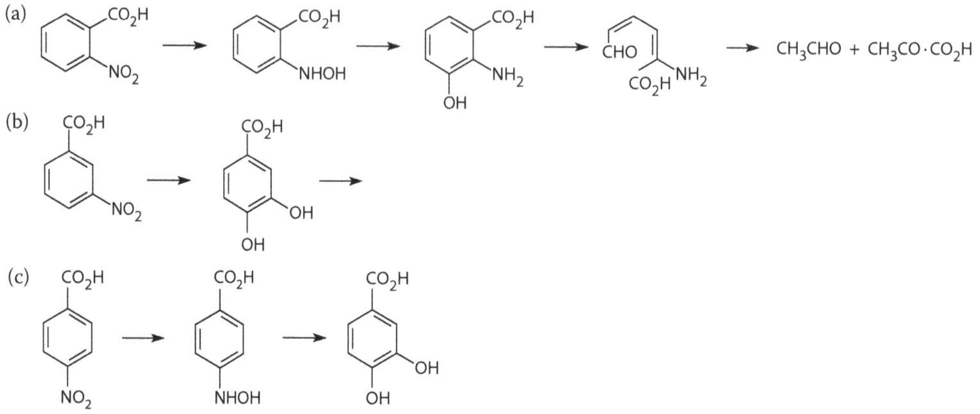

FIGURE 9.63 Degradation of (a) 2-nitrobenzoate, (b) 3-nitrobenzoate, and (c) 4-nitrobenzoate.

(3-hydroxyanthranilate), which was degraded by extradiol fission. Further reactions involved decarboxylation, deamination, and further decarboxylation to acetaldehyde and pyruvate (Figure 9.63a) (Muraki et al. 2003).

b. In *Arthrobacter protophormiae*, reduction to anthranilate took place, followed by the established pathway for its degradation (Chauhan and Jain 2000).

2. The degradation of 3-nitrobenzoate by *Pseudomonas* sp. strain JS51 and *Comamonas* sp. strain JS46 is initiated by dioxygenation to 3,4-dihydroxybenzoate with loss of nitrite (Nadeau and Spain 1995) (Figure 9.63b).

3. The degradation of 4-nitrobenzoate proceeds by a different pathway. During the degradation of 4-nitrobenzoate by *Comamonas acidovorans*, 4-nitroso and 4-hydroxylaminobenzoate were formed successively, and the latter was then metabolized to 3,4-dihydroxybenzoate with the elimination of NH_4^+ (Groenewegen et al. 1992) (Figure 9.63c). It should be noted that these cells were not adapted to growth with either 4-aminobenzoate or 4-hydroxy-benzoate that are alternative plausible intermediates. A comparable pathway is used by strains of *Pseudomonas* sp. for the degradation of 4-nitrotoluene that is initially oxidized to 4-nitrobenzoate via 4-nitrobenzyl alcohol and 4-nitrobenzaldehyde (Haigler and Spain 1993; Rhys-Williams et al. 1993; James and Williams 1998).

9.23 Nitrophenols

Nitrophenols are phytotoxic, and dinoseb (6-*sec*-butyl-2,4-dinitrophenol) has been used as a herbicide, while nitrophenols have been detected in rainwater and plausible mechanisms for their abiotic formation have been proposed (Kohler and Heeb 2003; Vione et al. 2005). The pathway for the degradation of phenols with a single nitro group depends on the position of the substituents, while that of phenols with two or more nitro groups is carried out by entirely different pathways involving reduction of the ring.

1. A strain of *Pseudomonas putida* was able to use both 2- and 3-nitrophenol as sources of carbon and nitrogen, with elimination of nitrite from the former and ammonium from the latter (Zeyer and Kearney 1984). The degradation of 2-nitrophenol by *Pseudomonas putida* B2 is carried out by monooxygenation and elimination of nitrite to produce catechol. The oxygenase had a broad substrate specificity (Zeyer and Kocher 1988), and might involve a quinone analogous to the monooxygenase-catalyzed loss of nitrite from 4-nitrophenol (Spain and Gibson 1991). Nitrated aromatic metabolites including 3-nitrotyrosine (van der Vliet et al. 1997) have been used as

markers for the physiological production of nitric oxide-derived oxidants, and the nature of the active species has been shown to be the NO_2 radical, which is produced by peroxidase oxidation of NO_2^- and H_2O_2 (Brennan et al. 2002). The degradation of 3-nitrotyrosine by *Variovorax paradoxus* strain JS171 and *Burkholderia* sp. strain JS165 has been examined (Nishino and Spain 2006). It involves reactions analogous to those for the degradation of tyrosine-deamination and decarboxylation to 4-hydroxy-3-nitrophenylacetate. This is converted to 3,4-dihydrophenylacetate (homoprotocatechuate) with loss of nitrite in a reaction (Figure 9.64) analogous to that for the degradation of 2-nitrophenol, which has already been noted.

2. For the degradation of 4-nitrophenol the pathway depends on the organism. Both involve monooxygenation to quinones at different stages that is analogous to that of 4-methyl-5-nitrocatechol produced from 2,4-dinitrotoluene and has already been described.

    a. In a strain of *Moraxella* sp., flavin-dependent oxygenation by a particulate enzyme produced successively benzoquinone and hydroquinone, which is degraded by dioxygenation to 3-oxoadipate (Spain and Gibson 1991) (Figure 9.65a). This is in contrast to the dioxygenase-mediated reaction in *Pseudomonas putida* B2 (Zeyer and Kocher 1988).

    b. In *Arthrobacter* sp. (Jain et al. 1994), *Bacillus sphaericus* JS905 (Kadiyala and Spain 1998) (Figure 9.65b), and in *Rhodococcus opacus* SAO 101 (Kitagawa et al. 2004), loss of nitrite from the initially formed 4-nitrocatechol produced 1,2,4-trihydroxy-benzene, which was further degraded by dioxygenation.

3. The degradation of 3-nitrophenol is different and is initiated in *Pseudomonas putida* strain B2 by partial reduction to the hydroxylamine (Meulenberg et al. 1996). This is followed by a Bamberger-type rearrangement to 2-aminohydroquinone catalyzed by a mutase before ring fission of the resulting 1,2,4-trihydroxybenzene (Meulenberg et al. 1996; Schenzle et al. 1997). The 3-hydroxylaminophenol mutase from *Ralstonia eutropha* JMP134 grown with 3-nitrophenol as nitrogen source has been purified (Schenzle et al. 1999). It is able to catalyze the rearrangement of a number of substituted aromatic hydroxylamines and the formation of both 2- and 4-hydroxyphenol from hydroxylaminobenzene, which is formally comparable to the classic Bamberger rearrangement.

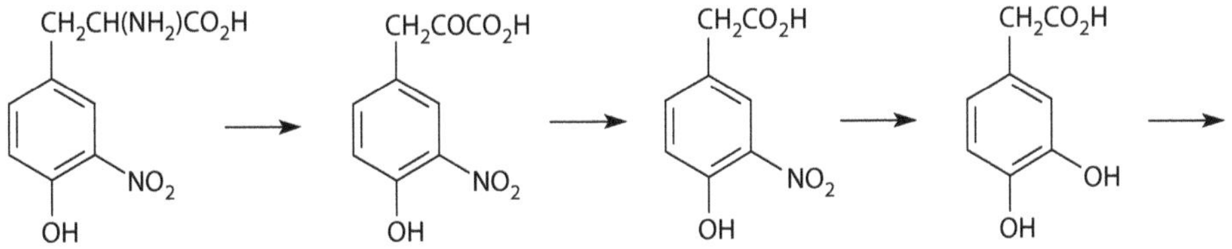

FIGURE 9.64   Degradation of 3-nitrotyrosine.

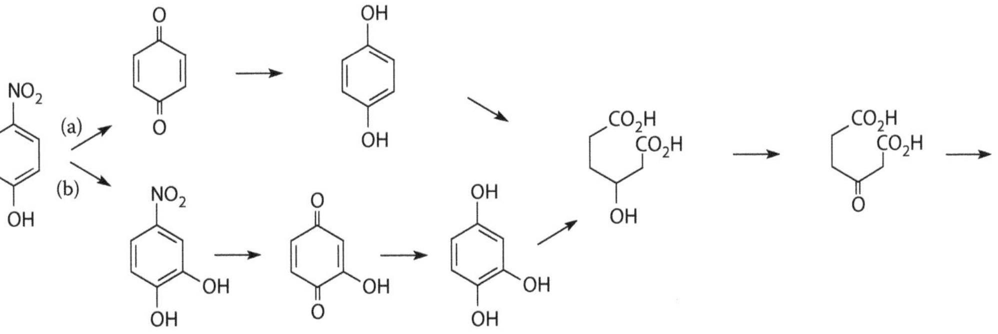

FIGURE 9.65   Degradation of 4-nitrophenol (a) by a strain of *Moraxella* sp. and (b) by *Arthrobacter* sp. and *Bacillus sphaericus* JS905. (Adapted from Jain RK, JH Dreisbach, JC Spain. 1994. *Appl Environ Microbiol* 60: 3030–3032.)

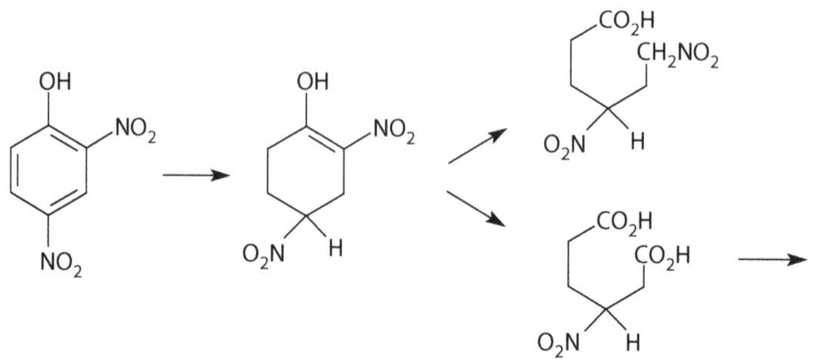

FIGURE 9.66 Degradation of 2,4-dinitrophenol.

Examination of the degradation of phenols containing several nitro groups has revealed that the elimination of nitrite did not proceed by any of the pathways already illustrated, and that terminal metabolites could be formed by reduction of the aromatic ring. For example, 4,6-dinitrohexanoate is produced from 2,4-dinitrophenol (Lenke et al. 1992), and 2,4,6-trinitrocyclohexanone from 2,4,6-trinitrophenol (Lenke and Knackmuss 1992). Depending on the strain, degradation of 2,4-dinitrophenol may produce 4,6-dinitrohexanoate or 3-nitroadipate, which is further degraded with loss of nitrite (Figure 9.66) (Blasco et al. 1999). This merits discussion in the separate paragraph that deals with Meisenheimer-type complexes.

9.23.1 Reduction of the Ring: Meisenheimer-Type Complexes

Meisenheimer-type complexes are produced by the reaction of arenes containing strongly electron-attracting substituents such as nitro groups with a nucleophile. In the biotransformation of 2,4,6-trinitrotoluene and 2,4,6-trinitrophenol the complex is the result of attack with hydride anions from the reductant.

9.23.1.1 Characterization

Reduction of the ring produces products with one or two additional hydrogen atoms that are termed the 3-monohydride or the 3,5-dihydride whose structures are normally written in the nitronate form with electrons delocalized over C_1, C_2, and C_6 (Figure 9.67a). *Rhodococcus erythropolis* produced the 3,5-dihydride and its protonated form by reduction of 2,4,6-trinitrotoluene that were characterized by ^{1}H and ^{13}C NMR (Vorbeck et al. 1998). In the yeast *Yarrowia lipolytica*, the 3-hydride, the 3,5-dihydride and their protonated isomers were the dominant products, and were characterized by MS and UV spectroscopy (Ziganchin et al. 2007). The analogous products from 2,4,6-trinitrophenol are normally written in the quinonoid form (Figure 9.67b) that has been characterized by ^{1}H-NMR from *Nocardioides* sp. strain CB 22-2 (Behrend et al. 1999) and from *Rhodococcus erythropolis* (Rieger et al. 1999).

9.23.1.2 Mechanisms

The degradation of 2,4,6-trinitrophenol by *Nocardioides simplex* FJ2-1A is initiated by reduction of the ring. This is carried out by two enzyme systems, an NADPH-dependent F_{420} reductase and a transferase that transfers hydride to the ring to produce the 3-hydride: this could be further reduced to the 3,5-dihydride that was a terminal product, or by loss of nitrite to produce 2,4-dinitrophenol that is reduced to 2,4-dinitrocyclohexanone that was hydrolyzed to 4,6-dinitrohexanoate (Ebert et al. 1999). Details were added from analysis of the genes involved in 2,4,6-trinitrophenol degradation by *Rhodococcus (opacus) erythropolis* HL PM-1 that made it possible to construct the pathway: NpdI was the transferase in the $F_{420}H_2$ production of the 3-hydride which underwent reduction by NpdC to the 3,5-dihydride from which nitrite was lost (Heiss et al. 2002). Regeneration of $F_{420}H_2$ by DADPH reduction was catalyzed by NpdG, and NpdH was proposed as involved in isomerization of the 3,5-dihydride and a protonated form.

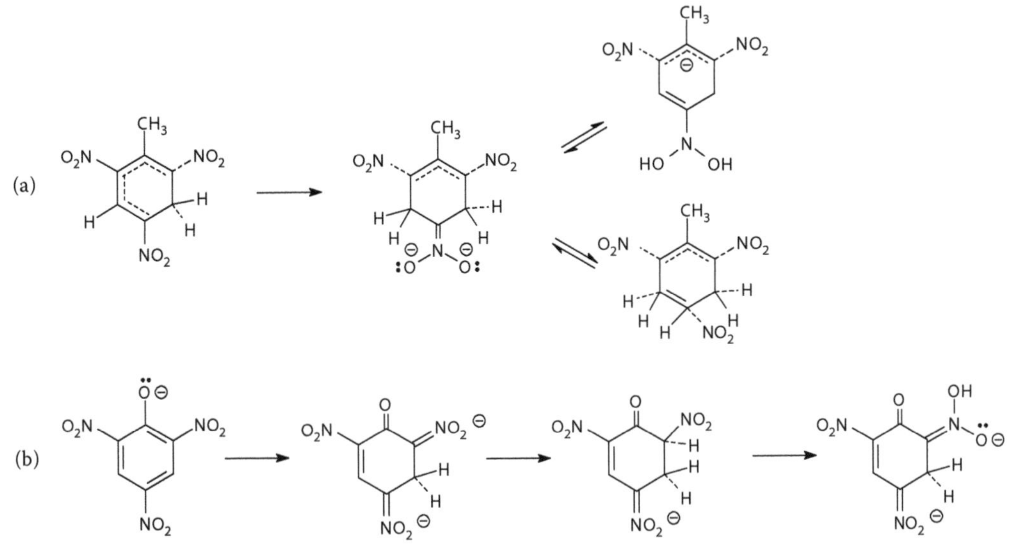

FIGURE 9.67   Formation of Meisenheimer complexes by *Rhodococcus erythropolis* from (a) 2,4,6-trinitrotoluene. (Redrawn from Vorbeck C et al. 1998. *Appl Environ Microbiol* 64: 246–252.) (b) 2,4,6-Trinitrophenol. (Redrawn from Rieger P-G et al. 1999. *J Bacteriol* 181: 1189–1195.)

9.23.1.3 Reactions

Elimination of nitrite is characteristic of these reactions that may be either degradations or complex biotransformations.

1. In the degradation of 2,4,6-trinitrophenol, nitrite could be eliminated from the *aci*-nitro—but not from the nitro form—of the dihydride Meisenheimer-type complex from *Rhodococcus opacus* HL PM-1 that has already been discussed The resulting 2,4-cyclohexanone then underwent hydrolysis to 4,6-dinitrohexanoate (Hofmann et al. 2004).

2. More complex biotransformations have been described: these involved reactions between Meisenheimer-type hydride complex and partly reduced nitroarenes concomitant with the loss of nitrite.

   a. Partial reduction of 2,4,6-trinitrotoluene produced the 2- and 4-hydroxylamines. Nucleophilic attack on the dihydride-Meisenheimer complex resulted in displacement of nitrite and the formation of tetranitro *N*-hydroxydiphenylamines that were reduced to the tetranitrodiphenylamines. The transformation of 2,4,6-trinitrotoluene was examined in *Pseudomonas fluorescens* strain I-C using a purified reductase XenB that is a flavoprotein oxidoreductase. Partial reduction produced the 2- and 4-hydroxylaminodinitro- and 2- and 4-aminodinitrobenzene from which nonenzymatic reactions resulted in the loss of nitrite and formation of tetranitrobiphenyls with an amino and two methyl substituents that were characterized by MS (Figure 9.68a) (Pak et al. 2000).

   b. An alternative reaction was studied using XenB from strains of *Pseudomonas putida* KT2440 in which two reactions took place: type I reduction of one nitro group to the 2- and 4-hydroxyaminodinitrobenzene and a type II reaction that produced a Meisenheimer dihydride. These reacted with the elimination of nitrite to produce methyltrinitro- or the dimethyltetranitro *N*-hydroxydiphenylamines that were reduced to the corresponding diphenylamines (Figure 9.68b) (van Dillewijn et al. 2008a) that were identified by MPLC-¹H NMR (Wittich et al. 2008).

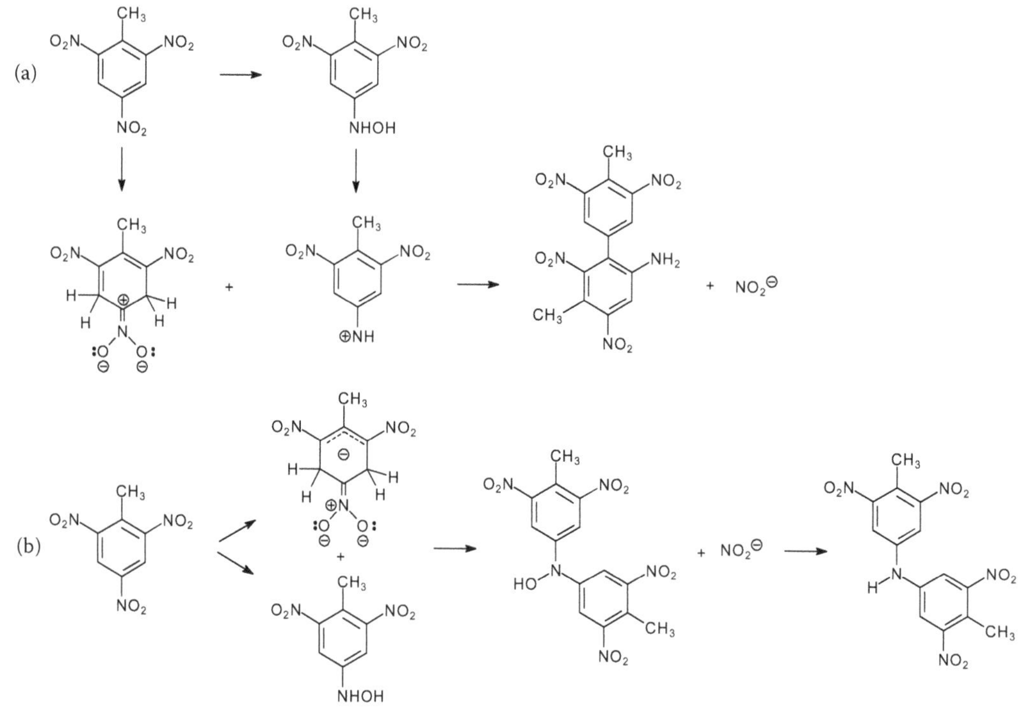

FIGURE 9.68 Reactions of Meisenheimer-type complexes: Transformation of (a) 2,4,6-trinitrotoluene by reductase XenB from *Pseudomonas fluorescens* with production of tetranitrobiphenyls. (Adapted from Pak JW et al. 2000. *Appl Environ Microbiol* 66: 4742–4750.) (b) 2,4,6-Trinitrotoluene by reductase XenB from *Pseudomonas putida* followed by condensation to produce *N*-hydroxydiphenylamines and diphenylamines. (Adapted from van Dillewijn P et al. 2008a. *Appl Environ Microbiol* 74: 6820–6823.)

9.24 References

An D, DT Gibson, JC Spain. 1994. Oxidative release of nitrite from 2-nitrotoluene by a three component enzyme system from *Pseudomonas* sp. strain JS42. *J Bacteriol* 176: 7462–7467.

Bayman P, GV Radkar. 1997. Transformation and tolerance of TNT (2,4,6-trinitrotoluene) by fungi. *Int Biodet Biorem* 39: 45–53.

Behrend C, K Heesche-Wagner. 1999. Formation of hydride-Meisenheimer complexes of picric acid (2,4,6-trinitrophenol) and 2,4-dinitrophenol during mineralization of picric acid by *Nocardiodes* sp. strain CB 22-2. *Appl Environ Microbiol* 65: 1372–1377.

Blasco R, E Moore, V Wray, D Pieper, K Timmis, F Castillo. 1999. 3-Nitroadipoate, a metabolic intermediate for mineralization of 2,4-dinitrophenol by a new strain of a *Rhodococcus* species. *J Bacteriol* 181: 149–152.

Brennan M-L et al. 2002. A tale of two controversies. Defining both the role of peroxidases in nitrotyrosine formation *in vivo* using eosinophil peroxidase and myeloperoxidase-deficient mice, and the nature of the peroxidase-generated reactive nitrogen species. *J Biol Chem* 277: 17415–17427.

Caballero A, A Esteve-Núñez, GJ Zylstra, JL Ramos. 2005. Assimilation of nitrogen from nitrite and trinitro-toluene in *Pseudomonas putida* JLR11. *J Bacteriol* 187: 396–399.

Cartwright NJ, RB Cain. 1959. Bacterial degradation of the nitrobenzoic acids. 2. Reduction of the nitro group. *Biochem J* 73: 305–314.

Chauhan A, RK Jain. 2000. Degradation of *o*-nitrobenzoate via anthranilic acid (*o*-aminobenzoate) by *Arthroacter protophormiae*: A plasmid-encoded new pathway. *Biochem Biophys Res Commun* 267: 236–244.

Dickel OD, H-J Knackmuss. 1991. Catabolism of 1,3-dinitrobenzene by *Rhodococcus* sp. QT-1. *Arch Microbiol* 157: 76–79.

Drzyzga O, A Schmidt, K-H Blotevogel. 1996. Cometabolic transformation and cleavage of nitrodiphenyl-amines by three newly isolated sulfate-reducing bacterial strains. *Appl Environ Microbiol* 62: 1710–1716.

Duque E, A Haidour, F Godoy, JL Ramos. 1993. Construction of a *Pseudomonas* hybrid strain that mineralizes 2,4,6-trinitrotoluene. *J Bacteriol* 175: 2278–2283.

Ebert S, P-G Rieger, H-J Knmackmuss. 1999. Function of coenzyme F_{420} in aerobic catabolism of 2,4,6-tri-nitrophenol and 2,4-dinitrophenol by *Nocardioides simplex* FJ2-1A. *J Bacteriol* 181: 2669–2694.

Esteve-Núñez A, A Caballero, JL Ramos. 2001. Biological degradation of 2,4,6-trinitrotoluene. *Microbiol Mol Biol Revs* 65: 335–352.

Esteve-Núñez A, G Lucchesi, B Phillip, B Schink, JL Ramos. 2000. Respiration of 2,4,6-trinitrotoluene by *Pseudomonas* sp. strain aw. *J Bacteriol* 182: 1352–1355.

Esteve-Núñez A, JL Ramos. 1998. Metabolism of 2,4,6-trinitrotoluene by *Pseudomonas* sp. JRL11. *Environ Sci Technol* 32: 3802–3808.

Fiorella PD, JC Spain. 1997. Transformation of 2,4,6-trinitrotoluene by *Pseudomonas pseudoalcaligenes* JS52. *Appl Environ Microbiol* 63: 2007–2015.

Fishman A, Y Tao, TK Wood. 2004. Toluene 3-monooxygenase of *Ralstonia pickettii* PKO1 is a para-hydroxylating enzyme. *J Bacteriol* 186: 3117–3123.

French CE, S Nicklin, NC Bruce. 1998. Aerobic degradation of 2,4,6-trinitrotoluene by *Enterobacter cloacae* PB2, and by pentaerythritol tetranitrate reductase. *Appl Environ Microbiol* 64: 2864–2868.

Groenewegen PEG, P Breeuwer, JMLM van Helvoort, AAM Langenhoff, FP de Vries, JAM de Bont. 1992. Novel degradative pathway of 4-nitrobenzoate in *Comamonas acidovorans* NBA-10. *J Gen Microbiol* 138: 1599–1605.

Haigler BE, JC Spain. 1993. Biodegradation of 4-nitrotoluene by *Pseudomonas* sp. strain 4NT. *Appl Environ Microbiol* 59: 2239–2243.

Haigler BE, WH Wallace, JC Spain. 1994. Biodegradation of 2-nitrotoluene by *Pseudomonas* sp. strain JS 42. *Appl Environ Microbiol* 60: 3466–3469.

Hallas LE, M Alexander. 1983. Microbial transformation of nitroaromatic compounds in sewage effluent. *Appl Environ Microbiol* 45: 1234–1241.

He Z, JC Spain. 1997. Studies of the catabolic pathway of degradation of nitrobenzene by *Pseudomonas pseu-doalcaligenes* JS45: removal of the amino group from 2-aminomuconic semialdehyde. *Appl Environ Microbiol* 63: 4839–4843.

He Z, JC Spain. 1998. A novel 2-aminomuconate deaminase in the nitrobenzene degradation pathways of *Pseudomonas pseudoalcaligenes* JS 45. *J Bacteriol* 180: 2502–2506.

Heiss G, KW Hofmann, N Trachtmann, DM Walters, P Rouvière, H-J Knackmuss. 2002. *npd* gene functions of *Rhodococcus* (*opacus*) *erythropolis* HL PM-1 in the initial steps of 2,4,6-trinitrophenol degradation. *Microbiology UK* 148: 799–806.

Hofmann KW, H-J Knackmuss, G Heiss. 2004. Nitrite elimination and hydrolytic ring cleavage in 2,4,6-trini-trophenol (picric acid) degradation. *Appl Environ Microbiol* 70: 2854–2860.

Huang S, PA Lindahl, C Wang, GN Bennett, FB Rudolph, JB Hughes. 2000. 2,4,6-trinitrotoluene reduction by car-bon monoxide dehydrogenase from *Clostridium thermoaceticum*. *Appl Environ Microbiol* 66: 1474–1478.

Hughes MA, MJ Baggs, J al-Dulayymi, MS Baird, PA Williams. 2002. Accumulation of 2-aminophenoxazine-3-one-7-carboxylate during growth of *Pseudomonas putida* TW3 on 4-nitro-substituted substrates requires 4-hydroxylaminbenzoate lyase (PnbB). *Appl Environ Microbiol* 68: 4965–4970.

Hughes JB, C Wang, K Yesland, A Richardson, R Bhadra, G Bennett, F Rudolph. 1998. Bamberger rearrange-ment during TNT metabolism by *Clostridium acetobutylicum*. *Environ Sci Technol* 32: 494–500.

Hughes MA, PA Williams. 2001. Cloning and characterization of the *pnb* genes, encoding enzymes for 4-nitrobenzoate catabolism in *Pseudomonas putida* TW3. *J Bacteriol* 183: 1225–1232.

Jain RK, JH Dreisbach, JC Spain. 1994. Biodegradation of *p*-nitrophenol via 1,2,4-benzenetriol by an *Arthrobacter* sp. *Appl Environ Microbiol* 60: 3030–3032.

James KD, PA Williams. 1998. *ntn* genes determining the early steps in the divergent catabolism of 4-nitrotoluene and toluene in *Pseudomonas* sp. strain TW3. *J Bacteriol* 180: 2043–2049.

Johnson GR, RK Jain, JC Spain. 2002. Origins of the 2,4-dinitrotolune pathway. *J Bacteriol* 184: 4219–4231.

Johnson GR, BF Smets, JC Spain. 2001. Oxidative transformation of aminodinitrotoluene isomers by multi-component dioxygenases. *Appl Environ Microbiol* 67: 5460–5466.

Ju K-S, RE Parales. 2006. Control of substrate specificity by active-site residues in nitrobenzene in nitrobenzene dioxygenase. *Appl Environ Microbiol* 72: 1817–1824.

Ju KS, RE Parales. 2010. Nitroaromatic compounds, from synthesis to biodegradation. *Microbiol Mol Biol Rev* 74: 250–272.

Kadiyala V, JC Spain. 1998. A two-component monoxygenase catalyzes both the hydroxylation of *p*-nitrophenol and the oxidative release of nitrite from 4-nitrocatechol in *Bacillus sphaericus* JS905. *Appl Environ Microbiol* 64: 2479–2484.

Kitagawa W, N Kimura, Y Kamagata. 2004. A novel *p*-nitrophenol degradation gene cluster from a Gram-positive bacterium *Rhodococcus opacus* SAO101. *J Bacteriol* 186: 4894–4902.

Kohler M, NV Heeb. 2003. Determination of nitrated phenolic compounds in rain by liquid chromatography/atmospheric pressure chemical ionization mass spectrometry. *Anal Chem* 75: 3115–3121.

Laanio TL, PC Kearney, DD Kaufman. 1973. Microbial metabolism of dinitramine. *Pest Biochem Physiol* 3: 271–277.

Lendenmann U, JC Spain. 1996. 2-Aminophenol 1,6-dioxygenase: A novel aromatic ring cleavage enzyme purified from *Pseudomonas pseudoalcaligenes* JS 45. *J Bacteriol* 178: 6227–6232.

Lenke H, H-J Knackmuss. 1992. Initial hydrogenation during catabolism of picric acid by *Rhodococcus erythropolis* HL 24-2. *Appl Environ Microbiol* 58: 2933–2937.

Lenke H, H-J Knackmuss. 1996. Initial hydrogenation and extensive reduction of substituted 2,4-dinitro-phenols. *Appl Environ Microbiol* 62: 784–790.

Lenke H, DH Pieper, C Bruhn, H-J Knackmuss. 1992. Degradation of 2,4-dinitrophenol by two *Rhodococcus erythropolis* strains, HL 24-1 and HL 24-2. *Appl Environ Microbiol* 58: 2928–2932.

Lessner DJ, GR Johnson, RE Parales, JC Spain, DT Gibson. 2002. Molecular characterization and substrate specificity of nitrobenzene dioxygenase from *Comamonas* sp. strain JS 765. *Appl Environ Microbiol* 68: 634–541.

Lewis TA, S Goszczynski, RL Crawford, RA Korus, W Adamassu. 1996. Products of anaerobic 2,4,6-trinitrotoluene (TNT) transformation by *Clostridium bifermentans*. *Appl Environ Microbiol* 62: 4669–4674.

McCormick NG, JH Cornell, AM Kaplan. 1978. Identification of biotransformation products from 2,4-dinitrotoluene. *Appl Environ Microbiol* 35: 945–948.

Meulenberg R, M Pepi, JAM de Bont. 1996. Degradation of 3-nitrophenol by *Pseudomonas putida* B2 occurs via 1,2,4-benzenetriol. *Biodegradation* 7: 303–311.

Muraki T, M Taki, Y Hasegawa, H Iwaki, PCK Lau. 2003. Prokaryotic homologues of the eukaryotic 3-hydroxy-anthranilate 3,4-dioxygenase and 2-amino-3-carboxymuconate-6-semialdehyde decarboxylase in the 2-nitrobenzoate degradation pathway of *Pseudomonas fluorescens* strain KU-7. *Appl Environ Microbiol* 69: 1564–1572.

Nadeau LJ, Z He, JC Spain. 2003. Bacterial conversion of hydroxylamino aromatic compounds by both lyase and mutase enzymes involves intramolecular transfer of hydroxyl groups. *Appl Environ Microbiol* 69: 2786–2793.

Nadeau IJ, JC Spain. 1995. The bacterial degradation of *m*-nitrobenzoic acid. *Appl Environ Microbiol* 61: 840–843.

Nishino SF, GC Paoli, JC Spain. 2000. Aerobic degradation of nitrotoluenes and pathway for bacterial degradation of 2,6-dinitrotoluene. *Appl Environ Microbiol* 66: 2139–2147.

Nishino SF, JC Spain. 1993. Degradation of nitrobenzene by a *Pseudomonas pseudoalcaligenes*. *Appl Environ Microbiol* 59: 2520–2525.

Nishino SF, JC Spain. 1995. Oxidative pathways for the degradation of nitrobenzene by *Comamonas* sp. strain JS 765. *Appl Environ Microbiol* 61: 2308–2313.

Nishino SF, JC Spain. 2006. Biodegradation of 3-nitrotyrosine by *Burkholderia* sp. strain JS165 and *Variovorax paradoxus* JS171. *Appl Environ Microbiol* 72: 1040–1044.

Noguera DR, DL Freedman. 1996. Reduction and acetylation of 2,4-dinitrotoluene by a *Pseudomonas aeruginosa* strain. *Appl Environ Microbiol* 62: 2257–2263.

Pak JW, Kl Knoke, DR Noguera, BG Fox, GH Chambliss. 2000. Transformation of 2,4,6-trinitrotoluene by purified xenobiotic reductase B from *Pseudomonas fluorescens* I-C. *Appl Environ Microbiol* 66: 4742–4750.

Parales JV, RE Parales, SM Resnick, DT Gibson. 1998. Enzyme specificity of 2-nitrotoluene 2,3-dioxygenase from *Pseudomonas* sp. strain JS42 is determined by the C-terminal region of the α subunit of the oxygenase component. *J Bacteriol* 180: 1194–1199.

Park H-S, H-S Kim. 2000. Identification and characterization of the nitrobenzene catabolic plasmids pNB1 and pNB2 in *Pseudomonas putida*. *J Bacteriol* 182: 573–580.

Purohit V, AK Basu. 2000. Mutagenicity of nitroaromatic compounds. *Chem Res Toxicol* 13: 673–692.

Qu Y, JC Spain. 2011. Molecular and biochemical characterization of the 5-nitroanthranilic acid degradation pathway in *Bradyrhizobium* sp. strain JS 329. *J Bacteriol* 193: 3057–3063.

Rafii F, W Franklin, RH Heflich, CE Cerniglia. 1991. Reduction of nitroaromatic compounds by anaerobic bacteria isolated from the human gastrointestinal tract. *Appl Environ Microbiol* 57: 962–968.

Rhys-Williams W, SC Taylor, PA Williams. 1993. A novel pathway for the catabolism of 4-nitrotoluene by *Pseudomonas*. *J Gen Microbiol* 139: 1967–1972.

Rieger P-G, V Sinnwell, A Preuβ, W Francke, H-J Knackmuss. 1999. Hydride-Meisenheimer complex formation and protonation as key reactions of 2,4,6-trinitrophenol biodegradation by *Rhodococcus erythropolis*. *J Bacteriol* 181: 1189–1195.

Sayama M, M Inoue, M-A Mori, Y Maruyama, H Kozuka. 1992. Bacterial metabolism of 2,6-dinitrotoluene with *Salmonella typhimurium* and mutagenicity of the metabolites of 2,6-dinitrotoluene and related compounds. *Xenobiotica* 22: 633–640.

Schenzle A, H Lenke, P Fischer, PA Williams, H-J Knackmuss. 1997. Catabolism of 3-nitrophenol by *Ralstonia eutropha* JMP 134. *Appl Environ Microbiol* 63: 1421–1427.

Schenzle A, H Lenke, JV Spain, H-J Knackmuss. 1999. 3-hydroxylaminophenol mutase from *Ralstonia eutropha* JMP134 catalyzes a Bamberger rearrangement. *J Bacteriol* 181: 1444–1450.

Somerville CC, SF Nishino, JC Spain. 1995. Purification and characterization of nitrobenzene nitroreductase from *Pseudomonas pseudoalcaligenes* JS45. *J Bacteriol* 177: 3837–3842.

Spain JC, DT Gibson. 1991. Pathway for degradation of *p*-nitrophenol in a *Moraxella* sp. *Appl Environ Microbiol* 57: 812–819.

Spanggord RJ, JC Spain, SF Nishino, KE Mortelmans. 1991. Biodegradation of 2,4-dinitrotoluene by a *Pseudomonas* sp. *Appl Environ Microbiol* 57: 3200–3205.

Spiess T, F Desiere, P Fischer, JC Spain, H-J Knackmuss, H Lenke. 1998. A new 4-nitrotoluene degradation pathway in a *Mycobacterium* strain. *Appl Environ Microbiol* 64: 446–452.

Stenuit B, L Eyers, R Rozenberg, J-L Habid-Jiwan, SN Agathos. 2006. Aerobic growth of *Escherichia coli* with 2,4,6-trinitrotoluene (TNT) as the sole nitrogen source and evidence of TNT denitration by whole cells and cell-free extracts. *Appl Environ Microbiol* 72: 7945–7948.

Suen W-C, BE Haigler, JC Spain. 1996. 2,4-Dinitrotoluene dioxygenase from *Burkholderia* sp strain DNT: Similarity to naphthalene dioxygenase. *J Bacteriol* 178: 4926–4934.

Suzuki H, Y Furusho, T Higashi, Y Ohnishi, S Horinouchi. 2006a. A novel *o*-aminophenol oxidase responsible for formation of the phenoxazinone chromophore of grixazone. *J Biol Chem* 281: 824–833.

Suzuki H, Y Ohnishi, Y Furusho, S Sakuda, S Horinouchi. 2006b. Novel benzene ring biosynthesis from C_3 and C_4 primary metabolites by two enzymes. *J Biol Chem* 281: 36944–36951.

Takenaka S, S Murakami, R Shinke, K Hatakeyama, H Yukuwa, K Aoki. 1997. Novel genes encoding 2-aminophenol 1,6-dioxygenase from *Pseudomonas* species AP-3 growing on 2-aminophenol and catalytic properties of the purified enzyme. *J Biol Chem* 272: 14727–14732.

Thorn KA, KR Kennedy. 2002. ^{15}N NMR investigation of the covalent binding of reduced TNT amines to soil humic acid, model compounds, and lignocellulose. *Environ Sci Technol* 36: 3787–3796.

Valli K, BJ Brock, DK Joshi, MH Gold. 1992. Degradation of 2,4-dinitrotoluene by the lignin-degrading fungus *Phanerochaete chrysosporium*. *Appl Environ Microbiol* 58: 221–228.

Valli K, MH Gold. 1991. Degradation of 2,4-dichlorophenol by the lignin-degrading fungus *Phanerochaete chrysosporium*. *J Bacteriol* 173: 345–352.

van der Vliet A, JP Eiserich, H Lliwell, BCE Cross. 1997. Formation of reactive nitrogen species during peroxidase-catalyzed oxidation of nitrite. *J Biol Chem* 272: 7617–7625.

van Dillewijn P, R-M Wittich, A Caballero, J-L Ramos. 2008a. Type II hydride transferases from different microorganisms yield nitrite and diarylamines from polynitroaromatic compounds. *Appl Environ Microbiol* 74: 6820–6823.

Vione D, V Maurino, C Minero, E Pelizzetti. 2005. Aqueous atmospheric chemistry: Formation of 2,4-nitrophenol upon nitration of 2-nitrophenol and 4-nitrophenol in solution. *Environ Sci Technol* 39: 7921–7931.

Vorbeck C, H Lenke, P Fischer, H-J Knackmuss. 1994. Identification of a hydride-Meisenheimer complex as a metabolite of 2,4,6-trinitrotoluene by a *Mycobacterium* strain. *J Bacteriol* 176: 932–934.

Vorbeck C, H Lenke, P Fischer, JC Spain, H-J Knackmuss. 1998. Initial reductive reactions in aerobic microbial metabolism of 2,4,6-trinitrotoluene. *Appl Environ Microbiol* 64: 246–252.

Watrous MM, S Clark, R Kutty, S Huang, FB Rudolph, JB Hughes, GN Bennett. 2003. 2,4,6-trinitrotoluene reduction by an Fe-only hydrogenase in *Clostridium acetobutylicum. J Bacteriol* 182: 5683–5691.

Williams RE, DA Rathbone, NS Scrutton, NC Bruce. 2004. Biotransformation of explosives by the Old Yellow Enzyme family of flavoproteins. *Appl Environ Microbiol* 70: 3566–3574.

Wittich R-M, A Haïdour, P van Dillewijn, J-L Ramos. 2008. OYE flavoprotein reductases initiated the condensation of TNT-derived intermediates to secondary diarylamines and nitrite. *Environ Sci Technol* 42: 734–739.

Wu J-F, C-Y Jiang, B-J Wang, Y-F Ma, Z-P Liu, S-J Liu. 2006. Novel partial reductive pathway for 4-chloronitrobenzene and nitrobenzene degradation in *Comamonas* sp. strain CNB-1. *Appl Environ Microbiol* 72: 1759–1765.

Zaripov SA, AV Naumov, JF Abdrakhmanova, AV Garusov, RP Naumova. 2002. Models of 2,4,6-trinitrotoluene (TNT) initial conversion by yeasts. *FEMS Microbiol Lett* 217: 213–217.

Zeyer J, HP Kocher. 1988. Purification and characterization of a bacterial nitrophenol oxygenase which converts *ortho*-nitrophenol to catechol and nitrite. *J Bacteriol* 170: 1789–1794.

Zeyer J, PC Kearney. 1984. Degradation of *o*-nitrophenol and *m*-nitrophenol by a *Pseudomonas putida. J Agric Food Chem* 32: 238–242.

Ziganchin AM, R Gerlach, T Borch, AV Naumov, RP Naumova. 2007. Production of eight different hydride complexes and nitrite release from 2,4,6-trinitrotoluene by *Yarrowia lipolytica. Appl Environ Microbiol* 73: 7989–7905.

Part 6: Azoarenes

Possibly the most significant discovery in the metabolism of aromatic azo compounds had implications that heralded the age of modern chemotherapy. It was shown that the bactericidal effect of the azo dye Prontosil *in vivo* was in fact due to the action of its transformation product, sulfanilamide, which is an antagonist of 4-aminobenzoate that is required for the synthesis of the vitamin folic acid. Indeed, this reduction is the typical reaction involved in the first stage of the biodegradation of aromatic azo compounds.

Aromatic azo compounds, many of which are sulfonated, are components of many commercially important dyes, colorants, and pigments, so that attention has been directed to their degradation and transformation. These compounds are often considered recalcitrant, although their transformation has been accomplished by reduction to amines with scission of the Ar–N=N–Ar bond to produce arylamines. The amines may then be degraded, for example, 6-aminonaphthalene-2-sulfonate by dioxygenation and ring fission to 5-aminosalicylate (Haug et al. 1991). This is then degraded by dioxygenation and deamination (Stolz and Knackmuss 1993) in a pathway analogous to that for the degradation of salicylate initiated by salicylate-5-hydroxylase (Zhou et al. 2002).

Although reduction of the azo groups may be carried out under both aerobic and anaerobic conditions, some of the complicating issues are noted later. Reduction is readily accomplished under anaerobic conditions (Haug et al. 1991), and the azoreductase and nitroreductase from *Clostridium perfringens* apparently involve the same protein (Rafii and Cerniglia 1993). However, bacterial azoreductases have also been purified from aerobic organisms including strains of *Pseudomonas* sp. adapted to grow at the expense of azo dyes (Zimmermann et al. 1984), and *Staphylococcus aureus* (Chen et al. 2005). The amines that result from these reductive transformations then enter well-established metabolic pathways for the degradation of anilines by oxidative deamination to catechol followed by ring fission (McClure and Venables 1986; Fuchs et al. 1991). However, it should be noted that aniline-4-sulfonate (sulfanilate) formed from sulfonated azo dyes may be excreted into the medium, or after further oxygenation, polymerized to terminal products (Kulla et al. 1983). However, sulfanilate is degraded by a mixed culture of *Hydrogenophaga palleroni* and *Agrobacterium radiobacter*, and proceeds via catechol-4-sulfonate followed by intradiol ring fission; sulfite is then eliminated from the muconolactone (Hammer et al. 1996). This pathway has been confirmed using an adapted strain of *Hydrogenophaga palleroni* strain S1 (Blümel et al. 1998). Although azoreductases have been described in many organisms, there are a number of unresolved issues. In *Sphingomonas xenophaga* BN6, quinoid redox mediators were involved and two of them, 1,2-naphthoquinones with 4-amino substituents, were identified during the interrupted degradation of naphthalene-2-sulfonate (Keck et al. 2002). The cells reduced these to hydroquinones that can accomplish reduction of the azo group outside the cells (Rau et al. 2002; Rau and Stolz 2003). The azoreductase from the aerobe *Xenophilus azovorans* KF46F is a protein lacking metal ions or cofactors, but is able to accept only a limited number of azo compounds of industrial relevance (Blümel et al. 2002), and a thermostable azoreductase from *Bacillus* sp. strain SF has been described (Maier et al. 2004).

A number of factors are, therefore, implicated in the reduction of aromatic azo compounds:

1. Limited reduction by whole cells may be the result of limited uptake of the substrates into the cells;
2. Flavin-dependent azo reductases may be laboratory artifacts, and do not play a significant role *in vivo* (Russ et al. 2000);
3. Mediators that have already been noted may play an essential role.

Reduction of the azo group in dyes with concomitant decolorization—though not necessarily degradation—has been observed in a number of organisms. These include the yeast *Issatchenkia occidentalis* under microaerophilic conditions in the presence of glucose or ethanol (Ramalho et al. 2004). In this organism, the 1-amino-2-naphthol that was produced could be used as a source of both carbon and nitrogen, whereas *N,N'*-dimethylaniline served only as a source of nitrogen. Decolorization of sulfonated azo dyes has also been accomplished with fungal peroxidases, and involves oxidative scission of the azo bond followed by a complex set of redox and hydrolytic reactions (Goszczynski et al. 1994). Although laccases have been proposed for the degradation of azo dyes, it has been shown that the enzyme from *Trametes villosa* produced metabolites that combined with the dyes to produce oligomeric products that retained the azo group, and were, therefore, unacceptable for discharge (Zille et al. 2005).

9.25 References

Blümel S, M Contzen, M Lutz, A Stolz, H-J Knackmnuss. 1998. Isolation of a bacterial strain with the ability to utilize the sulfonated azo compound 4-carboxy-4'-sulfoazobenzene as the sole source of carbon and energy. *Appl Environ Microbiol* 64: 2315–2317.

Blümel S, H-J Knackmuss, A Stolz. 2002. Molecular cloning and characterization of the gene coding for the aerobic azoreductase from *Xenophilus azovorans* KF46. *Appl Environ Microbiol* 68: 3948–3955.

Chen H, SL Hopper, CE Cerniglia. 2005. Biochemical and molecular characterization of an azoreductase from *Staphylococcus aureus*, a tetrametric NADPH-dependent flavoprotein. *Microbiology (UK)* 151: 1433–1441.

Fuchs K, A Schreiner, F Lingens. 1991. Degradation of 2-methylaniline and chlorinated isomers of 2-methyl-aniline by *Rhodococcus rhodochrous* strain CTM. *J Gen Microbiol* 137: 2033–2039.

Goszczynski S, A Paszczynski, MB Pasti-Grigsby, RL Crawford, DL Crawford. 1994. New pathways for degradation of sulfonated azo dyes by microbial peroxidases of *Phanerochaete chrysosporium* and *Streptomyces chromofuscus*. *J Bacteriol* 176: 1339–1347.

Hammer A, A Stolz, H-J Knackmuss. 1996. Purification and characterization of a novel type of protocatechuate 3,4-dioxygenase with the ability to oxidize 4-sulfocatechol. *Arch Microbiol* 166: 92–100.

Haug W, A Schmidt, B Nörtemann, DC Hempel, A Stolz, H-J Knackmuss. 1991. Mineralization of the sulfonated azo dye mordant Yellow 3 by a 6-aminonaphthalene-2-sulfonate-degrading bacterial consortium. *Appl Environ Microbiol* 57: 3144–3149.

Keck A, J Rau, T Reemtsma, R Mattes, A Stolz, J Klein. 2002. Identification of quinonoid redox mediators that are formed during the degradation of naphthalene-2-sulfonate by *Sphingomonas xenophaga* BN6. *Appl Environ Microbiol* 68: 4341–4349.

Kulla HG, F Klausener, U Meyer, B Lüdeke, T Leisinger. 1983. Interference of aromatic sulfo groups in the microbial degradation of the aza dyes Orange I and Orange II. *Arch Microbiol* 135: 1–7.

Maier J, A Kandelbauer, A Erlacher, A Cavaco-Paulo, GM Gübitz. 2004. A new alkali-thermostable azoreductase from *Bacillus* sp strain SF. *Appl Environ Microbiol* 70: 837–844.

McClure NC, WA Venables. 1986. Adaptation of *Pseudomonas putida* mt-2 to growth on aromatic amines. *J Gen Microbiol* 132: 2209–2218.

Rafii F, CE Cerniglia. 1993. Comparison of the azoreductase and nitroreductase from *Clostridium perfringens*. *Appl Environ Microbiol* 59: 1731–1734.

Ramalho PA, MH Cardoso, A Cavaco-Paulo, MT Ramalho. 2004. Characterization of azo reduction activity in a novel ascomycete yeast strain. *Appl Environ Microbiol* 70: 2279–2288.

Rau J, A Stolz. 2003. Oxygen-insensitive nitroreductases NfsA and NfsB of *Escherichia coli* function under anaerobic conditions as lawsone-dependent azo reductases. *Appl Environ Microbiol* 69: 3448–3455.

Rau J, H-J Knackmuss, A Stolz. 2002. Effects of different quinoid redox mediators on the anaerobic reduction of azo dyes by bacteria. *Environ Sci Technol* 36: 1497–1504.

Russ R, J Rau, A Stolz. 2000. The function of cytoplasmic flavin reductases in the reduction of azo dyes by bacteria. *Appl Environ Microbiol* 66: 1429–1434.

Stolz A, H-J Knackmuss. 1993. Bacterial metabolism of 5-aminosalicylic acid; enzymatic conversion to malate, pyruvate and ammonia. *J Gen Microbiol* 139: 1019–1025.

Zhou N-Y, J Al-Dulayymi, MS Baird, PA Williams. 2002. Salicylate 5-hydroxylase from *Ralstonia* sp. strain U2: A monooxygenase with close relationships to and shared electron transport proteins with naphthalene dioxygenase. *J Bacteriol* 184: 1547–1555.

Zille A, B Górnacka, A Rehorek, A Cavaco-Paulo. 2005. Degradation of azo dyes by *Trametes villosa* laccase over long periods of oxidative conditions. *Appl Environ Microbiol* 71: 6711–6718.

Zimmermann T, F Gasser, HG Kulla, T Leisinger. 1984. Comparison of two bacterial azoreductases acquired during adaptation to growth on azo dyes. *Arch Microbiol* 138: 37–43.

10

Heterocyclic Aromatic Compounds

Part 1: Azaarenes

10.1 Five-Membered Monocyclic Aza, Oxa, and Thiaarenes

There are large numbers of naturally occurring representatives, especially of pyrrole that include the important polypyrroles (porphyrins and corrins), and the nitropyrrole antibiotics such as pyrrolomycins and pyrroxamycin. Derivatives of furan have been used as fungicides and *N*-vinylpyrrolidone is an important monomer for the production of blood plasma extenders and for cosmetic applications. On account of the similarity in the pathways for the aerobic degradation of monocyclic furan, thiophene, and pyrrole, all of them are considered here. Anaerobic degradation of furans is discussed in Part 2 of this chapter.

10.2 Aerobic Conditions

10.2.1 Monocyclic Heteroarenes

The degradation of furan-2-carboxylate (Trudgill 1969), thiophene-2-carboxylate (Cripps 1973), and pyrrrole-2-carboxylate (Hormann and Andreesen 1991), proceeds by initial ring-hydroxylation rather than dioxygenation that is used for the analogs annelated with benzene rings. After ring fission, 2-oxoglutarate is produced from all these compounds (Figure 10.1), and this then enters central anabolic and catabolic pathways. A strain of *Xanthobacter tagetidis* is able to grow aerobically at the expense of furan-2-carboxylate, thiophene-2-carboxylate, thiophene-2-acetate, and pyrrole-2-carboxylate (Padden et al. 1997), and *Vibrio* sp. strain YC1 with some of them (Evans and Venables 1990). The initial steps in the aerobic degradation of furan 2-carboxylate by *Pseudomonas putida* F2 involved formation of the coenzyme-A ester followed by hydroxylation with production of 2-oxoglutarate (Kitcher et al. 1972), and analogously in *P. putida* Ful (Koenig and Andreesen 1990a). The enzyme has been purified and the involvement of a Mo-containing dehydrogenase was suggested (Koenig and Andreesen 1990a,b). Although formally analogous reactions occur in the degradation of pyrrole-2-carboxylate by *Arthrobacter* strain Pyl (Hormann and Andreesen 1991), the hydroxylation of pyrrole-2-carboxylate in a species of *Rhodococcus* is carried out by a two-component monooxygenase (Becker et al. 1997).

The degradation of histidine has been explored intensively in a number of organisms including *Pseudomonas aeruginosa*, *Pseudomonas fluorescens*, *Aerobacter (Enterobacter) aerogenes*, and *Bacillus subtilis*, and both the enzymology and the regulation of the pathway have been elucidated. Degradation is accomplished by four enzymes. It is initiated by the formation of urocanate and ammonia by the activity of a lyase, urocanate is then converted into 4-iminazolone-5-propionate in an unusual reaction catalyzed by histidine ammonia-lyase (HAL) (urocanase), followed by reduction and fission of the ring to glutamate and formamide (Kaminskas et al. 1970) (Figure 10.2). The mechanism of HAL has attracted considerable attention, and the crystal structure of the enzyme from *P. putida* has been examined: this revealed details of the active step and made it possible to propose a mechanism (Schwede et al.

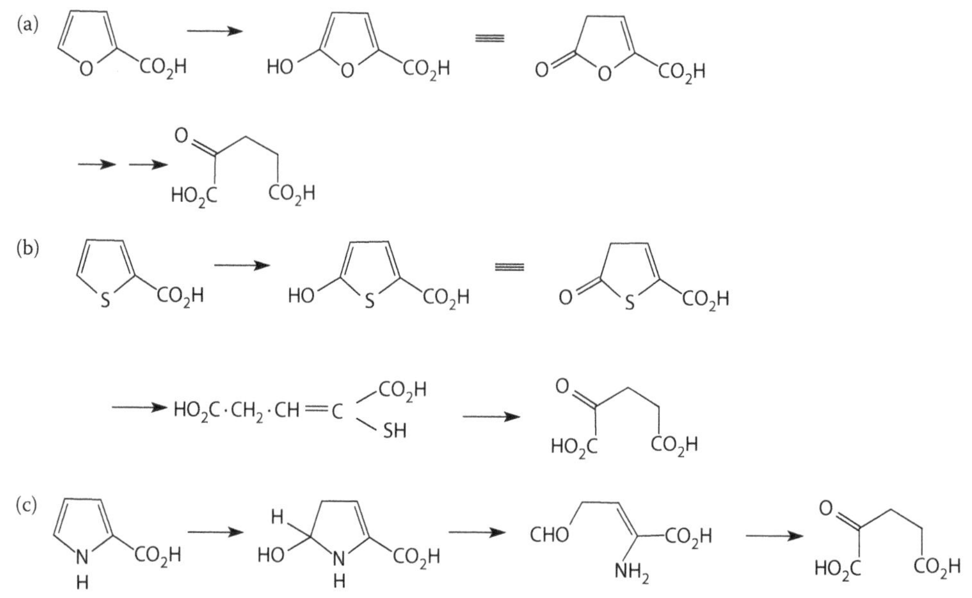

FIGURE 10.1 Aerobic degradation: (a) furan-2-carboxylate, (b) thiophene-2-carboxylate, (c) pyrrole-2-carboxylate.

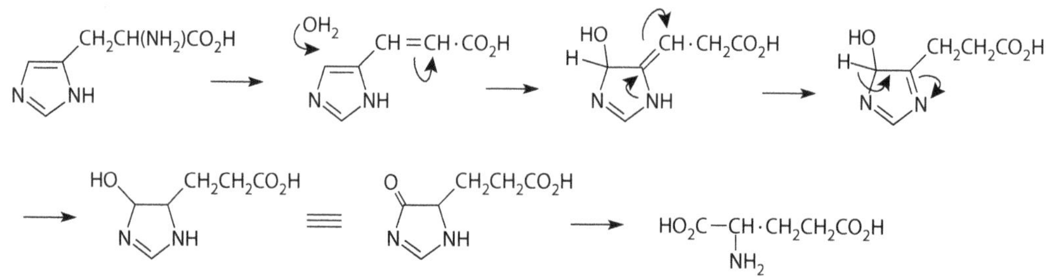

FIGURE 10.2 Aerobic degradation of histidine.

1999). In contrast, the degradation of iminazolacetate by *Pseudomonas* sp. strain ATCC 11299B is carried out by oxygenation with a flavoprotein monooxygenase (Maki et al. 1969).

The degradation of 2-nitroimidazole by *Mycobacterium* sp. strain JS330 is carried out by hydrolytic loss of nitrite to produce iminazol-2-one that served as a source of carbon and nitrogen (Qu and Spain 2011).

10.2.2 Benzpyrroles: Indole and Carbazole

Whereas the degradation of the carboxylates of the monocyclic furan, thiophene, and pyrrole is initiated by hydroxylation, degradation of their benzo analogs is often carried out by dioxygenation. The degradation of the analogs dibenzofuran and dibenzo-[1,4]-dioxin is discussed in Part 2 of this chapter.

10.3 Indole and 3-Alkylindoles

10.3.1 Aerobic Reactions

The aerobic degradation of indole has been described for strain In3 of *Alcaligenes* sp. and was carried out by the following steps: hydroxylation at C_3 produced indoxyl (3-hydroxyindole) that was oxidized to isatin and hydrolyzed to anthranilate. This was degraded by the gentisate pathway to maleylpyruvate

(Claus and Kutzner 1983). Parts of this pathway are also used for the degradation of indole-3-acetic acid (IAA) that is discussed later.

Several pathways have been demonstrated for the aerobic degradation of tryptophan.

10.3.1.1 Side-Chain Degradation

1. Degradation initiated by tryptophanase has been demonstrated in a number of bacteria (DeMoss and Moser 1969) including *Escherichia coli* (Morino and Snell 1967), *Paenibacillus (Bacillus) alvei* (Hoch and DeMoss 1972), *Enterobacter aerogenes* (Kawasaki et al. 1993), and the tryptophanase operon in *E. coli* and *Proteus vulgaris* are similar in structure and regulation (Kamath and Yanofsky 1992). The enzyme carries out an α,β-elimination to produce indole, pyruvate, and NH_4^+ (Figure 10.3a). The reaction is dependent on pyridoxal that binds to the ε-amino group of lysine and is initiated by enzymatic removal of the proton at the α-position of the resulting Schiff base (Vederas et al. 1978). Additionally, the enzyme is capable of carrying out other α,β-eliminations including that of L-cysteine to pyruvate, sulfide, and ammonia; and the synthesis of L-tryptophan from L-serine and indole (Newton et al. 1965).

2. In *P. fluorescens* ATCC 29574, oxidation produced a number of C_2-metabolites after initial transamination (Figure 10.3b) (Narumiya et al. 1979).

3. In *Pseudomonas pyrrocinia*, transformation was initiated by oxidation of the side chain to indole-3-acetate that was further degraded to indole-3-carboxylate before decarboxylation to indole (Figure 10.3c) (Lübbe et al. 1983).

4. Considerable attention has been given to the bacterial synthesis of the plant hormone IAA, and a range of pathways involve transformation of the tryptophan side chain. These substrates include indole-3-pyruvic acid (IPyA); indole-3-acetamide (IAM); indole-3-acetaldehyde (IAAld); and indole-3-acetonitrile (IAN). The IPyA and IAM pathways have been most thoroughly explored, and plant growth-promoting strains generally use the indole-3-pyruvic acid IPyA pathway whereas phytopathogenic strains use the IAM pathway. The IAM pathway is initiated by FAD-dependent monooxygenation by tryptophan 2-monooxygenase to indole-3-acetamide (Ralph et al. 2006) that is followed by hydrolysis. The synthesis and regulation of tryptophan 2-monooxygenase has been examined in *Pseudomonas savastanoi* in which IAA is obligatory for virulence against olive and oleander (Hutcheson and Kosuge 1985). In *Erwinia herbicola* (Brandl and Lindow 1996) and *Enterobacter cloacae* (Schütz et al. 2003), however,

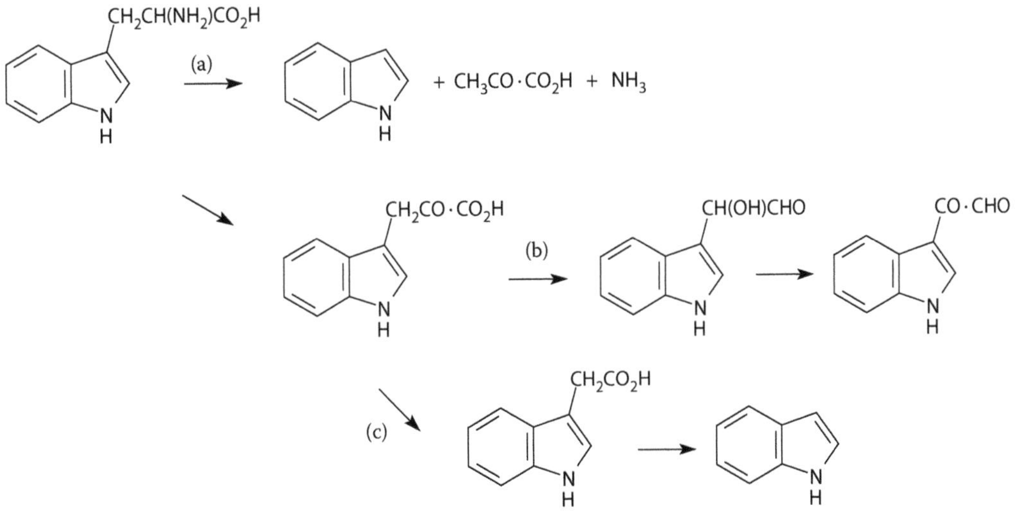

FIGURE 10.3 Pathways for degradation of L-tryptophan (a) by tryptophanase, (b) by dioxygenation to kynurenine, and (c) via side-chain oxidation and decarboxylation to indole.

the alternative pathway is used: this is initiated by transamination to indole-3-pyruvate that is decarboxylated to indole-3-acetaldehyde followed by oxidation to indole-3-acetate by an oxidase. There are some additional aspects that are worth noting:

a. The plant pathogen *Pseudomonas syringae* pv. *syringae* contains a nitrilase that can hydrolyze IAN to IAA (Howden et al. 2009);

b. The plant–root-associated *Azospirillum brasilense* is able to synthesize IAA by the IPyA pathway for which the phenylpyruvate decarboxylase has been characterized (Spaepen et al. 2007);

c. In *A. brasilense*, an additional route has been described that is independent of L-tryptophan when exogenous tryptophan is not available (Prinsen et al. 1993).

10.3.1.2 Ring Dioxygenation

The degradation of indole to anthranilate, putatively by dioxygenation, has been proposed for an undetermined Gram-positive organism (Figure 10.4) (Fujioka and Wada 1968), and this pathway has been found for the degradation of L-tryptophan and IAA.

The bacterial degradation of L-tryptophan is initiated by dioxygenation to *N*-formylkynurenine and hydrolysis to kynurenine. Further degradation may then take place by either of two pathways: (a) formation of anthranilate that is degraded by dioxygenation to catechol or (b) cyclization of *N*-formylkynurenine to 4-hydroxyquinoline-2-carboxylate (kynurenic acid) that is termed the quinoline pathway and is found in pseudomonads. The subsequent steps in the pathway have been studied in *P. fluorescens* ATCC 11299B (Taniuchi and Hayaishi 1963), and in a strain of *Aerococcus* sp. (Dagley and Johnson 1963). These reactions involved dioxygenation to 4-hydroxyquinoline-2-carboxylate-7,8-dihydrodiol followed by dehydrogenation, extradiol ring fission, and reduction of the α,β-unsaturated carboxylate (Figure 10.5a). In *Burkholderia cepacia* strain J2315, an alternative degradation of kynurenine involved monooxygenation to 3-hydroxykynurenine followed by hydrolysis to 3-hydroxyanthranilate, ring fission by 3-hydroxyanthranilate 3,4-dioxygenase, and production of 4-oxalocrotonate, and subsequently, pyruvate and acetaldehyde (Figure 10.5b) (Colabroy and Begley 2005).

The degradation of tryptophan in mammalian systems is initiated by the same reactions as far as kynurenine, which is hydroxylated to 3-hydroxykynurenine and 3-hydroxyanthranilate. This is followed by ring fission and nonenzymatic cyclization to pyridine 2,3-dicarboxylate (Malherbe et al. 1994). 3-Hydroxyanthranilate is also an intermediate during the bacterial degradation of 2-nitrobenzoate when it is degraded to acetaldehyde and pyruvate (Muraki et al. 2003). The transformation of both indole-2- and indole-3-carboxylate by dioxygenases produced isatin, and the dimeric coupling products indigo and indirubin (Eaton and Chapman 1995).

Although most attention has been devoted to the *synthesis* of the plant hormone IAA that has already been discussed, its *degradation* has been examined in a few bacteria. It was produced from tryptophan by *P. pyrrocinia* before further degradation of the side chain (Lübbe et al. 1983). IAA was used for the growth of *P. putida*, 1290, as a source of carbon, nitrogen, and energy, although no degradation pathway was proposed (Leveau and Lindow 2005). In *Bradyrhizobium japonicum* strain 110 degradation was initiated by oxidation to dioxindole-3-acetic acid followed by loss of acetate to isatin that was successively hydrolyzed to 2-aminophenyl glyoxylic acid and anthranilic acid (Figure 10.6) (Jensen et al. 1995). This pathway is consistent with the activities of two enzymes whose activity was demonstrated: isatin reductase to produce dioxindole and an isatin hydrolase to produce 2-aminophenyl-glyoxylic acid (Olesen and Jochimsen 1996). The genes for the degradation of IAA occur in the complex *iac* locus, and similar

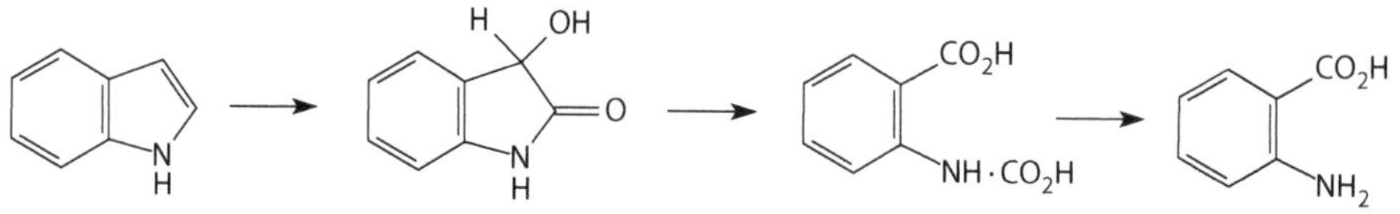

FIGURE 10.4 Biodegradation of indole.

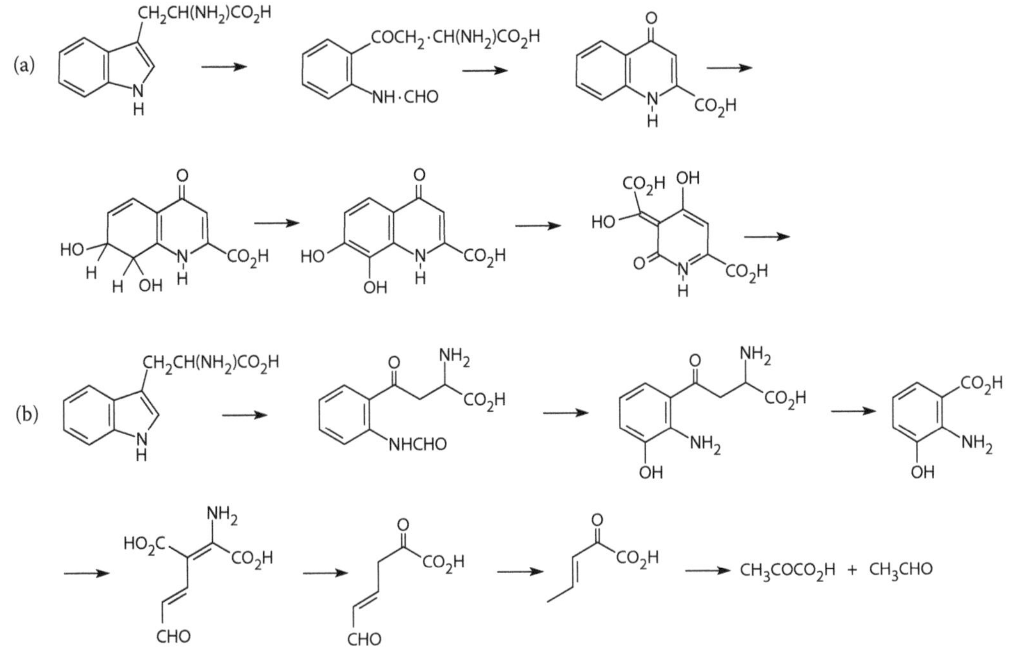

FIGURE 10.5 Pathways for degradation of L-tryptophan (a) via kynurenic acid and (b) via quinolone-2-carboxylate.

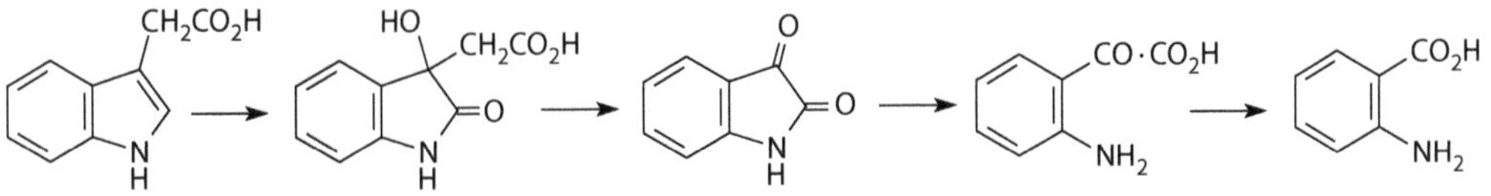

FIGURE 10.6 Pathway for degradation of indole-3-acetate.

genes have been found in *P. putida* GB-1, *Sphingomonas wittichii*, *Marinomonas* MWYL1, and *Rhodococcus* sp strain RHA1: all of these were able to grow on minimal medium supplemented with IAA (Leveau and Gerards 2008).

10.3.2 Anaerobic Reactions

Cells of *Lactobacillus* sp. strain 11201 grown anaerobically and induced with IAA were able to transform this to 3-methylindole (Honeyfield and Carlson 1990). Tryptophan was metabolized by *Clostridium drakei* and *Clostridium scatologenes* to IAA and, by decarboxylation to 3-methylindole (Whitehead et al. 2008). Indole can be oxidized by *Desulfobacterium indolicum* to CO_2 using sulfate as electron acceptor (Bak and Widdel 1986), and indole was transformed in this strain by hydroxylation to 2-hydroxyindole (oxindole) followed by dehydrogenation to isatin and hydrolysis to anthranilic acid (Johansen et al. 1996).

10.4 Carbazole

The transformation of carbazole has been examined, and naphthalene 1,2-dioxygenase activity was induced in *Pseudomonas* sp. strain NCIB 9816-4 and in *Beijerinckia* sp. strain B8/36 (Resnick et al. 1993). The 3-hydroxycarbazole that was formed could have resulted from two pathways: (a) from the initial production of *cis*-3,4-dihydro-3,4-dihydroxycarbazole followed by dehydration or (b) by monooxygenation that cannot be excluded since monooxygenase activity can be mediated by naphthalene 1,2-dioxygenase (Gibson et al. 1995).

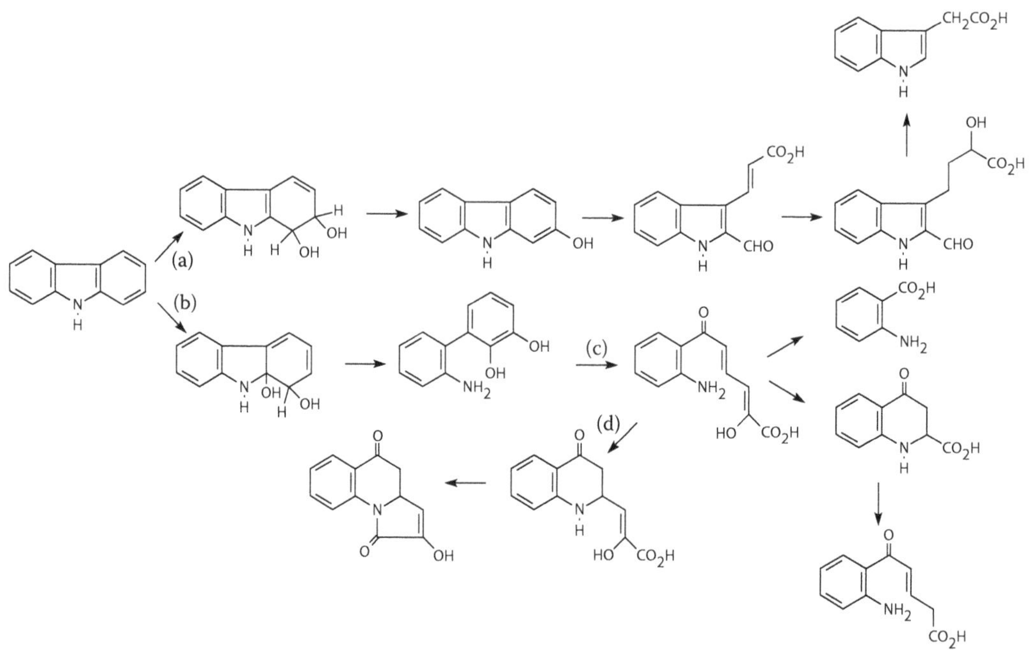

FIGURE 10.7 Degradation of carbazole by: (a) 1,2-dioxygenation, (b) 1,9a dioxygenation followed by (c) extradiol ring fission, and (d) production of quinolones.

The degradation of carbazole has been examined in a number of organisms. These include:

1. The Gram-negative *Pseudomonas* sp. strain LD2 (Gieg et al. 1996) and *Pseudomonas resinovorans* strain CA10 in which the genes occur on a megaplasmid (Nojiri et al. 2001), *Sphingomonas* sp. strain KA1 (Urata et al. 2006).
2. The Gram-positive *Nocardioides aromaticivorans* strain IC177 in which the genes occur in an operon (Inoue et al. 2006).

All use essentially the same pathway involving dioxygenation, ring fission, and hydrolysis. The degradation by *Pseudomonas* sp. strain LD2, which uses carbazole as sole source of carbon, nitrogen, and energy, revealed a complex set of intermediates (Gieg et al. 1996). Anthranilic acid and catechol were intermediates, together with a number of terminal transformation products including indole-3-acetate. These products may plausibly arise from 1,2-dioxygenation (Figure 10.7a) or from angular 1,9a-dioxygenation (Figure 10.7b). In both of them production of the dihydrodiol is apparently followed by elimination rather than dehydrogenation, and for angular 1,9-dioxygenation there are several branches in the subsequent pathways (Figure 10.7c,d). Carbazole 1,9a-dioxygenase, which produced biphenyl-2′-amino-2,3-diol, has been examined in *Pseudomonas resinovorans* strain CA10, and consisted of a terminal dioxygenase, a ferredoxin, and a ferredoxin reductase (Sato et al. 1997b), which were products of the genes *carAa*, *carAc*, and *carAd* (Nam et al. 2002, 2005). The extradiol fission enzyme and the enzyme that produced anthranilic acid by hydrolytic fission of the vinylogous 1:3-diketone were products of *carB* and *carC*, respectively (Sato et al. 1997a). The three-component dioxygenase (Nam et al. 2002) has a relaxed specificity that includes dioxygenation not only of these heterocyclic compounds but also of the carbocyclic naphthalene, biphenyl, anthracene, and fluoranthene (Nojiri et al. 1999).

10.4.1 Pyridine

Azaarenes are structural elements of coal and petroleum, and are found in products derived from them by pyrolysis and distillation. They are components of creosote, which has been widely used as a timber preservative.

10.4.2 Aerobic Conditions

Compared with monocyclic aromatic hydrocarbons and the five-membered azaarenes, the pathways used for the degradation of pyridines are less uniform, and this is consistent with the differences in electronic structure and thereby their chemical reactivity. For pyridines, both hydroxylation and dioxygenation that is typical of aromatic compounds have been observed, although these are often accompanied by reduction of one or more of the double bonds in the pyridine ring. Examples are used to illustrate the metabolic possibilities.

10.4.2.1 Reduction

Reduction may be either the initial reaction in degradation or take place at a later stage after hydroxylation:

1. Pyridine is degraded by two Gram-positive organisms that are postulated to initiate degradation by reduction rather than by hydroxylation (dehydrogenation), which is widely used by azaarenes. *Nocardia* sp. strain Z1 produced succinic semialdehyde by ring fission of 1,4-dihydropyridine between C2 and C3, whereas *Bacillus* sp. strain 4 produced glutaric dialdehyde by fission of 2,3-dihydropyridine after hydroxylation, and hydrolysis between N and C-2 (Figure 10.8) (Watson and Cain 1975). These metabolites are then incorporated into normal catabolic cycles. A strain of *Azoarcus* sp., which can degrade pyridine under both aerobic and denitrifying conditions, was presumed to use the first pathway (Rhee et al. 1997). Analogous pathways are used for the aerobic degradation of 3-methylpyridine by *Gordonia nitida* LKE31 (Lee et al. 2001) and in the initial step in one of the pathways for *N*-methylisonicotinate (Wright and Cain 1972), which is formed by photolysis of the herbicide paraquat.

2. The degradation of pyridoxal (vitamin B_6) by *Pseudomonas* sp. strain MA-1 involves 3-hydroxy-2-methylpyridine-5-carboxylate as an intermediate, which is then degraded by reduction of the pyridine ring followed by ring fission (Sundaram and Snell 1969; Nyns et al. 1969) (Figure 10.9).

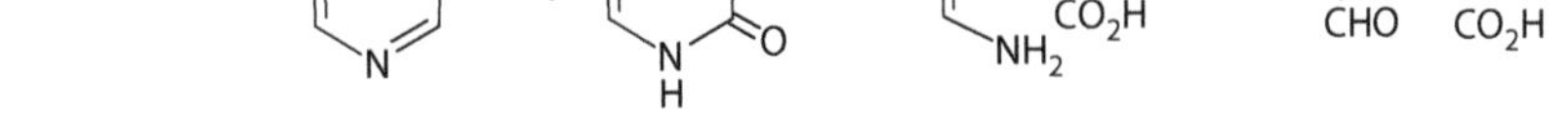

FIGURE 10.8 Reductive pathways in degradation of pyridine.

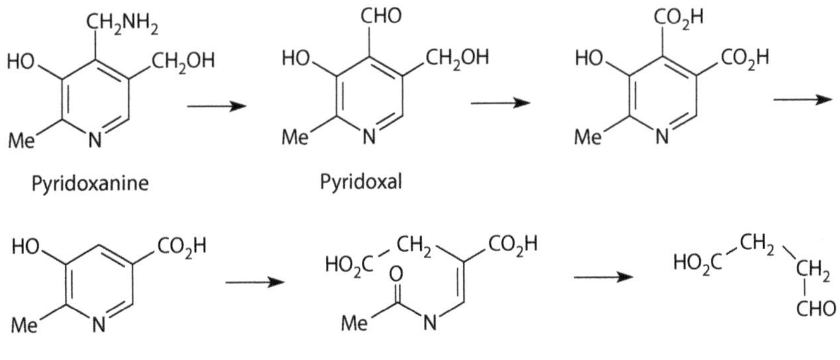

FIGURE 10.9 Degradation of pyridoxal.

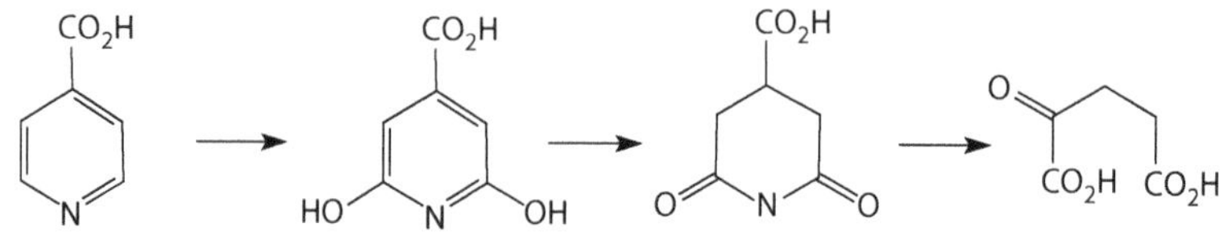

FIGURE 10.10 Biodegradation of pyridine-4-carboxylate.

The inducible oxygenase contains FAD and the formation of the product apparently includes a reductive step. Whereas experiments with $^{18}O_2$ showed that this metabolite contained two atoms of ^{18}O, some ^{18}O incorporation was also observed when $H_2^{18}O$ was used. This suggests that both dioxygenation and a hydroxylation mechanism might be operating (Sparrow et al. 1969).

3. The degradation of pyridine-4-carboxylate by *Mycobacterium* sp. strain INA1 takes place by successive hydroxylations before reduction and ring fission (Kretzer and Andreesen 1991) (Figure 10.10).

10.4.2.2 Hydroxylation

In contrast, degradation of 2- and 4-hydroxypyridines—which exists as pyridones—generally, proceeds by successive hydroxylations (dehydrogenations) to 2,3- or 2,6-dihydroxypyridines that are subsequently degraded by ring fission. Hydroxypyridines are also formed from pyridine carboxylates after hydroxylation and decarboxylation, and are then degraded to maleate (Figure 10.11). This pathway is used for the alternative degradation of *N*-methylisonicotinate (Orpin et al. 1972).

Both 2-hydroxy- and 3-hydroxypyridine are hydroxylated to 2,5-dihydroxypyridine by strains of *Achromobacter* sp. (Houghton and Cain 1972). These metabolites are probably, however, formed by different reactions; whereas 3-hydroxypyridine behaves as a true pyridine, addition of H_2O across the C_6-N_1 bond would produce the 2,5-dihydroxy compound; 2-hydroxypyridine is a cyclic amide and hydroxylation apparently occurs at the diagonal position. The degradation of 4-hydroxypyridine is also initiated by hydroxylation and is followed by dioxygenation before ring fission (Figure 10.12) (Watson et al. 1974).

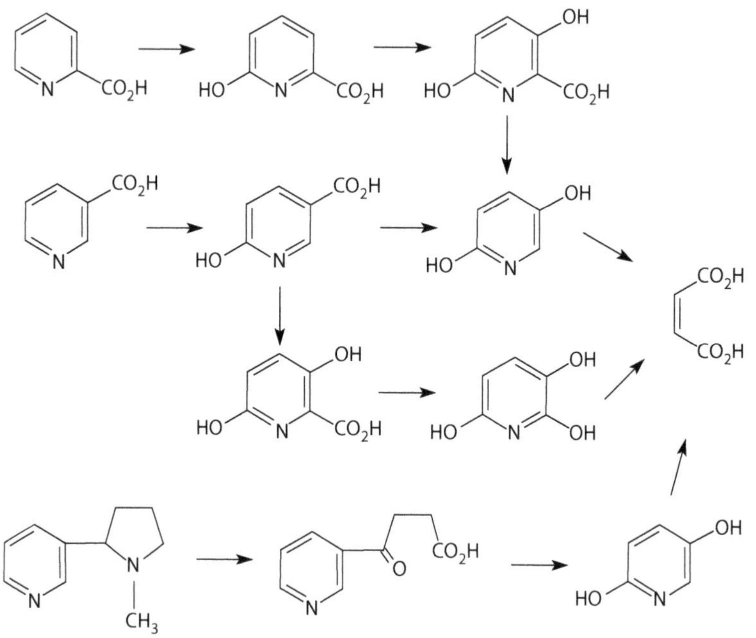

FIGURE 10.11 Interaction of pathways for degradation of pyridine-2-carboxylate, pyridine-3- carboxylate and nicotine.

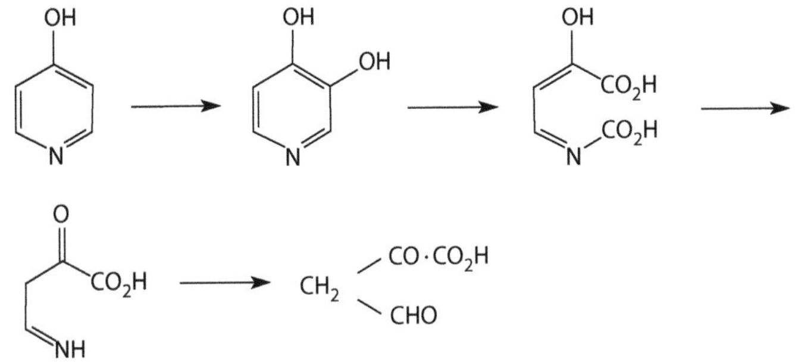

FIGURE 10.12 Biodegradation of 4-hydroxypyridine.

Hydroxylation is also involved in the degradation of all the pyridine carboxylates and the interrelations of these pathways are shown (Figure 10.11):

1. Pyridine-2-carboxylate is hydroxylated in *Arthrobacter picolinophilus* to 3,6-dihydroxypyridine-2-carboxylate (as for 2-hydroxypyridine) by a particulate enzyme that introduces oxygen from H_2O (Tate and Ensign 1974), has a molecular mass of 130 kDa, and contains Mo (Siegmund et al. 1990). Further degradation involves decarboxylation to 2,5-dihydroxypyridine before ring fission. This reaction has been examined using $^{18}O_2$ and $H_2^{18}O$ (Gauthier and Rittenberg 1971), and shown to involve the incorporation of both atoms of oxygen, one each into formate and maleamate (Figure 10.13).

2. Analogously, pyridine-3-carboxylate is hydroxylated to 6-hydroxypyridine 3-carboxylate, and then either to 2,5-dihydroxypyridine in *P. fluorescens* N-9 (Behrman and Stanier 1957a), or in a *Bacillus* sp. to the 2,6-dihydroxy carboxylate (Ensign and Rittenberg 1964). Hydroxylation to 2,6-dihydroxypyridine carboxylate in a strain of *Bacillus* sp. (now *B. niacinii*) was unambiguously shown using $H_2^{18}O$ to show that both atoms of the oxygen incorporated were derived from water (Hirschberg and Ensign 1971b). Pyridine-3-carboxylate and 6-hydroxypyridine-3-carboxylate hydroxylases were shown to contain two molecules of flavin and eight atoms of iron per molecule (Hirschberg and Ensign 1971a). The enzymes (nicotinate dehydrogenase and 6-hydroxynicotinate dehydrogenase) in *B. niacini* were separated, and each contained FAD, an [Fe–S] center, and probably Mo (Nagel and Andreesen 1990), and the ring fission is accomplished by a 5,6-dioxygenase (Gauthier and Rittenberg 1971). The pathway used for the further metabolism of 6-hydroxypyridine-3-carboxylate by *Azorhizobium caulinodans* involves reduction (Kitts et al. 1989).

3. Pyridine-4-carboxylate is hydroxylated by *Mycobacterium* sp. strain INA1 to 2,6-dihydroxypyridine-4-carboxylate. Two different hydroxylation enzymes were involved and were apparently Mo-dependent (Kretzer and Andreesen 1991). The formation of 2-oxoglutarate can, however, be rationalized equally as β-oxidation to hexahydropyridine-2,3,6-trione-4-carboxy-CoA ester followed by hydrolysis.

4. The degradation of 6-methylpyridine-3-carboxylate by an organism selected by enrichment with 6-methylpyridine-3-carboxylate takes place by selective hydroxylation at C-2, and the organism is also able to hydroxylate pyridine 3-carboxylate at the same position (Tinschert et al. 1997).

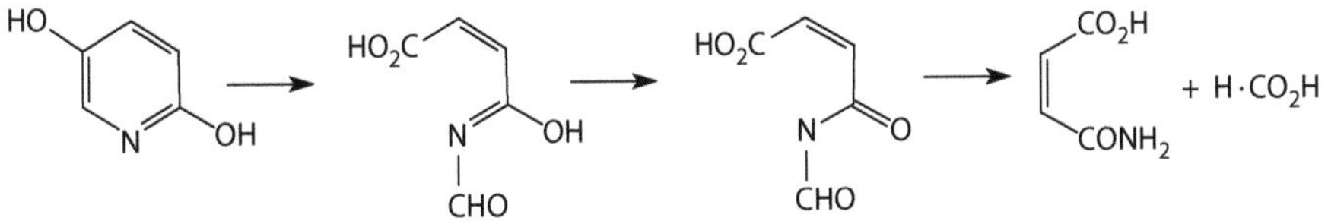

FIGURE 10.13 Biodegradation of 2,5-dihydroxypyridine.

5. Advantage has been taken of the electronic similarity of phthalates to pyridine dicarboxylates:

   a. Strains capable of degrading *o*-, *m*-, or *p*-phthalate could oxidize the analogous pyridine 2,3-, 2,6-, or 2,5-dicarboxylate (Taylor and Amador 1988).

   b. The degradation of *o*-phthalate is initiated by dioxygenation to produce the 4,5-dihydrodiol before dehydrogenation and decarboxylation to 3,4-dihydroxybenzoate. In addition, the dioxygenase is able to accept pyridine-2,3-dicarboxylate and pyridine-3,4-dicarboxylate (quinolinate) (Batie et al. 1987). There are two regions of DNA that encode the enzymes for *o*-phthalate degradation and the gene encoding quinolinate phosphoribosyl transferase is located between them, and insertional knockout mutants with elevated levels of this enzyme enhanced growth on *o*-phthalate (Chang and Zylstra 1999).

The degradation of nicotine has been examined extensively in *Arthrobacter nicotinovorans (oxydans)* in which it is mediated by a plasmid (Brandsch et al. 1982; Schenk et al. 1998). In strain P34, the first metabolite was 6-hydroxynicotine, and experiments with $^{18}O_2$ and $H_2^{18}O$ showed that the oxygen in the hydroxyl group was derived from H_2O (Hochstein and Dalton 1965). Nicotine dehydrogenase has a molecular mass of 120,000 and contains FAD, Mo, Fe, and acid-labile sulfur (Freudenberg et al. 1988). Degradation involves a sequence of reactions:

1. Hydroxylation of the *N*-methylpyrrolidine ring at the benzylic carbon atom by an FAD-containing oxidase (Dai et al. 1968), followed by fission of the resulting carbinolamine.

2. Hydrolysis of the β-oxoamide with the formation of γ-*N*-methylaminobutyrate and 2,6-dihydroxypyridine (Sachelaru et al. 2005).

3. Further hydroxylation to 2,3,6-trihydroxypyridine by an FAD, Mo, and [Fe–S] dehydrogenase (Baitsch et al. 2001).

4. Fission of the oxygen-labile 2,3,6-pyridine to maleamate (Figure 10.14) (Holmes et al. 1972; Baitsch et al. 2001).

A diazodiphenoquinone was produced as a chemical artifact (Knackmuss and Beckmann 1973), and there are several unresolved features of the degradation, including the mechanisms for production of 2,6-dihydroxypyridine and oxidative fission of the pyridine ring.

An alternative pathway has been explored in *P. putida* strain S16. Nicotine was converted by an oxidoreductase into methylamine and succinoylpyridine in which the ^{18}O-carbonyl is derived in consecutive reactions from two molecules of $H_2^{18}O$ (Tang et al. 2009). This was followed by hydroxylation to 2,5-dihydroxypyridine that underwent fission to maleamic and fumaric acids (Tang et al. 2008). The hydroxylation is formally analogous to the successive hydroxylation of pyridine-3-carboxylate to 6-hydroxypyridine 3-carboxylate and 2,5-dihydroxypyridine in *P. fluorescens* N-9 (Behrman and Stanier 1957a).

There has been interest in the fate of the products from photolysis of the herbicide paraquat (1,1′-dimethyl-4,4′-bipyridylium ion), in particular, *N*-methylisonicotinate. Its degradation has been examined in a strain of *Achromobacter* sp. and resembles one of the pathways for the degradation of pyridine. It is initiated by reduction of the ring, followed by ring fission and decarboxylation, with the production of formate, methylamine, and succinate (Figure 10.15a) (Wright and Cain 1972).

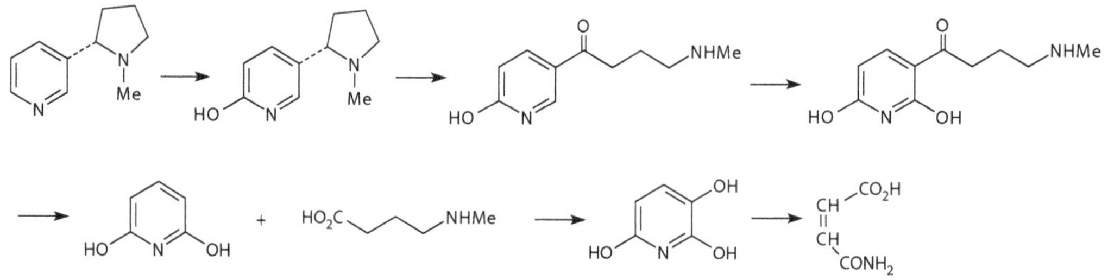

FIGURE 10.14 Aerobic degradation of nicotine. (Adapted from Baitsch D et al. 2001. *J Bacteriol* 183: 5262–5267.)

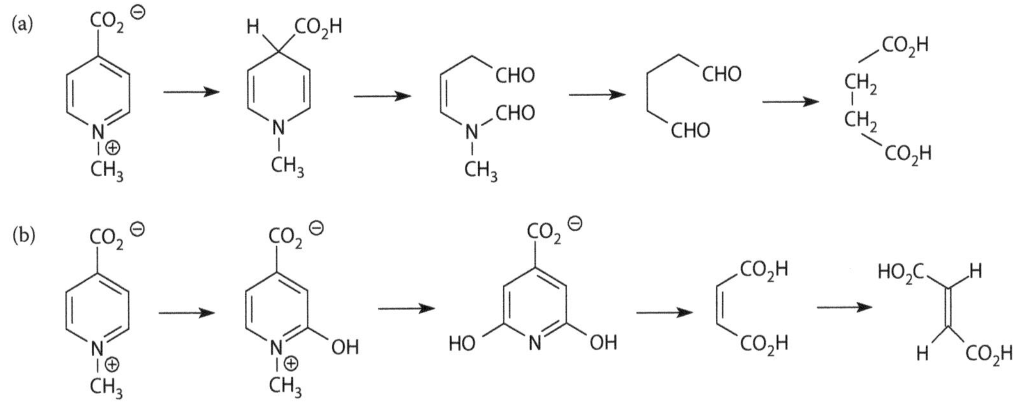

FIGURE 10.15 Degradation of 1-methylpyridinium-4-carboxylate. (a) Reduction and ring fission. (Adapted from Wright KA, RB Cain. 1972. *Biochem J* 128: 543–559.) (b) Successive hydroxylation to 2,6-dihydroxypyridine-4-carboxylate. (Adapted from Orpin CG, M Knight, WC Evans. 1972. *Biochem J* 127: 833–844.)

An alternative pathway in a Gram-positive organism involved hydroxylation at C-2 and loss of the methyl group, followed by further hydroxylation to 2,6-dihydroxypyridine-4-carboxylate that is degraded to maleate (Figure 10.15b) (Orpin et al. 1972).

10.4.2.3 Dioxygenation

Dioxygenation is illustrated by a few examples:

1. The degradation of pyridine by a species of *Bacillus* (Watson and Cain 1975) produces formate and succinate (Figure 10.16).
2. The degradation of 2,5-dihydroxypyridine has been examined using $^{18}O_2$ and $H_2^{18}O$ (Gauthier and Rittenberg 1971), and was shown to involve the incorporation of both atoms of oxygen, one each into formate and maleamate.
3. Degradation of the toxin mimosine, which is produced by *Leucaena glauca* and *Mimosa pudica*, is initiated by the formation of 3-hydroxypyrid-4-one, plausibly by the activity of an enzyme analogous to tryptophanase (Borthakur et al. 2003). The 3-hydroxypyrid-4-one is then degraded to formate and pyruvate by enzymes encoded by an extradiol dioxygenase (pydA) and a hydrolase (pydB) (Figure 10.17) (Awaya et al. 2005).

10.4.2.4 Halogenated Pyridines

Only a few investigations have examined the degradation of halogenated pyridines, some of which are the basis of herbicides:

1. Chlorpyrifos (*O,O*-diethyl-*O*-(3,5,6-trichlo-2-pyridyl)phosphorothioate) is hydrolyzed by *Enterobacter* sp. to 3,5,6-trichloropyridin-2-ol (Singh et al. 2004) whose degradation has been described (Feng et al. 1997).

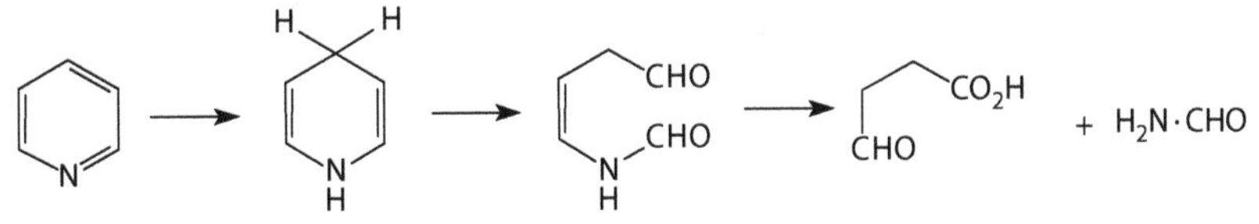

FIGURE 10.16 Biodegradation of pyridine.

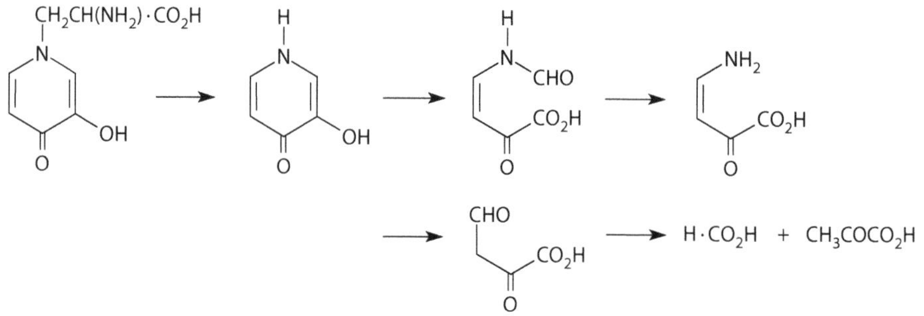

FIGURE 10.17   Degradation of mimosine.

2. 5-Chloro-2-hydroxypyridine-3-carboxylate is a terminal metabolite in the degradation of 3-chloroquinoline-8-carboxylate, but can be degraded by *Mycobacterium* sp. strain BA to chlorofumarate by reactions analogous to those described above for pyridine carboxylates (Figure 10.18) (Tibbles et al. 1989a).

3. Oxidation of the trichloromethyl group in 2-chloro-6-(trichloromethyl)-pyridine to the corresponding carboxylic acid (Vannelli and Hooper 1992) by *Nitrosomonas europaea* occurs at high oxygen concentrations during cooxidation of ammonia or hydrazine. In contrast, at low oxygen concentrations in the presence of hydrazine, reductive dechlorination to 2-chloro-6-dichloromethylpyridine takes place (Figure 10.19).

4. Although *P. fluorescens* strain N-9 can oxidize 5-fluoronicotinic acid, it is unable to grow at its expense. The initial reaction—analogous to that for nicotinic acid—is the formation of 5-fluoro-6-hydroxynicotinic acid (Behrman and Stanier 1957b). Although this was oxidized further, the products were not identifiable by the methods then available. Pyruvate was formed from nicotinic acid and it was suggested by analogy that fluoropyruvate could be produced from 5-fluoronicotinic acid, and thence fluoroacetate that would inhibit effective functioning of the TCA cycle as a source of carbon and energy. In contrast, the degradation of nicotinic acid produces maleic acid (Behrman

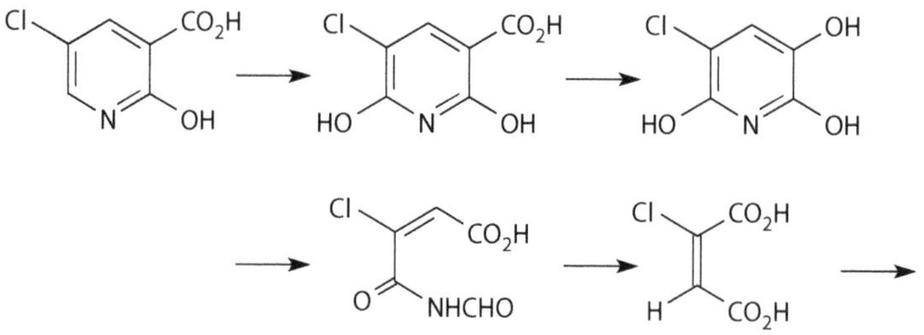

FIGURE 10.18   Degradation of 3-chloro-6-hydroxypyridine-5-carboxylate.

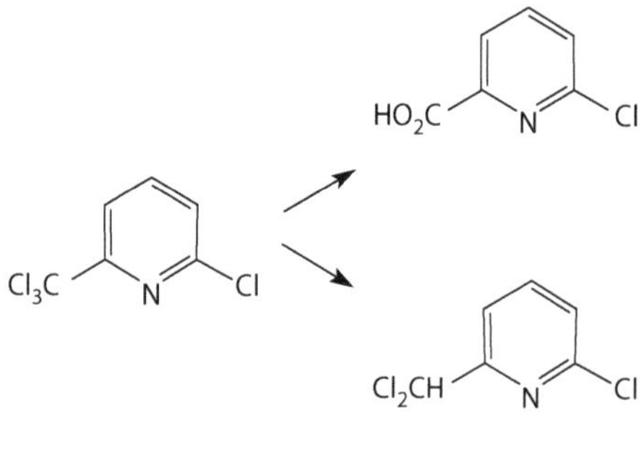

FIGURE 10.19   Metabolism of 2-chloro-6-trichloromethylpyridine.

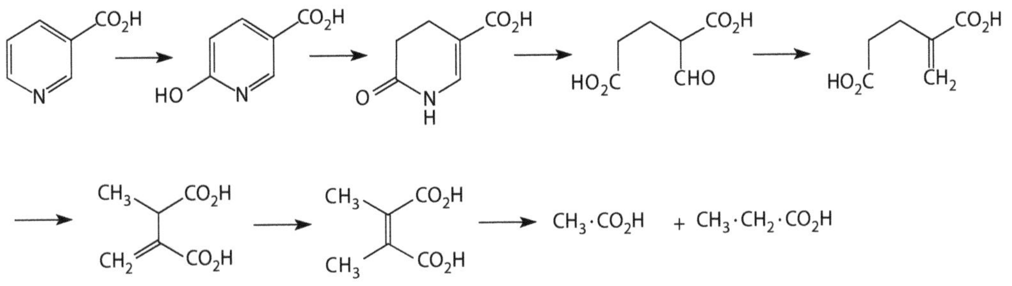

FIGURE 10.20 Anaerobic biodegradation of nicotinate.

and Stanier 1957a); and 5-fluorinicotinic acid would be expected to yield fluorofumarate whose degradation by fumarase is known to yield oxalacetate and fluoride (Marletta et al. 1982).

The partial anaerobic dechlorination of pentachloropyridine to a trichloropyridine of unestablished orientation by the dehalorespiring *Desulfitobacterium hafniense (frappieri)* PCP-1 has been demonstrated (Dennie et al. 1998).

10.4.3 Anaerobic Conditions

As for the aerobic degradation of pyridines, hydroxylation of the heterocyclic ring is a key reaction in the anaerobic degradation of azaarenes by clostridia. Whereas in *Clostridium barkeri*, the end products are carboxylic acids, CO_2, and ammonium, the anaerobic sulfate-reducing *Desulfococcus niacinii* degraded nicotinate completely to CO_2 (Imhoff-Stuckle and Pfennig 1983), although the details of the pathway remain incompletely resolved.

The metabolism of nicotinate has been extensively studied in clostridia and the details of the pathway (Figure 10.20) have been delineated in a series of studies (Kung and Stadtman 1971; Kung and Tsai 1971; Kung et al. 1971). Degradation is initiated by hydroxylation of the ring, and the level of nicotinic acid hydroxylase is substantially increased by the addition of selenite to the medium (Imhoff and Andreesen 1979). Nicotinate hydroxylase from *Cl. barkeri* contains molybdenum that is coordinated to selenium, which is essential for hydroxylase activity (Gladyshev et al. 1994). The most remarkable feature of the pathway is the mechanism whereby 2-methylene-glutarate is converted into methylitaconate by a coenzyme B_{12}-mediated reaction (Kung and Stadtman 1971).

10.4.4 Quinoline and Isoquinoline

It is appropriate to note that (a) 2- and 4-hydroxyquinolines generally exist as quinolones and (b) that the term hydroxylation is not strictly accurate, and the enzymes are preferably described as oxidoreductases. They are formally analogous to the dehydrogenases that are involved in the hydroxylation of purines, and the degradation of quinoline and its methyl homologs (quinaldines) has been extensively investigated. As a generalization, it may be stated that degradation is almost always initiated by hydroxylation at C_2 or (C_4) in the hetero ring, and the resulting hydroxyquinolines (quinolones) are then further oxygenated in the carbocyclic ring.

10.4.5 Bacterial Metabolism

10.4.5.1 Quinoline

Experiments with $H_2^{18}O$ using *P. putida* strain 86 (Bauder et al. 1990) showed that the oxygen incorporated into quinol-2-one originates from water. The oxidoreductases have been purified from a number of organisms that degrade quinoline or quinaldine (2-methylquinoline). These include *P. putida* strain 86 (Bauder et al. 1990), *Comamonas testosteroni* strain 63 (Schach et al. 1995), *Rhodococcus* sp. strain B1 (Peschke and Lingens 1991), *Arthrobacter* sp. strain Rü 61a (De Beyer and Lingens 1993), and *Agrobacterium* sp. strain 1B, which can degrade quinoline-4-carboxylate (Bauer and Lingens 1992). The oxidoreductases have a molecular mass of 300–360 kDa, and contain per molecule, eight atoms of Fe, eight atoms of acid-labile S, two atoms of Mo, and two molecules of FAD. The organic component of the molybdenum cofactor is generally molybdopterin cytosine dinucleotide (Hetterich et al. 1991; Schach et al. 1995).

Although fission of the quinolones formed by hydroxylation may sometimes be accomplished by hydrolysis, degradation generally involves hydroxylation of the hetero ring followed by dioxygenation and fission of the carbocyclic ring. The distinction between monooxygenation and dioxygenation in quinoline is, however, less clear than in the case for carbocyclic compounds. An illustrative example is provided by *P. putida* strain 86 in which 8-hydroxy-quinol-2-one is produced from quinol-2-one. The enzyme system consists of a reductase that transfers electrons from NADH and contains FAD and a [2Fe–2S] ferredoxin, and a high-molecular-weight oxygenase consisting of six identical subunits and six Rieske [2Fe–2S] clusters. It was suggested that this complex belonged to class IB oxygenases that include benzoate-1,2-dioxygenase. It is, therefore, possible that this apparent monooxygenase is, in fact, a dioxygenase introducing oxygen at the 8,8a-positions followed by elimination (Rosche et al. 1995). This is reminiscent of naphthalene dioxygenase in which both activities are catalyzed by the same enzyme (Gibson et al. 1995).

Mycobacterium gilvum that is able to degrade fluorene, fluoranthene, and pyrene (Boldrin et al. 1993) degraded 2,3-benzo-, 5,6-benzo-, and 7,8-benzoquinoline. 5,6-Benzoquinoline served as a source of both carbon and nitrogen, and 2-oxo-5,6-benzoquinoline was tentatively proposed as an intermediate in the degradation (Willumsen et al. 2001).

10.4.6 Hydroxylation

Degradation generally involves both hydroxylation and dioxygenation. For example, in organisms that degrade quinoline, initial hydroxylation that takes place at C_2 is followed in some organisms by dioxygenation. For example, *C. testosteroni* degradation of quinoline is carried out first by hydroxylation to quinol-2-one, and subsequently, by dioxygenation to the *cis*-5,6-dihydrodiol (Schach et al. 1995). Most dioxygenases for carbocyclic aromatic compounds are multicomponent enzymes, but in contrast the 2-hydroxyquinoline-5,6-dioxygenase from *C. testosteroni* is apparently a single-component enzyme. 4-Hydroxyquinoline-2-carboxylate (kynurenic acid) is an intermediate on one of the pathways for the metabolism of L-tryptophan. Its degradation by *P. fluorescens* ATCC 11299B involves dioxygenation to quinol-4-one-2-carboxylate-7,8-dihydrodiol followed by dehydrogenation, extradiol ring fission, and reduction of the α,β-unsaturated carboxylate, which has already been noted (Figure 10.5a) (Dagley and Johnson 1963; Taniuchi and Hayaishi 1963). The possibly ambiguous position of quinol-2-one 8-monooxygenase has already been noted.

The degradation of quinoline by *P. fluorescens* strain 3 and *P. putida* strain 86, and by a strain of *Rhodococcus* sp. has been investigated (Schwarz et al. 1989). The initial reaction was hydroxylation of the heterocyclic ring at C-2 for all the organisms examined, although subsequent reactions were different for the Gram-negative and Gram-positive organisms. In the pseudomonads, 8-hydroxyquinol-2-one was metabolized to 8-hydroxycoumarin by an unestablished reaction involving loss of NH_3 (Figure 10.21a). In the rhodococcus, however, hydroxylation initially produced the 6-hydroxy compound and, by further hydroxylation, the 5,6-dihydroxyquinol-2-one followed by ring fission by extradiol dioxygenation to the terminal metabolite (Figure 10.21b). The degradation of quinoline-4-carboxylate by *Microbacterium* sp. strain H2, *Agrobacterium* sp. strain 1B, and *Pimelobacter simplex* strains 4B and

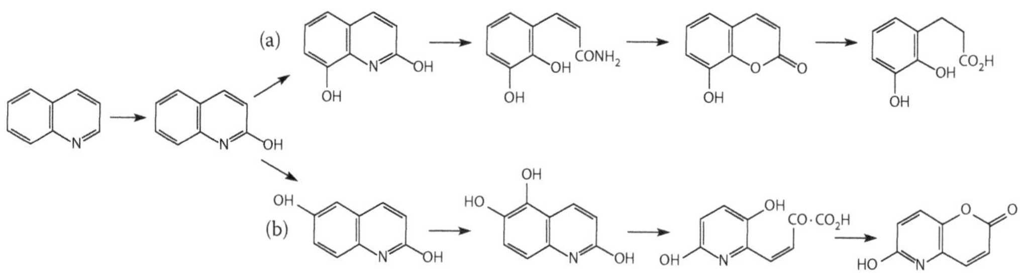

FIGURE 10.21 Degradation of quinoline. (a) Hydroxylation to 8-hydroxyquinoline and fission of ring A, (b) hydroxylation to 2,8-dihydroxyquinoline and fission of ring A.

5B follows an analogous pathway to produce 8-hydroxy-coumarin-4-carboxylate, which is then hydrolyzed and reduced (Schmidt et al. 1991).

The metabolism of 2-methylquinoline in *Arthrobacter* sp. strain Rü 61a is analogous, with the introduction of oxygen at C-4 (Hund et al. 1990). The enzymes—oxidoreductases that introduce oxygen into the pyridine rings in *Rhodococcus* sp. strain B1, *Arthrobacter* sp. strain Rü61a, and into quinoline in *P. putida* strain 86 are virtually identical. Like those already noted they have molecular masses of 300–320 kDa and contain Mo, Fe, FAD, and acid-labile sulfur (De Beyer and Lingens 1993). The enzymes from *C. testosteroni* for hydroxylation of quinoline to 2-hydroxylation (quinoline 2-oxidoreductase), and the dioxygenase responsible for the introduction of oxygen in the benzene ring (2-oxo-1,2-dihydroquinoline 5,6-dioxygenase) have been described (Schach et al. 1995). The hydroxylase showed similarities to the enzyme from the other quinoline-degrading organisms'—*P. putida* strain 86 and *Rhodococcus* sp. strain B1. In contrast, the degradation of quinol-4-one by *P. putida* strain 33/1 involves initial monooxygenation at C-3 (Block and Lingens 1992a), followed by ring fission that is mediated by a dioxygenase with the formation of anthranilate (Block and Lingens 1992b).

10.4.7 Dioxygenation

For quinoline and its methyl analogs, several types of dioxygenation have been found, and this may take place at the 5,6- or 7,8-positions, followed by fission of the carbocyclic ring (Figure 10.22). For example, a mutant strain of *P. putida* transformed quinoline to the *cis*-(5R,6S)- and *cis*-(7S,8R)-dihydrodiols, and quinoxaline to the *cis*-(5R,6S)-dihydrodiol (Boyd et al. 1992). Dioxygenation at the 7,8-position is involved in the degradation of (a) 3-chloroquinoline-8-carboxylate to 5-chloro-2-hydroxynicotinate (Tibbles et al. 1989b) and (b) 4-hydroxyquinoline-2-carboxylate (kynurenate), which is an intermediate in one of the degradation pathways of tryptophan. Ring fission of 3-hydroxyquinol-4-one is carried out by an unusual dioxygenation that is independent of metal or organic cofactors (Steiner et al. 2010), and incorporates both atoms of oxygen at C_2 and C_4, with formation of formyl anthranilate and carbon monoxide (Fischer et al. 1999). It was proposed that these belonged to the α/β hydrolase-fold superfamily functionally related to serine hydroxylases and a plausible mechanism was put forward (Fischer et al. 1999).

Fluoroquinolones are important antibiotics that have seen extensive use in veterinary medicine, since they are effective against both Gram-negative organisms including *Salmonella* spp and *Helicobacter* (*Campylobacter*) spp, and Gram-positive organisms. The degradation of the quinolone antibacterial drugs enrofloxacin (Wetzstein et al. 1999) and ciprofloxacin (Wetzstein et al. 1997) has been studied in the brown-rot fungus *Gloeophyllum striatum*. A wide range of metabolites was isolated, among them a series in which the 6-fluorine substituent was replaced by hydroxyl (Figure 10.23). The major metabolite produced from enrofloxacin by *Mucorramannianus* was the *N*-oxide of the piperazine ring distal from the hetero ring (Parshikov et al. 2000). The important issue of the photolability of aqueous solutions of these antibiotics has been discussed in Chapter 1.

10.4.7.1 Isoquinoline

The degradation of isoquinoline by *Alcaligenes faecalis* strain Pa and *Pseudomonas diminuta* strain 7 (Röger et al. 1990, 1995) is mediated by an oxidoreductase that produces 1,2-dihydroisoquinoline-1-one, followed by ring fission with the formation of *o*-phthalate and oxidation to 3,4-dihydroxybenzoate (Figure 10.24). The oxidoreductase has been purified and, like most typical azaarene oxidoreductases,

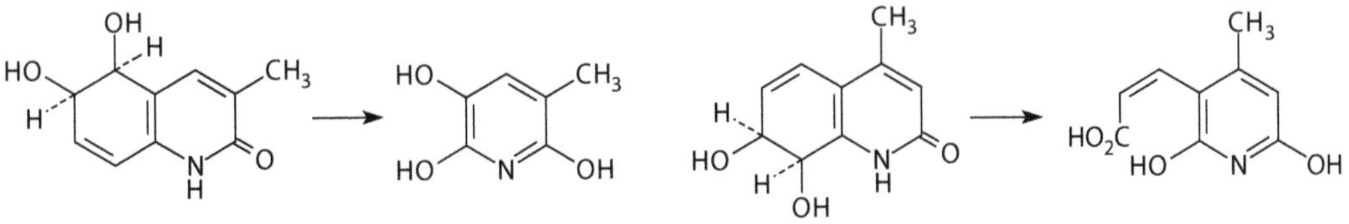

FIGURE 10.22 Degradation of methylquinolones.

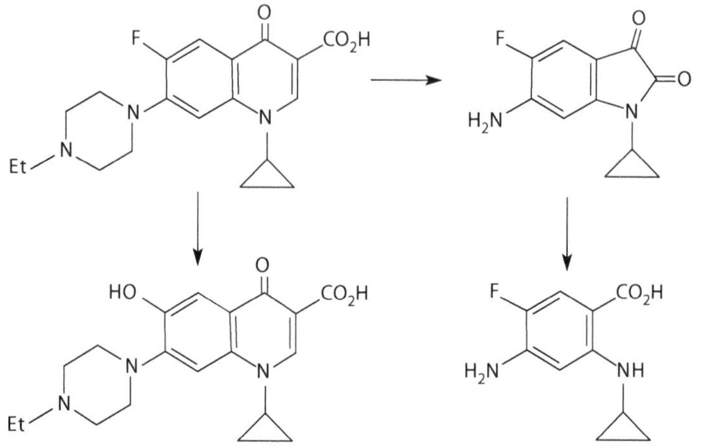

FIGURE 10.23 Transformation of enrofloxacin by *Gleophyllum striatum*.

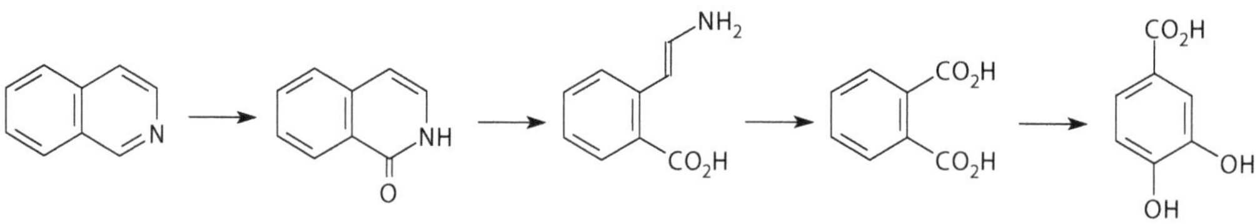

FIGURE 10.24 Degradation of isoquinoline. (Adapted from Röger P, G Bär, F Lingens. 1995. *FEMS Microbiol Lett* 129: 281–286.)

contained 0.85 g atoms of Mo per mole, 3.9 g atoms of Fe, and acid-labile S (Lehmann et al. 1994). The oxidoreductase had a rather limited specificity, but it was able to hydroxylate quinazoline (1,3-diazanaph-thalene) and phthalazine (2,3-diazanaphthalene) (Stephan et al. 1996). When the position adjacent to the nitrogen atom was substituted with a methyl group, hydroxylation by quinaldine oxidoreductase took place at the 4-position (De Beyer and Lingens 1993).

10.4.7.2 Pyrimidine and Related Compounds Excluding Purines

The degradation of pyrimidines can be carried out under both aerobic and anaerobic conditions.

1. The aerobic degradation of uracil (2,4-dihydroxypyrimidine) may take place by mono-oxygenation to barbituric acid (2,4,6-trihydroxypyrimidine) or, from thymine the analogous 5-methylbarbi-turic acid. Barbituric acid is then hydrolyzed to ureidomalonate by barbiturase (Soong et al. 2002), and subsequently to urea and malonate (Figure 10.25a) (Hayaishi and Kornberg 1952; Batt and Woods 1961).

2. The degradation of uracil by *E. coli* K-12 involves an alternative pathway: this produces 2 mol utilizable nitrogen and 1 mol 3-hydroxypropionate as waste per mol uracil. The pathway is initiated by RutA and RutF that carry out monooxygenation in the presence of NADH, FMN, and O_2 to produce peroxyureidoacrylate. This is hydrolyzed by RutB to carbamate and peroxy-acrylate, followed by reduction to aminoacrylate by RutC, and then via malonate semialdehyde by reduction to 3-hydroxypropionate by RutE (Figure 10.25b) (Kim et al. 2010).

3. The degradation of pyrimidines may also take place by reduction of the ring under both aerobic and anaerobic conditions. For uracil, the pathway is initiated by dihydropyrimidine dehydro-genase to 4,5-dihydroxyuracil, hydrolysis by dihydropyrimidinase to ureidopropionate and hydrolysis to alanine, CO_2 and NH_3. The analogous pathway for orotic acid produces *N*-carbamoyl-β-alanine that is hydrolyzed to aspartate (Xu and West 1992) (Figure 10.26).

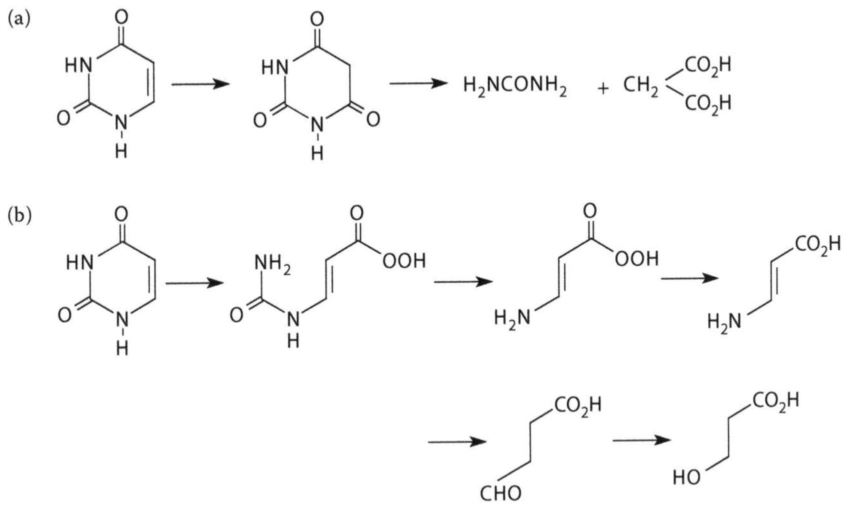

FIGURE 10.25 Degradation of uracil. (a) Reduction, (b) hydrolysis.

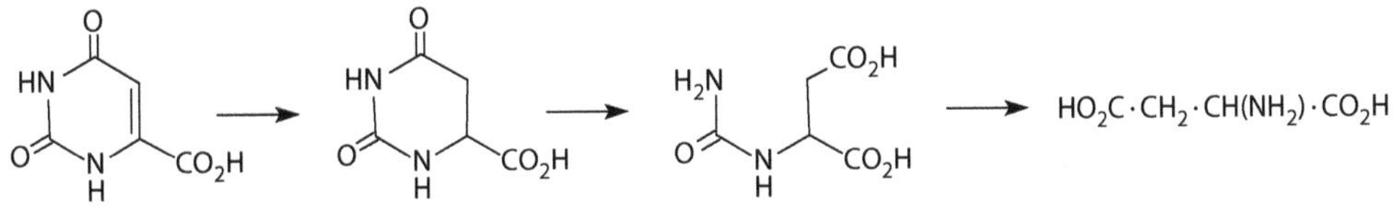

FIGURE 10.26 Reductive degradation of orotic acid.

The dihydropyrimidase that mediates the ring reduction of uracil and thymidine in *Pseudomonas stutzeri* (Xu and West 1994), and the corresponding reductase for orotase (Ogawa and Shimizu 1995) have been characterized. Although *E. coli* K-12 is unable to degrade uracil by the reductive pathway since it lacks the ring-fission enzyme, it has been shown that it contains an enzyme homologous to mammalian dihydropyrimidine 5,6-dehydrogenase that was an NAD-dependent iron-sulfur flavoenzyme (Hidese et al. 2011).

The degradation or transformation of only a few halogenated azaarenes has been examined under aerobic conditions:

1. 5-Fluoro-2′-deoxyuridine has been extensively used in studies of the mechanism of action of thymidylate synthase, and 5-fluorouracil is an anticancer drug that has provided a lead to the development of others. The metabolism of 5-fluorouracil by the ascomycete fungus *Nectria haematococca* has been studied using ^{19}F NMR (Parisot et al. 1991). α-fluoro-β-alanine (2-fluoro-3-aminopropionate) was produced (Figure 10.27), while 5-fluorouridine-5′-mono-, di-, and triphosphate were found in acid extracts of the mycelia, and the 2′- and 3′-monophosphates were recovered from RNA.

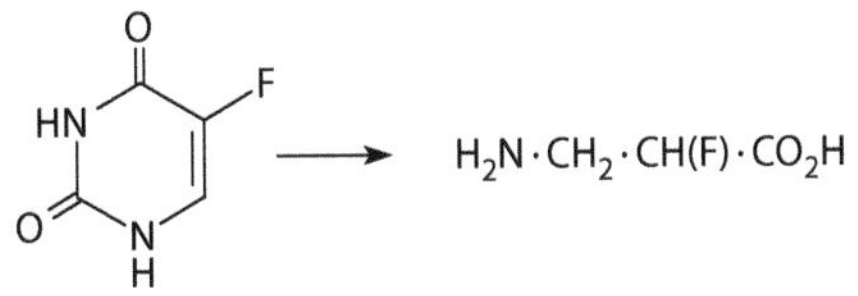

FIGURE 10.27 Degradation of 5-fluorouracil by *Nectria haematococca*.

2. An enzyme may also carry out gratuitous dehalogenation even though this is not its primary function:

   a. Thymidylate synthetase catalyzes the reductive methylation of 2′-deoxyuridylate to thymidylate using 5,10-methylene-5,6,7,8-tetrahydrofolate as both methyl donor and reductant. The enzyme can also dehalogenate 5-bromo- and 5-iododeoxyuridylate (Wataya and Santi 1975).

   b. Dihydropyrimidine dehydrogenase is a key enzyme in pyrimidine synthesis, and contains FAD, FMN, and nonheme iron (Lu et al. 1992). It is also able to catalyze the dehalogenation of 5-bromo- and 5-iodo-5,6-dihydrouracil to uracil (Porter 1994).

Unusual 2-oxoglutarate-dependent dioxygenations have been found in the yeast *Rhodotorula glutinis*:

1. Conversion of deoxyuridine by 1′-hydroxylation into uracil and ribonolactone (Stubbe 1985).
2. Thymine-7-hydroxylase catalyzed the oxidation of the 5-methyl group to 5-hydroxymethyl uracil (Wondrack et al. 1978).

10.4.8 Anaerobic Conditions

Reduction of the pyrimidine ring has been shown to be the first step in the degradation of orotic acid by *Clostridium* (*Zymobacterium*) *oroticum* (Lieberman and Kornberg 1953, 1954, 1955) (Figure 10.28a), and of uracil by *Cl. uracilicum* (Campbell 1957a–c, 1960) (Figure 10.28b), which closely follows the pathways for aerobic degradation.

10.4.8.1 Pyrazine

Substituted pyrazines are metabolites of species of the fungus *Aspergillus* and the bacterium *Bacillus*. Their degradation has seldom, however, been examined. Degradation by a strain of *Mycobacterium* sp. was plausibly postulated to proceed by hydroxylation at C-6 followed by ring fission with the elimination of NH_4^+ (Rappert et al. 2006). Analogous reactions have already been illustrated for the degradation of pyridines.

10.4.8.2 Isoalloxazine

Although simple quinoxalines have not been found as natural products, extensive investigation have been devoted to the biodegradation of the important isoalloxazine riboflavin (vitamin B_2). *Pseudomonas* sp. strain RF was obtained by enrichment with riboflavin and a number of degradation products have been

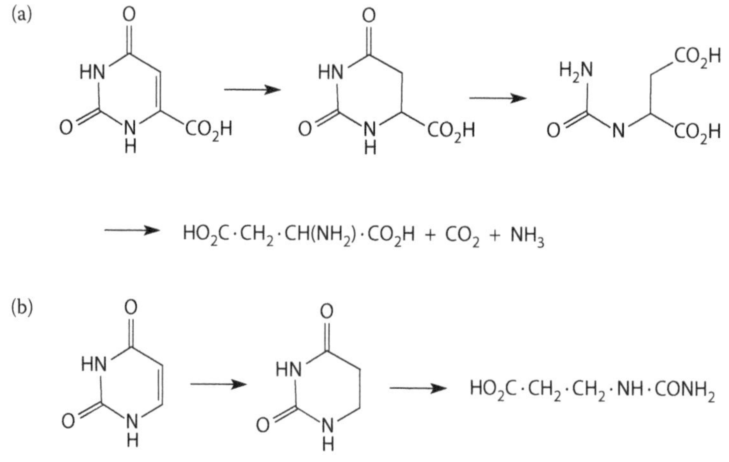

FIGURE 10.28 Anaerobic biodegradation of (a) orotic acid and (b) uracil.

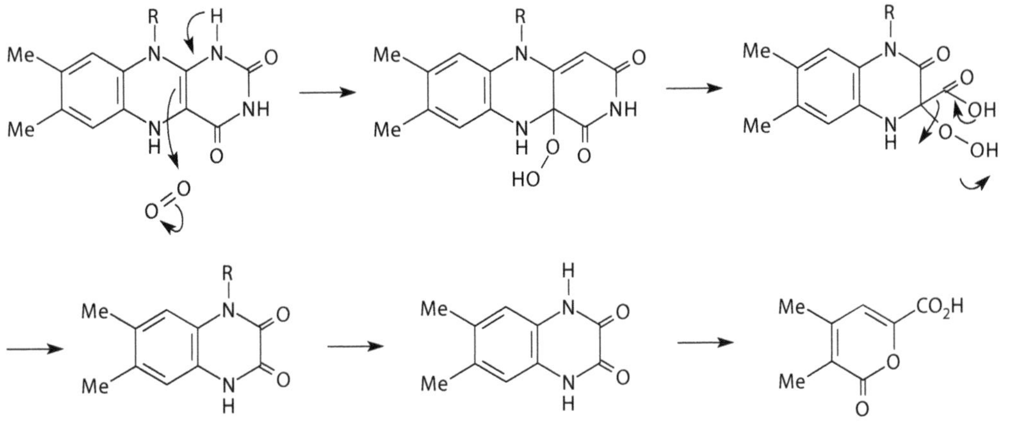

FIGURE 10.29 Aerobic degradation of riboflavin. (Adapted from Harkness DR, L Tsai, ER Stadtman. 1964. *Arch Biochem Biophys* 108: 323–333.)

characterized. These included oxamide that provided a clue to the pathway and identification of metabolites from labeled substrate (Harkness et al. 1964). The pathway for degradation may plausibly be rationalized on the basis of a number of successive steps: (a) monooxygenation, (b) oxidative loss of the side chain, (c) hydrolysis of the dioxoquinoxaline with loss of oxamide, and (d) fission of the benzenoid ring by dioxygenation to 3,4-dimethylpyrone-6-carboxylate (Figure 10.29).

10.5 Purines

10.5.1 Aerobic Degradation

Whereas the methylated xanthines that are constituents of coffee and cocoa are prepared from the seeds of the plants *Coffea robusta* and *Theobroma cacoa*, adenine and guanine occur universally as the 9β-ribofuranoside-5′(3′)-phosphates in ribonucleic acids and the analogous 2-deoxy-9β-ribofuranosides in deoxyribonucleic acids.

Although the anaerobic degradation of pyrimidines and purines has been extensively examined in a range of organisms, particularly in species of clostridia, aerobic degradation has been studied less often.

1. The aerobic metabolism of uric acid (2,6,8-trihydroxypurine) has been examined in pseudomonads and degradation has been shown to occur by alternative pathways:

    a. Fission of the pyrimidine ring catalyzed by uricase to produce allantoin and allantoic acid, which is then degraded to glyoxylate and urea (Canellakis and Cohen 1955; Bachrach 1957) (Figure 10.30a).

    b. Fission of the iminazolone ring to produce barbituric acid, alloxan, and alloxanic acid, which is produced by a benzylic acid rearrangement of the barbituric acid initially produced by monooxygenation (Figure 10.30b) (Hayaishi and Kornberg 1952).

2. The degradation of caffeine (1,3,7-trimethylxanthine) has been examined in *P. putida* strain 40 that could also oxidize 1,3-dimethylxanthine (theophylline) and 3,7-dimethylxanthine (theobromine) (Woolfolk 1975). A later study using *P. putida* strain CBB5 and theophylline, theobromine and caffeine illustrated partial. *N*-demethylation and the formation of 1- and 3-methyluric acid that were stable terminal metabolites (Yu et al. 2009). Oxidative demethylation with the production of formaldehyde and xanthine is a common feature (Glück and Lingens 1988), though not all methyl-xanthines are demethylated. *P. putida* CBB5 can degrade a range of *N*-methylated purines including caffeine, theobromine, theophylline and 3-methylxanthine,

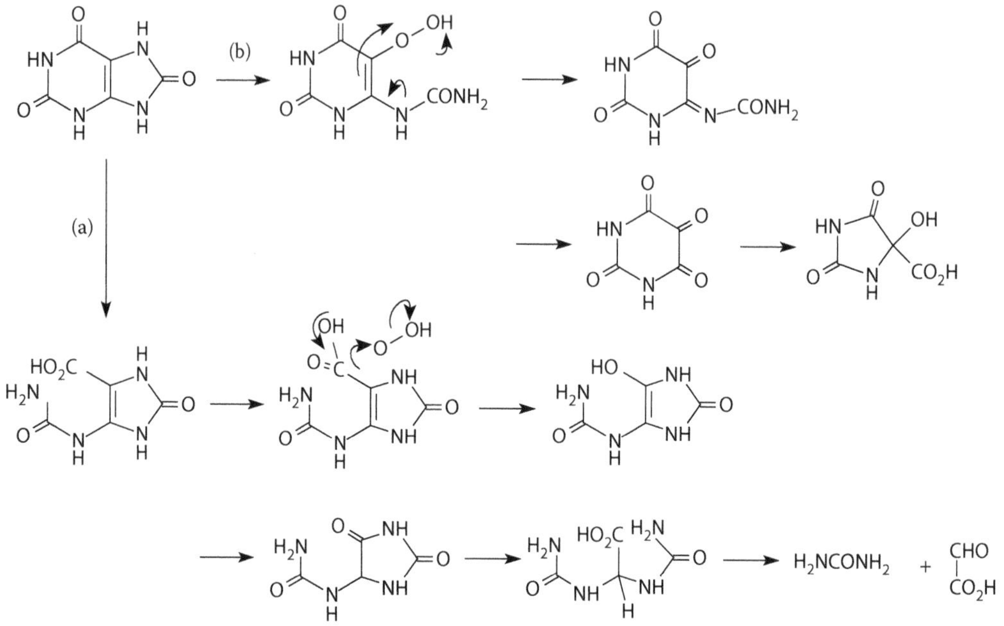

FIGURE 10.30 Aerobic degradation of uric acid. (a) Uricase-mediated fission of pyrimidine ring, (b) hydrolysis of imin-azolone ring. (Adapted from Schultz AC, P Nygaard, HH Saxild. 2001. *J Bacteriol* 183: 3293–3302.)

and the broad specificity of the *N*-demethylase has been characterized (Summers et al. 2011). It consisted of a two-subunit *N*-demethylase component containing a Rieske domain [2Fe–2S] and a reductase component with cytochrome *c* reductase activity. Caffeine can also be hydroxyl-ated at C_8 by a specific dehydrogenase in *Pseudomonas* sp. strain CBB1 with coenzyme Q_0 as the preferred electron acceptor (Yu et al. 2008).

3. The *puc* genes that encode the enzymes in the aerobic pathway for the degradation of hypoxan-thine have been determined for *B. subtilis* (Schultz et al. 2001). They encode enzymes that carry out the following sequences in which xanthine and hypoxanthine are central metabolites (Figure 10.31), produced by deamination of adenine and guanine:

Adenine (6-aminopurine) → hypoxanthine (6-hydroxypurine) → xanthine (2,6-dihydroxypu-rine) → uric acid (2,6,8-trihydroxypurine) → allantoin → allantoic acid → ureidoglycine → ureidoglycolic acid. In *B. subtilis*, the first two reactions are catalyzed by expression of xanthine dehydrogenase (*pucABCDE*), ring fission by urocanase (*pucLM*), and subsequent hydrolysis by *S*(+) allantoinase (*pucH*) and of allantoate by an amidohydrolase (*pucF*). It should be noted that in higher organisms the oxidation of urate to 5-hydroxyurate is carried out by an oxidase (Jung et al. 2006; Cendron et al. 2007).

4. A number of additional aspects of purine metabolism when they are used as sources of nitrogen by *Enterobacteriaceae* include the following.

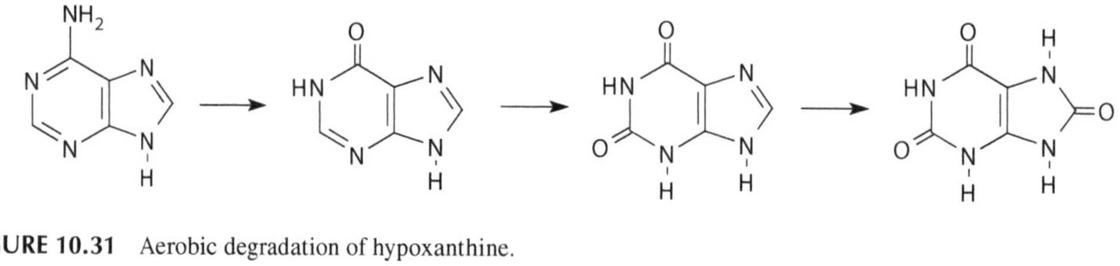

FIGURE 10.31 Aerobic degradation of hypoxanthine.

a. Purines are not generally used as nitrogen sources by *E. coli* and, although allantoate was putatively produced from adenine, neither NH_4^+ nor carbamoyl phosphate was produced (Xi et al. 2000). This is consistent with the use of allantoin as a nitrogen source under anaerobic conditions by *E. coli* that is not, however, able to use it as a growth substrate (Cusa et al. 1999).

b. Important metabolic aspects have emerged from the study of purine metabolism in *Klebsiella pneumoniae* (de la Riva et al. 2008) and *K. oxytoca* (Pope et al. 2009) that are able to use purines as sources of nitrogen. In *K. oxytoca*, the oxidation to 2,4,6-trihydroxypurine (uric acid) is mediated by a dioxygenase/reductase system whose genes have been identified. They encode a two-component Rieske nonheme iron, aromatizing hydroxylase (HpxD and HpxE) (Pope et al. 2009), and genetic analysis of *K. pneumoniae* (de la Riva et al. 2008) failed to find evidence for the existence of typical molybdenum-containing enzymes, and xanthine dehydrogenase appears, therefore, not to be involved in either of these bacteria. The pathway for degradation of guanine has been explored in *K. pneumoniae* (Guzmán et al. 2011). This is initiated by deamination (GuaD) to xanthine, and dehydrogenation (HpxDE) to urate that is oxidized to 5-hydroxyisourate by a flavoprotein monooxygenase (HpxO). This is followed by hydrolysis (HpxT), ring fission (HpxQ), decarboxylation to allantoin (HpxQ), and hydrolysis to allantoate by allantoinase (HpxB). These reactions occur in three clusters of genes that carry out: (i) hypoxanthine → allantoin, (ii) allantoin → allantoate, (iii) further catabolism of allantoate (Guzmán et al. 2011).

5. Yeast *Trichosporon adeninovorans* that was isolated by enrichment with adenine was able to grow at the expense of adenine, xanthine, and uric acid as source of carbon, nitrogen, and energy. The presence of uric acid oxidase and allantoinase in cells grown with adenine and uric acid was consistent with the pathways already described for bacteria (Middelhoven et al. 1984).

10.5.2 Anaerobic Degradation

Demonstration of the anaerobic degradation of purines belongs to the golden age of microbiology and was appropriately discovered in Beijerinck's laboratory in Delft. Liebert (1909) obtained a pure culture of an organism that was able to grow anaerobically with 2,6,8-trihydroxypurine (uric acid), which he named *Bacillus acidi urici* (Liebert 1909). (The organism is now known as *Clostridium acidurici*). Subsequently, several other purinolytic clostridia have been isolated, as well as anaerobic bacteria belonging to other groups.

10.5.3 Ring-Fission Reactions

Substantial effort has been devoted to elucidating the details of the anaerobic degradation of purines containing hydroxyl and amino groups. Although the most studied group of organisms are the nutritionally restricted clostridia, *Cl.*, *Cl. cylindrospermum*, and *Cl. purinilyticum* (Schieffer-Ullrich et al. 1984), attention should also be drawn to the nonspore-forming *Eubacterium angustum* (Beuscher and Andreesen 1984), *Peptostreptococcus barnesae* (Schiefer-Ullrich and Andreesen 1985), and *Methanococcus vannielii*, which can use purines as sources of nitrogen (DeMoll and Auffenberg 1993).

Barker and his colleagues (Barker 1961) carried out many of the basic investigations on the mechanisms of purine degradation by clostridia, and more recent developments have been presented (Dürre and Andreesen 1983). One of the significant findings was the selenium dependency in *Cl. purinilyticum* that was the result of its requirement for the synthesis of several critical enzymes—xanthine dehydrogenase, formate dehydrogenase, and glycine reductase. Selenium is required for the synthesis of active xanthine dehydrogenase in *Cl. acidiurici* and is a component of the purified enzyme that contained iron, acid-labile sulfur, FAD, and molybdenum in the ratios 7.7:7.5:1.7:1.8 (Wagner et al. 1984). Under conditions of selenium starvation in *Cl. purinilyticum*, uric acid is the central metabolite and is degraded by fission of the iminazole ring to produce 4,5-diaminouracil, which is then degraded to formate, acetate, glycine, and CO_2 (Dürre and Andreesen 1982). Under conditions of selenium sufficiency, xanthine is the central metabolite

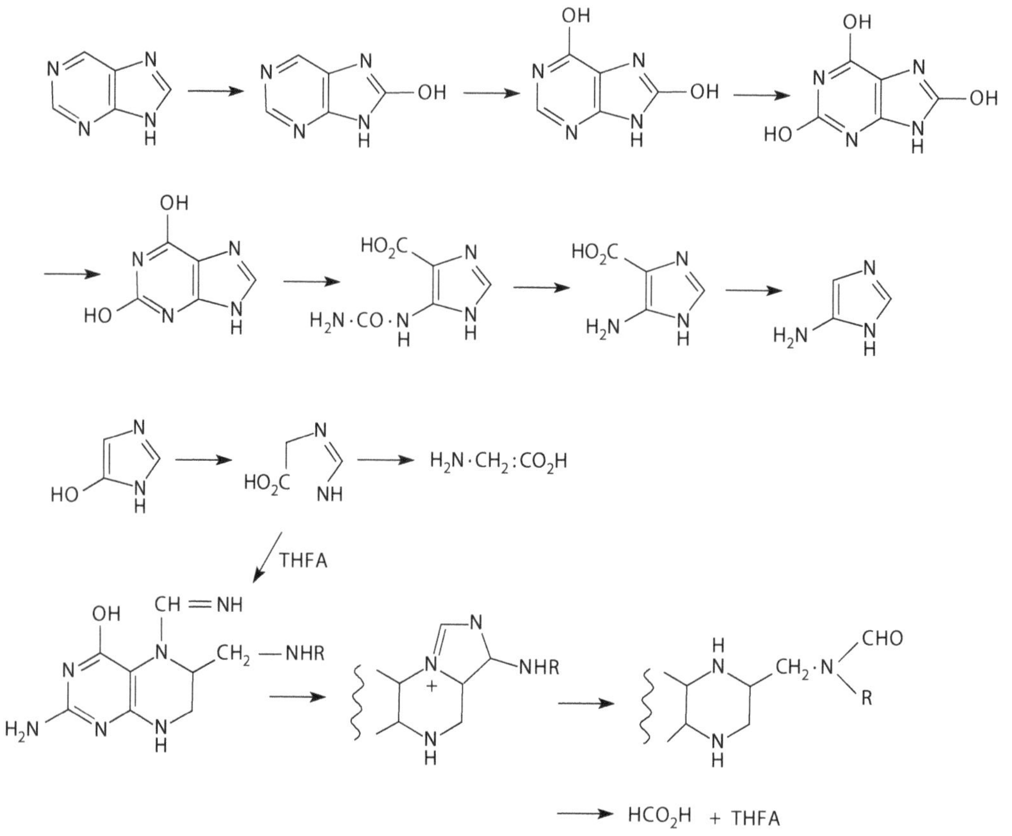

FIGURE 10.32   Anaerobic biodegradation of purine.

that is produced from all purines including uric acid, which is reductively transformed to xanthine (Dürre and Andreesen 1983).

The pathways for the degradation of purines containing amino and/or hydroxyl groups converge on the synthesis of xanthine (2,6-dihydroxypurine), and are followed by its degradation to formiminoglycine (Figure 10.32). This compound is then used for the synthesis of glycine and 5-formiminotetrahydrofolate whose further metabolism to formate results in the synthesis of ATP. The energy requirements of the cell are supplemented by the contribution of ATP produced during the reduction of glycine to acetate in an unusual reaction catalyzed by glycine reductase (Arkowitz and Abeles 1989). An essentially similar pathway is used by *M. vannielii* that can utilize a number of purines as nitrogen sources (DeMoll and Auffenberg 1993). There are some important additional issues:

1. The anaerobe *Peptococcus (Micrococcus) aerogenes* had a dehydrogenase that carried out specific hydroxylation at the 6-positions of 2- and 8-hydroxypurine, and was, therefore, distinct from xanthine dehydrogenase from which it could be separated (Woolfolk et al. 1970). It was also able to carry out dismutation of 2-hydroxypurine to xanthine (2,6-dihydroxypurine) and hypoxanthine (6-hydroxypurine).

2. Although it had been assumed that only hypoxanthine dehydrogenase is required for the conversion of hypoxanthine (6-hydroxypurine) into uric acid, in *Cl. purinolyticum*, two enzymes, both of which contain a selenium cofactor, are required. The enzymes differ in the molecular mass of their subunits, in their terminal amino acid sequences, in their kinetic parameters, and in their specific activities for purines (Self and Stadman 2000). Purine hydroxylase converts purine into hypoxanthine and xanthine (2,6-dihydroxypurine), which is then further hydroxylated to uric acid (2,6,8-trihydroxypurine) by xanthine dehydrogenase (Self 2002).

10.6 Triazines

10.6.1 1,3,5-Triazines

Derivatives of 1,3,5-triazine are important herbicides so that attention has been directed to their persistence particularly in the terrestrial environment. In some experiments when a growth substrate was supplied, they have been used as sources of nitrogen or sulfur:

1. 2-Chloro-4,6-diamino-1,3,5-triazine as a nitrogen source with lactate as carbon source (Grossenbacher et al. 1984) or 2-chloro-4-aminoethyl-6-amino-1,3,5-triazine as a nitrogen source and glycerol as carbon source (Cook and Hütter 1984).
2. 2-(1-Methylethyl)amino-4-hydroxy-6-methylthio-1,3,5-triazine as sulfur source and glucose as carbon source (Cook and Hütter 1982).

Degradation of substituted triazines is accomplished by a sequence of hydrolytic reactions (Jutzi et al. 1982; Mulbry 1994). Atrazine, which is a representative herbicide, is degraded to cyanuric acid (trihydroxytriazine), and ultimately to CO_2 and ammonia (Figure 10.33) (Mulbry 1994; Seffernick et al. 2002; Strong et al. 2002; Fruchey et al. 2003). The trihydroxy compound (cyanuric acid) is produced by three hydrolases encoded in *Pseudomonas* sp. strain ADP by AtzA, AtzB, and AtzC that do not occur on an operon. Biuret, which is the ring fission product of AtzD, is then hydrolyzed by AtzE to allophanate, and by AtzF to NH_4^+ and CO_2 (Cheng et al. 2005; Shapir et al. 2005).

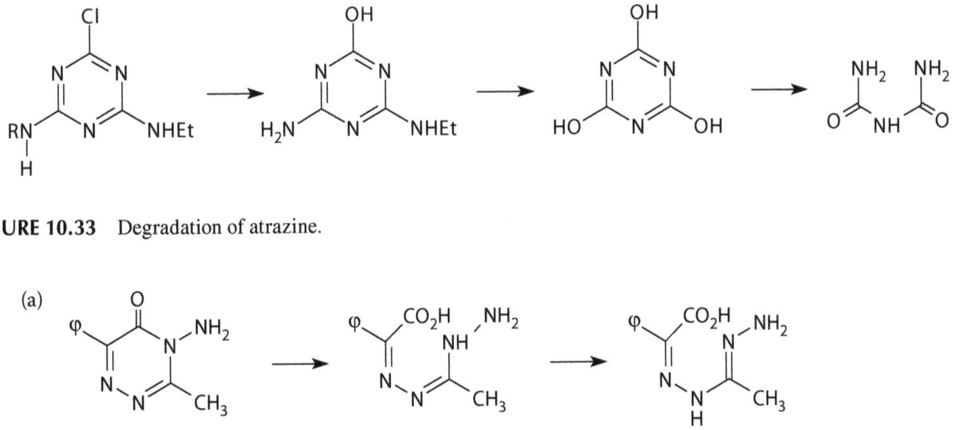

FIGURE 10.33 Degradation of atrazine.

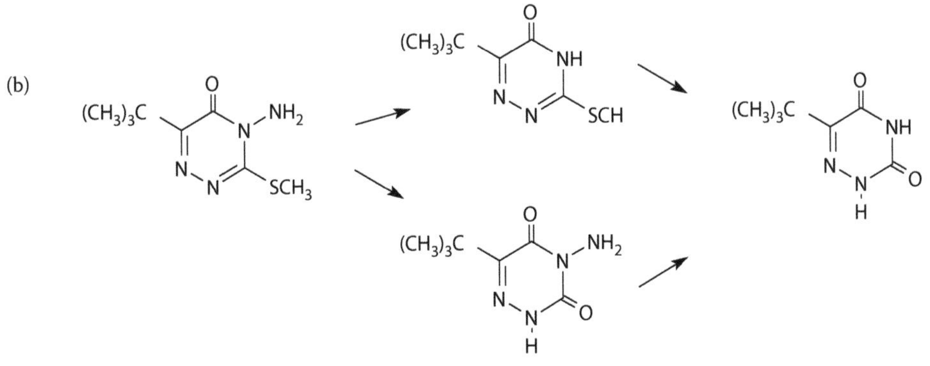

FIGURE 10.34 Aerobic degradation of 1,2,4-triazolone herbicides (a) metamitron and (b) metribuzin.

10.6.2 1,2,4-Triazines

There are a few important examples of 1,2,4-triazine herbicides. The 1,2,4-triazinone metamitron can be degraded to phenylglyoxalate by an *Arthrobacter* sp. (Figure 10.34a) (Engelhardt et al. 1982), whereas metribuzin undergoes mainly only transformation by reductive loss of the *N*-amino group and hydrolytic loss of the thiomethyl group (Figure 10.34b) (Lawrence et al. 1993).

10.7 References

Arkowitz RA, RH Abeles. 1989. Identification of acetyl phosphate as the product of clostridial glycine reductase: Evidence for an acyl enzyme intermediate. *Biochemistry* 28: 4639–4644.

Awaya JD, PM Fox, D Borthakur. 2005. pyd genes of *Rhizobium* sp, strain TAL1145 are required for degradation of 3-hydroxy-4-pyridone, an aromatic intermediate in mimosine metabolism. *J Bacteriol* 187: 4480–4487.

Bachrach U. 1957. The aerobic breakdown of uric acid by certain pseudomonads. *J Gen Microbiol* 17: 1–11.

Baitsch D, C Sandu, R Brandsch, GL Igloi. 2001. Gene cluster on pAO1 of *Arthrobacter nicotinovorans* involved in degradation of the plant alkaloid nicotine: cloning, purification, and characterization of 2,6-dihydroxypyridine 3-hydrolase. *J Bacteriol* 183: 5262–5267.

Bak F, F Widdel. 1986. Anaerobic degradation of indolic compounds by sulfate-reducing enrichment cultures, and description of *Desulfobacterium indolicm* gen. nov., sp. nov. *Arch Microbiol* 146: 170–176.

Barker HA. 1961. Fermentations of nitrogenous organic compounds, Vol 2, pp. 151–207. In *The Bacteria* (Eds. IC Gunsalus and R.Y Stanier), Academic Press, New York.

Batie CJ, E LaHaie, DP Ballou. 1987. Purification and characterization of phthalate oxygenase and phthalate oxygenase reductase from *Pseudomonas cepacia. J Biol Chem* 262: 1510–1518.

Batt RD, DD Woods. 1961. Decomposition of pyrimidines by *Nocardia corallina. J Gen Microbiol* 24: 207–224.

Bauder R, B Tshisuaka, F Lingens. 1990. Microbial metabolism of quinoline and related compounds VII. Quinoline oxidoreductase from *Pseudomonas putida*: A molybdenum-containing enzyme. *Biol Chem Hoppe-Seyler* 371: 1137–1144.

Bauer G, F Lingens. 1992. Microbial metabolism of quinoline and related compounds XV. Quinoline-4-carboxylic acid oxidoreductase from *Agrobacterum* spec. 1B: A molybdenum-containing enzyme. *Biol Chem Hoppe-Seyler* 370: 1183–1189.

Becker D, T Schräder, JR Andreesen. 1997. Two-component flavin-dependent pyrrole-2-carboxylate monooxygenase from *Rhodococcus* sp. *Eur J Biochem* 249: 739–747.

Behrman EJ, RY Stanier. 1957a. The bacterial oxidation of nicotinic acid. *J Biol Chem* 228: 923–945.

Behrman EJ, RY Stanier. 1957b. Observations on the oxidation of halogenated nicotinic acids. *J Biol Chem* 228: 947–953.

Beuscher HU, JR Andreesen. 1984. *Eubacterium angustum* sp. nov., a Gram-positive anaerobic, non-sporeforming, obligate purine fermenting organism. *Arch Microbiol* 140: 2–8.

Block DW, F Lingens. 1992a. Microbial metabolism of quinoline and related compounds XIII. Purification and properties of 1H-4-oxoquinoline monooxygenase from *Pseudomonas putida* strain 33/1. *Biol Chem Hoppe-Seyler* 373: 249–254.

Block DW, F Lingens. 1992b. Microbial metabolism of quinoline and related compounds XIV. Purification and properties of 1H-3-hydroxy-4-oxoquinoline oxygenase, a new extradiol cleavage enzyme from *Pseudomonas putida* strain 33/1. *Biol Chem Hoppe-Seyler* 370: 343–349.

Boldrin B, A Tiehm, C Fritzsche. 1993. Degradation of phenanthrene, fluorene, fluoranthene, and pyrene by a *Mycobacterium* sp. *Appl Environ Microbiol* 59: 1927–1930.

Borthakur D, M Soedarjo, PM Fox, DT Webb. 2003. The mid genes of *Rhizobium* sp. strain TAL1145 are required for degradation of mimosine into 3-hydroxy-4-pyridone and are inducible by mimosine. *Microbiology (UK)* 149: 537–546.

Boyd DR, ND Sharma, R Boyle, RAS McMordie, J Chima, H Dalton. 1992. A 1H NMR method for the determination of enantiomeric excess and absolute configuration of cis-dihydrodiol metabolites of polycyclic arenes and heteroarenes. *Tetrahedron Lett* 33: 1241–1244.

Brandl MT, SE LIndow. 1996. Cloning and characterization of the locus encoding an indolpyruvate decarboxylase involved in indole-3-acetic acid synthesis in *Erwinia herbicola*. *Appl Environ Microbiol* 62: 4121–4128.

Brandsch R, AE Hinkkanen, K Decker. 1982. Plasmid-mediated nicotine degradation in *Arthrobacter oxidans*. *Arch Microbiol* 132: 26–30.

Campbell LL. 1957a. Reductive degradation of pyrimidines I. The isolation and characterization of a uracil fermenting bacterium, *Clostridium uracilicum* NOV. SPEC. *J Bacteriol* 73: 220–224.

Campbell LL. 1957b. Reductive degradation of pyrimidines II. Mechanism of uracil degradation by *Clostridium uracilicum*. *J Bacteriol* 73: 225–229.

Campbell LL. 1957c. Reductive degradation of pyrimidines IV. Purification and properties of dihydrouracil hydrase. *J Biol Chem* 233: 1236–1240.

Campbell LL. 1960. Reductive degradation of pyrimidines V. Enzymatic conversion of N-carbamoyl-β-alanine to β-alanine, carbon dioxide, and ammonia. *J Biol Chem* 235: 2375–2378.

Canellakis ES, PP Cohen. 1955. The end-products and intermediates of uric acid oxidation by uricase. *J Biol Chem* 213: 385–395.

Cendron L, R Berni, C Folli, I Ramazzina, R Percudani, G Zanotti. 2007. The structure of 2-oxo-4-hydroxy-4-carboxy-5-ureidoimidazoline decarboxylase provides insight into the mechanism of uric acid degradation. *J Biol Chem* 282: 18182–18189.

Chang H-K, GJ Zylstra. 1999. Role of quinolinate phosphoribosyl transferase in degradation of phthalate by *Burkholderia cepacia* DBO1. *J Bacteriol* 181: 3069–3075.

Cheng G, N Shapir, MJ Sadowsky, LP Wackett. 2005. Allophanate hydrolase, not urease functions in bacterial cyanuric acid metabolism. *Appl Environ Microbiol* 71: 4437–4445.

Claus G, HJ Kutzner. 1983. Degradation of indole by *Alcaligenes* spec. *Syst Appl Microbiol* 4: 169–180.

Colabroy KL, TP Begley. 2005. Tryptophan catabolism: identification and characterization of a new degradative pathway. *J Bacteriol* 187: 7866–7869.

Cook AM, R Hütter. 1982. Ametryne and prometryne as sulfur sources for bacteria. *Appl Environ Microbiol* 43: 781–786.

Cook AM, R Hütter. 1984. Deethylsimazine: Bacterial dechlorination, deamination and complete degradation. *J Agric Food Chem* 32: 581–585.

Cripps RE. 1973. The microbial metabolism of thiophen-2-carboxylate. *Biochem J* 134: 353–366.

Cusa E, N Obradors, L Baldomá, J Badía, J Aguilar. 1999. Genetic analysis of a chromosomal region containing genes required for assimilation of allantoin nitrogen and linked glyoxylate metabolism in *Escherichia coli*. *J Bacteriol* 181: 7479–7484.

Dagley S, PA Johnson. 1963. Microbial oxidation of kynurenic, xanthurenic and picolinic acids. *Biochim Biophys Acta* 78: 577–587.

Dai VD, K Decker, H Sund. 1968. Purification and properties of L-6-hydroxynicotine oxidase. *Eur J Biochem* 4: 95–102.

Dash SS, SN Gummadi. 2006. Catabolic pathways and biotechnological applications of microbial caffeine degradation. *Biotechnol Lett* 28: 1993–2002.

De Beyer A, F Lingens. 1993. Microbial metabolism of quinoline and related compounds XVI. Quinaldine oxidoreductase from *Arthrobacter* spec. Rü 61a: a molybdenum-containing enzyme catalysing the hydroxylation at C-4 of the heterocycle. *Biol Chem Hoppe-Seyler* 374: 101–120.

De la Riva L, J Badia, J Aguilar, RA Bender, L Baldoma. 2008. The hpx genetic system for hypoxanthine assimilation as a nitrogen source in *Klebsiella pneumoniae*: gene organization and transcriptional regulation. *J Bacteriol* 190: 7892–7903.

DeMoll E, T Auffenberg. 1993. Purine metabolism in *Methanococcus vannielii*. *J Bacteriol* 175: 5754–5761.

DeMoss RD, K Moser. 1969. Tryptophananase in diverse bacterial species. *J Bacteriol* 98: 167–171.

Dennie D, I Gladu, F Lépine, R Villemur, J-G Bisaillon, R Beaudet. 1998. Spectrum of the reductive dehalogenation activity of *Desulfitobacterium frappieri* PCP-1. *Appl Environ Microbiol* 64: 4603–4606.

Dürre P, JR Andreesen. 1982. Anaerobic degradation of uric acid via pyrimidine derivatives by selenium-starved cells of *Clostridium purinolyticum*. *Arch Microbiol* 131: 255–260.

Dürre P, JR Andreesen. 1983. Purine and glycine metabolism by purinolytic clostridia. *J Bacteriol* 154: 192–199.

Eaton RW, PJ Chapman. 1995. Formation of indigo and related compounds from indolecarboxylic acids by aromatic acid-degrading bacteria: Chromogenic reactions for cloning genes encoding dioxygenases that act on aromatic acids. *J Bacteriol* 177: 6983–6988.

Engelhardt G, W Ziegler, PR Wallnöfer, HJ Jarcszyk, L Oehlmann. 1982. Degradation of the triazinone herbicide metamitron by *Arthrobacter* sp. DSM 20389. *J Agric Food Chem* 30: 278–282.

Ensign JC, SC Rittenberg. 1964. The pathway of nicotinic acid oxidation by a *Bacillus* species. *J Biol Chem* 239: 2285–2291.

Evans JS, WA Venables. 1990. Degradation of thiophene-2-carboxylate, furan-2-carboxylate, and pyrrole-2-carboxylate and other thiophene derivatives by the bacterium *Vibrio* YC1. *Appl Microbiol Biotechnol* 32: 715–720.

Feng Y, KD Racke, J-M Bollag. 1997. Isolation and characterization of a chlorinated-pyridinol-degrading bacterium. *Appl Environ Microbiol* 63: 4096–4098.

Fischer F, S Künne, S Fetzner. 1999. Bacterial 2,4-dioxygenases: new members of the α,β-hydrolase-fold superfamily of enzymes functionally related to serine hydrolases. *J Bacteriol* 181: 5725–5733.

Freudenberg W, K König, JR Andressen. 1988. Nicotine dehydrogenase from *Arthrobacter oxidans*: A molybdenum-containing hydroxylase. *FEMS Microbiol Lett* 52: 13–18.

Fruchey I, N Shapir, MJ Sadowsky, LP Wackett. 2003. On the origins of cyanuric acid hydrolase: purification, substrates, and prevalence of AtdzD from *Pseudomonas* sp. strain ADP. *Appl Environ Microbiol* 69: 3653–3657.

Fujioka M, H Wada. 1968. The bacterial oxidation of indole. *Biochim Biophys Acta* 158: 70–78.

Gauthier JJ, SC Rittenberg. 1971. The metabolism of nicotinic acid II. 2,5-dihydroxypyridine oxidation, product formation, and oxygen 18 incorporation. *J Biol Chem* 246: 3743–3748.

Gibson DT, SM Resnick, K Lee, JM Brand, DS Torok, LP Wackett, MJ Schocken, BE Haigler. 1995. Desaturation, dioxygenation, and monooxygenation reactions catalysed by naphthalene dioxygenase from *Pseudomonas* sp. strain 9816-4. *J Bacteriol* 177: 2615–2621.

Gieg LM, A Otter, PM Fedorak. 1996. Carbazole degradation by *Pseudomonas* sp. LD2: metabolic characteristics and identification of some metabolites. *Environ Sci Technol* 30: 575–585.

Gladyshev VN, SV Khangulov, TC Stadtman. 1994. Nicotinic acid hydroxylase from *Clostridium barkeri*: Electron paramagnetic resonance studies show that selenium is coordinated with molybdenum in the catalytically active selenium-dependent enzyme. *Proc Natl Acad Sci USA* 91: 232–236.

Glück M, F Lingens. 1988. Heteroxanthine demethylase, a new enzyme in the degradation of caffeine by *Pseudomonas putida*. *Appl Microbiol Biotechnol* 28: 59–62.

Grossenbacher H, C Horn, AM Cook, R Hütter. 1984. 2-chloro-4-amino-1,3,5-triazine-6(5H)-one: A new intermediate in the biodegradation of chlorinated s-triazines. *Appl Environ Microbiol* 48: 451–453.

Guzmán, K, J Badia, R Giménez, J Aguilar, L Baldoma. 2011. Transcriptional regulation of the gene cluster encoding allantoinase and guanine deaminase in *Klebsiella pneumoniae*. *J Bacteriol* 193: 2197–2207.

Harkness DR, L Tsai, ER Stadtman. 1964. Bacterial degradation of riboflavin V. Stoichiometry of riboflavin degradation to oxamide and other products, oxidation of C14-labeled intermediates and isolation of the pseudomonad effecting these transformations. *Arch Biochem Biophys* 108: 323–333.

Hayaishi O, A Kornberg. 1952. Metabolism of cytosine, thymidine, uracil and barbituric acid by bacterial enzymes. *J Biol Chem* 197:717–732.

Hetterich D, B Peschke, B Tshisuaka, F Lingens. 1991. Microbial metabolism of quinoline and related compounds. X. The molybdopterin cofactors of quinoline oxidoreductases from Pseudomonas putida 86 and *Rhodococcus* spec. B1 and of xanthine dehydrogenase from *Pseudomonas putida* 86. *Biol Chem Hoppe-Seyler* 372: 513–517.

Hidese R, H Mihara, T Kurihara, N Esaki. 2011. *Escherichia coli* dihydropyrimidine dehydrogenase is a novel NAD-dependent heterotetramer essential for the production of 5,6-dihydrouracil. *J Bacteriol* 193: 989–993.

Hirschberg R, JC Ensign. 1971a. Oxidation of nicotinic acid by a *Bacillus* species: source of oxygen atoms for hydroxylation of nicotinic acid and 6-hydroxynicotinic acid. *J Bacteriol* 108: 757–759.

Hirschberg R, JC Ensign. 1971b. Oxidation of nicotinic acid by a *Bacillus* species: purification and properties of nicotinic acid and 6-hydroxynicotinic acid hydroxylases. *J Bacteriol* 108: 751–756.

Hoch SO'N, RD DeMoss. 1972. Tryptophanase from *Bacillus alvei*. 1. Subunit structure. *J Biol Chem* 247: 1750–1756.

Hochstein LI, BP Dalton. 1965. The hydroxylation of nicotine: The origin of the hydroxyl oxygen. *Biochem Biophys Res Commun* 21: 644–648.

Holmes PE, SC Rittenberg, H-J Knackmuss. 1972. The bacterial oxidation of nicotine. VIII. Synthesis of 2,3,6-trihydroxypyridine and accumulation and partial characterization of the product of 2,6-dihydroxy-pyridine oxidation. *J Biol Chem* 247: 7628–7633.

Honeyfield DC, JR Carlson. 1990. Effect of indoleacetic acid and related indoles on *Lactobacillus* sp. strain 11201 growth, indoleacetic acid catabolism and 3-methylindole formation. *Appl Environ Microbiol* 56: 1373–1377.

Hormann K, JR Andreesen. 1991. A flavin-dependent oxygenase reaction initiates the degradation of pyrrole-2-carboxylate in *Arthrobacter* strain Py1 (DSM 6386). *Arch Microbiol* 157: 43–48.

Houghton C, RB Cain. 1972. Microbial metabolism of the pyridine ring. Formation of pyridinediols (dihydroxypyridines) as intermediates in the degradation of pyridine compounds by microorganisms. *Biochem J* 130: 879–893.

Howden AJM, A Rico, T Mantlaki, K Miguet, GM Preston. 2009. *Pseudomonas syingae* pv. syringae B728a hydrolyses indole-3-acetonitrile to the plant hormone indole-3-acetic acid. *Mol Plant Pathol* 10: 857–865.

Hund HK, A de Beyer, F Lingens. 1990. Microbial metabolism of quinoline and related compounds. VI. Degradation of quinaldine by *Arthrobacter* sp. *Biol Chem Hoppe-Seyler* 371: 1005–1008.

Hutcheson SW, T Kosuge. 1985. Regulation of 3-indolacetic acid production in *Pseudomonas savastanoi*. Purification and properties of trypyptophan 2-monooxygenase. *J Biol Chem* 260: 6281–6287.

Imhoff D, JR Andreesen. 1979. Nicotinic acid hydroxylase from *Clostridium barkeri*: Selenium-dependent formation of active enzyme. *FEMS Microbiol Lett* 5: 155–158.

Imhoff-Stuckle D, N Pfennig. 1983. Isolation and characterization of a nicotinic acid-degrading sulfate-reducing bacterium, *Desulfococcus niacinii* sp. nov. *Arch Microbiol* 136: 194–198.

Inoue K, H Habe, H Yamane, H Nojiri. 2006. Characterization of novel carbazole catabolism genes from Gram-positive carbazole degrader *Nocardioides aromaticivorans* IC177. *Appl Environ Microbiol* 72: 3321–3329.

Jensen JB, H Egsgaard, H van Onckelen, BU Jochimsen. 1995. Catabolism of indole-3-acetic acid and 4- and 5-chloroindole-3-acetic acid in *Bradyrhizobium japonicum*. *J Bacteriol* 177: 5762–5766.

Johansen SS, D Licht, E Arvin, H Mosbaek, AB Hansen. 1996. Metabolic pathways of quinoline, indole and their methylated analogues by *Desulfobacterium indolicum* (DSM 3383). *Appl Microbiol Biotechnol* 47: 292–300.

Jung D-K, Y Lee, SG Park, BC Park, G-H Kim, S Rhee. 2006. Structural and functional analysis of PucM, a hydrolase in the ureide pathway and a member of the transthyretin-related protein family. *Proc Natl Acad Sci USA* 103: 9790–9795.

Jutzi K, AM Cook, R Hütter. 1982. The degradative pathway of the s-triazine melamine. *Biochem J* 208: 679–684.

Kamath AV, C Yanofsky. 1992. Characterization of the tryptophanase operon in *Proteus vulgaris*. *J Biol Chem* 267: 19978–19985.

Kaminskas E, Y Kimhi, B Maganasik. 1970. Urocanase and N-formimino-L-glutamate formiminohydrolase of *Bacillus subtilis*, two enzymes of the histidine degradation pathway. *J Biol Chem* 245: 3536–3544.

Kawasaki K, A Yokota, S Oita, C Kobayashi, S Yoshikawa, S Kawamoto, S Takao, F Tomita. 1993. Cloning and characterization of a tryptophanse gene from *Enterobacter aerogenes* SM-18. *J Gen Microbiol* 139: 3275–3281.

Kim K-S, JG Pelton, WB Inwood, U Andersen, S Kustu, DE Wemmer. 2010. The Rut pathway for pyrimidine degradation: novel chemistry and toxicity problems. *J Bacteriol* 192: 4089–4120.

Kitcher JO, PW Trudgill, JS Rees. 1972. Purification and properties of 2-furoyl-coenzyme A hydroxylase from *Pseudomonas putida* F2. *Biochem J* 130: 121–132.

Kitts CL, LE Schaechter, RS Rabin, RA Ludwig. 1989. Identification of cyclic intermediates in *Azorhizobium caulinodans* nicotinate metabolism. *J Bacteriol* 171: 3406–3411.

Knackmuss H-J, W Beckmann. 1973. The structure of nicotine blue from *Arthrobacter oxidans*. *Arch Microbiol* 90: 167–169.

Koenig K, JR Andreesen. 1990a. Molybdenum involvement in aerobic degradation of 2-furoic acid by *Pseudomonas putida* Fu1. *Appl Environ Microbiol* 55: 1829–1834.

Koenig K, JR Andreesen. 1990b. Xanthine dehydrogenase and 2-furoyl-coenzyme A dehydrogenase from *Pseudomonas putida* Fu1: Two molybdenum-containing enzymes of novel structural composition. *J Bacteriol* 172: 5999–6009.

Kretzer A, JR Andreesen. 1991. A new pathway for isonicotinate degradation by *Mycobacterium* sp. INA1. *J Gen Microbiol* 137: 1073–1080.

Kung H-F, TC Stadtman. 1971. Nicotinic acid metabolism. VI. Purification and properties of α-methyleneglutarate mutase (B12-dependent) and methylitaconate isomerase. *J Biol Chem* 246: 3378–3388.

Kung H-F, L Tsai. 1971. Nicotinic acid metabolism. VII. Mechanisms of action of clostridial α-methyleneglutarate mutase (B12-dependent) and methylitaconate isomerase. *J Biol Chem* 246: 6436–6443.

Kung H-F, L Tsai, TC Stadtman. 1971. Nicotinic acid metabolism. VIII. Tracer studies on the intermediary roles of α-methyleneglutarate, methylitaconate, dimethymaleate and pyruvate. *J Biol Chem* 246: 6444–6451.

Lawrence JR, M Eldan, WC Sonzogni. 1993. Metribuzin and metabolites in Wisconsin (USA) well water. *Water Res* 27: 1263–1268.

Lee JL, S-K Rhee, S-T Lee. 2001. Degradation of 3-methylpyridine and 3-ethylpyridine by Gordonia nitida LE31. *Appl Environ Microbiol* 67: 4342–4345.

Lehmann M, B Tshisuaka, S Fetzner, P Röger, F Lingens. 1994. Purification and characterization of isoquinoline 1-oxidoreductase from *Pseudomonas diminuta* 7, a molybdenum-containing hydroxylase. *J Biol Chem* 269: 11254–11260.

Leveau JHL, S Gerards. 2008. Discovery of a bacterial gene cluster for catabolism of the plant hormone indole-3-acetic acid. *FEMS Microbiol Ecol* 65: 238–250.

Leveau JHL, SE Lindow. 2005. Utilization of the plant hormone indole-3-acetic acid for growth by *Pseudomonas putida* strain 1290. *Appl Environ Microbiol* 71: 2365–2371.

Lieberman I, A Kornberg. 1953. Enzymatic synthesis and breakdown of a pyrimidine, orotic acid I. Dihydroorotic dehydrogenase. *Biochim Biophys Acta* 12: 223–234.

Lieberman I, A Kornberg. 1954. Enzymatic synthesis and breakdown of a pyrimidine, orotic acid II. Dihydroorotic acid, ureidosuccinic acid, and 5-carboxymethylhydantoin. *J Biol Chem* 207: 911–924.

Lieberman I, A Kornberg. 1955. Enzymatic synthesis and breakdown of a pyrimidine, orotic acid. III. Ureidosuccinase. *J Biol Chem* 212: 909–920.

Liebert F. 1909. The decomposition of uric acid by bacteria. *Proc K Acad Ned Wetensch* 12: 54–64.

Lu Z-H, R Zhang, RB Diasio. 1992. Purification and characterization of dihydropyrimidine dehydrogenase from human liver. *J Biol Chem* 267: 17102–17109A.

Lübbe C, K-H van Pée, O Salcher, F Lingens. 1983. The metabolism of tryptophan and 7-chlorotryptophan in Pseudomonas pyrrocinia and *Pseudomonas aureofaciens*. *Z Physiol Chem* 364: 447–453.

Maki Y, S Yamamoto, M Nozaki, O Hayaishi. 1969. Studies on monooxygenases II. Crystallization and some properties of imidazole acetate monooxygenase. *J Biol Chem* 244: 2942–2950.

Malherbe P, C Köhler, M Da Prada, G Lang, V Kiefer, R Schwarcz, H-W Lahm, AM Cesura. 1994. Molecular cloning and functional expression of human 3-hydroxyanthranilic-acid dioxygenase. *J Biol Chem* 269: 13792–13979.

Marletta MA, K-F Chenng, C Walsh. 1982. Stereochemical studies on the hydration of monofluorofumarate and 2,3-difluorofumarate by fumarase. *Biochemistry* 21: 2637–2644.

Middelhoven WJ, MC Hoogkamer-Te Niet, NJ Kreger-Van Rij. 1984. *Trichosporon adeninovorans* sp. nov., a yeast species utilizing adenine, xanthine, uric acid, putrescine and primary n-alkylamines as the sole source of carbon, nitrogen and energy. *Antonie van Leeuwenhoek* 50: 369–378.

Morino Y, EE Snell. 1967. A kinetic study of the reaction mechanism of tryptophanase-catalyzed reactions. *J Biol Chem* 242: 2793–2799.

Mulbry WW. 1994. Purification and characterization of an inducible s-triazine hydrolase from *Rhodococcus corallinus* NRRL B-15444R. *Appl Environ Microbiol* 60: 613–618.

Muraki T, M Taki, Y Hasegawa, H Iwaki, PCK Lau. 2003. Prokaryotic homologues of the eukaryotic 3-hydroxyanthranilate 3,4-dioxygenase and 2-amino-3-carboxymuconate-6-semialdehyde decarboxylase in the 2-nitrobenzoate degradation pathway of *Pseudomonas fluorescens* strain KU-7. *Appl Environ Microbiol* 69: 1564–1572.

Nagel M, JR Andreesen. 1990. Purification and characterization of the molybdoenzymes nicotinate dehydrogenase and 6-hydroxynicotinate dehydrogenase from *Bacillus niacini*. *Arch Microbiol* 154: 605–613.

Nam J-W, H Nojiri, H Noguchi, H Uchimura, T Yoshida, H Habe, H Yamane, T Omori. 2002. Purification and characterization of carbazole 1,9a-dioxygenase, a three-component dioxygenase system of *Pseudomonas resinovorans* strain CA10. *Appl Environ Microbiol* 68: 5882–5890.

Nam JW, H Noguchi, Z Fujimoto, H Mizuno, Y Ashikawa, M Abo, S Fushinobu. et al. 2005. Crystal structure of the ferredoxin component of carbazole 1,9a-dioxygenase of *Pseudomonas resinovorans* strain CA10, a novel Rieske non-heme iron oxygenase system. *Proteins* 58: 779–789.

Narumiya S, K Takai, T Tokuyama, Y Noda, H Ushiro, O Hayaishi. 1979. A new metabolic pathway of tryptophan initiated by tryptophan side chain oxidase. *J Biol Chem* 254: 7007–7015.

Newton WA, Y Morino, EE Snell. 1965. Properties of crystalline tryptophanase. *J Biol Chem* 240: 1211–1218.

Nojiri H, H Sekiguchi, K Maeda, M Urata, S-I Nakai, T Yoshida, H Habe, T Omori. 2001. Genetic characterization and evolutionary implications of a car gene cluster in the carbazole degrader *Pseudomonas* sp. strain CA10. *J Bacteriol* 183: 3663–3679.

Nojiri H, J-W Nam, M Kosaka, K-I Morii, T Takemura, K Furihata, H Yamane, T Omori. 1999. Diverse oxygenations catalyzed by carbazole 1,9a-dioxygenase from *Pseudomonas* sp. strain CA 10. *J Bacteriol* 181: 3105–3113.

Nyns EJ, D Zach, EE Snell. 1969. The bacterial oxidation of vitamin B6. VIII. Enzymatic breakdown of α-(N-acetylaminomethylene)succinic acid. *J Biol Chem* 244: 2601–2605.

Ogawa J, S Shimizu. 1995. Purification and characterization of dihydroorotase from *Pseudomonas putida*. *Arch Microbiol* 164: 353–357.

Olesen MR, BU Jochimsen. 1996. Identification of enzymes involved in indole-3-acetic acid degradation. *Plant Soil* 186: 143–149.

Orpin CG, M Knight, WC Evans. 1972. The bacterial oxidation of N-methylisonicotinate, a photolytic product of paraquat. *Biochem J* 127: 833–844.

Padden AN, FA Rainey, DP Kelly, AP Wood. 1997. *Xanthobacter tagetidis* sp. nov., an organism associated with Tagetes species and able to grow on substituted thiophenes. *Int J Syst Bacteriol* 47: 394–401.

Parisot D, MC Malet-Martino, R Martino, P Crasnier. 1991. 19F nuclear magnetic resonance analysis of 5-fluorouracil metabolism in four differently pigmented strains of *Nectria haematococca*. *Appl Environ Microbiol* 57: 3605–3612.

Parshikov IA, JP Freeman, JO Lay, RD Beger, AJ Williams, JB Sutherland. 2000. Microbiological transformation of enrofloxacin by the fungus Mucor ramannianus. *Appl Environ Microbiol* 66: 2664–2667.

Peschke B, F Lingens. 1991. Microbial metabolism of quinoline and related compounds XII. Isolation and characterization of the quinoline oxidoreductase from *Rhodococcus* spec. B1 compared with the quinoline oxidoreductase from *Pseudomonas putida* 86. *Biol Chem Hoppe-Seyler* 370: 1081–1088.

Pope SD, L-L Chen, V Stewart. 2009. Purine utilization by *Klebsiella oxytoca* M5a1: Genes for ring-oxidizing and –opening enzymes. *J Bacteriol* 191: 1006–1017.

Porter DJT. 1994. Dehalogenating and NADPH-modifying activities of dihydropyrimidine dehydrogenase. *J Biol Chem* 269: 24177–24182.

Prinsen E, A Costacurta, K Michiels, J Vanderleyden, H Van Onckelen. 1993. Azospirillum brasilense indole-3-acetic acid biosynthesis: evidence for a non-tryptophan dependent pathway. *Mol Plant-Microbe Interact* 6: 609–615.

Qu Y, JC Spain. 2011. Catabolic pathway for 2-nitromidazole involves a novel nitrohydrolase that also confers drug resistance. *Environ Microbiol* 13: 1010–1017.

Ralph EC, MA Anderson, WW Cleland, PF Fitzpatrick. 2006. Mechanistic studies of the flavoenzyme tryptophan 2-monooxygenase: deuterium and 15N kinetic isotope effects on alanine oxidation by an L-amino acid oxidase. *Biochemistry* 45: 15844–15852.

Rappert S, KC Botsch, S Nagorny, W Francke, R Müller. 2006. Degradation of 2,3-diethyl-5-methylpyrazine by a newly discovered bacterium, *Mycobacterium* sp. Strain DM-11. *Appl Environ Microbiol* 72: 1437–1444.

Resnick SM, DS Torok, DT Gibson. 1993. Oxidation of carbazole to 3-hydroxycarbazole by naphthalene 1,2-dioxygenase and biphenyl 2,3-dioxygenase. *FEMS Microbiol Lett* 113: 297–302.

Rhee S-K, GM Lee, J-H Yoon, Y-H Park, H-S Bae, S-T Lee. 1997. Anaerobic and aerobic degradation of pyridine by a newly isolated denitrifying bacterium. *Appl Environ Microbiol* 63: 2578–2585.

Röger P, A Erben, F Lingens. 1990. Microbial metabolism of quinoline and related compounds IV. Degradation of isoquinoline by Alcaligenes faecalis Pa and Pseudomonas diminuta 7. *Biol Chem Hoppe-Seyler* 370: 1183–1189.

Röger P, G Bär, F Lingens. 1995. Two novel metabolites in the degradation pathway of isoquinoline by Pseudomonas diminuta 7. *FEMS Microbiol Lett* 129: 281–286.

Rosche B, B Shisuaka, S Getzner, F Lingens. 1995. 2-oxo-1,2-dihydroquinoline 8-monooxygenase, a two-component enzyme system from *Pseudomonas putida* 86. *J Biol Chem* 270: 17836–17842.

Sachelaru P, E Schiltz, GL Igloi, R Brandsch. 2005. An α/β-fold C–C bond hydrolase is involved in a central step of nicotine metabolism by *Arthrobacter nicotinovorans*. *J Bacteriol* 187: 8516–8519.

Sato S-I, J-W Nam, K Kasuga, H Nojiri, H Yamane, T Omori. 1997b. Identification and characterization of genes encoding carbazole 1,9a-dioxygenase in *Pseudomonas* sp. strain CA10. *J Bacteriol* 179: 4850–4858.

Sato, S-I, N Ouchiyama, T Kimura, H Nojiri, H Yamane, T Omori. 1997a. Cloning of gees involved in carbazole degradation of *Pseudomonas* sp. strain CA10: nucleotide sequences of genes and characterization of meta cleavage enzymes and hydrolase. *J Bacteriol* 179: 4841–4849.

Schach S, B Tshisuaka, S Fetzner, F Lingens. 1995. Quinoline 2-oxidoreductase and 2-oxo-1,2-dihydroquinoline 5,6-dioxygenase from *Comamonas testosteroni* 63. The first two enzymes in quinoline and 3-methylquinoline degradation. *Eur J Biochem* 232: 536–544.

Schenk S, A Hoelz, B Krauss, K Decker. 1998. Gene structures and properties of enzymes of the plasmid-encoded nicotine catabolism of *Arthrobacter nicotinovorans*. *J Mol Biol* 284: 1323–1339.

Schiefer-Ullrich H, JR Andreesen. 1985. *Peptostreptoccus barnesae* sp. nov, a Gram-positive, anaerobic, obligately purine utilizing coccus from chicken feces. *Arch Microbiol* 143: 26–31.

Schieffer-Ullrich H, R Wagner, P Dürre, JR Andreesen. 1984. Comparative studies on physiology and taxonomy of obligately purinolytic clostridia. *Arch Microbiol* 138: 345–353.

Schmidt M, P Röger, F Lingens. 1991. Microbial metabolism of quinoline and related compounds XI. Degradation of quinoline-4-carboxylic acid by Microbacterium sp. H2, Agrobacterium sp. 1B and Pimelobacter simplex 4B and 5B. *Biol Chem Hoppe-Seyler* 370: 1015–1020.

Schultz AC, P Nygaard, HH Saxild. 2001. Functional analysis of 14 genes that constitute the purine catabolic pathway in *Bacillus subtilis* and evidence for a novel regulon controlled by the PucR transcription activator. *J Bacteriol* 183: 3293–3302.

Schütz A, T Sandalova, S Ricagno, G Hübner, S König, G Schneider. 2003. Crystal structure of thiaminediphosphate-dependent indolpyruvate decarboxylase from *Enterobacter cloacae*, an enzyme involved in the biosynthesis of the plant hormone indol-3-acetic acid. *Eur J Biochem* 270: 2312–2321.

Schwarz G, R Bauder, M Speer, TO Rommel, F Lingens. 1989. Microbial metabolism of quinoline and related compounds II. Degradation of quinoline by *Pseudomonas fluorescens* 3, *Pseudomonas putida* 86 and *Rhodococcus* spec. B1. *Biol Chem Hoppe-Seyler* 370: 1183–1189.

Schwede TF, J Rétey, GE Schulz. 1999. Crystal structure of histidine ammonia-lyase revealed a novel polypeptide modification as the catalytic electrophile. *Biochemistry* 38: 5355–5361.

Seffernick JL, N Shapir, M Schoeb, G Johnson, MJ Sadowsky, LP Wackett. 2002. Enzymatic degradation of chlorodiamino-s-triazine. *Appl Environ Microbiol* 68: 4672–4675.

Self WT. 2002. Regulation of purine hydroxylase and xanthine dehydrogenase from Clostridium purinolyticum in response to purines, selenium and molybdenum. *J Bacteriol* 184: 2039–2044.

Self WT, TC Stadman. 2000. Selenium-dependent metabolism of purines: a selenium-dependent purine hydroxylase and xanthine dehydrogenase were purified from *Clostridium purinolyticum* and characterized. *Proc Natl Acad Sci USA* 97: 7208–7213.

Shapir N, MJ Sadowsky, LP Wackett. 2005. Purification and characterization of allophanate hydrolase/AtzF) from Pseudomonas sp. strain ADP. *J Bacteriol* 187: 3731–3738.

Siegmund I, K Koenig, JR Andreesen. 1990. Molybdenum involvement in aerobic degradation of picolinic acid by *Arthrobacter picolinophilus*. *FEMS Microbiol Lett* 67: 281–284.

Singh BK, A Walker, JAW Morgan, DJ Wright. 2004. Biodegradation of chloropyrifos by *Enterobacter* strain B-14 and its use in bioremediation of contaminated soils. *Appl Environ Microbiol* 70: 4855–4863.

Soong C-L, J Ogawa, E Sakuradani, S Shimizu. 2002. Barbiturase, a novel zinc-containing amidohydrolase involved in oxidative pyrimidine metabolism. *J Biol Chem* 277: 7051–7058.

Sparrow LG, PPK Ho, TK Sundaram, D Zach, EJ Nyns, EE Snell. 1969. The bacterial oxidation of vitamin B6 VII. Purification, properties, and mechanism of action of an oxygenase which cleaves the 3-hydroxypyridine ring. *J Biol Chem* 244: 2590–2660.

Spaepen S, W Versées, D Gocke, M Pohl, J Steyaert, J Vanderleyden. 2007. Characterization of phenylpyruvate decarboxylase, involved in auxin production of *Azospirillum brasilense*. *J Bacteriol* 189: 7626–7633.

Steiner RA, HJ Janssen, P Roversi, AJ Oakley, S Fetzner. 2010. Structural basis for cofactor-independent dioxygenation of N-heteroaromatic compounds at the α/β-hydrolase fold. *Proc Natl Acad Sci USA* 107: 657–662.

Stephan I, B Tshisuaka, S Fetzner, F Lingens. 1996. Quinaldine 4-oxidase from *Arthrobacter* sp. Rü61a, a versatile procaryotic molybdenum-containing hydroxylase active towards N-containing heterocyclic compounds and aromatic aldehydes. *Eur J Biochem* 236: 155–162.

Strong LC, C Rosendahl, G Johnson, MJ Sadowsky, LP Wackett. 2002. *Arthrobacter aurescens* TC1 metabolizes diverse s-triazine ring compounds. *Appl Environ Microbiol* 68: 5973–5980.

Stubbe J. 1985. Identification of two α-ketoglutarate-dependent dioxygenases in extracts of *Rhodotolula glutinis* catalyzing deoxyuridine hydroxylation. *J Biol Chem* 260: 9972–9975.

Summers RM, TM Louie, CL Yu, M Subramanian. 2011. Characterization of a broad-specificity non-heme iron N-demethylase from *Pseudomonas putida* CBB5 capable of utilizing several purine alkaloids as sole carbon and nitrogen source. *Microbiology UK* 147: 583–592.

Sundaram TK, EE Snell. 1969. The bacterial oxidation of vitamin B6. V. The enzymatic formation of pyridoxal and isopyridoxal from pyridoxine. *J Biol Chem* 244: 2577–2584.

Tang H, S Wang, L Ma, X Meng, Z Deng, D Zhang, C Ma, P Xu. 2008. A novel gene encoding 6-hydroxy-3-succinoylpyridine hydroxylase, involved in nicotine degradation by *Pseudomonas putida* strain S16. *Appl Environ Microbiol* 74: 1567–1574.

Tang H, S Wang, X Meng, L. Ma, S Wang, X He, G Wu, P Xu. 2009. Novel oxidoreductase-encoding gene involved in nicotine degradation by *Pseudomonas putida* strain S16. *Appl Environ Microbiol* 75: 772–778.

Taniuchi H, O Hayaishi. 1963. Studies on the metabolism of kynurenic acid. III. Enzymatic formation of 7,8-dihydroxykynurenic acid from kynurenic acid. *J Biol Chem* 238: 283–293.

Tate RL, JC Ensign. 1974. Picolinic acid hydroxylase of *Arthrobacter picolinophilus*. *Can J Microbiol* 20: 695–702.

Taylor BF, JA Amador. 1988. Metabolism of pyridine compounds by phthalate-degrading bacteria. *Appl Environ Microbiol* 54: 2342–2344.

Tibbles PE, R Müller, F Lingens. 1989a. Degradation of 5-chloro-2-hydroxynicotinic acid by *Mycobacterium* sp. BA. *Biol Chem Hoppe-Seyler* 370: 601–606.

Tibbles PE, R Müller, F Lingens. 1989b. Microbial metabolism of quinoline and related compounds. III. Degradation of 3-chloroquinoline-8-carboxylic acid by Pseudomonas spec. EKIII. *Biol Chem Hoppe-Seyler* 370: 1191–1196.

Tinschert A, A Kiener, K Heinzmann, and A Tschech. 1997. Isolation of new 6-methylnicotinic-acid-degrading bacteria, one of which catalyses the regioselective hydroxylation of nicotinic acid at position C2. *Arch Microbiol* 168: 355–361.

Trudgill PW. 1969. The metabolism of 2-furoic acid by Pseudomonas F2. *Biochem J* 113: 577–587.

Urata M, H Uchimura, H Noguchi, T Sakaguchi, T Takemura, K Eto, H Habe, T Omori, H Yamane, H Nojiri. 2006. Plasmid pCAR3 contains multiple gene sets involved in the conversion of carbazole to anthranilate. *Appl Environ Microbiol* 72: 3198–3205.

Vannelli T, AB Hooper. 1992. Oxidation of nitrapyrin to 6-chloropicolinic acid by the ammonia-oxidizing bacterium *Nitrosomonas europaea*. *Appl Environ Microbiol* 58: 2321–2325.

Vederas JC, E Schleicher, M-D Tsai, HG Floss. 1978. Stereochemistry and mechanism of reactions catalyzed by tryptophanase from *Escherichia coli*. *J Biol Chem* 253: 5350–5354.

Wagner R, R Cammack, JR Andreesen. 1984. Purification and characterization of xanthine dehydrogenase from *Clostridium acidiurici* grown in the presence of selenium. *Biochim Biophys Acta* 791: 63–74.

Wataya Y, DV Santi. 1975. Thymidylate synthetase catalyzed dehalogenation of 5-bromo- and 5-iodo-2′-deoxyuridylate. *Biochem Biophys Res Commun* 67: 818–823.

Watson K, RB Cain. 1975. Microbial metabolism of the pyridine ring. Metabolic pathways of pyridine biodegradation by soil bacteria. *Biochem J* 146: 157–172.

Watson GK, C Houghton, RB Cain. 1974. Microbial metabolism of the pyridine ring. The hydroxylation of 4-hydroxypyridine to pyridine-3,4-diol (3,4-dihydroxypyridine) by 4-hydroxypyridine-3-hydroxylase. *Biochem J* 140: 265–276.

Wetzstein H-G, N Schmeer, W Karl. 1997. Degradation of the fluoroquinolone enrofloxacin by the brown-rot fungus *Gloeophyllum striatum*: Identification of metabolites. *Appl Environ Microbiol* 63: 4272–4281.

Wetzstein H-G, M Stadler, H-V Tichy, A Dalhoff, W Karl. 1999. Degradation of ciprofloxacin by basidiomycetes and identification of metabolites generated by the brown rot fungus *Gloeophyllum striatum*. *Appl Environ Microbiol* 65: 1556–1563.

Whitehead TR, NP Price, HL Draka, MA Cotta. 2008. Catabolic pathway for the production of skatole and indoleacetic acid by the acetogen *Clostridium drakei*, *Clostridium scatologenes*, and swine manure. *Appl Environ Microbiol* 74: 1950–1953.

Willumsen PA, JK Nielsen, U Karlson. 2001. Degradation of phenanthrene-analogue azaarenes by *Mycobacterium gilvum*. *Appl Microbiol Biotechnol* 56: 539–544.

Wondrack LM, C-A Hsu, MT Abbott. 1978. Thymine-7-hydroxylase and pyrimidine deoxyribonucleoside 2′-hydroxylase activities in *Rhodotorula glutinis*. *J Biol Chem* 253: 6511–6515.

Woolfolk CA. 1975. Metabolism of N-methylpurines by a Pseudomonas putida strain isolated by enrichment on caffeine as the sole source of carbon and nitrogen. *J Bacteriol* 123: 1088–1106.

Woolfolk CA, BS Woolfolk, HR Whiteley. 1970. 2-oxypurine dehydrogenase from *Micrococcus aerogenes*. I. Isolation, specificity, and some chemical and physical properties. *J Biol Chem* 245: 3167–3178.

Wright KA, RB Cain. 1972. Microbial metabolism of pyridinium compounds. Metabolism of 4-carboxy-1-methylpyridinium chloride, a photolytic product of paraquat. *Biochem J* 128: 543–559.

Xi H, BL Schneider, L Reitzer. 2000. Purine catabolism in *Escherichia coli* and function of xanthine dehydrogenase in purine salvage. *J Bacteriol* 182: 5332–5341.

Xu G, TP West. 1992. Reductive catabolism of pyrimidine bases by *Pseudomonas steutzeri*. *J Gen Microbiol* 138: 2459–2463.

Xu G, TP West. 1994. Characterization of dihydropyrimidase from *Pseudomonas steutzeri*. *Arch Microbiol* 161: 70–74.

Yu CL, Y Kale, S Gopishetty, TM Louie, M Subramanian. 2008. A novel caffeine dehydrogenase in *Pseudomonas* sp. strain CBB1 oxidizes caffeine to trimethyluric acid. *J Bacteriol* 190: 772–776.

Yu CL, TM Louie, R Summers, Y Kale, S Gopishetty, M Subramanian. 2009. Two distinct pathways for metabolism of theophylline and caffeine are coexpressed in *Pseudomonas putida* CBB5. *J Bacteriol* 191: 4624–4632.

Part 2: Oxaarenes

Naturally occurring oxaarenes based on polycyclic pyrans encompass a plethora of structures including the plant polyphenols such as anthocyanins and α-tocopherol (vitamin E). Halogenated dibenzo[1,4]dioxins (dibenzo-*p*-dioxins) and dibenzofurans are formed both as by-products during the manufacture of chlorophenols, and from the incineration of organic matter in the presence of inorganic halides.

10.8 Aerobic Conditions

10.8.1 Monocyclic Oxaarenes

The aerobic degradation of furan-2-carboxylate has already been summarized in Part 1 of this chapter and emphasis is placed here on anaerobic pathways.

The degradation of furan-2-carboxaldehyde (furfural) has been examined in *Desulfovibrio furfuralis* and is initiated by dehydrogenation to furan-2-carboxylate followed by hydroxylation. This is hydrolyzed to 2-oxoglutarate and decarboxylated to succinic acid semialdehyde that is degraded to acetate (Figure 10.35a) (Folkerts et al. 1989). Furan-3-carboxylate is degraded by *Paracoccus denitrificans* strain MK33 under denitrifying conditions by the formation of the CoA ester and reduction at $\Delta^{4,5}$ before

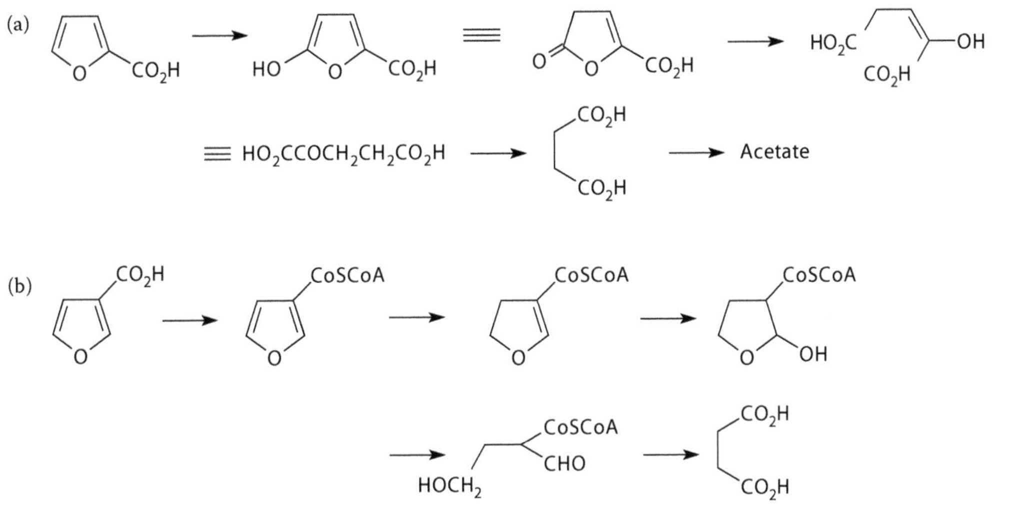

FIGURE 10.35 Anaerobic degradation of (a) furan-2-carboxylate, (b) furan-3-carboxylate.

hydration (Figure 10.35b) (Koenig and Andreesen 1991). This strain is also able to degrade furan-3-carboxylate by an analogous pathway.

10.8.2 Polycyclic Oxaarenes

10.8.2.1 Monooxygenation

Monooxygenation mediated by the cytochrome P450 system of *Streptomyces* has been observed in oxaarenes:

1. 2,2-Dimethyl-6,7-dimethoxychromene was transformed to an epoxide, which then underwent further reactions (Figure 10.36) (Sariaslani et al. 1987).
2. 7-Ethoxycoumarin was transformed initially to the 7-hydroxycompound, which was subsequently hydroxylated and partially *O*-methylated (Sariaslani et al. 1989).

10.8.2.2 Dioxygenation

It is appropriate to note briefly the degradation of the analogous halogenated diphenyl ethers, which have been used as agrochemicals and flame retardants. In *P. cepacia*, the degradation of diphenyl ether was initiated by dioxygenation and produced pyr-2-one-6-carboxylate as the end product (Figure 10.37) (Pfeifer et al. 1993). A strain of *Sphingomonas* strain SS33 was able to degrade diphenyl ether and grow

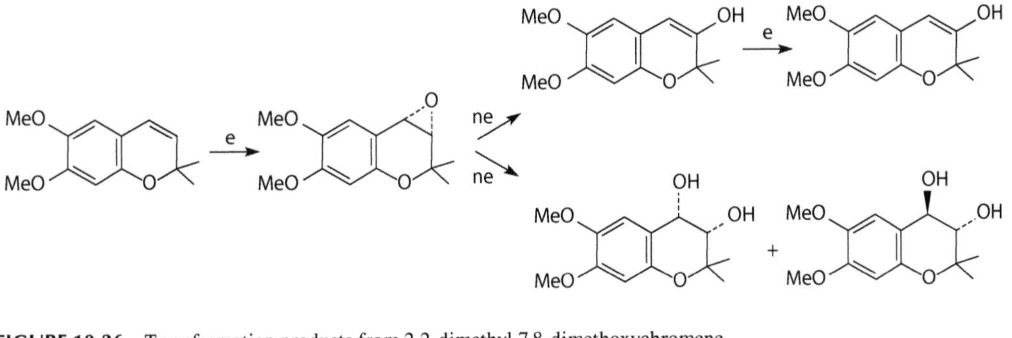

FIGURE 10.36 Transformation products from 2,2-dimethyl-7,8-dimethoxychromene.

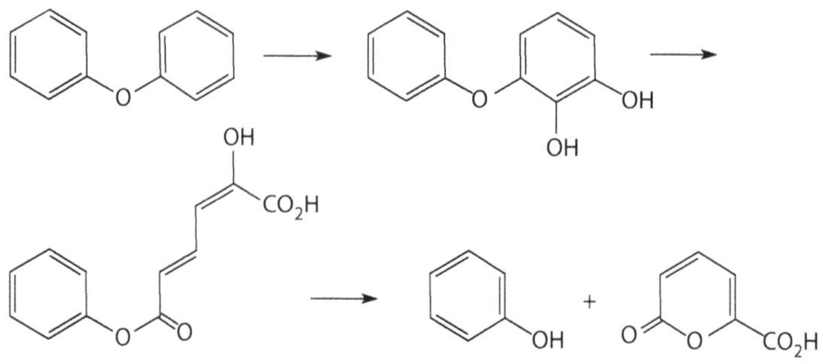

FIGURE 10.37 Biodegradation of diphenyl ether.

with 4-fluoro- and 4-chlorodiphenyl ether by dioxygenation to form 4-halogenated catechol and phenol (Schmidt et al. 1992).

There are two variants for dioxygenation of polycyclic oxaarenes. Dioxygenation of the benzenoid ring occurs most commonly in transformations, rather than in degradations that proceed by angular dioxygenation at the ring junction in a range of polycyclic oxaarenes. For example, benzofuran is oxygenated by *P. putida* strain UV4 to *cis*-(7S,6S)-benzofuran-dihydrodiol (the configuration at C_7 is determined from the Sequence Rule that gives priority to 7S) (Boyd et al. 1992), dibenzofuran by *Beijerinckia* sp. strain B8/36 to the *cis*-2,3-dihydrodiol (Cerniglia et al. 1979), and dibenzo[1,4]dioxin by *Pseudomonas* sp. strain NCIB 9816 to the *cis*-1,2-dihydrodiol, which was dehydrogenated to 1,2-dihydroxydibenzo[1,4] dioxin by naphthalene-grown cells (Klecka and Gibson 1979).

The dioxygenases that carry out angular dioxygenation have been characterized in a number of degradations, which include not only dibenzofuran, dibenzo[1,4]dioxin, and xanthone, but also carbazole. All of them involve angular dioxygenation between the heteroatom and the benzene ring. The properties of the three-component dioxygenases have been summarized (Armengaud et al. 1998), and the three-component carbazole 1,9a dioxygenase from *P. resinovorans* strain CA10 has been described (Nam et al. 2002). It has a relaxed specificity that includes dioxygenation not only of these heterocyclic compounds but also of the carbocyclic naphthalene, biphenyl, anthracene, and fluoranthene (Nojiri et al. 1999). The 3,4-dihydroxyxanthone dioxygenase (Chen and Tomasek 1991) that is used for the degradation of xanthone by a strain of *Arthrobacter* sp. (Figure 10.38) (Chen et al. 1986; Tomasek and Crawford 1986; Chen and Tomasek 1991) and the dioxygenase that carries out angular dioxygenation of dibenzofuran by a strain of *Sphingomonas* sp. have been characterized (Bünz and Cook 1993). A summary of the reactions involved in the biodegradation of dibenzo[1,4]dioxins, dibenzofurans, diphenyl ethers, and fluoren-9-one has been given (Wittich 1998). The degradation of 3-hydroxybenzofuran and 2-hydroxydibenzo[1,4] dioxin has been examined in *Sphingomonas* sp. strain RW1 and also involves angular dioxygenation. For both substrates, 1,2,4-trihydroxybenzene is produced from 3-hydroxybenzofuran by oxidative decarboxylation of 4-hydroxysalicylate (Armengaud et al. 1999).

10.8.2.3 Xanthone

The degradation of xanthone by *Arthrobacter* sp. strain GFB100 closely resembled the pathways used for carbocyclic compounds. In a yeast-supplemented medium, initial dioxygenation to the 3,4-dihydrodiol was followed by dehydrogenation to 3,4-dihydroxyxanthone and extradiol ring fission to a coumarin and subsequent production of 2,5-dihydroxybenzoate (Figure 10.38) (Tomasek and Crawford 1986; Chen and Tomasek 1991).

10.8.2.4 Dibenzofuran

The degradation of dibenzofuran by *Pseudomonas* sp. strain HH69 was initiated by dioxygenation with the formation of 2,2′,3-trihydroxybiphenyl that was further degraded to salicylate and several chroman-4-ones (Figure 10.39) (Fortnagel et al. 1990). The same strain (now designated as *Sphingomonas* sp.) is able

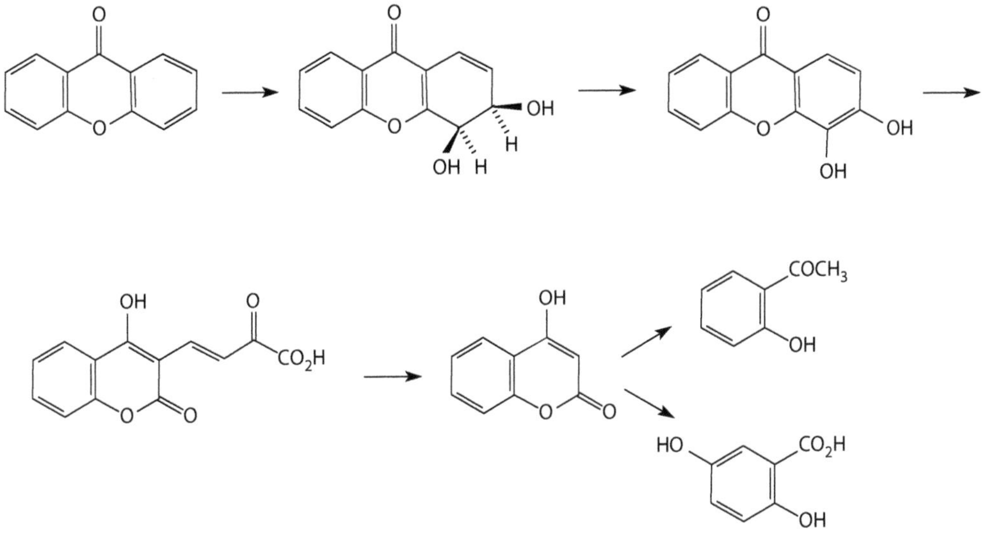

FIGURE 10.38 Biodegradation of xanthone.

to metabolize a range of substituted dibenzofurans (Harms et al. 1995). Other organisms that carry out the same initial reactions include *Brevibacterium* sp. strain DPO1361 (Strubel et al. 1991) and *Sphingomonas* sp. strain RW1 from which the 4,4a-dioxygenase has been isolated and characterized (Bünz and Cook 1993). Like the dioxygenases from carbocyclic compounds, the latter consists of four proteins, a reductase containing two isofunctional flavoproteins containing FAD that accepts electrons from NADH, a ferredoxin, and a terminal oxygenase containing a Rieske [2Fe–2S] center. It was suggested that the system

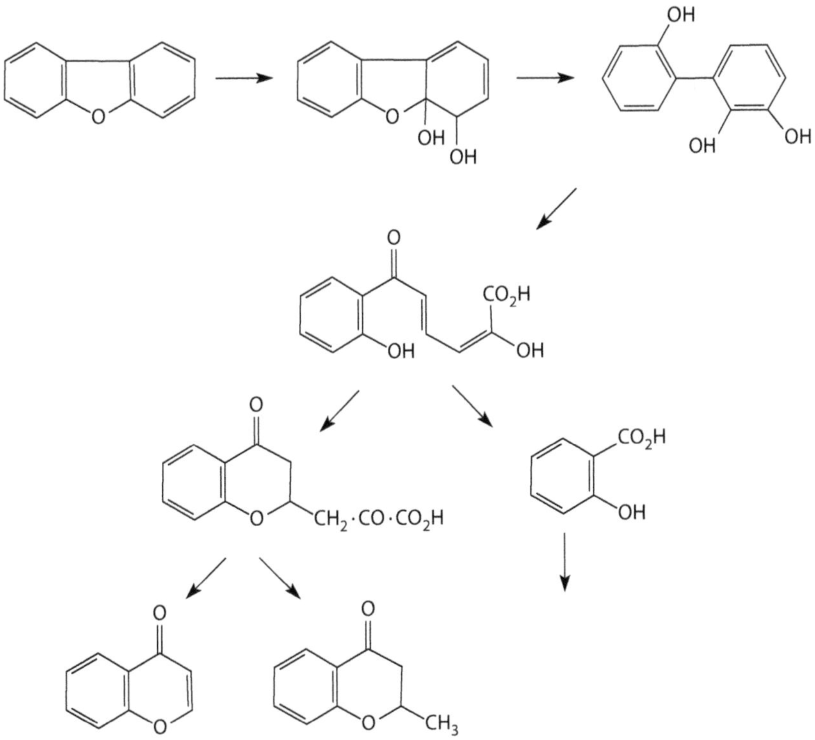

FIGURE 10.39 Pathways for degradation of dibenzofuran.

involves IIA dioxygenase. The dioxygenase genes in *Terrabacter* sp. strain YK3 are carried on a plasmid, and a phylogenetic analysis has compared the amino acid sequences of the gene products with those from a number of other dioxygenases (Iida et al. 2002). The key enzyme for the ring fission of 2,2′,3-trihydroxy-biphenyl has been biochemically and genetically analyzed in a strain of *Sphingomonas* sp. (Happe et al. 1993). Three extradiol dioxygenases are involved in the degradation of dibenzofuran by *Terrabacter* strain DPO360 (*Brevibacter* sp. strain DPO360)—two in ring fission of the intermediate 2,2′,3-trihydroxybiphenyl (BphC1 and BphC2) and a catechol 2,3-dioxygenase (Schmid et al. 1997).

10.8.2.5 Dibenzo[1,4]dioxin

The analogous 4,4a-dioxygenation of dibenzo[1,4]dioxin by *Sphingomonas* sp. strain RW1 produced 2,2′,3-trihydroxydiphenyl ether that was degraded by an extradiol dioxygenase (Figure 10.40) (Happe et al. 1993) to catechol that underwent both intra- and extradiol fission (Wittich et al. 1992). A strain of *Sphingomonas* sp. grown with diphenyl ether was able to oxidize dibenzo[1,4]dioxin, and the versatility of members of the genus has been associated with the scattering of the genes for the component proteins of the dioxygenase system around the genome of *Sphingomomas* sp. strain RW1, which can degrade a number of substituted dibenzo[1,4]dioxins (Armengaud et al. 1998) that are discussed in the next paragraph. Plausible pathways for the degradation of 3-hydroxybenzofuran and 2-hydroxybenzo[1,4]dioxin have been proposed (Armengaud et al. 1999).

10.8.2.6 Polyhalogenated Dibenzofurans, and Dibenzo[1,4]dioxins

10.8.2.6.1 Aerobic

Dioxygenation of chlorinated dibenzofurans and dibenzo[1,4] dioxins takes place at the angular positions analogously to those of the non-chlorinated analogues. *Sphingomonas* sp. strain RW1 produced 4,5-dichlorosalicylate from 2,3-dichlorodibenzofuran, and 6-chloro-2-methylchromenone from 2,8-dichlorodibenzofuran (Figure 10.41a) (Wilkes et al. 1996). The same strain transformed chlorinated dibenzo[1,4]dioxins to chlorocatechols with fission of the less highly chlorinated ring: 3,4,5-trichlorocatechol was produced from 1,2,3,-trichlorodibenzo[1,4]dioxin (Hong et al. 2002), and 3,4,5,6-tetrachlorocatechol from 1,2,3,4,7,8-hexachlorodibenzo[1,4]dioxin (Nam et al. 2006) (Figure 10.41b). The competitive angular dioxygenation at the two rings has been examined in a range of chlorinated dibenzo[1,4]dioxins (Habe et al. 2001).

10.8.2.6.2 Anaerobic

Chlorinated dioxins occur in atmospheric deposition (Koester and Hites 1992) and thereby enter the terrestrial environment and watercourses. In addition, they have been recovered from sediments contaminated with industrial discharge (Macdonald et al. 1992; Evers et al. 1993; Götz et al. 1993). Attention has

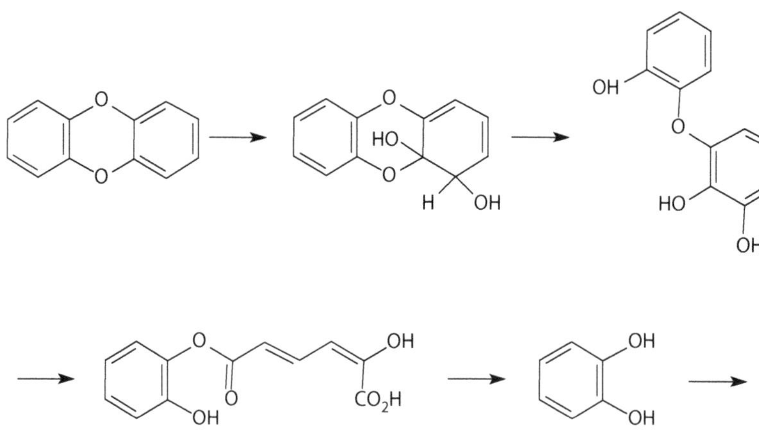

FIGURE 10.40 Degradation of 2,3-dichlorodibenzo[1,4]dioxin.

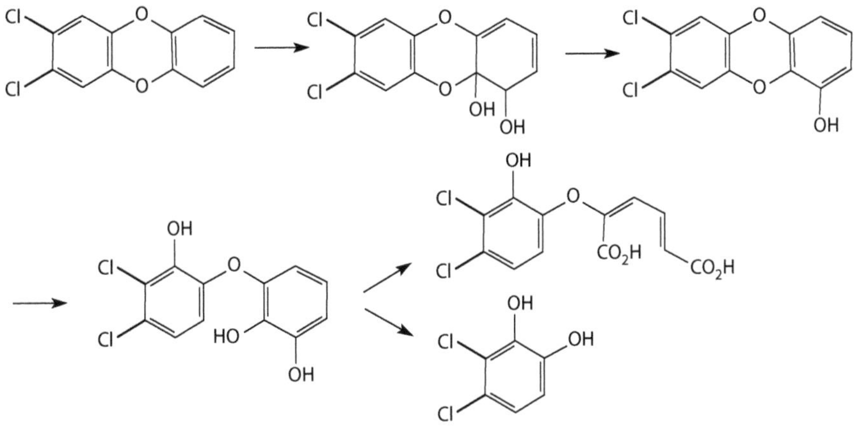

FIGURE 10.41 Degradation of 2,3-dichlorodibenzo[1,4]dioxin.

been, therefore, been directed to anaerobic dechlorination processes (Beurskens et al. 1995; Adriaens et al. 1995). Partial anaerobic dechlorination has been observed in sediment slurries: (a) the 1,2,3,4-tetrachloro compound produced predominantly the 1,3-dichloro compound (Beurskens et al. 1995; Ballerstedt et al. 1997), and (b) for compounds with 5–7 chlorine substituents, chlorine was removed from both the *peri* and the lateral positions (Barkovskii and Adriaens 1996). Dehalogenation has also been observed in microcosms where a combination of anaerobic and aerobic processes account for the total loss without the formation of intermediates (Yoshida et al. 2005), and it has been suggested that microbial dechlorination of cell-partitioned 2,3,7,8-tetrachlorodibenzo[1,4]dioxin in aged sediments was as effective as that in freshly spiked sediment (Barkovskii and Adriaens 1996). As for PCBs, the positive effect of brominated analogues in priming the anaerobic dechlorination of octachlorod-ibenzo[1,4]dioxin to the 2,3,7,8-congener in the presence of H_2 was induced by 2-bromodibenzo[1,4]dioxin (Albrecht et al. 1999). Dechlorination of chlorinated dibenzo[1,4]dioxins has been examined in *Dehaloccoides* sp. strain CBDB1, and successive dechlorinations were observed (Bunge et al. 2003):

1. 1,2,3,4-TetrachloroDBD → 1,2,3-TrichloroDBD → 2,3-DichloroDBD → 2-MonochloroDBD
2. 1,2,3,7,8-PentachloroDBD → 2,3,7,8-TetrachloroDBD → 2,3,7-TrichloroDBD → 2,7-BDB or 2,8-DichloroDBD.

It is worth noting that polybrominated dibenzofurans and dibenzo[1,4]dioxins can be produced by thermal reactions from polybrominated diphenyl ethers (Buser 1986) and other flame retardants (Thoma et al. 1986).

10.9 Fungal Reactions

The imperfect fungus *Paecilomyces liliacinus* is able to produce successively mon-, di-, and trihydroxyl-ated metabolites from biphenyl, and carry out *intradiol* ring fission of rings carrying adjacent hydroxyl groups (Gesell et al. 2001). Analogously, mono- and dihydroxylated metabolites are produced from dibenzofuran followed by *intradiol* fission of the ring with adjacent hydroxyl groups (Gesell et al. 2004). Fungal peroxidases that have already been noted in the context of degrading 2,4-dichlorophenol (Valli and Gold 1991) and 2,4-nitrotoluene (Valli et al. 1992b) are able to degrade 2,7-dichlorodibenzo[1,4]dioxin by a pathway completely different from those used by bacteria (Valli et al. 1992b) (Figure 10.42). The transformation of tetrachloro- to octachlorodibenzo[1,4]dioxins has been examined in low-nitrogen medium by *Phanerochaete sordida* YK-624 (Takada et al. 1996). All the compounds were extensively degraded, and the ring-fission of 2,3,7,8-tetrachloro-dibenzo[1,4]dioxin and octachlorodibenzo[1,4]dioxin produced 4,5-dichlorocatechol and tetrachlorocatechol, respectively.

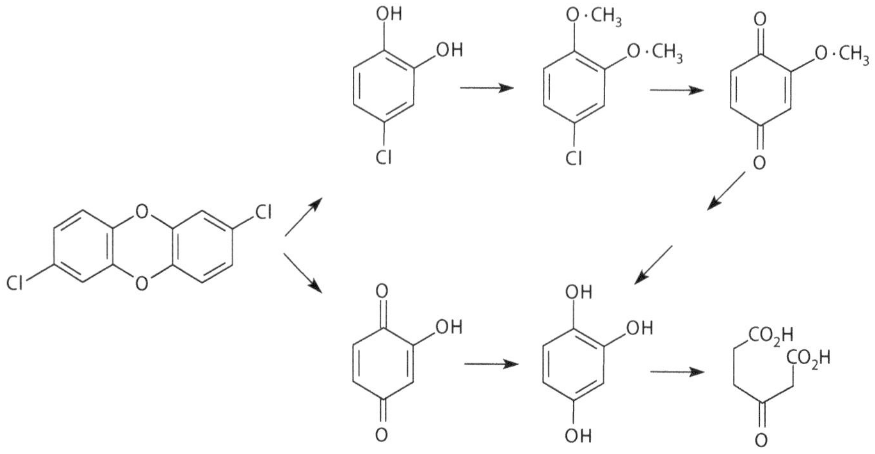

FIGURE 10.42 Fungal degradation of 2,7-dichlorodibenzo[1,4]dioxin.

10.9.1 Flavanoids, Isoflavanoids, and Related Compounds

Many higher plants synthesize flavanes, flavanones, flavones, and isoflavones with a wide range of structural complexity. They make a significant contribution to the food intake of both herbivores and humans, and they have aroused particular interest on account of their degradation by mammals that are mediated by intestinal bacteria. Most of them exist naturally as glycosides and these are readily hydrolyzed to the aglycones.

10.9.2 Reactions Carried Out by Fungi

1. The unusual dioxygenation of rutin by *Aspergillus flavus* to a depside of 3,4-dihydroxybenzoate and 2,4,6-trihydroxybenzoate (Figure 10.43) (Krishnamurty et al. 1970) is an unusual example of a dioxygenase synthesized by a eukaryotic microorganism (Krishnamurty and Simpson 1970). Experiments with $^{18}O_2$ showed that both atoms were incorporated into the depside, whereas with $H_2^{18}O$ no incorporation occurred into either the depside or into carbon monoxide. The enzyme in *Aspergillus japonicus* that brings about fission of ring C with the formation of carbon monoxide is a copper containing dioxygenase for which mechanisms have been proposed (Kooter et al. 2002; Steiner et al. 2002).

2. The transformation of representatives of flavanones and isoflavanones has been examined in a number of fungi and in *Streptomyces fulvissimus* (Ibrahim and Abul-Hajj 1990a). The principal reactions involved dehydrogenation, aromatic ring hydroxylation, and reductive ring scission of flavanones (Ibrahim and Abul-Hajj 1990b), although the mechanisms of these transformations remain unresolved. These reactions are illustrated by an overview of the

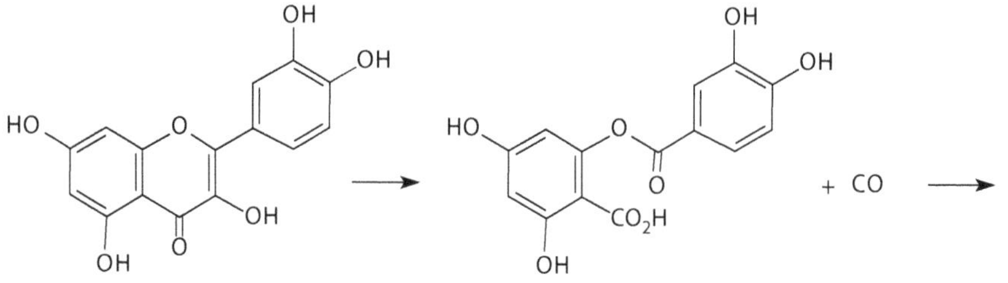

FIGURE 10.43 Transformation of rutin by dioxygenation.

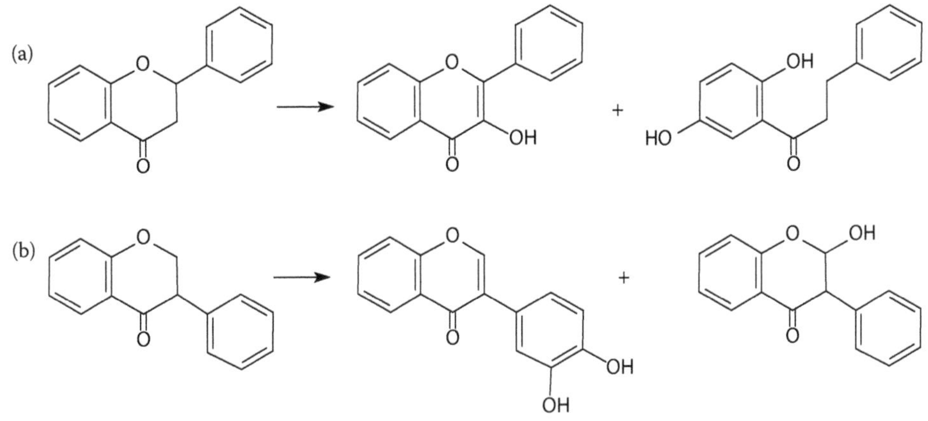

FIGURE 10.44 Transformation of (a) flavanone and (b) isoflavanone.

transformation of flavanone by *Aspergillus niger* NRRL 599 (Figure 10.44a) and of isoflavanone by *A. niger* X172 (Figure 10.44b).

3. The metabolism of oxazines and related compounds with both oxygen and nitrogen in the same ring has been explored for only a few compounds. Cyclic hydroxamates related to 2-hydroxy-benz- 1,4-oxazin-3-ones can serve as defense chemicals against fungal infection in the *Gramineae* including rye, maize, and wheat, though not in rice, barley, or oats (Chapter 2) (Niemeyer 1988). The biotransformation of 2-hydroxy-benz-1,4-oxazin-3-one and 2-benzoxazolinone has been examined in endophytic fungi isolated from *Aphelandra tetragona*. The transformations included hydrolysis to 2-aminophenol followed by: (a) oxidation to phenoxazin-3-ones and (b) acetylation followed by nitration to 2-hydroxy-3(5)-nitroacetanilides (Zikmundová et al. 2002).

10.9.3 Reactions Carried Out by Bacteria

10.9.3.1 Aerobic Conditions

The key reaction under both aerobic and anaerobic conditions is hydrolysis of the 1,3-dicarbonyl intermediate involving the keto form of phloroglucinol or resorcinol. In species of *Rhizobium* and *Brachyrhizobium*, the aerobic degradation of flavanoids (naringenin, quercetin, and luteolin) is initiated

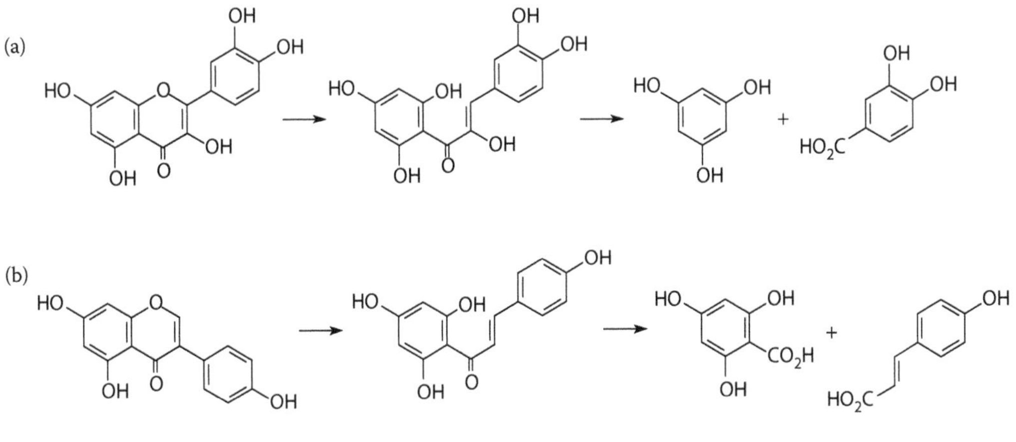

FIGURE 10.45 Aerobic transformation of (a) quercitin and (b) genistein.

by reductive fission of the C ring between C_2 and O followed by hydrolysis (Figure 10.45a) (Rao et al. 1991; Rao and Cooper 1994). In contrast, in the isoflavanoids (genistein and daidzein), reduction is accompanied by migration of the phenolic group from C_3 to C_2 (Figure 10.45b). An analogous pathway was followed in the degradation of quercetin by *P. putida* strain PML2. It was initiated by reductive loss of hydroxyl groups at C-3 and C-3′ followed by fission of the C-ring to form phloroglucinol and 3,4-dihydroxycinnamic acid that was degraded to 3,4-dihydroxybenzoate before further degradation (Pillai and Swarup 2002). The dioxygenation of 7,4′-dihydroxyisoflavone by biphenyl-2,3-dioxygenase from *Burkholderia* sp. strain LB400 was unusual, since it is followed by elimination to the phenol 7,2′,4′-trihydroxyisoflavone rather than by dehydrogenation (Seeger et al. 2003).

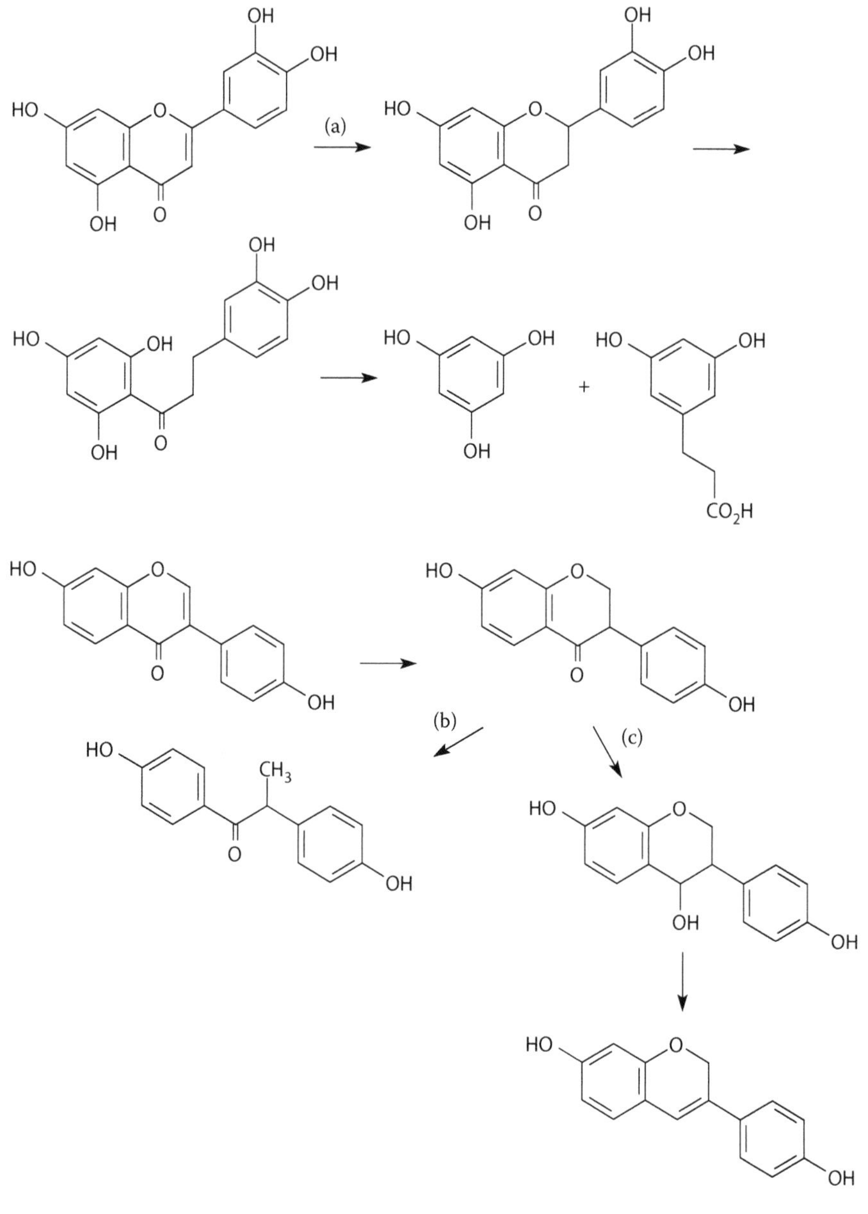

FIGURE 10.46 Anaerobic transformation of (a) quercitin (b,c) daidzein.

10.9.3.2 Anaerobic Conditions

The anaerobic degradation of flavanoids such as quercetin, naringin, and luteolin has been examined in a number of anaerobic bacteria including *Butyrovibrio* sp. (Cheng et al. 1971), *Clostridium* sp. (Winter et al. 1989), *Eubacterium ramulus* (Braune et al. 2001), and *Clostridium orbiscindens* (Schoefer et al. 2003). The initial reactions are enone reductions of the B ring followed by fission of the C_2–O bond and hydrolysis (Figure 10.46a). For isoflavanoids such as daidzein two reactions after reduction have been found (i) fission of the C_2–O bond in *Clostridium* sp. (Hur et al. 2002) (Figure 10.46b) or (ii) reduction of the ketone followed by dehydration to dehydroequol and a stereospecific reductive dehydroxylation to *S*-equol (Figure 10.46c) (Wang et al. 2005). The reduction of daidzein to the dihydrodaidzein was characterized in *Lactococcus* sp. strain 20-92 and it was proposed to be a flavin oxidoreductase belonging to the old yellow enzyme family (Shimada et al. 2010).

10.10 References

Adriaens P, Q Fu, D Grbic-Galic. 1995. Bioavailability and transformation of highly chlorinated dibenzo-p-dioxins and dibenzofurans in anaerobic soils and sediments. *Environ Sci Technol* 29: 2252–2260.

Albrecht ID, AL Barkowski, P Adriaens. 1999. Production and dechlorination of 2,3,7,8-tetrachlorodibenzo-p-dioxin in historically-contaminated estuarine sediments. *Environ Sci Technol* 33: 737–744.

Armengaud J, B Happe, KN Timmis. 1998. Genetic analysis of dioxin dioxygenase of *Sphingomonas* sp. strain RW1: catabolic genes dispersed on the genome. *J Bacteriol* 180: 3954–3966.

Armengaud J, KN Timmis, R-M Wittich. 1999. A functional 4-hydroxysalicylate/hydroxyquinol degradative pathway gene cluster is linked to the initial dibenzo-p-dioxin pathway genes in *Sphingomonas* sp. strain RW1. *J Bacteriol* 181: 3452–3461.

Ballerstedt H, A Kraus, U Lechner. 1997. Reductive dechlorination of 1,2,3,4-tetrachlorodibenzo-p-dioxin and its products by anaerobic mixed cultures from Saale River sediment. *Environ Sci Technol* 31: 1749–1753.

Barkovskii AI, P Adriaens. 1996. Microbial dechlorination of historically present and freshly spiked chlorinated dioxins and diversity of dioxin-dechlorinating populations *Appl Environ Microbiol* 62: 4556–4562.

Bedard DL, H van Dort, KA Deweerd. 1998. Brominated biphenyls prime extensive microbial reductive dehalogenation of Arochlor 1260 in Hoisatonic River sediment. *Appl Environ Microbiol* 64: 1786–1795.

Beurskens JEM, M Toussaint, J de Wolf, JMD van der Steen, PC Slot, LCM Commandeur, JR Parsons. 1995. Dehalogenation of chlorinated dioxins by an anaerobic consortium from sediment. *Environ Toxicol Chem* 14: 939–943.

Boyd DR, ND Sharma, R Boyle, RAS McMordie, J Chima, H Dalton. 1992. A 1H NMR method for the determination of enantiomeric excess and absolute configuration of cis-dihydrodiol metabolites of polycyclic arenes and heteroarenes. *Tetrahedron Lett* 33: 1241–1244.

Braune A, M Gütschow, W Engst, M Baut. 2001. Degradation of quercetin and luteolin by *Eubacterium ramulus*. *Appl Environ Microbiol* 67: 558–5567.

Bunge M, L Adrian, A Kraus, M Opel, WG Lorenz, JR Andreesen, H Görisch, U Lechner. 2003. Reductive dehalogenation of chlorinated dioxins by an anaerobic bacterium. *Nature* 421: 357–360.

Bünz PV, AM Cook. 1993. Dibenzofuran 4,4a-dioxygenase from Sphingomonas sp. strain RW1: Angular dioxygenation by a three-component system. *J Bacteriol* 175: 6467–6475.

Buser HR. 1986. Polybrominated dibenzofurans and dibenzo-p-dioxins: Thermal reaction products of polybrominated diphenyl ether flame retardants. *Environ Sci Technol* 20: 404–408.

Cerniglia CE, JC Morgan, DT Gibson. 1979. Bacterial and fungal oxidation of dibenzofuran. *Biochem J* 180: 175–185.

Chen C-M, PH Tomasek. 1991. 3,4-Dihydroxyxanthone dioxygenase from *Arthrobacter* sp. strain GFB 100. *Appl Environ Microbiol* 57: 2217–2222.

Chen C-M, PH Tomasek, RL Crawford. 1986. Initial reactions of xanthone biodegradation by an *Arthrobacter* sp. *J Bacteriol* 167: 818–827.

Cheng K-J, HG Krishnamurty, GA Jones, FJ Simpson. 1971. Identification of products produced by the anaerobic degradation of naringin by *Butyrivibrio* sp. C3. *Can J Microbiol* 17: 129–131.

Evers EHG, HJC Klamer, RWPM Laane, HAJ Govers. 1993. Polychlorinated dibenzo-p-dioxin and dibenzofuran residues in estuarine and coastal North Sea sediments. Sources and distribution. *Environ Toxicol Chem* 12: 1583–1598.

Folkerts M, U Ney, H Kneifel, E Stackenbrandt, EG Witte, H Förstel, SM Schoberth, H Sahm. 1989. *Desulfovibrio furfuralis* sp. nov., a furfural degrading strictly anaerobic bacterium. *Syst Appl Microbiol* 11: 161–169.

Fortnagel P, H Harms, R-M Wittich, S Krohn, H Meter, V Sinnwell, H Wilkes, W Francke. 1990. Metabolism of dibenzofuran by Pseudomonas sp. strain HH 69 and the mixed culture HH27. *Appl Environ Microbiol* 56: 1148–1156.

Gerecke AC, PC Hartmann, NV Heeb, H-P E Kohler, W Giger, P Schmid, M Zennegg, M Kohler. 2005. Anaerobic degradation of decabromodiphenyl ether. *Environ Sci Technol* 39: 1078–1083.

Gesell M, E Hammer, A Mikolasch. 2004. Oxidation and ring cleavage of dibenzofuran by the filamentous fungus *Paecilomyces lilacinus*. *Arch Microbiol* 182: 51–59.

Gesell M, E Hammer, M Specht, W Francke, F Schauer. 2001. Biotransformation of biphenyl by *Paecilomyces lilacinus* and characterization of ring cleavage products. *Appl Environ Microbiol* 67: 1551–1557.

Götz R, P Friesel, K Roch, O Päpke, M Ball, M, A Lis. 1993. Polychlorinated-p-dioxins PCDDs, dibenzofurans PCDFs, and other chlorinated compounds in the River Elbe: Results on bottom sediments and fresh sediments collected in sedimentation chambers. *Chemosphere* 27: 105–111.

Habe H, J-S Chung, J-H Lee, K Kasuga, T Yoshida, H Nojiri, T Omori. 2001. Degradation of chlorinated dibenzofurans and dibenzo-p-dioxins by two types of bacteria having angular dioxygenases with different features. *Appl Environ Microbiol* 67: 3610–3617.

Happe B, LD Eltis, H Poth, R Hedderich, KN Timmis. 1993. Characterization of 2,2′,3-trihydroxybiphenyl dioxygenase, an extradiol dioxygenase from the dibenzofuran- and dibenzo-p-dioxin-degrading bacterium *Sphingomonas* sp. strain RW1. *J Bacteriol* 175: 7313–7320.

Harms H, H Wilkes, R-M Wittich, P Fortnagel. 1995. Metabolism of hydroxydibenzofurans, methoxydibenzofurans, acetoxydibenzofurans, and nitrodibenzofurans by *Sphingomonas* sp. strain HH69. *Appl Environ Microbiol* 61: 2499–2505.

He J, KR Robrock, L Alvarez-Cohen. 2006. Microbial reductive debromination of polybrominated diphenyl ethers (PBDEs). *Environ Sci Technol* 40: 4429–4434.

Hites RA. 2004. Polybrominated diphenyl ethers in the environment and in people: A meta-analysis of concentrations. *Environ Sci Technol* 38: 945–956.

Hong H-B, Y-S Chang, I-H Nam, P Fortnagel, S Schmidt. 2002. Biotransformation of 2,7-dichloro- and 1,2,3,4-tetrachlorodibenzo-p-dioxin by *Sphingomonas wittichii* RW1. *Appl Environ Microbiol* 68: 2584–2588.

Hur H-G, JO Lay, RD Beger, JP Freeman, F Rafii. 2002. isolation of human intestinal bacteria metabolizing the natural isoflavone glycosides daidzin and genistin. *Arch Microbiol* 174: 422–428.

Ibrahim A-R S, YJ Abul-Hajj. 1990a. Microbiological transformation of flavone and isoflavone. *Xenobiotica* 20: 363–373.

Ibrahim A-R S, YJ Abul-Hajj. 1990b. Microbiological transformation of (±)-flavanone and (±)-isoflavanone. *J Nat Prod* 53: 644–656.

Iida T, Y Mukouzaka, K Nakamura, T Kudo. 2002. Plasmid-borne genes code for an angular dioxygenase involved in dibenzofuran degradation by *Terrabacter* sp. strain YK3. *Appl Environ Microbiol* 68: 3716–3723.

Klecka GM, DT Gibson. 1979. Metabolism of dibenzo[1,4]dioxan by a *Pseudomonas* species. *Biochem J* 180: 639–645.

Koenig K, JR Andreesen. 1991. Aerobic and anaerobic degradation of furan-3-carboxylate by *Paracoccus denitrificans* strain MK 33. *Arch Microbiol* 157: 70–75.

Koester CJ, RA Hites. 1992. Wet and dry deposition of chlorinated dioxins and furans. *Environ Sci Technol* 26: 1375–1382.

Kooter IM, RA Steiner, BW Dijkstra, PI van Noort, MR Egmond, M Huber. 2002. EPR characterization of the mononuclear Cu-containing *Aspergillus japonicus* quercetin 2,3-dioxygenase reveals dramatic changes upon anaerobic binding of substrates. *Eur J Biochem* 269: 2971–2979.

Krishnamurty HG, K-J Cheng, GA Jones, FJ Simpson, JE Watkin. 1970. Identification of products produced by the anaerobic degradation of rutin and related flavonoids by *Butyrivibrio* sp. C3. *Can J Microbiol* 16: 759–767.

Krishnamurty HG, FJ Simpson. 1970. Degradation of rutin by Aspergillus flavus. Studies with oxygen 18 on the action of a dioxygenase on quercetin. *J Biol Chem* 245: 1467–1471.

Lee LK, J He. 2010. Reductive debromination of polybrominated diphenyl ethers by anaerobic bacteria from soils and sediments. *Appl Environ Microbiol* 76: 794–802.

Macdonald RW, WJ Cretney, N Crewe, D Paton. 1992. A history of octachlorodibenzo-p-dioxin, 2,3,7,8-tetrachlorodibenzofuran, and 3,3′,4,4′-tetrachlorobiphenyl contamination in Howe Sound, British Columbia. *Environ Sci Technol* 26: 1544–1550.

Nam I-H, Y-M Kim, S Schmidt, Y-S Chang. 2006. Biotransformation of 1,2,3-tri- and 1,2,3,4,7,8-hexachlorodibenzo-p-dioxin by *Sphingomonas wittichii* strain RW1. *Appl Environ Microbiol* 72: 112–116.

Nam J-W, H Nojiri, H Noguchi, H Uchimura, T Yoshida, H Habe, H Yamane, T Omori. 2002. Purification and characterization of carbazole 1,9a-dioxygenase, a three-component dioxygenase system of *Pseudomonas resinovorans* strain CA10. *Appl Environ Microbiol* 68: 5882–5890.

Niemeyer HM. 1988. Hydroxamic acids (4-hydroxy-1,4-benzoxazin-3-ones), defence chemicals in the Gramineae. *Phytochemistry* 27: 3349–3358.

Nojiri H, J-W Nam, M Kosaka, K-I Morii, T Takemura, K Furihata, H Yamane, T Omori. 1999. Diverse oxygenations catalyzed by carbazole 1,9a-dioxygenase from *Pseudomonas* sp. strain CA 10. *J Bacteriol* 181: 3105–3113.

Pfeifer F, HG Trüper, J Klein, S Schacht. 1993. Degradation of diphenylether by *Pseudomonas cepacia* Et4: Enzymatic release of phenol from 2,3-dihydroxydiphenylether. *Arch Microbiol* 159: 323–329.

Phanerochaete sordida YK-624. *Appl Environ Microbiol* 62: 4323–4328.

Pillai BVS, Swarup. 2002. Elucidation of the flavonoid catabolism pathway in *Pseudomonas putida* PML2 by comparative metabolic profiling. *Appl Environ Microbiol* 68: 143–151.

Rao JR, E Cooper. 1994. Rhizobia catabolize nod gene-inducing flavonoids via C-ring fission mechanisms. *J Bacteriol* 176: 5409–5413.

Rao JR, ND Sharma, JTG Hamilton, DR Boyd, JE Cooper. 1991. Biotransformation of the pentahydroxy flavone quercetin by *Rhizobium loti* and *Bradyrhizobium* strains (Lotus). *Appl Environ Microbiol* 57: 1563–1565.

Robrock KR, P Korytár, L Alvarez-Cohen. 2008. Pathways for the anaerobic microbial debromination of polybrominated diphenyl ethers. *Environ Sci Technol* 42: 2845–2852.

Sariaslani FS, LR McGee, DW Ovenall. 1987. Microbial transformation of precocene II: Oxidative reactions by *Streptomyces griseus*. *Appl Environ Microbiol* 53: 1780–1784.

Sariaslani FS, MK Trower, SE Buchloz. 1989. Xenobiotic transformations by *Streptomyces griseus*. *Dev Ind Microbiol* 30: 161–171.

Schmid AS, B Rothe, J Altenbuchnerm, W Ludwig, K-H Engesser. 1997. Characterization of three distinct extradiol dioxygenases involved in mineralization of dibenzofuran by *Terrabacter* sp. strain DPO 360. *J Bacteriol* 179: 53–62.

Schmidt S, R-M Wittich, D Erdmann, H Wilkes, W Francke, P Fortnagel. 1992. Biodegradation of diphenyl ether and its monohalogenated derivatives by *Sphingomonas* sp. strain SS3. *Appl Environ Microbiol* 58: 2744–2750.

Schoefer L, R Mohan, A Schwiertz, A Braune, M Blaut. 2003. Anaerobic degradation of flavonoids by *Clostridium orbiscindens*. *Appl Environ Microbiol* 69: 5849–5854.

Seeger M, M González, B Cámara, L Mũnoz, E Ponce, L Mejías, C Mascayano, Y Vásquez, S Sepúlveda-Boza. 2003. Biotransformation of natural and synthetic isoflavanoids by two recombinant microbial enzymes. *Appl Environ Microbiol* 69: 5045–5050.

Shimada Y, S Yasuda, M Takahashi, T Hayashi, N Miyazawam, I Sato, Y Abiru, S Uchiyama, H Hishigaki. 2010. Cloning and ecpression of a novel NADP(H)-dependent daidzein reductase, an enzyme involved in the metabolism of daidzein, from equol-producing *Lactococcus* strain 20–92. *Appl Environ Microbiol* 76: 5892–5901.

Steiner RA, KH Kalk, BW Dijkstra. 2002. Anaerobic enzyme.substrate structures provide insight into the reaction mechanism of the copper-dependent quercetin 2,3-dioxygenase. *Proc Natl Acad USA* 99: 16625–16630.

Strubel V, K-H Engesser, P Fischer, H-J Knackmuss. 1991. 3-(2-hydroxyphenyl)catechol as substrate for proximal meta ring cleavage in dibenzofuran degradation by *Brevibacterium* sp. strain DPO 1361. *J Bacteriol* 173: 1932–1937.

Takada S, M Nakamura, T Matsueda, R Kondo, K Sakai. 1996. Degradation of polychlorinated dibenzo-p-dioxins and polychlorinated dibenzofurans by the white-rot fungus *Phanerochaete sordida* YK-624. *Appl Environ Microbiol* 62: 4322–4328.

Thoma H, S Rist, G Hauschulz, O Hutzinger. 1986. Polybrominated dibenzodioxins and—furans from the pyrolysis of some flame retardants. *Chemosphere* 15: 649–652.

Tokarz JA, M-Y Ahn, J Leng, TR Filley, L Nies. 2008. Reductive debromination of polybrominated diphenyl ethers in anaerobic sediment and a biomimetic system. *Environ Sci Technol* 42: 1157–1164.

Tomasek PH, RL Crawford. 1986. Initial reactions of xanthone biodegradation by an *Arthrobacter* sp. *J Bacteriol* 167: 818–827.

Valli K, BJ Brock, DK Joshi, MH Gold. 1992a. Degradation of 2,4-dinitrotoluene by the lignin-degrading fungus *Phanerochaete chrysosporium*. *Appl Environ Microbiol* 58: 221–228.

Valli K, MH Gold. 1991. Degradation of 2,4-dichlorophenol by the lignin-degrading fungus *Phanerochaete chrysosporium*. *J Bacteriol* 173: 345–352.

Valli K, H Warishi, MH Gold. 1992b. Degradation of 2,7-dichlorodibenzo-p-dioxin by the lignin-degrading basidiomycete *Phanerochaete chrysosporium*. *J Bacteriol* 174: 2131–2137.

Wang X-Y, H-G Hur, JH Lee, KT Kim, S-I Kim. 2005. Enantioselective synthesis of S-equol from dihydrodaidzein by a newly isolated anaerobic human intestinal bacterium. *Appl Environ Microbiol* 71: 214–219.

Wilkes H, RM Wittich, KN Timmis, P Fortnagel, W Francke. 1996. Degradation of chlorinated dibenzofurans and dibenzo-p-dioxins by *Sphingomonas* sp. strain RW1. *Appl Environ Microbiol* 62: 367–371.

Winter J, LH Moore, VR Dowell, VD Bokkenheuser. 1989. C-ring cleavage of flavonoids by human intestinal bacteria. *Appl Environ Microbiol* 55: 1203–1208.

Wittich R-M. 1998. Degradation of dioxin-like compounds by microorganisms. *Appl Microbiol Biotechnol* 49: 489–499.

Wittich R-M, H Wilkes, V Sinnwell, W Francke, P Fortnagel. 1992. Metabolism of dibenzo-p-dioxin by *Sphingomonas* sp. strain RW1. *Appl Environ Microbiol* 58: 1005–1010.

Yoshida N, N Takahashi, A Hiraishi. 2005. Phylogenetic characterization of polychlorinated-dioxin-dechlorinating microbial community by use of microcosm studies. *Appl Environ Microbiol* 71: 4325–4334.

Zikmundová M, K Drandarov, L Bigler, M Hesse, C Werner. 2002. Biotransformation of 2-benzoxazolinone and 2-hydroxy-1,4-benzoxazin-3-one by endophytic fungi isolated from *Aphelandra tetragona*. *Appl Environ Microbiol* 68: 4863–4870.

Part 3: Thiaarenes: Benzothiophenes, Dibenzothiophenes, and Benzothiazole

Thiaarenes are major components of crude oil and extensive effort has been devoted to microbial processes for their removal since their presence generates undesirable SO_x during incineration of fossil fuels. Attention has been directed to a number of organisms including species of *Rhodococcus*, *Corynebacterium*, and *Gordonia*.

10.11 Benzothiophene and Dibenzothiophene

Benzothiophene is isoelectronic with naphthalene, dibenzothiophene with anthracene, and benzothiazole with quinoline. This is reflected in their aerobic degradation that is initiated by dioxygenation. The diversity of pathways for the degradation of dibenzothiophene is illustrated by the following examples:

1. By analogy with anthracene, dioxygenation of dibenzothiophene produces 2-hydroxybenzothiophene-3-carboxaldehyde, and this may be transformed into a number of products including benzothiophene-2,3-dione and further into disulfides and thioindigo (Bressler and Fedorak 2001a,b).

2. Dioxygenation of one of the rings occurs, and after dehydrogenation to a catechol, ring fission takes place (Figure 10.47) (Kodama et al. 1973). This pathway is analogous to that used for the degradation of naphthalene and the isoelectronic anthracene.

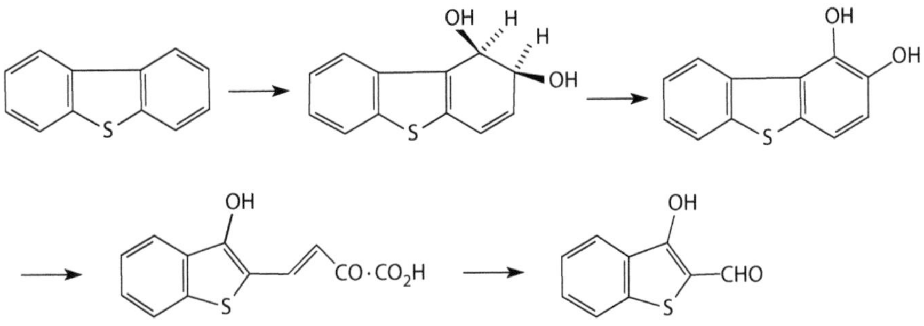

FIGURE 10.47 Alternative pathways for the biodegradation of dibenzothiophene.

3. Considerable attention has been directed to reactions in which sulfur is removed from dibenzothiophene:

   a. Successive oxidation at the sulfur atom may take place with the formation of the sulfoxide, followed by elimination of sulfite to yield either 2-hydroxybiphenyl (Omori et al. 1992; Rhee et al. 1998) or benzoate (Van Afferden et al. 1990) (Figure 10.48).

   b. Alternatively, when they serve as sources of sulfur, transformation with elimination of sulfite can be initiated in *Rhodococcus* sp. strain WU-K2R by oxidation at the sulfur atom to sulfones, followed by a complex series of reactions to produce 2-hydroxy-styrene and benzofuran (Kirimura et al. 2002).

   c. In *Rhodococcus erythropolis* strain D-1, four flavin reductases designated DszC, DszB, and DszA bring about analogous oxygenations of dibenzothiophene to form 2′-hydroxybiphenyl sulfinate that loses sulfite to produce 2-hydroxybiphenyl (Figure 10.49) (Oldfield et al. 1997; Matsubara et al. 2001).

   d. In *Rhodococus* sp. strain ECRD-1, degradation involves successive oxidation to the sulfoxide that is further degraded primarily via 2′hydroxybiphenyl-2-sulfonate to 2-hydroxybiphenyl and sulfate (Macpherson et al. 1998): the cyclized sultones and sultines were also isolated.

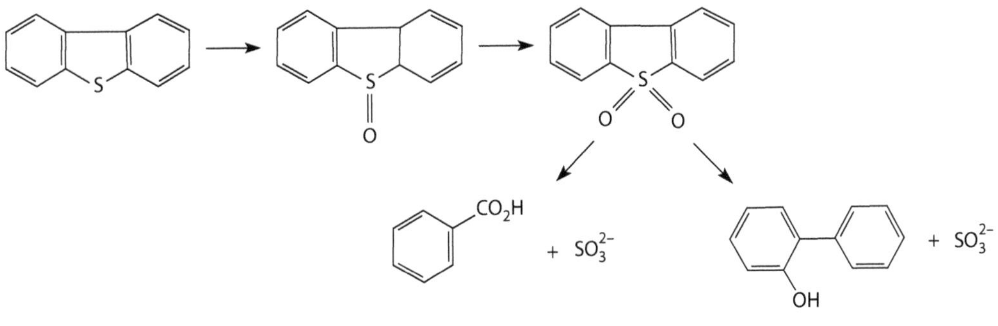

FIGURE 10.48 Alternative pathways for the biodegradation of dibenzothiophene.

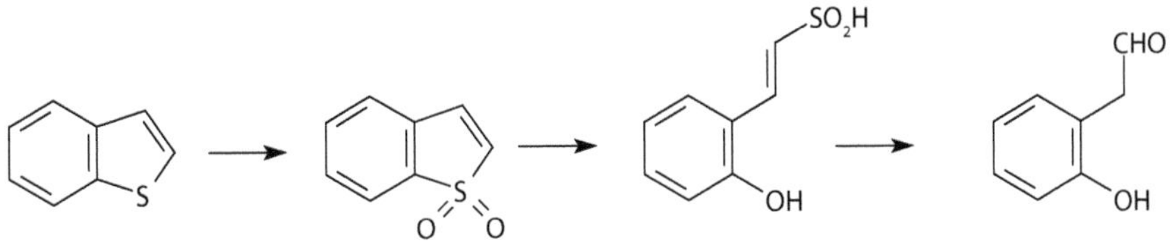

FIGURE 10.49 Degradation of benzothiophene.

e. In *Gordonia* sp. strain 213E, successive oxidation of benzothiophene to the sulfoxide is followed by fission of the thiophene ring to 2-(2′hydroxyphenyl)ethen-1-sulfinate that is converted into 2-(2′hydroxyphenyl)ethan-1-al (Gilbert et al. 1998) (Figure 10.50).

f. Benzothiophene is oxidized by strains of *Pseudomonas* sp. to both sulfoxide and sulfone (Kropp et al. 1994a), and the sulfoxide undergoes an abiotic reaction with the formation of benzo[*b*]naphtho[1,2-*d*]thiophene (Figure 10.51) (Kropp et al. 1994b).

4. The *Corynebacterium* sp. that utilizes dibenzothiophene as a sulfur source produced 2-hydroxy-biphenyl, and subsequently, nitrated thus using the nitrate in the growth medium to form two hydroxynitrobiphenyls (Omori et al. 1992) (Figure 10.52). This reaction is reminiscent of a similar one that takes place during the metabolism of 4-chlorobiphenyl (Figure 10.51b) (Sylvestre et al. 1982). These products are plausibly formed from arene oxide intermediate produced by monooxygenase systems and are discussed in Chapter 2.

Under anaerobic conditions dibenzothiophene is metabolized to a carboxybenzothiophene whose structure has not been established (Annweiler et al. 2001).

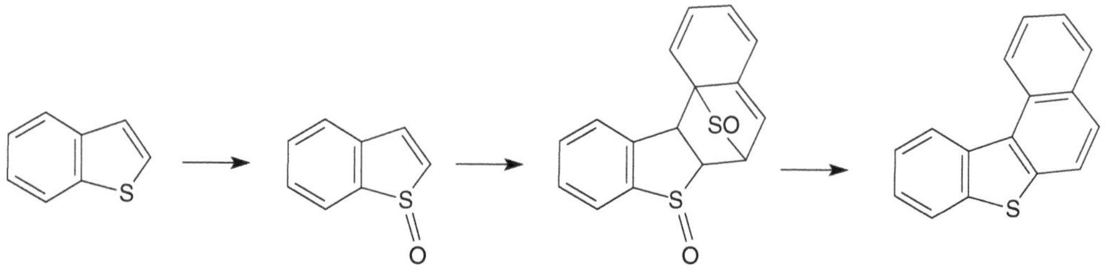

FIGURE 10.50 Transformation of benzothiophene involving microbial oxidation and chemical reaction of the sulfoxide.

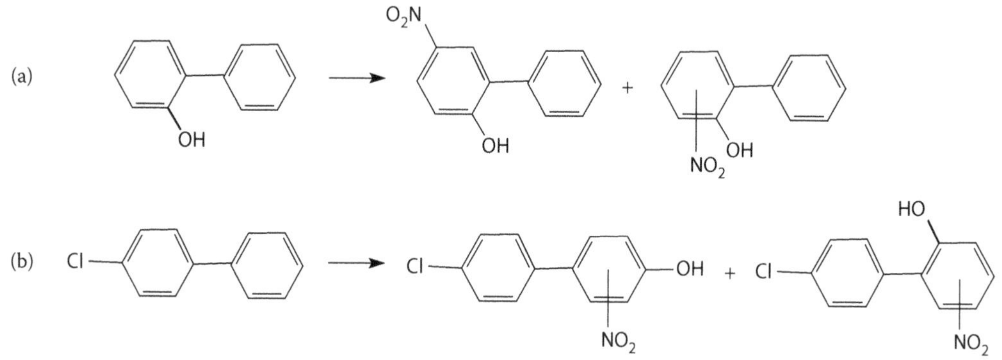

FIGURE 10.51 Formation of nitrohydroxybiphenyls during metabolism of (a) dibenzothiophene and (b) 4-chlorobiphenyl.

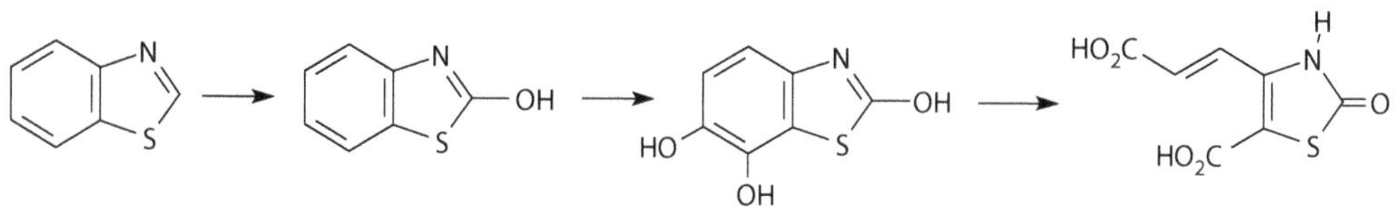

FIGURE 10.52 Aerobic degradation of benzothiazole. (Adapted from Sylvestre M et al. 1982. *Appl Environ Microbiol* 44: 871–877.)

10.12 Benzothiazole

2-Marcaptobenzothiazole is widely used as a rubber vulcanizer and as a corrosion inhibitor. The metabolism of benzothiazole by two strains of *Rhodococcus* involved hydroxylation to 2,6-dihydroxybenzothiazole (Besse et al. 2001). Degradation by a pyridine-degrading strain of *Rhodococcus pyridinovorans* proceeded by a pathway analogous to that for the degradation of the isoelectronic quinoline by *Rhodococcus* sp. strain B1 (Schwarz et al. 1989, Section 10.4.6, Figure 10.21b): hydroxylation at C_2 and C_6 followed by the formation of the 6,7-diol and dioxygenase intradiol fission to a thiazolone dicarboxylate (Haroune et al. 2002). Although biotransformation of 2-marcaptobenzothiazole by *Rhodococcus rhodochrous* may produce the *S*-methyl compound as a terminal metabolite, the aerobic biodegradation may also proceed by dioxygenation at the 6,7-positions followed by intradiol fission of the catechol to a mercaptothiazole dicarboxylic acid (Haroune et al. 2004).

10.13 References

Annweiler E, W Michaelis, RU Meckenstock. 2001. Anaerobic cometabolic conversion of benzothiophene by a sulfate-reducing enrichment culture and in a tar-oil-contaminated aquifer. *Appl Environ Microbiol* 67: 5077–5083.

Besse P, B Combourieu, G Boyse, M Sancelme, H de Wever, A-M Delort. 2001. Long-range ^{1}H-^{15}N heteronuclear shift correlation at natural abundance: a tool to study benzothiazole biodegradation by two *Rhodococcus* strains. *Appl Environ Microbiol* 67: 1412–1417.

Bressler DC, PM Fedorak. 2001a. Purification, stability, and mineralization of 3-hydroxy-2-formylbenzothiophene, a metabolite of dibenzothiophene. *Appl Environ Microbiol* 67: 821–826.

Bressler DC, PM Fedorak. 2001b. Identification of disulfides from the biodegradation of dibenzothiophene. *Appl Environ Microbiol* 67: 5084–5093.

Gilbert SC, J Morton, S Buchanan, C Oldfield, A McRoberts. 1998. Isolation of a unique benzothiophene-desulphurizing bacterium, *Gordona* sp. strain 213E (NCIMB 40816), and characterization of the desulphurization pathway. *Microbiology (UK)* 144: 2545–2553.

Haroune N, B Combourieu, P Besse, M Sancelme, A Kloepfer, T Reemtsma, H De Wever, A-M Delort. 2004. Metabolism of 2-mercaptobenzothiazole by *Rhodococcus rhodochrous*. *Appl Environ Microbiol* 70: 6315–6319.

Haroune N, B Combourieu, P Besse, M Sancelme, T Reemtsma, A Kloepfer, A Diab, JS Knapp, S Baumberg, A-M Delort. 2002. Benzothiazole degradation by *Rhodococcus pyridinovorans* strain PA: Evidence of a catechol 1,2-dioxygenase activity. *Appl Environ Microbiol* 68: 6114–6120.

Kirimura K, T Furuya, R Sato, Y Ishii, K Kino, S Usami. 2002. Biodesulfurizarion of naphthothiophene and benzothiophene through selective cleavage of carbon–sulfur bonds by *Rhodococcus* sp. strain WU-K2R. *Appl Environ Microbiol* 68: 3867–3872.

Kodama K, K Umehara, K Shikmizu, S Nakatani, Y Minoda, K Yamada. 1973. Identification of microbial products from dibenzothiophene and its proposed oxidation pathway. *Agric Biol Chem* 37: 45–50.

Kropp KG, JA Goncalves, JT Anderson, PM Fedorak PM. 1994a. Bacterial transformations of benzothiophene and methylbenzothiophenes. *Environ Sci Technol* 28: 1348–1356.

Kropp KG, JA Goncalves, JT Anderson, PM Fedorak. 1994b. Microbially mediated formation of benzonaphthothiophenes from benzo[*b*]thiophenes. *Appl Environ Microbiol* 60: 3624–3631.

Macpherson T, CW Greer, E Zhou, AM Jones, G Wisse, PCK Lau, B Sankey, MJ Grossman, J Hawari. 1998. Application of SPME/GC-MS to characterize metabolitres in the biodesulfurization of organosulur model compounds in bitumen. *Environ Sci Technol* 32: 421–426.

Matsubara T, T Ohshiro, Y Nishina, Y Izumi. 2001. Purification, characterization, and overexpression of flavin reductase involved in dibenzothiophene desulfurization by *Rhodococcus erythropolis* D-1. *Appl Environ Microbiol* 67: 1179–1184.

Oldfield C, O Pogrebinsky, J Simmonds, ES Olson, CF Kulpa. 1997. Elucidation of the metabolic pathway for dibenzothiophene desulphurization by *Rhodococcus* sp. strain IGTS8 (ATCC 53968). *Microbiology (UK)* 143: 2961–2973.

Omori T, L Monna, Y Saiki, T Kodama. 1992. Desulfurization of dibenzothiophene by *Corynebacterium* sp. strain SY1. *Appl Environ Microbiol* 58: 911–915.

Rhee S-K, JH Chang, YK Chang, HN Chang. 1998. Desulfurization of dibenzothiophene and diesel oils by a newly isolated *Gordona* strain CYKS1. *Appl Environ Microbiol* 64: 2327–2331.

Sylvestre M, R Massé, F Messier, J Fauteux, J-G Bisaillon, R Beaudet. 1982. Bacterial nitration of 4-chlorobiphenyl. *Appl Environ Microbiol* 44: 871–877.

Van Afferden M, S Schacht, J Klein, HG Trüper. 1990. Degradation of dibenzothiophene by *Brevibacterium* sp. DO. *Arch Microbiol* 153: 324–328.

11

Miscellaneous Compounds

Parts 1 through 5 of this chapter provide an outline of the reactions involved in the biodegradation of aliphatic esters, ethers, nitramines, phosphonates and sulfonates, and organic compounds of metals and metalloids.

Part 1: Carboxylate, Sulfate, Phosphate, and Nitrate Esters

Hydrolysis is the first step in the degradation of a number of important contaminants. These include o-phthalate esters used as plasticizers; organophosphate, organothiophosphate, and organodithiophosphate insecticides; phosphorofluoridates that have been considered as chemical warfare agents; aliphatic sulfates in surfactants; and polyester polyurethane (Akutsu et al. 1998). In addition, there are important naturally occurring phosphonates, phosphates, and pyrophosphates; polysaccharide sulfates; linear alkyl (C_{16}–C_{30}) sulfates; and sulfate conjugate esters of phenolic compounds such as tyrosine sulfate. The hydrolysis of a range of naturally occurring sulfate esters may make an important contribution to the sulfate present in aerobic soils (Fitzgerald 1976), quite apart from the anthropogenic contribution of SO_x.

11.1 Carboxylates

The hydrolases that depolymerize poly-3-hydroxyalkanoates and generally contain serine at the active site are discussed in Chapter 7. Phthalate esters are widespread contaminants and the dialkyl phthalates are hydrolyzed before degradation of the resulting phthalate (Eaton and Ribbons 1982), which is discussed in Chapter 8, Part 3. For dimethyl phthalate, dimethyl terephthalate, and dimethyl isophthalate only partial hydrolysis may take place (Li et al. 2005). Cocaine is hydrolyzed to benzoate and ecgonine methyl ester by a strain of *Pseudomonas maltophilia* (Britt et al. 1992).

11.2 Sulfates

Alkyl sulfates and alkylethoxy sulfates have been extensively used as detergents so that concern has been expressed over their biodegradability. A review (Cain 1981) covers the degradation of a wide range of surfactants including both of these groups, and one by White and Russell (1994) discusses in detail the biodegradation of alkyl sulfates including the enzymology and regulation.

Although sulfate is formed by hydrolysis of both alkyl and aryl sulfates, the pathway of degradation for aryl sulfates is controlled by the source of sulfur (Cook et al. 1999). The complex issues surrounding the hydrolysis of sulfate esters have been discussed (Kertesz 1999). Illustrative examples include:

1. *Pseudomonas putida* strain FLA was able to carry out hydrolysis of the sulfate ester 2-(2,4-dichlorophenoxy)ethyl sulfate, and displayed three primary alkylsulfates and three secondary sulfates although only one of these was apparently involved in hydrolysis of the herbicide substrate (Lillis et al. 1983).

2. *Pseudomonas* sp. strain AE-A was able to grow at the expense of either sodium dodecyl sulfate (SDS) or the secondary 2-butyloctyl sulfate (2BOS) (Ellis et al. 2002). Three alkyl sulfatases were produced: AP3 was SDS-inducible, specific for SDS and inactive on 2BOS, while AP2 was 2BOS-inducible, specific for 2BOS and inactive on SDS.

There are several metabolically distinct groups of sulfatases: Group I enzymes require post-translational modification of serine or cysteine residues to α-formyl-glycine that is discussed in detail in Chapter 3, Group II comprise the Fe(II) 2-oxoglutarate-dependent enzymes in *Pseudomonas putida* S-313 that produce sulfate and the corresponding aldehyde from primary sulfonates (Kahnert and Kertesz 2000), and Group III comprise a distinct group containing a binuclear Zn cluster enzymes that constituted a distinct member of the metallo-β-lactam family and carried out the degradation of dodecyl sulfate (Hagelueken et al. 2006). The arylsulfatases are discussed later.

The degradation or transformation of aliphatic sulfate esters can be carried out by a range of mechanisms.

1. Long-chain unbranched aliphatic sulfate esters are generally degraded by initial hydrolysis to sulfate and the alkanol, which is then degraded by conventional pathways. The alkylsulfatases show diverse specificity (Dodson and White 1983)—generally for sulfates with chain lengths ≥5—although organisms have been isolated that degrade short-chain (C_1–C_4) primary alkyl sulfates (White et al. 1987). *Pseudomonas* sp. strain C12B produces two primary (P1 and P2) and three secondary (S1, S2, and S3) alkylsulfatases, and hydrolysis by the P1 enzyme takes place by O–S bond fission. The S1 and S2 enzymes are constitutive, and for chiral compounds such as octan-2-yl sulfate, hydrolysis proceeds with inversion of configuration by cleavage of the alkyl–oxygen bond (Bartholomew et al. 1977). For example, hydrolysis of (*R*)-2-octyl sulfate by *Rhodococcus ruber* DSM 44541 proceeded with inversion to (*S*)-2-octanol (Pogorevc and Faber 2003).

2. In an alternative pathway for degradation, oxidation may precede elimination of sulfate. Examples include the degradation of propan-2-yl sulfate (Crescenzi et al. 1985) and of mono-methyl sulfate (Davies et al. 1990; Higgins et al. 1993) (Figure 11.1). Under sulfate limitation, *Pseudomonas putida* strain S-313 degraded alkyl sulfates (C_4–C_{12}) to sulfate and the corresponding aldehyde by a 2-ketoacid-dependent reaction in the presence of Fe^{2+}. Exceptionally, other 2-ketoacids could be used including 2-ketoglutarate, 2-ketovalerate, and 2-ketoadipate (Kahnert and Kertesz 2000).

3. For alkylethoxy sulfates, a greater range of possibilities exist including ether-cleavage reactions, while direct removal of sulfate may be of lesser significance (Hales et al. 1986). It should be noted, however, that an unusual reaction may occur simultaneously: chain elongation of the carboxylic acid. For example, during degradation of dodecyl sulfate, lipids containing 14, 16, and 18 carbon atoms were synthesized (Thomas and White 1989).

4. Transformations by monooxygenation rather than hydrolysis have been described for the organochlorine insecticides endosulfan and endosulfate in organisms that use them as sources of sulfur in the absence of inorganic sulfur. Hydroxylation at the position adjacent to the oxygen is mediated by a monooxygenase and a reduced flavin, and results in the production of sulfite from β-endosulfan by *Mycobacterium* sp. strain ESD (Sutherland et al. 2002). Analogous monooxygenation of α- and β-endosulfan, and endosulfate can be carried out by a species of *Arthrobacter* (Weir et al. 2006).

Aryl sulfates are widely synthesized from phenolic substrates and serve as a detoxification mechanism both for microorganisms and for fish. Hydrolysis of aryl esters takes place with fission of the O–S bond (Recksiek et al. 1998), whereas for aliphatic sulfates, fission of the C–O bond takes place. In addition, utilization of aryl sulfate may be mediated by an arylsulfo-transferase (Kahnert et al. 2000). As for alkyl

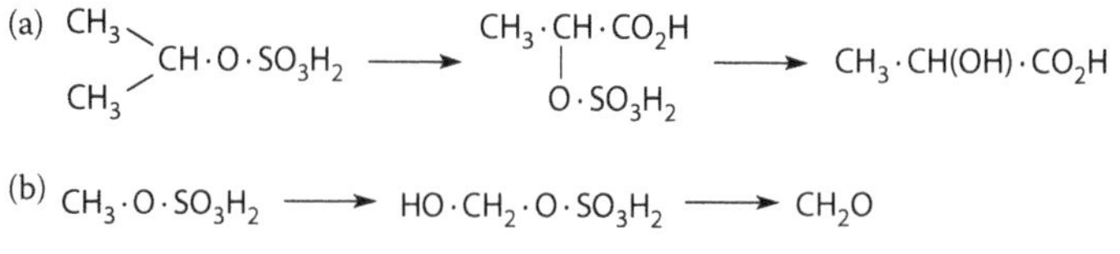

(a) CH_3\
$\quad\quad\quad$CH·O·SO$_3$H$_2$ $\longrightarrow$ $\begin{array}{c} CH_3\cdot CH\cdot CO_2H \\ | \\ O\cdot SO_3H_2 \end{array}$ $\longrightarrow$ $CH_3\cdot CH(OH)\cdot CO_2H$\
CH_3

(b) $CH_3\cdot O\cdot SO_3H_2$ $\longrightarrow$ $HO\cdot CH_2\cdot O\cdot SO_3H_2$ $\longrightarrow$ CH_2O

FIGURE 11.1 Degradation of (a) propan-2-yl sulfate and (b) methyl sulfate.

sulfates, sulfate is formed from aryl sulfates and, for aryl sulfates the pathway of degradation is controlled by the source of sulfur (Cook et al. 1999). The hydrolysis of aryl sulfates has traditionally been a useful taxonomic character in the genus *Mycobacterium* (Wayne et al. 1991). A positive result in the 3-d aryl sulfatase test based on the hydrolysis of phenolphthalein sulfate has been particularly valuable for distinguishing members of the rapidly growing *M. fortuitum* group, which are potentially pathogenic to man, whereas the slow-growing *M. tuberculosis* and *M. bovis* are negative even after 10 d. Regulation of the synthesis of tyrosine sulfate sulfohydrolase has been examined in a strain of *Comamonas terrigena* (Fitzgerald et al. 1979), and both inducible and constitutive forms of the enzyme exist. These are, however, apparently distinct from the aryl sulfate sulfohydrolase, which has found application in taxonomic classification.

11.3 Phosphates

Aryl phosphates and thiophosphates, and alkyl dithiophosphates are important agrochemicals, while phosphorofluoridates have been prepared as chemical warfare agents. Concern over the persistence and the biodegradability of organophosphate and organophosphorothioates, which are used as agrochemicals, has stimulated studies into their degradation. Considerable attention has been directed to biodegradation of all of them, and references may be found in Munnecke et al. (1982), DeFrank and White (2002), and Singh et al. (2004). The hydrolytic enzyme(s)—organophosphorus acid anhydrase (OPA)—are responsible for defluorination of phosphorofluoridates. These studies have revealed the widespread distribution of bacterial triesterases, whereas diesterases are less common, although an enzyme from *Delftia acidovorans* has been described (Tehara and Keasling 2003).

The first step in the degradation of phosphate and phosphorothioate esters is hydrolysis, and substantial effort has been directed to all groups. Investigations have also been directed to the use of their degradation products as a source of phosphate for the growth of bacteria, and a wide range of phosphates, dialkylphosphates, and phosphorothioates has, therefore, been examined as sources of phosphorus (Cook et al. 1978).

It is important to emphasize that the initial metabolites after hydrolysis may be both toxic and sometimes resistant to further degradation. Examples include nitrophenols, whose degradation is discussed in Chapter 9 and 3,5,6-trichloropyridin-2-ol (Feng et al. 1997), which is produced by the hydrolysis of chlorpyrifos (*O,O*-diethyl-*O*-[3,5,6-trichlo-2-pyridyl]phosphorothioate).

11.4 Nitrates

Compared with the fairly numerous investigations on the microbial degradation of carboxylic acid, sulfate and phosphate esters, data on the degradation of nitrate esters is limited to a few structural groups including glycerol trinitrate and pentaerythritol tetranitrate, and the pharmaceutical product isosorbide 2,5-dinitrate. Glycerol trinitrate (White et al. 1996; Marshall and White 2001), and pentaerythritol tetranitrate (French et al. 1996) are important explosives whose degradation has attracted the greatest attention. This has been clearly revealed in a review (White and Snape 1993) that summarized existing knowledge on the microbial degradation of nitrate esters.

Examples of degradation include glycerol trinitrate (Marshall and White 2001) and pentaerythritol tetranitrate (French et al. 1996), which are important explosives. Reactions involving glutathione transferases are important in eukaryotic microorganisms (White et al. 1996). In bacteria, however, the degradation of nitrate esters generally takes place by reduction with loss of nitrite (Figure 11.2), and pentaerythritol tetranitrate reductase, which is related to "old yellow enzyme" (French et al. 1996), and glycerol trinitrate

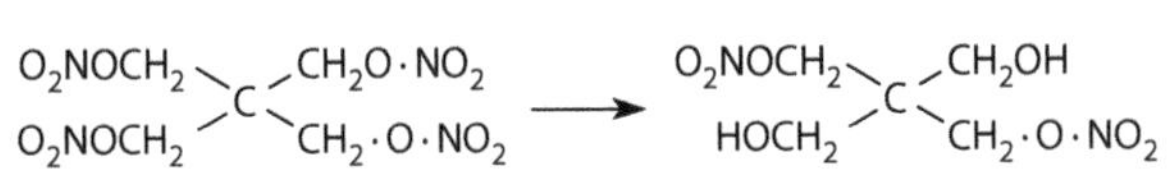

FIGURE 11.2 Reductive degradation of pentaerythritol tetranitrate.

reductase (Snape et al. 1997) have been purified. The reductase from *E. cloacae* is strongly inhibited by steroids and is capable of the reduction of cyclohex-2-ene-1-one (French et al. 1996).

The biotransformation of glycerol trinitrate by strains of *Bacillus thuringiensis/cereus* or *Enterobacter agglomerans* (Meng et al. 1995), by strains of *Pseudomonas* sp., and some *Enterobacteriaceae* (Blehert et al. 1997) involves the expected successive loss of nitrite by FMN-dependent reduction (Marshall et al. 2004) although removal of all the nitrate groups may not occur and the mononitrate is generally the terminal product. This has been found for the metabolism of glycerol trinitrate by *Arthrobacter* sp. strain JBH1 when glycerol is used for cell growth and 2-nitroglycerol is a terminal metabolite (Husserl et al. 2010). 2-Nitroglycerol can, however, be degraded in this strain by the activity of a kinase to produce 1-nitro-3-phosphoglycerol followed by loss of nitrite before entry into central metabolism (Husserl et al. 2012). On the other hand, a strain of *Rhodococcus* sp. is capable of removing all three nitrate groups (Marshall and White 2001). The biotransformation of pentaerythritol tetranitrate by *Enterobacter cloacae* proceeds comparably by metabolism of two hydroxymethyl groups produced by loss of nitrite to the aldehyde (Binks et al. 1996). 2-Ethylhexyl nitrate is a major additive to diesel fuel and was degraded by *Mycobacterium austroafricanum*. This involved conversion to 2-ethylpentan-1,6-diol, followed by dehydrogenation to the 1-carboxylate and β-oxidation to 3(hydroxymethyl)pentanoate. This was then cyclized to the terminal metabolite 4-ethyldihydrofuran-2(3*H*)-one (Nicolau et al. 2008).

In a medium containing glucose and ammonium nitrate, glyceryl trinitrate is degraded by *Penicillium corylophilum* to the di- and mononitrate before complete degradation. In contrast, the metabolism of glyceryl trinitrate by *Phanerochaete chrysosporium* involves the production of nitric oxide (Servent et al. 1991). Nitric oxide is produced during conversion of L-arginine into L-citrulline by a strain of *Nocardia* sp. (Chen and Rosazza 1995), and a summary of the mechanism, which includes both prokaryotic and mammalian systems, has been given (Stuehr et al. 2004).

11.5 References

Akutsu Y, T Nakajima-Kambe, N Nomura, T Nakahara. 1998. Purification and properties of a polyester polyurethane-degrading enzyme from *Comamonas acidovorans* TB-35. *Appl Environ Microbiol* 64: 62–67.

Bartholomew B, KS Dodgson, GWJ Matcham, DJ Shaw, GF White. 1977. A novel mechanism of enzymatic hydrolysis. Inversion of configuration and carbon–oxygen bond cleavage by secondary alkylsulphohydrolases from detergent-degrading micro-organisms. *Biochem J* 165: 575–580.

Binks PR, CE French, S Nicklin, NC Bruce. 1996. Degradation of pentaerythritol tetranitrate by *Enterobacter cloacae* PB2. *Appl Environ Microbiol* 62: 1214–1219.

Blehert DS, KL Knoke, BG Fox, GH Cambliss. 1997. Regioselectivity of nitroglycerine denitration by flavoprotein nitroester reductases purified from two *Pseudomonas* species. *J Bacteriol* 179: 6912–6920.

Britt AJ, NC Bruce, CR Lowe. 1992. Identification of a cocaine esterase in a strain of *Pseudomonas maltophilia*. *J Bacteriol* 174: 2087–2094.

Cain RB. 1981. Microbial degradation of surfactants and "builder" components. In *Microbial Degradation of Xenobiotics and Recalcitrant Compounds* (Eds. T Leisinger, AM Cook, R Hütter, and J Nüesch), pp. 325–370. Academic Press, London.

Chen Y, JPN Rosazza. 1995. Purification and characterization of nitric oxide synthase NOS$_{Noc}$ from a *Nocardia* sp. *J Bacteriol* 177: 5122–5128.

Cheng T-C, JJ Calomiris. 1996. A cloned bacterial enzyme for nerve agent decontamination. *Enz Microbiol Technol* 18: 597–601.

Cheng T-C, SP Harvey, GL Chen. 1996. Cloning and expression of a gene encoding a bacterial enzyme for decontamination of organophosphorus nerve agents and nucleotide sequence of the enzyme. *Appl Environ Microbiol* 62: 1636–1641.

Cook AM, CG Daughton, M Alexander. 1978. Phosphorus-containing pesticide breakdown products: Quantitative utilization as phosphorus sources by bacteria. *Appl Environ Microbiol* 36: 668–672.

Cook AM, H Laue, F Junker. 1999. Microbial desulfonation *FEMS Microbiol Rev* 22: 399–419.

Crescenzi AMV, KS Dodgson, GF White, WJ Payne. 1985. Initial oxidation and subsequent desulphation of propan-2-yl sulphate by *Pseudomonas syringae* strain GG. *J Gen Microbiol* 131: 469–477.

Davies I, GF White, WJ Payne. 1990. Oxygen-dependent desulphation of monomethyl sulphate by *Agrobacterium* sp. M3C. *Biodegradation* 1: 229–241.

DeFrank JJ. 1991. Organophosphorus cholinesterase inhibitors: Detoxification by microbial enzymes, pp. 165–180. In *Applications of Enzyme Biotechnology* (Eds. JW Kelly and TO Baldwin), Plenum Press, New York.

DeFrank JJ, WT Beaudry, T-C Cheng, SP Harvey, AN Stroup, LL Szafraniec. 1993. Screening of halophilic bacteria and *Alteromonas* species for organophosphorus hydrolyzing enzyme activity. *Chem-Biol Interact* 87: 141–148.

DeFrank JJ, T-C Cheng. 1991. Purification and properties of an organophosphorus acid anhydrase from a halophilic bacterial isolate. *J Bacteriol* 173: 1938–1943.

DeFrank JJ, WE White. 2002. Phosphorofluoridates: Biological activity and biodegradation. *Handbook Environ Chem* 3N: 295–343.

Dodson KS, GF White. 1983. Some microbial enzymes involved in the biodegradation of sulfated surfactants. In *Topics in Enzyme and Fermentation Technology* (Ed. A Wiseman), Vol. 7, pp. 90–155. Ellis-Horwood, Chichester.

Dumas DP, JR Wild, FM Raushel. 1989. Diisopropylfluorophosphate hydrolysis by an organophosphate anhydrase *from Pseudomonas diminuta*. *Biotechnol Appl Biochem* 11: 235–243.

Eaton RW, DW Ribbons. 1982. Metabolism of dibutylphthalate and phthalate by *Micrococcus* sp. strain 12B. *J Bacteriol* 151: 48–57.

Ellis AW, SG Hales, NGA Ur-Rehman, GF White. 2002. Novel alkylsulfatases required for biodegradation of the branched primary alkyl sulfate surfactant 2-butyloctyl sulfate. *Appl Environ Microbiol* 68: 31–36.

Feng Y, KD Racke, J-M Bollag. 1997. Isolation and characterization of a chlorinated-pyridinol-degrading bacterium. *Appl Environ Microbiol* 63: 4096–4098.

Fitzgerald JW. 1976. Sulfate ester formation and hydrolysis: A potentially important yet often ignored aspect of the sulfur cycle of aerobic soils. *Bacteriol Rev* 40: 698–721.

Fitzgerald JW, HW Maca, FA Rose. 1979. Physiological factors regulating tyrosine-sulphate sulphohydrolase activity in *Comamonas terrigena*: Occurrence of constitutive and inducible enzymes. *J Gen Microbiol* 111: 407–415.

French CE, S Nicklin, NC Bruce. 1996. Sequence and properties of pentaerythritol tetranitrate reductase from *Enterobacter cloacae* PB 2. *J Bacteriol* 178: 6623–6627.

Hagelueken G, TM Adams, L Wiehlmann, U Widow, H Kolmar, B Tümmler, DW Heinz, W-D Schubert. 2006. The crystal structure of SdsA1, an alkylsulfatase from *Pseudomonas aeruginosa* defines a third class of sulfatases. *Proc Natl Acad Sci USA* 103: 7631–7636.

Hales SG, GK Watson, KS Dodson, GF White. 1986. A comparative study of the biodegradation of the surfactant sodium dodecyltriethoxy sulphate by four detergent-degrading bacteria. *J Gen Microbiol* 132: 953–961.

Higgins TP, JR Snape, GF White. 1993. Comparison of pathways for biodegradation of monomethyl sulphate in *Agrobacterium* and *Hyphomicrobium* species. *J Gen Microbiol* 139: 2915–2920.

Husserl J, JB Hughes, JC Spain. 2012. Key enzymes enabling the growth of *Arthrobacter* sp. strain JBH1 with nitroglycerin as the sole source of carbon and energy. *Appl Environ Microbiol* 78: 3649–3655.

Husserl J, JC Spain, JB Hughes. 2010. Growth of *Arthrobacter* sp. strain JBH1 on nitroglycerin as sole source of carbon and nitrogen. *Appl Environ Microbiol* 76: 1689–1691.

Kahnert A, MA Kertesz. 2000. Characterization of a sulfur-regulated oxygenative alkylsulfatase from *Pseudomonas putida* S-313. *J Biol Chem* 275: 31661–31667.

Kahnert A, P Vermeli, C Wietek, P James, T Leisinger, MA Kertesz. 2000. The *ssu* locus plays a key role in organosulfur metabolism in *Pseudomonas putida* S-313. *J Bacteriol* 182: 2869–2878.

Kertesz MA. 1999. Riding the sulfur cycle—Metabolism of sulfonates and sulfate esters by Gram-negative bacteria. *FEMS Microbiol Rev* 24: 135–175.

Landis WG, JJ DeFrank. 1990. Enzymatic hydrolysis of toxic organofluorophosphate compounds. In *Biotechnology and Biodegradation* Vol. 4, pp. 183–201 (Eds. D Kamely, A Chakrabarty and GS Omenn). Gulf Publishing Company, Houston, USA.

Li J, J-D Gu, L Pan. 2005. Transformation of dimethyl phthalate, dimethyl isophthalate, and dimethylterephthalate by *Rhodococcus rubber* Sa and modeling the process using the modified Gompertz model. *Int Biodet Biodeg* 55: 223–232.

Lillis V, KS Dodson, GF White, WJ Payne. 1983. Initiation of activation of a preemergent herbicide by a novel alkylsulfatase of *Pseudomonas putida* FLA. *Appl Environ Microbiol* 46: 988–994.

Marshall SJ, D Krause, DK Blencowe, GF White. 2004. Characterization of glycerol trinitrate reductase (NerA) and the catalytic role of active-site residues. *J Bacteriol* 186: 1802–1810.

Marshall SJ, GF White. 2001. Complete denitration of nitroglycerin by bacteria isolated from a washwater soakaway. *Appl Environ Microbiol* 67: 2622–2626.

Meng M, W-Q Sun, LA Geelhaar, G Kumar, AR Patel, GF Payne, MK Speedie, JR Stacy. 1995. Denitration of glycerol trinitrate by resting cells and cell extracts of *Bacillus thuringiensis/cereus* and *Enterobacter agglomerans*. *Appl Environ Microbiol* 61: 2548–2553.

Munnecke DM. 1976. Enzymatic hydrolysis of organophosphate insecticides, a possible pesticide disposal method. *Appl Environ Microbiol* 32: 7–13.

Munnecke DM, LM Johnson, HW Talbot, S Barik. 1982. Microbial metabolism and enzymology of selected pesticides. In *Biodegradation and Detoxification of Environmental Pollutants* (Ed. AM Chrakrabarty), pp. 1–32. CRC Press, Boca Raton, FL.

Nicolau E, L Kerhoas, M Lettere, Y Jouanneau, R Marchal. 2008. Biodegradation of 2-ethylhexyl nitrate by *Mycobacterium austroafricanum* IFP 2173. *Appl Environ Microbiol* 74: 6187–6193.

Pogorevc M, K Faber. 2003. Purification and characterization of an inverting stero- and enantioselective *sec*-alkylsulfatase from the Gram-positive bacterium *Rhodoccus ruber* DSM 44541. *Appl Environ Microbiol* 69: 2810–2815.

Recksiek M, T Selmer, T Dierks, B Schmidt, K von Figura. 1998. Sulfatases, trapping of the sulfonated enzyme intermediate by substituting the active site formylglycine. *J Biol Chem* 273: 6096–6103.

Reid KA, JTG Hamilton, RD Bowden, D O'Hagan, L Dasaradhi, MR Amin, MR, DB Harper. 1995. Biosynthesis of fluorinated secondary metabolites by *Streptomyces cattleya*. *Microbiology (UK)* 141: 1385–1393.

Servent D, C Ducrorq, Y Henry, A Guissani, M Lenfant. 1991. Nitroglycerin metabolism by *Phanerochaete chrysosporium*: Evidence for nitric oxide and nitrite formation. *Biochim Biophys Acta* 1074: 320–325.

Singh BK, A Walker, JAW Morgan, DJ Wright. 2004. Biodegradation of chloropyrifos by *Enterobacter* strain B-14 and its use in bioremediation of contaminated soils. *Appl Environ Microbiol* 70: 4855–4863.

Snape JR, NA Walkley, AP Morby, S Nicklin, GF White. 1997. Purification, properties, and sequence of glycerol trinitrate reductase from *Agrobacterium radiobacter*. *J Bacteriol* 179: 7796–7802.

Stuehr DJ, J Santolini, Z-Q Wang, C-C Wei, S Adak. 2004. Update on mechanism and catalytic regulation in the NO synthases. *J Biol Chem* 36167–36170.

Sutherland TD, I Horne, RJ Russell, JG Oakeshott. 2002. Gene cloning and molecular characterization of a two-enzyme system catalyzing the oxidative detoxification of β-endosulfan. *Appl Environ Microbiol* 68: 6237–6245.

Tehara SK, JD Keasling. 2003. Gene cloning, purification, and characterization of a phosphodiesterase from *Delftia acidovorans*. *Appl Environ Microbiol* 69: 504–508.

Thomas ORT, GF White. 1989. Metabolic pathway for the biodegradation of sodium dodecyl sulfate by *Pseudomonas* spC12B. *Biotechnol Appl Biochem* 11: 318–327.

Wayne LG et al. 1991. Fourth report of the cooperative, open-ended study of slowly growing mycobacteria by the international working group on mycobacterial taxonomy. *Int J Syst Bacteriol* 41: 463–472.

Weir KM, TD Sutherland, I Horne, RJ Russell, JG Oakshott. 2006. A single monooxygenase, Ese, is involved in the metabolism of the organochlorines endosulfan and endosulfate in an *Arthrobacter* sp. *Appl Environ Microbiol* 72: 3524–3530.

White GF, JR Snape. 1993. Microbial cleavage of nitrate esters: Defusing the environment. *J Gen Microbiol* 139: 1947–1957.

White GF, JR Snape, S Nicklin. 1996. Bacterial biodegradation of glycerol trinitrate. *Int Biodet Biodeg* 38: 77–82.

White GF, KS Dodson, I Davies, PJ Matts, JP Shapleigh, WJ Payne. 1987. Bacterial utilisation of short-chain primary alkyl sulphate esters. *FEMS Microbiol Lett* 40: 173–177.

White GF, NJ Russell. 1994. Biodegradation of anionic surfactants and related molecules. In *Biochemistry of Microbial Degradation* (Ed. C Ratledge), pp. 143–177. Kluwer Academic Publishers, Dordrecht, The Netherlands.

Part 2: Ethers and Sulfides

The degradation of a structurally wide range of ethers under both aerobic and anaerobic conditions is covered in a review (White et al. 1996), which should be consulted for details, particularly of polyethers.

11.6 Aliphatic and Benzylic Ethers

Aliphatic and benzylic ethers are degraded by hydroxylation of the α-methylene group followed by scission of the ether bond with the formation of an aldehyde and an alkanol (White et al. 1996; Kim and Engesser 2004). In contrast, the degradation of 2-chloroethylvinyl ether by *Ancylobacter aquaticus* is initiated by a dehalogenase, although fission of the C–O–C bond is nonenzymatic (van den Wijngaard et al. 1993).

1. The monooxygenase from *Burkholderia cepacia* G4/PR1 in which the synthesis of toluene-2-monooxygenase is constitutive is able to degrade a number of ethers including diethyl ether and *n*-butyl methyl ether, though not *tert*-butyl methyl ether (Hur et al. 1997).

2. Methyl *tert*-butyl ether (MTBE) has been used as a gasoline additive, and concern has arisen over its biodegradability in view of its water solubility that facilitates dispersion in aquatic systems (Johnson et al. 2000). Although pure cultures have been isolated that are able to mineralize MTBE (Hanson et al. 1999; Hatzinger et al. 2001), there are a number of issues that determine its biotransformation or biodegradation, and several pathways have been proposed.

   a. Hydroxylase activity that is gratuitously induced by dicyclopropylketone and is involved in *n*-octane-grown cells of *Pseudomonas putida* Gpo1 was used to support hydroxylation in which *tert*-butanol was produced (Smith and Hyman 2004).

   b. Propane-grown cells of *Mycobacterium vaccae* JOB5 and a strain ENV425 obtained by propane enrichment transformed MTBE to *tert*-butanol or, by hydroxylation (monooxygenation) of the quaternary methyl group to 2-hydroxyisobutyric acid, which was not, however, used as a growth substrate for the organisms (Figure 11.3) (Steffan et al. 1997).

   c. *Mycobacterium austroafricanum* strain IFP 2012 transformed MTBE to *tert*-butanol, 2-methyl-1,2-propandiol, and 2-hydroxyisobutyrate that was degraded (Ferreira et al. 2006). A pathway for its degradation has been proposed that involved a cobalamin-dependent mutase which converted 2-hydroxyisobutyrate-CoA into 3-hydroxybutyryl-CoA (Rohwerder et al. 2006).

   d. *Methylilibium petroleiphilum* (Nakatsu et al. 2006) is a methylotroph that is able to degrade C_1 and C_2 substrates including MTBE when an enzyme MdpA closely homologous to alkane hydrolases was associated with its degradation—though not for *tert*-butanol (Schmidt et al. 2008), and the genetics of the pathway have been elucidated (Kane et al. 2007). On the other hand, it was proposed that degradation of MTBE in treatment systems involved a cytochrome P450 monooxygenase encoded by the *ethB* gene (Jechalke et al. 2011).

   e. As an alternative to degradation, elimination of tertiary alkanols to alkenes was observed with *Aquincola tertiarcarbonis* L108, *Methylibium petroleiphilum* PM1, and *Methylibium* sp. strain R-8: *tert*-butanol produced low concentrations of isobutene and *tert*-amyl alcohol formed a mixture of β- and γ-isoamylene (Schäfer et al. 2011). It was suggested tentatively that the desaturation was carried out by the alcohol monooxygenase MdpJ that has been found by transcriptome analysis of *M. petroleiphilum* exposed to methyl *tert*-butyl ether (Hristova et al. 2007).

3. Degradations of symmetrical long-chain dialkyl ethers are used to illustrate an entirely different metabolic pathway. The di-*n*-heptyl-, di-*n*-octyl-, di-*n*-nonyl-, and di-*n*-decyl ethers are degraded by a strain of *Acinetobacter* sp. to two different groups of metabolites (Figure 11.4):

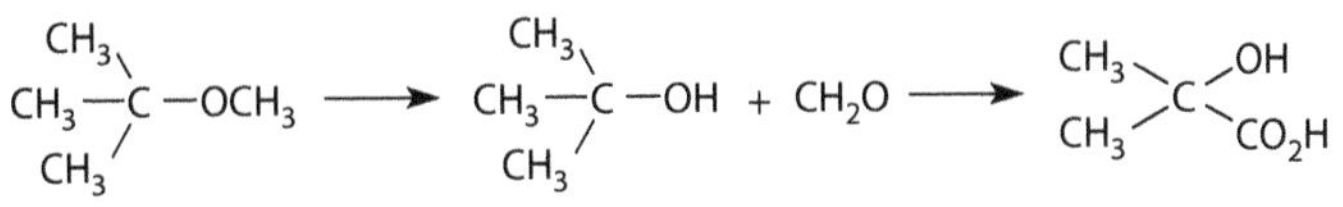

FIGURE 11.3 Degradation of methyl *tert*-butyl ether.

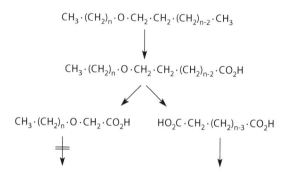

FIGURE 11.4   Biodegradation of di-*n*-heptyl ether by *Acinetobacter* sp.

a. To *n*-heptan-, *n*-octan-, *n*-nonan-, and *n*-decanol-1-acetic acids, which were not metabolized further.

b. To glutaric (C_5), adipic (C_6), pimelic (C_7), and suberic (C_8) acids, which served as sources of carbon and energy. These compounds were formed by terminal oxidation followed by an unusual oxidation at the carbon atom β to the ether bond (Modrzakowski and Finnerty 1980).

4. A less usual reaction is involved in the degradation of succinyloxyacetate by *Zoogloea* sp. (Peterson and Llaneza 1974), which is accomplished by a lyase that produced fumarate and glycolic acid.

5. Bis-(1-chloro-2-propyl)ether has two chiral centers and exists in (R,R)-, (S,S)-, and a *meso* form. It is degraded by *Rhodococcus* sp. with a preference for the (S,S) enantiomer with the intermediate formation of 1-chloro-propan-2-ol and chloroacetone (Garbe et al. 2006).

6. Although hydrolysis of alkyl sulfates by sulfatases is noted in Part 1 of this chapter, ether cleavage has been shown to be the major pathway for the degradation of dodecyltriethoxy sulfate (Hales et al. 1986).

7. Polyethylene glycol can be metabolized aerobically and used for the growth of several species of *Sphingomonas*. Aerobic degradation is generally initiated by oxidation, and an unusual flavoprotein dehydrogenase has been characterized from *Sphingopyxis* (*Sphingomonas*) *terrae* (Sugimoto et al. 2001). Details of the mechanism of degradation have not, however, been resolved (White et al. 1996). In contrast, anaerobic degradation involves rearrangement. This has been investigated in a variety of organisms including *Pelobacter venetianus* (Schink and Stieb 1983), an *Acetobacterium* sp. (Schramm and Schink 1991), *Desulfovibrio desulfuricans*, and *Bacteroides* sp. (Dwyer and Tiedje 1986). The initial product is acetaldehyde, which is formed in two stages by the action of a diol dehydratase and a polyethylene glycol acetaldehyde lyase (Figure 11.5), which is apparently found in all PGE-degrading anaerobic bacteria (Frings et al. 1992).

8. Alkylphenol polyethoxylates, which are widely used as nonionic surfactants, are partially degradable by oxidation and loss of ethoxyethyl groups. Concern has arisen, since, although the number of the ethoxy groups is reduced from about nine or more to two or three (Figure 11.6) (Maki et al. 1994; John and White 1998; Fenner et al. 2002), the metabolites are both appreciably persistent and toxic. Although the alkyl phenols may be formed from the complete oxidation of the polyethoxylate side chains, partially degraded metabolites may apparently be resistant to further degradation (Ball et al. 1989). The degradation of a highly branched

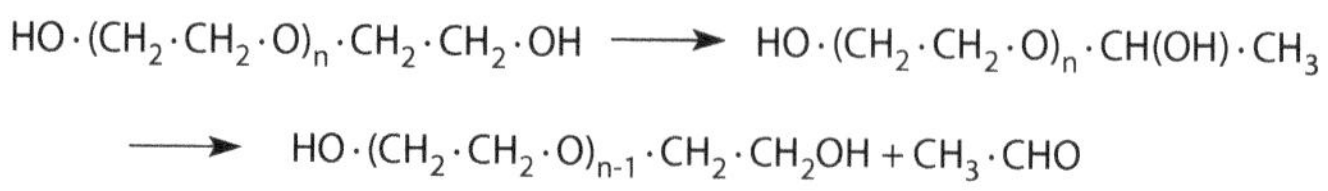

$$HO \cdot (CH_2 \cdot CH_2 \cdot O)_n \cdot CH_2 \cdot CH_2 \cdot OH \longrightarrow HO \cdot (CH_2 \cdot CH_2 \cdot O)_n \cdot CH(OH) \cdot CH_3$$

$$\longrightarrow HO \cdot (CH_2 \cdot CH_2 \cdot O)_{n-1} \cdot CH_2 \cdot CH_2OH + CH_3 \cdot CHO$$

FIGURE 11.5   Anaerobic biodegradation of polyethylene glycol.

$$\text{Ar}-\text{O}-(\text{CH}_2\text{CH}_2\text{O})_n-\text{CH}_2\text{CH}_2\text{OH} \longrightarrow \text{ArO}-(\text{CH}_2\text{CH}_2\text{O})_n-\text{CH}_2\text{CO}_2\text{H} \longrightarrow$$

$$\text{ArO}-(\text{CH}_2\text{CH}_2\text{O})_{n-1}-\text{CH}_2\text{CH}_2\text{OH} \longrightarrow \text{ArO}-(\text{CH}_2\text{CH}_2\text{O})_{n-1}-\text{CH}_2\text{CO}_2\text{H}$$

FIGURE 11.6 Partial degradation of alkylphenol polyethoxylates.

nonylphenol polyethoxylate by *Pseudomonas putida* isolated from activated sludge involved the loss of single ethoxylate groups as acetaldehyde until two ethoxylate residues remained (John and White 1998). The mechanism is reminiscent of that involved in the anaerobic degradation of polyethylene glycols.

9. The metabolism of diethyl ether has been studied in the fungus *Graphium* sp. strain ATCC 58400, which was able to use this as sole source of carbon and energy. When grown with *n*-butane, the fungus was able to transform—though not to degrade—*tert*-butyl methyl ether to *tert*-butanol and *tert*-butyl formate (Hardison et al. 1997). The former is biodegradable under aerobic, and some anaerobic, conditions (Bradley et al. 2002).

10. *Rhodococcus* sp. strain 219 (Bernhardt and Diekmann 1991) and *Pseudonocardia* sp. strain K1 (Thiemer et al. 2003) are able to grow at the expense of tetrahydrofuran. It was suggested that degradation was initiated by monooxygenation to 2-hydroxytetrahydrofuran. Mineralization of 1,4-dioxane by *Pseudonocardia dioxanivorans* strain CB1190 has been described (Parales et al. 1994), and this strain was able to use 1,4-dioxane, tetrahydrofuran, and was distinguished from *P. sulfidoxydans* and *P. hydrocarboxydans* additionally by its ability to use dinitrogen as nitrogen source (Mahendra and Alvarez-Cohen 2005). Although *Pseudonocardia* sp. strain ENV4789 that was isolated by enrichment with tetrahydrofuran was unable to grow with 1,4-dioxan, it produced 2-hydroxyethoxyacetate as a metabolite, analogous to that produced from tetrahydrofuran (Vainberg et al. 2006). In *Pseudonocardia dioxanivorans* the degradation of 1,4-dioxane is initiated by monooxygenation to 2-hydroxy-1,4-dioxane or the open-chain isomer 2-hydroxyethoxyacetaldehyde, both of which produced glyoxylic acid (Mahendra et al. 2007). Details of glyoxylate metabolism by the glyoxylate carboligase pathway to 2-phosphoglycerate and pyruvate have been established using analysis of ^{13}C-1,4-dioxane to establish the route for complete degradation (Grostern et al. 2012).

11. The fungus *Cordyceps sinensis* degraded 1,4-dioxan and other cyclic ethers including 1,3-dioxan and tetrahydrofuran. The pathway for 1,4-dioxane involved reductive formation of ethylene glycol, glycolate, and oxalate, although details of the mechanism were not presented (Nakamiya et al. 2005).

11.7 Aryl Ethers

11.7.1 Diaryl Ethers

Halogenated derivatives of diphenyl ether have been used as herbicides (Scalla et al. 1990) and flame retardants (references in Sellström et al. 1998), and also occur naturally (Voinov et al. 1991). Attention has, therefore, been directed to this class of compounds that formally includes dibenzofurans and dibenzo[1,4] dioxins, which are discussed in Chapter 10, Part 2. The degradation of diphenyl ether itself by *Pseudomonas cepacia* has been examined (Pfeifer et al. 1989, 1993) and yields 2-pyrone-6-carboxylate as a stable end product. This may be formed from the initially produced 2,3-dihydroxydiphenyl ether in a reaction formally analogous (Figure 11.7) to that, whereby 3-*O*-methylgallate is converted into 2-pyrone-4,6-dicarboxylate by 3,4-dihydroxybenzoate 4,5-dioxygenase in pseudomonads (Kersten et al. 1982). The degradation of diphenyl ether by a strain of *Sphingomonas* sp. strain SS33 took place with the fission of both rings (Schmidt et al. 1992), and cells grown with diphenyl ether were able to oxidize dibenzo[1,4] dioxin to 2-(2-hydroxyphenoxy)-*cis,cis*-muconate. After adaptation to growth with 4,4′-difluorodiphenyl ether, the organism grew with the chlorinated, but not the brominated analogs (Schmidt et al. 1993). In general, some degree of recalcitrance seems to be associated with halogenated diaryl ethers.

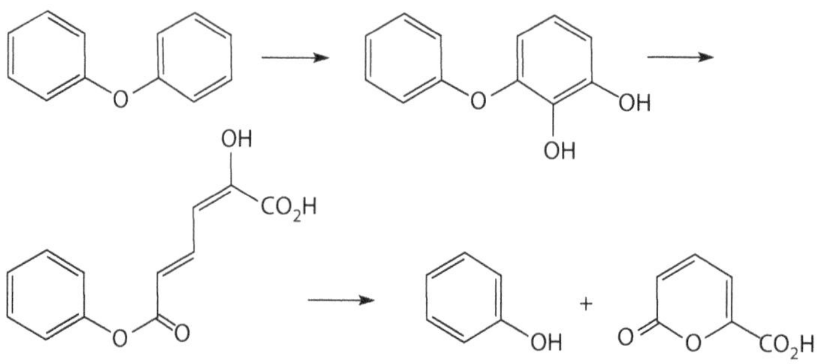

FIGURE 11.7   Biodegradation of diphenyl ether.

11.7.2 Brominated Diphenyl Ethers

Brominated diphenyl ethers have been widely used as flame retardants, and concern has been expressed over their persistence and the toxicity of some congeners. It is worth noting that polybrominated dibenzofurans and dibenzo[1,4]dioxins can be produced by thermal reactions from polybrominated diphenyl ethers (Buser 1986) and other flame retardants (Thoma et al. 1986). Brominated diphenyl ethers are now ubiquitous, and have been found in biota including humans, marine mammals, fish, and birds' eggs, and in samples of air and sediment (Hites 2004). In these, the predominant congeners were: BDE 47 (2,4,2′,4′-tetrabromo), BDE 99 (2,4,5,2′,4′-pentabromo), BDE 100 (2,4,6,2′,4′-pentabromo), BDE 153 (2,4,5,2′,4′,5′-hexabromo), and BDE 154 (2,4,5,2′,4′,6′-hexabromo).

The aerobic metabolism of brominated diphenyl ethers has been examined in several strains of bacteria including strains with an established ability to degrade PCBs. All the strains were able to bring about transformation of the congeners with ≤4 substituents, although *Rhodococcus jostii* RHA1 metabolized the pentabromo and *Burkholderia xenovorans* LB400 one of the hexabromo congeners. Whereas strain RHA1 released bromine stoichiometrically, strain LB400 converted monobromo BDE 3 to a hydroxylated derivative (Robrock et al. 2009).

In view of their recovery from sediments, considerable effort has been directed to their anaerobic debromination. This has been examined under a number of different conditions.

1. By analogy with the positive effect of brominated biphenyls in the dechlorination of PCBs (Bedard et al. 1998), the effect of primers in inducing anaerobic debromination has been examined. In the presence of the primers that included 4-bromobenzoate, 2,6-dibromo-biphenyl and decabromobiphenyl, the concentration of decabromodiphenyl ether BDE-209 decreased by 30% in 238 d. Two nona-brominated, and six octa-brominated congeners were produced in which debromination took place primarily at the *meta-* and *para-*positions (Gerecke et al. 2005).

2. Debromination was examined in sediments that were supplemented with tetrahydrofuran to facilitate solubilization, titanium citrate as reductant, and vitamin B_{12}. Within 24 h, decabromodiphenyl ether (BDE-209) was debrominated to BDE-99, BDE-119, BDE-47, and BDE-66, compared with the slow debromination after 3.5 years in their absence that produced a range of partly debrominated nona-, octa- hepta-, and hexa-BDEs in low concentrations (Tokarz et al. 2008).

3. The anaerobic debromination of brominated diphenyl ethers was examined in microcosms, and in sediment-free enrichment cultures. The substrates were supplied as solutions in trichloroethene or in nonane, and cultures were either amended with a range of organic hydrogen donors or in their absence. In a culture amended with acetate and H_2 that was the most active, nonabromo-congeners were unchanged after incubation for 42 d, whereas heptabromo- and octabromo- congeners were debrominated to tetrabromo-, pentabromo-, and hexabromo-congeners (Lee and He 2010).

4. a. In cells grown with trichloroethene, *Sulfurospirillum multivorans* debrominated decabromodiphenyl ether to octa and hepta congeners (He et al. 2006).

   b. Anaerobic debromination was examined in three cultures with established dehalogenation capabilities—a trichloroethene-enrichment containing *Dehalococcoides* spp., *Dehalobacter restrictus* PER-K23, and *Desulfitobacterium hafniense* PCP-1 (Robrock et al. 2008). All of them were able to carry out debromination of PBDE congeners, the common pattern involving the successive formation:

octa-197/octa-203/octa-196 → hepta-183 → hexa-153 → penta-99 → tetra-47

with preferential debromination at the *meta-* and *para-*positions that is also typical for PCBs and DBEs. Although debromination of the most highly brominated congeners was extremely slow, some were able to carry out the complete debromination of DBE-47. Not all dehalogenating bacteria were capable of this activity since *Desulfomonile tiedjei* DCB-1 lacked this activity.

11.7.3 Aryl–Alkyl Ethers

1. *Phenoxyalkanoates.* There has been considerable interest in the persistence of chlorinated phenoxyalkanoates—and particularly of phenoxyacetates and phenoxypropionates, which have been used as herbicides. This has, therefore, stimulated studies on the degradation of these aryl–alkyl ethers. Considerable effort has been directed to elucidating the subsequent steps that culminate in the fission of the aromatic ring and have been discussed in Chapter 9, Part 2. The first step in the degradation of phenoxyalkanoates is dealkylation to the corresponding phenol with the formation of glyoxylate from phenoxyacetates or acetoacetate and acetone from phenoxypropionates. This reaction is mediated by an α-oxoglutarate-dependent dioxygenase (Fukumori and Hausinger 1993), and it has been shown that one atom of this is incorporated into pyruvate and succinate in cell extracts of *Sphingomonas herbicidovorans* by using $^{18}O_2$ (Nickel et al. 1997). This strain is able to degrade both enantiomers of the racemate (*R,S*)-4-chloro-2-methylphenoxypropionate (mecoprop) (Zipper et al. 1996), and dioxygenases specific for the *R*- and *S*-enantiomers are induced by growth of cells with the respective enantiomer (Nickel et al. 1997). Enantiospecific dioxygenases involved in the degradation of 2-(2,4-dichlorophenoxy)propionate have been characterized in *Delftia acidovorans* (Schleinitz et al. 2004) and in *Sphingomonas herbicidovorans* (Müller et al. 2006). The resulting chlorophenols are then degraded by intradiol fission of 2,4-dichlorocatechol and established pathways (Müller et al. 2004).

2. *Aryl methyl ethers.* A great deal of attention has been directed to the demethylation of aryl methyl ethers on account of interest in the degradation of lignin and related compounds by both aerobic and anaerobic organisms.

   a. In aerobic bacteria, the degradation of vanillate, isovanillate, and syringate is initiated by de-*O*-methylation, and the resulting catechols are then degraded by extradiol dioxygenation (Kasai et al. 2004; Providenti et al. 2006). Three mechanisms for the demethylation of aryl *O*-methyl ethers have been identified, and are involved in the degradation of methoxylated benzoates.

      i. The methyl group is converted into CH_2O by a dioxygenase vanA and a reductase vanB in *Pseudomonas* sp. strain HR199 (Priefert et al. 1997) and *Acinetobacter* sp. (Segura et al. 1999). The sequences of van B have been divided into three clusters (Civolani et al. 2000).

      ii. A tetrahydrofolate (THF) pathway in which the methyl group is transferred to 5-CH_3-THF and then to 5,10-CH_2-THF. For example, in *Sphingomonas paucimobilis* strain SYK-6, the demethylase LigM is required for the degradation of both vanillate and the product of partial *O*-demethylation of syringate (Abe et al. 2005). In this strain, however, the initial de-*O*-methylation of 2,2′-dihydroxy-3,3′-dimethoxy-4,4′-dicarboxybiphenyl is mediated by an unstable oxygenase (Sonoki et al. 2000), so that both mechanisms operate in the same organism.

      iii. In the degradation of substrates such as syringate (Kasai et al. 2004) and 3,4,5-trimethoxybenzoate (Donnelly and Dagley 1980) from which 3-*O*-methylgallate is formed, the methyl group may be lost as methanol produced by hydrolysis of an intermediate methyl ester. This is discussed in Chapter 8.

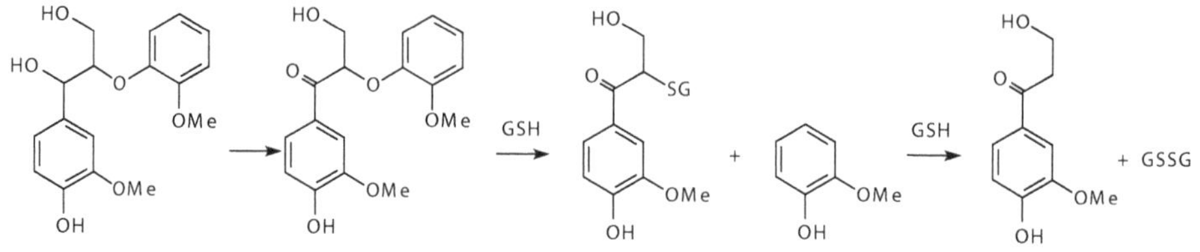

FIGURE 11.8 Biotransformation of a β-aryl ether.

b. Several mechanisms have been elucidated for anaerobic bacteria:

i. Different organisms are able to utilize the methyl of the methoxy group for growth, for example, the acetogenic *Acetobacterium woodii*, although this organism is unable to degrade the aromatic ring (Bache and Pfennig 1981). In *Clostridium pfennigii*, the methyl group may alternatively be converted into butyrate (Krumholz and Bryant 1985), and *Desulfotomaculum thermobenzoicum* is able to use methoxylated benzoates in the presence or absence of sulfate (Tasaki et al. 1992). The pathways of de-*O*-methylation have been elucidated in *A. woodii* (Berman and Frazer 1992), *Sporomusa ovata* (Stupperich and Konle 1993), *Acetobacterium dehalogenans* (Kaufmann et al. 1998), and *Moorella thermoacetica*, and all of them involve an intermediate THF. The *O*-demethylase of *M. thermoacetica* consists of three components: MtvB carries out the *O*-demethylation, MtvC carries out transfer of the methyl group to Co(III) and MtvA the transmethylation to THF with production of CH_3-THF that is then metabolized to acetate after incorporation of CO from CO_2 (Naidu and Ragsdale 2001).

ii. An alternative to de-*O*-methylation by homoacetogenic organisms has been explored in strains of *Desulfitobacterium*. In the presence of fumarate, rather than CO_2 as electron acceptor, vanillate and syringate were able to support growth with the production of 3,4-dihydroxybenzoate and 3,4,5-trihydroxybenzoate, while fumarate was reduced to succinate. A growing culture of the *Desulfitobacterium hafniense* strain DCB2 was also able to use vanillate or syringate as electron donors for the dechlorination of 3-chloro-4-hydroxyphenylacetate to 4-hydroxyphenylacetate (Neumann et al. 2004). During fumarate respiration of *Desulfitobacterium hafniense* strains DCB2 and PCE-S with phenyl methyl ethers, the methyl group was oxidized to CO_2. Although activities of the acetyl-CoA pathway were observed, this was not the major pathway, and it was suggested that a bifunctional CO dehydrogenase/acetyl-CoA synthase was used (Kreher et al. 2008).

iii. Methanethiol is produced by de-*O*-methylation in a few organisms. *Sporobacterium olearium* is able to use a wide range of hydroxylated and methoxylated aromatic compounds for growth (Mechichi et al. 1999). Methanethiol was produced from the methoxy groups and acetate and butyrate from the resulting 3,4,5-trihydroxylated compounds. Sulfide-dependent de-*O*-methylation has been demonstrated in *Parasporobacterium paucivorans* that is able to degrade syringate with the production of both methanethiol and dimethyl sulfide, while gallate that isproduced from 3,4,5-trimethoxybenzoate is degraded to butyrate and acetate(Lomans et al. 2001).

3. Fission of the aryl–O bond in an β-aryl ether by *Sphingomonas paucimobilis* SYK-6 involves the operation of four genes: *LigD*, a dehydrogenase; and three tandem-located glutathione *S*-transferase genes *LigE*, *LigF*, a β-etherase, and *LigG*, a glutathione lyase (Figure 11.8) (Masai et al. 2003).

11.8 References

Abe T, E Masai, K Miyauchi, Y Katayama, M Fukuda. 2005. A tetrahydrofolate-dependent *O*-demethylase, LigM, is crucial for catabolism of vanillate and syringate in *Sphingomonas paucimobilis* SYK-6. *J Bacteriol* 187: 2030–2037.

Adam W, F Heckel, CR Saha-Möller, M Taupp, J-M Meyer, P Schrier. 2005. Opposite enantioselectivities of two phenotypically and genotypically similar strains of *Pseudomonas frederiksbergensis* in bacterial whole-cell sulfoxidation. *Appl Environ Microbiol* 71: 2199–2202.

Ansede JH, PJ Pellechia, DC Yoch. 1999. Metabolism of acrylate to β-hydroxypropionate and its role in dimethylsulfoniopropionate lyase induction by a salt march sediment bacterium, *Alcaligenes faecalis* M3A. *Appl Environ Microbiol* 65: 5075–5081.

Ansede JH, PJ Pellechia, DC Yoch. 2001. Nuclear magnetic resonance analysis of [1-^{13}C]dimethylsulfoniopropionate (DMSP) and [1-^{13}C]acrylate metabolism by a DMSP lyase-producing marine isolate of the α-subclass proteobacteria. *Appl Environ Microbiol* 67: 3134–3139.

Bache R, N Pfennig. 1981. Selective isolation of *Acetobacterium woodii* on methoxylated aromatic acids and determination of growth yields. *Arch Microbiol* 130: 255–261.

Ball HA, M Reinhard, PL McCarty. 1989. Biotransformation of halogenatednonhalogenated octylphenol polyethoxylate residues under aerobic and anaerobic conditions. *Environ Sci Technol* 23: 951–961.

Bedard DL, H van Dort, KA Deweerd. 1998. Brominated biphenyls prime extensive microbial reductive dehalogenation of Arochlor 1260 in Hoisatonic River sediment. *Appl Environ Microbiol* 64: 1786–1795.

Berman MH, AC Frazer. 1992. Importance of tetrahydrofolate and ATP in the anaerobic *O*-demethylation reaction for phenylmethylethers. *Appl Environ Microbiol* 58: 925–931.

Bernhardt D, H Diekmann. 1991. Degradation of dioxane, tetrahydrofuran and other cyclic ethers by an environmental *Rhodococcus* strain. *Appl Microbiol Biotechnol* 36: 120–123.

Bradley PM, JE Landmeyer, FH Chapelle. 2002. TBA biodegradation in surface-water sediments under aerobic and anaerobic conditions. *Environ Sci Technol* 36: 4087–4090.

Buser HR. 1986. Polybrominated dibenzofurans and dibenzo-*p*-dioxins: Thermal reaction products of polybrominated diphenyl ether flame retardants. *Environ Sci Technol* 20: 404–408.

Charlson RJ, JE Lovelock, MO Andreae, SG Warren. 1987. Oceanic phytoplankton, atmospheric sulfur, cloud albedo and climate. *Nature (London)* 326: 655–661.

Civolani C, P Barghini, AR Roncetti, M Ruzzi, A Schiesser. 2000. Bioconversion of ferulic acid into vanillic acid by means of a vanillate-negative mutant of *Pseudomonas fluorescens* strain BF13. *Appl Environ Microbiol* 66: 2311–2317.

de Souza MP, DC Yoch. 1995a. Purification and characterization of dimethylsulfoniopropionate lyase from an *Alcaligenes*-like dimethyl sulfide-producing marine isolate. *Appl Environ Microbiol* 61: 21–26.

de Souza MP, DC Yoch. 1995b. Comparative physiology of dimethyl sulfide production by dimethylsulfoniopropionate lyase in *Pseudomonas doudoroffii* and *Alcaligenes* sp., strain M3A. *Appl Environ Microbiol* 61: 3986–3991.

Denger K, THM Smits, AM Cook. 2006. L-Cysteate sulpho-lyase, a widespread pyridoxal 5'-phosphate-coupled desulfonative enzyme purified from *Silicibacter pomeroyi* DSS-3. *Biochem J* 394: 657–664.

Dominy JE, CR Simmons, PA Karplus, AM Gehring, MH Stipanuk. 2006. Identification and characterization of bacterial cysteine dioxygenases: A new route of cysteine degradation for eubacteria. *J Bacteriol* 188: 5561–5569.

Donnelly MI, S Dagley. 1980. Production of methanol from aromatic acids by *Pseudomonas putida*. *J Bacteriol* 142: 916–924.

Dwyer DF, JM Tiedje. 1986. Metabolism of polyethylene glycol by two anaerobic bacteria, *Desulfovibrio desulfuricans* and a *Bacteroides* sp. *Appl Environ Microbiol* 52: 852–856.

Fenner K, C Kooijman, M Scheringer, K Hungerbühler. 2002. Including transformation products into the risk assessment for chemicals: The case of nonylphenol ethoxylate usage in Switzerland. *Environ Sci Technol* 36: 1147–1154.

Ferreira NL, D Labbé, F Monot, F Fayolle-Guichard, CW Greer. 2006. Genes involved in the methyl *tert*-butyl ether (MTBE) metabolic pathway of *Mycobacterium austroafricanum* IFP 2012. *Microbiology (UK)* 152: 1361–1374.

Frings J, E Schramm, B Schink. 1992. Enzymes involved in anaerobic polyethylene glycol degradation by *Pelobacter venetianus* and *Bacteroides* strain PG1. *Appl Environ Microbiol* 58: 2164–2167.

Fukumori F, RP Hausinger. 1993. *Alcaligenes eutrophus* JMP 134 "2,4-dichlorophenoxyacetate monooxygenase" is an a-ketoglutarate-dependent dioxygenase. *J Bacteriol* 175: 2083–2086.

Garbe L-A, M Moreno-Horn, R Tressi, H Görisch. 2006. Preferential attack of the (*S*)-configured ether-linked carbons in bis-(1-chloro-2-propyl)ether by *Rhodococcus* sp. strain DTB. *FEMS Microbiol Ecol* 55: 113–121.

Gerecke AC, PC Hartmann, NV Heeb, H-PE Kohler, W Giger, P Schmid, M Zennegg, M Kohler. 2005. Anaerobic degradation of decabromodiphenyl ether. *Environ Sci Technol* 39: 1078–1083.

González JM et al. 2003. *Silicibacter pomeroyi* sp. nov. and *Roseovarius nubinhibens* sp. nov., dimethylsulfonio-propionate-demethylating bacteria from marine environments. *Int J Syst Evol Microbiol* 53: 1261–1269.

Grostern A, CM Sales, W-Q Zhuang, O Erbilgin, L Alvarez-Cohen. 2012. Glyoxylate metabolism is a key feature of the metabolic degradation of 1,4-dioxane by *Pseudonocardia dioxanivorans* strain CB1190. *Appl Environ Microbiol* 78: 3298–3308.

Hales SG, GK Watson, KS Dodson, GF White. 1986. A comparative study of the biodegradation of the surfactant sodium dodecyltriethoxy sulphate by four detergent-degrading bacteria. *J Gen Microbiol* 132: 953–961.

Hanlon SP, RA Holt, GR Moore, AG McEwan. 1994. Isolation and characterization of a strain of *Rhodobacter sulfidophilus*: A bacterium which grows autotrophically with dimethylsulphide as electron donor. *Microbiology (UK)* 140: 1953–1958.

Hanson JR, CE Ackerman, KM Scow. 1999. Biodegradation of methyl *tert*-butyl ether by a bacterial pure culture. *Appl Environ Microbiol* 65: 4788–4792.

Hardison L, SS Curie, LM Ciuffeti, MR Hyman. 1997. Metabolism of diethyl ether and cometabolism of methyl *tert*-butyl ether by a filamentous fungus, a *Graphium* sp. *Appl Environ Microbiol* 63: 3059–3167.

Hatzinger PB, K McClay, S Vainberg, M Tugusheva, CW Condee, RJ Steffan. 2001. Biodegradation of methyl *tert*-butyl ether by a pure bacterial culture. *Appl Environ Microbiol* 67: 5601–5607.

He J, KR Robrock, L Alvarez-Cohen. 2006. Microbial reductive debromination of polybrominated diphenyl ethers (PBDEs). *Environ Sci Technol* 40: 4429–4434.

Hites RA. 2004. Polybrominated diphenyl ethers in the environment and in people: A meta-analysis of concentrations. *Environ Sci Technol* 38: 945–956.

Holland HL. 1988. Chiral sulfoxidation by biotransformation of organic sulfides. *Chem Rev* 88: 473–485.

Holland HL, M Carey, and S Kumaresan. 1993. Fungal biotransformation of organophosphines. *Xenobiotica* 23: 519–524.

Howard EC et al. 2006. Bacterial taxa that limit sulfur flux from the ocean. *Science* 314: 649–651.

Hristova KR, R Schmidt, AY Chakicherla, TC Legler, J Wu, PS Chain, KM Scow, SR Kane. 2007. Comparative transcriptome analysis of *Methylibium petroleiphilum* PM1 exposed to the fuel oxygenates methyl *tert*-butyl ether and ethanol. *Appl Environ Microbiol* 73: 7347–7357.

Hur H-G, LM Newman, LP Wackett, MJ Sadowsky. 1997. Toluene 2-monooxygenase-dependent growth of *Burkholderia cepacia* G4/PR1 on diethyl ether. *Appl Environ Microbiol* 63: 1606–1609.

Ince JE, CJ Knowles. 1986. Ethylene formation by cell-free extracts of *Escherichia coli*. *Arch Microbiol* 146: 151–158.

Jechalke S, M Rosell, PM Martínez-Lavanchy, P Pérez-Leiva, T Rohwerder, C Vogt, HH Richnow. 2011. Linking low-level stable isotope fractionation to expression of the cytochrome P450 monooxygenase-encoding *ethB* gene for elucidation of methyl *tert*-butyl ether biodegradation in aerated treatment pond systems. *Appl Environ Microbiol* 77: 1086–1096.

John DM, GF White. 1998. Mechanism for biotransformation of nonylphenol polyethoxylates to xenoestrogens in *Pseudomonas putida*. *J Bacteriol* 180: 4332–4338.

Johnson R, J Pankow, D Bender, C Price, J Zogorski. 2000. MTBE: To what extent will past releases contaminate community water supply wells? *Environ Sci Technol* 34: 210A–217A.

Jordan SL, AJ Kraczkiewicz-Dowjat, DP Kelly, AP Wood. 1995. Novel eubacteria able to grow on carbon disulfide. *Arch Microbiol* 163: 131–137.

Jordan SL, IR McDonald, AJ Kraczkiewicz-Dowjat, DP Kelly, FA Rainey, J-C Murrell, AP Wood. 1997. Autotrophic growth on carbon disulfide is a property of novel strains of *Paracoccus denitrificans*. *Arch Microbiol* 168: 225–236.

Kane SR, et al. 2007. Whole-genome analysis of the methyl tert-butyl ether-degrading beta-proteobacterium *Methylibium petroleiphilum*. *J Bacteriol* 189: 1931–1945.

Kasai K, E Masai, K Miyauchi, Y Katayama, M Fukuda. 2004. Characterization of the 3-*O*-methylgallate dioxygenase gene and evidence of multiple 3-*O*-methylgallate catabolic pathways in *Sphingomonas paucimobilis* SYK-6. *J Bacteriol* 186: 4951–4959.

Kaufmann F, G Wohlfarth, G Diekert. 1998. *O*-demethylase from *Acetobacterium dehalogenans* Substrate specificity and function of the participating proteins. *Eur J Biochem* 253: 706–711.

Kende H. 1989. Enzymes of ethylene biosynthesis. *Plant Physiol* 91: 1–4.

Kersten PJ, S Dagley, JW Whittaker, DM Arciero, JD Lipscomb. 1982. 2-pyrone-4,6-dicarboxylic acid, a catabolite of gallic acids in *Pseudomonas species*. *J Bacteriol* 152: 1154–1162.

Kim Y-H, K-H Engesser. 2004. Degradation of alkyl ethers, aralkyl ethers, and dibenzyl ether by *Rhodococcus* sp. strain DEE 5151, isolated from diethyl ether-containing enrichment cultures. *Appl Environ Microbiol* 70: 4398–4401.

Kredich NM, LJ Foote, BS Keenan. 1973. The stoichiometry and kinetics of the inducible cysteine desulfhydrase from *Salmonella typhimurium*. *J Biol Chem* 248: 6187–6196.

Kreher S, A Schilhabel, G Diekert. 2008. Enzymes involved in the anoxic utilization of phenyl methyl ethers by *Desulfitobacterium hafniense* DCB2 and *Desulfitobacterium hafniense* PCE-S. *Arch Microbiol* 190: 489–495.

Krumholz LR, MP Bryant. 1985. *Clostridium pfennigi* sp. nov. uses methoxyl groups of monobenzenoids and produces butyrate. *Appl Environ Microbiol* 35: 454–456.

Lee LK, J He. 2010. Reductive debromination of polybrominated diphenyl ethers by anaerobic bacteria from soils and sediments. *Appl Environ Microbiol* 76: 794–802.

Lomans BP, P Leijdekkers, J-P Wesselink, P Bakkes, A Pol, C van der Drift, HJP op den Camp. 2001. Obligate sulfide-dependent degradation of methoxylated aromatic compounds and formation of methanethiol and dimethyl sulfide by a freshwater sediment isolate, *Parasporobacterium paucivorans* gen. nov., sp. nov. *Appl Environ Microbiol* 67: 4017–4203.

Lomans BP, R Maas, R Luderer, HJP op den Camp, A Pol, C van der Drift, GD Vogels. 1999. Isolation and characterization of *Methanomethylovorans hollandica* gen. nov., sp. nov., isolated from freshwater sediment, a methylotrophic methanogen able to grow with dimethyl sulfide and methanethiol. *Appl Environ Microbiol* 65: 3641–3650.

Mahendra S, L Alvarez-Cohen. 2005. *Pseudonocardia dioxanivorans* sp. nov., a novel actinomycete that grows on 1,4-dioxane. *Int J Syst Evol Microbiol* 55: 593–598.

Mahendra S, CJ Petzold, EE Baidoo, JD Keasling, L Alvarez-Cohen. 2007. Identification of the intermediates of *in vivo* oxidation of 1,4-dioxane by monooxygenase-containing bacteria. *Environ Sci Technol* 41: 7330–7336.

Maki H, N Masuda, Y Fujiwara, M Ile, M Fujita. 1994. Degradation of alkylphenol ethoxylates by *Pseudomonas* sp. strain TR01. *Appl Environ Microbiol* 60: 2265–2271.

Mansouri S, AW Bunch. 1989. Bacterial synthesis from 2-oxo-4-thiobutyric acid and from methionine. *J Gen Microbiol* 135: 2819–2827.

Masai E, A Ichimura, Y Sato, K Miyauchi, Y Katayama, M Fukuda. 2003. Roles of enantioselective glutathione *S*-transferases in cleavage of β-aryl ether. *J Bacteriol* 185: 1768–1775.

Mechichi T, M Labat, J-L Garcia, P Thomas, BKC Patel. 1999. *Sporobacterium olearium* gen. nov., sp. nov., a new methanethiol-producing bacterium that degrades aromatic compounds, isolated from an olive mill wastewater treatment digester. *Int J Syst Bacteriol* 49: 1741–1748.

Miller KW. 1992. Reductive desulfurization of dibenzyldisulfide. *Appl Environ Microbiol* 58: 2176–2179.

Modrzakowski MC, WR Finnerty. 1980. Metabolism of symmetrical dialkyl ethers by *Acinetobacter* sp. HO1-N. *Arch Microbiol* 126: 285–290.

Müller TA, SM Byrde, C Werlen, JR van der Meer, H-P E Kohler. 2004. Genetic analysis of phenoxyalkanoic acid degradation in *Sphingomonas herbicidovorans* MH. *Appl Environ Microbiol* 70: 6066–6075.

Müller TA, T Fleischmann, JR van der Meer, H-P E Kohler. 2006. Purification and characterization of two enantioselective α-ketoglutarate-dependent dioxygenases, RdpA and SdpA, from *Sphingomonas herbicidovorans*. *Appl Environ Microbiol* 72: 4853–4861.

Nagasawa T, T Ishii, H Kumagai, H Yamada. 1985. D-cysteine desulfhydrase of *Escherichia coli*. *Eur J Biochem* 153: 541–551.

Nagasawa T, T Ishii, H Yamada. 1988. Physiological comparison of D-cycteine desulfhydrase of *Escherichia coli* with 3-chloro-D-alanine dehydrochlorinase of *Pseudomonas putida* CR 1-1. *Arch Microbiol* 149: 413–416.

Naidu D, SW Ragsdale. 2001. Characterization of a three-component vanillate *O*-demethylase from *Moorella thermoacetica*. *J Bacteriol* 183: 3276–3281.

Nakamiya K, S Hashimoto, H Ito, JS Edmonds. 2005. Degradation of 1,4-dioxane and cyclic ethers by an isolated fungus. *Appl Environ Microbiol* 71: 1254–1258.

Nakatsu CH, K Hristova, S Hanada, X-Y Meng, JR Hanson, KM Scow, Y Kamagata. 2006. *Methylibium petroleiphilum* gen. nov., sp. nov., a novel methyl *tert*-butyl ether-degrading methylotroph of the *Betaproteobacteria*. *Int J Syst Evol Microbiol* 56: 983–989.

Neumann A, T Engelmann, R Schmitz, Y Greiser, A Orthaus, G Diekert. 2004. Phenyl methyl ethers: Novel electron donors for respiratory growth of *Desulfitobacterium hafniense* and *Desulfitobacterium* sp. strain PCE-S. *Arch Microbiol* 181: 245–249.

Newton WA, Y Morino, EE Snell. 1965. Properties of crystalline tryptophanase. *J Biol Chem* 240: 1211–1218.

Nickel K, MJ-F Suter, H-PE Kohler. 1997. Involvement of two β-ketoglutarate-dependent dioxygenases in enantioselective degradation of (*R*)- and (*S*)-mecoprop by *Sphingomonas herbicidovorans* MH. *Appl Environ Microbiol* 63: 6674–6679.

Oremland RS, RS Kiene, I Mathrani, MJ Whiticar, DR Boone. 1989. Description of an estuarine methylotrophic methanogen which grows on dimethyl sulfide. *Appl Environ Microbiol* 55: 994–1002.

Parales RE, JE Adanus, N White, HD May. 1994. Degradation of 1,4-dioxane by an actinomycete in pure cultures. *Appl Environ Microbiol* 60: 4527–4530.

Peterson D, J Llaneza. 1974. Identification of a carbon–oxygen lyase activity cleaving the ether linkage in carboxymethyloxysuccinic acid. *Arch Biochem Biophys* 162: 135–146.

Priefert H, J Rabenborst, A Steinbüchek. 1997. Molecular characterization of genes of *Pseudomonas* sp. strain HR199 involved in bioconversion of vanillin to protocatechuate. *J Bacteriol* 179: 2595–2607.

Pfeifer F, S Schacht, J Klein, HG Trüper. 1989. Degradation of diphenyl ether by *Pseudomonas cepacia*. *Arch Microbiol* 152: 515–519.

Pfeifer F, HG Trüper, J Klein, S Schacht. 1993. Degradation of diphenylether by *Pseudomonas cepacia* Et4: Enzymatic release of phenol from 2,3-dihydroxydiphenylether. *Arch Microbiol* 159: 323–329.

Providenti MA, JM O'Brien, J Ruff, AM Cook, IB Lambert. 2006. Metabolism of isovanillate, vanillate, and veratrate by *Comamonas testosteroni* strain BR6020. *J Bacteriol* 188: 3862–3869.

Robrock KR, M Coelhan, DL Sedlak, L Alvarez-Cohen. 2009. Aerobic biotransformation of polybrominated diphenyl ethers (PBDEs) by bacterial isolates. *Environ Sci Technol* 43: 5705–5711.

Robrock KR, P Korytár, L Alvarez-Cohen. 2008. Pathways for the anaerobic microbial debromination of polybrominated diphenyl ethers. *Environ Sci Technol* 42: 2845–2852.

Rohwerder T, U Breuer, D Benndorf, U Lechner, RH Müller. 2006. The alkyl *tert*-butyl ether intermediate 2-hydroxyisobutyrate is degraded via a novel cobalamin-dependent mutase pathway. *Appl Environ Microbiol* 72: 4128–4135.

Rossol I, A Pühler. 1992. The *Corynebacterium glutamicum aecD* gene encodes a C–S lyase with alpha-beta-elimination activity that degrades aminoethylcysteine. *J Bacteriol* 174: 2968–2977.

Scalla R, M Matringe, J-M Camadroo, P Labbe. 1990. Recent advances in the mode of action of diphenyl ethers and related herbicides. *Z Naturforsch* 45c: 503–511.

Schink B, M Stieb. 1983. Fermentative degradation of polyethylene glycol by a strictly anaerobic, Gram-negative, nonsporeforming bacterium, *Pelobacter venetianus* sp. nov. *Appl Environ Microbiol* 45: 1905–1913.

Schleinitz KM, S Kleinsteuber, T Vallaeys, W Babel. 2004. Localization and characterization of two novel genes encoding stereospecific dioxygenases catalyzing 2(2,4-dichlorophenoxy)propionate cleavage in *Delftia acidovorans* MC1. *Appl Environ Microbiol* 70: 5357–5365.

Schmidt R, V Battaglia, K Scow, S Kane, KR Hristova. 2008. Involvement of a novel enzyme MdpA, in methyl *tert*-butyl ether degradation in *Methylibium petroleiphilum* PM1. *Appl Environ Microbiol* 74: 6631–6638.

Schmidt S, P Fortnagel, R-M Wittich. 1993. Biodegradation and transformation of 4,4'- and 2,4-dihalodiphenyl ethers by *Sphingomonas* sp. strain SS33. *Appl Environ Microbiol* 59: 3931–3933.

Schmidt S, R-M Wittich, D Erdmann, H Wilkes, W Francke, P Fortnagel. 1992. Biodegradation of diphenyl ether and its monohalogenated derivatives by *Sphingomonas* sp. strain SS3. *Appl Environ Microbiol* 58: 2744–2750.

Schramm E, B Schink. 1991. Ether-cleaving enzyme and diol dehydratase involved in anaerobic polyethylene glycol degradation by a new *Acetobacterium* sp. *Biodegradation* 2: 71–79.

Schäfer F, L Muzica, J Schuster, N Treuter, M Rosell, H Harms, RK Müller, T Rohwerder. 2011. Formation of alkenes via degradation of *tert*-alkyl ethers and alcohols, by *Aquinocola tertiarcarbonis* L108, *Methylibium spp. Appl Environ Microbiol* 77: 5981–5987.

Segura A, PV Bünz, DA D'Argenio, LN Ornston. 1999. Genetic analysis of a chromosomal region containing van A and van B, genes required for conversion of either ferulate or vanillate to protocatechuate in *Acinetobacter. J Bacteriol* 181: 3494–3504.

Sellström U, A Kierkegaard, C de Witt, B Jansson. 1998. Polybrominated diphenyl ethers and hexabromocyclododecane in sediment and fish from a Swedish river. *Environ Toxicol Chem* 17: 1065–1072.

Smith CA, MR Hyman. 2004. Oxidation of methyl *tert*-butyl ether by alkane hydroxylase in dicyclopropylketone-induced and *n*-octane-grown *Pseudomonas putida* Gpo1. *Appl Environ Microbiol* 70: 4544–4550.

Smith NA, DP Kelly. 1988. Isolation and physiological characterization of autotrophic sulphur bacteria oxidizing dimethyl disulphide as sole source of energy. *J Gen Microbiol* 134: 1407–1417.

Sonoki T, T Obi, S Kubota, M Higashi, E Masai, Y Katayama. 2000. Coexistence of two different O demethylation systems in lignin metabolism by *Sphingomonas paucimobilis* SYK-6: Cloning and sequencing of the lignin biphenyl-specific O-demthylase4 (LigX) gene. *Appl Environ Microbiol* 66: 2125–2132.

Steffan RJ, K McClay, S Vainberg, CW Condee, D Zhang. 1997. Biodegradation of the gasoline oxygenates methyl *tert*-butyl ether, ethyl *tert*-butyl ether, and amyl *tert*-butyl ether by propane-oxidizing bacteria. *Appl Environ Microbiol* 63: 4216–4222.

Stupperich E, R Konle. 1993. Corrinoid-dependent methyl transfer reactions are involved in methanol and 3,4-dimethoxybenmzoate metabolism by *Sporomusa ovata. Appl Environ Microbiol* 59: 3110–3116.

Sugimoto M, M Tanaba, M Hataya, S Enokibara, JA Duine, F Kawai. 2001. The first step in polyethylene glycol degradation by sphingomonads proceeds via a flavoprotein alcohol dehydrogenase containing flavin adenine dinucleotide. *J Bacteriol* 183: 6694–6698.

Suylen GMH, PJ Large, JP van Dijken, JG Kuenen. 1987. Methyl mercaptan oxidase, a key enzyme in the metabolism of methylated sulphur compounds by *Hyphomicrobium* EG. *J Gen Microbiol* 133: 2989–2997.

Suylen GMH, GC Stefess, JG Kuenen. 1986. Chemolithotrophic potential of a *Hyphomicrobium* species, capable of growth on methylated sulphur compounds. *Arch Microbiol* 146: 192–198.

Tasaki M, Y Kamagata, K Nakamura, E Mikami. 1992. Utilization of methoxylated benzoates and formation of intermediates by *Desulfotomaculum thermobenzoicum* in the presence or absence of sulfate. *Arch Microbiol* 157: 209–212.

Thiemer B, JR Andreesen, T Schrader. 2003. Cloning and characterization of a gene cluster involved in tetrahydrofuran degradation in *Pseudonocardia* sp. strain K1. *Arch Microbiol* 179: 266–277.

Thoma H, S Rist, G Hauschulz, O Hutzinger. 1986. Polybrominated dibenzodioxins and—Furans from the pyrolysis of some flame retardants. *Chemosphere* 15: 649–652.

Tokarz JA, M-Y Ahn, J Leng, TR Filley, L Nies. 2008. Reductive debromination of polybrominated diphenyl ethers in anaerobic sediment and a biomimetic system. *Environ Sci Technol* 42: 1157–1164.

Vainberg S, K McClay, H Masuda, D Root, C Condee, GJ Zylstra, RJ Steffan. 2006. Biodegradation of ether pollutants by *Pseudonocardia* sp. strain ENV478. *Appl Environ Microbiol* 72: 5218–5224.

van den Wijngaard AJ, J Prins, AJAC Smal, DB Janssen. 1993. Degradation of 2-chloroethylvinyl ether by *Ancylobacter aquaticus* AD25 and AD27. *Appl Environ Microbiol* 59: 2777–2783.

van der Maarel MJEC, M Jansen, R Haanstra, WG Meijer, TA Hansen. 1996b. Demethylation of dimethylsulfoniopropionate to 3-*S*-methylmercaptopropionate by marine sulfate-reducing bacteria. *Appl Environ Microbiol* 62: 3978–3984.

van der Maarel MJEC, S van Bergeijk, AF van Werkhoven, AM Laverman, WG Maijer, WT Stam, TA Hansen. 1996a. Cleavage of dimethylsulfoniopropionate and reduction of acrylate by *Desulfovibrio acrylicus* sp. nov. *Arch Microbiol* 166: 109–115.

van Hamme JD, PM Fedorak, JM Foght, MR Gray, HD Dettman. 2004. Use of novel fluorinated organosulfur compound to isolate bacteria capable of carbon–sulfur bond fission. *Appl Environ Microbiol* 70: 1487–1493.

Visscher PT, BF Taylor. 1993. A new mechanism for the aerobic catabolism of dimethyl sulfide. *Appl Environ Microbiol* 59: 3784–3789.

Visscher PT, BF Taylor. 1994. Demethylation of dimethylsulfoniopropionate to 3-mercaptopropionatre by an aerobic bacterium. *Appl Environ Microbiol* 60: 4617–4619.

Voinov VG, YuN El'kin, TA Kuznetsova, II Mal'tsev, VV Mikhailov, VA Sasunkevich. 1991. Use of mass spectrometry for the detection and identification of bromine-containing diphenyl ethers. *J Chromatogr* 586: 360–362.

Wagner C, ER Stadtman. 1962. Bacterial fermentation of dimethyl-β-propiothetin. *Arch Biochem Biophys* 98: 331–336.

White GF, NJ Russell, EC Tidswell. 1996. Bacterial scission of ether bonds. *Microbiol Rev* 60: 216–232.

Yoch D. 2002. Dimethylsulfoniopropionate: Its sources, role in the marine food web, and biological degradation to dimethylsulfide. *Appl Environ Microbiol* 68: 5804–5815.

Yoshida Y, Y Nakano, A Amano, M Yoshimura, H Fukamachi, T Oho, Y Koga. 2002. lcd from *Streptococcus anginosus* encodes a C–S lyase with α,β-elimination activity that degrades L-cysteine. *Microbiology (UK)* 148: 3961–3970.

Zipper C, K Nickel, W Angst, H-P E Kohler. 1996. Complete microbial degradation of both enantiomers of the chiral herbicide Mecoprop [(R,S)-2-(4-chloro-2-methylphenoxy)]propionic acid in an enantioselective manner by *Sphingomonas herbicidovorans* sp. nov. *Appl Environ Microbiol* 62: 4318–4322.

11.9 Sulfides, Disulfides, and Related Compounds

Interest in the possible persistence of aliphatic sulfides has arisen since they are produced in marine anaerobic sediments, and dimethylsulfide may be implicated in climate alteration (Charlson et al. 1987). Dimethylsulfoniopropionate is produced by marine algae as an osmolyte, and has aroused attention for several reasons. It can be the source of climatically active dimethylsulfide (Yoch 2002), so the role of specific bacteria has been considered in limiting its flux from the ocean and deflecting the products of its transformation into the microbial sulfur cycle (Howard et al. 2006).

Sulfides and related compounds may be degraded by a range of different pathways, which are illustrated in the following examples.

11.9.1 Introduction

Bacteria selected for growth with *bis*-(3-pentfluorophenylpropy)-sulfide as sulfur source were able to use dimethyl sulfoxide, dibenzyl sulfide, and some long-chain disulfides as sources of sulfur (van Hamme et al. 2004). Degradation took place by oxidation to the sulfone, scission of the C–S bond to an alkanol and an alkyl sulfinate that is degraded with loss of the sulfur that was used for growth. Sulfides and sulfur-containing amino acids may be degraded by a range of pathways that are illustrated in the following examples and involve either, or both, α,β-elimination and γ-scission.

11.9.1.1 Cysteine

11.9.1.1.1 Production of H_2S from Cysteine by α,β-Elimination

1. A great deal of attention has been directed to the conversion of L-cysteine to pyruvate, ammonia, and sulfide (Figure 11.9) and the desulfhydrase has been characterized from *Salmonella typhimurium* (Kredich et al. 1972), and from *Streptococcus anginosus* (Yoshida et al. 2002). In *S. typhimurium*, only a fraction of the pyruvate is recovered, since reaction with the intermediate aminoacrylate produces 2-methyl-2,4-thiazolidine-2-carboxylate (Kredich et al. 1973) and proteins encoding L-cysteine desulfhydrase activity have been found in *O*-acetylserine sulfhydrases (Awano et al. 2005). D-cysteine can also be degraded to pyruvate, sulfide, and ammonia by the desulfhydrase that has been found in enteric bacteria including *Klebsiella pneumoniae*, *Enterobacter cloacae*, and *Citrobacter freundii* (Nagasawa et al. 1985). The ability to degrade D-cysteine is significant, since D-cysteine is an inhibitor of *E. coli*, and protection against this is obtained by addition of the branched-chain aminoacids and the synthesis of D-cysteine desulfhydrase (Soutourina et al. 2001). The desulfhydrase is able to bring about elimination reactions from other D-cysteine derivatives, and from 3-chloro-D-alanine (Nagasawa et al. 1988), while pyruvate is also formed from L-cysteine by the activity of tryptophanase (Newton

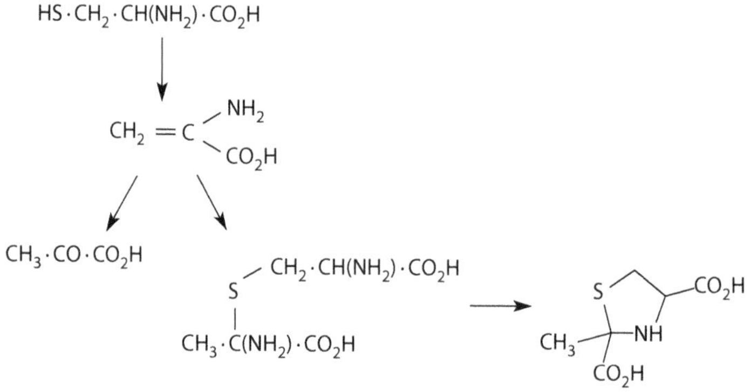

FIGURE 11.9 Elimination reactions during metabolism of cysteine.

et al. 1965). It is worth noting, however, that the formation of L-alanine and H_2Se (or elemental Se^0) from L-selenocysteine is carried out by a β-lyase containing pyridoxal 5′phosphate that has been characterized from *Citrobacter freundii* (Chocat et al. 1985). This activity has been found in a few other bacteria, fungi, and yeasts (Chocat et al. 1983).

2. This reaction can also be carried out by a totally different enzyme that, unlike those already noted and is not dependent on pyridoxal-5′-phosphate. Metabolism of cysteine to pyruvate, NH_3, H_2S by the desulfidase pathway that involved [4Fe–4S] in place of pyridoxal-5-phosphate for removal of the β-proton was originally described from the thermophilic methanogen *Methanocaldococcus jannaschii* (Tchong et al. 2005). On the basis of sequence comparison, it was found that the enzyme had a wide distribution in anaerobic organisms. This has been extended to the existence of the *cdaAB* operon in the red-mouth fish pathogen in *Yersinia ruckeri* in which CdsA was a permease. Sequence analysis revealed the occurrence of the desulfidase in a range of bacteria including the anaerobic foot-rot pathogen *Dichelobacter nodosus* for cattle, sheep and goats, and the facultative anaerobes *Y. enterocolitica*, *Klebsiella pneumoniae*, *Proteus mirabilis* (Méndez et al. 2011). In addition an important finding was the requirement of the *cdsAB* operon to establish full virulence in strains of *Y. ruckeri*.

It is appropriate to note here that a few other α,β-elimination reactions have been found to depend on the presence of [4Fe–4S] clusters, instead of pyridoxal 5′-phosphate. These involve [4Fe–4S] clusters for L-serine dehydratase in *Peptostreptococcus asaccharolyticus* that produces pyruvate (Hofmeister et al. 1997), and for serine deaminase in *Escherichia coli* that produces pyruvate and ammonia (Cicchillo et al. 2004). These reactions may be formally analogous to PLP-independent amino acid racemases in which the α-proton undergoes exchange by a direct 1:1 proton transfer mechanism involving pairs of amino acids, for example, lysine/histidine, two lysines, two cysteines, or two aspartates.

11.9.1.1.2 Dioxygenation and Desulfination

1. In a few bacteria, cysteine can exceptionally be dioxygenated to cysteine sulfinic acid from which pyruvate and sulfite are produced by a desulfinase (Dominy et al. 2006):

$$HS-CH_2-CH(NH_2)-CO_2H \rightarrow HO_2C-CH(NH_2)-CH_2-SO_2H$$
$$\rightarrow CH_3-CH(NH_2)-CO_2H + HSO_3^-$$

Phylogenetic analysis suggested, however, a much wider distribution of this activity. The γ-scission of cysteine sulfinate to alanine and sulfite can be carried out gratuitously by aspartate β-decarboxylase (Soda et al. 1964), by an enzyme that metabolizes selenocysteine to alanine and selenium, contains pyridoxal 5′phosphate and has been designated cysteine sulfinate desulfinase (Mihara et al. 1997).

2. Aerobic degradation of 3,3′-dithiodipropionate by *Advenella* (*Tetrathiobacter*) *mimigardefordensis* produced sulfite and propionyl CoA that entered the methylcitrate cycle (Wübbeler et al. 2008). Production of the 3-mercaptopropionate from the substrate was catalyzed by a dihydrolipoamide dehydrogenase (Wübbeler et al. 2010a), followed by dioxygenation to the sulfinate that was converted into the coenzyme A ester by a succinyl coenzyme A synthase that was purified and characterized by MS (Schürmann et al. 2011).

3. The degradation of the analogous 4,4′-dithiodibutyrate by *Rhodococcus erythropolis* followed a different pathway, although degradation was initiated by reduction to 4-mercaptobutyrate. This was, however followed by oxidation to 4-oxo-4-sulfanyl-butanoate and hydrolysis by a desulfhydrase to sulfide and succinate (Wübbeler et al. 2010b).

11.9.1.1.3 Degradation of Cysteate

This has already been discussed in the section dealing with aliphatic sulfonates, but is repeated here for convenience.

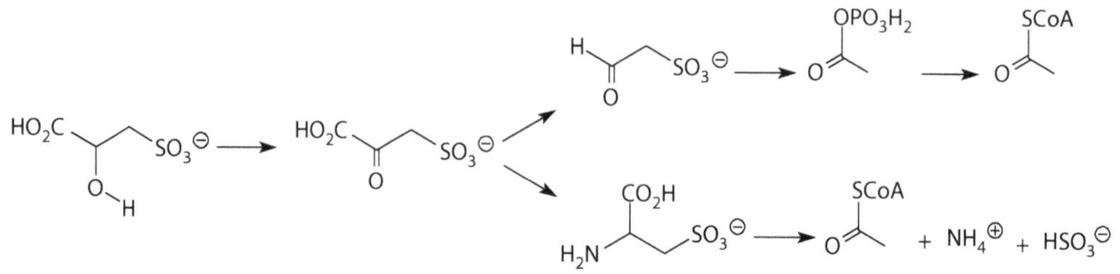

FIGURE 11.10   Production of acetyl-phosphate from 3-sulfolactate in *Roseovarius nubinhibensk.*

1. Degradation of cysteate (2-amino-3-sulfopropionate) by *Paracoccus pantotrophus* takes place by the formation of 3-sulfopyruvate and dehydrogenation to 3-sulfolactate that undergoes γ-scission to pyruvate and sulfite, mediated by a sulfolyase (Rein et al. 2005):

$$HSO_3-CH_2-CH(NH_2)-CO_2H \rightarrow HSO_3-CH_2-CO-CO_2H + NH_3$$
$$\rightarrow HSO_3-CH_2-CH(OH)-CO_2H \rightarrow CH_3-CO-CO_2H + HSO_3^-$$
$$\rightarrow CH_3-CO-CO_2H + NH_3 + HSO_3^-$$

2. In *Ruegeria (Silicibacter) pomeroyi*, L-cysteate sulfolyase brings about the conversion of L-cysteate directly to sulfite, ammonia, and pyruvate in a reaction involving pyridoxal 5′-phosphate (Denger et al. 2006).

3. In *Roseovarius nubinhibens*, however, the degradation pathway for 3-sulfolactate is bifurcated after dehydrogenation to 3-sulfopyruvate by the dehydrogenase SlcD (Denger et al. 2009).

    a. Production of cysteate by transamination, and thence by the activity of cysteate sulfolyase (CuyA) to pyruvate, ammonia and bisulfite,

    b. Production of acetyl-phosphate by the successive activities of two thiamine diphosphate-dependent enzymes—sulfopyruvate decarboxylase (ComDE), and sul foacetaldehyde acetyl transferase (Xsc), and thence of acetyl-CoA (Figure 11.10).

11.9.1.2 Elimination Reactions Carried Out by Lyases

1. A C–S lyase from *Corynebacterium glutamicum* is able to degrade aminoethylcysteine to pyruvate and cysteamine, and this activity is able to confer resistance to aminoethylcysteine (Rossol and Pühler 1992), and has also been found in *Streptococcus anginosus* (Yoshida et al. 2002).

2. There are two metabolic pathways for the degradation of cystathionine

$$HO_2C-CH(NH_2)-CH_2-CH_2-S-CH_2-CH(NH_2)-CO_2H$$

    a. The cystathionine β-lyase that produces homocysteine, pyruvate and ammonia β-lyase → $HO_2C-CH(NH_2)-CH_2-CH_2-SH + CH_3-CO-CO_2H + NH_3$ is encoded by the *malY* gene in *Escherichia coli*, and it has been characterized (Zdych et al. 1995) as a pyridoxal 5′phosphate-dependent enzyme with a rather low sequence homology to the C–S lyase from *Corynebacterium glutamicum* that has already been noted.

    b. Cystathionine γ-lyase catalyzes the α,γ-elimination of L-cystathionine to produce L-cysteine, 2-oxobutyrate, and ammonia:

$$\gamma\text{-lyase} \rightarrow HO_2C-CO-CH_2-CH_3 + HS-CH_2-CH(NH_2)-CO_2H + NH_3$$

    It has been purified from *Lactococcus lactis* subsp. *cremoris* SK11 (Bruinenberg et al. 1997), and is widely distributed in actinomycetes (Nagasawa et al. 1984).

3. Elimination is one of the pathways used for the degradation of dimethylsulfoniopropionate $HO_2C-CH_2-CH_2-S^+(CH_3)_2$ and has been demonstrated in a strain of *Clostridium* sp. (Wagner and Stadtman 1962):

$$\gamma\text{-lyase: } HO_2C-CH_2-CH_2-S^+(CH_3)_2 \rightarrow (CH_3)_2S + CH_2{=}CH-CO_2H$$
$$\rightarrow CH_3-CH_2-CO_2H$$

The enzyme has been purified from a strain of an *Alcaligenes*-like organism (de Souza and Yoch 1995), and the acrylate can be metabolized by *Alcaligenes* sp. strain M3A to β-hydroxypropionate (Ansede et al. 1999, 2001), and by *Desulfovibrio acrylicus* to propionate (van der Maarel et al. 1996a). The alternative de-*S*-methylation is noted here:

$$HO_2C-CH_2-CH_2-S^+(CH_3)_2 \rightarrow CH_3S-CH_2-CH_2-CO_2H \rightarrow HS-CH_2-CH_2-CO_2H.$$

11.9.1.3 Metabolism of Methionine

1. The formation of ethene from methionine by *Escherichia coli* takes place by an α,β-elimination reaction (Ince and Knowles 1986) (Figure 11.11), though considerable complexities in the control and regulation of this reaction have emerged (Mansouri and Bunch 1989). It should be noted that the synthesis of ethene in plants proceeds by an entirely different reaction via *S*-adenosyl methionine and 1-aminocyclopropane-1-carboxylate (Kende 1989).

2. L-methionine γ-lyase produces methanethiol, ammonia and 2-oxobutyrate from L-methionine by an α,γ-elimination:

$$\gamma\text{-lyase} \rightarrow CH_3-SH + HO_2C-CO-CH_2-CH_3$$

and the enzyme has been characterized from *Brevibacterium linens* strain BL2 (Dias and Weimer 1998). This reaction is probably a major source of volatile organosulfur compounds in this organism (Amarita et al. 2004). *Ferroplasma acidarmanus* has an obligate requirement for a minimum concentration of 100 mM sulfate and produced methanethiol from ^{35}S-labelled methionine, cysteine and sulfate, while ^{3}H-labelled methionine suggested that this was used as the source of the methyl group (Baumler et al. 2007).

3. In the methionine salvage pathway for *Klebsiella pneumoniae*, 5-thiomethyladenosine is converted into 2,3-dioxo-5-methylthio-1-phosphopentane from which it is transformed non-enzymatically to 2-oxo-4-thiomethylbutyrate and thence enzymatically to methionine, or by dioxygenation to 4-thiomethylbutyrate (Wray and Abeles 1995) (Figure 11.12).

11.9.1.4 Metabolism of Dimethyl Sulfide and Related Compounds

1. Dimethyldisulfide is degraded by autotrophic sulfur bacteria with the formation of sulfate and CO_2 that then enters the Benson–Calvin cycle (Smith and Kelly 1988). On the other hand,

$$CH_3\cdot S\cdot CH_2\cdot CH_2\cdot CH(NH_2)\cdot CO_2H \longrightarrow CH_3\cdot S\cdot CH_2\cdot CH_2\cdot CO\cdot CO_2H$$
$$\longrightarrow CH_2{=}CH_2 + CH_3SH + CO_2$$

FIGURE 11.11 Elimination reactions during metabolism of methionine.

$$(CH_3)_2S^+-CH_2-CH_2-CO_2^- \longrightarrow CH_3-S-CH_2-CH_2-CO_2H \longrightarrow HS-CH_2-CH_2-CO_2H$$

FIGURE 11.12 The methionine salvage pathway by dioxygenation of enediol to formate + CO.

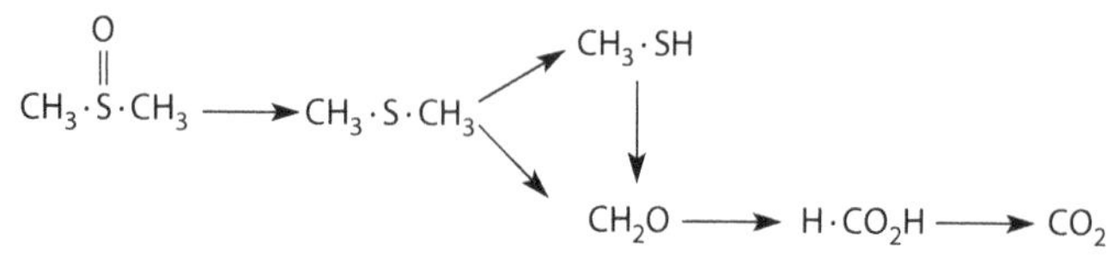

FIGURE 11.13 Biodegradation of dimethyl sulfide by *Hyphomicrobium sp.*

dimethyl sulfide and dimethyl sulfoxide are degraded by a strain of *Hyphomicrobium* sp. by pathways involving the formation from both carbon atoms of formaldehyde which subsequently enters the serine pathway (Suylen et al. 1986) (Figure 11.13). The key enzyme is methanethiol oxidase which converts methanethiol into formaldehyde, sulfide, and peroxide (Suylen et al. 1987). A strain of *Thiobacillus* sp. metabolizes dimethyl sulfide by an alternative pathway involving transfer of the methyl group probably to tetrahydrofolate by a cobalamin carrier (Visscher and Taylor 1993). Oxygen is not involved in the removal of the methyl groups so that the reaction may proceed anaerobically.

2. The metabolism of dimethyl sulfide has attracted attention in the context of climate change and its metabolism has been examined in a number of organisms generally with the production of sulfate. Substantial effort has been given to marine organisms belonging to the genus *Methylophaga* including *Methylophaga aminisulfidivorans* (Kim et al. 2007) that is able to utilize dimethylsulfide, methanol and all the methylamines, *Methylophaga sulfidovorans* can oxidize both sulfide and dimethylsulfide (de Zwart et al. 1996), and *Methylphaga thiooxidans* is able to degrade dimethylsulfide by demethylation to sulfide that was oxidized through thiosulfate to tetrathionate as a terminal metabolite (Boden et al. 2010). In addition to these, *Methanomethylovorans hollandica* is able to grow with both dimethyl sulfide and methanethiol, in addition to methanol and methylamines (Lomans et al. 1999).

3. The metabolism of dimethylsulfone has been described in several facultative methylotrophs (Borodina et al. 2002). It is initiated by the activities of constitutive membrane-associated reductases to produce successively dimethylsulfoxide and dimethylsulfide. This is metabolized to methylsulfide by a monooxygenase that has been purified from *Hyphomicrobium sulfonivorans* (Boden et al. 2011), followed by the activity of methanethiol oxidase to produce sulfide and formaldehyde.

4. Strains of facultatively heterotrophic and methylotrophic bacteria can use CS_2 as sole energy source, and under aerobic conditions also COS, dimethyl sulfide, dimethyl disulfide, and thioacetate (Jordan et al. 1995). It was proposed that the strains belonged to the genus *Thiobacillus* though they are clearly distinct from previously described species, and they have now been assigned to *Paracoccus denitrificans* (Jordan et al. 1997). *Thiomonas* strain WZW is able to grow at the expense of carbon disulfide, dimethylsulfide, dimethyldisulfide and thiosulfate, as well as growth on a complex medium containing glucose (Pol et al. 2007).

5. Anaerobic reduction of dimethyl sulfide with the production of methane (Oremland et al. 1989), and of dibenzyl disulfide to toluenethiol and finally toluene (Miller 1992) have been described.

   Strains provisionally assigned to *Desulfotomaculum* were able to degrade methyl sulfide and dimethyl sulfide, using sulfate, sulfite, or thiosulfate as electron acceptors with the production of sulfide and CO_2 (Tanimoto and Bak 1994). On the other hand, dimethyl sulfide that was used as an electron donor for the growth of *Rhodobacter sulfidophilus* was converted into dimethyl sulfoxide (Hanlon et al. 1994).

11.9.1.5 Transformation of Dimethylsulfoniopropionate

Dimethylsulfoniopropionate (DMSP) is synthesized from methionine and occurs in a ubiquitous range of marine algae from both coastal and oceanic sites (Yoch 2002). It can fulfill a number of functions including osmoregulation, and exists in both a soluble and an insoluble form which may be released during algal senescence or by zooplankton grazing (Wolfe et al. 1994). Degradation of DMSP has been examined in several bacteria which exemplify the basic mechanisms: demethylation to 3-mercaptomethylpropionate with release of methanethiol which may be utilized as sources of sulfur by marine phytoplankton—remarkably in preference to the much greater concentration of readily available sulfate in sea water (Kiene et al. 1999), or fission by a group of lyases to form dimethyl sulfide and acrylate or 3-hydroxypropionate. Whereas the genes encoding the lyases *dddD*, *dddL*, and *dddP* (Curson et al. 2008; Todd et al. 2009) have only a low frequency in oceanic surface water, the *dmdA* that encodes demethylation is abundant in surface water (Howard et al. 2008). Details of the metabolic pathways and the relevant genes and enzymes have been extensively explored.

11.9.1.5.1 Metabolic Pathways

11.9.1.5.1.1 Successive Demethylation Demethylation to 3-methylmercaptopropionate and methanethiol that is encoded by *dmdA* is abundant in marine metagenome sets and represents the dominant pathway for DMSP degradation (Howard et al. 2008).

$$(CH_3)_2S^+–CH_2–CH_2–CO_2^- \rightarrow CH_3–S–CH_2–CH_2–CO_2^- \rightarrow HS–CH_2–CH_2–CO_2^-$$

This has been established in a versatile methylotroph strain BIS-6 (Visscher and Taylor 1994), and the enzyme that has been purified from *Pelagibacter ubique* HTCC1062 and *Ruegeria (Silicibacter) pomeroyi* DSS-3 was strictly specific for dimethylsulfoniopropionate (Reisch et al. 2008). The methyl group is metabolized by the formation of 5-methyl-THF and successively to 10-formyl-THF (Howard et al. 2006), while the metabolism of 3-methylmercaptopropionate involves a complex of reactions. This is initiated by conversion into the CoA thioester by a ligase (DmbB), followed by dehydrogenation to methylthioacrylyl-CoA (DmbC), and hydration with the formation of methanethiol, acetaldehyde, release of CO_2 and free coenzyme A (DmdD) (Reisch et al. 2011b). These enzyme activities have been assayed in *R. pomeroyi* DSS-3 and *R. lacuscaerulensis* (Reisch et al. 2011a). In addition, the genomic database contained a number of bacteria that possessed *dmdB* and *dmdC* so that strains of *Burkholderia thailandensis*, *Pseudoalteromonas atlantica*, *Myxococcus xanthus*, and *Deinococcus radiodurans*—which are not considered to be marine—were able to produce methanethiol from 3-methylmercaptopropionate (Reisch et al. 2011a).

11.9.1.5.1.2 Scission by Lyase Activity Scission of DMSP to produce dimethyl sulfide and acrylate or 3-hydroxypropionate may be carried out by the activity of bacterial lyases. There are several different enzymes encoded by *dddL*, *dddP*, *dddQ*, *dddY*, *ddpW* that have been identified in *Roseovarius nubinhibens* (Kirkwood et al. 2010) and in *Ruegeria pomeroyi* DSS-3 (Todd et al. 2011, 2012), and it has emerged that the protein encoded by *dddQ* is a novel cupin-containing lyase (Todd et al. 2011). Whereas *S. pomeroyi* could grow at the expense of dimethylsulfoniopropionate and a number of putative degradation products including acrylate, *R. nubinhibens* was unable to grow with acrylate, and neither strain could utilize 3-mercaptopropionate (González et al. 2003).

Lyase activity has also been demonstrated in crude cell extracts of algae supplemented with DMSP: the prymnesiophyte *Phaeocystis* sp. (Stefels and Dijkhuizen 1996), and in several strains of the coccolithophore *Emiliania huxleyi* (Steinke et al. 1998). Although these algae may dominate algal blooms, there seems to be an apparent contradiction with the minor role of dimethylsulfide production by bacteria.

A strain of *Halomonas* sp. isolated from the phyllophane of the green macroalga *Ulva* sp. was able to use dimethylsulfoniopropionate or acrylate as sole carbon sources, while the formation of dimethylsulfide and 3-hydroxypropionate was mediated by *dddA* and *dddC* (Todd et al. 2009).

Some additional issues merit note:

1. The transformation of DMSP has been examined in strains of bacteria isolated from seawater that belonged to the *Roseobacter* group. Most strains were able to use DMSP as the sole source

of carbon, and during growth with glucose they were able to produce dimethylsulfide from DMSP, while some were in addition able to release methanethiol. Both the lyase and demethyl-ase pathways were found in the same strain and were therefore able to function simultaneously (González et al. 1999).

2. a. Location of DMSP lyase was examined in a strain LFR that had been isolated from Sargasso Sea surface water and belonged to the α subgroup of the Proteobacteria. Transport of DMSP into the cell took place before the formation of methylsulfide which is consistent with an intracellular location of the lyase (Ledyard and Dacey 1994).

   b. The uptake of DMSP was analyzed in strains of marine bacteria that were isolated from estuarine surface sediments and it was shown that in one strain DMSP was fissioned on the surface of lyase-induced cells, in the other DMSP was actively taken up by an inducible system and concentrated intracellularly before fission to dimethylsulfide and acrylate (Yoch et al. 1997).

   c. Acrylate is produced from DMSP by the activity of lyases and may serve as a growth sub-strate for bacteria. In strain LFR that belonged to the α subgroup of Proteobacteria, the kinetics of production from [^{13}C]-DMSP and transformation to 3-hydroxypropionate have been analyzed by [^{13}C]-NMR, although subsequent stages of the metabolism were not examined (Ansede et al. 2001).

   d. In anaerobic freshwater habitats, however, methyl sulfide (methanethiol) and dimethyl sul-fide can be produced by the reaction of sulfide with methoxylated aromatic compounds (Lomans et al. 2001), and the anaerobic *Methanomethylovorans hollandica* is able to grow with both dimethyl sulfide and methanethiol, in addition to methanol and methylamines (Lomans et al. 1999).

11.9.1.5.2 Anaerobic Transformation

In addition to these pathways, an alternative has been demonstrated in a strain of *Clostridium* that involved elimination with the formation of acrylate (Wagner and Stadtman 1962) and which could be reduced to propionate in *Desulfovibrio acrylicus* (van der Maarel et al. 1996a):

$$(CH_3)_2S^+-CH_2-CH_2-CO_2^- \rightarrow (CH_3)_2S + CH_2=CH-CO_2^- \rightarrow CH_3-CH_2-CO_2^-$$

The latter organism could use both sulfate and acrylate as electron acceptors, and a range of electron donors including lactate, succinate, ethanol, propanol, glycerol, glycine, and alanine. Demethylation to 3-*S*-methylmercaptopropionate has been observed in marine sulfate-reducing bacteria (van der Maarel et al. 1996b), and it is worth noting that *Methanosarcina barkeri* is able to use both dimethyl sulfide and 3-methylmercaptopropionate for methanogenesis (Tallant and Krzycki 1997).

11.9.1.5.3 Fungal Transformation

A fungus *Fusarium lateritium* isolated from the marine environment was able to carry out transforma-tion of dimethylsulfoniopropionate to dimethyl sulfide (Bacic and Yoch 1998).

11.9.1.6 Enantiomeric Oxidation of Sulfides

There has been considerable interest in the enantiomeric oxidation of sulfides to sulfoxides, and illustra-tive examples include the following.

1. Enantiomerically pure alkyl–aryl sulfoxides have been obtained by microbial oxidation of the corresponding sulfides (Holland 1988). Both *Corynebacterium equi* and fungi including *Aspergillus niger*, species of *Helminthosporium*, and *Mortierella isabellina* were effective, although the same fungi were not able to carry out enantiomeric-selective oxidation of ethyl-methylphenylphosphine due apparently to the intrusion of nonselective chemical autoxidation (Figure 11.14) (Holland et al. 1993).

$$(CH_3)_2\overset{+}{S} \cdot CH_2 \cdot CH_2 \cdot CO_2H \longrightarrow (CH_3)_2S + CH_2{=}CH \cdot CO_2H$$

FIGURE 11.14 Elimination reaction to produce acrylate from methionine.

2. The oxidation of a series of substituted phenylmethyl sulfides was examined in two strains of putatively the same organism—*Pseudomonas frederiksbergensis*. Two significant features emerged: (a) the enantioselectivity varied widely among the substrates and (b) one of the strains consistently produced sulfoxides with the *S* configuration, whereas the other produced those with the *R* configuration (Adam et al. 2005). This is unique among sulfoxide-producing strains.

3. The monooxygenase from a strain of *Rhodococcus* selected by enrichment was able to metabolize substituted phenylmethyl sulfides to the (*S*)-sulfoxides with a high enantioselectivity (Li et al. 2009). The exception was the 4-methoxyphenylmethyl sulfide that produced the (*R*)-sulfoxide.

11.9.2 Aerobic Metabolism

11.9.2.1 Thiosulfate and Polythionate Metabolism

A complex set of reactions interrelate the metabolism of thiosulfate $[^-S{-}SO_3^-]$, tetrathionate $[^-O_3S{-}S{-}S{-}SO_3^-]$, trithionate $[^-O_3S{-}S{-}SO_3^-]$, pentathionate $[^-O_3S{-}S{-}S{-}S{-}SO_3^-]$, and elemental sulfur $[S_8]$. These have been discussed in the context of alternative electron acceptors in Chapter 3, and their metabolism may involve either hydrolytic mechanisms:

$$S_3O_6^{2-} + H_2O \rightarrow S_2O_3^{2-} + SO_4^{2-} + 2H^+$$
$$S_4O_6^{2-} + H_2O \rightarrow S_2O_3^{2-} + SO_4^{2-} + S + 2H^+$$
$$S_5O_6^{2-} + H_2O \rightarrow S_2O_3^{2-} + SO_4^{2-} + 2S + 2H^+$$
$$\text{or redox reactions: } S_2O_3^{2-} + 2H^+ + 2e \rightarrow HS^- + HSO_3^-$$
$$S_4O_6^{2-} + 2e \rightarrow 2S_2O_3^{2-}$$

The sulfur oxidation system (Sox) involved in oxidation of reduced sulfur compounds carries out the following reactions (Friedrich et al. 2001; Sauvé et al. 2009):

$$\text{SoxAX: } SoxY{-}SH + {-}S{-}SO_3^- \rightarrow SoxY{-}S{-}S{-}SO_3^-$$
$$\text{SoxB: } SoxY{-}S{-}S{-}SO_3^- + H_2O \rightarrow SoxY{-}S{-}S^- + SO_4^{2-} + 2H^+$$
$$\text{SoxCD: } SoxY{-}S{-}S^- + 3H_2O \rightarrow SoxY{-}S{-}SO_3^- + 6H^+ + 6e$$
$$\text{SoxB: } SoxY{-}S{-}SO_3^- + H_2O \rightarrow SoxY{-}SH + SO_4^{-2} + H^+$$

Additional issues include the following:

1. Tetrathionate (Ttr) and thiosulfate reductase (PhS) in *Salmonella enterica*: Ttr reduces tetra- and trithionate—though not thiosulfate or elementary sulfur. Anaerobic growth on a basal medium supplemented with formate and elemental sulfur, takes place either by thiosulfate cycling or, possibly by a polysulfide reduction model (Hinsley and Berks 2002).

2. The archaeon *Acidianus ambivalens* is a chemolithotroph grows aerobically with S^0 by simultaneous oxygenation and disproportionation (Veith et al. 2011):

$$S^0 + O_2 + H_2O \rightarrow HSO_3^- + H^+; \; 3S^0 + 3H_2O \rightarrow 2H_2S + HSO_3^- + H^+, \text{ and tetrathionate}$$
hydrolase in cells grown aerobically with tetrathionate is an extracellular enzyme (Protze et al. 2011).

3. Tetrathionate and pentathionate hydrolase have been purified from cells of *Thiobacillus ferro-oxidans* grown with thiosulfate with the production of thiosulfate, sulfate, and sulfur (de Jong et al. 1997a), and the analogous tetrathionate hydrolase from *Thiobacillus acidophilus* (de Jong et al. 1997b).

4. Trithionate hydrolase from *Thiobacillus acidophilus* produced thiosulfate and sulfate (Meulenberg et al. 1992).

5. Even though thiosulfate may enhance growth, a carbon source is required.

Oxidation of thiosulfate to tetrathionate rather than sulfate in *Catenococcus thiocyclus* is carried out using organic carbon as source of carbon and energy, and the organism does not grow chemolithotrophically. Aerobic heterotrophic growth with carbohydrates can support anaerobic growth by fermentation and with fatty acids and some amino acids (Sorokin 1992). Analogously, *Bosea thiooxidans* oxidized thiosulfate using carbohydrates, acetate, succinate and malate, and amino acids for heterotrophic growth, and was incapable of lithotrophic growth (Das et al. 1996).

11.9.2.2 Metabolism of Elemental (S⁰)

Reviews regarding sulfur metabolism in *Thiobacillus* and *Paracoccus* (Kelly et al. 1997), and a brief discussion of sulfur dioxygenase are given in Chapter 3. Only a few salient features of the aerobic metabolism of elemental sulfur (S_8) are given here.

1. a. The oxidation of elemental sulfur in *Thiobacillus thiooxidans* produced sulfite (Suzuki et al. 1992), and the metabolism of a range of sulfur compounds has been examined in *Thiobacillus caldus* that is able to oxidize sulfur (S_8), thiosulfate, tetrathionate, sulfite, and sulfide (Hallberg et al. 1996). The use of inhibitors and uncouplers was used to define the regions in which the various reactions took place. Neither 2,4-dinitrophenol nor carbonyl cyanide *m*-chlorophenyl hydrazone (CCCP) had any effect on the oxidation of thiosulfate that, therefore, took place in the periplasm. On the other hand, they completely inhibited the oxidation of sulfur, sulfite, tetrathionate and sulfide that were metabolized in the cytoplasm, and *N*-ethylmaleamide inhibited oxidation of sulfur to sulfite whose further oxidation was inhibited by 2-heptyl-4-hydroxyquinoline-*N*-oxide. These results made it possible to establish the pathways for metabolism of all these compounds (Hallberg et al. 1996).

   b. The status of the genus *Thiomonas* has been clarified (Kelly et al. 2007) and, except for *Thiomonas cuprina*, all species that have been described were able to grow chemolithotrophically with sulfur, thiosulfate and tetrathionate, although optimal growth took place mixotrophically in complex media supplemented with reduced sulfur compounds (Moreira and Amils 1997; Kelly et al. 2007).

2. A number of bacteria can use thiosulfate as energy source for chemolithotrophic growth using oxygen as electron acceptor while the electrons released are used for the assimilation of CO_2. This has been extensively examined in the facultative lithoautotrophs *Paracoccus pantotrophus*, and *Starkey novella* in which thiosulfate is oxidized to sulfate. These are discussed in Chapter 3. Those with the metabolic potential that includes sulfur oxyanions as well as S⁰ are summarized in Table 11.1. In addition to these, the alkaliphiles *Thioalkalimicrobium* and *Thioalkalivibrio* are obligate chemolithotrophs that are able to use sulfide and thiosulfate and, for the latter polythionates (Sorokin et al. 2001).

3. Several species of the microaerophilic genus *Sulfurospirillum* such as *S. deleyianum*, *S. barnesii, and S. arsenophilum* are able to use S⁰ as electron acceptor and pyruvate, Fumarate, and formate as electron donors using acetate as the source of carbon and energy (Stolz et al. 1999).

4. Species of *Shewanella* are facultatively anaerobic bacteria that are generally isolated from freshwater or marine environments and are capable of using a range of compounds as electron acceptors which include oxygen, nitrate, fumarate, and metal cations. *Shewanella oneidensis*

TABLE 11.1

Organisms Able to Use Elementary S^0 and Sulfur Oxyanions

Substrates	Organism	Reference
S_8, $S_2O_3^{2-}$	*Sulfurimonas*	Takai et al. (2006b)
S_8, $S_2O_3^{2-}$, S^{2-}	*Thiovirga sulfuroxydans*	Ito et al. (2005)
S_8, $S_2O_3^{2-}$, $S_4O_6^{2-}$, S^{2-}	*Thiofaba tepidiphila*	Mori and Suzuki (2008)
S_8, $S_2O_3^{2-}$, $S_4O_6^{2-}$	*Thiomicrospira arctica* *Thiomicrospira psychrophila*	Knittel et al. (2005)
$S_2O_3^{2-}$, $S_4O_6^{2-}$	*Sulfurivirga caldicuralii*	Takai et al. (2006a)

(putrefaciens) is able to grow using the reduction of sulfur S^0 to sulfide in a medium using lactate as principal carbon source and casamino acids as a supplement (Moser and Nealson 1996). Anaerobic respiration at the expense of sulfur S^0 reduction has been observed for *Shewanella oneidensis* MR-1, and required the presence of a homologue of the *phsA* gene in *Salmonella enterica* serovar *typhimurium* LT2 (Burns and DiChristina 2009).

5. A range of hyperthermophilic organisms is able to reduce elemental sulfur to sulfide, many of them use H_2, and they include both archaea such as *Acidianus*, *Ignococcus*, *Pyrococcus* and methanogens, as well as bacteria such as *Wolinella* and *Aquifex* (Hedderich et al. 1999). Illustrative examples include the following.

   a. *Acidianus infernus* and *Ac. brierleyi* (previously designated as species of *Sulfolobus*) are facultative aerobes with the unusual ability to grow aerobically with S_8 to produce SO_4^{2-} as lithotrophs, and anaerobically by reduction with hydrogen of S_8 to S^{2-} (Segerer et al. 1986). In *Acidianus ambivalens*, although the sulfur reductase and the [Ni–Fe] hydrogenase could not be separated, the reductase reduced S^0 with hydrogenase in the presence of cytochrome *c* (Laska et al. 2003).

   b. Although strains of the strictly anaerobic hyperthermophilic genus *Ignicoccus* can use elemental S^0 as electron acceptor, they are unable to use any sulfur oxyanions as electron acceptor, and only H_2 as electron donor (Huber et al. 2000).

   c. *Pyrococcus furiosus* ferments carbohydrates to acetate, CO_2 and H_2 and, unlike other hyperthermophiles, it grows equally well with or without S^0 sulfur. It contains two hydrogenases, a heterotetrameric cytoplasmic soluble enzyme I, and a multimeric transmembrane enzyme II. The soluble enzyme has been characterized as a [Ni–Fe] enzyme (van Haaster et al. 2008), and the reduced membrane-bound after solubilization produced an EPR signal typical of the Ni center in mesophilic enzymes (Sapra et al. 2000). The soluble enzyme was able to catalyze both the production of H_2 and the reduction of S^0 to H_2S. This organism is able to produce H_2S by reduction of S^0 or polysulfide by several mechanisms.

      i. The hydrogenase that is involved in the reduction of H^+ is also able to carry out the reduction of S^0 and has appropriately been termed a sulfhydrogenase (Ma et al. 1993).

      ii. A sulfide dehydrogenase is able to catalyze the reduction of polysulfide to H_2S using NADPH as electron donor, and contains several [Fe–S] clusters (Ma and Adams 1994).

      iii. A reductase that is able to reduce sulfur is an oxidoreductase that has been characterized as a homodimeric flavoprotein (Schut et al. 2007).

6. A number of anaerobes have been isolated that are able to grow by linking the reduction of elemental sulfur to sulfide, generally with the oxidation of organic compounds such as acetate.

   a. The canonical reaction was described for *Desulfuromonas acetoxidans* that was able to couple the oxidation of acetate to the reduction of elemental sulfur S^0 to sulfide (Pfennig and Biebl 1976). Strains of *Desulfuromonas acetexigens* were able to use elemental sulfur or polysulfide and used exclusively acetate as carbon source (Finster et al. 1994), while *Desulfuromonas thiophila* was able to use in addition, pyruvate and succinate (Finster et al. 1997).

b. *Desulfuromonas svalbardensis* was able to conserve energy by the reduction of Fe(III) as well as S^0, and was able to use acetate and propionate as electron donors.

   Desulfuromonas ferrireducens could use formate and lactate as electron donors and Fe(III), Mn(IV) and S^0 as electron acceptors (Vandieken et al. 2006).

c. *Desulfocapsa sulfoexigens* was a chemolithotroph that could grow at the expense of elemental S^0 in the presence of ferrohydrite whose function was to remove sulfide as insoluble FeS. In the absence of ferrohydrite, when no growth occurred, disproportionation took place: $4S^0 + 4H_2O \rightarrow SO_4^{2-} + 3HS^- + 5H^+$, and an analogous reaction was observed with thiosulfate independently of the presence of ferrohydrite (Finster et al. 1998). Organic carbon was not required and carbon was obtained from HCO_3^-.

d. Thermophilic species of *Desulfurella kamchatkensis* obligately required sulfur S^0 as electron acceptor and used a range of substrates including acetate, pyruvate, palmitate, and stearate for growth, and *Desulfurella propionica* could use S^0 and thiosulfate as electron acceptors. In addition to acetate, propionate, and long-chain carboxylates could be used for growth (Miroshnichenko et al. 1998).

e. *Wolinella succinogenes* is able to grow anaerobically using formate or H_2 as electron donor and elemental sulfur or polysulfide as electron acceptor. The polysulfide is produced by the reaction of S_8 with sulfide produced by the organism (Krafft et al. 1992): $H_2 + S^0 \rightarrow HS^- + H^+$; $HCO_2^- + S^0 \rightarrow CO_2 + HS^-$. Liposomes contained formate dehydrogenase or hydrogenase, and polysulfide isolated from the cytoplasmic membrane catalyze the reaction in the presence of methyl-menaquinone-6. The hydrogenase and formate dehydrogenase are identical to those involved using fumarate respiration with hydrogen or formate, and their cytochrome *b* subunits are similar. The polysulfide reductase consists of three sub-units, containing the molybdopterin guanine dinucleotide, [Fe–S] clusters and an anchor protein that contains methyl-menaquinone-6 (Dietrich and Klimmek 2002).

f. *Aquifex aeolicus* is a hyperthermophilic chemolithoautrophic microaerophilic bacterium that is able to grow with H_2 in a medium containing S^0. The membrane-bound multienzyme complex that carries out sulfur reduction consists of a sulfur reductase connected by quinones and cytochrome b^{II} to hydrogenase II and via cytochrome b^I to hydrogenase. The reductase is able to accept sulfur, polysulfide, and tetrathionate, though not thiosulfate (Guiral et al. 2005).

It is worth drawing attention to some unusual metabolic reactions involving S^0.

1. The novel production of sulfate from elemental S^0 has been observed in several strains of sulfate-reducing bacteria specifically when Mn(IV) was provided as electron acceptor: these included *Desulfovibrio desulfuricans*, *Desulfobacterium autotrophicum*, *Desulfomicrobium baculatum*, and *Desulfuromonas acetoxidans*, whereas *Desulfobulbus propionicum* exceptionally carried this out using Fe(III) (Lovley and Phillips 1994).

2. *Thermotoga lettingae* is an anaerobic thermophilic organism that is able to grow at the expense of a range of substrates including methanol and the methylamines. During growth with methanol in the presence of elemental sulfur S^0, alanine was produced in addition to CO_2 and sulfide (Balk et al. 2002).

11.10 References

Adam W, F Heckel, CR Saha-Möller, M Taupp, J-M Meyer, P Schrier. 2005. Opposite enantioselectivities of two phenotypically and genotypically similar strains of *Pseudomonas frederiksbergensis* in bacterial whole-cell sulfoxidation. *Appl Environ Microbiol* 71: 2199–2202.

Amarita F, M Yvon, M Nardi, E Chambellon, J Delettre, P Bonnarme. 2004. Identification and functional analysis of the gene encoding methionine-γ-lyase in *Brevibacterium linens*. *Appl Environ Microbiol* 70: 7348–7354.

Ansede JH, PJ Pellechia, DC Yoch. 1999. Metabolism of acrylate to β-hydroxypropionate and its role in dimethylsulfoniopropionate lyase induction by a salt march sediment bacterium, *Alcaligenes faecalis* M3A. *Appl Environ Microbiol* 65: 5075–5081.

Ansede JH, PJ Pellechia, DC Yoch. 2001. Nuclear magnetic resonance analysis of [1-^{13}C]dimethylsulfoniopropionate (DMSP) and [1-^{13}C]acrylate metabolism by a DMSP lyase-producing marine isolate of the δ-subclass proteobacteria. *Appl Environ Microbiol* 67: 3134–3139.

Awano N, M Wada, H Mori, S Nakamori, H Takagi. 2005. Identification and functional analysis of *Escherichia coli* cysteine desulfhydrases. *Appl Environ Microbiol* 71: 4149–4152.

Bacic MK, DC Yoch. 1998. *In vivo* characterization of dimethylsulfoniopropiopnate lyase in the fungus *Fusarium lateritium. Appl Environ Microbiol* 64: 106–111.

Balk M, J Weijma, AJM Stams. 2002. *Thermotoga lettingae* sp. nov., a novel thermophilic, methanol-degrading bacterium isolated from a thermophilic anaerobic reactor. *Int J Syst Evol Microbiol* 52: 1361–1368.

Baumler DJ, K-F Hung, KC Jeong, CW Kaster. 2007. Production of methanethiol and volatile sulfur compounds by the archaeon "*Ferroplasma acidarmanus*". *Extremophiles* 11: 841–851.

Boden R, E Borodina, AP Wood, DP Kelly, JC Murrell, H Schäfer. 2011. Purification and characterization of dimethylsulfide monooxygenase from *Hyphomicrobium sulfonivorans. J Bacteriol* 193: 1250–1258.

Boden R, DP Kelly, JC Murrell, H Schäfer. 2010. Oxidation of dimethylsulfide to tetrathionate by *Methylophaga thiooxidans* sp. nov.: A new link in the sulfur cycle. *Environ Microbiol* 12: 2688–2699.

Borodina E, DP Kelly, P Schumann, FA Rainey, NL Ward-Rainey, AP Wood. 2002. Enzymes of dimethylsulfone metabolism and the phylogenetic characterization of the facultative methylotrophs *Arthrobacter sulfonivorans* sp. nov., *Arthrobacter methylotrophus* sp. nov., and *Hyphomicrobium sulfonivorans* sp. nov. *Arch Microbiol* 77: 173–183.

Bruinenberg PG, G de Roo, GY Limsowtin. 1997. Purification and characterization of cystathionine γ-lyase from *Lactococcus linens* subsp. *cremoris* SK11: Possible role in flavor compound formation during cheese maturation. *Appl Environ Microbiol* 63: 561–566.

Burns JL, TJ DiChristina. 2009. Anaerobic respiration of elemental sulfur and thiosulfate by *Shewanella oneidensis* MR-1 requires *psr*A, a homologue of the *phs*A gene of *Salmonella enterica* serovar *typhimurium* LT2. *Appl Environ Microbiol* 75: 5209–5217.

Charlson RJ, JE Lovelock, MO Andreae, SG Warren. 1987. Oceanic phytoplankton, atmospheric sulfur, cloud albedo and climate. *Nature* (London) 326: 655–661.

Chocat P, N Esaki, T Nakamura, H Tanaka, K Soda. 1983. Microbial distribution of selenocysteine lyase. *J Bacteriol* 156: 455–457.

Chocat P, N Esaki, K Tanizawa, K Nakamura, H Tanaka, K Soda. 1985. Purification and characterization of selenocysteine β-lyase from *Citrobacter freundii. J Bacteriol* 163: 669–676.

Cicchillo RM, MA Baker, EJ Schnitzer, EB Newman, C Krebs, SJ Booker. 2004. *Escherichia coli* L-serine deaminase requires a [4Fe–4S] cluster in catalysis. *J Biol Chem* 279: 32418–31425.

Curson ARJ, R Rogers, JD Todd, CA Brearley, AWB Johnson. 2008. Molecular genetic analysis of a dimethylsulfoniopropionate lyase that liberates the climate-changing gas dimethylsulfide in several marine α-proteobacteria and *Rhodobacter sphaeroides. Environ Microbiol* 10: 757–767; Correction 10: 1099.

Das SK, AK Mishra, BJ Tindall, FA Rainey, E Stackebrandt. 1996. Oxidation of thiosulfate by a new bacterium, *Bosea thiooxidans* (strain BI-42) gen. nov., sp nov.: Analysis of phylogeny based on chemotaxonomy and 16S ribosomal DNA sequencing. *Int J Syst Bacteriol* 46: 871–987.

de Jong GAH, W Hazeau, P Bos, JG Kuenen. 1997a. Polythionate degradation by tetrathionate hydrolase of *Thiobacillus ferrooxidans. Microbiology UK* 143: 499–504.

de Jong GAH, W Hazeau, P Bos, JG Kuenen. 1997b. Isolation of tetrathionate hydrolase from *Thiobacillus acidophilus. Eur J Biochem* 243: 678–683.

de Souza MP, D Yoch. 1995. Purification and characterization of dimethylsufoniopropionate lyase from an *Alcaligenes*-like dimethyl sulfide-producing marine isolate. *Appl Environ Microbiol* 61: 21–26.

de Zwart JMM, PN Nelisse, JG Kuenen. 1996. Isolation and characterization of *Methylophaga sulfidovorans* sp. nov.: An obligately methylotrophic, aerobic, dimethylsulfide oxidizing bacterium from a microbial mat. *FEMS Microbiol Ecol* 20: 261–270.

Denger K, J Mayer, M Buhmann, S Weinitschke, THM Smits, AM Cook. 2009. Bifurcated degradative pathway of 3-sulfolactate in *Roseovarius nubinhibens* ISM via sulfoacetaldehyde acetyltransferase and (*S*)-cysteate sulfolyase. *J Bacteriol* 191: 5648–5656.

Denger K, THM Smits, AM Cook. 2006. L-cysteate sulpho-lyase, a widespread pyridoxal 5′-phosphate-coupled desulfonative enzyme purified from *Silicibacter pomeroyi* DSS-3. *Biochem J* 394: 657–664.

Dias B, B Weimer. 1998. Purification and characterization of L-methionine γ-lyase from *Brevibacterium linens* BL2. *Appl Environ Microbiol* 64: 3327–3331.

Dietrich W, O Klimmek. 2002. The function of methyl-menaquinone-6 and polysulfide reductase membrane anchor (PsrC) in polysulfide respiration of *Wolinella succinogenes*. *Eur J Biochem* 269: 1086–1095.

Dominy JE, CR Simmons, PA Karplus, AM Gehring, MH Stipanuk. 2006. Identification and characterization of bacterial cysteine dioxygenases: A new route of cysteine degradation for eubacteria. *J Bacteriol* 188: 5561–5569.

Finster K, F Bak, N Pfennig. 1994. *Desulfuromonas acetexigens* sp. nov., a dissimilatory sulfur-reducing eubacterium from anoxic freshwater sediments. *Arch Microbiol* 161: 328–332.

Finster K, JD Coates, W Liesack, N Pfennig. 1997. *Desulfuromonas thiophila* sp. nov., a new obligately sulfur-reducing bacterium from anoxic freshwater sediment. *Int J Syst Bacteriol* 47: 754–758.

Finster K, W Liesack, B Thamdrup. 1998. Elemental sulfur and thiosulfate disproportionation by *Desulfocapsa sulfoexigens* sp. npv., a new anaerobic bacterium isolated from marine surface sediment. *Appl Environ Microbiol* 64: 119–125.

Friedrich CG, D Rother, F Bardischewsky, A Quenmeier, J Fischer. 2001. Oxidation of reduced inorganic sulfur compounds by bacteria: Emergence of a common mechanism? *Appl Environ Microbiol* 67: 2873–2882.

González JM et al. 2003. *Silicibacter pomeroyi* sp. nov. and *Roseovarius nubinhibens* sp. nov., dimethylsulfoniopropionate-demethylating bacteria from marine environments. *Int J Syst Evol Microbiol* 53: 1261–1269.

González JM, RP Kiene, MA Moran. 1999. Transformation of sulfur compounds by an abundant lineage of marine bacteria in the α-subclass of the class Proteobacteria. *Appl Environ Microbiol* 65: 3810–3819.

Guiral M, P Tron, C Aubert, A Gloter, C Iobbi-Nivol, M-T Giudici-Orticoni. 2005. A membrane-bound multienzyme, hydrogen-oxidizing, and sulfur-reducing complex from the hyperthermophilic bacterium *Aquifex aeolicus*. *J Biol Chem* 280: 42004–42015.

Hallberg KB, M Dopson, EB Lindström. 1996. Reduced sulfur compound oxidation by *Thiobacillus caldus*. *J Bacteriol* 178: 6–11.

Hanlon SP, RA Holt, GR Moore, AG McEwan. 1994. Isolation and characterization of a strain of *Rhodobacter sulfidophilus*: A bacterium which grows autotrophically with dimethylsulphide as electron donor. *Microbiology (UK)* 140: 1953–1958.

Hedderich R, O Klimmek, A Kröger, R Dirmeier, M Keller, KO Stetter. 1999. Anaerobic respiration with elemental sulfur and with disulfides. *FEMS Mictobiol Revs* 22: 353–381.

Hinsley AP, BC Berks. 2002. Specificity of regulatory pathways involved in the reduction of sulfur compounds by *Salmonella enterica*. *Microbiology UK* 148: 3631–3638.

Hofmeister AEM, S Textor, W Buckel. 1997. Cloning and expression of the two genes coding for L-serine dehydratase from *Peptostreptoccus asaccharolyticus*: Relationship of the iron–sulfur protein to both L-serine dehydratases from *Escherichia coli*. *J Bacteriol* 179: 4937–4941.

Holland HL. 1988. Chiral sulfoxidation by biotransformation of organic sulfides. *Chem Revs* 88: 473–485.

Holland HL, M Carey, and S Kumaresan. 1993. Fungal biotransformation of organophosphines. *Xenobiotica* 23: 519–524.

Howard EC , JR Henriksen, A Buchan, CR Reisch, H Bürgmann, R Welsh. 2006. Bacterial taxa that limit sulfur flux from the ocean. *Science* 314: 649–652.

Howard EC, S Sun, EJ Biers, MA Moran. 2008. Abundant and divers bacteria involved in DMSP degradation in marine surface waters. *Environ Microbiol* 10: 2397–2410.

Huber H, S Burggraf, T Mayer, I Wyschkony, R Rachel, KO Stetter. 2000. *Ignicoccus* gen. nov., a novel genus of hyperthermophilic, chemolithotrophic archaea, represented by two new species, *Ignicoccus islandicus* sp. nov., and *Ignicoccus pacificus* sp. nov. *Int J Syst Evol Microbiol* 50: 2093–21200.

Ince JE, CJ Knowles. 1986. Ethylene formation by cell-free extracts of *Escherichia coli*. *Arch Microbiol* 146: 151–158.

Ito T, K Sugita, I Yumoto, Y Nodasaka, S Okabe. 2005. *Thiovirga sulfuroxydans* gen. nov., sp. nov., a chemolithotrophic sulfur-oxidizing bacterium isolated from a microaerobic waste-water biofilm. *Int J Syst Evol Microbiol* 55: 1059–1064.

Jordan SL, AJ Kraczkiewicz-Dowjat, DP Kelly, AP Wood. 1995. Novel eubacteria able to grow on carbon disulfide. *Arch Microbiol* 163: 131–137.

Jordan SL, IR McDonald, AJ Kraczkiewicz-Dowjat, DP Kelly, FA Rainey, J-C Murrell AP Wood. 1997. Autotrophic growth on carbon disulfide is a property of novel strains of *Paracoccus denitrificans*. *Arch Microbiol* 168: 225–236.

Kelly DP, JK Shergill, WP Lu, AP Wood. 1997. Oxidative metabolism of inorganic sulfur compounds by bacteria. *Antonie van Leeuwenhoek* 71: 95–107.

Kelly DP, Y Uchino, H Huber, R Amils, AP Wood. 2007. Reassessment of the phylogenetic relationships of *Thiomonas cuprina*. *Int J Syst Evol Microbiol* 57: 2720–2724.

Kende H. 1989. Enzymes of ethylene biosynthesis. *Plant Physiol* 91: 1–4.

Kiene RP, LJ Linn, J González, MA Moran, JA Bruton. 1999. Dimethylsulfoniopropionate and methanethiol are important precursors of methionine and protein-sulfur in marine bacterioplankton. *Appl Environ Microbiol* 65: 4549–4558.

Kim, HG, NV Doronina, YA Trotsenko, SW Kim. 2007. Methylophaga aminisulfidivorans sp. nov., a restricted facultatively methylotrophic marine bacterium. *Int J Syst Evol Microbiol* 57: 2096–2101.

Kirkwood M, NE Le Brun, TD Todd, AWB Johnston. 2010. The *dddP* gene of *Roseovarius nubinhibens* encodes a novel lyase that cleaves dimethylsulfoniopropionate into acrylate plus dimethyl sulfide. *Microbiology UK* 156: 1900–1906.

Knittel K, J Kuever, A Meyerdierks, R Meinke, R Amann, T Brinkhoff. 2005. *Thiomicrospira arctica* sp. nov. and *Thiomicrospira psychrophila* sp. nov., psychrophilic, obligately chemolithoauthotrophic, sulphur-oxidizing bacteria isolated from marine Arctic sediments. *Int J Syst Evol Microbiol* 55: 781–786.

Krafft T, M Bokranz, O Klimmek, I Schröder, F Fahrenholz, E Kojro, A Kröger. 1992. Cloning and nucleotide sequence of the *psfA* gene *of Wolinella succinogenes* polysulfide reductase. *Eur J Biochem* 206: 503–510.

Kredich NM, BS Keenan, LJ Foote. 1972. The purification and subunit structure of cysteine desulfhydrase from *Salmonella typhimurium*. *J Biol Chem* 247: 7157–7162.

Kredich NM, LJ Foote, BS Keenan. 1973. The stoichiometry and kinetics of the inducible cysteine desulfhydrase from *Salmonella typhimurium*. *J Biol Chem* 248: 6187–6196.

Laska S, F Lottspeich, A Kletzin. 2003. Membrane-bound hydrogenase and sulfur reductase of the hyperthermophilic and acidophilic archaeon *Acidianus ambivalens*. *Microbiology (UK)* 149: 2357–2371.

Ledyard KM, JWH Dacey. 1994. Dimethylsulfide production from dimethylsulfoniopropionate by a marine bacterium. *Mar Ecol Prog Ser* 110: 95–103.

Li A.-T, J-D Zhang, J-H Xu, W-Y Lu, G-Q Lin. 2009. Isolation of *Rhodococcus* sp. strain ECU0066, a new sulfide monooxygenase-producing strain for asymmetric sulfoxidation. *App Environ Microbiol* 75: 551–556.

Lomans BP, P Leijdekkers, J-P Wesselink, P Bakkes, A Pol, C van der Drift, HJP op den Camp. 2001. Obligate sulfide-dependent degradation of methoxylated aromatic compounds and formation of methanethiol and dimethyl sulfide by a freshwater sediment isolate. *Parasporobacterium paucivorans* gen. nov., sp. nov. *Appl Environ Microbiol* 67: 4017–4203.

Lomans BP, R Maas, R Luderer, HJP op den Camp, A Pol, C van der Drift, GD Vogels. 1999. Isolation and characterization of *Methanomethylovorans hollandica* gen. nov., sp. nov., isolated from freshwater sediment, a methylotrophic methanogen able to grow with dimethyl sulfide and methanethiol. *Appl Environ Microbiol* 65: 3641–3650.

Lovley DR, EJP Phillips. 1994. Novel process for anaerobic sulfate production from elemental sulfur by sulfate-reducing bacteria. *Appl Environ Microbiol* 60: 2394–2399.

Ma K, MWW Adams. 1994. Sulfide dehydrogenase from the hyperthermophilic archaeon *Pyrococcus furiosus*: A new multifunctional enzyme involved in the reduction of elemental sulfur. *J Bacteriol* 176: 6509–8517.

Ma, K, RN Schicho, RM Kelly, MW Adams. 1993. Expand+ hydrogenase of the hyperthermophile *Pyrococcus furiosus* is an elemental sulfur reductase or sulfhydrogenase: Evidence for a sulfur-reducing hydrogenase ancestor. *Proc Natl Acad Sci USA* 90: 65341–5344.

Mansouri S, AW Bunch. 1989. Bacterial synthesis from 2-oxo-4-thiobutyric acid and from methionine. *J Gen Microbiol* 135: 2819–2827.

Méndez J, P Reimundo, D Pérez-Pascual, R Navais, E Gómez, JA Giijarro. 2011. A novel *cdsAB* operon is involved in the uptake of L-cysteine and participates in the pathogenesis of *Yersinia ruckeri*. *J Bacteriol* 193: 944–951.

Meulenberg R, JT Pronk, J Frank, W Hazeau, JG Kuenen. 1992. Purification and partial characterization of a thermostable trithionate hydrolase from the acidophilic sulfur oxidizer *Thiobacillus acidophilus*. *Eur J Biochem* 209: 367–374.

Mihara H, T Kurihara T Yoshimura, K Soda, N Esaki. 1997. Cysteine sulfinate desulfinase, a NIFS-like protein of *Escherichia coli* with selenocysteine lyase and cysteine desulfurase activities. *J Biol Chem* 272: 22417–22424.

Miller KW. 1992. Reductive desulfurization of dibenzyldisulfide. *Appl Environ Microbiol* 58: 2176–2179.

Miroshnichenko ML, FA Rainey, H Hippe, NA Chernyh, NA Kostrikina, EA Bonch-Osmolovskaya. 1998. *Desulfurella kamchatkensis* sp. nov. and *Desulfurella propionica* sp. nov., new sulfur-respiring thermophilic bacteria from Kamchatka thermal environments. *Int J Syst Bacteriol* 48: 475–479.

Moreira D, R Amils. 1997. Phylogeny of *Thiobacillus cuprinus* and other mixotrophic thiobacilli: Proposal for *Thiomonas* gen. nov. *Int J Syst Bacteriol* 47: 522–528.

Mori K, K-I Suzuki. 2008. *Thiofaba tepidiphila* gen. nov., sp. nov., a novel obligately chemolithoautotrophic, sulfur-oxidizing bacterium of the *Gammaproteobacteria* isolated from a hot spring. *Int J Syst Evol Microbiol* 58: 1885–1892.

Moser DP, KH Nealson. 1996. Growth of the facultatively anaerobe *Shewanella putrefaciens* by elemental sulphur reduction. *Appl Environ Microbiol* 62: 2100–2105.

Müller FB, TM Bandeiras, T Ulrich, M Teixeira, CM Gomes, A Kletzin. 2004. Coupling of the pathway of sulfur oxidation to dioxygen reduction: Characterization of a novel membrane-bound thiosulfate: Quinone oxidoreductase. *Mol Microbiol* 53: 1147–1160.

Nagasawa T, T Ishii, H Kumagai, H Yamada. 1985. D-cysteine desulfhydrase of *Escherichia coli*. *Eur J Biochem* 153: 541–551.

Nagasawa T, T Ishii, H Yamada. 1988. Physiological comparison of D-cysteine desulfhydrase of *Escherichia coli* with 3-chloro-D-alanine dehydrochlorinase of *Pseudomonas putida* CR 1-1. *Arch Microbiol* 149: 413–416.

Nagasawa T, H Kanaki, H Yamada. 1984. Cystathionine γ-lyase of *Streptomyces phaeochromogenes*. The occurrence of cystathionine γ-lyase in filamentous bacteria and its purification and characterization. *J Biol Chem* 259: 10393–10403.

Newton WA, Y Morino, EE Snell. 1965. Properties of crystalline tryptophanase. *J Biol Chem* 240: 1211–1218.

Oremland RS, RS Kiene, I Mathrani, MJ Whiticar, DR Boone. 1989. Description of an estuarine methylotrophic methanogen which grows on dimethyl sulfide. *Appl Environ Microbiol* 55: 994–1002.

Pfennig N, H Biebl. 1976. *Desulfuromonas acetoxidans* gen. nov. and sp. nov., a new anaerobic, sulfur-reducing, acetate oxidizing bacterium. *Arch Microbiol* 110: 3–12.

Pol A, C van der Drift, HJM Op den Camp. 2007. Isolation of a carbon disulfide utilizing *Thiomonas* sp. and its application in a trickling filter. *Appl Microbiol Biotechnol* 74: 439–446.

Protze J, F Müller, K Lauber, B Naß, R Mentele, F Lottspeich, A Kletzin. 2011. An extracellular tetrathionate hydrolase from the thermoacidophilic archaeon *Acidianus ambivalens* with an activity optimum at pH 1. *Front Microbiol* doi: 10.3389/fmicb.2011.00068.

Rein U, R Gueta, K Denger, J Ruff, K Hollemeyer, AM Cook. 2005. Dissimilation of cysteate via 3-sulfolactate sulfo-lyase and a sulfate exported in *Paracoccus pantotrophus* NKNCYSA. *Microbiology (UK)* 151: 737–747.

Reisch CR, MA Moran, WB Whitman. 2008. Dimethylsulfoniopropionate-dependent demethylase (DmdA) from *Pelagibacter ubique* and *Silicibacter pomeroyi*. *J Bacteriol* 190: 8018–8024.

Reisch CR, MA Moran, WB Whitman. 2011b. Bacterial catabolism of dimethylsulfoniopropionate. *Frontiers Microbiol* 2, doi: 10.3389/fmicb.2011.00172.

Reisch CR, MJ Stoudemayer, VA Varaljay, IJ Amster, MA Moran, WB Whitman. 2011a. Novel pathway for assimilation of dimethylsulfoniopropionate widespread in marine bacteria. *Nature* 47: 208–211.

Rossol I, A Pühler. 1992. The *Corynebacterium glutamicum aecD* gene encodes a C–S lyase with α,β-elimination activity that degrades aminoethylcysteine. *J Bacteriol* 174: 2968–2977.

Sapra R, MFJM Verhagen, MWW Adams. 2000. Purification and characterization of a membrane-bound hydrogenase from the hyperthermophilic archaeon *Pyrococcus furiosus*. *J Bacteriol* 182: 3423–3428.

Sauvé V, P Roversi, KL Leath, EF Garman, R Antrobus, SM Lea, BC Berks. 2009. Mechanism for the hydrolysis of a sulfur–sulfur bond based on the crystal structure of the thiosulfohydrolase SoxB. *J Biol Chem* 284: 21707–21718.

Schut GJ, SL Bridger, MWW Adams. 2007. Insights into the metabolism of elemental sulfur by the hyperthermophilic archaeon *Pyrococcus furiosus*: Characterization of a coenzyme A-dependent NAD(P)H sulfur oxidoreductase. *J Bacteriol* 189: 4431–4441.

Schürmann M, JH Wübbeler, J Grote, A Steinbüchel. 2011. Novel reaction of succinyl coenzyme A (succinyl-CoA) synthetase: Activation of 3-sulfinopropionate to 3-sulfinopropionyl-CoA in *Advenella mimigardefordensis* strain DPN7 during degradation of 3,3′-dithiodipropionic acid. *J Bacteriol* 193: 3078–3089.

Segerer A, A Neuner, JK Kristjansson, KO Stetter. 1986. *Acidianus infernus* gen. nov., sp. nov., and *Acidianus brierleyi* comb. nov.: Facultatively aerobic, extremely acidophilic thermophilic sulfur-metabolizing archaebacteria. *Int J Syst Evol Microbiol* 36: 559–564.

Smith NA, DP Kelly. 1988. Isolation and physiological characterization of autotrophic sulphur bacteria oxidizing dimethyl disulphide as sole source of energy. *J Gen Microbiol* 134: 1407–1417.

Soda K, A Novogrodsky, A Meister. 1964. Enzymatic desulfination of cysteine sulfinic acid. *Biochemistry* 3: 1450–1454.

Sorokin DY. 1992. *Catenococcus thiocyclus* gen. nov. sp. nov.—A new facultatively anaerobic bacterium from a new-shore sulphidic hydrothermal area. *J Gen Microbiol* 138: 2287–2292.

Sorokin DY, AM Lysenko, LL Mityuishina, TP Tourova, BE Jones, FA Rainey, LA Robertson, GJ Kuenen. 2001. *Thioalkalimicrobium aerophilum* gen. nov., sp. nov and *Thioalkalimicrobium sibericum* sp. nov., and *Thioalkalivibrio versutus* gen. nov., sp. nov., *Thioalkalivibrio nitratis* sp. nov. and *Thioalkalivibrio denitrificans* sp. nov., novel alkaliphilic and obligately chemolithoautotrophic sulfur-oxidizing bacteria from soda lakes. *Int J Syst Evol Microbiol* 51: 565–580.

Soutourina J, S Blanquet, P Plateau. 2001. Role of D-cysteine desulfhydrase in the adaptation of *Escherichia coli* to D-cysteine. *J Biol Chem* 276: 40864–40872.

Stefels J, L Dijlkhuizen. 1996. Characteristics of DMSP-lyase in *Phaeocystis* sp . (Prymnesiophyceae). *Mar Ecol Prog Ser* 131: 307–313.

Steinke M, GV Wolfe, GO Kirst. 1998. Partial characterization of dimethylsulfoniopropionate (DMSP) lyase isoenzymes in 6 strains of *Emiliania huxleyi*. *Mar Ecol Prog Ser* 175: 215–225.

Stolz JF, DJ Ellis, JS Blum, D Ahmann, DR Lovley, RS Oremland. 1999. *Sulfurospirillum barnesii* sp. nov. and *Sulfurospirillum arsenophilum* sp. nov., new members of the *Sulfurosopirillum* clade of the ε-Proteobacteria. *Int J Syst Bacteriol* 49: 1177–1180.

Suylen GMH, GC Stefess, JG Kuenen. 1986. Chemolithotrophic potential of a *Hyphomicrobium* species, capable of growth on methylated sulphur compounds. *Arch Microbiol* 146: 192–198.

Suylen GMH, PJ Large, JP van Dijken, JG Kuenen. 1987. Methyl mercaptan oxidase, a key enzyme in the metabolism of methylated sulphur compounds by *Hyphomicrobium* EG. *J Gen Microbiol* 133: 2989–2997.

Suzuki I, CW Chan, TL Takeuchi. 1992. Oxidation of elemental sulfur to sulfite by *Thiobacillus thioooxidans* cells. *Appl Environ Microbiol* 58: 38767–3769.

Takai K, M Miyazaki, T Nunoura, H Hirayama, H Oida, Y Furushima, H Yamamoto, K Horikoshi. 2006a. *Sulfurivirga caldicuralii* gen. nov., sp. nov., a novel microaerobic, thermophilic, thiosulfate-oxidizing chemolithotroph, isolated from a shallow marine hydrothermal system occurring in a coral reef, Japan. *Int J Syst Evol Microbiol* 56: 1921–1929.

Takai K, M Suzuki, S Nakagawa, M Miyazaki, Y Suzuki, F Inagaki, K Horikoshi. 2006b. *Sulfurimonas paralvinellae* sp. nov., a novel mesophilic, hydrogen- and sulfur-oxidizing chemolithoautotroph within the *Epsilonproteobacteria* isolated from a deep-sea hydrothermal vent polychaete nest, reclassification of *Thiomicrospira denitrificans* as *Sulfurimonas denitrificans* comb. nov. and emended description of the genus *Sulfurimonas*. *Int J Syst Evol Microbiol* 56: 1725–1733.

Tallant TC, JA Krzycki. 1997. Methylthiol: coenzyme M methyltransferase from *Methanosarcina barkeri*, an enzyme of methanogenesis from dimethylsulfide and methylmercaptopropionate. *J Bacteriol* 179: 6902–6911.

Tanimoto Y, F Bak. 1994. Anaerobic degradation of methyl mercaptan and dimethyl sulfide by newly isolated thermophilic sulfate-reducing bacteria. *Appl Environ Microbiol* 60: 2450–2455.

Tchong SI, H Xu, RH White. 2005. L-cysteine desulfidase: An [4Fe–4S] enzyme isolated from *Methanocaldococcus jannaschii* that catalyzes the breakdown of L-cysteine into pyruvate, ammonia, and sulfide. *Biochemistry* 44: 1659–1670.

Todd JD, ARJ Curson, CL Dupont, P Nicholson, AWB Johnston. 2009. The *dddP* gene, encoding a novel enzyme that converts dimethylsulfoniopropionate into dimethyl sulfide, is widespread in ocean metagenomes and marine bacteria and also occurs in some Ascomycete fungi. *Environ Microbiol* 11: 1376–1385.

Todd JD, ARJ Curson, M Kirkwood, MJ Sullivan, RT Green, AWB Johnston. 2011. DddQ, a novel, cupin-containing dimethylsulfoniopropionate lyase in marine roseobacters and in uncultured marine bacteria. *Environ Microbiol* 13: 427–438.

Todd JD, M Kirkwood, S Newton-Payne, AWB Johnston. 2012. DddW, a third DMSP lyase in a model *Roseobacter* marine bacterium, *Ruegeria pomeroyi* DSS-3. *ISME J* 6: 223–226.

van der Maarel MJEC, M Jansen, R Haanstra, WG Meijer, TA Hansen. 1996b. Demethylation of dimethylsulfoniopropionate to 3-*S*-methymercaptopropionate by marine sulfate-reducing bacteria. *Appl Environ Microbiol* 62: 3978–3984.

van der Maarel MJEC, S van Bergeijk, AF van Werkhoven, AM Laverman, WG Maijer, WT Stam, TA Hansen. 1996a. Cleavage of dimethylsulfoniopropionate and reduction of acrylate by *Desulfovibrio acrylicus* sp. nov. *Arch Microbiol* 166: 109–115.

van der Maarel P Quist, L Dijkhuizen, TA Hansen. 1993. Anaerobic degradation of dimethylsulfoniopropionate to 3-*S*-methylmercaptopropionate by a marine *Desulfobacterium* strain. *Arch Microbiol* 160: 411–412.

Vandieken V, M Mußmann, H Niemann, BB Jørgensen. 2006. *Desulfuromonas svalbardensis* sp. nov. and *Desulfuromusa ferrireducens* sp. nov., psychrophilic, Fe(III)-reducing bacteria isolated from Arctic sediments, Svalbard. *Int J Syst Evol Microbiol* 56: 1133–1139.

van Haaster DJ, PJ Silva, P-L Hagedoorn, JA Jongejan, WR Hagen. 2008. Reinvestigation of the steady-state kinetics and physiological function of the soluble NiFe-hydrogenase I of *Pyrococcus furiosus*. *J Bacteriol* 190: 1584–1587.

van Hamme JD, PM Fedorak, JM Foght, MR Gray, HD Dettman. 2004. Use of novel fluorinated organosulfur compound to isolate bacteria capable of carbon–sulfur bond fission. *Appl Environ Microbiol* 70: 1487–1493.

Veith A, T Urich, K Seyfarth, J Protze, C Frazão, A Kletzin. 2011. Substrate pathways and mechanisms of inhibition in the sulfur oxygenase reductase of *Acidianus ambivalens*. *Front Microbiol* doi: 10.3389/fmicb.2011.00037.

Visscher PT, BF Taylor. 1993. A new mechanism for the aerobic catabolism of dimethyl sulfide. *Appl Environ Microbiol* 59: 3784–3789.

Visscher PT, BF Taylor. 1994. Demethylation of dimethylsulfoniopropionate to 3-mercapropionate by an aerobic bacterium. *Appl Environ Microbiol* 60: 4617–4619.

Wagner C, ER Stadtman. 1962. Bacterial fermentation of dimethyl-β-propiothetin. *Arch Biochem Biophys* 98: 331–336.

Wolfe GW, EB Sherr, BF Sherr. 1994. Release and consumption of DMSP from *Emiliania huxleyi* during grazing by *Oxyrrhis marina*. *Mar Ecol Prog Rep* 111: 111–119.

Wray JW, RH Abeles. 1995. The methionine salvage pathway in *Klebsiella pneumoniae* and rat liver. *J Biol Chem* 270: 3147–3153.

Wübbeler JH, N Bruland, K Kretschmer, A Steinbüchel. 2008. Novel pathway for catabolism of the organic sulfur compound 3,3′-dithiodipropionic acid via 3-mercaptopropionic acid and 3-sulfinopropionic acid to propionyl-coenzyme A by the aerobic bacterium *Tetrathiobacter minigardefordensis*. *Appl Environ Microbiol* 74: 4028–4035.

Wübbeler JH, M Raberg, U Brandt, A Steinbüchel. 2010a. Dihydrolipoamide dehydrogenases of *Advenella mimigardefordensis* and *Ralstonia eutropha* catalyze cleavage of 3,3′-dithiodipropionic acid into 3-mercaptopropionic acid. *Appl Environ Microbiol* 76: 7023–7028.

Wübbeler JH, N Bruland, M Wozniczka, A Steinbüchel. 2010b. Biodegradation of the xenobiotic organic disulfide 4,4′-dithiodibutyric acid by *Rhodococcus erythropolis* strain M2 and comparison with the microbial utilization of 3,3′-dithiodipropionic acid and 3,3′-thiopropionic acid. *Microbiology UK* 156: 1221–1233.

Yoch D. 2002. Dimethylsulfoniopropionate: Its sources, role in the marine food web, and biological degradation to dimethylsulfide. *Appl Environ Microbiol* 68: 5804–5815.

Yoch DC, JH Ansede, KS Rabinowitz. 1997. Evidence for intracellular and extracellular dimethylsulfoniopropionate (DMSP) lyases and DMSP uptake sites in two species of marine bacteria. *Appl Environ Microbiol* 63: 3182–3188.

Yoshida Y, Y Nakano, A Amano, M Yoshimura, H Fukamachi, T Oho, T Kogas. 2002. *lcd* from *Streptococcus anginosus* encodes a C–S lyase with α,β-elimination activity that degrades L-cysteine. *Microbiology (UK)* 148: 3961–3971.

Zdych E, R Peist, J Reidl, W Boos. 1995. MalY of *Escherichia coli* is an enzyme with the activity of a β-C-S lyase (cystathionase). *J Bacteriol* 177: 5035–5039.

Part 3: Aliphatic Nitramines and Nitroalkanes

11.11 Nitramines

Aliphatic nitramines based on oligomers of $-CH_2-N-NO_2-$ have been used extensively as explosives. Hexahydro-1,3,5-trinitro-1,3,5-triazine (RDX) can be used as a source of nitrogen by a range of bacteria, including the aerobe *Stenotrophomonas maltophilia* strain PB1 (Binks et al. 1995), the facultative anaerobe *Klebsiella pneumoniae* (Zhou et al. 2002), and the anaerobe *Acetobacterium paludosum* (Sherburne et al. 2005). In *Rhodococcus* sp. this was accomplished by reduction of the ring, loss of nitrite, and formation of 4-nitro-2,4-diazabutanal as a terminal metabolite that retained one of the nitro groups (Figure 11.15) (Fournier et al. 2002; Seth-Smith et al. 2002). This metabolite is also produced from octahydro-1,3,5,7-tetranitro-1,3,5,7-tetrazocine (HMX) by *Phanerochaete chrysosporium* (Fournier et al. 2004), although it can be used as a nitrogen source by the facultative methylotroph *Methylobacterium* sp. strain JS178 (Fournier et al. 2005). RDX can be degraded via methylenedinitramine to CH_2O, CH_3OH, and CO_2 by *Klebsiella pneumoniae* strain SCZ-1 (Zhao et al. 2002). Two groups of bacteria that were able to degrade RDX and HMX were isolated from contaminated sediment: one was anaerobic and belonged to various genera, and the other was aerobic that were closely similar to species of *Shewanella* including the novel species *Shewanella sediminis* (Zhao et al. 2005), *Shewanella canadensis*, and *S. atlantica* (Zhao et al. 2007). The second group transformed RDX principally to the mononitroso compound and further to 4-nitro-2,4-diazabutanal and methylene dinatramine (Zhao et al. 2004). Although the use of RDX as a source of both carbon and nitrogen is less common, this has been observed in *Williamsia* sp. strain KTR4 and *Gordonia* sp. strain KTR9 (Thompson et al. 2005).

Denitration to 4-nitro-2,4-diazabutanal is carried out by two enzymes, XplA that is a cytochrome P450 in which a flavodoxin domain is fused to the C-terminal residue (Sabbadin et al. 2009), and XplB that serves as partner for the NADH-utilizing flavodoxin reductase. It has been shown in this strain that the *xplAB* genes occur on a 182-kilobase plasmid (Indest et al. 2010). Under anaerobic conditions, the formation of methylenedinitramine is followed by abiotic loss of formaldehyde. Bacterial degradation of the polyazapolycyclic-caged polynitramine (2,4,6,8,10,12-hexanitro-2,4,6,8,10,12-hexaazaisowurtzitane), which will probably displace RDX, has been accomplished when it is used as a source of nitrogen (Bhushan et al. 2003; Trott et al. 2003). Its degradation by salicylate 1-monooxygenase from *Pseudomonas* sp. strain ATCC 29352 involved successive loss of nitrite followed by hydrolytic reactions and abiotic reactions (Bhushan et al. 2004).

11.12 Nitroalkanes

These contain $C-NO_2$ bonds in contrast to the nitrate esters of, for example, glycerol and pentaerythritol with $O-NO_2$ bonds. Nitroalkanes have been used as solvents, and there is a range of naturally occurring nitroalkane derivatives: (a) 3-nitropropionic acid in species of *Aspergillus*, and as the glycoside that is widespread in leguminous plants (b) derivatives of 3-nitropropanol—miserotoxin that is a glycoside of nitropropanol, and the glucose esters cibarian (1,6-di-*O*-(3-nitropropanoyl)-β-D-glucopyranose) and karakin (1,2,6-tri-*O*-(3-nitropropanoyl)-β-D-glucopyranose) (Anderson et al.

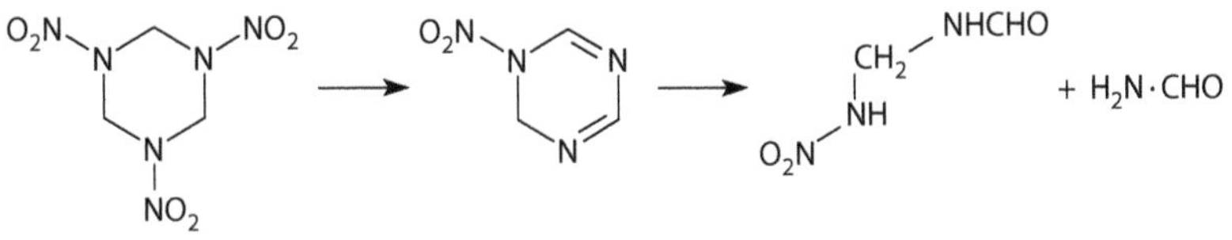

FIGURE 11.15 Degradation of RDX.

2005). They exert their toxicity by metabolism to 3-nitropropionic acid which is a suicide inactivator of succinate dehydrogenase (Alston et al. 1977). In addition, the *O*-methyl ethers of *aci*-nitro compounds such as enteromycin are produced by species of *Streptomyces*. Exceptionally, the aglycones can be used as electron acceptors by *Denitrobacterium detoxificans* during growing with lactate that is converted into acetate (Anderson et al. 2000). The structurally related aliphatic azoxy compounds are represented by the toxic macrozamin that occurs in plants as a glucoside, the antibiotic elaiomycin, and the *Streptomyces* nematicidal jietacin A. The metabolism of nitroalkanes has attracted considerable attention that is briefly summarized in the following examples, and the enzymology is discussed in Chapter 3.

11.12.1 Plants

1. The oxidation of 2-nitropropane that was carried out by an extract of pea plants produced acetone, nitrite and H_2O_2, and the oxidase was highly specific for this substrate (Little 1957).

2. An enzyme purified from leaf extracts of horse shoe vetch *Hippocrepis comosa* that produces large amounts of 3-nitropropionate degraded this to malonate semialdehyde, nitrite, nitrate, and H_2O_2 (Hipkin et al. 1999). It was a flavoprotein with one molecule of FMN in the subunit, was quite specific for 3-nitropropionate, and it was suggested that it was an oxidase.

11.12.2 Yeasts and Fungi

1. The metabolism of 2-nitropropane to acetone in the yeast *Hansenula mrakii* was studied (Kido et al. 1978b), and the intermolecular reaction (Figure 11.16) is formally analogous to the dioxygenation of propane (Babu and Brown 1984). The dioxygenase gene has been isolated and it was shown that it had a unique primary structure (Tchorzewski et al. 1994). The enzyme after expression in *Escherichia coli* was shown to require FMN, preferred primary nitronates, and reacted specifically with *aci*-nitroalkanes (nitronates) and, significantly not the neutral nitroalkanes (Mijatovic and Gadda 2008).

2. The enzyme in *Neurospora crassa* has been characterized, showed a preference for the nitronate, required FMN, and was less active with primary nitroalkanes. A mechanism was proposed to illustrate the reaction of two molecules of the nitronate with the superoxide anion (Gorlatova et al. 1998). This is consistent with a reaction between the nitronate and a flavosemiquinone that has emerged (Francis et al. 2005; Gadda and Francis 2010).

3. A monooxygenase (oxidase) from *Fusarium oxysporum* oxidized 1-nitro- and 2-nitropropane, and 1-nitrocyclohexane to the aldehyde (ketone) with the production of nitrite and H_2O_2 (Kido et al. 1978a). A later study with nitroethane showed that the anion reacted with FAD to produce 5-nitrobutyl-FAD in a sequence of two reactions (Gadda et al. 1997). The enzyme belonged to the acyl-coenzyme A dehydrogenase superfamily although it was unable to accept acyl-CoA substrates (Daubner et al. 2002).

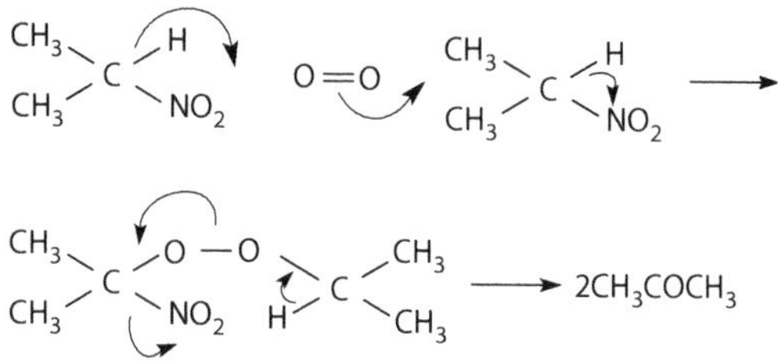

FIGURE 11.16 Metabolism of 2-nitropropane to acetone.

4. *Penicillium atrovenutum* excreted 3-nitropropionate into the medium, and the enzyme that catalyzed the metabolism to malonate semialdehyde, nitrite, nitrate, and H_2O_2 was a flavoprotein with FMN in the subunit, while the preferred substrate was the *aci*-nitronate (Porter and Bright 1987).

5. A protein was selected from the database as a hypothetical nitroalkane oxidase in the fungus *Podospora anserina* and the gene was expressed and characterized (Tormos et al. 2010). It was comparable to the enzyme from *Fusarium oxysporum* and shared the same active site triad, although it had a different substrate specificity.

11.12.3 Bacteria

1. The oxidation of nitroalkanes to the corresponding ketones has been reported in several strains of *Streptomyces* (Dhwale and Hornemann 1979), the nitroalkane-oxidizing gene has been cloned from a genomic library of *Streptomyces ansochromogenes*, and it was suggested that a superoxide anion was involved (Zhang et al. 2002).

2. Strains of bacteria were isolated after enrichment with 3-nitropropionate as sole source of carbon, nitrogen, and energy. It was proposed that degradation took place via the *aci*-nitronate that produced nitrite, nitrate and H_2O_2, and malonate semialdehyde before oxidative decarboxylation to acetyl-CoA (Nishino et al. 2010).

There is clearly some divergence of opinion on the nature of the enzyme that carries out oxygenation although the production of H_2O_2 would be consistent with the activity of an oxidase. There is substantial agreement that: (i) the enzyme is a flavoprotein—FMN or FAD, (ii) the isomeric *aci*-nitro-compounds (nitronates) are the preferred substrates, (iii) oxygen atoms are incorporated separately into two molecules of the substrate, (iv) the initial product is nitrite that can be oxidized to nitrate. Designation of the enzyme as nitronate monooxygenase is, therefore, appropriate (Gadda and Francis 2010).

The enantiomeric reduction of 2-nitro-1-phenylprop-1-ene has been studied in a range of Gram-positive organisms including strains of *Rhodococcus rhodochrous* (Sakai et al. 1985). The enantiomeric purity of the product depended on the strain used, the length of cultivation, and the maintenance of a low pH that is consistent with the later results of Meah and Massey (2000). It has been shown that an NADPH-linked reduction of α,β-unsaturated nitro compounds may also be accomplished by old yellow enzyme via the *aci*-nitro form (Meah and Massey 2000). This is formally analogous to the reduction and dismutation of cyclic enones by the same enzyme (Vaz et al. 1995), and the reductive fission of nitrate esters by an enzyme homologous to the old yellow enzyme from *Saccharomyces cerevisiae* (Snape et al. 1997).

The transformation of *N*-nitrosodimethylamine by *Pseudomonas mendocina* KR1 that has toluene-4-monooxygenase activity was initiated by monooxygenation to the *N*-nitro compound, which produced *N*-nitromethylamine and formaldehyde, presumably by hydroxylation of the methyl group (Fournier et al. 2006).

It is worth noting that, in spite of the prevalence of reductive removal of the nitro group from nitrate esters, hydrolytic fission with loss of nitrite has been observed when the nitro group was attached to an sp^2 carbon atom: (a) the degradation of 2-nitroimidazole to iminazolone by *Mycobacterium* sp. strain JS330 (Qu and Spain 2011) (b) loss of nitrite from the direct ring fission product from 5-nitrosalicylate by *Bradyrhizobium* sp. (Qu and Spain 2010).

11.13 References

Alston TA, L Mela, HJ Bright. 1977. 3-Nitropropionate, the toxic substance of *Indigofera*, is a suicide inactivator of succinate dehydrogenase. *Proc Natl Acad USA* 74: 3767–3771.

Anderson RC, W Majak, MA Rasmussen, TR Callaway, RC Beier, DJ Nisbet, MJ Allison. 2005. Toxicity and metabolism of the conjugates of 3-nitropropanol and 3-nitropropionioc acid in forage poisonous to livestock. *J Agric Food Chem* 53: 2344–2350.

Anderson RC, MA Rasmussen, NS Jensen, MJ Allison. 2000. *Denitrobacterium detoxificans* gen. nov., sp. nov., a ruminal bacterium that respires on nitrocompounds. *Int J Syst Evol Microbiol* 50: 633–638.

Babu JP, LR Brown. 1984. New type of oxygenase involved in the metabolism of propane and isobutane. *Appl Environ Microbiol* 48: 260–264.

Bhushan B, A Halasz, J Spain, J Hawari. 2004. Initial reaction(s) in biotransformation of CL-20 is catalyzed by salicylate 1-monooxygenase from *Pseudomonas* sp. strain ATCC 29352. *Appl Environ Microbiol* 70: 4040–4047.

Bhushan B, L Paquet, JC Spain, J Hawari. 2003. Biotransformation of 2,4,6,8,10,12-hexanitro-2,4,6,8,10,12-hexaazaisowurtzitane (CL-20) by denitrifying *Pseudomonas* sp. strain FA1. *Appl Environ Microbiol* 69: 5216–5221.

Binks PR, S Nicklin, NC Bruce. 1995. Degradation of hexahydro-1,3,5-trinitro-1,3,5-triazine (RDX) by *Stenotrophomonas maltophilia* PB1. *Appl Environ Microbiol* 61: 1318–1322.

Daubner SC, G Gadda, Mp Valley, PF Fitzpatrick. 2002. Cloning of nitroalkane oxidase from *Fusarium oxysporum* identifies a new member of the acyl-CoA dehydrogenase superfamily. *Proc Natl Acad Sci USA* 99: 2702–2707.

Dhawale MR, U Hornemann. 1979. Nitroalkane oxidation by *Streptomyces*. *J Bacteriol* 137: 916–924.

Fournier D, A Halasz, J Spain, P Fiurasek, J Hawari. 2002. Determination of key intermediates during biodegradation of hexahydro-1,3,5-trinitro-1,3,5-triazine with *Rhodococcus* sp. strain DN 22. *Appl Environ Microbiol* 68: 166–172.

Fournier D, A Halasz, S Thiboutot, G Ampleman, D Manno, J Hawari. 2004. Biodegradation of octahydro-1,3,5,7-tetranitro-1,3,5,7-tetrazocine (HMX) by *Phanerochaete chrysosporium*: New insight into the degradation pathway. *Environ Sci Technol* 38: 4130–4133.

Fournier D, J Hawari, SH Streger, K McClay, PB Hatzinger. 2006. Biotransformation of *N*-nitrosodimethylamine by *Pseudomonas mendocina* KR1. *Appl Environ Microbiol* 72: 6693–6698.

Fournier D, S Trott, J Hawari, J Spain. 2005. Metabolism of the aliphatic nitramine 4-nitro-2,4-diazabutranal by *Methylobacterium* sp. strain JS 178. *Appl Environ Microbiol* 71: 4199–4202.

Francis K, B Russell, G Gadda. 2005. Involvement of a flavosemiquinone in the enzymatic oxidation of nitroalkenes catalyzed by 2-nitropropane dioxygenase. *J Biol Chem* 280: 5195–5204.

Gadda G, RD Edmonson, DH Russell, PF Fitzpatrick. 1997. Identification of the naturally occurring flavin of nitroalkane oxidase from *Fusarium oxysporum* as a 5-nitrobutyl-FAD and conversion of the enzyme to the active FAD-containing form. *J Biol Chem* 272: 5563–5570.

Gadda G, K Francis. 2010. Nitronate monooxygenase, a model for anionic flavin semiquinone intermediates in oxidative catalysis. *Arch Biochem Biophys* 493: 53–61.

Gorlatova N, M Tchorzewski, T Kurihara, K Soda, N Esaki. 1998. Purification, characterization, and mechanism of a flavin mononucleotide-dependent 2-nitropropane dioxygenase from *Neurospora crassa*. *Appl Environ Microbiol* 64: 1029–1033.

Hipkin CR, MA Salem, D Simpson, SJ Wainwright. 1999. 3-Nitropropionic acid oxidase from horseshoe vetch (*Hippocrepis comosa*): A novel plant enzyme. *Biochem J* 340: 491–495.

Indest KJ, CM Jung, H-P Chen, D Hancock, C Flozizone, LD Eltis, FH Crocker. 2010. Functional characterization of pGKT2, a 182-kilobase plasmid containing the *xplAB* genes, which are involved in the degradation of hexahydro-1,3,5-trinitro-1,3,5-triazine by *Gordonia* sp. strain KTR9. *Appl Environ Microbiol* 76: 6329–6337.

Kido T, K Hashizume, K Soda. 1978a. Purification and properties of nitroalkane oxidase from *Fusarium oxysporium*. *J Bacteriol* 133: 53–58.

Kido T, K Sida, K Asada. 1978b. Properties of 2-nitropropane dioxygenase of *Hansenula mrakii*. *J BiolChem* 253: 226–232.

Little HN. 1957. The oxidation of 2-nitropropane by extracts of pea plants. *J Biol Chem* 229: 231–238.

Meah Y, V Massey. 2000. Old yellow enzyme: Stepwise reduction of nitroolefins and catalysis of acid-nitro tautomerization. *Proc Natl Acad Sci USA* 97: 10733–10738.

Mijatovic S, G Gadda. 2008. oxidation of alkyl nitronates catalyzed by 2-nitropropane dioxygenase from *Hansenula mrakii*. *Arch Biochem Biophys* 473: 61–68.

Nishino SF, KA Shin, RB Payne, JC Spain. 2010. Growth of bacteria on 3-nitropropionic acid as sole source of carbon, nitrogen and energy. *Appl Environ Microbiol* 76: 3590–3598.

Porter DJ, HJ Bright. 1987. Propionate-3-nitronate from *Penicillium atrovenetum* is a flavoprotein which initiates the autoxidation of its substrate by O_2. *J Biol Chem* 262: 14428–14434.

Qu Y, JC Spain. 2010. Catabolic pathway for 2-nitroimidazole involves a novel nitrohydrolase that also confers drug resistance. *Environ Microbiol* 13: 1010–1017.

Qu Y, JC Spain. 2011. Molecular and biochemical characterization of the 5-nitroanthranilic acid degradation pathway in *Bradyrhizobium* sp. strain JS 329. *J Bacteriol* 193: 3057–3063.

Sabbadin F, R Jackson, K Haider, G Tampi, JP Turkenburg, S Hart, NC Bruce, G Grogan. 2009. The 1.5-Å structure of the XplA-heme, an unusual cytochrome P450 heme domain that catalyzes reductive biotransformation of royal demolition explosive. *J Biol Chem* 284: 28467–28475.

Sakai K, A Nakazawa, K Kondo, H Ohta. 1985. Microbial hydrogenation of nitroolefins. *Agric Biol Chem* 49: 2231–2236.

Seth-Smith HMB, SJ Rosser, A Basran, ER Travis, ER Dabbs, S Nicklin, NC Bruce. 2002. Cloning, sequencing, and characterization of the hexahydro-1,3,5-trinitro-1,3,5-triazine degradation gene cluster from *Rhodococcus rhodochrous*. *Appl Environ Microbiol* 68: 4764–4771.

Sherburne LA, JD Shrout, PJJ Alvarez. 2005. Hexahydro-1,3,5-trinitro-1,3,5-triazine (RDX) degradation by *Acetobacterium paludosum*. *Biodegradation* 16: 539–547.

Snape JR, NA Walkley, AP Morby, S Nicklin, GF White. 1997. Purification, properties, and sequence of glycerol trinitrate reductase from *Agrobacterium radiobacter*. *J Bacteriol* 179: 7796–7802.

Tchorzewski M, T Kurihara, N Esaki, K Soda. 1994. Unique primary structure of 2-nitropropane dioxygenase from *Hansenula mrakii*. *Eur J Biochem* 226: 841–846.

Thompson KT, FH Crocker, HL Fredrickson. 2005. Mineralization of the cyclic nitramine explosive hexahydro-1,3,5-trinitro-1,3,5-triazine by *Gordonia* and *Williamsia* spp. *Appl Environ Microbiol* 71: 8265–8272.

Tormos JR, AB Taylor, SC Daubner, PJ Harts, PF Fitzpatrick. 2010. Identification of a hypothetical protein from *Podospora anserina* as a nitroalkane oxidase. *Biochemistry* 49: 5035–5041.

Trott S, SF Nishino, J Hawari, JC Spain. 2003. Biodegradation of the nitramine explosive CL-20. *Appl Environ Microbiol* 69: 1871–1874.

Vaz ADN, S Chakraborty, V Massey. 1995. Old yellow enzyme: Aromatization of cyclic enones and the mechanism of a novel dismutation reaction. *Biochemistry* 34: 4246–4256.

Zhang J, W Ma, H Tan. 2002. Cloning, expression and characterization of a gene encoding nitroalkane-oxidizing enzymes from *Streptomyces ansochromogenes*. *Eur J Biochem* 269: 6302–6307.

Zhao J-S, A Halasz, L Paquet, C Beaulieu, J Hawari. 2002. Biodegradation of hexahydro-1,3,5-trinitro-1,3,5-triazine and its mononitroso derivative hexahydro-1-nitroso-3,5-dinitro-1,3,5-triazine by *Klebsiella pneumoniae* strain SCZ-1 isolated from an anaerobic sludge. *Appl Environ Microbiol* 68: 5336–5341.

Zhao J-S, D Manno, C Beaulieu, L Paquet, J Hawari. 2005. *Shewanella sediminis* sp. nov., a novel Na⁺-requiring and hexahydro-1,3,5-trinitro-1,3,5-triazine-degrading bacterium from marine sediment. *Int J Syst Evol Microbiol* 55: 1511–1520.

Zhao J-S, D Manno, D Thiboutot, G Ampleman, J Hawari. 2007. *Shewanella canadensis* sp. nov. and *Shewanella atlantica* sp. nov., manganese dioxide- and hexahydro-1,3,5-trinitro-1,3,5-triazine-reducing, psychrophilic marine bacteria. *Int J Syst Evol Microbiol* 57: 2155–2162.

Zhao J-S, J Spain, S Thiboutot, G Ampleman, C Greer, J Hawari. 2004. Phylogeny of cyclic nitramines-degrading psychrophilic bacteria in marine sediment and their potential role in the natural attenuation of explosives. *FEMS Microbiol Ecol* 49: 349–357.

Part 4: Aliphatic Phosphonates and Sulfonates

11.14 Introduction

Organic phosphonates are both anthropogenic and biogenic. They have been incorporated into household detergents and synthesized as antiviral compounds (acyclic nucleoside phosphonates): they exist in naturally occurring compounds including antibiotics (phosphonomycin [fosfo- mycin], alaphosphin [alafosfalin], and plumbemycin), in the microbial herbicide bialaphos (phosphinothricin), and in complex cell components. Although alkyl sulfonates have been used as detergents and the aromatic analogs have a number of applications, there are also biogenic alkyl sulfonates including the metabolically important coenzyme M (2-mercaptoethanesulfonate), taurine (2-aminoethanesulfonate) derived from cysteine, and sulfonolipids. It has been shown that sulfonates, which are present in agricultural soil, are important for the survival of *Pseudomonas putida* in the soil and the rhizosphere (Mirleau et al.

2005). Both phosphonates and sulfonates can be degraded, and the mechanisms of their degradation generally depend on the absence of alternative sources of inorganic phosphorus or sulfur. There are only a few naturally occurring organic compounds containing boron, and these are borate complexes with tetradentate oxygen ligands containing substituted 1,2-dihydroxyethanes. Organic boronates with C–B bonds, particularly aryl boronates are, however, important intermediates in organic synthesis.

11.15 Phosphonates

Phosphonates can serve as sources of phosphorus, generally in the absence of inorganic phosphate (Cook et al. 1978), although only a single strain may not be able to use all the phosphonates that have been examined (Schowanek and Verstraete 1990). Degradation of phosphonates involves cleavage of the C–P bond with the formation of inorganic phosphate, and may be accomplished by a carbon–phosphorus lyase, although the enzymology and its regulation are extremely complex (Chen et al. 1990). The gene cluster required for the utilization of phosphonates is induced in *Escherichia coli* by phosphate limitation, and genetic evidence suggests a connection between the metabolism of phosphonates and phosphites. On the basis of this, the interesting suggestion has been made that there may exist a phosphorus redox cycle and that phosphorus is involved not only at the +5 oxidation level, but also at lower oxidation levels (Metcalf and Wanner 1991).

Two main pathways are used for the degradation of phosphonates that are initiated by a lyase or a hydrolase (Wanner 1994):

1. C–P lyase has a broad substrate specificity and, for example, dimethyl phosphonate is degraded to methane, methylphenyl phosphonate to benzene, and the degradation of the widely used herbicide glyphosate may follow alternative pathways both of which involve C–P fission.

2. The phosphonatase (hydrolase) pathway is less widely used, typically for the degradation of 2-aminoethylphosphonate via phosphonoacetaldehyde to acetaldehyde, and the degradation of phosphonoacetate that involves a specific hydrolase (Kulakova et al. 2001).

Different enzymatic activities have, however, been shown in species of *Campylobacter* in which these are exceptionally expressed in the presence of high concentrations of phosphate (Mendz et al. 2005). In enteric bacteria and pseudomonads, both or only one of the pathways may be expressed (Jiang et al. 1995), and further examples are given in the illustrations.

Examples of the various mechanisms available for the metabolism of phosphonates include the following:

1. In *B. cereus*, 2-aminoethylphosphonate is initially oxidized to 2-phosphonacetaldehyde (La Nauze and Rosenberg 1968) before cleavage of the C–P bond (La Nauze et al. 1970) (Figure 11.17). The degradation of 2-aminoethylphosphonate by *Pseudomonas putida* strain NG2 is carried out even in the presence of phosphate, and is mediated by pyruvate aminotransferase and phosphonoacetaldehyde hydrolase activities that are induced by 2-aminoethylphosphonate (Ternan and Quinn 1998). In contrast, in *Enterobacter aerogenes* strain IFO 12010, these activities are induced only under conditions of phosphate limitation.

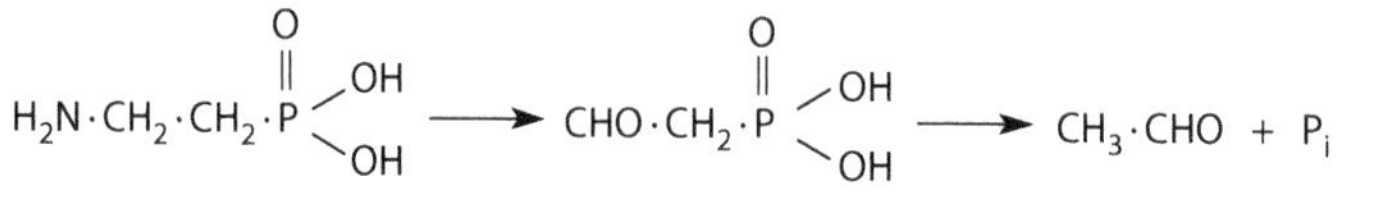

FIGURE 11.17 Biodegradation of 2-aminomethylphosphonate.

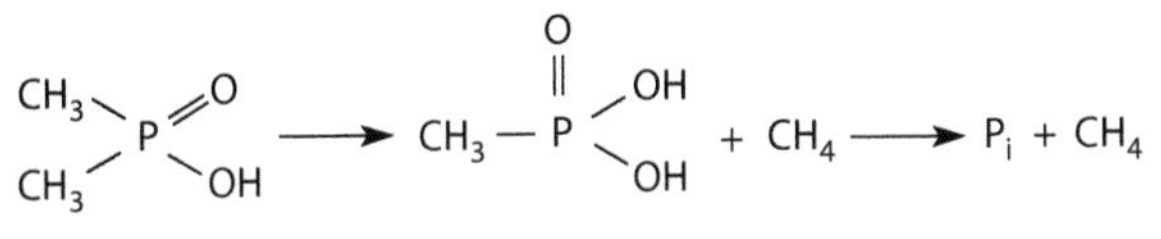

FIGURE 11.18 Reductive biodegradation of alkyl phosphonates and phosphites.

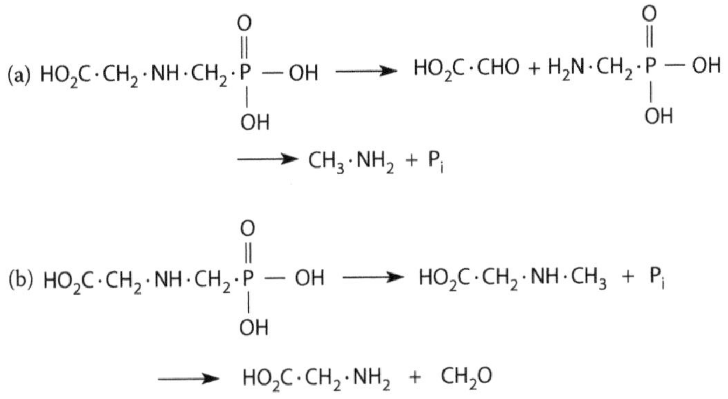

FIGURE 11.19 Alternative pathways for the biodegradation of glyphosate.

2. Reductive pathways have been observed in a number of degradations. *Klebsiella pneumoniae* metabolized methylphenylphosphonate to benzene (Cook et al. 1979). Alkyl phosphonates and phosphites are degraded by a number of bacteria including species of *Klyvera* and *Klebsiella* (Wackett et al. 1987a) (Figure 11.18), and by *Escherichia coli* (Wackett et al. 1987b) using a pathway in which the alkyl groups are reduced to alkanes.

3. In view of its importance as a herbicide, the degradation of glyphosate has been investigated in a number of organisms and two pathways have been elucidated, differing in the stage at which the C–P bond undergoes fission:

   a. Loss of a C_2 fragment—formally glyoxylate—with the formation of aminomethylphospho-nate (Pipke and Amrhein 1988), which may be further degraded by cleavage of the C–P bond to methylamine and phosphate (Jacob et al. 1988) (Figure 11.19a).

   b. Initial cleavage of the C–P bond with the formation of sarcosine, which is then metabolized to glycine (Pipke et al. 1987; Liu et al. 1991) (Figure 11.19b).

11.16 Sulfonates

Methane sulfonic acid is produced by tropospheric oxidation of methyl sulfides, and there are naturally occurring sulfonates including derivatives of taurine and of glucose-6-sulfonate (sulfoquinovose), which are widely distributed in the sulfonolipids of algae and some cyanobacteria. Sulfonates can be degraded by a range of mechanisms, both aerobically and anaerobically (Cook et al. 1999). Several are able to serve as sources of sulfur in the absence of sulfate.

1. a. Strains of methylotrophic bacteria degrade the primary methane sulfonate by initial oxidation to formaldehyde and sulfite (Kelly et al. 1994; Thompson et al. 1995; Baxter et al. 2002). The aerobic strains capable of degrading methanesulfonate have been assigned to the genera *Methylosulfonomonas* and *Marinosulfonomonas* (Holmes et al. 1997). Degradation by *Methylosulfonomonas methylovora* strain M2 is initiated by

monooxygenation (hydroxylation) by a multicomponent enzyme. One component is an electron transfer protein of which the larger subunit contains a Rieske [2Fe–2S] center, and both this and the small subunit show a high degree of homology with those of dioxygenase enzymes (De Marco et al. 1999). The degradation of C_2–C_{10} unsubstituted alkyl sulfonates also takes place by monooxygenation with the formation of sulfite and the corresponding aldehyde (Eichhorn et al. 1999).

b. An analogous pathway was found for the degradation of the secondary alkyl sulfonate sulfosuccinate by *Pseudomonas* sp. strain BS1 that could use this as a source of carbon. The sulfur was recovered as sulfate, and the oxalacetate produced by desulfonation was fed into the TCA cycle (Quick et al. 1994). The degradation of 3-mercaptosuccinate by *Variovorax paradoxus* B4 was initiated by successive oxygenation to sulfinosuccinate and sulfosuccinate followed by desulfonation to oxaloacetate and sulfite (Carbajal-Rodríguez et al. 2011).

2. a. Primary alkanesulfonates C_2–C_{10} can be used by *Escherichia coli* as sources of sulfur under sulfate or cysteine limitation that is carried out aerobically by a two-component monooxygenase dependent on the NAD(P)H reduction of FMN (Eichhorn et al. 1999). The corresponding aldehydes are produced with the formation of sulfite:

$R-CH_2-SO_3H \rightarrow RCHO + SO_3^{2-}$. This enzyme was inactive with methane sulfonate and taurine whose degradation by dioxygenation is described in the next paragraph.

b. *Bacillus subtilis* strain BD99 is able to use a range of aliphatic sulfonates C_1, C_2, C_4, C_6 and the organic sulfonate buffer MOPS as sources of sulfur using a monooxygenase to produce the aldehydes and sulfite. This involves an ABC transport system *ssuBAC*, a monooxygenase *ssuD*, *ygaN* without similarity to established proteins, and $FMNH_2$ provided by an NADH::FMN oxidoreductase. The synthesis of proteins encoded by *ssuBACD* was repressed by sulfate (van der Ploeg et al. 1998).

c. *Corynebacterium glutamicum* is able to use a wide range of aliphatic sulfonates and sulfonate esters as sources of sulfur. They include ethanesulfonate, octanesulfonate, ethyl methanesulfonate, taurine, the cyclic 1,4-butanesultone and 1,3-propanesultone, and organic sulfonate buffers including MOPS. Two gene clusters were assigned to sulfonate utilization, *ssuD1CBA* and *ssuI-seuABC-ssuD2* where the Ssu proteins were similar to those found in *Escherichia coli*, whereas the *sue* gene products were similar to the dibenzothiophene-degrading monooxygenase of *Rhodococcus*. The sulfonates were degraded by sulfonatases encoded by *ssuD1* and *ssuD2* and the monooxygenases depended on a putative reductase encoded by SsuI requiring the $FMNH_2$-dependent reductase that is produced from NAD(P)H (Koch et al. 2005).

3. Taurine is degraded aerobically either by a 2-ketoglutarate-dependent dioxygenation to sulfite and aminoacetaldehyde (Eichhorn et al. 1997; Ryle et al. 2003) (cf. degradation of 2,4-dichlorophenoxyacetate, Chapter 3) or by transamination and fission by a lyase that is also used anaerobically with the formation of acetate (Cook et al. 1999).

4. Sulfoacetaldehyde is a central metabolite in the degradation of a number of C_2 sulfonates (Cook and Denger 2002), including ethane-1,2-disulfonate by *Ralstonia* sp. strain EDS1 (Denger and Cook 2001), and taurine by dehydrogenation and transamination (references in Cook and Denger 2002). Degradation of sulfoacetaldehyde is carried out by sulfoacetaldehyde acetyl transferase, also termed sulfoacetaldehyde sulfolyase. The enzyme has been purified and characterized from cells of *Alcaligenes defragrans* grown with taurine using nitrate as electron acceptor. Its activity required thiamine pyrophosphate, phosphate, and Mg^{2+} (Ruff et al. 2003), and the reaction involves the C-2 anion of the thiazolium ring and results in the formation of sulfite and acetyl phosphate in a reaction that is formally analogous to the conversion of pyruvate into acetyl-CoA.

5. The aerobic degradation of L-cysteate by *Paracoccus pantotrophus* is carried out by deamination to 3-sulfolactate from which sulfite is lost with the formation of pyruvate (Cook et al.

2006). The activity of the lyase (3-L-sulfolactatesulfolyase) has been shown to involve pyridoxal 5'-phosphate, and the enzyme has been found in a number of organisms using L-cysteate either as a source of carbon or as an electron acceptor (Denger et al. 2006).

6. a. When lactate is supplied as the carbon source, a few aliphatic sulfonates such as 2-hydroxyethylsulfonate, alanine-3-sulfonate, and acetaldehyde-2-sulfonate are able to serve as sulfur sources and electron acceptors during the anaerobic growth of some sulfate-reducing bacteria (Lie et al. 1996). Several sulfite-reducing species of *Desulfito-bacterium* were able to use 2-hydroxyethanesulfonate as terminal electron acceptor producing acetate and sulfide (Lie et al. 1999). It is relevant to note that species of *Desulfitobacterium* are also able to use widely different terminal electron acceptors including chlorinated organic compounds, acrylate, and arsenate.

   b. Anaerobic degradation of alkylsulfonates has been demonstrated in a strain of *Bilophila wadsworthia* (*Desulfovibrio* sp.) strain GRZCYSA by different pathways:

      i. Fermentation of cysteate with the formation of acetate, NH_4^+, and equimolar amounts of sulfide and sulfate (Laue et al. 1997a).

      ii. Utilization of aminomethansulfonate and taurine with lactate as electron donor (Laue et al. 1997b).

7. A strain of *Rhodopseudomonas palustris*, which was isolated by enrichment with taurine, could use this as electron source, and as a source of sulfur and nitrogen during photoautotrophic growth with CO_2. Taurine was metabolized to sulfoacetaldehyde and acetyl phosphate by a pathway, which has already been noted (Novak et al. 2004).

It has already been noted (Chapter 9, Part 4) that the degradation of aromatic sulfonates when they are used as source of carbon involves dioxygenation, whereas when they serve as a source of sulfur in the absence of sulfate, degradation takes place by monooxygenation to produce the corresponding phenol and sulfite (Kertesz 1999).

11.17 Boronates

The C–B bond in boronates is fissioned by monooxygenases in both alkyl (Latham and Walsh 1986), and in aryl boronates including substituted phenylboronates and naphthylboronates with production of the corresponding phenols (Negrete-Raymond et al. 2003).

11.18 References

Baxter NJ, J Scanlan, P De Marco, AP Wood, JC Murrell. 2002. Duplicate copies of genes encoding methanesulfonate monooxygenase in *Marinosulfonomonas methylotrophica* strain TR3 and detection of methanesulfonate utilizers in the environment. *Appl Environ Microbiol* 68: 289–296.

Bedard DL, H van Dort, KA Deweerd. 1998. Brominated biphenyls prime extensive microbial reductive dehalogenation of Arochlor 1260 in Hoisatonic River sediment. *Appl Environ Microbiol* 64: 1786–1795.

Buser HR. 1986. Polybrominated dibenzofurans and dibenzo-*p*-dioxins: Thermal reaction products of polybrominated diphenyl ether flame retardants. *Environ Sci Technol* 20: 404–408.

Carbajal-Rodríguez I, N Stöveken, B Satola, JH Wübbeler, A Steinbüchel. 2011. Aerobic degradation of mercaptosuccinate by the Gram-negative bacterium *Variovorax paradoxus* strain B4. *J Bacteriol* 193: 527–539.

Chen C-M, Q-Z Zhuang, Z Zhu, BL Wanner, CT Walsh. 1990. Molecuar biology of carbon–phosphorus bond cleavage cloning and sequencing of the *phn* (*psiD*) genes involved in alkylphosphonate uptake and C–P lyase activity in *Escherichia coli*. *J Biol Chem* 265: 4461–4471.

Cook AM, CG Daughton, M Alexander. 1978. Phosphonate utilization by bacteria. *J Bacteriol* 133: 85–90.

Cook AM, CG Daughton, M Alexander. 1979. Benzene from bacterial cleavage of the carbon–phosphorus bond of phenylphosphonates. *Biochem J* 184: 453–455.

Cook AM, K Denger. 2002. Dissimilation of the C_2 sulfonates. *Arch Microbiol* 179: 1–6.

Cook AM, K Denger, THM Smits. 2006. Dissimilation of C_3-sulfonates. *Arch Microbiol* 185: 83–90.

Cook AM, H Laue, F Junker. 1999. Microbial desulfonation *FEMS Microbiol Rev* 22: 399–419.

De Marco P, P Moradas-Ferreira, TP Higgins, I McDonald, EM Kenna, JC Murrell. 1999. Molecular analysis of a novel methanesulfonic acid monooxygenase from the methylotroph *Methylosulfonomonas methylovora. J Bacteriol* 181: 2244–2251.

Denger K, AM Cook. 2001. Ethanedisulfonate is degraded via sulfoacetaldehyde in *Ralstonia* sp. strain EDS1. *Arch Microbiol* 176: 89–95.

Denger K, THM Smits, AM Cook. 2006. L-cysteate sulpho-lyase, a widespread pyridoxal 5′-phosphate-coupled desulfonative enzyme purified from *Silicibacter pomeroyi* DSS-3. *Biochem J* 394: 657–664.

Eichhorn E, JR van der Ploeg, MA Kertesz, T Leisinger. 1997. Characterization of α-ketoglutarate-dependent taurin dioxygenase from *Escherichia coli. J Biol Chem* 272: 23031–23036.

Eichhorn E, JR van der Ploeg, T Leisinger. 1999. Characterization of a two-component alkanesulfonate monooxygenase from *Escherichia coli. J Biol Chem* 274: 26639–26646.

Holmes AJ, DP Kelly, SC Baker, AS Thompson, P de Marco, EM Kenna, JC Murrell. 1997. *Methylosulfonomonas methylovora* gen. nov., sp. nov., and *Marinosulfonomonas methylotropha* gen. nov., sp. nov.: Novel methylotrophs able to grow on methansulfonic acid. *Arch Microbiol* 167: 46–53.

Jacob GS, JR Garbow, LE Hallas, NM Kimack, GN Kishore, J Schaefer. 1988. Metabolism of glyphosate in *Pseudomonas* sp strain LBr. *Appl Environ Microbiol* 54: 2953–2958.

Jiang W, WW Metcalf, K-S Lee, BL Wanner. 1995. Molecular cloning, mapping, and regulation of the pho regulon genes for phosphonate breakdown by the phosphonatase pathway of *Salmonella typhimurium* LT2. *J Bacteriol* 177: 6411–6421.

Kelly DP, SC Baker, J Trickett, M Davey, JC Murrell. 1994. Methanesulphonate utilization by a novel methylotrophic bacterium involves an unusual monooxygenase. *Microbiology (UK)* 140: 1419–1426.

Kertesz MA. 1999. Riding the sulfur cycle—Metabolism of sulfonates and sulfate esters by Gram-negative bacteria. *FEMS Microbiol Rev* 24: 135–175.

Koch DJ, C Rückert, DA Rey, A Mix, A Pühler, J Kalinowski. 2005. Role of the *ssu* and *seu* genes of *Corynebacterium glutamicum* ATCC 13032 in utilization of sulfonates and sulfonate esters as sulfur sources. *Appl Environ Microbiol* 71: 6104–6114.

Kulakova AN, LA Kulakov, NY Akulenko, VN Ksenzenko, JTG Hamilton, JP Quinn. 2001. Structural and functional analysis of the phosphonoacetate hydrolase (phnA) gene region in *Pseudomonas fluorescens* 23F. *J Bacteriol* 183: 3268–3275.

La Nauze JM, H Rosenberg. 1968. The identification of 2-phosphonoacetaldehyde as an intermediate in the degradation of 2-aminoethylphosphonate by *Bacillus cereus. Biochim Biophys Acta* 165: 438–447.

La Nauze JM, H Rosenberg, DC Shaw. 1970. The enzymatic cleavage of the carbon–phosphorus bond: Purification and properties of phosphonatase. *Biochim Biophys Acta* 212: 332–350.

Latham J, AC Walsh. 1986. Retention of configuration in oxidation of a chiral boronic acid by the flavoenzyme cyclohexanone oxygenase. *J Chem Soc Chem Commun* 527–528.

Laue H, K Denger, AM Cook. 1997a. Fermentation of cysteate by a sulfate-reducing bacterium. *Arch Microbiol* 168: 210–214.

Laue H, K Denger, AM Cook. 1997b. Taurine reduction in anaerobic respiration of *Bilophila wadsworthia* RZATAU. *Appl Environ Microbiol* 63: 2016–2021.

Lie TJ, T Pitta, ER Leadbetter, W Godchaux, JR Leadbetter. 1996. Sulfonates: Novel electron acceptors in anaerobic respiration. *Arch Microbiol* 166: 204–210.

Lie TJ, W Godchaux, JR Leadbetter. 1999. Sulfonates as terminal electron acceptors for growth of sulfite-reducing bacteria (*Desulfitobacterium* spp.) and sulfate-reducing bacteria: Effects of inhibitors of sulfidogenesis. *Appl Environ Microbiol* 65: 4611–4617.

Liu C-M, PA McLean, CC Sookdeo, FC Cannon. 1991. Degradation of the herbicide glyphosate by members of the family *Rhizobiaceae. Appl Environ Microbiol* 57: 1799–1804.

Mendz GL, F Megraud, V Korolik. 2005. Phosphonate catabolism by *Campylobacter* spp. *Arch Microbiol* 183: 113–120.

Metcalf WW, BL Wanner. 1991. Involvement of the *Escherichia coli phn* (*psiD*) gene cluster in assimilation of phosphorus in the form of phosphonates, phosphite, P_i esters, and P_i. *J Bacteriol* 173: 587–600.

Mirleau P, R Wogelius, A Smith, MA Kertesz. 2005. Importance of organosulfur utilization for survival of *Pseudomonas putida* in soil and rhizosphere. *Appl Environ Microbiol* 71: 6571–6577.

Negrete-Raymond AC, B Weder, LP Wackett. 2003. Catabolism of arylboronic acids by *Arthrobacter nicotinovorans* strain PBA. *Appl Environ Microbiol* 69: 4263–4267.

Novak, RT, RF Gritzer, ER Leadbetter, W Godchaux. 2004. Phototrophic utilization of taurine by the purple nonsulfur bacteria *Rhodopseudomonas palustris* and *Rhodobacter sphaeroides*. *Microbiology (UK)* 150: 1881–1891.

Pipke R, N Amrhein. 1988. Degradation of the phosphonate herbicide glyphosate by *Arthrobacter atrocyaneus* ATCC 13752. *Appl Environ Microbiol* 54: 1293–1296.

Pipke R, N Amrhein, GS Jacob, J Schaefer, GM Kishore. 1987. Metabolism of glyphosate in an *Arthrobacter* sp GLP-1. *Eur J Biochem* 165: 267–273.

Quick A, NJ Russell, SG Hales, GF White. 1994. Biodegradation of sulphosuccinate: Direct desulphonation of a secondary sulphonate. *Microbiology (UK)* 140: 2991–2998.

Ruff J, K Denger, AM Cook. 2003. Sulphoacetaldehyde acetyltransferase yields acetyl phosphate: Purification from *Alcaligenes defragrans* and gene clusters in taurine degradation. *Biochem J* 369: 275–285.

Ryle MJ, KD Koehntorp, A Liu, L Que, RP Hausinger. 2003. Interconversion of two oxidized forms of taurine/α-ketoglutarate dioxygenase, a non-heme iron hydroxylase: Evidence for bicarbonate binding. *Proc Natl Acad Sci USA* 100: 3790–3795.

Schowanek D, W Verstraete. 1990. Phosphonate utilization by bacterial cultures and enrichments from environmental samples. *Appl Environ Microbiol* 56: 895–903.

Ternan NG, JP Quinn. 1998. Phosphate starvation-independent 2-aminoethylphosphonic acid biodegradation in a newly isolated strain of *Pseudomonas putida*, NG2. *System Appl Microbiol* 21: 346–352.

Thompson AS, NJP Owens, JC Murrell. 1995. Isolation and characterization of methansulfonic acid-degrading bacteria from the marine environment. *Appl Environ Microbiol* 61: 2388–2393.

van der Ploeg JR, NJ Cummings, T Leisinger, IF Connerton. 1998. *Bacillus subtilis* genes for the utilization of sulfur from aliphatic sulfonates. *Microbiology (UK)* 144: 2555–2561.

Wackett LP, BL Wanner, CP Venditti, CT Walsh. 1987b. Involvement of the phosphate regulon and the *psiD* locus in carbon–phosphorus lyase activity of *Escherichi coli* K-12. *J Bacteriol* 169: 1753–1756.

Wackett LP, SL Shames, CP Venditti, CT Walsh. 1987a. Bacterial carbon–phosphorus lyase: Products, rates and regulation of phosphonic and phosphinic acid metabolism. *J Bacteriol* 169: 710–717.

Wanner BL. 1994. Molecular genetics of carbon–phosphorus bond cleavage in bacteria. *Biodegradation* 5: 175–184.

Part 5: Degradation of Organic Compounds of Metals and Metalloids

Although knowledge on the biodegradation of these compounds is sparse, a number of them are important in industrial processes. Formation of methylated derivatives may take place in metals and metalloids belonging to groups 15 and 16 of the periodic table, and a few of group 14. These have been discussed in a critical review (Thayer 2002) and in Chapter 3, Part 4, and they have been noted in the context of the bacterial resistance to metals and metalloids. Since carbon monoxide has been considered as an organic compound (Chapter 7, Part 1), it is consistent to make brief comments on metal carbonyls.

There are distinct structural types of organic compounds containing metals and metalloids. The first contain covalent carbon–metal bonds and are strictly organometallic compounds, for example, the alkylated compounds of Hg, Sn, and Pb, and of Li, Mg, and Al (and formerly Hg), which have been extensively used in laboratory organic synthesis, and $Al(C_2H_5)_3$ that is a component of the Ziegler–Natta catalyst for polymerization of alkenes. Considerable attention has been directed to double-bonded Fischer carbenes of Cr and W, the Schrock carbenes of Ta and Ti, and cyclic polyene ligands of Fe, Co, Cr, and U. Carbonyls of transition metals from groups 6 to 10 of the periodic table include both the monomeric compounds such as $Cr(CO)_6$, $Fe(CO)_5$, $Ni(CO)_4$ and those with two metal groups such as $Mn_2(CO)_{10}$ and

Co$_2$(CO)$_8$, which is used industrially for hydroformylation. Although their source has not been identified, it has been shown that volatile compounds from landfills contain carbonyls of Mo and W (Feldmann and Cullen 1997).

Metals may also be linked through an oxygen or nitrogen atom to form a stable metal complex without a carbon–metal bond. These include metal complexes of ethylenediamine tetraacetate (EDTA), diethylenetriamine pentaacetate (DTPA), or ethylenediamine tetramethylphosphonate (EDTMP). Metalloid compounds include antimonyl gluconate and bismuth salicylate.

Organic compounds of metals and metalloids are both anthropogenic and produced naturally as methylated compounds of which methylmercury has received considerable attention in view of its established toxicity. In addition, resistance to toxic metal cations and metalloid oxyanions by methylation is well established and is discussed in Chapter 3. Phenylmercury and organic tin compounds have been used as biocides; methylcyclopentadienyl manganese tricarbonyl has been proposed as a fuel additive. Methylarsonate and dimethylarsinate have been used as herbicides and insecticides, and several organic arsenic compounds have been examined as chemotherapeutic agents. These include phenylarsonic acids and arsenobenzene of which the 3,3′-diamino-4,4′-dihydroxy derivative known by its trade name Salvarsan occupied an important place in the history of chemotherapy. A range of organoarsenic compounds including arsenolipids and arsenocarbohydrates that are degraded to arsenobetaine is found in marine biota.

A number of the less common metals including Gd, Sm, In, Tc, Au, and the platinum metals have achieved importance in clinical medicine, and include the γ-emitters ^{99m}Tc, ^{111}In, and ^{153}Sm with short half-lives (6.0 h, 2.8 days, and 1.93 days, respectively). They are generally administered as organic complexes—the DTPA of ^{111}In in brain imaging and of Gd that is used in MRI; the EDTMP of ^{153}Sm; the *cis*-diammine-(1,1-cyclobutanedicarboxylato)platinum (carboplatin); and the gold complex of (Et$_3$P) ≡ Au–X, where X is 1-β-thiopenta-*O*-acetylglucose (auranofin) (Thompson and Orvig 2003). Little is known, however, of the ultimate fate of these metal complexes (Kümmerer and Helmers 2000).

Although the degradation or transformation of some organometallic compounds has been examined, only a few of the pathways and the enzymology have been unambiguously established. Only a brief account of the important issues is attempted, but coverage of essentially ecological aspects has not been attempted. Redox systems that are involved in bacterial reactions involving inorganic arsenic have been reviewed (Silver and Phung 2005), and reactions in which selenate and arsenate serve as electron acceptors for growth in the absence of oxygen have been discussed in Chapter 3, Part 2.

11.19 Tin

The aerobic transformation of tributyltin by bacteria takes place by successive loss of butyl groups to monobutyltin and eventually to inorganic tin (Kawai et al. 1998). A comparable degradation takes place with fungi and yeasts, but is complicated by the simultaneous methylation of the products (Errécalde et al. 1995). Triphenyltin chloride is degraded to diphenyltin and benzene (Inoue et al. 2000), apparently mediated by pyoverdine siderophores produced by the fluorescent pseudomonad (Inoue et al. 2003). The analogous pyochelin has been implicated in the degradation of triphenyltin chloride by *Pseudomonas aeruginosa* (Sun et al. 2006), and the role of the ferric complex of pyochelin in generating hydroxyl radicals has been demonstrated using EPR (Sun and Zhong 2006).

11.20 Lead

It has been shown using (1-^{14}C-ethyl) tetraethyl lead that this was biodegradable, and that the rate of mineralization was adversely affected by the presence of hydrocarbons that are generally simultaneous contaminants (Mulroy and Ou 1998).

11.21 Mercury

In Hg-resistant bacteria that are resistant to organic forms of Hg such as phenylmercuric acetate and methylmercury chloride, lyases are involved in the fission of the C–Hg to form Hg^{2+} and benzene or methane, and the enzyme has been partly purified (Schottel 1978). The Hg^{2+} may then be reduced to nontoxic Hg^0. The situation under anaerobic conditions for sulfate-reducing bacteria is complicated by the simultaneous occurrence of both methylation and demethylation in the same strain (Pak and Bartha 1998; Gilmour et al. 2011), plausibly by operation of the acetyl-CoA pathway (Choi et al. 1994; Ekstrom et al. 2003).

Under anaerobic conditions, demethylation, though not methylation, has been reported for a methanogen (Pak and Bartha 1998).

11.22 Arsenic

Tamaki and Frankenberger (1992) quote the degradation of methanearsonic acid to arsenate and CO_2 by aerobic bacteria, and growth at the expense of organoarsenic compounds has been reported. Growth with 2-aminoethylarsonic acid has been demonstrated and the products identified as alanine and arsonoacetaldehyde, the latter decomposing spontaneously by hydrolysis (Lacoste et al. 1992). Growth on arsenoacetate has also been reported without, however, identifying the enzyme responsible for scission of the C–As bond (Quinn and McMullen 1995), and strains of *Campylobacter* degraded both phenylphosphonate and phenylarsonate (Mendz et al. 2005). Dimethylselenide is degraded to CH_4 and CO_2 by a methylotrophic methanogen grown with dimethylsulfide (Oremland and Zehr 1986), and serves as a source of selenium for *Methanococcus voltae* during selenium limitation (Niess and Klein 2004).

11.23 References

Choi S-C, TT Chase, R Bartha. 1994. Metabolic pathways leading to mercury methylation in *Desulfovibrio desulfuricans* LS. *Appl Environ Microbiol* 60: 4072–4077.

Ekstrom EB, FM Morel, JM Benoit. 2003. Mercury methylation independent of the acetyl-coenzyme a pathway in sulfate-reducing bacteria. *Appl Environ Microbiol* 69: 5414–5422.

Errécalde O, M Astruc, G Maury, R Pinel. 1995. Biotransformation of butyltin compounds using pure strains of microorganisms. *Appl Organomet Chem* 9: 23–28.

Feldmann J, WR Cullen. 1997. Occurrence of volatile transition metal compounds in landfill has: Synthesis of molybdenum and tungsten carbonyls in the environment. *Environ Sci Technol* 31: 2125–2129.

Gilmour CC, DA Elios, AM Kucken, SD Brown, AV Palumbo, CW Schadt, JD Wall. 2011. Sulfate-reducing bacterium *Desulfovibrio desulfuricans* DN 132 as model for understanding bacterial mercury methylation. *Appl Environ Microbiol* 77: 3938–3951.

Inoue H, O Takimura, H Fuse, K Murakami, K Kamimura, Y Yamaoka. 2000. Degradation of triphenyltin by a fluorescent pseudomonad *Appl Environ Microbiol* 66: 3492–3498.

Inoue H, O Takimura, K Kawaguchi, T Nitoda, H Fuse, K Murakami, Y Yamaoka. 2003. Tin–carbon cleavage of organotin compounds by pyoverdine from *Pseudomonas chlororaphis*. *Appl Environ Microbiol* 69: 878–883.

Kawai S, Y Kurokawa, H Harino, M Fukushima. 1998. Degradation of tributyltin by a bacterial strain isolated from polluted river water. *Environ Pollut* 102: 259–263.

Kümmerer K, E Helmers. 2000. Hospital effluents as a source of gadolinium in the aquatic environment. *Environ Sci Technol* 34: 573–577.

Lacoste A-M, C Dumora, BRS Ali, E Neuzil, HBF Dixon. 1992. Utilization of 2-aminoethylarsonic acid in *Pseudomonas aeruginosa*. *J Gen Microbiol* 138: 1283–1287.

Mendz GL, F Mégraud, V Korolik. 2005. Phosphonate catabolism by *Campylobacter* spp. *Arch Microbiol* 183: 113–120.

Mulroy, PT, L-T Ou. 1998. Degradation of tetraethyllead during the degradation of leaded gasoline hydrocarbons in soil. *Environ Toxicol Chem* 17: 777–782.

Niess UM, A Klein. 2004. Dimethylselenide demethylation is an adaptaive response to selenium deprivation in the archaeon *Methanococcus voltae. J Bacteriol* 186: 3640–3648.

Oremland RS, JP Zehr. 1986. Formation of methane and carbon dioxide from dimethylselenide in anoxic sediments and by a methanogenic bacterium. *Appl Environ Microbiol* 52: 1031–1036.

Pak K-R, R Bartha. 1998. Mercury methylation and demethylation in anoxic lake sediments by strictly anaerobic bacteria. *Appl Environ Microbiol* 64: 1013–1017.

Quinn JP, G McMullan. 1995. Carbon–arsenic bond cleavage by a newly isolated Gram-negative bacterium strain ASV2. *Microbiology (UK)* 141: 721–727.

Schottel JL. 1978. The mercuric and organomercurial detoxifying enzymes from a plasmid-bearing strain of *Escherichia coli. J Biol Chem* 253: 4341–4349.

Silver S, LT Phung. 2005. Genes and enzymes in bacterial oxidation and reduction of inorganic arsenic. *Appl Environ Microbiol* 71: 599–608.

Sun G-X, J-J Zhong. 2006. Mechanism of augmentation of organotin decomposition by ferripyochelin: Formation of hydroxyl radical and organotin-iron ternary complex. *Appl Environ Microbiol* 72: 7264–7269.

Sun G-X, W-Q Zhou, J-J Zhong. 2006. Organotin decomposition by pyochelin secreted by *Pseudomonas aeruginosa* even in an iron-sufficient environment. *Appl Environ Microbiol* 72: 6411–6413.

Tamaki S, WT Frankenberger. 1992. Environmental biochemistry of arsenic. *Rev Environ Contam Toxicol* 124: 79–110.

Thayer JS. 2002. Biological methylation of less-studied elements. *Appl Organometal Chem* 16: 677–691.

Thompson KH, C Orvig. 2003. Boon and bane of metal ions in medicine. *Science* 300: 936–939.

Index

9 781032 099170